Universe

NINTH EDITION

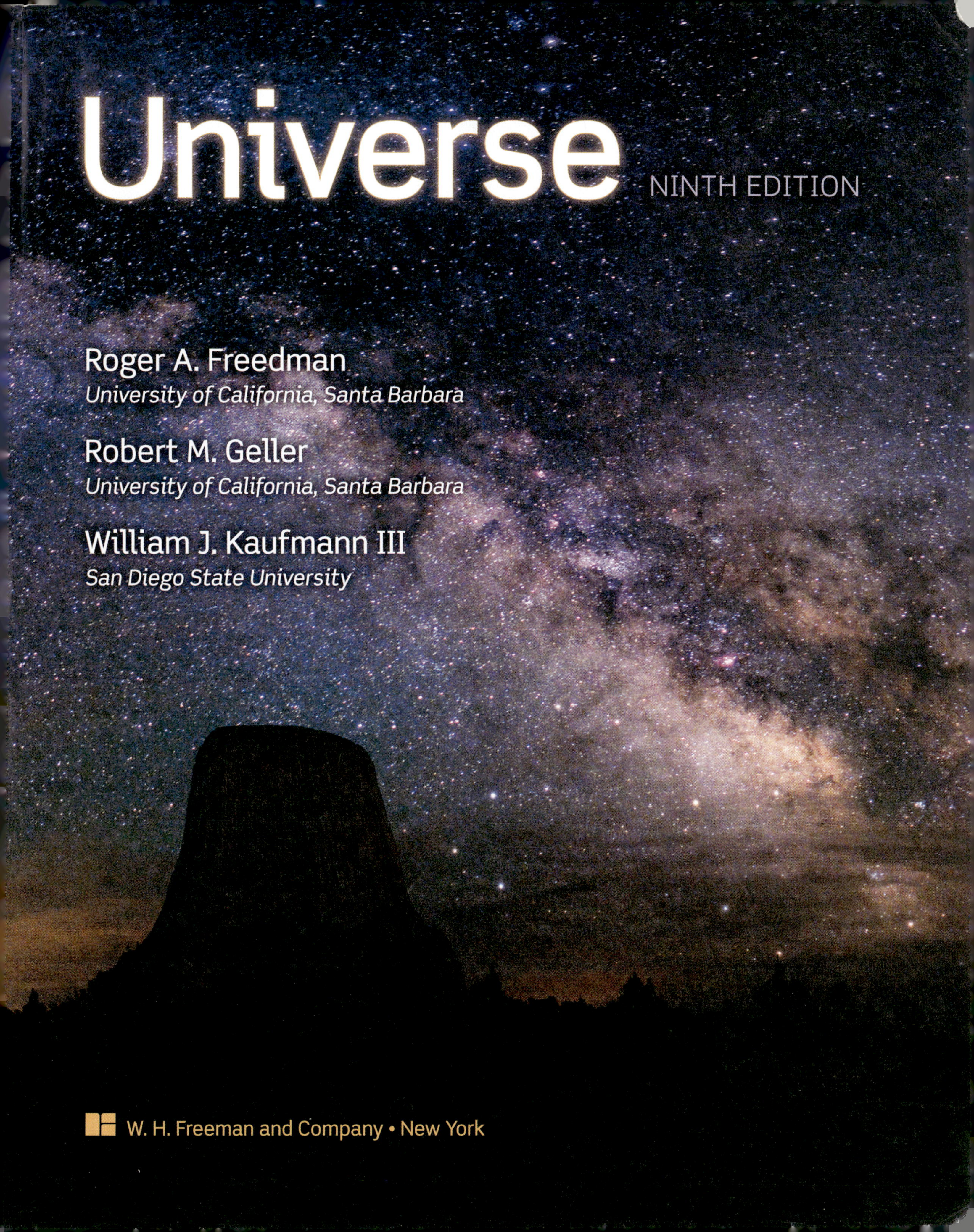

Universe

NINTH EDITION

Roger A. Freedman
University of California, Santa Barbara

Robert M. Geller
University of California, Santa Barbara

William J. Kaufmann III
San Diego State University

W. H. Freeman and Company • New York

Publisher: Clancy Marshall
Senior Acquisitions Editor: Anthony Palmiotto
Senior Development Editor: Kharissia Pettus
Assistant Editor: Anthony Petrites
Editorial Assistant: Brittany Murphy
Project Editor: Leigh Renhard
Media and Supplements Editor: Amy Thorne
Marketing Director: John Britch
Art Director: Diana Blume
Production Coordinator: Susan Wein
Senior Illustration Coordinator: Bill Page
Illustrations: Imagineering Media Services
Photo Editor: Bianca Moscatelli
Composition: Matrix Publishing Services
Printing and Binding: Quad/Graphics Versailles

Guest Essay Photo Credits
p. 20, Don Fukuda/National Optical Astronomy Observatories; p. 44, courtesy of James Randi; p. 63, courtesy of Mark Hollabaugh; p. 401, courtesy of Scott S. Sheppard; p. 763, courtesy of Lisa Plaxco

Library of Congress Control Number: 2009941774
ISBN-13: 978-1-4292-3153-4
ISBN-10: 1-4292-3153-X

Printed in the United States of America

Third printing

W. H. Freeman and Company
41 Madison Avenue
New York, NY 10010
Houndmills, Basingstoke RG21 6XS, England
www.whfreeman.com

To Lee Johnson Kaufmann and
Caroline Robillard-Freedman,
strong survivors

and to the memory of
PFC Richard Freedman, AUS
and S/Sgt. Ann Kazmierczak Freedman, WAC

About the Authors

Roger A. Freedman is on the faculty of the Department of Physics at the University of California, Santa Barbara. He grew up in San Diego, California and was an undergraduate at the University of California campuses in San Diego and Los Angeles. He did his doctoral research in nuclear theory and its astrophysical applications at Stanford University under the direction of Professor J. Dirk Walecka. Dr. Freedman came to UCSB in 1981 after three years of teaching and doing research at the University of Washington. Dr. Freedman holds a commercial pilot's license, and when not teaching or writing he can frequently be found flying with his wife, Caroline. He has flown across the United States and Canada.

Robert M. Geller teaches and conducts research in astrophysics at the University of California, Santa Barbara. He also obtained his Ph.D. at the University of California, Santa Barbara; his doctorial research was in observational cosmology under Professor Robert Antonucci. Using data from the Hubble Space Telescope, he is currently involved in a search for bursts of light that are predicted to occur when a supermassive black hole consumes a star. Dr. Geller also has a strong emphasis on education, and he received the Distinguished Teaching Award at UCSB in 2003. His hobbies include rock climbing, and he's also building an unusual telescope with lenses made of water.

William J. Kaufmann III was the author of the first four editions of *Universe*. Dr. Kaufmann earned his bachelor's degree *magna cum laude* in physics from Adelphi University in 1963, a master's degree in physics from Rutgers in 1965, and a Ph.D. in astrophysics from Indiana University in 1968. At 27, he became the youngest director of any major planetarium in the United States when he took the helm of the Griffith Observatory in Los Angeles. During his career, he also held positions at San Diego State University, UCLA, Caltech, and the University of Illinois. A prolific author, his many books include *Black Holes and Warped Spacetime, Relativity and Cosmology, Planets and Moons, Stars and Nebulas,* and *Galaxies and Quasars*. Dr. Kaufmann died in 1994.

Contents Overview

III Stars and Stellar Evolution

IV Galaxies and Cosmology

Contents

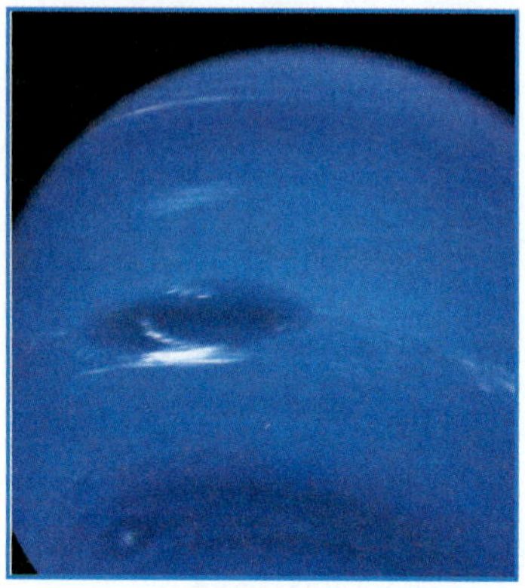

Preface

Astronomy is one of the most dynamic and exciting areas of modern science. Its recent discoveries often capture widespread interest, and its subjects—the planets, stars, and the universe—inspire awe and inquiry. From observations of previously unseen worlds in the outer depths of our solar system, to new evidence of how the most massive stars can collapse into black holes, astronomy provides a wide array of opportunities to show students the continuing process of science.

With so many potential concepts to explore, students can easily get overwhelmed. *Universe* has been developed to help students focus on and learn core concepts. Through clear, highlighted explanations of key topics, and multiple interactions with the material, the text fosters a deep, appreciative comprehension of the subject.

COSMIC CONNECTIONS figures summarize key ideas visually

Many students learn more through visual presentations than from reading long passages of text. To help these students, large figures, called **Cosmic Connections**, appear in most chapters in *Universe*. Cosmic Connections give an overview or summation of a particular important topic in a chapter. Subjects range from the variety of modern telescopes to the formation of the solar system; from the scale of distances in the universe to the life cycles of stars; and from the influence of gravitational tidal forces to the evolution of the universe after the Big Bang. Cosmic Connections also convey some of the excitement and adventure that lead students to study astronomy in the first place.

Enhanced, animated versions of the Cosmic Connections are available within the textbook's associated multimedia resources.

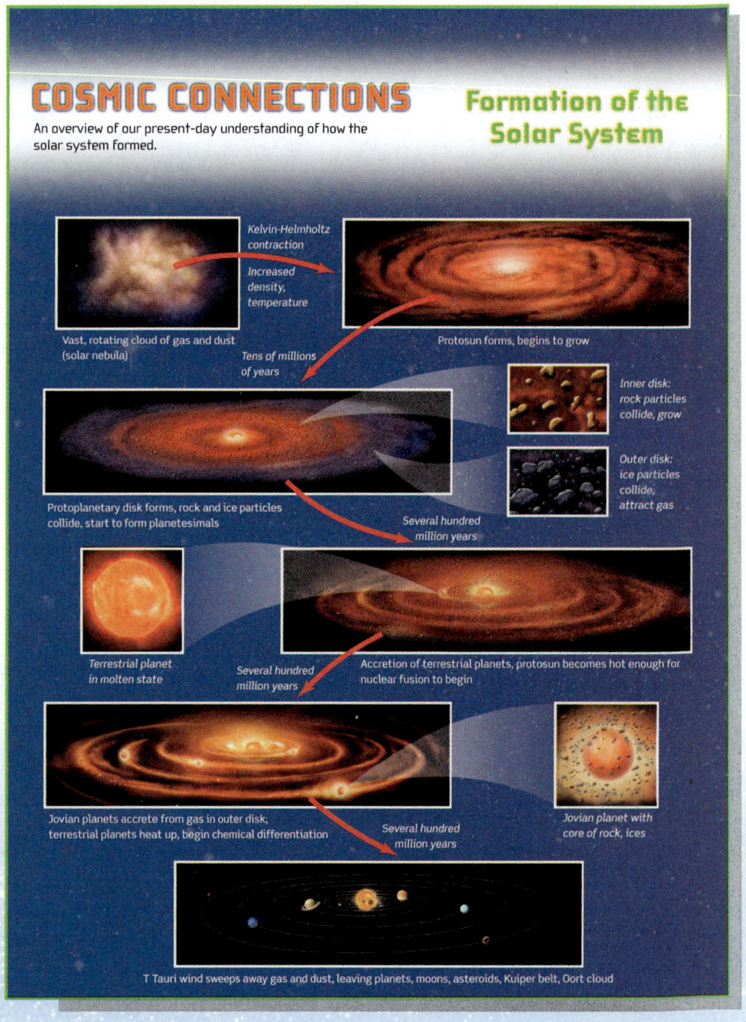

COSMIC CONNECTIONS
An overview of our present-day understanding of how the solar system formed.
Formation of the Solar System
Kelvin-Helmholtz contraction
Increased density, temperature
Vast, rotating cloud of gas and dust (solar nebula)
Protosun forms, begins to grow
Tens of millions of years
Inner disk: rock particles collide, grow
Outer disk: ice particles collide, attract gas
Protoplanetary disk forms, rock and ice particles collide, start to form planetesimals
Several hundred million years
Terrestrial planet in molten state
Several hundred million years
Accretion of terrestrial planets, protosun becomes hot enough for nuclear fusion to begin
Jovian planets accrete from gas in outer disk, terrestrial planets heat up, begin chemical differentiation
Several hundred million years
Jovian planet with core of rock, ices
T Tauri wind sweeps away gas and dust, leaving planets, moons, asteroids, Kuiper belt, Oort cloud

INTERACTIVE TUTORIALS guide students through core concepts at their own pace

Developed by prominent astronomy education researchers, the **Interactive Tutorials** present astronomy topics in a flexible multimedia environment. They take advantage of the best means to illustrate each topic, using a blend of text, review questions, animations, video, and quizzing to produce a thorough understanding that students can carry with them. The sixty tutorial topics were chosen after careful analysis of the most commonly taught subjects, with particular emphasis on which topics were most often misunderstood by students.

Each tutorial first outlines the learning goals and the introductory material related to the concept at hand. Students are then guided through a progression of deeper interaction with the material. They are encouraged to make decisions based on their interpretations of art and animations, and are coached through the topic by way of review questions with answer-specific feedback to correct their mistakes. The tutorials have been shown to increase student understanding and produce a meaningful, memorable learning experience.

To provide flexibility for instructors and students, the tutorials are a premium resource available through a number of online environments, including the **eBook, AstroPortal, WebAssign Premium,** and the **Companion Web site,** all described below.

10-3 The Moon has no global magnetic field but has a small core beneath a thick mantle

We saw in Section 7-6 and Section 7-7 that a small planet or satellite will have little internal heat, and hence is likely to have a mostly solid interior. Lacking substantial amounts of moving molten material in its interior, a small world is unlikely to generate a global magnetic field. As we will see, these general rules describe the Moon very well.

Compared to earthquakes, moonquakes are few and feeble—but they reveal key aspects of the Moon's interior

MOTIVATION STATEMENTS connect detailed discussion to the broader picture

As students read a science text, they often wonder "What is the key idea? How does it relate to the discussion on previous pages? What motivated scientists to make these observations and deductions?" Free-standing **Motivation Statements** address these questions; a statement is a brief sentence that shows students how a section's material fits in with the larger picture of astronomy presented in a chapter. Some statements deal with the scientific method. Others compare topics in one section with topics in another section or chapter; still other statements give a quick summary of a section's conclusion.

CAUTIONS confront misconceptions

Many people think that Earth is closer to the Sun in summer than in winter and that the phases of the Moon are caused by Earth's shadow falling on the Moon. However, these "common sense" ideas are incorrect (as explained in Chapters 2 and 3). Throughout *Universe*, paragraphs marked by the Caution icon alert the reader to such common conceptual pitfalls.

CAUTION! The images in Figure 6-18 are **false color** images: They do not represent the true color of the stars shown. False color is often used when the image is made using wavelengths that the eye cannot detect, as with the infrared images in Figure 6-18. A different use of false color is to indicate the relative brightness of different parts of the image, as in the infrared image of a person in Figure 5-10. Throughout this book, we'll always point out when false color is used in an image.

ANALOGIES make Astronomy Relatable to Students

When learning new astronomical ideas, it can be helpful to relate them to more familiar experiences on Earth. Throughout *Universe*, Analogy paragraphs make these connections. For example, the motions of the planets can be related to children on a merry-go-round, and the bending of light through a telescope lens is similar to the path of a car driving from firm ground onto sand.

ANALOGY An inflationary epoch can also account for the flatness of the universe. To see why, think about a small portion of Earth's surface, such as a small lake. For all practical purposes, it is impossible to detect Earth's curvature over such a small area, and a small lake looks flat. Similarly, the observable universe is such a tiny fraction of the entire inflated universe that any overall curvature in it is virtually undetectable (Figure 27-3). Like a small lake, the segment of space we can observe looks flat.

EXPLANATORY ART help students visualize concepts

Many **figures** in the text have balloon captions to help students interpret complex figures. Many **photos** also contain brief labels to point out the most important features.

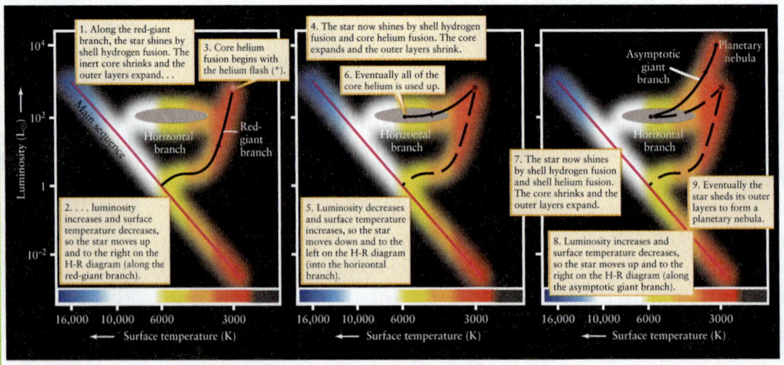

1. Along the red-giant branch, the star shines by shell hydrogen fusion. The inert core shrinks and the outer layers expand. . .
3. Core helium fusion begins with the helium flash (*).
Main sequence
Horizontal branch
Red-giant branch
2. . . . luminosity increases and surface temperature decreases, so the star moves up and to the right on the H-R diagram (along the red-giant branch).
Luminosity ($L_\odot$)
10^4
10^2
1
10^{-2}
16,000 10,000 6000 3000
Surface temperature (K)
4. The star now shines by shell hydrogen fusion and core helium fusion. The core expands and the outer layers shrink.
6. Eventually all of the core helium is used up.
5. Luminosity decreases and surface temperature increases, so the star moves down and to the left on the H-R diagram (into the horizontal branch).
Asymptotic giant branch
Planetary nebula
7. The star now shines by shell hydrogen fusion and shell helium fusion. The core shrinks and the outer layers expand.
9. Eventually the star sheds its outer layers to form a planetary nebula.
8. Luminosity increases and surface temperature decreases, so the star moves up and to the right on the H-R diagram (along the asymptotic giant branch).

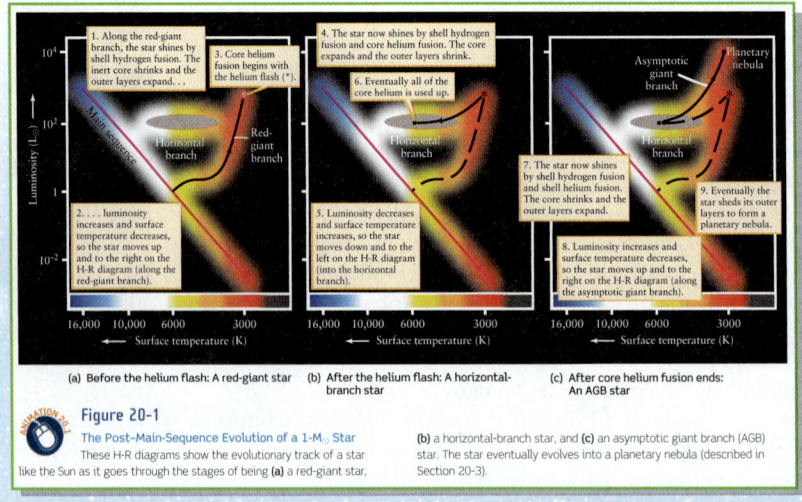

(a) Before the helium flash: A red-giant star (b) After the helium flash: A horizontal-branch star (c) After core helium fusion ends: An AGB star

ANIMATION 20.1

Figure 20-1

The Post-Main-Sequence Evolution of a 1-$M_\odot$ Star These H-R diagrams show the evolutionary track of a star like the Sun as it goes through the stages of being **(a)** a red-giant star, **(b)** a horizontal-branch star, and **(c)** an asymptotic giant branch (AGB) star. The star eventually evolves into a planetary nebula (described in Section 20-3).

WAVELENGTH TABS—
Astronomers observe the sky using many forms of light

Astronomers rely on special telescopes that are sensitive to nonvisible forms of light. To help students appreciate these different kinds of observations, all the images in *Universe* appear with wavelength tabs. The highlighted letter on each tab indicates whether the image was made with **R**adio waves, **I**nfrared radiation, **V**isible light, **U**ltraviolet light, **X** rays, or **G**amma rays.

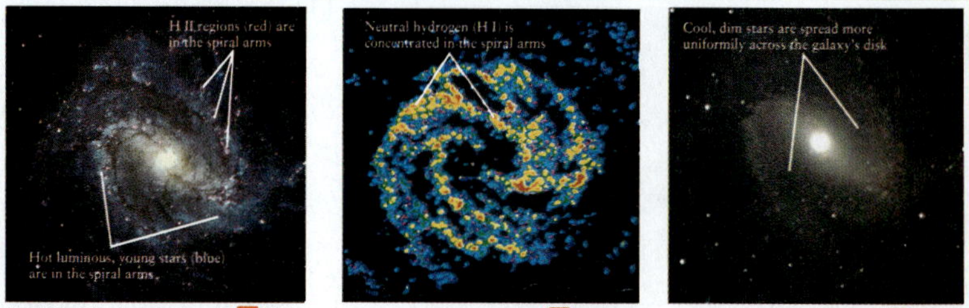

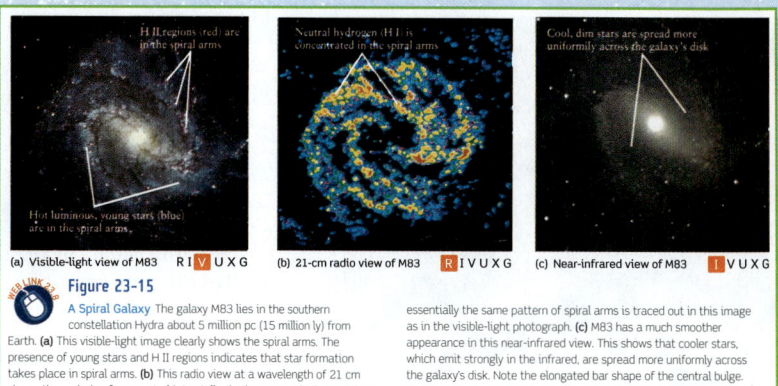

(a) Visible-light view of M83 R I V U X G (b) 21-cm radio view of M83 R I V U X G (c) Near-infrared view of M83 I V U X G

WEB LINK 23.8

Figure 23-15

A Spiral Galaxy The galaxy M83 lies in the southern constellation Hydra about 5 million pc (15 million ly) from Earth. **(a)** This visible-light image clearly shows the spiral arms. The presence of young stars and H II regions indicates that star formation takes place in spiral arms. **(b)** This radio view at a wavelength of 21 cm shows the emission from neutral interstellar hydrogen gas (H I). Note that essentially the same pattern of spiral arms is traced out in this image as in the visible-light photograph. **(c)** M83 has a much smoother appearance in this near-infrared view. This shows that cooler stars, which emit strongly in the infrared, are spread more uniformly across the galaxy's disk. Note the elongated bar shape of the central bulge. (a: Anglo-Australian Observatory; b: VLA, NRAO; c: S. Van Dyk/IPAC)

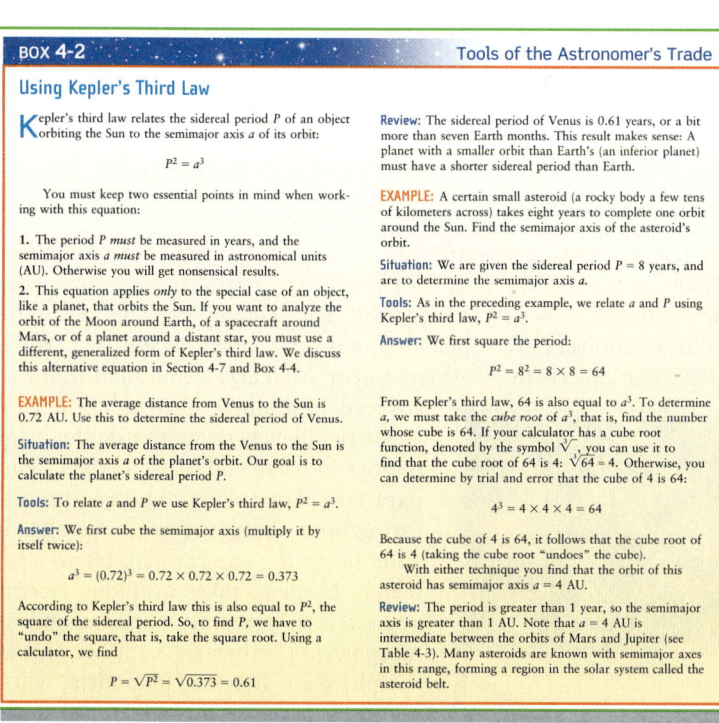

BOX 4-2 Tools of the Astronomer's Trade

Using Kepler's Third Law

Kepler's third law relates the sidereal period P of an object orbiting the Sun to the semimajor axis a of its orbit:

$$P^2 = a^3$$

You must keep two essential points in mind when working with this equation:

1. The period P *must* be measured in years, and the semimajor axis a *must* be measured in astronomical units (AU). Otherwise you will get nonsensical results.
2. This equation applies *only* to the special case of an object, like a planet, that orbits the Sun. If you want to analyze the orbit of the Moon around Earth, of a spacecraft around Mars, or of a planet around a distant star, you must use a different, generalized form of Kepler's third law. We discuss this alternative equation in Section 4-7 and Box 4-4.

EXAMPLE: The average distance from Venus to the Sun is 0.72 AU. Use this to determine the sidereal period of Venus.

Situation: The average distance from the Venus to the Sun is the semimajor axis a of the planet's orbit. Our goal is to calculate the planet's sidereal period P.

Tools: To relate a and P we use Kepler's third law, $P^2 = a^3$.

Answer: We first cube the semimajor axis (multiply it by itself twice):

$$a^3 = (0.72)^3 = 0.72 \times 0.72 \times 0.72 = 0.373$$

According to Kepler's third law this is also equal to P^2, the square of the sidereal period. So, to find P, we have to "undo" the square, that is, take the square root. Using a calculator, we find

$$P = \sqrt{P^2} = \sqrt{0.373} = 0.61$$

Review: The sidereal period of Venus is 0.61 years, or a bit more than seven Earth months. This result makes sense: A planet with a smaller orbit than Earth's (an inferior planet) must have a shorter sidereal period than Earth.

EXAMPLE: A certain small asteroid (a rocky body a few tens of kilometers across) takes eight years to complete one orbit around the Sun. Find the semimajor axis of the asteroid's orbit.

Situation: We are given the sidereal period P = 8 years, and are to determine the semimajor axis a.

Tools: As in the preceding example, we relate a and P using Kepler's third law, $P^2 = a^3$.

Answer: We first square the period:

$$P^2 = 8^2 = 8 \times 8 = 64$$

From Kepler's third law, 64 is also equal to a^3. To determine a, we must take the *cube root* of a^3, that is, find the number whose cube is 64. If your calculator has a cube root function, denoted by the symbol $\sqrt[3]{\ }$, you can use it to find that the cube root of 64 is 4: $\sqrt[3]{64} = 4$. Otherwise, you can determine by trial and error that the cube of 4 is 64:

$$4^3 = 4 \times 4 \times 4 = 64$$

Because the cube of 4 is 64, it follows that the cube root of 64 is 4 (taking the cube root "undoes" the cube).

With either technique you find that the orbit of this asteroid has semimajor axis a = 4 AU.

Review: The period is greater than 1 year, so the semimajor axis is greater than 1 AU. Note that a = 4 AU is intermediate between the orbits of Mars and Jupiter (see Table 4-3). Many asteroids are known with semimajor axes in this range, forming a region in the solar system called the asteroid belt.

TOOLS OF THE ASTRONOMER'S TRADE and S.T.A.R.—A problem-solving rubric

All the worked examples in *Universe* (found in boxes called **Tools of the Astronomer's Trade**) follow a logical and consistent sequence of steps called **S.T.A.R.**: Assess the **S**ituation, select the **T**ools, find the **A**nswer, and **R**eview the answer and explore its significance. Feedback from instructors shows that when students follow these steps in their own work, they more rapidly become skillful problem solvers.

ASTRONOMY DOWN TO EARTH reveals the universal applicability of the laws of nature

Astronomy Down to Earth boxes illustrate how the same principles astronomers use to explain celestial phenomena can also explain everyday behavior here on Earth, from the color of the sky to why diet soft drink cans float in water.

BOX 5-3 Astronomy Down to Earth

Photons at the Supermarket

A beam of light can be regarded as a stream of tiny packets of energy called photons. The Planck relationships $E = hc/\lambda$ and $E = h\nu$ can be used to relate the energy E carried by a photon to its wavelength λ and frequency ν.

As an example, the laser bar-code scanners used at stores and supermarkets emit orange-red light of wavelength 633 nm. To calculate the energy of a single photon of this light, we must first express the wavelength in meters. A nanometer (nm) is equal to 10^{-9} m, so the wavelength is

$$\lambda = (633 \text{ nm})\left(\frac{10^{-9}\text{ m}}{1 \text{ nm}}\right) = 633 \times 10^{-9}\text{ m} = 6.33 \times 10^{-7}\text{ m}$$

Then, using the Planck formula $E = hc/\lambda$, we find that the energy of a single photon is

$$E = \frac{hc}{\lambda} = \frac{(6.625 \times 10^{-34}\text{ J s})(3 \times 10^{8}\text{ m/s})}{6.33 \times 10^{-7}\text{ m}} = 3.14 \times 10^{-19}\text{ J}$$

This amount of energy is very small. The laser in a typical bar-code scanner emits 10^{-3} joule of light energy per second, so the number of photons emitted per second is

$$\frac{10^{-3}\text{ joule per second}}{3.14 \times 10^{-19}\text{ joule per photon}} = 3.2 \times 10^{15}\text{ photons per second}$$

This number is so large that the laser beam seems like a continuous flow of energy rather than a stream of little energy packets.

Figure 14-20

The Largest Trans-Neptunian Objects This artist's impression depicts Earth and the largest objects known beyond Neptune (as of 2009) to the same scale. The largest of these, Eris, has a diameter of about 2900 km versus 2274 km for Pluto. Note the differences in color among these objects and that three of them (Eris, Pluto, and Haumea) have satellites of their own. Table 7-4 in Section 7-5 lists the properties of these objects. (NASA; ESA; and A. Field, STScI)

Up-to-date information shows the cutting edge of astronomy

Universe is up to date with the latest astronomical discoveries, ideas, and images. These include:

- New material on using transits to study extrasolar planets (Chapter 8)
- Updated discussion of global climate change, including historical data on global temperatures and CO_2 concentrations (Chapter 9)
- New information on the history of Martian water from the Mars Exploration Rovers (Chapter 11)
- New images and preliminary interpretations from flybys of Mercury by the MESSENGER spacecraft (Chapter 11)
- Cassini observations of the moons of Saturn (Chapter 13)
- New discoveries about trans-Neptunian objects (Chapter 14 and Chapter 15)
- Updated information about hypernovae and the origin of gamma ray bursts (Chapter 20)
- Chandra X-Ray Observatory images revealing the presence of supermassive black holes in galaxies (Chapter 24)
- Observations of the infrared background due to the first stars (Chapter 27)

7

Comparative Planetology I: Our Solar System

Fifty years ago, astronomers knew precious little about the worlds that orbit the Sun. Even the best telescopes provided images of the planets that were frustratingly hazy and indistinct. Of asteroids, comets, and the satellites of the planets, we knew even less.

Today, our knowledge of the solar system has grown exponentially, due almost entirely to robotic spacecraft. (The illustration shows an artist's impression of one such robotic explorer, the *Mars Reconnaissance Orbiter* spacecraft, as it approached Mars in 2006.) Spacecraft have been sent to fly past all the planets at close range, revealing details unimagined by astronomers of an earlier generation. We have landed spacecraft on the Moon, Venus, and Mars and dropped a probe into the immense atmosphere of Jupiter. This is truly the golden age of solar system exploration.

In this chapter we paint in broad outline our present understanding of the solar system. We will see that the planets come in a variety of sizes and chemical compositions. There is also rich variety among the moons of the planets and among smaller bodies we call asteroids, comets, and trans-Neptunian objects. We will investigate the nature of craters on the Moon and other worlds of the solar system. And by exploring the magnetic fields

Learning Goals

By reading the sections of this chapter, you will learn

7-1 The important differences between the two broad categories of planets: terrestrial and Jovian

7-2 The similarities and differences among the large planetary satellites, including Earth's Moon

7-3 How the spectrum of sunlight reflected from a planet reveals the composition of its atmosphere and surface

7-4 Why some planets have atmospheres and others do not

11

Mercury, Venus, and Mars: Earthlike yet Unique

(Mercury: NASA; Venus: NASA/JPL, MIT, and USGS; Mars: NASA, J. Bell/Cornell University, and M. Wolff/SSI)

These images show the three planets that share the inner solar system with our Earth. All three have solid surfaces, and, in principle, a properly protected astronaut could stand on any of them. The differences between these three worlds, however, are pronounced.

Mercury is small, airless, and extensively cratered. It is also mysterious: Half of its surface has never been viewed at close range, and it has a perplexing and unexpected magnetic field. It also rotates in a manner unique in the solar system, spinning three times on its axis for every two orbits around the Sun.

Venus is nearly the same size as Earth, but it is shrouded by a perpetual cloud cover that hides its surface from view. To penetrate the clouds and learn what lies beneath, two advanced technologies were needed: radar, which allowed astronomers to "see" through the clouds, and unmanned spacecraft, which orbited Venus at close range and even landed on its surface. The false-color topographic map shown here reveals that Venus has no true continents but merely highlands (shown in red) that rise gently above the planet's lower-lying areas (shown in blue). We have also learned that Venus's atmosphere is thick and searing hot, and that its clouds contain droplets of corrosive sulfuric acid.

Mars has captivated the popular imagination like no other planet. But rather than being the abode of warlike aliens, Mars proves to be an enigmatic world. Some parts of its surface are drier than any desert on Earth, while other locations show evidence of having been underwater for extended periods. Our challenge is to understand how these three nominally Earthlike worlds evolved to be so unique and so different than our own Earth.

Learning Goals

By reading the sections of this chapter, you will learn

11-1 What astronomers have learned by observing the terrestrial planets from Earth

11-2 The radically different ways in which Mercury, Venus, and Mars rotate on their axes

11-3 The outstanding features of Mercury, and why its magnetic field came as a surprise

11-4 How the advent of the space age transformed our understanding of Venus and Mars

11-5 How geologic activity took a very different form on Venus than on Earth, and why it essentially stopped on Mars

11-6 The key differences among the atmospheres of Earth, Venus, and Mars

11-7 How the atmospheres of Earth, Venus, and Mars evolved to their present states

11-8 The evidence that there was once liquid water on Mars

11-9 What we know about the two small satellites of Mars

255

Planetary Themes: Comparative Planetology and System Approach

The two chapters on comparative planetology—Chapter 7, Our Solar System, and Chapter 8, The Origin of the Solar System—have been revised and updated with the latest information. They contain explanations of the official distinctions between planets and dwarf planets, as well as new discoveries about objects in our solar system. Mercury, Venus, and Mars are covered in a single Chapter 11 on the terrestrial planets to emphasize their similarities to Earth and to each other. Comparative planetology figures are also throughout the planetary coverage. In addition to the comparative planetology theme, *Universe* takes a unique *systems* approach to planetary surfaces and atmospheres: Each planet's surface and atmosphere are discussed in tandem, where other texts address them separately. This underscores their interrelationships and builds a deeper understanding of the planets' overall characteristics.

End-of-Chapter material provides students with opportunities to review and assess what they have learned and opportunities to extend their knowledge

Key Words

A list of **Key words** appears at the end of each chapter, along with the number of the page where each term is introduced.

Key Ideas

Students can get the most benefit from these brief chapter summaries by using them in conjunction with the notes they take while reading.

Questions

Items from these sections are designed to be assigned as homework or used as jumping-off points for class discussion. Some questions ask students to analyze images in the text or evaluate how the mass media portrays concepts in astronomy. Advanced questions are accompanied by a **Problem-Solving Tips and Tools** box to provide guidance. Web/eBook questions challenge students to work with animations and interactive modules on the *Universe* Web site or eBook (described below) or to research topics on the World Wide Web. Near the end of the book is a section of **Answers to Selected Questions.**

Key Words

active optics, p. 140
adaptive optics, p. 140
angular resolution, p. 139
baseline, p. 140
Cassegrain focus, p. 137
charge-coupled device (CCD), p. 142
chromatic aberration, p. 134
coma, p. 138
coudé focus, p. 137
diffraction, p. 139
diffraction grating, p. 143
eyepiece lens, p. 131
false color, p. 140
focal length, p. 130
focal plane, p. 131
focal point, p. 130
focus (of a lens or mirror), p. 130
imaging, p. 142
interferometry, p. 140
lens, p. 129
light-gathering power, p. 132
light pollution, p. 141
magnification (magnifying power), p. 132
medium (*plural* media), p. 130
Newtonian reflector, p. 136
objective lens, p. 131
objective mirror (primary mirror), p. 135
optical telescope, p. 129
optical window (in Earth's atmosphere), p. 147
photometry, p. 142
pixel, p. 142
prime focus, p. 136
radio telescope, p. 144
radio window (in Earth's atmosphere), p. 147
reflecting telescope (reflector), p. 135
reflection, p. 134
refracting telescope (refractor), p. 131
refraction, p. 130
seeing disk, p. 139
spectrograph, p. 143
spectroscopy, p. 143
spherical aberration, p. 137
very-long-baseline interferometry (VLBI), p. 146

Key Ideas

Refracting Telescopes: Refracting telescopes, or refractors, produce images by bending light rays as they pass through glass lenses.

• Chromatic aberration is an optical defect whereby light of different wavelengths is bent in different amounts by a lens.

• Glass impurities, chromatic aberration, opacity to certain wavelengths, and structural difficulties make it inadvisable to build extremely large refractors.

Reflecting Telescopes: Reflecting telescopes, or reflectors, produce images by reflecting light rays to a focus point from curved mirrors.

• Reflectors are not subject to most of the problems that limit the useful size of refractors.

Angular Resolution: A telescope's angular resolution, which indicates ability to see fine details, is limited by two key factors.

• Diffraction is an intrinsic property of light waves. Its effects can be minimized by using a larger objective lens or mirror.

• The blurring effects of atmospheric turbulence can be minimized by placing the telescope atop a tall mountain with very smooth air. They can be dramatically reduced by the use of adaptive optics and can be eliminated entirely by placing the telescope in orbit.

Charge-Coupled Devices: Sensitive light detectors called charge-coupled devices (CCDs) are often used at a telescope's focus to record faint images.

Spectrographs: A spectrograph uses a diffraction grating to form the spectrum of an astronomical object.

Radio Telescopes: Radio telescopes use large reflecting dishes to focus radio waves onto a detector.

• Very large dishes provide reasonably sharp radio images. Higher resolution is achieved with interferometry techniques that link smaller dishes together.

Transparency of Earth's Atmosphere: Earth's atmosphere absorbs much of the radiation that arrives from space.

• The atmosphere is transparent chiefly in two wavelength ranges known as the optical window and the radio window. A few wavelengths in the near-infrared also reach the ground.

Telescopes in Space: For observations at wavelengths to which Earth's atmosphere is opaque, astronomers depend on telescopes carried above the atmosphere by rockets or spacecraft.

• Satellite-based observatories provide new information about the universe and permit coordinated observation of the sky at all wavelengths.

Questions

Review Questions

1. Describe refraction and reflection. Explain how these processes enable astronomers to build telescopes.
2. Explain why a flat piece of glass does not bring light to a focus while a curved piece of glass can.
3. Explain why the light rays that enter a telescope from an astronomical object are essentially parallel.
4. With the aid of a diagram, describe a refracting telescope. Which dimensions of the telescope determine its light-gathering power? Which dimensions determine the magnification?
5. What is the purpose of a telescope eyepiece? What aspect of the eyepiece determines the magnification of the image? In what circumstances would the eyepiece not be used?
6. Do most professional astronomers actually look through their telescopes? Why or why not?
7. Quite often advertisements appear for telescopes that extol their magnifying power. Is this a good criterion for evaluating telescopes? Explain your answer.
8. What is chromatic aberration? For what kinds of telescopes does it occur? How can it be corrected?
9. With the aid of a diagram, describe a reflecting telescope. Describe four different ways in which an astronomer can access the focal plane.
10. Explain some of the disadvantages of refracting telescopes compared to reflecting telescopes.
11. What kind of telescope would you use if you wanted to take a color photograph entirely free of chromatic aberration? Explain your answer.
12. Explain why a Cassegrain reflector can be substantially shorter than a refractor of the same focal length.

NEWS SCAN

ASTRONOMY

Dangling a COROT

Space telescope aims to find more planets orbiting other stars BY ALEXANDER HELLEMANS

(from Alexander Hellemans, "Dangling a COROT," *Scientific American*, September 2007, 32)

This year is shaping up to be an important one in the search for planets around other stars. In April astronomers at the Geneva Observatory announced the discovery of the most Earth-like exoplanet yet located, the first rocky world beyond our solar system that could hold liquid water. Just 1.5 times the size of Earth and possessing five times its mass, the planet circles the red dwarf Gliese 581. More such announcements will likely come in the months to follow, as the first space observatory dedicated to hunting exoplanets, called COROT, begins full operation and researchers complete their calculations.

In hunting for other worlds, scientists have proved adept at wringing the most from ground-based observatories. The Geneva astronomers found the rocky, Earth-like planet by first detecting the slight wobbling in the motion of its parent star, as star and planet circled around their common center of gravity. Such wobbling betrays itself as tiny Doppler shifts in the spectral lines present in a star's light. Although the investigators determined the exoplanet's mass via the observed movement, they could not directly gauge its size, because it does not transit—that is, pass in between Earth and its parent star. Hence, the team had to derive the exoplanet's density from planet formation models, team member Stéphane Udry reports.

Even if the planet moved across its star, the team might not have been able to see it. Peering through Earth's turbulent atmosphere requires adaptive optics to correct for distortions, but the same technology also precludes precise measurements of the dimming of light when a planet passes in front of a star. Even with future giant telescopes up to 42 meters in diameter, the sensitivity for such photometric detection will remain limited. "It is difficult to combine transit search with adaptive optics," Udry explains. "Adaptive optics continuously disturbs the photometric calibration."

That is where space observatories come in, such as the $46-million COROT (COnvection ROtation and planetary Transits). In addition to studying the ripples on the surface of stars to glean information about their interior, the observatory looks at changes in stellar luminosity that might indicate an exoplanetary transit. COROT began taking measurements in February, and, once fully calibrated, its 27-centimeter-wide telescope will detect light variations as low as one part in 20,000—200 times the resolution of ground-based instruments. COROT will ultimately look at 120,000 stars. "It is a hit-or-miss search, but we need a large number of stars so as to increase the probability of discovering planets," explains Pierre Barge of the Astrophysics Laboratory of Marseille in France, who leads the COROT exoplanet working group.

Besides finding exoplanets, COROT will also be able to determine their sizes, because the amount of dimming during a planet's occultation is proportional to that planet's girth. Moreover, knowing its size pins down its density and hence confirms whether the planet is rocky or gaseous. Currently, astronomers know the diameter of only about 20 of the approximately 240 known exoplanets.

Other space observatories can use transits to discover details about any atmosphere that might enshroud an exoplanet. In the July 12 *Nature*, Giovanna Tinetti of the European Space Agency and University College London and her colleagues report the discovery of water on a planet circling a star 64 light-years away, based on infrared measurements taken by NASA's Spitzer Space Telescope. They found that the way the light is absorbed by the planet's atmosphere when the body passes in front of its parent star matches the absorption characteristics of water vapor. But the planet is a gas giant, about 15 percent larger than Jupiter, so it probably does not harbor life.

Unfortunately, none of today's orbiting observatories, not even the Hubble Space Telescope, can find water on smaller worlds. "They are not sensitive enough to go for the smaller rocky planets," Tinetti says. She expects that such detections will come within the reach of the James Webb Space Telescope, Hubble's successor, which will be launched in 2013.

Space observatories will not put their terrestrial counterparts out of business in the search for exoplanets. Ground instruments can detect Doppler shifting quite well and are needed to corroborate findings from space. COROT has already spied possible new exoplanets, but researchers will not announce results until later this year. "We have to do a certain number of checks," Barge says. If the checks pan out, expect to hear a lot more otherworldly news.

Alexander Hellemans is a science writer based in Antwerp, Belgium.

Scientific American Articles

Chosen by the textbook authors to illuminate core concepts in the text, these brief, relevant selections demonstrate the process of science and discovery. Revealing recent developments related to specific topics in the text, the articles also provide touchstones for classroom discussion.

Guest Essays

Several chapters in *Universe* end with essays written by scientists involved with some of the recent discoveries described in the text. These include essays by Scott Sheppard on Pluto and the Kuiper belt and Kevin Plaxco on astrobiology. For a full list of essays, see the Contents Overview on pages vii and viii.

A revised selection of Observing Projects

The CD-ROM that comes free (upon request) with this text contains the *Starry Night*™ planetarium software from Imaginova Inc. Every chapter of *Universe* includes one or more **Observing Projects** that use this software package. Many projects are accompanied by an **Observing Tips and Tools** box to provide help and guidance.

> **58.** Use the *Starry Night Enthusiast*™ program to plan observations of Cygnus X-1. Use the **Find** pane to **Centre** the view on Cygnus X-1. Select **View > Show Daylight** from the menu. (**a**) Using the time controls in the toolbar, determine when Cygnus X-1 rises and sets on today's date from your home location. (**b**) At approximately what time on today's date is Cygnus X-1 highest in the sky? Is tonight a good night for observing this star with a visible-light telescope? Would it be better placed in the sky for observation six months from now? Explain how you determined this.

Three versions of the text meet the needs of different instructors

In addition to the complete 28-chapter version of *Universe*, two shorter versions are also available. *Universe: The Solar System* includes Chapters 1-16 and 28; it omits the chapters on stars, stellar evolution, galaxies, and cosmology. *Universe: Stars and Galaxies* includes Chapters 1-8 and 16-28; it omits the detailed chapters on the solar system but includes the overview of the solar system in Chapters 7 and 8.

Multimedia

Universe was developed to more deeply integrate learning resources, study aids, and multimedia tools into the use of the textbook. A variety of teaching and learning options provide an array of choices for students and instructors in their use of these materials, which were created based on input and contributions from a large number of faculty.

Premium Multimedia Resources

The **Interactive Tutorials**, described in greater detail above, were developed by astronomy education researchers to address the core concepts in the introductory course.

Electronic Versions

Universe is offered in two electronic versions. One is an **Interactive eBook**, and the other is pdf-based **eBook from CourseSmart**. Students and faculty can obtain the text directly from the publisher or through the bookstore. These options are provided to offer students and instructors flexibility in their use of course materials.

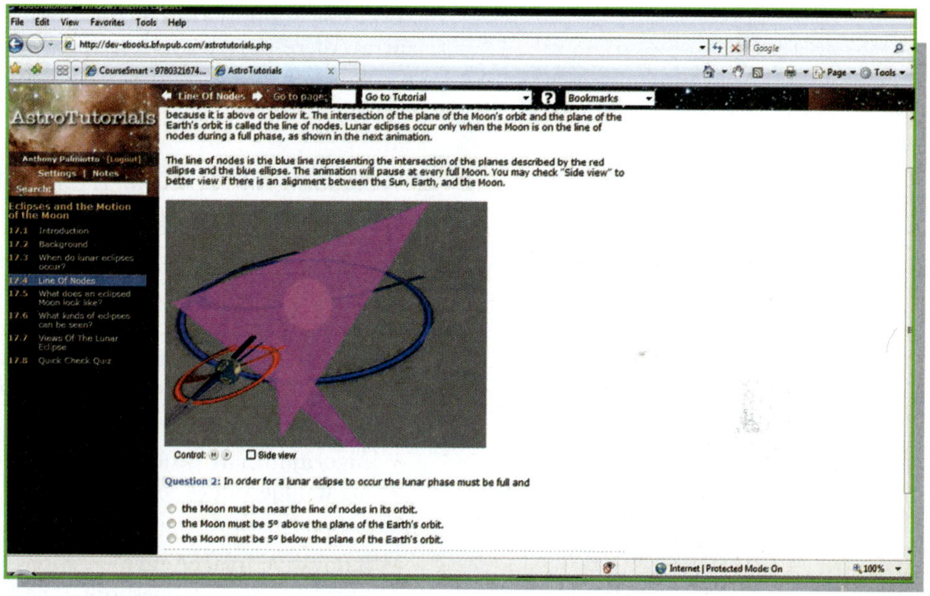

Interactive eBook

The **Interactive eBook** is a complete online version of the textbook with easy access to rich multimedia resources that support a complete student understanding. All text, graphics, tables, boxes, and end-of-chapter resources are included in the eBook, and the eBook provides instructors and students with powerful functionality to tailor their course resources to fit their needs.

- Allows seamless usage by students by using a page-to-page correlation between the online and printed versions. Each eBook page corresponds to a specific print-text page.
- Quick, intuitive navigation to any section of subsection
- Full text search, including the glossary and index
- Sticky-note feature allows users to place notes anywhere on a screen, and choose the note color for easy categorization
- "Top-note" feature allows users to place a prominent note at the top of the page to provide a more significant alert or reminder.
- Text highlighting, down to the level of individual phrases, in a variety of colors
- All instructor notes and highlights are delivered to their students' eBooks.
- Note subscription capability offers users the opportunity to incorporate another user's notes, particularly helpful if instructors or course coordinators prefer to align their notes.
- Instructor customizations can be saved from semester to semester.

Additional features for instructors

- Custom Chapter Selection: Instructors can customize the eBook to fit their syllabus by choosing only the chapters they'd like to cover; their students will then log in to this customized version of the eBook.
- Custom Content: Instructors notes can include text, Web links, images, and other materials, allowing instructors to place any content they choose within the textbook.
- Online quizzing: The online review quizzes from the Book Companion Site are integrated into the eBook.

CourseSmart eBook

Universe **CourseSmart eBook** offers the complete text in an easy-to-use, flexible format. Students have the choice to view the CourseSmart eBook online or to download it to their computer or a portable media player, such as an iPhone. To help students study and mirror the experience of a printed textbook, CourseSmart eBooks feature notetaking, highlighting, and bookmark features.

Online Learning Options

Universe supports instructors with a variety of online learning preferences. Its rich array of resources and platforms provides solutions according to each instructors' teaching method. Students can also access the resources through the **Companion Web Site.**

WebAssign Premium: Trusted Homework Management, Enhanced Learning Tools

For instructors interested in online homework management, **WebAssign Premium** features a time-tested, secure online environment already used by millions of students worldwide. Featuring algorithmic problem-generation and supported by a wealth of astronomy-specific learning tools, WebAssign Premium for *Universe* presents instructors with a powerful assignment manager and study environment. WebAssign Premium provides the following resources:

- **Algorithmically generated problems:** Students receive homework problems containing unique values for computation, encouraging them to work out the problems on their own.
- Innovative **ImageActive** questions, which allow students to directly interact with figures to answer questions. These problems are most often critical thinking questions, which ask the student to make connections across concepts and apply their knowledge.
- **Complete access to the interactive eBook**, from a live table of contents, as well as from relevant problem statements.
- **Personalized Study Plan:** This powerful study engine helps students improve their understanding by identifying areas where they need the most help, then delivering study aids to help them succeed. After taking an initial quiz, students receive a detailed study plan showing students the areas where they need to focus their attention; the study plan also presents animations, eBook sections, videos, and other resources tied to the concepts in need of further study.
- Links to Tutorials, eBook sections, animations, videos, and other interactive tools as hints and feedback to ensure a clearer understanding of the problems and the concepts they reinforce.

AstroPortal: Ease of use for instructors, success for students

AstroPortal can be used in introductory courses as student-driven compilation of study aids or a complete course management system, and anything in between.

- **Personalized Study Plan: Study > Assess > Review Targeted Concepts > Improve** This powerful study engine helps students improve their understanding by identifying areas where they need the most help, then delivering study aids to help them succeed. Students first take a quiz keyed to core concepts within a chapter of the text. The results of the quiz generate a detailed study plan showing students the areas where they need to focus their attention; the study plan also presents animations, eBook sections, videos, and other resources tied to the concepts in need of further study. Instructors can also access the results to determine student progress and adjust their lecture and other activities accordingly. The personalized study plan quizzes have been pre-made for each chapter of the text; instructors can edit the questions, assign the quizzes, and associate student activity with a grade.
- **Interactive eBook:** The online version of *Universe*, described in detail above, is included within AstroPortal and combines the text, all existing student media resources, and additional study features including highlighting, bookmarking, note-taking, Google™-style searching, in-text glossary definitions of terms and outlinks to Google.™
- **Assignments for Online Quizzing, Homework, and Self-Study.** Instructors can create and assign automatically-graded homework and quizzes from the complete test bank, which is pre-loaded in AstroPortal. It also includes Mastery Quizzes for each chapter. All quiz results feed directly into the instructor's gradebook
- ***Scientific American* Newsfeed:** To demonstrate the continued process of science and the exciting new developments in the field, the *Scientific American* Newsfeed delivers regularly updated material from the well-known magazine. Articles, podcasts, news briefs, and videos on subjects related to introductory geoscience are selected for inclusion by *Scientific American's* editors. The newsfeed provides several updates per week, and instructors can archive or assign the content they find most valuable.

Student Companion Web Site

The *Universe* Book Companion Web site, accessed through www.whfreeman.com/universe, provides a range of tools for student self-study and review. They include:

- All **60 Interactive Tutorials** provide multimedia-based learning experiences on the core concepts in the course.
- **Online Self-Study Quizzes** offer randomized questions and answers with instant feedback referring to specific sections in the text, to help students study, review, and prepare for exams. Instructors can access results through an online database or they can have them e-mailed directly to their accounts.
- **Animations** of key figures
- **NASA videos** of important processes and phenomena
- **Vocabulary and concept-review flashcards**
- **Interactive Drag and Drop Exercises,** based on text illustrations, help students grasp the vocabulary in context.

Starry Night™ Planetarium Software

***Starry Night*™** is a brilliantly realistic planetarium software package. It is designed for easy use by anyone with an interest in the night sky. See the sky from anywhere on Earth or lift off and visit any solar system body or any location up to 20,000 light years away. View 2,500,000 stars along with more than 170 deep-space objects like galaxies, star clusters, and nebulae. You can travel 15,000 years in time, check out the view from the International Space Station, and see planets up close from any one of their moons. Included are stunning OpenGL graphics. You can also print handy star charts to explore outside. The Starry Night™ CD is available at no extra charge with the text upon request; see the back cover for the ISBN.

Observing Projects Using Starry Night™ Planetarium Software

ISBN 1-4292-5886-1, T. Alan Clark and William J. F. Wilson, University of Calgary, and Marcel Bergman
Available for packaging with the text, and compatible with both PC and Mac, this book contains a variety of comprehensive lab activities for ***Starry Night*™** planetarium software.

Instructor's Resource CD-ROM

ISBN 1-4292-5872-1
To help instructors create lecture presentations, course Web sites, and other resources, this CD-ROM contains the following resources:

- All images from the text
- Instructor's Manual
- Test Bank
- Online resources from the text's companion Web site

Online Instructor's Manual

by Mark Hollabaugh, Normandale Community College
This manual includes solutions to all end-of-chapter problems, detailed chapter outlines, and classroom-tested teaching hints and strategies. The extensive resource guide covers student and instructor reading materials, audiovisual material, and discussion/paper topics.

Test Bank CD-ROM

Windows and Mac versions on one disc, ISBN: 1-4292-5888-8, by T. Alan Clark and William J. F. Wilson, University of Calgary, and Thomas Krause, Towson University
More than 3,500 multiple-choice questions are section-referenced. The easy-to-use CD-ROM version includes Windows and Mac versions on a single disc, in a format that lets you add, edit, re-sequence, and print questions to suit your needs.

Online Course Materials (WebCT, Blackboard)

As a service for adopters, we will provide content files in the appropriate online course format, including the instructor and student resources for this text. The files can be used as is or can be customized to fir specific needs. Course outlines, prebuilt quizzes, links, activities, and a whole array of materials are included.

PowerPoint Lecture Presentations

A set of online lecture presentations created in Powerpoint allows instructors to tailor their lectures to suit their own needs using images and notes from the textbook. These presentations are available on the instructor portion of the companion Web site.

Acknowledgements

We would like to thank my colleagues who have carefully scrutinized the manuscript of this and earlier editions. *Universe* is a stronger and better textbook because of their conscientious efforts:

Robert Antonucci, *University of California, Santa Barbara*
Ann Bragg, *Marietta College*
Richard Bowman, *Bridgewater College*
Robert Braunstein, *Northern Virginia Community College (Loudoun)*
Thomas Campbell, *Allen Community College*
Brian Carter, *Grossmont College*
Stephen Case, *Olivet Nazarene University*
Demian H. J. Cho, *Kenyon College*
Richard A. Christie, *Okanagan College*
Eric Collins, *California State University, Northridge*
George J. Corso, *DePaul University*
Jamie Day, *Transylvania University*
Steven Desch, *Arizona State University*
Peter Detterline, *Kutztown University*
James Dickinson, *Clackamas Community College*
Jeanne Digel, *Canada College*
Edward Dingler, *Southwest Virginia Community College*
David Duluk, *Los Angeles City College/Glendale Community College*
Jean W. Dupon, *Menlo College*
Stephanie Fawcett, *McKendree University*
Kent Fisher, *Columbus State Community College*
Juhan Frank, *Louisiana State University*
Tony George, *Columbia Basin College*
Erika Gibb, *University of Missouri - St. Louis*
Daniel Greenberger, *City College of New York*
Wayne K. Guinn, *Lon Morris College*
Joshua Gundersen, *University of Miami*
Hunt Guitar, *New Mexico State University*
Kathleen A. Harper, *Denison University*
Lynn Higgs, *University of Utah*
Melinda Hutson, *Portland Community College*
Douglas R. Ingram, *Texas Christian University*
Adam G. Jensen, *University of Colorado, Boulder*
Darell Johnson, *Missouri Western State University*
Michael Joner, *Brigham Young University*
Lauren Jones, *Denison University*
Steve Kawaler, *Iowa State University*
William Keel, *University of Alabama*
Charles Kerton, *Iowa State University*
Kevin Kimberlin, *Bradley* University
H. S. Krawczynski, *Washington University at St. Louis*
Lauren Likkel, *University of Wisconsin, Eau Claire*
Dennis Machnik, *Plymouth State University*
Franck Marchis, *University of California, Berkeley*
Anatoly Miroshnichenko, *University of North Carolina at Greensboro*
Michele M. Montgomery, University of Central Florida
Krishna Mukherjee, *Slippery Rock University*
Edward M. Murphy, *University of Virginia*
Donna Naples, *University of Pittsburgh*
Gerald H. Newsom, *The Ohio State University*
Brian Oetiker, *Sam Houston State University*
Ronald P. Olowin, *Saint Mary's College*
Michael J. O'Shea, *Kansas State University*
Fritz Osell, *Northern Oklahoma College*
J. Douglas Patterson, *Johnson County Community College*
Charles Peterson, *University of Missouri*
Ylva Pihlstrom, *University of New Mexico*
Richard Rand, *University of New Mexico*
Mike Reynolds, *Florida Community College*
Frederick Ringwald, *California State University, Fresno*
Todd Rimkus, *Marymount University*
Paul Robinson, *Westchester Community College*
Louis Rubbo, *Coastal Carolina University*
Kyla Scarborough, *Pittsburg State University*
Doug Showell, *University of Nebraska at Omaha*
Caroline Simpson, *Florida International University*
Glenn Spiczak, *University of Wisconsin - River Falls*
James D. Stickler, *Allegany College of Maryland*
Larry K. Smith, *Snow College*
Chris Taylor, *California State University, Sacramento*
Charles M. Telesco, University of Florida
Dale Trapp, *Concordia University, St. Paul*

We are particularly grateful to Kevin Plaxco and Scott Sheppard, whose essays for this edition expand its coverage. Their commentary on astrobiology and the controversy surrounding Pluto are valuable additions to the text.

Many others have participated in the preparation of this book and we thank them for their efforts. We are grateful for the guidance we have received from Anthony Palmiotto, our acquisitions editor. We're also grateful for the wide-ranging talents of our development editor on *Universe*, Kharissia Pettus, and for the careful and insightful contributions of her assistant, Anthony Petrites. Amy Thorne, the media and supplements editor, supervised the creation of the superb multimedia with support from Brittany Murphy. Special thanks go to Leigh Renhard, who skillfully kept the various versions on track and on target; and to Susan Wein for coordinating the vast array of resources needed to produce a book such as this. Diana Blume deserves full credit for the stunningly beautiful cover design as well as the handsome look and feel of the book. We also thank Bill Page and Imagineering Media Services for coordinating and producing the excellent artwork, and extend our gratitude for Bianca Moscatelli for finding just the right photographs. Louise Ketz deserves a medal for wading through the prose and copyediting it into proper English.

On a personal note, Roger Freedman would like to thank his father, Richard Freedman, for first cultivating an interest in space many years ago, and for his father's personal contributions to the exploration of the universe as an engineer for the Atlas and Centaur launch vehicle programs. Most of all, Roger thanks his charming wife, Caroline, for putting up with his long nights slaving over the computer!

Robert Geller would like to thank his parents for nurturing a love of science. Although some home appliances that were taken apart in the name of science were never put back together, he always had their support. Robert would also like to thank his wife Susanne for her support, even during her own efforts on the design and construction of the CMS detector for the Large Hadron Collider at CERN. He would also like to thank his seven-year-old daughter Zoe for her wonderful science questions each day at breakfast.

Although we have made a concerted effort to make this edition error-free, some mistakes may have crept in unbidden. We would appreciate hearing from anyone who finds an error or wishes to comment on the text.

Roger A. Freedman
Department of Physics
University of California, Santa Barbara
Santa Barbara CA 93106
airboy@physics.ucsb.edu

Robert M. Geller
Department of Physics
University of California, Santa Barbara
Santa Barbara CA 93106
rhmg@physics.ucsb.edu

To the Student

How To Get the Most From *Universe*

If you're like most students just opening this textbook, you're enrolled in one of the few science courses you'll take in college. As you study astronomy, you'll probably do relatively little reading compared to a literature or history course—at least in terms of the number of pages. But your readings will be packed with information, much of it new to you and (we hope) exciting. You can't read this textbook like a novel and expect to learn much from it. Don't worry, though. we wrote this book with you in mind. In this section, we'll suggest how *Universe* can help you succeed in your astronomy course, and take you on a guided tour of the book and media.

Apply these techniques to studying astronomy

- **Read before each lecture** You'll get the most out of your astronomy course if you read each chapter before hearing a lecture about its subject matter. That way, many of the topics will already be clear in your mind, and you'll understand the lecture better. You'll be able to spend more of your listening and note-taking time on the more challenging ideas presented in the lecture.
- **Take notes as you read and make use of office hours** Keep a notebook handy as you read, and write down the key points of each section so that you can review them later. If any parts of the section don't seem clear on first reading, make a note of them, too, including the page numbers. Once you've gone through the chapter, re-read it with special emphasis on the ideas that gave you trouble the first time. If you're still unsure after the lecture, consult your instructor, either during office hours or after class. Bring your notes with you so your instructor can see which concepts are giving you trouble. Once your instructor has helped clarify things for you, revise your notes so you'll remember your new-found insights. You'll end up with a chapter summary in your own words. This will be a tremendous help when studying for exams!
- **Make use of your fellow students** Many students find it useful to form study groups for astronomy. You can hash out challenging topics with each other and have a good time while you're doing it. But make sure that you write up your homework by yourself, because the penalties for copying or plagiarizing other students' work can be severe in the extreme. Some students find individual assistance useful too. If you think a tutor will be helpful, link up with one early. Getting a tutor late in the course, in the belief that you'll be able to catch up with what you missed earlier on, is almost always a lost cause.
- **Take advantage of the Web site and eBook** Take some time to explore the *Universe* Web site (www.whfreeman.com/universe). There you'll find review materials, animations, videos, interactive exercises, flashcards, and many other features keyed to chapters in *Universe.* All of these features are designed to help you learn and enjoy astronomy, so make sure to take full advantage of them.

  Your instructor may also have asked you to purchase the eBook version of this textbook (www.ebooks.bfwpub.com), which combines the complete content of the book and the Web site in a convenient online format.

- **Try astronomy for yourself with your star charts** At the back of this book, you'll find a set star charts for the each month of the year in the Northern Hemisphere. (For a set of Southern Hemisphere star charts, see the *Universe* Web site.) Star charts can get you started with your own observations of the universe. Hold the chart overhead in the same orientation as the compass points, with southern horizon toward the south and western horizon toward the west. (To save strain on your arms, you may want to cut these pages out of the book.) Depending on the version of this textbook that your instructor requested, this book may also include a CD-ROM with the easy-to-use *Starry Night Enthusiast*™ planetarium program, which you can use to view the sky on any date and time as seen from anywhere on Earth.

Here's the most important advice of all

We haven't mentioned the most important thing you should do when studying astronomy: Have fun! Of all the different kinds of scientists, astronomers are among the most excited about what they do and what they study. Let some of that excitement about the universe rub off on you, and you'll have a great time with this course and with this textbook.

In preparing this edition of *Universe*, we've tried very hard to make it the kind of textbook that a student like you will find useful. We're very interested in your comments and opinions! Please feel free to send us e-mail or write us, and we'll respond personally.

Best wishes for success in your studies!

Roger A. Freedman
Department of Physics
University of California, Santa Barbara
Santa Barbara CA 93106
airboy@physics.ucsb.edu

Robert M. Geller
Department of Physics
University of California, Santa Barbara
Santa Barbara CA 93106
rhmg@physics.ucsb.edu

1 Astronomy and the Universe

The night sky as seen from the European Southern Observatory in Chile. This picture is made from visible light, which is indicated by the highlighted V in the wavelength tab. (Hännes Heyer, ESO) R I **V** U X G

Imagine yourself in the desert on a clear, dark, moonless night, far from the glare of city lights. As you gaze upward, you see a panorama that no poet's words can truly describe and that no artist's brush could truly capture. Literally thousands of stars are scattered from horizon to horizon, many of them grouped into a luminous band called the Milky Way (which extends up and down across the middle of this photograph). As you watch, the entire spectacle swings slowly overhead from east to west as the night progresses.

For thousands of years people have looked up at the heavens and contemplated the universe. Like our ancestors, we find our thoughts turning to profound questions as we gaze at the stars. How was the universe created? Where did the Earth, Moon, and Sun come from? What are the planets and stars made of? And how do we fit in? What is our place in the cosmic scope of space and time?

Wondering about the universe is a key part of what makes us human. Our curiosity, our desire to explore and discover, and, most important, our ability to reason about what we have discovered are qualities that distinguish us from other animals. The study of the stars transcends all boundaries of culture, geography, and politics. In a literal sense, astronomy is a universal subject—its subject is the entire universe.

1-1 To understand the universe, astronomers use the laws of physics to construct testable theories and models

Astronomy has a rich heritage that dates back to the myths and legends of antiquity. Centuries ago, the heavens were thought to be populated with demons and heroes, gods and goddesses. Astronomical phenomena were explained as the result of supernatural forces and divine intervention.

The course of civilization was greatly affected by a profound realization: *The universe is comprehensible.* This awareness is one of the great gifts to come to us from ancient Greece. Greek astronomers discovered that by observing the heavens and carefully reasoning about what they saw, they could learn something about how the universe operates. For example, as we shall see in Chapter 3, they measured the size of Earth and were able to understand and predict eclipses without appealing to supernatural

Learning Goals

By reading the sections of this chapter, you will learn

1-1 What distinguishes the methods of science from other human activities

1-2 How exploring other planets provides insight into the origins of the solar system and the nature of our Earth

1-3 Stars have a life cycle—they form, evolve over millions or billions of years, and die

1-4 Stars are grouped into galaxies, which are found throughout the universe

1-5 How astronomers measure position and size of a celestial object

1-6 How to express very large or very small numbers in convenient notation

1-7 Why astronomers use different units to measure distances in space

1-8 What astronomy can tell us about our place in the universe forces. Modern science is a direct descendant of the intellectual endeavors of these ancient Greek pioneers.

TUTORIAL 1.1

The Scientific Method

Like art, music, or any other human creative activity, science makes use of intuition and experience. But the approach used by scientists to explore physical reality differs from other forms of intellectual endeavor in that it is based fundamentally on *observation, logic,* and *skepticism.* This approach, called the **scientific method,** requires that our ideas about the world around us be consistent with what we actually observe.

The scientific method goes something like this: A scientist trying to understand some observed phenomenon proposes a **hypothesis,** which is a collection of ideas that seems to explain what is observed. It is in developing hypotheses that scientists are at their most creative, imaginative, and intuitive. But their hypotheses must always agree with existing observations and experiments, because a discrepancy with what is observed implies that the hypothesis is wrong. (The exception is if the scientist thinks that the existing results are wrong and can give compelling evidence to show that they are wrong.) The scientist then uses logic to work out the implications of the hypothesis and to make predictions that can be tested. A hypothesis is on firm ground only after it has accurately forecast the results of new experiments or observations. (In practice, scientists typically go through these steps in a less linear fashion than we have described.)

Hypotheses, models, theories, and laws are essential parts of the scientific way of knowing

Scientists describe reality in terms of **models,** which are hypotheses that have withstood observational or experimental tests. A model tells us about the properties and behavior of some object or phenomenon. A familiar example is a model of the atom, which scientists picture as electrons orbiting a central nucleus. Another example, which we will encounter in Chapter 18, is a model that tells us about physical conditions (for example, temperature, pressure, and density) in the interior of the Sun (Figure 1-1). A well-developed model uses mathematics—one of the most powerful tools for logical thinking—to make detailed predictions. For example, a successful model of the Sun's interior should describe what the values of temperature, pressure, and density are at each depth within the Sun, as well as the relations between these quantities. For this reason, mathematics is one of the most important tools used by scientists.

A body of related hypotheses can be pieced together into a self-consistent description of nature called a **theory.** An example from Chapter 4 is the theory that the planets are held in their orbits around the Sun by the Sun's gravitational force (Figure 1-2). Without models and theories there is no understanding and no science, only collections of facts.

CAUTION! In everyday language the word "theory" is often used to mean an idea that looks good on paper, but has little to do with reality. In science, however, a good theory is one that explains reality very well and that can be applied to explain new observations. An excellent example is the theory of gravitation (Chapter 4), which was devised by the English scientist Isaac Newton in the late 1600s to explain the orbits of the six planets known at that time. When astronomers of later centuries discovered the planets Uranus and Neptune and the dwarf planet Pluto, they found that these planets also moved in accordance with Newton's theory. The same theory describes the motions of satellites around Earth as well as the orbits of planets around other stars (see Chapter 8).

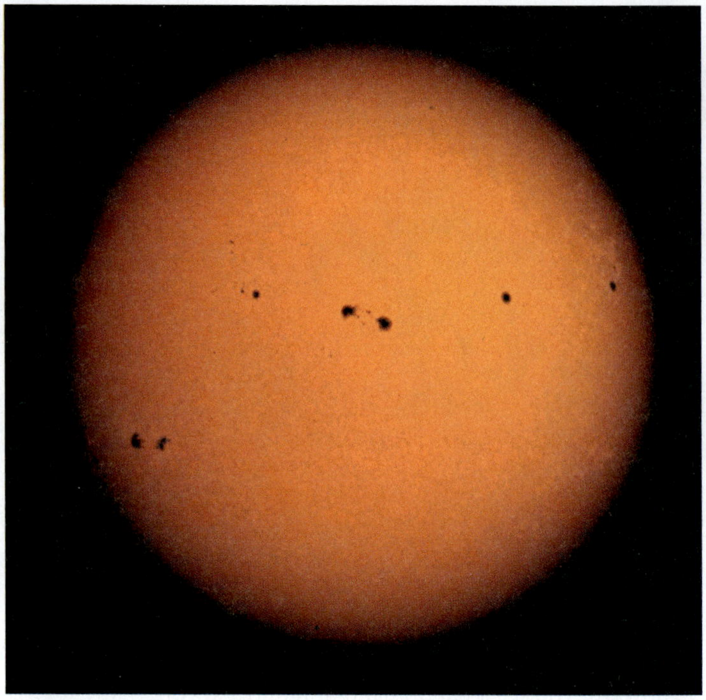

WEB LINK 1.1

Figure 1-1 R I V U X G

Our Star, the Sun The Sun is a typical star. Its diameter is about 1.39 million kilometers (roughly a million miles), and its surface temperature is about 5500°C (10,000°F). A detailed scientific model of the Sun tells us that it draws its energy from thermonuclear reactions occurring at its center, where the temperature is about 15 million degrees Celsius. (NSO/AURA/NSF)

An important part of a scientific theory is its ability to make predictions that can be tested by other scientists. If the predictions are verified by observation, that lends support to the theory and suggests that it might be correct. If the predictions are *not* verified, the theory needs to be modified or completely replaced. For example, an old theory held that the Sun and planets orbit around a stationary Earth. This theory led to certain predictions that could be checked by observation, as we will see in Chapter 4. In the early 1600s the Italian scientist Galileo Galilei used one of the first telescopes to show that these predictions were incorrect. As a result, the theory of a stationary Earth was rejected, eventually to be replaced by the modern picture shown in Figure 1-2 in which Earth and other planets orbit the Sun.

An idea that *cannot* be tested by observation or experiment does not qualify as a scientific theory. An example is the idea that there is a little man living in your refrigerator who turns the inside light on or off when you open and close the door. The little

Figure 1-2

Planets Orbiting the Sun An example of a scientific theory is the idea that Earth and planets orbit the Sun due to the Sun's gravitational attraction. This theory is universally accepted because it makes predictions that have been tested and confirmed by observation. (The Sun and planets are actually much smaller than this illustration would suggest.) (Detlev Van Ravenswaay/Science Photo Library)

man is invisible, weightless, and makes no sound, so you cannot detect his presence. While this is an amusing idea, it cannot be tested and so cannot be considered science.

Skepticism is an essential part of the scientific method. New hypotheses must be able to withstand the close scrutiny of other scientists. The more radical the hypothesis, the more skepticism and critical evaluation it will receive from the scientific community, because the general rule in science is that extraordinary claims require extraordinary evidence. That is why scientists as a rule do not accept claims that people have been abducted by aliens and taken aboard UFOs. The evidence presented for these claims is flimsy, secondhand, and unverifiable.

At the same time, scientists must be open-minded. They must be willing to discard long-held ideas if these ideas fail to agree with new observations and experiments, provided the new data have survived evaluation. (If an alien spacecraft really did land on Earth, scientists would be the first to accept that aliens existed—provided they could take a careful look at the spacecraft and its occupants.) That is why the validity of scientific knowledge is always temporary. As you go through this book, you will encounter many instances where new observations have transformed our understanding of Earth, the planets, the Sun and stars, and indeed the very structure of the universe.

Theories that accurately describe the workings of physical reality have a significant effect on civilization. For example, basing his conclusions in part on observations of how the planets orbit the Sun, Isaac Newton deduced a set of fundamental principles that describe how *all* objects move. These theoretical principles, which we will encounter in Chapter 4, work equally well on Earth as they do in the most distant corner of the universe. They represent our first complete, coherent description of the behavior of the physical universe. **Newtonian mechanics** had an immediate practical application in the construction of machines, buildings, and bridges. It is no coincidence that the Industrial Revolution followed hard on the heels of these theoretical and mathematical advances inspired by astronomy.

Newtonian mechanics and other physical theories have stood the test of time and been shown to have great and general validity. Proven theories of this kind are collectively referred to as the **laws of physics.** Astronomers use these laws to interpret and understand their observations of the universe. The laws governing light and its relationship to matter are of particular importance, because the only information we can gather about distant stars and galaxies is in the light that we receive from them. Using the physical laws that describe how objects absorb and emit light, astronomers have measured the temperature of the Sun and even learned what the Sun is made of. By analyzing starlight in the same way, they have discovered that our own Sun is a rather ordinary star and that the observable universe may contain billions of stars just like the Sun.

Technology in Science

An important part of science is the development of new tools for research and new techniques of observation. As an example, until fairly recently everything we knew about the distant universe was based on visible light. Astronomers would peer through telescopes to observe and analyze visible starlight. By the end of the nineteenth century, however, scientists had begun discovering forms of light invisible to the human eye: X rays, gamma rays, radio waves, microwaves, and ultraviolet and infrared radiation.

As we will see in Chapter 6, in recent years astronomers have constructed telescopes that can detect such nonvisible forms of light (Figure 1-3). These instruments give us views of the universe vastly different from anything our eyes can see. These new views have allowed us to see through the atmospheres of distant planets,

Figure 1-3 R I V U X G

A Telescope in Space Because it orbits outside Earth's atmosphere in the near-vacuum of space, the Hubble Space Telescope (HST) can detect not only visible light but also ultraviolet and near-infrared light coming from distant stars and galaxies. These forms of nonvisible light are absorbed by our atmosphere and hence are difficult or impossible to detect with a telescope on Earth's surface. This photo of HST was taken by the crew of the space shuttle *Columbia* after a servicing mission in 2002. (NASA)

to study the thin but incredibly violent gas that surrounds our Sun, and even to observe new solar systems being formed around distant stars. Aided by high-technology telescopes, today's astronomers carry on the program of careful observation and logical analysis begun thousands of years ago by their ancient Greek predecessors.

1-2 By exploring the planets, astronomers uncover clues about the formation of the solar system

The science of astronomy allows our intellects to voyage across the cosmos. We can think of three stages in this voyage: from Earth to other parts of the solar system, from the solar system to the stars, and from stars to galaxies and the grand scheme of the universe.

Studying planetary science gives us a better perspective on our own unique Earth

The star we call the Sun and all the celestial bodies that orbit the Sun—including Earth, the other planets, all their various moons, and smaller bodies such as asteroids and comets—make up the **solar system.** Since the 1960s a series of unmanned spacecraft has been sent to explore each of the planets (Figure 1-4). Using the remote "eyes" of such spacecraft, we have flown over Mercury's cratered surface, peered beneath Venus's poisonous cloud cover, and discovered enormous canyons and extinct volcanoes on Mars. We have found active volcanoes on a moon of Jupiter, probed the atmosphere of Saturn's moon Titan, seen the rings of Uranus up close, and looked down on the active atmosphere of Neptune.

Along with rocks brought back by the Apollo astronauts from the Moon (the only world beyond Earth visited by humans), new information from spacecraft has revolutionized our understanding of the origin and evolution of the solar system. We have come to realize that many of the planets and their satellites were shaped by collisions with other objects. Craters on the Moon and on many other worlds are the relics of innumerable impacts by bits of interplanetary rock. The Moon may itself be the result of a catastrophic collision between Earth and a planet-sized object shortly after the solar system was formed. Such a collision could have torn sufficient material from the primordial Earth to create the Moon.

The oldest objects found on Earth are **meteorites,** bits of interplanetary debris that sometimes fall to our planet's surface. By using radioactive age-dating techniques, scientists have found that all meteorites are 4.56 billion years old—older than any other rocks found on Earth or the Moon. The conclusion is that our entire solar system, including the Sun and planets, formed 4.56 billion years ago. The few thousand years of recorded human history is no more than the twinkling of an eye compared to the long history of our solar system.

The discoveries that we have made in our journeys across the solar system are directly relevant to the quality of human life on our own planet. Until recently, our understanding of geology, weather, and climate was based solely on data from Earth. Since the advent of space exploration, however, we have been able to

Figure 1-4

The Sun and Planets to Scale This montage of images from various spacecraft and ground-based telescopes shows the relative sizes of the planets and the Sun. The Sun is so large compared to the planets that only a portion of it fits into this illustration. The distances from the Sun to each planet are *not* shown to scale; the actual distance from the Sun to Earth, for instance, is 12,000 times greater than Earth's diameter. (Calvin J. Hamilton and NASA/JPL) compare and contrast other worlds with our own. This new knowledge gives us valuable insight into our origins, the nature of our planetary home, and the limits of our natural resources.

1-3 By studying stars and nebulae, astronomers discover how stars are born, grow old, and die

The nearest of all stars to Earth is the Sun. Although humans have used the Sun's warmth since the dawn of our species, it was only in the 1920s and 1930s that physicists figured out how the Sun shines. At the center of the Sun, thermonuclear reactions—so called because they require extremely high temperatures—convert hydrogen (the Sun's primary constituent) into helium. This violent process releases a vast amount of energy, which eventually makes its way to the Sun's surface and escapes as light (see Figure 1-1). All the stars you can see in the nighttime sky also shine by thermonuclear reactions (Figure 1-5). By 1950 physicists could reproduce such thermonuclear reactions here on Earth in the form of a hydrogen bomb (Figure 1-6). And in the future, converting hydrogen into helium might someday offer a cleaner method for the production of nuclear energy.

Figure 1-6 R I V U X G

A Thermonuclear Explosion A hydrogen bomb uses the same physical principle as the thermonuclear reactions at the Sun's center: the conversion of matter into energy by nuclear reactions. This thermonuclear detonation on October 31, 1952, had an energy output equivalent to 10.4 million tons of TNT. This is a mere ten-billionth of the amount of energy released by the Sun in one second. (Defense Nuclear Agency)

Figure 1-5 R I V U X G

Stars like Grains of Sand This Hubble Space Telescope image shows thousands of stars in the constellation Sagittarius. Each star shines because of thermonuclear reactions that release energy in its interior. Different colors indicate stars with different surface temperatures: stars with the hottest surfaces appear blue, while those with the coolest surfaces appear red. (The Hubble Heritage Team, AURA/STScI/NASA)

Thermonuclear reactions consume the material of which stars are made, which means that stars cannot last forever. Rather, they must form, evolve, and eventually die.

CAUTION! Astronomers often use biological terms such as "birth" and "death" to describe stages in the evolution of inanimate objects like stars. Keep in mind that such terms are used only as *analogies,* which help us visualize these stages. They are not to be taken literally!

The Life Stories of Stars

The rate at which stars emit energy in the form of light tells us how rapidly they are consuming their thermonuclear "fuel," and hence how long they can continue to shine before reaching the end of their life spans. More massive stars have more thermonuclear "fuel," but consume it at such a prodigious rate that they live out their lives in just a few million years. Less massive stars have less material to consume, but their thermonuclear reactions proceed so slowly that their life spans are measured in billions of years. (Our own star, the Sun, is in early middle age: it is 4.56 billion years old, with a lifetime of 12.5 billion years.)

While no astronomer can watch a single star go through all of its life stages, we have been able to piece together the life stories of stars by observing many different stars at different points in their life cycles. Important pieces of the puzzle have been discovered by studying huge clouds of interstellar gas, called **nebulae** (singular **nebula**), which are found scattered across the sky.

Within some nebulae, such as the Orion Nebula shown in Figure 1-7, stars are born from the material of the nebula itself. Other nebulae reveal what happens when thermonuclear reactions stop and a star dies. Some stars that are far more massive than the Sun end their lives with a spectacular detonation called a **supernova** (plural **supernovae**) that blows the star apart. The Crab Nebula (Figure 1-8) is a striking example of a remnant left behind by a supernova.

Studying the life cycles of stars is crucial for understanding our own origins

Dying stars can produce some of the strangest objects in the sky. Some dead stars become **pulsars,** which spin rapidly at rates of tens or hundreds of rotations per second. And some stars end their lives as almost inconceivably dense objects called **black holes,** whose gravity is so powerful that nothing—not even light—can escape. Even though a black hole itself emits essentially no radiation, a number of black holes have been discovered by Earth-orbiting telescopes. This is done by detecting the X rays emitted by gases falling toward a black hole.

During their death throes, stars return the gas of which they are made to interstellar space. (Figure 1-8 shows these expelled gases expanding away from the site of a supernova explosion.) This gas contains heavy elements—that is, elements heavier than hydrogen and helium—that were created during the star's lifetime by thermonuclear reactions in its interior. Interstellar space thus becomes enriched with newly manufactured atoms and molecules. The Sun and its planets were formed from interstellar material that was enriched in this way. This means that the atoms of iron and nickel that make up Earth, as well as the carbon in our bodies and the oxygen we breathe, were created deep inside ancient stars. By studying stars and their evolution, we are really studying our own origins.

WEB LINK 1.5

Figure 1-8 R I V U X G

The Crab Nebula—Wreckage of an Exploded Star When a dying star exploded in a supernova, it left behind this elegant funeral shroud of glowing gases blasted violently into space. A thousand years after the explosion these gases are still moving outward at about 1800 kilometers per second (roughly 4 million miles per hour). The Crab Nebula is 6500 light-years from Earth and about 13 light-years across. (NASA, ESA, J. Hester and A. Loll/Arizona State University)

WEB LINK 1.4

Figure 1-7 R I V U X G

The Orion Nebula—Birthplace of Stars This beautiful nebula is a stellar "nursery" where stars are formed out of the nebula's gas. Intense ultraviolet light from newborn stars excites the surrounding gas and causes it to glow. Many of the stars embedded in this nebula are less than a million years old, a brief interval in the lifetime of a typical star. The Orion Nebula is some 1500 light-years from Earth and is about 30 light-years across. (NASA, ESA, M. Robberto/STScI/ESA, and the Hubble Space Telescope Orion Treasury Project Team)

1-4 By observing galaxies, astronomers learn about the origin and fate of the universe

Stars are not spread uniformly across the universe but are grouped together in huge assemblages called **galaxies.** Galaxies come in a wide range of shapes and sizes. A typical galaxy, like the Milky Way, of which our Sun is part, contains several hundred billion stars. Some galaxies are much smaller, containing only a few million stars. Others are monstrosities that devour neighboring galaxies in a process called "galactic cannibalism."

WEB LINK 1.6

Figure 1-9 R I V U X G

A Galaxy This spectacular galaxy, called M63, contains about a hundred billion stars. M63 has a diameter of about 60,000 light-years and is located about 35 million light-years from Earth. Along this galaxy's spiral arms you can see a number of glowing clumps. Like the Orion Nebula in our own Milky Way Galaxy (see Figure 1-7), these are sites of active star formation. (Subaru Telescope, National Astronomical Observatory of Japan)

Our Milky Way Galaxy has arching spiral arms like those of the galaxy shown in Figure 1-9. These arms are particularly active sites of star formation. In recent years, astronomers have discovered a mysterious object at the center of the Milky Way with a mass millions of times greater than that of our Sun. It now seems certain that this curious object is an enormous black hole.

Some of the most intriguing galaxies appear to be in the throes of violent convulsions and are rapidly expelling matter. The centers of these strange galaxies, which may harbor even more massive black holes, are often powerful sources of X rays and radio waves.

Even more awesome sources of energy are found still deeper in space. At distances so great that their light takes billions of years to reach Earth, we find the mysterious **quasars.** Although quasars look like nearby stars (Figure 1-10), they are among the most distant and most luminous objects in the sky. A typical quasar shines with the brilliance of a hundred galaxies. Detailed observations of quasars imply that they draw their energy from material falling into enormous black holes.

Galaxies and the Expanding Universe

The motions of distant galaxies reveal that they are moving away from us and from each other. In other words, the universe is *expanding*. Extrapolating into the past, we learn that the universe must have been born from an incredibly dense state some 13.7 billion years ago. A variety of evidence indicates that at that moment—the beginning of time—the universe began with a cosmic explosion, known as the **Big Bang,** which occurred throughout all space.

The motions of distant galaxies motivate the ideas of the expanding universe and the Big Bang

Thanks to the combined efforts of astronomers and physicists, we are making steady advances in understanding the nature and history of the universe. This understanding may reveal the origin of some of the most basic properties of physical reality. Studying the most remote galaxies is also helping to answer questions about the ultimate fate of the universe. Such studies suggest that the expansion of the universe will continue forever, and is actually gaining speed.

LOOKING DEEPER 1.1

The work of unraveling the deepest mysteries of the universe requires specialized tools, including telescopes, spacecraft, and computers. But for many purposes the most useful device for studying the universe is the human brain itself. Our goal in this book is to help you use *your* brain to share in the excitement of scientific discovery.

In the remainder of this chapter we introduce some of the key concepts and mathematics that we will use in subsequent

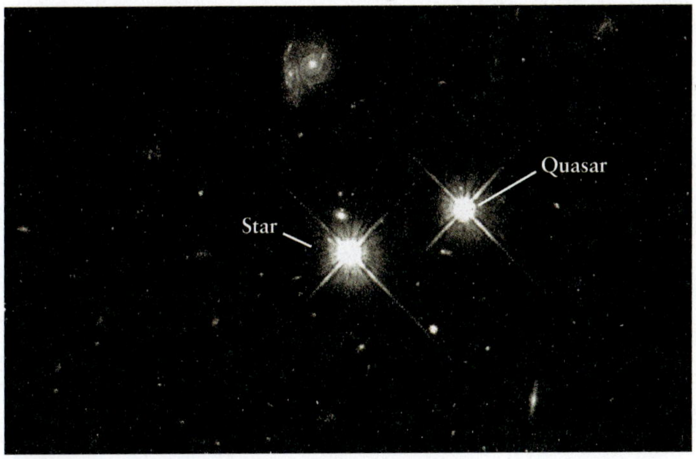

WEB LINK 1.7

Figure 1-10 R I V U X G

A Quasar The two bright starlike objects in this image look almost identical, but they are dramatically different. The object on the left is indeed a star that lies a few hundred light-years from Earth. But the "star" on the right is actually a quasar about 9 billion light-years away. To appear so bright even though they are so distant, quasars like this one must be some of the most luminous objects in the universe. The other objects in this image are galaxies like that in Figure 1-9. (Charles Steidel, California Institute of Technology; and NASA)

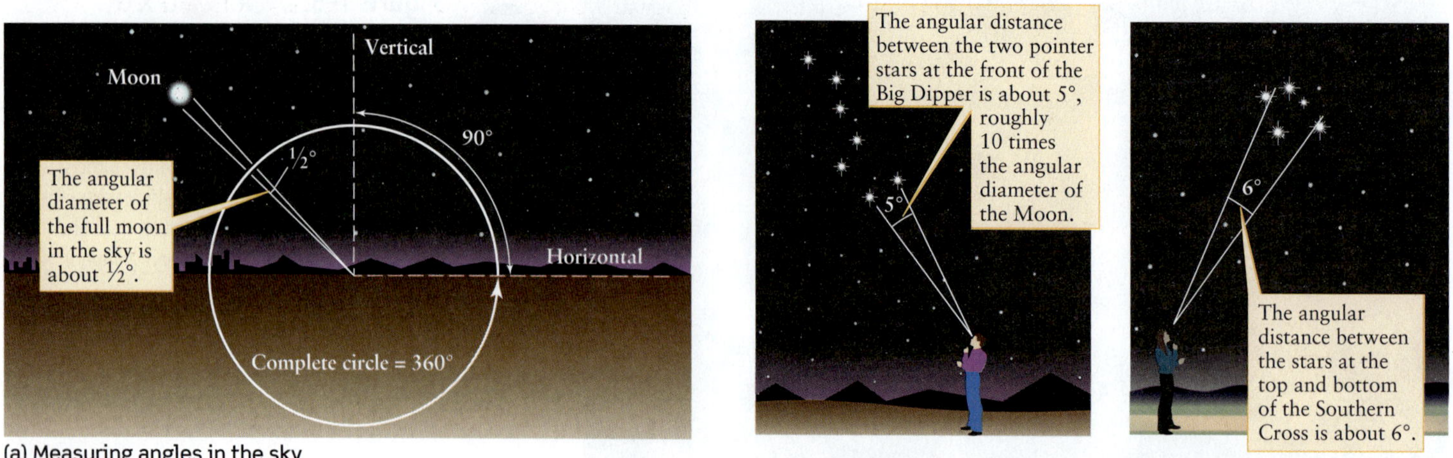

(a) Measuring angles in the sky

(b) Angular distances in the northern hemisphere

(c) Angular distances in the southern hemisphere

Figure 1-11

Angles and Angular Measure **(a)** Angles are measured in degrees (°). There are 360° in a complete circle and 90° in a right angle. For example, the angle between the vertical direction (directly above you) and the horizontal direction (toward the horizon) is 90°. The angular diameter of the full moon in the sky is about $\frac{1}{2}$°. **(b)** The seven bright stars that make up the Big Dipper can be seen from anywhere in the northern hemisphere. The angular distance between the two "pointer stars" at the front of the Big Dipper is about 5°. **(c)** The four bright stars that make up the Southern Cross can be seen from anywhere in the southern hemisphere. The angular distance between the stars at the top and bottom of the cross is about 6°.

chapters. Study these carefully, for you will use them over and over again throughout your own study of astronomy.

1-5 Astronomers use angles to denote the positions and apparent sizes of objects in the sky

Whether they study planets, stars, galaxies, or the very origins of the universe, astronomers must know where to point their telescopes. For this reason, an important part of astronomy is keeping track of the positions of objects in the sky. Angles and a system of angular measure are essential parts for this aspect of astronomy (Figure 1-11).

Angular measure is a tool that we will use throughout our study of astronomy

An **angle** is the opening between two lines that meet at a point. **Angular measure** describes the size of an angle exactly. A basic unit of angular measure is the **degree,** designated by the symbol °. A full circle is divided into 360°, and a right angle measures 90° (Figure 1-11*a*). As Figure 1-11*b* shows, if you draw lines from your eye to each of the two "pointer stars" in the Big Dipper, the angle between these lines—that is, the **angular distance** between these two stars—is about 5°. (In Chapter 2 we will see that these two stars "point" to Polaris, the North Star.) The angular distance between the stars that make up the top and bottom of the Southern Cross, which is visible from south of the equator, is about 6° (Figure 1-11*c*).

Astronomers also use angular measure to describe the apparent size of a celestial object—that is, how wide the object appears on the sky. For example, the angle covered by the diameter of the full moon is about $\frac{1}{2}$° (Figure 1-11*a*). We therefore say that the **angular diameter** (or **angular size**) of the Moon is $\frac{1}{2}$°. Alternatively, astronomers say that the Moon **subtends** an angle of $\frac{1}{2}$°. Ten full moons could fit side by side between the two pointer stars in the Big Dipper.

The average adult human hand held at arm's length provides a means of estimating angles, as Figure 1-12 shows. For example, a fist covers an angle of about 10°, whereas a finger tip is about 1° wide. You can use various segments of your index finger extended to arm's length to estimate angles a few degrees across.

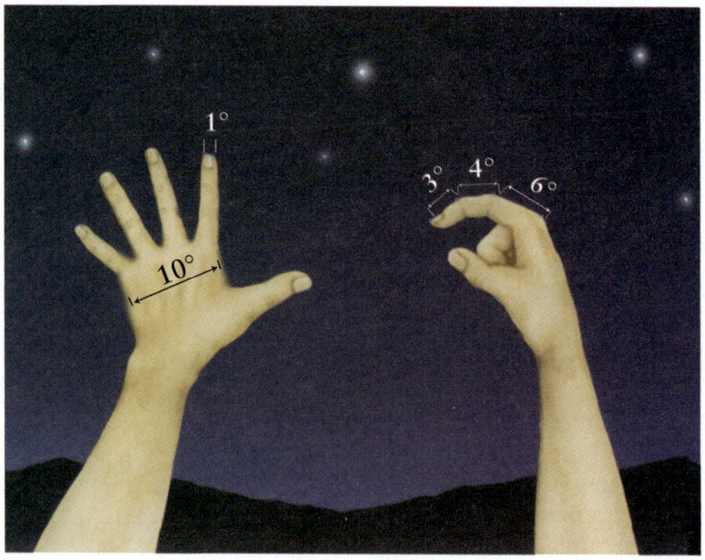

Figure 1-12

Estimating Angles with Your Hand The adult human hand extended to arm's length can be used to estimate angular distances and angular sizes in the sky.

To talk about smaller angles, we subdivide the degree into 60 **arcminutes** (also called minutes of arc), which is commonly abbreviated as 60 arcmin or 60′. An arcminute is further subdivided into 60 **arcseconds** (or seconds of arc), usually written as 60 arcsec or 60″. Thus,

$$1° = 60 \text{ arcmin} = 60'$$

$$1' = 60 \text{ arcsec} = 60''$$

For example, on January 1, 2007, the planet Saturn had an angular diameter of 19.6 arcsec as viewed from Earth. That is a convenient, precise statement of how big the planet appeared in Earth's sky on that date. (Because this angular diameter is so small, to the naked eye Saturn appears simply as a point of light. To see any detail on Saturn, such as the planet's rings, requires a telescope.)

If we know the angular size of an object as well as the distance to that object, we can determine the actual linear size of the object (measured in kilometers or miles, for example). Box 1-1 describes how this is done.

BOX 1-1 — Tools of the Astronomer's Trade

The Small-Angle Formula

You can estimate the angular sizes of objects in the sky with your hand and fingers (see Figure 1-12). Using rather more sophisticated equipment, astronomers can measure angular sizes to a fraction of an arcsecond. Keep in mind, however, that *angular* size is not the same as *actual* size. As an example, if you extend your arm while looking at a full moon, you can completely cover the Moon with your thumb. That's because from your perspective, your thumb has a larger angular size (that is, it subtends a larger angle) than the Moon. But the actual size of your thumb (about 2 centimeters) is much less than the actual diameter of the Moon (more than 3000 kilometers).

The accompanying figure shows how the angular size of an object is related to its linear size. Part **a** of the figure shows that for a given angular size, the more distant the object, the larger its actual size. For example, your finger tip held at arm's length covers the full moon, but the Moon is much farther away and is far larger in linear size. Part **b** shows that for a given linear size, the angular size decreases the farther away the object. This is why a car looks smaller and smaller as it drives away from you.

We can put these relationships together into a single mathematical expression called the small-angle formula. Suppose that an object subtends an angle α (the Greek letter alpha) and is at a distance d from the observer, as in part **c** of the figure. If the angle α is small, as is almost always the case for objects in the sky, the linear size (D) of the object is given by the following expression:

The small-angle formula

$$D = \frac{\alpha d}{206{,}265}$$

D = linear size of an object

α = angular size of the object, in arcsec

d = distance to the object

The number 206,265 is required in the formula and is equal to the number of arcseconds in a complete circle (that is, 360°) divided by the number 2π (the ratio of the circumference of a circle to that circle's radius).

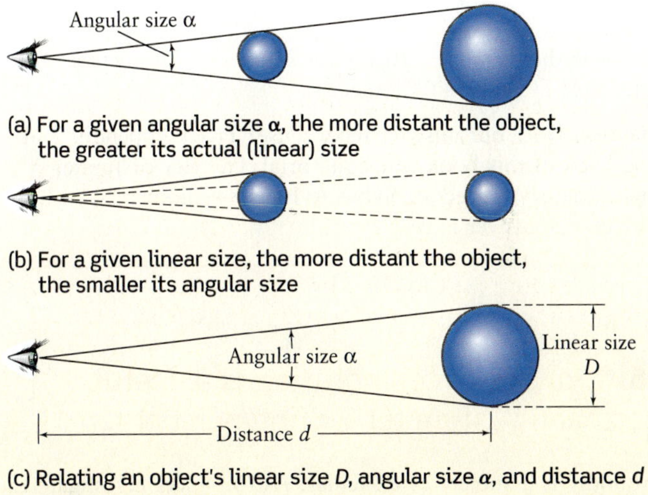

Angular size α
(a) For a given angular size α, the more distant the object, the greater its actual (linear) size
(b) For a given linear size, the more distant the object, the smaller its angular size
Angular size α
Linear size D
Distance d
(c) Relating an object's linear size D, angular size α, and distance d

(a) Two objects that have the same angular size may have different linear sizes if they are at different distances from the observer. **(b)** For an object of a given linear size, the angular size is smaller the farther the object is from the observer. **(c)** The small-angle formula relates the linear size D of an object to its angular size α and its distance d from the observer.

The following examples show two different ways to use the small-angle formula. In both examples we follow a four-step process: Evaluate the ***situation*** given in the example, decide which ***tools*** are needed to solve the problem, use those tools to find the ***answer*** to the problem, and ***review*** the result to see what it tells you. Throughout this book, we'll use these same four steps in *all* examples that require the use of formulas. We encourage you to follow this four-step process when solving problems for homework or exams. You can remember these steps by their acronym: *S.T.A.R.*

EXAMPLE: On December 11, 2006, Jupiter was 944 million kilometers from Earth and had an angular diameter of 31.2 arcsec. From this information, calculate the actual diameter of Jupiter in kilometers.

(continued on the next page)

BOX 1-1 *(continued)*

Situation: The astronomical object in this example is Jupiter, and we are given its distance d and its angular size α (the same as angular diameter). Our goal is to find Jupiter's diameter D.

Tools: The equation to use is the small-angle formula, which relates the quantities d, α, and D. Note that when using this formula, the angular size α *must* be expressed in arcseconds.

Answer: The small-angle formula as given is an equation for D. Plugging in the given values $\alpha = 31.2$ arcsec and $d = 944$ million kilometers,

$$D = \frac{31.2 \times 944{,}000{,}000 \text{ km}}{206{,}265} = 143{,}000 \text{ km}$$

Because the distance d to Jupiter is given in kilometers, the diameter D is also in kilometers.

Review: Does our answer make sense? From Appendix 2 at the back of this book, the equatorial diameter of Jupiter measured by spacecraft flybys is 142,984 kilometers, so our calculated answer is very close.

EXAMPLE: Under excellent conditions, a telescope on Earth can see details with an angular size as small as 1 arcsec. What is the greatest distance at which you could see details as small as 1.7 meters (the height of a typical person) under these conditions?

Situation: Now the object in question is a person, whose linear size D we are given. Our goal is to find the distance d at which the person has an angular size α equal to 1 arcsec.

Tools: Again we use the small-angle formula to relate d, α, and D.

Answer: We first rewrite the formula to solve for the distance d, then plug in the given values $D = 1.7$ m and $\alpha = 1$ arcsec:

$$d = \frac{206{,}265 D}{\alpha} = \frac{206{,}265 \times 1.7 \text{ m}}{1} = 350{,}000 \text{ m} = 350 \text{ km}$$

Review: This value is much less than the distance to the Moon, which is 384,000 kilometers. Thus, even the best telescope on Earth could not be used to see an astronaut walking on the surface of the Moon.

1-6 Powers-of-ten notation is a useful shorthand system for writing numbers

Astronomy is a subject of extremes. Astronomers investigate the largest structures in the universe, including galaxies and clusters of galaxies. But they must also study atoms and atomic nuclei, among the smallest objects in the universe, in order to explain how and why stars shine. They also study conditions ranging from the incredibly hot and dense centers of stars to the frigid near-vacuum of interstellar space. To describe such a wide range of phenomena, we need an equally wide range of both large and small numbers.

Learning powers-of-ten notation will help you deal with very large and very small numbers

Powers-of-Ten Notation: Large Numbers

Astronomers avoid such confusing terms as "a million billion billion" by using a standard shorthand system called **powers-of-ten notation.** All the cumbersome zeros that accompany a large number are consolidated into one term consisting of 10 followed by an **exponent,** which is written as a superscript. The exponent indicates how many zeros you would need to write out the long form of the number. Thus,

$10^0 = 1$ (one)

$10^1 = 10$ (ten)

$10^2 = 100$ (one hundred)

$10^3 = 1000$ (one thousand)

$10^4 = 10{,}000$ (ten thousand)

$10^6 = 1{,}000{,}000$ (one million)

$10^9 = 1{,}000{,}000{,}000$ (one billion)

$10^{12} = 1{,}000{,}000{,}000{,}000$ (one trillion)

and so forth. The exponent also tells you how many tens must be multiplied together to give the desired number, which is why the exponent is also called the **power of ten.** For example, ten thousand can be written as 10^4 ("ten to the fourth" or "ten to the fourth power") because $10^4 = 10 \times 10 \times 10 \times 10 = 10{,}000$.

In powers-of-ten notation, numbers are written as a figure between one and ten multiplied by the appropriate power of ten. The approximate distance between Earth and the Sun, for example, can be written as 1.5×10^8 kilometers (or 1.5×10^8 km for short). Once you get used to it, this is more convenient than writing "150,000,000 kilometers" or "one hundred and fifty million kilometers." (The same number could also be written as 15×10^7 or 0.15×10^9, but the preferred form is *always* to have the first figure be between 1 and 10.)

Calculators and Powers-of-Ten Notation

Most electronic calculators use a shorthand for powers-of-ten notation. To enter the number 1.5×10^8, you first enter 1.5, then press a key labeled "EXP" or "EE," then enter the exponent 8. (The EXP or EE key takes care of the "$\times$ 10" part of the expression.) The number will then appear on your calculator's display as "1.5 E 8," "1.5 8," or some variation of this; typically the "$\times$ 10" is not displayed as such. There are some variations from one kind of calculator to another, so you should spend a few minutes reading over your calculator's instruction manual to make sure you know the correct procedure for working with numbers in powers-of-ten notation. You will be using this notation continually in your study of astronomy, so this is time well spent.

CAUTION! Confusion can result from the way that calculators display powers-of-ten notation. Since 1.5×10^8 is displayed as "1.5 8" or "1.5 E 8," it is not uncommon to think that 1.5×10^8 is the same as 1.5^8. That is not correct, however; 1.5^8 is equal to 1.5 multiplied by itself 8 times, or 25.63, which is not even close to $150{,}000{,}000 = 1.5 \times 10^8$. Another, not uncommon, mistake is to write 1.5×10^8 as 15^8. If you are inclined to do this, perhaps you are thinking that you can multiply 1.5 by 10, then tack on the exponent later. Another process also does not work; 15^8 is equal to 15 multiplied by itself 8 times, or 2,562,890,625, which again is nowhere near 1.5×10^8. Reading over the manual for your calculator will help you to avoid these common errors.

Powers-of-Ten Notation: Small Numbers

You can use powers-of-ten notation for numbers that are less than one by using a minus sign in front of the exponent. A negative exponent tells you to *divide* by the appropriate number of tens. For example, 10^{-2} ("ten to the minus two") means to divide by 10 twice, so $10^{-2} = 1/10 \times 1/10 = 1/100 = 0.01$. This same idea tells us how to interpret other negative powers of ten:

$10^0 = 1$ (one)

$10^{-1} = 1/10 = 0.1$ (one tenth)

$10^{-2} = 1/10 \times 1/10 = 1/10^2 = 0.01$ (one hundredth)

$10^{-3} = 1/10 \times 1/10 \times 1/10 = 1/10^3 = 0.001$ (one thousandth)

$10^{-4} = 1/10 \times 1/10 \times 1/10 \times 1/10 = 1/10^4 = 0.0001$ (one ten-thousandth)

$10^{-6} = 1/10 \times 1/10 \times 1/10 \times 1/10 \times 1/10 \times 1/10 = 1/10^6 = 0.000001$ (one millionth)

$10^{-12} = 1/10 \times 1/10 \times 1/10 \times 1/10 \times 1/10 \times 1/10 \times 1/10 \times 1/10 \times 1/10 \times 1/10 \times 1/10 \times 1/10 = 1/10^{12} = 0.000000000001$ (one trillionth)

and so forth.

As these examples show, negative exponents tell you how many tenths must be multiplied together to give the desired number. For example, one ten-thousandth, or 0.0001, can be written as 10^{-4} ("ten to the minus four") because $10^{-4} = 1/10 \times 1/10 \times 1/10 \times 1/10 = 0.0001$.

A useful shortcut in converting a decimal to powers-of-ten notation is to notice where the decimal point is. For example, the decimal point in 0.0001 is four places to the left of the "1," so the exponent is -4, that is, $0.0001 = 10^{-4}$.

You can also use powers-of-ten notation to express a number like 0.00245, which is not a multiple of 1/10. For example, $0.00245 = 2.45 \times 0.001 = 2.45 \times 10^{-3}$. (Again, the standard for powers-of-ten notation is that the first figure is a number between one and ten.) This notation is particularly useful when dealing with very small numbers. A good example is the diameter of a hydrogen atom, which is much more convenient to state in powers-of-ten notation (1.1×10^{-10} meter, or 1.1×10^{-10} m) than as a decimal (0.00000000011 m) or a fraction (110 trillionths of a meter.)

Because it bypasses all the awkward zeros, powers-of-ten notation is ideal for describing the size of objects as small as atoms or as big as galaxies (Figure 1-13). Box 1-2 explains how powers-of-ten notation also makes it easy to multiply and divide numbers that are very large or very small.

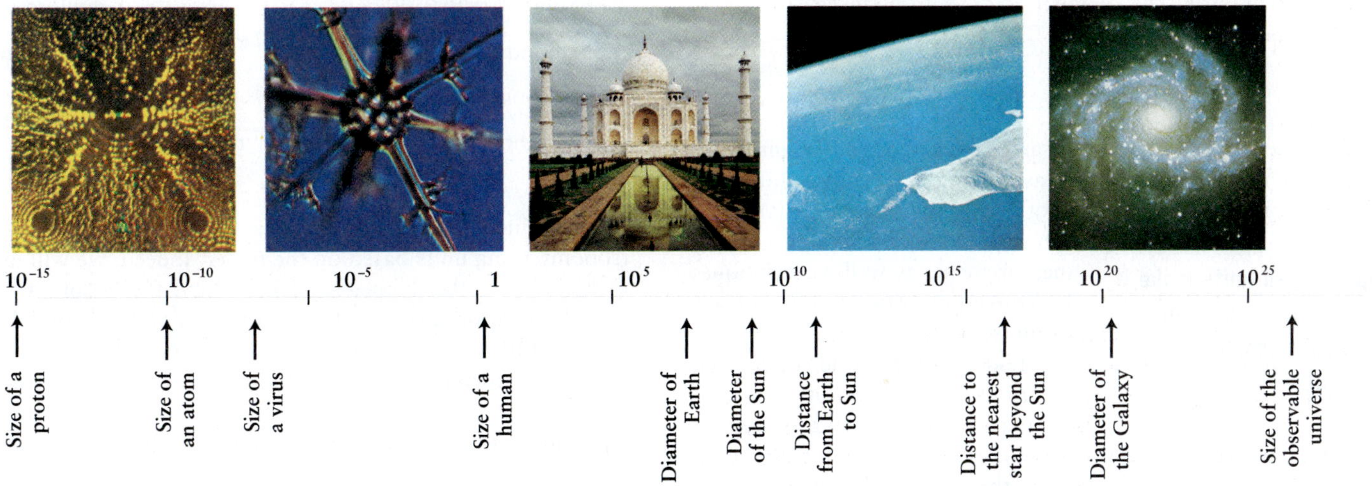

Figure 1-13

Examples of Powers-of-Ten Notation The scale gives the sizes of objects in meters, ranging from subatomic particles at the left to the entire observable universe on the right. The photograph at the left shows tungsten atoms, 10^{-10} meter in diameter. Second from left is the crystalline skeleton of a diatom (a single-celled organism), 10^{-4} meter (0.1 millimeter) in size. At the center is the Taj Mahal, about 60 meters tall and within reach of our unaided senses. On the right, looking across the Indian Ocean toward the south pole, we see the curvature of Earth, about 10^7 meters in diameter. At the far right is a galaxy, 10^{21} meters (100,000 light-years) in diameter. (Courtesy of Scientific American Books; NASA; and photograph by David Malin from the Anglo-Australian Observatory)

BOX 1-2 Tools of the Astronomer's Trade

Arithmetic with Powers-of-Ten Notation

Using powers-of-ten notation makes it easy to multiply numbers. For example, suppose you want to multiply 100 by 1000. If you use ordinary notation, you have to write a lot of zeros:

$$100 \times 1000 = 100{,}000 \text{ (one hundred thousand)}$$

By converting these numbers to powers-of-ten notation, we can write this same multiplication more compactly as

$$10^2 \times 10^3 = 10^5$$

Because $2 + 3 = 5$, we are led to the following general rule for *multiplying* numbers expressed in terms of powers of ten: Simply *add* the exponents.

EXAMPLE: $10^4 \times 10^3 = 10^{4+3} = 10^7$

To *divide* numbers expressed in terms of powers of ten, remember that $10^{-1} = 1/10$, $10^{-2} = 1/100$, and so on. The general rule for any exponent n is

$$10^{-n} = \frac{1}{10^n}$$

In other words, dividing by 10^n is the same as multiplying by 10^{-n}. To carry out a division, you first transform it into multiplication by changing the sign of the exponent, and then carry out the multiplication by adding the exponents.

EXAMPLE: $\frac{10^4}{10^6} = 10^4 \times 10^{-6} = 10^{4+(-6)} = 10^{4-6} = 10^{-2}$

Usually a computation involves numbers like 3.0×10^{10}, that is, an ordinary number multiplied by a factor of 10 with an exponent. In such cases, to perform multiplication or division, you can treat the numbers separately from the factors of 10^n.

EXAMPLE: We can redo the first numerical example from Box 1-1 in a straightforward manner by using exponents:

$$D = \frac{31.2 \times 944{,}000{,}000 \text{ km}}{206{,}265}$$

$$= \frac{3.12 \times 10 \times 9.44 \times 10^8}{2.06265 \times 10^5} \text{km}$$

$$= \frac{3.12 \times 9.44 \times 10^{1+8-5}}{2.06265} \text{km} = 14.3 \times 10^4 \text{ km}$$

$$= 1.43 \times 10 \times 10^4 \text{ km} = 1.43 \times 10^5 \text{ km}$$

1-7 Astronomical distances are often measured in astronomical units, light-years, or parsecs

Astronomers use many of the same units of measurement as do other scientists. They often measure lengths in meters (abbreviated m), masses in kilograms (kg), and time in seconds (s). (You can read more about these units of measurement, as well as techniques for converting between different sets of units, in Box 1-3.)

Specialized units make it easier to comprehend immense cosmic distances

Like other scientists, astronomers often find it useful to combine these units with powers of ten and create new units using prefixes. As an example, the number 1000 ($= 10^3$) is represented by the prefix "kilo," and so a distance of 1000 meters is the same as 1 kilometer (1 km). Here are some of the most common prefixes, with examples of how they are used:

one-billionth meter	$= 10^{-9}$ m	= 1 nanometer
one-millionth second	$= 10^{-6}$ s	= 1 microsecond
one-thousandth arcsecond	$= 10^{-3}$ arcsec	= 1 milliarcsecond
one-hundredth meter	$= 10^{-2}$ m	= 1 centimeter
one thousand meters	$= 10^3$ m	= 1 kilometer
one million tons	$= 10^6$ tons	= 1 megaton

In principle, we could express all sizes and distances in astronomy using units based on the meter. Indeed, we will use kilometers to give the diameters of Earth and the Moon, as well as the Earth-Moon distance. But, while a kilometer (roughly equal to three-fifths of a mile) is an easy distance for humans to visualize, a megameter (10^6 m) is not. For this reason, astronomers have devised units of measure that are more appropriate for the tremendous distances between the planets and the far greater distances between the stars.

When discussing distances across the solar system, astronomers use a unit of length called the **astronomical unit** (abbreviated **AU**). This is the average distance between Earth and the Sun:

$$1 \text{ AU} = 1.496 \times 10^8 \text{ km} = 92.96 \text{ million miles}$$

BOX 1-3 Tools of the Astronomer's Trade

Units of Length, Time, and Mass

To understand and appreciate the universe, we need to describe phenomena not only on the large scales of galaxies but also on the submicroscopic scale of the atom. Astronomers generally use units that are best suited to the topic at hand. For example, interstellar distances are conveniently expressed in either light-years or parsecs, whereas the diameters of the planets are more comfortably presented in kilometers.

Most scientists prefer to use a version of the metric system called the International System of Units, abbreviated **SI** (after the French name Système International). In SI units, length is measured in meters (m), time is measured in seconds (s), and mass (a measure of the amount of material in an object) is measured in kilograms (kg). How are these basic units related to other measures?

When discussing objects on a human scale, sizes and distances can also be expressed in millimeters (mm), centimeters (cm), and kilometers (km). These units of length are related to the meter as follows:

$$1 \text{ millimeter} = 0.001 \text{ m} = 10^{-3} \text{ m}$$

$$1 \text{ centimeter} = 0.01 \text{ m} = 10^{-2} \text{ m}$$

$$1 \text{ kilometer} = 1000 \text{ m} = 10^{3} \text{ m}$$

A useful set of conversions for the English system is

$$1 \text{ in} = 2.54 \text{ cm}$$

$$1 \text{ ft} = 0.3048 \text{ m}$$

$$1 \text{ mi} = 1.609 \text{ km}$$

Each of these equalities can also be written as a fraction equal to 1. For example, you can write

$$\frac{0.3048 \text{ m}}{1 \text{ ft}} = 1$$

Fractions like this are useful for converting a quantity from one set of units to another. For example, the *Saturn V* rocket used to send astronauts to the Moon stands about 363 ft tall. How can we convert this height to meters? The trick is to remember that a quantity does not change if you multiply it by 1. Expressing the number 1 by the fraction (0.3048 m)/(1 ft), we can write the height of the rocket as

$$363 \text{ ft} \times 1 = 363 \text{ ft} \times \frac{0.3048 \text{ m}}{1 \text{ ft}} = 111 \frac{\text{ft} \times \text{m}}{\text{ft}} = 111 \text{ m}$$

EXAMPLE: The diameter of Mars is 6794 km. Let's try expressing this in miles.

CAUTION! You can get into trouble if you are careless in applying the trick of taking the number whose units are to be converted and multiplying it by 1. For example, if we multiply the diameter by 1 expressed as (1.609 km)/(1 mi), we get

$$6794 \text{ km} \times 1 = 6794 \text{ km} \times \frac{1.609 \text{ km}}{1 \text{ mi}} = 10{,}930 \frac{\text{km}^2}{\text{mi}}$$

The unwanted units of km did not cancel, so this answer cannot be right. Furthermore, a mile is larger than a kilometer, so the diameter expressed in miles should be a smaller number than when expressed in kilometers.

The correct approach is to write the number 1 so that the unwanted units *will* cancel. The number we are starting with is in kilometers, so we must write the number 1 with kilometers in the denominator ("downstairs" in the fraction). Thus, we express 1 as (1 mi)/(1.609 km):

$$6794 \text{ km} \times 1 = 6794 \text{ km} \times 1 \text{ mi}/1.609 \text{ km}$$

$$= 4222 \text{ km} \times \text{mi/km} = 4222 \text{ mi}$$

Now the units of km cancel as they should, and the distance in miles is a smaller number than in kilometers (as it must be).

When discussing very small distances such as the size of an atom, astronomers often use the micrometer (μm) or the nanometer (nm). These are related to the meter as follows:

$$1 \text{ micrometer} = 1\ \mu\text{m} = 10^{-6} \text{ m}$$

$$1 \text{ nanometer} = 1 \text{ nm} = 10^{-9} \text{ m}$$

Thus, 1 μm = 10^3 nm. (Note that the micrometer is often called the micron.)

The basic unit of time is the second (s). It is related to other units of time as follows:

$$1 \text{ minute (min)} = 60 \text{ s}$$

$$1 \text{ hour (h)} = 3600 \text{ s}$$

$$1 \text{ day (d)} = 86{,}400 \text{ s}$$

$$1 \text{ year (y)} = 3.156 \times 10^{7} \text{ s}$$

(continued on the next page)

BOX 1-3 *(continued)*

In the SI system, speed is properly measured in meters per second (m/s). Quite commonly, however, speed is also expressed in km/s and mi/h:

$$1 \text{ km/s} = 10^3 \text{ m/s}$$

$$1 \text{ km/s} = 2237 \text{ mi/h}$$

$$1 \text{ mi/h} = 0.447 \text{ m/s}$$

$$1 \text{ mi/h} = 1.47 \text{ ft/s}$$

In addition to using kilograms, astronomers sometimes express mass in grams (g) and in solar masses ($M_\odot$), where the subscript $\odot$ is the symbol denoting the Sun. It is especially convenient to use solar masses when discussing the masses of stars and galaxies. These units are related to each other as follows:

$$1 \text{ kg} = 1000 \text{ g}$$

$$1 M_\odot = 1.99 \times 10^{30} \text{ kg}$$

CAUTION! You may be wondering why we have not given a conversion between kilograms and pounds (lb). The reason is that these units do not refer to the same physical quantity! A kilogram is a unit of *mass,* which is a measure of the amount of material in an object. By contrast, a pound is a unit of *weight,* which tells you how strongly gravity pulls on that object's material. Consider a person who weighs 110 pounds on Earth, corresponding to a mass of 50 kg. Gravity is only about one-sixth as strong on the Moon as it is on Earth, so on the Moon this person would weigh only one-sixth of 110 pounds, or about 18 pounds. But that person's mass of 50 kg is the same on the Moon; wherever you go in the universe, you take all of your material along with you. We will explore the relationship between mass and weight in Chapter 4.

Thus, the average distance between the Sun and Jupiter can be conveniently stated as 5.2 AU.

ANIMATION 1.1

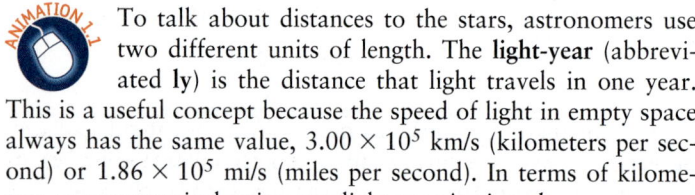

To talk about distances to the stars, astronomers use two different units of length. The **light-year** (abbreviated **ly**) is the distance that light travels in one year. This is a useful concept because the speed of light in empty space always has the same value, 3.00×10^5 km/s (kilometers per second) or 1.86×10^5 mi/s (miles per second). In terms of kilometers or astronomical units, one light-year is given by

$$1 \text{ ly} = 9.46 \times 10^{12} \text{ km} = 63{,}240 \text{ AU}$$

This distance is roughly equal to 6 trillion miles.

CAUTION! Keep in mind that despite its name, the light-year is a unit of distance and *not* a unit of time. As an example, Proxima Centauri, the nearest star other than the Sun, is a distance of 4.2 light-years from Earth. This means that light takes 4.2 years to travel to us from Proxima Centauri.

Physicists often measure interstellar distances in light-years because the speed of light is one of nature's most important numbers. But many astronomers prefer to use another unit of length, the *parsec,* because its definition is closely related to a method of measuring distances to the stars.

Imagine taking a journey far into space, beyond the orbits of the outer planets. As you look back toward the Sun, Earth's orbit subtends a smaller angle in the sky the farther you are from the Sun. As Figure 1-14 shows, the distance at which 1 AU subtends an angle of 1 arcsec is defined as 1 **parsec** (abbreviated **pc**):

$$1 \text{ pc} = 3.09 \times 10^{13} \text{ km} = 3.26 \text{ ly}$$

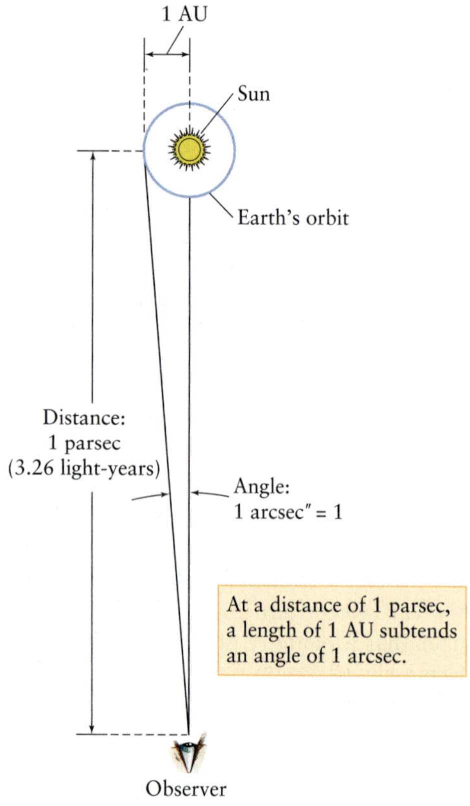

Figure 1-14

A Parsec The parsec, a unit of length commonly used by astronomers, is equal to 3.26 light-years. The parsec is defined as the distance at which 1 AU perpendicular to the observer's line of sight subtends an angle of 1 arcsec.

The distance from Earth to Proxima Centauri can be stated as 1.3 pc or as 4.2 ly. Whether you choose to use parsecs or light-years is a matter of personal taste.

For even greater distances, astronomers commonly use **kiloparsecs** and **megaparsecs** (abbreviated **kpc** and **Mpc**). As we saw before, these prefixes simply mean "thousand" and "million," respectively:

$$1 \text{ kiloparsec} = 1 \text{ kpc} = 1000 \text{ pc} = 10^3 \text{ pc}$$

$$1 \text{ megaparsec} = 1 \text{ Mpc} = 1{,}000{,}000 \text{ pc} = 10^6 \text{ pc}$$

For example, the distance from Earth to the center of our Milky Way Galaxy is about 8 kpc, and the galaxy shown in Figure 1-9 is about 11 Mpc away.

Some astronomers prefer to talk about thousands or millions of light-years rather than kiloparsecs and megaparsecs. Once again, the choice is a matter of personal taste. As a general rule, astronomers use whatever measuring sticks seem best suited for the issue at hand and do not restrict themselves to one system of measurement. For example, an astronomer might say that the supergiant star Antares has a diameter of 860 million kilometers and is located at a distance of 185 parsecs from Earth. The *Cosmic Connections* figure on page 16 shows where these different systems are useful.

1-8 Astronomy is an adventure of the human mind

An underlying theme of this book is that the universe is rational. It is not a hodgepodge of unrelated things behaving in unpredictable ways. Rather, we find strong evidence that fundamental laws of physics govern the nature of the universe and the behavior of everything in it. These unifying concepts enable us to explore realms far removed from our earthly experience. Thus, a scientist can do experiments in a laboratory to determine the properties of light or the behavior of atoms and then use this knowledge to investigate the structure of the universe.

Studying the universe benefits our lives on Earth

The discovery of fundamental laws of nature has had a profound influence on humanity. These laws have led to an immense number of practical applications that have fundamentally transformed commerce, medicine, entertainment, transportation, and other aspects of our lives. In particular, space technology has given us instant contact with any point on the globe through communication satellites, precise navigation to any point on Earth using signals from the satellites of the Global Positioning System (GPS), and accurate weather forecasts from meteorological satellites (Figure 1-15).

Figure 1-15 R I V U X G
A Hurricane Seen from Space This image of Hurricane Frances was made on September 2, 2004, by *GOES-12* (*Geostationary Operational Environmental Satellite 12*). A geostationary satellite like *GOES-12* orbits around Earth's equator every 24 hours, the same length of time it takes the planet to make a complete rotation. Hence, this satellite remains over the same spot on Earth, from which it can monitor the weather continuously. By tracking hurricanes from orbit, *GOES-12* makes it much easier for meteorologists to give early warning of these immense storms. The resulting savings in lives and property more than pay for the cost of the satellite. (NASA, NOAA)

As important as the applications of science are, the pursuit of scientific knowledge for its own sake is no less important. We are fortunate to live in an age in which this pursuit is in full flower. Just as explorers such as Columbus and Magellan discovered the true size of our planet in the fifteenth and sixteenth centuries, astronomers of the twenty-first century are exploring the universe to an extent that is unparalleled in human history. Indeed, even the voyages into space imagined by such great science fiction writers as Jules Verne and H. G. Wells pale in comparison to today's reality. Over a few short decades, humans have walked on the Moon, sent robot spacecraft to dig into the Martian soil and explore the satellites of Saturn, and used the most powerful telescopes ever built to probe the limits of the observable universe. Never before has so much been revealed in so short a time.

As you proceed through this book, you will learn about the tools that scientists use to explore the natural world, as well as what they observe with these tools. But, most important, you will see how astronomers build from their observations an understanding of the universe in which we live. It is this search for understanding that makes science more than merely a collection of data and elevates it to one of the great adventures of the human mind. It is an adventure that will continue as long as there are mysteries in the universe—an adventure we hope you will come to appreciate and share.

Key Words

Terms preceded by an asterisk (*) are discussed in the Boxes.

angle, p. 8
angular diameter (angular size), p. 8
angular distance, p. 8
angular measure, p. 8
arcminute (′, minute of arc), p. 9
arcsecond (″, second of arc), p. 9
astronomical unit (AU), p. 12

COSMIC CONNECTIONS Sizes in the Universe

Powers-of-ten notation provides a convenient way to express the sizes of astronomical objects and distances in space. This illustration suggests the immense distances within our solar system, the far greater distances between the stars within our Milky Way Galaxy, and the truly cosmic distances between galaxies.

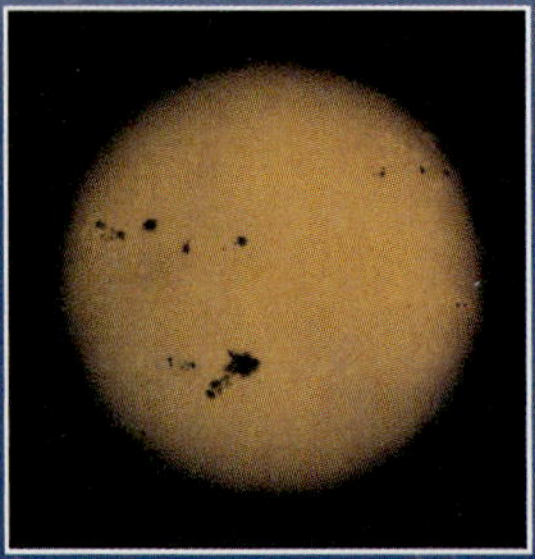

Sun: diameter = 1.39×10^6 km

The Sun, Earth, and other planets are members of our solar system

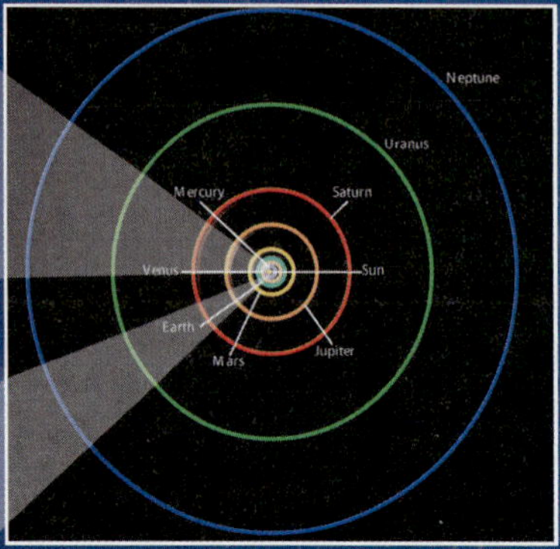

Diameter of Neptune's orbit: 60 AU
1 AU (astronomical unit) = 1.50×10^8 km = average Earth-Sun distance

Earth: diameter = 1.28×10^4 km

The Sun is a typical star. Typical distances between our neighboring stars = 1 to 5 ly
1 ly = distance that light travels in one year = 6.32×10^4 AU

Our Sun is one of more than 10^{11} stars in the Milky Way Galaxy. Distance from the center of the Milky Way to the Sun = 2.8×10^4 ly

Galaxies are grouped into clusters, which can be up to 10^7 ly across.

Each of the 1.6×10^6 dots in this map of the entire sky represents a relatively nearby galaxy. This is a tiny fraction of the number of galaxies in the observable universe.

Big Bang, p. 7
black hole, p. 6
degree (°), p. 8
exponent, p. 10
galaxy, p. 6
hypothesis, p. 2
kiloparsec (kpc), p. 15
laws of physics, p. 3
light-year (ly), p. 14
megaparsec (Mpc), p. 15
meteorite, p. 4
model, p. 2
nebula (*plural* nebulae), p. 6
Newtonian mechanics, p. 3
parsec (pc), p. 14
power of ten, p. 10
powers-of-ten notation, p. 10
pulsar, p. 6
quasar, p. 7
scientific method, p. 2
*SI units, p. 13
*small-angle formula, p. 9
solar system, p. 4
subtend (an angle), p. 8
supernova (*plural* supernovae), p. 6
theory, p. 2

Key Ideas

Astronomy, Science, and the Nature of the Universe: The universe is comprehensible. The scientific method is a procedure for formulating hypotheses about the universe. Hypotheses are tested by observation or experimentation in order to build consistent models or theories that accurately describe phenomena in nature.

• Observations of the heavens have helped scientists discover some of the fundamental laws of physics. The laws of physics are in turn used by astronomers to interpret their observations.

The Solar System: Exploration of the planets provides information about the origin and evolution of the solar system, as well as about the history and resources of Earth.

Stars and Nebulae: Studying the stars and nebulae helps us learn about the origin and history of the Sun and the solar system.

Galaxies: Observations of galaxies tell us about the origin and history of the universe.

Angular Measure: Astronomers use angles to denote the positions and sizes of objects in the sky. The size of an angle is measured in degrees, arcminutes, and arcseconds.

Powers-of-Ten Notation is a convenient shorthand system for writing numbers. It allows very large and very small numbers to be expressed in a compact form.

Units of Distance: Astronomers use a variety of distance units. These include the astronomical unit (the average distance from Earth to the Sun), the light-year (the distance that light travels in one year), and the parsec.

Questions

Review Questions

TUTORIAL 1.1

1. What is the difference between a hypothesis and a theory?
2. What is the difference between a theory and a law of physics?
3. How are scientific theories tested?
4. Describe the role that skepticism plays in science.
5. Describe one reason why it is useful to have telescopes in space.
6. What caused the craters on the Moon?
7. What are meteorites? Why are they important for understanding the history of the solar system?
8. What makes the Sun and stars shine?
9. What role do nebulae like the Orion Nebula play in the life stories of stars?
10. What is the difference between a solar system and a galaxy?
11. What are degrees, arcminutes, and arcseconds used for? What are the relationships among these units of measure?
12. How many arcseconds equal 1°?
13. With the aid of a diagram, explain what it means to say that the Moon subtends an angle of $\frac{1}{2}°$.
14. What is an exponent? How are exponents used in powers-of-ten notation?
15. What are the advantages of using powers-of-ten notation?
16. Write the following numbers using powers-of-ten notation: (a) ten million, (b) sixty thousand, (c) four one-thousandths, (d) thirty-eight billion, (e) your age in months.
17. How is an astronomical unit (AU) defined? Give an example of a situation in which this unit of measure would be convenient to use.
18. What is the advantage to the astronomer of using the light-year as a unit of distance?
19. What is a parsec? How is it related to a kiloparsec and to a megaparsec?
20. Give the word or phrase that corresponds to the following standard abbreviations: (a) km, (b) cm, (c) s, (d) km/s, (e) mi/h, (f) m, (g) m/s, (h) h, (i) y, (j) g, (k) kg. Which of these are units of speed? (*Hint:* You may have to refer to a dictionary. All of these abbreviations should be part of your working vocabulary.)
21. In the original (1977) *Star Wars* movie, Han Solo praises the speed of his spaceship by saying, "It's the ship that made the Kessel run in less than 12 parsecs!" Explain why this statement is obvious misinformation.
22. A reporter once described a light-year as "the time it takes light to reach us traveling at the speed of light." How would you correct this statement?

Advanced Questions

Questions preceded by an asterisk (*) involve topics discussed in the Boxes.

> **Problem-solving tips and tools**
>
> The small-angle formula, given in Box 1-1, relates the size of an astronomical object to the angle it subtends. Box 1-3 illustrates how to convert from one unit of measure to another. An object traveling at speed v for a time t covers a distance d given by $d = vt$; for example, a car traveling at 90 km/h (v) for 3 h (t) covers a distance d = (90 km/h) (3 h) = 270 km. Similarly, the time t required to cover a given distance d at speed v is $t = d/v$; for example, if d = 270 km and v = 90 km/h, then t = (270 km)/(90 km/h) = 3 h.

23. What is the meaning of the letters R I V U X G that appear under some of the figures in this chapter? Why in each case is one of the letters highlighted? (*Hint:* See the Preface that precedes Chapter 1.)

24. The diameter of the Sun is 1.4×10^{11} cm, and the distance to the nearest star, Proxima Centauri, is 4.2 ly. Suppose you want to build an exact scale model of the Sun and Proxima Centauri, and you are using a ball 30 cm in diameter to represent the Sun. In your scale model, how far away would Proxima Centauri be from the Sun? Give your answer in kilometers, using powers-of-ten notation.
25. How many Suns would it take, laid side by side, to reach the nearest star? Use powers-of-ten notation. (*Hint:* See the preceding question.)
26. A hydrogen atom has a radius of about 5×10^{-9} cm. The radius of the observable universe is about 14 billion light-years. How many times larger than a hydrogen atom is the observable universe? Use powers-of-ten notation.
27. The Sun's mass is 1.99×10^{30} kg, three-quarters of which is hydrogen. The mass of a hydrogen atom is 1.67×10^{-27} kg. How many hydrogen atoms does the Sun contain? Use powers-of-ten notation.
28. The average distance from Earth to the Sun is 1.496×10^{8} km. Express this distance (**a**) in light-years and (**b**) in parsecs. Use powers-of-ten notation. (**c**) Are light-years or parsecs useful units for describing distances of this size? Explain.
29. The speed of light is 3.00×10^{8} m/s. How long does it take light to travel from the Sun to Earth? Give your answer in seconds, using powers-of-ten notation. (*Hint:* See the preceding question.)
30. When the *Voyager 2* spacecraft sent back pictures of Neptune during its flyby of that planet in 1989, the spacecraft's radio signals traveled for 4 hours at the speed of light to reach Earth. How far away was the spacecraft? Give your answer in kilometers, using powers-of-ten notation. (*Hint:* See the preceding question.)
31. The star Altair is 5.15 pc from Earth. (**a**) What is the distance to Altair in kilometers? Use powers-of-ten notation. (**b**) How long does it take for light emanating from Altair to reach Earth? Give your answer in years. (*Hint:* You do not need to know the value of the speed of light.)
32. The age of the universe is about 13.7 billion years. What is this age in seconds? Use powers-of-ten notation.
*33. Explain where the number 206,265 in the small-angle formula comes from.
*34. At what distance would a person have to hold a European 2-euro coin (which has a diameter of about 2.6 cm) in order for the coin to subtend an angle of (**a**) 1°? (**b**) 1 arcmin? (**c**) 1 arcsec? Give your answers in meters.
*35. A person with good vision can see details that subtend an angle of as small as 1 arcminute. If two dark lines on an eye chart are 2 millimeters apart, how far can such a person be from the chart and still be able to tell that there are two distinct lines? Give your answer in meters.
*36. The average distance to the Moon is 384,000 km, and the Moon subtends an angle of ½°. Use this information to calculate the diameter of the Moon in kilometers.
*37. Suppose your telescope can give you a clear view of objects and features that subtend angles of at least 2 arcsec. What is the diameter in kilometers of the smallest crater you can see on the Moon? (*Hint:* See the preceding question.)
*38. On April 18, 2006, the planet Venus was a distance of 0.869 AU from Earth. The diameter of Venus is 12,104 km. What was the angular size of Venus as seen from Earth on April 18, 2006? Give your answer in arcminutes.
*39. (**a**) Use the information given in the caption to Figure 1-7 to determine the angular size of the Orion Nebula. Give your answer in degrees. (**b**) How does the angular diameter of the Orion Nebula compare to the angular diameter of the Moon?

Discussion Questions

40. Scientists assume that "reality is rational." Discuss what this means and the thinking behind it.
41. All scientific knowledge is inherently provisional. Discuss whether this is a weakness or a strength of the scientific method.
42. How do astronomical observations differ from those of other sciences?

Web/eBook Questions

WEB LINK 1.1

43. Use the links given in the *Universe* Web site or eBook, Chapter 1, to learn about the Orion Nebula (Figure 1-7). Can the nebula be seen with the naked eye? Does the nebula stand alone, or is it part of a larger cloud of interstellar material? What has been learned by examining the Orion Nebula with telescopes sensitive to infrared light?
44. Use the links given in the *Universe* Web site or eBook, Chapter 1, to learn more about the Crab Nebula (Figure 1-8). When did observers on Earth see the supernova that created this nebula? Does the nebula emit any radiation other than visible light? What kind of object is at the center of the nebula?

Activities

Observing Projects

45. On a dark, clear, moonless night, can you see the Milky Way from where you live? If so, briefly describe its appearance. If not, what seems to be interfering with your ability to see the Milky Way?
46. Look up at the sky on a clear, cloud-free night. Is the Moon in the sky? If so, does it interfere with your ability to see the fainter stars? Why do you suppose astronomers prefer to schedule their observations on nights when the Moon is not in the sky?
47. Look up at the sky on a clear, cloud-free night and note the positions of a few prominent stars relative to such reference markers as rooftops, telephone poles, and treetops. Also note the location from where you make your observations. A few hours later, return to that location and again note the positions of the same bright stars that you observed earlier. How have their positions changed? From these changes, can you deduce the general direction in which the stars appear to be moving?
48. If your book comes with a CD-ROM, use it to install the *Starry Night Enthusiast*™ planetarium software on your computer and run this program to determine when the Moon is

visible today from your location. You will see that the Moon can be seen in the daytime as well as at night. Note that the **Time Flow Rate** is set to **1x,** indicating that time is running forward at the normal rate. Set the **Time Flow Rate** to **1 minute** and find the time of moonset at your location. Determine which, if any, of the following planets are visible tonight: Mercury, Venus, Mars, Jupiter, and Saturn. Feel free to experiment with *Starry Night Enthusiast*™.

Operating Hints: (1) To see the sky from your actual location on Earth, select **Set Home Location . . .** in the **File** menu (on a Macintosh, this command is found under the **Starry Night Enthusiast 5.0** menu). Click the **List** tab in the Home Location dialog box; then select the name of your city or town and click the **Save As Home Location** button. You can always return to this starting screen by clicking the **Home** button in the toolbar. (2) To change your viewing direction, move the mouse in the view. When the mouse cursor appears as a little hand, hold down the mouse button (on a Windows computer, the left button) and drag the mouse; this action will move the sky and change the gaze direction. (3) Use the toolbar at the top of the main window to change the time and date appropriate to the display and to adjust the time flow rate, as explained above. (4) You can use the **Find . . .** command in the **Edit** menu or click the **Find** tab on the left side of the main view window to open the Find pane to locate specific planets or stars by name. (5) To learn about any object in the sky, point the cursor at the object. A panel of information about the chosen object will appear on the display. The position of this information panel and its content will depend upon the selected options. These options can be altered from the **Preferences** item in the **File** menu (on a Macintosh, the **Preferences** item is under the **Starry Night Enthusiast 5.0** menu). Select **Cursor Tracking (HUD)** in the left-hand edit box in the Preferences dialog. You can then select the information to be displayed and the position of this information panel. Another way to get information about an object is to position the mouse cursor over the object in the main view window and double-click. This will open the **Info** side-pane which includes specific information about the chosen object under several headings. You can expand and collapse the layers beneath these headings by clicking the + or − icons to the left of the heading label (▶ and ▼ icons on a Macintosh).

WEB LINK 1.8

You can find further information about the program and its many modes of operation in the *Starry Night Enthusiast*™ manual (in *Starry Night Enthusiast*™, select **User's Guide** in the **Help** menu).

49. Use the *Starry Night Enthusiast*™ program to investigate the Milky Way Galaxy. Select **Deep Space > Local Universe** in the **Favorites** menu. Click and hold the **minus (−)** symbol in the **Zoom** control on the toolbar to adjust the field of view to an appropriate value for this image of the Milky Way. A foreground image of astronaut's feet may be superimposed upon this view. Click on **View > Feet** to remove this foreground, if desired. The view shows the Milky Way near the center of the window against a background of distant galaxies as seen from a point in space 282,000 light-years from the Sun. To center the Milky Way in the view, position the mouse cursor over the Milky Way and click and hold the mouse button (on a two-button mouse, click the right button). Select **Centre** from the drop down menu that appears. (a) To view the Milky Way from different angles, hold down the **SHIFT** key on the keyboard. The cursor should turn into a small square surrounded by four small triangles. Then hold down the mouse button (the left button on a two-button mouse) and drag the mouse in order to change the viewing angle. This will be equivalent to your moving around the galaxy, viewing it against different backgrounds of both near and far galaxies. How would you describe the shape of the Milky Way? (b) You can zoom in and zoom out on this galaxy using the Zoom buttons at the right side of the toolbar. (c) Another way to zoom in on the view is to change the elevation of the viewing location. Click the **Find** tab at the left of the view window and double-click the entry for the Sun. This will center the view on the Sun. Now use the elevation buttons to the left of the **Home** button in the toolbar to move closer to the Sun. Move in until you can see the planets in their orbits around the Sun, then zoom back out until you can see the entire Milky Way Galaxy again. Are the Sun and planets located at the center of the Milky Way? How would you describe their location?

Collaborative Exercises

50. A scientific theory is fundamentally different than the everyday use of the word "theory." List and describe any three scientific theories of your choice and creatively imagine an additional three hypothetical theories that are not scientific. Briefly describe what is scientific and what is nonscientific about each of these theories.

51. Angular measure describes how far apart two objects appear to an observer. From where you are currently sitting, estimate the angular distance between the floor and the ceiling at the front of the room you are sitting in, the angular distance between the two people sitting closest to you, and the angular size of a clock or an exit sign on the wall. Draw sketches to illustrate each answer and describe how each of your answers would change if you were standing in the very center of the room.

52. Astronomers use powers of ten to describe the distances to objects. List an object or place that is located at very roughly each of the following distances from you: 10^{-2} m, 10^{0} m, 10^{1} m, 10^{3} m, 10^{7} m, 10^{10} m, and 10^{20} m.

Why Astronomy?

by Sandra M. Faber

As you study astronomy, you may ask, "Why am I studying this subject? What good is it for people in general and for me in particular?" Admittedly, astronomy does not offer the same practical benefits as other sciences, so how can it be important to your life?

On the most basic level, I think of astronomy as providing the ultimate background for human history. Recorded history goes back about 3000 years. For knowledge of the time before that, we consult archeologists and anthropologists about early human history and paleontologists, biologists, and geologists about the evolution of life and of our planet—altogether going back some five billion years. Astronomy tells us about the time before that, the ten billion years or so when the Sun, solar system, and Milky Way Galaxy formed, and even about the origin of the universe in the Big Bang. Knowledge of astronomy is part of a well-educated person's view of history.

Astronomy challenges our belief system and impels us to put our "philosophical house" in order. For example, the Bible says the world and everything in it were created in six days by the hand of God. However, according to the ancient Egyptians, Earth arose spontaneously from the infinite waters of the eternal universe, called Nun. Alaskan legends teach that the world was created by the conscious imaginings of a deity named Father Raven.

Modern astronomy, supported by physics and observations, differs from these stories of the creation of Earth. Astronomers believe that the Sun formed about five billion years ago by gravitational collapse from a dense cloud of interstellar gas and dust. At the same time, and over a period of several hundred thousand years, the planets condensed within the swirling solar nebula. Astronomers have actually seen young stars form in this way.

At issue here, really, is the question of how we are to gain information about the nature of the physical world—whether by revelation and intuition or by logic and observation. Where science stops and faith begins is a thorny issue for everyone, but particularly for astronomers—and for astronomy students.

Astronomy cultivates our notions about cosmic time and cosmic evolution. Given the short span of human life, it is all too easy to overlook that the universe is a dynamic place. This idea implies fragility—if something can change, it might even some day disappear. For instance, in another five billion years or so, the Sun will swell up and brighten to 1000 times its present luminosity, incinerating Earth in the process. This is far enough in the future that neither you nor I need to feel any personal responsibility for preparing to meet this challenge. However, other cosmic catastrophes will inevitably occur before then. Earth will be hit by a sizable piece of space debris—craters show that this happens every few million years or so. Enormous volcanic eruptions have occurred in the past and will certainly occur again. Another Ice Age is virtually certain to begin within the next 20,000 years, unless we first cook Earth ourselves by burning too much fossil fuel.

Such common notions as the inevitability of human progress, the desirability of endless economic growth, and Earth's ability to support its human population are all based on limited experience—they will probably not prove viable in the long run. Consequently, we must rethink who we are as a species and what is our proper activity on Earth. These long-term problems involve the whole human race and are vital to our survival and well-being. Astronomy is essential to developing a perspective on human existence and its relation to the cosmos.

Many astronomers believe that the ultimate, proper concept of "home" for the human race is our universe. It seems increasingly likely that a large number of other universes exist, with the vast majority incapable of harboring intelligent life as we know it. The parallel with Earth is striking. Among the solar system planets, only Earth can support human life. Among the great number of planets in our galaxy, only a small fraction may be such that we can call them home. The fraction of hospitable universes is likely to be smaller still. It seems, then, that our universe is the ultimate "home," a sanctuary in a vast sea of inhospitable universes.

I began this essay by talking about history and ended with issues that border on the ethical and religious. Astronomy is like that: It offers a modern-day version of Genesis—and of the Apocalypse, too. I hope that during this course you will be able to take time out to contemplate the broader implications of what you are studying. This is one of the rare opportunities in life to think about who you are and where you and the human race are going. Don't miss it.

Sandra M. Faber is professor of astronomy at the University of California, Santa Cruz, and astronomer at Lick Observatory. She chaired the now legendary group of astronomers called the Seven Samurai, who surveyed the nearest 400 elliptical galaxies and discovered a new mass concentration, called the Great Attractor. Dr. Faber received the Bok Prize from Harvard University in 1978, was elected to the National Academy of Sciences in 1985, and in 1986 won the coveted Dannie Heineman Prize from the American Astronomical Society, in recognition of her sustained body of especially influential astronomical research.

2

Knowing the Heavens

It is a clear night at the Gemini North Observatory atop Mauna Kea, a dormant volcano on the island of Hawaii. As you gaze toward the north, as in this time-exposure photograph, you find that the stars are not motionless. Rather, they move in counterclockwise circles around a fixed point above the northern horizon. Stars close to this point never dip below the horizon, while stars farther from the fixed point rise in the east and set in the west. These motions fade from view when the Sun rises in the east and illuminates the sky. The Sun, too, arcs across the sky in the same manner as the stars. At day's end, when the Sun sets in the west, the panorama of stars is revealed for yet another night.

These observations are at the heart of *naked-eye astronomy*—the sort that requires no equipment but human vision. Naked-eye astronomy cannot tell us what the Sun is made of or how far away the stars are. For such purposes we need tools such as the Gemini Telescope, housed within the dome shown in the photograph. But by studying naked-eye astronomy, you will learn the answers to equally profound questions such as why there are seasons, why the night sky is different at different times of year, and why the night sky looks different in Australia than in North America. In discovering the answers to these questions, you will learn how Earth moves through space and will begin to understand our true place in the cosmos.

R I V U X G

Earth's rotation makes stars appear to trace out circles in the sky. (Gemini Observatory)

2-1 Naked-eye astronomy had an important place in civilizations of the past

WEB LINK 2.1

Positional astronomy—the study of the positions of objects in the sky and how these positions change—has roots that extend far back in time. Four to five thousand years ago, the inhabitants of the British Isles erected stone structures, such as Stonehenge, that suggest a preoccupation with the motions of the sky. Alignments of these stones appear to show where the Sun rose and set at key times during the year. Stone works of a different sort but with a similar astronomical purpose

Learning Goals

By reading the sections of this chapter, you will learn

- 2-1 The importance of astronomy in ancient civilizations around the world
- 2-2 That regions of the sky are divided around groups of stars called constellations
- 2-3 How the sky changes from night to night
- 2-4 How astronomers locate objects in the sky
- 2-5 What causes the seasons
- 2-6 The effect of changes in the direction of Earth's axis of rotation
- 2-7 The role of astronomy in measuring time
- 2-8 How the modern calendar developed

Figure 2-1 R I V U X G

The Sun Dagger at Chaco Canyon On the first day of winter, rays of sunlight passing between stone slabs bracket a spiral stone carving, or petroglyph, at Chaco Canyon in New Mexico. A single band of light strikes the center of the spiral on the first day of summer. This astronomically aligned petroglyph and others were carved by the ancestral Puebloan culture between 850 and 1250 A.D. (Courtesy Karl Kernberger)

are found in America from the Southwestern United States to the Andes of Peru. The ancestral Puebloans (also called Anasazi) of modern-day New Mexico, Arizona, Utah, and Colorado created stone carvings that were illuminated by the Sun on the first days of summer or winter (Figure 2-1). At the Incan city of Machu Picchu in Peru, a narrow window carved in a rock 2 meters thick looks out on the sunrise on December 21 each year.

The astronomical knowledge of ancient peoples is the foundation of modern astronomy

Other ancient peoples designed buildings with astronomical orientations. The great Egyptian pyramids, built around 3000 B.C., are oriented north-south and east-west with an accuracy much better than 1 degree. Similar alignments are found in the grand tomb of Shih Huang Ti (259 B.C.–210 B.C.), the first emperor of China.

Evidence of a highly sophisticated understanding of astronomy can be found in the written records of the Mayan civilization of central America. Mayan astronomers deduced by observation that the apparent motions of the planet Venus follow a cycle that repeats every 584 days. They also developed a technique for calculating the position of Venus on different dates. The Maya believed that Venus was associated with war, so such calculations were important for choosing the most promising dates on which to attack an enemy.

These archaeological discoveries bear witness to an awareness of naked-eye astronomy by the peoples of many cultures. Many of the concepts of modern positional astronomy come to us from these ancients, including the idea of dividing the sky into constellations.

2-2 Eighty-eight constellations cover the entire sky

WEB LINK 2.2

Looking at the sky on a clear, dark night, you might think that you can see millions of stars. Actually, the unaided human eye can detect only about 6000 stars. Because half of the sky is below the horizon at any one time, you can see at most about 3000 stars. When ancient peoples looked at these thousands of stars, they imagined that groupings of stars traced out pictures in the sky. Astronomers still refer to these groupings, called **constellations** (from the Latin for "group of stars").

The constellations provide a convenient framework for stating the position of an object in the heavens

You may already be familiar with some of these pictures or patterns in the sky, such as the Big Dipper, which is actually part of the large constellation Ursa Major (the Great Bear). Many constellations, such as Orion in Figure 2-2, have names derived from the myths and legends of antiquity. Although some star groupings vaguely resemble the figures they are supposed to represent (see Figure 2-2*c*), most do not.

The term "constellation" has a broader definition in present-day astronomy. On modern star charts, the entire sky is divided into 88 regions, each of which is called a constellation. For example, the constellation Orion is now defined to be an irregular patch of sky whose borders are shown in Figure 2-2*b*. When astronomers refer to the "Great Nebula" M42 in Orion, they mean that as seen from Earth this nebula appears to be within Orion's

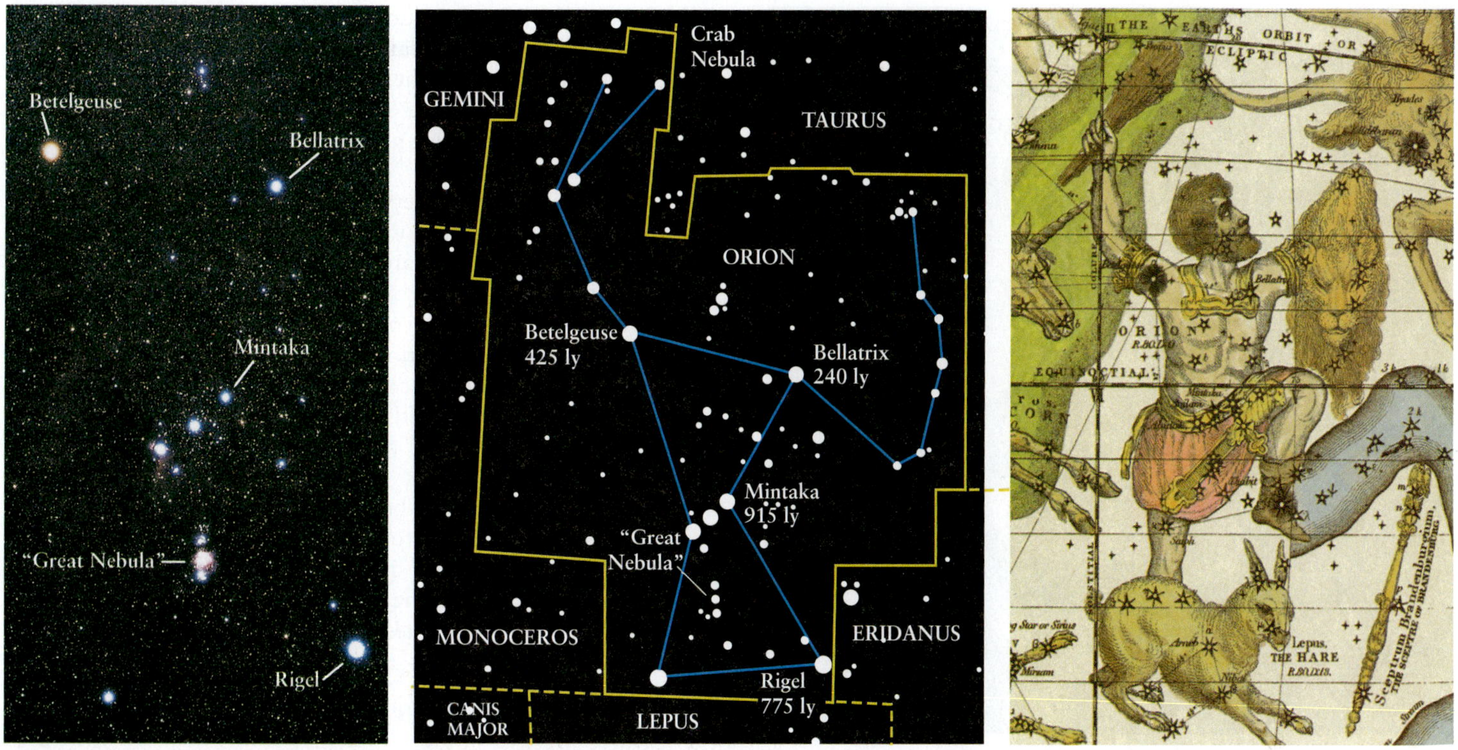

(a) (b) (c)

Figure 2-2 R I **V** U X G

Three Views of Orion The constellation Orion is easily seen on nights from December through March. **(a)** This photograph of Orion shows many more stars than can be seen with the naked eye. **(b)** A portion of a modern star atlas shows the distances in light-years (ly) to some of the stars in Orion. The yellow lines show the borders between Orion and its neighboring constellations (labeled in capitals). **(c)** This fanciful drawing from a star atlas published in 1835 shows Orion the Hunter as well as other celestial creatures. (a: Luke Dodd/Science Photo Library/Photo Researchers; c: Courtesy of Janus Publications)

patch of sky. Some constellations cover large areas of the sky (Ursa Major being one of the biggest) and others very small areas (Crux, the Southern Cross, being the smallest). But because the modern constellations cover the entire sky, every star lies in one constellation or another.

CAUTION! When you look at a constellation's star pattern, it is tempting to conclude that you are seeing a group of stars that are all relatively close together. In fact, most of these stars are nowhere near one another. As an example, Figure 2-2*b* shows the distances in light-years to four stars in Orion. Although Bellatrix (Arabic for "the Amazon") and Mintaka ("the belt") appear to be close to each other, Mintaka is actually more than 600 light-years farther away from us. The two stars only *appear* to be close because they are in nearly the same direction as seen from Earth. The same illusion often appears when you see an airliner's lights at night. It is very difficult to tell how far away a single bright light is, which is why you can mistake an airliner a few kilometers away for a star trillions of times more distant.

LOOKING DEEPER 2.1

The star names shown in Figure 2-2*b* are from the Arabic language. For example, Betelgeuse means "armpit," which makes sense when you look at the star atlas drawing in Figure 2-2*c*. Other types of names are also used for stars. For example, Betelgeuse is also known as α Orionis because it is the brightest star in Orion (α, or alpha, is the first letter in the Greek alphabet).

CAUTION! A number of unscrupulous commercial firms offer to name a star for you for a fee. The money that they charge you for this "service" is real, but the star names are not; none of these names are recognized by astronomers. If you want to use astronomy to commemorate your name or the name of a friend or relative, consider making a donation to your local planetarium or science museum. The money will be put to much better use!

2-3 The appearance of the sky changes during the course of the night and from one night to the next

Go outdoors soon after dark, find a spot away from bright lights, and note the patterns of stars in the sky. Do the same a few hours later. You will find that the entire pattern of stars (as well as the Moon, if it is visible) has shifted its position. New constellations will have risen above the eastern horizon, and some will have disappeared below the western horizon. If you look again before dawn, you will see that the stars that were just rising in the east

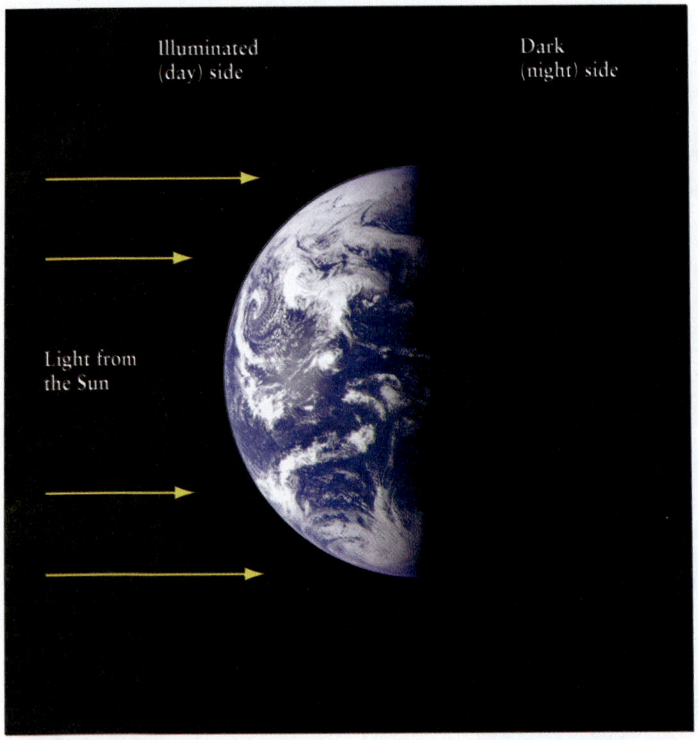

Figure 2-3 R I V U X G

Day and Night on Earth At any moment, half of Earth is illuminated by the Sun. As Earth rotates from west to east, your location moves from the dark (night) hemisphere into the illuminated (day) hemisphere and back again. This image was recorded in 1992 by the *Galileo* spacecraft as it was en route to Jupiter. (JPL/NASA)

when the night began are now low in the western sky. This daily motion, or **diurnal motion,** of the stars is apparent in time-exposure photographs (see the photograph that opens this chapter).

By understanding the motions of Earth through space, we can understand why the Sun and stars appear to move in the sky

If you repeat your observations on the following night, you will find that the motions of the sky are almost but not quite the same. The same constellations rise in the east and set in the west, but a few minutes earlier than on the previous night. If you look again after a month, the constellations visible at a given time of night (say, midnight) will be noticeably different, and after six months you will see an almost totally different set of constellations. Only after a year has passed will the night sky have the same appearance as when you began.

Why does the sky go through diurnal motion? Why do the constellations slowly shift from one night to the next? As we will see, the answer to the first question is that Earth *rotates* once a day around an axis from the north pole to the south pole, while the answer to the second question is that Earth also *revolves* once a year around the Sun.

Diurnal Motion and Earth's Rotation

To understand diurnal motion, note that at any given moment it is daytime on the half of Earth illuminated by the Sun and nighttime on the other half (Figure 2-3). Earth rotates from west to east, making one complete rotation every 24 hours, which is why there is a daily cycle of day and night. Because of this rotation, stars appear to us to rise in the east and set in the west, as do the Sun and Moon.

Figure 2-4 helps to further explain diurnal motion. It shows two views of Earth as seen from a point above the north pole.

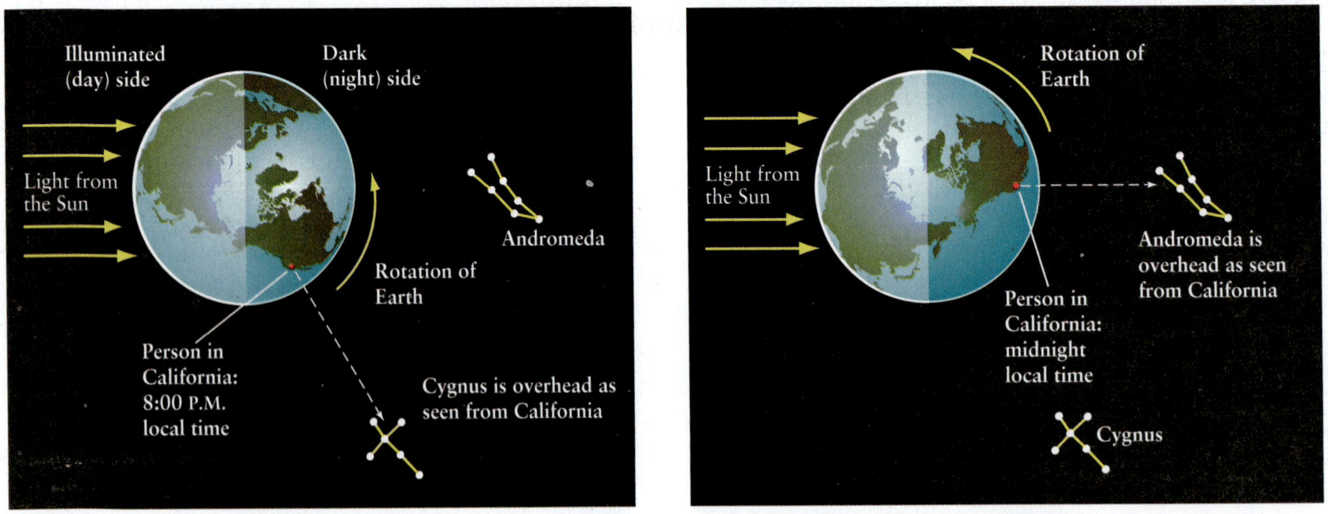

(a) Earth as seen from above the north pole

(b) 4 hours (one-sixth of a complete rotation) later

Figure 2-4

Why Diurnal Motion Happens The diurnal (daily) motion of the stars, the Sun, and the Moon is a consequence of Earth's rotation. **(a)** This drawing shows Earth from a vantage point above the north pole. In this drawing, for a person in California the local time is 8:00 P.M. and the constellation Cygnus is directly overhead. **(b)** Four hours later, Earth has made one-sixth of a complete rotation to the east. As seen from Earth, the entire sky appears to have rotated to the west by one-sixth of a complete rotation. It is now midnight in California, and the constellation directly over California is Andromeda.

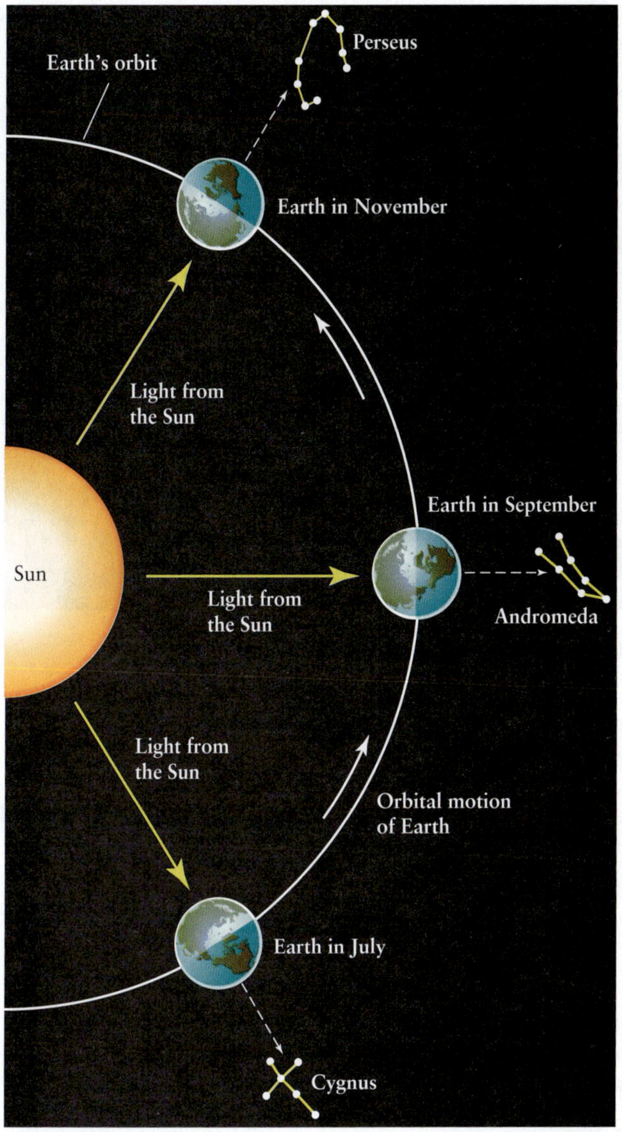

Figure 2-5

Why the Night Sky Changes During the Year As Earth orbits around the Sun, the nighttime side of Earth gradually turns toward different parts of the sky. Hence, the particular stars that you see in the night sky are different at different times of the year. This figure shows which constellation is overhead at midnight local time—when the Sun is on the opposite side of Earth from your location—during different months for observers at midnorthern latitudes (including the United States). If you want to view the constellation Andromeda, the best time of the year to do it is in late September, when Andromeda is nearly overhead at midnight.

At the instant shown in Figure 2-4*a*, it is day in Asia but night in most of North America and Europe. Figure 2-4*b* shows Earth 4 hours later. Four hours is one-sixth of a complete 24-hour day, so Earth has made one-sixth of a rotation between Figures 2-4*a* and 2-4*b*. Europe is now in the illuminated half of Earth (the Sun has risen in Europe), while Alaska has moved from the illuminated to the dark half of Earth (the Sun has set in Alaska). For a person in California, in Figure 2-4*a* the time is 8:00 P.M. and the constellation Cygnus (the Swan) is directly overhead. Four hours later, the constellation over California is Andromeda (named for a mythological princess). Because Earth rotates from west to east, it appears to us on Earth that the entire sky rotates around us in the opposite direction, from east to west.

Yearly Motion and Earth's Orbit

We described earlier that in addition to the diurnal motion of the sky, the constellations visible in the night sky also change slowly over the course of a year. This happens because Earth orbits, or revolves around, the Sun (Figure 2-5). Over the course of a year, Earth makes one complete orbit, and the darkened, nighttime side of Earth gradually turns toward different parts of the heavens. For example, as seen from the northern hemisphere, at midnight in late July the constellation Cygnus is close to overhead; at midnight in late September the constellation Andromeda is close to overhead; and at midnight in late November the constellation Perseus (commemorating a mythological hero) is close to overhead. If you follow a particular star on successive evenings, you will find that it rises approximately 4 minutes earlier each night, or 2 hours earlier each month.

Constellations and the Night Sky

Constellations can help you find your way around the sky. For example, if you live in the northern hemisphere, you can use the Big Dipper in Ursa Major to find the north direction by drawing a straight line through the two stars at the front of the Big Dipper's bowl (Figure 2-6). The first moderately bright star you come to is Polaris, also called the North Star because it is located almost directly over Earth's north pole. If you draw a line from

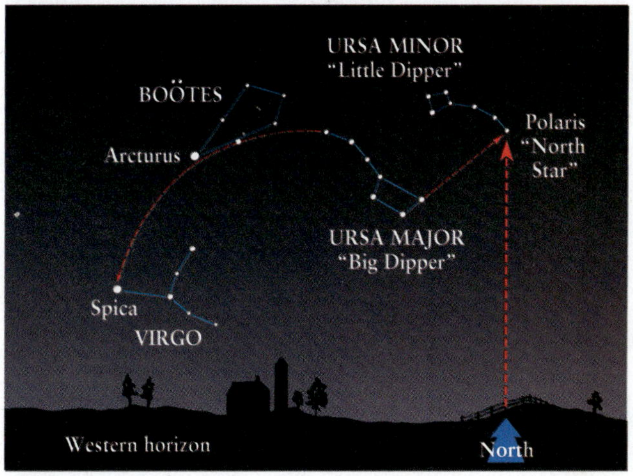

Figure 2-6

The Big Dipper as a Guide The North Star can be seen from anywhere in the northern hemisphere on any night of the year. This star chart shows how the Big Dipper can be used to point out the North Star as well as the brightest stars in two other constellations. The chart shows the sky at around 11 P.M. (daylight savings time) on August 1. Due to Earth's orbital motion around the Sun, you will see this same view at 1 A.M. on July 1 and at 9 P.M. on September 1. The angular distance from Polaris to Spica is 102°.

Polaris straight down to the horizon, you will find the north direction.

As Figure 2-6 shows, by following the handle of the Big Dipper you can locate the bright reddish star Arcturus in Boötes (the Shepherd) and the prominent bluish star Spica in Virgo (the Virgin). The saying "Follow the arc to Arcturus and speed to Spica" may help you remember these stars, which are conspicuous in the evening sky during the spring and summer.

During winter in the northern hemisphere, you can see some of the brightest stars in the sky. Many of them are in the vicinity of the "winter triangle" (Figure 2-7), which connects bright stars in the constellations of Orion (the Hunter), Canis Major (the Large Dog), and Canis Minor (the Small Dog).

A similar feature, the "summer triangle," graces the summer sky in the northern hemisphere. This triangle connects the brightest stars in Lyra (the Harp), Cygnus (the Swan), and Aquila (the Eagle) (Figure 2-8). A conspicuous portion of the Milky Way forms a beautiful background for these constellations, which are nearly overhead during the middle of summer at midnight.

A wonderful tool to help you find your way around the night sky is the planetarium program *Starry Night Enthusiast*™, which is on the CD-ROM that accompanies certain print copies of this book. (You can also obtain *Starry Night Enthusiast*™ separately.) In addition, at the end of this book you will find a set of selected star charts for the evening hours of all 12 months of the year. You may find stargazing an enjoyable experience, and *Starry Night Enthusiast*™ and the star charts will help you identify many well-known constellations.

Note that all the star charts in this section and at the end of this book are drawn for an observer in the northern hemisphere. If you live in the southern hemisphere, you can see constellations that are not visible from the northern hemisphere, and vice versa. In the next section we will see why this is so.

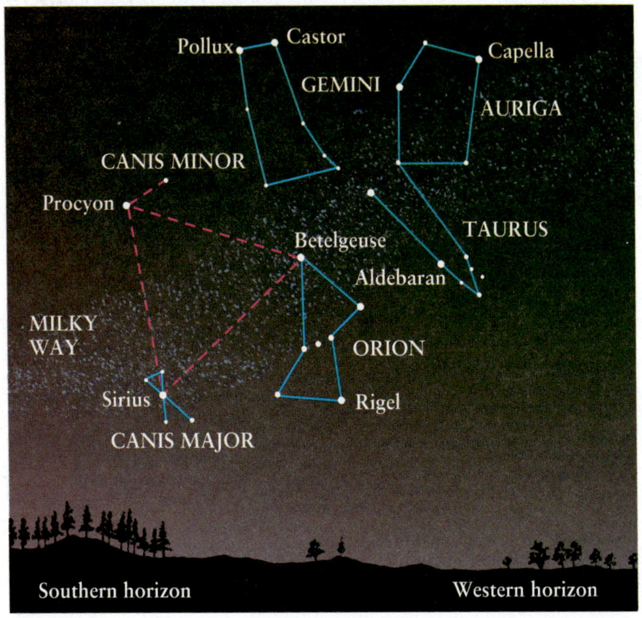

Figure 2-7

The "Winter Triangle" This star chart shows the view toward the southwest on a winter evening in the northern hemisphere (around midnight on January 1, 10 P.M. on February 1, or 8 P.M. on March 1). Three of the brightest stars in the sky make up the "winter triangle," which is about 26° on a side. In addition to the constellations involved in the triangle, the chart shows the prominent constellations Gemini (the Twins), Auriga (the Charioteer), and Taurus (the Bull).

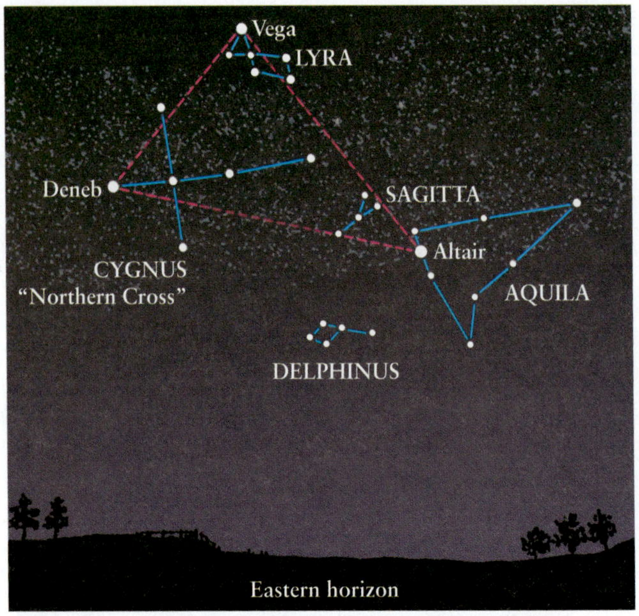

Figure 2-8

The "Summer Triangle" This star chart shows the eastern sky as it appears in the evening during spring and summer in the northern hemisphere (around 1 A.M. daylight savings time on June 1, around 11 P.M. on July 1, and around 9 P.M. on August 1). The angular distance from Deneb to Altair is about 38°. The constellations Sagitta (the Arrow) and Delphinus (the Dolphin) are much fainter than the three constellations that make up the triangle.

2-4 It is convenient to imagine that the stars are located on a celestial sphere

Many ancient societies believed that all the stars are the same distance from Earth. They imagined the stars to be bits of fire imbedded in the inner surface of an immense hollow sphere, called the **celestial sphere,** with Earth at its center. In this picture of the universe, Earth was fixed and did not rotate. Instead, the entire celestial sphere rotated once a day around Earth from east to west, thereby causing the diurnal motion of the sky. The picture of a rotating celestial sphere fit well with naked-eye observations, and for its time was a

The concept of the imaginary celestial sphere helps us visualize the motions of stars in the sky

useful model of how the universe works. (We discussed the role of models in science in Section 1-1).

Today's astronomers know that this simple model of the universe is not correct. Diurnal motion is due to the rotation of Earth, not the rest of the universe. Furthermore, as we learned when discussing the constellations in Section 2-2, the stars are not all at the same distance from Earth. Indeed, the stars that you can see with the naked eye range from 4.2 to more than 1000 light-years away, and telescopes allow us to see objects at distances of billions of light-years.

Thus, astronomers now recognize that the celestial sphere is an *imaginary* object that has no basis in physical reality. Nonetheless, the celestial sphere model remains a useful tool of positional astronomy. If we imagine, as did the ancients, that Earth is stationary and that the celestial sphere rotates around us, it is relatively easy to specify the directions to different objects in the sky and to visualize the motions of these objects.

Figure 2-9 depicts the celestial sphere, with Earth at its center. (A truly proportional drawing would show the celestial sphere as being millions of times larger than Earth.) We picture the stars as points of light that are fixed on the inner surface of the celestial sphere. If we project Earth's equator out into space, we obtain the **celestial equator.** The celestial equator divides the sky into northern and southern hemispheres, just as Earth's equator divides Earth into two hemispheres.

If we project Earth's north and south poles into space, we obtain the **north celestial pole** and the **south celestial pole.** The two celestial poles are where Earth's axis of rotation (extended out into space) intersects the celestial sphere (see Figure 2-9). The star Polaris is less than 1° away from the north celestial pole, which is why it is called the North Star or the Pole Star.

The point in the sky directly overhead an observer anywhere on Earth is called that observer's **zenith.** The zenith and celestial sphere are shown in Figure 2-10 for an observer located at 35° north latitude (that is, at a location on Earth's surface 35° north of the equator). The zenith is shown at the top of Figure 2-10, so Earth and the celestial sphere appear "tipped" compared to Figure 2-9. At any time, an observer can see only half of the celestial sphere; the other half is below the horizon, hidden by the body of Earth. The hidden half of the celestial sphere is darkly shaded in Figure 2-10.

Motions of the Celestial Sphere

For an observer anywhere in the northern hemisphere, including the observer in Figure 2-10, the north celestial pole is always

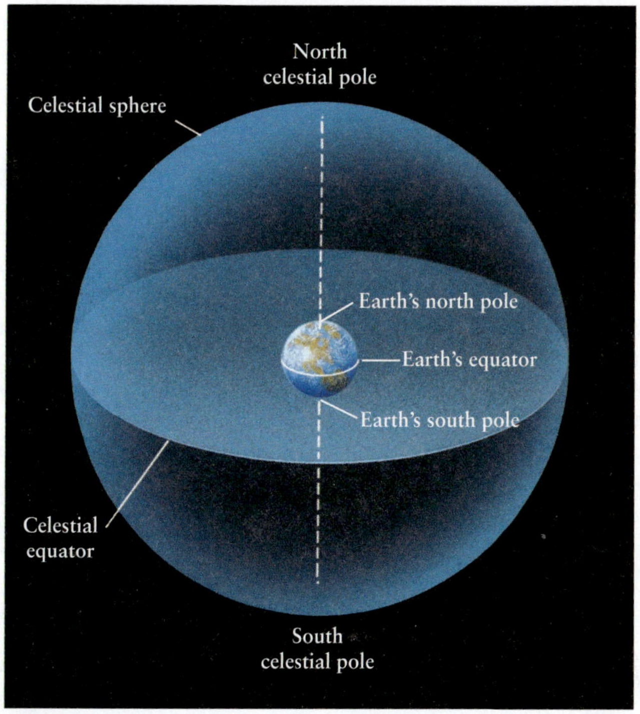

ANIMATION 2.1

Figure 2-9

The Celestial Sphere The celestial sphere is the apparent sphere of the sky. The view in this figure is from the outside of this (wholly imaginary) sphere. Earth is at the center of the celestial sphere, so our view is always of the *inside* of the sphere. The celestial equator and poles are the projections of Earth's equator and axis of rotation out into space. The celestial poles are therefore located directly over Earth's poles.

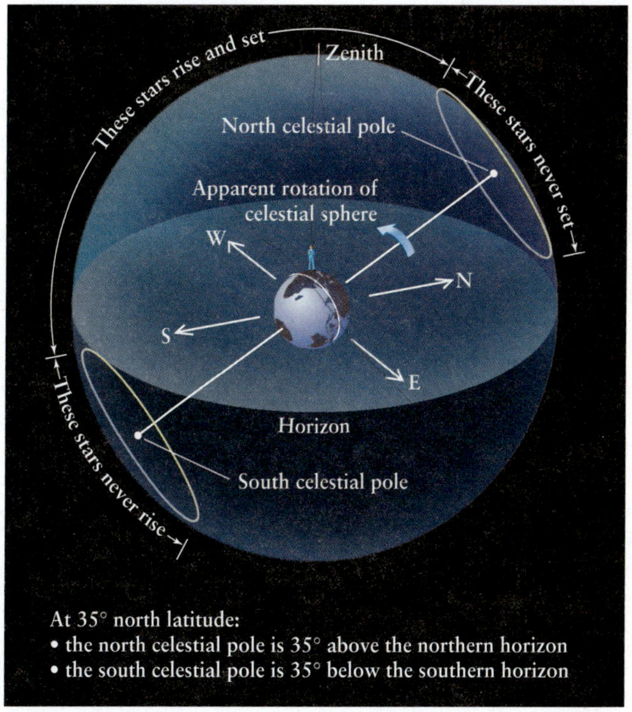

Figure 2-10

The View from 35° North Latitude To an observer at 35° north latitude (roughly the latitude of Los Angeles, Atlanta, Tel Aviv, and Tokyo), the north celestial pole is always 35° above the horizon. Stars within 35° of the *north* celestial pole are circumpolar; they trace out circles around the north celestial pole during the course of the night, and are always above the horizon on any night of the year. Stars within 35° of the *south* celestial pole are always below the horizon and can never be seen from this latitude. Stars that lie between these two extremes rise in the east and set in the west.

(a) At middle northern latitudes (b) At the north pole (c) At the equator

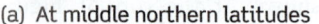

Figure 2-11 RIVUXG

The Apparent Motion of Stars at Different Latitudes As Earth rotates, stars appear to rotate around us along paths that are parallel to the celestial equator. **(a)** As shown in this long time exposure, at most locations on Earth the rising and setting motions are at an angle to the horizon that depends on the latitude. **(b)** At the north pole (latitude 90° north) the stars appear to move parallel to the horizon. **(c)** At the equator (latitude 0°) the stars rise and set along vertical paths. (a: David Miller/DMI)

above the horizon. As Earth turns from west to east, it appears to the observer that the celestial sphere turns from east to west. Stars sufficiently near the north celestial pole revolve around the pole, never rising or setting. Such stars are called **circumpolar.** For example, as seen from North America or Europe, Polaris is a circumpolar star and can be seen at any time of night on any night of the year. The photograph that opens this chapter shows the circular trails of stars around the north celestial pole as seen from Hawaii (at 20° north latitude). Stars near the south celestial pole revolve around that pole but always remain below the horizon of an observer in the northern hemisphere. Hence, these stars can never be seen by the observer in Figure 2-10. Stars between those two limits rise in the east and set in the west.

CAUTION! Keep in mind that which stars are circumpolar, which stars never rise, and which stars rise and set depends on the latitude from which you view the heavens. As an example, for an observer at 35° south latitude (roughly the latitude of Sydney, Cape Town, and Buenos Aires), the roles of the north and south celestial poles are the opposite of those shown in Figure 2-10: Objects close to the *south* celestial pole are circumpolar, that is, they revolve around that pole and never rise or set. For an observer in the southern hemisphere, stars close to the *north* celestial pole are always below the horizon and can never be seen. Hence, astronomers in Australia, South Africa, and Argentina never see the North Star but are able to see other stars that are forever hidden from North American or European observers.

For observers at most locations on Earth, stars rise in the east and set in the west at an angle to the horizon (Figure 2-11*a*). To see why this is so, notice that the rotation of the celestial sphere carries stars across the sky in paths that are parallel to the celestial equator. If you stand at the north pole, the north celestial pole is directly above you at the zenith (see Figure 2-9) and the celestial equator lies all around you at the horizon. Hence, as the celestial sphere rotates, the stars appear to move parallel to the horizon (Figure 2-11*b*). If instead you stand on the equator, the celestial equator passes from the eastern horizon through the zenith to the western horizon. The north and south celestial poles are 90° away from the celestial equator, so they lie on the northern and southern horizons, respectively. As the celestial sphere rotates around an axis from pole to pole, the stars rise and set straight up and down—that is, in a direction perpendicular to the horizon (Figure 2-11*c*). At any location on Earth between the equator and either pole, the rising and setting motions of the stars are at an angle intermediate between Figures 2-11*b* and 2-11*c*. The particular angle depends on the latitude.

Using the celestial equator and poles, we can define a coordinate system to specify the position of any star on the celestial sphere. As Box 2-1 describes, the most commonly used coordinate system uses two angles, *right ascension* and *declination,* that are analogous to longitude and latitude on Earth. These coordinates tell us in what direction we should look to see the star. To locate the star's true position in three-dimensional space, we must also know the distance to the star.

2-5 The seasons are caused by the tilt of Earth's axis of rotation

In addition to rotating on its axis every 24 hours, Earth revolves around the Sun—that is, it *orbits* the Sun—in about 365¼ days (see Figure 2-5). As we travel with Earth around its orbit, we experience the annual cycle of seasons. But why *are* there seasons? Furthermore, the seasons are opposite in the northern and southern hemispheres. For example, February is midwinter in North America but midsummer in Australia. Why should this be?

The celestial sphere concept helps us visualize how the Sun appears to move relative to the stars

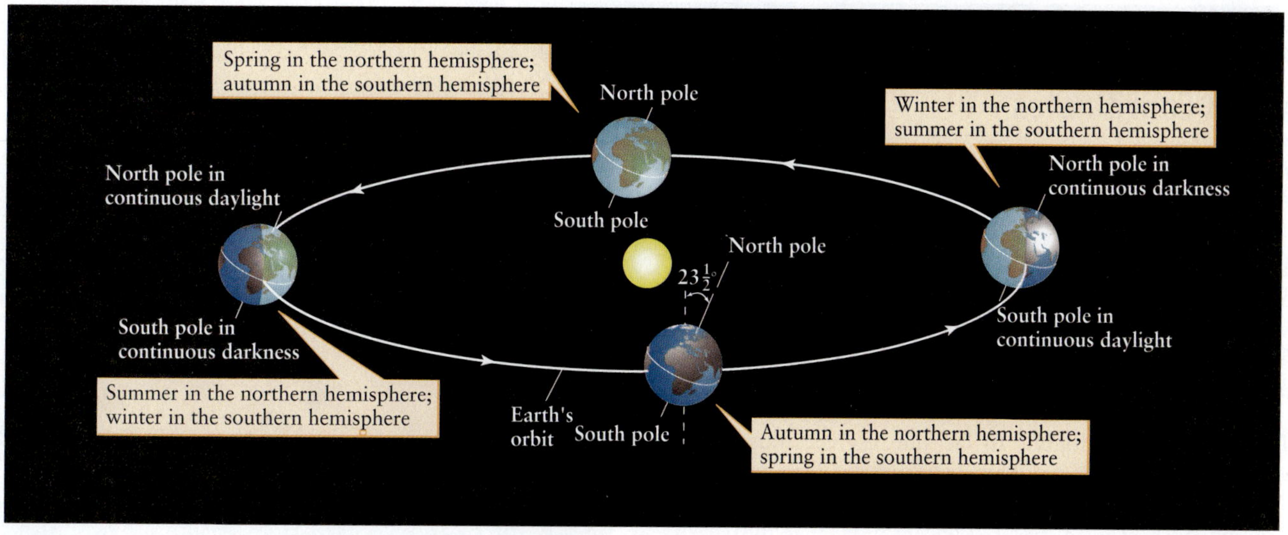

Figure 2-12

The Seasons Earth's axis of rotation is inclined $23\frac{1}{2}°$ away from the perpendicular to the plane of Earth's orbit. The north pole is aimed at the north celestial pole, near the star Polaris. Earth maintains this orientation as it orbits the Sun. Consequently, the amount of solar illumination and the number of daylight hours at any location on Earth vary in a regular pattern throughout the year. This is the origin of the seasons.

The Origin of the Seasons

The reason why we have seasons, and why they are different in different hemispheres, is that Earth's axis of rotation is not perpendicular to the plane of Earth's orbit. Instead, as Figure 2-12 shows, the axis is tilted about $23\frac{1}{2}°$ away from the perpendicular. Earth maintains this tilt as it orbits the Sun, with Earth's north pole pointing toward the north celestial pole. (This stability is a hallmark of all rotating objects. A top will not fall over as long as it is spinning, and the rotating wheels of a motorcycle help to keep the rider upright.)

During part of the year, when Earth is in the part of its orbit shown on the left side of Figure 2-12, the northern hemisphere is tilted toward the Sun. As Earth spins on its axis, a point in the northern hemisphere spends more than 12 hours in the sunlight. Thus, the days there are long and the nights are short, and it is summer in the northern hemisphere. The summer is hot not only because of the extended daylight hours but also because the Sun is high in the northern hemisphere's sky. As a result, sunlight strikes the ground at a nearly perpendicular angle that heats the ground efficiently (Figure 2-13*a*). During this same time of year in the southern hemisphere, the days are short and the nights are

(a) The Sun in summer

(b) The Sun in winter

Figure 2-13

Solar Energy in Summer and Winter At different times of the year, sunlight strikes the ground at different angles. **(a)** In summer, sunlight is concentrated and the days are also longer, which further increases the heating. **(b)** In winter the sunlight is less concentrated, the days are short, and little heating of the ground takes place. This accounts for the low temperatures in winter.

BOX 2-1 Tools of the Astronomer's Trade

Celestial Coordinates

Your *latitude* and *longitude* describe where on Earth's surface you are located. The latitude of your location denotes how far north or south of the equator you are, and the longitude of your location denotes how far west or east you are of an imaginary circle that runs from the north pole to the south pole through the Royal Observatory in Greenwich, England. In an analogous way, astronomers use coordinates called *declination* and *right ascension* to describe the position of a planet, star, or galaxy on the celestial sphere.

Declination is analogous to latitude. As the illustration shows, the **declination** of an object is its angular distance north or south of the celestial equator, measured along a circle passing through both celestial poles. Like latitude, it is measured in degrees, arcminutes, and arcseconds (see Section 1-5).

Right ascension is analogous to longitude. It is measured from a line that runs between the north and south celestial poles and passes through a point on the celestial equator called the *vernal equinox* (shown as a blue dot in the illustration). This point is one of two locations where the Sun crosses the celestial equator during its apparent annual motion, as we discuss in Section 2-5. In Earth's northern hemisphere, spring officially begins when the Sun reaches the vernal equinox in late March. The **right ascension** of an object is the angular distance from the vernal equinox eastward along the celestial equator to the circle used in measuring its declination (see illustration). Astronomers measure right ascension in *time* units (hours, minutes, and seconds), corresponding to the time required for the celestial sphere to rotate through this angle. For example, suppose there is a star at your zenith right now with right ascension $6^h\ 0^m\ 0^s$. Two hours and 30 minutes from now, there will be a different object at your zenith with right ascension $8^h\ 30^m\ 0^s$.

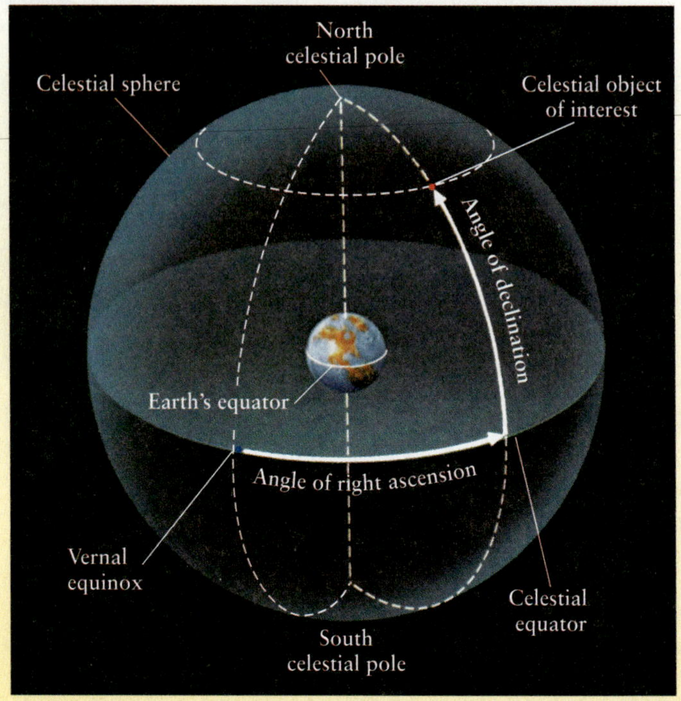

long, because a point in this hemisphere spends fewer than 12 hours a day in the sunlight. The Sun is low in the sky, so sunlight strikes the surface at a grazing angle that causes little heating (Figure 2-13*b*), and it is winter in the southern hemisphere.

Half a year later, Earth is in the part of its orbit shown on the right side of Figure 2-12. Now the situation is reversed, with winter in the northern hemisphere (which is now tilted away from the Sun) and summer in the southern hemisphere. During spring and autumn, the two hemispheres receive roughly equal amounts of illumination from the Sun, and daytime and nighttime are of roughly equal length everywhere on Earth.

CAUTION! A common misconception is that the seasons are caused by variations in the distance from Earth to the Sun. According to this idea, Earth is closer to the Sun in summer and farther away in winter. But in fact, Earth's orbit around the Sun is very nearly circular, and the Earth-Sun distance varies only about 3% over the course of a year. (Earth's orbit only *looks* elongated in Figure 2-12 because this illustration shows an oblique view.) We are slightly closer to the Sun in January than in July, but this small variation has little influence on the cycle of the seasons. Also, if the seasons were really caused by variations in the Earth-Sun distance, the seasons would be the same in both hemispheres!

How the Sun Moves on the Celestial Sphere

The plane of Earth's orbit around the Sun is called the **ecliptic plane** (Figure 2-14*a*). As a result of Earth's annual motion around the Sun, it appears to us (observing from Earth) that the Sun slowly changes its position on the celestial sphere over the course of a year. The circular path that the Sun appears to trace out against the background of stars is called the **ecliptic** (Figure 2-14*b*). The plane of this path is the same as the ecliptic plane. (The name *ecliptic* suggests that the path traced out by the Sun has something to do with eclipses. We will discuss the connection in Chapter 3.) Because there are $365\frac{1}{4}$ days in a year and 360° in a circle, the Sun appears to move along the ecliptic at a rate of about 1° per day. This motion is from west to east, that is, in the direction opposite to the apparent motion of the celestial sphere.

INTERACTIVE EXERCISE 2.1 The coordinates of the bright star Rigel for the year 2000 are R.A. = $5^h\ 14^m\ 32.2^s$, Decl. = $-8°\ 12'\ 06''$. (R.A. and Decl. are abbreviations for right ascension and declination.) A minus sign on the declination indicates that the star is south of the celestial equator; a plus sign (or no sign at all) indicates that an object is north of the celestial equator. As we discuss in Box 2-2, right ascension helps determine the best time to observe a particular object.

It is important to state the year for which a star's right ascension and declination are valid. This is so because of precession, which we discuss in Section 2-6.

EXAMPLE: What are the coordinates of a star that lies exactly halfway between the vernal equinox and the south celestial pole?

Situation: Our goal is to find the right ascension and declination of the star in question.

Tools: We use the definitions depicted in the figure.

Answer: Since the circle used to measure this star's declination passes through the vernal equinox, this star's right ascension is R.A. = $0^h\ 0^m\ 0^s$. The angle between the celestial equator and south celestial pole is $90°\ 0'\ 0''$, so the declination of this star is Decl. = $-45°\ 0'\ 0''$.

Review: The declination in this example is negative because the star is in the southern half of the celestial sphere.

EXAMPLE: At midnight local time you see a star with R.A. = $2^h\ 30^m\ 0^s$ at your zenith. When will you see a star at your zenith with R.A. = $21^h\ 0^m\ 0^s$?

Situation: If you held your finger stationary over a globe of Earth, the longitude of the point directly under your finger would change as you rotated the globe. In the same way, the right ascension of the point directly over your head (the zenith) changes as the celestial sphere rotates. We use this concept to determine the time in question.

Tools: We use the idea that a change in right ascension of 24^h corresponds to an elapsed time of 24 hours and a complete rotation of the celestial sphere.

Answer: The time required for the sky to rotate through the angle between the stars is the difference in their right ascensions: $21^h\ 0^m\ 0^s - 2^h\ 30^m\ 0^s = 18^h\ 30^m\ 0^s$. So the second star will be at your zenith $18\frac{1}{2}$ hours after the first one, or at 6:30 P.M. the following evening.

Review: Our answer was based on the idea that the celestial sphere makes *exactly* one complete rotation in 24 hours. If this were so, from one night to the next each star would be in exactly the same position at a given time. But because of the way that we customarily measure time, the celestial sphere makes slightly more than one complete rotation in 24 hours. (We explore the reasons for this in Box 2-2.) As a result, our answer is in error by about 3 minutes. For our purposes, this is a small enough error that we can ignore it.

ANALOGY Note that at the same time that the Sun is making its yearlong trip around the ecliptic, the entire celestial sphere is rotating around us once per day. You can envision the celestial sphere as a merry-go-round rotating clockwise, and the Sun as a restless child who is walking slowly around the merry-go-round's rim in the counterclockwise direction. During the time it takes the child to make a round trip, the merry-go-round rotates $365\frac{1}{4}$ times.

The ecliptic plane is *not* the same as the plane of Earth's equator, thanks to the $23\frac{1}{2}°$ tilt of Earth's rotation axis shown in Figure 2-12. As a result, the ecliptic and the celestial equator are inclined to each other by that same $23\frac{1}{2}°$ angle (Figure 2-15).

Equinoxes and Solstices

The ecliptic and the celestial equator intersect at only two points, which are exactly opposite each other on the celestial sphere. Each point is called an **equinox** (from the Latin for "equal night"), because when the Sun appears at either of these points, day and night are each about 12 hours long at all locations on Earth. The term "equinox" is also used to refer to the date on which the Sun passes through one of these special points on the ecliptic.

On about March 21 of each year, the Sun passes northward across the celestial equator at the **vernal equinox.** This marks the beginning of spring in the northern hemisphere ("vernal" is from the Latin for "spring"). On about September 22 the Sun moves southward across the celestial equator at the **autumnal equinox,** marking the moment when fall begins in the northern hemisphere. Since the seasons are opposite in the northern and southern hemispheres, for Australians, South Africans, and South Americans the vernal equinox actually marks the beginning of autumn. The names of the equinoxes come from astronomers of the past who lived north of the equator.

Between the vernal and autumnal equinoxes lie two other significant locations along the ecliptic. The point on the ecliptic farthest north of the celestial equator is called the **summer solstice.** "Solstice" is from the Latin for "solar standstill," and it is at the summer solstice that the Sun stops moving northward on the

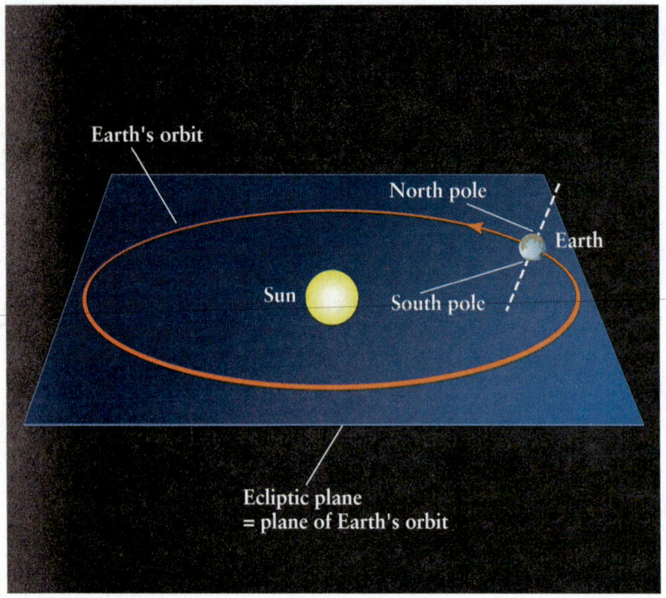

(a) In reality Earth orbits the Sun once a year

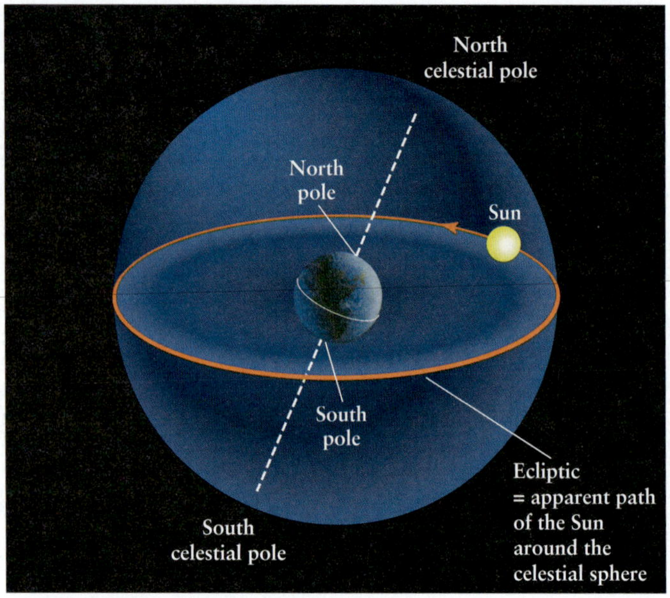

(b) It appears from Earth that the Sun travels around the celestial sphere once a year

Figure 2-14

The Ecliptic Plane and the Ecliptic **(a)** The ecliptic plane is the plane in which Earth moves around the Sun. **(b)** As seen from Earth, the Sun appears to move around the celestial sphere along a circular path called the ecliptic. Earth takes a year to complete one orbit around the Sun, so as seen by us the Sun takes a year to make a complete trip around the ecliptic.

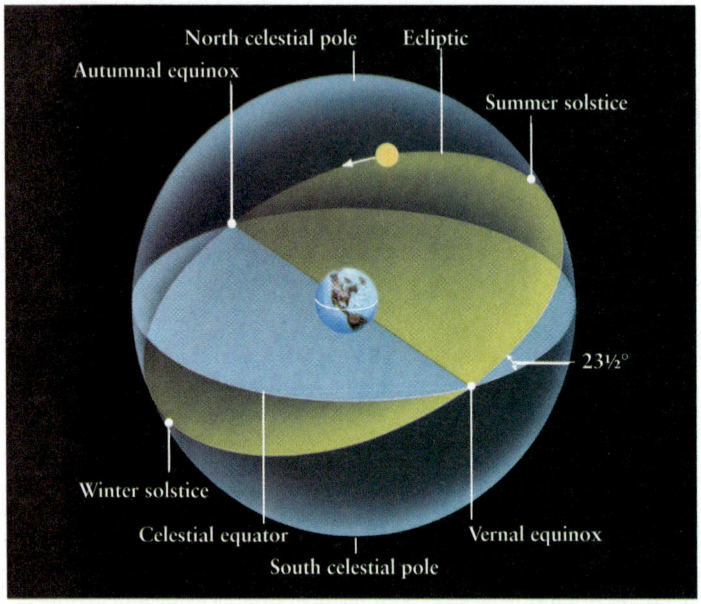

Figure 2-15

The Ecliptic, Equinoxes, and Solstices This illustration of the celestial sphere is similar to Figure 2-14*b*, but is drawn with the north celestial pole at the top and the celestial equator running through the middle. The ecliptic is inclined to the celestial equator by $23\frac{1}{2}°$ because of the tilt of Earth's axis of rotation. It intersects the celestial equator at two points, called equinoxes. The northernmost point on the ecliptic is the summer solstice, and the southernmost point is the winter solstice. The Sun is shown in its approximate position for August 1.

celestial sphere. At this point, the Sun is as far north of the celestial equator as it can get. It marks the location of the Sun at the moment summer begins in the northern hemisphere (about June 21). At the beginning of the northern hemisphere's winter (about December 21), the Sun is farthest south of the celestial equator at a point called the **winter solstice.**

Because the Sun's position on the celestial sphere varies slowly over the course of a year, its daily path across the sky (due to Earth's rotation) also varies with the seasons (Figure 2-16). On the first day of spring or the first day of fall, when the Sun is at one of the equinoxes, the Sun rises directly in the east and sets directly in the west.

When the northern hemisphere is tilted away from the Sun and it is winter in the northern hemisphere, the Sun rises in the southeast. Daylight lasts for fewer than 12 hours as the Sun skims low over the southern horizon and sets in the southwest. Northern hemisphere nights are longest when the Sun is at the winter solstice.

The closer you get to the north pole, the shorter the winter days and the longer the winter nights. In fact, anywhere within $23\frac{1}{2}°$ of the north pole (that is, north of latitude $90° - 23\frac{1}{2}° = 66\frac{1}{2}°$ N) the Sun is below the horizon for 24 continuous hours at least one day of the year. The circle around Earth at $66\frac{1}{2}°$ north latitude is called the **Arctic Circle** (Figure 2-17). The corresponding region around the south pole is bounded by the **Antarctic Circle** at $66\frac{1}{2}°$ south latitude. At the time of the winter solstice, explorers south of the Antarctic Circle enjoy "the midnight sun," or 24 hours of continuous daylight.

During summer in the northern hemisphere, when the northern hemisphere is tilted toward the Sun, the Sun rises in the north-

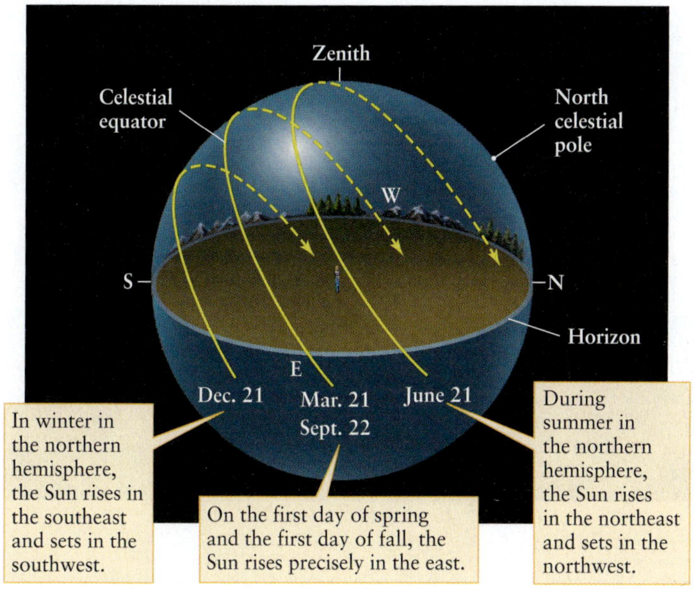

Figure 2-16

The Sun's Daily Path Across the Sky This drawing shows the apparent path of the Sun during the course of a day on four different dates. Like Figure 2-10, this drawing is for an observer at 35° north latitude.

east and sets in the northwest. The Sun is at its northernmost position at the summer solstice, giving the northern hemisphere the greatest number of daylight hours. At the summer solstice the Sun does not set at all north of the Arctic Circle (Figure 2-18) and does not rise at all south of the Antarctic Circle.

The variations of the seasons are much less pronounced close to the equator. Between the **Tropic of Capricorn** at $23\frac{1}{2}°$ south latitude and the **Tropic of Cancer** at $23\frac{1}{2}°$ north latitude, the Sun is directly overhead—that is, at the zenith—at high noon at least one day a year. Outside of the tropics, the Sun is never directly overhead, but is always either south of the zenith (as seen from locations north of the Tropic of Cancer) or north of the zenith (as seen from south of the Tropic of Capricorn).

2-6 The Moon helps to cause precession, a slow, conical motion of Earth's axis of rotation

The Moon is by far the brightest and most obvious naked-eye object in the nighttime sky. Like the Sun, the Moon slowly changes its position relative to the background stars; unlike the Sun, the Moon makes a complete trip around the celestial sphere in only about 4 weeks, or about a month. (The word "month" comes from the same Old English root as the word "moon.") Ancient astronomers realized that this motion occurs because the Moon orbits Earth in roughly 4 weeks. In 1 hour the Moon moves on the celestial sphere by about $\frac{1}{2}°$, or roughly its own angular size.

Precession causes the apparent positions of the stars to slowly change over the centuries

The Moon's path on the celestial sphere is never far from the Sun's path (that is, the ecliptic). This is because the plane of the Moon's orbit around Earth is inclined only slightly from the plane of Earth's orbit around the Sun (the ecliptic plane shown in Figure 2-14*a*). The Moon's path varies somewhat from one month

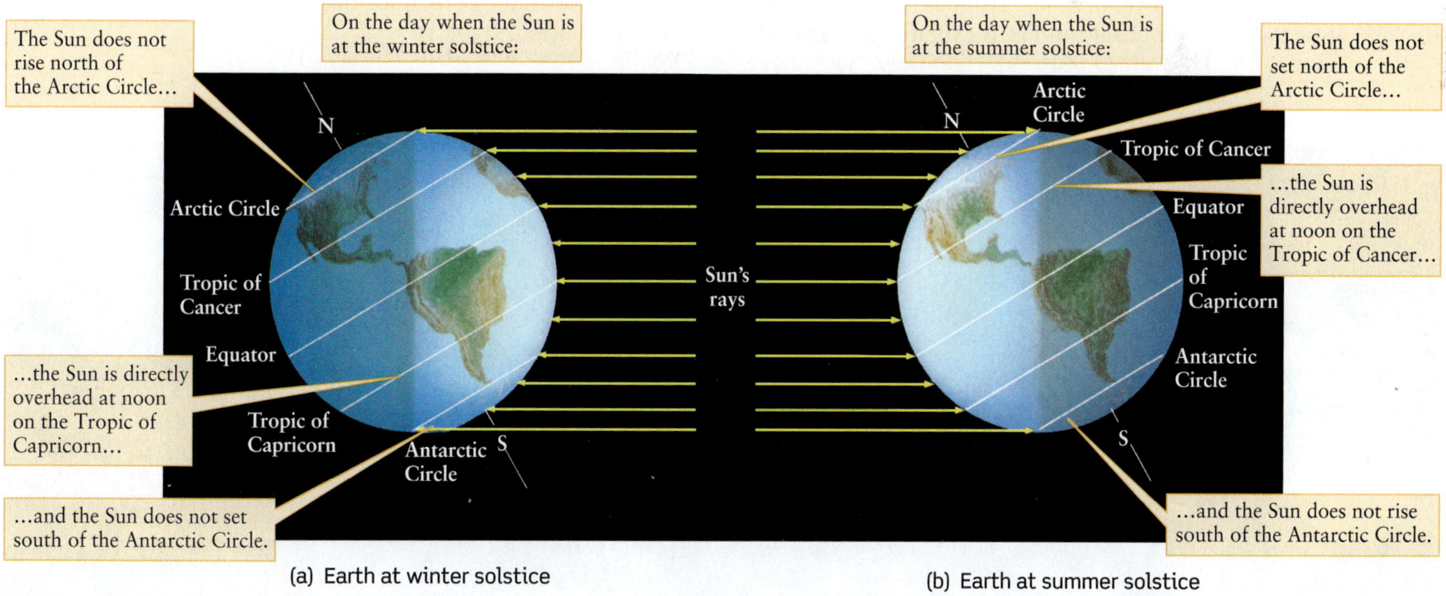

Figure 2-17

Tropics and Circles Four important latitudes on Earth are the Arctic Circle ($66\frac{1}{2}°$ north latitude), Tropic of Cancer ($23\frac{1}{2}°$ north latitude), Tropic of Capricorn ($23\frac{1}{2}°$ south latitude), and Antarctic Circle ($66\frac{1}{2}°$ south latitude). These drawings show the significance of these latitudes when the Sun is **(a)** at the winter solstice and **(b)** at the summer solstice.

Figure 2-18 R I V U X G

The Midnight Sun This time-lapse photograph was taken on July 19, 1985, at 69° north latitude in northeast Alaska. At this latitude, the Sun is above the horizon continuously (that is, it is circumpolar) from mid-May to the end of July. (Doug Plummer/Science Photo Library)

to the next, but always remains within a band called the **zodiac** that extends about 8° on either side of the ecliptic. Twelve famous constellations—Aries, Taurus, Gemini, Cancer, Leo, Virgo, Libra, Scorpius, Sagittarius, Capricornus, Aquarius, and Pisces—lie along the zodiac. The Moon is generally found in one of these 12 constellations. (Thanks to a redrawing of constellation boundaries in the mid-twentieth century, the zodiac actually passes through a thirteenth constellation—Ophiuchus, the Serpent Bearer—between Scorpius and Sagittarius.) As it moves along its orbit, the Moon appears north of the celestial equator for about two weeks and then south of the celestial equator for about the next two weeks. We will learn more about the Moon's motion, as well as why the Moon goes through phases, in Chapter 3.

The Moon not only moves around Earth but, in concert with the Sun, also causes a slow change in Earth's rotation. This is because both the Sun and the Moon exert a gravitational pull on Earth. We will learn much more about gravity in Chapter 4; for now, all we need is the idea that gravity is a universal attraction of matter for other matter.

The gravitational pull of the Sun and the Moon affects Earth's rotation because Earth is slightly fatter across the equator than it is from pole to pole: Its equatorial diameter is 43 kilometers (27 miles) larger than the diameter measured from pole to pole. Earth is therefore said to have an "equatorial bulge." Because of the gravitational pull of the Moon and the Sun on this bulge, the orientation of Earth's axis of rotation gradually changes.

Earth behaves somewhat like a spinning top, as illustrated in Figure 2-19. If the top were not spinning, gravity would pull the top over on its side. When the top is spinning, gravity causes the top's axis of rotation to trace out a circle, producing a motion called **precession.**

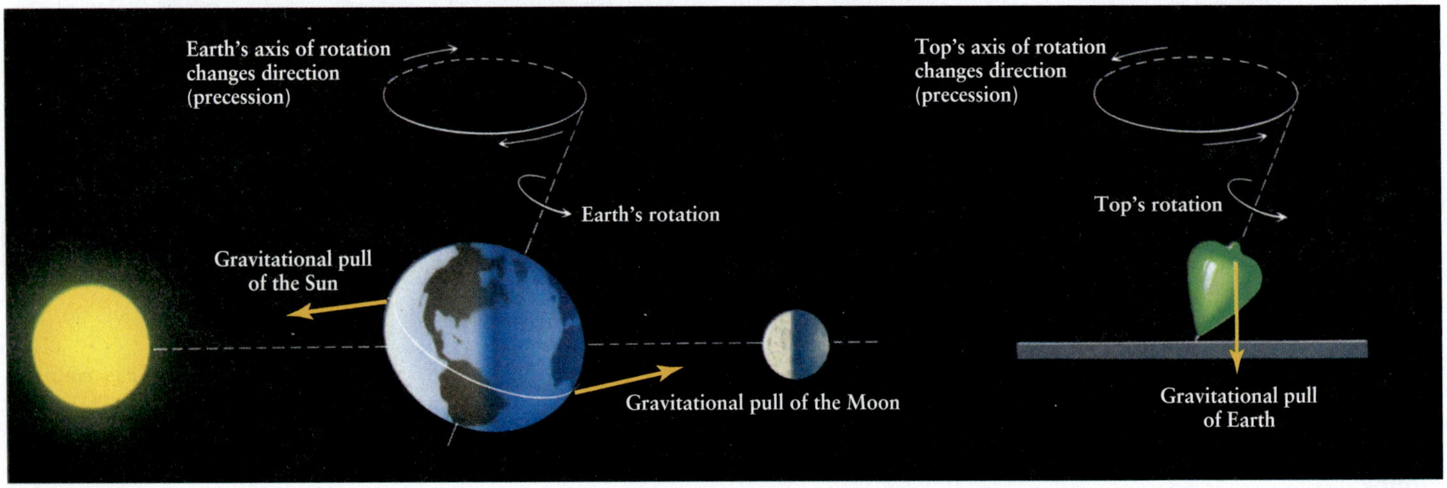

Figure 2-19

Precession Because Earth's rotation axis is tilted, the gravitational pull of the Moon and the Sun on Earth's equatorial bulge together cause Earth to precess. As Earth precesses, its axis of rotation slowly traces out a circle in the sky, like the shifting axis of a spinning top.

As the Sun and Moon move along the zodiac, each spends half its time north of Earth's equatorial bulge and half its time south of it. The gravitational pull of the Sun and Moon tugging on the equatorial bulge tries to twist Earth's axis of rotation to be perpendicular to the plane of the ecliptic. But because Earth is spinning, the combined actions of gravity and rotation cause Earth's axis to trace out a circle in the sky, much like what happens to the toy top. As the axis precesses, it remains tilted about $23\frac{1}{2}°$ to the perpendicular.

As Earth's axis of rotation slowly changes its orientation, the north and south celestial poles—which are the projections of that axis onto the celestial sphere—change their positions relative to the stars. At present, the north celestial pole lies within 1° of the star Polaris, which is why Polaris is the North Star. But 5000 years ago, the north celestial pole was closest to the star Thuban in the constellation of Draco (the Dragon). Thus, that star and not Polaris was the North Star. And 12,000 years from now, the North Star will be the bright star Vega in Lyra (the Harp). It takes 26,000 years for the north celestial pole to complete one full precessional circle around the sky (Figure 2-20). The south celestial pole executes a similar circle in the southern sky.

Precession also causes Earth's equatorial plane to change its orientation. Because this plane defines the location of the celestial equator in the sky, the celestial equator precesses as well. The intersections of the celestial equator and the ecliptic define the equinoxes (see Figure 2-15), so these key locations in the sky also shift slowly from year to year. For this reason, the precession of Earth is also called the **precession of the equinoxes.** The first person to detect the precession of the equinoxes, in the second century B.C., was the Greek astronomer Hipparchus, who compared his own observations with those of Babylonian astronomers three centuries earlier. Today, the vernal equinox is located in the constellation Pisces (the Fishes). Two thousand years ago, it was in Aries (the Ram). Around the year A.D. 2600, the vernal equinox will move into Aquarius (the Water Bearer).

CAUTION! Astrological terms like the "Age of Aquarius" involve boundaries in the sky that are not recognized by astronomers and are generally not even related to the positions of the constellations. For example, most astrologers would call a person born on March 21, 1988, an "Aries" because the Sun was supposedly in the direction of that constellation on March 21. But due to precession, the Sun was actually in the constellation Pisces on that date! Indeed, astrology is *not* a science at all, but merely a collection of superstitions and hokum. Its practitioners use some of the terminology of astronomy but reject the logical thinking that is at the heart of science. James Randi has more to say about astrology and other pseudosciences in his essay "Why Astrology Is Not Science" at the end of this chapter.

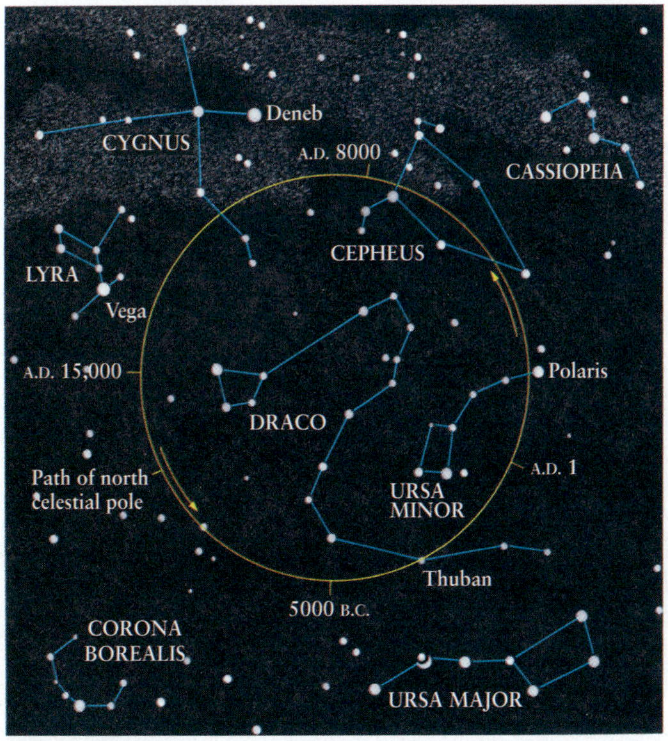

Figure 2-20

Precession and the Path of the North Celestial Pole As Earth precesses, the north celestial pole slowly traces out a circle among the northern constellations. At present, the north celestial pole is near the moderately bright star Polaris, which serves as the North Star. Twelve thousand years from now the bright star Vega will be the North Star.

The astronomer's system of locating heavenly bodies by their right ascension and declination, discussed in Box 2-1, is tied to the positions of the celestial equator and the vernal equinox. Because of precession, these positions are changing, and thus the coordinates of stars in the sky are also constantly changing. These changes are very small and gradual, but they add up over the years. To cope with this difficulty, astronomers always make note of the date (called the **epoch**) for which a particular set of coordinates is precisely correct. Consequently, star catalogs and star charts are periodically updated. Most current catalogs and star charts are prepared for the epoch 2000. The coordinates in these reference books, which are precise for January 1, 2000, will require very little correction over the next few decades.

2-7 Positional astronomy plays an important role in keeping track of time

Astronomers have traditionally been responsible for telling time. This is because we want the system of timekeeping used in everyday life to reflect the position of the Sun in the sky. Thousands of years ago, the sundial was invented to keep track of **apparent solar time.** To obtain more accurate measurements astronomers use the **meridian.** As Figure 2-21 shows, this is a north-south circle on the celestial sphere that passes through the zenith (the point directly overhead) and both celestial poles. *Local noon* is defined to be when the Sun crosses the **upper meridian,** which is the half of the meridian above the horizon. At *local midnight,* the Sun crosses the **lower**

Ancient scholars developed a system of timekeeping based on the Sun

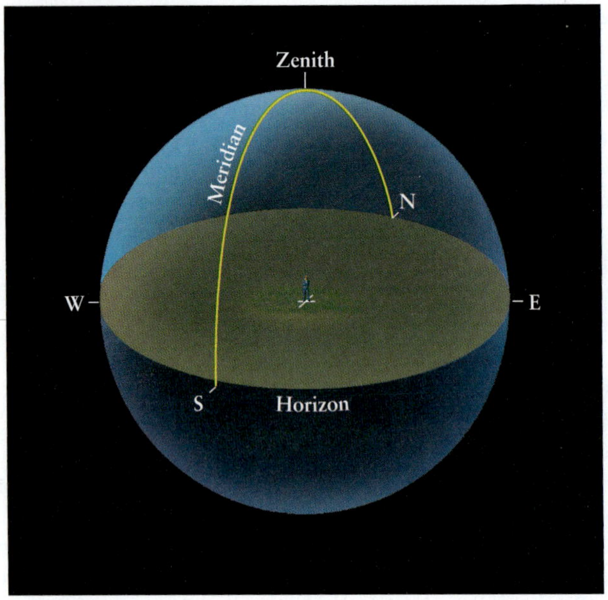

Figure 2-21

The Meridian The meridian is a circle on the celestial sphere that passes through the observer's zenith (the point directly overhead) and the north and south points on the observer's horizon. The passing of celestial objects across the meridian can be used to measure time. The upper meridian is the part above the horizon, and the lower meridian (not shown) is the part below the horizon.

meridian, the half of the meridian below the horizon; this crossing cannot be observed directly.

The crossing of the meridian by any object in the sky is called a **meridian transit** of that object. If the crossing occurs above the horizon, it is an *upper* meridian transit. An **apparent solar day** is the interval between two successive upper meridian transits of the Sun as observed from any fixed spot on Earth. Stated less formally, an apparent solar day is the time from one local noon to the next local noon, or from when the Sun is highest in the sky to when it is again highest in the sky.

The Sun as a Timekeeper

Unfortunately, the Sun is not a good timekeeper (Figure 2-22). The length of an apparent solar day (as measured by a device such as an hourglass) varies from one time of year to another. There are two main reasons why this is so, both having to do with the way in which Earth orbits the Sun.

The first reason is that Earth's orbit is not a perfect circle; rather, it is an ellipse, as Figure 2-22*a* shows in exaggerated form. As we will learn in Chapter 4, Earth moves more rapidly along its orbit when it is near the Sun than when it is farther away. Hence, the Sun appears to us to move more than 1° per day along the ecliptic in January, when Earth is nearest the Sun, and less than 1° per day in July, when Earth is farthest from the Sun. By itself, this effect would cause the apparent solar day to be longer in January than in July.

The second reason why the Sun is not a good timekeeper is the $23\frac{1}{2}°$ angle between the ecliptic and the celestial equator (see

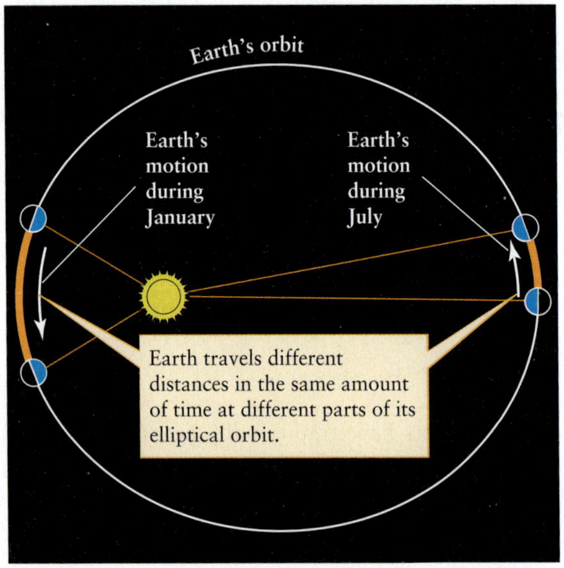

(a) A month's motion of the Earth along its orbit

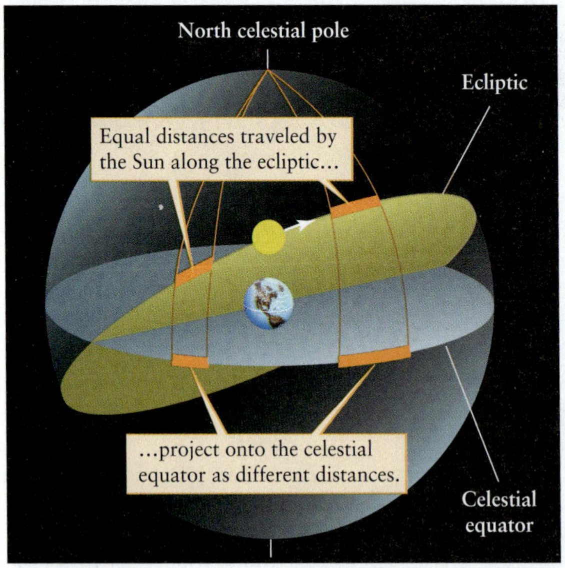

(b) A day's motion of the Sun along the ecliptic

Figure 2-22

Why the Sun Is a Poor Timekeeper There are two main reasons that the Sun is a poor timekeeper. **(a)** Earth's speed along its orbit varies during the year. It moves fastest when closest to the Sun in January and slowest when farthest from the Sun in July. Hence, the apparent speed of the Sun along the ecliptic is not constant. **(b)** Because of the tilt of Earth's rotation axis, the ecliptic is inclined with respect to the celestial equator. Therefore, the projection of the Sun's daily progress along the ecliptic onto the celestial equator (shown in blue) varies during the year. This causes further variations in the length of the apparent solar day.

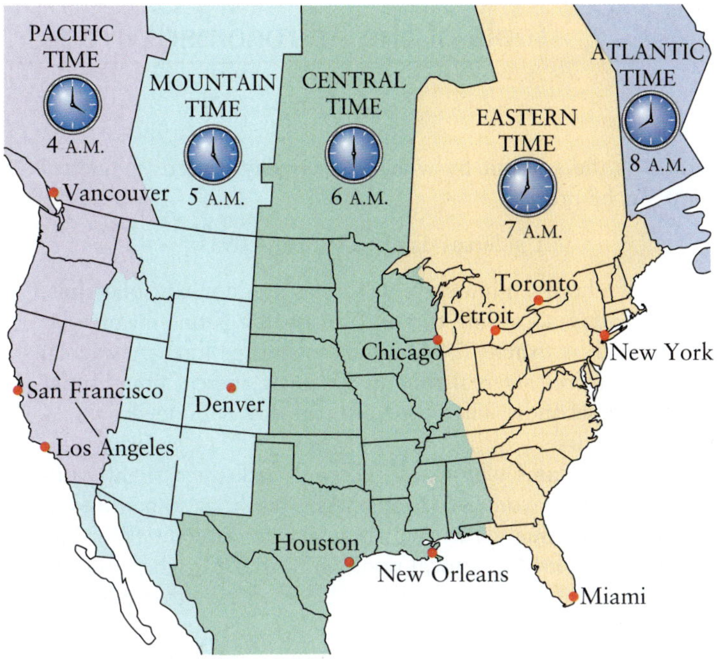

Figure 2-23

Time Zones in North America For convenience, Earth is divided into 24 time zones, generally centered on 15° intervals of longitude around the globe. There are four time zones across the continental United States, making for a 3-hour time difference between New York and California.

Figure 2-15). As Figure 2-22*b* shows, this causes a significant part of the Sun's apparent motion when near the equinoxes to be in a north-south direction. The net daily eastward progress in the sky is then somewhat foreshortened. At the summer and winter solstices, by contrast, the Sun's motion is parallel to the celestial equator. Thus, there is no comparable foreshortening around the beginning of summer or winter. This effect by itself would make the apparent solar day shorter in March and September than in June or December. Combining these effects with those due to Earth's noncircular orbit, we find that the length of the apparent solar day varies in a complicated fashion over the course of a year.

To avoid these difficulties, astronomers invented an imaginary object called the **mean sun** that moves along the celestial equator at a uniform rate. (In science and mathematics, "mean" is a synonym for "average.") The mean sun is sometimes slightly ahead of the real Sun in the sky, sometimes behind. As a result, mean solar time and apparent solar time can differ by as much as a quarter of an hour at certain times of the year.

Because the mean sun moves at a constant rate, it serves as a fine timekeeper. A **mean solar day** is the interval between successive upper meridian transits of the mean sun. It is exactly 24 hours long, the average length of an apparent solar day. One 24-hour day as measured by your alarm clock or wristwatch is a mean solar day.

WEB LINK 2.3 **Time zones** were invented for convenience in commerce, transportation, and communication. In a time zone, all clocks and watches are set to the mean solar time for a meridian of longitude that runs approximately through the center of the zone. Time zones around the world are generally centered on meridians of longitude at 15° intervals. In most cases, going from one time zone to the next requires you to change the time on your wristwatch by exactly 1 hour. The time zones for most of North America are shown in Figure 2-23.

In order to coordinate their observations with colleagues elsewhere around the globe, astronomers often keep track of time using Coordinated Universal Time, somewhat confusingly abbreviated UTC or UT. This is the time in a zone that includes Greenwich, England, a seaport just outside of London where the first internationally accepted time standard was kept. (UTC was formerly known as Greenwich Mean Time.) In North America, Eastern Standard Time (EST) is 5 hours different from UTC; 9:00 A.M. EST is 14:00 UTC. Coordinated Universal Time is also used by aviators and sailors, who regularly travel from one time zone to another.

Although it is natural to want our clocks and method of timekeeping to be related to the Sun, astronomers often use a system that is based on the apparent motion of the stars. This system, called **sidereal time**, is useful when aiming a telescope. Most observatories are therefore equipped with a clock that measures sidereal time, as discussed in Box 2-2.

2-8 Astronomical observations led to the development of the modern calendar

Just as the day is a natural unit of time based on Earth's rotation, the year is a natural unit of time based on Earth's revolution about the Sun. However, nature has not arranged things for our convenience. The year does not divide into exactly 365 whole days. Ancient astronomers realized that the length of a year is approximately $365\frac{1}{4}$ days, so the Roman emperor Julius Caesar established the system of "leap years" to account for this extra quarter of a day. By adding an extra day to the calendar every four years, he hoped to ensure that seasonal astronomical events, such as the beginning of spring, would occur on the same date year after year.

Our calendar is complex because a year does not contain a whole number of days

Caesar's system would have been perfect if the year were exactly $365\frac{1}{4}$ days long and if there were no precession. Unfortunately, this is not the case. To be more accurate, astronomers now use several different types of years. For example, the **sidereal year** is defined to be the time required for the Sun to return to the same position with respect to the stars. It is equal to 365.2564 mean solar days, or $365^d\,6^h\,9^m\,10^s$.

The sidereal year is the orbital period of Earth around the Sun, but it is *not* the year on which we base our calendar. Like Caesar, most people want annual events—in particular, the first days of the seasons—to fall on the same date each year. For example, we want the first day of spring to occur on March 21. But spring begins when the Sun is at the vernal equinox, and the vernal equinox moves slowly against the background stars because of precession. Therefore, to set up a calendar we use the **tropical**

BOX 2-2 Tools of the Astronomer's Trade

Sidereal Time

TUTORIAL 2.3

If you want to observe a particular object in the heavens, the ideal time to do so is when the object is high in the sky, on or close to the upper meridian. Timing observations to occur when an object is high in the sky minimizes the distorting effects of Earth's atmosphere, which increase as you view closer to the horizon. For astronomers who study the Sun, this means making observations at local noon, which is not too different from noon as determined using mean solar time. For astronomers who observe planets, stars, or galaxies, however, the optimum time to observe depends on the particular object to be studied. Given the location of a given object on the celestial sphere, when will that object be on the upper meridian?

To answer this question, astronomers use *sidereal time* rather than solar time. It is different from the time on your wristwatch. In fact, a *sidereal clock* and an ordinary clock even tick at different rates, because they are based on different astronomical objects. Ordinary clocks are related to the position of the Sun, while sidereal clocks are based on the position of the vernal equinox, the location from which right ascension is measured. (See Box 2-1 for a discussion of right ascension.)

Regardless of where the Sun is, midnight sidereal time at your location is defined to be when the vernal equinox crosses your upper meridian. (Like solar time, sidereal time depends on where you are on Earth.) A **sidereal day** is the time between two successive upper meridian passages of the vernal equinox. By contrast, an apparent solar day is the time between two successive upper meridian crossings of the Sun. The illustration shows why these two kinds of day are not equal. Because Earth orbits the Sun, Earth must make one complete rotation plus about 1° to get from one local solar noon to the next. This extra 1° of rotation corresponds to 4 minutes of time, which is the amount by which a solar day exceeds a sidereal day. To be precise:

$$1 \text{ sidereal day} = 23^h\ 56^m\ 4.091^s$$

where the hours, minutes, and seconds are in mean solar time.

One day according to your wristwatch is one mean solar day, which is exactly 24 hours of solar time long. A **sidereal clock** measures sidereal time in terms of sidereal hours, minutes, and seconds, where one sidereal day is divided into 24 sidereal hours.

This explains why a sidereal clock ticks at a slightly different rate than your wristwatch. As a result, at some times of the year a sidereal clock will show a very different time than an ordinary clock. (At local noon on March 21, when the Sun is at the vernal equinox, a sidereal clock will say that it is midnight. Do you see why?)

We can now answer the question in the opening paragraph. The vernal equinox, whose celestial coordinates are R.A. = $0^h\ 0^m\ 0^s$, Decl. = 0° 0′ 0″, crosses the upper meridian at midnight sidereal time (0:00). The autumnal equinox, which is on the opposite side of the celestial sphere at R.A. = $12^h\ 0^m\ 0^s$, Decl. = 0° 0′ 0″, crosses the upper meridian 12 sidereal hours later at noon sidereal time (12:00). As these examples illustrate, *any* object crosses the upper meridian when the sidereal time is equal to the object's right ascension. That is why astronomers measure right ascension in units of time rather than degrees, and why right ascension is always given in sidereal hours, minutes, and seconds.

EXAMPLE: Suppose you want to observe the bright star Spica (Figure 2-6), which has epoch 2000 coordinates R.A. = $13^h\ 25^m\ 11.6^s$, Decl. = –11° 9′ 41″. What is the best time to do this?

Situation: Our goal is to find the time when Spica passes through your upper meridian, where it can best be observed.

Tools: We use the idea that a celestial object is on your upper meridian when the sidereal time equals the object's right ascension.

Answer: Based on the given right ascension of Spica, it will be best placed for observation when the sidereal time at your location is about 13:25. (Note that sidereal time is measured using a 24-hour clock.)

Review: By itself, our answer doesn't tell you what time on your wristwatch (which measures mean solar time) is best for observing Spica. That's why most observatories are equipped with a sidereal clock.

While sidereal time is extremely useful in astronomy, mean solar time is still the best method of timekeeping for most earthbound purposes. All time measurements in this book are expressed in mean solar time unless otherwise stated.

ANIMATION 2.3

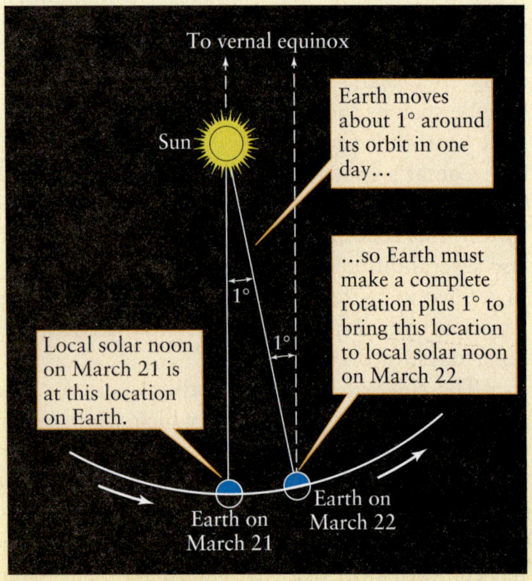
To vernal equinox
Sun
Earth moves about 1° around its orbit in one day...
...so Earth must make a complete rotation plus 1° to bring this location to local solar noon on March 22.
1°
1°
Local solar noon on March 21 is at this location on Earth.
Earth on March 21
Earth on March 22

(a) A month's motion of Earth along its orbit

year, which is equal to the time needed for the Sun to return to the vernal equinox. This period is equal to 365.2422 mean solar days, or $365^d\ 5^h\ 48^m\ 46^s$. Because of precession, the tropical year is 20 minutes and 24 seconds shorter than the sidereal year.

Caesar's assumption that the tropical year equals 365¼ days was off by 11 minutes and 14 seconds. This tiny error adds up to about three days every four centuries. Although Caesar's astronomical advisers were aware of the discrepancy, they felt that it was too small to matter. However, by the sixteenth century the first day of spring was occurring on March 11.

The Roman Catholic Church became concerned because Easter kept shifting to progressively earlier dates. To straighten things out, Pope Gregory XIII instituted a calendar reform in 1582. He began by dropping ten days (October 4, 1582, was followed by October 15, 1582), which brought the first day of spring back to March 21. Next, he modified Caesar's system of leap years.

Caesar had added February 29 to every calendar year that is evenly divisible by four. Thus, for example, 2004, 2008, and 2012 are all leap years with 366 days. But we have seen that this system produces an error of about three days every four centuries. To solve the problem, Pope Gregory decreed that only the century years evenly divisible by 400 should be leap years. For example, the years 1700, 1800, and 1900 (which would have been leap years according to Caesar) were not leap years in the improved Gregorian system, but the year 2000, which can be divided evenly by 400, *was* a leap year.

We use the Gregorian system today. It assumes that the year is 365.2425 mean solar days long, which is very close to the true length of the tropical year. In fact, the error is only one day in every 3300 years, which won't cause any problems for a long time.

Key Words

Terms preceded by an asterisk (*) are discussed in the Boxes.

Antarctic Circle, p. 32
apparent solar day, p. 36
apparent solar time, p. 35
Arctic Circle, p. 32
autumnal equinox, p. 31
celestial equator, p. 27
celestial sphere, p. 26
circumpolar, p. 28
constellation, p. 22
*declination, p. 30
diurnal motion, p. 24
ecliptic, p. 30
ecliptic plane, p. 30
epoch, p. 35
equinox, p. 31
lower meridian, p. 35
mean solar day, p. 37
mean sun, p. 37
meridian, p. 35
meridian transit, p. 36
north celestial pole, p. 27
positional astronomy, p. 21
precession, p. 34
precession of the equinoxes, p. 35
*right ascension, p. 29
*sidereal clock, p. 38
*sidereal day, p. 38
sidereal time, p. 37
sidereal year, p. 37
south celestial pole, p. 27
summer solstice, p. 31
time zone, p. 37
tropical year, p. 37
Tropic of Cancer, p. 33
Tropic of Capricorn, p. 33
upper meridian, p. 35
vernal equinox, p. 31
winter solstice, p. 32
zenith, p. 27
zodiac, p. 34

Key Ideas

Ideas preceded by an asterisk (*) are discussed in the Boxes.

Constellations and the Celestial Sphere: It is convenient to imagine the stars fixed to the celestial sphere with Earth at its center.

• The surface of the celestial sphere is divided into 88 regions called constellations.

Diurnal (Daily) Motion of the Celestial Sphere: The celestial sphere appears to rotate around Earth once in each 24-hour period. In fact, it is actually Earth that is rotating.

• The poles and equator of the celestial sphere are determined by extending the axis of rotation and the equatorial plane of Earth out to the celestial sphere.

• *The positions of objects on the celestial sphere are described by specifying their right ascension (in time units) and declination (in angular measure).

Seasons and the Tilt of Earth's Axis: Earth's axis of rotation is tilted at an angle of about 23½° from the perpendicular to the plane of Earth's orbit.

• The seasons are caused by the tilt of Earth's axis.

• Over the course of a year, the Sun appears to move around the celestial sphere along a path called the ecliptic. The ecliptic is inclined to the celestial equator by about 23½°.

• The ecliptic crosses the celestial equator at two points in the sky, the vernal and autumnal equinoxes. The northernmost point that the Sun reaches on the celestial sphere is the summer solstice, and the southernmost point is the winter solstice.

Precession: The orientation of Earth's axis of rotation changes slowly, a phenomenon called precession.

• Precession is caused by the gravitational pull of the Sun and Moon on Earth's equatorial bulge.

• Precession of Earth's axis causes the positions of the equinoxes and celestial poles to shift slowly.

• *Because the system of right ascension and declination is tied to the position of the vernal equinox, the date (or epoch) of observation must be specified when giving the position of an object in the sky.

Timekeeping: Astronomers use several different means of keeping time.

• Apparent solar time is based on the apparent motion of the Sun across the celestial sphere, which varies over the course of the year.

• Mean solar time is based on the motion of an imaginary mean sun along the celestial equator, which produces a uniform mean solar day of 24 hours. Ordinary watches and clocks measure mean solar time.

• *Sidereal time is based on the apparent motion of the celestial sphere.

The Calendar: The tropical year is the period between two passages of the Sun across the vernal equinox. Leap year corrections are needed because the tropical year is not exactly 365 days. The sidereal year is the actual orbital period of Earth.

Questions

Review Questions

1. Describe three structures or carvings made by past civilizations that show an understanding of astronomy.
2. How are constellations useful to astronomers? How many stars are not part of any constellation?
3. A fellow student tells you that only those stars in Figure 2-2*b* that are connected by blue lines are part of the constellation Orion. How would you respond?
4. Why are different stars overhead at 10:00 P.M. on a given night than two hours later at midnight? Why are different stars overhead at midnight on June 1 than at midnight on December 1?

TUTORIAL 2.1

5. What is the celestial sphere? Why is this ancient concept still useful today?
6. Imagine that someone suggests sending a spacecraft to land on the surface of the celestial sphere. How would you respond to such a suggestion?
7. What is the celestial equator? How is it related to Earth's equator? How are the north and south celestial poles related to Earth's axis of rotation?
8. Where would you have to look to see your zenith? Where on Earth would you have to be for the celestial equator to pass through your zenith? Where on Earth would you have to be for the south celestial pole to be at your zenith?
9. How many degrees is the angle from the horizon to the zenith? Does your answer depend on what point on the horizon you choose?
10. Why can't a person in Antarctica use the Big Dipper to find the north direction?
11. Is there any place on Earth where you could see the north celestial pole on the northern horizon? If so, where? Is there any place on Earth where you could see the north celestial pole on the western horizon? If so, where? Explain your answers.
12. How do the stars appear to move over the course of the night as seen from the north pole? As seen from the equator? Why are these two motions different?

TUTORIAL 2.2

13. Using a diagram, explain why the tilt of Earth's axis relative to Earth's orbit causes the seasons as we orbit the Sun.
14. Give two reasons why it's warmer in summer than in winter.
15. What is the ecliptic plane? What is the ecliptic?
16. Why is the ecliptic tilted with respect to the celestial equator? Does the Sun appear to move along the ecliptic, the celestial equator, or neither? By about how many degrees does the Sun appear to move on the celestial sphere each day?
17. Where on Earth do you have to be in order to see the north celestial pole directly overhead? What is the maximum possible elevation of the Sun above the horizon at that location? On what date can this maximum elevation be observed?
18. What are the vernal and the autumnal equinoxes? What are the summer and winter solstices? How are these four points related to the ecliptic and the celestial equator?
19. At what point on the horizon does the vernal equinox rise? Where on the horizon does it set? (*Hint:* See Figure 2-16.)
20. How does the daily path of the Sun across the sky change with the seasons? Why does it change?
21. Where on Earth do you have to be in order to see the Sun at the zenith? As seen from such a location, will the Sun be at the zenith every day? Explain your reasoning.
22. What is precession of the equinoxes? What causes it? How long does it take for the vernal equinox to move 1° along the ecliptic?
23. What is the (fictitious) mean sun? What path does it follow on the celestial sphere? Why is it a better timekeeper than the actual Sun in the sky?
24. Why is it convenient to divide Earth into time zones?
25. Why is the time given by a sundial not necessarily the same as the time on your wristwatch?
26. What is the difference between the sidereal year and the tropical year? Why are these two kinds of year slightly different in length? Why are calendars based on the tropical year?
27. When is the next leap year? Was 2000 a leap year? Will 2100 be a leap year?

Advanced Questions

Questions preceded by an asterisk (*) involve topics discussed in the Boxes.

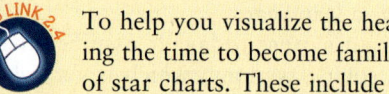

> **Problem-solving tips and tools**
>
> WEB LINK 2.4
>
> To help you visualize the heavens, it is worth taking the time to become familiar with various types of star charts. These include the simple star charts at the end of this book, the monthly star charts published in such magazines as *Sky & Telescope* and *Astronomy*, and the more detailed maps of the heavens found in star atlases.
>
> One of the best ways to understand the sky and its motions is to use the *Starry Night Enthusiast*™ computer program on the CD-ROM that accompanies certain printed copies of this book. This easy-to-use program allows you to view the sky on any date and at any time, as seen from any point on Earth, and to animate the sky to visualize its diurnal and annual motions.
>
> You may also find it useful to examine a planisphere, a device consisting of two rotatable disks. The bottom disk shows all the stars in the sky (for a particular latitude), and the top one is opaque with a transparent oval window through which only some of the stars can be seen. By rotating the top disk, you can immediately see which constellations are above the horizon at any time of the year. A planisphere is a convenient tool to carry with you when you are out observing the night sky.

28. On November 1 at 8:30 P.M. you look toward the eastern horizon and see the bright star Bellatrix (shown in Figure 2-2*b*) rising. At approximately what time will Bellatrix rise one week later, on November 8?
29. Figure 2-4 shows the situation on September 21, when Cygnus is highest in the sky at 8:00 P.M. local time and Andromeda is highest in the sky at midnight. But as Figure 2-5 shows, on

July 21 Cygnus is highest in the sky at midnight. On July 21, at approximately what local time is Andromeda highest in the sky? Explain your reasoning.

30. Figure 2-5 shows which constellations are high in the sky (for observers in the northern hemisphere) in the months of July, September, and November. From this figure, would you be able to see Perseus at midnight on May 15? Draw a picture to justify your answer.

31. Figure 2-6 shows the appearance of Polaris, the Little Dipper, and the Big Dipper at 11 P.M. (daylight savings time) on August 1. Sketch how these objects would appear on this same date at (**a**) 8 P.M. and (**b**) 2 A.M. Include the horizon in your sketches, and indicate the north direction.

32. Figure 2-6 shows the appearance of the sky near the North Star at 11 P.M. (daylight savings time) on August 1. Explain why the sky has this same appearance at 1 A.M. on July 1 and at 9 P.M. on September 1.

33. The time-exposure photograph that opens this chapter shows the trails made by individual stars as the celestial sphere appears to rotate around Earth. (**a**) For approximately what length of time was the camera shutter left open to take this photograph? (**b**) The stars in this photograph (taken in Hawaii, at roughly 20° north latitude) appear to rotate around one of the celestial poles. Which celestial pole is it? As seen from this location, do the stars move clockwise or counterclockwise around this celestial pole? (**c**) If you were at 20° south latitude, which celestial pole could you see? In which direction would you look to see it? As seen from this location, do the stars move clockwise or counterclockwise around this celestial pole?

34. (**a**) Redraw Figure 2-10 for an observer at the north pole. (*Hint:* The north celestial pole is directly above this observer.) (**b**) Redraw Figure 2-10 for an observer at the equator. (*Hint:* The celestial equator passes through this observer's zenith.) (**c**) Using Figure 2-10 and your drawings from (a) and (b), justify the following rule, long used by navigators: The latitude of an observer in the northern hemisphere is equal to the angle in the sky between that observer's horizon and the north celestial pole. (**d**) State the rule that corresponds to (c) for an observer in the southern hemisphere.

35. The photograph that opens this chapter was taken next to the Gemini North Observatory atop Mauna Kea in Hawaii. The telescope is at longitude 155° 28′ 09″ west and latitude 19° 49′ 26″ north. (**a**) By making measurements on the photograph, find the approximate angular width and angular height of the photo. (**b**) How far (in degrees, arcminutes, and arcseconds) from the south celestial pole can a star be and still be circumpolar as seen from the Gemini North Observatory?

36. The Gemini North Observatory shown in the photograph that opens this chapter is located in Hawaii, roughly 20° north of the equator. Its near-twin, the Gemini South Observatory, is located roughly 30° south of the equator in Chile. Why is it useful to have telescopes in both the northern and southern hemispheres?

37. Is there any place on Earth where all the visible stars are circumpolar? If so, where? Is there any place on Earth where none of the visible stars is circumpolar? If so, where? Explain your answers.

RIVUXG
(NASA/JPL)

38. The above image of Earth was made by the *Galileo* spacecraft while en route to Jupiter. South America is at the center of the image and Antarctica is at the bottom of the image. (**a**) In which month of the year was this image made? Explain your reasoning. (**b**) When this image was made, was Earth relatively close to the Sun or relatively distant from the Sun? Explain your reasoning.

39. Figure 2-16 shows the daily path of the Sun across the sky on March 21, June 21, September 22, and December 21 for an observer at 35° north latitude. Sketch drawings of this kind for (**a**) an observer at 35° south latitude; (**b**) an observer at the equator; and (**c**) an observer at the north pole.

40. Suppose that you live at a latitude of 40° N. What is the elevation (angle) of the Sun above the southern horizon at noon (**a**) at the time of the vernal equinox? (**b**) at the time of the winter solstice? Explain your reasoning. Include a drawing as part of your explanation.

41. In the northern hemisphere, houses are designed to have "southern exposure," that is, with the largest windows on the southern side of the house. But in the southern hemisphere houses are designed to have "northern exposure." Why are houses designed this way, and why is there a difference between the hemispheres?

42. The city of Mumbai (formerly Bombay) in India is 19° north of the equator. On how many days of the year, if any, is the Sun at the zenith at midday as seen from Mumbai? Explain your answer.

43. Ancient records show that 2000 years ago, the stars of the constellation Crux (the Southern Cross) were visible in the southern sky from Greece. Today, however, these stars cannot be seen from Greece. What accounts for this change?

44. The Great Pyramid at Giza has a tunnel that points toward the north celestial pole. At the time the pyramid was built, around 2600 B.C., toward which star did it point? Toward which star does this same tunnel point today? (See Figure 2-20.)

45. The photo on the next page shows a statue of the Greek god Atlas. The globe that Atlas is holding represents the celestial sphere, with depictions of several important constellations and the celestial equator. Although the statue dates from around 150 A.D., it has been proposed that the arrangement

R I V U X G

(Scala/Art Resource, NY)

of constellations depicts the sky as it was mapped in an early star atlas that dates from 129 B.C. Explain the reasoning that could lead to such a proposal.

46. Unlike western Europe, Imperial Russia did not use the revised calendar instituted by Pope Gregory XIII. Explain why the Russian Revolution, which started on November 7, 1917, according to the modern calendar, is called "the October revolution" in Russia. What was this date according to the Russian calendar at the time? Explain your answer.

*47. What is the right ascension of a star that is on the meridian at midnight at the time of the autumnal equinox? Explain.

*48. The coordinates on the celestial sphere of the summer solstice are R.A. = $6^h\ 0^m\ 0^s$, Decl. = $+23°\ 27'$. What are the right ascension and declination of the winter solstice? Explain your answer.

*49. Because 24 hours of right ascension takes you all the way around the celestial equator, $24^h = 360°$, what is the angle in the sky (measured in degrees) between a star with R.A. = $8^h\ 0^m\ 0^s$, Decl. = $0°\ 0'\ 0''$ and a second star with R.A. = $11^h\ 20^m\ 0^s$, Decl. = $0°\ 0'\ 0''$? Explain your answer.

*50. On a certain night, the first star in Advanced Question 49 passes through the zenith at 12:30 A.M. local time. At what time will the second star pass through the zenith? Explain your answer.

*51. At local noon on March 21, when the Sun is at the vernal equinox, a sidereal clock will say that it is midnight. Explain why.

*52. (a) What is the sidereal time when the vernal equinox rises? (b) On what date is the sidereal time nearly equal to the solar time? Explain your answer.

*53. How would the sidereal and solar days change (a) if Earth's rate of rotation increased, (b) if Earth's rate of rotation decreased, and (c) if Earth's rotation were retrograde (that is, if Earth rotated about its axis opposite to the direction in which it revolves about the Sun)?

Discussion Questions

WEB LINK 2.5

54. Examine a list of the 88 constellations. Are there any constellations whose names obviously date from modern times? Where are these constellations located? Why do you suppose they do not have archaic names?

55. Describe how the seasons would be different if Earth's axis of rotation, rather than having its present $23\frac{1}{2}°$ tilt, were tilted (a) by 0° or (b) by 90°.

56. In William Shakespeare's *Julius Caesar* (act 3, scene 1), Caesar says:

> ***But I am constant as the northern star,***
> ***Of whose true-fix'd and resting quality***
> ***There is no fellow in the firmament.***

Translate Caesar's statement about the "northern star" into modern astronomical language. Is the northern star truly "constant"? Was the northern star the same in Shakespeare's time (1564–1616) as it is today?

Web/eBook Questions

57. Search the World Wide Web for information about the national flags of Australia, New Zealand, and Brazil and the state flag of Alaska. Which stars are depicted on these flags? Explain any similarities or differences among these flags.

58. Some people say that on the date that the Sun is at the vernal equinox, and only on this date, you can stand a raw egg on end. Others say that there is nothing special about the vernal equinox, and that with patience you can stand a raw egg on end on any day of the year. Search the World Wide Web for information about this story and for hints about how to stand an egg on end. Use these hints to try the experiment yourself on a day when the Sun is *not* at the vernal equinox. What do you conclude about the connection between eggs and equinoxes?

WEB LINK 2.6

59. Use the U.S. Naval Observatory Web site to find the times of sunset and sunrise on (a) your next birthday and (b) the date this assignment is due. (c) Are the times the same for the two dates? Explain why or why not.

Activities

Observing Projects

Observing tips and tools

WEB LINK 2.7

Moonlight is so bright that it interferes with seeing the stars. For the best view of the constellations, do your observing when the Moon is below the horizon. You can find the times of moonrise and moonset in your local newspaper or on the World Wide Web. Each monthly issue of the magazines *Sky & Telescope* and *Astronomy* includes much additional observing information.

60. On a clear, cloud-free night, use the star charts at the end of this book to see how many constellations of the zodiac you can identify. Which ones were easy to find? Which were difficult? Are the zodiacal constellations the most prominent ones in the sky?

61. Examine the star charts that are published monthly in such popular astronomy magazines as *Sky & Telescope* and *As-*

tronomy. How do they differ from the star charts at the end of this book? On a clear, cloud-free night, use one of these star charts to locate the celestial equator and the ecliptic. Note the inclination of the Milky Way to the ecliptic and celestial equator. The Milky Way traces out the plane of our galaxy. What do your observations tell you about the orientation of Earth and its orbit relative to the galaxy's plane?

62. Suppose you wake up before dawn and want to see which constellations are in the sky. Explain how the star charts at the end of this book can be quite useful, even though chart times are given only for the evening hours. Which chart most closely depicts the sky at 4:00 A.M. on the morning that this assignment is due? Set your alarm clock for 4:00 A.M. to see if you are correct.

63. Use the ***Starry Night Enthusiast™*** program to observe the diurnal motion of the sky. Select **Viewing Location . . .** in the **Options** menu, click on the **List** tab, highlight the name of your town or city and click the **Set Location** button to display the sky where you live. a) In the northern hemisphere, press the "N" key to set the gaze direction to the northern sky. (If your location is in the southern hemisphere, press the "S" key to select your southern sky.) Select **Hide Daylight** under the **View** menu. In the toolbar, click on the **Time Flow Rate** control and set the discrete time step to **1 minute.** Then click on the **Play** button (a triangle that points to the right) to run time forward. (To reduce confusion, remove rapidly moving artificial Earth-orbiting satellites from this view by clicking on **View/Solar System** and turning off **Satellites**). Do the background stars appear to rotate clockwise or counterclockwise? Explain this observation in terms of Earth's rotation. Are any of the stars circumpolar, that is, do they stay above your horizon for the full 24 hours of a day? b) Now center your field of view on the southern horizon if you live in the northern hemisphere) (or the northern horizon, if you live in the southern hemisphere). Describe what you see. Are any of these stars circumpolar?

64. Use the ***Starry Night Enthusiast™*** program to observe the Sun's motion on the celestial sphere. Select **Guides > Atlas** in the **Favorites** menu to see the entire celestial sphere as if you were at the center of a transparent Earth. Open the **Find pane** and double-click on the entry for the **Sun** to center it in the view. a) In the toolbar at the top of the main window, set the **Time Flow Rate** to **1 day** and set time to move forward, using the **Play** button (a triangle that points to the right). Observe the Sun for a full year of simulated time. How does the Sun appear to move against the background stars? What path does it follow? Does it ever change direction? b) Open the **Options** pane, expand the **Constellations** layer and select the **Auto Identify, Boundaries and Labels** checkboxes to turn these constellation options on. In the toolbar, click on the **Now** button to return to your present time. Again, open the **Find** pane and double-click on the Sun to center it in your view. In which constellation is the Sun located today? Is this the same as the astrological sign for today's date? Explain your answer in terms of precession. (c) Set the discrete **Time Flow Rate** to **1 day** and click the **Play** button. Through which constellations does the Sun appear to pass over the course of a year?

65. Use the ***Starry Night Enthusiast™*** program to investigate the orbits of Earth and some of the other planets around the Sun from a position 4 AU above the North Pole of the Sun. Select **Viewing Location . . .** in the **Options** menu. In the Viewing Location dialog box, select **stationary location** in the **View from** drop box. In the edit boxes beneath the **Cartesian coordinates** label, enter "0" for the **X** and **Y** values and "4 AU" for the **Z** value. Then click the **Set Location** button. You are now at the above position in space. Open the **Find** pane and center the view upon the Sun from this point in space, 4 AU above the North Pole of the Sun. You will now be able to see the orbits of the four inner planets of the Solar System nearly face-on. To display Earth's orbit, click the checkbox to the right of the label **Earth** in the **Find** pane. a) Can you tell from this view that Earth's orbit is not a perfect circle, as we learned in Section 2-7? b) In the **Find** pane, click in the checkbox to the right of the labels for the other three inner planets (Mercury, Venus, and Mars) to display their orbits in the view. Of these four inner planets, which have orbits that are clearly noncircular? (You may want to use a ruler to measure the distance from the Sun to various points on each planet's orbit. If the distance is the same at all points then the orbit is circular; otherwise, it is noncircular.) c) Now change your view so that it is in the plane of Earth's orbit. Select **Viewing Location . . .** from the **Options** menu. In the Viewing Location dialog box, enter "0" for the **X** and **Z** Cartesian coordinates and enter "4 AU" for the **Y** coordinate. This view shows an edge-on view of the orbits of the four inner planets. Do the orbits of the other planets lie in the same plane as the plane of Earth's orbit (the ecliptic plane)? Which planet's orbit appears to be tilted the most from Earth's orbit? (You may need to use the **Zoom** controls at the right side of the toolbar to investigate this.) If you were on any of these other planets in the solar system, would the Sun appear to follow exactly the same path across the celestial sphere as is seen from Earth?

Why Astrology Is Not Science

by James Randi

I'm involved in the strange business of telling folks what they should already know. I meet audiences who believe in all sorts of impossible things, often despite their education and intelligence. My job is to explain how science differs from the unproven, illogical assumptions of pseudoscience—and why it matters. Perhaps my best example is the difference between astronomy and astrology.

Both astrology and astronomy arose from the wonders of the night sky, from the stars to comets, planets, the Sun, and the Moon. Surely, humans have long reasoned, there must be some meaning in their motions. Surely the Moon's effect on tides hints at hidden "causes" for strange events. *Judiciary* (literally "judging") astrology therefore attempted to foretell the future—our earthly future. To serve it, *horary* (literally "hourly") astrology carefully tracked the heavens.

It is the latter that has become astronomy. Thanks to its process of careful measurement and testing, we now understand more about the true nature of the starry universe than astrologers could ever have imagined. With the birth of a new science, astronomers had a logical framework based on physical causes and systematic observations.

Astrology remains a popular delusion. Far too many believe today that patterns in the sky govern our lives. They accept the vague tendencies and portents of seers who cast horoscopes. They shouldn't. Just a glance at the tenets of astrology provides ample evidence of its absurdity.

An individual is said to be born under a sign. To the astrologer, the Sun was located "in" that sign at the moment of birth. (Stars are not seen in the daytime, but no matter—a calculation tells where the Sun is.) Each sign takes its name from a constellation, a totally imaginary figure invented for our convenience in referring to stars. Different cultures have different mythical figures up there, and so different schools of astrology assign different meanings to the signs they use.

In the spirit of equal-opportunity swindling, astrologers divide up the year fairly, ignoring variations in the size of constellations. Since Libra is tiny, while Virgo is huge, they chop some of the sky off Virgo and add it—along with bits of Scorpio—to bring Libra up to size. The Sun could well be declared "in" Libra when it is actually outside that constellation.

It gets worse. Science constantly challenges itself and changes. The rules of astrology could not, although they were made up thousands of years ago, and since then the "fixed" stars have moved. In particular, precession of the equinoxes has shifted objects in the sky relative to our calendar. The constellations have changed but astrology has not. If you were born August 7, you are said to be a Leo, but the Sun that day was really in the same part of the sky as the constellation Cancer.

With a theory like this to back it up, we should not be surprised at the bottom line: *A pseudoscience does not work.* Test after test has checked its predictions, and the result is always the same. One such investigator is Shawn Carlson of the University of California, San Diego. As he put it in *Nature* magazine, astrology is "a hopeless cause." Johannes Kepler, the pioneering astronomer, himself cast horoscopes, but they are little remembered today. Owen Gingerich, a historian of science at Harvard, puts it well: Kepler was the astrologer who destroyed astrology.

Astronomy works—it works very well indeed—which isn't easy. Because we humans tend to find what we want in any body of data, it takes science's careful process of observation, creative insight, and critical thinking to understand and predict changes in nature. As I write, a transit of Ganymede is due next Thursday at 21:47:20. At exactly that time, the satellite of Jupiter will cross in front of its planet as seen from Earth, and yet most of us will never know it. Still other moons of Jupiter may hold fresh clues to the formation of our entire solar system and the conditions for life elsewhere.

For most people, astronomy has too little fantasy or money in it, and they will never experience the beauty in its predictions. The dedicated labors of generations of scientists have enabled us to perform a genuine wonder.

James ("the Amazing") Randi works tirelessly to expose trickery so that others can relish the greater wonder of science. As a magician, he has had his own television show and an enormous public following. As a lecturer, he addresses teachers, students, and others worldwide. His newsletter and column for *The Skeptic* are key resources for educators. His many books include *Flim-Flam!, The Faith Healers,* and *The Mask of Nostradamus,* about a legendary con man with secrets of his own.

Mr. Randi is the founder of the James Randi Educational Foundation, and his one million dollar prize for "the performance of any paranormal event . . . under proper observing conditions" has gone unclaimed for more than 25 years. An amateur archeologist and astronomer as well, he lives in Florida with several untalented parrots and the occasional visiting magus.

3

Eclipses and the Motion of the Moon

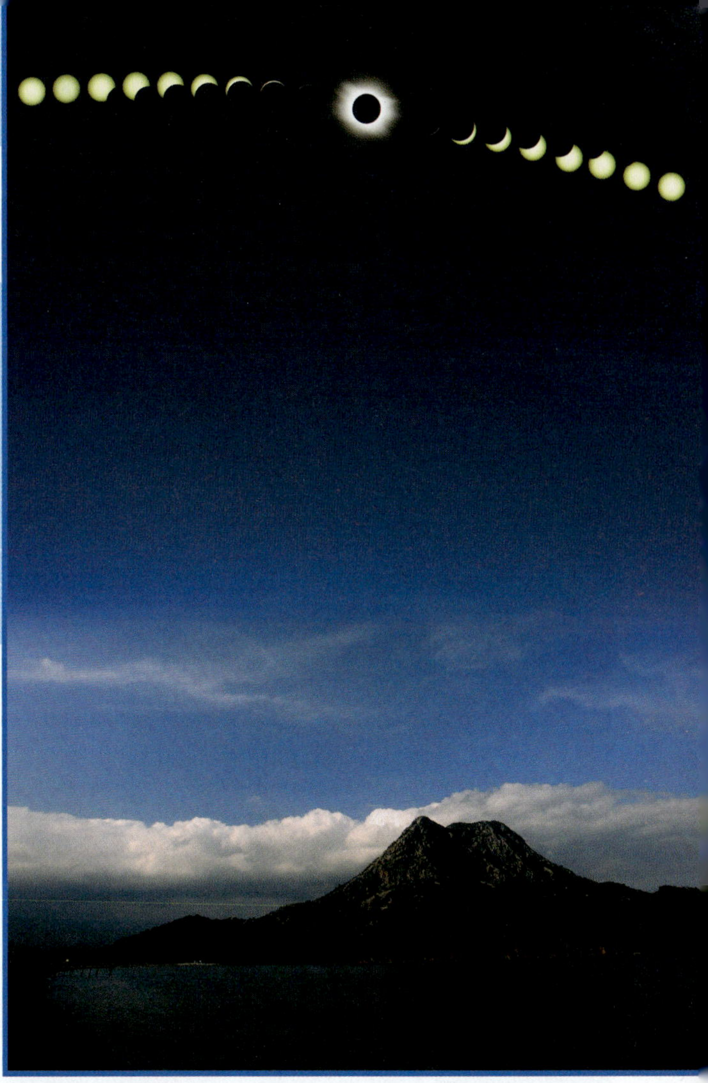

RIVUXG

The Sun in total eclipse, March 29, 2006. (Stefan Seip)

On March 29, 2006, a rare cosmic spectacle—a total solar eclipse—was visible along a narrow corridor that extended from the coast of Brazil through equatorial Africa and into central Asia. As shown in this digital composite taken in Turkey, the Moon slowly moved over the disk of the Sun. (Time flows from left to right in this image.) For a few brief minutes the Sun was totally covered, darkening the sky and revealing the Sun's thin outer atmosphere, or corona, which glows with an unearthly pearlescent light.

Such eclipses can be seen only on specific dates from special locations on Earth, so not everyone will ever see the Moon cover the Sun in this way. But anyone can find the Moon in the sky and observe how its appearance changes from night to night, from new moon to full moon and back again, and how the times when the Moon rises and sets differ noticeably from one night to the next.

In this chapter our subject is how the Moon moves as seen from Earth. We will explore why the Moon goes through a regular cycle of phases, and how the Moon's orbit around Earth leads to solar eclipses as well as lunar eclipses. We will also see how ancient astronomers used their observations of the Moon to determine the size and shape of Earth, as well as other features of the solar system. Thus, the Moon—which has always loomed large in the minds of poets, lovers, and dreamers—has also played a key role in the development of our modern picture of the universe.

3-1 The phases of the Moon are caused by its orbital motion

As seen from Earth, both the Sun and the Moon appear to move from west to east on the celestial sphere—that is, relative to the background of stars. They move at very different rates, however. The Sun takes one year to make a complete trip around the imaginary celestial sphere along the path we call the *ecliptic* (Section 2-5). By comparison, the Moon takes only about four weeks. In the past, these similar motions led people to believe that both the Sun and the

You can tell the Moon's position relative to Earth and the Sun by observing its phase

Learning Goals

By reading the sections of this chapter, you will learn

3-1 Why we see the Moon go through phases

3-2 Why we always see the same side of the Moon

3-3 The differences between lunar and solar eclipses

3-4 Why not all lunar eclipses are total eclipses

3-5 Why solar eclipses are visible only from certain special locations on Earth

3-6 How ancient Greek astronomers deduced the sizes of Earth, the Moon, and the Sun

Figure 3-1 R I **V** U X G

Earth and the Moon This picture of Earth and the Moon was taken in 1992 by the *Galileo* spacecraft on its way toward Jupiter. The Sun, which provides the illumination for both Earth and the Moon, was far to the right and out of the camera's field of view when this photograph was taken. (NASA/JPL)

Moon orbit around Earth. We now know that only the Moon orbits Earth, while the Earth-Moon system as a whole (Figure 3-1) orbits the Sun. (In Chapter 4 we will learn how this was discovered.)

One key difference between the Sun and the Moon is the nature of the light that we receive from them. The Sun emits its own light. So do the stars, which are objects like the Sun but much farther away, and so does an ordinary light bulb. By contrast, the light that we see from the Moon is reflected light. This is sunlight that has struck the Moon's surface, bounced off, and ended up in our eyes here on Earth.

CAUTION! You probably associate *reflection* with shiny objects like a mirror or the surface of a still lake. In science, however, the term refers to light bouncing off any object. You see most objects around you by reflected light. When you look at your hand, for example, you are seeing light from the Sun (or from a light fixture) that has been reflected from the skin of your hand and into your eye. In the same way, moonlight is really sunlight that has been reflected by the Moon's surface.

Understanding the Moon's Phases

Figure 3-1 shows both the Moon and Earth as seen from a spacecraft. When this image was recorded, the Sun was far off to the right. Hence, only the right-hand hemispheres of both worlds were illuminated by the Sun; the left-hand hemispheres were in darkness and are not visible in the picture. In the same way, when we view the Moon from Earth, we see only the half of the Moon that faces the Sun and is illuminated. However, not all of the illuminated half of the Moon is necessarily facing us. As the Moon moves around Earth, from one night to the next we see different amounts of the illuminated half of the Moon. These different appearances of the Moon are called **lunar phases.**

Figure 3-2 shows the relationship between the lunar phase visible from Earth and the position of the Moon in its orbit. For example, when the Moon is at position A, we see it in roughly the same direction in the sky as the Sun. Hence, the dark hemisphere of the Moon faces Earth. This phase, in which the Moon is barely visible, is called **new moon.** Since a new moon is near the Sun in the sky, it rises around sunrise and sets around sunset.

As the Moon continues around its orbit from position A in Figure 3-2, more of its illuminated half becomes exposed to our view. The result, shown at position B, is a phase called **waxing crescent moon** ("waxing" is a synonym for "increasing"). About a week after new moon, the Moon is at position C; we then see half of the Moon's illuminated hemisphere and half of the dark hemisphere. This phase is called **first quarter moon.**

As seen from Earth, a first quarter moon is one-quarter of the way around the celestial sphere from the Sun. It rises and sets about one-quarter of an Earth rotation, or six hours, after the Sun does: moonrise occurs around noon, and moonset occurs around midnight.

CAUTION! Despite the name, a first quarter moon appears to be *half* illuminated, not one-quarter illuminated! The name means that this phase is one-quarter of the way through the complete cycle of lunar phases.

About four days later, the Moon reaches position D in Figure 3-2. Still more of the illuminated hemisphere can now be seen from Earth, giving us the phase called **waxing gibbous moon** ("gibbous" is another word for "swollen"). When you look at the Moon in this phase, as in the waxing crescent and first quarter phases, the illuminated part of the Moon is toward the west. Two weeks after new moon, when the Moon stands opposite the Sun in the sky (position E), we see the fully illuminated hemisphere. This phase is called **full moon.** Because a full moon is opposite the Sun on the celestial sphere, it rises at sunset and sets at sunrise.

Over the following two weeks, we see less and less of the Moon's illuminated hemisphere as it continues along its orbit, and the Moon is said to be *waning* ("decreasing"). While the Moon is waning, its illuminated side is toward the east. The phases are called **waning gibbous moon** (position F), **third quarter moon** (position G, also called *last quarter moon*), and **waning crescent moon** (position H). A third quarter moon appears one-quarter of the way around the celestial sphere from the Sun, but on the opposite side of the celestial sphere from a first quarter

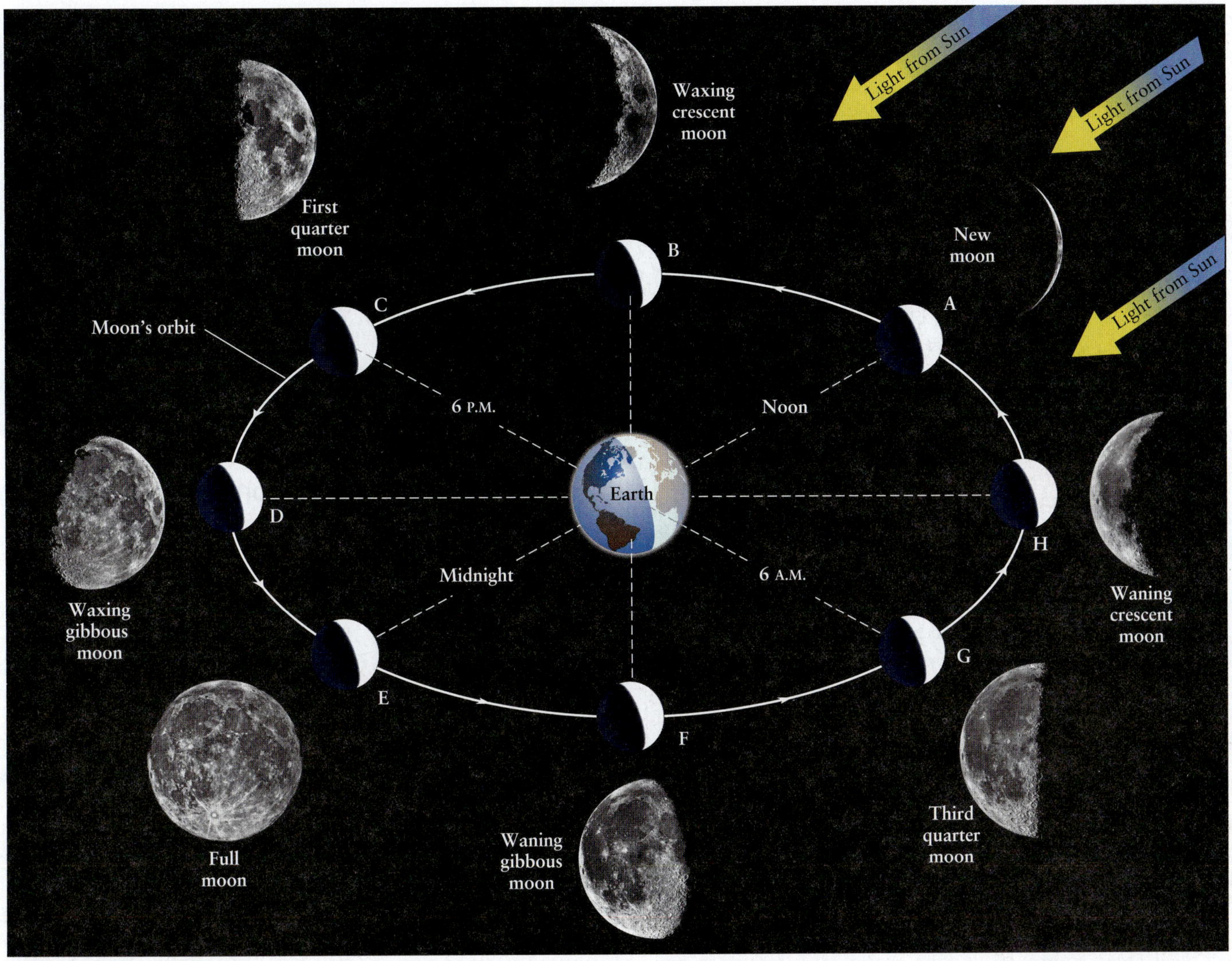

Figure 3-2

Why the Moon Goes Through Phases This figure shows the Moon at eight positions on its orbit, along with photographs of what the Moon looks like at each position as seen from Earth. The changes in phase occur because light from the Sun illuminates one half of the Moon, and as the Moon orbits Earth we see varying amounts of the Moon's illuminated half. It takes about $29\frac{1}{2}$ days for the Moon to go through a complete cycle of phases. (Photographs from Yerkes Observatory and Lick Observatory)

moon. Hence, a third quarter moon rises and sets about one-quarter Earth rotation, or 6 hours, *before* the Sun: moonrise is around midnight and moonset is around noon.

The Moon takes about four weeks to complete one orbit around Earth, so it likewise takes about four weeks for a complete cycle of phases from new moon to full moon and back to new moon. Since the Moon's position relative to the Sun on the celestial sphere is constantly changing, and since our system of timekeeping is based on the Sun (see Section 2-7), the times of moonrise and moonset are different on different nights. On average the Moon rises and sets about an hour later each night.

Figure 3-2 also explains why the Moon is often visible in the daytime, as shown in Figure 3-3. From any location on Earth, about half of the Moon's orbit is visible at any time. For example, if it is midnight at your location, you are in the middle of the dark side of Earth that faces away from the Sun. At that time you can easily see the Moon if it is at position C, D, E, F, or G. If it is midday at your location, you are in the middle of Earth's illuminated side, and the Moon will be easily visible if it is at position A, B, C, G, or H. (The Moon is so bright that it can be seen even against the bright blue sky.) You can see that the Moon is prominent in the midnight sky for about half of its orbit, and prominent in the midday sky for the other half.

CAUTION! A very common misconception about lunar phases is that they are caused by the shadow of *Earth* falling on the Moon. As Figure 3-2 shows, this is not the case at all. Instead, phases are simply the result of our seeing the illuminated half of the Moon at different angles as the Moon moves around its orbit. To help you better visualize how this works, Box 3-1 describes how you can simulate the cycle shown in Figure 3-2 using ordinary objects on Earth. (As we will learn in Section 3-3, Earth's shadow does indeed fall on the Moon on rare occasions. When this happens, we see a lunar eclipse.)

Figure 3-3 R I V U X G

The Moon During the Day The Moon can be seen during the daytime as well as at night. The time of day or night when it is visible depends on its phase. [Karl Beath/Gallo Images/Getty Images]

BOX 3-1 **Astronomy Down to Earth**

Phases and Shadows

Figure 3-2 shows how the relative positions of Earth, Moon, and Sun explain the phases of the Moon. You can visualize lunar phases more clearly by doing a simple experiment here on Earth. All you need are a small round object, such as an orange or a baseball, and a bright source of light, such as a street lamp or the Sun.

In this experiment, you play the role of an observer on Earth looking at the Moon, and the round object plays the role of the Moon. The light source plays the role of the Sun. Hold the object in your right hand with your right arm stretched straight out in front of you, with the object directly between you and the light source (position A in the accompanying illustration). In this orientation the illuminated half of the object faces away from you, like the Moon when it is in its new phase (position A in Figure 3-2).

Now, slowly turn your body to the left so that the object in your hand "orbits" around you (toward positions C, E, and G in the illustration). As you turn, more and more of the illuminated side of the "moon" in your hand becomes visible, and it goes through the same cycle of phases—waxing crescent, first quarter, and waxing gibbous—as does the real Moon. When you have rotated through half a turn so that the light source is directly behind you, you will be looking face on at the illuminated side of the object in your hand. This corresponds to a full moon (position E in Figure 3-2). Make sure your body does not cast a shadow on the "moon" in your hand—that would correspond to a lunar eclipse!

As you continue turning to the left, more of the unilluminated half of the object becomes visible as its phase moves through waning gibbous, third quarter, and waning crescent. When your body has rotated back to the same orientation that you were in originally, the unilluminated half of your handheld "moon" is again facing toward you, and its phase is again new. If you continue to rotate, the object in your hand repeats the cycle of "phases," just as the Moon does as it orbits around Earth.

The experiment works best when there is just one light source around. If there are several light sources, such as in a room with several lamps turned on, the different sources will create multiple shadows and it will be difficult to see the phases of your hand-held "moon." If you do the experiment outdoors using sunlight, you may find that it is best to perform it in the early morning or late afternoon when shadows are most pronounced and the Sun's rays are nearly horizontal.

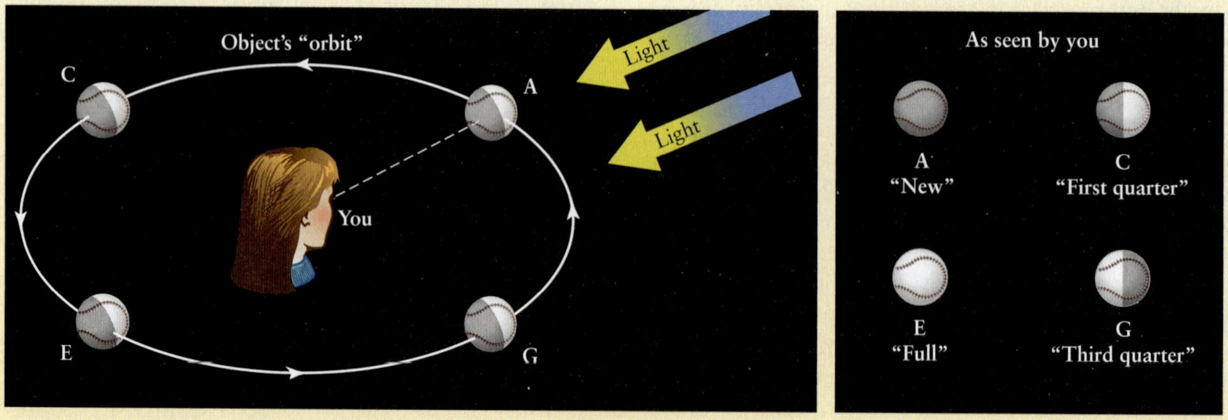

3-2 The Moon always keeps the same face toward Earth

Although the phase of the Moon is constantly changing, one aspect of its appearance remains the same: it always keeps essentially the same hemisphere, or face, toward Earth. Thus, you will always see the same craters and mountains on the Moon, no matter when you look at it; the only difference will be the angle at which these surface features are illuminated by the Sun. (You can verify this by carefully examining the photographs of the Moon in Figure 3-2.)

The Moon rotates in a special way: it spins exactly once per orbit

The Moon's Synchronous Rotation

Why is it that we only ever see one face of the Moon? You might think that it is because the Moon does not rotate (unlike Earth, which rotates around an axis that passes from its north pole to its south pole). To see that Earth must rotate, consider Figure 3-4. This figure shows Earth and the orbiting Moon from a vantage point far above Earth's north pole. In this figure two craters on the lunar surface have been colored, one in red and one in blue. If the Moon did not rotate on its axis, as in Figure 3-4*a*, at some times the red crater would be visible from Earth, while at other times the blue crater would be visible. Thus, we would see different parts of the lunar surface over time, which does not happen in reality.

In fact, the Moon always keeps the same face toward us because it *is* rotating, but in a very special way: It takes exactly as long to rotate on its axis as it does to make one orbit around Earth. This situation is called **synchronous rotation.** As Figure 3-4*b* shows, this keeps the crater shown in red always facing Earth, so that we always see the same face of the Moon. In Chapter 4 we will learn why the Moon's rotation and orbital motion are in step with each other.

An astronaut standing in the crater shown in red (Figure 3-4*b*) would spend two weeks (half of a lunar orbit) in darkness, or lunar nighttime, and the next two weeks in sunlight, or lunar daytime. Thus, as seen from the Moon, the Sun rises and sets, and no part of the Moon is perpetually in darkness. This means that there really is no "dark side of the Moon." The side of the Moon that constantly faces away from Earth is properly called the *far* side. The Sun rises and sets on the far side just as on the side toward Earth. Hence, the blue crater on the far side of the Moon in Figure 3-4*b* is in sunlight for half of each lunar orbit.

Sidereal and Synodic Months

The time for a complete lunar "day"—the same as the time that it takes the Moon to rotate once on its axis—is about four weeks. (Because the Moon's rotation is synchronous, it takes the same time for one complete lunar orbit.) It also takes about four weeks for the Moon to complete one cycle of its phases as seen from Earth. This regular cycle of phases inspired our ancestors to invent the concept of a month. For historical reasons, none of which

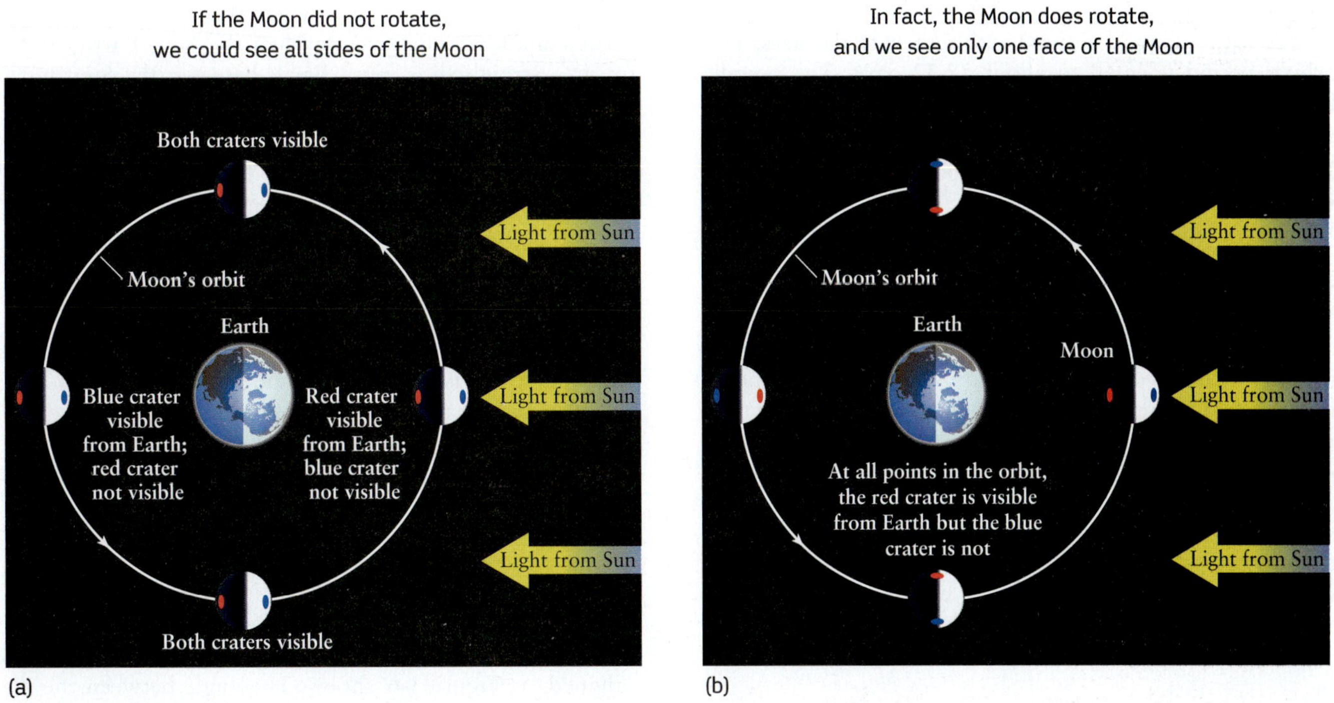

ANIMATION 3.2

Figure 3-4

The Moon's Rotation These diagrams show the Moon at four points in its orbit as viewed from high above Earth's north pole. **(a) The wrong model:** If the Moon did not rotate, then at various times the red crater would be visible from Earth while at other times the blue crater would be visible. Over a complete orbit, the entire surface of the Moon would be visible. **(b)** In reality, the Moon rotates on its north-south axis. Because the Moon makes one rotation in exactly the same time that it makes one orbit around Earth, we see only one face of the Moon.

ANIMATION 3.3

Figure 3-5

The Sidereal and Synodic Months The sidereal month is the time the Moon takes to complete one full revolution around Earth with respect to the background stars. However, because Earth is constantly moving along its orbit about the Sun, the Moon must travel through slightly more than 360° of its orbit to get from one new moon to the next. Thus, the synodic month—the time from one new moon to the next—is longer than the sidereal month.

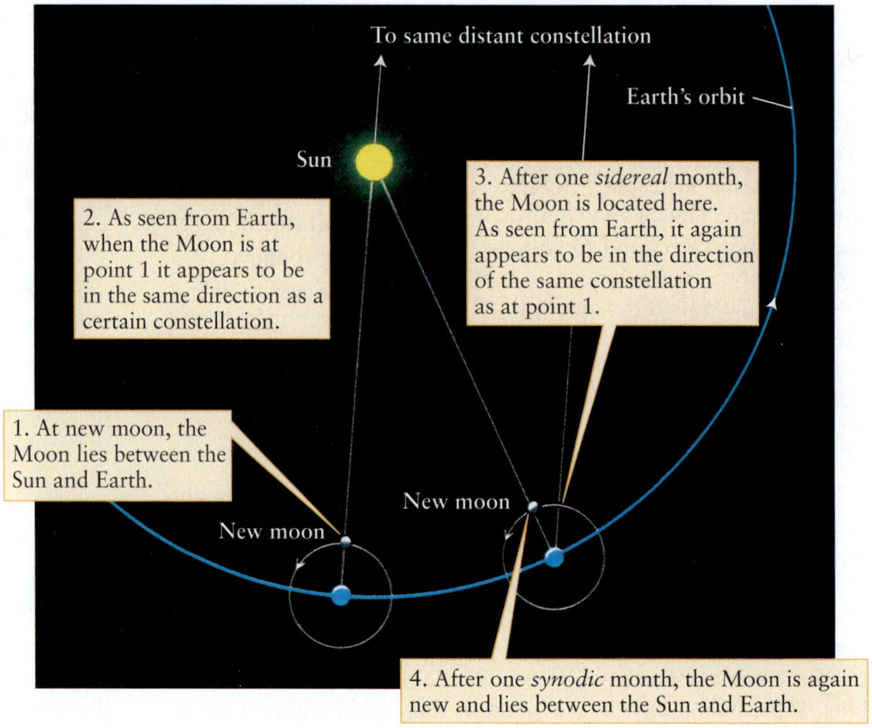

To same distant constellation
Earth's orbit
Sun
2. As seen from Earth, when the Moon is at point 1 it appears to be in the same direction as a certain constellation.
3. After one *sidereal* month, the Moon is located here. As seen from Earth, it again appears to be in the direction of the same constellation as at point 1.
1. At new moon, the Moon lies between the Sun and Earth.
New moon
New moon
4. After one *synodic* month, the Moon is again new and lies between the Sun and Earth.

has much to do with the heavens, the calendar we use today has months of differing lengths. Astronomers find it useful to define two other types of months, depending on whether the Moon's motion is measured relative to the stars or to the Sun. Neither corresponds exactly to the familiar months of the calendar.

The **sidereal month** is the time it takes the Moon to complete one full orbit of Earth, as measured with respect to the stars. This true orbital period is equal to about 27.32 days. The **synodic month,** or *lunar month*, is the time it takes the Moon to complete one cycle of phases (that is, from new moon to new moon or from full moon to full moon) and thus is measured with respect to the Sun rather than the stars. The length of the "day" on the Moon is a synodic month, not a sidereal month.

The synodic month is longer than the sidereal month because Earth is orbiting the Sun while the Moon goes through its phases. As Figure 3-5 shows, the Moon must travel *more* than 360° along its orbit to complete a cycle of phases (for example, from one new moon to the next). Because of this extra distance, the synodic month is equal to about 29.53 days, about two days longer than the sidereal month.

Both the sidereal month and synodic month vary somewhat from one orbit to another, the latter by as much as half a day. The reason is that the Sun's gravity sometimes causes the Moon to speed up or slow down slightly in its orbit, depending on the relative positions of the Sun, Moon, and Earth. Furthermore, the Moon's orbit changes slightly from one month to the next.

3-3 Eclipses occur only when the Sun and Moon are both on the line of nodes

TUTORIAL 3.2

From time to time the Sun, Earth, and Moon all happen to lie along a straight line. When this occurs, the shadow of Earth can fall on the Moon or the shadow of the Moon can fall on Earth. Such phenomena are called **eclipses.** They are perhaps the most dramatic astronomical events that can be seen with the naked eye.

The tilt of the Moon's orbit makes lunar and solar eclipses rare events

A **lunar eclipse** occurs when the Moon passes through Earth's shadow. This occurs when the Sun, Earth, and Moon are in a straight line, with Earth between the Sun and Moon so that the Moon is at full phase (position E in Figure 3-2). At this point in the Moon's orbit, the face of the Moon seen from Earth would normally be fully illuminated by the Sun. Instead, it appears quite dim because Earth casts a shadow on the Moon.

A **solar eclipse** occurs when Earth passes through the Moon's shadow. As seen from Earth, the Moon moves in front of the Sun. Once again, this can happen only when the Sun, Moon, and Earth are in a straight line. However, for a solar eclipse to occur, the Moon must be between Earth and the Sun. Therefore, a solar eclipse can occur only at new moon (position A in Figure 3-2).

CAUTION! Both new moon and full moon occur at intervals of $29\frac{1}{2}$ days. Hence, you might expect that there would be a solar eclipse every $29\frac{1}{2}$ days, followed by a lunar eclipse about two weeks (half a lunar orbit) later. But in fact, there are only a few solar eclipses and lunar eclipses per year. Solar and lunar eclipses are so infrequent because the plane of the Moon's orbit and the plane of Earth's orbit are not exactly aligned, as Figure 3-6 shows. The angle between the plane of Earth's orbit and the plane of the Moon's orbit is about 5°. Because of this tilt, new moon and full moon usually occur when the Moon is either above or below the plane of Earth's orbit. When the Moon is not in the plane of Earth's orbit, the Sun, Moon, and Earth cannot align perfectly, and an eclipse cannot occur.

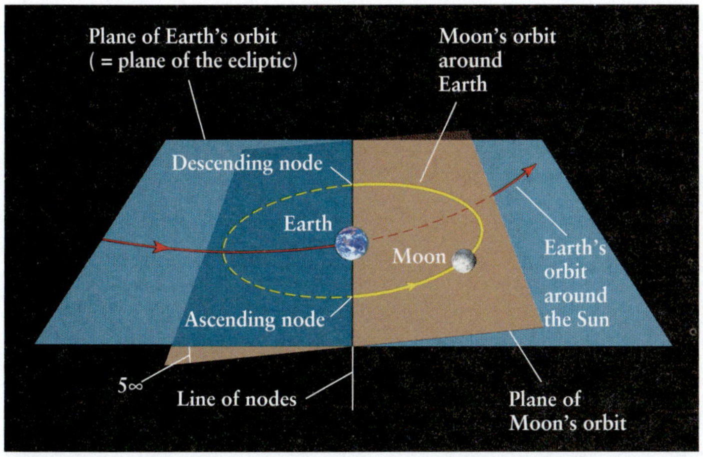

Figure 3-6

The Inclination of the Moon's Orbit This drawing shows the Moon's orbit around Earth (in yellow) and part of Earth's orbit around the Sun (in red). The plane of the Moon's orbit (shown in brown) is tilted by about 5° with respect to the plane of Earth's orbit, also called the plane of the ecliptic (shown in blue). These two planes intersect along a line called the line of nodes.

In order for the Sun, Earth, and Moon to be lined up for an eclipse, the Moon must lie in the same plane as Earth's orbit around the Sun. As we saw in Section 2-5, this plane is called the *ecliptic plane* because it is the same as the plane of the Sun's apparent path around the sky, or ecliptic (see Figure 2-14). Thus, when an eclipse occurs, the Moon appears from Earth to be on the ecliptic—which is how the ecliptic gets its name.

The planes of Earth's orbit and the Moon's orbit intersect along a line called the **line of nodes**, shown in Figure 3-6. The line of nodes passes through Earth and is pointed in a particular direction in space. Eclipses can occur only if the line of nodes is pointed toward the Sun—that is, if the Sun lies on or near the line of nodes—and if, at the same time, the Moon lies on or very near the line of nodes. Only then do the Sun, Earth, and Moon lie in a line straight enough for an eclipse to occur (Figure 3-7).

LOOKING DEEPER 3.1 Anyone who wants to predict eclipses must know the orientation of the line of nodes. But the line of nodes is gradually shifting because of the gravitational pull of the Sun on the Moon. As a result, the line of nodes rotates slowly westward. Astronomers calculate such details to fix the dates and times of upcoming eclipses.

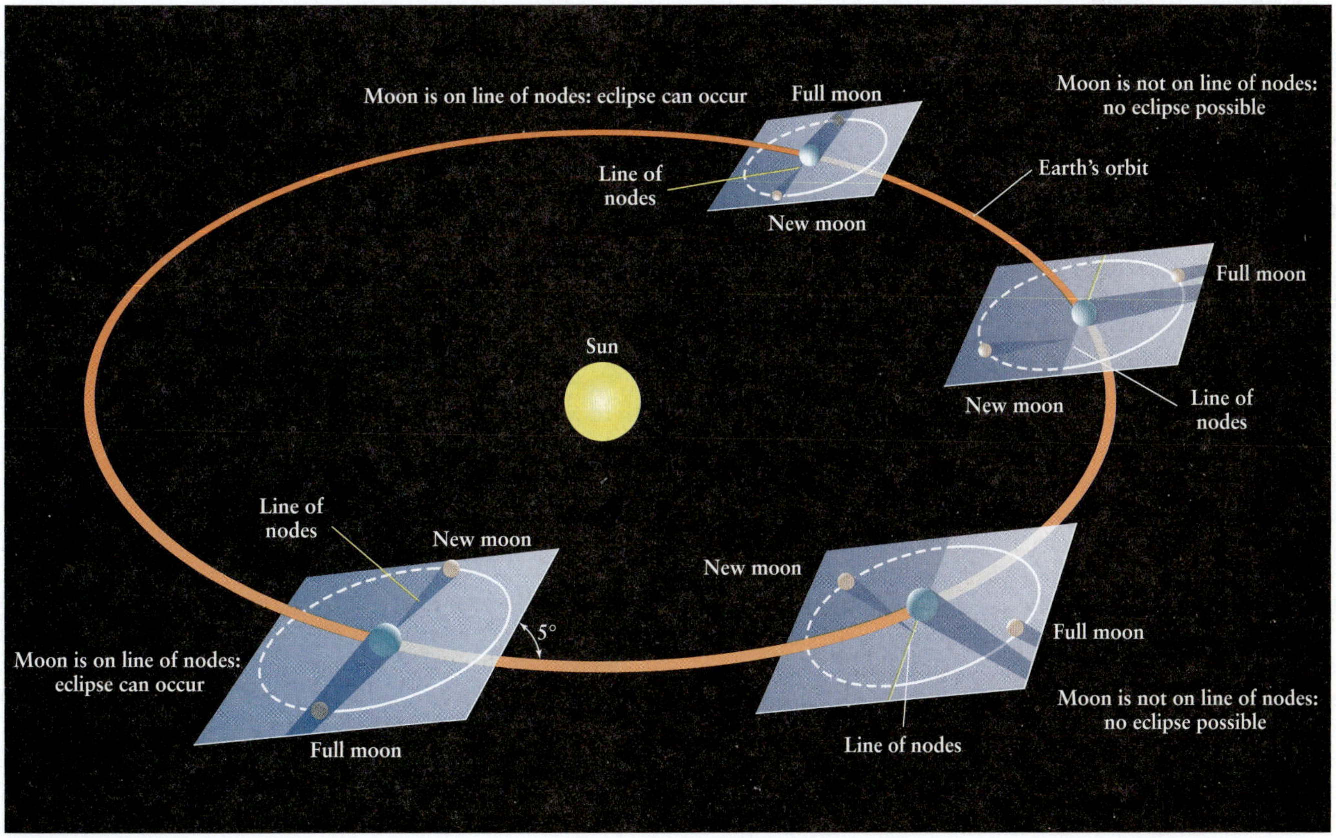

INTERACTIVE EXERCISE 3.2

Figure 3-7

Conditions for Eclipses Eclipses can take place only if the Sun and Moon are both very near to or on the line of nodes. Only then can the Sun, Earth, and Moon all lie along a straight line. A solar eclipse occurs only if the Moon is very near the line of nodes at new moon; a lunar eclipse occurs only if the Moon is very near the line of nodes at full moon. If the Sun and Moon are not near the line of nodes, the Moon's shadow cannot fall on Earth and Earth's shadow cannot fall on the Moon.

INTERACTIVE EXERCISE 3.3

Figure 3-8

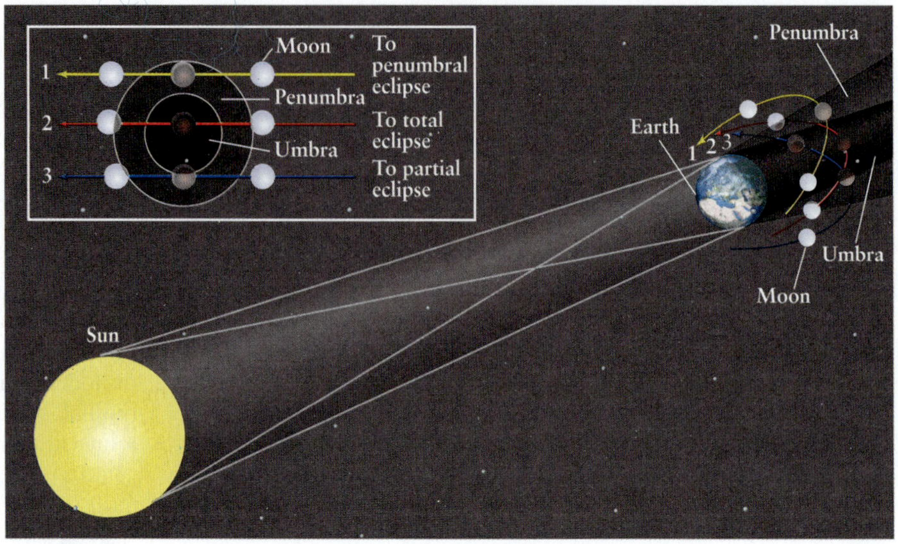

Three Types of Lunar Eclipse People on the nighttime side of Earth see a lunar eclipse when the Moon moves through Earth's shadow. In the umbra, the darkest part of the shadow, the Sun is completely covered by Earth. The penumbra is less dark because only part of the Sun is covered by Earth. The three paths show the motion of the Moon if the lunar eclipse is penumbral (Path 1, in yellow), total (Path 2, in red), or partial (Path 3, in blue). The inset shows these same paths, along with the umbra and penumbra, as viewed from Earth.

There are at least two—but never more than five—solar eclipses each year. The last year in which five solar eclipses occurred was 1935. The least number of eclipses possible (two solar, zero lunar) happened in 1969. Lunar eclipses occur just about as frequently as solar eclipses, but the maximum possible number of eclipses (lunar and solar combined) in a single year is seven.

3-4 The character of a lunar eclipse depends on the alignment of the Sun, Earth, and Moon

The character of a lunar eclipse depends on exactly how the Moon travels through Earth's shadow. As Figure 3-8 shows, the shadow of Earth has two distinct parts. Imagine an observer on the moon. In the **umbra**, the darkest part of the shadow, no portion of the Sun's surface can be seen. A portion of the Sun's surface is visible in the **penumbra**, which therefore is not quite as dark. Most people notice a lunar eclipse only if the Moon passes into Earth's umbra. As this umbral phase of the eclipse begins, a bite seems to be taken out of the Moon.

A lunar eclipse is most impressive when total, but can also be partial or penumbral

The inset in Figure 3-8 shows the different ways in which the Moon can pass into Earth's shadow. When the Moon passes through only Earth's penumbra (Path 1), we see a **penumbral eclipse.** During a penumbral eclipse, Earth blocks only part of the Sun's light and so none of the lunar surface is completely shaded. Because the Moon still looks full but only a little dimmer than usual, penumbral eclipses are easy to miss. If the Moon travels completely into the umbra (Path 2), a **total lunar eclipse** occurs. If only part of the Moon passes through the umbra (Path 3), we see a **partial lunar eclipse.**

If you were on the Moon during a total lunar eclipse, the Sun would be hidden behind Earth. But some sunlight would be visible through the thin ring of atmosphere around Earth, just as you can see sunlight through a person's hair if they stand with their head between your eyes and the Sun. As a result, a small amount of light reaches the Moon during a total lunar eclipse, and so the Moon does not completely disappear from the sky as seen from Earth. Most of the sunlight that passes through Earth's atmosphere is red, and thus the eclipsed Moon glows faintly in reddish hues, as Figure 3-9 shows. (We'll see in Chapter 5 how our atmosphere causes this reddish color.)

Lunar eclipses occur at full moon, when the Moon is directly opposite the Sun in the sky. Hence, a lunar eclipse can be seen at any place on Earth where the Sun is below the horizon (that is, where it is nighttime). A lunar eclipse has the maximum possible duration if the Moon travels directly through the center of the umbra. The Moon's speed through Earth's shadow is roughly 1 kilometer per second (3600 kilometers per hour, or 2280 miles per hour), which means that **totality**—the period when the Moon is completely within Earth's umbra—can last for as long as 1 hour and 42 minutes.

WEB LINK 3.2

On average, two or three lunar eclipses occur in a year. Table 3-1 lists all 12 lunar eclipses from 2009 to 2012. Of all lunar eclipses, roughly one-third are total, one-third are partial, and one-third are penumbral.

3-5 Solar eclipses also depend on the alignment of the Sun, Earth, and Moon

TUTORIAL 3.3

As seen from Earth, the angular diameter of the Moon is almost exactly the same as the angular diameter of the far larger but more distant Sun—about 0.5°. Thanks to this coincidence of nature, the Moon just "fits" over the Sun during a **total solar eclipse.**

Total Solar Eclipses

A total solar eclipse is a dramatic event. The sky begins to darken, the air temperature falls, and winds increase as the Moon gradually covers more and more of the Sun's disk. All nature responds: Birds go to roost,

The Sun's tenuous outer atmosphere is revealed during a total solar eclipse

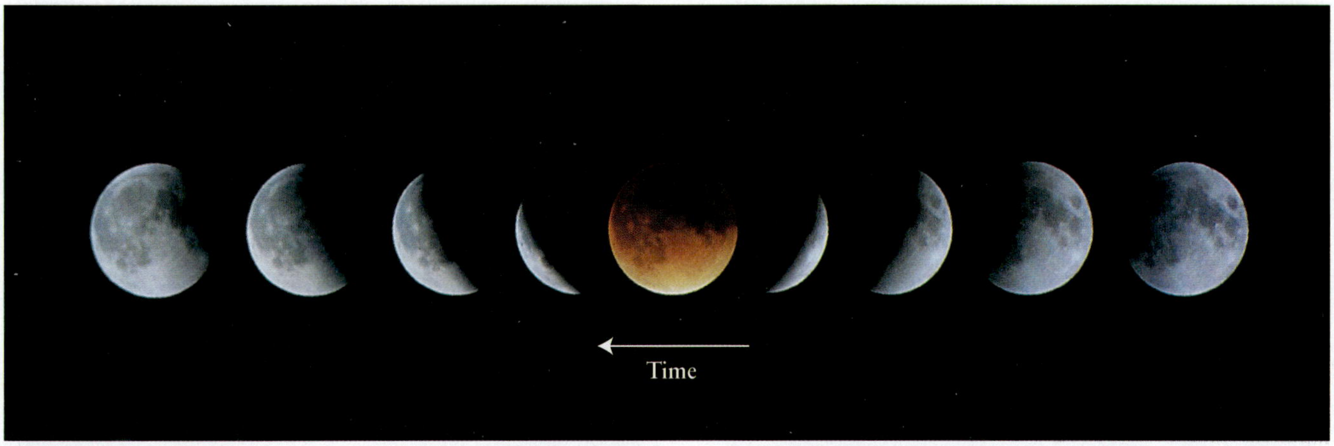

Figure 3-9 RIVUXG

A Total Lunar Eclipse This sequence of nine photographs was taken over a 3-hour period during the lunar eclipse of January 20, 2000. The sequence, which runs from right to left, shows the Moon moving through Earth's umbra. During the total phase of the eclipse (shown in the center), the Moon has a distinct reddish color. (Fred Espenak, NASA/Goddard Space Flight Center; ©2000 Fred Espenak, MrEclipse.com)

flowers close their petals, and crickets begin to chirp as if evening had arrived. As the last few rays of sunlight peek out from behind the edge of the Moon and the eclipse becomes total, the landscape around you is bathed in an eerie gray or, less frequently, in shimmering bands of light and dark. Finally, for a few minutes the Moon completely blocks out the dazzling solar disk and not much else (Figure 3-10*a*). The **solar corona**—the Sun's thin, hot outer atmosphere, which is normally too dim to be seen—blazes forth in the darkened daytime sky (Figure 3-10*b*). It is an awe-inspiring sight.

CAUTION! If you are fortunate enough to see a solar eclipse, keep in mind that the only time when it is safe to look at the Sun is during **totality,** when the solar disk is blocked by the Moon and only the solar corona is visible. Viewing this magnificent spectacle cannot harm you in any way. But you must *never* look directly at the Sun when even a portion of its intensely brilliant disk is exposed. *If you look directly at the Sun at any time without a special filter approved for solar viewing, you will suffer permanent eye damage or blindness.*

To see the remarkable spectacle of a total solar eclipse, you must be inside the darkest part of the Moon's shadow, also called the umbra, where the Moon completely blocks the Sun. Because the Sun and the Moon have nearly the same angular diameter as seen from Earth, only the tip of the Moon's umbra reaches Earth's

Table 3-1 Lunar Eclipses, 2009–2012

Date	Type	Where visible	Duration of totality (h = hours, m = minutes)
2009 February 9	Penumbral	Europe, Asia, Australia, Pacific, North America	—
2009 July 7	Penumbral	Australia, Pacific, Americas	—
2009 August 6	Penumbral	Americas, Europe, Africa, Asia	—
2009 December 31	Partial	Europe, Africa, Asia, Australia	—
2010 June 26	Partial	Asia, Australia, Pacific, Americas	—
2010 December 21	Total	Asia, Australia, Pacific, Americas, Europe	1h 13m
2011 June 15	Total	South America, Europe, Africa, Asia, Australia	1h 41m
2011 December 10	Total	Europe, Africa, Asia, Australia, Pacific, North America	52m
2012 June 4	Partial	Asia, Australia, Pacific, Americas	—
2012 November 28	Penumbral	Europe, eastern Africa, Asia, Australia, North America	—

Eclipse predictions by Fred Espenak, NASA/Goddard Space Flight Center. All dates are given in standard astronomical format: year, month, day.

(a)

(b)

Figure 3-10 R I V U X G

A Total Solar Eclipse **(a)** This photograph shows the total solar eclipse of August 11, 1999, as seen from Elâziğ, Turkey. The sky is so dark that the planet Venus can be seen to the left of the eclipsed Sun. **(b)** When the Moon completely covers the Sun's disk during a total eclipse, the faint solar corona is revealed. (Fred Espenak, MrEclipse.com)

surface (Figure 3-11). As Earth rotates, the tip of the umbra traces an **eclipse path** across Earth's surface. Only those locations within the eclipse path are treated to the spectacle of a total solar eclipse. The inset in Figure 3-11 shows the dark spot on Earth's surface produced by the Moon's umbra.

Partial Solar Eclipses

Immediately surrounding the Moon's umbra is the region of partial shadow called the penumbra. As seen from this area, the Sun's surface appears only partially covered by the Moon. During a solar eclipse, the Moon's penumbra covers a large portion of Earth's surface, and anyone standing inside the penumbra sees a **partial solar eclipse.** Such eclipses are much less interesting events than total solar eclipses, which is why astronomy enthusiasts strive to be inside the eclipse path. If you are within the eclipse path, you will see a partial eclipse before and after the brief period of totality (see the photograph that opens this chapter).

The width of the eclipse path depends primarily on the Earth-Moon distance during totality. The eclipse path is widest if the Moon happens to be at **perigee,** the point in its orbit nearest Earth. In this case the width of the eclipse path can be as great as 270 kilometers (170 miles). In most eclipses, however, the path is much narrower.

Annular Solar Eclipses

In some eclipses the Moon's umbra does not reach all the way to Earth's surface. This can happen if the Moon is at or near **apogee,** its farthest position from Earth. In this case, the Moon appears too small to cover the Sun completely. The result is a third type of solar eclipse, called an **annular eclipse.** During an annular eclipse, a thin ring of the Sun is seen around the edge of the Moon (Figure 3-12). The length of the Moon's umbra is nearly 5000 kilometers (3100 miles) less than the average distance between the Moon and Earth's surface. Thus, the Moon's shadow often fails to reach Earth even when the Sun, Moon, and Earth are properly aligned for an eclipse. Hence, annular eclipses are slightly more common—as well as far less dramatic—than total eclipses.

Even during a total eclipse, most people along the eclipse path observe totality for only a few moments. Earth's rotation, coupled with the orbital motion of the Moon, causes the umbra to race eastward along the eclipse path at speeds in excess of 1700 kilometers per hour (1060 miles per hour). Because of the umbra's high speed, totality never lasts for more than $7\frac{1}{2}$ minutes. In a typical total solar eclipse, the Sun-Moon-Earth alignment and the Earth-Moon distance are such that totality lasts much less than this maximum.

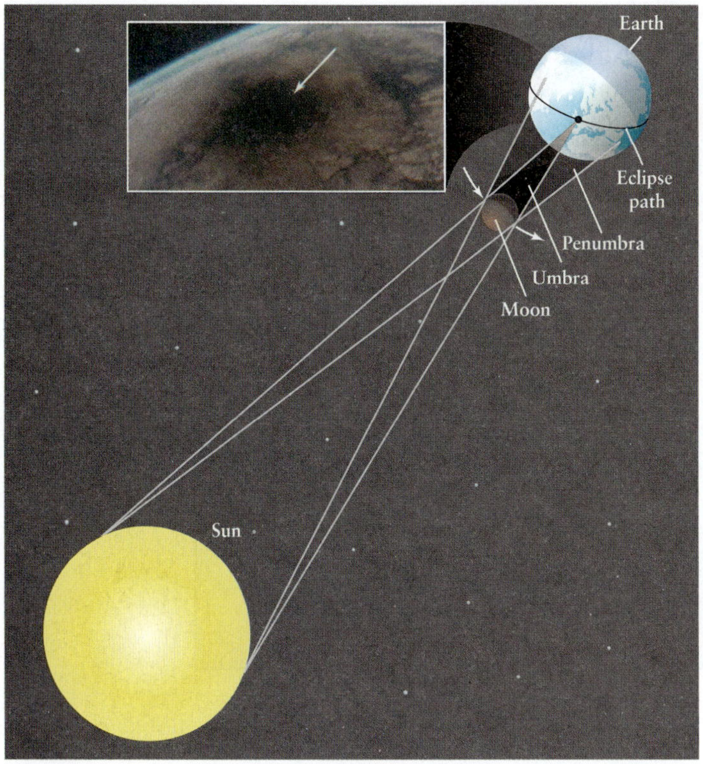

ANIMATION 3.4

Figure 3-11 R I V U X G

The Geometry of a Total Solar Eclipse During a total solar eclipse, the tip of the Moon's umbra reaches Earth's surface. As Earth and the Moon move along their orbits, this tip traces an eclipse path across Earth's surface. People within the eclipse path see a total solar eclipse as the tip moves over them. Anyone within the penumbra sees only a partial eclipse. The inset photograph was taken from the *Mir* space station during the August 11, 1999, total solar eclipse (the same eclipse shown in Figure 3-10). The tip of the umbra appears as a black spot on Earth's surface. At the time the photograph was taken, this spot was 105 km (65 mi) wide and was crossing the cloud-covered English Channel at 3000 km/h (1900 mi/h). (Photograph by Jean-Pierre Haigneré, Centre National d'Etudes Spatiales, France/GSFS)

Figure 3-12 R I V U X G

An Annular Solar Eclipse This composite of six photographs taken at sunrise in Costa Rica shows the progress of an annular eclipse of the Sun on December 24, 1973. (Five photographs were made of the Sun, plus one of the hills and sky.) Note that at mideclipse the limb, or outer edge, of the Sun is visible around the Moon. (Courtesy of Dennis di Cicco)

WEB LINK 3.2

The details of solar eclipses are calculated well in advance. They are published in such reference books as the *Astronomical Almanac* and are available on the World Wide Web. Figure 3-13 shows the eclipse paths for all total solar eclipses from 1997 to 2020. Table 3-2 lists all the total, annular, and partial eclipses from 2009 to 2012, including the maximum duration of totality for total eclipses.

Ancient astronomers achieved a limited ability to predict eclipses. In those times, religious and political leaders who were able to predict such awe-inspiring events as eclipses must have made a tremendous impression on their followers. One of three priceless manuscripts to survive the devastating Spanish Conquest shows that the Mayan astronomers of Mexico and Guatemala had a fairly reliable method for predicting eclipses. The great Greek astronomer Thales of Miletus is said to have predicted the famous eclipse of 585 B.C., which occurred during the middle of a war. The sight was so unnerving that the soldiers put down their arms and declared peace.

In retrospect, it seems that what ancient astronomers actually produced were eclipse "warnings" of various degrees of reliability rather than true predictions. Working with historical records, these astronomers generally sought to discover cycles and regularities from which future eclipses could be anticipated. Box 3-2 describes how you might produce eclipse warnings yourself.

3-6 Ancient astronomers measured the size of Earth and attempted to determine distances to the Sun and Moon

The prediction of eclipses was not the only problem attacked by ancient astronomers. More than 2000 years ago, centuries before sailors of Columbus's era crossed the oceans, Greek astronomers were fully aware that Earth is not flat. They had come to this conclusion using a combination of observation and logical

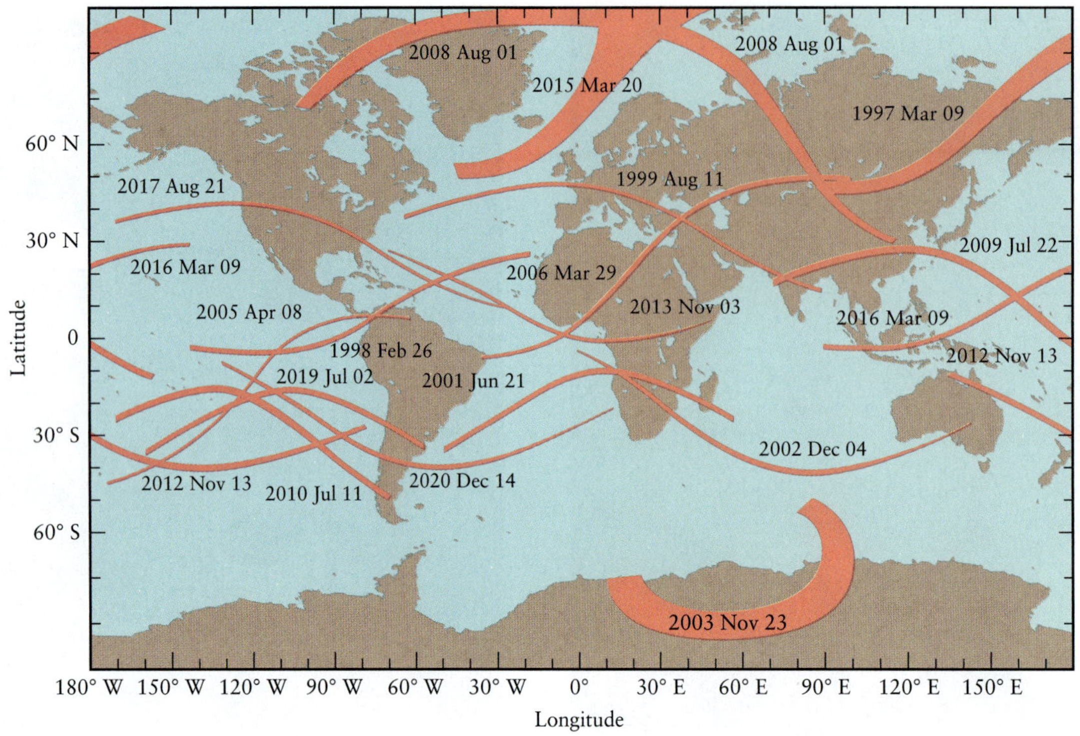

Figure 3-13

Eclipse Paths for Total Eclipses, 1997–2020 This map shows the eclipse paths for all 18 total solar eclipses occurring from 1997 through 2020. In each eclipse, the Moon's shadow travels along the eclipse path in a generally eastward direction across Earth's surface. (Courtesy of Fred Espenak, NASA/Goddard Space Flight Center)

deduction, much like modern scientists. The Greeks noted that during lunar eclipses, when the Moon passes through Earth's shadow, the edge of the shadow is always circular. Because a sphere is the only shape that always casts a circular shadow from any angle, they concluded that Earth is spherical.

The ideas of geometry made it possible for Greek scholars to estimate cosmic distances

Eratosthenes and the Size of Earth

Around 200 B.C., the Greek astronomer Eratosthenes devised a way to measure the circumference of the spherical Earth. It was known that on the date of the summer solstice (the first day of summer; see Section 2-5) in the town of Syene in Egypt, near present-day Aswan, the Sun shone directly down the vertical shafts of water wells. Hence, at local noon on that day, the Sun was at

Table 3-2 Solar Eclipses, 2009–2012

Date	Type	Where visible	Notes
2009 January 26	Annular	Southern Africa, Antarctica, southeast Asia, Australia	—
2009 July 22	Total	Eastern Asia, Pacific Ocean, Hawaii	Maximum duration of totality 6m 39s
2010 January 15	Annular	Africa, Asia	—
2010 July 11	Total	Pacific Ocean, South America	Maximum duration of totality 5m 20s
2011 January 4	Partial	Europe, Africa, central Asia	86% eclipsed
2011 June 1	Partial	Eastern Asia, northern North America, Iceland	60% eclipsed
2011 July 1	Partial	Indian Ocean	10% eclipsed
2011 November 25	Partial	Southern Africa, Antarctica, Australia, New Zealand	91% eclipsed
2012 May 20	Annular	Asia, Pacific, North America	—
2012 November 13	Total	Australia, New Zealand, southern South America	—

Eclipse predictions by Fred Espenak, NASA/Goddard Space Flight Center. All dates are given in standard astronomical format: year, month, day.

BOX 3-2 **Tools of the Astronomer's Trade**

Predicting Solar Eclipses

Suppose that you observe a solar eclipse in your hometown and want to figure out when you and your neighbors might see another eclipse. How would you begin?

First, remember that a solar eclipse can occur only if the line of nodes points toward the Sun at the same time that there is a new moon (see Figure 3-7). Second, you must know that it takes 29.53 days (one synodic month) to go from one new moon to the next. Because solar eclipses occur only during new moon, you must wait several whole lunar months for the proper alignment to occur again.

However, there is a complication: The line of nodes gradually shifts its position with respect to the background stars. It takes 346.6 days to move from one alignment of the line of nodes pointing toward the Sun to the next identical alignment. This period is called the **eclipse year.**

Therefore, to predict when you will see another solar eclipse, you need to know how many whole lunar months equal some whole number of eclipse years. This information will tell you how long you will have to wait for the next virtually identical alignment of the Sun, the Moon, and the line of nodes. By trial and error, you find that 223 lunar months is the same length of time as 19 eclipse years, because

$$223 \times 29.53 \text{ days} = 19 \times 346.6 \text{ days} = 6585 \text{ days}$$

This calculation is accurate to within a few hours. A more accurate calculation gives an interval, called the **saros,** that is about one-third day longer, or 6585.3 days (18 years, 11.3 days). Eclipses separated by the saros interval are said to form an *eclipse series.*

You might think that you and your neighbors would simply have to wait one full saros interval to go from one solar eclipse to the next. However, because of the extra one-third day, Earth will have rotated by an extra 120° (one-third of a complete rotation) when the next solar eclipse of a particular series occurs. The eclipse path will thus be one-third of the way around the world from you. Therefore, you must wait three full saros intervals (54 years, 34 days) before the eclipse path comes back around to your part of Earth. The illustration shows a series of six solar eclipse paths, each separated from the next by one saros interval.

There is evidence that ancient Babylonian astronomers knew about the saros interval. However, the discovery of the saros is more likely to have come from lunar eclipses than solar eclipses. If you are far from the eclipse path, there is a good chance that you could fail to notice a solar eclipse. Even if half the Sun is covered by the Moon, the remaining solar surface provides enough sunlight for the outdoor illumination not to be greatly diminished. By contrast, anyone on the nighttime side of Earth can see an eclipse of the Moon unless clouds block the view.

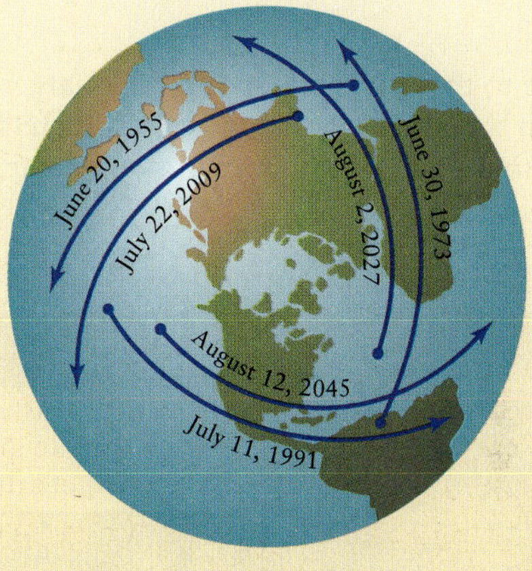

the zenith (see Section 2-4) as seen from Syene. Eratosthenes knew that the Sun never appeared at the zenith at his home in the Egyptian city of Alexandria, which is on the Mediterranean Sea almost due north of Syene. Rather, on the summer solstice in Alexandria, the position of the Sun at local noon was about 7° south of the zenith (Figure 3-14). This angle is about one-fiftieth of a complete circle, so he concluded that the distance from Alexandria to Syene must be about one-fiftieth of Earth's circumference.

In Eratosthenes's day, the distance from Alexandria to Syene was said to be 5000 stades. Therefore, Eratosthenes found Earth's circumference to be

$$50 \times 5000 \text{ stades} = 250{,}000 \text{ stades}$$

Unfortunately, no one today is sure of the exact length of the Greek unit called the stade. One guess is that the stade was about one-sixth of a kilometer, which would mean that Eratosthenes obtained a circumference for Earth of about 42,000 kilometers. This is remarkably close to the modern value of 40,000 kilometers.

Aristarchus and Distances in the Solar System

Eratosthenes was only one of several brilliant astronomers to emerge from the so-called Alexandrian school, which by his time had a distinguished tradition. One of the first Alexandrian astronomers, Aristarchus of Samos, had proposed a method of determining the relative distances to the Sun and Moon, perhaps as long ago as 280 B.C.

Aristarchus knew that the Sun, Moon, and Earth form a right triangle at the moment of first or third quarter moon, with the right angle at the location of the Moon (Figure 3-15). He estimated that, as seen from Earth, the angle between the Moon and the Sun at first and third quarters is 87°, or 3° less than a right angle. Using the rules of geometry, Aristarchus concluded that the Sun is about 20 times farther from us than is the Moon. We now know that Aristarchus erred in measuring angles and that the

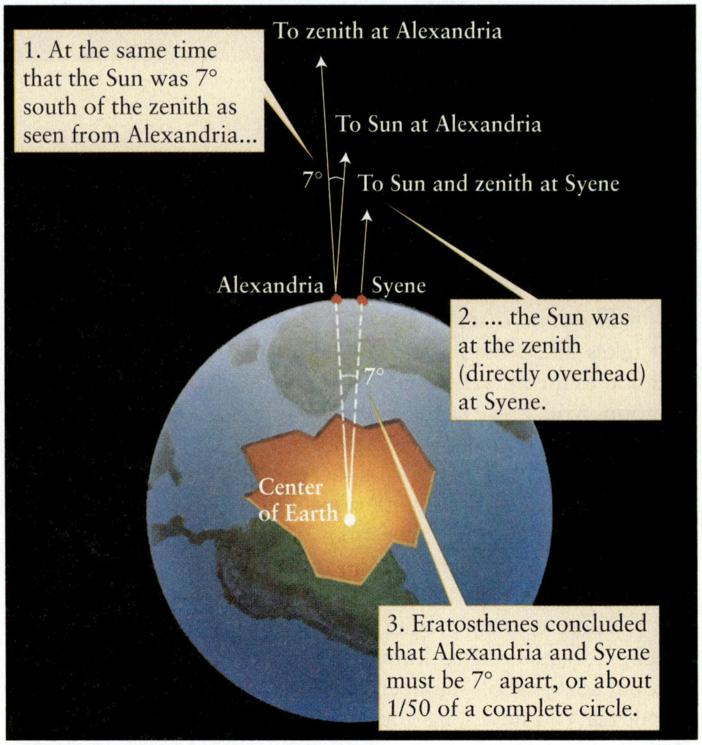

To Sun and zenith at Syene

2. ... the Sun was at the zenith (directly overhead) at Syene.

Figure 3-14

Eratosthenes's Method of Determining the Diameter of Earth Around 200 B.C., Eratosthenes used observations of the Sun's position at noon on the summer solstice to show that Alexandria and Syene were about 7° apart on the surface of Earth. This angle is about one-fiftieth of a circle, so the distance between Alexandria and Syene must be about one-fiftieth of Earth's circumference.

average distance to the Sun is about 390 times larger than the average distance to the Moon. It is nevertheless impressive that people were trying to measure distances across the solar system more than 2000 years ago.

Aristarchus also made an equally bold attempt to determine the relative sizes of the Earth, Moon, and Sun. From his observations of how long the Moon takes to move through Earth's shadow during a lunar eclipse, Aristarchus estimated the diameter of Earth

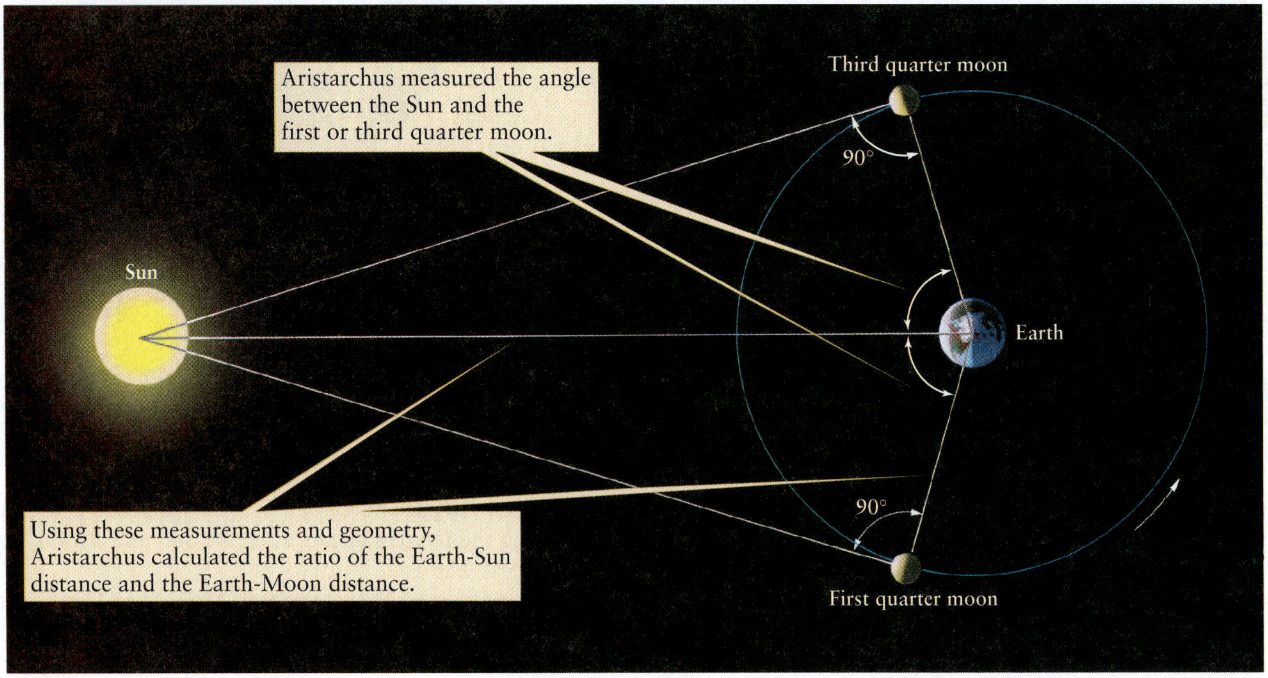

Aristarchus measured the angle between the Sun and the first or third quarter moon.
Third quarter moon
90°
Sun
Earth
90°
Using these measurements and geometry, Aristarchus calculated the ratio of the Earth-Sun distance and the Earth-Moon distance.
First quarter moon

Figure 3-15

Aristarchus's Method of Determining Distances to the Sun and Moon Aristarchus knew that the Sun, Moon, and Earth form a right triangle at first and third quarter phases. Using geometrical arguments, he calculated the relative lengths of the sides of these triangles, thereby obtaining the distances to the Sun and Moon.

Table 3-3 Comparison of Ancient and Modern Astronomical Measurements

	Ancient (km)	Modern (km)
Earth's diameter	13,000	12,756
Moon's diameter	4,300	3,476
Sun's diameter	9×10^4	1.39×10^6
Earth-Moon distance	4×10^5	3.84×10^5
Earth-Sun distance	10^7	1.50×10^8

to be about 3 times larger than the diameter of the Moon. To determine the diameter of the Sun, Aristarchus simply pointed out that the Sun and the Moon have the same angular size in the sky. Therefore, their diameters must be in the same proportion as their distances (see part *a* of the figure in Box 1-1). In other words, because Aristarchus thought the Sun to be 20 times farther from Earth than the Moon, he concluded that the Sun must be 20 times larger than the Moon. Once Eratosthenes had measured Earth's circumference, astronomers of the Alexandrian school could estimate the diameters of the Sun and Moon as well as their distances from Earth.

Table 3-3 summarizes some ancient and modern measurements of the sizes of Earth, the Moon, and the Sun and the distances between them. Some of these ancient measurements are far from the modern values. Yet the achievements of our ancestors still stand as impressive applications of observation and reasoning and important steps toward the development of the scientific method.

Key Words

Terms preceded by an asterisk (*) are discussed in the Boxes.

annular eclipse, p. 54
apogee, p. 54
eclipse, p. 50
eclipse path, p. 54
*eclipse year, p. 57
first quarter moon, p. 46
full moon, p. 46
line of nodes, p. 51
lunar eclipse, p. 50
lunar phases, p. 46
new moon, p. 46
partial lunar eclipse, p. 52
partial solar eclipse, p. 54
penumbra (*plural* penumbrae), p. 52
penumbral eclipse, p. 52
perigee, p. 54
*saros, p. 57
sidereal month, p. 50
solar corona, p. 53
solar eclipse, p. 50
synchronous rotation, p. 49
synodic month, p. 50
third quarter moon, p. 46
totality (lunar eclipse), p. 52
totality (solar eclipse), p. 53
total lunar eclipse, p. 52
total solar eclipse, p. 52
umbra (*plural* umbrae), p. 52
waning crescent moon, p. 46
waning gibbous moon, p. 46
waxing crescent moon, p. 46
waxing gibbous moon, p. 46

Key Ideas

Lunar Phases: The phases of the Moon occur because light from the Moon is actually reflected sunlight. As the relative positions of Earth, the Moon, and the Sun change, we see more or less of the illuminated half of the Moon.

Length of the Month: Two types of months are used in describing the motion of the Moon.

• With respect to the stars, the Moon completes one orbit around Earth in a sidereal month, averaging 27.32 days.

• The Moon completes one cycle of phases (one orbit around Earth with respect to the Sun) in a synodic month, averaging 29.53 days.

The Moon's Orbit: The plane of the Moon's orbit is tilted by about 5° from the plane of Earth's orbit, or ecliptic.

• The line of nodes is the line where the planes of the Moon's orbit and Earth's orbit intersect. The gravitational pull of the Sun gradually shifts the orientation of the line of nodes with respect to the stars.

Conditions for Eclipses: During a lunar eclipse, the Moon passes through Earth's shadow. During a solar eclipse, Earth passes through the Moon's shadow.

• Lunar eclipses occur at full moon, while solar eclipses occur at new moon.

• Either type of eclipse can occur only when the Sun and Moon are both on or very near the line of nodes. If this condition is not met, Earth's shadow cannot fall on the Moon and the Moon's shadow cannot fall on Earth.

Umbra and Penumbra: The shadow of an object has two parts: the umbra, within which the light source is completely blocked, and the penumbra, where the light source is only partially blocked.

Lunar Eclipses: Depending on the relative positions of the Sun, Moon, and Earth, lunar eclipses may be total (the Moon passes completely into Earth's umbra), partial (only part of the Moon passes into Earth's umbra), or penumbral (the Moon passes only into Earth's penumbra).

Solar Eclipses: Solar eclipses may be total, partial, or annular.

• During a total solar eclipse, the Moon's umbra traces out an eclipse path over Earth's surface as Earth rotates. Observers outside the eclipse path but within the penumbra see only a partial solar eclipse.

• During an annular eclipse, the umbra falls short of Earth, and the outer edge of the Sun's disk is visible around the Moon at mideclipse.

The Moon and Ancient Astronomers: Ancient astronomers such as Aristarchus and Eratosthenes made great progress in determining the sizes and relative distances of Earth, the Moon, and the Sun.

Questions

Review Questions

Questions preceded by an asterisk (*) involve topics discussed in the Boxes.

1. Explain the difference between sunlight and moonlight.
2. (a) Explain why the Moon exhibits phases. (b) A common misconception about the Moon's phases is that they are caused by Earth's shadow. Use Figure 3-2 to explain why this not correct.
3. How would the sequence and timing of lunar phases be affected if the Moon moved around its orbit (a) in the same di-

rection, but at twice the speed; (**b**) at the same speed, but in the opposite direction? Explain your answers.

4. At approximately what time does the Moon rise when it is (**a**) a new moon; (**b**) a first quarter moon; (**c**) a full moon; (**d**) a third quarter moon?
5. Astronomers sometimes refer to lunar phases in terms of the *age* of the Moon. This is the time that has elapsed since new moon phase. Thus, the age of a full moon is half of a $29\frac{1}{2}$-day synodic period, or approximately 15 days. Find the approximate age of (**a**) a waxing crescent moon; (**b**) a third quarter moon; (**c**) a waning gibbous moon.
6. If you lived on the Moon, would you see Earth go through phases? If so, would the sequence of phases be the same as those of the Moon as seen from Earth, or would the sequence be reversed? Explain using Figure 3-2.
7. Is the far side of the Moon (the side that can never be seen from Earth) the same as the dark side of the Moon? Explain.
8. (**a**) If you lived on the Moon, would you see the Sun rise and set, or would it always be in the same place in the sky? Explain. (**b**) Would you see Earth rise and set, or would it always be in the same place in the sky? Explain using Figure 3-4.
9. What is the difference between a sidereal month and a synodic month? Which is longer? Why?
10. On a certain date the Moon is in the direction of the constellation Gemini as seen from Earth. When will the Moon next be in the direction of Gemini: one sidereal month later, or one synodic month later? Explain your answer.

TUTORIAL 3.2

11. What is the difference between the umbra and the penumbra of a shadow?
12. Why doesn't a lunar eclipse occur at every full moon and a solar eclipse at every new moon?
13. What is the line of nodes? Why is it important to the subject of eclipses?
14. What is a penumbral eclipse of the Moon? Why do you suppose that it is easy to overlook such an eclipse?
15. Why is the duration of totality different for different total lunar eclipses, as shown in Table 3-1?
16. Can one ever observe an annular eclipse of the Moon? Why or why not?

TUTORIAL 3.3

17. If you were looking at Earth from the side of the Moon that faces Earth, what would you see during (**a**) a total lunar eclipse? (**b**) a total solar eclipse? Explain your answers.
18. If there is a total eclipse of the Sun in April, can there be a lunar eclipse three months later in July? Why or why not?
19. Which type of eclipse—lunar or solar—do you think most people on Earth have seen? Why?
20. How is an annular eclipse of the Sun different from a total eclipse of the Sun? What causes this difference?

*21. What is the saros? How did ancient astronomers use it to predict eclipses?

22. How did Eratosthenes measure the size of Earth?
23. How did Aristarchus try to estimate the distance from Earth to the Sun and Moon?
24. How did Aristarchus try to estimate the diameters of the Sun and Moon?

Advanced Questions

Problem-solving tips and tools

To estimate the average angular speed of the Moon along its orbit (that is, how many degrees around its orbit the Moon travels per day), divide 360° by the length of a sidereal month. It is helpful to know that the saros interval of 6585.3 days equals 18 years and $11\frac{1}{3}$ days if the interval includes 4 leap years, but is 18 years and $10\frac{1}{3}$ days if it includes 5 leap years.

25. The dividing line between the illuminated and unilluminated halves of the Moon is called the *terminator*. The terminator appears curved when there is a crescent or gibbous moon, but appears straight when there is a first quarter or third quarter moon (see Figure 3-2). Describe how you could use these facts to explain to a friend why lunar phases cannot be caused by Earth's shadow falling on the Moon.
26. What is the phase of the Moon if it rises at (**a**) midnight; (**b**) sunrise; (**c**) halfway between sunset and midnight; (**d**) halfway between noon and sunset? Explain your answers.
27. The Moon is highest in the sky when it crosses the meridian (see Figure 2-21), halfway between the time of moonrise and the time of moonset. At approximately what time does the Moon cross the meridian if it is (**a**) a new moon; (**b**) a first quarter moon; (**c**) a full moon; (**d**) a third quarter moon? Explain your answers.
28. The Moon is highest in the sky when it crosses the meridian (see Figure 2-21), halfway between the time of moonrise and the time of moonset. What is the phase of the Moon if it is highest in the sky at (**a**) midnight; (**b**) sunrise; (**c**) noon; (**d**) sunset? Explain your answers.
29. Suppose it is the first day of autumn in the northern hemisphere. What is the phase of the Moon if the Moon is located at (**a**) the vernal equinox? (**b**) the summer solstice? (**c**) the autumnal equinox? (**d**) the winter solstice? Explain your answers. (*Hint:* Make a drawing showing the relative positions of the Sun, Earth, and Moon. Compare with Figure 3-2.)

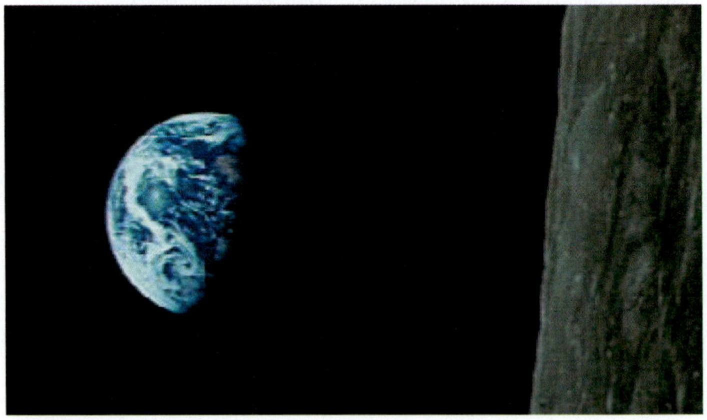

R I V U X G

(NASA/JSC)

30. The above photograph of Earth was taken by the crew of the *Apollo 8* spacecraft as they orbited the Moon. A portion of

the lunar surface is visible at the right-hand side of the photo. In this photo, Earth is oriented with its north pole approximately at the top. When this photo was taken, was the Moon waxing or waning as seen from Earth? Explain your answer with a diagram.

31. (a) The Moon moves noticeably on the celestial sphere over the space of a single night. To show this, calculate how long it takes the Moon to move through an angle equal to its own angular diameter ($1/2°$) against the background of stars. Give your answer in hours. (b) Through what angle (in degrees) does the Moon move during a 12-hour night? Can you notice an angle of this size? (*Hint:* See Figure 1-10.)

32. During an occultation, or "covering up," of Jupiter by the Moon, an astronomer notices that it takes the Moon's edge 90 seconds to cover Jupiter's disk completely. If the Moon's motion is assumed to be uniform and the occultation was "central" (that is, center over center), find the angular diameter of Jupiter. (*Hint:* Assume that Jupiter does not appear to move against the background of stars during this brief 90-second interval. You will need to convert the Moon's angular speed from degrees per day to arcseconds per second.)

33. The plane of the Moon's orbit is inclined at a 5° angle from the ecliptic, and the ecliptic is inclined at a $23\frac{1}{2}°$ angle from the celestial equator. Could the Moon ever appear at your zenith if you lived at (a) the equator; (b) the south pole? Explain your answers.

34. How many more sidereal months than synodic months are there in a year? Explain your answer.

35. Suppose Earth moved a little faster around the Sun, so that it took a bit less than one year to make a complete orbit. If the speed of the Moon's orbit around Earth were unchanged, would the length of the sidereal month be the same, longer, or shorter than it is now? What about the synodic month? Explain your answers.

36. If the Moon revolved about Earth in the same orbit but in the opposite direction, would the synodic month be longer or shorter than the sidereal month? Explain your reasoning.

37. One definition of a "blue moon" is the second full moon within the same calendar month. There is usually only one full moon within a calendar month, so the phrase "once in a blue moon" means "hardly ever." Why are blue moons so rare? Are there any months of the year in which it would be impossible to have two full moons? Explain your answer.

38. You are watching a lunar eclipse from some place on Earth's night side. Will you see the Moon enter Earth's shadow from the east or from the west? Explain your reasoning.

39. The total lunar eclipse of October 28, 2004, was visible from South America. The duration of totality was 1 hour, 21 minutes. Was this total eclipse also visible from Australia, on the opposite side of Earth? Explain your reasoning.

40. During a total solar eclipse, the Moon's umbra moves in a generally eastward direction across Earth's surface. Use a drawing like Figure 3-11 to explain why the motion is eastward, not westward.

41. A total solar eclipse was visible from Africa on March 29, 2006 (see Figure 3-10). Draw what the eclipse would have looked like as seen from France, to the north of the path of totality. Explain the reasoning behind your drawing.

42. Figures 3-11 and 3-13 show that the path of a total eclipse is quite narrow. Use this to explain why a glow is visible all around the horizon when you are viewing a solar eclipse during totality (see Figure 3-10*a*).

43. (a) Suppose the diameter of the Moon were doubled, but the orbit of the Moon remained the same. Would total solar eclipses be more common, less common, or just as common as they are now? Explain. (b) Suppose the diameter of the Moon were halved, but the orbit of the Moon remained the same. Explain why there would be *no* total solar eclipses.

44. Just as the distance from Earth to the Moon varies somewhat as the Moon orbits Earth, the distance from the Sun to Earth changes as Earth orbits the Sun. Earth is closest to the Sun at its *perihelion*; it is farthest from the Sun at its *aphelion*. In order for a total solar eclipse to have the maximum duration of totality, should Earth be at perihelion or aphelion? Assume that the Earth-Moon distance is the same in both situations. As part of your explanation, draw two pictures like Figure 3-11, one with Earth relatively close to the Sun and one with Earth relatively far from the Sun.

*45. On March 29, 2006, residents of northern Africa were treated to a total solar eclipse. (a) On what date and over what part of the world will the next total eclipse of that series occur? Explain. (b) On what date might you next expect a total eclipse of that series to be visible from northern Africa? Explain.

Discussion Questions

46. Describe the cycle of lunar phases that would be observed if the Moon moved around Earth in an orbit perpendicular to the plane of Earth's orbit. Would it be possible for both solar and lunar eclipses to occur under these circumstances? Explain your reasoning.

47. How would a lunar eclipse look if Earth had no atmosphere? Explain your reasoning.

48. In his 1885 novel *King Solomon's Mines*, H. Rider Haggard described a total solar eclipse that was seen in both South Africa and in the British Isles. Is such an eclipse possible? Why or why not?

49. Why do you suppose that total solar eclipse paths fall more frequently on oceans than on land? (You may find it useful to look at Figure 3-13.)

50. Examine Figure 3-13, which shows all of the total solar eclipses from 1997 to 2020. What are the chances that you might be able to travel to one of the eclipse paths? Do you think you might go through your entire life without ever seeing a total eclipse of the Sun?

Web/eBook Questions

51. Access the animation "The Moon's Phases" in Chapter 3 of the *Universe* Web site or eBook. This shows the Earth-Moon system as seen from a vantage point looking down onto the northpole. (a) Describe where you would be on the diagram if you are on the equator and the time is 6:00 P.M. (b) If it is 6:00 P.M. and you are standing on Earth's equator, would a third quarter moon be visible? Why or why not? If it would be visible, describe its appearance.

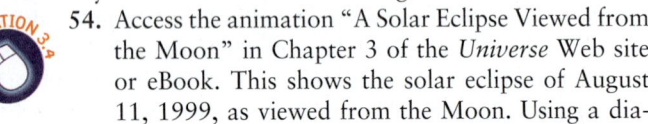

52. Search the World Wide Web for information about the next total lunar eclipse. Will the total phase of the eclipse be visible from your location? If not, will the penumbral phase be visible? Draw a picture showing the Sun, Earth, and Moon when the totality is at its maximum duration, and indicate your location on the drawing of Earth.
53. Search the World Wide Web for information about the next total solar eclipse. Through which major cities, if any, does the path of totality pass? What is the maximum duration of totality? At what location is this maximum duration observed? Will this eclipse be visible (even as a partial eclipse) from your location? Draw a picture showing the Sun, Earth, and Moon when the totality is at its maximum duration, and indicate your location on the drawing of Earth.
54. Access the animation "A Solar Eclipse Viewed from the Moon" in Chapter 3 of the *Universe* Web site or eBook. This shows the solar eclipse of August 11, 1999, as viewed from the Moon. Using a diagram, explain why the stars and the Moon's shadow move in the directions shown in this animation.

Activities

Observing Projects

55. Observe the Moon on each clear night over the course of a month. On each night, note the Moon's location among the constellations and record that location on a star chart that also shows the ecliptic. After a few weeks, your observations will begin to trace the Moon's orbit. Identify the orientation of the line of nodes by marking the points where the Moon's orbit and the ecliptic intersect. On what dates is the Sun near the nodes marked on your star chart? Compare these dates with the dates of the next solar and lunar eclipses.
56. It is quite possible that a lunar eclipse will occur while you are taking this course. Look up the date of the next lunar eclipse in Table 3-1, on the World Wide Web, or in the current issue of a reference such as the *Astronomical Almanac* or *Astronomical Phenomena*. Then make arrangements to observe this lunar eclipse. You can observe the eclipse with the naked eye, but binoculars or a small telescope will enhance your viewing experience. If the eclipse is partial or total, note the times at which the Moon enters and exits Earth's umbra. If the eclipse is penumbral, can you see any changes in the Moon's brightness as the eclipse progresses?
57. Use the *Starry Night Enthusiast*™ program to observe the motion of the Moon. (a) Display the entire celestial sphere, including the part below the horizon, by moving to the Atlas mode. You do this by selecting **Favorites > Guides > Atlas.** Here, you will see the sky, containing the background stars and the planets, overlaid by a coordinate grid. One axis, the **Right Ascension** axis, is the extension of Earth's equator on to the sky and is marked in hours along the **Celestial Equator.** At right angles to this equator are the **Declination** lines at constant Right Ascension, converging upon the North and South Celestial Poles. These poles are the extensions of the two ends of Earth's spin axis. You can use the Hand Tool to explore this coordinate system by moving your viewpoint around the sky. (Move the mouse while holding down the mouse button to achieve this motion.) Across this sky, inclined at an angle to the celestial equator, is the **Ecliptic,** or the path along which the Sun appears to move across our sky. This is the plane of Earth's orbit. (If this green line does not appear, open the **Options** pane and check that the **Ecliptic** is selected in the **Guides** layer.) Use the Hand Tool to move the sky around to find the **Moon,** which will be close to, but not on, the ecliptic plane. Once you have found the Moon, use the Hand tool to move the Moon to the right-hand side of the main window. On the toolbar across the top of the main window, click on the **Time Flow Rate** control (immediately to the right of the date and time display) and set the discrete time step to **1 sidereal day.** Then advance time in one-sidereal-day intervals by clicking on the **Step Time Forward** button (the icon consisting of a black vertical line and right-pointing triangle to the far right of the time controls). You will note that the background sky remains fixed, as expected when time moves ahead in sidereal- day intervals. How does the Moon appear to move against the background of stars? Does it ever change direction? (b) Use this **Step Time Forward** button to determine how many days elapse between successive times when the Moon is on the ecliptic. Then move forward in time to a date when the Moon is on the ecliptic and either full or new. What type of eclipse will occur on that date? Confirm your answer by comparing with Tables 3-1 and 3-2 or with lists of eclipses on the World Wide Web.
58. Use the *Starry Night Enthusiast*™ program to examine the Moon as seen from space. Select **Solar System > Inner Solar System** in the **Favorites** menu. Click the **Stop** button in the toolbar to stop time flow. Then, click on the **Find** tab and double-click on the entry for the Moon in the Find pane in order to center the view on the Moon. Close the Find pane and zoom in on the Moon by clicking and holding the mouse cursor on the **Decrease current elevation** button (the downward-pointing arrow to the left of the Home button in the toolbar) to approach the Moon until detail is visible on the lunar surface. You can now view the Moon from any angle by holding down the **Shift** key while holding down the mouse button (the left button on a two-button mouse) and dragging the mouse. This is equivalent to flying a spaceship around the Moon at a constant distance. (a) Use this technique to rotate the Moon and view it from different perspectives. How does the phase of the Moon change as you rotate it around? (*Hint:* Compare with Box 3-1.) (b) Rotate the Moon until you can also see the Sun and note particularly the Moon's phase when it is in front of the Sun. Explain how your observations show that the phases of the Moon cannot be caused by Earth's shadow falling on the Moon.

Collaborative Exercise

59. Using a bright light source at the center of a darkened room, or a flashlight, use your fist held at arm's length to demonstrate the difference between a full moon and a lunar eclipse. (Use yourself or a classmate as Earth.) How must your fist "orbit" Earth so that lunar eclipses do not happen at every full moon? Create a simple sketch to illustrate your answers.

Archaeoastronomy and Ethnoastronomy

by Mark Hollabaugh

Many years ago I read astronomer John Eddy's *National Geographic* article about the Wyoming Medicine Wheel. A few years later I visited this archaeological site in the Bighorn Mountains and started on the major preoccupation of my career.

Archaeoastronomy combines astronomy and archaeology. You may be familiar with sites such as Stonehenge or the Mayan ruins of the Yucatán. You may not know that there are many archaeological sites in the United States that demonstrate the remarkable understanding a people, usually known as the Anasazi, had about celestial motions (see Figure 2-1). Chaco Canyon, Hovenweep, and Chimney Rock in the Four Corners area of the Southwest preserve ruins from this ancient Pueblo culture.

Moonrise at Chimney Rock in southern Colorado provides a good example of the Anasazi's knowledge of the lunar cycles. If you watched the rising of the Moon for many, many years, you would discover that the Moon's northernmost rising point undergoes an 18.6-year cycle. Dr. McKim Malville of the University of Colorado discovered that the Anasazi who lived there knew of the lunar standstill cycle and watched the northernmost rising of the Moon between the twin rock pillars of Chimney Rock.

My own specialty is the ethnoastronomy of the Lakota, or Teton Sioux, who flourished on the Great Plains of what are now Nebraska, the Dakotas, and Wyoming. Ethnoastronomy combines ethnography with astronomy. As an ethnoastronomer, I am less concerned with physical evidence in the form of ruins and more interested in myths, legends, religious belief, and current practices. In my quest to understand the astronomical thinking and customs of the nineteenth-century Lakota, I have traveled to museums, archives, and libraries in Nebraska, Colorado, Wyoming, South Dakota, and North Dakota. I frequently visit the Pine Ridge and Rosebud Reservations in South Dakota. The Lakota, and other Plains Indians, had a rich tradition of understanding celestial motions and developed an even richer explanation of why things appear the way they do in the sky.

My first professional contribution to ethnoastronomy was in 1996 at the Fifth Oxford International Conference on Astronomy in Culture held in Santa Fe, New Mexico. I had noticed that images of eclipses often appeared in Lakota winter counts—their method of making a historic record of events. The great Leonid meteor shower of 1833 appears in almost every Plains Indian winter count. As I looked at the hides or in ledger books recording these winter counts, I wondered why the Sun, the Moon, and the stars are so common among the Lakota of 150 years ago. My curiosity led me to look deeper: What did the Lakota think about eclipses? Why does their central ritual, the Sun Dance, focus on the Sun? Why do so many legends involve the stars?

The Lakota observed lunar and solar eclipses. Perhaps they felt they had the power to restore the eclipsed Sun or Moon. In August 1869, an Indian agency physician in South Dakota told the Lakota there would be an eclipse. When the Sun disappeared from view, the Lakota began firing their guns in the air. In their minds, they were more powerful than the white doctor because the result of their action was to restore the Sun.

The Lakota used a lunar calendar. Their names for the months came from the world around them. October was the moon of falling leaves. In some years, there actually are 13 new moons, and they often called this extra month the "Lost Moon." The lunar calendar often dictated the timing of their sacred rites. Although the Lakota were never dogmatic about it, they preferred to hold their most important ceremony, the Sun Dance, at the time of the full moon in June, which is when the summer solstice occurs.

Why did the Lakota pay attention to the night sky? Lakota elder Ringing Shield's statement about Polaris, recorded in the late nineteenth century, provides a clue: "One star never moves and it is *wakan*. Other stars move in a circle about it. They are dancing in the dance circle." For the Lakota, a driving force in their culture was a quest to understand the nature of the sacred, or *wakan*. Anything hard to understand or different from the ordinary was *wakan*.

Reaching for the stars, as far away as they are, was a means for the Lakota to bring the incomprehensible universe a bit closer to Earth. Their goal was the same as what Dr. Sandra M. Faber says in her essay "Why Astronomy?" at the end of Chapter 1, "a perspective on human existence and its relation to the cosmos."

Mark Hollabaugh teaches physics and astronomy at Normandale Community College, in Bloomington, Minnesota. A graduate of St. Olaf College, he earned his Ph.D. in science education at the University of Minnesota. He has also taught at St. Olaf College, Augsburg College, and the U.S. Air Force Academy. As a boy he watched the dance of the northern lights and the flash of meteors. Meeting *Apollo 13* commander Jim Lovell and going for a ride in Roger Freedman's airplane are as close as he came to his dream of being an astronaut.

4

Gravitation and the Waltz of the Planets

RIVUXG

An astronaut orbits Earth attached to the International Space Station's manipulator arm. (NASA)

Sixty years ago the idea of humans orbiting Earth or sending spacecraft to other worlds was regarded as science fiction. Today science fiction has become commonplace reality. Literally thousands of artificial satellites orbit our planet to track weather, relay signals for communications and entertainment, and collect scientific data about Earth and the universe. Humans live and work in Earth orbit (as in the accompanying photograph), have ventured as far as the Moon, and have sent dozens of robotic spacecraft to explore all the planets of the solar system.

While we think of spaceflight as an innovation of the twentieth century, we can trace its origins to a series of scientific revolutions that began in the 1500s. The first of these revolutions overthrew the ancient idea that Earth is an immovable object at the center of the universe, around which the Sun, the Moon, and the planets move. In this chapter we will learn how Nicolaus Copernicus, Tycho Brahe, Johannes Kepler, and Galileo Galilei helped us understand that Earth is itself one of several planets orbiting the Sun.

We will learn, too, about Isaac Newton's revolutionary discovery of why the planets move in the way that they do. This was just one aspect of Newton's immense body of work, which included formulating the fundamental laws of physics and developing a precise mathematical description of the force of gravitation—the force that holds the planets in their orbits.

Gravitation proves to be a truly universal force: It guides spacecraft as they journey across the solar system, and keeps satellites and astronauts in their orbits. In this chapter you will learn more about this important force of nature, which plays a central role in all parts of astronomy.

4-1 Ancient astronomers invented geocentric models to explain planetary motions

Since the dawn of civilization, scholars have attempted to explain the nature of the universe. The ancient Greeks were the first to use the principle that still guides scientists today: *The universe can be described and understood logically*. For example, more than 2500 years ago Pythagoras and his followers put forth the idea that nature can be described with mathematics. About 200 years later, Aristotle asserted that the universe is governed by physical laws. One of the most important tasks before the scholars of ancient Greece was to create a model (see Section 1-1) to explain the motions of objects in the heavens.

Learning Goals

By reading the sections of this chapter, you will learn

4-1 How ancient astronomers attempted to explain the motions of the planets

4-2 What led Copernicus to a Sun-centered model of planetary motion

4-3 How Tycho's naked-eye observations of the sky revolutionized ideas about the heavens

4-4 How Kepler deduced the shapes of the orbits of the planets

4-5 How Galileo's pioneering observations with a telescope supported a Sun-centered model

4-6 The ideas behind Newton's laws, which govern the motion of all physical objects, including the planets

4-7 Why planets stay in their orbits and do not fall into the Sun

4-8 What causes ocean tides on Earth

The Greek Geocentric Model

Most Greek scholars assumed that the Sun, the Moon, the stars, and the planets revolve about a stationary Earth. A model of this kind, in which Earth is at the center of the universe, is called a **geocentric model.** Similar ideas were held by the scholars of ancient China.

The scholars of ancient Greece imagined that the planets follow an ornate combination of circular paths

Today we recognize that the stars are not merely points of light on an immense celestial sphere. But in fact this is how the ancient Greeks regarded the stars in their geocentric model of the universe. To explain the diurnal motions of the stars, they assumed that the celestial sphere was *real,* and that it rotated around the stationary Earth once a day.

The Sun and Moon both participated in this daily rotation of the sky, which explained their rising and setting motions. To explain why the Sun and Moon both move slowly with respect to the stars, the ancient Greeks imagined that both of these objects orbit around Earth.

ANALOGY Imagine a merry-go-round that rotates clockwise as seen from above, as in Figure 4-1*a*. As it rotates, two children walk slowly counterclockwise at different speeds around the merry-go-round's platform. Thus, the children rotate along with the merry-go-round and also change their positions with respect to the merry-go-round's wooden horses. This scene is analogous to the way the ancient Greeks pictured the motions of the stars, the Sun, and the Moon. In their model, the celestial sphere rotated to the west around a stationary Earth (Figure 4-1*b*). The stars rotate along with the celestial sphere just as the wooden horses rotate along with the merry-go-round in Figure 4-1*a*. The Sun and Moon are analogous to the two children; they both turn westward with the celestial sphere, making one complete turn each day, and also move slowly eastward at different speeds with respect to the stars.

The geocentric model of the heavens also had to explain the motions of the planets. The ancient Greeks and other cultures of that time knew of five planets: Mercury, Venus, Mars, Jupiter, and Saturn, each of which is a bright object in the night sky. For example, when Venus is at its maximum brilliancy, it is 16 times brighter than the brightest star. (By contrast, Uranus and Neptune are quite dim and were not discovered until after the invention of the telescope.)

Like the Sun and Moon, all of the planets rise in the east and set in the west once a day. And like the Sun and Moon, from night to night the planets slowly move on the celestial sphere, that is, with respect to the background of stars. However, the character of this motion on the celestial sphere is quite different for the planets. Both the Sun and the Moon always move from west to east on the celestial sphere, that is, opposite the direction in which the celestial sphere appears to rotate. The Sun follows the path called the ecliptic (see Section 2-5), while the Moon follows a path that is slightly inclined to the ecliptic (see Section 3-3). Furthermore, the Sun and the Moon each move at relatively constant speeds around the celestial sphere. (The Moon's speed is faster than that of the Sun: It travels all the way around the celestial sphere in about a month while the Sun takes an entire year.) The planets, too, appear to move along paths that are close to the ecliptic. The difference is that each of the planets appears to wander back and forth on the celestial sphere with varying speed. As an example, Figure 4-2 shows the wandering motion of Mars with respect to the background of stars during 2011 and 2012. (This figure shows that the name *planet* is well deserved; it comes from a Greek word meaning "wanderer.")

CAUTION! On a map of Earth with north at the top, west is to the left and east is to the right. Why, then, is *east* on the left and *west* on the right in Figure 4-2? The answer is that a map of Earth is a view looking downward at the ground from above, while a star map like Figure 4-2 is a view looking upward at the sky. If the constellation Leo in Figure 4-2 were directly overhead, Virgo would be toward the eastern horizon and Cancer would be toward the western horizon.

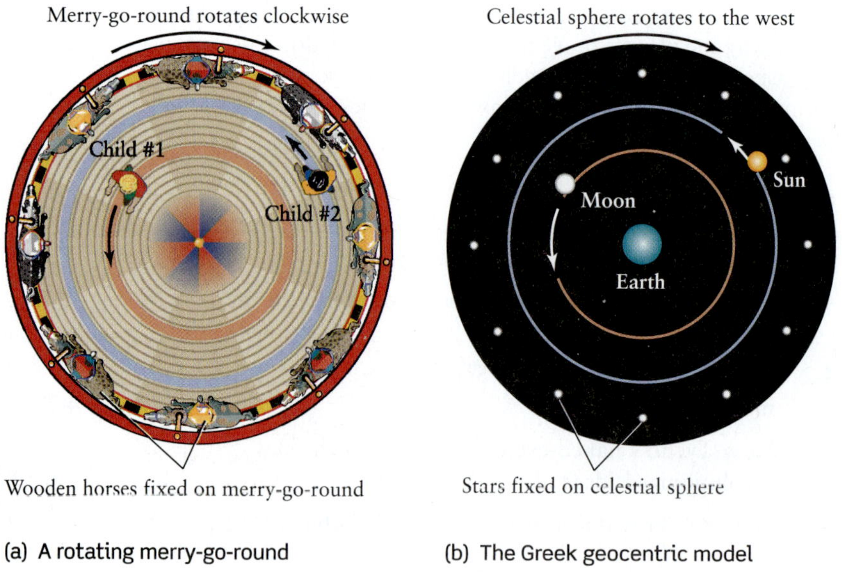

ANIMATION 4.1

Figure 4-1

A Merry-Go-Round Analogy (a) Two children walk at different speeds around a rotating merry-go-round with its wooden horses. **(b)** In an analogous way, the ancient Greeks imagined that the Sun and Moon move around the rotating celestial sphere with its fixed stars. Thus, the Sun and Moon move from east to west across the sky every day and also move slowly eastward from one night to the next relative to the background of stars.

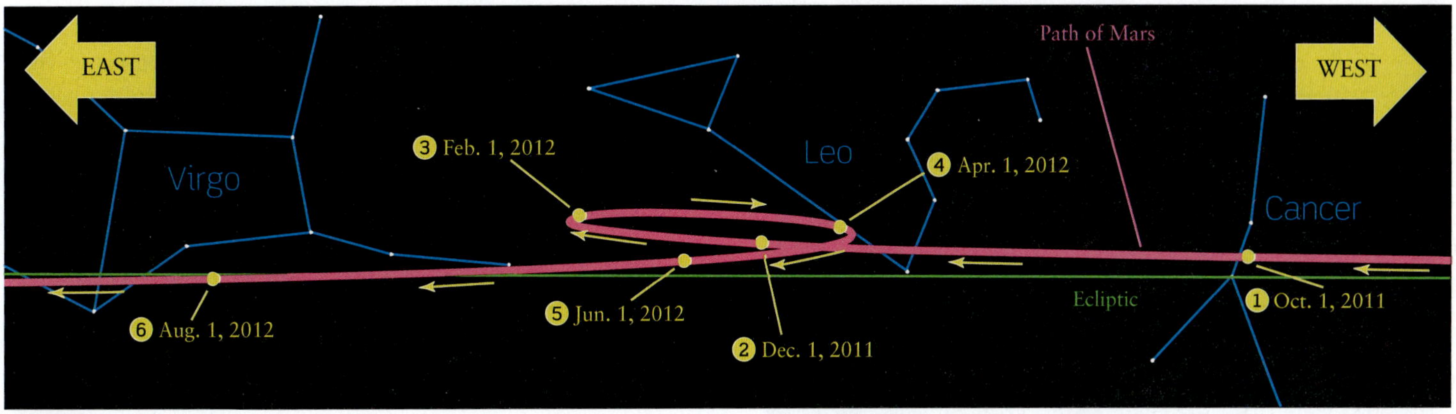

ANIMATION 4.2

Figure 4-2

The Path of Mars in 2011–2012 From October 2011 through August 2012, Mars will move across the zodiacal constellations Cancer, Leo, and Virgo. Mars's motion will be direct (from west to east, or from right to left in this figure) most of the time but will be retrograde (from east to west, or from left to right in this figure) during February and March 2011. Notice that the speed of Mars relative to the stars is not constant: The planet travels farther across the sky from October 1 to December 1 than it does from December 1 to February 1.

Most of the time planets move slowly eastward relative to the stars, just as the Sun and Moon do. This eastward progress is called **direct motion.** For example, Figure 4-2 shows that Mars will be in direct motion from October 2011 through January 2012 and from April through August 2012. Occasionally, however, the planet will seem to stop and then back up for several weeks or months. This occasional westward movement is called **retrograde motion.** Mars will undergo retrograde motion during February and March 2012 (see Figure 4-2), and will do so again about every $22\frac{1}{2}$ months. All the other planets go through retrograde motion, but at different intervals. In the language of the merry-go-round analogy in Figure 4-1, the Greeks imagined the planets as children walking around the rotating merry-go-round but who keep changing their minds about which direction to walk!

CAUTION! Whether a planet is in direct or retrograde motion, over the course of a single night you will see it rise in the east and set in the west. That's because both direct and retrograde motions are much slower than the apparent daily rotation of the sky. Hence, they are best detected by mapping the position of a planet against the background stars from night to night over a long period. Figure 4-2 is a map of just this sort.

The Ptolemaic System

Explaining the nonuniform motions of the five planets was one of the main challenges facing the astronomers of antiquity. The Greeks developed many theories to account for retrograde motion and the loops that the planets trace out against the background stars. One of the most successful and enduring models was originated by Apollonius of Perga and by Hipparchus in the second century B.C. and expanded upon by Ptolemy, the last of the great Greek astronomers, during the second century A.D. Figure 4-3*a* sketches the basic concept, usually called the **Ptolemaic system.** Each planet is assumed to move in a small circle called an **epicycle,** whose center in turn moves in a larger circle, called a **deferent,** which is centered approximately on Earth. Both the epicycle and deferent rotate in the same direction, shown as counterclockwise in Figure 4-3*a*.

As viewed from Earth, the epicycle moves eastward along the deferent. Most of the time the eastward motion of the planet on its epicycle adds to the eastward motion of the epicycle on the deferent (Figure 4-3*b*). Then the planet is seen to be in direct (eastward) motion against the background stars. However, when the planet is on the part of its epicycle nearest Earth, the motion of the planet along the epicycle is opposite to the motion of the epicycle along the deferent. The planet therefore appears to slow down and halt its usual eastward movement among the constellations, and actually goes backward in retrograde (westward) motion for a few weeks or months (Figure 4-3*c*). Thus, the concept of epicycles and deferents enabled Greek astronomers to explain the retrograde loops of the planets.

Using the wealth of astronomical data in the library at Alexandria, including records of planetary positions for hundreds of years, Ptolemy deduced the sizes and rotation rates of the epicycles and deferents needed to reproduce the recorded paths of the planets. After years of tedious work, Ptolemy assembled his calculations into 13 volumes, collectively called the *Almagest.* His work was used to predict the positions and paths of the Sun, Moon, and planets with unprecedented accuracy. In fact, the *Almagest* was so successful that it became the astronomer's bible, and for more than 1000 years, the Ptolemaic system endured as a useful description of the workings of the heavens.

A major problem posed by the Ptolemaic system was a philosophical one: It treated each planet independent of the others. There was no rule in the *Almagest* that related the size and rotation speed of one planet's epicycle and deferent to the corresponding sizes and speeds for other planets. This problem made the Ptolemaic system very unsatisfying to many Islamic and European astronomers of the Middle Ages. They felt that a correct model of the universe should be based on a simple set of underlying principles that applied to all of the planets.

WEB LINK 4.2

The idea that simple, straightforward explanations of phenomena are most likely to be correct is called **Occam's razor,** after William of Occam (or Ockham), the fourteenth-century English philosopher who first expressed it.

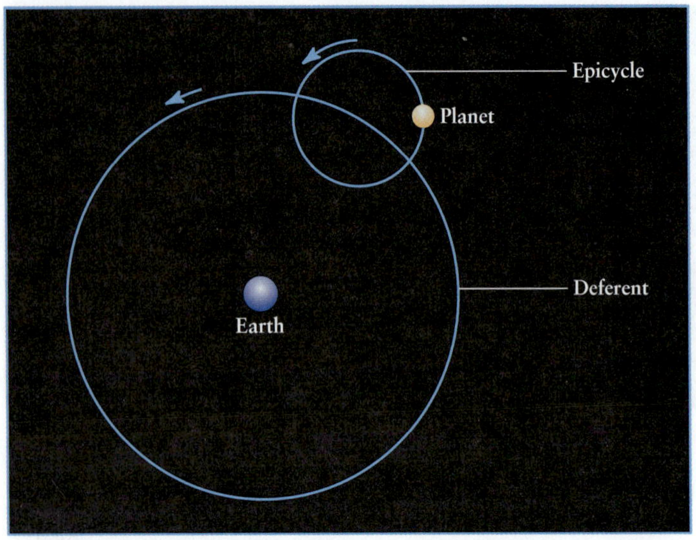

Epicycle
Planet
Earth
Deferent

(a) Planetary motion modeled as a combination of circular motions

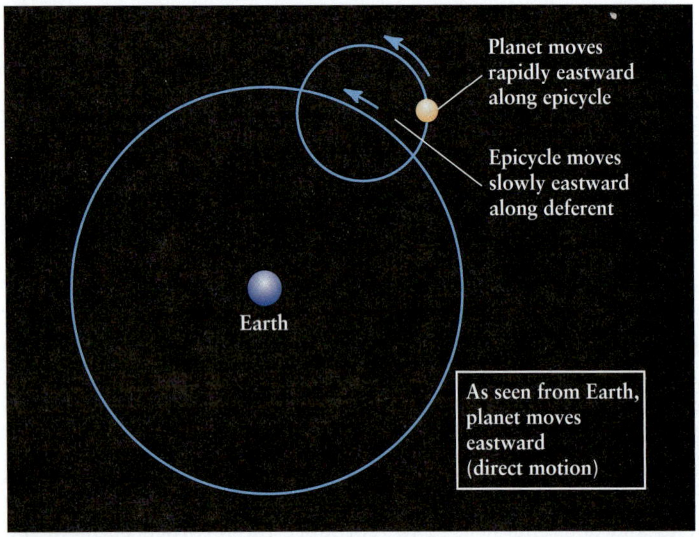

Planet moves rapidly eastward along epicycle
Epicycle moves slowly eastward along deferent
Earth
As seen from Earth, planet moves eastward (direct motion)

(b) Modeling direct motion

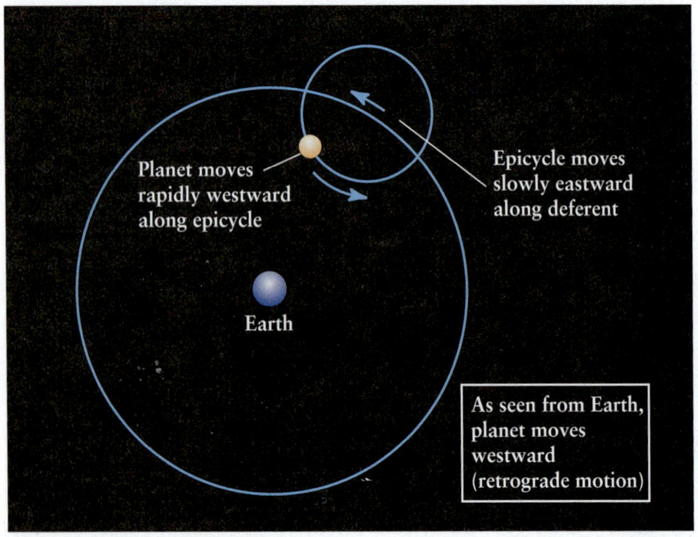

Planet moves rapidly westward along epicycle
Epicycle moves slowly eastward along deferent
Earth
As seen from Earth, planet moves westward (retrograde motion)

(c) Modeling retrograde motion

Figure 4-3

A Geocentric Explanation of Retrograde Motion **(a)** The ancient Greeks imagined that each planet moves along an epicycle, which in turn moves along a deferent centered approximately on Earth. The planet moves along the epicycle more rapidly than the epicycle moves along the deferent. **(b)** At most times the eastward motion of the planet on the epicycle adds to the eastward motion of the epicycle on the deferent. Then the planet moves eastward in direct motion as seen from Earth. **(c)** When the planet is on the inside of the deferent, its motion along the epicycle is westward. Because this motion is faster than the eastward motion of the epicycle on the deferent, the planet appears from Earth to be moving westward in retrograde motion.

(The "razor" refers to shaving extraneous details from an argument or explanation.) Although Occam's razor has no proof or verification, it appeals to the scientist's sense of beauty and elegance, and it has helped lead to the simple and powerful laws of nature that scientists use today. In the centuries that followed, Occam's razor would help motivate a new and revolutionary view of the universe.

4-2 Nicolaus Copernicus devised the first comprehensive heliocentric model

During the first half of the sixteenth century, a Polish lawyer, physician, canon of the church, and gifted mathematician named Nicolaus Copernicus (Figure 4-4) began to construct a new model of the universe. His model, which placed the Sun at the center, explained the motions of the planets in a more natural way than the Ptolemaic system. As we will see, it also helped lay the foundations of modern physical science. But Copernicus was not the first to conceive a Sun-centered model of planetary motion. In the third century B.C., the Greek astronomer Aristarchus suggested such a model as a way to explain retrograde motion.

A Heliocentric Model Explains Retrograde Motion

Imagine riding on a fast racehorse. As you pass a slowly walking pedestrian, he appears to move backward, even though he is traveling in the same direction as you and your horse. This sort of simple observation inspired Aristarchus to formulate a **heliocentric** (Sun-centered) **model** in which all the planets, including Earth, revolve about the Sun. Different planets take different lengths of time to complete an orbit, so from time to time one planet will overtake another, just as a fast-moving horse overtakes a person on foot. When Earth overtakes Mars, for example, Mars appears to move backward in retrograde motion, as Figure 4-5 shows. Thus, in the heliocentric picture, the occasional retrograde motion of a planet is merely the result of Earth's motion.

The retrograde motion of the planets is a result of our viewing the universe from a moving Earth

As we saw in Section 3-6, Aristarchus demonstrated that the Sun is bigger than Earth (see Table 3-3). This made it sensible to imagine Earth orbiting the larger Sun. He also imagined that Earth rotated on its axis once a day, which explained the daily

Figure 4-4

Nicolaus Copernicus (1473–1543) Copernicus was the first person to work out the details of a heliocentric system in which the planets, including Earth, orbit the Sun. (E. Lessing/Magnum)

rising and setting of the Sun, Moon, and planets and the diurnal motions of the stars. To explain why the apparent motions of the planets never take them far from the ecliptic, Aristarchus proposed that the orbits of Earth and all the planets must lie in nearly the same plane. (Recall from Section 2-5 that the ecliptic is the projection onto the celestial sphere of the plane of Earth's orbit.)

This heliocentric model is conceptually much simpler than an Earth-centered system, such as that of Ptolemy, with all its "circles upon circles." In Aristarchus's day, however, the idea of an orbiting, rotating Earth seemed inconceivable, given Earth's apparent stillness and immobility. Nearly 2000 years would pass before a heliocentric model found broad acceptance.

Copernicus and the Arrangement of the Planets

In the years after 1500, Copernicus came to realize that a heliocentric model has several advantages beyond providing a natural explanation of retrograde motion. In the Ptolemaic system, the arrangement of the planets—that is, which are close to Earth and which are far away—was chosen in large part by guesswork. But using a heliocentric model, Copernicus could determine the arrangement of the planets without ambiguity.

Copernicus realized that because Mercury and Venus are always observed fairly near the Sun in the sky, their orbits must be smaller than Earth's. Planets in such orbits are called **inferior planets** (Figure 4-6). The other visible planets—Mars, Jupiter, and Saturn—are sometimes seen on the side of the celestial sphere opposite the Sun, so these planets appear high above the horizon at midnight (when the Sun is far below the horizon). When this

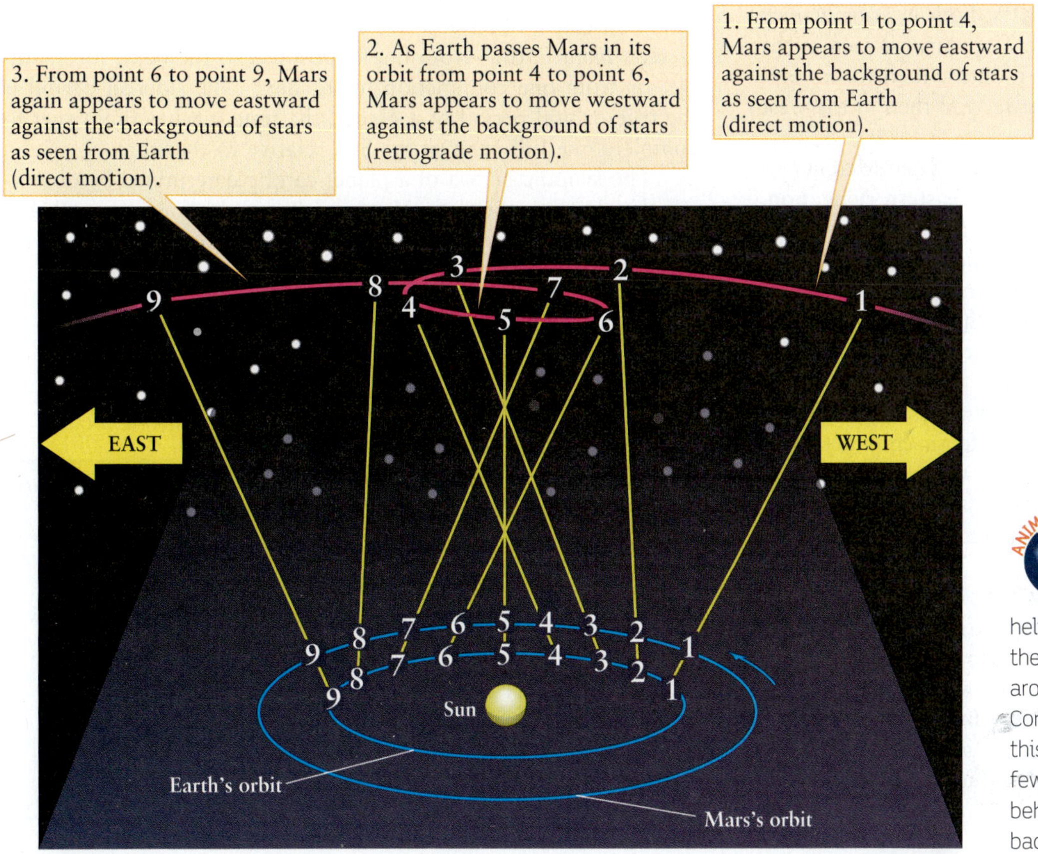

3. From point 6 to point 9, Mars again appears to move eastward against the background of stars as seen from Earth (direct motion).
2. As Earth passes Mars in its orbit from point 4 to point 6, Mars appears to move westward against the background of stars (retrograde motion).
1. From point 1 to point 4, Mars appears to move eastward against the background of stars as seen from Earth (direct motion).
EAST
WEST
Sun
Earth's orbit
Mars's orbit

Figure 4-5

A Heliocentric Explanation of Retrograde Motion In the heliocentric model of Aristarchus, Earth and the other planets orbit the Sun. Earth travels around the Sun more rapidly than Mars. Consequently, as Earth overtakes and passes this slower-moving planet, Mars appears for a few months (from points 4 through 6) to fall behind and move backward with respect to the background of stars.

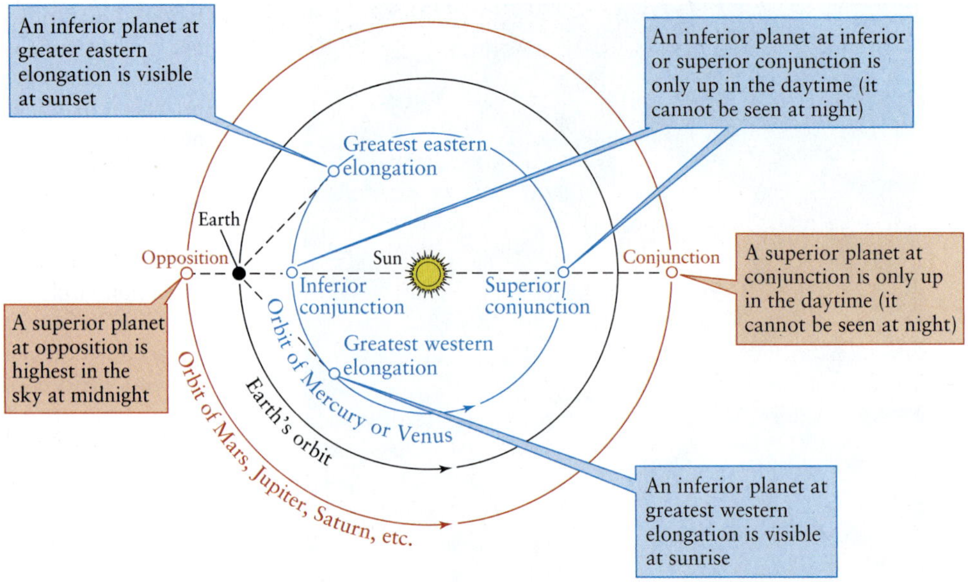
An inferior planet at greater eastern elongation is visible at sunset
An inferior planet at inferior or superior conjunction is only up in the daytime (it cannot be seen at night)
Greatest eastern elongation
Earth
Opposition
Sun
Inferior conjunction
Superior conjunction
Conjunction
A superior planet at conjunction is only up in the daytime (it cannot be seen at night)
A superior planet at opposition is highest in the sky at midnight
Greatest western elongation
Orbit of Mercury or Venus
Earth's orbit
Orbit of Mars, Jupiter, Saturn, etc.
An inferior planet at greatest western elongation is visible at sunrise

INTERACTIVE EXERCISE 4.1

Figure 4-6

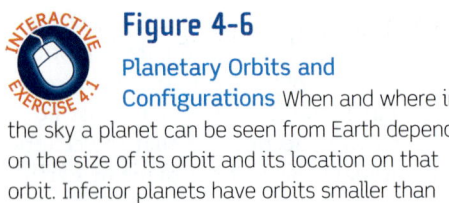

Planetary Orbits and Configurations When and where in the sky a planet can be seen from Earth depends on the size of its orbit and its location on that orbit. Inferior planets have orbits smaller than Earth's, while superior planets have orbits larger than Earth's. (Note that in this figure you are looking down onto the solar system from a point far above Earth's northern hemisphere.)

happens, Earth must lie between the Sun and these planets. Copernicus therefore concluded that the orbits of Mars, Jupiter, and Saturn must be larger than Earth's orbit. Hence, these planets are called **superior planets.**

Uranus and Neptune, as well as Pluto and a number of other small bodies called asteroids that also orbit the Sun, were discovered after the telescope was invented (and after the death of Copernicus). All of these can be seen at times in the midnight sky, so these also have orbits larger than Earth's.

The heliocentric model also explains why planets appear in different parts of the sky on different dates. Both inferior planets (Mercury and Venus) go through cycles: The planet is seen in the west after sunset for several weeks or months, then for several weeks or months in the east before sunrise, and then in the west after sunset again.

Figure 4-6 shows the reason for this cycle. When Mercury or Venus is visible after sunset, it is near **greatest eastern elongation.** (The angle between the Sun and a planet as viewed from Earth is called the planet's **elongation.**) The planet's position in the sky is as far east of the Sun as possible, so it appears above the western horizon after sunset (that is, to the east of the Sun) and is often called an "evening star." At **greatest western elongation,** Mercury or Venus is as far west of the Sun as it can possibly be. It then rises before the Sun, gracing the predawn sky as a "morning star" in the east. When Mercury or Venus is at **inferior conjunction,** it is between us and the Sun, and it moves from the evening sky into the morning sky over weeks or months. At **superior conjunction,** when the planet is on the opposite side of the Sun, it moves back into the evening sky.

A superior planet such as Mars, whose orbit is larger than Earth's, is best seen in the night sky when it is at **opposition.** At this point the planet is in the part of the sky opposite the Sun and is highest in the sky at midnight. This is also when the planet appears brightest, because it is closest to us. But when a superior planet like Mars is located behind the Sun at **conjunction,** it is above the horizon during the daytime and thus is not well placed for nighttime viewing.

Planetary Periods and Orbit Sizes

The Ptolemaic system has no simple rules relating the motion of one planet to another. But Copernicus showed that there *are* such rules in a heliocentric model. In particular, he found a correspondence between the time a planet takes to complete one orbit—that is, its **period**—and the size of the orbit.

Determining the period of a planet takes some care, because Earth, from which we must make the observations, is also moving. Realizing this, Copernicus was careful to distinguish between two different periods of each planet. The **synodic period** is the time that elapses between two successive identical configurations as seen from Earth—from one opposition to the next, for example, or from one conjunction to the next. The **sidereal period** is the true orbital period of a planet, the time it takes the planet to complete one full orbit of the Sun relative to the stars.

The synodic period of a planet can be determined by observing the sky, but the sidereal period has to be found by calculation. Copernicus figured out how to do this (Box 4-1). Table 4-1 shows the results for all of the planets.

Table 4-1 Synodic and Sidereal Periods of the Planets

Planet	Synodic period	Sidereal period
Mercury	116 days	88 days
Venus	584 days	225 days
Earth	—	1.0 year
Mars	780 days	1.9 years
Jupiter	399 days	11.9 years
Saturn	378 days	29.5 years
Uranus	370 days	84.1 years
Neptune	368 days	164.9 years

BOX 4-1 Tools of the Astronomer's Trade

Relating Synodic and Sidereal Periods

We can derive a mathematical formula that relates a planet's sidereal period (the time required for the planet to complete one orbit) to its synodic period (the time between two successive identical configurations). To start with, let's consider an inferior planet (Mercury or Venus) orbiting the Sun as shown in the accompanying figure. Let P be the planet's sidereal period, S the planet's synodic period, and E Earth's sidereal period or sidereal year (see Section 2-8), which Copernicus knew to be nearly $365\frac{1}{4}$ days.

The rate at which Earth moves around its orbit is the number of degrees around the orbit divided by the time to complete the orbit, or $360°/E$ (equal to a little less than 1° per day). Similarly, the rate at which the inferior planet moves along its orbit is $360°/P$.

During a given time interval, the angular distance that Earth moves around its orbit is its rate, $(360°/E)$, multiplied by the length of the time interval. Thus, during a time S, or one synodic period of the inferior planet, Earth covers an angular distance of $(360°/E)S$ around its orbit. In that same time, the inferior planet covers an angular distance of $(360°/P)S$. Note, however, that the inferior planet has gained one full lap on Earth, and hence has covered 360° more than Earth has (see the figure). Thus, $(360°/P)S = (360°/E)S + 360°$. Dividing each term of this equation by $360°S$ gives

For an inferior planet:

$$\frac{1}{P} = \frac{1}{E} + \frac{1}{S}$$

P = inferior planet's sidereal period

E = Earth's sidereal period = 1 year

S = inferior planet's synodic period

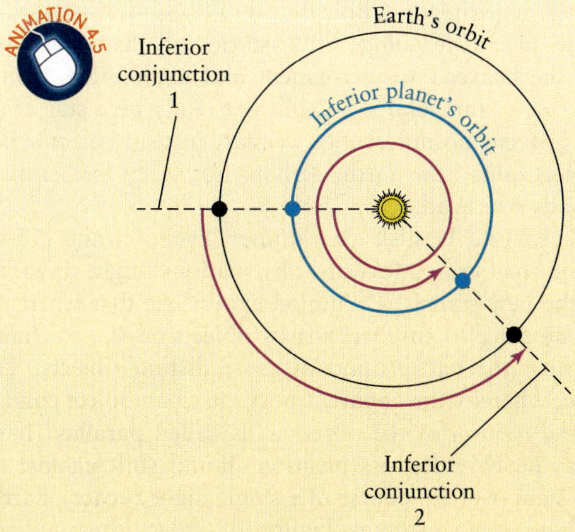

A similar analysis for a superior planet (for example, Mars, Jupiter, or Saturn) yields

For a superior planet:

$$\frac{1}{P} = \frac{1}{E} - \frac{1}{S}$$

P = superior planet's sidereal period

E = Earth's sidereal period = 1 year

S = superior planet's synodic period

Using these formulas, we can calculate a planet's sidereal period P from its synodic period S. Often astronomers express P, E, and S in terms of years, by which they mean Earth years of approximately 365.26 days.

EXAMPLE: Jupiter has an observed synodic period of 398.9 days, or 1.092 years. What is its sidereal period?

Situation: Our goal is to find the sidereal period of Jupiter, a superior planet.

Tools: Since Jupiter is a superior planet, we use the second of the two equations given above to determine the sidereal period P.

Answer: We are given Earth's sidereal period $E = 1$ year and Jupiter's synodic period $S = 1.092$ years. Using the equation $1/P = 1/E - 1/S$,

$$\frac{1}{P} = \frac{1}{1} - \frac{1}{1.092} = 0.08425,$$

so

$$P = \frac{1}{0.08425} = 11.87 \text{ years}$$

Review: Our answer means that it takes 11.87 years for Jupiter to complete one full orbit of the Sun. This sidereal period is greater than Earth's 1-year sidereal period because Jupiter's orbit is larger than Earth's orbit. Jupiter's synodic period of 1.092 years is the time from one opposition to the next, or the time that elapses from when Earth overtakes Jupiter to when it next overtakes Jupiter. This synodic period is so much shorter than the sidereal period because Jupiter moves quite slowly around its orbit. Earth overtakes it a little less often than once per Earth orbit, that is, at intervals of a little bit more than a year.

Table 4-2 Average Distances of the Planets from the Sun

Planet	Copernican value (AU*)	Modern value (AU)
Mercury	0.38	0.39
Venus	0.72	0.72
Earth	1.00	1.00
Mars	1.52	1.52
Jupiter	5.22	5.20
Saturn	9.07	9.55
Uranus	—	19.19
Neptune	—	30.07

**1 AU = 1 astronomical unit = average distance from Earth to the Sun.*

To find a relationship between the sidereal period of a planet and the size of its orbit, Copernicus still had to determine the relative distances of the planets from the Sun. He devised a straightforward geometric method of determining the relative distances of the planets from the Sun using trigonometry. His answers turned out to be remarkably close to the modern values, as shown in Table 4-2.

The distances in Table 4-2 are given in terms of the astronomical unit, which is the average distance from Earth to the Sun (Section 1-7). Copernicus did not know the precise value of this distance, so he could only determine the *relative* sizes of the orbits of the planets. One method used by modern astronomers to determine the astronomical unit is to measure the Earth-Venus distance very accurately using radar. At the same time, they measure the angle in the sky between Venus and the Sun, and then calculate the Earth-Sun distance using trigonometry. In this way, the astronomical unit is found to be 1.496×10^8 km (92.96 million miles). Once the astronomical unit is known, the average distances from the Sun to each of the planets in kilometers can be determined from Table 4-2. A table of these distances is given in Appendix 2.

By comparing Tables 4-1 and 4-2, you can see the unifying relationship between planetary orbits in the Copernican model: The farther a planet is from the Sun, the longer it takes to travel around its orbit (that is, the longer its sidereal period). That is so for *two* reasons: (1) the larger the orbit, the farther a planet must travel to complete an orbit; and (2) the larger the orbit, the slower a planet moves. For example, Mercury, with its small orbit, moves at an average speed of 47.9 km/s (107,000 mi/h). Saturn travels around its large orbit much more slowly, at an average speed of 9.64 km/s (21,600 mi/h). The older Ptolemaic model offers no such simple relations between the motions of different planets.

The Shapes of Orbits in the Copernican Model

At first, Copernicus assumed that Earth travels around the Sun along a circular path. He found that perfectly circular orbits could not accurately describe the paths of the other planets, so he had to add an epicycle to each planet. These epicycles were *not* to explain retrograde motion, which Copernicus realized was because of the differences in orbital speeds of different planets, as shown in Figure 4-5. Rather, the small epicycles helped Copernicus account for slight variations in each planet's speed along its orbit.

Even though he clung to the old notion that orbits must be made up of circles, Copernicus had shown that a heliocentric model could explain the motions of the planets. He compiled his ideas and calculations into a book entitled *De revolutionibus orbium coelestium* (On the Revolutions of the Celestial Spheres), which was published in 1543, the year of his death.

For several decades after Copernicus, most astronomers saw little reason to change their allegiance from the older geocentric model of Ptolemy. The predictions that the Copernican model makes for the apparent positions of the planets are, on average, no better or worse than those of the Ptolemaic model. The test of Occam's razor does not really favor either model, because both use a combination of circles to describe each planet's motion.

More concrete evidence was needed to convince scholars to abandon the old, comfortable idea of a stationary Earth at the center of the universe. The story of how this evidence was accumulated begins nearly 30 years after the death of Copernicus, when a young Danish astronomer pondered the nature of a new star in the heavens.

4-3 Tycho Brahe's astronomical observations disproved ancient ideas about the heavens

A supernova explosion and a comet revealed to Tycho that our universe is more dynamic than had been imagined

On November 11, 1572, a bright star suddenly appeared in the constellation of Cassiopeia. At first, it was even brighter than Venus, but then it began to grow dim. After 18 months, it faded from view. Modern astronomers recognize this event as a supernova explosion, the violent death of a massive star.

In the sixteenth century, however, the vast majority of scholars held with the ancient teachings of Aristotle and Plato, who had argued that the heavens are permanent and unalterable. Consequently, the "new star" of 1572 could not really be a star at all, because the heavens do not change; it must instead be some sort of bright object quite near Earth, perhaps not much farther away than the clouds overhead.

The 25-year-old Danish astronomer Tycho Brahe (1546–1601) realized that straightforward observations might reveal the distance to the new star. It is common experience that when you walk from one place to another, nearby objects appear to change position against the background of more distant objects. This phenomenon, whereby the apparent position of an object changes because of the motion of the observer, is called **parallax.** If the new star was nearby, then its position should shift against the background stars over the course of a single night because Earth's rotation changes our viewpoint. Figure 4-7 shows this predicted shift. (Actually, Tycho believed that the heavens rotate about the Earth, as in the Ptolemaic model, but the net effect is the same.)

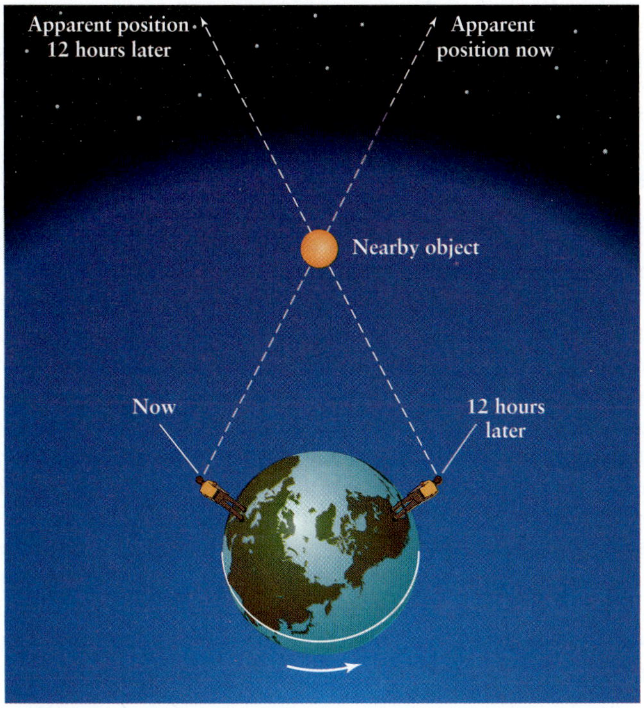

Figure 4-7

A Nearby Object Shows a Parallax Shift Tycho Brahe argued that if an object is near Earth, an observer would have to look in different directions to see that object over the course of a night and its position relative to the background stars would change. Tycho failed to measure such changes for a supernova in 1572 and a comet in 1577. He therefore concluded that these objects were far from Earth.

Tycho's careful observations failed to disclose any parallax. The farther away an object is, the less it appears to shift against the background as we change our viewpoint, and the smaller the parallax. Hence, the new star had to be quite far away, farther from Earth than anyone had imagined. This was the first evidence that the "unchanging" stars were in fact changeable.

Tycho also attempted to measure the parallax of a bright comet that appeared in 1577 and, again, found it too small to measure. Thus, the comet also had to be far beyond Earth. Furthermore, because the comet's position relative to the stars changed from night to night, its motion was more like a planet than a star. But most scholars of Tycho's time taught that the motions of the planets had existed in unchanging form since the beginning of the universe. If new objects such as comets could appear and disappear within the realm of the planets, the conventional notions of planetary motions needed to be revised.

Tycho's observations showed that the heavens are by no means pristine and unchanging. This discovery flew in the face of nearly 2000 years of astronomical thought. In support of these revolutionary observations, the king of Denmark financed the construction of two magnificent observatories for Tycho's use on the island of Hven, just off the Danish coast. The bequest for these two observatories—Uraniborg ("heavenly castle") and Stjerneborg ("star castle")—allowed Tycho to design and have built a set of astronomical instruments vastly superior in quality to any earlier instruments (Figure 4-8).

Tycho and the Positions of the Planets

With this state-of-the-art equipment, Tycho proceeded to measure the positions of stars and planets with unprecedented accuracy. In addition, he and his assistants were careful to make several observations of the same star or planet with different instruments, in order to identify any errors that might be caused by the instruments themselves. This painstaking approach to mapping the heavens revolutionized the practice of astronomy and is used by astronomers today.

A key goal of Tycho's observations during this period was to test the ideas Copernicus had proposed decades earlier about Earth going around the Sun. Tycho argued that if Earth was in motion, then nearby stars should appear to shift their positions with respect to background stars as we orbit the Sun. Tycho failed to detect any such parallax, and he concluded that Earth was at rest and the Copernican system was wrong.

On this point Tycho was in error, for nearby stars do in fact shift their positions as he had suggested. But even the nearest stars are so far away that the shifts in their positions are less than an arcsecond, too small to be seen with the naked eye. Tycho would have needed a telescope to detect the parallax that he was looking for, but the telescope was not invented until after his death in

WEB LINK 4.4

Figure 4-8

Tycho Brahe (1546–1601) Observing This contemporary illustration shows Tycho Brahe with some of the state-of-the-art measuring apparatus at Uraniborg, one of the two observatories that he built under the patronage of Frederik II of Denmark. (This magnificent observatory lacked a telescope, which had not yet been invented.) The data that Tycho collected were crucial to the development of astronomy in the years after his death. (Photo Researchers, Inc.)

WEB LINK 4.5

Figure 4-9

Johannes Kepler (1571–1630) By analyzing Tycho Brahe's detailed records of planetary positions, Kepler developed three general principles, called Kepler's laws, that describe how the planets move about the Sun. Kepler was the first to realize that the orbits of the planets are ellipses and not circles. (E. Lessing/Magnum)

1601. Indeed, the first accurate determination of stellar parallax was not made until 1838.

Although he remained convinced that Earth was at the center of the universe, Tycho nonetheless made a tremendous contribution toward putting the heliocentric model on a solid foundation. From 1576 to 1597, he used his instruments to make comprehensive measurements of the positions of the planets with an accuracy of 1 arcminute. These measurements are as well as can be done with the naked eye and were far superior to any earlier measurements. Within the reams of data that Tycho compiled lay the truth about the motions of the planets. The person who would extract this truth was a German mathematician who became Tycho's assistant in 1600, a year before the great astronomer's death. His name was Johannes Kepler (Figure 4-9).

4-4 Johannes Kepler proposed elliptical paths for the planets about the Sun

TUTORIAL 4.1

The task that Johannes Kepler took on at the beginning of the seventeenth century was to find a model of planetary motion that agreed completely with Tycho's extensive and very accurate observations of planetary positions. To do this, Kepler found that he had to break with an ancient prejudice about planetary motions.

Kepler's ideas apply not just to the planets, but to all orbiting celestial objects

Elliptical Orbits and Kepler's First Law

Astronomers had long assumed that heavenly objects move in circles, which were considered the most perfect and harmonious of all geometric shapes. They believed that if a perfect God resided in heaven along with the stars and planets, then the motions of these objects must be perfect too. Against this context, Kepler dared to try to explain planetary motions with noncircular curves. In particular, he found that he had the best success with a particular kind of curve called an **ellipse.**

You can draw an ellipse by using a loop of string, two thumbtacks, and a pencil, as shown in Figure 4-10*a*. Each thumbtack in the figure is at a **focus** (plural **foci**) of the ellipse; an ellipse has two foci. The longest diameter of an ellipse, called the **major axis,** passes through both foci. Half of that distance is called the **semimajor axis** and is usually designated by the letter *a*. A circle is a special case of an ellipse in which the two foci are at the same point (this corresponds to using only a single thumbtack in Figure 4-10*a*). The semimajor axis of a circle is equal to its radius.

By assuming that planetary orbits were ellipses, Kepler found, to his delight, that he could make his theoretical calculations match precisely to Tycho's observations. This important discovery, first published in 1609, is now called **Kepler's first law:**

The orbit of a planet about the Sun is an ellipse with the Sun at one focus.

The semimajor axis *a* of a planet's orbit is the average distance between the planet and the Sun.

CAUTION! The Sun is at one focus of a planet's elliptical orbit, but there is *nothing* at the other focus. This "empty focus" has geometrical significance, because it helps to define the shape of the ellipse, but plays no other role.

Ellipses come in different shapes, depending on the elongation of the ellipse. The shape of an ellipse is described by its **eccentricity,** designated by the letter *e*. The value of *e* can range from 0 (a circle) to just under 1 (nearly a straight line). The greater the eccentricity, the more elongated the ellipse. Figure 4-10*b* shows a few examples of ellipses with different eccentricities. Because a circle is a special case of an ellipse, it is possible to have a perfectly circular orbit. But all of the objects that orbit the Sun have orbits that are at least slightly elliptical. The most circular of any planetary orbit is that of Venus, with an eccentricity of just 0.007; Mercury's orbit has an eccentricity of 0.206, and a number of small bodies called comets move in very elongated orbits with eccentricities just less than 1.

Orbital Speeds and Kepler's Second Law

Once he knew the shape of a planet's orbit, Kepler was ready to describe exactly *how* it moves on that orbit. As a planet travels in an elliptical orbit, its distance from the Sun varies. Kepler realized that the speed of a planet also varies along its orbit. A planet moves most rapidly when it is nearest the Sun, at a point on its orbit called **perihelion.** Conversely, a planet moves most slowly when it is farthest from the Sun, at a point called **aphelion** (Figure 4-11).

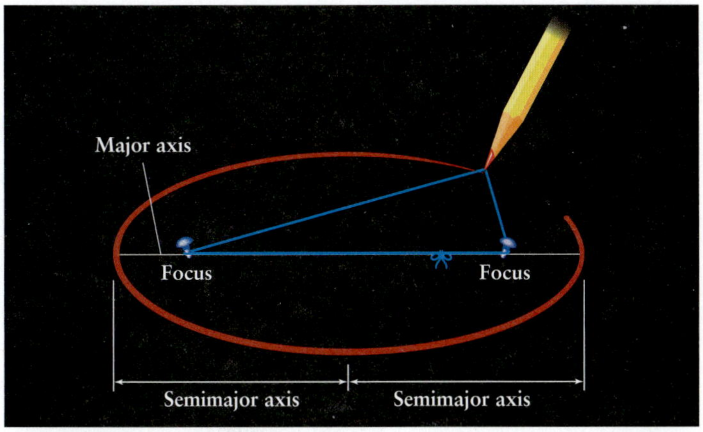

(a) The geometry of an ellipse

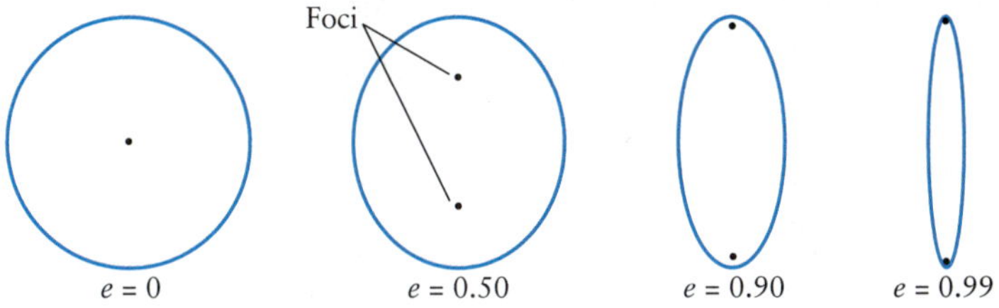

(b) Ellipses with different eccentricities

Figure 4-10

Ellipses (a) To draw an ellipse, use two thumbtacks to secure the ends of a piece of string, then use a pencil to pull the string taut. If you move the pencil while keeping the string taut, the pencil traces out an ellipse. The thumbtacks are located at the two foci of the ellipse. The major axis is the greatest distance across the ellipse; the semimajor axis is half of this distance. (b) A series of ellipses with the same major axis but different eccentricities. An ellipse can have any eccentricity from $e = 0$ (a circle) to just under $e = 1$ (virtually a straight line).

456-780-222-80

After much trial and error, Kepler found a way to describe just how a planet's speed varies as it moves along its orbit. Figure 4-11 illustrates this discovery, also published in 1609. Suppose that it takes 30 days for a planet to go from point *A* to point *B*. During that time, an imaginary line joining the Sun and the planet sweeps out a nearly triangular area. Kepler discovered that a line joining the Sun and the planet also sweeps out exactly the same area during any other 30-day interval. In other words, if the planet also takes 30 days to go from point *C* to point *D*, then the two shaded segments in Figure 4-11 are equal in area. **Kepler's second law** can be stated thus:

A line joining a planet and the Sun sweeps out equal areas in equal intervals of time.

This relationship is also called the **law of equal areas.** In the idealized case of a circular orbit, a planet would have to move at a constant speed around the orbit in order to satisfy Kepler's second law. In the actual case of an elliptical orbit, a planet speeds up when it's closer to the Sun, and slows down when it's further away.

ANALOGY An analogy for Kepler's second law is a twirling ice skater holding weights in each hand. If the skater moves the weights closer to her body by pulling her arms straight in, her rate of spin increases and the weights move faster; if she extends her arms so the weights move away from her body, her rate of spin decreases and the weights slow down. Just like the weights, a planet in an elliptical orbit travels at a higher speed when it moves closer to the Sun (toward perihelion) and travels at a lower speed when it moves away from the Sun (toward aphelion).

Orbital Periods and Kepler's Third Law

Kepler's second law describes how the speed of a given planet changes as it orbits the Sun. Kepler also deduced from Tycho's data a relationship that can be used to compare the motions of *different* planets. Published in 1618 and now called **Kepler's third law**, it states a relationship between the size of a planet's orbit and the time the planet takes to go once around the Sun:

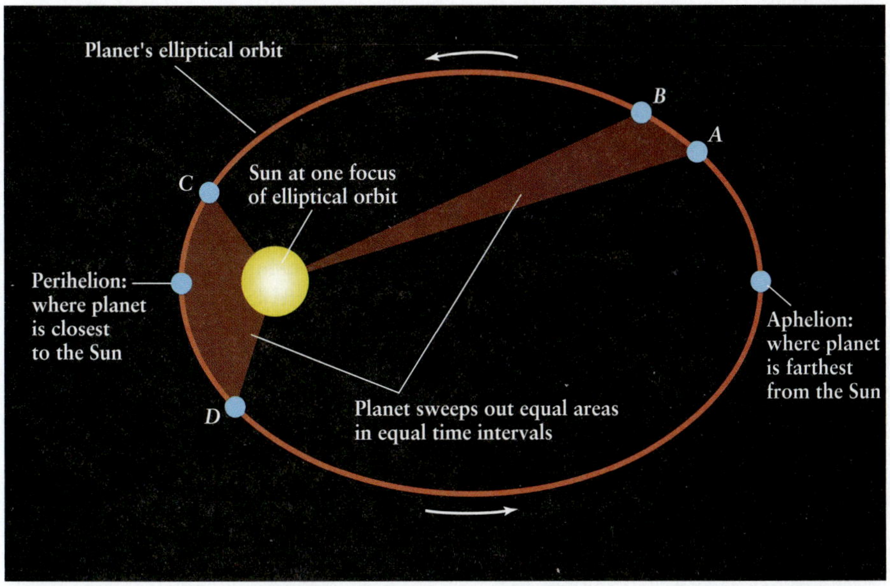

ANIMATION 4.6

Figure 4-11

Kepler's First and Second Laws According to Kepler's first law, a planet travels around the Sun along an elliptical orbit with the Sun at one focus. According to his second law, a planet moves fastest when closest to the Sun (at perihelion) and slowest when farthest from the Sun (at aphelion). As the planet moves, an imaginary line joining the planet and the Sun sweeps out equal areas in equal intervals of time (from *A* to *B* or from *C* to *D*). By using these laws in his calculations, Kepler found an excellent fit to the apparent motions of the planets.

BOX 4-2 Tools of the Astronomer's Trade

Using Kepler's Third Law

Kepler's third law relates the sidereal period P of an object orbiting the Sun to the semimajor axis a of its orbit:

$$P^2 = a^3$$

You must keep two essential points in mind when working with this equation:

1. The period P *must* be measured in years, and the semimajor axis a *must* be measured in astronomical units (AU). Otherwise you will get nonsensical results.
2. This equation applies *only* to the special case of an object, like a planet, that orbits the Sun. If you want to analyze the orbit of the Moon around Earth, of a spacecraft around Mars, or of a planet around a distant star, you must use a different, generalized form of Kepler's third law. We discuss this alternative equation in Section 4-7 and Box 4-4.

EXAMPLE: The average distance from Venus to the Sun is 0.72 AU. Use this to determine the sidereal period of Venus.

Situation: The average distance from the Venus to the Sun is the semimajor axis a of the planet's orbit. Our goal is to calculate the planet's sidereal period P.

Tools: To relate a and P we use Kepler's third law, $P^2 = a^3$.

Answer: We first cube the semimajor axis (multiply it by itself twice):

$$a^3 = (0.72)^3 = 0.72 \times 0.72 \times 0.72 = 0.373$$

According to Kepler's third law this is also equal to P^2, the square of the sidereal period. So, to find P, we have to "undo" the square, that is, take the square root. Using a calculator, we find

$$P = \sqrt{P^2} = \sqrt{0.373} = 0.61$$

Review: The sidereal period of Venus is 0.61 years, or a bit more than seven Earth months. This result makes sense: A planet with a smaller orbit than Earth's (an inferior planet) must have a shorter sidereal period than Earth.

EXAMPLE: A certain small asteroid (a rocky body a few tens of kilometers across) takes eight years to complete one orbit around the Sun. Find the semimajor axis of the asteroid's orbit.

Situation: We are given the sidereal period $P = 8$ years, and are to determine the semimajor axis a.

Tools: As in the preceding example, we relate a and P using Kepler's third law, $P^2 = a^3$.

Answer: We first square the period:

$$P^2 = 8^2 = 8 \times 8 = 64$$

From Kepler's third law, 64 is also equal to a^3. To determine a, we must take the *cube root* of a^3, that is, find the number whose cube is 64. If your calculator has a cube root function, denoted by the symbol $\sqrt[3]{\ }$, you can use it to find that the cube root of 64 is 4: $\sqrt[3]{64} = 4$. Otherwise, you can determine by trial and error that the cube of 4 is 64:

$$4^3 = 4 \times 4 \times 4 = 64$$

Because the cube of 4 is 64, it follows that the cube root of 64 is 4 (taking the cube root "undoes" the cube).

With either technique you find that the orbit of this asteroid has semimajor axis $a = 4$ AU.

Review: The period is greater than 1 year, so the semimajor axis is greater than 1 AU. Note that $a = 4$ AU is intermediate between the orbits of Mars and Jupiter (see Table 4-3). Many asteroids are known with semimajor axes in this range, forming a region in the solar system called the asteroid belt.

The square of the sidereal period of a planet is directly proportional to the cube of the semimajor axis of the orbit.

Kepler's third law says that the larger a planet's orbit—that is, the larger the semimajor axis, or average distance from the planet to the Sun—the longer the sidereal period, which is the time it takes the planet to complete an orbit. From Kepler's third law one can show that the larger the semimajor axis, the slower the average speed at which the planet moves around its orbit. (By contrast, Kepler's *second* law describes how the speed of a given planet is sometimes faster and sometimes slower than its average speed.) This qualitative relationship between orbital size and orbital speed is just what Aristarchus and Copernicus used to explain retrograde motion, as we saw in Section 4-2. Kepler's great contribution was to make this relationship a quantitative one.

It is useful to restate Kepler's third law as an equation. If a planet's sidereal period P is measured in years and the length of its semimajor axis a is measured in astronomical units (AU), where 1 AU is the average distance from Earth to the Sun (see Section 1-7), then Kepler's third law is

Kepler's third law

$$P^2 = a^3$$

P = planet's sidereal period, in years

a = planet's semimajor axis, in AU

Table 4-3 A Demonstration of Kepler's Third Law ($P^2 = a^3$)

Planet	Sidereal period P (years)	Semimajor axis a (AU)	P^2	a^3
Mercury	0.24	0.39	0.06	0.06
Venus	0.61	0.72	0.37	0.37
Earth	1.00	1.00	1.00	1.00
Mars	1.88	1.52	3.53	3.51
Jupiter	11.86	5.20	140.7	140.6
Saturn	29.46	9.55	867.9	871.0
Uranus	84.10	19.19	7,072	7,067
Neptune	164.86	30.07	27,180	27,190

Kepler's third law states that $P^2 = a^3$ for each of the planets. The last two columns of this table demonstrate that this relationship holds true to a very high level of accuracy.

If you know either the sidereal period of a planet or the semimajor axis of its orbit, you can find the other quantity using this equation. Box 4-2 gives some examples of how this is done.

We can verify Kepler's third law for all of the planets, including those that were discovered after Kepler's death, using data from Tables 4-1 and 4-2. If Kepler's third law is correct, for each planet the numerical values of P^2 and a^3 should be equal. This result is indeed true to very high accuracy, as Table 4-3 shows.

The Significance of Kepler's Laws

Kepler's laws are a landmark in the history of astronomy. They made it possible to calculate the motions of the planets with better accuracy than any geocentric model ever had, and they helped to justify the idea of a heliocentric model. Kepler's laws also pass the test of Occam's razor, for they are simpler in every way than the schemes of Ptolemy or Copernicus, both of which used a complicated combination of circles.

But the significance of Kepler's laws goes beyond understanding planetary orbits. These same laws are also obeyed by spacecraft orbiting Earth, by two stars revolving about each other in a binary star system, and even by galaxies in their orbits about each other. Throughout this book, we shall use Kepler's laws in a wide range of situations.

As impressive as Kepler's accomplishments were, he did not prove that the planets orbit the Sun, nor was he able to explain why planets move in accordance with his three laws. These advances were made by two other figures who loom large in the history of astronomy: Galileo Galilei and Isaac Newton.

4-5 Galileo's discoveries with a telescope strongly supported a heliocentric model

When Dutch opticians invented the telescope during the first decade of the seventeenth century, astronomy was changed forever. The scholar who used this new tool to amass convincing evidence that the planets orbit the Sun, not Earth, was the Italian mathematician and physical scientist Galileo Galilei (Figure 4-12).

Galileo used the cutting-edge technology of the 1600s to radically transform our picture of the universe

While Galileo did not invent the telescope, he was the first to point one of these new devices toward the sky and to publish his observations. Beginning in 1610, he saw sights of which no one had ever dreamed. He discovered mountains on the Moon, sunspots on the Sun, and the rings of Saturn, and he was the first to see that the Milky Way is not a featureless band of light but rather "a mass of innumerable stars."

The Phases of Venus

One of Galileo's most important discoveries with the telescope was that Venus exhibits phases like those of the Moon (Figure 4-13). Galileo also noticed that the apparent size of Venus as seen through his telescope was related to the planet's phase. Venus appears small at gibbous phase and largest at crescent phase. There is also a correlation between the phases of Venus and the planet's angular distance from the Sun.

Figure 4-14 shows that these relationships are entirely compatible with a heliocentric model in which Earth and Venus both go around the Sun. They are also completely *incompatible* with the Ptolemaic system, in which the Sun and Venus both orbit Earth.

WEB LINK 4.6

Figure 4-12

Galileo Galilei (1564–1642) Galileo was one of the first people to use a telescope to observe the heavens. He discovered craters on the Moon, sunspots on the Sun, the phases of Venus, and four moons orbiting Jupiter. His observations strongly suggested that Earth orbits the Sun, not vice versa. (Eric Lessing/Art Resource)

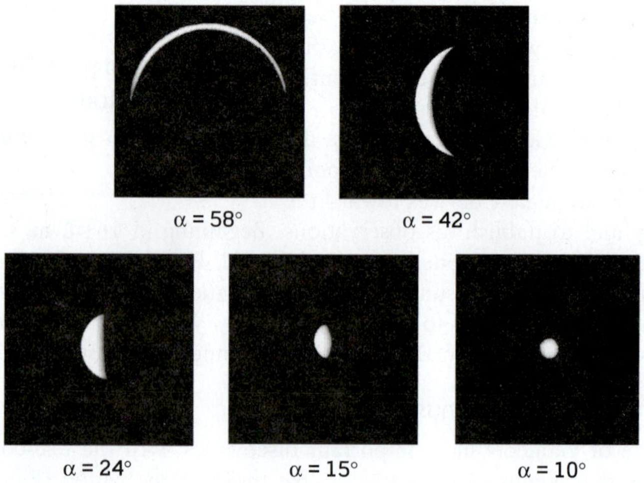
α = 58°
α = 42°
α = 24°
α = 15°
α = 10°

Figure 4-13 R I V U X G

The Phases of Venus This series of photographs shows how the appearance of Venus changes as it moves along its orbit. The number below each view is the angular diameter α of the planet in arcseconds. Venus has the largest angular diameter when it is a crescent, and the smallest angular diameter when it is gibbous (nearly full). (New Mexico State University Observatory)

To explain why Venus is never seen very far from the Sun, the Ptolemaic model had to assume that the deferents of Venus and of the Sun move together in lockstep, with the epicycle of Venus centered on a straight line between Earth and the Sun (Figure 4-15). In this model, Venus was never on the opposite side of the Sun from Earth, and so it could never have shown the gibbous phases that Galileo observed.

The Moons of Jupiter

Galileo also found more unexpected evidence for the ideas of Copernicus. In 1610 Galileo discovered four moons, now called the Galilean satellites, orbiting Jupiter (Figure 4-16). He realized that they were orbiting Jupiter because they appeared to move back and forth from one side of the planet to the other. Figure 4-17 shows confirming observations made by Jesuit observers in 1620. Astronomers soon realized that the larger the orbit of one of the moons around Jupiter, the slower that moon moves and the longer it takes that moon to travel around its orbit. These are the same relationships that Copernicus deduced for the motions of the planets around the Sun. Thus, the moons of Jupiter behave like a Copernican system in miniature.

Galileo's telescopic observations constituted the first fundamentally new astronomical data in almost 2000 years. Contradicting prevailing opinion and religious belief, his discoveries strongly suggested a heliocentric structure of the universe. The Roman Catholic Church, which was a powerful political force in Italy and whose

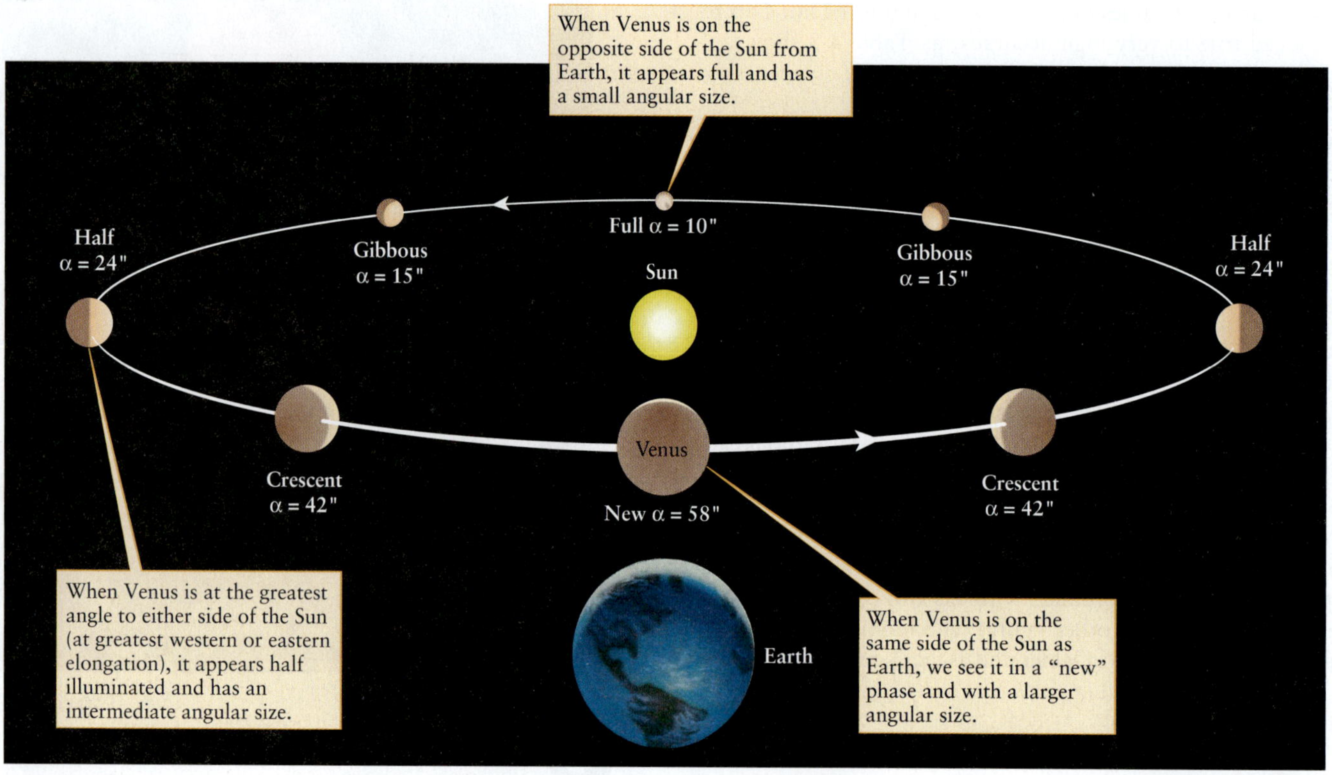
When Venus is on the opposite side of the Sun from Earth, it appears full and has a small angular size.
Full α = 10"
Half α = 24"
Gibbous α = 15"
Sun
Gibbous α = 15"
Half α = 24"
Venus
Crescent α = 42"
New α = 58"
Crescent α = 42"
When Venus is at the greatest angle to either side of the Sun (at greatest western or eastern elongation), it appears half illuminated and has an intermediate angular size.
Earth
When Venus is on the same side of the Sun as Earth, we see it in a "new" phase and with a larger angular size.

Figure 4-14

The Changing Appearance of Venus Explained in a Heliocentric Model A heliocentric model, in which Earth and Venus both orbit the Sun, provides a natural explanation for the changing appearance of Venus shown in Figure 4-13.

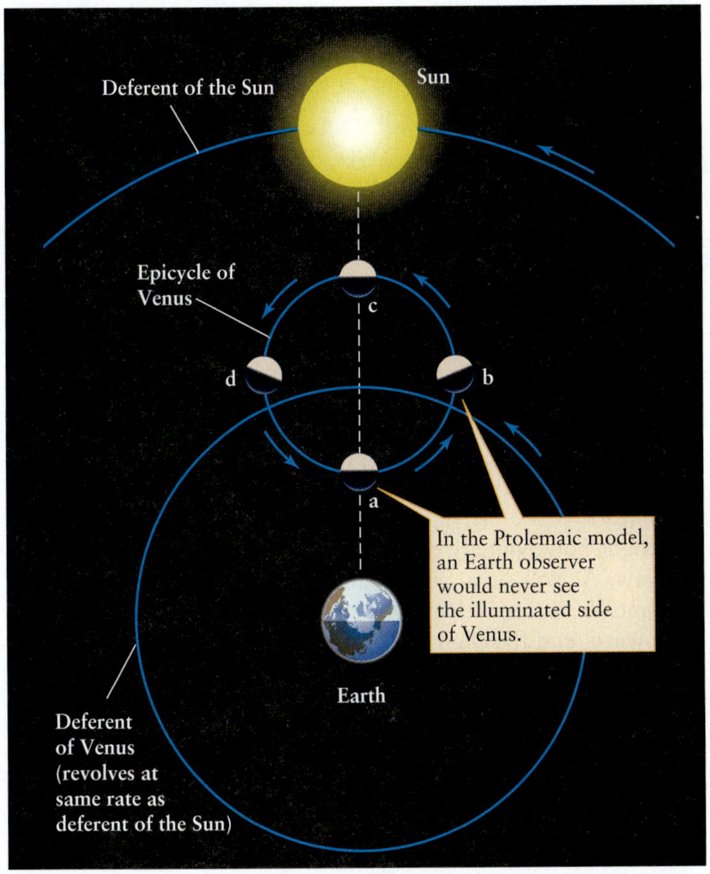

Figure 4-15

The Appearance of Venus in the Ptolemaic Model In the geocentric Ptolemaic model the deferents of Venus and the Sun rotate together, with the epicycle of Venus centered on a line (shown dashed) that connects the Sun and Earth. In this model an Earth observer would never see Venus as more than half illuminated. (At positions *a* and *c*, Venus appears in a "new" phase; at positions *b* and *d*, it appears as a crescent. Compare with Figure 3-2, which shows the phases of the Moon.) Because Galileo saw Venus in nearly fully illuminated phases, he concluded that the Ptolemaic model must be incorrect.

doctrine at the time placed Earth at the center of the universe, cautioned Galileo not to advocate a heliocentric model. He nonetheless persisted and was sentenced to spend the last years of his life under house arrest "for vehement suspicion of heresy." Nevertheless, there was no turning back. (The Roman Catholic Church lifted its ban against Galileo's heliocentric ideas in the 1700s.)

While Galileo's observations showed convincingly that the Ptolemaic model was entirely wrong and that a heliocentric model is the more nearly correct one, he was unable to provide a complete explanation of why Earth should orbit the Sun and not vice versa. The first person who was able to provide such an explanation was the Englishman Isaac Newton, born on Christmas Day of 1642, a dozen years after the death of Kepler and the same year that Galileo died. While Kepler and Galileo revolutionized our understanding of planetary motions, Newton's contribution was far greater: He deduced the basic laws that govern all motions on Earth as well as in the heavens.

Figure 4-16 RI**V**UXG

Jupiter and Its Largest Moons This photograph, taken by an amateur astronomer with a small telescope, shows the four Galilean satellites alongside an overexposed image of Jupiter. Each satellite is bright enough to be seen with the unaided eye, were it not overwhelmed by the glare of Jupiter. (Courtesy of C. Holmes)

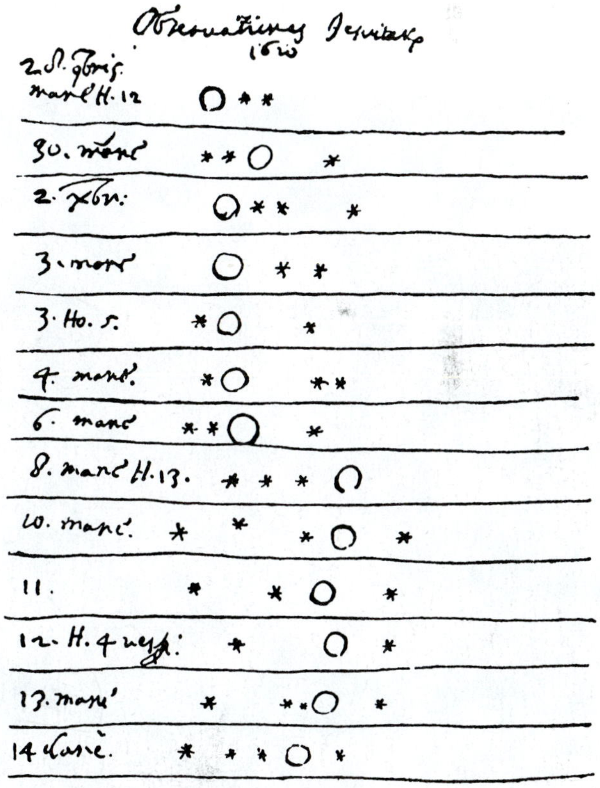

Figure 4-17

Early Observations of Jupiter's Moons In 1610 Galileo discovered four "stars" that move back and forth across Jupiter from one night to the next. He concluded that these are four moons that orbit Jupiter, much as our Moon orbits Earth. This drawing shows notations made by Jesuit observers on successive nights in 1620. The circle represents Jupiter and the stars its moons. Compare the drawing numbered 13 with the photograph in Figure 4-16. (Yerkes Observatory)

4-6 Isaac Newton formulated three laws that describe fundamental properties of physical reality

TUTORIAL 4.2

Until the mid-seventeenth century, virtually all attempts to describe the motions of the heavens were *empirical*, or based directly on data and observations. From Ptolemy to Kepler, astronomers would adjust their ideas and calculations by trial and error until they ended up with answers that agreed with observation.

The same laws of motion that hold sway on Earth apply throughout the universe

Isaac Newton (Figure 4-18) introduced a new approach. He began with three quite general statements, now called **Newton's laws of motion.** These laws, deduced from experimental observation, apply to all forces and all objects. Newton then showed that Kepler's three laws follow logically from these laws of motion and from a formula for the force of gravity that he derived from observation.

In other words, Kepler's laws are not just an empirical description of the motions of the planets, but a direct consequence of the fundamental laws of physical matter. Using this deeper insight into the nature of motions in the heavens, Newton and his successors were able to describe accurately not just the orbits of the planets but also the orbits of the Moon and comets.

WEB LINK 4.7

Figure 4-18

Isaac Newton (1642–1727) Using mathematical techniques that he devised, Isaac Newton formulated the law of universal gravitation and demonstrated that the planets orbit the Sun according to simple mechanical rules. (National Portrait Gallery, London)

Newton's First Law

Newton's laws of motion describe objects on Earth as well as in the heavens. Thus, we can understand each of these laws by considering the motions of objects around us. We begin with **Newton's first law of motion:**

An object remains at rest, or moves in a straight line at a constant speed, unless acted upon by a net outside force.

By **force** we mean any push or pull that acts on the object. An *outside* force is one that is exerted on the object by something other than the object itself. The net, or total, outside force is the combined effect of all of the individual outside forces that act on the object.

Right now, you are demonstrating the first part of Newton's first law. As you sit in your chair reading this passage, there are two outside forces acting on you: The force of gravity pulls you downward, and the chair pushes up on you. These two forces are of equal strength but of opposite direction, so their effects cancel—there is no *net* outside force. Hence, your body remains at rest as stated in Newton's first law. If you try to lift yourself out of your chair by grabbing your knees and pulling up, you will remain at rest because this force is not an outside force.

CAUTION! It is easy to confuse the *net* outside force on an object (central to Newton's first law), with *individual* outside forces on an object. If you want to make this book move across the floor in a straight line at a constant speed, you must continually push on it. You might therefore think that your push is a net outside force. But another force also acts on the book—the force of friction as the book rubs across the floor. The force of your push and the force of friction combine to make the *net* outside force. If you push the book at a constant speed, then the force of your push exactly balances the force of friction, so there is no net outside force. If you stop pushing, there will be nothing to balance the effects of friction. Then the friction will be a net outside force and the book will slow to a stop.

Newton's first law tells us that if no net outside force acts on a moving object, it can only move in a straight line and at a constant speed. This means that a net outside force *must* be acting on the planets since they don't move in straight lines, but instead move around elliptical paths. Another way to see that planetary orbits require a net outside force is that a planet would tend to fly off into space at a constant speed along a straight line if there were no net outside force acting on it. Because planets don't fly off, Newton concluded that a force *must* act continuously on the planets to keep them in their elliptical orbits.

Newton's Second Law

Newton's second law describes how the motion of an object *changes* if there is a net outside force acting on it. To appreciate Newton's second law, we must first understand three quantities that describe motion—speed, velocity, and acceleration.

Speed is a measure of how fast an object is moving. Speed and direction of motion together constitute an object's **velocity.** Compared with a car driving north at 100 km/h (62 mi/h), a car driving east at 100 km/h has the same speed but a different velocity. We can restate Newton's first law to say that an object has a constant velocity (its speed and direction of motion do not change) if no net outside force acts on the object.

Acceleration is the rate at which velocity changes. Because velocity involves both speed and direction, acceleration can result from changes in either. Contrary to popular use of the term, acceleration does not simply mean speeding up. A car is accelerating if it is speeding up, and it is also accelerating if it is slowing down or turning (that is, changing the direction in which it is moving).

You can verify these statements about acceleration if you think about the sensations of riding in a car. If the car is moving with a constant velocity (in a straight line at a constant speed), you feel the same, aside from vibrations, as if the car were not moving at all. But you can feel it when the car accelerates in any way: You feel thrown back in your seat if the car speeds up, thrown forward if the car slows down, and thrown sideways if the car changes direction in a tight turn. In Box 4-3 we discuss the reasons for these sensations, along with other applications of Newton's laws to everyday life.

An apple falling from a tree is a good example of acceleration that involves only an increase in speed. Initially, at the moment the stem breaks, the apple's speed is zero. After 1 second, its downward speed is 9.8 meters per second, or 9.8 m/s (32 feet per second, or 32 ft/s). After 2 seconds, the apple's speed is twice this, or 19.6 m/s. After 3 seconds, the speed is 29.4 m/s. Because the apple's speed increases by 9.8 m/s for each second of free fall, the rate of acceleration is 9.8 meters per second per second, or 9.8 m/s^2 (32 ft/s^2). Thus, Earth's gravity gives the apple a constant acceleration of 9.8 m/s^2 downward, toward the center of Earth.

A planet revolving about the Sun along a perfectly circular orbit is an example of acceleration that involves change of direction only. As the planet moves along its orbit, its speed remains

BOX 4-3 — Astronomy Down to Earth

Newton's Laws in Everyday Life

In our study of astronomy, we use Newton's three laws of motion to help us understand the motions of objects in the heavens. But you can see applications of Newton's laws every day in the world around you. By considering these everyday applications, we can gain insight into how Newton's laws apply to celestial events that are far removed from ordinary human experience.

Newton's *first* law, or principle of inertia, says that an object at rest naturally tends to remain at rest and that an object in motion naturally tends to remain in motion. This law explains the sensations that you feel when riding in an automobile. When you are waiting at a red light, your car and your body are both at rest. When the light turns green and you press on the gas pedal, the car accelerates forward but your body attempts to stay where it was. Hence, the seat of the accelerating car pushes forward into your body, and it feels as though you are being pushed back in your seat.

Once the car is up to cruising speed, your body wants to keep moving in a straight line at this cruising speed. If the car makes a sharp turn to the left, the right side of the car will move toward you. Thus, you will feel as though you are being thrown to the car's right side (the side on the outside of the turn). If you bring the car to a sudden stop by pressing on the brakes, your body will continue moving forward until the seat belt stops you. In this case, it feels as though you are being thrown toward the front of the car.

Newton's *second* law states that the net outside force on an object equals the product of the object's mass and its acceleration. You can accelerate a crumpled-up piece of paper to a pretty good speed by throwing it with a moderate force. But if you try to throw a heavy rock by using the same force, the acceleration will be much less because the rock has much more mass than the crumpled paper. Because of the smaller acceleration, the rock will leave your hand moving at only a slow speed.

Automobile airbags are based on the relationship between force and acceleration. It takes a large force to bring a fast-moving object suddenly to rest because this requires a large acceleration. In a collision, the driver of a car not equipped with airbags is jerked to a sudden stop and the large forces that act can cause major injuries. But if the car has airbags that deploy in an accident, the driver's body will slow down more gradually as it contacts the airbag, and the driver's acceleration will be less. (Remember that *acceleration* can refer to slowing down as well as to speeding up.) Hence, the force on the driver and the chance of injury will both be greatly reduced.

Newton's *third* law, the principle of action and reaction, explains how a car can accelerate at all. It is not correct to say that the engine pushes the car forward, because Newton's second law tells us that it takes a force acting from outside the car to make the car accelerate. Rather, the engine makes the wheels and tires turn, and the tires push backward on the ground. (You can see this backward force in action when a car drives through wet ground and sprays mud backward from the tires.) From Newton's third law, the ground must exert an equally large forward force on the car, and this is the force that pushes the car forward.

You use the same principles when you walk: You push backward on the ground with your foot, and the ground pushes forward on you. Icy pavement or a freshly waxed floor have greatly reduced friction. In these situations, your feet and the surface under you can exert only weak forces on each other, and it is much harder to walk.

constant. Nevertheless, the planet is continuously being accelerated because its direction of motion is continuously changing.

Newton's second law of motion says that in order to give an object an acceleration (that is, to change its velocity), a net outside force *must* act on the object. To be specific, this law says that the acceleration of an object is proportional to the net outside force acting on the object. That is, the harder you push on an object, the greater the resulting acceleration. This law can be succinctly stated as an equation. If a net outside force F acts on an object of mass m, the object will experience an acceleration a such that

Newton's second law

$$F = ma$$

F = net outside force on an object

m = mass of object

a = acceleration of object

The **mass** of an object is a measure of the total amount of material in the object. It is usually expressed in kilograms (kg) or grams (g). For example, the mass of the Sun is 2×10^{30} kg, the mass of a hydrogen atom is 1.7×10^{-27} kg, and the mass of an average adult is 75 kg. The Sun, a hydrogen atom, and a person have these masses regardless of where they happen to be in the universe.

CAUTION! It is important not to confuse the concepts of mass and weight. **Weight** is the force of gravity that acts on an object and, like any force, is usually expressed in pounds or newtons (1 newton = 0.225 pounds).

We can use Newton's second law to relate mass and weight. We have seen that the acceleration caused by Earth's gravity is 9.8 m/s^2. When a 50-kg swimmer falls from a diving board, the only outside force acting on her as she falls is her weight. Thus, from Newton's second law ($F = ma$), her weight is equal to her mass multiplied by the acceleration due to gravity:

$$50 \text{ kg} \times 9.8 \text{ m/s}^2 = 490 \text{ newtons} = 110 \text{ pounds}$$

Note that this answer is correct only when the swimmer is on Earth. She would weigh less on the Moon, where the pull of gravity is weaker, and more on Jupiter, where the gravitational pull is stronger. Floating deep in space, she would have no weight at all; she would be "weightless." Nevertheless, in all these circumstances, she would always have exactly the same mass, because mass is an inherent property of matter unaffected by details of the environment. Whenever we describe the properties of planets, stars, or galaxies, we speak of their masses, never of their weights.

We have seen that a planet is continually accelerating as it orbits the Sun. From Newton's second law, this means that there must be a net outside force that acts continually on each of the planets. As we will see in the next section, this force is the gravitational attraction of the Sun.

Newton's Third Law

The last of Newton's general laws of motion, called **Newton's third law of motion,** is the famous statement about action and reaction:

Whenever one object exerts a force on a second object, the second object exerts an equal and opposite force on the first object.

For example, if you weigh 110 pounds, when you are standing up you are pressing down on the floor with a force of 110 pounds. Newton's third law tells us that the floor is also pushing up against your feet with an equal force of 110 pounds. You can think of each of these forces as a reaction to the other force, which is the origin of the phrase "action and reaction."

Newton realized that because the Sun is exerting a force on each planet to keep it in orbit, each planet must also be exerting an equal and opposite force on the Sun. However, the planets are much less massive than the Sun (for example, Earth has only 1/300,000 of the Sun's mass). Therefore, although the Sun's force on a planet is the same as the planet's force on the Sun, the planet's much smaller mass gives it a much larger acceleration, according to Newton's second law. This is why the planets circle the Sun instead of vice versa. Thus, Newton's laws reveal the reason for our heliocentric solar system.

4-7 Newton's description of gravity accounts for Kepler's laws and explains the motions of the planets

TUTORIAL 4.3

Tie a ball to one end of a piece of string, hold the other end of the string in your hand, and whirl the ball around in a circle. As the ball "orbits" your hand, it is continuously accelerating because its velocity is changing. (Even if its speed is constant, its direction of motion is changing.) In accordance with Newton's second law, this can happen only if the ball is continuously acted on by an outside force—the pull of the string. The pull is directed along the string toward your hand. In the same way, Newton saw, the force that keeps a planet in orbit around the Sun is a pull that always acts toward the Sun. That pull is **gravity,** or **gravitational force.**

Newton's discovery about the forces that act on planets led him to suspect that the force of gravity pulling a falling apple straight down to the ground is fundamentally the same as the force on a planet that is always directed straight at the Sun. In other words, gravity is the force that shapes the orbits of the planets. What is more, he was able to determine how the force of gravity depends on the distance between the Sun and the planet. His result was a law of gravitation that could apply to the motion of distant planets as well as to the flight of a football on Earth. Using this law, Newton achieved the remarkable goal of deducing Kepler's laws from fundamental principles of nature.

Newton's law of gravitation is truly universal: It applies to falling apples as well as to planets and galaxies

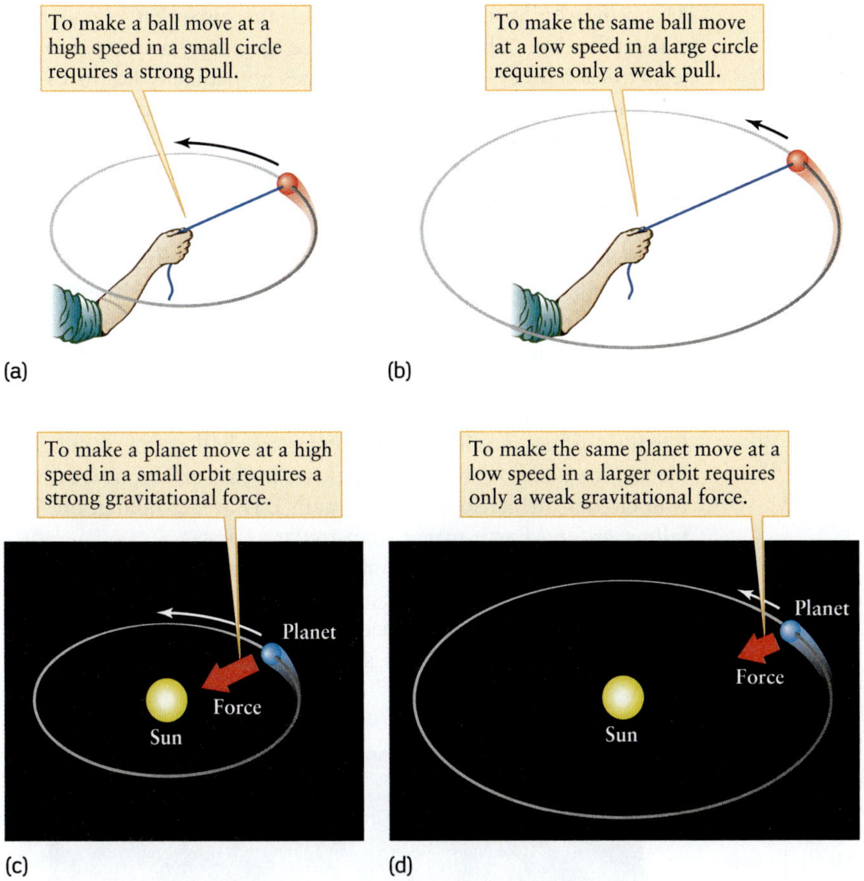
To make a ball move at a high speed in a small circle requires a strong pull.
To make the same ball move at a low speed in a large circle requires only a weak pull.
(a)
(b)
To make a planet move at a high speed in a small orbit requires a strong gravitational force.
To make the same planet move at a low speed in a larger orbit requires only a weak gravitational force.
Planet
Force
Sun
Planet
Force
Sun
(c)
(d)

Figure 4-19

An Orbit Analogy **(a)** To make a ball on a string move at high speed around a small circle, you have to exert a substantial pull on the string. **(b)** If you lengthen the string and make the same ball move at low speed around a large circle, much less pull is required. **(c), (d)** Similarly, a planet that orbits close to the Sun moves at high speed and requires a substantial gravitational force from the Sun, while a planet in a large orbit moves at low speed and requires less gravitational force to stay in orbit.

To see how Newton reasoned, think again about a ball attached to a string. If you use a short string, so that the ball orbits in a small circle, and whirl the ball around your hand at a high speed, you will find that you have to pull fairly hard on the string (Figure 4-19*a*). But if you use a longer string, so that the ball moves in a larger orbit, and if you make the ball orbit your hand at a slow speed, you only have to exert a light tug on the string (Figure 4-19*b*). The orbits of the planets behave in the same way; the larger the size of the orbit, the slower the planet's speed (Figure 4-19*c, d*). By analogy to the force of the string on the orbiting ball, Newton concluded that the force that attracts a planet toward the Sun must decrease with increasing distance between the Sun and the planet.

The Law of Universal Gravitation

Using his own three laws and Kepler's three laws, Newton succeeded in formulating a general statement that describes the nature of the gravitational force. Newton's **law of universal gravitation** is as follows:

Two objects attract each other with a force that is directly proportional to the mass of each object and inversely proportional to the square of the distance between them.

This law states that *any* two objects exert gravitational pulls on each other. Normally, you notice only the gravitational force that Earth exerts on you, otherwise known as your weight. In fact, you feel gravitational attractions to *all* the objects around you. For example, this book is exerting a gravitational force on you as you read it. But because the force exerted on you by this book is proportional to the book's mass, which is very small compared to Earth's mass, the force is too small to notice. (It can actually be measured with sensitive equipment.)

Consider two 1-kg objects separated by a distance of 1 meter. Newton's law of universal gravitation says that the force is directly proportional to the mass, so if we double the mass of one object to 2 kg, the force between the objects will double. If we double both masses so that we have two 2-kg objects separated by 1 meter, the force will be $2 \times 2 = 4$ times what it was originally (the force is directly proportional to the mass of *each* object). If we go back to two 1-kg masses, but double their separation to 2 meters, the force will be only one-quarter its original value. This is because the force is inversely proportional to the square of the distance: If we double the distance, the force is multiplied by a factor of

$$\frac{1}{2^2} = \frac{1}{4}$$

Newton's law of universal gravitation can be stated more succinctly as an equation. If two objects have masses m_1 and m_2 and are separated by a distance r, then the gravitational force F between these two objects is given by the following equation:

Newton's law of universal gravitation

$$F = G\left(\frac{m_1 m_2}{r^2}\right)$$

F = gravitational force between two objects

m_1 = mass of first object

m_2 = mass of second object

r = distance between objects

G = universal constant of gravitation

If the masses are measured in kilograms and the distance between them in meters, then the force is measured in newtons. In this formula, G is a number called the **universal constant of gravitation.** Laboratory experiments have yielded a value for G of

$$G = 6.67 \times 10^{-11} \text{ newton} \cdot \text{m}^2/\text{kg}^2$$

We can use Newton's law of universal gravitation to calculate the force with which any two objects attract each other. For example, to compute the gravitational force that the Sun exerts on Earth, we substitute values for Earth's mass ($m_1 = 5.98 \times 10^{24}$ kg), the Sun's mass ($m_2 = 1.99 \times 10^{30}$ kg), the distance between them ($r = 1$ AU $= 1.5 \times 10^{11}$ m), and the value of G into Newton's equation. We get

$$F_{\text{Sun-Earth}} = 6.67 \times 10^{-11}\left[\frac{(5.98 \times 10^{24}) \times (1.99 \times 10^{30})}{(1.50 \times 10^{11})^2}\right]$$
$$= 3.53 \times 10^{22} \text{ newtons}$$

If we calculate the force that Earth exerts on the Sun, we get exactly the same result. (Mathematically, we just let m_1 be the Sun's mass and m_2 be Earth's mass instead of the other way around. The product of the two numbers is the same, so the force is the same.) This is in accordance with Newton's third law: Any two objects exert *equal* gravitational forces on each other.

Your weight is just the gravitational force that Earth exerts on you, so we can calculate it using Newton's law of universal gravitation. Earth's mass is $m_1 = 5.98 \times 10^{24}$ kg, and the distance r to use is the distance between the *centers* of Earth and you. This distance is just the radius of Earth, which is $r = 6378$ km $= 6.378 \times 10^6$ m. If your mass is $m_2 = 50$ kg, your weight is

$$F_{\text{Earth-you}} = 6.67 \times 10^{-11}\left[\frac{(5.98 \times 10^{24}) \times (50)}{(6.378 \times 10^{6})^2}\right]$$
$$= 490 \text{ newtons}$$

This value is the same as the weight of a 50-kg person that we calculated in Section 4-6. This example shows that your weight would have a different value on a planet with a different mass m_1 and a different radius r.

Gravitational Force and Orbits

Because there is a gravitational force between any two objects, Newton concluded that gravity is also the force that keeps the Moon in orbit around Earth. It is also the force that keeps artificial satellites in orbit. But if the force of gravity attracts two objects to each other, why don't satellites immediately fall to Earth? Why doesn't the Moon fall into Earth? And, for that matter, why don't the planets fall into the Sun?

To see the answer, imagine (as Newton did) dropping a ball from a great height above Earth's surface, as in Figure 4-20. After you drop the ball, it, of course, falls straight down (path A in Figure 4-20). But if you *throw* the ball horizontally, it travels some distance across Earth's surface before hitting the ground (path B). If you throw the ball harder, it travels a greater distance (path C). If you could throw at just the right speed, the curvature of the ball's path will exactly match the curvature of Earth's surface (path E). Although Earth's gravity is making the ball fall, Earth's surface is falling away under the ball at the same rate. Hence, the ball does not get any closer to the surface, and the ball is in circular orbit. So the ball in path E is in fact falling, but it is falling *around* Earth rather than *toward* Earth.

A spacecraft is launched into orbit in just this way—by throwing it fast enough. The thrust of a rocket is used to give the spacecraft the necessary orbital speed. Once the spacecraft is in orbit, the rocket turns off and the spacecraft falls continually around Earth.

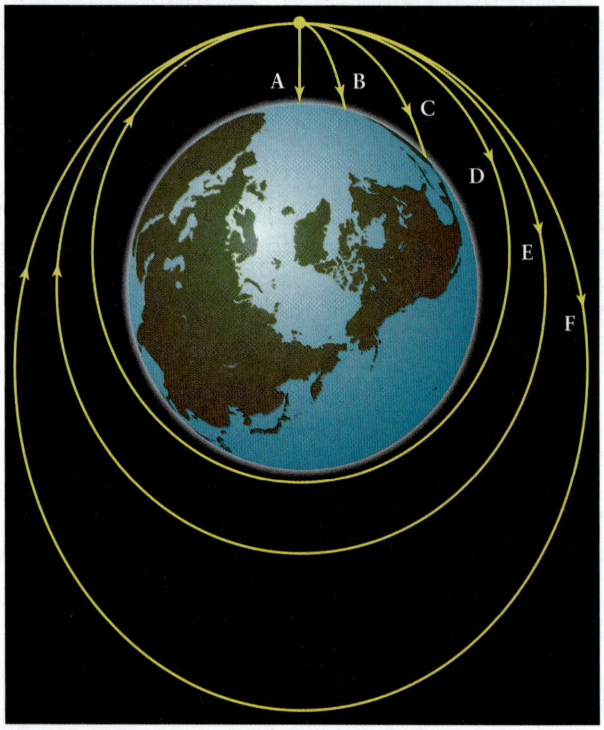

WEB LINK 4.8 ANIMATION 4.7

Figure 4-20

An Explanation of Orbits If a ball is dropped from a great height above Earth's surface, it falls straight down (A). If the ball is thrown with some horizontal speed, it follows a curved path before hitting the ground (B, C). If thrown with just the right speed (E), the ball goes into circular orbit; the ball's path curves but it never gets any closer to Earth's surface. If the ball is thrown with a speed that is slightly less (D) or slightly more (F) than the speed for a circular orbit, the ball's orbit is an ellipse.

CAUTION! An astronaut on board an orbiting spacecraft (like the one shown in the photograph that opens this chapter) feels "weightless." However, this is *not* because she is "beyond the pull of gravity." The astronaut is herself an independent satellite of Earth, and Earth's gravitational pull is what holds her in orbit. She feels "weightless" because she and her spacecraft are falling *together* around Earth, so there is nothing pushing her against any of the spacecraft walls. You feel the same "weightless" sensation whenever you are falling, such as when you jump off a diving board or ride the free-fall ride at an amusement park.

If the ball in Figure 4-20 is thrown with a slightly slower speed than that required for a circular orbit, its orbit will be an ellipse (path D). An elliptical orbit also results if instead the ball is thrown a bit too fast (path F). In this way, spacecraft can be placed into any desired orbit around Earth by adjusting the thrust of the rockets.

Just as the ball in Figure 4-20 will not fall to Earth if given enough speed, the Moon does not fall to Earth and the planets do not fall into the Sun. The planets acquired their initial speeds around the Sun when the solar system first formed 4.56 billion years ago. Figure 4-20 shows that a circular orbit is a very special case, so it is no surprise that the orbits of the planets are not precisely circular.

CAUTION! Orbiting satellites do sometimes fall out of orbit and crash back to Earth. When this happens, however, the real culprit is not gravity but air resistance. A satellite in a relatively low orbit is actually flying through the tenuous outer wisps of Earth's atmosphere. The resistance of the atmosphere slows the satellite and changes a circular orbit like E in Figure 4-20 to an elliptical one like D. As the satellite sinks to lower altitude, it encounters more air resistance and sinks even lower. Eventually, it either strikes Earth or burns up in flight due to air friction. By contrast, the Moon and planets orbit in the near-vacuum of interplanetary space. Hence, they are unaffected by this kind of air resistance, and their orbits are much more long-lasting.

Gravitation and Kepler's Laws

Using his three laws of motion and his law of gravity, Newton found that he could derive Kepler's three laws mathematically. Kepler's first law, concerning the elliptical shape of planetary orbits, proved to be a direct consequence of the $1/r^2$ factor in the law of universal gravitation. (Had the nature of gravity in our universe been different, so that this factor was given by a different function such as $1/r$ or $1/r^3$, elliptical orbits would not have been possible.) The law of equal areas, or Kepler's second law, turns out to be a consequence of the Sun's gravitational force on a planet being directed straight toward the Sun.

Newton also demonstrated that Kepler's third law follows logically from his law of gravity. Specifically, he proved that if two objects with masses m_1 and m_2 orbit each other, the period P of their orbit and the semimajor axis a of their orbit (that is, the average distance between the two objects) are related by an equation that we call **Newton's form of Kepler's third law:**

$$P^2 = \left[\frac{4\pi^2}{G(m_1 + m_2)}\right] a^3$$

Newton's form of Kepler's third law is valid whenever two objects orbit each other because of their mutual gravitational attraction (Figure 4-21). It is invaluable in the study of binary star systems, in which two stars orbit each other. If the orbital period

WEB LINK 4.9 **Figure 4-21** RI**V**UXG

In Orbit Around the Moon This photograph taken from the spacecraft *Columbia* shows the lunar lander *Eagle* after returning from the first human landing on the Moon in July 1969. Newton's form of Kepler's third law describes the orbit of a spacecraft around the Moon, as well as the orbit of the Moon around Earth (visible in the distance). (Michael Collins, *Apollo 11*, NASA)

P and semimajor axis a of the two stars in a binary system are known, astronomers can use this formula to calculate the sum $m_1 + m_2$ of the masses of the two stars. Within our own solar system, Newton's form of Kepler's third law makes it possible to learn about the masses of planets. By measuring the period and semimajor axis for a satellite, astronomers can determine the sum of the masses of the planet and the satellite. (The satellite can be a moon of the planet or a spacecraft that we place in orbit around the planet. Newton's laws apply in either case.) Box 4-4 gives an example of using Newton's form of Kepler's third law.

Newton also discovered new features of orbits around the Sun. For example, his equations soon led him to conclude that the orbit of an object around the Sun need not be an ellipse. It could be any one of a family of curves called conic sections.

A **conic section** is any curve that you get by cutting a cone with a plane, as shown in Figure 4-22. You can get circles and ellipses by slicing all the way through the cone. You can also get two types of open curves called **parabolas** and **hyperbolas.** If you were to throw the ball in Figure 4-20 with a fast enough speed, it would follow a parabolic or hyperbolic orbit and would fly off into space, never to return. Comets hurtling toward the Sun from the depths of space sometimes follow hyperbolic orbits.

BOX 4-4 Tools of the Astronomer's Trade

Newton's Form of Kepler's Third Law

Kepler's original statement of his third law, $P^2 = a^3$, is valid only for objects that orbit the Sun. (Box 4-2 shows how to use this equation.) But Newton's form of Kepler's third law is much more general: It can be used in *any* situation where two objects of masses m_1 and m_2 orbit each other. For example, Newton's form is the equation to use for a moon orbiting a planet or a satellite orbiting Earth. This equation is

Newton's form of Kepler's third law

$$P^2 = \left[\frac{4\pi^2}{G(m_1 + m_2)}\right]a^3$$

P = sidereal period of orbit, in seconds

a = semimajor axis of orbit, in meters

m_1 = mass of first object, in kilograms

m_2 = mass of second object, in kilograms

G = universal constant of gravitation = 6.67×10^{-11}

Notice that P, a, m_1, and m_2 *must* be expressed in these particular units. If you fail to use the correct units, your answer will be incorrect.

EXAMPLE: Io (pronounced "eye-oh") is one of the four large moons of Jupiter discovered by Galileo and shown in Figure 4-16. It orbits at a distance of 421,600 km from the center of Jupiter and has an orbital period of 1.77 days. Determine the combined mass of Jupiter and Io.

Situation: We are given Io's orbital period P and semimajor axis a (the distance from Io to the center of its circular orbit, which is at the center of Jupiter). Our goal is to find the sum of the masses of Jupiter (m_1) and Io (m_2).

Tools: Because this orbit is not around the Sun, we must use Newton's form of Kepler's third law to relate P and a. This relationship also involves m_1 and m_2, whose sum ($m_1 + m_2$) we are asked to find.

Answer: To solve for $m_1 + m_2$, we rewrite the equation in the form

$$m_1 + m_2 = \frac{4\pi^2 a^3}{GP^2}$$

To use this equation, we have to convert the distance a from kilometers to meters and convert the period P from days to seconds. There are 1000 meters in 1 kilometer and 86,400 seconds in 1 day, so

$$a = (421{,}600 \text{ km}) \times \frac{1000 \text{ m}}{1 \text{ km}} = 4.216 \times 10^8 \text{ m}$$

$$P = (1.77 \text{ days}) \times \frac{86{,}400 \text{ s}}{1 \text{ day}} = 1.529 \times 10^5 \text{ s}$$

We can now put these values and the value of G into the above equation:

$$m_1 + m_2 = \frac{4\pi^2(4.216 \times 10^8)^3}{(6.67 \times 10^{-11})(1.529 \times 10^5)^2} = 1.90 \times 10^{27} \text{ kg}$$

Review: Io is very much smaller than Jupiter, so its mass is only a small fraction of the mass of Jupiter. Thus, $m_1 + m_2$ is very nearly the mass of Jupiter alone. We conclude that Jupiter has a mass of 1.90×10^{27} kg, or about 300 times the mass of Earth. This technique can be used to determine the mass of any object that has a second, much smaller object orbiting it. Astronomers use this technique to find the masses of stars, black holes, and entire galaxies of stars.

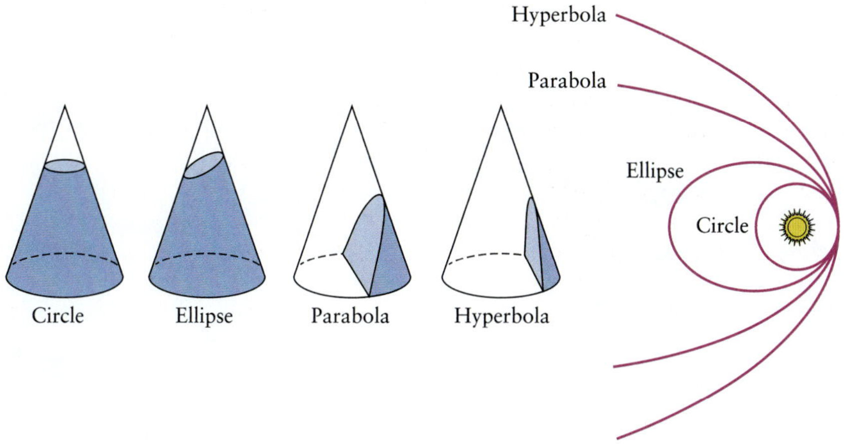

INTERACTIVE EXERCISE 4.2

Figure 4-22

Conic Sections A conic section is any one of a family of curves obtained by slicing a cone with a plane. The orbit of one object about another can be any one of these curves: a circle, an ellipse, a parabola, or a hyperbola.

The Triumph of Newtonian Mechanics

Newton's ideas turned out to be applicable to an incredibly wide range of situations. Using his laws of motion, Newton himself proved that Earth's axis of rotation must precess because of the gravitational pull of the Moon and the Sun on Earth's equatorial bulge (see Figure 2-19). In fact, all the details of the orbits of the planets and their satellites could be explained mathematically with a body of knowledge built on Newton's work that is today called **Newtonian mechanics.**

Not only could Newtonian mechanics explain a variety of known phenomena in detail, but it could also predict new phenomena. For example, one of Newton's friends, Edmund Halley, was intrigued by three similar historical records of a comet that had been sighted at intervals of 76 years. Assuming these records to be accounts of the same comet, Halley used Newton's methods to work out the details of the comet's orbit and predicted its return in 1758. It was first sighted on Christmas night of 1757, a fitting memorial to Newton's birthday. To this day the comet bears Halley's name (Figure 4-23).

Another dramatic success of Newton's ideas was their role in the discovery of the eighth planet from the Sun. The seventh planet, Uranus, was discovered accidentally by William Herschel in 1781 during a telescopic survey of the sky. Fifty years later, however, it was clear that Uranus was not following its predicted orbit. John Couch Adams in England and Urbain Le Verrier in France independently calculated that the gravitational pull of a yet unknown, more distant planet could explain the deviations of Uranus from its orbit. Le Verrier predicted that the planet would be found at a certain location in the constellation of Aquarius. A brief telescopic search on September 23, 1846, revealed the planet Neptune within 1° of the calculated position. Before it was sighted with a telescope, Neptune was actually predicted with pencil and paper.

Because it has been so successful in explaining and predicting many important phenomena, Newtonian mechanics has become the cornerstone of modern physical science. Even today, as we send astronauts into Earth orbit and spacecraft to the outer planets, Newton's equations are used to calculate the orbits and trajectories of these spacecraft. The *Cosmic Connections* figure on the next page summarizes some of the ways that gravity plays an important role on scales from apples to galaxies.

In the twentieth century, scientists found that Newton's laws do not apply in all situations. A new theory called *quantum mechanics* had to be developed to explain the behavior of matter on the very smallest of scales, such as within the atom and within the atomic nucleus. Albert Einstein developed the *theory of relativity* to explain what happens at very high speeds approaching the speed of light and in places where gravitational forces are very strong. For many purposes in astronomy, however, Newton's laws are as useful today as when Newton formulated them more than three centuries ago.

WEB LINK 4.10

Figure 4-23 R I **V** U X G

Comet Halley This most famous of all comets orbits the Sun with an average period of about 76 years. During the twentieth century, the comet passed near the Sun in 1910 and again in 1986 (when this photograph was taken). It will next be prominent in the sky in 2061. (Harvard College Observatory/Photo Researchers, Inc.)

COSMIC CONNECTIONS Universal Gravitation

Gravity is one of the fundamental forces of nature. We can see its effects here on Earth as well as in the farthest regions of the observable universe.

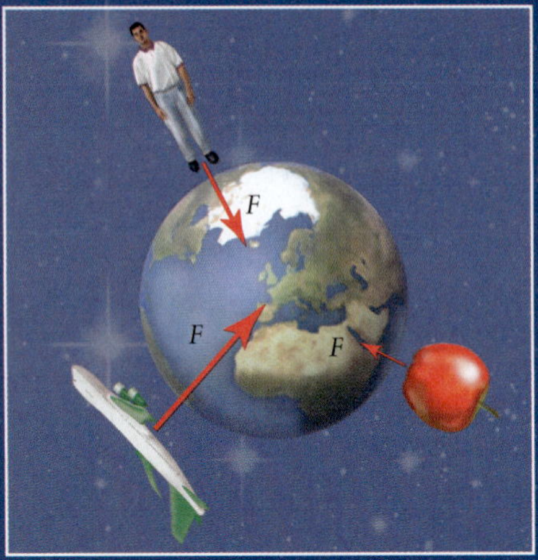

The weight of an ordinary object is just the gravitational force exerted on that object by Earth.

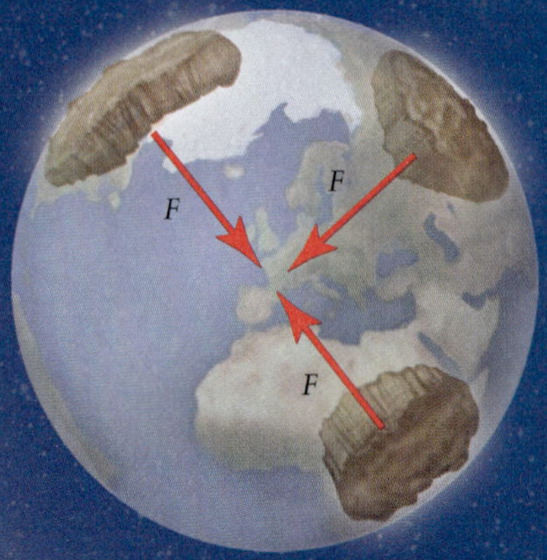

Earth is held together by the mutual gravitational attraction of its parts.

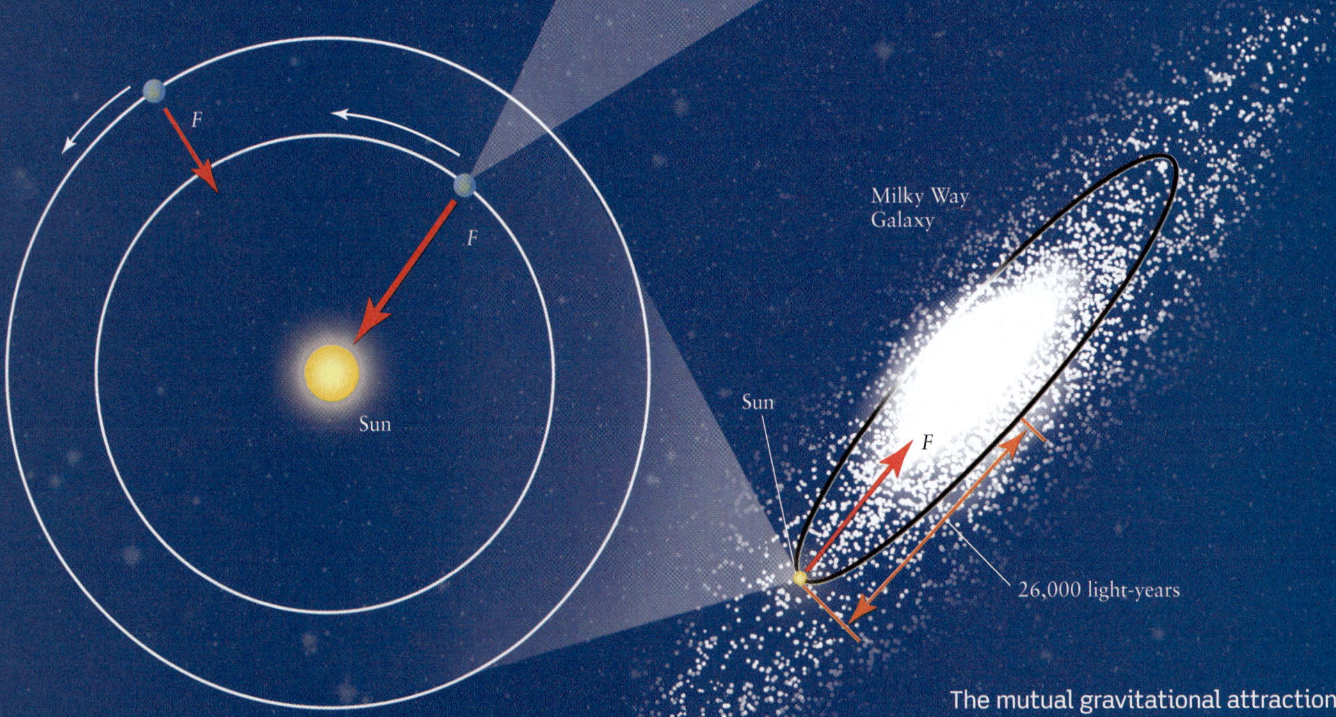

Gravitational forces exerted by the Sun keep the planets in their orbits. The farther a planet is from the Sun, the weaker the gravitational force that acts on the planet.

The mutual gravitational attraction of all the matter in the Milky Way Galaxy holds it together. The gravitational force of the Galaxy on our Sun and solar system holds us in an immense orbit around the galactic center.

4-8 Gravitational forces between Earth and the Moon produce tides

We have seen how Newtonian mechanics explains why the Moon stays in orbit around Earth. It also explains why there are ocean tides, as well as why the Moon always keeps the same face toward Earth. Both of these are the result of *tidal forces*—a consequence of gravity that deforms planets and reshapes galaxies.

Tidal forces reveal how gravitation can pull objects apart rather than drawing them together

Tidal forces are differences in the gravitational pull at different points in an object. As an illustration, imagine that three billiard balls are lined up in space at some distance from a planet, as in Figure 4-24*a*. According to Newton's law of universal gravitation, the force of attraction between two objects is greater the closer the two objects are to each other. Thus, the planet exerts more force on the 3-ball (in red) than on the 2-ball (in blue), and exerts more force on the 2-ball than on the 1-ball (in yellow). Now, imagine that the three balls are released and allowed to fall toward the planet. Figure 4-24*b* shows the situation a short time later. Because of the differences in gravitational pull, a short time later the 3-ball will have moved farther than the 2-ball, which will in turn have moved farther than the 1-ball. But now imagine that same motion from the perspective of the 2-ball. From this perspective, it appears as though the 3-ball is pulled toward the planet while the 1-ball is pushed away (Figure 4-24*c*). These apparent pushes and pulls are called tidal forces.

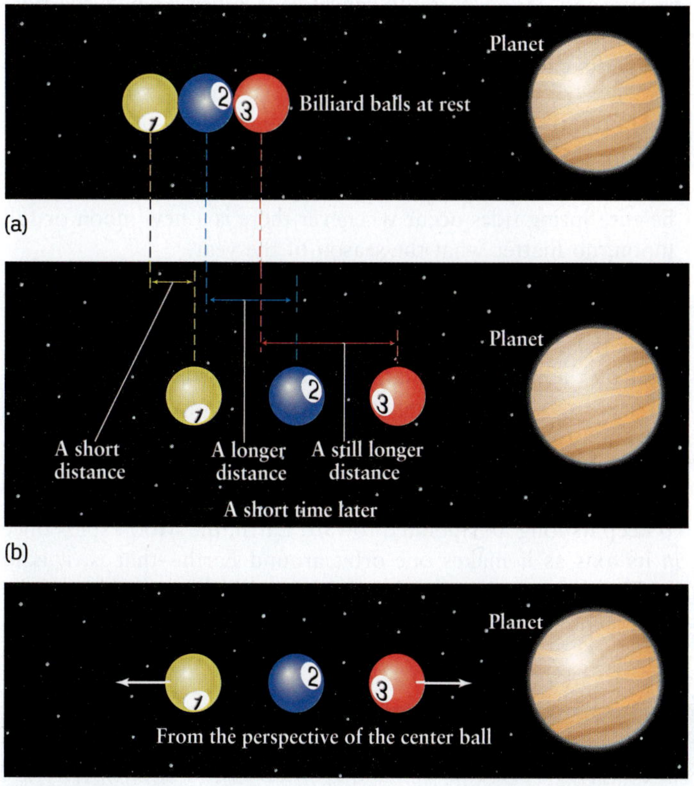

(a) (b) (c)

Figure 4-24

ANIMATION 4.8

The Origin of Tidal Forces **(a)** Imagine three identical billiard balls placed some distance from a planet and released. **(b)** The closer a ball is to the planet, the more gravitational force the planet exerts on it. Thus, a short time after the balls are released, the blue 2-ball has moved farther toward the planet than the yellow 1-ball, and the red 3-ball has moved farther still. **(c)** From the perspective of the 2-ball in the center, it appears that forces have pushed the 1-ball away from the planet and pulled the 3-ball toward the planet. These forces are called tidal forces.

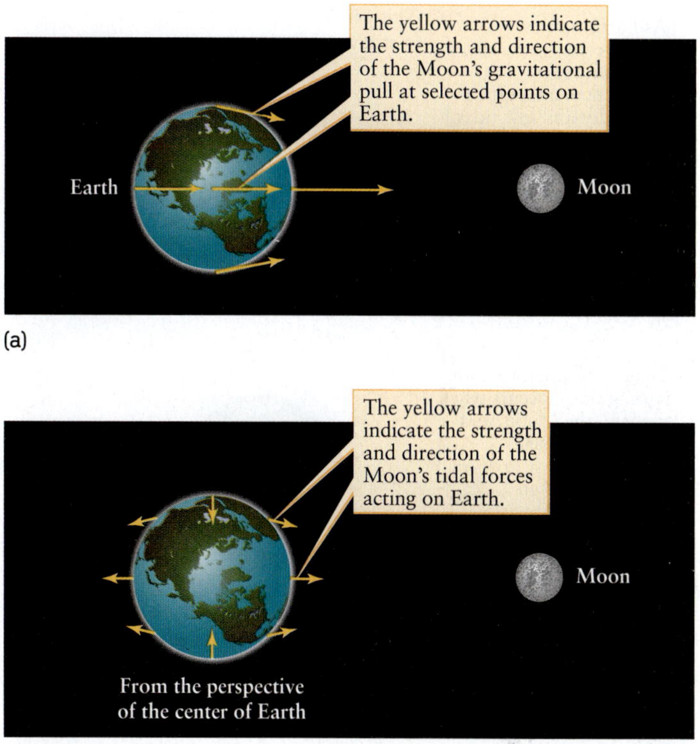

(a) (b)

Figure 4-25

Tidal Forces on the Earth **(a)** The Moon exerts different gravitational pulls at different locations on Earth. **(b)** At any location, the tidal force equals the Moon's gravitational pull at that location minus the gravitational pull of the Moon at the center of Earth. These tidal forces tend to deform Earth into a nonspherical shape.

Tidal Forces on the Earth

The Moon has a similar effect on Earth as the planet in Figure 4-24 has on the three billiard balls. The arrows in Figure 4-25*a* indicate the strength and direction of the gravitational force of the Moon at several locations on Earth. The side of Earth closest to the Moon feels a greater gravitational pull than Earth's center does, and the side of Earth that faces away from the Moon feels less gravitational pull than does Earth's center. This means that just as for the billiard balls in Figure 4-24, there are tidal forces acting on Earth (Figure 4-25*b*). These tidal forces exerted by the Moon try to elongate Earth along a line connecting the centers of

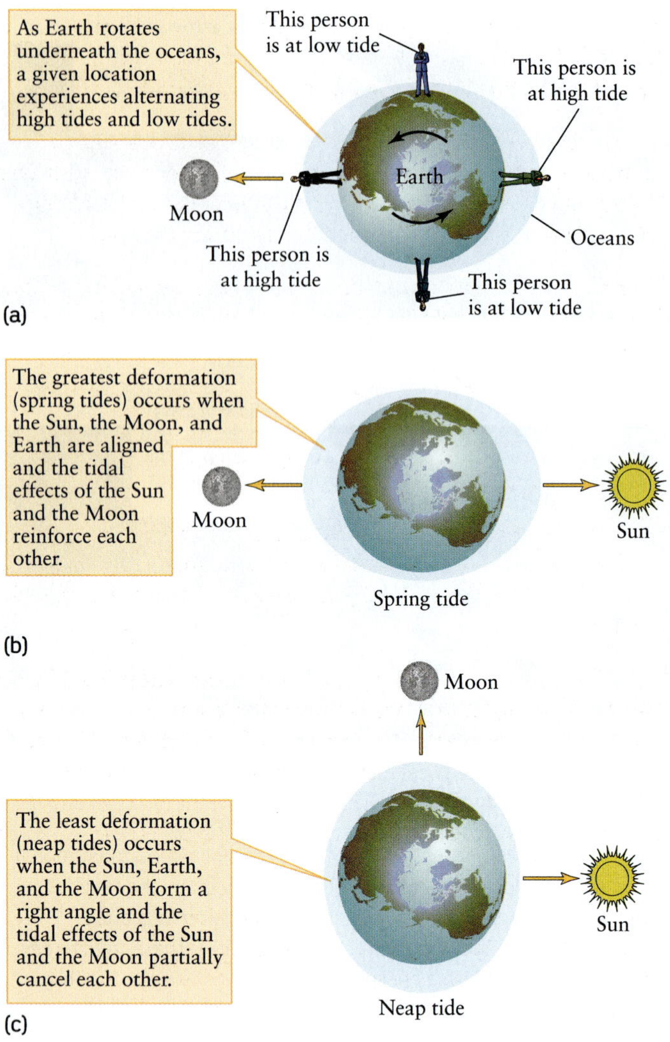
As Earth rotates underneath the oceans, a given location experiences alternating high tides and low tides.
This person is at low tide
This person is at high tide
Earth
Moon
Oceans
This person is at high tide
This person is at low tide
(a)
The greatest deformation (spring tides) occurs when the Sun, the Moon, and Earth are aligned and the tidal effects of the Sun and the Moon reinforce each other.
Moon
Sun
Spring tide
(b)
Moon
The least deformation (neap tides) occurs when the Sun, Earth, and the Moon form a right angle and the tidal effects of the Sun and the Moon partially cancel each other.
Sun
Neap tide
(c)

Figure 4-26

High and Low Tides **(a)** The gravitational forces of the Moon and the Sun deform Earth's oceans, giving rise to low and high tides. **(b), (c)** The strength of the tides depends on the relative positions of the Sun, the Moon, and Earth.

Earth and the Moon and try to squeeze Earth inward in the direction perpendicular to that line.

Because the body of Earth is largely rigid, it cannot deform very much in response to the tidal forces of the Moon. But the water in the oceans can and does deform into a football shape, as Figure 4-26*a* shows. As Earth rotates, a point on its surface goes from where the water is shallow to where the water is deep and back again. This is the origin of low and high ocean tides. (In this simplified description we have assumed that Earth is completely covered with water. The full story of the tides is much more complex, because the shapes of the continents and the effects of winds must also be taken into account.)

The Sun also exerts tidal forces on Earth's oceans. (The tidal effects of the Sun are about half as great as those of the Moon.) When the Sun, the Moon, and Earth are aligned, which happens at either new moon or full moon, the tidal effects of the Sun and Moon reinforce each other and the tidal distortion of the oceans is greatest. This produces large shifts in water level called **spring tides** (Figure 4-26*b*). At first quarter and last quarter, when the Sun and Moon form a right angle with Earth, the tidal effects of the Sun and Moon partially cancel each other. Hence, the tidal distortion of the oceans is the least pronounced, producing smaller tidal shifts called **neap tides** (Figure 4-26*c*).

CAUTION! Note that spring tides have nothing to do with the season of the year called spring. Instead, the name refers to the way that the ocean level "springs up" to a greater than normal height. Spring tides occur whenever there is a new moon or full moon, no matter what the season of the year.

Tidal Forces on the Moon and Beyond

Just as the Moon exerts tidal forces on Earth, Earth exerts tidal forces on the Moon. Soon after the Moon formed some 4.56 billion years ago, it was molten throughout its volume. Earth's tidal forces deformed the molten Moon into a slightly elongated shape, with the long axis of the Moon pointed toward Earth. The Moon retained this shape and orientation when it cooled and solidified. To keep its long axis pointed toward Earth, the Moon spins once on its axis as it makes one orbit around Earth—that is, it is in synchronous rotation (Section 3-2). Hence, the same side of the Moon always faces Earth, and this is the side that we see. For the

Figure 4-27 R I V U X G

Tidal Forces on a Galaxy For millions of years the galaxies NGC 2207 and IC 2163 have been moving ponderously past each other. The larger galaxy's tremendous tidal forces have drawn a streamer of material a hundred thousand light-years long out of IC 2163. If you lived on a planet orbiting a star within this streamer, you would have a magnificent view of both galaxies. NGC 2207 and IC 2163 are respectively 143,000 light-years and 101,000 light-years in diameter. Both galaxies are 114 million light-years away in the constellation Canis Major. (NASA and the Hubble Heritage Team, AURA/STScI)

IC 2163
NGC 2207
Stars and gas pulled out of IC 2163 by tidal forces from NGC 2207

same reason, most of the satellites in the solar system are in synchronous rotation, and thus always keep the same side facing their planet.

Tidal forces are also important on scales much larger than the solar system. Figure 4-27 shows two spiral galaxies, like the one in Figure 1-7, undergoing a near-collision. During the millions of years that this close encounter has been taking place, the tidal forces of the larger galaxy have pulled an immense streamer of stars and interstellar gas out of the smaller galaxy.

Many galaxies, including our own Milky Way Galaxy, show signs of having been disturbed at some time by tidal interactions with other galaxies. By their effect on the interstellar gas from which stars are formed, tidal interactions can actually trigger the birth of new stars. Our own Sun and solar system may have been formed as a result of tidal interactions of this kind. Hence, we may owe our very existence to tidal forces. In this and many other ways, the laws of motion and of universal gravitation shape our universe and our destinies.

Key Words

acceleration, p. 81
aphelion, p. 74
conic section, p. 86
conjunction, p. 70
deferent, p. 67
direct motion, p. 67
eccentricity, p. 74
ellipse, p. 74
elongation, p. 70
epicycle, p. 67
focus (of an ellipse; *plural* foci), p. 74
force, p. 80
geocentric model, p. 66
gravitational force, p. 82
gravity, p. 82
greatest eastern elongation, p. 70
greatest western elongation, p. 70
heliocentric model, p. 68
hyperbola, p. 86
inferior conjunction, p. 70
inferior planet, p. 69
Kepler's first law, p. 74
Kepler's second law, p. 75
Kepler's third law, p. 75
law of equal areas, p. 75
law of universal gravitation, p. 83
major axis (of an ellipse), p. 74
mass, p. 82
neap tides, p. 90
Newtonian mechanics, p. 87
Newton's first law of motion, p. 80
Newton's form of Kepler's third law, p. 85
Newton's second law of motion, p. 82
Newton's third law of motion, p. 82
Occam's razor, p. 67
opposition, p. 70
parabola, p. 86
parallax, p. 72
perihelion, p. 74
period (of a planet), p. 70
Ptolemaic system, p. 67
retrograde motion, p. 67
semimajor axis (of an ellipse), p. 74
sidereal period, p. 70
speed, p. 81
spring tides, p. 90
superior conjunction, p. 70
superior planet, p. 70
synodic period, p. 70
tidal forces, p. 89
universal constant of gravitation, p. 84
velocity, p. 81
weight, p. 82

Key Ideas

Apparent Motions of the Planets: Like the Sun and the Moon, the planets move on the celestial sphere with respect to the background of stars. Most of the time a planet moves eastward in direct motion, in the same direction as the Sun and the Moon, but from time to time it moves westward in retrograde motion.

The Ancient Geocentric Model: Ancient astronomers believed Earth to be at the center of the universe. They invented a complex system of epicycles and deferents to explain the direct and retrograde motions of the planets on the celestial sphere.

Copernicus's Heliocentric Model: Copernicus's heliocentric (Sun-centered) theory simplified the general explanation of planetary motions.

• In a heliocentric system, Earth is one of the planets orbiting the Sun.

• A planet undergoes retrograde motion as seen from Earth when Earth and the planet pass each other.

• The sidereal period of a planet, its true orbital period, is measured with respect to the stars. Its synodic period is measured with respect to Earth and the Sun (for example, from one opposition to the next).

Kepler's Improved Heliocentric Model and Elliptical Orbits: Copernicus thought that the orbits of the planets were combinations of circles. Using data collected by Tycho Brahe, Kepler deduced three laws of planetary motion: (1) the orbits are in fact ellipses; (2) a planet's speed varies as it moves around its elliptical orbit; and (3) the orbital period of a planet is related to the size of its orbit.

Evidence for the Heliocentric Model: The invention of the telescope led Galileo to new discoveries that supported a heliocentric model. These included his observations of the phases of Venus and of the motions of four moons around Jupiter.

Newton's Laws of Motion: Isaac Newton developed three principles, called the laws of motion, that apply to the motions of objects on Earth as well as in space. These are (1) the tendency of an object to maintain a constant velocity, (2) the relationship between the net outside force on an object and the object's acceleration, and (3) the principle of action and reaction. These laws and Newton's law of universal gravitation can be used to deduce Kepler's laws. They lead to extremely accurate descriptions of planetary motions.

• The mass of an object is a measure of the amount of matter in the object. Its weight is a measure of the force with which the gravity of some other object pulls on it.

• In general, the path of one object about another, such as that of a planet or comet about the Sun, is one of the curves called conic sections: circle, ellipse, parabola, or hyperbola.

Tidal Forces: Tidal forces are caused by differences in the gravitational pull that one object exerts on different parts of a second object.

• The tidal forces of the Moon and the Sun produce tides in Earth's oceans.

• The tidal forces of Earth have locked the Moon into synchronous rotation.

Questions

Review Questions

1. How did the ancient Greeks explain why the Sun and the Moon slowly change their positions relative to the background stars?
2. In what direction does a planet move relative to the stars when it is in direct motion? When it is in retrograde motion? How do these compare with the direction in which we see the Sun move relative to the stars?
3. (a) In what direction does a planet move relative to the horizon over the course of one night? (b) The answer to (a) is the same whether the planet is in direct motion or retrograde motion. What does this tell you about the speed at which planets move on the celestial sphere?
4. What is an epicycle? How is it important in Ptolemy's explanation of the retrograde motions of the planets?
5. What is the significance of Occam's razor as a tool for analyzing theories?
6. How did the models of Aristarchus and Copernicus explain the retrograde motion of the planets?
7. How did Copernicus determine that the orbits of Mercury and Venus must be smaller than Earth's orbit? How did he determine that the orbits of Mars, Jupiter, and Saturn must be larger than Earth's orbit?
8. At what configuration (for example, superior conjunction, greatest eastern elongation, and so on) would it be best to observe Mercury or Venus with an Earth-based telescope? At what configuration would it be best to observe Mars, Jupiter, or Saturn? Explain your answers.
9. Is it ever possible to see Mercury at midnight? Explain your answer.
10. Which planets can never be seen at opposition? Which planets can never be seen at inferior conjunction? Explain your answers.
11. What is the difference between the synodic period and the sidereal period of a planet?
12. What is parallax? What did Tycho Brahe conclude from his attempt to measure the parallax of a supernova and a comet?
13. What observations did Tycho Brahe make in an attempt to test the heliocentric model? What were his results? Explain why modern astronomers get different results.

TUTORIAL 4.1

14. What are the foci of an ellipse? If the Sun is at one focus of a planet's orbit, what is at the other focus?
15. What are Kepler's three laws? Why are they important?
16. At what point in a planet's elliptical orbit does it move fastest? At what point does it move slowest? At what point does it sweep out an area at the fastest rate?
17. A line joining the Sun and an asteroid is found to sweep out an area of 6.3 AU^2 during 2010. How much area is swept out during 2011? Over a period of five years?
18. The orbit of a spacecraft about the Sun has a perihelion distance of 0.1 AU and an aphelion distance of 0.4 AU. What is the semimajor axis of the spacecraft's orbit? What is its orbital period?
19. A comet with a period of 125 years moves in a highly elongated orbit about the Sun. At perihelion, the comet comes very close to the Sun's surface. What is the comet's average distance from the Sun? What is the farthest it can get from the Sun?
20. What observations did Galileo make that reinforced the heliocentric model? Why did these observations contradict the older model of Ptolemy? Why could these observations not have been made before Galileo's time?
21. Why does Venus have its largest angular diameter when it is new and its smallest angular diameter when it is full?

TUTORIAL 4.2

22. What are Newton's three laws? Give an everyday example of each law.
23. How much force do you have to exert on a 3-kg brick to give it an acceleration of 2 m/s^2? If you double this force, what is the brick's acceleration? Explain your answer.
24. What is the difference between weight and mass?
25. What is your weight in pounds and in newtons? What is your mass in kilograms?

TUTORIAL 4.3

26. Suppose that Earth were moved to a distance of 3.0 AU from the Sun. How much stronger or weaker would the Sun's gravitational pull be on Earth? Explain your answer.
27. How far would you have to go from Earth to be completely beyond the pull of its gravity? Explain your answer.
28. What are conic sections? In what way are they related to the orbits of planets in the solar system?
29. Why was the discovery of Neptune an important confirmation of Newton's law of universal gravitation?
30. What is a tidal force? How do tidal forces produce tides in Earth's oceans?
31. What is the difference between spring tides and neap tides?

Advanced Questions

Questions preceded by an asterisk (*) involve topics discussed in the Boxes.

> **Problem-solving tips and tools**
>
> Box 4-1 explains sidereal and synodic periods in detail. The semimajor axis of an ellipse is half the length of the long, or major, axis of the ellipse. For data about the planets and their satellites, see Appendices 1, 2, and 3 at the back of this book. If you want to calculate the gravitational force that you feel on the surface of a planet, the distance *r* to use is the planet's radius (the distance between you and the center of the planet). Boxes 4-2 and 4-4 show how to use Kepler's third law in its original form and in Newton's form.

32. Figure 4-2 shows the retrograde motion of Mars as seen from Earth. Sketch a similar figure that shows how Earth would appear to move against the background of stars during this same time period as seen by an observer on Mars.

*33. The synodic period of Mercury (an inferior planet) is 115.88 days. Calculate its sidereal period in days.

*34. Table 4-1 shows that the synodic period is *greater* than the sidereal period for Mercury, but the synodic period is *less* than the sidereal period for Jupiter. Draw diagrams like the one in Box 4-1 to explain why this is so.

*35. A general rule for superior planets is that the greater the average distance from the planet to the Sun, the more frequently that planet will be at opposition. Explain how this rule comes about.

36. In 2006, Mercury was at greatest western elongation on April 8, August 7, and November 25. It was at greatest eastern elongation on February 24, June 20, and October 17. Does Mercury take longer to go from eastern to western elongation, or vice versa? Explain why, using Figure 4-6.

37. Explain why the semimajor axis of a planet's orbit is equal to the average of the distance from the Sun to the planet at perihelion (the *perihelion distance*) and the distance from the Sun to the planet at aphelion (the *aphelion distance*).

38. A certain comet is 2 AU from the Sun at perihelion and 16 AU from the Sun at aphelion. (**a**) Find the semimajor axis of the comet's orbit. (**b**) Find the sidereal period of the orbit.

39. A comet orbits the Sun with a sidereal period of 64.0 years. (**a**) Find the semimajor axis of the orbit. (**b**) At aphelion, the comet is 31.5 AU from the Sun. How far is it from the Sun at perihelion?

40. One trajectory that can be used to send spacecraft from Earth to Mars is an elliptical orbit that has the Sun at one focus, its perihelion at Earth, and its aphelion at Mars. The spacecraft is launched from Earth and coasts along this ellipse until it reaches Mars, when a rocket is fired to either put the spacecraft into orbit around Mars or cause it to land on Mars. (**a**) Find the semimajor axis of the ellipse. (*Hint:* Draw a picture showing the Sun and the orbits of Earth, Mars, and the spacecraft. Treat the orbits of Earth and Mars as circles.) (**b**) Calculate how long (in days) such a one-way trip to Mars would take.

41. The mass of the Moon is 7.35×10^{22} kg, while that of Earth is 5.98×10^{24} kg. The average distance from the center of the Moon to the center of Earth is 384,400 km. What is the size of the gravitational force that Earth exerts on the Moon? What is the size of the gravitational force that the Moon exerts on Earth? How do your answers compare with the force between the Sun and Earth calculated in the text?

42. The mass of Saturn is approximately 100 times that of Earth, and the semimajor axis of Saturn's orbit is approximately 10 AU. To this approximation, how does the gravitational force that the Sun exerts on Saturn compare to the gravitational force that the Sun exerts on Earth? How do the accelerations of Saturn and Earth compare?

43. Suppose that you traveled to a planet with 4 times the mass and 4 times the diameter of Earth. Would you weigh more or less on that planet than on Earth? By what factor?

44. On Earth, a 50-kg astronaut weighs 490 newtons. What would she weigh if she landed on Jupiter's moon Callisto? What fraction is this of her weight on Earth? See Appendix 3 for relevant data about Callisto.

45. Imagine a planet like Earth orbiting a star with 4 times the mass of the Sun. If the semimajor axis of the planet's orbit is 1 AU, what would be the planet's sidereal period? (*Hint:* Use Newton's form of Kepler's third law. Compared with the case of Earth orbiting the Sun, by what factor has the quantity $m_1 + m_2$ changed? Has a changed? By what factor must P^2 change?)

46. A satellite is said to be in a "geosynchronous" orbit if it appears always to remain over the exact same spot on rotating Earth. (**a**) What is the period of this orbit? (**b**) At what distance from the center of Earth must such a satellite be placed into orbit? (*Hint:* Use Newton's form of Kepler's third law.) (**c**) Explain why the orbit must be in the plane of Earth's equator.

47. Figure 4-21 shows the lunar module *Eagle* in orbit around the Moon after completing the first successful lunar landing in July 1969. (The photograph was taken from the command module *Columbia,* in which the astronauts returned to Earth.) The spacecraft orbited 111 km above the surface of the Moon. Calculate the period of the spacecraft's orbit. See Appendix 3 for relevant data about the Moon.

*48. In Box 4-4 we analyze the orbit of Jupiter's moon Io. Look up information about the orbits of Jupiter's three other large moons (Europa, Ganymede, and Callisto) in Appendix 3. Demonstrate that these data are in agreement with Newton's form of Kepler's third law.

*49. Suppose a newly discovered asteroid is in a circular orbit with synodic period 1.25 years. The asteroid lies between the orbits of Mars and Jupiter. (**a**) Find the sidereal period of the orbit. (**b**) Find the distance from the asteroid to the Sun.

50. The average distance from the Moon to the center of Earth is 384,400 km, and the diameter of Earth is 12,756 km. Calculate the gravitational force that the Moon exerts (**a**) on a 1-kg rock at the point on Earth's surface closest to the Moon and (**b**) on a 1-kg rock at the point on Earth's surface farthest from the Moon. (**c**) Find the difference between the two forces you calculated in parts (a) and (b). This difference is the tidal force pulling these two rocks away from each other, like the 1-ball and 3-ball in Figure 4-24. Explain why tidal forces cause only a very small deformation of Earth.

Discussion Questions

51. Which planet would you expect to exhibit the greatest variation in apparent brightness as seen from Earth? Which planet would you expect to exhibit the greatest variation in angular diameter? Explain your answers.

52. Use two thumbtacks, a loop of string, and a pencil to draw several ellipses. Describe how the shape of an ellipse varies as the distance between the thumbtacks changes.

Web/eBook Questions

53. (**a**) Search the World Wide Web for information about Kepler. Before he realized that the planets move on elliptical paths, what other models of planetary motion did he consider? What was Kepler's idea of "the music of the spheres"? (**b**) Search the World Wide Web for information about Galileo. What were his contributions to physics? Which of Galileo's new ideas were later used by Newton to construct his laws of motion? (**c**) Search the World Wide Web for information about Newton. What were some of the contributions that he made to physics other than developing his laws of motion? What contributions did he make to mathematics?

54. **Monitoring the Retrograde Motion of Mars.** Watching Mars night after night reveals that it changes its position with respect to the background stars. To track its motion, access and view the animation "The Path of Mars in 2011–2012" in Chapter 4 of the *Universe* Web site or eBook. (**a**) Through which constellations does Mars move? (**b**) On approximately what date does Mars stop its direct (west-to-east) motion and begin its retrograde motion? (*Hint:* Use the "Stop" function on your animation controls.) (**c**) Over how many days does Mars move retrograde?

Activities

Observing Projects

55. It is quite probable that within a few weeks of your reading this chapter one of the planets will be near opposition or greatest eastern elongation, making it readily visible in the evening sky. Select a planet that is at or near such a configuration by searching the World Wide Web or by consulting a reference book, such as the current issue of the *Astronomical Almanac* or the pamphlet entitled *Astronomical Phenomena* (both published by the U.S. government). At that configuration, would you expect the planet to be moving rapidly or slowly from night to night against the background stars? Verify your expectations by observing the planet once a week for a month, recording your observations on a star chart.
56. If Jupiter happens to be visible in the evening sky, observe the planet with a small telescope on five consecutive clear nights. Record the positions of the four Galilean satellites by making nightly drawings, just as the Jesuit priests did in 1620 (see Figure 4-17). From your drawings, can you tell which moon orbits closest to Jupiter and which orbits farthest? Was there a night when you could see only three of the moons? What do you suppose happened to the fourth moon on that night?
57. If Venus happens to be visible in the evening sky, observe the planet with a small telescope once a week for a month. On each night, make a drawing of the crescent that you see. From your drawings, can you determine if the planet is nearer or farther from Earth than the Sun is? Do your drawings show any changes in the shape of the crescent from one week to the next? If so, can you deduce if Venus is coming toward us or moving away from us?
58. Use the *Starry Night Enthusiast*™ program to observe the moons of Jupiter. Display the entire celestial sphere (select **Guides > Atlas** in the **Favorites** menu) and center on the planet Jupiter (double-click the entry for Jupiter in the **Find** pane). Then use the zoom controls at the right-hand end of the toolbar (at the top of the main window) to adjust your view to about **20′ × 14′** so that you can see the planet and its four Galilean satellites. (Click on the + button or press the + key on the keyboard to zoom in and click on the − button or press the − key on the keyboard to zoom out.) This is equivalent to increasing the magnification of a telescope while looking at Jupiter. (**a**) Click on the **Time Flow Rate** control (immediately to the right of the date and time display) and set the discrete time step to **2 hours.** Using the **Step Forward** button (just to the right of the **Forward** button), observe and draw the positions of the moons relative to Jupiter at 2-hour intervals. (**b**) From your drawings, can you tell which moon orbits closest to Jupiter and which orbits farthest away? Explain your reasoning. (**c**) Are there times when only three of the satellites are visible? What happens to the fourth moon at those times?
59. Use the *Starry Night Enthusiast*™ program to observe the changing appearance of Mercury. Display the entire celestial sphere (select **Guides > Atlas** in the **Favorites** menu) and center on Mercury (double-click the entry for Mercury in the **Find** pane); then use the zoom controls at the right-hand end of the toolbar (at the top of the main window) to adjust your view so that you can clearly see details on the planet's surface. (Click on the + button to zoom in and on the − button to zoom out.) (**a**) Click on the **Time Flow Rate** control (immediately to the right of the date and time display) and set the discrete time step to **1 day.** Using the **Step Forward** button, observe and record the changes in Mercury's phase and apparent size from one day to the next. Run time forward for some time to see these changes more graphically. (**b**) Explain why the phase and apparent size change in the way that you observe.
60. Use the *Starry Night Enthusiast*™ program to observe retrograde motion. Display the entire celestial sphere (select **Guides > Atlas** in the **Favorites** menu) and center on Mars (double-click the entry for Mars in the **Find** pane). Set the **Date** (in the time and date display in the toolbar) to **January 1, 2010.** Then click on the **Time Flow Rate** control (immediately to the right of the date and time display) and set the discrete time step to **1 day.** Press and hold down the − key on the keyboard to zoom out to the maximum field of view. Then click on the **Forward** button (a triangle that points to the right) and watch the motion of Mars for two years of simulated time. Since you have centered Mars in the view, the sky appears to move but the relative motion of Mars against this sky is obvious. (**a**) For most of the two-year period, does Mars move generally to the left (eastward) or to the right (westward) on the celestial sphere? On what date during this period does this *direct* motion end, so that Mars appears to come to a momentary halt, and retrograde motion begins? On what date does *retrograde* motion end and direct motion resume? You may want to use the **Stop** button and the **Back** button (a triangle that points to the left) to help you to pin down the exact dates of these events. (**b**) You have been observing the motion of Mars as seen from Earth. To observe the motion of Earth as seen from Mars, locate yourself on the north pole of Mars. To do this, select **Viewing Location . . .** in the **Options** menu. In the Viewing Location dialog box that appears on the screen, change the pull-down menus next to the words "View from:" so that they read "the surface of" and "Mars." Click on **Latitude/Longitude** tab in the **Viewing Location** dialog box and change the Latitude to **90° N.** Then click on the **Set Location** button. You will now see a Martian horizon and a pink Martian sky, both of which will interfere with your view of Earth as seen from Mars. Remove the horizon by clicking the **Options** tab on the left of the main view to open the **Options** pane. Under the **Local View**

heading in the **Options** pane, click the checkbox to the left of the item labeled **Local Horizon.** Then, in the View menu, select **Hide Daylight** to remove the Martian sky. Now center the field of view on **Earth** (double-click the entry for Earth in the **Find** pane). Set the date to **January 1, 2010,** set the discrete time step to **1 day,** and click on the **Forward** button. As before, watch the motion for two years of simulated time. In which direction does Earth appear to move most of the time? On what date does its motion change from direct to retrograde? On what date does its motion change from retrograde back to direct? Are these roughly the same dates as you found in part (a)? (**c**) To understand the motions of Mars as seen from Earth and vice versa, observe the motion of the planets from a point above the solar system. Again, select **Viewing Location . . .** in the **Options** menu. In the Viewing Location dialog box that appears on the screen, select "stationary location" from the pull-down menu immediately to the right of the words "View from." Then type **0 au** in the **X:** box, **0 au** in the **Y:** box, and **5 au** in the **Z:** box, and click on the **Set Location** button. This will place you at a position 5 AU above the plane of the solar system. Then center the field of view on the Sun (double-click the entry for the Sun in the **Find** pane). Once again, set the date to **January 1, 2010,** set the discrete time step to **1 day,** and click on the **Forward** button. Watch the motions of the planets for two years of simulated time. On what date during this two-year period is Earth directly between Mars and the Sun? How does this date compare to the two dates you recorded in part (a) and the two dates you recorded in part (b)? Explain the significance of this. (*Hint:* see See Figure 4-5.)

5 The Nature of Light

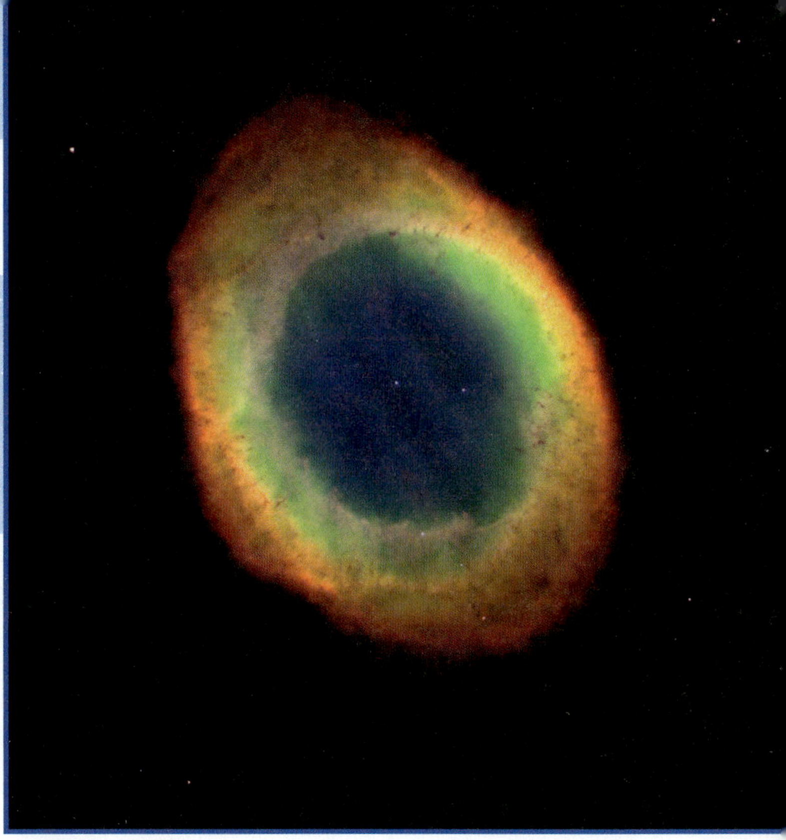

WEB LINK 5.1 RIVUXG

The Ring Nebula is a shell of glowing gases surrounding a dying star. The spectrum of the emitted light reveals which gases are present. (Hubble Heritage Team, AURA/STScI/NASA)

In the early 1800s, the French philosopher Auguste Comte argued that because the stars are so far away, humanity would never know their nature and composition. But the means to learn about the stars was already there for anyone to see—starlight. Just a few years after Comte's bold pronouncement, scientists began analyzing starlight to learn the very things that he had deemed unknowable.

We now know that atoms of each chemical element emit and absorb light at a unique set of wavelengths characteristic of that element alone. The red light in the accompanying image of a gas cloud in space is of a wavelength emitted by nitrogen and no other element; the particular green light in this image is unique to oxygen, and the particular blue light is unique to helium. The light from nearby planets, distant stars, and remote galaxies also has characteristic "fingerprints" that reveal the chemical composition of these celestial objects.

In this chapter we learn about the basic properties of light. Light has a dual nature: It has the properties of both waves and particles. The light emitted by an object depends upon the object's temperature; we can use this to determine the surface temperatures of stars. By studying the structure of atoms, we will learn why each element emits and absorbs light only at specific wavelengths and will see how astronomers determine what the atmospheres of planets and stars are made of. The motion of a light source also affects wavelengths, permitting us to deduce how fast stars and other objects are approaching or receding. These are but a few of the reasons why understanding light is a prerequisite to understanding the universe.

5-1 Light travels through empty space at a speed of 300,000 km/s

Galileo Galilei and Isaac Newton were among the first to ask basic questions about light. Does light travel instantaneously from one place to another, or does it move with a measurable speed? Whatever the nature of light, it does seem to travel swiftly from a source to our eyes. We see a distant event before we hear the accompanying sound. (For example, we see a flash of lightning before we hear the thunderclap.)

The speed of light in a vacuum is a universal constant: It has the same value everywhere in the cosmos

In the early 1600s, Galileo tried to measure the speed of light. He and an assistant stood at night on two hilltops a known distance apart, each holding a shuttered lantern. First, Galileo opened the shutter of

Learning Goals

By reading the sections of this chapter, you will learn

5-1 How we measure the speed of light

5-2 How we know that light is an electromagnetic wave

5-3 How an object's temperature is related to the radiation it emits

5-4 The relationship between an object's temperature and the amount of energy it emits

5-5 The evidence that light has both particle and wave aspects

5-6 How astronomers can detect an object's chemical composition by studying the light it emits

5-7 The quantum rules that govern the structure of an atom

5-8 The relationship between atomic structure and the light emitted by objects

5-9 How an object's motion affects the light we receive from that object his lantern; as soon as his assistant saw the flash of light, he opened his own. Galileo used his pulse as a timer to try to measure the time between opening his lantern and seeing the light from his assistant's lantern. From the distance and time, he hoped to compute the speed at which the light had traveled to the distant hilltop and back.

Galileo found that the measured time failed to increase noticeably, no matter how distant the assistant was stationed. Galileo therefore concluded that the speed of light is too high to be measured by slow human reactions. Thus, he was unable to tell whether or not light travels instantaneously.

The Speed of Light: Astronomical Measurements

The first evidence that light does *not* travel instantaneously was presented in 1676 by Olaus Rømer, a Danish astronomer. Rømer had been studying the orbits of the moons of Jupiter by carefully timing the moments when they passed into or out of Jupiter's shadow. To Rømer's surprise, the timing of these eclipses of Jupiter's moons seemed to depend on the relative positions of Jupiter and Earth. When Earth was far from Jupiter (that is, near conjunction; see Figure 4-6), the eclipses occurred several minutes later than when Earth was close to Jupiter (near opposition).

Rømer realized that this puzzling effect could be explained if light needs time to travel from Jupiter to Earth. When Earth is closest to Jupiter, the image of one of Jupiter's moons disappearing into Jupiter's shadow arrives at our telescopes a little sooner than it does when Jupiter and Earth are farther apart (Figure 5-1). The range of variation in the times at which such eclipses are observed is about 16.6 minutes, which Rømer interpreted as the length of time required for light to travel across the diameter of Earth's orbit (a distance of 2 AU). The size of Earth's orbit was not accurately known in Rømer's day, and he never actually calculated the speed of light. Today, using the modern value of 150 million kilometers for the astronomical unit, Rømer's method yields a value for the speed of light equal to roughly 300,000 km/s (186,000 mi/s).

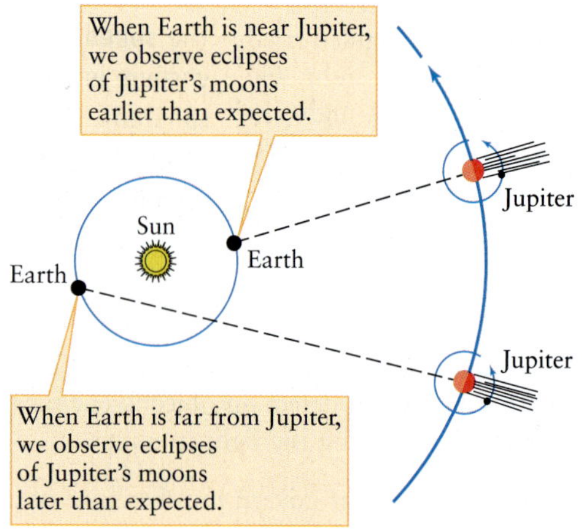
When Earth is near Jupiter, we observe eclipses of Jupiter's moons earlier than expected.
Sun
Earth
Earth
Jupiter
Jupiter
When Earth is far from Jupiter, we observe eclipses of Jupiter's moons later than expected.

Figure 5-1

Rømer's Evidence That Light Does Not Travel Instantaneously The timing of eclipses of Jupiter's moons as seen from Earth depends on the Earth-Jupiter distance. Rømer correctly attributed this effect to variations in the time required for light to travel from Jupiter to Earth.

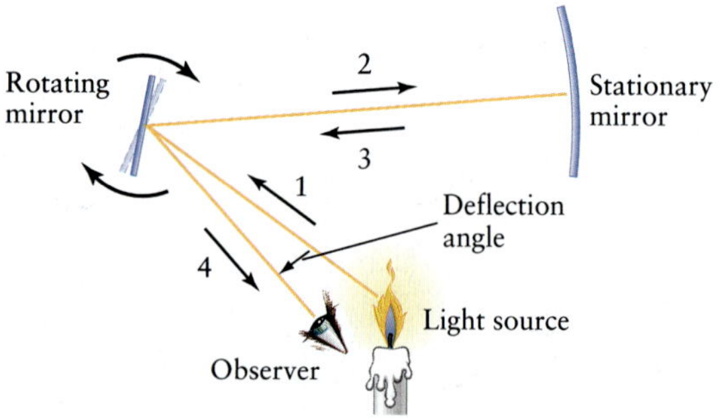
Rotating mirror
2
3
Stationary mirror
1
Deflection angle
4
Light source
Observer

Figure 5-2

The Fizeau-Foucault Method of Measuring the Speed of Light Light from a light source (1) is reflected off a rotating mirror to a stationary mirror (2) and from there back to the rotating mirror (3). The ray that reaches the observer (4) is deflected away from the path of the initial beam because the rotating mirror has moved slightly while the light was making the round trip. The speed of light is calculated from the deflection angle and the dimensions of the apparatus.

The Speed of Light: Measurements on Earth

Almost two centuries after Rømer, the speed of light was measured very precisely in an experiment carried out on Earth. In 1850, the French physicists Armand-Hippolyte Fizeau and Jean Foucault built the apparatus sketched in Figure 5-2. Light from a light source reflects from a rotating mirror toward a stationary mirror 20 meters away. The rotating mirror moves slightly while the light is making the round trip, so the returning light ray is deflected away from the source by a small angle. By measuring this angle and knowing the dimensions of their apparatus, Fizeau and Foucault could deduce the speed of light. Once again, the answer was very nearly 300,000 km/s.

The speed of light in a vacuum is usually designated by the letter c (from the Latin *celeritas,* meaning "speed"). The modern value is $c = 299{,}792.458$ km/s (186,282.397 mi/s). In most calculations you can use

$$c = 3.00 \times 10^5 \text{ km/s} = 3.00 \times 10^8 \text{ m/s}$$

The most convenient set of units to use for c is different in different situations. The value in kilometers per second (km/s) is often most useful when comparing c to the speeds of objects in space, while the value in meters per second (m/s) is preferred when doing calculations involving the wave nature of light (which we will discuss in Section 5-2).

CAUTION! Note that the quantity c is the speed of light *in a vacuum.* Light travels more slowly through air, water, glass, or any other transparent substance than it does in a vacuum. In our study of astronomy, however, we will almost always consider light traveling through the vacuum (or near-vacuum) of space.

The speed of light in empty space is one of the most important numbers in modern physical science. This value appears in many equations that describe atoms, gravity, electricity, and mag-

Figure 5-3

A Prism and a Spectrum When a beam of sunlight passes through a glass prism, the light is broken into a rainbow-colored band called a spectrum. The wavelengths of different colors of light are shown on the right (1 nm = 1 nanometer = 10^{-9} m).

netism. According to Einstein's special theory of relativity, nothing can travel faster than the speed of light.

5-2 Light is electromagnetic radiation and is characterized by its wavelength

TUTORIAL 5.1

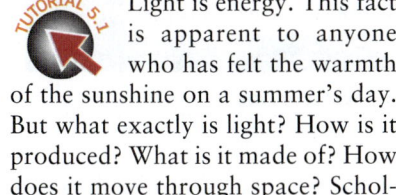

Light is energy. This fact is apparent to anyone who has felt the warmth of the sunshine on a summer's day. But what exactly is light? How is it produced? What is it made of? How does it move through space? Scholars have struggled with these questions throughout history.

Visible light, radio waves, and X rays are all the same type of wave: They differ only in their wavelength

Newton and the Nature of Color

The first major breakthrough in understanding light came from a simple experiment performed by Isaac Newton around 1670. Newton was familiar with what he called the "celebrated Phenomenon of Colours," in which a beam of sunlight passing through a glass prism spreads out into the colors of the rainbow (Figure 5-3). This rainbow is called a **spectrum** (plural **spectra**).

WEB LINK 5.2

Until Newton's time, it was thought that a prism somehow added colors to white light. To test this idea, Newton placed a second prism so that just one color of the spectrum passed through it (Figure 5-4). According to the old theory, this should have caused a further change in the color of the light. But Newton found that each color of the spectrum was unchanged by the second prism; red remained red, blue remained

Figure 5-4

Newton's Experiment on the Nature of Light In a crucial experiment, Newton took sunlight that had passed through a prism and sent it through a second prism. Between the two prisms was a screen with a hole in it that allowed only one color of the spectrum to pass through. This same color emerged from the second prism. Newton's experiment proved that prisms do not add color to light but merely bend different colors through different angles. It also showed that white light, such as sunlight, is actually a combination of all the colors that appear in its spectrum.

blue, and so on. He concluded that a prism merely separates colors and does not add color. Hence, the spectrum produced by the first prism shows that sunlight is a mixture of all the colors of the rainbow.

Newton suggested that light is composed of particles too small to detect individually. In 1678, however, the Dutch physicist and astronomer Christiaan Huygens proposed a rival explanation. He suggested that light travels in the form of waves rather than particles.

Young and the Wave Nature of Light

Around 1801, Thomas Young in England carried out an experiment that convincingly demonstrated the wavelike aspect of light. He passed a beam of light through two thin, parallel slits in an opaque screen, as shown in Figure 5-5*a*. On a white surface some distance beyond the slits, the light formed a pattern of alternating bright and dark bands. Young reasoned that if a beam of light was a stream of particles (as Newton had suggested), the two beams of light from the slits should simply form bright images of the slits on the white surface. The pattern of bright and dark bands he observed is just what would be expected, however, if light had wavelike properties. An analogy with water waves demonstrates why.

ANALOGY Imagine ocean waves pounding against a reef or breakwater that has two openings (Figure 5-5*b*). A pattern of ripples is formed on the other side of the barrier as the waves come through the two openings and interfere with each other. At certain points, wave crests arrive simultaneously from the two openings. These reinforce each other and produce high waves. At other points, a crest from one opening meets a trough from the other opening. These cancel each other out, leaving areas of still water. This process of combining two waves also takes place in Young's double-slit experiment: The bright bands are regions where waves from the two slits reinforce each other, while the dark bands appear where waves from the two slits cancel each other.

Maxwell and Light as an Electromagnetic Wave

The discovery of the wave nature of light posed some obvious questions. What exactly is "waving" in light? That is, what is it about light that goes up and down like water waves on the ocean? Because we can see light from the Sun, planets, and stars, light waves must be able to travel across empty space. Hence, whatever is "waving" cannot be any material substance. What, then, is it?

The answer came from a seemingly unlikely source—a comprehensive theory that described electricity and magnetism. Numerous experiments during the first half of the nineteenth century demonstrated an intimate connection between electric and magnetic forces. A central idea to emerge from these experiments is the concept of a *field,* an immaterial yet measurable disturbance of any region of space in which electric or magnetic forces are felt. Thus, an electric charge is surrounded by an electric field, and a magnet is surrounded by a magnetic field. Experiments in the early 1800s demonstrated that moving an electric charge produces a magnetic field; conversely, moving a magnet gives rise to an electric field.

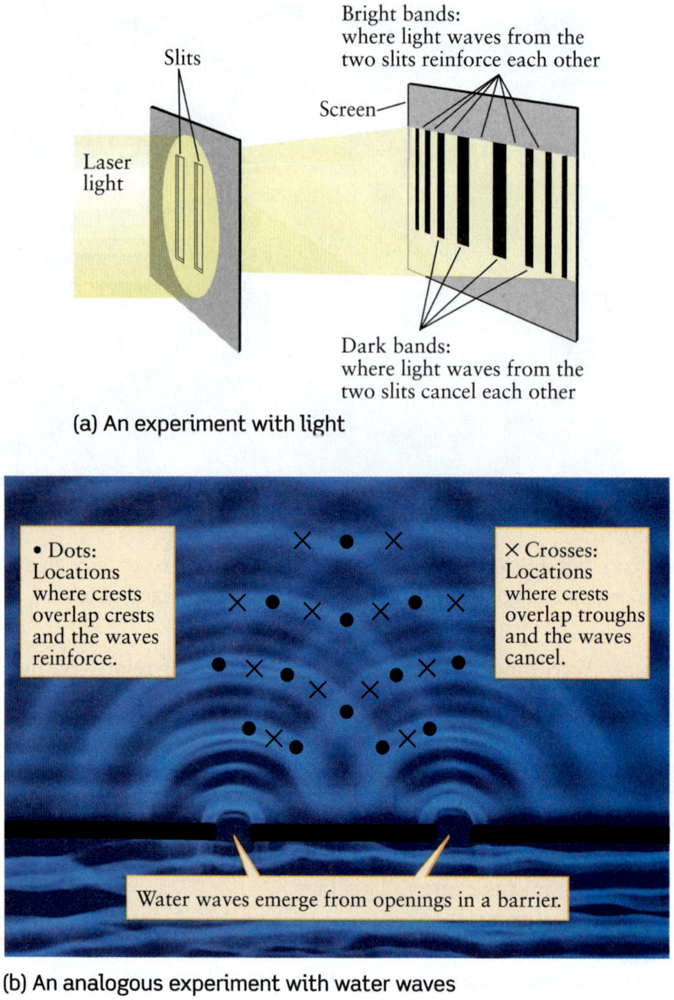

Figure 5-5

Young's Double-Slit Experiment (a) Thomas Young's classic double-slit experiment can easily be repeated in the modern laboratory by shining light from a laser onto two closely spaced parallel slits. Alternating dark and bright bands appear on a screen beyond the slits. **(b)** The intensity of light on the screen in (a) is analogous to the height of water waves that pass through a barrier with two openings. (The photograph shows this experiment with water waves in a small tank.) In certain locations, wave crests from both openings reinforce each other to produce extra high waves. At other locations a crest from one opening meets a trough from the other. The crest and trough cancel each other, producing still water. (Eric Schrempp/Photo Researchers)

In the 1860s, the Scottish mathematician and physicist James Clerk Maxwell succeeded in describing all the basic properties of electricity and magnetism in four equations. This mathematical achievement demonstrated that electric and magnetic forces are really two aspects of the same phenomenon, which we now call **electromagnetism.**

By combining his four equations, Maxwell showed that electric and magnetic fields should travel through space in the form of waves at a speed of 3.0×10^5 km/s—a value equal to the best available value for the speed of light. Maxwell's suggestion that

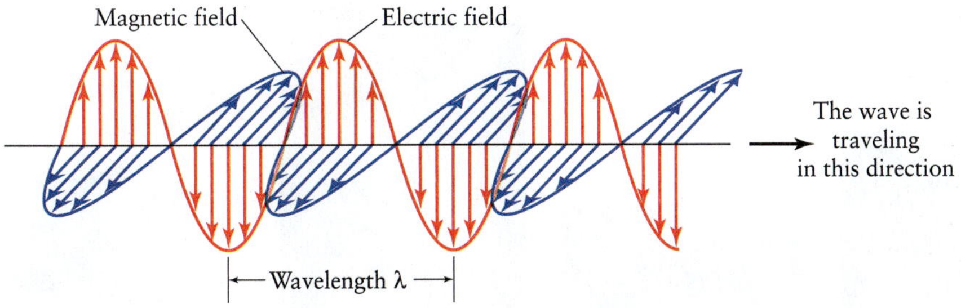

Figure 5-6

Electromagnetic Radiation All forms of light consist of oscillating electric and magnetic fields that move through space at a speed of 3.00×10^5 km/s = 3.00×10^8 m/s. This figure shows a "snapshot" of these fields at one instant. The distance between two successive crests, called the wavelength of the light, is usually designated by the Greek letter λ (lambda).

these waves do exist and are observed as light was soon confirmed by experiments. Because of its electric and magnetic properties, light is also called **electromagnetic radiation.**

CAUTION! You may associate the term *radiation* with radioactive materials like uranium, but this term refers to anything that radiates, or spreads away, from its source. For example, scientists sometimes refer to sound waves as "acoustic radiation." Radiation does not have to be related to radioactivity!

Electromagnetic radiation consists of oscillating electric and magnetic fields, as shown in Figure 5-6. The distance between two successive wave crests is called the **wavelength** of the light, usually designated by the Greek letter λ (lambda). No matter what the wavelength, electromagnetic radiation always travels at the same speed $c = 3.0 \times 10^5$ km/s $= 3.0 \times 10^8$ m/s in a vacuum.

More than a century elapsed between Newton's experiments with a prism and the confirmation of the wave nature of light. One reason for this delay is that **visible light,** the light to which the human eye is sensitive, has extremely short wavelengths—less than a thousandth of a millimeter—that are not easily detectable. To express such tiny distances conveniently, scientists use a unit of length called the **nanometer** (abbreviated nm), where 1 nm = 10^{-9} m. Experiments demonstrated that visible light has wavelengths covering the range from about 400 nm for violet light to about 700 nm for red light. Intermediate colors of the rainbow like yellow (550 nm) have intermediate wavelengths, as shown in Figure 5-7. (Some astronomers prefer to measure wavelengths in *angstroms.* One angstrom, abbreviated Å, is one-tenth of a nanometer: 1 Å = 0.1 nm = 10^{-10} m. In these units, the wavelengths of visible light extend from about 4000 Å to about 7000 Å. We will not use the angstrom unit in this book, however.)

Visible and Nonvisible Light

Maxwell's equations place no restrictions on the wavelength of electromagnetic radiation. Hence, electromagnetic waves could and should exist with wavelengths both longer and shorter than the 400–700 nm range of visible light. Consequently, researchers began to look for *invisible* forms of light. These are forms of electromagnetic radiation to which the cells of the human retina do not respond.

The first kind of invisible radiation to be discovered actually preceded Maxwell's work by more than a half century. Around 1800 the British astronomer William Herschel passed sunlight through a prism and held a thermometer just beyond the red end of the visible spectrum. The thermometer registered a temperature increase, indicating that it was being exposed to an invisible form of energy. This invisible energy, now called **infrared radiation,** was later realized to be electromagnetic radiation with wavelengths somewhat longer than those of visible light.

In experiments with electric sparks in 1888, the German physicist Heinrich Hertz succeeded in producing electromagnetic radiation with even longer wavelengths of a few centimeters or more. These are now known as **radio waves.** In 1895 another German physicist, Wilhelm Röntgen, invented a machine that produces electromagnetic radiation with wavelengths shorter than

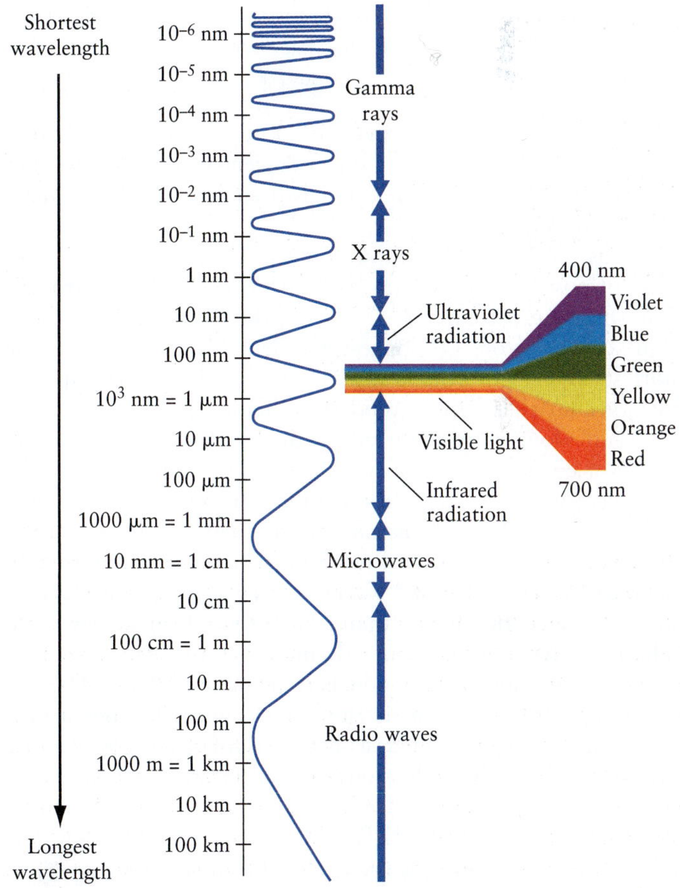

WEB LINK 5.3

Figure 5-7

The Electromagnetic Spectrum The full array of all types of electromagnetic radiation is called the electromagnetic spectrum. It extends from the longest-wavelength radio waves to the shortest-wavelength gamma rays. Visible light occupies only a tiny portion of the full electromagnetic spectrum.

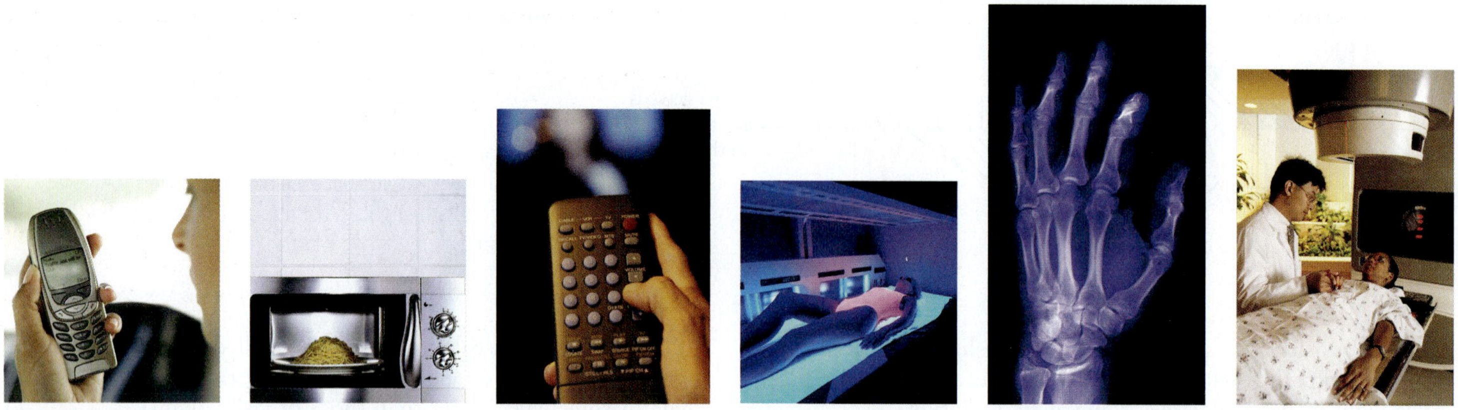

(a) Mobile phone: radio waves

(b) Microwave oven: microwaves

(c) TV remote: infrared light

(d) Tanning booth: ultraviolet light

(e) Medical imaging: X rays.

(f) Cancer radiotherapy: gamma rays

Figure 5-8

Uses of Nonvisible Electromagnetic Radiation (a) A mobile phone is actually a radio transmitter and receiver. The wavelengths used are in the range 16 to 36 cm. **(b)** A microwave oven produces radiation with a wavelength near 10 cm. The water in food absorbs this radiation, thus heating the food. **(c)** A remote control sends commands to a television using a beam of infrared light. **(d)** Ultraviolet radiation in moderation gives you a suntan, but in excess can cause sunburn or skin cancer. **(e)** X rays can penetrate through soft tissue but not through bone, which makes them useful for medical imaging. **(f)** Gamma rays destroy cancer cells by breaking their DNA molecules, making them unable to multiply. (Ian Britton, Royalty-Free/Corbis, Michael Porsche/Corbis, Bill Lush/Taxi/Getty, Neil McAllister/Alamy, Edward Kinsman/Photo Researchers, Inc., Will and Deni McIntyre/Science Photo Library)

10 nm, now known as **X rays.** The X-ray machines in modern medical and dental offices are direct descendants of Röntgen's invention. Over the years radiation has been discovered with many other wavelengths.

Thus, visible light occupies only a tiny fraction of the full range of possible wavelengths, collectively called the **electromagnetic spectrum.** As Figure 5-7 shows, the electromagnetic spectrum stretches from the longest-wavelength radio waves to the shortest-wavelength gamma rays. Figure 5-8 shows some applications of nonvisible light in modern technology.

On the long-wavelength side of the visible spectrum, infrared radiation covers the range from about 700 nm to 1 mm. Astronomers interested in infrared radiation often express wavelength in *micrometers* or *microns,* abbreviated μm, where 1 μm = 10^{-3} mm = 10^{-6} m. **Microwaves** have wavelengths from roughly 1 mm to 10 cm, while radio waves have even longer wavelengths.

At wavelengths shorter than those of visible light, **ultraviolet radiation** extends from about 400 nm down to 10 nm. Next are X rays, which have wavelengths between about 10 and 0.01 nm, and beyond them at even shorter wavelengths are **gamma rays.** Note that the rough boundaries between different types of radiation are simply arbitrary divisions in the electromagnetic spectrum.

Frequency and Wavelength

Astronomers who work with radio telescopes often prefer to speak of *frequency* rather than wavelength. The **frequency** of a wave is the number of wave crests that pass a given point in one second. Equivalently, it is the number of complete *cycles* of the wave that pass per second (a complete cycle is from one crest to the next). Frequency is usually denoted by the Greek letter ν (nu). The unit of frequency is the cycle per second, also called the *hertz* (abbreviated Hz) in honor of Heinrich Hertz, the physicist who first produced radio waves. For example, if 500 crests of a wave pass you in one second, the frequency of the wave is 500 cycles per second or 500 Hz.

In working with frequencies, it is often convenient to use the prefix *mega-* (meaning "million," or 10^6, and abbreviated M) or *kilo-* (meaning "thousand," or 10^3, and abbreviated k). For example, AM radio stations broadcast at frequencies between 535 and 1605 kHz (kilohertz), while FM radio stations broadcast at frequencies in the range from 88 to 108 MHz (megahertz).

The relationship between the frequency and wavelength of an electromagnetic wave is a simple one. Because light moves at a constant speed $c = 3 \times 10^8$ m/s, if the wavelength (distance from one crest to the next) is made shorter, the frequency must increase (more of those closely spaced crests pass you each second). The frequency ν of light is related to its wavelength λ by the equation

Frequency and wavelength of an electromagnetic wave

$$\nu = \frac{c}{\lambda}$$

ν = frequency of an electromagnetic wave (in Hz)

c = speed of light = 3×10^8 m/s

λ = wavelength of the wave (in meters)

That is, the frequency of a wave equals the wave speed divided by the wavelength.

For example, hydrogen atoms in space emit radio waves with a wavelength of 21.12 cm. To calculate the frequency of this radiation, we must first express the wavelength in meters rather than centimeters: $\lambda = 0.2112$ m. Then we can use the above formula to find the frequency ν:

$$\nu = \frac{c}{\lambda} = \frac{3 \times 10^8 \text{ m/s}}{0.2112 \text{ m}} = 1.42 \times 10^9 \text{ Hz} = 1420 \text{ MHz}$$

Visible light has a much shorter wavelength and higher frequency than radio waves. You can use the above formula to show that for yellow-orange light of wavelength 600 nm, the frequency is 5×10^{14} Hz or 500 *million* megahertz!

While Young's experiment (Figure 5-5) showed convincingly that light has wavelike aspects, it was discovered in the early 1900s that light *also* has some of the characteristics of a stream of particles and waves. We will explore light's dual nature in Section 5-5.

5-3 An opaque object emits electromagnetic radiation according to its temperature

To learn about objects in the heavens, astronomers study the character of the electromagnetic radiation coming from those objects. Such studies can be very revealing because different kinds of electromagnetic radiation are typically produced in different ways. As an example, on Earth the most common way to generate radio waves is to make an electric current oscillate back and forth (as is done in the broadcast antenna of a radio station). By contrast, X rays for medical and dental purposes are usually produced by bombarding atoms in a piece of metal with fast-moving particles extracted from within other atoms. Our own Sun emits radio waves from near its glowing surface and X rays from its corona (see the photo that opens Chapter 3). Hence, these observations indicate the presence of electric currents near the Sun's surface and of fast-moving particles in the Sun's outermost regions. (We will discuss the Sun at length in Chapter 18.)

As an object is heated, it glows more brightly and its peak color shifts to shorter wavelengths

Radiation from Heated Objects

TUTORIAL 5.2

The simplest and most common way to produce electromagnetic radiation, either on or off Earth, is to heat an object. The hot filament of wire inside an ordinary lightbulb emits white light, and a neon sign has a characteristic red glow because neon gas within the tube is heated by an electric current. In like fashion, almost all the visible light that we receive from space comes from hot objects like the Sun and the stars. The kind and amount of light emitted by a hot object tell us not only how hot it is but also about other properties of the object.

We can tell whether the hot object is made of relatively dense or relatively thin material. Consider the difference between a lightbulb and a neon sign. The dense, solid filament of a lightbulb makes white light, which is a mixture of all different visible wavelengths, while the thin, transparent neon gas produces light of a rather definite red color and, hence, a rather definite wavelength. For now we will concentrate our attention on the light produced by dense, opaque objects. (We will return to the light produced by gases in Section 5-6.) Even though the Sun and stars are gaseous, not solid, it turns out that they emit light with many of the same properties as light emitted by a hot, glowing, solid object.

Imagine a welder or blacksmith heating a bar of iron. As the bar becomes hot, it begins to glow deep red, as shown in Figure 5-9*a*. (You can see this same glow from the coils of a toaster or from an electric range turned on "high.") As the temperature rises further, the bar begins to give off a brighter orange light (Figure 5-9*b*). At still higher temperatures, it shines with a brilliant yellow light (Figure 5-9*c*). If the bar could be prevented from melt-

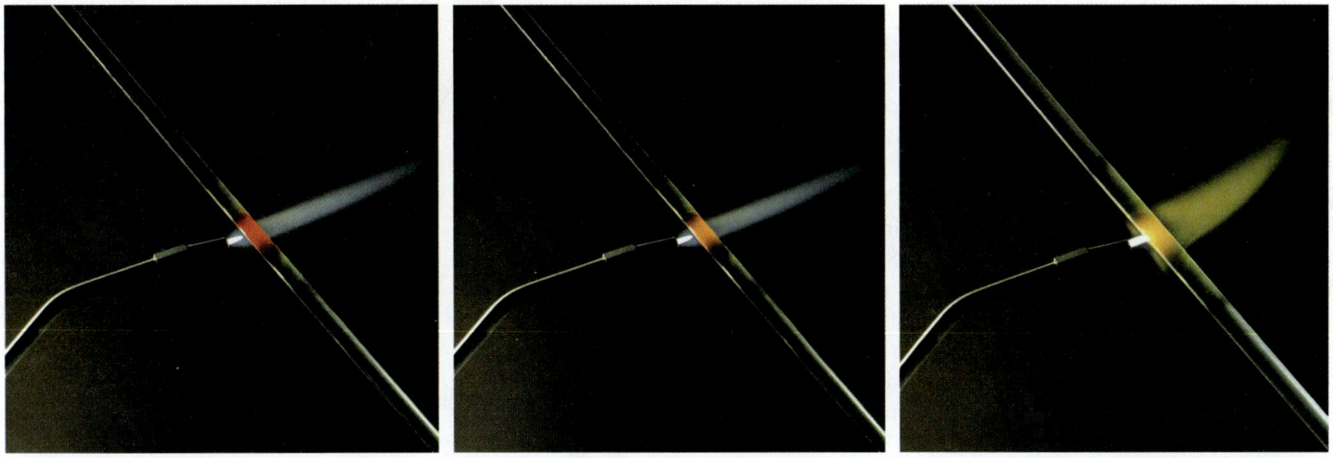

(a) Hot: glows deep red (b) Hotter: glows orange (c) Even hotter: glows yellow

Figure 5-9 R I V U X G

Heating a Bar of Iron This sequence of photographs shows how the appearance of a heated bar of iron changes with temperature. As the temperature increases, the bar glows more brightly because it radiates more energy. The color of the bar also changes because as the temperature goes up, the dominant wavelength of light emitted by the bar decreases. (©1984 Richard Megna/Fundamental Photographs)

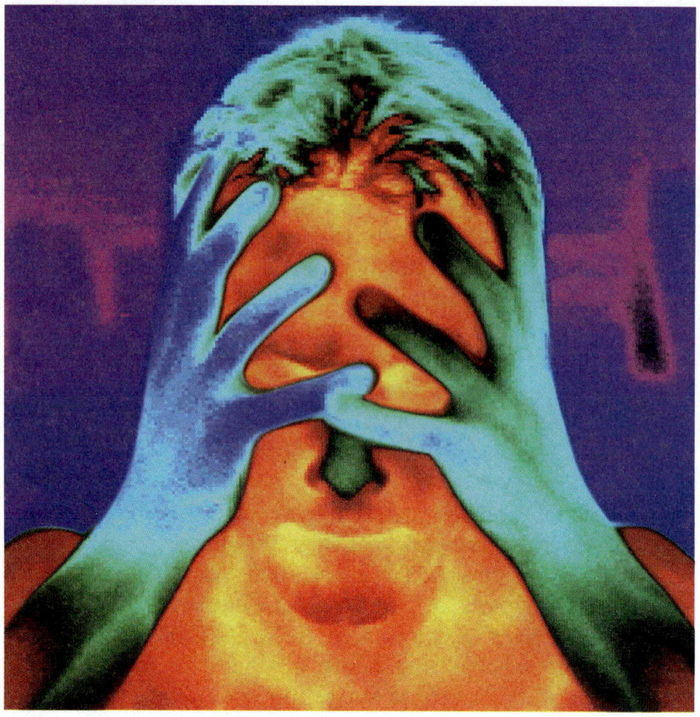

Figure 5-10 R I V U X G

An Infrared Portrait In this image made with a camera sensitive to infrared radiation, the different colors represent regions of different temperature. That this image is made from infrared radiation is indicated by the highlighted I in the wavelength tab. Red areas (like the man's face) are the warmest and emit the most infrared light, while blue-green areas (including the man's hands and hair) are at the lowest temperatures and emit the least radiation. (Dr. Arthur Tucker/Photo Researchers)

ing and vaporizing, at extremely high temperatures it would emit a dazzling blue-white light.

As this example shows, the amount of energy emitted by the hot, dense object and the dominant wavelength of the emitted radiation both depend on the temperature of the object. The hotter the object, the more energy it emits and the shorter the wavelength at which most of the energy is emitted. Colder objects emit relatively little energy, and this emission is primarily at long wavelengths.

These observations explain why you can't see in the dark. The temperatures of people, animals, and furniture are rather less than even that of the iron bar in Figure 5-9*a*. So, while these objects emit radiation even in a darkened room, most of this emission is at wavelengths greater than those of red light, in the infrared part of the spectrum (see Figure 5-7). Your eye is not sensitive to infrared, and you thus cannot see ordinary objects in a darkened room. But you can detect this radiation by using a camera that is sensitive to infrared light (Figure 5-10).

To better understand the relationship between the temperature of a dense object and the radiation it emits, it is helpful to know just what "temperature" means. The temperature of a substance is directly related to the average speed of the tiny **atoms**—the building blocks that come in distinct forms for each distinct chemical element—that make up the substance. (Typical atoms are about 10^{-10} m = 0.1 nm in diameter, or about 1/5000 as large as a typical wavelength of visible light.)

If something is hot, its atoms are moving at high speeds; if it is cold, its atoms are moving slowly. Scientists usually prefer to use the Kelvin temperature scale, on which temperature is measured in **kelvins** (K) upward from **absolute zero.** This is the coldest possible temperature, at which atoms move as slowly as possible (they can never quite stop completely). On the more familiar Celsius and Fahrenheit temperature scales, absolute zero (0 K) is −273°C and −460°F. Ordinary room temperature is 293 K, 20°C, or 68°F. Box 5-1 discusses the relationships among the Kelvin, Celsius, and Fahrenheit temperature scales.

Figure 5-11 depicts quantitatively how the radiation from a dense object depends on its Kelvin temperature. Each curve in this figure shows the intensity of light emitted at each wavelength by a dense object at a given temperature: 3000 K (the temperature at which molten gold boils), 6000K (the temperature of an iron-welding arc), and 12,000 K (a temperature found in special industrial furnaces). In other words, the curves show the spectrum of light emitted by such an object. At any temperature, a hot,

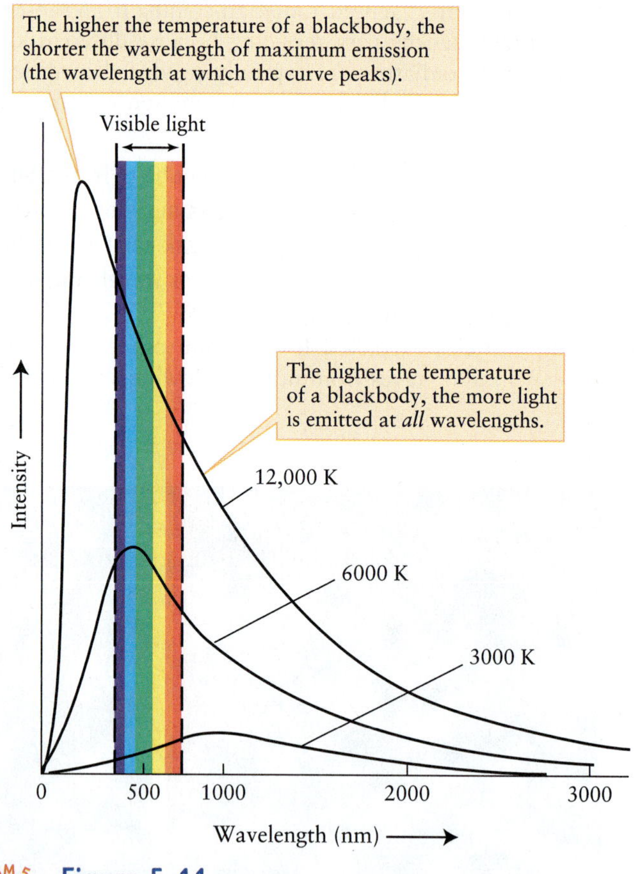
The higher the temperature of a blackbody, the shorter the wavelength of maximum emission (the wavelength at which the curve peaks).
Visible light
The higher the temperature of a blackbody, the more light is emitted at *all* wavelengths.
Intensity
12,000 K
6000 K
3000 K
0
500
1000
2000
3000
Wavelength (nm)

AIMM 5.1

Figure 5-11

Blackbody Curves Each of these curves shows the intensity of light at every wavelength that is emitted by a blackbody (an idealized case of a dense object) at a particular temperature. The rainbow-colored band shows the range of visible wavelengths. The vertical scale has been compressed so that all three curves can be seen; the peak intensity for the 12,000-K curve is actually about 1000 times greater than the peak intensity for the 3000-K curve.

BOX 5-1 Tools of the Astronomer's Trade

Temperatures and Temperature Scales

WEB LINK 5.4

Three temperature scales are in common use. Throughout most of the world, temperatures are expressed in **degrees Celsius** (°C). The Celsius temperature scale is based on the behavior of water, which freezes at 0°C and boils at 100°C at sea level on Earth. This scale is named after the Swedish astronomer Anders Celsius, who proposed it in 1742.

Astronomers usually prefer the Kelvin temperature scale. This is named after the nineteenth-century British physicist Lord Kelvin, who made many important contributions to our understanding of heat and temperature. Absolute zero, the temperature at which atomic motion is at the absolute minimum, is –273°C in the Celsius scale but 0 K in the Kelvin scale. Atomic motion cannot be any less than the minimum, so nothing can be colder than 0 K; hence, there are no negative temperatures on the Kelvin scale. Note that we do *not* use degree (°) with the Kelvin temperature scale.

A temperature expressed in kelvins is always equal to the temperature in degrees Celsius plus 273. On the Kelvin scale, water freezes at 273 K and boils at 373 K. Water must be heated through a change of 100 K or 100°C to go from its freezing point to its boiling point. Thus, the "size" of a kelvin is the same as the "size" of a Celsius degree. When considering temperature changes, measurements in kelvins and Celsius degrees are the same. For extremely high temperatures the Kelvin and Celsius scales are essentially the same: for example, the Sun's core temperature is either 1.55×10^7 K or 1.55×10^7 °C.

The now-archaic Fahrenheit scale, which expresses temperature in **degrees Fahrenheit** (°F), is used only in the United States. When the German physicist Gabriel Fahrenheit introduced this scale in the early 1700s, he intended 100°F to represent the temperature of a healthy human body. On the Fahrenheit scale, water freezes at 32°F and boils at 212°F. There are 180 Fahrenheit degrees between the freezing and boiling points of water, so a degree Fahrenheit is only 100/180 = 5/9 as large as either a Celsius degree or a kelvin.

Two simple equations allow you to convert a temperature from the Celsius scale to the Fahrenheit scale and from Fahrenheit to Celsius:

$$T_F = \frac{9}{5}\, T_C + 32$$

$$T_C = \frac{5}{9}\, (T_F - 32)$$

T_F = temperature in degrees Fahrenheit

T_C = temperature in degrees Celsius

EXAMPLE: A typical room temperature is 68°F. We can convert this to the Celsius scale using the second equation:

$$T_C = \frac{5}{9}\, (68 - 32) = 20°\text{C}$$

To convert this to the Kelvin scale, we simply add 273 to the Celsius temperature. Thus,

$$68° = 20°\text{C} = 293\ \text{K}$$

The diagram displays the relationships among these three temperature scales.

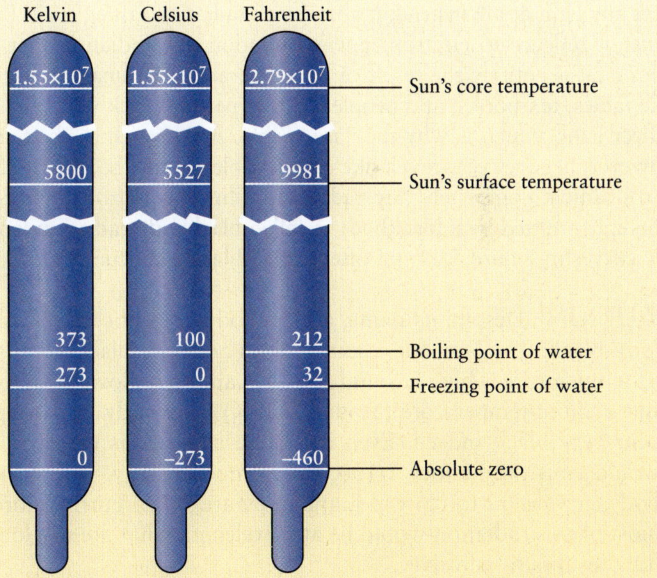

dense object emits at all wavelengths, so its spectrum is a smooth, continuous curve with no gaps in it.

The shape of the spectrum depends on temperature, however. An object at relatively low temperature (say, 3000 K) has a low curve, indicating a low intensity of radiation. The **wavelength of maximum emission,** at which the curve has its peak and the emission of energy is strongest, is at a long wavelength. The higher the temperature, the higher the curve (indicating greater intensity) and the shorter the wavelength of maximum emission.

Figure 5-11 shows that for a dense object at a temperature of 3000 K, the wavelength of maximum emission is around 1000 nm (1 μm). Because this wavelength corresponds to the infrared range and is well outside the visible range, you might think that you cannot see the radiation from an object at this temperature. In fact, the glow from such an object *is* visible; the curve shows that this object emits plenty of light within the visible range, as well as at even shorter wavelengths.

The 3000-K curve is quite a bit higher at the red end of the visible spectrum than at the violet end, so a dense object at this temperature will appear red in color. Similarly, the 12,000-K curve has its wavelength of maximum emission in the ultraviolet part of the spectrum, at a wavelength shorter than visible light. But such a hot, dense object also emits copious amounts of visible light (much more than at 6000 K or 3000 K, for which the curves are lower) and thus will have a very visible glow. The curve for this temperature is higher for blue light than for red light, and so the color of a dense object at 12,000 K is a brilliant blue or blue-white. These conclusions agree with the color changes of a heated rod shown in Figure 5-9. The same principles apply to stars: A star that looks blue, such as Bellatrix in the constellation Orion (see Figure 2-2*a*), has a high surface temperature, while a red star such as Betelgeuse (Figure 2-2*a*) has a relatively cool surface.

These observations lead to a general rule:

The higher an object's temperature, the more intensely the object emits electromagnetic radiation and the shorter the wavelength at which it emits most strongly.

We will make frequent use of this general rule to analyze the temperatures of celestial objects such as planets and stars.

The curves in Figure 5-11 are drawn for an idealized type of dense object called a **blackbody.** A perfect blackbody does not reflect any light at all; instead, it absorbs all radiation falling on it. Because it reflects no electromagnetic radiation, the radiation that it does emit is entirely the result of its temperature. Ordinary objects, like tables, textbooks, and people, are not perfect blackbodies; they reflect light, which is why they are visible. A star such as the Sun, however, behaves very much like a perfect blackbody, because it absorbs almost completely any radiation falling on it from outside. The light emitted by a blackbody is called **blackbody radiation,** and the curves in Figure 5-11 are often called **blackbody curves.**

CAUTION! Despite its name, a blackbody does not necessarily *look* black. The Sun, for instance, does not look black because its temperature is high (around 5800 K), and so it glows brightly. But a room-temperature (around 300 K) blackbody would appear very black indeed. Even if it were as large as the Sun, it would emit only about 1/100,000 as much energy. (Its blackbody curve is far too low to graph in Figure 5-11.) Furthermore, most of this radiation would be at wavelengths that are too long for our eyes to perceive.

Figure 5-12 shows the blackbody curve for a temperature of 5800 K. It also shows the intensity curve for light from the Sun, as measured from above Earth's atmosphere. (This is necessary because Earth's atmosphere absorbs certain wavelengths.) The peak of both curves is at a wavelength of about 500 nm, near the middle of the visible spectrum. Note how closely the observed intensity curve for the Sun matches the blackbody curve. This is a strong indication that the temperature of the Sun's glowing surface is about 5800 K—a temperature that we can measure across a distance of 150 million kilometers! The close correlation between blackbody curves and the observed intensity curves for most stars is a key reason why astronomers are interested in the physics of blackbody radiation.

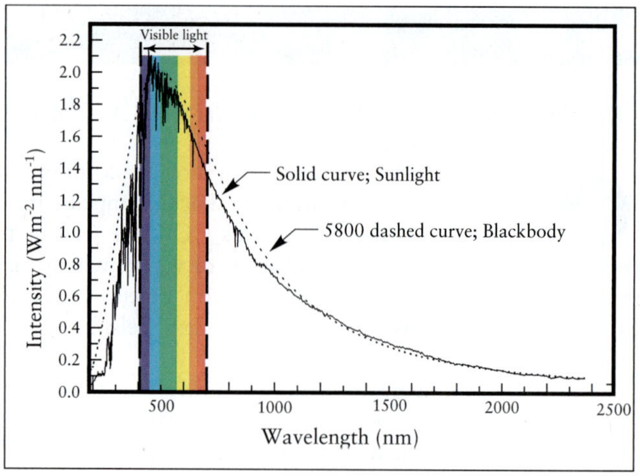

Figure 5-12

The Sun as a Blackbody This graph shows that the intensity of sunlight over a wide range of wavelengths (solid curve) is a remarkably close match to the intensity of radiation coming from a blackbody at a temperature of 5800 K (dashed curve). The measurements of the Sun's intensity were made above Earth's atmosphere (which absorbs and scatters certain wavelengths of sunlight). It's not surprising that the range of visible wavelengths includes the peak of the Sun's spectrum; the human eye evolved to take advantage of the most plentiful light available.

Blackbody radiation depends *only* on the temperature of the object emitting the radiation, not on the chemical composition of the object. The light emitted by molten gold at 2000 K is very nearly the same as that emitted by molten lead at 2000 K. Therefore, it might seem that analyzing the light from the Sun or from a star can tell astronomers the object's temperature but not what the star is made of. As Figure 5-12 shows, however, the intensity curve for the Sun (a typical star) is not precisely that of a blackbody. We will see later in this chapter that the *differences* between a star's spectrum and that of a blackbody allow us to determine the chemical composition of the star.

5-4 Wien's law and the Stefan-Boltzmann law are useful tools for analyzing glowing objects like stars

The mathematical formula that describes the blackbody curves in Figure 5-11 is a rather complicated one. But there are two simpler formulas for blackbody radiation that prove to be very useful in many branches of astronomy. They are used by astronomers who investigate the stars as well as by those who study the planets (which are dense, relatively cool objects that emit infrared radiation). One of these formulas relates the temperature of a blackbody to its wavelength of maximum emission,

Two simple mathematical formulas describing blackbodies are essential tools for studying the universe

and the other relates the temperature to the amount of energy that the blackbody emits. These formulas, which we will use throughout this book, restate in precise mathematical terms the qualitative relationships that we described in Section 5-3.

Wien's Law

Figure 5-11 shows that the higher the temperature (T) of a blackbody, the shorter its wavelength of maximum emission (λ_{max}). In 1893 the German physicist Wilhelm Wien used ideas about both heat and electromagnetism to make this relationship quantitative. The formula that he derived, which today is called **Wien's law,** is

Wien's law for a blackbody

$$\lambda_{max} = \frac{0.0029 \text{ K m}}{T}$$

λ_{max} = wavelength of maximum emission of the object (in meters)

T = temperature of the object (in kelvins)

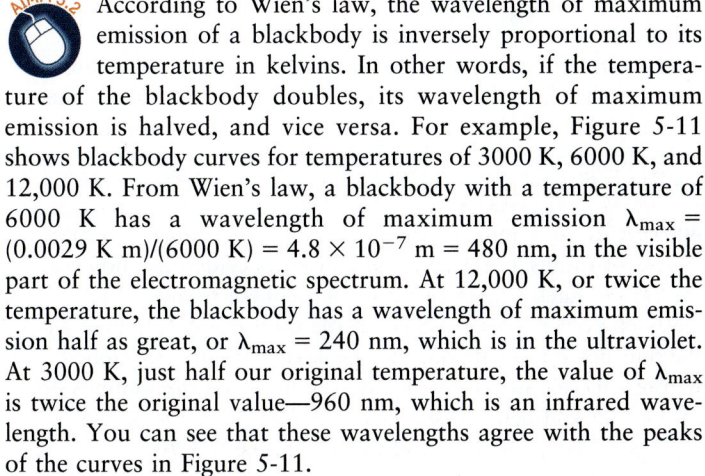

According to Wien's law, the wavelength of maximum emission of a blackbody is inversely proportional to its temperature in kelvins. In other words, if the temperature of the blackbody doubles, its wavelength of maximum emission is halved, and vice versa. For example, Figure 5-11 shows blackbody curves for temperatures of 3000 K, 6000 K, and 12,000 K. From Wien's law, a blackbody with a temperature of 6000 K has a wavelength of maximum emission $\lambda_{max} = (0.0029 \text{ K m})/(6000 \text{ K}) = 4.8 \times 10^{-7} \text{ m} = 480 \text{ nm}$, in the visible part of the electromagnetic spectrum. At 12,000 K, or twice the temperature, the blackbody has a wavelength of maximum emission half as great, or $\lambda_{max} = 240$ nm, which is in the ultraviolet. At 3000 K, just half our original temperature, the value of λ_{max} is twice the original value—960 nm, which is an infrared wavelength. You can see that these wavelengths agree with the peaks of the curves in Figure 5-11.

CAUTION! Remember that Wien's law involves the wavelength of maximum emission in *meters*. If you want to convert the wavelength to nanometers, you must multiply the wavelength in meters by $(10^9 \text{ nm})/(1 \text{ m})$.

Wien's law is very useful for determining the surface temperatures of stars. It is not necessary to know how far away the star is, how large it is, or how much energy it radiates into space. All we need to know is the dominant wavelength of the star's electromagnetic radiation.

The Stefan-Boltzmann Law

The other useful formula for the radiation from a blackbody involves the total amount of energy the blackbody radiates at all wavelengths. (By contrast, the curves in Figure 5-11 show how much energy a blackbody radiates at each individual wavelength.)

Energy is usually measured in **joules** (J), named after the nineteenth-century English physicist James Joule. A joule is the amount of energy contained in the motion of a 2-kilogram mass moving at a speed of 1 meter per second. The joule is a convenient unit of energy because it is closely related to the familiar **watt** (W): 1 watt is 1 joule per second, or 1 W = 1 J/s = 1 J s^{-1}. (The superscript -1 means you are dividing by that quantity.) For example, a 100-watt lightbulb uses energy at a rate of 100 joules per second, or 100 J/s. The energy content of food is also often measured in joules; in most of the world, diet soft drinks are labeled as "low joule" rather than "low calorie."

The amount of energy emitted by a blackbody depends both on its temperature and on its surface area. These characteristics make sense: A large burning log radiates much more heat than a burning match, even though the temperatures are the same. To consider the effects of temperature alone, it is convenient to look at the amount of energy emitted from each square meter of an object's surface in a second. This quantity is called the **energy flux** (F). Flux means "rate of flow," and thus F is a measure of how rapidly energy is flowing out of the object. It is measured in joules per square meter per second, usually written as $\text{J/m}^2\text{/s}$ or J m^{-2} s^{-1}. Alternatively, because 1 watt equals 1 joule per second, we can express flux in watts per square meter (W/m^2, or W m^{-2}).

The nineteenth-century Irish physicist David Tyndall performed the first careful measurements of the amount of radiation emitted by a blackbody. (He studied the light from a heated platinum wire, which behaves approximately like a blackbody.) By analyzing Tyndall's results, the Slovenian physicist Josef Stefan deduced in 1879 that the flux from a blackbody is proportional to the fourth power of the object's temperature (measured in kelvins). Five years after Stefan announced his law, Austrian physicist Ludwig Boltzmann showed how it could be derived mathematically from basic assumptions about atoms and molecules. For this reason, Stefan's law is commonly known as the **Stefan-Boltzmann law.** Written as an equation, the Stefan-Boltzmann law is

Stefan-Boltzmann law for a blackbody

$$F = \sigma T^4$$

F = energy flux, in joules per square meter of surface per second

σ = a constant = 5.67×10^{-8} W m^{-2} K^{-4}

T = object's temperature, in kelvins

The value of the constant σ (the Greek letter sigma) is known from laboratory experiments.

The Stefan-Boltzmann law says that if you double the temperature of an object (for example, from 300 K to 600 K), then the energy emitted from the object's surface each second increases by a factor of $2^4 = 16$. If you increase the temperature by a factor of 10 (for example, from 300 K to 3000 K), the rate of energy emission increases by a factor of $10^4 = 10{,}000$. Thus, a chunk of iron at room temperature (around 300 K) emits very little electromagnetic radiation (and essentially no visible light), but an iron bar heated to 3000 K glows quite intensely.

Box 5-2 gives several examples of applying Wien's law and the Stefan-Boltzmann law to typical astronomical problems.

BOX 5-2 Tools of the Astronomer's Trade

Using the Laws of Blackbody Radiation

The Sun and stars behave like nearly perfect blackbodies. Wien's law and the Stefan-Boltzmann law can therefore be used to relate the surface temperature of the Sun or a distant star to the energy flux and wavelength of maximum emission of its radiation. The following examples show how to do this.

EXAMPLE: The maximum intensity of sunlight is at a wavelength of roughly 500 nm = 5.0×10^{-7} m. Use this information to determine the surface temperature of the Sun.

Situation: We are given the Sun's wavelength of maximum emission λ_{max}, and our goal is to find the Sun's surface temperature, denoted by $T_\odot$. (The symbol $\odot$ is the standard astronomical symbol for the Sun.)

Tools: We use Wien's law to relate the values of λ_{max} and $T_\odot$.

Answer: As written, Wien's law tells how to find λ_{max} if we know the surface temperature. To find the surface temperature from λ_{max}, we first rearrange the formula, then substitute the value of λ_{max}:

$$T_\odot = \frac{0.0029 \text{ K m}}{\lambda_{max}} = \frac{0.0029 \text{ K m}}{5.0 \times 10^{-7} \text{ m}} = 5800 \text{ K}$$

Review: This temperature is very high by Earth standards, about the same as an iron-welding arc.

EXAMPLE: Using detectors above Earth's atmosphere, astronomers have measured the average flux of solar energy arriving at Earth. This value, called the **solar constant,** is equal to 1370 W m^{-2}. Use this information to calculate the Sun's surface temperature. (This calculation provides a check on our result from the preceding example.)

Situation: The solar constant is the flux of sunlight as measured at Earth. We want to use the value of the solar constant to calculate $T_\odot$.

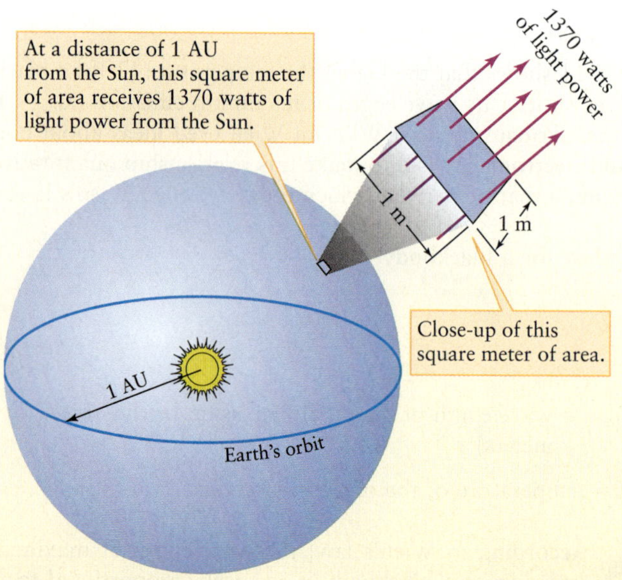
At a distance of 1 AU from the Sun, this square meter of area receives 1370 watts of light power from the Sun.
1370 watts of light power
1 m
1 m
Close-up of this square meter of area.
1 AU
Earth's orbit

Tools: It may seem that all we need is the Stefan-Boltzmann law, which relates flux to surface temperature. However, the quantity *F* in this law refers to the flux measured at the Sun's surface, *not* at Earth. Hence, we will first need to calculate *F* from the given information.

Answer: To determine the value of *F,* we first imagine a huge sphere of radius 1 AU with the Sun at its center, as shown in the figure. Each square meter of that sphere receives 1370 watts of power from the Sun, so the total energy radiated by the Sun per second is equal to the solar constant multiplied by the sphere's surface area. The result, called the **luminosity** of the Sun and denoted by the symbol $L_\odot$, is $L_\odot = 3.90 \times 10^{26}$ W. That is, in 1 second the Sun radiates 3.90×10^{26} joules of energy into space. Because we know the size of the Sun, we can compute the energy flux (energy emitted per square meter per second) at its surface. The radius of the Sun is $R_\odot = 6.96 \times 10^8$ m, and the Sun's surface area is $4\pi R_\odot^2$.

5-5 Light has properties of both waves and particles

At the end of the nineteenth century, physicists mounted a valiant effort to explain all the characteristics of blackbody radiation. To this end they constructed theories based on Maxwell's description of light as electromagnetic waves. But all such theories failed to explain the characteristic shapes of blackbody curves shown in Figure 5-11.

The Photon Hypothesis

In 1900, however, the German physicist Max Planck discovered that he could derive a formula that correctly described blackbody curves if he made certain assumptions. In 1905 the great German-born physicist Albert Einstein realized that these assumptions implied a radical new view of the nature of light. One tenet of this new view is that electromagnetic energy is emitted in discrete, particle-like packets, or light *quanta* (the plural of *quantum*, from a Latin word meaning "how much"). The second tenet is that the energy of each light quantum—today called a **photon**—is related to the wavelength of light: the greater the wavelength, the lower the energy of a photon associated with that wavelength. Thus, a photon of red light (wavelength λ = 700 nm) has less en-

The revolutionary photon concept is necessary to explain blackbody radiation and the photoelectric effect

Therefore, its energy flux $F_\odot$ is the Sun's luminosity (total energy emitted by the Sun per second) divided by the Sun's surface area (the number of square meters of surface):

$$F_\odot = \frac{L_\odot}{4\pi R_\odot^2} = \frac{3.90 \times 10^{26}\ \text{W}}{4\pi(6.96 \times 10^8\ \text{m})^2} = 6.41 \times 10^7\ \text{W m}^{-2}$$

Once we have the Sun's energy flux $F_\odot$ we can use the Stefan-Boltzmann law to find the Sun's surface temperature $T_\odot$:

$$T_\odot^4 = F_\odot/\sigma = 1.13 \times 10^{15}\ \text{K}^4$$

Taking the fourth root (the square root of the square root) of this value, we find the surface temperature of the Sun to be $T_\odot = 5800$ K.

Review: Our result for $T_\odot$ agrees with the value we computed in the previous example using Wien's law. Notice that the solar constant of 1370 W m^{-2} is very much less than $F_\odot$, the flux at the Sun's surface. By the time the Sun's radiation reaches Earth, it is spread over a greatly increased area.

EXAMPLE: Sirius, the brightest star in the night sky, has a surface temperature of about 10,000 K. Find the wavelength at which Sirius emits most intensely.

Situation: Our goal is to calculate the wavelength of maximum emission of Sirius (λ_{max}) from its surface temperature T.

Tools: We use Wien's law to relate the values of λ_{max} and T.

Answer: Using Wien's law,

$$\lambda_{max} = \frac{0.0029\ \text{K m}}{T} = \frac{0.0029\ \text{K m}}{10{,}000\ \text{K}}$$
$$= 2.9 \times 10^{-7}\ \text{m} = 290\ \text{nm}$$

Review: Our result shows that Sirius emits light most intensely in the ultraviolet. In the visible part of the spectrum, it emits more blue light than red light (like the curve for 12,000 K in Figure 5-11), so Sirius has a distinct blue color.

EXAMPLE: How does the energy flux from Sirius compare to the Sun's energy flux?

Situation: To compare the energy fluxes from the two stars, we want to find the *ratio* of the flux from Sirius to the flux from the Sun.

Tools: We use the Stefan-Boltzmann law to find the flux from Sirius and from the Sun, which from the preceding examples have surface temperatures 10,000 K and 5800 K, respectively.

Answer: For the Sun, the Stefan-Boltzmann law is $F_\odot = \sigma T_\odot^4$, and for Sirius we can likewise write $F_* = \sigma T_*^4$, where the subscripts $\odot$ and * refer to the Sun and Sirius, respectively. If we divide one equation by the other to find the ratio of fluxes, the Stefan-Boltzmann constants cancel out and we get

$$\frac{F_*}{F_\odot} = \frac{T_*^4}{T_\odot^4} = \frac{(10{,}000\ \text{K})^4}{(5800\ \text{K})^4} = \left(\frac{10{,}000}{5800}\right)^4 = 8.8$$

Review: Because Sirius has such a high surface temperature, each square meter of its surface emits 8.8 times more energy per second than a square meter of the Sun's relatively cool surface. Sirius is actually a larger star than the Sun, so it has more square meters of surface area and, hence, its *total* energy output is *more* than 8.8 times that of the Sun.

ergy than a photon of violet light (λ = 400 nm). In this picture, light has a dual personality; it behaves as a stream of particlelike photons, but each photon has wavelike properties. In this sense, the best answer to the question "Is light a wave or a stream of particles?" is "Yes!"

It was soon realized that the photon hypothesis explains more than just the detailed shape of blackbody curves. For example, it explains why only ultraviolet light causes suntans and sunburns. The reason is that tanning or burning involves a chemical reaction in the skin. High-energy, short-wavelength ultraviolet photons can trigger these reactions, but the lower-energy, longer-wavelength photons of visible light cannot. Similarly, normal photographic film is sensitive to visible light but not to infrared light; a long-wavelength infrared photon does not have enough energy to cause the chemical change that occurs when film is exposed to the higher-energy photons of visible light.

WEB LINK 5.5

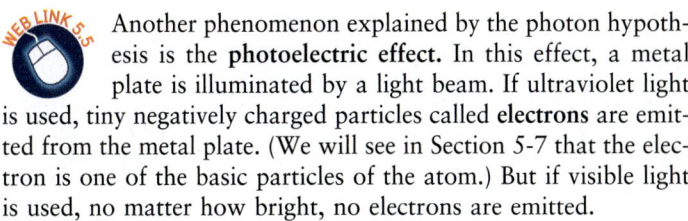

Another phenomenon explained by the photon hypothesis is the **photoelectric effect.** In this effect, a metal plate is illuminated by a light beam. If ultraviolet light is used, tiny negatively charged particles called **electrons** are emitted from the metal plate. (We will see in Section 5-7 that the electron is one of the basic particles of the atom.) But if visible light is used, no matter how bright, no electrons are emitted.

WEB LINK 5.6

Einstein explained this behavior by noting that a certain minimum amount of energy is required to remove an electron from the metal plate. The energy of a short-wavelength ultraviolet photon is greater than this minimum value, so an electron that absorbs a photon of ultraviolet light will have enough energy to escape from the plate. But an electron that absorbs a photon of visible light, with its longer wavelength and lower energy, does not gain enough energy to escape and so remains within the metal. Einstein and Planck both won Nobel prizes for their contributions to understanding the nature of light.

The Energy of a Photon

The relationship between the energy E of a single photon and the wavelength of the electromagnetic radiation can be expressed in a simple equation:

Energy of a photon (in terms of wavelength)

$$E = \frac{hc}{\lambda}$$

E = energy of a photon

h = Planck's constant

c = speed of light

λ = wavelength of light

The value of the constant h in this equation, now called *Planck's constant,* has been shown in laboratory experiments to be

$$h = 6.625 \times 10^{-34} \text{ J s}$$

The units of h are joules multiplied by seconds, called "joule-seconds" and abbreviated J s.

Because the value of h is so tiny, a single photon carries a very small amount of energy. For example, a photon of red light with wavelength 633 nm has an energy of only 3.14×10^{-19} J (Box 5-3), which is why we ordinarily do not notice that light comes in the form of photons. Even a dim light source emits so many photons per second that it seems to be radiating a continuous stream of energy.

The energies of photons are sometimes expressed in terms of a small unit of energy called the **electron volt** (eV). One electron volt is equal to 1.602×10^{-19} J, so a 633-nm photon has an energy of 1.96 eV. If energy is expressed in electron volts, Planck's constant is best expressed in electron volts multiplied by seconds, abbreviated eV s:

$$h = 4.135 \times 10^{-15} \text{ eV s}$$

Because the frequency ν of light is related to the wavelength λ by $\nu = c/\lambda$, we can rewrite the equation for the energy of a photon as

Energy of a photon (in terms of frequency)

$$E = h\nu$$

E = energy of a photon

h = Planck's constant

ν = frequency of light

The equations $E = hc/\lambda$ and $E = h\nu$ are together called **Planck's law.** Both equations express a relationship between a particlelike property of light (the energy E of a photon) and a wavelike property (the wavelength λ or frequency ν).

The photon picture of light is essential for understanding the detailed shapes of blackbody curves. As we will see, it also helps to explain how and why the spectra of the Sun and stars differ from those of perfect blackbodies.

BOX 5-3 — Astronomy Down to Earth

Photons at the Supermarket

A beam of light can be regarded as a stream of tiny packets of energy called photons. The Planck relationships $E = hc/\lambda$ and $E = h\nu$ can be used to relate the energy E carried by a photon to its wavelength λ and frequency ν.

As an example, the laser bar-code scanners used at stores and supermarkets emit orange-red light of wavelength 633 nm. To calculate the energy of a single photon of this light, we must first express the wavelength in meters. A nanometer (nm) is equal to 10^{-9} m, so the wavelength is

$$\lambda = (633 \text{ nm})\left(\frac{10^{-9}\text{ m}}{1 \text{ nm}}\right) = 633 \times 10^{-9} \text{ m} = 6.33 \times 10^{-7} \text{ m}$$

Then, using the Planck formula $E = hc/\lambda$, we find that the energy of a single photon is

$$E = \frac{hc}{\lambda} = \frac{(6.625 \times 10^{-34} \text{ J s})(3 \times 10^{8} \text{ m/s})}{6.33 \times 10^{-7} \text{ m}} = 3.14 \times 10^{-19} \text{ J}$$

This amount of energy is very small. The laser in a typical bar-code scanner emits 10^{-3} joule of light energy per second, so the number of photons emitted per second is

$$\frac{10^{-3} \text{ joule per second}}{3.14 \times 10^{-19} \text{ joule per photon}} = 3.2 \times 10^{15} \text{ photons per second}$$

This number is so large that the laser beam seems like a continuous flow of energy rather than a stream of little energy packets.

5-6 Each chemical element produces its own unique set of spectral lines

In 1814 the German master optician Joseph von Fraunhofer repeated the classic experiment of shining a beam of sunlight through a prism (see Figure 5-3). But this time Fraunhofer subjected the resulting rainbow-colored spectrum to intense magnification. To his surprise, he discovered that the solar spectrum contains hundreds of fine, dark lines, now called **spectral lines.** By contrast, if the light from a perfect blackbody were sent through a prism, it would produce a smooth, continuous spectrum with no dark lines. Fraunhofer counted more than 600 dark lines in the Sun's spectrum; today we know of more than 30,000. The photograph of the Sun's spectrum in Figure 5-13 shows hundreds of these spectral lines.

Spectroscopy is the key to determining the chemical composition of planets and stars

Spectral Analysis

Half a century later, chemists discovered that they could produce spectral lines in the laboratory and use these spectral lines to analyze what kinds of atoms different substances are made of. Chemists had long known that many substances emit distinctive colors when sprinkled into a flame. To facilitate study of these colors, around 1857 the German chemist Robert Bunsen invented a gas burner (today called a Bunsen burner) that produces a clean flame with no color of its own. Bunsen's colleague, the Prussian-born physicist Gustav Kirchhoff, suggested that the colored light produced by substances in a flame might best be studied by passing the light through a prism (Figure 5-14). The two scientists promptly discovered that the spectrum from a flame consists of a pattern of thin, bright spectral lines against a dark background. The same kind of spectrum is produced by heated gases such as neon or argon.

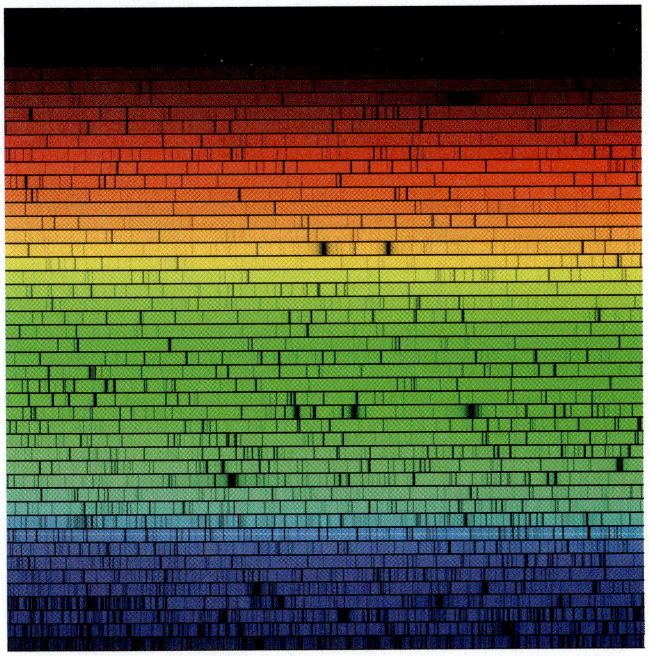

Figure 5-13 RI**V**UXG

The Sun's Spectrum Numerous dark spectral lines are seen in this image of the Sun's spectrum. The spectrum is spread out so much that it had to be cut into segments to fit on this page. (N. A. Sharp, NOAO/NSO/Kitt Peak FTS/AURA/NSF)

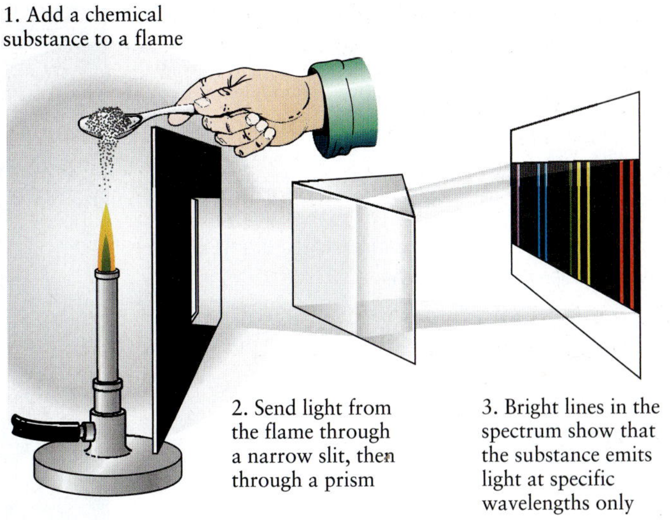
1. Add a chemical substance to a flame
2. Send light from the flame through a narrow slit, then through a prism
3. Bright lines in the spectrum show that the substance emits light at specific wavelengths only

Figure 5-14

The Kirchhoff-Bunsen Experiment In the mid-1850s, Gustav Kirchhoff and Robert Bunsen discovered that when a chemical substance is heated and vaporized, the spectrum of the emitted light exhibits a series of bright spectral lines. They also found that each chemical element produces its own characteristic pattern of spectral lines. (In an actual laboratory experiment, lenses would be needed to focus the image of the slit onto the screen.)

Kirchhoff and Bunsen then found that each chemical element produces its own unique pattern of spectral lines. Thus was born in 1859 the technique of **spectral analysis,** the identification of chemical substances by their unique patterns of spectral lines.

A chemical **element** is a fundamental substance that cannot be broken down into more basic chemicals. Some examples are hydrogen, oxygen, carbon, iron, gold, and silver. After Kirchhoff and Bunsen had recorded the prominent spectral lines of all the then-known elements, they soon began to discover other spectral lines in the spectra of vaporized mineral samples. In this way they discovered elements whose presence had never before been suspected. In 1860, Kirchhoff and Bunsen found a new line in the blue portion of the spectrum of a sample of mineral water. After isolating the previously unknown element responsible for making the line, they named it cesium (from the Latin *caesium,* "gray-blue"). The next year, a new line in the red portion of the spectrum of a mineral sample led them to discover the element rubidium (Latin *rubidium,* "red").

Spectral analysis even allowed the discovery of new elements outside Earth. During the solar eclipse of 1868, astronomers found a new spectral line in light coming from the hot gases at the upper surface of the Sun while the main body of the Sun was hidden by the Moon. This line was attributed to a new element

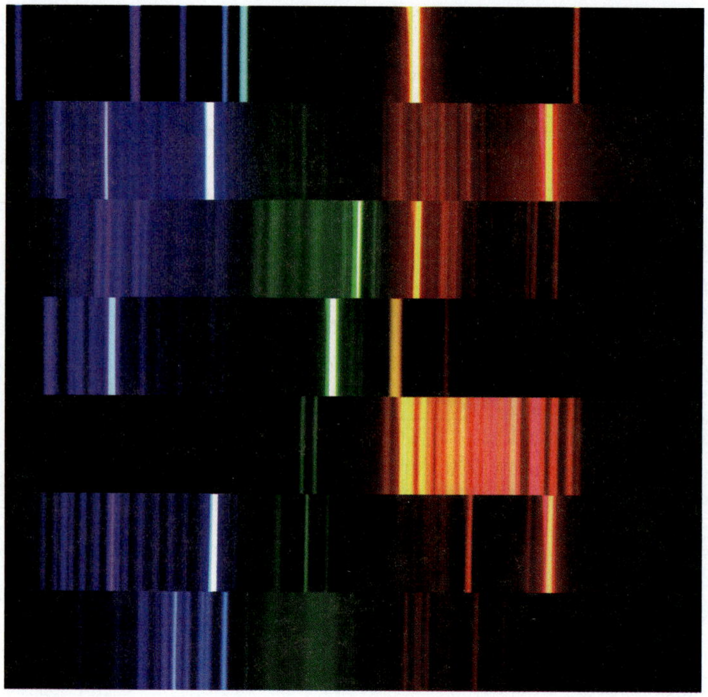

Figure 5-15 R I **V** U X G

Various Spectra These photographs show the spectra of different types of gases as measured in a laboratory on Earth. Each type of gas has a unique spectrum that is the same wherever in the universe the gas is found. Water vapor (H_2O) is a compound whose molecules are made up of hydrogen and oxygen atoms; the hydrogen molecule (H_2) is made up of two hydrogen atoms. (Ted Kinsman/Science Photo Library)

that was named helium (from the Greek *helios,* "sun"). Helium was not discovered on Earth until 1895, when it was found in gases obtained from a uranium mineral.

A sample of an element contains only a single type of atom; carbon contains only carbon atoms, helium contains only helium atoms, and so on. Atoms of the same or different elements can combine to form **molecules.** For example, two hydrogen atoms (symbol H) can combine with an oxygen atom (symbol O) to form a water molecule (symbol H_2O). Substances like water whose molecules include atoms of different elements are called **compounds.** Just as each type of atom has its own unique spectrum, so does each type of molecule. Figure 5-15 shows the spectra of several types of atoms and molecules.

Kirchhoff's Laws

The spectrum of the Sun, with its dark spectral lines superimposed on a bright background (see Figure 5-13), may seem to be unrelated to the spectra of bright lines against a dark background produced by substances in a flame (see Figure 5-14). But by the early 1860s, Kirchhoff's experiments had revealed a direct connection between these two types of spectra. His conclusions are summarized in three important statements about spectra that are today called **Kirchhoff's laws.** These laws, which are illustrated in Figure 5-16, are as follows:

Law 1 A hot opaque body, such as a perfect blackbody, or a hot, dense gas produces a **continuous spectrum**—a complete rainbow of colors without any spectral lines.

Law 2 A hot, transparent gas produces an **emission line spectrum**—a series of bright spectral lines against a dark background.

Law 3 A cool, transparent gas in front of a source of a continuous spectrum produces an **absorption line spectrum**—a series of dark spectral lines among the colors of the continuous spectrum. Furthermore, the dark lines in the absorption spectrum of a particular gas occur at exactly the *same* wavelengths as the bright lines in the emission spectrum of that same gas.

Kirchhoff's laws imply that if a beam of white light is passed through a gas, the atoms of the gas somehow extract light of very specific wavelengths from the white light. Hence, an observer who looks straight through the gas at the white-light source (the blackbody in Figure 5-16) will receive light whose spectrum has dark absorption lines superimposed on the continuous spectrum of the white light. The gas atoms then radiate light of precisely these same wavelengths in all directions. An observer at an oblique angle (that is, one who is not sighting directly through the cloud toward the blackbody) will receive only this light radiated by the gas cloud; the spectrum of this light is bright emission lines on a dark background.

CAUTION! Figure 5-16 shows that light can either pass through a cloud of gas or be absorbed by the gas. But there is also a third possibility: The light can simply bounce off the atoms or molecules that make up the gas, a phenomenon called **light scattering.** In other words, photons passing through a gas cloud can miss the gas atoms altogether, be swallowed whole by the atoms (absorption), or bounce off the atoms like billiard balls colliding (scattering). Box 5-4 describes how light scattering explains the blue color of the sky and the red color of sunsets.

Whether an emission line spectrum or an absorption line spectrum is observed from a gas cloud depends on the relative temperatures of the gas cloud and its background. Absorption lines are seen if the background is hotter than the gas, and emission lines are seen if the background is cooler.

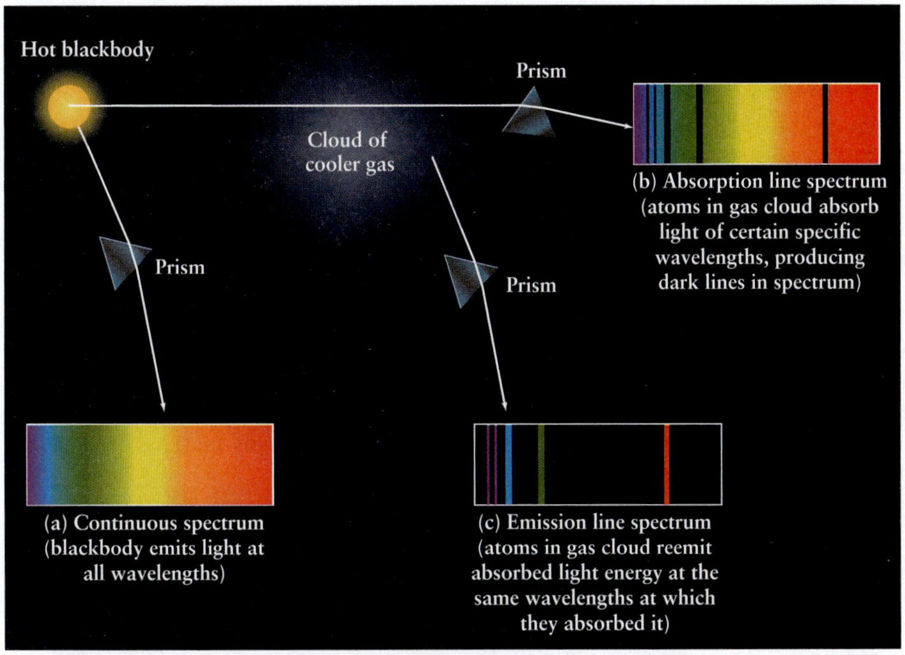
Hot blackbody
Prism
Cloud of cooler gas
(b) Absorption line spectrum (atoms in gas cloud absorb light of certain specific wavelengths, producing dark lines in spectrum)
Prism
Prism
(a) Continuous spectrum (blackbody emits light at all wavelengths)
(c) Emission line spectrum (atoms in gas cloud reemit absorbed light energy at the same wavelengths at which they absorbed it)

ANIMATION 5.1

Figure 5-16

Continuous, Absorption Line, and Emission Line Spectra A hot, opaque body (like a blackbody) emits a continuous spectrum of light (spectrum *a*). If this light is passed through a cloud of a cooler gas, the cloud absorbs light of certain specific wavelengths, and the spectrum of light that passes directly through the cloud has dark absorption lines (spectrum *b*). The cloud does not retain all the light energy that it absorbs but radiates it outward in all directions. The spectrum of this reradiated light contains bright emission lines (spectrum *c*) with exactly the same wavelengths as the dark absorption lines in spectrum *b*. The specific wavelengths observed depend on the chemical composition of the cloud.

For example, if sodium is placed in the flame of a Bunsen burner in a darkened room, the flame will emit a characteristic orange-yellow glow. (This same glow is produced if we use ordinary table salt, which is a compound of sodium and chlorine.) If we pass the light from the flame through a prism, it displays an emission line spectrum with two closely spaced spectral lines at wavelengths of 588.99 and 589.59 nm, in the orange-yellow part of the spectrum. We now turn on a lightbulb whose filament is hotter than the flame and shine the bulb's white light through the flame. The spectrum of this light after it passes through the flame's sodium vapor is the continuous spectrum from the lightbulb, but with two closely spaced *dark* lines at 588.99 and 589.59 nm. Thus, the chemical composition of the gas is revealed by either bright emission lines or dark absorption lines.

Spectroscopy

Spectroscopy is the systematic study of spectra and spectral lines. Spectral lines are tremendously important in astronomy, because they provide reliable evidence about the chemical composition of distant objects. As an example, the spectrum of the Sun shown in Figure 5-13 is an absorption line spectrum. The continuous spectrum comes from the hot surface of the Sun, which acts like a blackbody. The dark absorption lines are caused by this light passing through a cooler gas; this gas is the atmosphere that surrounds the Sun. Therefore, by identifying the spectral lines present in the solar spectrum, we can determine the chemical composition of the Sun's atmosphere.

Figure 5-17 shows both a portion of the Sun's absorption line spectrum and the emission line spectrum of iron vapor over the same wavelength range. This pattern of bright spectral lines in the lower spectrum is iron's own distinctive "fingerprint," which no other substance can imitate. Because some absorption lines in the Sun's spectrum coincide with the iron lines, some vaporized iron must exist in the Sun's atmosphere.

Spectroscopy can also help us analyze gas clouds in space, such as the nebula surrounding the star cluster NGC 346 shown

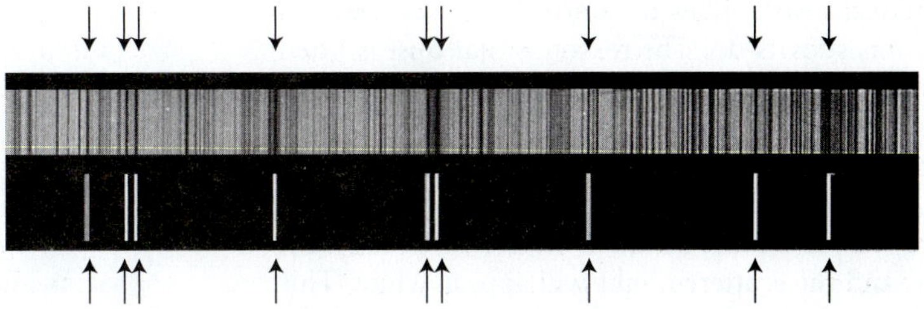
Absorption spectrum of the Sun
Emission spectrum of iron (in the laboratory on Earth)
For each emission line of iron, there is a corresponding absorption line in the solar spectrum; hence there must be iron in the Sun's atmosphere

Figure 5-17 R I **V** U X G

Iron in the Sun The upper part of this figure is a portion of the Sun's spectrum at violet wavelengths, showing numerous dark absorption lines. The lower part of the figure is a corresponding portion of the emission line spectrum of vaporized iron. The iron lines coincide with some of the solar lines, which proves that there is some iron (albeit a relatively small amount) in the Sun's atmosphere. (Carnegie Observatories)

BOX 5-4 Astronomy Down to Earth

Light Scattering

Light scattering is the process whereby photons bounce off particles in their path. These particles can be atoms, molecules, or clumps of molecules. You are reading these words using photons from the Sun or a lamp that bounced off the page—that is, were scattered by the particles that make up the page.

An important fact about light scattering is that very small particles—ones that are smaller than a wavelength of visible light—are quite effective at scattering short-wavelength photons of blue light, but less effective at scattering long-wavelength photons of red light. This fact explains a number of phenomena that you can see here on Earth.

The light that comes from the daytime sky is sunlight that has been scattered by the molecules that make up our atmosphere (see part *a* of the accompanying figure). Air molecules are less than 1 nm across, far smaller than the wavelength of visible light, so they scatter blue light more than red light—which is why the sky looks blue. Smoke particles are also quite small, which explains why the smoke from a cigarette or a fire has a bluish color.

Distant mountains often appear blue thanks to sunlight being scattered from the atmosphere between the mountains and your eyes. (The Blue Ridge Mountains, which extend from Pennsylvania to Georgia, and Australia's Blue Mountains derive their names from this effect.) Sunglasses often have a red or orange tint, which blocks out blue light. This cuts down on the amount of scattered light from the sky reaching your eyes and allows you to see distant objects more clearly.

Light scattering also explains why sunsets are red. The light from the Sun contains photons of all visible wavelengths, but as this light passes through our atmosphere the blue photons are scattered away from the straight-line path from the Sun to your eye. Red photons undergo relatively little scattering, so the Sun always looks a bit redder than it really is. When you look toward the setting sun, the sunlight that reaches your eye has had to pass through a relatively thick layer of atmosphere (part *b* of the accompanying figure). Hence, a large fraction of the blue light from the Sun has been scattered, and the Sun appears quite red.

The same effect also applies to sunrises, but sunrises seldom look as red as sunsets do. The reason is that dust is lifted into the atmosphere during the day by the wind (which is typically stronger in the daytime than at night), and dust particles in the atmosphere help to scatter even more blue light.

If the small particles that scatter light are sufficiently concentrated, there will be almost as much scattering of red light as of blue light, and the scattered light will appear white. This explains the white color of clouds, fog, and haze, in which the scattering particles are ice crystals or water droplets. Whole milk looks white because of light scattering from tiny fat globules; nonfat milk has only a very few of these globules and so has a slight bluish cast.

Light scattering has many applications to astronomy. For example, it explains why very distant stars in our Galaxy appear surprisingly red. The reason is that there are tiny dust particles in the space between the stars, and this dust scatters blue photons. By studying how much scattering takes place, astronomers have learned about the tenuous material that fills interstellar space.

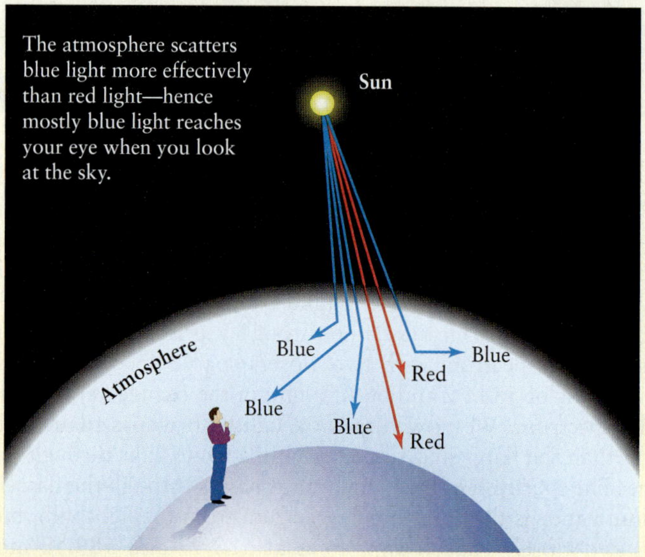

(a) Why the sky looks blue

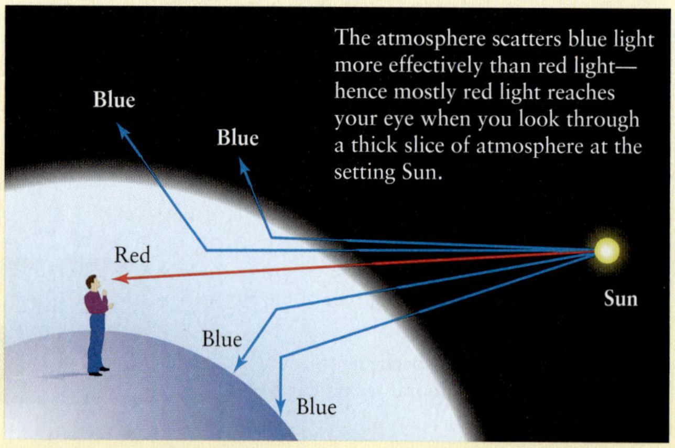

(b) Why the setting Sun looks red

WEB LINK 5.7

Figure 5-18 RI**V**UXG

Analyzing the Composition of a Distant Nebula The glowing gas cloud in this Hubble Space Telescope image lies 210,000 light-years away in the constellation Tucana (the Toucan). Hot stars within the nebula emit high-energy, ultraviolet photons, which are absorbed by the surrounding gas and heat the gas to high temperature. This heated gas produces light with an emission line spectrum. The particular wavelength of red light emitted by the nebula is 656 nm, characteristic of hydrogen gas. (NASA, ESA, and A. Nota, STScI/ESA)

in Figure 5-18. Such glowing clouds have emission line spectra, because we see them against the black background of space. The particular shade of red that dominates the color of this nebula is due to an emission line at a wavelength near 656 nm. This is one of the characteristic wavelengths emitted by hydrogen gas, so we can conclude that this nebula contains hydrogen. More detailed analyses of this kind show that hydrogen is the most common element in gaseous nebulae, and indeed in the universe as a whole. The spectra of other nebulae, such as the Ring Nebula shown in the image that opens this chapter, also reveal the presence of nitrogen, oxygen, helium, and other gases.

What is truly remarkable about spectroscopy is that it can determine chemical composition at any distance. The 656-nm red light produced by a sample of heated hydrogen gas on Earth (the bright red line in the hydrogen spectrum in Figure 5-15) is the same as that observed coming from the nebula shown in Figure 5-18, located about 210,000 light-years away. By using the basic principles outlined by Kirchhoff, astronomers have the tools to make chemical assays of objects that are almost inconceivably distant. Throughout this book we will see many examples of how astronomers use Kirchhoff's laws to determine the nature of celestial objects.

To make full use of Kirchhoff's laws, it is helpful to understand why they work. Why does an atom absorb light of only particular wavelengths? And why does it then emit light of only these same wavelengths? Maxwell's theory of electromagnetism (see Section 5-2) could not answer these questions. The answers did not come until early in the twentieth century, when scientists began to discover the structure and properties of atoms.

5-7 An atom consists of a small, dense nucleus surrounded by electrons

The first important clue about the internal structure of atoms came from an experiment conducted in 1910 by Ernest Rutherford, a gifted physicist from New Zealand. Rutherford and his colleagues at the University of Manchester in England had been investigating the recently discovered phenomenon of radioactivity. Certain radioactive elements, such as uranium and radium, were known to emit particles of various types. One type, the alpha particle, has about the same mass as a helium atom and is emitted from some radioactive substances with considerable speed.

To decode the information in the light from immense objects like stars and galaxies, we must understand the structure of atoms

In one series of experiments, Rutherford and his colleagues were using alpha particles as projectiles to probe the structure of solid matter. They directed a beam of these particles at a thin sheet of metal (Figure 5-19). Almost all the alpha particles passed through the metal sheet with little or no deflection from their straight-line paths. To the surprise of the experimenters, however, an occasional alpha particle bounced back from the metal sheet as though it had struck something quite dense. Rutherford later remarked, "It was almost as incredible as

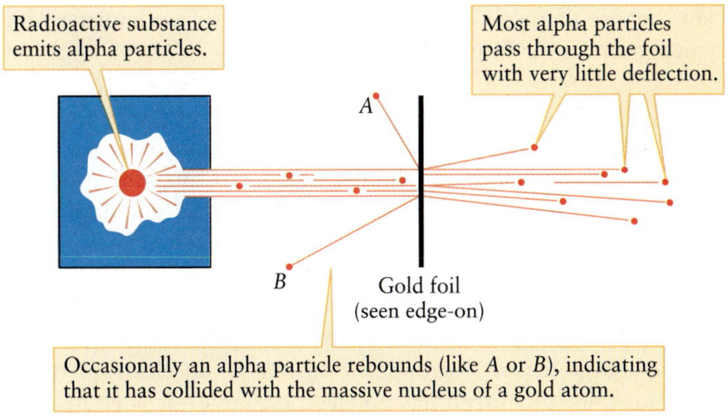

Figure 5-19

Rutherford's Experiment Alpha particles from a radioactive source are directed at a thin metal foil. This experiment provided the first evidence that the nuclei of atoms are relatively massive and compact.

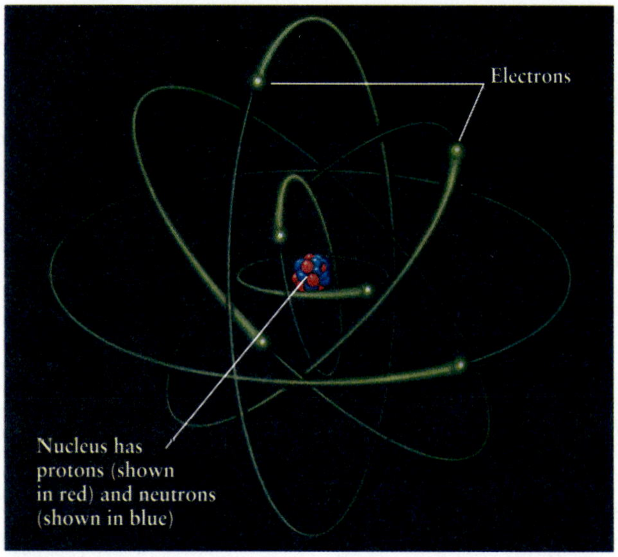

Figure 5-20

Rutherford's Model of the Atom In this model, electrons orbit the atom's nucleus, which contains most of the atom's mass. The nucleus contains two types of particles, protons and neutrons.

if you fired a fifteen-inch shell at a piece of tissue paper and it came back and hit you."

Rutherford concluded from this experiment that most of the mass of an atom is concentrated in a compact, massive lump of matter that occupies only a small part of the atom's volume. Most of the alpha particles pass freely through the nearly empty space that makes up most of the atom, but a few particles happen to strike the dense mass at the center of the atom and bounce back.

The Nucleus of an Atom

Rutherford proposed a new model for the structure of an atom, shown in Figure 5-20. According to this model, a massive, positively charged **nucleus** at the center of the atom is orbited by tiny, negatively charged electrons. Rutherford concluded that at least 99.98% of the mass of an atom must be concentrated in its nucleus, whose diameter is only about 10^{-14} m. (The diameter of a typical atom is far larger, about 10^{-10} m.)

ANALOGY To appreciate just how tiny the nucleus is, imagine expanding an atom by a factor of 10^{12} to a diameter of 100 meters, about the length of a football field. To this scale, the nucleus would be just a centimeter across—no larger than your thumbnail.

We know today that the nucleus of an atom contains two types of particles, **protons** and **neutrons.** A proton has a positive electric charge, equal and opposite to that of an electron. As its name suggests, a neutron has no electric charge—it is electrically neutral. As an example, an alpha particle (such as those Rutherford's team used) is actually a nucleus of the helium atom, with two protons and two neutrons. Protons and neutrons are held together in a nucleus by the so-called strong nuclear force, whose great strength overcomes the electric repulsion between the positively charged protons. A proton and a neutron have almost the same mass, 1.7×10^{-27} kg, and each has about 2000 times as much mass as an electron (9.1×10^{-31} kg). In an ordinary atom there are as many positive protons as there are negative electrons, so the atom has no net electric charge. Because the mass of the electron is so small, the mass of an atom is not much greater than the mass of its nucleus. That is why an alpha particle has nearly the same mass as an atom of helium.

While the solar system is held together by gravitational forces, atoms are held together by electrical forces. Opposites attract: The negative charges on the orbiting electrons are attracted to the positive charges on the protons in the nucleus. **Box 5-5** on page 118 describes more about the connection between the structure of atoms and the chemical and physical properties of substances made of those atoms.

Rutherford's experiments clarified the structure of the atom, but they did not explain how these tiny particles within the atom give rise to spectral lines. The task of reconciling Rutherford's atomic model with Kirchhoff's laws of spectral analysis was undertaken by the young Danish physicist Niels Bohr, who joined Rutherford's group at Manchester in 1912.

5-8 Spectral lines are produced when an electron jumps from one energy level to another within an atom

WEB LINK 5.8

Niels Bohr began his study of the connection between atomic spectra and atomic structure by trying to understand the structure of hydrogen, the simplest and lightest of the elements. (As we discussed in Section 5-6, hydrogen is also the most common element in the universe.) When Bohr was done, he had not only found a way to explain this atom's spectrum but had also found a justification for Kirchhoff's laws in terms of atomic physics.

Niels Bohr explained spectral lines with a radical new model of the atom

Hydrogen and the Balmer Series

The most common type of hydrogen atom consists of a single electron and a single proton. Hydrogen atoms have a simple visible-light spectrum consisting of a pattern of lines that begins at a wavelength of 656.3 nm and ends at 364.6 nm. The first spectral line is called H_α (H-alpha), the second spectral line is called H_β (H-beta), the third is H_γ (H-gamma), and so forth. (These are the bright lines in the spectrum of hydrogen shown in Figure 5-15. The fainter lines between these appear when hydrogen atoms form into molecules.) The closer you get to the short-wavelength end of the spectrum at 364.6 nm, the more spectral lines you see.

The regularity in this spectral pattern was described mathematically in 1885 by Johann Jakob Balmer, a Swiss schoolteacher. The spectral lines of hydrogen at visible wavelengths are today called **Balmer lines,** and the entire pattern from H_α onward is called the **Balmer series.** Eight Balmer lines are seen in the spectrum of the star shown in **Figure 5-21**. Stars in general, including

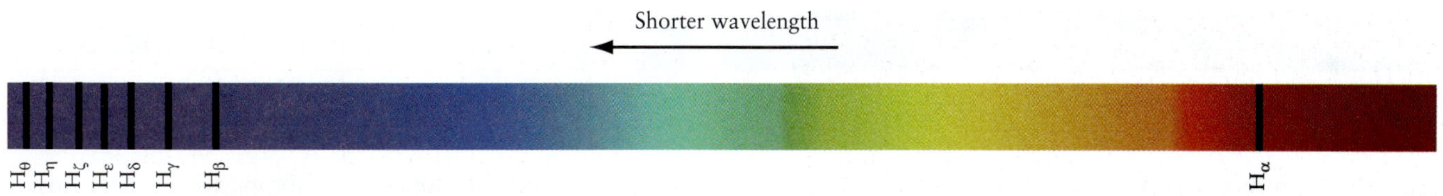

Figure 5-21 R I V U X G

Balmer Lines in the Spectrum of a Star This portion of the spectrum of the star Vega in the constellation Lyra (the Harp) shows eight Balmer lines, from H_α at 656.3 nm through H_θ (H-theta) at 388.9 nm. The series converges at 364.6 nm, slightly to the left of H_θ. Parts of this image were made using ultraviolet radiation, which is indicated by the highlighted U in the wavelength tab. (NOAO)

the Sun, have Balmer absorption lines in their spectra, which shows they have atmospheres that contain hydrogen.

Using trial and error, Balmer discovered a formula from which the wavelengths (λ) of hydrogen's spectral lines can be calculated. Balmer's formula is usually written

$$\frac{1}{\lambda} = R\left(\frac{1}{4} - \frac{1}{n^2}\right)$$

In this formula R is the *Rydberg constant* ($R = 1.097 \times 10^7 \text{ m}^{-1}$), named in honor of the Swedish spectroscopist Johannes Rydberg, and n can be any integer (whole number) greater than 2. To get the wavelength λ_α of the spectral line H_α, you first put $n = 3$ into Balmer's formula:

$$\frac{1}{\lambda_\alpha} = (1.097 \times 10^7 \text{ m}^{-1})\left(\frac{1}{4} - \frac{1}{3^2}\right) = 1.524 \times 10^6 \text{ m}^{-1}$$

Then take the reciprocal:

$$\lambda_\alpha = \frac{1}{1.524 \times 10^6 \text{ m}^{-1}} = 6.563 \times 10^{-7} \text{ m} = 656.3 \text{ nm}$$

To get the wavelength of H_β, use $n = 4$, and to get the wavelength of H_γ, use $n = 5$. If you use $n = \infty$ (the symbol ∞ stands for infinity), you get the short-wavelength end of the hydrogen spectrum at 364.6 nm. (Note that 1 divided by infinity equals zero.)

Bohr's Model of Hydrogen

Bohr realized that to fully understand the structure of the hydrogen atom, he had to be able to derive Balmer's formula using the laws of physics. He first made the rather wild assumption that the electron in a hydrogen atom can orbit the nucleus only in certain specific orbits. (This idea was a significant break with the ideas of Newton, in whose mechanics any orbit should be possible.) Figure 5-22 shows the four smallest of these **Bohr orbits,** labeled by the numbers $n = 1$, $n = 2$, $n = 3$, and so on.

Although confined to one of these allowed orbits while circling the nucleus, an electron can jump from one Bohr orbit to another. For an electron to jump, the hydrogen atom must gain or lose a specific amount of energy. The atom must absorb energy for the electron to go from an inner to an outer orbit; the atom must release energy for the electron to go from an outer to an inner orbit. As an example, Figure 5-23 shows an electron jumping between the $n = 2$ and $n = 3$ orbits of a hydrogen atom as the atom absorbs or emits an H_α photon.

When the electron jumps from one orbit to another, the energy of the photon that is emitted or absorbed equals the difference in energy between these two orbits. This energy difference, and hence the photon energy, is the same whether the jump is from a low orbit to a high orbit (Figure 5-23*a*) or from the high orbit back to the low one (Figure 5-23*b*). According to Planck and Einstein, if two photons have the same energy E, the relationship $E = hc/\lambda$ tells us that they must also have the same wavelength λ. It follows that if an atom can emit photons of a given energy and wavelength, it can also absorb photons of precisely the same energy and wavelength. Thus, Bohr's picture explains Kirchhoff's observation that atoms emit and absorb the same wavelengths of light.

The Bohr picture also helps us visualize what happens to produce an emission line spectrum. When a gas is heated, its atoms move around rapidly and can collide forcefully with each other. These energetic collisions excite the atoms' electrons into high orbits. The electrons then cascade back down to the innermost

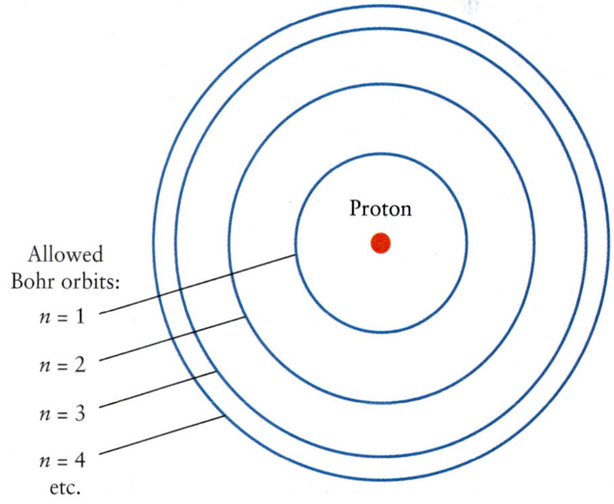

Figure 5-22

The Bohr Model of the Hydrogen Atom In this model, an electron circles the hydrogen nucleus (a proton) only in allowed orbits $n = 1, 2, 3,$ and so forth. The first four Bohr orbits are shown here. This figure is *not* drawn to scale; in the Bohr model, the $n = 2$, 3, and 4 orbits are respectively 4, 9, and 16 times larger than the $n = 1$ orbit.

BOX 5-5 Tools of the Astronomer's Trade

Atoms, the Periodic Table, and Isotopes

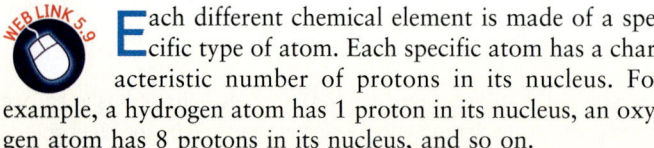

Each different chemical element is made of a specific type of atom. Each specific atom has a characteristic number of protons in its nucleus. For example, a hydrogen atom has 1 proton in its nucleus, an oxygen atom has 8 protons in its nucleus, and so on.

The number of protons in an atom's nucleus is the **atomic number** for that particular element. The chemical elements are most conveniently listed in the form of a **periodic table** (shown in the figure). Elements are arranged in the periodic table in order of increasing atomic number. With only a few exceptions, this sequence also corresponds to increasing average mass of the atoms of the elements. Thus, hydrogen (symbol H), with atomic number 1, is the lightest element. Iron (symbol Fe) has atomic number 26 and is a relatively heavy element.

All the elements listed in a single vertical column of the periodic table have similar chemical properties. For example, the elements in the far right column are all gases under the conditions of temperature and pressure found at Earth's surface, and they are all very reluctant to react chemically with other elements.

In addition to nearly 100 naturally occurring elements, the periodic table includes a number of artificially produced elements. Most of these elements are heavier than uranium (symbol U) and are highly radioactive, which means that they decay into lighter elements within a short time of being created in laboratory experiments. Scientists have succeeded in creating only a few atoms of elements 104 and above.

The number of protons in the nucleus of an atom determines which element that atom is. Nevertheless, the same element may have different numbers of neutrons in its nucleus. For example, oxygen (O) has atomic number 8, so every oxygen nucleus has exactly 8 protons. But oxygen nuclei can have 8, 9, or 10 neutrons. These three slightly different kinds of oxygen are called **isotopes.** The isotope with 8 neutrons is by far the most abundant variety. It is written as ^{16}O, or oxygen-16. The rarer isotopes with 9 and 10 neutrons are designated as ^{17}O and ^{18}O, respectively.

The superscript that precedes the chemical symbol for an element equals the total number of protons and neutrons in a nucleus of that particular isotope. For example, a nucleus of the most common isotope of iron, ^{56}Fe or iron-56, contains a total of 56 protons and neutrons. From the periodic table, the atomic number of iron is 26, so every iron atom has 26 protons in its nucleus. Therefore, the number of neutrons in an iron-56 nucleus is $56 - 26 = 30$. (Most nuclei have more neutrons than protons, especially in the case of the heaviest elements.)

It is extremely difficult to distinguish chemically between the various isotopes of a particular element. Ordinary chemical reactions involve only the electrons that orbit the atom, never the neutrons buried in its nucleus. But there are small differences in the wavelengths of the spectral lines for different isotopes of the same element. For example, the spectral line wavelengths of the hydrogen isotope ^{2}H are about 0.03% greater than the wavelengths for the most common hydrogen isotope, ^{1}H. Thus, different isotopes can be distinguished by careful spectroscopic analysis.

Isotopes are important in astronomy for a variety of reasons. By measuring the relative amounts of different isotopes of a given element in a Moon rock or meteorite, the age of that sample can be determined. The mixture of isotopes left behind when a star explodes into a supernova (see Section 1-3) tells astronomers about the processes that led to the explosion. And knowing the properties of different isotopes of hydrogen and helium is crucial to understanding the nuclear reactions that make the Sun shine. Look for these and other applications of the idea of isotopes in later chapters.

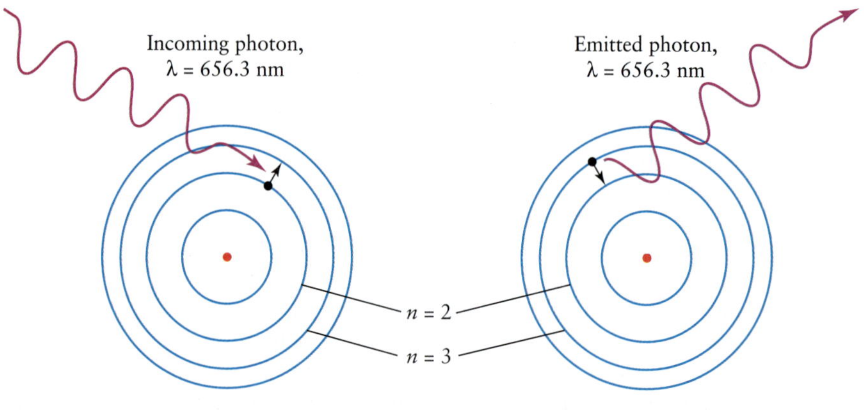

(a) Atom absorbs a 656.3-nm photon; absorbed energy causes electron to jump from the $n = 2$ orbit up the $n = 3$ orbit

(b) Electron falls from the $n = 3$ orbit to the $n = 2$ orbit; energy lost by atom goes into emitting a 656.3-nm photon

Figure 5-23

The Absorption and Emission of an H_α Photon This schematic diagram, drawn according to the Bohr model, shows what happens when a hydrogen atom **(a)** absorbs or **(b)** emits a photon whose wavelength is 656.3 nm.

Periodic Table of the Elements

1 H Hydrogen																	2 He Helium
3 Li Lithium	4 Be Beryllium											5 B Boron	6 C Carbon	7 N Nitrogen	8 O Oxygen	9 F Fluorine	10 Ne Neon
11 Na Sodium	12 Mg Magnesium											13 Al Aluminum	14 Si Silicon	15 P Phosphorus	16 S Sulfur	17 Cl Chlorine	18 Ar Argon
19 K Potassium	20 Ca Calcium	21 Sc Scandium	22 Ti Titanium	23 V Vanadium	24 Cr Chromium	25 Mn Manganese	26 Fe Iron	27 Co Cobalt	28 Ni Nickel	29 Cu Copper	30 Zn Zinc	31 Ga Gallium	32 Ge Germanium	33 As Arsenic	34 Se Selenium	35 Br Bromine	36 Kr Kryton
37 Rb Rubidium	38 Sr Strontium	39 Y Yttrium	40 Zr Zirconium	41 Nb Niobium	42 Mo Molybdenum	43 Tc Technetium	44 Ru Ruthenium	45 Rh Rhodium	46 Pd Palladium	47 Ag Silver	48 Cd Cadmium	49 In Indium	50 Sn Tin	51 Sb Antimony	52 Te Tellurium	53 I Iodine	54 Xe Xenon
55 Cs Cesium	56 Ba Barium	57 La Lanthanum	72 Hf Hafnium	73 Ta Tantaium	74 W Tungsten	75 Re Rhenium	76 Os Osmium	77 Ir Iridium	78 Pt Platinum	79 Au Gold	80 Hg Mercury	81 Tl Thallium	82 Pb Lead	83 Bi Bismuth	84 Po Polonium	85 At Astatine	86 Rn Radon
87 Fr Francium	88 Ra Radium	89 Ac Actinium	104 Rf Rutherfordium	105 Db Dubnium	106 Sg Seaborgium	107 Bh Bohrium	108 Hs Hassium	109 Mt Meitnerium	110 Ds Darmstadium	111 Rg Roentgenium							
			58 Ce Cerium	59 Pr Praseodymium	60 Nd Neodymium	61 Pm Promethium	62 Sm Samarium	63 Eu Europium	64 Gd Gadolinium	65 Tb Terbium	66 Dy Dysprosium	67 Ho Holmium	68 Er Erbium	69 Tm Thulium	70 Yb Ytterbium	71 Lu Lutetium	
			90 Th Thorium	91 Pa Protactinium	92 U Uranium	93 Np Neptunium	94 Pu Plutonium	95 Am Americium	96 Cm Curium	97 Bk Berkelium	98 Cf Californium	99 Es Einsteinium	100 Fm Fermium	101 Md Mendelevium	102 No Nobelium	103 Lr Lawrencium	

possible orbit, emitting photons whose energies are equal to the energy differences between different Bohr orbits. In this fashion, a hot gas produces an emission line spectrum with a variety of different wavelengths.

To produce an absorption line spectrum, begin with a relatively cool gas, so that the electrons in most of the atoms are in inner, low-energy orbits. If a beam of light with a continuous spectrum is shone through the gas, most wavelengths will pass through undisturbed. Only those photons will be absorbed whose energies are just right to excite an electron to an allowed outer orbit. Hence, only certain wavelengths will be absorbed, and dark lines will appear in the spectrum at those wavelengths.

Using his picture of allowed orbits and the formula $E = hc/\lambda$, Bohr was able to prove mathematically that the wavelength λ of the photon emitted or absorbed as an electron jumps between an inner orbit N and an outer orbit n is

Bohr formula for hydrogen wavelengths

$$\frac{1}{\lambda} = R\left(\frac{1}{N^2} - \frac{1}{n^2}\right)$$

N = number of inner orbit

n = number of outer orbit

R = Rydberg constant = $1.097 \times 10^7\ \text{m}^{-1}$

λ = wavelength (in meters) of emitted or absorbed photon

If Bohr let $N = 2$ in this formula, he got back the formula that Balmer discovered by trial and error. Hence, Bohr deduced the meaning of the Balmer series: All the Balmer lines are produced by electrons jumping between the second Bohr orbit

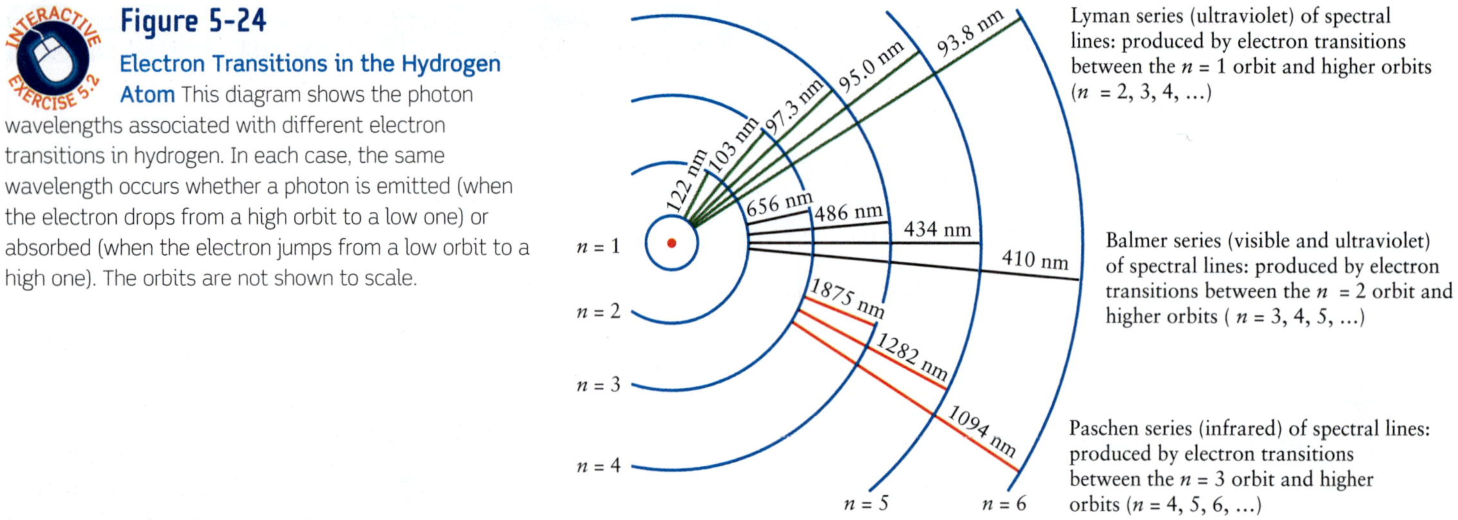

Figure 5-24

Electron Transitions in the Hydrogen Atom This diagram shows the photon wavelengths associated with different electron transitions in hydrogen. In each case, the same wavelength occurs whether a photon is emitted (when the electron drops from a high orbit to a low one) or absorbed (when the electron jumps from a low orbit to a high one). The orbits are not shown to scale.

($N = 2$) and higher orbits ($n = 3$, 4, 5, and so on). Remarkably, as part of his calculation Bohr was able to *derive* the value of the Rydberg constant in terms of Planck's constant, the speed of light, and the mass and electric charge of the electron. This gave particular credence to his radical model.

Bohr's formula also correctly predicts the wavelengths of other series of spectral lines that occur at nonvisible wavelengths (Figure 5-24). Using $N = 1$ gives the **Lyman series,** which is entirely in the ultraviolet. All the spectral lines in this series involve electron transitions between the lowest Bohr orbit and all higher orbits ($n = 2$, 3, 4, and so on). This pattern of spectral lines begins with L_α (Lyman alpha) at 122 nm and converges on L_∞ at 91.2 nm. Using $N = 3$ gives a series of infrared wavelengths called the **Paschen series.** This series, which involves transitions between the third Bohr orbit and all higher orbits, begins with P_α (Paschen alpha) at 1875 nm and converges on P_∞ at 822 nm. Additional series exist at still longer wavelengths.

Atomic Energy Levels

Today's view of the atom owes much to the Bohr model, but is different in certain ways. The modern picture is based on **quantum mechanics,** a branch of physics developed during the 1920s that deals with photons and subatomic particles. As a result of this work, physicists no longer picture electrons as moving in specific orbits about the nucleus. Instead, electrons are now known to have both particle and wave properties and are said to occupy only certain **energy levels** in the atom.

An extremely useful way of displaying the structure of an atom is with an **energy-level diagram.** Figure 5-25 shows such a diagram for hydrogen. The lowest energy level, called the **ground state,** corresponds to the $n = 1$ Bohr orbit. Higher energy levels, called **excited states,** correspond to successively larger Bohr orbits.

An electron can jump from the ground state up to the $n = 2$ level if the atom absorbs a Lyman-alpha photon with a wavelength of 122 nm. Such a photon has energy $E = hc/\lambda = 10.2$ eV (electron volts; see Section 5-5). That's why the energy level of $n = 2$ is shown in Figure 5-25 as having an energy 10.2 eV above that of the ground state (which is usually assigned a value of 0 eV). Similarly, the $n = 3$ level is 12.1 eV above the ground state, and so forth. Electrons can make transitions to higher energy levels by absorbing a photon or in a collision between atoms; they can make transitions to lower energy levels by emitting a photon.

On the energy-level diagram for hydrogen, the $n = \infty$ level has an energy of 13.6 eV. (This corresponds to an infinitely large orbit in the Bohr model.) If the electron is initially in the ground state and the atom absorbs a photon of any energy greater than 13.6 eV, the electron will be removed completely from the atom. This process is called **ionization.** A 13.6-eV photon has a wave-

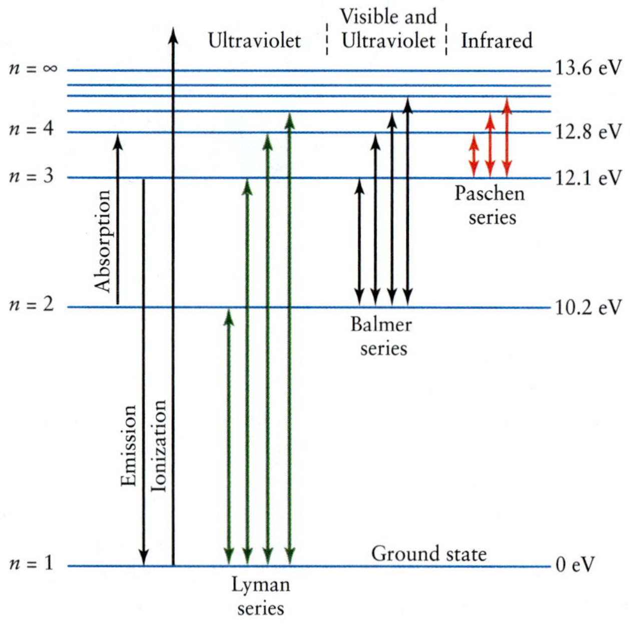

Figure 5-25

Energy-Level Diagram of Hydrogen A convenient way to display the structure of the hydrogen atom is in a diagram like this, which shows the allowed energy levels. The diagram shows a number of possible electron jumps, or transitions, between energy levels. An upward transition occurs when the atom absorbs a photon; a downward transition occurs when the atom emits a photon. (Compare with Figure 5-24.)

length of 91.2 nm, equal to the shortest wavelength in the ultraviolet Lyman series (L_∞). So any photon with a wavelength of 91.2 nm or less can ionize hydrogen. (The Planck formula $E = hc/\lambda$ tells us that the higher the photon energy, the shorter the wavelength.)

As an example, the gaseous nebula shown in Figure 5-18 surrounds a cluster of hot stars which produce copious amounts of ultraviolet photons with wavelengths less than 91.2 nm. Hydrogen atoms in the nebula that absorb these photons become ionized and lose their electrons. When the electrons recombine with the nuclei, they cascade down the energy levels to the ground state and emit visible light in the process. This process is what makes the nebula glow.

The Spectra of Other Elements

LOOKING DEEPER 5.1

The same basic principles that explain the hydrogen spectrum also apply to the atoms of other elements. Electrons in each kind of atom can be only in certain energy levels, so only photons of certain wavelengths can be emitted or absorbed. Because each kind of atom has its own unique arrangement of electron levels, the pattern of spectral lines is likewise unique to that particular type of atom (see the photograph that opens this chapter). These patterns are in general much more complicated than for the hydrogen atom. Hence, there is no simple relationship analogous to the Bohr formula that applies to the spectra of all atoms.

The idea of energy levels explains the emission line spectra and absorption line spectra of gases. But what about the continuous spectra produced by dense objects like the filament of a lightbulb or the coils of a toaster? These objects are made of atoms, so why don't they emit light with an emission line spectrum characteristic of the particular atoms of which they are made?

The reason is directly related to the difference between a gas on the one hand and a liquid or solid on the other. In a gas, atoms are widely separated and can emit photons without interference from other atoms. But in a liquid or a solid, atoms are so close that they almost touch, and thus these atoms interact strongly with each other. These interactions interfere with the process of emitting photons. As a result, the pattern of distinctive bright spectral lines that the atoms would emit in isolation becomes "smeared out" into a continuous spectrum.

ANALOGY Think of atoms as being like tuning forks. If you strike a single tuning fork, it produces a sound wave with a single clear frequency and wavelength, just as an isolated atom emits light of definite wavelengths. But if you shake a box packed full of tuning forks, you will hear a clanging noise that is a mixture of sounds of all different frequencies and wavelengths. This is directly analogous to the continuous spectrum of light emitted by a dense object with closely packed atoms.

With the work of such people as Planck, Einstein, Rutherford, and Bohr, the interchange between astronomy and physics came full circle. Modern physics was born when Newton set out to understand the motions of the planets. Two and a half centuries later, physicists in their laboratories probed the properties of light and the structure of atoms. Their labors had immediate applications in astronomy. Armed with this new understanding of light and matter, astronomers were able to probe in detail the chemical and physical properties of planets, stars, and galaxies.

5-9 The wavelength of a spectral line is affected by the relative motion between the source and the observer

In addition to telling us about temperature and chemical composition, the spectrum of a planet, star, or galaxy can also reveal something about that object's motion through space. This idea dates from 1842, when Christian Doppler, a professor of mathematics in Prague, pointed out that the observed wavelength of light must be affected by motion.

The Doppler effect makes it possible to tell whether astronomical objects are moving toward us or away from us

The Doppler Effect

In **Figure 5-26** a light source is moving from right to left; the circles represent the crests of waves emitted from the moving source at various positions. Each successive wave crest is emitted from a position slightly closer to the observer on the left, so she sees a shorter wavelength—the distance from one crest to the next—than she would if the source were stationary. All the lines in the spectrum

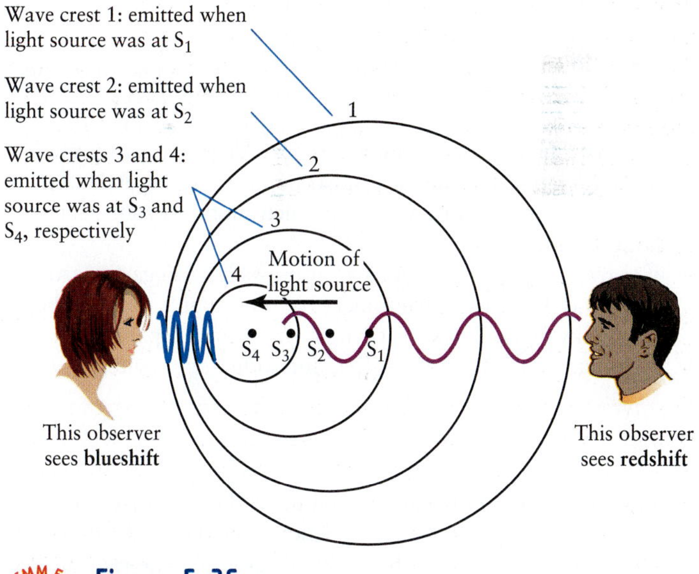

Wave crest 1: emitted when light source was at S_1
Wave crest 2: emitted when light source was at S_2
Wave crests 3 and 4: emitted when light source was at S_3 and S_4, respectively
1
2
3
4
Motion of light source
S_4 S_3 S_2 S_1
This observer sees **blueshift**
This observer sees **redshift**

AIMM 5.3

Figure 5-26

The Doppler Effect The wavelength of light is affected by motion between the light source and an observer. The light source shown here is moving, so wave crests 1, 2, etc., emitted when the source was at points S_1, S_2, etc., are crowded together in front of the source but are spread out behind it. Consequently, wavelengths are shortened (blueshifted) if the source is moving toward the observer and lengthened (redshifted) if the source is moving away from the observer. Motion perpendicular to an observer's line of sight does not affect wavelength.

BOX 5-6 Tools of the Astronomer's Trade

Applications of the Doppler Effect

Doppler's formula relates the radial velocity of an astronomical object to the wavelength shift of its spectral lines. Here are two examples that show how to use this remarkably powerful formula.

EXAMPLE: As measured in the laboratory, the prominent H_α spectral line of hydrogen has a wavelength $\lambda_0 = 656.285$ nm. But in the spectrum of the star Vega (Figure 5-21), this line has a wavelength $\lambda = 656.255$ nm. What can we conclude about the motion of Vega?

Situation: Our goal is to use the ideas of the Doppler effect to find the velocity of Vega toward or away from Earth.

Tools: We use the Doppler shift formula, $\Delta\lambda/\lambda_0 = v/c$, to determine Vega's velocity v.

Answer: The wavelength shift is

$$\Delta\lambda = \lambda - \lambda_0 = 656.255 \text{ nm} - 656.285 \text{ nm} = -0.030 \text{ nm}$$

The negative value means that we see the light from Vega shifted to shorter wavelengths—that is, there is a blueshift. (Note that the shift is very tiny and can be measured only using specialized equipment.) From the Doppler shift formula, the star's radial velocity is

$$v = c\frac{\Delta\lambda}{\lambda_0} = (3.00 \times 10^5 \text{ km/s})\left(\frac{-0.030 \text{ nm}}{656.285 \text{ nm}}\right) = -14 \text{ km/s}$$

Review: The minus sign indicates that Vega is coming toward us at 14 km/s. The star may also be moving perpendicular to the line from Earth to Vega, but such motion produces no Doppler shift.

By plotting the motions of stars such as Vega toward and away from us, astronomers have been able to learn how the Milky Way Galaxy (of which our Sun is a part) is rotating. From this knowledge, and aided by Newton's universal law of gravitation (see Section 4-7), they have made the surprising discovery that the Milky Way contains roughly 10 times more matter than had once been thought! The nature of this unseen *dark matter* is still a subject of debate.

EXAMPLE: In the radio region of the electromagnetic spectrum, hydrogen atoms emit and absorb photons with a wavelength of 21.12 cm, giving rise to a spectral feature commonly called the *21-centimeter line.* The galaxy NGC 3840 in the constellation Leo (the Lion) is receding from us at a speed of 7370 km/s, or about 2.5% of the speed of light. At what wavelength do we expect to detect the 21-cm line from this galaxy?

Situation: Given the velocity of NGC 3840 away from us, our goal is to find the wavelength as measured on Earth of the 21-centimeter line from this galaxy.

Tools: We use the Doppler shift formula to calculate the wavelength shift $\Delta\lambda$, then use this to find the wavelength λ measured on Earth.

Answer: The wavelength shift is

$$\Delta\lambda = \lambda_0\left(\frac{v}{c}\right) = (21.12 \text{ cm})\left(\frac{7370 \text{ km/s}}{3.00 \times 10^5 \text{ km/s}}\right) = 0.52 \text{ cm}$$

Therefore, we will detect the 21-cm line of hydrogen from this galaxy at a wavelength of

$$\lambda = \lambda_0 + \Delta\lambda = 21.12 \text{ cm} + 0.52 \text{ cm} = 21.64 \text{ cm}$$

Review: The 21-cm line has been redshifted to a longer wavelength because the galaxy is receding from us. In fact, most galaxies are receding from us. This observation is one of the key pieces of evidence that the universe is expanding, and has been doing so since the Big Bang that took place almost 14 billion years ago.

of an approaching source are shifted toward the short-wavelength (blue) end of the spectrum. This phenomenon is called a **blueshift.**

The source is receding from the observer on the right in Figure 5-26. The wave crests that reach him are stretched apart, so that he sees a longer wavelength than he would if the source were stationary. All the lines in the spectrum of a receding source are shifted toward the longer-wavelength (red) end of the spectrum, producing a **redshift.** In general, the effect of relative motion on wavelength is called the **Doppler effect.** Police radar guns use the Doppler effect to check for cars exceeding the speed limit: The radar gun sends a radio wave toward the car, and measures the wavelength shift of the reflected wave (and thus the speed of the car).

ANALOGY You have probably noticed a similar Doppler effect for sound waves. When a police car is approaching, the sound waves from its siren have a shorter wavelength and higher frequency than if the siren were at rest, and hence you hear a higher pitch. After the police car passes you and is moving away, you hear a lower pitch from the siren because the sound waves have a longer wavelength and a lower frequency.

Suppose that λ_0 is the wavelength of a particular spectral line from a light source that is not moving. It is the wavelength that you might look up in a reference book or determine in a laboratory experiment for this spectral line. If the source is moving, this particular spectral line is shifted to a different wavelength λ. The size of the wavelength shift is usually written as $\Delta\lambda$, where $\Delta\lambda = \lambda - \lambda_0$. Thus, $\Delta\lambda$ is the difference between the wavelength listed in reference books and the wavelength that you actually observe in the spectrum of a star or galaxy.

Doppler proved that the wavelength shift ($\Delta\lambda$) is governed by the following simple equation:

Doppler shift equation

$$\frac{\Delta\lambda}{\lambda_0} = \frac{v}{c}$$

$\Delta\lambda$ = wavelength shift

λ_0 = wavelength if source is not moving

v = velocity of the source measured along the line of sight

c = speed of light = 3.0×10^5 km/s

CAUTION! The capital Greek letter Δ (delta) is commonly used to denote a change in the value of a quantity. Thus, $\Delta\lambda$ is the change in the wavelength λ due to the Doppler effect. It is *not* equal to a quantity Δ multiplied by a second quantity λ!

Interpreting the Doppler Effect

The velocity determined from the Doppler effect is called **radial velocity,** because v is the component of the star's motion parallel to our line of sight, or along the "radius" drawn from Earth to the star. Of course, a sizable fraction of a star's motion may be perpendicular to our line of sight. The speed of this transverse movement across the sky does not affect wavelengths if the speed is small compared with c. Box 5-6 includes two examples of calculations with radial velocity using the Doppler formula.

CAUTION! The redshifts and blueshifts of stars visible to the naked eye, or even through a small telescope, are only a small fraction of a nanometer. These tiny wavelength changes are far too small to detect visually. (Astronomers were able to detect the tiny Doppler shifts of starlight only after they had developed highly sensitive equipment for measuring wavelengths. This was done around 1890, a half-century after Doppler's original proposal.) So, if you see a star with a red color, it means that the star really is red; it does *not* mean that it is moving rapidly away from us.

The Doppler effect is an important tool in astronomy because it uncovers basic information about the motions of planets, stars, and galaxies. For example, the rotation of the planet Venus was deduced from the Doppler shift of radar waves reflected from its surface. Small Doppler shifts in the spectrum of sunlight have shown that the entire Sun is vibrating like an immense gong. The back-and-forth Doppler shifting of the spectral lines of certain stars reveals that these stars are being orbited by unseen companions; from this astronomers have discovered planets around other stars and massive objects that may be black holes. Astronomers also use the Doppler effect along with Kepler's third law to measure the masses of galaxies. These are but a few examples of how Doppler's discovery has empowered astronomers in their quest to understand the universe.

In this chapter we have glimpsed how much can be learned by analyzing light from the heavens. To analyze this light, however, it is first necessary to collect as much of it as possible, because most light sources in space are very dim. Collecting the faint light from distant objects is a key purpose of telescopes. In the next chapter we will describe both how telescopes work and how they are used.

Key Words

Terms preceded by an asterisk (*) are discussed in the Boxes.

absolute zero, p. 104
absorption line spectrum, p. 112
atom, p. 104
*atomic number, p. 118
Balmer line, p. 116
Balmer series, p. 116
blackbody, p. 106
blackbody curve, p. 106
blackbody radiation, p. 106
blueshift, p. 122
Bohr orbits, p. 117
compound, p. 112
continuous spectrum, p. 112
*degrees Celsius, p. 105
*degrees Fahrenheit, p. 105
Doppler effect, p. 122
electromagnetic radiation, p. 101
electromagnetic spectrum, p. 102
electromagnetism, p. 100
electron, p. 109
electron volt, p. 110
element, p. 111
emission line spectrum, p. 112
energy flux, p. 107
energy level, p. 120
energy-level diagram, p. 120
excited state, p. 120
frequency, p. 102
gamma rays, p. 102
ground state, p. 120
infrared radiation, p. 101
ionization, p. 120
*isotope, p. 118
joule, p. 107
kelvin, p. 104
Kirchhoff's laws, p. 112
light scattering, p. 112
*luminosity, p. 108
Lyman series, p. 120
microwaves, p. 102
molecule, p. 112
nanometer, p. 101
neutron, p. 116
nucleus, p. 116
Paschen series, p. 120
*periodic table, p. 118
photoelectric effect, p. 109
photon, p. 108
Planck's law, p. 110
proton, p. 116
quantum mechanics, p. 120
radial velocity, p. 123
radio waves, p. 101
redshift, p. 122
*solar constant, p. 108
spectral analysis, p. 111
spectral line, p. 111
spectroscopy, p. 113
spectrum (*plural* spectra), p. 99
Stefan-Boltzmann law, p. 107
ultraviolet radiation, p. 102
visible light, p. 101
watt, p. 107
wavelength, p. 101
wavelength of maximum emission, p. 105
Wien's law, p. 107
X rays, p. 102

Key Ideas

The Nature of Light: Light is electromagnetic radiation. It has wavelike properties described by its wavelength λ and frequency ν, and travels through empty space at the constant speed $c = 3.0 \times 10^8$ m/s $= 3.0 \times 10^5$ km/s.

Blackbody Radiation: A blackbody is a hypothetical object that is a perfect absorber of electromagnetic radiation at all wavelengths. Stars closely approximate the behavior of blackbodies, as do other hot, dense objects.

• The intensities of radiation emitted at various wavelengths by a blackbody at a given temperature are shown by a blackbody curve.

• Wien's law states that the dominant wavelength at which a blackbody emits electromagnetic radiation is inversely proportional to the Kelvin temperature of the object: λ_{max} (in meters) = (0.0029 K m)/T.

• The Stefan-Boltzmann law states that a blackbody radiates electromagnetic waves with a total energy flux F directly proportional to the fourth power of the Kelvin temperature T of the object: $F = \sigma T^4$.

Photons: An explanation of blackbody curves led to the discovery that light has particlelike properties. The particles of light are called photons.

• Planck's law relates the energy E of a photon to its frequency ν or wavelength λ: $E = h\nu = hc/\lambda$, where h is Planck's constant.

Kirchhoff's Laws: Kirchhoff's three laws of spectral analysis describe conditions under which different kinds of spectra are produced.

• A hot, dense object such as a blackbody emits a continuous spectrum covering all wavelengths.

• A hot, transparent gas produces a spectrum that contains bright (emission) lines.

• A cool, transparent gas in front of a light source that itself has a continuous spectrum produces dark (absorption) lines in the continuous spectrum.

Atomic Structure: An atom has a small dense nucleus composed of protons and neutrons. The nucleus is surrounded by electrons that occupy only certain orbits or energy levels.

• When an electron jumps from one energy level to another, it emits or absorbs a photon of appropriate energy (and hence of a specific wavelength).

• The spectral lines of a particular element correspond to the various electron transitions between energy levels in atoms of that element.

• Bohr's model of the atom correctly predicts the wavelengths of hydrogen's spectral lines.

The Doppler Shift: The Doppler shift enables us to determine the radial velocity of a light source from the displacement of its spectral lines.

• The spectral lines of an approaching light source are shifted toward short wavelengths (a blueshift); the spectral lines of a receding light source are shifted toward long wavelengths (a redshift).

• The size of a wavelength shift is proportional to the radial velocity of the light source relative to the observer.

Questions

Review Questions

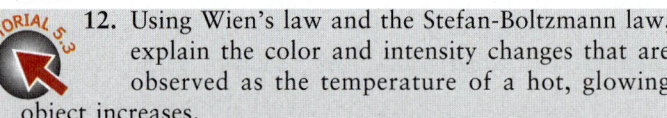

1. When Jupiter is undergoing retrograde motion as seen from Earth, would you expect the eclipses of Jupiter's moons to occur several minutes early, several minutes late, or neither? Explain your answer.
2. Approximately how many times around Earth could a beam of light travel in one second?
3. How long does it take light to travel from the Sun to Earth, a distance of 1.50×10^8 km?
4. How did Newton show that a prism breaks white light into its component colors, but does not add any color to the light?
5. For each of the following wavelengths, state whether it is in the radio, microwave, infrared, visible, ultraviolet, X-ray, or gamma-ray portion of the electromagnetic spectrum. Explain your reasoning. (**a**) 2.6 μm, (**b**) 34 m, (**c**) 0.54 nm, (**d**) 0.0032 nm, (**e**) 0.620 μm, (**f**) 310 nm, (**g**) 0.012 m
6. What is meant by the frequency of light? How is frequency related to wavelength?
7. A cellular phone is actually a radio transmitter and receiver. You receive an incoming call in the form of a radio wave of frequency 880.65 MHz. What is the wavelength (in meters) of this wave?
8. A light source emits infrared radiation at a wavelength of 1150 nm. What is the frequency of this radiation?
9. (**a**) What is a blackbody? (**b**) In what way is a blackbody black? (**c**) If a blackbody is black, how can it emit light? (**d**) If you were to shine a flashlight beam on a perfect blackbody, what would happen to the light?
10. Why do astronomers find it convenient to use the Kelvin temperature scale in their work rather than the Celsius or Fahrenheit scale?
11. Explain why astronomers are interested in blackbody radiation.
12. Using Wien's law and the Stefan-Boltzmann law, explain the color and intensity changes that are observed as the temperature of a hot, glowing object increases.
13. If you double the Kelvin temperature of a hot piece of steel, how much more energy will it radiate per second?
14. The bright star Bellatrix in the constellation Orion has a surface temperature of 21,500 K. What is its wavelength of maximum emission in nanometers? What color is this star?
15. The bright star Antares in the constellation Scorpius (the Scorpion) emits the greatest intensity of radiation at a wavelength of 853 nm. What is the surface temperature of Antares? What color is this star?
16. (**a**) Describe an experiment in which light behaves like a wave. (**b**) Describe an experiment in which light behaves like a particle.
17. How is the energy of a photon related to its wavelength? What kind of photons carry the most energy? What kind of photons carry the least energy?
18. To emit the same amount of light energy per second, which must emit more photons per second: a source of red light, or a source of blue light? Explain your answer.
19. Explain how we know that atoms have massive, compact nuclei.
20. (**a**) Describe the spectrum of hydrogen at visible wavelengths. (**b**) Explain how Bohr's model of the atom accounts for the Balmer lines.

21. Why do different elements display different patterns of lines in their spectra?
22. What is the Doppler effect? Why is it important to astronomers?
23. If you see a blue star, what does its color tell you about how the star is moving through space? Explain your answer.

Advanced Questions

Questions preceded by an asterisk (*) involve topics discussed in the Boxes.

Problem-solving tips and tools

You can find formulas in Box 5-1 for converting between temperature scales. Box 5-2 discusses how a star's radius, luminosity, and surface temperature are related. Box 5-3 shows how to use Planck's law to calculate the energy of a photon. To learn how to do calculations using the Doppler effect, see Box 5-6.

24. Your normal body temperature is 98.6°F. What kind of radiation do you predominantly emit? At what wavelength (in nm) do you emit the most radiation?
25. What is the temperature of the Sun's surface in degrees Fahrenheit?
26. What wavelength of electromagnetic radiation is emitted with greatest intensity by this book? To what region of the electromagnetic spectrum does this wavelength correspond?
27. Black holes are objects whose gravity is so strong that not even an object moving at the speed of light can escape from their surface. Hence, black holes do not themselves emit light. But it is possible to detect radiation from material falling *toward* a black hole. Calculations suggest that as this matter falls, it is compressed and heated to temperatures around 10^6 K. Calculate the wavelength of maximum emission for this temperature. In what part of the electromagnetic spectrum does this wavelength lie?

*28. Use the value of the solar constant given in Box 5-2 and the distance from Earth to the Sun to calculate the luminosity of the Sun.

*29. The star Alpha Lupi (the brightest in the constellation Lupus, the Wolf) has a surface temperature of 21,600 K. How much more energy is emitted each second from each square meter of the surface of Alpha Lupi than from each square meter of the Sun's surface?

*30. Jupiter's moon Io has an active volcano named Pele whose temperature can be as high as 320°C. (**a**) What is the wavelength of maximum emission for the volcano at this temperature? In what part of the electromagnetic spectrum is this? (**b**) The average temperature of Io's surface is −150°C. Compared with a square meter of surface at this temperature, how much more energy is emitted per second from each square meter of Pele's surface?

*31. The bright star Sirius in the constellation of Canis Major (the Large Dog) has a radius of 1.67 $R_\odot$ and a luminosity of 25 $L_\odot$. (**a**) Use this information to calculate the energy flux at the surface of Sirius. (**b**) Use your answer in part (a) to calculate the surface temperature of Sirius. How does your answer compare to the value given in Box 5-2?

32. In Figure 5-13 you can see two distinct dark lines at the boundary between the orange and yellow parts of the Sun's spectrum (in the center of the third colored band from the top of the figure). The wavelengths of these dark lines are 588.99 and 589.59 nm. What do you conclude from this about the chemical composition of the Sun's atmosphere? (*Hint:* See Section 5-6.)
33. Instruments on board balloons and spacecraft detect 511-keV photons coming from the direction of the center of our Galaxy. (The prefix k means *kilo*, or thousand, so 1 keV = 10^3 eV.) What is the wavelength of these photons? To what part of the electromagnetic spectrum do these photons belong?
34. (**a**) Calculate the wavelength of P_δ (P-delta), the fourth wavelength in the Paschen series. (**b**) Draw a schematic diagram of the hydrogen atom and indicate the electron transition that gives rise to this spectral line. (**c**) In what part of the electromagnetic spectrum does this wavelength lie?
35. (**a**) Calculate the wavelength of H_η (H-eta), the spectral line for an electron transition between the $n = 7$ and $n = 2$ orbits of hydrogen. (**b**) In what part of the electromagnetic spectrum does this wavelength lie? Use this to explain why Figure 5-21 is labeled R I V U X G.
36. Certain interstellar clouds contain a very cold, very thin gas of hydrogen atoms. Ultraviolet radiation with any wavelength shorter than 91.2 nm cannot pass through this gas; instead, it is absorbed. Explain why.
37. (**a**) Can a hydrogen atom in the ground state absorb an H-alpha (H_α) photon? Explain why or why not. (**b**) Can a hydrogen atom in the $n = 2$ state absorb a Lyman-alpha (L_α) photon? Explain why or why not.
38. An imaginary atom has just 3 energy levels: 0 eV, 1 eV, and 3 eV. Draw an energy-level diagram for this atom. Show all possible transitions between these energy levels. For each transition, determine the photon energy and the photon wavelength. Which transitions involve the emission or absorption of visible light?
39. The star cluster NGC 346 and nebula shown in Figure 5-18 are located within the Small Magellanic Cloud (SMC), a small galaxy that orbits our Milky Way Galaxy. The SMC and the stars and gas within it are moving away from us at 158 km/s. At what wavelength does the red H_α line of hydrogen (which causes the color of the nebula) appear in the nebula's spectrum?
40. The wavelength of H_β in the spectrum of the star Megrez in the Big Dipper (part of the constellation Ursa Major, the Great Bear) is 486.112 nm. Laboratory measurements demonstrate that the normal wavelength of this spectral line is 486.133 nm. Is the star coming toward us or moving away from us? At what speed?
41. You are given a traffic ticket for going through a red light (wavelength 700 nm). You tell the police officer that because you were approaching the light, the Doppler effect caused a blueshift that made the light appear green (wavelength 500 nm). How fast would you have had to be going for this to be true? Would the speeding ticket be justified? Explain your answer.

Discussion Questions

42. The equation that relates the frequency, wavelength, and speed of a light wave, $\nu = c/\lambda$, can be rewritten as $c = \nu\lambda$. A friend who has studied mathematics but not much astronomy or physics might look at this equation and say: "This equation tells me that the higher the frequency ν, the greater the wave speed c. Since visible light has a higher frequency than radio waves, this means that visible light goes faster than radio waves." How would you respond to your friend?
43. (**a**) If you could see ultraviolet radiation, how might the night sky appear different? Would ordinary objects appear different in the daytime? (**b**) What differences might there be in the appearance of the night sky and in the appearance of ordinary objects in the daytime if you could see infrared radiation?
44. The accompanying visible-light image shows the star cluster NGC 3293 in the constellation Carina (the Ship's Keel). What can you say about the surface temperatures of most of the bright stars in this cluster? In what part of the electromagnetic spectrum do these stars emit most intensely? Are your eyes sensitive to this type of radiation? If not, how is it possible to see these stars at all? There is at least one bright star in this cluster with a distinctly different color from the others; what can you conclude about its surface temperature?

RIVUXG

(David Malin/Anglo-Australian Observatory)

45. The human eye is most sensitive over the same wavelength range at which the Sun emits the greatest intensity of radiation. Suppose creatures were to evolve on a planet orbiting a star somewhat hotter than the Sun. To what wavelengths would their vision most likely be sensitive?
46. Why do you suppose that ultraviolet light can cause skin cancer but ordinary visible light does not?

Web/eBook Questions

47. Search the World Wide Web for information about rainbows. Why do rainbows form? Why do they appear as circular arcs? Why can you see different colors?

Activities

Observing Projects

48. Turn on an electric stove or toaster oven and carefully observe the heating elements as they warm up. Relate your observations to Wien's law and the Stefan-Boltzmann law.
49. Obtain a glass prism (or a diffraction grating, which is probably more readily available and is discussed in the next chapter) and look through it at various light sources, such as an ordinary incandescent light, a neon sign, and a mercury vapor street lamp. **Do not look at the Sun! Looking directly at the Sun causes permanent eye damage or blindness.** Do you have any trouble seeing spectra? What do you have to do to see a spectrum? Describe the differences in the spectra of the various light sources you observed.
50. Use the *Starry Night Enthusiast*™ program to examine some distant celestial objects. First display the entire celestial sphere (select **Guides > Atlas** in the **Favorites** menu) and ensure that deep space objects are displayed by opening **View > Deep Space** and clicking on **Messier Objects** and **Bright NGC Objects.** You can now search for objects (i), (ii), and (iii) listed below. Click the **Find** tab at the left of the main view window to open the Find pane, click on the magnifying glass icon at the left of the edit box at the top of the Find pane and select **Search All** from the menu, and then type the name of the object in the edit box followed by the **Enter (Return)** key. The object will be centered in the view. For each object, use the zoom controls at the right-hand end of the Toolbar (at the top of the main window) to adjust your view until you can see the object in detail. For each object, state whether it has a continuous spectrum, an absorption line spectrum, or an emission line spectrum, and explain your reasoning. (i) The Lagoon Nebula in Sagittarius. (*Hint:* See Figure 5-18.) (With a field of view of about 6° × 4°, you can compare and contrast the appearance of the Lagoon Nebula with the Trifid Nebula just to the north of it.) (ii) M31, the great galaxy in the constellation Andromeda. (*Hint:* The light coming from this galaxy is the combined light of hundreds of billions of individual stars.) (ii) The Moon. (*Hint:* Recall from Section 3-1 that moonlight is simply reflected sunlight.)
51. Use the *Starry Night Enthusiast*™ program to examine the temperatures of several relatively nearby stars. First display the entire celestial sphere (select **Guides > Atlas** in the **Favorites** menu). You can now search for each of the stars listed below. Open the **Find** pane, click on the magnifying glass icon at the left side of the edit box at the top of the Find pane, select Star from the menu that appears, type the name of the star in the edit box and click the **Enter (Return)** key. (i) Altair; (ii) Procyon; (iii) Epsilon Indi; (iv) Tau Ceti; (v) Epsilon Eridani; (vi) Lalande 21185. Information for each star

can then be found by clicking on the **Info** tab at the far left of the *Starry Night Enthusiast*™ window. For each star, record its temperature (listed in the **Info** pane under **Other Data**). Then answer the following questions. (**a**) Which of the stars have a longer wavelength of maximum emission λ_{max} than the Sun? Which of the stars have a shorter λ_{max} than the Sun? (**b**) Which of the stars has a reddish color?

52. Use the *Starry Night Enthusiast*™ program to explore the colors of galaxies. Select **Deep Space > Virgo Cluster** in the **Favorites** menu. The Virgo cluster is a grouping of more than 2000 galaxies extending across about 9 million light-years, located about 50 million light-years from Earth. Move the cursor over one of the larger galaxies in the view (for example, the galaxy named "The Eyes" near the center of the screen) and double-click over the image to center that galaxy in the view. You can zoom in and zoom out on the cluster using the elevation buttons (an upward-pointing triangle and a downward-pointing triangle) to the left of the Home button in the toolbar. You can also rotate the cluster by moving the mouse while holding down both the mouse button (left button on a two-button mouse) and the **Shift** key on the keyboard. (**a**) Zoom in and rotate the image to examine several different galaxies. Are all galaxies the same color? Do all galaxies contain stars of the same surface temperature? (Recall from Section 1-4 that galaxies are assemblages of stars.) (**b**) Examine several spiral galaxies like the one shown in Figure 1-7. What are the dominant colors of the inner regions and outer regions of these galaxies? What can you conclude about the surface temperatures of the brightest stars in the inner regions of spiral galaxies compared to the surface temperatures of those in the outer regions?

Collaborative Exercise

53. The Doppler effect describes how relative motion impacts wavelength. With a classmate, stand up and demonstrate each of the following: (**a**) a blueshifted source for a stationary observer; (**b**) a stationary source and an observer detecting a redshift; and (**c**) a source and an observer both moving in the same direction, but the observer is detecting a redshift. Create simple sketches to illustrate what you and your classmate did.

6 Optics and Telescopes

R I **V** U X G   R **I** V U X G

WEB LINK 6.1 Modern telescopes can view the universe in every range of electromagnetic radiation, although some must be placed above Earth's atmosphere. (Visible: NOAO; Infrared: NASA/JPL-Caltech/C. Engelbracht, University of Arizona)

There is literally more to the universe than meets the eye. As seen through a small telescope, the galaxy M82 (shown in the accompanying image) appears as a bright patch that glows with the light of its billions of stars. But when observed with a telescope sensitive to infrared light—with wavelengths longer than your eye can detect—M82 displays an immense halo that extends for tens of thousands of light-years. The spectrum of this halo reveals it to be composed of tiny dust particles, which are ejected by newly formed stars. The halo's tremendous size shows that stars are forming in M82 at a far greater rate than within our own Galaxy.

These observations are just one example of the tremendous importance of telescopes to astronomy. Whether a telescope detects visible or nonvisible light, its fundamental purpose is the same—to gather more light than can the unaided eye. Telescopes are used to gather the feeble light from distant objects to make bright, sharp images. Telescopes also produce finely detailed spectra of objects in space. These spectra reveal the chemical composition of nearby planets as well as of distant galaxies, and help astronomers understand the nature and evolution of astronomical objects of all kinds.

As the accompanying images of the galaxy M82 show, telescopes for nonvisible light reveal otherwise hidden aspects of the universe. In addition to infrared telescopes, radio telescopes have mapped out the structure of our Milky Way Galaxy; ultraviolet telescopes have revealed the workings of the Sun's outer atmosphere; and gamma-ray telescopes have detected the most powerful explosions in the universe. The telescope, in all its variations, is by far astronomers' most useful tool for collecting data about the universe.

6-1 A refracting telescope uses a lens to concentrate incoming light at a focus

The **optical telescope**—that is, a telescope designed for use with visible light—was invented in the Netherlands in the early seventeenth century. Soon after, Galileo used one of these new inventions for his groundbreaking astronomical observations (see Section 4-5). These first telescopes used carefully shaped pieces of glass, or **lenses,** to make distant objects appear larger and brighter. Telescopes of this same basic design are used today by many amateur astronomers. To understand telescopes of this kind, we need to understand how lenses work.

Refracting telescopes gave humans the first close-up views of the Moon and planets

Refraction of Light

All lenses, including those used in telescopes, make use of the same physical principle: *Light travels at a slower speed in a dense substance.* Thus, although the speed of light in a vacuum is 3.00×10^8 m/s, its speed in glass is less than 2×10^8 m/s. Just as a woman's walking pace slows suddenly when she walks from a boardwalk onto a sandy beach, so light slows abruptly as it enters a piece of glass. The same woman easily resumes her original

Learning Goals

By reading the sections of this chapter, you will learn

6-1 How a refracting telescope uses a lens to form an image

6-2 How a reflecting telescope uses a curved mirror to form an image

6-3 How a telescope's size and Earth's atmosphere limit the sharpness of a telescopic image

6-4 How electronic light detectors have revolutionized astronomy

6-5 How telescopes are used to obtain spectra of astronomical objects

6-6 The advantages of using telescopes that detect radio waves from space

6-7 The advantages of placing telescopes in Earth orbit

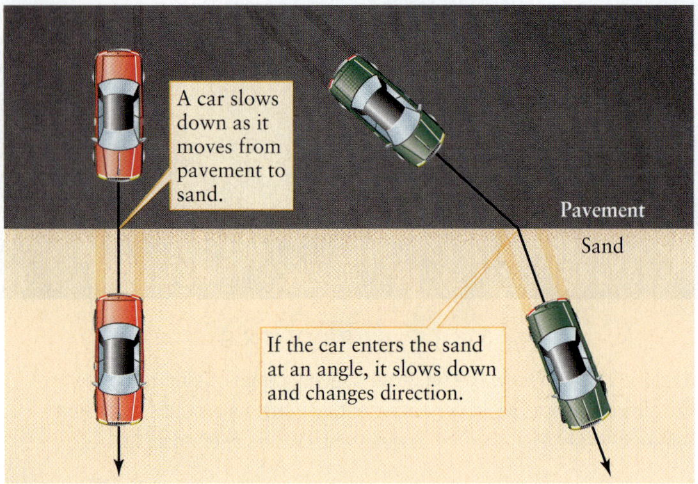

(a) How cars behave

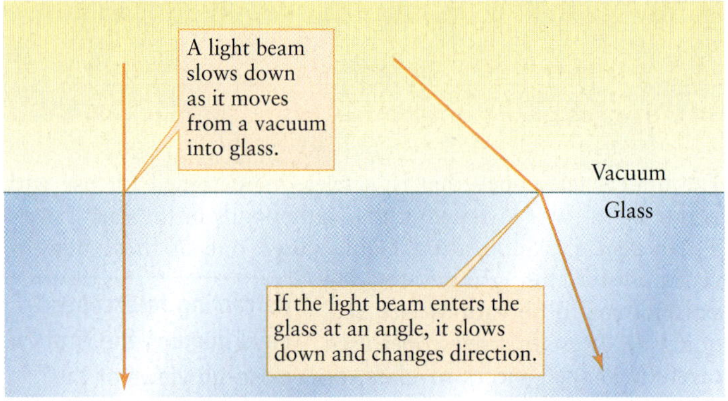

A light beam slows down as it moves from a vacuum into glass.
Vacuum
Glass
If the light beam enters the glass at an angle, it slows down and changes direction.

(b) How light beams behave

Figure 6-1

Refraction (a) When a car drives from smooth pavement into soft sand, it slows down. If it enters the sand at an angle, the front wheel on one side feels the drag of the sand before the other wheel, causing the car to veer to the side and change direction. (b) Similarly, light slows down when it passes from a vacuum into glass and changes direction if it enters the glass at an angle.

pace when she steps back onto the boardwalk; in the same way, light resumes its original speed upon exiting the glass.

A material through which light travels is called a **medium** (*plural* **media**). As a beam of light passes from one transparent medium into another—say, from air into glass, or from glass back into air—the direction of the light can change. This phenomenon, called **refraction,** is caused by the change in the speed of light.

ANALOGY Imagine driving a car from a smooth pavement onto a sandy beach (Figure 6-1*a*). If the car approaches the beach head-on, it slows down when it enters the sand but keeps moving straight ahead. If the car approaches the beach at an angle, however, one of the front wheels will be slowed by the sand before the other is, and the car will veer from its original direction. In the same way, a beam of light changes direction when it enters a piece of glass at an angle (Figure 6-1*b*).

Figure 6-2*a* shows the refraction of a beam of light passing through a piece of flat glass. As the beam enters the upper surface of the glass, refraction takes place, and the beam is bent to a direction more nearly perpendicular to the surface of the glass. As the beam exits from the glass back into the surrounding air, a second refraction takes place, and the beam bends in the opposite sense. (The amount of bending depends on the speed of light in the glass, so different kinds of glass produce slightly different amounts of refraction.) Because the two surfaces of the glass are parallel, the beam emerges from the glass traveling in the same direction in which it entered.

Lenses and Refracting Telescopes

Something more useful happens if the glass is curved into a convex shape (one that is fatter in the middle than at the edges), like the lens in Figure 6-2*b*. When a beam of light rays passes through the lens, refraction causes all the rays to converge at a point called the **focus.** If the light rays entering the lens are all parallel, the focus occurs at a special point called the **focal point.** The distance from the lens to the focal point is called the **focal length** of the lens.

The case of parallel light rays, shown in Figure 6-2*b*, is not merely a theoretical ideal. The stars are so far away that light rays

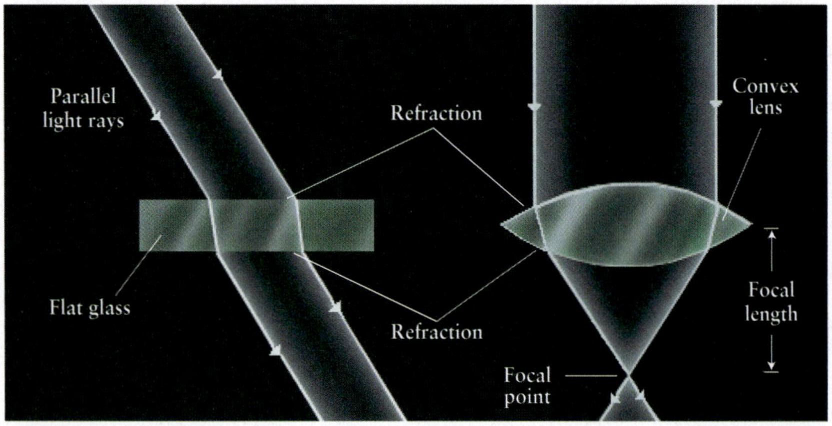

(a) (b)

Figure 6-2

Refraction and Lenses (a) Refraction is the change in direction of a light ray when it passes into or out of a transparent medium such as glass. When light rays pass through a flat piece of glass, the two refractions bend the rays in opposite directions. There is no overall change in the direction in which the light travels. (b) If the glass is in the shape of a convex lens, parallel light rays converge to a focus at a special point called the focal point. The distance from the lens to the focal point is called the focal length of the lens.

from them are essentially parallel, as Figure 6-3 shows. Consequently, a lens always focuses light from an astronomical object to the focal point. If the object has a very small angular size, like a distant star, all the light entering the lens from that object converges onto the focal point. The resulting image is just a single bright dot.

However, if the object is *extended*—that is, has a relatively large angular size, like the Moon or a planet—then light coming from each point on the object is brought to a focus at its own individual point. The result is an extended image that lies in the **focal plane** of the lens (Figure 6-4), which is a plane that includes the focal point. You can use an ordinary magnifying glass in this way to make an image of the Sun on the ground.

To use a lens to make a photograph of an astronomical object, you would place a piece of film or an electronic detector in the focal plane. An ordinary film camera or digital camera works in the same way for taking photographs of relatively close objects here on Earth. Most observations for astronomical research are done in this way to produce a permanent record of the observation. Furthermore, modern image-recording technology is far more sensitive than the human eye, as we will see in Section 6-4. But many amateur astronomers want to view the image visually, and so they add a second lens to magnify the image formed in the focal plane. Such an arrangement of two lenses is called a **refracting telescope**, or **refractor** (Figure 6-5). The large-diameter, long-focal-length lens at the front of the telescope, called the **objective lens,** forms the image; the smaller, shorter-focal-length lens at the rear of the telescope, called the **eyepiece lens,** magnifies the image for the observer.

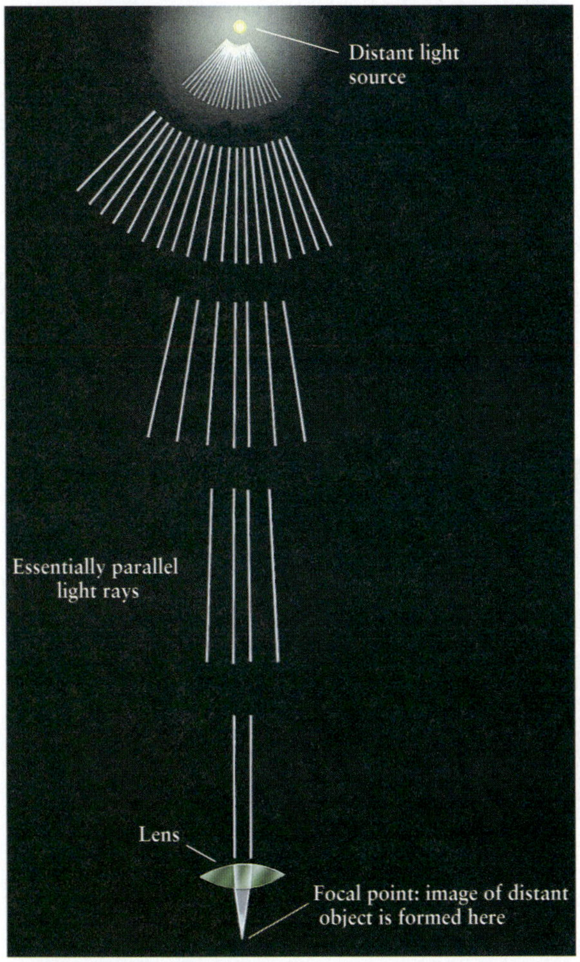

Figure 6-3

Light Rays from Distant Objects Are Parallel Light rays travel away in all directions from an ordinary light source. If a lens is located very far from the light source, only a few of the light rays will enter the lens, and these rays will be essentially parallel. This observation is why we drew parallel rays entering the lens in Figure 6-2*b*.

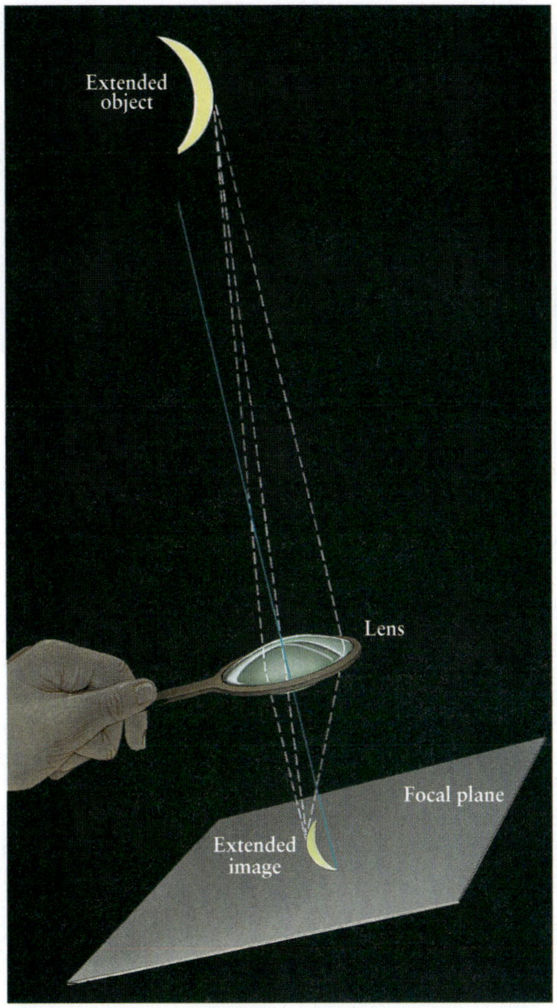

Figure 6-4

A Lens Creates an Extended Image of an Extended Object Light coming from each point on an extended object passes through a lens and produces an image of that point. All of these tiny images put together make an extended image of the entire object. The image of a very distant object is formed in a plane called the focal plane. The distance from the lens to the focal plane is called the focal length.

Light-Gathering Power

In addition to the focal length, the other important dimension of the objective lens of a refractor is the diameter. Compared with

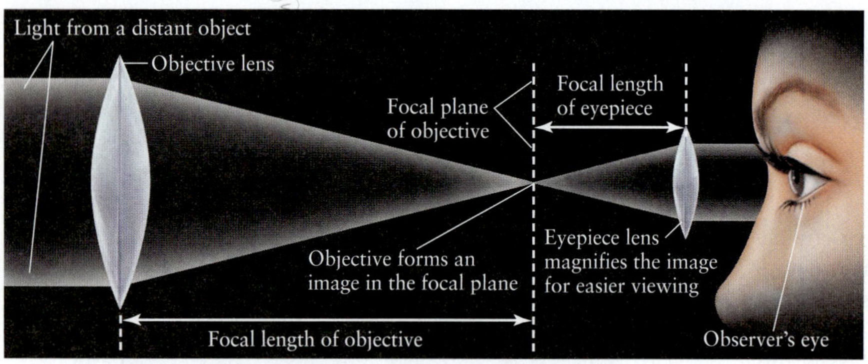

ANIMATION 6.1

Figure 6-5

A Refracting Telescope A refracting telescope consists of a large-diameter objective lens with a long focal length and a small eyepiece lens of short focal length. The eyepiece lens magnifies the image formed by the objective lens in its focal plane (shown as a dashed line). To take a photograph, the eyepiece is removed and the film or electronic detector is placed in the focal plane.

a small-diameter lens, a large-diameter lens captures more light, produces brighter images, and allows astronomers to detect fainter objects. (For the same reason, the iris of your eye opens when you go into a darkened room to allow you to see dimly lit objects.)

The **light-gathering power** of a telescope is directly proportional to the area of the objective lens, which in turn is proportional to the square of the lens diameter (Figure 6-6). Thus, if you double the diameter of the lens, the light-gathering power increases by a factor of $2^2 = 2 \times 2 = 4$. **Box 6-1** describes how to compare the light-gathering power of different telescopes.

Because light-gathering power is so important for seeing faint objects, the lens diameter is almost always given in describing a telescope. For example, the Lick telescope on Mount Hamilton in California is a 90-cm refractor, which means that it is a refracting telescope whose objective lens is 90 cm in diameter. By comparison, Galileo's telescope of 1610 was a 3-cm refractor. The Lick telescope has an objective lens 30 times larger in diameter, and so has $30 \times 30 = 900$ times the light-gathering power of Galileo's instrument.

Magnification

In addition to their light-gathering power, telescopes are useful because they magnify distant objects. As an example, the angular diameter of the Moon as viewed with the naked eye is about 0.5°. But when Galileo viewed the Moon through his telescope, its apparent angular diameter was 10°, large enough so that he could identify craters and mountain ranges. The **magnification, or magnifying power,** of a telescope is the ratio of an object's angular diameter seen through the telescope to its naked-eye angular diameter. Thus, the magnification of Galileo's telescope was $10°/0.5° = 20$ times, usually written as 20×.

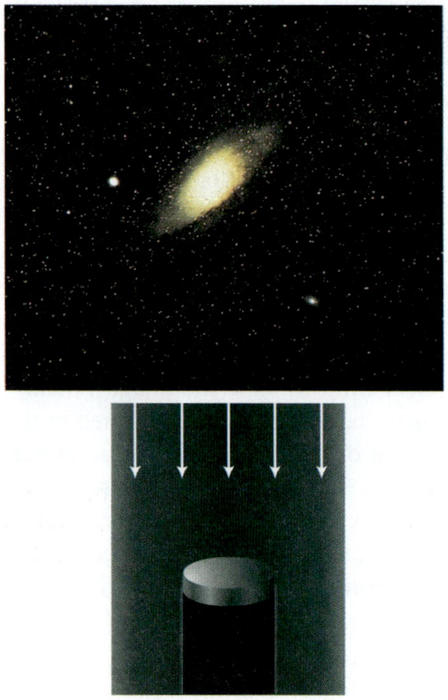
Small-diameter objective lens: dimmer image, less detail

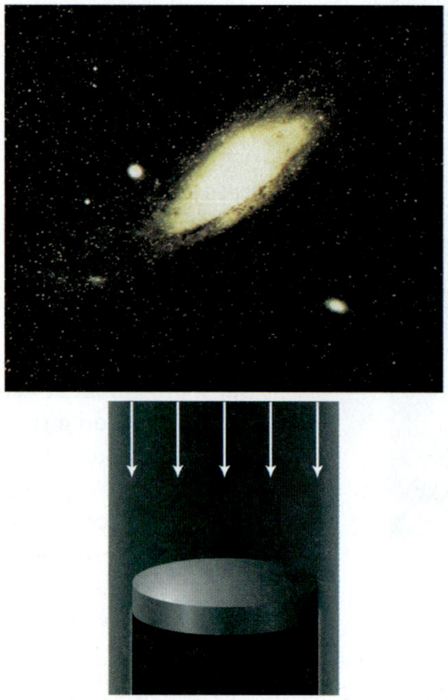
Large-diameter objective lens: brighter image, more detail

Figure 6-6 R I V U X G

Light-Gathering Power These two photographs of the galaxy M31 in Andromeda were taken using the same exposure time and at the same magnification, but with two different telescopes with objective lenses of different diameters. The right-hand photograph is brighter and shows more detail because it was made using the large-diameter lens, which intercepts more starlight than a small-diameter lens. This same principle applies to telescopes that use curved mirrors rather than lenses to collect light (see Section 6-2). (Association of Universities for Research in Astronomy)

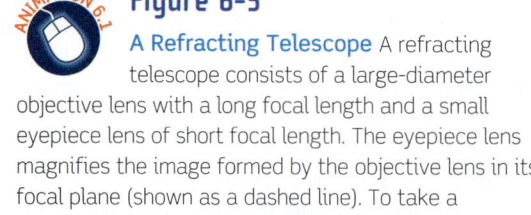

BOX 6-1 Tools of the Astronomer's Trade

Magnification and Light-Gathering Power

The magnification of a telescope is equal to the focal length of the objective divided by the focal length of the eyepiece. Telescopic eyepieces are usually interchangeable, so the magnification of a telescope can be changed by using eyepieces of different focal lengths.

EXAMPLE: A small refracting telescope has an objective of focal length 120 cm. If the eyepiece has a focal length of 4.0 cm, what is the magnification of the telescope?

Situation: We are given the focal lengths of the telescope's objective and eyepiece lenses. Our goal is to calculate the magnification provided by this combination of lenses.

Tools: We use the relationship that the magnification equals the focal length of the objective (120 cm) divided by the focal length of the eyepiece (4.0 cm).

Answer: Using this relationship,

$$\text{Magnification} = \frac{120\text{ cm}}{4.0\text{ cm}} = 30 \text{ (usually written as } 30\times)$$

Review: A magnification of 30× means that as viewed through this telescope, a large lunar crater that subtends an angle of 1 arcminute to the naked eye will appear to subtend an angle 30 times greater, or 30 arcminutes (one-half of a degree). This magnification makes the details of the crater much easier to see.

If a 2.0-cm-focal-length eyepiece is used instead, the magnification will be (120 cm)/(2.0 cm) = 60×. The shorter the focal length of the eyepiece, the greater the magnification.

The light-gathering power of a telescope depends on the diameter of the objective lens; it does not depend on the focal length. The light-gathering power is proportional to the square of the diameter. As an example, a fully dark adapted human eye has a pupil diameter of about 5 mm. By comparison, a small telescope whose objective lens is 5 cm in diameter has 10 times the diameter and $10^2 = 100$ times the light-gathering power of the eye. (Recall that there are 10 mm in 1 cm.) Hence, this telescope allows you to see objects 100 times fainter than you can see without a telescope.

EXAMPLE: The same relationships apply to reflecting telescopes, discussed in Section 6-2. Each of the two Keck telescopes on Mauna Kea in Hawaii (discussed in Section 6-3; see Figure 6-16) uses a concave mirror 10 m in diameter to bring starlight to a focus. How many times greater is the light-gathering power of either Keck telescope compared to that of the human eye?

Situation: We are given the diameters of the pupil of the human eye (5 mm) and of the mirror of either Keck telescope (10 m). Our goal is to compare the light-gathering powers of these two optical instruments.

Tools: We use the relationship that light-gathering power is proportional to the square of the diameter of the area that gathers light. Hence, the *ratio* of the light-gathering powers is equal to the square of the ratio of the diameters.

Answer: We first calculate the ratio of the diameter of the Keck mirror to the diameter of the pupil. To determine this ratio, we must first express both diameters in the same units. Because there are 1000 mm in 1 meter, the diameter of the Keck mirror can be expressed as

$$10\text{ m} \times \frac{1000\text{ mm}}{1\text{ m}} = 10{,}000\text{ mm}$$

Thus, the light-gathering power of either of the Keck telescopes is greater than that of the human eye by a factor of

$$\frac{(10{,}000\text{ mm})^2}{(5\text{ mm})^2} = (2000)^2 = 4 \times 10^6 = 4{,}000{,}000$$

Review: Either Keck telescope can gather *4 million* times as much light as a human eye. When it comes to light-gathering power, the bigger the telescope, the better!

INTERACTIVE EXERCISE 6.1

The magnification of a refracting telescope depends on the focal lengths of both of its lenses:

$$\text{Magnification} = \frac{\text{focal length of objective lens}}{\text{focal length of eyepiece lens}}$$

This formula shows that using a long-focal-length objective lens with a short-focal-length eyepiece gives a large magnification. Box 6-1 illustrates how this formula is used.

CAUTION! Many people think that the primary purpose of a telescope is to magnify images. But in fact magnification is *not* the most important aspect of a telescope. The reason is that there is a limit to how sharp any astronomical image can be, due either to the blurring caused by Earth's atmosphere or to fundamental limitations imposed by the nature of light itself. (We will describe these effects in more detail in Section 6-3.) Magnifying a blurred image may make it look bigger but will not make it any clearer. Thus, beyond a certain point, there is nothing to be gained by further magnification. Astronomers put much more store in the light-gathering power of a telescope than in its magnification. Greater light-gathering power means brighter images, which makes it easier to see faint details.

Disadvantages of Refracting Telescopes

If you were to build a telescope like that in Figure 6-5 using only the instructions given so far, you would probably be disappointed with the results. The problem is that a lens bends different colors of light through different angles, just as a prism does (recall Figure 5-3). As a result, different colors do not focus at the same point, and stars viewed through a telescope that uses a simple lens are surrounded by fuzzy, rainbow-colored halos. Figure 6-7*a* shows this optical defect, called **chromatic aberration.**

One way to correct for chromatic aberration is to use an objective lens that is not just a single piece of glass. Different types of glass can be manufactured by adding small amounts of chemicals to the glass when it is molten. Because of these chemicals, the speed of light varies slightly from one kind of glass to another, and the refractive properties vary as well. If a thin lens is mounted just behind the main objective lens of a telescope, as shown in Figure 6-7*b*, and if the telescope designer carefully chooses two different kinds of glass for these two lenses, different colors of light can be brought to a focus at the same point.

Chromatic aberration is only the most severe of a host of optical problems that must be solved in designing a high-quality refracting telescope. Master opticians of the nineteenth century devoted their careers to solving these problems, and several magnificent refractors were constructed in the late 1800s (Figure 6-8).

Unfortunately, there are several negative aspects of refractors that even the finest optician cannot overcome:

1. Because faint light must readily pass through the objective lens, the glass from which the lens is made must be totally free of defects, such as the bubbles that frequently form when molten glass is poured into a mold. Such defect-free glass is extremely expensive.

2. Glass is opaque to certain kinds of light. Ultraviolet light is absorbed almost completely, and even visible light is dimmed substantially as it passes through the thick slab of glass that makes up the objective lens.

3. It is impossible to produce a large lens that is entirely free of chromatic aberration.

4. Because a lens can be supported only around its edges, a large lens tends to sag and distort under its own weight. This distortion has adverse effects on the image clarity.

For these reasons and more, few major refractors have been built since the beginning of the twentieth century. Instead, astronomers have avoided all of the limitations of refractors by building telescopes that use a mirror instead of a lens to form an image.

6-2 A reflecting telescope uses a mirror to concentrate incoming light at a focus

Almost all modern telescopes form an image using the principle of **reflection.** To understand reflection, imagine drawing a dashed line perpendicular to the surface of a flat mirror at the point where a light ray strikes the mirror (Figure 6-9*a*). The angle *i* between the *incident* (arriving) light ray and the perpendicular is al-

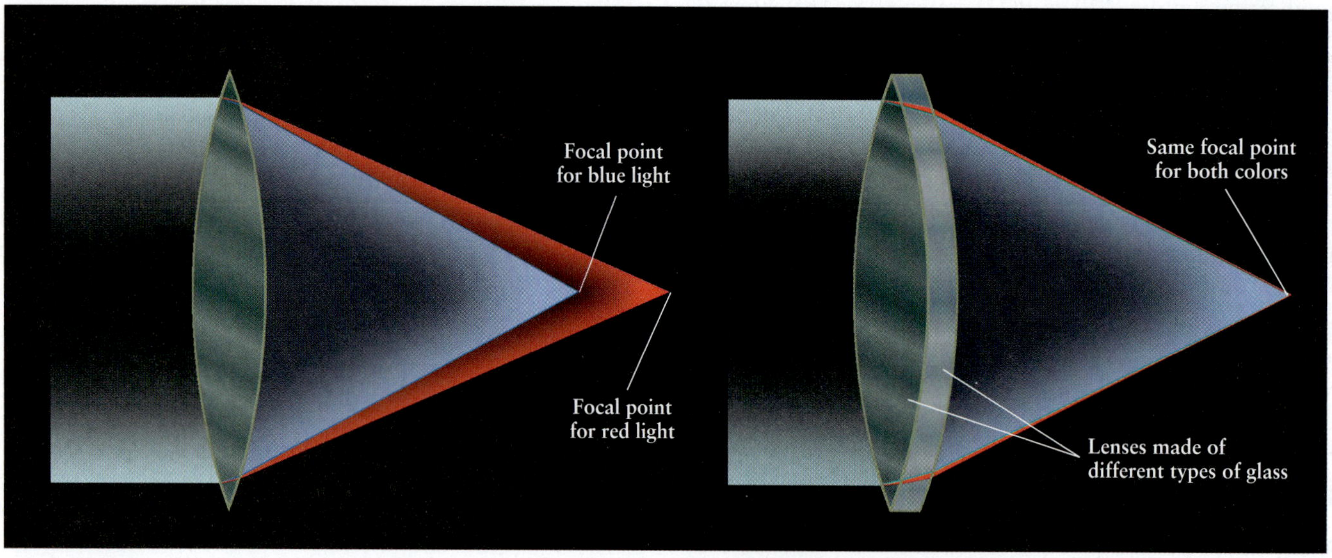

(a) The problem: chromatic aberration

(b) The solution: use two lenses

Figure 6-7

Chromatic Aberration (a) A single lens suffers from a defect called chromatic aberration, in which different colors of light are brought to a focus at different distances from the lens. **(b)** This problem can be corrected by adding a second lens made from a different kind of glass.

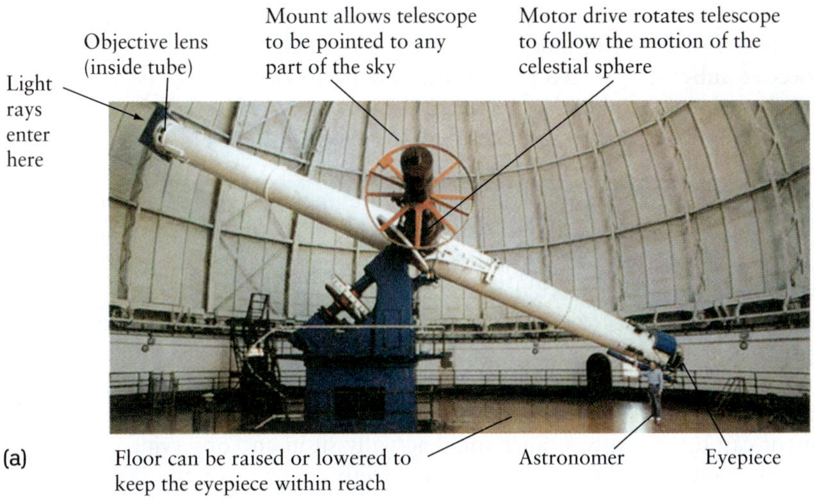

(a)

(b)

Figure 6-8 R I V U X G

A Large Refracting Telescope **(a)** This giant refractor, built in 1897, is housed at Yerkes Observatory near Chicago. The telescope tube is 19.5 m (64 ft) long; it has to be this long because the focal length of the objective is just under 19.5 m (see Figure 6-5). As Earth rotates, the motor drive rotates the telescope in the opposite direction in order to keep the object being studied within the telescope's field of view. **(b)** This historical photograph shows the astronomer George van Biesbrock with the objective lens of the Yerkes refractor. This lens, the largest ever made, is 102 cm (40 in.) in diameter. (Yerkes Observatory)

ways equal to the angle *r* between the *reflected* ray and the perpendicular.

In 1663, the Scottish mathematician James Gregory first proposed a telescope using reflection from a concave mirror—one that is fatter at the edges than at the middle. Such a mirror makes parallel light rays converge to a focus (Figure 6-9*b*). The distance between the reflecting surface and the focus is the focal length of the mirror. A telescope that uses a curved mirror to make an image of a distant object is called a **reflecting telescope,** or **reflector.** Using terminology similar to that used for refractors, the mirror that forms the image is called the **objective mirror** or **primary mirror.**

All of the largest professional telescopes—and most amateur telescopes—are reflecting telescopes

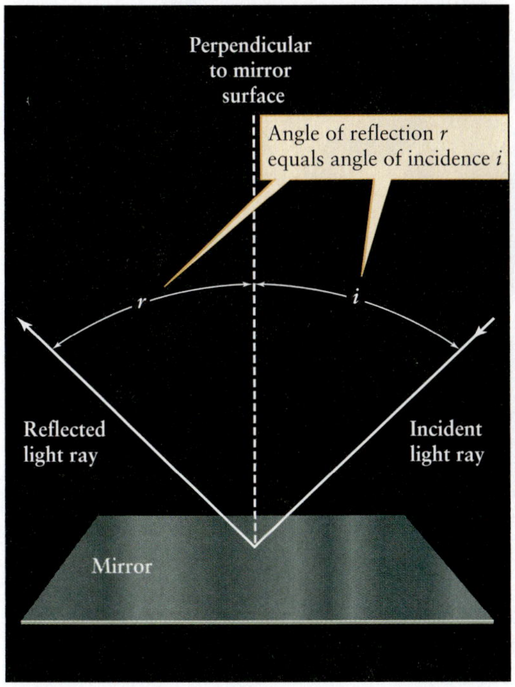

(a)

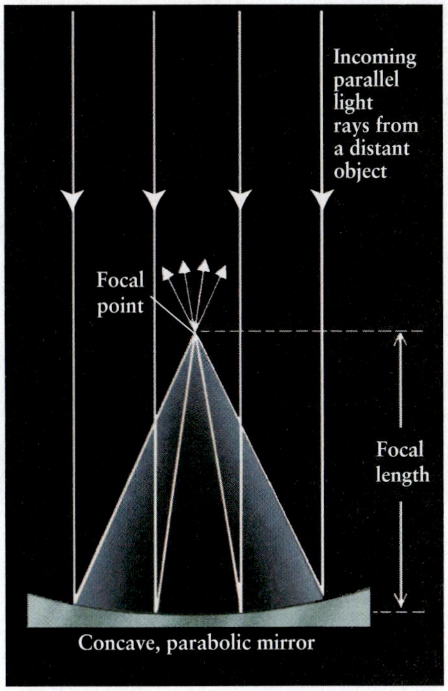

(b)

Figure 6-9

Reflection **(a)** The angle at which a beam of light approaches a mirror, called the angle of incidence (*i*), is always equal to the angle at which the beam is reflected from the mirror, called the angle of reflection (*r*). **(b)** A concave mirror causes parallel light rays to converge to a focus at the focal point. The distance between the mirror and the focal point is the focal length of the mirror.

To make a reflector, an optician grinds and polishes a large slab of glass into the appropriate concave shape. The glass is then coated with silver, aluminum, or a similar highly reflective substance. Because light reflects off the surface of the glass rather than passing through it, defects within the glass—which would have very negative consequences for the objective lens of a refracting telescope—have no effect on the optical quality of a reflecting telescope.

Another advantage of reflectors is that they do not suffer from the chromatic aberration that plagues refractors. This is because reflection is not affected by the wavelength of the incoming light, so all wavelengths are reflected to the same focus. (A small amount of chromatic aberration may arise if the image is viewed using an eyepiece lens.) Furthermore, the mirror can be fully supported by braces on its back, so that a large, heavy mirror can be mounted without much danger of breakage or surface distortion.

Designs for Reflecting Telescopes

Although a reflecting telescope has many advantages over a refractor, the arrangement shown in Figure 6-9*a* is not ideal. One problem is that the focal point is in front of the objective mirror. If you try to view the image formed at the focal point, your head will block part or all of the light from reaching the mirror.

To get around this problem, in 1668 Isaac Newton simply placed a small, flat mirror at a 45° angle in front of the focal point, as sketched in Figure 6-10*a*. This secondary mirror deflects the light rays to one side, where Newton placed an eyepiece lens to magnify the image. A reflecting telescope with this optical design is appropriately called a **Newtonian reflector** (Figure 6-10*b*). The magnifying power of a Newtonian reflector is calculated in the same way as for a refractor: The focal length of the objective mirror is divided by the focal length of the eyepiece (see Box 6-1).

Later astronomers modified Newton's original design. The objective mirrors of some modern reflectors are so large that an astronomer could actually sit in an "observing cage" at the undeflected focal point directly in front of the objective mirror. (In practice, riding in this cage on a winter's night is a remarkably cold and uncomfortable experience.) This arrangement is called a **prime focus** (Figure 6-11*a*). It usually provides the highest-quality image, because there is no need for a secondary mirror (which might have imperfections).

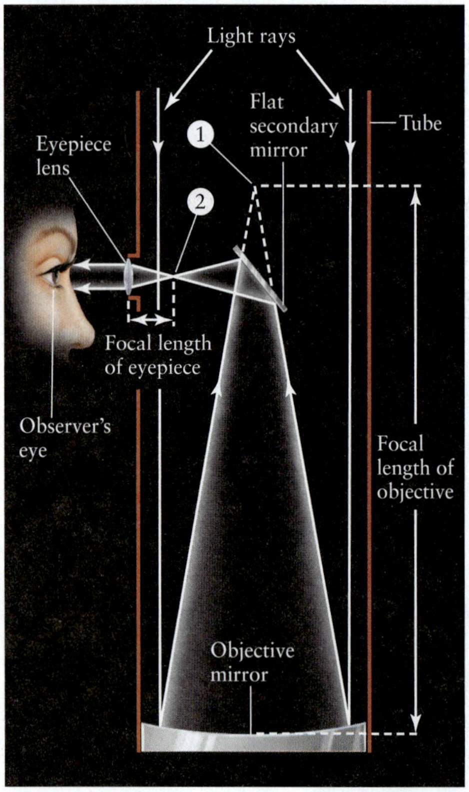

(a)

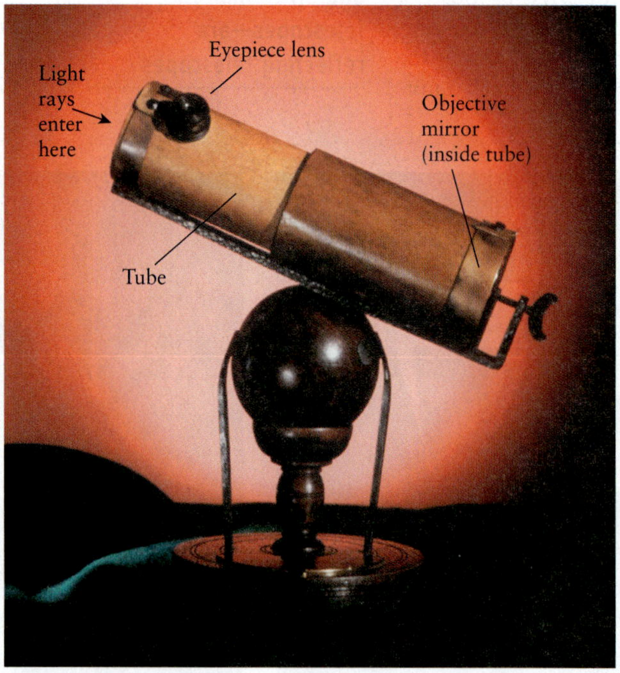

(b)

Figure 6-10 R I V U X G

A Newtonian Telescope (a) In a Newtonian telescope, the image made by the objective is moved from point 1 to point 2 by means of a flat mirror called the secondary. An eyepiece magnifies this image, just as for a refracting telescope (Figure 6-5). **(b)** This is a replica of a Newtonian telescope built by Isaac Newton in 1672. The objective mirror is 3 cm (1.3 inches) in diameter and the magnification is 40× (Royal Greenwich Observatory/Science Photo Library)

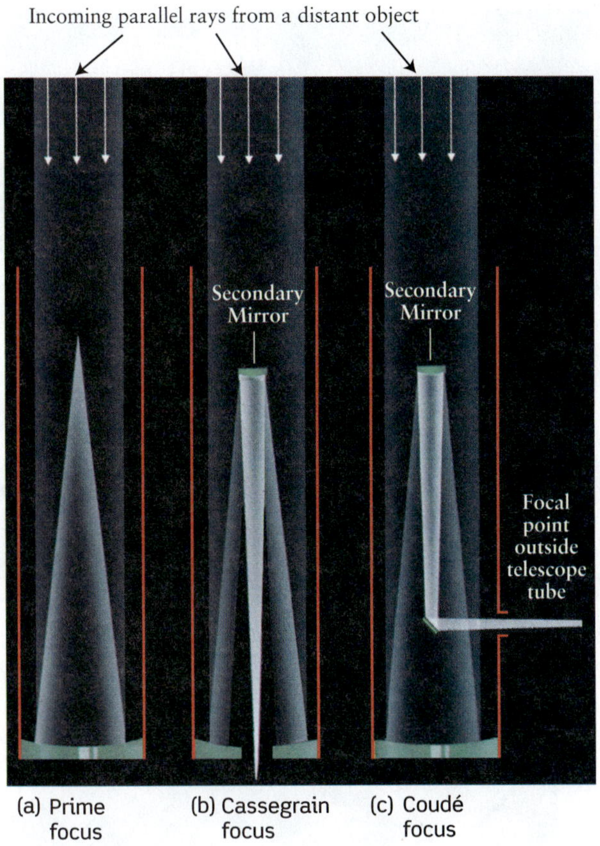

ANIMATION 6.2

Figure 6-11

Designs for Reflecting Telescopes Three common optical designs for reflecting telescopes are shown here. **(a)** Prime focus is used only on some large telescopes; an observer or instrument is placed directly at the focal point, within the barrel of the telescope. **(b)** The Cassegrain focus is used on reflecting telescopes of all sizes, from 90-mm (3.5-in) reflectors used by amateur astronomers to giant research telescopes on Mauna Kea (see Figure 6-14). **(c)** The coudé focus is useful when large and heavy optical apparatus is to be used at the focal point. Light reflects off the objective mirror to a secondary mirror, then back down to a third, angled mirror. With this arrangement, the heavy apparatus at the focal point does not have to move when the telescope is repositioned.

Another popular optical design, called a **Cassegrain focus** after the French contemporary of Newton who first proposed it, also has a convenient, accessible focal point. A hole is drilled directly through the center of the primary mirror, and a convex secondary mirror placed in front of the original focal point reflects the light rays back through the hole (Figure 6-11*b*).

A fourth design is useful when there is optical equipment too heavy or bulky to mount directly on the telescope. Instead, a series of mirrors channels the light rays away from the telescope to a remote focal point where the equipment is located. This design is called a **coudé focus,** from a French word meaning "bent like an elbow" (Figure 6-11*c*).

CAUTION! You might think that the secondary mirror in the Newtonian design and the Cassegrain and coudé designs shown in Figure 6-11 would cause a black spot or hole in the center of the telescope image. But this spot does not form. The reason is that light from every part of the object lands on every part of the primary, objective mirror. Hence, any portion of the mirror can itself produce an image of the distant object, as Figure 6-12 shows. The only effect of the secondary mirror is that it prevents part of the light from reaching the objective mirror, which reduces somewhat the light-gathering power of the telescope.

Spherical Aberration

A reflecting telescope must be designed to minimize a defect called **spherical aberration** (Figure 6-13). At issue is the precise shape of a mirror's concave surface. A spherical surface is easy to grind and polish, but different parts of a spherical mirror focus light onto slightly different spots (Figure 6-13*a*). This results in a fuzzy image.

One common way to eliminate spherical aberration is to polish the mirror's surface to a parabolic shape, because a parabola reflects parallel light rays to a common focus (Figure 6-13*b*). Unfortunately, the astronomer then no longer has a wide-angle view. Furthermore, unlike spherical mirrors, parabolic mirrors suffer

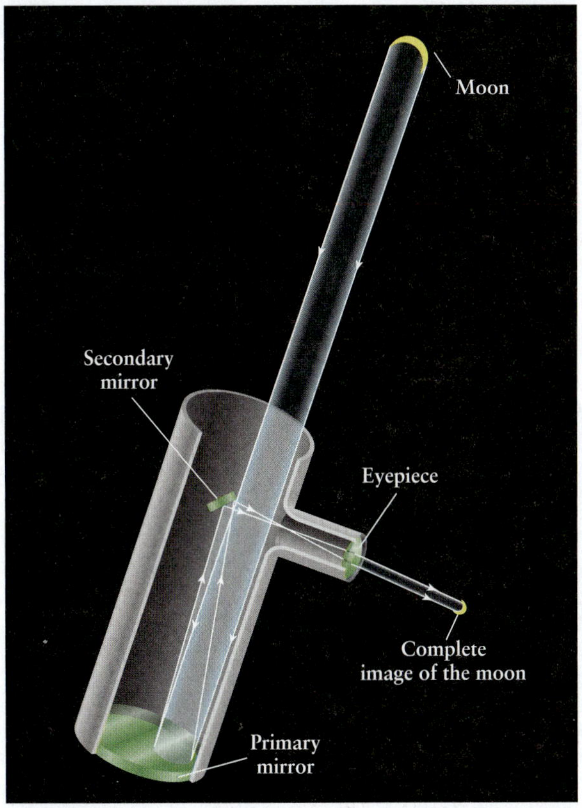

Figure 6-12

The Secondary Mirror Does Not Cause a Hole in the Image This illustration shows how even a small portion of the primary (objective) mirror of a reflecting telescope can make a complete image of the Moon. Thus, the secondary mirror does not cause a black spot or hole in the image. (It does, however, make the image a bit dimmer by reducing the total amount of light that reaches the primary mirror.)

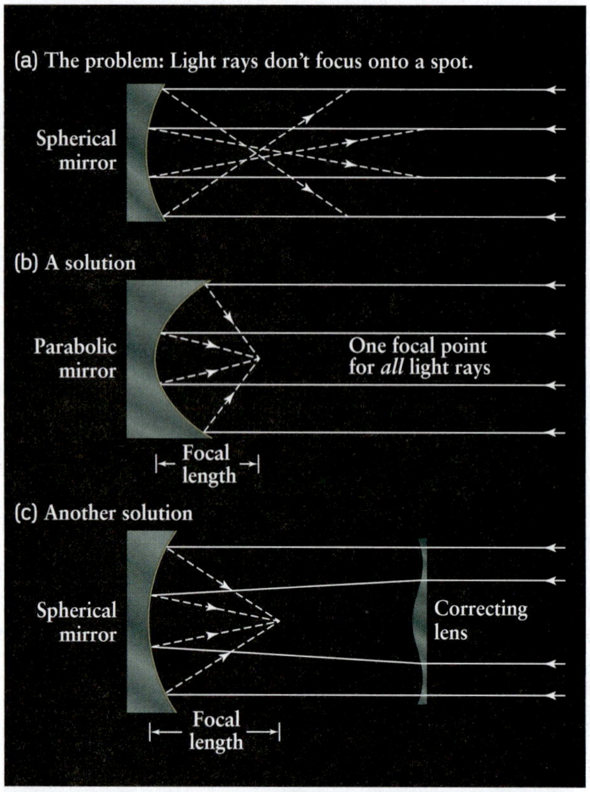

Figure 6-13

Spherical Aberration (a) Different parts of a spherically concave mirror reflect light to slightly different points. This effect, called spherical aberration, causes image blurring. This difficulty can be corrected by either (b) using a parabolic mirror or (c) using a correcting lens in front of the mirror.

from a defect called **coma,** wherein star images far from the center of the field of view are elongated to look like tiny teardrops. A different approach is to use a spherical mirror, thus minimizing coma, and to place a thin correcting lens at the front of the telescope to eliminate spherical aberration (Figure 6-13*c*). This approach is only used on relatively small reflecting telescopes for amateur astronomers.

The Largest Reflectors

WEB LINK 6.2

There are over a dozen optical reflectors in operation with primary mirrors between 8 meters (26.2 feet) and 11 meters (36.1 feet) in diameter. Figure 6-14*a* shows the objective mirror of one of the four Very Large Telescope (VLT) units in Chile, and Figure 6-14*b* shows the objective and secondary mirrors of the Gemini North telescope in Hawaii. A near-twin of Gemini North, called Gemini South, is in Cerro Pachón, Chile. These twins allow astronomers to observe both the northern and southern parts of the celestial sphere with essentially the same state-of-the-art instrument. Two other "twins" are the side-by-side 8.4-m objective mirrors of the Large Binocular Telescope in Arizona. Combining the light from these two mirrors gives double the light-gathering power, equivalent to a single 11.8-m mirror.

(a) A large objective mirror

(b) A large Cassegrain telescope

Figure 6-14 R I V U X G

Reflecting Telescopes (a) This photograph shows technicians preparing an objective mirror 8.2 meters in diameter for the European Southern Observatory in Chile. The mirror was ground to a curved shape with a remarkable precision of 8.5 nanometers. (b) This view of the Gemini North telescope shows its 8.1-meter objective mirror (1). Light incident on this mirror is reflected toward the 1.0-meter secondary mirror (2), then through the hole in the objective mirror (3) to the Cassegrain focus (see Figure 6-11*b*). (a: SAGEM; b: NOAO/AURA/NSF)

Several other reflectors around the world have objective mirrors between 3 and 6 meters in diameter, and dozens of smaller but still powerful telescopes have mirrors in the range of 1 to 3 meters. There are thousands of professional astronomers, each of whom has several ongoing research projects, and thus the demand for all of these telescopes is high. On any night of the year, nearly every research telescope in the world is being used to explore the universe.

6-3 Telescope images are degraded by the blurring effects of the atmosphere and by light pollution

TUTORIAL 6.1

In addition to providing a brighter image, a large telescope also helps achieve a second major goal: It produces star images that are sharp and crisp. A quantity called **angular resolution** gauges how well fine details can be seen. Poor angular resolution causes star images to be fuzzy and blurred together.

To determine the angular resolution of a telescope, pick out two adjacent stars whose separate images are just barely discernible (Figure 6-15). The angle θ (the Greek letter theta) between these stars is the telescope's angular resolution; the *smaller* that angle, the finer the details that can be seen and the sharper the image.

When you are asked to read the letters on an eye chart, what's being measured is the angular resolution of your eye. If you have 20/20 vision, the angular resolution θ of your eye is about 1 arcminute, or 60 arcseconds. (You may want to review the definitions of these angular measures in Section 1-5.) Hence, with the naked eye it is impossible to distinguish two stars less than 1 arcminute apart or to see details on the Moon with an angular size smaller than this. All the planets have angular sizes (as seen from Earth) of 1 arcminute or less, which is why they appear as featureless points of light to the naked eye.

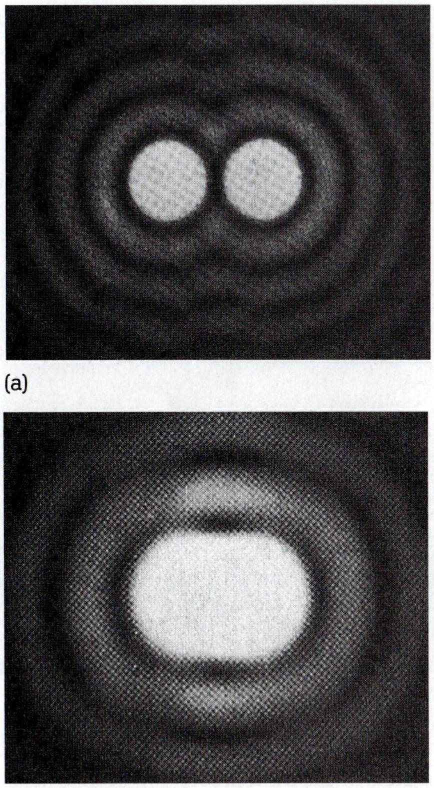

Two light sources with angular separation greater than angular resolution of telescope: Two sources easily distinguished
(a)
Light sources moved closer so that angular separation equals angular resolution of telescope: Just barely possible to tell that there are two sources
(b)

Figure 6-15 R I V U X G

Angular Resolution The angular resolution of a telescope indicates the sharpness of the telescope's images. **(a)** This telescope view shows two sources of light whose angular separation is greater than the angular resolution. **(b)** The light sources have been moved together so that their angular separation is equal to the angular resolution. If the sources were moved any closer together, the telescope image would show them as a single source.

Limits to Angular Resolution

One factor limiting angular resolution is **diffraction,** which is the tendency of light waves to spread out when they are confined to a small area like the lens or mirror of a telescope. (A rough analogy is the way water exiting a garden hose sprays out in a wider angle when you cover part of the end of the hose with your thumb.) As a result of diffraction, a narrow beam of light tends to spread out within a telescope's optics, thus blurring the image. If diffraction were the only limit, the angular resolution of a telescope would be given by the formula

Diffraction-limited angular resolution

$$\theta = 2.5 \times 10^5 \frac{\lambda}{D}$$

θ = diffraction-limited angular resolution of a telescope, in arcseconds

λ = wavelength of light, in meters

D = diameter of telescope objective, in meters

For a given wavelength of light, using a telescope with an objective of *larger* diameter D *reduces* the amount of diffraction and makes the angular resolution θ *smaller* (and hence better). For example, with red light with wavelength 640 nm, or 6.4×10^{-7} m, the diffraction-limited resolution of an 8-meter telescope (see Figure 6-15) would be

$$\theta = (2.5 \times 10^5)\,\frac{6.4 \times 10^{-7}\ \text{m}}{8\ \text{m}} = 0.02\ \text{arcsec}$$

In practice, however, ordinary optical telescopes cannot achieve such fine angular resolution. The problem is that turbulence in the air causes star images to jiggle around and twinkle. Even through the largest telescopes, a star still looks like a tiny blob rather than a pinpoint of light. A measure of the limit that atmospheric turbulence places on a telescope's resolution is called the **seeing disk.** This disk is the angular diameter of a star's image broadened by turbulence. The size of the seeing disk varies from one observatory site to another and from one night to another. At the observatories on Kitt Peak in Arizona and Cerro Tololo in Chile, the seeing disk is typically around 1 arcsec. Some

Figure 6-16 RIVUXG
The Telescopes of Mauna Kea The summit of Mauna Kea—an extinct Hawaiian volcano that reaches more than 4100 m (13,400 ft) above the waters of the Pacific—has nighttime skies that are unusually clear, still, and dark. To take advantage of these superb viewing conditions, Mauna Kea has become the home of many powerful telescopes. (Richard J. Wainscoat, University of Hawaii)

of the very best conditions in the world can be found at the observatories atop Mauna Kea in Hawaii, where the seeing disk is often as small as 0.5 arcsec. These great conditions are one reason why so many telescopes have been built there (Figure 6-16).

Active Optics and Adaptive Optics

In many cases the angular resolution of a telescope is even worse than the limit imposed by the seeing disk. This occurs if the objective mirror deforms even slightly due to variations in air temperature or flexing of the telescope mount. To combat this, many large telescopes are equipped with an **active optics** system. Such a system adjusts the mirror shape every few seconds to help keep the telescope in optimum focus and properly aimed at its target.

Changing the mirror shape is also at the heart of a more refined technique called **adaptive optics.** The goal of this technique is to compensate for atmospheric turbulence, so that the angular resolution can be smaller than the size of the seeing disk and can even approach the theoretical limit set by diffraction. Turbulence causes the image of a star to "dance" around erratically. In an adaptive optics system, sensors monitor this dancing motion 10 to 100 times per second, and a powerful computer rapidly calculates the mirror shape needed to compensate. Fast-acting mechanical devices called *actuators* then deform the mirror accordingly, at a much faster rate than in an active optics system. In some adaptive optics systems, the actuators deform a small secondary mirror rather than the large objective mirror.

Adaptive optics produces sharper images by "undoing" atmospheric turbulence

One difficulty with adaptive optics is that a fairly bright star must be in or near the field of the telescope's view to serve as a "target" for the sensors that track atmospheric turbulence. This is seldom the case, since the field of view of most telescopes is rather narrow. Astronomers get around this limitation by shining a laser beam toward a spot in the sky near the object to be observed (Figure 6-17). The laser beam causes atoms in the upper atmosphere to glow, making an artificial "star." The light that comes down to Earth from this "star" travels through the same part of our atmosphere as the light from the object being observed, so its image in the telescope will "dance" around in the same erratic way as the image of a real star.

Figure 6-18 shows the dramatic improvement in angular resolution possible with adaptive optics. Images made with adaptive optics are nearly as sharp as if the telescope were in the vacuum of space, where there is no atmospheric distortion whatsoever and the only limit on angular resolution is diffraction. A number of large telescopes are now being used with adaptive optics systems.

CAUTION! The images in Figure 6-18 are **false color** images: They do not represent the true color of the stars shown. False color is often used when the image is made using wavelengths that the eye cannot detect, as with the infrared images in Figure 6-18. A different use of false color is to indicate the relative brightness of different parts of the image, as in the infrared image of a person in Figure 5-10. Throughout this book, we'll always point out when false color is used in an image.

Interferometry

Several large observatories are developing a technique called **interferometry** that promises to further improve the angular resolution of telescopes. The idea is to have two widely separated telescopes observe the same object simultaneously, then use fiber optic cables to "pipe" the light signals from each telescope to a central location where they "interfere" or blend together. This method makes the combined signal sharp and clear. The effective resolution of such a combination of telescopes is equivalent to that of one giant telescope with a diameter equal to the **baseline,** or distance between the two telescopes. For example, the Keck I

Figure 6-17 R I V U X G

Creating an Artificial "Star" A laser beam shines upward from Yepun, an 8.2-meter telescope at the European Southern Observatory in the Atacama Desert of Chile. (Figure 6-14*a* shows the objective mirror for this telescope.) The beam strikes sodium atoms that lie about 90 km (56 miles) above Earth's surface, causing them to glow and make an artificial "star." Tracking the twinkling of this "star" makes it possible to undo the effects of atmospheric turbulence on telescope images. (European Southern Observatory)

and Keck II telescopes atop Mauna Kea (Figure 6-16) are 85 meters apart, so when used as an interferometer the angular resolution is the same as a single 85-meter telescope.

Interferometry has been used for many years with radio telescopes (which we will discuss in Section 6-6), but is still under development with telescopes for visible light or infrared wavelengths. Astronomers are devoting a great deal of effort to this development because the potential rewards are great. For example, the Keck I and II telescopes used together should give an angular resolution as small as 0.005 arcsec, which corresponds to being able to read the bottom row on an eye chart 36 km (22 miles) away!

Light Pollution

WEB LINK 6.4

Light from city street lamps and from buildings also degrades telescope images. This **light pollution** illuminates the sky, making it more difficult to see the stars. You can appreciate the problem if you have ever looked at the night sky from a major city. Only a few of the very brightest stars can be seen, as against the thousands that can be seen with the naked eye in the desert or the mountains. To avoid light pollution, observatories are built in remote locations far from any city lights.

Unfortunately, the expansion of cities has brought light pollution to observatories that in former times had none. As an example, the growth of Tucson, Arizona, has had deleterious effects on observations at the nearby Kitt Peak National Observatory. Efforts have been made to have cities adopt light fixtures that provide safe illumination for their citizens but produce little light pollution. These efforts have met with only mixed success.

One factor over which astronomers have absolutely no control is the weather. Optical telescopes cannot see through clouds, so it is important to build observatories where the weather is usually good. One advantage of mountaintop observatories such as Mauna Kea is that most clouds form at altitudes below the observatory, giving astronomers a better chance of having clear skies.

In many ways the best location for a telescope is in orbit around Earth, where it is unaffected by weather, light pollution, or atmospheric turbulence. We will discuss orbiting telescopes in Section 6-7.

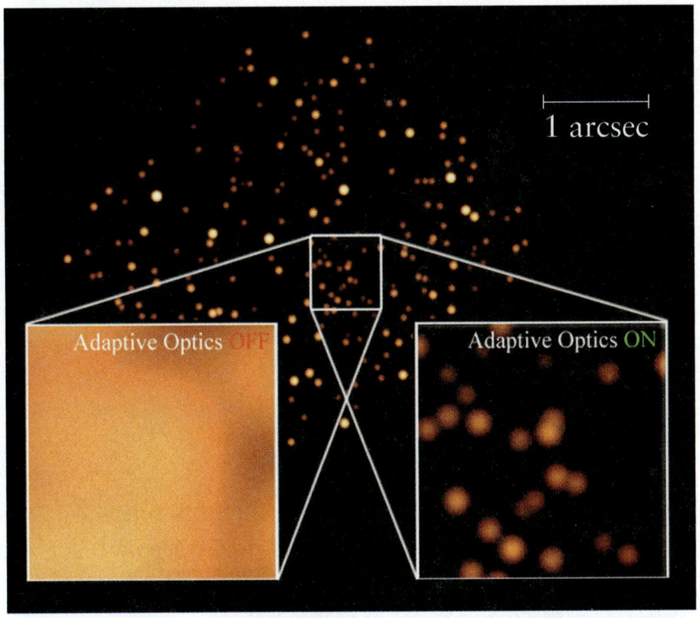

Figure 6-18 R I V U X G

Using Adaptive Optics to "Unblur" Telescope Images The two false-color, inset images show the same 1-arcsecond-wide region of the sky at infrared wavelengths as observed with the 10.0-m Keck II telescope on Mauna Kea (see Figure 6-16). Without adaptive optics, it is impossible to distinguish individual stars in this region. With adaptive optics turned on, more than two dozen stars can be distinguished. (UCLA Galactic Center Group)

6-4 A charge-coupled device is commonly used to record the image at a telescope's focus

TUTORIAL 6.2

Telescopes provide astronomers with detailed pictures of distant objects. The task of recording these pictures is called **imaging.**

Astronomical imaging really began in the nineteenth century with the invention of photography. It was soon realized that this new invention was a boon to astronomy. By taking long exposures with a camera mounted at the focus of a telescope, an astronomer can record features too faint to be seen by simply looking through the telescope. Such long exposures can reveal details in galaxies, star clusters, and nebulae that would not be visible to an astronomer looking through a telescope. Indeed, most large, modern telescopes do not have eyepieces at all.

The same technology that makes digital cameras possible has revolutionized astronomy

Unfortunately, photographic film is not a very efficient light detector. Only about 1 out of every 50 photons striking photographic film triggers the chemical reaction needed to produce an image. Thus, roughly 98% of the light falling onto photographic film is wasted.

Charge-Coupled Devices

The most sensitive light detector currently available to astronomers is the **charge-coupled device** (CCD). At the heart of a CCD is a semiconductor wafer divided into an array of small light-sensitive squares called picture elements or, more commonly, **pixels** (Figure 6-19*a*). For example, each of the 40 CCDs shown in Figure 6-19*a* has more than 9.4 million pixels arranged in 2048 rows by 4608 columns. They have about a thousand times more pixels per square centimeter than on a typical computer screen, which means that a CCD of this type can record very fine image details. CCDs with smaller numbers of pixels are used in digital cameras, scanners, and fax machines.

When an image from a telescope is focused on the CCD, an electric charge builds up in each pixel in proportion to the number of photons falling on that pixel. When the exposure is finished, the amount of charge on each pixel is read into a computer, where the resulting image can be stored in digital form and either viewed on a monitor or printed out. Compared with photographic film, CCDs are some 35 times more sensitive to light (they commonly respond to 70% of the light falling on them, versus 2% for film), can record much finer details, and respond more uniformly to light of different colors. Figures 6-19*b*, 6-19*c*, and 6-19*d* show the dramatic difference between photographic and CCD images. The great sensitivity of CCDs also makes them useful for **photometry,** which is the measurement of the brightnesses of stars and other astronomical objects.

In the modern world of CCD astronomy, astronomers need no longer spend the night in the unheated dome of a telescope. Instead, they operate the telescope electronically from a separate control room, where the electronic CCD images can be viewed on a computer monitor. The control room need not even be adjacent to the telescope. Although the Keck I and II

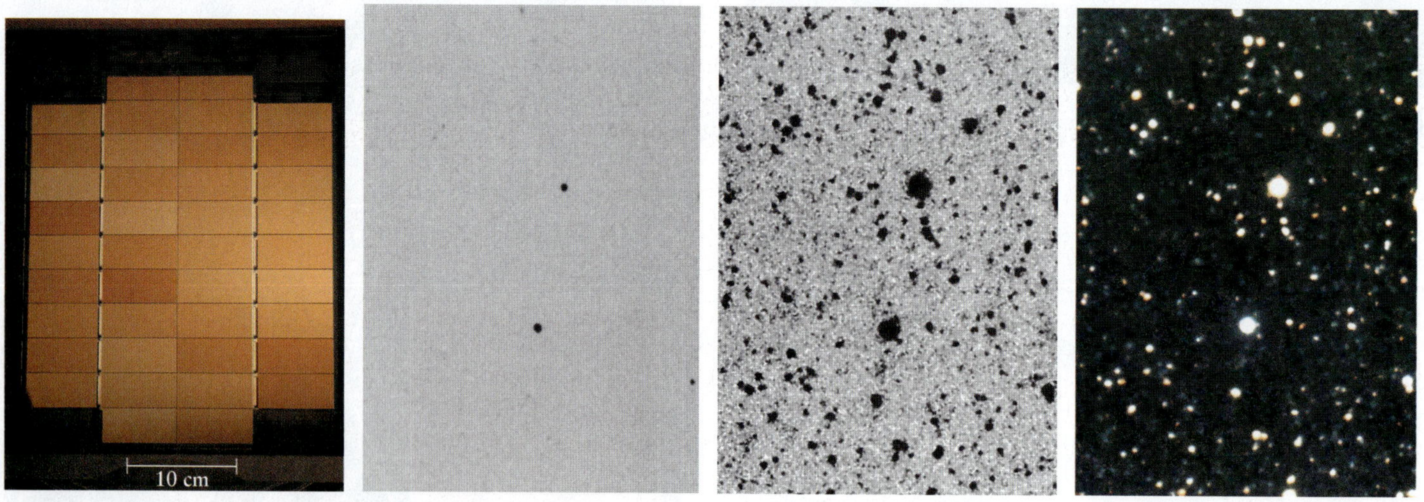

(a) A mosaic of 40 charge-coupled devices (CCDs)

(b) An image made with photographic film

(c) An image of the same region of the sky made with a CCD

(d) Combining several CCD images made with different color filters

WEB LINK 6.5

Figure 6-19 R I V U X G

Charge-Coupled Devices (CCDs) and Imaging (a) The 40 CCDs in this mosaic are used to record the light gathered by the Canada-France-Hawaii Telescope (see Figure 6-16). After an exposure, data from each of the 377 million light-sensitive pixels in the mosaic is transferred to a waiting computer. **(b)** This negative print (black stars and white sky) shows a portion of the sky as imaged with a 4-meter telescope and photographic film. **(c)** This negative image of the same region of the sky was made with the same telescope, but with the photographic film replaced by a CCD. Many more stars and galaxies are visible. **(d)** To produce this color positive view of the same region, a series of CCD images were made using different color filters. These were then combined using a computer image-processing program.

(a: J. C. Cullandre/Canada-France-Hawaii Telescope; b, c, d: Patrick Seitzer, National Optical Astronomy Observatories) telescopes (see Figure 6-16) are at an altitude of 4100 m (13,500 feet), astronomers can now make observations from a facility elsewhere on the island of Hawaii that is much closer to sea level. This saves the laborious drive to the summit of Mauna Kea and eliminates the need for astronomers to acclimate to the high altitude.

Most of the images that you will see in this book were made with CCDs. Because of their extraordinary sensitivity and their ability to be used in conjunction with computers, CCDs have attained a role of central importance in astronomy.

6-5 Spectrographs record the spectra of astronomical objects

We saw in Section 5-6 how the spectrum of an astronomical object provides a tremendous amount of information about that object, including its chemical composition and temperature. So, measuring spectra, or **spectroscopy**, is one of the most important uses of telescopes. Indeed, some telescopes are designed solely for measuring the spectra of distant, faint objects; they are never used for imaging.

The spectrum of a planet, star, or galaxy can reveal more about its nature than an image

Spectrographs and Diffraction Gratings

An essential tool of spectroscopy is the **spectrograph,** a device that records spectra. This optical device is mounted at the focus of a telescope. Figure 6-20*a* shows one design for a spectrograph, in which a **diffraction grating** is used to form the spectrum of a planet, star, or galaxy. A diffraction grating is a piece of glass on which thousands of very regularly spaced parallel lines have been cut. Some of the finest diffraction gratings have more than 10,000 lines per centimeter, which are usually cut by drawing a diamond back and forth across the glass. When light is shone on a diffraction grating, a spectrum is produced by the way in which light waves leaving different parts of the grating interfere with each other. (This same effect produces the rainbow of colors you see reflected from a compact disc or DVD, as shown in Figure 6-20*b*. Information is stored on the disc in a series of closely spaced pits, which act as a diffraction grating.)

Older types of spectrographs used a prism rather than a diffraction grating to form a spectrum. Using a prism had several drawbacks. A prism does not disperse the colors of the rainbow evenly: Blue and violet portions of the spectrum are spread out more than the red portion. In addition, because the blue and violet wavelengths must pass through more of the prism's glass than do the red wavelengths (examine Figure 5-3), light is absorbed unevenly across the spectrum. Indeed, a glass prism is opaque to near-ultraviolet light. For these reasons, diffraction gratings are preferred in modern spectrographs.

In Figure 6-20*a* the spectrum of a planet, star, or galaxy formed by the diffraction grating is recorded on a CCD. Light from a hot gas (such as helium, neon, argon, iron, or a combination of these) is then focused on the spectrograph slit. The result is a *comparison spectrum* alongside the spectrum of the celestial object under study. The wavelengths of the bright spectral lines

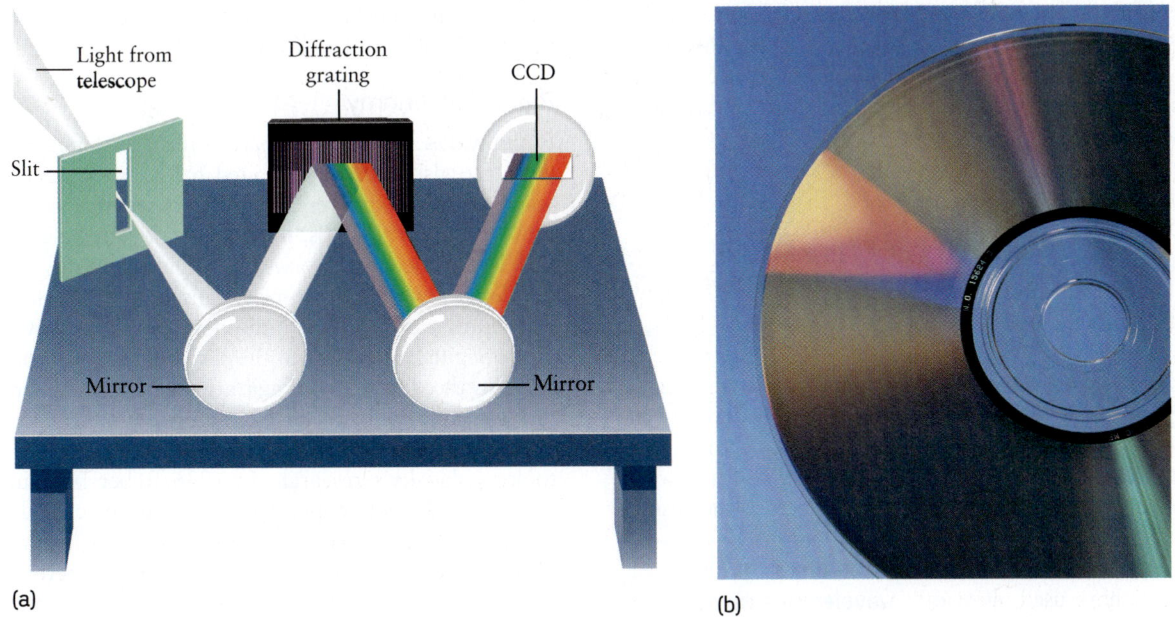

Figure 6-20 R I V U X G

A Grating Spectrograph **(a)** This optical device uses a diffraction grating to break up the light from a source into a spectrum. The spectrum is then recorded on a CCD. **(b)** A diffraction grating has a large number of parallel lines in its surface that reflect light of different colors in different direction. A compact disc, which stores information in a series of closely spaced pits, reflects light in a similar way. (Dale E. Boyer/Photo Researchers)

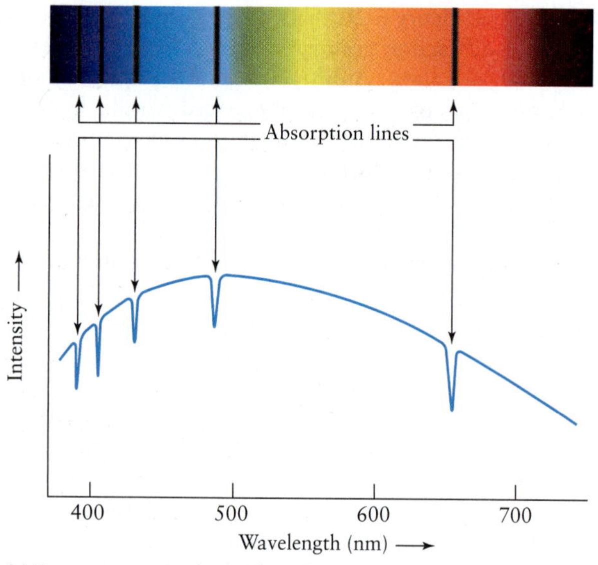

(a) Two representations of an absorption line spectrum

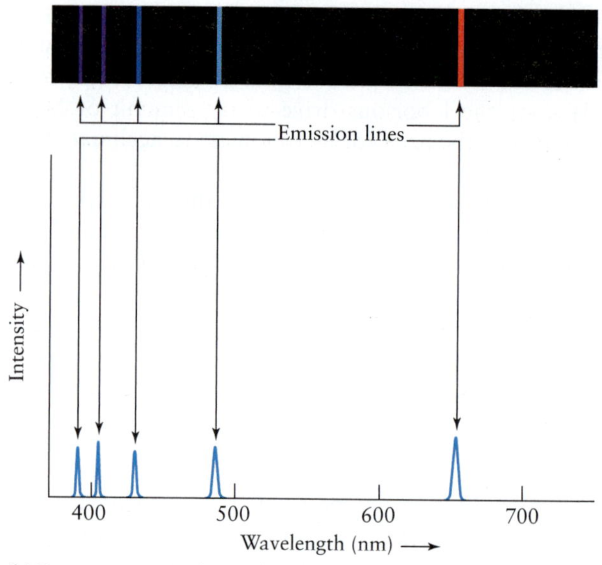

(b) Two representations of an emission line spectrum

Figure 6-21

Two Ways to Represent Spectra When a CCD is placed at the focus of a spectrograph, it records the rainbow-colored spectrum. A computer program can be used to convert the recorded data into a graph of intensity versus wavelength. **(a)** Absorption lines appear as dips on such a graph, while **(b)** emission lines appear as peaks. The dark absorption lines and bright emission lines in this example are the Balmer lines of hydrogen (see Section 5-8).

of the comparison spectrum are known from laboratory experiments and can therefore serve as reference markers. (See Figure 5-17, which shows the spectrum of the Sun and a comparison spectrum of iron.)

When the exposure is finished, electronic equipment measures the charge that has accumulated in each pixel. These data are used to graph light intensity against wavelength. Dark absorption lines in the spectrum appear as depressions or valleys on the graph, while bright emission lines appear as peaks. Figure 6-21 compares two ways of exhibiting spectra with absorption lines and emission lines. Later in this book we shall see spectra presented in both these ways.

6-6 A radio telescope uses a large concave dish to reflect radio waves to a focus

For thousands of years, all the information that astronomers gathered about the universe was based on ordinary visible light. In the twentieth century, however, astronomers first began to explore the nonvisible electromagnetic radiation coming from astronomical objects. In this way they have discovered aspects of the cosmos that are forever hidden to optical telescopes.

Observing at radio wavelengths reveals aspects of the universe hidden from ordinary telescopes

Astronomers have used ultraviolet light to map the outer regions of the Sun and the clouds of Venus, and used infrared radiation to see new stars and perhaps new planetary systems in the process of formation. By detecting radio waves from Jupiter and Saturn, they have mapped the intense magnetic fields that surround those giant planets; by detecting curious bursts of X rays from space, they have learned about the utterly alien conditions in the vicinity of a black hole. It is no exaggeration to say that today's astronomers learn as much about the universe using telescopes for nonvisible wavelengths as they do using visible light.

Radio Astronomy

Radio waves were the first part of the electromagnetic spectrum beyond the visible to be exploited for astronomy. The use of radio waves is a result of a research project seemingly unrelated to astronomy. In the early 1930s, Karl Jansky, a young electrical engineer at Bell Telephone Laboratories, was trying to locate what was causing interference with the then-new transatlantic radio link. By 1932, he realized that one kind of radio noise is strongest when the constellation Sagittarius is high in the sky. The center of our Galaxy is located in the direction of Sagittarius, and Jansky concluded that he was detecting radio waves from an astronomical source.

At first only Grote Reber, a radio engineer living in Illinois, took up Jansky's research. In 1936 Reber built in his backyard the first **radio telescope**, a radio-wave detector dedicated to astronomy. He modeled his design after an ordinary reflecting telescope, with a parabolic metal "dish" (reflecting antenna) measuring 10 m (31 ft) in diameter and a radio receiver at the focal point of the dish.

Reber spent the years from 1938 to 1944 mapping radio emissions from the sky at wavelengths of 1.9 m and 0.63 m. He found radio waves coming from the entire Milky Way, with the greatest emission from the center of the Galaxy. These results, together with the development of improved radio technology during World War II, encouraged the growth of radio astronomy and the construction of new radio telescopes around the world. Radio observatories are as common today as major optical observatories.

Modern Radio Telescopes

Like Reber's prototype, a typical modern radio telescope has a large parabolic dish (Figure 6-22). An antenna tuned to the desired frequency is located at the focus (like the prime focus design for optical reflecting telescopes shown in Fig. 6-11*a*). The incoming signal is relayed from the antenna to amplifiers and recording instruments, typically located in a room at the base of the telescope's pier.

CAUTION! The radio telescope in Figure 6-22 looks like a radar dish but is used in a different way. In radar, the dish is used to send out a narrow beam of radio waves. If this beam encounters an object like an airplane, some of the radio waves will be reflected back to the radar dish and detected by a receiver at the focus of the dish. Thus, a radar dish looks for radio waves *reflected* by distant objects. By contrast, a radio telescope is designed to receive radio waves *emitted* by objects in space.

Many radio telescope dishes, like the one in Figure 6-22, have visible gaps in them like a wire mesh. This does not affect their reflecting power because the holes are much smaller than the wavelengths of the radio waves. The same idea is used in the design of microwave ovens. The glass window in the oven door would allow the microwaves to leak out, so the window is covered by a metal screen with small holes. These holes are much smaller than the 12.2-cm (4.8-in.) wavelength of the microwaves, so the screen reflects the microwaves back into the oven.

Figure 6-22 R I V U X G

A Radio Telescope The dish of the Parkes radio telescope in New South Wales, Australia, is 64 m (210 ft) in diameter. Radio waves reflected from the dish are brought to a focus and collected by an antenna at the focal point. (David Nunuk/Photo Researchers)

Radio Telescopes: Limits to Angular Resolution

One great drawback of early radio telescopes was their very poor angular resolution. Recall from Section 6-3 that angular resolution is the smallest angular separation between two stars that can just barely be distinguished as separate objects. Unlike visible light, radio waves are only slightly affected by turbulence in the atmosphere, so the limitation on the angular resolution of a radio telescope is diffraction. The problem is that diffraction-limited angular resolution is directly proportional to the wavelength being observed: The longer the wavelength, the larger (and hence worse) the angular resolution and the fuzzier the image. (See the formula for angular resolution θ in Section 6-3.) As an example, a 1-m radio telescope detecting radio waves of 5-cm wavelength has 100,000 times poorer angular resolution than a 1-m optical telescope. Because radio radiation has very long wavelengths, small radio telescopes can produce only blurry, indistinct images.

VIDEO 6.1 A very large radio telescope can produce a somewhat sharper radio image, because as the diameter of the telescope increases, the angular resolution decreases. In other words, the bigger the dish, the better the resolution. For this reason, most modern radio telescopes have dishes more than 30 m (100 ft) in diameter. A large dish is also useful for increasing light-gathering power, because radio signals from astronomical objects are typically very weak in comparison with the intensity of visible light. But even the largest single radio dish in existence, the 305-m (1000-ft) Arecibo radio telescope in Puerto Rico, cannot come close to the resolution of the best optical instruments.

To improve angular resolution at radio wavelengths, astronomers combine observations from two or more widely separated telescopes using the interferometry technique that we described in Section 6-3. Using radio waves is much easier than for visible or infrared light because radio signals can be carried over electrical wires. Consequently, two radio telescopes observing the same astronomical object can be hooked together, even if they are separated by many kilometers. As for visible light, the resulting angular resolution is that of a single telescope whose diameter is equal to the baseline, or separation between the telescopes.

One of the largest arrangements of radio telescopes for interferometry is the Very Large Array (VLA), located in the desert near Socorro, New Mexico (Figure 6-23). The VLA consists of 27 parabolic dishes, each 25 m (82 ft) in diameter. These 27 telescopes are arranged along the arms of a gigantic Y that covers an area 27 km (17 mi) in diameter. By pointing all 27 telescopes at the same object and combining the 27 radio signals, this system can produce radio views of the sky with an angular resolution as small as 0.05 arcsec, comparable to that of the very best optical telescopes.

Figure 6-23 R I V U X G

The Very Large Array (VLA) The 27 radio telescopes of the VLA in central New Mexico are arranged along the arms of a Y. The north arm of the array is 19 km long; the southwest and southeast arms are each 21 km long. By spreading the telescopes out along the legs and combining the signals received, the VLA can give the same angular resolution as a single dish many kilometers in radius. (Courtesy of NRAO/AUI)

Dramatically better angular resolution can be obtained by combining the signals from radio telescopes at different observatories thousands of kilometers apart. This technique is called **very-long-baseline interferometry (VLBI).** VLBI is used by a system called the Very Long Baseline Array (VLBA), which consists of ten 25-meter dishes at different locations between Hawaii and the Caribbean. Although the ten dishes are not physically connected, they are all used to observe the same object at the same time. The data from each telescope are recorded electronically and processed later. By carefully synchronizing the ten recorded signals, they can be combined just as if the telescopes had been linked together during the observation. With the VLBA, features smaller than 0.001 arcsec can be distinguished at radio wavelengths. This angular resolution is 100 times better than a large optical telescope with adaptive optics.

Even better angular resolution can be obtained by adding radio telescopes in space; the baseline is then the distance from the VLBA to the orbiting telescope. The first such space radio telescope, the Japanese HALCA spacecraft, operated from 1997 to 2003. Orbiting at a maximum distance of 21,400 km (13,300 mi) above Earth's surface, HALCA gave a baseline three times longer—and thus an angular resolution three times better—than can be obtained with Earthbound telescopes alone. Successors to HALCA may be placed in orbit by 2013.

Figure 6-24 shows how optical and radio images of the same object can give different and complementary kinds of information. The visible-light image of Saturn (Figure 6-24*a*) shows clouds in the planet's atmosphere and the structure of the rings. Like the visible light from the Moon, the light used to make this image is just reflected sunlight. By contrast, the false-color radio image of Saturn (Figure 6-24*b*) is a record of waves *emitted* by the planet and its rings. Analyzing such images provides information about the structure of Saturn's atmosphere and rings that could never be obtained from a visible-light image such as Figure 6-24*a*.

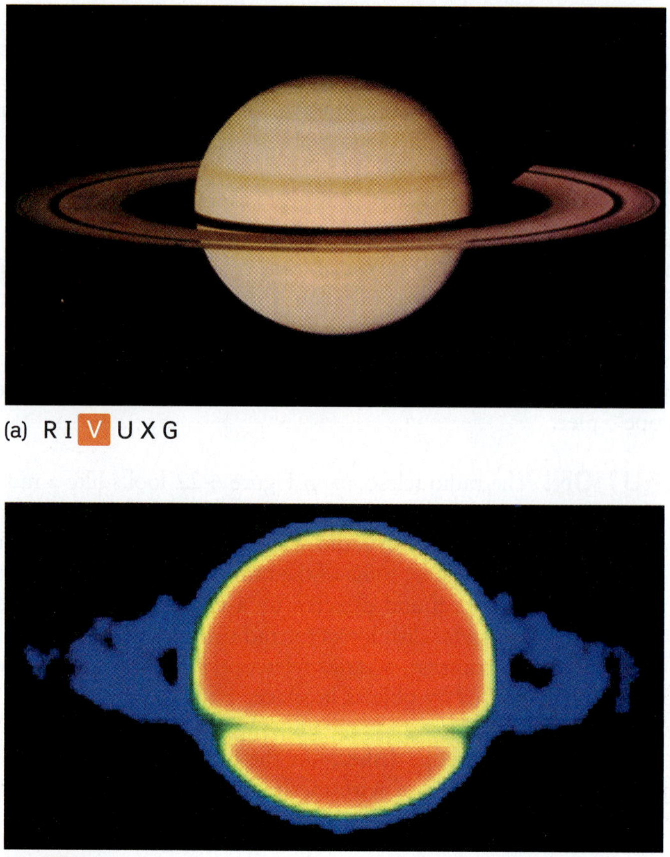

(a) R I V U X G

(b) R I V U X G

Figure 6-24

Optical and Radio Views of Saturn **(a)** This picture was taken by a spacecraft 18 million kilometers from Saturn. The view was produced by sunlight reflecting from the planet's cloudtops and rings. **(b)** The image is made by using radio waves, which is indicated by the highlighted R in the wavelength tab. This VLA image shows radio emission from Saturn at a wavelength of 2 cm. In this false-color image, the most intense radio emission is shown in red, the least intense in blue. Yellow and green represent intermediate levels of radio intensity; black indicates no detectable radio emission. Note the radio "shadow" caused by Saturn's rings where they lie in front of the planet. (a: NASA; b: Image courtesy of NRAO/AUI)

6-7 Telescopes in orbit around Earth detect radiation that does not penetrate the atmosphere

The many successes of radio astronomy show the value of observations at nonvisible wavelengths. But Earth's atmosphere is opaque to many wavelengths.

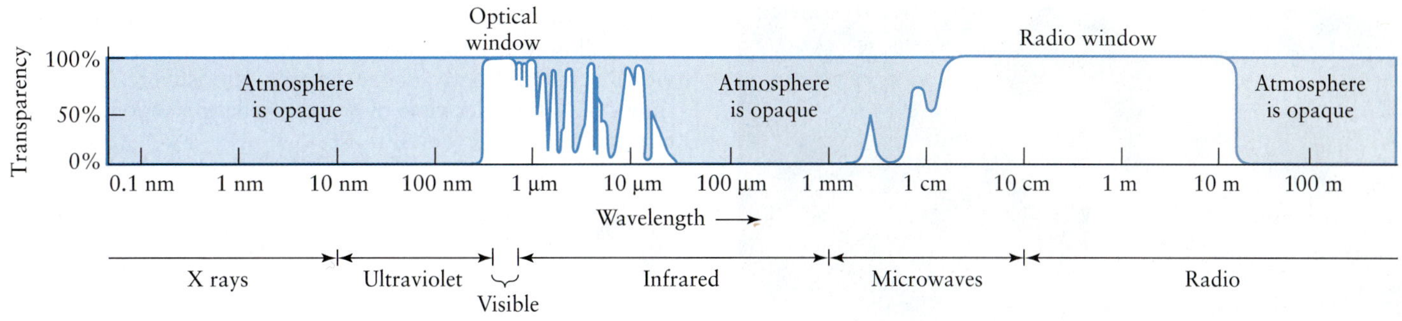

Figure 6-25

The Transparency of Earth's Atmosphere This graph shows the percentage of radiation that can penetrate Earth's atmosphere at different wavelengths. Regions in which the curve is high are called "windows," because the atmosphere is relatively transparent at those wavelengths. There are also three wavelength ranges in which the atmosphere is opaque and the curve is near zero: at wavelengths less than about 290 nm, which are absorbed by atmospheric oxygen and nitrogen; between the optical and radio windows, due to absorption by water vapor and carbon dioxide; and at wavelengths longer than about 20 m, which are reflected back into space by ionized gases in the upper atmosphere.

Space telescopes make it possible to study the universe across the entire electromagnetic spectrum

Other than visible light and radio waves, very little radiation from space manages to penetrate the air we breathe. To overcome this, astronomers have placed a variety of telescopes in orbit.

Figure 6-25 shows the transparency of Earth's atmosphere to different wavelengths of electromagnetic radiation. The atmosphere is most transparent in two wavelength regions, the **optical window** (which includes the entire visible spectrum) and the **radio window** (which includes part, but not all, of the radio spectrum). There are also several relatively transparent regions at infrared wavelengths between 1 and 40 μm. Infrared radiation within these wavelength intervals can penetrate Earth's atmosphere somewhat and can be detected with ground-based telescopes. This wavelength range is called the *near-infrared,* because it lies just beyond the red end of the visible spectrum.

Infrared Astronomy

Water vapor is the main absorber of infrared radiation from space, which is why infrared observatories are located at sites with exceptionally low humidity. The site must also be at high altitude to get above as much of the atmosphere's water vapor as possible. One site that meets both criteria is the summit of Mauna Kea in Hawaii, shown in Figure 6-16. (The complete lack of vegetation on the summit attests to its extreme dryness.) Some of the telescopes on Mauna Kea are designed exclusively for detecting infrared radiation. Others, such as the Keck I and Keck II telescopes, are used for both visible and near-infrared observations (see Figure 6-18).

WEB LINK 6.8 Even at the elevation of Mauna Kea, water vapor in the atmosphere restricts the kinds of infrared observations that astronomers can make. This situation can be improved by carrying telescopes on board high-altitude balloons or aircraft. But the ultimate solution is to place a telescope in Earth's orbit and radio its data back to astronomers on the ground. The first such orbiting infrared observatory, the Infrared Astronomical Satellite (IRAS), was launched in 1983. During its nine-month mission, IRAS used its 57-cm (22-in.) telescope to map almost the entire sky at wavelengths from 12 to 100 μm.

The IRAS data revealed the presence of dust disks around nearby stars. Planets are thought to coalesce from disks of this kind, so this was the first concrete (if indirect) evidence that there might be planets orbiting other stars. The dust that IRAS detected is warm enough to emit infrared radiation but too cold to emit much visible light, so it remained undetected by ordinary optical telescopes. IRAS also discovered distant, ultraluminous galaxies that emit almost all their radiation at infrared wavelengths.

In 1995 the Infrared Space Observatory (ISO), a more advanced 60-cm reflector with better light detectors, was launched into orbit by the European Space Agency. During its $2\frac{1}{2}$-year mission, ISO made a number of groundbreaking observations of very distant galaxies and of the thin, cold material between the stars of our own Galaxy. Like IRAS, ISO had to be cooled by liquid helium to temperatures just a few degrees above absolute zero. Had this not been done, the infrared blackbody radiation from the telescope itself would have outshone the infrared radiation from astronomical objects. The ISO mission came to an end when the last of the helium evaporated into space.

WEB LINK 6.9 At the time of this writing the largest orbiting infrared observatory is the Spitzer Space Telescope, an 85-cm infrared telescope designed to survey the infrared sky with unprecedented resolution (Figure 6-26). Placed in orbit in 2003, the Spitzer Space Telescope is being used to study the formation of new stars, probe the inner structure of galaxies, and examine the first galaxies that formed after the Big Bang. The images that open this chapter illustrate how Spitzer can reveal aspects of astronomical objects that are hidden from visible-light telescopes.

Figure 6-26 R I V U X G

The Spitzer Space Telescope Launched in 2003, this is the largest infrared telescope ever placed in space. Its 85-cm objective mirror and three science instruments, kept cold by 360 liters of liquid helium, enable the Spitzer Space Telescope to observe the universe at wavelengths from 3 to 180 μm. The background to this illustration of the telescope is a false-color Spitzer image of the Milky Way and the constellation Orion at 100 μm, showing emission from warm interstellar dust. (NASA/JPL-Caltech)

Ultraviolet Astronomy

Astronomers are also very interested in observing at ultraviolet wavelengths. These observations can reveal a great deal about hot stars, ionized clouds of gas between the stars, and the Sun's high-temperature corona (see Section 3-5), all of which emit copious amounts of ultraviolet light. The spectrum of ultraviolet sunlight reflected from a planet can also reveal the composition of the planet's atmosphere. However, Earth's atmosphere is opaque to ultraviolet light except for the narrow *near-ultraviolet* range, which extends from about 400 nm (the violet end of the visible spectrum) down to 300 nm.

To see shorter-wavelength *far-ultraviolet* light, astronomers must again make their observations from space. The first ultraviolet telescope was placed in orbit in 1962, and several others have since followed it into space. Small rockets have also lifted ultraviolet cameras briefly above Earth's atmosphere. Figure 6-27 shows an ultraviolet view of the constellation Orion, along with infrared and visible views.

WEB LINK 6.10 The Far Ultraviolet Spectroscopic Explorer (FUSE), which went into orbit in 1999, specializes in measuring spectra at wavelengths from 90 to 120 nm. Highly ionized oxygen atoms, which can exist only in an extremely high-temperature gas, have a characteristic spectral line in this range. By looking for this spectral line in various parts of the sky, FUSE has confirmed that our Milky Way Galaxy (Section 1-4) is surrounded by an immense "halo" of gas at temperatures in excess of 200,000 K. Only an ultraviolet telescope could have detected this "halo," which is thought to have been produced by exploding stars called supernovae (Section 1-3).

The Hubble Space Telescope

Infrared and ultraviolet satellites give excellent views of the heavens at selected wavelengths. But since the 1940s astronomers had dreamed of having one large telescope that could be operated at any wavelength from the near-infrared through the visible range and out into the ultraviolet. This is the mission of the Hubble Space Telescope (HST), which was placed in a 600-km-high orbit by the space shuttle *Discovery* in 1990 (Figure 6-28). HST has a 2.4-m (7.9-ft) objective mirror and was designed to observe at wavelengths from 115 nm to 1 μm. Like most ground-based telescopes, HST uses a CCD to record images. (In fact, the development of HST helped drive advances in CCD technology.) The images are then radioed back to Earth in digital form.

The great promise of HST was that from its vantage point high above the atmosphere, its angular resolution would be limited only by diffraction. But soon after HST was placed in orbit, astronomers discovered that a manufacturing error had caused the telescope's objective mirror to suffer from spherical aberration. The mirror should have been able to concentrate 70% of a star's light into an image with an angular diameter of 0.1 arcsec. Instead, only 20% of the light was focused into this small area. The remainder was smeared out over an area about 1 arcsec wide, giving images little better than those achieved at major ground-based observatories.

On an interim basis, astronomers used only the 20% of incoming starlight that was properly focused and, with computer processing, discarded the remaining poorly focused 80%. This was practical only for brighter objects on which astronomers could afford to waste light. But many of the observing projects scheduled for HST involved extremely dim galaxies and nebulae.

These problems were resolved by a second space shuttle mission in 1993. Astronauts installed a set of small secondary mirrors whose curvature exactly compensated for the error in curvature of the primary mirror. Once these were in place, HST was able to make truly sharp images of extremely faint objects. Astronomers have used the repaired HST to make discoveries about the nature of planets, the evolution of stars, the inner

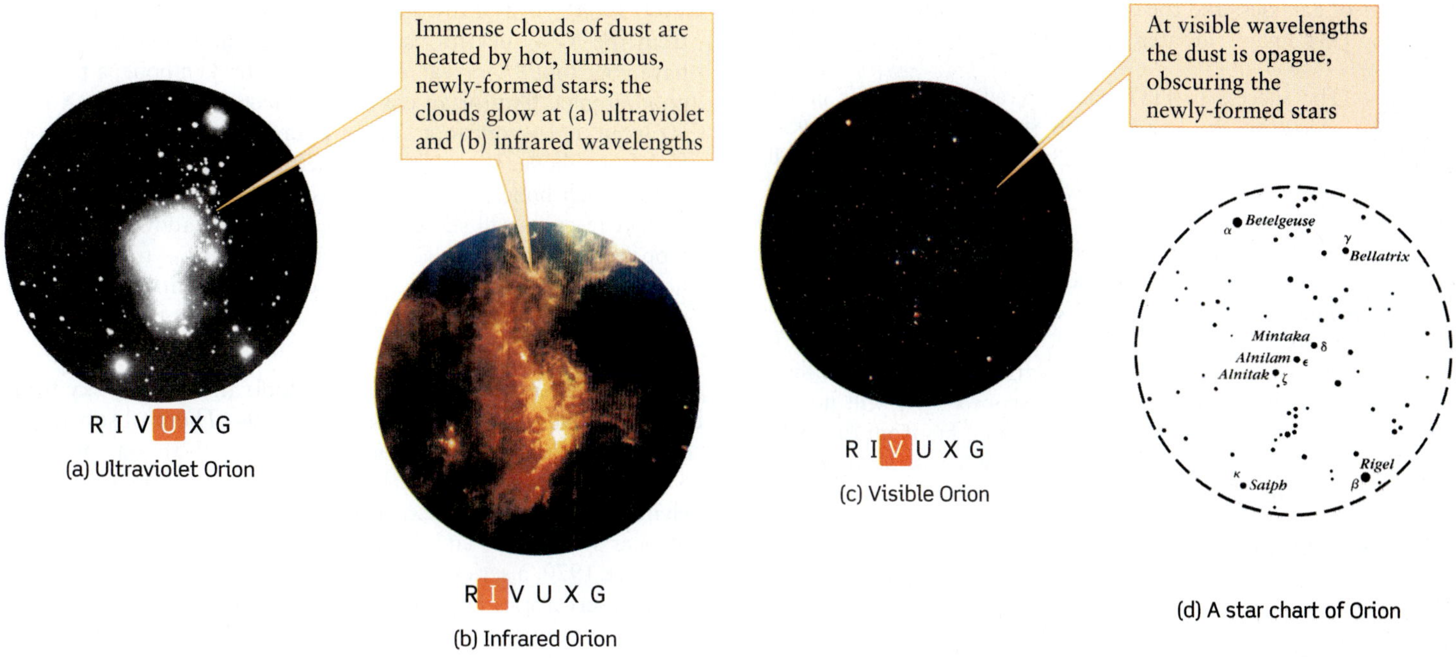

Figure 6-27

Orion Seen at Ultraviolet, Infrared, and Visible Wavelengths **(a)** An ultraviolet view of the constellation of Orion was obtained during a brief rocket flight in 1975. This 100-s exposure covers the wavelength range 125–200 nm. **(b)** The false-color view from the Infrared Astronomical Satellite displays emission at different wavelengths in different colors: red for 100-μm radiation, green for 60-μm radiation, and blue for 12-μm radiation. Compare these images with **(c)** an ordinary visible-light photograph and (d) a star chart of Orion. (a: G. R. Carruthers, Naval Research Laboratory; b: NASA; c: R. C. Mitchell, Central Washington University)

WEB LINK 6.11

Figure 6-28 R I V U X G

The Hubble Space Telescope The largest telescope yet placed in orbit, HST is a joint project of NASA and the European Space Agency (ESA). HST has helped discover new moons of Pluto, probed the formation of stars, and found evidence that the universe is now expanding at a faster rate than several billion years ago. This photograph was taken from the space shuttle *Discovery* during a 1997 mission to service HST. (NASA) workings of galaxies, and the expansion of the universe. You will see many HST images in later chapters.

WEB LINK 6.12

The success of HST has inspired plans for its larger successor, the James Webb Space Telescope or JWST (Figure 6-29). Planned for a 2013 launch, JWST will observe at visible and infrared wavelengths from 600 nm to 28 μm. With its 6.5-m objective mirror—2.5 times the diameter of the HST objective mirror, with six times the light-gathering power—JWST will study faint objects such as planetary systems forming around other stars and galaxies near the limit of the observable universe. Unlike HST, which is in a relatively low-altitude orbit around Earth, JWST will orbit the Sun some 1.5 million km beyond Earth. In this orbit the telescope's view will not be blocked by Earth. Furthermore, by remaining far from the radiant heat of Earth it will be easier to keep JWST at the very cold temperatures required by its infrared detectors.

X-Ray Astronomy

Space telescopes have also made it possible to explore objects whose temperatures reach the almost inconceivable values of 10^6 to 10^8 K. Atoms in such a high-temperature gas move so fast that when they collide, they emit X-ray photons of very high energy and very short wavelengths less than 10 nm. X-ray telescopes designed to detect these photons must be placed in orbit, since Earth's atmosphere is totally opaque at these wavelengths.

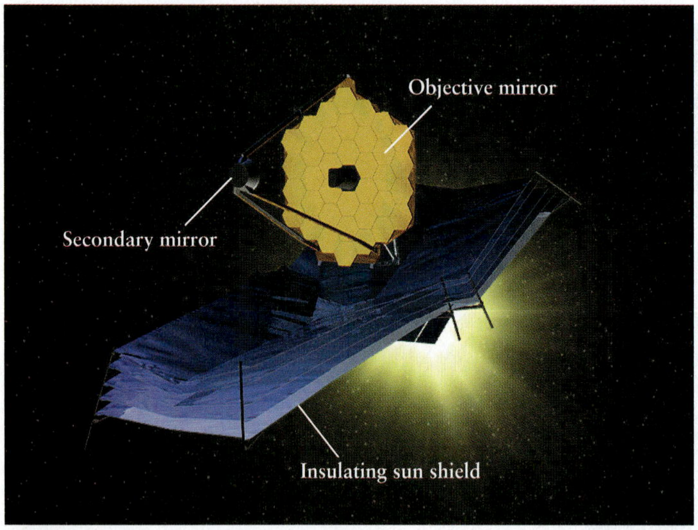

ANIMATION 6.3

Figure 6-29

The James Webb Space Telescope (JWST) The successor to the Hubble Space Telescope, JWST will have an objective mirror 6.5 m (21 ft) in diameter. Like many Earthbound telescopes, JWST will be of Cassegrain design (see Figures 6-11*b* and 6-14*b*). Rather than using liquid helium to keep the telescope and instruments at the low temperatures needed to observe at infrared wavelengths, JWST will keep cool using a multilayer sunshield the size of two tennis courts. (Courtesy of Northrop Grumman Space Technology)

CAUTION! X-ray telescopes work on a very different principle from the X-ray devices used in medicine and dentistry. If you have your foot "X rayed" to check for a broken bone, a piece of photographic film (or an electronic detector) sensitive to X rays is placed under your foot and an X-ray beam is directed at your foot from above. The radiation penetrates through soft tissue but not through bone, so the bones cast an "X-ray shadow" on the film. A fracture will show as a break in the shadow. X-ray telescopes, by contrast, do *not* send beams of X rays toward astronomical objects in an attempt to see inside them. Rather, these telescopes detect X rays that the objects emit on their own.

Astronomers got their first quick look at the X-ray sky from brief rocket flights during the late 1940s. These observations confirmed that the Sun's corona (see Section 3-5) is a source of X rays, and must therefore be at a temperature of millions of kelvins. In 1962 a rocket experiment revealed that objects beyond the solar system also emit X rays.

Since 1970, a series of increasingly sensitive and sophisticated X-ray observatories have been placed in orbit, including NASA's Einstein Observatory, the European Space Agency's Exosat, and the German-British-American ROSAT. These telescopes have shown that other stars also have high-temperature coronae, and have found hot, X-ray-emitting gas clouds so immense that hundreds of galaxies fit inside them. They also discovered unusual stars that emit X rays in erratic bursts. These bursts are now thought to be coming from heated gas swirling around a small but massive object—possibly a black hole.

X-ray astronomy took a quantum leap forward in 1999 with the launch of NASA's Chandra X-ray Observatory and the European Space Agency's XMM-Newton. Named for the Indian-American Nobelaureate astrophysicist Subrahmanyan Chandrasekhar, Chandra can view the X-ray sky with an angular resolution of 0.5 arcsec (Figure 6-30*a*). This is comparable to the best ground-based optical telescopes, and more than a thousand times better than the resolution of the first orbiting X-ray telescope. Chandra can also measure X-ray spectra 100 times more precisely than any previous spacecraft and can detect variations in X-ray emissions on time scales as short as 16 microseconds. This latter capability is essential for understanding how X-ray bursts are produced around black holes.

XMM-Newton (for *X*-ray *M*ulti-mirror *M*ission) is actually three X-ray telescopes that all point in the same direction (Figure 6-30*b*). Their combined light-gathering power is five times greater than that of Chandra, which makes XMM-Newton able to observe fainter objects. (For reasons of economy, the mirrors were not ground as precisely as those on Chandra, so the angular resolution of XMM-Newton is only about 6 arcseconds.) It also carries a small but highly capable telescope for ultraviolet and visible observations. Hot X-ray sources are usually accompanied by cooler material that radiates at these longer wavelengths, so XMM-Newton can observe these hot and cool regions simultaneously.

Gamma-Ray Astronomy

Gamma rays, the shortest-wavelength photons of all, help us to understand phenomena even more energetic than those that pro-

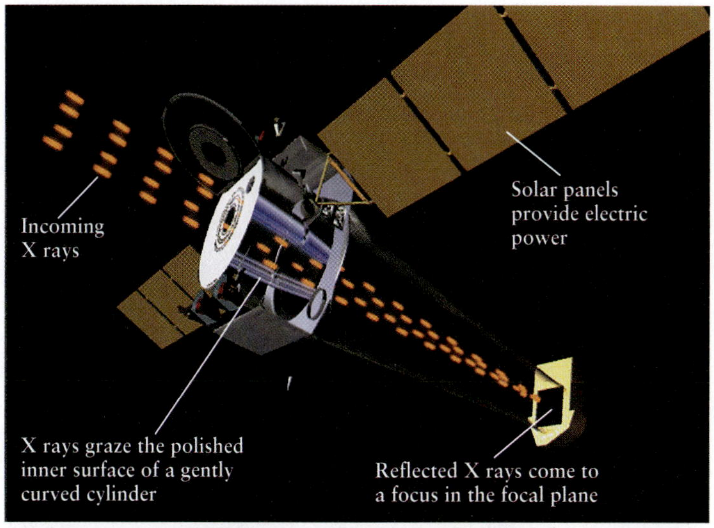

(a) Chandra X-ray Observatory

(b) XMM-Newton

WEB LINK 6.13

Figure 6-30

Two Orbiting X-Ray Observatories **(a)** X rays are absorbed by ordinary mirrors like those used in optical reflectors, but they can be reflected if they graze the mirror surface at a very shallow angle. In the Chandra X-ray Observatory, X rays are focused in this way onto a focal plane 10 m (33 ft) behind the mirror. **(b)** XMM-Newton is about the same size as Chandra, and its three X-ray telescopes form images in the same way. (a: NASA/Chandra X-ray Observatory Center/Smithsonian Astrophysical Observatory; b: D. Ducros/European Space Agency)

duce X rays. As an example, when a massive star explodes into a supernova, it produces radioactive atomic nuclei that are strewn across interstellar space. Observing the gamma rays emitted by these nuclei helps astronomers understand the nature of supernova explosions.

Like X rays, gamma rays do not penetrate Earth's atmosphere, so space telescopes are required. One of the most important gamma-ray telescopes placed in orbit to date is the Compton Gamma Ray Observatory (CGRO), shown in Figure 6-31. One particularly important task for CGRO was the study of gamma-ray bursts, which are brief, unpredictable, and very intense flashes of gamma rays that are found in all parts of the sky. By analyzing data from CGRO and other orbiting observatories, astronomers have shown that the sources of these gamma-ray bursts are billions of light-years away. For these bursts to be visible across such great distances, their sources must be among the most energetic objects in the universe. By combining these gamma-ray observations with images made by optical telescopes, astronomers have found that at least some of the gamma-ray bursts emanate from stars that explode catastrophically. Since the CGRO mission ended in 2000, more advanced gamma-ray telescopes, including the European Space Agency's INTEGRAL (International Gamma-Ray Astrophysics Laboratory), have taken its place.

The *Cosmic Connections* figure shows the wavelengths at which Earth-orbiting telescopes are particularly useful, as well as summarizing the design of refracting and reflecting Earth-based telescopes. The advantages and benefits of Earth-orbiting observatories cannot be overemphasized. We are no longer limited to the narrow ranges of whatever wavelengths manage to leak through our shimmering, hazy atmosphere (Figure 6-32). For the first time, we are really *seeing* the universe.

WEB LINK 6.14

Figure 6-31 R I **V** U X G

The Compton Gamma Ray Observatory (CGRO) This photograph shows the Compton Observatory being deployed from the space shuttle *Atlantis* in 1991. Named in honor of Arthur Holly Compton, an American scientist who made important discoveries about gamma rays, CGRO carried four different gamma-ray detectors. When its mission ended in 2000, CGRO disintegrated as it reentered our atmosphere. (NASA)

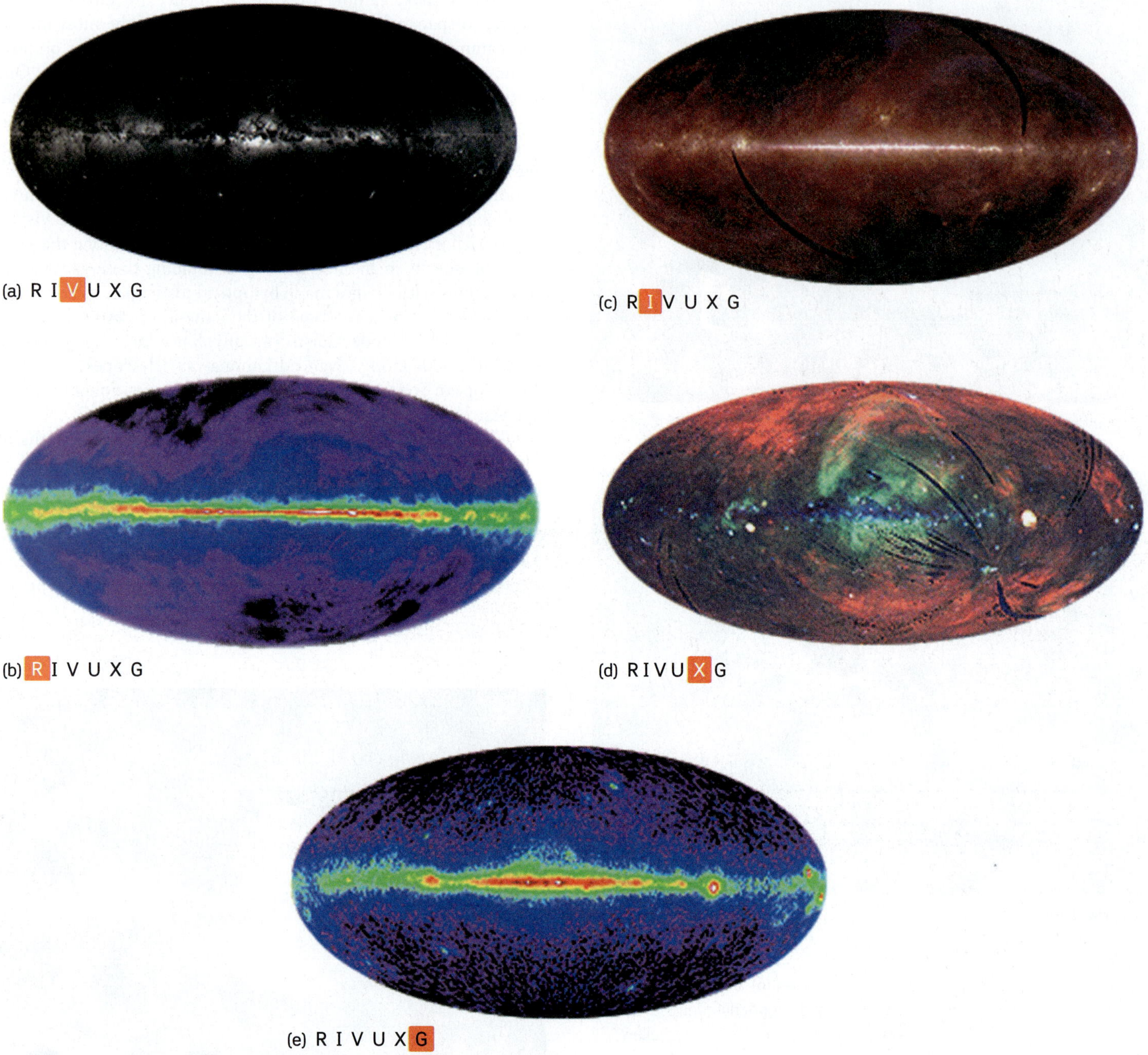

WEB LINK 6.15

Figure 6-32

The Entire Sky at Five Wavelength Ranges These five views show the entire sky at visible, radio, infrared, X-ray, and gamma-ray wavelengths. The entire celestial sphere is mapped onto an oval, with the Milky Way stretching horizontally across the center. The black crescents in the infrared and X-ray images are where data are missing. (GSFC/NASA) **(a)** In the visible view the constellation Orion is at the right, Sagittarius in the middle, and Cygnus toward the left. Many of the dark areas along the Milky Way are locations where interstellar dust is sufficiently thick to block visible light. **(b)** The radio view shows the sky at a wavelength of 21 cm. This wavelength is emitted by hydrogen atoms in interstellar space. The brightest regions (shown in red) are in the plane of the Milky Way, where the hydrogen is most concentrated. **(c)** The infrared view from IRAS shows emission at 100 μm, 60 μm, and 12 μm. Most of the emission is from dust particles in the plane of the Milky Way that have been warmed by starlight. **(d)** The image is made by using X-rays, which is indicated by the highlighted X in the wavelength tab. The X-ray view from ROSAT shows wavelengths of 0.8 nm (blue), 1.7 nm (green), and 5.0 nm (red), corresponding to photon energies of 1500, 750, and 250 eV. Extremely high temperature gas emits these X rays. The white regions, which emit strongly at all X-ray wavelengths, are remnants of supernovae. **(e)** The image is made by using gamma rays, which is indicated by the highlighted G in the wavelength tab. The gamma-ray view from the Compton Gamma Ray Observatory includes all wavelengths less than about 1.2×10^{-5} nm (photon energies greater than 10^8 eV). The diffuse radiation from the Milky Way is emitted when fast-moving subatomic particles collide with the nuclei of atoms in interstellar gas clouds. The bright spots above and below the Milky Way are distant, extremely energetic galaxies.

COSMIC CONNECTIONS

Telescopes Across the EM Spectrum

Today, telescopes can view the universe in every range of electromagnetic radiation, although some must be above Earth's atmosphere to receive radiation without interference.

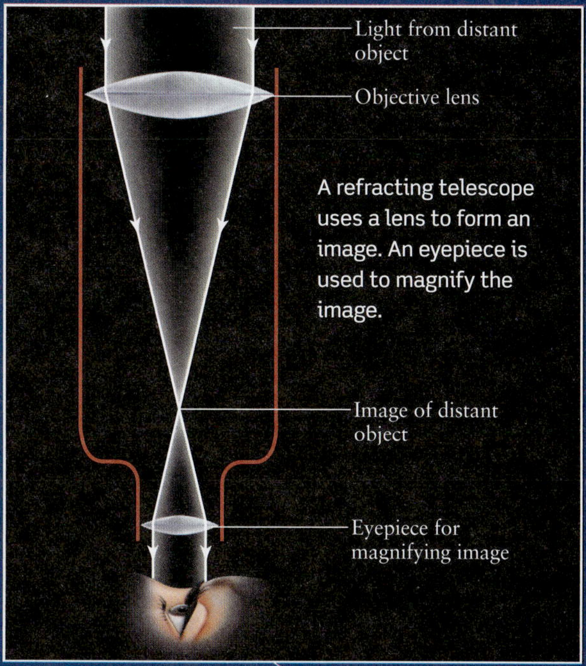

Light from distant object
Objective lens
A refracting telescope uses a lens to form an image. An eyepiece is used to magnify the image.
Image of distant object
Eyepiece for magnifying image

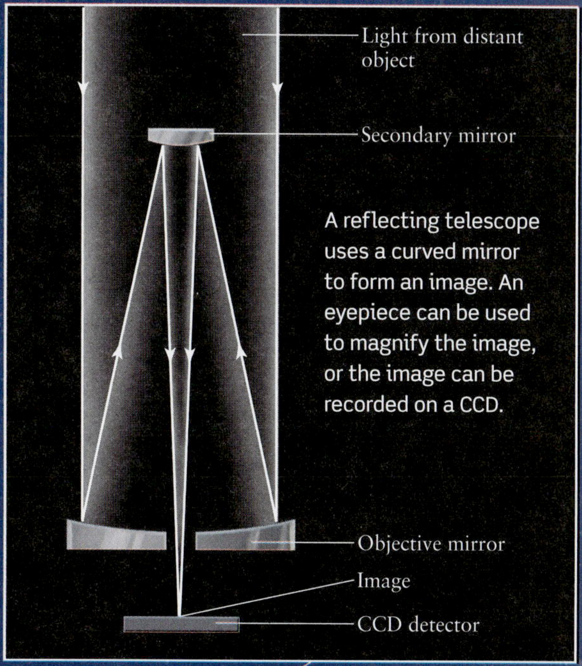

Light from distant object
Secondary mirror
A reflecting telescope uses a curved mirror to form an image. An eyepiece can be used to magnify the image, or the image can be recorded on a CCD.
Objective mirror
Image
CCD detector

Earth's atmosphere is transparent to visible wavelengths, so visible-light telescopes (refracting or reflecting) can be used from Earth's surface.
Infrared telescopes are best placed in orbit since most infrared wavelengths do not penetrate the atmosphere.
Infrared telescope
Ultraviolet telescope
Gamma-ray, x-ray, and ultraviolet telescopes must be placed in orbit because these wavelengths do not penetrate the atmosphere.
X-ray telescope
Radio telescope
The atmosphere is transparent to radio waves, so radio telescopes can be used from Earth's surface.
Gamma-ray telescope

Key Words

active optics, p. 140
adaptive optics, p. 140
angular resolution, p. 139
baseline, p. 140
Cassegrain focus, p. 137
charge-coupled device (CCD), p. 142
chromatic aberration, p. 134
coma, p. 138
coudé focus, p. 137
diffraction, p. 139
diffraction grating, p. 143
eyepiece lens, p. 131
false color, p. 140
focal length, p. 130
focal plane, p. 131
focal point, p. 130
focus (of a lens or mirror), p. 130
imaging, p. 142
interferometry, p. 140
lens, p. 129
light-gathering power, p. 132
light pollution, p. 141
magnification (magnifying power), p. 132
medium (*plural* media), p. 130
Newtonian reflector, p. 136
objective lens, p. 131
objective mirror (primary mirror), p. 135
optical telescope, p. 129
optical window (in Earth's atmosphere), p. 147
photometry, p. 142
pixel, p. 142
prime focus, p. 136
radio telescope, p. 144
radio window (in Earth's atmosphere), p. 147
reflecting telescope (reflector), p. 135
reflection, p. 134
refracting telescope (refractor), p. 131
refraction, p. 130
seeing disk, p. 139
spectrograph, p. 143
spectroscopy, p. 143
spherical aberration, p. 137
very-long-baseline interferometry (VLBI), p. 146

Key Ideas

Refracting Telescopes: Refracting telescopes, or refractors, produce images by bending light rays as they pass through glass lenses.

• Chromatic aberration is an optical defect whereby light of different wavelengths is bent in different amounts by a lens.

• Glass impurities, chromatic aberration, opacity to certain wavelengths, and structural difficulties make it inadvisable to build extremely large refractors.

Reflecting Telescopes: Reflecting telescopes, or reflectors, produce images by reflecting light rays to a focus point from curved mirrors.

• Reflectors are not subject to most of the problems that limit the useful size of refractors.

Angular Resolution: A telescope's angular resolution, which indicates ability to see fine details, is limited by two key factors.

• Diffraction is an intrinsic property of light waves. Its effects can be minimized by using a larger objective lens or mirror.

• The blurring effects of atmospheric turbulence can be minimized by placing the telescope atop a tall mountain with very smooth air. They can be dramatically reduced by the use of adaptive optics and can be eliminated entirely by placing the telescope in orbit.

Charge-Coupled Devices: Sensitive light detectors called charge-coupled devices (CCDs) are often used at a telescope's focus to record faint images.

Spectrographs: A spectrograph uses a diffraction grating to form the spectrum of an astronomical object.

Radio Telescopes: Radio telescopes use large reflecting dishes to focus radio waves onto a detector.

• Very large dishes provide reasonably sharp radio images. Higher resolution is achieved with interferometry techniques that link smaller dishes together.

Transparency of Earth's Atmosphere: Earth's atmosphere absorbs much of the radiation that arrives from space.

• The atmosphere is transparent chiefly in two wavelength ranges known as the optical window and the radio window. A few wavelengths in the near-infrared also reach the ground.

Telescopes in Space: For observations at wavelengths to which Earth's atmosphere is opaque, astronomers depend on telescopes carried above the atmosphere by rockets or spacecraft.

• Satellite-based observatories provide new information about the universe and permit coordinated observation of the sky at all wavelengths.

Questions

Review Questions

1. Describe refraction and reflection. Explain how these processes enable astronomers to build telescopes.
2. Explain why a flat piece of glass does not bring light to a focus while a curved piece of glass can.
3. Explain why the light rays that enter a telescope from an astronomical object are essentially parallel.
4. With the aid of a diagram, describe a refracting telescope. Which dimensions of the telescope determine its light-gathering power? Which dimensions determine the magnification?
5. What is the purpose of a telescope eyepiece? What aspect of the eyepiece determines the magnification of the image? In what circumstances would the eyepiece not be used?
6. Do most professional astronomers actually look through their telescopes? Why or why not?
7. Quite often advertisements appear for telescopes that extol their magnifying power. Is this a good criterion for evaluating telescopes? Explain your answer.
8. What is chromatic aberration? For what kinds of telescopes does it occur? How can it be corrected?
9. With the aid of a diagram, describe a reflecting telescope. Describe four different ways in which an astronomer can access the focal plane.
10. Explain some of the disadvantages of refracting telescopes compared to reflecting telescopes.
11. What kind of telescope would you use if you wanted to take a color photograph entirely free of chromatic aberration? Explain your answer.
12. Explain why a Cassegrain reflector can be substantially shorter than a refractor of the same focal length.

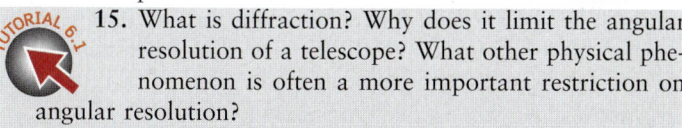

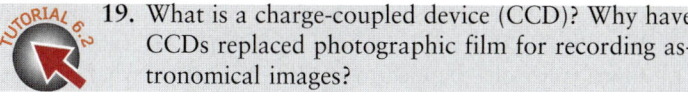

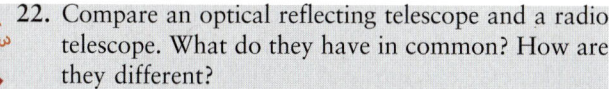

13. No major observatory has a Newtonian reflector as its primary instrument, whereas Newtonian reflectors are extremely popular among amateur astronomers. Explain why this is so.
14. What is spherical aberration? How can it be corrected?
15. What is diffraction? Why does it limit the angular resolution of a telescope? What other physical phenomenon is often a more important restriction on angular resolution?
16. What is active optics? What is adaptive optics? Why are they useful? Would either of these be a good feature to include on a telescope to be placed in orbit?
17. Explain why combining the light from two or more optical telescopes can give dramatically improved angular resolution.
18. What is light pollution? What effects does it have on the operation of telescopes? What can be done to minimize these effects?
19. What is a charge-coupled device (CCD)? Why have CCDs replaced photographic film for recording astronomical images?
20. What is a spectrograph? Why do many astronomers regard it as the most important device that can be attached to a telescope?
21. What are the advantages of using a diffraction grating rather than a prism in a spectrograph?
22. Compare an optical reflecting telescope and a radio telescope. What do they have in common? How are they different?
23. Why can radio astronomers make observations at any time during the day, whereas optical astronomers are mostly limited to observing at night? (*Hint:* Does your radio work any better or worse in the daytime than at night?)
24. Why are radio telescopes so large? Why does a single radio telescope have poorer angular resolution than a large optical telescope? How can the resolution be improved by making simultaneous observations with several radio telescopes?
25. What are the optical window and the radio window? Why isn't there an X-ray window or an ultraviolet window?
26. Why is it necessary to keep an infrared telescope at a very low temperature?
27. How are the images made by an X-ray telescope different from those made by a medical X-ray machine?
28. Why must astronomers use satellites and Earth-orbiting observatories to study the heavens at X-ray and gamma-ray wavelengths?

Advanced Questions

> **Problem-solving tips and tools**
>
> You may find it useful to review the small-angle formula discussed in Box 1-1. The area of a circle is proportional to the square of its diameter. Data on the planets can be found in the appendices at the end of this book. Section 5-2 discusses the relationship between frequency and wavelength. Box 6-1 gives examples of how to calculate magnifying power and light-gathering power.

29. Show by means of a diagram why the image formed by a simple refracting telescope is upside down.
30. Ordinary photographs made with a telephoto lens make distant objects appear close. How does the focal length of a telephoto lens compare with that of a normal lens? Explain your reasoning.
31. The observing cage in which an astronomer can sit at the prime focus of the 5-m telescope on Palomar Mountain is about 1 m in diameter. Calculate what fraction of the incoming starlight is blocked by the cage.
32. (**a**) Compare the light-gathering power of the Keck I 10.0-m telescope with that of the Hubble Space Telescope (HST), which has a 2.4-m objective mirror. (**b**) What advantages does Keck I have over HST? What advantages does HST have over Keck I?
33. Suppose your Newtonian reflector has an objective mirror 20 cm (8 in.) in diameter with a focal length of 2 m. What magnification do you get with eyepieces whose focal lengths are (**a**) 9 mm, (**b**) 20 mm, and (**c**) 55 mm? (**d**) What is the telescope's diffraction-limited angular resolution when used with orange light of wavelength 600 nm? (**e**) Would it be possible to achieve this angular resolution if you took the telescope to the summit of Mauna Kea? Why or why not?
34. Several groups of astronomers are making plans for large ground-based telescopes. (**a**) What would be the diffraction-limited angular resolution of a telescope with a 40-meter objective mirror? Assume that yellow light with wavelength 550 nm is used. (**b**) Suppose this telescope is placed atop Mauna Kea. How will the actual angular resolution of the telescope compare to that of the 10-meter Keck I telescope? Assume that adaptive optics is not used.
35. The Hobby-Eberly Telescope (HET) at the McDonald Observatory in Texas has a spherical mirror, which is the least expensive shape to grind. Consequently, the telescope has spherical aberration. Explain why this doesn't affect the usefulness of HET for spectroscopy. (The telescope is not used for imaging.)
36. The four largest moons of Jupiter are roughly the same size as our Moon and are about 628 million (6.28×10^8) kilometers from Earth at opposition. What is the size in kilometers of the smallest surface features that the Hubble Space Telescope (resolution of 0.1 arcsec) can detect? How does this compare with the smallest features that can be seen on the Moon with the unaided human eye (resolution of 1 arcmin)?
37. The Hubble Space Telescope (HST) has been used to observe the galaxy M100, some 70 million light-years from Earth. (**a**) If the angular resolution of the HST image is 0.1 arcsec, what is the diameter in light-years of the smallest detail that can be discerned in the image? (**b**) At what distance would a U.S. dime (diameter 1.8 cm) have an angular size of 0.1 arcsec? Give your answer in kilometers.
38. At its closest to Earth, Pluto is 28.6 AU from Earth. Can the Hubble Space Telescope distinguish any features on Pluto? Justify your answer using calculations.
39. The Institute of Space and Astronautical Science in Japan proposes to place a radio telescope into an even higher orbit than the HALCA telescope. Using this telescope in concert with ground-based radio-telescopes, baselines as long as

25,000 km may be obtainable. Astronomers want to use this combination to study radio emission at a frequency of 43 GHz from the molecule silicon monoxide, which is found in the interstellar clouds from which stars form. (1 GHz = 1 gigahertz = 10^9 Hz.) (**a**) What is the wavelength of this emission? (**b**) Taking the baseline to be the effective diameter of this radio-telescope array, what angular resolution can be achieved?

40. The mission of the Submillimeter Wave Astronomy Satellite (SWAS), launched in 1998, was to investigate interstellar clouds within which stars form. One of the frequencies at which it observed these clouds is 557 GHz (1 GHz = 1 gigahertz = 10^9 Hz), characteristic of the emission from interstellar water molecules. (**a**) What is the wavelength (in meters) of this emission? In what part of the electromagnetic spectrum is this? (**b**) Why was it necessary to use a satellite for these observations? (**c**) SWAS had an angular resolution of 4 arcminutes. What was the diameter of its primary mirror?

41. To search for ionized oxygen gas surrounding our Milky Way Galaxy, astronomers aimed the ultraviolet telescope of the FUSE spacecraft at a distant galaxy far beyond the Milky Way. They then looked for an ultraviolet spectral line of ionized oxygen in that galaxy's spectrum. Were they looking for an emission line or an absorption line? Explain.

42. A sufficiently thick interstellar cloud of cool gas can absorb low-energy X rays but is transparent to high-energy X rays and gamma rays. Explain why both part *b* and part *d* of Figure 6-32 reveal the presence of cool gas in the Milky Way. Could you infer the presence of this gas from the visible-light image in Figure 6-32*a*? Explain.

Discussion Questions

43. If you were in charge of selecting a site for a new observatory, what factors would you consider important?

44. Discuss the advantages and disadvantages of using a small telescope in Earth's orbit versus a large telescope on a mountaintop.

Web/e-Book Questions

45. Several telescope manufacturers build telescopes with a design called a "Schmidt-Cassegrain." These use a correcting lens in an arrangement like that shown in Figure 6-13*c*. Consult advertisements on the World Wide Web to see the appearance of these telescopes and find out their cost. Why do you suppose they are very popular among amateur astronomers?

46. The Large Zenith Telescope (LZT) in British Columbia, Canada, uses a 6.0-m *liquid* mirror made of mercury. Use the World Wide Web to investigate this technology. How can a liquid metal be formed into the necessary shape for a telescope mirror? What are the advantages of a liquid mirror? What are the disadvantages?

47. Three of the telescopes shown in Figure 6-16—the James Clerk Maxwell Telescope (JCMT), the Caltech Submillimeter Observatory (CSO), and the Submillimeter Array (SMA)—are designed to detect radiation with wavelengths close to 1 mm. Search for current information about JCMT, CSO, and SMA on the World Wide Web. What kinds of celestial objects emit radiation at these wavelengths? What can astronomers see using JCMT, CSO, and SMA that cannot be observed at other wavelengths? Why is it important that they be at high altitude? How large are the primary mirrors used in JCMT, CSO, and SMA? What are the differences among the three telescopes? Which can be used in the daytime? What recent discoveries have been made using JCMT, CSO, or SMA?

48. In 2003 an ultraviolet telescope called GALEX (Galaxy Evolution Explorer) was placed into orbit. Use the World Wide Web to learn about GALEX and its mission. What aspects of galaxies was GALEX designed to investigate? Why is it important to make these observations using ultraviolet wavelengths?

49. At the time of this writing, NASA's plans for the end of the Hubble Space Telescope's mission were uncertain. Consult the Space Telescope Science Institute Web site to learn about plans for HST's final years of operation. Are future space shuttle missions planned to service HST? If so, what changes will be made to HST on such missions? What will become of HST at the end of its mission lifetime?

50. NASA and the European Space Agency (ESA) have plans to launch a number of advanced space telescopes. These include Herschel, the Wide-Field Infrared Survey Explorer (WISE), the Gamma-ray Large Area Space Telescope (GLAST), the Space Interferometry Mission (SIM), and the X-ray Evolving Universe Spectroscopy (XEUS) mission. Search the World Wide Web for information about at least two of these. What are the scientific goals of these projects? What is unique about each telescope? What advantages would they have over existing ground-based or orbiting telescopes? What kind of orbit will each telescope be in? When will each be launched and placed in operation?

Activities

Observing Projects

51. Obtain a telescope during the daytime along with several eyepieces of various focal lengths. If you can determine the telescope's focal length, calculate the magnifying powers of the eyepieces. Focus the telescope on some familiar object, such as a distant lamppost or tree. **DO NOT FOCUS ON THE SUN! Looking directly at the Sun can cause blindness.** Describe the image you see through the telescope. Is it upside down? How does the image move as you slowly and gently shift the telescope left and right or up and down? Examine the eyepieces, noting their focal lengths. By changing the eyepieces, examine the distant object under different magnifications. How do the field of view and the quality of the image change as you go from low power to high power?

52. On a clear night, view the Moon, a planet, and a star through a telescope using eyepieces of various focal lengths and known magnifying powers. (To determine the locations in the sky of the Moon and planets, you may want to use the *Starry Night Enthusiast*™ program on the CD-ROM that comes with certain printed copies of this book. You may also want to consult such magazines as *Sky & Telescope* and *Astronomy* or their Web sites.) In what way does the image seem to degrade as you view with increasingly higher magnification? Do you see any chromatic aberration? If so, with which object and which eyepiece is it most noticeable?

53. Many towns and cities have amateur astronomy clubs. If you are so inclined, attend a "star party" hosted by your local club. People who bring their telescopes to such gatherings are delighted to show you their instruments and take you on a telescopic tour of the heavens. Such an experience can lead to a very enjoyable, lifelong hobby.

54. The field of view of a typical small telescope for amateur astronomers is about 30 arcminutes, or 30′. (Many research telescopes have much smaller fields of view. For instance, the widest field of view available to the Hubble Space Telescope is only 3.4 arcminutes.) Use the *Starry Night Enthusiast*™ program to simulate the view that such a telescope provides of various celestial objects. First display the entire celestial sphere (select **Guides > Atlas** in the **Favorites** menu). Open the **Find** pane (select **Find . . .** under the **Edit** menu or click the **Find** tab on the left side of the view window). Click on the magnifying glass icon at the left-hand side of the edit box at the top of the Find pane and select the **Orbiting Objects** item from the drop-down menu that appears. This will bring up a list of Solar System objects in the Find pane. For each of the following objects, (i) the Moon, (ii) Jupiter, and (iii) Saturn, double-click on the name of the object in the Find pane to center it in the view. Use the zoom controls (+ and − buttons) at the right-hand end of the toolbar to adjust the field of view to approximately 30 arcminutes. (You can display an indication of this specific field of view on the sky by opening the **FOV** (**Field of View**) pane on the left side of the window and clicking on **Other > 30 Arcminutes.**) Describe how much detail you can see on each object. Open the Find pane again. Click on the magnifying glass icon on the left-hand side of the edit box at the top of the Find pane and select Messier Objects from the drop-down menu that appears. From the list of Messier Objects that populates the Find pane, double-click the entry for "M31 (Andromeda Galaxy)" to center this object in the view. Again, use the zoom controls to adjust the field of view to approximately 30 arcminutes. Describe how much detail you can see for this object.

55. Use the *Starry Night Enthusiast*™ program to explore the concept of angular resolution. Click the **Find** tab to the left of the main view window to open the Find pane. Click on the magnifying glass icon at the left-hand side of the edit box at the top of the Find pane and select the **Orbiting Objects** item from the drop-down menu that appears. This will bring up a list of Solar System objects in the Find pane. Double-click the entry labeled **The Moon.** You can zoom in and zoom out using the **Zoom** buttons at the right side of the toolbar. You can also rotate the Moon by putting the mouse cursor over the image, holding down the mouse button and the **Shift** key on the keyboard, and moving the mouse. (On a two-button mouse, hold down the left mouse button.) (**a**) What is the size of the smallest detail that you can see? (You will have to make measurements on the screen using a ruler and compare it to the diameter of the Moon, which is 3476 km.) (**b**) The angular resolution of the Hubble Space Telescope (HST) is 0.1 arcsec. How far away from the Moon could HST be and still be able to resolve details as small as you determined in part (a)? Give your answer in kilometers and in astronomical units (AU).

Collaborative Exercises

56. Stand up and have everyone in your group join hands, making as large a circle as possible. If a telescope mirror were built as big as your circle, what would be its diameter? What would be your telescope's diffraction-limited angular resolution for blue light? Would atmospheric turbulence have a noticeable effect on the angular resolution?

57. Are there enough students in your class to stand and join hands and make two large circles that recreate the sizes of the two Keck telescopes? Explain how you determined your answer.

7

Comparative Planetology I: Our Solar System

WEB LINK 7.1

March 10, 2006: The *Mars Reconnaissance Orbiter* spacecraft arrives at Mars (artist's impression). (JPL/NASA)

Fifty years ago, astronomers knew precious little about the worlds that orbit the Sun. Even the best telescopes provided images of the planets that were frustratingly hazy and indistinct. Of asteroids, comets, and the satellites (or moons) of the planets, we knew even less.

Today, our knowledge of the solar system has grown exponentially, due almost entirely to robotic spacecraft. (The illustration shows an artist's impression of one such robotic explorer, the *Mars Reconnaissance Orbiter* spacecraft, as it approached Mars in 2006.) Spacecraft have been sent to fly past all the planets at close range, revealing details unimagined by astronomers of an earlier generation. We have landed spacecraft on the Moon, Venus, and Mars and dropped a probe into the immense atmosphere of Jupiter. This is truly the golden age of solar system exploration.

In this chapter we paint in broad outline our present understanding of the solar system. We will see that the planets come in a variety of sizes and chemical compositions. A rich variety also exists among the moons (or satellites) of the planets and among smaller bodies we call asteroids, comets, and trans-Neptunian objects. We will investigate the nature of craters on the Moon and other worlds of the solar system. And by exploring the magnetic fields of planets, we will be able to peer inside Earth and other worlds and learn about their interior compositions.

An important reason to study the solar system is to search for our own origins. In Chapter 8 we will see how astronomers have used evidence from the present-day solar system to understand how the Sun, Earth, and the other planets formed some four and a half billion years ago, and how they have evolved since then. But for now, we invite you to join us on a guided tour of the worlds that orbit our Sun.

7-1 The solar system has two broad categories of planets: Earthlike and Jupiterlike

Each of the planets that orbit the Sun is unique. Only Earth has liquid water and an atmosphere breathable by humans; only Venus has a perpetual cloud layer made of sulfuric acid droplets; and only Jupiter has immense storm systems that persist for centuries. But there are also striking similarities among planets.

Learning Goals

By reading the sections of this chapter, you will learn

7-1 The important differences between the two broad categories of planets: terrestrial and Jovian

7-2 The similarities and differences among the large planetary satellites, including Earth's Moon

7-3 How the spectrum of sunlight reflected from a planet reveals the composition of its atmosphere and surface

7-4 Why some planets have atmospheres and others do not

7-5 The categories of the many small bodies that also orbit the Sun

7-6 How craters on a planet or satellite reveal the age of its surface and the nature of its interior

7-7 Why a planet's magnetic field indicates a fluid interior in motion

7-8 How the diversity of the solar system is a result of its origin and evolution

Volcanoes are found not only on Earth but also on Venus and Mars; rings encircle Jupiter, Saturn, Uranus, and Neptune; and impact craters dot the surfaces of Mercury, Venus, Earth, and Mars, showing that all of these planets have been bombarded by interplanetary debris.

We can understand the most important similarities and differences among the planets by comparing their orbits, masses, and diameters

How can we make sense of the many similarities and differences among the planets? An important step is to organize our knowledge of the planets in a systematic way. We can organize this information in two ways. First, we can contrast the orbits of different planets around the Sun; and second, we can compare the planets' physical properties such as diameter, mass, average density, and chemical composition.

Comparing the Planets: Orbits

TUTORIAL 7.1

The planets fall naturally into two classes according to the sizes of their orbits. As Figure 7-1 shows, the orbits of the four inner planets (Mercury, Venus, Earth, and Mars) are crowded in close to the Sun. In contrast, the orbits of the next four planets (Jupiter, Saturn, Uranus, and Neptune) are widely spaced at great distances from the Sun. Table 7-1 lists the orbital characteristics of these eight planets.

CAUTION! While Figure 7-1 shows the orbits of the planets, it does not show the planets themselves. The reason is simple: If Jupiter, the largest of the planets, were to be drawn to the same scale as the rest of this figure, it would be a dot just 0.0002 cm across—about $^1/_{300}$ of the width of a human hair and far too small to be seen without a microscope. The planets themselves are *very* small compared to the distances between them. Indeed, while an airliner traveling at 1000 km/h (620 mi/h) can fly around Earth in less than two days, at this speed it would take 17 *years* to fly from Earth to the Sun. The solar system is a very large and very empty place!

Most of the planets have orbits that are nearly circular. As we learned in Section 4-4, Kepler discovered in the seventeenth

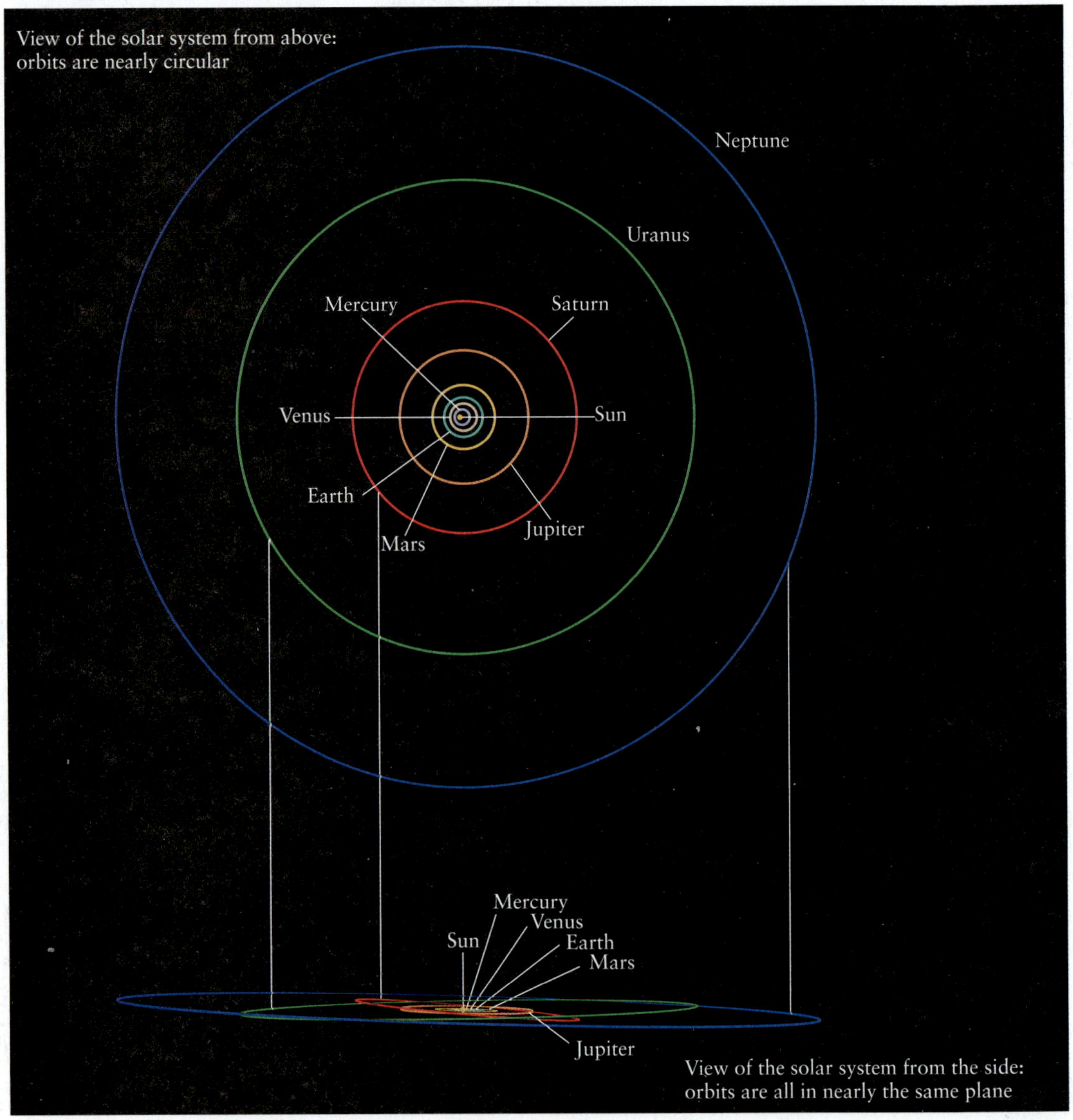
View of the solar system from above:
orbits are nearly circular
Neptune
Uranus
Mercury
Saturn
Venus
Sun
Earth
Jupiter
Mars
Mercury
Venus
Sun
Earth
Mars
Jupiter
View of the solar system from the side:
orbits are all in nearly the same plane

ANIMATION 7.1

Figure 7-1
The Solar System to Scale This scale drawing shows the orbits of the planets around the Sun. The four inner planets are crowded in close to the Sun, while the four outer planets orbit the Sun at much greater distances. On the scale of this drawing, the planets themselves would be much smaller than the diameter of a human hair and too small to see.

WEB LINK 7.2

Table 7-1 Characteristics of the Planets

	The Inner (Terrestrial) Planets			
	Mercury	Venus	Earth	Mars
Average distance from the Sun (10^6 km)	57.9	108.2	149.6	227.9
Average distance from the Sun (AU)	0.387	0.723	1.000	1.524
Orbital period (years)	0.241	0.615	1.000	1.88
Orbital eccentricity	0.206	0.007	0.017	0.093
Inclination of orbit to the ecliptic	7.00°	3.39°	0.00°	1.85°
Equatorial diameter (km)	4880	12,104	12,756	6794
Equatorial diameter (Earth = 1)	0.383	0.949	1.000	0.533
Mass (kg)	3.302×10^{23}	4.868×10^{24}	5.974×10^{24}	6.418×10^{23}
Mass (Earth = 1)	0.0553	0.8149	1.0000	0.1074
Average density (kg/m^3)	5430	5243	5515	3934

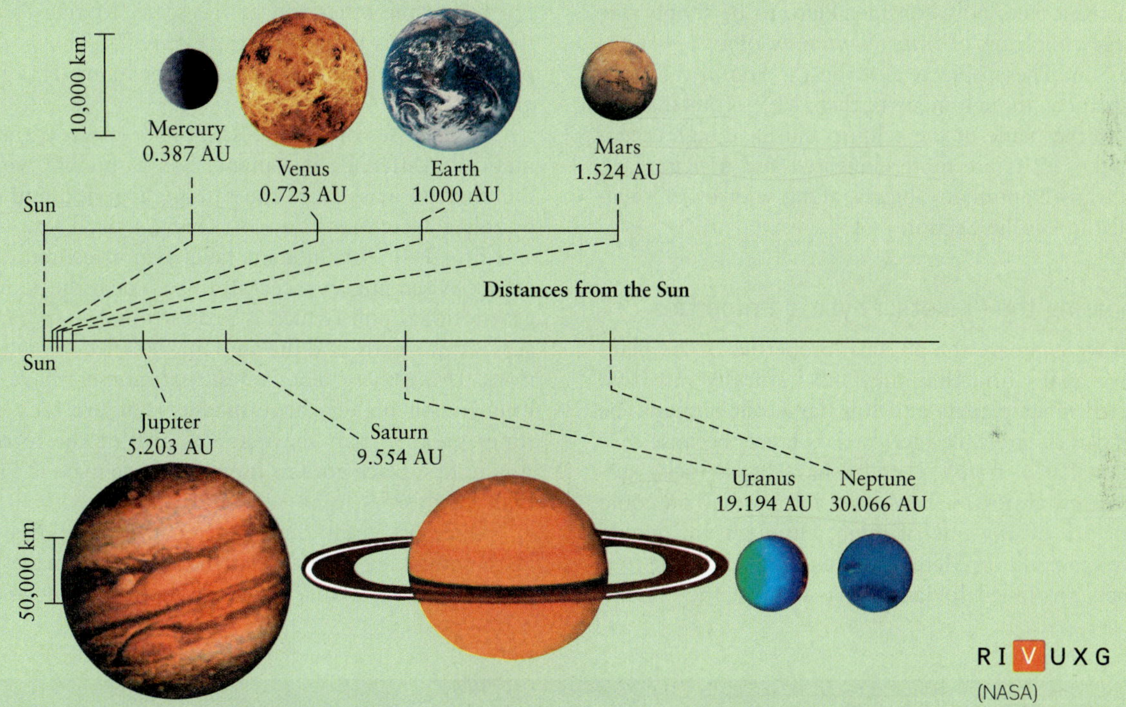

(NASA)

	The Outer (Jovian) Planets			
	Jupiter	Saturn	Uranus	Neptune
Average distance from the Sun (10^6 km)	778.3	1429	2871	4498
Average distance from the Sun (AU)	5.203	9.554	19.194	30.066
Orbital period (years)	11.86	29.46	84.10	164.86
Orbital eccentricity	0.048	0.053	0.043	0.010
Inclination of orbit to the ecliptic	1.30°	2.48°	0.77°	1.77°
Equatorial diameter (km)	142,984	120,536	51,118	49,528
Equatorial diameter (Earth = 1)	11.209	9.449	4.007	3.883
Mass (kg)	1.899×10^{27}	5.685×10^{26}	8.682×10^{25}	1.024×10^{26}
Mass (Earth = 1)	317.8	95.16	14.53	17.15
Average density (kg/m^3)	1326	687	1318	1638

century that these orbits are actually ellipses. Astronomers denote the elongation of an ellipse by its *eccentricity* (see Figure 4-10*b*). The eccentricity of a circle is zero, and indeed most of the eight planets (with the notable exception of Mercury) have orbital eccentricities that are very close to zero.

If you could observe the solar system from a point several astronomical units (AU) above Earth's north pole, you would see that all the planets orbit the Sun in the same counterclockwise direction. Furthermore, the orbits of the eight planets all lie in nearly the same plane. In other words, these orbits are inclined at only slight angles to the plane of the ecliptic, which is the plane of Earth's orbit around the Sun (see Section 2-5). What is more, the plane of the Sun's equator is very closely aligned with the orbital planes of the planets. As we will see in Chapter 8, these near-alignments are not a coincidence. They provide important clues about the origin of the solar system.

Not included in Figure 7-1 or Table 7-1 is Pluto, which is in an even larger orbit than Neptune. Until the late 1990s, Pluto was generally regarded as the ninth planet. But in light of recent discoveries many astronomers now consider Pluto to be simply one member of a large collection of *trans-Neptunian objects* that orbit far from the Sun. These objects all orbit the Sun in the same direction as the planets, though many of them have orbits that are steeply inclined to the plane of the ecliptic and have high eccentricities (that is, the orbits are quite elongated and noncircular). We will discuss trans-Neptunian objects, along with other small bodies that orbit the Sun, in Section 7-5.

Comparing the Planets: Physical Properties

TUTORIAL 7.2

When we compare the physical properties of the planets, we again find that they fall naturally into two classes—four small inner planets and four large outer ones. The four small inner planets are called **terrestrial planets** because they resemble Earth (in Latin, *terra*). They all have hard, rocky surfaces with mountains, craters, valleys, and volcanoes. You could stand on the surface of any one of them, although you would need a protective spacesuit on Mercury, Venus, or Mars. The four large outer planets are called **Jovian planets** because they resemble Jupiter. (Jove was another name for the Roman god Jupiter.) An attempt to land a spacecraft on the surface of any of the Jovian planets would be futile, because the materials of which these planets are made are mostly gaseous or liquid. The visible "surface" features of a Jovian planet are actually cloud formations in the planet's atmosphere. The photographs in Figure 7-2 show the distinctive appearances of the two classes of planets.

The most apparent difference between the terrestrial and Jovian planets is their *diameters*. You can compute the diameter of a planet if you know its angular diameter as seen from Earth and its distance from Earth. For example, on March 16, 2007, Venus was 1.97×10^8 km from Earth and had an angular diameter of 12.7 arcsec. Using the small-angle formula from Box 1-1, we can calculate the diameter of Venus to be 12,100 km (7520 mi). Similar calculations demonstrate that Earth, with its diameter of about 12,756 km (7926 mi), is the largest of the four inner, terrestrial planets. In sharp contrast, the four outer, Jovian planets are much larger than the terrestrial planets. First place goes to Jupiter, whose equatorial diameter is more than 11 times that of Earth. On the other end of the scale, Mercury's diameter is less than two-fifths that of Earth. Figure 7-2 shows the Sun and the planets drawn to the same scale. The diameters of the planets are given in Table 7-1.

The *masses* of the terrestrial and Jovian planets are also dramatically different. If a planet has a satellite, you can calculate the planet's mass from the satellite's period and semimajor axis by using Newton's form of Kepler's third law (see Section 4-7 and Box 4-4). Astronomers have also measured the mass of each planet by sending a spacecraft to pass near the planet. The planet's gravitational pull (which is proportional to its mass) deflects the spacecraft's path, and the amount of deflection tells us the planet's mass. Using these techniques, astronomers have found that the four Jovian planets have masses that are tens or hundreds of times greater than the mass of any of the terrestrial planets. Again, first place goes to Jupiter, whose mass is 318 times greater than Earth's.

Once we know the diameter and mass of a planet, we can learn something about what that planet is made of. The trick is to calculate the planet's **average density**, or mass divided by vol-

Figure 7-2 RIVUXG

The Planets to Scale This figure shows the planets from Mercury to Neptune to the same scale. The four terrestrial planets have orbits nearest the Sun, and the Jovian planets are the next four planets from the Sun. (Calvin J. Hamilton and NASA/JPL)

BOX 7-1 Astronomy Down to Earth

Average Density

Average density—the mass of an object divided by that object's volume—is a useful quantity for describing the differences between planets in our solar system. This same quantity has many applications here on Earth.

A rock tossed into a lake sinks to the bottom, while an air bubble produced at the bottom of a lake (for example, by the air tanks of a scuba diver) rises to the top. These are examples of a general principle: An object sinks in a fluid if its average density is greater than that of the fluid, but rises if its average density is less than that of the fluid. The average density of water is 1000 kg/m^3, which is why a typical rock (with an average density of about 3000 kg/m^3) sinks, while an air bubble (average density of about 1.2 kg/m^3) rises.

At many summer barbecues, cans of soft drinks are kept cold by putting them in a container full of ice. When the ice melts, the cans of diet soda always rise to the top, while the cans of regular soda sink to the bottom. Why is this? The average density of a can of diet soda—which includes water, flavoring, artificial sweetener, and the trapped gas that makes the drink fizzy—is slightly less than the density of water, and so the can floats. A can of regular soda contains sugar instead of artificial sweetener, and the sugar is a bit heavier than the sweetener. The extra weight is just enough to make the average density of a can of regular soda slightly more than that of water, making the can sink. (You can test these statements for yourself by putting unopened cans of diet soda and regular soda in a sink or bathtub full of water.)

The concept of average density provides geologists with important clues about the early history of Earth. The average density of surface rocks on Earth, about 3000 kg/m^3, is less than Earth's average density of 5515 kg/m^3. The simplest explanation is that in the ancient past, Earth was completely molten throughout its volume, so that low-density materials rose to the surface and high-density materials sank deep into Earth's interior in a process called *chemical differentiation.* This series of events also suggests that Earth's core must be made of relatively dense materials, such as iron and nickel. A tremendous amount of other geological evidence has convinced scientists that this picture is correct.

ume, measured in kilograms per cubic meter (kg/m^3). The average density of any substance depends in part on that substance's composition. For example, air near sea level on Earth has an average density of 1.2 kg/m^3, water's average density is 1000 kg/m^3, and a piece of concrete has an average density of 2000 kg/m^3. Box 7-1 describes some applications of the idea of average density to everyday phenomena on Earth.

The four inner, terrestrial planets have very high average densities (see Table 7-1); the average density of Earth, for example, is 5515 kg/m^3. By contrast, a typical rock found on Earth's surface has a lower average density, about 3000 kg/m^3. Thus, Earth must contain a large amount of material that is denser than rock. This information provides our first clue that terrestrial planets have dense iron cores.

In sharp contrast, the outer, Jovian planets have quite low densities. Saturn has an average density less than that of water. This information strongly suggests that the giant outer planets are composed primarily of light elements such as hydrogen and helium. All four Jovian planets probably have large cores of mixed rock and highly compressed water buried beneath low-density outer layers tens of thousands of kilometers thick.

We can conclude that the following general rule applies to the planets:

The terrestrial planets are made of rocky materials and have dense iron cores, which gives these planets high average densities. The Jovian planets are composed primarily of light elements such as hydrogen and helium, which gives these planets low average densities.

7-2 Seven large satellites are almost as big as the terrestrial planets

All the planets except Mercury and Venus have moons. More than 160 satellites are known: Earth has one (the Moon), Mars has two, Jupiter has at least 63, Saturn at least 61, Uranus at least 27, and Neptune at least 13. Dozens of other small satellites probably remain to be discovered as our telescope technology continues to improve. Like the terrestrial planets, all of the satellites of the planets have solid surfaces.

The various moons of the planets are not simply copies of Earth's Moon

You can see that there is a striking difference between the terrestrial planets, with few or no satellites, and the Jovian planets, each of which has so many moons that it resembles a solar system in miniature. In Chapter 8 we will explore this evidence (as well as other evidence) that the Jovian planets formed in a manner similar to the solar system as a whole, but on a smaller scale.

Of the known satellites, seven are roughly as big as the planet Mercury. Table 7-2 lists these satellites and shows them to the same scale. Note that Earth's Moon and Jupiter's satellites Io and Europa have relatively high average densities, indicating that these moons are made primarily of rocky materials. By contrast, the average densities of Ganymede, Callisto, Titan, and Triton are all relatively low. Planetary scientists conclude that the interiors of these four moons also contain substantial amounts of water ice, which is less dense than rock. (In Section 7-4 we

WEB LINK 7.3

Table 7-2 The Seven Giant Satellites

	Moon	Io	Europa	Ganymede	Callisto	Titan	Triton
Parent planet	Earth	Jupiter	Jupiter	Jupiter	Jupiter	Saturn	Neptune
Diameter (km)	3476	3642	3130	5268	4806	5150	2706
Mass (kg)	7.35×10^{22}	8.93×10^{22}	4.80×10^{22}	1.48×10^{23}	1.08×10^{23}	1.34×10^{23}	2.15×10^{22}
Average density (kg/m^3)	3340	3530	2970	1940	1850	1880	2050
Substantial atmosphere?	No	No	No	No	No	Yes	No

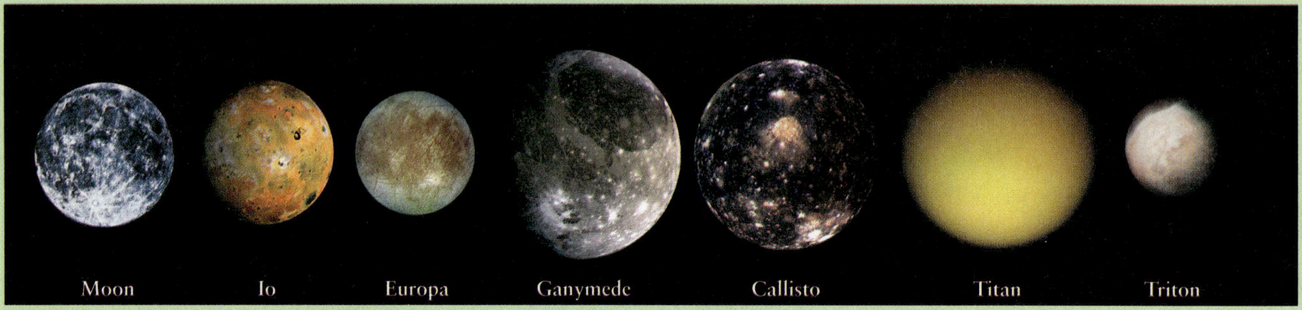

R I V U X G
(NASA/JPL/Space Science Institute)

will learn about types of frozen "ice" made of substances other than water.)

CAUTION! Water ice may seem like a poor material for building a satellite, since the ice you find in your freezer can easily be cracked or crushed. But under high pressure, such as is found in the interior of a large satellite, water ice becomes as rigid as rock. (It also becomes denser than the ice found in ice cubes, although not as dense as rock.) Note that water ice is an important constituent only for satellites in the outer solar system, where the Sun is far away and temperatures are very low. For example, the surface temperature of Titan is a frigid 95 K (−178°C = −288°F). In Section 7-4 we will learn more about the importance of temperature in determining the composition of a planet or satellite.

The satellites listed in Table 7-2 are actually unusually large. Most of the known satellites have diameters less than 2000 km, and many are irregularly shaped and just a few kilometers across.

Interplanetary spacecraft have made many surprising and fascinating discoveries about the satellites of the solar system. We now know that Jupiter's satellite Io is the most geologically active world in the solar system, with numerous geyserlike volcanoes that continually belch forth sulfur-rich compounds. The fractured surface of Europa, another of Jupiter's large satellites, suggests that a worldwide ocean of liquid water may lie beneath its icy surface. Saturn's largest satellite, Titan, has an atmosphere nearly twice as dense as Earth's atmosphere, with a perpetual haze layer that gives it a featureless appearance (see the photographs in Table 7-2).

7-3 Spectroscopy reveals the chemical composition of the planets

As we have seen, the average densities of the planets and satellites give us a crude measure of their **chemical compositions**—that is, what substances they are made of. For example, the low average density of the Moon (3340 kg/m^3) compared with Earth (5515 kg/m^3) tells us that the Moon contains relatively little iron or other dense metals. But to truly understand the nature of the planets and satellites, we need to know their chemical compositions in much greater detail than we can learn from average density alone.

The light we receive from a planet or satellite is reflected sunlight—but with revealing differences in its spectrum

The most accurate way to determine chemical composition is by directly analyzing samples taken from a planet's atmosphere and soil. Unfortunately, of all the planets and satellites, we have such direct information only for Earth and the three worlds on which spacecraft have landed—Venus, the Moon, and Mars. In all other cases, astronomers must analyze sunlight reflected from the distant planets and their satellites. To do that, astronomers bring to bear one of their most powerful tools, **spectroscopy,** the systematic study of spectra and spectral lines. (We discussed spectroscopy in Sections 5-6 and 6-5.)

Determining Atmospheric Composition

Spectroscopy is a sensitive probe of the composition of a planet's *atmosphere*. If a planet has an atmosphere, then sunlight reflected from that planet must have passed through its atmosphere. During this passage, some of the wavelengths of sunlight will have been absorbed. Hence, the spectrum of this reflected sunlight will have dark absorption lines. Astronomers look at the particular wavelengths absorbed and the amount of light absorbed at those wavelengths. Both of these depend on the kinds of chemicals present in the planet's atmosphere and the abundance of those chemicals.

For example, astronomers have used spectroscopy to analyze the atmosphere of Saturn's largest satellite, Titan (see Figure 7-3*a* and Table 7-2). The graph in Figure 7-3*b* shows the spectrum of visible sunlight reflected from Titan. (We first saw this method of displaying spectra in Figure 6-21.) The dips in this curve of intensity versus wavelength represent absorption lines. However, not all of these absorption lines are produced in the atmosphere of Titan (Figure 7-3*c*). Before reaching Titan, light from the Sun's glowing surface must pass through the Sun's hydrogen-rich atmosphere. This produces the hydrogen absorption line in Figure 7-3*b* at a wavelength of 656 nm. After being reflected from Titan, the light must pass through Earth's atmosphere before reaching the telescope; this is where the oxygen absorption line in Figure 7-3*b* is produced. Only the two dips near 620 nm and 730 nm are caused by gases in Titan's atmosphere.

These two absorption lines are caused not by individual atoms in the atmosphere of Titan but by atoms combined to form molecules. (We introduced the idea of molecules in Section 5-6.) Molecules, like atoms, also produce unique patterns of lines in the spectra of astronomical objects. The absorption lines in Figure 7-3*b* indicate the presence in Titan's atmosphere of methane molecules (CH_4, a molecule made of one carbon atom and four hydrogen atoms). Titan must be a curious place indeed, because on Earth, methane is a rather rare substance that is the primary ingredient in natural gas! When we examine other planets and satellites with atmospheres, we find that all of their spectra have absorption lines of molecules of various types.

In addition to visible-light measurements such as those shown in Figure 7-3*b*, it is very useful to study the *infrared* and *ultraviolet* spectra of planetary atmospheres. Many molecules have much stronger spectral lines in these nonvisible wavelength bands than in the visible. As an example, the ultraviolet spectrum of Titan shows that nitrogen molecules (N_2) are the dominant constituent of Titan's atmosphere. Furthermore, Titan's infrared spectrum

(a) Saturn's satellite Titan

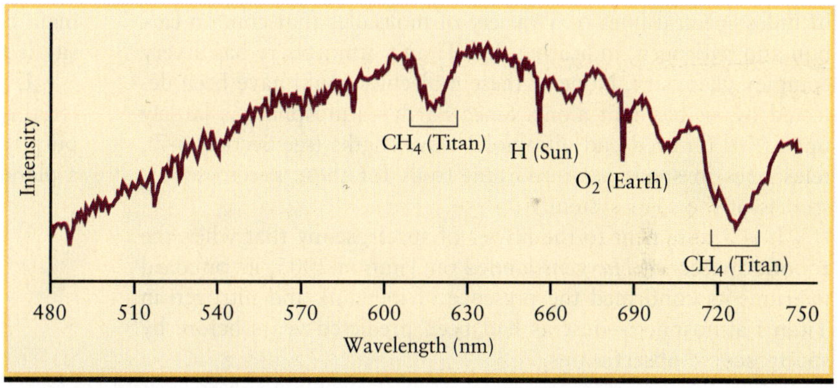

(b) The spectrum of sunlight reflected from Titan

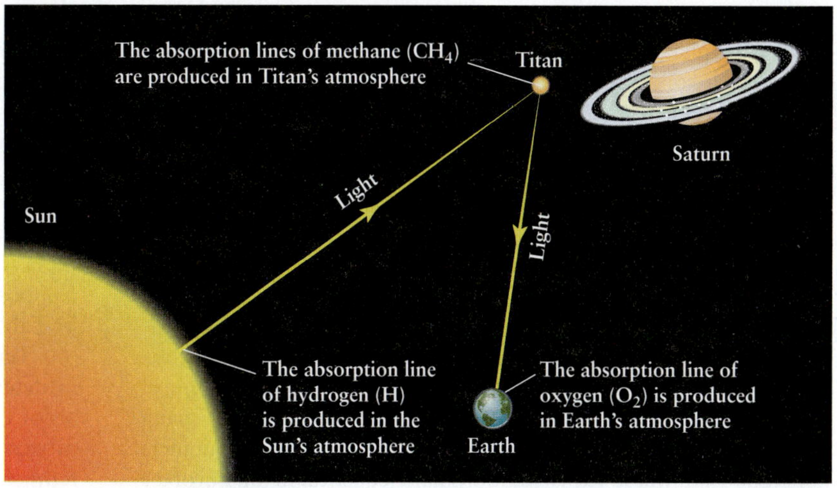

(c) Interpreting Titan's spectrum

Figure 7-3 R I V U X G

Analyzing a Satellite's Atmosphere through its Spectrum **(a)** Titan is the only satellite in the solar system with a substantial atmosphere. **(b)** The dips in the spectrum of sunlight reflected from Titan are due to absorption by hydrogen atoms (H), oxygen molecules (O_2), and methane molecules (CH_4). Of these, only methane is actually present in Titan's atmosphere. **(c)** This illustration shows the path of the light that reaches us from Titan. To interpret the spectrum of this light as shown in (b), astronomers must account for the absorption that takes place in the atmospheres of the Sun and Earth. (a: NASA/JPL/Space Science Institute)

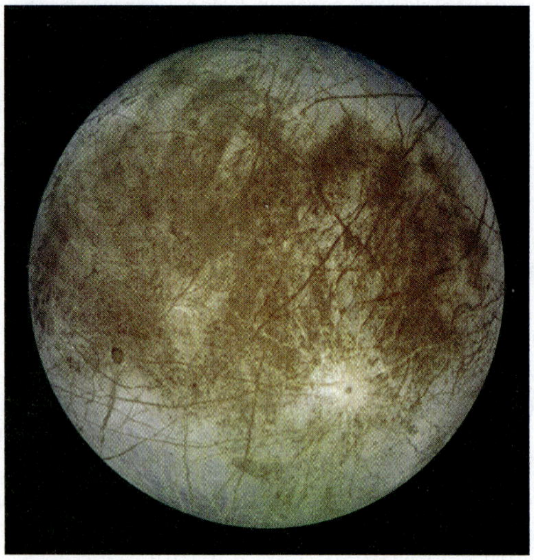

(a) Jupiter's moon Europa R I V U X G

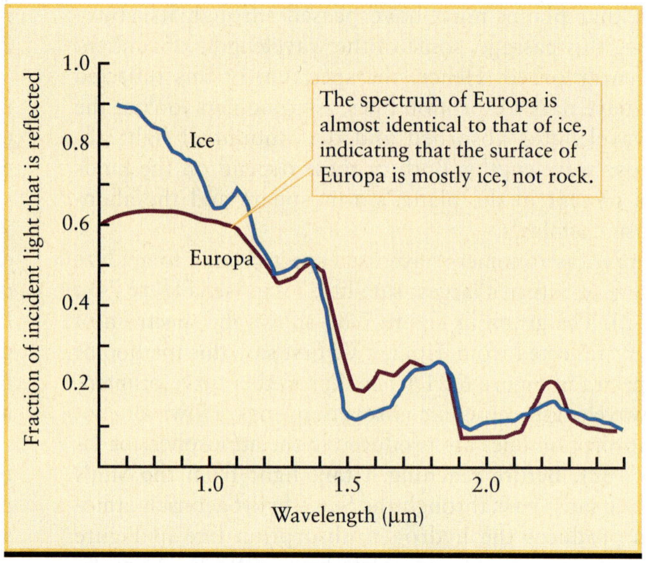
1.0
0.8
0.6
0.4
0.2
Fraction of incident light that is reflected
Ice
Europa
The spectrum of Europa is almost identical to that of ice, indicating that the surface of Europa is mostly ice, not rock.
1.0
1.5
2.0
Wavelength (μm)

(b) The spectum of light reflected from Europa

Figure 7-4

Analyzing a Satellite's Surface from Its Spectrum

(a) Unlike Titan (Figure 7-3*a*), Jupiter's satellite Europa has no atmosphere. **(b)** Infrared light from the Sun that is reflected from the surface of Europa has almost exactly the same spectrum as sunlight reflected from ordinary water ice.

includes spectral lines of a variety of molecules that contain carbon and hydrogen, indicating that Titan's atmosphere has a very complex chemistry. None of these molecules could have been detected by visible light alone. Since Earth's atmosphere is largely opaque to infrared and ultraviolet wavelengths (see Section 6-7), telescopes in space are important tools for these spectroscopic studies of the solar system.

It is a testament to the power of spectroscopy that when the robotic spacecraft *Huygens* landed on Titan in 2005, its onboard instruments confirmed the presence of methane and nitrogen in Titan's atmosphere—just as had been predicted years before by spectroscopic observations.

Determining Surface Composition

Spectroscopy can also provide useful information about the *solid surfaces* of planets and satellites without atmospheres. When light shines on a solid surface, some wavelengths are absorbed while others are reflected. (For example, a plant leaf absorbs red and violet light but reflects green light—which is why leaves look green.) Unlike a gas, a solid illuminated by sunlight does not produce sharp, definite spectral lines. Instead, only broad absorption features appear in the spectrum. By comparing such a spectrum with the spectra of samples of different substances on Earth, astronomers can infer the chemical composition of the surface of a planet or satellite.

As an example, Figure 7-4*a* shows Jupiter's satellite Europa (see Table 7-2), and Figure 7-4*b* shows the infrared spectrum of light reflected from the surface of Europa. Because this spectrum is so close to that of water ice—that is, frozen water—astronomers conclude that water ice is the dominant constituent of Europa's surface. (We saw in Section 7-2 that water ice cannot be the dominant constituent of Europa's *interior,* because this satellite's density is too high.)

Unfortunately, spectroscopy tells us little about what the material is like just below the surface of a satellite or planet. For this purpose, there is simply no substitute for sending a spacecraft to a planet and examining its surface directly.

7-4 The Jovian planets are made of lighter elements than the terrestrial planets

Spectroscopic observations from Earth and spacecraft show that the outer layers of the Jovian planets are composed primarily of the lightest gases, hydrogen and helium (see Box 5-5). In contrast, chemical analysis of soil samples from Venus, Earth, and Mars demonstrate that the terrestrial planets are made mostly of heavier elements, such as iron, oxygen, silicon, magnesium, nickel, and sulfur. Spacecraft images such as Figure 7-5 and Figure 7-6 only hint at these striking differences in chemical composition, which are summarized in Table 7-3.

Differences in distance from the Sun, and hence in temperature, explain many distinctions between terrestrial and Jovian planets

Temperature plays a major role in determining whether the materials of which planets are made exist as solids, liquids, or gases. Hydrogen (H_2) and helium (He) are gaseous except at extremely low temperatures and extraordinarily high pressures. By contrast, rock-forming substances such as iron and silicon are solids except

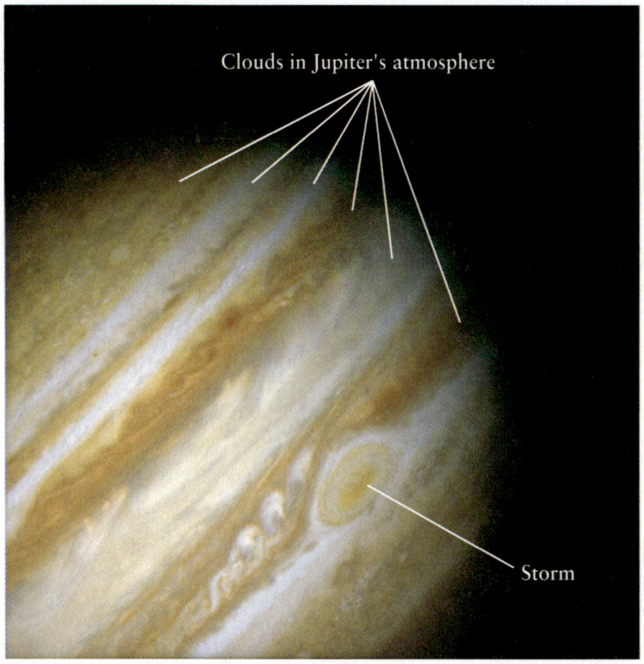

Figure 7-5 R I V U X G

A Jovian Planet This Hubble Space Telescope image gives a detailed view of Jupiter's cloudtops. Jupiter is composed mostly of the lightest elements, hydrogen and helium, which are colorless; the colors in the atmosphere are caused by trace amounts of other substances. The giant storm at lower right, called the Great Red Spot, has been raging for more than 300 years. (Space Telescope Science Institute/JPL/NASA)

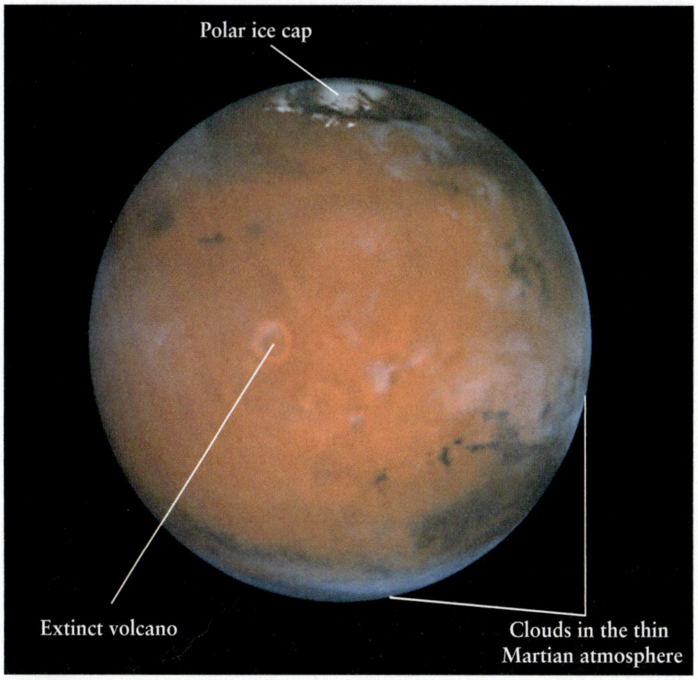

Figure 7-6 R I V U X G

A Terrestrial Planet Mars is composed mostly of heavy elements such as iron, oxygen, silicon, magnesium, nickel, and sulfur. The planet's red surface can be seen clearly in this Hubble Space Telescope image because the Martian atmosphere is thin and nearly cloudless. Olympus Mons, the extinct volcano to the left of center, is nearly 3 times the height of Mount Everest. (Space Telescope Science Institute/JPL/NASA)

at temperatures well above 1000 K. (You may want to review the discussion of temperature scales in Box 5-1.) Between these two extremes are substances such as water (H_2O), carbon dioxide (CO_2), methane (CH_4), and ammonia (NH_3), which solidify at low temperatures (from below 100 to 300 K) into solids called **ices.** (In astronomy, frozen water is just one kind of "ice.") At somewhat higher temperatures, they can exist as liquids or gases. For example, clouds of ammonia ice crystals are found in the cold upper atmosphere of Jupiter, but within Jupiter's warmer interior, ammonia exists primarily as a liquid.

CAUTION! The Jovian planets are sometimes called "gas giants." It is true that their primary constituents, including hydrogen, helium, ammonia, and methane, are gases under normal conditions on Earth. But in the interiors of these planets, pressures are so high that these substances are *liquids,* not gases. The Jovian planets might be better described as "liquid giants"!

As you might expect, a planet's surface temperature is related to its distance from the Sun. The four inner planets are quite warm. For example, midday temperatures on Mercury may climb to 700 K (= 427°C = 801°F), and during midsummer on Mars, it is sometimes as warm as 290 K (= 17°C = 63°F). The outer planets, which receive much less solar radiation, are cooler. Typical temperatures range from about 125 K (= –148°C = –234°F) in Jupiter's upper atmosphere to about 55 K (= –218°C = –360°F) at the tops of Neptune's clouds.

How Temperature Affects Atmospheres

The higher surface temperatures of the terrestrial planets help to explain the following observation: The atmospheres of the terrestrial planets contain virtually *no* hydrogen molecules or helium atoms.

Table 7-3 Comparing Terrestrial and Jovian Planets

	Terrestrial Planets	Jovian Planets
Distance from the Sun	Less than 2 AU	More than 5 AU
Size	Small	Large
Composition	Mostly rocky materials containing iron, oxygen, silicon, magnesium, nickel, and sulfur	Mostly light elements such as hydrogen and helium
Density	High	Low

BOX 7-2 Tools of the Astronomer's Trade

Kinetic Energy, Temperature, and Whether Planets Have Atmospheres

A moving object possesses energy as a result of its motion. The faster it moves, the more energy it has. Energy of this type is called **kinetic energy.** If an object of mass m is moving with a speed v, its kinetic energy E_k is given by

Kinetic energy

$$E_k = \frac{1}{2} mv^2$$

E_k = kinetic energy of an object

m = mass of object

v = speed of object

This expression for kinetic energy is valid for all objects, both big and small, from atoms and molecules to planets and stars, as long as their speeds are slow in relation to the speed of light. If the mass is expressed in kilograms and the speed in meters per second, the kinetic energy is expressed in joules (J).

EXAMPLE: An automobile of mass 1000 kg driving at a typical freeway speed of 30 m/s (= 108 km/h = 67 mi/h) has a kinetic energy of

$$E_k = \frac{1}{2}(1000 \times 30^2) = 450{,}000 \text{ J} = 4.5 \times 10^5 \text{ J}$$

Consider a gas, such as the atmosphere of a star or planet. Some of the gas atoms or molecules will be moving slowly, with little kinetic energy, while others will be moving faster and have more kinetic energy. The temperature of the gas is a direct measure of the *average* amount of kinetic energy per atom or molecule. The hotter the gas, the faster atoms or molecules move, on average, and the greater the average kinetic energy of an atom or molecule.

If the gas temperature is sufficiently high, typically several thousand kelvins, molecules move so fast that when they collide with one another, the energy of the collision can break the molecules apart into their constituent atoms. Thus, the Sun's atmosphere, where the temperature is 5800 K, consists primarily of individual hydrogen atoms rather than hydrogen molecules. By contrast, the hydrogen atoms in Earth's atmosphere (temperature 290 K) are combined with oxygen atoms into molecules of water vapor (H_2O).

The physics of gases tells us that in a gas of temperature T (in kelvins), the average kinetic energy of an atom or molecule is

Kinetic energy of a gas atom or molecule

$$E_k = \frac{3}{2} kT$$

E_k = average kinetic energy of a gas atom or molecule, in joules

$k = 1.38 \times 10^{-23}$ J/K

T = temperature of gas, in kelvins

The quantity k is called the Boltzmann constant. Note that the higher the gas temperature, the greater the average kinetic energy of an atom or molecule of the gas. This average kinetic energy becomes zero at absolute zero, or $T = 0$, the temperature at which molecular motion is at a minimum.

At a given temperature, all kinds of atoms and molecules will have the same average kinetic energy. But the average *speed* of a given kind of atom or molecule depends on the particle's mass. To see this dependence on mass, note that the average kinetic energy of a gas atom or molecule can be written in two equivalent ways:

$$E_k = \frac{1}{2} mv^2 = \frac{3}{2} kT$$

where v represents the average speed of an atom or molecule in a gas with temperature T. Rearranging this equation, we obtain

Average speed of a gas atom or molecule

$$v = \sqrt{\frac{3kT}{m}}$$

v = average speed of a gas atom or molecule, in m/s

$k = 1.38 \times 10^{-23}$ J/K

T = temperature of gas, in kelvins

m = mass of the atom or molecule, in kilograms

For a given gas temperature, the greater the mass of a given type of gas atom or molecule, the slower its average

Instead, the atmospheres of Venus, Earth, and Mars are composed of heavier molecules such as nitrogen (N_2, 14 times more massive than a hydrogen molecule), oxygen (O_2, 16 times more massive), and carbon dioxide (22 times more massive). To understand the connection between surface temperature and the absence of hydrogen and helium, we need to know a few basic facts about gases.

The temperature of a gas is directly related to the speeds at which the atoms or molecules of the gas move: The higher the gas temperature, the greater the speed of its atoms or molecules. Furthermore, for a given temperature, lightweight atoms and molecules move more rapidly than heavy ones. On the four inner, terrestrial planets, where atmospheric temperatures are high, low-speed. (The value of v given by this equation is actually slightly higher than the average speed of the atoms or molecules in the gas, but it is close enough for our purposes here. If you are studying physics, you may know that v is actually the root-mean-square speed.)

EXAMPLE: What is the average speed of the oxygen molecules that you breathe at a room temperature of 20°C (= 68°F)?

Situation: We are given the temperature of a gas and are asked to find the average speed of the gas molecules.

Tools: We use the relationship $v = \sqrt{3kT/m}$, where T is the gas temperature in kelvins and m is the mass of a single oxygen molecule in kilograms.

Answer: To use the equation to calculate the average speed v, we must express the temperature T in kelvins (K) rather than degrees Celsius (°C). As we learned in Box 5-1, we do this by adding 273 to the Celsius temperature, so 20°C becomes (20 + 273) = 293 K. The mass m of an oxygen molecule is not given, but from a reference book you can find that the mass of an oxygen *atom* is 2.66×10^{-26} kg. The mass of an oxygen molecule (O_2) is twice the mass of an oxygen atom, or $2(2.66 \times 10^{-26}\ \text{kg}) = 5.32 \times 10^{-26}$ kg. Thus, the average speed of an oxygen molecule in 20°C air is

$$v = \sqrt{\frac{3(1.38 \times 10^{-23})(293)}{5.32 \times 10^{-26}}} = 478\ \text{m/s} = 0.478\ \text{km/s}$$

Review: This speed is about 1700 kilometers per hour, or about 1100 miles per hour. Hence, atoms and molecules move rapidly in even a moderate-temperature gas.

In some situations, atoms and molecules in a gas may be moving so fast that they can overcome the attractive force of a planet's gravity and escape into interplanetary space. The minimum speed that an object at a planet's surface must have in order to permanently leave the planet is called the planet's **escape speed.** The escape speed for a planet of mass M and radius R is given by

$$v_{\text{escape}} = \sqrt{\frac{2GM}{R}}$$

where $G = 6.67 \times 10^{-11}\ \text{N m}^2/\text{kg}^2$ is the universal constant of gravitation.

The accompanying table gives the escape speed for the Sun, the planets, and the Moon. For example, to escape Earth, a cannon ball would have to be shot with an implausible speed greater than 11.2 km/s (25,100 mi/h).

A good rule of thumb is that a planet can retain a gas if the escape speed is at least 6 times greater than the average speed of the molecules in the gas. (Some molecules are moving slower than average, and others are moving faster, but very few are moving more than 6 times faster than average.) In such a case, very few molecules will be moving fast enough to escape from the planet's gravity.

EXAMPLE: Consider Earth's atmosphere. We saw that the average speed of oxygen molecules is 0.478 km/s at room temperature. The escape speed from Earth (11.2 km/s) is much more than 6 times the average speed of the oxygen molecules, so Earth has no trouble keeping oxygen in its atmosphere.

A similar calculation for hydrogen molecules (H_2) gives a different result, however. At 293 K, the average speed of a hydrogen molecule is 1.9 km/s. Six times this speed is 11.4 km/s, which is slightly higher than the escape speed from Earth. Thus, Earth does not retain hydrogen in its atmosphere. Any hydrogen released into the air slowly leaks away into space. On Jupiter, by contrast, the escape speed is so high that even the lightest gases such as hydrogen are retained in its atmosphere. But on Mercury the escape speed is low and the temperature high (so that gas molecules move faster), so Mercury cannot retain any significant atmosphere at all.

Object	Escape speed (km/s)
Sun	618
Mercury	4.3
Venus	10.4
Earth	11.2
Moon	2.4
Mars	5.0
Jupiter	59.5
Saturn	35.5
Uranus	21.3
Neptune	23.5

mass hydrogen molecules and helium atoms move so swiftly that they can escape from the relatively weak gravity of these planets. Hence, the atmospheres that surround the terrestrial planets are composed primarily of more massive, slower-moving molecules such as CO_2, N_2, O_2, and water vapor (H_2O). On the four Jovian planets, low temperatures and relatively strong gravity prevent even lightweight hydrogen and helium gases from escaping into space, and so their atmospheres are much more extensive. The combined mass of Jupiter's atmosphere, for example, is about a million (10^6) times greater than that of Earth's atmosphere. This is comparable to the entire mass of Earth! Box 7-2 describes more about the ability of a planet's gravity to retain gases.

7-5 Small chunks of rock and ice also orbit the Sun

In addition to the eight planets, many smaller objects orbit the Sun. These objects fall into three broad categories: asteroids, which are rocky objects found in the inner solar system; trans-Neptunian objects, which are found beyond Neptune in the outer solar system and contain both rock and ice; and comets, which are mixtures of rock and ice that originate in the outer solar system but can venture close to the Sun.

The smaller bodies of the solar system contain important clues about its origin and evolution

Asteroids

Within the orbit of Jupiter are hundreds of thousands of rocky objects called **asteroids.** There is no sharp dividing line between planets and asteroids, which is why asteroids are also called **minor planets.** The largest asteroid, Ceres, has a diameter of about 900 km. The next largest, Pallas and Vesta, are each about 500 km in diameter. Still smaller ones, like the asteroid shown in close-up in Figure 7-7, are more numerous. Hundreds of thousands of kilometer-sized asteroids are known, and there are probably hundreds of thousands more that are boulder-sized or smaller. All of these objects orbit the Sun in the same direction as the planets.

Most (although not all) asteroids orbit the Sun at distances of 2 to 3.5 AU. This region of the solar system between the orbits of Mars and Jupiter is called the **asteroid belt.**

CAUTION! One common misconception about asteroids is that they are the remnants of an ancient planet that somehow broke apart or exploded, like the fictional planet Krypton in the comic book adventures of Superman. In fact, the combined mass of the asteroids is less than that of the Moon, and they were probably never part of any planet-sized body. The early solar system is thought to have been filled with asteroidlike objects, most of which were incorporated into the planets. The "leftover" objects that missed out on this process make up our present-day population of asteroids.

Trans-Neptunian Objects

While asteroids are the most important small bodies in the inner solar system, the outer solar system is the realm of the **trans-Neptunian objects.** As the name suggests, these are small bodies whose orbits lie beyond the orbit of Neptune. The first of these to be discovered (1930) was Pluto, with a diameter of only 2274 km. This is larger than any asteroid, but smaller than any planet or any of the satellites listed in Table 7-2. The orbit of Pluto has a greater semimajor axis (39.54 AU), is more steeply inclined to the ecliptic (17.15°), and has a greater eccentricity (0.250) than that of any of the planets (Figure 7-8). In fact, Pluto's noncircular orbit sometimes takes it nearer the Sun than Neptune. (Happily, the orbits of Neptune and Pluto are such that they will never collide.) Pluto's density is only 2000 kg/m^3, about the same as Neptune's moon Triton, shown in Table 7-2. Hence, its composition is thought to be a mixture of about 70% rock and 30% ice.

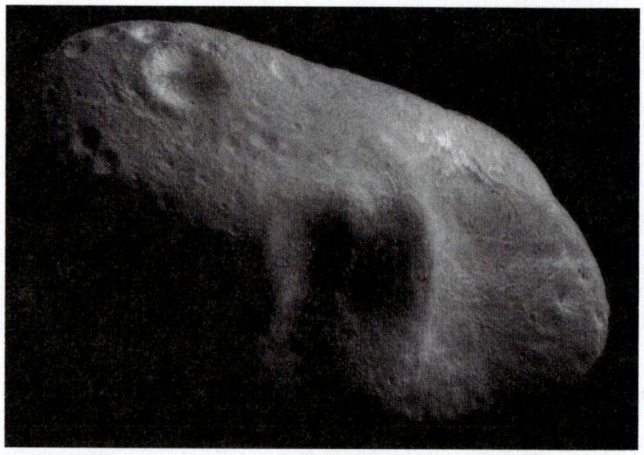

WEB LINK 7.7

Figure 7-7 RI**V**UXG

An Asteroid The asteroid shown in this image, 433 Eros, is only 33 km (21 mi) long and 13 km (8 mi) wide—about the same size as the island of Manhattan. Because Eros is so small, its gravity is too weak to have pulled it into a spherical shape. This image was taken in March 2000 by *NEAR Shoemaker,* the first spacecraft to orbit around and land on an asteroid. (*NEAR* Project, Johns Hopkins University, Applied Physics Laboratory, and NASA)

Since 1992 astronomers have discovered more than 900 other trans-Neptunian objects, and are discovering more each year. Like asteroids, all trans-Neptunian objects orbit the Sun in the same direction as the planets. A handful of trans-Neptunian objects are comparable in size to Pluto; at least one, Eris, is even larger than Pluto, as well as being in an orbit that is much larger, more steeply inclined, and more eccentric (Figure 7-8). Table 7-4 lists the seven largest trans-Neptunian objects known as of this writing. (Note that Charon is actually a satellite of Pluto.)

Just as most asteroids lie in the asteroid belt, most trans-Neptunian objects orbit within a band called the **Kuiper belt** (pronounced "ki-per") that extends from 30 AU to 50 AU from the Sun and is centered on the plane of the ecliptic. Like asteroids, many more trans-Neptunian objects remain to be discovered: astronomers estimate that there are 35,000 or more such objects with diameters greater than 100 km. If so, the combined mass of all trans-Neptunian objects is comparable to the mass of Jupiter, and is several hundred times greater than the combined mass of all the asteroids found in the inner solar system.

Like asteroids, trans-Neptunian objects are thought to be debris left over from the formation of the solar system. In the inner regions of the solar system, rocky fragments have been able to endure continuous exposure to the Sun's heat, but any ice originally present would have evaporated. Far from the Sun, ice has survived for billions of years. Thus, debris in the solar system naturally divides into two families (asteroids and trans-Neptunian objects), which can be arranged according to distance from the Sun, just like the two categories of planets (terrestrial and Jovian).

Comets

Two objects in the Kuiper belt can collide if their orbits cross each other. When this happens, a fragment a few kilometers across can be knocked off one of the colliding objects and be diverted into an elongated orbit that brings it close to the Sun. Such small

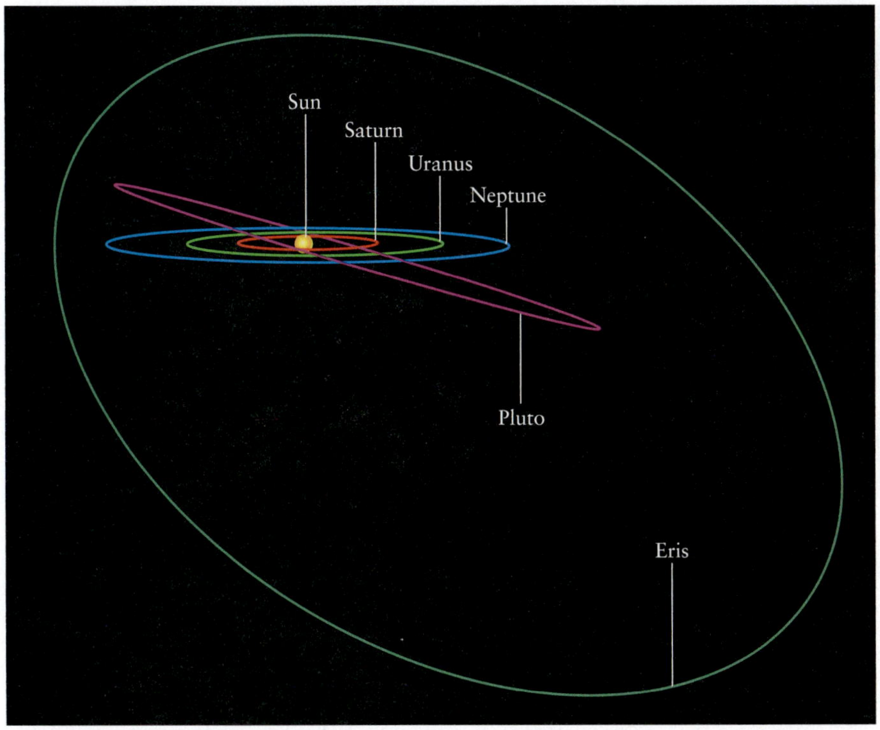

Figure 7-8

Trans-Neptunian Objects Pluto and Eris are the two largest trans-Neptunian objects, small worlds of rock and ice that orbit beyond Neptune. Unlike the orbits of the planets, the orbits of these two objects are steeply inclined to the ecliptic: Pluto's orbit is tilted by about 17°, and that of Eris is tilted by 44°.

WEB LINK 7.8

Table 7-4 Seven Large Trans-Neptunian Objects

	Quaoar	Haumea	Sedna	Makemake	Pluto	Charon (satellite of Pluto)	Eris
Average distance from the Sun (AU)	43.54	43.34	489	45.71	39.54	39.54	67.67
Orbital period (years)	287	285	10,800	309	248.6	248.6	557
Orbital eccentricity	0.035	0.189	0.844	0.155	0.250	0.250	0.442
Inclination of orbit to the ecliptic	8.0°	28.2°	11.9°	29.0°	17.15°	17.15°	44.2°
Approximate diameter (km)	1250	1500	1600	1800	2274	1190	2900

R I V U X G

(Images of Pluto and Charon: Alan Stern, Southwest Research Institute; Marc Buie, Lowell Observatory; NASA; and ESA)

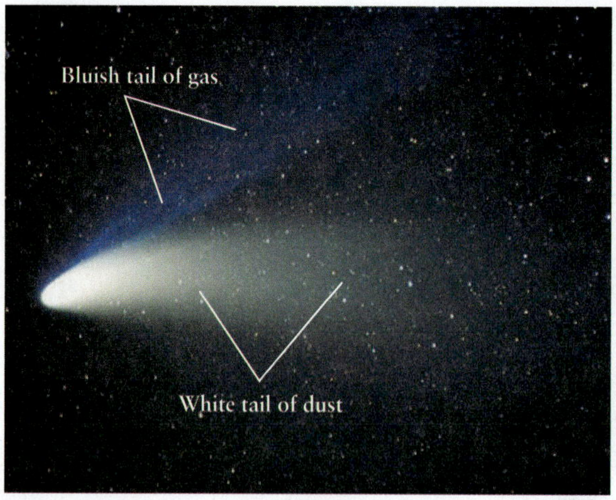

WEB LINK 7.9

Figure 7-9 RIVUXG

A Comet This photograph shows Comet Hale-Bopp as it appeared in April 1997. The solid part of a comet like this is a chunk of dirty ice a few tens of kilometers in diameter. When a comet passes near the Sun, solar radiation vaporizes some of the icy material, forming a bluish tail of gas and a white tail of dust. Both tails can extend for tens of millions of kilometers. (Richard J. Wainscoat, University of Hawaii)

objects, each a combination of rock and ice, are called **comets.** When a comet comes close enough to the Sun, the Sun's radiation vaporizes some of the comet's ices, producing long flowing tails of gas and dust particles (Figure 7-9). Astronomers deduce the composition of comets by studying the spectra of these tails.

CAUTION! Science-fiction movies and television programs sometimes show comets tearing across the night sky like a rocket, which would be a pretty impressive sight. However, comets do not zoom across the sky. Like the planets, comets orbit the Sun. And like the planets, comets move hardly at all against the background of stars over the course of a single night (see Section 4-1). If you are lucky enough to see a bright comet, it will not zoom dramatically from horizon to horizon. Instead, it will seem to hang majestically among the stars, so you can admire it at your leisure.

Some comets appear to originate from locations far beyond the Kuiper belt. The source of these is thought to be a swarm of comets that forms a spherical "halo" around the solar system called the **Oort comet cloud** (also known as Oort cloud). This hypothesized "halo" extends to 50,000 AU from the Sun (about one-fifth of the way to the nearest other star). Because the Oort cloud is so distant, it has not yet been possible to detect objects in the Oort cloud directly.

7-6 Cratering on planets and satellites is the result of impacts from interplanetary debris

One of the great challenges in studying planets and satellites is how to determine their internal structures. Are they solid or liquid inside? If there is liquid in the interiors, is the liquid calm or in agitated motion? Because planets are opaque, we cannot see directly into their interiors to answer these questions. But we can gather important clues about the interiors of terrestrial planets and satellites by studying the extent to which their surfaces are covered with craters (Figure 7-10). To see how information about the interiors is gathered, we first need to understand where craters come from.

Scientists study craters on a planet or satellite to learn the age and geologic history of the surface

The Origin of Craters

The planets orbit the Sun in roughly circular orbits. But many asteroids and comets are in more elongated orbits. Such an elongated orbit can put these small objects on a collision course with a planet or satellite. If the object collides with a Jovian planet, it is swallowed up by the planet's thick atmosphere. (Astronomers actually saw an event of this kind in 1994, when a comet crashed into Jupiter.) But if the object collides with the solid surface of a terrestrial planet or a satellite, the result is an **impact crater** (see Figure 7-10). Such impact craters, found throughout the solar system, offer stark evidence of these violent collisions.

The easiest way to view impact craters is to examine the Moon through a telescope or binoculars. Some 30,000 lunar craters are visible, with diameters ranging from 1 km to several hundred km. Close-up photographs from lunar orbit have revealed millions of craters too small to be seen from Earth (Figure 7-10*a*). These smaller craters are thought to have been caused by impacts of relatively small objects called **meteoroids,** which range in size from a few hundred meters across to the size of a pebble or smaller. Meteoroids are the result of collisions between asteroids, whose orbits sometimes cross. The chunks of rock that result from these collisions go into independent orbits around the Sun, which can lead them to collide with the Moon or another world.

When German astronomer Franz Gruithuisen proposed in 1824 that lunar craters were the result of impacts, a major sticking point was the observation that nearly all craters are circular. If craters were merely gouged out by high-speed rocks, a rock striking the Moon in any direction except straight downward would have created a noncircular crater. A century after Gruithuisen, it was realized that a meteoroid colliding with the Moon generates a shock wave in the lunar surface that spreads out from the point of impact. Such a shock wave produces a circular crater no matter what direction the meteoroid was moving. (In a similar way, the craters made by artillery shells are almost always circular.) Many of the larger lunar craters also have a central peak, which is characteristic of a high-speed impact (see Figure 7-10*a*). Craters made by other processes, such as volcanic action, would not have central peaks of this sort.

Comparing Cratering on Different Worlds

Not all planets and satellites show the same amount of cratering. The Moon is heavily cratered over its entire surface, with craters on top of craters, as shown in Figure 7-10*a*. On Earth, by contrast, craters are very rare. Geologists have identified fewer than 200 impact craters on our planet (Figure 7-10*b*). Our understanding is that both Earth and the Moon formed at nearly the same time and have been bombarded at comparable rates over their

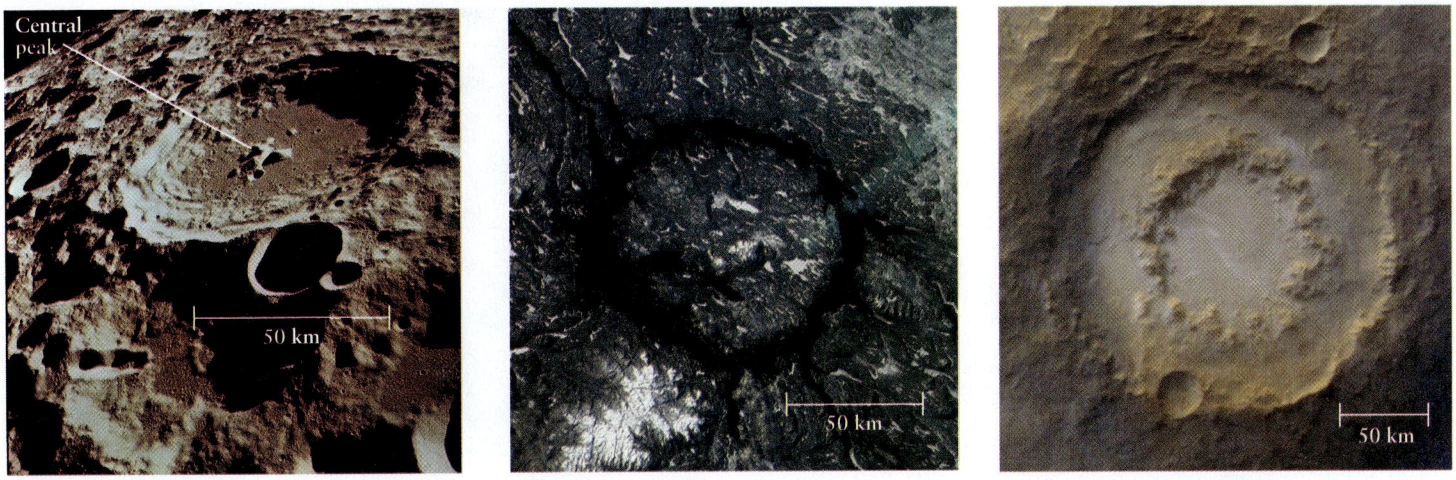

(a) A crater on the Moon (b) A crater on Earth (c) A crater on Mars

Figure 7-10 R I V U X G

Impact Craters These images, all taken from spacecraft, show impact craters on three different worlds. **(a)** The Moon's surface has craters of all sizes. The large crater near the middle of this image is about 80 km (50 miles) in diameter, equal to the length of San Francisco Bay. **(b)** Manicouagan Reservoir in Quebec is the relic of a crater formed by an impact more than 200 million years ago. The crater was eroded over the ages by the advance and retreat of glaciers, leaving a ring lake 100 km (60 miles) across. **(c)** Lowell Crater in the southern highlands of Mars is 201 km (125 miles) across. Like the image of the Moon in part (a), there are craters on top of craters. Note the light-colored frost formed by condensation of carbon dioxide from the Martian atmosphere. (a: NASA; b: JSC/NASA; c: NASA/JPL/MSSS)

histories. Why, then, are craters so much rarer on Earth than on the Moon?

The answer is that Earth is a *geologically active* planet: the continents slowly change their positions over eons, new material flows onto the surface from the interior (as occurs in a volcanic eruption), and old surface material is pushed back into the interior (as occurs off the coast of Chile, where the ocean bottom is slowly being pushed beneath the South American continent). These processes, coupled with erosion from wind and water, cause craters on Earth to be erased over time. The few craters found on Earth today must be relatively recent, since there has not yet been time to erase them.

The Moon, by contrast, is geologically *inactive*. There are no volcanoes and no motion of continents (and, indeed, no continents). Furthermore, the Moon has neither oceans nor an atmosphere, so there is no erosion as we know it on Earth. With none of the processes that tend to erase craters on Earth, the Moon's surface remains pockmarked with the scars of billions of years of impacts.

In order for a planet to be geologically active, its interior must be at least partially molten. This partially molten state is necessary so that continents can slide around on the underlying molten material and so that molten lava can come to the surface, as in a volcanic eruption. Hence, we would expect geologically inactive (and hence heavily cratered) worlds like the Moon to have less molten material in their interiors than does Earth. Investigations of these inactive worlds bear this out. But *why* is the Moon's interior less molten than Earth's?

ANALOGY To see one simple answer to this question, notice that a large turkey or roast taken from the oven will stay warm inside for hours, but a single meatball will cool off much more rapidly. The reason is that the meatball has more surface area relative to its volume, and so it can more easily lose heat to its surroundings. A planet or satellite also tends to cool down as it emits electromagnetic radiation into space (see Section 5-3); the smaller the planet or satellite, the greater its surface area relative to its volume, and the more readily it can radiate away heat (Figure 7-11). Both Earth and the Moon were probably completely molten when they first formed, but because the Moon (diameter 3476 km) is so much smaller than Earth (diameter 12,756 km), it has lost much of its internal heat and has a much more solid interior.

Cratering Measures Geologic Activity

By considering these differences between Earth and the Moon, we have uncovered a general rule for worlds with solid surfaces:

The smaller the terrestrial world, the less internal heat it is likely to have retained, and, thus, the less geologic activity it will display on its surface. The less geologically active the world, the older and hence more heavily cratered its surface.

This rule means that we can use the amount of cratering visible on a planet or satellite to estimate the age of its surface and how geologically active it is. As an example, Mercury has a heavily cratered surface, which means that the surface is very old. This geologically inactive surface is in agreement with Mercury being the smallest of the terrestrial planets (see Table 7-1). Due to its small size, it has lost the internal heat required to sustain geologic activity. On Venus, by comparison, there are only about a thousand craters larger than a few kilometers in diameter, many more than have been found on Earth but only a small fraction of the number on the Moon or Mercury. Venus is only slightly smaller than Earth, and it has enough internal heat to power the geologic activity required to erase most of its impact craters.

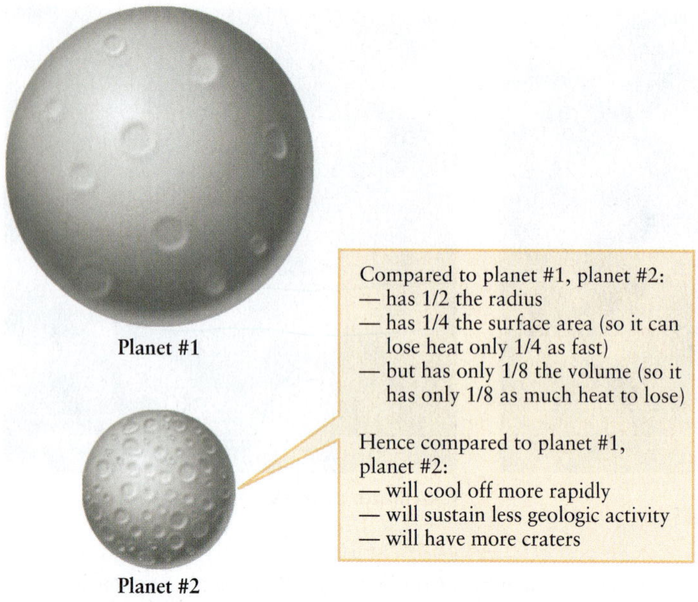
Planet #1
Compared to planet #1, planet #2:
— has 1/2 the radius
— has 1/4 the surface area (so it can lose heat only 1/4 as fast)
— but has only 1/8 the volume (so it has only 1/8 as much heat to lose)
Hence compared to planet #1, planet #2:
— will cool off more rapidly
— will sustain less geologic activity
— will have more craters
Planet #2

Figure 7-11

Planet Size and Cratering Of these two hypothetical planets, the smaller one (#2) has less volume and less internal heat, as well as less surface area from which to radiate heat into space. But the *ratio* of surface area to volume is greater for the smaller planet. Hence, the smaller planet will lose heat faster, have a colder interior, and be less geologically active. It will also have a more heavily cratered surface, since it takes geologic activity to erase craters.

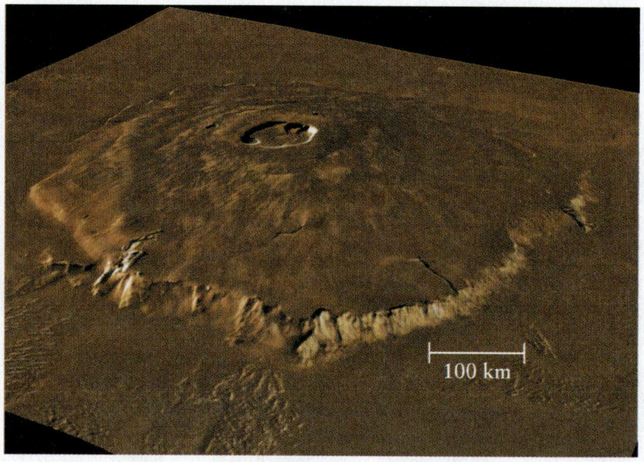

Figure 7-12 R I V U X G

A Martian Volcano Olympus Mons is the largest of the inactive volcanoes of Mars and the largest volcano in the solar system. The base of Olympus Mons measures 600 km (370 mi) in diameter, and the scarps (cliffs) that surround the base are 6 km (4 mi) high. The caldera, or volcanic crater, at the summit is approximately 70 km across, large enough to contain the state of Rhode Island. This view was creating by combining a number of images taken from Mars orbit. (© Calvin J. Hamilton)

Mars is an unusual case, in that extensive cratering (Figure 7-10*c*) is found only in the higher terrain; the lowlands of Mars are remarkably smooth and free of craters. Thus, it follows that the Martian highlands are quite old, while the lowlands have a younger surface from which most craters have been erased. Considering the planet as a whole, the amount of cratering on Mars is intermediate between that on Mercury and Earth. This agrees with our general rule, because Mars is intermediate in size between Mercury and Earth. The interior of Mars was once hotter and more molten than it is now, so that geologic processes were able to erase some of the impact craters. A key piece of evidence that supports this picture is that Mars has a number of immense volcanoes (Figure 7-12). These volcanoes were active when Mars was young, but as this relatively small planet cooled down and its interior solidified, the supply of molten material to the volcanoes from the Martian interior was cut off. As a result, all of the volcanoes of Mars are now inactive.

As for all rules, there are limitations and exceptions to the rule relating a world's size to its geologic activity. One limitation is that the four terrestrial planets all have slightly different compositions, which affects the types and extent of geologic activity that can take place on their surfaces. The different compositions also complicate the relationship between the number of craters and the age of the surface. An important exception to our rule is Jupiter's satellite Io, which, despite its small size, is the most volcanic world in the solar system (see Section 7-2). Something must be supplying Io with energy to keep its interior hot; this energy comes from Jupiter, which exerts powerful tidal forces on Io as it moves in a relatively small orbit around its planet. These tidal forces cause Io to flex like a ball of clay being kneaded between your fingers, and this flexing heats up the satellite's interior. But despite these limitations and exceptions, the relationships between a world's size, internal heat, geologic activity, and amount of cratering are powerful tools for understanding the terrestrial planets and satellites.

7-7 A planet with a magnetic field indicates a fluid interior in motion

The amount of impact cratering on a terrestrial planet or satellite provides indirect evidence about whether the planet or satellite has a molten interior. But another, more direct tool for probing the interior of *any* planet or satellite is an ordinary compass, which senses the magnetic field outside the planet or satellite. Magnetic field measurements prove to be an extremely powerful way to investigate the internal structure of a world without having to actually dig into its interior. To illustrate how this works, consider the behavior of a compass on Earth.

By studying the magnetic field of a planet or satellite, scientists can learn about that world's interior motions

The needle of a compass on Earth points north because it aligns with Earth's *magnetic field.* Such fields arise whenever electrically charged particles are in motion. For example, a loop of wire carrying an electric current generates a magnetic field in the space around it. The magnetic field that surrounds an ordinary bar magnet (Figure 7-13*a*) is created by the motions of negatively charged electrons within the iron atoms of which the magnet is made. Earth's magnetic field is similar to that of a bar magnet, as Figure 7-13*b* shows. The consensus among geologists

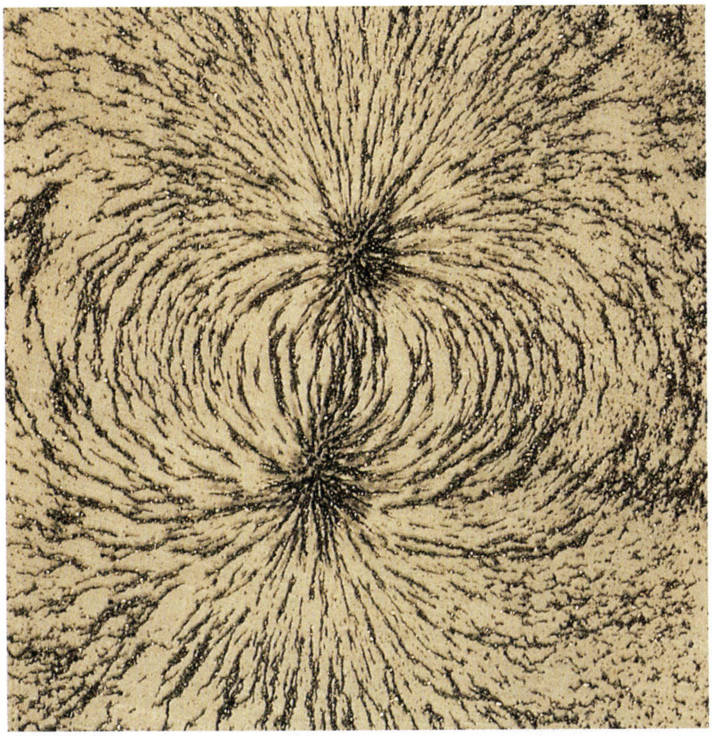

(a)

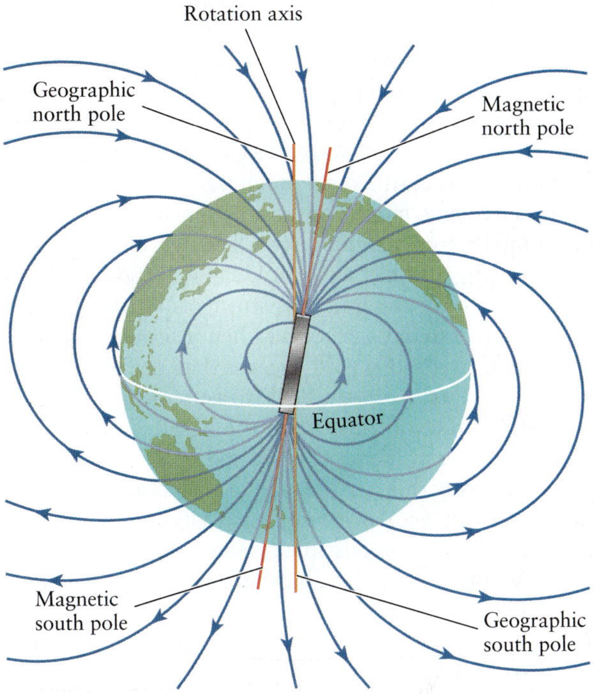

(b)

Figure 7-13

The Magnetic Fields of a Bar Magnet and of Earth (a) This picture was made by placing a piece of paper on top of a bar magnet, then spreading iron filings on the paper. The pattern of the filings show the magnetic field lines, which appear to stream from one of the magnet's poles to the other. **(b)** Earth's magnetic field lines have a similar pattern. Although Earth's field is produced in a different way—by electric currents in the liquid portion of our planet's interior—the field is much the same as if there were a giant bar magnet inside Earth. This "bar magnet" is not exactly aligned with Earth's rotation axis, which is why the magnetic north and south poles are not at the same locations as the true, or geographic, poles. A compass needle points toward the north magnetic pole, not the true north pole. (a: Jules Bucher/Photo Researchers)

is that this magnetic field is caused by the motion of the liquid portions of Earth's interior. Because this molten material (mostly iron) conducts electricity, these motions give rise to electric currents, which in turn produce Earth's magnetic field. Our planet's rotation helps to sustain these motions and hence the magnetic field. This process for producing a magnetic field is called a **dynamo.**

CAUTION! While Earth's magnetic field is similar to that of a giant bar magnet, you should not take this picture too literally. Earth is *not* simply a magnetized ball of iron. In an iron bar magnet, the electrons of different atoms orbit their nuclei in the same general direction, so that the magnetic fields generated by individual atoms add together to form a single, strong field. But at temperatures above 770°C (1418°F = 1043 K), the orientations of the electron orbits become randomized. The fields of individual atoms tend to cancel each other out, and the iron loses its magnetism. Geological evidence shows that almost all of Earth's interior is hotter than 770°C, so the iron there cannot be extensively magnetized. The correct picture is that Earth acts as a dynamo: The liquid iron carries electric currents, and these currents create Earth's magnetic field.

If a planet or satellite has a mostly solid interior, then the dynamo mechanism cannot work: Material in the interior cannot flow, there are no electric currents, and the planet or satellite does not generate a magnetic field. One example of this is the Moon. As we saw in Section 7-6, the extensive cratering of the lunar surface indicates that the Moon has no geologic activity and must therefore have a mostly solid interior. Measurements made during the *Apollo* missions, in which 12 humans visited the lunar surface between 1969 and 1972, showed that the present-day Moon indeed has no global magnetic field. However, careful magnetic measurements of lunar rocks returned by the *Apollo* astronauts indicate that the Moon *did* have a weak magnetic field when the rocks solidified. These rocks, like the rest of the lunar surface, are very old. Hence, in the distant past the Moon may have had a small amount of molten iron in its interior that acted as a dynamo. This material presumably solidified at least partially as the Moon cooled, so that the lunar magnetic field disappeared.

We have now identified another general rule about planets and satellites:

A planet or satellite with a global magnetic field has liquid material in its interior that conducts electricity and is in motion, generating the magnetic field.

Thus, by studying the magnetic field of a planet or satellite, we can learn about that world's interior. So, many spacecraft carry devices called **magnetometers** to measure magnetic fields. Magnetometers are often placed on a long boom extending outward from the body of the spacecraft (Figure 7-14). The boom isolates them from the magnetic fields produced by electric currents in the spacecraft's own circuitry.

Measurements made with magnetometers on spacecraft have led to a number of striking discoveries. For example, it has been found that Mercury has a planetwide magnetic field like Earth's, although it is only about 1% as strong as Earth's field. Mercury has a heavily cratered surface and hence little or no geologic activity, which by itself would suggest that the planet's interior is mostly solid. The magnetic field measurements show that *some* of Mercury's interior must be in the liquid state to act as a dynamo. By contrast, Venus has no measurable planetwide magnetic field, even though the paucity of craters on its surface indicates the presence of geologic activity and hence a hot interior of the planet. One possible reason for the lack of a magnetic field on Venus is that the planet turns on its axis very slowly, taking 243 days for a complete rotation. Because of this slow rotation, the fluid material within the planet is hardly agitated at all, and so may not move in the fashion that generates a magnetic field.

Like Venus, Mars has no planetwide magnetic field. But the magnetometer aboard the *Mars Global Surveyor* spacecraft, which went into orbit around Mars in 1997, found magnetized regions in the cratered highlands of the Martian southern hemisphere (Figure 7-15). As we saw in Section 7-6, these portions of the Martian surface are very old, and so would have formed when the planet's interior was still hot and molten. Electric currents in the flowing molten material could then produce a planetwide magnetic field. As surface material cooled and solidified, it became magnetized by the planetwide field, and this material has retained its magnetization over the eons. *Mars Global Surveyor* has not found magnetized areas in the younger terrain of the lowlands. Hence, the planetwide Martian magnetic field must have shut off by the time the lowland terrain formed.

The most intense planetary magnetic field in the solar system is that of Jupiter: at the tops of Jupiter's clouds, the magnetic field is about 14 times stronger than the field at Earth's surface. It is thought that Jupiter's field, like Earth's, is produced by a dynamo acting deep within the planet's interior. Unlike Earth, however, Jupiter is composed primarily of hydrogen and helium, not substances like iron that conduct electricity. How, then, can Jupiter have a dynamo that generates such strong magnetic fields?

WEB LINK 7.12

To answer this question, recall from Section 5-8 that a hydrogen atom consists of a single proton orbited by a single electron. Deep inside Jupiter, the pressure is so great and hydrogen atoms are squeezed so close together that electrons can hop from one atom to another. This hopping motion creates an electric current, just as the ordered movement of electrons in the copper wires of a flashlight constitutes an electric current. In other words, the highly compressed hydrogen deep inside Jupiter behaves like an electrically conducting metal; thus, it is called **liquid metallic hydrogen.**

Laboratory experiments show that hydrogen becomes a liquid metal when the pressure is more than about 1.4 million times ordinary atmospheric pressure on Earth. Recent calculations suggest that this transition occurs about 7000 km below Jupiter's cloudtops. Most of the planet's enormous bulk lies below this level, so there is a tremendous amount of liquid metallic hydrogen within Jupiter. Since Jupiter rotates rapidly—a "day" on

WEB LINK 7.11

Figure 7-14

Probing the Magnetic Field of Saturn This illustration depicts the *Cassini* spacecraft as it entered orbit around Saturn on July 1, 2004. In addition to telescopes for observing Saturn and its satellites, *Cassini* carries a magnetometer for exploring Saturn's magnetic field. The magnetometer is located on the long boom that extends down and to the right from the body of the spacecraft. (The glow at the end of the boom is the reflection of the Sun.) (JPL/NASA)

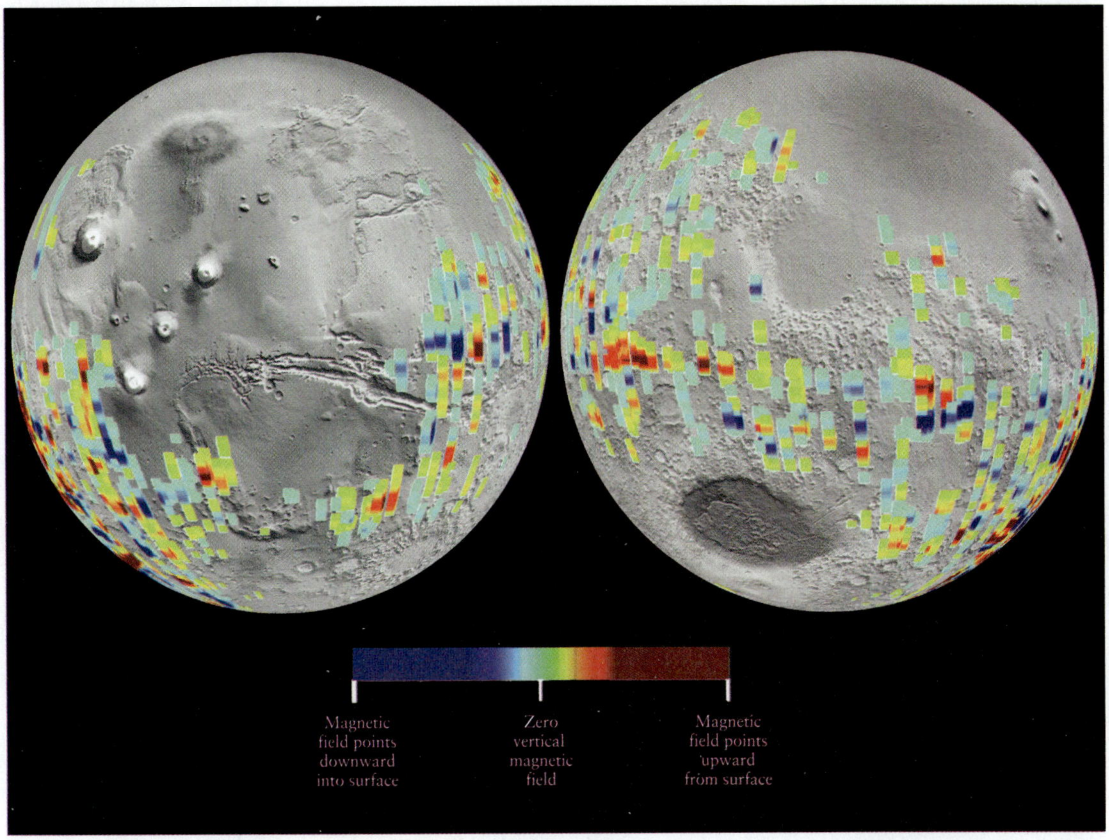

Figure 7-15

Relic Magnetism on Mars Portions of the highlands of Mars became magnetized during the period early in Martian history when the planet had an extensive magnetic field. The planetwide field has long since disappeared, but certain surface regions remained magnetized. The false colors in this illustration indicate the direction and strength of the field at different locations on Mars superimposed on images of the two Martian hemispheres. (MGS Magnetometer Team, GSFC/NASA)

Jupiter is just less than 10 hours long—this liquid metal moves rapidly, generating the planet's powerful magnetic field. Saturn also has a magnetic field produced by dynamo action in liquid metallic hydrogen. (The field is weaker than Jupiter's because Saturn is a smaller planet with less internal pressure, so there is less of the liquid metal available.)

Uranus and Neptune also have magnetic fields, but they cannot be produced in the same way: Because these planets are relatively small, the internal pressure is not great enough to turn liquid hydrogen into a metal. Instead, it is thought that both Uranus and Neptune have large amounts of liquid water in their interiors and that this water has molecules of ammonia and other substances dissolved in it. (The fluid used for washing windows has a similar chemical composition.) Under the pressures found in this interior water, the dissolved molecules lose one or more electrons and become electrically charged (that is, they become ionized; see Section 5-8). Water is a good conductor of electricity when it has such electrically charged molecules dissolved in it, and electric currents in this fluid are probably the source of the magnetic fields of Uranus and Neptune.

7-8 The diversity of the solar system is a result of its origin and evolution

Our brief tour of the solar system has revealed its almost dizzying variety. No two planets are alike, satellites come in all sizes, the extent of cratering varies from one terrestrial planet to another, and the magnetic fields of different planets vary dramatically in their strength and in how they are produced. (The *Cosmic Connections* figure that closes this chapter summarizes these properties of the planets.) All of this variety leads us to a simple yet profound question: *Why* are the planets and satellites of the solar system so different from each other?

The similarities and differences among the planets can be logically explained by a model of the solar system's origin and evolution

Among humans, the differences from one individual to another result from heredity (the genetic traits passed on from an individual's parents) and environment (the circumstances under which the individual

COSMIC CONNECTIONS

Characteristics of the Planets

The Inner (Terrestrial) Planets Close to the Sun - Small diameter, small mass - High density

	Mercury	Venus	Earth	Mars
Average distance from the Sun (AU)	0.387	0.723	1.000	1.524
Equatorial diameter (Earth = 1)	0.383	0.949	1.000	0.533
Mass (Earth = 1)	0.0553	0.8149	1.0000	0.1074
Average density (kg/m^3)	5430	5243	5515	3934

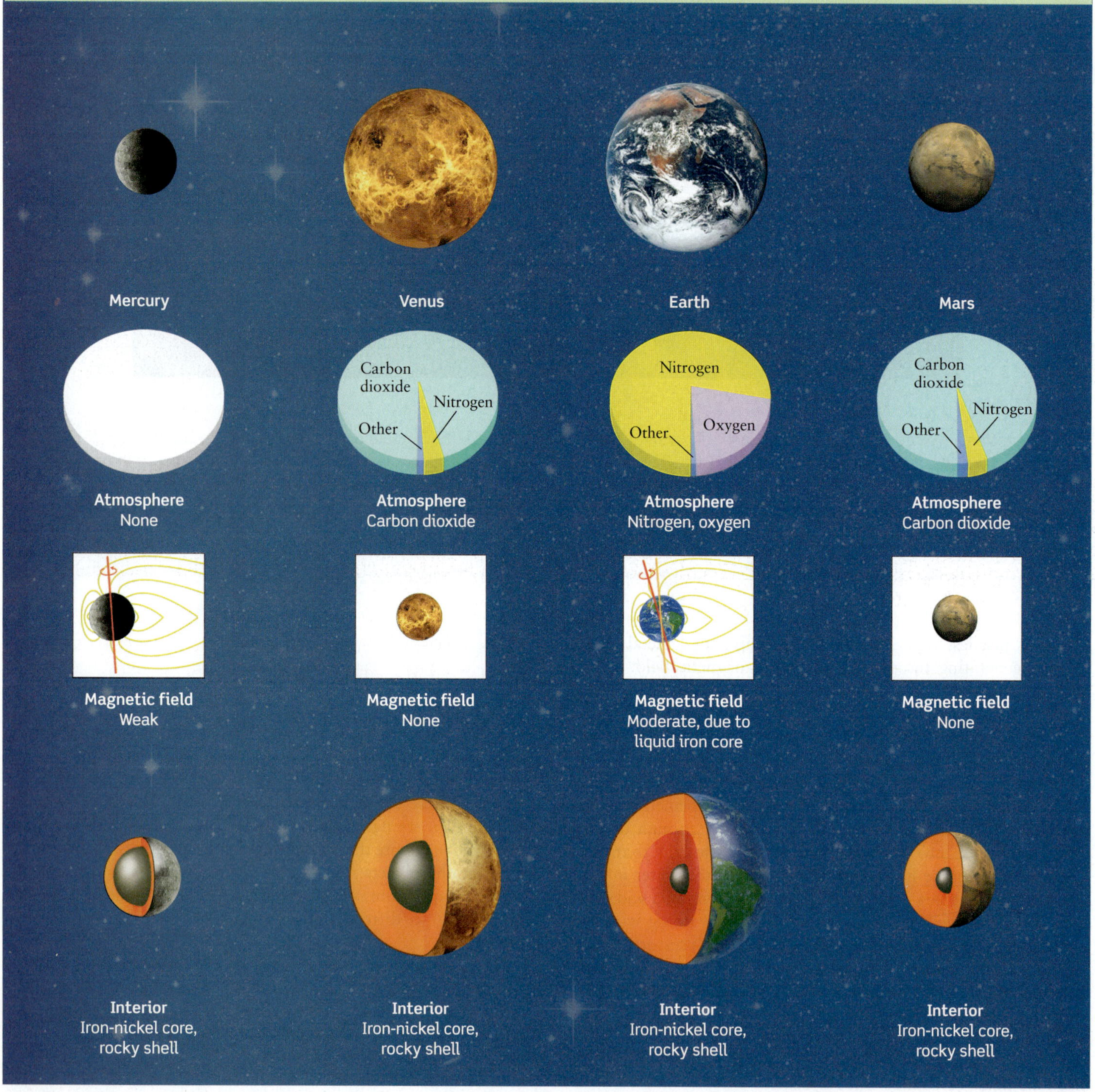

The Outer (Jovian) Planets   Far from the Sun - Large diameter, large mass - Low density

	Jupiter	Saturn	Uranus	Neptune
Average distance from the Sun (AU)	5.203	9.554	19.194	30.066
Equatorial diameter (Earth = 1)	11.209	9.449	4.007	3.883
Mass (Earth = 1)	317.8	95.16	14.53	17.15
Average density (kg/m^3)	1326	687	1318	1638

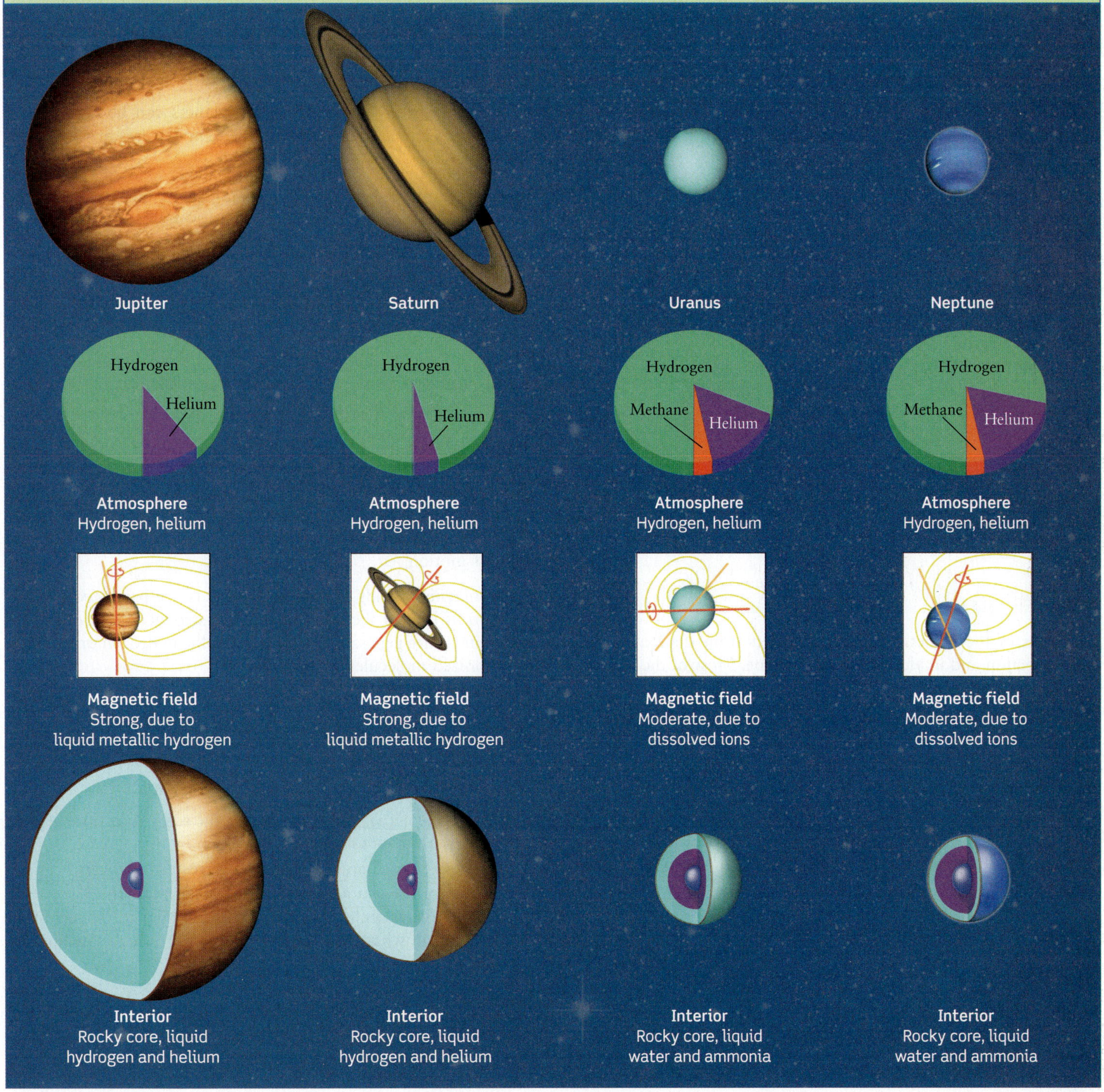

matures to an adult). As we will find in the following chapter, much the same is true for the worlds of the solar system.

In Chapter 8 we will see evidence that the entire solar system shares a common "heredity," in that the planets, satellites, comets, asteroids, and the Sun itself formed from the same cloud of interstellar gas and dust. The composition of this cloud was shaped by cosmic processes, including nuclear reactions that took place within stars that died long before our solar system was formed. We will see how different planets formed in different environments depending on their distance from the Sun and will discover how these environmental variations gave rise to the planets and satellites of our present-day solar system. And we will see how we can test these ideas of solar system origin and evolution by studying planetary systems orbiting other stars.

Our journey through the solar system is just beginning. In this chapter we have explored space to examine the variety of the present-day solar system; in Chapter 8 we will journey through time to see how our solar system came to be.

Key Words

Terms preceded by an asterisk (*) are discussed in the Boxes.

asteroid, p. 170
asteroid belt, p. 170
average density, p. 162
chemical composition, p. 164
comet, p. 172
dynamo, p. 175
*escape speed, p. 169
ices, p. 167
impact crater, p. 172
Jovian planet, p. 162
*kinetic energy, p. 168
Kuiper belt, p. 170
liquid metallic hydrogen, p. 176
magnetometer, p. 176
meteoroid, p. 172
minor planet, p. 170
Oort cloud, p. 172
spectroscopy, p. 164
terrestrial planet, p. 162
trans-Neptunian object, p. 170

Key Ideas

Properties of the Planets: All of the planets orbit the Sun in the same direction and in almost the same plane. Most of the planets have nearly circular orbits.

• The four inner planets are called terrestrial planets. They are relatively small (with diameters of 5000 to 13,000 km), have high average densities (4000 to 5500 kg/m^3), and are composed primarily of rocky materials.

• The four giant outer planets are called Jovian planets. They have large diameters (50,000 to 143,000 km) and low average densities (700 to 1700 kg/m^3) and are composed primarily of light elements such as hydrogen and helium.

Satellites and Small Bodies in the Solar System: Besides the planets, the solar system includes satellites of the planets, asteroids, comets, and trans-Neptunian objects.

• Seven large planetary satellites (one of which is the Moon) are comparable in size to the planet Mercury. The remaining satellites of the solar system are much smaller.

• Asteroids are small, rocky objects, while comets and trans-Neptunian objects are made of ice and rock. All are remnants left over from the formation of the planets.

• Most asteroids are found in the asteroid belt between the orbits of Mars and Jupiter, and most trans-Neptunian objects lie in the Kuiper belt outside the orbit of Neptune. Pluto is one of the largest members of the Kuiper belt.

Spectroscopy and the Composition of the Planets: Spectroscopy, the study of spectra, provides information about the chemical composition of objects in the solar system.

• The spectrum of a planet or satellite with an atmosphere reveals the atmosphere's composition. If there is no atmosphere, the spectrum indicates the composition of the surface.

• The substances that make up the planets can be classified as gases, ices, or rock, depending on the temperatures at which they solidify.

Impact Craters: When an asteroid, comet, or meteoroid collides with the surface of a terrestrial planet or satellite, the result is an impact crater.

• Geologic activity renews the surface and erases craters, so a terrestrial world with extensive cratering has an old surface and little or no geologic activity.

• Because geologic activity is powered by internal heat, and smaller worlds lose heat more rapidly, as a general rule smaller terrestrial worlds are more extensively cratered.

Magnetic Fields and Planetary Interiors: Planetary magnetic fields are produced by the motion of electrically conducting liquids inside the planet. This mechanism is called a dynamo. If a planet has no magnetic field, that is evidence that there is little such liquid material in the planet's interior or that the liquid is not in a state of motion.

• The magnetic fields of terrestrial planets are produced by metals such as iron in the liquid state. The stronger fields of the Jovian planets are generated by liquid metallic hydrogen or by water with ionized molecules dissolved in it.

Questions

Review Questions

1. Do all the planets orbit the Sun in the same direction? Are all of the orbits circular?
2. What are the characteristics of a terrestrial planet?
3. What are the characteristics of a Jovian planet?
4. In what ways are the largest satellites similar to the terrestrial planets? In what ways are they different? Which satellites are largest?

TUTORIAL 7.2

5. What is meant by the average density of a planet? What does the average density of a planet tell us?
6. What are the differences in chemical composition between the terrestrial and Jovian planets?

7. The absorption lines in the spectrum of a planet or satellite do not necessarily indicate the composition of the planet or satellite's atmosphere. Why not?
8. Why are hydrogen and helium abundant in the atmospheres of the Jovian planets but present in only small amounts in Earth's atmosphere?
9. What is an asteroid? What is a trans-Neptunian object? In what ways are these minor members of the solar system like or unlike the planets?
10. What are the asteroid belt, the Kuiper belt, and the Oort cloud? Where are they located? How do the objects found in these three regions compare?
11. In what ways is Pluto similar to a terrestrial planet? In what ways is it different?
12. What is the connection between comets and the Kuiper belt? Between comets and the Oort cloud?
13. What is one piece of evidence that impact craters are actually caused by impacts?
14. What is the relationship between the extent to which a planet or satellite is cratered and the amount of geologic activity on that planet or satellite?
15. How do we know that the surface of Venus is older than Earth's surface but younger than the Moon's surface?
16. Why do smaller worlds retain less of their internal heat?
17. How does the size of a terrestrial planet influence the amount of cratering on the planet's surface?
18. How is the magnetic field of a planet different from that of a bar magnet? Why is a large planet more likely to have a magnetic field than a small planet?
19. Could you use a compass to find your way around Venus? Why or why not?
20. If Mars has no planetwide magnetic field, why does it have magnetized regions on its surface?
21. What is liquid metallic hydrogen? Why is it found only in the interiors of certain planets?

Advanced Questions

Questions preceded by an asterisk (*) involve topics discussed in the Boxes.

Problem-solving tips and tools

The volume of a sphere of radius r is $4\pi r^3/3$, and the surface area of a sphere of radius r is $4\pi r^2$. The surface area of a circle of radius r is πr^2. The average density of an object is its mass divided by its volume. To calculate escape speeds, you will need to review Box 7-2. Be sure to use the same system of units (meters, seconds, kilograms) in all your calculations involving escape speeds, orbital speeds, and masses. Appendix 6 gives conversion factors between different sets of units, and Box 5-1 has formulas relating various temperature scales.

22. Mars has two small satellites, Phobos and Deimos. Phobos circles Mars once every 0.31891 day at an average altitude of 5980 km above the planet's surface. The diameter of Mars is 6794 km. Using this information, calculate the mass and average density of Mars.
23. Figure 7-3 shows the spectrum of Saturn's largest satellite, Titan. Can you think of a way that astronomers can tell which absorption lines are due to Titan's atmosphere and which are due to the atmospheres of the Sun and Earth? Explain.

*24. (a) Find the mass of a hypothetical spherical asteroid 2 km in diameter and composed of rock with average density 2500 kg/m^3. (b) Find the speed required to escape from the surface of this asteroid. (c) A typical jogging speed is 3 m/s. What would happen to an astronaut who decided to go for a jog on this asteroid?

*25. The hypothetical asteroid described in Question 24 strikes Earth with a speed of 25 km/s. (a) What is the kinetic energy of the asteroid at the moment of impact? (b) How does this energy compare with that released by a 20-kiloton nuclear weapon, like the device that destroyed Hiroshima, Japan, on August 6, 1945? (*Hint:* 1 kiloton of TNT releases 4.2×10^{12} joules of energy.)

*26. Suppose a spacecraft landed on Jupiter's moon Europa (see Table 7-2), which moves around Jupiter in an orbit of radius 670,900 km. After collecting samples from the satellite's surface, the spacecraft prepares to return to Earth. (a) Calculate the escape speed from Europa. (b) Calculate the escape speed from Jupiter at the distance of Europa's orbit. (c) In order to begin its homeward journey, the spacecraft must leave Europa with a speed greater than either your answer to (a) or your answer to (b). Explain why.

*27. A hydrogen atom has a mass of 1.673×10^{-27} kg, and the temperature of the Sun's surface is 5800 K. What is the average speed of hydrogen atoms at the Sun's surface?

*28. The Sun's mass is 1.989×10^{30} kg, and its radius is 6.96×10^8 m. (a) Calculate the escape speed from the Sun's surface. (b) Using your answer to Question 27, explain why the Sun has lost very little hydrogen over its entire 4.56-billion-year history.

*29. Saturn's satellite Titan has an appreciable atmosphere, yet Jupiter's satellite Ganymede—which is about the same size and mass as Titan—has no atmosphere. Explain why there is a difference.

30. The distance from the asteroid 433 Eros (Figure 7-7) to the Sun varies between 1.13 and 1.78 AU. (a) Find the period of Eros's orbit. (b) Does Eros lie in the asteroid belt? How can you tell?
31. Imagine a trans-Neptunian object with roughly the same mass as Earth but located 50 AU from the Sun. (a) What do you think this object would be made of? Explain your reasoning. (b) On the basis of this speculation, assume a reasonable density for this object and calculate its diameter. How many times bigger or smaller than Earth would it be?
32. Consider a hypothetical trans-Neptunian object located 100 AU from the Sun. (a) What would be the orbital period (in years) of this object? (b) There are 360 degrees in a circle, and 60 arcminutes in a degree. How long would it take this object to move 1 arcminute across the sky? (c) Trans-Neptunian objects are discovered by looking for "stars" that move on the celestial sphere. Use your answer from part (b)

to explain why these discoveries require patience. (**d**) Discovering trans-Neptunian objects also requires large telescopes equipped with sensitive detectors. Explain why.

33. The surfaces of Mercury, the Moon, and Mars are riddled with craters formed by the impact of space debris. Many of these craters are billions of years old. By contrast, there are only a few conspicuous craters on Earth's surface, and these are generally less than 500 million years old. What do you suppose explains the difference?

34. During the period of most intense bombardment by space debris, a new 1-km-radius crater formed somewhere on the Moon about once per century. During this same period, what was the probability that such a crater would be created within 1 km of a certain location on the Moon during a 100-year period? During a 10^6-year period? (*Hint:* If you drop a coin onto a checkerboard, the probability that the coin will land on any particular one of the board's 64 squares is $^1/_{64}$.)

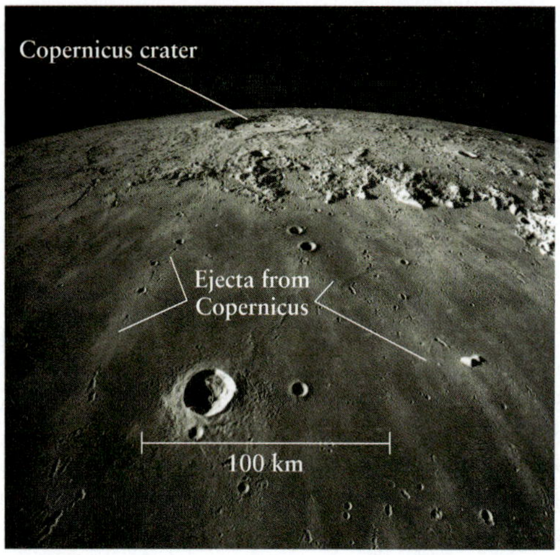

R I V U X G

(Courtesy of USRA)

35. When an impact crater is formed, material (called *ejecta*) is sprayed outward from the impact. (The accompanying photograph of the Moon shows light-colored ejecta extending outward from the crater Copernicus.) While ejecta are found surrounding the craters on Mercury, they do not extend as far from the craters as do ejecta on the Moon. Explain why, using the difference in surface gravity between the Moon (surface gravity = 0.17 that on Earth) and Mercury (surface gravity = 0.38 that on Earth).

36. Mercury rotates once on its axis every 58.646 days, compared to 1 day for Earth. Use this information to argue why Mercury's magnetic field should be much smaller than Earth's.

37. Suppose *Mars Global Surveyor* had discovered magnetized regions in the lowlands of Mars. How would this discovery have affected our understanding of the evolution of the Martian interior?

38. Liquid metallic hydrogen is the source of the magnetic fields of Jupiter and Saturn. Explain why liquid metallic hydrogen cannot be the source of Earth's magnetic field.

Discussion Questions

***39.** There are no asteroids with an atmosphere. Discuss why not.

40. The *Galileo* spacecraft that orbited Jupiter from 1995 to 2003 discovered that Ganymede (Table 7-2) has a magnetic field twice as strong as that of Mercury. Does this discovery surprise you? Why or why not?

Web/eBook Questions

41. Search the World Wide Web for information about impact craters on Earth. Where is the largest crater located? How old is it estimated to be? Which crater is closest to where you live?

42. Determining Terrestrial Planet Orbital Periods. Access the animation "Planetary Orbits" in Chapter 7 of the *Universe* Web site or eBook. Focus on the motions of the inner planets at the last half of the animation. Using the stop and start buttons, determine how many days it takes Mars, Venus, and Mercury to orbit the Sun once if Earth takes approximately 365 days.

Activities

Observing Projects

43. Use a telescope or binoculars to observe craters on the Moon. Make a drawing of the Moon, indicating the smallest and largest craters that you can see. Can you estimate their sizes? For comparison, the Moon as a whole has a diameter of 3476 km. *Hint:* You can see craters most distinctly when the Moon is near first quarter or third quarter (see Figure 3-2). At these phases, the Sun casts long shadows across the portion of the Moon in the center of your field of view, making the variations in elevation between the rims and centers of craters easy to identify. You can determine the phase of the Moon by looking at a calendar or the weather page of the newspaper, by using the *Starry Night Enthusiast*™ program, or on the World Wide Web.

44. Use the *Starry Night Enthusiast*™ program to examine magnified images of the terrestrial planets Mercury, Venus, Earth, and Mars and the asteroid Ceres. In the **Favourites** menu, under the **Solar System** submenu, select the desired planet. This will place you in the position of an astronaut orbiting above the surface of the planet. The astronaut's spacesuit and feet are shown in the foreground in this view but can be removed by clicking on **View > Feet.** To show the planet as we see it from Earth, we need to show its atmosphere. Select **Solar System > Planets-Moons . . .** in the **Options** menu. In the Planets-Moons Options dialog that pops up, click on the **Show atmosphere** checkbox to turn this option on and then click the **OK** button. The mouse icon will change to the **location scroller** when moved over the planet. You can use this scroller to rotate the image to see different views of the planet. This is equivalent to flying around the planet at a fixed distance. Follow the above steps to examine each planet and asteroid from different viewpoints and describe each planet's appearance. From what you observe in each case, is there any way of knowing whether you are looking at a planet's surface or at complete cloud cover over the planet? Which planet or planets have clouds? Which planet or asteroid shows the heaviest cratering? Which of these planets show evidence of liquid water?

45. Use the *Starry Night Enthusiast*™ program to examine the Jovian planets Jupiter, Saturn, Uranus, and Neptune. Select each of these planets from the **Solar System** submenu in the **Favourites** menu. If you desire, you can remove the image of the astronaut's feet by selecting **Feet** in the **View** menu. Position the mouse cursor over the planet and click and drag the image to examine the planet from different views. Describe each planet's appearance. Which has the greatest color contrast in its cloud tops? Which has the least color contrast? What can you say about the thickness of Saturn's rings compared to their diameter?

8 Comparative Planetology II: The Origin of Our Solar System

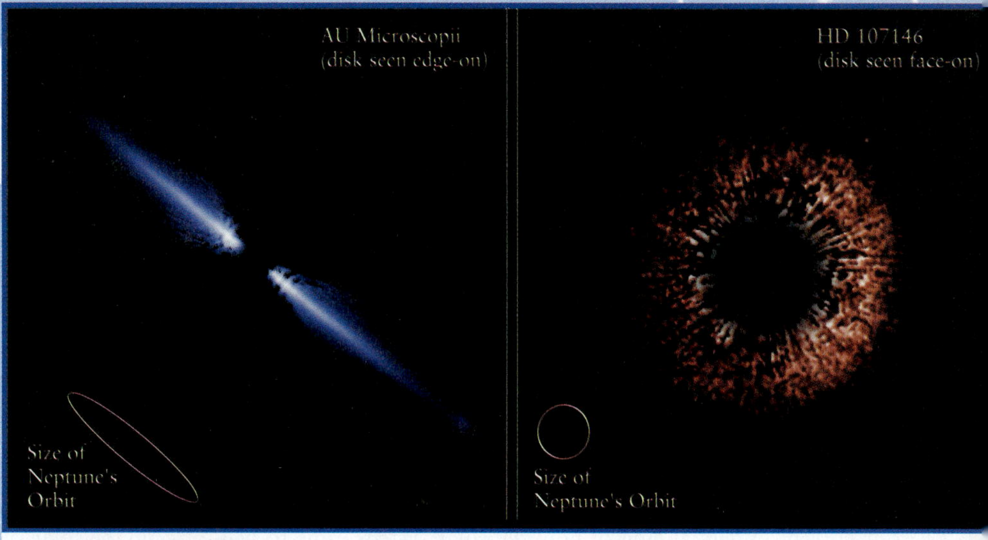

WEB LINK 8.1

R I V U X G

Planets are thought to form within the disks surrounding young stars such as these. Neptune's orbit, shown for scale, is about 60 AU across. (NASA, ESA, D. R. Ardila (JHU), D. A. Golimowski (JHU), J. E. Krist (STScI/JPL), M. Clampin (NASA/GSFC), J. P. Williams (UH/IfA), J. P. Blakeslee (JHU), H. C. Ford (JHU), G. F. Hartig (STScI), G. D. Illingworth (UCO-Lick) and the ACS Science Team)

What did our solar system look like before the planets were fully formed? The answer may lie in these remarkable images from the Hubble Space Telescope. Each image shows an immense disk of gas and dust centered on a young star. (In each image the light from the star itself was blocked out within the telescope to make the rather faint disk more visible.) Astronomers strongly suspect that in its infancy our own Sun was surrounded by a disk of this kind and that the planets coalesced from it.

In this chapter we will examine the evidence that led astronomers to this picture of the origin of the planets. We will see how the abundances of different chemical elements in the solar system indicate that the Sun and planets formed from a thin cloud of interstellar matter some 4.56 billion years ago, an age determined by measuring the radioactivity of meteorites. We will learn how the nature of planetary orbits gives important clues to what happened as this cloud contracted and how evidence from meteorites reveals the chaotic conditions that existed within the cloud. And we will see how this cloud eventually evolved into the solar system that we see today.

In the past few years astronomers have been able to test this picture of planetary formation by examining disks around young stars (like the ones in the accompanying images). Most remarkably of all, they have discovered planets in orbit around dozens of other stars. These recent observations provide valuable information about how our own system of planets came to be.

8-1 Any model of solar system origins must explain the present-day Sun and planets

How did the Sun and planets form? In other words, where did the solar system come from? This question has tantalized astronomers for centuries. Our goal in this chapter is to examine

Learning Goals

By reading the sections of this chapter, you will learn

8-1 The key characteristics of the solar system that must be explained by any theory of its origins

8-2 How the abundances of chemical elements in the solar system and beyond explain the sizes of the planets

8-3 How we can determine the age of the solar system by measuring abundances of radioactive elements

8-4 Why scientists think the Sun and planets all formed from a cloud called the solar nebula

8-5 How the solar nebula model explains the formation of the terrestrial planets

8-6 Two competing models for the origin of the Jovian planets

8-7 How astronomers test the solar nebula model by observing planets around other stars

Table 8-1 Three Key Properties of Our Solar System

Any theory of the origin of the solar system must be able to account for these properties of the planets.

Property 1: *Sizes and compositions of terrestrial planets versus Jovian planets*	The terrestrial planets, which are composed primarily of rocky substances, are relatively small, while the Jovian planets, which are composed primarily of hydrogen and helium, are relatively large (see Sections 7-1 and 7-4).
Property 2: *Directions and orientations of planetary orbits*	All of the planets orbit the Sun in the same direction, and all of their orbits are in nearly the same plane (see Section 7-1).
Property 3: *Sizes of terrestrial planet orbits versus Jovian planet orbits*	The terrestrial planets orbit close to the Sun, while the Jovian planets orbit far from the Sun (see Section 7-1).

our current understanding of how the solar system came to be—that is, our current best *theory* of the origin of the solar system.

Recall from Section 1-1 that a theory is not merely a set of wild speculations, but a self-consistent collection of ideas that must pass the test of providing an accurate description of the real world. Since no humans were present to witness the formation of the planets, scientists must base their theories of solar system origins on their observations of the present-day solar system. (In an analogous way, paleontologists base their understanding of the lives of dinosaurs on the evidence provided by fossils that have survived to the present day.) In so doing, they are following the steps of the scientific method that we described in Section 1-1.

What key attributes of the solar system should guide us in building a theory of solar system origins? Among the many properties of the planets that we discussed in Chapter 7, three of the most important are listed in Table 8-1.

Any theory that attempts to describe the origin of the solar system must be able to explain how these attributes came to be. We begin by considering what Property 1 tells us; we will return to Properties 2 and 3 and the orbits of the planets later in this chapter.

8-2 The cosmic abundances of the chemical elements are the result of how stars evolve

The small sizes of the terrestrial planets compared to the Jovian planets (Property 1 in Table 8-1) suggest that some chemical elements are quite common in our solar system, while others are quite rare. The tremendous masses of the Jovian planets—Jupiter alone has more mass than all of the other planets combined—means that the elements of which they are made, primarily hydrogen and helium, are very abundant. The Sun, too, is made almost entirely of hydrogen and helium. Its average density of 1410 kg/m^3 is in the same range as the densities of the Jovian planets (see Table 7-1), and its absorption spectrum (see Figure 5-12) shows the dominance of hydrogen and helium in the Sun's atmosphere. Hydrogen, the most abundant element, makes up nearly three-quarters of the combined mass of the Sun and planets. Helium is the second most abundant element. Together, hydrogen and helium account for about 98% of the mass of all the material in the solar system. All of the other chemical elements are relatively rare; combined, they make up the remaining 2% (Figure 8-1).

The terrestrial planets are small because they are made of less abundant elements

The dominance of hydrogen and helium is not merely a characteristic of our local part of the universe. By analyzing the spectra of stars and galaxies, astronomers have found essentially the same pattern of chemical abundances out to the farthest distance attainable by the most powerful telescopes. Hence, the vast majority of the atoms in the universe are hydrogen and helium atoms. The elements that make up the bulk of Earth—mostly iron, oxygen, and silicon—are relatively rare in the universe as a whole, as are the elements of which living organisms are made—carbon, oxygen, nitrogen, and phosphorus, among others. (You may find

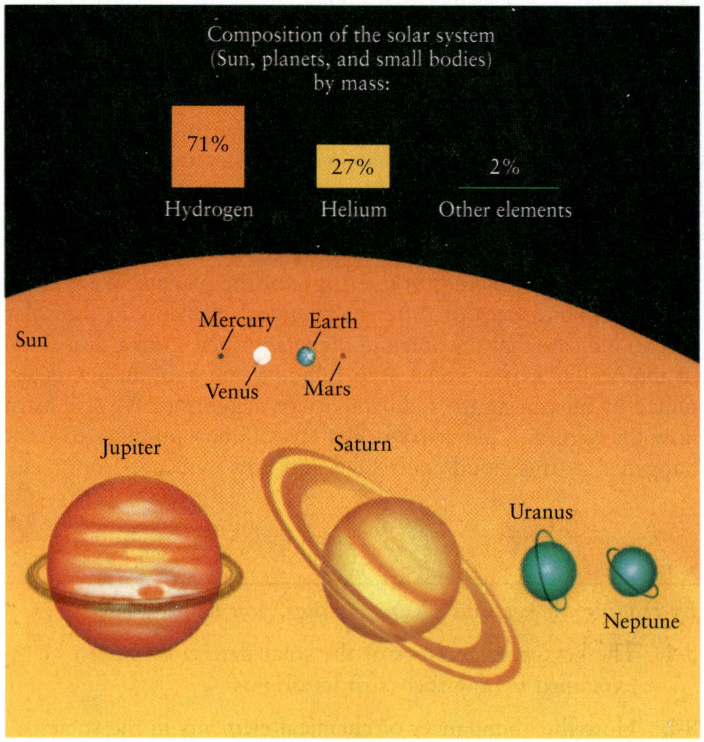

Figure 8-1

Composition of the Solar System Hydrogen and helium make up almost all of the mass of our solar system. Other elements such as carbon, oxygen, nitrogen, iron, gold, and uranium constitute only 2% of the total mass.

Antares, an aging star, ejects gas and dust.

Figure 8-2 R I V U X G

A Mature Star Ejecting Gas and Dust The star Antares is shedding material from its outer layers, forming a thin cloud around the star. We can see the cloud because some of the ejected material has condensed into tiny grains of dust that reflect the star's light. (Dust particles in the air around you reflect light in the same way, which is why you can see them within a shaft of sunlight in a darkened room). Antares lies some 600 light-years from Earth in the constellation Scorpio. (David Malin/Anglo-Australian Observatory)

it useful to review the periodic table of the elements, described in Box 5-5.)

The Origin of the Elements and Cosmic "Recycling"

There is a good reason for this overwhelming abundance of hydrogen and helium. A wealth of evidence has led astronomers to conclude that the universe began some 13.7 billion years ago with a violent event called the Big Bang (see Section 1-4). Only the lightest elements—hydrogen and helium, as well as tiny amounts of lithium and perhaps beryllium—emerged from the enormously high temperatures following this cosmic event. All the heavier elements were manufactured by stars later, either by thermonuclear fusion reactions deep in their interiors or by the violent explosions that mark the end of massive stars. Were it not for these processes that take place only in stars, there would be no heavy elements in the universe, no planet like our Earth, and no humans to contemplate the nature of the cosmos.

Because our solar system contains heavy elements, it must be that at least some of its material was once inside other stars. But how did this material become available to help build our solar system? The answer is that near the ends of their lives, stars cast much of their matter back out into space. For most stars this process is a comparatively gentle one, in which a star's outer layers are gradually expelled. Figure 8-2 shows a star losing material in this fashion. This ejected material appears as the cloudy region, or **nebulosity** (from *nubes*, Latin for "cloud"), that surrounds the star and is illuminated by it. A few stars eject matter much more dramatically at the very end of their lives, in a spectacular detonation called a *supernova*, which blows the star apart (see Figure 1-6).

No matter how it escapes, the ejected material contains heavy elements dredged up from the star's interior, where they were formed. This material becomes part of the **interstellar medium,** a tenuous collection of gas and dust that pervades the spaces between the stars. As different stars die, they increasingly enrich the interstellar medium with heavy elements. Observations show that new stars form as condensations in the interstellar medium (Figure 8-3). Thus, these new stars have an adequate supply of heavy elements from which to develop a system of planets, satellites, comets, and asteroids. Our own solar system must have formed from enriched material in just this way. Thus, our solar system contains "recycled" material that was produced long ago inside a now-dead star. This "recycled" material includes all of the carbon in your body, all of the oxygen that you breathe, and all of the iron and silicon in the soil beneath your feet.

The Abundances of the Elements

Stars create different heavy elements in different amounts. For example, oxygen (as well as carbon, silicon, and iron) is readily produced in the interiors of massive stars, whereas gold (as well as silver, platinum, and uranium) is created only under special

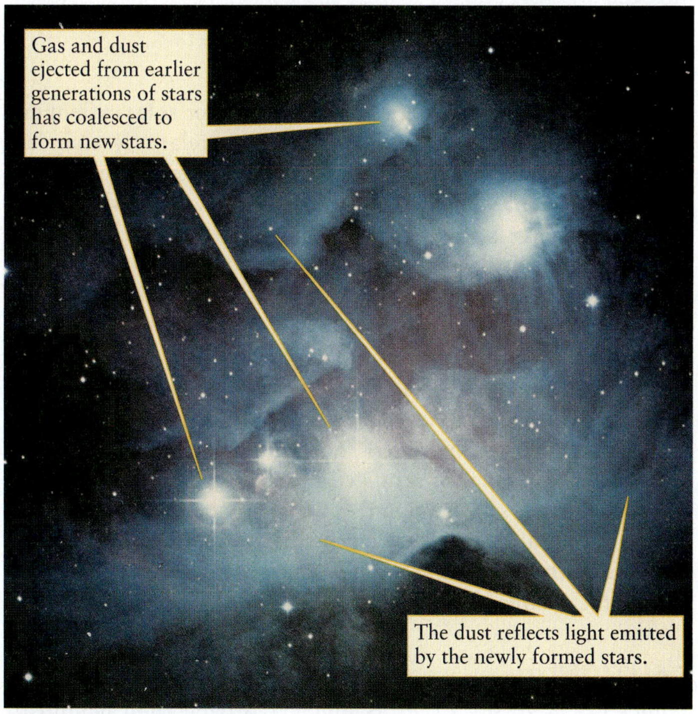

Gas and dust ejected from earlier generations of stars has coalesced to form new stars.
The dust reflects light emitted by the newly formed stars.

Figure 8-3 R I V U X G

New Stars Forming from Gas and Dust Unlike Figure 8-2, which depicts an old star that is ejecting material into space, this image shows young stars in the constellation Orion (the Hunter) that have only recently formed from a cloud of gas and dust. The bluish, wispy appearance of the cloud (called NGC 1973-1975-1977) is caused by starlight reflecting off interstellar dust grains within the cloud (see Box 5-4). The grains are made of heavy elements produced by earlier generations of stars. (David Malin/Anglo-Australian Observatory)

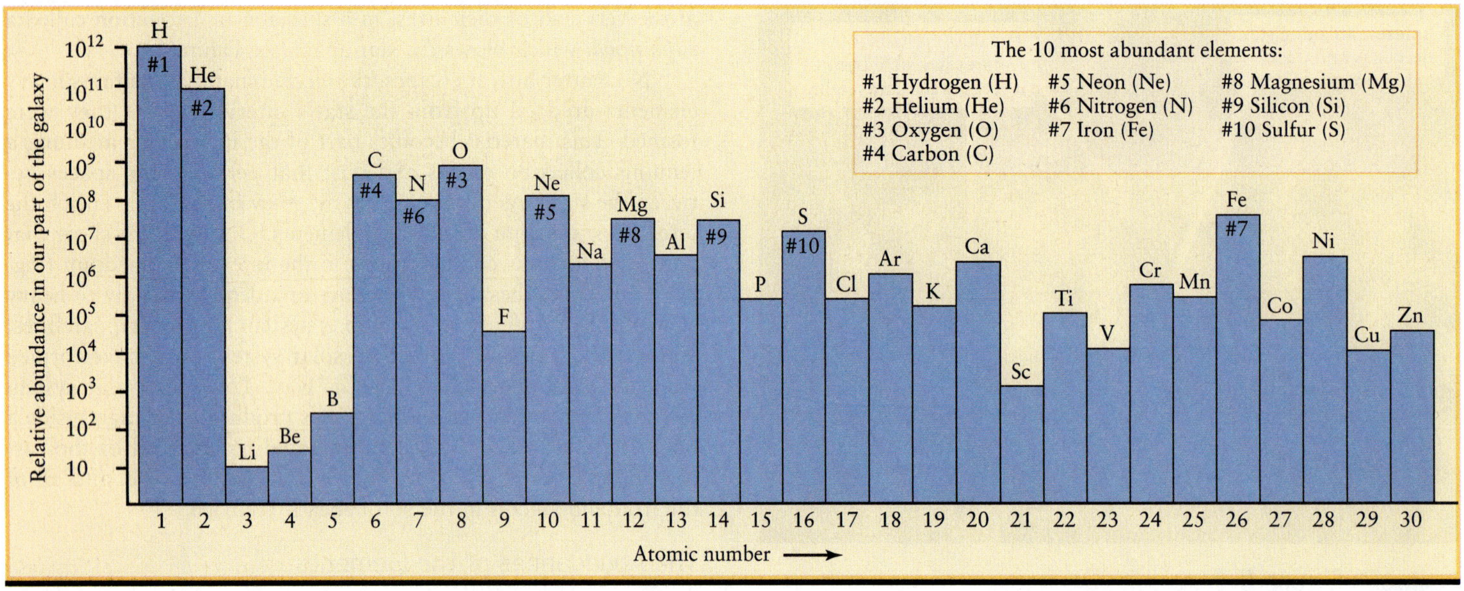

Figure 8-4

Abundances of the Lighter Elements This graph shows the abundances in our part of the Galaxy of the 30 lightest elements (listed in order of increasing atomic number) compared to a value of 10^{12} for hydrogen. The inset lists the 10 most abundant of these elements, which are also indicated in the graph. Notice that the vertical scale is not linear; each division on the scale corresponds to a tenfold increase in abundance. All elements heavier than zinc (Zn) have abundances of fewer than 1000 atoms per 10^{12} atoms of hydrogen.

circumstances. Consequently, gold is rare in our solar system and in the universe as a whole, while oxygen is relatively abundant (although still much less abundant than hydrogen or helium).

A convenient way to express the relative abundances of the various elements is to say how many atoms of a particular element are found for every trillion (10^{12}) hydrogen atoms. For example, for every 10^{12} hydrogen atoms in space, there are about 100 billion (10^{11}) helium atoms. From spectral analysis of stars and chemical analysis of Earth rocks, Moon rocks, and bits of interplanetary debris called meteorites, scientists have determined the relative abundances of the elements in our part of the Milky Way Galaxy today. Figure 8-4 shows the relative abundances of the 30 lightest elements, arranged in order of their **atomic numbers.** An element's atomic number is the number of protons in the nucleus of an atom of that element. It is also equal to the number of electrons orbiting the nucleus (see Box 5-5). In general, the greater the atomic number of an atom, the greater its mass.

CAUTION! Figure 8-1 shows that our solar system has 2.6 times (71% versus 27%) as much hydrogen than helium by *mass,* while Figure 8-4 shows that there is about 10 times as much hydrogen than helium by *number of atoms*. The explanation of this seeming inconsistency is that a helium atom has about 4 times the mass of a hydrogen atom, which makes helium more important on a per-mass basis than on a per-number basis.

The inset in Figure 8-4 lists the 10 most abundant elements. Note that even oxygen (chemical symbol O), the third most abundant element, is quite rare relative to hydrogen (H) and helium (He): There are only 8.5×10^8 oxygen atoms for each 10^{12} hydrogen atoms and each 10^{11} helium atoms. Expressed another way, for each oxygen atom in our region of the Milky Way Galaxy, there are about 1200 hydrogen atoms and 120 helium atoms.

In addition to the 10 most abundant elements listed in Figure 8-3, five elements are moderately abundant: sodium (Na), aluminum (Al), argon (Ar), calcium (Ca), and nickel (Ni). These elements have abundances in the range of 10^6 to 10^7 relative to the standard 10^{12} hydrogen atoms. Most of the other elements are much rarer. For example, for every 10^{12} hydrogen atoms in the solar system, there are only six atoms of gold.

The small cosmic abundances of elements other than hydrogen and helium help to explain why the terrestrial planets are so small (Property 1 in Table 8-1). Because the heavier elements required to make a terrestrial planet are rare, only relatively small planets can form out of them. By contrast, hydrogen and helium are so abundant that it was possible for these elements to form large Jovian planets.

8-3 The abundances of radioactive elements reveal the solar system's age

The heavy elements can tell us even more about the solar system: They also help us determine its age. The particular heavy elements that provide us with this information are *radioactive*. Their atomic nuclei are unstable because they contain too many protons or too many neutrons. A radioactive nucleus therefore ejects particles until it becomes stable. In doing so, a nucleus may change from one element to another. Physicists refer to this transmutation as **radioactive decay.** For example, a radioactive form of the element rubidium (atomic number 37) decays into the element strontium (atomic number 38) when one of the neutrons in the rubidium nucleus decays into a proton and an electron (which is ejected from the nucleus).

Our solar system, which formed 9 billion years after the Big Bang, is a relative newcomer in the universe

Experiment shows that each type of radioactive nucleus decays at its own characteristic rate, which can be measured in the laboratory. This observation is the key to a technique called **radioactive dating,** which is used to determine the ages of rocks. If a rock contained a certain amount of radioactive rubidium when it first formed, over time more and more of the atoms of rubidium within the rock will decay into strontium atoms. The ratio of the number of strontium atoms the rock contains to the number of rubidium atoms it contains then gives a measure of the age of the rock. Box 8-1 describes radioactive dating in more detail.

Dating the Solar System

Scientists have applied techniques of radioactive dating to rocks taken from all over Earth. The results show that most rocks are tens or hundreds of millions of years old, but that some rocks are as much as 4 billion (4×10^9) years old. These results confirm that geologic processes have produced new surface material over Earth's history, as we concluded from the small number of impact craters found on Earth (see Section 7-6). They also show that Earth is at least 4×10^9 years old.

Radioactive dating has also been applied to rock samples brought back from the Moon by the *Apollo* astronauts. The oldest *Apollo* specimen, collected from one of the most heavily cratered and hence most ancient regions of the Moon, is 4.5×10^9 years old. But the oldest rocks found anywhere in the solar system are **meteorites,** bits of interplanetary debris that survive passing through Earth's atmosphere and land on our planet's surface (Figure 8-5). Radioactive dating of meteorites reveals that they are all nearly the *same* age, about 4.56 billion years old. The absence of any younger or older meteorites indicates that these are all remnants of objects that formed at the same time in the early solar system. We conclude that the age of the oldest meteorites, about 4.56×10^9 years, is the age of the solar system itself. Note that this almost inconceivably long span of time is only about one-third of the current age of the universe, 13.7×10^9 years.

Thus, by studying the abundances of radioactive elements, we are led to a remarkable insight: Some 4.56 billion years ago, a collection of hydrogen, helium, and heavy elements came together to form the Sun and all of the objects that orbit around it. All of those heavy elements, including the carbon atoms in your body and the oxygen atoms that you breathe, were created and cast off by stars that lived and died long before our solar system formed, during the first 9 billion years of the universe's existence. We are literally made of star dust.

Figure 8-5 R I V U X G

A Meteorite Although it resembles an ordinary Earth rock, this is actually a meteorite that fell from space. The proof of its extraterrestrial origin is the meteorite's surface, which shows evidence of having been melted by air friction as it entered our atmosphere at 40,000 km/h (25,000 mi/h). Meteorites are the oldest objects in the solar system. (Ted Kinsman/Photo Researchers, Inc.)

8-4 The Sun and planets formed from a solar nebula

We have seen how processes in the Big Bang and within ancient stars produced the raw ingredients of our solar system. But given these ingredients, how did they combine to make the Sun and planets? Astronomers have developed a variety of models for the origin of the solar system. The test of these models is whether they explain the properties of the present-day system of Sun and planets.

Astronomers see young stars that may be forming planets today in the same way that our solar system did billions of years ago

The Failed Tidal Hypothesis

Any model of the origin of the solar system must explain why all the planets orbit the Sun in the same direction and in nearly the same plane (Property 2 in Table 8-1). One model that was devised explicitly to address this issue was the *tidal hypothesis,* proposed in the early 1900s. As we saw in Section 4-8, two nearby planets, stars, or galaxies exert tidal forces on each other that cause the objects to elongate. In the tidal hypothesis, another star happened to pass close by the Sun, and the star's tidal forces drew a long filament out of the Sun. The filament material would then go into orbit around the Sun, and all of it would naturally orbit in the same direction and in the same plane. From this filament the planets would condense. However, it was shown in the 1930s that tidal forces strong enough to pull a filament out of the Sun would cause the filament to disperse before it could condense into planets. Hence, the tidal hypothesis cannot be correct.

BOX 8-1 Tools of the Astronomers Trade

Radioactive Dating

How old are the rocks found on Earth and other planets? Are rocks found at different locations the same age or of different ages? How old are meteorites? Questions like these are important to scientists who wish to reconstruct the history of our solar system. But simply looking at a rock cannot tell us whether it was formed a thousand years, a million years, or a billion years ago. Fortunately, most rocks contain trace amounts of radioactive elements such as uranium. By measuring the relative abundances of various radioactive isotopes and their decay products within a rock, scientists can determine the rock's age.

As we saw in Box 5-5, every atom of a particular element has the same number of protons in its nucleus. However, different isotopes of the same element have different numbers of neutrons in their nuclei. For example, the common isotopes of uranium are ^{235}U and ^{238}U. Each isotope of uranium has 92 protons in its nucleus (correspondingly, uranium is element 92 in the periodic table; see Box 5-5). However, a ^{235}U nucleus contains 143 neutrons, whereas a ^{238}U nucleus has 146 neutrons.

A radioactive nucleus with too many protons or too many neutrons is unstable; to become stable, it *decays* by ejecting particles until it becomes stable. If the number of protons (the atomic number) changes in this process, the nucleus changes from one element to another.

Some radioactive isotopes decay rapidly, while others decay slowly. Physicists find it convenient to talk about the decay rate in terms of an isotope's **half-life.** The half-life of an isotope is the time interval in which one-half of the nuclei decay. For example, the half-life of ^{238}U is 4.5 billion (4.5×10^9) years. Uranium's half-life means that if you start out with 1 kg of ^{238}U, after 4.5 billion years, you will have only $\frac{1}{2}$ kg of ^{238}U remaining; the other $\frac{1}{2}$ kg will have turned into other elements. If you wait another half-life, so that a total of 9.0 billion years has elapsed, only $\frac{1}{4}$ kg of ^{238}U—one-half of one-half of the original amount—will remain. Several isotopes useful for determining the ages of rocks are listed in the accompanying table.

To see how geologists date rocks, consider the slow conversion of radioactive rubidium (^{87}Rb) into strontium (^{87}Sr). (The periodic table in Box 5-5 shows that the atomic numbers for these elements are 37 for rubidium and 38 for strontium, so in the decay a neutron is transformed into a proton. In this process an electron is ejected from the nucleus.) Over the years, the amount of ^{87}Rb in a rock decreases, while the amount of ^{87}Sr increases. Because the ^{87}Sr appears in the rock due to radioactive decay, this isotope is called *radiogenic.* Dating the rock is not simply a matter of measuring its ratio of rubidium to strontium, however, because the rock already had some strontium in it when it was formed. Geologists must therefore determine how much fresh strontium came from the decay of rubidium after the rock's formation.

To make this determination, geologists use as a reference another isotope of strontium whose concentration has remained constant. In this case, they use ^{86}Sr, which is stable and is not created by radioactive decay; it is said to be *nonradiogenic.* Dating a rock thus entails comparing the ratio of radiogenic and nonradiogenic strontium ($^{87}Sr/^{86}Sr$) in the rock to the ratio of radioactive rubidium to nonradiogenic strontium ($^{87}Rb/^{86}Sr$). Because the half-life for converting ^{87}Rb into ^{87}Sr is known, the rock's age can then be calculated from these ratios (see the table).

Radioactive isotopes decay with the same half-life no matter where in the universe they are found. Hence, scientists have used the same techniques to determine the ages of rocks from the Moon and of meteorites.

Original Radioactive Isotope	Final Stable Isotope	Half-Life (Years)	Range of Ages that Can Be Determined (Years)
Rubidium (^{87}Rb)	Strontium (^{87}Sr)	47.0 billion	10 million–4.56 billion
Uranium (^{238}U)	Lead (^{206}Pb)	4.5 billion	10 million–4.56 billion
Potassium (^{40}K)	Argon (^{40}Ar)	1.3 billion	50,000–4.56 billion
Carbon (^{14}C)	Nitrogen (^{14}N)	5730	100–70,000

The Successful Nebular Hypothesis

An entirely different model is now thought to describe the most likely series of events that led to our present solar system (Figure 8-6). The central idea of this model dates to the late 1700s, when the German philosopher Immanuel Kant and the French scientist Pierre-Simon de Laplace turned their attention to the manner in which the planets orbit the Sun. Both concluded that the arrangement of the orbits—all in the same direction and in nearly the same plane—could not be mere coincidence. To explain the orbits, Kant and Laplace independently proposed that our entire solar system, including the Sun as well as all of its planets and satellites, formed from a vast, rotating cloud of gas and dust called the **solar nebula** (Figure 8-6*a*). This model is called the **nebular hypothesis.**

The consensus among today's astronomers is that Kant and Laplace were exactly right. In the modern version of the nebular hypothesis, at the outset the solar nebula was similar in character to the nebulosity shown in Figure 8-3 and had a mass somewhat greater than that of our present-day Sun.

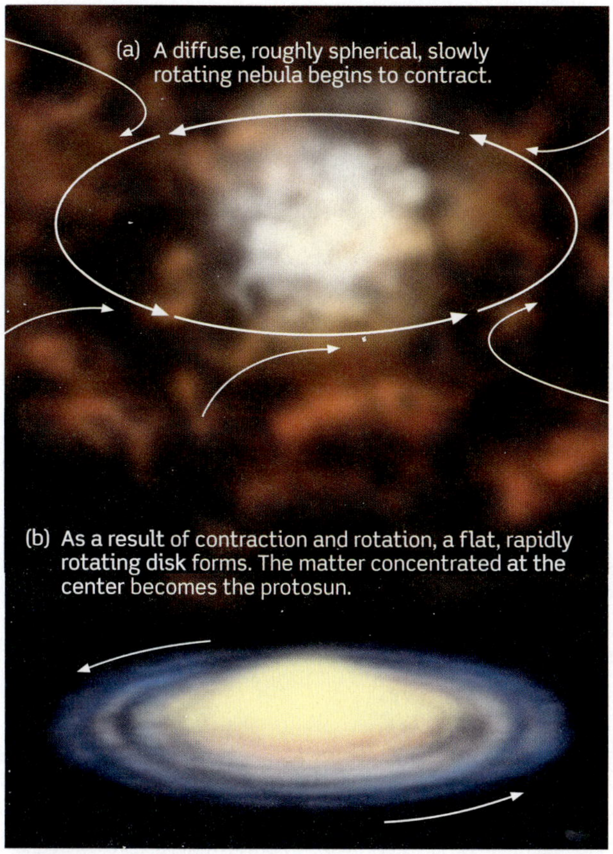

Figure 8-6

The Birth of the Solar System **(a)** A cloud of interstellar gas and dust begins to contract because of its own gravity. **(b)** As the cloud flattens and spins more rapidly around its rotation axis, a central condensation develops that evolves into a glowing protosun. The planets will form out of the surrounding disk of gas and dust.

Each part of the nebula exerted a gravitational attraction on the other parts, and these mutual gravitational pulls tended to make the nebula contract. As it contracted, the greatest concentration of matter occurred at the center of the nebula, forming a relatively dense region called the **protosun.** As its name suggests, this part of the solar nebula eventually developed into the Sun. The planets formed from the much sparser material in the outer regions of the solar nebula. Indeed, the mass of all the planets together is only 0.1% of the Sun's mass.

Evolution of the Protosun

When you drop a ball, the gravitational attraction of Earth makes the ball fall faster and faster as it falls; in the same way, material falling inward toward the protosun would have gained speed. As this fast-moving material ran into the protosun, the energy of the collision was converted into thermal energy, causing the temperature deep inside the solar nebula to climb. This process, in which the gravitational energy of a contracting gas cloud is converted into thermal energy, is called **Kelvin-Helmholtz contraction,** after the nineteenth-century physicists who first described it.

As the newly created protosun continued to contract and become denser, its temperature continued to climb as well. After about 10^5 (100,000) years, the protosun's surface temperature stabilized at about 6000 K, but the temperature in its interior kept increasing to ever higher values as the central regions of the protosun became denser and denser. Eventually, after perhaps 10^7 (10 million) years had passed since the solar nebula first began to contract, the gas at the center of the protosun reached a density of about 10^5 kg/m^3 (a hundred times denser than water) and a temperature of a few million kelvins (that is, a few times 10^6 K). Under these extreme conditions, nuclear reactions that convert hydrogen into helium began in the protosun's interior. When these nuclear reactions began, the pressure increased due to the energy released by these reactions. When the pressure stopped further contraction, a true star was born. Nuclear reactions continue to the present day in the interior of the Sun and are the source of all the energy that the Sun radiates into space.

The Protoplanetary Disk

If the solar nebula had not been rotating at all, everything would have fallen directly into the protosun, leaving nothing behind to form the planets. Instead, the solar nebula must have had an overall slight rotation, which caused its evolution to follow a different path. As the slowly rotating nebula collapsed inward, it would naturally have tended to rotate faster. This relationship between the size of an object and its rotation speed is an example of a general principle called the **conservation of angular momentum.**

ANALOGY Figure skaters make use of the conservation of angular momentum. When a spinning skater pulls her arms and legs in close to her body, the rate at which she spins automatically increases (Figure 8-7). Even if you are not a figure skater,

(a) (b)

Figure 8-7 R I V U X G

Conservation of Angular Momentum A figure skater who **(a)** spins slowly with her limbs extended will naturally speed up when **(b)** she pulls her limbs in. In the same way, the solar nebula spun more rapidly as its material contracted toward the center of the nebula. (AP Photo/Amy Sancetta)

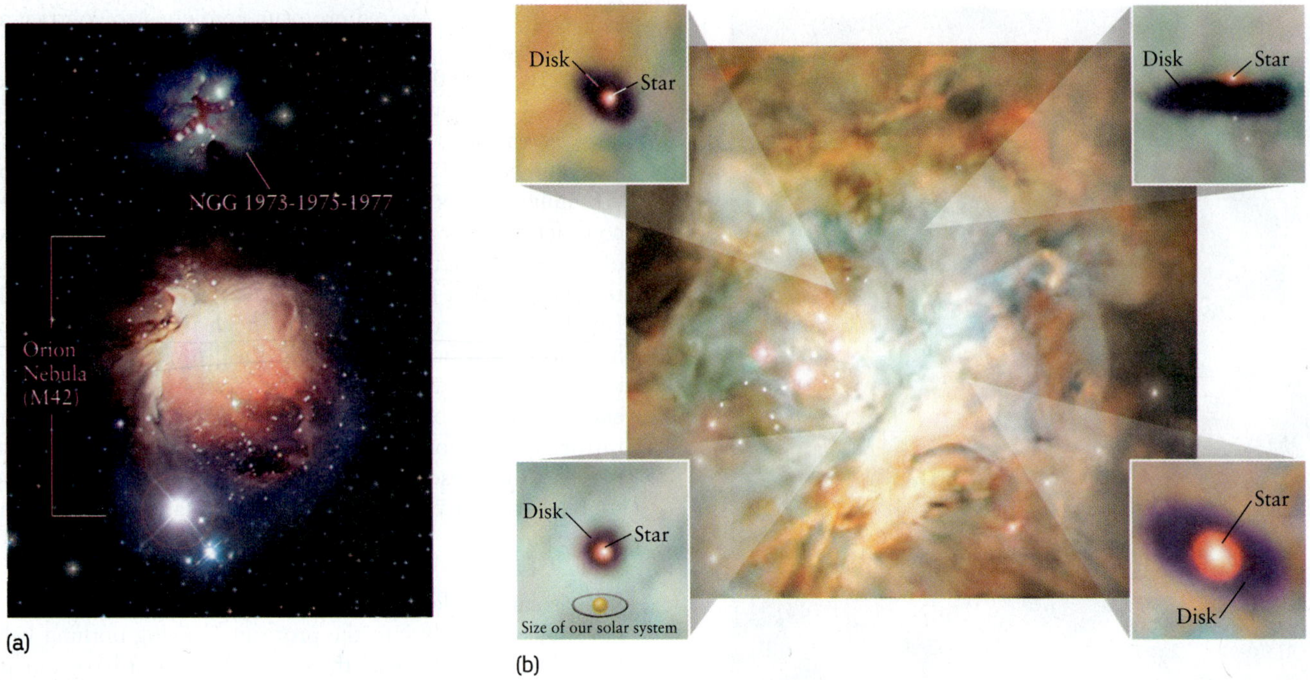

VIDEO 8.1 **Figure 8-8** RI**V**UXG
Protoplanetary Disks (a) The Orion Nebula is a star-forming region located some 1500 light-years from Earth. It is the middle "star" in Orion's "sword" (see Figure 2-2a). The smaller, bluish nebula is the object shown in Figure 8-3. **(b)** This view of the center of the Orion Nebula is a mosaic of Hubble Space Telescope images. The four insets are false-color close-ups of four protoplanetary disks that lie within the nebula. A young, recently formed star is at the center of each disk. (The disk at upper right is seen nearly edge-on.) The inset at the lower left shows the size of our own solar system for comparison. (a: Anglo-Australian Observatory image by David Malin; b: C. R. O'Dell and S. K. Wong, Rice University; NASA)

you can demonstrate this by sitting on an office chair. Sit with your arms outstretched and hold a weight, like a brick or a full water bottle, in either hand. Now use your feet to start your body and the chair rotating, lift your feet off the ground, and then pull your arms inward. Your rotation will speed up quite noticeably.

As the solar nebula began to rotate more rapidly, it also tended to flatten out (Figure 8-6*b*)—but why? From the perspective of a particle rotating along with the nebula, it felt as though there were a force pushing the particle away from the nebula's axis of rotation. (Likewise, passengers on a spinning carnival ride seem to feel a force pushing them outward and away from the ride's axis of rotation.) This apparent force was directed opposite to the inward pull of gravity, and so it tended to slow the contraction of material toward the nebula's rotation axis. But there was no such effect opposing contraction in a direction parallel to the rotation axis. Some 10^5 (100,000) years after the solar nebula first began to contract, it had developed the structure shown in Figure 8-6*b*, with a rotating, flattened disk surrounding what will become the protosun. This disk is called the **protoplanetary disk** or **proplyd,** since planets formed from its material. This model explains why their orbits all lie in essentially the same plane and why they all orbit the Sun in the same direction.

There were no humans to observe these processes taking place during the formation of the solar system. But Earth astronomers have seen disks of material surrounding other stars that formed only recently. These, too, are called protoplanetary disks, because it is thought that planets can eventually form from these disks around other stars. Hence, these disks are planetary systems that are still "under construction." By studying these disks around other stars, astronomers are able to examine what our solar nebula may have been like some 4.56×10^9 years ago.

WEB LINK 8.5 Figure 8-8 shows a number of protoplanetary disks in the Orion Nebula, a region of active star formation. A star is visible at the center of each disk, which reinforces the idea that our Sun began to shine before the planets were fully formed. (The images that open this chapter show even more detailed views of disks surrounding young stars.) A study of 110 young stars in the Orion Nebula detected protoplanetary disks around 56 of them, which suggests that systems of planets may form around a substantial fraction of stars. Later in this chapter we will see direct evidence for planets that have formed around stars other than the Sun.

8-5 The terrestrial planets formed by the accretion of planetesimals

We have seen how the solar nebula would have contracted to form a young Sun with a protoplanetary disk rotating around it. But how did the material in this disk form into planets? And why are terrestrial planets in the inner solar system, while the giant Jovian planets are in the outer solar system (Property 3 in Table 8-1)? In this section and the next we will see how the nebular hypothesis provides answers to these questions.

Rocky planets formed in the inner solar nebula as a consequence of the high temperatures close to the protosun

Temperatures in the Solar Nebula

To understand how the planets, asteroids, and comets formed, we must look at the conditions that prevailed within the solar nebula. The density of material in the part of the nebula outside the protosun was rather low, as was the pressure of the nebula's gas. If the pressure is sufficiently low, a substance cannot remain in the liquid state, but must end up as either a solid or a gas. For a given pressure, what determines whether a certain substance is a solid or a gas is its **condensation temperature.** If the temperature of a substance is above its condensation temperature, the substance is a gas; if the temperature is below the condensation temperature, the substance solidifies into tiny specks of dust or snowflakes. You can often see similar behavior on a cold morning. The air temperature can be above the condensation temperature of water, while the cold windows of parked cars may have temperatures below the condensation temperature. Thus, water molecules in the air remain as a gas (water vapor) but form solid ice particles (frost) on the car windows.

Substances such as water (H_2O), methane (CH_4), and ammonia (NH_3) have low condensation temperatures, ranging from 100 to 300 K. Rock-forming substances have much higher condensation temperatures, in the range from 1300 to 1600 K. The gas cloud from which the solar system formed had an initial temperature near 50 K, so all of these substances could have existed in solid form. Thus, the solar nebula would have been populated by an abundance of small ice particles and solid dust grains like the one shown in Figure 8-9. (Recall from Section 7-4 that "ice" can refer to frozen CO_2, CH_4, or NH_3, as well as frozen water.) But hydrogen and helium, the most abundant elements in the solar nebula, have condensation temperatures so near absolute zero that these substances always existed as gases during the creation of the solar system. You can best visualize the initial state of the solar nebula as a thin gas of hydrogen and helium strewn with tiny dust particles.

This state of affairs changed as the central part of the solar nebula underwent Kelvin-Helmholtz contraction to form the protosun. During this phase, the protosun was actually quite a bit more luminous than the present-day Sun, and it heated the innermost part of the solar nebula to temperatures above 2000 K. Meanwhile, temperatures in the outermost regions of the solar nebula remained below 50 K.

Figure 8-10 shows the probable temperature distribution in the solar nebula at this stage in the formation of our solar system. In the inner regions of the solar nebula, water, methane, and ammonia were vaporized by the high temperatures. Only materials with high condensation temperatures could have remained solid. Of these materials, iron, silicon, magnesium, and sulfur were particularly abundant, followed closely by aluminum, calcium, and nickel. (Most of these elements were present in the form of oxides, which are chemical compounds containing oxygen. These compounds also have high condensation temperatures.) In contrast, ice particles and ice-coated dust grains were able to survive in the cooler, outer portions of the solar nebula.

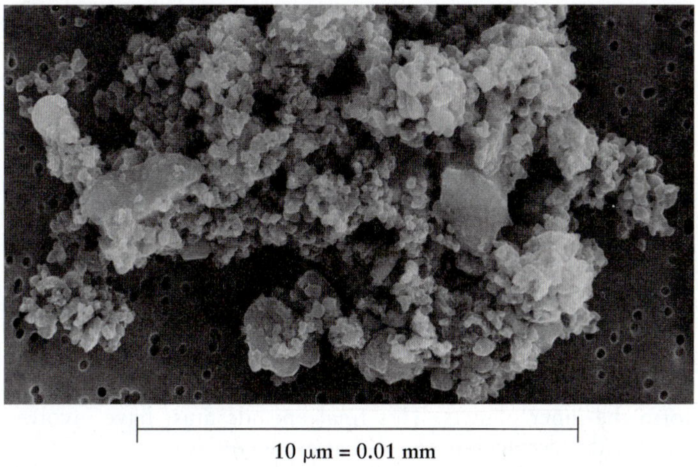

Figure 8-9

WEB LINK 8.6

A Grain of Cosmic Dust This highly magnified image shows a microscopic dust grain that came from interplanetary space. It entered Earth's upper atmosphere and was collected by a high-flying aircraft. Dust grains of this sort are abundant in star-forming regions like that shown in Figure 8-3. These tiny grains were also abundant in the solar nebula and served as the building blocks of the planets. (Donald Brownlee, University of Washington)

Planetesimals, Protoplanets, and Terrestrial Planets

ANIMATION 8.1

In the inner part of the solar nebula, the grains of high-condensation-temperature materials would have

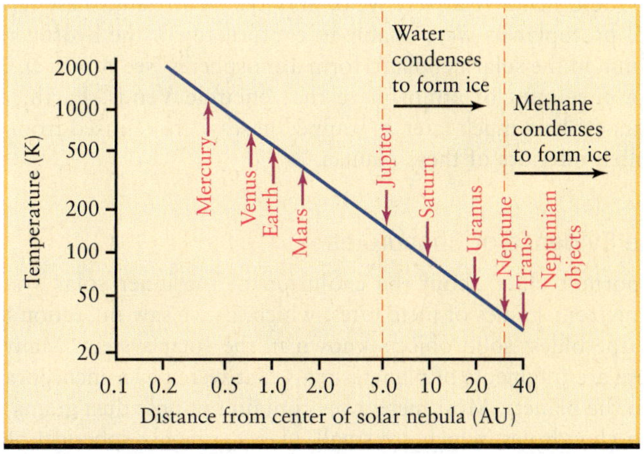

Figure 8-10

Temperature Distribution in the Solar Nebula This graph shows how temperatures probably varied across the solar nebula as the planets were forming. Note the general decline in temperature with increasing distance from the center of the nebula. Beyond 5 AU from the center of the nebula, temperatures were low enough for water to condense and form ice; beyond 30 AU, methane (CH_4) could also condense into ice.

collided and merged into small chunks. Initially, electric forces—that is, chemical bonds—held these chunks together, in the same way that chemical bonds hold an ordinary rock together. Over a few million years, these chunks coalesced into roughly 10^9 asteroidlike objects called **planetesimals,** with diameters of a kilometer or so. These planetesimals were large enough to be held together by the gravitational attraction of the different small chunks within the planetesimal. During the next stage, gravitational attraction between the planetesimals caused them to collide and accumulate into still-larger objects called **protoplanets,** which were roughly the size and mass of our Moon. This accumulation of material to form larger and larger objects is called **accretion.** During the final stage, these Moon-sized protoplanets collided to form the inner planets. This final episode must have involved some truly spectacular, world-shattering collisions.

In the inner solar nebula only materials with high condensation temperatures could form dust grains and hence protoplanets, so the result was a set of planets made predominantly of materials such as iron, silicon, magnesium, and nickel, and their oxides. This makeup is just the composition of the present-day terrestrial planets.

At first the material that coalesced to form protoplanets in the inner solar nebula remained largely in solid form, despite the high temperatures close to the protosun. But as the protoplanets grew, they were heated by the violent impacts of planetesimals as well as the energy released from the decay of radioactive elements such as uranium, and this heat caused melting. Thus, the terrestrial planets began their existence as spheres of at least partially molten rocky materials. Material was free to move within these molten spheres, so the denser, iron-rich minerals sank to the centers of the planets while the less dense silicon-rich minerals floated to their surfaces. This process is called **chemical differentiation** (see Box 7-1). In this way the terrestrial planets developed their dense iron cores. We saw in Section 7-7 that this electrically conductive iron gives rise to the magnetic fields of the terrestrial planets.

Because the materials that went into the terrestrial planets are relatively scarce, these planets ended up having relatively small mass and hence relatively weak gravity. As a result, these terrestrial protoplanets were unable to capture any of the hydrogen or helium in the solar nebula to form atmospheres (see Box 7-2). The thin envelopes of atmosphere that encircle Venus, Earth, and Mars evolved much later as trapped gases were released from the molten interiors of these planets.

The Evidence of Chondrules

Important clues about the evolution of the inner solar system come from studies of meteorites, which, as we saw in Section 8-3, are the oldest solid objects known in the solar system. Many of these are fragments of planetesimals that were never incorporated into the planets. Most meteorites contain not only dust grains but also **chondrules,** which are small, glassy, roughly spherical blobs (Figure 8-11). (The "ch" in "chondrule" is pronounced as in "chord.") Liquids tend to form spherical drops (like drops of water), so the shape of chondrules shows that they were once molten.

Attempts to produce chondrules in the laboratory show that the blobs must have melted suddenly, then solidified over the space of only an hour or so. This means they could not have been melted by the temperature of the inner solar nebula, which would have remained high for hundreds of thousands of years after the formation of the protosun. Instead, chondrules must have been produced by sudden, high-energy events that quickly melted material and then permitted it to cool. These events were probably shock waves propagating through the gas of the solar nebula, like the sonic booms produced by an airplane traveling faster than sound. The sheer number of chondrules suggests that such shock waves occurred throughout the inner solar nebula over a period of more than 10^6 years. The words "cloud" and "nebula" may suggest material moving in gentle orbits around the protosun; the existence of chondrules suggests that conditions in the inner solar nebula were at times quite violent.

WEB LINK 8.7

Figure 8-11 R I V U X G

Chondrules This is a cross-section of a meteorite that landed in Chihuahua, Mexico in 1969. The light-colored spots are individual chondrules that range in size from a few millimeters to a few tenths of a millimeter. Many of the chondrules are coated with dust grains like the one shown in Figure 8-8. (The R. N. Hartman Collection)

Simulating the Formation of Planets

Astronomers use computer simulations to learn how the inner planets could have formed from planetesimals. A computer is programmed to simulate a large number of particles circling a newborn sun along orbits dictated by Newtonian mechanics. As the simulation proceeds, the particles coalesce to form larger objects, which in turn collide to form planets. By performing a variety of simulations, each beginning with somewhat different numbers of planetesimals in different orbits, it is possible to see what kinds of planetary systems would have been created under different initial conditions. Such studies demonstrate that a wide range of initial conditions ultimately lead to basically the same result in the

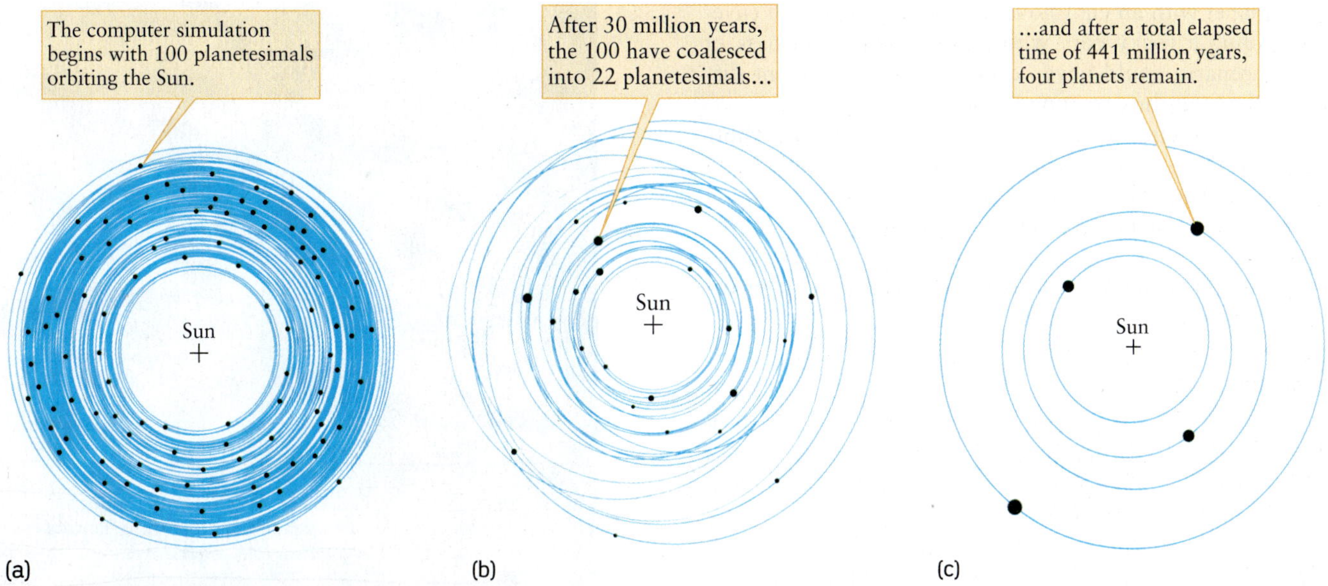
The computer simulation begins with 100 planetesimals orbiting the Sun.
After 30 million years, the 100 have coalesced into 22 planetesimals...
...and after a total elapsed time of 441 million years, four planets remain.
Sun
Sun
Sun
(a)
(b)
(c)

Figure 8-12

Accretion of the Terrestrial Planets These three drawings show the results of a computer simulation of the formation of the inner planets. In this simulation, the inner planets were essentially formed after 150 million years. (Adapted from George W. Wetherill)

inner solar system: Accretion continues for roughly 10^8 (100 million) years and typically forms four or five terrestrial planets with orbits between 0.3 and 1.6 AU from the Sun.

Figure 8-12 shows one particular computer simulation. The calculations began with 100 planetesimals, each having a mass of 1.2×10^{23} kg. This choice ensures that the total mass (1.2×10^{25} kg) equals the combined mass of the four terrestrial planets (Mercury through Mars) plus their satellites. The initial orbits of planetesimals are inclined to each other by angles of less than 5° to simulate a thin layer of particles orbiting the protosun.

After an elapsed time simulating 30 million (3×10^7) years, the 100 original planetesimals have merged into 22 protoplanets (Figure 8-9*b*). After 79 million (7.9×10^7) years, these protoplanets have combined into 11 larger protoplanets. Nearly another 100 million (10^8) years elapse before further mergers leave just six growing protoplanets. Finally, Figure 8-9*c* shows four planets following nearly circular orbits after a total elapsed time of 441 million (4.41×10^8) years. In this particular simulation, the fourth planet from the Sun ends up being the most massive; in our own solar system, the third planet (Earth) is the most massive of the terrestrial planets. Note that due to collisions, the four planets in the simulation end up in orbits that are nearly circular, just like the orbits of most of the planets in our solar system.

8-6 Gases in the outer solar nebula formed the Jovian planets, whose gravitational influence gave rise to the small bodies of the solar system

We have seen how the high temperatures in the inner solar nebula led to the formation of the small, rocky terrestrial planets. To explain the very different properties of the Jovian planets, we need to consider the conditions that prevailed in the relatively cool outer regions of the solar nebula. Two leading models explain how the Jovian planets could have formed there.

Low temperatures in the outer solar nebula made it possible for planets to grow to titanic size

The Core Accretion Model

Like the inner planets, the outer planets may have begun to form by the accretion of planetesimals. The key difference is that ices as well as rocky grains were able to survive in the outer regions of the solar nebula, where temperatures were relatively low (see Figure 8-10). The elements of which ices are made are much more abundant than those that form rocky grains. Thus, much more solid material would have been available to form planetesimals in the outer solar nebula than in the inner part. As a result, solid objects several times larger than any of the terrestrial planets could have formed in the outer solar nebula. Each such object, made up of a mixture of ices and rocky material, could have become the core of a Jovian planet and served as a "seed" around which the rest of the planet eventually grew.

Thanks to the lower temperatures in the outer solar system, gas atoms (principally hydrogen and helium) were moving relatively slowly and so could more easily be captured by the strong gravity of these massive cores (see Box 7-2). Thus, the core of a Jovian protoplanet could have captured an envelope of gas as it continued to grow by accretion. This picture is called the **core accretion model** for the origin of the Jovian planets.

Calculations based on the core accretion model suggest that both rocky materials and gas slowly accumulated for several million years, until the masses of the core and the envelope became equal. From that critical moment on, the envelope pulled in all the gas it could get, dramatically increasing the protoplanet's mass and size. This runaway growth of the protoplanet would have continued until all the available gas was used up. The result was a huge planet with an enormously thick, hydrogen-rich envelope surrounding a rocky core with 5 to 10 times the mass of Earth. This scenario could have occurred at four different distances from the Sun, thus creating the four Jovian planets.

In the core accretion model, Uranus and Neptune probably did not form at their present locations, respectively about 19 and 30 AU from the Sun. The solar nebula was too sparse at those distances to allow these planets to have grown to the present-day sizes. Instead, it is thought that Uranus and Neptune formed between 4 and 10 AU from the Sun, but were flung into larger orbits by gravitational interactions with Jupiter.

The Disk Instability Model

An alternative model for the origin of the Jovian planets, the **disk instability model,** suggests that they formed directly from the gas of the solar nebula. If the gas of the outer solar nebula was not smooth but clumpy, a sufficiently large and massive clump of gas could begin to collapse inward all on its own, like the protosun but on a smaller scale. Such a large clump would attract other gas, forming a very large planet of hydrogen and helium within just a few hundreds or thousands of years.

In the disk instability model the rocky core of a Jovian planet is *not* the original nucleus around which the planet grew; rather, it is the result of planetesimals and icy dust grains that were drawn gravitationally into the planet and then settled at the planet's center. Another difference is that in the disk instability model it is possible for Uranus and Neptune to have formed in their present-day orbits, since they would have grown much more rapidly than would be possible in the core accretion model.

Planetary scientists are actively researching whether the core accretion model or the disk instability model is the more correct description of the formation of the Jovian planets. In either case, we can explain why the terrestrial planets formed in the inner solar system while the Jovian planets formed in the outer solar system. Figure 8-13 summarizes our overall picture of the formation of the terrestrial and Jovian planets.

The Origin of Jovian Satellites and Small Bodies

Like the terrestrial planets, the Jovian planets were quite a bit hotter during their formation than they are today. A heated gas expands, so these gas-rich planets must also have been much larger than at present. As each planet cooled and contracted, it would have formed a disk like a solar nebula in miniature (see Figure 8-6). Many of the satellites of the Jovian planets, including those shown in Table 7-2, are thought to have formed from ice particles and dust grains within these disks.

Jupiter, the most massive of the planets, was almost certainly the first to form. This may explain why there are asteroids, trans-Neptunian objects, and comets (see Section 7-5). Jupiter's tremendous mass would have exerted strong gravitational forces on any nearby planetesimals. Some would have been sent crashing into the Sun, while others would have been sent crashing into the terrestrial planets to form impact craters (see Section 7-6). Still others would have been ejected completely from the solar system. Only a relatively few rocky planetesimals survived to produce the present-day population of asteroids. Indeed, the mass of all known asteroids combined is less than the mass of the Moon.

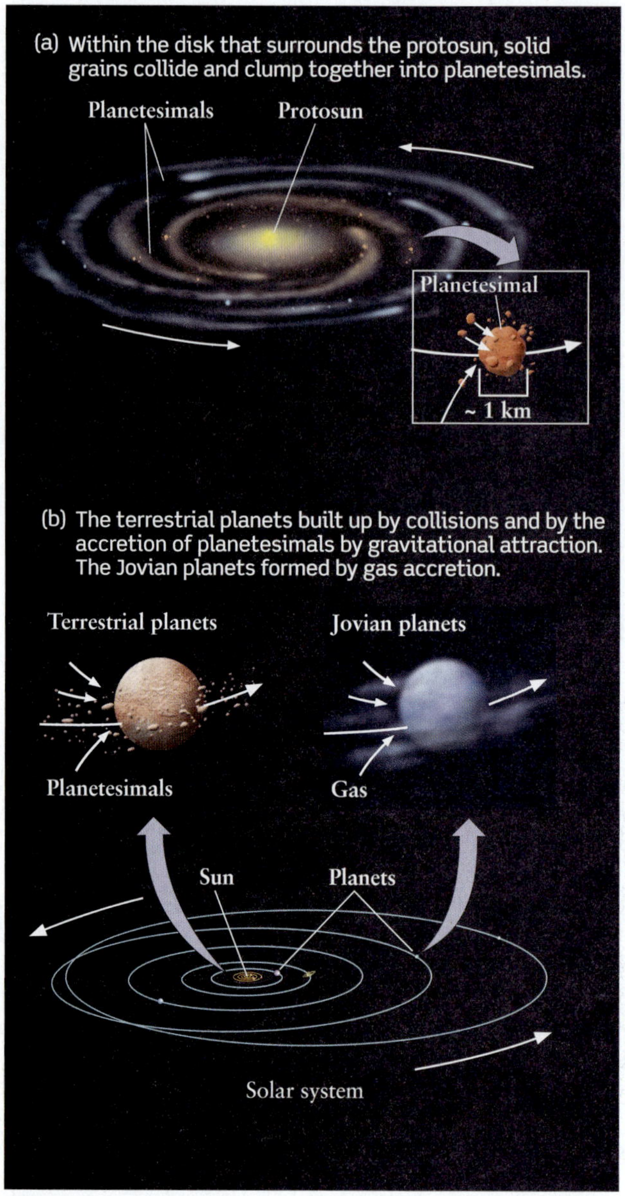
(a) Within the disk that surrounds the protosun, solid grains collide and clump together into planetesimals.
Planetesimals
Protosun
Planetesimal
~ 1 km
(b) The terrestrial planets built up by collisions and by the accretion of planetesimals by gravitational attraction. The Jovian planets formed by gas accretion.
Terrestrial planets
Jovian planets
Planetesimals
Gas
Sun
Planets
Solar system

ANIMATION 8.2

Figure 8-13

Terrestrial Versus Jovian Planet Formation

(a) Planetesimals about 1 km in size formed in the solar nebula from small dust grains sticking together. **(b)** Planetesimals in the inner solar system grouped together to form the terrestrial planets, as in Figure 8-12. In the outer solar system, the Jovian planets may have begun as terrestrial-like planets that accumulated massive envelopes of hydrogen and helium. Alternatively, the Jovian planets may have formed directly from the gas of the solar nebula.

ANIMATION 8.3

By contrast, trans-Neptunian objects (including Pluto and the other objects shown in Table 7-4) began as icy planetesimals beyond the orbit of Jupiter. The strong gravitational forces from all of the Jovian planets flung these into orbits even further from the Sun, forming the immense ring of the Kuiper belt (Figure 8-14). The images that open this chapter show two young stars surrounded by dusty disks, one seen edge-on and

Figure 8-14

The Kuiper Belt: A Dusty Debris Ring The gravitational influence of the Jovian planets pushed small, icy objects to the outer reaches of the solar system beyond Neptune. The result shown in this artist's conception is the Kuiper belt, a ring populated by trans-Neptunian objects like Pluto, icy planetesimals, and dust particles. (NASA/JPL-Caltech/T. Pyle, SSC)

the other seen face-on, that resemble the Kuiper belt. As we saw in Section 7-5, the combined mass of all trans-Neptunian objects is hundreds of times larger than that of the asteroid belt, in part because the materials that form ices are much more abundant than those that form rock.

Some of the smaller icy objects are thought to have been pushed as far as 50,000 AU from the Sun, forming a spherical "halo" around the solar system called the Oort cloud. From time to time one of the smaller chunks of ice and rock from either the Kuiper belt or the Oort cloud is deflected into the inner solar system. If one of these chunks comes close enough to the Sun, it begins to evaporate, producing a visible tail and appearing as a comet (see Figure 7-9).

The Final Stages of Solar System Evolution

While the planets, satellites, asteroids, and comets were forming, the protosun was also evolving into a full-fledged star with nuclear reactions occurring in its core (see Section 1-3). The time required for this to occur was about 10^8 years, roughly the same as that required for the formation of the terrestrial planets. Before the onset of nuclear reactions, however, the young Sun probably expelled a substantial portion of its mass into space (Figure 8-15).

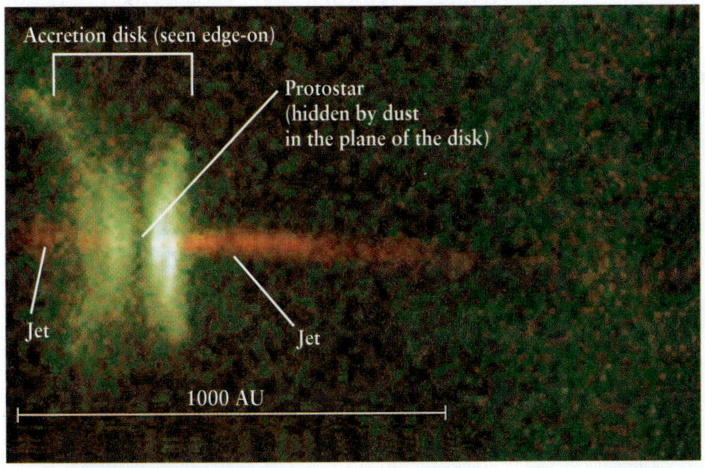

(a) Jets from a young star

(b) Winds from young stars

Figure 8-15 R I V U X G

Jets and Winds from Young Stars **(a)** This protostar in the constellation Taurus (the Bull) is ejecting matter in two immense jets directed perpendicular to the plane of the accretion disk. Red denotes light emitted by hot ionized gas in the jet, while green denotes starlight scattered by dust particles in the disk. **(b)** An outpouring of particles and radiation from the surfaces of these young stars has carved out a cavity in the surrounding dusty material. The stars lie within the Trifid Nebula in the constellation Sagittarius (the Archer). The scale of this image is about 100 times greater than that of image (a). (a: C. Burrows, the WFPC-2 Investigation Definition Team, and NASA; b: David Malin/Anglo-Australian Observatory)

Magnetic fields within the solar nebula would have funneled a portion of the nebula's mass into oppositely directed **jets** along the rotation axis of the nebula. Figure 8-15*a* shows one such jet emanating from a young star.

In addition, instabilities within the young Sun would have caused it to eject its tenuous outermost layers into space. This brief but intense burst of mass loss, observed in many young stars across the sky, is called a **T Tauri wind,** after the star in the constellation Taurus (the Bull) where it was first identified (see Figure 8-15*b*). Each of the protoplanetary disks in Figure 8-8 has a T Tauri star at its center. (The present-day Sun also loses mass from its outer layers in the form of high-speed electrons and protons, a flow called the **solar wind.** But this wind is minuscule compared with a T Tauri wind, which causes a star to lose mass 10^6 to 10^7 times faster than in the solar wind.)

With the passage of time, the combined effects of jets, the T Tauri wind, and accretion onto the planets would have swept the solar system nearly clean of gas and dust. With no more interplanetary material to gather up, the planets would have stabilized at roughly their present-day sizes, and the formation of the solar system would have been complete.

The *Cosmic Connections* figure summarizes our modern-day picture of the origin of our solar system. Prior to 1995 the only fully formed planetary system to which we could apply this model was our own. As we will see in the next section, astronomers can now further test this model on an ever-growing number of planets known to orbit other stars.

8-7 A variety of observational techniques reveal planets around other stars

If planets formed around our Sun, have they formed around other stars? That is, are there **extrasolar planets** orbiting stars other than the Sun? Our model for the formation of the planets would seem to suggest so. This model is based on the laws of physics and chemistry, which to the best of our understanding are the same throughout the universe. The discovery of a set of planets orbiting another star, with terrestrial planets in orbit close to the star and Jovian planets orbiting farther away, would be a tremendous vindication of our theory of solar system formation. It would also tell us that our planetary system is not unique in the universe. Because at least one planet in our solar system—Earth—has the ability to support life, perhaps other planetary systems could also harbor living organisms.

Several stars are known to have systems of two or more planets, but none of these resemble our own solar system

Since 1995 astronomers have in fact discovered over 340 planets orbiting stars other than the Sun. However, none of these extrasolar planets is a terrestrial planet. Most of them are more massive than Jupiter, and some are in eccentric, noncircular orbits quite unlike planetary orbits in our solar system. To appreciate how remarkable these discoveries are, we must look at the process that astronomers go through to search for extrasolar planets.

Searching for Extrasolar Planets

It is very difficult to make direct observations of planets orbiting other stars. The problem is that planets are small and dim compared with stars; at visible wavelengths, the Sun is 10^9 times brighter than Jupiter and 10^{10} times brighter than Earth. A hypothetical planet orbiting a distant star, even a planet 10 times larger than Jupiter, would be lost in the star's glare as seen through even the largest telescope on Earth.

Instead, indirect methods are used to search for extrasolar planets. One very powerful method is to search for stars that appear to "wobble" (Figure 8-16). If a star has a planet, it is not

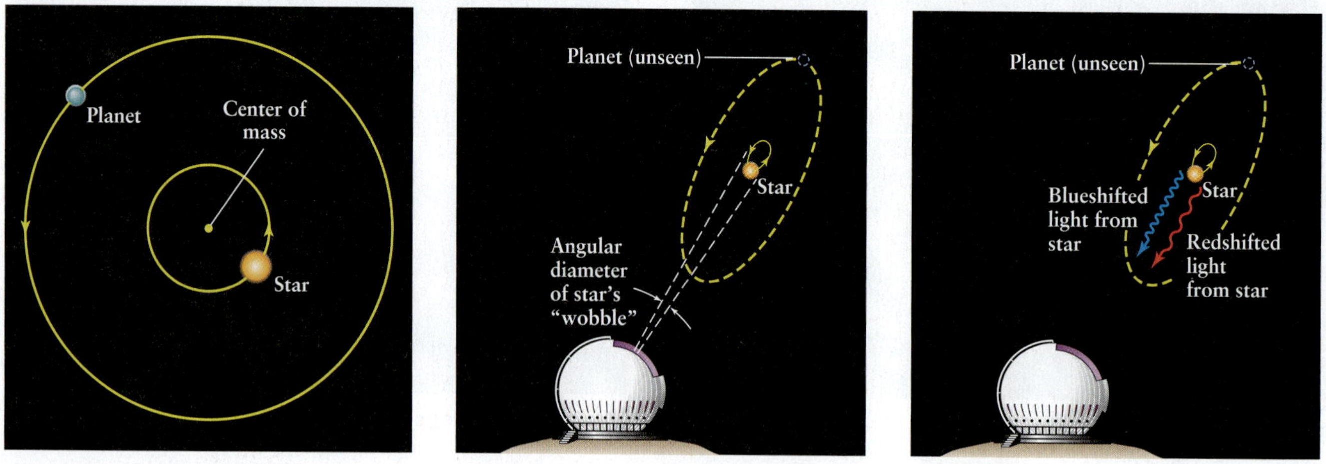

(a) A star and its planet (b) The astrometric method (c) The radial velocity method

Figure 8-16

Detecting a Planet by Measuring Its Parent Star's Motion
(a) A planet and its star both orbit around their common center of mass, always staying on opposite sides of this point. Even if the planet cannot be seen, its presence can be inferred if the star's motion can be detected. **(b)** The astrometric method of detecting the unseen planet involves making direct measurements of the star's orbital motion. **(c)** In the radial velocity method, astronomers measure the Doppler shift of the star's spectrum as it moves alternately toward and away from Earth. The amount of Doppler shift determines the size of the star's orbit, which in turn tells us about the unseen planet's orbit.

COSMIC CONNECTIONS

Formation of the Solar System

An overview of our present-day understanding of how the solar system formed.

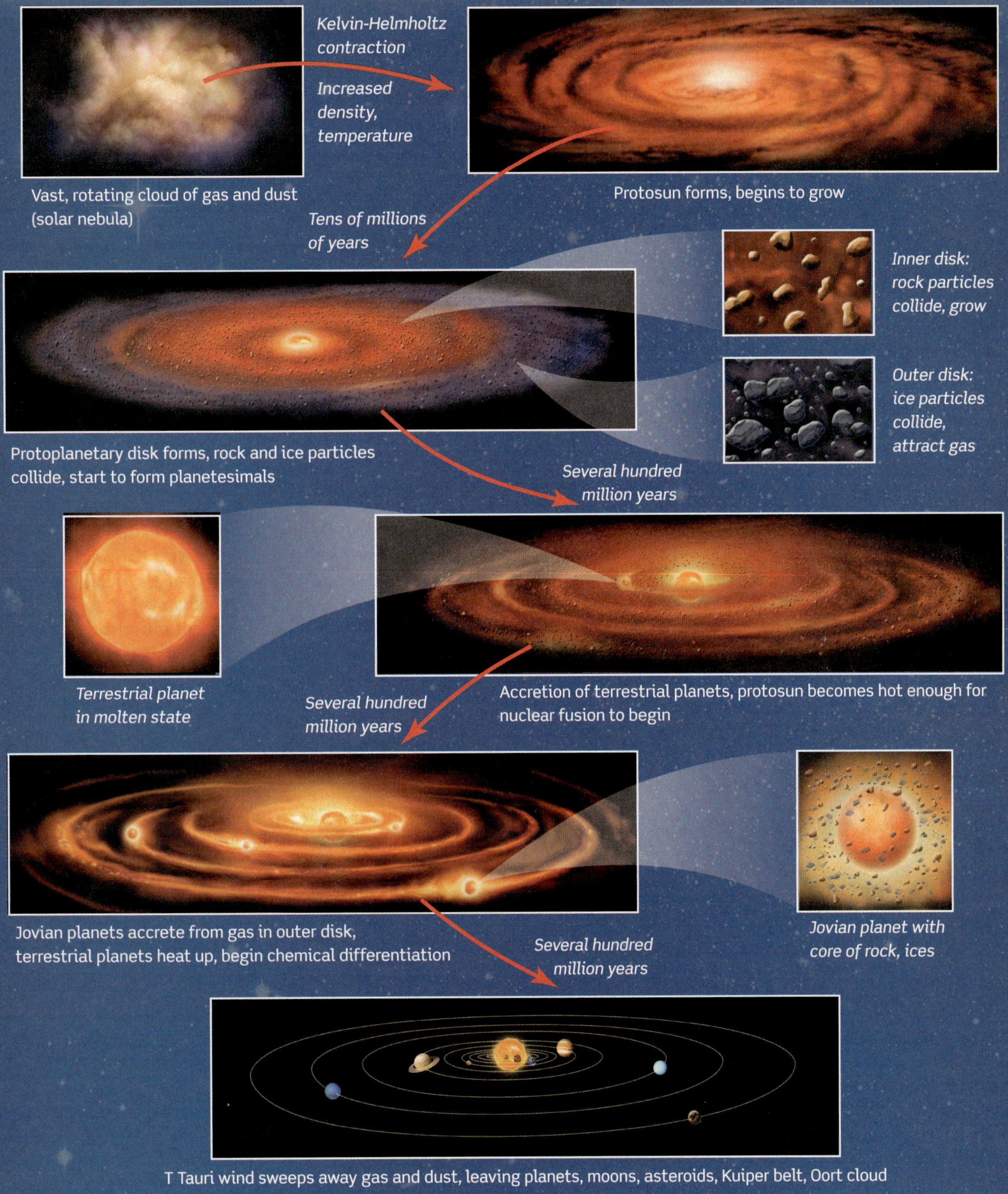

Kelvin-Helmholtz contraction
Increased density, temperature
Vast, rotating cloud of gas and dust (solar nebula)
Protosun forms, begins to grow
Tens of millions of years
Inner disk: rock particles collide, grow
Outer disk: ice particles collide, attract gas
Protoplanetary disk forms, rock and ice particles collide, start to form planetesimals
Several hundred million years
Terrestrial planet in molten state
Several hundred million years
Accretion of terrestrial planets, protosun becomes hot enough for nuclear fusion to begin
Jovian planet with core of rock, ices
Jovian planets accrete from gas in outer disk, terrestrial planets heat up, begin chemical differentiation
Several hundred million years
T Tauri wind sweeps away gas and dust, leaving planets, moons, asteroids, Kuiper belt, Oort cloud

quite correct to say that the planet orbits the star. Rather, both the planet and the star move in elliptical orbits around a point called the **center of mass.** Imagine the planet and the star as sitting at opposite ends of a very long seesaw; the center of mass is the point where you would have to place the fulcrum in order to make the seesaw balance. Because of the star's much greater mass, the center of mass is much closer to the star than to the planet. Thus, while the planet may move in an orbit that is hundreds of millions of kilometers across, the star will move in a much smaller orbit (Figure 8-16*a*).

As an example, the Sun and Jupiter both orbit their common center of mass with an orbital period of 11.86 years. (Jupiter has more mass than the other eight planets put together, so it is a reasonable approximation to consider the Sun's wobble as being due to Jupiter alone.) Jupiter's orbit has a semimajor axis of 7.78×10^8 km, while the Sun's orbit has a much smaller semimajor axis of 742,000 km. The Sun's radius is 696,000 km, so the Sun slowly wobbles around a point not far outside its surface. If astronomers elsewhere in the Galaxy could detect the Sun's wobbling motion, they could tell that there was a large planet (Jupiter) orbiting the Sun. They could even determine the planet's mass and the size of its orbit, even though the planet itself was unseen.

Detecting the wobble of other stars is not an easy task. One approach to the problem, called the **astrometric method,** involves making very precise measurements of a star's position in the sky relative to other stars. The goal is to find stars whose positions change in a cyclic way (Figure 8-16*b*). The measurements must be made with very high accuracy (0.001 arcsec or better) and, ideally, over a long enough time to span an entire orbital period of the star's motion. Because this technique is so challenging to put into practice, it has not yet led to any confirmed detections of extrasolar planets.

ANIMATION 8.5

A different approach to the problem is the **radial velocity method** (Figure 8-16*c*). This method is based on the Doppler effect, which we described in Section 5-9. A wobbling star will alternately move away from and toward Earth. The star's movement will cause the dark absorption lines in the star's spectrum (see Figure 5-13) to change their wavelengths in a periodic fashion. When the star is moving away from us, its spectrum will undergo a redshift to longer wavelengths. When the star is approaching, there will be a blueshift of the spectrum to shorter wavelengths. These wavelength shifts are very small because the star's motion around its orbit is quite slow. As an example, the Sun moves around its small orbit at only 12.5 m/s (45 km/h, or 28 mi/h). If the Sun were moving directly toward an observer at this speed, the hydrogen absorption line at a wavelength of 656 nm in the Sun's spectrum would be shifted by only 2.6×10^{-5} nm, or about 1 part in 25 million. Detecting these tiny shifts requires extraordinarily careful measurements and painstaking data analysis.

In 1995, Michel Mayor and Didier Queloz of the Geneva Observatory in Switzerland announced that by using the radial velocity method, they had discovered a planet orbiting the star 51 Pegasi, which is 47.9 light-years from the Sun in the constellation Pegasus (the Winged Horse). Their results were soon confirmed by the team of Geoff Marcy of San Francisco State University and Paul Butler of the University of California, Berkeley, using observations made at the University of California's Lick Observatory. For the first time, solid evidence had been found for a planet orbiting a star like our own Sun. Astronomers have since used the radial velocity method to discover planets orbiting more than 160 other stars at distances up to 600 light-years from the Sun. In at least 20 cases the star's "wobble" is too complex to have been caused by a single planet, so in these cases the star has a multiple-planet system. Figure 8-17 depicts a selection of these extrasolar planetary discoveries.

Extrasolar Planets: Surprising Masses and Orbits

Most of the extrasolar planets discovered by the radial velocity method have masses comparable with or larger than that of Saturn, and thus are presumably Jovian planets made primarily of hydrogen and helium. (This is difficult to confirm directly, because the spectra of the planets are very faint.) According to the picture we presented in Section 8-5, such Jovian planets would be expected to orbit relatively far from their stars, where temperatures were low enough to allow the buildup of a massive envelope of hydrogen and helium gas. But as Figure 8-17 shows, many extrasolar planets are in fact found orbiting very *close* to their stars. For example, the planet orbiting 51 Pegasi has a mass at least 0.46 times that of Jupiter but orbits only 0.052 AU from its star with an orbital period of just 4.23 days. In our own solar system, this orbit would lie well inside the orbit of Mercury!

Another surprising result is that many of the extrasolar planets found so far have orbits with very large eccentricities (see Figure 4-10*b*). As an example, the planet around the star 16 Cygni B has an orbital eccentricity of 0.67; its distance from the star varies between 0.55 AU and 2.79 AU. This eccentricity is quite unlike planetary orbits in our own solar system, where no planet has an orbital eccentricity greater than 0.25.

Do these observations mean that our picture of how planets form is incorrect? If Jupiterlike extrasolar planets such as that orbiting 51 Pegasi formed close to their stars, the mechanism of their formation must have been very different from that which operated in our solar system.

But another possibility is that extrasolar planets actually formed at large distances from their stars and have migrated inward since their formation. If enough gas and dust remain in a disk around a star after its planets form, interactions between the disk material and an orbiting planet will cause the planet to lose energy and to spiral inward toward the star around which it orbits. The planet's inward migration can eventually stop because of subtle gravitational effects from the disk or from the star. Gravitational interactions between the planet and the disk, or between planets in the same planetary system, could also have forced an extrasolar planet into a highly eccentric, noncircular orbit.

ANIMATION 8.6

One piece of evidence that young planets may spiral inward toward their parent stars is the spectrum of the star HD 82943, which has at least two planets orbiting it. The spectrum shows that this star's atmosphere contains a rare form of lithium that is found in planets, but which is destroyed in stars by nuclear reactions within 30 million years. The presence of this exotic form of lithium, known as ^{6}Li, means that in the past HD 82943 was orbited by at least one other planet.

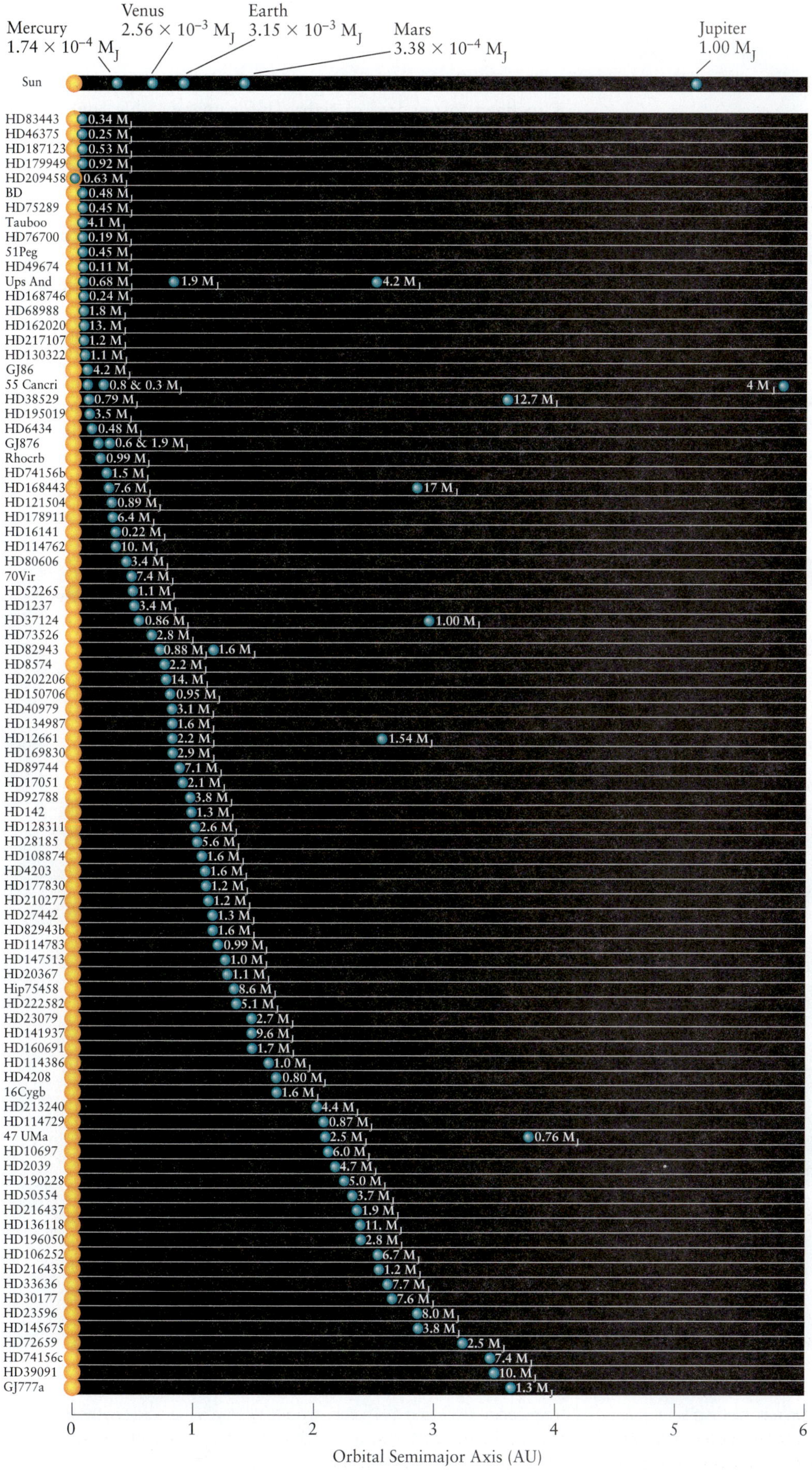

WEB LINK 8.9

Figure 8-17

A Selection of Extrasolar Planets This figure summarizes what we know about planets orbiting a number of other stars. The star name is given at the left of each line. Each planet is shown at its average distance from its star (equal to the semimajor axis of its orbit). The mass of each planet—actually a lower limit—is given as a multiple of Jupiter's mass (M_J), equal to 318 Earth masses. Comparison with our own solar system (at the top of the figure) shows how closely many these extrasolar planets orbit their stars. (Adapted from the California and Carnegie Planet Search)

This planet vanished when it spiraled too close to the star, was vaporized, and was swallowed whole.

Are all stars equally likely to have planets? The answer appears to be no: Only about 5% of the stars surveyed with the radial velocity method show evidence of planets. However, the percentage is 10% to 20% for stars in which heavy elements are at least as abundant as in the Sun. The conclusion is that heavy elements appear to play an important role in the evolution of planets from an interstellar cloud. This agrees with the idea discussed in Section 8-5 that planetesimals (which are composed of heavy elements) are the building blocks of planets.

Analyzing Extrasolar Planets with the Transit Method

The radial velocity method gives us only a partial picture of what extrasolar planets are like. It cannot give us precise values for a planet's mass, only a lower limit. (An exact determination of the mass would require knowing how the plane of the planet's orbit is inclined to our line of sight. Unfortunately, this angle is not known because the planet itself is unseen.) Furthermore, the radial velocity method by itself cannot tell us the diameter of a planet or what the planet is made of.

A technique called the **transit method** makes it possible to fill in these blanks about the properties of certain extrasolar planets. This method looks for the rare situation in which a planet comes between us and its parent star, an event called a **transit** (Figure 8-18). As in a partial solar eclipse (Section 3-5), this causes a small but measurable dimming of the star's light. If a transit is seen, the orbit must be nearly edge-on to our line of sight. Knowing the orientation of the orbit, the information obtained by radial velocity measurements of the star tells us the true mass (not just a lower limit) of the orbiting planet. About 5% of all known extrasolar planets undergo transits of their parent

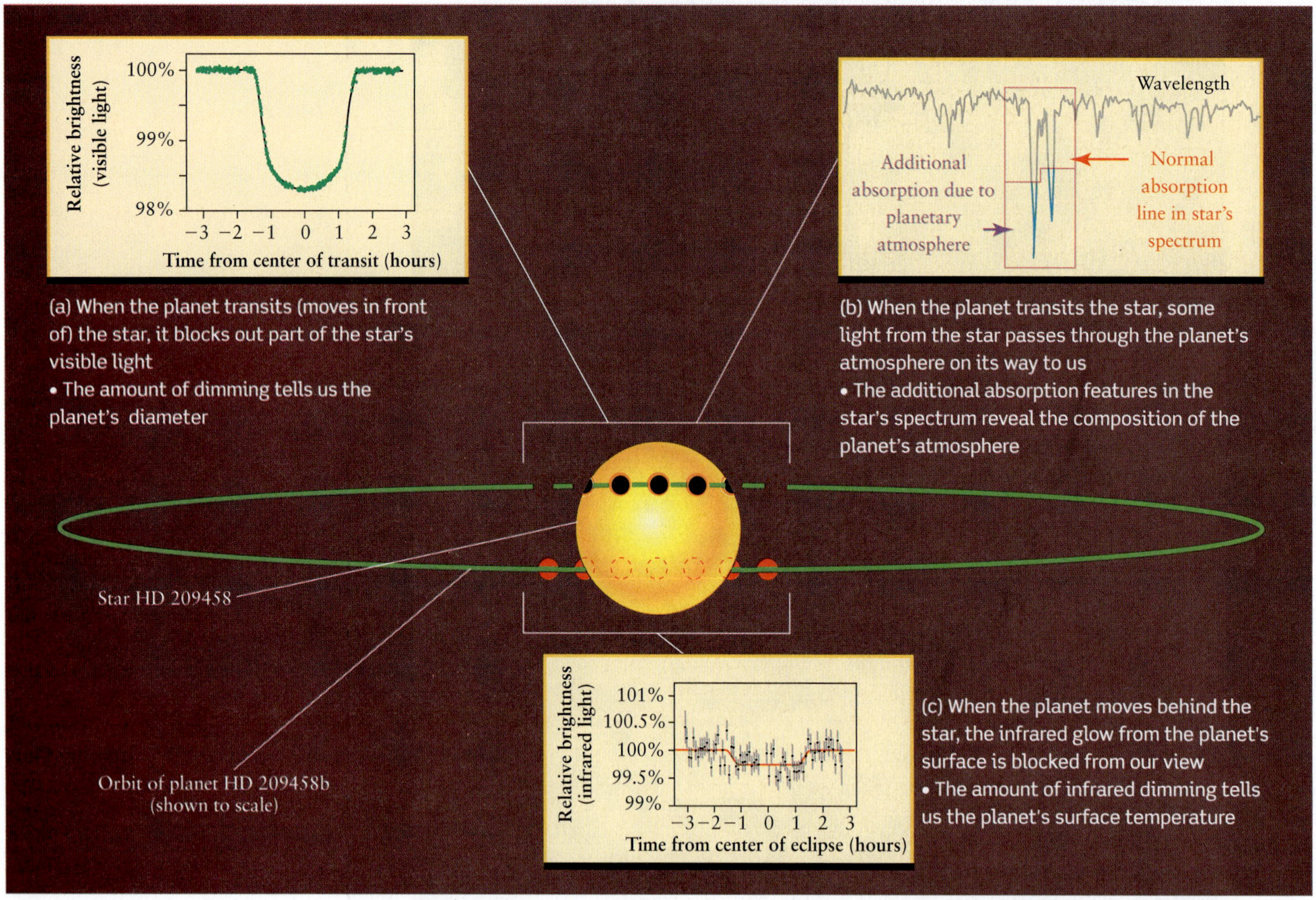
Relative brightness (visible light)
100%
99%
98%
−3 −2 −1 0 1 2 3
Time from center of transit (hours)
(a) When the planet transits (moves in front of) the star, it blocks out part of the star's visible light
• The amount of dimming tells us the planet's diameter
Wavelength
Additional absorption due to planetary atmosphere
Normal absorption line in star's spectrum
(b) When the planet transits the star, some light from the star passes through the planet's atmosphere on its way to us
• The additional absorption features in the star's spectrum reveal the composition of the planet's atmosphere
Star HD 209458
Orbit of planet HD 209458b (shown to scale)
Relative brightness (infrared light)
101%
100.5%
100%
99.5%
99%
−3 −2 −1 0 1 2 3
Time from center of eclipse (hours)
(c) When the planet moves behind the star, the infrared glow from the planet's surface is blocked from our view
• The amount of infrared dimming tells us the planet's surface temperature

Figure 8-18

A Transiting Extrasolar Planet If the orbit of an extrasolar planet is nearly edge-on to our line of sight, like the planet that orbits the star HD 209458, we can learn about the planet's **(a)** diameter, **(b)** atmospheric composition, and **(c)** surface temperature. (S. Seager and C. Reed, *Sky and Telescope*; H. Knutson, D. Charbonneau, R. W. Noyes (Harvard-Smithsonian CfA), T. M. Brown (HAO/NCAR), and R. L. Gilliland (STScI); A. Feild (STScI); NASA/JPL-Caltech/D. Charbonneau, Harvard-Smithsonian CfA)

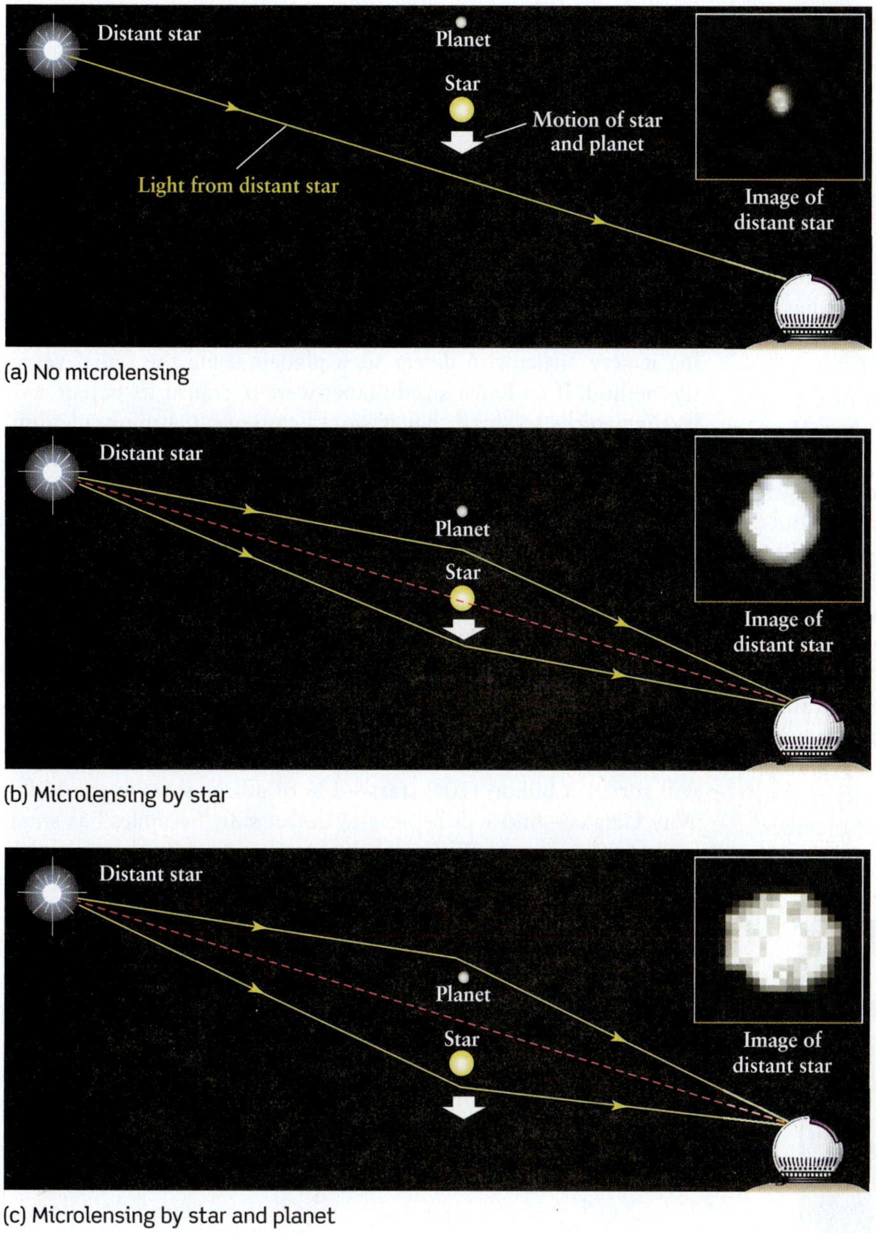

ANIMATION 8.8

Figure 8-19

Microlensing Reveals an Extrasolar Planet

(a) A star with a planet drifts across the line of sight between a more distant star and a telescope on Earth. **(b)** The gravity of the closer star bends the light rays from the distant star, focusing the distant star's light and making it appear brighter. **(c)** The gravity of the planet causes a second increase in the distant star's brightness.

1) diameter
2) composition of atmosphere
3) radiation emitted

star, and all of these planets turn out to have masses comparable to that of Jupiter. These results reinforce the idea that other stars can end up with planets much like those in our solar system.

ANIMATION 8.8

There are three other benefits of the transit method. One, the amount by which the star is dimmed during the transit depends on how large the planet is, and so tells us the planet's diameter (Figure 8-18*a*). Two, during the transit the star's light passes through the planet's atmosphere, where certain wavelengths are absorbed by the atmospheric gases. This absorption affects the spectrum of starlight that we measure and so allows us to determine the composition of the planet's atmosphere (Figure 8-18*b*). Three, with an infrared telescope it's possible to detect a slight dimming when the planet goes *behind* the star (Figure 8-18*c*). That is because the planet emits infrared radiation due to its own temperature, and this radiation is blocked when the planet is behind the star. Measuring the amount of dimming tells us the amount of radiation emitted by the planet, which in turn tells us the planet's surface temperature.

One example of what the transit method can tell us is about the planet that orbits the star HD 209458, which is 153 light-years from the Sun. The mass of this planet is 0.69 that of Jupiter, but its diameter is 1.32 times larger than that of Jupiter. This low-mass, large-volume planet thus has only one-quarter the density of Jupiter. It is also very hot: Because the planet orbits just 0.047 AU from the star, its surface temperature is a torrid 1130 K (860°C, or 1570°F). Observations of the spectrum during a transit reveal the presence of hydrogen, carbon, oxygen, and sodium in a planetary atmosphere that is literally evaporating away in the intense starlight. Discoveries such as these give insight into the exotic circumstances that can be found in other planetary systems.

Other Techniques for Discovering Extrasolar Planets

In 2004 astronomers began to use a property of space discovered by Albert Einstein as a tool for detecting extrasolar planets. Einstein's general theory of relativity makes the remarkable statement that a star's gravity can deflect the path of a light beam just as it deflects the path of a planet or spacecraft. If a star drifts through the line of sight between Earth and a more distant star, the closer star's gravity acts like a lens that focuses the more distant star's light. Such **microlensing** causes the distant star's image as seen in a telescope to become brighter. If the closer star has a planet, the planet's gravity will cause a secondary brightening whose magnitude depends on the planet's mass (Figure 8-19). Using this technique astronomers have detected a handful of extrasolar planets at distances up to 21,000 light-years, and expect to find many more.

Since 2004 astronomers have also been able to record images of a few extrasolar planets. At infrared wavelengths a star outshines a Jupiter-sized planet by only about 100 to 1, compared to 10^9 to 1 at visible wavelengths. Hence, by using an infrared telescope aided by adaptive optics (Section 6-3) it is possible to resolve both a star and its planet if the two are sufficiently far apart. Figure 8-20 shows an infrared image of the star 2M1207 and its planet. Unlike planets detected with the radial velocity or transit method, it is impractical to watch this planet go through a complete orbit: It is so far from its parent star that the orbital period is more than a thousand years. Instead, astronomers verify that it is a planet by confirming that the star and planet move together through space, as they must do if they are held together by their mutual gravitational attraction. While it is not possible to measure the planet's mass directly, the spectrum and brightness of the planet tell astronomers its surface temperature and diameter.

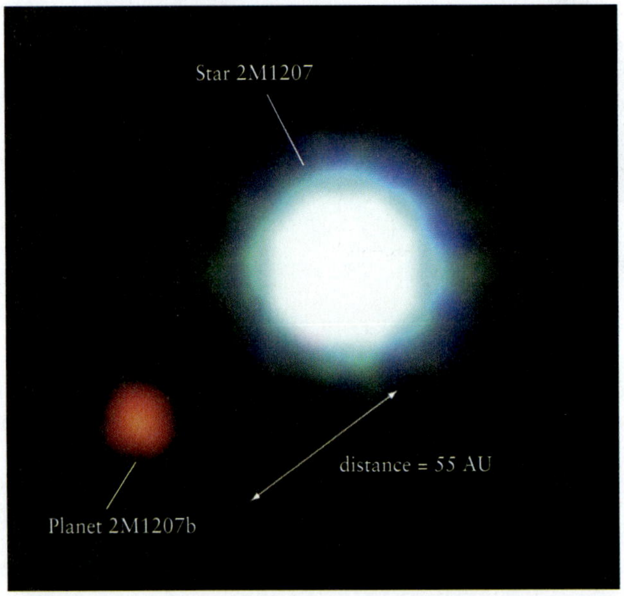

Figure 8-20 R I V U X G

Imaging an Extrasolar Planet This infrared image from the European Southern Observatory shows the star 2M1207 and a planet with about 1.5 times the diameter of Jupiter. First observed in 2004, this extrasolar planet was the first to be visible in a telescopic image. 2M1207 and its planet lie about 170 light-years from the Sun in the constellation Hydra (the Water Snake). (ESO/VLT/NACO)

The Search for Earth-sized Planets

If there are planets orbiting other stars, are there any that resemble our own Earth? As of this writing we do not know the answer to this question. An Earth-sized planet has such weak gravity that it can produce only a tiny "wobble" in its parent star, making it very difficult to detect such planets using the radial velocity method. If an Earth-sized planet were to transit its parent star, its diameter is so small that it would cause only a miniscule dimming of the star's light. For these reasons, as of this writing, astronomers have been unable to detect extrasolar planets less massive than 7.5 Earth masses.

In the near future, however, orbiting telescopes being planned by both NASA and the European Space Agency (ESA) will search for terrestrial planets around other stars. One such orbiting telescope, launched in March 2009, is NASA's Kepler, which uses a one-meter telescope and sensitive detectors to look for transits of Earth-sized planets within a narrow patch of sky.

An even broader search for extrasolar planets will be carried out by ESA's Gaia mission, scheduled for launch in 2011. Gaia will survey a billion (10^9) stars—1% of all the stars in the Milky Way Galaxy—and will be able to detect star "wobbles" as small as 2×10^{-6} arcsec. With such high-precision data the astrometric method (Figure 8-16*b*) will finally come into its own. Astronomers estimate that Gaia will find from 10,000 to 50,000 Jupiter-sized planets. If our solar system is fairly typical, a substantial number of these planets should move around their parent stars in orbits like those of the Jovian planets around the Sun, with semimajor axes greater than 3 AU and low eccentricities. If Gaia finds such planetary systems that resemble our own, they will be natural places to look for Earth-sized planets. The next few decades may tell whether our own solar system is an exception or just one of a host of similar systems throughout the Galaxy.

Key Words

Term preceded by an asterisk (*) is discussed in Box 8-1.

accretion, p. 194
astrometric method (for detecting extrasolar planets), p. 200
atomic number, p. 188
center of mass, p. 200
chemical differentiation, p. 194
chondrule, p. 194
condensation temperature, p. 193
conservation of angular momentum, p. 191
core accretion model, p. 195
disk instability model, p. 196
extrasolar planet, p. 198
*half-life, p. 190
interstellar medium, p. 187
jets, p. 198
Kelvin-Helmholtz contraction, p. 191
meteorite, p. 189
microlensing, p. 204
nebulosity, p. 187
nebular hypothesis, p. 190
planetesimal, p. 194
protoplanet, p. 194

protoplanetary disk (proplyd), p. 192
protosun, p. 191
radial velocity method (for detecting extrasolar planets), p. 200
radioactive dating, p. 189
radioactive decay, p. 189
solar nebula, p. 190
solar wind, p. 198
T Tauri wind, p. 198
transit, p. 202
transit method (for detecting extrasolar planets), p. 202

Key Ideas

Models of Solar System Formation: The most successful model of the origin of the solar system is called the nebular hypothesis. According to this hypothesis, the solar system formed from a cloud of interstellar material called the solar nebula. This occurred 4.56 billion years ago (as determined by radioactive dating).

The Solar Nebula and Its Evolution: The chemical composition of the solar nebula, by mass, was 98% hydrogen and helium (elements that formed shortly after the beginning of the universe) and 2% heavier elements (produced much later in the centers of stars, and cast into space when the stars died). The heavier elements were in the form of ice and dust particles.

• The nebula flattened into a disk in which all the material orbited the center in the same direction, just as do the present-day planets.

Formation of the Planets and Sun: The terrestrial planets, the Jovian planets, and the Sun followed different pathways to formation.

• The four terrestrial planets formed through the accretion of dust particles into planetesimals, then into larger protoplanets.

• In the core accretion model, the four Jovian planets began as rocky protoplanetary cores, similar in character to the terrestrial planets. Gas then accreted onto these cores in a runaway fashion.

• In the alternative disk instability model, the Jovian planets formed directly from the gases of the solar nebula. In this model the cores formed from planetesimals falling into the planets.

• The Sun formed by gravitational contraction of the center of the nebula. After about 10^8 years, temperatures at the protosun's center became high enough to ignite nuclear reactions that convert hydrogen into helium, thus forming a true star.

Extrasolar Planets: Astronomers have discovered planets orbiting other stars.

• Most of these planets are detected by the "wobble" of the stars around which they orbit.

• A small but growing number of extrasolar planets have been discovered by the transit method, by microlensing, and direct imaging.

• Most of the extrasolar planets discovered to date are quite massive and have orbits that are very different from planets in our solar system.

Questions

Review Questions

1. Describe three properties of the solar system that are thought to be a result of how the solar system formed.
2. The graphite in your pencil is a form of carbon. Where were these carbon atoms formed?
3. What is the interstellar medium? How does it become enriched over time with heavy elements?
4. What is the evidence that other stars existed before our Sun was formed?
5. Why are terrestrial planets smaller than Jovian planets?
6. How do radioactive elements make it possible to determine the age of the solar system? What are the oldest objects that have been found in the solar system?
7. What is the tidal hypothesis? What aspect of the solar system was it designed to explain? Why was this hypothesis rejected?
8. What is the nebular hypothesis? Why is this hypothesis accepted?
9. What was the protosun? What caused it to shine? Into what did it evolve?
10. Why is it thought that a disk appeared in the solar nebula?
11. What are proplyds? What do they tell us about the plausibility of our model of the solar system's origin?
12. What is meant by a substance's condensation temperature? What role did condensation temperatures play in the formation of the planets?
13. What is a planetesimal? How did planetesimals give rise to the terrestrial planets?
14. (**a**) What is meant by accretion? (**b**) Why are the terrestrial planets denser at their centers than at their surfaces?
15. If hydrogen and helium account for 98% of the mass of all the atoms in the universe, why aren't Earth and the Moon composed primarily of these two gases?
16. What is a chondrule? How do we know they were not formed by the ambient heat of the solar nebula?
17. Why did the terrestrial planets form close to the Sun while the Jovian planets formed far from the Sun?
18. What are the competing models of how the Jovian planets formed?
19. Explain how our current understanding of the formation of the solar system can account for the following characteristics of the solar system: (**a**) All planetary orbits lie in nearly the same plane. (**b**) All planetary orbits are nearly circular. (**c**) The planets orbit the Sun in the same direction in which the Sun itself rotates.
20. Why is the combined mass of all the asteroids so small? Why is the combined mass of all trans-Neptunian objects so much greater than that for the asteroids?
21. Explain why most of the satellites of Jupiter orbit that planet in the same direction that Jupiter rotates.
22. What is the radial velocity method used to detect planets orbiting other stars? Why is it difficult to use this method to detect planets like Earth?
23. Summarize the differences between the planets of our solar system and those found orbiting other stars.

24. Is there evidence that planets have fallen into their parent stars? Explain your answer.
25. What does it mean for a planet to transit a star? What can we learn from such events?
26. What is microlensing? How does it enable astronomers to discover extrasolar planets?
27. A 1999 news story about the discovery of three planets orbiting the star Upsilon Andromedae ("Ups And" in Figure 8-17) stated that "the newly discovered galaxy, with three large planets orbiting a star known as Upsilon Andromedae, is 44 light-years away from Earth." What is wrong with this statement?

Advanced Questions

Questions preceded by an asterisk (*) involve topics discussed in Box 4-4 or Box 8-1.

Problem-solving tips and tools

The volume of a disk of radius r and thickness t is $\pi r^2 t$. Box 1-1 explains the relationship between the angular size of an object and its actual size. An object moving at speed v for a time t travels a distance $d = vt$; Appendix 6 includes conversion factors between different units of length and time. To calculate the mass of 70 Virginis or the orbital period of the planet around 2M1207, review Box 4-4. Section 5-4 describes the properties of blackbody radiation.

28. Figure 8-4 shows that carbon, nitrogen, and oxygen are among the most abundant elements (after hydrogen and helium). In our solar system, the atoms of these elements are found primarily in the molecules CH_4 (methane), NH_3 (ammonia), and H_2O (water). Explain why you suppose this is.
29. (a) If Earth had retained hydrogen and helium in the same proportion to the heavier elements that exist elsewhere in the universe, what would its mass be? Give your answer as a multiple of Earth's actual mass. Explain your reasoning. (b) How does your answer to (a) compare with the mass of Jupiter, which is 318 Earth masses? (c) Based on your answer to (b), would you expect Jupiter's rocky core to be larger, smaller, or the same size as Earth? Explain your reasoning.

*30. If you start with 0.80 kg of radioactive potassium (^{40}K), how much will remain after 1.3 billion years? After 2.6 billion years? After 3.9 billion years? How long would you have to wait until there was *no* ^{40}K remaining?

*31. Three-quarters of the radioactive potassium (^{40}K) originally contained in a certain volcanic rock has decayed into argon (^{40}Ar). How long ago did this rock form?

32. Suppose you were to use the Hubble Space Telescope to monitor one of the protoplanetary disks shown in Figure 8-8*b*. Over the course of 10 years, would you expect to see planets forming within the disk? Why or why not?
33. The protoplanetary disk at the upper right of Figure 8-8*b* is seen edge-on. The diameter of the disk is about 700 AU. (a) Make measurements on this image to determine the thickness of the disk in AU. (b) Explain why the disk will continue to flatten as time goes by.

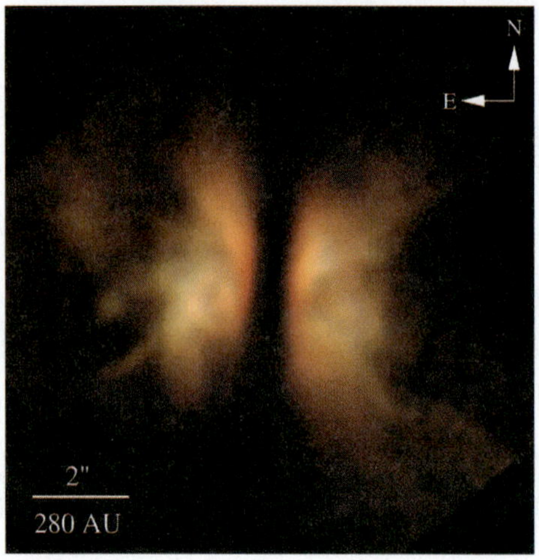

R I V U X G

(Courtesy of D. Padgett and W. Brandner, IPAC/Caltech; K. Stapelfeldt, JPL; and NASA)

34. The accompanying infrared image shows IRAS 04302+2247, a young star that is still surrounded by a disk of gas and dust. The scale bar at the lower left of the image shows that at the distance of IRAS 04302+2247, an angular size of 2 arcseconds corresponds to a linear size of 280 AU. Use this information to find the distance to IRAS 04302+2247.
35. The image accompanying Question 34 shows a dark, opaque disk of material surrounding the young star IRAS 04302+2247. The disk is edge-on to our line of sight, so it appears as a dark band running vertically across this image. The material to the left and right of this band is still falling onto the disk. (a) Make measurements on this image to determine the diameter of the disk in AU. Use the scale bar at the lower left of this image. (b) If the thickness of the disk is 50 AU, find its volume in cubic meters. (c) The total mass of the disk is perhaps 2×10^{28} kg (0.01 of the mass of the Sun). How many atoms are in the disk? Assume that the disk is all hydrogen. A single hydrogen atom has a mass of 1.673×10^{-27} kg. (d) Find the number of atoms per cubic meter in the disk. Is the disk material thick or thin compared to the air that you breathe, which contains about 5.4×10^{25} atoms per cubic meter?
36. If the material in the jet shown in Figure 8-15*a* is moving at 200 km/s, how long ago was the material at the far right-hand end of the jet ejected from the star? Give your answer in years. (You will need to make measurements on the image.)

*37. The planet discovered orbiting the star 70 Virginis ("70Vir" in Figure 8-17), 59 light-years from Earth, moves in an orbit with semimajor axis 0.48 AU and eccentricity 0.40. The period of the orbit is 116.7 days. Find the mass of 70 Virginis. Compare your answer with the mass of the Sun. (*Hint:* The planet has far less mass than the star.)

*38. Because of the presence of Jupiter, the Sun moves in a small orbit of radius 742,000 km with a period of 11.86 years. (a) Calculate the Sun's orbital speed in meters per second. (b) An astronomer on a hypothetical planet orbiting the star

Vega, 25 light-years from the Sun, wants to use the astrometric method to search for planets orbiting the Sun. What would be the angular diameter of the Sun's orbit as seen by this alien astronomer? Would the Sun's motion be discernible if the alien astronomer could measure positions to an accuracy of 0.001 arcsec? (c) Repeat part (b), but now let the astronomer be located on a hypothetical planet in the Pleiades star cluster, 360 light-years from the Sun. Would the Sun's motion be discernible to this astronomer?

39. (a) Figure 8-18*c* shows how astronomers determine that the planet of HD 209458 has a surface temperature of 1130 K. Treating the planet as a blackbody, calculate the wavelength at which it emits most strongly. (b) The star HD 209458 itself has a surface temperature of 6030 K. Calculate its wavelength of maximum emission, assuming it to be a blackbody. (c) If a high-resolution telescope were to be used in an attempt to record an image of the planet orbiting HD 209458, would it be better for the telescope to use visible or infrared light? Explain your reasoning.

*40. (a) The star 2M1207 shown in Figure 8-20 is 170 light-years from Earth. Find the angular distance between this star and its planet as seen from Earth. Express your answer in arcseconds. (b) The mass of 2M1207 is 0.025 that of the Sun; the mass of the planet is very much smaller. Calculate the orbital period of the planet, assuming that the distance between the star and planet shown in Figure 8-20 is the semimajor axis of the orbit. Is it possible that an astronomer could observe a complete orbit in one lifetime?

Discussion Questions

41. Propose an explanation why the Jovian planets are orbited by terrestrial-like satellites.
42. Suppose that a planetary system is now forming around some protostar in the sky. In what ways might this planetary system turn out to be similar to or different from our own solar system? Explain your reasoning.
43. Suppose astronomers discovered a planetary system in which the planets orbit a star along randomly inclined orbits. How might a theory for the formation of that planetary system differ from that for our own?

Web/eBook Questions

44. Search the World Wide Web for information about recent observations of protoplanetary disks. What insights have astronomers gained from these observations? Is there any evidence that planets have formed within these disks?
45. In 2000, extrasolar planets with masses comparable to that of Saturn were first detected around the stars HD 16141 (also called 79 Ceti) and HD 46375. Search the World Wide Web for information about these "lightweight" planets. Do these planets move around their stars in the same kind of orbit as Saturn follows around the Sun? Why do you suppose this is? How does the discovery of these planets reinforce the model of planet formation described in this chapter?
46. In 2006 a planet called XO-1b was discovered using the transit method. Search the World Wide Web for information about this planet and how it was discovered. What unusual kind of telescope was used to make this discovery? Have other extrasolar planets been discovered using the same kind of telescope?

Activities

Observing Projects

47. Use the *Starry Night Enthusiast*™ program to investigate stars that have planets orbiting them. First display the entire celestial sphere (select **Guides > Atlas** in the **Favourites** menu). Then use the **Find** pane to find and center each of the stars listed below. To do this, click the magnifying glass icon on the left side of the edit box at the top of the Find pane and select **Star** from the drop-down menu; then type the name of the star in the edit box and press the Enter or Return key on the keyboard. Click on the **Info** tab on the left-hand side of the *Starry Night Enthusiast*™ window for full information about the star. For each star, record the luminosity of the star (a measure of the star's total light output). How far from Earth is each star? Which stars are more luminous than the Sun? Which are less luminous? How do you think these differences would have affected temperatures in the nebula in which each star's planets formed (see Figure 8-10)? (i) 47 Ursae Majoris; (ii) 51 Pegasi; (iii) 70 Virginis; (iv) Rho Coronae Borealis.
48. Use the *Starry Night Enthusiast*™ program to examine stars that have planets. Select **Stars > ExtraSolar Planets** in the **Favourites** menu. In the star map that appears, each circled star has one or more planets. (You can zoom in and zoom out using the buttons at the right side of the toolbar. Click the **Info** tab at the left of the main window to open the Info pane and open the **Other Data** panel. Then click on a circled star to learn more about its properties. Note that the information given for each star includes the *apparent magnitude*, which is a measure of how bright each star appears as seen from Earth. Apparent magnitude uses a "backwards" scale: The greater the value of the apparent magnitude, the *dimmer* the star. Most of the brighter stars you can see with the naked eye from Earth have apparent magnitudes between 0 and 1, while the dimmest star you can see from a dark location has apparent magnitude 6. Are most of the circled stars visible to the naked eye? List at least two stars that are visible, and include their apparent magnitudes.

ASTRONOMY

Dangling a COROT

Space telescope aims to find more planets orbiting other stars BY ALEXANDER HELLEMANS

(from Alexander Hellemans, "Dangling a COROT," *Scientific American*, September 2007, 32)

This year is shaping up to be an important one in the search for planets around other stars. In April astronomers at the Geneva Observatory announced the discovery of the most Earth-like exoplanet yet located, the first rocky world beyond our solar system that could hold liquid water. Just 1.5 times the size of Earth and possessing five times its mass, the planet circles the red dwarf Gliese 581. More such announcements will likely come in the months to follow, as the first space observatory dedicated to hunting exoplanets, called COROT, begins full operation and researchers complete their calculations.

In hunting for other worlds, scientists have proved adept at wringing the most from ground-based observatories. The Geneva astronomers found the rocky, Earth-like planet by first detecting the slight wobbling in the motion of its parent star, as star and planet circled around their common center of gravity. Such wobbling betrays itself as tiny Doppler shifts in the spectral lines present in a star's light. Although the investigators determined the exoplanet's mass via the observed movement, they could not directly gauge its size, because it does not transit—that is, pass in between Earth and its parent star. Hence, the team had to derive the exoplanet's density from planet formation models, team member Stéphane Udry reports.

Even if the planet moved across its star, the team might not have been able to see it. Peering through Earth's turbulent atmosphere requires adaptive optics to correct for distortions, but the same technology also precludes precise measurements of the dimming of light when a planet passes in front of a star. Even with future giant telescopes up to 42 meters in diameter, the sensitivity for such photometric detection will remain limited. "It is difficult to combine transit search with adaptive optics," Udry explains. "Adaptive optics continuously disturbs the photometric calibration."

That is where space observatories come in, such as the $46-million COROT (COnvection ROtation and planetary Transits). In addition to studying the ripples on the surface of stars to glean information about their interior, the observatory looks at changes in stellar luminosity that might indicate an exoplanetary transit. COROT began taking measurements in February, and, once fully calibrated, its 27-centimeter-wide telescope will detect light variations as low as one part in 20,000—200 times the resolution of ground-based instruments. COROT will ultimately look at 120,000 stars. "It is a hit-or-miss search, but we need a large number of stars so as to increase the probability of discovering planets," explains Pierre Barge of the Astrophysics Laboratory of Marseille in France, who leads the COROT exoplanet working group.

Besides finding exoplanets, COROT will also be able to determine their sizes, because the amount of dimming during a planet's occultation is proportional to that planet's girth. Moreover, knowing its size pins down its density and hence confirms whether the planet is rocky or gaseous. Currently, astronomers know the diameter of only about 20 of the approximately 240 known exoplanets.

Other space observatories can use transits to discover details about any atmosphere that might enshroud an exoplanet. In the July 12 *Nature*, Giovanna Tinetti of the European Space Agency and University College London and her colleagues report the discovery of water on a planet circling a star 64 light-years away, based on infrared measurements taken by NASA's Spitzer Space Telescope. They found that the way the light is absorbed by the planet's atmosphere when the body passes in front of its parent star matches the absorption characteristics of water vapor. But the planet is a gas giant, about 15 percent larger than Jupiter, so it probably does not harbor life.

Unfortunately, none of today's orbiting observatories, not even the Hubble Space Telescope, can find water on smaller worlds. "They are not sensitive enough to go for the smaller rocky planets," Tinetti says. She expects that such detections will come within the reach of the James Webb Space Telescope, Hubble's successor, which will be launched in 2013.

Space observatories will not put their terrestrial counterparts out of business in the search for exoplanets. Ground instruments can detect Doppler shifting quite well and are needed to corroborate findings from space. COROT has already spied possible new exoplanets, but researchers will not announce results until later this year. "We have to do a certain number of checks," Barge says. If the checks pan out, expect to hear a lot more otherworldly news.

Alexander Hellemans is a science writer based in Antwerp, Belgium.

9 The Living Earth

RIVUXG

The two hemispheres of Earth. (NASA)

WEB LINK 9.1

When astronauts made the first journeys from Earth to the Moon between 1968 and 1972, they often reported that our planet is the most beautiful sight visible from space. Of all the worlds of the solar system, only Earth has the distinctive green color of vegetation, and only Earth has liquid water on its surface. Compared with the Moon, whose lifeless, dry surface has been ravaged by billions of years of impacts by interplanetary debris, Earth seems an inviting and tranquil place.

Yet the appearance of tranquility is deceiving. The seemingly solid surface of the planet is in a state of slow but constant motion, driven by the flow of molten rock within Earth's interior. Sunlight provides Earth's atmosphere with the energy that drives our planet's sometimes violent weather, including all the cloud patterns visible in the images shown here.

Unseen in these images are immense clouds of subatomic particles that wander around the outside of the planet in two giant belts, held in thrall by Earth's magnetic field. And equally unseen are trace gases in Earth's atmosphere whose abundance may determine the future of our climate, and on which may depend the survival of entire species.

In this chapter our goal is to learn about Earth's components—its dynamic oceans and atmosphere, its ever-changing surface, and its hot, active interior—and how they interact to make up our planetary home. In later chapters we will use this knowledge as a point of reference for studying the Moon and the other planets.

9-1 Earth's atmosphere, oceans, and surface are extraordinarily active

VIDEO 9.1

The crew of an alien spacecraft exploring our solar system might overlook Earth altogether. Although the largest of the terrestrial planets, with a mass greater than that of Mercury, Venus, and Mars put together, Earth is far smaller than any of the giant Jovian planets (see Table 7-1 and Table 9-1). But a more careful inspection would reveal Earth to be unique among all the planets that orbit the Sun.

Dynamic Earth: Water, Air, and Land

Unlike the arid surfaces of Venus and Mars, Earth's surface is very wet. Indeed, nearly 71% of Earth's surface is covered with water (Figure 9-1). Water is also locked into the chemical structure of many Earth rocks, including those found in the driest deserts. Furthermore, Earth's liquid water is in constant motion. The oceans ebb and flow with the tide, streams and rivers flow down to the sea, and storms whip lake waters into a frenzy.

To understand our dynamic planet, we must understand the energy sources that power its activity

Earth's atmosphere is also in a state of perpetual activity. Winds blow at all altitudes, with speeds and directions that change from one hour to the next. The atmosphere is at its most dynamic

Learning Goals

By reading the sections of this chapter, you will learn

9-1 What powers the motions of Earth's atmosphere, oceans, and land surfaces

9-2 How scientists have deduced the layered structure of our planet's interior

9-3 The evidence that the continents are being continuously moved and reshaped

9-4 How life on Earth is protected from subatomic particles emitted by the Sun

9-5 How the evolution of life has transformed Earth's atmosphere

9-6 What causes the patterns of weather in our atmosphere

9-7 How human civilization is causing dramatic and adverse changes to our planetary habitat

Table 9-1 Earth Data

Average distance from Sun:	1.000 AU = 1.496×10^8 km
Maximum distance from Sun:	1.017 AU = 1.521×10^8 km
Minimum distance from Sun:	0.983 AU = 1.471×10^8 km
Eccentricity of orbit:	0.017
Average orbital speed:	29.79 km/s
Orbital period:	365.256 days
Rotation period:	23.9345 hours
Inclination of equator to orbit:	23.45°
Diameter (equatorial):	12,756 km
Mass:	5.974×10^{24} kg
Average density:	5515 kg/m^3
Escape speed:	11.2 km/s
Albedo:	0.31
Surface temperature range	Maximum: 60°C = 140°F = 333 K Mean: 14°C = 57°F = 287 K Minimum: −90°C = −130°F = 183 K
Atmospheric composition (by number of molecules):	78.08% nitrogen (N_2) 20.95% oxygen (O_2) 0.035% carbon dioxide (CO_2) about 1% water vapor

WEB LINK 9.2

(NASA)

when water evaporates to form clouds, then returns to the surface as rain or snow (Figure 9-2). While other worlds of the solar system have atmospheres, only Earth's contains the oxygen that animals (including humans) need to sustain life.

The combined effects of water and wind cause erosion of mountains and beaches. But these are by no means the only forces reshaping the face of our planet. All across Earth we find evidence that the surface has been twisted, deformed, and folded (Figure 9-3). What is more, new material is continually being added to Earth's surface as lava pours forth from volcanoes and from immense cracks in the ocean floors. At the same time, other geologic activity slowly pushes old surface material back into our planet's interior. These processes and others work together to renew and refresh our planet's exterior. Thus, although Earth is some 4.56 billion (4.56×10^9) years old, much of its surface is a few hundred million (a few times 10^8) years old (Figure 9-4). (We saw in Section 8-2 how scientists use radioactive dating to determine the ages of rocks and of Earth as a whole.)

Earth's Energy Sources

TUTORIAL 9.1

What powers all of this activity in Earth's oceans, atmosphere, and surface? There are three energy sources: radiation from the Sun, the tidal effects of the Moon, and Earth's own internal heat (Table 9-2).

Figure 9-1 R I **V** U X G

Earth's Dynamic Oceans Nearly three-quarters of Earth's surface is covered with water, a substance that is essential to the existence of life. In contrast, there is no liquid water at all on Mercury, Venus, Mars, or the Moon. (Farley Lewis/Photo Researchers)

Figure 9-2 R I V U X G

Earth's Dynamic Atmosphere This space shuttle image shows thunderstorm clouds over the African nation of Zaire. The tops of thunderstorm clouds frequently reach altitudes of 10,000 m (33,000 ft) or higher. At any given time, nearly 2000 thunderstorms are in progress over Earth's surface. (JSC/NASA)

2. Over millions of years, layers of sediments built up over that rock. The most recent layer—the top—is about 250 million years old.
1. The rocks at the bottom of the Grand Canyon are 1.7-2.0 billion years old.

Figure 9-4 R I V U X G

Old and Young Rocks in the Grand Canyon The ages of rocks in Arizona's Grand Canyon demonstrate that geologic processes take place over very long time scales. (John Wang/PhotoDisc/Getty Images)

Figure 9-3 R I V U X G

Earth's Dynamic Surface These tan-colored ridges, or hogbacks, in Colorado's Rocky Mountains were once layers of sediment at the bottom of an ancient body of water. Forces within Earth folded this terrain and rotated the layers into a vertical orientation. The layers were revealed when wind and rain eroded away the surrounding material. (Tom Till/DRK)

Table 9-2 Earth's Energy Sources

Activity	Energy Sources
Motion of water in oceans, lakes, rivers	Solar energy, tidal forces
Motion of the atmosphere	Solar energy
Reshaping of surface	Earth's internal heat
Life	Solar energy (a few species that live on the ocean floor make use of Earth's internal heat)

1) albedo- reflecting sunlight back to space
2) Earth emits radiation

The Sun is the principal source of energy for the atmosphere. Earth's surface is warmed by sunlight, which in turn warms the air next to the surface. Hot air is less dense than cool air and so tends to rise (see the discussion of density in Box 7-1). As the air rises, it transfers heat to its surroundings. As a result, the rising air cools and becomes denser. It then sinks downward to be heated again, and the process starts over. This up-and-down motion is called **convection,** and the overall pattern of circulation is called a **convection current.** (You can see convection currents in action by heating water on a stove, as Figure 9-5 shows.) In Section 9-6 we will see how Earth's rotation modifies the up-and-down motion of the atmosphere to produce a pattern of global circulation.

Solar energy also powers atmospheric activity by evaporating water from the surface. The energy in the water vapor is released when it condenses to form water droplets, like those that make up clouds. In a typical thunderstorm (see Figure 9-2), some 5×10^8 kg of water vapor is lifted to great heights. The amount of energy released when this water condenses is as much as a city of 100,000 people uses in a month!

Solar energy also helps to power the oceans. Warm water from near the equator moves toward the poles, while cold polar water returns toward the equator. As we saw in Section 4-8, however, there is back-and-forth motion of the oceans due to the tidal forces of the Moon and Sun. Sometimes these two influences can reinforce each other, as when a storm (caused by solar energy) reaches a coastline at high tide (caused by tidal forces), producing waves strong enough to seriously erode beaches and sea cliffs.

Neither solar energy nor tidal forces can explain the reshaping of Earth's surface suggested by Figure 9-3. Rather, all geologic activity is powered by heat flowing from the interior of Earth itself. Earth formed by collisions among planetesimals (see Section 8-4), and these collisions heated the body of Earth. The interior remains hot to this day. An additional source of heat is the decay of radioactive elements such as uranium and thorium deep inside Earth.

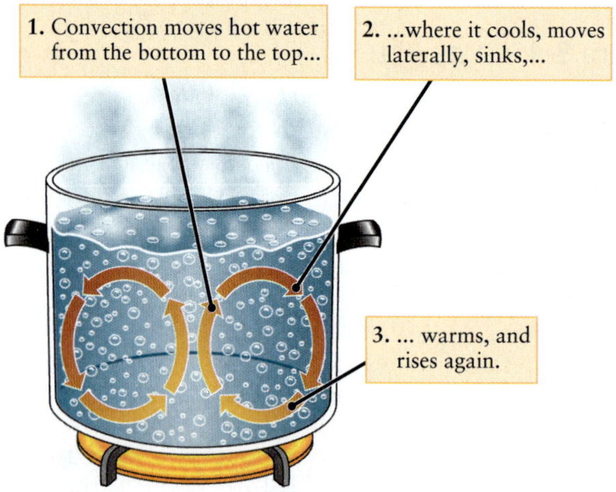

1. Convection moves hot water from the bottom to the top...
2. ...where it cools, moves laterally, sinks,...
3. ... warms, and rises again.

Figure 9-5

Convection in the Kitchen Heat supplied to the bottom of a pot warms the water there, making the water expand and lowering its average density. This low-density water rises and transfers heat to its cooler surroundings. The water that began at the bottom thus cools down, becomes denser, and sinks back to the bottom to repeat the process. (Adapted from F. Press, R. Siever, T. Grotzinger, and T. H. Jordan, *Understanding Earth,* 4th ed., W. H. Freeman, 2004)

The heat flow from Earth's interior to its surface is minuscule—just 1/6000 as great as the flow of energy we receive from the Sun—but it has a profound effect on the face of our planet. In Sections 9-2 and 9-3 we will explore the interior structure of Earth and discover the connection between geologic activity on Earth's surface and heat coming from its interior.

Earth's Surface Temperature and the Greenhouse Effect

Since so little heat comes from inside our planet, the average surface temperature of Earth depends almost entirely on the amount of energy that reaches us from the Sun in the form of electromagnetic radiation. (In an analogous way, whether you feel warm or cool outdoors on a summer day depends on whether you are in the sunlight and receiving lots of solar energy or in the shade and receiving little of this energy.)

If Earth did nothing but *absorb* radiation from the Sun, it would get hotter and hotter until the surface temperature became high enough to melt rock. Happily, there are two reasons why this does not happen. One is that the clouds, snow, ice, and sand *reflect* about 31% of the incoming sunlight back into space. The fraction of incoming sunlight that a planet reflects is called its **albedo** (from the Latin for "whiteness"); thus Earth's albedo is about 0.31. Thus, only 69% of the incoming solar energy is absorbed by Earth. A second reason is that Earth also *emits* radiation into space because of its temperature, in accordance with the laws that describe heated dense objects (see Section 5-3). Earth's average surface temperature is nearly constant, which means that on the whole it is neither gaining nor losing energy. Thus, the rate at which Earth emits energy into space must equal the rate at which it absorbs energy from the Sun.

To better understand this balance between absorbed and emitted radiation, remember that Wien's law tells us that the wavelength at which such an object emits most strongly (λ_{max}) is inversely proportional to its temperature (T) on the Kelvin scale (see Section 5-4 and Box 5-2). For example, the Sun's surface temperature is about 5800 K, and sunlight has its greatest intensity at a wavelength λ_{max} of 500 nm, in the middle of the visible spectrum. Earth's average surface temperature of 287 K is far lower than the Sun's, so Earth radiates most strongly at longer wavelengths in the infrared portion of the electromagnetic spectrum. The Stefan-Boltzmann law tells us that temperature also determines the *amount* of radiation that Earth emits: The higher the temperature, the more energy it radiates.

Given the amount of energy reaching us from the Sun each second as well as Earth's albedo, we can calculate the amount of solar energy that Earth should *absorb* each second. Since this amount must equal the amount of electromagnetic energy that Earth *emits* each second, which in turn depends on Earth's average surface temperature, we can calculate what Earth's average surface temperature should be. The result is a very chilly 254 K (−19°C = −2°F), so cold that oceans and lakes around the world should be frozen over. In fact, Earth's actual average surface temperature is 287 K (14°C = 57°F). What is wrong with our model? Why is Earth warmer than we would expect?

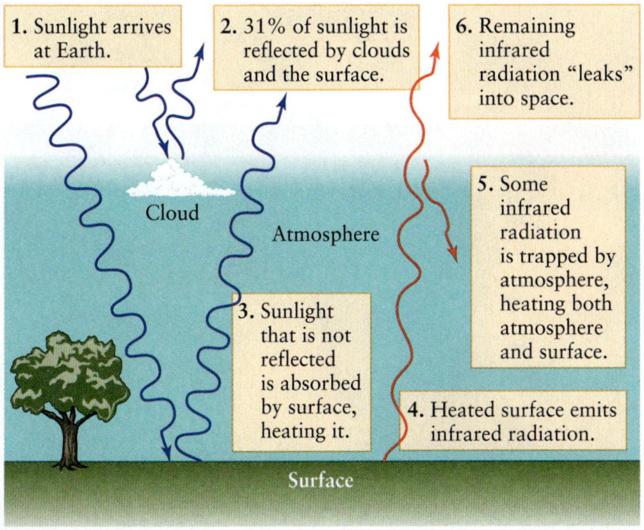
1. Sunlight arrives at Earth.
2. 31% of sunlight is reflected by clouds and the surface.
6. Remaining infrared radiation "leaks" into space.
Cloud
Atmosphere
5. Some infrared radiation is trapped by atmosphere, heating both atmosphere and surface.
3. Sunlight that is not reflected is absorbed by surface, heating it.
4. Heated surface emits infrared radiation.
Surface

ANIMATION 9.1

Figure 9-6

The Greenhouse Effect Sunlight warms Earth's surface, which due to its temperature emits infrared radiation. Much of this radiation is absorbed by atmospheric water vapor and carbon dioxide, helping to raise the average temperature of the surface. Some infrared radiation does penetrate the atmosphere and leaks into space. In a state of equilibrium, the rate at which Earth loses energy to space in this way is equal to the rate at which it absorbs energy from the Sun.

The explanation for this discrepancy is called the **greenhouse effect:** Our atmosphere prevents some of the radiation emitted by Earth's surface from escaping into space. Certain gases in our atmosphere called **greenhouse gases,** among them water vapor and carbon dioxide, are transparent to visible light but not to infrared radiation. Consequently, visible sunlight has no trouble entering our atmosphere and warming the surface. But the infrared radiation coming from the heated surface is partially trapped by the atmosphere, thus raising the temperatures of both the atmosphere and the surface. As the surface and atmosphere become hotter, they both emit more infrared radiation, part of which is able to escape into space. The temperature levels off when the amount of infrared energy that escapes just balances the amount of solar energy reaching the surface (Figure 9-6). The result is that our planet's surface is some 33°C (59°F) warmer than it would be without the greenhouse effect, and water remains unfrozen over most of Earth.

The warming caused by the greenhouse effect gives our planet the moderate temperatures needed for the existence of life. For over a century, however, our technological civilization has been adding greenhouse gases to the atmosphere at an unprecedented rate. As we will see in Section 9-7, the likely consequences to our climate are extremely grave.

9-2 Studies of earthquakes reveal Earth's layered interior structure

TUTORIAL 9.2

The kinds of rocks found on and near Earth's surface provide an important clue about our planet's interior. The densities of typical surface rocks are around 3000 kg/m^3, but the average density of Earth as a whole (that is, its mass divided by its volume) is 5515 kg/m^3. The interior of Earth must therefore be composed of substances much denser than those found near the surface. But what is this substance? Why is Earth's interior more dense than its crust? And is Earth's interior solid like rock or molten like lava?

Like an exotic dessert, our planet's interior has layers of both solid and liquid material

An Iron-Rich Planet

Iron (chemical symbol Fe) is a good candidate for the substance that makes up most of Earth's interior for two reasons. First, iron atoms are quite massive (a typical iron atom has 56 times the mass of a hydrogen atom), and second, iron is relatively abundant. (Figure 8-4 shows that it is the seventh most abundant element in our part of the Milky Way Galaxy.) Other elements such as lead and uranium have more massive atoms, but these elements are quite rare. Hence, the solar nebula could not have had enough of these massive atoms to create Earth's dense interior. Furthermore, iron is common in meteoroids that strike Earth, which suggests that it was abundant in the planetesimals from which Earth formed.

Earth was almost certainly molten throughout its volume soon after its formation, about 4.56×10^9 years ago. Energy released by the violent impacts of numerous meteoroids and asteroids and by the decay of radioactive isotopes likely melted the solid material collected from the earlier planetesimals. Gravity caused abundant, dense iron to sink toward Earth's center, forcing less dense material to the surface. Figure 9-7*a* shows this process of *chemical differentiation.* (We discussed chemical differentiation in Section 8-4.) The result was a planet with the layered structure shown in Figure 9-7*b*—a central **core** composed of almost pure iron, surrounded by a **mantle** of dense, iron-rich minerals. The mantle, in turn, is surrounded by a thin **crust** of relatively light silicon-rich minerals. We live on the surface of this crust.

Seismic Waves as Earth Probes

How do we know that this layered structure is correct? The challenge in testing this model is that Earth's interior is as inaccessible as the most distant galaxies in space. The deepest wells go down only a few kilometers, barely penetrating the surface of our planet. Despite these difficulties, geologists have learned basic properties of Earth's interior by studying earthquakes and the seismic waves that they produce.

WEB LINK 9.3

Over the centuries, stresses build up in Earth's crust. Occasionally, these stresses are relieved with a sudden motion called an **earthquake.** Most earthquakes occur deep within Earth's crust. The point on Earth's surface directly over an earthquake's location is called the **epicenter.**

Earthquakes produce three different kinds of **seismic waves,** which travel around or through Earth in different ways and at different speeds. Geologists use sensitive instruments called **seismographs** to detect and record these vibratory motions. The first type of wave, which is analogous to water waves on the ocean, causes the rolling motion that people feel around an epicenter. These waves are called **surface waves** because they

Figure 9-7

Chemical Differentiation and Earth's Internal Structure **(a)** The newly formed Earth was molten throughout its volume. Dense materials such as iron (shown in orange) sank toward the center, while low-density materials (shown in blue) rose toward the surface. **(b)** The present-day Earth is no longer completely molten inside. A dense, solid iron core is surrounded by a less dense liquid core and an even less dense mantle. The crust, which includes the continents and ocean floors, is the least dense of all; it floats atop the mantle like the skin that forms on the surface of a cooling cup of cocoa.

1) Surface waves
2) P waves - longitudinal
3) S-wave - transverse

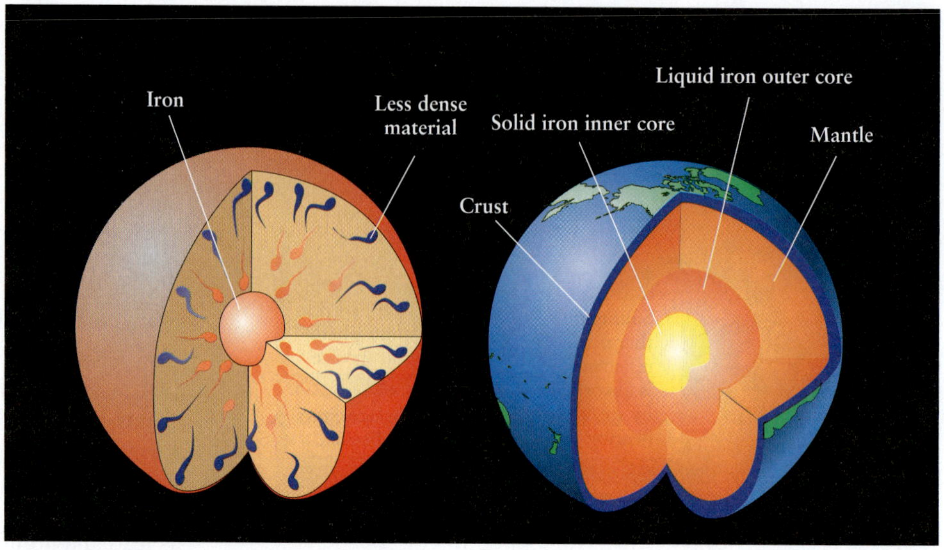

(a) During differentiation, iron sank to the center and less dense material floated upward

(b) As a result of differentiation, Earth has the layered structure that we see today

travel only over Earth's surface. The two other kinds of waves, called **P waves** (for "primary") and **S waves** (for "secondary"), travel through the interior of Earth. P waves are called *longitudinal* waves because their oscillations are parallel to the direction of wave motion, like a spring that is alternately pushed and pulled. In contrast, S waves are called *transverse* waves because their vibrations are perpendicular to the direction in which the waves move. S waves are analogous to waves produced by a person shaking a rope up and down (Figure 9-8).

What makes seismic waves useful for learning about Earth's interior is that they do not travel in straight lines. Instead, the paths that they follow through the body of Earth are bent because of the varying density and composition of Earth's interior. We saw in Section 6-1 that light waves behave in a very similar way. Just as light waves bend, or refract, when they pass from air into glass or vice versa (see Figure 6-2), seismic waves refract as they pass through different parts of Earth's interior. By studying how the paths of these waves bend, geologists can map out the general interior structure of Earth.

One key observation about seismic waves and how they bend has to do with the differences between S and P waves. When an earthquake occurs, seismographs relatively close to the epicenter record both S and P waves, but seismometers on the opposite side of Earth record only P waves. The absence of S waves was first explained in 1906 by British geologist Richard Dixon Oldham, who noted that transverse vibrations such as S waves cannot travel far through liquids. Oldham therefore concluded that our planet has a molten core. Furthermore, there is a region in which neither S waves nor P waves from an earthquake can be detected (Figure 9-9). This "shadow zone" results from the specific way in which P waves are refracted at the boundary between the solid mantle and the molten core. By measuring the size of the shadow zone, geologists have concluded that the radius of the molten core is about 3500 km (2200 mi), about 55% of our planet's overall radius but about double the radius of the Moon (1738 km = 1080 mi).

As the quality and sensitivity of seismographs improved, geologists discovered faint traces of P waves in an earthquake's shadow zone. In 1936, the Danish seismologist Inge Lehmann explained that some of the P waves passing through Earth are deflected into the shadow zone by a small, solid **inner core** at the center of our planet. The radius of this inner core is about 1300 km (800 mi).

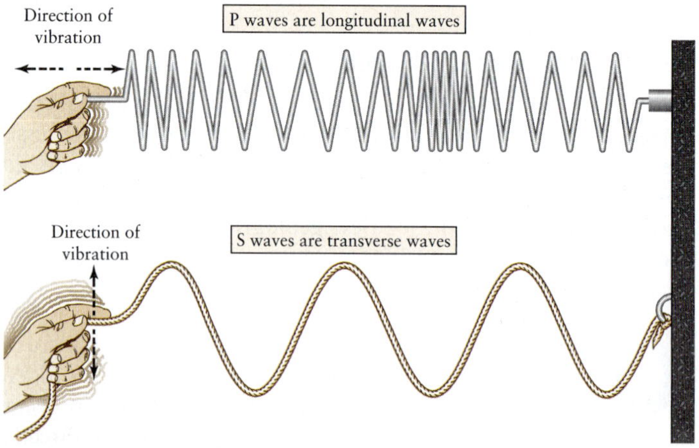

ANIMATION 9.2

Figure 9-8

Seismic Waves Earthquakes produce two kinds of waves that travel through the body of our planet. One kind, called P waves, are longitudinal waves. They are analogous to those produced by pushing a spring in and out. The other kind, S waves, are transverse waves analogous to the waves produced by shaking a rope up and down.

Earth's Major Layers

LOOKING DEEPER 9.1

The seismic evidence reveals that our planet has a curious internal structure—a liquid **outer core** sandwiched between a solid inner core and a solid mantle. Table 9-3 summarizes this structure. To understand

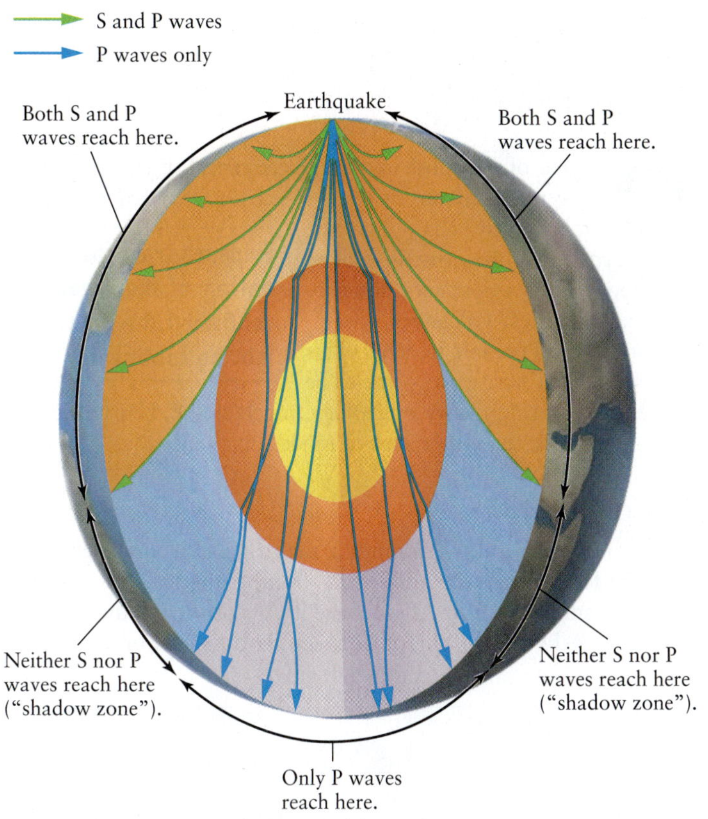
S and P waves
P waves only
Earthquake
Both S and P waves reach here.
Both S and P waves reach here.
Neither S nor P waves reach here ("shadow zone").
Neither S nor P waves reach here ("shadow zone").
Only P waves reach here.

Figure 9-9

Earth's Internal Structure and the Paths of Seismic Waves Seismic waves follow curved paths because of differences in the density and composition of the material in Earth's interior. The paths curve gradually where there are gradual changes in density and composition. Sharp bends occur only where there is an abrupt change from one kind of material to another, such as at the boundary between the outer core and the mantle. Only P waves can pass through Earth's liquid outer core.

this arrangement, we must look at how temperature and pressure inside Earth affect the melting point of rock.

Both temperature and pressure increase with increasing depth below Earth's surface. The temperature of Earth's interior rises steadily from about 14°C on the surface to nearly 5000°C at our planet's center (Figure 9-10).

Earth's outermost layer, the crust, is only about 5 to 35 km thick. It is composed of rocks for which the **melting point,** or temperature at which the rock changes from solid to liquid, is far higher than the temperatures actually found in the crust. Thus, the crust is solid.

Earth's mantle, which extends to a depth of about 2900 km (1800 mi), is largely composed of substances rich in iron and magnesium. On Earth's surface, specimens of these substances have melting points slightly over 1000°C. However, the melting point of a substance depends on the pressure to which it is subjected—the higher the pressure, the higher the melting point. As Figure 9-10 shows, the actual temperatures throughout most of the mantle are less than the melting point of all the substances of which the mantle is made. Hence, the mantle is primarily solid. However, the upper levels of the mantle—called the **asthenosphere** (from the Greek *asthenia,* meaning "weakness")—are able to flow slowly and are therefore referred to as being **plastic.**

CAUTION! It may seem strange to think of a plastic material as one that is able to flow. Normally we think of plastic objects as being hard and solid, like the plastic out of which compact discs and bicycle helmets are made. But to make these objects, the plastic material is heated so that it can flow into a mold, then cooled so that it solidifies. Strictly, the material is only "plastic" when it is able to flow. Maple syrup and tree sap are two other substances that flow at warm temperatures but become solid when cooled.

At the boundary between the mantle and the outer core, there is an abrupt change in chemical composition, from silicon-rich materials to almost-pure iron with a small admixture of nickel. Because this nearly pure iron has a lower melting point than the iron-rich materials, the melting curve in Figure 9-10 changes abruptly as it crosses from the mantle to the outer core and remains. As a result, the melting point is less than the temperature at depths of about 2900 to 5100 km (1800 to 3200 mi), and the outer core is liquid.

At depths greater than about 5100 km, the pressure is more than 10^{11} newtons per square meter. This pressure is about 10^6 times ordinary atmospheric pressure, or about 10^4 tons per square inch. Because the melting point of the iron-nickel mixture under this pressure is higher than the actual temperature (see Figure 9-10), Earth's inner core is solid.

Table 9-3 Earth's Internal Structure

Region	Depth Below Surface (km)	Distance From Center (km)	Average Density (kg/m^3)
Crust (solid)	0–5 (under oceans) 0–35 (under continents)	6343–6378	3500
Mantle (plastic, solid)	from bottom of crust to 2900	3500–6343	3500–5500
Outer core (liquid)	2900–5100	1300–3500	10,000–12,000
Inner core (solid)	5100–6400	0–1300	13,000

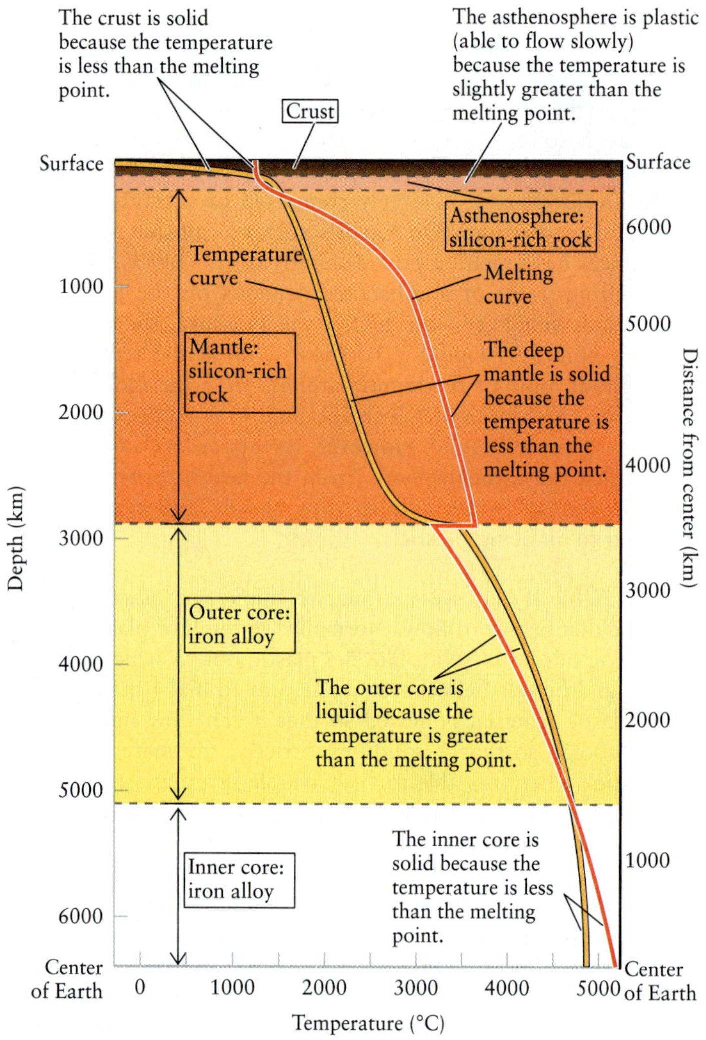
The crust is solid because the temperature is less than the melting point.
Crust
The asthenosphere is plastic (able to flow slowly) because the temperature is slightly greater than the melting point.
Surface
Surface
Asthenosphere: silicon-rich rock
Temperature curve
Melting curve
Mantle: silicon-rich rock
The deep mantle is solid because the temperature is less than the melting point.
Outer core: iron alloy
The outer core is liquid because the temperature is greater than the melting point.
Inner core: iron alloy
The inner core is solid because the temperature is less than the melting point.
Depth (km)
1000
2000
3000
4000
5000
6000
Center of Earth
Distance from center (km)
6000
5000
4000
3000
2000
1000
Center of Earth
0
1000
2000
3000
4000
5000
Temperature (°C)

Figure 9-10

Temperature and Melting Point of Rock Inside Earth The temperature (yellow curve) rises steadily from Earth's surface to its center. The melting point of Earth's material (red curve) is also shown on this graph. Where the temperature is below the melting point, as in the mantle and inner core, the material is solid; where the temperature is above the melting point, as in the outer core, the material is liquid. (Adapted from T. Grotzinger, T. H. Jordan, F. Press, and R. Siever, *Understanding Earth*, 5th ed., W. H. Freeman, 2007)

LOOKING DEEPER 9.2

Until the 1980s, geologists knew little more about the inner core than that it is solid and dense. Since then, evidence has accumulated that suggests the inner core is a single **crystal** of iron—that is, iron atoms arranged in orderly rows, like carbon atoms are arranged within a diamond. If so, the inner core is the largest crystalline object anywhere in the solar system. Furthermore, there are strong indications that the inner core is rotating at a slightly faster rate than the rest of Earth! These remarkable discoveries suggest that even more surprises may lurk deep within Earth's interior.

Heat naturally flows from where the temperature is high to where it is low. Figure 9-10 thus explains why heat flows from the center of Earth outward. We will see in the next section how this heat flow acts as the "engine" that powers our planet's geologic activity.

9-3 Plate movement produces earthquakes, mountain ranges, and volcanoes that shape Earth's surface

WEB LINK 9.4

One of the most important geological discoveries of the twentieth century was the realization that Earth's crust is constantly changing. We have learned that the crust is divided into huge **plates** whose motions produce earthquakes, volcanoes, mountain ranges, and oceanic trenches. This picture of Earth's crust has come to be the central unifying theory of geology, much as the theory of evolution has become the centerpiece of modern biology.

Evidence from the bottom of the ocean confirmed the theory of moving continents

Continental Drift

Anyone who carefully examines a map of Earth might come up with the idea of moving continents. South America, for example, would fit snugly against Africa were it not for the Atlantic Ocean. As Figure 9-11 shows, the fit between land masses on either side

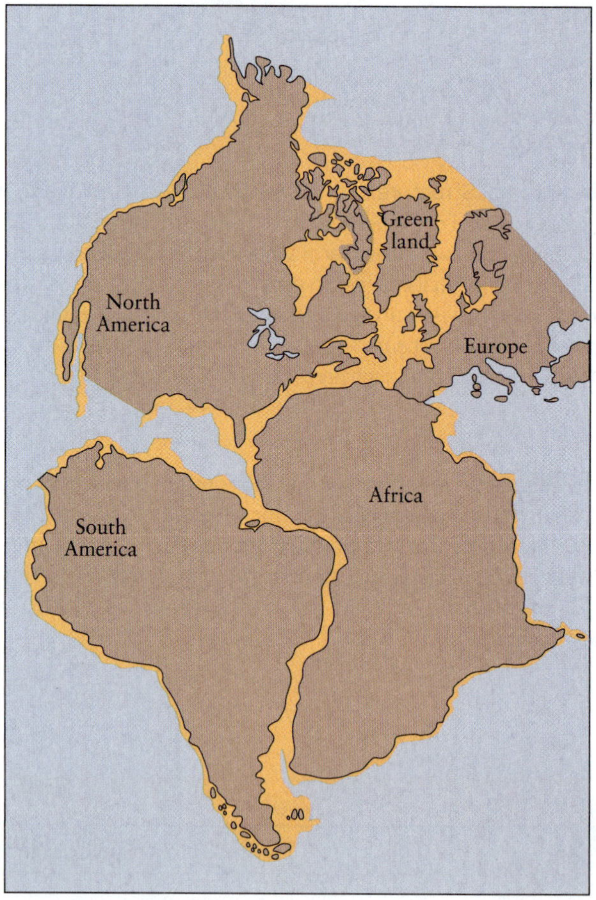
Greenland
North America
Europe
Africa
South America

Figure 9-11

Fitting the Continents Together Africa, Europe, Greenland, North America, and South America fit together remarkably well. The fit is especially convincing if the edges of the continental shelves (shown in yellow) are used, rather than today's shorelines. This strongly suggests that these continents were in fact joined together at some point in the past. (Adapted from P. M. Hurley)

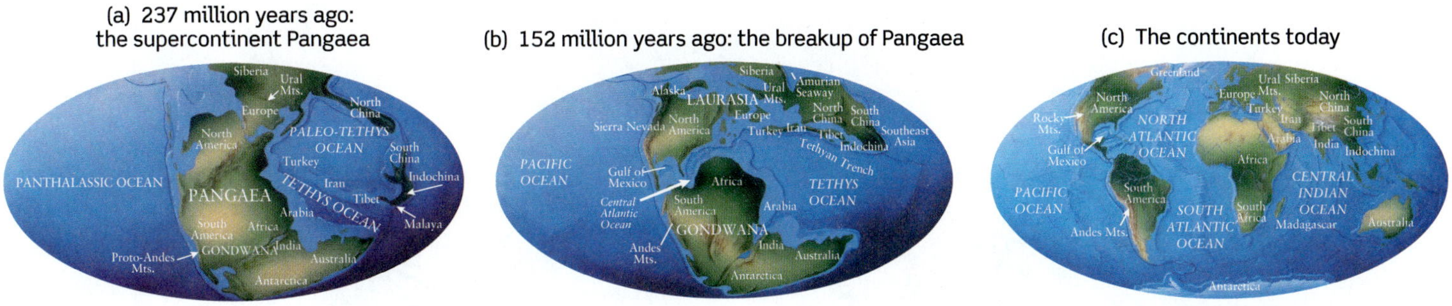

Figure 9-12

The Breakup of the Supercontinent Pangaea **(a)** The shapes of the continents led Alfred Wegener to conclude that more than 200 million (2×10^8) years ago, the continents were merged into a single supercontinent, which he called Pangaea. **(b)** Pangaea first split into two smaller land masses, Laurasia and Gondwana. **(c)** Over millions of years, the continents moved to their present-day locations. Among the evidence confirming this picture are nearly identical rock formations 200 million years in age that today are thousands of kilometers apart but would have been side by side on Pangaea. (Adapted from F. Press, R. Siever, T. Grotzinger, and T. H. Jordan, *Understanding Earth,* 4th ed., W. H. Freeman, 2004)

of the Atlantic Ocean is quite remarkable. This observation inspired the German meteorologist Alfred Wegener to advocate "continental drift"—the idea that the continents on either side of the Atlantic Ocean have simply drifted apart. After much research, in 1915 Wegener published the theory that there had originally been a single gigantic supercontinent, which he called Pangaea (meaning "all lands"), that began to break up and drift apart roughly 200 million years ago. (No humans were present to witness the break-up; the earliest humans did not appear on Earth until about 1.9 million years ago. At the time that Pangaea broke up, during what geologists refer to as the early Jurassic period, small dinosaurs were the dominant land animals.)

Other geologists refined this theory, arguing that Pangaea must have first split into two smaller supercontinents, which they called Laurasia and Gondwana, separated by what they called the Tethys Ocean. Gondwana later split into Africa and South America, with Laurasia dividing to become North America and Eurasia. According to this theory, the Mediterranean Sea is a surviving remnant of the ancient Tethys Ocean (Figure 9-12).

Initially, most geologists ridiculed Wegener's ideas. Although it was generally accepted that the continents do "float" on the denser, somewhat plastic mantle beneath them, few geologists could accept the idea that entire continents could move around Earth at speeds as great as several centimeters per year. The "continental drifters" could not explain what forces could be shoving the massive continents around.

Plate Tectonics

WEB LINK 9.5

Beginning in the mid-1950s, however, geologists found evidence that material is being forced upward to the crust from deep within Earth. Bruce C. Heezen of Columbia University and his colleagues began discovering long and completely submerged mountain ranges on the ocean floors, such as the Mid-Atlantic Ridge, which stretches all the way from Iceland to Antarctica (Figure 9-13). During the 1960s, Harry Hess of Princeton University and Robert Dietz of the University of California again carefully examined the floor of the Atlantic Ocean. They concluded that rock from Earth's mantle is being melted and then forced upward along the Mid-Atlantic Ridge, which is in essence a long chain of underwater volcanoes.

The upwelling of new material from the mantle to the crust forces the existing crusts apart, causing **seafloor spreading.** For example, the floor of the Atlantic Ocean to the east of the Mid-Atlantic Ridge is moving eastward and the floor to the west is moving westward. By explaining what fills in the gap between continents as they move apart, seafloor spreading helps to fill out the

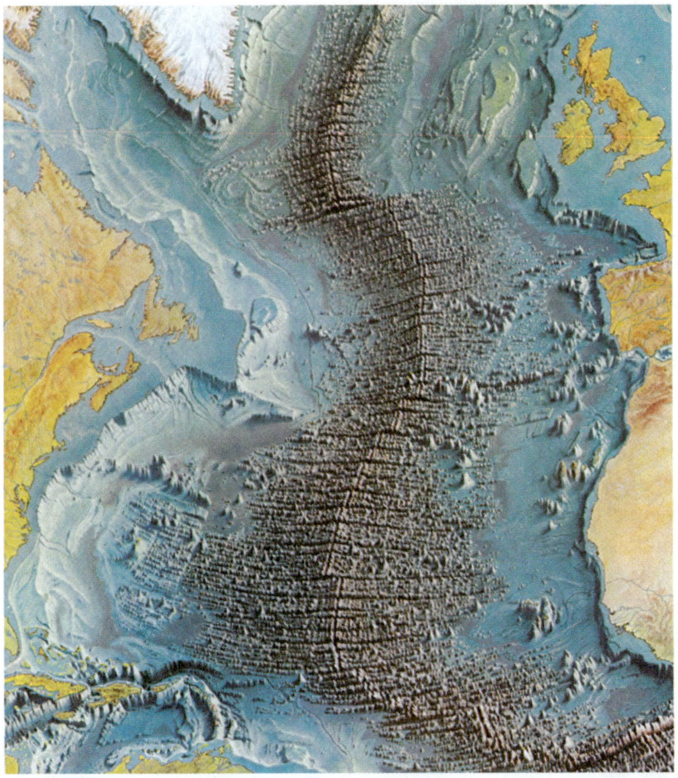

Figure 9-13

The Mid-Atlantic Ridge This artist's rendition shows the Mid-Atlantic Ridge, an immense mountain ridge that rises up from the floor of the North Atlantic Ocean. It is caused by lava seeping up from Earth's interior along a rift that extends from Iceland to Antarctica. (Courtesy of M. Tharp and B. C. Heezen)

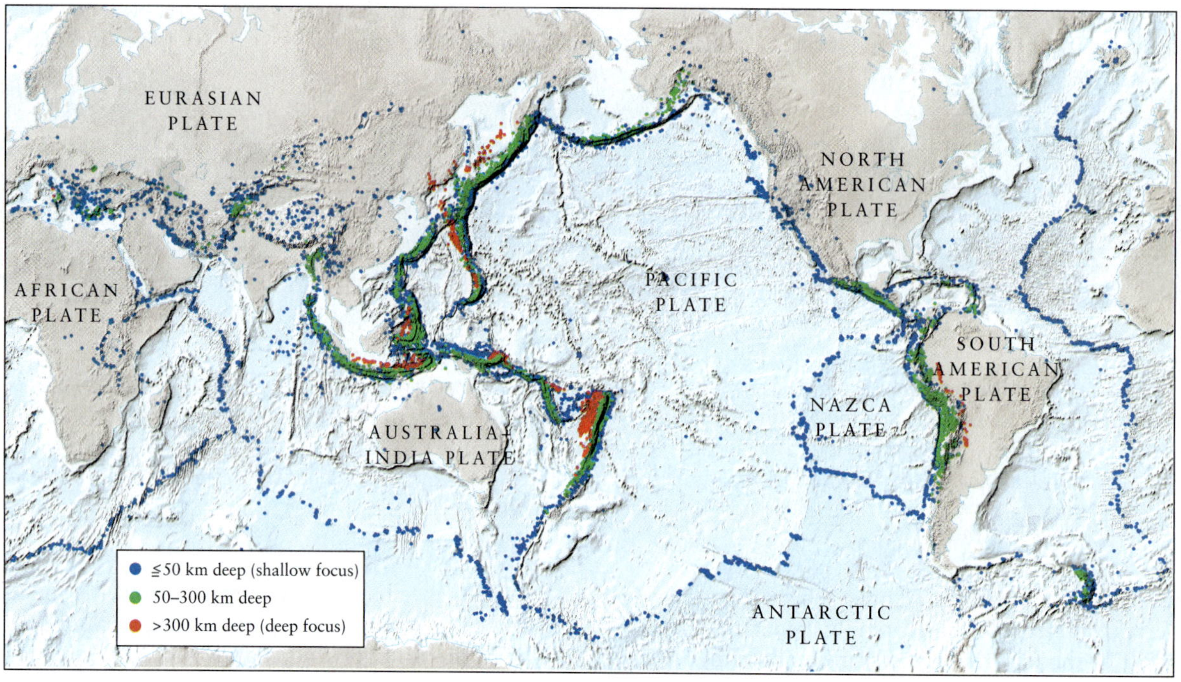

ANIMATION 9.3

Figure 9-14

Earth's Major Plates The boundaries of Earth's major plates are the scenes of violent seismic and geologic activity. Most earthquakes occur where plates separate, collide, or rub together. Plate boundaries are therefore easily identified simply by plotting earthquake epicenters (shown here as dots) on a map. The colors of the dots indicate the depths at which the earthquakes originate. (Data from Harvard CMT catalog; plot by M. Boettcher and T. Jordan.)

theories of continental drift. Because of the seafloor spreading from the Mid-Atlantic Ridge, South America and Africa are moving apart at a speed of roughly 3 cm per year. Working backward, these two continents would have been next to each other some 200 million years ago—just as Wegener suggested (see Figure 9-11).

In the early 1960s, geologists began to find additional evidence supporting the existence of large, moving plates. Thus was born the modern theory of crustal motion, which came to be known as **plate tectonics** (from the Greek *tekton,* meaning "builder").

Geologists today realize that earthquakes tend to occur at the boundaries of Earth's crustal plates, where the plates are colliding, separating, or rubbing against each other. The boundaries of the plates therefore stand out clearly when the epicenters of earthquakes are plotted on a map, as in Figure 9-14.

WEB LINK 9.6

The vast majority of volcanoes also occur at plate boundaries. Figure 9-15 shows a volcanic eruption at the boundary between the Pacific and Eurasian plates.

Convection and Plate Motion

ANIMATION 9.4

What makes the plates move? The answer is twofold. First, heat flows outward from Earth's hot core to its cool crust, and second, this heat flows through the mantle by convection, the same process that takes place in Earth's atmosphere or in a pot of boiling water.

Figure 9-5 shows that convection occurs when a fluid is heated from below. The heat that drives convection in Earth's outer layers actually comes from very far below, at the boundary between the outer and inner cores. As material deep in the liquid core cools and solidifies to join the solid portion of the core, it releases the energy needed to heat the overlying mantle and cause

Figure 9-15 R I **V** U X G

An Erupting Volcano On September 30, 1994, a volcano erupted at the boundary of the Pacific and Eurasian plates on Russia's Kamchatka Peninsula. This space shuttle photograph shows an immense cloud of ash belching forth from this volcano. The cloud reached an altitude of 18 km (60,000 ft) above sea level and was carried by the winds for at least 1000 km from the volcano. (*STS-68,* NASA)

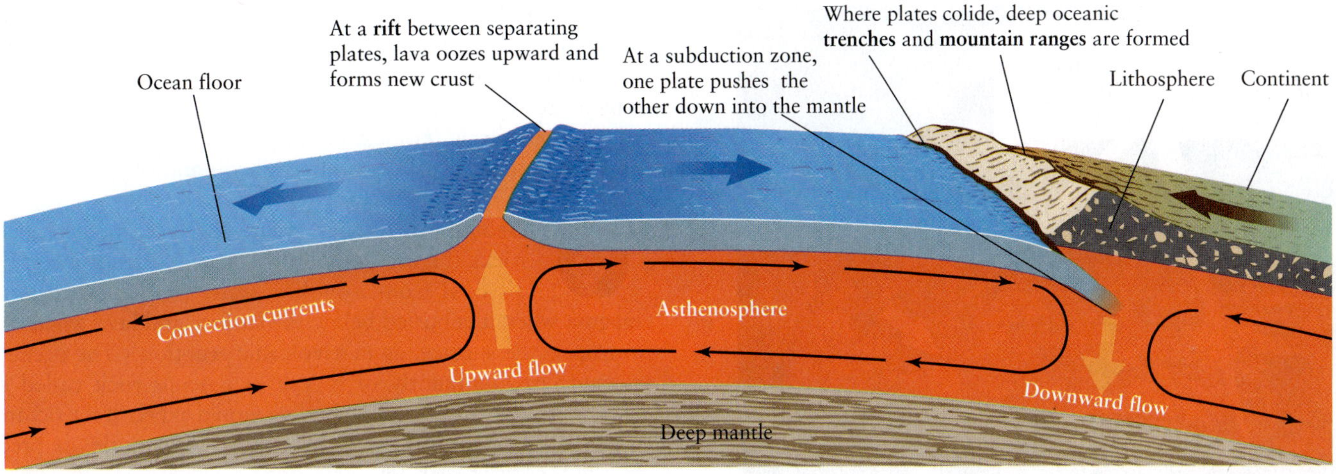
At a rift between separating plates, lava oozes upward and forms new crust
Where plates colide, deep oceanic trenches and mountain ranges are formed
At a subduction zone, one plate pushes the other down into the mantle
Ocean floor
Lithosphere
Continent
Convection currents
Asthenosphere
Upward flow
Downward flow
Deep mantle

INTERACTIVE EXERCISE 9.1

Figure 9-16

The Mechanism of Plate Tectonics Convection currents in the asthenosphere, the soft upper layer of the mantle, are responsible for pushing around rigid, low-density crustal plates. New crust forms in oceanic rifts, where lava oozes upward between separating plates. Mountain ranges and deep oceanic trenches are formed where plates collide.

convection. (In Section 9-1 we described a similar process in Earth's atmosphere: When water vapor in the air cools and forms liquid drops, it releases the energy that powers thunderstorms.)

The asthenosphere, or upper level of the mantle, is hot and soft enough to permit an oozing, plastic flow. Atop the asthenosphere is a rigid layer, called the **lithosphere** (from the Greek word for "rock.") The lithosphere is divided into plates that ride along the convection currents of the asthenosphere. The crust is simply the uppermost layer of the lithosphere, with a somewhat different chemical composition than the lithosphere's lower regions.

Figure 9-16 shows how convection causes plate movement. Molten subsurface rock seeps upward along **oceanic rifts,** where plates are separating. The Mid-Atlantic Ridge, shown in Figure 9-13, is an oceanic rift. Where plates collide, cool crustal material from one of the plates sinks back down into the mantle along a **subduction zone.** One such subduction zone is found along the west coast of South America, where the oceanic Nazca plate is being subducted into the mantle under the continental South American plate at a relatively speedy 10 centimeters per year. As the material from the subducted plate sinks, it pulls the rest of its plate along with it, thus helping to keep the plates in motion. New material is added to the crust from the mantle at the oceanic rifts and is "recycled" back into the mantle at the subduction zones. In this way the total amount of crust remains essentially the same.

ANIMATION 9.5 ANIMATION 9.6

The boundaries between plates are the sites of some of the most impressive geological activity on our planet. Great mountain ranges, such as the Sierras and Cascades along the western coast of North America and the Andes along South America's west coast, are thrust up by ongoing collisions between continental plates and the plates of the ocean floor. Subduction zones, where old crust is drawn back down into the mantle, are typically the locations of deep oceanic trenches, such as the Peru-Chile Trench off the west coast of South America. Figure 9-17 and Figure 9-18 show two well-known geographic features that resulted from tectonic activity at plate boundaries.

Plate Tectonics and the Varieties of Rocks

Plate tectonics also helps to explain geology on the scale of individual rocks and minerals. These are composed of chemical

The Red Sea is formed by the spreading apart of the African plate (at lower left) and the Arabian plate (at upper right)
Saudi Arabia
Gulf of Aqaba
Sinai Peninsula
Egypt
Gulf of Suez
Red Sea

ANIMATION 9.7

Figure 9-17 R I V U X G

The Separation of Two Plates The plates that carry Africa and Arabia are moving apart, leaving a great rift that has been flooded to form the Red Sea. This view from orbit shows the northern Red Sea, which splits into the Gulf of Suez and the Gulf of Aqaba. (*Gemini 12,* NASA)

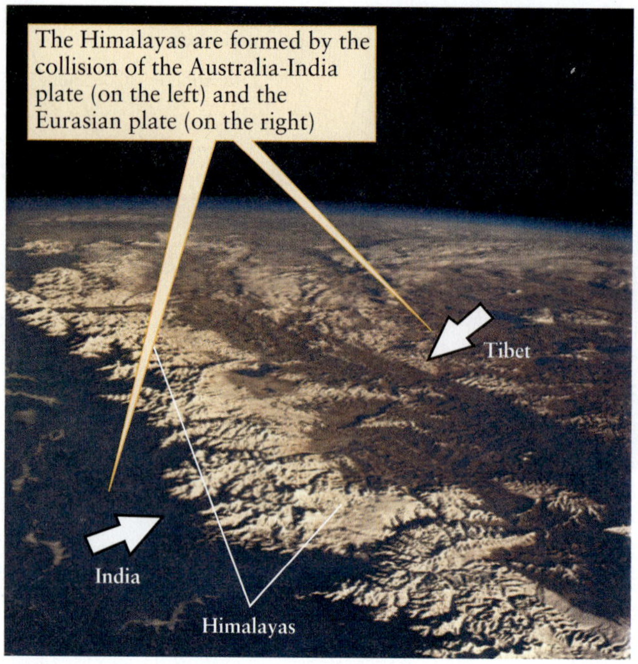
The Himalayas are formed by the collision of the Australia-India plate (on the left) and the Eurasian plate (on the right)
Tibet
India
Himalayas

Figure 9-18 RIVUXG

The Collision of Two Plates The plates that carry India and China are colliding. Instead of one plate being subducted beneath the other, both plates are pushed upward, forming the Himalayas. Mount Everest is one of the snow-covered peaks near the center of this photograph taken from orbit. (*Apollo 7*, NASA)

elements (see the periodic table in Box 5-5), but individual elements are rarely found in a pure state. The exceptions include diamonds, which are crystals of carbon, as well as nuggets of gold, silver, and copper. Somewhat more common are solids composed of a particular chemical combination of elements called a *compound*. An example is feldspar, a crystalline material made up of potassium, aluminum, silicon, and oxygen atoms. Diamond and feldspar are both **minerals**—naturally occurring solids composed of a single element or compound. Most common of all are **rocks,** which are solid parts of Earth's crust that are composed of one or more minerals. For example, granite contains both feldspar and quartz (a mineral containing silicon and oxygen).

Geologic processes create three major categories of rocks (Figure 9-19). **Igneous rocks** result when minerals cool from a molten state. (Molten rock is called **magma** when it is buried below the surface and **lava** when it flows out upon the surface, as in a volcanic eruption.) The ocean floor is made predominantly of a type of igneous rock called basalt (Figure 9-19*a*). This is just what we would expect if the ocean floor is produced by material welling up from the mantle, as predicted by the theory of seafloor spreading.

Sedimentary rocks are produced by the action of wind, water, or ice. For example, winds can pile up layer upon layer of sand grains. Other minerals present amid the sand can gradually cement the grains together to produce sandstone (Figure 9-19*b*). Minerals that precipitate out of the oceans can cover the ocean floor with layers of a sedimentary rock called limestone. The motion of tectonic plates can move such rocks to places far from where they were formed. This motion explains why sedimentary rock from the floor of an ancient ocean can now be found in central Alaska and as hogback ridges in Colorado (see Figure 9-3).

Sometimes igneous or sedimentary rocks become buried far beneath the surface, where they are subjected to enormous pressure and high temperatures. These severe conditions change the structure of the rocks, producing **metamorphic rock** (Figure 9-19*c*). The presence of metamorphic rocks at Earth's surface tells us that tectonic activity sometimes lifts up material from deep within the crust. This activity can happen when two plates collide, as shown in Figure 9-18. The boundaries of the colliding plates are thrust upward to form a new chain of mountains, and buried metamorphic rocks are brought to the surface.

(a) Igneous rock (basalt) (b) Sedimentary rock (sandstone) (c) Metamorphic rocks (marble, schist)

Figure 9-19 RIVUXG

Igneous, Sedimentary, and Metamorphic Rocks **(a)** Igneous rocks such as basalt are created when molten materials solidify. This example contains iron-rich minerals that give it its dark color. **(b)** Sedimentary rocks such as sandstone are typically formed when loose particles of soil or sand are fused into rock by the presence of other minerals, which act as a cement. **(c)** Metamorphic rocks are produced when igneous or sedimentary rocks are subjected to high temperatures and pressures deep within Earth's crust. Marble (left) is formed from sedimentary limestone; schist (right) is formed from igneous rock. (W. J. Kaufmann III; specimens courtesy Mineral Museum, California Division of Mines)

The Cycle of Supercontinents

Plate tectonic theory offers insight into geology on the largest of scales, that of an entire supercontinent. In recent years, geologists have uncovered evidence that points to a whole succession of supercontinents that once broke apart and then reassembled. Pangaea is only the most recent supercontinent in this cycle, which repeats about every 500 million (5×10^8) years. As a result, intense episodes of mountain building have occurred at roughly 500-million-year intervals.

Apparently, a supercontinent sows the seeds of its own destruction because it blocks the flow of heat from Earth's interior. As soon as a supercontinent forms, temperatures beneath it rise, much as they do when a lid is placed on a heated cooking pot. As heat accumulates, the lithosphere domes upward and cracks. Molten rock from the overheated asthenosphere wells up to fill the resulting fractures, which continue to widen as pieces of the fragmenting supercontinent move apart.

It can take a very long time for the heat trapped under a supercontinent to escape. Although Pangaea broke apart some 200 million years ago (see Figure 9-11), the mantle under its former location is still hot and is still trying to rise upward. As a result, Africa—which lies close to the center of this mass of rising material—sits several tens of meters higher than the other continents.

The changes wrought by plate tectonics are very slow on the scale of a human lifetime, but they are very rapid in comparison with the age of Earth. For example, the period over which Pangaea broke into Laurasia and Gondwana was only about 0.4% of Earth's age of 4.56×10^9 years. (To put this in perspective, 0.4% of a human lifetime is about 4 months.) The lesson of plate tectonics is that the seemingly permanent face of Earth is in fact dynamic and ever-changing.

9-4 Earth's magnetic field produces a magnetosphere that traps particles from the solar wind

Like the motions of tectonic plates, Earth's magnetic field results from our planet's internal heat

We saw in Section 7-7 that Earth's magnetic field provides evidence that our planet's interior is partially molten. Our understanding is that the magnetic field is produced by the motion of iron-rich material in the liquid outer core. Because this material is a good conductor of electricity, its motions generate a magnetic field in a mechanism called a *dynamo*.

The Source of Earth's Magnetic Field

The same energy source that powers plate tectonics is also responsible for Earth's dynamo. As we saw in Section 9-3, energy is released when material from our planet's liquid core cools and solidifies onto the solid core. This released energy stirs up the motions of the liquid core, and these motions generate a magnetic field. (While there is also fluid motion in the mantle, it does not produce an appreciable magnetic field: The silicon-rich rocks that comprise the mantle are poor conductors of electricity, and their motions are a million times slower than those in the outer core.)

LOOKING DEEPER 9.3

Studies of ancient rocks reinforce the idea that our planet's magnetism is due to fluid material in motion. When iron-bearing lava cools and solidifies to form igneous rock, it becomes magnetized in the direction of Earth's magnetic field. By analyzing samples of igneous rock of different ages from around the world, geologists have found that Earth's magnetic field actually flips over and reverses direction on an irregular schedule ranging from tens of thousands to hundreds of thousands of years. As an example, lava that solidified 30,000 years ago is magnetized in the opposite direction to lava that has solidified recently. Therefore, 30,000 years ago a compass needle would have pointed south, not north! If Earth was a permanent magnet, like the small magnets used to attach notes to refrigerators, it would be hard to imagine how its magnetic field could spontaneously reverse direction. But computer simulations show that fields produced by moving fluids in Earth's outer core can indeed change direction from time to time.

The Magnetosphere

Earth's magnetic field has important effects far above the atmosphere, where it interacts dramatically with charged particles from the Sun. This *solar wind* is a flow of mostly protons and electrons that streams constantly outward from the Sun's upper atmosphere. Near Earth, the particles in the solar wind move at speeds of roughly 450 km/s, or about a million miles per hour. Because this is considerably faster than sound waves can travel in the very thin gas between the planets, the solar wind is said to be *supersonic*. (Because the gas between the planets is so thin, interplanetary sound waves carry too little energy to be heard by astronauts.)

If Earth had no magnetic field, we would be continually bombarded by the solar wind. But our planet does have a magnetic field, and the forces that this field can exert on charged particles are strong enough to deflect them away from us. The region of space around a planet in which the motion of charged particles is dominated by the planet's magnetic field is called planet's **magnetosphere.** Figure 9-20 is a scale drawing of Earth's magnetosphere, which was discovered in the late 1950s by the first satellites placed in orbit.

When the supersonic particles in the solar wind first encounter Earth's magnetic field, they abruptly slow to subsonic speeds. The boundary where this sudden decrease in velocity occurs is called a **shock wave.** Still closer to Earth lies another boundary, called the **magnetopause,** where the outward magnetic pressure of Earth's field is exactly counterbalanced by the impinging pressure of the solar wind. Most of the particles of the solar wind are deflected around the magnetopause, just as water is deflected to either side of the bow of a ship.

Some charged particles of the solar wind manage to leak through the magnetopause. When they do, they are trapped by Earth's magnetic field in two huge, doughnut-shaped rings around Earth called the **Van Allen belts.** These belts were discovered in 1958 during the flight of the first successful U.S. Earth-orbiting satellite. They are named after the physicist James Van Allen, who

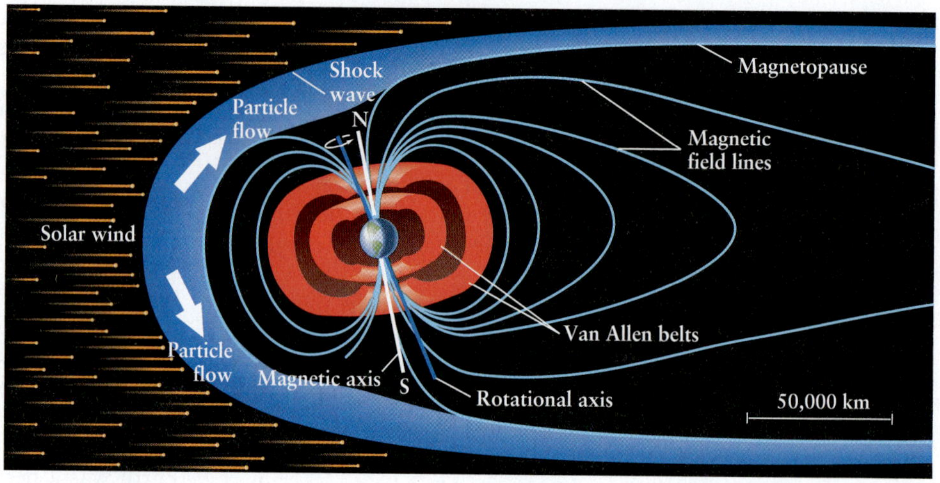

INTERACTIVE EXERCISE 9.2

Figure 9-20

Earth's Magnetosphere Earth's magnetic field carves out a cavity in the solar wind, shown here in cross section. A shock wave marks the boundary where the supersonic solar wind is abruptly slowed to subsonic speeds. Most of the particles of the solar wind are deflected around Earth in a turbulent region (colored blue in this drawing). Earth's magnetic field also traps some charged particles in two huge, doughnut-shaped rings called the Van Allen belts (shown in red). This figure shows only a slice through the Van Allen belts.

insisted that the satellite carry a Geiger counter to detect charged particles. The inner Van Allen belt, which extends over altitudes of about 2000 to 5000 km, contains mostly protons. The outer Van Allen belt, about 6000 km thick, is centered at an altitude of about 16,000 km above Earth's surface and contains mostly electrons.

Aurorae

Sometimes the Van Allen belts become overloaded with particles. The particles then leak through the magnetic fields at their weakest points and cascade down into Earth's upper atmosphere, usually in a ring-shaped pattern (Figure 9-21*a*). As these high-speed charged particles collide with atoms in the upper atmosphere, they excite the atoms to high energy levels. The atoms then emit visible light as they drop down to their ground states, like the excited gas atoms in a neon light (see Section 5-8). The result is a beautiful, shimmering display called an **aurora** (*plural* **aurorae**). These are also called the **northern lights (aurora borealis)** or **southern lights (aurora australis)**, depending on the hemisphere in which the phenomenon is observed. Figures 9-21*b* and 9-21*c* show the aurorae as seen from orbit and from Earth's surface.

Occasionally, a violent event on the Sun's surface sends a particularly intense burst of protons and electrons toward Earth. The resulting auroral display can be exceptionally bright and can often be seen over a wide range of latitudes. Such events also disturb radio transmissions and can damage communication satellites and transmission lines.

It is remarkable that Earth's magnetosphere, including its vast belts of charged particles, was entirely unknown until a few decades ago. Such discoveries remind us of how little we truly understand and how much remains to be learned even about our own planet.

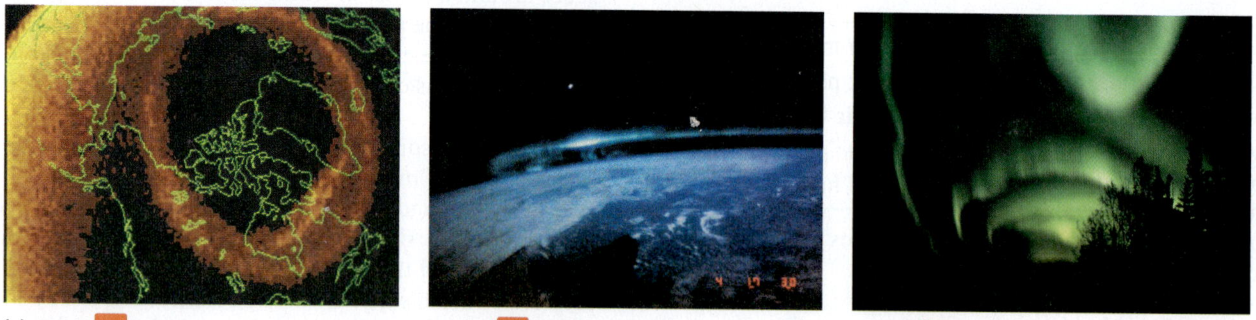

(a) R I V **U** X G (b) R I **V** U X G (c) R I **V** U X G

VIDEO 9.2

Figure 9-21

The Aurora An increased flow of charged particles from the Sun can overload the Van Allen belts and cascade toward Earth, producing aurorae. **(a)** This false-color ultraviolet image from the *Dynamics Explorer 1* spacecraft shows a glowing oval of auroral emission about 4500 km in diameter centered on the north magnetic pole. **(b)** This photograph shows the aurora australis over Antarctica as seen from an orbiting space shuttle. **(c)** This view from Alaska shows aurorae at their typical altitudes of 100 to 400 km. The green color shows that the light is emitted by excited oxygen atoms in the upper atmosphere. (a: Courtesy of L. A. Frank and J. D. Craven, University of Iowa; b: MTU/Geological & Mining Engineering & Sciences; c: J. Finch/Photo Researchers, Inc.)

9-5 Earth's atmosphere has changed substantially over our planet's history

Both plate tectonics and Earth's magnetic field paint a picture of a planet with a dynamic, evolving surface and interior. Our atmosphere, too, has changed and evolved substantially over Earth's 4.56-billion-year history. As we will see, this evolution explains why we have an atmosphere totally unlike that of any other world in the solar system.

The composition of our atmosphere has been dramatically altered by the evolution of life

Earth's Early Atmosphere

When Earth first formed by accretion of planetesimals (see Section 8-4), gases were probably trapped within Earth's interior in the same proportions that they were present in the solar nebula. But since the early Earth was hot enough to be molten throughout its volume, most of these trapped gases were released. As we discussed in Section 7-4 and Box 7-2, Earth's gravity was too weak to prevent hydrogen and helium—the two most common kinds of atoms in the universe, but also the least massive—from leaking away into space. The atmosphere that remained still contained substantial amounts of hydrogen, but in the form of relatively massive molecules of water vapor (H_2O) made by combining two atoms of hydrogen with one atom of oxygen, the third most common element in our part of the Milky Way Galaxy (see Figure 8-4). In fact, water vapor was probably the dominant constituent of the early atmosphere, which is thought to have been about 100 times denser than our present-day atmosphere.

Water vapor is one of the greenhouse gases that we discussed in Section 9-1. It traps infrared radiation from Earth's surface, and its presence in the early atmosphere helped to sustain the high temperature of the surface. But if water vapor had been the *only* constituent of the atmosphere, you would not be reading these words! The reason is that as the molten Earth inevitably cooled, water vapor in the atmosphere condensed into raindrops and fell to Earth to form the oceans. As water vapor was lost from the atmosphere, its contribution to the greenhouse effect weakened, which would have made the surface temperature drop even lower. A further cooling factor is that the young Sun was only about 70% as bright as it is today. The net result is that the entire Earth should have been frozen over! Ice is so reflective that Earth would have remained frozen even as the Sun aged and became more luminous. Liquid water is absolutely essential to all living creatures, so life as we know it would probably never have evolved on such an icy Earth.

In fact, the first living organisms appeared on Earth within 400 million years after the planet first formed. Hence, Earth could not have been frozen over for very long, if at all. What saved our planet from a perpetual deep freeze was the presence of carbon, the fourth most common of the elements. When combined with oxygen, carbon forms carbon dioxide (CO_2), a greenhouse gas that remains gaseous at low temperatures. Carbon dioxide would have been released into the atmosphere by volcanic activity, a process called **outgassing** (see Figure 9-15). It would also have been added to the atmosphere by carbon-rich meteorites striking the planet's surface. Once CO_2 was in the atmosphere, its greenhouse effect raised the planet's temperature and melted the ice. Some of the water evaporated into the atmosphere, further enhancing Earth's greenhouse effect.

Fewer than one in 2500 of the molecules in today's atmosphere is carbon dioxide. But CO_2 must have been thousands of times more abundant in the early Earth's atmosphere in order to melt the global ice sheath. Unlike hydrogen or helium, this excess CO_2 was not lost into space; instead, it has been trapped in rocks. Carbon dioxide dissolves in rainwater and falls into the oceans, where it combines with other substances to form a class of minerals called *carbonates.* (Limestone and marble are examples of carbonate-bearing rock.) These form sediments on the ocean floor, which are eventually recycled into the crust by subduction. As an example, marble (Figure 9-19*c*) is a metamorphic rock formed deep within the crust from limestone, a carbonate-rich sedimentary rock.

This process removed most of the CO_2 from the atmosphere within the first billion (10^9) years after the formation of Earth. Although this weakened the greenhouse effect, temperatures remained warm as the Sun's brightness increased. The small amounts of atmospheric CO_2 that remain today are the result of a balance between volcanic activity (which releases CO_2 into the atmosphere) and the formation of carbonates (which removes CO_2). (As we will see in Section 9-7, human activity is upsetting this balance.)

Life's Impact on Earth's Atmosphere

The appearance of life on Earth set into motion a radical transformation of the atmosphere. Early single-cell organisms converted energy from sunlight into chemical energy using **photosynthesis,** a chemical process that consumes CO_2 and water and releases oxygen (O_2). Oxygen molecules are very reactive, so originally most of the O_2 produced by photosynthesis combined with other substances to form minerals called oxides. (Evidence for this can be found in rock formations of various ages. The oldest rocks have very low oxide content, while oxides are prevalent in rocks that formed after the appearance of organisms that used photosynthesis.) But as life proliferated, the amount of photosynthesis increased dramatically. Eventually so much oxygen was being produced that it could not all be absorbed to form oxides, and O_2 began to accumulate in the atmosphere. Figure 9-22 shows how the amount of O_2 in the atmosphere has increased over the history of Earth.

About 2 billion (2×10^9) years ago, a new type of life evolved to take advantage of the newly abundant oxygen. These new organisms produce energy by consuming oxygen and releasing carbon dioxide—a process called **respiration** that is used by all modern animals, including humans. Such organisms thrived because photosynthetic plants continued to add even more oxygen to the atmosphere.

Several hundred million years ago, the number of oxygen molecules in the atmosphere stabilized at 21% of the total, the same as the present-day value. This value represents a balance between the release of oxygen from plants by photosynthesis and the intake of oxygen by respiration. Thus, the abundance of oxygen is regulated almost exclusively by the presence of life—a situation that has no parallel anywhere else in the solar system.

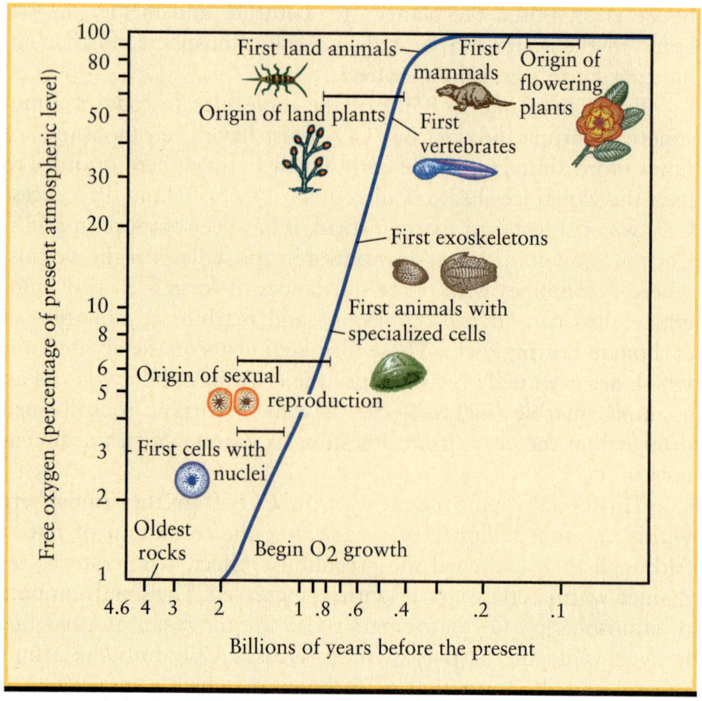

Figure 9-22

The Increase in Atmospheric Oxygen This graph shows how the amount of oxygen in the atmosphere (expressed as a percentage of its present-day value) has evolved with time. Note that the atmosphere contained essentially no oxygen until about 2 billion years ago. (Adapted from Preston Cloud, "The Biosphere," *Scientific American*, September 1983, p. 176)

The most numerous molecules in our atmosphere are nitrogen (N_2), which make up 78% of the total. Nitrogen molecules too, are a consequence of the presence of life on Earth. Certain bacteria extract oxygen from minerals called nitrates, and in the process liberate nitrogen into the atmosphere. The amount of atmospheric nitrogen is kept in check by lightning. The energy in a lightning flash causes atmospheric nitrogen and oxygen to combine into nitrogen oxides, which dissolve in rainwater, fall into the oceans, and form nitrates.

Comparing Atmospheres: Earth, Venus, and Mars

WEB LINK 9.7

Table 9-4 shows the dramatic differences between Earth's atmosphere and those of Venus and Mars. (Mercury, the other terrestrial planet, is too small and has too little gravity to hold an appreciable atmosphere.) The greater intensity of sunlight on Venus caused higher temperatures, which boiled any liquid water and made it impossible for CO_2 to be taken out of the atmosphere and put back into rocks. Venus's atmosphere thus became far denser than our own and rich in greenhouse gases. The result was a very strong greenhouse effect that raised temperatures on Venus to their present value of about 460°C (733 K = 855°F).

Just the opposite happened on Mars, where sunlight is less than half as intense as on Earth. The lower temperatures drove CO_2 from the atmosphere into Martian rocks, and froze all of the water beneath the planet's surface. The atmosphere that remains on Mars has a similar composition to that of Venus but is less than 1/10,000 as dense. On neither Venus nor Mars was life able to blossom and transform the atmosphere as it did here on Earth.

Table 9-4 Chemical Compositions of Three Planetary Atmospheres

	Venus	Earth	Mars
Nitrogen (N_2)	3.5%	78.08%	2.7%
Oxygen (O_2)	almost zero	20.95%	almost zero
Carbon dioxide (CO_2)	96.5%	0.035%	95.3%
Water vapor (H_2O)	0.003%	about 1%	0.03%
Other gases	almost zero	almost zero	2%

We have thus uncovered a general rule about the terrestrial planets:

The closer a terrestrial planet is to the Sun, the more important the greenhouse effect in that planet's atmosphere. The stronger the greenhouse effect, the higher the planet's surface temperature.

This rule suggests that if Earth had formed a bit closer to or farther from the Sun, temperatures on our planet might have been too high or too low for life ever to evolve. Thus, our very existence is a result of Earth's special position in the solar system.

9-6 Like Earth's interior, our atmosphere has a layered structure

While life has shaped our atmosphere's chemical composition, the structure of the atmosphere is controlled by the influence of sunlight. Scientists describe the structure of any atmosphere in terms of two properties: temperature and atmospheric pressure, which vary with altitude.

Circulation in our atmosphere results from convection and Earth's rotation

Pressure , Temperature, and Convection in the Atmosphere

LOOKING DEEPER 9.1

Atmospheric pressure at any height in the atmosphere is caused by the weight of all the air above that height. The average atmospheric pressure at sea level is defined to be one **atmosphere** (1 atm), equal to 1.01×10^5 N/m^2 or 14.7 pounds per square inch. As you go up in the atmosphere, there is less air above you to weigh down on you. Hence, atmospheric pressure decreases smoothly with increasing altitude.

Unlike atmospheric pressure, temperature varies with altitude in a complex way. Figure 9-23 shows that temperature decreases with increasing altitude in some layers of the atmosphere, but in other layers actually *increases* with increasing altitude. These differences result from the individual ways in which each layer is heated.

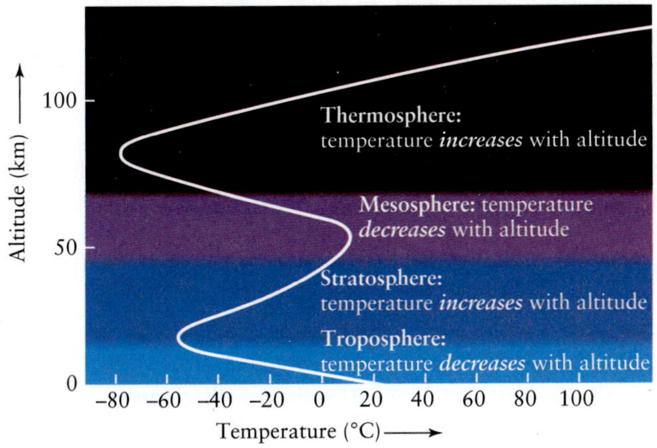

Figure 9-23

Temperature Profile of Earth's Atmosphere This graph shows how the temperature in Earth's atmosphere varies with altitude. In the troposphere and mesosphere, temperature decreases with increasing altitude; in the stratosphere and thermosphere, temperature actually increases with increasing altitude.

ANIMATION 9.10

The lowest layer, called the **troposphere,** extends from the surface to an average altitude of 12 km (roughly 7.5 miles, or 39,000 ft). It is heated only indirectly by the Sun. Sunlight warms Earth's surface, which heats the lower part of the troposphere. By contrast, the upper part of the troposphere remains at cooler temperatures. This vertical temperature variation causes convection currents that move up and down through the troposphere (see Section 8-1). All of Earth's weather is a consequence of this convection.

Convection on a grand scale is caused by the temperature difference between Earth's equator and its poles. If Earth did not rotate, heated air near the equator would rise upward and flow at high altitude toward the poles. There it would cool and sink to lower altitudes, at which it would flow back to the equator. However, Earth's rotation breaks up this simple convection pattern into a series of **convection cells.** In these cells, air flows east and west as well as vertically and in a north-south direction. The structure of these cells explains why the prevailing winds blow in different directions at different latitudes (Figure 9-24).

Upper Layers of the Atmosphere

Almost all the oxygen in the troposphere is in the form of O_2, a molecule made of two oxygen atoms. But in the **stratosphere,** which extends from about 12 to 50 km (about 7.5 to 31 mi) above the surface, an appreciable amount of oxygen is in the form of **ozone,** a molecule made of *three* oxygen atoms (O_3). Unlike O_2, ozone is very efficient at absorbing ultraviolet radiation from the Sun, which means that the stratosphere can directly absorb solar energy. The result is that the temperature actually increases as you move upward in the stratosphere. Convection requires that the temperature must decrease, not increase, with increasing altitude, so there are essentially no convection currents in the stratosphere.

Above the stratosphere lies the **mesosphere.** Very little ozone is found here, so solar ultraviolet radiation is not absorbed within the mesosphere, and atmospheric temperature again declines with increasing altitude.

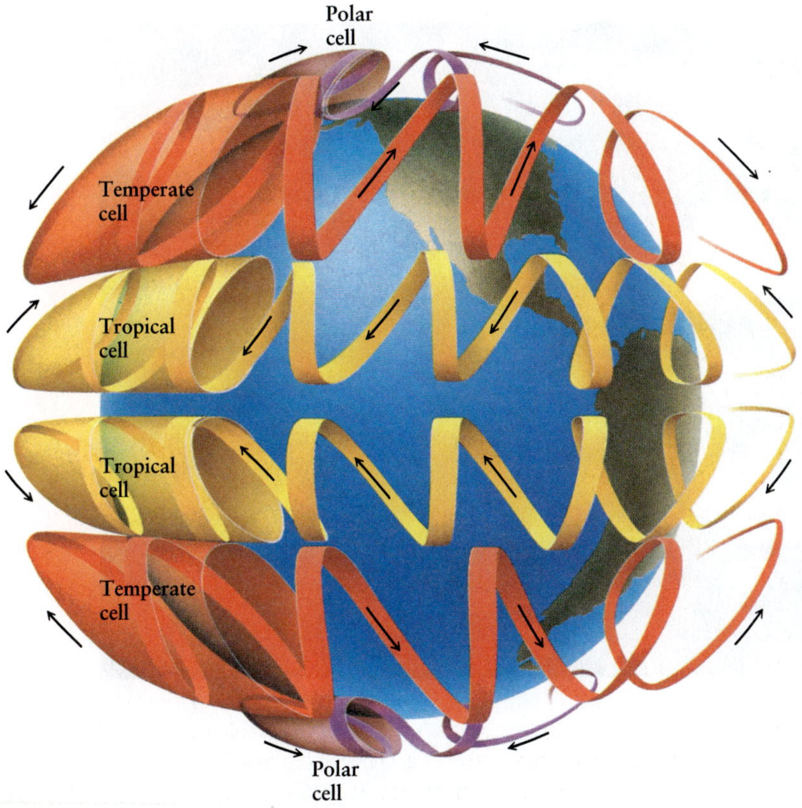

Figure 9-24

Circulation Patterns in Earth's Atmosphere The dominant circulation in our atmosphere consists of six convection cells, three in the northern hemisphere and three in the southern hemisphere. In the northern temperate region (including the continental United States), the prevailing winds at the surface are from the southwest toward the northeast. Farther south, within the northern tropical region (for example, Hawaii), the prevailing surface winds are from northeast to southwest.

LOOKING DEEPER 9.4

The temperature of the mesosphere reaches a minimum of about −75°C (= −103°F = 198 K) at an altitude of about 80 km (50 mi). This minimum marks the bottom of the atmosphere's thinnest and uppermost layer, the **thermosphere,** in which temperature once again rises with increasing altitude. This temperature increase is not due to the presence of ozone, because in this very low-density region oxygen and nitrogen are found as individual atoms rather than in molecules. Instead, the thermosphere is heated because these isolated atoms absorb very-short-wavelength solar ultraviolet radiation (which oxygen and nitrogen molecules cannot absorb).

CAUTION! At altitudes near 300 km the temperature of the thermosphere is about 1000°C (1800°F). This altitude is near the altitude at which the space shuttle and satellites orbit Earth. Nonetheless, a satellite in orbit does *not* risk being burned up as it moves through the thermosphere. The reason is that the thermosphere is far less dense than the atmosphere at sea level. The high temperature simply means that an average atom in the thermosphere is moving very fast (see Box 7-2). But because the thermosphere is so thin (only about 10^{-11} as dense as the air at sea level), these fast-moving atoms are few and far between. Hence, the thermosphere contains very little energy. Nearly all the heat that an orbiting satellite receives is from sunlight, not the thermosphere.

The *Cosmic Connections* figure compares the layered structure of our atmosphere with that of Earth's interior.

9-7 A burgeoning human population is profoundly altering Earth's biosphere

One of the extraordinary characteristics of Earth is that it is covered with life, from the floors of the oceans to the tops of mountains and from frigid polar caps to blistering deserts. So far in this chapter we have hinted at how our planet and living organisms interact: The greenhouse effect has given Earth a suitable temperature for the evolution of life (Section 9-1), and over billions of years that evolution has transformed the chemical composition of the atmosphere (Section 9-5). Let's explore in greater depth how Earth and the organisms that live on it, especially humans, affect each other.

Global warming and its consequences pose a major challenge to our civilization

The Biosphere and Natural Climate Variation

All life on Earth subsists in a relatively thin layer called the **biosphere,** which includes the oceans, the lowest few kilometers of the troposphere, and the crust to a depth of almost 3 kilometers. Figure 9-25 is a portrait of Earth's biosphere based on NASA satellite data. The biosphere, which has taken billions of years to evolve to its present state, is a delicate, highly complex system in which plants and animals depend on each other for their mutual survival.

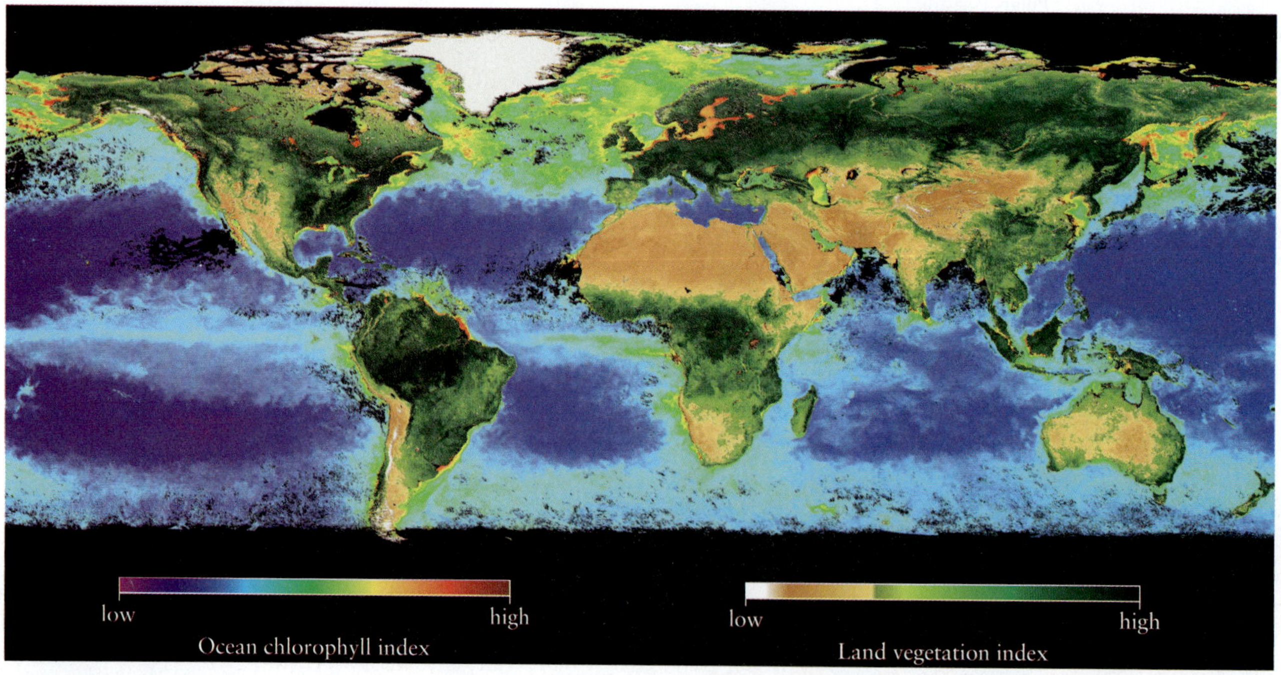

WEB LINK 9.8

Figure 9-25

Earth's Biosphere This image, based on data from the *SeaWIFS* spacecraft, shows the distribution of plant life over Earth's surface. The ocean colors show where free-floating microscopic plants called phytoplankton are found. (NASA Visible Earth)

COSMIC CONNECTIONS

Comparing Earth's Atmosphere and Interior

Earth's atmosphere (which is a gas) and Earth's interior (which is partly solid, partly liquid) both decrease in density and pressure as you go farther away from Earth's center.

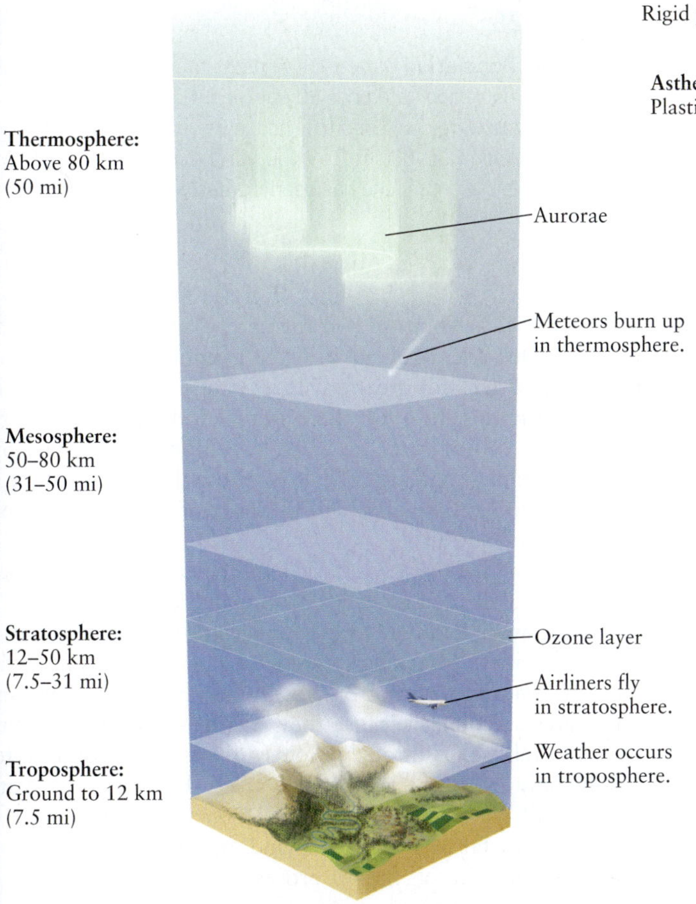

Structure of Earth's Atmosphere

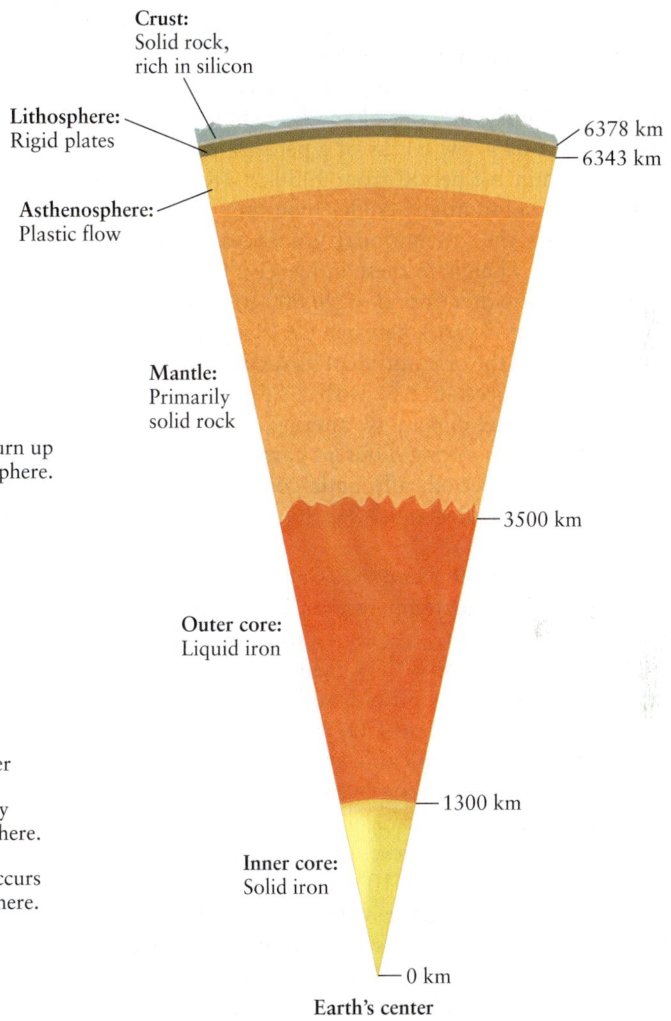

Structure of Earth's interior

WEB LINK 9.9

The state of the biosphere depends crucially on the temperature of the oceans and atmosphere. Even small temperature changes can have dramatic consequences. An example that recurs every three to seven years is the El Niño phenomenon, in which temperatures at the surface of the equatorial Pacific Ocean rise by 2 to 3°C. Ordinarily, water from the cold depths of the ocean is able to well upward, bringing with it nutrients that are used by microscopic marine organisms called phytoplankton that live near the surface (see Figure 9-25). But during an El Niño, the warm surface water suppresses this upwelling, and the phytoplankton starve. This wreaks havoc on organisms such as mollusks that feed on phytoplankton, on the fish that feed on the mollusks, and on the birds and mammals that eat the fish. During the 1982–1983 El Niño, a quarter of the adult sea lions off the Peruvian coast starved, along with all of their pups.

Many different factors can change the surface temperature of our planet. One is that the amount of energy radiated by the Sun can vary up or down by a few tenths of a percent. Reduced solar brightness may explain the period from 1450 to 1850 when European winters were substantially colder than they are today.

Other factors are the gravitational influences of the Moon and the other planets. Thanks to these influences, the eccentricity of Earth's orbit varies with a period of 90,000 to 100,000 years, the tilt of its rotation axis varies between 22.1° and 24.5° with a 40,000-year period, and the orientation of its rotation axis changes due to precession (see Section 2-6) with a 26,000-year period. These variations can affect climate by altering the amount of solar energy that heats Earth during different parts of the year. They help explain why Earth periodically undergoes an extended period of low temperatures called an *ice age,* the last of which ended about 11,000 years ago.

As we saw in Sections 9-1 and 9-5, one of the most important factors affecting global temperatures is the abundance of greenhouse gases such as CO_2. Geologic processes can alter this abundance, either by removing CO_2 from the atmosphere (as happens when fresh rock is uplifted and exposed to the air, where it can absorb atmospheric CO_2 to form carbonates) or by supplying new CO_2 (as in volcanic eruptions like that shown in Figure 9-15). From time to time in our planet's history, natural events have caused dramatic increases in the amount of greenhouse gases in the atmosphere. One such event may have taken place 251 million (2.51×10^8) years ago, when Siberia went through a period of intense volcanic activity. (The evidence for this is a layer of igneous rock in Siberia that covers an area about the size of Europe. Like the rock shown in Figure 9-19*a*, this layer—which radioactive dating reveals to be 2.51×10^8 years old—is solidified lava.) The tremendous amounts of CO_2 released by this activity would have elevated the global temperature by several degrees. Remarkably, the fossil record reveals that 95% of all species on Earth became extinct at this same time. The coincidence of these two events suggests that greenhouse-induced warming can have catastrophic effects on life.

Human Effects on the Biosphere: Deforestation

Our species is having an increasing effect on the biosphere because our population is skyrocketing. Figure 9-26 shows the sharp rise in the human population that began in the late 1700s with the Industrial Revolution and the spread of modern ideas about hygiene. This rise accelerated in the twentieth century thanks to medical and technological advances ranging from antibiotics to high-yield grains. In 1960 there were 3 billion people on Earth; in 1975, 4 billion; and in 1999, 6 billion. Projections by the United Nations Population Division show that there will be more than 8 billion people on Earth by the year 2030 and more than 9 billion by 2050.

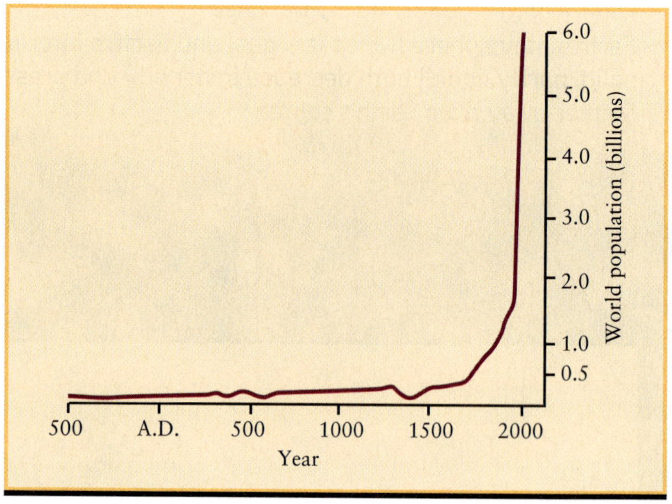

Figure 9-26

The Human Population Data and estimates from the U.S. Bureau of the Census, the Population Reference Bureau, and the United Nations Population Fund were combined to produce this graph showing the human population from 500 B.C. to 2000 A.D. The population began to rise in the eighteenth century and has been increasing at an astonishing rate since 1900.

Every human being has basic requirements: food, clothing, and housing. We all need fuel for cooking and heating. To meet these demands, we cut down forests, cultivate grasslands, and build sprawling cities. A striking example of this activity is occurring in the Amazon rain forest of Brazil. Tropical rain forests are vital to our planet's ecology because they absorb significant amounts of CO_2 and release O_2 through photosynthesis. Although rain forests occupy only 7% of the world's land areas, they are home to at least 50% of all plant and animal species on Earth. Nevertheless, to make way for farms and grazing land, people simply cut down the trees and set them on fire—a process called slash-and-burn. Such deforestation, along with extensive lumbering operations, is occurring in Malaysia, Indonesia, and Papua New Guinea. The rain forests that once thrived in Central America, India, and the western coast of Africa are almost gone (Figure 9-27).

Human Effects on the Biosphere: The Ozone Layer

Human activity is also having potentially disastrous effects on the upper atmosphere. Certain chemicals released into the air—in particular chlorofluorocarbons (CFCs), which have been used in refrigeration and electronics, and methyl bromide, which is used in fumigation—are destroying the ozone in the stratosphere. We saw in Section 9-6 that ozone molecules absorb ultraviolet

Figure 9-27 RIVUXG

The Deforestation of Amazonia The Amazon, the world's largest rain forest, is being destroyed at a rate of 20,000 square kilometers per year in order to provide land for grazing and farming and as a source for lumber. About 80% of the logging is being carried out illegally. (Martin Wendler/Okapia/Photo Researchers)

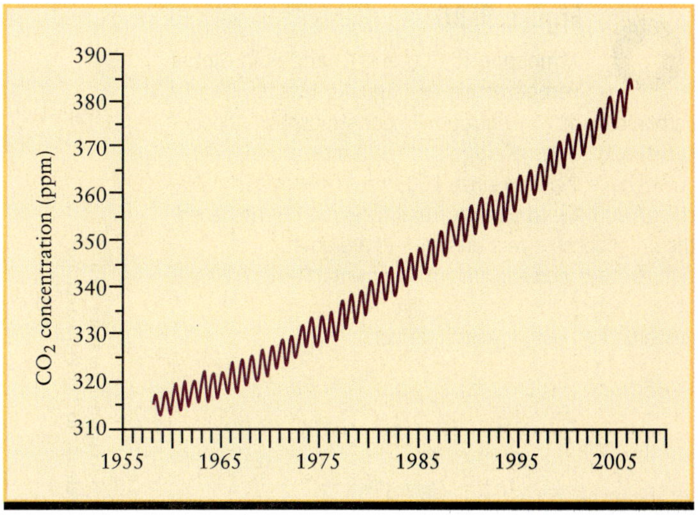

Figure 9-29

Atmospheric Carbon Dioxide Is Increasing This graph shows measurements of atmospheric carbon dioxide in parts per million (ppm). The sawtooth pattern results from plants absorbing more carbon dioxide during the spring and summer. The CO_2 concentration in the atmosphere has increased by 21% since continuous observations started in 1958. (NOAA/Scripps Institution of Oceanography)

radiation from the Sun. Without this high-altitude **ozone layer,** solar ultraviolet radiation would beat down on Earth's surface with greatly increased intensity. Such radiation breaks apart most of the delicate molecules that form living tissue. Hence, a complete loss of the ozone layer would lead to a catastrophic ecological disaster.

In 1985 scientists discovered an **ozone hole,** a region with an abnormally low concentration of ozone, over Antarctica. Since then this hole has generally expanded from one year to the next (Figure 9-28). Smaller but still serious effects have been observed in the stratosphere above other parts of Earth. As a result, there has been a worldwide increase in the number of deaths due to skin cancer caused by solar ultraviolet radiation. By international agreement, CFCs are being replaced by compounds that do not deplete stratospheric ozone, and sunlight naturally produces more ozone in the stratosphere. But even in the best of circumstances, the damage to the ozone layer is not expected to heal for many decades.

Human Effects on the Biosphere: Global Warming and Climate Change

The most troubling influence of human affairs on the biosphere is a consequence of burning fossil fuels (petroleum and coal) in automobiles, airplanes, and power plants as well as burning forests and brushland for agriculture and cooking (Figure 9-27). This burning releases carbon dioxide into Earth's atmosphere—and we are adding CO_2 to the atmosphere faster than plants and geological processes can extract it. Figure 9-29 shows how the

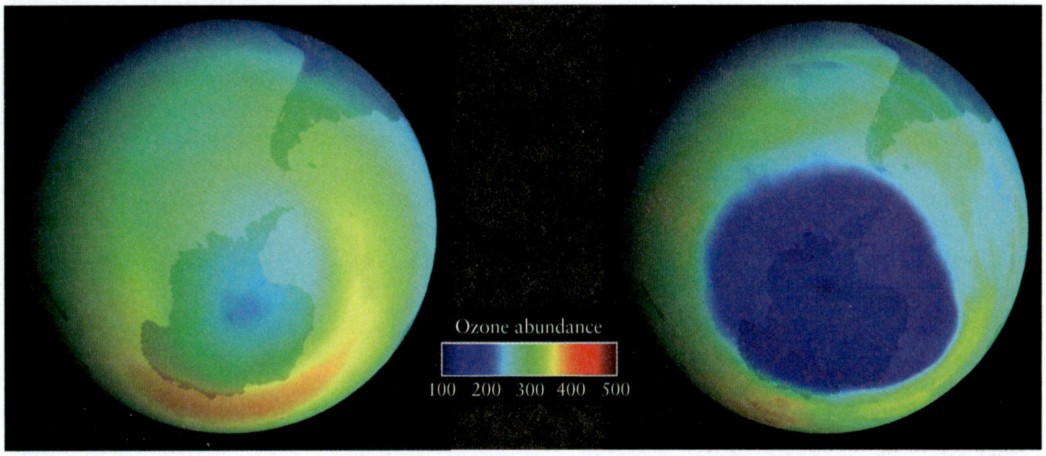

WEB LINK 9.10

Figure 9-28

The Antarctic Ozone Hole These two false-color images show that there was a net decrease of 50% in stratospheric ozone over Antarctica between October 1979 and September 2003. The amount of ozone at midlatitudes, where most of the human population lives, decreased by 10 to 20% over the same period. (GSFC/NASA)

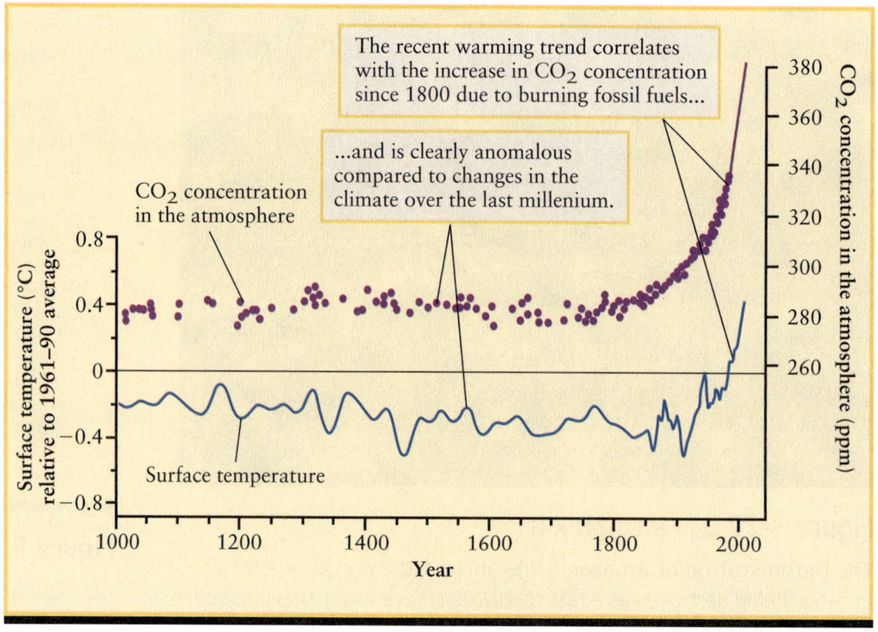

Figure 9-30

Atmospheric CO_2 and Changes in Global Temperature This figure shows how the carbon dioxide concentration in our atmosphere (upper curve) and Earth's average surface temperature (lower curve) have changed since 1000 A.D. The increase in CO_2 since 1800 due to burning fossil fuels has strengthened the greenhouse effect and caused a dramatic temperature increase. (Intergovernmental Panel on Climate Change and Hadley Centre for Climate Prediction and Research, U.K. Meteorological Office)

carbon dioxide content of the atmosphere has increased since 1958, when scientists began to measure this quantity on an ongoing basis.

To put the values shown in Figure 9-29 into perspective, we need to know the atmospheric CO_2 concentration in earlier eras. Scientists have learned this by analyzing air bubbles trapped at various depths in the ice that blankets the Antarctic and Greenland. Each winter a new ice layer is deposited, so the depth of a bubble indicates the year in which it was trapped. Figure 9-30 uses data obtained in this way to show how the atmospheric CO_2 concentration has varied since 1000 A.D. While there has been some natural variation in the concentration, its value has skyrocketed since the beginning of the Industrial Revolution around 1800. Data from older, deeper bubbles of trapped air show that in the 650,000 years before the Industrial Revolution, the CO_2 concentration was never greater than 300 parts per million. The present-day CO_2 concentration is greater than this by 25%, and has grown to its present elevated level in just over half a century. If there are no changes in our energy consumption habits, by 2050 the atmospheric CO_2 concentration will be greater than 600 parts per million.

Increasing atmospheric CO_2 is of concern because this strengthens the greenhouse effect and raises Earth's average surface temperature. Figure 9-30 also shows how this average temperature has varied since 1000 A.D. (Like the CO_2 data, the temperature data from past centuries comes from analyzing trapped air bubbles. Scientists measure the relative amounts of two different types of atmospheric oxygen called ^{16}O and ^{18}O, described in Box 5-5, in a bubble. The ratio of these is a sensitive measure of the temperature at that time the air was trapped inside the ice.) The recent dramatic increase in atmospheric CO_2 concentration has produced an equally dramatic increase in the average surface temperature. This temperature increase is called **global warming.** Other explanations for global warming have been proposed, such as changes in the Sun's brightness, but only greenhouse gases produced by human activity can explain the steep temperature increase shown in Figure 9-30. These gases include methane (CH_4) and nitrous oxide (N_2O), which are released in relatively small amounts by agriculture and industry but which are far more effective greenhouse gases than CO_2.

The effects of global warming can be seen around the world. Nine of the 10 warmest years on record have occurred since 1995, and the years since 2000 have seen increasing numbers of droughts, water shortages, and unprecedented heat waves. Glaciers worldwide are receding, the size of the ice cap around the North Pole has decreased by 20% since 1979, and a portion of the Antarctic ice shelf has broken off (Figure 9-31).

Figure 9-31 RIVUXG

A Melting Antarctic Ice Shelf Global warming caused the Larsen B ice shelf to break up in early 2002. This ice shelf, which was about the size of Rhode Island, is thought to have been part of the Antarctic coast for the past 12,000 years. (NASA/GSFC/LaRC/JPL, MIST Team)

Unfortunately, global warming is predicted to intensify in the decades to come. The UN Intergovernmental Panel on Climate Change predicts that if nothing is done to decrease the rate at which we add greenhouse gases to our atmosphere, the average surface temperature will continue to rise by an additional 1.4 to 5.8°C during the twenty-first century. What is worse, the temperature increase is predicted to be greater at the poles than at the equator. The global pattern of atmospheric circulation (Figure 9-24) depends on the temperature difference between the warm equator and cold poles, so this entire pattern will be affected. The same is true for the circulation patterns in the oceans. As a result, temperatures will rise in some regions and decline in others and the patterns of rainfall will be substantially altered. Agriculture depends on rainfall, so these changes in rainfall patterns can cause major disruptions in the world food supply. Studies suggest that the climate changes caused by a 3°C increase in the average surface temperature would cause a worldwide drop in cereal crops of 20 to 400 million tons, putting 400 million more people at risk of hunger.

The melting of polar ice due to global warming poses an additional risk to our civilization. When floating ice like that found near the North Pole melts, the ocean level remains the same. (You can see this by examining a glass of water with an ice cube floating in it. The water level doesn't change when the ice melts.) But the ocean level rises when ice on land melts and runs off into the sea. The Greenland ice cap has been melting at an accelerating rate since 2000 and has the potential to raise sea levels by half a meter or more. Low-lying coastal communities such as Boston and New Orleans, as well as river cities such as London, will have a greater risk of catastrophic flooding. Some island nations of Pacific will disappear completely beneath the waves.

While global warming is an unintended consequence of human activity, the solution to global warming will require concerted and thoughtful action. Global warming cannot be stopped completely: Even if we were to immediately halt all production of greenhouse gases, the average surface temperature would increase an additional 2°C by 2100 thanks to the natural inertia of Earth's climate system. Instead, our goal is to minimize the effects of global warming by changing the ways in which we produce energy, making choices about how to decrease our requirements for energy, and finding ways to remove CO_2 from the atmosphere and trap it in the oceans or beneath our planet's surface. Confronting global warming is perhaps the greatest challenge to face our civilization in the twenty-first century.

Key Words

albedo, p. 212
asthenosphere, p. 215
atmosphere (atm), p. 224
atmospheric pressure, p. 224
aurora (*plural* aurorae), p. 222
aurora australis, p. 222
aurora borealis, p. 222
biosphere, p. 226
convection, p. 210
convection cell, p. 225
convection current, p. 210
core, p. 213
crust (of Earth), p. 213
crystal, p. 216
earthquake, p. 213
epicenter, p. 213
global warming, p. 230
greenhouse effect, p. 213
greenhouse gas, p. 213
igneous rock, p. 220
inner core (of Earth), p. 214
lava, p. 220
lithosphere, p. 219
magma, p. 220
magnetopause, p. 221
magnetosphere, p. 221
mantle, p. 213
melting point, p. 215
mesosphere, p. 225
metamorphic rock, p. 220
mineral, p. 220
northern lights, p. 222
oceanic rift, p. 219
outer core (of Earth), p. 214
outgassing, p. 223
ozone, p. 225
ozone layer, p. 229
ozone hole, p. 229
P wave, p. 214
photosynthesis, p. 223
plastic, p. 215
plate (lithospheric), p. 216
plate tectonics, p. 218
respiration, p. 223
rock, p. 220
S wave, p. 214
seafloor spreading, p. 217
sedimentary rock, p. 220
seismic wave, p. 213
seismograph, p. 213
shock wave, p. 221
southern lights, p. 222
stratosphere, p. 225
subduction zone, p. 219
surface wave, p. 214
thermosphere, p. 226
troposphere, p. 225
Van Allen belts, p. 221

Key Ideas

Earth's Energy Sources: All activity in Earth's atmosphere, oceans, and surface is powered by three sources of energy.

• Solar energy is the energy source for the atmosphere. In the greenhouse effect, some of this energy is trapped by infrared-absorbing gases in the atmosphere, raising Earth's surface temperature.

• Tidal forces from the Moon and Sun help to power the motion of the oceans.

• The internal heat of Earth is the energy source for geologic activity.

Earth's Interior: Studies of seismic waves (vibrations produced by earthquakes) show that Earth has a small, solid inner core surrounded by a liquid outer core. The outer core is surrounded by the dense mantle, which in turn is surrounded by the thin low-density crust.

• Seismologists deduce Earth's interior structure by studying how longitudinal P waves and transverse S waves travel through Earth's interior.

• Earth's inner and outer cores are composed of almost pure iron with some nickel mixed in. The mantle is composed of iron-rich minerals.

• Both temperature and pressure steadily increase with depth inside Earth.

Plate Tectonics: Earth's crust and a small part of its upper mantle form a rigid layer called the lithosphere. The lithosphere is divided into huge plates that move about over the plastic layer called the asthenosphere in the upper mantle.

• Plate tectonics, or movement of the plates, is driven by convection within the asthenosphere. Molten material wells up at oceanic rifts, producing seafloor spreading, and is returned to the asthenosphere in subduction zones. As one end of a plate is subducted back into the asthenosphere, it helps to pull the rest of the plate along.

• Plate tectonics is responsible for most of the major features of Earth's surface, including mountain ranges, volcanoes, and the shapes of the continents and oceans.

• Plate tectonics is involved in the formation of the three major categories of rocks: igneous rocks (cooled from molten material), sedimentary rocks (formed by the action of wind, water, and ice), and metamorphic rocks (altered in the solid state by extreme heat and pressure).

Earth's Magnetic Field and Magnetosphere: Electric currents in the liquid outer core generate a magnetic field. This magnetic field produces a magnetosphere that surrounds Earth and blocks the solar wind from hitting the atmosphere.

• A bow-shaped shock wave, where the supersonic solar wind is abruptly slowed to subsonic speeds, marks the outer boundary of the magnetosphere.

• Most of the particles of the solar wind are deflected around Earth by the magnetosphere.

• Some charged particles from the solar wind are trapped in two huge, doughnut-shaped rings called the Van Allen belts. An excess of these particles can initiate an auroral display.

Earth's Atmosphere: Earth's atmosphere differs from those of the other terrestrial planets in its chemical composition, circulation pattern, and temperature profile.

• Earth's atmosphere evolved from being mostly water vapor to being rich in carbon dioxide. A strong greenhouse effect kept Earth warm enough for water to remain liquid and to permit the evolution of life.

• The appearance of photosynthetic living organisms led to our present atmospheric composition, about four-fifths nitrogen and one-fifth oxygen.

• Earth's atmosphere is divided into layers called the troposphere, stratosphere, mesosphere, and thermosphere. Ozone molecules in the stratosphere absorb ultraviolet light.

• Because of Earth's rapid rotation, the circulation in its atmosphere is complex, with three circulation cells in each hemisphere.

The Biosphere: Human activity is changing Earth's biosphere, on which all living organisms depend.

• Industrial chemicals released into the atmosphere have damaged the ozone layer in the stratosphere.

• Deforestation and the burning of fossil fuels are increasing the greenhouse effect in our atmosphere and warming the planet. This can lead to destructive changes in the climate.

Questions

Review Questions

1. Describe the various ways in which Earth is unique among the planets of our solar system.
2. Describe the various ways in which Earth's surface is reshaped over time.
3. Why are typical rocks found on Earth's surface much younger than Earth itself?
4. What is convection? What causes convection in Earth's atmosphere?
5. Describe how energy is transferred from Earth's surface to the atmosphere by both convection and radiation.
6. If heat flows to Earth's surface from both the Sun and Earth's interior, why do we say that the motions of the atmosphere are powered by the Sun?
7. How does solar energy help power the motion of water in Earth's oceans?
8. How does the greenhouse effect influence the temperature of the atmosphere? Which properties of greenhouse gases in the atmosphere cause this effect?
9. How do we know that Earth was once entirely molten?
10. What are the different types of seismic waves? Why are seismic waves useful for probing Earth's interior structure?
11. Describe the interior structure of Earth.
12. What is a plastic material? Which parts of Earth's interior are described as being plastic?
13. The deepest wells and mines go down only a few kilometers. What, then, is the evidence that iron is abundant in Earth's core? That Earth's outer core is molten but the inner core is solid?
14. The inner core of Earth is at a higher temperature than the outer core. Why, then, is the inner core solid and the outer core molten instead of the other way around?
15. Describe the process of plate tectonics. Give specific examples of geographic features created by plate tectonics.
16. Explain how convection in Earth's interior drives the process of plate tectonics.
17. Why do you suppose that active volcanoes, such as Mount St. Helens in Washington State, are usually located in mountain ranges that border on subduction zones?
18. What is the difference between a rock and a mineral?
19. What are the differences among igneous, sedimentary, and metamorphic rocks? What do these rocks tell us about the sites at which they are found?
20. Why do geologists suspect that Pangaea was the most recent in a succession of supercontinents?
21. Why do geologists think that Earth's magnetic field is produced in the liquid outer core rather than in the mantle?
22. Describe Earth's magnetosphere. If Earth did not have a magnetic field, do you think aurorae would be more common or less common than they are today?
23. What are the Van Allen belts?
24. What phenomena tended to make the young Earth freeze over? What other phenomena prevented this from happening?
25. Summarize the history of Earth's atmosphere. What role has biological activity played in this evolution?
26. Ozone and carbon dioxide and ozone each make up only a fraction of a percent of our atmosphere. Why, then, should we be concerned about small increases or decreases in the atmospheric abundance of these gases?
27. How are scientists able to measure what the atmospheric CO_2 concentration and average surface temperature were in the distant past?

28. Does global warming increase the surface temperature of all parts of Earth by equal amounts or by different amounts? What consequences does this have?

Advanced Questions

Problem-solving tips and tools

You can find most of Earth data that you need in Tables 9-1 and 9-3. You will need to know that a sphere has surface area $4\pi r^2$ and volume $4\pi r^3/3$, where r is the sphere's radius. You may have to consult an atlas to examine the geography of the South Atlantic. Also, remember that the average density of an object is its mass divided by its volume. Sections 5-3 and 5-4 and Box 5-2 describe how to solve problems involving blackbody radiation, Section 2-6 discusses precession, and Section 4-4 describes the properties of elliptical orbits.

29. The total power in sunlight that reaches the top of our atmosphere is 1.75×10^{17} W. (**a**) How many watts of power are reflected back into space due to Earth's albedo? (**b**) If Earth had no atmosphere, all of the solar power that was not reflected would be absorbed by Earth's surface. In equilibrium, the heated surface would act as a blackbody that radiates as much power as it absorbs from the Sun. How much power would the entire Earth radiate? (**c**) How much power would one square meter of the surface radiate? (**d**) What would be the average temperature of the surface? Give your answer in both the Kelvin and Celsius scales. (**e**) Why is Earth's actual average temperature higher than the value you calculated in (d)?
30. On average, the temperature beneath Earth's crust increases at a rate of 20°C per kilometer. At what depth would water boil? (Assume the surface temperature is 20°C and ignore the effect of the pressure of overlying rock on the boiling point of water.)
31. What fractions of Earth's total volume are occupied by the core, the mantle, and the crust?
32. What fraction of the total mass of Earth lies within the inner core?
33. (**a**) Using the mass and diameter of Earth listed in Table 9-1, verify that the average density of Earth is 5500 kg/m^3. (**b**) Assuming that the average density of material in Earth's mantle is about 3500 kg/m^3, what must the average density of the core be? Is your answer consistent with the values given in Table 9-3?
34. Identical fossils of the reptile *Mesosaurus,* which lived 300 million years ago, are found in eastern South America and western Africa and nowhere else in the world. Explain how these fossils provide evidence for the theory of plate tectonics.
35. Measurements of the sea floor show that the Eurasian and North American plates have moved 60 km apart in the past 3.3 million years. How far apart (in millimeters) do they move in one year? (By comparison, your fingernails grow at a rate of about 50 mm/year.)
36. The oldest rocks found on the continents are about 4 billion years old. By contrast, the oldest rocks found on the ocean floor are only about 200 million years old. Explain why there is such a large difference in ages.
37. It was stated in Section 7-7 that iron loses its magnetism at temperatures above 770°C. Use this fact and Figure 9-10 to explain why Earth's magnetic field cannot be due to a magnetized solid core, but must instead be caused by the motion of electrically conducting material in the liquid core.
38. Most auroral displays (like those shown in Figure 9-21*c*) have a green color dominated by emission from oxygen atoms at a wavelength of 557.7 nm. (**a**) What minimum energy (in electron volts) must be imparted to an oxygen atom to make it emit this wavelength? (**b**) Why is your answer in (a) a *minimum* value?
39. Describe how the present-day atmosphere and surface temperature of Earth might be different (**a**) if carbon dioxide had never been released into the atmosphere; (**b**) if carbon dioxide had been released, but life had never evolved on Earth.
40. The photograph below shows the soil at Daspoort Tunnel near Pretoria, South Africa. The whitish layer that extends from lower left to upper right is 2.2 billion years old. Its color is due to a lack of iron oxide. More recent soils typically contain iron oxide and have a darker color. Explain what this tells us about the history of Earth's atmosphere.

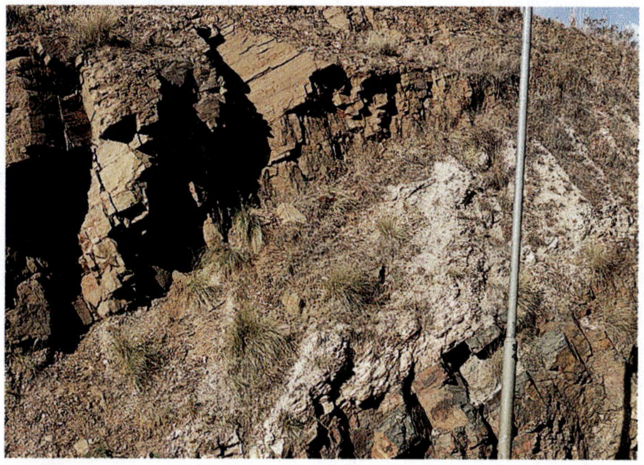

R I V U X G

(Courtesy of H. D. Holland)

41. Earth's atmospheric pressure decreases by a factor of one-half for every 5.5-km increase in altitude above sea level. Construct a plot of pressure versus altitude, assuming the pressure at sea level is one atmosphere (1 atm). Discuss the characteristics of your graph. At what altitude is the atmospheric pressure equal to 0.001 atm?
42. Earth is at perihelion on January 3 and at aphelion on July 4. Because of precession, in 15,000 A.D. the amount of sunlight in summer will be more than at present in the northern hemisphere but less than at present in the southern hemisphere. Explain why.
43. Antarctica has an area of 13 million square kilometers and is covered by an icecap that varies in thickness from 300 meters near the coast to 1800 meters in the interior. Estimate

the volume of this icecap. Assuming that water and ice have roughly the same density, estimate the amount by which the water level of the world's oceans would rise if Antarctica's ice were to melt completely (see Figure 9-31). What portions of Earth's surface would be inundated by such a deluge?

Discussion Questions

44. The human population on Earth is currently doubling about every 30 years. Describe the various pressures placed on Earth by uncontrolled human population growth. Can such growth continue indefinitely? If not, what natural and human controls might arise to curb this growth? It has been suggested that overpopulation problems could be solved by colonizing the Moon or Mars. Do you think this is a reasonable solution? Explain your answer.
45. In order to alleviate global warming, it will be necessary to dramatically reduce the amount of carbon dioxide that we release into the atmosphere by burning petroleum. What changes in technology and society do you think will be needed to bring this about?

Web/eBook Questions

46. Search the World Wide Web for information about Rodinia, the supercontinent that predated Pangaea. When did this supercontinent form? When did it break apart? Do we have any evidence of supercontinents that predated Rodinia? Why or why not?
47. Because of plate tectonics, the arrangement of continents in the future will be different from today. Search the World Wide Web for information about "Pangaea Ultima," a supercontinent that may form in the distant future. When is it predicted to form? How will it compare to the supercontinent that existed 200 million years ago (see Figure 9-12*a*)?
48. Use the World Wide Web to investigate the current status of the Antarctic ozone hole. Is the situation getting better or worse? Is there a comparable hole over the North Pole? Why do most scientists blame the chemicals called CFCs for the existence of the ozone hole?
49. Use the World Wide Web to research predictions of Earth's future surface temperature. What are some predictions for the surface temperature in 2050? In 2100? What effects may the increased temperature have on human health?
50. The Kyoto Protocol is an agreement made by a number of nations around the world to reduce their production of greenhouse gases. Use the World Wide Web to investigate the status of the Kyoto Protocol. How many nations have signed the protocol? What fraction of the world's greenhouse gas production comes from these nations? Have any developed nations failed to sign? How effective has the protocol been?
51. **Observing Mountain Range Formation.** Access the two animations "The Collision of Two Plates: South America" and "The Collision of Two Plates: The Himalayas" in Chapter 9 of the *Universe* Web site or eBook. (**a**) In which case are the mountains formed by tectonic uplift? In which case are the mountains formed by volcanoes from rising lava? (**b**) For each animation, describe which plate is moving in which direction.

Activities

Observing Project

52. Use the *Starry Night Enthusiast*™ program to view Earth from space. In the **Favorites** menu, select **Solar System > Earth** to place yourself in the position of an astronaut flying outside his space-ship at about 5800 km above Earth's surface on September 26, 2002. [You may need to click once on a **Zoom** button to show Earth against the background stars. You may also see the astronaut's spacesuit and feet in the foreground. This foreground can be removed by clicking off **Feet** in the **View** menu.] Click and hold down and move the mouse to use the **Location Scroller** to rotate the cloud-enshrouded Earth beneath you, allowing you to view different parts of Earth. You can change the **Time** to view different parts of this surface in daylight. (**a**) Is there any evidence for the presence of an atmosphere on Earth? Explain your answer. Remove the cloud cover by opening **Options > Solar System,** clicking on **Planets-Moons** to display the Planets-Moons Options panel and turning off **Show atmosphere.** It is also helpful to remove all Earth-orbiting satellites by clicking on **View > Solar System > Satellites.** (**b**) Can you see any evidence of plate tectonics as you rotate Earth again? Locate the position of your home and zoom in on it, using the Zoom tool on the right of the toolbar to set the Field of View to about 16° × 11°. Again, use the location scroller to move around on Earth. (**c**) Can you see any evidence of the presence of life or of man-made objects? Right-click on Earth and click on Google Map to examine regions of Earth in great detail on these images from an orbiting satellite. You can attempt to locate your own home on these maps. As an exercise, find the country of Panama, zoom in progressively upon the Panama Canal and search for ships traversing this incredible waterway. (**d**) What do you conclude about the importance of sending spacecraft to planets to explore their surfaces at close range? Close the internet connection to Google Map. In *Starry Night Enthusiast*™ open the **Favorites** menu, select **Solar System,** and click on **Earth** to return to the initial time and position above Earth. Rotate Earth to place eastern Brazil, on the eastern tip of South America, near to the center of the window. At the time of this observation this area is near the terminator (the dividing line between the day and night sides of Earth). (**e**) As seen from eastern Brazil, is the sun rising or setting? Explain your answer.

10 Our Barren Moon

RIVUXG

Walking on the lunar surface, April 1972. (John Young, *Apollo 16*, NASA/Science Photo Library)

Other than Earth itself, the most familiar world in the solar system is the Moon. It is so near to us that some of its surface features are visible to the naked eye. But the Moon is also a strange and alien place, with dramatic differences from our own Earth. It has no substantial atmosphere, no global magnetic field, and no liquid water of any kind. On the airless Moon, the sky is black even at midday. And unlike the geologically active Earth, the Moon has a surface that changes very little with time. When *Apollo 16* astronauts Charlie Duke (shown in the photograph) and John Young explored the region of the lunar highlands known as the Descartes Formation, the youngest rocks that they found were virtually the same age as the oldest rocks found on Earth.

Because the Moon has changed so little over the eons, it has preserved the early history of the inner solar system and the terrestrial planets. In this chapter we will see evidence that the Moon's formation and early evolution was violent and chaotic. The *Apollo* missions revealed that the Moon may have formed some 4.5 billion years ago as the result of a titanic collision between the young Earth and a rogue protoplanet the size of Mars. The young Moon was then pelted incessantly for hundreds of millions of years by large chunks of interplanetary debris, producing a cratered landscape. We will discover that the largest of these impacts formed immense circular basins that then flooded with molten lava. You can see the solidified lava as dark patches that cover much of the face of the Moon. In later chapters we will find other evidence that catastrophic collisions have played an important role in shaping the solar system.

10-1 The Moon's airless, dry surface is covered with plains and craters

Compared with the distances between planets, which are measured in tens or hundreds of millions of kilometers, the average distance from Earth to the Moon is just 384,400 km (238,900 mi). Despite being so close, the Moon has an angular diameter of just $\frac{1}{2}°$, which tells us that it is a rather small world. The diameter of the Moon is 3476 km (2160 mi), just 27% of Earth's diameter and about the same as the distance from Las Vegas to Philadelphia or from London to Cairo (Figure 10-1). Table 10-1 lists some basic data about the Moon.

Motions of the Earth-Moon System

Although the Moon has a small mass (just 1.23% of Earth's mass), Newton's third law tells us that the Moon exerts just as much gravitational force on Earth as Earth does on the Moon (see Section 4-6). Hence, it is not strictly correct to say that the Moon orbits Earth; rather, they both orbit around a point between their centers called the

Learning Goals

By reading the sections of this chapter, you will learn

10-1 The nature of the Moon's surface

10-2 The story of human exploration of the Moon

10-3 How we have learned about the Moon's interior

10-4 How Moon rocks differ from rocks on Earth

10-5 Why scientists think the Moon formed as the result of a violent collision between worlds

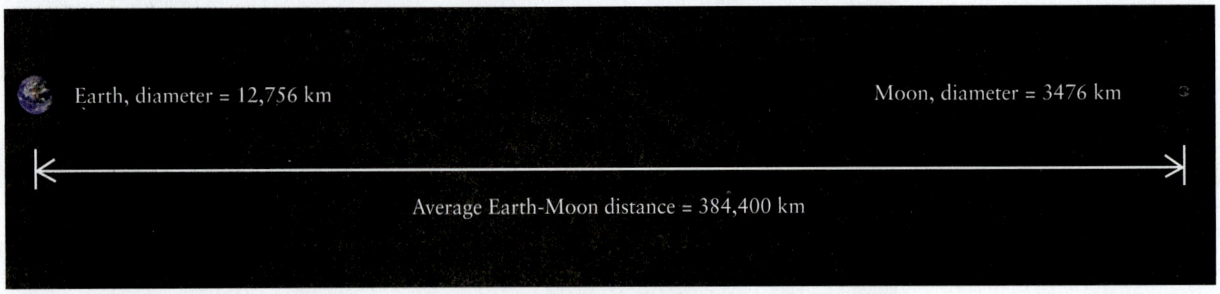

(a) The Earth-Moon system

(b) Earth and the Moon to scale, shown 10 times larger than in part (a)

Figure 10-1 R I V U X G

Comparing Earth and the Moon (a) The Moon is a bit more than one-quarter the diameter of Earth. The average Earth-Moon distance is about 30 times Earth's diameter. **(b)** Earth has blue oceans, an atmosphere streaked with white clouds, and continents continually being reshaped by plate tectonics. By contrast, the Moon has no oceans, no atmosphere, and no evidence of plate tectonics. (a: NASA; b: John Sanford, Science Photo Library)

center of mass of Earth-Moon system. (Because Earth is so much more massive than the Moon, the center of mass is close to Earth's center: Indeed, it actually lies within Earth's interior.) The center of mass then follows an elliptical orbit around the Sun (Figure 10-2*a*). In a similar way, a spinning wrench sliding along a table actually rotates around its center of mass (Figure 10-2*b*).

The Moon's Rotation and Libration

Even without binoculars or a telescope, you can easily see that dark gray and light gray areas cover vast expanses of the Moon's rocky surface. If you observe the Moon over many nights, you will see the Moon go through its phases as the dividing line between the illuminated and darkened portion of the Moon (the **terminator**) shifts. But you will always see the same pattern of dark and light areas, because the Moon keeps the same side turned toward Earth; you never see the **far side** (or back side) of the Moon. We saw in Section 3-2 that this is the result of the Moon's **synchronous rotation:** It takes precisely the same amount of time for the Moon to rotate on its axis (one lunar "day") as it does to complete one orbit around Earth.

ANIMATION 10.1

With patience, however, you can see slightly more than half the lunar surface, because the Moon appears to wobble slightly over the course of an orbit. It seems to rock back and forth around its north-south axis and to nod up and down in a north-south direction. This apparent periodic wobbling, called **libration,** permits us to view 59% of our satellite's surface.

LOOKING DEEPER 10.1

We stress that the Moon only *appears* to be wobbling. In fact, its rotation axis keeps nearly the same orientation in space as the Moon goes around its orbit. The reason why the Moon seems to rock back and forth is that its orbit around Earth is slightly elliptical. As a result, the Moon's orbital motion does not keep pace with its rotation at all points around the orbit. Furthermore, because the Moon's rotation axis is not exactly perpendicular to the plane of its orbit, it appears to us to nod up and down.

An Airless World

Clouds always cover some portion of Earth's surface. However, clouds are never seen on the Moon (see Figure 10-1), because the Moon is too small a world to have an atmosphere. Its surface gravity is low, only about one-sixth as great as that of Earth, and thus any gas molecules can easily escape into space (see Box 7-2). Because there is no atmosphere to scatter sunlight, the daytime sky on the Moon appears as black as the nighttime sky on Earth (see the photograph that opens this chapter).

The absence of an atmosphere means that the Moon can have no liquid water on its surface. Here on Earth, water molecules are kept in the liquid state by the pressure of the atmosphere pushing down on them. To see what happens to liquid water on the

Table 10-1 Moon Data

Distance from Earth (center to center):	Average: 384,400 km = 238,900 mi Maximum (apogee): 405,500 km Minimum (perigee): 363,300 km
Eccentricity of orbit:	0.0549
Average orbital speed:	3680 km/h
Sidereal period (relative to fixed stars):	27.322 days
Synodic period (new moon to new moon):	29.531 days
Inclination of lunar equator to orbit:	6.68°
Inclination of orbit to ecliptic:	5.15°
Diameter (equatorial):	3476 km = 2160 mi = 0.272 Earth diameter
Mass:	7.349×10^{22} kg = 0.0123 Earth mass
Average density:	3344 kg/m^3
Escape speed:	2.4 km/s
Surface gravity (Earth = 1):	0.17
Albedo:	0.11
Average surface temperatures:	Day: 130°C = 266°F = 403 K Night: −180°C = −292°F = 93 K
Atmosphere:	Essentially none

WEB LINK 10.2

RIVUXG

(Larry Landolfi, Science Photo Library)

airless Moon, consider what happens when the pressure of the atmosphere is reduced. The air pressure in Denver, Colorado (elevation 1650 m, or 5400 ft), is only 83% as great as at sea level. Therefore, in Denver, less energy has to be added to the molecules in a pot of water to make them evaporate, and water boils at only 95°C (= 368 K = 203°F), as compared with 100°C (= 373 K = 212°F) at sea level. (This temperature dependence on pressure is why some foods require longer cooking times at high elevations, where the boiling water used for cooking is not as hot as at sea level.) On the Moon, the atmospheric pressure is essentially

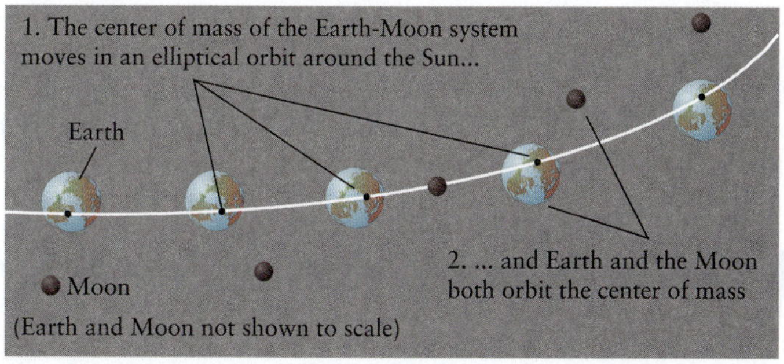

1. The center of mass of the Earth-Moon system moves in an elliptical orbit around the Sun...
Earth
Moon
2. ... and Earth and the Moon both orbit the center of mass
(Earth and Moon not shown to scale)

(a)

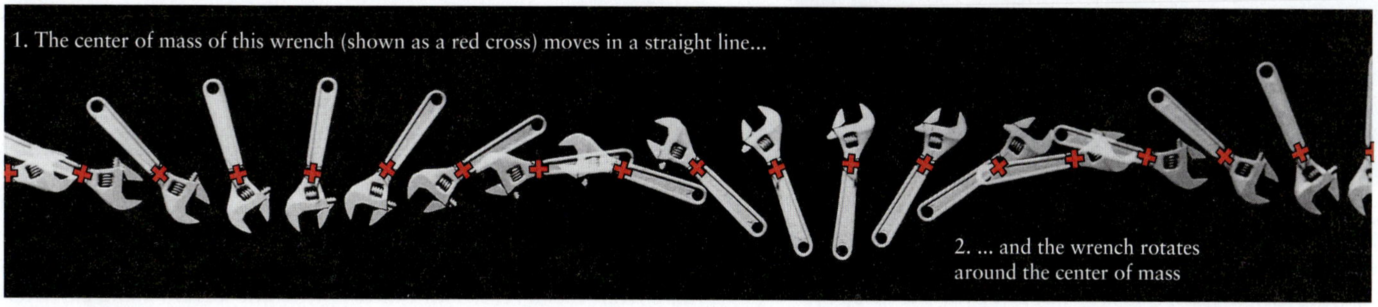

1. The center of mass of this wrench (shown as a red cross) moves in a straight line...
2. ... and the wrench rotates around the center of mass

(b)

Figure 10-2 RIVUXG

Motion of the Earth-Moon System (a) Earth and the Moon both orbit around their center of mass, which in turn follows an elliptical orbit around the Sun. **(b)** The illustration in part (a) is analogous to this time-lapse image of a spinning wrench sliding across a table.

(b: Berenice Abbott/Photo Researchers)

zero and the boiling temperature is lower than the temperature of the lunar surface. Under these conditions, water can exist as a solid (ice) or a vapor but not as a liquid.

CAUTION! If astronauts were to venture out onto the lunar surface without wearing spacesuits, the lack of atmosphere would *not* cause their blood to boil. That's because our skin is strong enough to contain bodily fluids. In the same way, if we place an orange in a sealed container and remove all the air, the orange retains all of its fluid and does not explode. It even tastes the same afterward! The reasons why astronauts wore pressurized spacesuits on the Moon (see the photograph that opens this chapter) were to provide oxygen for breathing, to regulate body temperature, and to protect against ultraviolet radiation from the Sun. Since the Moon has no atmosphere, and thus no ozone layer like that of Earth to screen out ultraviolet wavelengths (Section 9-7), an unprotected astronaut would suffer a sunburn after just 10 seconds of exposure!

Lunar Craters

Since the Moon has no atmosphere to obscure its surface, you can get a clear view of several different types of lunar features with a small telescope or binoculars. These include *craters,* the dark-colored *maria,* and the light-colored *lunar highlands* or *terrae* (Figure 10-3). Taken together, these different features show the striking differences between the surfaces of Earth and the Moon.

With an Earth-based telescope, some 30,000 **craters** are visible, with diameters ranging from 1 km to several hundred km. Close-up photographs from lunar orbit have revealed millions of craters too small to be seen from Earth. Following a tradition established in the seventeenth century, the most prominent craters are named after famous philosophers, mathematicians, and scientists, such as Plato, Aristotle, Pythagoras, Copernicus, and Kepler. Virtually all lunar craters, both large and small, are the result of the Moon's having been bombarded by meteoritic material. For this reason, they are also called **impact craters** (see Section 7-6). One piece of evidence that lunar craters are caused by impacts is that many large craters have a central peak, with a shape that could only result from high-speed impacts (see Figure 7-10*a*).

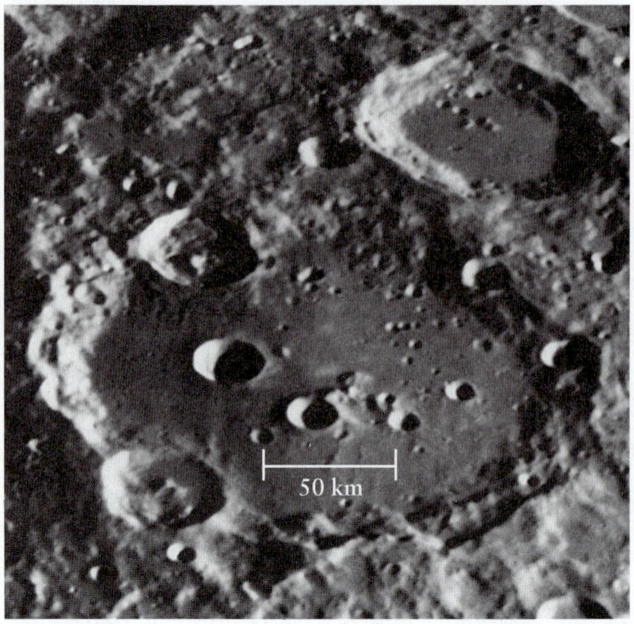

Figure 10-4 R I **V** U X G

The Crater Clavius This photograph from the 5-m (200-in) Palomar telescope shows one of the largest craters on the Moon. Clavius has a diameter of 232 km (144 mi) and a depth of 4.9 km (16,000 ft), measured from the crater's floor to the top of the surrounding rim. Later impacts formed the smaller craters inside Clavius. (Palomar Observatory)

Figure 10-3 R I **V** U X G

The Moon as Seen from Earth The key features of the lunar surface can be seen with binoculars, a small telescope, or even the naked eye. (Larry Landolfi, Science Photo Library)

WEB LINK 10.3

A meteoroid striking the lunar surface generates a shock wave over the surface outward from the point of impact. Such shock waves can spread over an area much larger than the size of the meteoroid that generated the wave. Very large craters more than 100 km across, such as the crater Clavius shown in Figure 10-4, were probably created by the impact of a fast-moving piece of rock only a few kilometers in radius.

Maria: Dark and Relatively Young

The large dark areas on the lunar surface shown in Figure 10-3 are called **maria** (pronounced "MAR-ee-uh"). They form a pattern that looks crudely like a human face, sometimes called "the man in the Moon." The singular form of maria is **mare** (pronounced "MAR-ay"), which means "sea" in Latin. These terms date from the seventeenth century, when observers using early telescopes thought these dark features were large bodies of water.

They gave the maria fanciful and romantic names such as Mare Tranquillitatis (Sea of Tranquillity), Mare Nubium (Sea of Clouds), Mare Nectaris (Sea of Nectar), Mare Serenitatis (Sea of Serenity), and Mare Imbrium (Sea of Showers). While modern astronomers still use these names, they now understand that the maria are not bodies of water (which, as we have seen, could not exist on the airless Moon) but rather the remains of huge lava flows. The maria get their distinctive appearance from the dark color of the solidified lava.

On the whole, the maria have far fewer craters than the surrounding terrain. Craters are caused by meteoritic bombardment, so the maria have not been exposed to that bombardment for as long as the surrounding terrain. Hence, the maria must be relatively young, and the lava flows that formed them must have occurred at a later stage in the Moon's geologic history (Figure 10-5). The craters that are found in the maria were caused by the relatively few impacts that have occurred since the maria solidified (Figure 10-6).

Although they are much larger than craters, maria also tend to be circular in shape (see Figure 10-5). This suggests that, like craters, these depressions in the lunar surface were also created by impacts. It is thought that very large meteoroids or asteroids some tens of kilometers across struck the Moon, forming basins. These depressions subsequently flooded with lava that flowed out from the Moon's interior through cracks in the lunar crust. The solidified lava formed the maria that we see today. The *Cosmic Connections* figure depicts our understanding of how lunar craters and maria formed.

An impact large enough to create a mare basin must have thrust upward tremendous amounts of material around the impact site. This explains the mountain ranges that curve in circular arcs around mare basins. Analysis of lunar rocks as well as observations made from lunar orbit have validated this picture of the origin of lunar mountains.

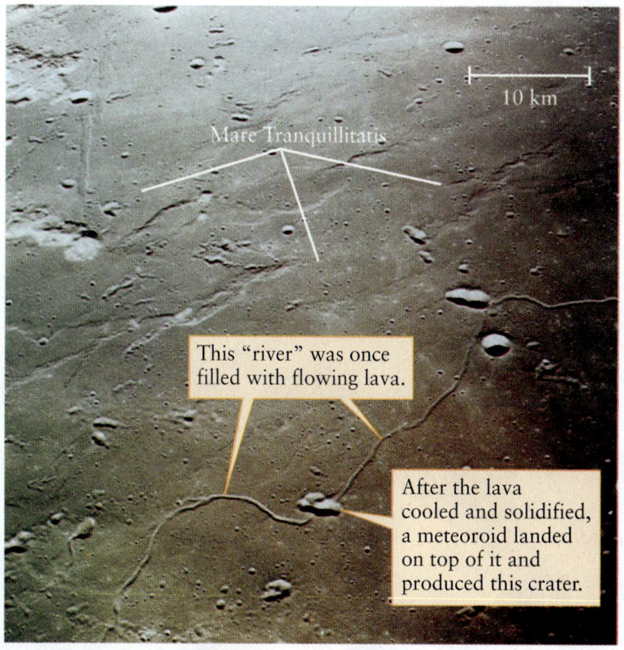

Figure 10-6 R I **V** U X G

Details of Mare Tranquillitatis This color photograph from lunar orbit reveals numerous tiny craters in the surface of a typical mare. The curving "river," or rille, was carved out by flowing lava that later solidified; similar features are found in volcanic areas on Earth. The *Apollo 11* landing site is near the center of the top edge of this image. (*Apollo 10*, NASA)

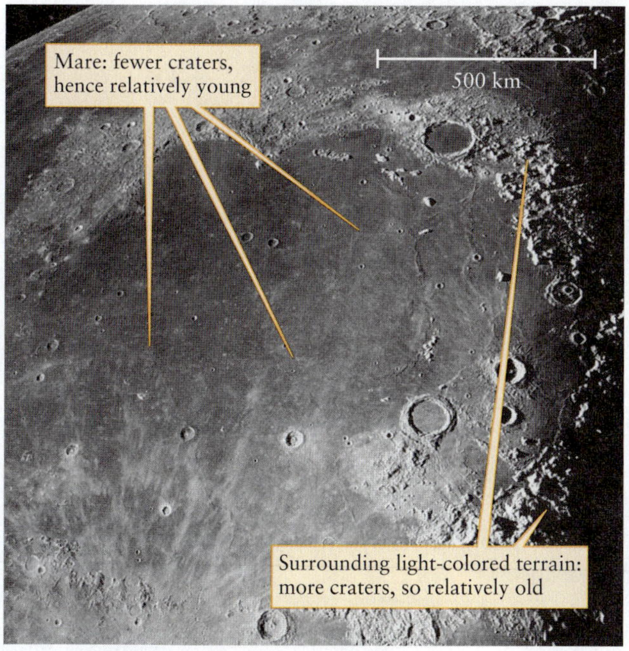

Figure 10-5 R I **V** U X G

Mare Imbrium from Earth Mare Imbrium is the largest of 14 dark plains that dominate the Earth-facing side of the Moon. Its diameter of 1100 km (700 mi) is about the same as the distance from Chicago to Philadelphia or from London to Rome. The maria formed after the surrounding light-colored terrain, so they have not been exposed to meteoritic bombardment for as long and hence have fewer craters. (Carnegie Observatories)

Lunar Highlands: Light-Colored and Ancient

The light-colored terrain that surrounds the maria is called **terrae,** from the Latin for "lands." (Seventeenth-century astronomers thought that this terrain was the dry land that surrounded lunar oceans.) Detailed measurements by astronauts in lunar orbit demonstrated that the maria on the Moon's Earth-facing side are 2 to 5 km below the average lunar elevation, while the terrae extend to several kilometers above the average elevation. For this reason terrae are often called **lunar highlands.** The flat, low-lying, dark gray maria cover only 15% of the lunar surface; the remaining 85% is light gray lunar highlands. Since the lunar highlands are more heavily cratered than the maria, we can conclude that the highlands are older.

WEB LINK 10.4

Remarkably, when the Moon's far side was photographed for the first time by the Soviet *Luna 3* spacecraft in 1959, scientists were surprised to find almost

COSMIC CONNECTIONS

The Formation of Craters and Maria on the Moon

The physical features of lunar craters and maria show that they were formed by meteoroids impacting the Moon's surface at high speed. The crater labeled in the photograph, called Aristillus after a Greek astronomer who lived around 300 B.C., is 55 km (34 mi) across and 3600 m (12,000 ft) deep. (Photograph: Lunar and Planetary Institute/Universities Space Research Association)

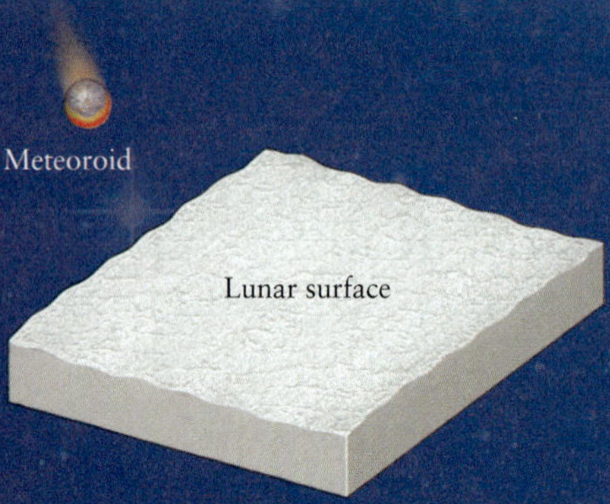

1. A meteoroid approaches the lunar surface.

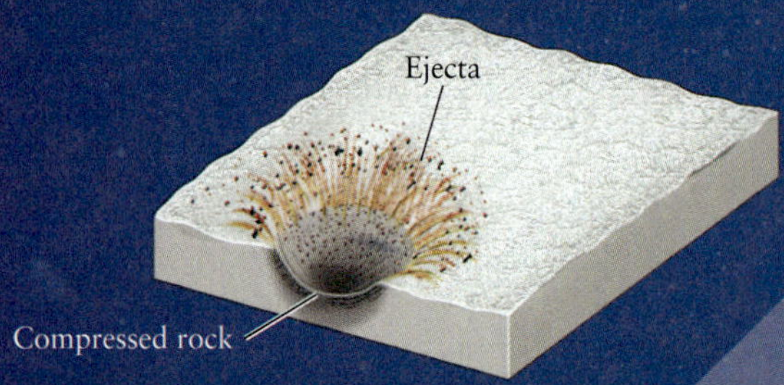

2. When the meteoroid strikes, it ejects rocks and boulders and produces a shock wave in the lunar surface.

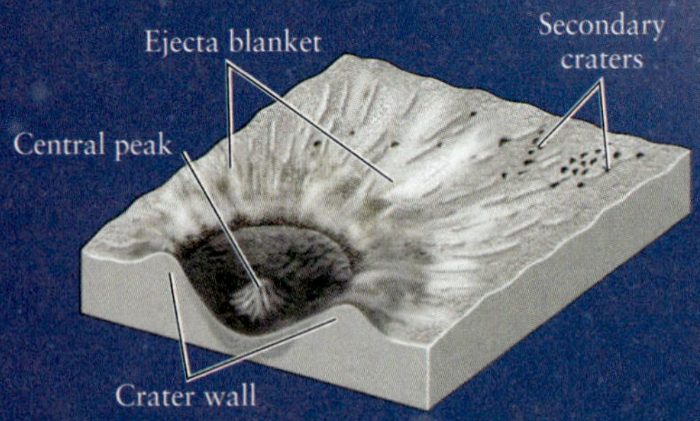

3. What remains is a crater with a central peak. The ejected material forms a blanket around the crater and can cause secondary craters.

R I V U X G

Lava fills crater and solidifies to form the dark mare floor.

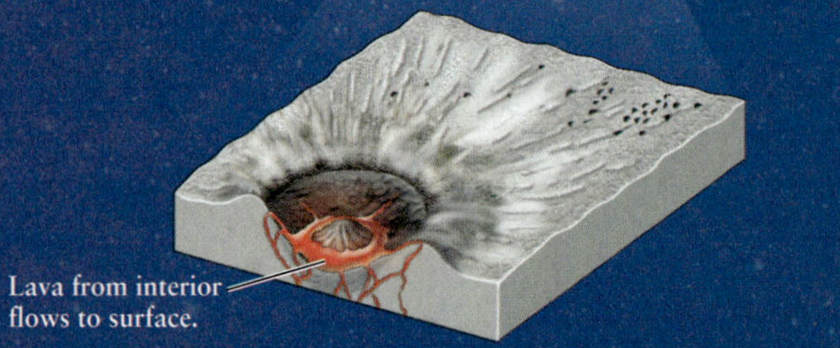

4. A very large meteoroid can crack the lunar crust, forming a mare.

no maria there. Thus, lunar highlands predominate on the side of the Moon that faces away from Earth (Figure 10-7). Presumably, the crust is thicker on the far side, so even the most massive impacts there were unable to crack through the crust and let lava flood onto the surface. The Moon is not unique in having two hemispheres with different crustal thicknesses: Earth's northern hemisphere has much more continental crust than its southern hemisphere, and the southern hemisphere of Mars has a thicker crust than the northern hemisphere.

Although the Moon's far side is deficient in maria compared to the near side, the highlands on the far side are as heavily pockmarked by impacts as those on the near side. The most prominent feature on the far side is the South Pole–Aitken Basin, the largest known impact crater in the solar system. This basin is about 2500 km (1600 mi) across—about the same size as the continent of Europe—and 12 km (7 mi) deep.

A World Without Plate Tectonics

Something is entirely missing from our description of the Moon's surface—any mention of plate tectonics. Indeed, no evidence for plate tectonics can be found on the Moon. Recall from Section 9-3 that Earth's major mountain ranges were created by collisions between lithospheric plates. In contrast, it appears that the Moon is a *one-plate world:* Its entire lithosphere is a single, solid plate. There are no rifts where plates are moving apart, no mountain ranges of the sort formed by collisions between plates, and no subduction zones. Thus, while lunar mountain ranges bear the names of terrestrial ranges, such as the Alps, Carpathians, and Apennines, they are entirely different in character: they were formed by material ejected from impact sites, not by plate tectonics. Geologically, the Moon is a dead world.

In contrast to Earth and its active geology, the Moon's surface is geologically dead

In Section 7-6 we discussed the general rule that a world with abundant craters has an old, geologically inactive surface. The Moon is a prime example of this rule. Without an atmosphere to cause erosion and without plate tectonics, the surface of the Moon changes only very slowly. That is why the Moon's craters, many of which are now known to be billions of years old, remain visible today. On Earth, by contrast, the continents and seafloor are continually being reshaped by seafloor spreading, mountain building, and subduction. Thus, any craters that were formed by impacts on the ancient Earth's surface have long since been erased. Fewer than 200 impact craters exist on Earth today; these were formed within the last few hundred million years and have not yet been erased.

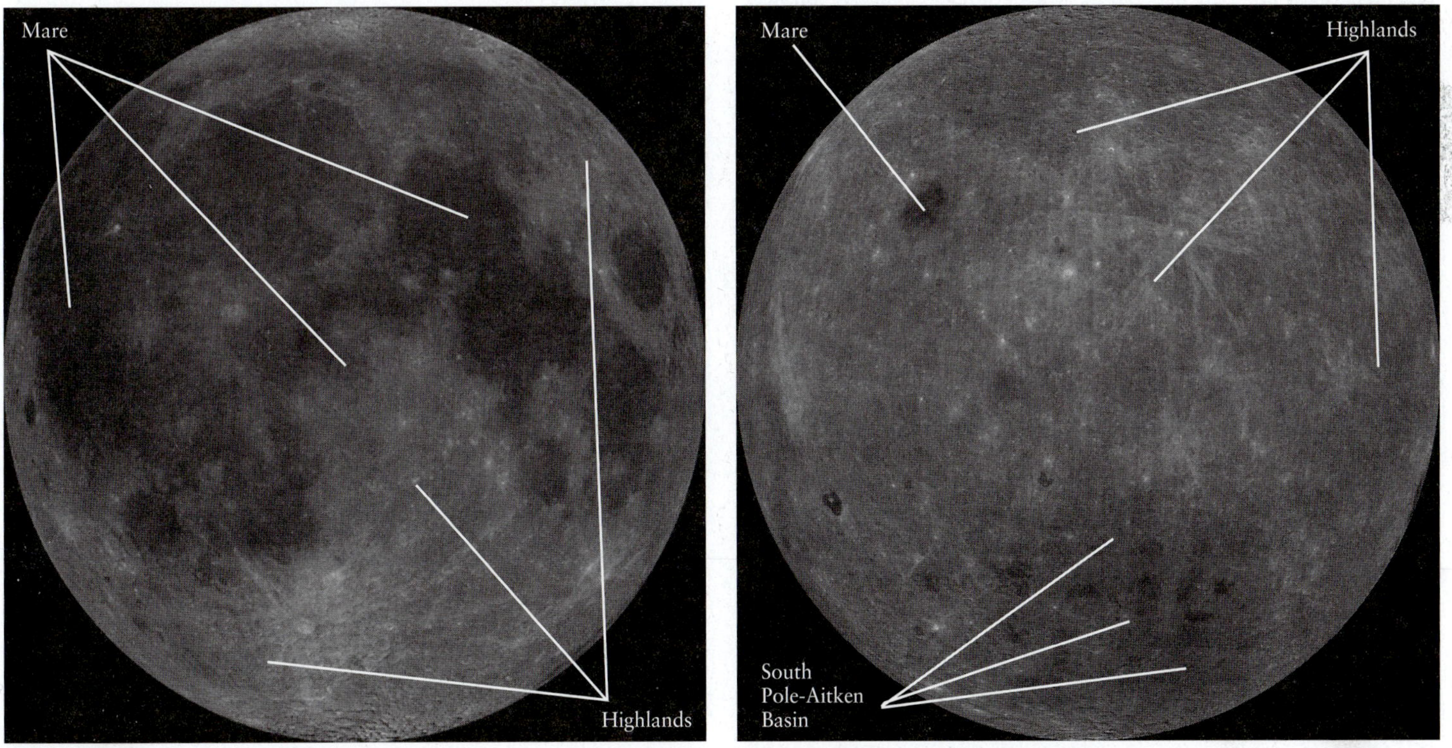

Near side: highlands and maria

Far side: highlands but almost no maria

VIDEO 10.1

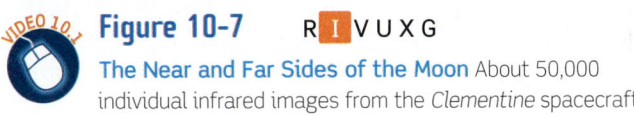

Figure 10-7 R **I** V U X G

The Near and Far Sides of the Moon About 50,000 individual infrared images from the *Clementine* spacecraft were combined to create these views of the near (Earth-facing) side and far side of the Moon. (NASA)

10-2 Human exploration of the Moon in the 1960s and 1970s has been continued by robotic spacecraft

For thousands of years, storytellers have invented fabulous tales of voyages to the Moon. The seventeenth-century French satirist Cyrano de Bergerac wrote of a spacecraft whose motive power was provided by spheres of morning dew. (After all, reasoned Cyrano, dew rises with the morning sun, so a spacecraft would rise along with it.) In his imaginative tale *Somnium,* Cyrano's contemporary Johannes Kepler (the same Kepler who deduced the laws of planetary motion) used the astronomical knowledge of his time to envision what it would be like to walk on the Moon's surface. In 1865 the French author Jules Verne published *From the Earth to the Moon,* a story of a spacecraft sent to the Moon from a launching site in Florida.

Almost exactly a hundred years later, Florida was indeed the starting point of the first voyages made by humans to other worlds. Between 1969 and 1972, 12 astronauts walked on the Moon, and fantasy became magnificent reality.

The Moon is the only world other than Earth to be visited by humans

The First Spacecraft to the Moon

As in Verne's story, politics had much to do with the motivation for going to the Moon. But in fact, there were excellent scientific reasons to do so. Because Earth is a geologically active planet, typical terrestrial surface rocks are only a few hundred million years old, which is only a small fraction of Earth's age. Thus, all traces of Earth's origins have been erased. By contrast, rocks on the geologically inactive surface of the Moon have been essentially undisturbed for billions of years. Samples of lunar rocks have helped answer many questions about the Moon and have shed light on the origin of Earth and the birth of the entire solar system.

Lunar missions began in 1959, when the Soviet Union sent three small spacecraft—*Lunas 1, 2,* and *3*—toward the Moon. American attempts to reach the Moon began in the early 1960s with Project Ranger. Equipped with six television cameras, the *Ranger* spacecraft transmitted close-up views of the lunar surface taken in the final few minutes before they crashed onto the Moon. These images showed far more detail than even the best Earth-based telescope.

In 1966 and 1967, five *Lunar Orbiter* spacecraft were placed in orbit around the Moon and made a high-resolution photographic survey of 99.5% of the lunar surface. Some of these photographs show features as small as 1 m across and were essential in helping NASA scientists choose suitable landing sites for the astronauts.

Even after the *Ranger* missions, it was unclear whether the lunar surface was a safe place to land. One leading theory was that billions of years of impacts by meteoroids had ground the Moon's surface to a fine powder. Perhaps a spacecraft attempting to land would simply sink into a quicksandlike sea of dust! Thus, before a manned landing on the Moon could be attempted, it was necessary to try a soft landing of an unmanned spacecraft. Thus, the purpose of the *Surveyor* program was to land unmanned spacecraft on the Moon. Between June 1966 and January 1968, five *Surveyors* successfully touched down on the Moon, sending back pictures and data directly from the lunar surface. These missions demonstrated convincingly that the Moon's surface was as solid as that of Earth. Figure 10-8 shows an astronaut visiting *Surveyor 3,* two and a half years after it had landed on the Moon.

The *Apollo* and *Luna* Missions

The first of six manned lunar landings took place on July 20, 1969, when the *Apollo 11* lunar module *Eagle* set down in Mare Tranquillitatis (see Figure 10-6). Astronauts Neil Armstrong and Edwin "Buzz" Aldrin were the first humans to set foot on the Moon. *Apollo 12* also landed in a mare, Oceanus Procellarum (Ocean of Storms). The *Apollo 13* mission experienced a nearly catastrophic inflight explosion that made it impossible for it to land on the Moon. Fortunately, titanic efforts by the astronauts and ground personnel brought it safely home. The remaining four missions, *Apollo 14* through *Apollo 17,* were made in progressively more challenging terrain, chosen to permit the exploration of a wide variety of geologically interesting features. The first era of human exploration of the Moon culminated in December 1972, when *Apollo 17* landed in

Figure 10-8 R I V U X G

Visiting an Unmanned Pioneer *Surveyor 3* landed on the Moon in April 1967. It was visited in November 1969 by *Apollo 12* astronaut Alan Bean, who took pieces of the spacecraft back to Earth for analysis. (The *Apollo 12* lunar module is visible in the distance.) The retrieved pieces showed evidence of damage by micrometeoroid impacts, demonstrating that small bits of space debris are still colliding with the Moon. (Pete Conrad, *Apollo 12,* NASA) a rugged region of the lunar highlands. Figure 10-9 shows one of the *Apollo* bases.

WEB LINK 10.8 In a curious postscript to the *Apollo* missions, in the 1990s an "urban legend" arose that the Moon landings had never taken place and were merely an elaborate hoax by NASA. While amusing, this legend does not hold up under even the slightest scientific scrutiny. For example, believers in the "Moon hoax" cannot explain why the rocks brought back from the Moon are substantially different in their chemistry from Earth rocks, as we will see later in this chapter. It makes as much sense to ask "Do you believe that humans actually went to the Moon?" as it does to ask "Do you believe there was a First World War?" or "Do you believe there is such a place as Antarctica?"

Between 1966 and 1976, a series of unmanned Soviet spacecraft also orbited the Moon and soft-landed on its surface. The first soft landing, by *Luna 9,* came four months before the U.S. *Surveyor 1* arrived. The first lunar satellite, *Luna 10,* orbited the Moon four months before *Lunar Orbiter 1*. In the 1970s, robotic spacecraft named *Luna 17* and *Luna 21* landed vehicles that roamed the lunar surface, and *Lunas 16, 20,* and *24* brought samples of rock back to Earth. The *Luna* samples complement those returned by the *Apollo* astronauts because they were taken from different parts of the Moon's surface.

Recent Lunar Exploration

WEB LINK 10.9 Unmanned spacecraft continued the exploration of the Moon in the 1990s and 2000s. In 1994 a small spacecraft named *Clementine* spent more than two months observing the Moon from lunar orbit. Although originally intended to test lightweight imaging systems for future defense satellites, *Clementine* proved to be a remarkable tool for lunar exploration. It carried an array of very high-resolution cameras sensitive to ultraviolet and infrared light as well as the visible spectrum (see Figure 10-7). Different types of materials on the lunar surface reflect different wavelengths, so analysis of *Clementine* images made in different parts of the electromagnetic spectrum has revealed the composition of large areas of the Moon's surface (Figure 10-10). The task of analyzing the Moon's surface composition was carried out in even greater depth by the European Space Agency's *SMART-1* spacecraft, which orbited the Moon from 2004 to 2006 and carried even more sensitive instruments than *Clementine*.

One remarkable result of the *Clementine* mission was evidence for a patch of water ice tens or hundreds of kilometers across at the Moon's south pole, in low-lying locations where sunlight never reaches and where temperatures are low enough to prevent ice from evaporating. Radio waves beamed from *Clementine* toward these areas were detected by radio telescopes on Earth and showed the characteristic signature of reflection from ice.

WEB LINK 10.10 Data from the *Lunar Prospector* spacecraft, which went into orbit around the Moon in January 1998, reinforced the *Clementine* findings and indicated that even more ice might lie in deep craters at the Moon's north pole. The ice may have been deposited by comets (see Section 7-5) that crashed into the lunar surface.

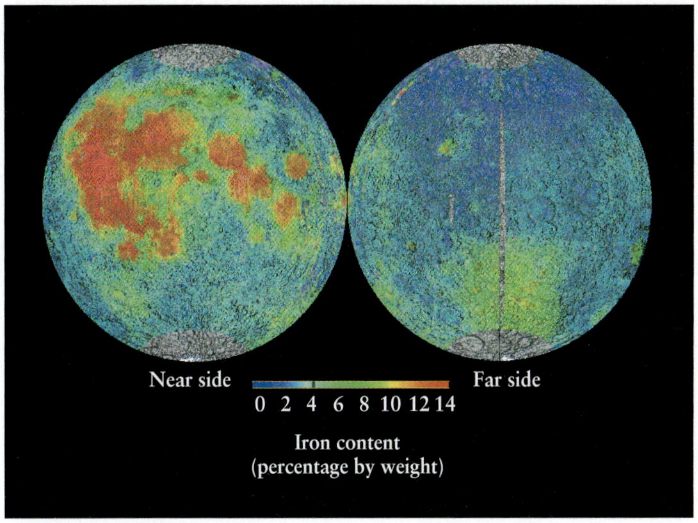

Figure 10-10

Iron on the Moon Images at different wavelengths from the *Clementine* spacecraft were used to make this map of the concentration of iron on the lunar surface. The areas of highest concentration (red) coincide closely with the maria on the near side (compare Figure 10-7), confirming that the maria formed from iron-rich lavas. The lowest iron concentration (blue) is found in the lunar highlands. (No data were collected in the gray areas.) The large green region of intermediate iron concentration on the far side is the South Pole–Aitken Basin (see Figure 10-7). The impact that created this basin may have excavated all the way to the Moon's mantle, in which iron is more abundant than in the crust. (JSC/NASA)

VIDEO 10.3 VIDEO 10.4 **Figure 10-9** R I V U X G

The *Apollo 15* Base This photograph shows astronaut James Irwin and the *Apollo 15* landing site at the foot of the Apennine Mountains near the eastern edge of Mare Imbrium. The hill in the background is about 5 km away and rises about 3 km high. At the right is a Lunar Rover, an electric-powered vehicle used by astronauts to explore a greater radius around the landing site. The last three *Apollo* missions carried Lunar Rovers. (Dave Scott, *Apollo 15,* NASA)

One complication is that the *Lunar Prospector* observations show only that there is an excess of hydrogen atoms at the lunar poles. These could be trapped within water molecules (H_2O) or within other, more exotic minerals. At the end of the *Lunar Prospector* mission in July 1999, the spacecraft was crashed into a crater at the lunar south pole in the hope that a plume of water vapor might rise from the impact site. Although more than a dozen large Earth-based telescopes watched the Moon during the impact, no such plume was observed, nor was any debris of any kind. Radar observations made from Earth in 2003 suggest that if there is ice at the lunar poles, it is in sheets no more than a few centimeters thick. A new generation of European, Japanese, Indian, Chinese, and U.S. unmanned spacecraft may help provide a definitive answer about ice at the Moon's poles.

10-3 The Moon has no global magnetic field but has a small core beneath a thick mantle

We saw in Section 7-6 and Section 7-7 that a small planet or satellite will have little internal heat, and hence is likely to have a mostly solid interior. Lacking substantial amounts of moving molten material in its interior, a small world is unlikely to generate a global magnetic field. As we will see, these general rules describe the Moon very well.

Compared to earthquakes, moonquakes are few and feeble—but they reveal key aspects of the Moon's interior

The Moon's Interior: The Core

The Moon landings gave scientists an unprecedented opportunity to explore an alien world. The *Apollo* spacecraft were therefore packed with an assortment of apparatus that the astronauts deployed around the landing sites (see Figure 10-9). For example, all the missions carried magnetometers, which confirmed that the present-day Moon has no global magnetic field. However, careful magnetic measurements of lunar rocks returned by the *Apollo* astronauts indicated that the Moon did have a weak magnetic field when the rocks solidified billions of years ago. Hence, the Moon must originally have had a small, molten, iron-rich core. (We saw in Section 7-7 that Earth's magnetic field is generated by fluid motions in its interior.) The core presumably solidified at least partially as the Moon cooled, so that the lunar magnetic field disappeared.

WEB LINK 10.11

More recently, scientists used the *Lunar Prospector* to probe the character of the Moon's core. The spacecraft made extensive measurements of lunar gravity and of how the Moon responds when it passes through Earth's magnetosphere (Section 9-4). When combined with detailed observations of how Earth's tidal forces affect the motions of the Moon, the data indicate that the present-day Moon has a partially liquid, iron-rich core about 700 km (435 mi) in diameter. While 32% of Earth's mass is in its core, the core of the Moon contains only 2% to 3% of the lunar mass.

The Moon's Interior: The Mantle

The *Apollo* astronauts also set up seismometers on the Moon's surface, which made it possible for scientists to investigate the Moon's interior using seismic waves just as is done for Earth (Section 9-2). It was found that the Moon exhibits far less seismic activity than Earth. Roughly 3000 **moonquakes** were detected per year, whereas a similar seismometer on Earth would record hundreds of thousands of earthquakes per year. Furthermore, typical moonquakes are very weak. A major terrestrial earthquake is in the range from 6 to 8 on the Richter scale, while a major moonquake measures 0.5 to 1.5 on that scale and would go unnoticed by a person standing near the epicenter.

Analysis of the feeble moonquakes reveals that most originate 600 to 800 km below the surface, deeper than most earthquakes. On Earth, the deepest earthquakes mark the boundary between the solid lithosphere and the plastic asthenosphere, because the lithosphere is brittle enough to fracture and produce seismic waves, whereas the deeper rock of the asthenosphere oozes and flows rather than cracks. We may conclude that the Moon's lithosphere is about 800 km thick (Figure 10-11). By contrast, Earth's lithosphere is only 50 to 100 km thick.

The lunar asthenosphere extends down to the Moon's relatively small iron-rich core, which is more than 1400 km below the lunar surface. The asthenosphere and the lower part of the lithosphere are presumably composed of relatively dense iron-rich rocks. The uppermost part of the lithosphere is a lower-density crust, with an average thickness of about 60 km on the Earth-facing side and up to 100 km on the far side. For comparison, the thickness of Earth's crust ranges from 5 km under the oceans to about 30 km under major mountain ranges on the continents

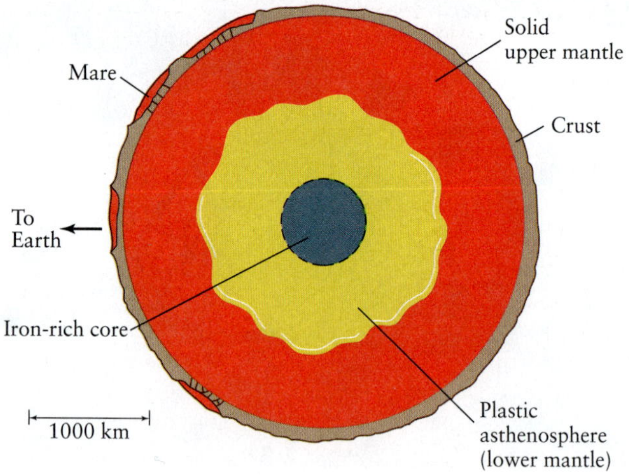

Figure 10-11

The Internal Structure of the Moon Like Earth, the Moon has a crust, a mantle, and a core. The lunar crust has an average thickness of about 60 km on the Earth-facing side but about 100 km on the far side. The crust and solid upper mantle form a lithosphere about 800 km thick. The plastic (nonrigid) asthenosphere extends all the way to the base of the mantle. The iron-rich core is roughly 700 km in diameter. (Compare Figure 9-9.)

(see Table 9-3). This slightly lopsided distribution of mass between the near and far sides of the Moon helps to keep the Moon in synchronous rotation, so that we always see the same side of the Moon (see Section 4-8).

On the whole, the Moon's interior is much more rigid and less fluid than the interior of Earth. Plate tectonics requires that there be fluid motion just below a planet's surface, so it is not surprising that there is no evidence for plate tectonics on the lunar surface.

The Origin of Moonquakes

Earthquakes tend to occur at the boundaries between tectonic plates, where plates move past each other, collide, or undergo subduction (see Figure 9-14). But there are no plate motions on the Moon. What, then, causes moonquakes? The answer turns out to be Earth's tidal forces. Just as the tidal forces of the Sun and Moon deform Earth's oceans and give rise to ocean tides (see Figure 4-26), Earth's tidal forces deform the solid body of the Moon. The amount of the deformation changes as the Moon moves around its elliptical orbit and the Earth-Moon distance varies. As a result, the body of the Moon flexes slightly, triggering moonquakes. The tidal stresses are greatest when the Moon is nearest Earth—that is, at perigee (see Section 3-5)—which is just when the *Apollo* seismometers reported the most moonquakes. Box 10-1 has more information about how the tidal stresses on the Moon depend on its distance from Earth.

Without tectonics, and without the erosion caused by an atmosphere or oceans, the only changes occurring today on the lunar surface are those due to meteoroid impacts. These impacts were also monitored by the seismometers left on the Moon by the *Apollo* astronauts. These devices, which remained in operation for several years, were sensitive enough to detect a hit by a grapefruit-sized meteoroid anywhere on the lunar surface. The data show that every year the Moon is struck by 80 to 150 meteoroids having masses between 0.1 and 1000 kg (roughly ¼ lb to 1 ton).

10-4 Lunar rocks reveal a geologic history quite unlike that of Earth

WEB LINK 10.12

One of the most important scientific goals of the *Apollo* missions was to learn as much as possible about the lunar surface. Although only one of the 12 astronauts who visited the Moon was a professional geologist, all of the astronauts received intensive training in field geology, and they were in constant communication with planetary scientists on Earth as they explored the lunar surface.

Weathering by Meteoroids

Although no wind or rain disturbs the lunar surface, it is "weathered" over the ages by the impact of tiny meteoroids

The Moon has no atmosphere or oceans to cause "weathering" of the surface. But the *Apollo* astronauts found that the lunar surface has undergone a kind of erosion due to billions of years of relentless meteoritic bombardment. This bombardment has pulverized the surface rocks into a layer of fine powder and rock fragments called the **regolith**, from the Greek for "blanket of stone" (Figure 10-12). This layer varies in depth from about 2 to 20 m. Just as the rough-surfaced acoustic tile used in ceilings absorbs sound waves, the regolith absorbs most of the sunlight that falls on it. This absorption of light helps to account for the Moon's very low albedo (see Table 10-1).

Moon Rocks

In addition to their observations of six different landing sites, the *Apollo* astronauts brought back 2415 individual samples of lunar material totaling 382 kg (842 lb). In addition, the unmanned Soviet spacecraft *Luna 16, 20,* and *24* returned 300 g (⅔ lb) from three other sites on the Moon. This collection of lunar material has provided important information about the early history of the Moon that could have been obtained in no other way.

All of the lunar samples are igneous rocks; there are no metamorphic or sedimentary rocks (see Section 9-3, especially Figure 9-19). The presence of only igneous rocks suggests that most or all of the Moon's surface was once molten. Indeed, Moon rocks are composed mostly of the same minerals that are found in volcanic rocks on Earth.

Astronauts who visited the maria discovered that these dark regions of the Moon are covered with basaltic rock quite similar to the dark-colored rocks formed by lava from volcanoes on

Figure 10-12 R I V U X G

The Regolith Billions of years of bombardment by space debris have pulverized the uppermost layer of the Moon's surface into powdered rock. This layer, called the regolith, is utterly bone-dry. It nonetheless sticks together like wet sand, as shown by the sharp outline of an astronaut's bootprint. (*Apollo 11,* NASA)

BOX 10-1 Tools of the Astronomer's Trade

Calculating Tidal Forces

The gravitational force between any two objects decreases with increasing distance between the objects. This principle leads to a simple formula for estimating the tidal force that Earth exerts on parts of the Moon.

Consider two small objects, each of mass m, on opposite sides of the Moon. Because the two objects are at different distances from Earth's center, Earth exerts different forces F_{near} and F_{far} on them (see the accompanying illustration). The consequence of this difference is that the two objects tend to pull away from each other and away from the center of the Moon (see Section 4-8, especially Figure 4-23). The *tidal force* on these two objects, which is the force that tends to pull them apart, is the difference between the forces on the individual objects:

$$F_{tidal} = F_{near} - F_{far}$$

Both F_{near} and F_{far} are inversely proportional to the square of the distance from Earth (see Section 4-7). If this distance were doubled, both forces would decrease to $\frac{1}{2}^2 = \frac{1}{4}$ of their original values. But the tidal force F_{tidal}, which is the *difference* between these two forces, decreases even more rapidly with distance: It is inversely proportional to the *cube* of the distance from Earth to the Moon. That is, doubling the Earth-Moon distance decreases the tidal force to $\frac{1}{2}^3 = \frac{1}{8}$ of its original value. If we use the symbols M_{Earth} for Earth's mass (5.974×10^{24} kg), d for the Moon's diameter, r for the center-to-center distance from Earth to the Moon, and G for the universal constant of gravitation (6.67×10^{-11} newton m^2/kg^2), the tidal force is given approximately by the formula

$$F_{tidal} = \frac{2GM_{Earth}md}{r^3}$$

EXAMPLE: Calculate the tidal force for two 1-kg lunar rocks at the locations shown in the figure.

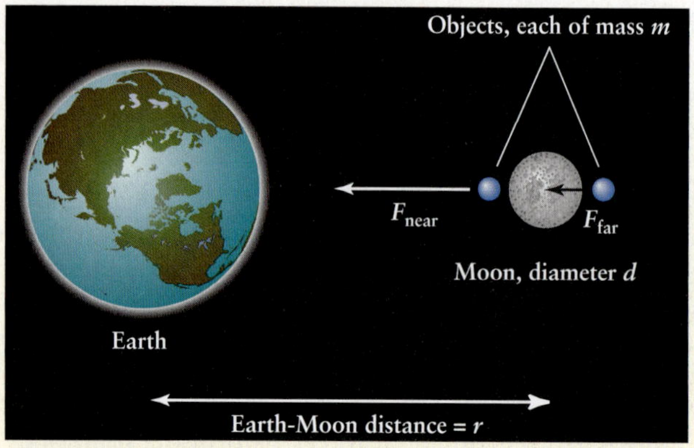

Situation: We are asked to calculate a tidal force. We know the masses of Earth (M_{Earth}) and of each of the two rocks ($m = 1$ kg). The Earth-Moon distance (r) and the diameter of the Moon (d, equal to the separation between the two rocks) are given in Table 10-1.

Tools: We use the equation given above for the tidal force F_{tidal}.

Answer: To use the given values in the equation, we must first convert the values for the average Earth-Moon distance r and the Moon's diameter d from kilometers to meters:

$$d = 3476 \text{ km} \times \frac{10^3 \text{ m}}{1 \text{ km}} = 3.476 \times 10^6 \text{ m}$$

$$r = 384{,}400 \text{ km} \times \frac{10^3 \text{ m}}{1 \text{ km}} = 3.844 \times 10^8 \text{ m}$$

The tidal force is then

$$F_{tidal} = \frac{2(6.67 \times 10^{-11})(5.974 \times 10^{24})(1)(3.476 \times 10^6)}{(3.844 \times 10^8)^3}$$
$$= 4.88 \times 10^{-5} \text{ newton}$$

Hawaii and Iceland. The rock of these low-lying lunar plains is called **mare basalt** (Figure 10-13).

In contrast to the dark maria, the lunar highlands are composed of a light-colored rock called **anorthosite** (Figure 10-14). On Earth, anorthositic rock is found only in such very old mountain ranges as the Adirondacks in the eastern United States. In comparison to the mare basalts, which have more of the heavier elements like iron, manganese, and titanium, anorthosite is rich in silicon, calcium, and aluminum. Anorthosite is therefore less dense than basalt. During the period when the Moon's surface was molten, the less dense anorthosite rose to the top, forming the terrae that make up the majority of the present-day lunar surface.

Although anorthosite is the dominant rock type in the lunar highlands, most of the rock samples collected there were not pure anorthosite but rather **impact breccias.** These rocks are composites of different rock types that were broken apart, mixed, and fused together by a series of meteoritic impacts (Figure 10-15). The

Review: To put our result into perspective, let's compare it to the *weight* of a 1-kg rock on the Moon. The weight of an object is equal to its mass multiplied by the acceleration due to gravity (see Section 4-6). This acceleration is equal to 9.8 m/s^2 on Earth, but is only 0.17 as great on the Moon. So, the weight of a 1-kg rock on the Moon is

$$0.17 \times 1 \text{ kg} \times 9.8 \text{ m/s}^2 = 1.7 \text{ newtons}$$

Thus, the tidal force on lunar rocks is *much* less than their weight, which is a good thing. If the tidal force (which tries to pull a rock away from the Moon's center) were greater than the weight (which pulls it toward the Moon's center), loose objects would levitate off the Moon's surface and fly into space!

The tidal force of Earth on the Moon deforms the solid body of the Moon, so that it acquires *tidal bulges* on its near and far sides. The mass of each bulge must be proportional to the tidal force that lifts lunar material away from the Moon's center. So, the above formula tells us that the mass of each bulge is inversely proportional to the cube of the Earth-Moon distance r:

$$m_{\text{bulge}} = \frac{A}{r^3}$$

(Determining the value of the constant A requires knowing how easily the Moon deforms, which is a complicated problem beyond our scope.) We can then use our formula for the tidal force F_{tidal} to tell us the *net* tidal force on the Moon. To make this calculation, we think of the objects of mass m on opposite sides of the Moon as being the tidal bulges. So, we substitute m_{bulge} for m in the tidal force formula:

$$F_{\text{tidal-net}} = \frac{2GM_{\text{Earth}}m_{\text{bulge}}d}{r^3} = \frac{2GM_{\text{Earth}}}{r^3}\frac{A}{r^3}d = \frac{2GM_{\text{Earth}}Ad}{r^6}$$

Thus, the net tidal force on the Moon is inversely proportional to the *sixth* power of the Earth-Moon distance. Even a small decrease in this distance can cause a substantial increase in the net tidal force.

EXAMPLE: Compare the net tidal force on the Moon at perigee (when it is closest to Earth) and at apogee (when it is farthest away).

Situation: We are asked to *compare* the tidal forces on the Moon at two different distances from Earth. To make this comparison, we will find the *ratio* of the tidal force at perigee to the tidal force at apogee.

Tools: We use the equation given above for the net tidal force on the Moon.

Answer: The Earth-Moon distances at perigee and apogee are given in Table 10-1. If we take the ratio of the tidal forces at these two distances, the constants (including the unknown quantity A) cancel out:

$$\frac{F_{\text{tidal-net-perigee}}}{F_{\text{tidal-net-apogee}}} = \frac{2GM_{\text{Earth}}Ad/r^6{}_{\text{perigee}}}{2GM_{\text{Earth}}Ad/r^6{}_{\text{apogee}}} = \left(\frac{r_{\text{apogee}}}{r_{\text{perigee}}}\right)^6 = \left(\frac{405{,}500 \text{ km}}{363{,}000 \text{ km}}\right)^6 = (1.116)^6 = 1.93$$

Review: Our result tells us that the net tidal force at perigee is 1.93 times as great (that is, almost double) as at apogee. Tidal forces on the Moon help create the stresses that generate moonquakes, so it is not surprising that these quakes are more frequent at perigee than at apogee.

prevalence of breccias is evidence that the lunar highlands have undergone eons of bombardment from space.

Every rock found on Earth, even those from the driest desert, has some water locked into its mineral structure. But lunar rocks are totally dry. With the exception of ice at the lunar poles (see Section 10-2), there is no evidence that water has ever existed on the Moon. Because the Moon lacks both an atmosphere and liquid water, it is not surprising that no traces of life have ever been found there.

By carefully measuring the abundances of trace amounts of radioactive elements in lunar samples and applying the principles of radioactive dating (see Section 8-2), geologists have determined that lunar rocks formed more than 3 billion years ago. Of these ancient rocks, however, anorthosite is much older than the mare basalts. Typical anorthositic specimens from the highlands are between 4.0 and 4.3 billion years old. All these extremely ancient specimens are probably samples of the Moon's original crust. In sharp contrast, all the mare basalts are between 3.1 and 3.8 bil-

Figure 10-13 R I V U X G

Mare Basalt This 1.53-kg (3.38-lb) specimen of mare basalt was brought back by the crew of *Apollo 15*. Gas must have dissolved under pressure in the lava from which this rock solidified. When the lava reached the airless lunar surface, the pressure on the lava dropped and the gas formed bubbles. Some of the bubbles were frozen in place as the rock cooled, forming the holes on its surface. (Figure 9-19*a* shows a sample of basalt from Earth.) (NASA)

lion years old. With no plate tectonics to bring fresh material to the surface, no new anorthosites or basalts have formed on the Moon for more than 3 billion years.

The History of Lunar Cratering

By correlating the ages of Moon rocks with the density of craters at the sites where the rocks were collected, geologists have found that the rate of impacts on the Moon has changed over the ages. The ancient, heavily cratered lunar highlands are evidence of an intense bombardment that dominated the Moon's early history. For nearly a billion years, rocky debris left over from the formation of the planets rained down upon the Moon's young surface. As Figure 10-16 shows, this barrage extended from 4.56 billion years ago, when the Moon's surface solidified, until about 3.8 billion years ago. At its peak, this bombardment from space would have produced a new crater of about 1 km radius somewhere on the Moon about once per century. (If 100 years seems like a long time interval, remember that we are talking about a bombardment that lasted hundreds of millions of years and produced millions of craters over that time.)

The rate of impacts should have tapered off as meteoroids and planetesimals were swept up by the newly formed planets. In fact, radioactive dating shows that the impact rate spiked upward between 4.0 and 3.8 billion years ago (see Figure 10-16). One proposed explanation is that the orbits of Jupiter and Saturn changed slightly but suddenly about 4.0 billion years ago during the process of settling into their present-day orbits. Slightly different orbits would have changed the gravitational forces that these planets exerted on the asteroid belt and could have disturbed the orbits of many asteroids, sending them careening toward the inner solar system.

Whatever the explanation for the final epoch of heavy bombardment, major impacts during this period gouged out the mare basins. Between 3.8 and 3.1 billion years ago iron-rich magma oozed up from the Moon's still-molten mantle, flooding the mare basins with lava that solidified to form basalt (see the *Cosmic Connections* figure).

The relative absence of craters in the maria (see Figure 10-5) tells us that the impact rate has been quite low since the lava solidified. During these past 3.1 billion years the Moon's surface has changed very little.

Figure 10-14 R I V U X G

Anorthosite The light-colored lunar terrae (highlands) are composed of this ancient type of rock, which is thought to be the material of the original lunar crust. Lunar anorthosites vary in color from dark gray to white; this sample from the *Apollo 16* mission is a medium gray. It measures 18 × 16 × 7 cm. (NASA)

Figure 10-15 R I V U X G

Impact Breccias These rocks are evidence of the Moon's long history of bombardment from space. Parts of the Moon's original crust were shattered and strewn across the surface by meteoritic impacts. Later impacts compressed and heated this debris, welding it into the type of rock shown here. Such impact breccias are rare on Earth but abundant on the Moon. (NASA)

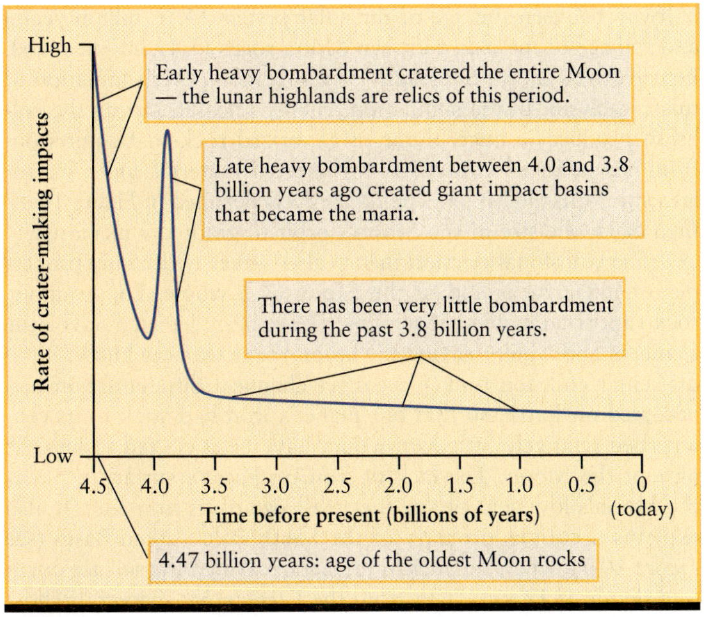

Figure 10-16

INTERACTIVE EXERCISE 10.1

The Rate of Crater Formation on the Moon This graph shows the rate at which impact craters formed over the Moon's history. The impact rate today is only about 1/10,000 as great as during the most intense bombardment. (Adapted from T. Grotzinger, T. H. Jordan, F. Press, and R. Siever, *Understanding Earth*, 5th ed., W. H. Freeman, 2007)

10-5 The Moon probably formed from debris cast into space when a huge protoplanet struck the young Earth

TUTORIAL 10.3

The lunar rocks collected by the *Apollo* and *Luna* missions have provided essential information about the history of the Moon. In particular, they have helped astronomers come to a consensus about one of the most important questions in lunar science: Where did the Moon come from?

The Receding Moon

Seventy years before the first spacecraft landed on the Moon, the British astronomer George Darwin (second son of the famous evolutionist Charles Darwin) deduced that the Moon must be slowly moving away from Earth. He began with the notion that the Moon's tidal forces elongate the oceans into a football shape (see Section 4-8, especially Figure 4-26). However, the long axis of this "football" does not point precisely at the Moon. Earth spins on its axis more rapidly than the Moon revolves around Earth, and this rapid rotation carries the tidal bulge about 10° ahead of a line connecting Earth and the Moon (Figure 10-17). This misaligned bulge produces a small but steady gravitational force that tugs the Moon forward and lifts it into an ever larger orbit around Earth. In other words, Darwin concluded that the Moon must be spiraling away from Earth.

It became possible to test Darwin's conclusions with the aid of a simple device that the *Apollo 11, 14,* and *15* and *Luna 21* missions left on the Moon—a set of reflectors, similar to the orange and red ones found in automobile taillights. Since 1969, scientists on Earth have been firing pulses of laser light at these reflectors and measuring very accurately the length of time it takes each pulse to return to Earth. Knowing the speed of light, they can use this data to calculate the distance to the Moon to an accuracy of just 3 centimeters. They have found that the Moon is moving away from Earth at a very gradual rate of 3.8 cm per year—which is just what Darwin predicted.

As the Moon moves away from Earth, its sidereal period is getting longer in accordance with Kepler's third law (see Section 4-7). At the same time, friction between the oceans and the body of Earth is gradually slowing Earth's rotation. The length of Earth's day is therefore increasing, by approximately 0.002 second per century.

Theories of the Moon's Origins

These observations mean that in the distant past, Earth was spinning faster than it is now and the Moon was much closer. Darwin theorized that the early Earth may have been spinning so fast that a large fraction of its mass tore away, and that this fraction coalesced to form the Moon. This is called the **fission theory** of the Moon's origin. Prior to the *Apollo* program, two other theories were in competition with the fission theory. The **capture theory** posits that the Moon was formed elsewhere in the solar system and then drawn into orbit about Earth by gravitational forces. By contrast, the **co-creation theory** proposes that Earth and the Moon were formed at the same time but separately.

One fact used to support the fission theory was that the Moon's average density (3344 kg/m^3) is comparable to that of Earth's outer layers, as would be expected if the Moon had been

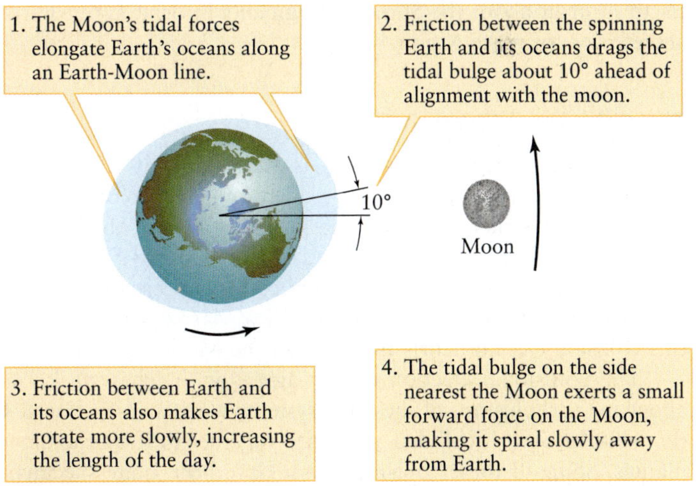

Figure 10-17

INTERACTIVE EXERCISE 10.2

The Moon's Tidal Recession Earth's rapid rotation drags the tidal bulge of the oceans about 10° ahead of a direct alignment with the Moon. The bulge on the side nearest the Moon exerts more gravitational force than the other, more distant bulge. The net effect is a small forward force on the Moon that makes it spiral slowly away from Earth.

flung out of a rapidly rotating proto-Earth. However, the fission theory predicts that lunar and terrestrial rocks should be similar in composition. In fact, samples returned from the Moon by the *Apollo* and *Luna* missions show that they are significantly different.

Volatile elements (such as potassium and sodium) boil at relatively low temperatures under 900°C, whereas **refractory elements** (such as titanium, calcium, and aluminum) boil at 1400°C or higher. Compared with terrestrial rocks, the lunar rocks have slightly greater proportions of the refractory elements and slightly lesser proportions of the volatile elements. This ratio suggests that the Moon formed from material that was heated to a higher temperature than that from which Earth formed. Some of the volatile elements boiled away, leaving the young moon relatively enriched in the refractory elements.

LOOKING DEEPER 10.2

Proponents of the capture theory used these differences to argue that the Moon formed elsewhere and was later captured by Earth. They also noted that the plane of the Moon's orbit is close to the plane of the ecliptic, as would be expected if the Moon had originally been in orbit about the Sun but was later captured by Earth.

There are difficulties with the capture theory, however. If Earth did capture the Moon intact, then a host of conditions must have been satisfied entirely by chance. The Moon must have coasted to within 50,000 km of Earth at exactly the right speed to leave a solar orbit and must have gone into orbit around Earth without actually hitting our planet.

Proponents of the co-creation theory argued that it is easier for a planet to capture swarms of tiny rocks. In this theory the Moon formed from just such rocky debris. Great numbers of rock fragments orbited the newborn Sun in the plane of the ecliptic along with the protoplanets. Heat generated in collisions could have baked the volatile elements out of these smaller rock fragments. Then, just as planetesimals accreted to form the proto-Earth in orbit about the Sun, the fragments in orbit about Earth accreted to form the Moon.

The Collisional Ejection Theory

The present scientific consensus is that *none* of these three theories—fission, capture, and co-creation—is correct. Instead, the evidence points toward an idea proposed in 1975 by William Hartmann and Donald Davis and independently by Alastair Cameron and William Ward. In this **collisional ejection theory**, the proto-Earth was struck off-center by a Mars-sized object, and this collision ejected debris from which the Moon formed.

The collisional ejection theory agrees with what we know about the early history of the solar system. We saw in Section 8-4 that smaller objects collided and fused together to form the inner planets. Some of these collisions must have been quite spectacular, especially near the end of planet formation, when most of the mass of the inner solar system was contained in a few dozen large protoplanets. The collisional ejection theory proposes that one such protoplanet struck the proto-Earth about 4.51 billion years ago, about halfway between the age of the solar system (4.56 billion years) and the age of the oldest known Moon rocks (4.47 billion years). Figure 10-18 shows the results of a supercomputer simulation of this cataclysm. In the simulation, energy released during the collision produces a huge plume of vaporized rock that squirts out from the point of impact. As this ejected material cools, it coalesces to form the Moon. The tidal effects depicted in Figure 10-17 then cause the size of the Moon's orbit to gradually increase.

The difference in chemical composition between lunar rocks and Earth rocks helps to constrain theories of the Moon's origin

The collisional ejection theory also agrees with many properties of lunar rocks and of the Moon as a whole. For example, rock vaporized by the impact would have been depleted of volatile elements and water, leaving the moon rocks we now know. If the off-center collision took place after chemical differentiation had occurred on Earth, so that our planet's iron had sunk to its center, then relatively little iron would have been ejected to become part of the Moon. The lack of iron on Earth's surface explains the Moon's low density and the small size of its iron core. It also explains a curious property of the South Pole–Aitken Basin (see Figure 10-7), where an ancient impact excavated the surface down to a depth of 12 km. Data from the *Clementine* spacecraft show that iron concentration in this basin is only 10% (see Figure 10-10), far less than the 20–30% concentration at a corresponding depth below Earth's surface.

An Overview of the Moon's History

We can now summarize the entire geologic history of the Moon. The surface of the newborn Moon was probably molten for many years, both from heat released during the impact of rock fragments falling into the young satellite and from the decay of short-lived radioactive isotopes. As the Moon gradually cooled, low-density lava floating on the Moon's surface began to solidify into the anorthositic crust that exists today. The barrage of rock fragments that ended about 3.8 billion years ago produced the craters that cover the lunar highlands.

During the relatively brief period of heavy bombardment between 4.0 and 3.8 billion years ago (see Figure 10-16), more than a dozen asteroid-size objects, each measuring at least 100 km across, struck the Moon and blasted out the vast mare basins. Meanwhile, heat from the decay of long-lived radioactive elements like uranium and thorium began to melt the inside of the Moon. From 3.8 to 3.1 billion years ago, great floods of molten rock gushed up from the lunar interior, filling the impact basins and creating the basaltic maria we see today.

Very little has happened on the Moon since those ancient times. A few relatively fresh craters have been formed, but the astronauts visited a world that has remained largely unchanged for more than 3 billion years. During that same period on Earth, by contrast, the continents have been totally transformed time and time again (see Section 9-3).

Many questions and mysteries still remain. The six American and three Soviet lunar landings have brought back samples from only nine locations, barely scratching the lunar surface. We still know very little about the Moon's far side and poles. Are there really vast ice deposits at the poles? How much of the Moon's interior is molten? Just how large is its iron core? How old are the youngest lunar rocks? Did lava flows occur over western Oceanus Procellarum only 2 billion years ago, as crater densities there sug-

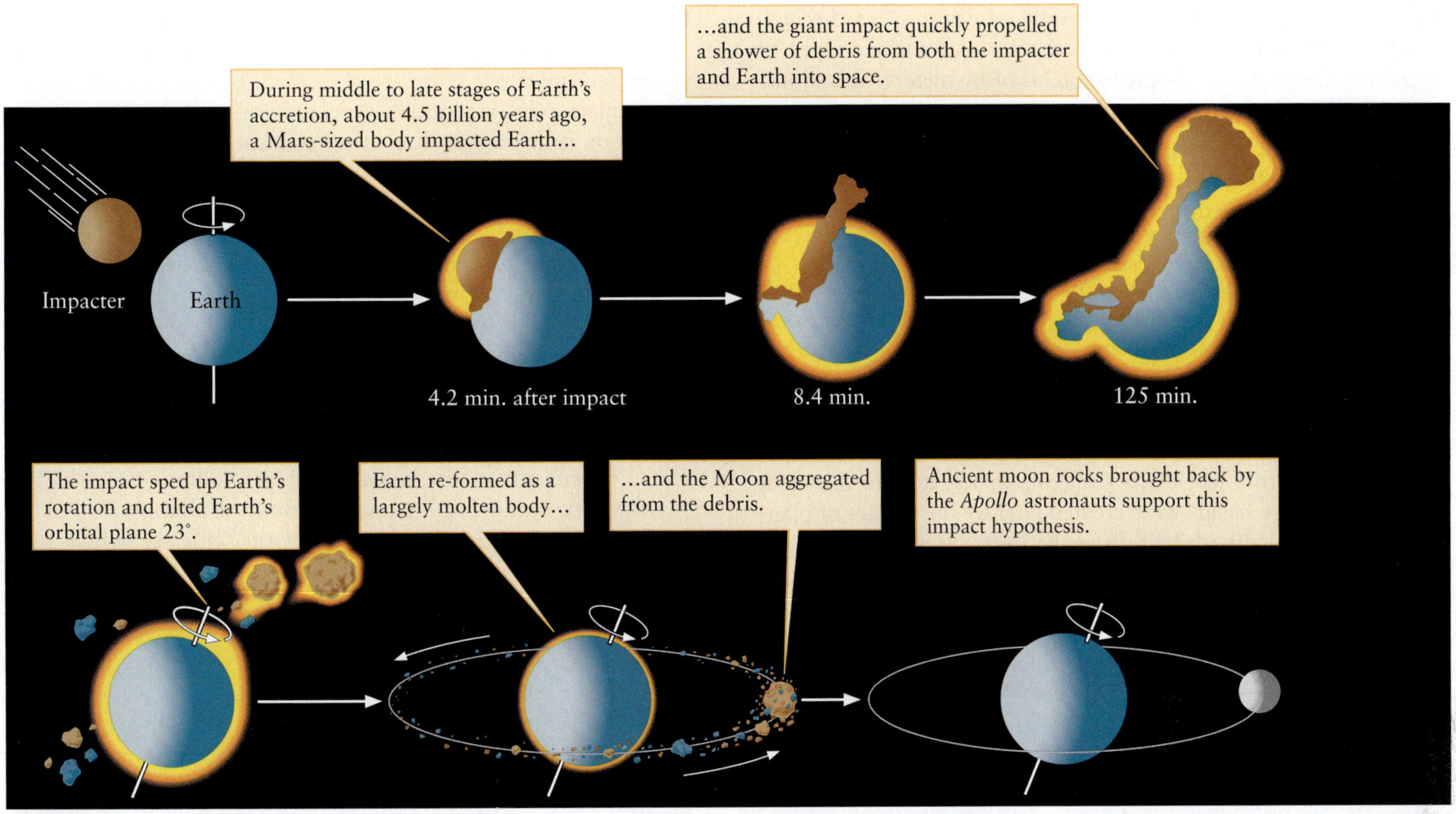

During middle to late stages of Earth's accretion, about 4.5 billion years ago, a Mars-sized body impacted Earth...
...and the giant impact quickly propelled a shower of debris from both the impacter and Earth into space.
Impacter
Earth
4.2 min. after impact
8.4 min.
125 min.
The impact sped up Earth's rotation and tilted Earth's orbital plane 23°.
Earth re-formed as a largely molten body...
...and the Moon aggregated from the debris.
Ancient moon rocks brought back by the Apollo astronauts support this impact hypothesis.

ANIMATION 10.2

Figure 10-18

The Formation of the Moon This figure shows how the Moon could have formed in the aftermath of a collision between a Mars-sized protoplanet and the proto-Earth. (Adapted from T. Grotzinger, T. H. Jordan, F. Press, and R. Siever, *Understanding Earth*, 5th ed., W. H. Freeman, 2007)

gest? Could there still be active volcanoes on the Moon? There is still much to be learned by exploring the Moon.

Key Words

anorthosite, p. 246
capture theory (of Moon's formation), p. 249
center of mass, p. 236
co-creation theory (of Moon's formation), p. 249
collisional ejection theory (of Moon's formation), p. 250
crater, p. 238
far side (of the Moon), p. 236
fission theory (of Moon's formation), p. 249
impact breccia, p. 246
impact crater, p. 238
libration, p. 236
lunar highlands, p. 239
mare (*plural* maria), p. 238
mare basalt, p. 246
moonquake, p. 244
refractory element, p. 250
regolith, p. 245
synchronous rotation, p. 238
terminator, p. 236
terrae, p. 239
volatile element, p. 250

Key Ideas

Appearance of the Moon: The Earth-facing side of the Moon displays light-colored, heavily cratered highlands and dark-colored, smooth-surfaced maria. The Moon's far side has almost no maria.

• Virtually all lunar craters were caused by space debris striking the surface. There is no evidence of plate tectonic activity on the Moon.

Internal Structure of the Moon: Much of our knowledge about the Moon has come from human exploration in the 1960s and early 1970s and from more recent observations by unmanned spacecraft.

• Analysis of seismic waves and other data indicates that the Moon has a crust thicker than that of Earth (and thickest on the far side of the Moon), a mantle with a thickness equal to about 80% of the Moon's radius, and a small iron core.

• The Moon's lithosphere is far thicker than that of Earth. The lunar asthenosphere probably extends from the base of the lithosphere to the core.

• The Moon has no global magnetic field today, although it had a weak magnetic field billions of years ago.

Geologic History of the Moon: The anorthositic crust exposed in the highlands was formed between 4.3 and 4.0 billion years ago. An era of heavy bombardment formed the maria basins between 4.0 and 3.8 billion years ago, and the mare basalts solidified between 3.8 and 3.1 billion years ago.

• The Moon's surface has undergone very little change over the past 3 billion years.

• Meteoroid impacts have been the only significant "weathering" agent on the Moon. The Moon's regolith, or surface layer of powdered and fractured rock, was formed by meteoritic action.

• All of the lunar rock samples are igneous rocks formed largely of minerals found in terrestrial rocks. The lunar rocks contain no water and also differ from terrestrial rocks in being relatively enriched in the refractory elements and depleted in the volatile elements.

Origin of the Moon: The collisional ejection theory of the Moon's origin holds that the proto-Earth was struck by a Mars-sized protoplanet and that debris from this collision coalesced to form the Moon. This theory successfully explains most properties of the Moon.

• The Moon was molten in its early stages, and the anorthositic crust solidified from low-density magma that floated to the lunar surface. The mare basins were created later by the impact of planetesimals and filled with lava from the lunar interior.

• Tidal interactions between Earth and the Moon are slowing Earth's rotation and pushing the Moon away from Earth.

Questions

Review Questions

TUTORIAL 10.1

1. Is it correct to say that the Moon orbits Earth? If not, what is a more correct description?
2. If the Moon always keeps the same face toward Earth, how is it possible for Earth observers to see more than half of the Moon's surface?
3. Why does the sky look black on the Moon even during daytime?
4. Why is it impossible for liquid water to exist on the surface of the Moon?
5. Describe two reasons why astronauts needed to wear spacesuits on the lunar surface.
6. Describe the kinds of features that can be seen on the Moon with a small telescope.
7. Are impact craters on the Moon the same size as the meteoroids that made the impact? Explain your answer.
8. Describe the differences between the maria and the lunar highlands. Which kind of terrain covers more of the Moon's surface? Which kind of terrain is more heavily cratered? Which kind of terrain was formed later in the Moon's history? How do we know?
9. Describe the differences between the near and far sides of the Moon. What is thought to be the likely explanation for these differences?
10. What does it mean to say the Moon is a "one-plate world"? What is the evidence for this statement?

TUTORIAL 10.2

11. Why was it necessary to send unmanned spacecraft to land on the Moon before sending humans there?
12. What is the evidence that ice exists at the lunar poles? Is this evidence definitive?
13. Why was it useful for the *Apollo* astronauts to bring magnetometers and seismometers to the Moon?
14. Could you use a magnetic compass to navigate on the Moon? Why or why not?
15. Describe the evidence that (a) the Moon has a more solid interior than Earth and (b) the Moon's interior is not completely solid.
16. Explain why moonquakes occur more frequently when the Moon is at perigee than at other locations along its orbit.
17. Why is Earth geologically active while the Moon is not?
18. What is the regolith? What causes its powdery character?
19. Why are there no sedimentary rocks on the Moon?
20. On the basis of moon rocks brought back by the astronauts, explain why the maria are dark-colored but the lunar highlands are light-colored.
21. Briefly describe the main differences and similarities between Moon rocks and Earth rocks.
22. Rocks found on the Moon are between 3.1 and 4.47 billion years old. By contrast, the majority of Earth's surface is made of oceanic crust that is less than 200 million years old, and the very oldest Earth rocks are about 4 billion years old. If Earth and the Moon are essentially the same age, why is there such a disparity in the ages of rocks on the two worlds?

TUTORIAL 10.3

23. If Earth's tidal bulge pointed directly toward the Moon, would the Moon still be receding from Earth? Explain your answer.
24. Why do most scientists favor the collisional ejection theory of the Moon's formation?
25. Some people who supported the fission theory proposed that the Pacific Ocean basin is the scar left when the Moon pulled away from Earth. Explain why this idea is probably wrong.

Advanced Questions

Questions preceded by an asterisk (*) involve topics discussed in Box 10-1.

> **Problem-solving tips and tools**
>
> Recall that the average density of an object is its mass divided by its volume. The volume of a sphere is $4\pi r^3/3$, where r is the sphere's radius. The surface area of a sphere of radius r is $4\pi r^2$, while the surface area of a circle of radius r is πr^2. Recall also that the acceleration of gravity on Earth's surface is 9.8 m/s^2. You may find it useful to know that a 1-pound (1-lb) weight presses down on Earth's surface with a force of 4.448 newtons. You might want to review Newton's universal law of gravitation in Section 4-7. The time to travel a certain distance is equal to the distance traveled divided by the speed of motion. Consult Table 10-1 and the Appendices for any additional data.

26. Suppose two worlds (say, a planet and its satellite) have masses m_1 and m_2, and the center-to-center distance between the worlds is r. The distance d_{cm} from the center of world 1 to the center of mass of the system of two worlds is given by the formula

$$d_c = \frac{m_2 r}{m_1 + m_2}$$

(a) Suppose world 1 is Earth and world 2 is the Moon. If Earth and Moon are at their average center-to-center distance, find the distance from the center of Earth to the center of mass of the Earth-Moon system. (**b**) Is the Earth-Moon system's center of mass within Earth? How far below Earth's surface is it located? (**c**) Find the distance from the center of the Sun (mass 1.989×10^{30} kg) to the center of mass of the Sun-Earth system. How does this compare to the diameter of the Sun? Is it a good approximation to say that Earth orbits around the center of the Sun?

27. If you view the Moon through a telescope, you will find that details of its craters and mountains are more visible when the Moon is near first quarter phase or third quarter phase than when it is at full phase. Explain why.
28. In a whimsical moment during the *Apollo 14* mission, astronaut Alan Shepard hit two golf balls over the lunar surface. Give two reasons why they traveled much farther than golf balls do on Earth.
29. Temperature variations between day and night are much more severe on the Moon than on Earth. Explain why.
30. How much would an 80-kg person weigh on the Moon? How much does that person weigh on Earth?
31. Using the diameter and mass of the Moon given in Table 10-1, verify that the Moon's average density is about 3344 kg/m^3. Explain why this average density implies that the Moon's interior contains much less iron than the interior of Earth.
*32. In Box 10-1 we calculated the tidal force that Earth exerts on two 1-kg rocks located on the near and far sides of the Moon. We assumed that the Earth-Moon distance was equal to its average value. Repeat this calculation (**a**) for the Moon at perigee and (**b**) for the Moon at apogee. (**c**) What is the ratio of the tidal force on the rocks at perigee to the tidal force at apogee?
33. The youngest lunar anorthosites are 4.0 billion years old, and the youngest mare basalts are 3.1 billion years old. Would you expect to find any impact breccias on the Moon that formed less than 3.1 billion years ago? Explain your answer.
34. In the maria, the lunar regolith is about 2 to 8 meters deep. In the lunar highlands, by contrast, it may be more than 15 meters deep. Explain how the different ages of the maria and highlands can account for these differences.
35. The mare basalts are volcanic rock. Is it likely that active volcanoes exist anywhere on the Moon today? Explain.
36. Calculate the round-trip travel time for a pulse of laser light that is fired from a point on Earth nearest the Moon, hits a reflector at the point on the Moon nearest Earth, and returns to its point of origin. Assume that Earth and Moon are at their average separation from each other.
37. Before the *Apollo* missions to the Moon, there were two diametrically opposite schools of thought about the history of lunar geology. The "cold moon" theory held that all lunar surface features were the result of impacts. The most violent impacts melted the surface rock, which then solidified to form the maria. The opposite "hot moon" theory held that all lunar features, including maria, mountains, and craters, were the result of volcanic activity. Explain how lunar rock samples show that neither of these theories is entirely correct.
*38. When the Moon originally coalesced, it may have been only one-tenth as far from Earth as it is now. (**a**) When the Moon first coalesced, was Earth's tidal force strong enough to lift rocks off the lunar surface? Explain. (**b**) Compared with the net tidal force that Earth exerts on the Moon today, how many times larger was the net tidal force on the newly coalesced Moon? (This strong tidal force kept the one axis of the Moon oriented toward Earth, and the Moon kept that orientation after it solidified.)

Discussion Questions

39. Comment on the idea that without the presence of the Moon in our sky, astronomy would have developed far more slowly.
40. No *Apollo* mission landed on the far side of the Moon. Why do you suppose this was? What would have been the scientific benefits of a mission to the far side?
41. NASA is planning a new series of manned missions to the Moon. Compare the advantages and disadvantages of exploring the Moon with astronauts as opposed to using mobile, unmanned instrument packages.
42. Describe how you would empirically test the idea that human behavior is related to the phases of the Moon. What problems are inherent in such testing?
43. How would our theories of the Moon's history have been affected if astronauts had discovered sedimentary rock on the Moon?
44. Imagine that you are planning a lunar landing mission. What type of landing site would you select? Where might you land to search for evidence of recent volcanic activity?

Web/eBook Questions

45. In 2005 the *SMART-1* spacecraft detected calcium on the lunar surface. Search the World Wide Web for information about the *SMART-1* mission and this discovery. How was the presence of calcium detected? What does this tell astronomers about the origin of the Moon?
46. **Determining the Size of the Planetesimal That Formed the Moon.** Access the animation "The Formation of the Moon" in Chapter 10 of the *Universe* Web site or eBook. By making measurements on the screen with a ruler, determine how many times larger the proto-Earth is than the impacting planetesimal. If Mars is about 50% the size of Earth, how does the planetesimal compare in size with present-day Mars?

WEB LINK 10.13

Observing tips and tools

If you do not have access to a telescope, you can learn a lot by observing the Moon through binoculars. Note that the Moon will appear right-side-up through binoculars but inverted through a telescope; if you are using a map of the Moon to aid your observations, you will need to take this into account. Inexpensive maps of the Moon can be purchased from most good bookstores or educational supply stores. You can determine the phase of the Moon either by looking at a calendar (most of which tell you the dates of new moon, first quarter, full moon, and third quarter), by checking the weather page of your newspaper, by consulting the current issue of *Sky & Telescope* or *Astronomy* magazine, by using the *Starry Night Enthusiast*™ program, or by using the World Wide Web.

Activities

Observing Projects

47. Observe the Moon through a telescope every few nights over a period of two weeks between new moon and full moon. Make sketches of various surface features, such as craters, mountain ranges, and maria. How does the appearance of these features change with the Moon's phase? Which features are most easily seen at a low angle of illumination? Which features show up best with the Sun nearly overhead?
48. Use the *Starry Night Enthusiast*™ program to observe the changing appearance of the Moon. Display the entire celestial sphere by selecting **Guides > Atlas** in the **Favorites** menu. Open the **Find** pane by clicking the **Find** tab at the top of the left border of the view window. Click the menu button (the blue colored button with a downward- pointing arrowhead) at the left of the entry in the list for the **Moon** and select the **Magnify** option. This will center a magnified image of the Moon in the view as seen from the center of a transparent Earth. Close the **Find** pane by clicking its tab. Click on the down arrow to the right of the **Time Flow Rate** control and select **hours** from the drop-down menu. Then click on the **Run Time Forward** button (a triangle that points to the right like a Play button). (**a**) Describe how the phase of the Moon changes over time. (**b**) Look carefully at features near the left-hand and right-hand limbs (edges) of the Moon. Are these features always at the same position relative to the limb? Explain in terms of libration.
49. Use the *Starry Night Enthusiast*™ program to observe the apparent change in size of the Moon as seen from the surface of Earth. Click the **Home** button in the toolbar. Stop the time flow by clicking the **Stop** time button (the button with a square icon in the **Time Flow Rate** section of the toolbar). Click each field of the **Time** and **Date** display pane in the toolbar and use the keyboard to set the time to 12:00:00 A.M. and the date to October 16, 2006. You can now set up the view of the Moon as if you were observing from a transparent Earth in continuous darkness. Open the **Options** pane by clicking its tab on the left border of the view window. In the **Local View** layer of the Options pane, uncheck the boxes beside the **Daylight** and **Local Horizon** options. Open the **Find** pane and click the menu button at the left of the entry for the Moon and select **Magnify** from the menu to display a stationary image of the magnified Moon in the center of the view. Notice the Moon's phase. (**a**) Select **Options > Solar System > Planets-Moons...** from the menu. In the Planets-Moons dialog box, click the slide control next to the **Show dark side** label near the top of the window and adjust the control all the way to the right (**Brighter**) side of the scale. Then click the **OK** button to close the dialog box. Note that the image of the Moon now appears full because *Starry Night Enthusiast*™ has artificially brightened the dark side of the image of the Moon, effectively removing the appearance of the Moon's phases. (**b**) Use the + button in the Zoom section of the toolbar to adjust the field of view to about **55′ × 45′**. *Starry Night Enthusiast*™ can display a reference field of view (FOV) upon this sky. Open the **FOV** pane by clicking its tab. Select the **30 Arcminutes** option. Then click the FOV tab to close the pane. (**c**) The final view is of the Moon, its dark side artificially brightened, as it would be seen from your home location if Earth were airless and transparent, surrounded by a yellow FOV circle 30 arcminutes in diameter. Note the size of the Moon relative to this reference circle. Set the **Time Flow Rate** in the toolbar to 1 minute. **Run Time Forward** for least 24 hours and observe the apparent size of the Moon relative to the reference circle. (**c**) Note that the apparent size of the Moon changes somewhat over the course of a day (of simulated time). Explain in terms of Earth's rotation. (*Hint:* In this view, *Starry Night Enthusiast*™ has made Earth transparent, so you can see objects that would normally be below the horizon. As Earth rotates, your observing location on the surface is carried along and your distance from the Moon changes.) (**d**) Change the **Time Flow Rate** to **1 day** and again click the **Run Time Forward** button. Does the apparent size of the Moon always stay the same, or does it vary? Explain what this tells you about the shape of the Moon's orbit.
50. Use the *Starry Night Enthusiast*™ program to examine the Moon. Select **Solar System > Moon** from the **Favorites** menu. (If desired, remove the image of the astronaut by clicking on **Feet** in the **View** menu) You can rotate the image of the Moon by placing the mouse cursor over the image, holding down the mouse button, and moving the mouse. (On a two-button mouse, hold down the left mouse button.) (**a**) From what you can see in the image, what evidence can you find that the Moon is geologically inactive? Explain. (**b**) Spreading outward from some of the largest craters on the Moon are straight lines of light-colored material called rays that were caused by material ejected outward by the impact that caused the crater. Rotate the Moon around to see the entire illuminated surface. Can you find any rays on the Moon? Zoom in on the Moon to examine various surface features such as craters and mountain ranges. Estimate the length of several rays extending from craters by measuring their length on the screen with a ruler and comparing them to the diameter of the Moon, which is about 30 minutes of arc when seen from Earth.

Collaborative Exercise

51. The image of Crater Clavius in Figure 10-4 reveals numerous craters. Using the idea that the Moon's landscape can only be changed by impacts, make a rough sketch showing ten of the largest craters and label them from oldest (those that showed up first) to youngest (the most recent ones). Explain your reasoning and any uncertainties.

11

Mercury, Venus, and Mars: Earthlike yet Unique

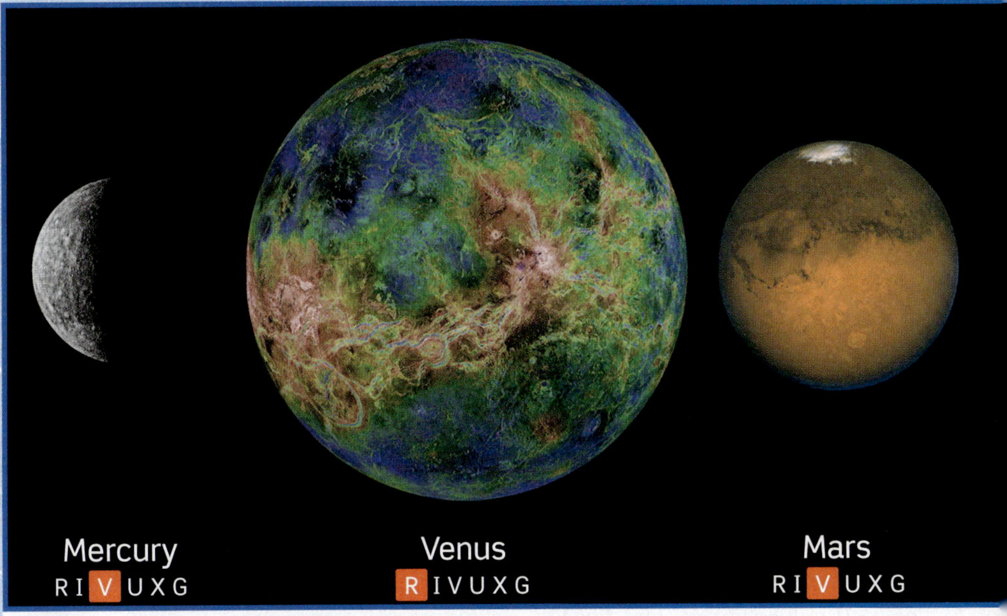

WEB LINK 11.1 VIDEO 11.1

(Mercury: NASA; Venus: NASA/JPL, MIT, and USGS; Mars: NASA, J. Bell/Cornell University, and M. Wolff/SSI)

These images above show the three planets that share the inner solar system with our Earth. All three have solid surfaces, and, in principle, a properly protected astronaut could stand on any of them. The differences between these three worlds, however, are pronounced.

Mercury is small, airless, and extensively cratered. It is also mysterious: Half of its surface has only been viewed recently, and it has a perplexing and unexpected magnetic field. It also rotates in a manner unique in the solar system, spinning three times on its axis for every two orbits around the Sun.

Venus is nearly the same size as Earth, but it is shrouded by a perpetual cloud cover that hides its surface from view. To penetrate the clouds and learn what lies beneath, two advanced technologies were needed: radar, which allowed astronomers to "see" through the clouds, and unmanned spacecraft, which orbited Venus at close range and even landed on its surface. The false-color topographic map shown in the photo above reveals that Venus has no true continents but merely highlands (shown in red) that rise gently above the planet's lower-lying areas (shown in blue). We have also learned that Venus's atmosphere is thick and searing hot, and that its clouds contain droplets of corrosive sulfuric acid.

Mars has captivated the popular imagination like no other planet. But rather than being the abode of warlike aliens, Mars proves to be an enigmatic world. Some parts of its surface are drier than any desert on Earth, while other locations show evidence of having been underwater for extended periods. Our challenge is to understand how these three nominally Earthlike worlds evolved to be so unique and so different than our own Earth.

Learning Goals

By reading the sections of this chapter, you will learn

11-1 What astronomers have learned by observing the terrestrial planets from Earth

11-2 The radically different ways in which Mercury, Venus, and Mars rotate on their axes

11-3 The outstanding features of Mercury, and why its magnetic field came as a surprise

11-4 How the advent of the space age transformed our understanding of Venus and Mars

11-5 How geologic activity took a very different form on Venus than on Earth, and why it essentially stopped on Mars

11-6 The key differences among the atmospheres of Earth, Venus, and Mars

11-7 How the atmospheres of Earth, Venus, and Mars evolved to their present states

11-8 The evidence that there was once liquid water on Mars

11-9 What we know about the two small satellites of Mars

11-1 Mercury, Venus, and Mars can all be seen with the naked eye

At various times Mercury, Venus, and Mars are among the brightest objects in the sky. At their greatest brilliance, each of these planets appears brighter than any star, which is why they have been known since ancient times and why their motions played a role in the religious beliefs of many ancient cultures.

Naked-Eye Observations of Mercury and Venus

Viewing Mercury is difficult because it is so close to the Sun; Venus is easier to see but covered with clouds; Mars is best seen at opposition

Mercury and Venus are *inferior* planets whose orbits around the Sun are smaller than Earth's; that is, each moves in an orbit with a semimajor axis a less than 1 astronomical unit (AU). For Mercury, a = 0.387 AU (57.9 million kilometers, or 36.0 million miles), and for Venus, a = 0.723 AU (108 million kilometers, or 67.2 million miles). As a result, Mercury and Venus never appear very far from the Sun in our sky (Figure 11-1). Table 11-1 and Table 11-2 summarize some basic information about these planets and their orbits.

We get the best view of either Mercury or Venus when it appears as far from the Sun in the sky as it can be, at its greatest eastern or western elongation (Section 4-2). On dates near its **greatest eastern elongation,** Mercury or Venus appears after sunset over the western horizon as an "evening star." On dates near its **greatest western elongation,** Mercury or Venus appears as a "morning star" that rises before the Sun in the eastern sky.

CAUTION! Remember that "greatest *western* elongation" and "greatest *eastern* elongation" refer to a planet's position in the sky relative to the Sun. When Mercury or Venus is at greatest western elongation, it is west of the Sun in the sky and has already risen at dawn when the Sun is rising in the east. When Mercury or Venus is at greatest eastern elongation, it is east of the Sun in the sky and is still above the horizon when the Sun sets in the west.

WEB LINK 11.5

Because Mercury's orbit is so close to the Sun, its maximum elongation is only 28° and it is quite difficult to observe. (It is said that Copernicus never saw it.) The celestial sphere appears to rotate at 15° per hour (360° divided by 24 hours), so Mercury never rises more than 2 hours before sunrise or sets more than 2 hours after sunset. Unfortunately, Mercury's orbit is very elliptical (its eccentricity of 0.206 is greater than that of any other planet) and is inclined to the ecliptic—the plane of Earth's orbit around the Sun—by about 7°. As a result, at greatest elongation Mercury is often much less than 28° from the horizon at sunset or sunrise. Hence, some elongations are more favorable for viewing Mercury than others, as Figure 11-2 shows. A total of six or seven greatest elongations (both eastern and western) occur each year, but usually only two of these will be favorable for viewing the planet.

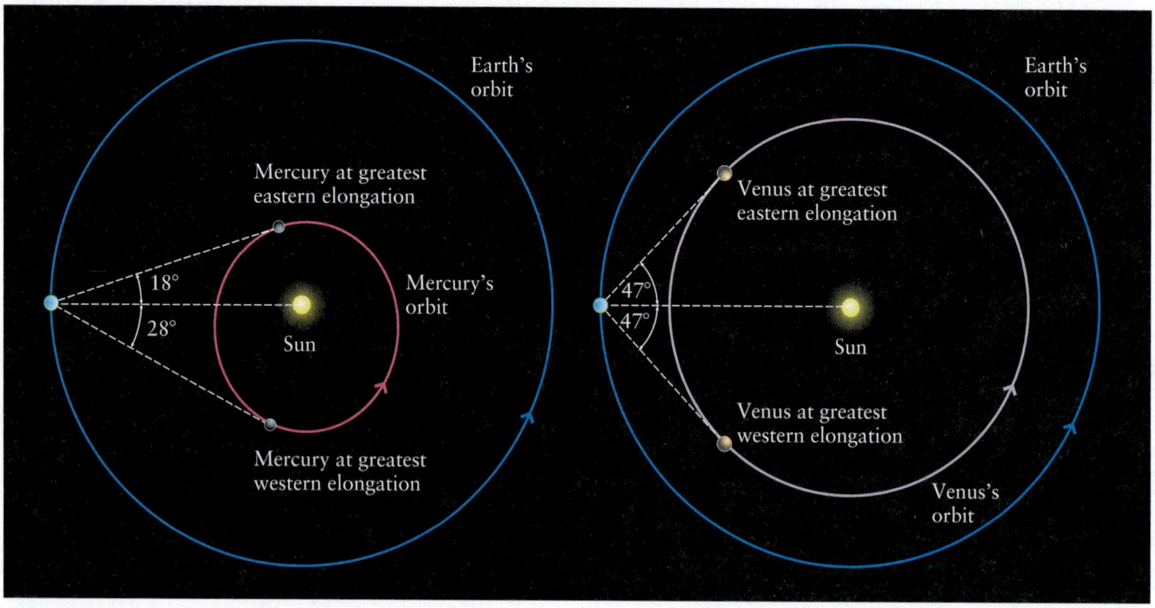

ANIMATION 11.1

Figure 11-1

The Orbits of Mercury and Venus Mercury moves around the Sun every 88 days in a rather eccentric orbit. As seen from Earth, the angle between Mercury and the Sun at greatest eastern or western elongation can be as large as 28° (when Mercury is near aphelion) or as small as 18° (near perihelion). By contrast, Venus follows a larger, nearly circular orbit with a 224.7-day period. The angle between Venus and the Sun at eastern or western elongation is 47°.

Table 11-1 Mercury Data

Average distance from the Sun:	0.387 AU = 5.79×10^7 km
Maximum distance from the Sun:	0.467 AU = 6.98×10^7 km
Minimum distance from the Sun:	0.307 AU = 4.60×10^7 km
Eccentricity of orbit:	0.206
Average orbital speed:	47.9 km/s
Orbital period:	87.969 days
Rotation period:	58.646 days
Inclination of equator to orbit:	0.5°
Inclination of orbit to ecliptic:	7° 00′ 16″
Diameter (equatorial):	4880 km = 0.383 Earth diameter
Mass:	3.302×10^{23} kg = 0.0553 Earth mass
Average density:	5430 kg/m^3
Escape speed:	4.3 km/s
Surface gravity (Earth = 1):	0.38
Albedo:	0.12
Average surface temperatures:	Day: 350°C = 662°F = 623 K Night: −170°C = − 274°F = 103 K
Atmosphere:	Essentially none

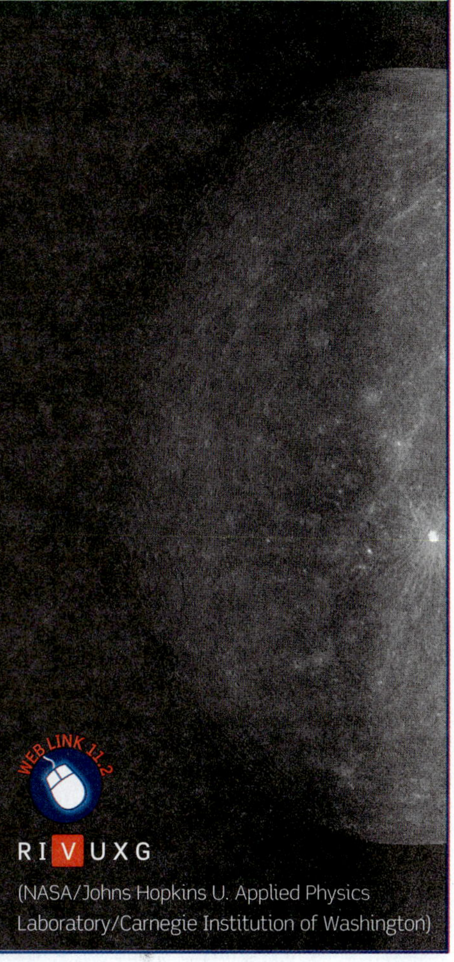

(NASA/Johns Hopkins U. Applied Physics Laboratory/Carnegie Institution of Washington)

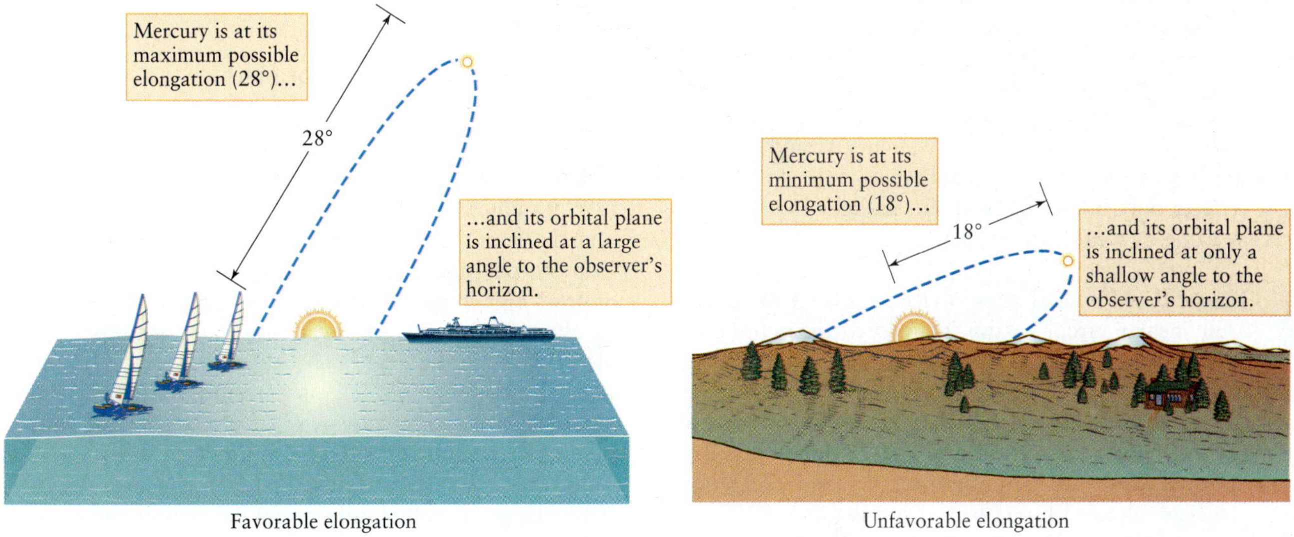

Figure 11-2

Favorable Versus Unfavorable Elongations The tilt of Earth's axis, the inclination and eccentricity of Mercury's orbit, and the latitude of the observer combine to make an elongation either favorable (left illustration) or unfavorable (right illustration) for viewing Mercury.

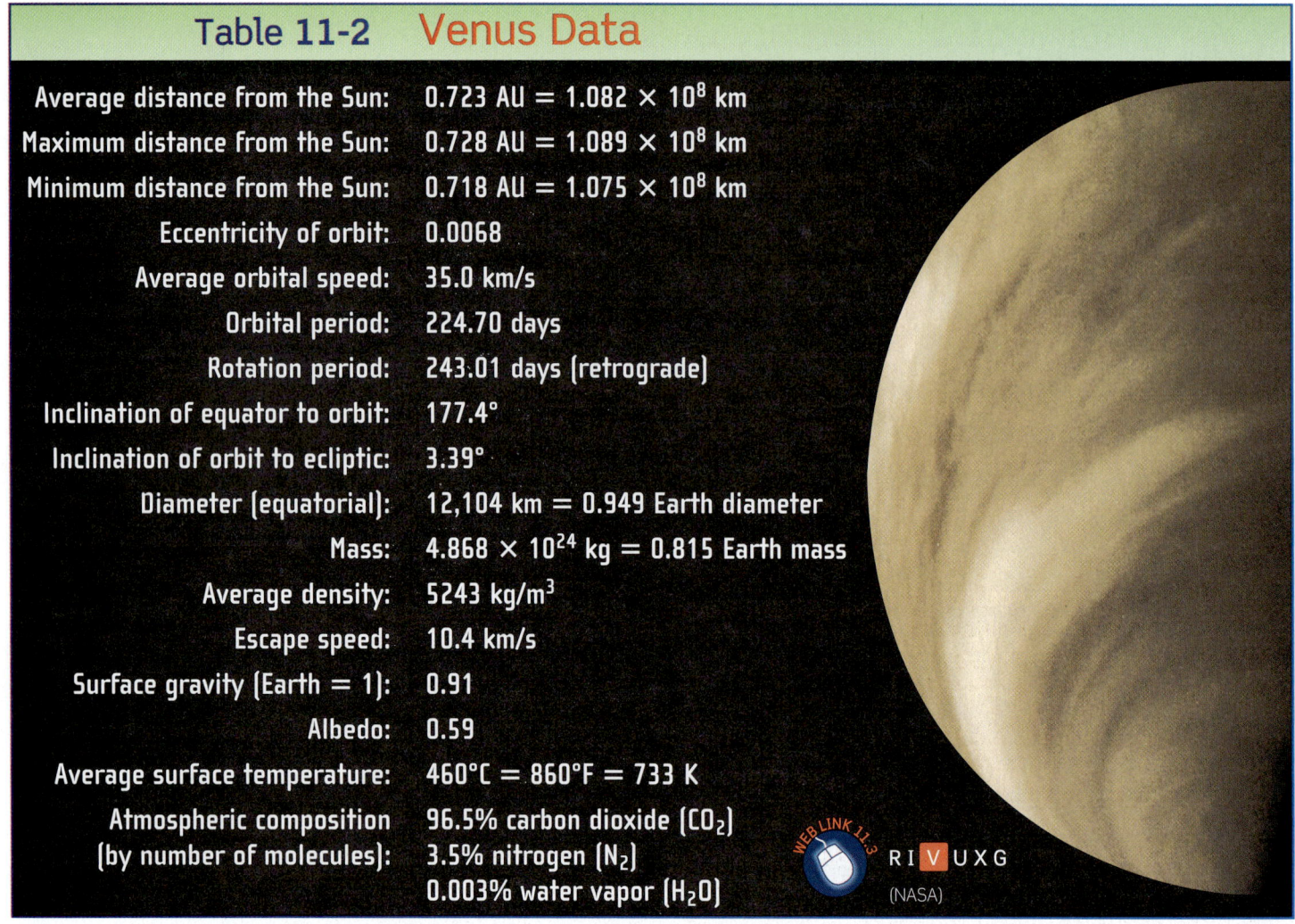

Table 11-2 Venus Data

Average distance from the Sun:	0.723 AU = 1.082×10^8 km
Maximum distance from the Sun:	0.728 AU = 1.089×10^8 km
Minimum distance from the Sun:	0.718 AU = 1.075×10^8 km
Eccentricity of orbit:	0.0068
Average orbital speed:	35.0 km/s
Orbital period:	224.70 days
Rotation period:	243.01 days (retrograde)
Inclination of equator to orbit:	177.4°
Inclination of orbit to ecliptic:	3.39°
Diameter (equatorial):	12,104 km = 0.949 Earth diameter
Mass:	4.868×10^{24} kg = 0.815 Earth mass
Average density:	5243 kg/m^3
Escape speed:	10.4 km/s
Surface gravity (Earth = 1):	0.91
Albedo:	0.59
Average surface temperature:	460°C = 860°F = 733 K
Atmospheric composition (by number of molecules):	96.5% carbon dioxide (CO_2) 3.5% nitrogen (N_2) 0.003% water vapor (H_2O)

WEB LINK 11.3 RIVUXG (NASA)

Like all the planets, Mercury shines by reflected sunlight. Because it orbits so close to the Sun, it is exposed to sunlight that on average is about 7 times as intense as on Earth. Furthermore, Mercury is never more than 1.5 AU from Earth. These factors explain why Mercury can have up to twice the brightness of Sirius, the brightest star in the nighttime sky. Mercury would be even brighter if its surface did not have a rather low albedo of 0.12 (that is, it reflects only about 12% of the sunlight that falls on it—about the same as weathered asphalt).

WEB LINK 11.6 The orbit of Venus is about twice the size of Mercury's orbit, nearly circular, and inclined to the ecliptic by only 3.39°. As a result, Venus can be seen about 47° away from the Sun and all elongations are nearly equally favorable. At greatest eastern elongation, which occurs at intervals of about 19 months, Venus sets in the west about 3 hours after sunset. About 5 months later Venus is at greatest western elongation and rises in the east about 3 hours before sunrise.

At its greatest brilliance, Venus is 16 times brighter than Sirius and is outshone only by the Sun and the Moon. Venus is more brilliant than Mercury because it is relatively large (more than double the diameter of Mercury and almost the same size as Earth) and has a much higher albedo of 0.59 (it reflects 59% of the sunlight that falls on it).

Naked-Eye Observations of Mars

WEB LINK 11.7 Unlike Mercury and Venus, Mars is a *superior* planet with an orbit larger than Earth's orbit. As Table 11-3 shows, its orbital semimajor axis is a = 1.524 AU (227.9 million kilometers, or 141.6 million miles). Hence, Mars and the Sun are sometimes on opposite sides of the sky as seen from Earth—that is, Mars is at opposition (see Figure 4-6)—and appears high in our night sky. Such oppositions occur at intervals of about 780 days. However, some oppositions provide better views of Mars than others because Mars has a noticeably elongated orbit (Figure 11-3). As a result, we get the best view of Mars when it is simultaneously at opposition and near the perihelion of its elliptical orbit, a configuration called a **favorable opposition.** The Earth-Mars distance can then be as small as 0.37 AU, or 56 million kilometers (35 million miles), and the angular diameter of Mars can be as large as 25.1 arcsec—about the same as a person's head seen at a distance of 1.6 km (1 mi). Under these conditions, Mars appears in the nighttime sky as a brilliant red object $3\frac{1}{2}$ times brighter than Sirius.

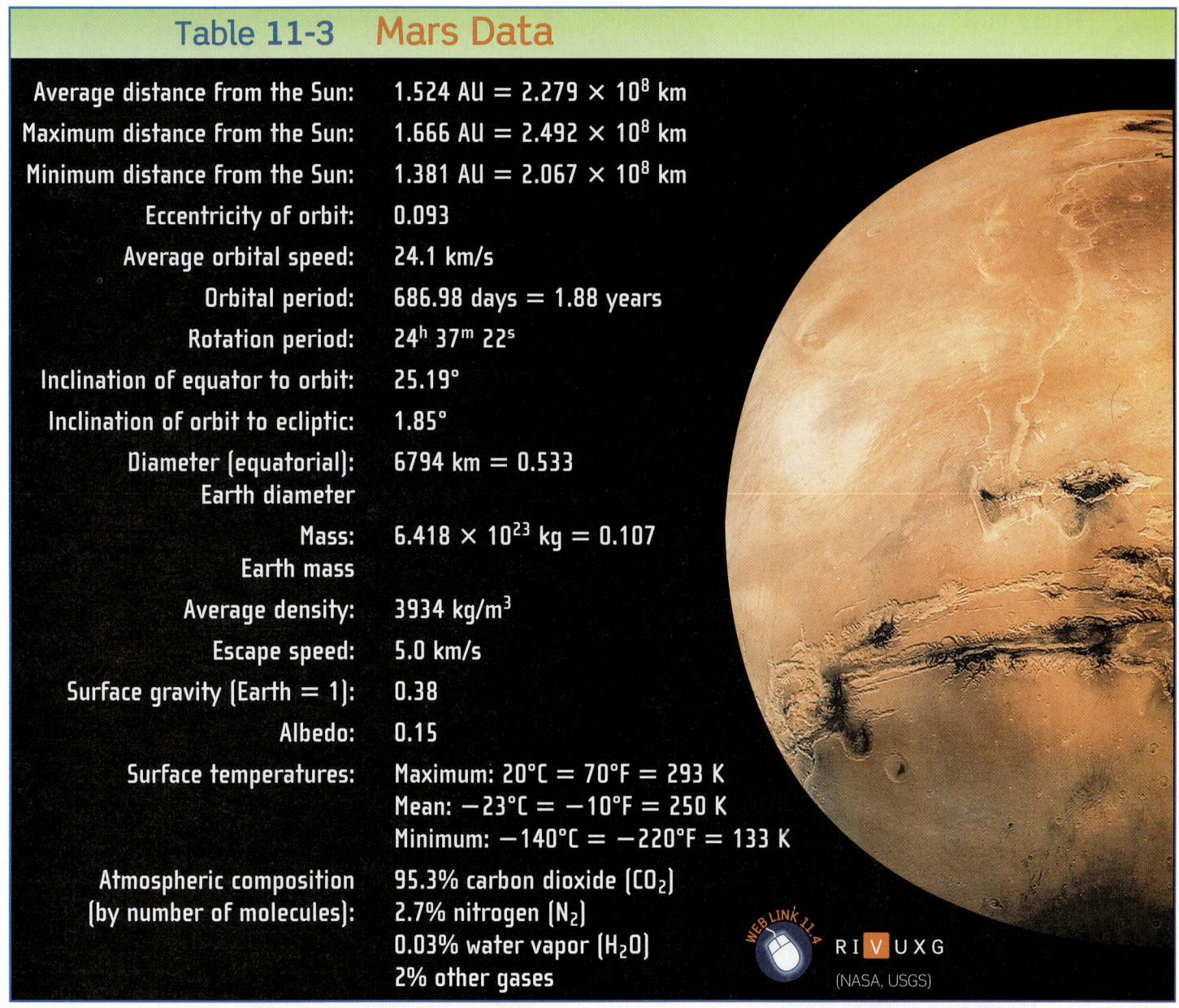

Table 11-3 Mars Data

Average distance from the Sun:	1.524 AU = 2.279×10^8 km
Maximum distance from the Sun:	1.666 AU = 2.492×10^8 km
Minimum distance from the Sun:	1.381 AU = 2.067×10^8 km
Eccentricity of orbit:	0.093
Average orbital speed:	24.1 km/s
Orbital period:	686.98 days = 1.88 years
Rotation period:	$24^h\ 37^m\ 22^s$
Inclination of equator to orbit:	25.19°
Inclination of orbit to ecliptic:	1.85°
Diameter (equatorial):	6794 km = 0.533 Earth diameter
Mass:	6.418×10^{23} kg = 0.107 Earth mass
Average density:	3934 kg/m^3
Escape speed:	5.0 km/s
Surface gravity (Earth = 1):	0.38
Albedo:	0.15
Surface temperatures:	Maximum: 20°C = 70°F = 293 K Mean: −23°C = −10°F = 250 K Minimum: −140°C = −220°F = 133 K
Atmospheric composition (by number of molecules):	95.3% carbon dioxide (CO_2) 2.7% nitrogen (N_2) 0.03% water vapor (H_2O) 2% other gases

WEB LINK 11.4

R I V U X G

(NASA, USGS)

Unfortunately, the time from one favorable opposition to the next is about 15 years. Hence, astronomers take advantage of all oppositions, even the not-so-favorable ones, to observe Mars. When Mars is at opposition but near aphelion the Earth-Mars distance can be as large as 0.68 AU, or 101 million kilometers (63 million miles).

Telescopic Views of Mercury, Venus, and Mars

Naked-eye observations of Mercury are best made at dusk or dawn, but the best telescopic views are obtained in the daytime when the planet is high in the sky, far above the degrading atmospheric effects near the horizon. A yellow filter can eliminate much of the blue light from our sky. The photograph in Figure 11-4*a*, which is among the finest Earth-based views of Mercury, was taken during daytime.

While an Earth-based telescope reveals some surface details on Mercury, Venus appears almost completely featureless, with neither continents nor mountains (see Figure 4-13). Astronomers soon realized that they were not seeing the true surface of the planet. Rather, Venus must be covered by a thick, unbroken layer of clouds. A cloud layer would also explain why Venus reflects such a large fraction of the sunlight that falls on it. Direct evidence that Venus has a thick atmosphere came when nineteenth-century astronomers observed Venus near the time of inferior conjunction. At inferior conjunction, Venus lies most nearly between Earth and the Sun, so that we see the planet lit from behind. As Figure 11-4*b* shows, sunlight is scattered by Venus's atmosphere, producing a luminescent ring around the planet that would not otherwise be present. No such ring is seen around Mercury at inferior conjunction, which indicates that Mercury has no appreciable atmosphere.

ANIMATION 11.2

Figure 11-3

The Orbits of Earth and Mars The best times to observe Mars are at opposition, when Mars is on the side of Earth opposite the Sun. The red and blue dots connected by dashed lines show the positions of Mars and Earth at various oppositions. The months shown around the Sun refer to the time of year when Earth is at each position around its orbit.

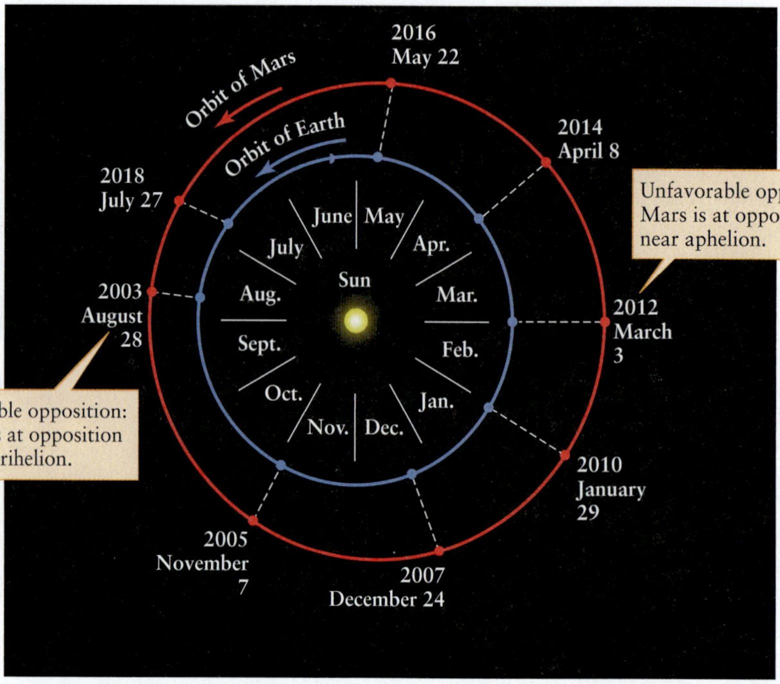

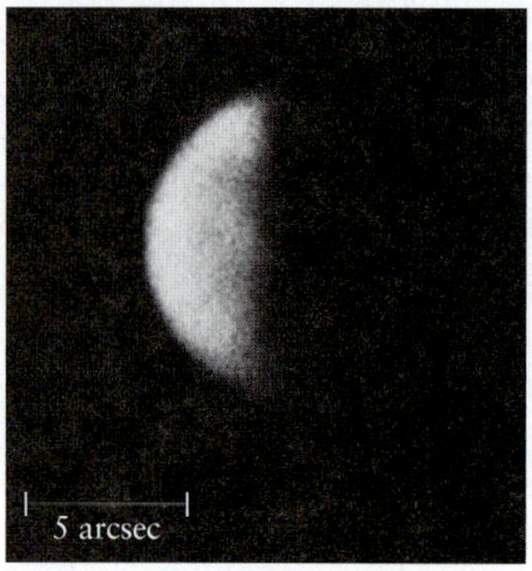

(a) Mercury at greatest elongation

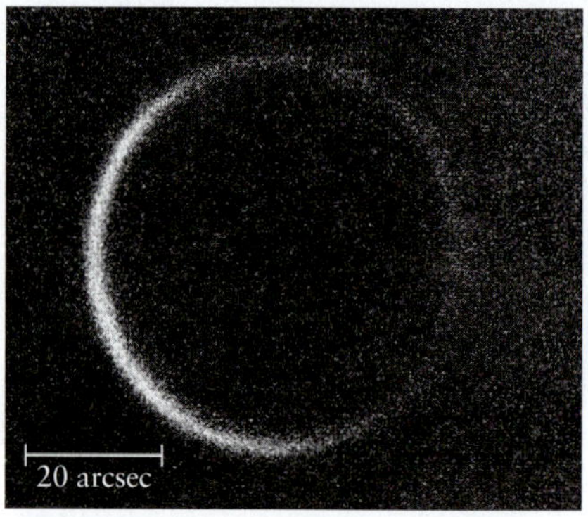

(b) Venus at inferior conjunction

WEB LINK 11.8

Figure 11-4 R I V U X G

Earth-Based Views of Mercury, Venus, and Mars **(a)** Hazy markings are faintly visible in this photograph of Mercury, one of the best ever produced with an Earth-based telescope. Note that Mercury goes through phases like Venus (see Figure 4-13). **(b)** Venus at inferior conjunction shows its atmosphere as a faint ring surrounding the planet. This photograph was processed to reveal the ring, which is too faint to show in Figure 4-13. **(c)** A 25-cm (10-in) telescope with a CCD detector (see Section 6-4) was used to make this image of Mars. Compare this with the Hubble Space Telescope image that opens this chapter, which was taken during the same month in 2003 and which shows the same face of Mars. (a: New Mexico State University Observatory; b: Lowell Observatory; c: Courtesy of Sheldon Faworski and Donald Parker)

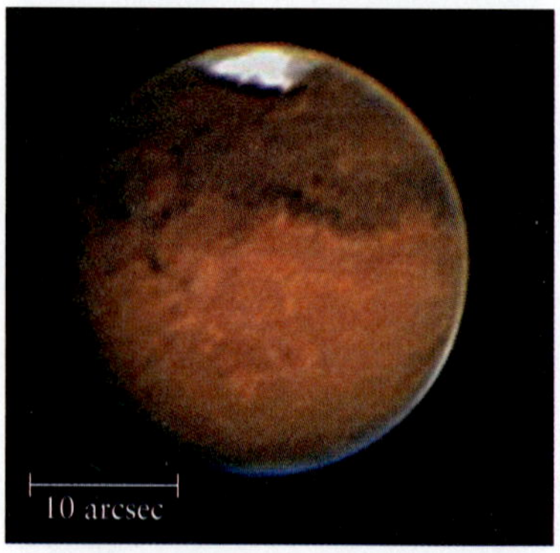

(c) Mars at opposition

Unlike cloud-shrouded Venus, Mars has a thin, almost cloudless atmosphere that permits a clear view of the surface. Under the right conditions, a relatively small telescope can reveal substantial detail on the Martian surface. Figure 11-4*c* shows an image of Mars during the very favorable 2003 opposition, when Mars was closer to Earth than it had been in several thousand years. The thin strip of blue and orange around the right-hand edge of Mars in this image is caused by the Martian atmosphere. This image also shows that different parts of the Martian surface have different colors. Most striking are the whitish ice caps at the Martian poles, which bear a striking superficial resemblance to the Arctic and Antarctic regions on Earth.

CAUTION! Although Mars appears bright red to Earth observers, its surface is actually reddish-brown. A sunlit brown surface, if surrounded by the blackness of space, is perceived by the human eye as having a red color.

11-2 While Mars rotates much like Earth, Mercury's rotation is coupled to its orbital motion and Venus's rotation is slow and retrograde

We have seen that the surface of Mars is relatively easy to observe with visible-light telescopes. By contrast, only a few faint, hazy markings can be seen on Mercury's surface, and the surface of Venus is perpetually hidden by clouds. These differences explain why the rotation of Mars has been well understood since the mid-seventeenth century, while it was not until the 1960s that astronomers accurately measured the rotation of Mercury and Venus. As we will see, Mars rotates in a very Earthlike way, while Mercury and Venus rotate like no other objects in the solar system.

Radar technology revealed the curious rotation of Mercury and Venus

The Rotation of Mars

WEB LINK 11.9

The first reliable record of surface features on Mars was made by the Dutch scientist Christiaan Huygens, who observed the planet during the opposition of 1659. He identified a prominent dark feature that we now call Syrtis Major. After observing this feature for several weeks, Huygens concluded that the rotation period of Mars is approximately 24 hours, the same as Earth.

In 1666, the Italian astronomer Gian Domenico Cassini made more-refined observations which showed that a Martian day is 37.5 minutes longer than an Earth day. More than a century later, the German-born English astronomer William Herschel found that Mars's axis of rotation is not perpendicular to the plane of the planet's orbit, but is tilted by about 25° away from the perpendicular. This tilt is very close to Earth's $23\frac{1}{2}$° tilt (see Figure 2-12). This striking coincidence means that Mars experiences Earthlike seasons, with opposite seasons in the northern and southern Martian hemispheres (see Section 2-5). Because Mars takes nearly 2 (Earth) years to orbit the Sun, the Martian seasons last nearly twice as long as on Earth.

The Challenge of Observing Mercury's Rotation

During the 1880s, the Italian astronomer Giovanni Schiaparelli attempted to make the first map of Mercury. Unfortunately, Schiaparelli's telescopic views of Mercury were so vague and indistinct that he made an understandable but major error, which went uncorrected for more than half a century. He erroneously concluded that Mercury always keeps the same side facing the Sun.

Many objects in our solar system are in synchronous rotation, so that their rotation period equals their period of revolution—a situation also called **1-to-1 spin-orbit coupling.** We saw in Section 4-8 how Earth's tidal forces keep the Moon in synchronous rotation, so that it always keeps the same side toward Earth. Tidal forces also keep the two moons of Mars with the same side facing their parent planet, and likewise many of the satellites of Jupiter and Saturn.

It had been suggested as early as 1865 that the Sun's tidal forces would keep Mercury in synchronous rotation. But this is not how Mercury rotates. The first clue about Mercury's true rotation period came in 1962, when astronomers detected radio radiation coming from the innermost planet.

As we learned in Sections 5-3 and 5-4, every dense object (such as a planet) emits electromagnetic radiation with a continuous spectrum. The dominant wavelength of this radiation depends on the object's temperature. Only at absolute zero does a dense object emit no radiation at all.

If Mercury exhibited synchronous rotation, one side of the planet would remain in perpetual, frigid darkness, quite near absolute zero. To test this idea, in 1962 the radio astronomer W. E. Howard and his colleagues at the University of Michigan began monitoring radiation from Mercury. (If the planet had one side that was very cold, most of the emission from that side should be at quite long wavelengths. Hence, a radio telescope was used for these observations.) They were surprised to discover that the temperature on the planet's nighttime side is about 100 K (= −173°C = −280°F), not nearly as cold as expected. These observations were in direct contradiction to the idea of synchronous rotation.

One idea offered to explain Howard's observations was that Mercury rotates synchronously, but that winds in the planet's atmosphere carried warmth from the daytime side of the planet around to the nighttime side. But, in fact, there are no winds on Mercury. Its daytime temperature is too high and its gravity too weak to retain any substantial atmosphere (see Box 7-2).

The definitive observation was made in 1965, when Rolf B. Dyce and Gordon H. Pettengill used the giant 1000-ft radio telescope at the Arecibo Observatory in Puerto Rico to bounce powerful radar pulses off Mercury. The outgoing radiation consisted of microwaves of a very specific wavelength. In the reflected signal echoed back from the planet, the wavelength had shifted as a result of the Doppler effect (see Section 5-9, especially Figure 5-26). As Mercury rotates, one side of the planet approaches Earth, while the other side recedes from Earth. Microwaves reflected from the planet's approaching side were shortened in wavelength, whereas those from its receding side were lengthened (**Figure 11-5**). Thus, the radar pulse went out at one specific wavelength, but it came back spread over a small wavelength

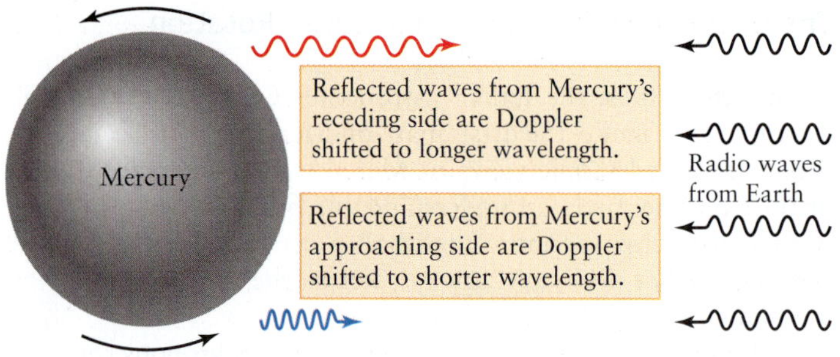

Figure 11-5

Measuring Mercury's Rotation Period As Mercury rotates, one side of the planet moves away from Earth and the other side toward Earth. If radio waves of a single wavelength are beamed toward Mercury, waves reflected from these two sides will be Doppler shifted to longer and shorter wavelengths, respectively. By measuring the wavelength shift of the reflected radiation, astronomers deduced how rapidly Mercury rotates and thus determined its rotation period.

range. From the width of this wavelength range, Dyce and Pettengill deduced that Mercury's rotation period is approximately 58.6 days.

Spin-Orbit Coupling and the Curious Rotation of Mercury

Giuseppe Colombo, an Italian physicist with a long-standing interest in Mercury, was intrigued by Dyce and Pettingill's results. Colombo noted that their result for the rotation period is very close to two-thirds of Mercury's sidereal period of 87.969 days:

$$2/3\ (87.969\ \text{days}) = 58.646\ \text{days}$$

Colombo therefore boldly speculated that Mercury's true rotation period is exactly 58.646 days, or 58 days and $15\frac{1}{2}$ hours. He realized that this figure would mean that Mercury is locked into a **3-to-2 spin-orbit coupling:** The planet makes *three* complete rotations on its axis for every *two* complete orbits around the Sun. No other planet or satellite in the solar system has this curious relationship between its rotation and its orbital motion.

Figure 11-6 shows how gravitational forces from the Sun cause Mercury's 3-to-2 spin-orbit coupling. The force of gravity decreases with increasing distance, which explains how the Moon is able to raise a tidal bulge in Earth's oceans (see Section 4-8). Mercury has no oceans, but it has a natural bulge of its own; thanks to the Sun's tidal forces, the planet is slightly elongated along one axis. The stronger gravitational force that the Sun exerts on the near side of the planet tends to twist the long axis to point toward the Sun, as Figure 11-6*a* shows. Indeed, Mercury's long axis would always point toward the Sun if its orbit were circular or nearly so. In this case the same side of Mercury would always face the Sun, and there would be synchronous rotation (Figure 11-6*b*). But Mercury's orbit has a rather high eccentricity, and the twisting effect on Mercury decreases rapidly as the planet moves away from the Sun. As a result, Mercury's long axis points toward the Sun only at perihelion (Figure 11-6*c*). From one perihelion to the next, alternate sides of Mercury face the Sun.

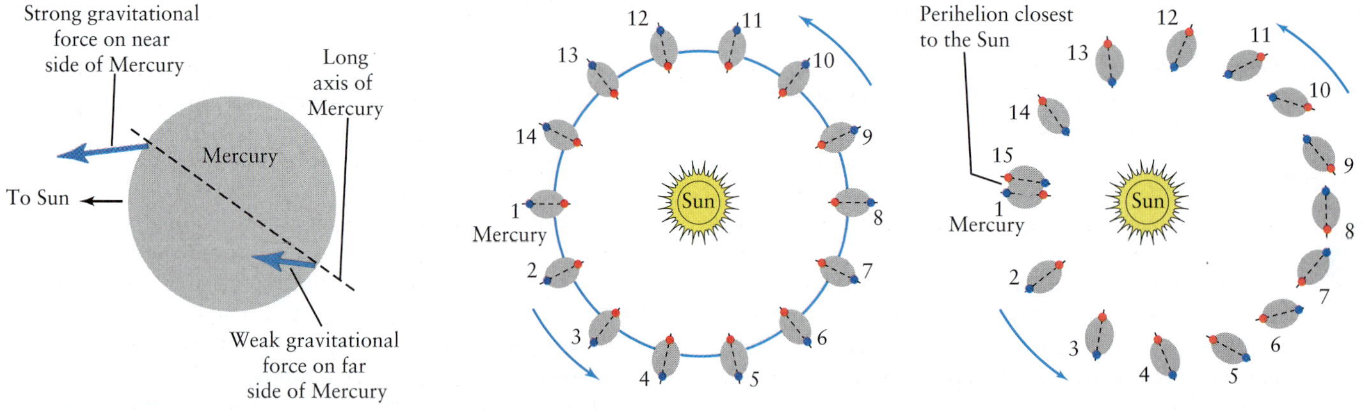

(a) Mercury is slightly elongated: The different gravitational forces on Mercury's two ends tend to twist the long axis to point toward the Sun.

(b) If Mercury were in a circular orbit, its long axis would always point toward the Sun: Mercury would be in synchronous rotation (1-to-1 spin-orbit coupling).

(c) In fact Mercury is in an elliptical orbit, and its long axis only points toward the Sun at perihelion: Mercury spins on its axis 1½imes during each complete orbit (3-to-2 spin-orbit coupling).

Figure 11-6

Mercury's Spin-Orbit Coupling (a) The Sun's gravitational force on the near side of Mercury's long axis is greater than the force on the far side. This tends to twist the long axis to point toward the Sun. **(b)** If Mercury were in a circular orbit, the twisting effect shown in (a) would always keep the same side of Mercury (shown by a red dot) facing the Sun. **(c)** Because Mercury's orbit is rather elongated, its rotation is more complex: The end of Mercury's long axis marked by a red dot faces the Sun at one perihelion (point 1), but at the next perihelion (point 15) the opposite end—marked by a blue dot—faces the Sun.

Because of the 3-to-2 spin-orbit coupling, the average time from sunrise to sunset on Mercury is just equal to its orbital period, or just under 88 days. This period helps explain the tremendous difference between daytime and nighttime temperatures on Mercury. Not only is sunlight about 7 times more intense on Mercury than on Earth (because of the smaller size of Mercury's orbit), but that sunlight has almost 3 Earth months to heat up the surface. As a result, daytime temperatures at the equator are high enough—about 430°C (roughly 700 K, or 800°F)—to melt lead. (In comparison, a typical kitchen oven reaches only about 300°C or 550°F.)

The time from sunset to the next sunrise is also 88 days, and the surface thus has almost 3 Earth months of darkness during which to cool down. Therefore, nighttime temperatures reach to below −170°C (about 100 K, or −270°F), which is cold enough to freeze carbon dioxide and methane. The surface of Mercury is truly inhospitable!

While the Sun moves slowly across Mercury's sky, it does not always move from east to west as it does as seen from Earth. The reason is that Mercury's speed along its elliptical orbit varies substantially, in accordance with Kepler's second law (see Section 4-4, especially Figure 4-11). It moves fastest (59 km/s) at perihelion and slowest (39 km/s) at aphelion.

As seen from Mercury's surface, the Sun rises in the east and sets in the west, just as it does on Earth. When Mercury is near perihelion, however, the planet's rapid motion along its orbit outpaces its leisurely rotation about its axis (one rotation in 58.646 days). Hence, the usual east-west movement of the Sun across Mercury's sky is interrupted, and the Sun actually stops and moves backward (from west to east) for a few Earth days. If you were standing on Mercury watching a sunset occurring at perihelion, the Sun would not simply set. It would dip below the western horizon and then come back up, only to set a second time a Mercury day or two later.

Observing Venus's Rotation: Penetrating the Clouds with Radio

Venus's perpetual cloud cover makes it impossible to measure the planet's rotation using visible-light telescopes. (It is possible to use visible-light telescopes to track features in the planet's cloud cover. But winds in the atmosphere cause such features to move relative to Venus's surface, so this approach does not give a direct measurement of how the planet itself rotates.) To penetrate the planet's perpetual cloud cover, astronomers applied to Venus the same technique of using radio waves and the Doppler effect that revealed the rotation of Mercury.

Clouds may contain gas, dust, haze, water droplets, or other small particles. In general, electromagnetic radiation can pass easily through such a cloud only if the wavelength is large compared to the size of the particles. We then say that the cloud is *transparent* to the radiation. For example, clouds in Earth's atmosphere are made of water droplets with an average diameter of about 20 μm (2×10^{-5} m, or 20,000 nm). Visible light has wavelengths between 400 to 700 nm, which is less than the droplet size. Hence, visible light cannot easily pass through such clouds, which is why cloudy days are darker than sunny days. But radio waves, with wavelengths of 0.1 m or more, and microwaves, with wavelengths from 10^{-3} m to 0.1 m, can pass through such a cloud with ease. So, your mobile phone, radio, or television works just as well on a cloudy or foggy day as on a clear one.

Like clouds in Earth's atmosphere, the clouds of Venus are transparent to radio waves and microwaves. Beginning in the early 1960s, advances in radio telescope technology made it possible to send microwave radiation to Venus and detect the waves reflected from its surface. Just as for Mercury (see Figure 11-5), the Doppler shift of waves reflected from the two sides of Venus reveal the speed and direction of the planet's rotation.

Venus: Where the Sun Rises in the West

These measurements show that Venus rotates very slowly; the sidereal rotation period of the planet is 243.01 days, even longer than the planet's 224.7-day orbital period. To an imaginary inhabitant of Venus who could somehow see through the cloud cover, stars would move across the sky at a rate of only $1\frac{1}{2}°$ per Earth day. As seen from Earth, by contrast, stars move $1\frac{1}{2}°$ across the sky in just 6 minutes.

Not only does Venus rotate slowly, it also rotates in an unusual direction. Most of the planets in the solar system spin on their axes in **prograde rotation.** Prograde means "forward," and planets with prograde rotation spin on their axes in the same direction in which they orbit the Sun (Figure 11-7*a*). As seen from the surface of a planet with prograde rotation, such as Earth, the Sun and stars rise in the east and set in the west.

Venus is an exception to the rule: It spins in **retrograde rotation.** That is, the direction in which it spins on its axis is backward compared to the direction in which it orbits the Sun, as Figure 11-7*b* shows. (Recall that "retrograde" means "backward.") If an observer on the surface of Venus could see through the thick layer of clouds, the Sun and stars would rise in the west and set in the east!

CAUTION! Be careful not to confuse retrograde *rotation* with retrograde *motion*. When Venus is near inferior conjunction, so that it is passing Earth on its smaller, faster orbit around the Sun, it appears to us to move from east to west on the celestial sphere from one night to the next. This apparent motion is called retrograde motion, because it is backward to the apparent west-to-east motion that Venus displays most of the time. (You may want to review the discussion of retrograde motion in Sections 4-1 and 4-2.) By contrast, the retrograde rotation of Venus means that it spins on its axis from east to west, rather than west to east like Earth.

Venus's retrograde rotation adds to the puzzle of our solar system's origins. If you could view our solar system from a great distance above Earth's north pole, you would see all the planets orbiting the Sun counterclockwise. Closer examination would reveal that the Sun and most of the planets also *rotate* counterclockwise on their axes; the exceptions are Venus and Uranus. Most of the satellites of the planets also move counterclockwise along their orbits and rotate in the same direction. Thus, the rotation of Venus is not just opposite to its orbital motion; it is opposite to most of the orbital and rotational motions in the solar system!

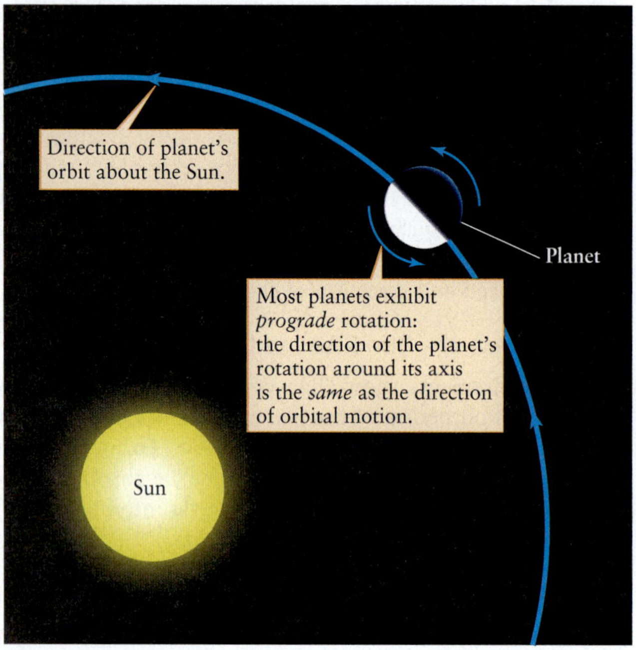

(a) Prograde rotation

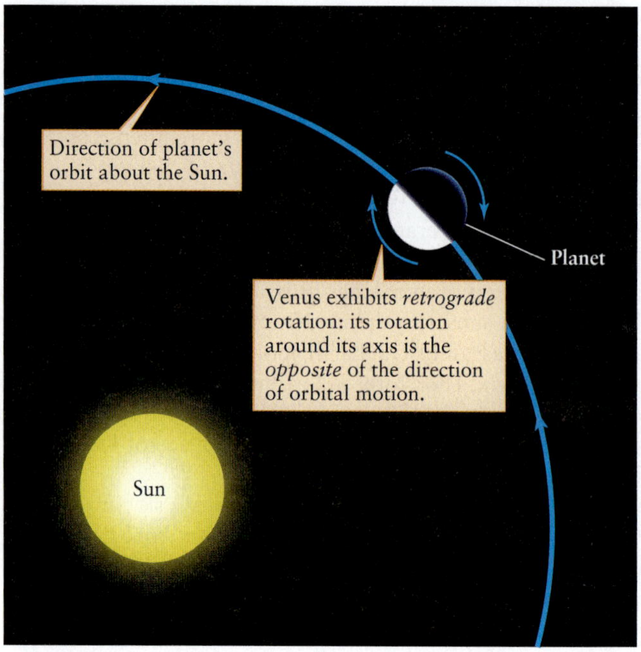

(b) Retrograde rotation

ANIMATION 11.3

Figure 11-7

Prograde and Retrograde Rotation (a) If you could view the solar system from a point several astronomical units above the Earth's north pole, you would see all of the planets orbiting the Sun in a counterclockwise direction. Most of the planets also rotate counterclockwise on their axes in prograde (forward) rotation. (b) Venus is an exception; from your perspective high above the plane of the solar system, you would see Venus rotating clockwise in retrograde (backward) rotation.

As we saw in Section 8-4, the Sun and planets formed from a solar nebula that was initially rotating. (Had it not been rotating, all of the material in the nebula would have fallen inward toward the protosun, leaving nothing to form the planets.) Because the planets formed from the material of this rotating cloud, they tend to rotate on their axes in that same direction. It is difficult to imagine how Venus and Uranus could have bucked this trend. One theory suggests that the tidal forces exerted by the Sun tended to elongate Venus's atmosphere toward the Sun, much as the Moon's tidal forces elongate Earth's oceans toward the Moon (see Section 4-8, especially Figure 4-25). Friction between the atmosphere and Venus's surface then acted to slow the planet's rotation. By themselves, these tidal effects would put Venus into synchronous prograde rotation. But in addition, the Sun's heat on the illuminated side of Venus generates strong winds in the planet's atmosphere, and the frictional forces of these winds drive Venus into slow retrograde rotation.

Venus's rotation has other puzzling aspects. The length of an apparent solar day on Venus—that is, from when the Sun is highest in the sky to when it is again highest in the sky (see Section 2-7)—is 116.8 Earth days. Remarkably, this is almost exactly one-fifth as long as the planet's 584-day synodic period. (We saw in Section 4-2 that this is the time between successive inferior conjunctions, when Venus and Earth are closest together.) Is this curious ratio the result of gravitational interactions between Venus and Earth? Or is it sheer coincidence? No one knows the answer.

11-3 Mercury is cratered like the Moon but has a surprising magnetic field

Imagine an alien astronomer observing Earth from Mercury, with the same telescope technology available to our own astronomers. Such an astronomer would be unable to resolve any features on Earth less than a few hundred kilometers across. It would be impossible for her to learn about Earth's mountain ranges or volcanoes. Earth-based astronomers face the same limitations in studying Mercury. To truly understand Mercury, it is essential to study this world at close range using spacecraft.

Spacecraft observations at close range were necessary to reveal Mercury's unusual properties

Mercury spacecraft: Mariner 10 and MESSENGER

LOOKING DEEPER 11.1

The first spacecraft to provide detailed knowledge about Mercury was *Mariner 10*, which flew close to Mercury on three occasions in 1974 and 1975. In 2008, two brief flybys produced the first images of Mercury from the spacecraft *MESSENGER* (MErcury Surface, Space ENvironment, GEochemistry, and Ranging). *MESSENGER* can record features as small as 18 m—a vast improvement over *Mariner 10*, which can record features only as small as 1 km. **Figure 11-8** shows a real-color image of Mercury from *MESSENGER*.

The *Mariner 10* images show the only hemisphere of Mercury's surface visible during the flybys. *MESSENGER*'s two flybys have already revealed 80% of the previously hidden side. As of this writing, analysis of the new data has just begun. In 2011, *MESSENGER* is scheduled to go into orbit around Mercury (not just fly by), and much more data will help unravel Mercury's mysteries.

Mercury's Cratered Surface

As *Mariner 10* closed in on Mercury, scientists were struck by the Moonlike pictures appearing on their television monitors. It was obvious that Mercury, like the Moon, is a barren, heavily cratered world, with no evidence for plate tectonics. These surface features are what we would expect based on Mercury's small size (see Table 7-1 and the figure that opens this chapter). As we saw in Section 7-6, a small world is expected to have little internal heat, and, hence, little or no geologic activity and an ancient, cratered surface (see Figure 7-11). But craters are not the only features of Mercury's surface; there are also gently rolling plains, long, meandering cliffs, and an unusual sort of jumbled terrain. As we will see, Mercury's distinctive surface shows that this planet is *not* merely a somewhat larger version of the Moon.

Figure 11-9 shows a typical close-up view of Mercury's surface. The consensus among astronomers is that most of the craters on both Mercury and the Moon were produced during the first 700 million years after the planets formed. Debris remaining after planet formation rained down on these young worlds, gouging out most of the craters we see today. We saw in Section 10-4 that the strongest evidence for this chronology comes from analysis and dating of Moon rocks. No spacecraft has landed on Mercury, so we are not able to make the same kind of direct analysis of rocks from the planet's surface.

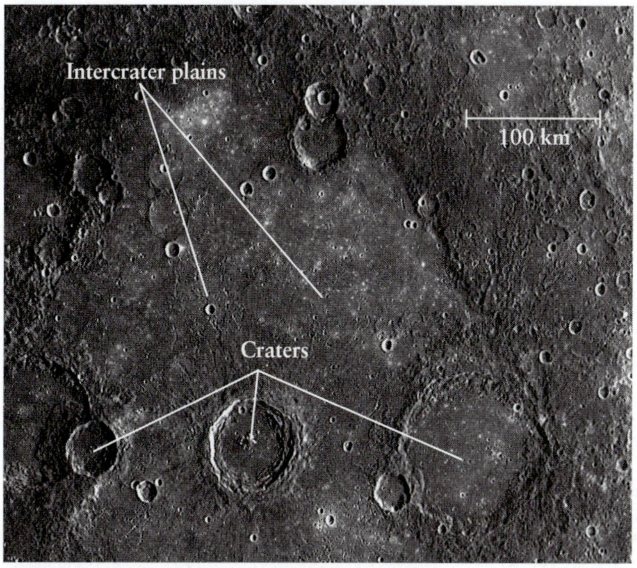

VIDEO 11.2 **Figure 11-9** RI V UXG

Mercurian Craters and Plains *MESSENGER* took this true-color image of a region near Mercury's equator at a range of 2800 km (1700 mi). Lava flows formed the plains inside and to the west of the crater on the right. (NASA/Johns Hopkins U. Applied Physics Laboratory/Carnegie Institution of Washington)

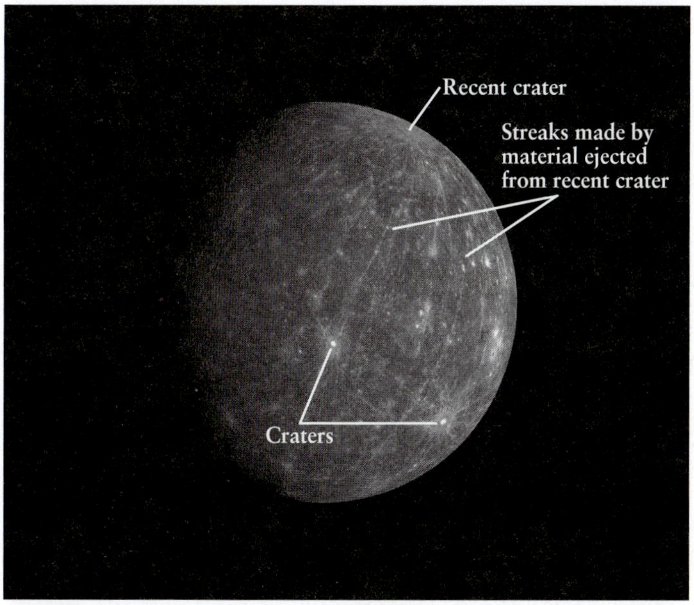

WEB LINK 11.10 **Figure 11-8** RI V UXG

***MESSENGER* Image of Mercury** This real-color image shows the natural gray color of Mercury's surface. The impact that formed the crater near the top of the image was so powerful that it ejected material to nearly the opposite end of the planet. Seen first in 2008, are the long streaks, or rays, from the northern hemisphere into the south. (NASA/Johns Hopkins U. Applied Physics Laboratory/Carnegie Institution of Washington)

Unlike the Moon, Mercury's surface has extensive low-lying plains (see Figure 11-9). These large, smooth areas are about 2 km lower than the cratered terrain. These probably have a similar origin to the lunar maria, which were produced by extensive lava flows between 3.1 and 3.8 billion years ago (see Section 10-4). Primordial lava flows probably also explain the plains of Mercury. As large meteoroids punctured the planet's thin, newly formed crust, lava welled up from the molten interior to flood low-lying areas.

From the number of craters that pit them, Mercury's plains appear to have been formed near the end of the era of heavy bombardment, a little more than 3.8 billion years ago. Mercury's plains are therefore older than most of the lunar maria. Analysis of images made at different wavelengths also shows that the material in Mercury's plains has a lower iron content than the lunar maria. This is presumably why the plains of Mercury do not have the dark color of the Moon's maria.

Scarps: Evidence of a Wrinkled Planet

Images also reveal numerous long cliffs, called **scarps**, meandering across Mercury's surface (Figure 11-10). Some scarps rise as much as 3 km (2 mi) above the surrounding plains and are 20 to 500 km long. These cliffs probably formed as the planet cooled and contracted.

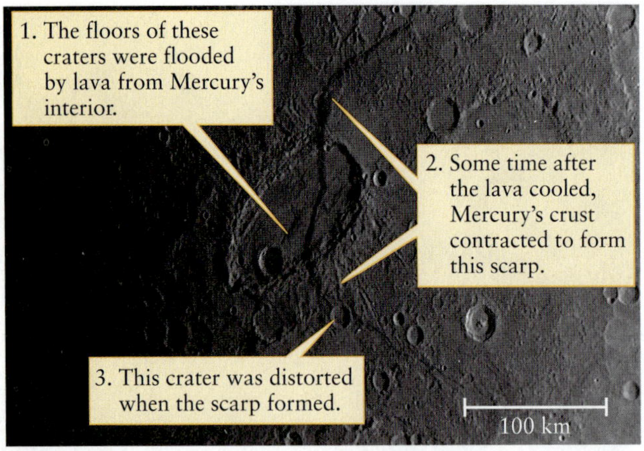

1. The floors of these craters were flooded by lava from Mercury's interior.
2. Some time after the lava cooled, Mercury's crust contracted to form this scarp.
3. This crater was distorted when the scarp formed.
100 km

ANIMATION 11.4

Figure 11-10 R I V U X G

A Scarp A long cliff, or scarp, called Beagle Rupes runs across this *MESSENGER* image. This scrap is almost a kilometer high and extends for more than 600 km (370 mi) across a region near Mercury's equator. (NASA/Johns Hopkins U. Applied Physics Laboratory/Carnegie Institution of Washington)

ANALOGY The shrinkage of Mercury's crust is probably a result of a general property of most materials: They shrink when cooled and expand when heated. The lid on a jar of jam makes a useful analogy. When the jar is kept cold in the refrigerator, the metal lid contracts more than the glass jar. The shrinking lid makes a tight seal, but also makes the lid hard to unscrew. You can loosen the lid by running hot water over it; this makes the metal lid expand more than the glass jar, helping to break the seal.

If there had still been molten material beneath Mercury's surface when this contraction took place, lava would have leaked onto the surface around the scarps. However, it does not appear that lava has flowed. Hence, the scarps must have formed relatively late in Mercury's history, after the lava flows had ended and after the planet had solidified to a substantial depth beneath the surface. With the exception of the scarps, there are no features on Mercury that resemble the boundaries of tectonic plates. Thus, we can regard Mercury's crust, like that of the Moon, as a single plate.

Fire and Ice: The Caloris Basin and Mercury's Poles

The most impressive feature discovered by *Mariner 10* was a huge impact scar called the Caloris Basin (*calor* is the Latin word for "heat"). The Sun is directly over the Caloris Basin during alternating perihelion passages, and thus it's the hottest place on the planet once every 176 days. Different chemical compositions influence the enhanced-colors of the Caloris Basin, in Figure 11-11. The human eye would see mostly gray as in the other Mercury figures. The bright orange spots near the rim are thought to indicate volcanic material, and the blue areas suggest an above-average titanium oxide content. Material appearing blue is thought to orginate a few kilometers below the surface and was exposed through impact events.

The Caloris Basin, which measures 1500 km (960 mi) in diameter, is both filled with and surrounded by smooth lava plains. The basin was probably gouged out by the impact of a large meteoroid that penetrated the planet's crust, allowing lava to flood out onto the surface and fill the basin. Because relatively few craters pockmark these lava flows, the Caloris impact must be relatively young. It must have occurred toward the end of the cratermaking period that dominated the first 700 million years of our solar system's history.

The impact that created the huge Caloris Basin must have been so violent that it shook the entire planet. On the side of Mercury opposite the Caloris Basin, *Mariner 10* discovered a consequence of this impact—a jumbled, hilly region covering about 500,000 square kilometers, about twice the size of the state of Wyoming. Geologists theorize that seismic waves from the Caloris impact became focused as they passed through Mercury. As this concentrated seismic energy reached the far surface of the planet, it deformed the surface and created jumbled terrain. (This same process may also have taken place on the Moon: There is a region of chaotic hills on the side of the Moon directly opposite from the large impact basin called Mare Orientale.)

WEB LINK 11.11

In the decades since the *Mariner 10* mission, astronomers have continued to study Mercury with both visible-light and radio telescopes. Radar observations have detected patches at Mercury's north and south poles that are unusually effective at reflecting radio waves. It has been suggested that these patches may be regions of water ice deep within craters where the Sun's rays never reach. (We saw in Section 10-2 that ice may also exist in shadowed regions at the Moon's

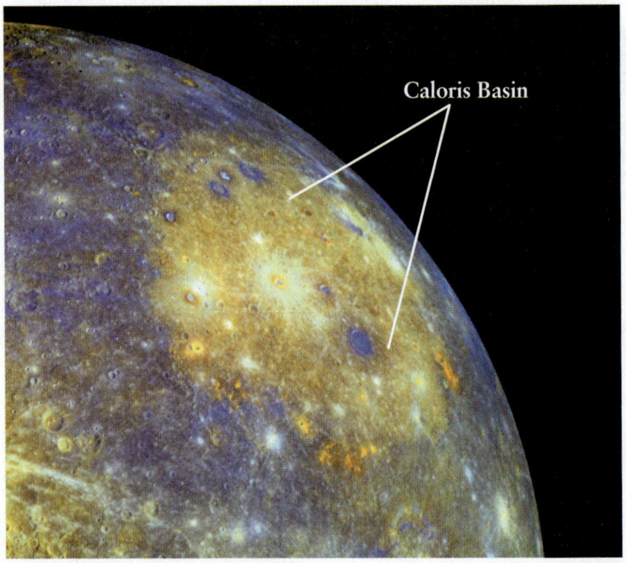

Figure 11-11 R I V U X G

The Caloris Basin About 4 billion years ago an impact on Mercury created an immense basin, shown in orange in this enhanced-color *MESSENGER* image. The crater is about 1500 km (960 mi) in diameter. The impact fractured the surface extensively, forming several concentric chains of mountains. The mountains in the outermost ring are up to 2 km (6500 ft) high. Craters inside the basin (shown in blue) were formed later by smaller impacts. (NASA/Johns Hopkins U. Applied Physics Laboratory/Arizona State U./Carnegie Institution of Washington)

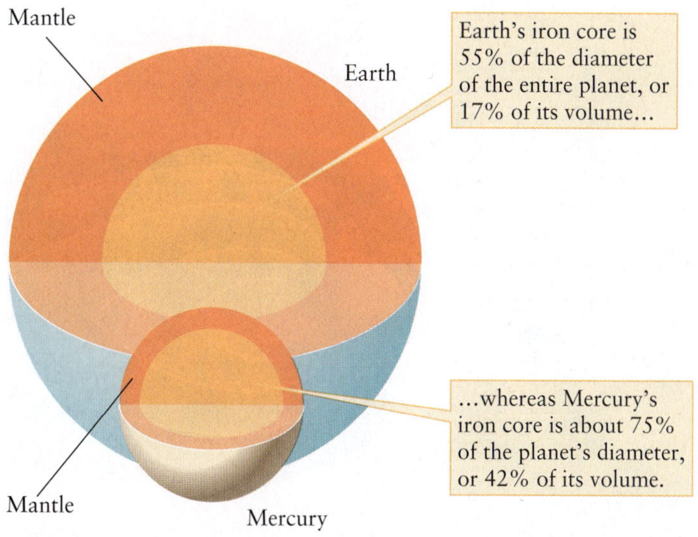

INTERACTIVE EXERCISE 11.1

Figure 11-12

The Internal Structures of Mercury and Earth Compared to Earth's core, Mercury's iron-rich core takes up a much larger fraction of the planet's volume. Indeed, Mercury is the most iron-rich planet in the solar system. Surrounding Mercury's core is a 600-km-thick rocky mantle.

poles.) If this is correct, then Mercury is truly a world of extremes, with perpetual ice caps at one or both poles but with midday temperatures at the equator that are high enough to melt lead.

Mercury's Interior and Iron Core

By measuring how much *Mariner 10* was deflected by Mercury's gravity, scientists were able to make a precise determination of Mercury's mass. Given the mass and the diameter of Mercury, they could then calculate that the planet has an average density of 5430 kg/m^3. This value is quite similar to Earth's average density of 5515 kg/m^3. We saw in Section 9-2 that typical rocks from Earth's surface have a density of only about 3000 kg/m^3 because they are composed primarily of lightweight, mineral-forming elements. The higher average density of our planet is caused by Earth's iron core. By studying how Earth vibrates during earthquakes, geologists have deduced that our planet's iron core occupies about 17% of Earth's volume.

Pressures inside a planet increase the density of rock by squeezing its atoms closer together. Because Earth is 18 times more massive than Mercury, the weight of this larger mass compresses Earth's core much more than Mercury's core. If Earth were uncompressed, its density would be 4400 kg/m^3, whereas Mercury's uncompressed density would still be a hefty 5300 kg/m^3. This higher uncompressed density means that Mercury has a *larger* proportion of iron than Earth: An iron core must occupy about 42% of Mercury's volume. Mercury is therefore the most iron-rich planet in the solar system. Figure 11-12 is a scale drawing of the interior structures of Mercury and Earth.

Several theories have been proposed to account for Mercury's high iron content. According to one theory, the inner regions of the primordial solar nebula were so warm that only those substances with high condensation temperatures—like iron-rich minerals—could have condensed into solids. Another theory suggests that a brief episode of very powerful solar winds could have stripped Mercury of its low-density mantle shortly after the Sun formed. A third possibility is that during the final stages of planet formation, Mercury was struck by a large planetesimal. Supercomputer simulations show that this cataclysmic collision would have ejected much of the lighter mantle, leaving a disproportionate amount of iron to reaccumulate to form the planet we see today (Figure 11-13).

Clues About the Core: Mercury's Magnetic Field

An important clue to the structure of Mercury's iron core came from the *Mariner 10* magnetometers, which discovered that Mercury has a magnetic field similar to that of Earth but only about

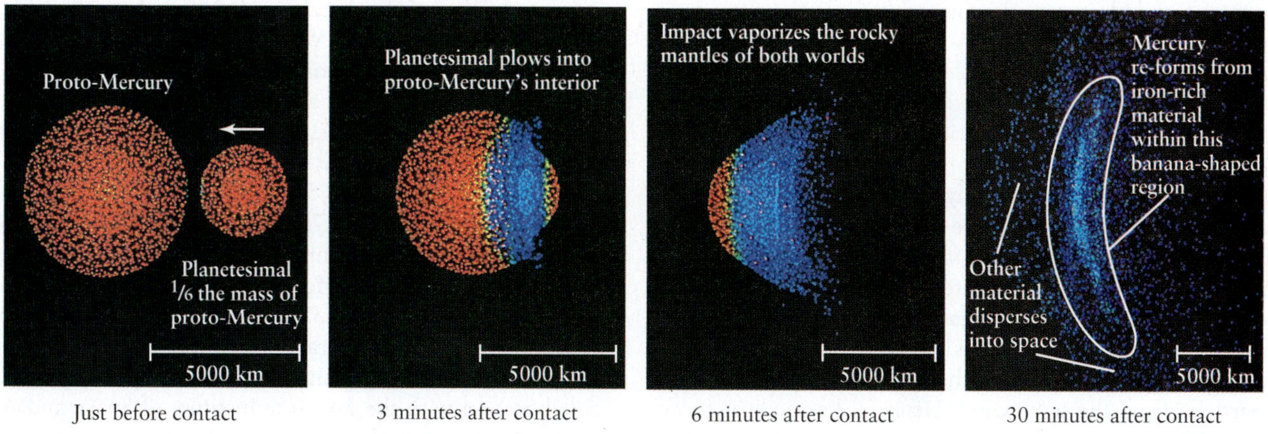

Figure 11-13

Stripping Mercury's Mantle by a Collision To account for Mercury's high iron content, one theory proposes that a collision with a planet-sized object stripped Mercury of most of its rocky mantle. These four images show a supercomputer simulation of a head-on collision between proto-Mercury and a planetesimal one-sixth its mass. (Courtesy of W. Benz, A. G. W. Cameron, and W. Slattery)

1% as strong. We saw in Section 7-7 that for a planet to have a global magnetic field, it must have moving liquid material in its interior that conducts electricity. This moving material acts as a dynamo and generates the magnetic field. Hence, we conclude that (1) at least part of Mercury's core must be in a liquid state, and (2) there must be a source of energy to make material flow within this liquid region in order to produce electric currents. The second requirement is evidence that Mercury, like Earth, also has a *solid* part of its core: As material deep in the liquid core cools and solidifies to join the solid portion of the core, it releases the energy needed to stir up the motions of the remaining liquid material. Mercury's magnetic field therefore suggests that its core has the same general structure as Earth's core (recall Figure 9-9).

Mercury's magnetic field comes as something of a surprise. The ancient, cratered surface of Mercury is evidence of a lack of geologic activity, which in turn shows that Mercury must have lost the internal heat that powers such activity. Again, we would expect that such a small planet would lose much of its internal heat (see Section 7-6). But the existence of a global magnetic field shows that Mercury is still partially molten inside, which is evidence that the planet has some internal heat. Thus, the evidence from cratering and from the global magnetic field contradict each other.

Two other observations further complicate our understanding of Mercury's interior. One is that the amount of electrically conducting molten material that remains within Mercury is thought to be too little to account for even the relatively weak global magnetic field. Another is that a planet must rotate fairly rapidly in order to stir its molten interior into the kind of motion that generates a magnetic field. But as we saw in Section 11-2, Mercury rotates very *slowly:* It takes 58.646 days to spin on its axis. We are forced to conclude that we still have much to learn about the interior of Mercury.

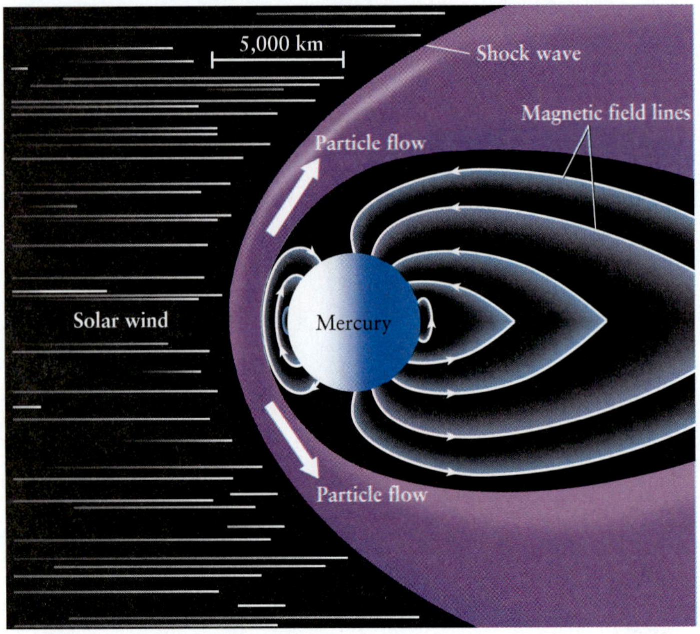

ANIMATION 11.5

Figure 11-14

Mercury's Magnetosphere Mercury's weak magnetic field has only 1% the strength of Earth's field, but just strong enough to carve out a cavity in the solar wind, preventing the impinging particles from striking the planet's surface directly. Most of the particles of the solar wind are deflected around the planet in a turbulent region colored purple in this scale drawing. (Compare this with Earth's much larger magnetosphere, shown in Figure 9-20.)

Mercury's Magnetosphere

Whatever the origin of Mercury's magnetic field, we know that it interacts with the solar wind much as Earth's field does. (**Figure 11-14**). The solar wind is a constant stream of charged particles (mostly protons and electrons) flowing away from the outer layers of the Sun's upper atmosphere. A planet's magnetic field can repel and deflect the impinging particles, thereby forming an elongated "cavity" in the solar wind called a magnetosphere. (We discussed Earth's magnetosphere in Section 9-4.)

When the particles in the solar wind first encounter Mercury's magnetic field, they are abruptly slowed, producing a shock wave that marks the boundary where this sudden decrease in velocity occurs. Most of the particles from the solar wind are deflected around the planet, just as water is deflected to either side of the bow of a ship. Mercury's magnetosphere thus prevents the solar wind from reaching the planet's surface. Because Mercury's magnetic field is much weaker than that of Earth, it is not able to capture particles from the solar wind. Hence, it has no structures like the Van Allen belts that surround Earth (recall Figure 9-20). During *MESSENGER*'s first flyby, the interaction between the solar wind and Mercury's magnetosphere was relatively mild. However, during the second flyby *MESSENGER* detected the solar wind interacting with Mercury's magnetosphere at ten times the rate as the solar wind interacts with Earth's magnetosphere during its most active phases.

Future Exploration of Mercury

Much about Mercury still remains unknown. What is the chemical composition of the surface rocks? Was the surface of Mercury once completely molten? What caused Mercury to have such a high iron content? What portion of the planet's core is liquid, and what portion is solid? Is there indeed ice perpetually frozen at the poles? Learning the answers will require extensive observations of Mercury by a spacecraft at close range.

WEB LINK 11.12

The *MESSENGER* spacecraft has a third flyby scheduled for September 29, 2009, before it enters a steady orbit around Mercury in March of 2011. A wealth of data from *MESSENGER* should answer some questions and will inevitably raise more. Another Mercury mission, *BepiColombo,* is being planned by the European Space Agency in collaboration with the Japanese Institute of Space and Astronautical Science. *BepiColombo* is actually two spacecraft: an orbiter that will study the planet itself and a smaller orbiter that will explore Mercury's magnetosphere. If launched as scheduled in 2013, *BepiColombo* will reach Mercury in 2019, revealing even more about one of the most enigmatic worlds of the solar system.

11-4 The first missions to Venus and Mars demolished decades of speculations about those planets

Prior to the 1960s, scientists developed a number of intriguing ideas about Venus and Mars. While these ideas were consistent with observations made using Earth-based telescopes, they needed to be radically revised with the advent of spacecraft data.

Speculations About Venus: A Lush Tropical Paradise?

Venus was once thought to be steamy and tropical, and Mars was thought to have vegetation that varied with the seasons

Venus is bathed in more intense sunlight than Earth because it is closer to the Sun. If Venus had no atmosphere, and if its surface had an albedo like that of Mercury or the Moon, heating by the Sun would bring its surface temperature to around 45°C (113°F)—comparable to that found in the hottest regions on Earth. What was unclear until the 1960s was how much effect Venus's atmosphere has on the planet's surface temperature.

We saw in Section 9-1 that gases such as water vapor (H_2O) and carbon dioxide (CO_2) in Earth's atmosphere trap some of the infrared radiation that is emitted from our planet's surface. This trapped radiation elevates the temperatures of both the atmosphere and its surface, a phenomenon called the greenhouse effect. In 1932, Walter S. Adams and Theodore Dunham Jr. at the Mount Wilson Observatory found absorption lines of CO_2 in the spectrum of sunlight reflected from Venus, indicating that carbon dioxide is also present in the Venusian atmosphere. (Section 7-3 describes how astronomers use spectroscopy to determine the chemical compositions of atmospheres.) Thus, Venus's surface, like Earth's, should be warmed by the greenhouse effect.

However, Venus's perpetual cloud cover reflects back into space a substantial amount of the solar energy reaching the planet. This effect by itself acts to cool Venus's surface, just as clouds in our atmosphere can make overcast days cooler than sunny days. Some scientists suspected that this cooling kept the greenhouse effect in check, leaving Venus with surface temperatures below the boiling point of water. Venus might then have warm oceans and possibly even life in the form of tropical vegetation. But if the greenhouse effect were strong enough, surface temperatures could be so high that any liquid water would boil away. Then Venus would have a dry, desertlike surface with no oceans, lakes, or rivers.

Close-Up Observations of Venus: Revealing a Broiled Planet

This scientific controversy was resolved in 1962 when the unmanned U.S. spacecraft *Mariner 2* (Figure 11-15) made the first close flyby of Venus. As we learned in Sections 5-3 and 5-4, a dense object (like a planet's surface) emits radiation whose intensity and spectrum depend on the temperature of the object. *Mariner 2* carried instruments that measured radiation coming from the planet at two microwave wavelengths, 1.35 cm and 1.9 cm. Venus's atmosphere is transparent to both these wavelengths, so *Mariner 2* saw radiation that had been emitted by the planet's surface and then passed through the atmosphere like visible light through glass.

WEB LINK 11.13

Figure 11-15

The *Mariner 2* Spacecraft *Mariner 2* was the first of all spacecraft from Earth to make a successful flyby of another planet. After coming within 34,773 km of Venus on December 14, 1962, it went into a permanent orbit around the Sun. (GSFC/NASA)

From the amount of radiation that *Mariner 2* detected at these two wavelengths, astronomers concluded that the surface temperature of Venus was more than 400°C—well above the boiling point of water and higher than even daytime temperatures on Mercury. Thus, there cannot be any liquid water on the planet's surface. Water vapor absorbs microwaves at 1.35 cm, so if there were substantial amounts of water vapor in Venus's atmosphere, it would have blocked this wavelength from reaching the detectors on board *Mariner 2*. In fact, the spacecraft did detect strong 1.35-cm emissions from Venus. Thus, the atmosphere, too, must be all but devoid of water.

This picture of Venus as a dry, hellishly hot world was confirmed in 1970 by the Soviet spacecraft *Venera 7*, which survived a descent through the Venusian atmosphere and managed to transmit data for a few seconds directly from the planet's surface. *Venera 7* and other successful Soviet landers during the early 1970s finally determined the surface temperature to be a nearly constant 460°C (860°F). Thus, Venus has an extraordinarily strong greenhouse effect, and the planet's surface is no one's idea of a tropical paradise.

Speculations About Mars: A World With Plant Life and Canals?

As we saw in Section 11-2, Mars's axis of rotation is tilted by almost the same angle as Earth's axis, and so Mars goes through seasons much like those on Earth. Observations with Earth-based telescopes show that the colors of Mars (see Figure 11-4*c*) vary with the seasons: During spring and summer in a Martian hemisphere, the polar cap in that hemisphere shrinks and the dark markings (which sometimes look greenish) become very distinct. Half a Martian year later, with the approach of fall and winter, the dark markings fade and the polar cap grows. Could these seasonal variations mean that there is vegetation on Mars that changes seasonally, just as on Earth? And if there is plant life on Mars, might there not also be intelligent beings?

Such speculations were fueled by the Italian astronomer Giovanni Schiaparelli, who observed Mars during the favorable opposition of 1877. He reported that during moments of "good seeing," when the air is exceptionally calm and clear, he was able to see 40 straight-line features crisscrossing the Martian surface (Figure 11-16). He called these dark linear features *canali,* an Italian word for "channels," which was soon mistranslated into English as "canals." The alleged discovery of canals implied that there were intelligent creatures on Mars capable of substantial engineering feats. This speculation helped motivate the American millionaire Percival Lowell to finance a major new observatory near Flagstaff, Arizona, primarily to study Mars. By the end of the nineteenth century, Lowell had reported observations of 160 Martian canals.

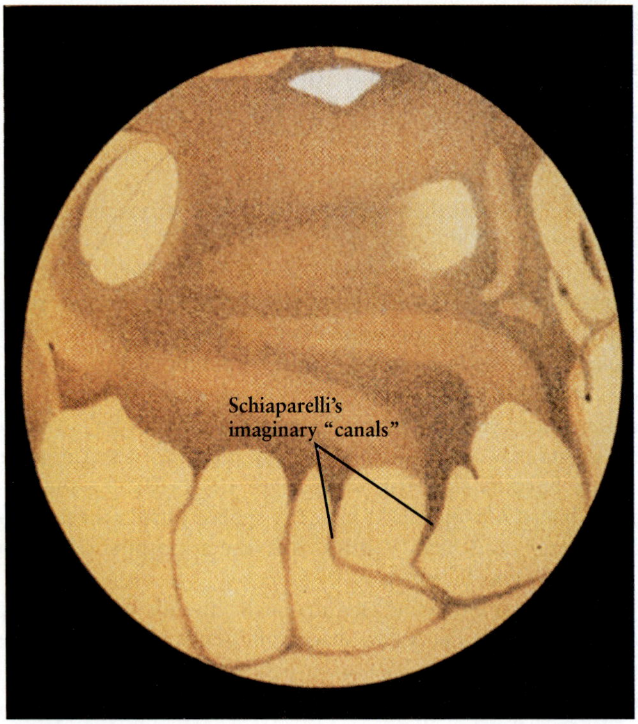

Figure 11-16

The Mirage of the Martian Canals Giovanni Schiaparelli studied the Martian surface using a 20-cm (8-in) telescope, the same size used by many amateur astronomers today. He recorded his observations in drawings like this one, which shows a network of linear features that Percival Lowell and others interpreted as irrigation canals. Later observations with larger telescopes revealed that the "canals" were mere illusions. (Michael Hoskin, ed., *The Cambridge Illustrated History of Astronomy,* Cambridge University Press, 1997, p. 286. Illustration by G. V. Schiaparelli. Courtesy Institute of Astronomy, University of Cambridge, UK)

Although many astronomers observed Mars, not all saw the canals. In 1894, the American astronomer Edward Barnard, working at Lick Observatory in California, complained that "to save my soul I can't believe in the canals as Schiaparelli draws them." But objections by Barnard and others did little to sway the proponents of the canals.

As the nineteenth century drew to a close, speculation about Mars grew more and more fanciful. Perhaps the reddish-brown color of the planet meant that Mars was a desert world, and perhaps the Martian canals were an enormous planetwide irrigation network. From these ideas, it was a small leap to envision Mars as a dying world with canals carrying scarce water from melting polar caps to farmlands near the equator. The terrible plight of the Martian race formed the basis of inventive science fiction by Edgar Rice Burroughs, Ray Bradbury, and many others. It also led to the notion that the Martians might be a warlike race who schemed to invade the Earth for its abundant resources. (In Roman mythology, Mars was the god of war.) Stories of alien invasions, from H. G. Wells's 1898 novel *The War of the Worlds* down to the present day, all owe their existence to the *canali* of Schiaparelli.

Observations of Mars: Revealing a Windblown, Cratered World

However, during the first decades of the twentieth century the Greek astronomer Eugène Antoniadi observed Mars with a state-of-the-art telescope and found that the "canals" were actually unconnected dark spots. There was no evidence whatsoever of linear canals. Why, then, were Schiaparelli, Lowell, and others so certain that the canals were there? One reason is that their observations were made from the Earth's surface, at the bottom of an often turbulent atmosphere that blurs images seen through a telescope. A second reason is that the human eye and brain can readily be deceived. The brain can connect together two dark streaks on the Martian surface to give the perception of a single "canal." In this way, astronomers such as Schiaparelli and Lowell deluded themselves.

WEB LINK 11.14

The other speculation, that Mars was covered with vegetation, was conclusively ruled out several decades later. In the 1960s, three American spacecraft flew past Mars and sent back the first close-up pictures of the planet's surface. (Figure 11-17 shows even better images from a subsequent mission to Mars.) These images showed that the dark surface markings are just different-colored terrain.

Why did many Earth-based observers report that dark areas on Mars had the greenish hue of vegetation? The probable explanation is a quirk of human color vision. When a neutral gray area is viewed next to a bright red or brown area, the eye perceives the gray to have a blue-green color.

The seasonal variations in color can be explained by winds in the Martian atmosphere that blow in different directions in different seasons. These seasonal winds blow fine dust across the Martian surface, producing planetwide dust storms that can be seen from Earth (Figure 11-18). The motion of dust covers some parts of the Martian terrain and exposes others. Thus, Earth observers see large areas of the planet vary from dark to light with

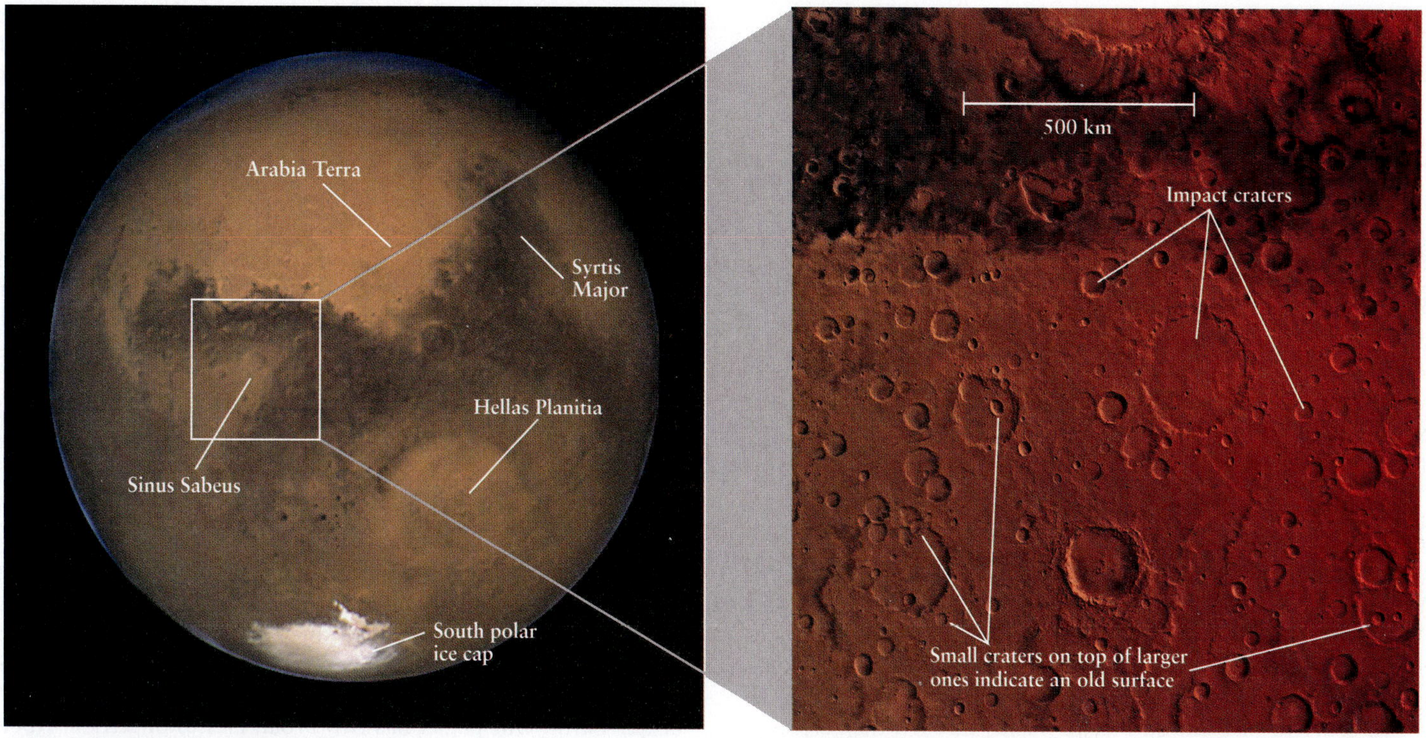

(a) Mars from the Hubble Space Telescope

(b) Closeup of Sinus Sabeus region

Figure 11-17 R I **V** U X G

Martian Craters **(a)** This image, made during the favorable opposition of 2003, shows cratered regions on the side of Mars opposite to that in the image that opens this chapter. Arabia Terra is dotted with numerous flat-bottom craters. Syrtis Major was first identified by Christiaan Huygens in 1659. A single titanic impact carved out Hellas Planitia, which is 5 times the size of Texas. **(b)** This mosaic of images from the *Viking Orbiter 1* and *2* spacecraft shows an extensively cratered region located south of the Martian equator. (a: NASA; J. Bell, Cornell University; and M. Wolff, SSI; b: USGS)

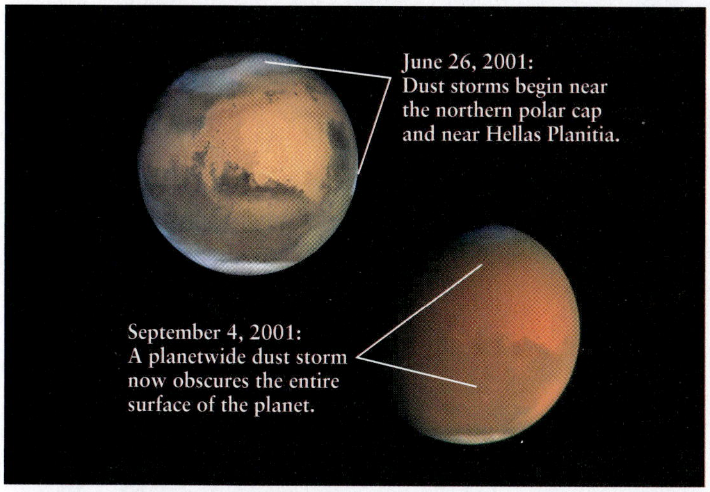

Figure 11-18 R I **V** U X G

A Martian Dust Storm Dust storms on Mars can expand to cover the entire planet. These Hubble Space Telescope images show a particularly intense storm that took place in 2001, when it was spring in the Martian northern hemisphere and autumn in the southern hemisphere. (NASA; J. Bell, Cornell University; M. Wolff, SSI; and the Hubble Heritage Team, STScI/AURA)

the passage of the seasons, mimicking the color changes that would be expected from vegetation.

The most remarkable discovery of the missions to Mars in the 1960s was that the planet's surface is pockmarked with craters, as Figure 11-17*b* and Figure 7-10*c* show. (Although many of the Martian craters are quite large, they had escaped detection from Earth because the thin Martian atmosphere obscures the planet's surface somewhat. That is why even Hubble Space Telescope images like Figure 11-17*a* do not show Martian craters distinctly, and why it took close-up images made by spacecraft near Mars to reveal these craters.) During that same decade of the 1960s, scientists were learning that most of the craters on the Moon were formed by interplanetary debris that struck the lunar surface more than 3 billion years ago. The Martian craters, too, are the result of long-ago impacts from space. Since so many craters survive to the present day, at least part of the Martian surface must be extremely ancient (see Section 7-6).

The episode of the Martian canals and Martian vegetation was not the first time that a number of scientists came to very wrong conclusions. Like any human activity, science will always involve human error and human frailties. But scientific investigation is fundamentally self-correcting, because new observations and experiments can always call into question the scientific theories of the past.

11-5 Both Venus and Mars have volcanoes and craters—but no Earthlike plate tectonics

TUTORIAL 11.1

From a distance, Venus and Mars appear radically different: Venus is nearly the size of Earth and has a thick atmosphere, while Mars is much smaller and has only a thin atmosphere. Yet spacecraft observations reveal that these two worlds have many surface features in common, including volcanoes and impact craters. By comparing the surfaces of these two worlds to Earth we can gain insight into the similarities and differences between the three largest terrestrial planets.

Venus has no plate tectonics because its crust is too thin, while on Mars the crust is too thick

Observing Venus and Mars from Orbit

WEB LINK 11.15

To make a detailed study of a planet's surface, a spacecraft that simply flies by the planet will not suffice. Instead, it is necessary to place a spacecraft in orbit around the planet. Since the 1970s, a number of spacecraft have been placed in Venus orbit and Mars orbit.

In order to map the surface of Venus through the perpetual cloud layer, several of the Venus orbiters carried radar devices. A beam of microwave radiation from the orbiter easily penetrates Venus's clouds and reflects off the planet's surface; a receiver on the orbiter then detects the reflected beam. Different types of terrain reflect microwaves more or less efficiently, so astronomers are able to construct a map of the Venusian surface by analyzing the reflected radiation.

ANIMATION 11.6

Magellan, the most recent spacecraft to orbit Venus, also carried a radar altimeter that bounced microwaves directly off the ground directly below the spacecraft. By measuring the time delay of the reflected waves, astronomers can determine the height and depth of Venus's terrain.

By contrast, a Mars orbiter can use more conventional telescopes to view the Martian surface through that planet's thin atmosphere. Instead of using a radar altimeter to map surface elevations, the *Mars Global Surveyor* spacecraft (which entered Mars's orbit in 1997) used a laser beam for the same purpose.

The Topography of Venus and Mars

Figure 11-19 and **Figure 11-20** show topographic maps of Venus and Mars derived using radar and laser altimeters, respectively. (Note that the scale of these two maps is different. The diameter of Mars is only 56% as large as the diameter of Venus.) The topographies of both worlds differ in important ways from that of our Earth. Our planet has two broad classes of terrain: About 71% of the surface is oceanic crust and about 27% is continental crust that rises above the ocean floors by about 4–6 km on av-

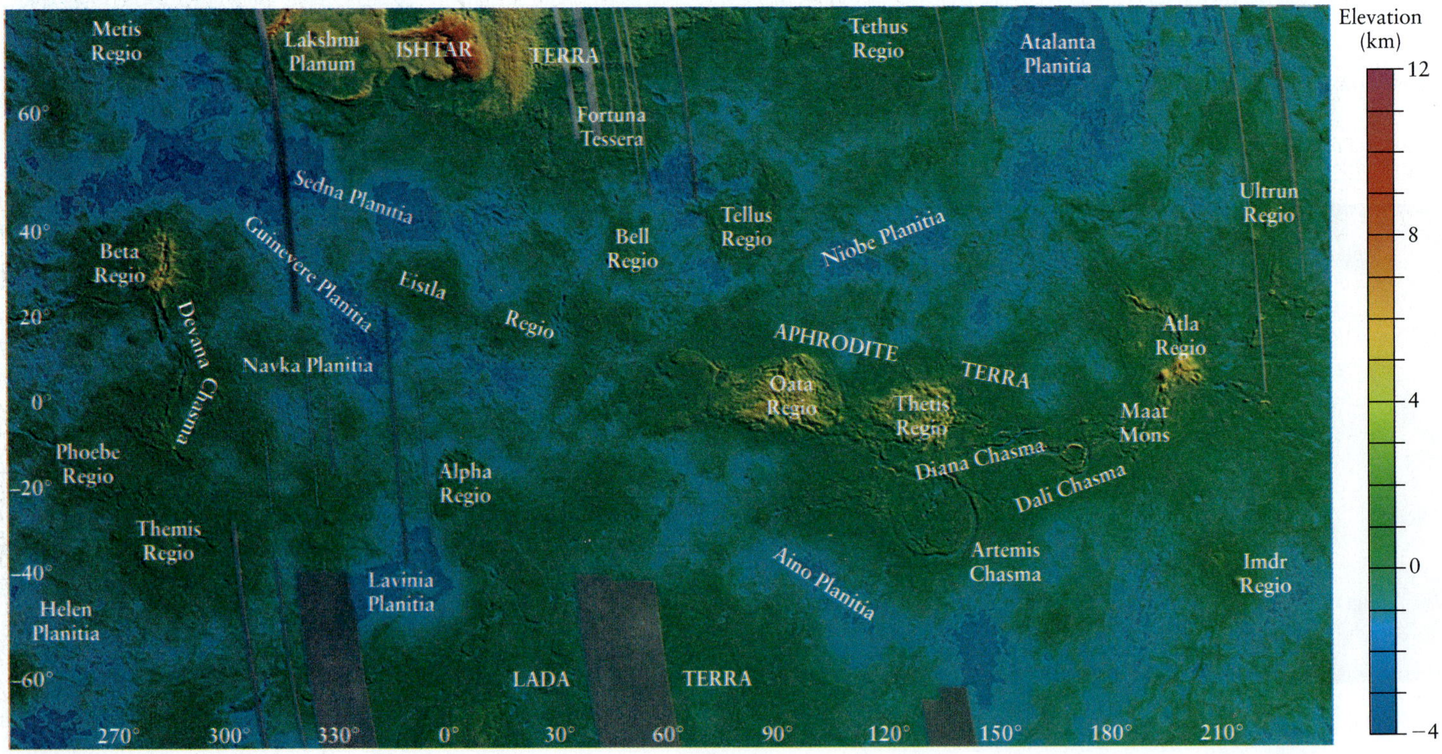

Figure 11-19 R I V U X G

A Topographic Map of Venus Radar altimeter measurements by *Magellan* were used to produce this topographic map of Venus. Color indicates elevations above (positive numbers) or below (negative numbers) the planet's average radius. (The blue areas are *not* oceans!) Gray areas were not mapped by *Magellan.* Flat plains of volcanic origin cover most of the planet's surface, with only a few continent-like highlands. (Peter Ford, MIT; NASA/JPL)

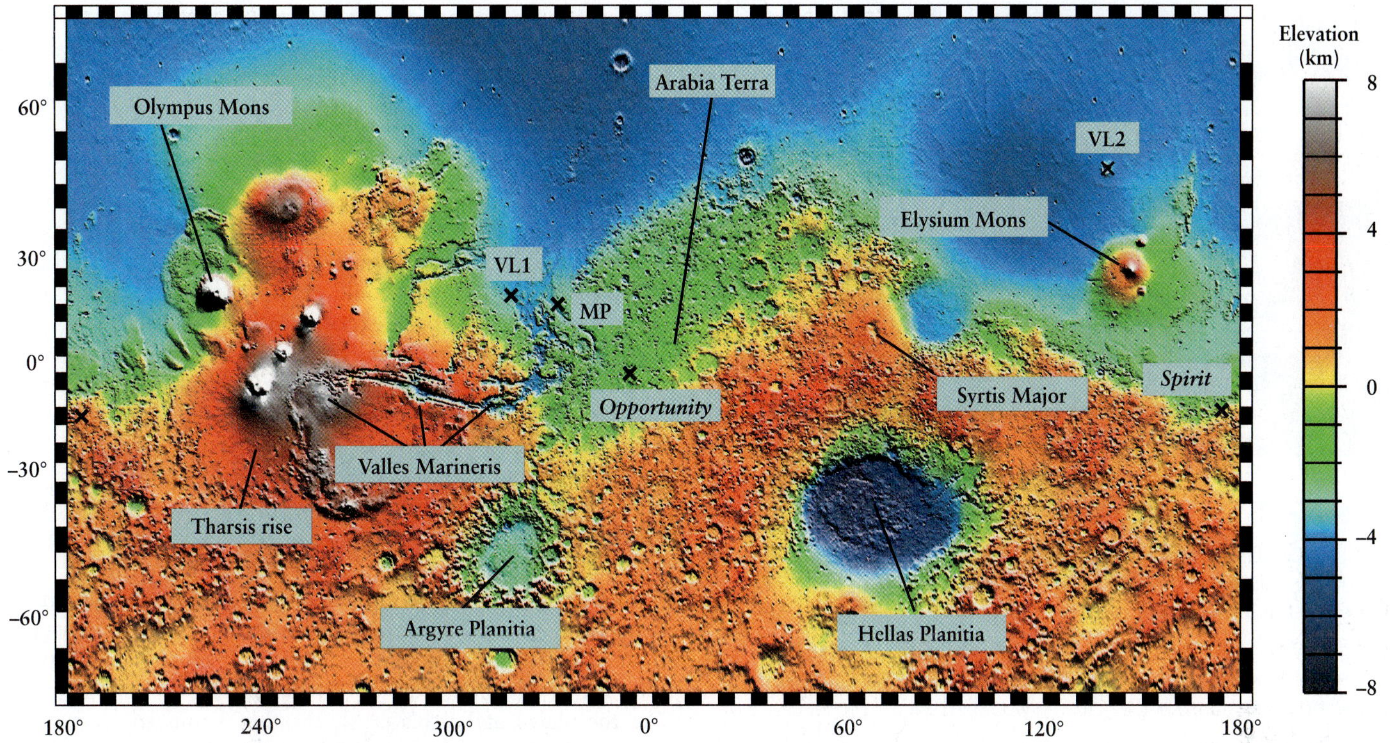

Figure 11-20 R I **V** U X G

A Topographic Map of Mars This map was generated from measurements made by the laser altimeter on board the *Mars Global Surveyor* spacecraft. As in Figure 11-19, color indicates elevations above or below the planet's average radius. Most of the southern hemisphere lies several kilometers above the northern hemisphere, with the exception of the immense impact feature called Hellas Planitia. The landing sites for *Viking Landers 1* and *2* (VL1 and VL2), *Mars Pathfinder* (MP), and the *Mars Exploration Rovers* (*Spirit* and *Opportunity*) are each marked with an X. (MOLA Science Team, NASA/GSFC)

erage. On Venus, by contrast, about 60% of the terrain lies within 500 m of the average elevation, with only a few localized highlands (shown in yellow and red in Figure 11-19). Mars is different from both Earth and Venus: Rather than having continents scattered among low-lying ocean floors, all of the high terrain on Mars (shown in red and orange in Figure 11-20) is in the southern hemisphere. Hence, planetary scientists refer to Mars as having **northern lowlands** and **southern highlands.** The implication is that the Martian crust is about 5 km thicker in the southern hemisphere than in the northern hemisphere, a situation called the **crustal dichotomy.** In this sense Mars resembles the Moon, which has a thicker crust on the far side than on the side that faces Earth (see Section 10-1).

Tectonics on Venus: A Light, Flaky Crust

Before radar maps like Figure 11-19 were available, scientists wondered whether Venus had plate tectonics like those that have remolded the face of Earth. Venus is only slightly smaller than Earth, and should have retained enough heat to sustain a molten interior and the convection currents that drive tectonic activity on Earth (see Section 9-3, especially Figure 9-16). If this were the case, then the same tectonic effects might also have shaped the surface of Venus. As we saw in Section 9-3, Earth's hard outer shell, or lithosphere, is broken into about a dozen large plates that slowly shuffle across the globe. Long mountain ranges, like the Mid-Atlantic Ridge (see Figure 9-13), are created where fresh magma wells up from Earth's interior to push the plates apart.

Radar images from *Magellan* show *no* evidence of Earthlike tectonics on Venus. On Earth, long chains of volcanic mountains (like the Cascades in North America or the Andes in South America) form along plate boundaries where subduction is taking place. Mountainous features on Venus, by contrast, do not appear in chains. There are also no structures like Earth's Mid-Atlantic Ridge, which suggests that there is no seafloor spreading on Venus. With no subduction or seafloor spreading, there have been only limited horizontal displacement of Venus's lithosphere. Thus, like the Moon (see Section 10-1) and Mercury (see Section 11-3), Venus has a one-plate crust.

Unlike the Moon, however, Venus has had local, small-scale deformations and reshaping of the surface. One piece of evidence for this is that roughly a fifth of Venus's surface is covered by folded and faulted ridges. Further evidence comes from close-up *Magellan* images that show that Venus has about a thousand craters larger than a few kilometers in diameter, many more than have been found on Earth but only a small fraction of the number on the Moon or Mercury. We saw in Section 7-6 that the number of impact craters is a clue to the age of a planet's surface. Such craters formed at a rapid rate during the early history of the solar system, when considerable interplanetary debris still orbited

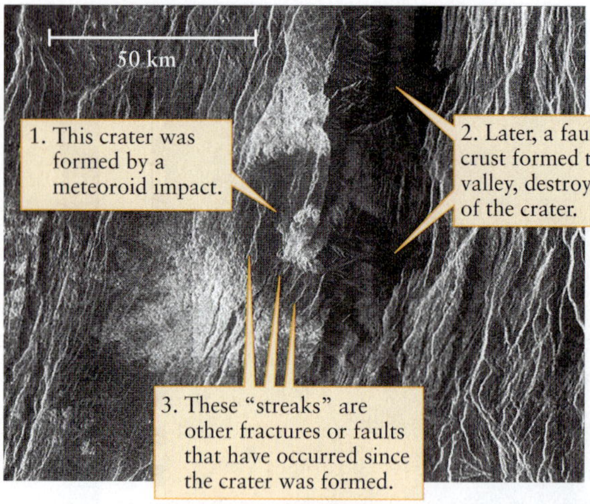

2. Later, a fault in the crust formed this deep valley, destroying half of the crater.
3. These "streaks" are other fractures or faults that have occurred since the crater was formed.

Figure 11-21 R I V U X G

A Partially Obliterated Crater on Venus This *Magellan* image shows how the right half of an old impact crater 37 km (20 mi) in diameter was erased by a fault in the crust. This crater lies in Beta Regio (see the left side of Figure 11-19). Most features on Venus are named for women in history and legend; this crater commemorates Emily Greene Balch, an American economist and sociologist who won the 1946 Nobel Peace Prize. (NASA/JPL)

the Sun, and have formed at a much slower rate since then. Consequently, the more craters a planet has, the older its surface. The number of craters on Venus indicates that the Venusian surface is roughly 500 million years old. This is about twice the age of Earth's surface but much younger than the surfaces of the Moon or Mercury, each of which is billions of years old. No doubt Venus was more heavily cratered in its youth, but localized activity in its crust has erased the older craters (Figure 11-21).

Surprisingly, Venus's craters are uniformly scattered across the planet's surface. We would expect that older regions on the surface—which have been exposed to bombardment for a longer time—would be more heavily cratered, while younger regions would be relatively free of craters. For example, the ancient highlands on the Moon are much more heavily cratered than the younger maria (see Section 10-4). Because such variations are not found on Venus, scientists conclude that the entire surface of the planet has essentially the *same* age. This is very different from Earth, where geological formations of widely different ages can be found.

One model that can explain these features suggests that the convection currents in Venus's interior are actually more vigorous than inside Earth, but that the Venusian crust is much thinner than the continental crust on Earth. Rather than sliding around like the plates of Earth's crust, the thin Venusian crust stays in roughly the same place but undergoes wrinkling and flaking (Figure 11-22). Hence, this model is called **flake tectonics.** Earth, too, may have displayed flake tectonics billions of years ago when its interior was hotter.

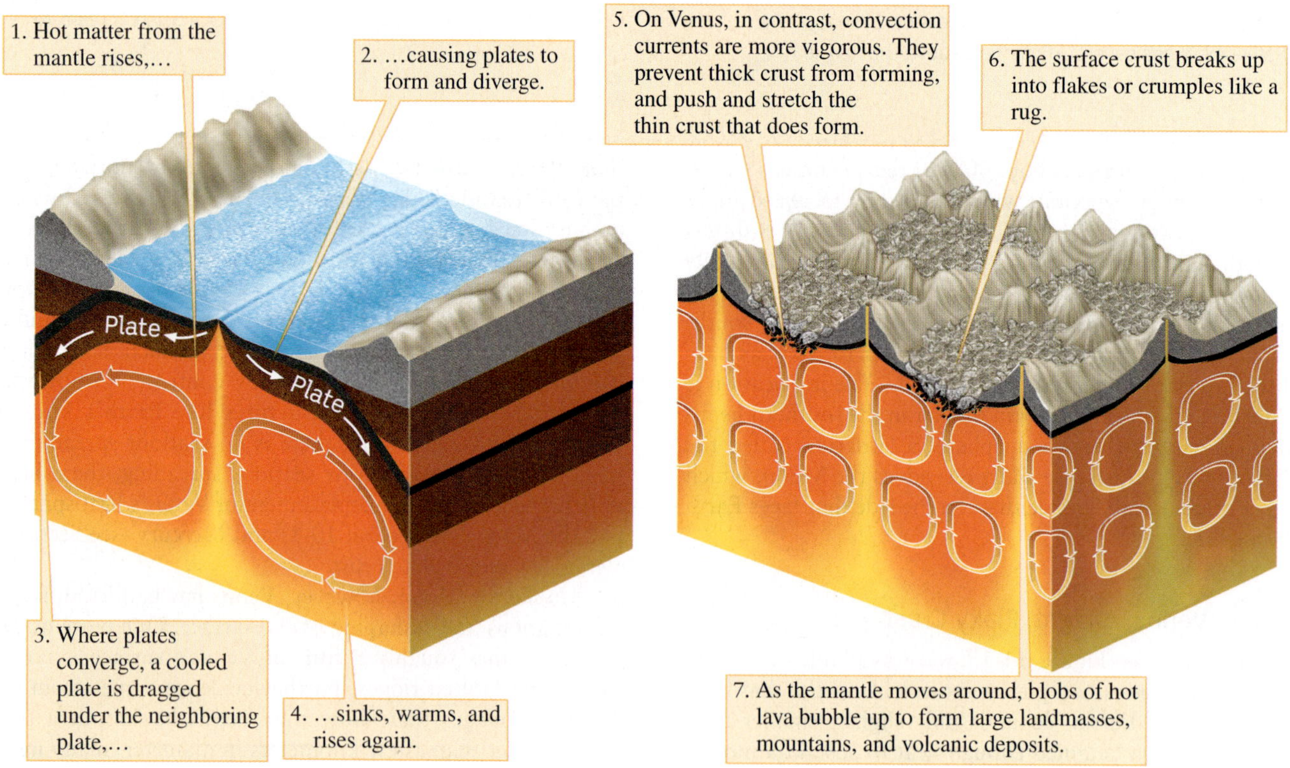

1. Hot matter from the mantle rises,...
2. ...causing plates to form and diverge.
Plate
Plate
3. Where plates converge, a cooled plate is dragged under the neighboring plate,...
4. ...sinks, warms, and rises again.
5. On Venus, in contrast, convection currents are more vigorous. They prevent thick crust from forming, and push and stretch the thin crust that does form.
6. The surface crust breaks up into flakes or crumples like a rug.
7. As the mantle moves around, blobs of hot lava bubble up to form large landmasses, mountains, and volcanic deposits.

Figure 11-22

Plate Tectonics Versus Flake Tectonics This illustration shows the difference between plate tectonics on Earth and the model of flake tectonics on Venus. (Adapted from T. Grotzinger, T. H. Jordan, F. Press, and R. Siever, *Understanding Earth,* 5th ed., W. H. Freeman, 2007)

LOOKING DEEPER 11.2

Although Venus almost certainly has molten material in its interior, it has no planetwide magnetic field. As we discussed in Section 7-7, Venus may have no magnetic field because it rotates too slowly to generate the kind of internal motions that would produce a magnetic field. With no such field, Venus has no magnetosphere. The planet is nonetheless shielded from the solar wind by ions in its upper atmosphere.

Tectonics on Mars: A Thick, Rigid Crust

Like Venus, Mars lacks the global network of ridges and subduction zones that plate tectonics has produced on Earth. Hence, the entire crust of Mars makes up a single tectonic plate, as is the case on the Moon, Mercury, and Venus. The explanation is quite different from that for Venus, however: Because Mars is a much smaller world than Earth or Venus, the outer layers of the red planet have cooled more extensively. Thus, Mars lacks plate tectonics because its crust is too thick for one part of the crust to be subducted beneath another. We see that for a terrestrial planet to have plate tectonics, the crust must not be too thin (like Venus) or too thick (like Mars), but just right (like Earth).

This idea has been verified by carefully monitoring the motion of a spacecraft orbiting Mars. If there is a concentration of mass (such as a thicker crust) at one location on the planet, gravitational attraction will make the spacecraft speed up as it approaches the concentration and slow down as it moves away. A team of scientists headed by David Smith of NASA's Goddard Space Flight Center and Maria Zuber of MIT analyzed the orbit of *Mars Global Surveyor* in just this way. They found that unlike Earth's crust, which varies in thickness from 5 to 35 km, the Martian crust is about 40 km thick under the northern lowlands but about 70 km thick under the southern highlands. Both regions of the Martian crust are too thick to undergo subduction, making plate tectonics impossible.

Although there is no plate motion on Mars, there are features on the planet's surface that indicate there has been substantial motion of material in the Martian mantle. The Tharsis rise, visible on the left-hand side of Figure 11-20, is a dome-shaped bulge that has been lifted 5 to 6 km above the planet's average elevation. Apparently, a massive plume of magma once welled upward from a hot spot under this region. East of the Tharsis rise, a vast chasm runs roughly parallel to the Martian equator (**Figure 11-23**). If this canyon were located on Earth, it would stretch from Los Angeles to New York. In honor of the *Mariner 9* spacecraft that first revealed its presence in 1971, this chasm has been named Valles Marineris.

Valles Marineris has heavily fractured terrain at its western end near the Tharsis rise. At its eastern end, by contrast, it is dominated by ancient cratered terrain. Many geologists suspect that

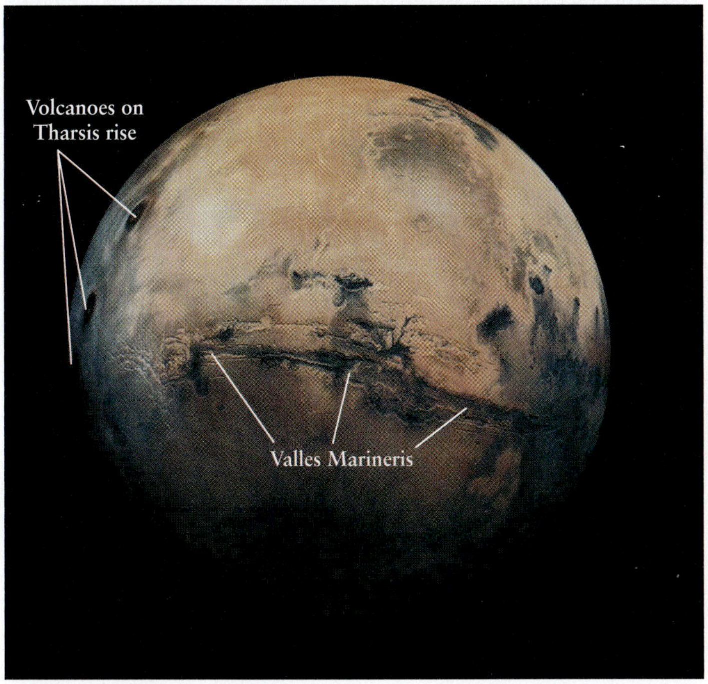

(a) Mars and Valles Marineris

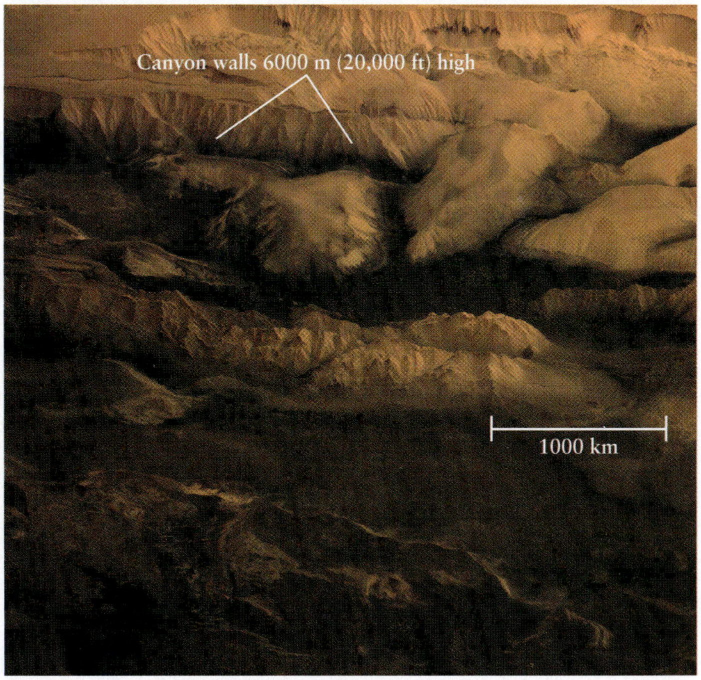

(b) The central region of Valles Marineris

VIDEO 11.3 VIDEO 11.4

Figure 11-23 R I V U X G

Valles Marineris (a) This mosaic of *Viking Orbiter* images shows the huge rift valley of Valles Marineris, which extends from west to east for more than 4000 km (2500 mi) and is 600 km (400 mi) wide at its center. Its deepest part is 8 km (5 mi) beneath the surrounding plateau. At its western end is the Tharsis rise. Compare this image with Figure 11-20. **(b)** This perspective image from the *Mars Express* spacecraft shows what you would see from a point high above the central part of Valles Marineris. (a: USGS/NASA; b: ESA/DLR/FU Berlin, G. Neukum)

Valles Marineris was caused by the same upwelling of material that formed the Tharsis rise. As the Martian surface bulged upward at Tharsis, there would have been tremendous stresses on the crust, which would have caused extensive fracturing. Thus, Valles Marineris may be a **rift valley**, a feature created when a planet's crust breaks apart along a fault line. Rift valleys are found on Earth; two examples are the Red Sea (see Figure 9-17) and the Rhine River valley in Europe. Other, smaller rifts in the Martian crust are found all around the Tharsis rise. All of these features are very old, however, and it is thought that there has been little geologic activity on Mars for billions of years.

Additional evidence that may point to ancient geologic activity on Mars is the crustal dichotomy between the northern lowlands and the southern highlands. The southern highlands are much more heavily cratered than either Earth or Venus, though less so than the Moon, while the northern lowlands are remarkably smooth and free of craters. (You can see this difference in Figure 11-20.) Since most cratering occurred early in the history of the solar system, this implies that the northern lowlands are relatively young, while the more heavily cratered southern highlands are relatively old. One proposed explanation is that older craters in the northern lowlands were erased by tectonic activity that took place long ago when the Martian crust had not yet cooled to its present thickness. Other hypotheses are that older craters could have been erased by a giant asteroid impact or by lava flows.

A final piece of evidence about ancient Mars is the existence of locally magnetized areas in the old southern highlands. These areas presumably solidified during an era in which the interior of Mars was much hotter, and moving, molten material in the interior developed a magnetic field. Today, by contrast, Mars has no planetwide magnetic field. There is probably still molten material in the core, but the Martian core is thought to contain substantial amounts of sulfur in addition to iron. While temperatures in the core should be high enough to melt sulfur and sulfur compounds, the electrical properties of these substances are such that their motions do not generate the electric currents needed to produce a magnetic field.

Volcanoes on Venus and Mars

Radar images of Venus and visible-light images of Mars show that both planets have a number of large volcanoes (Figure 11-24). *Magellan* observed more than 1600 major volcanoes and volcanic features on Venus, two of which are shown in Figure 11-24*a*. Both of these volcanoes have gently sloping sides. A volcano with this characteristic is called a **shield volcano**, because in profile it resembles an ancient Greek warrior's shield lying on the ground. Martian volcanoes are less numerous than those on Venus, but they are also shield volcanoes; the largest of these, Olympus Mons, is the largest volcano in the solar system (Figure 11-24*b*). Olympus Mons rises 24 km (15 mi) above the surrounding plains. By comparison, the highest volcano on Earth, Mauna Loa in the Hawaiian Islands, has a summit only 8 km (5 mi) above the ocean floor.

Most volcanoes on Earth are found near the boundaries of tectonic plates, where subducted material becomes molten magma and rises upward to erupt from the surface. This cannot explain

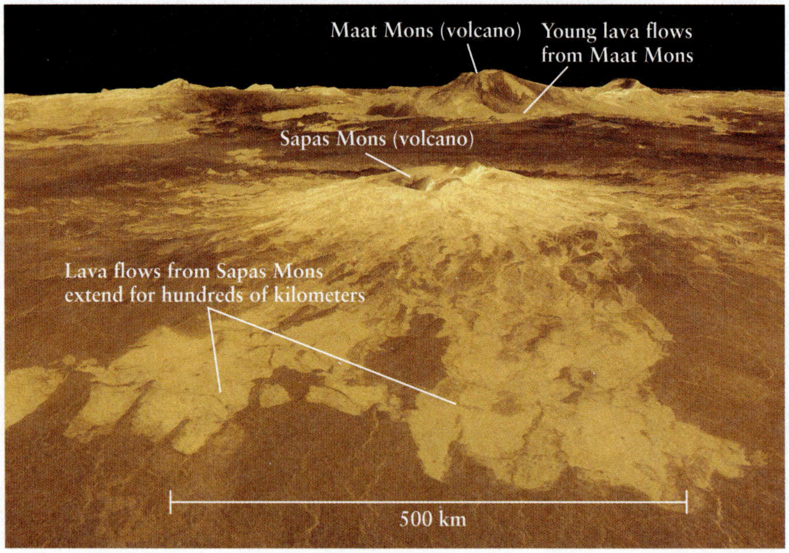

(a) Volcanoes and lava flows on Venus R I V U X G

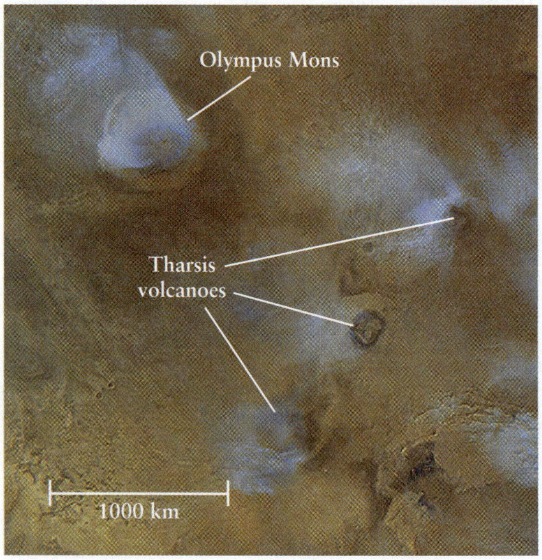

(b) Cloud-topped volcanoes on Mars R I V U X G

Figure 11-24

Volcanoes on Venus and Mars **(a)** The false color in this computer-generated perspective view suggests the actual color of sunlight that penetrates Venus's thick clouds. The brighter color of the extensive lava flows indicates that they reflect radio waves more strongly. To emphasize the gently sloping volcanoes, the vertical scale has been exaggerated 10 times. **(b)** The volcanoes of Mars also have gently sloping sides. In this view looking down from Mars orbit you can see bluish clouds topping the summits of the volcanoes. These clouds, made of water ice crystals, form on most Martian afternoons. (a: NASA, JPL Multimission Image Processing Laboratory; b: NASA/JPL/Malin Space Science Systems)

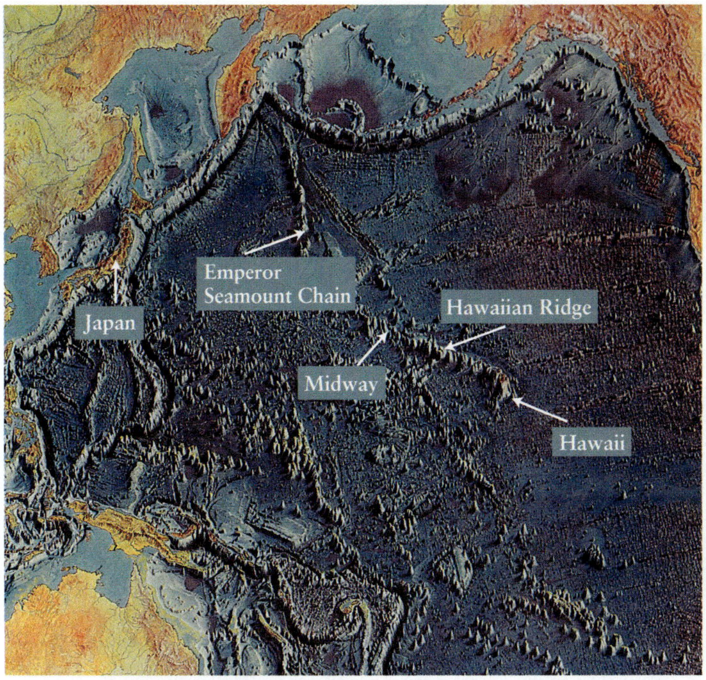

Figure 11-25

Hot-Spot Volcanoes on Earth A hot spot under the Pacific plate has remained essentially stationary for 70 million years while the plate has moved some 6000 km to the northwest. The upwelling magma has thus produced a long chain of volcanoes. The Hawaiian Islands are the newest of these; the oldest, the Emperor Seamount Chain, have eroded so much that they no longer protrude above the ocean surface.

the volcanoes of Venus and Mars, since there are no subduction zones on those planets. Instead, Venusian and Martian volcanoes probably formed by **hot-spot volcanism.** In this process, magma wells upward from a hot spot in a planet's mantle, elevating the overlying surface and producing a shield volcano.

On Earth, hot-spot volcanism is the origin of the Hawaiian Islands. These islands are part of a long chain of shield volcanoes that formed in the middle of the Pacific tectonic plate as that plate moved over a long-lived hot spot (Figure 11-25). On Venus and Mars, by contrast, the absence of plate tectonics means that the crust remains stationary over a hot spot. On Mars, a single hot spot under Olympus Mons probably pumped magma upward through the same vent for millions of years, producing one giant volcano rather than a long chain of smaller ones. The Tharsis rise and its volcanoes (see Figure 11-20, Figure 11-23*a*, and Figure 11-24*b*) may have formed from the same hot spot as gave rise to Olympus Mons; a different hot spot on the opposite side of Mars produced a smaller bulge centered on the volcano Elysium Mons, shown near the right-hand side in Figure 11-20. The same process of hot-spot volcanism presumably gave rise to large shield volcanoes on Venus like those shown in Figure 11-24*a*.

Volcanic Activity on Venus

About 80% of the surface of Venus is composed of flat plains of volcanic origin. In other words, essentially the entire planet is covered with lava! This observation shows the tremendous importance of volcanic activity in Venusian geology.

To verify the volcanic nature of the Venusian surface, it is necessary to visit the surface and examine rock samples. Figure 11-26 is a panoramic view taken in 1981 by the Soviet spacecraft *Venera 13,* one of 10 unmanned spacecraft that the Soviet Union landed successfully on the surface of Venus. Russian scientists hypothesize that this region was covered with a thin layer of lava that fractured upon cooling to create the rounded, interlocking shapes seen in the photograph. This hypothesis agrees with information obtained from chemical analyses of surface material made by the spacecraft's instruments. These analyses indicate that the surface composition is similar to lava rocks called basalt, which are common on Earth (see Figure 9-19*a*) and in the maria of the Moon (see Figure 10-13). The results from *Venera 13* and other landers are consistent with the picture of Venus as a world whose surface and atmosphere have been shaped by volcanic activity.

Most of the volcanoes on Venus are probably inactive at present, just as is the case with most volcanoes on Earth. But *Magellan* found evidence of recent volcanic activity, some of which may be continuing today. The key to estimating the amount of recent volcanic activity is that the radar reflectivity of volcanic materials depends on whether the material is relatively fresh or relatively old. By mapping these reflectivity variations on Venus, *Magellan* found many areas with young lava flows. Some of the youngest material found by *Magellan* caps the volcano Maat Mons (see Figure 11-24*a*). Geologists estimate that the topmost material is no more than 10 million years old and could be much younger. The presence of such young lava flows suggests that Venus, like Earth, has some present-day volcanic activity.

Another piece of evidence for ongoing volcanic activity on Venus comes from the planet's atmosphere. An erupting volcano on Earth, like the one shown in Figure 9-15, ejects substantial amounts of sulfur dioxide, sulfuric acid, and other sulfur compounds into the air. Many of these substances are highly reactive and short-lived, forming sulfate compounds that become part of the planet's surface rocks. For these substances to be relatively abundant in a planet's atmosphere, they must be constantly replenished by new eruptions. Sulfur compounds make up about 0.015% of the Venusian atmosphere, compared to less than 0.0001% of Earth's atmosphere. This evidence suggests that ongoing volcanic eruptions on Venus are ejecting sulfur compounds into the atmosphere to sustain the high sulfur content.

Volcanic Activity on Mars

Spectroscopic observations from Mars's orbit confirm that the planet's rocks and sands are made almost entirely of the three minerals feldspar, pyroxene, and olivine. These are the components of basalt, or solidified lava. Thus, Mars, like Venus, had a volcanic past.

Unlike lava flows on Venus, however, most of the lava flows on Mars have impact craters on them. These craters suggest that most Martian lava flows are very old and that most of the volcanoes on the red planet are no longer active. This is what we would expect from a small planet whose crust has cooled and solidified to a greater depth than on Earth, making

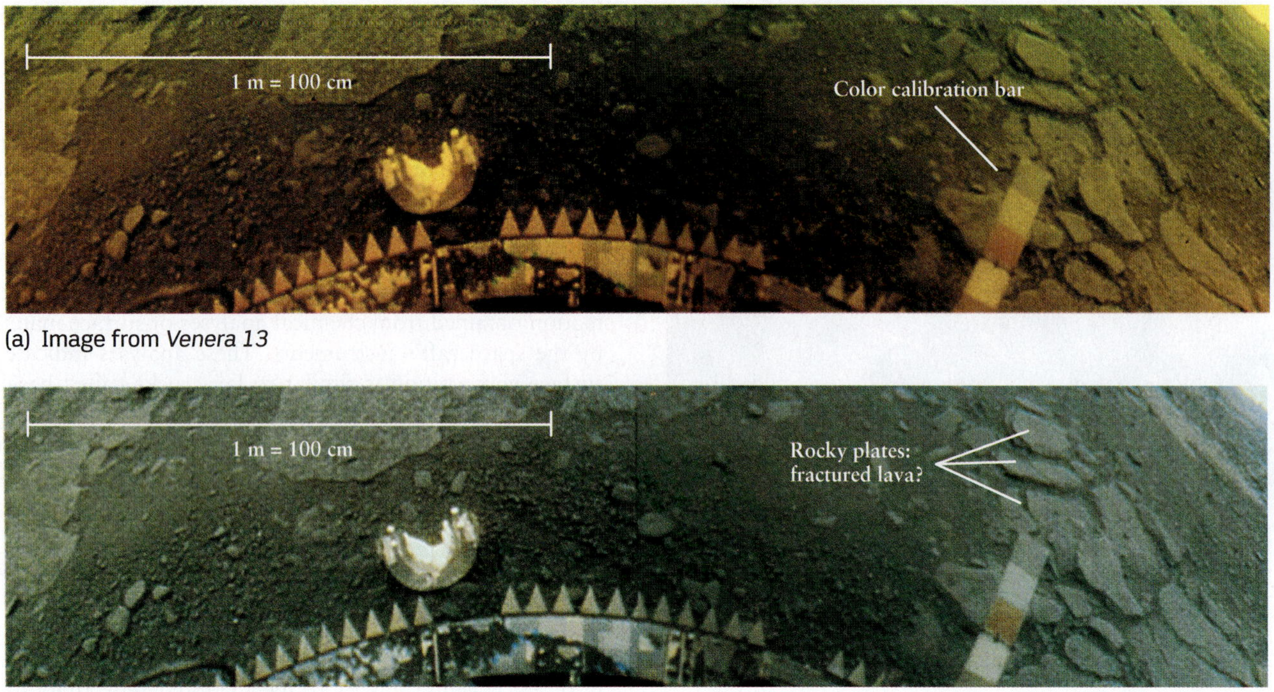

(a) Image from *Venera 13*

(b) Color-corrected image

Figure 11-26 R I V U X G

A Venusian Landscape (a) This wide-angle color photograph from *Venera 13* shows the rocky surface of Venus. The thick atmosphere absorbs the blue component of sunlight, giving the image an orange tint: The stripes on the spacecraft's color calibration bar that appear yellow are actually white in color. (b) This color-corrected view shows that the rocks are actually gray in color. The rocky plates covering the ground may be fractured segments of a thin layer of lava. (Courtesy of C. M. Pieters and the Russian Academy of Sciences)

it difficult for magma to travel from the Martian mantle to the surface.

However, a few Martian lava flows are crater-free, which suggests that they are only a few million years old. If volcanoes erupted on Mars within the past few million years, are they erupting now? Scientists have used infrared telescopes to search for telltale hot spots on the Martian surface, but have yet to discover any. Perhaps volcanism on Mars is rare but not yet wholly extinct.

Spacecraft that have landed on Mars have taught us a great deal about the geology and history of the red planet. Before we explore the Martian surface in detail, however, it is useful to examine the unique atmospheres of both Mars and Venus.

11-6 The dense atmosphere of Venus and the thin Martian atmosphere are dramatically different but have similar chemical compositions

We have seen that the three large terrestrial planets—Earth, Venus, and Mars—display very different styles of geology. As we will discover, these differences help to explain why each of these three worlds has a distinctively different atmosphere.

To explore a planet's atmosphere it is necessary to send a spacecraft to make measurements as it descends through that atmosphere. Both Soviet and American spacecraft have done this for Venus (see Figure 11-26), and five U.S. spacecraft have successfully landed on Mars. The results of these missions are summarized in Figure 11-27, which shows how pressure and temperature vary with altitude in the atmospheres of Earth, Venus, and Mars.

On Venus, sulfuric acid rain evaporates before reaching the ground; on Mars, it snows frozen carbon dioxide

The Venusian Atmosphere: Dense, Hot, and Corrosive

Figure 11-27*b* depicts just how dense the Venusian atmosphere is. At the surface, the pressure is 90 atmospheres. This pressure is 90 times greater than the average air pressure at sea level on Earth and is about the same as the water pressure at a depth of about 1 kilometer below the surface of Earth's oceans. The density of the atmosphere at the surface of Venus is likewise high, more than 50 times greater than the sea-level density of our atmosphere. The atmosphere is so massive that once heated by the Sun, it retains its heat throughout the long Venusian night. As a result, temperatures on the day and night sides of Venus are almost identical.

The composition of atmosphere explains why Venus's surface temperatures are so high: 96.5% of the molecules in Venus's atmosphere are carbon dioxide, with nitrogen (N_2) making up most of the remaining 3.5% (see Table 9-4 for a comparison of the atmospheres of Venus, Earth, and Mars). Because Venus's atmosphere is so thick, and because most of the atmosphere is the

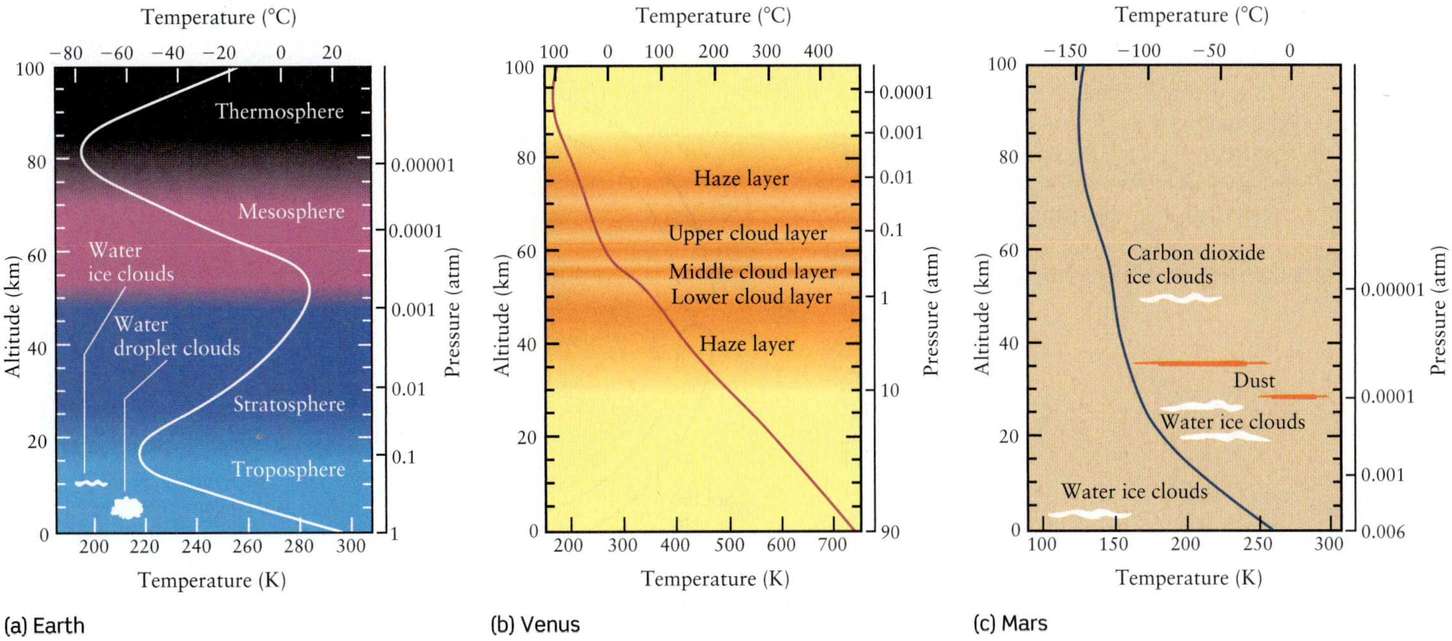

Figure 11-27

Atmospheres of the Terrestrial Planets In each graph the curve shows how temperature varies with altitude from 0 to 100 km above the planet's surface. The scale on the right-hand side of each graph shows how pressure varies with altitude. **(a)** Clouds in Earth's atmosphere are seldom found above 12 km (40,000 ft). **(b)** Venus's perpetual cloud layers lie at much higher altitudes. The atmosphere is so dense that the pressure 50 km above the Venusian surface is 1 atm, the same as at sea level on Earth. **(c)** By contrast, the Martian atmosphere is so thin that the surface pressure is the same as the pressure at an altitude of 35 km on Earth. Wispy clouds can be found at extreme altitudes.

greenhouse gas CO_2, the greenhouse effect (Section 9-1) has run wild. On Earth the greenhouse gases H_2O and CO_2 together make up only about 1% of our relatively sparse atmosphere, and the greenhouse effect has elevated the surface temperature by an additional 33°C (59°F). But on Venus, the dense shroud of CO_2 traps infrared radiation from the surface so effectively that it has raised the surface temperature by more than an additional 400°C (720°F).

Soviet and American spacecraft also discovered that Venus's clouds are primarily confined to three high-altitude layers. An upper cloud layer lies at altitudes between 68 and 58 km, a denser and more opaque cloud layer from 58 to 52 km, and an even more dense and opaque layer between 52 and 48 km. Above and below the clouds are 20-km-thick layers of haze. Below the lowest haze layer, the atmosphere is remarkably clear all the way down to the surface of Venus.

We saw in Section 11-5 that sulfur, which is not found in any appreciable amount in our atmosphere, plays an important role in the Venusian atmosphere. Sulfur combines with other elements to form gases such as sulfur dioxide (SO_2) and hydrogen sulfide (H_2S), along with sulfuric acid (H_2SO_4), the same acid used in automobile batteries. While Earth's clouds are composed of water droplets, Venusian clouds contain almost no water. Instead, they are composed of droplets of concentrated, corrosive sulfuric acid. Due to the high temperatures on Venus, these droplets never rain down on the planet's surface; they simply evaporate at high altitude. (You can see a similar effect on Earth. On a hot day in the desert of the American Southwest, streamers of rain called *virga* appear out of the bottoms of clouds but evaporate before reaching the ground.)

Convection in the Venusian Atmosphere

We saw in Section 9-6 that Earth's atmosphere is in continuous motion: Hot gases rise and cooler gases sink, forming convection cells in the atmosphere. The same is true on Venus, but with some important differences. On Venus, gases in the equatorial regions are warmed by the Sun, then rise upward and travel in the upper cloud layer toward the cooler polar regions. At the polar latitudes, the cooled gases sink to the lower cloud layer, in which they are transported back toward the equator. The result is two huge convection cells, one in the northern hemisphere and another in the southern hemisphere (Figure 11-28*a*). These cells are almost entirely contained within the main cloud layers shown in Figure 11-27*b*. The circulation is so effective at transporting heat around Venus's atmosphere that there is almost no temperature difference between the planet's equator and its poles. By contrast, Earth's atmosphere has a more complicated convection pattern (see Figure 11-28*b* and Figure 9-24) because Earth's rotation—which is far more rapid than Venus's—distorts the convection cells.

In addition to the north-south motions due to convection, high-altitude winds with speeds of 350 km/h (220 mi/h) blow from east to west, the same direction as Venus's retrograde rotation. Thus, the upper atmosphere rotates around the planet once every 4 days. The fast-moving winds stretch out the convection cells and produce V-shaped, chevronlike patterns in the clouds (Figure 11-29). In a similar way, Earth's rotation stretches out the convection cells in our atmosphere to produce the complicated circulation patterns shown in Figure 9-24.

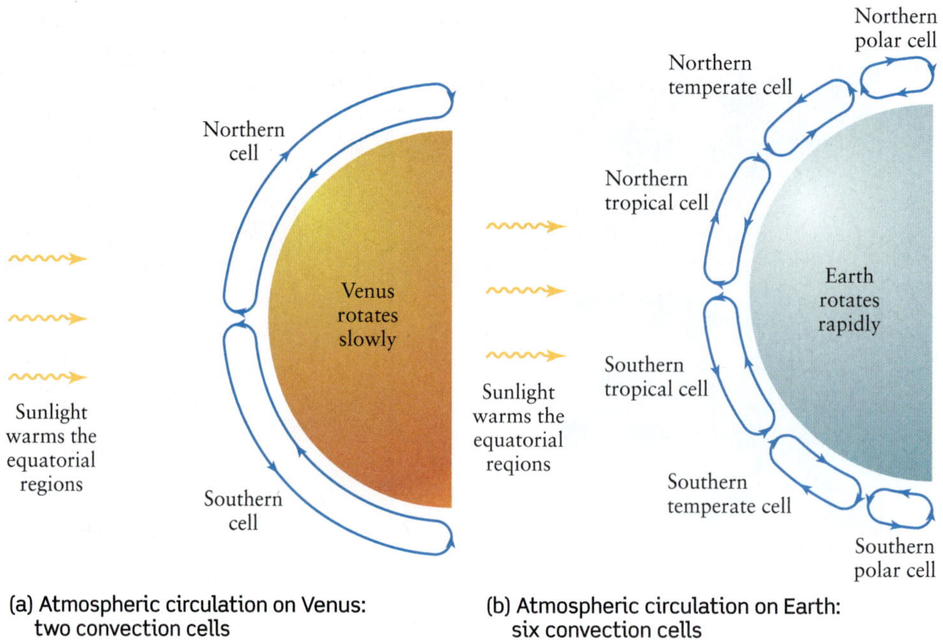

Figure 11-28

Atmospheric Circulation on Venus and Earth Solar heating causes convection in the atmospheres of both Venus and Earth, with warm air rising at the equator and cold air descending at the poles. **(a)** Because Venus rotates very slowly, it has little effect on the circulation. **(b)** Earth's rapid rotation distorts the atmospheric circulation into a more complex pattern (compare Figure 9-24).

On Earth, friction between the atmosphere and the ground causes wind speeds at the surface to be much less than at high altitude. The same effect occurs on Venus: The greatest wind speed measured by spacecraft on Venus's surface is only about 5 km/h (3 mi/h). Thus, only slight breezes disturb the crushing pressures and infernal temperatures found on Venus's dry, lifeless surface.

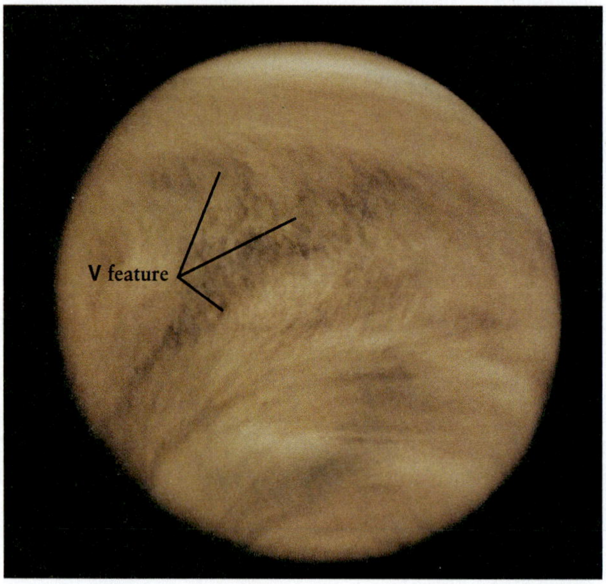

Figure 11-29 R I V U X G

Venus's Cloud Patterns This false-color image of Venus was made at ultraviolet wavelengths, at which the planet's atmospheric markings stand out best. The dark V feature is produced by the rapid motion of clouds around the planet's equator. The clouds move in the same retrograde direction as the rotation of the planet itself but at a much greater speed. (GSFC/NASA)

The Martian Atmosphere: Not a Drop to Drink

As Figure 11-27*c* shows, the atmosphere of Mars is very cold and thin compared to the atmosphere of either Earth or Venus: The surface pressure is a mere 0.006 atmosphere. Yet its chemical composition is very close to that of Venus: The Martian atmosphere is 95.3% carbon dioxide (versus 96.5% for Venus) and 2.7% nitrogen (versus 3.5% for Venus). The predominance of carbon dioxide means that Mars, like Earth and Venus, is warmed by the greenhouse effect. However, the Martian atmosphere is so thin that the greenhouse effect is very weak and warms the Martian surface by only 5°C (versus 33°C on Earth). While the perpetual cloud cover on Venus maintains a steady surface temperature, the thin atmosphere on Mars provides very little thermal insulation. While daytime highs on Mars can be as high as 20°C (68°F), at night the temperature can plummet to −140°C (−220°F).

Water vapor makes up about 0.03% of the Martian atmosphere (versus about 1% for Earth), and this vapor can form clouds. Figure 11-24*b* shows clouds that form near the tops of Martian volcanoes when air moves up the volcanic slopes. The rising air cools until the water vapor condenses into ice crystals. The same process occurs on Earth, but the clouds that form in this way are made of droplets of liquid water rather than ice crystals. Why is there a difference?

The explanation is that liquid water cannot exist anywhere on the Martian surface or in the Martian atmosphere. Water is liquid over only a limited temperature range: If the temperature is too low, water becomes ice, and if the temperature is too high, it becomes water vapor. What determines this temperature range is the atmospheric pressure above a body of water or around a water drop. If the pressure is very low, molecules easily escape from the liquid's surface, causing the water to vaporize. Thus, at low pressures, water more easily becomes water vapor. The average surface temperature on Mars is only about 250 K (−23°C, or

−10°F), and the average pressure is only 0.006 atmosphere. With this combination of temperature and pressure, water can exist as a solid (ice) and as a gas (water vapor) but not as a liquid. (You can see this same situation inside a freezer, where water vapor swirls around over ice cubes.) Hence, it never rains on Mars and there are no bodies of standing water anywhere on the planet.

In order to keep water on Mars in the liquid state, it would be necessary to increase both the temperature (to keep water from freezing) and the pressure (to keep the liquid water from evaporating). Later in this chapter we will see evidence that higher temperatures and pressures did exist on Mars billions of years ago, and that there was once liquid water on the surface.

The Martian Poles: Freezing the Atmosphere

Figure 11-27*c* shows that at very high altitudes above the Martian surface, atmospheric carbon dioxide can freeze into crystals and form clouds. (Frozen carbon dioxide can be made on Earth and is called "dry ice.") Such frozen carbon dioxide is also found in larger quantities in ice caps at the north and south poles of Mars, one of which you can see in Figure 11-4*c*. Although no spacecraft has successfully landed on the ice caps, their composition can be measured from orbit by comparing the spectrum sunlight reflected from the ice caps with the known reflection spectrum of frozen carbon dioxide. (Figure 7-4 shows another application of this idea.)

The size of the ice caps varies substantially with the seasons (Figure 11-30). When it is winter at one of the Martian poles, temperatures there are so low that atmospheric carbon dioxide freezes and the ice cap grows. (Think of it: It gets so cold at the Martian poles that the atmosphere freezes!) With the coming of spring, much of the carbon dioxide returns to the gaseous state and the ice cap shrinks. With the arrival of summer, however, the rate of shrinkage slows abruptly and a **residual ice cap** remains solid throughout the summer. Scientists had long suspected that the north and south residual polar caps contain a large quantity of frozen water that is less easily evaporated. The European *Mars Express* spacecraft confirmed this idea in 2004. It used its infrared cameras to peer through the carbon dioxide layer at the south polar ice cap and detect the characteristic spectrum of water ice. Because we do not know the thickness of the layer of water ice, the amount of frozen water stored in the Martian polar caps is still unknown.

The wintertime freezing of atmospheric carbon dioxide was measured by the *Viking Lander 1* and *Viking Lander 2* spacecraft, which made the first successful landings on Mars in 1976. Both of these spacecraft touched down in the northern hemisphere of Mars when it was spring in that hemisphere and autumn in the southern hemisphere. (You can see the landing sites in Figure 11-20.) Figure 11-31*a* shows the springtime view from *Viking Lander 1*. Initially, both spacecraft measured pressures around 0.008 atmosphere. After only a few weeks on the Martian surface, however, atmospheric pressure at both landing sites was dropping steadily. Mars seemed to be rapidly losing its atmosphere, and some scientists joked that soon all the air would be gone!

The actual explanation was that winter was coming to the southern hemisphere, causing carbon dioxide to freeze in that hemisphere's atmosphere and fall as dry-ice "snow." The formation of this "snow" removed gas from the Martian atmosphere, thereby lowering the atmospheric pressure across the planet. Several months later, when spring came to the southern hemisphere, the dry-ice snow evaporated rapidly and the atmospheric pressure returned to its prewinter levels. The pressure dropped again in the southern hemisphere's summer, because it is then winter in the northern hemisphere, and dry-ice snow condenses at northern latitudes (Figure 11-31*b*). Therefore, both temperature and atmospheric pressure change significantly with the Martian seasons.

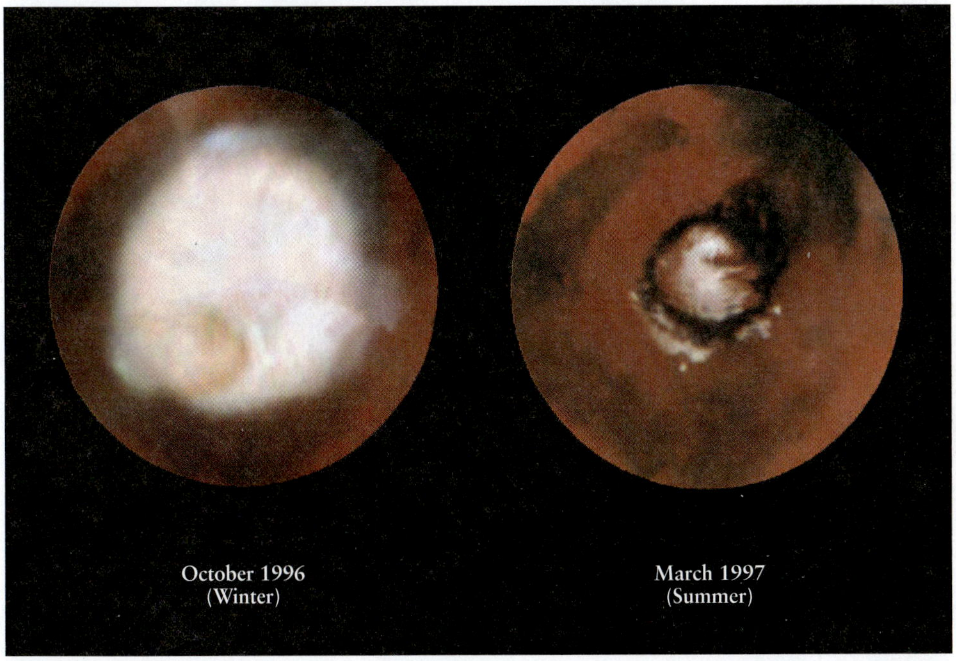

Figure 11-30 R I V U X G

Changing Seasons at the Martian North Pole
During the Martian winter, the temperature drops so low that carbon dioxide freezes out of the Martian atmosphere. A thin coating of carbon dioxide frost covers a broad region around Mars's north pole. During summer in the northern hemisphere, the ice cap is much smaller because carbon dioxide goes back into the atmosphere. The shrinking of the ice cap exposes a ring of dark sand dunes around the Martian north pole. (S. Lee/J. Bell/M. Wolff/Space Science Institute/NASA)

(a) A view from *Viking Lander 1*

(b) A wintertime view from *Viking Lander 2*

WEB LINK 11.16

Figure 11-31 R I V U X G

Seasons on the Martian Surface (a) This view from *Viking Lander 1* shows rocks that resemble volcanic rocks on Earth (see Figure 9-19*a*). They are thought to be part of an ancient lava flow that was broken apart by crater-forming asteroid impacts. Fine-grained debris has formed sand dunes. **(b)** This picture was taken during midwinter at the *Viking Lander 2* site. Freezing carbon dioxide adheres to water-ice crystals and dust grains in the atmosphere, causing them to fall to the ground and coat the surface with frost. This frost lasted for about a hundred days. (a: Dr. Edwin Bell II/NSSDC/GSFC/NASA; b: NASA/JPL)

Martian Dust

VIDEO 11.5

Figure 11-31*a* shows that the landscape around the *Viking Lander 1* is covered with a fine-grained dust. The dust particles have piled up in places to form drifts that resemble sand dunes on Earth. The dust particles are much smaller than ordinary household dust on Earth: They are only about 1 μm (10^{-6} m) across, about the size of the particles in cigarette smoke. The rocks at the *Viking Lander 2* site are also covered with dust, though it does not pile up in dunes. More recent missions to Mars have also found dust around their landing sites.

Because Martian dust is so fine, it can be carried aloft and spread over the surface by Martian winds even though the atmosphere is so tenuous. (A wind of 100 km/h, or 60 mi/h, on Mars is only about as strong as an Earth breeze of 10 km/h or 6 mi/h.) Windblown dust can sometimes be seen from Earth as a yellow haze that obscures Martian surface features. Figure 11-18 shows a dust storm that covered the entire planet, triggered by the flow of carbon dioxide evaporating from the north polar ice cap with the coming of northern spring. On Earth, stormy weather means cloudy skies, wind, and falling rain; on Mars, where no raindrop falls, stormy weather means wind and dust.

Dust also plays a role in the ordinary daily weather on Mars. Each afternoon, parcels of warm air rise from the heated surface and form whirlwinds called **dust devils.** (A similar phenomenon with the same name occurs in dry or desert terrain on Earth.) Martian dust devils can reach an altitude of 6 km (20,000 ft) and help spread dust particles across the planet's surface. All three landers measured changes in air pressure as dust devils swept past. Often several of these would pass in a single afternoon. Dust devils are large enough to be seen by spacecraft orbiting Mars (Figure 11-32). As the sun sets and afternoon turns to evening, the dust devils subside until the following day.

Airborne dust also affects the color of the Martian sky, shown in Figure 11-31. Earth's largely dust-free sky is blue because molecules in our atmosphere scatter blue sunlight more effectively than other colors (see Box 5-4). On Mars, however, dust particles in the air do a good job of scattering all visible wavelengths, which by itself would make the sky appear white. But, in addition, the dust particles contain a mineral called magnetite, which absorbs blue light. The result is that the Martian sky has a yellowish color reminiscent of butterscotch.

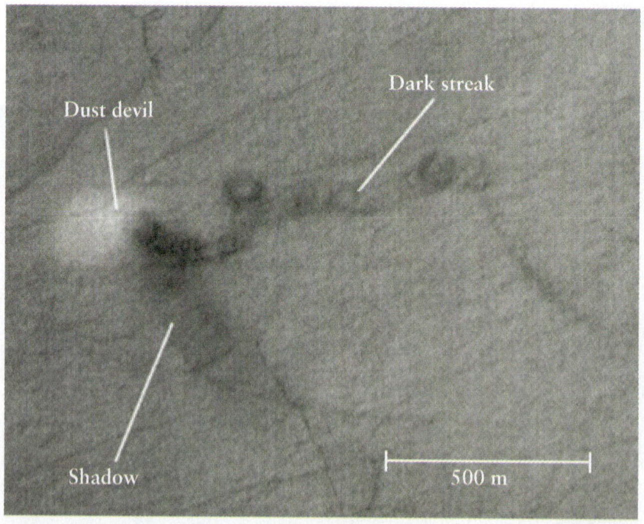

VIDEO 11.6

Figure 11-32 R I V U X G

A Martian Dust Devil This *Mars Global Surveyor* image shows a dust devil as seen from almost directly above. This tower of swirling air and dust casts a long shadow in the afternoon sun. The dust devil had been moving from right to left before the picture was taken, leaving a dark, curlicue-shaped trail in its wake. (NASA/JPL/Malin Space Science Systems)

11-7 The atmospheres of Venus and Mars were very different billions of years ago

TUTORIAL 11.2

Why do Earth, Venus, and Mars have such dramatically different atmospheres? As we will see, the original atmospheres of all three worlds were essentially the same. However, each atmosphere evolved in a unique way determined by the planet's size and distance from the Sun. The *Cosmic Connections* figure summarizes the processes that led to the present-day atmospheres of the three large terrestrial planets.

The early atmospheres of Earth, Venus, and Mars were predominantly water vapor and carbon dioxide

The Origin of Atmospheres on Earth, Venus, and Mars

The original atmospheres of Earth, Venus, and Mars all derived from gases that were emitted, or *outgassed,* from volcanoes (Figure 11-33). The gases released by present-day Earth volcanoes are predominantly water vapor (H_2O), carbon dioxide (CO_2), sulfur dioxide (SO_2), and nitrogen (N_2). These gases should therefore have been important parts of the original atmospheres of all three planets. Much of the water on all three planets is thought to have come from impacts from comets, icy bodies from the outer solar system (see Section 7-5 and Section 8-5).

Figure 11-33 R I V U X G

A Volcanic Eruption on Earth The eruption of Mount St. Helens in Washington state on May 18, 1980 released a plume of ash and gas 40 km (25 mi) high. Eruptions of this kind on Earth, Venus, and Mars probably gave rise to those planets' original atmospheres. (USGS)

In fact, water probably once dominated the atmospheres of all three planets. Venus and Earth are similar in size and mass, so Venusian volcanoes may well have outgassed as much water vapor as on Earth, and both planets would have had about the same number of comets strike their surfaces. Studies of how stars evolve suggest that the early Sun was only about 70% as luminous as it is now, so the temperature in Venus's early atmosphere must have been quite a bit lower. Water vapor may actually have been able to liquefy and form oceans on Venus. Mars is smaller than either Earth or Venus, but its large volcanoes could also have emitted substantial amounts of water vapor. With a thicker atmosphere and hence greater pressure, liquid water could also have existed on Mars.

The Evolution of Earth's Atmosphere

On the present-day Earth most of the water is in the oceans. Nitrogen, which is not very reactive, is still in the atmosphere. By contrast, carbon dioxide dissolves in water, which can fall as rain; as a result, rain removed most of the H_2O and CO_2 out of our planet's atmosphere long ago. The dissolved CO_2 reacted with rocks in rivers and streams, and the residue was ultimately deposited on the ocean floors, where it turned into carbonate rocks, such as limestone. Consequently, most of Earth's carbon dioxide is tied up in Earth's crust, and only about 1 in every 3000 molecules in our atmosphere is a CO_2 molecule. If Earth became as hot as Venus, much of its CO_2 would be boiled out of the oceans and baked out of the crust, and our planet would soon develop a thick, oppressive carbon dioxide atmosphere much like that of Venus.

The small amount of carbon dioxide in our atmosphere is sustained in part by plate tectonics. Tectonic activity causes carbonate rocks to cycle through volcanoes, where they are heated and forced to liberate their trapped CO_2. This liberated gas then rejoins the atmosphere. Without volcanoes, rainfall would remove all of the CO_2 from our atmosphere in only a few thousand years. This lack of CO_2 would dramatically reduce the greenhouse effect, our planet's surface temperature would drop precipitously, and the oceans would freeze. Plate tectonics is thus essential for maintaining our planet's livable temperatures.

The oxygen (O_2) in our atmosphere is the result of photosynthesis by plant life, which absorbs CO_2 and releases O_2. The net result is an atmosphere that is predominantly N_2 and O_2, with enough pressure and a high enough temperature (thanks to the greenhouse effect) for water to remain a liquid—and hence for life, which requires liquid water, to exist.

The Evolution of the Venusian Atmosphere: A Runaway Greenhouse Effect

When Venus was young and had liquid water, the amount of atmospheric CO_2 was kept in check much as on Earth today: CO_2 was released by volcanoes, then dissolved in the oceans and bound up in carbonate rocks. Enough of the liquid water would have vaporized to create a thick cover of water vapor clouds. Since water vapor is a greenhouse gas, this humid atmosphere—perhaps denser than Earth's present-day atmosphere, but far less dense than the atmosphere that surrounds Venus today—would

COSMIC CONNECTIONS

Evolution of Terrestrial Atmospheres

Earth, Venus, and Mars formed with similar original atmospheres. However, these atmospheres changed dramatically over time due to factors such as planetary size and distance from the Sun.

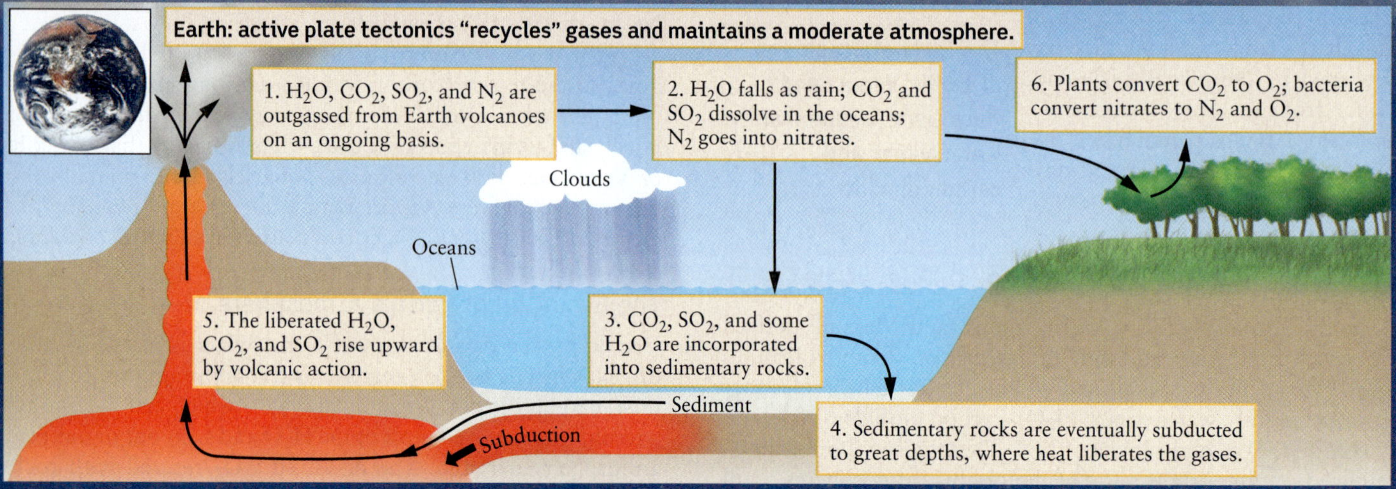

Earth: active plate tectonics "recycles" gases and maintains a moderate atmosphere.
1. H_2O, CO_2, SO_2, and N_2 are outgassed from Earth volcanoes on an ongoing basis.
2. H_2O falls as rain; CO_2 and SO_2 dissolve in the oceans; N_2 goes into nitrates.
3. CO_2, SO_2, and some H_2O are incorporated into sedimentary rocks.
4. Sedimentary rocks are eventually subducted to great depths, where heat liberates the gases.
5. The liberated H_2O, CO_2, and SO_2 rise upward by volcanic action.
6. Plants convert CO_2 to O_2; bacteria convert nitrates to N_2 and O_2.
Clouds
Oceans
Sediment
Subduction

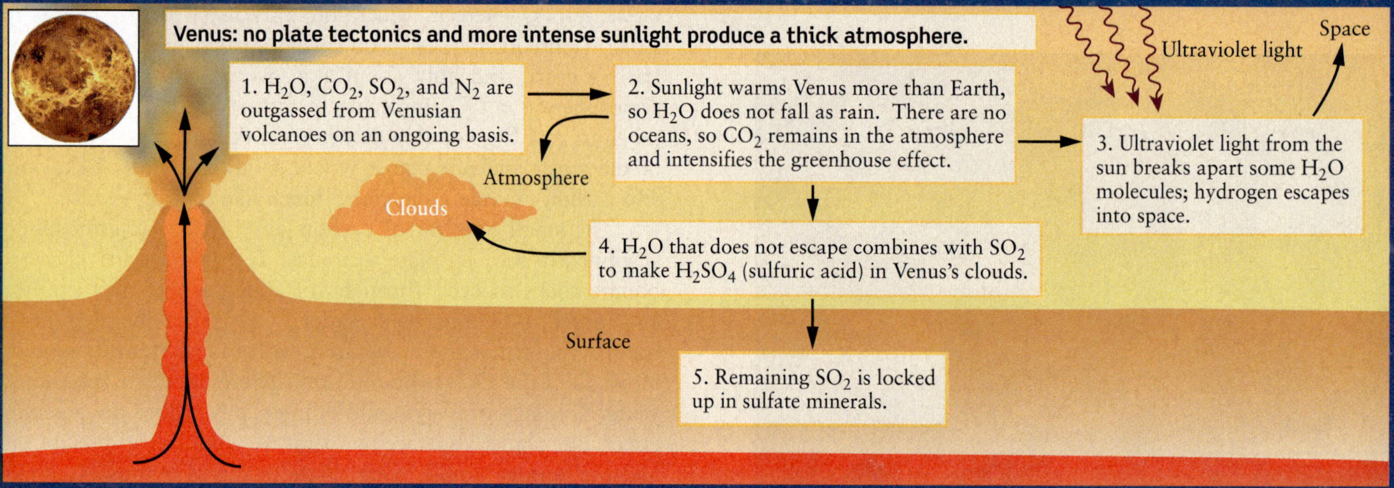

Venus: no plate tectonics and more intense sunlight produce a thick atmosphere.
1. H_2O, CO_2, SO_2, and N_2 are outgassed from Venusian volcanoes on an ongoing basis.
2. Sunlight warms Venus more than Earth, so H_2O does not fall as rain. There are no oceans, so CO_2 remains in the atmosphere and intensifies the greenhouse effect.
3. Ultraviolet light from the sun breaks apart some H_2O molecules; hydrogen escapes into space.
4. H_2O that does not escape combines with SO_2 to make H_2SO_4 (sulfuric acid) in Venus's clouds.
5. Remaining SO_2 is locked up in sulfate minerals.
Ultraviolet light
Space
Atmosphere
Clouds
Surface

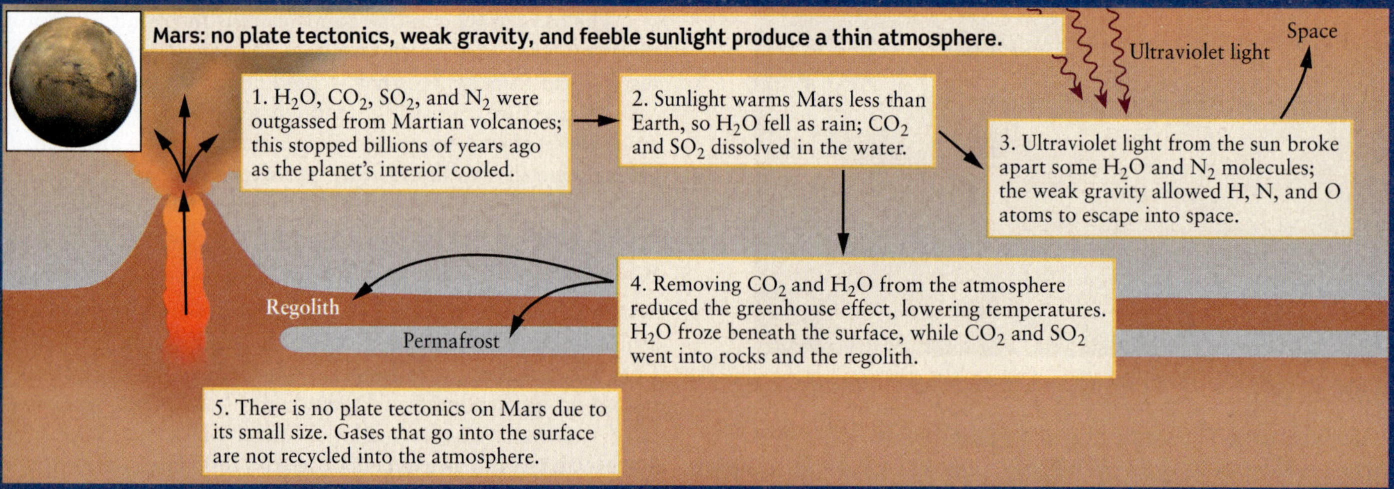

Mars: no plate tectonics, weak gravity, and feeble sunlight produce a thin atmosphere.
1. H_2O, CO_2, SO_2, and N_2 were outgassed from Martian volcanoes; this stopped billions of years ago as the planet's interior cooled.
2. Sunlight warms Mars less than Earth, so H_2O fell as rain; CO_2 and SO_2 dissolved in the water.
3. Ultraviolet light from the sun broke apart some H_2O and N_2 molecules; the weak gravity allowed H, N, and O atoms to escape into space.
4. Removing CO_2 and H_2O from the atmosphere reduced the greenhouse effect, lowering temperatures. H_2O froze beneath the surface, while CO_2 and SO_2 went into rocks and the regolith.
5. There is no plate tectonics on Mars due to its small size. Gases that go into the surface are not recycled into the atmosphere.
Ultraviolet light
Space
Regolith
Permafrost

have efficiently trapped heat from the Sun. At first, this would have had little effect on the oceans of Venus. Although the temperature would have climbed above 100°C, the boiling point of water at sea level on Earth, the added atmospheric pressure from water vapor would have kept the water in Venus's oceans in the liquid state.

This hot and humid state of affairs may have persisted for several hundred million years. But as the Sun's energy output slowly increased over time, the temperature at the surface would eventually have risen above 374°C (647 K, or 705°F). Above this temperature, no matter what the atmospheric pressure, Venus's oceans would have begun to evaporate. The added water vapor in the atmosphere would have increased the greenhouse effect, making the temperature even higher and causing the oceans to evaporate faster. This temperature increase would have added even more water vapor to the atmosphere, further intensifying the greenhouse effect and making the temperature climb higher still. This is an example of a **runaway greenhouse effect,** in which an increase in temperature causes a further increase in temperature, and so on.

ANALOGY A runaway greenhouse effect is like a house in which the thermostat has accidentally been connected backward. If the temperature in such a house gets above the value set on the thermostat, the heater comes on and makes the house even hotter.

Once Venus's oceans disappeared, so did the mechanism for removing carbon dioxide from the atmosphere. With no oceans to dissolve it, outgassed CO_2 began to accumulate in the atmosphere, making the greenhouse effect "run away" even faster. Temperatures eventually became high enough to "bake out" any CO_2 that was trapped in carbonate rocks. This liberated carbon dioxide formed the thick atmosphere of present-day Venus.

Sulfur dioxide (SO_2) is also a greenhouse gas released by volcanoes. Although present in far smaller amounts than carbon dioxide, it would have contributed to Venus's rising temperatures. Like CO_2, it dissolves in water, a process that helps moderate the amount of sulfur dioxide in our atmosphere. But when the oceans evaporated on Venus, the amount of atmospheric SO_2 would have increased.

The temperature in Venus's atmosphere would not have increased indefinitely. In time, solar ultraviolet radiation striking molecules of water vapor—the dominant cause of the runaway greenhouse effect—would have broken them into hydrogen and oxygen atoms. The lightweight hydrogen atoms would have then escaped into space (see Box 7-2 for a discussion of why lightweight atoms can escape a planet more easily than heavy atoms). The remaining atoms of oxygen, which is one of the most chemically active elements, would have readily combined with other substances in Venus's atmosphere. Eventually, almost all of the water vapor would have been irretrievably lost from Venus's atmosphere. With all the water vapor gone, and essentially all the carbon dioxide removed from surface rocks, the greenhouse effect would no longer have "run away," and the rising temperature would have leveled off.

Today, the infrared-absorbing properties of CO_2 have stabilized Venus's surface temperature at its present value of 460°C. Only minuscule amounts of water vapor—about 30 parts per million, or 0.003%—remain in the atmosphere. As on Earth, volcanic outgassing might still add small amounts of water vapor to the atmosphere, along with carbon dioxide and sulfur dioxide. But on Venus, these water molecules either combine with sulfur dioxide to form sulfuric acid clouds or break apart due to solar ultraviolet radiation. The great irony is that this state of affairs is the direct result of an earlier Venusian atmosphere that was predominantly water vapor!

The Evolution of the Martian Atmosphere: A Runaway Icehouse Effect

Mars probably had a thicker atmosphere 4 billion years ago. Thanks to Mars's greater distance from the Sun and hence less intense sunlight, temperatures would have been lower than on young Earth, and any water in the atmosphere would more easily have fallen as rain or snow. This precipitation would have washed much of the planet's carbon dioxide from its atmosphere, perhaps creating carbonate minerals in which the CO_2 is today chemically bound. Measurements from Mars orbit show only small amounts of carbonate materials on the surface, suggesting that the amount of atmospheric CO_2 that rained out was small. Hence, even the original Martian atmosphere was relatively thin, though thicker than the present-day atmosphere.

Because Mars is so small, it cooled early in its history and volcanic activity came to an end. Thus, any solid carbonates were not recycled through volcanoes as they are on Earth. The depletion of carbon dioxide from the Martian atmosphere into the surface would therefore have been permanent.

As the amount of atmospheric CO_2 declined, the greenhouse effect on Mars would have weakened and temperatures begun to fall. This temperature decrease would have caused more water vapor to condense into rain or snow and fall to the surface, taking even more CO_2 with them and further weakening the greenhouse effect. Thus, a decrease in temperature would have caused a further decrease in temperature—a phenomenon sometimes called a **runaway icehouse effect.** (This is the reverse of the runaway greenhouse effect that has taken place on Venus.) Ultimately, both water vapor and most of the carbon dioxide would have been removed from the Martian atmosphere. With only a very thin CO_2 atmosphere remaining, surface temperatures on Mars eventually stabilized at their present frigid values.

As both water vapor and carbon dioxide became depleted, ultraviolet light from the Sun could penetrate the thinning Martian atmosphere to strip it of nitrogen. Nitrogen molecules (N_2) normally do not have enough thermal energy to escape from Mars, but they can acquire that energy from ultraviolet photons, which break the molecules in two. Ultraviolet photons can also split carbon dioxide and water molecules, giving their atoms enough energy to escape. Indeed, in 1971 the Soviet *Mars 2* spacecraft found a stream of oxygen and hydrogen atoms (from the breakup of water molecules) escaping into space. Oxygen atoms that did not escape into space could have combined with iron-bearing minerals in the surface, forming rustlike compounds. Such compounds have a characteristic reddish-brown color and may be responsible for the overall color of the planet.

Was the early Martian atmosphere thick enough and warm enough for water to remain as a liquid on the planet's surface? As we will see in the next section, scientists have found evidence on Mars that the answer is a qualified yes.

11-8 Rovers have found evidence of ancient Martian water

Just as ancient fossils on Earth provide evidence of how life has evolved on our planet, ancient rocks and geological formations help us understand how our planet's surface and atmosphere have evolved. (For example, the oldest rocks on Earth have a chemical structure which shows that they formed in a time when there was little oxygen in the atmosphere and hence before the appearance of photosynthetic planet life.) Planetary scientists who want to learn about the history of Venus and Mars are therefore very interested in looking at ancient rocks on those worlds.

Parts of Mars are far drier than a bone, while others had standing water for extended periods

On Venus we cannot look very far into the past, since the planet's volcanic activity and flake tectonics have erased features more than about 500 million years old (see Section 11-5). On Mars, by contrast, we expect to find rocks that are billions of years old: The thin atmosphere causes very little erosion, there has been little recent volcanic activity to cover ancient rocks, and there is no subduction to drag old surface features back into the planet's interior. (The heavy cratering of the southern highlands bears testament to the great age of much of the Martian surface.) In a quest to find these ancient rocks and learn about the early history of Mars, scientists have sent spacecraft to land on the red planet and explore it at close range.

Landing on Mars

As of this writing (2009) six spacecraft have successfully landed on the Martian surface. *Viking Lander 1* and *Viking Lander 2*, which we first discussed in Section 11-6, touched down in 1976. They had no capability to move around on the surface, and so each *Viking Lander* could only study objects within reach of the scoop at the end of its mechanical arm (Figure 11-34*a*). Figure 11-31 shows the view from the two *Viking Lander* sites.

Three more recent unmanned explorers, *Mars Pathfinder* (landed 1997) and two *Mars Exploration Rovers* (landed 2004), were equipped with electrically driven wheels that allow travel over the surface (Figure 11-34*b*). Unlike the *Viking Landers,* these rovers did not have a heavy and expensive rocket engine to lower it gently to the Martian surface. Instead, the spacecraft was surrounded by airbags like those used in automobiles. Had there been anyone to watch *Mars Pathfinder* or one of the *Mars Exploration Rovers* land, they would have seen an oversized beach ball hit the surface at 50 km/h (30 mi/h), then bounce for a kilometer before finally rolling to a stop. Unharmed by its wild ride, the spacecraft then deflated its airbags and began to study the geology of Mars.

Figure 11-20 shows the landing sites of the *Mars Pathfinder* rover (named *Sojourner*) and the two *Mars Exploration Rovers* (named *Spirit* and *Opportunity*). Following directions from Earth, *Sojourner* spent three months exploring the rocks around its landing site. The *Mars Exploration Rovers* were also designed for a three-month lifetime, but both were still functioning more than 5 years after landing. As of this writing (mid-2009), *Opportunity* has traveled more than 16 km over the Martian surface and con-

(a) A *Viking Lander*

(b) Two generations of rovers

WEBLINK 11.17

Figure 11-34 RIVUXG

Martian Explorers **(a)** Each *Viking Lander* is about 2 m (6 ft) tall from the base of its footpads to the top of its dish antenna, used for sending data to an orbiter, which in turn relayed the information back to Earth. (This photograph shows a full-size replica under test on Earth.) **(b)** The technicians in this photograph pose with one of the *Mars Exploration Rovers* and with a replica of its predecessor, the *Mars Pathfinder* rover. (a: NASA; b: NASA/JPL) tinues to collect data. However, *Spirit*, has recently become stuck in soft sand. *Spirit* continues to collect data, but it's not clear if the rover will ever free itself.

In addition to cameras, all five landers carried a device called an X-ray spectrometer to measure the chemical composition of rock samples. *Spirit* and *Opportunity* also carried a tool designed to grind holes into rocks, thus revealing parts of the rock's interior that have not been weathered by exposure to the Martian atmosphere.

The Martian Regolith

The *Viking Landers* were the first to reveal the nature of the Martian regolith, or surface material. The scoop on the extendable arm (Figure 11-34*a*) was used to dig into the regolith and retrieve samples for analysis. Bits of the regolith clung to a magnet mounted on the scoop, indicating that the regolith contains iron. The X-ray spectrometers showed that samples collected at both *Viking Lander* sites were rich in iron, silicon, and sulfur.

The *Viking Landers* each carried a miniature chemistry laboratory for analyzing the regolith. Curiously, when a small amount of water was added to a sample of regolith, the sample began to fizz and release oxygen. This experiment indicated that the regolith contains unstable chemicals called peroxides and superoxides, which break down in the presence of water to release oxygen gas.

The unstable chemistry of the Martian regolith probably comes from solar ultraviolet radiation that beats down on the planet's surface. (Unlike Earth, Mars has no ozone layer to screen out ultraviolet light.) Ultraviolet photons easily break apart molecules of carbon dioxide (CO_2) and water vapor (H_2O) by knocking off oxygen atoms, which then become loosely attached to chemicals in the regolith. Ultraviolet photons also produce highly reactive ozone (O_3) and hydrogen peroxide (H_2O_2), which also become incorporated in the regolith.

Among the most important scientific goals of the *Viking Landers* was to search for evidence of life on Mars. Although it was clear that Mars has neither civilizations nor vast fields of plants, it still seemed possible that simple microorganisms could have evolved on Mars during its early brief epoch of Earthlike climate. If so, perhaps some species had survived to the present day. To test this idea, the *Viking Landers* looked for biologically significant chemical reactions in samples of the Martian regolith, but they failed to detect any reactions of this type. Perhaps life never existed on Mars, or perhaps it existed once but was destroyed by the presence of peroxides and superoxides in the regolith.

On Earth, hydrogen peroxide is commonly used as an antiseptic. When poured on a wound, it fizzes and froths as it reacts with organic material and destroys germs. Thus, the *Viking Landers* may have failed to detect any biochemistry on Mars because the regolith has literally been sterilized! Hence, we cannot yet tell whether life ever existed on Mars, or whether it might yet exist elsewhere on or beneath the planet's surface. (In Chapter 28 we discuss the *Viking Lander* experiments in detail, as well as the controversial proposal that fossil Martian microorganisms may have traveled to Earth aboard meteorites.)

Exploring Martian Rocks

Images made from Mars orbit show a number of features that suggest water flowed there in the past (Figure 11-35). To investigate these, *Mars Pathfinder* landed in an area of the northern lowlands called Ares Valles, which appears to be an ancient flood plain. Figure 11-35*b* shows a portion of this plain. Flood waters can carry rocks of all kinds great distances from their original locations, so scientists expected that this landing site (Figure 11-36) would show great geologic diversity. They were not disappointed.

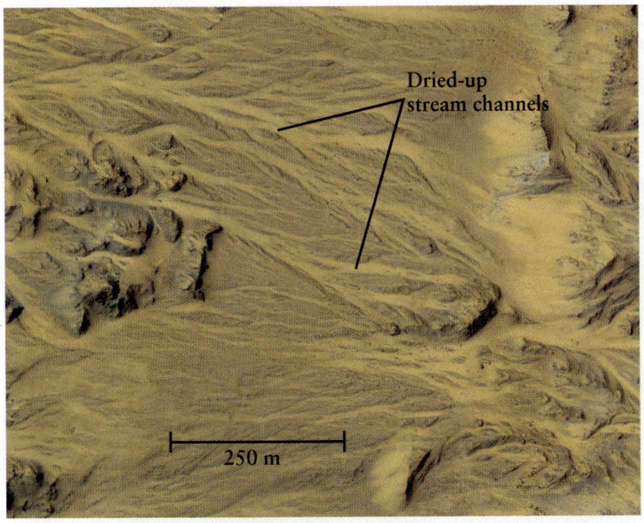

(a)

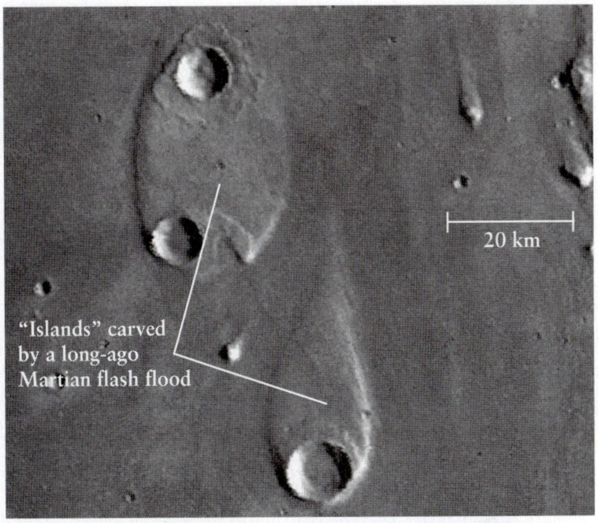

(b)

Figure 11-35 R I V U X G

Sign of Ancient Martian Water **(a)** This *Mars Reconnaissance Orbiter* image shows patterns from flowing water almost identical to those caused by rainfall in deserts on Earth. No rain falls in today's sparse Martian atmosphere, so these must date from an earlier era when the atmosphere was thicker. **(b)** These features rise above the floor of Ares Valles. They were carved out by a torrent of water that flowed from the bottom of the image toward the top. Similar flood-carved islands are found on Earth in eastern Washington state. (a: NASA/JPL/U. of Arizona; b: NASA/USGS)

Figure 11-36 R I **V** U X G

Roving the Martian Surface This view from the stationary part of *Mars Pathfinder* shows the rover, called *Sojourner,* deploying its X-ray spectrometer against a rock named Moe. Scientists gave other rocks whimsical names such as Jiminy Cricket and Pooh Bear. (NASA/JPL)

Many of the rocks appeared to be igneous rock produced by volcanic action. Other rocks, which resemble the impact breccias found in the ancient highlands of the Moon (see Figure 10-15), appear to have come from areas that were struck by crater-forming meteoroids. Still others have a layered structure like sedimentary rocks on Earth (see Figure 9-19*b*). Such rocks form gradually at the bottom of bodies of water, suggesting that liquid water was stable in at least some regions of Mars for a substantial period of time.

Searching for Martian Water

The two *Mars Exploration Rovers* also landed in sites that were suspected to have been under water in the distant past. *Spirit* touched down in Gusev Crater, a large impact basin. A valley opens into this crater, resembling a river that runs into a lake. The destination for *Opportunity* was the flat plain of Sinus Meridianii, where orbiting spectrometers had spotted evidence of hematite—a type of iron oxide that, on Earth, tends to form in wet locations.

Contrary to expectations, most of the rocks encountered by *Spirit* show *no* evidence that they have been exposed to water. The volcanic rocks found in and around Gusev Crater (Figure 11-37*a*) are composed of the minerals olivine and pyroxene, which break down if they are exposed to liquid water. In a few locations *Spirit* found deposits of sulfur salts, which can be formed by the action of water (Figure 11-37*b*). However, olivine and pyroxene were still found among the sulfur salts. While it remains a possibility that Gusev Crater was once a river-fed lake, the results from *Spirit* show that for the past billion years this region has been drier than any desert on Earth.

The results from *Opportunity* were entirely different. This rover discovered outcrops of sedimentary rocks on the plains of Sinus Meridianii (Figure 11-38). These rocks are heavily laden with sulfur compounds, strongly suggesting that they were produced by the evaporation of sulfur-rich water. *Opportunity* also found millimeter-sized spheres of hematite that, like analogous spheres found on Earth, probably formed from iron-rich water circulating through sediment. These observations suggest Sinus

(a) Volcanic rocks

(b) Sulfur salts

Figure 11-37 R I **V** U X G

Dry Mars At the *Spirit* landing site, both **(a)** the dusty volcanic rocks that dominate the site and **(b)** these light-colored sulfur salts contain minerals that break apart upon contact with liquid water. Hence, this area must have been dry for billions of years. (a: NASA/JPL-Caltech/Cornell; b: NASA/JPL/Cornell)

On Earth, layered rocks like these form by sedimentation in a body of water.
These bluish spheres are made of hematite. On Earth this mineral forms in wet sediments.

Figure 11-38 RIVUXG
Wet Mars This false-color image shows two signatures of water found by *Opportunity*. Both the sedimentary rock (exposed by erosion) and the millimeter-sized hematite spheres shown in blue require standing water to form. So, this area must have been under water for an extended period. (NASA/JPL/Cornell)

Meridianii, which is among the flattest places known in the solar system, was once a lake bed. It is not known how long ago this area was under water, but the density of craters found there suggests that the surface formed 3 to 4 billion years ago.

When combined with observations from Mars orbit, the *Spirit* and *Opportunity* results show that ancient Mars was a very diverse planet. Most of the planet was probably very dry, like Gusev Crater, but there were isolated areas like Sinus Meridianii that had liquid water for some period of time. It remains a challenge to planetary scientists to explain why liquid water appeared on Mars only in certain places and at certain times.

Where Is the Water Now?

Whatever liquid water once coursed over parts of the Martian surface is presumably now frozen, either at the polar ice caps or beneath the planet's surface. To check this idea, scientists used a device on board the *Mars Odyssey* orbiter to measure the result of cosmic rays hitting Mars.

Cosmic rays are fast-moving subatomic particles that enter our solar system from interstellar space. When these particles strike and penetrate into the Martian soil, they collide with the nuclei of atoms in the soil and knock out neutrons. (Recall from Section 5-7 that atomic nuclei contain both positively charged protons and electrically neutral neutrons.) If the ejected neutrons originate within a meter or so of the surface, they can penetrate through the regolith and escape into space. However, if there are water molecules (H_2O) in the surface, the nuclei of the hydrogen atoms in these molecules will absorb the neutrons and prevent them from escaping. An instrument on board *Mars Odyssey* detects neutrons coming from Mars; where the instrument finds a *deficiency* of neutrons is where the soil contains hydrogen and, presumably, water.

The results from the *Mars Odyssey* neutron measurements show that there is abundant water beneath the surface at both Martian poles (Figure 11-39). But they also show a surprising amount of subsurface water in two regions near the equator, Arabia Terra (see Figure 11-17*a* and Figure 11-20) and another region on the opposite side of the planet. One possible explanation is that about a million years ago, gravitational forces from other planets caused a temporary change in the tilt of Mars's rotation axis. The ice caps were then exposed to more direct sunlight, causing some of the water to evaporate and eventually refreeze elsewhere on the planet.

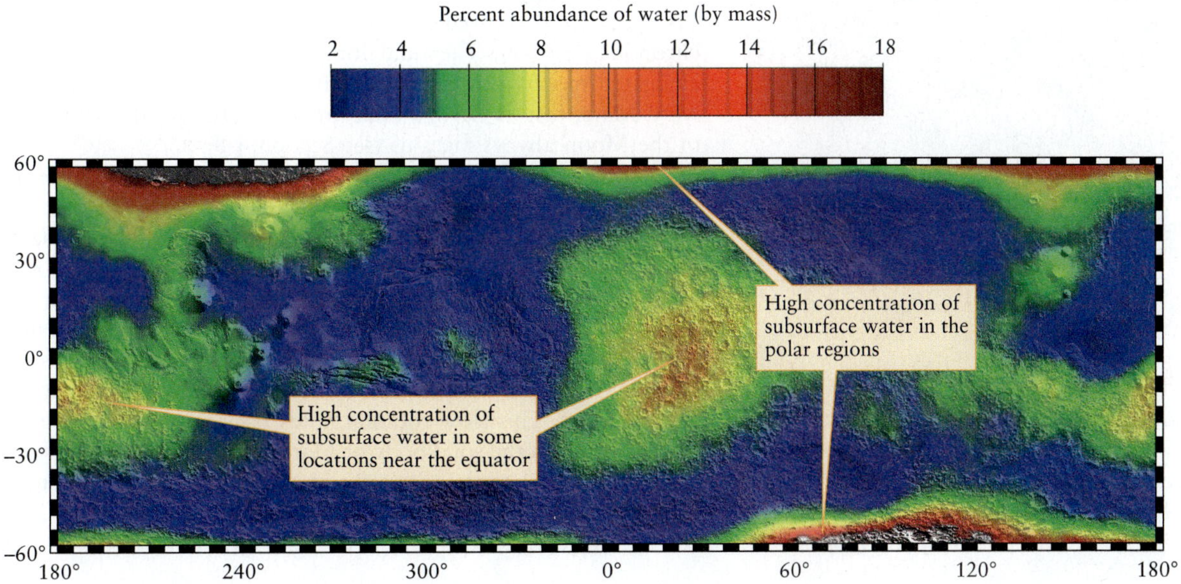

Figure 11-39
Water Beneath the Martian Surface This map of the abundance of water (probably in the form of ice) just below the surface of Mars is based on data collected from orbit by the *Mars Odyssey* spacecraft. These measurements can only reveal the presence of water or water ice to a depth of about a meter; there may be much more water at greater depths. (LANL)

There are other surprising features on Mars related to the presence of water. A number of *Mars Global Surveyor* images show gullies apparently carved by water flowing down the walls of pits or craters (Figure 11-40). These gullies could have formed when water trapped underground—where the pressure is greater and water can remain a liquid—seeped onto the surface. The gullies appear to be geologically young, so it is possible that even today some liquid water survives below the Martian surface. Alternatively, these gullies may have formed during periods when the Martian climate was colder than usual and the slopes of craters were covered with a layer of dust and snow. Melting snow on the underside of this layer could have carved out gullies, which were later exposed when the climate warmed and the snow evaporated.

The total amount of water on Mars is not known. But by examining flood channels at various locations around Mars, Michael Carr of the U.S. Geological Survey has estimated that there is enough water to cover the planet to a depth of 500 m (1600 ft). (By comparison, Earth has enough water to cover our planet to a depth of 2700 m, or 8900 ft.) Thus, while the total amount of water on Mars is less than on Earth, there may be enough to have once formed lakes or seas—one of which may have covered the plains of Sinus Meridianii and produced the sedimentary rocks shown in Figure 11-38. Future missions to Mars will delve more deeply into the past climate and geology of the red planet.

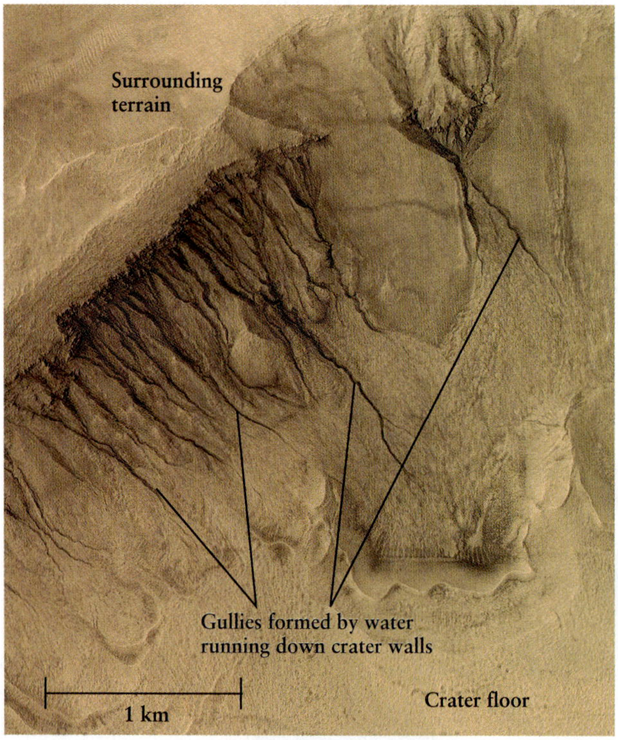

Figure 11-40 R I **V** U X G

Martian Gullies When the *Mars Global Surveyor* spacecraft looked straight down into this crater in the Martian southern hemisphere, it saw a series of gullies along the crater wall. These gullies may have formed by subsurface water seeping out to the surface, or by the melting of snow that fell on crater walls. (NASA/JPL/Malin Space Science Systems)

11-9 The two Martian moons resemble asteroids

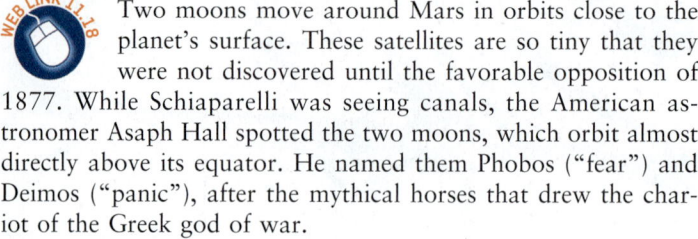

Two moons move around Mars in orbits close to the planet's surface. These satellites are so tiny that they were not discovered until the favorable opposition of 1877. While Schiaparelli was seeing canals, the American astronomer Asaph Hall spotted the two moons, which orbit almost directly above its equator. He named them Phobos ("fear") and Deimos ("panic"), after the mythical horses that drew the chariot of the Greek god of war.

The Orbits of Phobos and Deimos

Phobos is the inner and larger of the Martian moons. It circles Mars in just 7 hours and 39 minutes at an average distance of only 9378 km (5828 mi) from the center of Mars. (Recall that our Moon orbits at an average distance of 384,400 km from Earth's center.) This orbital period is much shorter than the Martian day; hence, Phobos rises in the *west* and gallops across the sky in only $5\frac{1}{2}$ hours, as viewed by an observer near the Martian equator. During this time, Phobos appears several times brighter in the Martian sky than Venus does from Earth.

As seen from Mars, Deimos rises in the east but Phobos rises in the west

Deimos, which is farther from Mars and somewhat smaller than Phobos, appears about as bright from Mars as Venus does from Earth. Deimos orbits at an average distance of 23,460 km (14,580 mi) from the center of Mars, slightly beyond the right distance for a synchronous orbit—an orbit in which a satellite seems to hover above a single location on the planet's equator. As seen from the Martian surface, Deimos rises in the east and takes about 3 full days to creep slowly from one horizon to the other. (Figure 11-41*a*).

Images from spacecraft in Mars's orbit reveal Phobos and Deimos to be jagged and heavily cratered (Figure 11-41*b*). Both moons are somewhat elongated from a spherical shape: Phobos is approximately 28 by 23 by 20 km in size, while Deimos is slightly smaller, roughly 16 by 12 by 10 km. Observations from the *Viking Orbiters* showed that Phobos and Deimos are both in synchronous rotation. The tidal forces that Mars exerts on these elongated moons cause them to always keep the same sides toward Mars, just as Earth's tidal forces ensure that the same side of the Moon always faces us (see Section 4-8).

We saw in Section 10-5 that the Moon raises a tidal bulge on Earth. Because Earth rotates on its axis more rapidly than the Moon orbits Earth, this bulge is dragged ahead of a line connecting Earth and the Moon (see Figure 10-17). The gravitational force of this bulge pulls the Moon into an ever larger orbit. In the same way, and for the same reason, Deimos is slowly moving away from Mars. But for fast-moving Phobos, the situation is reversed: Mars rotates on its axis more slowly than Phobos moves around Mars. This slow rotation holds back the tidal bulge that Phobos raises in the solid body of Mars, and this bulge thus is *behind* a line connecting the centers of Mars and Phobos. (Imagine that the bulge in Figure 10-17 is tilted clockwise rather than counterclockwise.) This bulge exerts a gravitational force on

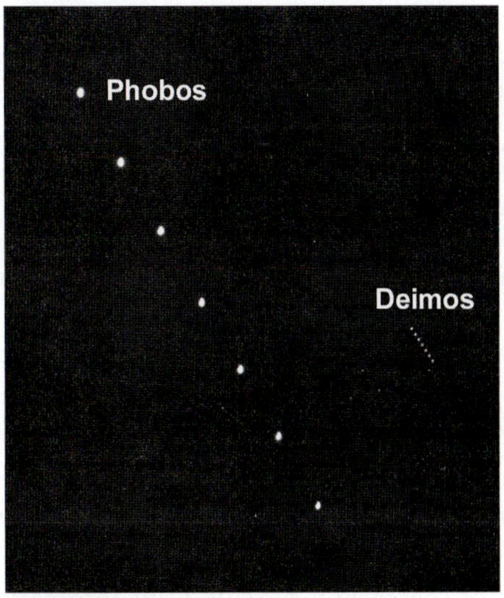

(a) As seen from Mars

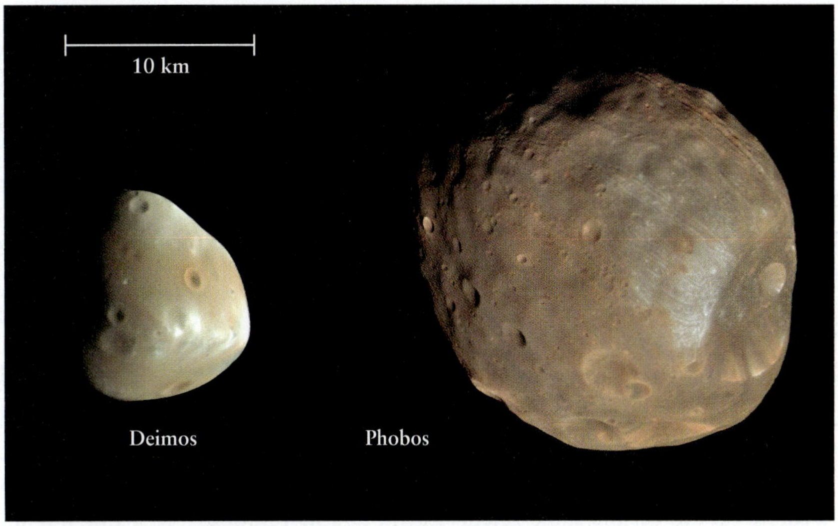

(b) As seen from Mars orbit

Figure 11-41 R I V U X G

The Moons of Mars (a) The *Spirit* rover recorded seven images of the Martian night sky at 15-minute intervals to produce this composite view of the motions of Deimos and Phobos. (b) Two *Mars Reconnaissance Orbiter* images have been combined to show Deimos and Phobos and their heavily cratered surfaces to the same scale. The gravity on these moons is only about 1/1000 of Earth's, and not strong enough to shape them into spheres. (a: NASA/JPL/Cornell/Texas A&M; b: NASA/JPL Caltech/University of Arizona)

Phobos that pulls it back, slowing the moon's motion and making it slowly spiral inward toward Mars. This inward spiral will cease in about 40 million years when Phobos smashes into the Martian surface.

The Nature of the Martian Moons

The origin of the Martian moons is unknown. One idea is that they are asteroids that Mars captured from the nearby asteroid belt (see Section 7-5). Like Phobos and Deimos, many asteroids reflect very little sunlight (typically less than 10%) because of a high carbon content. Phobos and Deimos also have very low densities (1900 kg/m^3 and 1760 kg/m^3, respectively), which are also characteristic of carbon-rich asteroids. Perhaps two of these asteroids wandered close enough to Mars to become permanently trapped by the planet's gravitational field. Alternatively, Phobos and Deimos may have formed in orbit around Mars out of debris left over from the formation of the planets. Some of this debris might have come from the asteroid belt, giving these two moons an asteroidlike character. Future measurements of the composition of the Martian moons—perhaps by landing spacecraft on their surfaces—may shed light on their origin.

Future Exploration of Mars

The exploration of Mars and its two moons has really just begun. Recent discoveries from orbiters and landers have only whetted our appetite to know more about Mars. Fortunately, every 25 months Earth and Mars are ideally placed for sending a spacecraft from one planet to another with the minimum thrust. For the next several years, one or more spacecraft will be launched toward Mars at each of these opportunities. One of these spacecraft may land on Mars, collect samples with an advanced rover, and return those samples to Earth. Detailed analysis of Martian samples in laboratories on Earth could answer many open questions about the red planet, its formation, and its evolution.

It is ironic that at the end of the nineteenth century, the English novelist H. G. Wells envisioned a fleet of Martian spaceships coming to Earth on a mission of conquest. In the twenty-first century, it is we on Earth who are sending robotic spacecraft on a peaceful scientific "invasion" of Mars.

Key Words

1-to-1 spin-orbit coupling, p. 261
3-to-2 spin-orbit coupling, p. 262
cosmic rays, p. 289
crustal dichotomy, p. 273
dust devil, p. 282
favorable opposition, p. 258
flake tectonics, p. 274
greatest eastern elongation, p. 256
greatest western elongation, p. 256
hot-spot volcanism, p. 277
northern lowlands, p. 273
prograde rotation, p. 263
residual polar cap, p. 281
retrograde rotation, p. 263
rift valley, p. 276
runaway greenhouse effect, p. 285
runaway icehouse effect, p. 285
scarp, p. 265
shield volcano, p. 276
southern highlands, p. 273

Key Ideas

Motions of Mercury, Venus, and Mars in Earth's Sky: Mercury and Venus can be seen in the morning or evening sky only, while it is possible to see Mars at any time of night depending on its position in its orbit.

• At their greatest eastern and western elongations, Mercury is only 28° from the Sun and Venus is only 47° from the Sun.

• The best Earth-based views of Mars are obtained at favorable oppositions, when Mars is simultaneously at opposition and near perihelion.

Rotation of Mercury, Venus, and Mars: Poor telescopic views of Mercury's surface led to the mistaken impression that the planet always keeps the same face toward the Sun (1-to-1 spin-orbit coupling).

• Radio and radar observations revealed that Mercury in fact has 3-to-2 spin-orbit coupling: The planet rotates on its axis three times every two orbits.

• Venus rotates slowly in a retrograde direction. Its rotation period is longer than its orbital period.

• Mars rotates at almost the same rate as Earth, and its rotation axis is tilted by almost the same angle as Earth's axis.

Mercury's Surface, Interior, and Magnetic Field: The Mercurian surface is pocked with craters, but there are extensive smooth plains between these craters.

• Long cliffs called scarps meander across the surface of Mercury. These probably formed as the planet's crust cooled, solidified, and shrank.

• Mercury has an iron core with a diameter equal to about $^3/_4$ of the planet's diameter. By contrast, the diameter of Earth's core is only slightly more than $^1/_2$ of Earth's diameter.

• Mercury has a weak magnetic field, which indicates that at least part of the iron core is liquid.

Comparing Venus and Mars: Most of the surface of Venus is at about the same elevation, with just a few elevated regions. On Mars, the southern highlands rise several kilometers above the northern lowlands.

• Venus has a thick atmosphere and a volcanically active surface. Mars has a very thin atmosphere and little or no current volcanism.

• There is no evidence of plate tectonics on either Venus or Mars. On Venus, there is vigorous convection in the planet's interior, but the crust is too thin to move around in plates; instead, it wrinkles and flakes. On Mars, the planet's smaller size means the crust has cooled and become too thick to undergo subduction.

• Volcanoes on both Venus and Mars were produced by hot spots in the planet's interior.

• The entire Venusian surface is about 500 million years old and has relatively few craters. By contrast, most of the Martian surface is cratered and is probably billions of years old. The southern highlands on Mars are the most heavily cratered and hence the oldest part of the planet's surface.

The Atmospheres of Venus and Mars: Both planetary atmospheres are over 95% carbon dioxide, with a few percent of nitrogen.

• The pressure at the surface of Venus is about 90 atmospheres. The greenhouse effect is very strong, which raises the surface temperature to 460°C. The pressure at the surface of Mars is only 0.006 atmosphere, and the greenhouse effect is very weak.

• The permanent high-altitude clouds on Venus are made primarily of sulfuric acid. By contrast, the few clouds in the Martian atmosphere are composed of water ice and carbon dioxide ice.

• The circulation of the Venusian atmosphere is dominated by two huge convection currents in the cloud layers, one in the northern hemisphere and one in the southern hemisphere. The upper cloud layers of the Venusian atmosphere move rapidly around the planet in a retrograde direction, with a period of only about 4 Earth days.

• Weather on Mars is dominated by the north and south flow of carbon dioxide from pole to pole with the changing seasons. This can trigger planetwide dust storms.

Evolution of Atmospheres: Earth, Venus and Mars all began with relatively thick atmospheres of carbon dioxide, water vapor, and sulfur dioxide.

• On Earth, most of the carbon dioxide went into carbonate rocks and most of the water into the oceans. Ongoing plate tectonics recycles atmospheric gases through the crust.

• On Venus, more intense sunlight and the absence of plate tectonics led to a thick carbon dioxide atmosphere and a runaway greenhouse effect.

• On Mars, a runaway icehouse effect resulted from weaker sunlight and the absence of plate tectonics.

Water on Mars: Liquid water cannot exist on present-day Mars because the atmosphere is too thin and cold. But there is evidence for frozen water at the polar ice caps and beneath the surface of the regolith.

• Geological evidence from unmanned rovers shows that much of the Martian surface has been dry for billions of years, but some regions had substantial amounts of water.

The Moons of Mars: Mars has two small, football-shaped satellites that move in orbits close to the surface of the planet. They may be captured asteroids or may have formed in orbit around Mars out of solar system debris.

Questions

Review Questions

1. Why is it impossible to see Mercury or Venus in the sky at midnight?
2. Why are naked-eye observations of Mercury best made at dusk or dawn, while telescopic observations are best made during the day?
3. Why can't you see any surface features on Mercury when it is closest to Earth?
4. Why is it best to view Mars near opposition? Why are some oppositions better than others?
5. Explain why Mars can best be viewed from Earth while it is undergoing retrograde motion.
6. Explain why the photograph in Figure 11-4*b* must have been made during the daytime.
7. In his 1964 science fiction story "The Coldest Place," author Larry Niven described the "dark side" of Mercury as the coldest place in the solar system. What assumption did he

make about the rotation of Mercury? Did this assumption turn out to be correct?

8. What is 3-to-2 spin-orbit coupling? How is the rotation period of an object exhibiting 3-to-2 spin-orbit coupling related to its orbital period? What aspects of Mercury's orbit cause it to exhibit 3-to-2 spin-orbit coupling? What telescopic observations proved this?
9. Why was it so difficult to determine the rate and direction of Venus's rotation? How were these finally determined? What is one proposed explanation for the slow, retrograde rotation of Venus?
10. Explain why Mercury does not have a substantial atmosphere.
11. What kind of surface features are found on Mercury? How do they compare to surface features on the Moon? Why are they probably much older than most surface features on Earth?
12. How do we know that the scarps on Mercury are younger than the lava flows? How can you tell that the scarp in Figure 11-10 is younger than the vertically distorted crater at the center of the figure?
13. If Mercury is the closest planet to the Sun and has such a high average surface temperature, how is it possible that ice might exist on its surface?
14. Why do astronomers think that Mercury has a very large iron core?
15. Why is it surprising that Mercury has a global magnetic field? Why does the 58.646-day rotation period of Mercury imply that the planet can have only a weak magnetic field?
16. How is Mercury's magnetosphere similar to that of Earth? How is it different? Why do you suppose Mercury does not have Van Allen belts?
17. Why was it difficult to determine Venus's surface temperature from Earth? How was this finally determined?
18. The *Mariner 2* spacecraft did not enter Venus's atmosphere, but it was nonetheless able to determine that the atmosphere is very dry. How was this done?
19. Why did Earth observers report that they had seen straight-line features ("canals") on Mars? How did the seasonal winds trick them into thinking that Mars had vegetation?
20. Do Venus and Mars have continents like those on Earth?
21. What is the Martian crustal dichotomy? What is the evidence that the southern highlands are older than the northern lowlands?
22. What is the evidence that the surface of Venus is only about 500 million years old?
23. What is flake tectonics? Why does Venus exhibit flake tectonics rather than plate tectonics?
24. What geologic features (or lack thereof) on Mars have convinced scientists that plate tectonics did not significantly shape the Martian surface?
25. What geologic processes are thought to have created Valles Marineris?
26. Compare the volcanoes of Venus, Earth, and Mars. Cite evidence that hot-spot volcanism is or was active on all three worlds.
27. Describe the evidence that there has been recent volcanic activity (**a**) on Venus and (**b**) on Mars.
28. Suppose all of Venus's volcanic activity suddenly stopped. (**a**) How would this affect Venus's clouds? (**b**) How would this affect the overall Venusian environment?
29. Why are the patterns of convection in the Venusian atmosphere so different from those in our atmosphere?
30. Why is it impossible for liquid water to exist on Mars today? If liquid water existed on Mars in the past, what must have been different then?
31. Why does the atmospheric pressure on Mars vary with the seasons? What is the relationship between this pressure variation and Martian dust storms?
32. Why is it reasonable to assume that the primordial atmospheres of Earth, Venus, and Mars were roughly the same?
33. Carbon dioxide accounts for about 95% of the present-day atmospheres of both Mars and Venus. Why, then, is there a strong greenhouse effect on Venus but only a weak greenhouse effect on Mars?
34. (**a**) What is a runaway greenhouse effect? (**b**) What is a runaway icehouse effect?
35. What is a dust devil? Why would you feel much less breeze from a Martian dust devil than from a dust devil on Earth?
36. (**a**) Why is Mars red? (**b**) Why is the Martian sky the color of butterscotch?
37. Were the *Viking Landers* able to determine whether life currently exists on Mars or whether it once existed there? Why or why not?
38. (**a**) The *Spirit* rover found the minerals olivine and pyroxene at its landing site on Mars. Explain how this shows that there has been no liquid water at that site for billions of years. (**b**) What evidence did the *Opportunity* rover find at its landing site to suggest that liquid water had once been present there?
39. How was the *Mars Odyssey* spacecraft able to detect water beneath the Martian surface without landing on the planet?
40. A full moon on Earth is bright enough to cast shadows. As seen from the Martian surface, would you expect a full Phobos or full Deimos to cast shadows? Why or why not?

Advanced Questions

Problem-solving tips and tools

We discussed the relationship between angular distance and linear distance in Box 1-1 and the idea of angular resolution in Section 6-3. You may need to refresh your memory about Kepler's third law, described in Section 4-4. You may also need to review the form of Kepler's third law that explicitly includes mass (see Section 4-7 and Box 4-4). Box 4-1 describes the relationship between a planet's sidereal orbital period and its synodic period (the time from one inferior conjunction to the next). You should recall that Wien's law (Section 5-4) relates the temperature of a blackbody to λ_{max}, its wavelength of maximum emission. Section 5-9 and Box 5-6 explain the Doppler effect and how to do calculations using it. The linear speed of a point on a planet's equator is the planet's circumference divided by its rotation period; recall that the circumference of a circle of radius r is $2\pi r$. The volume of a sphere of radius r is $4\pi r^3/3$. The speed of light is given in Appendix 7.

41. Figure 11-1 shows Mercury with a greatest eastern elongation of 18° and a greatest western elongation of 28°. On November 25, 2006, Mercury was at a greatest western elongation of 20°. Was Mercury at perihelion, aphelion, or some other point on its orbit? Explain your answer.
42. Venus takes 440 days to move from greatest western elongation to greatest eastern elongation, but it needs only 144 days to go from greatest eastern elongation to greatest western elongation. With the aid of a diagram like Figure 11-1, explain why.
43. Venus's sidereal rotation period is 243.01 days and its orbital period is 224.70 days. Use these data to prove that a solar day on Venus lasts 116.8 days. (*Hint:* Develop a formula relating Venus's solar day to its sidereal rotation period and orbital period similar to the first formula in Box 4-1.)
44. In Section 11-2 we described the relationship between the length of Venus's synodic period and the length of an apparent solar day on Venus. Using this relationship and a diagram, explain why at each inferior conjunction the same side of Venus is turned toward Earth.
45. This time-lapse photograph was taken on May 7, 2003, during a *solar transit* of Mercury. Over a period of 5 hours and 19 minutes, Mercury appeared to move across the face of the Sun. Such solar transits of Mercury occur 13 or 14 times each century; they do *not* happen each time that Mercury is at inferior conjunction. Explain why not. (*Hint*: For a solar transit to occur, the Sun, Mercury, and Earth must be in a nearly perfect alignment. Does the orbit of Mercury lie in the plane of the ecliptic?)

R I V U X G

(Dominique Dierick)

46. Find the largest angular size that Mercury can have as seen from Earth. In order for Mercury to have this apparent size, at what point in its orbit must it be?
47. (a) Suppose you have a telescope with an angular resolution of 1 arcsec. What is the size (in kilometers) of the smallest feature you could have seen on the Martian surface during the opposition of 2005, when Mars was 0.464 AU from Earth? (b) Suppose you had access to the Hubble Space Telescope, which has an angular resolution of 0.1 arcsec. What is the size (in kilometers) of the smallest feature you could have seen on Mars with the HST during the 2005 opposition?
48. For a planet to appear to the naked eye as a disk rather than as a point of light, its angular size would have to be 1 arcmin, or 60 arcsec. (This is the same as the angular separation between lines in the bottom row of an optometrist's eye chart.) (a) How close would you have to be to Mars in order to see it as a disk with the naked eye? Does Mars ever get this close to Earth? (b) Would Earth ever be visible as a disk to an astronaut on Mars? Would she be able to see Earth and the Moon separately, or would they always appear as a single object? Explain your answers.
49. Using Figure 11-3, explain why oppositions of Mars are most favorable when they occur in August.
50. Imagine that you are part of the scientific team monitoring a spacecraft that has landed on Mars. At 5:00 P.M. in your control room on Earth, the spacecraft reports that the Sun is highest in the sky as seen from its location on Mars. When the Sun is next at its highest point as seen from the spacecraft, what time will it be in the control room on Earth?
51. (a) Mercury has a 58.646-day rotation period. What is the speed at which a point on the planet's equator moves due to this rotation? (*Hint:* Remember that speed is distance divided by time. What distance does a point on Mercury's equator travel as the planet makes one rotation?) (b) Use your answer to (a) to answer the following: As a result of rotation, what difference in wavelength is observed for a radio wave of wavelength 12.5 cm (such as is actually used in radar studies of Mercury) emitted from either the approaching or receding edge of the planet?
52. The orbital period of *Mariner 10* is twice that of Mercury. Use this fact to calculate the length of the semimajor axis of the spacecraft's orbit.
53. Much of the area shown in Figure 11-10 is pockmarked with small impact craters. By contrast, there are very few small craters within the large crater at the upper right of this figure. Explain why this suggests that the interior of this crater was flooded with lava while the area outside the crater was not.
54. Consider the idea that Mercury has a solid iron-bearing mantle that is permanently magnetized like a giant bar magnet. Using the fact that iron demagnetizes at temperatures above 770°C, present an argument against this explanation of Mercury's magnetic field.
55. (a) At what wavelength does Venus's surface emit the most radiation? (b) Do astronomers have telescopes that can detect this radiation? (c) Why can't we use such telescopes to view the planet's surface?
56. The *Mariner 2* spacecraft detected more microwave radiation when its instruments looked at the center of Venus's disk than when it looked at the edge, or limb, of the planet. (This effect is called *limb darkening.*) Explain how these observations show that the microwaves are emitted by the planet's surface rather than its atmosphere.
57. In the classic Ray Bradbury science-fiction story "All Summer in a Day," human colonists on Venus are subjected to con-

tinuous rainfall except for one day every few years when the clouds part and the Sun comes out for an hour or so. Discuss how our understanding of Venus's atmosphere has evolved since this story was first published in 1954.

58. Suppose that Venus had no atmosphere at all. How would the albedo of Venus then compare with that of Mercury or the Moon? Explain your answer.
59. A hypothetical planet has an atmosphere that is opaque to visible light but transparent to infrared radiation. How would this affect the planet's surface temperature? Contrast and compare this hypothetical planet's atmosphere with the greenhouse effect in Venus's atmosphere.
60. Earth's northern hemisphere is 39% land and 61% water, while its southern hemisphere is only 19% land and 81% water. Thus, the southern hemisphere could also be called the "water hemisphere." The Moon also has two distinct hemispheres, the near side (which has a number of maria) and the far side (which has almost none). How are these hemispheric differences on Earth and on the Moon similar to the Martian crustal dichotomy? How are they different?
61. For a group of properly attired astronauts equipped with oxygen tanks, a climb to the summit of Olympus Mons would actually be a relatively easy (albeit long) hike rather than a true mountain climb. Give two reasons why.
62. On Mars, the difference in elevation between the highest point (the summit of Olympus Mons) and the lowest point (the bottom of the Hellas Planitia basin) is 30 km. On Earth, the corresponding elevation difference (from the peak of Mount Everest to the bottom of the deepest ocean) is only 20 km. Discuss why the maximum elevation difference is so much greater on Mars.
63. The *Mars Global Surveyor* (MGS) spacecraft is in a nearly circular orbit around Mars with an orbital period of 117 minutes. (**a**) Using the data in Table 11-3, find the radius of the orbit. (**b**) What is the average altitude of MGS above the Martian surface? (**c**) The orbit of MGS passes over the north and south poles of Mars. Explain how this makes it possible for the spacecraft to observe the entire surface of the planet.
64. The elevations in Figure 11-20 were measured using an instrument called MOLA (Mars Orbiter Laser Altimeter) on board *Mars Global Surveyor*. MOLA fires a laser beam downward, then measures how long it takes for the beam to return to the spacecraft after reflecting off the surface. Suppose MOLA measures this round-trip time for the laser beam to reflect off the summit of Olympus Mons, as well as the round-trip time to reflect off the bottom of Hellas Planitia (see Question 61). Which round-trip time is longer? How much longer is it? Do you need to know the distance from *Mars Global Surveyor* to the surface to answer this question?
65. (**a**) The Grand Canyon in Arizona was formed over 15 to 20 million years by the flowing waters of the Colorado River, as well as by rain and wind. Contrast this formation scenario to that of Valles Marineris on Mars. (**b**) Valles Marineris is sometimes called "the Grand Canyon of Mars." Is this an appropriate description? Why or why not?
66. Water has a density of 1000 kg/m^3, so a column of water n meters tall and 1 meter square at its base has a mass of $n \times$ 1000 kg. On either Earth or Venus, which have nearly the same surface gravity, a mass of 1 kg weighs about 9.8 newtons (2.2 lb). Calculate how deep you would have to descend into Earth's oceans for the pressure to equal the atmospheric pressure on Venus's surface, 90 atm or 9×10^6 newtons per square meter. Give your answer in meters.
67. Marine organisms produce sulfur-bearing compounds, some of which escape from the oceans into Earth's atmosphere. (These compounds are largely responsible for the characteristic smell of the sea.) Even more sulfurous gases are injected into our atmosphere by the burning of sulfur-rich fossil fuels, such as coal, in electric power plants. Both of these processes add more sulfur compounds to the atmosphere than do volcanic eruptions. On lifeless Venus, by contrast, volcanoes are the only source for sulfurous atmospheric gases. Why, then, are sulfur compounds so much rarer in our atmosphere than in the Venusian atmosphere?
68. The classic 1950 science-fiction movie *Rocketship X-M* shows astronauts on the Martian surface with oxygen masks for breathing but wearing ordinary clothing. Would this be a sensible choice of apparel for a walk on Mars? Why or why not?
69. Suppose that the only information you had about Mars was the images of the surface in Figure 11-31. Describe at least two ways that you could tell from these images that Mars has an atmosphere.
70. Although the *Viking Lander 1* and *Viking Lander 2* landing sites are 6500 km apart and have different geologic histories, the chemical compositions of the dust at both sites are nearly identical. (**a**) What does this suggest about the ability of the Martian winds to transport dust particles? (**b**) Would you expect that larger particles such as pebbles would also have identical chemical compositions at the two *Viking Lander* sites? Why or why not?
71. Why do you suppose that Phobos and Deimos are not round like our Moon?
72. The orbit of Phobos has a semimajor axis of 9378 km. Use this information and the orbital period given in the text to calculate the mass of Mars. How does your answer compare with the mass of Mars given in Table 11-3?
73. Calculate the angular sizes of Phobos and Deimos as they pass overhead, as seen by an observer standing on the Martian equator. How do these sizes compare with that of the Moon seen from Earth's surface? Would Phobos and Deimos appear as impressive in the Martian sky as the Moon does in our sky?
74. You are to put a spacecraft into a synchronous circular orbit around the Martian equator, so that its orbital period is equal to the planet's rotation period. Such a spacecraft would always be over the same part of the Martian surface. (**a**) Find the radius of the orbit and the altitude of the spacecraft above the Martian surface. (**b**) Suppose Mars had a third moon that was in a synchronous orbit. Would tidal forces make this moon tend to move toward Mars, away from Mars, or neither? Explain your answer.

Discussion Questions

75. Before about 350 B.C., the ancient Greeks did not realize that Mercury seen in the morning sky (which they called Apollo) and seen in the evening sky (which they called Hermes) were

actually the same planet. Discuss why you think it took some time to realize this.

76. If you were planning a new mission to Mercury, what features and observations would be of particular interest to you?
77. What evidence do we have that the surface features on Mercury were not formed during recent geological history?
78. Describe the apparent motion of the Sun during a "day" on Venus relative to (**a**) the horizon and (**b**) the background stars. (Assume that you can see through the cloud cover.)
79. If you could examine rock samples from the surface of Venus, would you expect them to be the same as rock samples from Earth? Would you expect to find igneous, sedimentary, and metamorphic rocks like those found on Earth (see Section 9-3)? Explain your answers.
80. In 1978 the *Pioneer Venus Orbiter* spacecraft arrived at Venus. It carried an ultraviolet spectrometer to measure the chemical composition of the Venusian atmosphere. This instrument recorded unexpectedly high levels of sulfur dioxide and sulfuric acid, which steadily declined over the next several years. Discuss how this observation suggests that volcanic eruptions occurred on Venus not long before *Pioneer Venus Orbiter* arrived there.
81. The total cost of the *Mars Global Surveyor* mission was about $154 million. (To put this number into perspective, in 2000 the U.S. Mint spent about $40 million to advertise its new $1 coin, which failed to be accepted by the public. Several recent Hollywood movies have had larger budgets than *Mars Global Surveyor*.) Does this expenditure seem reasonable to you? Why or why not?

Web/eBook Questions

ANIMATION 11.1

82. **Elongations of Mercury.** Access the animation "Elongations of Mercury" in Chapter 11 of the *Universe* Web site or eBook. (**a**) View the animation and notice the dates of the greatest eastern and greatest western elongations. Which time interval is greater: from a greatest eastern elongation to a greatest western elongation, or vice versa? (**b**) Based on what you observe in the animation, draw a diagram to explain your answer to the question in (a).
83. Just as Mercury can pass in front of the Sun as seen from Earth (see Question 45), so can Venus. Transits of Venus are quite rare. The dates of the only transits in the twenty-first century are June 8, 2004, and June 6, 2012; the next ones will occur in 2117 and 2125, A number of European astronomers traveled to Asia and the Pacific islands to observe the transits of Venus in 1761 and 1769. Search the World Wide Web for information about these expeditions. Why were these events of such interest to astronomers? How definitive were the results of these observations?
84. Search the World Wide Web for information about possible manned missions to Mars. How long might such a mission take? How expensive would such a project be? What would be the advantages of a manned mission compared to an unmanned one?

ANIMATION 11.2

85. **Conjunctions of Mars.** Access and view the animation "The Orbits of Earth and Mars" in Chapter 11 of the *Universe* Web site or eBook. (**a**) The animation highlights three dates when Mars is in opposition, so that Earth lies directly between Mars and the Sun. By using the "Stop" and "Play" buttons in the animation, find two times during the animation when Mars is in *conjunction,* so that the Sun lies directly between Mars and Earth (see Figure 4-6). For each conjunction, make a drawing showing the positions of the Sun, Earth, and Mars, and record the month and year when the conjunction occurs. (*Hint:* See Figure 11-3.) (**b**) When Mars is in conjunction, at approximately what time of day does it rise as seen from Earth? At what time of day does it set? Is Mars suitably placed for telescopic observation when in conjunction?

Activities

Observing Projects

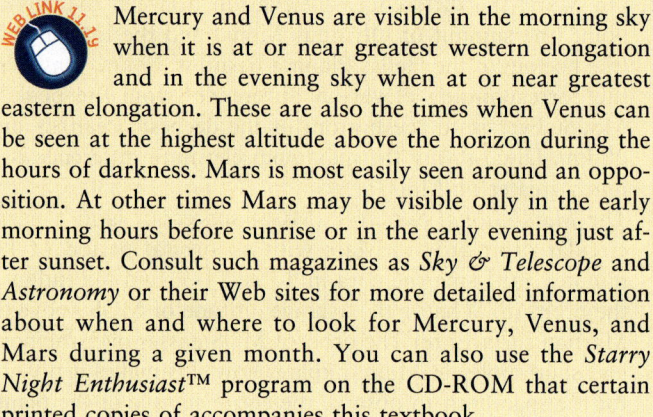

Observing tips and tools

WEB LINK 11.19

Mercury and Venus are visible in the morning sky when it is at or near greatest western elongation and in the evening sky when at or near greatest eastern elongation. These are also the times when Venus can be seen at the highest altitude above the horizon during the hours of darkness. Mars is most easily seen around an opposition. At other times Mars may be visible only in the early morning hours before sunrise or in the early evening just after sunset. Consult such magazines as *Sky & Telescope* and *Astronomy* or their Web sites for more detailed information about when and where to look for Mercury, Venus, and Mars during a given month. You can also use the *Starry Night Enthusiast*™ program on the CD-ROM that certain printed copies of accompanies this textbook.

WEB LINK 11.5

86. Refer to the *Universe* Web site or eBook for a link to a Web site that calculates the dates of upcoming greatest elongations of Mercury. Consult such magazines as *Sky & Telescope* and *Astronomy,* or the Web sites for these magazines, to determine if any of these greatest elongations is going to be a favorable one. If so, make plans to be one of those rare individuals who has actually seen the innermost planet of the solar system. Set aside several evenings (or mornings) around the date of the favorable elongation to reduce the chances of being "clouded out." Select an observing site that has a clear, unobstructed view of the horizon where the Sun sets (or rises). If possible, make arrangements to have a telescope at your disposal. Search for the planet on the dates you have selected, and make a drawing of its appearance through your telescope.
87. *This observing project should be performed only under the direct supervision of an astronomer who knows how to point a telescope safely at Mercury.* Make arrangements to view

Mercury during broad daylight. This is best done by visiting an observatory where the coordinates (right ascension and declination) of Mercury's position can be used to point the telescope. **DO NOT LOOK AT THE SUN! Looking directly at the Sun can cause blindness.**

88. Refer to the *Universe* Web site or eBook for a link to a Web site that calculates the dates of greatest elongations of Venus. If Venus is near a greatest elongation, view the planet through a telescope. Make a sketch of the planet's appearance. From your sketch, can you determine if Venus is closer to us or farther from us than the Sun?

89. Using a small telescope, observe Venus once a week for a month and make a sketch of the planet's appearance on each occasion. From your sketches, can you determine whether Venus is approaching us or moving away from us?

90. *This observing project should be performed only under the direct supervision of an astronomer who knows how to point a telescope safely at Venus.* Make arrangements to view Venus during broad daylight. This is best done by visiting an observatory where the coordinates (right ascension and declination) of Venus's position can be used to point the telescope. **DO NOT LOOK AT THE SUN! Looking directly at the Sun can cause blindness.**

91. If Mars is suitably placed for observation, arrange to view the planet through a telescope. Draw a picture of what you see. What magnifying power seems to give you the best image? Can you distinguish any surface features? Can you see a polar cap or dark markings? If not, can you offer an explanation for Mars's bland appearance?

92. Use the *Starry Night Enthusiast*™ program to observe solar transits of Mercury (see Question 45) when the planet passes in front of the Sun as viewed from Earth. Display the entire celestial sphere (select **Guides > Atlas** in the **Favorites** menu). Open the **Find** pane and click the menu button for Mercury and select **Centre** from the menu. Using the controls at the right-hand end of the Toolbar, zoom in until the field of view is about 1° × 1°. (**a**) In the toolbar, stop time flow and set the date and time to November 8, 2009, at 12:00:00 A.M. Set the **Time Flow Rate** to **1 hour.** Step backward or forward through time using the single-step buttons (the leftmost and the rightmost buttons) and record the times at which the solar transit begins and ends. You can zoom in to a field of view of about 11 × 8 arc minutes centered upon Mercury and change the **Time Flow Rate** to **1 minute** and use the **Time Flow** controls in the toolbar to move back and forth in time to increase the accuracy of your measurement. What is the total duration of the solar transit? (**b**) Set the date and time to May 9, 2016, at 12:00:00 A.M. Again, step backward or forward through time, record the times when the solar transit begins and ends, and find the total duration of the solar transit. (**c**) The maximum duration of a solar transit of Mercury is 9 hours. Explain why the transits you observed in (a) and (b) last a substantially shorter time.

93. Use the *Starry Night Enthusiast*™ program to examine Mercury. Select **Favorites > Solar System > Mercury** from the menu. Select **View > Feet** from the menu to remove the image of the astronaut's spacesuit from the view. Select **Options > Solar System > Planets-Moons. . .** from the menu to allow you to examine the complete surface of the planet. In the Planets-Moons Options dialog box, slide the control next to the label **Show dark side** all the way to the right (**Brighter**) end of the scale. Click **OK** to close the dialog box. Use the **Zoom** controls in the toolbar to zoom in and out on the view. Rotate the image of Mercury by placing the mouse cursor over the image of the planet, holding down the mouse button, and moving the mouse. (On a two-button mouse, hold down the left mouse button.) As you explore the surface of the planet Mercury, estimate the diameter of the largest craters by measuring their size on the screen with a ruler and comparing to the diameter of Mercury (see Table 11-1). (Note that the rosette patterns surrounding both planetary poles are an artifact of the technique used to produce these images).

94. Use the *Starry Night Enthusiast*™ program to observe solar transits of Venus (see Question 83). Display the entire celestial sphere (select **Guides > Atlas** in the **Favorites** menu). Open the **Find** pane and click the menu button for Venus (the downward-pointing blue arrow to the left of Venus) in the list. Select **Centre** from the menu. Use the zoom controls in the toolbar to adjust the field of view to about 1° × 1°. (**a**) In the toolbar, **Stop time flow** and then set the **Time** and **Date** to June 8, 2004, at 12:00:00 A.M. Set the **Time Flow Rate** to **1 hour.** Step backward or forward through time using the single-step buttons (the leftmost and the rightmost buttons) and record the times at which the solar transit begins and ends, changing the Time Flow Rate to 1 minute and the field of view to increase the accuracy of your measurement, as necessary. What is the total duration of the solar transit? (**b**) The ecliptic appears in *Starry Night Enthusiast*™ as a green line. During the transit, is Venus precisely on the ecliptic? If not, about how far off is it? (*Hint:* The Sun has an angular diameter of about 30 arcmin.) (**c**) Repeat parts (a) and (b) for the solar transit of Venus on June 6, 2012.

95. Use the *Starry Night Enthusiast*™ program to observe the apparent motion of Venus on the celestial sphere. Display the entire celestial sphere (select **Guides > Atlas** in the **Favorites** menu). Open the **Find** pane and click the menu button in the list to the left of the label for the Sun. Select **Centre** from the menu that appears. Using the controls at the right-hand end of the toolbar, zoom out until the field of view is 100°. **Stop** Time Flow and in the toolbar, set the date and time to January 1, 2007, at 12:00:00 A.M. and the **Time Flow Rate** to **1 day.** (**a**) Use the **Run Time Forward** and **Stop time** buttons to find the first date after January 1, 2007, when Venus is as far to the right of the Sun as possible, and the first date after January 1, 2007, when Venus is as far to the left of the Sun as possible. What is your interpretation of these two dates and how would you label them? (**b**) Set the date to December 1, 2007, and start the animation by clicking on the **Run Time Forward** button. Based on your observations, explain why Venus has neither a greatest western elongation nor a greatest eastern elongation during 2008.

96. Use the *Starry Night Enthusiast*™ program to compare the orbits of Venus and Earth. Select **Options > Viewing Location. . .** from the menu. In the Viewing Location dialog box, set the **View from** to a **position moving with the Sun** and

choose the option **Above orbital plane.** Then click the **Set Location** button to close the dialog. Click on and hold the **Increase current elevation** button in the Viewing Location section of the toolbar until the distance shown in the Viewing Location display pane is approximately **1.5 au** from the Sun. Open the **Find** pane and click both of the checkboxes on either side of the listing for Earth and for Venus. This labels the two planets and draws their orbits in the view. Close the Find pane. **Stop** time flow and set the time and the date to 0:00:00 UT on August 19, 2007 AD. You can zoom in and zoom out on these two planets and their orbits using the buttons in the **Zoom** section of the toolbar. You can also rotate the solar system by holding down the **Shift** key and then holding down the mouse button and moving the mouse. (On a two-button mouse, hold down the left mouse button.) Are the orbits of Venus and of Earth in the same plane? At the time shown in the image, is Venus nearest to inferior conjunction, superior conjunction, greatest eastern elongation, or greatest western elongation as seen from Earth? Explain your answers. Rotate your view to look down upon the orbits from above the pole of the Sun. Are the orbits of Earth and Venus circular?

97. Use the *Starry Night Enthusiast*™ program to observe the appearance of Mars. Select **Favorites > Guides > Atlas** from the menu. Select **View > Celestial Grid** from the menu to turn this option off. Open the **Find** pane and click the menu button for **Mars** and choose **Centre** from the menu. Close the Find pane and then use the Zoom controls in the toolbar to set a field of view of approximately 58″ × 40″. (**a**) Set the **Time Flow Rate** to **1 hour** and then run **Time Forward.** Describe what you see. (**b**) **Stop** time flow. Change the **Time Flow Rate** to **1 lunar month. Run Time Forward** again. Describe what you see. Using a diagram like Figure 4-6, explain the changes in the apparent size of the planet. (**c**) **Stop** time flow and **zoom out** to a field of view of approximately **2′ × 1′**. Change the time and date in the toolbar to 12:00:00 A.M. on August 28, 2003, to see Mars during a very favourable opposition. You will see Mars and its two moons, Phobos and Deimos. In the toolbar, set the **Time Flow Rate** to **1 minute.** Record the date and time in the display, and note the position of Phobos (the inner moon). Click the **Run Time Forward** and single time step button (the rightmost time control button) to advance time until Phobos returns to approximately the same position relative to Mars. Record the date and time in the display. From your observations, what is the orbital period of Phobos? How does your result compare with the orbital period given in Appendix 3? (**d**) Repeat part (c) for Deimos (the outer moon).

98. Use the *Starry Night Enthusiast*™ program to compare the orbits of Mars and Earth. **Select Options > Viewing Location. . .** from the menu. In the Viewing Location dialog box, set the **View from** to a **position moving with** the **Sun** and choose the option **Above orbital plane** and click the **Set Location** button to close the dialog. Click on and hold the **Increase current elevation** button in the Viewing Location section of the toolbar until the distance shown in the Viewing Location display pane is approximately **3.5 au** from the Sun. Open the **Find** pane and click the two checkboxes on either side of the listings for **Earth** and for **Mars** to label these two planets and draw their orbits in the view. Close the Find pane. You can zoom in and zoom out using the buttons in the **Zoom** section of the toolbar. You can also rotate the solar system by holding down the **Shift** key and then holding down the mouse button and moving the mouse. (On a two-button mouse, hold down the left mouse button.) Stop time flow and set the time and date to 0:00:00 UT on October 16, 2005 AD. (**a**) Is Mars a few weeks before an opposition or a few weeks after an opposition at this time on this date? (*Hint:* Arrowheads on the orbits show you which way each planet moves around its orbit.) (**b**) Are the orbits of Mars and of Earth in the same plane? Use your answer to explain the shape of the apparent path of Mars on the celestial sphere in 2011–2012, shown in Figure 4-2.

Collaborative Exercises

99. Figures 11-6*b* and 11-6*c* show a planet in synchronous rotation and Mercury with a 3-to-2 spin-orbit coupling, respectively. Stand up and demonstrate how planets move in each of these rotations by "orbiting" around a stationary classmate who represents our Sun. How would Mercury's motion be different if it had a 4-to-2 spin-orbit coupling instead?

100. In the nineteenth century, French mathematician and astronomer U. J. J. Le Verrier led a failed search for a hypothetical planet named Vulcan (after the mythical blacksmith of the gods) orbiting closer to our Sun than Mercury. If Vulcan had an orbit with the same eccentricity as Mercury's orbit but only one-half the size, what would have been its maximum eastern and western elongations?

101. Figure 11-7 shows prograde and retrograde rotation. Stand up and demonstrate how Venus rotates as it orbits our Sun, using a classmate as our stationary Sun. How is this different from how Earth rotates in its orbit?

102. The image of Mars that opens this chapter is from the Hubble Space Telescope. Draw a circle on your paper roughly 5 cm in diameter and, taking turns, have each person in your group sketch a different region of Mars. How is your collaborative sketch different than Schiaparelli's drawing shown in Figure 11-16?

Father of Spirit and Opportunity

With the success of twin rovers on the Red Planet, Steven W. Squyres and his team are showing how to conduct robotic missions–and setting the stage for human exploration BY DAVID APPELL

(from David Appell, "Father of Spirit and Opportunity," *Scientific American*, October 2004, 44, 46)

It's 8 A.M. at NASA's Jet Propulsion Laboratory, and the Mars Opportunity rover team is gathering in a conference room for its daily science kickoff meeting. For "Martian sol 149"—the 149th day on the Red Planet since the start of the rover missions—it is assembling the minute-by-minute plan of what the rover will do: A little spectroscopy. "Ratting" a few rocks—that is, drilling them with the rover's Rock Abrasion Tool, or RAT. "I'm interested in knowing if this stuff is red or not," says Steven W. Squyres of the rocks that the rover is currently rolling over.

A professor of astronomy at Cornell University, Squyres, 48, is the principal investigator for the Mars Exploration Team, which consists of 170 members. He is responsible for all the scientific activities of both the Opportunity and Spirit rovers, leading colleague John Grotzinger to liken him to a "flea on a hot griddle," with his hands in everything.

On this late June day in 2004, the Opportunity team is facing a critical decision about whether to go farther down into Endurance Crater, a 20-meter-deep hole near the landing site on the Meridiani Planum. The team members have been eyeing the crater for weeks. But first they have to figure out if Opportunity can negotiate an inclined rock step right in front of it that drops 30 centimeters along a 25-degree incline. The step is slightly above the rover's design limits, but everyone on the team wants to do it if they can.

"Welcome to the monster truck convention," jokes Opportunity mission manager Matt Wallace at a 10 A.M. traversability meeting. A test rover at the JPL Mars yard successfully went up and down a mock-up of the step; slight slippage and a minor wheelie were the only hitches. After some discussion, the engineering team decides to go ahead and send Opportunity down the step. They will drill several RAT holes on the way in case the rover cannot climb back up, Squyres says.

Departing from the way it normally does business, NASA entrusted the science package of the Mars rovers to a single individual. And unlike many earlier principal investigators, Squyres was heavily involved in the design and engineering of the rovers. In fact, he has been working on them in one form or another since 1987, when he developed the panoramic camera now in use on the rovers. In 1992 he assembled the RAT as well as spectrometers to measure thermal emissions and to detect iron and other chemicals in rocks. Then he spent the next few years writing proposals for the rover project. NASA selected Squyres and his team in 1997, and the Mars Exploration Rover (MER) project came together in its final form in July 2000 [see "The Spirit of Exploration," by George Musser; *Scientific American,* March].

The Spirit and Opportunity landers have proved to be the most complex robotic mission NASA has ever attempted. They have performed beautifully, operating for at least twice their projected lifetimes. Barring mechanical failure—and Spirit has lately been dragging around a bad wheel—the rovers' lives are limited by the buildup of dust on their solar arrays, which cuts the electricity being generated. (Over both rovers, the power loss is about 0.15 percent per sol, slightly less than the 0.18 percent loss per sol of Pathfinder, the rover that landed on Mars in 1997.) With Mars now entering winter, the arrays are taking longer to recharge the batteries, which will most likely result in extended sleep cycles for the vehicles. But there is hope the rovers will survive the Martian winter.

The Mars rovers are about 20 light-minutes from Earth, too far for any type of real-time joystick control. So they are programmed to run one Martian day at a time between code uploads. In the afternoon Squyres sits with the computer jocks who translate the agreedon strategy into the precise series of wheel and arm movements that will accomplish the tasks. It is tight, detailed code that calls for an in-depth understanding of the vehicles—the kind of difficult day-to-day engineering that Squyres has lived through since the rovers landed in January 2004. Although the rovers have performed beyond anyone's expectations, some small problems have arisen: a heating-element switch remains stuck open on Opportunity, and Spirit had software problems just after emerging from its landing cocoon.

But neither problem has detracted from the science that's been done. "It's fair to say that the rovers would certainly have not been as successful as they are and possibly would never even have happened if it wasn't for Steve," says Jim Bell, lead scientist for the panoramic cameras on both rovers

continued

and a close colleague of Squyres' at Cornell. To him, Squyres has set the example for those leading future missions. You not only "have to be a top-notch scientist," Bell states, "but you also have to be willing to get head over heels into the design of the instruments."

The most important discovery is the evidence for great amounts of water at Meridiani Planum. Opportunity detected, for instance, sulfate salts and hematite concretions—small, grayish, iron-containing spherules that scientists have been calling "blueberries." And as Squyres had wondered, shavings of the ratted rock in the crater were indeed brick-red, typical of hematite when it is pulverized. Sulfate and hematite are left in rocks by water, so they suggest that Opportunity is working on what was once the shoreline of a salty sea, although clues gathered so far do not indicate how long, or how long ago, liquid water covered the area. "Not only did we find evidence for a habitable environment at Meridiani," Squyres remarks, "but we've got these wonderful geologic deposits—sulfates and in particular the hematite concretions—that are very good at preserving evidence of whether there was interesting organic chemistry, whether there was life."

It's unlikely the rovers themselves will directly discover signs of life, though. Instead they are laying the groundwork for a sample-return mission, by robots or someday by humans. Squyres is all in favor of a manned mission to Mars. "I'm a huge fan of sending robots to Mars, obviously—that's what I do for a living. But even I believe that the best exploration, the most comprehensive, the most inspiring exploration is going to be conducted by humans."

Although some argue that NASA's many successful robotic missions prove that costly human flights are unnecessary, Squyres doesn't buy it. "I think people who would point to the successes of these two rovers as evidence that you don't need to send humans to Mars are missing the point entirely. I view our rovers not as competitors to humans but as precursors. And they're precursors in the sense of telling us more about the Martian surface and what it's like, what it takes to walk across it, build on it, launch from it, that kind of thing." A human base on Meridiani Planum, now being explored by Squyres and his MER team, would be a good place to start.

David Appell is based in Newmarket, N.H.

12

Jupiter and Saturn: Lords of the Planets

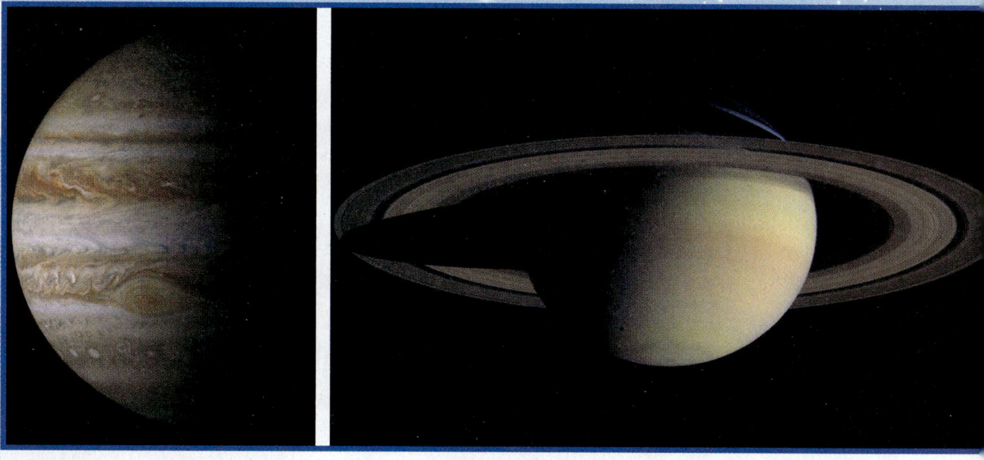

WEB LINK 12.1

R I V U X G

Jupiter (left) and Saturn (right) to scale. (NASA/JPL/Space Science Institute)

Among the most remarkable sights in our solar system are the colorful, turbulent atmosphere of Jupiter and the ethereal beauty of Saturn's rings. Both of these giant worlds dwarf our own planet: More than 1200 Earths would fit inside Jupiter's immense bulk, and more than 700 Earths inside Saturn. Unlike Earth, both Jupiter and Saturn are composed primarily of the lightweight elements hydrogen and helium, in abundances very similar to those in the Sun.

As these images show, the rapid rotations of both Jupiter and Saturn stretch their weather systems into colorful bands that extend completely around each planet. Both planets also have intense magnetic fields that are generated in their interiors, where hydrogen is so highly compressed that it becomes a metal.

Saturn is not merely a miniature version of Jupiter, however. Its muted colors and more flattened shape are clues that Saturn's atmosphere and interior have important differences from those of Jupiter. The most striking difference is Saturn's elaborate system of rings, composed of countless numbers of icy fragments orbiting in the plane of the planet's equator. The rings display a complex and elegant structure, which is shaped by subtle gravitational influences from Saturn's retinue of moons. Jupiter, too, has rings, but they are made of dark, dustlike particles that reflect little light. These systems of rings, along with the distinctive properties of the planets themselves, make Jupiter and Saturn highlights of our tour of the solar system.

12-1 Jupiter and Saturn are the most massive planets in the solar system

Jupiter and Saturn are respectively the largest and second largest of the planets, and the largest objects in the solar system other

Learning Goals

By reading the sections of this chapter, you will learn

12-1 What observations from Earth reveal about Jupiter and Saturn

12-2 How Jupiter and Saturn rotate differently from terrestrial planets like Earth

12-3 The nature of the immense storms seen in the clouds of Jupiter and Saturn

12-4 How the internal heat of Jupiter and Saturn drives activity in their atmospheres

12-5 What the *Galileo* space probe revealed about Jupiter's atmosphere

12-6 How the shapes of Jupiter and Saturn indicate the sizes of their rocky cores

12-7 How Jupiter and Saturn's intense magnetic fields are produced by an exotic form of hydrogen

12-8 The overall structure and appearance of Saturn's ring system

12-9 What kinds of particles form the rings of Jupiter and Saturn

12-10 How spacecraft observations revealed the intricate structure of Saturn's rings

12-11 How Saturn's rings are affected by the presence of several small satellites

Table 12-1 Jupiter Data

Average distance from the Sun:	5.203 AU = 7.783×10^8 km
Maximum distance from the Sun:	5.455 AU = 8.160×10^8 km
Minimum distance from the Sun:	4.950 AU = 7.406×10^8 km
Eccentricity of orbit:	0.048
Average orbital speed:	13.1 km/s
Orbital period:	11.86 years
Rotation period:	$9^h\ 50^m\ 28^s$ (equatorial) $9^h\ 55^m\ 29^s$ (internal)
Inclination of equator to orbit:	3.12°
Inclination of orbit to ecliptic:	1.30°
Diameter:	142,984 km = 11.209 Earth diameters (equatorial) 133,708 km = 10.482 Earth diameters (polar)
Mass:	1.899×10^{27} kg = 317.8 Earth masses
Average density:	1326 kg/m^3
Escape speed:	60.2 km/s
Surface gravity (Earth = 1):	2.36
Albedo:	0.44
Average temperature at cloudtops:	−108°C = −162°F = 165 K
Atmospheric composition (by number of molecules):	86.2% hydrogen (H_2), 13.6% helium (He), 0.2% methane (CH_4), ammonia (NH_3), water vapor (H_2O), and other gases

(NASA/JPL)

than the Sun itself. Astronomers have known for centuries about the huge diameters and immense masses of these two planets. Given the distance to each planet and their angular sizes, they used the small-angle formula (Box 1-1) to calculate that Jupiter is about 11 times larger in diameter than Earth, while Saturn's diameter is about 9 times larger than Earth's (see Figure 7-2). By observing the orbits of Jupiter's four large moons and Saturn's large moon Titan (see Table 7-2) and applying Newton's form of Kepler's third law (see Section 4-7 and Box 4-4), astronomers also determined that Jupiter and Saturn are, respectively, 318 times and 95 times more massive than Earth. In fact, Jupiter has $2\frac{1}{2}$ times the combined mass of all the other planets, satellites, asteroids, meteoroids, and comets in the solar system. A visitor from interstellar space might well describe our solar system as the Sun, Jupiter, and some debris! Table 12-1 and Table 12-2 list some basic data about Jupiter and Saturn.

Observing Jupiter and Saturn

As for any planet whose orbit lies outside Earth's orbit, the best time to observe Jupiter or Saturn is when they are at opposition. At opposition, Jupiter can appear nearly 3 times brighter than Sirius, the brightest star in the sky. Only the Moon and Venus

Table 12-2 Saturn Data

Average distance from the Sun:	9.572 AU = 1.432×10^9 km
Maximum distance from the Sun:	10.081 AU = 1.508×10^9 km
Minimum distance from the Sun:	9.063 AU = 1.356×10^9 km
Eccentricity of orbit:	0.053
Average orbital speed:	9.64 km/s
Orbital period:	29.37 years
Rotation period:	$10^h\ 13^m\ 59^s$ (equatorial) $10^h\ 39^m\ 25^s$ (internal)
Inclination of equator to orbit:	26.73°
Inclination of orbit to ecliptic:	2.48°
Diameter:	120,536 km = 9.449 Earth diameters (equatorial) 108,728 km = 8.523 Earth diameters (polar)
Mass:	5.685×10^{26} kg = 95.16 Earth masses
Average density:	687 kg/m^3
Escape speed:	35.5 km/s
Surface gravity (Earth = 1):	0.92
Albedo:	0.46
Average temperature at cloudtops:	−180°C = −292°F = 93 K
Atmospheric composition (by number of molecules):	96.3% hydrogen (H_2), 3.3% helium (He), 0.4% methane (CH_4), ammonia (NH_3), water vapor (H_2O), and other gases

RIVUXG (STScI/Hubble Heritage Team)

can outshine Jupiter at opposition. Saturn's brightness at opposition is only about 1/7 that of Jupiter, but it still outshines all of the stars except Sirius and Canopus.

Through a telescope, Jupiter at opposition presents a disk nearly 50 arcsec in diameter, approximately twice the angular diameter of Mars under the most favorable conditions (Figure 12-1*a*). Because Jupiter takes almost a dozen Earth years to orbit the Sun, it appears to meander slowly across the 12 constellations of the zodiac at the rate of approximately one constellation per year. Successive oppositions occur at intervals of about 13 months.

Saturn's orbital period is even longer, more than 29 Earth years, so it moves even more slowly across the zodiac. If Saturn did not move at all, successive oppositions would occur exactly one Earth year apart; thanks to its slow motion, Saturn's oppositions occur at intervals of about one year and two weeks. At opposition Saturn appears as a disk about 20 arcsec in diameter, with a dramatic difference from Jupiter: Saturn is surrounded by a magnificent system of rings (Figure 12-1*b*). Later in this chapter we will examine these rings in detail.

Even a small telescope reveals colored bands on Jupiter and rings around Saturn

(a)

(b)

Figure 12-1 R I V U X G

Jupiter and Saturn as Viewed from Earth **(a)** The disk of Jupiter at opposition appears about $2\frac{1}{2}$ times larger than **(b)** the disk of Saturn at opposition. Both planets display dark and light bands, though these are fainter on Saturn. (a: Courtesy of Stephen Larson; b: NASA)

Dark Belts and Light Zones

As seen through an Earth-based telescope, both Jupiter and Saturn display colorful bands that extend around each planet parallel to its equator (Figure 12-1). More detailed close-up views of Jupiter from a spacecraft (Figure 12-2*a*) show alternating dark and light bands parallel to Jupiter's equator in subtle tones of red, orange, brown, and yellow. The dark, reddish bands are called **belts,** and the light-colored bands are called **zones.** The image of Saturn in Figure 12-2*b* shows similar features, but with colors that are much less pronounced. (We will see in Section 12-4 how differences between the atmospheres of Jupiter and Saturn can explain Saturn's washed-out appearance.)

In addition to these conspicuous stripes, a huge, red-orange oval called the **Great Red Spot** is often visible in Jupiter's southern hemisphere. This remarkable feature was first seen by the English scientist Robert Hooke in 1664 but may be much older. It appears to be an extraordinarily long-lived storm in the planet's

(a) Jupiter

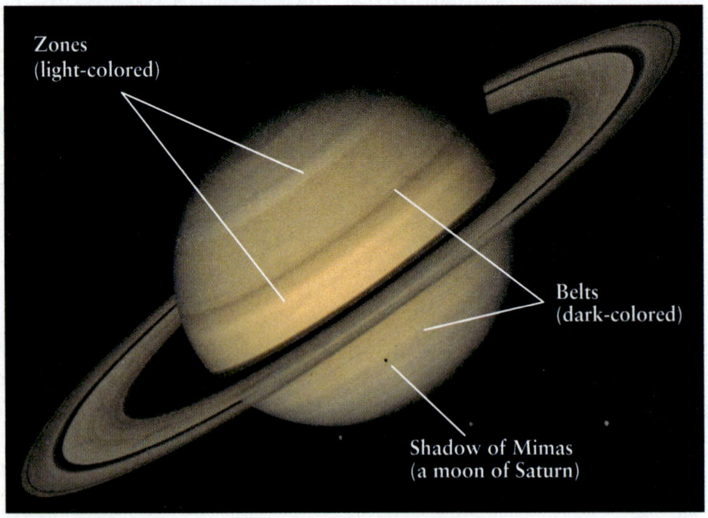

(b) Saturn

VIDEO 12.1 VIDEO 12.2

Figure 12-2 R I V U X G

Jupiter and Saturn as Viewed from Space **(a)** This view of Jupiter is a composite of four images made by the *Cassini* spacecraft as it flew past Jupiter in 2000. **(b)** Images from the *Voyager 2* spacecraft as it flew past Saturn in 1981 were combined to show the planet in approximately natural color. (a: NASA/JPL/University of Arizona; b: USGS/NASA/JPL) dynamic atmosphere. Saturn has no long-lived storm systems of this kind. However, many careful observers have reported smaller spots and blemishes in the atmospheres of both Jupiter and Saturn that last for only a few weeks or months.

12-2 Unlike the terrestrial planets, Jupiter and Saturn exhibit differential rotation

Observations of features like the Great Red Spot and smaller storms allow astronomers to determine how rapidly Jupiter and Saturn rotate. At its equator, Jupiter completes a full rotation in only 9 hours, 50 minutes, and 28 seconds, making it not only the largest and most massive planet in the solar system but also the one with the fastest rotation. However, Jupiter rotates in a strikingly different way from Earth, the Moon, Mercury, Venus, or Mars.

Differential Rotation

If Jupiter were a solid body like a terrestrial planet (or, for that matter, a billiard ball), all parts of Jupiter's surface would rotate through one complete circle in this same amount of time (Figure 12-3*a*). But by watching features in Jupiter's cloud cover, Gian Domenico Cassini discovered in 1690 that the polar regions of the planet rotate a little more slowly than do the equatorial regions. (You may recall this Italian astronomer from Section 11-2 as the gifted observer who first determined Mars's rate of rotation.) Near the poles, the rotation period of Jupiter's atmosphere is about 9 hours, 55 minutes, and 41 seconds. Saturn, too, has a longer rotation period near its poles (10 hours, 39 minutes, and 24 seconds) than at its equator (10 hours, 13 minutes, and 59 seconds).

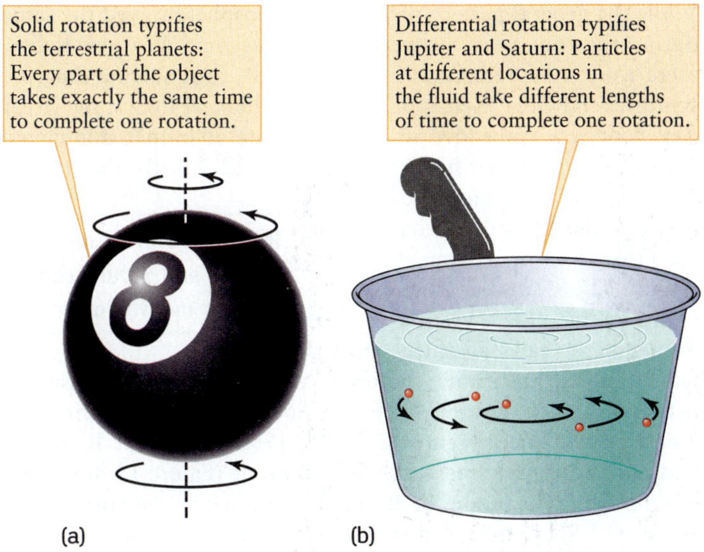

Solid rotation typifies the terrestrial planets: Every part of the object takes exactly the same time to complete one rotation.
Differential rotation typifies Jupiter and Saturn: Particles at different locations in the fluid take different lengths of time to complete one rotation.
(a)
(b)

Figure 12-3

Solid Rotation versus Differential Rotation **(a)** All parts of a solid object rotate together, but **(b)** a rotating fluid displays differential rotation. To see differential rotation, put some grains of sand, bread crumbs, or other small particles in a pot of water. Stir the water with a spoon to start it rotating, then take out the spoon. The particles near the center of the pot take less time to make a complete rotation than those away from the center.

ANALOGY You can see this kind of rotation, called **differential rotation,** in the kitchen. As you stir the water in a pot, different parts of the liquid take different amounts of time to make one "rotation" around the center of the pot (Figure 12-3*b*). Differential rotation shows that neither Jupiter nor Saturn can be solid throughout their volumes: They must be at least partially fluid, like water in a pot.

The Compositions of Jupiter and Saturn

If Jupiter and Saturn have partially fluid interiors, they cannot be made of the rocky materials that constitute the terrestrial planets. An important clue to the compositions of Jupiter and Saturn are their average densities, which are only 1326 kg/m^3 for Jupiter and 687 kg/m^3 for Saturn. (By comparison, Earth's average density is 5515 kg/m^3.) To explain these low average densities, Rupert Wildt of the University of Göttingen in Germany suggested in the 1930s that Jupiter and Saturn are composed mostly of hydrogen and helium atoms—the two lightest elements in the universe—held together by their mutual gravitational attraction to form a planet. Wildt was motivated in part by his observations of prominent absorption lines of methane and ammonia in Jupiter's spectrum. (We saw in Section 7-3 how spectroscopy plays an important role in understanding the planets.) A molecule of methane (CH_4) contains four hydrogen atoms, and a molecule of ammonia (NH_3) contains three. The presence of these hydrogen-rich molecules was strong, but indirect, evidence of abundant hydrogen in Jupiter's atmosphere.

Direct evidence for hydrogen and helium in Jupiter's atmosphere, however, was slow in coming. The problem was that neither gas produces prominent spectral lines in the visible sunlight reflected from the planet. To show the presence of these elements conclusively, astronomers had to look for spectral lines in the ultraviolet part of the spectrum. These lines are very difficult to measure from Earth, because almost no ultraviolet light penetrates our atmosphere (see Figure 6-25). Astronomers first detected the weak spectral lines of hydrogen molecules in Jupiter's spectrum in 1960. The presence of helium on Jupiter and Saturn was finally confirmed in the 1970s and 1980s, when spacecraft first flew past these planets and measured their hydrogen spectra in detail. (Collisions between helium and hydrogen atoms cause small but measurable changes in the hydrogen spectrum, which is what the spacecraft instruments detected.)

Spacecraft observations were needed to determine the compositions of Jupiter and Saturn

Today we know that the chemical composition of Jupiter's atmosphere is 86.2% hydrogen molecules (H_2), 13.6% helium atoms, and 0.2% methane, ammonia, water vapor, and other gases. The percentages in terms of *mass* are somewhat different because a helium atom is twice as massive as a hydrogen molecule. Hence, by mass, Jupiter's atmosphere is approximately 75% hydrogen, 24% helium, and 1% other substances, quite similar to that of the Sun. We will see evidence in Section 12-6 that Jupiter has a large rocky core made of heavier elements. It is estimated that the breakdown by mass of the planet as a whole (atmosphere plus interior) is approximately 71% hydrogen, 24% helium, and 5% all heavier elements.

Saturn's Missing Helium

Like Jupiter, Saturn is thought to have a large rocky core. But unlike Jupiter, data from Earth-based telescopes and spacecraft show that the atmosphere of Saturn has a serious helium deficiency: Its chemical composition is 96.3% hydrogen molecules, 3.3% helium, and 0.4% other substances (by mass, 92% hydrogen, 6% helium, and 2% other substances). Saturn's atmosphere is a puzzle because Jupiter and Saturn are thought to have formed in similar ways from the gases of the solar nebula (see Section 8-4), and so both planets (and the Sun) should have essentially the same abundances of hydrogen and helium. So where did Saturn's helium go?

The explanation may be simply that Saturn is smaller than Jupiter, and as a result Saturn probably cooled more rapidly. (We saw in Section 7-6 why a small world cools down faster than a large one.) This cooling would have triggered a process analogous to the way rain develops here on Earth. When the air is cool enough, humidity in Earth's atmosphere condenses into raindrops that fall to the ground. On Saturn, however, it is droplets of liquid helium that condense within the planet's cold, hydrogen-rich outer layers. In this scenario, helium is deficient in Saturn's upper atmosphere simply because it has fallen farther down into the planet. By contrast, Jupiter's helium has not rained out because its upper atmosphere is warmer and the helium does not form droplets.

ANALOGY An analogy to helium "rainfall" within Saturn is what happens when you try to sweeten tea by adding sugar. If the tea is cold, the sugar does not dissolve well and tends to sink to the bottom of the glass even if you stir the tea with a spoon. But if the tea is hot, the sugar dissolves with only a little stirring. In the same way, it is thought that the descending helium droplets once again dissolve in hydrogen once they reach the warmer depths of Saturn's interior.

In this scenario, Jupiter and Saturn both have about the same overall chemical composition. But Saturn's smaller mass, less than a third that of Jupiter, means that there is less gravitational force tending to compress its hydrogen and helium. This lack of compression explains why Saturn's density is only about half that of Jupiter, and is in fact the lowest of any planet in the solar system.

CAUTION! Because Jupiter and Saturn are almost entirely hydrogen and helium, it would be impossible to land a spacecraft on either planet. An astronaut foolish enough to try would notice the hydrogen and helium around the spacecraft becoming denser, the temperature rising, and the pressure increasing as the spacecraft descended. But the hydrogen and helium would never solidify into a surface on which the spacecraft could touch down. Long before reaching the planet's rocky core, the pressure of the hydrogen and helium would reach such unimaginably high levels that any spacecraft, even one made of the strongest known materials, would be crushed.

12-3 Spacecraft images show remarkable activity in the clouds of Jupiter and Saturn

Most of our detailed understanding of Jupiter and Saturn comes from a series of robotic spacecraft that have examined these remarkable planets at close range. They found striking evidence of stable, large-scale weather patterns in both planets' atmospheres, as well as evidence of dynamic changes on smaller scales.

Spacecraft to Jupiter and Saturn

WEB LINK 12.6

The first several spacecraft to visit Jupiter and Saturn each made a single flyby of the planet. *Pioneer 10* flew past Jupiter in December 1973; it was followed a year later by the nearly identical *Pioneer 11,* which went on to make the first-ever flyby of Saturn in 1979. Also in 1979, another pair of spacecraft, *Voyager 1* and *Voyager 2,* sailed past Jupiter. These spacecraft sent back spectacular close-up color pictures of Jupiter's dynamic atmosphere. Both *Voyagers* subsequently flew past Saturn.

The first spacecraft to go into Jupiter orbit was *Galileo,* which carried out an extensive program of observations from 1995 to 2003. The *Cassini* spacecraft went into orbit around Saturn in 2004. A cooperative project of NASA, the European Space Agency, and the Italian Space Agency, this spacecraft is named for astronomer Gian Domenico Cassini. On the way to its destination it viewed Jupiter at close range, recording detailed images such as Figure 12-2*a* and the image of Jupiter that opens this chapter.

Observing Jupiter's Dynamic Atmosphere

While the general pattern of Jupiter's atmosphere stayed the same during the four years between the *Pioneer* and *Voyager* flybys, there were some remarkable changes in the area surrounding the centuries-old Great Red Spot. During the *Pioneer* flybys, the Great Red Spot was embedded in a broad white zone that dominated the planet's southern hemisphere (Figure 12-4*a*). By the time of the *Voyager* missions, a dark belt had broadened and encroached on the Great Red Spot from the north (Figure 12-4*b*). In the same way that colliding weather systems in our atmosphere can produce strong winds and turbulent air, the interaction in Jupiter's atmosphere between the belt and the Great Red Spot embroiled the entire region in turbulence (Figure 12-5). By 1995, the Great Red Spot was once again centered within a white zone (Figure 12-4*c*); when *Cassini* flew past Jupiter in 2000, the dark belt to the north of the Great Red Spot embroiled the entire region in turbulence.

Immense storms on Jupiter and Saturn can last for months or years

Over the past three centuries, Earth-based observers have reported many long-term variations in the Great Red Spot's size and color. At its largest, it measured 40,000 by 14,000 km—so large that three Earths could fit side by side across it. At other times (as in 1976 and 1977), the spot almost faded from view. During the *Voyager* flybys of 1979, the Great Red Spot was comparable in size to Earth (see Figure 12-5).

Clouds at different levels in Jupiter's atmosphere reflect different wavelengths of infrared light. Using this effect, astronomers

The Great Red Spot lies within a white zone.
A dark belt has encroached on the Great Red Spot.
The Great Red Spot again lies within a white zone.
Brown ovals
White ovals

(a) *Pioneer 11*, December 1974 (b) *Voyager 2*, July 1979 (c) HST, February 1995

Figure 12-4 R I V U X G

Jupiter's Changing Appearance These images from **(a)**, **(b)** spacecraft and **(c)** the Hubble Space Telescope show major changes in the planet's upper atmosphere over a 20-year period. (a and b: NASA/JPL; c: Reta Beebe and Amy Simon, New Mexico State University, and NASA)

used an infrared telescope on board the *Galileo* spacecraft to help clarify the vertical structure of the Great Red Spot. Most of the Great Red Spot is made of clouds at relatively high altitudes, surrounded by a collar of very low-level clouds about 50 km (30 miles, or 160,000 feet) below the high clouds at the center of the spot. This same kind of structure is seen in high-pressure areas in Earth's atmosphere, although on a very much smaller scale.

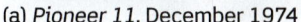

Cloud motions in and around the Great Red Spot reveal that the spot rotates counterclockwise with a period of about six days. Furthermore, winds to the north of the spot blow to the west, and winds south of the spot move toward the east. The circulation around the Great Red Spot is thus like a wheel spinning between two oppositely moving surfaces (see Figure 12-5). Weather patterns on Earth tend to change character and eventually dissipate when they move between plains and mountains or between land and sea. But because no solid surface or ocean underlies Jupiter's clouds, no such changes can occur for the Great Red Spot—which may explain how this wind pattern has survived for at least three centuries.

Other persistent features in Jupiter's atmosphere are the **white ovals.** Several white ovals are visible in Figure 12-4*c*. As in the Great Red Spot, wind flow in white ovals is counterclockwise. White ovals are also apparently long-lived; Earth-based observers have reported seeing them in the same location since 1938.

Most of the white ovals are observed in Jupiter's southern hemisphere, whereas **brown ovals** are more common in Jupiter's northern hemisphere. Brown ovals appear dark in a visible-light image like Figure 12-4*c*, but they appear bright in an infrared image. For this reason, brown ovals are understood to be holes in Jupiter's cloud cover. They permit us to see into the depths of the Jovian atmosphere, where the temperature is higher and the atmosphere emits infrared light more strongly. White ovals, by contrast, have relatively low temperatures. They are areas with cold, high-altitude clouds that block our view of the lower levels of the atmosphere.

Between 1998 and 2000 three white ovals merged in Jupiter's southern hemisphere. The merger of these ovals, each of which had been observed for 60 years, led to the formation of a massive storm about half the size of the Great Red Spot (Figure 12-6). Time will tell whether this new feature is as long-lived as its larger cousin.

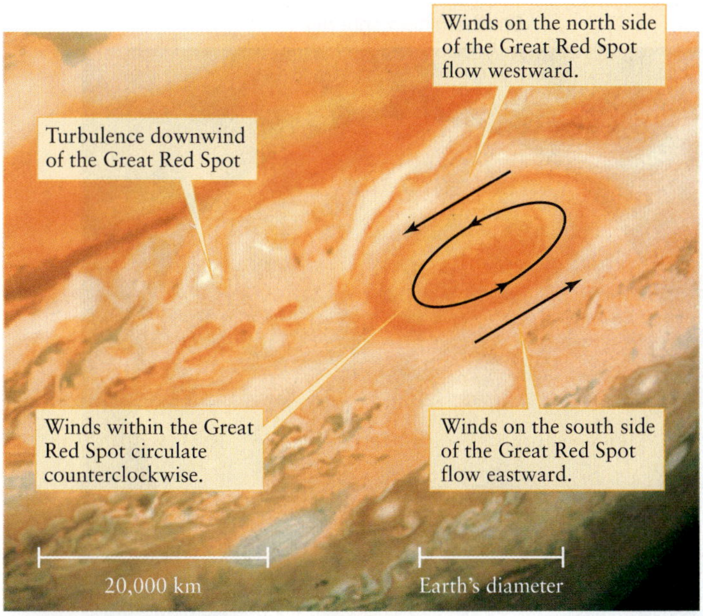
Winds on the north side of the Great Red Spot flow westward.
Turbulence downwind of the Great Red Spot
Winds within the Great Red Spot circulate counterclockwise.
Winds on the south side of the Great Red Spot flow eastward.
20,000 km
Earth's diameter

VIDEO 12.3

Figure 12-5 R I V U X G

Circulation around the Great Red Spot This 1979 image from *Voyager 2* shows atmospheric turbulence and the direction of winds around the Great Red Spot (compare Figure 12-4*b*, taken at approximately the same time). Winds within the Great Red Spot itself make it spin counterclockwise, completing a full revolution in about six days. (NASA/JPL; adapted from A. P. Ingersoll)

Figure 12-6 R I V U X G

Jupiter's "Red Spot Jr." This 2006 image of Jupiter and a new storm that formed between 1998 and 2000 was made using adaptive optics on the Gemini North telescope in Hawaii (see Section 6-3). Because Jupiter displays differential rotation (see Section 12-2), the two storms travel around Jupiter at different rates; they are not always side-by-side as shown here. (Gemini Observatory ALTAIR Adaptive Optics image/Chris Go)

Regularities in Jupiter's Atmosphere

The *Voyager*, *Galileo*, and *Cassini* images might suggest a state of incomprehensible turmoil, but, surprisingly, there is also great regularity in the Jovian atmosphere. The views in Figure 12-7 (assembled from a large number of *Cassini* images) show how Jupiter would look to someone located directly over either the planet's north pole or its south pole. Note the regular spacing of such cloud features as ripples, plumes, colored wisps, and white ovals in the southern hemisphere. These regularities are probably the result of stable, large-scale weather patterns that encircle Jupiter.

Storms on Saturn

While Saturn has belts and zones like Jupiter, it has no storm systems as long-lived as Jupiter's Great Red Spot. But about every 30 years—roughly the orbital period of Saturn—Earth-based observers have reported seeing storms in Saturn's clouds that last for

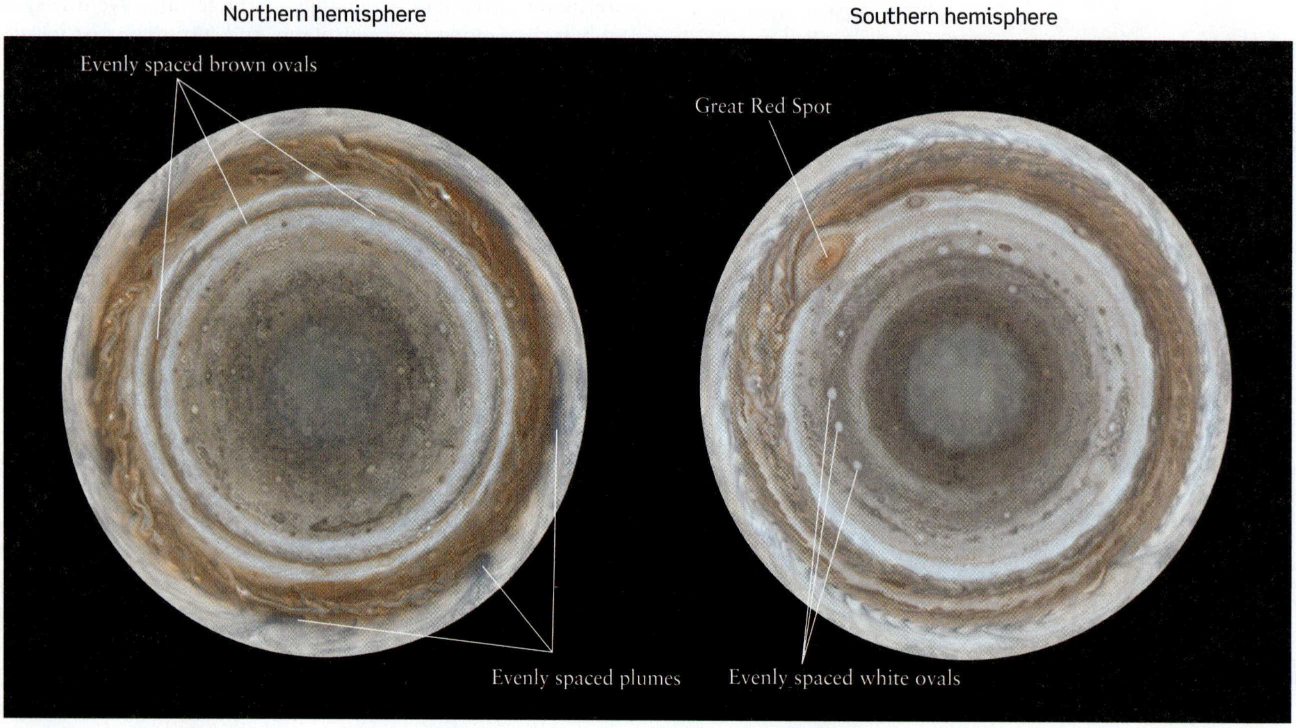

Figure 12-7 R I V U X G

Jupiter's Northern and Southern Hemispheres *Cassini* images were combined and computer processed to construct these views that look straight down onto Jupiter's north and south poles. Various cloud features are evenly spaced in longitude. (NASA/JPL/Space Science Institute)

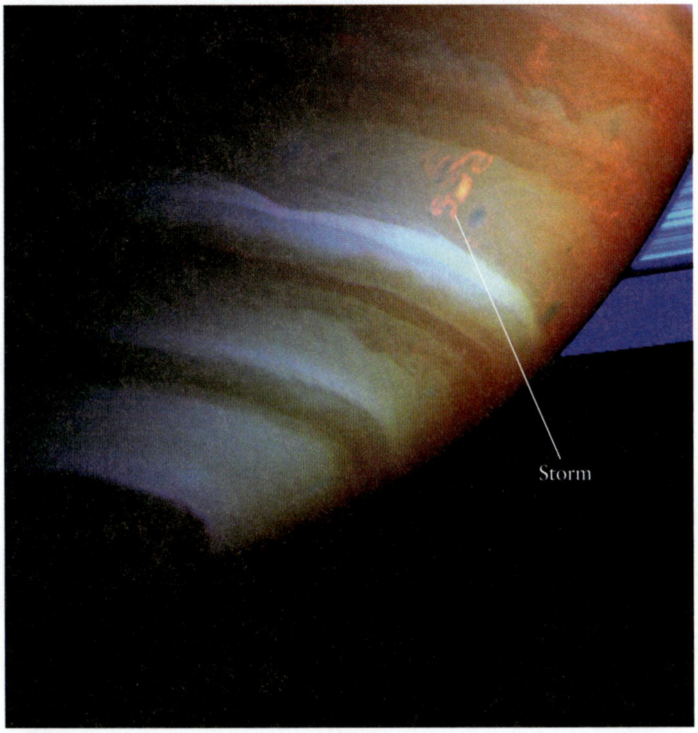

VIDEO 12.5

Figure 12-8 RIVUXG

A New Storm on Saturn This infrared image from the *Cassini* spacecraft shows a storm the size of Earth that appeared in Saturn's southern hemisphere in 2004. *Cassini* has observed several storms in this same region of Saturn. (NASA/JPL/Space Science Institute)

several days or even months. Figure 12-8 shows one such storm that appeared in Saturn's southern hemisphere in 2004.

VIDEO 12.4

Saturn's storms are thought to form when warm atmospheric gases rise upward and cool, causing gaseous ammonia to solidify into crystals and form white clouds. On Earth, similar rapid lifting of air occurs within thunderstorms and can cause water droplets to solidify into hailstones. Earth thunderstorms produce electrical discharges (lightning) that generate radio waves, which you hear as "static" on an AM radio. Radio receivers on board *Cassini* have recorded similar "static" emitted by storms on Saturn, which strongly reinforces the idea that these storms are like thunderstorms on Earth.

12-4 The internal heat of Jupiter and Saturn has a major effect on the planets' atmospheres

TUTORIAL 12.1

Weather patterns on Earth are the result of the motions of air masses. Winter storms in the midwestern United States occur when cold, moist air moves southward from Canada, and tropical storms in the southwest United States are caused by the northeastward motion of hot, moist air from Hawaii. All such motions, as well as the overall global circulation of our atmosphere, are powered by the Sun. Sunlight is absorbed by Earth's surface, and the heated surface in turn warms the atmosphere and stirs it into motion (see Sections 9-1 and 9-5). The energy of sunlight also powers the motions of the atmospheres of Venus and Mars (see Section 11-6). On both Jupiter and Saturn, however, atmospheric motions are powered both by solar energy and by the internal energy of the planet.

Jupiter and Saturn: Radiating Energy into Space

In the late 1960s, astronomers using Earth-based telescopes made the remarkable discovery that Jupiter emits more energy in the form of infrared radiation than it absorbs from sunlight—in fact, about twice as much. (By comparison, the internal heat that Earth radiates into space is only about 0.005% of what it absorbs from the Sun.) The excess heat presently escaping from Jupiter is thought to be energy left over from the formation of the planet 4.56 billion years ago. As gases from the solar nebula fell into the protoplanet, vast amounts of gravitational energy were converted into thermal energy that heated the planet's interior. Jupiter has been slowly radiating this energy into space ever since. Because Jupiter is so large, it has retained substantial thermal energy even after billions of years, and so still has plenty of energy to emit as infrared radiation.

Saturn, too, radiates into space more energy than it receives as sunlight. Because Saturn is smaller than Jupiter, it should have begun with less internal heat trapped inside, and it should have radiated that heat away more rapidly. Hence, we would expect Saturn to be radiating very little energy today. In fact, when we take Saturn's smaller mass into account, we find that Saturn releases about 25% *more* energy from its interior on a per-kilogram basis than does Jupiter. The explanation for this seeming paradox may be that helium is raining out of Saturn's upper atmosphere, as we described in Section 12-2. The helium condenses into droplets at cold upper altitudes, but generates heat at lower altitudes due to friction between the falling droplets and the surrounding gases. This heat eventually escapes from the planet's surface as infrared radiation. This "raining out" of helium from Saturn's upper layers is calculated to have begun 2 billion years ago. The amount of thermal energy released by this process adequately accounts for the extra heat radiated by Saturn since that time.

Heat naturally flows from a hot place to a colder place, never the other way around. Hence, in order for heat to flow upward through the atmospheres of Jupiter and Saturn and radiate out into space, the temperature must be warmer deep inside the atmosphere than it is at the cloudtops. Infrared measurements have confirmed that the temperature within the atmospheres of both planets does indeed increase with increasing depth (Figure 12-9).

ANALOGY You can understand the statement that "heat naturally flows from a hot place to a colder place" by visualizing an ice cube placed on a hot frying pan. Experience tells you that the hot frying pan cools down a little while the ice warms up and melts. If heat were to flow the other way, the ice would get colder and the frying pan would get hotter—which never happens in the real world.

A Convection Paradox

When a fluid is warm at the bottom and cool at the top, like water being warmed in a pot, one effective way for energy to be

Figure 12-9

The Upper Atmospheres of Jupiter and Saturn The black curves in these graphs show temperature versus altitude in each atmosphere, as well as the probable arrangements of the cloud layers. Zero altitude in each atmosphere is chosen to be the point where the pressure is 100 millibars, or one-tenth of Earth's atmospheric pressure. Beneath both planets' cloud layers, the atmosphere is composed almost entirely of hydrogen and helium. (Adapted from A. P. Ingersoll)

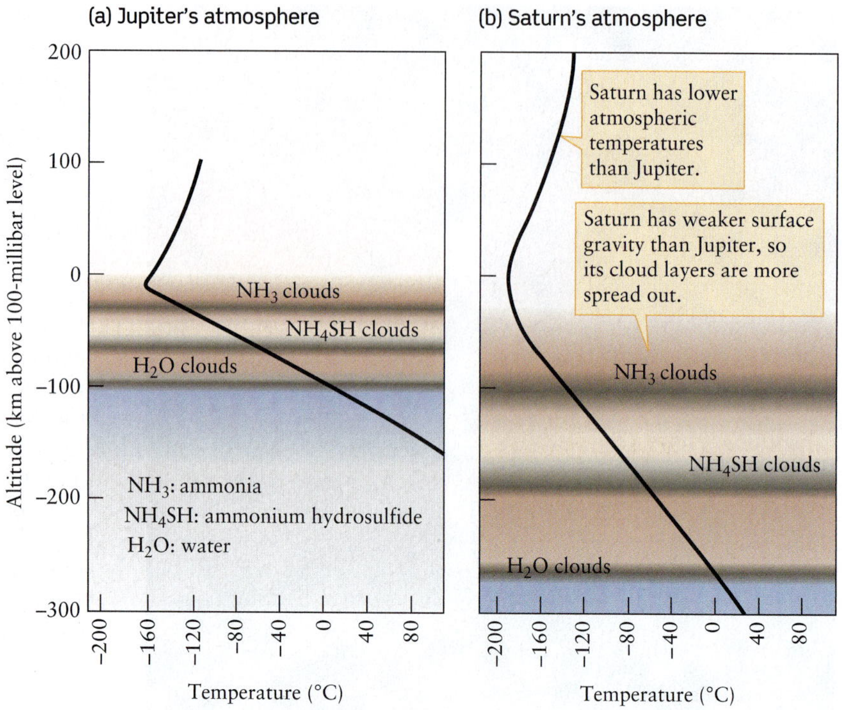

distributed through the fluid is by the up-and-down motion called *convection*. (We introduced the idea of convection in Section 9-1; see, in particular, Figure 9-5.) For many years it was thought that Jupiter's light-colored zones are regions where warm gas from low levels is rising and cooling, forming high-altitude clouds, while the dark-colored belts are regions where cool high-altitude gas is descending and being heated. However, this picture has been called into question by images made by *Cassini* during its 2000–2001 flyby of Jupiter, which showed details as small as 58 km (36 mi) across. The *Cassini* images revealed numerous white cloud cells welling upward in Jupiter's atmosphere, almost all of which appeared in the dark belts.

One interpretation of these observations is that air is actually rising in the *belts* and sinking in the *zones*. Another suggestion is that air is rising in the zones and producing clouds, but these are hidden from our view by a high-altitude layer of opaque white ammonia crystals. In this model the belts are regions in which air is descending except in the localized regions where the clouds seen by *Cassini* are found.

One thing that is well known about Jupiter's atmosphere is that the belts are regions in which we can see into the atmosphere's lower levels. You can see this in Figure 12-10, which compares visible-light and infrared images of Jupiter. The cold Jovian clouds emit radiation primarily at infrared wavelengths, with an intensity that depends on the cloud temperature (see Section 5-4) and hence on the depth of the cloud within the atmosphere (see Figure 12-9*a*). The infrared image in Figure 12-10*b* shows stronger emission from the belts, where we are looking at lower-lying, warmer levels of Jupiter's atmosphere.

The Fastest Winds in the Solar System

VIDEO 12.6

In addition to the vertical motion of gases within the belts and zones, the very rapid rotation of both Jupiter and Saturn creates a global pattern of eastward and westward **zonal winds.** Wind speeds on Jupiter can exceed 500 km/h (300 mi/h). Earth's atmospheric circulation has a similar pattern of eastward and westward flow (see Figure 9-24), but with slower wind speeds. The faster winds on Jupiter presumably result from the planet's more rapid rotation as well as the substantial flow of heat from the planet's interior.

The zonal winds on Jupiter are generally strongest at the boundaries between belts and zones. Within a given belt or zone, the wind reverses direction between the northern and southern boundaries. As Figure 12-5 shows, such reversals in wind direction are associated with circulating storms such as the Great Red Spot.

Saturn rotates at about the same rate as Jupiter, but its interior releases less heat and it receives only a quarter as much energy in sunlight as does Jupiter. With less energy available to power the motions of Saturn's atmosphere, we would expect its zonal winds to be slower than Jupiter's. Surprisingly, Saturn's winds are *faster!* In 1980 the *Voyager* spacecraft measured wind speeds near Saturn's equator that approach 1800 km/h (1100 mi/h), approximately two-thirds the speed of sound in Saturn's atmosphere and faster than winds on any other planet. Since then the equatorial winds seem to have abated somewhat: Between 1996 and 2004, astronomers using the Hubble Space Telescope showed that the wind speed had slowed to about 1000 km/h, and in 2005 *Cassini* measured wind speeds of about 1350 km/h at the equator. (These changes may be due to seasonal variations or changes in Saturn's clouds.) Why Saturn's winds should be even faster than Jupiter's is not yet completely understood.

Jupiter and Saturn's Cloud Layers

From spectroscopic observations and calculations of atmospheric temperature and pressure, scientists conclude that both Jupiter and Saturn have three main cloud layers of differing chemical

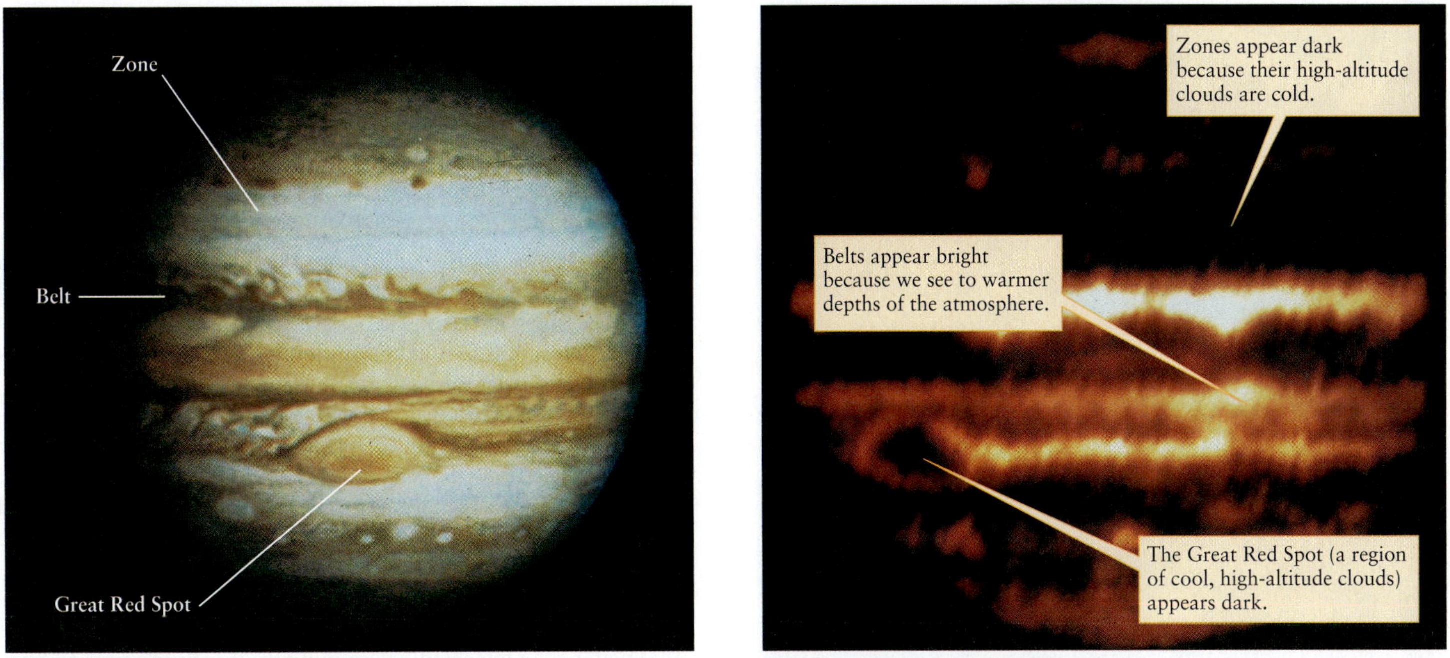

Zone
Belt
Great Red Spot
Zones appear dark because their high-altitude clouds are cold.
Belts appear bright because we see to warmer depths of the atmosphere.
The Great Red Spot (a region of cool, high-altitude clouds) appears dark.

(a) Visible-light image R I V U X G

(b) Infrared image R I V U X G

Figure 12-10

Jupiter's Warm Belts and Cool Zones **(a)** This visible-light image of Jupiter was made by the *Voyager 1* spacecraft. **(b)** Bright and dark areas in this Earth-based infrared image, taken at the same time as the image in (a), correspond to high and low temperatures, respectively. (NASA/JPL)

composition. The uppermost cloud layer is composed of crystals of frozen ammonia. Deeper in the atmosphere, ammonia (NH_3) and hydrogen sulfide (H_2S)—a compound of hydrogen and sulfur—combine to produce ammonium hydrosulfide (NH_4SH) crystals. At even greater depths, the clouds are composed of crystals of frozen water.

Jupiter's strong surface gravity compresses these cloud layers into a region just 75 km deep in the planet's upper atmosphere (Figure 12-9*a*). But Saturn has a smaller mass and hence weaker surface gravity, so the atmosphere is less compressed and the same three cloud layers are spread out over a range of nearly 300 km (Figure 12-9*b*). The colors of Saturn's clouds are less dramatic than Jupiter's (see Figure 12-2) because deeper cloud layers are partially obscured by the hazy atmosphere above them.

The colors of clouds on Jupiter and Saturn depend on the temperatures of the clouds and, therefore, on the depth of the clouds within the atmosphere. Brown clouds are the warmest and are thus the deepest layers that we can see. Whitish clouds form the next layer up, followed by red clouds in the highest layer. The whitish zones on each planet are therefore somewhat higher than the brownish belts, while the red clouds in Jupiter's Great Red Spot are among the highest found anywhere on that planet.

Despite all the data from various spacecraft, we still do not know what gives the clouds of Jupiter and Saturn their colors. The crystals of NH_3, NH_4SH, and water ice in the three main cloud layers are all white. Thus, other chemicals must cause the browns, reds, and oranges. Certain molecules closely related to ammonium hydrosulfide, which can form into long chains that have a yellow-brown color, may play a role. Compounds of sulfur or phosphorus, which can assume many different colors, depending on their temperature, might also be involved. The Sun's ultraviolet radiation may induce the chemical reactions that produce these colorful compounds.

12-5 A space probe has explored Jupiter's deep atmosphere

By observing Jupiter's and Saturn's cloud layers from a distance, we have been able to learn a great deal about the atmospheres beneath the clouds. But there is no substitute for making measurements on site, and for many years scientists planned to send a spacecraft to explore deep into Jupiter's atmosphere. While still 81.52 million kilometers (50.66 million miles) from Jupiter, the *Galileo* spacecraft released the *Galileo Probe,* a cone-shaped body about the size of an office desk. While *Galileo* itself later fired its rockets to place itself into an orbit around Jupiter, the *Galileo Probe* continued on a course that led it on December 7, 1995, to a point in Jupiter's clouds just north of the planet's equator.

Exploring Jupiter's Atmosphere

A heat shield protected the *Galileo Probe* as air friction slowed its descent speed from 171,000 km/h (106,000 mi/h) to 40 km/h (25 mi/h) in just 3 minutes. The spacecraft then deployed a parachute and floated down through the atmosphere (Figure 12-11). For the next hour, the probe observed its surroundings and radioed its findings back to the main *Galileo* spacecraft, which in turn radioed them to Earth. The mission ceased at a point some 200 km (120 mi) below Jupiter's upper cloud layer, where the

Figure 12-11

The *Galileo Probe* Enters Jupiter's Atmosphere This artist's impression shows the *Galileo Probe* descending under a parachute through Jupiter's clouds. The jettisoned heat shield (shown to the right of the probe) protected the probe during its initial high-speed entry into the atmosphere. The *Galileo Probe* returned data for 58 minutes before it was crushed and melted by the pressure and temperature of the atmosphere. (NASA)

tremendous pressure (24 atmospheres) and high temperature (152°C = 305°F) finally overwhelmed the probe's electronics.

Although the *Galileo Probe* did not have a camera, it did carry a variety of instruments that made several new discoveries. A radio-emissions detector found evidence for lightning discharges that, while less frequent than on Earth, are individually much stronger than lightning bolts in our atmosphere. Other measurements showed that Jupiter's winds, which are brisk in the atmosphere above the clouds, are even stronger beneath the clouds. The *Galileo Probe* measured a nearly constant wind speed of 650 km/h (400 mi/h) throughout its descent. This shows convincingly that the energy source for the winds is Jupiter's interior heat. If the winds were driven primarily by solar heating, as is the case on Earth, the wind speed would have decreased with increasing depth.

A central purpose of the *Galileo Probe* mission was to test the three-layer cloud model depicted in Figure 12-9*a* by making direct measurements of Jupiter's clouds. But the probe saw only traces of the NH_3 and NH_4SH cloud layers and found no sign at all of the low-lying water clouds. One explanation of this surprising result is that the probe by sheer chance entered a **hot spot,** an unusually warm and cloud-free part of Jupiter's atmosphere. Indeed, observations from Earth showed strong infrared emission from the probe entry site, as would be expected where a break in the cloud cover allows a view of Jupiter's warm interior (see Figure 12-10*b*).

The *Galileo Probe* also made measurements of the chemical composition of Jupiter's atmosphere. These data revealed that the relative proportions of hydrogen and helium are almost exactly the same as in the Sun, in line with the accepted picture of how Jupiter formed from the solar nebula (see Section 8-4). But a number of heavy elements, including carbon, nitrogen, and sulfur, were found to be significantly more abundant on Jupiter than in the Sun, which was an expected result. Over the past several billion years, Jupiter's strong gravity should have pulled in many pieces of interplanetary debris, thus building up a small but appreciable abundance of heavy elements.

Water is a major constituent of small bodies in the outer solar system and thus should likewise be present inside Jupiter in noticeable quantities. Surprisingly, the amount of water actually detected by the *Galileo Probe* was less than half the expected amount. One proposed explanation for this disparity is that the hot spot where the probe entered the atmosphere was also unusually dry. If this explanation is correct, Jupiter's hot spots are analogous to hot, dry deserts on Earth. An alternative idea is that more water than expected has rained out of Jupiter's upper atmosphere and coalesced deep within the planet. For now, the enigma of Jupiter's missing water remains unresolved.

Noble Gases and the Origin of Jupiter

Another surprising and potentially important result from the *Galileo Probe* concerns three elements, argon (Ar), krypton (Kr), and xenon (Xe). These elements are three of the so-called **noble gases,** which do not combine with other atoms to form molecules. (The noble gases, which also include helium, appear in the rightmost column of the periodic table in Box 5-5.) All three of these elements appear in tiny amounts in Earth's atmosphere and in the Sun, and so must also have been present in the solar nebula. But the *Galileo Probe* found that argon, krypton, and xenon are about 3 times as abundant in Jupiter's atmosphere as in the Sun's.

If these elements were incorporated into Jupiter directly from the gases of the solar nebula, they should be equally as abundant in Jupiter as in the Sun (which formed from the gas at the center of the solar nebula; see Section 8-3). So, the excess amounts of argon, krypton, and xenon must have entered Jupiter in the form of solid planetesimals, just as did carbon, nitrogen, and sulfur. But at Jupiter's distance from the Sun, the presumed temperature of the solar nebula was too high to permit argon, krypton, or xenon to solidify. How, then, did the excess amounts of these noble gases become part of Jupiter?

Measuring trace amounts of rare gases on Jupiter provides clues about the origins of the planets

Several hypotheses have been proposed to explain these results. One idea is that icy bodies, including solid argon, krypton, and xenon, formed in an interstellar cloud even before the cloud collapsed to form the solar nebula. A second notion is that the solar nebula may actually have been colder than previously thought. If either of these ideas proves to be correct, it would mean that Jovian planets can form closer to their stars than had heretofore been thought. This would help to explain why astronomers find massive, presumably Jovian planets orbiting close to stars other than the Sun (see Section 8-5).

A third hypothesis proposes that our models of the solar nebula are correct, but that Jupiter originally formed much farther from the Sun. Out in the cold, remote reaches of the solar nebula, argon, krypton, and xenon could have solidified into icy particles and fallen into the young Jupiter, thus enriching it in those elements. Interactions between Jupiter and the material of the solar nebula could then have forced the planet to spiral inward, eventually reaching its present distance from the Sun. As we saw in Section 8-5, a model of this kind has been proposed to explain why extrasolar planets orbit so close to their stars. More research will be needed to decide which of these hypotheses is closest to the truth.

The story of the *Galileo Probe* exemplifies two of the most remarkable aspects of scientific research. First, when new scientific observations are made, the results often pose many new questions at the same time that they provide answers to some old ones. And second, it sometimes happens that investigating what might seem to be trivial details—such as the abundances of trace elements in Jupiter's atmosphere—can reveal clues about truly grand issues, such as how the planets formed.

12-6 The oblateness of Jupiter and Saturn reveals their rocky cores

The *Galileo Probe* penetrated only a few hundred kilometers into Jupiter's atmosphere before being crushed. To learn about Jupiter's structure at greater depths as well as that of Saturn, astronomers must use indirect clues. The shapes of Jupiter and Saturn are important clues, since they indicate the size of the rocky cores at the centers of the planets.

Oblateness and Rotation

Even a casual glance through a small telescope shows that Jupiter and Saturn are not spherical but slightly flattened or **oblate.** (You can see this in Figures 12-1 and 12-2.) The diameter across Jupiter's equator (142,980 km) is 6.5% larger than its diameter from pole to pole (133,700 km). Thus, Jupiter is said to have an **oblateness** of 6.5%, or 0.065. Saturn has an even larger oblateness of 9.8%, or 0.098, making it the most oblate of all the planets. By comparison, the oblateness of Earth is just 0.34%, or 0.0034.

If Jupiter and Saturn did not rotate, both would be perfect spheres. A massive, nonrotating object naturally settles into a spherical shape so that every atom on its surface experiences the same gravitational pull aimed directly at the object's center. Because Jupiter and Saturn do rotate, however, the body of each planet tends to fly outward and away from the axis of rotation.

ANALOGY You can demonstrate this effect for yourself. Stand with your arms hanging limp at your sides, then spin yourself around. Your arms will naturally tend to fly outward, away from the vertical axis of rotation of your body. Likewise, if you drive your car around a circular road, you feel thrown toward the outside of the circle; the car as a whole is rotating around the center of the circle, and you tend to move away from the rotation axis. (We discussed the physical principles behind this in Box 4-3.) For Jupiter and Saturn, this same effect deforms the planets into their nonspherical, oblate shapes.

Modeling Jupiter and Saturn's Interiors

The oblateness of a planet depends both on its rotation rate and on how the planet's mass is distributed over its volume. The challenge to planetary scientists is to create a model of a planet's mass distribution that gives the right value for the oblateness. One such model for Jupiter suggests that 2.6% of its mass is concentrated in a dense, rocky core (Figure 12-12*a*). Although this core has about 8 times the mass of Earth, the crushing weight of the remaining bulk of Jupiter compresses it to a diameter of just 11,000 km, slightly smaller than Earth's diameter of 12,800 km. The pressure at the center of the core is estimated to be about 70 million atmospheres, and the central temperature is probably about 22,000 K. By contrast, the temperature at Jupiter's cloudtops is only 165 K.

If Jupiter formed by accretion of gases onto a rocky "seed," the present-day core is presumably that very seed. However, additional amounts of rocky material were presumably added later

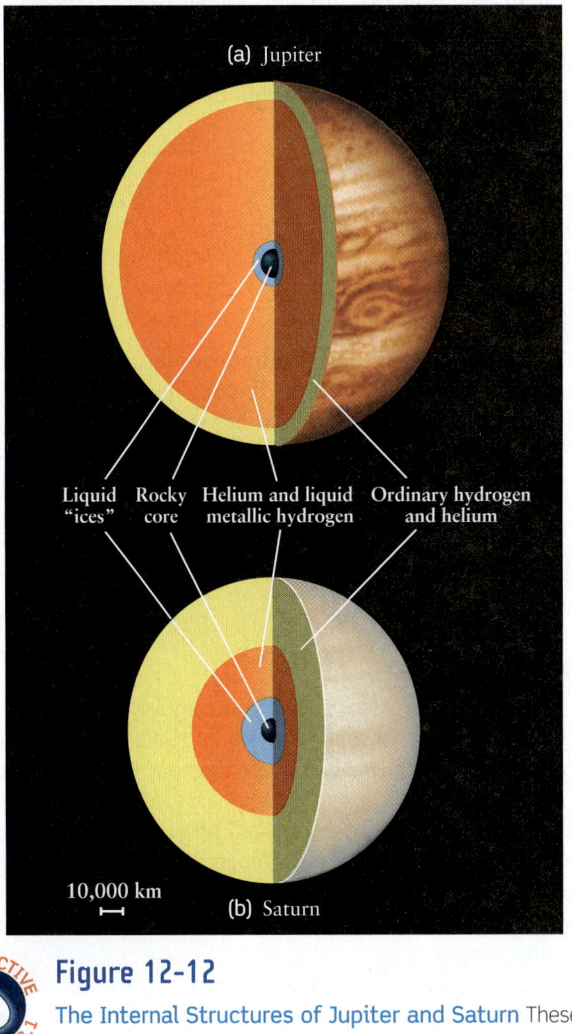

INTERACTIVE EXERCISE 12.1

Figure 12-12

The Internal Structures of Jupiter and Saturn These diagrams of the interiors of Jupiter and Saturn are drawn to the same scale. Each planet's rocky core is surrounded by an outer core of liquid "ices," a layer of helium and liquid metallic hydrogen, and a layer of helium and ordinary molecular hydrogen (H_2). Saturn's rocky core contains a larger fraction of the planet's mass than does Jupiter's, while Saturn has a smaller volume of liquid metallic hydrogen than does Jupiter.

by meteoritic material that fell into the planet and sank to the center. (If Jupiter formed directly from the gases of the solar nebula, this additional material makes up the present-day core.) Material from icy planetesimals, too, would have sunk deep within Jupiter and added to the core. In the model we have been describing, these "ices"—principally water (H_2O), methane (CH_4), ammonia (NH_3), and other molecules made by chemical reactions among these substances—form a layer some 3000 km thick that surrounds the rocky core. (Because these substances are less dense than rock, they "float" on top of the rocky core.) Despite the high pressures in this layer, the temperature is so high that the "ices" are probably in a liquid state.

Saturn rotates at about the same rate as Jupiter but has less mass, and therefore less gravity to pull its material inward. Hence, Saturn's rotation should cause material at its equator to bulge outward more than on Jupiter, giving Saturn a greater oblateness. In fact, if Saturn and Jupiter had the same internal structure, we would actually expect Saturn to be even more oblate than it really is. We can account for this discrepancy if Saturn has a different mass distribution than Jupiter. Detailed calculations suggest that about 10% of Saturn's mass is contained in its rocky core, compared to 2.6% for Jupiter's rocky core (see Figure 12-12*b*). As for Jupiter, the rocky core of Saturn is probably surrounded by an outer core of liquid "ices."

12-7 Metallic hydrogen inside Jupiter and Saturn endows the planets with strong magnetic fields

The oblateness of Jupiter and Saturn gives information about the size of the planets' rocky cores. It does not, however, provide much insight into the planets' bulk between their cores and their colorful cloudtops. Most of our knowledge of these parts of the interiors of Jupiter and Saturn comes from observations with radio telescopes. By using these telescopes to detect radio waves from Jupiter and Saturn, astronomers have found evidence of electric currents generated within both planets' hydrogen-rich interiors.

Radio Emissions from Jupiter

Astronomers began discovering radio emissions coming from Jupiter in the 1950s. A small portion of these emissions is **thermal radiation,** with the sort of spectrum that would be expected from an opaque object due to its temperature. (We discussed the electromagnetic radiation from heated objects in Sections 5-3 and 5-4.) However, most of the radio energy emitted from Jupiter is **nonthermal radiation,** which has a very different sort of spectrum from radiation from a heated object and is found in two distinct wavelength ranges.

At wavelengths of a few meters, Jupiter emits sporadic bursts of radio waves. This **decametric** (10-meter) **radiation** is probably caused by electrical discharges associated with powerful electric currents in Jupiter's ionosphere. These discharges result from electromagnetic interactions between Jupiter and its large satellite Io. At shorter wavelengths of a few tenths of a meter, Jupiter emits a nearly constant stream of electromagnetic radiation. The properties of this **decimetric** (tenth-meter) **radiation** show that it is emitted by electrons moving through a strong magnetic field at speeds comparable to the speed of light. The magnetic field deflects the electrons into spiral-shaped trajectories, and the spiraling electrons radiate energy like miniature radio antennas. Energy produced in this fashion is called **synchrotron radiation.** Astronomers now realize that synchrotron radiation is very important in a wide range of situations throughout the universe, including radio emissions from entire galaxies.

Jupiter's Magnetic Field

If some of Jupiter's radiation is indeed synchrotron radiation, then Jupiter must have a magnetic field. Measurements by the *Pioneer* and *Voyager* spacecraft verified the presence of a magnetic field and indicated that the field at Jupiter's equator is 14 times stronger than at Earth's equator. Like Earth's magnetic field, Jupiter's field is thought to be generated by motions of an electrically conducting fluid in the planet's interior. But wholly unlike Earth, the moving fluid within Jupiter is an exotic form of hydrogen called **liquid metallic hydrogen.**

As we discussed in Section 7-7, hydrogen becomes a liquid metal only when the pressure exceeds about 1.4 million atmospheres. This pressure occurs at depths more than about 7000 km below Jupiter's cloudtops. Thus, the internal structure of Jupiter consists of four distinct regions, as shown in Figure 12-12*a*: a rocky core, a roughly 3000-km thick layer of liquid "ices," a layer of helium and liquid metallic hydrogen about 56,000 km thick, and a layer of helium and ordinary hydrogen about 7000 km thick. The colorful cloud patterns that we can see through telescopes are located in the outermost 100 km of the outer layer.

Figure 12-12*a* shows that much of the enormous bulk of Jupiter is composed of liquid metallic hydrogen. Jupiter's rapid rotation sets the liquid metallic hydrogen into motion, giving rise to a powerful dynamo 20,000 times stronger than Earth's and generating an intense magnetic field.

Saturn's Magnetic Field

Saturn also has a substantial magnetic field, but it pales by Jupiter standards: Saturn's internal dynamo is only 3% as strong as Jupiter's. (To be fair, it is still 600 times stronger than Earth's dynamo.) This difference between the planets suggests that even though Saturn rotates nearly as rapidly as Jupiter, there must be far less liquid metallic hydrogen within Saturn to be stirred up by the rotation and thereby generate a magnetic field. This conclusion makes sense: Compared to Jupiter, Saturn has less mass, less gravity, and less internal pressure to compress ordinary hydrogen into liquid metallic hydrogen. As a consequence, as Figure 12-12 shows, liquid metallic hydrogen probably makes up only a relatively small fraction of Saturn's bulk.

Jupiter's Magnetosphere

Jupiter's strong magnetic field surrounds the planet with a magnetosphere so large that it envelops the orbits of many of its moons. The *Pioneer* and *Voyager* spacecraft, which carried instruments to detect charged particles and magnetic fields, revealed the awesome dimensions of the Jovian magnetosphere. The shock wave that surrounds Jupiter's magnetosphere is nearly 30 million

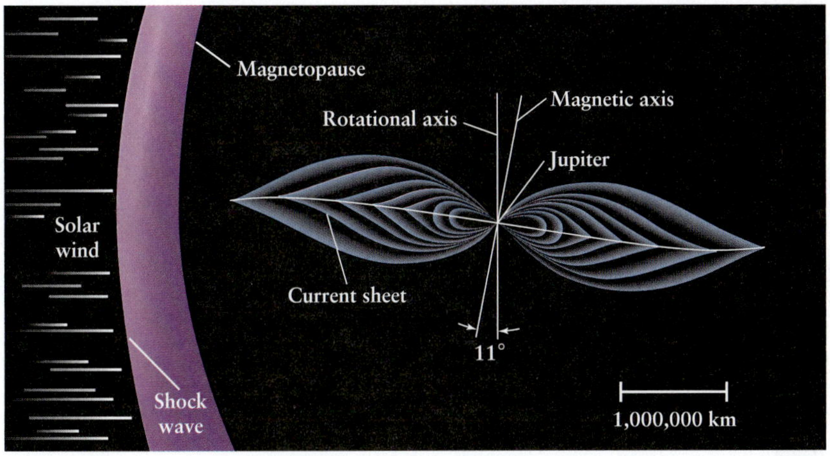

INTERACTIVE EXERCISE 12.2

Figure 12-13

Jupiter's Magnetosphere The shock wave marks where the supersonic solar wind is abruptly slowed to subsonic speeds. Most solar wind particles are deflected around Jupiter in the turbulent region (shown in purple) between the shock wave and magnetopause (where the pressure within the magnetosphere balances the pressure of the solar wind). Particles trapped inside Jupiter's magnetosphere are spread out into a vast current sheet by the planet's rapid rotation.

kilometers across. If you could see Jupiter's magnetosphere from Earth, it would cover an area in the sky 16 times larger than the Moon. Figure 12-13 shows the structure of Jupiter's magnetosphere. (Compare to Earth's much smaller magnetosphere, shown in Figure 9-20.)

Both Jupiter and Saturn have magnetospheres that are millions of kilometers across

The outer boundary of Jupiter's magnetosphere (on the left in Figure 12-13) lies at distances ranging from 3 to 7 million kilometers (2 to 4 million miles) above the planet. The "downstream" side of the magnetosphere (on the right in Figure 12-13) extends for more than a billion kilometers (650 million miles), even beyond the orbit of Saturn.

The inner regions of Earth's magnetosphere are dominated by two huge Van Allen belts (recall Figure 9-20), which are filled with charged particles. In a similar manner, Jupiter entraps immense quantities of charged particles in belts of its own. An astronaut venturing into these belts would quickly receive a lethal dose of radiation. In addition, Jupiter's rapid rotation spews the charged particles out into a huge **current sheet,** which lies close to the plane of Jupiter's magnetic equator (see Figure 12-13). Jupiter's magnetic axis is inclined 11° from the planet's axis of rotation, and the orientation of Jupiter's magnetic field is the reverse of Earth's magnetic field—a compass would point toward the south on Jupiter.

Figure 12-14 shows radio emissions from charged particles in the densest regions of Jupiter's magnetosphere. These emissions vary slightly with a period of 9 hours, 55 minutes, and 30 seconds, as Jupiter's rotation changes the angle at which we view its magnetic field. Because the magnetic field is anchored deep within the planet, this variation reveals Jupiter's **internal rotation period.** This period, indicative of the rotation of the bulk of the planet's mass, is slightly slower than the atmospheric rotation at the equator (period 9 hours, 50 minutes, 28 seconds) but about the same as the atmospheric rotation at the poles (period 9 hours, 55 minutes, 41 seconds). The faster motion of the atmosphere at the equator relative to the interior is driven by Jupiter's internal heat.

The *Voyager* spacecraft discovered that the inner regions of Jupiter's magnetosphere contain a hot, gaslike mixture of charged particles called a **plasma.** A plasma is formed when a gas is heated to such extremely high temperatures that electrons are torn off the atoms of the gas. As a result, a plasma is a mixture of positively charged ions and negatively charged electrons. The plasma that envelops Jupiter consists primarily of electrons and protons, with some ions of helium, sulfur, and oxygen. A major source of these particles is Jupiter's volcanically active moon Io (see Table 7-2), which lies deep within the magnetosphere. Io ejects a ton of material into Jupiter's magnetosphere each second from its surface and volcanic plumes. These charged particles are caught up in Jupiter's rapidly rotating magnetic field and are accelerated to high speeds.

The fast-moving particles in Jupiter's plasma exert a substantial pressure that holds off the solar wind. The *Voyager* data suggest that the pressure balance between the solar wind and the hot plasma inside the Jovian magnetosphere is precarious. A gust in the solar wind can blow away some of the plasma, at which point the magnetosphere deflates rapidly to as little as one-half its

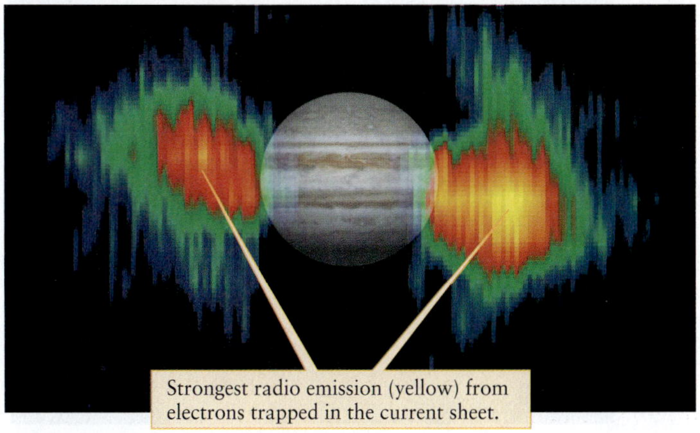

Strongest radio emission (yellow) from electrons trapped in the current sheet.

Figure 12-14 R I V U X G

A Radio View of Jupiter Instruments on board the *Cassini* spacecraft recorded this false-color map of decimetric emission from Jupiter at a wavelength of 2.2 cm. The emission comes from a region elongated parallel to the planet's magnetic equator, which is tilted relative to its geographic equator. A visible-light image of Jupiter is superimposed on the radio image. (NASA/JPL) original size. However, additional electrons and ions accelerated by Jupiter's rotating magnetic field soon replenish the plasma, and the magnetosphere expands again.

Saturn's Magnetosphere

Saturn's magnetosphere is similar to Jupiter's, but is only about 10 to 20% as large and contains far fewer charged particles than Jupiter's. There are two reasons for this deficiency. First, Saturn lacks a highly volcanic moon like Io to inject particles into the magnetosphere. Second, and more important, many charged particles are absorbed by the icy particles that make up Saturn's rings (see Section 12-10). The charged particles that do exist in Saturn's magnetosphere are concentrated in radiation belts similar to the Van Allen belts in Earth's magnetosphere.

Cassini measurements show, however, that the moons of Saturn do influence the magnetosphere. The inner magnetosphere is dominated by water and atomic oxygen, material that is thought to come from the planet's icy moons (which we will discuss in Chapter 13), as well as from Saturn's rings (which we will learn about later in this chapter).

WEB LINK 12.8

As for Jupiter, the motion of Saturn's magnetic field allows us to track the planet's internal rotation. *Voyager* measurements in 1980 showed Saturn is similar to Jupiter in that its internal rotation period (10 hours, 39 minutes, and 25 seconds) is slower than the rotation period of the visible surface at the equator (10 hours, 13 minutes, and 59 seconds). Surprisingly, when *Cassini* repeated these measurements in 2003–2004, it found that the magnetosphere's rotation had slowed down: The new rotation period is 10 hours, 45 minutes, and 45 seconds, more than 6 minutes longer than in 1980. It seems unlikely that the rotation of the planet's interior could have slowed so much in such a short time. Instead, the magnetic field inside Saturn may have slipped and reattached itself at higher latitudes, where the differentially rotating planet spins at a slower rate. The reason for this slippage is not known.

Aurorae on Jupiter and Saturn

On Earth, particles trapped in our planet's magnetic field stream onto the north and south magnetic poles, creating the aurora (see Section 9-4). The same effect takes place on both Jupiter and Saturn. Figure 12-15 shows images from the Hubble Space Telescope that reveal glowing auroral rings centered on the north and south magnetic poles of Jupiter and of Saturn. As on Earth, the emission comes from a region high in each planet's atmosphere, hundreds of kilometers above the cloudtops.

(a) Jupiter R I V U X G

(b) Saturn R I V U X G

Figure 12-15

Aurorae on Jupiter and Saturn On both Jupiter and Saturn, charged particles from the magnetosphere are funneled onto the planet's magnetic poles. When the particles collide with and excite molecules in the upper atmosphere, the molecules emit ultraviolet light. (The ultraviolet images in (a) are superimposed on a visible-light image of Jupiter.) (a: J. Clarke, University of Michigan, and NASA; b: J. Trauger, JPL, and NASA)

12-8 Earth-based observations reveal three broad rings encircling Saturn

TUTORIAL 12.2

Compared to Saturn, Jupiter is unquestionably larger, more massive, more colorful, and more dynamic in both its atmosphere and magnetosphere. But as seen through even a small telescope, Saturn stands out as perhaps the most beautiful of all the worlds of the solar system, thanks to its majestic rings (Figure 12-16). As we will see in Section 12-9, Jupiter also has rings, but they are so faint that they were not discovered until spacecraft could observe Jupiter at close range.

Discovering Saturn's Rings

In 1610, Galileo became the first person to see Saturn through a telescope. He saw few details, but he did notice two puzzling lumps protruding from opposite edges of the planet's disk. Curiously, these lumps disappeared in 1612, only to reappear in 1613. Other observers saw similar appearances and disappearances over the next several decades.

In 1655, the Dutch astronomer Christiaan Huygens began to observe Saturn with a better telescope than was available to any of his predecessors. (We discussed in Section 11-2 how Huygens used this same telescope to observe Mars.) On the basis of his observations, Huygens suggested that Saturn was surrounded by a thin, flattened ring. At times this ring was edge-on as viewed from Earth, making it almost impossible to see. At other times Earth observers viewed Saturn from an angle either above or below the plane of the ring, and the ring was visible, as in Figure 12-1. (The lumps that Galileo saw were the parts of the ring to either side of Saturn, blurred by the poor resolution of his rather small telescope.) Astronomers confirmed this brilliant deduction over the next several years as they watched the ring's appearance change just as Huygens had predicted.

As the quality of telescopes improved, astronomers realized that Saturn's "ring" is actually a *system* of rings, as Figure 12-16 shows. In 1675, the Italian astronomer Cassini discovered a dark division in the ring. The **Cassini division** is an apparent gap about 4500 kilometers wide that separates the outer **A ring** from the brighter **B ring** closer to the planet. In the mid-1800s, astronomers managed to identify the faint **C ring,** or *crepe* ring, that lies just inside the B ring.

Saturn and its rings are best seen when the planet is at or near opposition. A modest telescope gives a good view of the A and B rings, but a large telescope and excellent observing conditions are needed to see the C ring.

How the Rings Appear from Earth

Earth-based views of the Saturnian ring system change as Saturn slowly orbits the Sun. Huygens was the first to understand that this change occurs because the rings lie in the plane of Saturn's equator, and this plane is tilted 27° from the plane of Saturn's orbit. As Saturn orbits the Sun, its rotation axis and the plane of its equator maintain the same orientation in space. (The same is true for Earth; see Figure 2-12.) Hence, over the course of a Saturnian year, an Earth-based observer views the rings from various angles (Figure 12-17). At certain times Saturn's north pole is tilted toward Earth and the observer looks "down" on the "top side" of the rings. Half a Saturnian year later, Saturn's south pole is tilted toward us and the "underside" of the rings is exposed to our Earth-based view.

As seen from Earth, Saturn's rings change their orientation as the planet orbits the Sun

When our line of sight to Saturn is in the plane of the rings, the rings are viewed edge-on and seem to disappear entirely (see the images at the upper left and lower right corners of Figure 12-17).

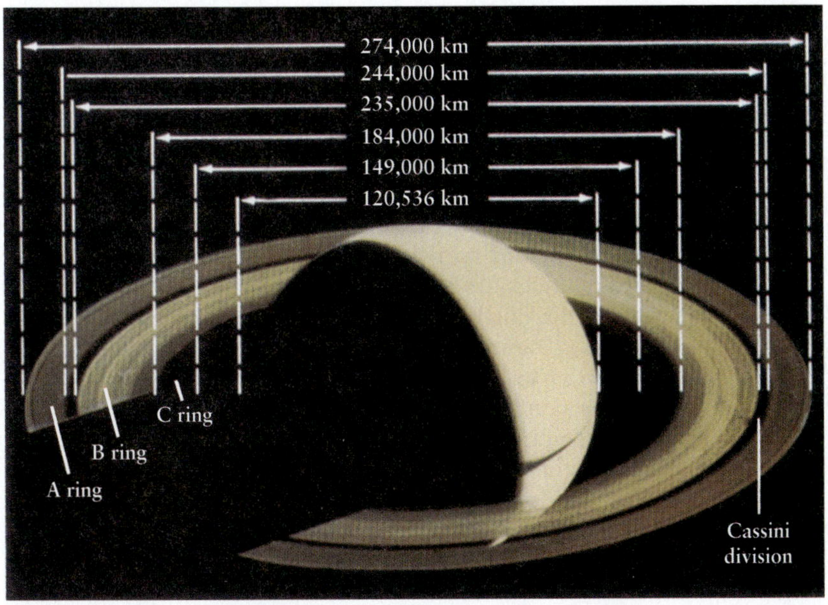

Figure 12-16 R I V U X G

Saturn's System of Rings This *Voyager 1* image shows many details of Saturn's rings. Saturn's equatorial diameter is labeled, as are the diameters of the inner and outer edges of the rings. The C ring is so faint that it is almost invisible in this view. Saturn is visible through the rings (look near the bottom of the image), which shows that the rings are not solid. (NASA/JPL)

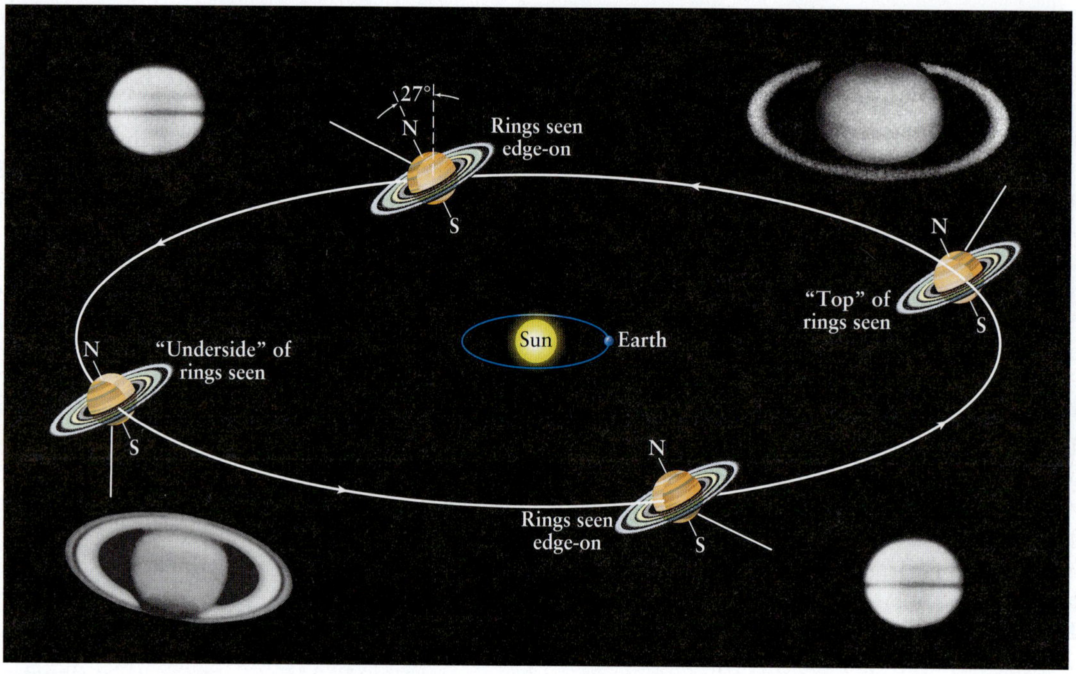

ANIMATION 12.1

Figure 12-17 R I V U X G

The Changing Appearance of Saturn's Rings Saturn's rings are aligned with its equator, which is tilted 27° from Saturn's orbital plane. As Saturn moves around its orbit, the rings maintain the same orientation in space, so observers on Earth see the rings at various angles. The accompanying Earth-based photographs show Saturn at various points in its orbit. Note that the rings seem to disappear entirely when viewed edge-on, which occurs about every 15 years. (Lowell Observatory)

This disappearance indicates that the rings are very thin. In fact, they are thought to be only a few tens of meters thick. In proportion to their diameter, Saturn's rings are thousands of times thinner than the sheets of paper used to print this book.

The last edge-on presentation of Saturn's rings of the 20th century was in 1995–1996; the first two of the 21st century are in 2008–2009 and in 2025. From 1996 to 2008, we Earth observers see the "underside" of the rings (as in the image at the lower left corner of Figure 12-17); from 2009 until 2025 we see the "top" of the rings.

12-9 Saturn's rings are composed of numerous icy fragments, while Jupiter's rings are made of small rocky particles

Astronomers have long known that Saturn's rings could not possibly be solid sheets of matter. In 1857, the Scottish physicist James Clerk Maxwell (whom we last encountered in Section 5-2) proved theoretically that if the rings were solid, differences in Saturn's gravitational pull on different parts of the rings would cause the rings to tear apart. He concluded that Saturn's rings are made of "an indefinite number of unconnected particles."

Ring Particles

In 1895, James Keeler at the Allegheny Observatory in Pittsburgh became the first to confirm by observation that the rings are not rigid. He made this conclusion by photographing the spectrum of sunlight reflected from Saturn's rings. As the rings orbit Saturn, the spectral lines from the side approaching us are blueshifted by the Doppler effect (recall Section 5-9, especially Figure 5-26). At the same time, spectral lines from the receding side of the rings are redshifted. Keeler noted that the size of the wavelength shift increased inward across the rings—the closer to the planet, the greater the shift. This variation in Doppler shift proved that the inner portions of Saturn's rings are moving around the planet more rapidly than the outer portions. Indeed, the orbital speeds across the rings are in complete agreement with Kepler's third law: The square of the orbital period about Saturn at any place in the rings is proportional to the cube of the distance from Saturn's center (see Section 4-7 and Box 4-4). This result is exactly what would be expected if the rings consisted of numerous tiny "moonlets," or **ring particles**, each individually circling Saturn.

Saturn's rings are quite bright; they reflect 80% of the sunlight that falls on them. (By comparison, Saturn itself reflects 46% of incoming sunlight.) Astronomers therefore long suspected that the ring particles are made of ice and ice-coated rock. This hunch was confirmed in the 1970s, when the American astronomers Gerard P. Kuiper and Carl Pilcher identified absorption features of frozen water in the rings' near-infrared spectrum. The *Voyager* and *Cassini* spacecraft have made even more detailed infrared measurements that indicate the temperature of the rings ranges from −180°C (−290°F) in the sunshine to less than −200°C (−330°F) in Saturn's shadow. Water ice is in no danger of melting or evaporating at these temperatures.

To determine the sizes of the particles that make up Saturn's rings, astronomers analyzed the radio signals received from a spacecraft as it passes behind the rings. How easily radio waves can travel through the rings depends on the relationship between the wavelength and the particle size. The results show that most of the particles range in size from pebble-sized fragments about 1 cm in diameter to chunks about 5 m across, the size of large boulders. Most abundant are snowball-sized particles about 10 cm in diameter.

The Roche Limit

All of Saturn's material may be ancient debris that failed to accrete (fall together) into satellites. The total amount of material in the rings is quite small. If Saturn's entire ring system were compressed together to make a satellite, it would be no more than 100 km (60 mi) in diameter. But, in fact, the ring particles are so close to Saturn that they will never be able to form moons.

To see why, imagine a collection of small particles orbiting a planet. Gravitational attraction between neighboring particles tends to pull the particles together. However, because the various particles are at differing distances from the parent planet, they also experience different amounts of gravitational pull from the planet. This difference in gravitational pull is a **tidal force** that tends to keep the particles separated. (We discussed tidal forces in detail in Section 4-8. You may want to review that section, and in particular Figure 4-24.)

Tidal forces prevent the material of Saturn's rings from coalescing into a moon

The closer a pair of particles is to the planet, the greater the tidal force that tries to pull the pair apart. At a certain distance from the planet's center, called the **Roche limit,** the disruptive tidal force is just as strong as the gravitational force between the particles. (The concept of this limit was developed in the mid-1800s by the French mathematician Edouard Roche.) Inside the Roche limit, the tidal force overwhelms the gravitational pull between neighboring particles, and these particles cannot accrete to form a larger body. Instead, they tend to spread out into a ring around the planet. (The *Cosmic Connections* figure on the next page depicts this process.) Indeed, most of Saturn's system of rings visible in Figure 12-16 lies within the planet's Roche limit.

All large planetary satellites are found outside their planet's Roche limit. If any large satellite were to come inside its planet's Roche limit, the planet's tidal forces would cause the satellite to break up into fragments. We will see in Chapter 14 that such a catastrophic tidal disruption may be the eventual fate of Neptune's large satellite, Triton.

CAUTION! It may seem that it would be impossible for any object to hold together inside a planet's Roche limit. But the ring particles inside Saturn's Roche limit survive and do not break apart. The reason is that the Roche limit applies only to objects held together by the *gravitational* attraction of each part of the object for the other parts. By contrast, the forces that hold a rock or a ball of ice together are *chemical* bonds between the object's atoms and molecules. These chemical forces are much stronger than the disruptive tidal force of a nearby planet, so the rock or ball of ice does not break apart. In the same way, people walking around on Earth's surface (which is inside Earth's Roche limit) are in no danger of coming apart, because we are held together by comparatively strong chemical forces rather than gravity.

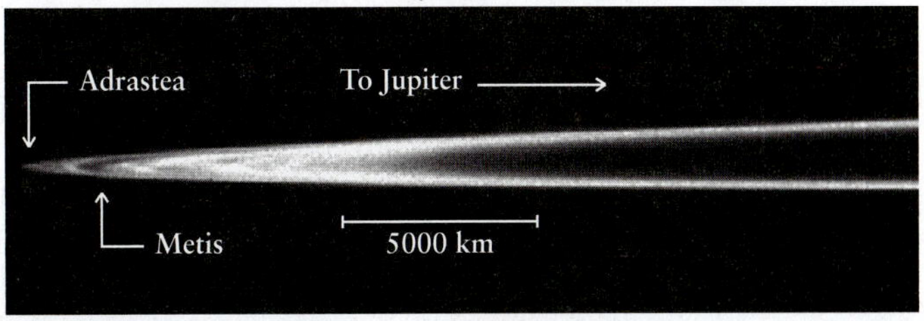

VIDEO 12.7

Figure 12-18 R I V U X G

Jupiter's Main Ring This *Galileo* image shows Jupiter's main ring almost edge-on. The ring's outer radius, about 129,000 km from the center of Jupiter, is close to the orbit of Adrastea, the second closest (after Metis) of Jupiter's moons. Not shown is an even larger and even more tenuous pair of "gossamer rings," one of which extends out to 181,000 km and the other out to 222,000 km. (NASA/JPL; Cornell University)

Jupiter's Ring and the Roche Limit

When the *Voyager 1* spacecraft flew past Jupiter in 1979, it trained its cameras not just on the planet but also on the space around the planet's equator. It discovered that Jupiter, too, has a system of rings that lies within its Roche limit (Figure 12-18). These rings differ from Saturn's in two important ways. First, Jupiter's rings are composed of tiny particles of rock with an average size of only about 1 μm (= 0.001 mm = 10^{-6} m) and that reflect less than 5% of the sunlight that falls on them. Second, there is very little material in the rings of Jupiter, less than 1/100,000 (10^{-5}) the amount of material in Saturn's rings. As a result, Jupiter's rings are extremely faint, which explains why their presence was first revealed by a spacecraft rather than an Earth-based telescope. The ring particles are thought to originate from meteorite impacts on Jupiter's four small, inner satellites, two of which are visible in Figure 12-18.

We will see in Chapter 14 that the rings of Uranus and Neptune are also made of many individual particles orbiting inside each planet's Roche limit. Like the rings of Jupiter, these rings are quite dim and difficult to see from Earth.

12-10 Saturn's rings consist of thousands of narrow, closely spaced ringlets

The photograph in Figure 12-1*b*, which was made in 1974, indicates what astronomers understood about the structure of Saturn's rings in the mid-1970s. Each of the A, B, and C rings appeared to be rather uniform, with little or no evidence of any internal structure. Within a few years, however, close-up observations from spacecraft revealed the true complexity of the rings, as well as their chemical composition.

COSMIC CONNECTIONS

Planetary Rings and the Roche Limit

If a small moon wanders too close to a planet, tidal forces tear the moon into smaller particles. These particles form a ring around the planet. The critical distance from the planet at which ring formation happens is called the Roche limit. This limit helps us understand why the rings of Jupiter and Saturn are made of small particles, as are the rings of Uranus and Neptune (discussed in Chapter 14).

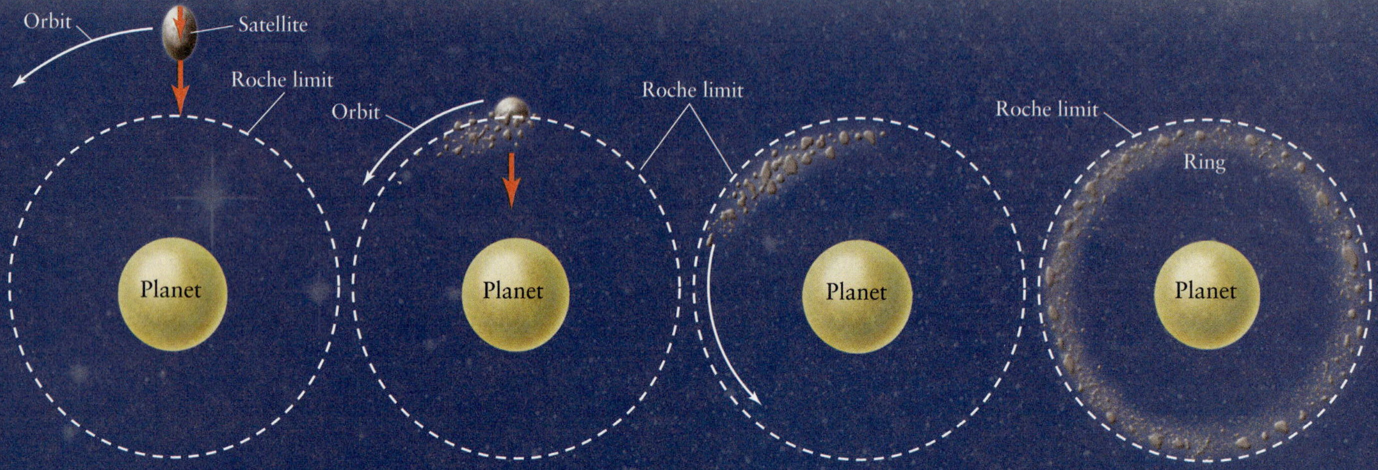

If a satellite is outside the planet's Roche limit, the tidal forces exerted by the planet may deform the satellite but will not tear it apart.

If the satellite crosses the planet's Roche limit, the tidal forces overwhelm the gravitational forces that hold the satellite together and the satellite fragments.

Fragments that are closer to the planet orbit faster (in accordance with Kepler's third law). This spreads the fragments around the planet...

...and the result is a ring.

Jupiter, Saturn, Uranus, and Neptune all have systems of rings that lie mostly within the Roche limit. This diagram shows each of the four ring systems scaled to the radius of the planet.

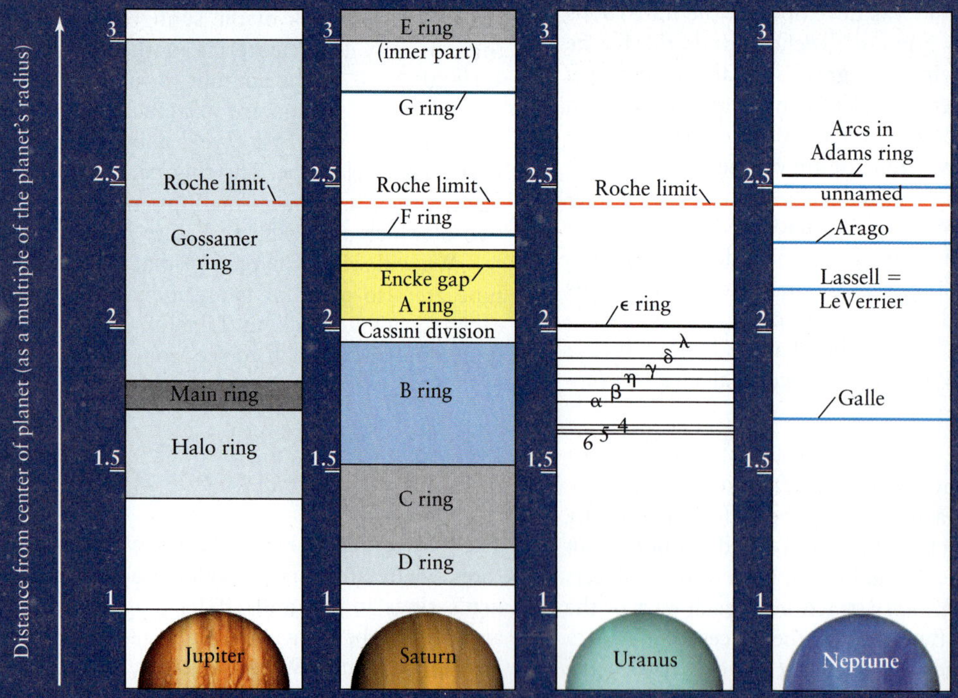

A planet's Roche limit is about 2.4 times the radius of the planet, provided the material orbiting the planet has the same density as the planet itself. For denser material the Roche limit is closer to the planet.

Spacecraft Views of Saturn's Rings

Pioneer 11, Voyager 1, and *Voyager 2* recorded images of Saturn's rings during their flybys in 1979, 1980, and 1981 respectively. *Pioneer 11* had only a relatively limited capability to make images, but cameras on board the two *Voyager* spacecraft sent back a number of pictures showing the detailed structure of Saturn's rings. Since entering Saturn orbit in 2004, the *Cassini* spacecraft has provided scientists with even higher-resolution images of the rings.

Figure 12-19 shows a *Voyager 1* image. Some of the features seen in this image were expected, including the **Encke gap,** a 270-km-wide division in the outer A ring observed by the German astronomer Johann Franz Encke in 1838. But to the amazement of scientists, images like the one in Figure 12-19 revealed that the broad A, B, and C rings are not uniform at all but instead consist of hundreds upon hundreds of closely spaced bands or **ringlets.** These ringlets are arrangements of ring particles that have evolved from the combined gravitational forces of neighboring particles, of Saturn's moons, and of the planet itself.

Pioneer 11 first detected the narrow **F ring,** which you can see in Figure 12-19. It is only about 100 km wide and lies 4000 km beyond the outer edge of the A ring. Close-up views from the *Voyagers* and *Cassini* show that the F ring is made up of several narrow ringlets. Small moons that orbit close to the F ring exert gravitational forces on the ring particles, thus warping and deforming these ringlets (Figure 12-20).

Exploring Saturn's Ring Particles

Spacecraft have done more than show Saturn's rings in greater detail; they have also viewed the rings from perspectives not possible from Earth. Through Earth-based telescopes, we can see only the sunlit side of the rings. From this perspective, the B ring appears very bright, the A ring moderately bright, the C ring dim, and the Cassini division dark (see Figure 12-19). The fraction of sunlight reflected back toward the Sun (the *albedo*) is directly related to the concentration and size of the particles in the ring. The B ring is bright because it has a high concentration of relatively large, icy, reflective particles, whereas the darker Cassini division has a lower concentration of such particles.

The *Voyagers* and *Cassini* have expanded our knowledge of the ring particles by imaging the shaded side of the rings, which cannot be seen from Earth. Figure 12-21 shows such an image. Because the spacecraft was looking back toward the Sun, this picture shows sunlight that has passed through the rings. The B ring looks darkest in this image because little sunlight gets through its dense collection of particles. If the Cassini division were completely empty, it would look black, because we would then see through it to the blackness of empty space. But the Cassini division looks *bright* in Figure 12-21. This shows that the Cassini division is not empty but contains a relatively small

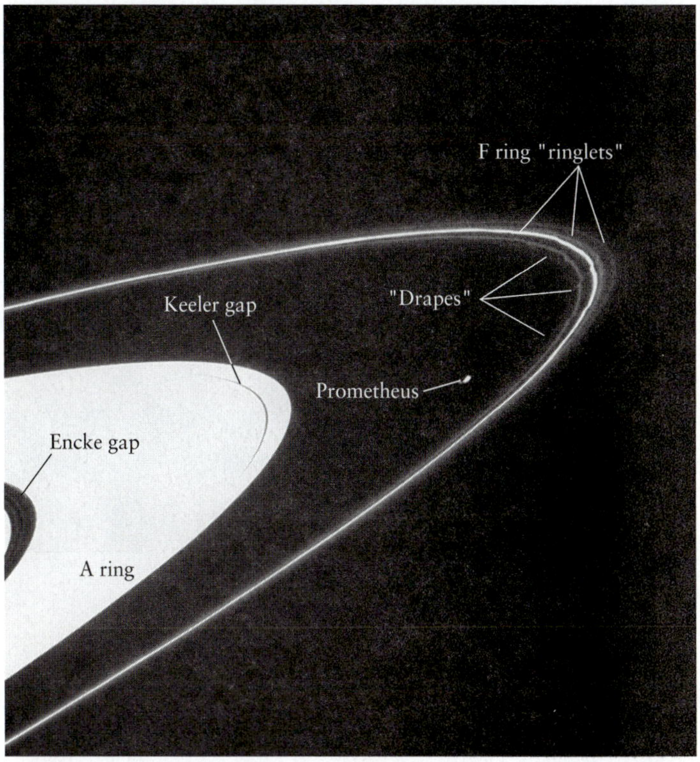

Figure 12-20 RIVUXG

VIDEO 12.8

Details of Saturn's F Ring This *Cassini* image shows Saturn's F ring (compare Figure 12-19). The bright central ringlet is about 50 km wide. As the small satellite Prometheus orbits just inside the F ring, its gravitational pull causes disturbances called "drapes" in the ringlets. The narrowness of the F ring is the result of gravitational forces exerted by Prometheus and another small satellite called Pandora (see Section 12-11). This image also shows two divisions in the A ring, the Encke gap and the Keeler gap. (NASA/JPL/Space Science Institute)

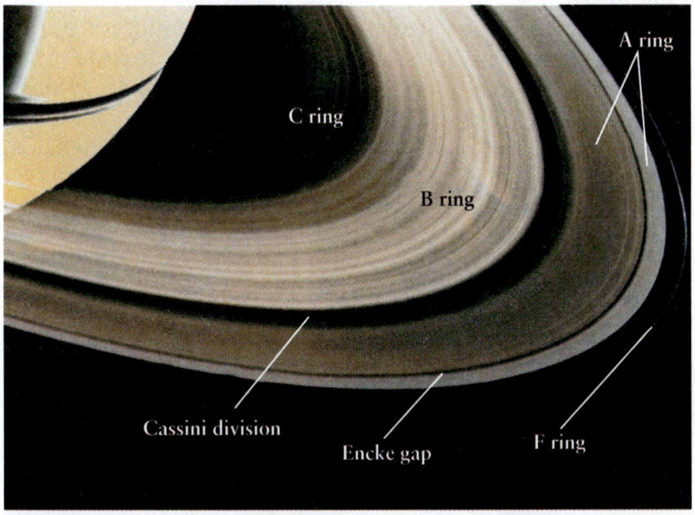

Figure 12-19 RIVUXG

Details of Saturn's Rings The colors in this *Voyager 1* image have been enhanced to emphasize small differences between different portions of the rings. The broad Cassini division is clearly visible, as is the narrow Encke gap in the outer A ring. The very thin F ring lies just beyond the outer edge of the A ring. (NASA/JPL)

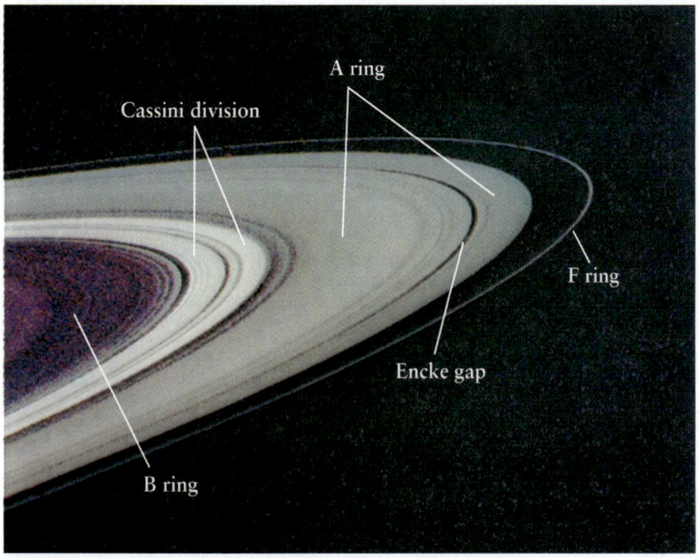

Figure 12-21 R I V U X G

The View from the Far Side of Saturn's Rings This false-color view of the side of the rings away from the Sun was taken by *Voyager 1*. The Cassini division appears white, not black like the empty gap between the A and F rings. Hence, the Cassini division cannot be empty, but must contain a number of relatively small particles that scatter sunlight like dust motes. (NASA/JPL)

number of particles. (In the same way, dust particles in the air in your room become visible when a shaft of sunlight passes through them.)

The process by which light bounces off particles in its path is called **light scattering.** The proportion of light scattered in various directions depends upon both the size of the particles and the wavelength of the light. (See Box 5-4 for examples of light scattering on Earth.) Because visible light has a much smaller wavelength than radio waves, light scattering allows scientists to measure the sizes of particles too small to detect using the radio technique described in Section 12-10. As an example, by measuring the amount of light scattered from the rings at various wavelengths and from different angles as the two *Voyager* spacecraft sped past Saturn, scientists showed that the F ring contains a substantial number of tiny particles about 1 μm (= 10^{-6} m) in diameter. This size is about the same as the size of the particles found in smoke.

There are also subtle differences in color from one ring to the next, as the computer-enhanced image in Figure 12-22 shows. Although the main chemical constituent of the ring particles is frozen water, trace amounts of other chemicals—perhaps coating the surfaces of the ice particles—are probably responsible for the different colors. These trace chemicals have not yet been identified.

The color variations in Figure 12-22 suggest that the icy particles do not migrate substantially from one ringlet to another. Had such migration taken place, the color differences would have been smeared out over time. The color differences may also indicate that different sorts of material were added to the rings at different times. In this scenario, the rings did not all form at the same time as Saturn but were added to over an extended period. New ring material could have come from small satellites that shattered after being hit by a stray asteroid or comet.

Different colors in the rings reveal variations in chemical composition

Discovering New Ring Systems

In addition to revealing new details about the A, B, C, and F rings, the *Voyager* cameras also discovered three new ring systems: the D, E, and G rings. The drawing in Figure 12-23 shows the layout of all of Saturn's known rings along with the orbits of some of Saturn's satellites. The **D ring** is Saturn's innermost ring system. It consists of a series of extremely faint ringlets located between the inner edge of the C ring and the Saturnian cloudtops. The **E ring** and the **G ring** both lie far from the planet, well beyond the narrow F ring. Both of these outer ring systems are extremely faint, fuzzy, and tenuous. Each lacks the ringlet structure so prominent in the main ring systems. The E ring encloses the orbit of Enceladus, one of Saturn's icy satellites. Some scientists suspect that water geysers on Enceladus are the source of ice particles in the E ring, much as Io's volcanoes supply material to Jupiter's magnetosphere (see Section 12-8).

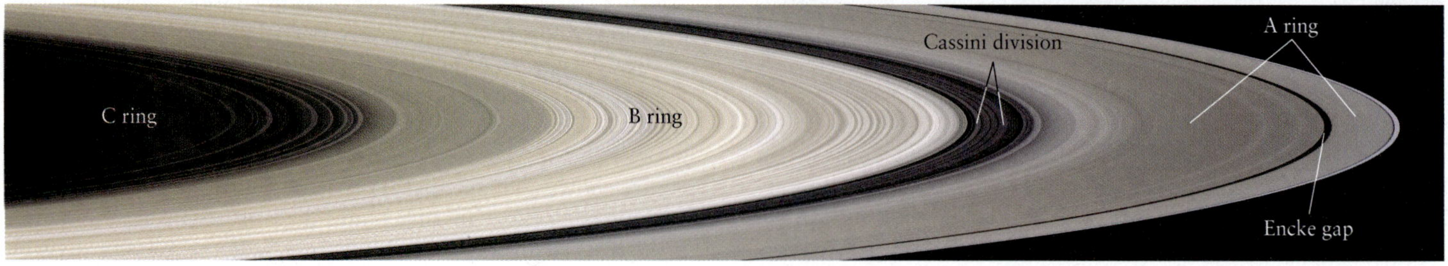

Figure 12-22 R I V U X G

Color Variations in Saturn's Rings This mosaic of *Cassini* images shows the rings in natural color. The color variations are indicative of slight differences in chemical composition among particles in different parts of the rings. (NASA/JPL/Space Science Institute)

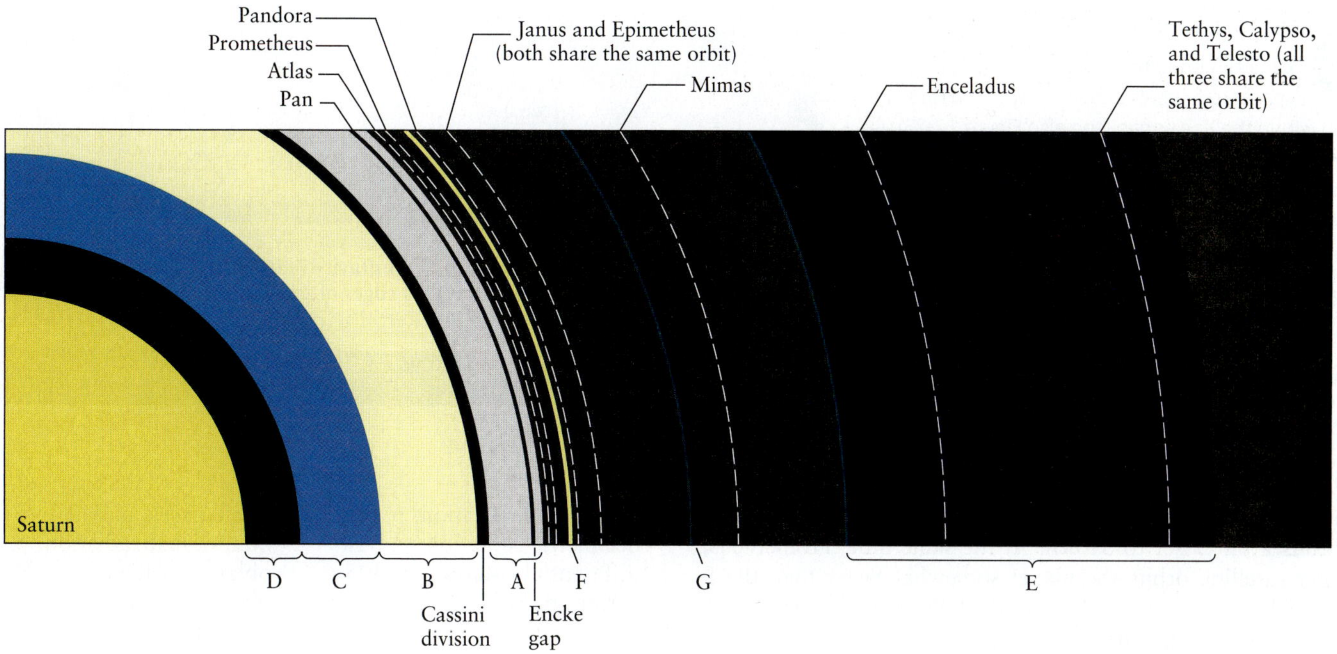

Figure 12-23

INTERACTIVE EXERCISE 12.3

The Arrangement of Saturn's Rings This scale drawing shows the location of Saturn's rings along with the orbits of the 11 inner satellites. Only the A, B, and C rings can readily be seen from Earth. The remaining rings are very faint, and their existence was only confirmed by spacecraft flybys.

12-11 Saturn's inner satellites affect the appearance and structure of its rings

Why do Saturn's rings have such a complex structure? The answer is that, like any object in the universe, the particles that make up the rings are affected by the force of gravity. Saturn's gravitational pull keeps the ring particles in orbit around the planet. But the satellites of Saturn also exert gravitational forces on the rings. These forces can shape the orbits of the ring particles and help give rise to the rings' structure.

Resonance and Shepherd Satellites

Astronomers have long known that one of Saturn's moons, Mimas, has a gravitational effect on its ring system. Mimas is a moderate-sized satellite that orbits Saturn every 22.6 hours. According to Kepler's third law, ring particles in the Cassini division should orbit Saturn approximately every 11.3 hours. Consequently, a given set of ring particles in the Cassini division will find itself between Saturn and Mimas on every second orbit (that is, every 22.6 hours). In other words, particles in the Cassini division are in a 2-to-1 resonance with Mimas. Because of these repeated alignments, the combined gravitational forces of Saturn and Mimas cause small particles to deviate from their original orbits, sweeping them out of the Cassini division. This helps explain why the Cassini division is relatively empty (although not completely so, as Figure 12-21 shows).

While Mimas tends to spread ring particles apart, other satellites exert gravitational forces that herd particles together. The *Voyager* spacecraft discovered two such satellites orbiting on either side of the narrow F ring (Figure 12-24). Pandora, the outer of the two satellites, moves around Saturn at a slightly slower speed than the particles in the ring. As ring particles pass Pandora, they experience a tiny gravitational tug that slows them

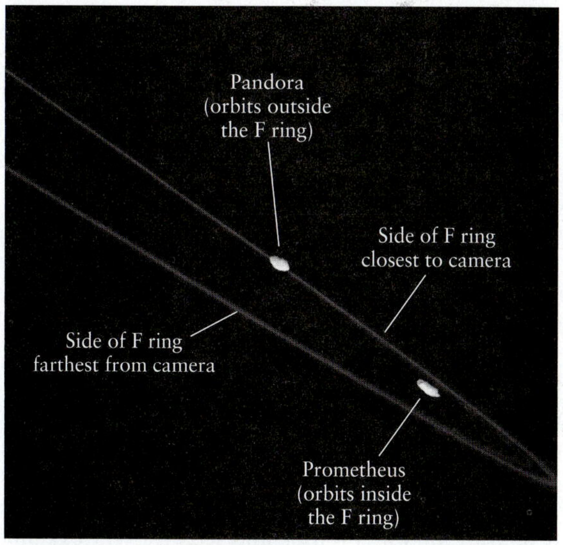

Figure 12-24 R I V U X G

Saturn's F Ring and Its Two Shepherds This *Cassini* image shows the tiny satellites Prometheus (102 km across) and Pandora (84 km across). (Compare with Figure 12-20.) These orbit Saturn on either side of its F ring, shepherding the ring particles into a narrow band. The satellites lap each other every 25 days. (NASA/JPL/Space Science Institute)

VIDEO 12.9

Figure 12-25 RI**V**UXG

Daphnis and the Keeler Gap This *Cassini* spacecraft shows Saturn's tiny moon Daphnis orbiting within the Keeler gap (see also Figure 12-20). Although it is just 7 km across, Daphnis exerts enough gravitational force to keep the 42-km wide gap open and induce wavelike disturbances in the two edges of the gap. (NASA/JPL/Space Science Institute)

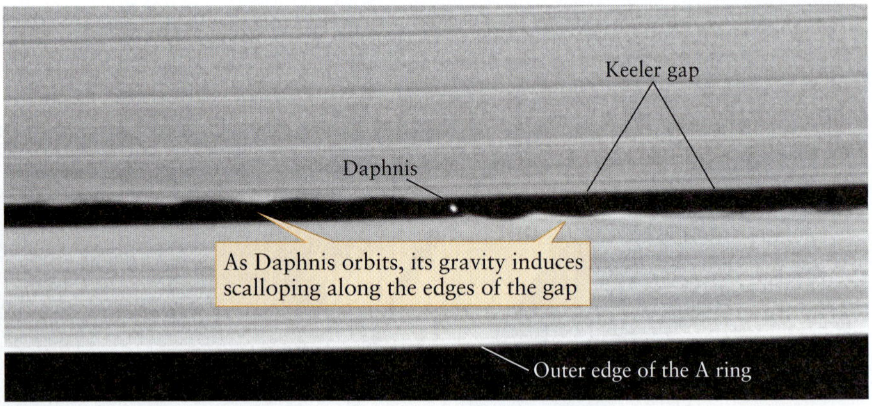

Keeler gap
Daphnis
As Daphnis orbits, its gravity induces scalloping along the edges of the gap
Outer edge of the A ring

down slightly. Hence, these particles lose a little energy and fall into orbits a bit closer to Saturn. At the same time, Prometheus, the inner satellite, orbits the planet somewhat faster than the F ring particles. The gravitational pull of Prometheus makes the ring particles speed up a little, giving them some extra energy and nudging them into slightly higher orbits. The competition between the pulls exerted by these two satellites focuses the F ring particles into a well-defined, narrow band.

Because of their confining influence, Prometheus and Pandora are called **shepherd satellites.** We will see in Chapter 14 that Uranus is encircled by narrow rings that are likewise kept confined by shepherd satellites.

An effect similar to shepherding explains two divisions in the A ring, the Encke gap and the even narrower Keeler gap (Figure 12-20). In 2005, *Cassini* images revealed a miniature moon, Daphnis, which actually orbits inside the Keeler gap (Figure 12-25). In a fashion similar to the F ring's shepherds, Daphnis's gravitational forces cause ring particles outside its orbit to move farther out and cause those inside its orbit to move farther inward toward Saturn. In this way, Daphnis helps to maintain a gap in the rings that is 42 km (26 mi) wide, many times wider than Daphnis's 7-km (4-mi) diameter. In a similar way, a 20-km wide moon called Pan orbits within and maintains the 270-km wide Encke gap. Both of these moons cause wavelike disturbances in the ring material on either side of the gap through which they travel (see Figure 12-25).

> **Gravitational effects carve gaps in the rings and focus ring particles into narrow ringlets**

In addition to their dynamic ring systems, Jupiter and Saturn together have an exotic collection of moons. These include Saturn's moon Titan, with a thick atmosphere rich in hydrocarbons; Jupiter's satellite Io, the most volcanic world in the solar system; and Saturn's satellite Enceladus, which has exotic geysers that spew a mixture of water vapor and ice. We will examine these alien worlds in Chapter 13.

Key Words

A ring, p. 317
B ring, p. 317
belts, p. 304
brown oval, p. 307
C ring, p. 317
Cassini division, p. 317
current sheet, p. 315
D ring, p. 322
decametric radiation, p. 314
decimetric radiation, p. 314
differential rotation, p. 305
E ring, p. 322
Encke gap, p. 321
F ring, p. 321
G ring, p. 322
Great Red Spot, p. 304
hot spot, p. 312
internal rotation period, p. 315
light scattering, p. 322
liquid metallic hydrogen, p. 314
noble gases, p. 312
nonthermal radiation, p. 314
oblate, oblateness, p. 313
plasma, p. 315
ring particles, p. 318
ringlets, p. 321
Roche limit, p. 319
shepherd satellite, p. 324
synchrotron radiation, p. 314
thermal radiation, p. 314
tidal force, p. 319
white oval, p. 307
zonal winds, p. 310
zones, p. 304

Key Ideas

Composition and Structure: Jupiter and Saturn are both much larger than Earth. Each is composed of 71% hydrogen, 24% helium, and 5% all other elements by mass. Both planets have a higher percentage of heavy elements than does the Sun.

• Jupiter probably has a rocky core several times more massive than Earth. The core is surrounded by a layer of liquid "ices" (water, ammonia, methane, and associated compounds). On top of this ice layer is a layer of helium and liquid metallic hydrogen and an outermost layer composed primarily of ordinary hydrogen and helium. All of Jupiter's visible features are near the top of this outermost layer.

• Saturn's internal structure is similar to that of Jupiter, but its core makes up a larger fraction of its volume and its liquid metallic hydrogen mantle is shallower than that of Jupiter.

• Jupiter and Saturn both rotate so rapidly that the planets are noticeably flattened. The rotation of both planets' interiors is revealed by variations in the radio emissions from the planets.

• Both Jupiter and Saturn emit more energy than they receive from the Sun. Presumably both planets are still cooling.

Atmospheres: The visible "surfaces" of Jupiter and Saturn are actually the tops of their clouds. The rapid rotation of the planets twists the clouds into dark belts and light zones that run parallel to the equator. Strong zonal winds run along the belts and zones.

• The outer layers of both planets' atmospheres show differential rotation: The equatorial regions rotate slightly faster than the polar regions. For both Jupiter and Saturn, the polar rotation rate is nearly the same as the internal rotation rate.

• The colored ovals visible in the Jovian atmosphere represent gigantic storms. Some, such as the Great Red Spot, are quite stable and persist for many years. Storms in Saturn's atmosphere seem to be shorter-lived.

• There are presumed to be three cloud layers in the atmospheres of Jupiter and Saturn. The reasons for the distinctive colors of these different layers are not yet known. The cloud layers in Saturn's atmosphere are spread out over a greater range of altitude than those of Jupiter, giving Saturn a more washed-out appearance.

• Saturn's atmosphere contains less helium than Jupiter's atmosphere. This lower abundance may be the result of helium raining downward into the planet. Helium "rainfall" may also account for Saturn's surprisingly strong heat output.

Magnetic Fields and Magnetospheres: Jupiter has a strong magnetic field created by currents in the metallic hydrogen layer. Its huge magnetosphere contains a vast current sheet of electrically charged particles. Charged particles in the densest portions of Jupiter's magnetosphere emit synchrotron radiation at radio wavelengths.

• The Jovian magnetosphere encloses a low-density plasma of charged particles. The magnetosphere exists in a delicate balance between pressures from the plasma and from the solar wind. When this balance is disturbed, the size of the magnetosphere fluctuates drastically.

• Saturn's magnetic field and magnetosphere are much less extensive than Jupiter's.

Rings: Saturn is circled by a system of thin, broad rings lying in the plane of the planet's equator. This system is tilted away from the plane of Saturn's orbit, which causes the rings to be seen at various angles by an Earth-based observer over the course of a Saturnian year.

Structure of the Rings: Saturn has three major, broad rings (A, B, and C) that can be seen from Earth. Other, fainter rings were found by the *Voyager* spacecraft.

• The principal rings of Saturn are composed of numerous particles of ice and ice-coated rock ranging in size from a few micrometers to about 10 m. Most of the rings exist inside the Roche limit of Saturn, where disruptive tidal forces are stronger than the gravitational forces attracting the ring particles to each other.

• Each of Saturn's major rings is composed of a great many narrow ringlets. The faint F ring, which is just outside the A ring, is kept narrow by the gravitational pull of shepherd satellites.

• Jupiter's faint rings are composed of a relatively small amount of small, dark, rocky particles that reflect very little light.

Questions

Review Questions

1. Mars passes closer to Earth than Jupiter does, but with an Earth-based telescope it is easier to see details on Jupiter than on Mars. Why is this?
2. Saturn is the most distant of the planets visible without a telescope. Is there any way we could infer this from naked-eye observations? Explain your answer. (*Hint:* Think about how Saturn's position on the celestial sphere must change over the course of weeks or months.)
3. As seen from Earth, does Jupiter or Saturn undergo retrograde motion more frequently? Explain your answer.
4. TUTORIAL 12.1 In what ways are the motions of Jupiter's atmosphere like the motion of water stirred in a pot (see Figure 12-3*b*)? In what ways are they different?
5. Is the chemical composition of Jupiter as a whole the same as that of its atmosphere? Explain any differences.
6. Astronomers can detect the presence of hydrogen in stars by looking for the characteristic absorption lines of hydrogen in the star's visible spectrum (Figure 5-21). They can also detect hydrogen in glowing gas clouds by looking for hydrogen's characteristic emission lines (Figure 5-18). Explain why neither of these techniques helped Earth-based astronomers to detect hydrogen in Jupiter's atmosphere.
7. On a warm, humid day, water vapor remains in the atmosphere. But if the temperature drops suddenly, the water vapor forms droplets, clouds appear, and it begins to rain. Relate this observation to why there is relatively little helium in Saturn's atmosphere compared to the atmosphere of Jupiter.
8. What would happen if you tried to land a spacecraft on the surface of Jupiter?
9. What are the belts and zones in the atmospheres of Jupiter and Saturn? Is the Great Red Spot more like a belt or a zone? Explain your answer.
10. Give one possible explanation why weather systems on Jupiter are longer-lived than weather systems on Earth.
11. What are white ovals and brown ovals? What can we infer about them from infrared observations?
12. Compare and contrast the source of energy for motions in Earth's atmosphere with the energy source for motions in the atmospheres of Jupiter and Saturn.
13. Both Jupiter and Saturn emit more energy than they receive from the Sun in the form of sunlight. Compare the internal energy sources of the two planets that produce this emission.
14. What observations from the *Cassini* spacecraft contradict the accepted picture of convection in Jupiter's atmosphere?
15. Compare the atmospheres of Jupiter and Saturn. Why does Saturn's atmosphere look "washed out" in comparison to that of Jupiter?
16. Which data from the *Galileo Probe* were in agreement with astronomers' predictions? Which data were surprising?
17. Fewer than one in every 10^5 atoms in Jupiter's atmosphere is an argon atom, and fewer than one in 10^8 is an atom of krypton or xenon. If these atoms are so rare, why are scientists concerned about them? How do the abundances of these elements in Jupiter's atmosphere compare to the abundances in the Sun? What hypotheses have been offered to explain these observations?
18. Why is Jupiter oblate? What do astronomers learn from the value of Jupiter's oblateness?
19. What is liquid metallic hydrogen? What is its significance for Jupiter?
20. Describe the internal structures of Jupiter and Saturn, and compare them with the internal structure of Earth.

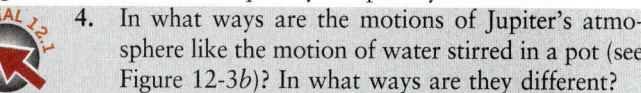

21. Explain why Saturn is more oblate than Jupiter, even though Saturn rotates more slowly.
22. If Jupiter does not have any observable solid surface and its atmosphere rotates differentially, how are astronomers able to determine the planet's internal rotation rate?
23. Compare and contrast Jupiter's magnetosphere with the magnetosphere of a terrestrial planet like Earth. Why is the size of the Jovian magnetosphere highly variable, while that of Earth's magnetosphere is not?
24. What is a plasma? Where are plasmas found in the vicinity of Jupiter?
25. Why is Saturn's magnetosphere less extensive than Jupiter's?
26. What observations of Saturn's rings proved that they are not solid?
27. If Saturn's rings are not solid, why do they look solid when viewed through a telescope?
28. Although the *Voyager* and *Cassini* spacecraft did not collect any samples of Saturn's ring particles, measurements from these spacecraft allowed scientists to determine the sizes of the particles. Explain how this was done.
29. The Space Shuttle and other spacecraft orbit Earth well within Earth's Roche limit. Explain why these spacecraft are not torn apart by tidal forces.
30. How do Jupiter's rings differ from those of Saturn?
31. Describe the structure of Saturn's rings. What evidence is there that ring particles do not migrate significantly between ringlets?
32. During the planning stages for the *Pioneer 11* mission, when relatively little was known about Saturn's rings, it was proposed to have the spacecraft fly through the Cassini division. Why would this have been a bad idea?
33. What is the relationship between Saturn's satellite Mimas and the Cassini division?
34. Why is the term "shepherd satellite" appropriate for the objects so named? Explain how a shepherd satellite operates.

Advanced Questions

The question preceded by an asterisk (*) involves topics discussed in Box 7-2.

> **Problem-solving tips and tools**
>
> Box 4-4 describes how to use Newton's form of Kepler's third law. Newton's universal law of gravitation, discussed in Section 4-7, is the basic equation from which you can calculate a planet's surface gravity. Box 7-2 discusses escape speed, and Sections 5-3 and 5-4 discuss the properties of thermal radiation (including the Stefan-Boltzmann law, which relates the temperature of a body to the amount of thermal radiation that it emits).

35. The angular diameter of Jupiter at opposition varies little from one opposition to the next. By contrast, the angular diameter of Mars at opposition is quite variable. Explain why there is a difference between these two planets.
36. Jupiter was at opposition on June 5, 2007. On that date Jupiter appeared to be in the constellation Ophiuchus. Approximately when will Jupiter next be at opposition in this same region of the celestial sphere? Explain your answer.
37. Jupiter's equatorial diameter and the rotation period at Jupiter's equator are both given in Table 12-1. Use these data to calculate the speed at which an object at the cloudtops along Jupiter's equator moves around the center of the planet.
38. Using orbital data for a Jovian satellite of your choice (see Appendix 3), calculate the mass of Jupiter. How does your answer compare with the mass quoted in Table 12-1?
39. It has been claimed that Saturn would float if one had a large enough bathtub. Using the mass and size of Saturn given in Table 12-2, confirm that the planet's average density is about 690 kg/m^3, and comment on this somewhat fanciful claim.
40. Roughly speaking, Jupiter's composition (by mass) is three-quarters hydrogen and one-quarter helium. The mass of a single hydrogen atom is given in Appendix 7; the mass of a single helium atom is about 4 times greater. Use these numbers to calculate how many hydrogen atoms and how many helium atoms there are in Jupiter.
41. Use the information given in Section 12-3 to estimate the wind velocities in the Great Red Spot, which rotates with a period of about 6 days.
42. An astronaut floating above Saturn's cloudtops would see a blue sky, even though Saturn's atmosphere has a very different chemical composition than Earth's. Explain why. (*Hint:* See Box 5-4.)
43. If Jupiter emitted just as much energy per second (as infrared radiation) as it receives from the Sun, the average temperature of the planet's cloudtops would be about 107 K. Given that Jupiter actually emits twice this much energy per second, calculate what the average temperature must actually be.
44. (**a**) From Figure 12-9*a*, by how much does the temperature increase as you descend from the 100-millibar level in Jupiter's atmosphere to an altitude 100 km below that level? (**b**) Use Figure 12-9*b* to answer the same question for Saturn's atmosphere. (**c**) In Earth's troposphere (see Section 9-5), the air temperature increases by 6.4°C for each kilometer that you descend. In which planet's atmosphere—Earth, Jupiter, or Saturn—does the temperature increase most rapidly with decreasing altitude?

*45. Consider a hypothetical future spacecraft that would float, suspended from a balloon, for extended periods in Jupiter's upper atmosphere. If we want this spacecraft to return to Earth after completing its mission, calculate the speed at which the spacecraft's rocket motor would have to accelerate it in order to escape Jupiter's gravitational pull. Compare with the escape speed from Earth, equal to 11.2 km/s.

46. The *Galileo Probe* had a mass of 339 kg. On Earth, its weight (the gravitational force exerted on it by Earth) was 3320 newtons, or 747 lb. What was the gravitational force that Jupiter exerted on the *Galileo Probe* when it entered Jupiter's clouds?
47. From the information given in Section 12-6, calculate the average density of Jupiter's rocky core. How does this compare with the average density of Earth? With the average density of Earth's solid inner core? (See Table 9-1 and Table 9-2 for data about Earth.)

48. In the outermost part of Jupiter's outer layer (shown in yellow in the upper part of Figure 12-12), hydrogen is principally in the form of molecules (H_2). Deep within the liquid metallic hydrogen layer (shown in orange in Figure 12-12), hydrogen is in the form of single atoms. Recent laboratory experiments suggest that there is a gradual transition between these two states, and that the transition layer overlaps the boundary between the ordinary hydrogen and liquid metallic hydrogen layers. Use this information to redraw the upper part of Figure 12-12 and to label the following regions in Jupiter's interior: (i) ordinary (nonmetallic) hydrogen molecules; (ii) nonmetallic hydrogen with a mixture of atoms and molecules; (iii) liquid metallic hydrogen with a mixture of atoms and molecules; (iv) liquid metallic hydrogen atoms.
49. When Saturn is at different points in its orbit, we see different aspects of its rings because the planet has a 27° tilt. If the tilt angle were different, would it be possible to see the upper and lower sides of the rings at all points in Saturn's orbit? If so, what would the tilt angle have to be? Explain your answers.
50. As seen from Earth, the intervals between successive edge-on presentations of Saturn's rings alternate between about 13 years, 9 months, and about 15 years, 9 months. Why do you think these two intervals are not equal?
51. (**a**) Use Newton's form of Kepler's third law to calculate the orbital periods of particles at the outer edge of Saturn's A ring and at the inner edge of the B ring. (**b**) Saturn's rings orbit in the same direction as Saturn's rotation. If you were floating along with the cloudtops at Saturn's equator, would the outer edge of the A ring and the inner edge of the B ring appear to move in the same or opposite directions? Explain.

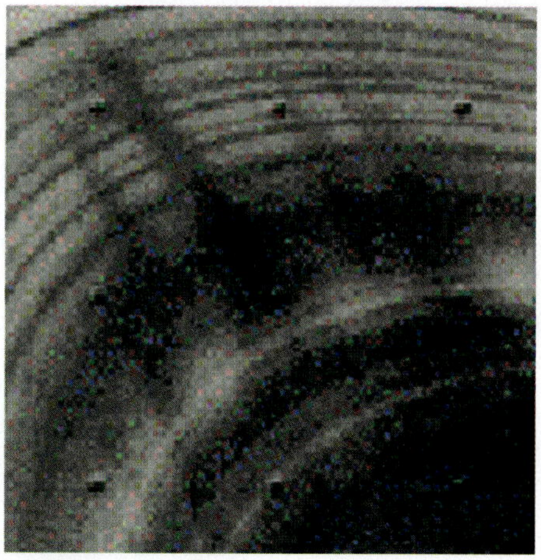

VIDEO 12.10 RIVUXG

(NASA/JPL)

52. The above *Voyager 2* close-up image of Saturn's rings shows a number of dark, straight features called *spokes*. As these features orbit around Saturn, they tend to retain their shape like the rigid spokes on a rotating bicycle wheel. (The black dots were added by the *Voyager* camera system to help scientists calibrate the electronic image.) The spokes rotate at the same rate as Saturn's magnetic field and are thought to be clouds of tiny, electrically charged particles kept in orbit by magnetic forces. Explain why the spokes could *not* maintain their shape if they were kept in orbit by gravitational forces alone.
53. The Cassini division involves a 2-to-1 resonance with Mimas. Does the location of the Encke gap—133,500 km from Saturn's center—correspond to a resonance with one of the other satellites? If so, which one? (See Appendix 3.)

Discussion Questions

54. Describe some of the semipermanent features in Jupiter's atmosphere. Compare and contrast these long-lived features with some of the transient phenomena seen in Jupiter's clouds.
55. Suppose you were asked to design a mission to Jupiter involving an unmanned airplanelike vehicle that would spend many days (months?) flying through the Jovian clouds. What observations, measurements, and analyses should this aircraft be prepared to make? What dangers might the aircraft encounter, and what design problems would you have to overcome?
56. What sort of experiment or space mission would you design in order to establish definitively whether Jupiter has a rocky core?
57. The classic science-fiction films *2001: A Space Odyssey* and *2010: The Year We Make Contact* both involve manned spacecraft in orbit around Jupiter. What kinds of observations could humans make on such a mission that cannot be made by robotic spacecraft? What would be the risks associated with such a mission? Do you think that a manned Jupiter mission would be as worthwhile as a manned mission to Mars? Explain your answers.
58. Suppose that Saturn were somehow moved to an orbit around the Sun with semimajor axis 1 AU, the same as Earth's. Discuss what long-term effects this would have on the planet and its rings.

Web/eBook Questions

59. On Jupiter, the noble gases argon, krypton, and xenon provide important clues about Jupiter's past. On Earth, xenon is used in electronic strobe lamps because it emits a very white light when excited by an electric current. Argon, by contrast, is one of the gases used to fill ordinary incandescent lightbulbs. Search the World Wide Web for information about why argon is used in this way. Why do premium, long-life lightbulbs use krypton rather than argon?
60. Search the World Wide Web, especially the Web sites for NASA's Jet Propulsion Laboratory and the European Space Agency, for information about the current status of the *Cassini* mission. What recent discoveries has *Cassini* made? What ideas are being considered for the *Cassini* extended mission, to begin in 2008?
61. The two *Voyager* spacecraft were launched from Earth along a trajectory that took them directly to Jupiter. The force of Jupiter's gravity then gave the two spacecraft a "kick" that helped push them onward to Saturn. The much larger *Cassini* spacecraft, by contrast, was first launched on a trajectory that took it past Venus. Search the World Wide Web, especially the Web sites for NASA's Jet Propulsion Laboratory and the European Space Agency, for information about the trajectory that *Cassini* took through the solar system to reach

Saturn. Explain why this trajectory was so different from that of the *Voyagers*.

62. **The Rotation Rate of Saturn.** Access and view the video "Saturn from the Hubble Space Telescope" in Chapter 12 of the *Universe* Web site or eBook. The total amount of time that actually elapses in this video is 42.6 hours. Using this information, identify and follow an atmospheric feature and determine the rotation period of Saturn. How does your answer compare with the value given in Table 12-2?

Activities

Observing Projects

Observing tips and tools

Like Mars, Jupiter and Saturn are most easily seen around opposition. (The dates of Jupiter's and Saturn's oppositions are available on the World Wide Web.) Each planet is visible in the night sky for several months before or after its opposition. At other times, Jupiter and Saturn may be visible in either the predawn morning sky or the early evening sky. Consult such magazines as *Sky & Telescope* and *Astronomy* or their Web sites for more detailed information about when and where to look for Jupiter during a given month. You can also use the *Starry Night Enthusiast*™ program on the CD-ROM that accompanies selected copies of this textbook. A relatively small telescope with an objective diameter of 15 to 20 cm (6 to 8 inches), used with a medium-power eyepiece to give a magnification of 25X or so, should enable you to see some of the dark belts on Jupiter and the rings of Saturn.

63. If Jupiter is visible in the night sky, make arrangements to view the planet through a telescope. What magnifying power seems to give you the best view? Draw a picture of what you see. Can you see any belts and zones? How many? Can you see the Great Red Spot?
64. Make arrangements to view Jupiter's Great Red Spot through a telescope. Consult the *Sky & Telescope* Web site, which lists the times when the center of the Great Red Spot passes across Jupiter's central meridian. The Great Red Spot is well placed for viewing for 50 minutes before and after this happens. You will need a refractor with an objective of at least 15 cm (6 in.) diameter or a reflector with an objective of at least 20 cm (8 in.) diameter. Using a pale blue or green filter can increase the color contrast and make the spot more visible. For other useful hints, see the article "Tracking Jupiter's Great Red Spot" by Alan MacRobert (*Sky & Telescope,* September 1997).
65. View Saturn through a small telescope. Make a sketch of what you see. Estimate the angle at which the rings are tilted to your line of sight. Can you see the Cassini division? Can you see any belts or zones in Saturn's clouds? Is there a faint, starlike object near Saturn that might be Saturn's satellite Titan? What observations could you perform to test whether the starlike object is a Saturnian satellite?
66. Use the *Starry Night Enthusiast*™ program to observe the appearance of Jupiter. Select **Favorites > Guides > Atlas** from the menu. Open the **Find** pane and click the menu button for **Jupiter** and select **Magnify** from the drop-down menu. **Stop** time flow and, in the toolbar, set the date and time to **March 4, 2004, at 12:00:00 A.M.** to see Jupiter at opposition. Set the **Time Flow Rate** to **1 minute.** Then **Run Time Forward.** Use the Time Flow controls to determine the rotation period of the planet. (*Hint:* You may want to track the motion of an easily recognizable feature, such as the Great Red Spot.) How does your answer compare with the rotation period given in Table 12-1?
67. Use the *Starry Night Enthusiast*™ program to observe the appearance of Saturn. Select **Favorites > Guides > Atlas** from the menu. Open the **Find** pane and click the menu button for **Saturn** and select **Magnify** from the drop-down menu. **Stop** time flow. Set the **Time Flow Rate** to 1 lunar month. Then **Run Time Forward.** Describe how Saturn's appearance changes over time. Explain what causes these changes.
68. Use the *Starry Night Enthusiast*™ program to examine Jupiter and Saturn. (**a**) Select **Favorites > Solar System > Jupiter** from the menu. Select **View > Feet** to remove the image of the spacesuit. You can zoom in and zoom out on the view and you can also rotate Jupiter by putting the mouse cursor over the image, holding down the mouse button, and moving the mouse. (On a two-button mouse, hold down the left mouse button.) How many white ovals and brown ovals can you count in Jupiter's southern hemisphere (the hemisphere in which the Great Red Spot is located)? How many white ovals and brown ovals can you count in the northern hemisphere? What general rule can you state about the abundance of these storms in the two hemispheres? (**b**) Select **Favorites > Solar System > Saturn** from the menu. Select **View > Feet** to remove the image of the spacesuit. You can zoom in and zoom out on the view and you can also rotate Saturn by placing the mouse cursor over the image, holding down the mouse button, and moving the mouse. (On a two-button mouse, hold down the left mouse button.) Rotate Saturn so that you are looking straight down on the plane of the rings. Draw a copy of what you see and label the different rings and divisions. (Compare Figure 12-23.)

Collaborative Exercises

69. Using a ruler with millimeter markings on five various images of Jupiter in the text, Figures 12-1*a*, 12-2*a*, and 12-4, determine the ratio of the longest width of the Great Red Spot to the full diameter of Jupiter. Each group member should measure a different image and all values should be averaged.
70. The text provides different years that spacecraft have flown by Jupiter and Saturn. List these dates and create a time line by listing one important event that was occurring on Earth during each of those years.
71. If the largest circle you can draw on a piece of paper represents the largest diameter of Saturn's rings, about how large would Saturn be if scaled appropriately? Which item in a group member's backpack is closest to this size?

13

Jupiter and Saturn's Satellites of Fire and Ice

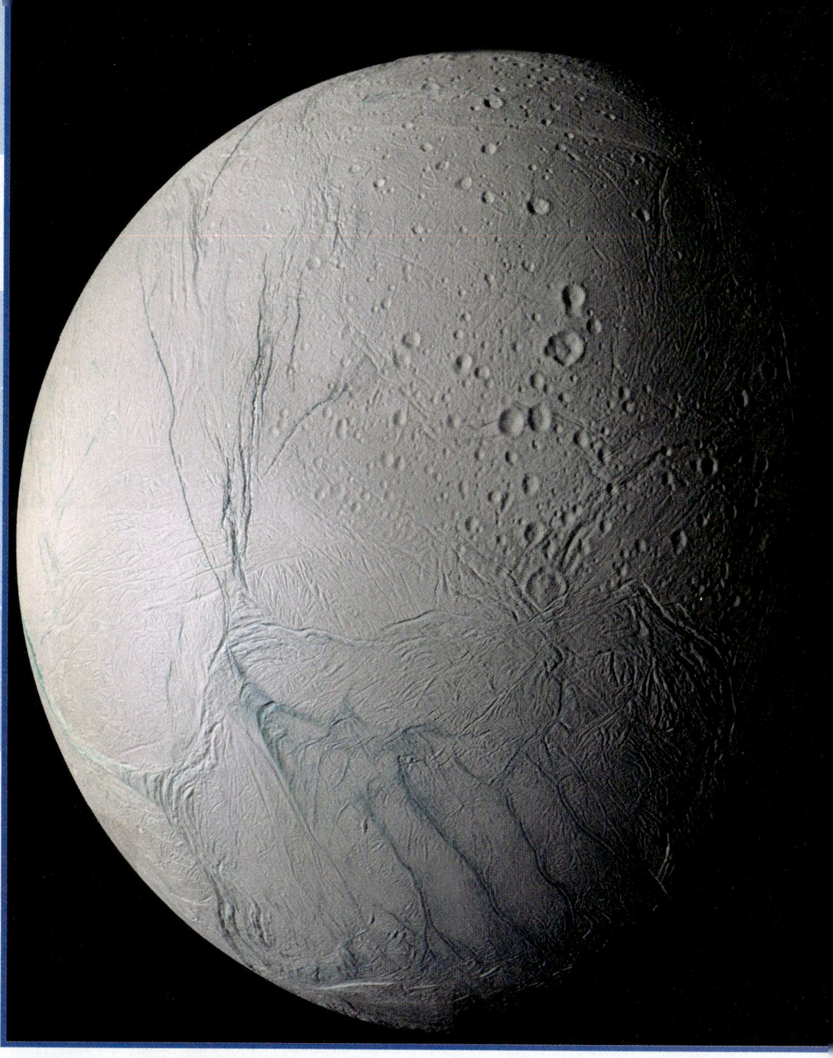

R I V U X G

A color-enhanced view of Saturn's satellite (or moon) Enceladus. (NASA/JPL/Space Science Institute)

The delicately colored world shown here is neither the creation of an abstract artist nor the product of a science fiction writer's imagination. It is Enceladus, a satellite, or moon of Saturn whose icy surface has a split personality: One hemisphere is ancient and covered with impact craters, while the other hemisphere is nearly crater-free and crisscrossed by stress faults and fractures. The latter hemisphere is actually the site of active volcanoes that eject icy particles rather than lava.

Enceladus is not the only satellite of the giant planets to display exotic geologic activity. Io, a large satellite of Jupiter discovered by Galileo in 1610, is the most geologically active world in the solar system. Its surface is covered with colorful sulfur compounds deposited by dozens of active volcanoes. Europa, another of Jupiter's large satellites, has no volcanoes but does have a weirdly cracked surface of ice. Beneath Europa's ice there is probably a worldwide ocean, within which there could possibly be single-celled organisms.

No less exotic is Saturn's large satellite Titan, which is enveloped by a thick, nitrogen-rich atmosphere from which liquid hydrocarbons fall as rain. In this chapter we will explore these and other intriguing aspects of Jupiter and Saturn's extensive collection of satellites.

13-1 Jupiter's Galilean satellites are easily seen with Earth-based telescopes

When Galileo trained his telescope on Jupiter in January 1610, he became the first person to observe satellites orbiting another planet (see Section 4-5). He

Learning Goals

By reading the sections of this chapter, you will learn

- 13-1 What the Galilean satellites are and how they orbit Jupiter
- 13-2 The similarities and differences among the Galilean satellites
- 13-3 How the Galilean satellites formed
- 13-4 Why Io is the most volcanically active world in the solar system
- 13-5 How Io interacts with Jupiter's magnetic field
- 13-6 The evidence that Europa may have an ocean beneath its surface
- 13-7 The kinds of geologic activity found on Ganymede and Callisto
- 13-8 The nature of Titan's thick, hydrocarbon-rich atmosphere
- 13-9 Why most of Jupiter's moons orbit in the "wrong" direction
- 13-10 What powers the volcanoes on Saturn's icy moon Enceladus

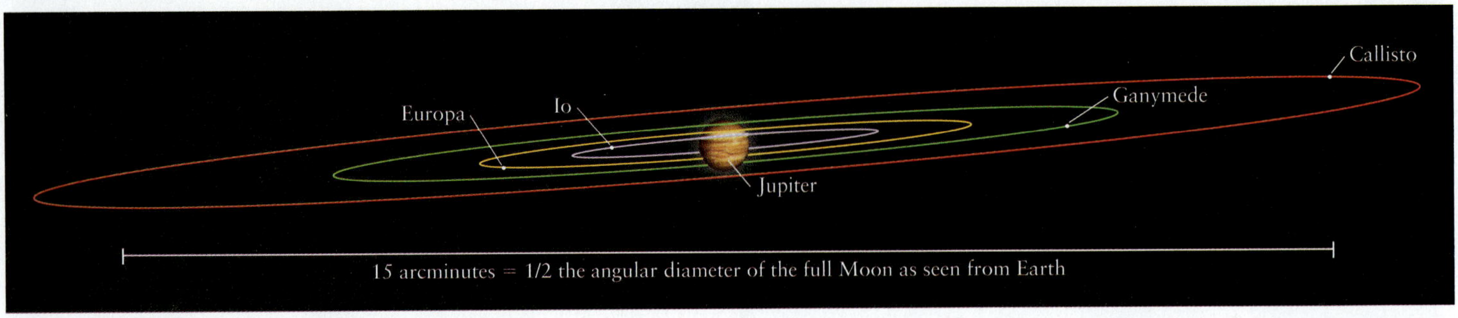

Figure 13-1

The Orbits of the Galilean Satellites This illustration shows the orbits of the four Galilean satellites as seen from Earth. All four orbits lie in nearly the same plane as Jupiter's equator. The apparent angular size and orientation of the orbits depends on the relative positions of Earth and Jupiter.

called them the Medicean Stars, in order to curry the favor of a wealthy Florentine patron of the arts and sciences. We now call them the **Galilean satellites.** Individually, they are named Io, Europa, Ganymede, and Callisto, after four mythical lovers and companions of the god whom the Greeks called Zeus and the Romans called Jupiter.

When viewed through an Earth-based telescope, the Galilean satellites look like pinpoints of light (see Figure 4-16). All four satellites orbit Jupiter in nearly the same plane as the planet's equator (Figure 13-1). From Earth, we always see that plane nearly edge-on, so the satellites appear to move back and forth relative to Jupiter (see Figure 4-17). The orbital periods are fairly short, ranging from 1.8 (Earth) days for Io to 16.7 days for Callisto, so we can easily see the satellites move from one night to the next and even during a single night. These motions led Galileo to realize that he was seeing objects orbiting around Jupiter, in much the same way that Copernicus said that planets move around the Sun (see Section 4-2).

Each of the four Galilean satellites is bright enough to be visible to the naked eye. Why, then, did Galileo need a telescope to discover them? The reason is that as seen from Earth, the angular separation between Jupiter and the satellites is quite small—never more than 10 arcminutes for Callisto and even less for the other three Galilean satellites. To the naked eye, these satellites are lost in the overwhelming glare of Jupiter. But a small telescope or even binoculars increases the apparent angular separation by enough to make the Galilean satellites visible.

Synchronous Rotation

The brightness of each satellite varies slightly as it orbits Jupiter, because the satellites also spin on their axes as they orbit Jupiter, and dark and light areas on their surfaces are alternately exposed to and hidden from our view. Remarkably, each Galilean satellite goes through one complete cycle of brightness during one orbital period. This observation tells us that each satellite is in *synchronous rotation,* so that it rotates exactly once on its axis during each trip around its orbit (see Figure 3-4*b*). As we saw in Section 4-8, our Moon's synchronous rotation is the result of gravitational forces exerted on the Moon by Earth. Likewise, Jupiter's gravitational forces keep the Galilean satellites in synchronous rotation.

ANIMATION 13.1

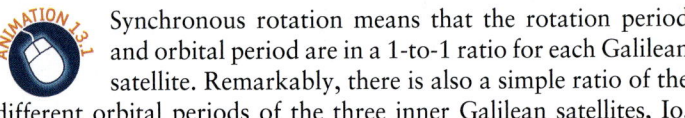

Synchronous rotation means that the rotation period and orbital period are in a 1-to-1 ratio for each Galilean satellite. Remarkably, there is also a simple ratio of the different orbital periods of the three inner Galilean satellites, Io, Europa, and Ganymede. During the 7.155 days that Ganymede takes to complete one orbit around Jupiter, Europa makes two orbits and Io makes four orbits. Thus, the orbital periods of Io, Europa, and Ganymede are in the ratio 1:2:4, which you can verify from the data in Table 13-1.

This special relationship among the satellites' orbits is maintained by the gravitational forces that they exert on one another. Those forces can be quite strong, because the three inner Galilean satellites pass relatively close to one another; at their closest approach, Io and Europa are separated by only two-thirds the distance from Earth to the Moon. Such a close approach occurs once for every two of Io's orbits, so Europa's gravitational pull acts on Io in a rhythmic way. Indeed, Io, Europa, and Ganymede all exert rhythmic gravitational tugs on one another. Just as a drummer's rhythm keeps musicians on the same beat, this gravitational rhythm maintains the simple ratio of orbital periods among the satellites.

Gravitational interactions between Io, Europa, and Ganymede induce a rhythmic relationship among their orbits

By contrast, Callisto orbits at a relatively large distance from the other three large satellites. Hence, the gravitational forces on Callisto from the other satellites are relatively weak, and there is no simple relationship between Callisto's orbital period and the period of Io, Europa, or Ganymede.

Transits, Eclipses, and Occultations

Because the orbital planes of the Galilean satellites are nearly edge-on to our line of sight, we see these satellites undergoing transits, eclipses, and occultations. In a *transit,* a satellite passes between us and Jupiter, and we see the satellite's shadow as a black dot against the planet's colorful cloudtops. (Figure 12-2*a* shows Europa's shadow on Jupiter.) In an *eclipse,* one of the satellites disappears and then reappears as it passes into and out of Jupiter's enormous shadow. In an **occultation,** a satellite passes

Table 13-1 Jupiter's Galilean Satellites Compared with the Moon, Mercury, and Mars

	Average distance from Jupiter (km)	Orbital period (days)	Diameter (km)	Mass		Average density	
				(kg)	(Moon = 1)	(kg/m³)	Albedo
Io	421,600	1.769	3642	8.932×10^{22}	1.22	3529	0.63
Europa	670,900	3.551	3120	4.791×10^{22}	0. 65	3018	0.64
Ganymede	1,070,000	7.155	5268	1.482×10^{23}	2.02	1936	0.43
Callisto	1,883,000	16.689	4800	1.077×10^{23}	1.47	1851	0.17
Moon	—	—	3476	7.349×10^{22}	1.00	3344	0.11
Mercury	—	—	4880	3.302×10^{23}	4.49	5430	0.12
Mars	—	—	6794	6.419×10^{23}	8.73	3934	0.15

Note: Jupiter is shown to the same scale as the distances of the satellites from Jupiter. Compared to this scale, the images of the satellites themselves have been enlarged 74×.

WEB LINK 13.2 RIVUXG

(NASA/JPL)

completely behind Jupiter, and the satellite is temporarily blocked from our Earth-based view. (The word *occultation* comes from a Latin verb meaning "to cover.")

Before spacecraft first ventured to Jupiter, astronomers used eclipses to estimate the diameters of the Galilean satellites. When a satellite emerges from Jupiter's shadow, it does not blink on instantly. Instead, there is a brief interval during which the satellite gets progressively brighter as more of its surface is exposed to sunlight. The satellite's diameter is calculated from the duration of this interval and the satellite's orbital speed (which is known from Kepler's laws).

Occultations can also be used to determine the diameter of a satellite by measuring the time it takes to disappear behind Jupiter or to reappear from behind the planet. In addition, the Galilean satellites occasionally occult one another, because all four orbit Jupiter in the plane of the planet's equator. Every six years, when Earth passes through this equatorial plane, there is a period of a few months during which Earth observers can see mutual occultations of the satellites. Once again, timing the occultations enables astronomers to calculate the diameters of the satellites.

As Table 13-1 shows, the smallest Galilean satellite (Europa) is slightly smaller than our Moon; the largest satellite (Ganymede) is larger than Mercury and more than three-quarters the size of Mars. Thus, the Galilean satellites truly are worlds in their own right.

13-2 Data from spacecraft reveal the unique properties of the Galilean satellites

WEB LINK 13.3

Even the finest images made with Earth-based telescopes have revealed relatively few details about the Galilean satellites. Almost everything we know about these satellites has come from observations made at close range by spacecraft.

The first close-range observations of the Galilean satellites were made by the *Pioneer 10* and *Pioneer 11* spacecraft as they flew past Jupiter in 1973 and 1974. The images from these missions were of relatively low resolution, however, and much more information has come from three subsequent missions. *Voyager 1* and *Voyager 2* flew past Jupiter in 1979 and recorded tens of thousands of images, one of which is shown in Figure 13-2. The *Galileo* spacecraft, which made 35 orbits around Jupiter between 1995 and 2003, carried out an even more extensive investigation of Jupiter's satellites.

Measuring Satellite Densities

One key goal of these missions was to determine the densities of the Galilean satellites to very high accuracy. Given the density of a planet or satellite, astronomers can draw conclusions about its chemical composition and internal structure. To find the densities, scientists first determined the satellite masses. They did this by measuring how the gravity of Io, Europa, Ganymede, and Callisto deflected the trajectories of the *Voyager* and *Galileo* spacecraft. Given the masses and diameters of the satellites, they then calculated each satellite's average density (mass divided by volume). Table 13-1 lists our current knowledge about the sizes and masses of the Galilean satellites.

Of the four Galilean satellites, Europa proves to be the least massive: Its mass is only two-thirds that of our Moon. Ganymede is by far the most massive of the four, with more than double the mass of our Moon. In fact, Ganymede is the most massive satellite anywhere in the solar system. In second place is Callisto, with about $1\frac{1}{2}$ times the Moon's mass.

However, Ganymede and Callisto are *not* merely larger versions of our Moon. The reason is that both Ganymede and Callisto have very low average densities of less than 2000 kg/m^3. By comparison, typical rocks in Earth's crust have densities around 3000 kg/m^3, and the average density of our Moon is 3344 kg/m^3. The low densities of Ganymede and Callisto mean that these satellites cannot be composed primarily of rock. Instead, they are probably roughly equal parts of rock and water ice. Water ice is substantially less dense than rock, but it becomes as rigid as rock under high pressure (such as is found in the interiors of the Galilean satellites). Furthermore, water molecules are relatively common in the solar system. Enough water would have been available in the early solar system to make up a substantial portion of a large satellite such as Ganymede or Callisto.

Three of the Galilean satellites are made of a mixture of ice and rock; only Io is ice-free

Water ice cannot be a major constituent of the two inner Galilean satellites, however. The innermost satellite, Io, has the highest average density, 3529 kg/m^3, slightly greater than the density of our Moon. The next satellite out, Europa, has an average density of 3018 kg/m^3. Both of these values are close to the densities of typical rocks in Earth's crust. Hence, it is reasonable to suppose that both Io and Europa are made primarily of rocky material.

The most definitive evidence for water ice on Jupiter's satellites has come from spectroscopic observations. The spectra of sunlight reflected from Europa, Ganymede, and Callisto show absorption at the infrared wavelengths characteristic of water ice molecules. You can see this in Figure 7-4, which compares the spectrum of Europa to the spectrum of ice. Europa's density shows that it is composed mostly of rock, so its ice must be limited to the satellite's outer regions. Among the Galilean satellites, only Io shows no trace of water ice on its surface or in its interior.

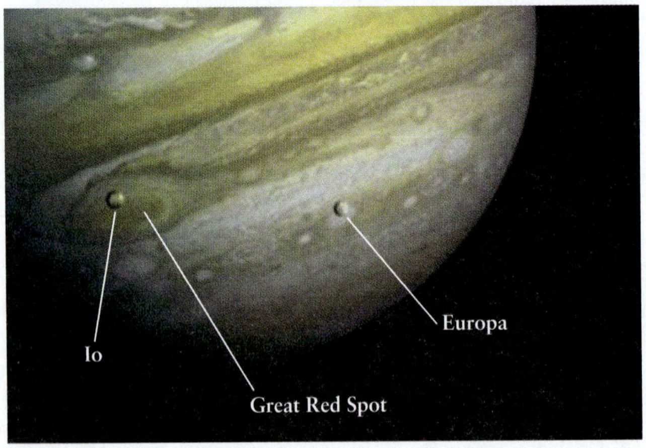

Figure 13-2 R I V U X G

Io and Europa from *Voyager 1* *Voyager 1* recorded this image of Jupiter and the Galilean satellites Io and Europa in February 1979, when it was 20 million kilometers (12.4 million miles) from the giant planet. The satellites are actually smaller than some of Jupiter's cloud features, such as the Great Red Spot (see Section 14-3). *Voyager 1* and *Voyager 2* also recorded many images of the Galilean satellites at closer range. (NASA/JPL)

13-3 The Galilean satellites formed like a solar system in miniature

As Table 13-1 shows, the more distant a Galilean satellite is from Jupiter, the greater the proportion of ice within the satellite and the lower its average density. But why should this be? An important clue is that the average densities of the planets show a similar trend: Moving outward from the Sun, the average density of the inner six planets steadily decreases, from more than 5000 kg/m^3 for Mercury to less than 1000 kg/m^3 for Saturn (recall Table 7-1). The best explanation for this similarity is that the Galilean satellites formed around Jupiter in much the same way that the planets formed around the Sun, although on a much smaller scale.

The differences in composition among the Galilean satellites mimic those among planets orbiting the Sun

We saw in Section 8-4 that dust grains in the solar nebula played an important role in planet formation. In the inner parts of the nebula, close to the glowing protosun, only dense, rocky grains were able to survive. These grains accumulated over time into the dense, rocky inner planets. But in the cold outer reaches of the nebula, dust grains were able to retain icy coatings of low-density material like water and ammonia. Hence, when the Jovian planets formed in the outer solar nebula and incorporated these ice-coated grains, the planets ended up with both rock and ices in their cores as well as substantial amounts of ammonia and water throughout their interiors and atmospheres (see Section 14-1 and Section 14-6).

As Jupiter coalesced, the gas accumulating around it formed a rotating "Jovian nebula." The central part of this nebula became the huge envelope of hydrogen and helium that makes up most of Jupiter's bulk. But in the outer parts of this nebula, dust grains could have accreted to form small solid bodies. These grew to become the Galilean satellites.

Although the "Jovian nebula" was far from the protosun, not all of it was cold. Jupiter, like the protosun, must have emitted substantial amounts of radiation due to Kelvin-Helmholtz contraction, which we discussed in Section 8-4. (We saw in Section 12-4 that even today, Jupiter emits more energy due to its internal heat than it absorbs from sunlight.) Hence, temperatures very close to Jupiter must have been substantially higher than at locations farther from the planet. Calculations show that only rocky material would condense at the orbital distances of Io and Europa, but frozen water could be retained and incorporated into satellites at the distances of Ganymede and Callisto. In this way Jupiter ended up with two distinct classes of Galilean satellites (Figure 13-3).

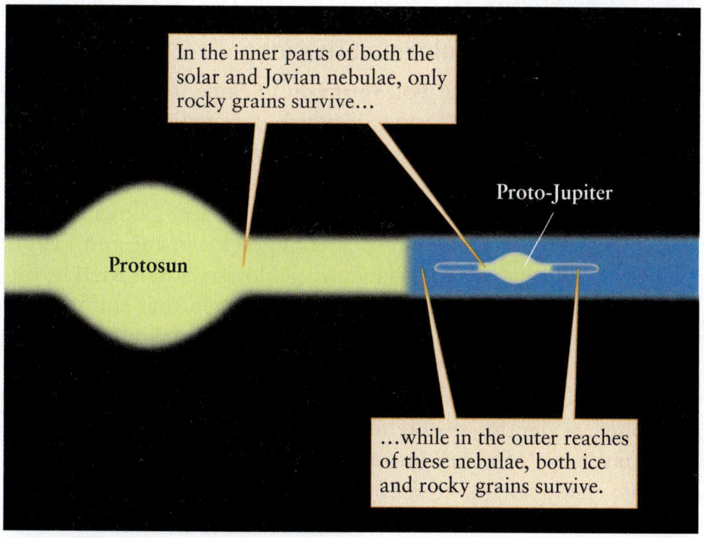

In the inner parts of both the solar and Jovian nebulae, only rocky grains survive...
Proto-Jupiter
Protosun
...while in the outer reaches of these nebulae, both ice and rocky grains survive.

Figure 13-3

Formation of the Galilean Satellites Heat from the protosun made it impossible for icy grains to survive within the innermost 2.5 to 4 AU of the solar nebula. In the same way, Jupiter's heat evaporated any icy grains that were too close to the center of the "Jovian nebula." Hence, the two inner Galilean satellites were formed primarily from rock, while the outer two incorporated both rock and ice.

The analogy between the solar nebula and the "Jovian nebula" is not exact. Unlike the Sun, Jupiter can be thought of as a "failed star." Its internal temperatures and pressures never became high enough to ignite nuclear reactions that convert hydrogen into helium. (Jupiter's mass would have had to be about 80 times larger for these reactions to have begun.) Furthermore, the icy worlds that formed in the outer region of the "Jovian nebula"—namely, Ganymede and Callisto—were of relatively small mass. Hence, they were unable to attract gas from the nebula and become Jovian planets in their own right. Nonetheless, it is remarkable how the main processes of "planet" formation seem to have occurred twice in our solar system: once in the solar nebula around the Sun, and once in microcosm in the gas cloud around Jupiter.

The diverse densities of the Galilean satellites suggest that these four worlds may be equally diverse in their geology. This conclusion turns out to be entirely correct. Because each of the Galilean satellites is unique, we devote the next several sections to each satellite in turn.

13-4 Io is covered with colorful sulfur compounds ejected from active volcanoes

WEB LINK 13.4

Before the two *Voyager* spacecraft flew past Jupiter, the nature of Io was largely unknown. On the basis of Io's size and density, however, it was expected that Io would be a world much like our own Moon—geologically dead, with little internal heat available to power tectonic or volcanic activity (see Section 7-6). Hence, it was thought that Io's surface would be extensively cratered, because there would have been little geologic activity to erase those craters over the satellite's history.

Io proved to be utterly different from these naive predictions. On March 5, 1979, *Voyager 1* came within 21,000 kilometers (13,000 miles) of Io and began sending back a series of bizarre and unexpected pictures of the satellite (Figure 13-4). These images showed that Io has *no* impact craters at all! Instead, the surface is pockmarked by irregularly shaped pits and is blotched with color.

The scientists were seeing confirmation of a prediction published in the journal *Science* just three days before *Voyager 1* flew past Io. In this article, Stanton Peale of the University of California, Santa Barbara, and Patrick Cassen and Ray Reynolds of NASA's Ames Research Center reported their conclusion that Io's interior must be kept hot by Jupiter's tidal forces, and that this should lead to intense volcanic activity.

Tidal Heating on Io

As we saw in Section 4-8, tidal forces are differences in the gravitational pull on different parts of a planet or satellite. These forces tend to deform the shape of the planet or satellite (see Figure 4-25). Io, for example, is somewhat distorted from a spherical shape by tidal forces from Jupiter. If Io were in a perfectly circular orbit so that it traveled around Jupiter at a steady speed, Io's rotation would always be in lockstep with its orbital motion. In this case, Io would be in perfect synchronous rotation, with its long axis always pointed directly toward Jupiter. But as Io moves

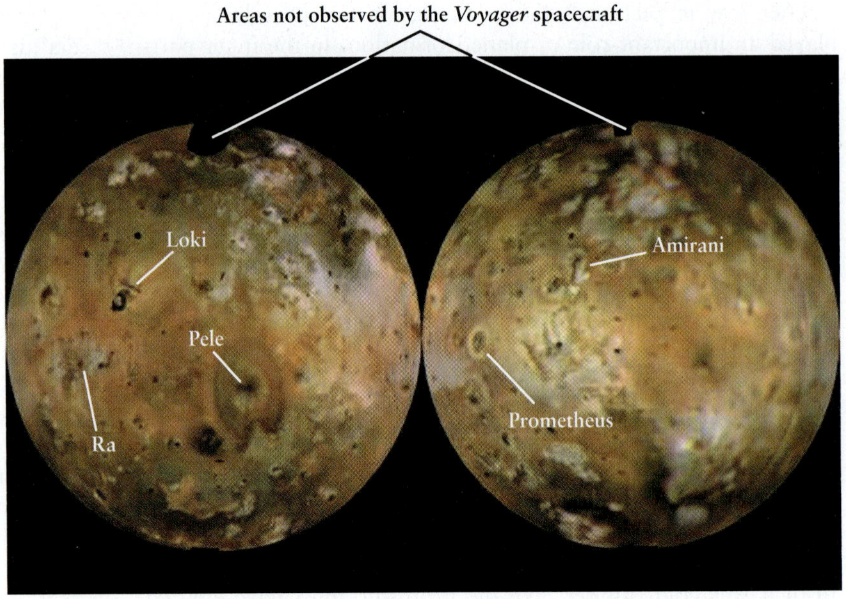

VIDEO 13.1 **Figure 13-4** RI**V**UXG

Io These mosaics of Io's two hemispheres were built up from individual *Voyager* images. The extraordinary colors are probably caused by deposits of sulfur and sulfur compounds ejected from Io's numerous volcanoes. The labels show several of these volcanoes, which are named for sun gods and fire gods of different cultures. (NASA/JPL)

around its orbit, Europa and Ganymede exert gravitational tugs on it in a regular, rhythmic fashion, thanks to the 1:2:4 ratio among the orbital periods of these three satellites. These tugs distort Io's orbit into an ellipse, and so its speed varies as it moves around its orbit. Consequently, Io's long axis "nods" back and forth by about ½° as seen from Jupiter. Due to this "nodding," the tidal stresses that Jupiter exerts on Io vary rhythmically as Io moves around its orbit. These varying tidal stresses alternately squeeze and flex Io. Just as a ball of clay or bread dough gets warm as you knead it between your fingers, this squeezing and flexing causes **tidal heating** of Io's interior. (Earth's Moon also "nods" as it orbits our planet, but experiences almost no tidal heating because Earth's tidal forces are very weak compared to those of massive Jupiter.)

The theory of tidal forces helped predict Io's volcanic nature

Tidal heating adds energy to Io at a rate of about 10^{14} watts, equivalent to 24 tons of TNT exploding every second. As this energy makes its way to the satellite's surface, it provides about 2.5 watts of power to each square meter of Io's surface. By comparison, the average global heat flow through Earth's crust is 0.06 watts per square meter. Only in volcanically active areas on Earth do we find heat flows that even come close to Io's average. Thus, Peale, Cassen, and Reynolds predicted "widespread and recurrent surface volcanism" on Io.

Io's Active Volcanoes

No one expected that *Voyager 1* would obtain images of erupting volcanoes on Io. After all, a spacecraft making a single trip past Earth would be highly unlikely to catch a large volcano actually erupting. But, in fact, *Voyager 1* images of Io revealed *eight* different giant plumes of gas from volcanic eruptions. (Figure 13-5*a* shows two of these.) More recently, the *Galileo* spacecraft, which came as close as 200 km to Io's surface during its eight years in orbit around Jupiter, returned detailed images of several such plumes (Figure 13-5*b*). These observations resoundingly confirm the Peale-Cassen-Reynolds prediction: Io is by far the most volcanic world in the solar system.

Io's volcanic plumes rise to astonishing heights of 70 to 280 km above Io's surface. To reach such altitudes, the material must emerge from volcanic vents with speeds between 300 and 1000 m/s (1100 to 3600 km/h, or 700 to 2200 mi/h). Even the most violent volcanoes on Earth have eruption speeds of only around 100 m/s (360 km/h, or 220 mi/h). Scientists thus began to suspect that Io's volcanic activity must be fundamentally different from volcanism here on Earth.

The Nature of Io's Volcanic Eruptions

An important clue about volcanism on Io came from the infrared spectrometers aboard *Voyager 1,* which detected abundant sulfur and sulfur dioxide in Io's volcanic plumes. This observation led to the idea that the plumes are actually more like geysers than volcanic eruptions. In a geyser on Earth, water seeps down to volcanically heated rocks, where it changes to steam and erupts explosively through a vent. Planetary geologists Susan Kieffer, Eugene Shoemaker, and Bradford Smith suggested that sulfur dioxide rather than water could be the principal propulsive agent driving volcanic plumes on Io. Sulfur dioxide is a solid at the frigid temperatures found on most of Io's surface, but it should be molten at depths of only a few kilometers. Just as the explosive conversion of water into steam produces a geyser on Earth, the conversion of liquid sulfur dioxide into a high-pressure gas could result in eruption velocities of up to 1000 m/s.

VIDEO 13.2 Io's dramatic coloration (see Figure 13-4) is probably due to sulfur and sulfur dioxide, which are ejected in volcanic plumes and later fall back to the surface. Sulfur is normally bright yellow, which explains the dominant color of Io's surface. But if sulfur is heated and suddenly cooled, as would happen if it were ejected from a volcanic vent and allowed to fall to the surface, it can assume a range of colors from orange

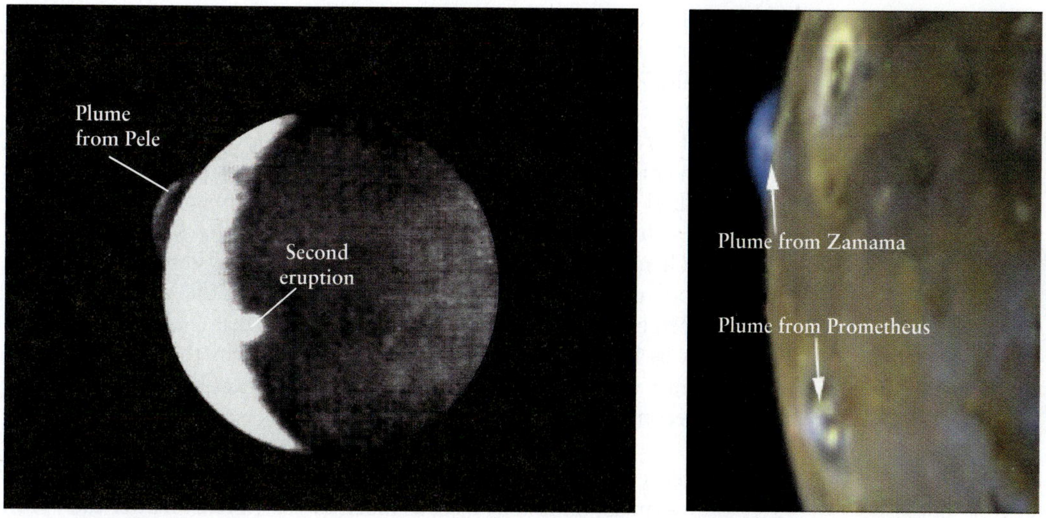

(a) *Voyager 1*, March 1979

(b) *Galileo*, November 1997

Figure 13-5 R I V U X G

Volcanic Plumes on Io **(a)** Io's volcanic eruptions were first discovered on this *Voyager 1* image. The plume at upper left, which extends 260 km (150 mi) above Io's surface, came from the volcano Pele (see Figure 13-4). **(b)** This *Galileo* image shows plumes from the volcanoes Prometheus (see Figure 13-4) and Zamama. The plumes rise to about 100 km (60 mi) above Io's surface and are about 250 km (150 mi) wide. (a: NASA; b: Planetary Image Research Laboratory/University of Arizona/JPL/NASA)

and red to black. Indeed, these colors are commonly found around active volcanic vents (Figure 13-6). Whitish surface deposits (examine Figure 13-4 and Figure 13-6*a*), by contrast, are probably due to sulfur dioxide (SO_2). Volcanic vents on Earth commonly discharge SO_2 in the form of an acrid gas. But on Io, when hot SO_2 gas is released by an eruption into the cold vacuum of space, it crystallizes into white snowflakes. This sulfur dioxide "snow" then rains back onto Io's surface.

The plumes are not the whole story of volcanism on Io, however. The sources of the plumes appear in *Voyager* images as black spots 10 to 50 km in diameter, which are actually volcanic vents (see Figure 13-4). Many of these black spots, which cover 5% of Io's surface, are surrounded by dark lava flows. Figure 13-7 is a *Galileo* image of one of Io's most active volcanic regions, where in 1999 lava was found spouting to altitudes of thousands of meters along a fissure 40 km (25 mi) in length. The glow from this lava fountain was so intense that it was visible with Earth-based telescopes. Thus, Io has two distinct styles of volcanic activity: unique geyserlike plumes, and lava flows that are more dramatic versions of volcanic flows on Earth.

While sulfur and sulfur compounds are the key constituents of Io's volcanic plumes, the lava flows must be made of something else. Infrared measurements from the *Galileo* spacecraft (Figure 13-8) show that lava flows on Io are at temperatures of 1700 to 2000 K (about 1450 to 1750°C, or 2600 to 3150°F). At these temperatures sulfur could not remain molten but would evaporate almost instantly. Io's lavas are also unlikely to have the same chemical composition as typical lavas on Earth, which have temperatures of only 1300 to 1450 K. Instead, lavas on Io are probably **ultramafic lavas.** These are enriched in magnesium and iron, which give the lava a higher melting temperature.

Solidified ultramafic lavas are found on Earth, but primarily in lava beds that formed billions of years ago when Earth's interior was much hotter than today. The presence of molten

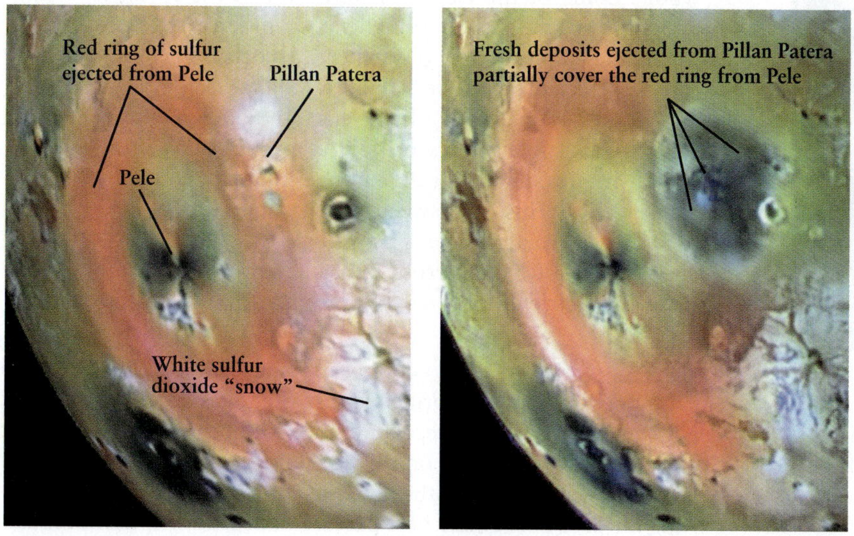

(a) (b)

WEB LINK 13.5

Figure 13-6 R I V U X G

Rapid Changes on Io **(a)** This *Galileo* image shows a portion of Io's southern hemisphere. Sulfur ejected from the active volcanic vent Pele produced a red ring that resembles paint sprayed from a can. **(b)** A few months later, a volcanic plume erupted from a vent called Pillan Patera, spraying dark material over an area some 400 km (250 mi) in diameter (about the size of Arizona). Other than Earth, Io is the only world in the solar system that shows such noticeable changes over short time spans. (NASA/JPL)

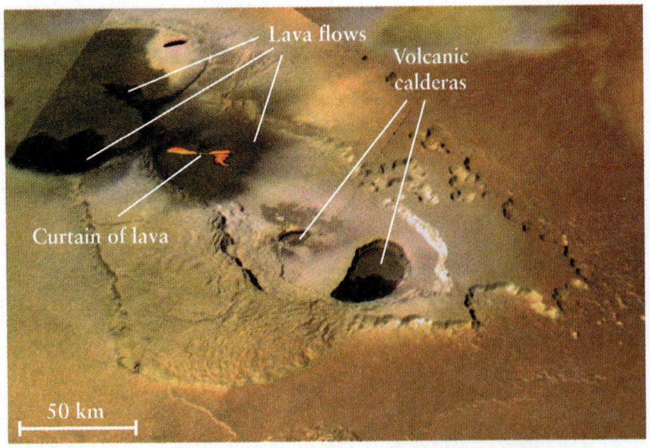

Figure 13-7 R I V U X G

Io's Lava Flows and a Curtain of Fire This mosaic of *Galileo* spacecraft images shows a nested chain of volcanic calderas (pits) with black lava flows. The red streak at upper left is a fissure in the surface about 40 km (25 mi) long, through which lava is erupting upward in a vertical sheet or curtain. The erupting lava reaches the amazing height of 1500 m (5000 ft). (University of Arizona/JPL/NASA)

ultramafic lava on Io suggests that its interior, too, is substantially hotter than that of the present-day Earth. One model proposes that Io has a 100-km (60-mi) thick crust floating atop a worldwide ocean of liquid magma 800 km (500 mi) deep. (As we saw in Section 9-2, the mantle that lies underneath Earth's crust is a plastic solid rather than a true liquid.) The global extent of this "magma ocean" would explain why volcanic activity is found at all points on Io's surface, rather than being concentrated in pockets as on Earth (see Section 9-3). It could also explain the origin of Io's mountains, which reach heights up to 10 km (30,000 ft). In this model, these mountains are blocks of Io's crust that have tilted before sinking into the depths of the magma ocean.

Io's worldwide volcanic activity is remarkably persistent. When *Voyager 2* flew through the Jovian system in July 1979, four months after *Voyager 1,* almost all of the volcanic plumes seen by *Voyager 1* were still active. And when *Galileo* first viewed Io in 1996, about half of the volcanoes seen by the *Voyager* spacecraft were still ejecting material, while other, new volcanoes had become active.

Io may have as many as 300 active volcanoes, each of which ejects an estimated 10,000 tons of material per second. Altogether, volcanism on Io may eject as much as 10^{13} tons of matter each year. This amount of material is sufficient to cover the satellite's entire surface to a depth of 1 meter in a century, or to cover an area of 1000 square kilometers in a few weeks (see Figure 13-6). Thanks to this continual "repaving" of the surface, there are probably no long-lived features on Io, and any impact craters are quickly obliterated.

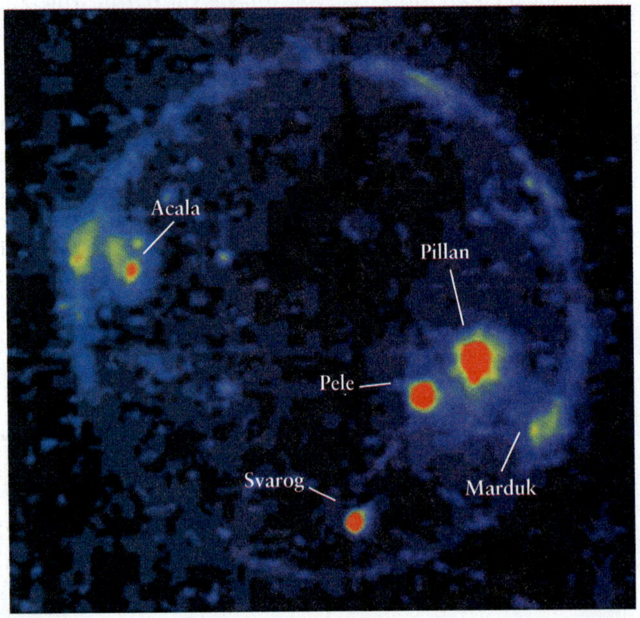

WEB LINK 13.6

Figure 13-8 R I V U X G

Io's Glowing Volcanoes This false-color *Galileo* image (made while Io was in the darkness of Jupiter's shadow) shows the infrared and visible glow from several volcanic regions, including Pele and Pillan (see Figure 13-6). These regions are substantially hotter than typical lava flows on Earth. (Planetary Image Research Laboratory/University of Arizona/JPL/NASA)

13-5 Jupiter's magnetic field makes electric currents flow through Io

Most of the material ejected from Io's volcanoes falls back onto the satellite's surface. But some material goes on a remarkable journey into space, by virtue of Io's location deep within Jupiter's magnetosphere.

The Io Torus

The Jovian magnetosphere contains charged particles, some of which collide with Io and its volcanic plumes. The impact of these collisions knocks ions out of the plumes and off the surface, and these ions become part of Jupiter's magnetosphere. Indeed, material from Io is the main source for Jupiter's magnetosphere as a whole. Some of the ejected material forms the **Io torus,** a huge doughnut-shaped ring of ionized gas, or plasma (see Section 12-7), that circles Jupiter at the distance of Io's orbit. The plasma's glow can be detected from Earth (**Figure 13-9*a***).

Jupiter and Io together act like an immense electric generator

Jupiter's magnetic field has other remarkable effects on Io. As Jupiter rotates, its magnetic field rotates with it. This field sweeps over Io at high speed and generates a voltage of 400,000 volts across the satellite, causing 5 million amperes of electric current to flow through Io.

ANALOGY Whenever you go shopping with a credit card, you use the same physical principle that generates an electric current within Io. The credit card's number is imprinted as a magnetic code in the dark stripe on the back of the card. The salesperson who swipes your card through the card reader is actually moving the card's magnetic field past a coil of wire within the reader. This action generates within the coil an electric current that carries the same coded information about your credit card number.

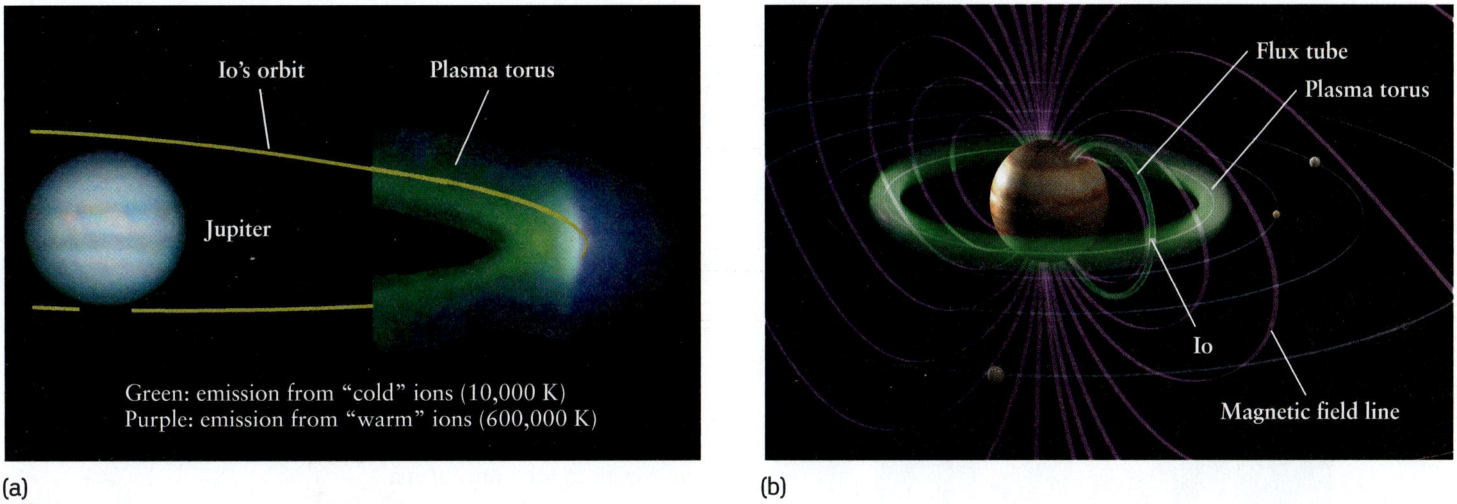

Figure 13-9 R I V U X G

The Io Torus **(a)** An Earth-based telescope recorded this false-color infrared image of Io's plasma torus. Because of Jupiter's glare, only the outer edge of the torus could be photographed; an artist added the visible-light picture of Jupiter and the yellow line indicating the rest of the torus. **(b)** A flux tube (a region of concentrated magnetic field) carries plasma between Jupiter and Io, forming an immense electric circuit. (a: Courtesy of J. Trauger; b: Alfred T. Kamajian and Torrence V. Johnson, "The Galileo Mission to Jupiter and Its Moons," *Scientific American*, February 2000, p. 44)

That information is transmitted to the credit card company, which (it is hoped) approves your purchase. An electric generator at a power plant works in the same way. A coil of wire is moved through a strong magnetic field, which makes current flow in the coil. This current is delivered to transmission lines and eventually to your home.

In fact, a current flows not only through Io but also through the sea of charged particles in Jupiter's magnetosphere and through Jupiter's atmosphere. This forms a gigantic, oval-shaped electric circuit that connects Io and Jupiter in the same way that wires connect the battery and the lightbulb in a flashlight (Figure 13-9*b*). Jupiter's aurora, which we discussed in Section 12-7, is particularly strong at the locations where this current strikes the upper atmosphere of Jupiter.

Part of the electric current that flows between Io and Jupiter is made up of electrons that spiral around Jupiter's magnetic field lines. As they spiral, the electrons act like miniature antennas and emit radio waves. This is the source of Jupiter's bursts of decametric radio radiation, which we discussed in Section 12-7.

Probing Io's Interior

The electric current within and near Io also creates a weak magnetic field, which was first detected by *Voyager 1*. Remarkably, measurements made by *Galileo* during its close approaches to Io suggest that the magnetic field near Io may be too strong—comparable to that of Mercury, although weaker than that of Earth—to be generated by Io's electric current alone. This result has led scientists to speculate that Io generates its own magnetic field through the motions of molten material in its interior, just as Earth does (see Section 7-7 and Section 9-4). If these speculations are correct, Io is the smallest world in the solar system to generate its own magnetic field.

If Io has undergone enough tidal heating to melt its interior, chemical differentiation must have taken place and the satellite should have a dense core. (We described chemical differentiation in Box 7-1 and Section 8-4.) To test for a dense core, scientists measured how Io's gravity deflected the trajectory of *Galileo* as the spacecraft flew past. From these measurements, they could determine not only Io's mass but also the satellite's oblateness (how much it deviates from a spherical shape because of its rotation). The amount of oblateness indicates the size of the core; the greater the fraction of the satellite's mass contained in its core, the less oblate the satellite will be. The *Galileo* observations suggest that Io has a dense core composed of iron and iron sulfide (FeS), with a radius of about 900 km (about half the satellite's overall radius). Surrounding the core is a mantle of partially molten rock, on top of which is Io's visible crust.

13-6 Europa is covered with a smooth layer of ice that may cover a worldwide ocean

Europa, the second of the Galilean satellites, is the smoothest body in the solar system. There are no mountains and no surface features greater than a few hundred meters high (Figure 13-10). There are almost no craters, indicating a young surface that has been reprocessed by geologic activity. The dominant surface feature is a worldwide network of stripes and cracks (Figure 13-11). Like Io, Europa is an exception to the general rule that a small world should be cratered and geologically dead (see Section 7-6).

Europa's Surface: Icy but Active

Neither *Voyager 1* nor *Voyager 2* flew very near Europa, so most of our knowledge of this satellite comes from the close passes

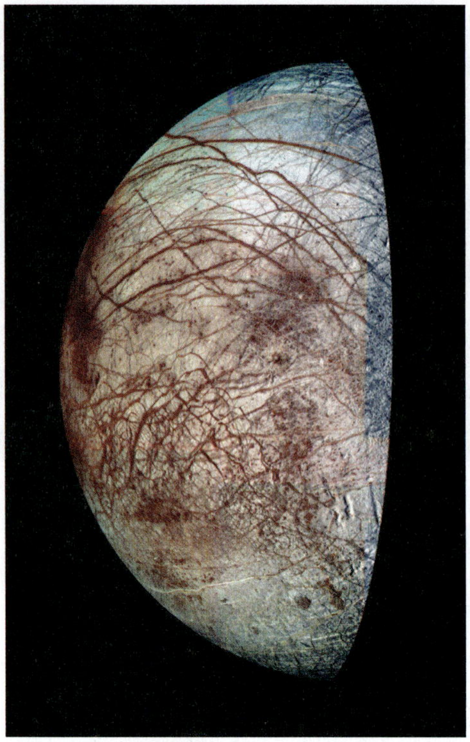

Figure 13-10 R I V U X G

Europa Dark lines crisscross Europa's smooth, icy surface in this false-color composite of visible and infrared images from *Galileo.* These lines are fractures in Europa's crust that can be as much as 20 to 40 km (12 to 25 mi) wide. Only a few impact craters are visible on Europa, which indicates that this satellite has a very young surface on which all older craters have been erased. (NASA/JPL/University of Arizona)

made by *Galileo*. (Both Figure 13-10 and Figure 13-11 are *Galileo* images.) But even before these spacecraft visited Jupiter, spectroscopic observations from Earth indicated that Europa's surface is almost pure frozen water (see Figure 7-4). This observation was confirmed by instruments on board *Galileo,* which showed that Europa's infrared spectrum is a close match to that of a thin layer of fine-grained water ice frost on top of a surface of pure water ice. (The brown areas in Figure 13-10 show where the icy surface contains deposits of rocky material from meteoritic impacts, from Europa's interior, or from a combination of these sources.)

Geology on Europa is based on solid ice and liquid water, not rock and molten magma

The purity of Europa's ice suggests that water is somehow brought upward from the moon's interior to the surface, where it solidifies to make a fresh, smooth layer of ice. Indeed, some *Galileo* images show what appear to be lava flows on Europa's surface, although the "lava" in this case is mostly ice. This idea helps to explain why Europa has very few craters (any old ones have simply been covered up) and why its surface is so smooth. Europa's surface may thus represent a water-and-ice version of plate tectonics.

Although Europa's surface is almost pure water ice, keep in mind that Europa is *not* merely a giant ice ball. The satellite's density shows that rocky material makes up about 85 to 90% of Europa's mass. Hence, only a small fraction of the mass, about 10 to 15%, is water ice. Because the surface is icy, we can conclude that the rocky material is found within Europa's interior.

Europa is too small to have retained much of the internal heat that it had when it first formed. But there must be internal heat nonetheless to power the geologic processes that erase craters and bring fresh water to Europa's surface. What keeps Europa's interior warm? The most likely answer, just as for Io, is heating by Jupiter's tidal forces. The rhythmic gravitational tugs exerted by Io and Ganymede on Europa deform its orbit into an ellipse; the varying speed of Europa as it goes around its orbit causes Europa to "nod" back and forth, causing variations in tidal stresses that make Europa flex. But because Europa is farther from Jupiter, tidal effects on Europa are only about one-fourth as strong as those on Io, which may explain why no ongoing volcanic activity has yet been seen on Europa.

Some features on Europa's surface, such as the fracture patterns shown in Figure 13-11, may be the direct result of the crust being stretched and compressed by tidal flexing. Figure 13-12 shows other features, such as networks of ridges and a young, very smooth circular area, that were probably caused by the internal heat that tidal flexing generates. The rich variety of terrain depicted in Figure 13-12, with stress ridges going in every direction, shows that Europa has a complex geologic history.

Among the unique structures found on Europa's surface are **ice rafts.** The area shown in Figure 13-13*a* was apparently subjected to folding, producing the same kind of linear features as those in Figure 13-12. But a later tectonic disturbance broke the surface into small chunks of crust a few kilometers across, which then "rafted" into new positions. A similar sort of rafting hap-

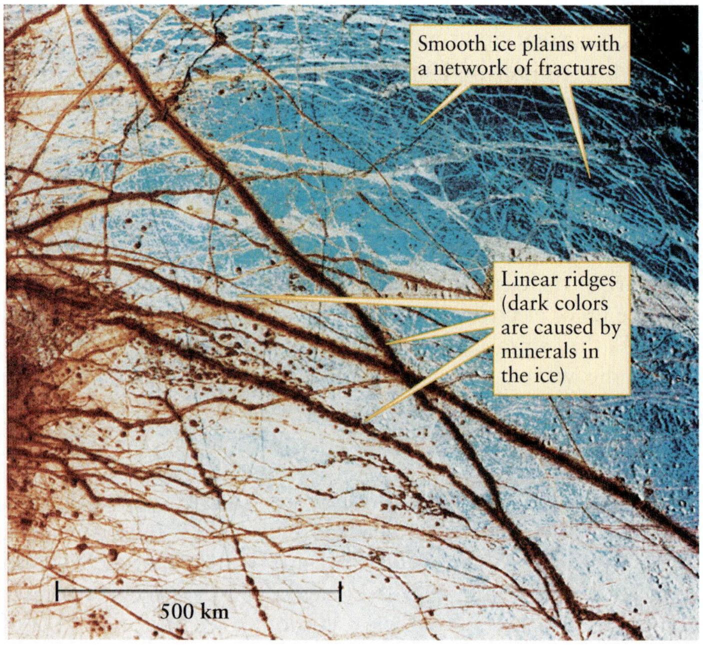

Figure 13-11 R I V U X G

Europa's Fractured Crust False colors in this *Galileo* composite image emphasize the difference between the linear ridges and the surrounding plains. The smooth ice plains (shown in blue) are the basic terrain found on Io. (NASA/JPL)

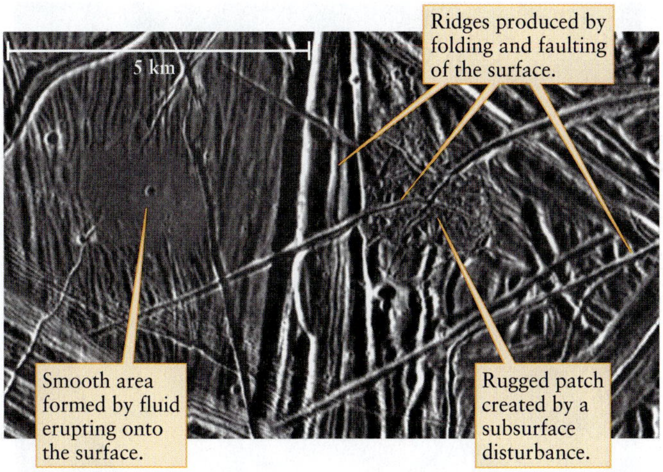

Figure 13-12 R I V U X G

Europa in Close-up This high-resolution *Galileo* image shows a network of overlapping ridges, part of which has been erased to leave a smooth area and part of which has been jumbled into a rugged patch of terrain. Europa's interior must be warm enough to power this complex geologic activity. (NASA/JPL)

pens in Earth's Arctic Ocean every spring, when the winter's accumulation of surface ice breaks up into drifting ice floes (Figure 13-13*b*). The existence of such structures on Europa strongly suggests that there is a subsurface layer of liquid water or soft ice over which the ice rafts can slide with little resistance.

An Underground Ocean?

The ridges, faults, ice rafts, and other features strongly suggest that Europa has substantial amounts of internal heat. This heat could prevent water from freezing beneath Europa's surface, creating a worldwide ocean beneath the crust. If this picture is correct, geologic processes on Europa may be an exotic version of those on Earth, with the roles of solid rock and molten magma being played by solid ice and liquid water.

A future spacecraft may use radar to penetrate through Europa's icy crust to search for definitive proof of liquid water beneath the surface. But magnetic field measurements made by the *Galileo* spacecraft have already provided some key evidence favoring this picture. Unlike Earth or Jupiter, Europa does not seem to create a steady magnetic field of its own. But as Europa moves through Jupiter's intense magnetic field, electric currents are induced within the satellite's interior, just as they are within Io (see Section 13-5), and these currents generate a weak but measurable field. (The strength and direction of this induced field vary as Europa moves through different parts of the Jovian magnetosphere, which would not be the case if Europa generated its field by itself.)

To explain these observations, there must be an electrically conducting fluid beneath Europa's crust—a perfect description of an underground ocean of water with dissolved minerals. (Pure water is a very poor conductor of electricity, so other substances must be present to make the water conducting.) If some of this water should penetrate upward to Europa's surface through cracks in the crust, it would vaporize and spread the dissolved minerals across the terrain. These minerals could explain the reddish-brown colors in Figure 13-13*a*.

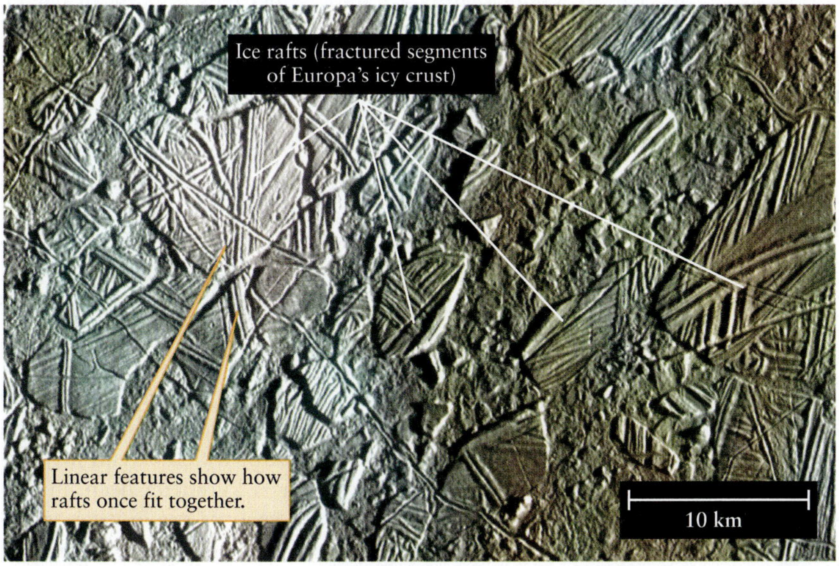

(a) Ice rafts on Europa

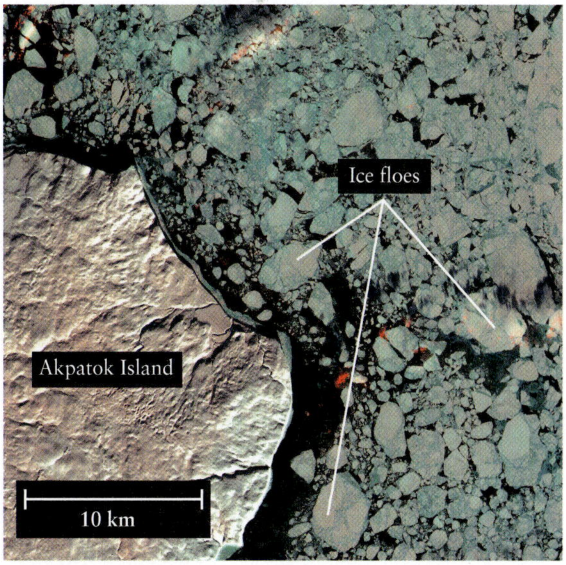

(b) Ice floes on Earth

Figure 13-13 R I V U X G

Moving Ice on Europa and Earth **(a)** Some time after a series of ridges formed in this region of Europa's surface, the icy crust broke into "rafts" that were moved around by an underlying liquid or plastic layer. The colors in this *Galileo* image may be due to minerals that were released from beneath the surface after the crust broke apart. **(b)** Europa's ice rafts are analogous to ice floes created when pack ice breaks up, as in this spacecraft view of part of the Canadian arctic. (a: NASA/JPL; b: USGS and NASA)

By combining measurements of Europa's induced magnetic field, gravitational pull, and oblateness, scientists conclude that Europa's outermost 100 to 200 km are ice and water. (It is not clear how much of this water is liquid and how much solid.) Within this outer shell is a rocky mantle surrounding a metallic core some 600 km (400 mi) in radius.

Remarkably, Europa also has an extremely thin atmosphere of oxygen. (By Earth standards, this atmosphere would qualify as a near-vacuum.) We saw in Section 9-5 that oxygen in Earth's atmosphere is produced by plants through photosynthesis. But Europa's oxygen atmosphere is probably the result of ions from Jupiter's magnetosphere striking the satellite's icy surface. These collisions break apart water molecules, liberating atoms of hydrogen (which escape into space) and oxygen.

The existence of a warm, subsurface ocean on Europa, if proved, would make Europa the only world in the solar system other than Earth on which there is liquid water. Liquid water on Europa would have dramatic implications. On Earth, water and warmth are essentials for the existence of life. Perhaps single-celled organisms have evolved in the water beneath Europa's crust, where they would use dissolved minerals and organic compounds as food sources. In light of this possibility, NASA has taken steps to prevent biological contamination of Europa. At the end of the *Galileo* mission in 2003, the spacecraft (which may have carried traces of organisms from Earth) was sent to burn up in Jupiter's atmosphere, rather than remaining in orbit where it might someday crash into Europa. An appropriately sterilized spacecraft may one day visit Europa and search for evidence of life within this exotic satellite.

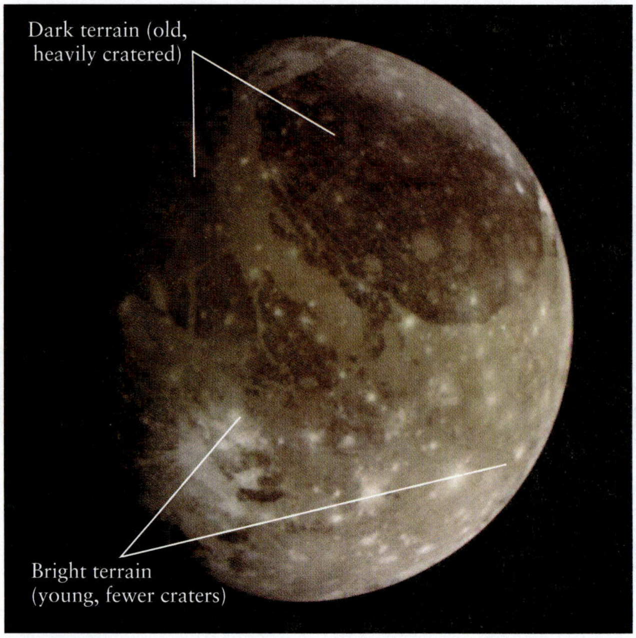

VIDEO 13.4

Figure 13-14 R I **V** U X G

Ganymede Two distinct types of terrain—one dark, heavily cratered, and hence old, the other bright, less cratered, and hence younger—are visible in this *Galileo* image of the hemisphere of Ganymede that always faces away from Jupiter. Craters in general appear bright, suggesting that the impacts that formed the craters excavated the surface to reveal ice underneath. (NASA/JPL)

13-7 Liquid water may also lie beneath the cratered surfaces of Ganymede and Callisto

Both Ganymede and Callisto show evidence of past geologic activity

Unlike Io and Europa, the two outer Galilean satellites have cratered surfaces. In this respect, Ganymede and Callisto bear a superficial resemblance to our own Moon (see the illustration with Table 13-1). But unlike the Moon's craters, the craters on both Ganymede and Callisto are made primarily of ice rather than rock, and Ganymede has a number of surface features that indicate a geologically active past.

The Two Faces of Ganymede's Surface

WEB LINK 13.10

Ganymede is the largest satellite in our solar system and is even larger than the planet Mercury. Like our Moon, Ganymede has two distinct kinds of terrain, called **dark terrain** and **bright terrain** (Figure 13-14). On the Moon, the dark maria are younger than the light-colored lunar highlands (see Section 10-1). On Ganymede, by contrast, the dark terrain is older, as indicated by its higher density of craters. The bright terrain is much less cratered and therefore younger. Because ice is more reflective than rock, even Ganymede's dark terrain is substantially brighter than the lunar surface.

Voyager and *Galileo* images show noticeable differences between young and old impact craters on Ganymede. The youngest craters are surrounded by bright rays of freshly exposed ice that has been excavated by the impact (see the lower left of Figure 13-14). Older craters are darker, perhaps because of chemical processes triggered by exposure to sunlight.

The close-up *Galileo* image in Figure 13-15 shows the contrast between dark and bright terrain. The area on the left-hand side of this image is strewn with large and small craters, reinforcing the idea that the dark terrain is billions of years old. The dark terrain is also marked by long, deep furrows, which are deformations of Ganymede's crust that have partially erased some of the oldest craters. Since other craters lie on top of the furrows, the stresses that created the furrows must have occurred long ago. Since geologic activity is a result of a planet or satellite's internal heat, this shows that Ganymede must have had a warm interior in the distant past.

Linear features are also seen in the bright terrain on the right in Figure 13-15. The *Voyager* missions discovered that the bright terrain is marked by long grooves, some of which extend for hundreds of kilometers and are as much as a kilometer deep. Images from *Galileo* show that some adjacent grooves run in different directions, indicating that the bright terrain has been subjected to a variety of tectonic stresses. *Galileo* also found a number of small craters overlying the grooves, some of which are visible in Figure 13-15. The density of craters shows that the bright terrain, while younger than the dark terrain, is at least a billion years old.

After the *Voyager* missions, it was thought that the bright terrain represented fresh ice that had flooded through cracks in

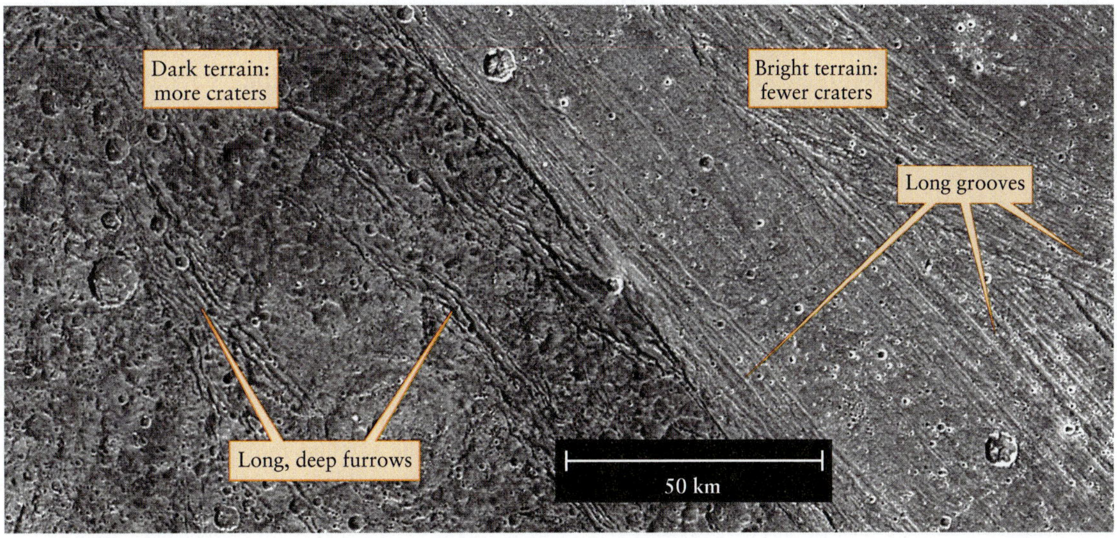

Figure 13-15 R I V U X G

Evidence for Geologic Activity on Ganymede Stresses in Ganymede's crust have produced linear features that extend for hundreds of kilometers across both the dark and bright terrain. These features must be quite ancient, since numerous impact craters have formed on top of them. (NASA)

Ganymede's crust, analogous to the maria on our Moon (see Section 10-1). However, high-resolution *Galileo* images do not show the kinds of surface features that would be expected if this terrain was volcanic in origin. Yet there are several low-lying areas within the bright terrain that appear to have been flooded by a watery fluid that subsequently froze. These low-lying areas may be *rift valleys* where the crust has been pulled apart, like Valles Marineris on Mars (see Section 11-5). In order to flood the floors of these valleys, there must have been liquid water beneath Ganymede's surface when the valleys formed. This is additional evidence that Ganymede must have had a warm interior in the past, and that this internal warmth persisted at least until the era when the bright terrain formed a billion years ago. The persistence of internal warmth is rather surprising, because Ganymede is too small to have retained much of the heat left over from its formation or from radioactive elements, and it orbits too far from Jupiter to have any substantial tidal heating.

Ganymede's Interior and Bizarre Magnetic Field

An equally great surprise from *Galileo* observations was the discovery that Ganymede has its own magnetic field, and that this field is twice as strong as that of Mercury. This field is sufficiently strong to trap charged particles, giving Ganymede its own "mini-magnetosphere" within Jupiter's much larger magnetosphere. The presence of a magnetic field shows that electrically conducting material must be in motion within Ganymede to act as a dynamo, which means that the satellite must have substantial internal heat even today. A warm interior causes differentiation, and *Galileo* measurements indeed showed that Ganymede is highly differentiated. It has a metallic core some 500 km in radius, surrounded by a rocky mantle and by an outer shell of ice some 800 km thick.

An even more surprising result is that Ganymede's magnetic field varies somewhat as it orbits Jupiter, and that these variations are correlated with Ganymede's location within Jupiter's magnetic field. One explanation for this variation is that Ganymede has a layer of salty, electrically conducting liquid water some 170 km beneath its surface. As the satellite moves through Jupiter's magnetic field, electric currents are induced in this layer just as happens within Europa (see Section 13-6), and these currents generate a weak field that adds to the stronger field produced by Ganymede's dynamo.

A possible explanation for Ganymede's internal heat is that gravitational forces from the other Galilean satellites may have affected its orbit. While Ganymede's present-day orbit around Jupiter is quite circular, calculations show that the orbit could have been more elliptical in the past. Just as for Io and Europa, such an elliptical orbit would have given rise to substantial tidal heating of Ganymede's interior. It may have retained a substantial amount of that heat down to the present day.

Callisto's Perplexing Surface

WEB LINK 13.11

In marked contrast to Io, Europa, and Ganymede is Callisto, Jupiter's outermost Galilean satellite. Images from *Voyager* showed that Callisto has numerous impact craters scattered over an icy crust (Figure 13-16). Callisto's ice is not as reflective as that on Europa or Ganymede, however; the surface appears to be covered with some sort of dark mineral deposit.

CAUTION! Figure 13-16, as well as the images that accompany Table 13-1, shows that Callisto is the darkest of the Galilean satellites. Nonetheless, its surface is actually more than twice as reflective as that of our own Moon. Callisto's surface may be made of "dirty" ice, but even dirty ice reflects more light than the gray rock found on the Moon. If you could somehow replace our own Moon with Callisto, a full moon as seen from Earth would be 4 times brighter than it is now (thanks also in part to Callisto's larger size).

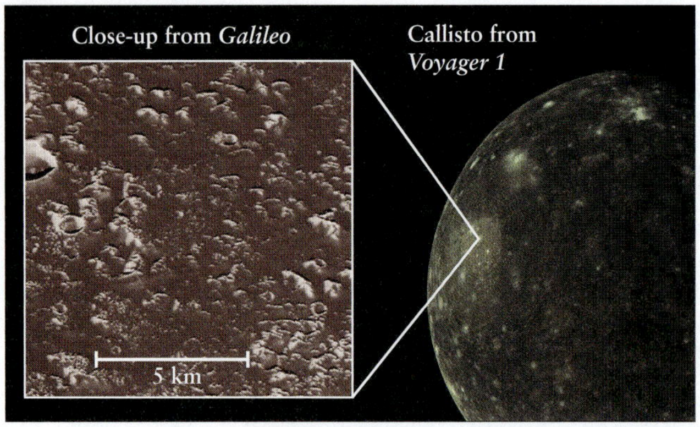

Figure 13-16 R I **V** U X G

Callisto Numerous craters pockmark Callisto's icy surface. The inset shows a portion of a huge impact basin called Valhalla. Most of the very smallest craters in this close-up view have been completely obliterated, and the terrain between the craters is blanketed by dark, dusty material. (Left: Arizona State University/JPL/NASA; right: JPL/NASA)

The *Voyager* images led scientists to conclude that Callisto is a dead world, with an ancient surface that has never been reshaped by geologic processes. There is no simple relationship between Callisto's orbital period and those of the other Galilean satellites, so it would seem that Callisto never experienced any tidal heating and hence was unable to power any geologic processes at its surface. But once the *Galileo* spacecraft went into orbit around Jupiter and began returning high-resolution images, scientists realized that Callisto's nature is not so simple. In fact, Callisto proves to be the most perplexing of the Galilean satellites.

One curious discovery from *Galileo* is that while Callisto has numerous large craters, there are very few with diameters less than 1 km. Such small craters are abundant on Ganymede, which has probably had the same impact history as Callisto. The lack of small craters on Callisto implies that most of these craters have been eroded away. How this erosion could have happened is not known. Another surprising result is that Callisto's surface is covered by a blanket of dark, dusty material (see the inset in Figure 13-16). Where this material came from, and how it came to be distributed across the satellite's surface, is not understood.

Callisto's Interior: Hot or Cold?

Callisto's most unexpected feature is that, like Europa and Ganymede, it appears to have an induced magnetic field that varies as the satellite moves through different parts of Jupiter's magnetosphere. Callisto must have a layer of electrically conducting material in order to generate such a field. One model suggests that this material is in the form of a subsurface ocean like Europa's but only about 10 km (6 mi) deep. However, without tidal heating, Callisto's interior is probably so cold that a layer of liquid water would quickly freeze. One proposed explanation is that Callisto's ocean is a mixture of water and a form of "antifreeze." In cold climates on Earth, antifreeze is added to the coolant in automobile radiators to lower the liquid's freezing temperature and prevent it from solidifying. Ammonia in Callisto's proposed subsurface ocean could play the same role.

The presence of a liquid layer beneath Callisto's surface is difficult to understand. Even with ammonia acting as an antifreeze, this layer cannot be too cold or it would solidify. This argues for a relatively *warm* interior. But *Galileo* measurements suggest that Callisto's interior is not highly differentiated. The lack of differenti-

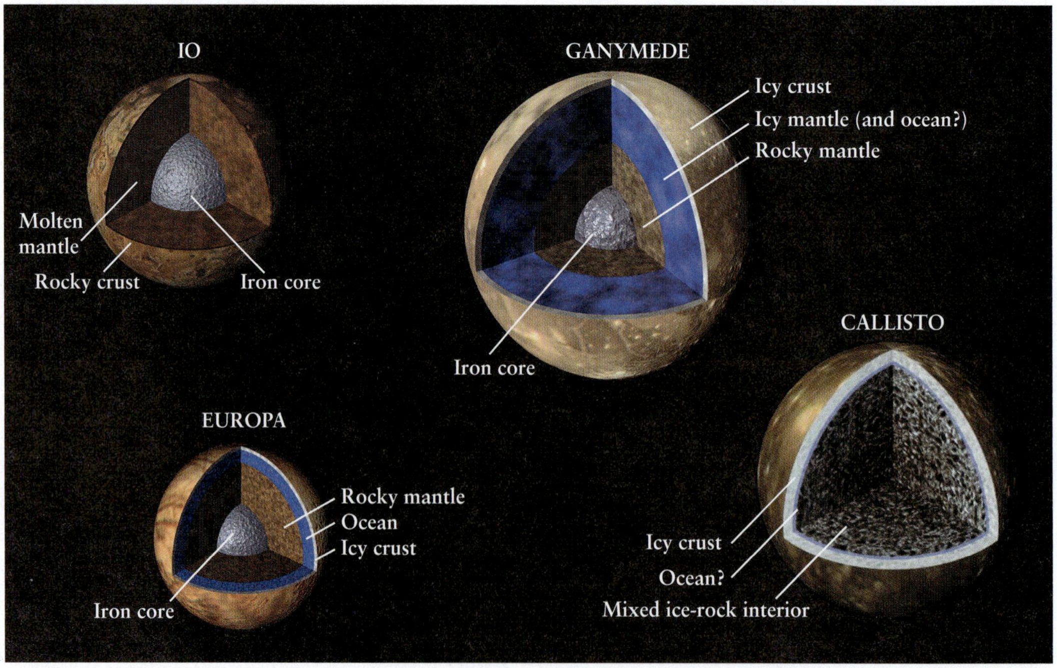

INTERACTIVE EXERCISE 13.1

Figure 13-17

Interiors of the Galilean Satellites These cross-sectional diagrams show the probable internal structures of the four Galilean satellites of Jupiter, based on information from the *Galileo* mission. (NASA/JPL) ation implies that the interior is *cold,* and may never have been heated enough to undergo complete chemical differentiation. The full story of Callisto's interior, it would seem, has yet to be told.

Although Ganymede and Callisto have many differences, one feature they share is an extremely thin and tenuous atmosphere. The molecules that make up Ganymede's atmosphere are oxygen (O_2) and ozone (O_3). These molecules are thought to be released from gas bubbles trapped in Ganymede's surface ice. On Callisto, by contrast, the atmosphere is composed of carbon dioxide (CO_2). This gas may have evaporated from Callisto's poles, where temperatures are low enough to form CO_2 ice.

The diagrams in Figure 13-17 summarize the probable interior structures of the Galilean satellites. They demonstrate the remarkable variety of these terrestrial worlds.

13-8 Titan has a thick atmosphere and streams carved by liquid hydrocarbons

WEB LINK 13.12 TUTORIAL 13.2

Unlike Jupiter, Saturn has only one large satellite that is comparable in size to our own Moon. This satellite, Titan, was discovered by Christiaan Huygens in 1665. Titan's diameter of 5150 km makes it second in size among satellites to Ganymede (Section 13.7). Like Ganymede, Titan has a low density that suggests it is made of a mixture of ice and rock. (See Table 7-2 for a comparison between Titan, the Moon, and the Galilean satellites.) But unlike Ganymede or any other satellite in the solar system, Titan has a thick atmosphere with a unique chemical composition.

Titan's Atmosphere

By the early 1900s, scientists had begun to suspect that Titan (Figure 13-18) might have an atmosphere, because it is cool enough and massive enough to retain heavy gases. (Box 7-2 explains the criteria for a planet or satellite to retain an atmosphere.) These suspicions were confirmed in 1944, when Gerard Kuiper discovered the spectral lines of methane in sunlight reflected from Titan. (We discussed Titan's spectrum and its interpretation in Section 7-3.)

Because of its atmosphere, Titan was a primary target for the *Voyager* missions. To the chagrin of mission scientists and engineers, however, *Voyager 1* spent hour after precious hour of its limited, preprogrammed observation time sending back featureless images like the one at the center of Figure 13-18. Titan's atmosphere is so thick—about 200 km (120 mi) deep, or about 10 times the depth of Earth's troposphere—that it blocked *Voyager*'s view of the surface. The haze surrounding Titan is so dense that little sunlight penetrates to the ground; noon on Titan is only about 1/1000 as bright as noon on Earth, though still 350 times brighter than night on Earth under a full moon.

By examining how the radio signal from *Voyager 1* was affected by passing through Titan's atmosphere, scientists inferred that the atmospheric pressure at Titan's surface is 50% greater than the atmospheric pressure at sea level on Earth. Titan's surface gravity is weaker than that of Earth, so to produce this pressure Titan must have considerably more gas in its atmosphere than Earth. Indeed, calculations show that about 10 times more gas (by mass) lies above a square meter of Titan's surface than above a square meter of Earth.

Voyager data show that more than 95% of Titan's atmosphere is nitrogen. Most of this nitrogen probably came from ammonia (NH_3), a compound of nitrogen and hydrogen that is quite common in the outer solar system. Ammonia is easily broken into hydrogen and nitrogen atoms by the Sun's ultraviolet radiation. Titan's gravity is too weak to retain hydrogen atoms (see Box 7-2), so these atoms escape into space and leave the more massive nitrogen atoms behind.

The remainder of Titan's atmosphere is predominantly methane (CH_4), which is the principal component of the "natural gas" used on Earth as fuel. The interaction of methane with ultraviolet light from the Sun produces a variety of other carbon-hydrogen compounds, or **hydrocarbons.** For example, the *Voyager* infrared spectrometers detected small amounts of ethane (C_2H_6),

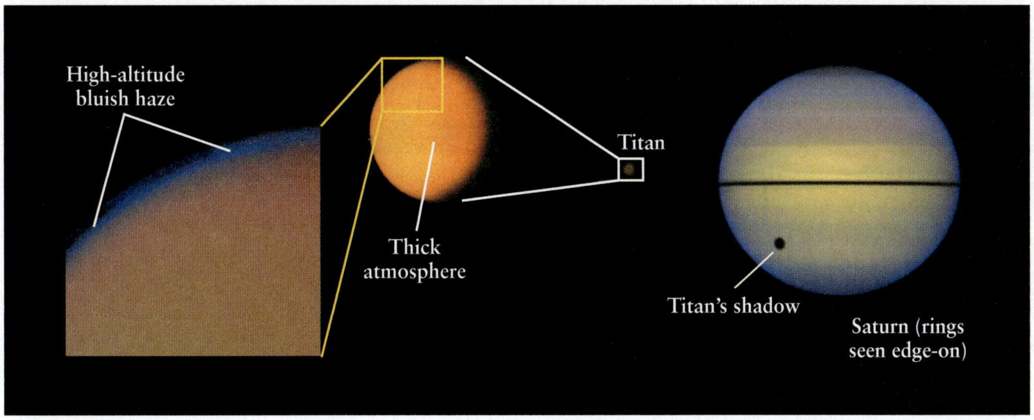

Figure 13-18 R I V U X G

Titan The Hubble Space Telescope image at far right shows Titan and the shadow it casts on Saturn's clouds. The *Voyager* image in the middle shows how Titan's atmosphere gives it a nearly featureless appearance. At far left is a color-enhanced *Voyager* image that shows a haze extending well above the satellite's visible edge, or limb. (Left: JPL/NASA; center: NASA; right: Erich Karkoschka, LPL/STScI/NASA)

acetylene (C_2H_2), ethylene (C_2H_4), and propane (C_3H_8) in Titan's atmosphere.

On Earth, water can be a gas (like water vapor in the atmosphere), a liquid (as in the oceans), or a solid (as in the polar ice caps). On Titan, the atmospheric pressure is so high and the surface temperature is so low—about 95 K (–178°C, or –288°F)—that any water is frozen solid. But conditions on Titan are just right for methane and ethane to exist as gases, liquids, or solids, which raises the tantalizing possibility that Titan could have methane rain and lakes of liquid ethane.

A faint mist of methane rain falls continually on Titan

Nitrogen in Titan's atmosphere can combine with airborne hydrocarbons to produce other compounds, such as hydrogen cyanide (HCN). Hydrogen cyanide, along with other molecules, can join together in long, repeating molecular chains to form substances called **polymers.** Droplets of some polymers remain suspended in the atmosphere, forming an **aerosol.** (Common aerosols on Earth include fog, mist, and paint sprayed from a can.) The polymers in Titan's airborne aerosol are thought to be responsible for the reddish-brown color seen in Figure 13-18.

Exploring Titan with *Cassini* and *Huygens*

WEBLINK 13.13

The designers of the *Cassini* mission to Saturn made the exploration of Titan a central goal. The Cassini spacecraft made 44 close flybys of Titan during its primary mission from 2004 to 2008, and will make 21 more flybys during a two-year extended mission. Titan's atmosphere is more transparent to infrared wavelengths than to visible light, so *Cassini* has used its infrared telescope to obtain the first detailed images of the satellite's surface (Figure 13-19*a*). Most of the surface is light-colored, but a swath of dark terrain extends around most of Titan's equator.

Cassini has used its atmosphere-penetrating radar to map the dark terrain. (Like infrared light, radio waves can pass through Titan's atmosphere.) These images reveal a series of long, parallel lines of sand dunes about 1 to 2 km apart. These are aligned west to east, in the same direction that winds blow at Titan's equator, and so presumably formed by wind action. Unlike Earth sand, which is made of small particles of silicate rock, sand on Titan is probably small particles of water ice combined with polymers that have fallen from Titan's atmosphere. These polymers give the dark terrain its color.

The infrared telescope aboard *Cassini* also detected a surface feature that appears to be a volcano. The presence of volcanism on Titan can explain why there is so much methane in the satellite's atmosphere. Methane molecules are broken apart by ultraviolet photons from the Sun, and the hydrogen atoms escape into space. This process would destroy all of Titan's methane within 10 million years unless a fresh supply is added to the atmosphere by volcanic activity. Volcanic activity can also explain the presence of very bright isolated patches on Titan's surface (see Figure

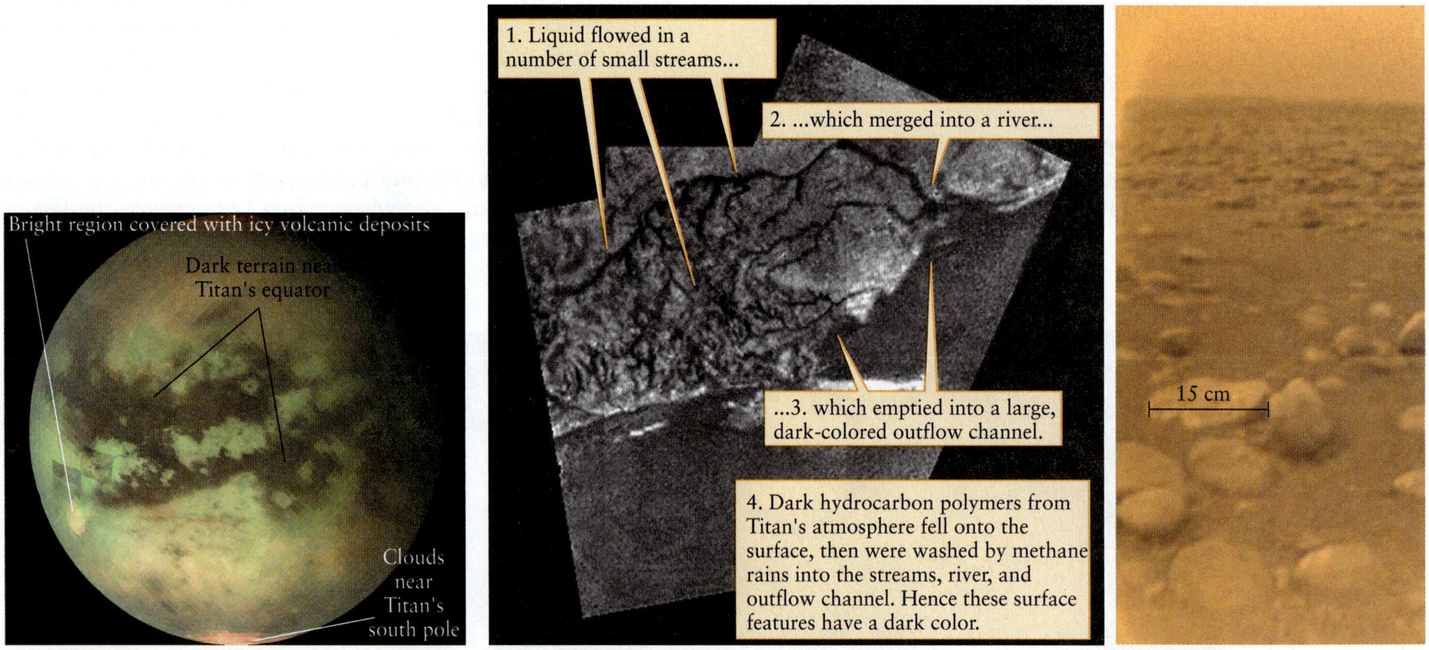

Bright region covered with icy volcanic deposits
Dark terrain near Titan's equator
Clouds near Titan's south pole
1. Liquid flowed in a number of small streams...
2. ...which merged into a river...
...3. which emptied into a large, dark-colored outflow channel.
4. Dark hydrocarbon polymers from Titan's atmosphere fell onto the surface, then were washed by methane rains into the streams, river, and outflow channel. Hence these surface features have a dark color.
15 cm

(a) R I V U X G (b) R I V U X G (c) R I V U X G

VIDEO 13.5 VIDEO 13.6

Figure 13-19

Beneath Titan's Clouds (a) This mosaic of Titan was constructed from *Cassini* images made at visible and infrared wavelengths during a close flyby in December 2005. **(b)** The *Huygens* lander recorded this view of features on Titan carved by flowing liquid. It is too cold on Titan for water to be a liquid, so these streams and river must have carried liquid methane or ethane. **(c)** This nearly true-color *Huygens* image shows the view from Titan's surface. The 15-cm (6-in.) wide "rock" is about 85 cm (33 in.) from the camera. Unlike Earth rocks, Titan "rocks" are chunks of water ice. (NASA/JPL/ESA/University of Arizona)

13-19*a*). A volcanic eruption would outgas not just methane but also water vapor and carbon dioxide. These gases would quickly freeze onto the surface, making a patch of highly reflective ice.

The most remarkable images of Titan's surface have come from *Huygens,* a small lander that *Cassini* carried on its journey from Earth and released before entering Saturn orbit. Named for the discoverer of Titan, *Huygens* entered Titan's atmosphere on January 14, 2005. The lander took $2\frac{1}{2}$ hours to descend to the surface under a parachute, during which time it made detailed images of the terrain (see Figure 13-19*b*). After touchdown, *Huygens* continued to return data for another 70 minutes (see Figure 13-19*c*) before its batteries succumbed to Titan's low temperatures.

Huygens images such as Figure 13-19*b* show that liquids have indeed flowed on Titan like water on Earth. Yet none of the images made by the lander during its descent or after touchdown showed any evidence of standing liquid. There was nonetheless evidence of recent methane rainfall at the *Huygens* landing site. A lander instrument designed to measure atmospheric composition was equipped with a heater for its gas inlet. During the three minutes it took to warm this inlet, the measured methane concentration jumped by 30% and then remained steady. The explanation is that the ground was soaked with liquid methane that had fallen as rain and which evaporated as it was warmed by the heater. *Huygens* measurements indicate that the average annual methane rainfall on Titan is about 5 cm (2 in), about the same as the amount of water rain that falls each year at Death Valley in California. The rivers and outflow channels that *Huygens* observed during its descent may date from an earlier period in Titan's history when methane was more abundant and rainfall was more intense.

Before the *Cassini* mission scientists expected to find lakes of liquid ethane across Titan's surface. By 2009, analysis of *Cassini* data indicate such lakes near Titan's north and south poles, with fewer lakes in the south (Figure 13-20). It is currently winter in the northern hemisphere, so temperatures are lowest there. Scientists also expected to find extensive clouds of ethane in Titan's atmosphere; these too have been found mostly near the north pole. Over the course of a Titan year (which, like that of its parent planet Saturn, lasts 29.37 years) the supply of liquid ethane may migrate with the seasons from pole to pole, much as frozen carbon dioxide migrates seasonally on Mars (see Section 11-6).

Impact craters are also conspicuously absent on Titan: Only a handful of craters have been observed by *Cassini*. The majority of craters may have been eroded by wind or rainfall, or covered by material ejected from volcanoes.

Titan's Geologic History

We have seen that Titan must have been volcanically active to provide its atmosphere with methane. But what has powered the volcanic activity? Titan is too small to have retained the internal heat from its formation. Furthermore, although Titan's rotation is synchronous like that of Io, its orbit around Saturn is so circular that it cannot be heated by tidal flexing in the way that Io is (see Section 13-4).

One explanation has been proposed by planetary scientists Gabriel Tobie, Jonathan Lunine, and Christophe Sotin. In their model, Titan developed a solid rocky core, a mantle of liquid water, and an icy crust soon after it formed some 4.5 billion years ago. The internal heat left over from formation facilitated volcanic activity that outgassed Titan's original complement of atmospheric methane molecules. Some of this methane was reabsorbed into the interior; the rest was destroyed by ultraviolet light from the Sun within the first billion years. From 3.5 billion to 2 billion years ago, Titan's atmosphere was essentially free of methane.

Over time, the rocky core increased in temperature due to the decay of radioactive elements like uranium. (These elements are concentrated in Titan's core to a greater extent than on Earth, which has rocky substances throughout its volume.) About 2 billion years ago, the core became warm enough to partially melt, and convection began in the melted portion. Convection allowed heat to flow from the core into the mantle, triggering a second episode of methane outgassing. This outgassing, too, eventually came to an end, and methane once again disappeared from Titan's atmosphere.

The third and final release of methane began about 500 million years ago, as the last of Titan's internal heat from radioactivity caused convection to take place in the satellite's ice crust. The amount of methane released during the current episode is enough to sustain the gas in the atmosphere, but not enough to provide extensive lakes of methane or ethane.

Within the next several hundred million years, methane outgassing on Titan will cease for good. In time, the hydrocarbon haze will disappear, the skies will clear, and the surface of Titan will be revealed to whatever outside observers may gaze upon it in that distant era.

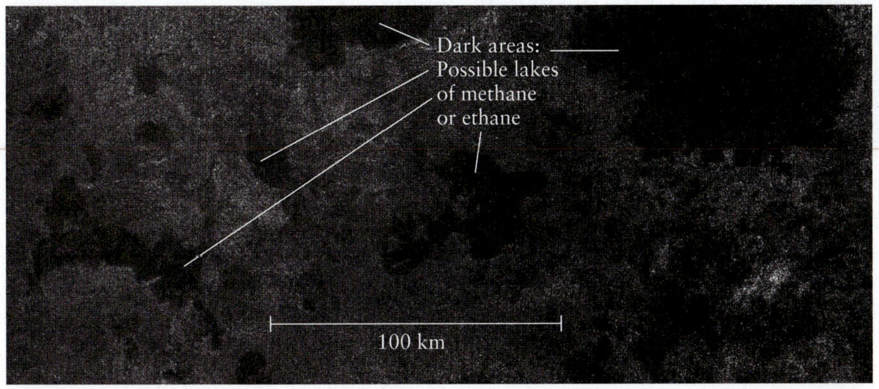

Figure 13-20 R I V U X G

Hydrocarbon Seas on Titan? *Cassini* used its onboard radar to record this view of a region near Titan's north pole. The dark regions show where the surface does not reflect radio waves at all, which would be the case if these regions were lakes filled with liquid methane or liquid ethane. In support of this idea, many of the dark regions appear to have liquid-carved inflow or outflow channels. (NASA/JPL)

13-9 Jupiter has dozens of small satellites that have different origins

WEB LINK 13.14

Besides the four Galilean satellites, Jupiter has at least 59 other small satellites. Each is named for a mythological character associated with Jupiter. Appendix 3 gives details of some of the satellites' orbits. It is helpful to sort Jupiter's 63 known satellites (as of 2009) into three groups according to their sizes and orbits.

Four planet-sized satellites, Io, Europa, Ganymede, and Callisto, can be thought of as four of the terrestrial worlds of the solar system. All are comparable in size to the planet Mercury (see Table 13-1). The semimajor axes of their orbits range from 5.9 to 26 times the radius of Jupiter. (By comparison, the distance from Earth to the Moon is about 60 times Earth's radius.)

Four small inner satellites, Metis, Adrastea, Amalthea, and Thebe, have orbital semimajor axes between 1.8 and 3.1 times the radius of Jupiter. As Figure 13-21 shows, these satellites have irregular shapes, much like a potato (or an asteroid). The largest of the inner satellites, Amalthea, measures about 270 by 150 km (170 by 95 mi)—about 10 times larger than the moons of Mars, but only about one-tenth the size of the Galilean satellites. Discovered in 1892, Amalthea has a distinct reddish color due to sulfur that Jupiter's magnetosphere removed from Io's volcanic plumes, then deposited onto Amalthea's surface. The three other inner satellites are even smaller than Amalthea. They were first discovered in images from *Voyager 1* and *Voyager 2.*

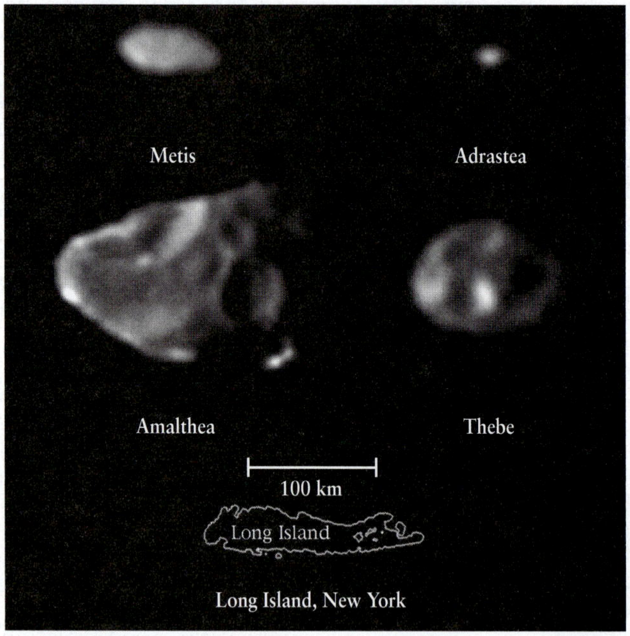

Figure 13-21 R I **V** U X G

Jupiter's Four Inner Satellites First observed in 1892, Amalthea was the last satellite of any planet to be discovered without the aid of photography. Metis, Adrastea, and Thebe, by contrast, are so small that they were not discovered until 1979. These images from the *Galileo* spacecraft are the first to show these three satellites as anything more than points of light. (NASA/JPL; Cornell University)

Fifty-five small outer satellites have diameters from 184 km for the largest to about 1 km for the smallest. All orbit far beyond Callisto, with semimajor axes that range from about 100 to 360 times the radius of Jupiter. (Twenty-three of these satellites were discovered in 2003 alone, and even more may be discovered as telescope technology continues to evolve.)

The Galilean satellites and the four inner satellites all orbit Jupiter in the plane of the planet's equator. (The same is true of Jupiter's rings, discussed in Section 12-9.) Furthermore, these orbits are all **prograde orbits,** which means that these objects orbit Jupiter in the same direction as Jupiter's rotation. We would expect orbits in the same direction if these satellites formed from the same primordial, rotating cloud as Jupiter itself. (In a similar way, the planets all orbit the Sun in nearly the same plane and in the same direction as the Sun's rotation. As we saw in Section 8-4, this reinforces the idea that the Sun and planets formed from the same solar nebula.)

Only a few of Jupiter's satellites orbit the planet in the same direction as Jupiter rotates

In contrast, the 55 outer satellites all circle Jupiter along orbits that are inclined at steep angles to the planet's equatorial plane. These satellites are thought not to have formed along with Jupiter; rather, they are probably wayward asteroids captured by Jupiter's powerful gravitational field. One piece of evidence for this is that 48 of the outer satellites are in **retrograde orbits:** They orbit Jupiter in a direction opposite the planet's rotation. Calculations show that it is easier for Jupiter to capture a satellite into a retrograde orbit than a prograde one.

13-10 The icy surfaces of Saturn's six moderate-sized moons provide clues to their histories

Saturn, too, has an extensive collection of satellites: 61 were known as of 2009. As we did for Jupiter's satellites in Section 13-9, we can subdivide the satellites of Saturn into three categories based on their sizes and their orbits. (Appendix 3 has information about the orbits of all of Saturn's satellites.) We will see that while Saturn does not have a set of large moons like the Galilean satellites of Jupiter, it does have a remarkably diverse collection of moderate-sized satellites.

Saturn's Retinue of Moons

One planet-sized satellite, Titan, is actually one of the terrestrial worlds of our solar system. With its diameter of 5150 km, Titan is intermediate in size between Mercury and Mars. Its prograde orbit has a semimajor axis of just over 20 times the radius of Saturn and lies in nearly the same plane as Saturn's equator and rings.

Six moderate-sized satellites were all discovered before 1800. These six satellites have properties and diameters that lead us to group them in pairs: Mimas and Enceladus (400 to 500 km in diameter), Tethys and Dione (roughly 1000 km in diameter), and Rhea and Iapetus (about 1500 km in diameter). The semimajor axes of their orbits range from 3.1 to 59 times the radius of

Saturn. Like Titan, all of their orbits are prograde and close to the plane of Saturn's equator.

Fifty-four small satellites have sizes from 3 to 266 km. Some are in prograde orbits as small as 2.2 times the radius of Saturn, and lie within or close to the rings (see Figure 12-25). These satellites may be jagged fragments of ice and rock produced by impacts and collisions. The shepherd satellites that shape Saturn's F ring (see Figure 12-24) may be such fragments. Others are in retrograde orbits with semimajor axes as large as 380 times Saturn's radius, and may be captured asteroids.

Six Strange Satellites

The most curious and mysterious of these worlds may be Saturn's six moderate-sized satellites (Figure 13-22). These six satellites have very low average densities of less than 1400 kg/m^3. This implies that unlike the satellites of the inner solar system or the Galilean satellites of Jupiter, they are composed primarily of ices (mostly frozen water and ammonia) with very little rock. The motions of these mid-sized satellites are not surprising: All six orbit the planet in the plane of Saturn's equator and in the direction of the planet's rotation, and all six exhibit synchronous rotation. What is surprising is the curious variety of surface features that the *Voyager* and *Cassini* spacecraft discovered on these satellites.

Mimas, the innermost of the six middle-sized moons, has the heavily cratered surface we would expect on such a small world. Its most remarkable feature is a crater so large (Figure 13-22*a*) that the impact that formed it must have come close to shattering Mimas into fragments.

In contrast to Mimas's heavily cratered surface, its neighbor and near twin, Enceladus, has extensive crater-free regions (Figure 13-22*b*). Some form of geologic activity has resurfaced these areas within the past 100 million years, and may still be under way. Geologic activity may also explain why Enceladus has an amazingly high albedo of 0.95, making it the most reflective large object in our solar system. Some process has freed Enceladus's icy surface of rock and dust, whose presence on the other icy satellites is the likeliest explanation for their lower albedos. Most remarkably, *Cassini* discovered that Enceladus is

Although just 500 km across, Enceladus has powerful volcanoes at its south pole

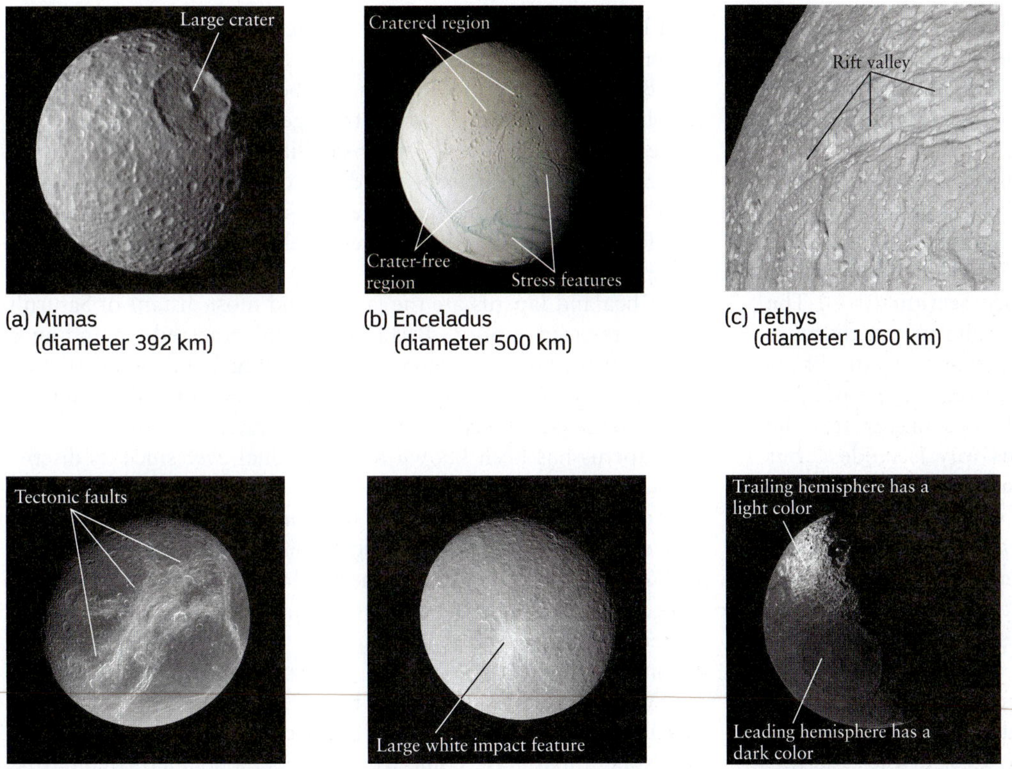

(a) Mimas (diameter 392 km)

(b) Enceladus (diameter 500 km)

(c) Tethys (diameter 1060 km)

(d) Dione (diameter 1120 km)

(e) Rhea (diameter 1530 km)

(f) Iapetus (diameter 1460 km)

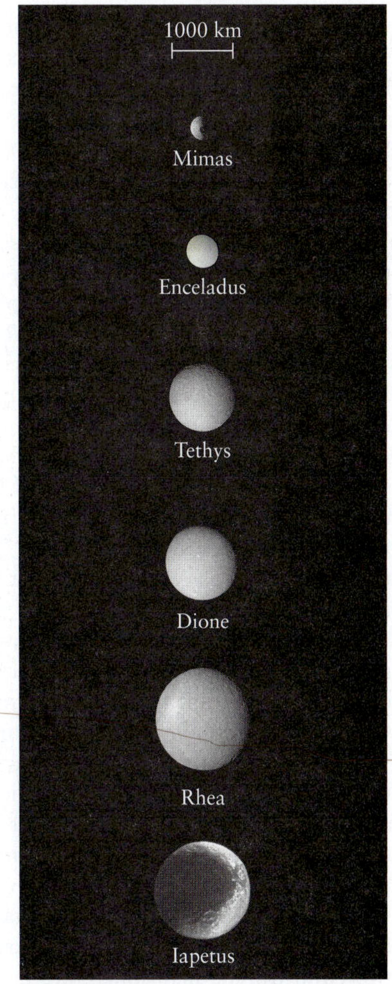

(g) Satellites to scale

Figure 13-22 RIVUXG

Saturn's Six Mid-Sized Satellites (a)–(f) These images from *Cassini* show the variety of surface features found on Saturn's six moderate-sized satellites. The satellites depicted in parts (a)–(f) are *not* shown to scale. **(g)** These images show the relative sizes of Saturn's moderate-sized satellites. (NASA/JPL/Space Science Institute)

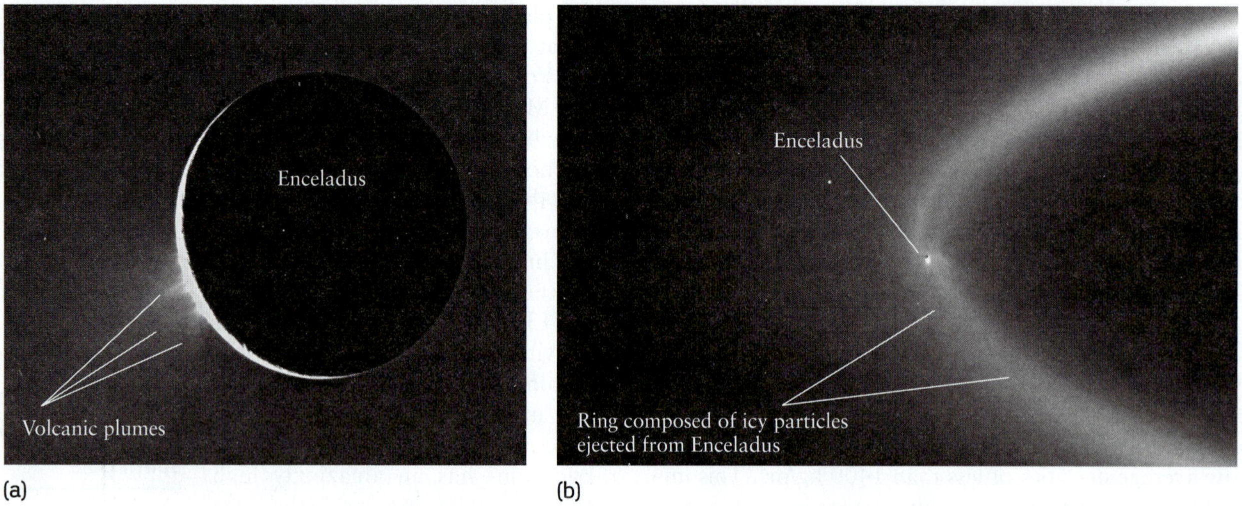

Figure 13-23 R I **V** U X G

Volcanoes on Enceladus **(a)** This *Cassini* image shows plumes of icy particles being ejected from ice volcanoes near the south pole of Enceladus. The faint plumes are best seen when they are backlit by the Sun, so Enceladus appears as a crescent. **(b)** A wide-angle backlit view from *Cassini* shows how particles ejected from Enceladus become part of Saturn's rings. This structure is actually part of the E ring, within which Enceladus orbits (see Figure 12-21). (NASA/JPL/Space Science Institute)

volcanically active, though with ice taking the place of lava. Figure 13-23*a* shows several distinct plumes ejecting icy particles from near the satellite's south pole, the same region where the surface is laced with bluish stress cracks (see Figure 13-22*b*). Some of the particles are ejected with such force that they escape Enceladus altogether and go into orbit around Saturn, forming a tenuous ring (Figure 13-23*b*).

What could be powering the geologic activity on Enceladus? One guess is tidal heating, which provides the internal heat for the inner Galilean satellites of Jupiter (see Section 13-4). The satellite Dione orbits Saturn in 65.7 hours, almost exactly twice Enceladus's orbital period of 32.9 hours, and the 1:2 ratio of orbital periods should set Enceladus up to be caught in a tidal tug-of-war between Saturn and Dione. Calculations suggest that the amount of tidal heating produced in this way is modest, but should be enough to melt the ices of Enceladus at least part of the time. Curiously, there is also a 1:2 ratio between the orbital periods of Mimas and Saturn's moon Tethys, but the cratered surface of Mimas shows no sign that this moon was ever tidally heated.

Tethys and Dione are the next-largest pair of Saturn's six moderate-sized satellites. Most of Tethys, like Mimas, is heavily cratered. But Tethys also has an immense rift valley called Ithaca Chasma that extends for 2000 km, or about $\frac{3}{4}$ of the way around Tethys (Figure 13-22*c*). This rift may have formed when the water inside Tethys cooled and solidified, or it may be the result of a shock wave that traveled through Tethys after an immense asteroid impact. (A large impact crater lies on the side of Tethys opposite Ithaca Chasma.) Dione, slightly larger than Tethys, has a strange surface dichotomy: The surfaces of its leading hemisphere (that is, the hemisphere facing toward its direction of orbital motion) and trailing hemisphere are quite different. Dione's leading hemisphere is heavily cratered, but its trailing hemisphere is covered by a network of bright stress cracks caused by some sort of tectonic activity (Figure 13-22*d*).

Because Tethys and Dione have about double the diameter of Enceladus, the energy to power their geologic activity might have come from within. These larger satellites retain heat more effectively, including heat from the decay of radioactive materials in each satellite's interior. Such materials are scarce within the icy satellites of Saturn, but only relatively little heat is needed to power ice volcanism.

Rhea and Iapetus are the largest and most distant of Saturn's moderate-sized moons. Rhea is heavily cratered, including a bright, fresh-looking crater on its leading hemisphere (Figure 13-22*e*). This crater may be the result of a more recent impact than those that formed most of Rhea's craters.

Iapetus has been known to be unusual ever since its discovery in 1671: Unlike the other satellites, its brightness varies greatly as it orbits Saturn. Images from *Cassini* show extreme differences between the leading and trailing hemispheres of Iapetus (Figure 13-22*f*). Most of the leading hemisphere is as black as asphalt (albedo = 0.05), but the trailing hemisphere is highly reflective (albedo = 0.50), like the surfaces of the other moderate-sized moons.

What could account for the two-sided nature of Iapetus? One possibility is that the dark material covering Iapetus's leading hemisphere may have come from Phoebe, a smaller satellite that orbits about $3\frac{1}{2}$ times farther from Saturn than Iapetus. Phoebe moves in a retrograde direction along an orbit tilted well away from the plane of Saturn's equator. This motion leads scientists to suspect that Phoebe is a captured asteroid, and its dark surface does indeed resemble the surfaces of a particular class of carbon-rich asteroids. Bits and pieces of this dark charcoal-like material drifting toward Saturn may have been swept up onto the leading hemisphere of Iapetus. However, a sticking point with this scenario is that the dark material on Iapetus is reddish in color, while Phoebe's surface is not. An alternative explanation is that a large portion of the leading hemisphere of Iapetus was coated with methane ice, which darkened when exposed to solar radiation.

Key Words

aerosol, p. 344
bright terrain (Ganymede), p. 340
dark terrain (Ganymede), p. 340
Galilean satellites, p. 330
hydrocarbon, p. 343
ice rafts (Europa), p. 338
Io torus, p. 336
occultation, p. 330
polymer, p. 344
prograde orbit, p. 346
retrograde orbit, p. 346
tidal heating, p. 334
ultramafic lava, p. 335

Key Ideas

Nature of the Galilean Satellites: The four Galilean satellites orbit Jupiter in the plane of its equator. All are in synchronous rotation.

• The orbital periods of the three innermost Galilean satellites, Io, Europa, and Ganymede, are in the ratio 1:2:4.

• The two innermost Galilean satellites, Io and Europa, have roughly the same size and density as our Moon. They are composed principally of rocky material. The two outermost Galilean satellites, Ganymede and Callisto, are roughly the size of Mercury. Lower in density than either the Moon or Mercury, they are made of roughly equal parts ice and rock.

• The Galilean satellites probably formed in a similar fashion to our solar system but on a smaller scale.

Io: Io is covered with a colorful layer of sulfur compounds deposited by frequent explosive eruptions from volcanic vents. These eruptions resemble terrestrial geysers.

• The energy to heat Io's interior and produce the satellite's volcanic activity comes from tidal forces that flex the satellite. This tidal flexing is aided by the 1:2:4 ratio of orbital periods among the inner three Galilean satellites.

• The Io torus is a ring of electrically charged particles circling Jupiter at the distance of Io's orbit. Interactions between this ring and Jupiter's magnetic field produce strong radio emissions. Io may also have a magnetic field of its own.

Europa: While composed primarily of rock, Europa is covered with a smooth layer of water ice.

• The surface has hardly any craters, indicating a geologically active history. Other indications are a worldwide network of long cracks and ice rafts that indicate a subsurface layer of liquid water or soft ice. As for Io, tidal heating is responsible for Europa's internal heat.

• An ocean may lie beneath Europa's frozen surface. Minerals dissolved in this ocean may explain Europa's induced magnetic field.

Ganymede: Two types of terrain are found on the icy surface of Ganymede: areas of dark, ancient, heavily cratered surface and regions of heavily grooved, lighter-colored, younger terrain.

• Ganymede is highly differentiated, and probably has a metallic core. It has a surprisingly strong magnetic field and a magnetosphere of its own.

• While there is at present little tidal heating of Ganymede, it may have been heated in this fashion in the past. An induced magnetic field suggests that it, too, has a layer of liquid water beneath the surface.

Callisto: Callisto has a heavily cratered crust of water ice. The surface shows little sign of geologic activity, because there was never any significant tidal heating of Callisto. However, some unknown processes have erased the smallest craters and blanketed the surface with a dark, dusty substance.

• Magnetic field data seem to suggest that Callisto has a shallow subsurface ocean.

Titan: The largest Saturnian satellite, Titan, is a terrestrial world with a dense nitrogen atmosphere. A variety of hydrocarbons are produced there by the interaction of sunlight with methane. These compounds form an aerosol layer in Titan's atmosphere and fall as a gentle rain on the surface.

• Titan's surface shows that liquid hydrocarbons have flowed over its surface, forming streams, rivers, and outflow channels. Very little of this liquid appears to be present on Titan's surface today.

Other Satellites: As of 2009, Jupiter has a total of 63 known satellites and Saturn has a total of 61.

• In addition to the Galilean satellites, Jupiter has four small inner satellites that lie inside Io's orbit. Like the Galilean satellites, these orbit in the plane of Jupiter's equator. The remaining satellites are small and move in much larger orbits that are noticeably inclined to the plane of Jupiter's equator. Many of these orbit in the direction opposite to Jupiter's rotation.

• In addition to Titan, six moderate-sized moons circle Saturn in regular orbits: Mimas, Enceladus, Tethys, Dione, Rhea, and Iapetus. They are probably composed largely of ice, but their surface features and histories vary significantly. The other, smaller moons include shepherd satellites that control the shapes of Saturn's rings and captured asteroids in large retrograde orbits.

Questions

Review Questions

1. If you observed the Galilean satellites through a telescope for a single night, could you notice their motions around Jupiter?
2. Why can't the Galilean satellites be seen with the naked eye?
3. During the time it takes Ganymede to complete one orbit, how many orbits do Io and Europa complete?
4. In what ways does the system of Galilean satellites resemble our solar system? In what ways is it different?
5. No spacecraft from Earth has ever landed on any of the Galilean satellites. How, then, can we know anything about the chemical compositions of these satellites?

6. In what ways did the formation of the Galilean satellites mimic the formation of the planets? In what ways were the two formation processes different?
7. In the classic science-fiction film *2010: The Year We Make Contact,* an alien intelligence causes Jupiter to contract so much that nuclear reactions begin at its center. As a result, Jupiter becomes a star like the Sun. Is this possible in principle? Explain your answer.
8. All of the Galilean satellites orbit Jupiter in the same direction. Furthermore, the planes of their orbits all lie within 0.5° of Jupiter's equatorial plane. Explain why these observations are consistent with the idea that the Galilean satellites formed from a "Jovian nebula."
9. What is the source of energy that powers Io's volcanoes? How is it related to the orbits of Io and the other Galilean satellites?
10. Io has no impact craters on its surface, while our Moon is covered with craters. What is the explanation for this difference?
11. Despite all the gases released from its interior by volcanic activity, Io does not possess a thick atmosphere. Explain why not.
12. Long before the *Voyager* flybys, Earth-based astronomers reported that Io appeared brighter than usual for the few hours after it emerged from Jupiter's shadow. From what we know about the material ejected from Io's volcanoes, suggest an explanation for this brief brightening of Io.
13. How do lavas on Io differ from typical lavas found on Earth? What does this difference tell us about Io's interior?
14. What accounts for the different colors found on Io's surface?
15. What is the Io torus? What is its source?
16. What is the origin of the electric current that flows through Io?
17. How was the *Galileo* spacecraft used to determine the internal structure of Io and the other Galilean satellites?
18. What surface features on Europa provide evidence for geologic activity?
19. What is the evidence for an ocean of liquid water beneath Europa's icy surface? What is the evidence that substances other than water are dissolved in this ocean?
20. What aspects of Europa lead scientists to speculate that life may exist there? What precautions have been taken to prevent contaminating Europa with Earth organisms?
21. Why is ice an important constituent of Ganymede and Callisto, but not of Earth's Moon?
22. How do scientists know that the dark terrain on Ganymede is younger than the bright terrain?
23. In what ways is Ganymede like our own Moon? In what ways is it different? What are the reasons for the differences?
24. Why were scientists surprised to learn that Ganymede has a magnetic field? What does this field tell us about Ganymede's history?
25. What leads scientists to suspect that there is liquid water beneath Ganymede's surface?
26. Why are numerous impact craters found on Ganymede and Callisto but not on Io or Europa?
27. Describe the surprising aspects of Callisto's surface and interior that were revealed by the *Galileo* spacecraft. Why did these come as a surprise?
28. Compare and contrast the surface features of the four Galilean satellites. Discuss their relative geological activity and the evolution of these four satellites.
29. The larger the orbit of a Galilean satellite, the less geologic activity that satellite has. Explain why.
30. Explain how the 1:2:4 ratio of the orbital periods of Io, Europa, and Ganymede is related to the geologic activity on these satellites.
31. Describe Titan's atmosphere. What effect has the Sun's ultraviolet radiation had on Titan's atmosphere?
32. In what ways do hydrocarbons on Titan behave like water on Earth?
33. Why does the presence of methane in Titan's atmosphere imply that Titan has had recent volcanic activity?
34. What is the evidence that there were liquid hydrocarbons on Titan in the past? That there are liquid hydrocarbons now?
35. How would you account for the existence of the satellites of Jupiter other than the Galilean ones?
36. Which of Saturn's moderate-sized satellites show evidence of geologic activity? What might be the energy source for this activity?
37. Explain why debris from Phoebe would be expected to pile up only on the leading hemisphere of Iapetus. (*Hint:* How do the orbits of these two satellites compare? How does the orbital motion of debris falling slowly inward toward Saturn compare with the orbital motion of Iapetus?)
38. Saturn's equator is tilted by 27° from the ecliptic, while Jupiter's equator is tilted by only 3°. Use these data to explain why we see fewer transits, eclipses, and occultations of Saturn's satellites than of the Galilean satellites.

Advanced Questions

Questions preceded by an asterisk (*) involve topics discussed in Box 1-1 or Box 7-2.

Problem-solving tips and tools

Newton's form of Kepler's third law (Box 4-4) relates the masses of two objects in orbit around each other to the period and size of the orbit. The small-angle formula is discussed in Box 1-1. The best seeing conditions on Earth give a limiting angular resolution of $\frac{1}{4}$ arcsec. Because the orbits of the Galilean satellites are almost perfect circles, you can easily calculate the orbital speeds of these satellites from the data listed in Table 13-1. Data about Jupiter itself are given in Table 12-1 and data about all of the satellites of Jupiter and Saturn can be found in Appendix 3. For a discussion of escape speed and how planets retain their atmospheres, see Box 7-2.

39. Using the orbital data in Table 13-1, demonstrate that the Galilean satellites obey Kepler's third law.
*40. What is the size of the smallest feature you should be able to see on a Galilean satellite through a large telescope under conditions of excellent seeing when Jupiter is near opposition? How does this compare with the best *Galileo* images, which have resolutions around 25 meters?
41. Using the diameter of Io (3642 km) as a scale, estimate the height to which the plume of Pele rises above the surface of Io in Figure 13-5*a*. (You will need to make measurements on this figure using a ruler.) Compare your answer to the value given in the figure caption.

42. Explain why the volcanic plumes in Figure 13-5*b* have a bluish color. (*Hint:* See Box 5-4.)
43. Jupiter, its magnetic field, and the charged particles that are trapped in the magnetosphere all rotate together once every 10 hours. Io takes 1.77 days to complete one orbit. Using a diagram, explain why particles from Jupiter's magnetosphere hit Io primarily from behind (that is, on the side of Io that trails as it orbits the planet).
44. Assuming material is ejected from Io into Jupiter's magnetosphere at the rate of 1 ton per second (1000 kg/s), how long will it be before Io loses 10% of its mass? How does your answer compare with the age of the solar system?
45. Volcanic eruptions from Io add about 1 ton of material to Jupiter's magnetosphere every second. Based on the information given in Section 13-4, what fraction of the material ejected from Io's volcanoes goes into Jupiter's magnetosphere?
46. How long does it take for Ganymede to entirely enter or entirely leave Jupiter's shadow? Assume that the shadow has a sharp edge.
47. Figure 13-15 shows that Ganymede's dark terrain has both more craters and larger craters than the bright terrain. Explain what this tells you about the sizes of meteoroids present in the solar system in the distant past and in the more recent past.
48. Put a few cubes of ice in a glass, then fill the glass with warm water to make the ice cubes crack. The fracture lines in each cube are clearly visible because they reflect light. Use this observation to suggest why Ganymede's icy bright terrain, which may have been badly fractured by the stresses that produced the grooves (Figure 13-15), is 25% more reflective than the dark terrain.
*49. Find the escape speed on Titan. What is the limiting molecular weight of gases that could be retained by Titan's gravity? (*Hint:* Use the ideas presented in Box 7-2 and assume an average atmospheric temperature of 95 K.)
50. Many of the gases in the atmosphere of Titan, such as methane, ethane, and acetylene, are highly flammable. Why, then, doesn't Titan's atmosphere catch fire? (*Hint:* What gas in our atmosphere is needed to make wood, coal, or gasoline burn?)
51. Suppose Earth's Moon were removed and replaced in its orbit by Titan. What changes would you expect to occur in Titan's atmosphere? Would solar eclipses be more or less common as seen from Earth? Explain your answers.
52. At infrared wavelengths, the Hubble Space Telescope can see details on Titan's surface as small as 580 km (360 mi) across. Determine the angular resolution of the Hubble Space Telescope using infrared light. If visible light is used, is the angular resolution better, worse, or the same? Explain your answer.
53. (a) To an observer on Enceladus, what is the time interval between successive oppositions of Dione? Explain your answer. (b) As seen from Enceladus, what is the angular diameter of Dione at opposition? How does this compare to the angular diameter of the Moon as seen from Earth (about $\frac{1}{2}^\circ$)?

Discussion Questions

54. If you could replace our Moon with Io, and if Io could maintain its present amount of volcanic activity, what changes would this cause in our nighttime sky? Do you think that Io could in fact remain volcanically active in this case? Why or why not?
55. Speculate on the possibility that Europa, Ganymede, or Callisto might harbor some sort of life. Explain your reasoning.
56. Comment on the suggestion that Titan may harbor life-forms.
57. Jupiter's satellite Io and Saturn's satellite Enceladus are both geologically active, and both are in 2-to-1 resonances with other satellites. However, the amount of geologic activity of Enceladus is far less than on Io. Discuss some possible reasons for this difference.
58. Imagine that you are in charge of planning a successor to the *Cassini* spacecraft to further explore the Saturnian system. In your opinion, which satellites in the system should be examined more closely? What data should be collected? What kinds of questions should the new mission attempt to answer?

Web/eBook Questions

59. Various spacecraft missions have been proposed to explore Europa in greater detail. Search the World Wide Web for information about these. How would these missions test for the presence of an ocean beneath Europa's surface?
60. Twenty-one new satellites of Jupiter were discovered in 2003. Search the World Wide Web for information about how these satellites were discovered. How was it determined that these satellites are actually in orbit around Jupiter?

VIDEO 13.4

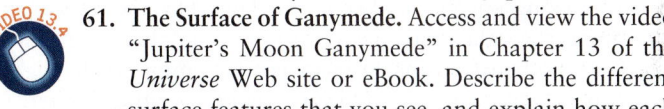

61. **The Surface of Ganymede.** Access and view the video "Jupiter's Moon Ganymede" in Chapter 13 of the *Universe* Web site or eBook. Describe the different surface features that you see, and explain how each type of feature was formed.

Activities

Observing Projects

Observing tips and tools

You can easily find the apparent positions of the Galilean satellites for any date and time using the *Starry Night Enthusiast*™ software on the CD-ROM that accompanies certain printed copies of this textbook. For even more detailed information about satellite positions, consult the "Satellites of Jupiter" section in the *Astronomical Almanac* for the current year. If your goal is to view Saturn's satellites, consult the section entitled "Satellites of Saturn" in the *Astronomical Almanac* for the current year. This includes a diagram showing the orbits of Mimas, Enceladus, Tethys, Dione, Rhea, Titan, and Hyperion. Plan your observing session by looking up the dates and times of the most recent greatest eastern elongations of the various satellites. You will have to convert from universal time (UT), also known as Greenwich Mean Time, to your local time zone. Then, using the tick marks along the orbits in the diagram, estimate the positions of the satellites relative to Saturn at the time you will be at the telescope. Another useful resource is the "Celestial Calendar" section of *Sky & Telescope*. During months when Saturn is visible in the night sky, this section of the magazine includes a chart of Saturn's satellites.

62. Observe Jupiter through a pair of binoculars. Can you see all four Galilean satellites? Make a drawing of what you observe. If you look again after an hour or two, can you see any changes?
63. Observe Jupiter through a small telescope on three or four consecutive nights. Make a drawing each night showing the positions of the Galilean satellites relative to Jupiter. Record the time and date of each observation. Consult the sources listed above in the "Observing tips and tools" to see if you can identify the satellites by name.
64. If you have access to a moderately large telescope, make arrangements to observe several of Saturn's satellites. At the telescope, you should have no trouble identifying Titan. Tethys, Dione, and Rhea are about one-sixth as bright as Titan and should be the next easiest satellites to find. Can you confidently identify any of the other satellites?
65. Use the *Starry Night Enthusiast*™ program to observe the Galilean satellites of Jupiter. Open the **Favorites pane** and click on **Guides > Atlas** to display the entire celestial sphere. Open the **Find** pane and double-click the entry for Jupiter to center this planet in the view. Using the controls at the right-hand end of the toolbar, zoom in to a field of view of approximately **13′ × 9′**. Stop time flow and in the toolbar, set the date and time to March 4, 2004, at 12:00:00 A.M to see Jupiter at opposition. Set the **Time Flow Rate** to 20 minutes by clicking on the number in the Time Flow Rate box and setting the value with the keyboard. Then click on the Run Time Forward button (a triangle that points to the right). You will see the four Galilean satellites orbiting Jupiter. (If these moons appear to move too quickly, adjust the Time Flow Rate to 10 minutes.) (**a**) Are all four satellites ever on the same side of Jupiter? (**b**) Observe the satellites passing in front of and behind Jupiter and look for their shadows upon the planet. (Zoom in as needed). Explain how your observations tell you that all four satellites orbit Jupiter in the same direction.
66. Use the *Starry Night Enthusiast*™ program to view Saturn from its satellite Enceladus. First click the **Home** button in the toolbar. Select **Options > Viewing Location . . .** from the menu. In the dialog window that appears at the top of the Viewing Location, set the drop-down box next to the "**View from:**" label to read "**the surface of**" and set the second drop-down box to read "**Enceladus**" (listed under Saturn). In the list of locations that appears under the **List** tab on Enceladus, scroll to "**Ahmad.**" Click on this entry to highlight it and then click the **Set Location** button. To see Saturn from this location on Enceladus, center on Saturn by double-clicking the entry for this planet in the **Find** pane. (**a**) Stop the time flow and set the Time Flow Rate to **1 minute** in the toolbar, then click on the **Run Time Forward** button (a triangle that points to the right). How do the stars appear to move as seen from this location on Enceladus? How does Saturn appear to move? What do these observations tell you about the relationship between the orbital period and rotation period of Enceladus? (**b**) Set the Time Flow Rate to **1 hour.** By stepping forward through time using the rightmost single-step button, determine how much time elapses from when Saturn appears full from this location to when it next appears full. Explain why this is the same as the orbital period of Enceladus, and compare this to the value of the orbital period given in Appendix 3.
67. Use the *Starry Night Enthusiast*™ program to examine the satellites of Saturn. Open the **Favorites** pane and select **Solar System > Saturn.** Remove the astronaut's feet from this view by clicking on **View > Feet.** In this view you can rotate Saturn by placing the mouse cursor over the image of the planet, holding down the mouse button, and moving the mouse (on a two-button mouse, hold down the left mouse button). Use this technique to rotate Saturn so that you are viewing the rings edge-on. Then use the elevation controls in the toolbar (the buttons to the left of the Home button) to move closer to and further away from Saturn. This allows you to identify the satellites of Saturn, since they will appear to move whereas the distant stars will remain stationary as you move with respect to the planet. Alternately, you can reduce the confusion in identifying these moons by clicking on **View > Stars > Stars** to remove the stars from the view. You should be able to see at least eight satellites. Which satellites are these? (Move the mouse to center the cursor on a satellite and its name will appear. If you cannot see all eight, try moving further from Saturn (increase the viewing location elevation with the elevation button showing the Up arrowhead). (**b**) The plane of Saturn's rings is the same as the plane of Saturn's equator. Which satellites appear to be the farthest from this plane?

Collaborative Exercises

68. Imagine that scientists are proposing to send a robotic lander to Jupiter's satellite Callisto. Create a 100-word written proposal describing a robotic lander mission to another of the Galilean satellites, explaining why your group found it to be the most interesting and why the government should allocate the money for your alternative project. In your proposal, be sure to demonstrate your knowledge of Callisto and the other satellite.
69. From the data and the accompanying images for Table 13-1, "The Galilean Satellites Compared to the Moon, Mercury, and Mars," use someone's shoe to represent the 150,000 km diameter of Jupiter and determine about how many "shoes" away would each of the Galilean satellites be from Jupiter.

14

Uranus, Neptune, Pluto, and the Kuiper Belt: Remote Worlds

WEB LINK 14.1

RIVUXG

Earth, Uranus, Neptune, Pluto, and Eris to scale. (Alan Stern, Southwest Research Institute; Marc Buie, Lowell Observatory; NASA; and ESA)

Beyond Saturn, in the cold, dark recesses of the solar system, orbit three worlds that have long been shrouded in mystery. These worlds—Uranus, Neptune, and Pluto (all shown here to the same scale as Earth)—are so distant, so dimly lit by the Sun, and so slow in their motion against the stars that they were unknown to ancient astronomers and were discovered only after the invention of the telescope. Even then, little was known about Uranus and Neptune until *Voyager 2* flew past these planets during the 1980s.

Surrounded by a system of small moons and thin, dark rings, Uranus is tipped on its side so that its axis of rotation lies nearly in its orbital plane. This remarkable orientation suggests that Uranus may have been the victim of a staggering impact by a massive planetesimal. Neptune is, at first glance, a denser, more massive version of Uranus, but it is a far more active world. It has an internal energy source that Uranus seemingly lacks, as well as atmospheric bands and storm activity resembling those on Jupiter.

Neptune also has dark rings, small, icy moons, and an intriguing large satellite, Triton, which is nearly devoid of impact craters and has geysers that squirt nitrogen-rich vapors. Triton's retrograde orbit suggests that this strange world may have been gravitationally captured by Neptune.

Close cousins of Triton are Pluto and thousands of other trans-Neptunian objects, among the most remote members of the solar system. These objects harbor many mysteries, in part because they have not yet been visited by spacecraft.

Learning Goals

By reading the sections of this chapter, you will learn

14-1 How Uranus and Neptune were discovered

14-2 The unusual properties of the orbit and atmosphere of Uranus

14-3 What gives Neptune's clouds and atmosphere their distinctive appearance

14-4 The internal structures of Uranus and Neptune

14-5 The unique orientations of the magnetic fields of Uranus and Neptune

14-6 Why the rings of Uranus and Neptune are hard to see

14-7 What could have powered geologic activity on Uranus's moderate-sized moons

14-8 Why Neptune's satellite Triton is destined to be torn apart

14-9 How Pluto came to be discovered

14-10 What shapes the orbits of Pluto and the thousands of other objects that orbit beyond Neptune

Table 14-1 Uranus Data

Average distance from the Sun:	19.194 AU = 2.871×10^9 km
Maximum distance from the Sun:	20.017 AU = 2.995×10^9 km
Minimum distance from the Sun:	18.371 AU = 2.748×10^9 km
Eccentricity of orbit:	0.0429
Average orbital speed:	6.83 km/s
Orbital period:	84.099 years
Rotation period (internal):	17.24 hours
Inclination of equator to orbit:	97.86°
Inclination of orbit to ecliptic:	0.77°
Diameter:	51,118 km = 4.007 Earth diameters (equatorial)
Mass:	8.682×10^{25} kg = 14.53 Earth masses
Average density:	1318 kg/m^3
Escape speed:	21.3 km/s
Surface gravity (Earth = 1):	0.90
Albedo:	0.56
Average temperature at cloudtops:	−218°C = −360°F = 55 K
Atmospheric composition (by number of molecules):	82.5% hydrogen (H_2), 15.2% helium (He), 2.3% methane (CH_4)

WEB LINK 14.2

R I V U X G

(NASA/JPL)

TUTORIAL 14.1

14-1 Uranus was discovered by chance, but Neptune's existence was predicted by applying Newtonian mechanics

Before the eighteenth century, only six planets were known to orbit the Sun: Mercury, Venus, Earth, Mars, Jupiter, and Saturn. Another planet was discovered by William Herschel, a German-born musician who moved to England in 1757 and became fascinated by astronomy.

Discovering the "Georgian Star"

Using a telescope that he built himself, Herschel was systematically surveying the sky on March 13, 1781, when he noticed a faint, fuzzy object that he first thought to be a distant comet. By the end of 1781, however, his observations had revealed that the object's orbit was relatively circular and was larger than Saturn's orbit. Comets, by contrast, can normally be seen only when they follow elliptical orbits that bring them much closer to the Sun.

Herschel had discovered the seventh planet from the Sun. In doing so, he had doubled the radius of the known solar system from 9.5 AU (the semimajor axis of Saturn's orbit) to 19.2 AU (the distance from the Sun to Uranus). Herschel originally named his discovery Georgium Sidus (Latin for "Georgian star") in honor of the reigning monarch, George III. The name Uranus—in Greek mythology, the personification of Heaven—came into currency only some decades later.

Although Herschel received the credit for discovering Uranus, he was by no means the first person to have seen it. At opposition, Uranus is just barely bright enough to be seen with the naked eye under good observing conditions, so it was probably seen by the ancients. Many other astronomers with telescopes had

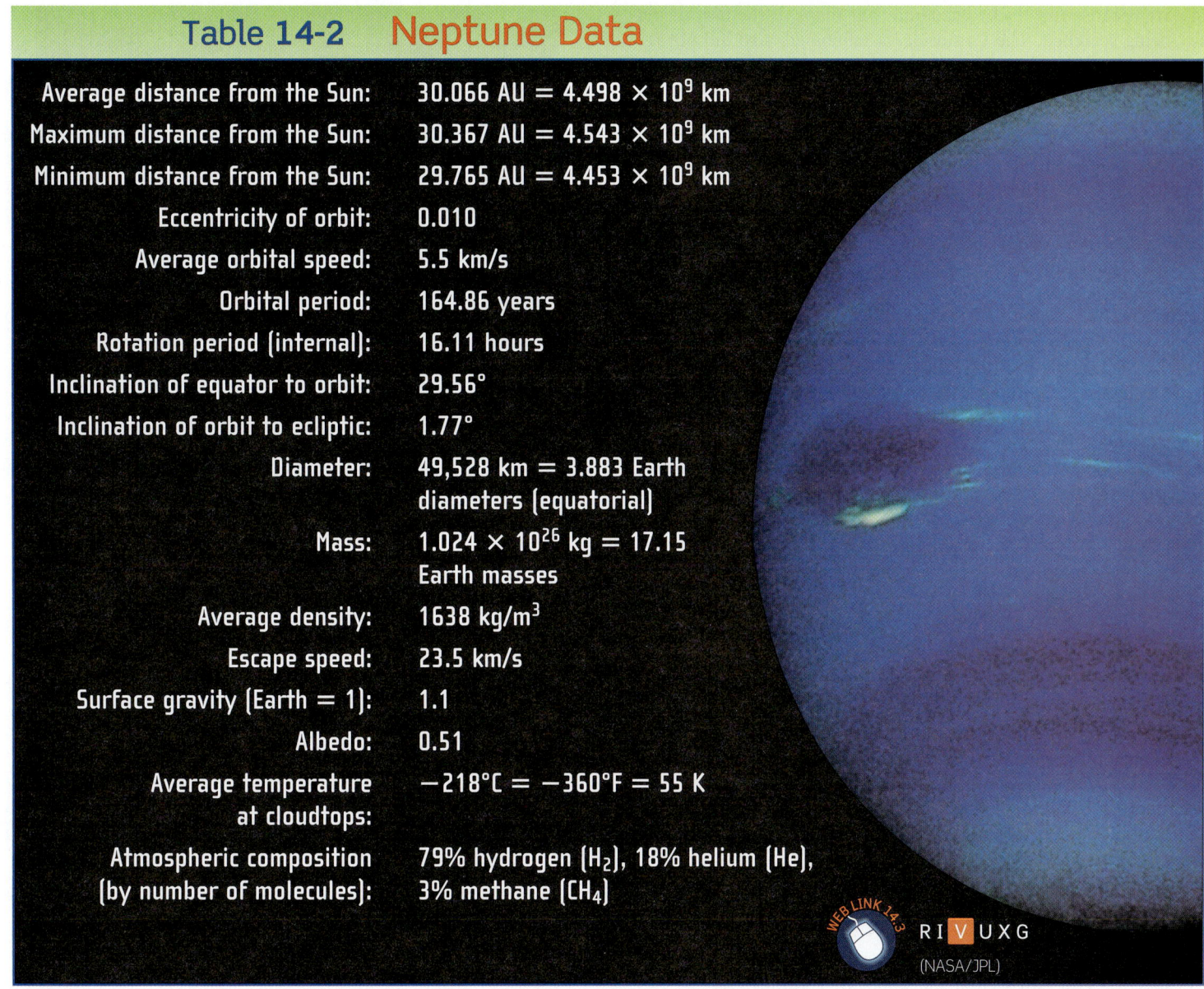

Table 14-2 Neptune Data

Average distance from the Sun:	30.066 AU = 4.498×10^9 km
Maximum distance from the Sun:	30.367 AU = 4.543×10^9 km
Minimum distance from the Sun:	29.765 AU = 4.453×10^9 km
Eccentricity of orbit:	0.010
Average orbital speed:	5.5 km/s
Orbital period:	164.86 years
Rotation period (internal):	16.11 hours
Inclination of equator to orbit:	29.56°
Inclination of orbit to ecliptic:	1.77°
Diameter:	49,528 km = 3.883 Earth diameters (equatorial)
Mass:	1.024×10^{26} kg = 17.15 Earth masses
Average density:	1638 kg/m^3
Escape speed:	23.5 km/s
Surface gravity (Earth = 1):	1.1
Albedo:	0.51
Average temperature at cloudtops:	−218°C = −360°F = 55 K
Atmospheric composition (by number of molecules):	79% hydrogen (H_2), 18% helium (He), 3% methane (CH_4)

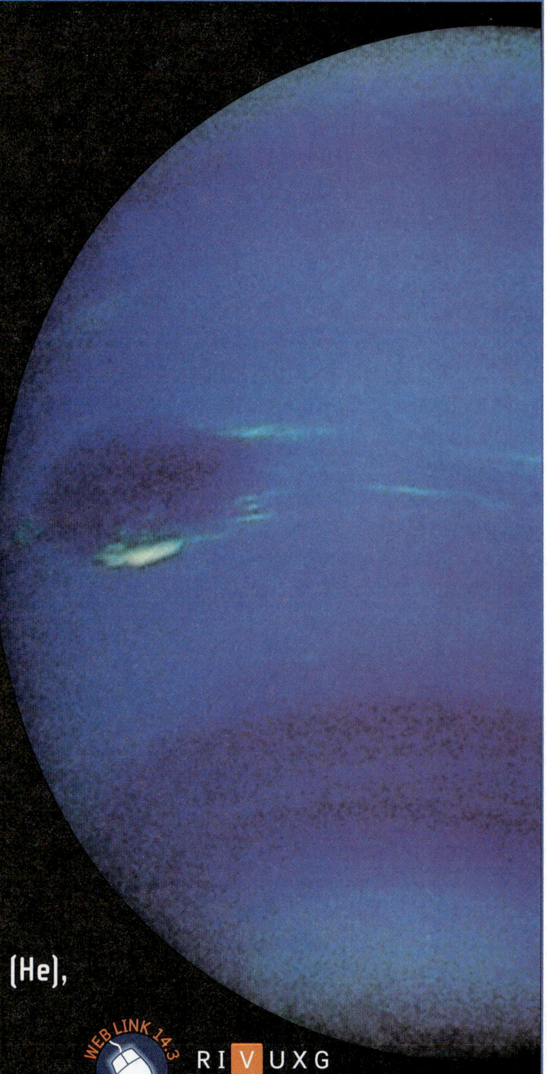

WEB LINK 14.3

RIVUXG (NASA/JPL)

sighted this planet before Herschel; it is plotted on at least 20 star charts drawn between 1690 and 1781. But all these other observers mistook Uranus for a dim star. Herschel was the first to track its motion relative to the stars and recognize it as a planet. Tracking Uranus was no small task, because Uranus moves very slowly on the celestial sphere, just over 4° in the space of a year (compared to about 12° for Saturn and about 35° for Jupiter).

The Discovery of Neptune

It was by carefully tracking Uranus's slow motions that astronomers were led to discover Neptune. By the beginning of the nineteenth century, it had become painfully clear to astronomers that they could not accurately predict the orbit of Uranus using Newtonian mechanics. By 1830 the discrepancy between the planet's predicted and observed positions had become large enough (2 arcmin) that some scientists suspected that Newton's law of gravitation might not be accurate at great distances from the Sun.

By the mid-1840s, two scientists working independently—the French astronomer Urbain Jean Joseph Le Verrier and the English mathematician John Couch Adams—were exploring an earlier and sounder suggestion. Perhaps the gravitational pull of an as yet undiscovered planet was causing Uranus to deviate slightly from its predicted orbit. Calculations by both scientists concluded that Uranus had indeed caught up with and had passed a more distant planet. Uranus had accelerated slightly as it approached the unknown planet, then decelerated slightly as it receded from the planet.

Inspired by Le Verrier's results, astronomers at Cambridge University Observatory undertook a six-week search for the proposed new planet in the summer of 1846. They were unsuccessful, in part because they lacked accurate star maps of the part of the sky being searched. The Cambridge astronomers also did not receive supporting information from Adams in a timely manner.

Meanwhile, Le Verrier wrote to Johann Gottfried Galle at the Berlin Observatory with detailed predictions of where to search for the new planet. Galle received the letter on September 23, 1846. That very night, aided by more complete star maps and after just a half-hour of searching, Galle and Heinrich d'Arrest located an uncharted star with the expected brightness in the predicted location. Subsequent observations confirmed that this new "star" showed a planetlike motion with respect to other stars.

Le Verrier proposed that the planet be called Neptune. After years of debate between English and French astronomers, the credit for its discovery came to be divided between Adams and Le Verrier. Neptune is the only planet whose existence was revealed by calculation rather than chance discovery.

At opposition, Neptune can be bright enough to be visible through small telescopes. The first person thus to have seen it was probably Galileo. His drawings from January 1613, when he was using his telescope to observe Jupiter and its four large satellites, show a "star" less than 1 arcmin from Neptune's location. Galileo even noted in his observation log that on one night this star seemed to have moved in relation to the other stars. But Galileo would have been hard pressed to identify Neptune as a planet, because its motion against the background stars is so slow (just over 2° per year).

Observing Uranus and Neptune

Through a large, modern telescope, both Uranus and Neptune are dim, uninspiring sights. Each planet appears as a hazy, featureless disk with a faint greenish-blue tinge. Although Uranus and Neptune are both about 4 times larger in diameter than Earth, they are so distant that their angular diameters as seen from Earth are tiny—no more than 4 arcsec for Uranus and just over 2 arcsec for Neptune. To an Earth-based observer, Uranus is roughly the size of a golf ball seen at a distance of 1 kilometer.

Uranus and Neptune are in such large orbits that they move very slowly on the celestial sphere

From 2007 through 2011, Uranus and Neptune are less than 40° apart in the sky in the adjacent constellations of Pisces, Aquarius, and Capricornus (the Sea Goat). During these years, the two planets are at opposition in either August or September, which are thus the best months to view them with a telescope. Table 14-1 and Table 14-2 give basic data about Uranus and Neptune, respectively.

14-2 Uranus is nearly featureless and has an unusually tilted axis of rotation

Scientists had hoped that *Voyager 2* would reveal cloud patterns in Uranus's atmosphere when it flew past the planet in January 1986. But even images recorded at close range showed Uranus to be remarkably featureless (Figure 14-1). Faint cloud markings became visible in images of Uranus only after extreme computer enhancement (Figure 14-2).

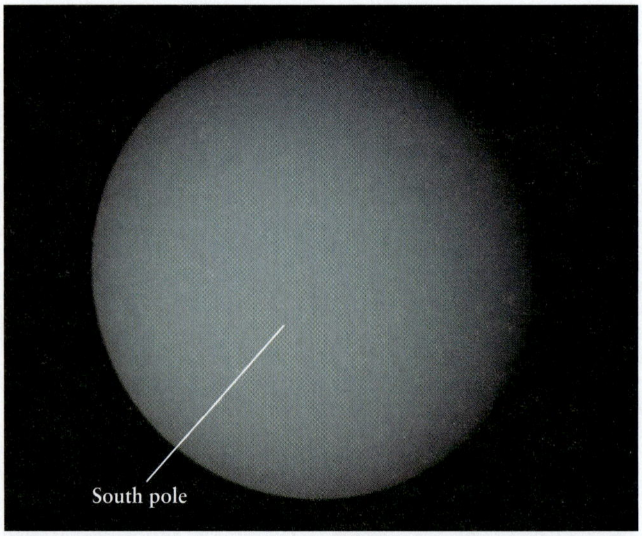

Figure 14-1 R I V U X G

Uranus from *Voyager 2* This image looks nearly straight down onto Uranus's south pole, which was pointed almost directly at the Sun when *Voyager 2* flew past in 1986. None of the *Voyager 2* images of Uranus show any pronounced cloud patterns. The color is due to methane in the planet's atmosphere, which absorbs red light but reflects green and blue. (JPL/NASA)

Uranus's Atmosphere

Voyager 2 data confirmed that the Uranian atmosphere is dominated by hydrogen (82.5%) and helium (15.2%), similar to the atmospheres of Jupiter and Saturn. Uranus differs, however, in that 2.3% of its atmosphere is methane (CH_4), which is 5 to 10 times the percentage found on Jupiter and Saturn. In fact, Uranus has a higher percentage of all heavy elements—including carbon atoms, which are found in molecules of methane—than Jupiter and Saturn. (In Section 14-4 we will investigate how this could have come about.)

Methane preferentially absorbs the longer wavelengths of visible light, so sunlight reflected from Uranus's upper atmosphere is depleted of its reds and yellows. Less reds and yellows in the reflected light gives the planet its distinct greenish-blue appearance. As on Saturn's moon Titan (see Section 13-8), ultraviolet light from the Sun turns some of the methane gas into a hydrocarbon haze, making it difficult to see the lower levels of the atmosphere.

Ammonia (NH_3), which constitutes 0.01 to 0.03% of the atmospheres of Jupiter and Saturn, is almost completely absent from the Uranian atmosphere. The reason is that Uranus is colder than Jupiter or Saturn: The temperature in its upper atmosphere is only 55 K (−218°C = −360°F). Ammonia freezes at these very low temperatures, so any ammonia has long since precipitated out of the atmosphere and into the planet's interior. For the same reason, Uranus's atmosphere is also lacking in water. Hence, the substances that make up the clouds on Jupiter and Saturn—ammonia, ammonium hydrosulfide (NH_4SH), and water—are not available in Uranus's atmosphere. This helps to explain the bland, uniform appearance of the planet shown in Figure 14-1.

The few clouds found on Uranus are made primarily of methane, which condenses into droplets only if the pressure is suf-

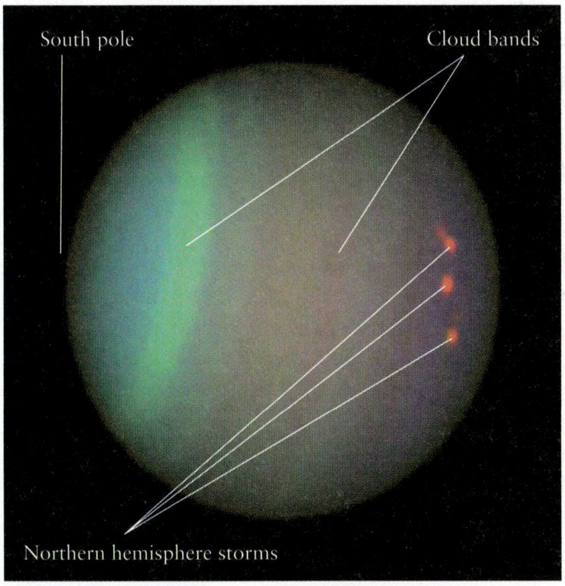

WEB LINK 14.5

Figure 14-2 R I V U X G

Uranus from the Hubble Space Telescope Images made at ultraviolet, visible, and infrared wavelengths were combined and enhanced to give this false-color view of cloud features on Uranus. These images were captured in August 2004 (18 years after the image in Figure 14-1), during springtime in Uranus's northern hemisphere. (NASA and Erich Karkoschka, University of Arizona)

ficiently high. Hence, methane clouds form only at lower levels within the atmosphere, where they are difficult to see.

An Oddly Tilted Planet

The rotation period of Uranus's atmosphere is about 16 hours. Like Jupiter and Saturn, Uranus rotates differentially, so this period depends on the latitude and can be measured by tracking the motions of clouds. To determine the rotation period for the underlying body of the planet, scientists looked to Uranus's magnetic field, which is presumably anchored in the planet's interior, or at least in the deeper and denser layers of its atmosphere. Data from *Voyager 2* indicate that Uranus's internal period of rotation is 17.24 hours.

Voyager 2 also confirmed that Uranus's rotation axis is tilted in a unique and bizarre way. Herschel found the first evidence of this in 1787, when he discovered two moons orbiting Uranus in a plane that is almost perpendicular to the plane of the planet's orbit around the Sun. Because the large moons of Jupiter and Saturn were known to orbit in the same plane as their planet's equator and in the same direction as their planet's rotation, it was thought that the same must be true for the moons of Uranus. Thus, Uranus's equator must be almost perpendicular to the plane of its orbit, and its rotation axis must lie very nearly in that plane (Figure 14-3).

Careful measurement shows that Uranus's axis of rotation is tilted by 98°, as compared to $23\frac{1}{2}$° for Earth (compare Figure 14-3 with Figure 2-12). A tilt angle greater than 90° means that

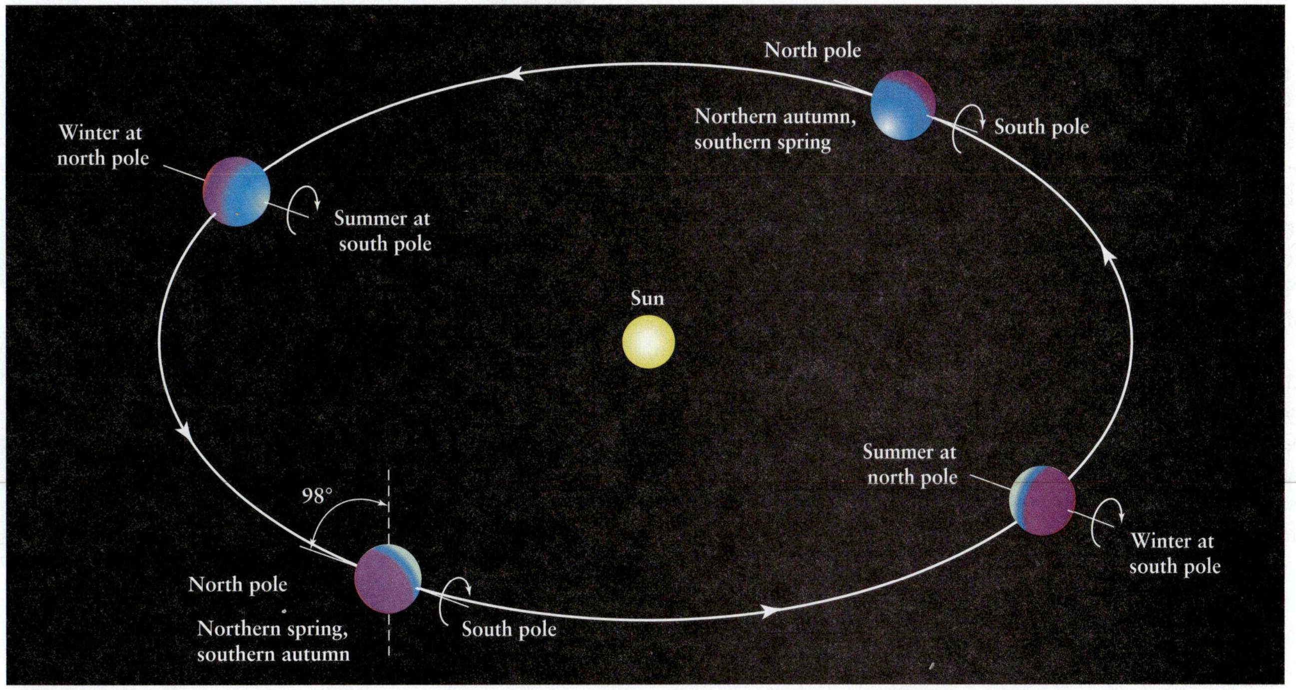

ANIMATION 14.1

Figure 14-3

Exaggerated Seasons on Uranus For most planets, the rotation axis is roughly perpendicular to the plane of the planet's orbit around the Sun. But for Uranus the rotation axis is tilted by 98° from the perpendicular. This tilt causes severely exaggerated seasons. For example, during midsummer at Uranus's south pole, the Sun appears nearly overhead for many Earth years, while the planet's northern regions are in continuous darkness. Half an orbit later, the seasons are reversed.

Uranus exhibits retrograde (backward) rotation like that of Venus, shown in Figure 11-7*b*. Astronomers suspect that Uranus might have acquired its large tilt angle billions of years ago, when another massive body collided with Uranus while the planet was still forming.

The radical tilt of its axis means that as Uranus moves along its 84-year orbit, its north and south poles alternately point toward or away from the Sun. This tilt produces highly exaggerated seasonal changes. For example, when *Voyager 2* flew by in 1986, Uranus's south pole was pointed toward the Sun. Most of the planet's southern hemisphere was bathed in continuous sunlight, while most of the northern hemisphere was subjected to a continuous, frigid winter night. But over the following quarter of a Uranian year later (21 of our years), sunlight has gradually been returning to the northern hemisphere, triggering immense storms there. Figure 14-2 shows some of these storms, which are much more visible at infrared wavelengths than with visible light.

Atmospheric Motions on Uranus

By following the motions of clouds and storm systems on Uranus, scientists find that the planet's winds flow to the east—that is, in the same direction as the planet's rotation—at northern and southern latitudes, but to the west near the equator. This wind pattern is quite unlike the situation on Jupiter and Saturn, where the zonal winds alternate direction many times between the north and south poles (see Section 12-4). The fastest Uranian winds (about 700 km/h, or 440 mi/h) are found at the equator.

We saw in Section 12-4 that the internal heat of Jupiter and Saturn plays a major role in driving atmospheric activity on those worlds. Uranus is different: It appears to have little or no internal source of thermal energy. Measurements to date show that unlike Jupiter or Saturn, Uranus radiates into space only as much energy as it receives from the Sun. With only feeble sunlight to provide energy to its atmosphere, and little internal thermal energy, Uranus lacks the dramatic wind and cloud dynamics found on Jupiter and Saturn. It's not known why Uranus lacks a significant source of internal thermal energy.

Unlike the other Jovian planets, Uranus's interior heat has no effect on its atmosphere's motions

Although Uranus's equatorial region was receiving little sunlight at the time of the *Voyager 2* flyby, the atmospheric temperature there (about 55 K = −218°C = −359°F) was not too different from that at the sunlit pole. Heat must therefore be efficiently transported from the poles to the equator. This north-south heat transport, which is perpendicular to the wind flow, may have mixed and homogenized the atmosphere to make Uranus nearly featureless.

14-3 Neptune is a cold, bluish world with Jupiterlike atmospheric features

VIDEO 14.1

At first glance, Neptune appears to be the twin of Uranus. (See the image that opens this chapter, and compare Tables 14-1 and 14-2.) But these two planets are by no means identical. While Neptune and Uranus have almost the same diameter, Neptune is 18% more massive. Neptune's axis of rotation also has a more modest $29\frac{1}{2}$° tilt. When *Voyager 2* flew past Neptune in August 1989, it revealed that the planet has a more active and dynamic atmosphere than Uranus. This activity suggests that Neptune, unlike Uranus, has a powerful source of energy in its interior.

Neptune's Atmosphere

The *Voyager 2* data showed that Neptune has essentially the same atmospheric composition as Uranus: 79% hydrogen, 18% helium, 3% methane, and almost no ammonia or water vapor. As for Uranus, the presence of methane gives Neptune a characteristic bluish-green color. The temperature in the upper atmosphere is also the same as on Uranus, about 55 K. That these temperatures should be so similar, even though Neptune is much farther from the Sun, is further evidence that Neptune has a strong internal heat source.

Unlike Uranus, however, Neptune has clearly visible cloud patterns in its atmosphere. At the time that *Voyager 2* flew past Neptune, the most prominent feature in the planet's atmosphere was a giant storm called the Great Dark Spot (Figure 14-4). The Great Dark Spot had a number of similarities to Jupiter's Great Red Spot (see Sections 12-1 and 12-3). The storms

Unlike its close relative Uranus, Neptune has truly immense storms

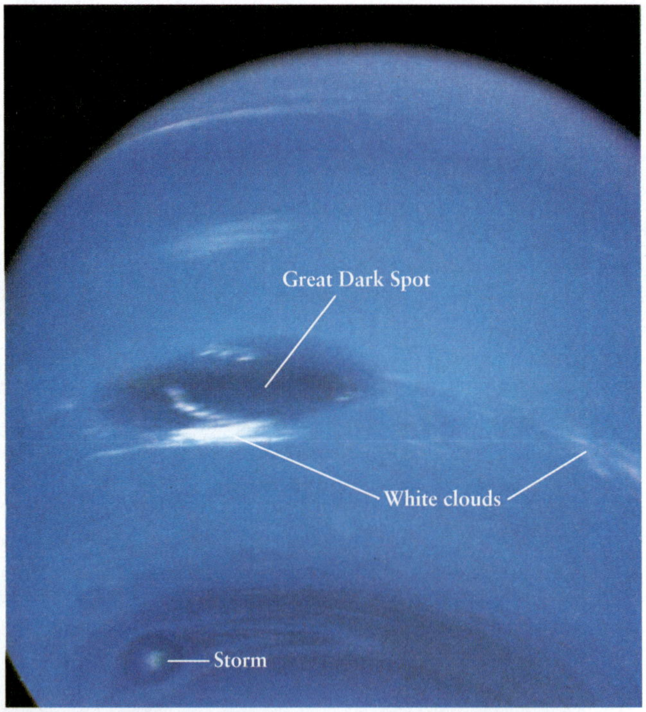

WEB LINK 14.6 VIDEO 14.2

Figure 14-4 RIVUXG
Neptune from *Voyager 2* When this picture of Neptune's southern hemisphere was taken in 1989, the Great Dark Spot measured about 12,000 by 8000 km, comparable in size to the Earth. (A smaller storm appears at the lower left.) The white clouds are thought to be composed of crystals of methane ice. The color contrast in this image has been exaggerated to emphasize the differences between dark and light areas. (NASA/JPL)

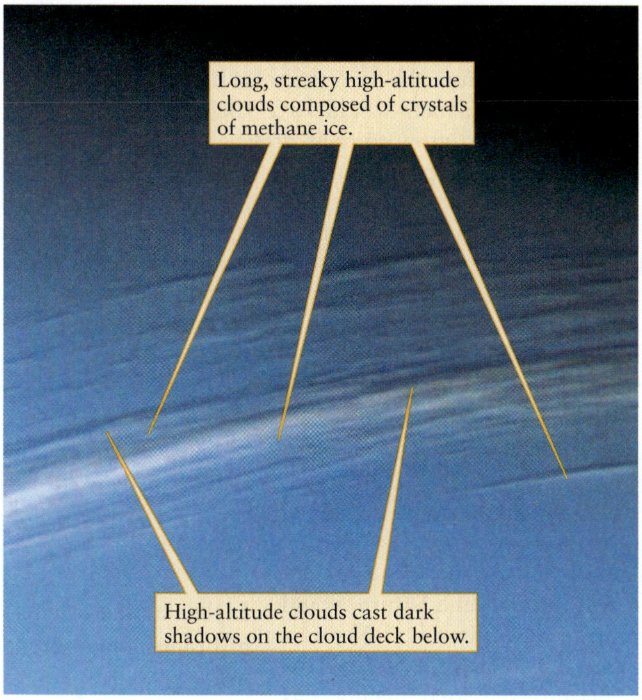
Long, streaky high-altitude clouds composed of crystals of methane ice.
High-altitude clouds cast dark shadows on the cloud deck below.

Figure 14-5 R I V U X G

Cirrus Clouds over Neptune *Voyager 2* recorded this image of clouds near Neptune's terminator (the border between day and night on the planet). Like wispy, high-altitude cirrus clouds in Earth's atmosphere, these clouds are thought to be made of ice crystals. The difference is that Neptune's cirrus clouds are probably methane ice, not water ice as on Earth. (NASA/JPL)

on both planets were comparable in size to Earth's diameter, both appeared at about the same latitude in the southern hemisphere, and the winds in both storms circulated in a counterclockwise direction (see Figure 12-5). But Neptune's Great Dark Spot appears not to have been as long-lived as the Great Red Spot on Jupiter. When the Hubble Space Telescope first viewed Neptune in 1994, the Great Dark Spot had disappeared. Another dark storm appeared in 1995 in Neptune's northern hemisphere.

Voyager 2 also saw a few conspicuous whitish clouds on Neptune. These clouds are thought to be produced when winds carry methane gas into the cool, upper atmosphere, where it condenses into crystals of methane ice. *Voyager 2* images show these high-altitude clouds casting shadows onto lower levels of Neptune's atmosphere (Figure 14-5). Images from the Hubble Space Telescope also show the presence of high-altitude clouds (Figure 14-6).

Neptune's Internal Heat

Due to its greater distance from the Sun, Neptune receives less than half as much energy from the Sun as Uranus. But with less solar energy available to power atmospheric motions, why are there high-altitude clouds and huge, dark storms on Neptune but not on Uranus? At least part of the answer is probably that Neptune is still slowly contracting, thus converting gravitational energy into thermal energy that heats the planet's core. (The same is true for Jupiter and Saturn, as we saw in Section 12-4.) The evidence for this contraction is that Neptune, like Jupiter and Saturn but unlike Uranus, emits more energy than it receives from the Sun. The combination of a warm interior and a cold outer atmosphere can cause convection in Neptune's atmosphere, producing the up-and-down motion of gases that generates clouds and storms. Neptune also resembles Jupiter in having faint belts and zones parallel to the planet's equator (see Figure 14-6).

Like those on Uranus, most of Neptune's clouds are probably made of droplets of liquid methane. Because these droplets form fairly deep within the atmosphere, the clouds are more difficult to see than the ones on Jupiter. Hence, Neptune's belts and zones are less pronounced than Jupiter's, although more so than those on Uranus (thanks to the extra cloud-building energy from Neptune's interior). As described above, Neptune's high-altitude clouds (see Figure 14-5) are probably made of *frozen* methane.

Although Neptune displays more evidence of up-and-down motion in its atmosphere than Uranus, the global pattern of east and west winds is almost identical on the two planets. This similarity is rather strange. The two planets are heated very differently by the Sun, thanks to their different distances from the Sun and the different tilts of their axes of rotation, so we might have expected the wind patterns on Uranus and Neptune also to be different. Perhaps the explanation of these wind patterns will involve understanding how heat is transported not only within the atmospheres of Uranus and Neptune but within their interiors as well.

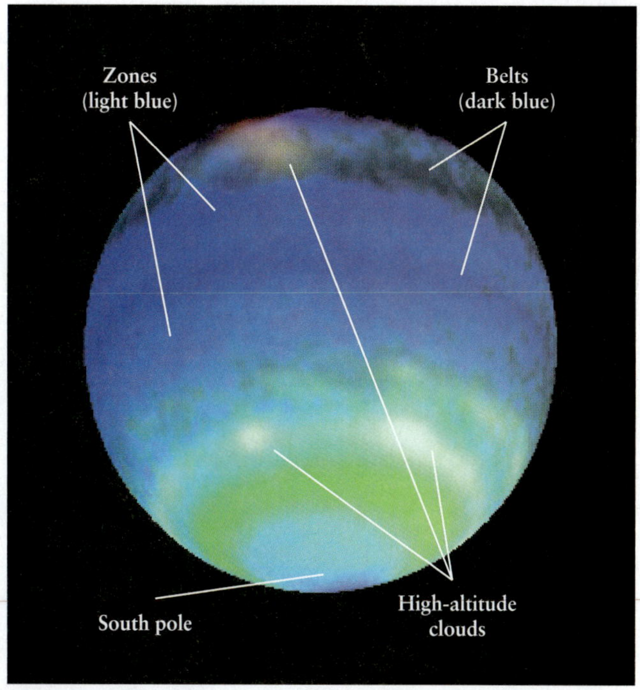
Zones (light blue)
Belts (dark blue)
South pole
High-altitude clouds

WEB LINK 14.7

Figure 14-6 R I V U X G

Neptune's Banded Structure This enhanced-color Hubble Space Telescope image shows Neptune's belts and zones. White areas denote high-altitude clouds; the very highest clouds (near the top of the image) are shown in yellow-red. The green belt near the south pole is a region where the atmosphere absorbs blue light, perhaps indicating a different chemical composition there. (Lawrence Sromovsky, University of Wisconsin-Madison; STScI/NASA)

14-4 Uranus and Neptune contain a higher proportion of heavy elements than Jupiter and Saturn

At first glance, Uranus and Neptune might seem to be simply smaller and less massive versions of Jupiter or Saturn (see Table 7-1). But like many first impressions, this one is misleading because it fails to take into account what lies beneath the surface. We saw in Section 14-2 that the interiors of both Jupiter and Saturn are composed primarily of hydrogen and helium, in nearly the same abundance as the Sun. But Uranus and Neptune must have a different composition, because their average densities are too high.

Uranus and Neptune: Curiously Dense

If Uranus and Neptune had solar abundances of the elements, the smaller masses of these planets would produce less gravitational compression, and Uranus and Neptune would both have lower average densities than Jupiter or Saturn. In fact, however, Uranus and Neptune have average densities (1320 kg/m^3 and 1640 kg/m^3, respectively) that are comparable to or greater than those of Jupiter (1330 kg/m^3) or Saturn (690 kg/m^3). Therefore, we can conclude that Uranus and Neptune contain greater proportions of the heavier elements than do Jupiter or Saturn.

This picture is not what we might expect. According to our discussion in Section 8-4 of how the solar system formed, hydrogen and helium should be relatively more abundant as distance from the vaporizing heat of the Sun increases. But Uranus and Neptune contain a greater percentage of heavy elements, and, therefore, a *smaller* percentage of hydrogen and helium, than Jupiter and Saturn.

A related problem is how to explain the masses of Uranus and Neptune. At the locations of Uranus (19 AU from the Sun) and Neptune (30 AU from the Sun) the solar nebula was probably so sparse that these planets would have taken tens of hundreds of millions of years to grow to their present sizes around a core of icy planetesimals. But protoplanetary disks observed around other stars (see Section 8-3) do not appear to survive for that long before they dissipate. In other words, the problem is not how to explain why Uranus and Neptune are smaller than Jupiter and Saturn; it is how to explain why Uranus and Neptune should exist at all!

Models of How Uranus and Neptune Formed

In one model for the origin of Uranus and Neptune, the planets formed in denser regions of the solar nebula between 4 and 10 AU from the Sun. In this region they could have grown rapidly to their present sizes. Had the planets remained at these locations, they could eventually have accumulated enough hydrogen and helium to become as large as Jupiter or Saturn. But before that could happen, gravitational interactions with Jupiter and Saturn would have deflected Uranus and Neptune outward into their present large orbits. Because the solar nebula was very sparse at those greater distances, Uranus and Neptune stopped growing and remained at the sizes we see today.

Explaining the origins of Uranus and Neptune poses a challenge to planetary scientists

An alternative idea is that Uranus and Neptune formed in their present locations, not around a core of planetesimals but directly from the gases of the solar nebula. (We introduced this idea in Section 8-4.) Once a ball of gas formed, icy particles within the ball would have settled to its center, forming a solid core. This

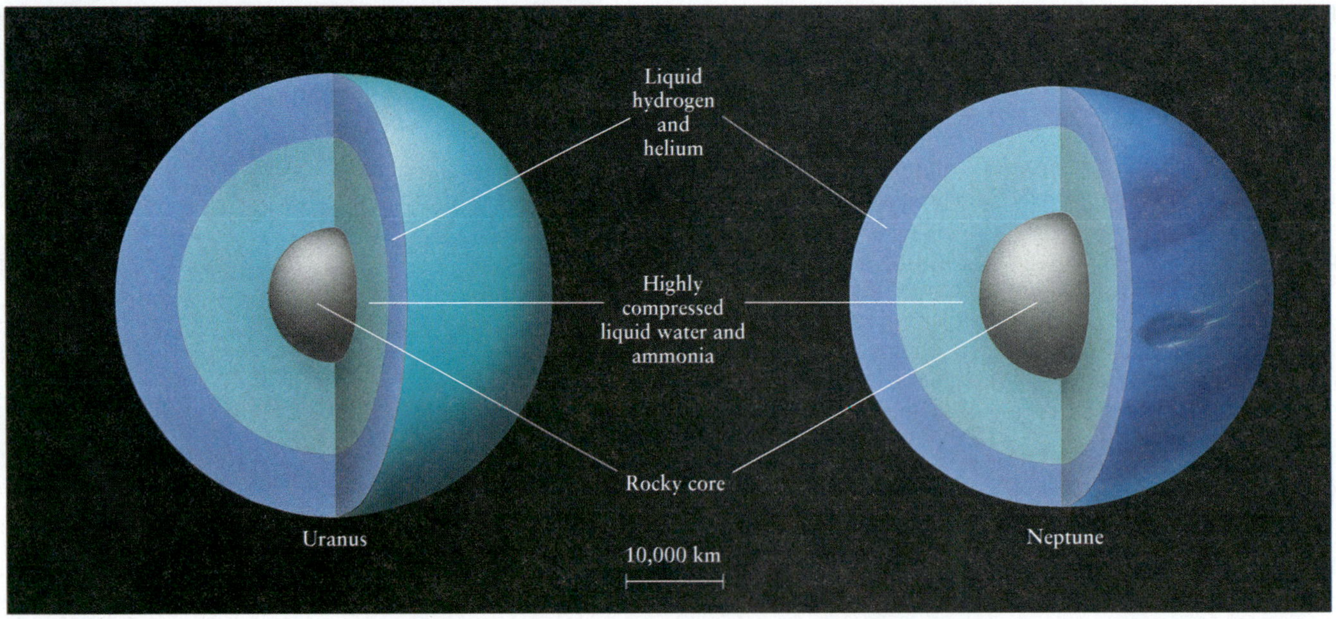

Figure 14-7

The Internal Structures of Uranus and Neptune In the model shown here, both Uranus and Neptune have a rocky core (resembling a terrestrial planet), a mantle of liquid water with ammonia dissolved in it, and an outer layer of liquid molecular hydrogen and liquid helium. The atmospheres are very thin shells on top of the hydrogen-helium layer. Uranus and Neptune have about the same diameter, but since Neptune is more massive it may have a somewhat larger core.

process could have formed Uranus and Neptune in only a few hundred years. In this model, Uranus and Neptune would actually have ended up larger than they are today, with substantially more hydrogen and helium. But if there happened to be a hot star relatively near the solar nebula, it would have emitted substantial amounts of ultraviolet radiation (see Section 5-3 and Figure 5-11). Ultraviolet photons have high energy (see Section 5-5), and a shower of them could have stripped Uranus and Neptune of many of their lightweight hydrogen and helium atoms, leaving shrunken planets with a larger percentage of heavy elements. (In this picture, Jupiter and Saturn were protected from the star's ultraviolet radiation because they were shrouded by the thicker clouds of material in their part of the solar nebula.) More research will be needed to better determine the origins of Uranus and Neptune.

Figure 14-7 shows one model for the present-day internal structures of Uranus and Neptune. (Compare this to the internal structures of Jupiter and Saturn, shown in Figure 12-12.) In this model each planet has a rocky core roughly the size of Earth, although more massive. Each planet's core is surrounded by a mantle of liquid water and ammonia. (This mixture means that the mantle is chemically similar to household window cleaning fluid.) Around the mantle is a layer of liquid molecular hydrogen and liquid helium, with a small percentage of liquid methane. This layer is relatively shallow compared to those on Jupiter and Saturn, and the pressure is not high enough to convert the liquid hydrogen into liquid metallic hydrogen.

14-5 The magnetic fields of both Uranus and Neptune are oriented at unusual angles

Astronomers were quite surprised by the data sent back from *Voyager 2*'s magnetometer as the spacecraft sped past Uranus and Neptune. These data, as well as radio emissions from charged particles in their magnetospheres, showed that the magnetic fields of both Uranus and Neptune are tilted at steep angles to their axes of rotation. Uranus's **magnetic axis,** the line connecting its north and south magnetic poles, is inclined by 59° from its axis of rotation; Neptune's magnetic axis is tilted by 47°. By contrast, the magnetic axes of Earth, Jupiter, and Saturn are all nearly aligned with their rotation axes; the angle between their magnetic and rotation axes is 12° or less (Figure 14-8). Scientists were also surprised to find that the magnetic fields of Uranus and Neptune are offset from the centers of the planets.

CAUTION! The drawing of Earth's magnetic field at the far left of Figure 14-8 may seem to be mislabeled, because it shows the *south* pole of a magnet at Earth's *north* pole. But, in fact, this is correct, as you can understand by thinking about how magnets work. On a magnet that is free to swivel, like the magnetized needle in a compass, the "north pole" is called that because it tends to point north on Earth. Likewise, a compass needle's south pole points toward the south on Earth. Furthermore, opposite magnetic poles attract. If you take two magnets and put them next to each other, they try to align themselves so that one magnet's north pole is next to the other magnet's south pole. Now, if you think of a compass needle as one magnet and the entire Earth as the other magnet, it makes sense that the compass needle's north pole is being drawn toward a magnetic *south* pole—which happens to be located near Earth's geographic north pole. Earth's magnetic pole nearest its geographic North Pole is called the "magnetic north pole." Note that the magnets drawn inside Jupiter, Saturn, and Neptune are oriented opposite to Earth's. On any of these planets, the north pole of a compass needle would point south, not north!

Explaining Misaligned Magnetism

Why are the magnetic axes and axes of rotation of Uranus and Neptune so badly misaligned? And why are the magnetic fields offset from the centers of the planets? One possibility is that their

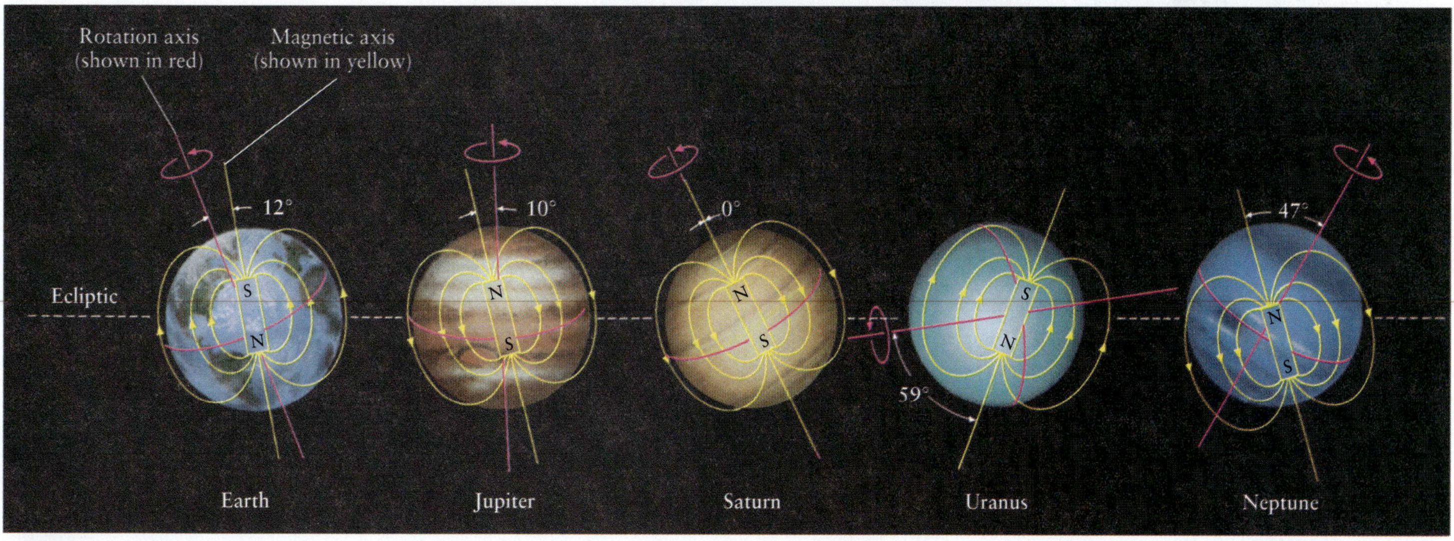

INTERACTIVE EXERCISE 14.1

Figure 14-8

The Magnetic Fields of Five Planets This illustration shows how the magnetic fields of Earth, Jupiter, Saturn, Uranus, and Neptune are tilted relative to their rotation axes (shown in red). For both Uranus and Neptune, the magnetic fields are offset from the planet's center and steeply inclined to the rotation axis.

magnetic fields might be undergoing a reversal; geological data show that Earth's magnetic field has switched north to south and back again many times in the past. Another possibility is that the misalignments resulted from catastrophic collisions with planet-sized bodies. As we discussed in Section 14-2, the tilt of Uranus's rotation axis and its system of moons suggest that such collisions occurred long ago. As we will see in Section 14-7, Neptune may have gravitationally captured its largest moon, Triton, but no one knows if that incident was responsible for offsetting Neptune's magnetic axis.

Because neither Uranus nor Neptune is massive enough to compress hydrogen to a metallic state, their magnetic fields cannot be generated in the same way as those of Jupiter and Saturn. Instead, under the high pressures found in the watery mantles of Uranus and Neptune, dissolved molecules such as ammonia lose one or more electrons and become electrically charged (that is, they become *ionized*; see Section 5-8). Water is a good conductor of electricity when it has such electrically charged molecules dissolved in it, and electric currents in this fluid are probably the source of the magnetic fields of Uranus and Neptune.

Unlike any other planets, Uranus and Neptune have magnetic fields caused by ionized ammonia

The *Cosmic Connections* figure summarizes the key properties of Uranus and Neptune and how they compare with those of Jupiter and Saturn.

14-6 Uranus and Neptune each has a system of thin, dark rings

On March 10, 1977, Uranus was scheduled to move in front of a faint star, as seen from the Indian Ocean. A team of astronomers headed by James L. Elliot of Cornell University observed this event, called an **occultation,** from a NASA airplane equipped with a telescope. They hoped that by measuring how long the star was hidden, they could accurately measure Uranus's size. In addition, by carefully measuring how the starlight faded when Uranus passed in front of the star, they planned to deduce important properties of Uranus's upper atmosphere.

The Surprising Rings of Uranus

To everyone's surprise, the background star briefly blinked on and off several times just before the star passed behind Uranus and again immediately after (Figure 14-9). The astronomers

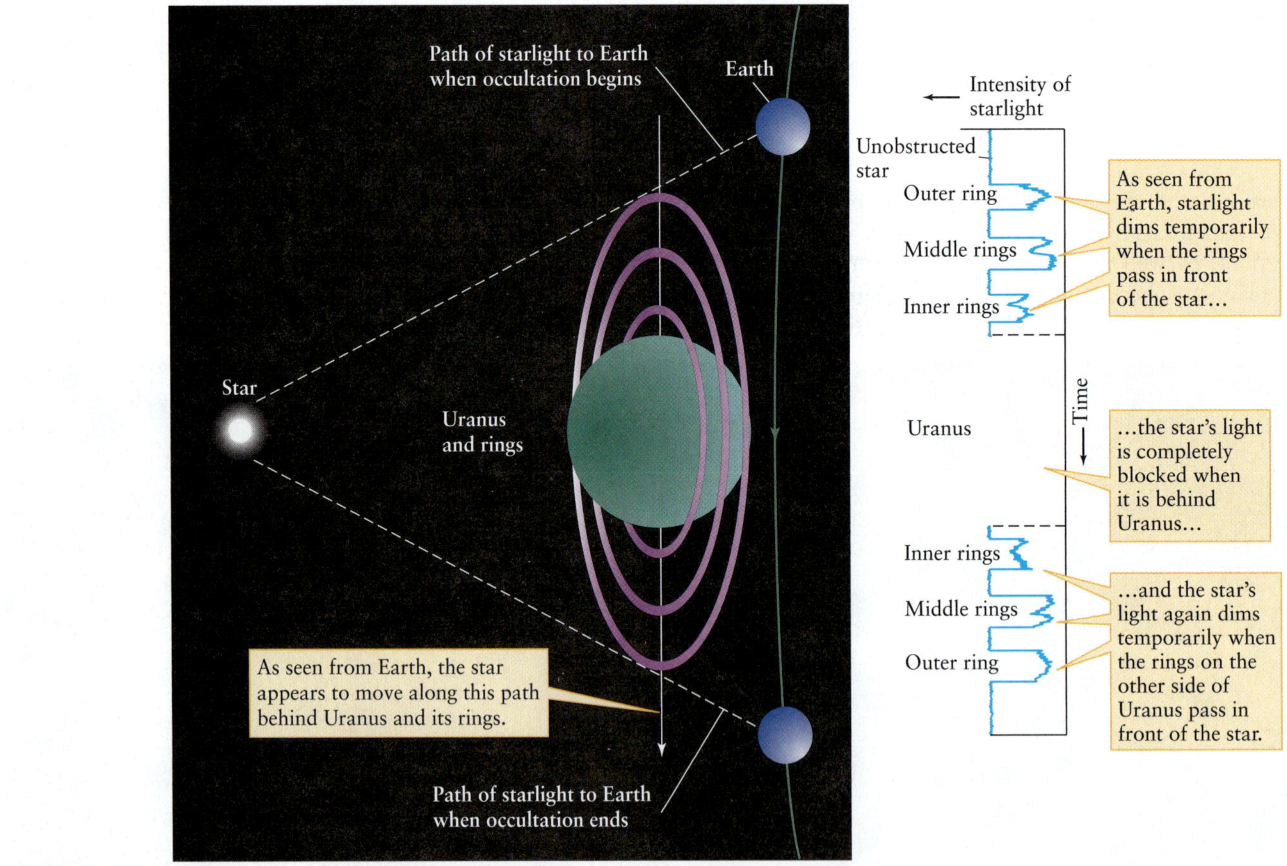

Path of starlight to Earth when occultation begins
Earth
Star
Uranus and rings
As seen from Earth, the star appears to move along this path behind Uranus and its rings.
Path of starlight to Earth when occultation ends
Intensity of starlight
Unobstructed star
Outer ring
Middle rings
Inner rings
Uranus
Time
Inner rings
Middle rings
Outer ring
As seen from Earth, starlight dims temporarily when the rings pass in front of the star...
...the star's light is completely blocked when it is behind Uranus...
...and the star's light again dims temporarily when the rings on the other side of Uranus pass in front of the star.
ANIMATION 14.2

Figure 14-9
How the Rings of Uranus Were Discovered As seen from Earth, Uranus occasionally appears to move in front of a distant star. Such an event is called an occultation. The star's light is completely blocked when it is behind the planet. But before and after the occultation by Uranus, the starlight dims temporarily as the rings pass in front of the star.

COSMIC CONNECTIONS

The Outer Planets: A Comparison

Uranus and Neptune are not simply smaller versions of Jupiter and Saturn. This table summarizes the key differences among the four Jovian planets.

	Interior	Surface	Rings	Atmosphere	Magnetic Field
Jupiter	Terrestrial core, liquid metallic hydrogen shell, liquid hydrogen mantle	No solid surface, atmosphere gradually thickens to liquid state, belt and zone structure, hurricane-like features	Yes	Primarily H, He	19,000× Earth's total field; at its cloud layer, 14× stronger than Earth's surface field
Saturn	Similar to Jupiter, with bigger terrestrial core and less metallic hydrogen	No solid surface, less distinct belt and zone structure than Jupiter	Yes	Primarily H, He	570× Earth's total field; at its cloud layer, $^2/_3$× Earth's surface field
Uranus	Terrestrial core, liquid water shell, liquid hydrogen and helium mantle	No solid surface, weak belt and zone system, hurricane-like features, color from methane absorption of red, orange, yellow	Yes	Primarily H, He, some CH_4	50× Earth's total field; at its cloud layer, 0.7× Earth's surface field
Neptune	Similar to Uranus	Similar to Uranus	Yes	Primarily H, He, some CH_4	35× Earth's total field; at its cloud layer, 0.4× Earth's surface field

For detailed comparisons between planets, see Appendices 1 and 2.

* To see the orientations of these magnetic fields relative to the rotation axes of the planets, see Figure 14-8.

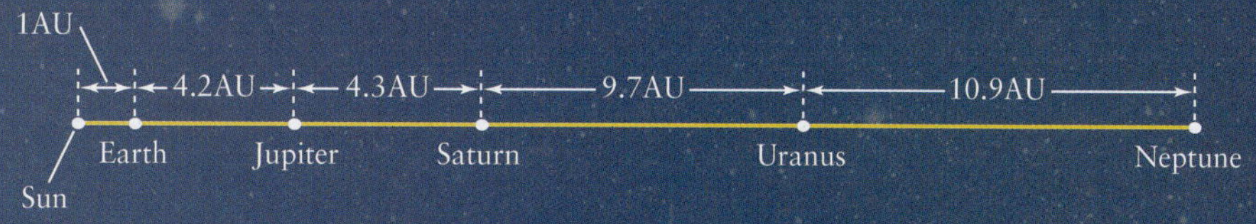

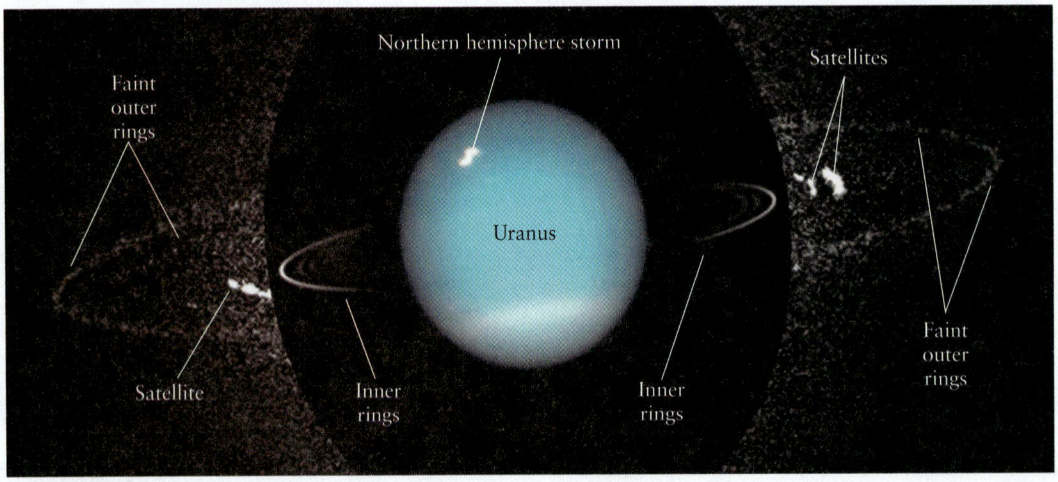

Figure 14-10 R I V U X G

Uranus's Rings from the Hubble Space Telescope Several Hubble Space Telescope images were assembled into this view of Uranus's rings. Relatively short exposure times reveal the planet and the inner rings, while much longer exposure times were needed to reveal the outer rings. During these long exposures several of Uranus's satellites moved noticeably, leaving bright trails on the image. (NASA; ESA; and M. Showalter, SETI Institute)

concluded that Uranus must be surrounded by a series of nine narrow rings. In addition to these nine rings, *Voyager 2* discovered two others that lie even closer to the cloudtops of Uranus. Two other extremely faint rings, much larger in diameter than any of the others, were found in 2005 using the Hubble Space Telescope (Figure 14-10).

Like the planet itself, the rings of Uranus were discovered by accident

Unlike Saturn's rings, the rings of Uranus are dark and narrow: Most are less than 10 km wide. Typical particles in Saturn's rings are chunks of reflective ice the size of snowballs (a few centimeters across), but typical particles in Uranus's rings are 0.1 to 10 meters in size and are no more reflective than lumps of coal. It is not surprising that these narrow, dark rings escaped detection for so long. Figure 14-11 is a *Voyager 2* image of the rings from the side of the planet away from the Sun, where light scattering from exceptionally small ring particles makes them more visible. (We discussed light scattering in Section 5-6 and Box 5-4.) Figure 14-10 shows the Sun-facing side of the rings.

The Uranian rings are so dark because sunlight at Uranus is only one-quarter as intense as at Saturn. As a result, Uranus's ring particles are so cold that they can retain methane ice. Scientists speculate that eons of impacts by electrons trapped in the magnetosphere of Uranus have converted this methane ice into dark carbon compounds. This **radiation darkening** can account for the low reflectivity of the rings.

Uranus's major rings are located less than 2 Uranian radii from the planet's center, well within the planet's Roche limit (see Section 12-9). Some sort of mechanism, possibly one involving shepherd satellites, efficiently confines particles to their narrow orbits. (In Section 12-11 we discussed how shepherd satellites help keep planetary rings narrow.) *Voyager 2* searched for shepherd satellites but found only two. The others may be so small and dark that they have simply escaped detection.

The two faint rings discovered in 2005 (see Figure 14-10) are different: They lie well *outside* Uranus's Roche limit. The outer of these two rings owes its existence to a miniature satellite called Mab (named for a diminutive fairy in English folklore) that orbits within this ring. Meteorites colliding with Mab knock dust off the satellite's surface. Since Mab is so small (just 36 km in diameter) and hence has little gravity, the ejected dust escapes from

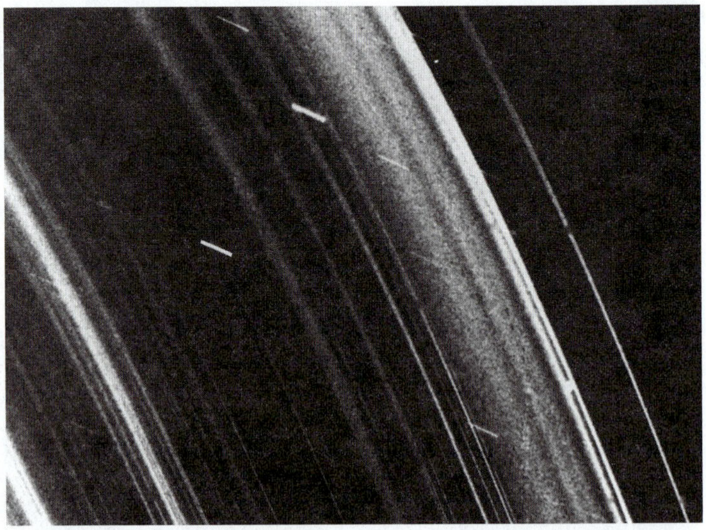

Figure 14-11 R I V U X G

The Shaded Side of Uranus's Rings This view from *Voyager 2*, taken when the spacecraft was in Uranus's shadow, looks back toward the Sun. Numerous fine-grained dust particles gleam in the spaces between the main rings. This dust is probably debris from collisions between larger particles in the main rings. The short horizontal streaks are star images blurred by the spacecraft's motion during the exposure. (NASA/JPL)

VIDEO 14.3

Figure 14-12 R I V U X G

Neptune's Rings The two main rings of Neptune and a faint, inner ring can easily be seen in this composite of two *Voyager 2* images. There is also a faint sheet of particles whose outer edge is located between the two main rings and extends inward toward the planet. The overexposed image of Neptune itself has been blocked out. (NASA)

the satellite and goes into orbit around Uranus, forming a ring. The inner of the two faint rings may be formed in the same way; however, no small satellite has yet been found within this ring.

The Decaying Rings of Neptune

Like Uranus, Neptune is surrounded by a system of thin, dark rings that were first detected in stellar occultations. Figure 14-12 is a *Voyager 2* image of Neptune's rings. The ring particles vary in size from a few micrometers ($1\ \mu\text{m} = 10^{-6}$ m) to about 10 meters. As for Uranus's rings, the particles that make up Neptune's rings reflect very little light because they have undergone radiation darkening.

In 2002 and 2003, a team led by Imke de Pater of the University of California, Berkeley, used the 10-meter Keck telescope in Hawaii to observe in detail the rings of Neptune. Remarkably, they found that all of the rings had become fainter since the *Voyager 2* flyby in 1989. Apparently, the rings are losing particles faster then new ones are being added. If the rings continue to decay at the same rate, one of them may vanish completely within a century. More research into the nature of Neptune's rings is needed to explain this curious and rapid decay.

14-7 Some of Uranus's satellites show evidence of past tidal heating

Before the *Voyager 2* flyby of Uranus, five moderate-sized satellites—Titania, Oberon, Ariel, Umbriel, and Miranda—were known to orbit the planet (Figure 14-13). Most are named after sprites and spirits in Shakespeare's plays. They range in diameter from about 1600 km (1000 mi) for Titania and Oberon to less than 500 km (300 mi) for Miranda. All these moons have average densities around 1500 kg/m^3, which is consistent with a mixture of ice and rock.

Voyager 2 discovered 11 other small Uranian satellites, most of which have diameters of less than 100 km (60 mi); 11 more were found using ground-based telescopes and the Hubble Space Telescope between 1997 and 2005. Only Jupiter and Saturn have more known satellites than Uranus. Figure 14-14 shows several of the smaller satellites. Appendix 3 summarizes information about all 27 known moons of Uranus.

Uranus's Unusual Moderate-Sized Satellites

Umbriel and Oberon both appear to be geologically dead worlds, with surfaces dominated by impact craters. By contrast, Ariel's surface appears to have been cracked at some time in the past, allowing some sort of ice lava to flood low-lying areas. A similar process appears to have taken place on Titania. This geologic activity may be due to a combination of the satellites' internal heat and tidal heating like that which powers volcanism on Jupiter's satellite Io (see Section 13-4). Io's tidal heating is only possible because of the 1:2:4 ratio of the orbital periods of Io, Europa, and Ganymede. While there are no such simple ratios between the

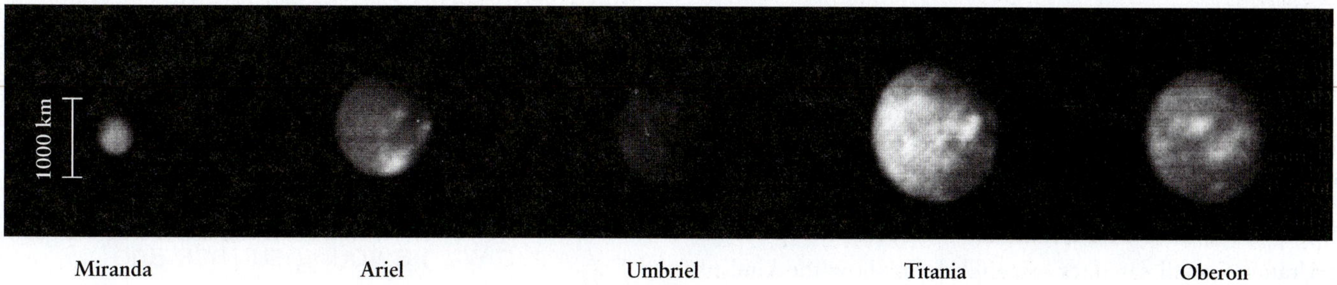

INTERACTIVE EXERCISE 14.2

Figure 14-13 R I V U X G

Uranus's Principal Satellites This "family portrait" (a montage of five *Voyager 2* images) shows Uranus's five largest moons to the same scale and correctly displays their respective reflectivities. (The darkest satellite, Umbriel, is actually more reflective than Earth's Moon.) All five satellites have grayish surfaces, with only slight variations in color. (NASA/JPL)

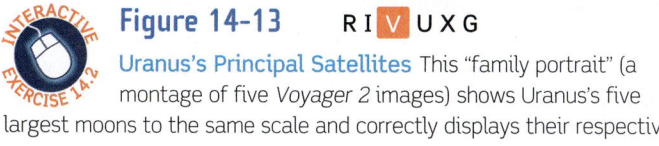

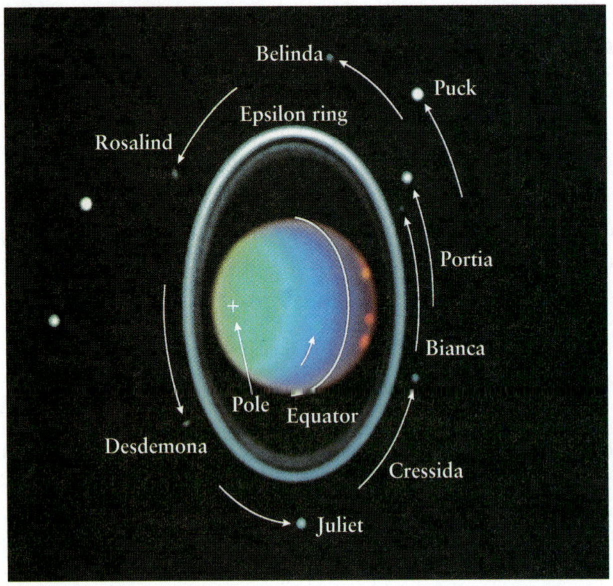

Figure 14-14 R I V U X G

Uranus's Rings and Small Satellites This false-color infrared image from the Hubble Space Telescope shows eight of Uranus's satellites, all of which were discovered by *Voyager 2* when it flew past Uranus in 1986. They all lie within 86,000 km of the planet's center (only about one-fifth of the distance from Earth to our Moon) and all are less than 160 km (100 mi) in diameter. The arcs show how far each satellite moves around its orbit in 90 minutes. (Erich Karkoschka, University of Arizona; and NASA)

Figure 14-15 R I V U X G

Miranda This composite of *Voyager 2* images shows that part of Miranda's surface is ancient and heavily cratered, while other parts are dominated by parallel networks of valleys and ridges. At the very bottom of the image—where a "bite" seems to have been taken out of Miranda—is a range of enormous cliffs that jut upward to an elevation of 20 km, twice as high as Mount Everest. (NASA/JPL)

present-day orbital periods of Uranus's satellites, there may have been in the past. If so, tidal heating could have helped reshape the surfaces of some of the satellites.

Unique among Uranus's satellites is Miranda, which has a landscape unlike that of any other world in the solar system. Much of the surface is heavily cratered, as we would expect for a satellite only 470 km in diameter, but several regions have unusual and dramatic topography (Figure 14-15). Detailed analysis of Miranda's geology suggests that this satellite's orbital period was once in a whole-number ratio with that of more massive Umbriel or Ariel. The resulting tidal heating melted Miranda's interior, causing dense rocks in some locations on the surface to settle toward the satellite's center as blocks of less dense ice were forced upward toward the surface, thus creating Miranda's resurfaced terrain. Tidal heating must have ceased before this process could run its full course, which explains why some ancient, heavily cratered regions remain on Miranda's surface.

Uranus's satellite Miranda has cliffs twice as high as Mount Everest

Uranus's Perplexing Small Satellites

While Uranus's small satellites are unlikely to show the kind of geology found in Miranda, some of them move in curious orbits. Eight of the nine small satellites in large orbits beyond Oberon are in retrograde orbits, moving around Uranus opposite to the direction in which Uranus rotates. Many of the small outer satellites of Jupiter and Saturn have retrograde orbits and are probably captured asteroids (see Section 13-9 and Section 13-10); the same is probably true of the outer satellites of Uranus.

The 13 satellites that orbit closest to Uranus (inside the orbit of Miranda) are all in prograde orbits, so they travel around Uranus in the same direction as the planet rotates. However, when Mark Showalter and Jack Lissauer compared 11 of these satellites' orbits in 2005 with their orbits in 1994, they found surprisingly large differences. (The other two inner satellites were only discovered in 2003, so it is not known whether their orbits underwent similar changes.) These satellites can pass within a few thousand to a few hundred kilometers from each other, so they can exert strong gravitational forces on each other. Over time these forces can modify their orbits. Computer simulations of these gravitational interactions indicate that the inner satellites might begin colliding with each other within a few million years. If these inner satellites are in such unstable orbits, have they really been in orbit around Uranus since it formed more than 4.5 billion years ago? Or did Uranus somehow acquire these satellites in the relatively recent past? More research will be needed to answer these questions.

14-8 Neptune's satellite Triton is an icy world with a young surface and a tenuous atmosphere

Neptune has 13 known satellites, listed in Appendix 3. They are named for mythological beings related to bodies of water. (Neptune itself is named for the Roman god of the sea.) Most of these worlds are small, icy bodies, probably similar to the smaller satellites of Uranus. The one striking exception is Triton, Neptune's largest satellite (see Table 7-2). In many ways Triton is quite unlike any other world in the solar system.

Triton's Backward Orbit and Young Surface

Like many of the small satellites of Jupiter, Saturn, Uranus, and Neptune, Triton is in a retrograde orbit, so that it moves around Neptune opposite to the direction of Neptune's rotation. Furthermore, the plane of Triton's orbit is inclined by 23° from the plane of Neptune's equator. It is difficult to imagine how a satellite might form out of the same cloud of material as a planet and end up orbiting in a direction opposite the planet's rotation and in such a tilted plane. Hence, Triton probably formed elsewhere in the solar system, collided long ago with a now-vanished satellite of Neptune, and was captured by Neptune's gravity. With a diameter of 2706 km, a bit smaller than our Moon but much larger than any other satellite in a retrograde orbit, Triton is certainly the largest captured satellite.

Figure 14-16 shows the icy, reflective surface of Triton as imaged by *Voyager 2*. There is a conspicuous absence of large craters, which immediately tells us that Triton has a young surface on which the scars of ancient impacts have largely been erased by tectonic activity. There are areas that resemble frozen lakes and may be the calderas of extinct ice volcanoes. Some of Triton's surface features resemble the long cracks seen on Europa (Section 13-6) and Ganymede (Section 13-7). Still other features are unique to Triton. For example, in the upper portion of Figure 14-16 you can see dimpled, wrinkled terrain that resembles the skin of a cantaloupe.

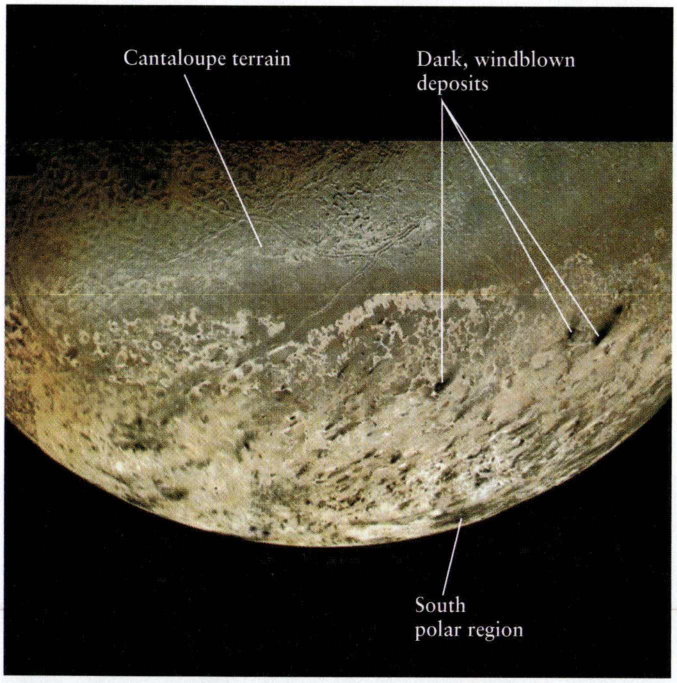

Figure 14-16 RIVUXG

Triton Several high-resolution *Voyager 2* images were combined to create this mosaic. The pinkish material surrounding Triton's south polar region is probably nitrogen frost. Some of this presumably evaporates when summer comes to the south pole; the northward flow of the evaporated gas may cause the dark surface markings. Farther north is a brown area of "cantaloupe terrain" that resembles the skin of a melon. (NASA/JPL)

Triton's tectonically active history is probably related to its having been captured into orbit around Neptune. After its capture, Triton most likely started off in a highly elliptical orbit, but today the satellite's orbit is quite circular. The satellite's original elliptical orbit would have been made circular by tidal forces exerted on Triton by Neptune's gravity. These tidal forces would also have stretched and flexed Triton, causing enough tidal heating to melt much of the interior. The resulting volcanic activity (with lavas made of ice rather than molten rock) would have obliterated Triton's original surface features, including craters.

There may still be warmth in Triton's interior today. *Voyager 2* observed plumes of dark material being ejected from the surface to a height of 8 km (5 mi). These plumes may have been generated from a hot spot far below Triton's surface, similar to geysers on Earth. Alternatively, the energy source for the plumes may be sunlight that warms the surface, producing subsurface pockets of gas and creating fissures in the icy surface through which the gas can escape.

Triton's surface temperature is only 38 K (−235°C = −391°F), the lowest of any world yet visited by spacecraft. This temperature is low enough to solidify nitrogen, and indeed the spectrum of sunlight reflected from Triton's surface shows absorption lines due to nitrogen ice as well as methane ice. But Triton is also warm enough to allow some nitrogen to evaporate from the surface, like the steam that rises from ice cubes when you first take them out of the freezer. *Voyager 2* confirmed that Triton has a very thin nitrogen atmosphere with a surface pressure of only 1.6×10^{-5} atmosphere, about the same as at an altitude of 100 km above Earth's surface. Despite its thinness, Triton's atmosphere has noticeable effects. *Voyager 2* saw areas on Triton's surface where dark material has been blown downwind by a steady breeze (see Figure 14-16). It also observed dark material ejected from the geyserlike plumes being carried as far as 150 km by high-altitude winds.

Tidal Forces and Triton's Doom

Just as tidal forces presumably played a large role in Triton's history, they also determine its future. Triton raises a tidal bulge on Neptune, just as our Moon distorts Earth (recall Figure 10-17). In the case of the Earth-Moon system, the gravitational pull of Earth's tidal bulge causes the Moon to spiral away from Earth. But because Triton's orbit is retrograde, the tidal bulge on Neptune exerts a force on Triton that makes the satellite slow down rather than speed up. (In Figure 10-17, imagine that the moon is orbiting toward the bottom of the figure rather than toward the top.) This is causing Triton to spiral gradually in toward Neptune. In approximately 100 million years, Triton will be inside Neptune's Roche limit, and the satellite will eventually be torn to pieces by tidal forces. When this happens, the planet will develop a spec-

Within 100 million years Triton will be torn apart by Neptune's tidal forces

tacular ring system—overshadowing by far its present-day set of narrow rings—as rock fragments gradually spread out along Triton's former orbit.

Prior to *Voyager 2,* only one other satellite was known to orbit Neptune. Nereid, which was first sighted in 1949, is in a prograde orbit. Hence, it orbits Neptune in the direction opposite to Triton. Nereid also has the most eccentric orbit of any satellite in the solar system; its distance from Neptune varies from 1.4 million to 9.7 million kilometers. One possible explanation is that when Triton was captured by Neptune's gravity, the interplay of gravitational forces exerted on Nereid by both Neptune and Triton moved Nereid from a relatively circular orbit (like those of Neptune's other, smaller moons) into its present elliptical one.

14-9 Pluto is smaller than any planet

Speculations about worlds beyond Neptune date back to the late 1800s, when a few astronomers suggested that Neptune's orbit was being perturbed by an unknown object. Encouraged by the fame of Adams and Le Verrier, several people set out to discover "Planet X." Two prosperous Boston gentlemen, William Pickering and Percival Lowell, were prominent in this effort. (Lowell also enthusiastically promoted the idea of canals on Mars; see Section 11-4.) Modern calculations show that there are, in fact, no unaccounted perturbations of Neptune's orbit. It is thus not surprising that no orbiting body was found at the positions predicted by Pickering, Lowell, and others. Yet the search continued.

The Discovery of Pluto

WEB LINK 14.8

Before he died in 1916, Lowell urged that a special camera with a wide field of view be constructed to help search for Planet X. After many delays, the camera was finished in 1929 and installed at the Lowell Observatory in Flagstaff, Arizona, where a young astronomer, Clyde W. Tombaugh, had recently joined the project. On February 18, 1930, Tombaugh finally discovered the long-sought object. It was disappointingly faint—a thousand times dimmer than the dimmest object visible with the naked eye and 250 times dimmer than Neptune at opposition—and presented no discernible disk. The new world, originally labeled a planet, was named for Pluto, the mythological god of the underworld. In 2006, Pluto was classified as a **dwarf planet**, which is an object orbiting the Sun (but not a satellite) with enough mass to gravitationally pull itself into a spherical shape, yet not enough gravity to clear out surrounding planetesimals.

Pluto's orbit (see Figure 14-17) about the Sun is more elliptical (eccentricity 0.2501) and more steeply inclined to the plane of the ecliptic (17.15°) than the orbit of any of the planets. In fact, Pluto's orbit is so eccentric that it is sometimes closer to the Sun than Neptune is. This was the case from 1979 until 1999; indeed, when Pluto was at perihelion in 1989, it was more than 10^8 km closer to the Sun than was Neptune. Happily, the orbits of Neptune and Pluto are such that the two worlds will never collide.

Pluto is so far away that its diameter subtends only a very small angle of 0.15 arcsec. Hence, it is extraordinarily difficult to resolve any of its surface features. But by observing Pluto with the Hubble Space Telescope over the course of a solar day on Pluto (6.3872 Earth days) and using computer image processing, Alan Stern and Marc Buie generated the maps of Pluto's surface shown in Figure 14-18. These maps show bright polar ice caps as well as regions of different reflectivity near Pluto's equator. Observations of Pluto's rotation confirm that its rotation axis is tipped by more than 90°, so that Pluto has retrograde rotation like Uranus.

Pluto's Satellites

In 1978, while examining some photographs of Pluto, James W. Christy of the U.S. Naval Observatory noticed that the image of Pluto on a photographic plate appeared slightly elongated, as though Pluto had a lump on one side. He promptly examined a number of other photographs of Pluto, and found a series that showed the lump moving clockwise around Pluto with a period of about 6 days.

Christy concluded that the lump was actually a satellite of Pluto. He proposed that the newly discovered moon be christened Charon (pronounced KAR-en), after the mythical boatman who ferried souls across the River Styx to Hades, the domain ruled by Pluto. (Christy also chose the name because of its similarity to Charlene, his wife's name.) The average distance between Charon

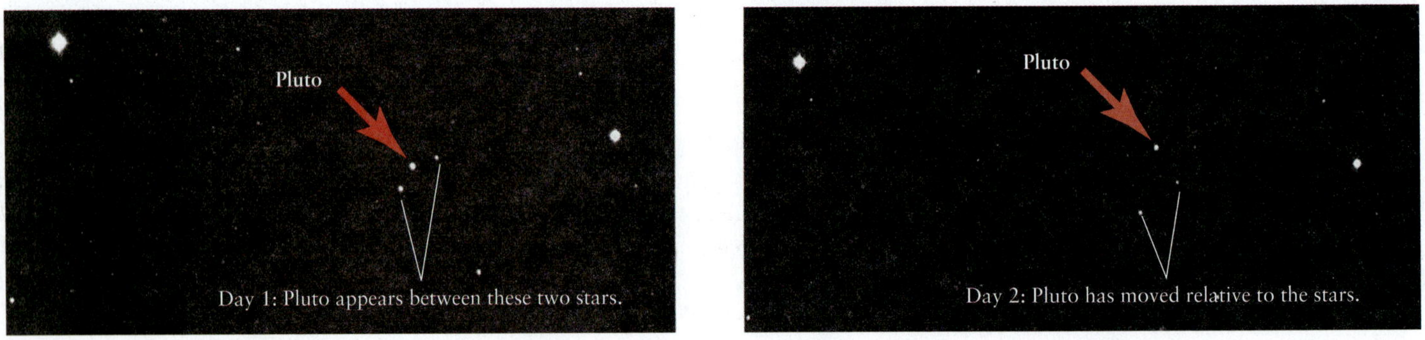

Figure 14-17 RIVUXG

Pluto's Motion across the Sky Pluto was discovered in 1930 by searching for a dim, starlike object that moves slowly in relation to the background stars. These two photographs were taken one day apart. Even when its apparent motion on the celestial sphere is fastest, Pluto moves only about 1 arcmin per day relative to the stars. (Lick Observatory)

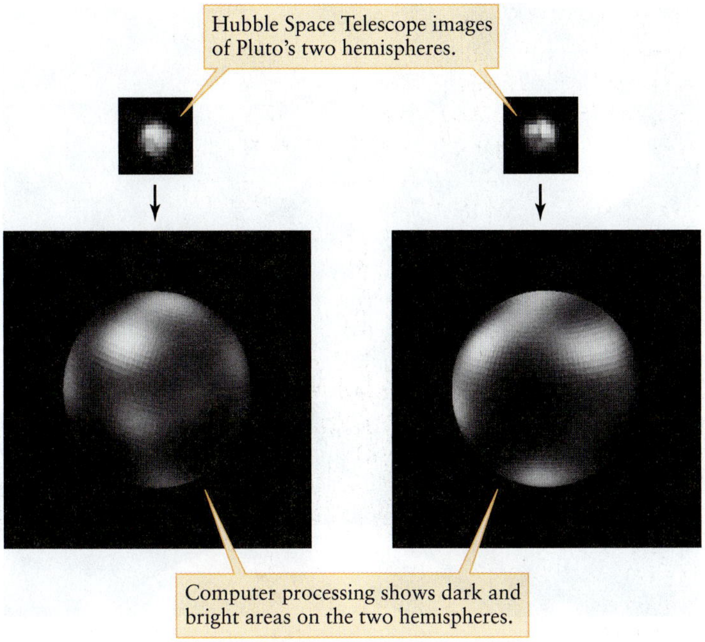
Hubble Space Telescope images of Pluto's two hemispheres.
Computer processing shows dark and bright areas on the two hemispheres.

Figure 14-18 R I V U X G

Pluto's Surface The bright regions at the top and bottom of each of Pluto's hemispheres may be polar ice caps. The bright regions nearer the equator may be impact basins where more reflective subsurface ice has been exposed. (Alan Stern, Southwest Research Institute; Marc Buie, Lowell Observatory; NASA; and ESA)

and Pluto is a scant 19,640 km, less than 5% of the distance between Earth and our Moon. In 2005 a team of astronomers using the Hubble Space Telescope discovered two additional, smaller satellites that orbit Pluto at two to three times the distance of Charon. The best pictures of Pluto and its moons have been obtained using the Keck telescope atop Mauna Kea in Hawaii (Figure 14-19).

Observations show that Charon's orbital period of 6.3872 days is the same as the rotational period of Pluto *and* the rotational period of Charon. In other words, both Pluto and Charon rotate synchronously with their orbital motion, and so both always keep the same face toward each other. As seen from the Charon-facing side of Pluto, the satellite remains perpetually suspended above the horizon. Likewise, Pluto neither rises nor sets as seen from Charon.

Pluto and its satellite Charon both rotate in lockstep, always keeping the same faces toward each other

Soon after the discovery of Charon, astronomers witnessed an alignment that occurs only once every 124 years. From 1985 through 1990, Charon's orbital plane appeared nearly edge-on as seen from Earth, allowing astronomers to view mutual eclipses of Pluto and its moon. As the bodies passed in front of each other, their combined brightness diminished in ways that revealed their sizes and surface characteristics. Data from the eclipses combined with subsequent measurements give Pluto's diameter as about 2274 km and Charon's as about 1190 km. For comparison, our Moon's diameter (3476 km) nearly equals the diameters of Pluto and Charon added together. The brightness variations during eclipses suggest that Charon has a bright southern polar cap.

The average densities of Pluto and Charon, at about 2000 kg/m^3, are essentially the same as that of Triton (2070 kg/m^3). All three worlds are therefore probably composed of a mixture of rock and ice, as we might expect for small bodies that formed in the cold outer reaches of the solar nebula.

Pluto's spectrum shows absorption lines of various solid ices that cover it's surface, including nitrogen (N_2), methane (CH_4), and carbon monoxide (CO). Stellar occultation measurements have shown that Pluto, like Triton, has a very thin atmosphere. Exposed to a daytime temperature of around 40 K, N_2 and CO ices turn to gas more easily than frozen CH_4. For this reason, most of Pluto's tenuous atmosphere probably consists of N_2 and CO. Charon's weaker gravity has apparently allowed most of the N_2, CH_4, and CO to escape into space; only water ice is found on its surface.

The Origin of Pluto's Satellites

Pluto and Charon are remarkably like each other in mass, size, and density. Throughout the solar system, planets are many times larger and more massive than any of their satellites. The excep-

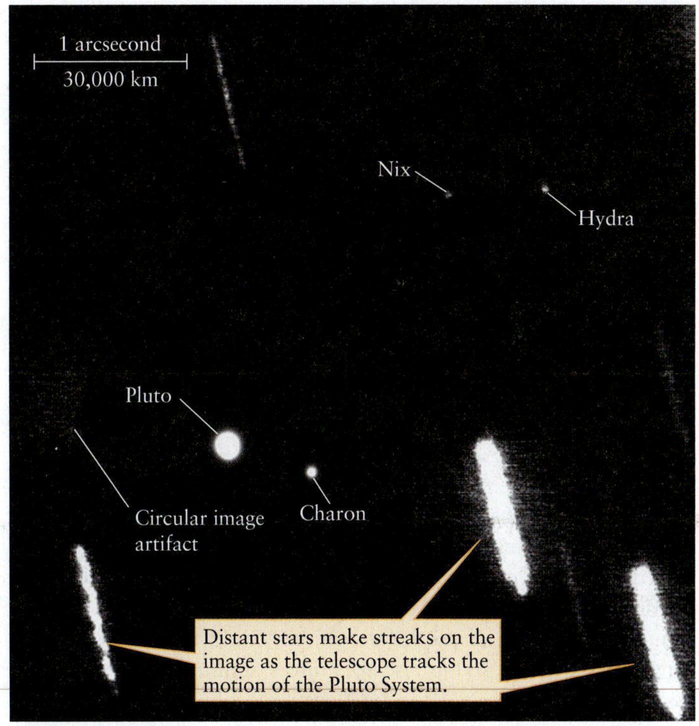
1 arcsecond
30,000 km
Nix
Hydra
Pluto
Circular image artifact
Charon
Distant stars make streaks on the image as the telescope tracks the motion of the Pluto System.

VIDEO 14.7 VIDEO 14.8

Figure 14-19 R I V U X G

Pluto and Its Satellites This composite image from the Keck II telescope atop Mauna Kea shows Pluto, its large satellite Charon (discovered 1978), and the two small satellites Nix and Hydra discovered in 2005. Nix is named for the mythological goddess of the night who was the mother of Charon; Hydra in mythology was a nine-headed, poisonous serpent who guarded the entrance to Hades. (David Tholen, Institute for Astronomy, University of Hawaii) tional similarities between Pluto (a dwarf planet) and Charon suggest to some astronomers that this binary system probably formed when Pluto collided with a similar body. Perhaps chunks of matter were stripped from the second body, leaving behind a mass, now called Charon, that was captured into orbit by Pluto's gravity. This same collision probably also left behind the small satellites Nix and Hydra shown in Figure 14-19.

Supporting this collision hypothesis is the observation that all three satellites have almost the same color and reflection spectrum, suggesting a common origin. Pluto, by contrast, has a redder color than its satellites, which is consistent with the idea that Pluto and the satellites formed in different ways.

For this scenario to be feasible, there must have been many Plutolike objects in the outer regions of the solar system. Astronomers estimate that in order for a collision or close encounter between two of them to have occurred at least once since the solar system formed some 4.56 billion years ago, there must have been at least a thousand Plutos. As we will see, astronomers have begun to discover this population of Plutolike objects in the dark recesses of the solar system beyond Neptune.

Figure 14-20

The Largest Trans-Neptunian Objects This artist's impression depicts Earth and the largest objects known beyond Neptune (as of 2009) to the same scale. The largest of these, Eris, has a diameter of about 2900 km versus 2274 km for Pluto. Note the differences in color among these objects and that three of them (Eris, Pluto, and Haumea) have satellites of their own. Table 7-4 in Section 7-5 lists the properties of these objects. (NASA; ESA; and A. Field, STScI)

14-10 Pluto is just one of thousands of icy objects that orbit the Sun beyond Neptune

For many years astronomers attempted to find other worlds beyond Neptune using the same technique used to discover Pluto (see Figure 14-17). The first to succeed were David Jewitt and Jane Luu, who in 1992 found an object with a semimajor axis of 42 AU and a diameter estimated to be only 240 km. This object, named 1992 QB1, has a reddish color like Pluto, possibly because frozen methane has been degraded by eons of radiation exposure. As of this writing (2009) about 1100 **trans-Neptunian objects**—icy worlds whose orbits have semimajor axes larger than that of Neptune—have been discovered. Most of these are relatively small like 1992 QB1, but a number are larger than Charon and one is larger than Pluto itself (Figure 14-20). Thus, Pluto is by no means unique: We can best regard it as a particularly large (but not the largest) trans-Neptunian object rather than a planet.

With improvements in the sensitivity of telescopes, new objects beyond Neptune are being discovered at a rapid pace; over 150 new trans-Neptunian objects were found since January 2005. Based on these observations, it is thought that there could be 35,000 or more objects beyond Neptune with diameters greater than 100 km. One of the larger objects could have collided with Pluto in the past, giving rise to Pluto's retinue of satellites (see Section 14-9). At least 29 trans-Neptunian objects have satellites, which suggests that such collisions have taken place many times.

The Kuiper Belt and Beyond

WEB LINK 14.9

Most of the trans-Neptunian objects lie within the **Kuiper belt,** which extends from about 30 to 50 AU from the Sun and is relatively close to the ecliptic. When the solar system was young, a large number of icy planetesimals formed in the region beyond Jupiter. Over time, the gravitational forces of the massive Jovian planets deflected most of these planetesimals beyond Neptune's orbit, concentrating them into a belt centered on the plane of the ecliptic. Most of the trans-Neptunian objects within the belt are in orbits that are only slightly inclined to the ecliptic; these objects are thought to have formed beyond Neptune and to be in roughly their original orbits. Other objects such as Makemake and Haumea (see Figure 14-20 and Table 7-4) are in orbits that are inclined by about 30° from the ecliptic. These objects are thought to have been pushed into their steeply inclined orbits by gravitational interactions with Neptune.

The processes that gave rise to the Kuiper belt in our solar system also appear to have taken place around other stars. Figure 14-21 shows a disk of material surrounding the young star HD 139664. This disk has dimensions comparable to our Kuiper belt. A number of other young stars have been found with disks of this same type.

There are relatively few members of the Kuiper belt between the orbits of Neptune and Pluto. Remarkably, there are about 100 objects that have nearly the same semimajor axis as Pluto. These so-called **plutinos,** which include Pluto itself, have the property that they complete nearly two orbits around the Sun in the same time that Neptune completes three orbits. The plutinos thus experience rhythmic gravitational pulls from Neptune, and these pulls keep them in their orbits. (In Section 13-1 we saw a similar relationship among the orbital periods of Jupiter's satellites Io, Europa, and Ganymede, though the ratio of their orbital periods is 1:2:4 rather

Neptune's gravity shapes the orbits of many icy worlds, including Pluto

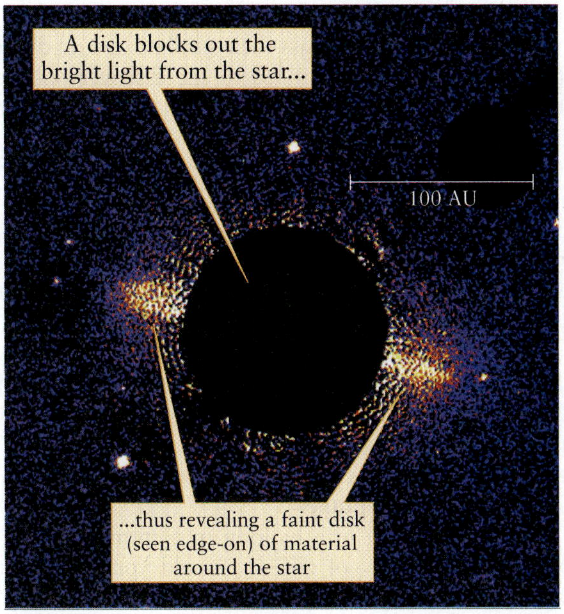
A disk blocks out the bright light from the star...
100 AU
...thus revealing a faint disk (seen edge-on) of material around the star

Figure 14-21 R I V U X G

A "Kuiper Belt" Around Another Star The star HD 139664 is only slightly more massive than the Sun but is thought to be just 300 million years old. This Hubble Space Telescope image shows a ring of material around HD 139664 that is similar in size to the Kuiper belt in our solar system. HD 139664 is 57 light-years (17 parsecs) from Earth in the constellation Lupus (the Wolf). (NASA; ESA; and P. Kalas/University of California, Berkeley)

than the 2:3 ratio for Neptune and the plutinos.) Most Kuiper belt objects orbit at distances beyond the plutinos but within about 50 AU from the Sun, at which distance an object completes one orbit for every two orbits of Neptune. At this distance the rhythmic gravitational forces of Neptune appear to pull small objects inward, giving the Kuiper belt a relatively sharp outer edge.

ANIMATION 14.3

Two examples of trans-Neptunian objects that are not members of the Kuiper belt are shown in Figure 14-20. Eris, the largest known trans-Neptunian object, has a semimajor axis of more than 67 AU and an orbital eccentricity of 0.442. This orbit takes Eris from inside the orbit of Pluto to far beyond the Kuiper belt. An even more extreme case is Sedna: Its orbital semimajor axis is 489 AU, giving it an orbital period of more than 10,000 years, and the orbital eccentricity has the remarkably high value of 0.844. It is not well understood how Sedna could have been moved into such an immense and elongated orbit.

WEB LINK 14.10

Excitement about the worlds beyond Neptune has motivated the development of a spacecraft called *New Horizons*. Launched in 2006, *New Horizons* swung by Jupiter in 2007 and will make the first-ever flyby of Pluto and Charon in 2015. It will then be aimed toward one or two other targets in the Kuiper belt. The choice of targets will not be made until a year before the Pluto-Charon flyby, giving astronomers extra time to discover new members of the Kuiper belt that might be suitable destinations. The high-resolution images and other data to be returned by *New Horizons* promise to revolutionize our understanding of these most remote members of the solar system.

Key Words

Kuiper belt, p. 370
magnetic axis, p. 361
occultation, p. 362
plutino, p. 370
radiation darkening, p. 364
trans-Neptunian object, p. 370
dwarf planet, p. 368

Key Ideas

Discovery of the Outer Planets: Uranus was discovered by chance, while Neptune was discovered at a location predicted by applying Newtonian mechanics.

Atmospheres of Uranus and Neptune: Both Uranus and Neptune have atmospheres composed primarily of hydrogen, helium, and a few percent methane.

• Methane absorbs red light, giving Uranus and Neptune their greenish-blue color.

• No white ammonia clouds are seen on Uranus or Neptune. Presumably the low temperatures have caused almost all the ammonia to precipitate into the interiors of the planets. All of these planets' clouds are composed of methane.

• Much more cloud activity is seen on Neptune than on Uranus. This is because Uranus lacks a substantial internal heat source.

Interiors and Magnetic Fields of Uranus and Neptune: Both Uranus and Neptune may have a rocky core surrounded by a mantle of water and ammonia. Electric currents in these mantles may generate the magnetic fields of the planets.

• The magnetic axes of both Uranus and Neptune are steeply inclined from their axes of rotation. The magnetic and rotational axes of all the other planets are more nearly parallel. The magnetic fields of Uranus and Neptune are also offset from the centers of the planets.

Uranus's Unusual Rotation: Uranus's axis of rotation lies nearly in the plane of its orbit, producing greatly exaggerated seasonal changes on the planet.

• This unusual orientation may be the result of a collision with a planetlike object early in the history of our solar system. Such a collision could have knocked Uranus on its side.

Ring Systems of Uranus and Neptune: Uranus and Neptune are both surrounded by systems of thin, dark rings. The low reflectivity of the ring particles may be due to radiation-darkened methane ice.

Satellites of Uranus and Neptune: Uranus has five satellites similar to the moderate-sized moons of Saturn, plus at least 22 more small satellites. Neptune has 13 satellites, one of which (Triton) is comparable in size to our Moon or the Galilean satellites of Jupiter.

• Triton has a young, icy surface indicative of tectonic activity. The energy for this activity may have been provided by tidal heating that occurred when Triton was captured by Neptune's gravity into a retrograde orbit.

• Triton has a tenuous nitrogen atmosphere.

Worlds Beyond Neptune: Pluto and its moon, Charon, move together in a highly elliptical orbit steeply inclined to the plane of the ecliptic.

• More than a thousand icy worlds have been discovered beyond Neptune. Pluto and Charon are part of this population.

• Most trans-Neptunian objects lie in a band called the Kuiper belt that extends from 30 to 50 AU from the Sun. Neptune's gravity shapes the orbits of objects within the Kuiper belt.

Questions

Review Questions

TUTORIAL 14.1

1. Could astronomers in antiquity have seen Uranus? If so, why was it not recognized as a planet?
2. Why do you suppose that the discovery of Neptune is rated as one of the great triumphs of science, whereas the discoveries of Uranus and Pluto are not?
3. Why is it so difficult to see features in the atmosphere of Uranus?
4. (**a**) Draw a figure like Figure 14-3, and indicate on it where Uranus was in 1986 and 2004. Explain your reasoning. (*Hint:* See Figure 14-1 and Figure 14-2.) (**b**) In approximately what year will the Sun next be highest in the sky as seen from Uranus's south pole? Explain your reasoning.
5. Why do you suppose the tilt of Uranus's rotation axis was deduced from the orbits of its satellites and not by observing the rotation of the planet itself?
6. Describe the seasons on Uranus. In what ways are the Uranian seasons different from those on Earth?

TUTORIAL 14.2

7. Explain the statement "Methane is to Uranus's atmosphere as water is to Earth's atmosphere."
8. A number of storms in the Uranian atmosphere can be seen in Figure 14-2, but none are visible in Figure 14-1. How can you account for the difference?
9. Why are Uranus and Neptune distinctly blue-green in color, while Jupiter or Saturn are not?
10. How does the energy source for Uranus's atmospheric motions differ from those from Jupiter, Saturn, and Neptune?
11. Why are fewer white clouds seen on Uranus and Neptune than on Jupiter and Saturn?
12. Why do Uranus and Neptune have higher densities than Jupiter and Saturn?
13. Discuss some competing explanations of why Uranus and Neptune are substantially smaller than Jupiter and Saturn.
14. How do the orientations of Uranus's and Neptune's magnetic axes differ from those of other planets?
15. Briefly describe the evidence supporting the idea that Uranus was struck by a large planetlike object several billion years ago.
16. Compare the rings that surround Jupiter, Saturn, Uranus, and Neptune. Briefly discuss their similarities and differences.
17. The 1977 occultation that led to the discovery of Uranus's rings was visible from the Indian Ocean. Explain why it could not be seen from other parts of Earth's night side.
18. As *Voyager 2* flew past Uranus, it produced images only of the southern hemispheres of the planet's satellites. Why do you suppose this was?
19. Why do astronomers think that the energy needed to resurface parts of Miranda came from tidal heating rather than the satellite's own internal heat?
20. Using the data in Appendix 3, explain why Uranus's satellites Caliban and Sycorax (both discovered in 1997) were probably captured from space rather than having formed at the same time as the planet itself.
21. Briefly describe the evidence supporting the idea that Triton was captured by Neptune.
22. If you were floating in a balloon in Neptune's upper atmosphere, in what part of the sky would you see Triton rise? Explain your reasoning.
23. Why is it reasonable to suppose that Neptune will someday be surrounded by a broad system of rings, perhaps similar to those that surround Saturn?
24. How can astronomers distinguish a faint solar system object like Pluto from background stars within the same field of view?
25. What is the evidence that the three moons of Pluto have a common origin?
26. How do the presence of Pluto's moons suggest that there must be other worlds beyond Neptune?
27. Why are there a large number of objects with the same semimajor axis as Pluto?
28. Are there any trans-Neptunian objects that are not members of the Kuiper belt? Are there any members of the Kuiper belt that are not trans-Neptunian objects? Explain your answers.

Advanced Questions

Questions preceded by an asterisk (*) involve topics discussed in Box 1-1.

> **Problem-solving tips and tools**
>
> See Box 1-1 for the small-angle formula. Recall that the volume of a sphere of radius r is $4\pi r^3/3$. Section 4-7 discusses the gravitational force between two objects and Newton's form of Kepler's third law. You will find discussions of the original form of Kepler's third law in Section 4-4, tidal forces in Section 4-8, Wien's law for blackbody radiation in Section 5-4, and the transparency of the Earth's atmosphere to various wavelengths of light in Section 6-7.

29. For which configuration of the Sun, Uranus, and Neptune is the gravitational force of Neptune on Uranus at a maximum? For this configuration, calculate the gravitational force exerted by the Sun on Uranus and by Neptune on Uranus. Then

calculate the fraction by which the sunward gravitational pull on Uranus is reduced by Neptune at that configuration. Based on your calculations, do you expect that Neptune has a relatively large or relatively small effect on Uranus's orbit?

30. At certain points in its orbit, a stellar occultation by Uranus would *not* reveal the existence of the rings. What points are those? How often does this circumstance arise? Explain using a diagram.

31. According to one model for the internal structure of Uranus, the rocky core and the surrounding shell of water and methane ices together make up 80% of the planet's mass. This interior region extends from the center of Uranus to about 70% of the planet's radius. (**a**) Find the average density of this interior region. (**b**) How does your answer to (a) compare with the average density of Uranus as a whole? Is this what you would expect? Why?

32. Uranus's epsilon (ε) ring has a radius of 51,150 km. (**a**) How long does it take a particle in the ε ring to make one complete orbit of Uranus? (**b**) If you were riding on one of the particles in the ε ring and watching a cloud near Uranus's equator, would the cloud appear to move eastward or westward as Uranus rotates? Explain your answer.

33. Suppose you were standing on Pluto. Describe the motions of Charon relative to the Sun, the stars, and your own horizon. Would you ever be able to see a total eclipse of the Sun? (*Hint:* You will need to calculate the angles subtended by Charon and by the Sun as seen by an observer on Pluto.) In what circumstances would you *never* see Charon?

34. It is thought that Pluto's tenuous atmosphere may become even thinner as the planet moves toward aphelion (which it will reach in 2113), then regain its present density as it again moves toward perihelion. Why should this be?

35. The brightness of sunlight is inversely proportional to the square of the distance from the Sun. For example, at a distance of 4 AU from the Sun, sunlight is only $(^1/_4)^2 = {}^1/_{16} = 0.0625$ as bright as at 1 AU. Compared with the brightness of sunlight on Earth, what is its brightness (**a**) on Pluto at perihelion (29.649 AU from the Sun) and (**b**) on Pluto at aphelion (49.425 AU from the Sun)? (**c**) How much brighter is it on Pluto at perihelion compared with aphelion? (Even this brightness is quite low. Noon on Pluto is about as dim as it is on Earth a half hour after sunset on a moonless night.)

***36.** If Earth-based telescopes can resolve angles down to 0.25 arcsec, how large could a trans-Neptunian object be at Pluto's average distance from the Sun and still not present a resolvable disk?

37. The observations of Pluto shown in Figure 14-19 were made using blue and ultraviolet light. What advantages does this have over observations made with red or infrared light?

***38.** Calculate the maximum angular separation between Pluto and Charon as seen from Earth. Assume that Pluto is at perihelion (29.649 AU from the Sun) and that Pluto is at opposition as seen from Earth.

39. Presumably Pluto and Charon raise tidal bulges on each other. Explain why the average distance between Pluto and Charon is probably constant, rather than increasing like the Earth-Moon distance or decreasing like the Neptune-Triton distance. Include a diagram like Figure 10-17 as part of your answer.

40. (**a**) Find the semimajor axis of the orbit of an object whose period is 3/2 of the orbital period of Neptune. How does your result compare to the semimajor axis of Pluto's orbit? (**b**) A number of Kuiper belt objects called plutinos have been discovered with the same orbital period and hence the same semimajor axis as Pluto. Explain how these objects can avoid colliding with Pluto.

41. Find the semimajor axis of the orbit of an object whose period is twice the orbital period of Neptune. How does your result compare to the outer limit of the Kuiper belt?

42. Suppose you wanted to search for trans-Neptunian objects. Why might it be advantageous to do your observations at infrared rather than visible wavelengths? (*Hint:* At visible wavelengths, the light we see from planets is reflected sunlight. At what wavelengths would you expect distant planets to *emit* their own light most strongly? Use Wien's law to calculate the wavelength range best suited for your search.) Could such observations be done at an observatory on Earth's surface? Explain your answer.

43. The *New Horizons* spacecraft swung by Jupiter to get a boost from that planet's gravity, enabling it to reach Pluto relatively quickly. To see what would happen if this technique were not used, consider a spacecraft trajectory that is an elliptical orbit around the Sun. The perihelion of this orbit is at 1 AU from the Sun (at Earth) and the aphelion is at 30 AU (at Pluto's position). Calculate how long it would take a spacecraft in this orbit to make the one-way trip from Earth to Pluto. Based on the information in Section 14-10, how much time is saved by making a swing by Jupiter instead?

Discussion Questions

44. Discuss the evidence presented by the outer planets that suggests that catastrophic impacts of planetlike objects occurred during the early history of our solar system.

45. Some scientists are discussing the possibility of placing spacecraft in orbit about Uranus and Neptune. What kinds of data should be collected, and what questions would you like to see answered by these missions?

46. If Triton had been formed along with Neptune rather than having been captured, would you expect it to be in a prograde or retrograde orbit? Would you expect the satellite to show signs of tectonic activity? Explain your answers.

47. Would you expect the surfaces of Pluto and Charon to be heavily cratered? Explain why or why not.

48. Imagine that you are in charge of planning the *New Horizons* flyby of Pluto and Charon. In your opinion, what data should be collected and what kinds of questions should the mission attempt to answer?

49. In 2006 the International Astronomical Union changed Pluto's designation from *planet* to *dwarf planet*. One criterion that Pluto failed to meet was that a planet must have "cleared the neighborhood" around its orbit. In what sense has Pluto not done so? In what sense have the eight planets (Mercury through Uranus) cleared their neighborhoods? Do you agree with this criterion?

Web/eBook Questions

50. **Miranda.** Access and view the video "Uranus's Moon Miranda" in Chapter 14 of the *Universe* Web site or eBook. Discuss some of the challenges that would be involved in launching a spacecraft from Earth to land on the surface of Miranda.
51. Charon was discovered by an astronomer at the U.S. Naval Observatory. Why do you suppose the U.S. Navy carries out work in astronomy? Search the World Wide Web for the answer.
52. Search the World Wide Web for a list of trans-Neptunian objects. What are the largest and smallest objects of this sort that have so far been found, and how large are they? Have any objects larger than Eris been found?

Activities

Observing Projects

> **Observing tips and tools**
>
> During the period 2007–2011, Uranus will be at opposition in September, Neptune in August, and Pluto in June. You can find Uranus and Neptune with binoculars if you know where to look (a good star chart is essential), but Pluto is so dim that it can be a challenge to spot even with a 25-cm (10-inch) telescope. Each year, star charts that enable you to find these planets are printed in the issue of *Sky & Telescope* for the month in which each planet is first visible in the nighttime sky. You can also locate the outer planets using the *Starry Night Enthusiast*™ program on the CD-ROM that accompanies certain printed copies of this textbook.

53. Make arrangements to view Uranus through a telescope. The planet is best seen at or near opposition. Use a star chart at the telescope to find the planet. Are you certain that you have found Uranus? Can you see a disk? What is its color?
54. If you have access to a large telescope, make arrangements to view Neptune. Like Uranus, Neptune is best seen at or near opposition and can most easily be found using a star chart. Can you see a disk? What is its color?
55. If you have access to a large telescope (at least 25 cm in diameter), make arrangements to view Pluto. Using the star chart from *Sky & Telescope* referred to above, view the part of the sky where Pluto is expected to be seen and make a careful sketch of all of the stars that you see. Repeat this process on a later night. Can you identify the "star" that has moved?
56. Use the *Starry Night Enthusiast*™ program to observe the five large satellites of Uranus. Open the **Favorites pane** and click on **Guides > Atlas** to display the entire celestial sphere. Open the **Find** pane and double-click the entry for Uranus to center this planet in the view. (Clicking once on the **Space** bar will speed up this centering). You can reduce the confusion in this view by removing the background stars by clicking on **View > Stars > Stars** and by ensuring that the celestial grid is removed by clicking on **View > Celestial Grid.** Using the controls at the right-hand end of the toolbar, zoom in to a field of view of about $2' \times 1'$. In the toolbar, set the year to 1986 and the **Time Flow Rate** to 1 hour. Then click on the **Run Time Forward** button, the right-pointing triangle on the toolbar. You can scroll on and off the labels for the moons by clicking on **Labels > Planets-Moons.** (**a**) Describe how the satellites move, and relate your observations to Kepler's third law (see Sections 4-4 and 4-7). (**b**) Set the year to 2007 and again click on the **Run Time Forward** button. How do the orbits look different than in (a)? Explain any differences.
57. Use the *Starry Night Enthusiast*™ program to examine the satellites of Uranus. (**a**) Select **Solar System > Uranus** from the **Favorites** menu. Remove the image of the astronaut's spacesuit by clicking on **View > Feet** in the menu and remove the background stars by selecting **View > Stars > Stars** from the menu. Use the Elevation buttons in the **Viewing Location** section of the toolbar to change the distance from the planet to about 0.004 AU. You should now be able to see at least five satellites of the planet Uranus. Which satellites are these? Select **Label > Planets-Moons** from the menu to confirm your identification of these satellites. (**b**) You can rotate the image of the planet and its moons by holding down the **Shift** key while clicking the mouse button and moving the mouse. Use this technique to rotate Uranus until you are looking at the plane of the satellites' orbits edge-on. Do all of the satellites appear to lie in the same plane? (To display the orbits of each of the moons, open the **Find** pane, expand the layer for Uranus, and click in the right-hand box next to each moon.) How do you imagine that this plane relates to the plane of Uranus's equator? Why do you suspect that this is so?
58. Use the *Starry Night Enthusiast*™ program to observe Neptune and its satellites Triton and Nereid. Display the entire celestial sphere (select **Guides > Atlas** in the **Favorites** menu). Open the **Find** pane and double-click on the entry for Neptune to center this planet in the view. In the Find Pane, expand the layer for Neptune and then click the box to the left of the entries for Triton and Nereid to label them in the view. To reduce confusion, remove the background stars and the coordinate grid by selecting **Stars > Stars** and **> Celestial Grid** under the View menu. Use the controls at the right-hand end of the toolbar to zoom in to a field of view of about $9' \times 6'$. In the toolbar, set the Time Flow Rate to 1 hour and click on the **Run Time Forward** button (a triangle that points to the right). (**a**) The satellite you see orbiting close to Neptune is Triton. In which direction (clockwise or counterclockwise) does it orbit Neptune? (**b**) Nereid is in a rather elongated orbit that takes it far from Neptune. You can use another viewpoint to show the relative positions and orientations of the orbits of these moons more clearly. In the **Options** menu, open the **Viewing Location** pane and, in the **View from** boxes, select **position hovering over** and **Neptune.** On the map that appears, you will note that you will be hovering over Neptune's equator. To choose this location, click on **Set Location.**

Increase your elevation to **0.05 a.u.** above Neptune by clicking and holding the **up** arrow in the **Viewing Location.** Again, you can remove the background stars by clicking on **View > Stars > Stars.** Open the **Find** pane and expand the list of moons under Neptune. You can now label **Triton** and **Nereid** and display their orbits by clicking in the boxes to the left and the right of their names respectively. Set the Time Flow Rate to 1 day and click on the **Run Time Forward** button. In which direction (clockwise or counterclockwise) does Nereid orbit Neptune? When it is closest to Neptune, does Nereid approach closer to Uranus than does Triton? You can view these orbits from different perspectives by holding down the Shift key while moving the mouse over Neptune to tilt this planetary system to different angles. (Note: You are at a point directly above a moon's orbit when its orbital track appears uniformly bright.) You can reduce the elevation above Neptune and set the Time Flow Rate to 10 minutes to show the planet's rotation and see the different clouds and dark spots on its surface. What is the planet's rotation period?

59. Use the *Starry Night Enthusiast*™ program to observe Pluto and Charon. First select **Options > Viewing Location** from the menu. At the top of the Viewing Location dialog box, select **position hovering over** and **Pluto** in the drop boxes. Then click on the center of the map of Pluto that appears in the dialog window and click the **Set Location** button. Use the elevation buttons in the toolbar to increase the distance from the surface of Pluto to about **35,000 km.** Use the **Location scroller** (hold down the Shift key while holding down the mouse button and moving the mouse, to rotate the view around Pluto. In the toolbar, set the **Time Flow Rate** to 1 hour, then click on the **Run Time Forward** button (a triangle that points to the right). (**a**) Estimate Charon's orbital period. (**b**) By following a spot on Pluto's surface, estimate Pluto's rotation period. How does it compare to your answer in part (a)? (**c**) Select **Options > Viewing Location** from the menu and set the dropdown boxes at the top of the Viewing Location dialog window to read **position moving with** and **Pluto.** Then select the **Above orbital plane** option and click the **Set Location** button. Open the **Find** pane and double-click the entry for the **Sun** to center the Sun in the view. Set the Time Flow Rate to 1 year, and click on the **Run Time Forward** button to see the apparent motion of the Sun as seen from Pluto. Observe the motion for several centuries of simulated time. Does the Sun always appear to move at the same speed? Use the properties of Pluto's orbit to explain why or why not.

Collaborative Exercise

60. Sir William Herschel, a British astronomer, discovered Uranus in 1781 and named it Georgium Sidus, after the reigning monarch, George III. What name might Uranus have been given in 1781 if an astronomer in your country had discovered it? Why? What if it had been discovered in your country in 1881? In 1991?

15 Vagabonds of the Solar System

WEB LINK 15.1

R I V U X G

Comet Hale-Bopp, the great comet of 1997. (Courtesy of Johnny Horne)

On the Tuesday night after Easter in the year 1066, a strange new star appeared in the European sky. Seemingly trailing fire, this "star" hung in the night sky for weeks. Some, including the English king Harold, may have taken it as an ill omen; others, such as Harold's rival William, Duke of Normandy, may have regarded it as a sign of good fortune. Perhaps they were both right, because by the end of the year Harold was dead and William was installed on the English throne.

Neither man ever knew that he was actually seeing the trail of a city-sized ball of ice and dust, part of which evaporated as it rounded the Sun to produce a long tail. They were seeing a comet—remarkably, the same Comet Halley that last appeared in 1986 and will next be seen in 2061. Although we now understand that comets are natural phenomena rather than supernatural omens, they still have the power to awe and inspire us, as did Comet Hale-Bopp (shown here) in 1997.

While comets are made of ice and dust, asteroids are rocky bits of the solar nebula that never formed into a full-sized planet. Some asteroids break into fragments, as do some comets, and some of these fragments fall to Earth as meteorites. On extraordinarily rare occasions, an entire asteroid strikes Earth. Such an asteroid impact may have led to the extinction of the dinosaurs some 65 million years ago. Thus, these minor members of our solar system can have major consequences for our planet.

15-1 A search for a planet between Mars and Jupiter led to the discovery of asteroids

TUTORIAL 15.1

After William Herschel's discovery of Uranus in 1781 (which we described in Section 14-1), many astronomers began to wonder if there were other, as yet undiscovered, planets. If these planets were too dim to be seen by the naked eye, they might still be visible through telescopes. These planets would presumably be found close to the plane of the ecliptic, because all the other planets orbit the Sun in or near that plane (see Section 7-1). But how far from the Sun might these additional planets be found?

The Hunt for the "Missing Planet"

LOOKING DEEPER 15.1

Astronomers in the late eighteenth century had a simple rule of thumb, called the *Titius-Bode law,* which

Learning Goals

By reading the sections of this chapter, you will learn

15-1 What led to the discovery of the asteroids

15-2 Why the asteroids never formed into a planet

15-3 What asteroids look like

15-4 How an asteroid led to the demise of the dinosaurs

15-5 What meteorites tell us about the nature of asteroids

15-6 What meteorites may reveal about the origin of the solar system

15-7 What comets are and why they have tails

15-8 How comets originate from the outer solar system

15-9 The connection between comets and meteor showers

relates the sizes of planetary orbits. The law states that from one planet to the next, the semimajor axis of the orbit increases by a factor between approximately 1.4 and 2. For example, the semimajor axis of Mercury's orbit is 0.39 AU. Venus's orbit has a semimajor axis of 0.72 AU, which is larger by a factor of (0.72)/(0.39) = 1.85. (Unlike Newton's laws, the Titius-Bode "law" is not a fundamental law of nature. It is probably just a reflection of how our solar system happened to form.)

The glaring exception is Jupiter (semimajor axis 5.20 AU), which is more than *3 times* farther from the Sun than Mars (semimajor axis 1.52 AU). Table 7-1 depicts this large space between the orbits of Mars and Jupiter. If the rule of thumb is correct, astronomers reasoned, there should be a "missing planet" with an orbit about 1.4 to 2 times larger than that of Mars. This planet should therefore have a semimajor axis between 2 and 3 AU.

Six German astronomers, who jokingly called themselves the "Celestial Police," organized an international group to begin a careful search for this missing planet. Before their search got under way, however, surprising news reached them from Giuseppe Piazzi, a Sicilian astronomer. Piazzi had been carefully mapping faint stars in the constellation of Taurus when on January 1, 1801, the first night of the nineteenth century, he noticed a dim, previously uncharted star. This star's position shifted slightly over the next several nights. Suspecting that he might have found the "missing planet," Piazzi excitedly wrote to Johann Bode, the director of the Berlin Observatory and a member of the Celestial Police.

Unfortunately, Piazzi's letter did not reach Bode until late March. By that time, Earth had moved around its orbit so that Piazzi's object appeared too near the Sun to be visible in the night sky. With no way of knowing where to look after it emerged from the Sun's glare, astronomers feared that Piazzi's object might have been lost.

Upon hearing of this dilemma, the brilliant young German mathematician Karl Friedrich Gauss took up the challenge. He developed a general method of computing an object's orbit from only three separate measurements of its position on the celestial sphere. (With slight modifications, this same method is used by astronomers today.) In November 1801, Gauss predicted that Piazzi's object would be found in the predawn sky in the constellation of Virgo. And indeed Piazzi's object was sighted again on December 31, 1801, only a short distance from the position Gauss had calculated. Piazzi named the object Ceres (pronounced SEE-reez), after the patron goddess of Sicily in Roman mythology.

Ceres orbits the Sun once every 4.6 years at an average distance of 2.77 AU, just where astronomers had expected to find the missing planet. But Ceres is very small; its equatorial diameter is a mere 974.6 km (Figure 15-1). (The pole-to-pole diameter is even smaller, just 909.4 km.) Hence, Ceres reflects only a little sunlight, which is why it cannot be seen with the naked eye even at opposition; it can be seen with binoculars, but it looks like just another faint star. Because of its small size, it is known today as a **minor planet**, or **asteroid.**

On March 28, 1802, Heinrich Olbers discovered another faint, starlike object that moved against the background stars. He called it Pallas, after the Greek goddess of wisdom. Like Ceres, Pallas orbits the Sun every 4.6 years at an average distance of 2.77 AU, but its orbit is more steeply inclined from the plane of the ecliptic and is somewhat more eccentric. Pallas is even smaller and dimmer than Ceres, with an estimated diameter of only 522 km. Obviously, Pallas was also not the missing planet.

Figure 15-1 R I V U X G

Ceres Compared with Earth and Moon A drawing of Ceres, the largest of the asteroids and the first one discovered, is shown here to the same scale as Earth and the Moon. Ceres is too small to be considered a planet, and it cannot be regarded as a moon because it does not orbit any other body. To denote their status, asteroids are also called minor planets. (NASA)

Did the missing planet even exist? Some astronomers speculated that perhaps there had once been such a full-size planet, but it had somehow broken apart or exploded to produce a population of asteroids orbiting between Mars and Jupiter. The search was on to discover this population. Two more such asteroids were discovered in the next few years, Juno in 1804 and Vesta in 1807. Several hundred more were discovered beginning in the mid-1800s, by which time telescopic equipment and techniques had improved.

An Abundance of Asteroids

The next major breakthrough came in 1891, when the German astronomer Max Wolf began using photographic techniques to search for asteroids. Before this, asteroids had to be painstakingly discovered by scrutinizing the skies for faint, uncharted, starlike objects whose positions move slightly from one night to the next. With photography, an astronomer simply aims a camera-equipped telescope at the stars and takes a long exposure. If an asteroid happens to be in the field of view, its orbital motion during the long exposure leaves a distinctive trail on the photographic plate (Figure 15-2). Using this technique, Wolf alone discovered 228 asteroids.

Today, more than 300,000 asteroids have been discovered, of which 100,000 have been studied well enough so that their orbits are known. About 5000 more are discovered every month, some by amateur astronomers. After the discoverer reports a new find to the Minor Planet Center of the Smithsonian Astrophysical Observatory, the new asteroid is given

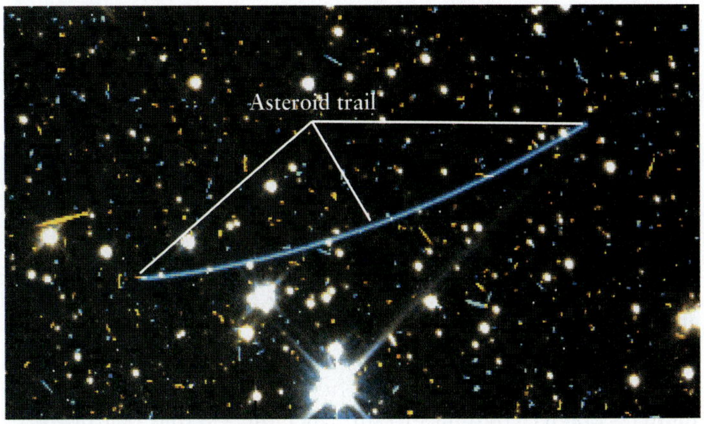

Figure 15-2 R I V U X G

The Trail of an Asteroid Telescopes used to photograph the stars are motorized to follow the apparent motion of the celestial sphere. Because asteroids orbit the Sun, their positions change with respect to the stars, and they leave blurred trails on time exposure images of the stars. This bluish asteroid trail was recorded quite by accident by the Hubble Space Telescope. (R. Evans and K. Stapelfeldt, Jet Propulsion Laboratory; and NASA)

a provisional designation. For example, asteroid 1980 JE was the fifth ("E") to be discovered during the second half of May (the tenth half-month of the year, "J") 1980. The same scheme for provisional designations is also used for small objects found in the outer solar system beyond Neptune, or trans-Neptunian objects (see Section 14-10).

If the asteroid is located again on at least four succeeding oppositions—a process that can take decades—the asteroid is assigned an official sequential number (Ceres is 1, Pallas is 2, and so forth), and the discoverer is given the privilege of suggesting a name for the asteroid. The suggested name must then be approved by the International Astronomical Union. For example, the asteroid 1980 JE was officially named 3834 Zappafrank (after the American musician Frank Zappa) in 1994.

Ceres—or, in modern nomenclature, 1 Ceres—is unquestionably the largest asteroid; it has about 25% of the mass of all the asteroids combined. In recognition of its size, 1 Ceres is also called a dwarf planet, a designation that it shares with Pluto and three other objects as of 2009 (see Section 14-9).

Besides 1 Ceres, only two asteroids—2 Pallas, and 4 Vesta—have diameters greater than 300 km. Thirty other asteroids have diameters between 200 and 300 km, and 200 more are bigger than 100 km across. The vast majority of asteroids are less than 1 km across.

CAUTION! Be careful not to confuse asteroids with the trans-Neptunian objects that we discussed in Section 14-9. Asteroids are found in the inner solar system, are made primarily of rocky materials, and are generally quite small. By contrast, trans-Neptunian objects are found primarily in the Kuiper belt beyond Neptune, are a mixture of ices and rock, and range in size from a few kilometers across to 2900 km in diameter—more than 3 times the diameter of 1 Ceres. The combined mass of all trans-Neptunian objects is not known, but is probably thousands of times greater than the combined mass of all asteroids. Asteroids make up only a tiny fraction of the total mass of the solar system!

The Asteroid Belt

Like Ceres, Pallas, Vesta, and Juno, most asteroids orbit the Sun at distances between 2 and 3.5 AU. This region of our solar system between the orbits of Mars and Jupiter is called the **asteroid belt** (Figure 15-3).

CAUTION! You may wonder how spacecraft such as *Voyager 1, Voyager 2, Galileo,* and *Cassini* were able to cross the asteroid belt to reach Jupiter or Saturn. Aren't asteroids hazards to space navigation, as they are sometimes depicted in science-fiction movies? Wouldn't these spacecraft have been likely to run into an asteroid? Happily, the answer to both questions is no! While it is true that there are more than 10^5 asteroids, they are spread over a belt with a total area (as seen from above the plane of the ecliptic) of about 10^{17} square kilometers—about 10^8 times greater than Earth's entire surface area. Hence, the average distance between asteroids in the ecliptic plane is about 10^6 kilometers, about twice the distance between Earth and Moon. Furthermore, many asteroids have orbits that are tilted out of the ecliptic. The *Galileo* spacecraft did pass within a few thousand kilometers of two asteroids, but these passes were intentional and required that the spacecraft be carefully aimed.

Science-fiction movies to the contrary, asteroids are not hazards to space navigation

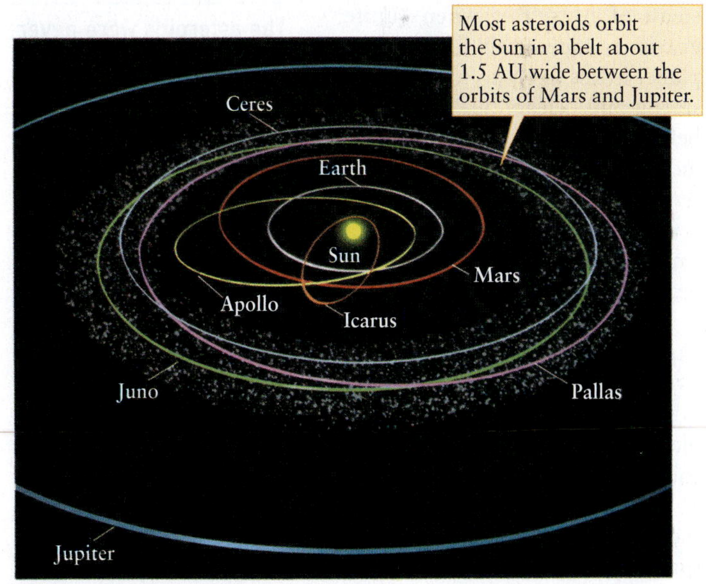

Figure 15-3

The Asteroid Belt Most asteroids, including the large asteroids 1 Ceres, 2 Pallas, and 3 Juno, have roughly circular orbits that lie within the asteroid belt. By contrast, some asteroids, such as 1862 Apollo and 1566 Icarus, have eccentric orbits that carry them inside the orbit of Earth.

What of the nineteenth-century notion of a missing planet? This idea has long since been discarded, because the combined matter of all the asteroids (including an estimate for those not yet officially known) would produce an object barely 1500 km in diameter. This combined matter is considerably smaller than anything that could be considered a missing planet. We now understand that asteroids are actually debris left over from the formation of the solar system.

15-2 Jupiter's gravity helped shape the asteroid belt

Where did the asteroid belt come from? Why did asteroids, rather than a planet, form in the solar nebula (see Section 8-5) in the region between Mars and Jupiter? Important insights into these questions come from supercomputer simulations of the formation of the terrestrial planets, like the one depicted in Figure 8-12 but more elaborate.

Jupiter and the Formation of the Asteroid Belt

A typical simulation of the inner solar system starts off with a billion (10^9) or so planetesimals, each with a mass of 10^{15} kg or more, so that their combined mass equals that of the terrestrial planets. The computer then follows these planetesimals as they collide and accrete to form the planets. By selecting different speeds and positions for the planetesimals at the start of the simulation, a scientist can study alternative scenarios of the development of the inner solar system.

If the effects of Jupiter's gravity are not included in a simulation, an Earth-sized planet usually forms in the asteroid belt, giving us five terrestrial planets instead of four. But when Jupiter's gravity is added, this fifth terrestrial planet is less likely to form. Jupiter's strong pull "clears out" the asteroid belt by disrupting the orbits of planetesimals in the belt, ejecting most of them from the solar system altogether. As a result, the asteroid belt becomes depleted of planetesimals before a planet has a chance to form. The few planetesimals that remain in the simulation—only a small fraction of those that originally orbited between Mars and Jupiter—become the asteroids that we see today.

The asteroids were never part of a larger planet

Jupiter's gravitational influence, however, probably cannot explain why asteroid orbits have the wide variety of semimajor axes, eccentricities, and inclinations to the ecliptic shown in Figure 15-3. Even with the influence of Jupiter, collisions between planetesimals in the solar nebula would have left the asteroids in roughly circular orbits close to the ecliptic plane, like the orbits of the planets. Therefore, something else must have "stirred up" the asteroids to put them into their current state. Recent analyses suggest that one or more Mars-sized objects—that is, terrestrial planets intermediate in size between Earth and the Moon—did in fact form within the asteroid belt. Planetesimals that passed within close range of these Mars-sized objects would have experienced strong gravitational forces, and these forces could have deflected the planetesimals into the eccentric or inclined orbits that many asteroids follow today.

Mars-sized objects would also have helped clear out the asteroid belt by deflecting planetesimals into orbits that crossed Jupiter's orbit. Once a planetesimal wandered close to Jupiter, it could be ejected from the solar system by the giant planet's gravitational force.

In this scheme, the Mars-sized objects disappeared when gravitational forces from Jupiter either accelerated them out of the solar system entirely or caused them to slow down and fall into the Sun. A Mars-sized object colliding with Earth is thought to have produced the Moon (see Figure 10-18). Perhaps this object was an "escapee" from the asteroid belt.

Kirkwood Gaps

Jupiter's gravity continues to influence the asteroid belt down to the present day. The American astronomer Daniel Kirkwood found the first evidence for this in 1867, when he discovered gaps in the asteroid belt. These features, today called **Kirkwood gaps,** can best be seen in a histogram like Figure 15-4, which shows asteroid orbital periods. The gaps in this histogram show regions where there are relatively few asteroids. Curiously, these gaps occur for asteroid orbits whose periods are simple fractions (such as 1/3, 2/5, 3/7, and 1/2) of Jupiter's orbital period.

To understand why the Kirkwood gaps exist, imagine an asteroid within the belt that circles the Sun once every 5.93 years, exactly half of Jupiter's orbital period. On every second trip around the Sun, the asteroid finds itself lined up between Jupiter

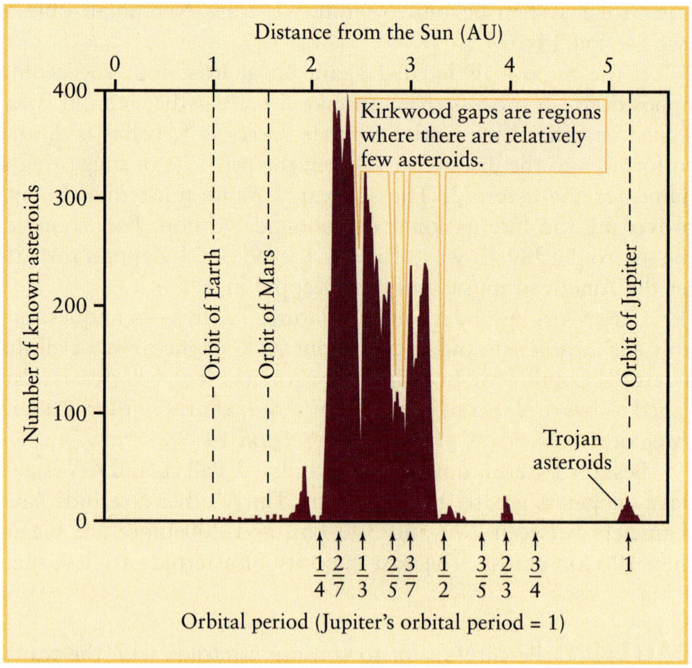

INTERACTIVE EXERCISE 15.1

Figure 15-4

The Kirkwood Gaps This graph displays the numbers of asteroids at various distances from the Sun. Very few asteroids are found in orbits whose orbital periods are equal to the simple fractions 1/3, 2/5, 3/7, and 1/2 of Jupiter's orbital period. The Trojan asteroids (at far right) follow the same orbit as Jupiter and thus have the same orbital period. We describe these in Section 15-4.

and the Sun, always at the same location and with the same orientation. Because of these repeated alignments, called a *2-to-1 resonance,* Jupiter's gravity deflects the asteroid from its original 5.93-year orbit, ultimately ejecting it from the asteroid belt. Similar resonances help clear out the Cassini division in Saturn's rings (see Section 12-11), drive the tidal heating of Jupiter's moon Io (see Section 13-4), and give an outer edge to the Kuiper belt beyond Neptune (see Section 14-10). According to Kepler's third law, a period of 5.93 years corresponds to a semimajor axis of 3.28 AU. Because of Jupiter's gravity, almost no asteroids orbit the Sun at this average distance.

Another Kirkwood gap corresponds to an orbital period of one-third Jupiter's period, or 3.95 years (a *3-to-1 resonance*). Additional gaps exist for other simple ratios between the periods of asteroids and Jupiter. Any Mars-sized bodies that existed in the asteroid belt were probably ejected after wandering into one of the Kirkwood gaps.

15-3 Astronomers use a variety of techniques to study asteroids

Although the asteroid belt is mostly empty space, collisions between asteroids should occur from time to time. Asteroids move in orbits with a variety of different eccentricities and inclinations (see Figure 15-3), and some of these orbits intersect each other. A collision between asteroids must be an awe-inspiring event. Typical collision speeds are estimated to be from 1 to 5 km/s (3600 to 18,000 km/h, or 2000 to 11,000 mi/h), which is more than sufficient to shatter rock. Recent observations show that these titanic impacts play an important role in determining the nature of asteroids.

Many asteroids have been battered by collisions with other asteroids

Observing Asteroids with Telescopes, Radar, and Spacecraft

State-of-the-art optical telescopes have good enough resolution to reveal surface details on large asteroids such as 1 Ceres, 2 Pallas, and 4 Vesta (Figure 15-5). These asteroids resemble miniature terrestrial planets. They have enough mass, and hence enough gravity, to pull themselves into roughly spherical shapes. These large asteroids are likely to have extensively cratered surfaces due to collisions with other asteroids in the asteroid belt.

Optical telescopes reveal much less detail about the smaller asteroids. But by studying how the brightness of an asteroid changes as it rotates, astronomers can infer the asteroid's shape. An elongated asteroid appears dimmest when its long axis points toward the Sun so that it presents less of its surface to reflect sunlight. By contrast, a rotating spherical asteroid will have a more constant brightness. Such observations show that most smaller asteroids are not spherical at all. Instead, they retain the odd shapes produced by previous collisions with other asteroids.

An important recent innovation in studying asteroids has been the use of radar. An intense radio beam is sent toward an asteroid, and the reflected signal is detected and analyzed to determine the asteroid's shape (Figure 15-6). The amount of the radio signal that is reflected also gives clues about the texture of the surface.

VIDEO 15.1 WEB LINK 15.6 The most powerful technique for studying asteroids is to send a spacecraft to make observations at close range. The Jupiter-bound *Galileo* spacecraft gave us our first close-up view of an asteroid

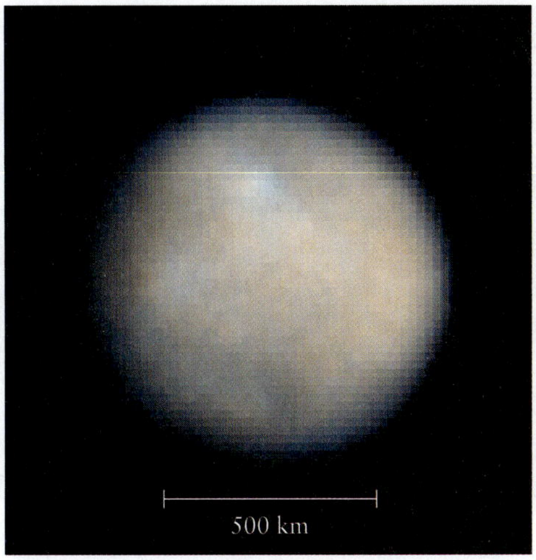

WEB LINK 15.4 **Figure 15-5** R I V U X G

A Hubble Space Telescope View of a Large Asteroid With an equatorial diameter of 974.6 km (605.6 mi), 1 Ceres is the largest of all the asteroids. Its very nearly spherical shape suggests that it was once molten throughout its volume, in which case it probably underwent chemical differentiation like the terrestrial planets (see Box 7-1 and Section 8-5). (NASA; ESA; J. Parker, SwRI; P. Thomas, Cornell U., L. McFadden, U. of Maryland; and M. Mutchler and Z. Levay, STScI)

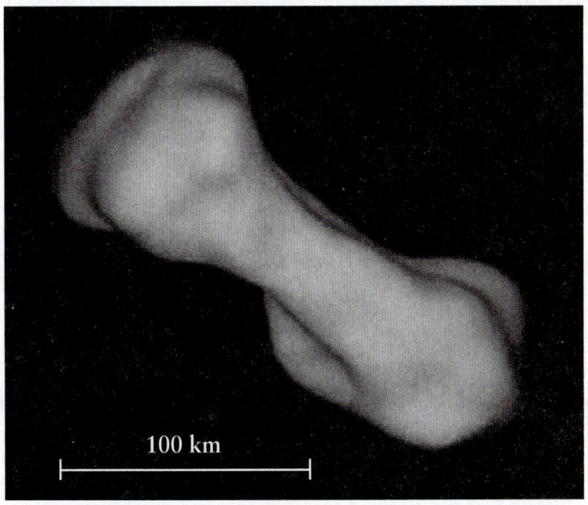

WEB LINK 15.5 **Figure 15-6** R I V U X G

A Radar View of a Medium-Sized Asteroid This radar image reveals the curious dog-bone shape of the asteroid 216 Kleopatra. The 300-m Arecibo radio telescope in Puerto Rico was used both to send radio waves toward Kleopatra and to detect the waves that were reflected back. Scientists then created this image by computer analysis of the reflected waves. (Arecibo Observatory, JPL/NASA)

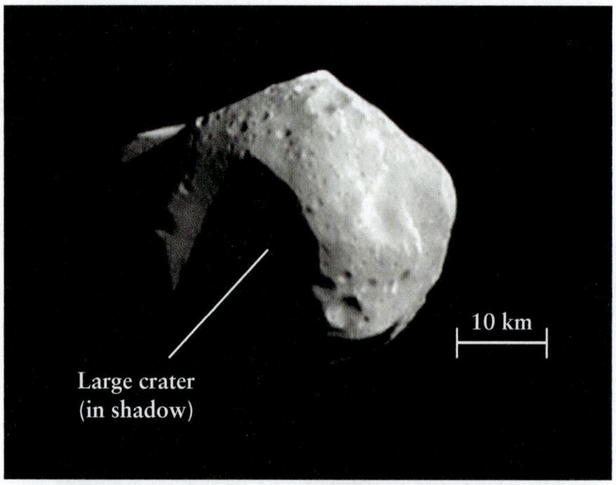

WEB LINK 15.7

Figure 15-7 R I V U X G

A Spacecraft View of a Small Asteroid The asteroid 253 Mathilde has an albedo of only 0.04, making it half as reflective as a charcoal briquette. It has an irregular shape, with dimensions 66 × 48 × 46 km. The partially shadowed crater indicated by the label is about 20 km across and 10 km deep. The *NEAR Shoemaker* spacecraft imaged Mathilde at close range during its close flyby in June 1997. (Johns Hopkins University, Applied Physics Laboratory)

in 1991, when it flew past 951 Gaspra. Two years later *Galileo* made a close pass by 243 Ida. The *NEAR Shoemaker* spacecraft made a close approach to 253 Mathilde in 1997 (Figure 15-7). It later went into orbit around 433 Eros (see Figure 7-7) and touched down on Eros in 2001. It was the first spacecraft to orbit and land on an asteroid. A number of other spacecraft have since made close flybys of asteroids.

One of the most ambitious asteroid missions to date is *Hayabusa* ("Falcon"), a project of the Japan Aerospace Exploration Agency. In 2005 it touched down briefly on the surface of 25143 Itokawa, where it was to have gathered small samples of the asteroid's surface. Scientists will not know whether this attempt was successful until *Hayabusa* returns to Earth in 2010. In 2007 the *Dawn* spacecraft was scheduled to be launched on a mission that would place it in orbit around 4 Vesta in 2011 and in orbit around 1 Ceres in 2015.

Collisions and the Nature of Asteroids

What have spacecraft and Earth-based radar told us about the asteroids? To appreciate the answer, we should first consider what scientists expected to find with these new tools. A few years ago, the consensus among planetary scientists was that the smaller asteroids were solid, rocky objects. The surfaces were presumed to be cratered by impacts, but only relatively small craters were expected. The reason is that an impact forceful enough to make a very large crater would probably fracture a rocky asteroid into two or more smaller asteroids.

The truth proves to be rather different. It is now suspected that most asteroids larger than about a kilometer are not entirely solid, but are actually composed of several pieces. An extreme example is Mathilde (see Figure 15-7). This asteroid has such a low density (1300 kg/m^3) that it cannot be made of solid rock, which typically has densities of 2500 kg/m^3 or higher. Instead, it is probably a "rubble pile" of small fragments that fit together loosely. Billions of years of impacts have totally shattered this asteroid. But the impacts were sufficiently gentle that the resulting fragments drifted away relatively slowly, only to be pulled back onto the asteroid by gravitational attraction.

Another surprising discovery is that some asteroids have extremely large craters, comparable in size to the asteroid itself. An example is the immense crater on Mathilde, shown in Figure 15-7. The object that formed this crater probably collided with Mathilde at a speed of 1 to 5 km/s (3600 to 18,000 km/h, or 2000 to 11,000 mi/h), which is more than sufficient to shatter rock. Why didn't this collision break the asteroid completely apart, sending the pieces flying in all directions? The explanation may be that the asteroid had already been broken into a loose collection of fragments before this major impact took place. Such a fragmented asteroid can more easily absorb the energy of a collision than can a solid, rocky one.

ANALOGY If you fire a rifle bullet at a wine glass, the glass will shatter into tiny pieces. But if you fire the same bullet into a bag full of sand (which has much the same chemical composition as glass), the main damage will be a hole in the bag. In a similar way, an asteroid that is a collection of small pieces can survive a major impact more easily than can a solid asteroid.

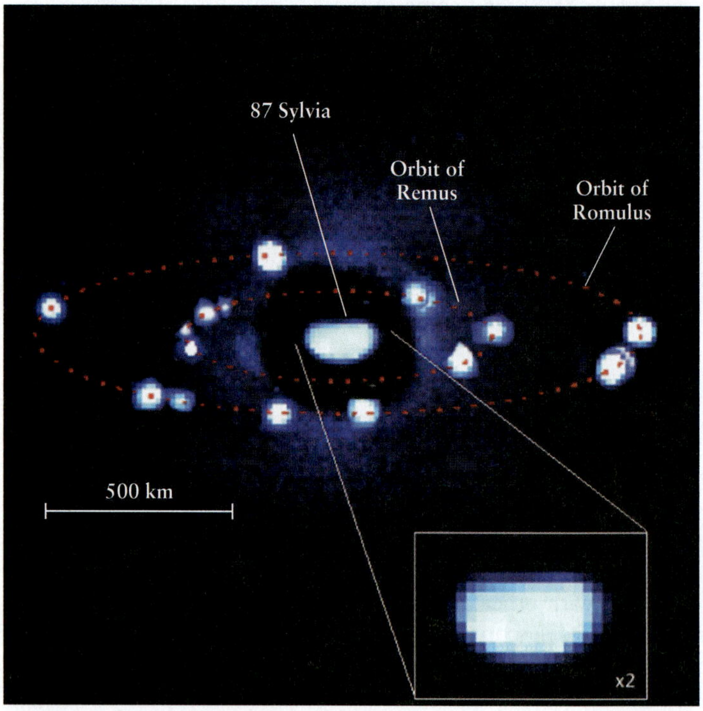

Figure 15-8 R I V U X G

An Asteroid with Two Satellites Although it is only about 280 km across, the potato-shaped asteroid 87 Sylvia has two satellites of its own. The outer satellite, Romulus, is 18 km across; the inner satellite, Remus, is just 7 km across. This image is actually a composite of 9 observations made with the adaptive optics system at the Yepun telescope of the European Southern Observatory (see Figure 6-17). (F. Marchis, U. of California; P. Descamps, D. Hestroffer, and J. Berthier, Observatoire de Paris; and ESO)

In some cases a collision that produces a rubble-pile asteroid can eject a piece that goes into orbit around the asteroid. This may explain why a handful of asteroids are known to have satellites. There is even one known case of an asteroid with two satellites, like a miniature planetary system (Figure 15-8).

If a collision is sufficiently energetic, it may shatter an asteroid permanently. Evidence for this was first pointed out in 1918 by the Japanese astronomer Kiyotsugu Hirayama. He drew attention to families of asteroids that share nearly identical orbits. Each of these groupings, now called **Hirayama families**, presumably resulted from a parent asteroid that was broken into fragments by a high-speed, high-energy collision with another asteroid.

15-4 Asteroids are found outside the asteroid belt—and have struck Earth

While the majority of asteroids lie within the asteroid belt, there are some notable exceptions. A few actually share the same orbit as Jupiter. Other asteroids venture well inside the orbit of Mars, and can pose a serious threat to life on Earth.

The Lagrange Points and Trojan Asteroids

We saw in Section 15-3 that Jupiter's gravitational pull by itself depletes certain orbits in the asteroid belt, forming the Kirkwood gaps. But the gravitational forces of the Sun and Jupiter work together to *capture* asteroids at two locations outside the asteroid belt called the **stable Lagrange points.** One of these points is one-sixth of the way around Jupiter's orbit ahead of the planet, while the other point is the same distance behind the planet (Figure 15-9*a*). The French mathematician Joseph Louis Lagrange predicted the existence of these points in 1772; asteroids were first discovered there in 1906.

The asteroids trapped at Jupiter's Lagrange points are called **Trojan asteroids,** named individually after heroes of the Trojan War. More than 1600 Trojan asteroids have been catalogued so far (see Figure 15-9*b*). (Four objects have also been found at the Lagrange points of Neptune. These are icy trans-Neptunian objects rather than rocky asteroids.)

Near-Earth Objects and Impacts from Space

WEB LINK 15.8

Some asteroids have highly elliptical orbits that bring them into the inner regions of the solar system. Asteroids that cross Mars's orbit, or whose orbits lie completely within that of Mars, are called **near-Earth objects** or **NEOs.** These are shown in red in Figure 15-9*b*.

Occasionally, a near-Earth object may pass relatively close to Earth. On October 30, 1937, Hermes passed Earth at a distance of 900,000 km (560,000 mi), only a little more than twice the distance to the Moon. One of the closest calls to date occurred on December 9, 1994, when an asteroid called 1994 XM1 passed within a mere 105,000 km (60,000 mi) of Earth.

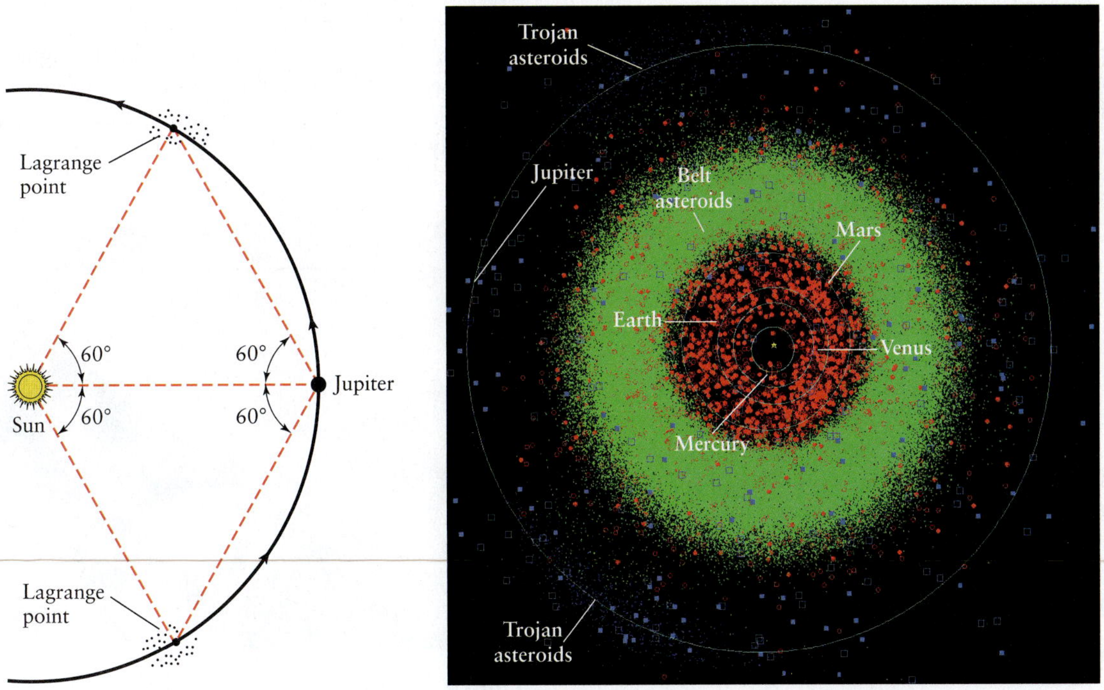

(a) A clump of asteroids is found at each of Jupiter's Lagrange points

(b) A map of all asteroids within Jupiter's orbit

Figure 15-9

Asteroids Outside the Belt (a) The combined gravitational forces of Jupiter and the Sun tend to trap asteroids near the two stable Lagrange points along Jupiter's orbit. Note that the Sun, Jupiter, and either one of these points lie at the vertices of an equilateral triangle. **(b)** This plot shows the actual positions of all known asteroids at Jupiter's orbit or closer. Green dots denote belt asteroids, deep blue squares denote Trojan asteroids, and red circles denote asteroids that come within 1.3 AU of the Sun. Comets are filled and unfilled light-blue squares. (b: Minor Planet Center)

WEB LINK 15.9

Figure 15-10 RI**V**UXG

The Barringer Crater This 1.2-km-($\frac{3}{4}$-mi-) wide crater was formed by a meteoroid impact about 50,000 years ago. Fewer than 200 other impact craters between 3 and 150 km across have been identified on Earth. None of these is more than 500 million years old, because the reshaping of Earth's surface by erosion and by tectonic processes tends to erase craters over time. (Meteor Crater Enterprises)

Because 1994 XM1 is only about 10 meters across (about the size of a house), even if it had been aimed directly at Earth it would probably have broken up in the atmosphere before reaching the surface.

More than 4100 near-Earth objects are known, and there may be thousands more that we have not yet detected. Fortunately, space is a big place and Earth is a rather small target, so asteroids strike Earth only very rarely.

However, collisions between asteroids produce numerous smaller chunks of rock, and many of these do eventually rain down on the terrestrial planets. Fortunately for us, the majority of these fragments, usually called **meteoroids,** are quite small. But on rare occasions a large fragment collides with our planet. When such a collision takes place, the result is an impact crater whose diameter depends on the mass and speed of the impinging object.

A relatively young, pristine terrestrial impact crater is the Barringer Crater near Winslow, Arizona (Figure 15-10). This crater was formed 50,000 years ago when an iron-rich object approximately 50 m across struck the ground at a speed of more than 11 km/s (40,000 km/h, or 25,000 mi/h). The resulting blast was equivalent to the detonation of a 20-megaton hydrogen bomb and left a crater 1.2 km ($\frac{3}{4}$ mi) wide and 200 m (650 ft) deep at its center. (Figure 7-10*b* shows a much larger but much older crater in Canada.)

Iridium and the Demise of the Dinosaurs

Iron is an important constituent of asteroids and meteoroids. Another element, iridium, is common in iron-rich minerals but rare in other types of rocks. Measurements of iridium in Earth's crust can thus tell us the rate at which meteoritic material has been deposited on Earth over the ages. The geologist Walter Alvarez and his physicist father, Luis Alvarez, from the University of California, Berkeley, made such measurements in the late 1970s.

Working at a site of exposed marine limestone in the Apennine Mountains in Italy, the Alvarez team discovered an exceptionally high abundance of iridium in a dark-colored layer of clay between limestone strata (Figure 15-11). Since this discovery in 1979, a comparable layer of iridium-rich material has been uncovered at numerous sites around the world. In every case, geological dating reveals that this apparently worldwide layer of iridium-rich clay was deposited about 65 million years ago.

An asteroid impact probably ended the age of dinosaurs 65 million years ago

Paleontologists were quick to realize the significance of this date, because it was 65 million years ago that all the dinosaurs rather suddenly became extinct. In fact, more than 75% of all the species on Earth disappeared within a relatively brief span of time.

The Alvarez discovery suggests a startling explanation for this dramatic mass extinction: Perhaps an asteroid hit Earth at that time. An asteroid 10 km in diameter slamming into our planet would have produced a fiery plume that ignited tremendous wildfires, killing untold numbers of animals and consuming much of Earth's vegetation. The dust thrown up from the impact and soot from wildfires hung in the atmosphere, blackening the sky for months after the impact. The dust eventually drifted to the ground, forming a worldwide layer of iridium-rich clay. (Evidence for this picture is that a layer of soot 65 million years old is also found at various sites around the world. From the thickness of this layer, scientists calculate that the wildfires produced nearly 70 *billion* tons of soot.) Temperatures would have plummeted due to the blocking out of sunlight by dust and soot, only to increase later due to global warming brought on by greenhouse gases (see Section 9-1 and Section 9-7) released by the wildfires. The *Cosmic Connections* figure depicts the sequence of events that may have followed the asteroid impact.

Figure 15-11 RI**V**UXG

Iridium-Rich Clay This photograph of strata in Italy's Apennine Mountains shows a dark-colored layer of iridium-rich clay. This clay is sandwiched between older layers of white limestone (lower right) and younger layers of grayish limestone (upper left). The iridium-rich layer is thought to be the result of an asteroid impact 65 million years ago. The coin is the size of a U.S. quarter. (Courtesy of W. Alvarez)

COSMIC CONNECTIONS

A Killer Asteroid

There is compelling evidence that 65 million years ago Earth was struck by an asteroid about 10 km in diameter. This impact triggered a biological cataclysm that wiped out 75 percent of all plant and animal species on our planet. (From "The Day the World Burned" by D. A. Kring and D. D. Durda, *Scientific American*, December 2003; art by Chris Butler)

Late Cretaceous swamps and rivers in North America had a mix of coniferous, broad-leaved evergreen, and deciduous trees. They formed canopied forests and open woodlands with understories of ferns, aquatic plants and flowering shrubs.

The Chicxulub Impact occurred in a shallow sea and immediately lofted rocky, molten and vaporous debris into the atmosphere. The bulk of the debris rained down on nearby continental regions, but much of it rose all the way into space.

The vapor-rich plume of material expanded to envelop Earth. As material in that plume fell back to the ground, it streaked through the atmosphere like trillions of meteors, heating it in some places by hundreds of degrees.

After fires had ravaged the landscape, only a few stark trunks and skeletons remained. Soot from the fires and dust from the impact slowly settled to the ground. Sunlight was dramatically, if not totally, attenuated for months.

The postimpact environment was less diverse. Ferns and algae were the first to recover. Plant species in swamps and swamp margins generally survived better than species in other types of ecosystems. Conifers fared particularly badly.

Shrubs took advantage of the vacant landscape and began to cover it. Species pollinated by the wind did better than those that relied on insects. Trees began to grow, but it took years for forest canopies to rebuild. The recovery time is uncertain.

Tiny rodentlike creatures capable of ferreting out seeds and nuts were among the animals that managed to survive this catastrophe, setting the stage for the rise of mammals in the eons that followed. As the dominant mammals on the planet, we may owe our very existence to an ancient asteroid.

WEB LINK 15.10 In 1992 a team of geologists suggested that this asteroid crashed into a site in Mexico. They based this conclusion on glassy debris and violently shocked grains of rock ejected from the 180-km-diameter Chicxulub Crater on the Yucatán Peninsula. Using radioactive dating techniques (see Section 8-3 and Box 8-1), the scientists found that the asteroid struck 64.98 million years ago, in remarkable agreement with the age of the iridium-rich layer of clay.

Some geologists and paleontologists are not convinced that an asteroid impact led to the extinction of the dinosaurs. But many scientists agree that this hypothesis fits the available evidence better than any other explanation that has been offered so far.

Tunguska and the Threat of Asteroid Impact

On June 30, 1908, a spectacular explosion occurred over the Tunguska region of Siberia that released about 10^{15} joules of energy, equivalent to a nuclear detonation of several hundred kilotons. The blast knocked a man off his porch some 60 km away and could be heard more than 1000 km away. Millions of tons of dust were injected into the atmosphere, darkening the sky as far away as California.

Preoccupied with political upheaval and World War I, Russia did not send a scientific expedition to the site until 1927. Researchers found that trees had been seared and felled radially outward in an area about 50 km in diameter (Figure 15-12). There was no clear evidence of a crater. In fact, the trees at "ground zero" were left standing upright, although they were completely stripped of branches and leaves.

Several teams of modern astronomers have argued that the Tunguska explosion was caused by an asteroid traveling at supersonic speed. They arrived at this conclusion after assessing the effects of various impactor sizes, speeds, and compositions. Even data from above-ground nuclear detonations of the 1940s and 1950s were worked into the calculations. The Tunguska event is well-matched by an asteroid about 80 m (260 ft) in diameter entering Earth's atmosphere at 22 km/s (80,000 km/h = 50,000 mph).

WEB LINK 15.11 Large objects, such as that which formed the Chicxulub Crater and may have led to the demise of the dinosaurs, occasionally strike Earth with the destructive force of as much as a million megatons of TNT. An asteroid 1 or 2 kilometers in diameter, striking Earth at 30 km/s, would destroy an area the size of California. Such an impact would throw more than 10^{13} kg of microscopic dust particles into the atmosphere, which would decrease the amount of sunlight reaching Earth's surface by enough to threaten the health of the world's agriculture. The loss of a year's crops could lead to the demise of a quarter of the world's population and would place our civilization—although perhaps not the survival of our species—in grave jeopardy.

Figure 15-12 R I V U X G

Aftermath of the Tunguska Event In 1908 a stony asteroid with a mass of about 10^8 kilograms entered Earth's atmosphere over the Tunguska region of Siberia. Its passage through the atmosphere made a blazing trail in the sky some 800 km long. The asteroid apparently exploded before reaching the surface, blowing down trees for hundreds of kilometers around "ground zero." (Courtesy of Sovfoto)

The good news is that such large objects strike Earth *very* infrequently. An asteroid 1 km in diameter might be expected to hit Earth only once every 300,000 years. Statistically, the probability that an average person will die because of such an event is about 1 in 20,000, or only about 1% as great as the probability of dying in an automobile accident. While the threat of asteroid or comet impact exists, a catastrophic impact is very unlikely to occur in our lifetimes.

15-5 Meteorites are classified as stones, stony irons, or irons, depending on their composition

TUTORIAL 15.2 Asteroids can only be studied at close range using spacecraft. But from time to time pieces of asteroids fall to Earth where they can be examined by scientists. These asteroid fragments are called *meteorites*.

Meteoroids, Meteors, and Meteorites

A *meteoroid*, like an asteroid, is a chunk of rock in space. There is no official dividing line between meteoroids and asteroids, but the term *asteroid* is generally applied only to objects larger than 50 m across.

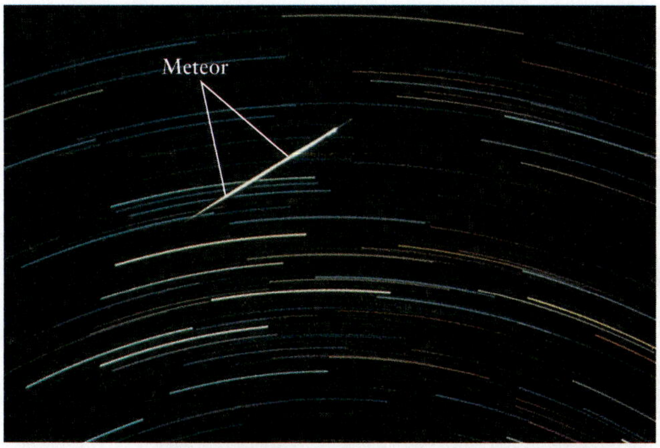

Figure 15-13 R I V U X G

A Meteor A meteor is produced when a meteoroid—a piece of interplanetary rock or dust—strikes Earth's atmosphere at high speed (typically 5 to 30 km/s). The heat of air friction makes the meteoroid glow as it disintegrates. Most meteoroids burn up completely at altitudes of 100 km (60 mi) or so. This long exposure (notice the star trails) shows an exceptionally bright meteor called a fireball. (Courtesy of R. A. Oriti)

A **meteor** is the brief flash of light (sometimes called a *shooting star*) that is visible at night when a meteoroid enters Earth's atmosphere (Figure 15-13). Frictional heat is generated as the meteoroid plunges through the atmosphere, leaving a fiery trail across the night sky.

WEB LINK 15.12

If a piece of rock survives its fiery descent through the atmosphere and reaches the ground, it is called a **meteorite.** As incredible as it may seem, an estimated total of 300 tons of extraterrestrial matter falls on Earth each day. Most of this matter is dust rather than meteorites, however, and is hardly noticeable. It is fortunate that large meteorite falls are rather rare events, because these fast-moving rock fragments can cause substantial damage when they strike the ground (Figure 15-14). Because they are so rare, meteorites are prized by scientists and collectors alike. Almost all meteorites are asteroid fragments and thus provide information about the chemical composition of asteroids. A very few meteorites have been identified as pieces of the Moon or Mars that were blasted off the surfaces of those worlds by asteroid impacts and eventually landed on Earth.

Studying meteorites reveals the properties of the asteroids from which they came

People have been finding specimens of meteorites for thousands of years, and descriptions of them appear in ancient Chinese, Greek, and Roman literature. Many civilizations have regarded meteorites as objects of veneration. The sacred Black Stone, enshrined at the Ka'aba in the Great Mosque at Mecca, may be a relic of a meteorite impact.

The extraterrestrial origin of meteorites was hotly debated until as late as the eighteenth century. Upon hearing a lecture by two Yale professors, President Thomas Jefferson is said to have remarked, "I could more easily believe that two Yankee professors could lie than that stones could fall from Heaven." Although several falling meteorites had been widely witnessed and specimens from them had been collected, many scientists were reluctant to accept the idea that rocks could fall to Earth from outer space. Conclusive evidence for the extraterrestrial origin of meteorites came on April 26, 1803, when fragments pelted the French town of L'Aigle. This event was analyzed by the noted physicist Jean-Baptiste Biot, and his conclusions helped finally to convince scientists that meteorites were indeed extraterrestrial.

Types of Meteorites

Meteorites are classified into three broad categories: stones, stony irons, and irons. As their name suggests, **stony meteorites,** or **stones,** look like ordinary rocks at first glance, but they are sometimes covered with a **fusion crust** (Figure 15-15*a*). This crust is produced by the melting of the meteorite's outer layers during its fiery descent through the atmosphere. When a stony meteorite is cut in two and polished, tiny flecks of iron can sometimes be found in the rock (Figure 15-15*b*).

Although stony meteorites account for about 95% of all the meteoritic material that falls to Earth, they are the most difficult meteorite specimens to find. If they lie undiscovered and become exposed to the weather for a few years, they look almost indistinguishable from common terrestrial rocks. (The asteroid that caused the Tunguska explosion, described in Section 15-4, was probably of stony composition—which would account for the lack of obvious meteoritic debris on the ground.) Meteorites with high iron content can be found much more easily, because they can be located with a metal detector. Consequently, iron and stony iron meteorites dominate most museum collections.

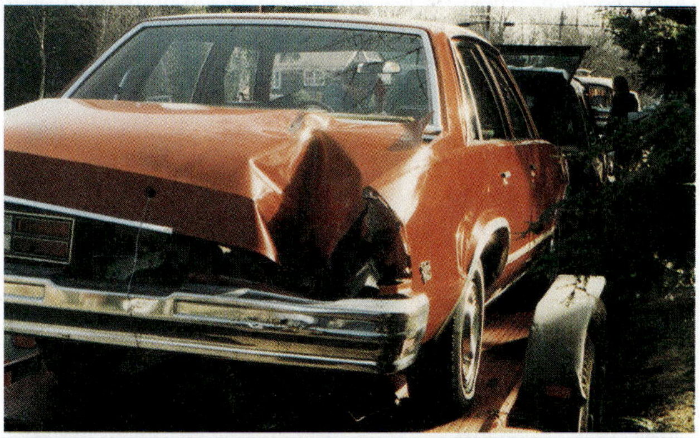

Figure 15-14 R I V U X G

A Meteorite "Fender-Bender" On the evening of October 9, 1992, Michelle Knapp of Peekskill, New York, heard a noise from outside that sounded like a car crash. She discovered that the trunk of her car had been smashed by a 12-kilogram (27-pound) meteorite, which was lying beside the car and was still warm to the touch. Meteorite damage is not covered by automobile insurance, but Ms. Knapp was offered several tens of thousands of dollars for the meteorite (and the car). (R. A. Langheinrich Meteorites, Ilion, N.Y.)

Many stony meteorites are coated with dark fusion crusts...

(a)

...but when cut and polished they reveal tiny specks of iron in the rock.

(b)

Figure 15-15 R I V U X G

Stony Meteorites Of all meteorites that fall on Earth, 95% are stones. **(a)** Many freshly discovered specimens, like the one shown here, are coated with dark fusion crusts. This particular stone fell in Texas. **(b)** Some stony meteorites are found to contain tiny specks of iron mixed in the rock. This specimen was discovered in California. (From the collection of R. A. Oriti)

Stony iron meteorites consist of roughly equal amounts of rock and iron. Figure 15-16, for example, shows the mineral olivine suspended in a matrix of iron. Only about 1% of the meteorites that fall to Earth are stony irons.

Iron meteorites (Figure 15-17), or **irons,** account for about 4% of the material that falls on Earth. Iron meteorites usually contain no stone, but many contain from 10% to 20% nickel. Before humans began extracting iron from ores around 2000 B.C., the only source of the metal was from iron meteorites. Because these are so rare, iron was regarded as a precious metal like gold or silver.

Figure 15-16 R I V U X G

A Stony Iron Meteorite Stony irons account for about 1% of all the meteorites that fall to Earth. This particular specimen, a variety of stony iron called a pallasite, fell in Chile. (Chip Clark)

In 1808, Count Alois von Widmanstätten, director of the Imperial Porcelain Works in Vienna, discovered a conclusive test for the most common type of iron meteorite. About 75% of all iron meteorites have a unique crystalline structure. These **Widmanstätten patterns** (pronounced VIT-mahn-shtetten) become visible when the meteorite is cut, polished, and briefly dipped into a dilute solution of acid (Figure 15-17*b*). Nickel-iron crystals of this type can form only if the molten metal cools slowly over many millions of years. Therefore, Widmanstätten patterns are never found in counterfeit meteorites, or "meteorwrongs."

Meteorites and the Early History of Asteroids

Widmanstätten patterns suggest that some asteroids were originally at least partly molten inside and remained partly molten long after they were formed. As soon as an asteroid accreted from planetesimals some 4.56 billion years ago, rapid decay of short-lived radioactive isotopes could have heated the asteroid's interior to temperatures above the melting point of rock. If the asteroid was large enough—at least 200 to 400 km in diameter—to insulate the interior and prevent cooling, the interior would have remained molten over the next few million years and chemical differentiation would have occurred (see Box 7-1 and Section 8-5). Iron and nickel would have sunk toward the asteroid's center, forcing less dense rock upward toward the asteroid's surface.

Like a terrestrial planet, such a **differentiated asteroid** would have a distinct core and crust. After a differentiated asteroid cooled and its core solidified, collisions with other asteroids would have fragmented the parent body into meteoroids. Iron meteorites are therefore specimens from a differentiated asteroid's core, while stony irons come from regions between a differentiated asteroid's core and its crust.

Unlike irons and stony irons, stony meteorites may or may not come from differentiated asteroids. Smaller asteroids would

Iron meteorites are composed of nickel-iron minerals and are characterized by a surface covered with depressions...

(a)

...and when cut and polished, by interlocking crystals in a Widmanstätten pattern.

(b)

Figure 15-17 R I V U X G

Iron Meteorites **(a)** The surface of a typical iron is covered with small depressions caused by ablation (removal by melting) during its high-speed descent through the atmosphere. Note the one-cent coin for scale. **(b)** When cut, polished, and etched with a weak acid solution, most iron meteorites exhibit interlocking crystals in designs called Widmanstätten patterns. Both of these iron meteorites were found in Australia. (From the collection of R. A. Oriti)

not have retained their internal heat long enough for chemical differentiation to occur. Some stony meteorites are fragments of such **undifferentiated asteroids** (also called *primitive* asteroids), while others are relics of the crust of differentiated asteroids.

15-6 Some meteorites retain traces of the early solar system

A class of rare stony meteorites called **carbonaceous chondrites** shows no evidence of ever having been melted as part of asteroids. As suggested by their name, carbonaceous chondrites contain substantial amounts of carbon and carbon compounds, including complex organic molecules and as much as 20% water bound into the minerals. These compounds would have been broken down and the water driven out if these meteorites had been subjected to heating and melting. Carbonaceous chondrites may therefore be samples of the original material from which our solar system was created. The asteroid 253 Mathilde, shown in Figure 15-7, has a very dark gray color and the same sort of spectrum as a carbonaceous chondrite. It, too, is likely composed of material that predates the formation of the solar system.

Amino acids, the building blocks of proteins upon which terrestrial life is based, are among the organic compounds occasionally found inside carbonaceous chondrites. Interstellar organic material has thus probably been falling on our planet since its formation. Some scientists suspect that carbonaceous chondrites may have played a role in the origin of life on Earth.

Shortly after midnight on February 8, 1969, the night sky around Chihuahua, Mexico, was illuminated by a brilliant blue-white light moving across the heavens. As the light crossed the sky, it disintegrated in a spectacular, noisy explosion that dropped thousands of rocks and pebbles over terrified onlookers. Within hours, teams of scientists were on their way to collect specimens of a carbonaceous chondrite, collectively named the Allende meteorite after the locality (Figure 15-18).

Specimens were scattered in an elongated ellipse approximately 50 km long by 10 km wide. Most fragments were coated with a fusion crust, but minerals immediately beneath the crust showed no signs of damage. Surface material is peeled away as it becomes heated during flight through the atmosphere; this process forms the fusion crust. Because the heat has little time to penetrate the meteorite's interior, compounds there are left intact.

One of the most striking discoveries to come from the Allende meteorite was evidence suggesting that a nearby star exploded into a **supernova** about 4.6 billion years ago. In a supernova, a massive star reaches the end of its life cycle and blows itself apart in a cataclysm that hurls matter outward at tremendous speeds (see Figure 1-8). During this detonation, violent collisions between nuclei produce a host of radioactive elements, including ^{26}Al, a radioactive isotope of aluminum.

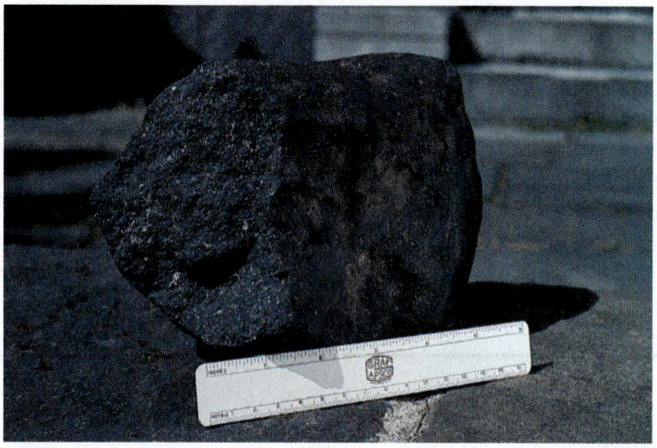

Figure 15-18 R I V U X G

A Piece of the Allende Meteorite This carbonaceous chondrite fell near Chihuahua, Mexico, in February 1969. The meteorite's dark color is due to its high abundance of carbon. Radioactive age-dating indicates that this meteorite is 4.56 billion years old, suggesting that this meteorite is a specimen of primitive planetary material that predates the formation of the planets. The ruler is 15 cm (6 in) long. (Courtesy of J. A. Wood)

Researchers found clear evidence for the former presence of ^{26}Al in the Allende meteorite: Chemical analyses revealed a high abundance of a stable isotope of magnesium (^{26}Mg), which is produced by the radioactive decay of ^{26}Al. Some astronomers interpret this as evidence for a supernova in our vicinity at about the time the Sun was born. Indeed, by compressing interstellar gas and dust, the supernova's shock wave may have triggered the birth of our solar system.

15-7 A comet is a chunk of ice and dust that partially vaporizes as it passes near the Sun

Just as heat from the protosun produced two classes of planets, the terrestrial and the Jovian, two main types of small bodies formed in the solar system. Near the Sun, interplanetary debris consists of the rocky objects called asteroids or meteoroids. Far from the Sun, where temperatures in the early solar system were low enough to permit ices of water, methane, ammonia, and carbon dioxide to form, interplanetary debris took the form of loose collections of ices and small rocky particles. These objects are like icy versions of "rubble pile" asteroids such as 253 Mathilde, which we described in Section 15-3. The Harvard astronomer Fred Whipple, who devised this model in 1950, called them "icy conglomerates" or "dirty snowballs." Some of these "snowballs" are trans-Neptunian objects that orbit forever in the outer regions of the solar system. Some of them, however, are in noncircular orbits that take them inward toward the Sun, where the ice begins to evaporate. These objects are known as **comets.**

The Structure of a Comet

Kuiper belt objects and asteroids travel around the Sun along roughly circular orbits that lie close to the plane of the ecliptic. In sharp contrast, comets travel around the Sun along highly elliptical orbits inclined at random angles to the ecliptic. Comets are just a few kilometers across, so when they are far from the Sun they are difficult to spot with even a large telescope. As a comet approaches the Sun, however, solar heat begins to vaporize the comet's ices, liberating gases as well as dust particles. The liberated gases begin to glow, producing a fuzzy, luminous ball called a **coma** that is typically 1 million km in diameter. Some of these luminous gases stream outward into a long, flowing **tail.** This tail, which can be more than 100 million kilometers in length—comparable to the distance from Earth to the Sun—is one of the most awesome sights that can be seen in the night sky (see Figure 15-19 and the image that opens this chapter).

Many comets are discovered when they are still far from the Sun, long before their tails grow to full size. A number of these discoveries are made by amateur astronomers, many of whom use only binoculars or small telescopes to aid in their search.

Figure 15-20 depicts the structure of a comet. The solid part of the comet, from which the coma and tail emanate, is called the **nucleus.** It is a mixture of ice

Figure 15-19 R I **V** U X G

Comet Hyakutake Using binoculars, Japanese amateur astronomer Yuji Hyakutake first noticed this comet on the morning of January 30, 1996. Two months later, Comet Hyakutake came within 0.1 AU (15 million km, or 9 million mi) of Earth and became one of the brightest comets of the twentieth century. This image was made in April 1996, when the comet's tail extended more than 30° across the sky (more than 60 times the diameter of the full moon). The stars appear as short streaks because the camera followed the comet's motion during the exposure. (Courtesy of Gary Goodman)

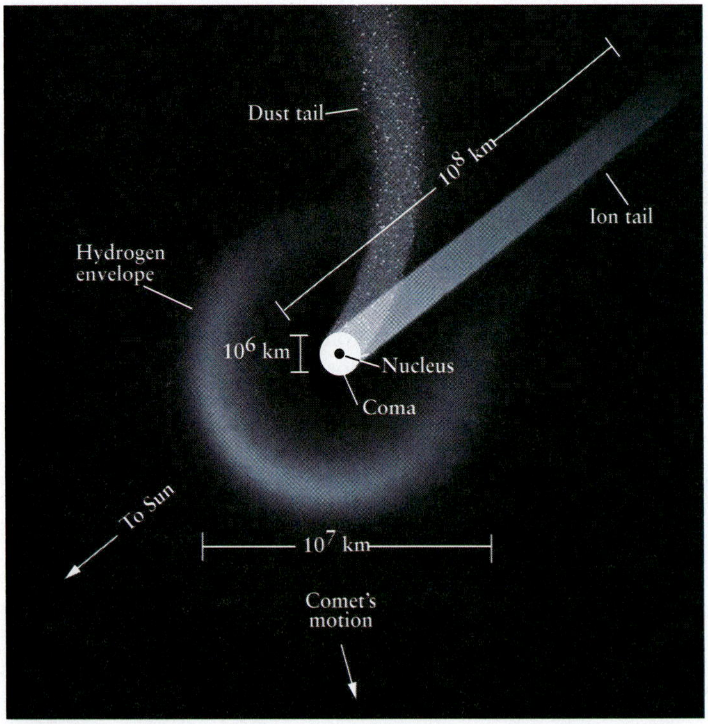

Figure 15-20

The Structure of a Comet The solid part of a comet is the nucleus. It is typically about 10 km in diameter. The coma typically measures 1 million kilometers in diameter, and the hydrogen envelope usually spans 10 million kilometers. A comet's tail can be as long as 1 AU.

Figure 15-21 RI**V**UXG
Comet Nuclei (a) This image from the European Space Agency spacecraft *Giotto* shows the dark, potato-shaped nucleus of Comet Halley in March 1986. Sunlight from the left illuminates half of the nucleus, which is about 15 km long and 8 km wide. Two bright jets of dust extend from the nucleus toward the Sun. (b) The *Deep Impact* spacecraft recorded this image of the nucleus of Comet Tempel 1 in July 2005. This nucleus, which measures 7.6 by 4.9 km, has a variety of different terrain on its surface. (a: Max-Planck-Institut für Aeronomie; b: NASA/JPL/U. of Maryland)

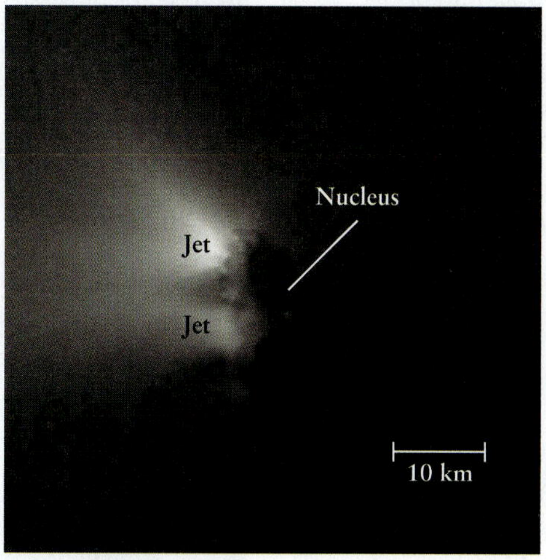

(a) Comet Halley

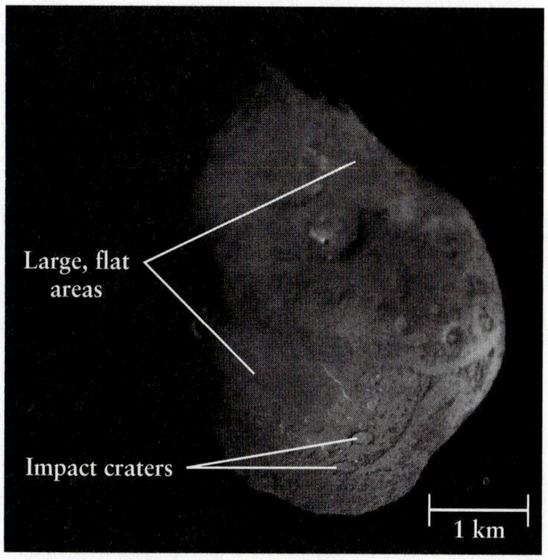

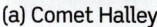

(b) Comet Tempel 1

and dust that typically measures a few kilometers across. Before 1986, no one had seen the nucleus of a comet. It is small, dim, and buried in the glare of the coma. In that year, the first close-up pictures of a comet's nucleus were obtained by a spacecraft that flew past Comet Halley (Figure 15-21*a*). Comet Halley's potato-shaped nucleus is darker than coal, reflecting only about 4% of the light that strikes it. This dark color is probably caused by a layer of carbon-rich compounds and dust that is left behind as the comet's ice evaporates.

Several 15-km-long jets of dust particles were seen emanating from bright areas on Halley's nucleus. These jets seem to be active only when exposed to the Sun. The bright areas, which probably cover about 10% of the surface of the nucleus, are presumably places where the dark layer covering the nucleus is particularly thin. When exposed to the Sun, these areas evaporate rapidly, producing the jets. When the nucleus's rotation brings the bright areas into darkness, away from the Sun, the jets shut off.

VIDEO 15.2

A few other comet nuclei have been imaged at close range by spacecraft. The best images to date have come from the *Deep Impact* spacecraft, which flew to within 500 km of Comet Tempel 1 in 2005 (Figure 15-21*b*). As it approached, *Deep Impact* launched a 372-kg copper projectile that slammed into the surface of the nucleus at 37,000 km/h (23,000 mi/h). The collision excavated about 11,000 tons of material from the nucleus. The character of the ejected material confirmed that the nucleus of Tempel 1 is a "rubble pile" held together not by chemical forces between its component parts (like a chocolate chip cookie) but by the gravitational forces between those parts.

The *Deep Impact* mission also measured the mass and hence the density of the comet nucleus. The very low density of 600 kg/m^3 (60% of the density of water) shows that the nucleus cannot be solid. It is best regarded as a porous jumble of rock, fine dust, and ice.

An even more ambitious mission to explore a comet is the European Space Agency's *Rosetta*. This spacecraft was lunched in 2004 and is scheduled to fly past at least one asteroid on its way to intercept Comet 67P/Churyumov-Gerasimenko in 2014. If all goes well, *Rosetta* will go into orbit around the comet nucleus and release a small probe that will actually land on the nucleus.

Comet Envelopes and Comet Tails

A part of a comet that is not visible to the human eye is the **hydrogen envelope,** a huge sphere of tenuous gas surrounding the nucleus. This hydrogen comes from water molecules (H_2O) that escape from the comet's evaporating ice and then break apart when they absorb ultraviolet photons from the Sun. The hydrogen atoms also absorb solar ultraviolet photons, which excite the atoms from the ground state into an excited state (see Section 5-8). When the atoms return to the ground state, they emit ultraviolet photons.

Unfortunately for astronomers, these emissions from a comet's hydrogen envelope cannot be detected by Earth-based telescopes because our atmosphere is largely opaque to ultraviolet light (see Section 6-7). Instead, cameras above Earth's atmosphere must be used. Figure 15-22 shows two views of the same comet, one as it appeared in visible light to Earth-based observers and one as photographed by an ultraviolet camera aboard a rocket. The ultraviolet image shows the enormous extent of the hydrogen envelope, which can span 10 million kilometers.

As the diagram in Figure 15-20 suggests, comet tails always point away from the Sun. This observation is true regardless of the direction of the comet's motion (Figure 15-23). The implication that something from the Sun was "blowing" the comet's gases radially outward led Ludwig Biermann to predict the existence of a *solar wind,* a stream of particles rushing away from the Sun (see Section 8-5). A decade later, in 1962, Biermann's prediction was confirmed when the solar wind was detected for the first time by instruments on the *Mariner 2* spacecraft (see Figure 11-15).

A comet's tail does not stream behind it; instead, it points generally away from the Sun

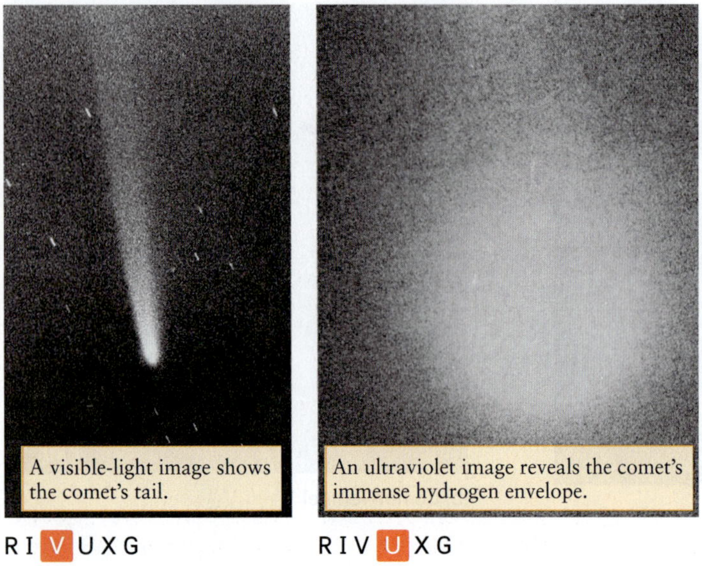

Figure 15-22

A Comet and Its Hydrogen Envelope These visible-light and ultraviolet images of Comet Kohoutek, which made an appearance in 1974, are reproduced to the same scale. The hydrogen envelope is only visible in the ultraviolet image. (Johns Hopkins University, Naval Research Laboratory)

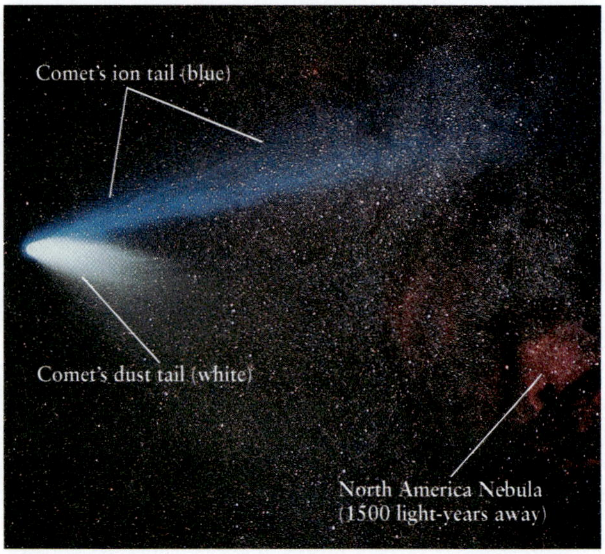

Figure 15-24 R I V U X G

WEB LINK 15.16

The Two Tails of Comet Hale-Bopp A comet's dust tail is white because it reflects sunlight, while the molecules in the ion tail emit their own light with a characteristic blue color. When this picture was taken on March 8, 1997, the ion tail extended more than 10° across the sky. The red object to the right is the North America Nebula, a star-forming region some 1500 light-years beyond the solar system. (Courtesy of Tony and Daphne Hallas Astrophotos)

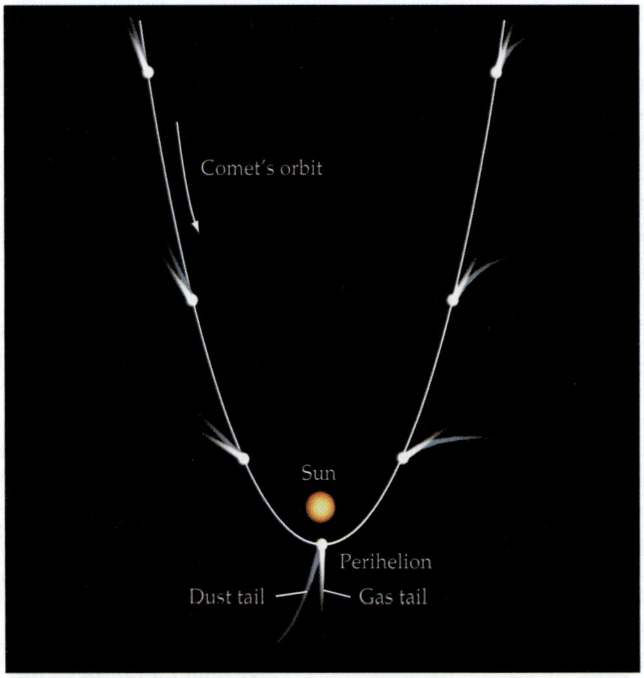

Figure 15-23

The Orbit and Tail of a Comet The solar wind and the pressure of sunlight blow a comet's dust particles and ionized atoms away from the Sun. Consequently, a comet's tail points generally away from the Sun. In particular, the tail does *not* always stream behind the nucleus. At the upper right of this figure, the comet is moving upward along its orbit and is literally chasing its own tail.

In fact, the Sun usually produces *two* comet tails—an **ion tail** and a **dust tail** (Figure 15-24). Ionized atoms and molecules—that is, atoms and molecules missing one or more electrons—are swept directly away from the Sun by the solar wind to form the relatively straight ion tail. The distinct blue color of the ion tail is caused by emissions from carbon-bearing molecules such as CN and C_2. The dust tail is formed when photons of light strike dust particles freed from the evaporating nucleus. (Figure 8-9 shows a dust particle of this type.) Light exerts a pressure on any object that absorbs or reflects it. This pressure, called **radiation pressure,** is quite weak but is strong enough to make fine-grained dust particles in a comet's coma drift away from the comet, thus producing a dust tail. The solar wind has less of an effect on dust particles than on (much smaller) ions, so the dust tail ends up being curved rather than straight. On rare occasions a comet is oriented in such a way that its dust tail appears to stick out in front of the comet (Figure 15-25).

After the comet passes perihelion, it recedes from the Sun back into the cold regions of the outer solar system. The ices stop vaporizing, the coma and tail disappear, and the comet goes back into an inert state—until the next time its orbit takes it toward the inner solar system. An object in an elliptical orbit spends most of its time far from the Sun, so it is only during a relatively brief period before and after perihelion that a comet can have a prominent tail. An example is Comet Halley, which orbits the Sun along a highly elliptical path that stretches from just inside Earth's orbit to slightly beyond the orbit of Neptune (Figure 15-26). The comet's tail is visible to the naked eye or with binoc-

Figure 15-25 R I V U X G

The Antitail of Comet Hale-Bopp In January 1998, nine months after passing perihelion, Comet Hale-Bopp was oriented in such a way that the end of its arched dust tail looked like a spike sticking out of the comet's head. This spike is called an "antitail" because it appears to protrude in front of the comet. (European Southern Observatory)

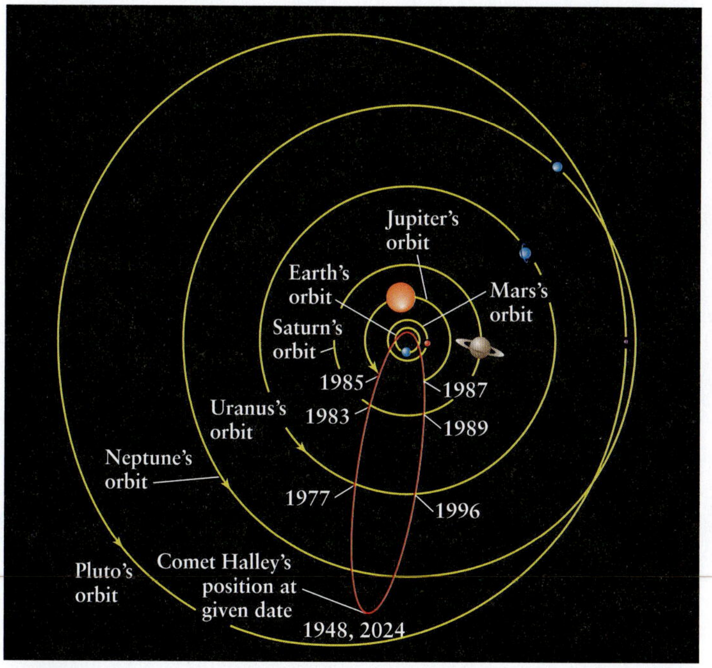

ANIMATION 15.1

Figure 15-26

Comet Halley's Eccentric Orbit Like most comets, Comet Halley has an elongated orbit and spends most of its time far from the Sun. Figure 4-23 shows the comet's tail, which appears only when the comet is close to perihelion. Comet Halley has been observed at intervals of about 76 years—the period of its orbit—since 88 B.C. (see Section 4-7).

ulars only during a few months around perihelion, which last occurred in 1986 and will happen again in 2061.

Other comets have orbits that are larger and more elongated, with even longer periods. The orbit of Comet Hyakutake (see Figure 15-19) takes it to a distance of about 2000 AU from the Sun, 50 times the size of Pluto's orbit, with an orbital period of around 100,000 years!

15-8 Comets originate either from the Kuiper belt or from a vast cloud in near-interstellar space

Both comets and Kuiper belt objects are mixtures of ice and rock that must have formed in the outer solar system. (Had they formed in the warm inner solar system, they would not have included ices.) But most Kuiper belt objects have relatively circular orbits that lie close to the plane of the ecliptic, while comets have very elongated orbits that are often steeply inclined to the ecliptic. What gives comets their distinctive orbits? As we will see, the answer to this question depends on whether the comet's orbital period is relatively short or relatively long.

Jupiter-Family Comets and the Kuiper Belt

As we saw in Section 14-10, the gravitational effects of Neptune shape the orbits of objects in the **Kuiper belt.** Most of these objects are found between the orbit of Pluto (where an object makes two orbits for every three orbits of Neptune) and about 50 AU from the Sun (where an object makes one orbit for every two orbits of Neptune).

However, collisions between Kuiper belt objects can break off relatively small chunks. Gravitational perturbations from Neptune can occasionally launch one of these chunks into a highly elliptical orbit that takes it close to the Sun. When one of these chunks approaches the Sun, it develops a visible tail and appears as a comet. These are called **Jupiter-family comets** because their orbits tend to be influenced strongly by Jupiter's gravitational pull. Jupiter-family comets orbit the Sun in fewer than 20 years. One example is Comet Tempel 1 (see Figure 15-21*b*), which has an orbital period of 5.5 years.

There may be tens of thousands of objects in the Kuiper belt that could eventually be perturbed into an elongated orbit and become comets. It is thought that a few of these refugees from the Kuiper belt have been captured by the gravitational pull of Saturn, becoming that planet's small outer satellites (see Section 13-10).

Intermediate-Period Comets, Long-Period Comets, and the Oort Cloud

The majority of comets are **intermediate-period comets,** with orbital periods between 20 and 200 years, and **long-period comets,** which take more than 200 years to complete one orbit of the Sun. Comet Halley, shown in Figure 15-21*a*, has an orbital period of 76 years and so is an intermediate-period comet; Comet Hyakutake (see Figure 15-19) is a long-period comet that takes more than 100,000 years to complete one orbit.

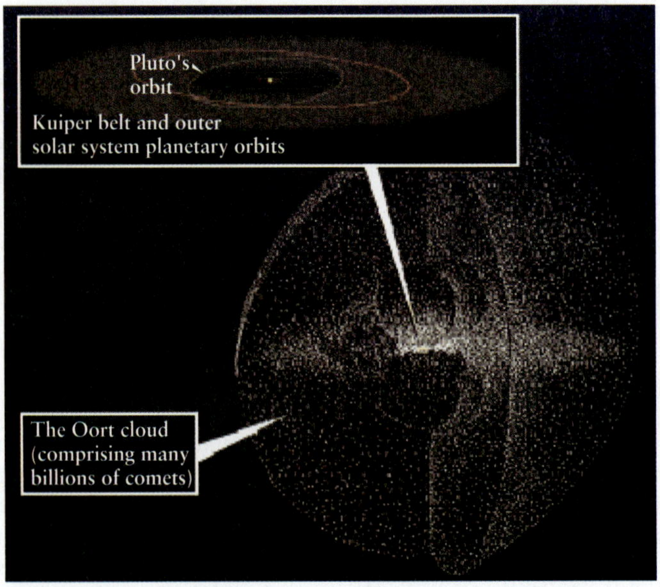
Pluto's orbit
Kuiper belt and outer solar system planetary orbits
The Oort cloud (comprising many billions of comets)

Figure 15-27

The Kuiper Belt and the Oort Cloud Objects in the Kuiper belt orbit close to the plane of the ecliptic. The Oort cloud, by contrast, is a spherical distribution of icy objects whose orbits are randomly oriented relative to the ecliptic. (NASA and A. Field/Space Telescope Science Institute)

Long-period comets travel along extremely elongated orbits and consequently spend most of their time at distances of roughly 10^4 to 10^5 AU from the Sun, or about one-fifth of the way to the nearest star. These orbits extend far beyond the Kuiper belt. Furthermore, intermediate-period and long-period comets have orbital planes that are often steeply inclined to the plane of the ecliptic. (Comet Halley's orbital plane is inclined by 162° to the ecliptic, which means that it orbits the Sun in the opposite direction to the planets.)

The best explanation for these observations is that intermediate-period and long-period comets come from a reservoir that extends from the Kuiper belt to some 50,000 AU from the Sun. This reservoir, first hypothesized by Dutch astronomer Jan Oort in 1950 and now called the **Oort cloud,** does not lie in or near the ecliptic; rather, it is a spherical distribution centered on the Sun (Figure 15-27). This explains why many intermediate-period and long-period comets have steeply inclined orbits.

Because astronomers discover long-period comets at the rate of about one per month, it is reasonable to suppose that there is an enormous population of comets in the Oort cloud. Estimates of the number of "dirty snowballs" in the Oort cloud range as high as 5 trillion (5×10^{12}). Only such a large reservoir of comet nuclei would explain why we see so many long-period comets, even though each one takes several million years to travel once around its orbit. Because the Oort cloud is so distant, it has not yet been possible to detect objects in the Oort cloud directly.

The Oort cloud was probably created 4.56 billion years ago from numerous icy planetesimals that orbited the Sun in the vicinity of the newly formed Jovian planets. During near-collisions with the giant planets, many of these chunks of ice and dust were catapulted by gravity into highly elliptical orbits, in much the same way that *Pioneer* and *Voyager* spacecraft were flung far from the Sun during their flybys of the Jovian planets. Gravitational perturbations from nearby stars could have tilted the planes of the orbits in all directions, giving the Oort cloud its spherical shape.

Changing a Comet's Orbit

The distinction between long-period, intermediate-period, and Jupiter-family comets can be blurred by the effects of gravitational perturbations. During a return trip toward the Sun, an encounter with a Jovian planet may force a comet into a much larger orbit. Alternatively, such an encounter can move a long-period comet into a smaller orbit (Figure 15-28). Several comets have been perturbed in this way into orbits that always remain within the inner solar system. One such is Comet Tempel 1 (see Figure 15-21*b*), whose orbit extends from the orbit of Mars out to the orbit of Jupiter.

Our description of cometary origins suggests that different comets are of different ages. Computer simulations suggest that many Jupiter-family comets were produced no more than a few hundred million years ago by collisions between larger icy objects in the Kuiper belt. Comets emanating from the Oort cloud may

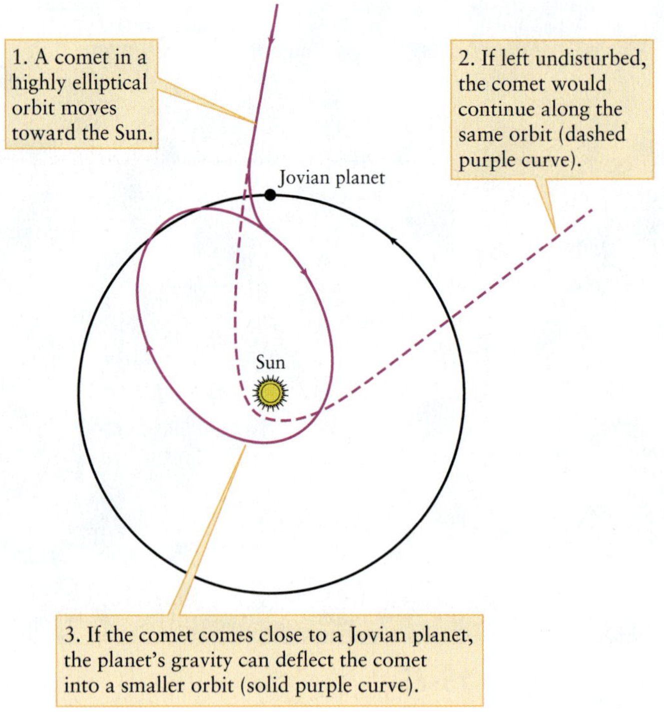
1. A comet in a highly elliptical orbit moves toward the Sun.
2. If left undisturbed, the comet would continue along the same orbit (dashed purple curve).
Jovian planet
Sun
3. If the comet comes close to a Jovian planet, the planet's gravity can deflect the comet into a smaller orbit (solid purple curve).

Figure 15-28

Transforming a Comet's Orbit The gravitational force of a planet can deflect a comet from a highly elliptical orbit into a less elliptical one. In some cases a comet ends up with an orbit that keeps it within the inner solar system.

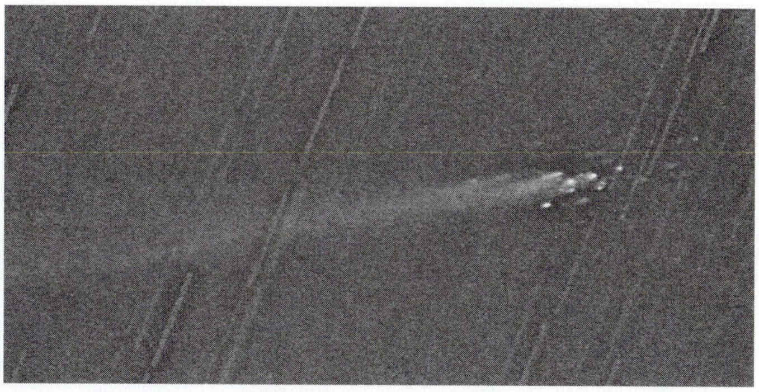

Figure 15-29 R I V U X G

The Fragmentation of Comet LINEAR The Lincoln Near Earth Asteroid Research (LINEAR) project discovered this comet in 1999. As it passed the perihelion of its orbit in July 2000, Comet LINEAR broke into more than a dozen small fragments. These continue to orbit the Sun along the same trajectory that the comet followed prior to its breakup. This image was made using the Very Large Telescope (see Figure 6-18). (European Southern Observatory)

date from the very early solar system, but may have originated at somewhat different times at different distances from the center of the solar nebula before being catapulted into the Oort cloud. This variety of ages is one reason why comets are of particular interest to planetary scientists: Comets carry information about the history of the cold, outer regions of the solar system, but are brought into the inner solar system where we can examine them at closer range using telescopes or spacecraft.

Comet Breakup and Meteor Showers

Because they evaporate away part of their mass each time they pass near the Sun, comets cannot last forever. A typical comet may lose about 0.5% to 1% of its ice each time it passes near the Sun. Hence, the ice completely vaporizes after about 100 or 200 perihelion passages, leaving only a swarm of dust and pebbles. Astronomers have observed some comet nuclei in the process of fragmenting (Figure 15-29).

Comets eventually break apart, and their fragments give rise to meteor showers

WEB LINK 15.17

One remarkable comet, called Shoemaker-Levy 9, broke apart for a different reason. As the comet swung by Jupiter in July 1992, the giant planet's tidal forces tore the nucleus into more than 20 fragments. Jupiter's gravity then deflected the trajectories of these fragments so that they plummeted into the planet's atmosphere.

As a comet's nucleus evaporates, residual dust and rock fragments form a **meteoritic swarm,** a loose collection of debris that continues to circle the Sun along the comet's orbit (Figure 15-30). If Earth's orbit happens to pass through this swarm, a **meteor shower** is seen as the dust particles strike Earth's upper atmosphere.

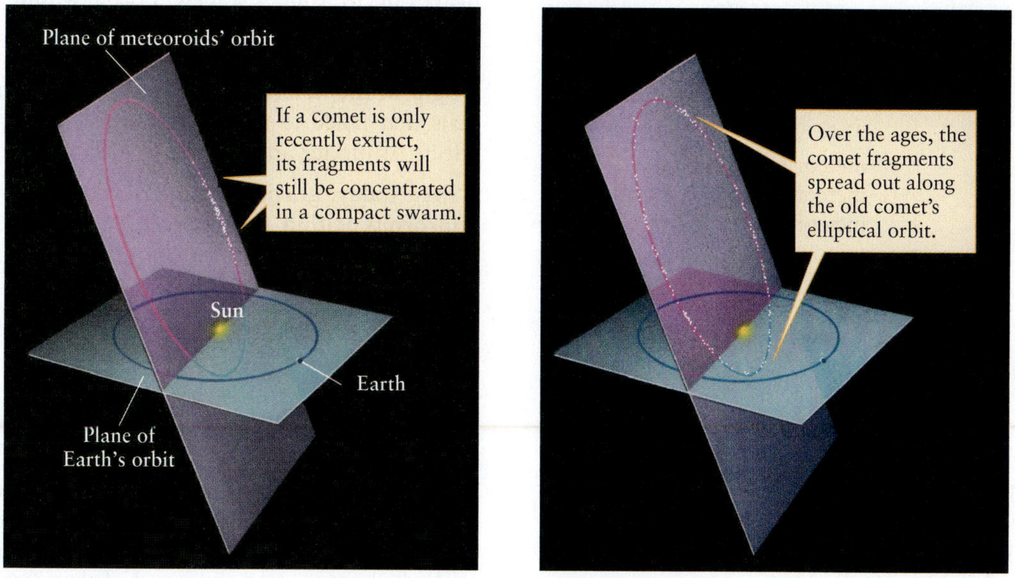

(a) (b)

Figure 15-30

Meteoritic Swarms Rock fragments and dust from "burned out" comets continue to circle the Sun. **(a)** The most spectacular meteor showers occur when Earth passes through compact swarms of debris from recently extinct comets. **(b)** Meteor showers associated with older comets are more predictable, because Earth passes through the evenly distributed swarm on each trip around the Sun.

LOOKING DEEPER 15.3

Table 15-1 Prominent Yearly Meteor Showers

Shower name	Date of maximum intensity*	Typical hourly rate	Average speed (km/s)	Radiant constellation
Quadrantids	January 3	40	40	Boötes
Lyrids	April 22	15	50	Lyra
Eta Aquarids	May 4	20	64	Aquarius
Delta Aquarids	July 30	20	40	Aquarius
Perseids	August 12	50	60	Perseus
Orionids	October 21	20	66	Orion
Taurids	November 4	15	30	Taurus
Leonids	November 16	15	70	Leo
Geminids	December 13	50	35	Gemini
Ursids	December 22	15	35	Ursa Minor

**The date of maximum intensity is the best time to observe a particular shower, although good displays can often be seen a day or two before or after the maximum. The typical hourly rate is given for an observer under optimum viewing conditions. The average speed refers to how fast the meteoroids are moving when they strike the atmosphere.*

Nearly a dozen meteor showers can be seen each year (Table 15-1). Note that the **radiants** for these showers—that is, the places among the stars from which the meteors appear to come—are not confined to the constellations of the zodiac. This means that the swarms do not all orbit in the plane of the ecliptic. The reason is that the orbits of their parent comets were inclined at random angles to the plane of the ecliptic (see Figure 15-30).

Key Words

amino acids, p. 389
asteroid, p. 378
asteroid belt, p. 379
carbonaceous chondrite, p. 389
coma (of a comet), p. 390
comet, p. 390
differentiated asteroid, p. 388
dust tail, p. 392
fusion crust, p. 387
Hirayama family, p. 383
hydrogen envelope, p. 391
intermediate-period comet, p. 393
iron meteorite (iron), p. 388
ion tail, p. 392
Jupiter-family comet, p. 393
Kirkwood gaps, p. 380
Kuiper belt, p. 393
long-period comet, p. 393
meteor, p. 387
meteor shower, p. 395
meteorite, p. 387
meteoritic swarm, p. 395
meteoroid, p. 384
minor planet, p. 378
near-Earth object (NEO), p. 383
nucleus (of a comet), p. 390
Oort cloud, p. 394
radiant (of a meteor shower), p. 396
radiation pressure, p. 392
stable Lagrange points, p. 383
stony iron meteorite, p. 388
stony meteorite (stone), p. 387
supernova, p. 389
tail (of a comet), p. 390
Trojan asteroid, p. 383
undifferentiated asteroid, p. 389
Widmanstätten patterns, p. 388

Key Ideas

Discovery of the Asteroids: Astronomers first discovered the asteroids while searching for a "missing planet."

• Thousands of asteroids with diameters ranging from a few kilometers up to 1000 kilometers orbit within the asteroid belt between the orbits of Mars and Jupiter.

Origin of the Asteroids: The asteroids are the relics of planetesimals that failed to accrete into a full-sized planet, thanks to the effects of Jupiter and other Mars-sized objects.

• Even today, gravitational perturbations by Jupiter deplete certain orbits within the asteroid belt. The resulting gaps, called Kirkwood gaps, occur at simple fractions of Jupiter's orbital period.

• Jupiter's gravity also captures asteroids in two locations, called Lagrangian points, along Jupiter's orbit.

Asteroid Collisions: Asteroids undergo collisions with each other, causing them to break up into smaller fragments.

• Some asteroids, called near-Earth objects, move in elliptical orbits that cross the orbits of Mars and Earth. If such an asteroid strikes Earth, it forms an impact crater whose diameter depends on both the mass and the speed of the asteroid.

• An asteroid struck Earth 65 million years ago, possibly causing the extinction of the dinosaurs and many other species.

Meteoroids, Meteors, and Meteorites: Small rocks in space are called meteoroids. If a meteoroid enters Earth's atmosphere, it produces a fiery trail called a meteor. If part of the object survives the fall, the fragment that reaches Earth's surface is called a meteorite.

• Meteorites are grouped into three major classes, according to composition: iron, stony iron, and stony meteorites. Irons and stony irons are fragments of the core of an asteroid that was large enough and hot enough to have undergone chemical differentiation, just like a terrestrial planet. Some stony meteorites come from the crust of such differentiated meteorites, while others are fragments of small asteroids that never underwent differentiation.

• Rare stony meteorites called carbonaceous chondrites may be relatively unmodified material from the solar nebula. These meteorites often contain organic material and may have played a role in the origin of life on Earth.

• Analysis of isotopes in certain meteorites suggests that a nearby supernova may have triggered the formation of the solar system 4.56 billion years ago.

Comets: A comet is a chunk of ice with imbedded rock fragments that generally moves in a highly elliptical orbit about the Sun.

• As a comet approaches the Sun, its icy nucleus develops a luminous coma, surrounded by a vast hydrogen envelope. An ion tail and a dust tail extend from the comet, pushed away from the Sun by the solar wind and radiation pressure, respectively.

Origin and Fate of Comets: Comets are thought to originate from two regions, the Kuiper belt and the Oort cloud.

• The Kuiper belt lies in the plane of the ecliptic at distances between 30 and 50 AU from the Sun. It is thought to contain many tens of thousands of comet nuclei. Many Jupiter-family comets probably come from the Kuiper belt, and hundreds of larger objects have been observed in the Kuiper belt.

• The Oort cloud contains billions of comet nuclei in a spherical distribution that extends out to 50,000 AU from the Sun. Intermediate-period and long-period comets are thought to originate in the Oort cloud. As yet no objects in the Oort cloud have been detected directly.

• Fragments of "burned out" comets produce meteoritic swarms. A meteor shower is seen when Earth passes through a meteoritic swarm.

Questions

Review Questions

1. What led astronomers to suspect that there were members of the solar system that orbit between Mars and Jupiter?
2. How did the first asteroid come to be discovered? What role did theoretical calculations play in confirming the discovery? How did this discovery differ from what astronomers had expected to find?
3. How do modern astronomers discover new asteroids?
4. What are the differences between asteroids and trans-Neptunian objects?
5. Describe the asteroid belt. Does it lie completely within the plane of the ecliptic? What are its inner and outer radii?
6. If Jupiter was not present in our solar system, would the asteroid belt exist? Why or why not?
7. What are Kirkwood gaps? What causes them?
8. Compare the explanation of the Kirkwood gaps in the asteroid belt to the way in which Saturn's moons help produce divisions in that planet's rings (see Section 12-11).
9. How is it possible to tell that some asteroids are nonspherical even though we do not have images of those asteroids?
10. What is the evidence that some asteroids are made of a loose conglomeration of smaller pieces?
11. The asteroid 243 Ida, which was viewed by the *Galileo* spacecraft, is a member of a Hirayama family. Discuss what this tells us about the history of this asteroid. Where might you look to find other members of the same family?
12. What are the Trojan asteroids, and where are they located? What holds them in this location?
13. What are near-Earth objects? What is the evidence that Earth has been struck by these objects?
14. What is the evidence that an asteroid impact could have contributed to the demise of the dinosaurs?
15. What is the difference between a meteoroid, a meteor, and a meteorite?
16. Is there anywhere on Earth where you might find large numbers of stony meteorites that are not significantly weathered? If so, where? If not, why not?
17. Scientists can tell that certain meteorites came from the interior of an asteroid rather than from its outer layers. Explain how this is done.
18. Why are some asteroids differentiated while others are not?
19. Suppose you found a rock you suspect might be a meteorite. Describe some of the things you could do to determine whether it was a meteorite or a "meteorwrong."
20. What is the evidence that carbonaceous chondrites are essentially unaltered relics of the early solar system? What do they suggest about how the solar system may have formed?
21. With the aid of a drawing, describe the structure of a comet.
22. Why is the phrase "dirty snowball" an appropriate characterization of a comet's nucleus?
23. What did scientists learn about the structure of comet nuclei from the *Deep Impact* mission?
24. Why do the ion tail and dust tail of a comet point in different directions?
25. Why do comets have prominent tails for only a short time during each orbit?
26. Why is it that Jupiter and Saturn can be seen in the night sky every year, while seeing specific comets such as Halley and Hyakutake is a once-in-a-lifetime event?
27. What is the relationship between the Kuiper belt and comets?
28. What is the Oort cloud? How might it be related to planetesimals left over from the formation of the solar system?
29. Why are comets more likely to break apart at perihelion than at aphelion?
30. Why do astronomers think that meteorites come from asteroids, while meteor showers are related to comets?

31. Why are asteroids, meteorites, and comets all of special interest to astronomers who want to understand the early history and subsequent evolution of the solar system?

Advanced Questions

Questions preceded by an asterisk (*) involve topics discussed in Box 1-1 or 7-2.

Problem-solving tips and tools

The small-angle formula is described in Box 1-1. We discussed retrograde motion in Section 4-1 and described its causes in Section 4-2. You will need to use Kepler's third law, described in Section 4-4 and Box 4-2, in some of the problems below. Box 7-2 discusses the concept of escape speed. A spherical object of radius r intercepts an amount of sunlight proportional to its cross-sectional area, equal to πr^2. The volume of a sphere of radius r is $4\pi r^3/3$.

32. When Olbers discovered Pallas in March 1802, the asteroid was moving from east to west relative to the stars. At what time of night was Pallas highest in the sky over Olbers's observatory? Explain your reasoning.

33. Consider the Kirkwood gap whose orbital period is two-fifths of Jupiter's 11.86-year period. Calculate the distance from the Sun to this gap. Does your answer agree with Figure 15-4?

*34. When the image in Figure 15-5 was made, the asteroid 1 Ceres was 1.63 AU, or 2.44×10^8 km, from the Hubble Space Telescope. (**a**) What was the angular size of the asteroid as a whole? (**b**) You can see individual pixels in the image shown in Figure 15-5. Using a ruler and the scale bar in the figure, determine how many kilometers on the surface of 1 Ceres are contained in the width of one pixel. (**c**) What is the angular width of each pixel in arcseconds?

35. Suppose that a binary asteroid (two asteroids orbiting each other) is observed in which one member is 16 times brighter than the other. Suppose that both members have the same albedo and that the larger of the two is 120 km in diameter. What is the diameter of the other member?

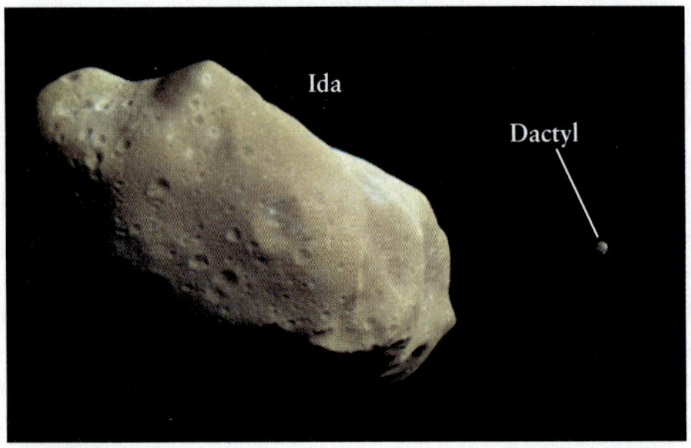

RI**V**UXG

(JPL/NASA)

36. The accompanying image from the *Galileo* spacecraft shows the asteroid 243 Ida, which has dimensions $56 \times 24 \times 21$ km. *Galileo* discovered a tiny moon called Dactyl, just $1.6 \times 1.4 \times 1.2$ km in size, which orbits Ida at a distance of about 100 km. (In Greek mythology, the Dactyli were beings who lived on the slopes of Mount Ida.) Describe a scenario that could explain how Ida came to have a moon.

*37. Assume that Ida's tiny moon Dactyl (see Question 36) has a density of 2500 kg/m^3. (**a**) Calculate the mass of Dactyl in kilograms. For simplicity, assume that Dactyl is a sphere 1.4 km in diameter. (**b**) Calculate the escape speed from the surface of Dactyl. If you were an astronaut standing on Dactyl's surface, could you throw a baseball straight up so that it would never come down? Professional baseball pitchers can throw at speeds around 40 m/s (140 km/h, or 90 mi/h); your throwing speed is probably a bit less.

38. Imagine that you are an astronaut standing on the surface of a Trojan asteroid. How will you see the phase of Jupiter change with the passage of time? How will you see Jupiter move relative to the distant stars? Explain your answers.

39. Use the percentages of stones, irons, and stony iron meteorites that fall to Earth to estimate what fraction of their parent asteroids' interior volume consisted of an iron core. Assume that the percentages of stones and irons that fall to Earth indicate the fractions of a parent asteroid's interior volume occupied by rock and iron, respectively. How valid do you think this assumption is?

40. On March 8, 1997, Comet Hale-Bopp was 1.39 AU from Earth and 1.00 AU from the Sun. Use this information and that given in the caption to Figure 15-24 to estimate the length of the comet's ion tail on that date. Give your answer in kilometers and astronomical units.

41. *Sun-grazing comets* come so close to the Sun that their perihelion distances are essentially zero. Find the orbital periods of Sun-grazing comets whose aphelion distances are (**a**) 100 AU, (**b**) 1000 AU, (**c**) 10,000 AU, and (**d**) 100,000 AU. Assuming that these comets can survive only a hundred perihelion passages, calculate their lifetimes. (*Hint:* Remember that the semimajor axis of an orbit is one-half the length of the orbit's long axis.)

42. Comets are generally brighter a few weeks after passing perihelion than a few weeks before passing perihelion. Explain why might this be. (*Hint:* Water, including water ice, does an excellent job of retaining heat.)

43. The hydrogen envelopes of comets are especially bright at an ultraviolet wavelength of 122 nm. Use Figure 5-24 to explain why.

44. A *very* crude model of a typical comet nucleus is a cube of ice (density 1000 kg/m^3) 10 km on a side. (**a**) What is the mass of this nucleus? (**b**) Suppose 1% of the mass of the nucleus evaporates away to form the comet's tail. Suppose further that the tail is 100 million (10^8) km long and 1 million (10^6) km wide. Estimate the average density of the tail (in kg/m^3). For comparison, the density of the air you breathe is about 1.2 kg/m^3. (**c**) In 1910 Earth actually passed through the tail of Comet Halley. At the time there was some concern among the general public that this could have deleterious effects on human health. Was this concern justified? Why or why not?

45. For many years it was thought that the Tunguska event was caused by a comet striking Earth. This idea was rejected because a small comet would have broken up too high in the atmosphere to cause significant damage on the ground. Explain why, using your knowledge of a comet's structure.

Discussion Questions

46. Discuss the idea that 1 Ceres should be regarded as the smallest dwarf planet rather than the largest asteroid. What are the advantages of this scheme? What are the disadvantages?
47. From the abundance of craters on the Moon and Mercury, we know that numerous asteroids and meteoroids struck the inner planets early in the history of our solar system. Is it reasonable to suppose that numerous comets also pelted the planets 3.5 to 4.5 billion years ago? Speculate about the effects of such a cometary bombardment, especially with regard to the evolution of the primordial atmospheres on the terrestrial planets.
48. In the 1998 movie *Armageddon,* an asteroid "the size of Texas" is on a collision course with Earth. The asteroid is first discovered by astronomers just 18 days prior to impact. To avert disaster, a team of astronauts blasts the asteroid into two pieces just 4 hours before impact. Discuss the plausibility of this scenario. (*Hint:* On average, the state of Texas extends for about 750 km from north to south and from east to west. How does this compare with the size of the largest known asteroids?)
49. Suppose astronomers discover that a near-Earth object the size of 1994 XM1 is on a collision course with Earth. Describe what humanity could do within the framework of present technology to counter such a catastrophe.

Web/eBook Questions

50. A NASA spacecraft called *Dawn* is intended to go into orbit around two asteroids, 1 Ceres and 4 Vesta. Search the World Wide Web for information about this mission. Why were these two particular asteroids selected for study? What types of observations will the spacecraft make?
51. Search the World Wide Web to find out why some scientists disagree with the idea that a tremendous impact led to the demise of the dinosaurs. (They do not dispute that the impact took place, only what its consequences were.) What are their arguments? From what you learn, what is your opinion?
52. Several scientific research programs are dedicated to the search for near-Earth objects (NEOs), especially those that might someday strike our planet. Search the World Wide Web for information about at least one of these programs. How does this program search for NEOs? How many NEOs has this program discovered? Will any of these pose a threat in the near future?

VIDEO 15.3

53. **Estimating the Speed of a Comet** Access and view the video "Two Comets and an Active Sun" in Chapter 15 of the *Universe* Web site or eBook. (a) Why don't the comets reappear after passing the Sun? (b) The white circle shows the size of the Sun, which has the diameter 1.39×10^6 km. Using this to set the scale, step through the video and measure how the position of one of the comets changes. Use your measurements and the times displayed in the video to estimate the comet's speed in km/h and km/s. (Assume that the comet moved in the same plane as that shown in the video.) As part of your answer, explain the technique and calculations you used. (c) How does your answer in (b) compare with the orbital speed of Mercury, the innermost and fastest-moving planet (see Table 11-1)? Why is there a difference? (d) If the comet's motion was not in the same plane as that shown in the video, was its actual speed more or less than your estimate in (b)? Explain.

Activities

Observing Projects

WEB LINK 15.18

Observing tips and tools

Meteors: Informative details concerning upcoming meteor showers appear on the Web sites for *Sky & Telescope* and *Astronomy* magazines.

Comets: Because astronomers discover dozens of comets each year, there is usually a comet visible somewhere in the sky. Unfortunately, they are often quite dim, and you will need to have access to a moderately large telescope (at least 35 cm, or 14 in.). You can get up-to-date information from the Minor Planet Center Web site. Also, if there is an especially bright comet in the sky, useful information about it might be found at the Web sites for *Sky & Telescope* and *Astronomy* magazines.

Asteroids: At opposition, some of the largest asteroids are bright enough to be seen through a modest telescope. Check the Minor Planet Center Web site to see if any bright asteroids are near opposition. If so, check the current issue as well as the most recent January issue of *Sky & Telescope* magazine for a star chart showing the asteroid's path among the constellations. You will need such a chart to distinguish the asteroid from background stars. Also, you can locate Ceres, Pallas, Juno, and Vesta using the *Starry Night Enthusiast™* program on the CD-ROM that accompanies certain printed copies of this textbook.

54. Make arrangements to view a meteor shower. Table 15-1 lists the dates of major meteor showers. Choose a shower that will occur near the time of a new moon. Set your alarm clock for the early morning hours (1 to 3 A.M.). Get comfortable in a reclining chair or lie on your back so that you can view a large portion of the sky. Record how long you observe, how many meteors you see, and what location in the sky they seem to come from. How well does your observed hourly rate agree with published estimates, such as those in Table 15-1? Is the radiant of the meteor shower apparent from your observations?
55. If a comet is visible with a telescope at your disposal, make arrangements to view it. Can you distinguish the comet from background stars? Can you see its coma? Can you see a tail?
56. Make arrangements to view an asteroid. Observe the asteroid on at least two occasions, separated by a few days. On

each night, draw a star chart of the objects in your telescope's field of view. Has the position of one starlike object shifted between observing sessions? Does the position of the moving object agree with the path plotted on published star charts? Do you feel confident that you have in fact seen the asteroid?

57. Use the *Starry Night Enthusiast*™ program to observe two asteroids. Display the entire celestial sphere by selecting **Guides > Atlas** in the **Favorites** menu. Center on **Ceres** by typing its name in the **Search All Databases** box in the **Find** pane and label it by clicking in the box to the left of its name as it appears in the Find list. In the toolbar, set the Time Flow Rate to 1 day. Then click the Run Time Forward button. (**a**) Watch Ceres for at least two years of simulated time. Describe how Ceres moves. How can you tell that Ceres orbits the Sun in the same direction as the planets? Does Ceres remain as close to the ecliptic (shown as a solid green line) as the planets do? (If you do not see a green line representing the ecliptic, make sure that the item **The Ecliptic** is checked in the View menu.) (**b**) Click on the **Stop** button and use the **Find** pane to locate and center the view on Pallas. Again click on the Run Time Forward button, and watch how Pallas moves for at least two years of simulated time. How does the motion of Pallas differ from that of Ceres? How are the two motions similar? Which asteroid's orbit is more steeply inclined to the plane of the Earth's orbit? How can you tell?

58. Use the *Starry Night Enthusiast*™ program to study the motion of a comet. First set up the field of view so that you are observing the inner solar system from a distance (select **Solar System > Inner Solar system** in the **Favorites** menu). In the toolbar, click on the **Stop** button to halt the animation, and then set the date to January 1, 1995, and the time step to 1 day. Select **View > Solar System > Asteroids** in the menu to remove the asteroids from the view. Open the **Find** pane and center on Comet Hyakutake by typing "*Hyakutake*" in the **Search All Databases** box and then pressing the **Enter** key. Use the Zoom controls to decrease the field of view to about 25° × 17°. Then click on the **Run Time Forward** button. (**a**) Watch the motion of Comet Hyakutake for at least two years of simulated time. Describe what you see. Is the comet's orbit in about the same plane as the orbits of the inner planets, or is it steeply inclined to that plane? (You can tilt the plane of the solar system by holding down the **Shift** key while clicking on and moving the mouse to investigate this off-ecliptic motion.) How does the comet's speed vary as it moves along its orbit? During which part of the orbit is the tail visible? In what direction does the tail point? (**b**) Click on the **Stop** button to halt the animation, and set up the field of view so that you are observing from the center of a transparent Earth by selecting **Guides > Atlas** in the **Favorites** menu. Set the date to **January 1, 1995**, and the **Time Flow Rate** to **1 day**, and again center on Comet Hyakutake. Use the controls at the right-hand end of the toolbar to zoom out as far as possible. Then click on the **Run Time Forward** button and watch the comet's motion for at least two years of simulated time. Describe the motion, and explain why it is more complicated than the motion you observed in part (a). (**c**) Stop the animation, set the date to today's date, set the **Time Flow Rate** to **1 month** ("lunar m."), and restart the animation. Comet Hyakutake is currently moving almost directly away from the Sun and so, as seen from the Sun, its position on the celestial sphere should not change. Is this what you see in *Starry Night Enthusiast*™? Explain any differences. (*Hint:* You are observing from Earth, not the Sun.)

59. Use the *Starry Night Enthusiast*™ program to examine Comet Halley as it would have been seen during its last visit to the Sun and the inner solar system. Display the entire celestial sphere as seen from the center of a transparent Earth by clicking on **Guides > Atlas** in the **Favorites** pane. **Stop** time flow and set the date to **February 28, 1986,** and the time to midnight, **12:00:00 A.M.** Center on Halley's Comet by using **Edit > Find** and entering the name ***Halle**y* in the **Search** box. Set the **Time Flow Rate** to **6 hours.** From this view at Earth's center with daylight turned off, you will note that the Sun is to the left of the view, on the ecliptic plane, as expected. (**a**) Based on the direction of the comet's tail, can you tell in which direction the comet is moving at this instant in time? Step time forward by a few steps to check your prediction. (**b**) Did Halley's Comet move in the direction that you predicted in (a) above? Run time backward to about January 7, 1986, and then run time forward and backward again several times to observe Halley's comet and particularly the direction of its tail during this encounter with the Sun. (**c**) In what direction does the tail of the comet point? Explain why.

Pluto and the Kuiper Belt

by Scott Sheppard

The "what is a planet" controversy began with the discovery of Ceres and several other objects between Mars and Jupiter in the early 1800s. At first, Ceres and then Pallas, Juno, and Vesta were considered planets, even though they were orders of magnitude less massive than the other known planets and all had similar orbits. The definition of a planet at this time was simple. Any object in orbit about the Sun that did not show cometary effects was a planet.

By the mid 1800s, the discovery of more and more objects between Mars and Jupiter saw the words "asteroid" and "minor planet" start to be used in order to signify these objects as an ensemble of bodies. That is, astronomers felt a major planet should be a rare and unique thing. The major planet club thus was left with eight members: Mercury, Venus, Earth, Mars, Jupiter, Saturn, Uranus, and Neptune.

The planet controversy arose once again in 1930 with the discovery of Pluto, where it was quickly regarded as the ninth planet. During the ensuing years, new observations downgraded Pluto's size from being near that of Earth to being smaller than our Moon. Its orbit also proved to be peculiar because it was highly inclined and eccentric, having it even crossing Neptune's orbit. Its uniqueness kept Pluto in the planet club with relatively few detractors until 1992, when the first Kuiper belt object was discovered. As it became apparent that hundreds of thousands of Kuiper belt objects existed on stable orbits just beyond Neptune, Pluto's planet status was becoming less and less certain. With the discovery of Eris (2003 UB313) in 2003, an object larger than Pluto, Pluto's last unique claim of being the largest Kuiper belt object was lost. Thus, the planet controversy had finally reached a point where a major planet's minimum size must be precisely defined.

No matter how you count them, there are no longer nine major planets in our solar system. The same situation of the main belt asteroids in the mid 1800's is now apparent with Pluto, Eris, and the rest of the Kuiper belt objects. Although the largest Kuiper belt objects are significantly bigger than the largest main belt asteroids, they are still an ensemble of bodies with a continuous size distribution and similar formation and evolution histories. The reason Pluto's status as a planet is a controversy now is simply because it took 62 years between the discovery of Pluto and the next Kuiper belt object in 1992. Within that time span Pluto became known as the ninth planet through several generations and was ingrained in society as such.

Classification is not that important to scientists but it is the understanding of how the object formed and its history. Even so, any classification should have a scientific base and not be based on tradition, as seems to be the case for Pluto in the recent past. Eris as well as Pluto and the rest of the Kuiper belt objects had very similar histories and formation scenarios and thus they should be considered as an ensemble of objects just like the main belt asteroids between Mars and Jupiter. As technology advances, our knowledge and thus understanding advances and so should our classification.

Several minimum-size planet definitions were debated recently. The first is that anything larger than a certain minimum size is a planet. The chosen size would likely be arbitrary and may be a nice round number like 1000 km or the size of Pluto. This stems from the traditional belief that Pluto has been considered a planet in the past and should continue as such. This would give us ten current planets including Eris. A second possible definition is that anything that is spherical should be a planet. This definition is based on physics in which the gravitational force of an object is large enough to overcome any other forces in determining the overall shape of the object. Determining if an object meets this criteria or not is complicated. This definition would also significantly change the status quo by increasing the number of known planets by several factors overnight. A third definition is that any object with a unique orbit and gravitationally dominates its local orbital environment should be a planet. This definition is the most scientifically based, in that objects with similar formation and evolutionary histories would be grouped together. Using this definition would give us eight known major planets, dropping Pluto from the list.

A final proposal as to what is a planet takes into account that the current known major planets themselves are remarkably different and can be split into individual categories themselves.

(continued on the next page)

Scott S. Sheppard is an astronomer at the Carnegie Institution of Washington in the Department of Terrestrial Magnetism located in Washington, D.C. His main research currently involves discovering small bodies in our solar system to understand their dynamical and physical properties that help constrain solar system and planet formation. He has discovered several small moons around all the giant planets, been a pioneer in the discovery of Neptune Trojans, as well as found tens of Kuiper Belt objects.

Pluto and the Kuiper Belt *(continued)*

Mercury, Venus, Earth and Mars are terrestrial planets whose compositions are dominated by rock. Jupiter and Saturn are gas giant planets dominated by their hydrogen and helium envelopes. Uranus and Neptune are ice giant planets dominated by gases other than hydrogen and helium. The trans-Neptunian objects or "ice dwarf planets," as some are calling them, are probably composed of large amounts of volatiles such as methane ice and water ice.

The International Astronomical Union (IAU) recently choose to use a combination of the second and third definitions above to define a planet. That is, a planet is an object that is both spherical and has a unique orbit in which it is gravitationally dominant. An object that only satisfies definition two above and not three is now being called a dwarf planet, of which Ceres, Pluto and Eris qualify, as well as several more objects in the Kuiper belt. In reality, it's the public that must accept this definition, and we will only know if that is the case a few generations from now.

Defining the word "planet" is like defining what an ocean is. At first it appears to be a simple term, but is very hard to precisely define. Further planet discoveries will be made and there are sure to be borderline cases. This is just the way nature is; it does not have little bins to nicely classify objects, but there is usually a continuous array of objects. The important thing to take away from all of this is to understand this array of objects by using our rapidly expanding scientific knowledge and understanding our place in the solar system.

16

Our Star, the Sun

RIVUXG

A composite view of the Sun (at ultraviolet wavelengths) and an upheaval in the Sun's outer atmosphere, or corona (at visible wavelengths). (*SOHO*/LASCO/EIT/ESA/NASA)

The Sun is by far the brightest object in the sky. By earthly standards, the temperature of its glowing surface is remarkably high, about 5800 K. Yet there are regions of the Sun that reach far higher temperatures of tens of thousands or even millions of kelvins. Gases at such temperatures emit ultraviolet light, which makes them prominent in the accompanying image from an ultraviolet telescope in space. Some of the hottest and most energetic regions on the Sun spawn immense disturbances and can propel solar material across space to reach Earth and other planets.

In recent decades, we have learned that the Sun shines because at its core hundreds of millions of tons of hydrogen are converted to helium every second. We have confirmed this picture by detecting the by-products of this transmutation—strange, ethereal particles called neutrinos—streaming outward from the Sun into space. We have discovered that the Sun has a surprisingly violent atmosphere, with a host of features such as sunspots whose numbers rise and fall on a predictable 11-year cycle. By studying the Sun's vibrations, we have begun to probe beneath its surface into hitherto unexplored realms. And we have just begun to investigate how changes in the Sun's activity can affect Earth's environment as well as our technological society.

16-1 The Sun's energy is generated by thermonuclear reactions in its core

TUTORIAL 16.1

The Sun is the largest member of the solar system. It has almost a thousand times more mass than all the planets, moons, asteroids, comets, and meteoroids put together. But the Sun is also a star. In fact, it is a remarkably typical star, with a mass, size, surface temperature, and chemical composition that are roughly midway between the extremes exhibited by the myriad other stars in the heavens. Table 16-1 lists essential data about the Sun.

Learning Goals

By reading the sections of this chapter, you will learn

16-1 The source of the Sun's heat and light

16-2 How scientists model the Sun's internal structure

16-3 How the Sun's vibrations reveal what lies beneath its glowing surface

16-4 How scientists are able to probe the Sun's energy-generating core

16-5 Why the gaseous Sun appears to have a sharp outer edge

16-6 Why the upper regions of the solar atmosphere have an emission spectrum

16-7 The relationship between the Sun's corona and the solar wind

16-8 The nature of sunspots

16-9 The connection between sunspots and the Sun's magnetic field

16-10 How magnetic reconnection can power immense solar eruptions

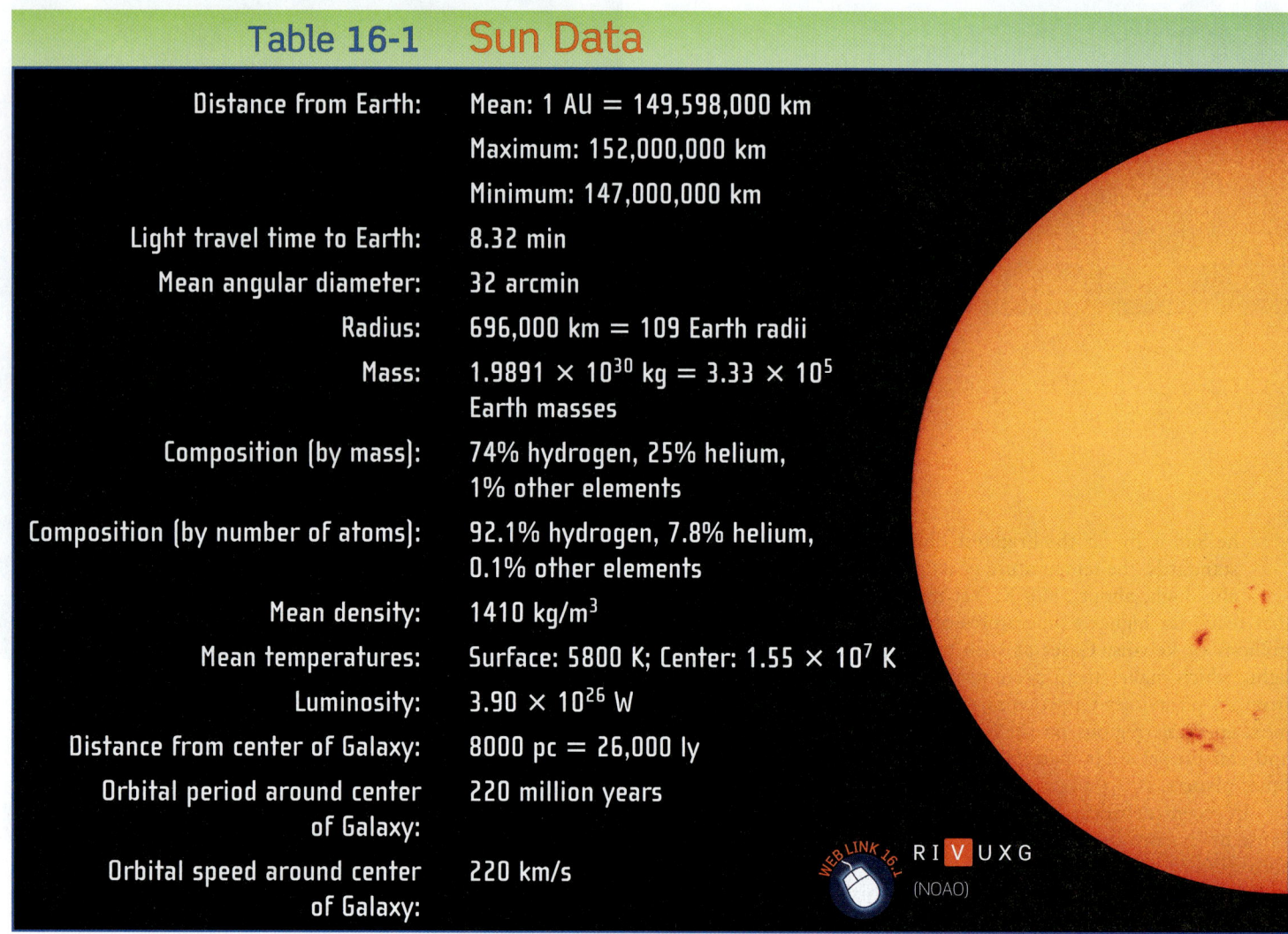

Table 16-1 Sun Data

Distance from Earth:	Mean: 1 AU = 149,598,000 km Maximum: 152,000,000 km Minimum: 147,000,000 km
Light travel time to Earth:	8.32 min
Mean angular diameter:	32 arcmin
Radius:	696,000 km = 109 Earth radii
Mass:	1.9891×10^{30} kg = 3.33×10^{5} Earth masses
Composition (by mass):	74% hydrogen, 25% helium, 1% other elements
Composition (by number of atoms):	92.1% hydrogen, 7.8% helium, 0.1% other elements
Mean density:	1410 kg/m^3
Mean temperatures:	Surface: 5800 K; Center: 1.55×10^{7} K
Luminosity:	3.90×10^{26} W
Distance from center of Galaxy:	8000 pc = 26,000 ly
Orbital period around center of Galaxy:	220 million years
Orbital speed around center of Galaxy:	220 km/s

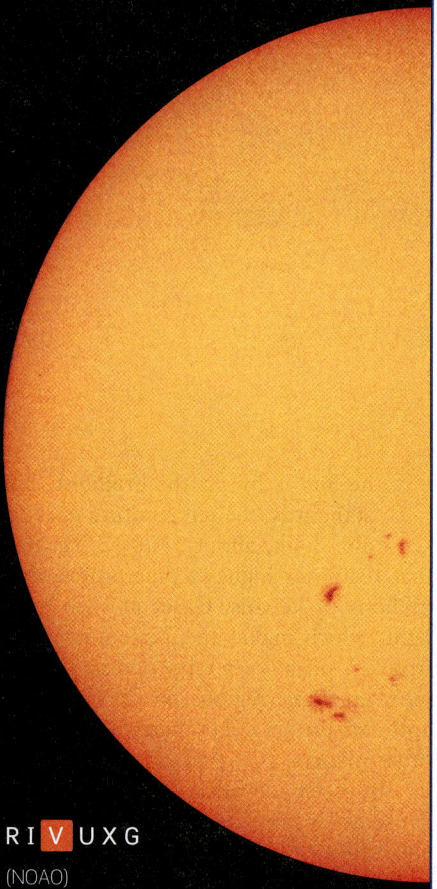

WEB LINK 16.1

RIVUXG

(NOAO)

Solar Energy

For most people, what matters most about the Sun is the energy that it radiates into space. Without the Sun's warming rays, our atmosphere and oceans would freeze into an icy layer coating a desperately cold planet, and life on Earth would be impossible. To understand why we are here, we must understand the nature of the Sun.

Why is the Sun such an important source of energy? An important part of the answer is that the Sun has a far higher surface temperature than any of the planets or moons. The Sun's spectrum is close to that of an idealized blackbody with a temperature of 5800 K (see Sections 5-3 and 5-4, especially Figure 5-12). Thanks to this high temperature, each square meter of the Sun's surface emits a tremendous amount of radiation, principally at visible wavelengths. Indeed, the Sun is the only object in the solar system that emits substantial amounts of visible light. The light that we see from the Moon and planets is actually sunlight that struck those worlds and was reflected toward Earth.

The Sun's size also helps us explain its tremendous energy output. Because the Sun is so large, the total number of square meters of radiating surface—that is, its surface area—is immense. Hence, the total amount of energy emitted by the Sun each second, called its **luminosity**, is very large indeed or about 3.9×10^{26} watts, (3.9×10^{26} joules of energy emitted per second.) (We discussed the relation among the Sun's surface temperature, radius, and luminosity in Box 5-2.) Astronomers denote the Sun's luminosity by the symbol $L_{\odot}$. A circle with a dot in the center is the astronomical symbol for the Sun and was also used by ancient astrologers.

The Source of the Sun's Energy: Early Ideas

These ideas lead us to a more fundamental question: What keeps the Sun's visible surface so hot? Or, put another way, what is the fundamental source of the tremendous energies that the Sun radiates into space? For centuries, this question was one of the greatest mysteries in science. The mystery deepened in the nineteenth century, when geologists and biologists found convincing evidence that life has existed on Earth for at least several hundred million years. (We now know that Earth is 4.56 billion years old and that life had existed on it for most of its history.) Since life as we know it depends crucially on sunlight, the Sun must be as

Figure 16-1 R I V U X G

The Sun The Sun's visible surface has a temperature of about 5800 K. At this temperature, all solids and liquids vaporize to form gases. It was only in the twentieth century that scientists discovered what has kept the Sun so hot for billions of years: the thermonuclear fusion of hydrogen nuclei in the Sun's core. (Jeremy Woodhouse/PhotoDisc)

old. The source of the Sun's energy posed a severe problem for physicists. What source of energy could have kept the Sun shining for so long (**Figure 16-1**)?

One attempt to explain solar energy was made in the mid-1800s by the English physicist Lord Kelvin (for whom the temperature scale is named) and the German scientist Hermann von Helmholtz. They argued that the tremendous weight of the Sun's outer layers should cause the Sun to contract gradually, compressing its interior gases. Whenever a gas is compressed, its temperature rises. (You can demonstrate this with a bicycle pump: As you pump air into a tire, the temperature of the air increases and the pump becomes warm to the touch.) Kelvin and Helmholtz thus suggested that gravitational contraction could cause the Sun's gases to become hot enough to radiate energy out into space.

This process, called *Kelvin-Helmholtz contraction,* actually does occur during the earliest stages of the birth of a star like the Sun (see Section 8-4). But Kelvin-Helmholtz contraction cannot be the major source of the Sun's energy today. If it were, the Sun would have had to be much larger in the relatively recent past. Helmholtz's own calculations showed that the Sun could have started its initial collapse from the solar nebula no more than about 25 million years ago. But the geological and fossil record shows that Earth is far older than that, and so the Sun must be as well. Hence, this model of a Sun that shines because it shrinks cannot be correct.

On Earth, a common way to produce heat and light is by burning fuel, such as a log in a fireplace or coal in a power plant. Is it possible that a similar process explains the energy released by the Sun? The answer is no, because this process could not continue for a long enough time to explain the age of Earth. The chemical reactions involved in burning release roughly 10^{-19} joule of energy per atom. Therefore, the number of atoms that would have to undergo chemical reactions each second to generate the Sun's luminosity of 3.9×10^{26} joules per second is approximately

$$\frac{3.9 \times 10^{26} \text{ joules per second}}{10^{-19} \text{ joule per atom}} = 3.9 \times 10^{45} \text{ atoms per second}$$

From its mass and chemical composition, we know that the Sun contains about 10^{57} atoms. Thus, the length of time that would be required to consume the entire Sun by burning is

$$\frac{10^{57} \text{ atoms}}{3.9 \times 10^{45} \text{ atoms per second}} = 3 \times 10^{11} \text{ seconds}$$

There are about 3×10^{7} seconds in a year. Hence, in this model, the Sun would burn itself out in a mere 10,000 (10^4) years! This period of time is far shorter than the known age of Earth, so chemical reactions also cannot explain how the Sun shines.

The Source of the Sun's Energy: Discovering Thermonuclear Fusion

The source of the Sun's luminosity could be explained if there were a process that was like burning but released much more energy per atom. Then the rate at which atoms would have to be consumed would be far less, and the lifetime of the Sun could be long enough to be consistent with the known age of Earth. Albert Einstein discovered the key to such a process in 1905. According to his *special theory of relativity,* a quantity m of mass can in principle be converted into an amount of energy E according to a now-famous equation:

Einstein's mass-energy equation

$$E = mc^2$$

m = quantity of mass, in kg

c = speed of light = 3×10^8 m/s

E = amount of energy into which the mass can be converted, in joules

The speed of light c is a large number, so c^2 is huge. Therefore, a small amount of matter can release an awesome amount of energy.

Inspired by Einstein's ideas, astronomers began to wonder if the Sun's energy output might come from the conversion of matter into energy. The Sun's low density of 1410 kg/m^3 indicates that it must be made of the very lightest atoms, primarily hydrogen and helium. In the 1920s, the British astronomer Arthur Eddington showed that temperatures near the center of the Sun must be so high that atoms become completely ionized. Hence, at the Sun's center we expect to find hydrogen nuclei and electrons flying around independent of each other.

Ideas from relativity and nuclear physics led to an understanding of how the Sun shines

Another British astronomer, Robert Atkinson, suggested that under these conditions hydrogen nuclei could fuse together to produce helium nuclei in a *nuclear reaction* that transforms a tiny amount of mass

BOX 16-1 Tools of the Astronomer's Trade

Converting Mass into Energy

The *Cosmic Connections* figure shows the steps involved in the thermonuclear fusion of hydrogen at the Sun's center. In these steps, four protons are converted into a single nucleus of ^{4}He, an isotope of helium with two protons and two neutrons. (As we saw in Box 5-5, different isotopes of the same element have the same number of protons but different numbers of neutrons.) The reaction depicted in the *Cosmic Connections: The Proton-Proton Chain* Step 1 also produces a neutral, nearly massless particle called the *neutrino*. Neutrinos respond hardly at all to ordinary matter, so they travel almost unimpeded through the Sun's massive bulk. Hence, the energy that neutrinos carry is quickly lost into space. This loss is not great, however, because the neutrinos carry relatively little energy. (See Section 16-4 for more about these curious particles.)

Most of the energy released by thermonuclear fusion appears in the form of gamma-ray photons. The energy of these photons remains trapped within the Sun for a long time, thus maintaining the Sun's intense internal heat. Some gamma-ray photons are produced by the reaction shown as Step 2 in the *Cosmic Connections* figure. Others appear when an electron in the Sun's interior annihilates a positively charged electron, or **positron,** which is a by-product of the reaction shown in Step 1 in the *Cosmic Connections* figure. An electron and a positron are respectively matter and antimatter, and they convert entirely into energy when they meet. (You may have thought that "antimatter" was pure science fiction. In fact, tremendous amounts of antimatter are being created and annihilated in the Sun as you read these words.)

We can summarize the thermonuclear fusion of hydrogen as follows:

$$4\ {}^1\text{H} \rightarrow {}^4\text{He} + \text{neutrinos} + \text{gamma-ray photons}$$

To calculate how much energy is released in this process, we use Einstein's mass-energy formula: The energy released is equal to the amount of mass consumed multiplied by c^2, where c is the speed of light. To see how much mass is consumed, we compare the combined mass of four hydrogen atoms (the ingredients) to the mass of one helium atom (the product):

$$\begin{aligned} 4 \text{ hydrogen atoms} &= 6.693 \times 10^{-27} \text{ kg} \\ -1 \text{ helium atom} &= 6.645 \times 10^{-27} \text{ kg} \\ \hline \text{Mass lost} &= 0.048 \times 10^{-27} \text{ kg} \end{aligned}$$

Thus, a small fraction (0.7%) of the mass of the hydrogen going into the nuclear reaction does not show up in the mass of the helium. This lost mass is converted into an amount of energy $E = mc^2$:

$$\begin{aligned} E = mc^2 &= (0.048 \times 10^{-27} \text{ kg})(3 \times 10^8 \text{ m/s})^2 \\ &= 4.3 \times 10^{-12} \text{ joule} \end{aligned}$$

This amount of energy is released by the formation of a single helium atom. It would light a 10-watt lightbulb for almost one-half of a trillionth of a second.

EXAMPLE: How much energy is released when 1 kg of hydrogen is converted to helium?

Situation: We are given the initial mass of hydrogen. We know that a fraction of the mass is lost when the hydrogen undergoes fusion to make helium; our goal is to find the quantity of energy into which this lost mass is transformed.

Tools: We use the equation $E = mc^2$ and the result that 0.7% of the mass is lost when hydrogen is converted into hydrogen.

Answer: When 1 kilogram of hydrogen is converted to helium, the amount of mass lost is 0.7% of 1 kg, or 0.007 kg. (This means that 0.993 kg of helium is produced.) Using Einstein's equation, we find that this missing 0.007 kg of matter is transformed into an amount of energy equal to

$$E = mc^2 = (0.007 \text{ kg})(3 \times 10^8 \text{ m/s})^2 = 6.3 \times 10^{14} \text{ joules}$$

Review: The energy released by the fusion of 1 kilogram of hydrogen is the same as that released by burning *20,000 metric tons* (2×10^7 kg) of coal! Hydrogen fusion is a *much* more efficient energy source than ordinary burning.

The Sun's luminosity is 3.9×10^{26} joules per second. To generate this much power, hydrogen must be consumed at a rate of

$$\frac{3.9 \times 10^{26} \text{ joules per second}}{6.3 \times 10^{14} \text{ joules per kilogram}} = 6 \times 10^{11} \text{ kilograms per second}$$

That is, the Sun converts 600 million metric tons of hydrogen into helium every second.

into a large amount of energy. Experiments in the laboratory using individual nuclei show that such reactions can indeed take place. The process of converting hydrogen into helium is called **hydrogen fusion.** (It is also sometimes called *hydrogen burning,* even though nothing is actually burned in the conventional sense. Ordinary burning involves chemical reactions that rearrange the outer electrons of atoms but have no effect on the atoms' nuclei.) Hydrogen fusion provides the devastating energy released in a hydrogen bomb (see Figure 1-6).

The fusing together of nuclei is also called **thermonuclear fusion,** because it can take place only at extremely high temperatures. The reason is that all nuclei have a positive electric charge

COSMIC CONNECTIONS

The Proton-Proton Chain

The most common form of hydrogen fusion in the Sun involves three steps, each of which releases energy.

Hydrogen fusion in the Sun usually takes place in a sequence of steps called the proton-proton chain. Each of these steps releases energy that heats the Sun and gives it its luminosity.

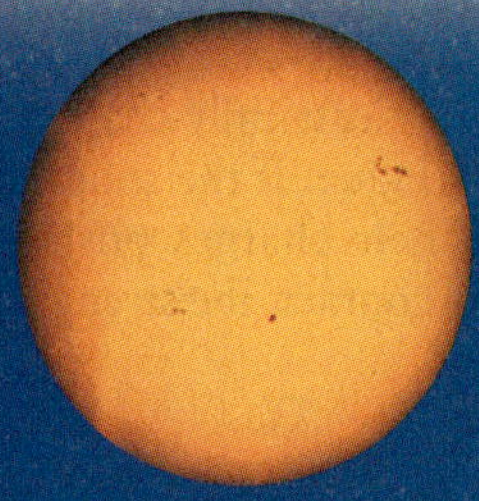

STEP 1

(a) Two protons (hydrogen nuclei, ^{1}H) collide.

(b) One of the protons changes into a neutron (shown in blue). The proton and neutron form a hydrogen isotope (^{2}H).

(c) One byproduct of converting a proton to a neutron is a neutral, nearly massless neutrino (ν), which escapes from the Sun.

(d) The other byproduct of converting a proton to a neutron is a positively charged electron, or positron (e^{+}). This encounters an ordinary electron (e^{-}), annihilating both particles and converting them into gamma-ray photons (γ). The energy of these photons goes into sustaining the Sun's internal heat.

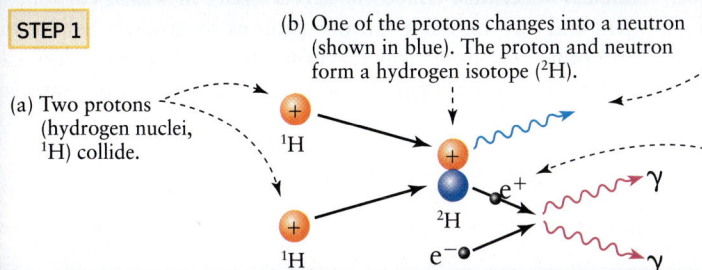

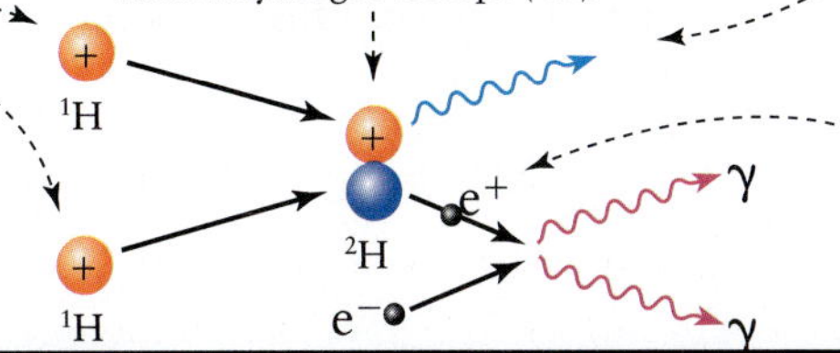

STEP 2

(a) The ^{2}H nucleus produced in Step 1 collides with a third proton (^{1}H).

(b) The result of the collision is a helium isotope (^{3}He) with two protons and one neutron.

(c) This nuclear reaction releases another gamma-ray photon (γ). Its energy also goes into sustaining the internal heat of the Sun.

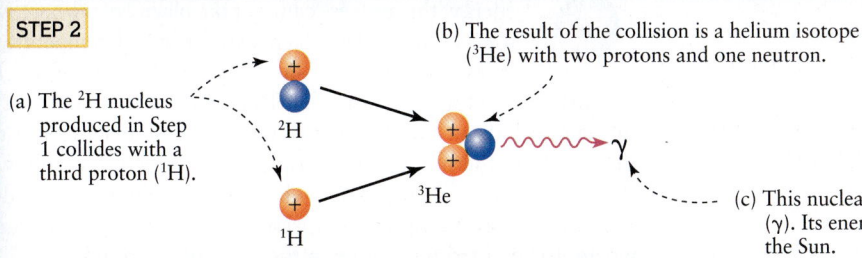

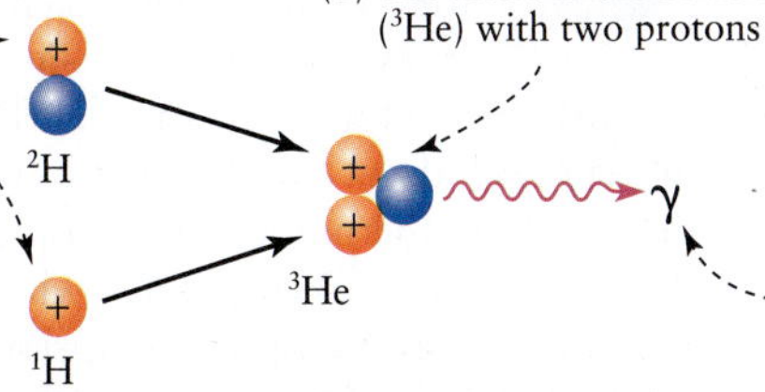

STEP 3

(a) The ^{3}He nucleus produced in Step 2 collides with another ^{3}He nucleus produced from three other protons.

(b) Two protons and two neutrons from the two ^{3}He nuclei rearrange themselves into a different helium isotope (^{4}He).

(c) The two remaining protons are released. The energy of their motion contributes to the Sun's internal heat.

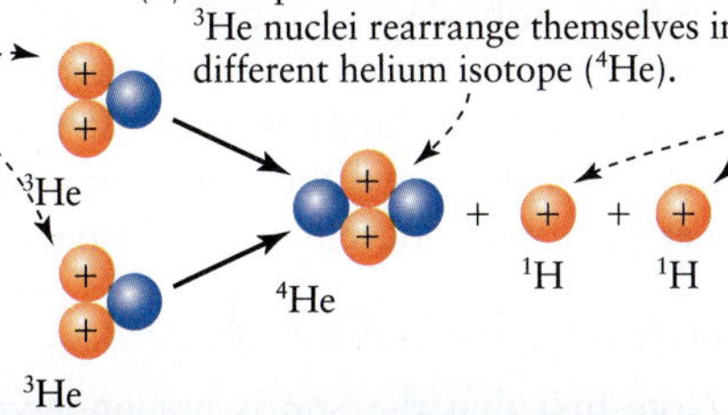

(d) Six ^{1}H nuclei went into producing the two ^{3}He nuclei, which combine to make one ^{4}He nucleus. Since two of the original ^{1}H nuclei are returned to their original state, we can summarize the three steps as:

$4\,{}^{1}\text{H} \rightarrow {}^{4}\text{He}$ + energy

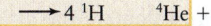

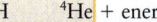

ANIMATION 16.1

Hydrogen fusion also takes place in all of the stars visible to the naked eye. (Fusion follows a different sequence of steps in the most massive stars, but the net result is the same.)

and so tend to repel one another. But in the extreme heat and pressure at the Sun's center, hydrogen nuclei (protons) are moving so fast that they can overcome their electric repulsion and actually touch one another. When that happens, thermonuclear fusion can take place.

ANALOGY You can think of protons as tiny electrically charged spheres that are coated with a very powerful glue. If the spheres are not touching, the repulsion between their charges pushes them apart. But if the spheres are forced into contact, the strength of the glue "fuses" them together.

CAUTION! Be careful not to confuse thermonuclear fusion with the similar-sounding process of *nuclear fission*. In nuclear fusion, energy is released by joining or fusing together nuclei of lightweight atoms such as hydrogen. In nuclear fission, by contrast, the nuclei of very massive atoms such as uranium or plutonium release energy by fragmenting into smaller nuclei. Nuclear power plants produce energy using fission, not fusion. (Generating power using fusion has been a goal of researchers for decades, but no one has yet devised a commercially viable way to do this.)

Converting Hydrogen to Helium

We learned in Section 5-8 that the nucleus of a hydrogen atom (H) consists of a single proton. The nucleus of a helium atom (He) consists of two protons and two neutrons. In the nuclear process that Atkinson described, four hydrogen nuclei combine to form one helium nucleus, with a concurrent release of energy:

$$4\ \mathrm{H} \rightarrow \mathrm{He} + \text{energy}$$

In several separate reactions, two of the four protons are changed into neutrons, and eventually combine with the remaining protons to produce a helium nucleus. This sequence of reactions is called the **proton-proton chain.** The *Cosmic Connections* figure depicts the proton-proton chain in detail.

Each time this process takes place, a small fraction (0.7%) of the initial combined mass of the hydrogen nuclei does not show up in the final mass of the helium nucleus. This "lost" mass is converted into energy. Box 16-1 describes how to use Einstein's mass-energy equation to calculate the amount of energy released.

CAUTION! You may have heard the idea that mass is always conserved (that is, it is neither created nor destroyed), or that energy is always conserved in a reaction. Einstein's ideas show that neither of these statements is quite correct, because mass can be converted into energy and vice versa. A more accurate statement is that the total amount of mass *plus* energy is conserved. Hence, the destruction of mass in the Sun does not violate any laws of nature.

For every four hydrogen nuclei converted into a helium nucleus, 4.3×10^{-12} joule of energy is released. This amount of energy may seem tiny, but it is about 10^7 times larger than the amount of energy released in a typical chemical reaction, such as occurs in ordinary burning. Thus, thermonuclear fusion explains how the Sun could have been shining for billions of years.

To produce the Sun's luminosity of 3.9×10^{26} joules per second, 6×10^{11} kg (600 million metric tons) of hydrogen must be converted into helium each second. This rate is prodigious, but there is a literally astronomical amount of hydrogen in the Sun. In particular, the Sun's core contains enough hydrogen to have been giving off energy at the present rate for as long as the solar system has existed, about 4.56 billion years, and to continue doing so for more than 6 billion years into the future.

LOOKING DEEPER 16.1

The proton-proton chain is also the energy source for many of the stars in the sky. In stars with central temperatures that are much hotter than that of the Sun, however, hydrogen fusion proceeds according to a different set of nuclear reactions, called the **CNO cycle,** in which carbon, nitrogen, and oxygen nuclei absorb protons to produce helium nuclei. Still other thermonuclear reactions, such as helium fusion, carbon fusion, and oxygen fusion, occur late in the lives of many stars.

16-2 A theoretical model of the Sun shows how energy gets from its center to its surface

TUTORIAL 16.2

While thermonuclear fusion is the source of the Sun's energy, this process cannot take place everywhere within the Sun. As we have seen, extremely high temperatures—in excess of 10^7 K—are required for atomic nuclei to fuse together to form larger nuclei. The temperature of the Sun's visible surface, about 5800 K, is too low for these reactions to occur there. Hence, thermonuclear fusion can be taking place only within the Sun's interior. But precisely where does it take place? And how does the energy produced by fusion make its way to the surface, where it is emitted into space in the form of photons?

To answer these questions, we must understand conditions in the Sun's interior. Ideally, we would send an exploratory spacecraft to probe deep into the Sun; in practice, the Sun's intense heat would vaporize even the sturdiest spacecraft. Instead, astronomers use the laws of physics to construct a theoretical model of the Sun. (We discussed the use of models in science in Section 1-1.) Let's see what ingredients go into building a model of this kind.

Hydrostatic Equilibrium

Note first that the Sun is not undergoing any dramatic changes. The Sun is not exploding or collapsing, nor is it significantly heating or cooling. The Sun is thus said to be in both *hydrostatic equilibrium* and *thermal equilibrium*.

To understand what is meant by **hydrostatic equilibrium,** imagine a slab of material in the solar interior (Figure 16-2*a*). In equilibrium, the slab on average will move neither up nor down. (In fact, there are upward and downward motions of material inside the Sun, but these motions average out.) Equilibrium is maintained by a balance among three forces that act on this slab:

1. The downward pressure of the layers of solar material above the slab.

2. The upward pressure of the hot gases beneath the slab.

3. The slab's weight—that is, the downward gravitational pull it feels from the rest of the Sun.

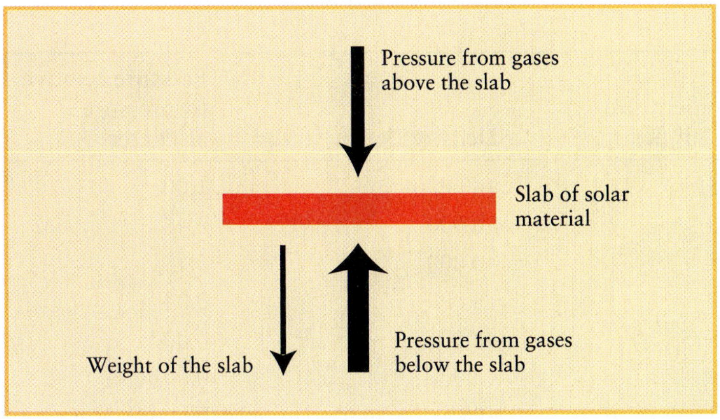

(a) Material inside the sun is in hydrostatic equilibrium, so forces balance

(b) A fish floating in water is in hydrostatic equilibrium, so forces balance

Figure 16-2

Hydrostatic Equilibrium (a) Material in the Sun's interior tends to move neither up nor down. The upward forces on a slab of solar material (due to pressure of gases below the slab) must balance the downward forces (due to the slab's weight and the pressure of gases above the slab). Hence, the pressure must increase with increasing depth. **(b)** The same principle applies to a fish floating in water. In equilibrium, the forces balance and the fish neither rises nor sinks. (Ken Usami/PhotoDisc)

The pressure from below must balance both the slab's weight and the pressure from above. Hence, the pressure below the slab must be greater than that above the slab. In other words, pressure has to increase with increasing depth. For the same reason, pressure increases as you dive deeper into the ocean (Figure 16-2*b*) or as you move toward lower altitudes in our atmosphere.

Hydrostatic equilibrium also tells us about the density of the slab. If the slab is too dense, its weight will be too large and it will sink; if the density is too low, the slab will rise. To prevent this, the density of solar material must have a certain value at each depth within the solar interior. (The same principle applies to objects that float beneath the surface of the ocean. Scuba divers wear weight belts to increase their average density so that they will neither rise nor sink but will stay submerged at the same level.)

Thermal Equilibrium

Another consideration is that the Sun's interior is so hot that it is completely gaseous. Gases compress and become more dense when you apply greater pressure to them, so density must increase along with pressure as you go to greater depths within the Sun. Furthermore, when you compress a gas, its temperature tends to increase, so the temperature must also increase as you move toward the Sun's center.

While the temperature in the solar interior is different at different depths, the temperature at each depth remains constant in time. This principle is called **thermal equilibrium.** For the Sun to be in thermal equilibrium, all the energy generated by thermonuclear reactions in the Sun's core must be transported to the Sun's glowing surface, where it can be radiated into space. If too much energy flowed from the core to the surface to be radiated away, the Sun's interior would cool down; alternatively, the Sun's interior would heat up if too little energy flowed to the surface.

Transporting Energy Outward from the Sun's Core

But exactly how is energy transported from the Sun's center to its surface? There are three methods of energy transport: *conduction, convection,* and *radiative diffusion.* Only the last two are important inside the Sun.

If you heat one end of a metal bar with a blowtorch, energy flows to the other end of the bar so that it too becomes warm. The efficiency of this method of energy transport, called **conduction,** varies significantly from one substance to another. For example, metal is a good conductor of heat, but plastic is not (which is why metal cooking pots often have plastic handles). Conduction is not an efficient means of energy transport in substances with low average densities, including the gases inside stars like the Sun.

Inside stars like our Sun, energy moves from center to surface by two other means: convection and radiative diffusion. **Convection** is the circulation of fluids—gases or liquids—between hot and cool regions. Hot gases (with lower density) rise toward a star's surface, while cool gases (with higher density) sink back down toward the star's center. This physical movement of gases transports heat energy outward in a star, just as the physical movement of water boiling in a pot transports energy from the bottom of the pot (where the heat is applied) to the cooler water at the surface.

In **radiative diffusion,** photons created in the thermonuclear inferno at a star's center diffuse outward toward the star's surface. Individual photons are absorbed and reemitted by atoms and electrons inside the star. The overall result is an outward migration from the hot core, where photons are constantly created, toward the cooler surface, where they escape into space.

Modeling the Sun

To construct a model of a star like the Sun, astrophysicists express the ideas of hydrostatic equilibrium, thermal equilibrium, and energy transport as a set of equations. To ensure that the model applies to the particular star under study, they also make

Table 16-2 A Theoretical Model of the Sun

Distance from the Sun's center (solar radii)	Fraction of luminosity	Fraction of mass	Temperature ($\times 10^6$ K)	Density (kg/m^3)	Pressure relative to pressure at center
0.0	0.00	0.00	15.5	160,000	1.00
0.1	0.42	0.07	13.0	90,000	0.46
0.2	0.94	0.35	9.5	40,000	0.15
0.3	1.00	0.64	6.7	13,000	0.04
0.4	1.00	0.85	4.8	4,000	0.007
0.5	1.00	0.94	3.4	1,000	0.001
0.6	1.00	0.98	2.2	400	0.0003
0.7	1.00	0.99	1.2	80	4×10^{-5}
0.8	1.00	1.00	0.7	20	5×10^{-6}
0.9	1.00	1.00	0.3	2	3×10^{-7}
1.0	1.00	1.00	0.006	0.00030	4×10^{-13}

Note: The distance from the Sun's center is expressed as a fraction of the Sun's radius ($R_\odot$). Thus, 0.0 is at the center of the Sun and 1.0 is at the surface. The fraction of luminosity is that portion of the Sun's total luminosity produced within each distance from the center; this is equal to 1.00 for distances of 0.25 $R_\odot$ or more, which means that all of the Sun's nuclear reactions occur within 0.25 solar radius from the Sun's center. The fraction of mass is that portion of the Sun's total mass lying within each distance from the Sun's center. The pressure is expressed as a fraction of the pressure at the center of the Sun.

use of astronomical observations of the star's surface. (For example, to construct a model of the Sun, they use the data that the Sun's surface temperature is 5800 K, its luminosity is 3.9×10^{26} W, and the gas pressure and density at the surface are almost zero.) The astrophysicists then use a computer to solve their set of equations and calculate conditions layer by layer in toward the star's center. The result is a model of how temperature, pressure, and density increase with increasing depth below the star's surface.

Table 16-2 and Figure 16-3 show a theoretical model of the Sun that was calculated in just this way. Different models of the Sun use slightly different assumptions, but all models give essen-

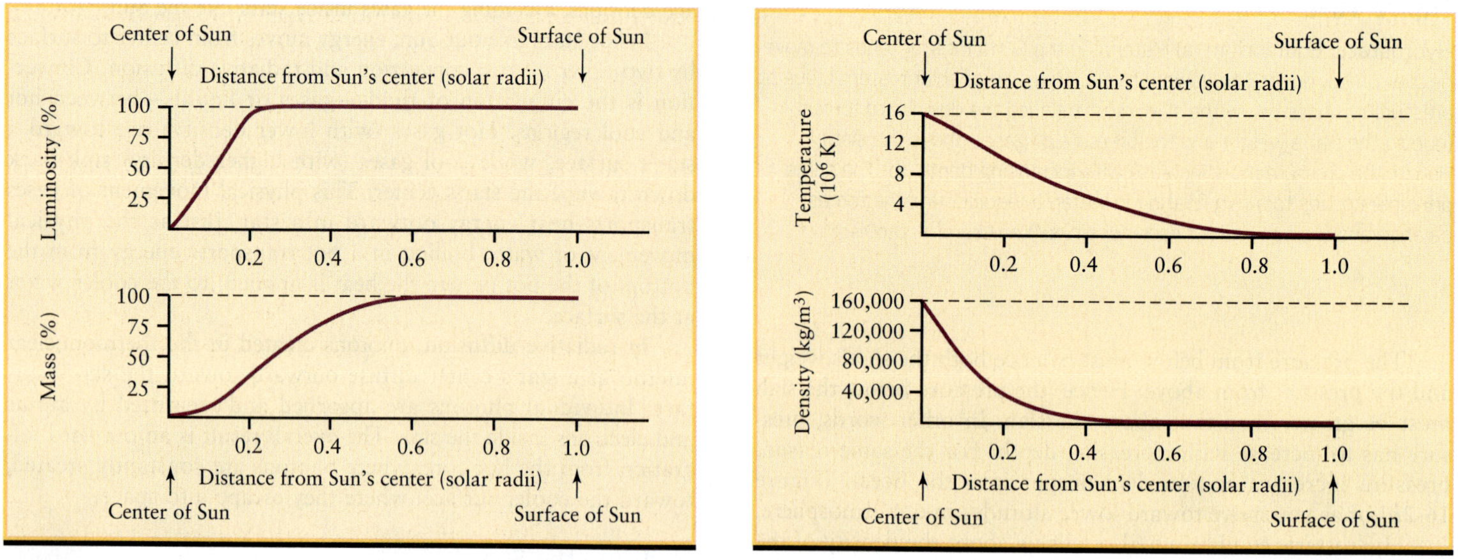

INTERACTIVE EXERCISE 16.1

Figure 16-3

A Theoretical Model of the Sun's Interior These graphs depict what percentage of the Sun's total luminosity is produced within each distance from the center (upper left), what percentage of the total mass lies within each distance from the center (lower left), the temperature at each distance (upper right), and the density at each distance (lower right). (See Table 16-2 for a numerical version of this model.)

tially the same results as those shown here. From such computer models we have learned that at the Sun's center the density is 160,000 kg/m^3 (14 times the density of lead!), the temperature is 1.55×10^7 K, and the pressure is 3.4×10^{11} atm. (One atmosphere, or 1 atm, is the average atmospheric pressure at sea level on Earth.)

Table 16-2 and Figure 16-3 show that the solar luminosity rises to 100% at about one-quarter of the way from the Sun's center to its surface. In other words, the Sun's energy production occurs within a volume that extends out only to 0.25 $R_\odot$. (The symbol $R_\odot$ denotes the solar radius, or radius of the Sun as a whole, equal to 696,000 km.) Outside 0.25 $R_\odot$, the density and temperature are too low for thermonuclear reactions to take place. Also note that 94% of the total mass of the Sun is found within the inner 0.5 $R_\odot$. Hence, the outer 0.5 $R_\odot$ contains only a relatively small amount of material.

How energy flows from the Sun's center toward its surface depends on how easily photons move through the gas. If the solar gases are comparatively transparent, photons can travel moderate distances before being scattered or absorbed, and energy is thus transported by radiative diffusion. If the gases are comparatively *opaque*, photons are frequently scattered or absorbed and can't easily get through the gas. In an opaque gas, heat builds up and convection then becomes the most efficient means of energy transport. The gases start to churn, with hot lower-density moving upward and cooler gas sinking downward.

From the center of the Sun out to about 0.71 $R_\odot$, energy is transported by radiative diffusion. Hence, this region is called the **radiative zone.** Beyond about 0.71 $R_\odot$, the temperature is low enough (a mere 2×10^6 K or so) for electrons and hydrogen nuclei to join into hydrogen atoms. These atoms are very effective at absorbing photons, much more so than free electrons or nuclei, and this absorption chokes off the outward flow of photons. Therefore, beyond about 0.71 $R_\odot$, radiative diffusion is not an effective way to transport energy. Instead, convection dominates the energy flow in this outer region, which is why it is called the **convective zone.** Figure 16-4 shows these aspects of the Sun's internal structure.

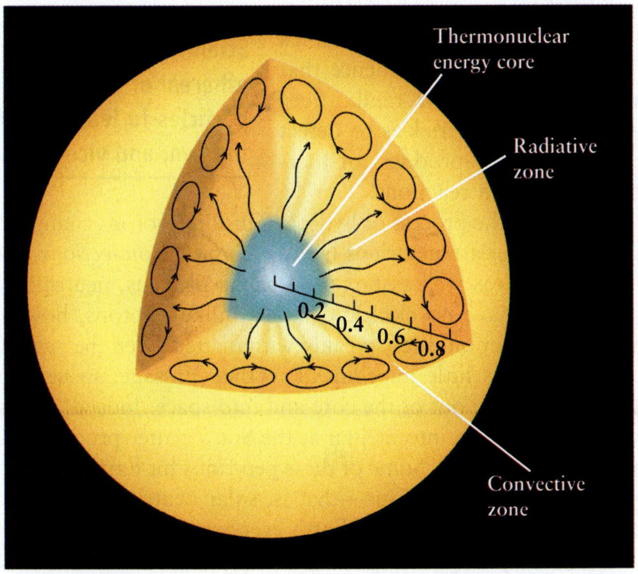

Figure 16-4

The Sun's Internal Structure Thermonuclear reactions occur in the Sun's core, which extends out to a distance of 0.25 $R_\odot$ from the center. Energy is transported outward, via radiative diffusion, to a distance of about 0.71 $R_\odot$. In the outer layers between 0.71 $R_\odot$ and 1.00 $R_\odot$, energy flows outward by convection.

Although energy travels through the radiative zone in the form of photons, the photons have a difficult time of it. Table 16-2 shows that the material in this zone is extremely dense, so photons from the Sun's core take a long time to diffuse through the radiative zone. As a result, it takes approximately 170,000 years for energy created at the Sun's center to travel 696,000 km to the solar surface and finally escape as sunlight. The energy flows outward at an average rate of 50 centimeters per hour, or about 20 times slower than a snail's pace.

Once the energy escapes from the Sun, it travels much faster—at the speed of light. Thus, solar energy that reaches you today took only 8 minutes to travel the 150 million kilometers from the Sun's surface to Earth. But this energy was actually produced by thermonuclear reactions that took place in the Sun's core hundreds of thousands of years ago.

The sunlight that reaches Earth today results from thermonuclear reactions that took place about 170,000 years ago

16-3 Astronomers probe the solar interior using the Sun's own vibrations

WEB LINK 16.2

We have described how astrophysicists construct models of the Sun. But since we cannot see into the Sun's opaque interior, how can we check these models to see if they are accurate? What is needed is a technique for probing the Sun's interior. A very powerful technique of just this kind involves measuring vibrations of the Sun as a whole. This field of solar research is called **helioseismology.**

Vibrations are a useful tool for examining the hidden interiors of all kinds of objects. Food shoppers test whether melons are ripe by tapping on them and listening to the vibrations. Geologists can determine the structure of Earth's interior by using seismographs to record vibrations during earthquakes.

Although there are no true "sunquakes," the Sun does vibrate at a variety of frequencies, somewhat like a ringing bell. These vibrations were first noticed in 1960 by Robert Leighton of the California Institute of Technology, who made high-precision Doppler shift observations of the solar surface. These measurements revealed that parts of the Sun's surface move up and down about 10 meters every 5 minutes. Since the mid-1970s, several astronomers have reported slower vibrations, having periods ranging from 20 to 160 minutes. The detection of extremely slow vibrations has inspired astronomers to organize networks of telescopes around and in orbit above Earth to monitor the Sun's vibrations on a continuous basis.

WEB LINK 16.3

The vibrations of the Sun's surface can be compared with sound waves. If you could somehow survive within the Sun's outermost layers, you would first notice a deafening roar, somewhat like a jet engine, produced by turbulence in the Sun's gases. But superimposed on this noise would be a variety of nearly pure tones. You would need greatly enhanced hearing to detect these tones, however; the strongest has a frequency of just 0.003 hertz, 13 octaves below the lowest frequency audible to humans. (Recall from Section 5-2 that one hertz is one oscillation per second.)

In 1970, Roger Ulrich, at UCLA, pointed out that sound waves moving upward from the solar interior would be reflected back inward after reaching the solar surface. However, as a reflected sound wave descends back into the Sun, the increasing density and pressure bend the wave severely, turning it around and aiming it back toward the solar surface. In other words, sound waves bounce back and forth between the solar surface and layers deep within the Sun. These sound waves can reinforce each other if their wavelength is the right size, just as sound waves of a particular wavelength resonate inside an organ pipe.

The Sun oscillates in millions of ways as a result of waves resonating in its interior. Figure 16-5 is a computer-generated illustration of one such mode of vibration. Helioseismologists can deduce information about the solar interior from measurements of these oscillations. For example, they have been able to set limits on the amount of helium in the Sun's core and convective zone and to determine the thickness of the transition region between the radiative zone and convective zone. They have also found that the convective zone is thicker than previously thought.

The Sun's oscillations resemble those inside an organ pipe, but are much more complex

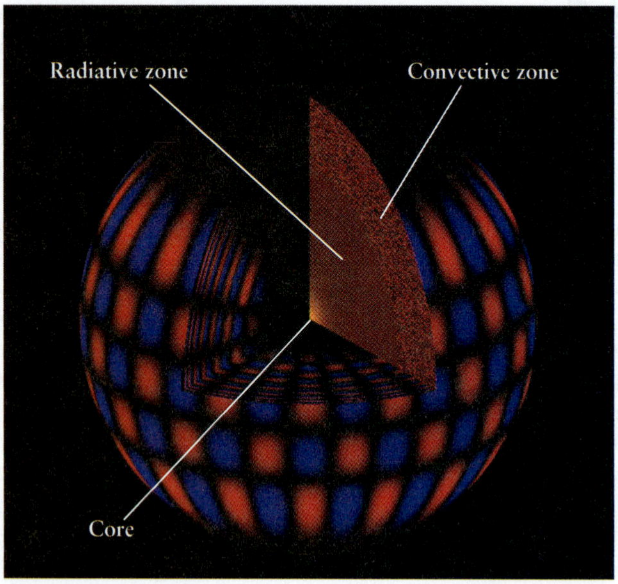

Figure 16-5

A Sound Wave Resonating in the Sun This computer-generated image shows one of the millions of ways in which the Sun's interior vibrates. The regions that are moving outward are colored blue, those moving inward, red. As the cutaway shows, these oscillations are thought to extend into the Sun's radiative zone (compare Figure 16-4). (National Solar Observatory)

16-4 Neutrinos reveal information about the Sun's core—and have surprises of their own

We have seen circumstantial evidence that thermonuclear fusion is the source of the Sun's power. To be certain, however, we need more definitive evidence. How can we show that thermonuclear fusion really is taking place in the Sun's core?

What Sunlight Cannot Tell Us

Although the light energy that we receive from the Sun originates in the core, it provides few clues about conditions there. The problem is that this energy has changed form repeatedly during its passage from the core: It appeared first as photons diffusing through the radiative zone, then as heat transported through the outer layers by convection, and then again as photons emitted from the Sun's glowing surface. As a result of these transformations, much of the information that the Sun's radiated energy once carried about conditions in the core has been lost.

ANALOGY If you make a photocopy of a photocopy of a photocopy of an original document, the final result may be so blurred as to be unreadable. In an analogous way, because solar energy is transformed many times while en route through the Sun, the story it could tell us about the Sun's core is hopelessly blurred.

Helioseismology also cannot reveal what is happening in the core. The vibrations that astronomers see on the Sun's surface do not penetrate that far into the interior.

Solar Neutrinos

Happily, there is a way for scientists to learn about conditions in the Sun's core and to get direct evidence that thermonuclear fusion really does happen there. The trick is to detect the subatomic by-products of thermonuclear fusion reactions.

Scientists use the most ethereal of subatomic particles to learn about the Sun, and vice versa

As part of the process of hydrogen fusion, protons change into neutrons and release **neutrinos** (see the *Cosmic Connections* figure in Section 16-1 as well as Box 16-1). Like photons, neutrinos are particles that have no electric charge. Unlike photons, however, neutrinos interact only very weakly with matter. Even the vast bulk of the Sun offers little impediment to their passage, so neutrinos must be streaming out of the core and into space. Indeed, the conversion of hydrogen into helium at the Sun's center produces 10^{38} neutrinos each second. Some of these neutrinos hit Earth and about 10^{14} neutrinos from the Sun—that is, **solar neutrinos**—must pass through each square meter of Earth every second.

If it were possible to detect these solar neutrinos, we would have direct evidence that thermonuclear reactions really do take place in the Sun's core. Beginning in the 1960s, scientists began to build neutrino detectors for precisely this purpose.

The challenge is that neutrinos are exceedingly difficult to detect. Just as neutrinos pass unimpeded through the Sun, they also pass through Earth almost as if it were not there. We stress the word "almost," because neutrinos can and do interact with matter, albeit infrequently.

On rare occasions a neutrino will strike a neutron and convert it into a proton. This effect was the basis of the original solar neutrino detector, designed and built by Raymond Davis of the Brookhaven National Laboratory in the 1960s. This device used 100,000 gallons of perchloroethylene (C_2Cl_4), a fluid used in dry cleaning, in a huge tank buried deep underground. Most of the solar neutrinos that entered Davis's tank passed right through it with no effect whatsoever. But occasionally a neutrino struck the nucleus of one of the chlorine atoms (^{37}Cl) in the cleaning fluid and converted one of its neutrons into a proton, creating a radioactive atom of argon (^{37}Ar).

The rate at which argon is produced is related to the neutrino flux—that is, the number of neutrinos from the Sun arriving at Earth per square meter per second. By counting the number of newly created argon atoms, Davis was able to determine the neutrino flux from the Sun. (Other subatomic particles besides neutrinos can also induce reactions that create radioactive atoms. By placing the experiment deep underground, however, the body of Earth absorbs essentially all such particles—with the exception of neutrinos.)

The Solar Neutrino Problem

Davis and his collaborators found that solar neutrinos created one radioactive argon atom in the tank every three days. But this rate corresponded to only one-third of the neutrino flux predicted from standard models of the Sun. This troubling discrepancy between theory and observation, called the **solar neutrino problem,** motivated scientists around the world to conduct further experiments to measure solar neutrinos.

One key question was whether the neutrinos that Davis had detected had really come from the Sun. (The Davis experiment had no way to determine the direction from which neutrinos had entered the tank of cleaning fluid.) The direction was resolved by an experiment in Japan called Kamiokande, which was designed by the physicist Masatoshi Koshiba. A large underground tank containing 3000 tons of water was surrounded by 1100 light detectors. From time to time, a high-energy solar neutrino struck an electron in one of the water molecules, dislodging it and sending it flying like a pin hit by a bowling ball. The recoiling electron produced a streaking flash of light, which was sensed by the detectors. By analyzing the flashes, scientists could tell the direction from which the neutrinos were coming and confirmed that they emanated from the Sun. These results in the late 1980s gave direct evidence that thermonuclear fusion is indeed occurring in the Sun's core. (Davis and Koshiba both received the 2002 Nobel Prize in Physics for their pioneering research on solar neutrinos.)

Like Davis's experiment, however, Koshiba and his colleagues at Kamiokande detected only a fraction of the expected flux of neutrinos. Where, then, were the missing solar neutrinos?

One proposed solution had to do with the energy of the detected neutrinos. The vast majority of neutrinos from the Sun are created during the first step in the proton-proton chain, in which two protons combine to form a heavy isotope of hydrogen (see the *Cosmic Connections* figure in Section 16-1). But these neutrinos have too little energy to convert chlorine into argon. Both Davis's and Koshiba's experiments responded only to high-energy neutrinos produced by reactions that occur only part of the time near the end of the proton-proton chain. (The *Cosmic Connections* figure in Section 16-1 does not show these reactions.) Could it be that the discrepancy between theory and observation would go away if the flux of low-energy neutrinos could be measured?

To test this idea, two teams of physicists constructed neutrino detectors that used several tons of gallium (a liquid metal) rather than cleaning fluid. Low-energy neutrinos convert gallium (^{71}Ga) into a radioactive isotope of germanium (^{71}Ge). By chemically separating the germanium from the gallium and counting the radioactive atoms, the physicists were able to measure the flux of low-energy solar neutrinos. These experiments—GALLEX in Italy and SAGE (Soviet-American Gallium Experiment) in Russia—detected only 50% to 60% of the expected neutrino flux. Hence, the solar neutrino problem was a discrepancy between theory and observation for neutrinos of all energies.

Another proposed solution to the neutrino problem was that the Sun's core is cooler than predicted by solar models. If the Sun's central temperature were only 10% less than the current estimate, fewer neutrinos would be produced and the neutrino flux would agree with experiments. However, a lower central temperature would cause other obvious features, such as the Sun's size and surface temperature, to be different from what we observe.

Finding the Missing Neutrinos

Only very recently has the solution to the neutrino problem been found. The answer lies not in how neutrinos are produced, but rather in what happens to them between the Sun's core and detectors on Earth. Motivated by the lack of observed solar neutrinos, physicists have found that there are actually three types of neutrinos. Only one of these types is produced in the Sun, and it is only this type that can be detected by the experiments we have described. But if some of the solar neutrinos change *in flight* into a different type of neutrino, the detectors in these experiments would record only a fraction of the total neutrino flux. This effect, where one type of neutrino can spontaneously transform into another type, is called **neutrino oscillation.** In June 1998, scientists at the Super-Kamiokande neutrino observatory (a larger and more sensitive device than Kamiokande) revealed evidence that neutrino oscillation does indeed take place.

The best confirmation of this idea has come from the Sudbury Neutrino Observatory (SNO) in Canada. Like Kamiokande and Super-Kamiokande, SNO uses a large tank of water placed deep underground (Figure 16-6). But unlike those earlier experiments, SNO can detect all of the three types of neutrinos. It detects neurinos by using *heavy water*. In ordinary, or "light," water, each hydrogen atom in the H_2O molecule has a solitary proton as its nucleus. In heavy water, by contrast, each of the hydrogen nuclei has a nucleus made up of a proton and a neutron. (This isotope ^2H is shown in the *Cosmic Connections* figure in Section 16-1.) If a high-energy solar neutrino of any type passes through SNO's tank of heavy water, it can knock the neutron out

Figure 16-6 R I V U X G

The Sudbury Neutrino Observatory Under Construction The transparent acrylic sphere holds 1000 tons of heavy water. Any of the three types of solar neutrino produces a flash of light when it interacts with the heavy water. The flash is sensed by 9600 light detectors surrounding the tank. (The detectors were not all installed when this photograph was taken.) (Photo courtesy of SNO)

of one of the ^2H nuclei. The ejected neutron can then be captured by another nucleus, and this capture releases energy that manifests itself as a tiny burst of light. As in Kamiokande, detectors around SNO's water tank record these light flashes.

Scientists using SNO have found that the combined flux of all three types of neutrinos coming from the Sun is *equal* to the theoretical prediction. Together with the results from earlier neutrino experiments, this result strongly suggests that the Sun is indeed producing neutrinos at the predicted rate as a by-product of thermonuclear reactions. But before these neutrinos can reach Earth, about two-thirds of them undergo spontaneous oscillation and change their type. Thus, there is really no solar neutrino problem (no lack of neutrinos)—scientists merely needed the right kind of detectors to observe all the neutrinos, including the ones that transformed in flight.

The story of the solar neutrino problem illustrates how two different branches of science—in this case, studies of the solar interior and investigations of subatomic particles—can sometimes interact, to the mutual benefit of both. While there is still much we do not understand about the Sun and about neutrinos, a new generation of neutrino detectors in Japan, Canada, and elsewhere promises to further our knowledge of these exotic realms of astronomy and physics.

16-5 The photosphere is the lowest of three main layers in the Sun's atmosphere

Although the Sun's core is hidden from our direct view, we can easily see sunlight coming from the high-temperature gases that make up the Sun's atmosphere. These outermost layers of the Sun prove to be the sites of truly dramatic activity, much of which has a direct impact on our planet. By studying these layers, we gain further insight into the character of the Sun as a whole.

Observing the Photosphere

A visible-light photograph like Figure 16-7 makes it appear that the Sun has a definite surface. This surface is actually an illusion; the Sun is gaseous throughout its volume because of its high internal temperature, and the gases simply become less and less dense as you move farther away from the Sun's center.

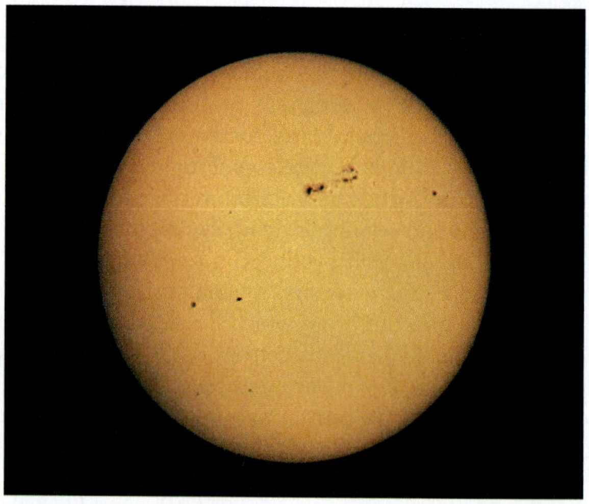

Figure 16-7 R I V U X G

The Photosphere The photosphere is the layer in the solar atmosphere from which the Sun's visible light is emitted. Note that the Sun appears darker around its limb, or edge; here we are seeing the upper photosphere, which is relatively cool and thus glows less brightly. (The dark sunspots, which we discuss in Section 16-8, are also relatively cool regions.) (Celestron International)

Why, then, does the Sun appear to have a sharp, well-defined surface? The reason is that essentially all of the Sun's visible light emanates from a single, thin layer of gas called the **photosphere** ("sphere of light"). Just as you can see only a certain distance through Earth's atmosphere before objects vanish in the haze, we can see only about 400 km into the photosphere. This distance is so small compared with the Sun's radius of 696,000 km that the photosphere appears to be a definite surface.

The photosphere is actually the lowest of the three layers that together constitute the solar atmosphere. Above it are the *chromosphere* and the *corona,* both of which are transparent to visible light. We can see them only using special techniques, which we discuss later in this chapter. Everything below the photosphere is called the *solar interior.*

The Lesson of Limb Darkening

The photosphere is heated from below by energy streaming outward from the solar interior. Hence, temperature should decrease as you go upward in the photosphere, just as in the solar interior (see Table 16-2 and Figure 16-3). We know this is the case because the photosphere appears darker around the edge, or *limb,* of the Sun than it does toward the center of the solar disk, an effect called **limb darkening** (examine Figure 16-7). This happens because when we look near the Sun's limb, we do not see as deeply into the photosphere as we do when we look near the center of the disk (Figure 16-8). The high-altitude gas we observe at the limb is not as hot and thus does not glow as brightly as the deeper, hotter gas seen near the disk center.

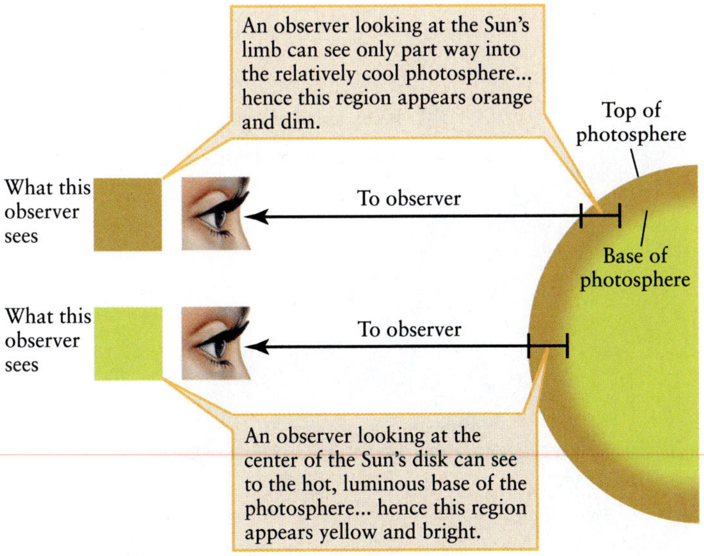

An observer looking at the Sun's limb can see only part way into the relatively cool photosphere... hence this region appears orange and dim.
Top of photosphere
What this observer sees
To observer
Base of photosphere
What this observer sees
To observer
An observer looking at the center of the Sun's disk can see to the hot, luminous base of the photosphere... hence this region appears yellow and bright.

Figure 16-8

The Origin of Limb Darkening Light from the Sun's limb and light from the center of its disk both travel about the same straight-line distance through the photosphere to reach us. Because of the Sun's curved shape, light from the limb comes from a greater height within the photosphere, where the temperature is lower and the gases glow less brightly. Hence, the limb appears darker and more orange.

The spectrum of the Sun's photosphere confirms how its temperature varies with altitude. As we saw in Section 16-1, the photosphere shines like a nearly perfect blackbody with an average temperature of about 5800 K. However, superimposed on this spectrum are many dark absorption lines (see Figure 5-12). As discussed in Section 5-6, we see an *absorption line spectrum* of this sort whenever we view a hot, glowing object through a relatively cool gas. In this case, the hot object is the lower part of the photosphere; the cooler gas is in the upper part of the photosphere, where the temperature declines to about 4400 K. All the absorption lines in the Sun's spectrum are produced in this relatively cool layer, as atoms selectively absorb photons of various wavelengths streaming outward from the hotter layers below.

CAUTION! You may find it hard to think of 4400 K as "cool." But keep in mind that the ratio of 4400 K to 5800 K, the temperature in the lower photosphere, is the same as the ratio of the temperature on a Siberian winter night to that of a typical day in Hawaii.

Granules and Supergranules in the Photosphere

We can learn still more about the photosphere by examining it with a telescope—but only when using special dark filters to prevent eye damage. ***Looking directly at the Sun without the correct filter, whether with the naked eye or with a telescope, can cause permanent blindness!*** Under good observing conditions, astronomers using such filter-equipped telescopes can often see a blotchy pattern in the photosphere, called **granulation** (Figure 16-9). Each light-colored **granule** measures about 1000 km (600 mi) across—equal in size to the areas of Texas and Oklahoma combined—and is surrounded by a darkish boundary. The difference in brightness between the center and the edge of a granule corresponds to a temperature drop of about 300 K.

Granulation is caused by convection of the gas in the photosphere. The inset in Figure 16-9 shows how hot gas from lower levels rises upward in granules, cools off, spills over the edges of the granules, and then plunges back down into the Sun. Convection can occur only if the gas is heated from below, like a pot of water being heated on a stove (see Section 16-2). Along with limb darkening and the Sun's absorption line spectrum, granulation shows that the upper part of the photosphere must be cooler than the lower part.

Like water boiling on a stove, the photosphere bubbles with convection cells

Time-lapse photography reveals more of the photosphere's dynamic activity. Granules form, disappear, and reform in cycles lasting only a few minutes. At any one time, about 4 million granules cover the solar surface.

Superimposed on the pattern of granulation are even larger convection cells called **supergranules** (Figure 16-10). As in granules, gases rise upward in the middle of a supergranule, move horizontally outward toward its edge, and descend back into the Sun. The difference is that a typical supergranule is about 35,000 km in diameter, large enough to enclose several hundred granules. This large-scale convection moves at only about 0.4 km/s (1400 km/h, or 900 mi/h), about one-tenth the speed of gases churning in a regular-sized granule. A given supergranule lasts about a day.

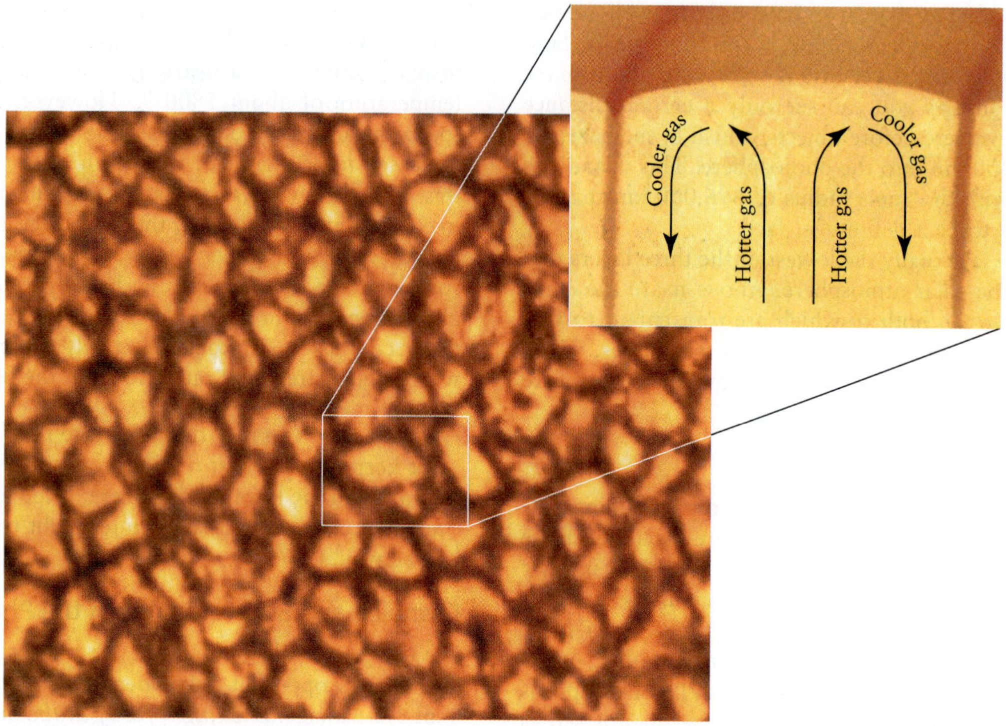

Figure 16-9 R I V U X G

VIDEO 16.1 VIDEO 16.2

Solar Granulation High-resolution photographs of the Sun's surface reveal a blotchy pattern called granulation. Granules are convection cells about 1000 km (600 mi) wide in the Sun's photosphere. **Inset:** Rising hot gas produces bright granules. Cooler gas sinks downward along the boundaries between granules; this gas glows less brightly, giving the boundaries their dark appearance. This convective motion transports heat from the Sun's interior outward to the solar atmosphere. (MSFC/NASA; inset: Goran Scharmer, Lund Observatory)

ANALOGY Similar patterns of large-scale and small-scale convection can be found in Earth's atmosphere. On the large scale, air rises gradually at a low-pressure area, then sinks gradually at a high-pressure area, which might be hundreds of kilometers away. This pattern is analogous to the flow in a supergranule. Thunderstorms in our atmosphere are small but intense convection cells within which air moves rapidly up and down. Like granules, they last only a relatively short time before they dissipate.

The Photosphere: Hot, Thin, and Opaque

Although the photosphere is a very active place, it actually contains relatively little material. Careful examination of the spectrum shows that it has a density of only about 10^{-4} kg/m^3,

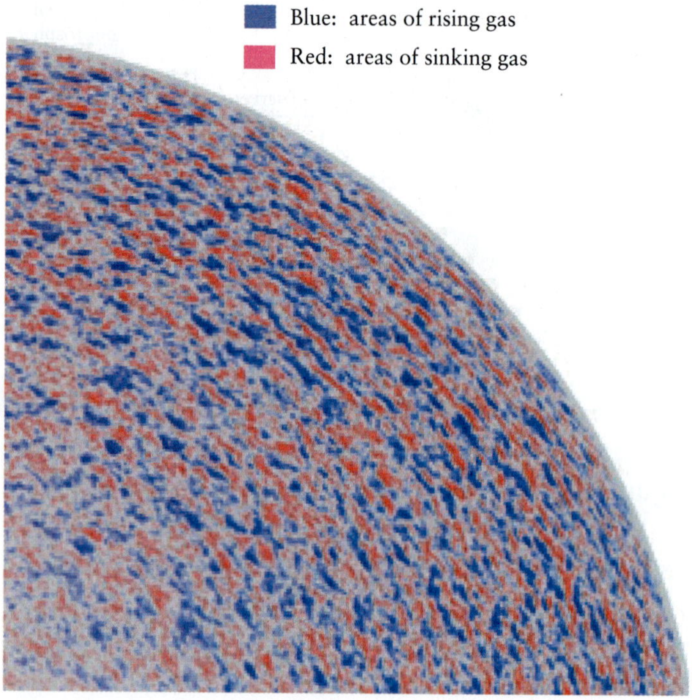

Figure 16-10 R I V U X G

Supergranules and Large-Scale Convection Supergranules display relatively little contrast between their center and edges, so they are hard to observe in ordinary images. But they can be seen in a false-color Doppler image like this one. Light from gas that is approaching us (that is, rising) is shifted toward shorter wavelengths, while light from receding gas (that is, descending) is shifted toward longer wavelengths (see Section 5-9). (David Hathaway, MSFC/NASA) roughly 0.01% the density of Earth's atmosphere at sea level. The photosphere is made primarily of hydrogen and helium, the most abundant elements in the solar system (see Figure 8-3).

Despite being such a thin gas, the photosphere is surprisingly opaque to visible light. If it were not so opaque, we could see into the Sun's interior to a depth of hundreds of thousands of kilometers, instead of the mere 400 km that we can see down into the photosphere. What makes the photosphere so opaque is that its hydrogen atoms sometimes acquire an extra electron, becoming **negative hydrogen ions.** The extra electron is only loosely attached and can be dislodged if it absorbs a photon of any visible wavelength. Hence, negative hydrogen ions are very efficient light absorbers, and there are enough of these light-absorbing ions in the photosphere to make it quite opaque. Because it is so opaque, the photosphere's spectrum is close to that of an ideal blackbody.

16-6 Spikes of rising gas extend through the Sun's chromosphere

An ordinary visible-light image such as Figure 16-7 gives the impression that the Sun ends at the top of the photosphere. But during a total solar eclipse, the Moon blocks the photosphere from our view, revealing a glowing, pinkish layer of gas above the photosphere (Figure 16-11). This layer is the tenuous **chromosphere** ("sphere of color"), the second of the three major levels in the Sun's atmosphere. The chromosphere is only about one ten-thousandth (10^{-4}) as dense as the photosphere, or about 10^{-8} as dense as our own atmosphere. No wonder it is normally invisible!

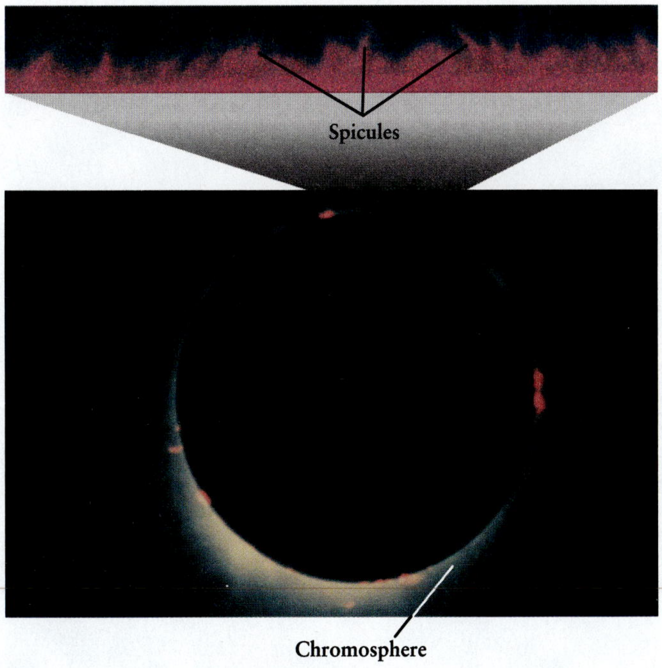

Figure 16-11 R I V U X G

The Chromosphere During a total solar eclipse, the Sun's glowing chromosphere can be seen around the edge of the Moon. It appears pinkish because its hot gases emit light at only certain discrete wavelengths, principally the H_α emission of hydrogen at a red wavelength of 656.3 nm. The expanded area above shows spicules, jets of chromospheric gas that surge upward into the Sun's outer atmosphere. (NOAO)

Comparing the Chromosphere and Photosphere

Unlike the photosphere, which has an absorption line spectrum, the chromosphere has a spectrum dominated by emission lines. An emission line spectrum is produced by the atoms of a hot, thin gas (see Section 5-6 and Section 5-8). As their electrons fall from higher to lower energy levels, the atoms emit photons.

One of the strongest emission lines in the chromosphere's spectrum is the H_α line at 656.3 nm, which is emitted by a hydrogen atom when its single electron falls from the $n = 3$ level to the $n = 2$ level (recall Figure 5-21*b*). This wavelength is in the red part of the spectrum, which gives the chromosphere its characteristic pinkish color. The spectrum also contains emission lines of singly ionized calcium, as well as lines due to ionized helium and ionized metals. In fact, helium was originally discovered in the chromospheric spectrum in 1868, almost 30 years before helium gas was first isolated on Earth.

Analysis of the chromospheric spectrum shows that temperature *increases* with increasing height in the chromosphere. This trend is just the opposite of the situation in the photosphere, where temperature decreases with increasing height. The temperature is about 4400 K at the top of the photosphere; 2000 km higher, at the top of the chromosphere, the temperature is nearly 25,000 K. This temperature difference is very surprising, since temperatures should decrease as you move away from the Sun's interior. In Section 16-10 we will see how solar scientists explain this seeming paradox.

WEB LINK 16.8

The photospheric spectrum is dominated by absorption lines at certain wavelengths, while the spectrum of the chromosphere has emission lines at these same wavelengths. In other words, the photosphere appears dark at the wavelengths at which the chromosphere emits most strongly, such as the H_α wavelength of 656.3 nm. By viewing the Sun through a special filter that is transparent to light only at the wavelength of H_α, astronomers can screen out light from the photosphere and make the chromosphere visible. (The same technique can be used with other wavelengths at which the chromosphere emits strongly, including nonvisible wavelengths.) Filters make it possible to see the chromosphere at any time, not just during a solar eclipse.

Spicules

The top photograph in Figure 16-11 is a high-resolution image of the Sun's chromosphere taken through an H_α filter. This image shows numerous vertical spikes, which are actually jets of rising gas called **spicules.** A typical spicule lasts just 15 minutes or so: It rises at the rate of about 20 km/s (72,000 km/h, or 45,000 mi/h), can reach a height of several thousand kilometers, and then collapses and fades away (Figure 16-12). Approximately 300,000 spicules exist at any one time, covering about 1% of the Sun's surface.

Jets of gas thousands of kilometers in height rise through the chromosphere

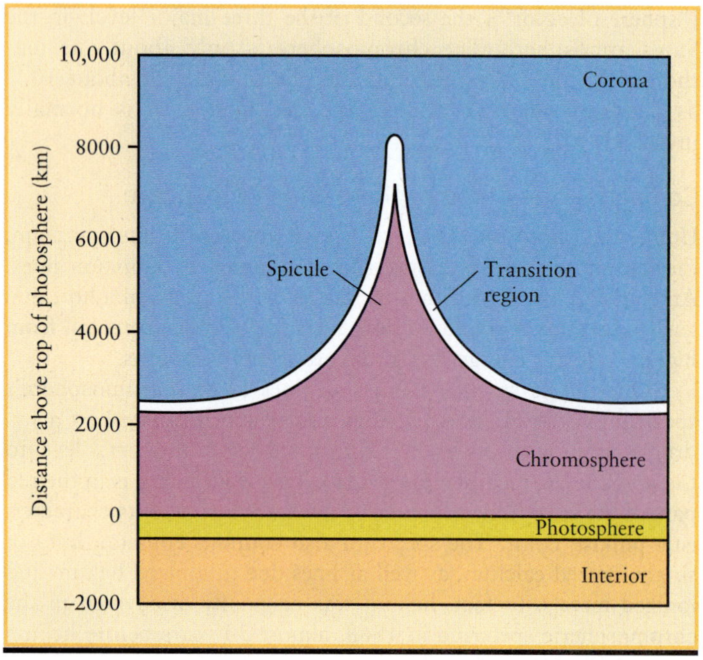

INTERACTIVE EXERCISE 16.2

Figure 16-12

The Solar Atmosphere This schematic diagram shows the three layers of the solar atmosphere. The lowest, the photosphere, is about 400 km thick. The chromosphere extends about 2000 km higher, with spicules jutting up to nearly 10,000 km above the photosphere. Above a transition region is the Sun's outermost layer, the corona, which we discuss in Section 16-7. It extends many millions of kilometers out into space. (Adapted from J. A. Eddy)

Spicules are generally located directly above the edges of supergranules (see Figure 16-10). This result is surprising, because chromospheric gases are rising in a spicule while photospheric gases are *descending* at the edge of a supergranule. What, then, is pulling gases upward to form spicules? As we will see in Section 16-10, the answer proves to be the Sun's intense magnetic field. But before we delve into how this happens, let us complete our tour of the solar atmosphere by exploring its outermost, least dense, most dynamic, and most bizarre layer—the region called the corona.

16-7 The corona ejects mass into space to form the solar wind

The **corona,** or outermost region of the Sun's atmosphere, begins at the top of the chromosphere. It extends out to a distance of several million kilometers. Despite its tremendous extent, the corona is only about one-millionth (10^{-6}) as bright as the photosphere—no brighter than the full moon. Hence, the corona can be viewed only when the light from the photosphere is blocked out, either by use of a specially designed telescope or during a total solar eclipse.

Figure 16-13 is an exceptionally detailed photograph of the Sun's corona taken during a solar eclipse. It shows that the corona is not merely a spherical shell of gas surrounding the Sun. Rather, numerous streamers extend in different directions far above the solar surface. The shapes of these streamers vary on time scales of days or weeks. (For another view of the corona during a solar eclipse, see Figure 3-10*b*.)

Comparing the Corona, Chromosphere, and Photosphere

Like the chromosphere that lies below it, the corona has an emission line spectrum characteristic of a hot, thin gas. When the spectrum of the corona was first measured in the nineteenth century, astronomers found a number of emission lines at wavelengths that had never been seen in the laboratory. Their explanation was that the corona contained elements that had not yet been detected on Earth. However, laboratory experiments in the 1930s revealed that these unusual emission lines were in fact caused by the same atoms found elsewhere in the universe—but in highly ionized states. For example, a prominent green line at 530.3 nm is caused by highly ionized iron atoms, each of which has been stripped of 13 of its 26 electrons. In order to strip that many electrons from atoms, temperatures in the corona must reach 2 million kelvins (2×10^6 K) or even higher—far greater than the temperatures in the chromosphere. Figure 16-14 shows how temperature in both the chromosphere and corona varies with altitude.

CAUTION! The corona is actually not very "hot"—that is, it contains very little thermal energy. The reason is that the corona is nearly a vacuum. In the corona there are only about 10^{11} atoms

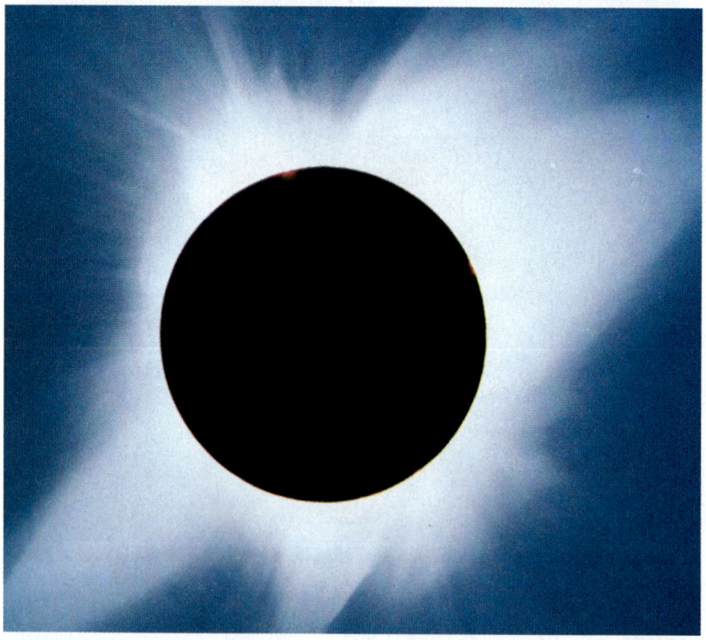

Figure 16-13 R I V U X G

The Solar Corona This striking photograph of the corona was taken during the total solar eclipse of July 11, 1991. Numerous streamers extend for millions of kilometers above the solar surface. The unearthly light of the corona is one of the most extraordinary aspects of experiencing a solar eclipse. (Courtesy of R. Christen and M. Christen, Astro-Physics, Inc.)

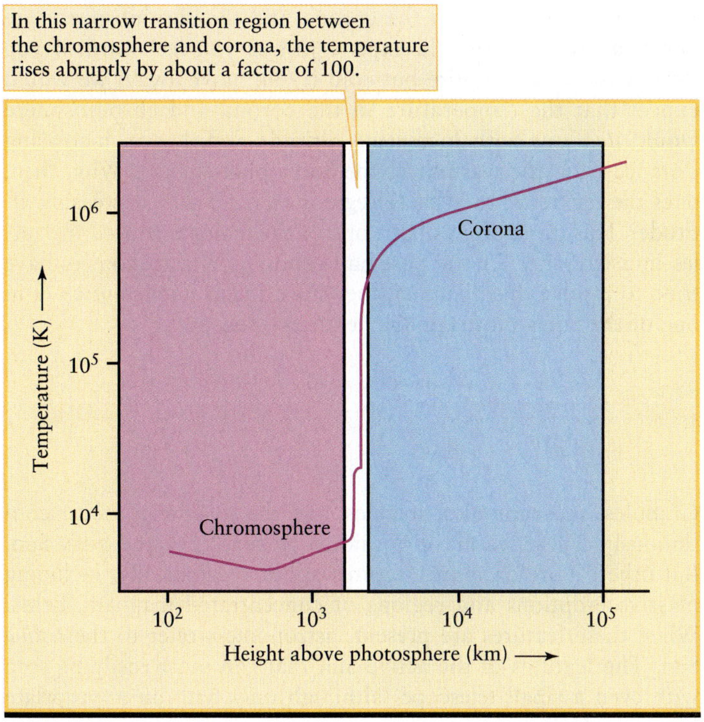
In this narrow transition region between the chromosphere and corona, the temperature rises abruptly by about a factor of 100.
Corona
Chromosphere
Temperature (K)
10^6
10^5
10^4
10^2
10^3
10^4
10^5
Height above photosphere (km)

Figure 16-14

Temperatures in the Sun's Upper Atmosphere This graph shows how temperature varies with altitude in the Sun's chromosphere and corona and in the narrow transition region between them. In order to show a large range of values, both the vertical and horizontal scales are nonlinear. (Adapted from A. Gabriel)

per cubic meter, compared with about 10^{23} atoms per cubic meter in the Sun's photosphere and about 10^{25} atoms per cubic meter in the air that we breathe. Because of the corona's high temperature, the atoms there are moving at very high speeds. But because there are so few atoms in the corona, the total amount of energy in these moving atoms (a measure of how "hot" the gas is) is rather low. If you flew a spaceship into the corona, you would have to worry about becoming overheated by the intense light coming from the photosphere, but you would notice hardly any heating from the corona's ultra-thin gas.

ANALOGY The situation in the corona is similar to that inside a conventional oven that is being used for baking. Both the walls of the oven and the air inside the oven are at the same high temperature, but the air contains very few atoms and thus carries little energy. If you put your hand in the oven momentarily, the lion's share of the heat you feel is radiation from the oven walls.

The low density of the corona explains why it is so dim compared with the photosphere. In general, the higher the temperature of a gas, the brighter it glows. But because there are so few atoms in the corona, the net amount of light that it emits is very feeble compared with the light from the much cooler, but also much denser, photosphere.

The Solar Wind and Coronal Holes

Earth's gravity keeps our atmosphere from escaping into space. In the same way, the Sun's powerful gravitational attraction keeps most of the gases of the photosphere, chromosphere, and corona from escaping. But the corona's high temperature means that its atoms and ions are moving at very high speeds, around a million kilometers per hour. As a result, some of the coronal gas can and does escape. This outflow of gas, which we first encountered in Section 8-5, is called the **solar wind.**

Each second the Sun ejects about a million tons (10^9 kg) of material into the solar wind. But the Sun is so massive that, even over its entire lifetime, it will eject only a few tenths of a percent of its total mass. The solar wind is composed almost entirely of electrons and nuclei of hydrogen and helium. About 0.1% of the solar wind is made up of ions of more massive atoms, such as silicon, sulfur, calcium, chromium, nickel, iron, and argon. The aurorae seen at far northern or southern latitudes on Earth are produced when electrons and ions from the solar wind enter our upper atmosphere.

Special telescopes enable astronomers to see the origin of the solar wind. To appreciate what sort of telescopes are needed, note that because the temperature of the coronal gas is so high, ions in the corona are moving very fast (see Box 7-2). When ions collide, the energy of the impact is so great that the ion's electrons are boosted to very high energy levels. As the electrons fall back to lower levels, they emit high-energy photons in the ultraviolet and X-ray portions of the spectrum—wavelengths at which the photosphere and chromosphere are relatively dim. Hence, telescopes sensitive to these short wavelengths are ideal for studying the corona and the flow of the solar wind.

WEB LINK 16.9 The Earth's atmosphere blocks out most ultraviolet light and X rays, so telescopes for these wavelengths must be placed on board spacecraft (see Section 6-7, especially Figure 6-25). Figure 16-15 shows an ultraviolet view of the corona from the *SOHO* spacecraft (*Solar and Heliospheric Observatory*), a joint project of the European Space Agency (ESA) and NASA.

Figure 16-15 reveals that the corona is not uniform in temperature or density. The densest, highest-temperature regions appear bright, while the thinner, lower-temperature regions are dark. Note the large dark area, called a **coronal hole** because it's almost devoid of luminous gas. Particles streaming away from the Sun can most easily flow outward through these particularly thin regions. Therefore, it is thought that coronal holes are the main corridors through which particles of the solar wind escape from the Sun.

Unlike the lower levels of the Sun's atmosphere, the corona has immense holes that shift and reshape

WEB LINK 16.10 Evidence in favor of this picture has come from the *Ulysses* spacecraft, another joint ESA/NASA mission. In 1994 and 1995, *Ulysses* became the first spacecraft to fly over the Sun's north and south poles, where there are apparently permanent coronal holes. The spacecraft indeed measured a stronger solar wind emanating from these holes.

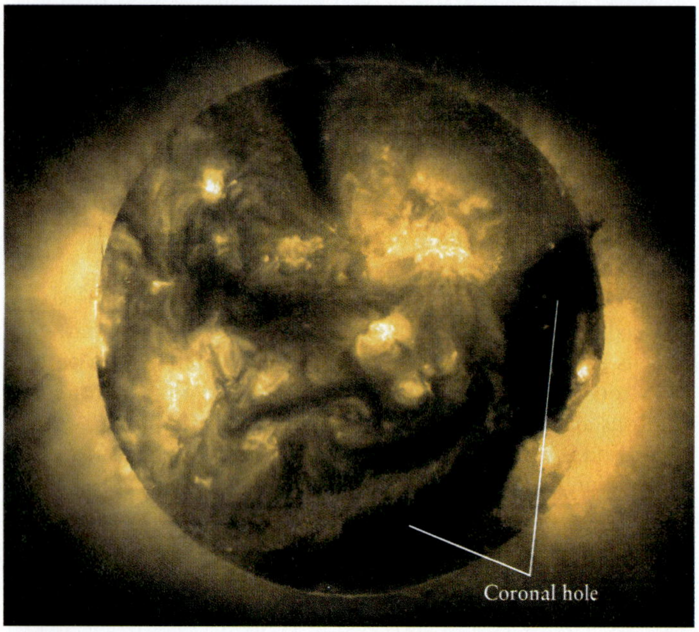

Figure 16-15 R I V **U** X G

The Ultraviolet Corona The *SOHO* spacecraft recorded this false-color ultraviolet view of the solar corona. The dark feature running across the Sun's disk from the top is a coronal hole, a region where the coronal gases are thinner than elsewhere. Such holes are often the source of strong gusts in the solar wind. (*SOHO*/EIT/ESA/NASA)

The temperatures in the corona and the chromosphere are not at all what we would expect. Just as you feel warm if you stand close to a campfire but cold if you move away, we would expect that the temperature in the corona and chromosphere would *decrease* with increasing altitude and, hence, increasing distance from the warmth of the Sun's photosphere. Why, then, does the temperature in these regions *increase* with increasing altitude? This question has been one of the major unsolved mysteries in astronomy for the past half-century. As astronomers have tried to resolve this dilemma, they have found important clues in one of the Sun's most familiar features—sunspots.

16-8 Sunspots are low-temperature regions in the photosphere

Granules, supergranules, spicules, and the solar wind occur continuously. These features are said to be aspects of the *quiet* Sun. But other, more dramatic features appear periodically, including massive eruptions and regions of concentrated magnetic fields. When these features are present, astronomers refer to the *active* Sun. The features of the active Sun that can most easily be seen with even a small telescope (although only with an appropriate filter attached) are sunspots.

Observing Sunspots

Sunspots are irregularly shaped dark regions in the photosphere. Sometimes sunspots appear in isolation (Figure 16-16*a*), but frequently they are found in sunspot groups (Figure 16-16*b*; see also Figure 16-7). Although sunspots vary greatly in size, typical ones

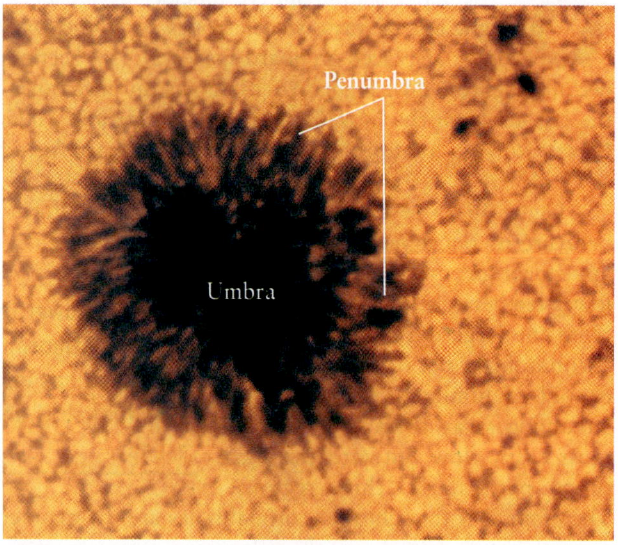

(a)

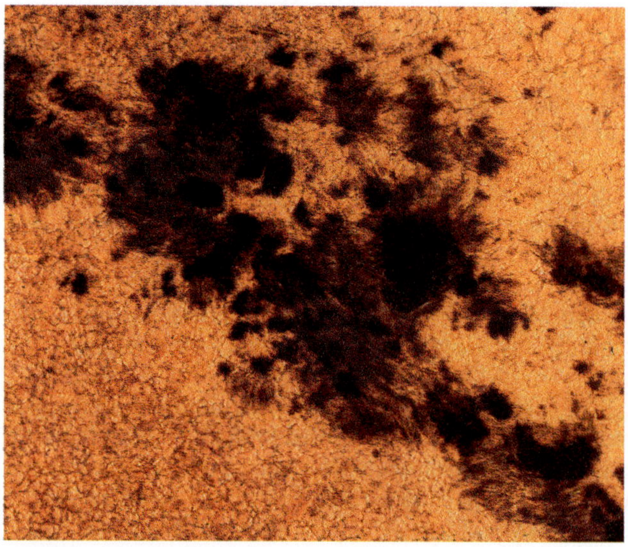

(b)

Figure 16-16 R I **V** U X G

Sunspots (a) This high-resolution photograph of the photosphere shows a mature sunspot. The dark center of the spot is called the umbra. It is bordered by the penumbra, which is less dark and has a featherlike appearance. **(b)** In this view of a typical sunspot group, several sunspots are close enough to overlap. In both images you can see granulation in the surrounding, undisturbed photosphere. (NOAO) measure a few tens of thousands of kilometers across—comparable to the diameter of Earth. Sunspots are not permanent features of the photosphere but last between a few hours and a few months.

Each sunspot has a dark central core, called the *umbra*, and a brighter border called the *penumbra*. We used these same terms in Section 3-4 to refer to different parts of Earth's or the Moon's shadow. But a sunspot is not a shadow; it's a region in the photosphere where the temperature is relatively low, which makes it appear darker than its surroundings. If the surrounding photosphere is blocked from view, a sunspot's umbra appears red and the penumbra appears orange. As we saw in Section 5-4, Wien's law relates the color of a blackbody (which depends on the wavelength at which it emits the most light) to the blackbody's temperature. The colors of a sunspot indicate that the temperature of the umbra is typically 4300 K and that of the penumbra is typically 5000 K. While high by earthly standards, these temperatures are quite a bit lower than the average photospheric temperature of 5800 K.

The Stefan-Boltzmann law (see Section 5-4) tells us that the energy flux from a blackbody is proportional to the fourth power of its temperature. This law lets us compare the amounts of light energy emitted by a square meter of a sunspot's umbra and by a square meter of undisturbed photosphere. The ratio is

$$\frac{\text{flux from umbra}}{\text{flux from photosphere}} = \left(\frac{4300\text{ K}}{5800\text{ K}}\right)^4 = 0.30$$

That is, the umbra emits only 30% as much light as an equally large patch of undisturbed photosphere, which is why sunspots appear so dark.

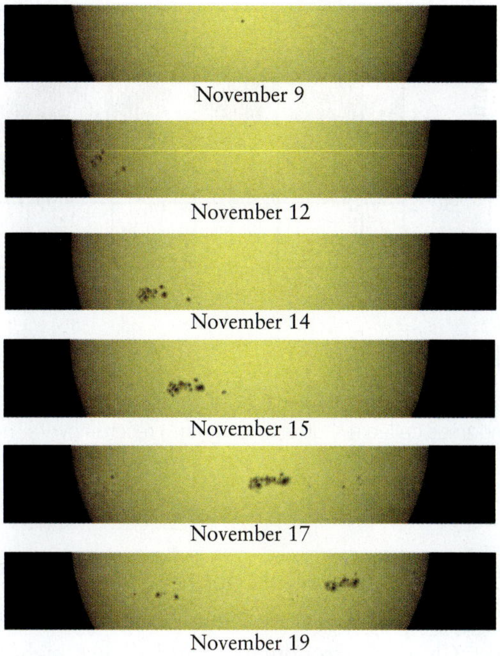

VIDEO 16.5

Figure 16-17 R I **V** U X G

Tracking the Sun's Rotation with Sunspots This series of photographs taken in 1999 shows the rotation of the Sun. By observing the same group of sunspots from one day to the next, Galileo found that the Sun rotates once in about four weeks. (The equatorial regions of the Sun actually rotate somewhat faster than the polar regions.) Notice how the sunspot group shown here changed its shape. (The Carnegie Observatories)

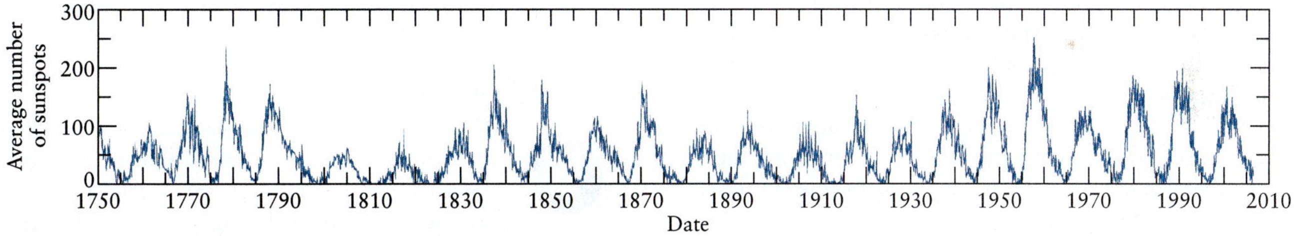

(a)

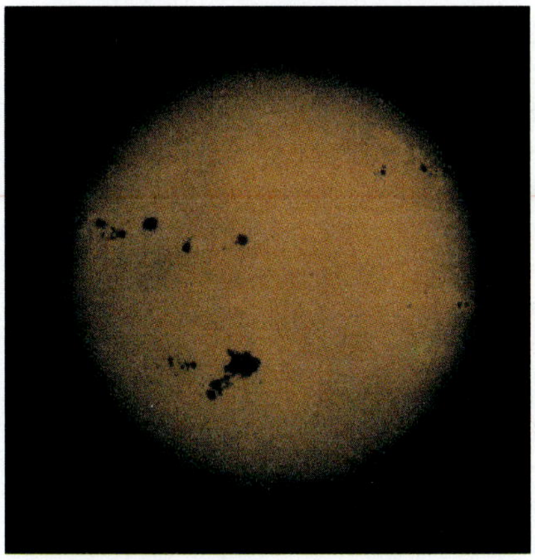

(b) Near sunspot maximum

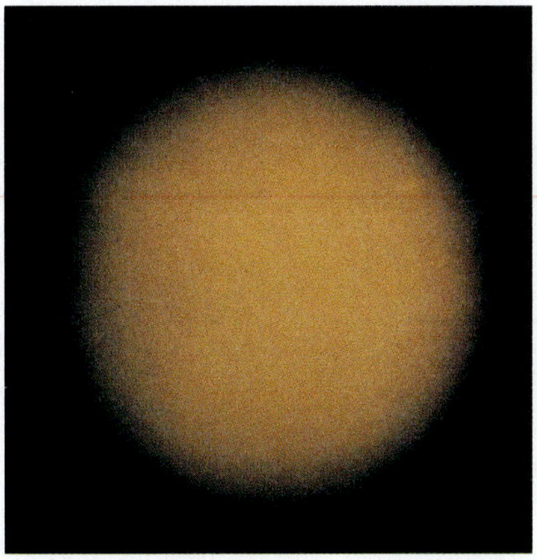

(c) Near sunspot minimum

WEB LINK 16.11

Figure 16-18 R I **V** U X G

The Sunspot Cycle (a) The number of sunspots on the Sun varies with a period of about 11 years. The most recent sunspot maximum occurred in 2000. **(b)** This photograph, taken near sunspot maximum in 1989, shows a number of sunspots and large sunspot groups. The sunspot group visible near the bottom of the Sun's disk has about the same diameter as the planet Jupiter. **(c)** Near sunspot minimum, as in this 1986 photograph, essentially no sunspots are visible. (NOAO)

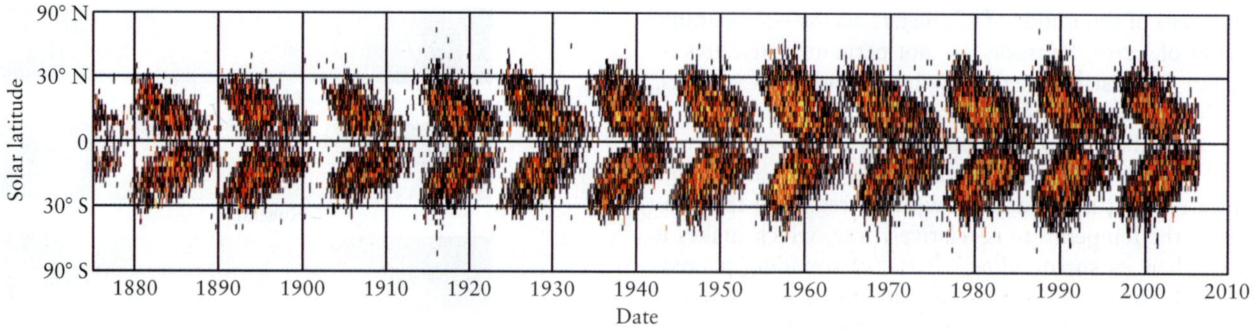

Figure 16-19

Variations in the Average Latitude of Sunspots The dots in this graph (sometimes called a "butterfly diagram") record how far north or south of the Sun's equator sunspots were observed. At the beginning of each sunspot cycle, most sunspots are found near latitudes 30° north or south. As the cycle goes on, sunspots typically form closer to the equator. (NASA Marshall Space Flight Center)

Sunspots and the Sun's Rotation

Occasionally, a sunspot group is large enough to be seen without a telescope. Chinese astronomers recorded such sightings 2000 years ago, and huge sunspot groups visible to the naked eye (with an appropriate filter) were seen in 1989 and 2003. But it was not until Galileo introduced the telescope into astronomy (see Section 4-5) that anyone was able to examine sunspots in detail.

Galileo discovered that he could determine the Sun's rotation rate by tracking sunspots as they moved across the solar disk (Figure 16-17). He found that the Sun rotates once in about four weeks. A typical sunspot group lasts about two months, so a specific one can be followed for two solar rotations.

Further observations by the British astronomer Richard Carrington in 1859 demonstrated that the Sun does not rotate as a rigid body. Instead, the equatorial regions rotate more rapidly than the polar regions. This phenomenon is known as **differential rotation.** Thus, while a sunspot near the solar equator takes only 25 days to go once around the Sun, a sunspot at 30° north or south of the equator takes $27\frac{1}{2}$ days. The rotation period at 75° north or south is about 33 days, while near the poles it may be as long as 35 days.

The Sunspot Cycle

The average number of sunspots on the Sun is not constant, but varies in a predictable **sunspot cycle** (Figure 16-18*a*). This phenomenon was first reported by the German astronomer Heinrich Schwabe in 1843 after many years of observing. As Figure 16-18*a* shows, the average number of sunspots varies with a period of about 11 years. A period of exceptionally many sunspots is a **sunspot maximum** (Figure 16-18*b*), as occurred in 1979, 1989, and 2000. Conversely, the Sun is almost devoid of sunspots at a **sunspot minimum** (Figure 16-18*c*), as occurred in 1976, 1986, 1996 and 2008. This most recent minimum in sunspot activity is ongoing as of this writing (2009), and is the lowest sunspot activity in over 50 years.

The number of sunspots increases and decreases on an 11-year cycle

The locations of sunspots also vary with the same 11-year sunspot cycle. At the beginning of a cycle, just after a sunspot minimum, sunspots first appear at latitudes around 30° north and south of the solar equator (Figure 16-19). Over the succeeding years, the sunspots occur closer and closer to the equator.

16-9 Sunspots are produced by a 22-year cycle in the Sun's magnetic field

TUTORIAL 16.3

Why should the number of sunspots vary with an 11-year cycle? Why should their average latitude vary over the course of a cycle? And why should sunspots exist at all? The first step toward answering these questions came in 1908, when the American astronomer George Ellery Hale discovered that sunspots are associated with intense magnetic fields on the Sun.

Probing Solar Magnetism

LOOKING DEEPER 16.2

When Hale focused a spectroscope on sunlight coming from a sunspot, he found that many spectral lines appear to be split into several closely spaced lines (Figure 16-20). This "splitting" of spectral lines is called the **Zeeman effect,** after the Dutch physicist Pieter Zeeman, who first observed it in his laboratory in 1896. Zeeman showed that a spectral line splits when the atoms are subjected to an intense magnetic field. The more intense the magnetic field, the wider the separation of the split lines.

Hale's discovery showed that sunspots are places where the hot gases of the photosphere are bathed in a concentrated magnetic field. Many of the atoms of the Sun's atmosphere are ionized due to the high temperature. The solar atmosphere is thus a special type of gas called a **plasma,** in which electrically charged ions and electrons can move freely. Like any moving, electrically charged objects, they can be deflected by magnetic fields. Figure 16-21 shows how a magnetic field in the laboratory bends a beam of fast-moving electrons into a curved trajectory. Similarly, the paths of moving ions and electrons in the photosphere are deflected by the Sun's magnetic field. In particular, magnetic forces act on the hot plasma that rises from the Sun's interior due to convection. Where the magnetic field

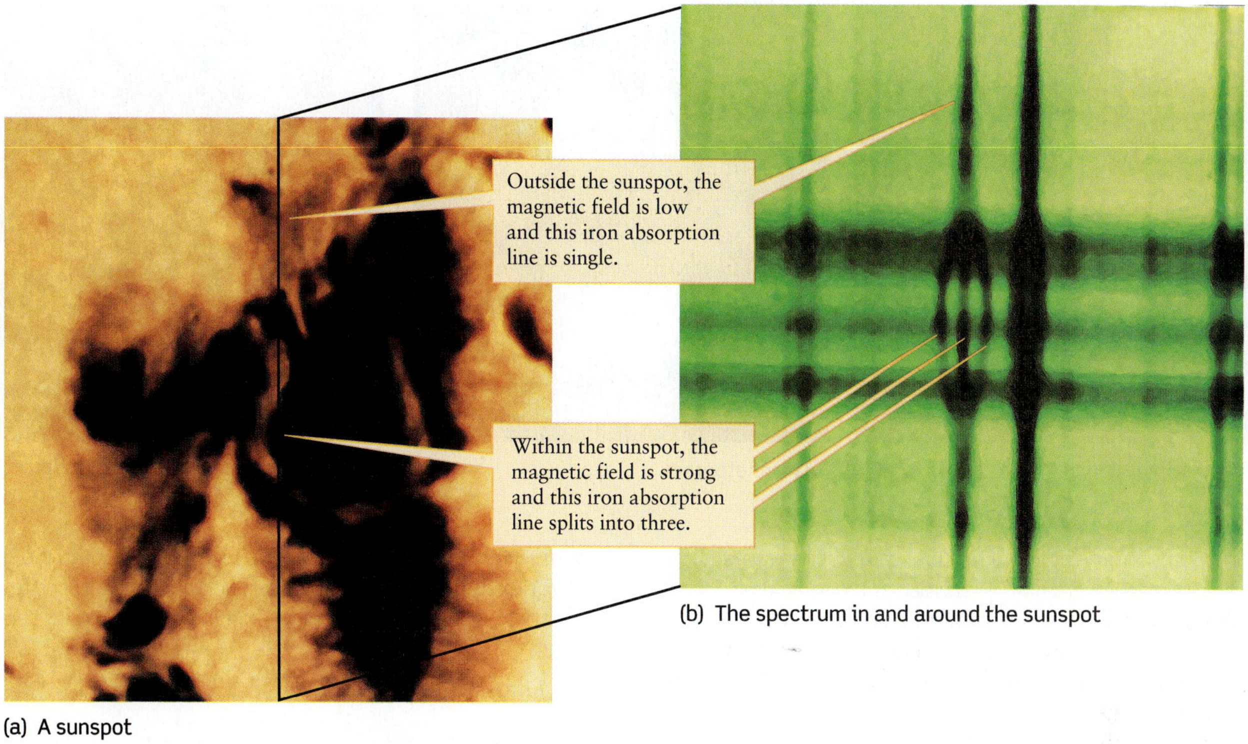

Outside the sunspot, the magnetic field is low and this iron absorption line is single.
Within the sunspot, the magnetic field is strong and this iron absorption line splits into three.
(a) A sunspot
(b) The spectrum in and around the sunspot

Figure 16-20 R I V U X G

Sunspots Have Strong Magnetic Fields **(a)** A black line in this image of a sunspot shows where the slit of a spectrograph was aimed. **(b)** This is a portion of the resulting spectrum, including a dark absorption line caused by iron atoms in the photosphere. The splitting of this line by the sunspot's magnetic field can be used to calculate the field strength. Typical sunspot magnetic fields are over 5000 times stronger than Earth's field at its north and south poles. (NOAO)

is particularly strong, these forces push the hot plasma away. The result is a localized region where the gas is relatively cool and thus glows less brightly—in other words, a sunspot.

To get a fuller picture of the Sun's magnetic fields, astronomers take images of the Sun at two wavelengths, one just less than and one just greater than the wavelength of a magnetically split spectral line. From the difference between these two images, they can construct a picture called a **magnetogram,** which displays the magnetic fields in the solar atmosphere. Figure 16-22*a* is an ordinary white-light photograph of the Sun taken at the same time as the magnetogram in Figure 16-22*b*. In the magnetogram, dark blue indicates areas of the photosphere with one magnetic polarity (north), and yellow indicates areas with the opposite (south) magnetic polarity. This image shows that many sunspot groups have roughly comparable areas covered by north

Figure 16-21 R I V U X G

Magnetic Fields Deflect Moving, Electrically Charged Objects In this laboratory experiment, a beam of negatively charged electrons (shown by a blue arc) is aimed straight upward from the center of the apparatus. The entire apparatus is inside a large magnet, and the magnetic field deflects the beam into a curved path. (Courtesy of Central Scientific Company)

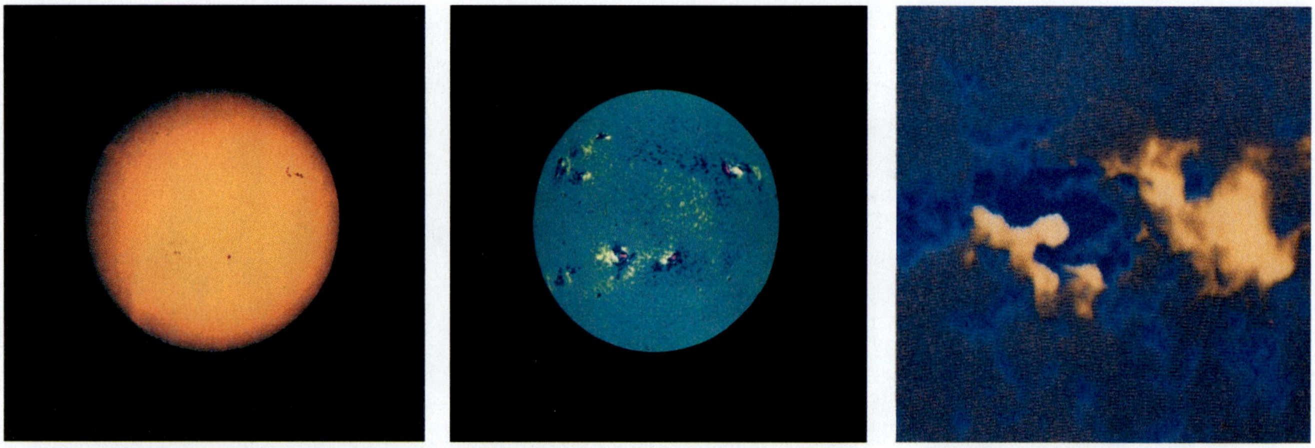

(a) Visible-light image (b) Magnetogram (c) Magnetogram of a sunspot group

Figure 16-22 R I V U X G

Mapping the Sun's Magnetic Field **(a)** This visible-light image and **(b)** this false-color magnetogram were recorded at the same time. Dark blue and yellow areas in the magnetogram have north and south magnetic polarity, respectively; blue-green regions have weak magnetic fields. The highly magnetized regions in (b) correlate with the sunspots in (a). **(c)** The two ends of this large sunspot group have opposite magnetic polarities (colored blue and yellow), like the ends of a giant bar magnet. (NOAO)

and south magnetic polarities (see also Figure 16-22*c*). Thus, a sunspot group resembles a giant bar magnet, with a north magnetic pole at one end and a south magnetic pole at the other.

If different sunspot groups were unrelated to one another, their magnetic poles would be randomly oriented, like a bunch of compass needles all pointing in random directions. As Hale discovered, however, there is a striking regularity in the magnetization of sunspot groups. As a given sunspot group moves with the Sun's rotation, the sunspots in front are called the "preceding members" of the group. The spots that follow behind are referred to as the "following members." Hale compared the sunspot groups in the two solar hemispheres, north or south of the Sun's equator. He found that the preceding members in one solar hemisphere all have the same magnetic polarity, while the preceding members in the other hemisphere have the opposite polarity. Furthermore, in the hemisphere where the Sun has its north magnetic pole, the preceding members of all sunspot groups have north magnetic polarity. In the opposite hemisphere, where the Sun has its south magnetic pole, the preceding members all have south magnetic polarity.

Along with his colleague Seth B. Nicholson, Hale also discovered that the Sun's polarity pattern completely reverses itself every 11 years—the same interval as the time from one solar maximum to the next. The hemisphere that has preceding north magnetic poles during one 11-year sunspot cycle will have preceding south magnetic poles during the next 11-year cycle, and vice versa. The north and south magnetic poles of the Sun itself also reverse every 11 years. Thus, the Sun's magnetic pattern repeats itself only after two sunspot cycles, which is why astronomers speak of a **22-year solar cycle.**

The Magnetic-Dynamo Model

In 1960, the American astronomer Horace Babcock proposed a description that seems to account for many features of this 22-year solar cycle. Babcock's scenario, called a **magnetic-dynamo model,** makes use of two basic properties of the Sun's photosphere—differential rotation and convection. Differential rotation causes the magnetic field in the photosphere to become wrapped around the Sun (**Figure 16-23**). As a result, the magnetic field becomes concentrated at certain latitudes on either side of the solar equator. Convection in the photosphere creates tangles in the concentrated magnetic field, and "kinks" erupt through the solar surface. Sunspots appear where the magnetic field protrudes through the photosphere. The theory suggests that sunspots should appear first at northern and southern latitudes and later form nearer to the equator, which is just what is observed (see Figure 16-19). Note also that as shown on the far right in Figure 16-23, the preceding member of a sunspot group has the same polarity (N or S) as the Sun's magnetic pole in that hemisphere, which is just as Hale observed.

Differential rotation eventually undoes the twisted magnetic field. The preceding members of sunspot groups move toward the Sun's equator, while the following members migrate toward the poles. Because the preceding members from the two hemispheres have opposite magnetic polarities, their magnetic fields cancel each other out when they meet at the equator. The following members in each hemisphere have the opposite polarity to the Sun's pole in that hemisphere; hence, when they converge on the pole, the following members first cancel out and then reverse the Sun's overall magnetic field. The fields are now completely relaxed. Once again, differential rotation begins to twist the Sun's magnetic field, but now with all magnetic polarities reversed. In this way, Babcock's model helps to explain the change in field direction every 11 years.

The Sun's differential rotation makes the magnetic field twist like a rubber band

Recent discoveries in helioseismology (Section 16-3) offer new insights into the Sun's magnetic field. By comparing the

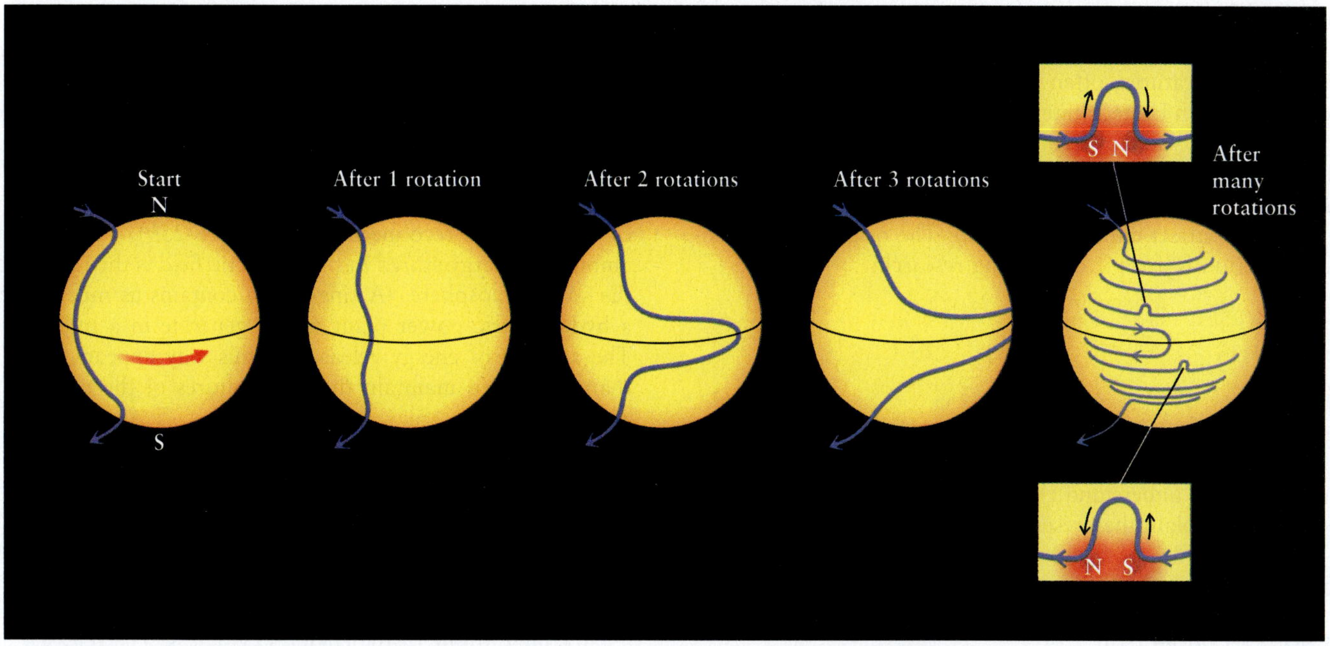

Figure 16-23

Babcock's Magnetic Dynamo Model Magnetic field lines tend to move along with the plasma in the Sun's outer layers. Because the Sun rotates faster at the equator than near the poles, a field line that starts off running from the Sun's north magnetic pole (N) to its south magnetic pole (S) ends up wrapped around the Sun like twine wrapped around a ball. The insets on the far right show how sunspot groups appear where the concentrated magnetic field rises through the photosphere.

speeds of sound waves that travel with and against the Sun's rotation, helioseismologists have been able to determine the Sun's rotation rate at different depths and latitudes. As shown in Figure 16-24, the Sun's surface pattern of differential rotation persists through the convective zone. Farther in, within the radiative zone, the Sun seems to rotate like a rigid object with a period of 27 days at all latitudes. Astronomers suspect that the Sun's magnetic field originates in a relatively thin layer where the radiative and convective zones meet and slide past each other due to their different rotation rates.

One dilemma about sunspots is that compressed magnetic fields tend to push themselves apart, which means that sunspots should dissipate rather quickly. Yet observations show that sunspots can persist for many weeks. The resolution of this paradox may have been found using helioseismology (see Section 16-3). Analysis of the vibrations of the Sun around sunspots shows that beneath the surface of the photosphere, the gases surrounding each sunspot are circulating at high speed—rather like a hurricane as large as Earth. The circulation of charged gases around the magnetic field holds the fields in place, thus stabilizing the sunspot.

Much about sunspots and solar activity remains mysterious. There are perplexing irregularities in the solar cycle. For example, the overall reversal of the Sun's magnetic field is often piecemeal and haphazard. One pole may reverse polarity long before the other. For several weeks the Sun's surface may have two north magnetic poles and no south magnetic pole at all.

What is more, there seem to be times when all traces of sunspots and the sunspot cycle vanish for many years. For

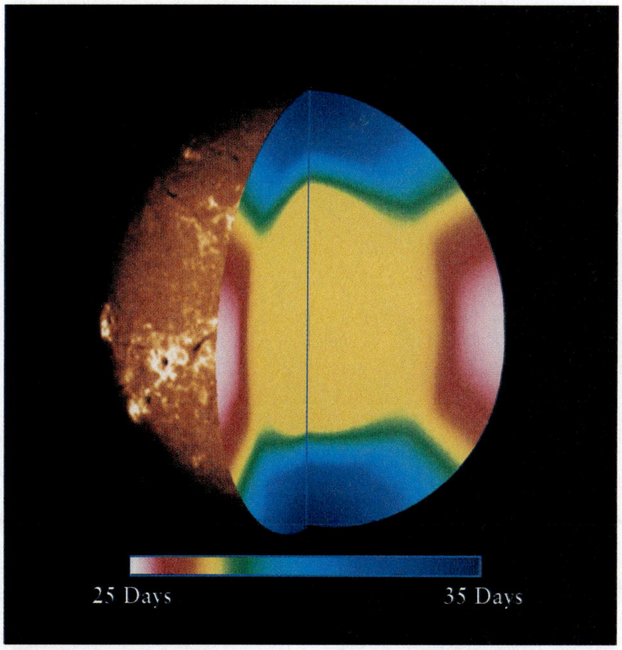

WEB LINK 16.12

Figure 16-24

Rotation of the Solar Interior This cutaway picture of the Sun shows how the solar rotation period (shown by different colors) varies with depth and latitude. The surface and the convective zone have differential rotation (a short period at the equator and longer periods near the poles). Deeper within the Sun, the radiative zone seems to rotate like a rigid sphere. (Courtesy of K. Libbrecht, Big Bear Solar Observatory) example, virtually no sunspots were seen from 1645 through 1715. Curiously, during these same years Europe experienced record low temperatures, often referred to as the Little Ice Age, whereas the western United States was subjected to severe drought. By contrast, there was apparently a period of increased sunspot activity during the eleventh and twelfth centuries, during which Earth was warmer than it is today. Thus, variations in solar activity appear to affect climates on Earth. The origin of this Sun-Earth connection is a topic of ongoing research.

16-10 The Sun's magnetic field heats the corona, produces flares, and causes massive eruptions

Astronomers now understand that the Sun's magnetic field does more than just explain the presence of sunspots. It is also responsible for the existence of spicules, as well as a host of other dramatic phenomena in the chromosphere and corona.

Magnetic Reconnection

In a plasma, magnetic field lines and the material of the plasma tend to move together. Their moving together means that as convection pushes material toward the edge of a supergranule, it pushes magnetic field lines as well. The result is that vertical magnetic field lines pile up around a supergranule. Plasma that "sticks" to these magnetic field lines thus ends up lifted upward, forming a spicule (see Figures 16-12 and 16-13).

The tendency of plasma to follow the Sun's magnetic field can also explain why the temperature of the chromosphere and corona is so high. Spacecraft observations show magnetic field arches extending tens of thousands of kilometers into the corona, with streamers of electrically charged particles moving along each arch (Figure 16-25*a*). If the magnetic fields of two arches come into proximity, their magnetic fields can rearrange in a process called **magnetic reconnection** (Figure 16-25*b*). The tremendous amount of energy stored in the magnetic field is then released into the solar atmosphere. (A single arch contains as much energy as a hydroelectric power plant would generate in a million years.) The amount of energy released in this way appears to be more than enough to maintain the temperatures of the chromosphere and corona.

ANALOGY The idea that a magnetic field can heat gases has applications on Earth as well as on the Sun. In an automobile engine's ignition system an electric current is set up in a coil of wire, which produces a magnetic field. When the current is shut off, the magnetic field collapses and its energy is directed to a spark plug in one of the engine's cylinders. The released energy heats the mixture of air and gasoline around the plug, causing the mixture to ignite. This drives the piston in that cylinder and makes the automobile go.

VIDEO 16.6

Magnetic heating can also explain why the parts of the corona that lie on top of sunspots are often the most prominent in ultraviolet images. (Some examples are the

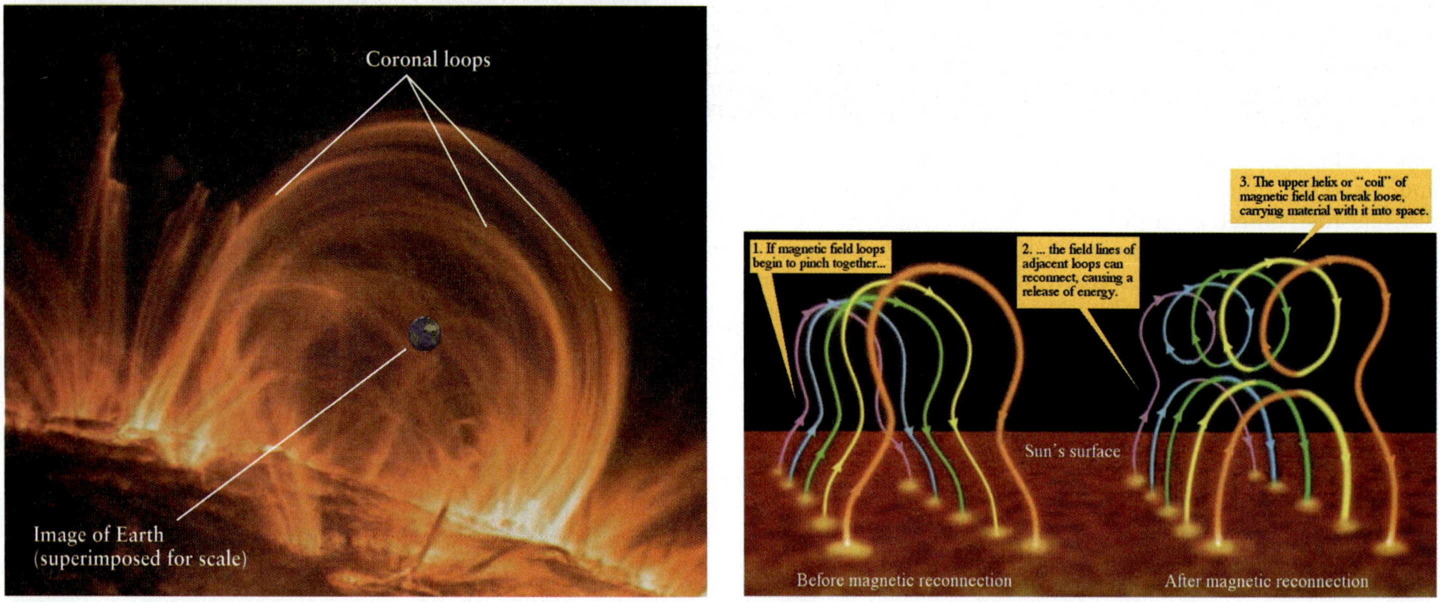

(a) (b)

WEB LINK 16.13

Figure 16-25 R I V U X G

Magnetic Arches and Magnetic Reconnection **(a)** This false-color ultraviolet image from the *TRACE* spacecraft (*Transition Region and Coronal Explorer*) shows magnetic field loops suspended high above the solar surface. The loops are made visible by the glowing gases trapped within them. **(b)** When the magnetic fields in these loops change their arrangement, a tremendous amount of energy is released and solar material can be ejected upward. (a: Stanford-Lockheed Institute for Space Research; *TRACE;* and NASA)

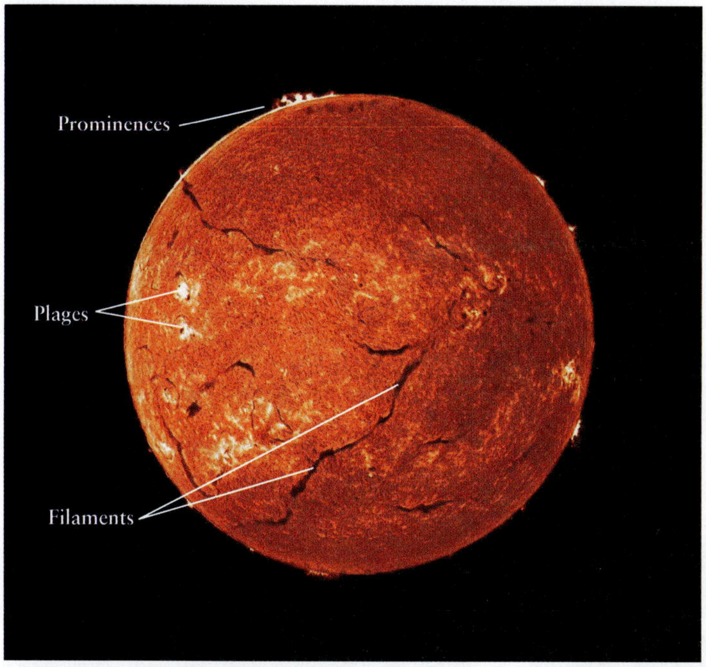

Figure 16-26 R I **V** U X G

The Active Sun Seen through an H_α Filter This image was made using a red filter that only passes light at a wavelength of 656 nm. The spectrum of the photosphere has an absorption line at this wavelength and so appears dark. Hence, this filter reveals the photosphere and corona. Prominences, plages, and filaments are associated with strong magnetic fields. (NASA)

bright regions in Figure 16-15.) The intense magnetic field of the sunspots helps trap and compress hot coronal gas, giving it such a high temperature that it emits copious amounts of high-energy ultraviolet photons and even more energetic X-ray photons.

Prominences, Flares, ands Coronal Mass Ejections

Spicules and coronal heating occur even when the Sun is quiet. But magnetic fields can also explain many aspects of the active Sun in addition to sunspots. Figure 16-26 is an image of the chromosphere made with an H_α filter during a sunspot maximum. The bright areas are called **plages** (from the French word for "beach"). These plages are bright, hot regions in the chromosphere that tend to form just before the appearance of new sunspots. They are probably created by magnetic fields that push upward from the Sun's interior, compressing and heating a portion of the chromosphere. The dark streaks, called **filaments,** are relatively cool and dense parts of the chromosphere that have been pulled along with magnetic field lines as they arch to high altitudes.

When seen from the side, so that they are viewed against the dark background of space, filaments appear as bright, arching columns of gas called **prominences** (Figure 16-27). They can extend for tens of thousands of kilometers above the photosphere. Some prominences last for only a few hours, while others persist for many months. The most energetic prominences break free of the magnetic fields that confined them and burst into space.

VIDEO 16.7

Violent, eruptive events on the Sun, called **solar flares,** occur in complex sunspot groups. Within only a few minutes, temperatures in a compact region may soar to 5×10^6 K and vast quantities of particles and radiation—including as much material as is in the prominence shown in Figure 16-27—are blasted out into space. These eruptions can also cause disturbances that spread outward in the solar atmosphere, like the ripples that appear when you drop a rock into a pond.

WEB LINK 16.14

The most energetic flares carry as much as 10^{30} joules of energy, equivalent to 10^{14} one-megaton nuclear weapons being exploded at once! However, the energy of a solar flare does not come from thermonuclear fusion in the solar atmosphere; instead, it appears to be released from the intense magnetic field around a sunspot group.

A solar flare can have as much energy as 100 trillion nuclear bombs

As energetic as solar flares are, they are dwarfed by **coronal mass ejections.** One such event is shown in the image that opens this chapter; Figure 16-28*a* shows another. In a coronal mass ejection, more than 10^{12} kilograms (a billion tons) of high-temperature coronal gas is blasted into space at speeds of hundreds of kilometers per second. A typical coronal mass ejection lasts a few hours. These explosive events seem to be related to large-scale alterations in the Sun's magnetic field, like the magnetic

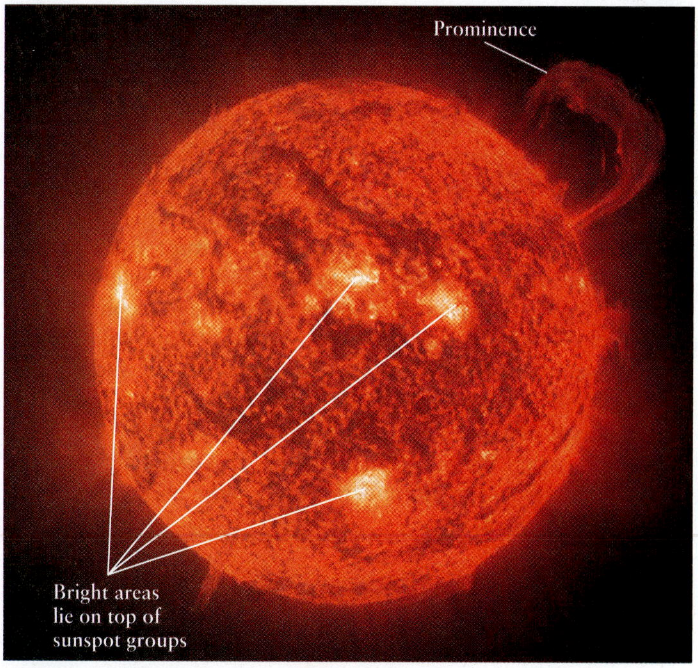

Figure 16-27 R I V **U** X G

A Solar Prominence A huge prominence arches above the solar surface in this ultraviolet image from the *SOHO* spacecraft. The image was recorded using light at a wavelength of 30.4 nm, emitted by singly ionized helium atoms at a temperature of about 60,000 K. By comparison, the material within the arches in Figure 16-25 reaches temperatures in excess of 2×10^6 K. (*SOHO*/EIT/ESA/NASA)

Figure 16-28 R I V **U** **X** G

A Coronal Mass Ejection **(a)** *SOHO* recorded this coronal mass ejection in an X-ray image. (The image of the Sun itself was made at ultraviolet wavelengths.) **(b)** Artist's illustration: Within two to four days the fastest-moving ejected material reaches a distance of 1 AU from the Sun. Most particles are deflected by Earth's magnetosphere, but some are able to reach Earth. (The ejection shown in (a) was not aimed toward Earth and did not affect us.) (*SOHO*/EIT/LASCO/ESA/NASA)

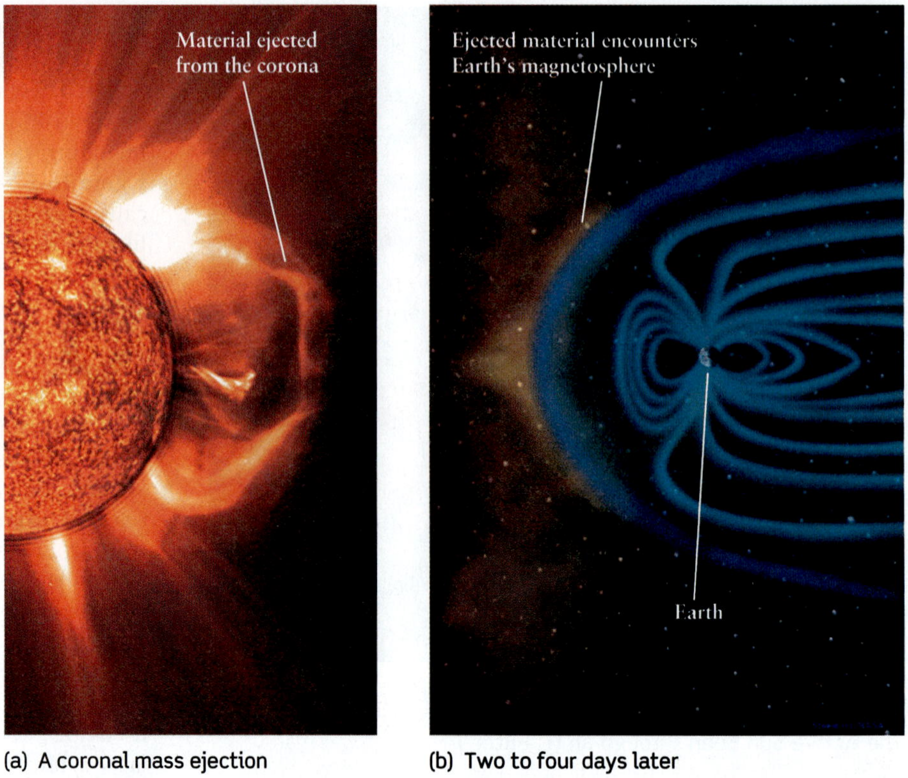

reconnection shown in Figure 16-25*b*. Coronal mass ejections occur every few months; smaller eruptions may occur almost daily.

If a solar flare or coronal mass ejection happens to be aimed toward Earth, a stream of high-energy electrons and nuclei reaches us a few days later (Figure 16-28*b*). When this plasma arrives, it can interfere with satellites, pose a health hazard to astronauts in orbit, and disrupt electrical and communications equipment on Earth's surface. Telescopes on Earth and on board spacecraft now monitor the Sun continuously to provide warnings of dangerous levels of solar particles.

The numbers of plages, filaments, solar flares, and coronal mass ejections all vary with the same 11-year cycle as sunspots. But unlike sunspots, coronal mass ejections never completely cease, even when the Sun is at its quietest. Astronomers are devoting substantial effort to understanding these and other aspects of our dynamic Sun.

Key Words

The term preceded by an asterisk (*) is discussed in Box 16-1.

22-year solar cycle, p. 424
chromosphere, p. 417
CNO cycle, p. 408
conduction, p. 409
convection, p. 409
convective zone, p. 411
corona, p. 418
coronal hole, p. 419
coronal mass ejection, p. 427
differential rotation, p. 422
filament, p. 427
granulation, p. 415
granule, p. 415
helioseismology, p. 411
hydrogen fusion, p. 406
hydrostatic equilibrium, p. 408
limb darkening, p. 415
luminosity (of the Sun), p. 404
magnetic-dynamo model, p. 424
magnetogram, p. 423
magnetic reconnection, p. 426
negative hydrogen ion, p. 417
neutrino, p. 412
neutrino oscillation, p. 413
photosphere, p. 415
plage, p. 427
plasma, p. 422
*positron, p. 406
prominence, p. 427
proton-proton chain, p. 408
radiative diffusion, p. 409
radiative zone, p. 411
solar flare, p. 427
solar neutrino, p. 412
solar neutrino problem, p. 413
solar wind, p. 419
spicule, p. 417
sunspot, p. 420
sunspot cycle, p. 422
sunspot maximum, p. 422
sunspot minimum, p. 422
supergranule, p. 415
thermal equilibrium, p. 409
thermonuclear fusion, p. 406
Zeeman effect, p. 422

Key Ideas

Hydrogen Fusion in the Sun's Core: The Sun's energy is produced by hydrogen fusion, a sequence of thermonuclear reactions in which four hydrogen nuclei combine to produce a single helium nucleus.

• The energy released in a nuclear reaction corresponds to a slight reduction of mass according to Einstein's equation $E = mc^2$.

• Thermonuclear fusion occurs only at very high temperatures; for example, hydrogen fusion occurs only at temperatures in excess of about 10^7 K. In the Sun, fusion occurs only in the dense, hot core.

Models of the Sun's Interior: A theoretical description of a star's interior can be calculated using the laws of physics.

• The standard model of the Sun suggests that hydrogen fusion takes place in a core extending from the Sun's center to about 0.25 solar radius.

• The core is surrounded by a radiative zone extending to about 0.71 solar radius. In this zone, energy travels outward through radiative diffusion.

• The radiative zone is surrounded by a rather opaque convective zone of gas at relatively low temperature and pressure. In this zone, energy travels outward primarily through convection.

Solar Neutrinos and Helioseismology: Conditions in the solar interior can be inferred from measurements of solar neutrinos and of solar vibrations.

• Neutrinos emitted in thermonuclear reactions in the Sun's core have been detected, but in smaller numbers than expected. Recent neutrino experiments explain why this is so.

• Helioseismology is the study of how the Sun vibrates. These vibrations have been used to infer pressures, densities, chemical compositions, and rotation rates within the Sun.

INTERACTIVE EXERCISE 16.3

The Sun's Atmosphere: The Sun's atmosphere has three main layers: the photosphere, the chromosphere, and the corona. Everything below the solar atmosphere is called the solar interior.

• The visible surface of the Sun, the photosphere, is the lowest layer in the solar atmosphere. Its spectrum is similar to that of a blackbody at a temperature of 5800 K. Convection in the photosphere produces granules.

• Above the photosphere is a layer of less dense but higher-temperature gases called the chromosphere. Spicules extend upward from the photosphere into the chromosphere along the boundaries of supergranules.

• The outermost layer of the solar atmosphere, the corona, is made of very high-temperature gases at extremely low density. Activity in the corona includes coronal mass ejections and coronal holes. The solar corona blends into the solar wind at great distances from the Sun.

The Active Sun: The Sun's surface features vary in an 11-year cycle. This is related to a 22-year cycle in which the surface magnetic field increases, decreases, and then increases again with the opposite polarity.

• Sunspots are relatively cool regions produced by local concentrations of the Sun's magnetic field. The average number of sunspots increases and decreases in a regular cycle of approximately 11 years, with reversed magnetic polarities from one 11-year cycle to the next. Two such cycles make up the 22-year solar cycle.

• The magnetic-dynamo model suggests that many features of the solar cycle are due to changes in the Sun's magnetic field. These changes are caused by convection and the Sun's differential rotation.

• A solar flare is a brief eruption of hot, ionized gases from a sunspot group. A coronal mass ejection is a much larger eruption that involves immense amounts of gas from the corona.

Questions

Review Questions

1. What is meant by the luminosity of the Sun?
2. What is Kelvin-Helmholtz contraction? Why is it ruled out as a source of the present-day Sun's energy?

TUTORIAL 16.1

3. Why is it impossible for the burning of substances like coal to be the source of the Sun's energy?
4. What is hydrogen fusion? Why is hydrogen fusion fundamentally unlike the burning of a log in a fireplace?
5. If the electric force between protons were somehow made stronger, what effect would this have on the temperature required for thermonuclear fusion to take place?
6. Why do thermonuclear reactions occur only in the Sun's core, not in its outer regions?
7. Describe how the net result of the reactions shown in the *Cosmic Connections* figure (Section 16-1) is the conversion of four protons into a single helium nucleus. What other particles are produced in this process? How many of each particle are produced?

TUTORIAL 16.2

8. Give an everyday example of hydrostatic equilibrium. Give an example of thermal equilibrium. Explain how these equilibrium conditions apply to each example.
9. If thermonuclear fusion in the Sun were suddenly to stop, what would eventually happen to the overall radius of the Sun? Justify your answer using the ideas of hydrostatic equilibrium and thermal equilibrium.
10. Give some everyday examples of conduction, convection, and radiative diffusion.
11. Describe the Sun's interior. Include references to the main physical processes that occur at various depths within the Sun.
12. Suppose thermonuclear fusion in the Sun stopped abruptly. Would the intensity of sunlight decrease just as abruptly? Why or why not?
13. Explain how studying the oscillations of the Sun's surface can give important, detailed information about physical conditions deep within the Sun.
14. What is a neutrino? Why is it useful to study neutrinos coming from the Sun? What do they tell us that cannot be learned from other avenues of research?
15. Unlike all other types of telescopes, neutrino detectors are placed deep underground. Why?
16. What was the solar neutrino problem? What solution to this problem was suggested by the results from the Sudbury Neutrino Observatory?
17. Describe the dangers in attempting to observe the Sun. How have astronomers learned to circumvent these observational problems?
18. Briefly describe the three layers that make up the Sun's atmosphere. In what ways do they differ from each other?
19. What is solar granulation? Describe how convection gives rise to granules.
20. High-resolution spectroscopy of the photosphere reveals that absorption lines are blueshifted in the spectrum of the central, bright regions of granules but are redshifted in the spec-

trum of the dark boundaries between granules. Explain how these observations show that granulation is due to convection.

21. What is the difference between granules and supergranules?
22. What are spicules? Where are they found? How can you observe them? What causes them?
23. How do astronomers know that the temperature of the corona is so high?

TUTORIAL 16.3

24. How do astronomers know when the next sunspot maximum and minimum will occur?
25. Why do astronomers say that the solar cycle is really 22 years long, even though the number of sunspots varies over an 11-year period?
26. Explain how the magnetic-dynamo model accounts for the solar cycle.
27. Describe one explanation for why the corona has a higher temperature than the chromosphere.
28. Why should solar flares and coronal mass ejections be a concern for businesses that use telecommunication satellites?

Advanced Questions

Questions preceded by an asterisk (*) involve the topic discussed in Box 16-1.

Problem-solving tips and tools

You may have to review Wien's law and the Stefan-Boltzmann law, which are the subjects of Section 5-4. Section 5-5 discusses the properties of photons. As we described in Box 5-2, you can simplify calculations by taking ratios, such as the ratio of the flux from a sunspot to the flux from the undisturbed photosphere. When you do this, all the cumbersome constants cancel out. Figure 5-7 shows the various parts of the electromagnetic spectrum. We introduced the Doppler effect in Section 5-9 and Box 5-6. For information about the planets, see Table 7-1.

29. Calculate how much energy would be released if each of the following masses were converted *entirely* into their equivalent energy: **(a)** a carbon atom with a mass of 2×10^{-26} kg, **(b)** 1 kilogram, and **(c)** a planet as massive as Earth (6×10^{24} kg).
30. Use the luminosity of the Sun (given in Table 16-1) and the answers to the previous question to calculate how long the Sun must shine in order to release an amount of energy equal to that produced by the complete mass-to-energy conversion of **(a)** a carbon atom, **(b)** 1 kilogram, and **(c)** Earth.
31. Assuming that the current rate of hydrogen fusion in the Sun remains constant, what fraction of the Sun's mass will be converted into helium over the next 5 billion years? How will this affect the overall chemical composition of the Sun?
32. **(a)** Estimate how many kilograms of hydrogen the Sun has consumed over the past 4.56 billion years, and estimate the amount of mass that the Sun has lost as a result. Assume that the Sun's luminosity has remained constant during that time. **(b)** In fact, the Sun's luminosity when it first formed was only about 70% of its present value. With this in mind, explain whether your answers to part (a) are an overestimate or an underestimate.

*33. To convert one kilogram of hydrogen (^{1}H) into helium (^{4}He) as described in Box 16-1, you must start with 1.5 kg of hydrogen. Explain why, and explain what happens to the other 0.5 kg. (*Hint:* How many ^{1}H nuclei are used to make the two ^{3}He nuclei shown in the *Cosmic Connections* figure in Section 16-1? How many of these ^{1}H nuclei end up being incorporated into the ^{4}He nucleus shown in the figure?)

*34. **(a)** A positron has the same mass as an electron (see Appendix 7). Calculate the amount of energy released by the annihilation of an electron and positron. **(b)** The products of this annihilation are two photons, each of equal energy. Calculate the wavelength of each photon, and confirm from Figure 5-7 that this wavelength is the gamma-ray range.

35. Sirius is the brightest star in the night sky. It has a luminosity of 23.5 $L_\odot$, that is, it is 23.5 times as luminous as the Sun and burns hydrogen at a rate 23.5 times greater than the Sun. How many kilograms of hydrogen does Sirius convert into helium each second?
36. (Refer to the preceding question.) Sirius has 2.3 times the mass of the Sun. Do you expect that the lifetime of Sirius will be longer, shorter, or the same length as that of the Sun? Explain your reasoning.
37. **(a)** If the Sun were not in a state of hydrostatic equilibrium, would its diameter remain the same? Explain your reasoning. **(b)** If the Sun were not in a state of thermal equilibrium, would its luminosity remain the same? What about its surface temperature? Explain your reasoning.
38. Using the mass and size of the Sun given in Table 16-1, verify that the average density of the Sun is 1410 kg/m^3. Compare your answer with the average densities of the Jovian planets.
39. Use the data in Table 16-2 to calculate the average density of material within 0.1 $R_\odot$ of the center of the Sun. (You will need to use the mass and radius of the Sun as a whole, given in Table 16-1.) Explain why your answer is not the same as the density at 0.1 $R_\odot$ given in Table 16-2.
40. In a typical solar oscillation, the Sun's surface moves up or down at a maximum speed of 0.1 m/s. An astronomer sets out to measure this speed by detecting the Doppler shift of an absorption line of iron with wavelength 557.6099 nm. What is the maximum wavelength shift that she will observe?
41. Explain why the results from the Sudbury Neutrino Observatory (SNO) only provide an answer to the solar neutrino problem for relatively high-energy neutrinos. (*Hint:* Can SNO detect solar neutrinos of all energies?)
42. The amount of energy required to dislodge the extra electron from a negative hydrogen ion is 1.2×10^{-19} J. **(a)** The extra electron can be dislodged if the ion absorbs a photon of sufficiently short wavelength. (Recall from Section 5-5 that the higher the energy of a photon, the shorter its wavelength.) Find the longest wavelength (in nm) that can accomplish this. **(b)** In what part of the electromagnetic spectrum does this wavelength lie? **(c)** Would a photon of visible light be able to dislodge the extra electron? Explain. **(d)** Explain why the photosphere, which contains negative hydrogen ions, is quite opaque to visible light but is less opaque to light with wavelengths longer than the value you calculated in (a).

43. Astronomers often use an H_α filter to view the chromosphere. Explain why this can also be accomplished with filters that are transparent only to the wavelengths of the H and K lines of ionized calcium. (*Hint:* The H and K lines are dark lines in the spectrum of the photosphere.)
44. Calculate the wavelengths at which the photosphere, chromosphere, and corona emit the most radiation. Explain how the results of your calculations suggest the best way to observe these regions of the solar atmosphere. (*Hint:* Treat each part of the atmosphere as a perfect blackbody. Assume average temperatures of 50,000 K and 1.5×10^6 K for the chromosphere and corona, respectively.)
45. On November 15, 1999, the planet Mercury passed in front of the Sun as seen from Earth. The TRACE spacecraft made these time-lapse images of this event using ultraviolet light (top) and visible light (bottom). (Mercury moved from left to right in these images. The time between successive views of Mercury is 6 to 9 minutes.) Explain why the Sun appears somewhat larger in the ultraviolet image than in the visible-light image.

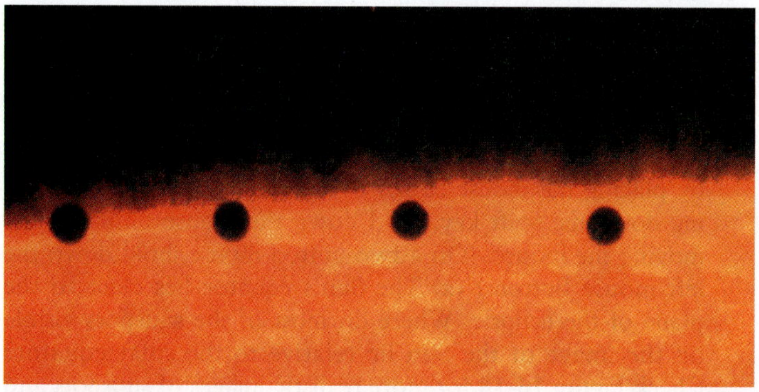

RIVUXG

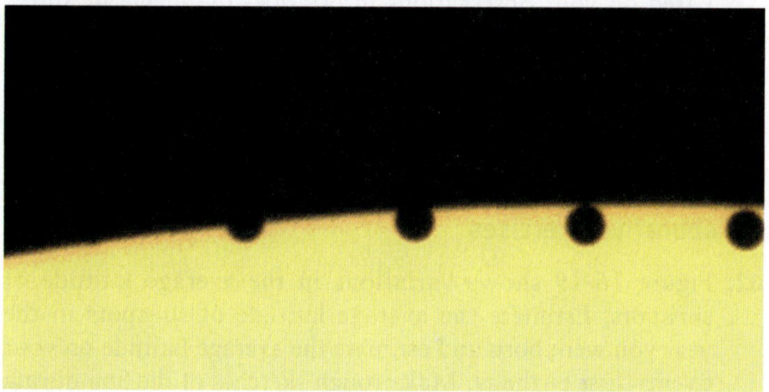

RIVUXG

(K. Schrijver, Stanford-Lockheed Institute for Space Research, TRACE, and NASA)

46. The moving images on a television set that uses a tube (as opposed to a LCD or plasma flat-screen TV) are made by a fast-moving electron beam that sweeps over the back of the screen. Explain why placing a strong magnet next to the screen distorts the picture. (*Caution:* Don't try this with your television! You can do permanent damage to your set.)
47. Find the wavelength of maximum emission of the umbra of a sunspot and the wavelength of maximum emission of a sunspot's penumbra. In what part of the electromagnetic spectrum do these wavelengths lie?
48. (**a**) Find the ratio of the energy flux from a patch of a sunspot's penumbra to the energy flux from an equally large patch of undisturbed photosphere. Which patch is brighter? (**b**) Find the ratio of the energy flux from a patch of a sunspot's penumbra to the energy flux from an equally large patch of umbra. Again, which patch is brighter?
49. Suppose that you want to determine the Sun's rotation rate by observing its sunspots. Is it necessary to take Earth's orbital motion into account? Why or why not?
50. (**a**) Using a ruler to make measurements of Figure 16-25, determine how far the arches in that figure extend above the Sun's surface. The diameter of Earth is 12,756 km. (**b**) In Figure 16-25 (an ultraviolet image) the photosphere appears dark compared to the arches. Explain why.
51. The amount of visible light emitted by the Sun varies only a little over the 11-year sunspot cycle. But the amount of X rays emitted by the Sun can be 10 times greater at solar maximum than at solar minimum. Explain why these two types of radiation should be so different in their variability.

Discussion Questions

52. Discuss the extent to which cultures around the world have worshiped the Sun as a deity throughout history. Why do you suppose there has been such widespread veneration?
53. In the movie *Star Trek IV: The Voyage Home,* the starship *Enterprise* flies on a trajectory that passes close to the Sun's surface. What features should a real spaceship have to survive such a flight? Why?
54. Discuss some of the difficulties in correlating solar activity with changes in Earth's climate.
55. Describe some of the advantages and disadvantages of observing the Sun (**a**) from space and (**b**) from Earth's south pole. What kinds of phenomena and issues might solar astronomers want to explore from these locations?

Web/eBook Questions

56. Search the World Wide Web for information about features in the solar atmosphere called *sigmoids.* What are they? What causes them? How might they provide a way to predict coronal mass ejections?
57. **Determining the Lifetime of a Solar Granule.** Access and view the video "Granules on the Sun's Surface" in Chapter 16 of the *Universe* Web site or eBook. Your task is to determine the approximate lifetime of a solar granule on the photosphere. Select an area, then slowly and rhythmically repeat "start, stop, start, stop" until you can consistently predict the appearance and disappearance of granules. While keeping your rhythm, move to a different area of the video and continue monitoring the appearance and disappearance of granules. When you are confident you have the timing right, move your eyes (or use a partner) to look at the clock shown in the video. Determine the length of time between the appearance and disappearance of the granules and record your answer.

Activities

Observing Projects

Observing tips and tools

At the risk of repeating ourselves, we remind you to ***never look directly at the Sun, because it can easily cause permanent blindness.*** You can view the Sun safely without a telescope just by using two pieces of white cardboard. First, use a pin to poke a small hole in one piece of cardboard; this will be your "lens," and the other piece of cardboard will be your "viewing screen." Hold the "lens" piece of cardboard so that it is face-on to the Sun and sunlight can pass through the hole. With your other hand, hold the "viewing screen" so that the sunlight from the "lens" falls on it. Adjust the distance between the two pieces of cardboard so that you see a sharp image of the Sun on the "viewing screen." This image is perfectly safe to view. It is actually possible to see sunspots with this low-tech apparatus.

For a better view, use a telescope with a solar filter that fits on the front of the telescope. A standard solar filter is a piece of glass coated with a thin layer of metal to give it a mirrorlike appearance. This coating reflects almost all the sunlight that falls on it, so that only a tiny, safe amount of sunlight enters the telescope. An H_α filter, which looks like a red piece of glass, keeps the light at a safe level by admitting only a very narrow range of wavelengths. (Filters that fit on the back of the telescope are *not* recommended. The telescope focuses concentrated sunlight on such a filter, heating it and making it susceptible to cracking—and if the filter cracks when you are looking through it, your eye will be ruined instantly and permanently.)

To use a telescope with a solar filter, first aim the telescope away from the Sun, then put on the filter. Keep the lens cap on the telescope's secondary wide-angle "finder scope" (if it has one), because the heat of sunlight can fry the finder scope's optics. Next, aim the telescope toward the Sun, using the telescope's shadow to judge when you are pointed in the right direction. You can then safely look through the telescope's eyepiece. When you are done, make sure you point the telescope away from the Sun before removing the filter and storing the telescope.

Note that the amount of solar activity that you can see (sunspots, filaments, flares, prominences, and so on) will depend on where the Sun is in its 11-year sunspot cycle.

58. Use a telescope with a solar filter to observe the surface of the Sun. Do you see any sunspots? Sketch their appearance. Can you distinguish between the umbrae and penumbrae of the sunspots? Can you see limb darkening? Can you see any granulation?
59. If you have access to an H_α filter attached to a telescope especially designed for viewing the Sun safely, use this instrument to examine the solar surface. How does the appearance of the Sun differ from that in white light? What do sunspots look like in H_α? Can you see any prominences? Can you see any filaments? Are the filaments in the H_α image near any sunspots seen in white light? (Note that the amount of activity that you see will be much greater at some times during the solar cycle than at others.)
60. Use the *Starry Night Enthusiast*™ program to measure the Sun's rotation. Display the entire celestial sphere by selecting **Guides > Atlas** in the **Favorites** menu and center on the Sun by double-clicking on **Sun** in the **Find** pane. Using the controls at the right-hand end of the toolbar, zoom in until you can see details on the Sun's surface clearly. In the toolbar, set the **Time Flow Rate** to 1 day. Using the time forward and backward buttons in the toolbar, step through enough time to determine the rotation period of the Sun. Which part of the actual Sun's surface rotates at the rate shown in *Starry Night Enthusiast*™? (Note: The program does not show the Sun's differential rotation.)
61. Use the *Starry Night Enthusiast*™ program to examine the Sun. Open the **Favorites** pane and double-click on **Solar System > Inner Solar System** to display the inner planets surrounding the Sun. Stop **Time Advance** and use the **down** arrow in the toolbar under **Viewing Location** to approach to within about 0.015 AU of the Sun. You can rotate the Sun by placing the mouse cursor over the image and, while holding down the **Shift** key, hold down the mouse button while moving the mouse. (On a two-button mouse, hold down the left mouse button.) **(a)** The Sun's equator lies close to the plane of the ecliptic. Where do most of the sunspots visible on the image of the Sun lie relative to the solar equator? Check Figure 16-18 for more realistic images of sunspots and Figure 16-19 for the latitude distribution of sunspots. **(b)** Based on your observations in (a), does the image in *Starry Night Enthusiast*™ show the Sun near the beginning, middle, or end of the 11-year sunspot cycle? Explain your reasoning. You can see current solar images from both ground and space-based solar telescopes by opening the LiveSky pane if you have an Internet connection on your computer.

Collaborative Exercises

62. Figure 16-19 shows variations in the average latitude of sunspots. Estimate the average latitude of sunspots in the year you were born and estimate the average latitude on your twenty-first birthday. Make rough sketches of the Sun during those years to illustrate your answers.
63. Create a diagram showing a sketch of how limb darkening on the Sun would look different if the Sun had either a thicker or thinner photosphere. Be sure to include a caption explaining your diagram.
64. Solar granules, shown in Figure 16-9, are about 1000 km across. What city is about that distance away from where you are right now? What city is that distance from the birthplace of each group member?
65. Magnetic arches in the corona are shown in Figure 16-25*a*. How many Earths high are these arches, and how many Earths could fit inside one arch?

17 The Nature of the Stars

R I V U X G

Some stars in this cluster (called M39) are distinctly blue in color, while others are yellow or red. (Heidi Schweiker/NOAO/AURA/NSF)

WEB LINK 17.1

To the unaided eye, the night sky is spangled with several thousand stars, each appearing as a bright pinpoint of light. With a pair of binoculars, you can see some 10,000 other, fainter stars; with a 15-cm (6-in) telescope, the total rises to more than 2 million. Astronomers now know that there are in excess of 100 billion (10^{11}) stars in our Milky Way Galaxy alone.

But what are these distant pinpoints? To the great thinkers of ancient Greece, the stars were bits of light embedded in a vast sphere with Earth at the center. They thought the stars were composed of a mysterious "fifth element," quite unlike anything found on Earth.

Today, we know that the stars are made of the same chemical elements found on Earth. We know their sizes, their temperatures, their masses, and something of their internal structures. We understand, too, why the stars in the accompanying image come in a range of beautiful colors: Blue stars have high surface temperatures, while the surface temperatures of red and yellow stars are relatively low.

How have we learned these things? How can we know the nature of the stars, objects so distant that their light takes years or centuries to reach us? In this chapter, we will learn about the measurements and calculations that astronomers make to determine the properties of stars. We will also take a first look at the Hertzsprung-Russell diagram, an important tool that helps astronomers systematize the wealth of available information about the stars. In later chapters, we will use this diagram to help us understand how stars are born, evolve, and eventually die.

17-1 Careful measurements of the parallaxes of stars reveal their distances

WEB LINK 17.2

The vast majority of stars are objects very much like the Sun. This understanding followed from the discovery that the stars are tremendously far from us, at distances so great that their light takes years to reach us. Because the stars at night are clearly visible to the naked eye despite these

Learning Goals

By reading the sections of this chapter, you will learn

17-1 How we can measure the distances to the stars

17-2 How we measure a star's brightness and luminosity

17-3 The magnitude scale for brightness and luminosity

17-4 How a star's color indicates its temperature

17-5 How a star's spectrum reveals its chemical composition

17-6 How we can determine the sizes of stars

17-7 How H-R diagrams summarize our knowledge of the stars

17-8 How we can deduce a star's size from its spectrum

17-9 How we can use binary stars to measure the masses of stars

17-10 How we can learn about binary stars in very close orbits

17-11 What eclipsing binaries are and what they tell us about the sizes of stars

huge distances, it must be that the **luminosity** of the stars—that is, how much energy they emit into space per second—is comparable to or greater than that of the Sun. Just as for the Sun, the only explanation for such tremendous luminosities is that thermonuclear reactions are occurring within the stars (see Section 16-1).

Clearly, then, it's important to know how distant the stars are. But how do we measure these distances? You might think these distances are determined by comparing how bright different stars appear. Perhaps the star Betelgeuse in the constellation Orion appears bright because it is relatively close, while the dimmer and less conspicuous star Polaris (the North Star, in the constellation Ursa Minor) is farther away.

But this line of reasoning is incorrect: Polaris is actually closer to us than Betelgeuse! How bright a star appears is *not* a good indicator of its distance. If you see a light on a darkened road, it could be a motorcycle headlight a kilometer away or a person holding a flashlight just a few meters away. In the same way, a bright star might be extremely far away but have an unusually high luminosity, and a dim star might be relatively close but have a rather low luminosity. Astronomers must use other techniques to determine the distances to the stars.

Parallax and the Distances to the Stars

TUTORIAL 17.1

The most straightforward way of measuring stellar distances uses an effect called **parallax,** which is the apparent displacement of an object because of a change in the observer's point of view (Figure 17-1). To see how parallax works, hold your arm out straight in front of you. Now look at the hand on your outstretched arm, first with your left eye closed, then with your right eye closed. When you close one eye and open the other, your hand appears to shift back and forth against the background of more distant objects.

When you are at position B, the tree appears to be in front of this mountain.
When you are at position A, the tree appears to be in front of this mountain.
Position A
Position B

Figure 17-1

Parallax Imagine looking at some nearby object (a tree) against a distant background (mountains). When you move from one location to another, the nearby object appears to shift with respect to the distant background scenery. This familiar phenomenon is called parallax.

The closer the object you are viewing, the greater the parallax shift. To see this increased shift, repeat the experiment with your hand held closer to your face. Your brain analyzes such parallax shifts constantly as it compares the images from your left and right eyes, and in this way determines the distances to objects around you. This analysis is the basis for depth perception.

You measure distances around you by comparing the images from your left and right eyes—we find the distances to stars using the same principle

To measure the distance to a star, astronomers measure the parallax shift of the star using two points of view that are as far apart as possible—at opposite sides of Earth's orbit. The direction from Earth to a nearby star changes as our planet orbits the Sun, and the nearby star appears to move back and forth against the background of more distant stars (Figure 17-2). This motion is called **stellar parallax.** The parallax (p) of a star is equal to half the angle through which the star's apparent position shifts as Earth moves from one side of its orbit to the other. The larger the parallax p, the smaller the distance d to the star (compare Figure 17-2*a* with Figure 17-2*b*).

It is convenient to measure the distance d in **parsecs.** A star with a parallax angle of 1 second of arc (p = 1 arcsec) is at a distance of 1 parsec (d = 1 pc). (The word "parsec" is a contraction of the phrase "the distance at which a star has a *par*allax of one arc*sec*ond." Recall from Section 1-7 that 1 parsec equals 3.26 light-years, 3.09×10^{13} km, or 206,265 AU; see Figure 1-14.) If the angle p is measured in arcseconds, then the distance d to the star in parsecs is given by the following equation:

Relation between a star's distance and its parallax

$$d = \frac{1}{p}$$

d = distance to a star, in parsecs

p = parallax angle of that star, in arcseconds

This simple relationship between parallax and distance in parsecs is one of the main reasons that astronomers usually measure cosmic distances in parsecs rather than light-years. For example, a star whose parallax is $p = 0.1$ arcsec is at a distance $d = 1/(0.1) = 10$ parsecs from Earth. Barnard's star, named for the American astronomer Edward E. Barnard, has a parallax of 0.547 arcsec. Hence, the distance to this star is

$$d = \frac{1}{p} = \frac{1}{0.547} = 1.83 \text{ pc}$$

Because 1 parsec is 3.26 light-years, this distance can also be expressed as

$$d = 1.83 \text{ pc} \times \frac{3.26 \text{ ly}}{1 \text{ pc}} = 5.97 \text{ ly}$$

All known stars have parallax angles less than one arcsecond. In other words, the closest star is more than 1 parsec away. Such

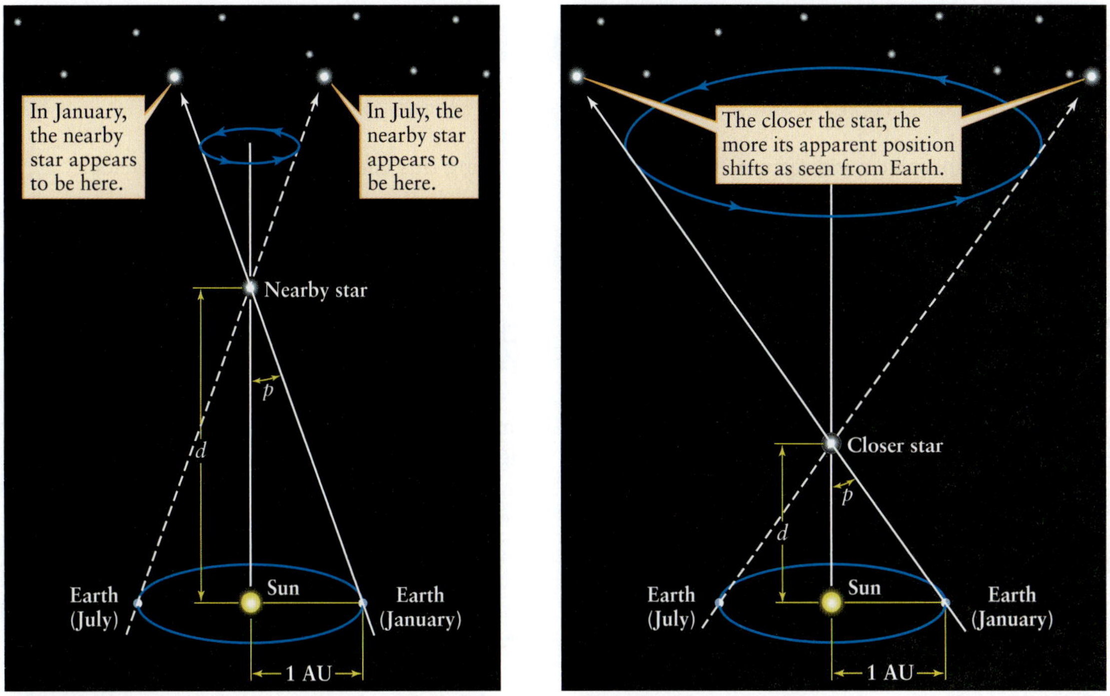
In January, the nearby star appears to be here.
In July, the nearby star appears to be here.
Nearby star
p
d
Earth (July)
Sun
Earth (January)
1 AU
The closer the star, the more its apparent position shifts as seen from Earth.
Closer star
p
d
Earth (July)
Sun
Earth (January)
1 AU

(a) Parallax of a nearby star

(b) Parallax of an even closer star

ANIMATION 17.1

Figure 17-2

Stellar Parallax (a) As Earth orbits the Sun, a nearby star appears to shift its position against the background of distant stars. The parallax (p) of the star is equal to the angular radius of Earth's orbit as seen from the star. (b) The closer the star is to us, the greater the parallax angle p. The distance d to the star (in parsecs) is equal to the reciprocal of the parallax angle p (in arcseconds): $d = 1/p$.

small parallax angles are difficult to detect, so it was not until 1838 that the first successful parallax measurements were made by the German astronomer and mathematician Friedrich Wilhelm Bessel. He found the parallax angle of the star 61 Cygni to be just 1/3 arcsec—equal to the angular diameter of a dime at a distance of 11 kilometers, or 7 miles. He thus determined that this star is about 3 pc from Earth. (Modern measurements give a slightly smaller parallax angle, which means that 61 Cygni is actually more than 3 pc away.) The star Proxima Centauri has the largest known parallax angle, 0.772 arcsec, and hence is the closest known star (other than the Sun); its distance is 1/(0.772) = 1.30 pc.

WEB LINK 17.3 AIMM 17.1

Appendix 4 at the back of this book lists all the stars within 4 pc of the Sun, as determined by parallax measurements. Most of these stars are far too dim to be seen with the naked eye, which is why their names are probably unfamiliar to you. By contrast, the majority of the familiar, bright stars in the nighttime sky (listed in Appendix 5) are so far away that their parallaxes cannot be measured from Earth's surface. They appear bright not because they are close, but because they are far more luminous than the Sun. The brightest stars in the sky are *not* necessarily the nearest stars!

Measuring Parallax from Space

Parallax angles smaller than about 0.01 arcsec are extremely difficult to measure from Earth, in part because of the blurring effects of the atmosphere. Therefore, the parallax method used with ground-based telescopes can give fairly reliable distances only for stars nearer than about 1/0.01 = 100 pc. But an observatory in space is unhampered by the atmosphere. Observations made from spacecraft therefore permit astronomers to measure even smaller parallax angles and thus determine the distances to more remote stars.

WEB LINK 17.4

In 1989 the European Space Agency (ESA) launched the satellite *Hipparcos*, an acronym for *H*igh *P*recision *Par*allax *Co*llecting *S*atellite (and a commemoration of the ancient Greek astronomer Hipparchus, who created one of the first star charts). Over more than three years of observations, the telescope aboard *Hipparcos* was used to measure the parallaxes of 118,000 stars, some with an accuracy of 0.001 arcsecond. This telescope has enabled astronomers to determine stellar distances out to several hundred parsecs, and with much greater precision than has been possible with ground-based observations. In the years to come, astronomers will increasingly turn to space-based observations to determine stellar distances.

Unfortunately, most of the stars in the Galaxy are so far away that their parallax angles are too small to measure even with an orbiting telescope. Later in this chapter, we will discuss a technique that can be used to find the distances to these more remote stars. In Chapters 24 and 26 we will learn about other techniques that astronomers use to determine the much larger distances to galaxies beyond the Milky Way. These techniques also

BOX 17-1 Tools of the Astronomer's Trade

Stellar Motions

Stars can move through space in any direction. The **space velocity** of a star describes how fast and in what direction it is moving. As the accompanying figure shows, a star's space velocity v can be broken into components parallel and perpendicular to our line of sight.

The component perpendicular to our line of sight—that is, across the plane of the sky—is called the star's **tangential velocity** (v_t). To determine it, astronomers must know the distance to a star (d) and its **proper motion** (μ, the Greek letter mu), which is the number of arcseconds that the star appears to move per year on the celestial sphere. Proper motion does not repeat itself yearly, so it can be distinguished from the apparent back-and-forth motion due to parallax. In terms of a star's distance and proper motion, its tangential velocity (in km/s) is

$$v_t = 4.74\mu d$$

where μ is in arcseconds per year and d is in parsecs. For example, Barnard's star (Figure 17-3) has a proper motion of 10.358 arcseconds per year and a distance of 1.83 pc. Hence, its tangential velocity is

$$v_t = 4.74(10.358)(1.83) = 89.8 \text{ km/s}$$

The component of a star's motion parallel to our line of sight—that is, either directly toward us or directly away from us—is its **radial velocity** (v_r). It can be determined from measurements of the Doppler shifts of the star's spectral lines (see Section 5-9 and Box 5-6). If a star is approaching us, the wavelengths of all of its spectral lines are decreased (blueshifted); if the star is receding from us, the wavelengths are increased (redshifted). The radial velocity v_r is related to the wavelength shift by the equation

$$\frac{\lambda - \lambda_0}{\lambda_0} = \frac{v_r}{c}$$

In this equation, λ is the wavelength of light coming from the star, λ_0 is what the wavelength would be if the star were not moving, and c is the speed of light. As an illustration, a particular spectral line of iron in the spectrum of Barnard's star has a wavelength (λ) of 516.445 nm. As measured in a laboratory on Earth, the same spectral line has a wavelength (λ_0) of 516.629 nm. Thus, for Barnard's star, our equation becomes

$$\frac{516.445 \text{ nm} - 516.629 \text{ nm}}{516.629 \text{ nm}} = -0.000356 = \frac{v_r}{c}$$

Solving this equation for the radial velocity v_r, we find

$$v_r = (-0.000356)\, c = (-0.000356)(3.00 \times 10^5 \text{ km/s})$$
$$= -107 \text{ km/s}$$

The minus sign means that Barnard's star is moving toward us. You can check this interpretation by noting that the wavelength λ = 516.445 nm received from Barnard's star is less than the laboratory wavelength λ_0 = 516.629 nm; hence, the light from the star is blueshifted, which indeed means that the star is approaching. If the star were receding, its light would be redshifted, and its radial velocity would be positive.

The illustration shows that the tangential velocity and radial velocity form two sides of a right triangle. The long side (hypotenuse) of this triangle is the space velocity (v). From the Pythagorean theorem, the space velocity is

$$v = \sqrt{v_t^2 + v_r^2}$$

For Barnard's star, the space velocity is

$$v = \sqrt{(89.4 \text{ km/s})^2 + (-107 \text{ km/s})^2} = 140 \text{ km/s}$$

Therefore, Barnard's star is moving through space at a speed of 140 km/s (503,000 km/h, or 312,000 mi/h) relative to the Sun.

Determining the space velocities of stars is essential for understanding the structure of the Galaxy. Studies show that the stars in our local neighborhood are moving in wide orbits around the center of the Galaxy, which lies some 8000 pc (26,000 light-years) away in the direction of the constellation Sagittarius (the Archer). While many of the orbits are roughly circular and lie in nearly the same plane, others are highly elliptical or steeply inclined to the galactic plane. We will see in Chapter 23 how the orbits of stars and gas clouds reveal the Galaxy's spiral structure.

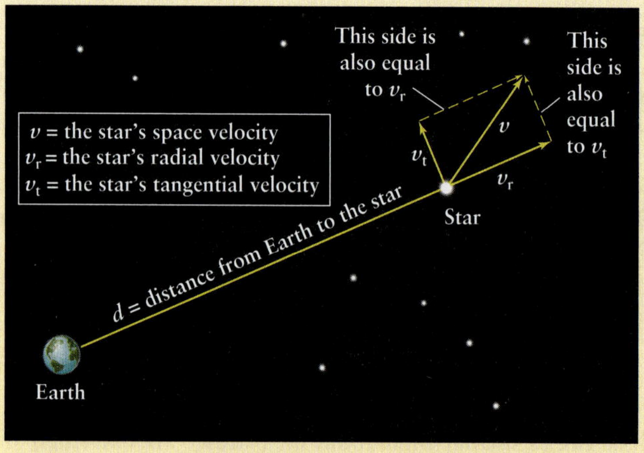

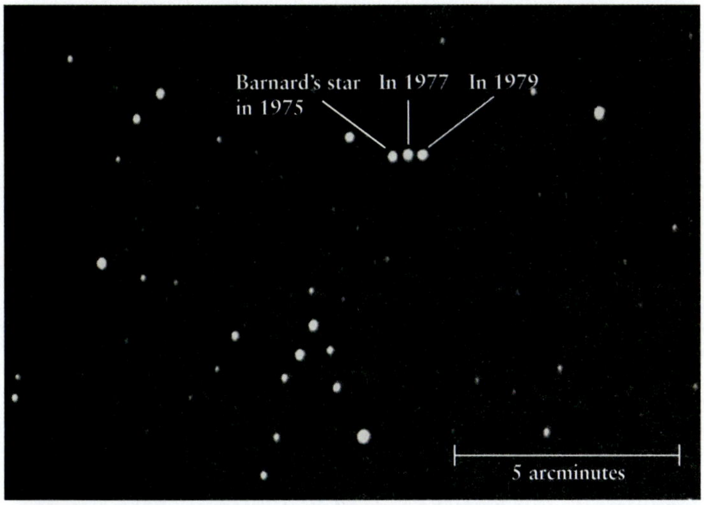

WEB LINK 17.5

Figure 17-3 RI**V**UXG

The Motion of Barnard's Star Three photographs taken over a four-year period were combined to show the motion of Barnard's star, which lies 1.82 parsecs away in the constellation Ophiuchus. Over this time interval, Barnard's star moved more than 41 arcseconds on the celestial sphere (about 0.69 arcminutes, or 0.012°), more than any other star. (John Sanford/Science Photo Library)

help us understand the overall size, age, and structure of the universe.

The Importance of Parallax Measurements

Because it can be used only on relatively close stars, stellar parallax might seem to be of limited usefulness. But parallax measurements are the cornerstone for all other methods of finding the distances to remote objects. These other methods require a precise and accurate knowledge of the distances to nearby stars, as determined by stellar parallax. Hence, any inaccuracies in the parallax angles for nearby stars can translate into substantial errors in measurement for the whole universe. To minimize these errors astronomers are continually trying to perfect their parallax-measuring techniques.

Stellar parallax is an *apparent* motion of stars caused by Earth's orbital motion around the Sun. But stars are not fixed objects and actually do move through space. As a result, stars change their positions on the celestial sphere (Figure 17-3), and they move either toward or away from the Sun. These motions are sufficiently slow, however, that changes in the positions of the stars are hardly noticeable over a human lifetime. Box 17-1 describes how astronomers study these motions and what insights they gain from these studies.

17-2 If a star's distance is known, its luminosity can be determined from its apparent brightness

TUTORIAL 17.2

All the stars you can see in the nighttime sky shine by thermonuclear fusion, just as the Sun does (see Section 16-1). But they are by no means merely identical copies of the Sun. Stars differ in their luminosity (L), the amount of light energy they emit each second. Luminosity is usually measured either in watts (1 watt, or 1 W, is 1 joule per second) or as a multiple of the Sun's luminosity ($L_\odot$, equal to 3.90×10^{26} W). Most stars are less luminous than the Sun, but some blaze forth with a million times the Sun's luminosity. Knowing a star's luminosity is essential for determining the star's history, its present-day internal structure, and its future evolution.

Luminosity, Apparent Brightness, and the Inverse-Square Law

Apparent brightness is a measure of how faint a star looks to us; luminosity is a measure of the star's total light output

To determine the luminosity of a star, we first note that as light energy moves away from its source, it spreads out over increasingly larger regions of space. Imagine a sphere of radius d centered on the light source, as in Figure 17-4. The amount of energy that passes each second through a square meter of the sphere's surface area is the total luminosity of the source (L) divided by the total surface area of the sphere (equal to $4\pi d^2$). This quantity is called the **apparent brightness** of the light, or just **brightness** (b), because how bright a light source appears depends on how much light energy per second enters through the area of a light detector (such as your eye). Apparent brightness is measured in watts per square meter (W/m^2). Written in the form of an equation, the relationship between apparent brightness and luminosity is

Inverse-square law relating apparent brightness and luminosity

$$b = \frac{L}{4\pi d^2}$$

b = apparent brightness of a star's light, in W/m^2

L = star's luminosity, in watts

d = distance to star, in meters

This relationship is called the **inverse-square law,** because the apparent brightness of light that an observer can see or measure is inversely proportional to the square of the observer's distance (d) from the source. If you double your distance from a light source, its radiation is spread out over an area 4 times larger, so the apparent brightness you see is decreased by a factor of 4. Similarly, at triple the distance, the apparent brightness is 1/9 as great (see Figure 17-4).

We can apply the inverse-square law to the Sun, which is 1.50×10^{11} m from Earth. Its apparent brightness ($b_\odot$) is

$$b_\odot = \frac{3.90 \times 10^{26}\ \text{W}}{4\pi(1.50 \times 10^{11}\ \text{m})^2} = 1370\ \text{W/m}^2$$

That is, a solar panel with an area of 1 square meter receives 1370 watts of power from the Sun.

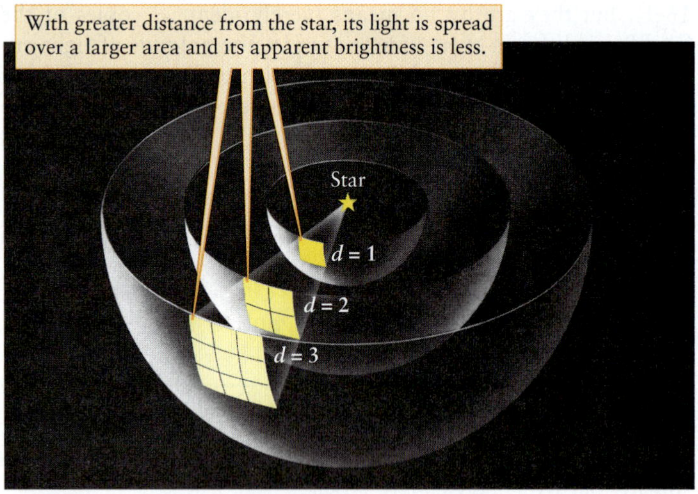
With greater distance from the star, its light is spread over a larger area and its apparent brightness is less.
Star
d = 1
d = 2
d = 3

ANIMATION 17.2

Figure 17-4

The Inverse-Square Law Radiation from a light source illuminates an area that increases as the square of the distance from the source. Hence, the apparent brightness decreases as the square of the distance. The brightness at $d = 2$ is $1/(2^2) = 1/4$ of the brightness at $d = 1$, and the brightness at $d = 3$ is $1/(3^2) = 1/9$ of that at $d = 1$.

Astronomers measure the apparent brightness of a star using a telescope with an attached light-sensitive instrument, similar to the light meter in a camera that determines the proper exposure. Measuring a star's apparent brightness is called **photometry.**

Calculating a Star's Luminosity

The inverse-square law says that we can find a star's luminosity if we know its distance and its apparent brightness. For convenience, this law can be expressed in a somewhat different form. We first rearrange the above equation:

$$L = 4\pi d^2 b$$

We then apply this equation to the Sun. That is, we write a similar equation relating the Sun's luminosity ($L_\odot$), the distance from Earth to the Sun ($d_\odot$, equal to 1 AU), and the Sun's apparent brightness ($b_\odot$):

$$L_\odot = 4\pi d_\odot^2 b_\odot$$

If we take the ratio of these two equations, the unpleasant factor of 4π drops out and we are left with the following:

Determining a star's luminosity from its apparent brightness

$$\frac{L}{L_\odot} = \left(\frac{d}{d_\odot}\right)^2 \frac{b}{b_\odot}$$

$L/L_\odot$ = ratio of the star's luminosity to the Sun's luminosity

$d/d_\odot$ = ratio of the star's distance to the Earth-Sun distance

$b/b_\odot$ = ratio of the star's apparent brightness to the Sun's apparent brightness

We need to know just two things to find a star's luminosity: the distance to a star as compared to the Earth-Sun distance (the ratio $d/d_\odot$), and how that star's apparent brightness compares to that of the Sun (the ratio $b/b_\odot$). Then we can use the above equation to find how luminous that star is compared to the Sun (the ratio $L/L_\odot$).

In other words, this equation gives us a general rule relating the luminosity, distance, and apparent brightness of a star:

We can determine the luminosity of a star from its distance and apparent brightness. For a given distance, the brighter the star, the more luminous that star must be. For a given apparent brightness, the more distant the star, the more luminous it must be to be seen at that distance.

Box 17-2 shows how to use the above equation to determine the luminosity of the nearby star ε (epsilon) Eridani, the fifth brightest star in the constellation Eridanus (named for a river in Greek mythology). Parallax measurements indicate that ε Eridani is 3.23 pc away, and photometry shows that the star appears only 6.73×10^{-13} as bright as the Sun. Using the above equation, we find that ε Eridani has only 0.30 times the luminosity of the Sun.

The Stellar Population

Calculations of this kind show that stars come in a wide variety of different luminosities, with values that range from about $10^6\ L_\odot$ (a million times the Sun's luminosity) to only about $10^{-4}\ L_\odot$ (a mere ten-thousandth of the Sun's light output). The most luminous star emits roughly 10^{10} times more energy each second than the least luminous! (To put this number in perspective, about 10^{10} human beings have lived on Earth since our species first evolved.)

As stars go, our Sun is neither extremely luminous nor extremely dim; it is a rather ordinary, garden-variety star. It is somewhat more luminous than most stars, however. Of more than 30 stars within 4 pc of the Sun (see Appendix 4), only three (α Centauri, Sirius, and Procyon) have a greater luminosity than the Sun.

To better characterize a typical population of stars, astronomers count the stars out to a certain distance from the Sun and plot the number of stars that have different luminosities. The resulting graph is called the **luminosity function.** Figure 17-5 on page 440 shows the luminosity function for stars in our part of the Milky Way Galaxy. The curve declines very steeply for the most luminous stars toward the left side of the graph, indicating that they are quite rare. For example, this graph shows that stars like the Sun are about 10,000 times more common than stars like Spica (which has a luminosity of 2100 $L_\odot$).

The exact shape of the curve in Figure 17-5 applies only to the vicinity of the Sun and similar regions in our Milky Way Galaxy. Other locations have somewhat different luminosity functions. In stellar populations in general, however, low-luminosity stars are much more common than high-luminosity ones.

BOX 17-2 Tools of the Astronomer's Trade

Luminosity, Distance, and Apparent Brightness

The inverse-square law (Section 17-2) relates a star's luminosity, distance, and apparent brightness to the corresponding quantities for the Sun:

$$\frac{L}{L_\odot} = \left(\frac{d}{d_\odot}\right)^2 \frac{b}{b_\odot}$$

We can use a similar equation to relate the luminosities, distances, and apparent brightnesses of *any* two stars, which we call star 1 and star 2:

$$\frac{L_1}{L_2} = \left(\frac{d_1}{d_2}\right)^2 \frac{b_1}{b_2}$$

EXAMPLE: The star ε (epsilon) Eridani is 3.23 pc from Earth. As seen from Earth, this star appears only 6.73×10^{-13} as bright as the Sun. What is the luminosity of ε Eridani compared with that of the Sun?

Situation: We are given the distance to ε Eridani (d = 3.23 pc) and this star's brightness compared to that of the Sun ($b/b_\odot = 6.73 \times 10^{-13}$). Our goal is to find the ratio of the luminosity of ε Eridani to that of the Sun, that is, the quantity $L/L_\odot$.

Tools: Since we are asked to compare this star to the Sun, we use the first of the two equations given above, $L/L_\odot = (d/d_\odot)^2(b/b_\odot)$, to solve for $L/L_\odot$.

Answer: Our equation requires the ratio of the star's distance to the Sun's distance, $d/d_\odot$. The distance from Earth to the Sun is $d_\odot = 1$ AU. To calculate the ratio $d/d_\odot$, we must express both distances in the same units. There are 206,265 AU in 1 parsec, so we can write the distance to ε Eridani as $d = (3.23 \text{ pc})(206{,}265 \text{ AU/pc}) = 6.66 \times 10^5$ AU. Hence, the ratio of distances is $d/d_\odot = (6.66 \times 10^5 \text{ AU})/(1 \text{ AU}) = 6.66 \times 10^5$. Then we find that the ratio of the luminosity of ε Eridani (L) to the Sun's luminosity ($L_\odot$) is

$$\frac{L}{L_\odot} = \left(\frac{d}{d_\odot}\right)^2 \frac{b}{b_\odot} = (6.66 \times 10^5)^2 \times (6.73 \times 10^{-13}) = 0.30$$

Review: This result means that ε Eridani is only 0.30 as luminous as the Sun; that is, its power output is only 30% as great.

EXAMPLE: Suppose star 1 is at half the distance of star 2 (that is, $d_1/d_2 = \frac{1}{2}$) and that star 1 appears twice as bright as star 2 (that is, $b_1/b_2 = 2$). How do the luminosities of these two stars compare?

Situation: For these two stars, we are given the ratio of distances (d_1/d_2) and the ratio of apparent brightnesses (b_1/b_2). Our goal is to find the ratio of their luminosities (L_1/L_2).

Tools: Since we are comparing two stars, neither of which is the Sun, we use the second of the two equations above: $L_1/L_2 = (d_1/d_2)^2(b_1/b_2)$.

Answer: Plugging values into our equation, we find

$$\frac{L_1}{L_2} = \left(\frac{d_1}{d_2}\right)^2 \frac{b_1}{b_2} = \left(\frac{1}{2}\right)^2 \times 2 = \frac{1}{2}$$

Review: This result says that star 1 has only one-half the luminosity of star 2. Despite this, star 1 appears brighter than star 2 because it is closer to us.

The two equations above are also useful in the method of *spectroscopic parallax,* which we discuss in Section 17-8. It turns out that a star's luminosity can be determined simply by analyzing the star's spectrum. If the star's apparent brightness is also known, the star's distance can be calculated. The inverse-square law can be rewritten as an expression for the ratio of the star's distance from Earth (d) to the Earth-Sun distance ($d_\odot$):

$$\frac{d}{d_\odot} = \sqrt{\frac{(L/L_\odot)}{(b/b_\odot)}}$$

We can also use this formula as a relation between the properties of any two stars, 1 and 2:

$$\frac{d_1}{d_2} = \sqrt{\frac{(L_1/L_2)}{(b_1/b_2)}}$$

EXAMPLE: The star Pleione in the constellation Taurus is 190 times as luminous as the Sun but appears only 3.19×10^{-13} as bright as the Sun. How far is Pleione from Earth?

Situation: We are told the ratio of Pleione's luminosity to that of the Sun ($L/L_\odot = 190$) and the ratio of their apparent brightnesses ($b/b_\odot = 3.19 \times 10^{-13}$). Our goal is to find the distance d from Earth to Pleione.

Tools: Since we are comparing Pleione to the Sun, we use the first of the two equations above.

Answer: Our equation tells us the ratio of the Earth-Pleione distance to the Earth-Sun distance:

$$\frac{d}{d_\odot} = \sqrt{\frac{(L/L_\odot)}{(b/b_\odot)}} = \sqrt{\frac{190}{3.19 \times 10^{-13}}}$$

$$= \sqrt{5.95 \times 10^{14}} = 2.44 \times 10^7$$

(continued on the next page)

BOX 17-2 *(continued)*

Hence, the distance from Earth to Pleione is 2.44×10^7 times greater than the distance from Earth to the Sun. The Sun-Earth distance is $d_\odot = 1$ AU and 206,265 AU = 1 pc, so we can express the star's distance as $d = (2.44 \times 10^7 \text{ AU}) \times (1 \text{ pc}/206{,}265 \text{ AU}) = 118$ pc.

Review: We can check our result by comparing it with the above example about the star ε Eridani. Pleione has a much greater luminosity than ε Eridani (190 times the Sun's luminosity versus 0.30 times), but Pleione appears dimmer than ε Eridani (3.19×10^{-13} times as bright as the Sun compared to 6.73×10^{-13} times). For this to be true, Pleione must be much farther away from Earth than is ε Eridani. This is just what our results show: $d = 118$ pc for Pleione compared to $d = 3.23$ pc for ε Eridani.

EXAMPLE: The star δ (delta) Cephei, which lies 300 pc from Earth, is thousands of times more luminous than the Sun. Thanks to this great luminosity, stars like δ Cephei can be seen in galaxies millions of parsecs away. As an example, the Hubble Space Telescope has detected stars like δ Cephei within the galaxy NGC 3351, which lies in the direction of the constellation Leo. These stars appear only 9×10^{-10} as bright as δ Cephei. What is the distance to NGC 3351?

Situation: To determine the distance we want, we need to find the distance to a star within NGC 3351. We are told that certain stars within this galaxy are identical to δ Cephei but appear only 9×10^{-10} as bright.

Tools: We use the equation $d_1/d_2 = \sqrt{(L_1/L_2)/(b_1/b_2)}$ to relate two stars, one within NGC 3351 (call this star 1) and the identical star δ Cephei (star 2). Our goal is to find d_1.

Answer: Since the two stars are identical, they have the same luminosity ($L_1 = L_2$, or $L_1/L_2 = 1$). The brightness ratio is $b_1/b_2 = 9 \times 10^{-10}$, so our equation tells us that

$$\frac{d_1}{d_2} = \sqrt{\frac{(L_1/L_2)}{(b_1/b_2)}} = \sqrt{\frac{1}{9 \times 10^{-10}}} = \sqrt{1.1 \times 10^9} = 33{,}000$$

Hence, NGC 3351 is 33,000 times farther away than δ Cephei, which is 300 pc from Earth. The distance from Earth to NGC 3351 is therefore (33,000)(300 pc) = 10^7 pc, or 10 megaparsecs (10 Mpc).

Review: This example illustrates one technique that astronomers use to measure extremely large distances. We will learn more about stars like δ Cephei in Chapter 19, and in Chapter 24 we will explore further how they are used to determine the distances to remote galaxies.

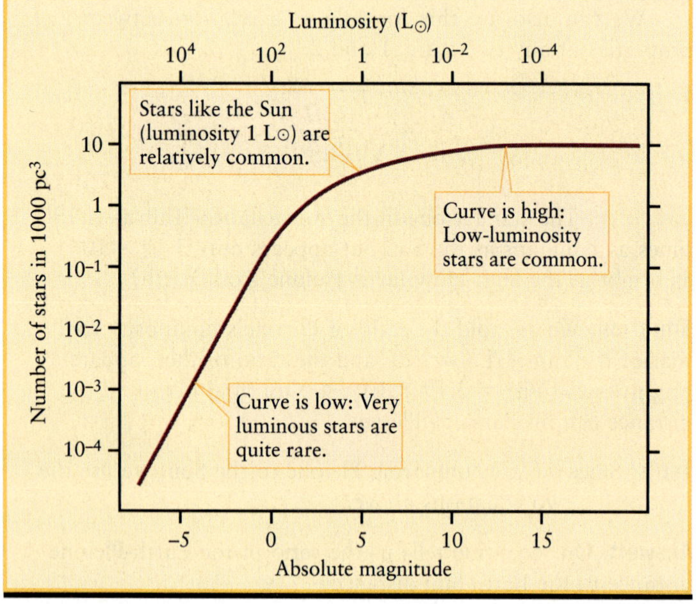

Figure 17-5

The Luminosity Function This graph shows how many stars of a given luminosity lie within a representative 1000 cubic-parsec volume. The scale at the bottom of the graph shows absolute magnitude, an alternative measure of a star's luminosity (described in Section 17-3). (Adapted from J. Bahcall and R. Soneira)

17-3 Astronomers often use the magnitude scale to denote brightness

Because astronomy is among the most ancient of sciences, some of the tools used by modern astronomers are actually many centuries old. One such tool is the **magnitude scale,** which astronomers frequently use to denote the brightness of stars. This scale was introduced in the second century B.C. by the Greek astronomer Hipparchus, who called the brightest stars first-magnitude stars. Stars about half as bright as first-magnitude stars were called second-magnitude stars, and so forth, down to sixth-magnitude stars, the dimmest ones he could see. After telescopes came into use, astronomers extended Hipparchus's magnitude scale to include even dimmer stars.

Apparent Magnitudes

The magnitudes in Hipparchus's scale are properly called **apparent magnitudes,** because they describe how bright an object *appears* to an Earth-based observer. Apparent magnitude is directly related to apparent brightness.

CAUTION! The magnitude scale can be confusing because it works "backward." Keep in mind that the *greater* the apparent magnitude, the *dimmer* the star. A star of apparent magnitude +3 (a third-magnitude star) is dimmer than a star of apparent magnitude +2 (a second-magnitude star).

In the nineteenth century, astronomers developed better techniques for measuring the light energy arriving from a star. These measurements showed that a first-magnitude star is about 100 times brighter than a sixth-magnitude star. In other words, it would take 100 stars of magnitude +6 to provide as much light energy as we receive from a single star of magnitude +1. To make computations easier, the magnitude scale was redefined so that a magnitude difference of 5 corresponds exactly to a factor of 100 in brightness. A magnitude difference of 1 corresponds to a factor of 2.512 in brightness, because

$$2.512 \times 2.512 \times 2.512 \times 2.512 \times 2.512 = (2.512)^5 = 100$$

Thus, it takes 2.512 third-magnitude stars to provide as much light as we receive from a single second-magnitude star.

Figure 17-6 illustrates the modern apparent magnitude scale. The dimmest stars visible through a pair of binoculars have an apparent magnitude of +10, and the dimmest stars that can be photographed in a one-hour exposure with the Keck telescopes (see Section 6-2) or the Hubble Space Telescope have apparent magnitude +30. Modern astronomers also use negative numbers to extend Hipparchus's scale to include very bright objects. For example, Sirius, the brightest star in the sky, has an apparent magnitude of −1.43. The Sun, the brightest object in the sky, has an apparent magnitude of −26.7.

Absolute Magnitudes

Apparent magnitude is a measure of a star's apparent brightness as seen from Earth. A related quantity that measures a star's true energy output—that is, its luminosity—is called **absolute magnitude.** This is the apparent magnitude a star would have if it were located exactly 10 parsecs from Earth.

Apparent magnitude measures a star's brightness; absolute magnitude measures its luminosity

ANALOGY If you wanted to compare the light output of two different lightbulbs, you would naturally place them side by side so that both bulbs were the same distance from you. In the absolute magnitude scale, we imagine doing the same thing with stars to compare their luminosities.

If the Sun were moved to a distance of 10 parsecs from Earth, it would have an apparent magnitude of +4.8. The absolute magnitude of the Sun is thus +4.8. The absolute magnitudes of the

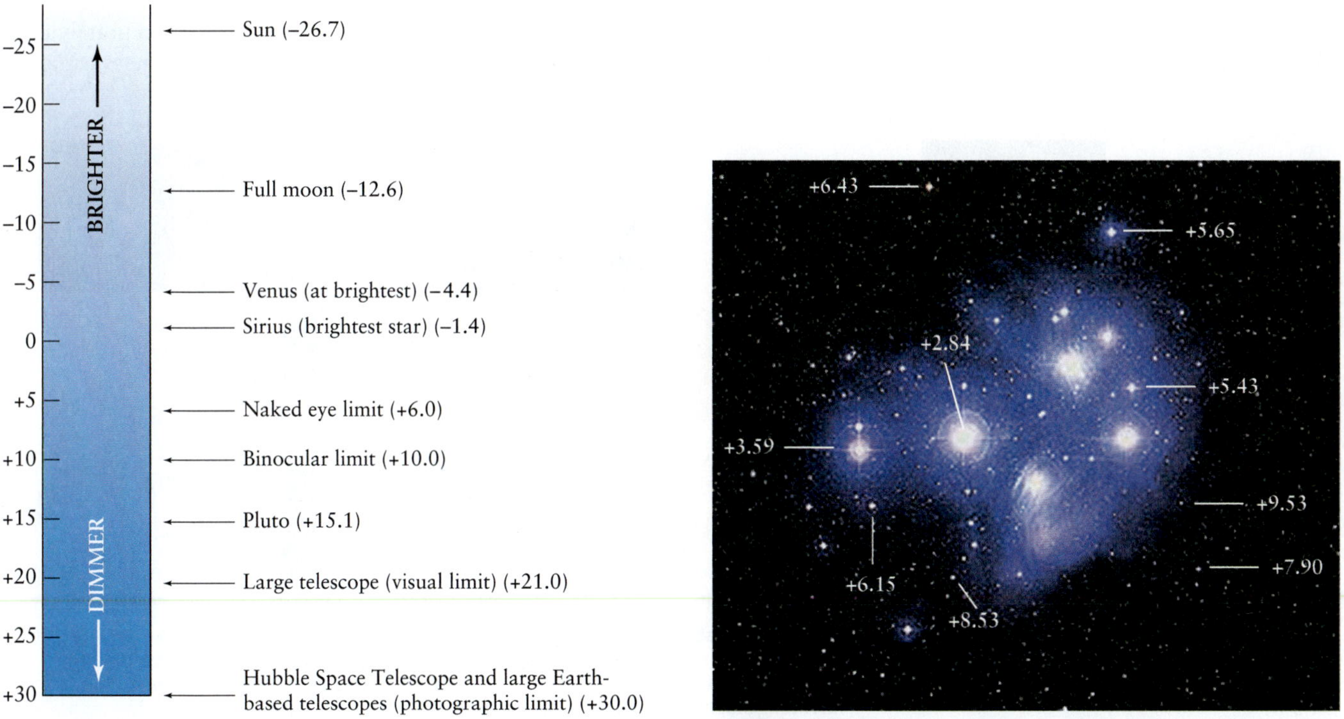

(a) Some apparent magnitudes

(b) Apparent magnitudes of stars in the Pleiades

RIVUXG

WEB LINK 17.6

Figure 17-6

The Apparent Magnitude Scale (a) Astronomers denote the apparent brightness of objects in the sky by their apparent magnitudes. The greater the apparent magnitude, the dimmer the object. **(b)** This photograph of the Pleiades cluster, located about 120 pc away in the constellation Taurus, shows the apparent magnitudes of some of its stars. Most are too faint to be seen by the naked eye. (David Malin/Anglo-Australian Observatory) stars range from approximately +15 for the least luminous to −10 for the most luminous. (*Note:* Like apparent magnitudes, absolute magnitudes work "backward": The *greater* the absolute magnitude, the *less luminous* the star.) The Sun's absolute magnitude is about in the middle of this range.

We saw in Section 17-2 that we can calculate the luminosity of a star if we know its distance and apparent brightness. There is a mathematical relationship between absolute magnitude and luminosity, which astronomers use to convert one to the other as they see fit. It is also possible to rewrite the inverse-square law, which we introduced in Section 17-2, as a mathematical relationship that allows you to calculate a star's absolute magnitude (a measure of its luminosity) from its distance and apparent magnitude (a measure of its apparent brightness). Box 17-3 describes these relationships and how to use them.

Because the "backward" magnitude scales can be confusing, we will use them only occasionally in this book. We will usually speak of a star's luminosity rather than its absolute magnitude and will describe a star's appearance in terms of apparent brightness rather than apparent magnitude. But if you go on to study more about astronomy, you will undoubtedly make frequent use of apparent magnitude and absolute magnitude.

17-4 A star's color depends on its surface temperature

WEB LINK 17.7

The image that opens this chapter shows that stars come in different colors. You can see these colors even with the naked eye. For example, you can easily see the red color of Betelgeuse, (the star in the "armpit" of the constellation Orion), and the blue tint of Bellatrix at Orion's other "shoulder" (see Figure 2-2). Colors are most evident for the brightest stars, because human color vision works poorly at low light levels.

CAUTION! It's true that the light from a star will appear redshifted if the star is moving away from you and blueshifted if it's moving toward you. But for even the fastest stars, these color shifts are so tiny that it takes sensitive instruments to measure them. The red color of Betelgeuse and the blue color of Bellatrix are not due to their motions; they are the actual colors of the stars.

Color and Temperature

We saw in Section 5-3 that a star's color is directly related to its surface temperature. The intensity of light from a relatively cool star peaks at long wavelengths, making the star look red (Figure 17-7*a*). A hot star's intensity curve peaks at shorter wavelengths, so the star looks blue (Figure 17-7*c*). For a star with an intermediate temperature, such as the Sun, the intensity peak is near the middle of the visible spectrum. The human visual system interprets an object with this spectrum of wavelengths as yellowish in color (Figure 17-7*b*). This leads to an important general rule about star colors and surface temperatures:

Red stars are relatively cold, with low surface temperatures; blue stars are relatively hot, with high surface temperatures.

Figure 17-7 shows that astronomers can accurately determine the surface temperature of a star by carefully measuring its color.

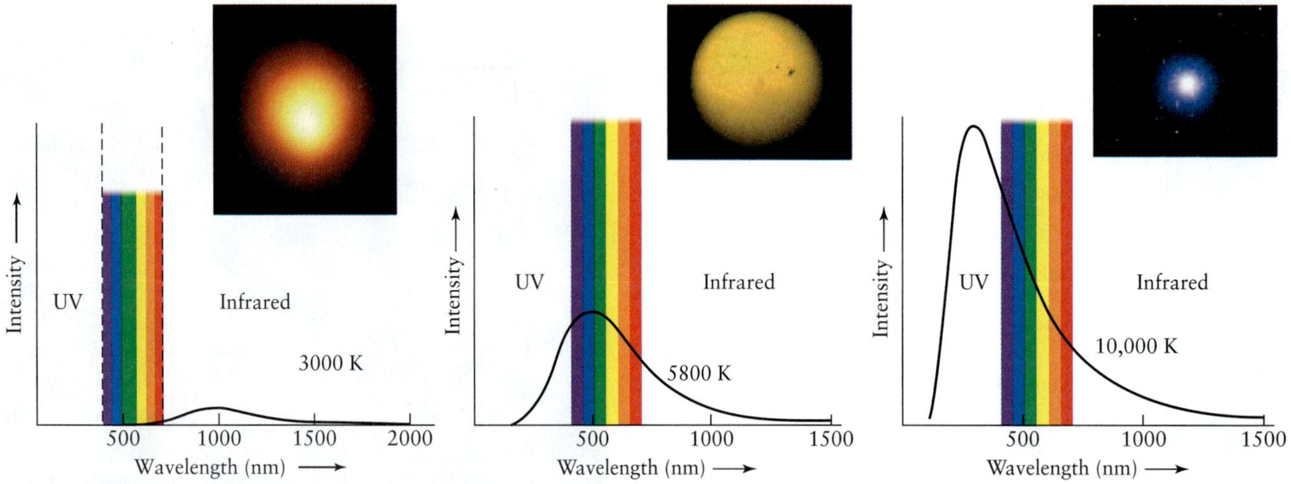

(a) A **cool star** with surface temperature 3000 K emits much more red light than blue light, and so appears red.

(b) A **warmer star** with surface temperature 5800 K (like the Sun) emits similar amounts of all visible wavelengths, and so appears yellow-white.

(c) A **hot star** with surface temperature 10,000 K emits much more blue light than red light, and so appears blue.

Figure 17-7

Temperature and Color These graphs show the intensity of light emitted by three hypothetical stars plotted against wavelength (compare with Figure 5-11). The rainbow band indicates the range of visible wavelengths. The star's apparent color depends on whether the intensity curve has larger values at the short-wavelength or long-wavelength end of the visible spectrum. The insets show stars of about these surface temperatures. UV stands for ultraviolet, which extends from 10 to 400 nm. See Figure 3-4 for more on wavelengths of the spectrum. (Inset a: Andrea Dupree/Harvard-Smithsonian CFA, Ronald Gilliland/STScI, NASA and ESA; inset b: NSO/AURA/NSF; inset c: Till Credner, Allthesky.com)

BOX 17-3 Tools of the Astronomer's Trade

Apparent Magnitude and Absolute Magnitude

Astronomers commonly express a star's apparent brightness in terms of apparent magnitude (denoted by a lowercase m), and the star's luminosity in terms of absolute magnitude (denoted by a capital M). While we do not use these quantities extensively in this book, it is useful to know a few simple relationships involving them.

Consider two stars, labeled 1 and 2, with apparent magnitudes m_1 and m_2 and brightnesses b_1 and b_2, respectively. The *ratio* of their apparent brightnesses (b_1/b_2) corresponds to a *difference* in their apparent magnitudes ($m_2 - m_1$). As we learned in Section 17-3, each step in magnitude corresponds to a factor of 2.512 in brightness; we receive 2.512 times more energy per square meter per second from a third-magnitude star than from a fourth-magnitude star. This idea was used to construct the following table:

Apparent magnitude difference ($m_2 - m_1$)	Ratio of apparent brightness (b_1/b_2)
1	2.512
2	$(2.512)^2 = 6.31$
3	$(2.512)^3 = 15.85$
4	$(2.512)^4 = 39.82$
5	$(2.512)^5 = 100$
10	$(2.512)^{10} = 10^4$
15	$(2.512)^{15} = 10^6$
20	$(2.512)^{20} = 10^8$

A simple equation relates the difference between two stars' apparent magnitudes to the ratio of their brightnesses:

Magnitude difference related to brightness ratio

$$m_2 - m_1 = 2.5 \log\left(\frac{b_1}{b_2}\right)$$

m_1, m_2 = apparent magnitudes of stars 1 and 2

b_1, b_2 = apparent brightnesses of stars 1 and 2

In this equation, $\log (b_1/b_2)$ is the logarithm of the brightness ratio. The logarithm of $1000 = 10^3$ is 3, the logarithm of $10 = 10^1$ is 1, and the logarithm of $1 = 10^0$ is 0.

EXAMPLE: At their most brilliant, Venus has a magnitude of about −4 and Mercury has a magnitude of about −2. How many times brighter are these planets than the dimmest stars visible to the naked eye, with a magnitude of +6?

Situation: In each case we want to find a ratio of two apparent brightnesses (the brightness of Venus or Mercury compared to that of the dimmest naked-eye stars).

Tools: In each case we will convert a *difference* in apparent magnitude between the planet and the naked-eye star into a *ratio* of their brightnesses.

Answer: The magnitude difference between Venus and the dimmest stars visible to the naked eye is $+6 - (-4) = 10$. From the table, this difference corresponds to a brightness ratio of $(2.512)^{10} = 10^4 = 10{,}000$, so Venus at it most brilliant is 10,000 times brighter than the dimmest naked-eye stars.

The magnitude difference between Mercury and the dimmest naked-eye stars is $+4 - (-4) = 8$. While this value is not in the table, you can see that the corresponding ratio of brightnesses is $(2.512)^8 = (2.512)^{5+3} = (2.512)^5 \times (2.512)^3$. From the table, $(2.512)^5 = 100$ and $(2.512)^3 = 15.85$, so the ratio of brightnesses is $100 \times 15.85 = 1585$. Hence, Mercury at its most brilliant is 1585 times brighter than the dimmest stars visible to the naked eye.

Review: Can you show that when at their most brilliant, Venus is 6.31 times brighter than Mercury? (*Hint:* No multiplication or division is required—just notice the difference in apparent magnitude between Venus and Mercury, and consult the table.)

EXAMPLE: The variable star RR Lyrae in the constellation Lyra (the Harp) periodically doubles its light output. By how much does its apparent magnitude change?

Situation: We are given a ratio of two brightnesses (the star at its maximum is twice as bright as at its minimum). Our goal is to find the corresponding difference in apparent magnitude.

Tools: We let 1 denote the star at its maximum brightness and 2 denote the same star at its dimmest, so the ratio of brightnesses is $b_1/b_2 = 2$. We then use the equation $m_2 - m_1 = 2.5 \log (b_1/b_2)$ to solve for the apparent magnitude difference $m_2 - m_1$.

Answer: Using a calculator, we find $m_2 - m_1 = 2.5 \log (2) = 2.5 \times 0.30 = 0.75$. RR Lyrae therefore varies periodically in brightness by 0.75 magnitude.

Review: Our answer means that at its dimmest, RR Lyrae has an apparent magnitude m_2 that is 0.75 *greater* than its apparent magnitude m_1 when it is brightest. (Remember that a greater value of apparent magnitude means the star is dimmer, not brighter!)

(continued on the next page)

BOX 17-3 (continued)

The inverse-square law relating a star's apparent brightness and luminosity can be rewritten in terms of the star's apparent magnitude (m), absolute magnitude (M), and distance from Earth (d). This can be expressed as an equation:

Relation between a star's apparent magnitude and absolute magnitude

$$m - M = 5 \log d - 5$$

m = star's apparent magnitude

M = star's absolute magnitude

d = distance from Earth to the star in parsecs

In this expression $m - M$ is called the **distance modulus,** and log d means the logarithm of the distance d in parsecs. For convenience, the following table gives the values of the distance d corresponding to different values of $m - M$.

Distance modulus $m - M$	Distance d (pc)
−4	1.6
−3	2.5
−2	4.0
−1	6.3
0	10
1	16
2	25
3	40
4	63
5	100
10	10^3
15	10^4
20	10^5

This table shows that if a star is less than 10 pc away, its distance modulus $m - M$ is negative. That is, its apparent magnitude (m) is less than its absolute magnitude (M). If the star is more than 10 pc away, $m - M$ is positive and m is greater than M. As an example, the star ε (epsilon) Indi, which is in the direction of the southern constellation Indus, has apparent magnitude $m = +4.7$. It is 3.6 pc away, which is less than 10 pc, so its apparent magnitude is less than its absolute magnitude.

EXAMPLE: Find the absolute magnitude of ε Indi.

Situation: We are given the distance to ε Indi ($d = 3.6$ pc) and its apparent magnitude ($m = +4.7$). Our goal is to find the star's absolute magnitude M.

Tools: We use the formula $m - M = 5 \log d - 5$ to solve for M.

Answer: Since $d = 3.6$ pc, we use a calculator to find log $d = \log 3.6 = 0.56$. Therefore the star's distance modulus is $m - M = 5(0.56) - 5 = -2.2$, and the star's absolute magnitude is $M = m - (-2.2) = +4.7 + 2.2 = +6.9$.

Review: As a check on our calculations, note that this star's distance modulus $m - M = -2.2$ is less than zero, as it should be for a star less than 10 pc away. Note that our Sun has absolute magnitude +4.8; ε Indi has a greater absolute magnitude, so it is less luminous than the Sun.

EXAMPLE: Suppose you were viewing the Sun from a planet orbiting another star 100 pc away. Could you see it without using a telescope?

Situation: We learned in the preceding examples that the Sun has absolute magnitude $M = +4.8$ and that the dimmest stars visible to the naked eye have apparent magnitude $m = +6$. Our goal is to determine whether the Sun would be visible to the naked eye at a distance of 100 pc.

Tools: We use the relationship $m - M = 5 \log d - 5$ to find the Sun's apparent magnitude at $d = 100$ pc. If this is greater than +6, the Sun would not be visible at that distance. (Remember that the greater the apparent magnitude, the dimmer the star.)

Answer: From the table, at $d = 100$ pc the distance modulus is $m - M = 5$. So, as seen from this distant planet, the Sun's apparent magnitude would be $m = M + 5 = +4.8 + 5 = +9.8$. This is greater than the naked-eye limit $m = +6$, so the Sun could not be seen.

Review: The Sun is by far the brightest object in Earth's sky. But our result tells us that to an inhabitant of a planetary system 100 pc away—a rather small distance in a galaxy that is thousands of parsecs across—our own Sun would be just another insignificant star, visible only through binoculars or a telescope.

The magnitude system is also used by astronomers to express the colors of stars as seen through different filters, as we describe in Section 17-4. For example, rather than quantifying a star's color by the color ratio b_V/b_B (a star's apparent brightness as seen through a V filter divided by the brightness through a B filter), astronomers commonly use the *color index* B–V, which is the difference in the star's apparent magnitude as measured with these two filters. We will not use this system in this book, however (but see Advanced Questions 53 and 54).

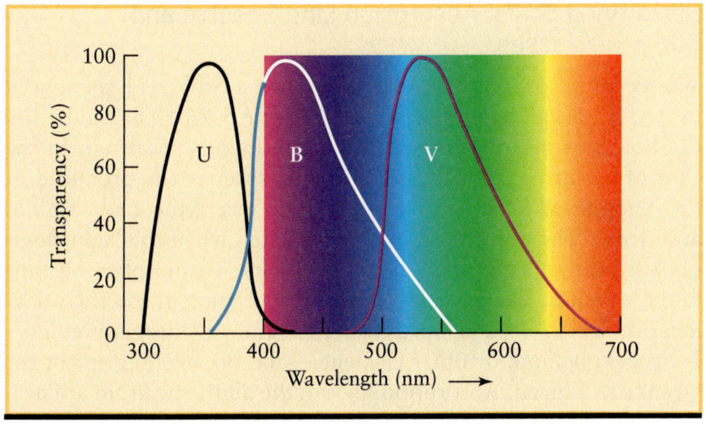

Figure 17-8

U, B, and V Filters This graph shows the wavelengths to which the standard filters are transparent. The U filter is transparent to near-ultraviolet light. The B filter is transparent to violet, blue, and green light, while the V filter is transparent to green and yellow light. By measuring the apparent brightness of a star with each of these filters and comparing the results, an astronomer can determine the star's surface temperature.

Astronomers use a set of filters in their telescopes to measure the surface temperatures of stars

To measure color, the star's light is collected by a telescope and passed through various color-filters. For example, a red filter passes red light while blocking other wavelengths. The filtered light is then collected by a light-sensitive device such as a CCD (see Section 6-4). The process is then repeated with each of the filters in the set. The star's image will have a different brightness through each colored filter, and by comparing these brightnesses astronomers can find the wavelength at which the star's intensity curve has its peak—and hence the star's temperature.

UBV Photometry

Let's look at this procedure in more detail. The most commonly used filters are called U, B, and V, and the technique that uses them is called **UBV photometry.** Each filter is transparent to a different band of wavelengths: the ultraviolet (U), the blue (B), and the yellow-green (V, for visual) region in and around the visible spectrum (Figure 17-8). The transparency of the V filter mimics the sensitivity of the human eye.

To determine a star's temperature using UBV photometry, the astronomer first measures the star's brightness through each of the filters individually. This gives three apparent brightnesses for the star, designated b_U, b_B, and b_V. The astronomer then compares the intensity of starlight in neighboring wavelength bands by taking the ratios of these brightnesses: b_V/b_B and b_B/b_U. Table 17-1 gives values for these **color ratios** for several stars with different surface temperatures.

If a star is very hot, its radiation is skewed toward short, ultraviolet wavelengths as in Figure 17-7*c*. This makes the star dim through the V filter, brighter through the B filter, and brightest through the U filter. Hence, for a hot star b_V is less than b_B, which in turn is less than b_U, and the ratios b_V/b_B and b_B/b_U are both less than 1. One such star is Bellatrix (see Table 17-1), which has a surface temperature of 21,500 K.

In contrast, if a star is cool, its radiation peaks at long wavelengths as in Figure 17-7*a*. Such a star appears brightest through the V filter, dimmer through the B filter, and dimmest through the U filter (see Figure 17-8). In other words, for a cool star b_V is greater than b_B, which in turn is greater than b_U. Hence, the ratios b_V/b_B and b_B/b_U will both be greater than 1. The star Betelgeuse (surface temperature 3500 K) is an example.

You can see these differences between hot and cool stars in parts *a* and *c* of Figure 6-27, which show the constellation Orion at ultraviolet wavelengths (a bit shorter than those transmitted by the U filter) and at visible wavelengths that approximate the transmission of a V filter. The hot star Bellatrix is brighter in the ultraviolet image (Figure 6-27*a*) than at visible wavelengths (Figure 6-27*c*). (Figure 6-27*d* shows the names of the stars.) The situation is reversed for the cool star Betelgeuse: It is bright at visible wavelengths, but at ultraviolet wavelengths it is too dim to show up in the image.

Figure 17-9 graphs the relationship between a star's b_V/b_B color ratio and its temperature. If you know the value of the b_V/b_B color ratio for a given star, you can use this graph to find the star's surface temperature. As an example, for the Sun b_V/b_B

Table 17-1 Colors of Selected Stars

Star	Surface temperature (K)	b_V/b_B	b_B/b_U	Apparent color
Bellatrix (γ Orionis)	21,500	0.81	0.45	Blue
Regulus (α Leonis)	12,000	0.90	0.72	Blue-white
Sirius (α Canis Majoris)	9400	1.00	0.96	Blue-white
Megrez (δ Ursae Majoris)	8630	1.07	1.07	White
Altair (α Aquilae)	7800	1.23	1.08	Yellow-white
Sun	5800	1.87	1.17	Yellow-white
Aldebaran (α Tauri)	4000	4.12	5.76	Orange
Betelgeuse (α Orionis)	3500	5.55	6.66	Red

Source: J.-C. Mermilliod, B. Hauck, and M. Mermilliod, University of Lausanne

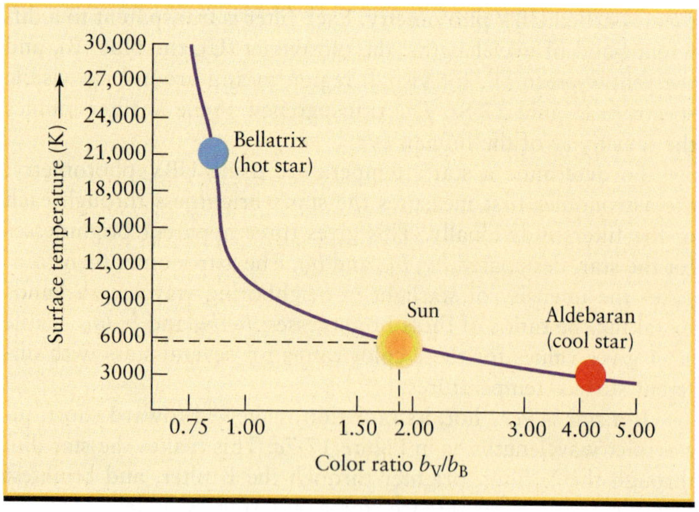

Figure 17-9

Temperature, Color, and Color Ratio The b_V/b_B color ratio is the ratio of a star's apparent brightnesses through a V filter and through a B filter. This ratio is small for hot, blue stars but large for cool, red stars. After measuring a star's brightness with the B and V filters, an astronomer can estimate the star's surface temperature from a graph like this one.

equals 1.87, which corresponds to a surface temperature of 5800 K.

CAUTION! As we will see in Chapter 18, tiny dust particles that pervade interstellar space cause distant stars to appear redder than they really are. (In the same way, particles in Earth's atmosphere make the setting Sun look redder; see Box 5-4.) Astronomers must take this reddening into account whenever they attempt to determine a star's surface temperature from its color ratios. A star's spectrum provides a more precise measure of a star's surface temperature, as we will see next. But it is quicker and easier to observe a star's colors with a set of U, B, and V filters than it is to take the star's spectrum with a spectrograph.

17-5 The spectra of stars reveal their chemical compositions as well as their surface temperatures

We have seen how the color of a star's light helps astronomers determine its surface temperature. To determine the other properties of a star, astronomers must analyze the spectrum of its light in more detail. This technique of *stellar spectroscopy* began in 1817 when Joseph Fraunhofer, a German instrument maker, attached a spectroscope to a telescope and pointed it toward the stars. Fraunhofer had earlier observed that the Sun has an absorption line spectrum—that is, a continuous spectrum with dark absorption lines (see Section 5-6). He found that stars have the same kind of spectra, which reinforces the idea that our Sun is a rather typical star. But Fraunhofer also found that the pattern of absorption lines is different for different stars.

Classifying Stars: Absorption Line Spectra and Spectral Classes

We see an absorption line spectrum when a cool gas lies between us and a hot, glowing object (recall Figure 5-16). The light from the hot, glowing object itself has a continuous spectrum. In the case of a star, light with a continuous spectrum is produced at low-lying levels of the star's atmosphere where the gases are hot and dense. The absorption lines are created when this light flows outward through the upper layers of the star's atmosphere. Atoms in these cooler, less dense layers absorb radiation at specific wavelengths, which depend on the specific kinds of atoms present—hydrogen, helium, or other elements—and on whether or not the atoms are ionized. Absorption lines in the Sun's spectrum are produced in this same way (see Section 16-5).

WEB LINK 17.8

Some stars have spectra in which the Balmer absorption lines of hydrogen are prominent. But in the spectra of other stars, including the Sun, the Balmer lines are nearly absent and the dominant absorption lines are those of heavier elements such as calcium, iron, and sodium. Still other stellar spectra are dominated by broad absorption lines caused by molecules, such as titanium oxide, rather than single atoms. To cope with this diversity, astronomers group similar-appearing stellar spectra into **spectral classes.** In a popular classification scheme that emerged in the late 1890s, a star was assigned a letter from A through O according to the strength or weakness of the hydrogen Balmer lines in the star's spectrum.

Nineteenth-century science could not explain why or how the spectral lines of a particular chemical are affected by the temperature and density of the gas. Nevertheless, a team of astronomers at the Harvard College Observatory forged ahead with a monumental project of examining the spectra of hundreds of thousands of stars. Their goal was to develop a system of spectral classification in which all spectral features, not just Balmer lines, change smoothly from one spectral class to the next.

Deciphering the information in starlight took the painstaking work of generations of astronomers

The Harvard project was financed by the estate of Henry Draper, a wealthy New York physician and amateur astronomer who in 1872 became the first person to photograph stellar absorption lines. Researchers on the project included Edward C. Pickering, Williamina Fleming, Antonia Maury, and Annie Jump Cannon (Figure 17-10). As a result of their efforts, many of the original A-through-O classes were dropped and others were consolidated. The remaining spectral classes were reordered in the sequence **OBAFGKM.** You can remember this sequence with the mnemonic: "Oh, *Be A* Fine Girl (or Guy), Kiss *Me*!"

Refining the Classification: Spectral Types

Cannon refined the original OBAFGKM sequence into smaller steps called **spectral types.** These steps are indicated by attaching an integer from 0 through 9 to the original letter. For example, the spectral class F includes spectral types F0, F1, F2, . . . , F8, F9, which are followed by the spectral types G0, G1, G2, . . . , G8, G9, and so on.

WEB LINK 17.9

Figure 17-10 RI**V**UXG

Classifying the Spectra of the Stars The modern scheme of classifying stars by their spectra was developed at the Harvard College Observatory in the late nineteenth century. A team of women astronomers led by Edward C. Pickering and Williamina Fleming (standing) analyzed hundreds of thousands of spectra. Social conventions of the time prevented most women astronomers from using research telescopes or receiving salaries comparable to men's. (Harvard College Observatory)

Figure 17-11 shows representative spectra of several spectral types. The strengths of spectral lines change gradually from one spectral type to the next. For example, the Balmer absorption lines of hydrogen become increasingly prominent as you go from spectral type B0 to A0. From A0 onward through the F and G classes, the hydrogen lines weaken and almost fade from view. The Sun, whose spectrum is dominated by calcium and iron, is a G2 star.

The Harvard project culminated in the *Henry Draper Catalogue,* published between 1918 and 1924. It listed 225,300 stars, each of which Cannon had personally classified. Meanwhile, physicists had been making important discoveries about the structure of atoms. Ernest Rutherford had shown that atoms have nuclei (recall Figure 5-19), and Niels Bohr made the remarkable hypothesis that electrons circle atomic nuclei along discrete orbits (see Figure 5-22). These advances gave scientists the conceptual and mathematical tools needed to understand stellar spectra.

In the 1920s, the Harvard astronomer Cecilia Payne and the Indian physicist Meghnad Saha demonstrated that the OBAFGKM spectral sequence is actually a sequence in temperature. The hottest stars are O stars. Their absorption lines can occur only if these stars have surface temperatures above 25,000 K. M stars are the coolest stars. The spectral features of M stars are consistent with stellar surface temperatures of about 3000 K.

Why Surface Temperature Affects Stellar Spectra

To see why the appearance of a star's spectrum is profoundly affected by the star's surface temperature, consider the Balmer lines of hydrogen. Hydrogen is by far the most abundant element in the universe, accounting for about three-quarters of the mass of a typical star. Yet the Balmer lines do not necessarily show up in a star's spectrum. As we saw in Section 5-8, Balmer absorption lines are produced when an electron in the $n = 2$ orbit of hydrogen is lifted into a higher orbit by absorbing a photon with the right amount of energy (see Figure 5-24). If the star is much hotter than 10,000 K, the photons pouring out of the star's interior have such high energy that they easily knock electrons out of hydrogen atoms in the star's atmosphere. This process ionizes the gas. With its only electron torn away, a hydrogen atom cannot produce absorption lines. Hence, the Balmer lines will be relatively weak in the spectra of such hot stars, such as the hot O and B2 stars in Figure 17-11.

Conversely, if the star's atmosphere is much cooler than 10,000 K, almost all the hydrogen atoms are in the lowest ($n = 1$) energy state. Most of the photons passing through the star's atmosphere possess too little energy to boost electrons up from the $n = 1$ to the $n = 2$ orbit of the hydrogen atoms. Hence, very few of these atoms will have electrons in the $n = 2$ orbit, and only these few can absorb the photons characteristic of the Balmer lines. As a result, these lines are nearly absent from the spectrum of a cool star. (You can see this in the spectra of the cool M0 and M2 stars in Figure 17-11.)

For the Balmer lines to be prominent in a star's spectrum, the star must be hot enough to excite the electrons out of the ground state but not so hot that all the hydrogen atoms become ionized. A stellar surface temperature of about 9000 K produces the strongest hydrogen lines; this is the case for the stars of spectral types A0 and A5 in Figure 17-11.

LOOKING DEEPER 17.1

Every other type of atom or molecule also has a characteristic temperature range in which it produces prominent absorption lines in the observable part of the spectrum. Figure 17-12 shows the relative strengths of absorption lines produced by different chemicals. By measuring the details of these lines in a given star's spectrum, astronomers can accurately determine that star's surface temperature.

For example, the spectral lines of neutral (that is, un-ionized) helium are strong around 25,000 K. At this temperature, photons have enough energy to excite helium atoms without tearing away the electrons altogether. In stars hotter than about 30,000 K, helium atoms become singly ionized, that is, they lose one of their two electrons. The remaining electron produces a set of spectral lines that is recognizably different from those of neutral helium. Hence, when the spectral lines of singly ionized helium appear in a star's spectrum, we know that the star's surface temperature is greater than 30,000 K.

Astronomers use the term **metals** to refer to all elements other than hydrogen and helium. This idiosyncratic use of the term "metal" is quite different from the definition used by chemists and other scientists. To a chemist, sodium and iron are metals but carbon and oxygen are not; to an astronomer, all of these substances are metals. In this terminology, metals dominate the

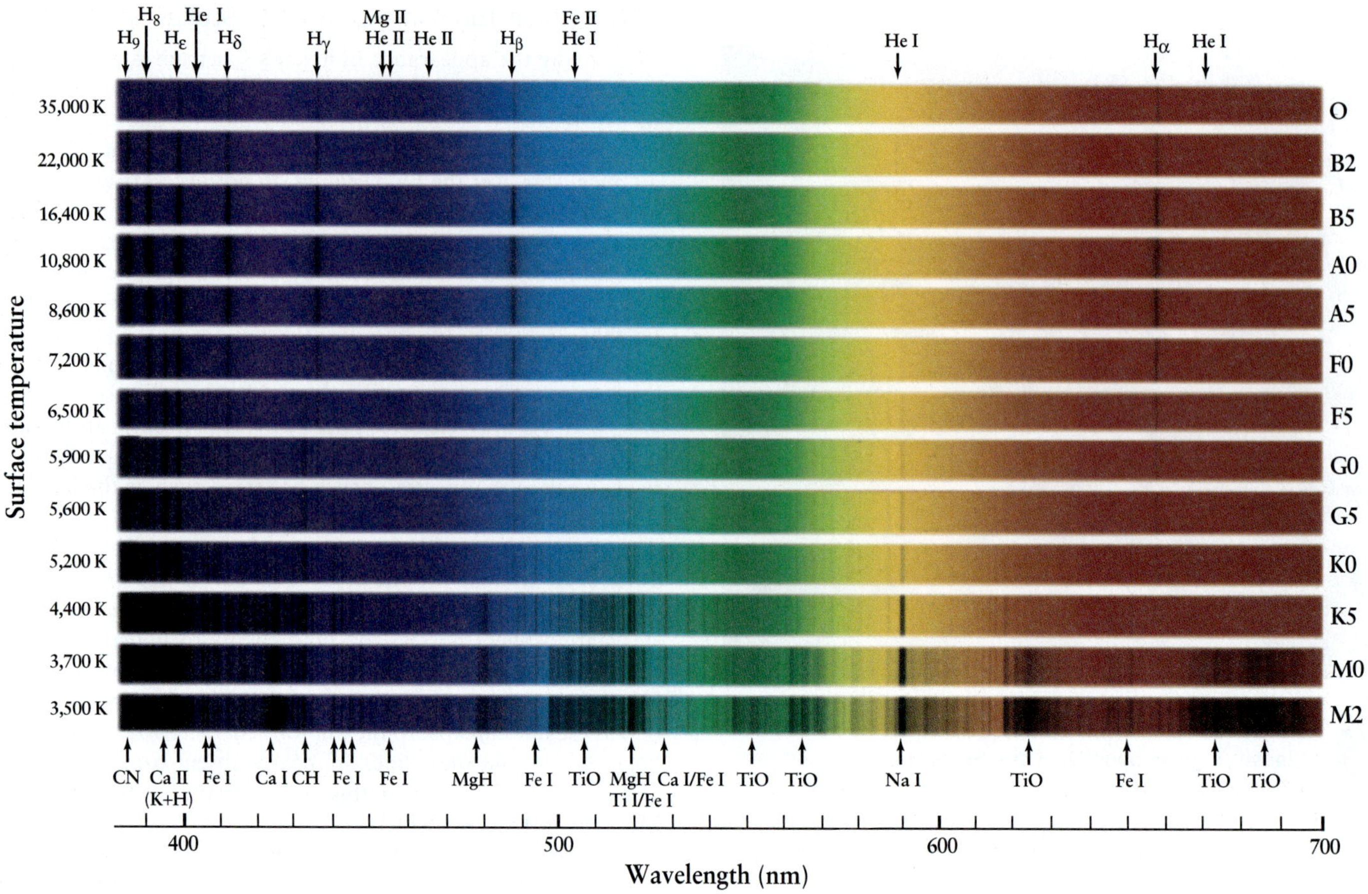

Figure 17-11 R I V U X G

Principal Types of Stellar Spectra Stars of different spectral classes and different surface temperatures have spectra dominated by different absorption lines. Notice how the Balmer lines of hydrogen (H_α, H_β, H_γ, and H_δ) are strongest for hot stars of spectral class A, while absorption lines due to calcium (Ca) are strongest in cooler K and M stars. The spectra of M stars also have broad, dark bands caused by molecules of titanium oxide (TiO), which can only exist at relatively low temperatures. A roman numeral after a chemical symbol shows whether the absorption line is caused by un-ionized atoms (roman numeral I) or by atoms that have lost one electron (roman numeral II). (R. Bell, University of Maryland, and M. Briley, University of Wisconsin at Oshkosh)

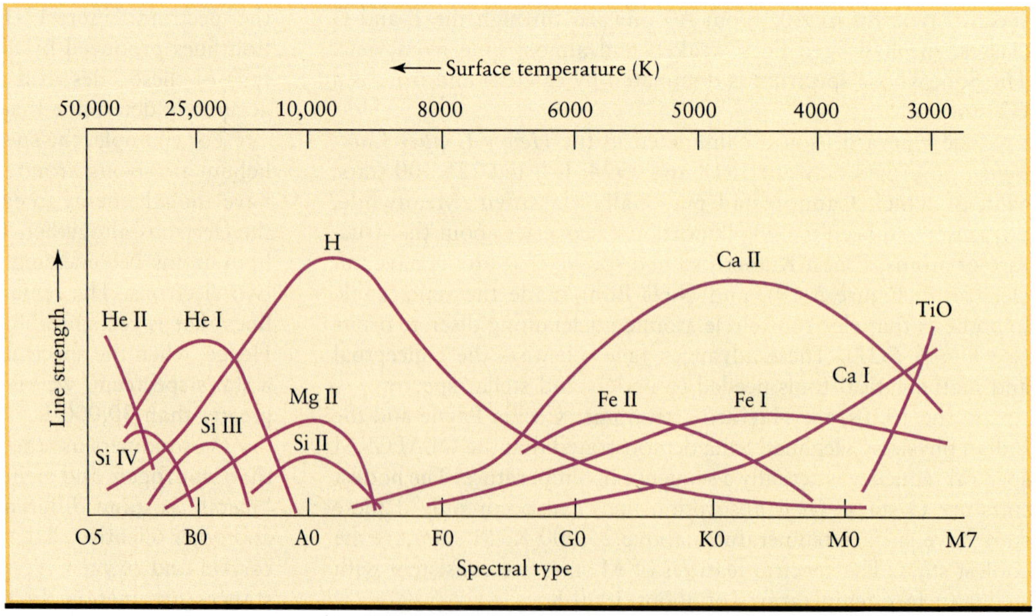

Figure 17-12

The Strengths of Absorption Lines

Each curve in this graph peaks at the stellar surface temperature for which that chemical's absorption line is strongest. For example, hydrogen (H) absorption lines are strongest in A stars with surface temperatures near 10,000 K. Roman numeral I denotes neutral, un-ionized atoms; II, III, and IV denote atoms that are singly, doubly, or triply ionized (that is, have lost one, two, or three electrons).

spectra of stars cooler than 10,000 K. Ionized metals are prominent for surface temperatures between 6000 and 8000 K, while neutral metals are strongest between approximately 5500 and 4000 K.

Below 4000 K, certain atoms in a star's atmosphere combine to form molecules. (At higher temperatures atoms move so fast that when they collide, they bounce off each other rather than "sticking together" to form molecules.) As these molecules vibrate and rotate, they produce bands of spectral lines that dominate the star's spectrum. Most noticeable are the lines of titanium oxide (TiO), which are strongest for surface temperatures of about 3000 K.

Spectral Classes for Brown Dwarfs

WEB LINK 17.10

Since 1995 astronomers have found a number of stars with surface temperatures even lower than those of spectral class M. Strictly speaking, these are not stars but **brown dwarfs,** which we introduced in Section 8-6. Brown dwarfs are too small to sustain thermonuclear fusion in their cores. Instead, these "substars" glow primarily from the heat released by Kelvin-Helmholtz contraction, which we described in Section 16-1. (They do undergo fusion reactions for a brief period during their evolution.) Brown dwarfs are so cold that they are best observed with infrared telescopes (see Figure 17-13). Such observations reveal that brown dwarf spectra have a rich variety of absorption lines due to molecules. Some of these molecules actually form into solid grains in a brown dwarf's atmosphere.

To describe brown dwarf spectra, astronomers have defined two new spectral classes, L and T. Thus, the modern spectral sequence of stars and brown dwarfs from hottest to coldest surface temperature is OBAFGKMLT. (Can you think of a new mnemonic that includes L and T?) For example, Figure 17-13 shows a star of spectral class K and a brown dwarf of spectral class T. Table 17-2 summarizes the relationship between the temperature and spectra of stars and brown dwarfs.

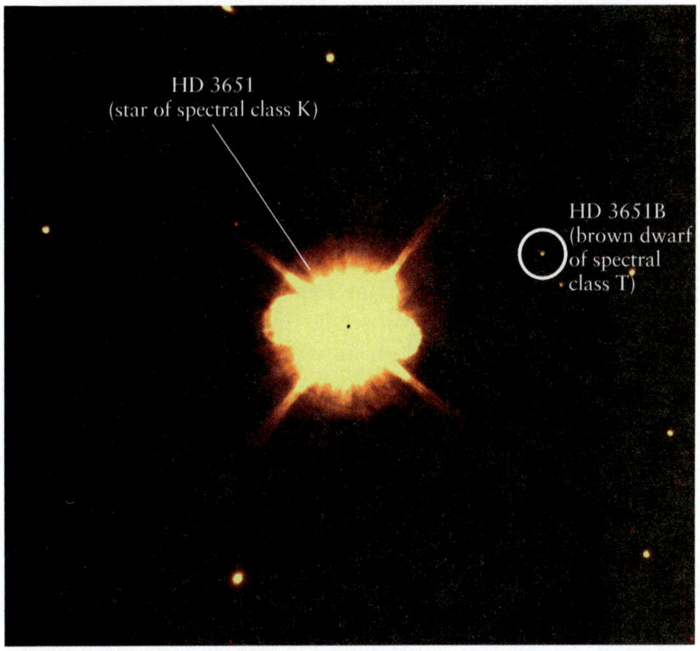

Figure 17-13 R I V U X G

An Infrared Image of Brown Dwarf HD 3651B The star HD 3561 is of spectral class K, with a surface temperature of about 5200 K. ("HD" refers to the *Henry Draper Catalogue.*) In 2006 it was discovered that HD 3651 is orbited by a brown dwarf named HD 3651B with a surface temperature between 800 and 900 K and a luminosity just 1/300,000 that of the Sun. The brown dwarf emits most of its light at infrared wavelengths, so an infrared telescope was used to record this image. The hotter and more luminous star HD 3651 is greatly overexposed in this image and appears much larger than its true size. HD 3651 and HD 3651B are both 11 pc (36 ly) from Earth in the constellation Pisces (the Fish); the other stars in this image are much farther away. (M. Mugrauer and R. Neuhäuser, U. of Jena; A. Seifahrt, ESO; and T. Mazeh, Tel Aviv U.)

Table 17-2 The Spectral Sequence

Spectral class	Color	Temperature (K)	Spectral lines	Examples
O	Blue-violet	30,000–50,000	Ionized atoms, especially helium	Naos (ζ Puppis), Mintaka (δ Orionis)
B	Blue-white	11,000–30,000	Neutral helium, some hydrogen	Spica (α Virginis), Rigel (β Orionis)
A	White	7500–11,000	Strong hydrogen, some ionized metals	Sirius (α Canis Majoris), Vega (α Lyrae)
F	Yellow-white	5900–7500	Hydrogen and ionized metals such as calcium and iron	Canopus (α Carinae), Procyon (α Canis Minoris)
G	Yellow	5200–5900	Both neutral and ionized metals, especially ionized calcium	Sun, Capella (α Aurigae)
K	Orange	3900–5200	Neutral metals	Arcturus (α Boötis), Aldebaran (α Tauri)
M	Red-orange	2500–3900	Strong titanium oxide and some neutral calcium	Antares (α Scorpii), Betelgeuse (α Orionis)
L	Red	1300–2500	Neutral potassium, rubidium, and cesium, and metal hydrides	Brown dwarf Teide 1
T	Red	below 1300	Strong neutral potassium and some water (H_2O)	Brown dwarfs Gliese 229B, HD 3651B

When the effects of temperature are accounted for, astronomers find that *all* stars have essentially the same chemical composition. We can state the results as a general rule:

By mass, almost all stars (including the Sun) and brown dwarfs are about three-quarters hydrogen, one-quarter helium, and 1% or less metals.

Our Sun is about 1% metals by mass, as are most of the stars you can see with the naked eye. But some stars have an even lower percentage of metals. We will see in Chapter 19 that these seemingly minor differences tell an important tale about the life stories of stars.

17-6 Stars come in a wide variety of sizes

With even the best telescopes, stars appear as nothing more than bright points of light. On a photograph or CCD image, brighter stars appear larger than dim ones (see Figures 17-3, 17-6*b*, and 17-13), but these apparent sizes give no indication of the star's actual size. To determine the size of a star, astronomers combine information about its luminosity (determined from its distance and apparent brightness) and its surface temperature (determined from its spectral type). In this way, they find that some stars are quite a bit smaller than the Sun, while others are a thousand times larger.

A star's radius can be calculated if we know its luminosity and surface temperature

Calculating the Radii of Stars

The key to finding a star's radius from its luminosity and surface temperature is the Stefan-Boltzmann law (see Section 5-4). This law says that the amount of energy radiated per second from a square meter of a blackbody's surface—that is, the energy flux (F)—is proportional to the fourth power of the temperature of that surface (T), as given by the equation $F = \sigma T^4$. This equation applies very well to stars, whose spectra are quite similar to that of a perfect blackbody. (Absorption lines, while important for determining the star's chemical composition and surface temperature, make only relatively small modifications to a star's blackbody spectrum.)

A star's luminosity is the amount of energy emitted per second from its entire surface. This quantity equals the energy flux F multiplied by the total number of square meters on the star's surface (that is, the star's surface area). We expect that most stars are nearly spherical, like the Sun, so we can use the formula for the surface area of a sphere. The formula is $4\pi R^2$, where R is the

BOX 17-4 Tools of the Astronomer's Trade

Stellar Radii, Luminosities, and Surface Temperatures

Because stars emit light in almost exactly the same fashion as blackbodies, we can use the Stefan-Boltzmann law to relate a star's luminosity (L), surface temperature (T), and radius (R). The relevant equation is

$$L = 4\pi R^2 \sigma T^4$$

As written, this equation involves the Stefan-Boltzmann constant σ, which is equal to 5.67×10^{-8} W m^{-2} K^{-4}. In many calculations, it is more convenient to relate everything to the Sun, which is a typical star. Specifically, for the Sun we have $L_\odot = 4\pi R_\odot^2 \sigma T_\odot^4$, where $L_\odot$ is the Sun's luminosity, $R_\odot$ is the Sun's radius, and $T_\odot$ is the Sun's surface temperature (equal to 5800 K). Dividing the general equation for L by this specific equation for the Sun, we obtain

$$\frac{L}{L_\odot} = \left(\frac{R}{R_\odot}\right)^2 \left(\frac{T}{T_\odot}\right)^4$$

This formula is easier to use because the constant σ has cancelled out. We can also rearrange terms to arrive at a useful equation for the radius (R) of a star:

Radius of a star related to its luminosity and surface temperature

$$\frac{R}{R_\odot} = \left(\frac{T_\odot}{T}\right)^2 \sqrt{\frac{L}{L_\odot}}$$

$R/R_\odot$ = ratio of the star's radius to the Sun's radius

$T_\odot/T$ = ratio of the Sun's surface temperature to the star's surface temperature

$L/L_\odot$ = ratio of the star's luminosity to the Sun's luminosity

EXAMPLE: The bright reddish star Betelgeuse in the constellation Orion (see Figure 2-2) is 60,000 times more luminous than the Sun and has a surface temperature of 3500 K. What is its radius?

Situation: We are given the star's luminosity $L = 60{,}000\ L_\odot$ and its surface temperature $T = 3500$ K. Our goal is to find the star's radius R.

star's radius (the distance from its center to its surface). Multiplying together the formulas for energy flux and surface area, we can write the star's luminosity as follows:

Relationship between a star's luminosity, radius, and surface temperature

$$L = 4\pi R^2 \sigma T^4$$

L = star's luminosity, in watts

R = star's radius, in meters

σ = Stefan-Boltzmann constant = 5.67×10^{-8} W m^{-2} K^{-4}

T = star's surface temperature, in kelvins

This equation says that a relatively cool star (low surface temperature T), for which the energy flux is quite low, can nonetheless be very luminous if it has a large enough radius R. Alternatively, a relatively hot star (large T) can have a very low luminosity if the star has only a little surface area (small R).

Box 17-4 describes how to use the above equation to calculate a star's radius if its luminosity and surface temperature are known. We can express the idea behind these calculations in terms of the following general rule:

We can determine the radius of a star from its luminosity and surface temperature. For a given luminosity, the greater the surface temperature, the smaller the radius must be. For a given surface temperature, the greater the luminosity, the larger the radius must be.

ANALOGY In a similar way, a roaring campfire can emit more light than a welder's torch. The campfire is at a lower temperature than the torch, but has a much larger surface area from which it emits light.

The Range of Stellar Radii

Using this general rule as shown in Box 17-4, astronomers find that stars come in a wide range of sizes. The smallest stars visible through ordinary telescopes, called *white dwarfs,* are about the same size as Earth. Although their surface temperatures can be very high (25,000 K or more), white dwarfs have so little surface area that their luminosities are very low (less than 0.01 $L_\odot$). The largest stars, called *supergiants,* are a thousand times larger in radius than the Sun and 10^5 times larger than white dwarfs. If our own Sun were replaced by one of these supergiants, Earth's orbit would lie completely inside the star!

Figure 17-14 summarizes how astronomers determine the distance from Earth, luminosity, surface temperature, chemical

Tools: We use the above equation to find the ratio of the star's radius to the radius of the Sun, $R/R_\odot$. Note that we also know the Sun's surface temperature, $T_\odot = 5800$ K.

Answer: Substituting these data into the above equation, we get

$$\frac{R}{R_\odot} = \left(\frac{5800\text{ K}}{3500\text{ K}}\right)^2 \sqrt{6 \times 10^4} = 670$$

Review: Our result tells us that Betelgeuse's radius is 670 times larger than that of the Sun. The Sun's radius is 6.96×10^5 km, so we can also express the radius of Betelgeuse as $(670)(6.96 \times 10^5\text{ km}) = 4.7 \times 10^8$ km, which is more than 3 AU. If Betelgeuse were located at the center of our solar system, it would extend beyond the orbit of Mars!

EXAMPLE: Sirius, the brightest star in the sky, is actually two stars orbiting each other (a binary star). The less luminous star, Sirius B, is a white dwarf that is too dim to see with the naked eye. Its luminosity is 0.0025 $L_\odot$ and its surface temperature is 10,000 K. How large is Sirius B compared to Earth?

Situation: Again we are asked to find a star's radius from its luminosity and surface temperature.

Tools: We use the same equation as in the preceding example.

Answer: The ratio of the radius of Sirius B to the Sun's radius is

$$\frac{R}{R_\odot} = \left(\frac{5800\text{ K}}{10{,}000\text{ K}}\right)^2 \sqrt{0.0025} = 0.017$$

Since the Sun's radius is $R_\odot = 6.96 \times 10^5$ km, the radius of Sirius B is $(0.017)(6.96 \times 10^5\text{ km}) = 12{,}000$ km. From Table 7-1, Earth's radius (half its diameter) is 6378 km. Hence, this star is only about twice the radius of Earth.

Review: Sirius B's radius would be large for a terrestrial planet, but it is minuscule for a star. The name *dwarf* is well deserved!

The radii of some stars have been measured with other techniques (see Section 17-11). These other methods yield values consistent with those calculated by the methods we have just described.

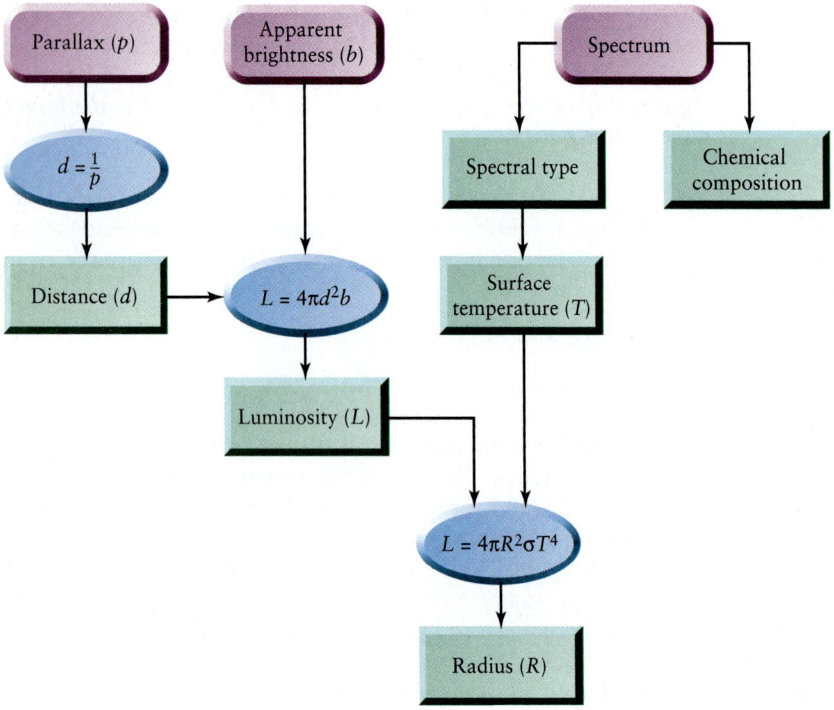

INTERACTIVE EXERCISE 17.1

Figure 17-14

Finding Key Properties of a Nearby Star This flowchart shows how astronomers determine the properties of a relatively nearby star (one close enough that its parallax can be measured). The rounded purple boxes show the measurements that must be made of the star, the blue ovals show the key equations that are used (from Sections 17-2, 17-5, and 17-6), and the green rectangles show the inferred properties of the stars. A different procedure is followed for more distant stars (see Section 17-8, especially Figure 17-17).

composition, and radius of a star close enough to us so that its parallax can be measured. Remarkably, all of these properties can be deduced from just a few measured quantities: the star's parallax angle, apparent brightness, and spectrum.

17-7 Hertzsprung-Russell (H-R) diagrams reveal the different kinds of stars

Astronomers have collected a wealth of data about the stars, but merely having tables of numerical data is not enough. Like all scientists, astronomers want to analyze their data to look for trends and underlying principles. One of the best ways to look for trends in any set of data, whether it comes from astronomy, finance, medicine, or meteorology, is to create a graph showing how one quantity depends on another. For example, investors consult graphs of stock market values versus dates, and weather forecasters make graphs of temperature versus altitude to determine whether thunderstorms are likely to form. Astronomers have found that a particular graph of stellar properties shows that stars fall naturally into just a few categories. This graph, one of the most important in all astronomy, will in later chapters help us understand how stars form, evolve, and eventually die.

H-R Diagrams

Which properties of stars should we include in a graph? Most stars have about the same chemical composition, but two properties of stars—their luminosities and surface temperatures—differ substantially from one star to another. Stars also come in a wide range of radii, but a star's radius is a secondary property that can be found from the luminosity and surface temperature (as we saw in Section 17-6 and Box 17-4). We also relegate the positions and space velocities of stars to secondary importance. (In a similar way, a physician is more interested in your weight and blood pressure than in where you live or how fast you drive.) We can then ask the following question: What do we learn when we graph the luminosities of stars versus their surface temperatures?

A graph of luminosity versus surface temperature reveals that stars fall into a few basic categories

The first answer to this question was given in 1911 by the Danish astronomer Ejnar Hertzsprung. He pointed out that a regular pattern appears when the absolute magnitudes of stars (which measure their luminosities) are plotted against their colors (which measure their surface temperatures). Two years later, the American astronomer Henry Norris Russell independently discovered a similar regularity in a graph using spectral types (another measure of surface temperature) instead of colors. In recognition of their originators, graphs of this kind are today known as **Hertzsprung-Russell diagrams,** or **H-R diagrams** (Figure 17-15).

Figure 17-15*a* is a typical Hertzsprung-Russell diagram. Each dot represents a star whose spectral type and luminosity have been determined. The most luminous stars are near the top of the diagram, the least luminous stars near the bottom. Hot stars of spectral classes O and B are toward the left side of the graph and cool stars of spectral class M are toward the right.

CAUTION! You are probably accustomed to graphs in which the numbers on the horizontal axis increase as you move to the right. (For example, the business section of a newspaper includes a graph of stock market values versus dates, with later

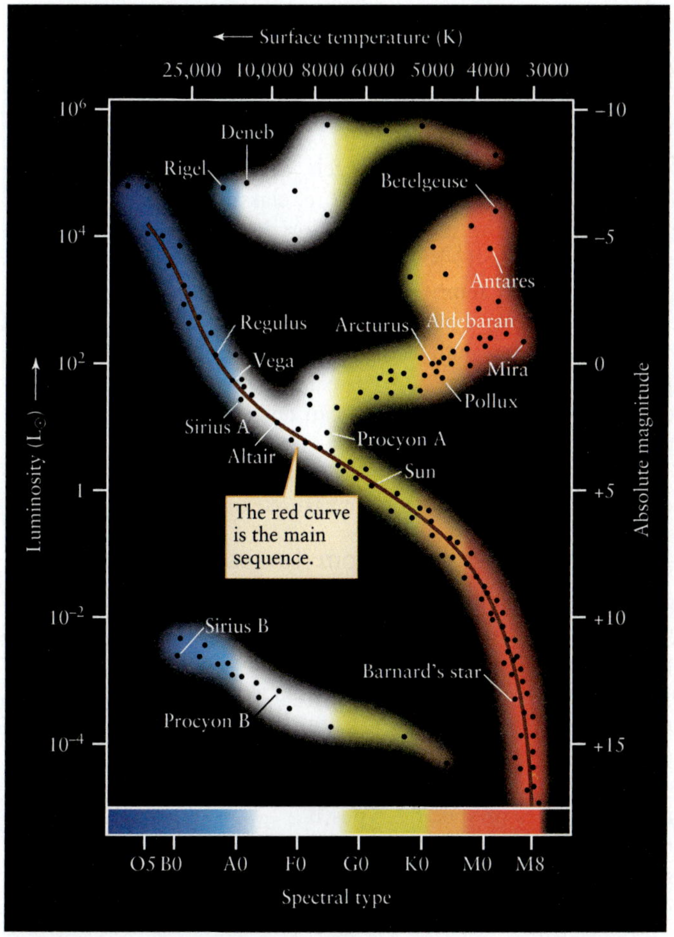

(a) A Hertzsprung-Russell (H-R) diagram

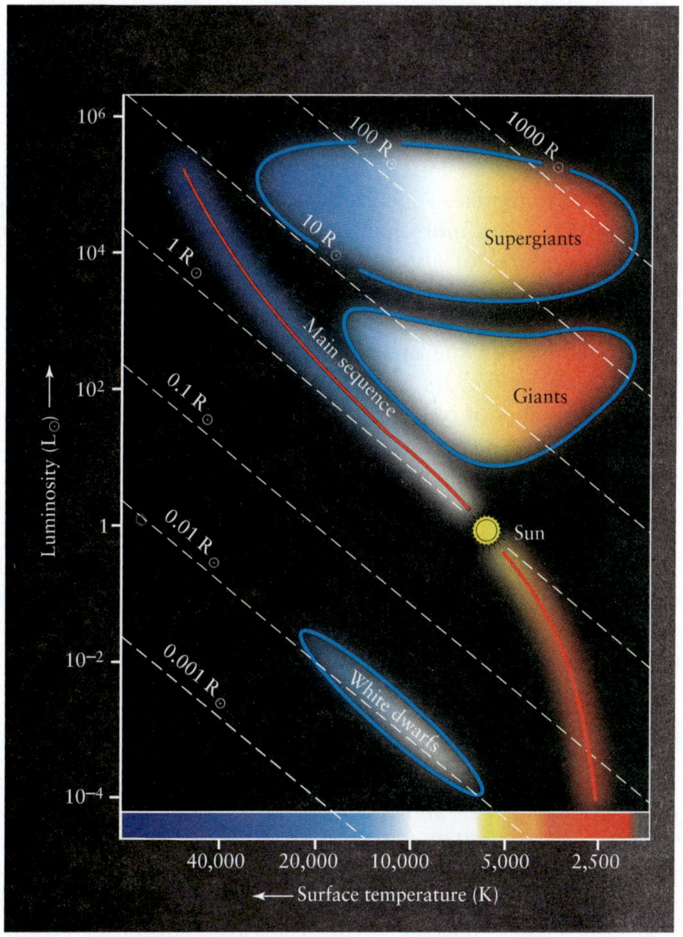

(b) The sizes of stars on an H-R diagram

INTERACTIVE EXERCISE 17.2

Figure 17-15

Hertzsprung-Russell (H-R) Diagrams On an H-R diagram, the luminosities (or absolute magnitudes) of stars are plotted against their spectral types (or surface temperatures). **(a)** The data points are grouped in just a few regions on the graph, showing that luminosity and spectral type are correlated. Most stars lie along the red curve called the main sequence. Giants like Arcturus as well as supergiants like Rigel and Betelgeuse are above the main sequence, and white dwarfs like Sirius B are below it. **(b)** The blue curves on this H-R diagram enclose the regions of the diagram in which different types of stars are found. The dashed diagonal lines indicate different stellar radii. For a given stellar radius, as the surface temperature increases (that is, moving from right to left in the diagram), the star glows more intensely and the luminosity increases (that is, moving upward in the diagram). Note that the Sun is intermediate in luminosity, surface temperature, and radius.

dates to the right of earlier ones.) But on an H-R diagram the temperature scale on the horizontal axis increases toward the *left*. This practice stems from the original diagrams of Hertzsprung and Russell, who placed hot O stars on the left and cool M stars on the right. This arrangement is a tradition that no one has seriously tried to change.

Star Varieties: Main-Sequence Stars, Giants, Supergiants, White Dwarfs, and Brown Dwarfs

The most striking feature of the H-R diagram is that the data points are not scattered randomly over the graph but are grouped in a few distinct regions. The luminosities and surface temperatures of stars do *not* have random values; instead, these two quantities are related!

The band stretching diagonally across the H-R diagram includes about 90% of the stars in the night sky. This band, called the **main sequence,** extends from the hot, luminous, blue stars in the upper left corner of the diagram to the cool, dim, red stars in the lower right corner. A star whose properties place it in this region of an H-R diagram is called a **main-sequence star.** The Sun (spectral type G2, luminosity 1 $L_\odot$, absolute magnitude +4.8) is such a star. We will find that all main-sequence stars are like the Sun in that *hydrogen fusion*—thermonuclear reactions that convert hydrogen into helium (see Section 16-1)—is taking place in their cores.

The upper right side of the H-R diagram shows a second major grouping of data points. Stars represented by these points are both luminous and cool. From the Stefan-Boltzmann law, we know that a cool star radiates much less light per unit of surface area than a hot star. In order for these stars to be as luminous as they are, they must be huge (see Section 17-6), and so they are called **giants.** These stars are around 10 to 100 times larger than the Sun. You can see this size difference in Figure 17-15*b,* which is an H-R diagram to which dashed lines have been added to represent stellar radii. Most giant stars are around 100 to 1000 times more luminous than the Sun and have surface temperatures of about 3000 to 6000 K. Cooler members of this class of stars (those with surface temperatures from about 3000 to 4000 K) are often called **red giants** because they appear reddish. In the image that opens this chapter, the yellowish stars, as well as the red star just left of center, are red giants. A number of red giants can easily be seen with the naked eye, including Aldebaran in the constellation Taurus and Arcturus in Boötes.

A few rare stars are considerably bigger and brighter than typical red giants, with radii up to 1000 $R_\odot$. Appropriately enough, these superluminous stars are called **supergiants.** Betelgeuse in Orion (see Box 17-4) and Antares in Scorpius are two supergiants you can find in the nighttime sky. Together, giants and supergiants make up about 1% of the stars in the sky.

Both giants and supergiants have thermonuclear reactions occurring in their interiors, but the character of those reactions and where in the star they occur can be quite different than for a main-sequence star like the Sun. We will study these stars in more detail in Chapters 21 and 22.

The remaining 9% of stars form a distinct grouping of data points toward the lower left corner of the Hertzsprung-Russell diagram. Although these stars are hot, their luminosities are quite low; hence, they must be small. They are appropriately called **white dwarfs.** These stars, which are so dim that they can be seen only with a telescope, are approximately the same size as Earth. As we will learn in Chapter 20, no thermonuclear reactions take place within white dwarf stars. Rather, like embers left from a fire, they are the still-glowing remnants of what were once giant stars.

By contrast, *brown* dwarfs (which lie at the extreme lower right of the main sequence, off the bottom and right-hand edge of Figure 17-15*a* or Figure 17-15*b*) are objects that will never become stars. They are comparable in radius to the planet Jupiter (that is, intermediate in size between Earth and the Sun; see Figure 7-2). The study of brown dwarfs is still in its infancy, but it appears that there may be twice as many brown dwarfs as there are "real" stars.

ANALOGY You can think of white dwarfs as "has-been" stars whose days of glory have passed. In this analogy, a brown dwarf is a "never-will-be."

The existence of fundamentally different types of stars is the first important lesson to come from the H-R diagram. In later chapters we will find that these different types represent various stages in the lives of stars. We will use the H-R diagram as an essential tool for understanding how stars evolve.

17-8 Details of a star's spectrum reveal whether it is a giant, a white dwarf, or a main-sequence star

A star's surface temperature largely determines which lines are prominent in its spectrum. Therefore, classifying stars by spectral type is essentially the same as categorizing them by surface temperature. But as the H-R diagram in Figure 17-15*b* shows, stars of the same surface temperature can have very different luminosities. As an example, a star with surface temperature 5800 K could be a white dwarf, a main-sequence star, a giant, or a supergiant, depending on its luminosity. By examining the details of a star's spectrum, however, astronomers can determine to which of these categories a star belongs. This gives astronomers a tool to determine the distances to stars millions of parsecs away, far beyond the maximum distance that can be measured using stellar parallax.

Determining a Star's Size from Its Spectrum

Figure 17-16 compares the spectra of two stars of the same spectral type but different luminosity (and hence different size): a B8 supergiant and a B8 main-sequence star. Note that the Balmer lines of hydrogen are narrow in the spectrum of the very large, very luminous supergiant but quite broad in the spectrum of the small, less luminous main-sequence star. In general, for stars of spectral types B through F, the larger and more luminous the star, the narrower its hydrogen lines.

The smaller a star and the denser its atmosphere, the broader the absorption lines in its spectrum

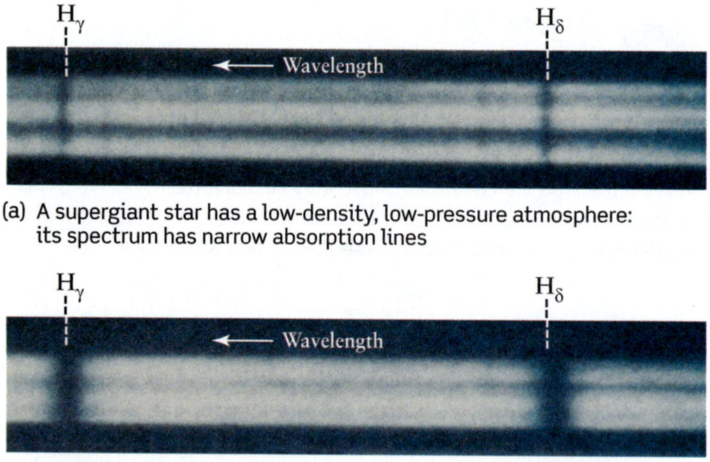

(a) A supergiant star has a low-density, low-pressure atmosphere: its spectrum has narrow absorption lines

(b) A main-sequence star has a denser, higher-pressure atmosphere: its spectrum has broad absorption lines

Figure 17-16 R I V U X G

How a Star's Size Affects Its Spectrum These are the spectra of two stars of the same spectral type (B8) and surface temperature (13,400 K) but different radii and luminosities: **(a)** the B8 supergiant Rigel (luminosity 58,000 $L_\odot$) in Orion, and **(b)** the B8 main-sequence star Algol (luminosity 100 $L_\odot$) in Perseus. (From W. W. Morgan, P. C. Keenan, and E. Kellman, *An Atlas of Stellar Spectra*)

Fundamentally, these differences between stars of different luminosity are due to differences between the stars' atmospheres, where absorption lines are produced. Hydrogen lines in particular are affected by the density and pressure of the gas in a star's atmosphere. The higher the density and pressure, the more frequently hydrogen atoms collide and interact with other atoms and ions in the atmosphere. These collisions shift the energy levels in the hydrogen atoms and thus broaden the hydrogen spectral lines.

In the atmosphere of a luminous giant star, the density and pressure are quite low because the star's mass is spread over a huge volume. Atoms and ions in the atmosphere are relatively far apart; hence, collisions between them are sufficiently infrequent that hydrogen atoms can produce narrow Balmer lines. A main-sequence star, however, is much more compact than a giant or supergiant. In the denser atmosphere of a main-sequence star, frequent interatomic collisions perturb the energy levels in the hydrogen atoms, thereby producing broader Balmer lines.

Luminosity Classes

In the 1930s, W. W. Morgan and P. C. Keenan of the Yerkes Observatory of the University of Chicago developed a system of **luminosity classes** based upon the subtle differences in spectral lines. When these luminosity classes are plotted on an H-R diagram (Figure 17-17), they provide a useful subdivision of the star types in the upper right of the diagram. Luminosity classes Ia and Ib are composed of supergiants; luminosity class V includes all the main-sequence stars. The intermediate classes distinguish giant stars of various luminosities. Note that for stars of a given surface temperature (that is, a given spectral type), the *higher* the number of the luminosity class, the *lower* the star's luminosity.

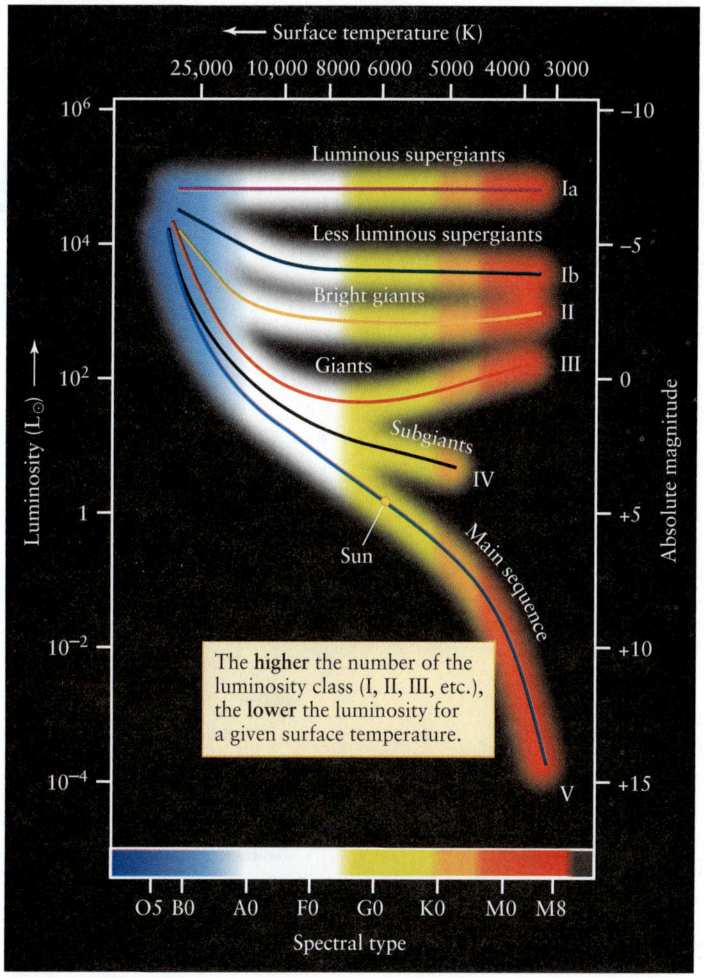

Figure 17-17

Luminosity Classes The H-R diagram is divided into regions corresponding to stars of different luminosity classes. (White dwarfs do not have their own luminosity class.) A star's spectrum reveals both its spectral type and its luminosity class; from these, the star's luminosity can be determined.

As we will see in Chapters 19 and 20, different luminosity classes represent different stages in the evolution of a star. White dwarfs are not always given a luminosity class of their own; as we mentioned in Section 17-7, they represent a final stage in stellar evolution in which no thermonuclear reactions take place.

Astronomers commonly use a shorthand description that combines a star's spectral type and its luminosity class. For example, the Sun is said to be a G2 V star. The spectral type indicates the star's surface temperature, and the luminosity class indicates its luminosity. Thus, an astronomer knows immediately that any G2 V star is a main-sequence star with a luminosity of about 1 $L_\odot$ and a surface temperature of about 5800 K. Similarly, a description of Aldebaran as a K5 III star tells an astronomer that it is a red giant with a luminosity of around 370 $L_\odot$ and a surface temperature of about 4000 K.

Spectroscopic Parallax

A star's spectral type and luminosity class, combined with the information on the H-R diagram, enable astronomers to estimate the star's distance from Earth. As an example, consider the star Pleione in the constellation Taurus. Its spectrum reveals Pleione to be a B8 V star (a hot, blue, main-sequence star, like the one in Figure 17-16*b*). Using Figure 17-17, we can read off that such a star's luminosity is 190 $L_\odot$. Given the star's luminosity and its apparent brightness—in the case of Pleione, 3.9×10^{-13} of the apparent brightness of the Sun—we can use the inverse-square law to determine its distance from Earth. The mathematical details are worked out in Box 17-2.

This method for determining distance, in which the luminosity of a star is found using spectroscopy, is called **spectroscopic parallax.** Figure 17-18 summarizes the method of spectroscopic parallax.

CAUTION! The name "spectroscopic parallax" is a bit misleading, because no parallax angle is involved! The idea is that measuring the star's spectrum takes the place of measuring its parallax as a way to find the star's distance. A better name for this method, although not the one used by astronomers, would be "spectroscopic distance determination."

Spectroscopic parallax is an incredibly powerful technique. No matter how remote a star is, this technique allows astronomers

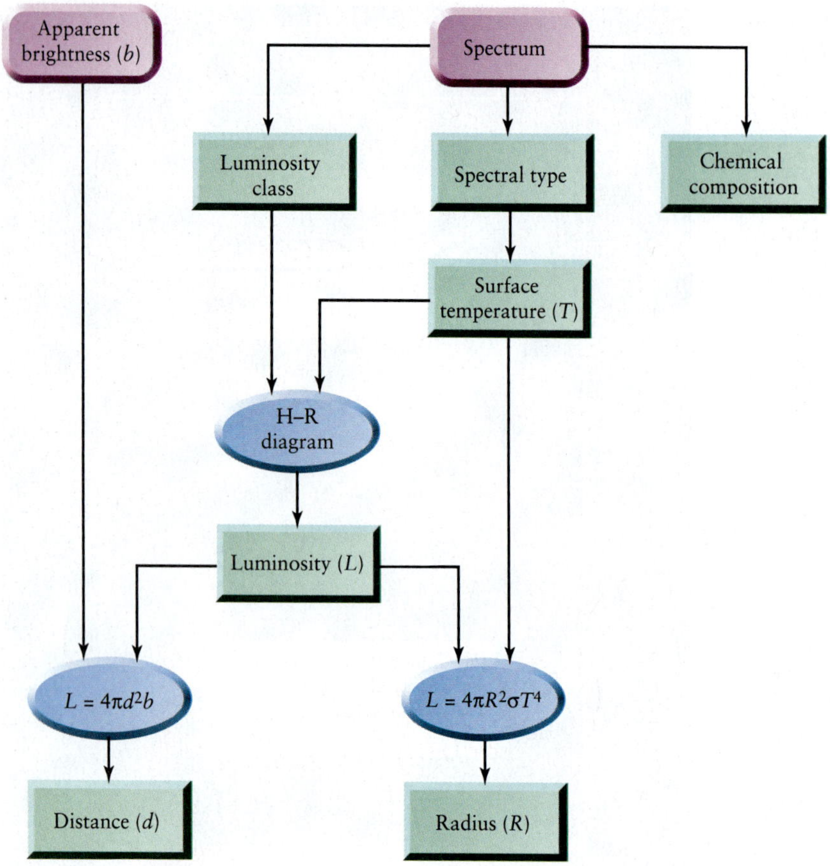

Figure 17-18

The Method of Spectroscopic Parallax If a star is too far away, its parallax angle is too small to allow a direct determination of its distance. This flowchart shows how astronomers deduce the properties of such a distant star. Note that the H-R diagram plays a central role in determining the star's luminosity from its spectral type and luminosity class. Just as for nearby stars (see Figure 17-14), the star's chemical composition is determined from its spectrum, and the star's radius is calculated from the luminosity and surface temperature.

to determine its distance, provided only that its spectrum and apparent brightness can be measured. Box 17-2 gives an example of how spectroscopic parallax has been used to find the distance to stars in other galaxies tens of millions of parsecs away. By contrast, we saw in Section 17-1 that "real" stellar parallaxes can be measured only for stars within a few hundred parsecs.

Unfortunately, spectroscopic parallax has its limitations; distances to individual stars determined using this method often have errors greater than 10%. The reason is that the luminosity classes shown in Figure 17-17 are not thin lines on the H-R diagram but are moderately broad bands. Hence, even if a star's spectral type and luminosity class are known, there is still some uncertainty in the luminosity that we read off an H-R diagram. Nonetheless, spectroscopic parallax is often the only means that an astronomer has to estimate the distance to remote stars.

What has been left out of this discussion is *why* different stars have different spectral types and luminosities. One key factor, as we shall see, turns out to be the mass of the star.

17-9 Observing binary star systems reveals the masses of stars

We now know something about the sizes, temperatures, and luminosities of stars. To complete our picture of the physical properties of stars, we need to know their masses. In this section, we will see that stars come in a wide range of masses. We will also discover an important relationship between the mass and luminosity of main-sequence stars. This relationship is crucial to understanding why some main-sequence stars are hot and luminous, while others are cool and dim. It will also help us understand what happens to a star as it ages and evolves.

For main-sequence stars, there is a direct correlation between mass and luminosity

Determining the masses of stars is not trivial, however. The problem is that there is no practical, direct way to measure the mass of an isolated star. Fortunately for astronomers, about half of the visible stars in the night sky are not isolated individuals. Instead, they are *multiple-star systems,* in which two or more stars orbit each other. By carefully observing the motions of these stars, astronomers can glean important information about their masses.

Binary Stars

A pair of stars located at nearly the same position in the night sky is called a **double star.** The Anglo-German astronomer William Herschel made the first organized search for such pairs. Between 1782 and 1821, he published three catalogs listing more than 800 double stars. Late in the nineteenth century, his son, John Herschel, discovered 10,000 more doubles. Some of these double stars are **optical double stars,** which are two stars that lie along nearly the same line of sight but are actually at very different distances from us. But many double stars are true **binary stars,** or

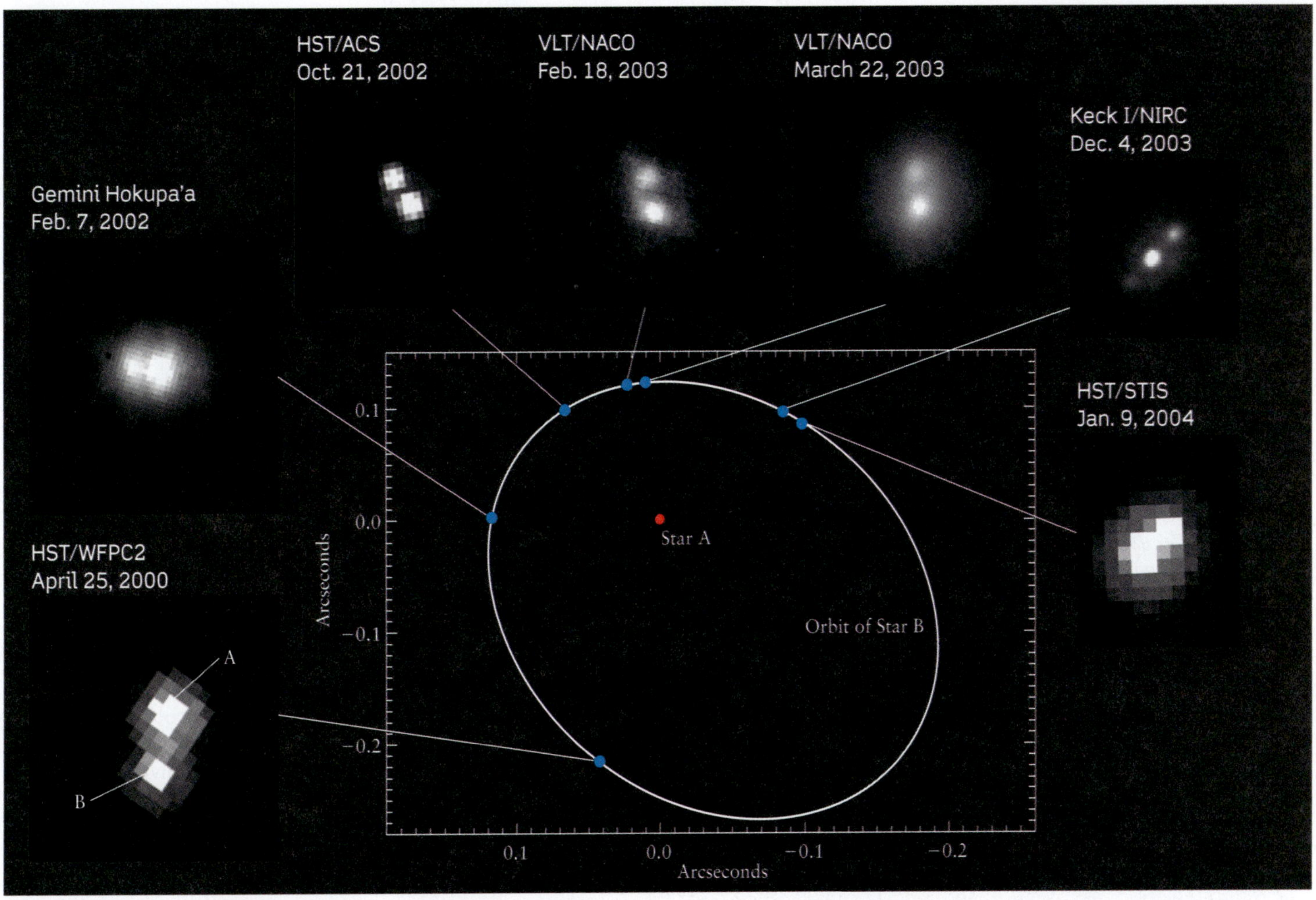

Figure 17-19 R I V U X G

A Binary Star System As seen from Earth, the two stars that make up the binary system called 2MASSW J0746425+2000321 are separated by less than 1/3 arcsecond. The images surrounding the center diagram show the relative positions of the two stars over a four-year period. These images were made by the Hubble Space Telescope (HST), the European Southern Observatory's Very Large Telescope (VLT), and Keck I and Gemini North in Hawaii (see Figure 6-16). For simplicity, the diagram shows one star as remaining stationary; in reality, both stars move around their common center of mass. (H. Bouy et al., MPE and ESO)

binaries—pairs of stars that actually orbit each other. Figure 17-19 shows an example of this orbital motion.

When astronomers can actually see the two stars orbiting each other, a binary is called a **visual binary.** By observing the binary over an extended period, astronomers can plot the orbit that one star appears to describe around the other, as shown in the center diagram in Figure 17-19.

In fact, *both* stars in a binary system are in motion. They orbit each other because of their mutual gravitational attraction, and their orbital motions obey Kepler's third law as formulated by Isaac Newton (see Section 4-7 and Box 4-4). This law can be written as follows:

Kepler's third law for binary star systems

$$M_1 + M_2 = \frac{a^3}{P^2}$$

M_1, M_2 = masses of two stars in binary system, in solar masses

a = semimajor axis of one star's orbit around the other, in AU

P = orbital period, in years

Here a is the semimajor axis of the elliptical orbit that one star appears to describe around the other, plotted as in the center diagram in Figure 17-19. As this equation indicates, if we can measure this semimajor axis (a) and the orbital period (P), we can learn something about the masses of the two stars.

In principle, the orbital period of a visual binary is easy to determine. All you have to do is see how long it takes for the two stars to revolve once about each other. The two stars shown in Figure 17-19 are relatively close, about 2.5 AU on average, and their orbital period is only 10 years. Many binary systems have

The center of mass of
the system of two children
is nearer to the more
massive child.
Fulcrum

(a) A "binary system" of two children

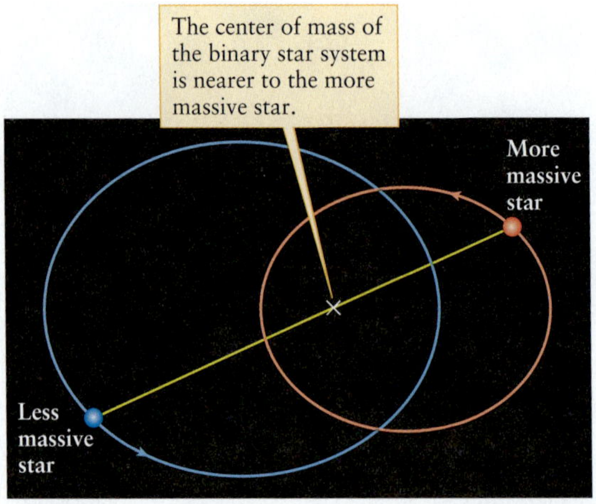
The center of mass of
the binary star system
is nearer to the more
massive star.
More
massive
star
Less
massive
star

(b) A binary star system

ANIMATION 17.3

Figure 17-20

Center of Mass in a Binary Star System (a) A seesaw balances if the fulcrum is at the center of mass of the two children. **(b)** The members of a binary star system orbit around the center of mass of the two stars. Although their elliptical orbits cross each other, the two stars are always on opposite sides of the center of mass and thus never collide.

much larger separations, however, and the period may be so long that more than one astronomer's lifetime is needed to complete the observations.

Determining the semimajor axis of an orbit can also be a challenge. The *angular* separation between the stars can be determined by observation. To convert this angle into a physical distance between the stars, we need to know the distance between the binary and Earth. This distance can be found from parallax measurements or by using spectroscopic parallax. The astronomer must also take into account how the orbit is tilted to our line of sight.

Once both P and a have been determined, Kepler's third law can be used to calculate $M_1 + M_2$, the sum of the masses of the two stars in the binary system. But this analysis tells us nothing about the *individual* masses of the two stars. To obtain these masses, more information about the motions of the two stars is needed.

Each of the two stars in a binary system actually moves in an elliptical orbit about the **center of mass** of the system. Imagine two children sitting on opposite ends of a seesaw (Figure 17-20*a*). For the seesaw to balance properly, they must position themselves so that their center of mass—an imaginary point that lies along a line connecting their two bodies—is at the fulcrum, or pivot point of the seesaw. If the two children have the same mass, the center of mass lies midway between them, and they should sit equal distances from the fulcrum. If their masses are different, the center of mass is closer to the heavier child.

Just as the seesaw naturally balances at its center of mass, the two stars that make up a binary system naturally orbit around their center of mass (Figure 17-20*b*). The center of mass always lies along the line connecting the two stars and is closer to the more massive star.

The center of mass of a visual binary is located by plotting the separate orbits of the two stars, as in Figure 17-20*b*, using the background stars as reference points. The center of mass lies at the common focus of the two elliptical orbits. Comparing the relative sizes of the two orbits around the center of mass yields the ratio of the two stars' masses, M_1/M_2. The sum $M_1 + M_2$ is already known from Kepler's third law, so the individual masses of the two stars can then be determined.

Main-Sequence Masses and the Mass-Luminosity Relation

Years of careful, patient observations of binaries have slowly yielded the masses of many stars. As the data accumulated, an important trend began to emerge: For main-sequence stars, there is a direct correlation between mass and luminosity. The more massive a main-sequence star, the more luminous it is. Figure 17-21 depicts this **mass-luminosity relation** as a graph. The range of stellar masses extends from less than 0.1 of a solar mass to more than 50 solar masses. The Sun's mass lies between these extremes.

The *Cosmic Connections* figure on the next page depicts the mass-luminosity relation for main-sequence stars on an H-R diagram. This figure shows the main sequence on an H-R diagram is a progression in mass as well as in luminosity and surface temperature. The hot, bright, bluish stars in the upper left corner of an H-R diagram are the most massive main-sequence stars. Likewise, the dim, cool, reddish stars in the lower right corner of an H-R diagram are the least massive. Main-sequence stars of intermediate temperature and luminosity also have intermediate masses.

The mass of a main-sequence star also helps determine its radius. Referring back to Figure 17-15*b*, we see that if we move along the main sequence from low luminosity to high luminosity, the radius of the star increases. Thus, we have the following general rule for main-sequence stars:

The greater the mass of a main-sequence star, the greater its luminosity, its surface temperature, and its radius.

COSMIC CONNECTIONS

The Main Sequence and Masses

The main sequence is an arrangement of stars according to their mass. The most massive main-sequence stars have the greatest luminosity, greatest radius, and greatest surface temperature. These characteristics are consequences of the behavior of thermonuclear reactions at the core of a main-sequence star.

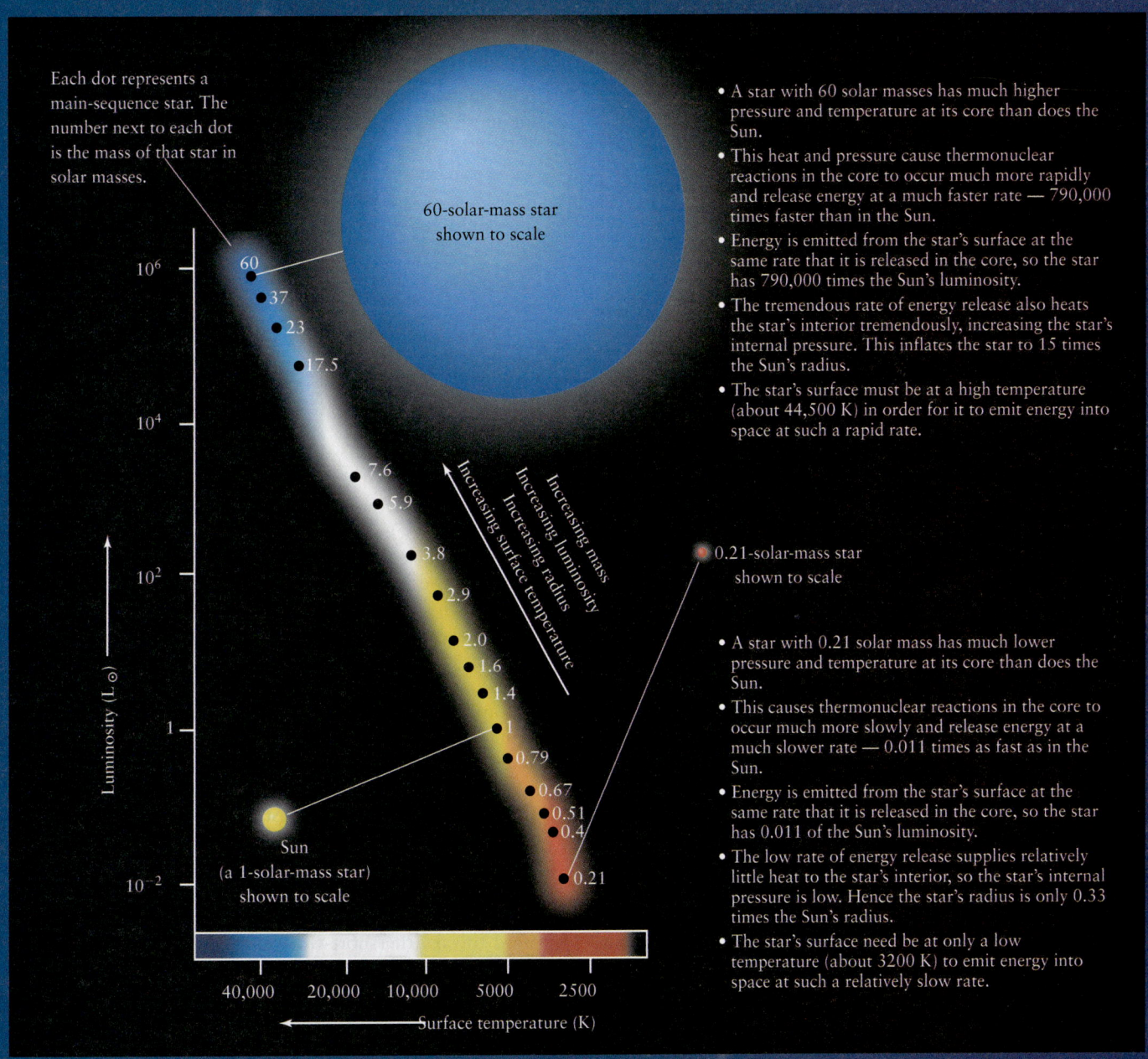

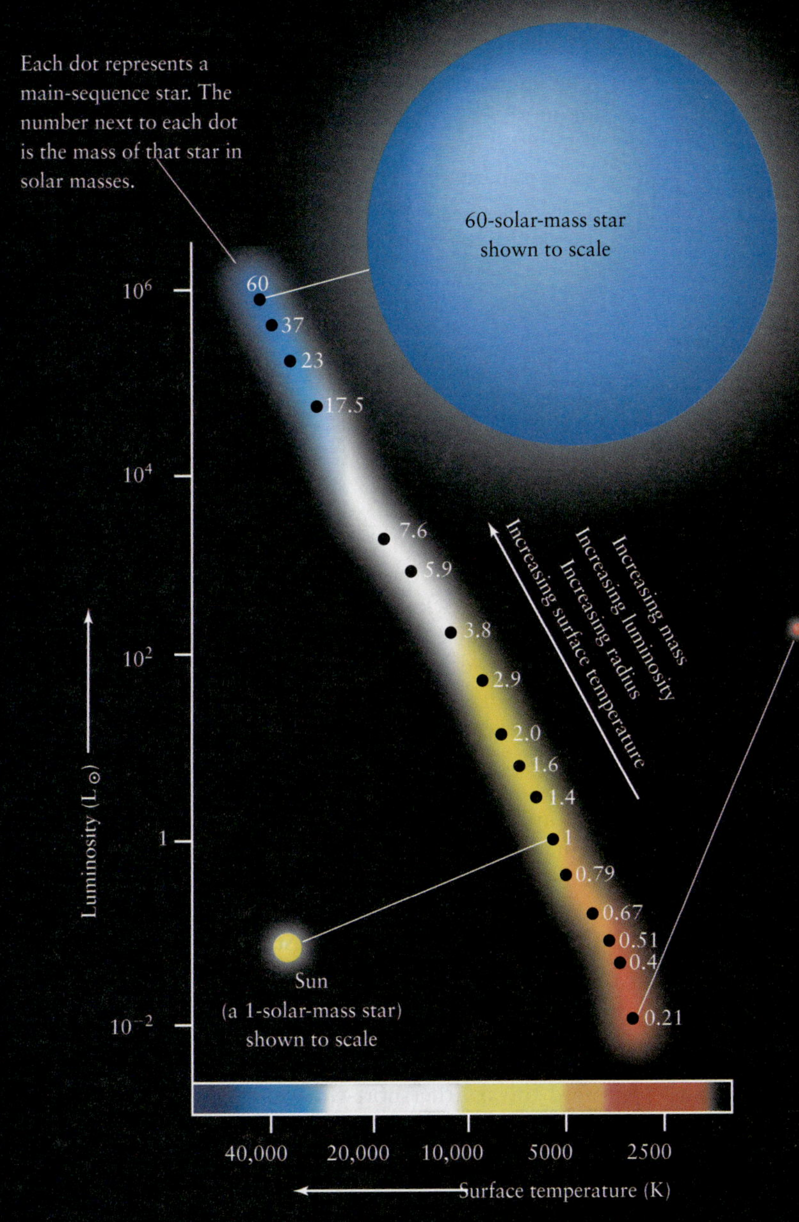

- A star with 60 solar masses has much higher pressure and temperature at its core than does the Sun.
- This heat and pressure cause thermonuclear reactions in the core to occur much more rapidly and release energy at a much faster rate — 790,000 times faster than in the Sun.
- Energy is emitted from the star's surface at the same rate that it is released in the core, so the star has 790,000 times the Sun's luminosity.
- The tremendous rate of energy release also heats the star's interior tremendously, increasing the star's internal pressure. This inflates the star to 15 times the Sun's radius.
- The star's surface must be at a high temperature (about 44,500 K) in order for it to emit energy into space at such a rapid rate.

- A star with 0.21 solar mass has much lower pressure and temperature at its core than does the Sun.
- This causes thermonuclear reactions in the core to occur much more slowly and release energy at a much slower rate — 0.011 times as fast as in the Sun.
- Energy is emitted from the star's surface at the same rate that it is released in the core, so the star has 0.011 of the Sun's luminosity.
- The low rate of energy release supplies relatively little heat to the star's interior, so the star's internal pressure is low. Hence the star's radius is only 0.33 times the Sun's radius.
- The star's surface need be at only a low temperature (about 3200 K) to emit energy into space at such a relatively slow rate.

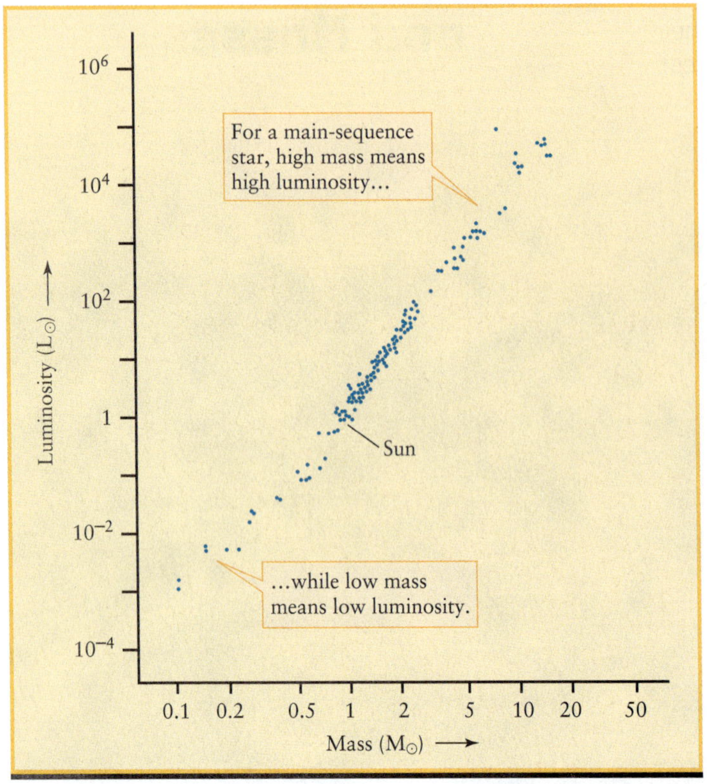

Figure 17-21

The Mass-Luminosity Relation For main-sequence stars, there is a direct correlation between mass and luminosity—the more massive a star, the more luminous it is. A main-sequence star of mass 10 $M_\odot$ (that is, 10 times the Sun's mass) has roughly 3000 times the Sun's luminosity (3000 $L_\odot$); one with 0.1 $M_\odot$ has a luminosity of only about 0.001 $L_\odot$.

Mass and Main-Sequence Stars

Why is mass the controlling factor in determining the properties of a main-sequence star? The answer is that all main-sequence stars are objects like the Sun, with essentially the same chemical composition as the Sun but with different masses. Like the Sun, all main-sequence stars shine because thermonuclear reactions at their cores convert hydrogen to helium and release energy. The greater the total mass of the star, the greater the pressure and temperature at the core, the more rapidly thermonuclear reactions take place in the core, and the greater the energy output—that is, the luminosity—of the star. In other words, the greater the mass of a main-sequence star, the greater its luminosity. This statement is just the mass-luminosity relation, which we can now recognize as a natural consequence of the nature of main-sequence stars.

Like the Sun, main-sequence stars are in a state of both hydrostatic equilibrium and thermal equilibrium. Calculations using models of a main-sequence star's interior (like the solar models we discussed in Section 16-2) show that to maintain equilibrium, a more massive star must have a larger radius and a higher surface temperature. This result is just what we see when we plot the curve of the main sequence on an H-R diagram (see Figure 17-15*b*). As you move up the main sequence from less massive stars (at the lower right in the H-R diagram) to more massive stars (at the upper left), the radius and surface temperature both increase.

Calculations using hydrostatic and thermal equilibrium also show that if a star's mass is less than about 0.08$M_\odot$, the core pressure and temperature are too low for thermonuclear reactions to take place. The "star" is then a brown dwarf. Brown dwarfs also follow a mass-luminosity relation: The greater the mass, the faster the brown dwarf contracts because of its own gravity, the more rapidly it radiates energy into space, and, hence, the more luminous the brown dwarf is.

CAUTION! The mass-luminosity relation we have discussed applies to main-sequence stars only. There are *no* simple mass-luminosity relations for giant, supergiant, or white dwarf stars. Why these stars lie where they do on an H-R diagram will become apparent when we study the evolution of stars in Chapters 19 and 20. We will find that main-sequence stars evolve into giant and supergiant stars, and that some of these eventually end their lives as white dwarfs.

17-10 Spectroscopy makes it possible to study binary systems in which the two stars are close together

We have described how the masses of stars can be determined from observations of visual binaries, in which the two stars can be distinguished from each other. But if the two stars in a binary system are too close together, the images of the two stars can blend to produce the semblance of a single star. Happily, in many cases we can use spectroscopy to decide whether a seemingly single star is in fact a binary system. Spectroscopic observations of binaries provide additional useful information about star masses.

Some binaries are discovered when the spectrum of a star shows incongruous spectral lines. For example, the spectrum of what appears to be a single star may include both strong hydrogen lines (characteristic of a type A star) and strong absorption bands of titanium oxide (typical of a type M star). Because a single star cannot have the differing physical properties of these two spectral types, such a star must actually be a binary system that is too far away for us to resolve its individual stars. A binary system detected in this way is called a **spectrum binary.**

Other binary systems can be detected using the Doppler effect. If a star is moving toward Earth, its spectral lines are displaced toward the short-wavelength (blue) end of the spectrum. Conversely, the spectral lines of a star moving away from us are shifted toward the long-wavelength (red) end of the spectrum. The upper portion of Figure 17-22 applies these ideas to a hypothetical binary star system with an orbital plane that is edge-on to our line of sight.

As the two stars move around their orbits, they periodically approach and recede from us. Hence, the spectral lines of the two stars are alternately blueshifted and redshifted. The two stars in this hypothetical system are so close together that they appear through a telescope as a single star with a single spectrum.

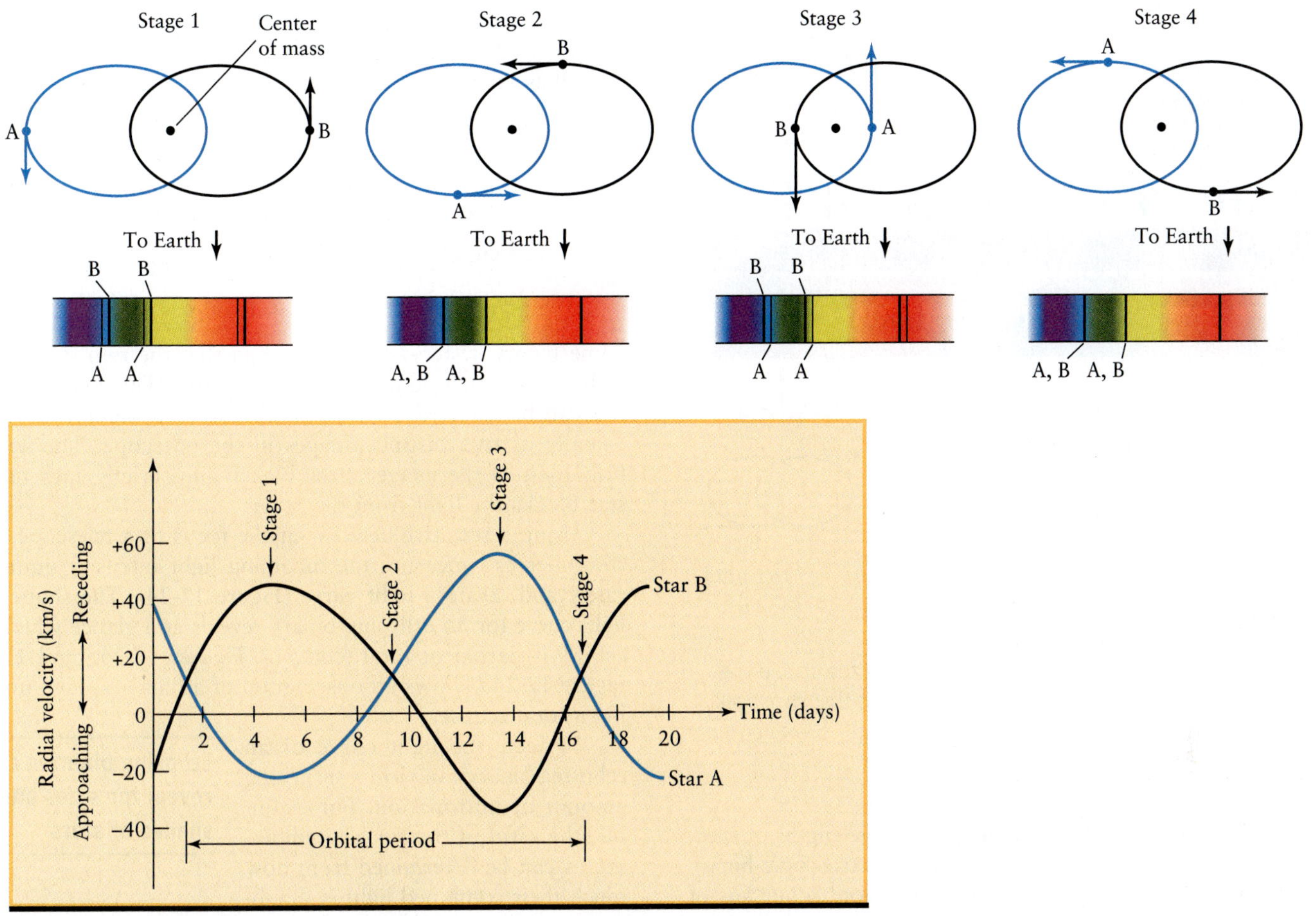

Figure 17-22

Radial Velocity Curves The lower graph displays the radial velocity curves of the binary system HD 171978. The drawings at the top indicate the positions of the stars (labeled A and B) and the spectra of the binary at four selected moments (stages 1, 2, 3, and 4) during an orbital period. Note that at stages 1 and 3, the Doppler effect splits apart the absorption lines from stars A and B; compare with Figure 17-23. The entire binary star system is moving away from us at 12 km/s, which is why the entire pattern of radial velocity curves is displaced upward from the zero-velocity line.

Because one star shows a blueshift while the other is showing a redshift, the spectral lines of the binary system appear to split apart and rejoin periodically. Stars whose binary character is revealed by such shifting spectral lines are called **spectroscopic binaries.**

Exploring Spectroscopic Binary Stars

Stars in close binary systems move so rapidly that we can detect their motion using the Doppler effect

To analyze a spectroscopic binary, astronomers measure the wavelength shift of each star's spectral lines and use the Doppler shift formula (introduced in Section 5-9 and Box 5-6) to determine the **radial velocity** of each star—that is, how fast and in what direction it is moving along our line of sight. The lower portion of Figure 17-22 shows a graph of the radial velocity versus time, called a **radial velocity curve,** for the binary system HD 171978. Each of the two stars alternately approaches and recedes as it orbits around the center of mass. The pattern of the curves repeats every 15 days, which is the orbital period of the binary.

Figure 17-23 shows two spectra of the spectroscopic binary κ (kappa) Arietis taken a few days apart. In Figure 17-23*a,* two sets of spectral lines are visible, offset slightly in opposite directions from the normal positions of these lines. This corresponds to stage 1 or stage 3 in Figure 17-22; one of the orbiting stars is moving toward Earth and has its spectral lines blueshifted, and the other star is moving away from Earth and has its lines redshifted. A few days later, the stars have progressed along their orbits so that neither star is moving toward or away from Earth, corresponding to stage 2 or stage 4 in Figure 17-22. At this time there are no Doppler shifts, and the spectral lines of both stars are at the same positions. That is why only one set of spectral lines appears in Figure 17-23*b*.

It is important to emphasize that the Doppler effect applies only to motion along the line of sight. Motion perpendicular to

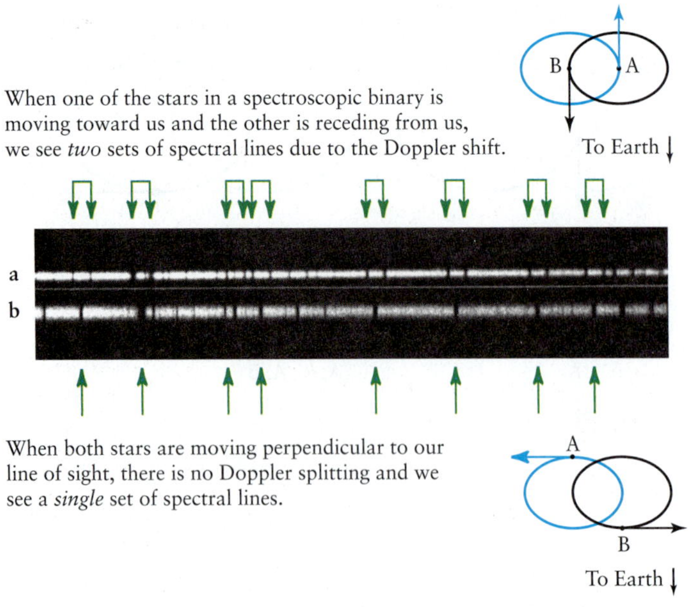

When one of the stars in a spectroscopic binary is moving toward us and the other is receding from us, we see *two* sets of spectral lines due to the Doppler shift.
B
A
To Earth
a
b
When both stars are moving perpendicular to our line of sight, there is no Doppler splitting and we see a *single* set of spectral lines.
A
B
To Earth

Figure 17-23 R I V U X G

A Spectroscopic Binary The visible-light spectrum of the double-line spectroscopic binary κ (kappa) Arietis has spectral lines that shift back and forth as the two stars revolve about each other. (Lick Observatory)

the line of sight does not affect the observed wavelengths of spectral lines. Hence, the ideal orientation for a spectroscopic binary is to have the stars orbit in a plane that is edge-on to our line of sight. (By contrast, a *visual* binary is best observed if the orbital plane is face-on to our line of sight.) For the Doppler shifts to be noticeable, the orbital speeds of the two stars should be at least a few kilometers per second.

The binaries depicted in Figures 17-22 and 17-23 are called *double-line* spectroscopic binaries, because the spectral lines of both stars in the binary system can be seen. Most spectroscopic binaries, however, are *single-line* spectroscopic binaries, in which one of the stars is so dim that its spectral lines cannot be detected. The star is obviously a binary, however, because its spectral lines shift back and forth, thereby revealing the orbital motions of two stars about their center of mass.

As for visual binaries, spectroscopic binaries allow astronomers to learn about stellar masses. From a radial velocity curve, one can find the *ratio* of the masses of the two stars in a binary. The *sum* of the masses is related to the orbital speeds of the two stars by Kepler's laws and Newtonian mechanics. If both the ratio of the masses and their sum are known, the individual masses can be determined using algebra. However, determining the sum of the masses requires that we know how the binary orbits are tilted from our line of sight. This is because the Doppler shifts reveal only the radial velocities of the stars rather than their true orbital speeds. This tilt is often impossible to determine, because we cannot see the individual stars in the binary. Thus, the masses of stars in spectroscopic binaries tend to be uncertain.

There is one important case in which we can determine the orbital tilt of a spectroscopic binary. If the two stars are observed to eclipse each other periodically, then we must be viewing the orbit nearly edge-on. As we will see next, individual stellar masses—as well as other useful data—can be determined if a spectroscopic binary also happens to be such an *eclipsing* binary.

17-11 Light curves of eclipsing binaries provide detailed information about the two stars

Some binary systems are oriented so that the two stars periodically eclipse each other as seen from Earth. These **eclipsing binaries** can be detected even when the two stars cannot be resolved visually as two distinct images in the telescope. The apparent brightness of the image of the binary dims briefly each time one star blocks the light from the other.

Using a sensitive detector at the focus of a telescope, an astronomer can measure the incoming light intensity quite accurately and create a **light curve** (Figure 17-24). The shape of the light curve for an eclipsing binary reveals at a glance whether the eclipse is partial or total (compare Figures 17-24*a* and 17-24*b*). Figure 17-24*d* shows an observation of a binary system undergoing a total eclipse.

Eclipsing binaries can reveal the sizes and shapes of stars

In fact, the light curve of an eclipsing binary can yield a surprising amount of information. For example, the ratio of the surface temperatures can be determined from how much their combined light is diminished when the stars eclipse each other. Also, the duration of a mutual eclipse tells astronomers about the relative sizes of the stars and their orbits.

If the eclipsing binary is also a double-line spectroscopic binary, an astronomer can calculate the mass and radius of each star from the light curves and the velocity curves. Unfortunately, very few binary stars are of this ideal type. Stellar radii determined in this way agree well with the values found using the Stefan-Boltzmann law, as described in Section 17-6.

The shape of a light curve can reveal many additional details about a binary system. In some binaries, for example, the gravitational pull of one star distorts the other, much as the Moon distorts Earth's oceans in producing tides (see Figure 4-26). Figure 17-24*c* shows how such tidal distortion gives the light curve a different shape than in Figure 17-24*b*.

Information about stellar atmospheres can also be derived from light curves. Suppose that one star of a binary is a luminous main-sequence star and the other is a bloated red giant. By observing exactly how the light from the bright main-sequence star is gradually cut off as it moves behind the edge of the red giant during the beginning of an eclipse, astronomers can infer the pressure and density in the upper atmosphere of the red giant.

Binary systems are tremendously important because they enable astronomers to measure stellar masses as well as other key properties of stars. In the next several chapters, we will use this information to help us piece together the story of *stellar evolution*—how stars are born, evolve, and eventually die.

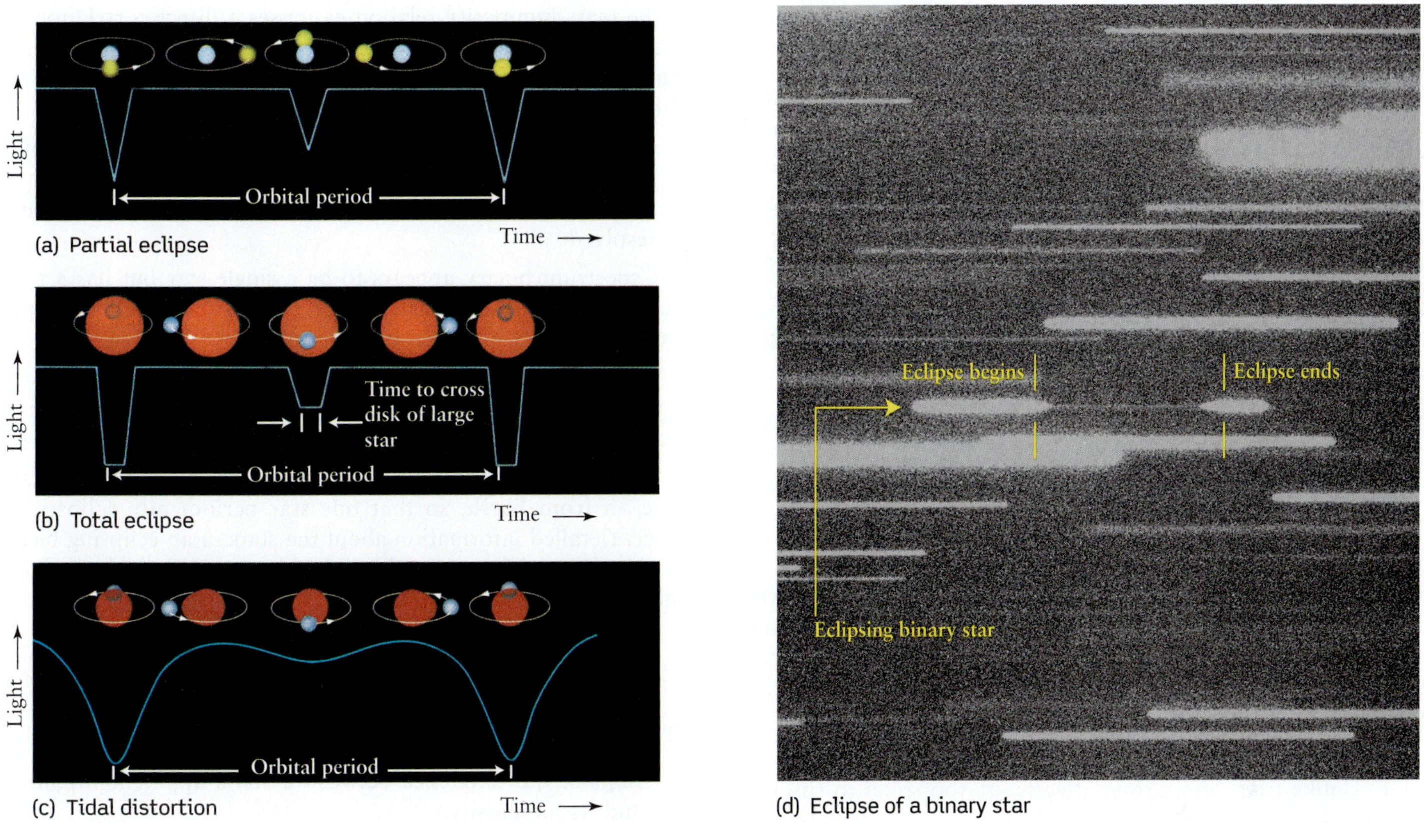

Figure 17-24 R I **V** U X G

Representative Light Curves of Eclipsing Binaries **(a), (b), (c)** The shape of the light curve of an eclipsing binary can reveal many details about the two stars that make up the binary. **(d)** This image shows the binary star NN Serpens (indicated by the arrow) undergoing a total eclipse. The telescope was moved during the exposure so that the sky drifted slowly from left to right across the field of view. During the 10.5-minute duration of the eclipse, the dimmer star of the binary system (an M6 main-sequence star) passed in front of the other; more luminous star (a white dwarf). The binary became so dim that it almost disappeared. (European Southern Observatory)

Key Words

Terms preceded by an asterisk (*) are discussed in the Boxes.

absolute magnitude, p. 441
apparent brightness (brightness), p. 437
apparent magnitude, p. 440
binary star (binary), p. 456
brown dwarf, p. 449
center of mass, p. 458
color ratio, p. 445
*distance modulus, p. 444
double star, p. 456
eclipsing binary, p. 462
giant, p. 454
Hertzsprung-Russell diagram (H-R diagram), p. 452
inverse-square law, p. 437
light curve, p. 462
luminosity, p. 434
luminosity class, p. 455
luminosity function, p. 438
magnitude scale, p. 440
main sequence, p. 453
main-sequence star, p. 453
mass-luminosity relation, p. 458
metals, p. 447
OBAFGKM, p. 446
optical double star, p. 456
parallax, p. 434
parsec, p. 434
photometry, p. 438
*proper motion, p. 436
radial velocity, p. 461
radial velocity curve, p. 461
red giant, p. 454
*space velocity, p. 436
spectral classes, p. 446
spectral types, p. 446
spectroscopic binary, p. 461
spectroscopic parallax, p. 455
spectrum binary, p. 460
stellar parallax, p. 434
supergiant, p. 454
*tangential velocity, p. 436
UBV photometry, p. 445
visual binary, p. 457
white dwarf, p. 454

Key Ideas

Measuring Distances to Nearby Stars: Distances to the nearer stars can be determined by parallax, the apparent shift of a star against the background stars observed as Earth moves along its orbit.

• Parallax measurements made from orbit, above the blurring effects of the atmosphere, are much more accurate than those made with Earth-based telescopes.

• Stellar parallaxes can only be measured for stars within a few hundred parsecs.

The Inverse-Square Law: A star's luminosity (total light output), apparent brightness, and distance from Earth are related by the inverse-square law. If any two of these quantities are known, the third can be calculated.

The Population of Stars: Stars of relatively low luminosity are more common than more luminous stars. Our own Sun is a rather average star of intermediate luminosity.

The Magnitude Scale: The apparent magnitude scale is an alternative way to measure a star's apparent brightness.

• The absolute magnitude of a star is the apparent magnitude it would have if viewed from a distance of 10 parsecs. A version of the inverse-square law relates a star's absolute magnitude, apparent magnitude, and distance.

Photometry and Color Ratios: Photometry measures the apparent brightness of a star. The color ratios of a star are the ratios of brightness values obtained through different standard filters, such as the U, B, and V filters. These ratios are a measure of the star's surface temperature.

Spectral Types: Stars are classified into spectral types (subdivisions of the spectral classes O, B, A, F, G, K, and M), based on the major patterns of spectral lines in their spectra. The spectral class and type of a star is directly related to its surface temperature: O stars are the hottest and M stars are the coolest.

• Most brown dwarfs are in even cooler spectral classes called L and T. Unlike true stars, brown dwarfs are too small to sustain thermonuclear fusion.

Hertzsprung-Russell Diagram: The Hertzsprung-Russell (H-R) diagram is a graph plotting the absolute magnitudes of stars against their spectral types—or, equivalently, their luminosities against surface temperatures.

• The positions on the H-R diagram of most stars are along the main sequence, a band that extends from high luminosity and high surface temperature to low luminosity and low surface temperature.

• On the H-R diagram, giant and supergiant stars lie above the main sequence, while white dwarfs are below the main sequence.

• By carefully examining a star's spectral lines, astronomers can determine whether that star is a main-sequence star, giant, supergiant, or white dwarf. Using the H-R diagram and the inverse-square law, the star's luminosity and distance can be found without measuring its stellar parallax.

Binary Stars: Binary stars, in which two stars are held in orbit around each other by their mutual gravitational attraction, are surprisingly common. Those that can be resolved into two distinct star images by a telescope are called visual binaries.

• Each of the two stars in a binary system moves in an elliptical orbit about the center of mass of the system.

• Binary stars are important because they allow astronomers to determine the masses of the two stars in a binary system. The masses can be computed from measurements of the orbital period and orbital dimensions of the system.

Mass-Luminosity Relation for Main-Sequence Stars: Main-sequence stars are stars like the Sun but with different masses.

• The mass-luminosity relation expresses a direct correlation between mass and luminosity for main-sequence stars. The greater the mass of a main-sequence star, the greater its luminosity (and also the greater its radius and surface temperature).

Spectroscopic Observations of Binary Stars: Some binaries can be detected and analyzed, even though the system may be so distant or the two stars so close together that the two star images cannot be resolved.

• A spectrum binary appears to be a single star but has a spectrum with the absorption lines for two distinctly different spectral types.

• A spectroscopic binary has spectral lines that shift back and forth in wavelength. This is caused by the Doppler effect, as the orbits of the stars carry them first toward and then away from Earth.

• An eclipsing binary is a system whose orbits are viewed nearly edge-on from Earth, so that one star periodically eclipses the other. Detailed information about the stars in an eclipsing binary can be obtained from a study of the binary's radial velocity curve and its light curve.

Questions

Review Questions

1. Explain the difference between a star's apparent brightness and its luminosity.

TUTORIAL 17.1

2. Describe how the parallax method of finding a star's distance is similar to binocular (two-eye) vision in humans.
3. Why does it take at least six months to make a measurement of a star's parallax?
4. Why are measurements of stellar parallax difficult to make? What are the advantages of making these measurements from orbit?

TUTORIAL 17.2

5. What is the inverse-square law? Use it to explain why an ordinary lightbulb can appear brighter than a star, even though the lightbulb emits far less light energy per second.
6. Briefly describe how you would determine the luminosity of a nearby star. Of what value is knowing the luminosity of various stars?
7. Which are more common, stars more luminous than the Sun or stars less luminous than the Sun?
8. Why is the magnitude scale called a "backward" scale? What is the difference between apparent magnitude and absolute magnitude?
9. The star Zubenelgenubi (from the Arabic for "scorpion's southern claw") has apparent magnitude +2.75, while the star Sulafat (Arabic for "tortoise") has apparent magnitude +3.25. Which star appears brighter? From this information alone, what can you conclude about the luminosities of these stars? Explain your answer.
10. Explain why the color ratios of a star are related to the star's surface temperature.
11. Would it be possible for a star to appear bright when viewed through a U filter or a V filter, but dim when viewed through a B filter? Explain your answer.

12. Which gives a more accurate measure of a star's surface temperature, its color ratios or its spectral lines? Explain.

13. Menkalinan (Arabic for "shoulder of the rein-holder") is an A2 star in the constellation Auriga (the Charioteer). What is its spectral class? What is its spectral type? Which gives a more precise description of the spectrum of Menkalinan?

14. What are the most prominent absorption lines you would expect to find in the spectrum of a star with a surface temperature of (**a**) 35,000 K, (**b**) 2800 K, and (**c**) 5800 K (like the Sun)? Briefly describe why these stars have such different spectra even though they have essentially the same chemical composition.

15. A fellow student expresses the opinion that since the Sun's spectrum has only weak absorption lines of hydrogen, this element cannot be a major constituent of the Sun. How would you enlighten this person?

16. If a red star and a blue star both have the same radius and both are the same distance from Earth, which one looks brighter in the night sky? Explain why.

17. If a red star and a blue star both appear equally bright and both are the same distance from Earth, which one has the larger radius? Explain why.

18. If a red star and a blue star both have the same radius and both appear equally bright, which one is farther from Earth? Explain why.

19. Sketch a Hertzsprung-Russell diagram. Indicate the regions on your diagram occupied by (**a**) main-sequence stars, (**b**) red giants, (**c**) supergiants, (**d**) white dwarfs, and (**e**) the Sun.

20. Most of the bright stars in the night sky (see Appendix 5) are giants and supergiants. How can this be, if giants and supergiants make up only 1% of the population of stars?

21. Explain why the dashed lines in Figure 17-15*b* slope down and to the right.

22. Some giant and supergiant stars are of the same spectral type (G2) as the Sun. What aspects of the spectrum of a G2 star would you concentrate on to determine the star's luminosity class? Explain what you would look for.

23. Briefly describe how you would determine the distance to a star whose parallax is too small to measure.

24. What information about stars do astronomers learn from binary systems that cannot be learned in any other way? What measurements do they make of binary systems to garner this information?

25. Suppose that you want to determine the temperature, diameter, and luminosity of an isolated star (not a member of a binary system). Which of these physical quantities require you to know the distance to the star? Explain your answer.

26. What is the mass-luminosity relation? Does it apply to stars of all kinds?

27. Use Figure 17-21 to (**a**) estimate the mass of a main-sequence star that is 1000 times as luminous as the Sun, and (**b**) estimate the luminosity of a main-sequence star whose mass is one-fifth that of the Sun. Explain your answers.

28. Which is more massive, a red main-sequence star or a blue main-sequence star? Which has the greater radius? Explain your answers.

29. How do white dwarfs differ from brown dwarfs? Which are more massive? Which are larger in radius? Which are denser?

30. Sketch the radial velocity curves of a binary consisting of two identical stars moving in circular orbits that are (**a**) perpendicular to and (**b**) parallel to our line of sight.

31. Give two reasons why a visual binary star is unlikely to also be a spectroscopic binary star.

32. Sketch the light curve of an eclipsing binary consisting of two identical stars in highly elongated orbits oriented so that (**a**) their major axes are pointed toward Earth and (**b**) their major axes are perpendicular to our line of sight.

Advanced Questions

Questions preceded by an asterisk (*) involve topics discussed in the Boxes.

Problem-solving tips and tools

Look carefully at the worked examples in Boxes 17-1, 17-2, 17-3, and 17-4 before attempting these exercises. For data on the planets, see Table 7-1 or Appendices 1 and 2 at the back of this book. Remember that a telescope's light-gathering power is proportional to the area of its objective or primary mirror. The volume of a sphere of radius r is $4\pi r^3/3$. Make use of the H-R diagrams in this chapter to answer questions involving spectroscopic parallax. As Box 17-3 shows, some of the problems concerning magnitudes may require facility with logarithms.

33. Find the average distance from the Sun to Neptune in parsecs. Compared to Neptune, how many times farther away from the Sun is Proxima Centauri?

34. Suppose that a dim star were located 2 million AU from the Sun. Find (**a**) the distance to the star in parsecs and (**b**) the parallax angle of the star. Would this angle be measurable with present-day techniques?

35. The star GJ 1156 has a parallax angle of 0.153 arcsec. How far away is the star?

***36.** Kapteyn's star (named after the Dutch astronomer who found it) has a parallax of 0.255 arcsec, a proper motion of 8.67 arcsec per year, and a radial velocity of +246 km/s. (**a**) What is the star's tangential velocity? (**b**) What is the star's actual speed relative to the Sun? (**c**) Is Kapteyn's star moving toward the Sun or away from the Sun? Explain.

***37.** How far away is a star that has a proper motion of 0.08 arcseconds per year and a tangential velocity of 40 km/s? For a star at this distance, what would its tangential velocity have to be in order for it to exhibit the same proper motion as Barnard's star (see Box 17-1)?

***38.** The space velocity of a certain star is 120 km/s and its radial velocity is 72 km/s. Find the star's tangential velocity.

***39.** In the spectrum of a particular star, the Balmer line H_α has a wavelength of 656.15 nm. The laboratory value for the wavelength of H_α is 656.28 nm. (**a**) Find the star's radial velocity. (**b**) Is this star approaching us or moving away? Explain your answer. (**c**) Find the wavelength at which you would expect to find H_α in the spectrum of this star, given that the laboratory wavelength of H_α is 486.13 nm. (**d**) Do your answers depend on the distance from the Sun to this star? Why or why not?

*40. Derive the equation given in Box 17-1 relating proper motion and tangential velocity. (*Hint:* See Box 1-1.)

41. How much dimmer does the Sun appear from Neptune than from Earth? (*Hint:* The average distance between a planet and the Sun equals the semimajor axis of the planet's orbit.)

42. Stars *A* and *B* are both equally bright as seen from Earth, but *A* is 120 pc away while *B* is 24 pc away. Which star has the greater luminosity? How many times greater is it?

43. Stars *C* and *D* both have the same luminosity, but *C* is 32 pc from Earth while *D* is 128 pc from Earth. Which star appears brighter as seen from Earth? How many times brighter is it?

44. Suppose two stars have the same apparent brightness, but one star is 8 times farther away than the other. What is the ratio of their luminosities? Which one is more luminous, the closer star or the farther star?

45. The *solar constant,* equal to 1370 W/m^2, is the amount of light energy from the Sun that falls on 1 square meter of Earth's surface in 1 second (see Section 17-2). What would the distance between Earth and the Sun have to be in order for the solar constant to be 1 watt per square meter (1 W/m^2)?

46. The star Procyon in Canis Minor (the Small Dog) is a prominent star in the winter sky, with an apparent brightness 1.3×10^{-11} that of the Sun. It is also one of the nearest stars, being only 3.50 parsecs from Earth. What is the luminosity of Procyon? Express your answer as a multiple of the Sun's luminosity.

47. The star HIP 92403 (also called Ross 154) is only 2.97 parsecs from Earth but can be seen only with a telescope, because it is 60 times dimmer than the dimmest star visible to the unaided eye. How close to us would this star have to be in order for it to be visible without a telescope? Give your answer in parsecs and in AU. Compare with the semimajor axis of Pluto's orbit around the Sun.

*48. The star HIP 72509 has an apparent magnitude of +12.1 and a parallax angle of 0.222 arcsecond. (**a**) Determine its absolute magnitude. (**b**) Find the approximate ratio of the luminosity of HIP 72509 to the Sun's luminosity.

*49. Suppose you can just barely see a twelfth-magnitude star through an amateur's 6-inch telescope. What is the magnitude of the dimmest star you could see through a 60-inch telescope?

*50. A certain type of variable star is known to have an average absolute magnitude of 0.0. Such stars are observed in a particular star cluster to have an average apparent magnitude of +14.0. What is the distance to that star cluster?

*51. (**a**) Find the absolute magnitudes of the brightest and dimmest of the labeled stars in Figure 17-6*b*. Assume that all of these stars are 110 pc from Earth. (**b**) If a star in the Pleiades cluster is just bright enough to be seen from Earth with the naked eye, what is its absolute magnitude? Is such a star more or less luminous than the Sun? Explain.

52. (**a**) On a copy of Figure 17-8, sketch the intensity curve for a blackbody at a temperature of 3000 K. Note that this figure shows a smaller wavelength range than Figure 17-7*a*. (**b**) Repeat part (a) for a blackbody at 12,000 K (see Figure 17-7*c*). (**c**) Use your sketches from parts (a) and (b) to explain why the color ratios b_V/b_B and b_B/b_U are less than 1 for very hot stars but greater than 1 for very cool stars.

*53. Astronomers usually express a star's color using apparent magnitudes. The star's apparent magnitude as viewed through a B filter is called m_B, and its apparent magnitude as viewed through a V filter is m_V. The difference $m_B - m_V$ is called the *B–V color index* ("B minus V"). Is the B–V color index positive or negative for very hot stars? What about very cool stars? Explain your answers.

*54. (See Question 53.) The B–V color index is related to the color ratio b_V/b_B by the equation

$$m_B = m_V = 2.5 \log\left(\frac{b_V}{b_B}\right)$$

(**a**) Explain why this equation is correct. (**b**) Use the data in Table 17-1 to calculate the B–V color indices for Bellatrix, the Sun, and Betelgeuse. From your results, describe a simple rule that relates the value of the B–V color index to a star's color.

55. The bright star Rigel in the constellation Orion has a surface temperature about 1.6 times that of the Sun. Its luminosity is about 64,000 $L_\odot$. What is Rigel's radius compared to the radius of the Sun?

56. (See Figure 17-12.) What temperature and spectral classification would you give to a star with equal line strengths of hydrogen (H) and neutral helium (He I)? Explain your answer.

57. The Sun's surface temperature is 5800 K. Using Figure 17-12, arrange the following absorption lines in the Sun's spectrum from the strongest to the weakest, and explain your reasoning: (i) neutral calcium; (ii) singly ionized calcium; (iii) neutral iron; (iv) singly ionized iron.

58. Star *P* has one-half the radius of star *Q*. Stars *P* and *Q* have surface temperatures 4000 K and 8000 K, respectively. Which star has the greater luminosity? How many times greater is it?

59. Star *X* has 12 times the luminosity of star *Y*. Stars *X* and *Y* have surface temperatures 3500 K and 7800 K, respectively. Which star has the larger radius? How many times larger is it?

60. Suppose a star experiences an outburst in which its surface temperature doubles but its average density (its mass divided by its volume) decreases by a factor of 8. The mass of the star stays the same. By what factors do the star's radius and luminosity change?

61. The Sun experiences solar flares (see Section 16-10). The amount of energy radiated by even the strongest solar flare is not enough to have an appreciable effect on the Sun's luminosity. But when a flare of the same size occurs on a main-sequence star of spectral class M, the star's brightness can increase by as much as a factor of 2. Why should there be an appreciable increase in brightness for a main-sequence M star but not for the Sun?

62. The bright star Zubeneschmali (β Librae) is of spectral type B8 and has a luminosity of 130 $L_\odot$. What is the star's approximate surface temperature? How does its radius compare to that of the Sun?

63. Castor (α Geminorum) is an A1 V star with an apparent brightness of 4.4×10^{-12} that of the Sun. Determine the approximate distance from Earth to Castor (in parsecs).

64. A brown dwarf called CoD–33°7795 B has a luminosity of $0.0025 L_{\odot}$. It has a relatively high surface temperature of 2550 K, which suggests that it is very young and has not yet had time to cool down by emitting radiation. (**a**) What is this brown dwarf's spectral class? (**b**) Find the radius of CoD–33°7795 B. Express your answer in terms of the Sun's radius and in kilometers. How does this compare to the radius of Jupiter? Is the name "dwarf" justified?

65. The star HD 3651 shown in Figure 17-13 has a mass of $0.79 M_{\odot}$. Its brown dwarf companion, HD 3651B, has about 40 times the mass of Jupiter. The average distance between the two stars is about 480 AU. How long does it take the two stars to complete one orbit around each other?

66. The visual binary 70 Ophiuchi (see the accompanying figure) has a period of 87.7 years. The parallax of 70 Ophiuchi is 0.2 arcsec, and the apparent length of the semimajor axis as seen through a telescope is 4.5 arcsec. (**a**) What is the distance to 70 Ophiuchi in parsecs? (**b**) What is the actual length of the semimajor axis in AU? (**c**) What is the sum of the masses of the two stars? Give your answer in solar masses.

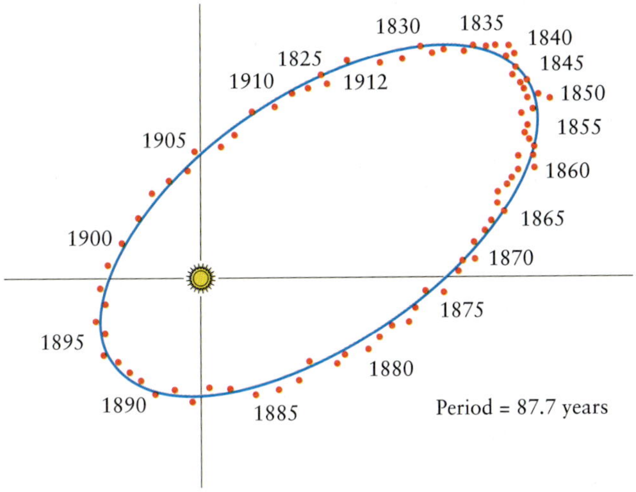

67. An astronomer observing a binary star finds that one of the stars orbits the other once every 5 years at a distance of 10 AU. (**a**) Find the sum of the masses of the two stars. (**b**) If the mass ratio of the system is $M_1/M_2 = 0.25$, find the individual masses of the stars. Give your answers in terms of the mass of the Sun.

Discussion Questions

68. From its orbit around Earth, the *Hipparcos* satellite could measure stellar parallax angles with acceptable accuracy only if the angles were larger than about 0.002 arcsec. Discuss the advantages or disadvantages of making parallax measurements from a satellite in a large solar orbit, say at the distance of Jupiter from the Sun. If this satellite can also measure parallax angles of 0.002 arcsec, what is the distance of the most remote stars that can be accurately determined? How much bigger a volume of space would be covered compared to Earth-based observations? How many more stars would you expect to be contained in that volume?

*69. As seen from the starship *Enterprise* in the *Star Trek* television series and movies, stars appear to move across the sky due to the starship's motion. How fast would the *Enterprise* have to move in order for a star 1 pc away to appear to move 1° per second? (*Hint:* The speed of the star as seen from the *Enterprise* is the same as the speed of the *Enterprise* relative to the star.) How does this compare with the speed of light? Do you think the stars appear to move as seen from an orbiting space shuttle, which moves at about 8 km/s?

70. It is desirable to be able to measure the radial velocity of stars (using the Doppler effect) to an accuracy of 1 km/s or better. One complication is that radial velocities refer to the motion of the star relative to the Sun, while the observations are made using a telescope on Earth. Is it important to take into account the motion of Earth around the Sun? Is it important to take into account Earth's rotational motion? To answer this question, you will have to calculate Earth's orbital speed and the speed of a point on Earth's equator (the part of Earth's surface that moves at the greatest speed because of the planet's rotation). If one or both of these effects are of importance, how do you suppose astronomers compensate for them?

Web/eBook Questions

71. Search the World Wide Web for information about *Gaia,* a European Space Agency (ESA) spacecraft planned to extend the work carried out by *Hipparcos*. When is the spacecraft planned to be launched? How will *Gaia* compare to *Hipparcos*? For how many more stars will it be able to measure parallaxes? What other types of research will it carry out?

72. Search the World Wide Web for recent discoveries about brown dwarfs. Are all brown dwarfs found orbiting normal stars, or are they also found orbiting other brown dwarfs? Are any found in isolation (that is, not part of a binary system)? The Sun experiences flares (see Section 16-10), as do other normal stars; is there any evidence that brown dwarfs also experience flares? If so, is there anything unusual about these flares?

Activities

Observing Projects

Observing tips and tools

Even through a telescope, the colors of stars are sometimes subtle and difficult to see. To give your eye the best chance of seeing color, use the "averted vision" trick: When looking through the telescope eyepiece, direct your vision a little to one side of the star you are most interested in. This places the light from that star on a more sensitive part of your eye's retina.

73. The table below lists five well-known red stars. It includes their right ascension and declination (celestial coordinates de-

scribed in Box 2-1), apparent magnitudes, and color ratios. As their apparent magnitudes indicate, all these stars are somewhat variable. Observe at least two of these stars both by eye and through a small telescope. Is the reddish color of the stars readily apparent, especially in contrast to neighboring stars? (The Jesuit priest and astronomer Angelo Secchi named Y Canum Venaticorum "La Superba," and μ Cephei is often called William Herschel's "Garnet Star.")

Star	Right ascension	Declination	Apparent magnitude	b_V/b_B
Betelgeuse	5^h 55.2^m	+7° 24′	0.4–1.3	5.5
Y Canum Venaticorum	12 45.1	+45 26	5.5–6.0	10.4
Antares	16 29.4	−26 26	0.9–1.8	5.4
μ Cephei	21 43.5	+58 47	3.6–5.1	8.7
TX Piscium	23 46.4	+3 29	5.3–5.8	11.0

Note: The right ascensions and declinations are given for epoch 2000.

74. The table of double stars shown below includes vivid examples of contrasting star colors. The table lists the angular separation between the stars of each double. Observe at least four of these double stars through a telescope. Use the spectral types listed to estimate the difference in surface temperature of the stars in each pair you observe. Does the double with the greatest difference in temperature seem to present the greatest color contrast? From what you see through the telescope and on what you know about the H-R diagram, explain why all the cool stars (spectral types K and M) listed are probably giants or supergiants.

75. Observe the eclipsing binary Algol (β Persei), using nearby stars to judge its brightness during the course of an eclipse. Algol has an orbital period of 2.87 days, and, with the onset of primary eclipse, its apparent magnitude drops from 2.1 to 3.4. It remains this faint for about 2 hours. The entire eclipse, from start to finish, takes about 10 hours. Consult the "Celestial Calendar" section of the current issue of *Sky & Telescope* for the predicted dates and times of the minima of Algol. Note that the schedule is given in Universal Time (the same as Greenwich Mean Time), so you will have to convert the time to that of your own time zone. Algol is normally the second brightest star in the constellation of Perseus. Because of its position on the celestial sphere (R.A. = 3^h 08.2^m, Decl. = 40° 57′), Algol is readily visible from northern latitudes during the fall and winter months.

76. Use the *Starry Night Enthusiast*™ program to investigate the brightest stars. Click on **Home** to show the sky as seen from your location. Set the date to today's date and the time to midnight(12:00:00 A.M.). Open the **Options** pane, expand the **Constellations** layer and click the **Boundaries** and **Labels** options to turn these displays on. You will now see the boundaries of the constellations. Open the **Info** pane and expand the **Position in Space** and **Other Data** lists. (**a**) Scroll around the sky and identify at least five of the brighter stars (shown as larger dots) and click on them to reveal relevant data in the **Info** pane. Make a list of these stars and record **Luminosity** and **Distance from Sun** from the **Info** pane. Which stars did you select? In which constellation does each of these stars lie? Which of these stars are listed in Appendix 5? Of these, which is the most luminous? Which is the most distant? (**b**) Set the date to six months from today, and again set the time to 12:00:00 A.M. Which of the stars that you selected in part (a) are visible? (You can use the **Find** pane to attempt to locate your selected stars.) Which are not? Explain why the passage of six months should make a difference.

77. Use the *Starry Night Enthusiast*™ program to examine the nearby stars. Click on **Favorites > Stars > Local Neighborhood** and **Stop time.** Select **View > Feet** to hide the spacesuit image. Center this view upon the Sun by opening the **Find** pane and double-clicking on Sun. You are now 16.41 light years from the Sun, looking at the labeled nearby stars. **Increase current elevation** to about 70,000 light-years using the button on the toolbar below the Viewing Location box (an upward-pointing triangle) to see these nearby stars within the Milky Way Galaxy. You can rotate the galaxy by placing the mouse cursor over the image and holding down the **Shift** key while holding down the mouse button and moving the mouse. (On a two-button mouse, hold down the left mouse button).

Star	Right ascension	Declination	Apparent magnitudes	Angular separation (arcseconds)	Spectral types
55 Piscium	0^h 39.9^m	+21° 26′	5.4 and 8.7	6.5	K0 and F3
γ Andromedae	2 03.9	+42 20	2.3 and 4.8	9.8	K3 and A0
32 Eridani	3 54.3	−2 57	4.8 and 6.1	6.8	G5 and A2
ι Cancri	8 46.7	+28 46	4.2 and 6.6	30.5	G5 and A5
γ Leonis	10 20.0	+19 51	2.2 and 3.5	4.4	K0 and G7
24 Coma Berenicis	12 35.1	+18 23	5.2 and 6.7	20.3	K0 and A3
ν Boötis	14 45.0	+27 04	2.5 and 4.9	2.8	K0 and A0
α Herculis	17 14.6	+14 23	3.5 and 5.4	4.7	M5 and G5
59 Serpentis	18 27.2	+0 12	5.3 and 7.6	3.8	G0 and A6
β Cygni	19 30.7	+27 58	3.1 and 5.1	34.3	K3 and B8
δ Cephei	22 29.2	+58 25	4* and 7.5	20.4	F5 and A0

Note: The right ascensions and declinations are given for epoch 2000.

**The brighter star in the δ Cephei binary system is a variable star of approximately the fourth magnitude.*

Decrease current elevation to a distance of about 100 light-years from the Sun to return to the solar neighborhood. Again, you can rotate this swarm of stars by holding down the **Shift** key while holding down the mouse button and moving the mouse. Open the **Info** pane. If you click the mouse while the cursor is over a star, you will see the star's apparent magnitude as seen from Earth in the **Other Data** layer and its distance from the Sun in the **Position in Space** layer of the **Info** pane. (**a**) Select at least 5 stars within 50 light-years of the Sun and note their names, apparent magnitudes, luminosities, and distances from the Sun in a list. Which of these stars would be visible from Earth with the naked eye from a dark location? Which are visible with the naked eye from a brightly lit city? (*Hint:* The naked eye can see stars as faint as apparent magnitude $m = +6$ from a dark location, but only as faint as $m = +4$ from an inner city.) (**b**) **Increase current elevation** once more to about 1000 light-years from Earth and locate at least 5 stars that are further than 500 light-years from the Sun, making a list of these stars, their names, apparent magnitudes, luminosities and distances from the Sun. Which of these stars are visible from Earth with the naked eye from a dark location? Are the naked-eye stars more likely to be giants or supergiants, or are they more likely to be main-sequence stars? Explain your answer.

18

The Birth of Stars

RI V UXG

WEB LINK 18.1

A region of star formation about 1400 pc (4000 ly) from Earth in the southern constellation Ara (the Altar). (European Southern Observatory)

The stars that illuminate our nights seem eternal and unchanging. But this permanence is an illusion. Each of the stars visible to the naked eye shines due to thermonuclear reactions and has only a finite amount of fuel available for these reactions. Hence, stars cannot last forever: They form from material in interstellar space, evolve over millions or billions of years, and eventually die. In this chapter our concern is with how stars are born and become part of the main sequence.

Stars form within cold, dark clouds of gas and dust that are scattered abundantly throughout our Galaxy. One such cloud appears as a dark area on the far right-hand side of the photograph at the top of this page. Perhaps a dark cloud like this encounters one of the Galaxy's spiral arms, or perhaps a supernova detonates nearby. From the shock of events like these, the cloud begins to contract under the pull of gravity, forming protostars—the fragments that will one day become stars. As a protostar develops, its internal pressure builds and its temperature rises. In time, hydrogen fusion begins, and a star is born. The hottest, bluest, and brightest young stars, like those in the accompanying image, emit ultraviolet radiation that excites the surrounding interstellar gas. The result is a beautiful glowing nebula, which typically has the red color characteristic of excited hydrogen (as shown in the photograph).

In Chapters 19 and 20 we will see how stars mature and grow old. Some even blow themselves apart in death throes that enrich interstellar space with the material for future generations of stars. Thus, like the mythical phoenix, new stars arise from the ashes of the old.

18-1 Understanding how stars evolve requires observation as well as ideas from physics

Over the past several decades, astronomers have labored to develop an understanding of **stellar evolution**, that is, how stars are born, live their lives, and finally die. Our own Sun provides evidence that stars are not permanent. The energy radiated by the Sun comes from thermonuclear reactions in its core, which consume 6×10^{11} kg of hydrogen each second and convert it into helium (see Section 16-1). While the amount of hydrogen in the Sun's core is vast, it is not infinite; therefore, the Sun cannot always have been shining, nor can it continue to shine forever. The same is true for all other main-sequence stars, which are fundamentally the same kinds of objects as the Sun but with different masses (see Section 17-9). Thus, stars must have a beginning as well as an end.

Stars consume the material of which they are made, and so cannot last forever

Learning Goals

By reading the sections of this chapter, you will learn

18-1 How astronomers have pieced together the story of stellar evolution

18-2 What interstellar nebulae are and what they are made of

18-3 What happens as a star begins to form

18-4 The stages of growth from young protostars to main-sequence stars

18-5 How stars gain and lose mass during their growth

18-6 What insights star clusters add to our understanding of stellar evolution

18-7 Where new stars form within galaxies

18-8 How the death of old stars can trigger the birth of new stars

Stars last very much longer than the lifetime of any astronomer—indeed, far longer than the entire history of human civilization. Thus, it is impossible to watch a single star go through its formation, evolution, and eventual demise. Rather, astronomers have to piece together the evolutionary history of stars by studying different stars at different stages in their life cycles.

ANALOGY To see the magnitude of this task, imagine that you are a biologist from another planet who sets out to understand the life cycles of human beings. You send a spacecraft to fly above Earth and photograph humans in action. Unfortunately, the spacecraft fails after collecting only 20 seconds of data, but during that time its sophisticated equipment sends back observations of thousands of different humans. From this brief snapshot of life on Earth—only 10^{-8} (a hundred-millionth) of a typical human lifetime—how would you decide which were the young humans and which were the older ones? Without a look inside our bodies to see the biological processes that shape our lives, could you tell how humans are born and how they die? And how could you deduce the various biological changes that humans undergo as they age?

Astronomers, too, have data spanning only a tiny fraction of any star's lifetime. A star like the Sun can last for about 10^{10} years, whereas astronomers have been observing stars in detail for only about a century—as in our analogy, roughly 10^{-8} of the life span of a typical star. Astronomers are also frustrated by being unable to see the interiors of stars. For example, we cannot see the thermonuclear reactions that convert hydrogen into helium. But astronomers have an advantage over the biologist in our story. Unlike humans, stars are made of relatively simple substances, primarily hydrogen and helium, that are found almost exclusively in the form of gases. Of the three phases of matter—gas, liquid, and solid—gases are by far the simplest to understand.

Astronomers use our understanding of gases to build theoretical models of the interiors of stars, like the model of the Sun we saw in Section 16-2. Models help to complete the story of stellar evolution. In fact, like all great dramas, the story of stellar evolution can be regarded as a struggle between two opposing and unyielding forces: Gravity continually tries to make a star shrink, while the star's internal pressure tends to make the star expand. When these two opposing forces are in balance, the star is in a state of hydrostatic equilibrium (see Figure 16-2).

But what happens when changes within the star cause either pressure or gravity to predominate? The star must then either expand or contract until it reaches a new equilibrium. In the process, it will change not only in size but also in luminosity and color.

In the following chapters, we will find that giant and supergiant stars are the result of pressure gaining the upper hand over gravity. Both giants and supergiants turn out to be aging stars that have become tremendously luminous and ballooned to hundreds or thousands of times their previous size. White dwarfs, by contrast, are the result of the balance tipping in gravity's favor. These dwarfs are even older stars that have collapsed to a fraction of the size they had while on the main sequence. In this chapter, however, we will see how the opposing influences of gravity and pressure explain the birth of stars. We start our journey within the diffuse clouds of gas and dust that permeate our galaxy.

18-2 Interstellar gas and dust pervade the galaxy

Different types of nebulae emit, absorb, or reflect light

Where do stars come from? As we saw in Section 8-4, our Sun condensed from a solar nebula, a collection of gas and dust in interstellar space. Observations suggest that other stars originate in a similar way (see Figure 8-8). To understand the formation of stars, we must first understand the nature of the interstellar matter from which the stars form.

Nebulae and the Interstellar Medium

At first glance, the space between the stars seems to be empty. On closer inspection, we find that it is filled with a thin gas laced with microscopic dust particles. This combination of gas and dust is called the **interstellar medium.** Evidence we'll discuss for this medium includes interstellar clouds of various types, curious lines in the spectra of binary star systems, and an apparent dimming and reddening of distant stars.

You can see evidence for the interstellar medium with the naked eye. Look carefully at the constellation Orion (Figure 18-1*a*), visible on winter nights in the northern hemisphere and summer nights in the southern hemisphere. While most of the stars in the constellation appear as sharply defined points of light, the middle "star" in Orion's sword has a fuzzy appearance. This fuzziness becomes more obvious with binoculars or a telescope. As Figure 18-1*b* shows, this "star" is actually not a star at all, but the Orion Nebula—a cloud in interstellar space. Any interstellar cloud is called a **nebula** (plural **nebulae**) or **nebulosity.**

Emission Nebulae: Clouds of Excited Gas

The Orion Nebula emits its own light, with the characteristic emission line spectrum of a hot, thin gas. For this reason it is called an **emission nebula.** Many emission nebulae can be seen with a small telescope. Figure 18-2 shows some of these nebulae in a different part of the constellation Orion. Emission nebulae are direct evidence of gas atoms in the interstellar medium.

Typical emission nebulae have masses that range from about 100 to about 10,000 solar masses. Because this mass is spread over a huge volume that is light-years across, the density is quite low by Earth standards, only a few thousand hydrogen atoms per cubic centimeter. (By comparison, the air you are breathing contains more than 10^{19} atoms per cm^3.)

Emission nebulae are found near hot, luminous stars of spectral types O and B. Such stars emit copious amounts of ultraviolet radiation. When atoms in the nearby interstellar gas absorb these energetic ultraviolet photons, the atoms become ionized. Indeed, emission nebulae are composed primarily of ionized hydrogen atoms, that is, free protons (hydrogen nuclei) and electrons. Astronomers use the notation H I for neutral, un-ionized hydrogen atoms and H II for ionized hydrogen atoms, which is why emission nebulae are also called **H II regions.**

H II regions emit visible light when some of the free protons and electrons get back together to form hydrogen atoms, a process

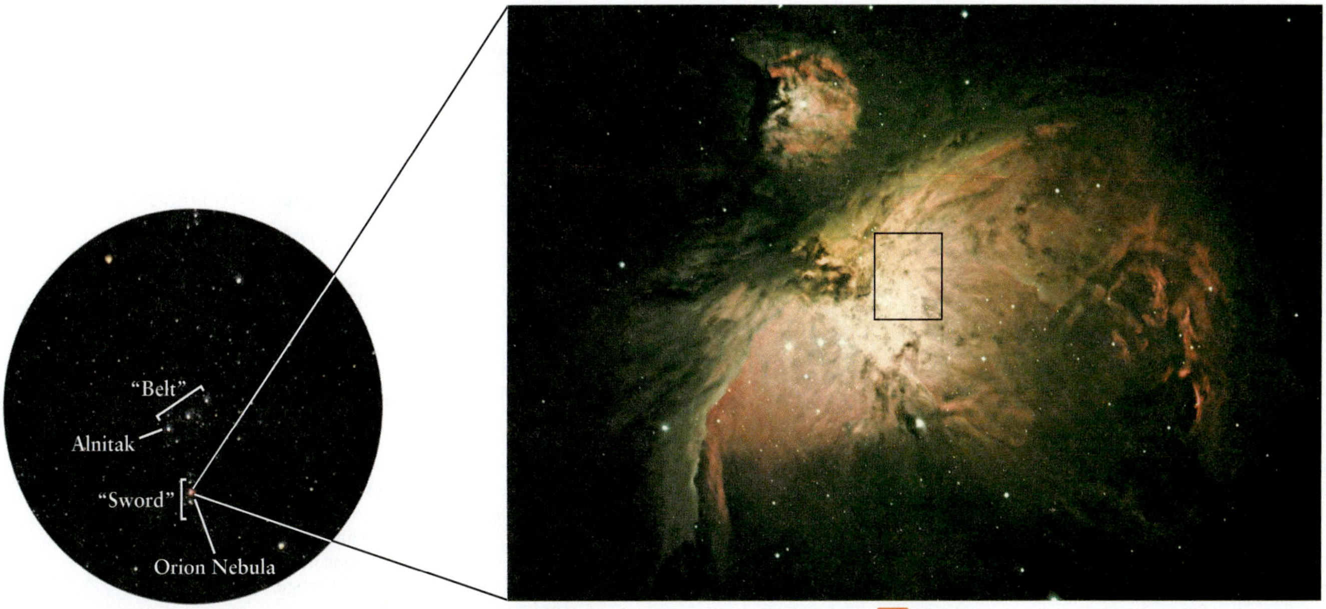

(a) A wide-angle view of Orion

(b) A closeup of the Orion Nebula R I **V** U X G

Figure 18-1

The Orion Nebula (a) The middle "star" of the three that make up Orion's sword is actually an interstellar cloud called the Orion Nebula. **(b)** The nebula is about 450 pc (1500 ly) from Earth and contains about 300 solar masses of material. Within the area shown by the box are four hot, massive stars called the Trapezium. They produce the ultraviolet light that makes the nebula glow. (a: R. C. Mitchell, Central Washington University; b: Anglo-Australian Observatory)

called **recombination** (Figure 18-3). When an atom forms by recombination, the electron is typically captured into a high-energy orbit. As the electron cascades downward through the atom's energy levels toward the ground state, the atom emits photons with lower energies and longer wavelengths than the photons that originally caused the ionization. Particularly important is the transition from $n = 3$ to $n = 2$. It produces H_α photons with a wavelength of 656 nm, in the red portion of the visible spectrum (see Section 5-8, especially Figure 5-23*b*). These photons give H II regions their distinctive reddish color.

For each high-energy, ultraviolet photon absorbed by a hydrogen atom to ionize it, several photons of lower energy are emitted when a proton and electron recombine. As Box 18-1 describes, a similar effect takes place in a fluorescent lightbulb. In this sense, H II regions are immense fluorescent light fixtures!

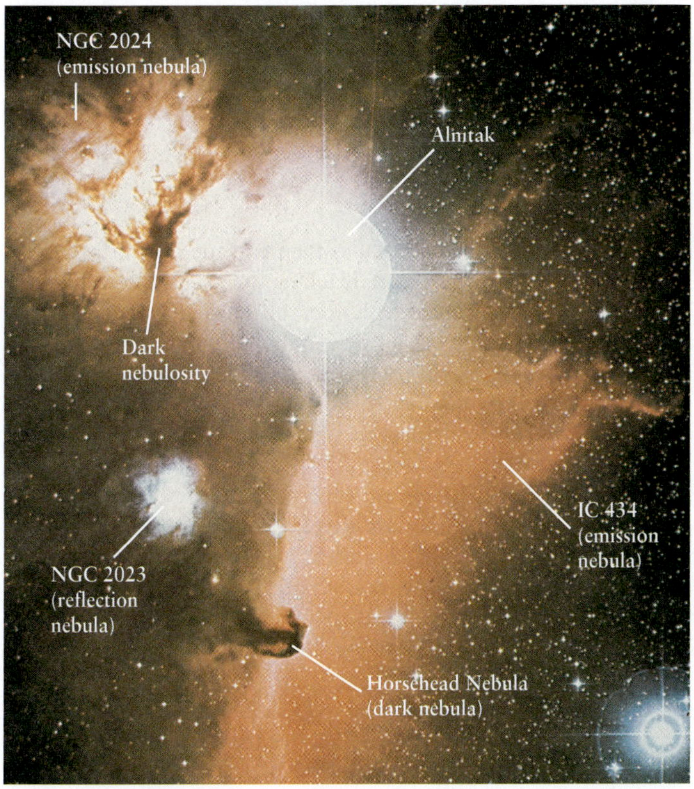

Figure 18-2 R I **V** U X G

Emission, Reflection, and Dark Nebulae in Orion A variety of different nebulae appear in the sky around Alnitak, the easternmost star in Orion's belt (see Figure 18-1a). All the nebulae lie approximately 500 pc (1600 ly) from Earth. They are actually nowhere near Alnitak, which is only 250 pc (820 ly) distant. This photograph shows an area of the sky about 1.5° across. (Royal Observatory, Edinburgh)

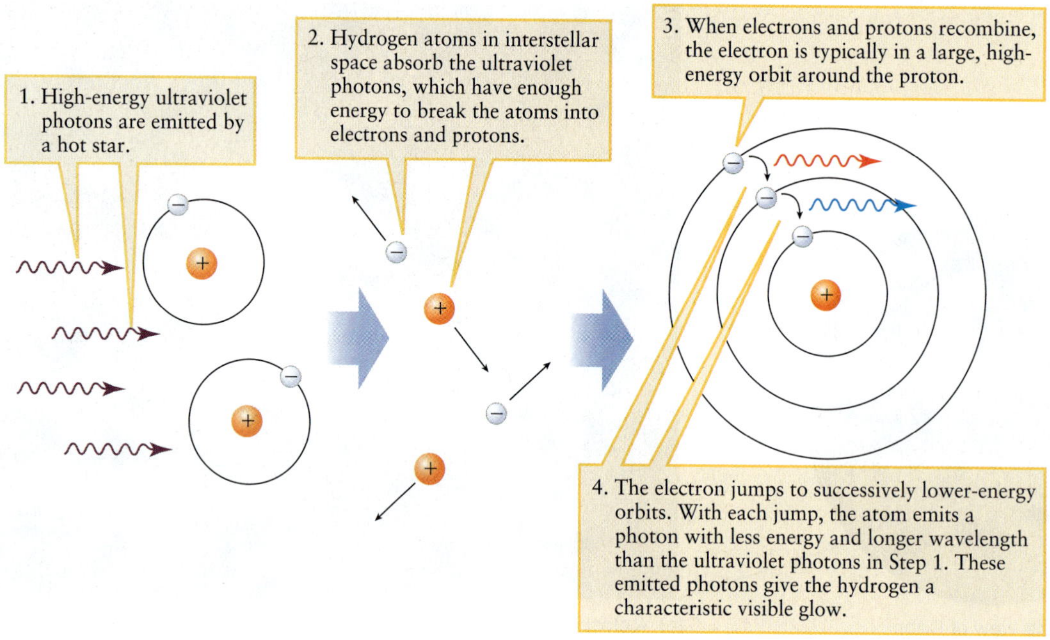

1. High-energy ultraviolet photons are emitted by a hot star.
2. Hydrogen atoms in interstellar space absorb the ultraviolet photons, which have enough energy to break the atoms into electrons and protons.
3. When electrons and protons recombine, the electron is typically in a large, high-energy orbit around the proton.
4. The electron jumps to successively lower-energy orbits. With each jump, the atom emits a photon with less energy and longer wavelength than the ultraviolet photons in Step 1. These emitted photons give the hydrogen a characteristic visible glow.

Figure 18-3

Ionization and Recombination The characteristic glow of emission nebulae (like those shown in Figure 18-1 and Figure 18-2) comes from gas atoms that are excited by ultraviolet radiation from nearby hot stars.

Further evidence of interstellar gas comes from the spectra of binary star systems. As the two stars that make up a binary system orbit their common center of mass, their spectral lines shift back and forth (see Figure 17-23). But certain calcium and sodium lines are found to remain at fixed wavelengths. These **stationary absorption lines** are therefore not associated with the binary star. Instead, they must be caused by interstellar gas between us and the binary system.

BOX 18-1 — Astronomy Down to Earth

Fluorescent Lights

The light that comes from glowing interstellar clouds is, quite literally, otherworldly. But the same principles that explain how such clouds emit light are also at the heart of light phenomena that we see here on Earth.

A fluorescent lamp produces light in a manner not too different from an emission nebula (H II region). In both cases, the physical effect is called **fluorescence:** High-energy ultraviolet photons are absorbed, and the absorbed energy is reradiated as lower-energy photons of visible light.

Within the glass tube of a fluorescent lamp is a small amount of the element mercury. When you turn on the lamp, an electric current passes through the tube, vaporizing the mercury and exciting its atoms. This excited mercury vapor radiates light with an emission-line spectrum, including lines in the ultraviolet. The white fluorescent coating on the inside of the glass tube absorbs these ultraviolet photons, exciting electrons in the coating's molecules to high energy levels.

The electrons in the molecules then cascade down through a number of lower levels before reaching the ground state. During this cascade, visible-light photons of many different wavelengths are emitted, giving an essentially continuous spectrum and a very white light. (By comparison, the hydrogen atoms in an H II region emit at only certain discrete wavelengths, because the spectrum of hydrogen is much simpler than that of the fluorescent coating's molecules. Another difference is that the molecules in the fluorescent tube never become ionized.)

Many common materials display fluorescence. Among them are teeth, fingernails, and certain minerals. When illuminated with ultraviolet light, these materials glow with a blue or green color. (Most natural history museums and science museums have an exhibit showing fluorescent minerals.) Laundry detergent also contains fluorescent material. After washing, your laundry absorbs ultraviolet light from the Sun and glows faintly, making it appear "whiter than white."

Dark Nebulae and Reflection Nebulae: Abodes of Dust

In addition to the presence of gas atoms in H II regions, Figure 18-2 also shows two kinds of evidence for larger bits of matter, called **dust grains,** in the interstellar medium. These dust grains make their appearance in *dark nebulae* and *reflection nebulae.*

A **dark nebula** is so opaque that it blocks any visible light coming from stars that lie behind it. One such dark nebula, called the Horsehead Nebula for its shape, protrudes in front of one of the H II regions in Figure 18-2. Figure 18-4 shows another dark nebula, called Barnard 86. These nebulae have a relatively dense concentration of microscopic dust grains, which scatter and absorb light much more efficiently than single atoms. Typical dark nebulae have temperatures between 10 K to 100 K, which is low enough for hydrogen atoms to form molecules. Such nebulae contain from 10^4 to 10^9 particles (atoms, molecules, and dust grains) per cubic centimeter. Although tenuous by Earth standards, dark nebulae are large enough—typically many light-years deep—that they block the passage of light. In the same way, a sufficient depth of haze or smoke in our atmosphere can make it impossible to see distant mountains.

The other evidence for dust in Figure 18-2 is the bluish haze, labeled NGC 2023, surrounding the star immediately above and to the left of the Horsehead Nebula. Figure 18-5 shows a similar haze around a different set of stars. A haze of this kind, called a **reflection nebula,** is caused by fine grains of dust in a lower concentration than that found in dark nebulae. The light we see coming from the nebula is starlight that has been scattered and reflected by these dust grains. The grains are only about 500 nm across, no larger than a typical wavelength of visible light, and they scatter short-wavelength blue light more efficiently than long-wavelength red light. Hence, reflection nebulae have a characteristic blue color. Box 5-4 explains how a similar process—the scattering of sunlight in our atmosphere—gives rise to the blue color of the sky.

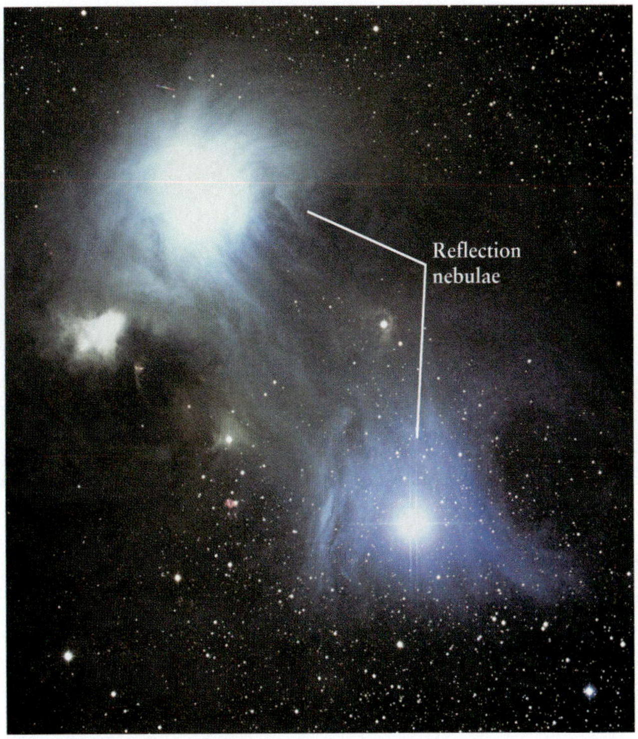

Figure 18-5 R I V U X G

Reflection Nebulae Wispy reflection nebulae called NGC 6726-27-29 surround several stars in the constellation Corona Australis (the Southern Crown). Unlike emission nebulae, reflection nebulae do not emit their own light, but scatter and reflect light from the stars that they surround. This scattered starlight is quite blue in color. The region shown here is about 23 arcminutes across. (David Malin/Anglo-Australian Observatory)

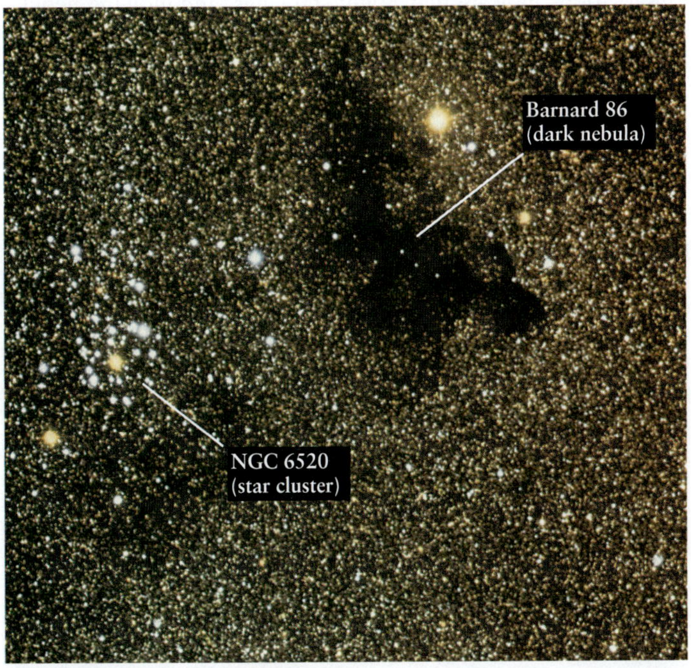

Figure 18-4 R I V U X G

A Dark Nebula When first discovered in the late 1700s, dark nebulae were thought to be "holes in the heavens" where very few stars are present. In fact, they are opaque regions that block out light from the stars beyond them. The few stars that appear to be within Barnard 86 lie between us and the nebula. Barnard 86 is in the constellation Sagittarius and has an angular diameter of 4 arcminutes, about 1/7 the angular diameter of the full moon. (Anglo-Australian Observatory)

Interstellar Extinction and Reddening

In the 1930s, the American astronomer Robert Trumpler discovered two other convincing pieces of evidence for the existence of interstellar matter—*interstellar extinction* and *interstellar reddening.* While studying the brightness and distances of certain star clusters, Trumpler noticed that remote clusters seem to be dimmer than would be expected from their distance alone. His observations demonstrated that the intensity of light from remote stars is reduced as the light passes through material in interstellar space. This process is called **interstellar extinction.** (In the same way, the headlights of oncoming cars appear dimmer when there is smoke or fog in the air.) The light from remote stars is also reddened as it passes through the interstellar medium, because the blue

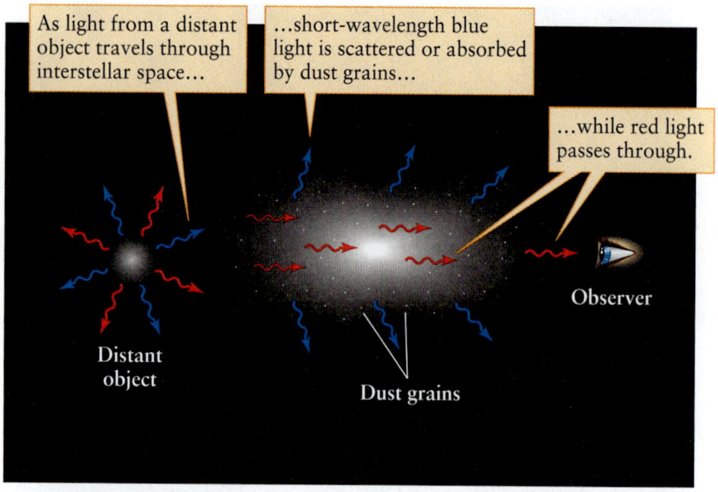

(a) How dust causes interstellar reddening

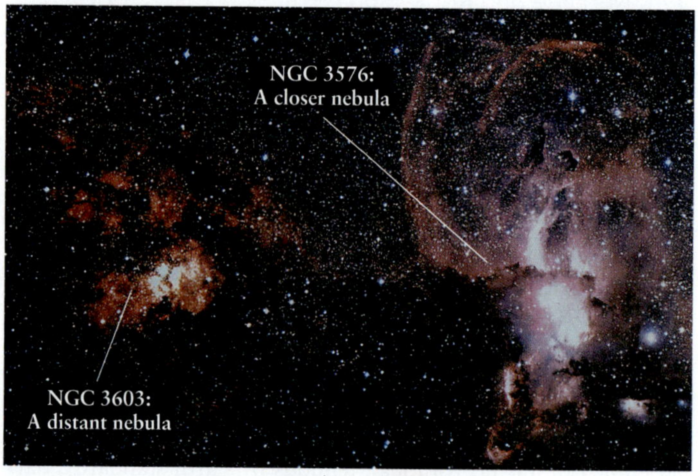

(b) Reddening depends on distance R I V U X G

Figure 18-6

Interstellar Reddening (a) Dust grains in interstellar space scatter or absorb blue light more than red light. Thus light from a distant object appears redder than it really is. (b) The emission nebulae NGC 3603 and NGC 3576 are different distances from Earth. Light from the more distant nebula must pass through more interstellar dust to reach us, so more interstellar reddening occurs and NGC 3603 is a ruddier shade of red. The two nebulae are about 1° apart in the sky. (Anglo-Australian Observatory)

component of their starlight is scattered and absorbed by interstellar dust. This effect, shown in Figure 18-6, is called **interstellar reddening.** The same effect makes the setting Sun look red (see Box 5-4).

CAUTION! It is important to understand the distinction between interstellar reddening and the Doppler effect. If an object is moving away from us, the Doppler effect *shifts* all of that object's light toward longer wavelengths (a redshift). Interstellar reddening, by contrast, makes objects appear red not by shifting wavelengths but by filtering out shorter blue wavelengths. The effect is the same as if we looked at an object through red-colored glasses, which let red light pass but block out blue light. The one similarity between Doppler shifts and interstellar reddening is that neither has any discernible effect on what you see with the naked eye. Both of these effects cause only subtle color changes that your eye (which does a poor job of seeing colors in faint objects like stars) cannot detect. If you see a star with a red color, the reason is *not* interstellar reddening or the Doppler effect; it is because the star really is red due to its low surface temperature (see Figure 17-7a).

Interstellar Gas and Dust in Spiral Galaxies

Observations indicate that interstellar gas and dust are largely confined to the plane of the Milky Way—that faint, hazy band of stars you can see stretching across the sky on a dark, moonless night. Figure 18-7 shows glowing emission nebulae and dark lanes of dust along the Milky Way.

As we will see in Chapter 23, the band of the Milky Way is actually our inside view of our Galaxy, which is a flat, disk-shaped collection of several hundred billion stars about 50,000 pc (160,000 ly) in diameter. The Sun is located about 8000 pc (26,000 ly) from our Galaxy's center. Astronomers know these dimensions because they can map our Galaxy from the locations of bright stars and nebulae and also by using radio telescopes. Observations indicate that bright stars, gas, and glowing nebulae are mostly located within a few hundred parsecs of the midplane of our Galaxy, form somewhat of a "pancake" shape, and are concentrated along arching spiral arms that wind outward from the Galaxy's center. If we could view our Galaxy from a great distance, it would look somewhat like one of the galaxies shown in Figure 18-8.

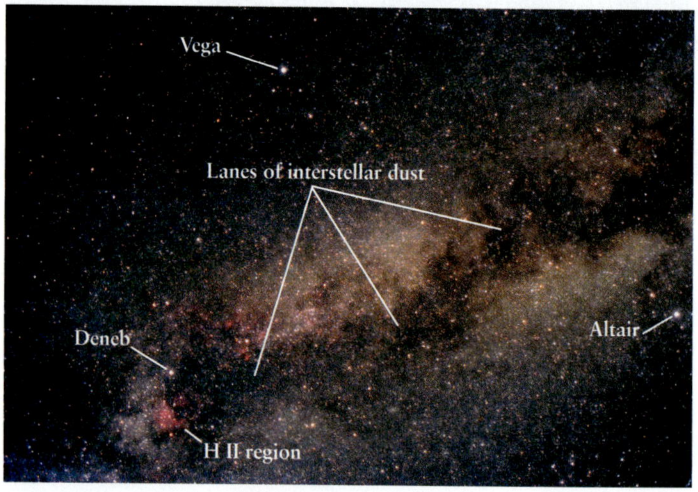

Figure 18-7 R I V U X G

Gas and Dust in the Milky Way Glowing gas clouds (emission nebulae or H II regions) and dark, dusty regions are concentrated close to the midplane of the Milky Way Galaxy, of which our Sun is part. This wide-angle photograph also shows the three bright stars that make up the "summer triangle" (see Figure 2-8). (© Jerry Lodriguss/Photo Researchers).

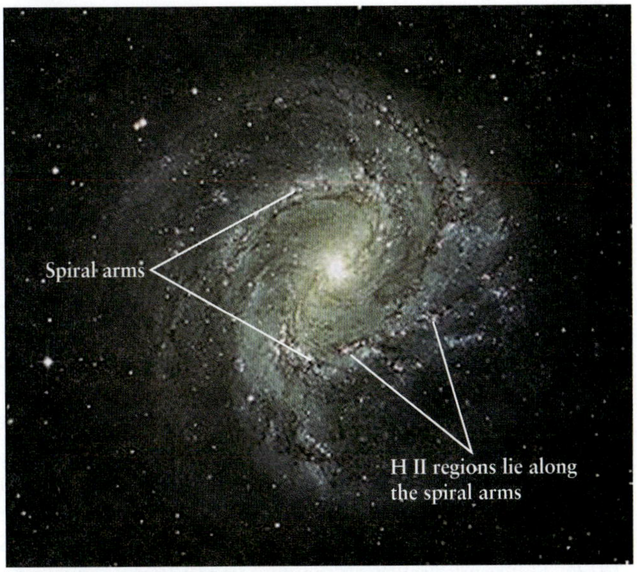

(a) We see spiral galaxy M83 nearly face-on

(b) We see spiral galaxy NGC 891 nearly edge-on

WEB LINK 18.4

Figure 18-8 R I V U X G

Spiral Galaxies Spiral galaxies, like our own Milky Way Galaxy, consist of stars, gas, and dust that are largely confined to a flattened, rotating disk. **(a)** This face-on view of M83 shows luminous stars and H II regions along the spiral arms. **(b)** This edge-on view of NGC 891 shows a dark band caused by dust in this galaxy's interstellar medium. Although in different parts of the sky, both galaxies are about 7 million pc (23 million ly) from Earth and have angular diameters of about 13 arcminutes. (a: David Malin/Anglo-Australian Observatory; b: Instituto de Astrofísica de Canarias/Royal Greenwich Observatory/David Malin)

Interstellar gas and dust are the raw material from which stars are made. The disk of our Galaxy, where most of this matter is concentrated, is therefore the site of ongoing star formation. Our next goal is to examine the steps by which stars are formed.

18-3 Protostars form in cold, dark nebulae

TUTORIAL 18.1

How do stars form in the interstellar medium? In this section, we discuss how a contracting cloud gives birth to a clump called a *protostar*—a future main-sequence star.

Protostars: Initial Gravitational Collapse

In order for interstellar material to condense and form a star, the force of gravity—which tends to draw interstellar material together—must overwhelm the internal pressure pushing the material apart. This means that stars will most easily form in regions where the interstellar material is relatively dense, so that atoms and dust grains are close together and gravitational attraction is enhanced.

To assist star formation, the pressure of the interstellar medium should be relatively low. This means that the star forming region of the interstellar medium should be as cold as possible, because the pressure of a gas goes down as the gas temperature decreases. (Conversely, increasing the temperature makes the pressure increase. This relationship is why the air pressure inside automobile tires is higher after the auto has been driven a while and the tires have warmed up.)

The only parts of the interstellar medium with high enough density and low enough temperature for stars to form are the dark nebulae. Many of these nebulae, such as the Horsehead Nebula (see Figure 18-2) and Barnard 86 (see Figure 18-4), were discovered and catalogued around 1900 by Edward Barnard and are known as **Barnard objects.** (The Horsehead Nebula is also known as Barnard 33.) Other relatively small, nearly spherical dark nebulae are known as **Bok globules,** after the Dutch-American astronomer Bart Bok, who first called attention to them during the 1940s (Figure 18-9). A Bok globule resembles the inner core of a Barnard object with the outer, less dense portions stripped away. The density of the gas and dust within a Barnard object or a Bok globule is indeed quite high by cosmic standards, in the range from 100 to 10,000 particles per cm^3; by comparison, most of the interstellar medium contains only 0.1 to 20 particles per cm^3. Radio emissions from molecules within these clouds indicate that their internal temperatures are very low, only about 10 K.

A typical Barnard object contains a few thousand solar masses of gas and dust spread over a volume roughly 10 pc (30 ly) across; a typical Bok globule is about one-tenth as large. The chemical composition of this material is the standard "cosmic abundance" of about 74% (by mass) hydrogen, 25% helium, and 1% heavier elements (review Figure 8-4). Within these clouds, the densest portions can contract under their own mutual gravitational attraction and form clumps called **protostars.** Each protostar will eventually evolve into a main-sequence star. Because dark nebulae contain many solar masses of material, it is possible for a large number of protostars to form

In the first stage of star formation, a protostar coalesces and contracts due to the mutual gravitational attraction of its parts

Figure 18-9 R I V U X G

Bok Globules The dark blobs in this photograph of a glowing H II region are clouds of dust and gas called Bok globules. A typical Bok globule is a parsec or less in size and contains from one to a thousand solar masses of material. The Bok globules and H II region in this image are part of a much larger star-forming region called NGC 281, which lies about 9500 ly (2900 pc) from Earth in the constellation Cassiopeia. The image shows an area about 8.8 ly (2.7 pc) across. (NASA, ESA, and The Hubble Heritage Team (STScI/AURA))

out of a single such nebula. Thus, we can think of dark nebulae as "stellar nurseries."

The Evolution of a Protostar

In the 1950s, astrophysicists such as Louis Henyey in the United States and C. Hayashi in Japan performed calculations that enabled them to describe the earliest stages of a protostar. At first, a protostar is merely a cool blob of gas several times larger than our solar system. The pressure inside the protostar is too low to support all this cool gas against the mutual gravitational attraction of its parts, and so the protostar collapses. As the protostar collapses, gravitational energy is converted into thermal energy, making the gases heat up and start glowing. (We discussed this process, called Kelvin-Helmholtz contraction, in Section 8-4 and Section 16-1.) Energy from the interior of the protostar is transported outward by convection, warming its surface.

After only a few thousand years of gravitational contraction, the protostar's surface temperature reaches 2000 to 3000 K. At this point the protostar is still quite large, so its glowing gases produce substantial luminosity. (The greater a star's radius and surface temperature, the greater its luminosity. You can review the details in Section 17-6 and Box 17-4.) For example, after only 1000 years of contraction, a protostar of 1 solar mass ($1\ M_\odot$) is 20 times larger in radius than the Sun and has 100 times the Sun's luminosity. Unlike the Sun, however, the luminosity of a young protostar is not the result of thermonuclear fusion, because the protostar's core is not yet hot enough for fusion reactions to begin. Instead, the radiated energy comes exclusively from the heating of the protostar as it contracts.

In order to determine the conditions inside a contracting protostar, astrophysicists use computers to solve equations similar to those used for calculating the structure of the Sun (see Section 18-2). The results tell how the protostar's luminosity and surface temperature change at various stages in its contraction. This information, when plotted on a Hertzsprung-Russell diagram, provides a protostar's **evolutionary track.** The track shows us how the protostar's appearance changes because of changes in its interior. These theoretical tracks agree quite well with actual observations of protostars.

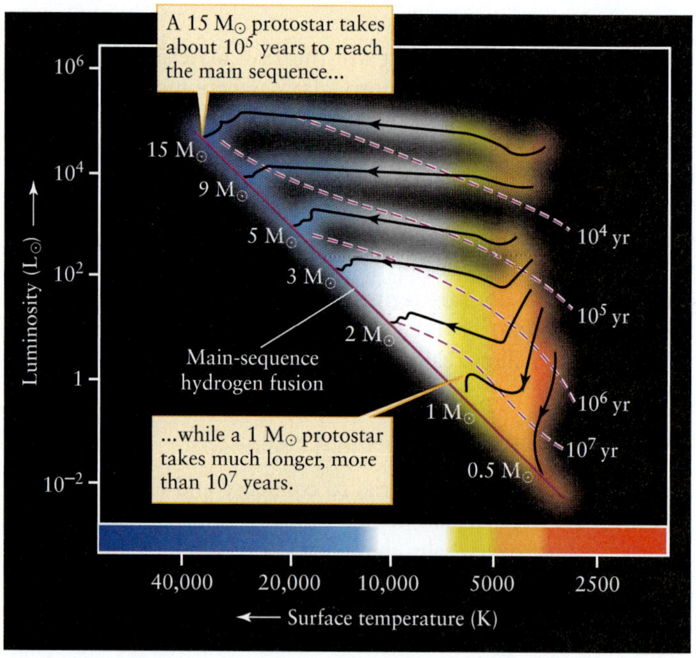

Figure 18-10

Pre-Main-Sequence Evolutionary Tracks on an H-R Diagram As a protostar evolves, its luminosity and surface temperature both change. The tracks shown here depict these changes for protostars of seven different masses. Each dashed red line shows the age of a protostar when its evolutionary track crosses that line. (We will see in Section 18-5 that protostars lose quite a bit of mass as they evolve: The mass shown for each track is the value when the protostar finally settles down as a main-sequence star.)

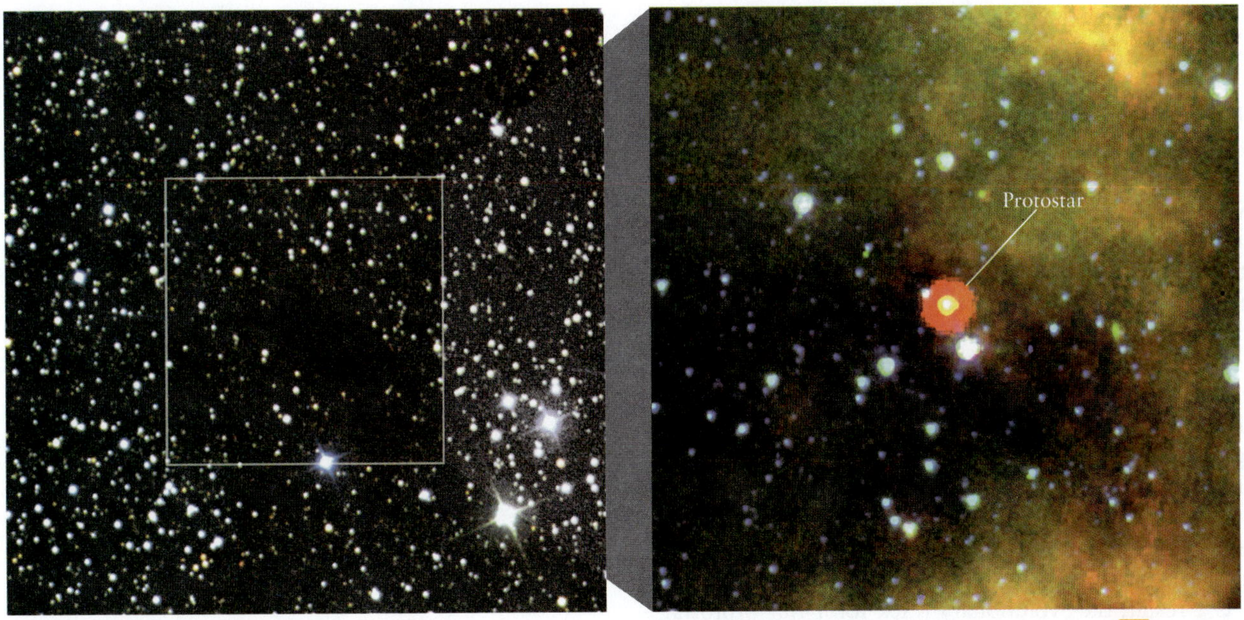

(a) A dark nebula R I V U X G

(b) A hidden protostar within the dark nebula R I V U X G

Figure 18-11

Revealing a Hidden Protostar **(a)** This visible-light view shows a dark nebula called L1014 in the constellation Cygnus (the Swan). No stars are visible within the nebula. **(b)** The Spitzer Space Telescope was used to make this false-color infrared image of the outlined area in (a). The bright red-yellow spot is a protostar within the dark nebula. (a: Deep Sky Survey; b: NASA/JPL-Caltech/N. Evans, Univ. of Texas at Austin)

CAUTION! When astronomers refer to a star "following an evolutionary track" or "moving on the H-R diagram," they mean that the star's luminosity, surface temperature, or both change. Hence, the point that represents the star on an H-R diagram changes its position. This change is completely unrelated to how the star physically moves through space!

Figure 18-10 shows evolutionary tracks for protostars of seven different masses, ranging from 0.5 $M_\odot$ to 15 $M_\odot$. Because protostars are relatively cool when they begin to shine at visible wavelengths, these tracks all begin near the right (low-temperature) side of the H-R diagram. The subsequent evolution is somewhat different, however, depending on the protostar's mass. Note that the greater a star's mass, the more rapidly it moves along its evolutionary track: a 15-$M_\odot$ protostar takes only about 10^5 years to reach the main sequence, while a 1-$M_\odot$ protostar takes over a hundred times longer (more than 10^7 years).

Observing Protostars

Observing the evolution of a protostar can be quite a challenge. Indeed, while Figure 18-10 shows that young protostars are much more luminous than the Sun, it is quite unlikely that you have ever seen a protostar shining in the night sky. The reason is that protostars form within clouds that contain substantial amounts of interstellar dust. The dust in a protostar's immediate surroundings, called its **cocoon nebula,** absorbs the vast amounts of visible light emitted by the protostar and makes it very hard to detect using visible wavelengths.

Protostars can be seen, however, using infrared wavelengths. Because a cocoon nebula absorbs so much energy from the protostar that it surrounds, the nebula becomes heated to a few hundred kelvins. The warmed dust then reradiates its thermal energy at infrared wavelengths, to which the cocoon nebula is relatively transparent. So, by using infrared telescopes, astronomers can see protostars within the "stellar nursery" of a dark nebula.

Figure 18-11 shows visible-light and infrared views of one such stellar nursery. The visible-light view (Figure 18-11*a*) shows a dark, dusty nebula that appears completely opaque. The infrared image (Figure 18-11*b*) allows us to see through the dust, revealing a newly formed protostar within the dark nebula. The properties of this protostar agree well with the theoretical ideas outlined in this section.

18-4 Protostars evolve into main-sequence stars

Evolutionary tracks in Figure 18-10 show how a protostar matures into a star as its gases contract. The details of this evolution depend on the star's mass. To follow these details, you should keep in mind the basic principle that a star's luminosity, radius, and surface temperature are intimately related: luminosity is proportional to the square of radius and to the fourth power of surface temperature (see Section 17-6).

The course of a protostar's evolution depends on its mass

A One-Solar-Mass Protostar

For a protostar with the same mass as our Sun (1 $M_\odot$), the outer layers are cool and quite opaque (for the same reasons that the Sun's photosphere is opaque; see Section 16-5). This means that energy released from the shrinking inner layers in the form of radiation cannot reach the surface. Instead, energy flows outward by the slower and less effective method of convection. Hence, the surface temperature of the contracting protostar stays roughly constant, the luminosity decreases as the radius decreases, and the evolutionary track for 1-$M_\odot$ protostar moves downward on the H-R diagram in Figure 18-10.

Although its surface temperature changes relatively little, the internal temperature of the shrinking protostar increases. After a time, the interior becomes ionized, which makes it less opaque. Energy is then conveyed outward by radiation in the interior and by convection in the opaque outer layers, just as in the present-day Sun (see Section 16-2, especially Figure 16-4). This makes it easier for energy to escape from the protostar, so the luminosity—the rate at which energy is emitted from the protostar's surface—increases. As a result, the evolutionary track for 1-$M_\odot$ protostar in Figure 18-10 bends upward (higher luminosity) and to the left (higher surface temperature, caused by the increased energy flow).

In time, the 1-$M_\odot$ protostar's interior temperature reaches a few million kelvins, hot enough for thermonuclear reactions to begin converting hydrogen into helium. As we saw in Section 16-1 and Box 16-1, these reactions release enormous amounts of energy. Eventually, these reactions provide enough heat and internal pressure to stop the star's gravitational contraction, and hydrostatic equilibrium is reached. The protostar's evolutionary track has now led it to the main sequence, and the protostar has become a full-fledged main-sequence star.

High-Mass and Low-Mass Protostars

More massive protostars evolve a bit differently. If its mass is more than about 4 $M_\odot$, a protostar contracts and heats more rapidly, and hydrogen fusion begins quite early. As a result, the luminosity quickly stabilizes at nearly its final value, while the surface temperature continues to increase as the star shrinks. Thus, the evolutionary tracks of massive protostars traverse the H-R diagram roughly horizontally (signifying approximately constant luminosity) in the direction from right to left (from low to high surface temperature). You can see this most easily for the 9-$M_\odot$ and 15-$M_\odot$ evolutionary tracks in Figure 18-10.

Greater mass means greater pressure and temperature in the interior, which means that a massive star has an even larger temperature difference between its core and its outer layers than the Sun. This difference causes convection deep in the interior of a massive star (**Figure 18-12**). By contrast, a massive star's outer layers are of such low density that energy flows through them more easily by radiation than by convection. Therefore, main-sequence stars with masses more than about 4 $M_\odot$ have convective interiors but radiative outer layers (Figure 18-12*a*). By contrast, less massive main-sequence stars such as the Sun have radiative interiors and convective outer layers (Figure 18-12*b*).

The internal structure is also different for main-sequence stars of very low mass. When such a star forms from a protostar,

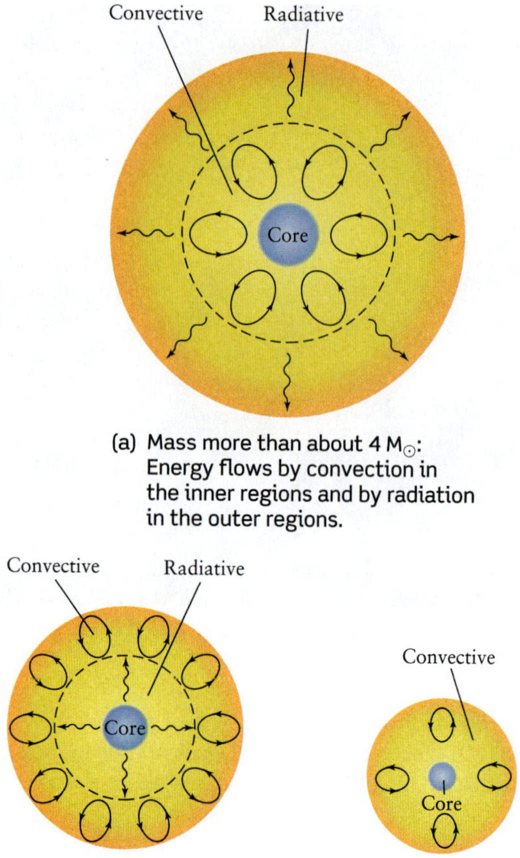

(a) Mass more than about 4 $M_\odot$: Energy flows by convection in the inner regions and by radiation in the outer regions.

(b) Mass between about 4 $M_\odot$ and 0.4 $M_\odot$: Energy flows by radiation in the inner regions and by convection in the outer regions.

(c) Mass less than 0.4 $M_\odot$: Energy flows by convection throughout the star's interior.

INTERACTIVE EXERCISE 18.1

Figure 18-12

Main-Sequence Stars of Different Masses Stellar models show that when a protostar evolves into a main-sequence star, its internal structure depends on its mass. *Note:* The three stars shown here are *not* drawn to scale. Compared with a 1-$M_\odot$ main-sequence star like that shown in (b), a 6-$M_\odot$ main-sequence star like that in (a) has more than 4 times the radius, and a 0.2-$M_\odot$ main-sequence star like that in (c) has only one-third the radius.

the interior temperature is never high enough to fully ionize the interior. The interior remains too opaque for radiation to flow efficiently, so energy is transported by convection throughout the volume of the star (Figure 18-12*c*).

Arriving on the Main Sequence

All of the protostar evolutionary tracks in Figure 18-10 end on the main sequence. The main sequence therefore represents stars in which thermonuclear reactions are converting hydrogen into helium. For most stars, these thermonuclear reactions are part of a stable situation. For example, our Sun will remain on or very near the main sequence, quietly fusing hydrogen into helium at its core, for a total of some 10^{10} years. The point along the main sequence where each evolutionary track ends depends on the star's mass. The most massive stars are the most luminous and their evolutionary tracks end at the upper left of the main sequence, while the least massive stars are the least luminous and their evolutionary tracks end at the lower right of the main sequence. The connection between a main-sequence star's mass and luminosity should be familiar because this relationship is just the mass-luminosity relation (recall the *Cosmic Connections* figure in Section 17-9).

The theory of how protostars evolve helps explain why the main sequence has both an upper mass limit and a lower mass limit. As we saw in Section 17-5, protostars less massive than about 0.08 $M_{\odot}$ can never develop the necessary pressure and temperature to start hydrogen fusion in their cores. Instead, such "failed stars" end up as *brown dwarfs,* which shine faintly by Kelvin-Helmholtz contraction (see Figure 17-13).

Protostars with masses greater than about 200 solar masses also do not become main-sequence stars. Such a protostar rapidly becomes very luminous, resulting in tremendous internal pressures. This pressure is so great that it overwhelms the effects of gravity, expelling the outer layers into space and disrupting the star. Main-sequence stars therefore have masses between about 0.08 and 200 $M_{\odot}$, although the high-mass stars are extremely rare.

CAUTION! Two words of caution are in order here. First, while the evolutionary tracks of protostars begin in the red giant region of the H-R diagram (the upper right), protostars are *not* red giants. As we will see in Chapter 19, red giant stars represent a stage in the evolution of stars that comes *after* being a main-sequence star. Second, it is worth remembering that stars live out most of their lives on the main sequence, after only a relatively brief period as protostars. A 15-$M_{\odot}$ protostar takes only 20,000 years to become a main-sequence star, and a 1-$M_{\odot}$ protostar takes about 2×10^7 years. By contrast, the Sun has been a main-sequence star for about 4.56×10^9 years. By astronomical standards, pre–main-sequence stars are quite transitory.

18-5 During the birth process, stars both gain and lose mass

After reading the previous section, you may think that a main-sequence star forms simply by collapsing inward. In fact, much of the material of a cold, dark nebula is ejected into space and never incorporated into stars. As it is ejected, this material may help sweep away the dust surrounding a young star, making the star observable at visible wavelengths.

Before settling down, protostars eject more than half of their initial mass into space

T Tauri Stars

Mass ejection into space is a hallmark of **T Tauri stars.** These objects are protostars with emission lines as well as absorption lines in their spectra and whose luminosity can change irregularly on time scales of a few days. The namesake of this class of stars, T Tauri, is a protostar in the constellation Taurus (the Bull).

T Tauri stars have masses less than about 3 $M_{\odot}$ and ages around 10^6 years, so on an H-R diagram such as Figure 18-10 they appear above the right-hand end of the main sequence. The emission lines show that these protostars are surrounded by a thin, hot gas. The Doppler shifts of these emission lines suggest that the protostars eject gas at speeds around 80 km/s (300,000 km/h, or 180,000 mi/h).

On average, T Tauri stars eject about 10^{-8} to 10^{-7} solar masses of material per year. This amount may seem small, but by comparison the present-day Sun loses only about 10^{-14} $M_{\odot}$ per year. The T Tauri phase of a protostar may last 10^7 years or so, during which time the protostar may eject roughly a solar mass of material. Thus, the mass of the final main-sequence star is quite a bit less than that of the cloud of gas and dust from which the star originated. (The stellar masses shown in Figure 18-10 are those of the final, main-sequence stars.)

Young stars that are more massive than about 3 $M_{\odot}$ do not vary in luminosity like T Tauri stars. They do lose mass, however, because the pressure of radiation at their surfaces is so strong that it blows gas into space. One place this can be seen is in the Omega Nebula (Figure 18-13*a*), where new stars are being formed (Figure 18-13*b*). The most massive of these stars eject gases with such high temperatures that they emit X rays (Figure 18-13*c*).

Bipolar Outflows and Herbig-Haro Objects

ANIMATION 18.1

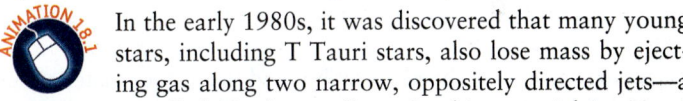

In the early 1980s, it was discovered that many young stars, including T Tauri stars, also lose mass by ejecting gas along two narrow, oppositely directed jets—a phenomenon called **bipolar outflow.** As this material is ejected into space at speeds of several hundred kilometers per second, it collides with the surrounding interstellar medium and produces knots of hot, ionized gas that glow with an emission-line spectrum. These glowing knots are called **Herbig-Haro objects** after the two astronomers, George Herbig in the United States and Guillermo Haro in Mexico, who discovered them independently.

Figure 18-14 is a Hubble Space Telescope image of the Herbig-Haro objects HH 1 and HH 2, which are produced by the two jets from a single young star in the constellation Orion. Herbig-Haro objects like these change noticeably in position, size, shape, and brightness from year to year, indicating the dynamic character of bipolar outflows.

Observations suggest that most protostars eject material in the form of jets at some point during their evolution. These bipolar outflows are very short-lived by astronomical standards, a mere 10^4 to 10^5 years, but they are so energetic that they typically eject into space more mass than ends up in the final protostar.

Accretion Disks

Protostars slowly add mass to themselves at the same time that they rapidly eject it into space. In fact, the two processes are related. As a protostar's nebula contracts, it spins faster and flattens into a disk with the protostar itself at the center. The same flattening took place in the solar nebula, from which the Sun and planets formed (see Section 8-4). Particles orbiting the protostar within this disk collide with each other, causing them to lose energy, spiral inward onto the protostar, and add to the protostar's

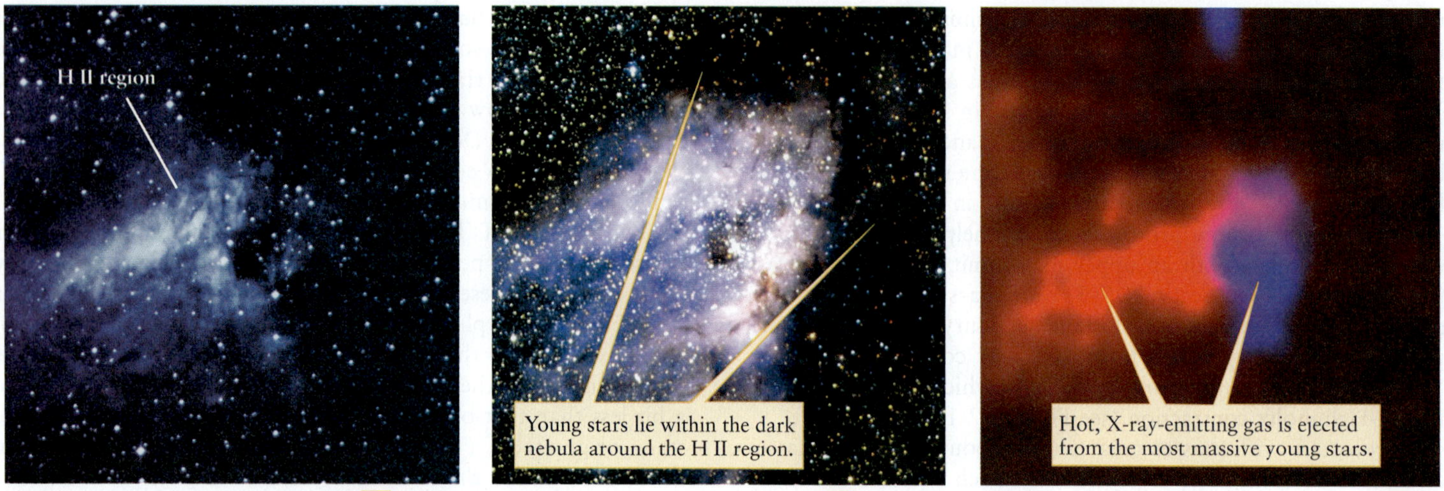
H II region
Young stars lie within the dark nebula around the H II region.
Hot, X-ray-emitting gas is ejected from the most massive young stars.

(a) Visible-light image R I V U X G (b) False-color infrared image R I V U X G (c) False-color X-ray image R I V U X G

WEB LINK 18.8

Figure 18-13

Mass Loss from Young, Massive Stars (a) The Omega Nebula, also known as M17, is a region of star formation in the constellation Sagittarius about 1700 pc (5500 ly) from Earth. **(b)** This infrared image allows us to see through dust, revealing recently formed stars that cannot be seen in (a). **(c)** The most massive young stars eject copious amounts of hot gas. Red indicates X-ray emission from gas at a temperature of 1.5×10^6 K; blue indicates even hotter gas at a temperature of 7×10^6 K. Astronomers do not see such X-ray emission from the Orion Nebula (Figure 18-1), which has many young stars but very few massive ones. (a: Palomar Observatory DSS; b: 2MASS/UMass/IPAC-Caltech/NASA/NSF; c: NASA/CXC/PSU/L. Townsley et al.)

mass. This process is called **accretion,** and the disk of material being added to the protostar in this way is called a **circumstellar accretion disk.** Figure 18-15 is an edge-on view of a circumstellar accretion disk, showing two oppositely directed jets emanating from a point at or near the center of the disk (where the protostar is located).

What causes some of the material in the disk to be blasted outward in a pair of jets? One model involves the magnetic field of the dark nebula in which the star forms (Figure 18-16). As material in the circumstellar accretion disk falls inward, it drags the magnetic field lines along with it. (We saw in Section 16-9 how a similar mechanism may explain the Sun's 22-year cycle.) Parts of the disk at different distances from the central protostar orbit at different speeds, and this can twist the magnetic field lines into two helix shapes, one on each side of the disk. The helices then act as channels that guide infalling material away from the protostar, forming two opposing jets.

Many astronomers suspect that interactions among the protostar, the accretion disk, and the jets help to slow the protostar's rotation. If so, this would explain why main-sequence stars generally spin much more slowly than protostars of the same final mass.

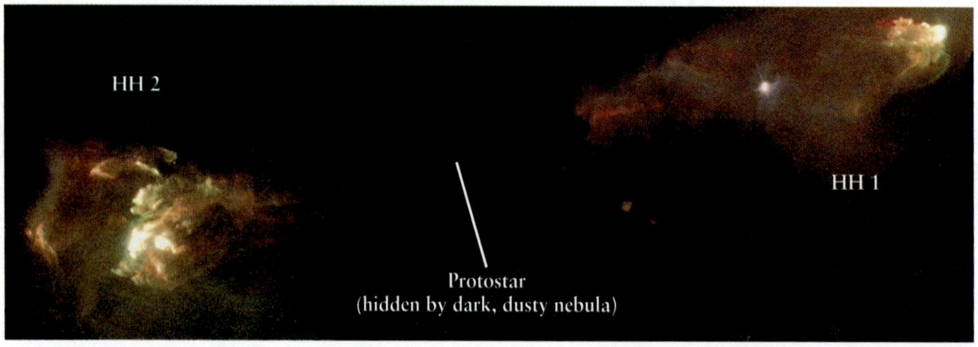
HH 2
HH 1
Protostar
(hidden by dark, dusty nebula)

WEB LINK 18.9

Figure 18-14 R I V U X G

Bipolar Outflow and Herbig-Haro Objects The two bright knots of glowing, ionized gas called HH 1 and HH 2 are Herbig-Haro objects. They are created when fast-moving gas ejected from a protostar slams into the surrounding interstellar medium, heating the gas to high temperature. HH 1 and HH 2 are 0.34 parsec (1.1 light-year) apart and lie 470 pc (1500 ly) from Earth in the constellation Orion. (J. Hester, the WFPC-2 Investigation Definition Team, and NASA)

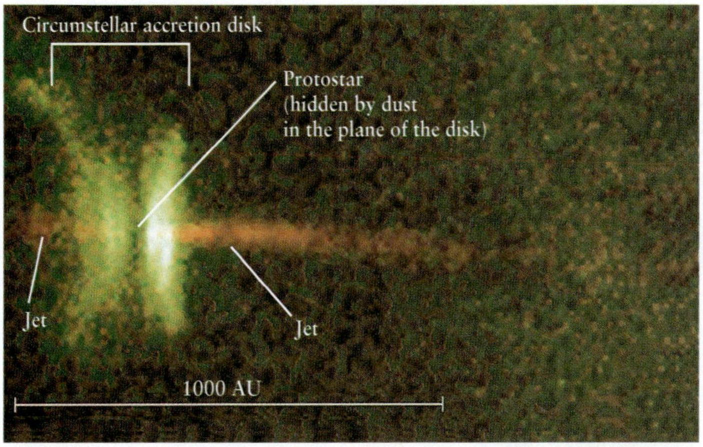
Circumstellar accretion disk
Protostar
(hidden by dust
in the plane of the disk)
Jet
Jet
1000 AU

Figure 18-15 RI V UXG

WEB LINK 18.10

A Circumstellar Accretion Disk and Jets This false-color image shows a star surrounded by an accretion disk, which we see nearly edge-on. Red denotes emission from ionized gas, while green denotes starlight scattered from dust particles in the disk. The midplane of the accretion disk is so dusty and opaque that it appears dark. Two oppositely directed jets flow away from the star, perpendicular to the disk and along the disk's rotation axis. This star lies 140 pc (460 ly) from Earth. (C. Burrows, the WFPC-2 Investigation Definition Team, and NASA)

ANIMATION 18.2

In the 1990s, astronomers using the Hubble Space Telescope discovered many examples of disks around newly formed stars in the Orion Nebula (see Figure 18-1), one of the most prominent star-forming regions in the northern sky. Figure 8-8 shows a number of these **protoplanetary disks,** or **proplyds,** that surround young stars within the nebula. As the name suggests, protoplanetary disks are thought to contain the material from which planets form around stars. They are what remains of a circumstellar accretion disk after much of the material has either fallen onto the star or been ejected by bipolar outflows.

Not all stars are thought to form protoplanetary disks; the exceptions probably include stars with masses in excess of about 3 $M_\odot$, as well as many stars in binary systems. But surveys of the Orion Nebula show that disks are found around most young, low-mass stars. Thus, disk formation may be a natural stage in the birth of many stars.

18-6 Young star clusters give insight into star formation and evolution

TUTORIAL 18.2

Dark nebulae contain tens or hundreds of solar masses of gas and dust, enough to form many stars. As a consequence, these nebulae tend to form groups or **clusters** of young stars. One such cluster is M16, shown in Figure 18-17; another is NGC 6520, visible in Figure 18-4.

Star Clusters as Evolutionary Laboratories

In addition to being objects of great natural beauty, star clusters give us a unique way to compare the evolution of different stars. That's because clusters typically include stars with a range of different masses, all of which began to form out of the parent nebula at roughly the same time.

ANALOGY A foot race is a useful way to compare the performance of sprinters because all the competitors start the race simultaneously. A young star cluster gives us the same kind of opportunity to compare the evolution of stars of different masses that all began to form roughly simultaneously. Unlike a foot

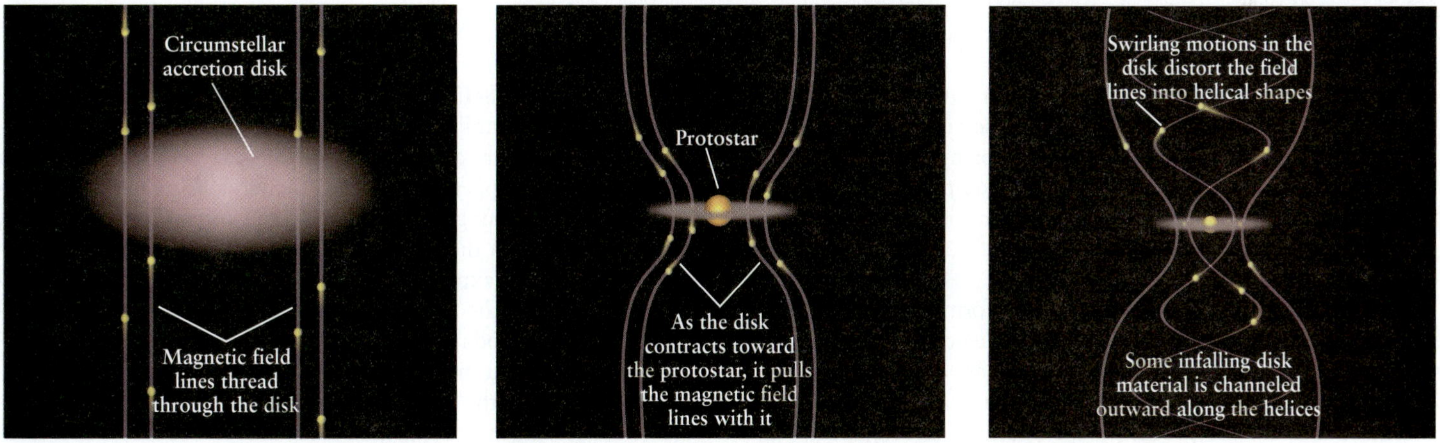
Circumstellar
accretion disk
Magnetic field
lines thread
through the disk
Protostar
As the disk
contracts toward
the protostar, it pulls
the magnetic field
lines with it
Swirling motions in the
disk distort the field
lines into helical shapes
Some infalling disk
material is channeled
outward along the helices
(a)
(b)
(c)

Figure 18-16

A Magnetic Model for Bipolar Outflow **(a)** Observations suggest that circumstellar accretion disks are threaded by magnetic field lines, as shown here. **(b), (c)** The contraction and rotation of the disk make the magnetic field lines distort and twist into helices. These helices steer some of the disk material into jets that stream perpendicular to the plane of the disk, as in Figure 18-15. (Adapted from Alfred T. Kamajian/Thomas P. Ray, "Fountain of Youth: Early Days in the Life of a Star," *Scientific American*, August 2000)

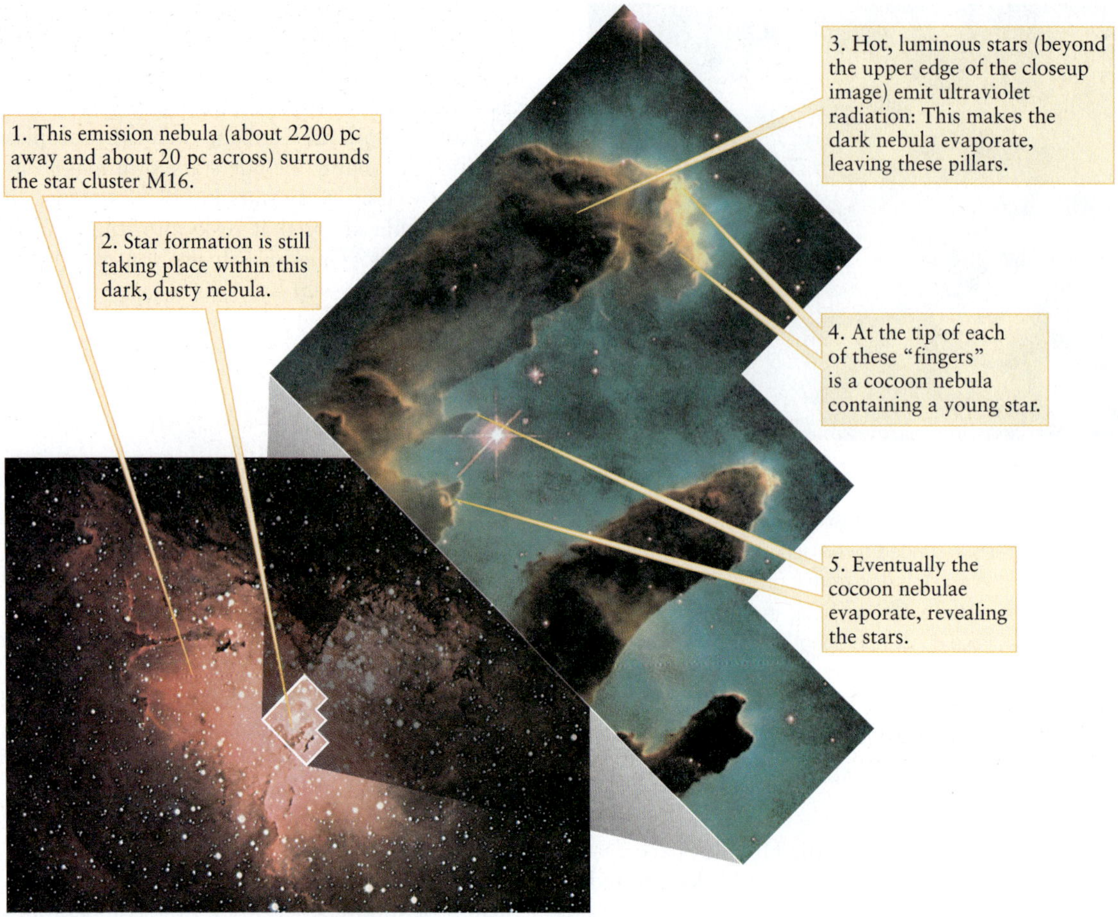

1. This emission nebula (about 2200 pc away and about 20 pc across) surrounds the star cluster M16.
2. Star formation is still taking place within this dark, dusty nebula.
3. Hot, luminous stars (beyond the upper edge of the closeup image) emit ultraviolet radiation: This makes the dark nebula evaporate, leaving these pillars.
4. At the tip of each of these "fingers" is a cocoon nebula containing a young star.
5. Eventually the cocoon nebulae evaporate, revealing the stars.

WEB LINK 18.11 ANIMATION 18.3

Figure 18-17 RIVUXG

A Star Cluster with an H II Region The star cluster M16 is thought to be no more than 800,000 years old, and star formation is still taking place within adjacent dark, dusty globules. The inset shows three dense, cold pillars of gas and dust silhouetted against the glowing background of the red emission nebula (called the Eagle Nebula for its shape). The pillar at the upper left extends about 0.3 parsec (1 light-year) from base to tip, and each of its "fingers" is somewhat broader than our entire solar system. (Anglo-Australian Observatory; J. Hester and P. Scowen, Arizona State University; NASA)

race, however, the entire "race" of stellar evolution in a single cluster happens too slowly for us to observe; as Figure 18-10 shows, protostars take many thousands or millions of years to evolve significantly. Instead, we must compare different star clusters at various stages in their evolution to piece together the history of star formation in a cluster.

All the stars in a cluster may begin to form nearly simultaneously, but they do not all become main-sequence stars at the same time. As you can see from their evolutionary tracks (see Figure 18-10), high-mass stars evolve more rapidly than low-mass stars. The more massive the protostar, the sooner it develops the central pressures and temperatures needed for steady hydrogen fusion to begin, thus joining the main sequence.

Upon reaching the main sequence, *high-mass* protostars become hot, ultraluminous stars of spectral types O and B. As we saw in Section 18-2, these types of stars have ultraviolet radiation that ionize the surrounding interstellar medium to produce an H II region. Figure 18-17 shows such an H II region, called the Eagle Nebula, surrounding the young star cluster M16. A few hundred thousand years ago, this region of space would have had a far less dramatic appearance. It was then a dark nebula, with protostars just beginning to form. Over the intervening millennia, mass ejection from these evolving protostars swept away the obscuring dust. The exposed young, hot stars heated the relatively thin remnants of the original dark nebula, creating the H II region that we see today.

When the most massive protostars to form out of a dark nebula have reached the main sequence, other *low-mass* protostars are still evolving nearby within their dusty cocoons. The evolution of these low-mass stars can be disturbed by their more massive neighbors. As an example, the inset in Figure 18-17 is a close-up of part of the Eagle Nebula. Within these opaque pillars of cold gas and dust, protostars are still forming. At the same time, however, the pillars are being eroded by intense ultraviolet light from hot, massive stars that have already shed their cocoons.

(a) The star cluster NGC 2264 R I V U X G

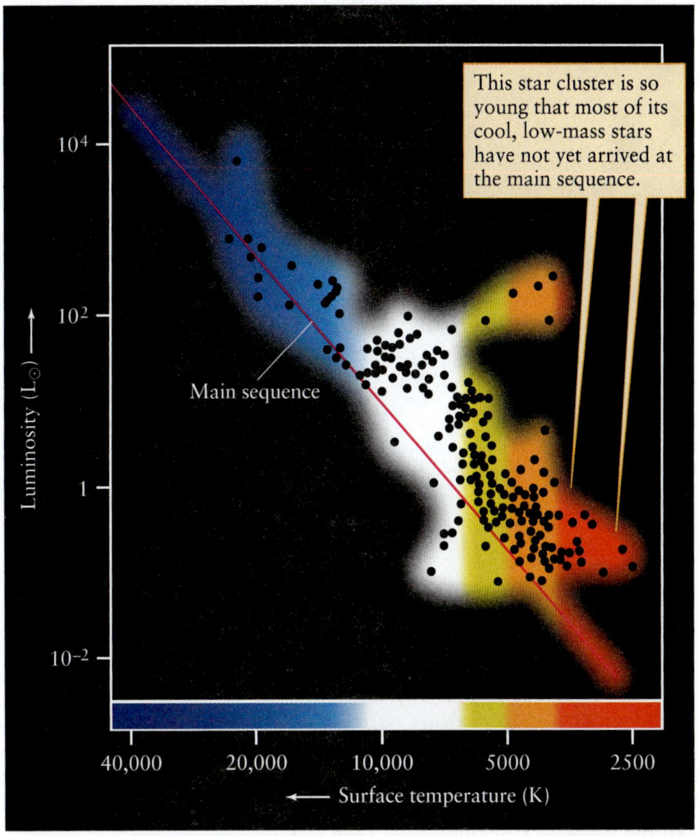
This star cluster is so young that most of its cool, low-mass stars have not yet arrived at the main sequence.
Main sequence
Luminosity ($L_\odot$)
10^4
10^2
1
10^{-2}
40,000
20,000
10,000
5000
2500
Surface temperature (K)

(b) An H-R diagram of the stars in NGC 2264

Figure 18-18

A Young Star Cluster and Its H-R Diagram (a) This photograph shows an H II region and the young star cluster NGC 2264 in the constellation Monoceros (the Unicorn). It lies about 800 pc (2600 ly) from Earth. **(b)** Each dot plotted on this H-R diagram represents a star in NGC 2264 whose luminosity and surface temperature have been determined. This star cluster probably started forming only 2 million years ago. (Anglo-Australian Observatory)

As each pillar evaporates, the embryonic stars within have their surrounding material stripped away prematurely, limiting the total mass that these stars can accrete.

Analyzing Young Clusters Using H-R Diagrams

Star clusters tell us still more about how high-mass and low-mass stars evolve. Figure 18-18*a* shows the young star cluster NGC 2264 and its associated emission nebula. Astronomers have measured each star's apparent brightness and color ratio. Knowing the distance to the cluster, they have deduced the luminosities and surface temperatures of the stars (see Section 17-2 and Section 17-4). Figure 18-18*b* shows all these stars on an H-R diagram. Note that the hottest and most massive stars, with surface temperatures around 20,000 K, are on the main sequence. Stars cooler than about 10,000 K, however, have not yet quite arrived at the main sequence. These are less massive stars in the final stages of pre–main-sequence contraction and are just now beginning to ignite thermonuclear reactions at their centers. To find the ages of these stars, we can compare Figure 18-18*b* with the theoretical calculations of protostar evolution in Figure 18-10. It turns out that this particular cluster is probably about 2 million years old.

The H-R diagram of a young cluster reveals how much time has elapsed since its stars began to form

Figure 18-19*a* shows another young star cluster called the Pleiades. The photograph shows gas that must once have formed an H II region around this cluster and has dissipated into interstellar space, leaving only traces of dusty material that forms reflection nebulae around the cluster's stars. This implies that the Pleiades must be older than NGC 2264, the cluster in Figure 18-18*a*, which is still surrounded by an H II region. The H-R diagram for the Pleiades in Figure 18-19*b* bears out this idea. In contrast to the H-R diagram for NGC 2264, nearly all the stars in the Pleiades are on the main sequence. The cluster's age is about 50 million years, which is how long it takes for the least massive stars to finally begin hydrogen fusion in their cores.

CAUTION! Note that the data points for the most massive stars in the Pleiades (at the upper left of the H-R diagram in Figure

(a) The Pleiades star cluster R I V U X G

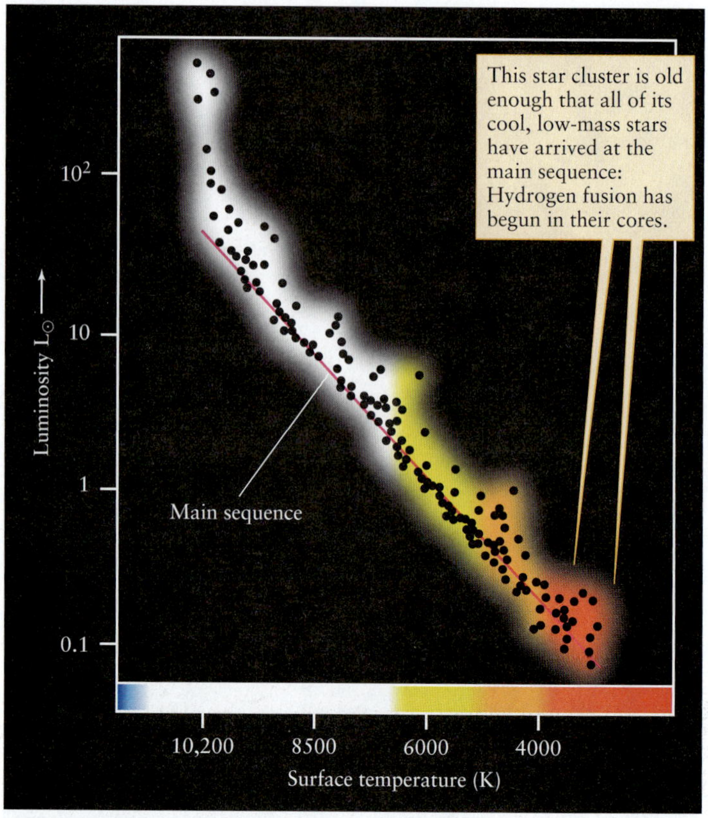
This star cluster is old enough that all of its cool, low-mass stars have arrived at the main sequence: Hydrogen fusion has begun in their cores.
Luminosity $L_\odot$ ⟶
10^2
10
1
0.1
Main sequence
10,200
8500
6000
4000
Surface temperature (K)

(b) An H-R diagram of the stars in the Pleiades

Figure 18-19

The Pleiades and Its H-R Diagram (a) The Pleiades star cluster is 117 pc (380 ly) from Earth in the constellation Taurus, and can be seen with the naked eye. (b) Each dot plotted on this H-R diagram represents a star in the Pleiades whose luminosity and surface temperature have been measured. (*Note:* The scales on this H-R diagram are different from those in Figure 18-18*b*.) The Pleiades is about 50 million (5×10^7) years old. (Anglo-Australian Observatory)

18-19*b*) lie above the main sequence. This is *not* because these stars have yet to arrive at the main sequence. Rather, these stars were the first members of the cluster to arrive at the main sequence some time ago and are now the first members to leave it. They have used up the hydrogen in their cores, so the steady process of core hydrogen fusion that characterizes main-sequence stars cannot continue. In Chapter 19 we will see why massive stars spend a rather short time as main-sequence stars and will study what happens to stars after the main-sequence phase of their lives.

WEB LINK 18.12 A loose collection of stars such as NGC 2264 or the Pleiades is referred to as an **open cluster** (or *galactic cluster,* since such clusters are usually found in the plane of the Milky Way Galaxy). Open clusters possess barely enough mass to hold themselves together by gravitation. Occasionally, a star moving faster than average will escape, or "evaporate," from an open cluster. Indeed, by the time the stars are a few billion years old, they may be so widely separated that a cluster no longer exists.

If a group of stars is gravitationally unbound from the very beginning—that is, if the stars are moving away from one another so rapidly that gravitational forces cannot keep them together—then the group is called a **stellar association.** Because young stellar associations are typically dominated by luminous O and B main-sequence stars, they are also called **OB associations.** The image that opens this chapter shows part of an OB association in the southern constellation Ara (the Altar).

18-7 Star birth can begin in giant molecular clouds

We have seen that star formation takes place within dark nebulae. But where within our Galaxy are these dark nebulae found? Does star formation take place everywhere within the Milky Way, or only in certain special locations? The answers to such questions can enhance our understanding of star formation and of the nature of our home Galaxy.

Exploring the Interstellar Medium at Millimeter Wavelengths

Dark nebulae are a challenge to locate simply because they *are* dark—they do not emit visible light. Nearby dark nebulae can be seen silhouetted against background stars or H II regions (see Figure 18-2), but sufficiently distant dark nebulae are impossible to see in contrast with background visible light because of interstellar extinction from dust grains. They can, however, be detected using longer-wavelength radiation that can pass unaffected through interstellar dust. In fact, dark nebulae actually emit radiation at millimeter wavelengths.

Such emission takes place because in the cold depths of interstellar space, atoms combine to form molecules. The laws of quantum mechanics predict that just as electrons within atoms can occupy only certain specific energy levels (see Section 5-8), molecules can vibrate and rotate only at certain specific rates. When a molecule goes from one vibrational state or rotational state to another, it either emits or absorbs a photon. (In the same way, an atom emits or absorbs a photon as an electron jumps from one energy level to another.) Most molecules are strong emitters of radiation with wavelengths of around 1 to 10 millimeters (mm). Consequently, observations with radio telescopes tuned to millimeter wavelengths make it possible to detect interstellar molecules of different types. More than 100 different kinds of molecules have so far been discovered in interstellar space, and the list is constantly growing.

Observing the Galaxy at millimeter wavelengths reveals the cold gas that spawns new stars

Hydrogen is by far the most abundant element in the universe. Unfortunately, in cold nebulae much of it is in a molecular form (H_2) that is difficult to detect. The reason is that the hydrogen molecule is symmetric, with two atoms of equal mass joined together, and such molecules do not emit many photons at radio frequencies. In contrast, asymmetric molecules that consist of two atoms of unequal mass joined together, such as carbon monoxide (CO), are easily detectable at radio frequencies. When a carbon monoxide molecule makes a transition from one rate of rotation to another, it emits a photon at a wavelength of 2.6 mm or shorter.

The ratio of carbon monoxide to hydrogen in interstellar space is reasonably constant: For every CO molecule, there are about 10,000 H_2 molecules. As a result, carbon monoxide is an excellent "tracer" for molecular hydrogen gas. Wherever astronomers detect strong emission from CO, they know molecular hydrogen gas must be abundant.

Giant Molecular Clouds

The first systematic surveys of our Galaxy looking for 2.6-mm CO radiation were undertaken in 1974 by the American astronomers Philip Solomon and Nicholas Scoville. In mapping the locations of CO emission, they discovered huge clouds, now called **giant molecular clouds,** that must contain enormous amounts of hydrogen. These clouds have masses in the range of 10^5 to 2×10^6 solar masses and diameters that range from about 15 to 100 pc (50 to 300 ly). Inside one of these clouds, there are about 200 hydrogen molecules per cubic centimeter. This density is several thousand times greater than the average density of matter in the disk of our Galaxy, yet only 10^{-17} as dense as the air we breathe. Astronomers now estimate that our Galaxy contains about 5000 of these enormous clouds.

Figure 18-20 is a map of radio emissions from carbon monoxide in the constellations Orion and Monoceros. Note the extensive areas of the sky covered by giant molecular clouds. This part of the sky is of particular interest because it includes several star-forming regions. By comparing the radio map with the star chart overlay, you can see that the areas where CO emission is strongest, and, thus, where giant molecular clouds are densest, are sites of star formation. Therefore, giant molecular clouds are associated with the formation of stars. Particularly dense regions within these clouds form dark nebulae, and within these stars are born.

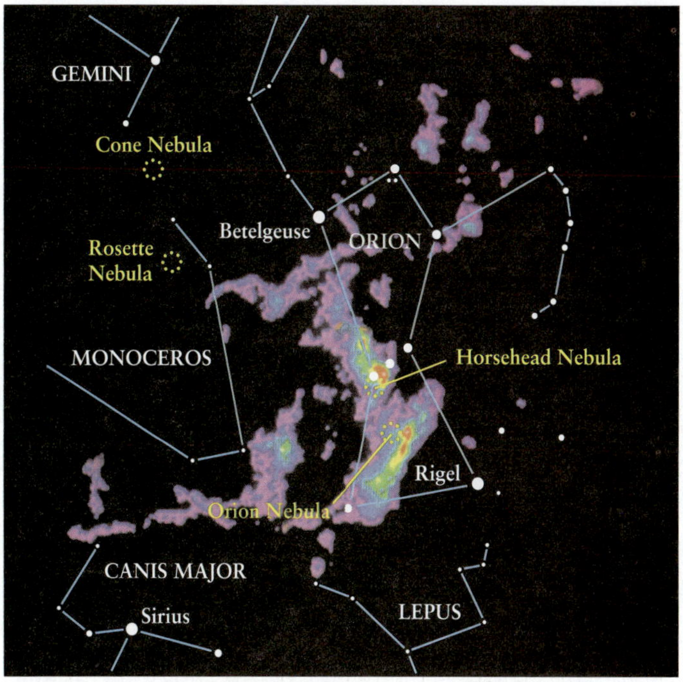

Figure 18-20 R I V U X G

Mapping Molecular Clouds A radio telescope was tuned to a wavelength of 2.6 mm to detect emissions from carbon monoxide (CO) molecules in the constellations Orion and Monoceros. The result was this false-color map, which shows a 35° × 40° section of the sky. The Orion and Horsehead star-forming nebulae are located at sites of intense CO emission (shown in red and yellow), indicating the presence of a particularly dense molecular cloud at these sites of star formation. The molecular cloud is much thinner at the positions of the Cone and Rosette nebulae, where star formation is less intense. (Courtesy of R. Maddalena, M. Morris, J. Moscowitz, and P. Thaddeus)

By using CO emissions to map out giant molecular clouds, astronomers can find the locations in our Galaxy where star formation occurs. These investigations reveal that molecular clouds clearly outline our Galaxy's spiral arms, as Figure 18-21 shows. These clouds lie roughly 1000 pc (3000 ly) apart and are strung along the spiral arms like beads on a string. This arrangement resembles the spacing of H II regions along the arms of other spiral galaxies, such as the galaxy shown in Figure 18-8*a*. The presence of both molecular clouds and H II regions shows that spiral arms are sites of ongoing star formation.

Star Formation in Spiral Arms

In Chapter 23 we will learn that spiral arms are locations where matter "piles up" temporarily as it orbits the center of the Galaxy.

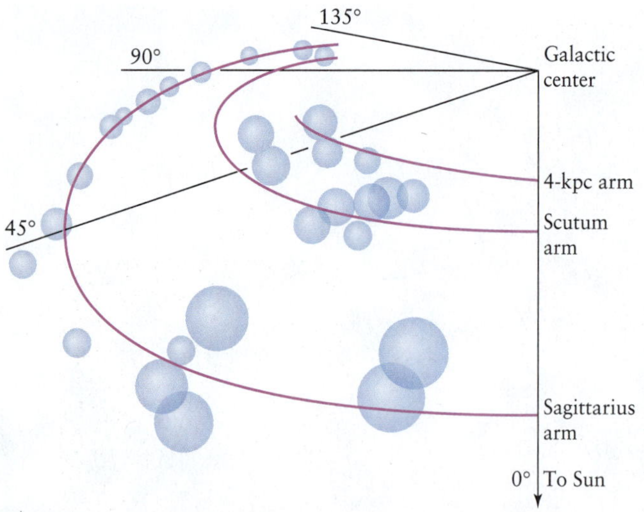

Figure 18-21

Giant Molecular Clouds in the Milky Way This perspective drawing shows the locations of giant molecular clouds in an inner part of our Galaxy as seen from a vantage point above the Sun. These clouds lie primarily along the Galaxy's spiral arms, shown by red arcs. The distance from the Sun to the galactic center is about 8000 pc (26,000 ly). (Adapted from T. M. Dame and colleagues)

Figure 18-22 R I V U X G

A Star-Forming Bubble Radiation and winds from the hot, young O and B stars at the center of this Spitzer Space Telescope image have carved out a bubble about 20 pc (70 ly) in diameter in the surrounding gas and dust. The material around the surface of the bubble has been compressed and heated, making the dust glow at the infrared wavelengths used to record this image. The compressed material is so dense that new stars have formed within that material. This glowing cloud, called RCW 79, lies about 5300 pc (17,200 ly) from Earth in the constellation Centaurus. (NASA; JPL-Caltech; and E. Churchwell, University of Wisconsin-Madison)

You can think of matter in a spiral arm as analogous to a freeway traffic jam. Just as cars are squeezed close together when they enter a traffic jam, a giant molecular cloud is compressed when it passes through a spiral arm. When this happens, vigorous star formation begins in the cloud's densest regions.

As soon as massive O and B stars form, they emit ultraviolet light that ionizes the surrounding hydrogen, and an H II region is born. An H II region is thus a small, bright "hot spot" in a giant molecular cloud. An example is the Orion Nebula, shown in Figure 20-1*b*. Four hot, luminous O and B stars at the heart of the nebula produce the ionizing radiation that makes the surrounding gases glow. The Orion Nebula is embedded on the edge of a giant molecular cloud whose mass is estimated at 500,000 $M_\odot$. The H II regions in Figure 18-2 are located at a different point on the edge of the same molecular cloud, some 25 pc (80 ly) from the Orion Nebula.

Once star formation has begun and an H II region has formed, the massive O and B stars at the core of the H II region induce star formation in the rest of the giant molecular cloud. Ultraviolet radiation and vigorous stellar winds from the O and B stars carve out a cavity in the cloud, and the H II region, heated by the stars, expands into it. These winds travel faster than the speed of sound in the gas—that is, they are **supersonic.** Just as an airplane creates a shock wave (a sonic boom) if it flies faster than sound waves in our atmosphere, a shock wave forms where the expanding H II region pushes at supersonic speed into the rest of the giant molecular cloud. This shock wave compresses the gas through which it passes, stimulating more star birth (Figure 18-22).

Newborn O and B stars further expand the H II region into the giant molecular cloud. Meanwhile, the older O and B stars, which were left behind, begin to disperse (Figure 18-23). In this way, an OB association "eats into" a giant molecular cloud, leaving stars in its wake.

18-8 Supernovae compress the interstellar medium and can trigger star birth

Spiral arms are not the only mechanism for triggering the birth of stars. Presumably, anything that compresses interstellar clouds will do the job. The most dramatic is a *supernova,* caused by the violent death of a massive star after it has left the main sequence. As we will see in Chapter 20, the core of the doomed star collapses suddenly, releasing vast quantities of particles and energy that blow the star apart. The star's outer layers are blasted into space at speeds of several thousand kilometers per second.

Supernova Remnants and Star Formation

Astronomers have found many nebulae across the sky that are the shredded funeral shrouds of these dead stars. Such nebulae, like the one shown in Figure 18-24, are known as **supernova remnants.**

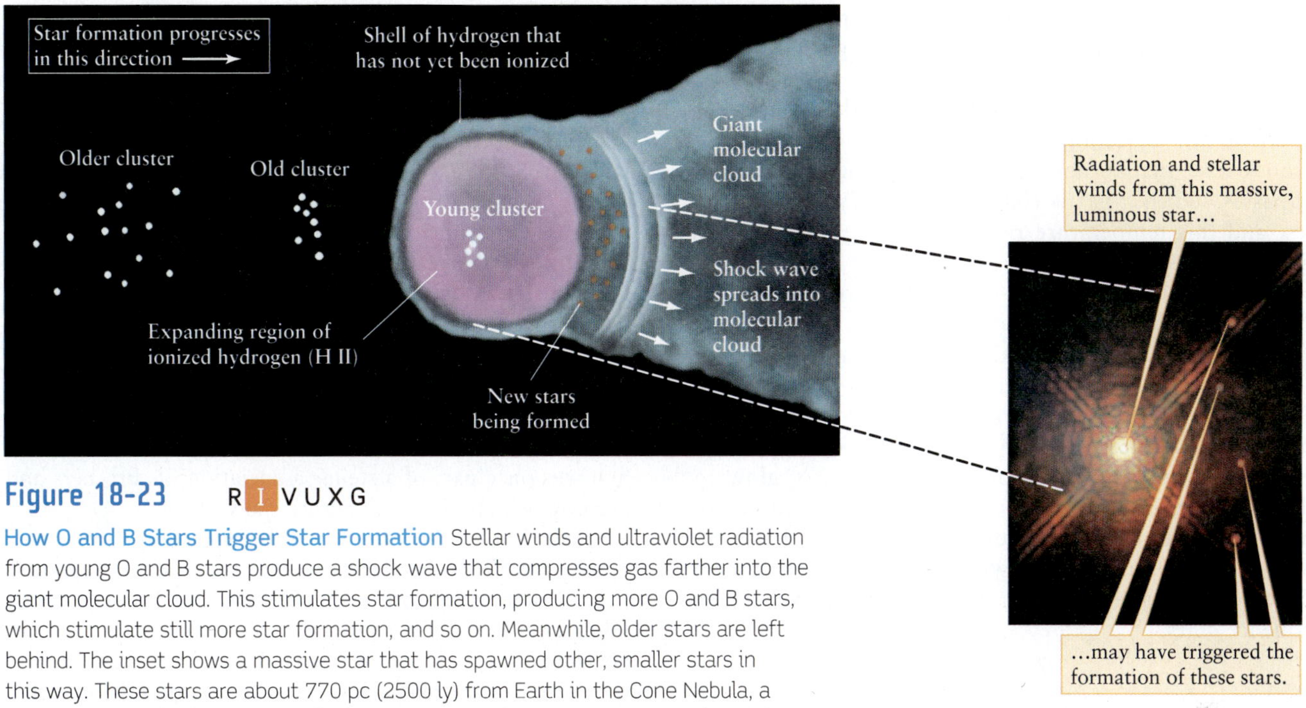

Figure 18-23 R I V U X G

How O and B Stars Trigger Star Formation Stellar winds and ultraviolet radiation from young O and B stars produce a shock wave that compresses gas farther into the giant molecular cloud. This stimulates star formation, producing more O and B stars, which stimulate still more star formation, and so on. Meanwhile, older stars are left behind. The inset shows a massive star that has spawned other, smaller stars in this way. These stars are about 770 pc (2500 ly) from Earth in the Cone Nebula, a star-forming region in the constellation Monoceros. The younger stars are just 0.04 to 0.08 ly (2500 to 5000 AU) from the central star. (Adapted from C. Lada, L. Blitz, and B. Elmegreen; inset: R. Thompson, M. Rieke, G, Schneider, and NASA)

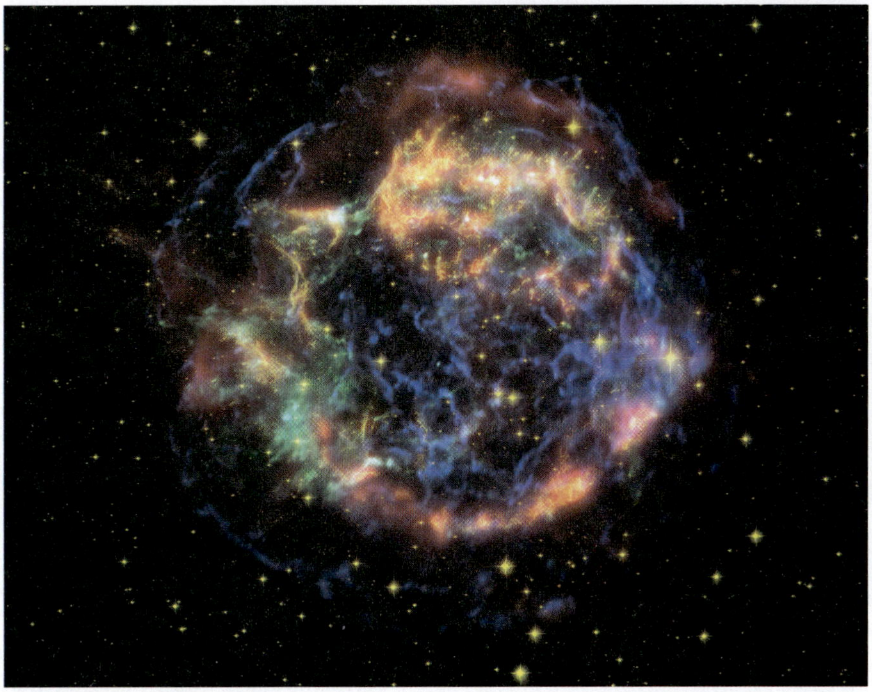

ANIMATION 18.4

Figure 18-24 R I V U X G

A Supernova Remnant This composite image shows Cassiopeia A, the remnant of a supernova that occurred about 3000 pc (10,000 ly) from Earth. In the roughly 300 years since the supernova explosion, a shock wave has expanded about 3 pc (10 ly) outward in all directions from the explosion site. The shock wave has warmed interstellar dust to a temperature of about 300 K (Spitzer Space Telescope infrared image in red), and has heated interstellar gases to temperatures that range from 10^4 K (Hubble Space Telescope visible-light image in yellow) to 10^7 K (Chandra X-ray Observatory X-ray image in green and blue). (NASA; JPL-Caltech; and O. Krause, Steward Observatory)

Many supernova remnants have a distinctly circular or arched appearance, as would be expected for an expanding shell of gas. This wall of gas is typically moving away from the dead star faster than sound waves can travel through the interstellar medium. As we saw in Section 18-7, such supersonic motion produces a shock wave that abruptly compresses the medium through which it passes. When a gas is compressed rapidly, its temperature rises, and this temperature rise causes the gas to glow as shown in Figure 18-24.

When the expanding shell of a supernova remnant slams into an interstellar cloud, it squeezes the cloud, stimulating star birth. This kind of star birth is taking place in the stellar association seen in Figure 18-25. This stellar nursery is located along a luminous arc of gas about 30 pc (100 ly) in length that is presumably the remnant of an ancient supernova explosion. In fact, this arc is part of an almost complete ring of glowing gas with a diameter of about 60 pc (200 ly). Spectroscopic observations of stars along this arc reveal substantial T Tauri activity. This activity results from newborn stars undergoing mass loss in their final stages of contraction before they become main-sequence stars.

ANIMATION 18.5

Figure 18-25 R I V U X G

The Canis Major R1 Association This luminous arc of gas, about 30 pc (100 ly) long, is studded with numerous young stars. Both the luminous arc and the young stars can be traced to the same source, a supernova explosion. The shock wave from the supernova explosion is exciting the gas and making it glow; the same shock wave also compresses the interstellar medium through which it passes, triggering star formation. (Courtesy of H. Vehrenberg)

Supernovae produce a variety of atomic nuclei, including some that are not produced in any other way. These nuclei are dispersed into space by the explosion. Some of these telltale nuclei have been discovered in meteorites that have fallen to Earth. Since meteorites formed very early in the history of our solar system (see Section 8-3), this suggests that a supernova occurred nearby when our solar system was very young. Some astronomers have used this idea to propose that the Sun was once a member of a loose stellar association created by a supernova. Individual stellar motions soon carry the stars of such an association in various directions away from their birthplaces. About 4.56 billion years have passed since the birth of our star, so if the Sun was once part of a stellar association, its brothers and sisters are now widely scattered across the Galaxy.

Our Sun may have formed in association with a number of other stars

Many other processes can also trigger star formation. For example, a collision between two interstellar clouds can create new stars. Compression occurs at the interface between the two colliding clouds and vigorous star formation follows. Similarly, stellar winds from a group of O and B stars may exert strong enough pressure on interstellar clouds to cause compression, followed by star formation. (This process is similar to the one depicted in Figure 18-23.)

Star Birth in Perspective

Our understanding of star birth has improved dramatically in recent years, primarily through infrared- and millimeter-wavelength observations. The *Cosmic Connections* figure on the next page summarizes our present state of knowledge about the formation of stars.

Nevertheless, many puzzles and mysteries remain. One problem is that different modes of star birth tend to produce different percentages of different kinds of stars. For example, the passage of a spiral arm through a giant molecular cloud tends to produce an abundance of massive O and B stars. In contrast, the shock wave from a supernova seems to produce fewer O and B stars but many more of the less massive A, F, G, and K stars. We do not yet know why this is so.

Despite these unanswered questions, it is now clear that star birth involves mechanisms on a colossal scale, from the deaths of massive stars to the rotation of an entire galaxy. In many respects, we have just begun to appreciate these cosmic processes. The study of cold, dark stellar nurseries will be an active and exciting area of astronomical research for many years to come.

In the chapters that follow, we will learn that the interstellar medium is both the birthplace of new stars and a dumping ground for dying stars. At the end of its life, a star can shed most of its mass in an outburst that enriches interstellar space with new chemical elements. The interstellar medium is therefore both nursery and graveyard. Because of this intimate relationship with stars, the interstellar medium evolves as successive generations of stars live out their lives. Understanding the details of this cosmic symbiosis is one of the challenges of modern astronomy.

COSMIC CONNECTIONS

How Stars Are Born

If a clump of interstellar matter is cold and dense enough, it will begin to collapse thanks to the mutual gravitational attraction of its parts. If the clump is massive enough, it will evolve into a main-sequence star through the sequence of events shown here.

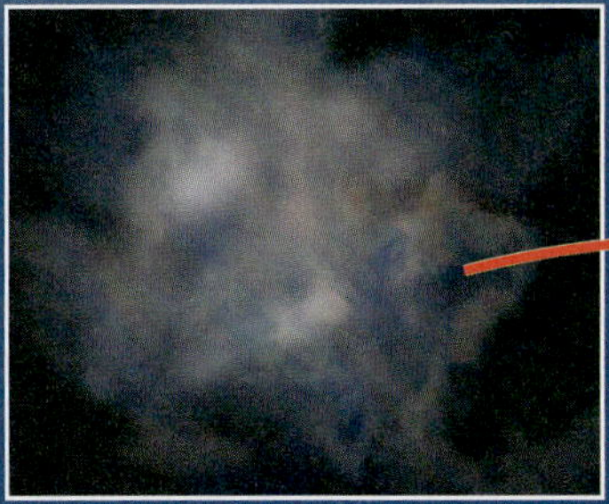

In this cold, dark nebula, gas atoms and dust particles move so slowly that gravity can draw them together.

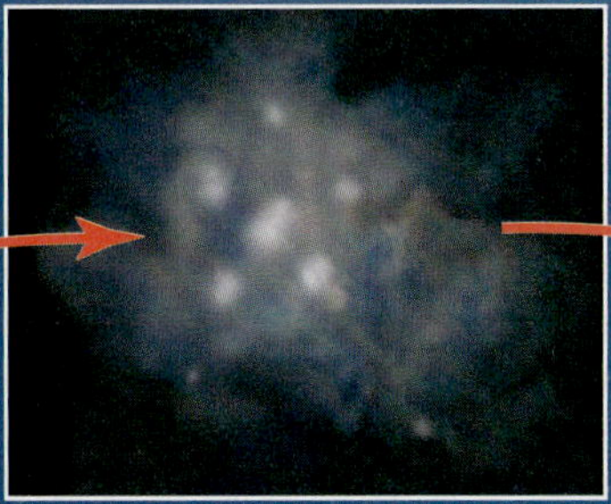

Gas and dust begin to condense into clumps, forming the cores of protostars.

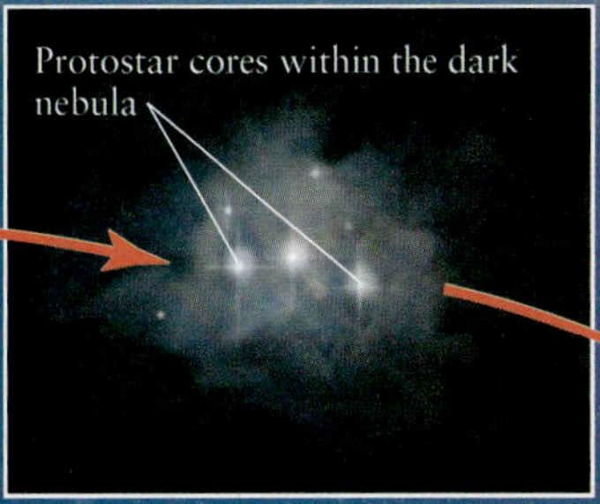

As the cores condense, their density and temperature both increase.

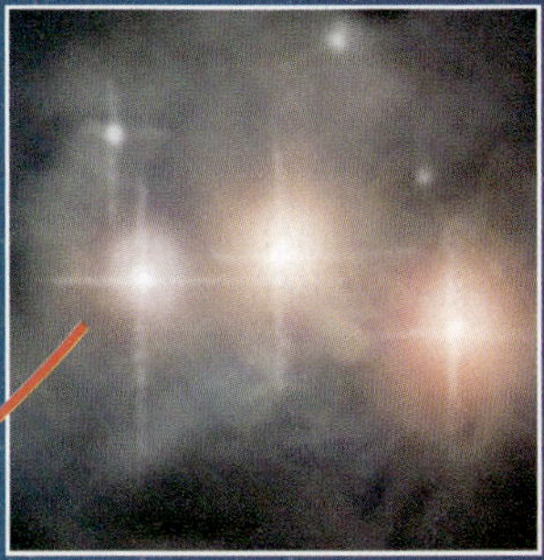

As the protostars continue to heat up and accrete matter from the nebula, they begin to glow due to their increasing temperature.

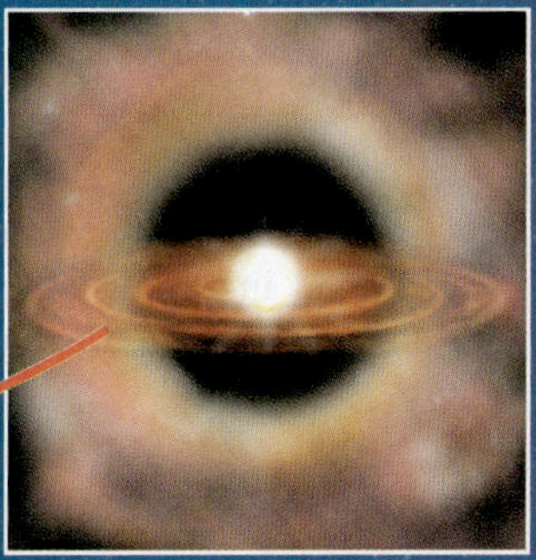

Once the temperature at the center of a protostar becomes sufficiently high, thermonuclear fusion of hydrogen into helium begins. The mass that is continuing to fall onto the star forms an accretion disk.

In the T Tauri stage, the young star ejects mass into space in a bipolar outflow. A stellar wind blows away the remaining parts of the nebula that surround the star, exposing the star to space.

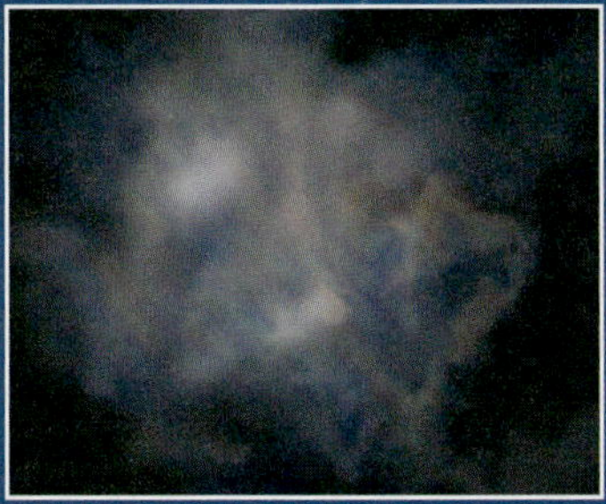

The ejected mass can induce a shock wave in the surrounding interstellar material, triggering the formation of additional stars.

Processes that cause the star to lose or gain mass come to an end, and the star stabilizes as a main-sequence star in hydrostatic equilibrium. The remnants of the accretion disk may remain as a protoplanetary disk, from which a system of planets may form around the star.

Key Words

The term preceded by an asterisk (*) is discussed in Box 18-1.

accretion, p. 482
Barnard object, p. 477
bipolar outflow, p. 481
Bok globule, p. 477
circumstellar accretion disk, p. 482
cluster (of stars), p. 483
cocoon nebula, p. 479
dark nebula, p. 475
dust grains, p. 475
emission nebula, p. 472
evolutionary track, p. 478
*fluorescence, p. 474
giant molecular cloud, p. 487
H II region, p. 472
Herbig-Haro object, p. 481
interstellar extinction, p. 475
interstellar medium, p. 472
interstellar reddening, p. 476
nebula (*plural* nebulae), p. 472
nebulosity, p. 472
OB association, p. 486
open cluster, p. 486
protoplanetary disk (proplyd), p. 483
protostar, p. 477
recombination, p. 473
reflection nebula, p. 475
stationary absorption lines, p. 474
stellar association, p. 486
stellar evolution, p. 471
supernova remnant, p. 488
supersonic, p. 488
T Tauri star, p. 481

Key Ideas

Stellar Evolution: Because stars shine by thermonuclear reactions, they have a finite life span. The theory of stellar evolution describes how stars form and change during that life span.

The Interstellar Medium: Interstellar gas and dust, which make up the interstellar medium, are concentrated in the disk of the Galaxy. Clouds within the interstellar medium are called nebulae.

• Dark nebulae are so dense that they are opaque. They appear as dark blots against a background of distant stars.

• Emission nebulae, or H II regions, are glowing, ionized clouds of gas. Emission nebulae are powered by ultraviolet light that they absorb from nearby hot stars.

• Reflection nebulae are produced when starlight is reflected from dust grains in the interstellar medium, producing a characteristic bluish glow.

Protostars: Star formation begins in dense, cold nebulae, where gravitational attraction causes a clump of material to condense into a protostar.

• As a protostar grows by the gravitational accretion of gases, Kelvin-Helmholtz contraction causes it to heat and begin glowing. Its relatively low temperature and high luminosity place it in the upper right region on an H-R diagram.

• Further evolution of a protostar causes it to move toward the main sequence on the H-R diagram. When its core temperatures become high enough to ignite steady hydrogen burning, it becomes a main-sequence star.

• The more massive the protostar, the more rapidly it evolves.

Mass Loss by Protostars: In the final stages of pre–main-sequence contraction, when thermonuclear reactions are about to begin in its core, a protostar may eject large amounts of gas into space.

• Low-mass stars that vigorously eject gas are called T Tauri stars.

• A circumstellar accretion disk provides material that a young star ejects as jets. Clumps of glowing gas called Herbig-Haro objects are sometimes found along these jets and at their ends.

Star Clusters: Newborn stars may form an open or galactic cluster. Stars are held together in such a cluster by gravity. Occasionally a star moving more rapidly than average will escape, or "evaporate," from such a cluster.

• A stellar association is a group of newborn stars that are moving apart so rapidly that their gravitational attraction for one another cannot pull them into orbit about one another.

O and B Stars and Their Relation to H II Regions: The most massive protostars to form out of a dark nebula rapidly become main sequence O and B stars. They emit strong ultraviolet radiation that ionizes hydrogen in the surrounding cloud, thus creating the reddish emission nebulae called H II regions.

• Ultraviolet radiation and stellar winds from the O and B stars at the core of an H II region create shock waves that move outward through the gas cloud, compressing the gas and triggering the formation of more protostars.

Giant Molecular Clouds: The spiral arms of our Galaxy are laced with giant molecular clouds, immense nebulae so cold that their constituent atoms can form into molecules.

• Star-forming regions appear when a giant molecular cloud is compressed. This can be caused by the cloud's passage through one of the spiral arms of our Galaxy, by a supernova explosion, or by other mechanisms.

Questions

Review Questions

1. If no one has ever seen a star go through the complete formation process, how are we able to understand how stars form?
2. Why is it more difficult to observe the life cycles of stars than the life cycles of planets or animals?
3. If an interstellar medium fills the space between the stars, how is that we are able to see the stars at all?
4. Summarize the evidence that interstellar space contains (**a**) gas and (**b**) dust.
5. What are H II regions? Near what kinds of stars are they found? Why do only these stars give rise to H II regions?
6. What are stationary absorption lines? In what sort of spectra are they seen? How do they give evidence for the existence of the interstellar medium?
7. In Figure 18-2, what makes the Horsehead Nebula dark? What makes IC 434 glow?
8. Why is the daytime sky blue? Why are distant mountains purple? Why is the Sun red when seen near the horizon at sunrise or sunset? In what ways are your answers analogous to the explanations for the bluish color of reflection nebulae and the process of interstellar reddening?

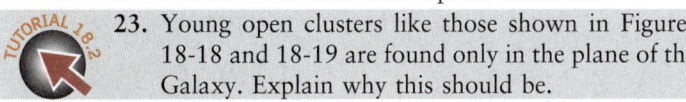

9. To see the constellation Coma Berenices (Berenice's Hair) you must look perpendicular to the plane of the Milky Way. By contrast, the Milky Way passes through the constellation Cassiopeia (named for a mythical queen). Would you expect H II regions to be more abundant in Coma Berenices or in Cassiopeia? Explain your reasoning.
10. The interior of a dark nebula is billions of times less dense than the air that you breathe. How, then, are dark nebulae able to block out starlight?
11. Why are low temperatures necessary in order for protostars to form inside dark nebulae?
12. Compare and contrast Barnard objects and Bok globules. How many Sun-sized stars could you make out of a Barnard object? Out of a Bok globule?
13. Describe the energy source that causes a protostar to shine. How does this source differ from the energy source inside a main-sequence star?
14. What is an evolutionary track? How can evolutionary tracks help us interpret the H-R diagram?
15. What happens inside a protostar to slow and eventually halt its gravitational contraction?
16. Why are the evolutionary tracks of high-mass stars different from those of low-mass stars? For which kind of star is the evolution more rapid? Why?
17. Why are protostars more easily seen with an infrared telescope than with a visible-light telescope?
18. In what ways is the internal structure of a $1\text{-}M_\odot$ main-sequence star different from that of a $5\text{-}M_\odot$ main-sequence star? From that of a $0.5\text{-}M_\odot$ main-sequence star? What features are common to all these stars?
19. What sets the limits on the maximum and minimum masses of a main-sequence star?
20. What are T Tauri stars? How do we know that they eject matter at high speed? How does their rate of mass loss compare to that of the Sun?
21. What are Herbig-Haro objects? Why are they often found in pairs?
22. Why do disks form around contracting protostars? What is the connection between disks and bipolar outflows?
23. Young open clusters like those shown in Figures 18-18 and 18-19 are found only in the plane of the Galaxy. Explain why this should be.
24. Why are observations at millimeter wavelengths so much more useful in exploring interstellar clouds than observations at visible wavelengths?
25. What are giant molecular clouds? What role do these clouds play in the birth of stars?
26. Giant molecular clouds are among the largest objects in our Galaxy. Why, then, were they discovered only relatively recently?
27. Consider the following stages in the evolution of a young star cluster: (i) H II region; (ii) dark nebula; (iii) formation of O and B stars; (iv) giant molecular cloud. Put these stages in the correct chronological order and discuss how they are related.
28. Briefly describe four mechanisms that compress the interstellar medium and trigger star formation.

Advanced Questions

Problem-solving tips and tools

You may find it helpful to review Box 17-4, which describes the relationship among a star's luminosity, radius, and surface temperature. The small-angle formula is described in Box 1-1. Orbital periods are described by Kepler's third law, which we discussed in Boxes 4-2 and 4-4. Remember that the Stefan-Boltzmann law (Box 5-2) relates the temperature of a blackbody to its energy flux. Remember, too, that the volume of a sphere of radius r is $4\pi r^3/3$.

29. If you looked at the spectrum of a reflection nebula, would you see absorption lines, emission lines, or no lines? Explain your answer. As part of your explanation, describe how the spectrum demonstrates that the light was reflected from nearby stars.
30. In the direction of a particular star cluster, interstellar extinction allows only 15% of a star's light to pass through each kiloparsec (1000 pc) of the interstellar medium. If the star cluster is 3.0 kiloparsecs away, what percentage of its photons survive the trip to Earth?
31. The visible-light photograph below shows the Trifid Nebula in the constellation Sagittarius. Label the following features on this photograph: (**a**) reflection nebulae (and the star or stars whose light is being reflected); (**b**) dark nebulae; (**c**) H II regions; (**d**) regions where star formation may be occurring. Explain how you identified each feature.

R I V U X G

(Anglo-Australian Observatory)

32. Find the density (in atoms per cubic centimeter) of a Bok globule having a radius of 1 light-year and a mass of 100 $M_\odot$. How does your result compare with the density of a typical H II region, between 80 and 600 atoms per cm^3? (Assume that the globule is made purely of hydrogen atoms.)
33. The *Becklin-Neugebauer object* is a newly formed star within the Orion Nebula. It is substantially more luminous than the other newly formed stars in that nebula. Assuming that all these stars began the process of formation of the same time, what can you conclude about the mass of the Becklin-Neugebauer object compared with those of the other newly formed stars? Does your conclusion depend on whether or not the stars have reached the main sequence? Explain your reasoning.
34. The two false-color images below show a portion of the Trifid Nebula (see Question 31). The reddish-orange view is a false-color infrared image, while the bluish picture (shown to the same scale) was made with visible light. Explain why the dark streaks in the visible-light image appear bright in the infrared image.
35. At one stage during its birth, the protosun had a luminosity of 1000 $L_\odot$ and a surface temperature of about 1000 K. At this time, what was its radius? Express your answer in three ways: as a multiple of the Sun's present-day radius, in kilometers, and in astronomical units.
36. A newly formed protostar and a red giant are both located in the same region on the H-R diagram. Explain how you could distinguish between these two.
37. (**a**) Determine the radius of the circumstellar accretion disk in Figure 18-15. (You will need to measure this image with a ruler. Note the scale bar in this figure.) Give your answer in astronomical units and in kilometers. (**b**) Assume that the young star at the center of this disk has a mass of 1 $M_\odot$. What is the orbital period (in years) of a particle at the outer edge of the disk? (**c**) Using your ruler again, determine the length of the jet that extends to the right of the circumstellar disk in Figure 18-15. At a speed of 200 km/s, how long does it take gas to traverse the entire visible length of the jet?
38. The star cluster NGC 2264 (Figure 18-18) contains numerous T Tauri stars, while the Pleiades (Figure 18-19) contains none. Explain why there is a difference.
39. The concentration or abundance of ethyl alcohol in a typical molecular cloud is about 1 molecule per 10^8 cubic meters. What volume of such a cloud would contain enough alcohol to make a martini (about 10 grams of alcohol)? A molecule of ethyl alcohol has 46 times the mass of a hydrogen atom (that is, ethyl alcohol has a molecular weight of 46).
40. From the information given in the caption to Figure 18-24, calculate the angular diameter in arcminutes of Cassiopeia A as seen from Earth.
41. From the information given in the caption for Figure 18-24, calculate the average speed at which the shock wave has spread away from the site of the supernova explosion. Give your answer in kilometers per second and as a fraction of the speed of light. (*Hint:* There are 3.16×10^7 seconds in a year and the speed of light is 3.00×10^5 km/s.)

Discussion Questions

42. Some science-fiction movies show stars suddenly becoming dramatically brighter when they are "born" (that is, when thermonuclear fusion reactions begin in their cores). Discuss whether this is a reasonable depiction.
43. Suppose that the electrons in hydrogen atoms were not as strongly attracted to the nuclei of those atoms, so that these atoms were easier to ionize. What consequences might this have for the internal structure of main-sequence stars? Explain your reasoning.
44. What do you think would happen if our solar system were to pass through a giant molecular cloud? Do you think Earth has ever passed through such clouds?
45. Many of the molecules found in giant molecular clouds are *organic* molecules (that is, they contain carbon). Speculate about the possibility of life-forms and biological processes occurring in giant molecular clouds. In what ways might the conditions existing in giant molecular clouds favor or hinder biological evolution?

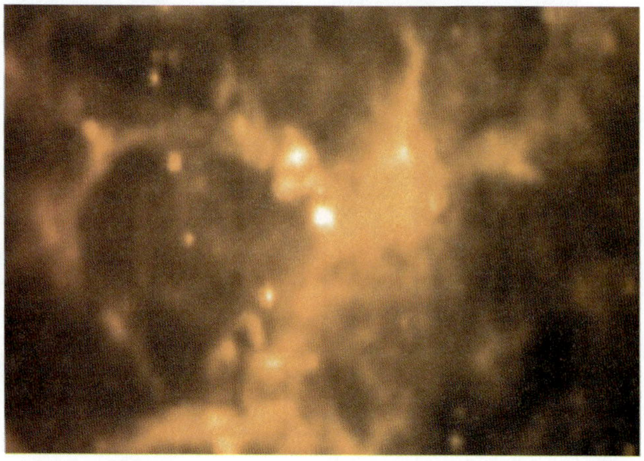

R I V U X G

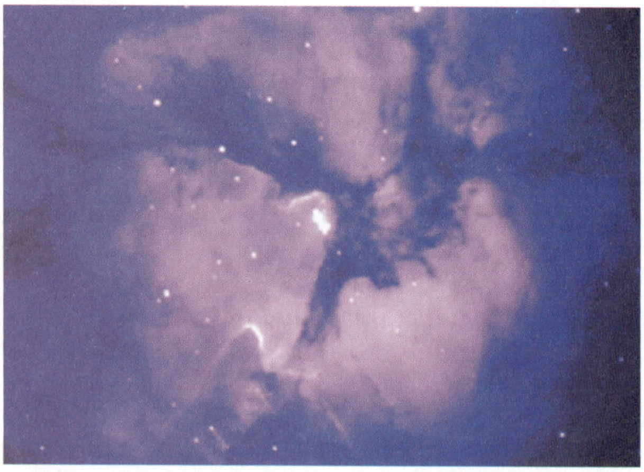

R I V U X G

(ESA/ISO, ISOCAM, and J. Cernicharo et al.; IAC, Observatorio del Teide, Tenerife)

46. Speculate on why a shock wave from a supernova seems to produce relatively few high-mass O and B stars, compared to the lower-mass A, F, G, and K stars.

Web/eBook Questions

47. In recent years astronomers have been able to learn about the character of the interstellar medium in the vicinity of the Sun. Search the World Wide Web for information about aspects of the nearby interstellar medium, including features called the Local Interstellar Cloud and the Local Bubble. How do astronomers study the nearby interstellar medium? What makes these studies difficult? Is the interstellar medium relatively uniform in our neighborhood, or is it clumpy? If the latter, is our solar system in a relatively thin or thick part of the interstellar medium? How is our solar system moving through the interstellar medium?

48. Search the World Wide Web for recent discoveries about how brown dwarfs form. Do they tend to form in the same locations as "real" stars? Do they form in relatively small or relatively large numbers compared to "real" stars? What techniques are used to make these discoveries?

49. **Measuring a Stellar Jet.** Access the animation "A Stellar Jet in the Trifid Nebula" in Chapter 18 of the *Universe* Web site or eBook. (**a**) The Trifid Nebula as a whole has an angular diameter of 28 arcmin. By stepping through the animation, estimate the angular size of the stellar jet shown at the end of the animation. (**b**) The Trifid Nebula is about 2800 pc (9000 ly) from Earth. Estimate the length of the jet in light-years. (**c**) If gas in the jet travels at 200 km/s, how long does it take to traverse the length of the jet? Give your answer in years. (*Hint:* There are 3.16×10^7 seconds in a year.)

Activities

Observing Projects

Observing tips and tools

After looking at the beautiful color photographs of nebulae in this chapter, you may find the view through a telescope a bit disappointing at first, but fear not. You can see a great deal with even a small telescope. To get the best view of a dim nebula using a telescope, use the same "averted vision" trick we described in "Observing tips and tools" in Chapter 17: If you direct your vision a little to one side of the object that you are looking at, the light from that object will go onto a more sensitive part of the retina.

50. Use a telescope to observe at least two of the H II regions listed in the following table. In each case, can you guess which stars are probably responsible for the ionizing radiation that causes the nebula to glow? Can you see any obscuration or silhouetted features that suggest the presence of interstellar dust? Draw a picture of what you see through the telescope and compare it with a photograph of the object. Take note of which portions of the nebula were not visible through your telescope.

Nebula	Right ascension	Declination
M42 (Orion)	$5^h\ 35.4^m$	$-5°\ 27'$
M43	$5\ 35.6^m$	$-5\ 16$
M20 (Trifid)	$18\ 02.6^m$	$-23\ 02$
M8 (Lagoon)	$18\ 03.8^m$	$-24\ 23$
M17 (Omega)	$18\ 20.8^m$	$-16\ 11$

Note: The right ascensions and declinations are given for epoch 2000.

51. On an exceptionally clear, moonless night, use a telescope to observe at least one of the dark nebulae listed in the following table. These nebulae are very difficult to find, because they are recognizable only by the absence of stars in an otherwise starry part of the sky. Are you confident that you actually saw the dark nebula? Does the pattern of background stars suggest a particular shape to the nebula?

Nebula	Right ascension	Declination
Barnard 72 (the Snake)	$17^h\ 23.5^m$	$-23°\ 38'$
Barnard 86	18 02.7	$-27\ 50$
Barnard 133	19 06.1	$-6\ 50$
Barnard 142 and 143	19 40.7	$-10\ 57$

Note: The right ascensions and declinations are given for epoch 2000.

52. A few fine objects cover such large regions of the sky that they are best seen with binoculars. If you have access to a high-quality pair of binoculars, observe the North American Nebula in Cygnus and the Pipe Nebula in Ophiuchus. Both nebulae are quite faint, so you should attempt to observe them only on an exceptionally dark, clear, moonless night. The North America Nebula is a cloud of glowing hydrogen gas located about 3° east of Deneb, the brightest star in Cygnus. While searching for the North America Nebula, you may glimpse another diffuse H II region, the Pelican Nebula, located about 2° southeast of Deneb. The Pipe Nebula is a 7°-long, meandering, dark nebula to the south and to the east of the star θ (theta) Ophiuchi, which is in a section of Ophiuchus that extends southward between the constellations of Scorpius and Sagittarius. Located about 12° east of the bright red star Antares, you can locate θ Ophiuchi using the *Starry Night Enthusiast*™ software on the CD-ROM that accompanies certain printed copies of this book.

53. Use the *Starry Night Enthusiast*™ program to investigate a star-forming region. Use the **Find . . .** command in the **Edit** menu to find and center on M20 (the Trifid Nebula, shown in the figure that accompanies Question 31) as seen from your location. Zoom out as far as possible using the **Zoom** controls at the right-hand end of the toolbar. Set the **Time** appropriately and adjust the **Month** and **Day** in the **Date** to answer the following questions. (*Hint:* You may want to remove daylight and display the local meridian to provide precise answers.) (**a**) On what day is M20 highest in the sky at noon? Explain how you determined this. (**b**) On what day is M20 highest in the sky at midnight, so that it is best placed

for observing with a telescope? Explain how you determined this.

54. Use the *Starry Night Enthusiast*™ program to examine the Milky Way Galaxy. Open the **Favorites** pane and click on **Stars > Sun in Milky Way** to display our Galaxy from a position 0.150 million light-years above the galactic plane. (You can remove the astronaut's feet from this view if desired by clicking on **View > Feet.**) You can zoom in or out on the Galaxy using the + and − buttons at the upper right end of the toolbar. You can move the Galaxy by holding down the mouse button while moving the mouse. You can also rotate the Galaxy by putting the mouse cursor over the image and holding down the **Shift** key while holding down the mouse button and moving the mouse. (**a**) You can identify H II regions by their characteristic magenta color. Describe where in the Galaxy you find these. Are most found in the inner part of the Galaxy or in its outer regions? (**b**) Where do you find dark lanes of dust—in the inner part of the Galaxy or in its outer regions? Do you see any connection between the locations of dust and of H II regions? If there is a connection, what do you think causes it? If there is not a connection, why is this the case? You can examine the location of our Galaxy in relation to neighboring galaxies by turning the Milky Way edge-on and by increasing the distance from Earth using the **up** key below the Viewing Location on the toolbar.

Collaborative Exercises

55. Imagine that your group walks into a store that specializes in selling antique clothing. Prepare a list of observable characteristics that you would look for to distinguish which items were from the early, middle, and late twentieth century. Also, write a paragraph that specifically describes how this task is similar to how astronomers understand the evolution of stars.
56. Consider advertisement signs visible at night in your community and provide specific examples of ones that are examples of the three different types of nebulae that astronomers observe and study. If an example doesn't exist in your community, creatively design an advertisement sign that could serve as an example.
57. The pre–main-sequence evolutionary tracks shown in Figure 18-10 describe the tracks of seven protostars of different masses. Imagine a new sort of H-R diagram that plots a human male's increasing age versus decreasing hair density on the head instead of increasing luminosity versus decreasing temperature. Create and carefully label a sketch of this imaginary HaiR diagram showing both the majority of the U.S. male population and a few oddities. Finally, draw a line that clearly labels your sketch to show how a typical male undergoing male-pattern baldness might slowly change position on the HaiR diagram over the course of a human life span.

19

Stellar Evolution: On and After the Main Sequence

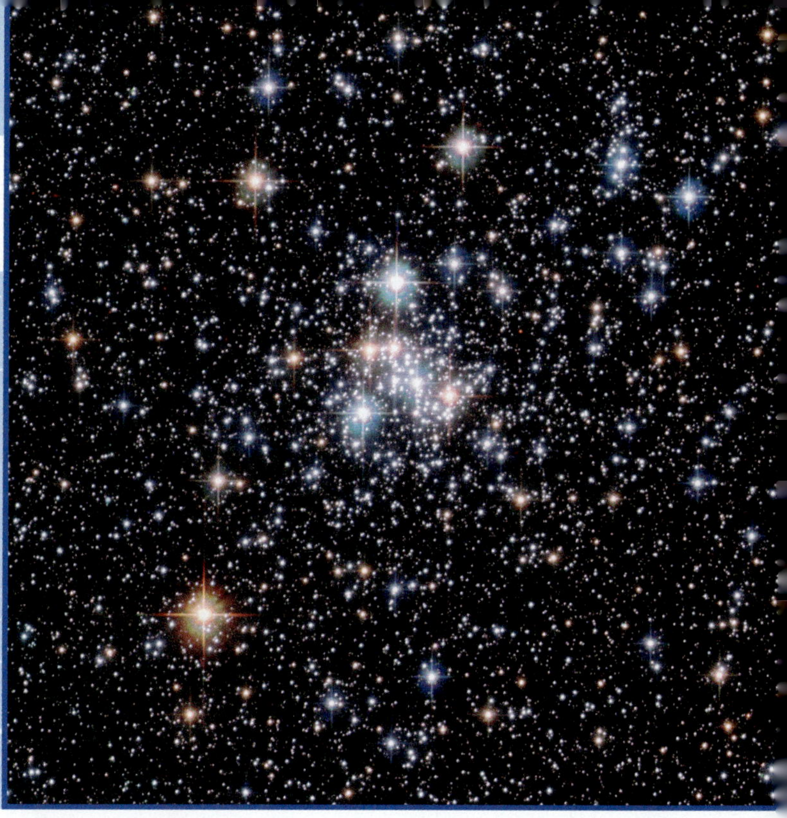

RIVUXG

The red stars in this image of open cluster NGC 290 are red giants, a late stage in stellar evolution. (ESA/NASA/Edward W. Olszewski, U. of Arizona)

Imagine a world like Earth, but orbiting a star more than 100 times larger and 2000 times more luminous than our Sun. Bathed in the star's intense light, the surface of this world is utterly dry, airless, and hot enough to melt iron. If you could somehow survive on the daytime side of this world, you would see the star filling almost the entire sky.

This bizarre planet is not a creation of science fiction—it is our own Earth some 7.6 billion years from now. The bloated star is our own Sun, which in that remote era will have become a red-giant star.

In this chapter, we'll learn how a main-sequence star evolves into a red giant when all the hydrogen in its core is consumed. The star's core contracts and heats up, but its outer layers expand and cool. In the hot, compressed core, helium fusion becomes a new energy source. The more massive a star, the more rapidly it consumes its core's hydrogen and the sooner it evolves into a giant.

The interiors of stars are hidden from our direct view, so much of the story in this chapter is based on theory. We back up those calculations with observations of star clusters, which contain stars of different masses but roughly the same age. (An example is the cluster shown here, many of whose stars have evolved into luminous red giants.) Other observations show that some red-giant stars actually pulsate, and that stars can evolve along very different paths if they are part of a binary star system.

19-1 During a star's main-sequence lifetime, it expands and becomes more luminous

In their cores, main-sequence stars are all fundamentally alike. As we saw in Section 18-4, it is in their cores that all such stars convert hydrogen into helium by thermonuclear reactions. This process is called **core hydrogen fusion.** The total time that a star will spend fusing hydrogen into helium in its core, and thus the total time that it will spend as a main-sequence star, is called its **main-sequence lifetime.** For our Sun, the main-sequence lifetime is about 12 billion (1.2×10^{10}) years. Hydrogen

Over the past 4.56 billion years, thermonuclear reactions have caused an accumulation of helium in our Sun's core

Learning Goals

By reading the sections of this chapter, you will learn

19-1 How a main-sequence star changes as it converts hydrogen to helium

19-2 What happens to a star when it runs out of hydrogen fuel

19-3 How aging stars can initiate a second stage of thermonuclear fusion

19-4 How H-R diagrams for star clusters reveal the later stages in the evolution of stars

19-5 The two kinds of stellar populations and their significance

19-6 Why some aging stars pulsate and vary in luminosity

19-7 How stars in a binary system can evolve very differently from single, isolated stars

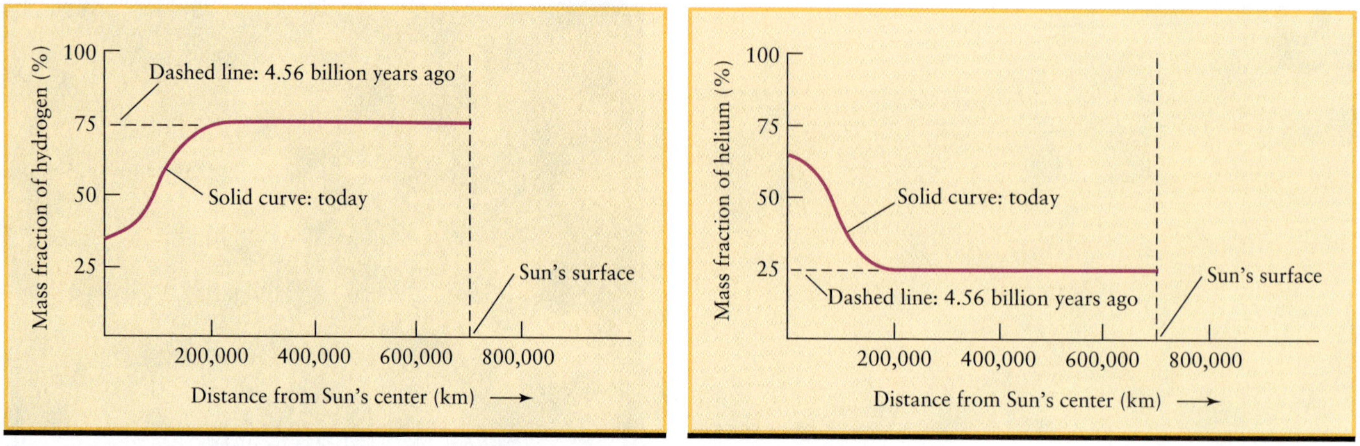

(a) Hydrogen in the Sun's interior

(b) Helium in the Sun's interior

Figure 19-1

Changes in the Sun's Chemical Composition These graphs show the percentage by mass of **(a)** hydrogen and **(b)** helium at different points within the Sun's interior. The dashed horizontal lines show that these percentages were the same throughout the Sun's volume when it first formed. As the solid curves show, over the past 4.56×10^9 years, thermonuclear reactions at the core have depleted hydrogen in the core and increased the amount of helium in the core.

fusion has been going on in the Sun's core for the past 4.56 billion (4.56×10^9) years, so our Sun is less than halfway through its main-sequence lifetime.

What happens to a star like the Sun after the core hydrogen has been used up, so that it is no longer a main-sequence star? As we will see, it expands dramatically to become a red giant. To understand why this happens, it is useful to first look at how a star evolves *during* its main-sequence lifetime. The nature of that evolution depends on whether its mass is less than or greater than about 0.4 $M_\odot$.

Main-Sequence Stars of 0.4 $M_\odot$ or Greater: Consuming Core Hydrogen

A protostar becomes a main-sequence star when steady hydrogen fusion begins in its core and it achieves *hydrostatic equilibrium*—a balance between the inward force of gravity and the outward pressure produced by hydrogen fusion (see Section 16-2 and Section 18-4). Such a freshly formed main-sequence star is called a **zero-age main-sequence star.**

We make the distinction between "main sequence" and "zero-age main sequence" because a star undergoes noticeable changes in luminosity, surface temperature, and radius during its main-sequence lifetime. These changes are a result of core hydrogen fusion, which alters the chemical composition of the core. As an example, when our Sun first formed, its composition was the same at all points throughout its volume: by mass, about 74% hydrogen, 25% helium, and 1% heavy elements. But as Figure 19-1 shows, the Sun's core now contains a greater mass of helium than of hydrogen. (There is still enough hydrogen in the Sun's core for another 7 billion years or so of core hydrogen fusion.)

CAUTION! Although the outer layers of the Sun are also predominantly hydrogen, there are two reasons why this hydrogen cannot undergo fusion. The first reason is that while the temperature and pressure in the core are high enough for thermonuclear reactions to take place, the temperatures and pressure in the outer layers are not. The second reason is that there is no flow of material between the Sun's core and outer layers, so the hydrogen in the outer layers cannot move into the hot, high-pressure core to undergo fusion. The same is true for main-sequence stars with masses of about 0.4 $M_\odot$ or greater. (We will see below that the outer layers *can* undergo fusion in main-sequence stars with a mass less than about 0.4 $M_\odot$.)

Thanks to core hydrogen fusion, the total number of atomic nuclei in a star's core decreases with time: In each reaction, four hydrogen nuclei are converted to a single helium nucleus (see the *Cosmic Connections* figure in Section 16-1, as well as Box 16-1). With fewer particles bouncing around to provide the core's internal pressure, the core contracts slightly under the weight of the star's outer layers. Compression makes the core denser and increases its temperature. (Box 19-1 gives some everyday examples of how the temperature of a gas changes when it compresses or expands.) As a result of these changes in density and temperature, the pressure in the compressed core is actually higher than before.

As the star's core shrinks, its outer layers expand and shine more brightly. Here's why: As the core's density and temperature increase, hydrogen nuclei in the core collide with one another more frequently, and the rate of core hydrogen fusion increases. Hence, the star's luminosity increases. The radius of the star as a whole also increases slightly, because increased core pressure pushes outward on the star's outer layers. The star's surface temperature changes as well, because it is related to the luminosity and radius (see Section 17-6 and Box 17-4). As an example, theoretical calculations indicate that over the past 4.56×10^9 years, our Sun has become 40% more luminous, grown in radius by 6%, and increased in surface temperature by 300 K (Figure 19-2).

As a main-sequence star ages and evolves, the increase in energy outflow from its core also heats the material immediately surrounding the core. As a result, hydrogen fusion can begin in

BOX 19-1 Astronomy Down to Earth

Compressing and Expanding Gases

As a star evolves, various parts of the star either contract or expand. When this happens, the gases behave in much the same way as gases here on Earth when they are forced to compress or allowed to expand.

When a gas is compressed, its temperature rises. You know this by personal experience if you have ever had to inflate a bicycle tire with a hand pump. As you pump, the compressed air gets warm and makes the pump warm to the touch. The same effect happens on a larger scale in southern California during Santa Ana winds or downwind from the Rocky Mountains when there are Chinook winds. Both of these strong winds blow from the mountains down to the lowlands. Even though the mountain air is cold, the winds that reach low elevations can be very hot. (Chinook winds have been known to raise the temperature by as much as 27°C, or 49°F, in only 2 minutes!) The explanation is compression. Air blown downhill by the winds is compressed by the greater air pressure at lower altitudes, and this compression raises the temperature of the air.

Expanding gases tend to drop in temperature. When you open a bottle of carbonated beverage, the gases trapped in the bottle expand and cool down. The cooling can be so great that a little cloud forms within the neck of the bottle. Clouds form in the atmosphere in the same way. Rising air cools as it goes to higher altitudes, where the pressure is lower, and the cooling makes water in the air condense into droplets.

Here's an experiment you can do to feel the cooling of expanding gases. Your breath is actually quite warm, as you can feel if you open your mouth wide, hold the back of your hand next to your mouth, and exhale. But if you bring your lips together to form an "o" and again blow on your hand, your breath feels cool. In the second case, your exhaled breath has to expand as it passes between your lips to the outside, which makes its temperature drop.

this surrounding material. By tapping this fresh supply of hydrogen, a star manages to eke out a few million more years of main-sequence existence.

Main-Sequence Stars of Less than 0.4 $M_\odot$: Consuming All Their Hydrogen

The story is somewhat different for the least massive main-sequence stars, with masses between 0.08 $M_\odot$ (the minimum mass for sustained thermonuclear reactions to take place in a star's core) and about 0.4 $M_\odot$. These stars, of spectral class M, are called **red dwarfs** because they are small in size and have a red color due to their low surface temperature. They are also very numerous; about 85% of all stars in the Milky Way Galaxy are red dwarfs.

In a red dwarf, helium does *not* accumulate in the core to the same extent as in the Sun's core. The reason is that in a red dwarf there are convection cells of rising and falling gas that extend throughout the star's volume and penetrate into the core (see Figure 18-12*c*). These convection cells drag helium outward from the core and replace it with hydrogen from the outer layers (Figure 19-3). The fresh hydrogen can undergo thermonuclear fusion that releases energy and makes additional helium. This helium is then dragged out of the core by convection and replaced by even more hydrogen from the red dwarf's outer layers.

As a consequence, over a red dwarf's main-sequence lifetime essentially all of the star's hydrogen can be consumed and converted to helium. The core temperature and pressure in a red dwarf is less than in the Sun, so thermonuclear reactions happen more slowly than in our Sun. Calculations indicate that it takes hundreds of billions of years for a red dwarf to convert all of its hydrogen completely to helium. The present age of the universe is only 13.7 billion years, so there has not yet been time for any red dwarfs to become pure helium.

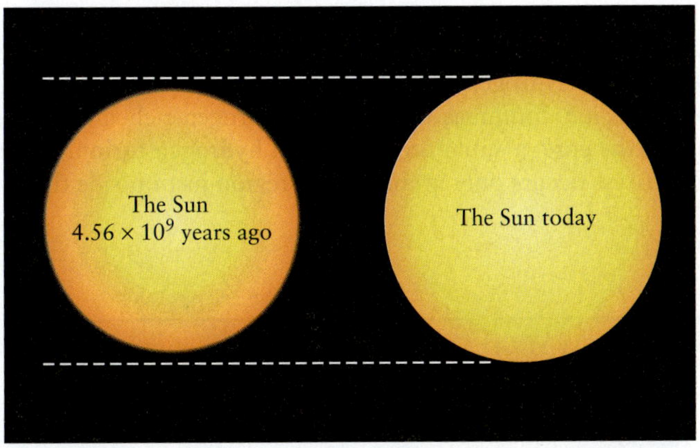

Figure 19-2

The Zero-Age Sun and Today's Sun Over the past 4.56×10^9 years, much of the hydrogen in the Sun's core has been converted into helium, the core has contracted a bit, and the Sun's luminosity has gone up by about 40%. These changes in the core have made the Sun's outer layers expand in radius by 6% and increased the surface temperature from 5500 K to 5800 K.

A Star's Mass Determines Its Main-Sequence Lifetime

The main-sequence lifetime of a star depends critically on its mass. As Table 19-1 shows, massive stars have short main-sequence lifetimes because they are also very luminous (see Section 17-9, and particularly the *Cosmic Connections* figure for Chapter 17). To emit energy so rapidly, these stars must be

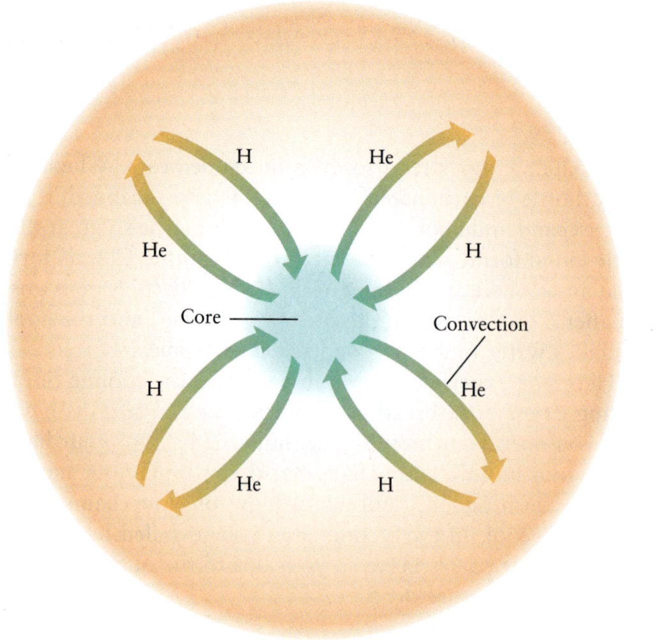

Figure 19-3

A Fully Convective Red Dwarf In a red dwarf—a main-sequence star with less than about 0.4 solar masses—helium (He) created in the core by thermonuclear reactions is carried to the star's outer layers by convection. Convection also brings fresh hydrogen (H) from the outer layers into the core. This process continues until the entire star is helium.

depleting the hydrogen in their cores at a prodigious rate. Hence, even though a massive O or B main-sequence star contains much more hydrogen fuel in its core than is in the entire volume of a red dwarf of spectral class M, the O or B star exhausts its hydrogen much sooner. High-mass O and B stars gobble up the available hydrogen fuel in only a few million years, while red dwarf stars of very low mass take hundreds of billions of years to use up their hydrogen. Thus, a main-sequence star's mass determines not only its luminosity and spectral type, but also how long it can remain a main-sequence star (see Box 19-2 for details).

We saw in Section 18-4 how more-massive stars evolve more quickly through the protostar phase to become main-sequence stars (see Figure 18-10). In general, the more massive the star, the more rapidly it goes through *all* the phases of its life. Still, most of the stars we are able to detect are in their main-sequence phase, because this phase lasts so much longer than other luminous phases.

In the remainder of this chapter we will look at the luminous phases that can take place after the end of a star's main-sequence lifetime. (In Chapters 20, 21, and 22 we will explore the final phases of a star's existence, when it ceases to have an appreciable luminosity.)

19-2 When core hydrogen fusion ceases, a main-sequence star like the Sun becomes a red giant

TUTORIAL 19.1

Like so many properties of stars, what happens at the end of a star's main-sequence lifetime depends on its mass. If the star is a red dwarf of less than about 0.4 $M_\odot$, after hundreds of billions of years the star has converted all of its hydrogen to helium. It is possible for helium to undergo thermonuclear fusion, but this requires temperatures and pressures far higher than those found within a red dwarf. Thus, this red dwarf will end its life as an inert ball of helium, which has no further nuclear reactions, but still glows due to its internal heat. As it radiates energy into space, it slowly cools and shrinks. This slow, quiet demise is the ultimate fate of the 85% of stars in the Milky Way that are red dwarfs. (As we have seen, there has not yet been time in the history of the universe for any red dwarf to reach this final stage in its evolution.)

What is the fate of stars more massive than about 0.4 $M_\odot$, including the Sun? As we will see, the late stages of their evolution are far more dramatic. Studying these stages will give us insight into the fate of our solar system and of life on Earth.

Stars of 0.4 $M_\odot$ or Greater: From Main-Sequence Star to Red Giant

When a star of at least 0.4 solar masses reaches the end of its main-sequence lifetime, all of the hydrogen in its core has been used up and hydrogen fusion ceases there. In this new stage, hydrogen fusion continues only in the hydrogen-rich material just outside the core, a situation called **shell hydrogen fusion.** At first, this process occurs only in the hottest region just outside the core,

Table 19-1 Approximate Main-Sequence Lifetimes

Mass ($M_\odot$)	Surface temperature (K)	Spectral class	Luminosity ($L_\odot$)	Main-sequence lifetime (10^6 years)
25	35,000	O	80,000	4
15	30,000	B	10,000	15
3	11,000	A	60	800
1.5	7000	F	5	4500
1.0	6000	G	1	12,000
0.75	5000	K	0.5	25,000
0.50	4000	M	0.03	700,000

The main-sequence lifetimes were estimated using the relationship $t \propto 1/M^{2.5}$ *(see Box 19-2).*

BOX 19-2 Tools of the Astronomer's Trade

Main-Sequence Lifetimes

Hydrogen fusion converts a portion of a star's mass into energy. We can use Einstein's famous equation relating mass and energy to calculate how long a star will remain on the main sequence.

Suppose that M is the mass of a star and f is the fraction of the star's mass that is converted into energy by hydrogen fusion. During its main-sequence lifetime, the total energy E supplied by the hydrogen fusion can be expressed as

$$E = fMc^2$$

In this equation c is the speed of light.

This energy from hydrogen fusion is released gradually over millions or billions of years. If L is the star's luminosity (energy released per unit time) and t is the star's main-sequence lifetime (the total time over which the hydrogen fusion occurs), then we can write

$$L = \frac{E}{t}$$

(Actually, this equation is only an approximation. A star's luminosity is not quite constant over its entire main-sequence lifetime. But the variations are not important for our purposes.) We can rewrite this equation as

$$E = Lt$$

From this equation and $E = fMc^2$, we see that

$$Lt = fMc^2$$

We can rearrange this equation as

$$t = \frac{fMc^2}{L}$$

Thus, a star's lifetime on the main sequence is proportional to its mass (M) divided by its luminosity (L). Using the symbol $\propto$ to denote "is proportional to," we write

$$t \propto \frac{M}{L}$$

We can carry this analysis further by recalling that main-sequence stars obey the mass-luminosity relation (see Section 17-9, especially the *Cosmic Connections* figure). The distribution of data on the graph in the *Cosmic Connections* figure in Section 17-9 tells us that a star's luminosity is roughly proportional to the 3.5 power of its mass:

$$L \propto M^{3.5}$$

Substituting this relationship into the previous proportionality, we find that

$$t \propto \frac{M}{M^{3.5}} = \frac{1}{M^{2.5}} = \frac{1}{M^2\sqrt{M}}$$

This approximate relationship can be used to obtain rough estimates of how long a star will remain on the main sequence. It is often convenient to relate these estimates to the Sun (a typical 1-$M_\odot$ star), which will spend 1.2×10^{10} years on the main sequence.

EXAMPLE: How long will a star whose mass is 4 $M_\odot$ remain on the main sequence?

Situation: Given the mass of a star, we are asked to determine its main-sequence lifetime.

Tools: We use the relationship $t \propto 1/M^{2.5}$.

Answer: The star has 4 times the mass of the Sun, so it will be on the main sequence for approximately

$$\frac{1}{4^{2.5}} = \frac{1}{4^2\sqrt{4}} = \frac{1}{32} \text{ times the Sun's main-sequence lifetime}$$

Thus, a 4-$M_\odot$ main-sequence star will fuse hydrogen in its core for about $(1/32) \times 1.2 \times 10^{10}$ years, or about 4×10^8 (400 million) years.

Review: Our result makes sense: A star more massive than the Sun must have a shorter main-sequence lifetime.

---

where the hydrogen fuel has not yet been exhausted. Outside this region, no fusion reactions take place.

Strangely enough, the end of the core hydrogen fusion process leads to an *increase* in the core's temperature. Here's why: When thermonuclear reactions first cease in the core, nothing remains to generate heat there. Hence, the core starts to cool and the pressure in the core starts to decrease. This pressure decrease allows the star's core to again compress under the weight of the outer layers. As the core contracts, its temperature again increases, and heat begins to flow outward from the core even though no nuclear reactions are taking place there. (Technically, gravitational energy is converted into thermal energy, as in Kelvin-Helmholtz contraction; see Section 8-4 and Section 16-1).

In the transition from main-sequence star to red giant, the star's core contracts while its outer layers expand

This new flow of heat warms the gases around the core, increasing the rate of shell hydrogen fusion and making the shell eat further outward into the surrounding matter. Helium produced by reactions in the shell falls down onto the core, which continues

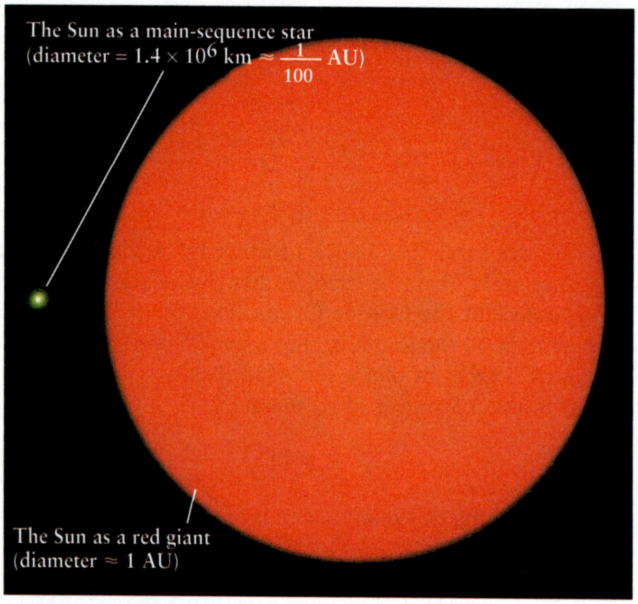

(a) The Sun today and as a red giant

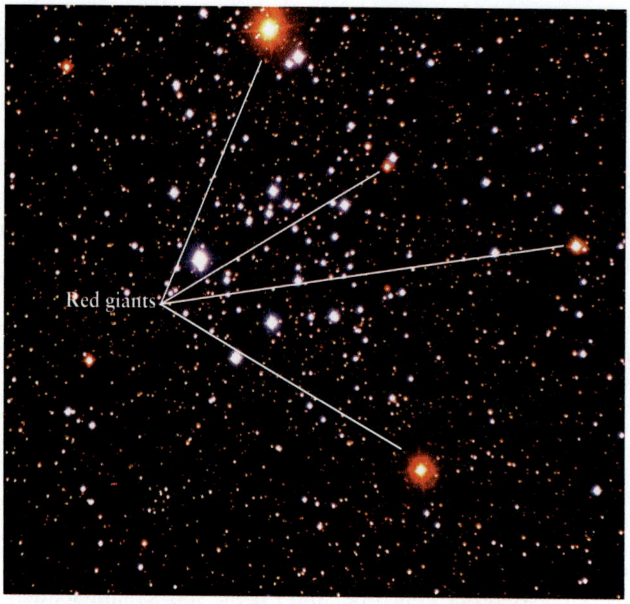

(b) Red giant stars in the star cluster M50 R I V U X G

WEB LINK 19.2

Figure 19-4

Red Giants (a) The present-day Sun produces energy in a hydrogen-fusing core about 100,000 km in diameter. Some 7.6 billion years from now, when the Sun becomes a red giant, its energy source will be a shell only about 30,000 km in diameter within which hydrogen fusion will take place at a furious rate. The Sun's luminosity will be about 2000 times greater than today, and the increased luminosity will make the Sun's outer layers expand to approximately 100 times their present size. (b) This composite of visible and infrared images shows bright red giant stars in the open cluster M50 in the constellation Monoceros (the Unicorn). (T. Credner and S. Kohle, Calar Alto Observatory)

to contract and heat up as it gains mass. Over the course of hundreds of millions of years, the core of a 1-$M_\odot$ star compresses to about one-third of its original radius, while its central temperature increases from about 15 million (1.5×10^7) K to about 100 million (10^8) K.

During this post–main-sequence phase, the star's outer layers expand just as dramatically as the core contracts. As the hydrogen-fusing shell works its way outward, egged on by heat from the contracting core, the star's luminosity increases substantially. This increases the star's internal pressure and makes the star's outer layers expand to many times the original radius. This tremendous expansion causes those outer layers to cool down, and the star's surface temperature drops (see Box 19-1). Once the temperature of the star's bloated surface falls to about 3500 K, the gases glow with a reddish hue, in accordance with Wien's law (see Figure 17-7*a*). The star is then appropriately called a **red giant** (Figure 19-4). Thus, we see that red-giant stars are former main-sequence stars that have evolved into a new stage of existence. We can summarize these observations as a general rule:

Stars join the main sequence when they begin hydrogen fusion in their cores. They leave the main sequence and become giant stars when the core hydrogen is depleted.

Red-giant stars undergo substantial **mass loss** because of their large diameters and correspondingly weak surface gravity. This makes it relatively easy for gases to escape from the red giant into space. Mass loss can be detected in a star's spectrum, because gas escaping from a red giant toward a telescope on Earth produces narrow absorption lines that are slightly blueshifted by the Doppler effect (review Figure 5-26). Typical observed blueshifts correspond to a speed of about 10 km/s. A typical red giant loses roughly 10^{-7} $M_\odot$ of matter per year. For comparison, the Sun's present-day mass loss rate is only 10^{-14} $M_\odot$ per year. Hence, an evolving star loses a substantial amount of mass as it becomes a red giant. Figure 19-5 shows a star losing mass in this way.

The Distant Future of Our Solar System

We can use these ideas to peer into the future of our planet and our solar system. The Sun's luminosity will continue to increase as it goes through its main-sequence lifetime, and the temperature of Earth will increase with it. One and a half billion years from now Earth's average surface temperature will be 50°C (122°F). By 3½ billion years from now the surface temperature of Earth will exceed the boiling temperature of water. All the oceans will boil away, and Earth will become utterly incapable of supporting life. These increasingly hostile conditions will pose the ultimate challenge to whatever intelligent beings might inhabit Earth in the distant future.

About 7 billion years from now, our Sun will finish converting hydrogen into helium at its core. As the Sun's core contracts, its atmosphere will expand to envelop Mercury and perhaps reach to the orbit of Venus. Roughly 700 million years after leaving the main sequence, our red-giant Sun will have swollen to a diameter

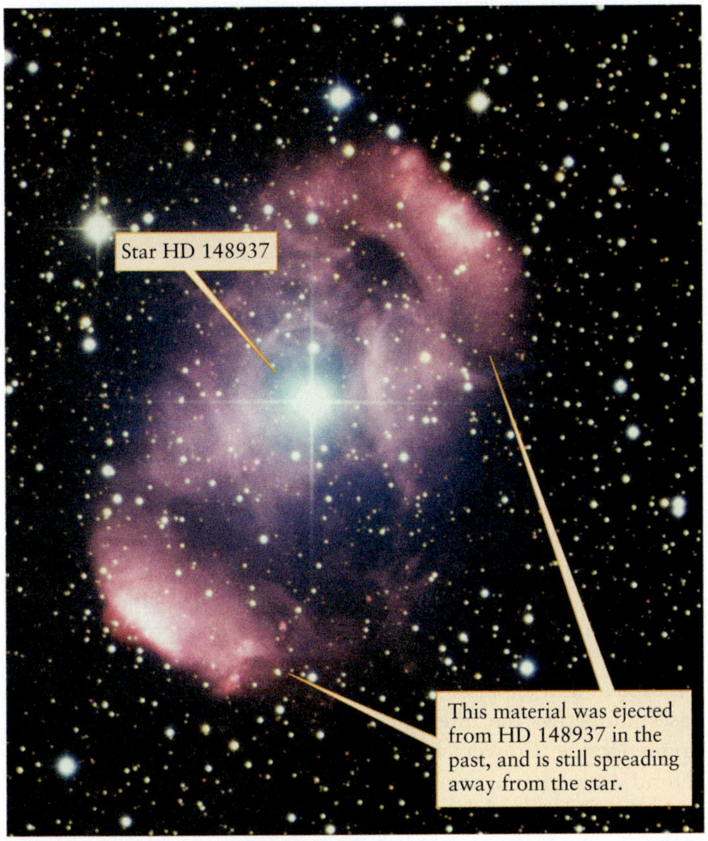

Star HD 148937
This material was ejected from HD 148937 in the past, and is still spreading away from the star.

Figure 19-5 R I V U X G

A Mass-Loss Star As stars age and become giant stars, they expand tremendously and shed matter into space. This star, HD 148937, is losing matter at a high rate. Other strong outbursts in the past ejected the clouds that surround HD 148937. These clouds absorb ultraviolet radiation from the star, which excites the atoms in the clouds and causes them to glow. The characteristic red color of the clouds reveals the presence of hydrogen (see Section 5-6) that was ejected from the star's outer layers. (David Malin, Anglo-Australian Observatory)

of about 1 AU—roughly 100 times larger than its present size—and its surface temperature will have dropped to about 3500 K. Shell hydrogen fusion will proceed at such a furious rate that our star will shine with the brightness of 2000 present-day Suns. Some of the inner planets will be vaporized, and the thick atmospheres of the outer planets will evaporate away to reveal tiny, rocky cores. Thus, in its later years, the aging Sun may destroy the planets that have accompanied it since its birth.

19-3 Fusion of helium into carbon and oxygen begins at the center of a red giant

When a star with a mass greater than 0.4 $M_\odot$ first changes from a main-sequence star (Figure 19-6*a*) to a red giant (Figure 19-6*b*), its hydrogen-fusing shell surrounds a small, compact core of almost pure helium. In a red giant of moderately low mass, which the Sun will become 7 billion years from now, the dense helium core is about twice the diameter of Earth. Most of this core helium was *produced* by thermonuclear reactions during the star's main-sequence lifetime; during the red-giant era, this helium will *undergo* thermonuclear reactions.

Core Helium Fusion

Helium, the "ash" of hydrogen fusion, is a potential nuclear fuel: **Helium fusion,** the thermonuclear fusion of helium nuclei to make heavier nuclei, releases energy. But this reaction cannot take place within the core of our present-day Sun because the temperature there is too low. Each helium nucleus contains two protons, so it has twice the positive electric charge of a hydrogen nucleus, and there is a much stronger electric repulsion between two helium nuclei than between two hydrogen nuclei. For helium nuclei to overcome this repulsion and get close enough to fuse together, they must be moving at very high speeds, which means that the temperature of the helium gas must be very high. (For more on the relationship between the temperature of a gas and the speed of atoms in the gas, see Box 7-2.)

When a star first becomes a red giant, the temperature of the contracted helium core is still too low for helium nuclei to fuse. But as the hydrogen-fusing shell adds mass to the helium core, the core contracts even more, further increasing the star's central temperature. When the central temperature finally reaches 100 million (10^8) K, **core helium fusion**—that is, thermonuclear fusion of helium in the core—begins. As a result, the aging star again has a central energy source for the first time since leaving the main sequence (Figure 19-6*c*).

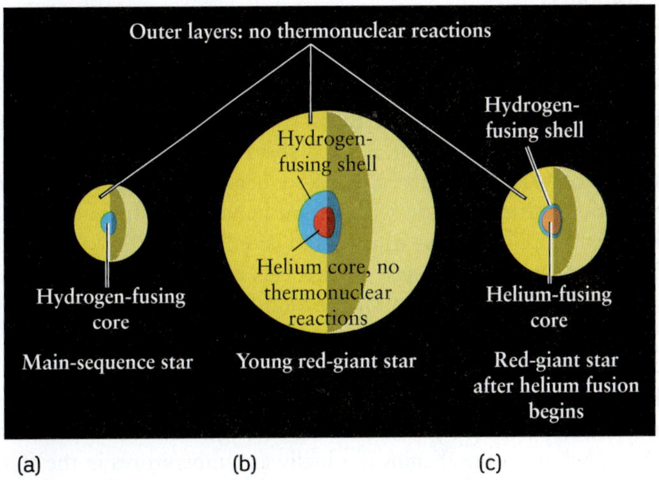

INTERACTIVE EXERCISE 19.1

Figure 19-6

Stages in the Evolution of a Star with More than 0.4 Solar Masses **(a)** During the star's main-sequence lifetime, hydrogen is converted into helium in the star's core. **(b)** When the core hydrogen is exhausted, hydrogen fusion continues in a shell, and the star expands to become a red giant. **(c)** When the temperature in the red giant's core becomes high enough because of contraction, core helium fusion begins (right). (These three pictures are *not* drawn to scale. The star is about 100 times larger in its red-giant phase than in its main-sequence phase, then shrinks somewhat when core helium fusion begins.)

Helium fusion occurs in two steps. First, two helium nuclei combine to form a beryllium nucleus:

$$^{4}\text{He} + {}^{4}\text{He} \rightarrow {}^{8}\text{Be}$$

This particular beryllium isotope, which has four protons and four neutrons, is very unstable and breaks into two helium nuclei soon after it forms. However, in the star's dense core a third helium nucleus may strike the ^{8}Be nucleus before it has a chance to fall apart. Such a collision creates a stable isotope of carbon and releases energy in the form of a gamma-ray photon (γ):

$$^{8}\text{Be} + {}^{4}\text{He} \rightarrow {}^{12}\text{C} + \gamma$$

This process of fusing three helium nuclei to form a carbon nucleus is called the **triple alpha process,** because helium nuclei (^{4}He) are also called **alpha particles** by nuclear physicists. Some of the carbon nuclei created in this process can fuse with an additional helium nucleus to produce a stable isotope of oxygen and release more energy:

$$^{12}\text{C} + {}^{4}\text{He} \rightarrow {}^{16}\text{O} + \gamma$$

Thus, both carbon and oxygen make up the "ash" of helium fusion. The *Cosmic Connections* figure summarizes the reactions involved in helium fusion.

It is interesting to note that ^{12}C and ^{16}O are the most common isotopes of carbon and oxygen, respectively; the vast majority of the carbon atoms in your body are ^{12}C, and almost all the oxygen you breathe is ^{16}O. We will discuss the significance of this in Section 19-5.

The second step in the triple alpha process and the process of oxygen formation both release energy. The onset of these reactions reestablishes thermal equilibrium and prevents any further gravitational contraction of the star's core. A mature red giant fuses helium in its core for a much shorter time than it spent fusing hydrogen in its core as a main-sequence star. For example, in the distant future the Sun will sustain helium core fusion for only about 100 million years—this period is only about 1% of the time that hydrogen fusion occurs. (While this is going on, hydrogen fusion is still continuing in a shell around the core.)

The Helium Flash and Electron Degeneracy

How helium fusion begins at a red giant's center depends on the mass of the star. In high-mass red giants (greater than about 2 to 3 $M_{\odot}$), helium fusion begins gradually as temperatures in the star's core approach 10^{8} K. In red giants with a mass less than about 2 to 3 $M_{\odot}$, helium fusion begins explosively and suddenly, in what is called the **helium flash.** Table 19-2 summarizes these differences.

Table 19-2 How Helium Core Fusion Begins in Different Red Giants

Mass of star	Onset of helium fusion in core
More than about 0.4 but less than 2–3 solar masses	Explosive (helium flash)
More than 2–3 solar masses	Gradual

Stars with less than about 0.4 solar masses do not become red giants (see Section 19-2).

The helium flash occurs because of unusual conditions that develop in the core of a moderately low-mass star as it becomes a red giant. To appreciate these conditions we must first understand how an ordinary gas behaves. Then we can explore how the densely packed electrons at the star's center alter this behavior.

When a gas is compressed, it usually becomes denser and warmer. To describe this process, scientists use the convenient concept of an **ideal gas,** which has a simple relationship between pressure, temperature, and density. Specifically, the pressure exerted by an ideal gas is directly proportional to both the density and the temperature of the gas. Many real gases actually behave like an ideal gas over a wide range of temperatures and densities.

Under most circumstances, the gases inside a star act like an ideal gas. If the gas expands, it cools down, and if it is compressed, it heats up (see Box 19-1). This behavior serves as a safety valve, ensuring that the star remains in thermal equilibrium (see Section 16-2). For example, if the rate of thermonuclear reactions in the star's core should increase, the additional energy releases heat and expands the core. This expansion cools the core's gases and slows the rate of thermonuclear reactions back to the original value. Conversely, if the rate of thermonuclear reactions should decrease, the core will cool down and compress under the pressure of the overlying layers. The compression of the core will make its temperature increase, thus speeding up the thermonuclear reactions and returning them to their original rate.

In a red giant with a mass between about 0.4 $M_{\odot}$ and 2-3 $M_{\odot}$, however, the core behaves very differently from an ideal gas. The core must be compressed tremendously in order to become hot enough for helium fusion to begin. At these extreme pressures and temperatures, the atoms are completely ionized, and most of the core consists of nuclei and detached electrons. Eventually, the free electrons become so closely crowded that a limit to further compression is reached, as predicted by a remarkable law of quantum mechanics called the **Pauli exclusion principle.** Formulated in 1925 by the Austrian physicist Wolfgang Pauli, this principle states that two electrons cannot simultaneously occupy the same quantum state. A quantum state is a particular set of circumstances concerning locations and speeds that are available to a particle. In the submicroscopic world of electrons, the Pauli exclusion principle is analogous to saying that you can't have two things in the same place at the same time.

Before helium fusion begins in a red giant's core, the helium gas there behaves like a metal

Just before the onset of helium fusion, the electrons in the core of a low-mass star are so closely crowded together that any further compression would violate the Pauli exclusion principle. Because the electrons cannot be squeezed any closer together, they produce a powerful pressure that resists further core contraction.

This phenomenon, in which closely packed particles resist compression as a consequence of the Pauli exclusion principle, is called **degeneracy.** Astronomers say that the electrons in the helium-rich core of a low-mass red giant are "degenerate," and

COSMIC CONNECTIONS

Helium Fusion In A Red Giant

A star becomes a red giant after the fusion of hydrogen into helium in its core has come to an end. As the red giant's core shrinks and heats up, a new cycle of reactions can occur that create the even heavier elements carbon and oxygen.

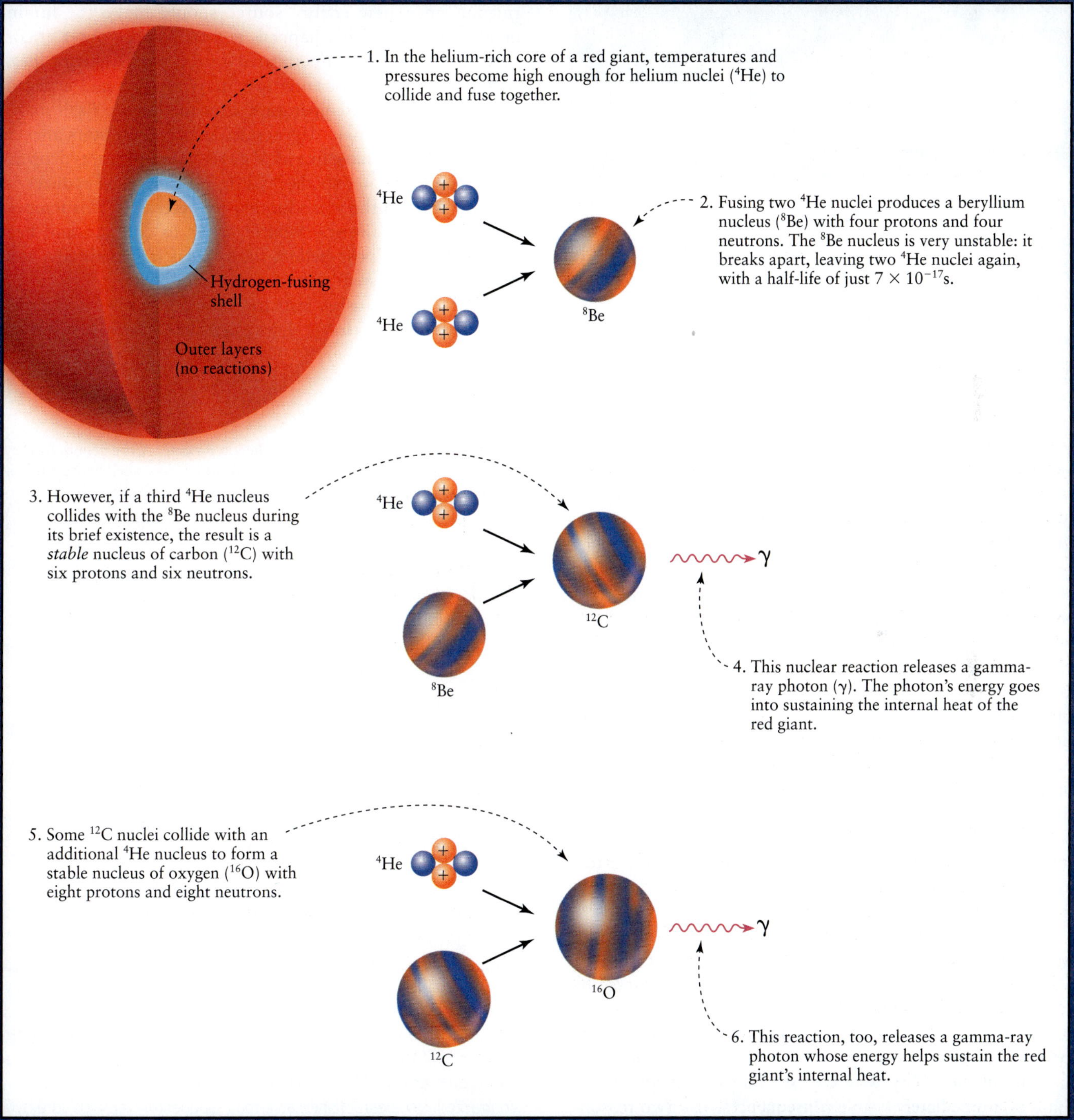

1. In the helium-rich core of a red giant, temperatures and pressures become high enough for helium nuclei (^{4}He) to collide and fuse together.
Hydrogen-fusing shell
Outer layers (no reactions)
^{4}He
^{4}He
^{8}Be
2. Fusing two ^{4}He nuclei produces a beryllium nucleus (^{8}Be) with four protons and four neutrons. The ^{8}Be nucleus is very unstable: it breaks apart, leaving two ^{4}He nuclei again, with a half-life of just 7×10^{-17}s.
3. However, if a third ^{4}He nucleus collides with the ^{8}Be nucleus during its brief existence, the result is a *stable* nucleus of carbon (^{12}C) with six protons and six neutrons.
^{4}He
^{12}C
^{8}Be
γ
4. This nuclear reaction releases a gamma-ray photon (γ). The photon's energy goes into sustaining the internal heat of the red giant.
5. Some ^{12}C nuclei collide with an additional ^{4}He nucleus to form a stable nucleus of oxygen (^{16}O) with eight protons and eight neutrons.
^{4}He
^{16}O
^{12}C
γ
6. This reaction, too, releases a gamma-ray photon whose energy helps sustain the red giant's internal heat.

Figure 19-7 RI**V**UXG

Degenerate Electrons The electrons in an ordinary piece of metal, like the chrome grille on this classic car, are so close together that they are affected by the Pauli exclusion principle. The resulting degenerate electron pressure helps make metals strong and difficult to compress. A more powerful version of this same effect happens inside the cores of low-mass red-giant stars. (Santokh Kochar/PhotoDisc)

that the core is supported by **degenerate-electron pressure.** This degenerate pressure, unlike the pressure of an ideal gas, does not depend on temperature. Remarkably, you can find degenerate electrons on Earth in an ordinary piece of metal (Figure 19-7).

When the temperature in the core of a low-mass red giant reaches the high level required for the triple alpha process, energy begins to be released. The helium heats up, which makes the triple alpha process happen even faster. However, the pressure provided by the degenerate electrons is independent of the temperature, so the pressure does not change. Without the "safety valve" of increasing pressure, the star's core cannot expand and cool. The rising temperature causes the helium to fuse at an ever-increasing rate, producing the helium flash.

Eventually, the temperature becomes so high that the electrons in the core are no longer degenerate. The electrons then behave like an ideal gas and the star's core expands, terminating the helium flash. These events occur so rapidly that the helium flash is over in seconds, after which the star's core settles down to a steady rate of helium fusion.

CAUTION! The term "helium flash" might give you the impression that a star emits a sudden flash of light when the helium flash occurs. If this were true, it would be an incredible sight. During the brief time interval when the helium flash occurs, the helium-fusing core is 10^{11} times more luminous than the present-day Sun, which is similar to the total luminosity of all the stars in the Milky Way Galaxy! But, in fact, the helium flash has no immediately visible consequences—for two reasons. First, much of the energy released during the helium flash goes into heating the core and terminating the degenerate state of the electrons. Second, the energy that does escape the core is largely absorbed by the star's outer layers, which are quite opaque (just like the Sun's present-day interior; see Section 16-2). Therefore, the explosive drama of the helium flash takes place where it cannot be seen directly.

The Continuing Evolution of a Red Giant

Whether a helium flash occurs or not, the onset of core helium fusion actually causes a *decrease* in the luminosity of the star. This decrease is the opposite of what you might expect—after all, turning on a new energy source should make the luminosity greater, not less. What happens is that after the onset of core helium fusion, a star's superheated core expands like an ideal gas. (If the star is of sufficiently low mass to have had a degenerate core, the increased temperature after the helium flash makes the core too hot to remain degenerate. Hence, these stars also end up with cores that behave like ideal gases.) Temperatures drop around the expanding core, so the hydrogen-fusing shell reduces its energy output and the star's luminosity decreases. This temperature decrease allows the star's outer layers to contract and heat up. Consequently, a post–helium-flash star is less luminous, hotter at the surface, and smaller than a red giant.

Core helium fusion lasts for only a relatively short time. Calculations suggest that a 1-$M_\odot$ star like the Sun sustains core

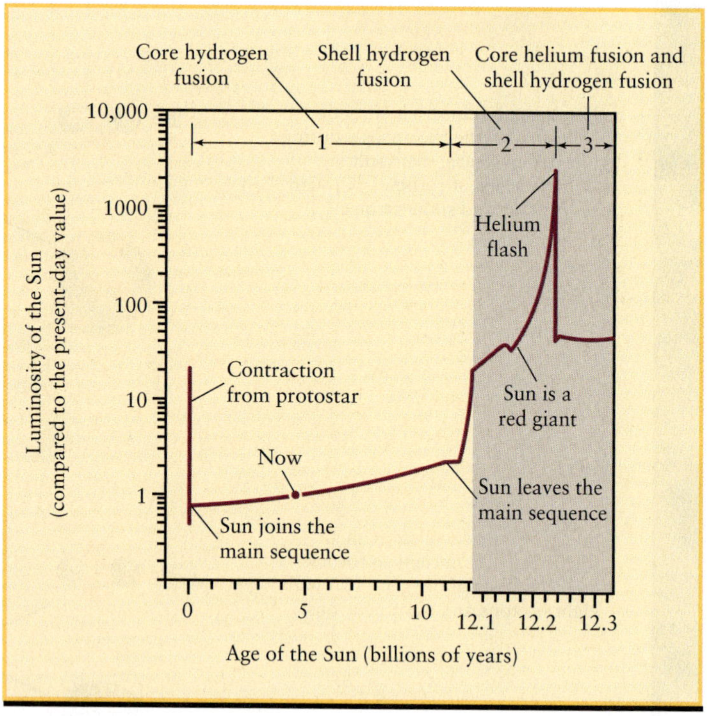

Figure 19-8

Stages in the Evolution of the Sun This diagram shows how the luminosity of the Sun (a 1-$M_\odot$ star) changes over time. The Sun began as a protostar whose luminosity decreased rapidly as the protostar contracted. Once established as a main-sequence star with core hydrogen fusion, the Sun's luminosity increases slowly over billions of years. The post-main-sequence evolution is much more rapid, so a different time scale is used in the right-hand portion of the graph. (Adapted from Mark A. Garlick, based on calculations by I.-Juliana Sackmann and Kathleen E. Kramer) hydrogen fusion for about 12 billion (1.2×10^{10}) years, followed by about 250 million (2.5×10^{8}) years of shell hydrogen fusion leading up to the helium flash. After the helium flash, such a star can fuse helium in its core (while simultaneously fusing hydrogen in a shell around the core) for only 100 million (10^{8}) years, a mere 1% of its main-sequence lifetime. Figure 19-8 summarizes these evolutionary stages in the life of a 1-$M_\odot$ star. In Chapter 20 we will take up the story of what happens after a star has consumed all the helium in its core.

Here is the story of post–main-sequence evolution in its briefest form: Before the beginning of core helium fusion, the star's core compresses and the outer layers expand, and just after core helium fusion begins, the core expands and the outer layers compress. We will see in Chapter 20 that this behavior, in which the inner and outer regions of the star change in opposite ways, occurs again and again in the final stages of a star's evolution.

19-4 H-R diagrams and observations of star clusters reveal how red giants evolve

TUTORIAL 19.2

To see how stars evolve during and after their main-sequence lifetimes, it is helpful to follow them on a Hertzsprung-Russell (H-R) diagram. On such a diagram, zero-age main-sequence stars lie along a line called the **zero-age main sequence**, or **ZAMS** (Figure 19-9*a*). These stars have just emerged from their protostar stage, are steadily fusing hydrogen into helium in their cores, and have attained hydrostatic equilibrium. With the passage of time, hydrogen in a main-sequence star's core is converted to helium, the luminosity slowly increases, the star slowly expands, and the star's position on the H-R diagram inches away from the ZAMS. As a result, the main sequence on an H-R diagram is a fairly broad band rather than a narrow line (Figure 19-9*b*).

Post-Main-Sequence Evolution on an H-R Diagram

The dashed line in Figure 19-9*a* denotes stars whose cores have been exhausted of hydrogen and in which core hydrogen fusion has ceased. These stars have reached the ends of their main-sequence lifetimes. From there, the points representing high-mass stars (3 $M_\odot$, 5 $M_\odot$, and 9 $M_\odot$) move rapidly from left to right across the H-R diagram. This means that, although the star's surface temperature is decreasing, its surface area is increasing at a rate that keeps its overall luminosity roughly constant. During this transition, the star's core contracts and its outer layers expand as energy flows outward from the hydrogen-fusing shell.

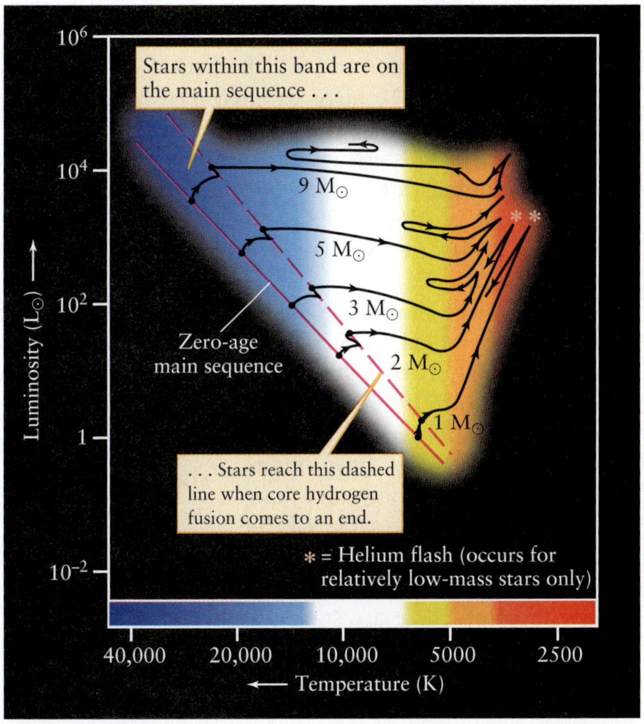

(a) Post-main-sequence evolutionary tracks of five stars with different mass

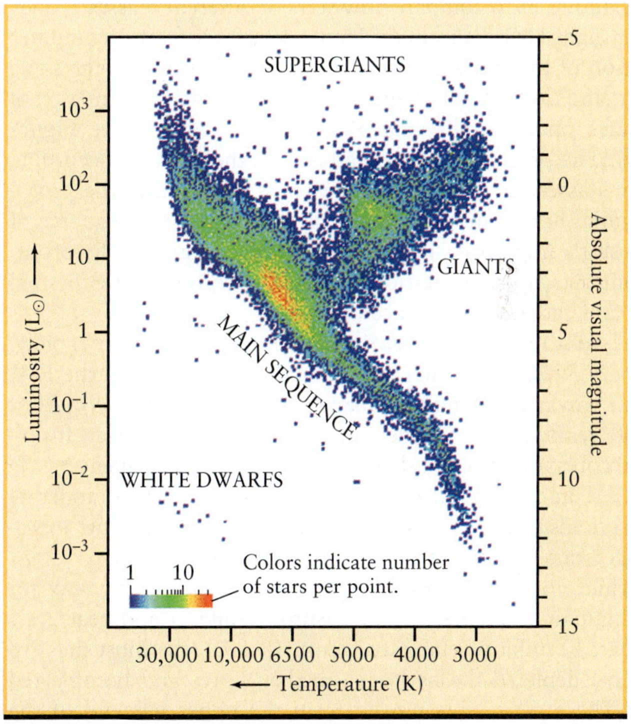

(b) H-R diagram of 20,853 stars—note the width of the main sequence

WEB LINK 19.3

Figure 19-9

H-R Diagrams of Stellar Evolution on and off the Main Sequence **(a)** The two lowest-mass stars shown here (1 $M_\odot$ and 2 $M_\odot$) undergo a helium flash at their centers, as shown by the asterisks. In the high-mass stars, core helium fusion ignites more gradually where the evolutionary tracks make a sharp downward turn in the red-giant region on the right hand side of the H-R diagram. **(b)** Data from the *Hipparcos* satellite (see Section 17-1) was used to create this H-R diagram. The thickness of the main sequence is due in large part to stars evolving during their main-sequence lifetimes. (a: Adapted from I. Iben; b: Adapted from M. A. C. Perryman)

Just before core helium fusion begins, the evolutionary tracks of high-mass stars turn upward in the red-giant region of the H-R diagram (to the upper right of the main sequence). After core helium fusion begins, however, the cores of these stars expand, the outer layers contract, and the evolutionary tracks back away from these temporary peak luminosities. The tracks then wander back and forth in the red-giant region while the stars readjust to their new energy sources.

Figure 19-9*a* also shows the evolutionary tracks of two stars of moderately low mass (1 $M_\odot$ and 2 $M_\odot$). The onset of core helium fusion in these stars occurs with a helium flash, indicated by the red asterisks in the figure. As we saw in the previous section, after the helium flash, these stars shrink and become less luminous. The decrease in size is proportionately greater than the decrease in luminosity, and so the surface temperatures increase. Hence, after the helium flash, the evolutionary tracks for the 1-$M_\odot$ and 2-$M_\odot$ stars move down and to the left.

A Simulated Star Cluster: Tracking 4½ Billion Years of Stellar Evolution

We can summarize our understanding of stellar evolution from birth through the onset of helium fusion by following the evolution of a hypothetical cluster of stars. We saw in Section 18-6 that the stars that make up a cluster all begin to form at essentially the same time but have different initial masses. Hence, studying star clusters allows us to compare how stars of different masses evolve.

The eight H-R diagrams in Figure 19-10 are from a computer simulation of the evolution of 100 stars that all form at the same moment and differ only in initial mass. All 100 stars begin as cool protostars on the right side of the H-R diagram (see Figure 19-10*a*). The protostars are spread out on the diagram according to their masses, and the greater the mass, the greater the protostar's initial luminosity. As we saw in Section 18-3, the source of a protostar's luminosity is its gravitational energy. As the protostar contracts, this gravitational energy is converted to thermal energy and radiated into space.

The most massive protostars contract and heat up very rapidly. After only 5000 years, they have already moved across the H-R diagram toward the main sequence (see Figure 19-10*b*). After 100,000 years, these massive stars have ignited hydrogen fusion in their cores and have settled down on the main sequence as O stars (see Figure 19-10*c*). After 3 million years, stars of moderate mass have also ignited core hydrogen fusion and become main-sequence stars of spectral classes B and A (see Figure 19-10*d*). Meanwhile, low-mass protostars continue to inch their way toward the main sequence as they leisurely contract and heat up.

After 30 million years (see Figure 19-10*e*), the most massive stars have depleted the hydrogen in their cores and become red giants. These stars have moved from the upper left end of the main sequence to the upper right corner of the H-R diagram. (This simulation follows stars only to the red-giant stage, after which they are simply deleted from the diagram.) Intermediate-mass stars lie on the main sequence, while the lowest-mass stars are still in the protostar stage and lie above the main sequence.

After 66 million years (see Figure 19-10*f*), even the lowest-mass protostars have finally ignited core hydrogen fusion and have settled down on the main sequence as cool, dim, M stars. These lowest-mass stars can continue to fuse hydrogen in their cores for hundreds of billions of years.

In the final two H-R diagrams (Figures 19-10*g* and 19-10*h*), the main sequence stars get "peeled" or "eaten away" from the upper left to the lower right as stars exhaust their core supplies of hydrogen and evolve into red giants. The stars that leave the main sequence between Figure 19-10*g* and Figure 19-10*h* have masses between about 1 $M_\odot$ and 3 $M_\odot$ and undergo the helium flash in their cores.

For all stars in this simulation, the giant stage lasts only a brief time compared to the star's main-sequence lifetime. Compared to a 1-$M_\odot$ star (see Figure 19-8), a more massive star has a shorter main-sequence lifetime *and* spends a shorter time as a giant star. Thus, at any given time, only a small fraction of the stellar population is passing through the giant stage. Hence, most of the stars we can see through telescopes are main-sequence stars. As an example, of the stars within 4.00 pc (13.05 ly) of the Sun listed in Appendix 4, only one—Procyon A—is presently evolving from a main-sequence star into a giant. Two other nearby stars are white dwarfs, an even later stage in stellar evolution that we will discuss in Chapter 20. (By contrast, most of the *brightest* stars listed in Appendix 5 are giants and supergiants. Although they make up only a small fraction of the stellar population, these stars stand out due to their extreme luminosity.)

Real Star Clusters: Cluster Ages and Turnoff Points

WEB LINK 19.4

The evolution of the hypothetical cluster displayed in Figure 19-10 helps us interpret what we see in actual star clusters. We can observe the early stages of stellar evolution in **open clusters,** which typically contain a few hundred to a few thousand stars. Many open clusters are just a few million years old, so their H-R diagrams resemble Figure 19-10*d*, 19-10*e*, or 19-10*f* (see Section 18-6, especially Figure 18-18 and Figure 18-19).

Figure 19-11 shows two open clusters of different ages. The nearer cluster, called M35, must be relatively young because it contains several dozen luminous, blue, high-mass main-sequence stars. These stars lie in the upper part of the main sequence on an H-R diagram. They have main-sequence lifetimes of only a few hundred million years, so M35 can be no older than that. Some of the most luminous stars in M35 are red or yellow in color; these are stars that ended their main-sequence lifetimes some time ago and have evolved into red giants. The H-R diagram for this cluster resembles Figure 19-10*g*.

As a cluster ages and its stars end their main-sequence lifetimes, the cluster's color changes from blue to red

There are no high-mass blue main-sequence stars at all in NGC 2158, the more distant cluster shown in Figure 19-11. Any such stars that were once in NGC 2158 have long since come to the end of their main-sequence lifetimes. As a result, the main sequence in this cluster has been "eaten away" more than that of M35, leaving only stars that are yellow or red in color. This tells us that NGC 2158 must be older than M35 (compare Figures

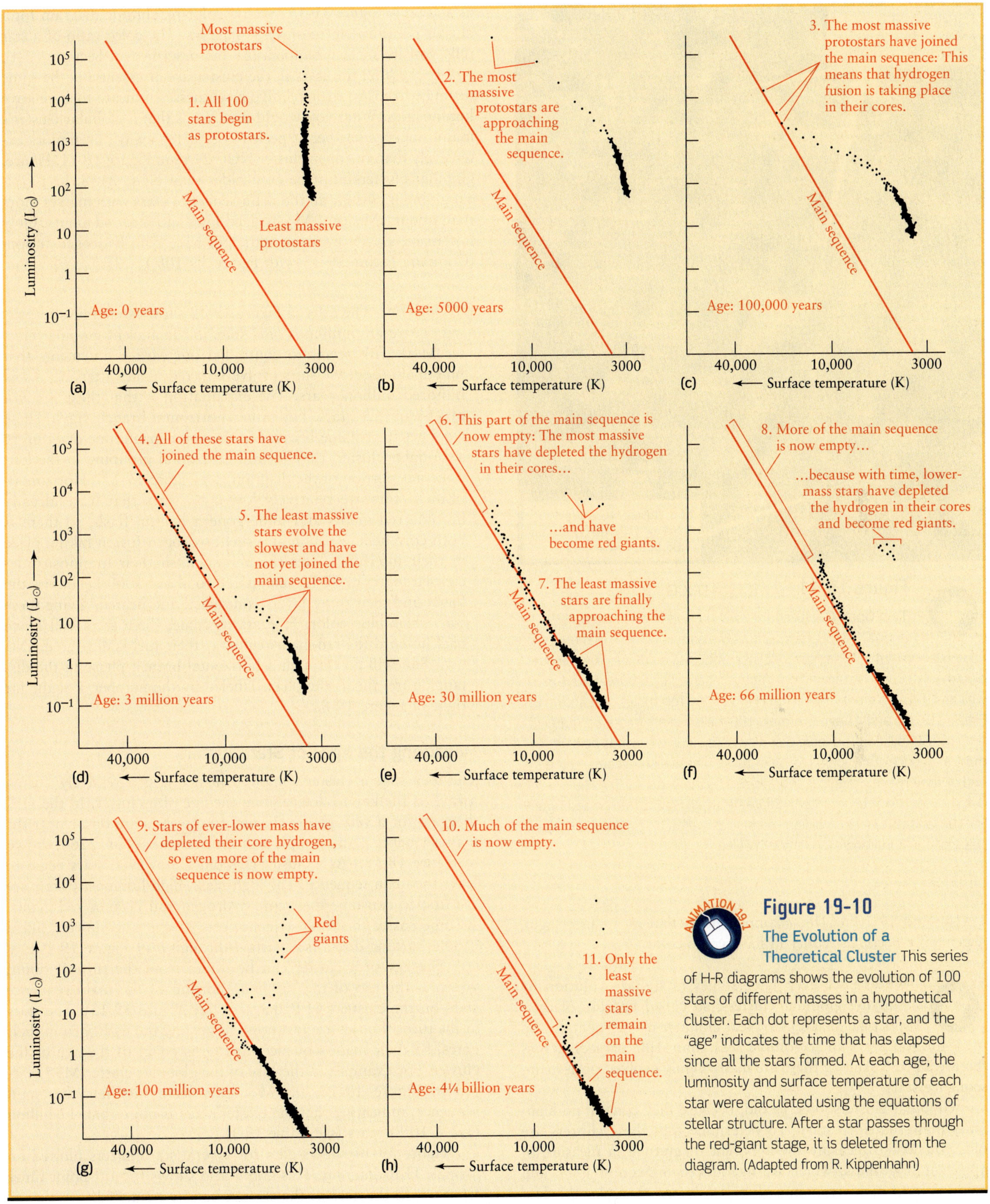
Most massive protostars
1. All 100 stars begin as protostars.
Main sequence
Least massive protostars
Age: 0 years
2. The most massive protostars are approaching the main sequence.
Age: 5000 years
3. The most massive protostars have joined the main sequence: This means that hydrogen fusion is taking place in their cores.
Age: 100,000 years
4. All of these stars have joined the main sequence.
5. The least massive stars evolve the slowest and have not yet joined the main sequence.
Age: 3 million years
6. This part of the main sequence is now empty: The most massive stars have depleted the hydrogen in their cores...
...and have become red giants.
7. The least massive stars are finally approaching the main sequence.
Age: 30 million years
8. More of the main sequence is now empty...
...because with time, lower-mass stars have depleted the hydrogen in their cores and become red giants.
Age: 66 million years
9. Stars of ever-lower mass have depleted their core hydrogen, so even more of the main sequence is now empty.
Red giants
Age: 100 million years
10. Much of the main sequence is now empty.
11. Only the least massive stars remain on the main sequence.
Age: 4¼ billion years
Luminosity ($L_\odot$)
10^5
10^4
10^3
10^2
10
1
10^{-1}
40,000
10,000
3000
Surface temperature (K)
(a)
(b)
(c)
(d)
(e)
(f)
(g)
(h)
ANIMATION 19.1

Figure 19-10

The Evolution of a Theoretical Cluster This series of H-R diagrams shows the evolution of 100 stars of different masses in a hypothetical cluster. Each dot represents a star, and the "age" indicates the time that has elapsed since all the stars formed. At each age, the luminosity and surface temperature of each star were calculated using the equations of stellar structure. After a star passes through the red-giant stage, it is deleted from the diagram. (Adapted from R. Kippenhahn)

Figure 19-11 R I V U X G

Two Open Clusters The two clusters in this image, M35 and NGC 2158, lie in almost the same direction in the constellation Gemini. The nearer cluster, M35, has a number of luminous blue main-sequence stars with surface temperatures around 10,000 K as well as a few red giants. Hence its H-R diagram resembles that shown in Figure 19-10*g*, and its age is around 100 million years (more accurately, around 150 million). The more distant cluster, NGC 2158, has *no* blue main-sequence stars; long ago, all of these massive stars came to the end of their main-sequence lifetimes and became giants. The H-R diagram for NGC 2158 is intermediate between Figure 19-10*g* and Figure 19-10*h*, and its age (1.05 billion years) is as well. (Canada-France-Hawaii Telescope; J.-C. Cuillandre, CFHT; and Coelum)

19-10*g* and 19-10*h*). This example shows that as a cluster ages, it generally becomes redder in its average color.

We can see even later stages in stellar evolution by studying **globular clusters,** so called because of their spherical shape. A typical globular cluster contains up to 1 million stars in a volume less than 100 parsecs across (Figure 19-12). Among these are many highly evolved post–main-sequence stars.

Globular clusters must be old, because they contain no high-mass main-sequence stars. To determine that these clusters are old, you would measure the apparent magnitude (a measure of apparent brightness, which we introduced in Section 17-3) and color ratio of many stars in a globular cluster, then plot the data as shown in Figure 19-13. Such a **color-magnitude diagram** for a cluster is equivalent to an H-R diagram. The color ratio of a star tells you its surface temperature (as described in Section 17-4), and because all the stars in the cluster are at essentially the same distance from us, their relative brightnesses indicate their relative luminosities. What you would discover is that a globular cluster's main sequence has been "peeled" or "eaten away" even more extensively than the open cluster NGC 2158 in Figure 19-11. Hence, globular clusters must be even older than NGC 2158. In a typical globular cluster, all the main-sequence stars with masses more than about $1M_\odot$ or $2\ M_\odot$ evolved long ago into red giants. Only low-mass, slowly evolving stars still have core hydrogen fusion. (Compare Figure 19-13 with Figure 19-10*h*.)

CAUTION! The inset in Figure 19-12 shows something surprising: There are luminous *blue* stars in the ancient globular cluster M10. This seems to contradict our earlier statements that blue main-sequence stars evolve into red giants after just a few hundred million years. The explanation is that these are *not* main-sequence stars, but rather **horizontal-branch stars.** These stars get their name because in the color-magnitude diagram of a globular cluster, they form a horizontal grouping in the left-of-center portion of the diagram (see Figure 19-13). Horizontal-branch stars are relatively low-mass stars that have already become red giants and undergone a helium flash, so there is both core helium fusion and shell hydrogen fusion taking place in their interiors. After the helium flash their luminosity decreased to about $50\ L_\odot$ (compared to about $1000\ L_\odot$ before the flash) and their outer layers contracted and heated, giving these stars their blue color. In years to come, these stars will move back toward the red-giant region as their fuel is devoured. Our own Sun will go through a horizontal-branch phase in the distant future; this is the phase labeled by the number 3 at the far right of Figure 19-8.

Measuring the Ages of Star Clusters

The idea that a cluster's main sequence is progressively "eaten away" is the key to determining the age of a cluster. In the H-R diagram for a very young cluster, all the stars are on or near the main sequence. (An example is the open cluster NGC 2264, shown in Figure 18-18.) As a cluster gets older, however, stars begin to leave the main sequence. The high-mass, high-luminosity stars are the first to consume their core hydrogen and become red giants. As time passes the main sequence gets shorter and shorter, like a candle burning down (see parts *d* through *h* of Figure 19-10).

The age of a cluster can be found from the **turnoff point,** which is the top of the surviving portion of the main sequence stars on the cluster's H-R diagram (see Figure 19-13). The stars at the turnoff point are just now exhausting the hydrogen in their cores, so their main-sequence lifetime is equal to the age of the cluster. For example, in the case of the globular cluster M55 plotted in Figure 19-13, $0.8\text{-}M_\odot$ stars have just left the main sequence, indicating that the cluster's age is more than 12 billion (1.2×10^{10}) years (see Table 19-1).

Figure 19-14 shows data for several star clusters plotted on a single H-R diagram. This graph also shows turnoff-point times from which the ages of the clusters can be estimated.

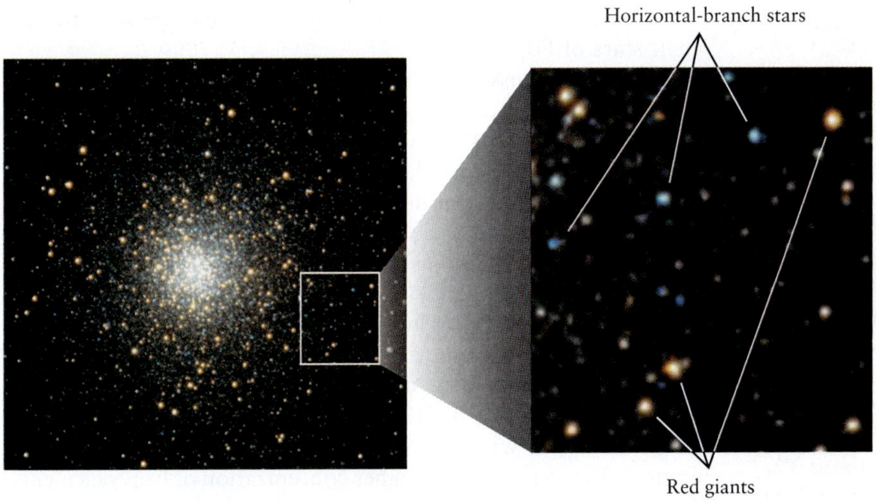

Figure 19-12 R I V U X G

A Globular Cluster This cluster, called M10, contains a few hundred thousand stars within a diameter of only 20 pc (70 ly). It lies approximately 5000 pc (16,000 ly) from Earth in the constellation Ophiuchus (the Serpent Holder). Most of the stars in this image are either red giants or blue, horizontal-branch stars with both core helium fusion and shell hydrogen fusion. (T. Credner and S. Kohle, Astronomical Institutes of the University of Bonn)

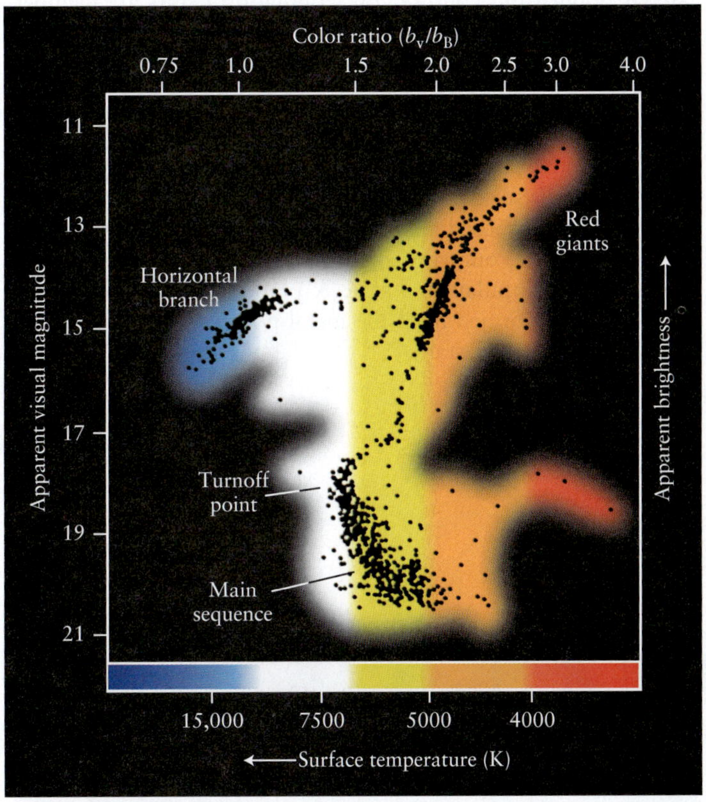

Figure 19-13

A Color-Magnitude Diagram of a Globular Cluster Each dot in this diagram represents the apparent visual magnitude (a measure of the brightness as seen through a V filter) and surface temperature (as measured by the color ratio b_V/b_B) of a star in the globular cluster M55 in Sagittarius. Because all the stars in M55 are at essentially the same distance from Earth (about 6000 pc or 20,000 ly), their apparent visual magnitudes are a direct measure of their luminosities. Note that the upper half of the main sequence is missing. (Adapted from D. Schade, D. VandenBerg, and F. Hartwick)

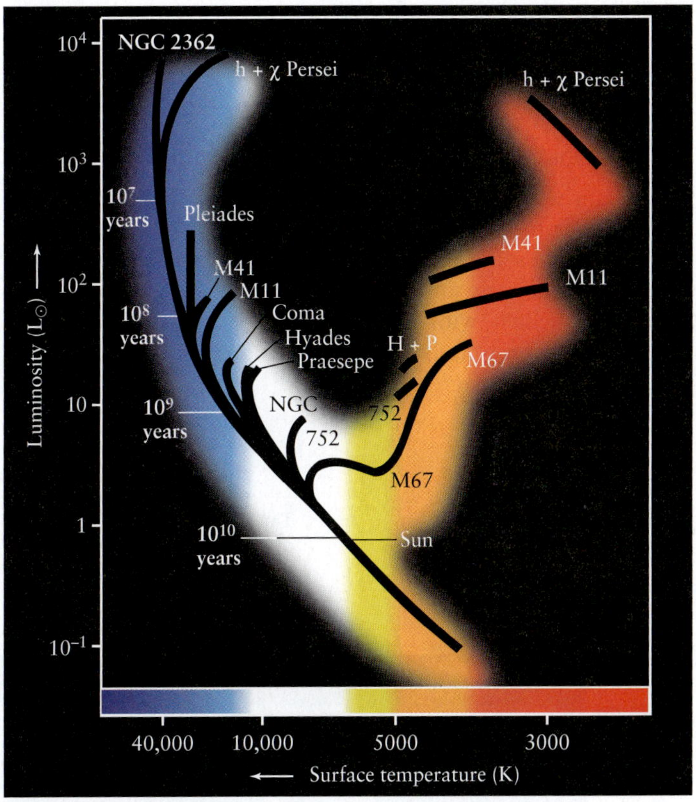

Figure 19-14

An H-R Diagram for Open Star Clusters The black bands indicate where stars from various open clusters fall on the H-R diagram. The age of a cluster can be estimated from the location of the cluster's turnoff point, where the cluster's most massive stars are just now leaving the main sequence. The times for these turnoff points are listed alongside the main sequence. For example, the Pleiades cluster turnoff point is near the 10^8-year point, so this cluster is about 10^8 years old. (Adapted from A. Sandage)

19-5 Stellar evolution has produced two distinct populations of stars

Studies of star clusters reveal a curious difference between the youngest and oldest stars in our Galaxy. Stars in the youngest clusters (those with most of their main sequences still intact) are said to be **metal rich,** because their spectra contain many prominent spectral lines of heavy elements. (Recall from Section 17-5 that astronomers use the term "metal" to denote any element other than hydrogen and helium, which are the two lightest elements.) Such stars are also called **Population I stars.** The Sun is a relatively young, metal-rich, Population I star.

By contrast, the spectra of stars in the oldest clusters show only weak lines of heavy elements. These ancient stars are thus said to be **metal poor,** because heavy elements are only about 3% as abundant in these stars as in the Sun. They are also called **Population II stars.** The stars in globular clusters are metal poor, Population II stars. Figure 19-15 shows the difference in spectra between a metal-poor, Population II star and the Sun (a metal-rich, Population I star).

CAUTION! Note that "metal rich" and "metal poor" are relative terms. In even the most metal-rich star known, metals make up just a few percent of the total mass of the star.

Stellar Populations and the Origin of Heavy Elements

To explain why there are two distinct populations of stars, we must go back to the Big Bang, the explosive origin of the universe that took place some 13.7 billion years ago. As we will discuss in Chapter 27, the early universe consisted almost exclusively of hydrogen and helium, with almost no heavy elements (metals). The first stars to form were likewise metal poor. The least massive of these stars have survived to the present day and are now the ancient stars of Population II.

The more massive of the original stars evolved more rapidly and no longer shine. But as these stars evolved, helium fusion in their cores produced metals—carbon and oxygen. In the most massive stars, as we will learn in Chapter 21, further thermonuclear reactions produced even heavier elements. As these massive original stars aged and died, they expelled their metal-enriched gases into space. (The star shown in Figure 19-5 is going through such a mass-loss phase late in its life.) This expelled material joined the interstellar medium and was eventually incorporated into a second generation of stars that have a higher concentration of heavy elements. These metal-rich members of the second stellar generation are the Population I stars, of which our Sun is an example.

Stars like the Sun contain material that was processed through an earlier generation of stars

CAUTION! Be careful not to let the designations of the two stellar populations confuse you. Population *I* stars are members of a *second* stellar generation, while Population *II* stars belong to an older *first* generation.

The relatively high concentration of heavy elements in the Sun means that the solar nebula, from which both the Sun and planets formed (see Section 8-4), must likewise have been metal rich. Earth is composed almost entirely of heavy elements, as are our bodies. Thus, our very existence is intimately linked to the Sun's being a Population I star. A planet like Earth probably could not have formed from the metal-poor gases that went into making Population II stars.

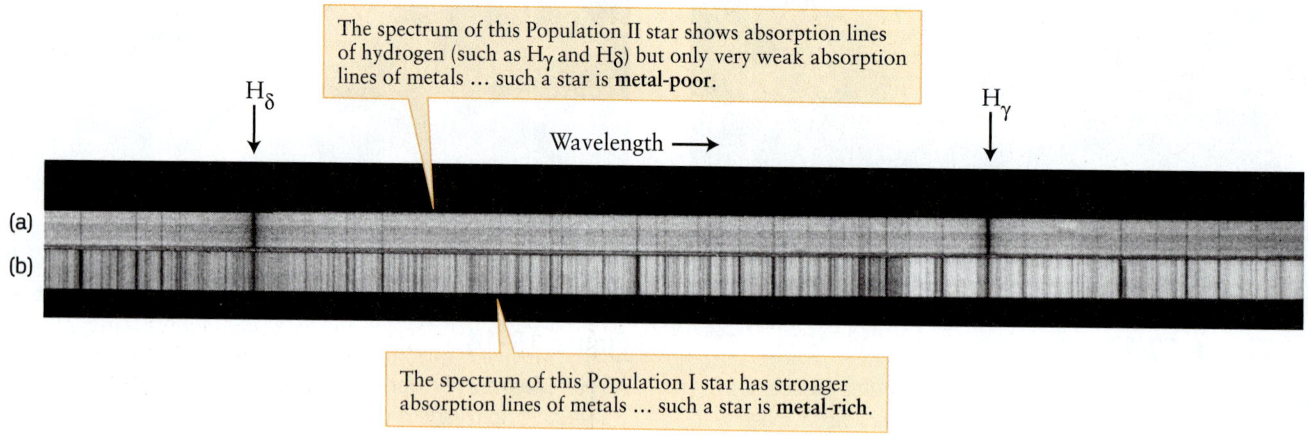

The spectrum of this Population II star shows absorption lines of hydrogen (such as Hγ and Hδ) but only very weak absorption lines of metals ... such a star is metal-poor.
Hδ
Hγ
Wavelength →
(a)
(b)
The spectrum of this Population I star has stronger absorption lines of metals ... such a star is metal-rich.

Figure 19-15 R I V U X G

Spectra of a Metal-Poor Star and a Metal-Rich Star The abundance of metals (elements heavier than hydrogen and helium) in a star can be inferred from its spectrum. These spectra compare **(a)** a metal-poor, Population II star and **(b)** a metal-rich, Population I star (the Sun) of the same surface temperature. We described the hydrogen absorption lines H_γ (wavelength 434 nm) and H_δ (wavelength 410 nm) in Section 5-8. (Lick Observatory)

The concept of two stellar populations provides insight into our own origins. Recall from Section 19-3 that helium fusion in red-giant stars produces the same isotopes of carbon (^{12}C) and oxygen (^{16}O) that are found most commonly on Earth. The reason is that Earth's carbon and oxygen atoms, including all of those in your body, actually *were* produced by helium fusion. These reactions occurred billions of years ago within an earlier generation of stars that died and gave up their atoms to the interstellar medium—the same atoms that later became part of our solar system, our planet, and our bodies. We are literally children of the stars.

19-6 Many mature stars pulsate

ANIMATION 19.2

We saw in Section 16-3 that the surface of our Sun vibrates in and out, although by only a small amount. But other stars undergo substantial changes in size, alternately swelling and shrinking. As these stars pulsate, they also vary dramatically in brightness. We now understand that these **pulsating variable stars** are actually evolved, post–main-sequence stars.

Long-Period Variables

Pulsating variable stars were first discovered in 1595 by David Fabricius, a Dutch minister and amateur astronomer. He noticed that the star o (omicron) Ceti is sometimes bright enough to be easily seen with the naked eye but at other times fades to invisibility (Figure 19-16). By 1660, astronomers realized that these brightness variations repeated with a period of 332 days. Seventeenth-century astronomers were so enthralled by this variable star that they renamed it Mira ("wonderful").

Mira is an example of a class of pulsating stars called **long-period variables.** These stars are cool red giants that vary in brightness by a factor of 100 or more over a period of months or years. With surface temperatures of about 3500 K and average luminosities that range from about 10 to 10,000 $L_\odot$, they occupy the upper right side of the H-R diagram (Figure 19-17). Some, like Mira, are periodic, but others are irregular. Many eject large amounts of gas and dust into space.

Astronomers do not fully understand why some cool red giants become long-period variables. It is difficult to calculate accurate stellar models to describe such huge stars with extended, tenuous atmospheres.

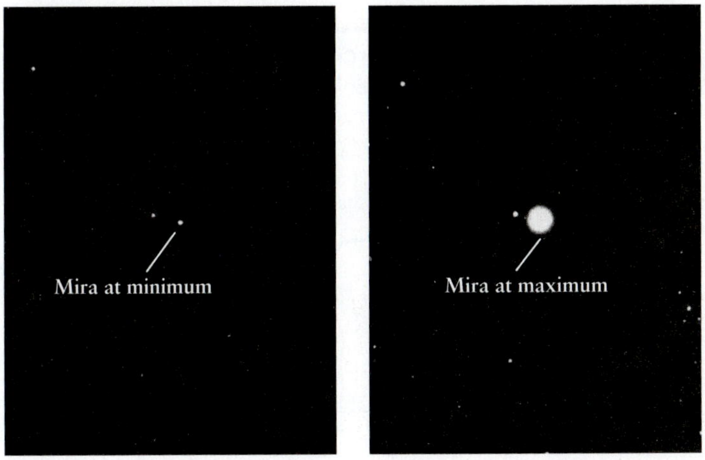

(a) (b)

WEB LINK 19.9

Figure 19-16 RIVUXG

Mira—A Long-Period Variable Star Mira, or o (omicron) Ceti, is a variable star whose luminosity varies with a 332-day period. At its dimmest, as in **(a)** (photographed in December 1961), Mira is less than 1% as bright as when it is at maximum, as in **(b)** (January 1965). These brightness variations occur because Mira pulsates. (Lowell Observatory)

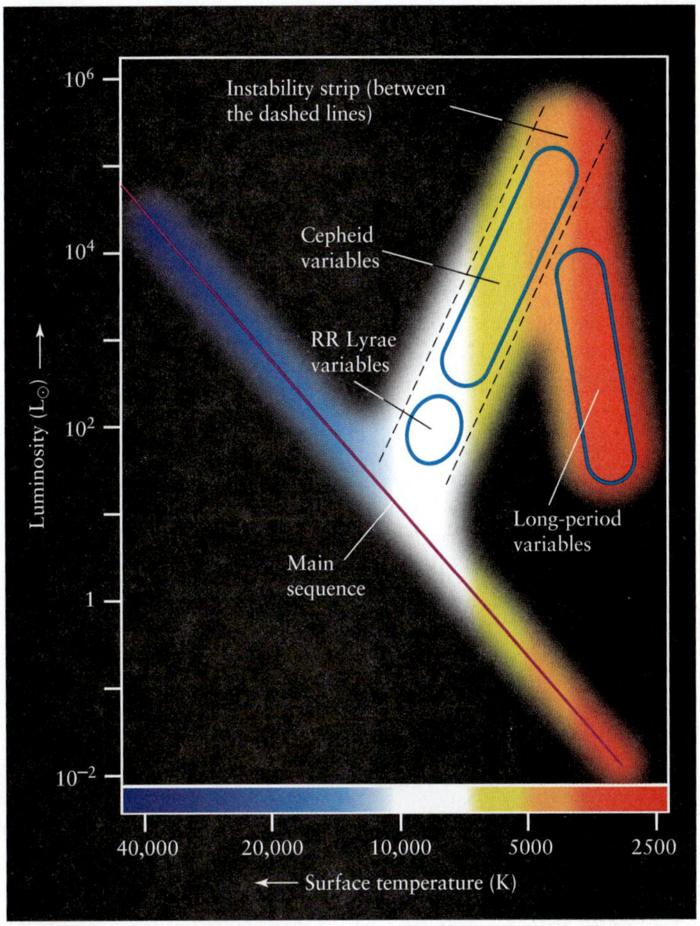

INTERACTIVE EXERCISE 19.2

Figure 19-17

Variable Stars on the H-R Diagram Pulsating variable stars are found in the upper right of the H-R diagram. Long-period variables like Mira are cool red giant stars that pulsate slowly, changing their brightness in a semi-regular fashion over months or years. Cepheid variables and RR Lyrae variables are located in the instability strip, which lies between the main sequence and the red-giant region. A star passing through this strip along its evolutionary track becomes unstable and pulsates.

Cepheid Variables

Astronomers have a much better understanding of other pulsating stars, called **Cepheid variables,** or simply Cepheids. A Cepheid variable is recognized by the characteristic way in which its light output varies—rapid brightening followed by gradual dimming. They are named for δ (delta) Cephei, an example of this type of star discovered in 1784 by John Goodricke, a deaf, mute, 19-year-old English amateur astronomer. He found that at its most brilliant, δ Cephei is 2.3 times as bright as at its dimmest. The cycle of brightness variations repeats every 5.4 days. (Sadly, Goodricke paid for his discoveries with his life; he caught pneumonia while making his nightly observations and died before his twenty-second birthday.) The surface temperatures and luminosities of the Cepheid variables place them in the upper middle of the H-R diagram (see Figure 19-17).

By studying variable stars, astronomers gain insight into late stages of stellar evolution

After core helium fusion begins, mature stars move across the middle of the H-R diagram. Figure 19-9*a* shows the evolutionary tracks of high-mass stars crisscrossing the H-R diagram. Post–helium-flash stars of moderate mass also cross the middle of the H-R diagram between the red-giant region and the horizontal branch.

During these transitions across the H-R diagram, a star can become unstable and pulsate. In fact, there is a region on the H-R diagram between the upper main sequence and the red-giant branch called the **instability strip** (see Figure 19-17). When an evolving star passes through this region, the star pulsates and its brightness varies periodically. Figure 19-18*a* shows the brightness variations of δ Cephei, which lies within the instability strip.

A Cepheid variable brightens and fades because the star's outer envelope cyclically expands and contracts. The first to observe this was the Russian astronomer Aristarkh Belopol'skii, who noticed in 1894 that spectral lines in the spectrum of δ Cephei shift back and forth with the same 5.4-day period as that of the magnitude variations. From the Doppler effect, we can

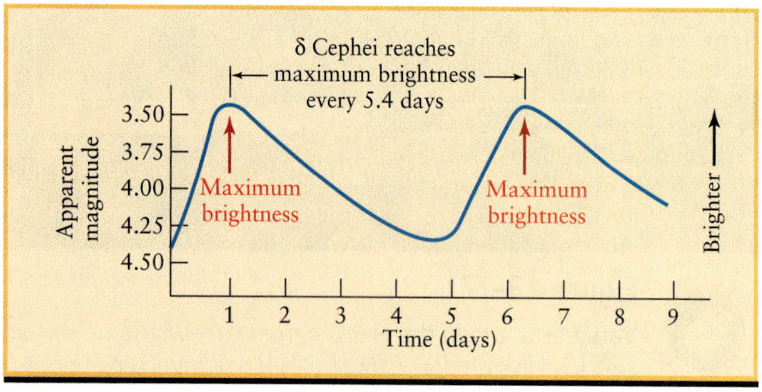

(a) The light curve of δ Cephei (a graph of brightness versus time)

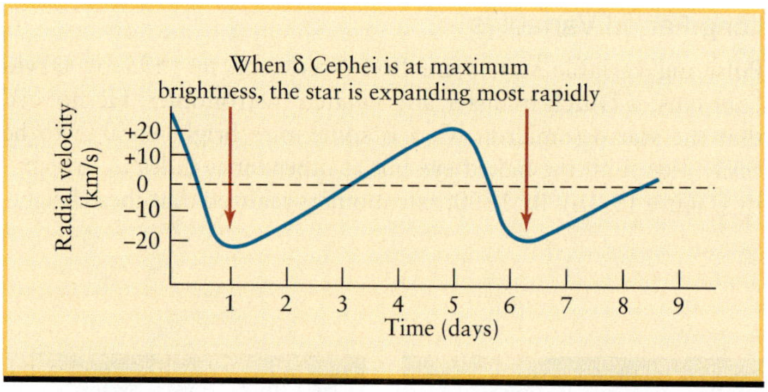

(b) Radial velocity versus time for δ Cephei
(positive: star is contracting; negative: star is expanding)

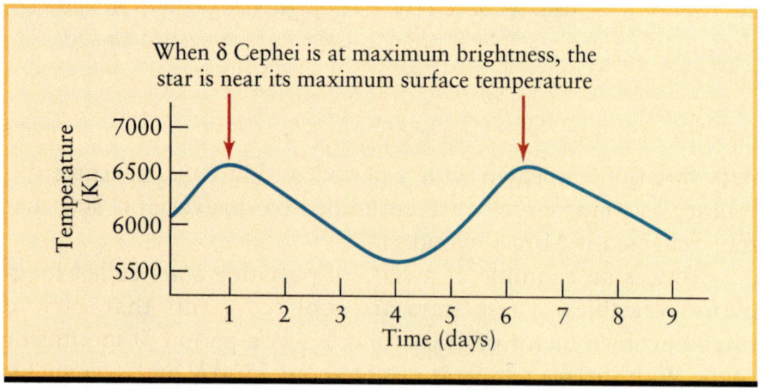

(c) Surface temperature versus time for δ Cephei

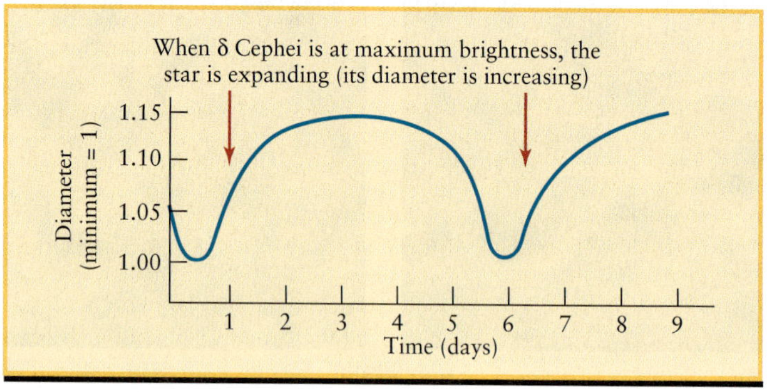

(d) Diameter versus time for δ Cephei

WEB LINK 19.10

Figure 19-18

δ Cephei—A Pulsating Star **(a)** As δ Cephei pulsates, it brightens quickly (the light curve moves upward sharply) but fades more slowly (the curve declines more gently). The increases and decreases in brightness are nearly in step with variations in **(b),** the star's radial velocity (positive when the star contracts and the surface moves away from us, negative when the star expands and the surface approaches us), as well as in **(c),** the star's surface temperature. **(d)** The star is still expanding when it is at its brightest and hottest (compare with parts a and b).

translate these wavelength shifts into radial velocities and draw a velocity curve (Figure 19-18*b*). Negative velocities mean that the star's surface is expanding toward us; positive velocities mean that the star's surface is receding. Note that the light curve and velocity curve are mirror images of each other. The star is brighter than average while it is expanding and dimmer than average while contracting.

When a Cepheid variable pulsates, the star's surface oscillates up and down like a spring. During these cyclical expansions and contractions, the star's gases alternately heat up and cool down. Figure 19-18*c* shows the resulting changes in the star's surface temperature. Figure 19-18*d* graphs the periodic changes in the star's diameter.

Just as a bouncing ball eventually comes to rest, a pulsating star would soon stop pulsating without something to keep its oscillations going. In 1914, the British astronomer Arthur Eddington suggested that a Cepheid pulsates because the star is more opaque when compressed than when expanded. When the star is compressed, trapped heat increases the internal pressure, which pushes the star's surface outward. When the star expands, the heat escapes, the internal pressure drops, and the star's surface falls inward.

In the 1960s, the American astronomer John Cox followed up on Eddington's idea and proved that helium is what keeps Cepheids pulsating. Normally, when a star's helium is compressed, the gas increases in temperature and becomes more transparent. But in certain layers near the star's surface, compression may ionize helium (remove one of its electrons) instead of raising its temperature. Ionized helium gas is quite opaque, so these layers effectively trap heat and make the star expand, as Eddington suggested. This expansion cools the outer layers and makes the helium ions recombine with electrons, which makes the gas more transparent and releases the trapped energy. The star's surface then falls inward, recompressing the helium, and the cycle begins all over again.

CAUTION! In our discussion of the behavior of gases (see Box 19-1, Section 19-1, and Section 19-3) we saw that a gas cools when it expands and heats up when it is compressed. Hence, you would expect that the gases in a pulsating star like δ Cephei would reach their maximum temperature when the star is at its smallest diameter, so that the gases are most compressed. The hotter the gas, the more brightly it glows, so δ Cephei should also have its maximum brightness when its diameter is smallest. But Figure 19-18 shows that the star's brightness and the temperature of the gases at the surface reach their maximum values when the star is expanding, some time *after* the star has contracted to its smallest diameter. How can this be? The explanation is again related to how opaque the gases are inside the star. The rate at which energy is emitted from the central regions of the star is indeed greatest when the star is at its minimum diameter, but the opaque gases in the star's outer layers impede the flow of energy to the surface. Hence, δ Cephei reaches its maximum brightness and maximum surface temperature about half a day after the star is at its smallest size.

Cepheid variables are important because they have two properties that allow astronomers to determine the distances to very remote objects. First, Cepheids can be seen even at distances of millions of parsecs, because they are very luminous, ranging from a few hundred times solar luminosity to more than $10^4\ L_\odot$. Second, there is a direct relationship between a Cepheid's period and its average luminosity: The dimmest Cepheid variables pulsate rapidly, with periods of 1 to 2 days, while the most luminous Cepheids pulsate with much slower periods of about 100 days.

Figure 19-19 shows this **period-luminosity relation.** By measuring the period of a distant Cepheid's brightness variations and using a graph like Figure 19-19, an astronomer can determine the star's luminosity. By also measuring the star's apparent brightness, the distance to the Cepheid can be found by using the inverse-square law (see Section 17-2). By applying the period-luminosity relation in this way to Cepheids in other galaxies, astronomers have been able to calculate the distances to those galaxies with great accuracy. (Box 17-2 gives an example of such a calculation.) As we will see in Chapters 24 and 26, such measurements play an important role in determining the overall size and structure of the universe.

How a Cepheid pulsates depends on the amount of heavy elements in the star's outer layers, because even trace amounts of these elements can have a large effect on how opaque the stellar gases are. Hence, Cepheids are classified according to their metal content. If the star is a metal-rich, Population I star, it is called a **Type I Cepheid;** if it is a metal-poor, Population II star, it is called

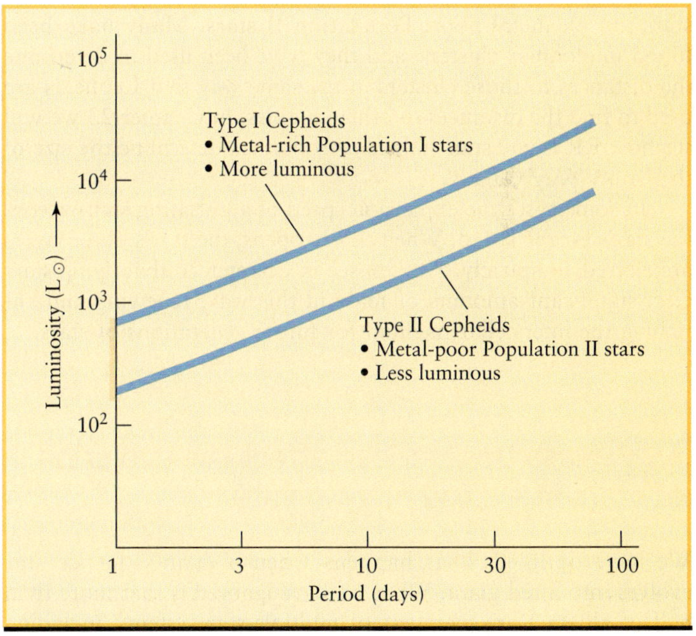

WEB LINK 19.11

Figure 19-19

Period-Luminosity Relations for Cepheids The greater the average luminosity of a Cepheid variable, the longer its period and the slower its pulsations. Note that there are actually two distinct period-luminosity relations—one for Type I Cepheids and one for the less luminous Type II Cepheids. (Adapted from H. C. Arp) a **Type II Cepheid.** As Figure 19-19 shows, these two types of Cepheids exhibit different period-luminosity relations. In order to know which period-luminosity relation to apply to a given Cepheid, an astronomer must determine the star's metal content from its spectrum (see Figure 19-15).

The evolutionary tracks of mature, high-mass stars pass back and forth through the upper end of the instability strip on the H-R diagram. These stars become Cepheids when helium ionization occurs at just the right depth to drive the pulsations. For stars on the high-temperature (left) side of the instability strip, helium ionization occurs too close to the surface and involves only an insignificant fraction of the star's mass. For stars on the cool (right) side of the instability strip, convection in the star's outer layers prevents the storage of the energy needed to drive the pulsations. Thus, Cepheids exist only in a narrow temperature range on the H-R diagram.

RR Lyrae Variables

Stars of lower mass do not become Cepheids. Instead, after leaving the main sequence, becoming red giants, and undergoing the helium flash, their evolutionary tracks pass through the lower end of the instability strip as they move along the horizontal branch. Some of these stars become **RR Lyrae variables,** named for their prototype in the constellation Lyra (the Harp). RR Lyrae variables all have periods shorter than one day and roughly the same average luminosity as horizontal-branch stars, about 100 $L_{\odot}$. In fact, the RR Lyrae region of the instability strip (see Figure 19-17) is actually a segment of the horizontal branch. RR Lyrae stars are all metal-poor, Population II stars. Many have been found in globular clusters, and they have been used to determine the distances to those clusters in the same way that Cepheids are used to find the distances to other galaxies. In Chapter 23 we will see how RR Lyrae stars helped astronomers determine the size of the Milky Way Galaxy.

In some cases the expansion speed of a pulsating star exceeds the star's escape speed. When this happens, the star's outer layers are ejected completely. We will see in Chapter 20 that dying stars eject significant amounts of mass in this way, renewing and enriching the interstellar medium for future generations of stars.

19-7 Mass transfer can affect the evolution of stars in a close binary system

We have outlined what happens when a main-sequence star evolves into a red giant. What we have ignored is that more than half of all stars are members of multiple-star systems, including binaries. If the stars in such a system are widely separated, the individual stars follow the same course of evolution as if they were isolated. In a **close binary,** however, when one star expands to become a red giant, its outer layers can be gravitationally captured by the nearby companion star. In other words, a bloated red giant in a close binary system can dump gas onto its companion, a process called **mass transfer.**

Roche Lobes and Lagrangian Points

Our modern understanding of mass transfer in close binaries is based on the work of the French mathematician Edouard Roche. In the mid-1800s, Roche studied how rotation and mutual tidal interaction affect the stars in a binary system. Tidal forces cause the two stars in a close binary to keep the same sides facing each other, just as our Moon keeps its same side facing Earth (see Section 4-8). But because stars are gaseous, not solid, rotation and tidal forces can have significant effects on their shapes.

If the stars in a binary system are sufficiently close, tidal forces can pull gases off one star and onto the other

In widely separated binaries, the stars are so far apart that tidal effects are small, and, therefore, the stars are nearly perfect spheres. In close binaries, where the separation between the stars is not much greater than their sizes, tidal effects are strong, causing the stars to be somewhat egg-shaped.

Roche discovered a mathematical surface that marks the gravitational domain of each star in a close binary. (This surface is not a real physical one, like the surface of a balloon, but a mathematical construct.) Figure 19-20*a* shows the outline of this surface as a dashed line. The two halves of this surface, each of which encloses one of the stars, are known as **Roche lobes.** The more massive star is always located inside the larger Roche lobe. If gas from a star leaks over its Roche lobe, it is no longer bound by gravity to that star. This escaped gas is free either to fall onto the companion star or to escape from the binary system.

The point where the two Roche lobes touch, called the **inner Lagrangian point,** is a kind of balance point between the two stars in a binary. It is here that the effects of gravity and rotation cancel each other. When mass transfer occurs in a close binary, gases flow through the inner Lagrangian point from one star to the other.

In many binaries, the stars are so far apart that even during their red-giant stages the surfaces of the stars remain well inside their Roche lobes. As a result, little mass transfer can occur and each star lives out its life as if it were single and isolated. A binary system of this kind is called a **detached binary** (Figure 19-20*a*).

However, if the two stars are close enough, when one star expands to become a red giant, it may fill or overflow its Roche lobe. Such a system is called a **semidetached binary** (Figure 19-20*b*). If both stars fill their Roche lobes, the two stars actually touch and the system is called a **contact binary** (Figure 19-20*c*). It is quite unlikely, however, that both stars exactly fill their Roche lobes at the same time. (This would only be the case if the two stars had identical masses, so that they both evolved at exactly the same rate.) It is more likely that they overflow their Roche lobes, giving rise to a common envelope of gas. Such a system is called an **overcontact binary** (Figure 19-20*d*).

Observations of Mass Transfer

The binary star system Algol (from an Arabic term for "demon") provided the first clear evidence of mass transfer in close binaries. Also called β (beta) Persei, Algol can easily be seen with the naked

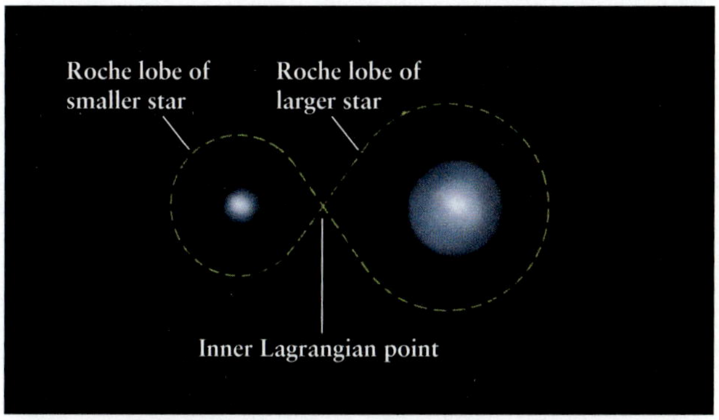

(a) Detached binary: Neither star fills its Roche lobe.

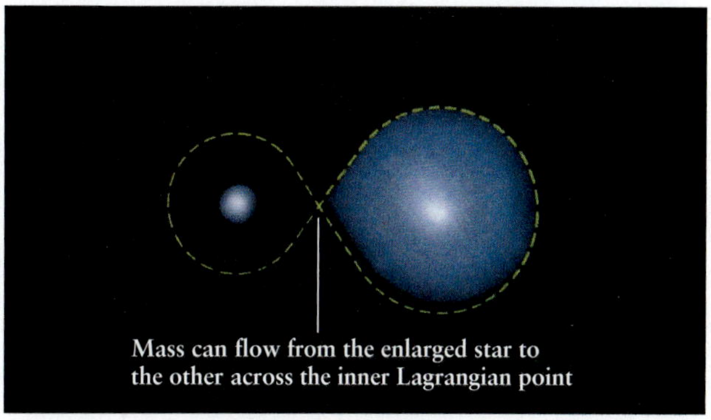

(b) Semi-detached binary: One star fills its Roche lobe.

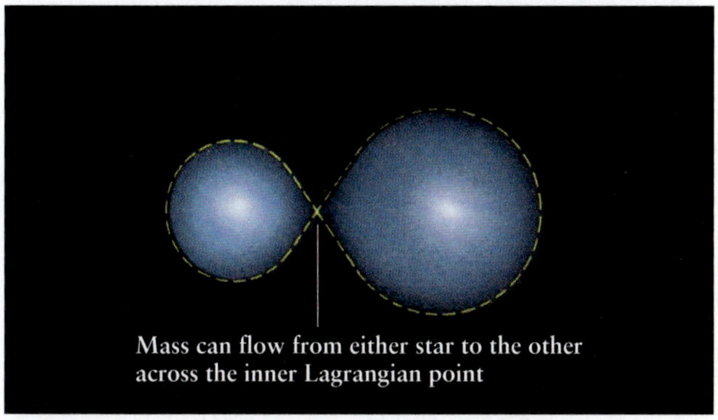

(c) Contact binary: Both stars fill their Roche lobes.

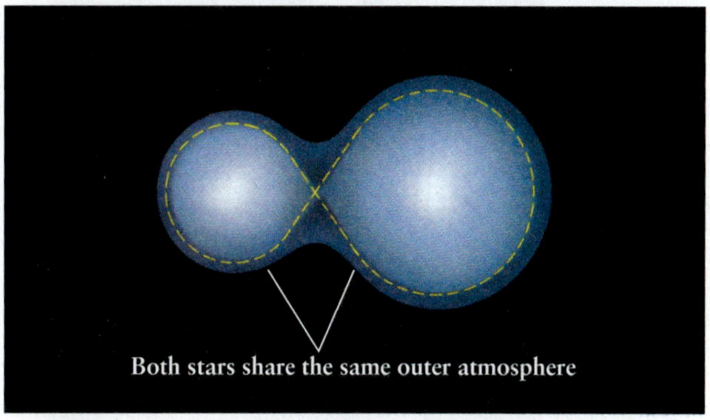

(d) Overcontact binary: Both stars overfill their Roche lobes.

INTERACTIVE EXERCISE 19.3

Figure 19-20

Close Binary Star Systems The gravitational domain of a star in a close binary system is called its Roche lobe. The two Roche lobes meet at the inner Lagrangian point. The sizes of the stars relative to their Roche lobes determine whether the system is **(a)** a detached binary, **(b)** a semidetached binary, **(c)** a contact binary, or **(d)** an overcontact binary.

eye. Ancient astronomers knew that Algol varies periodically in brightness by a factor of more than 2. In 1782, John Goodricke (the discoverer of δ Cephei's variability) first suggested that these brightness variations take place because Algol is an *eclipsing* binary. (We discussed this type of binary in Section 17-11.) The orbital plane of the two stars that make up the binary system happens to be nearly edge-on to our line of sight, so one star periodically eclipses the other. Algol's light curve (Figure 19-21*a*) and spectrum show that Goodricke's brilliant hypothesis is correct, and that Algol is a semidetached binary. The detached star (on the right in Figure 19-21*a*) is a luminous blue main-sequence star, while its less massive companion is a dimmer red giant that fills its Roche lobe.

CAUTION! According to stellar evolution theory, the more massive a star, the more rapidly it should evolve. Since the two stars in a binary system form simultaneously and thus are the same age, the more massive star should become a red giant before the less massive one. But in Algol and similar binaries, the *more* massive star (on the right in Figure 19-21*a*) is still on the main sequence, whereas the *less* massive star (on the left in Figure 19-21*a*) has evolved to become a red giant. How can we explain this apparent contradiction? The answer is that the red giant in Algol-type binaries was *originally* the more massive star. As it left the main sequence to become a red giant, this star expanded until it overflowed its Roche lobe and dumped gas onto its originally less massive companion. Because of the resulting mass transfer, that companion (which is still on the main sequence) became the more massive star.

Mass transfer is also important in another class of semidetached binaries, called β (beta) Lyrae variables, after their prototype in the constellation Lyra. As with Algol, the less massive star in β Lyrae (on the left in Figure 19-21*b*) fills its Roche lobe. Unlike Algol, however, the more massive detached star (on the right in Figure 19-21*b*) is the dimmer of the two stars. Apparently, this detached star is enveloped in a rotating *accretion disk* of gas captured from its bloated companion. This disk partially blocks the light coming from the detached star, making it appear dimmer.

What is the fate of an Algol or β Lyrae system? If the detached star is massive enough, it will evolve rapidly, expanding to also fill its Roche lobe. The result will be an overcontact binary in which the two stars share the gases of their outer layers. Such binaries are sometimes called W Ursae Majoris stars, after the prototype of this class (Figure 19-21*c*).

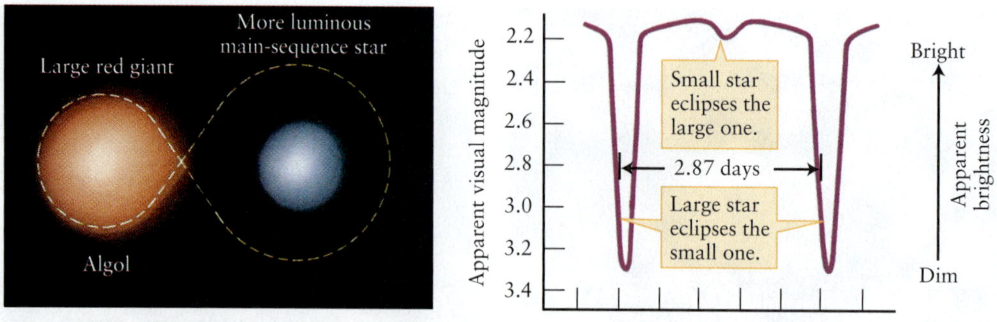

(a) A semidetached binary

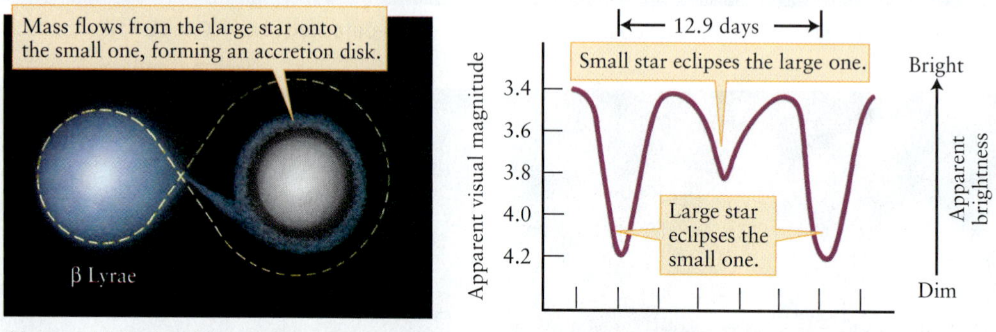

(b) A semidetached binary with mass transfer

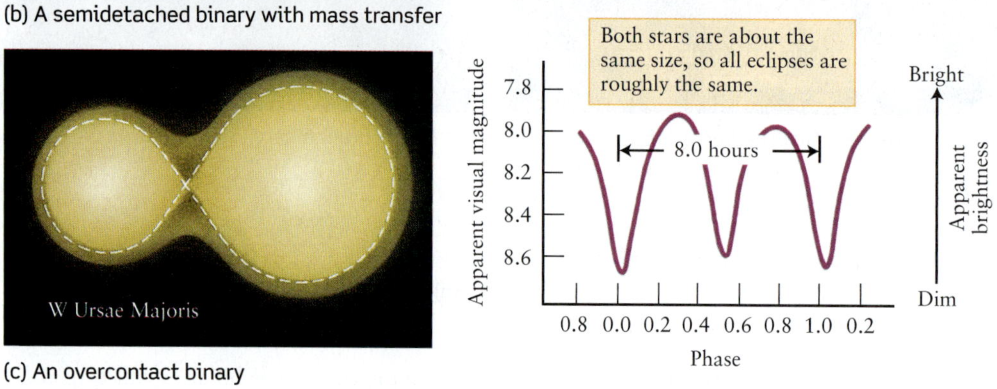

(c) An overcontact binary

WEB LINK 19.12

Figure 19-21

Three Eclipsing Binaries Compare the light curves of these three eclipsing binaries to those in Figure 17-24. **(a)** Algol is a semidetached binary. The deep eclipse occurs when the large red-giant star blocks the light from the smaller but more luminous main-sequence star. (Compare Figure 17-24*b*.) **(b)** β Lyrae's light curve is also at its lowest when the larger star completely eclipses the smaller one. Half an orbital period later, the smaller star partially eclipses the larger one, making a shallower dip in the light curve. **(c)** W Ursae Majoris is an overcontact binary in which both stars overfill their Roche lobes. The extremely short period of this binary indicates that the two stars are very close to each other.

Mass transfer can also continue even after nuclear reactions cease in one of the stars in a close binary. In Chapters 20 and 21 we will see how mass transfer onto different types of dead stars called *white dwarfs*, *neutron stars*, and *black holes* can produce some of the most unusual and dramatic objects in the sky.

Key Words

alpha particle, p. 504
Cepheid variable, p. 514
close binary, p. 516
color-magnitude diagram, p. 510
contact binary, p. 516
core helium fusion, p. 503
core hydrogen fusion, p. 497
degeneracy, p. 504
degenerate-electron pressure, p. 506
detached binary, p. 516
globular cluster, p. 510
helium flash, p. 504
helium fusion, p. 503
horizontal-branch star, p. 510
ideal gas, p. 504
inner Lagrangian point, p. 516
instability strip, p. 514
long-period variable, p. 513
main-sequence lifetime, p. 497
mass loss, p. 502
mass transfer, p. 516
metal-poor star, p. 512
metal-rich star, p. 512
open cluster, p. 508
overcontact binary, p. 516
Pauli exclusion principle, p. 504
period-luminosity relation, p. 515
Population I and Population II stars, p. 512
pulsating variable star, p. 513
red dwarf, p. 499
red giant, p. 502

Roche lobe, p. 516
RR Lyrae variable, p. 516
semidetached binary, p. 516
shell hydrogen fusion, p. 500
triple alpha process, p. 504
turnoff point, p. 510
Type I and Type II Cepheids, p. 515–516
zero-age main sequence (ZAMS), p. 507
zero-age main-sequence star, p. 498

Key Ideas

The Main-Sequence Lifetime: The duration of a star's main-sequence lifetime depends on the amount of hydrogen available to be consumed in the star's core and the rate at which this hydrogen is consumed.

• The more massive a star, the shorter is its main-sequence lifetime. The Sun has been a main-sequence star for about 4.56 billion years and should remain one for about another 7 billion years.

• During a star's main-sequence lifetime, the star expands somewhat and undergoes a modest increase in luminosity.

• If a star's mass is greater than about 0.4 $M_\odot$, only the hydrogen present in the core can undergo thermonuclear fusion during the star's main-sequence lifetime. If the star is a red dwarf with a mass less than about 0.4 $M_\odot$, over time convection brings all of the star's hydrogen to the core where it can undergo fusion.

Becoming a Red Giant: Core hydrogen fusion ceases when the hydrogen has been exhausted in the core of a main-sequence star with mass greater than about 0.4 $M_\odot$. This leaves a core of nearly pure helium surrounded by a shell through which hydrogen fusion works its way outward in the star. The core shrinks and becomes hotter, while the star's outer layers expand and cool. The result is a red giant star.

• As a star becomes a red giant, its evolutionary track moves rapidly from the main sequence to the red-giant region of the H-R diagram. The more massive the star, the more rapidly this evolution takes place.

Helium Fusion: When the central temperature of a red giant reaches about 100 million K, helium fusion begins in the core. This process, also called the triple alpha process, converts helium to carbon and oxygen.

• In a more massive red giant, helium fusion begins gradually; in a less massive red giant, it begins suddenly, in a process called the helium flash.

• After the helium flash, a low-mass star moves quickly from the red-giant region of the H-R diagram to the horizontal branch.

Star Clusters and Stellar Populations: The age of a star cluster can be estimated by plotting its stars on an H-R diagram.

• The cluster's age is equal to the age of the main-sequence stars at the turnoff point (the upper end of the remaining main sequence).

• As a cluster ages, the main sequence is "eaten away" from the upper left as stars of progressively smaller mass evolve into red giants.

• Relatively young Population I stars are metal rich; ancient Population II stars are metal poor. The metals (heavy elements) in Population I stars were manufactured by thermonuclear reactions in an earlier generation of Population II stars, then ejected into space and incorporated into a later stellar generation.

Pulsating Variable Stars: When a star's evolutionary track carries it through a region in the H-R diagram called the instability strip, the star becomes unstable and begins to pulsate.

• Cepheid variables are high-mass pulsating variables. There is a direct relationship between their periods of pulsation and their luminosities.

• RR Lyrae variables are low-mass, metal-poor pulsating variables with short periods.

• Long-period variable stars also pulsate but in a fashion that is less well understood.

Close Binary Systems: Mass transfer in a close binary system occurs when one star in a close binary overflows its Roche lobe. Gas flowing from one star to the other passes across the inner Lagrangian point. This mass transfer can affect the evolutionary history of the stars that make up the binary system.

Questions

Review Questions

1. How does the chemical composition of the present-day Sun's core compare to the core's composition when the Sun formed? What caused the change?
2. Which regions of the Sun are denser today than they were a billion years ago? Which regions are less dense? What has caused these changes?
3. What is a red dwarf? How are thermonuclear reactions in the core of a red dwarf able to consume hydrogen from the star's outer layers?
4. Why do high-mass main-sequence stars have shorter lifetimes than those of lower mass?
5. On what grounds are astronomers able to say that the Sun has about 7×10^9 years remaining in its main-sequence stage?
6. What will happen inside the Sun 7 billion years from now, when it begins to mature into a red giant?
7. Explain why Earth is expected to become inhospitable to life long before the Sun becomes a red giant.
8. Explain how it is possible for the core of a red giant to contract at the same time that its outer layers expand.
9. Why does helium fusion require much higher temperatures than hydrogen fusion?
10. How is a degenerate gas different from ordinary gases?
11. What is the helium flash? Why does it happen in some stars but not in others?
12. Why does a star's luminosity decrease after helium fusion begins in its core?
13. What does it mean when an astronomer says that a star "moves" from one place to another on an H-R diagram?
14. Explain why the majority of the stars visible through telescopes are main-sequence stars.
15. On an H-R diagram, main-sequence stars do not lie along a single narrow line but are spread out over a band (see Figure 19-9*b*). On the basis of how stars evolve during their main-sequence lifetimes, explain why this should be so.

16. Explain how and why the turnoff point on the H-R diagram of a cluster is related to the cluster's age.
17. There is a good deal of evidence that our universe is about 13.7 billion years old (see Chapter 24). Explain why no main-sequence stars of spectral class M have yet evolved into red-giant stars.
18. How do astronomers know that globular clusters are made of old stars?
19. Red-giant stars appear more pronounced in composites of infrared images and visible-light images, like those in Figure 19-4*b* and Figure 19-12. Explain why.
20. The horizontal-branch stars in Figure 19-12 appear blue. (**a**) Explain why this is consistent with the color-magnitude diagram shown in Figure 19-13. (**b**) All horizontal-branch stars were once red giants. Explain what happened to these stars to change their color.
21. What is the difference between Population I and Population II stars? In what sense can the stars of one population be regarded as the "children" of the other population?
22. Both diamonds and graphite (the material used in pencils to make marks on paper) are crystalline forms of carbon. Most of the carbon atoms in these substances have nuclei with 6 protons and 6 neutrons (^{12}C). Where did these nuclei come from?
23. Why do astronomers attribute the observed Doppler shifts of a Cepheid variable to pulsation, rather than to some other cause, such as orbital motion?
24. Why do Cepheid stars pulsate? Why are these stars important to astronomers who study galaxies beyond the Milky Way?
25. What is a Roche lobe? What is the inner Lagrangian point? Why are Roche lobes important in close binary star systems?
26. What is the difference between a detached binary, a semidetached binary, a contact binary, and an overcontact binary?
27. Massive main-sequence stars turn into red giants before less massive stars. Why, then, is the more massive star in Algol a main-sequence star and the less massive star a red giant?

Advanced Questions

Questions preceded by an asterisk (*) involve topics discussed in Box 7-2, the Boxes in Chapter 17, or the Boxes in this chapter.

Problem-solving tips and tools

Recall from Section 16-1 that 6×10^{11} kg of hydrogen is converted into helium each second at the Sun's center. Recall also that you must use absolute (Kelvin) temperatures when using the Stefan-Boltzmann law. You may find it helpful to review the discussion of apparent magnitude, absolute magnitude, and luminosity in Box 17-3. Section 17-4 discusses the connection between the surface temperatures and colors of stars. Newton's form of Kepler's third law (see Section 17-9) describes the orbits of stars in binary systems. Box 7-2 gives the formula for escape speed and for the average speeds of gas molecules.

28. The radius of the Sun has increased over the past several billion years. Over the same time period, the size of the Moon's orbit around Earth has also increased. A few billion years ago, were annular eclipses of the Sun (see Figure 3-12) more or less common than they are today? Explain your answer.
29. The Sun has increased in radius by 6% over the past 4.56 billion years. Its present-day radius is 696,000 km. What was its radius 4.56 billion years ago? (*Hint:* The answer is *not* 654,000 km.)

*30. Calculate the escape speed from (**a**) the surface of the present-day Sun and (**b**) the surface of the Sun when it becomes a red giant, with essentially the same mass as today but with a radius that is 100 times larger. (**c**) Explain how your results show that a red-giant star can lose mass more easily than a main-sequence star.

*31. Calculate the average speed of a hydrogen atom (mass 1.67×10^{-27} kg) (**a**) in the atmosphere of the present-day Sun, with temperature 5800 K, and (**b**) in the atmosphere of a 1-$M_\odot$ red giant, with temperature 3500 K. (**c**) Compare your results with the escape speeds that you calculated in Question 27. Use this comparison to discuss how well the present-day Sun and a 1-$M_\odot$ red giant can retain hydrogen in their atmospheres.

32. Use the value of the Sun's luminosity (3.90×10^{26} watts, or 3.90×10^{26} joules per second) to calculate what mass of hydrogen the Sun will convert into helium during its entire main-sequence lifetime of 1.2×10^{10} years. (Assume that the Sun's luminosity remains nearly constant during the entire 1.2×10^{10} years.) What fraction does this represent of the total mass of hydrogen that was originally in the Sun?
33. (**a**) The main-sequence stars Sirius (spectral type A1), Vega (A0), Spica (B1), Fomalhaut (A3), and Regulus (B7) are among the 20 brightest stars in the sky. Explain how you can tell that all these stars are younger than the Sun. (**b**) The third-brightest star in the sky, although it can be seen only south of 29° north latitude, is α (alpha) Centauri A. It is a main-sequence star of spectral type G2, the same as the Sun. Can you tell from this whether α Centauri A is younger than the Sun, the same age, or older? Explain your reasoning.
34. Using the same horizontal and vertical scales as in Figure 19-9*a*, make points on an H-R diagram for each of the stars listed in Table 19-1. Label each point with the star's mass and its main-sequence lifetime. Which of these stars will remain on the main sequence after 10^9 years? After 10^{11} years?

*35. Explain why the quantity f in Box 19-2 has a different value for stars with masses less than 0.4 $M_\odot$ than for stars with masses greater than 0.4 $M_\odot$. In which case does f have a greater value?

*36. Calculate the main-sequence lifetimes of (**a**) a 9-$M_\odot$ star and (**b**) a 0.25-$M_\odot$ star. Compare these lifetimes with that of the Sun.

*37. The earliest fossil records indicate that life appeared on Earth about a billion years after the formation of the solar system. What is the most mass that a star could have in order that its lifetime on the main sequence is long enough to permit life to form on one or more of its planets? Assume that the evolutionary processes would be similar to those that occurred on Earth.

38. As a red giant, the Sun's luminosity will be about 2000 times greater than it is now, so the amount of solar energy falling on Earth will increase to 2000 times its present-day value. Hence, to maintain thermal equilibrium, each square meter of Earth's surface will have to radiate 2000 times as much energy into space as it does now. Use the Stefan-Boltzmann law to determine what Earth's surface temperature will be under these conditions. (*Hint:* The present-day Earth has an average surface temperature of 14°C.)

39. When the Sun becomes a red giant, its luminosity will be about 2000 times greater than it is today. Assuming that this luminosity is caused *only* by fusion of the Sun's remaining hydrogen, calculate how long our star will be a red giant. (In fact, only a fraction of the remaining hydrogen will be consumed, and the luminosity will vary over time as shown in Figure 19-8.)

40. What observations would you make of a star to determine whether its primary source of energy is hydrogen fusion or helium fusion?

41. The star whose spectrum is shown in Figure 19-15*a* has a lower percentage of heavy elements than the Sun, whose spectrum is shown in Figure 19-15*b*. Hence, the star in Figure 19-15*a* has a higher percentage of hydrogen. Why, then, isn't the H_δ absorption line of hydrogen noticeably darker for the star in Figure 19-15*a*?

42. Would you expect the color of a Cepheid variable star (see Figure 19-18) to change during the star's oscillation period? If not, why not? If so, describe why the color should change, and describe the color changes you would expect to see during an oscillation period.

43. The brightness of a certain Cepheid variable star increases and decreases with a period of 10 days. (**a**) What must this star's luminosity be if its spectrum has strong absorption lines of hydrogen and helium, but no strong absorption lines of heavy elements? (**b**) Repeat part (a) for the case in which the star's spectrum also has strong absorption lines of heavy elements.

44. The star X Arietis is an RR Lyrae variable. Its apparent brightness varies between 2.0×10^{-15} and 4.9×10^{-15} that of the Sun with a period of 0.65 day. Interstellar extinction dims the star by 37%. Approximately how far away is the star?

45. The apparent brightness of δ Cephei (a Type I Cepheid variable) varies with a period of 5.4 days. Its average apparent brightness is 5.1×10^{-13} that of the Sun. Approximately how far away is δ Cephei? (Ignore interstellar extinction.)

46. Suppose you find a binary star system in which the more massive star is a red giant and the less massive star is a main-sequence star. Would you expect that mass transfer between the stars has played an important role in the evolution of these stars? Explain your reasoning.

47. The larger star in the Algol binary system (see Figure 19-21*a*) is of spectral class K, while the smaller star is of spectral class B. Discuss how the color of Algol changes as seen through a small telescope (through which Algol appears as a single star). What is the color during a deep eclipse, when the large star eclipses the small one? What is the color when the small star eclipses the large one?

48. Suppose the detached star in β Lyrae (Figure 19-21*b*) did not have an accretion disk. Would the deeper dips in the light curve be deeper, shallower, or about the same? What about the shallower dip? Explain your answers.

49. The two stars that make up the overcontact binary W Ursae Majoris (Figure 19-21*c*) have estimated masses of $0.99\ M_\odot$ and $0.62\ M_\odot$. (**a**) Find the average separation between the two stars. Give your answer in kilometers. (**b**) The radii of the two stars are estimated to be $1.14\ R_\odot$ and $0.83\ R_\odot$. Show that these values and your result in part (a) are consistent with the statement that this is an overcontact binary.

50. The stars that make up the binary system W Ursae Majoris (see Figure 19-21*c*) have particularly strong magnetic fields. Explain how astronomers could have discovered this. (*Hint:* See Section 16-9.)

51. Consult recent issues of *Sky & Telescope* and *Astronomy* to find out when Mira will next reach maximum brightness. Look up the star's location in the sky using the *Starry Night Enthusiast*™ program on the CD-ROM that comes with selected printed copies of this book. (Use the **Find . . .** command in the **Edit** menu to search for Omicron Ceti.) Why is it unlikely that you will be able to observe Mira at maximum brightness?

Discussion Questions

52. Eventually the Sun's luminosity will increase to the point where Earth can no longer sustain life. Discuss what measures a future civilization might take to preserve itself from such a calamity.

53. The half-life of the ^{8}Be nucleus, 2.6×10^{-16} second, is the average time that elapses before this unstable nucleus decays into two alpha particles. How would the universe be different if instead the ^{8}Be half-life were zero? How would the universe be different if the ^{8}Be nucleus were stable and did not decay?

54. Discuss how H-R diagrams of star clusters could be used to set limits on the age of the universe. Could they be used to set lower limits on the age? Could they be used to set upper limits? Explain your reasoning.

Web/eBook Questions

55. Suppose that an oxygen nucleus (^{16}O) were fused with a helium nucleus (^{4}He). What element would be formed? Look up the relative abundance of this element in, for example, the *Handbook of Chemistry and Physics* or on the World Wide Web. Based on the abundance, comment on whether such a process is likely. (*Hint:* See Figure 8-4.)

56. Although Polaris, the North Star, is a Cepheid variable, it pulsates in a somewhat different way than other Cepheids. Search the World Wide Web for information about this star's pulsations and how they have been measured by astronomers at the U.S. Naval Observatory. How does Polaris pulsate? How does this differ from other Cepheids?

57. **Observing Stellar Evolution.** Step through the animation "The Hertzsprung-Russell Diagram and Stellar Evolution" in Chapter 19 of the *Universe* Web site or eBook. Use this animation to answer

the following questions. (**a**) How does a 1-$M_\odot$ star move on the H-R diagram during its first 4.56 billion (4560 million) years of existence? Compare this with the discussion in Section 19-1 of how the Sun has evolved over the past 4.56 billion years. (**b**) What is the zero-age spectral class of a 2-$M_\odot$ star? At what age does such a star evolve into a red giant of spectral class K? (**c**) What is the approximate zero-age luminosity of a 1.3-$M_\odot$ star? What is its approximate luminosity when it becomes a red giant? (**d**) Suppose a star cluster has no main-sequence stars of spectral classes O or B. What is the approximate age of the cluster? (**e**) Approximately how long do the most massive stars of spectral class B live before leaving the main sequence? What about the most massive stars of spectral class F?

Activities

Observing Projects

Observing tips and tools

An excellent resource for learning how to observe variable stars is the Web site of the American Association of Variable Star Observers. A wealth of data about specific variable stars can be found on the *Sky & Telescope* Web site and in the three volumes of *Burnham's Celestial Handbook: An Observer's Guide to the Universe Beyond the Solar System* (Dover, 1978). This book also provides useful information about observing star clusters.

58. Observe several of the red giants and supergiants listed below with the naked eye and through a telescope. (Note that γ Andromedae, α Tauri, and ε Pegasi are all multiple-star systems. The spectral type and luminosity class refer to the brightest star in these systems.) You can locate these stars in the sky using the *Starry Night Enthusiast*™ program on the CD-ROM that comes with selected printed copies of this book. Is the reddish color of these stars apparent when they are compared with neighboring stars?

Star	Spectral type	R.A.	Decl.
Almach (γ Andromedae)	K3 II	2^h 03.9^m	+42° 20′
Aldebaran (α Tauri)	K5 III	4 35.9	+16 31
Betelgeuse (α Orionis)	M2 I	5 55.2	+07 24
Arcturus (α Boötis)	K2 III	14 15.7	+19 11
Antares (α Scorpii)	M1 I	16 29.5	−26 26
Eltanin (γ Draconis)	K5 III	17 56.7	+51 29
Enif (ε Pegasi)	K2 I	21 44.2	+09 52

Note: The right ascensions and declinations are given for epoch 2000.

59. Several of the open clusters referred to in Figure 19-14 can be seen with a good pair of binoculars. Observe as many of these clusters as you can, using both a telescope and binoculars. (Some are ac-tually so large that they will not fit in the field of view of many telescopes.) You can locate these clusters in the sky using the *Starry Night Enthusiast*™ program on the CD-ROM that comes with selected printed copies of this book. Note that in the following table, the M prefix refers to the Messier Catalog, NGC refers to the New General Catalogue, and Mel refers to the Melotte Catalog. In making your own observations, note the overall distribution of stars in each cluster. Which clusters are seen better through binoculars than through a telescope? Which clusters can you see with the naked eye?

Star cluster	Constellation	R.A.	Decl.
h Persei (NGC 869)	Perseus	2^h 19.0^m	+57° 09′
χ Persei (NGC 884)	Perseus	2 22.4	+57 07
Pleiades (M45)	Taurus	3 47.0	+24 07
Hyades (Mel 25)	Taurus	4 27	+16 00
Praesepe (M44)	Cancer	8 40.1	+19 59
Coma (Mel 111)	Coma Berenices	12 25	+26 00
Wild Duck (M11)	Scutum	18 51.1	−06 16

Note: The right ascensions and declinations are given for epoch 2000.

60. There are many beautiful globular clusters scattered around the sky that can be easily seen with a small telescope. Several of the brightest and nearest globulars are listed below. You can locate these clusters in the sky using the *Starry Night Enthusiast*™ program on the CD-ROM that comes with selected printed copies of this book. Observe as many of these globular clusters as you can. Can you distinguish individual stars toward the center of each cluster? Do you notice any differences in the overall distribution of stars between clusters?

Star cluster	Constellation	R.A.	Decl.
M3 (NGC 5272)	Canes Venatici	13^h 42.2^m	+28° 23′
M5 (NGC 5904)	Serpens	15 18.6	+2 05
M4 (NGC 6121)	Scorpius	16 23.6	−26 32
M13 (NGC 6205)	Hercules	16 41.7	+36 28
M12 (NGC 6218)	Ophiuchus	16 47.2	−1 57
M28 (NGC 6626)	Sagittarius	18 24.5	−24 52
M22 (NGC 6656)	Sagittarius	18 36.4	−23 54
M55 (NGC 6809)	Sagittarius	19 40.0	−30 58
M15 (NGC 7078)	Pegasus	21 30.0	+12 10

Note: The right ascensions and declinations are given for epoch 2000.

61. Use the *Starry Night Enthusiast*™ program to view some of the objects described in this chapter. Click the **Home** button in the toolbar. **Stop time** flow and change the **Time** in the toolbar to **12:00:00 A.M.** Use the **Find** pane to locate the giant star Aldebaran, the open cluster M44, and the globular cluster M12. (**a**) Which of these objects are visible from your home location at midnight tonight? If the object is visible from your home location at midnight, click the menu button next to the object's name in the Find pane and select the **Show Info** option. (**b**) For each object that is visible, use the Zoom controls at the upper right of the toolbar to get the best view. Describe the appearance of the object. (**c**) For each object that is visible, in which direction of the compass would you have to look at midnight to see it (that is, what is its *azimuth*)? How far above the horizon would you have to look (that is, what is its *altitude*)? (Look under the **Position in Sky** layer of the **Info** pane for this information.)

62. Use the ***Starry Night Enthusiast*™** program to look for signs of stellar evolution in our Milky Way Galaxy. In the menu, select **Favorites > Stars > Sun in Milky Way.** You can zoom in or out usingthe buttons at the right of the toolbar. You can move your viewpoint around the Galaxy by holding down the **Shift** key and the mouse button as you drag the mouse. (**a**) What is the color of the central part of the Galaxy? Based on the color of this region, do you expect that there are many massive main-sequence stars there? Would you expect to find many young stars there? Explain your reasoning. (**b**) Repeat part (a) for the outer regions of the Galaxy.

Collaborative Exercises

63. The inverse relationship between a star's mass and its main-sequence lifetime is sometimes likened to automobiles in that the more massive vehicles, such as commercial semi–tractor-trailer trucks, need to consume significantly more fuel to travel at highway speeds than more lightweight and economical vehicles. As a group, create a table called "Maximum Vehicle Driving Distances," much like Table 19-1, "Main-Sequence Lifetimes," by making estimates for any five vehicles of your groups' choosing. The table's column headings should be (1) vehicle make and model; (2) estimated gas tank size; (3) cost to fill tank; (4) estimated mileage (in miles per gallon); and (5) number of miles driven on a single fill-up.

64. Consider Figure 19-19, showing the period-luminosity relation for Cepheids. If the length of time it has been since someone in your group last purchased milk or juice at the grocery store was identical to the pulsation period of a Type I Cepheid, how much longer a pulsation period would a Type II Cepheid need in order to reach the same luminosity as the Type I Cepheid in this time frame?

65. Figure 19-18*a* shows a light curve of apparent magnitude versus time in days for δ Cephei—a pulsating star that reaches maximum brightness every 5.4 days. Create a new sketch of apparent magnitude versus time in days showing three different stars: (1) δ Cephei; (2) a slightly smaller pulsating star; and (3) a slightly larger pulsating star, all of which have about the same total change in apparent magnitude.

20 Stellar Evolution: The Deaths of Stars

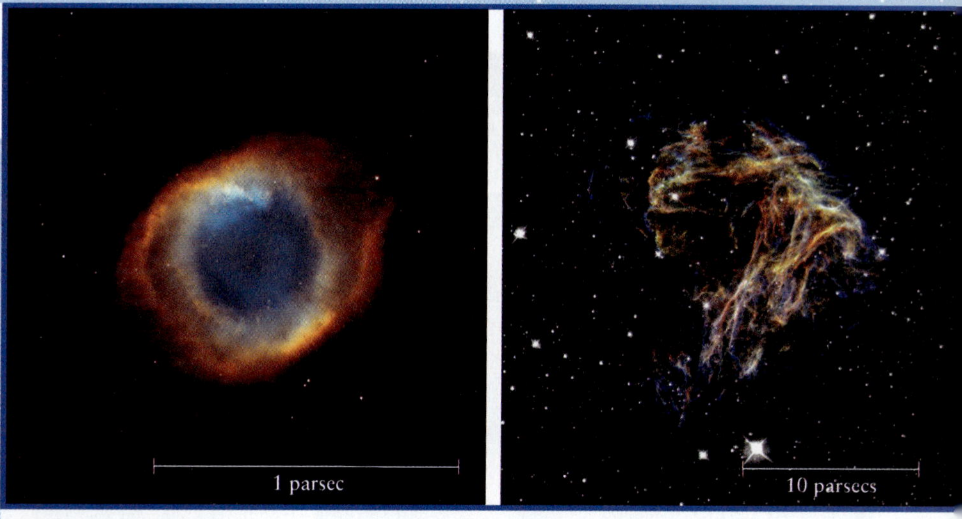

(a) A planetary nebula (b) A supernova remnant

WEB LINK 20.1

R I **V** U X G

Left: The planetary nebula NGC 7293. (the Helix Nebula) Right: The supernova remnant LMC N49. (NASA, NOAO, ESA, the Hubble Helix Nebula Team, M. Meixner/STScI, and T. A. Rector/NRAO; NASA and the Hubble Heritage Team, STScI/AURA)

When a star of 0.4 solar mass or more reaches the end of its main-sequence lifetime and becomes a red giant, it has a compressed core and a bloated atmosphere. Finally, the star devours its remaining nuclear fuel and begins to die. As we'll learn in this chapter, the character of the star's death depends crucially on the value of its mass.

A star of relatively low mass—such as our own Sun—ends its evolution by gently expelling its outer layers into space. These ejected gases form a glowing cloud called a *planetary nebula* such as the one shown here in the left-hand image. The burned-out core that remains is called a *white dwarf.*

In contrast, a high-mass star ends its life in almost inconceivable violence. At the end of its short life, the core of such a star collapses suddenly, which triggers a powerful *supernova* explosion that can be as luminous as an entire galaxy of stars. A white dwarf, too, can become a supernova if it accretes gas from a companion star in a close binary system.

Thermonuclear reactions in supernovae produce a wide variety of heavy elements, which are ejected into the interstellar medium. (The supernova remnant shown here in the right-hand image is rich in these elements.) Such heavy elements are essential building blocks for terrestrial worlds like our Earth. Thus, the deaths of massive stars can provide the seeds for planets orbiting succeeding generations of stars.

20-1 Stars of between 0.4 and 4 solar masses go through two distinct red-giant stages

TUTORIAL 20.1

All main-sequence stars convert hydrogen to helium in their cores in a series of energy-releasing thermonuclear reactions. As we saw in Section 19-1, convection within a low-mass main-sequence star—a so-called *red dwarf* with a mass between 0.08 and 0.4 $M_{\odot}$—will eventually bring all of the star's hydrogen into the core. Over hundreds of billions of years a red dwarf evolves into an inert ball of helium. Convection is less important in main-sequence stars with masses greater than 0.4 $M_{\odot}$, so these stars are able to consume only the hydrogen that is present within the core. These stars of greater mass then leave the main sequence. Let's examine what happens next for a star of moderately low mass, between 0.4 and 4 $M_{\odot}$. One example for such a star is our own Sun, with a mass of 1 $M_{\odot}$. We'll begin by

Learning Goals

By reading the sections of this chapter, you will learn

20-1 What kinds of thermonuclear reactions occur inside a star of moderately low mass as it ages

20-2 How evolving stars disperse carbon into the interstellar medium

20-3 How stars of moderately low mass eventually die

20-4 The nature of white dwarfs and how they are formed

20-5 What kinds of reactions occur inside a high-mass star as it ages

20-6 How high-mass stars explode and die

20-7 Why supernova SN 1987A was both important and unusual

20-8 What role neutrinos play in the death of a massive star

20-9 How white dwarfs in close binary systems can explode

20-10 What remains after a supernova explosion

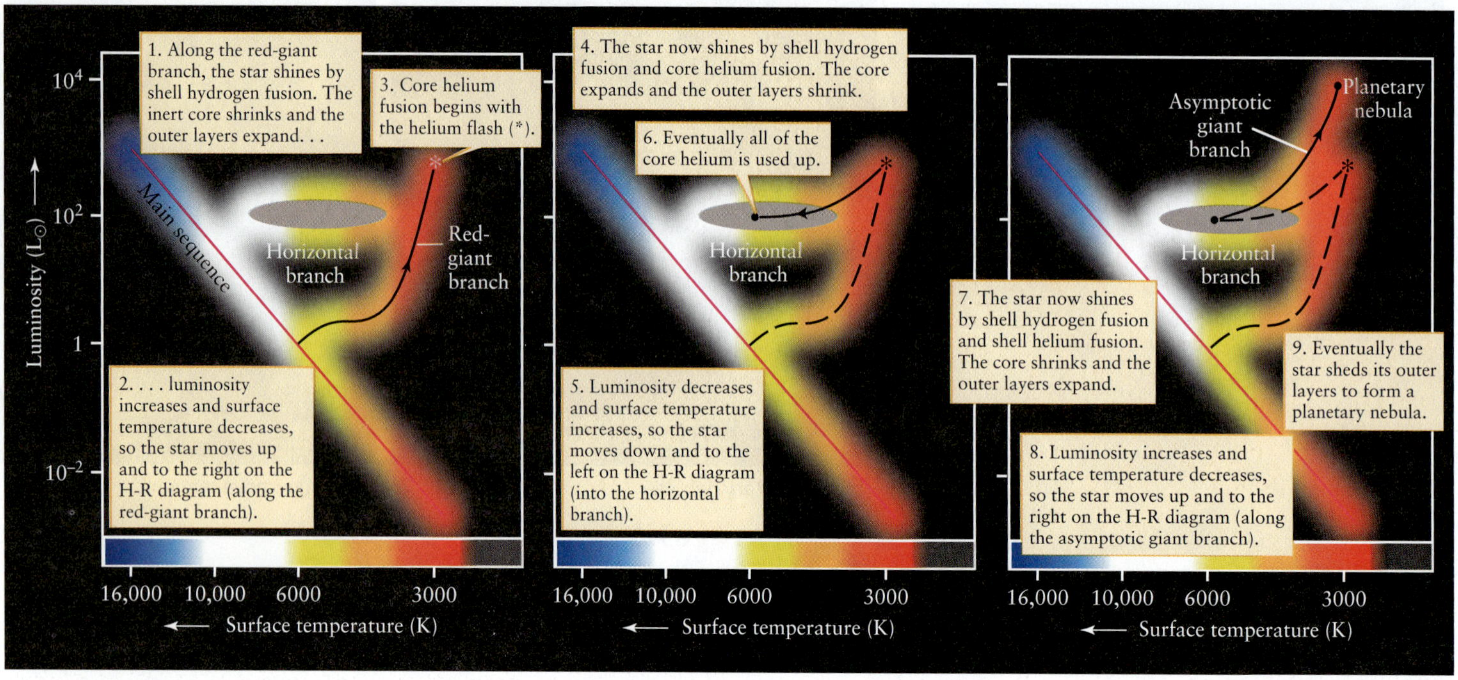
1. Along the red-giant branch, the star shines by shell hydrogen fusion. The inert core shrinks and the outer layers expand. . .
2. . . . luminosity increases and surface temperature decreases, so the star moves up and to the right on the H-R diagram (along the red-giant branch).
3. Core helium fusion begins with the helium flash (*).
4. The star now shines by shell hydrogen fusion and core helium fusion. The core expands and the outer layers shrink.
5. Luminosity decreases and surface temperature increases, so the star moves down and to the left on the H-R diagram (into the horizontal branch).
6. Eventually all of the core helium is used up.
7. The star now shines by shell hydrogen fusion and shell helium fusion. The core shrinks and the outer layers expand.
8. Luminosity increases and surface temperature decreases, so the star moves up and to the right on the H-R diagram (along the asymptotic giant branch).
9. Eventually the star sheds its outer layers to form a planetary nebula.
Main sequence
Horizontal branch
Red-giant branch
Asymptotic giant branch
Planetary nebula
Luminosity (L⊙)
10^4
10^2
1
10^-2
16,000 10,000 6000 3000
Surface temperature (K)

(a) Before the helium flash: A red-giant star (b) After the helium flash: A horizontal-branch star (c) After core helium fusion ends: An AGB star

ANIMATION 20.1

Figure 20-1

The Post-Main-Sequence Evolution of a 1-$M_\odot$ Star These H-R diagrams show the evolutionary track of a star like the Sun as it goes through the stages of being **(a)** a red-giant star, **(b)** a horizontal-branch star, and **(c)** an asymptotic giant branch (AGB) star. The star eventually evolves into a planetary nebula (described in Section 20-3).

reviewing what we learned in Chapter 19 about the first stages of post–main-sequence evolution for such a star. (Later in this chapter we'll study the evolution of more massive stars.)

The Red-Giant and Horizontal-Branch Stages: A Review

We can describe a star's post–main-sequence evolution using an evolutionary track on a H-R diagram. Figure 20-1 shows the track for a 1-$M_\odot$ star like the Sun. Once core hydrogen fusion ceases, the core shrinks, heating the surrounding hydrogen and triggering shell hydrogen fusion. The new outpouring of energy causes the star's outer layers to expand and cool, and the star becomes a red giant. As the luminosity increases and the surface temperature drops, the post–main-sequence star moves up and to the right along the **red-giant branch** on an H-R diagram (Figure 20-1*a*).

Next, the helium-rich core of the star shrinks and heats until eventually **core helium fusion** begins. This second post–main-sequence stage begins gradually in stars more massive than about 2–3 $M_\odot$, but for less massive stars it comes suddenly—in a *helium flash*. During core helium fusion, the surrounding hydrogen-fusing shell still provides most of the red giant's luminosity.

As we learned in Section 19-3, the core expands when core helium fusion begins, which makes the core cool down a bit. (We saw in Box 19-1 that letting a gas expand tends to lower its temperature, while compressing a gas tends to increase its temperature.) The cooling of the core also cools the surrounding hydrogen-fusing shell, so that the shell releases energy more slowly. Hence, the luminosity goes down a bit after core helium fusion begins.

The slower rate of energy release also lets the star's outer layers contract. As they contract, they heat up, so the star's surface temperature increases and its evolutionary track moves to the left on the H-R diagram in Figure 20-1*b*. The luminosity changes relatively little during this stage, so the evolutionary track moves almost horizontally, along a path called the **horizontal branch.** Horizontal-branch stars have helium-fusing cores surrounded by hydrogen-fusing shells. Figure 19-12 shows horizontal-branch stars in a globular cluster, and Figure 19-8 shows the evolution of the luminosity of a 1-$M_\odot$ star up to this point in its history.

AGB Stars: The Second Red-Giant Stage

Helium fusion produces nuclei of carbon and oxygen. After about a hundred million (10^8) years of core helium fusion, essentially all the helium in the core of a 1-$M_\odot$ star has been converted into carbon and oxygen, and the fusion of helium in the core ceases. (This corresponds to the right-hand end of the graph in Figure 19-8.) Without thermonuclear reactions to maintain the core's internal pressure, the core again contracts, until it is stopped by degenerate-electron pressure (described in Section 19-3). This contraction releases heat into the surrounding helium-rich gases, and

Stars like the Sun go through a second red-giant phase, during which helium fusion takes place in a shell around an inert core

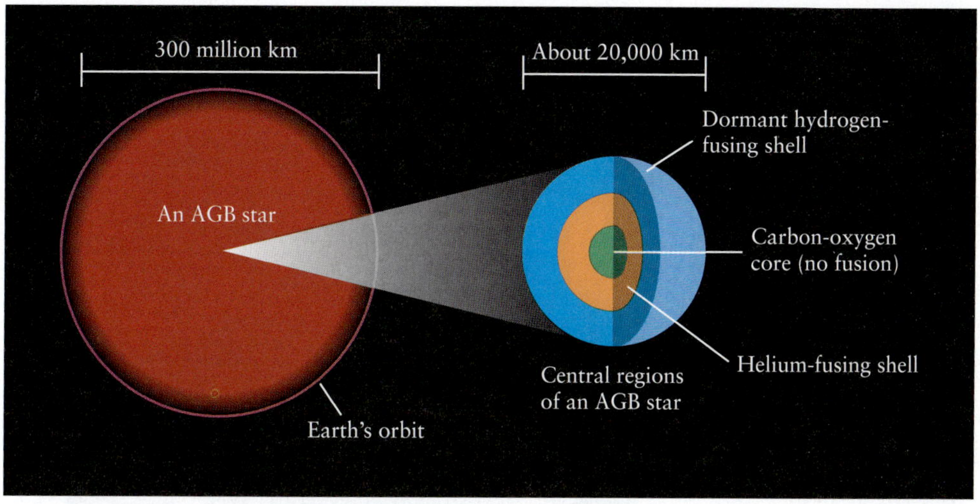

Figure 20-2

The Structure of an Old, Moderately Low-Mass AGB Star Near the end of its life, a star like the Sun becomes an immense, red, asymptotic giant branch (AGB) star. The star's inert core, active helium-fusing shell, and dormant hydrogen-fusing shell are all contained within a volume roughly the size of Earth. Thermonuclear reactions in the helium-fusing shell are so rapid that the star's luminosity is thousands of times that of the present-day Sun. (The relative sizes of the shells in the star's interior are not shown to scale.)

a new stage of helium fusion begins in a thin shell around the core. This process is called **shell helium fusion.**

History now repeats itself—the star enters a *second* red-giant phase. A star first becomes a red giant at the end of its main-sequence lifetime, when the outpouring of energy from shell hydrogen fusion makes the star's outer layers expand and cool. In the same way, the outpouring of energy from shell *helium* fusion causes the outer layers to expand again. The low-mass star ascends into the red-giant region of the H-R diagram for a second time (Figure 20-1*c*), but now with even greater luminosity than during its first red-giant phase.

Stars in this second red-giant phase are commonly called **asymptotic giant branch stars,** or **AGB stars,** and their evolutionary tracks follow what is called the **asymptotic giant branch.** (*Asymptotic* means "approaching"; the name means that a star on the asymptotic giant branch approaches the red-giant branch from the left on an H-R diagram.)

When a low-mass star first becomes an AGB star, it consists of an inert, degenerate carbon-oxygen core and a helium-fusing shell, both inside a hydrogen-fusing shell, all within a volume not much larger than Earth. This small, dense central region is surrounded by an enormous hydrogen-rich envelope about as big as Earth's orbit around the Sun. After a while, the expansion of the star's outer layers causes the hydrogen-fusing shell to also expand and cool, and thermonuclear reactions in this shell temporarily cease. This leaves the aging star's structure as shown in Figure 20-2.

We saw in Section 19-1 that the more massive a star, the shorter the amount of time it remains on the main sequence. Similarly, the greater the star's mass, the more rapidly it goes through the stages of post–main-sequence evolution. Hence, we can see all of these stages by studying star clusters, which contain stars that are all the same age but that have a range of masses (see Section 19-4). Figure 20-3 shows a color-magnitude diagram for the globular cluster M55, which is at least 13 billion years old. The least massive stars in this cluster are still on the main sequence. Progressively more massive stars have evolved to the red giant branch, the horizontal branch, and the asymptotic giant branch.

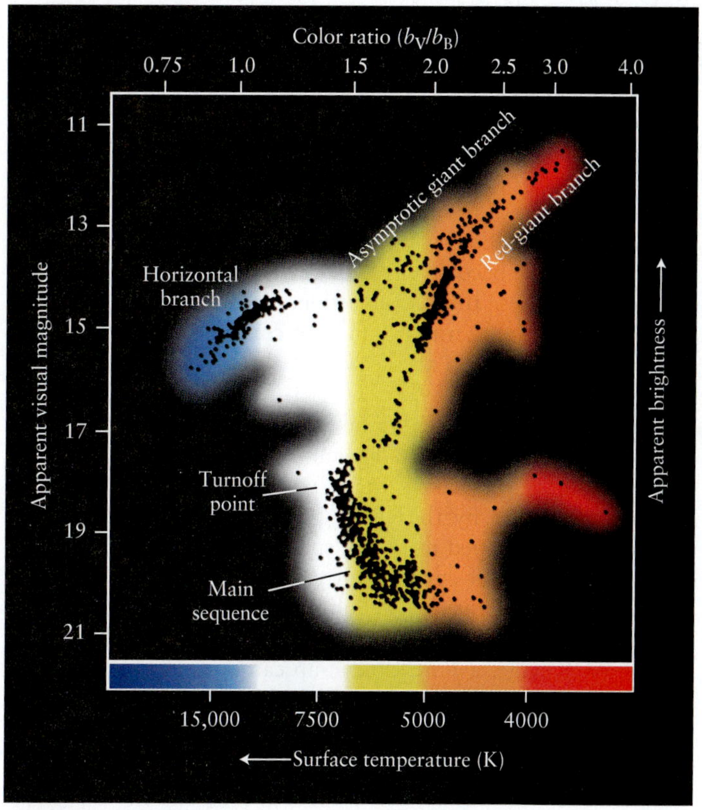

WEB LINK 20.2

Figure 20-3

Stellar Evolution in a Globular Cluster In the old globular cluster M55, stars with masses less than about 0.8 $M_\odot$ are still on the main sequence, converting hydrogen into helium in their cores. Slightly more massive stars have consumed their core hydrogen and are ascending the red-giant branch; even more massive stars have begun helium core fusion and are found on the horizontal branch. The most massive stars (which still have less than 4 $M_\odot$) have consumed all the helium in their cores and are ascending the asymptotic giant branch. (Compare with Figure 21-11.) (Adapted from D. Schade, D. VandenBerg, and F. Hartwick)

A 1-M$_\odot$ AGB star can reach a maximum luminosity of nearly 10^4 L$_\odot$, as compared with approximately 10^3 L$_\odot$ when it reached the helium flash and a relatively paltry 1 L$_\odot$ during its main-sequence lifetime. When the Sun becomes an AGB star some 7.8 billion years from now, this tremendous increase in luminosity will cause Mars and the Jovian planets to largely evaporate away. The Sun's bloated outer layers will reach to Earth's orbit. Mercury and perhaps Venus will simply be swallowed whole.

20-2 Dredge-ups bring the products of nuclear fusion to a giant star's surface

As we saw in Section 16-2, energy is transported outward from a star's core by one of two processes—radiative diffusion or convection. The first is the passage of energy in the form of electromagnetic radiation, and it dominates only when a star's gases are relatively transparent. The second involves up-and-down movement of the star's gases. Convection plays a very important role in giant stars, and it helps supply the cosmos with the elements essential to life.

Convection, Dredge-ups, and Carbon Stars

ANIMATION 20.2

In the Sun, convection dominates only the outer layers, from around 0.71 solar radius (measured from the center of the Sun) up to the photosphere (recall Figure 16-4). During the final stages of a star's life, however, the convective zone can become so broad that it extends down to the star's core. At these times, convection can "dredge up" the heavy elements produced in and around the core by thermonuclear fusion, transporting them all the way to the star's surface.

LOOKING DEEPER 20.1

The *first* **dredge-up** takes place after core hydrogen fusion stops, when the star becomes a red giant for the first time. Convection dips so deeply into the star that material processed by the CNO cycle of hydrogen fusion (see Section 16-1) is carried up to the star's surface, changing the relative abundances of carbon, nitrogen, and oxygen. A *second* dredge-up occurs after core helium fusion ceases, further altering the abundances of carbon, nitrogen, and oxygen. Still later, during the AGB stage, a *third* dredge-up can occur if the star has a mass greater than about 2 M$_\odot$. This third dredge-up transports large amounts of freshly synthesized carbon to the star's surface, and the star's spectrum thus exhibits prominent absorption bands of carbon-rich molecules like C_2, CH, and CN. For this reason, an AGB star that has undergone a third dredge-up is called a **carbon star.**

All AGB stars have very strong stellar winds that cause them to lose mass at very high rates, up to 10^{-4} M$_\odot$ per year (a thousand times greater than that of a red giant, and 10^{10} times greater than the rate at which our present-day Sun loses mass). The surface temperature of AGB stars is relatively low, around 3000 K, so any ejected carbon-rich molecules can condense to form tiny grains of soot. Indeed, carbon stars are commonly found to be obscured in sooty cocoons of ejected matter (Figure 20-4).

The carbon that forms the basis of all life on Earth was ejected billions of years ago from giant stars

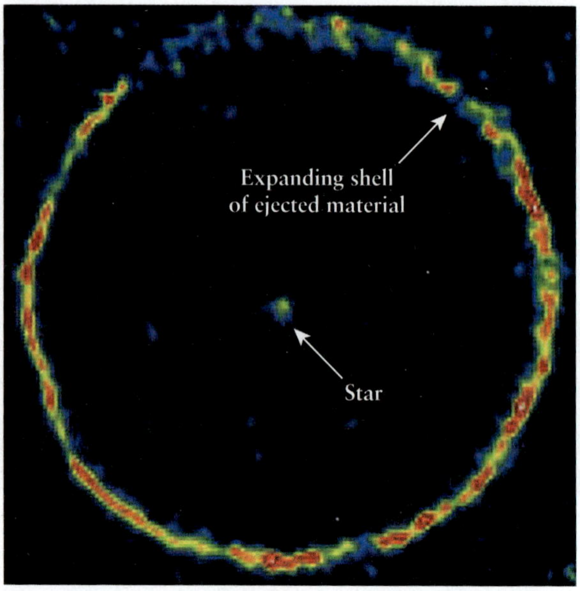

WEB LINK 20.3

Figure 20-4 R I V U X G

A Carbon Star TT Cygni is an AGB star in the constellation Cygnus that ejects some of its carbon-rich outer layers into space. Some of the ejected carbon combines with oxygen to form molecules of carbon monoxide (CO), whose emissions can be detected with a radio telescope. This radio image shows the CO emissions from a shell of material that TT Cygni ejected some 7000 years ago. Over that time, the shell has expanded to a diameter of about $\frac{1}{2}$ light-year. (H. Olofsson, Stockholm Observatory, et al./NASA)

Carbon stars are important because they enrich the interstellar medium with carbon and some nitrogen and oxygen. The triple alpha process that occurs in helium fusion is the *only* way that carbon can be made, and carbon stars are the primary avenue by which this element is dispersed into interstellar space. Indeed, most of the carbon in your body was produced long ago inside a star by the triple alpha process (see Section 19-3). This carbon was later dredged up to the star's surface and ejected into space. Some 4.56 billion years ago a clump of the interstellar medium which contained this carbon coalesced into the solar nebula from which our Earth—and all of the life on it—eventually formed. In this sense you can think of your body as containing "recycled" material—substances that were once in the heart of a star that formed and evolved long before our solar system existed.

20-3 Stars of moderately low mass die by gently ejecting their outer layers, creating planetary nebulae

For a star that began with a moderately low mass (between about 0.4 and 4 M$_\odot$), the AGB stage in its evolution is a dramatic turning point. Before this stage, a star loses mass only gradually through steady stellar winds. But as it evolves during its AGB stage, a star divests itself completely of its outer layers. The aging star undergoes a series of bursts in luminosity, and in each burst it ejects a shell of material into space. (The shell around the AGB star TT Cygni, shown in Figure 20-4, was probably created in this way.) Eventually, all that remains of a low-mass star is a fiercely hot, exposed core, surrounded by glowing shells of ejected gas. This late stage in the life of a star is called a **planetary nebula.** The left-hand image on the opening page of this chapter shows one such planetary nebula, called the Ring Nebula for its shape.

An aging AGB star casts off much of the mass that it possesses and makes a beautiful glowing nebula.

Making a Planetary Nebula

To understand how an AGB star can eject its outer layers in shells, consider the internal structure of such a star as shown in Figure 20-2. As the helium in the helium-fusing shell is used up, the pressure that holds up the dormant hydrogen-fusing shell decreases. Hence, the dormant hydrogen shell contracts and heats up, and hydrogen fusion begins anew. This revitalized hydrogen fusion creates helium, which rains downward onto the temporarily dormant helium-fusing shell. As the helium shell gains mass, it shrinks and heats up. When the temperature of the helium shell reaches a certain critical value, it reignites in a **helium shell flash** that is similar to (but less intense than) the helium flash that occurred earlier in the evolution of a low-mass star (see Section 19-3). The released energy pushes the hydrogen-fusing shell outward, making it cool off, so that hydrogen fusion ceases and this shell again becomes dormant. The process then starts over again.

When a helium shell flash occurs, the luminosity of an AGB star increases substantially in a relatively short-lived burst called a **thermal pulse.** Figure 20-5, which is based on a theoretical calculation of the evolution of a 1-$M_\odot$ star, shows that thermal pulses begin when the star is about 12.365 billion years old. The calculations predict that thermal pulses occur at ever-shorter intervals of about 100,000 years.

During these thermal pulses, the dying star's outer layers can separate completely from its carbon-oxygen core. As the ejected material expands into space, dust grains condense out of the cooling gases. Radiation pressure from the star's hot, burned-out core acts on the specks of dust, propelling them further outward, and the star sheds its outer layers altogether. In this way an aging 1-$M_\odot$ star loses as much as 40% of its mass. More massive stars eject even greater fractions of their original mass.

As a dying star ejects its outer layers, the star's hot core becomes exposed. With a surface temperature of about 100,000 K, this exposed core emits ultraviolet radiation intense enough to ionize and excite the expanding shell of ejected gases. These gases therefore glow and emit visible light through the process of fluorescence (see Box 18-1), producing a planetary nebula like those shown in Figure 20-6.

CAUTION! Despite their name, planetary nebulae have nothing to do with planets. This misleading term was introduced in the nineteenth century because these glowing objects looked like distant Jovian planets—a small colored blur—when viewed through the small telescopes then available. The difference between planets and planetary nebulae first became obvious with the advent of spectroscopy: Planets have *absorption* line spectra (see Section 7-3), but the excited gases of planetary nebulae have *emission* line spectra.

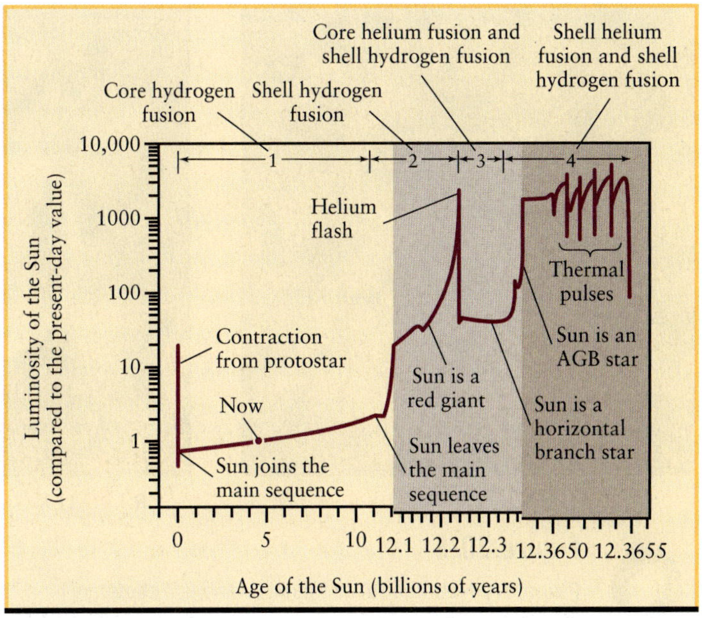

Figure 20-5

Further Stages in the Evolution of the Sun This diagram, which shows how the luminosity of the Sun (a 1-$M_\odot$ star) changes over time, is an extension of Figure 19-8. We use different scales for the final stages because the evolution is so rapid. During the AGB stage there are brief periods of runaway helium fusion, causing spikes in luminosity called thermal pulses. (Adapted from Mark A. Garlick, based on calculations by I.-Juliana Sackmann and Kathleen E. Kramer)

The Properties of Planetary Nebulae

Planetary nebulae are quite common. Astronomers estimate that there are 20,000 to 50,000 planetary nebulae in our Galaxy alone. Many planetary nebulae, such as those in Figure 20-6, are more or less spherical in shape. This shape is a result of the symmetrical way in which the gases were ejected. But if the rate of expansion is not the same in all directions, the resulting nebula can take on an hourglass or dumbbell appearance (Figure 20-7).

Spectroscopic observations of planetary nebulae show emission lines of ionized hydrogen, oxygen, and nitrogen. From the Doppler shifts of these lines, astronomers have concluded that the expanding shell of gas moves outward from a dying star at speeds from 10 to 30 km/s. For a shell expanding at such speeds to have attained the typical diameter of a planetary nebula, about 1 light-year, it must have begun expanding about 10,000 years ago. Thus, by astronomical standards, the planetary nebulae we see today were created only very recently.

We do not observe planetary nebulae that are more than about 50,000 years old. After this length of time, the shell has spread out so far from the cooling central star that its gases cease to glow and simply fade from view. The nebula's gases then mix with the surrounding interstellar medium.

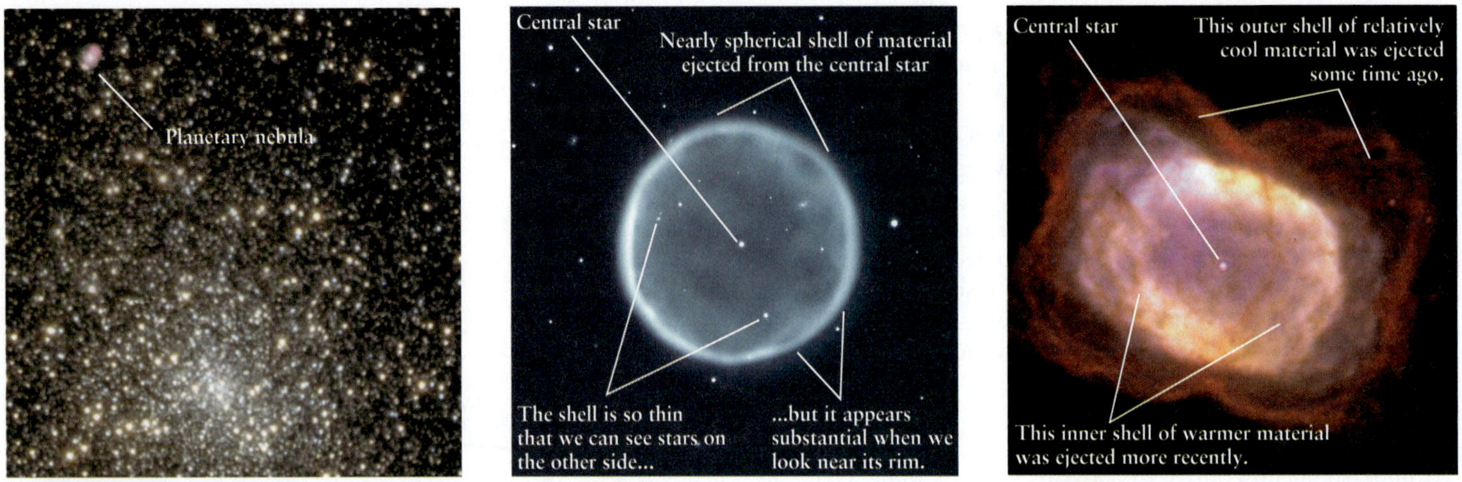

The shell is so thin that we can see stars on the other side...
...but it appears substantial when we look near its rim.

This outer shell of relatively cool material was ejected some time ago.
This inner shell of warmer material was ejected more recently.

R I V U X G   R I V U X G   R I V U X G

Figure 20-6

Planetary Nebulae (a) The pinkish blob is a planetary nebula surrounding a star in the globular cluster M15, about 10,000 pc (33,000 ly) from Earth in the constellation Pegasus. **(b)** The planetary nebula Abell 39 lies about 2200 pc (7000 ly) from Earth in the constellation Hercules. The almost perfectly spherical shell that comprises the nebula is about 1.5 pc (5 ly) in diameter; the thickness of the shell is only about 0.1 pc (0.3 ly). **(c)** This infrared image of the planetary nebula NGC 7027 suggests a more complex evolutionary history than that of Abell 39. NGC 7027 is about 900 pc (3000 ly) from Earth in the constellation Cygnus and is roughly 14,000 AU across. (a: NASA/Hubble Heritage Team, STScI/AURA; b: WIYN/NOAO/NSF; c: William B. Latter, SIRTF Science Center/Caltech, and NASA)

Astronomers estimate that all the planetary nebulae in the Galaxy return a total of about 5 $M_\odot$ to the interstellar medium each year. This amount is about 15% of all the matter expelled by all the various sorts of stars in the Galaxy each year. Because this contribution is so significant, and because the ejected material includes heavier elements (metals) manufactured within a nebula's central star, planetary nebulae play an important role in the chemical evolution of the Galaxy as a whole.

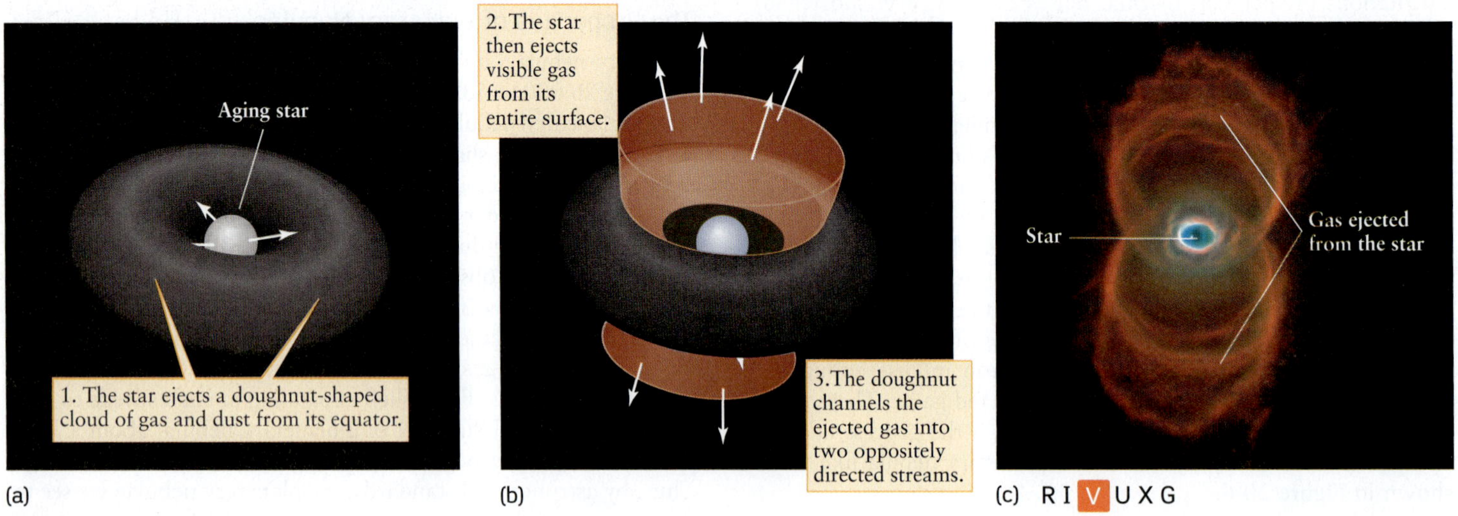

Aging star
1. The star ejects a doughnut-shaped cloud of gas and dust from its equator.
2. The star then ejects visible gas from its entire surface.
3.The doughnut channels the ejected gas into two oppositely directed streams.
Star
Gas ejected from the star

(a) (b) (c) R I V U X G

Figure 20-7

Making an Elongated Planetary Nebula (a), (b) These illustrations show one proposed explanation for why many planetary nebulae have an elongated shape. **(c)** The planetary nebula MyCn18, shown here in false color, may have acquired its elongated shape in this way. It lies some 2500 pc (8000 ly) from Earth in the constellation Musca (the Fly). (R. Sahai and J. Trauger, Jet Propulsion Laboratory; the WFPC-2 Science Team; and NASA)

20-4 The burned-out core of a moderately low-mass star cools and contracts until it becomes a white dwarf

We have seen that after a moderately low-mass star (from about 0.4 to about 4 solar masses) consumes all the hydrogen in its core, it is able to ignite thermonuclear reactions that convert helium to carbon and oxygen. Given sufficiently high temperature and pressure, carbon and oxygen can also undergo fusion reactions that release energy. But for such a moderately low-mass star, the core temperature and pressure never reach the extremely high values needed for these reactions to take place. Instead, as we have seen, the process of mass ejection just strips away the star's outer layers and leaves behind the hot carbon-oxygen core. With no thermonuclear reactions taking place, the core simply cools down like a dying ember. Such a burnt-out relic of a star's former glory is called a **white dwarf.** Such white dwarfs prove to have exotic physical properties that are wholly unlike any object found on Earth.

CAUTION! Unfortunately, the word *dwarf* is used in astronomy for several very different kinds of small objects. Here's a review of the three kinds that we have encountered so far in this book. A *white* dwarf is the relic that remains at the very end of the evolution of a star of initial mass between about 0.4 $M_\odot$ and 4 $M_\odot$. Thermonuclear reactions are no longer taking place in its interior; it emits light simply because it is still hot. A *red* dwarf, discussed in Section 19-1, is a cool main-sequence star with a mass between about 0.08 $M_\odot$ and 0.4 $M_\odot$. The energy emitted by a red dwarf in the form of light comes from its core, where fusion reactions convert hydrogen into helium. Finally, a *brown* dwarf (see Section 8-6 and Section 17-5) is an object like a main-sequence star but with a mass less than about 0.08 $M_\odot$. Because its mass is so small, its internal pressure and temperature are too low to sustain thermonuclear reactions. Instead, a brown dwarf emits light because it is slowly contracting, a process that releases energy (see Section 16-1). White dwarfs are comparable in size to Earth (see Section 17-7); by contrast, brown dwarfs are larger than the planet Jupiter, and red dwarfs are even larger.

Properties of White Dwarfs

You might think that without thermonuclear reactions to provide internal heat and pressure, a white dwarf should keep on shrinking under the influence of its own gravity as it cools. Actually, however, a cooling white dwarf maintains its size, because the burnt-out stellar core is so dense that most of its electrons are degenerate (see Section 17-3). Thus, degenerate-electron pressure supports the star against further collapse. This pressure does not depend on temperature, so it continues to hold up the star even as the white dwarf cools and its temperature drops.

A white dwarf is kept from collapsing by the pressure of its degenerate electrons

Many white dwarfs are found in the solar neighborhood, but all are too faint to be seen with the naked eye. One of the first white dwarfs to be discovered is a companion to Sirius, the brightest star in the night sky. In 1844 the German astronomer Friedrich Bessel noticed that Sirius was moving back and forth slightly, as if it was being orbited by an unseen object. This companion, designated Sirius B (Figure 20-8), was first glimpsed in 1862 by the American astronomer Alvan Clark. Recent Hubble Space Telescope observations at ultraviolet wavelengths, where hot white dwarfs emit most of their light, show that the surface temperature of Sirius B is 25,200 K. (By contrast, the main-sequence star Sirius A has a surface temperature of 10,500 K, while the Sun's surface temperature is a relatively frosty 5800 K.)

Observations of white dwarfs in binary systems like Sirius allow astronomers to determine the mass, radius, and density of these stars (see Sections 17-9, 17-10, and 17-11). Such observations show that the density of the degenerate matter in a white dwarf is typically 10^9 kg/m^3 (a million times denser than water). A teaspoonful of white dwarf matter brought to Earth would weigh nearly 5.5 tons—as much as an elephant!

The Mass-Radius Relation for White Dwarfs

As we learned in Section 17-3, degenerate matter has a very different relationship between its pressure, density, and temperature

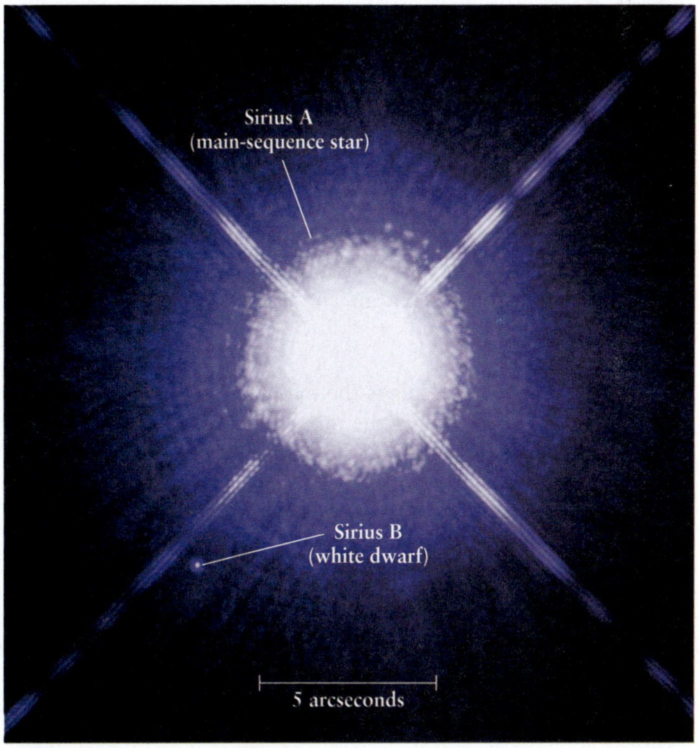

Figure 20-8 R I **V** U X G

Sirius A and Its White Dwarf Companion Sirius, the brightest-appearing star in the sky, is actually a binary star: The secondary star, called Sirius B, is a white dwarf. In this Hubble Space Telescope image, Sirius B is almost obscured by the glare of the overexposed primary star, Sirius A, which is about 10^4 times more luminous than Sirius B. The halo and rays around Sirius A are the result of optical effects within the telescope. (NASA; H. E. Bond and E. Nelan, STScI; M. Barstow and M. Burleigh, U. of Leicester; and J. B. Holberg, U. of Arizona)

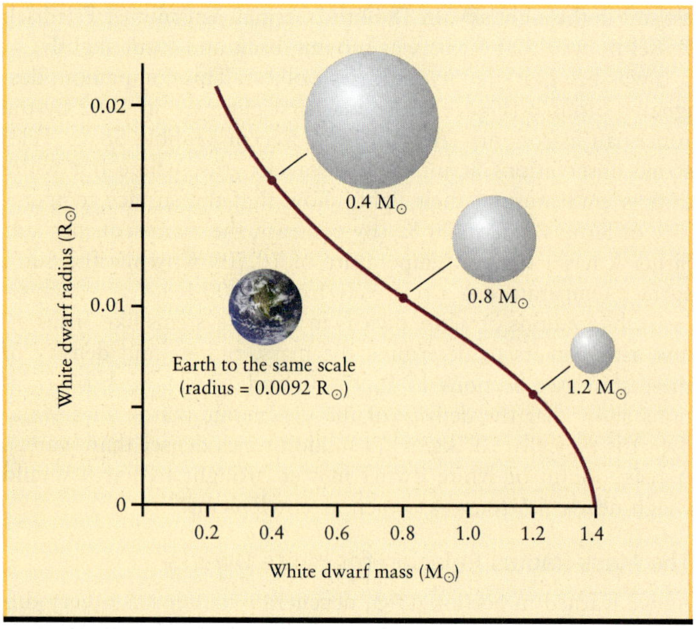

Figure 20-9

The Mass-Radius Relationship for White Dwarfs The more massive a white dwarf is, the smaller its radius. (The drawings of white dwarfs of different mass are drawn to the same scale as the image of Earth.) This unusual relationship is a result of the degenerate-electron pressure that supports the star. The maximum mass of a white dwarf, called the Chandrasekhar limit, is 1.4 $M_\odot$.

than that of ordinary gases. Consequently, white dwarf stars have an unusual **mass-radius relation:** The more massive a white dwarf star, the *smaller* it is.

WEB LINK 20.6

Figure 20-9 displays the mass-radius relation for white dwarfs. Note that the more degenerate matter you pile onto a white dwarf, the smaller it becomes. However, there is a limit to how much pressure degenerate electrons can produce. As a result, there is an upper limit to the mass that a white dwarf can have. This maximum mass is called the **Chandrasekhar limit,** after the Indian-American scientist Subrahmanyan Chandrasekhar, who pioneered theoretical studies of white dwarfs in the 1930s. (The orbiting Chandra X-ray Observatory, described in Section 6-7, is named in his honor.) The Chandrasekhar limit is equal to 1.4 $M_\odot$, meaning that all white dwarfs must have masses less than 1.4 $M_\odot$.

The material inside a white dwarf consists mostly of ionized carbon and oxygen atoms floating in a sea of degenerate electrons. As the dead star cools, the carbon and oxygen ions slow down, and electric forces between the ions begin to prevail over the random thermal motions. About 5×10^9 years after the star first becomes a white dwarf, when its luminosity has dropped to about 10^{-4} $L_\odot$ and its surface temperature is a mere 4000 K, the ions no longer move freely. Instead, they arrange themselves in orderly rows, like an immense crystal lattice. From this time on, you could say that the star is "solid." The degenerate electrons move around freely in this crystalline material, just as electrons move freely through an electrically conducting metal like copper or silver. A diamond is also crystalline (carbon ordered in a crystal lattice), so a cool carbon-oxygen white dwarf resembles an immense spherical diamond!

From Red Giant to Planetary Nebula to White Dwarf

Figure 20-10 shows the evolutionary tracks followed by three burned-out stellar cores as they pass through the planetary nebula stage and become white dwarfs. When these three stars were red giants, they had masses of 0.8, 1.5, and 3.0 $M_\odot$. Mass ejection strips these dying stars of up to 60% of their matter. During their final spasms, the luminosity and surface temperature of these stars change quite rapidly. The points representing these stars on an H-R diagram race along their evolutionary tracks, sometimes executing loops corresponding to thermal pulses (see

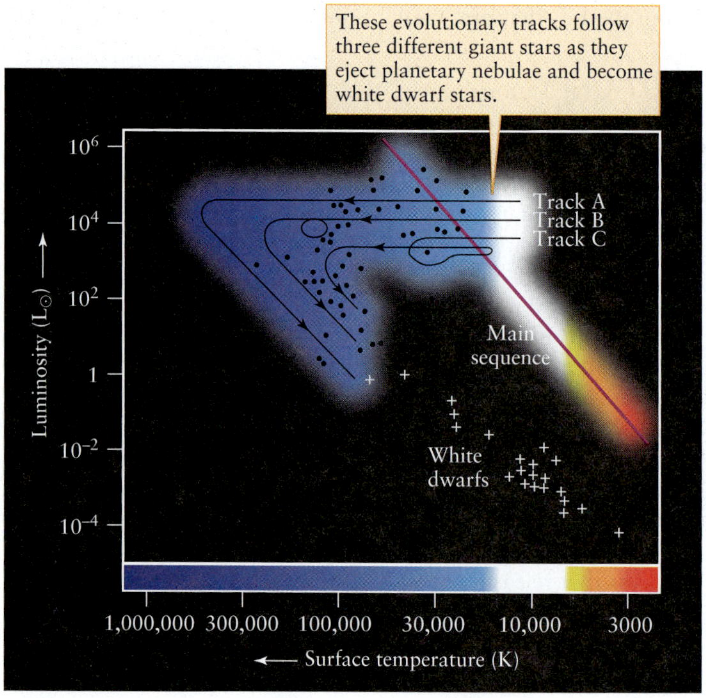

Evolutionary track	Mass ($M_\odot$) Giant star	Ejected nebula	White dwarf
A	3.0	1.8	1.2
B	1.5	0.7	0.8
C	0.8	0.2	0.6

Figure 20-10

Evolution from Giants to White Dwarfs This H-R diagram shows the evolutionary tracks of three low-mass giant stars as they eject planetary nebulae. The table gives the extent of mass loss in each case. The dots represent the central stars of planetary nebulae whose surface temperatures and luminosities have been determined; the crosses represent white dwarfs of known temperature and luminosity. (Adapted from B. Paczynski)

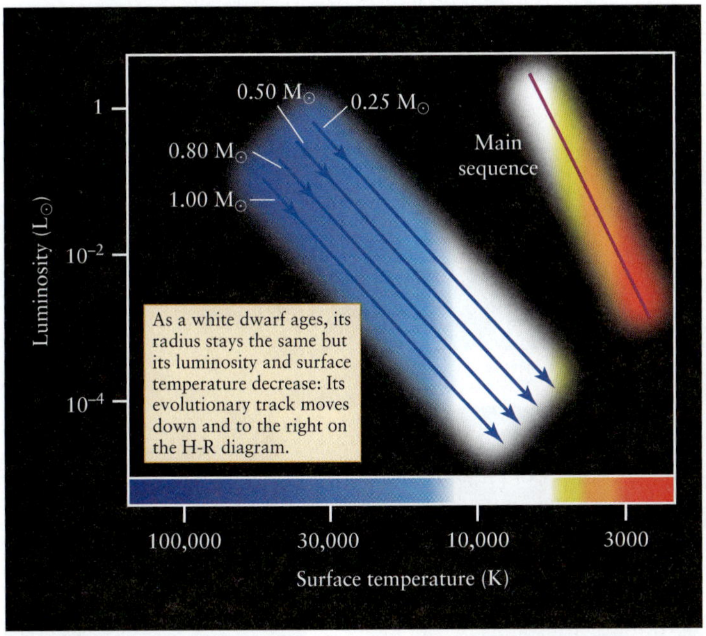

Figure 20-11

White Dwarf "Cooling Curves" As white dwarf stars radiate their internal energy into space, they become dimmer and cooler. The blue lines show the evolutionary tracks of four white dwarfs of different mass: The more massive a white dwarf, the smaller and hence fainter it is. Compare these "cooling curves" with the lines of constant radius in Figure 17-15*b*.

Track B and Track C in Figure 20-10). Finally, as the ejected nebulae fade and the stellar cores cool, the evolutionary tracks of these dying stars take a sharp turn toward the white dwarf region of the H-R diagram. As the table accompanying Figure 20-10 shows, the final white dwarf has only a fraction of the mass of the giant star from which it evolved.

Although a white dwarf maintains the same size as it cools, its luminosity and surface temperature both decrease with time. Consequently, the evolutionary tracks of aging white dwarfs point toward the lower right corner of the H-R diagram. You can see this in Figure 20-10; Figure 20-11 shows it in more detail. The energy that the white dwarf radiates into space comes only from the star's internal heat, which is a relic from the white dwarf's past existence as a stellar core. Over billions of years, white dwarfs grow dimmer and dimmer as their surface temperatures drop toward absolute zero.

After ejecting much of its mass into space, our own Sun will eventually evolve into a white dwarf star about the size of Earth and with perhaps one-tenth of its present luminosity. It will become even dimmer as it cools. After 5 billion years as a white dwarf, the Sun will radiate with no more than one ten-thousandth of its present brilliance. With the passage of eons, our Sun will simply fade into obscurity. The *Cosmic Connections* figure summarizes the full evolutionary cycle of a 1-$M_\odot$ star like the Sun, from its birth as a main-sequence star to its demise as a white dwarf.

20-5 High-mass stars create heavy elements in their cores

During the entire lifetime of a low-mass red dwarf star (with an initial mass less than about 0.4 $M_\odot$), the only thermonuclear reaction that takes place is the fusion of hydrogen nuclei to form helium nuclei. In stars with initial masses from about 0.4 $M_\odot$ to about 4 $M_\odot$, a second kind of thermonuclear reaction takes place—helium fusion. The heaviest elements manufactured by helium fusion are carbon and oxygen.

The life story of a *high-mass* star (with an initial, zero-age mass greater than about 4 $M_\odot$) begins with these same reactions. But theoretical calculations show that high-mass stars can also go through several additional stages of thermonuclear reactions involving the fusion of carbon, oxygen, and other heavy nuclei. As a result, high-mass stars end their lives quite differently from low-mass stars.

Heavy-Element Fusion in Massive Stars

Why is fusion of heavy nuclei possible only in a high-mass star? The reason is that heavy nuclei have large electric charges: For example, a nucleus of carbon has 6 positively charged protons and hence 6 times the charge of a hydrogen nucleus (which has a single proton). This means that there are strong repulsive electric forces that tend to keep these nuclei apart. Only at the great speeds associated with extremely high temperatures can the nuclei travel fast enough to overcome their mutual electric repulsion and fuse together. To produce these very high temperatures at a star's center, the pressure must also be very high. Hence, the star must have a very large mass, because only such a star has strong enough gravity trying to pull it together and thus strong enough pressure at its center.

As we discussed in Section 19-2, when a main-sequence star with a mass greater than about 0.4 $M_\odot$ uses up its core hydrogen, it begins shell hydrogen fusion and enters a red-giant phase. Such a star then begins core helium fusion when the core temperature becomes high enough. The differences between moderately low-mass (from about 0.4 $M_\odot$ to about 4 $M_\odot$) and high-mass stars (more than about 4 $M_\odot$) become pronounced after helium core fusion ends, when the core is composed primarily of carbon and oxygen.

Let us consider how the late stages in the evolution of a high-mass star differ from those of a low-mass star. In low-mass stars, as we saw in Section 20-4, the carbon-oxygen core eventually becomes exposed and becomes a white dwarf. But in stars whose overall mass is more than about 4 $M_\odot$, the carbon-oxygen core is more massive than the Chandrasekhar limit of 1.4 $M_\odot$, so degenerate-electron pressure cannot prevent the core from contracting and heating. Hence, a high-mass star is able to enter a new round of core thermonuclear reactions. When the central temperature of such a high-mass star reaches 600 million kelvins (6×10^8 K), the first of the new thermonuclear reactions, **carbon fusion**, begins. Carbon fusion consumes carbon nuclei (^{12}C, with 6 protons in each nucleus) and produces oxygen (^{16}O, 8 protons), neon (^{20}Ne, 10 protons), sodium (^{23}Na, 11 protons), and magnesium (^{23}Mg and ^{24}Mg, each with 12 protons).

If a star has an even larger main-sequence mass of about 8 $M_\odot$ or so (before mass ejection), even more thermonuclear

COSMIC CONNECTIONS

Our Sun: The Next Eight Billion Years

The Sun is presently less than halfway through its lifetime as a main-sequence star. The H-R diagram and cross-sections on this page summarize the dramatic changes that will take place when the Sun's main-sequence lifetime comes to an end.

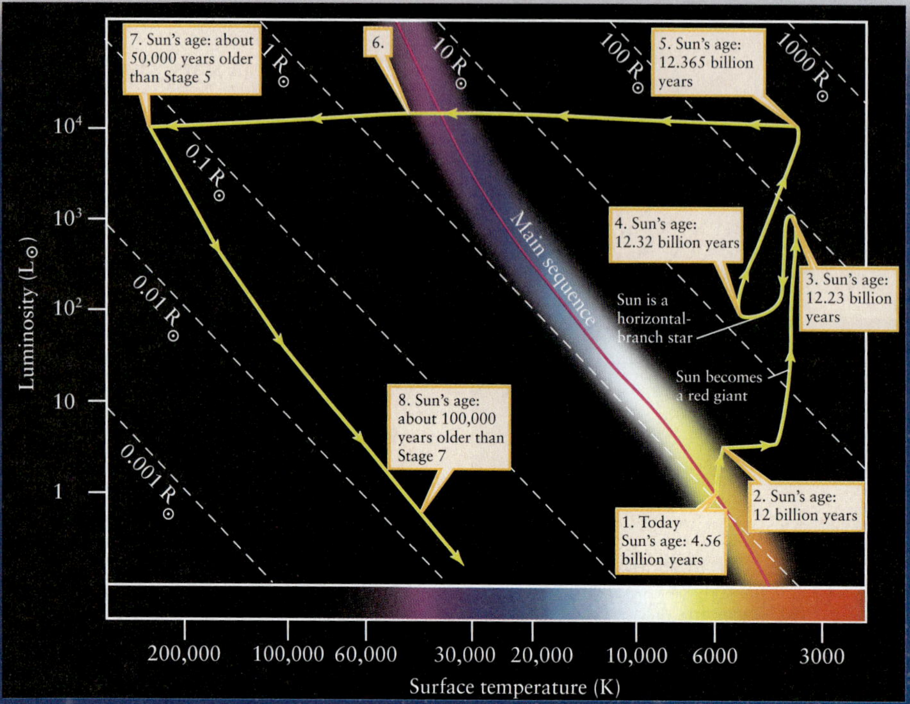

NOTE: The illustrations below do *not* show the dramatic changes in the Sun's radius as it evolves. The sizes of the various layers are not shown to scale.

1. On the main sequence

The present-day Sun is a main-sequence star – in its core, hydrogen fuses to produce helium.

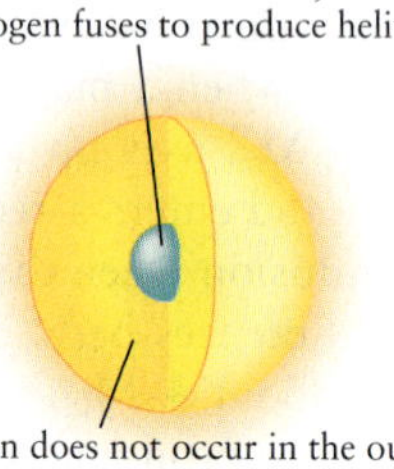

Fusion does not occur in the outer layers (which contain predominantly hydrogen and helium).

2. Becoming a red giant

At the end of the Sun's main-sequence lifetime, fusion stops in the core (which has been converted to helium).

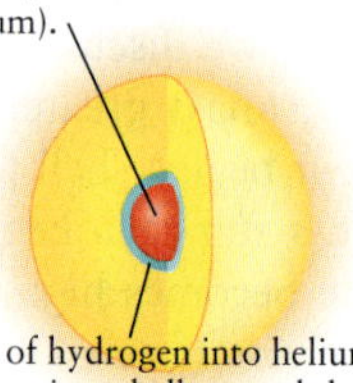

Fusion of hydrogen into helium continues in a shell around the core. The core shrinks, accelerating the fusion reactions in the shell and making the outer layers expand and cool.

3. The helium flash

As the core contracts and heats, the core helium begins to fuse to make carbon and oxygen. The core expands and the rate of energy release slows.

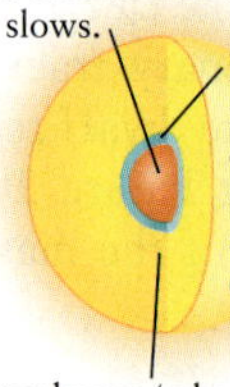

The outer layers (where there are still no fusion reactions) contract and get hotter due to the slower rate of energy release.

4. Beginning the second red giant phase

Once the core helium is consumed, what remains is an inert core of carbon and oxygen. The core again shrinks and gets hotter.

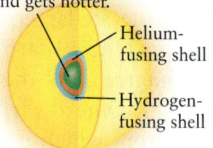

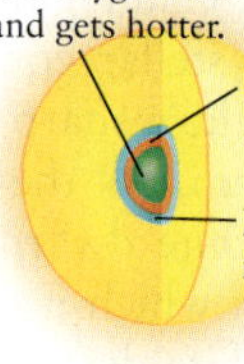

The shrinkage of the core again accelerates fusion reactions in the shells, making the inert outer layers expand and cool.

5. The Sun reaches its maximum size

Inert carbon-oxygen core

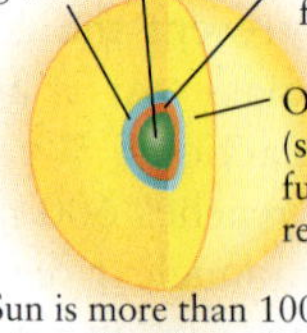

The Sun is more than 100 times larger in radius than when it was a main-sequence star. Part of the outer layers escapes into space in a stellar wind.

6. A planetary nebula

Thermal pulses cause spikes in luminosity that eject the star's outer layers.

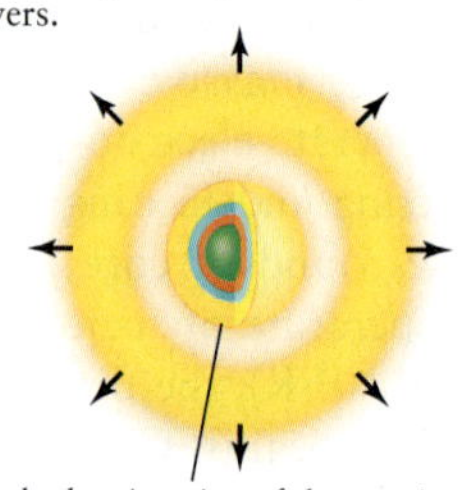

As the hot interior of the star is exposed, we observe an increase in the star's surface temperature.

7. The end of nuclear reactions

With the outer layers gone, the pressure on the shells around the core is too little to sustain nuclear reactions.

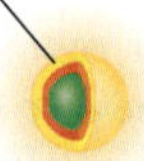

The star still glows intensely because of its high temperature. As energy is lost in the form of electromagnetic radiation, the star slowly cools.

8. A white dwarf

The core is now a white dwarf star, and the former shells around the core become its thin atmosphere.

The carbon-oxygen interior of the white dwarf is degenerate, so it does not contract as it cools. Hence the white dwarf's radius no longer changes.

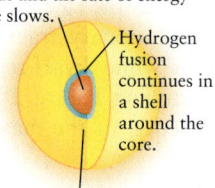

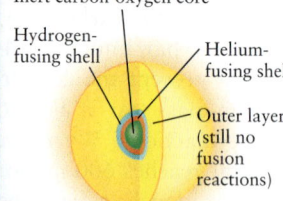

Hydrogen and helium, no fusion | Hydrogen fusion producing helium | Helium, no fusion | Helium fusion producing carbon and oxygen | Carbon and oxygen, no fusion

Figure 20-12 R I V U X G

Mass Loss from a Supergiant Star At the heart of this nebulosity, called SMC N76, lies a supergiant star with a mass of at least 18 $M_\odot$. This star is losing mass at a rapid rate in a strong stellar wind. As this wind collides with the surrounding interstellar gas and dust, it creates the "bubble" shown here. SMC N76, which has an angular diameter of 130 arcsec, lies within the Small Magellanic Cloud, a small galaxy that orbits our Milky Way. It is about 60,600 pc (198,000 ly) distant. (Y. Nazé, G. Rauw, J. Manfroid, and J.-M. Vreux, Liège Institute; Y.-H. Chu, U. of Illinois; and ESO)

reactions can take place. After the cessation of carbon fusion, the core will again contract, and the star's central temperature can rise to 1 billion kelvins (10^9 K). At this temperature **neon fusion** begins. This uses up the neon accumulated from carbon fusion and further increases the concentrations of oxygen and magnesium in the star's core.

After neon fusion ends, the core will again contract, and **oxygen fusion** will begin when the central temperature of the star reaches about 1.5 billion kelvins (1.5×10^9 K). The principal product of oxygen fusion is silicon (^{28}Si, 14 protons). Once oxygen fusion is over, the core will contract yet again. If the central temperature reaches about 2.7 billion kelvins (2.7×10^9 K), **silicon fusion** begins, producing a variety of nuclei from sulfur (^{32}S, 16 protons) to iron (^{56}Fe, 26 protons) and nickel (^{56}Ni, 28 protons). While all of this fusion is going on in the star's interior, at the surface the star is losing mass at a rapid rate (Figure 20-12).

As a high-mass star consumes increasingly heavier nuclei, the thermonuclear reactions produce a wider variety of products. For example, oxygen fusion produces not only silicon but also magnesium (^{24}Mg, with 12 protons), phosphorus (^{31}P, with 15 protons), and sulfur (^{31}S and ^{32}S, each with 16 protons). Some thermonuclear reactions that create heavy elements also release neutrons. A neutron is like a proton except that it carries no electric charge. Therefore, neutrons are not repelled by positively charged nuclei, and so can easily collide and combine with them. This absorption of neutrons by nuclei, called **neutron capture,** creates many elements and isotopes that are not produced directly in fusion reactions.

Each stage of thermonuclear reactions in a high-mass star helps to trigger the succeeding stage. In each stage, when the star exhausts a given variety of nuclear fuel in its core, gravitational contraction takes the core to ever-higher densities and temperatures, thereby igniting the "ash" of the previous fusion stage—and possibly the outlying shell of unburned fuel as well.

Supergiant Stars and Their Evolution

The increasing density and temperature of the core make each successive thermonuclear reaction more rapid than the one that preceded it. As an example, Table 20-1 shows a theoretical calculation of the evolutionary stages for a star with a zero-age mass of 25 $M_\odot$. This calculation indicates that carbon fusion in such a star lasts for 600 years, neon fusion for 1 year, and oxygen fusion for only 6 months. The last, and briefest, stage of nuclear reactions is silicon fusion. The entire core supply of silicon in a 25-$M_\odot$ star is used up in only one day!

Table 20-1 Evolutionary Stages of a 25-$M_\odot$ Star

Stage	Core temperature (K)	Core density (kg/m^3)	Duration of stage
Hydrogen fusion	4×10^7	5×10^3	7×10^6 years
Helium fusion	2×10^8	7×10^5	7×10^5 years
Carbon fusion	6×10^8	2×10^8	600 years
Neon fusion	1.2×10^9	4×10^9	1 year
Oxygen fusion	1.5×10^9	10^{10}	6 months
Silicon fusion	2.7×10^9	3×10^{10}	1 day
Core collapse	5.4×10^9	3×10^{12}	¼ second
Core bounce	2.3×10^{10}	4×10^{15}	milliseconds
Explosive (supernova)	about 10^9	varies	10 seconds

Based on calculations by Stanford Woosley (University of California, Santa Cruz) and Thomas Weaver (Lawrence Livermore National Laboratory).

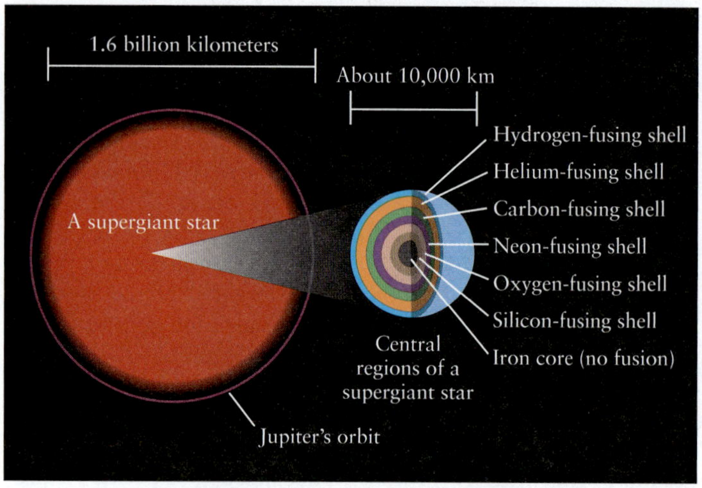

Figure 20-13

The Structure of an Old High-Mass Star Near the end of its life, a star with an initial mass greater than about 8 $M_\odot$ becomes a red supergiant. The star's overall size can be as large as Jupiter's orbit around the Sun. The stars energy comes from a series of concentric fusing shells, all combined within a volume roughly the same size as Earth. Thermonuclear reactions do not occur within the iron core, because fusion reactions that involve iron absorb energy rather than release it.

Each stage of core fusion in a high-mass star generates a new shell of material around the core. After several such stages, the internal structure of a truly massive star—say, 25 to 30 $M_\odot$ or greater—resembles that of an onion (see Figure 20-13). Because thermonuclear reactions can take place simultaneously in several shells, energy is released at such a rapid rate that the star's outer layers expand tremendously. The result is a **supergiant** star, whose luminosity and radius are much larger than those of a giant (see Section 17-7).

A star of 8 or more solar masses evolves into a supergiant 100 times (or more) larger than the Sun

Several of the brightest stars in the sky are supergiants, including Betelgeuse and Rigel in the constellation Orion and Antares in the constellation Scorpius. (Figure 17-15 shows the locations of these stars on an H-R diagram.) They appear bright not because they are particularly close, but because they are extraordinarily luminous.

A supergiant star cannot keep adding shells to its "onion" structure forever, because the sequence of thermonuclear reactions cannot go on indefinitely. In order for an element to serve as a thermonuclear fuel, energy must be given off when its nuclei collide and fuse. This released energy is a result of the strong nuclear force of attraction that draws nucleons (neutrons and protons) together. However, protons also repel one another by the weaker electric force. As a result of this electric repulsion, adding extra protons to nuclei larger than iron, which has 26 protons, requires an *input* of energy rather than causing energy to be released. Nuclei of this size or larger cannot act as fuel for thermonuclear reactions. Hence, the sequence of fusion stages ends with silicon fusion. One of the products of silicon fusion is iron, and the result is a star with an iron-rich core in which no thermonuclear reactions take place (see Figure 20-13).

Shell fusion in the layers surrounding the iron-rich core consumes the star's remaining reserves of fuel. At this stage the entire energy-producing region of the star is contained in a volume no bigger than Earth, some 10^6 times smaller in radius than the overall size of the star. This state of affairs will soon come to an end, because the buildup of an inert, iron-rich core signals the impending violent death of a massive supergiant star.

20-6 High-mass stars violently blow apart in core-collapse supernova explosions

Our present understanding is that all stars of about 8 $M_\odot$ or less divest most of their mass in the form of planetary nebulae. The burned-out core that remains settles down to become a white dwarf star. But the truly massive stars—stellar heavyweights that begin their lives with more than 8 solar masses of material—do not pass through a planetary nebula phase. Instead, they die in spectacular *core-collapse supernova explosions.*

The Violent End of a High-Mass Star

To understand what happens in a core-collapse supernova explosion, we must look deep inside a massive star at the end of its life. Of course, we cannot do this in actuality, because the interiors of stars are opaque. But astronomers have developed theoretical models based on what we know about the behavior of gases and atomic nuclei. The description that follows, while largely theoretical, describes our observations of supernovae fairly well. And, as we will see in Section 20-8, a special kind of "telescope" has allowed us to glimpse the interior of at least one relatively nearby supernova.

The core of an aging, massive star gets progressively hotter as it contracts to ignite successive stages of thermonuclear fusion (see Stage 1 in Figure 20-14). Wien's law (Section 5-4) and Planck's law (Section 5-5) together tell us that as the temperature of an object like a star increases, so does the energy of the photons it emits. When the temperature in the core of a massive star reaches a few hundred million kelvins, the photons are energetic enough to initiate a host of nuclear reactions that create neutrinos. These neutrinos, which carry off energy, escape from the star's core, just as solar neutrinos flow freely out of the Sun (see Section 16-4).

To compensate for the energy drained by the neutrinos, the star must provide energy either by consuming more thermonuclear fuel, by contracting, or both. But when the star's core is converted into iron, no more energy-producing thermonuclear reactions are possible, and the only source of energy is contraction and rapid heating (see Stage 2 in Figure 20-14).

By forming a dense core of iron, a massive star sows the seeds of its own destruction

Once a star with an original mass of about 8 $M_\odot$ or more develops an iron-rich core, the core contracts very rapidly, so that the core temperature skyrockets to 5×10^9 K within a tenth of a second. The gamma-ray photons emitted by the intensely hot core have so much energy that when they collide with iron nuclei, they begin to break the iron nuclei down into much smaller helium nuclei (^{4}He). This process is called

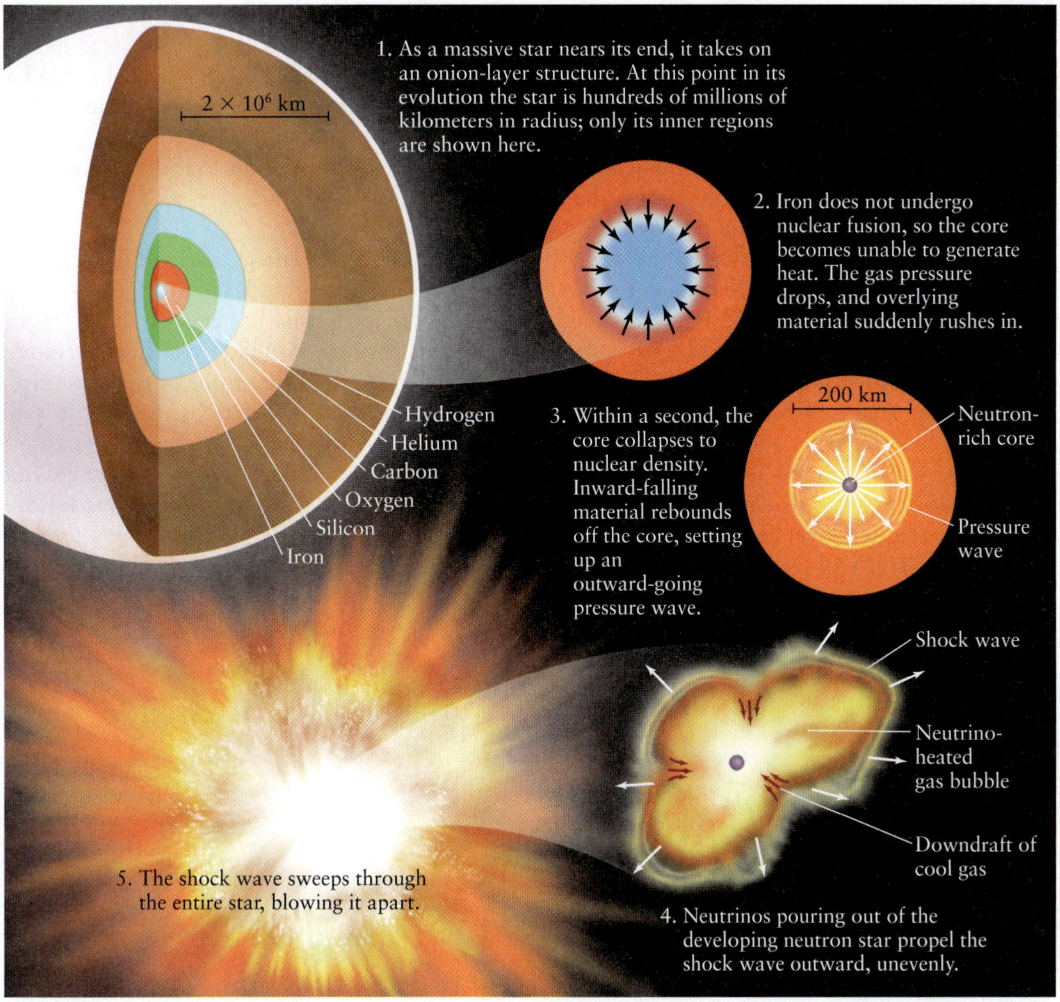

1. As a massive star nears its end, it takes on an onion-layer structure. At this point in its evolution the star is hundreds of millions of kilometers in radius; only its inner regions are shown here.
2 × 10⁶ km
2. Iron does not undergo nuclear fusion, so the core becomes unable to generate heat. The gas pressure drops, and overlying material suddenly rushes in.
Hydrogen
Helium
Carbon
Oxygen
Silicon
Iron
3. Within a second, the core collapses to nuclear density. Inward-falling material rebounds off the core, setting up an outward-going pressure wave.
200 km
Neutron-rich core
Pressure wave
Shock wave
Neutrino-heated gas bubble
Downdraft of cool gas
5. The shock wave sweeps through the entire star, blowing it apart.
4. Neutrinos pouring out of the developing neutron star propel the shock wave outward, unevenly.

Figure 20-14

A Core-Collapse Supernova This series of illustrations depicts our understanding of the last day in the life of a star of more than about 8 $M_\odot$. (Illustration by Don Dixon, adapted from Wolfgang Hillebrandt, Hans-Thomas Janka, and Ewald Müller, "How to Blow Up a Star," *Scientific American*, October 2006)

photodisintegration. As Table 20-1 shows, it takes a high-mass star millions of years and several stages of thermonuclear reactions to build up an iron core; within a fraction of a second, photodisintegration undoes the result of those millions of years of reactions.

Within another tenth of a second, the core becomes so dense that the negatively charged electrons within it are forced to combine with the positively charged protons to produce electrically neutral neutrons. This process also releases a flood of neutrinos, denoted by the Greek letter ν (nu):

$$e^- + p \rightarrow n + \nu$$

Although neutrinos interact only very weakly with matter (see Section 16-4), the core is now so dense that even neutrinos cannot escape from it immediately. But because these neutrinos carry away a substantial amount of energy when they do escape from the core, the core cools down and condenses even further.

At about 0.25 seconds after its rapid contraction begins, the core is less than 20 km in diameter and its density is in excess of 4×10^{17} kg/m^3. This is **nuclear density,** the density with which neutrons and protons are packed together inside nuclei. (If Earth were compressed to this density, it would be only 300 meters, or 1000 feet, in diameter.)

Matter at nuclear density or higher is extraordinarily difficult to compress. Thus, when the density of the neutron-rich core begins to exceed nuclear density, the core suddenly becomes very stiff and rigid. The core's contraction comes to a sudden halt, and the innermost part of the core actually bounces back and expands somewhat. This *core bounce* sends a powerful wave of pressure, like an unimaginably intense sound wave, outward into the outer core (see Stage 3 in Figure 20-14).

During this critical stage, the cooling of the core has caused the pressure to decrease profoundly in the regions surrounding the core. Without pressure to hold it up against gravity, the material from these regions plunges inward at speeds up to 15% of the speed of light. When this inward-moving material crashes down onto the rigid core, it encounters the outward-moving pressure wave. In just a fraction of a second, the material that fell onto the core begins to move back out toward the star's surface, propelled in part by the flood of neutrinos trying to escape from the star's core.

Supercomputer simulations of this complex process show that if the pressure wave were to spread outward at precisely the same speed in all directions, its energy would be absorbed by the gas around the core and the wave would fizzle out. But when the simulations allow for the presence of convection and turbulence in the dying star's gases, the result is quite different: The material surrounding the core behaves more like water boiling furiously in a heated pot. Rising bubbles of superheated gases deliver extra energy to the pressure wave, sustaining it and making it accelerate as it plows outward through the doomed star's outer layers. The wave soon reaches a speed greater than the speed of sound waves in the star's outer layers. When this happens, the wave becomes a *shock wave,* like the sonic boom produced by a supersonic airplane (see Stage 4 in Figure 20-14).

After a few hours, the shock wave reaches the star's surface, by which time the star's outer layers have begun to lift away from the core. When the star's outer layers thin out sufficiently, a portion of this energy escapes in a torrent of light (see Stage 5 in Figure 20-14). The star has become a **supernova** (plural **supernovae**). Specifically, what we have described is the formation of a **core-collapse supernova.** We use this term because, as we will see in Section 20-9, it is possible for supernova explosions to occur that do not involve the collapse of the core of a massive star.

CAUTION! The energy released in a core-collapse supernova is an incomprehensibly large 10^{46} joules—a hundred times more energy than the Sun has emitted due to thermonuclear reactions over its entire 4.56-billion year history. However, it is important to recognize that the source of the supernova's energy release is *not* thermonuclear reactions. Rather, it is the *gravitational* energy released by the collapse of the core and by the inward fall of the star's outer layers. (You release gravitational energy when you fall off a diving board, and this released energy goes into making a big splash in the swimming pool.) The energy released by the collapse of the core reappears in the form of neutrinos;

(a) A simulated supernova $5\frac{1}{2}$ hours after the core "bounce" (red = hydrogen, green = helium, turquoise and blue = carbon, oxygen, silicon, and iron)

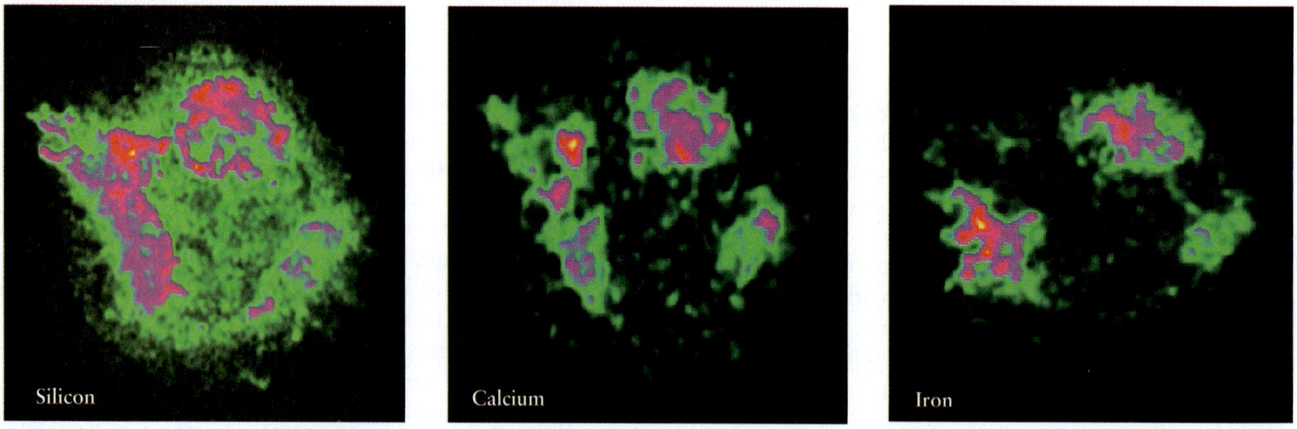

(b) Material was ejected in "blobs" from the supernova that produced the Cassiopeia A supernova remnant R I V U X G

ANIMATION 20.3 WEB LINK 20.8

Figure 20-15

Turbulence in a Core-Collapse Supernova

(a) This image from a supercomputer simulation show a cross section of a massive star several hours into the supernova explosion. The colors show the turbulent mixing of material from the star's inner regions (turquoise and blue) with hydrogen and helium from the outer layers (green and red). **(b)** Turbulence causes material to be ejected from the supernova in irregular "blobs," as shown by these images of the supernova remnant Cassiopeia A. Each image was made using an X-ray wavelength emitted by a particular element. (Figure 18-24 shows a false-color image of Cassiopeia A made using visible, infrared, and X-ray wavelengths.) (a: Konstantinos Kifonidis, Max-Planck-Institut für Astrophysik; b: U. Hwang et al., NASA/GSFC) the fall of the outer layers provides the energy to power the nuclear reactions that generate the supernova's electromagnetic radiation. The amount of energy release from the supernova is so great because the star is so massive; hence, the amount of material that falls inward is immense, it falls a great distance, and is acted on by a strong gravitational pull as it falls.

The Wreckage of a Core-Collapse Supernova

Supercomputer simulations provide many insights into the violent, complex, and rapidly changing conditions deep inside a star as it is torn apart by a supernova explosion (Figure 20-15). For example, Figure 20-15*a* shows a snapshot of the interior of a high-mass star $5\frac{1}{2}$ hours after the stiffening of the core. The simulation predicts turbulent swirls and eddies that grow behind the shock wave as it moves outward from the star's core. Evidence in favor of such turbulence comes from images of the remnants of long-ago supernovae (Figure 20-15*b*). Such images show that material is ejected from the supernova not in uniform shells but in irregular clumps. These clumps are just what would be expected from a turbulent explosion. (We discuss supernova remnants further in Section 20-10.)

Detailed computer calculations suggest that a 25-$M_\odot$ star ejects about 96% of its material to the interstellar medium for use in producing future generations of stars. Less massive stars eject a smaller percentage of their mass into space when they become core-collapse supernovae.

Before this material is ejected into space, it is compressed so much by the passage of the shock wave through the star's outer layers that a new wave of thermonuclear reactions sets in. These reactions can produce many more chemical elements, including elements heavier than iron. Reactions of this kind require a tremendous input of energy, and thus cannot take place during the star's pre-supernova lifetime.

The energy-rich environment of a supernova shock wave is almost the only place in the universe where such heavy elements as zinc, silver, tin, gold, mercury, lead, and uranium can be produced. (In Chapter 21 we will see another, even more exotic mechanism for producing these heaviest elements.) Remarkably, all of these elements are found on Earth. Hence, some of the material that makes up our solar system, our Earth, and our bodies must long ago have been part of a star that lived, evolved, and died as a supernova.

20-7 In 1987 a nearby supernova gave us a close-up look at the death of a massive star

Supernovae have peak luminosities as great as 10^9 $L_\odot$, rivaling the light output of an entire galaxy. These large luminosities make it possible to see supernovae in galaxies far beyond our own Milky Way Galaxy, and indeed hundreds of these distant supernovae are observed each year (Figure 20-16). In a handful of cases, images made before the explosion have allowed astronomers to identify the star that subsequently exploded into a supernova, called the **progenitor star.** For example, the progenitor star pointed to in Figure 20-16*b* was a red supergiant star whose internal structure probably resembled that shown in cross section in Figure 20-13.

One frustrating aspect of these distant supernovae is simply that they *are* distant, and so cannot be studied in as much detail as astronomers would like. But one recent and unusually close supernova has provided astronomers with a unique opportunity to check the theoretical ideas presented in Section 20-6.

A Supernova in the Galaxy Next Door

On February 23, 1987, a supernova was discovered in the Large Magellanic Cloud (LMC), a companion galaxy to our Milky Way some 51,500 pc (168,000 ly) from Earth. The supernova, designated SN 1987A because it was the first discovered that year, occurred near an enormous H II region in the LMC called the

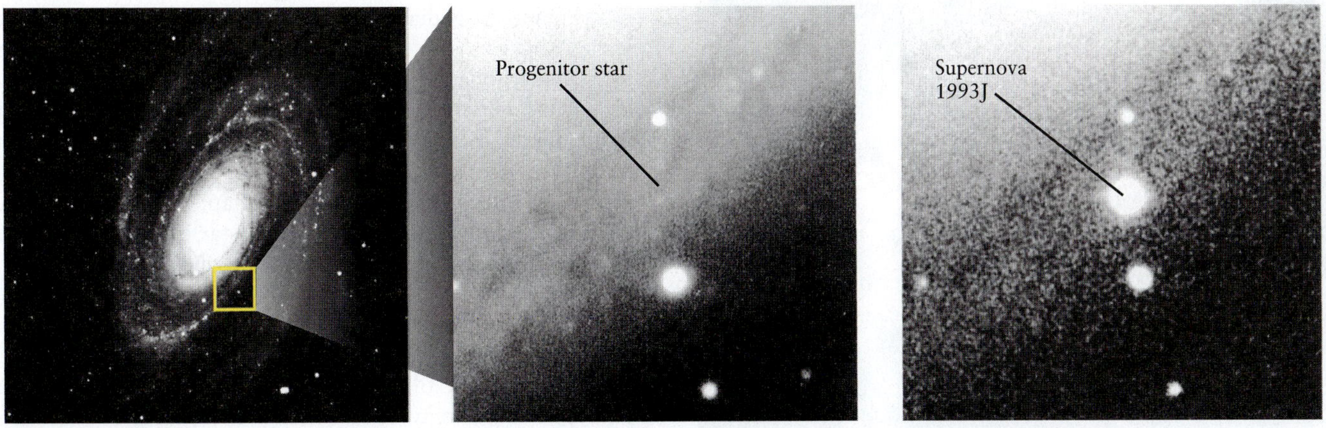

(a) Spiral galaxy M81 (b) Before the explosion (c) After the explosion

WEB LINK 20.9

Figure 20-16 R I V U X G

A Supernova in a Distant Galaxy On the night of March 28, 1993, Francisco Garcia Diaz, a Spanish amateur astronomer, discovered supernova SN 1993J in the galaxy M81 in Ursa Major. **(a)** M81 lies some 3.6 million pc (12 million ly) from Earth. Its angular size is about half that of the full moon. **(b)** The progenitor star that later exploded into SN 1993J was a K0 red supergiant. **(c)** This image shows the same part of the sky as (b). Like SN 1987A, SN 1993J resulted from the core collapse and subsequent explosion of a massive star. (a: Palomar Observatory; b, c: D. Jones and E. Telles, Isaac Newton Telescope)

Tarantula Nebula (Figure 20-17). The supernova was so bright that observers in the southern hemisphere could see it without a telescope.

Supernova 1987A was the first nearby supernova to be seen since the invention of the telescope

Such a bright supernova is a rare event. In the last thousand years, only five other supernovae—in 1006, 1054, 1181, 1572, and 1604—have been bright enough to be seen with the naked eye, and all of these occurred in our own Milky Way Galaxy. (As we described in Section 4-3, the supernova of 1572 had a major influence on Tycho Brahe's ideas about the heavens.) Although outside our Galaxy, SN 1987A occurred relatively close to us in a part of the heavens that is obscured only slightly by the Milky Way's interstellar dust. Furthermore, not long after SN 1987A appeared, several new orbiting telescopes were placed into service. These enabled astronomers to study the supernova's evolution with unprecedented resolution and in wavelength ranges not accessible from Earth's surface (see Section 6-7). As a result, SN 1987A has given astronomers an unprecedented view of the violent death of a massive star.

The light from a supernova such as SN 1987A does not all come in a single brief flash; the outer layers continue to glow as they expand into space. For the first 20 days after the detonation of SN 1987A, its glow was powered primarily by the tremendous heat that the shock wave deposited in the star's outer layers. As the expanding gases cooled, the light energy began to be provided by a different source—the decay of radioactive isotopes of cobalt, nickel, and titanium produced in the supernova explosion.

Astronomers have been able to pinpoint the specific isotopes involved because different radioactive nuclei emit gamma rays of different wavelengths when they decay. These emissions have been detected by orbiting gamma-ray telescopes (see Figure 6-31). Thanks to these radioactive decays, the brightness of SN 1987A actually *increased* for the first 85 days after the detonation, then settled into a slow decline as the radioactive isotopes were used up. The supernova remained visible to the naked eye for several months after the detonation.

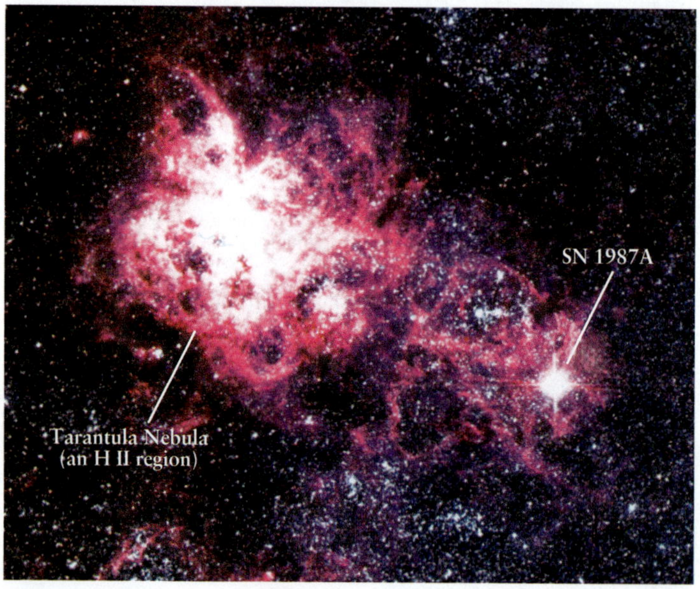

Figure 20-17 R I **V** U X G

Supernova 1987A This photograph, taken soon after the discovery of SN 1987A, shows a portion of the Large Magellanic Cloud that includes the supernova and a huge H II region called the Tarantula Nebula. Although it was 51,000 pc from Earth, SN 1987A was bright enough to be seen without a telescope. (European Southern Observatory)

Why SN 1987A Was Unusual

Ideally, SN 1987A would have confirmed the theories of astronomers about typical supernovae. But SN 1987A was *not* typical. Its luminosity peaked at roughly 10^8 $L_\odot$, only a tenth of the maximum luminosity observed for other, more distant supernovae. Fortunately, the doomed star had been observed prior to becoming a supernova, and these observations helped explain why SN 1987A was an exceptional case.

Figure 20-13 suggests that a massive star should be in a red supergiant stage when the iron core collapses and the star becomes a supernova. As we have mentioned, this was indeed the case for the progenitor star shown in Figure 20-16*b*, and is also the case for other supernovae seen in distant galaxies. The progenitor star of SN 1987A was indeed a high-mass star: Its estimated main sequence mass was about 20 $M_\odot$, although by the time it exploded—some 10^7 years after it first formed—it probably had shed a few solar masses. However, this progenitor star was identified not as a red supergiant, but as a *blue* B3 I supergiant (Figure 20-18).

The explanation of this seeming contradiction is that stars in the Large Magellanic Cloud, including the progenitor of SN 1987A, are Population II stars with a very low percentage of metals—that is, elements heavier than hydrogen and helium (see Section 19-5). A small difference in the amount of metals present can affect whether a star's interior is relatively transparent or opaque, just as a small amount of dirt can make a window difficult to see through. As a result, a high-mass Population II star follows a somewhat different evolutionary track than does a Population I star (with a greater percentage of metals) of the same mass. On an H-R diagram, an aging high-mass star of Population I goes directly from the main sequence to the red supergiant region at the upper right of the diagram. By contrast, the track for an equally massive Population II star wanders from left to right and back again across the top of the H-R diagram as the star alternates between being a hot, blue supergiant (on the left in an H-R diagram) and a cool red supergiant (on the right in an H-R diagram). Apparently the progenitor star of SN 1987A developed an iron core and became a supernova when it was in the blue supergiant stage.

A massive Population II star like the SN 1987A progenitor changes dramatically in size as it alternates between being a red supergiant and a blue supergiant. When in a blue supergiant phase, its radius may be less than 1/10 as large as when it is in a red supergiant phase. Hence, the progenitor star was relatively small when its core collapsed (though its radius was still more than 10 times that of our Sun). This means that the star's outer layers were close to the core and thus held more strongly by the core's gravitational attraction. When the detonation occurred, a relatively large fraction of the shock wave's energy had to be used against this gravitational attraction to push the outer layers into

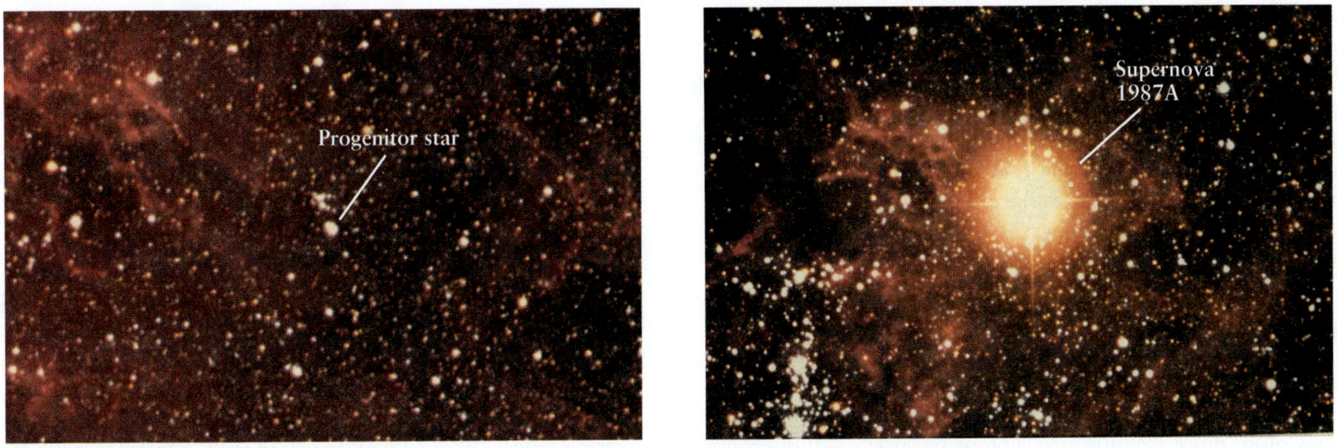

(a) Before the star exploded

(b) After the star exploded

WEB LINK 20.10

Figure 20-18 R I V U X G

SN 1987A—Before and After **(a)** This photograph shows a small section of the Large Magellanic Cloud as it appeared before the explosion of SN 1987A. The supernova's progenitor star was a B3 blue supergiant. **(b)** This image shows a somewhat larger region of the sky a few days after the supernova exploded into brightness. (Anglo-Australian Observatory)

space. Hence, the amount of shock wave energy available to be converted into light was smaller than for most supernovae. This explains why SN 1987A was only a tenth as bright as an exploding red supergiant would have been.

The Aftermath of SN 1987A

Three and a half years after SN 1987A exploded, astronomers used the newly launched Hubble Space Telescope to obtain a picture of the supernova. To their surprise, the image showed a ring of glowing gas around the exploded star. After the optics of the Hubble Space Telescope were repaired in 1994 (see Section 6-7), SN 1987A was observed again and a set of *three* glowing rings was revealed (Figure 20-19*a*).

These rings are relics of a hydrogen-rich outer atmosphere that was ejected by gentle stellar winds from the progenitor star when it was a red supergiant, about 20,000 years ago. This

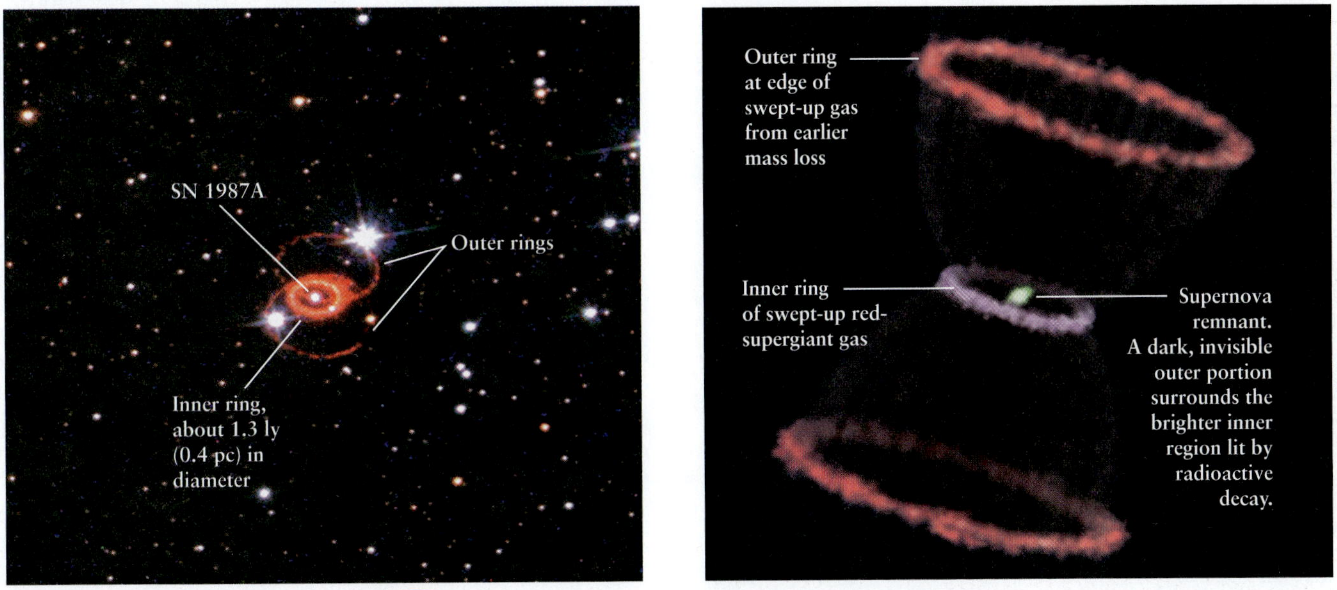

(a) Supernova 1987A seen in 1996 R I V U X G

(b) An explanation of the rings

ANIMATION 20.4

Figure 20-19

SN 1987A and Its "Three-Ring Circus" **(a)** This true-color view from the Hubble Space Telescope shows three bright rings around SN 1987A. **(b)** This drawing shows the probable origin of the rings. A wind from the progenitor star, shown in Figure 20-18*a*, formed an hourglass-shaped shell surrounding the star. (Compare with Figure 20-7.) Ultraviolet light from the supernova explosion ionized ring-shaped regions in the shell, causing them to glow. (Robert Kirshner and Peter Challis, Harvard-Smithsonian Center for Astrophysics; STScI) diffuse gas expanded in a hourglass shape (Figure 20-19*b*), because it was blocked from expanding around the star's equator either by a preexisting ring of gas or by the orbit of an as yet unseen companion star. (Figure 20-7 shows a similar model used to explain the shapes of certain planetary nebulae.) The outer rings in Figure 20-19*a* are parts of the hourglass that were ionized by the initial flash of ultraviolet radiation from the supernova; as electrons recombine with the ions, the rings emit visible-light photons. (We described this process of *recombination* in Section 18-2.)

WEB LINK 20.11 ANIMATION 20.5 By the early years of the twenty-first century, the shock wave from the supernova was beginning to collide with the "waist" of the hourglass shown in Figure 20-19*b*. This collision is making the hourglass glow more brightly in visible wavelengths—though not enough, unfortunately, to make the supernova again visible to the naked eye—and emit copious radiation at X-ray and ultraviolet wavelengths.

By studying this cosmic collision, astronomers hope to learn more about the shock wave and the supernova explosion that spawned it. They hope as well to learn about the matter in the inner ring, which will give us insight into the stellar winds that blew from the progenitor star thousands of years ago. Because SN 1987A provides a unique laboratory for studying the evolution of a supernova, astronomers will monitor it carefully for decades to come.

20-8 Neutrinos emanate from supernovae like SN 1987A

In addition to electromagnetic radiation, supernovae also emit a brief but intense burst of neutrinos from their collapsing cores. In fact, theory suggests that *most* of the energy released by the exploding star is in the form of neutrinos. If it were possible to detect the flood of neutrinos from an exploding star, astronomers would have direct evidence of the nuclear processes that occur within the star during its final seconds before becoming a supernova.

Observations of SN 1987A confirm that a core-collapse supernova emits most of its energy in the form of neutrinos

Unfortunately, detecting neutrinos is difficult because under most conditions matter is transparent to neutrinos (see Section 16-4). Consequently, when supernova neutrinos encounter Earth, almost all of them pass completely through the planet as if it were not there. The challenge to scientists is to detect the tiny fraction of neutrinos that *do* interact with the matter through which they pass.

Neutrino Telescopes

During the 1980s, two "neutrino telescopes" designed uniquely for detecting supernova neutrinos went into operation—the Kamiokande detector in Japan (a joint project of the University of Tokyo and the University of Pennsylvania), and the IMB detector (a collaboration of the University of California, Irvine, the University of Michigan, and Brookhaven National Laboratory). Both detectors consisted of large tanks containing thousands of tons of water. On the rare occasion when a neutrino collided with one of the water molecules, it produced a brief flash of light. Any light flash was recorded by photomultiplier tubes lining the walls of the tank. Because other types of subatomic particles besides neutrinos could also produce similar light flashes, the detectors were placed deep underground so that hundreds of meters of Earth would screen out almost all particles except neutrinos. (The more recent Sudbury Neutrino Observatory, shown in Figure 16-6, has a similar design.)

What causes the light flashes? And how do we know whether a neutrino comes from a supernova? The key is that a single supernova neutrino carries a relatively large amount of energy, typically 20 MeV or more. (We introduced the unit of energy called the electron volt, or eV, in Section 5-5. One MeV is equal to 10^6 electron volts. Only the most energetic nuclear reactions produce particles with energies of more than 1 MeV.) If such a high-energy neutrino hits a proton in the water-filled tank of a neutrino telescope, the collision produces a positron (see Box 16-1). The positron then recoils at a speed greater than the speed of light in water, which is 2.3×10^5 km/s. (Such motion does not violate the ultimate speed limit in the universe, 3×10^5 km/s, which is the speed of light in a *vacuum*.)

Just as an airplane that flies faster than sound produces a shock wave (a sonic boom), a positron that moves through a substance such as water faster than the speed of light in that substance produces a shock wave of light. This shock wave is called **Cerenkov radiation,** after the Russian physicist Pavel A. Cerenkov, who first observed it in 1934. It is this radiation that is detected by the photomultiplier tubes that line the detector walls.

By measuring the properties of the Cerenkov radiation from a recoiling positron, scientists can determine the positron's energy and, therefore, the energy of the neutrino that created the positron. Determining the energy allows them to tell the difference between the high-energy neutrinos from supernovae and neutrinos from the Sun, which typically have energies of 1 MeV or less.

Neutrinos from SN 1987A

Both the Kamiokande and IMB detectors were operational on February 23, 1987, when SN 1987A was first observed in the Large Magellanic Cloud. Soon afterward, the physicists working with these detectors excitedly reported that they had detected Cerenkov flashes from a 12-second burst of neutrinos that reached Earth 3 hours before astronomers saw the light from the exploding star. Only a few neutrinos were seen: The Kamiokande detector saw flashes from 11 neutrinos at about the same time that 8 were recorded by the IMB detector. But when the physicists factored in the sensitivity of their detectors, they calculated that Kamiokande and IMB had actually been exposed to a torrent of more than 10^{16} neutrinos.

Given the flux of neutrinos measured by the detectors and the distance of 168,000 light-years from SN 1987A to Earth, physicists used the inverse-square law to determine the total number of neutrinos that had been emitted from the supernova. (This law applies to neutrinos just as it does to electromagnetic radiation; see Section 17-2.) They found that over a 10-second period, SN 1987A emitted 10^{58} neutrinos with a total energy of 10^{46} joules. This energy is more than 100 times as much energy as the Sun has emitted in its entire history and more than 100 times the amount of energy that the supernova emitted in the form of electromagnetic radiation. Indeed, for a few seconds the supernova's neutrino luminosity—that is, the *rate* at which it emitted energy in the form of neutrinos—was 10 times greater than the total luminosity in electromagnetic radiation of all of the stars in the observable universe! Such comparisons give a hint of the incomprehensible violence with which a supernova explodes.

Why did the neutrinos from SN 1987A arrive 3 hours *before* the first light was seen? As we saw in Section 20-6, neutrinos are produced when thermonuclear reactions cease in the core of a massive star and the core collapses. These neutrinos encounter little delay as they pass through the volume of the star. The tremendous increase in the star's light output, by contrast, occurs only when the shock wave reaches the star's outermost layers (which are thin enough to allow light to pass through them). It took 3 hours for this shock wave to travel outward from the star's core to its surface, by which time the neutrino burst was already billions of kilometers beyond the dying star. Traveling toward Earth for the next 168,000 years, the neutrinos that would eventually produce light flashes in Kamiokande and IMB remained in front of the photons emitted from the star's surface, and so the neutrinos were detected before the supernova's light. Thus, the neutrino data from SN 1987A gave astronomers direct confirmation of theoretical ideas about how supernova explosions take place.

Kamiokande and IMB have both been replaced by a new generation of neutrino telescopes (see Section 16-4). While these new detectors are intended primarily to observe neutrinos from the Sun, they are also fully capable of measuring neutrino bursts from nearby supernovae. Astronomers have identified a number of supergiant stars in our Galaxy that are likely to explode into supernovae, among them the bright red supergiant Betelgeuse in the constellation Orion (see Figure 2-2 and Figure 6-27).

Unfortunately, astronomers do not yet know how to predict precisely when such stars will explode into supernovae. It may be many thousands of years before Betelgeuse explodes. Then again, it could happen tomorrow. If it does, neutrino telescopes will be ready to record the collapse of its massive core. However, you won't need a telescope to observe this spectacular event, as it would outshine the Moon!

20-9 White dwarfs in close binary systems can also become supernovae

Astronomers discover dozens of supernovae in distant galaxies every year, but not all of these are the result of massive stars dying violently. A totally different type of supernova occurs when a white dwarf star in a binary system blows itself completely apart. The first clue that two entirely distinct chains of events could produce supernovae was rather subtle: Some supernovae have prominent hydrogen emission lines in their spectra but others do not.

In just a few seconds, a thermonuclear supernova completely destroys an entire white dwarf star

Types of Supernovae

Supernovae with hydrogen emission lines, called **Type II supernovae,** are core-collapse supernovae of the sort we described in Section 20-6. They are caused by the death of highly evolved massive stars that still have ample hydrogen in their atmospheres when they explode. When the star explodes, the hydrogen atoms are excited and glow prominently, producing hydrogen emission lines. SN 1987A (the topic of Sections 20-7 and 20-8) and SN 1993J (shown in Figure 20-16*c*) were both Type II supernovae.

Hydrogen lines are missing in the spectrum of a **Type I supernova,** which tells us that little or no hydrogen is left in the debris from the explosion. Type I supernovae are further divided into three important subclasses. **Type Ia supernovae** have spectra that include a strong absorption line of ionized silicon. **Type Ib** and **Type Ic supernovae** both lack the ionized silicon line. The difference between them is that the spectra of Type Ib supernovae have a strong helium absorption line, while those of Type Ic supernovae do not. Figure 20-20 shows these different supernova spectra.

Astronomers suspect that Type Ib and Ic supernovae are caused by core collapse in dying massive stars, just like Type II supernovae. The difference is that the progenitor stars of Type Ib and Ic supernovae have been stripped of their outer layers before they explode. A star can lose its outer layers to a strong stellar wind (see Figure 20-12) or, if it is part of a close binary system, by transferring mass to its companion star (see Figure 19-21*b*). If enough mass remains for the star's core to collapse, the star dies as a Type Ib supernova. Because the outer layers of hydrogen are absent, the supernova's spectrum exhibits no hydrogen lines but many helium lines (Figure 20-20*b*). Type Ic supernovae have apparently undergone even more mass loss prior to their explosion; their spectra show that they have lost much of their helium as well as their hydrogen (Figure 20-20*c*).

A key piece of evidence that Type II, Type Ib, and Type Ic supernovae all begin as massive stars is that all three types are found only near sites of recent star formation. The life span of a massive star from its formation to its explosive death as a supernova is only about 10^7 years, less time than it took our own Sun to condense from a protostar to a main-sequence star (see Figure 18-10) and a mere blink of an eye on the time scale of stellar evolution. Because massive stars live such a short time, it makes sense that they should meet their demise very close to where they were formed.

Type Ia Supernovae: Detonating a White Dwarf

Type Ia supernovae, by contrast, are found even in galaxies where there is no ongoing star formation. Hence, they are probably *not* the death throes of massive supergiant stars. Instead, Type Ia supernovae are thought to result from the thermonuclear explosion of a white dwarf star. This may sound contradictory, because we saw in Section 20-4 that white dwarf stars have no thermonuclear reactions going on in their interiors. But these reactions *can* occur if a carbon-oxygen–rich white dwarf is in a close, semidetached binary system with a red giant star (see Figure 19-20*b* and Figure 20-20*a*).

ANIMATION 20.6 ANIMATION 20.7

Figure 20-21 shows the likely series of events that lead to a Type Ia supernova. Stage 1 in this figure shows a close binary system in which both stars have less than 4 $M_{\odot}$. The more massive star on the left evolves more rapidly than its less massive companion and eventually becomes a white dwarf. As the companion evolves and its outer layers expand, it overflows its Roche lobe

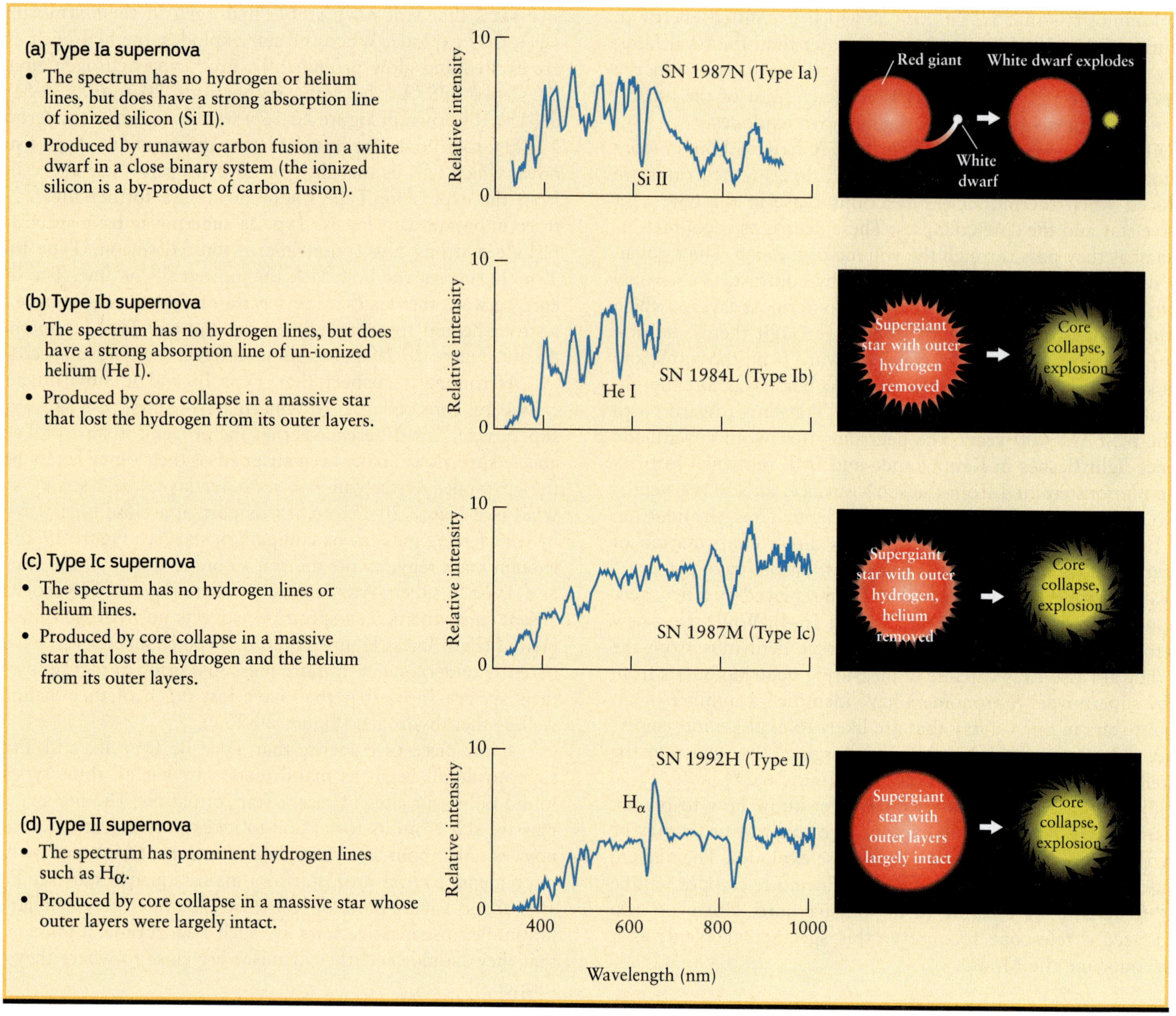
(a) Type Ia supernova
• The spectrum has no hydrogen or helium lines, but does have a strong absorption line of ionized silicon (Si II).
• Produced by runaway carbon fusion in a white dwarf in a close binary system (the ionized silicon is a by-product of carbon fusion).
SN 1987N (Type Ia)
Si II
Red giant
White dwarf explodes
White dwarf
(b) Type Ib supernova
• The spectrum has no hydrogen lines, but does have a strong absorption line of un-ionized helium (He I).
• Produced by core collapse in a massive star that lost the hydrogen from its outer layers.
SN 1984L (Type Ib)
He I
Supergiant star with outer hydrogen removed
Core collapse, explosion
(c) Type Ic supernova
• The spectrum has no hydrogen lines or helium lines.
• Produced by core collapse in a massive star that lost the hydrogen and the helium from its outer layers.
SN 1987M (Type Ic)
Supergiant star with outer hydrogen, helium removed
Core collapse, explosion
(d) Type II supernova
• The spectrum has prominent hydrogen lines such as Hα.
• Produced by core collapse in a massive star whose outer layers were largely intact.
SN 1992H (Type II)
Hα
Supergiant star with outer layers largely intact
Core collapse, explosion
Relative intensity
10
0
400
600
800
1000
Wavelength (nm)

Figure 20-20

Supernova Types These illustrations show the characteristic spectra and the probable origins of supernovae of **(a)** Type Ia, **(b)** Type Ib, **(c)** Type Ic, and **(d)** Type II. (Spectra courtesy of Alexei V. Filippenko, University of California, Berkeley)

and dumps gas from its outer layers onto the white dwarf (see Stage 2 in Figure 20-21). When the total mass of the white dwarf approaches the Chandrasekhar limit, the increased pressure applied to the white dwarf's interior causes carbon fusion to begin there (Stage 3 in Figure 20-21). Hence, the interior temperature of the white dwarf increases.

If the white dwarf were made of ordinary matter, the temperature increase would cause a further increase in pressure, the white dwarf would expand and cool, and the carbon-fusing reactions would abate. But because the white dwarf is composed of degenerate matter, this "safety valve" between temperature and pressure does not operate. Instead, the increased temperature just makes the reactions proceed at an ever-increasing rate, in a catastrophic runaway process reminiscent of the helium flash in low-mass stars (Stage 4 in Figure 20-21). The reaction spreads rapidly outward from the white dwarf's center, with its leading edge (called the *flame front*) being propelled by convection and turbulence in a manner analogous to what happens to the shock wave in a core-collapse supernova. Within seconds the white dwarf blows apart, dispersing 100% of its mass into space (Stage 5 in Figure 20-21).

Before exploding, the white dwarf contained primarily carbon and oxygen and almost no hydrogen or helium, which explains the absence of hydrogen and helium lines in the spectrum

1. The more massive member of a pair of sunlike stars exhausts its fuel and turns into a white dwarf star.
White dwarf
Companion star
2. The white dwarf sucks in gas from its companion, eventually reaching a critical mass.
Helium
3. A "flame"–a runaway nuclear reaction–ignites in the turbulent core of the dwarf.
Carbon, oxygen
Core
4. The flame spreads outward, converting carbon (^{12}C) and oxygen (^{16}O) to radioactive nickel (^{56}Ni).
Flame front
Nickel
5. Within a few seconds, the dwarf has been completely destroyed. Over the following weeks, the radioactive nickel decays, powering the bright glow of the debris.

Figure 20-21

A Type Ia Supernova This series of illustrations depicts our understanding of how a white dwarf in a close binary system can undergo a sudden nuclear detonation that destroys it completely. Such a cataclysmic event is called a Type Ia supernova or thermonuclear supernova. (Illustration by Don Dixon, adapted from Wolfgang Hillebrandt, Hans-Thomas Janka, and Ewald Müller, "How to Blow Up a Star," *Scientific American*, October 2006)

of the resulting supernova. Silicon is a by-product of the carbon-fusing reaction and gives rise to the silicon absorption line characteristic of Type Ia supernovae.

CAUTION! Different types of supernovae have fundamentally different energy sources. We saw in Section 20-6 that core-collapse supernovae (which we have now sorted into Types II, Ib, and Ic) are powered by *gravitational* energy released as the star's iron-rich core and outer layers fall inward. Type Ia supernovae, by contrast, are powered by *nuclear* energy released in the explosive thermonuclear fusion of a white dwarf star. For this reason we also use the term **thermonuclear supernova** to refer to a Type Ia supernova. While Type Ia supernovae typically emit more energy in the form of electromagnetic radiation than supernovae of other types, they do not emit copious numbers of neutrinos because there is no core collapse. If we include the energy emitted in the form of neutrinos, the most luminous supernovae by far are those of Type II.

The Decay of a Supernova: Light Curves

In addition to the differences in their spectra, different types of supernovae can be distinguished by their light curves (Figure 20-22). All supernovae begin with a sudden rise in brightness that occurs in less than a day. After reaching peak luminosity, Type Ia, Ib, and Ic supernovae settle into a steady, gradual decline in luminosity. (An example is the supernova of 1006, which is thought to have been of Type Ia. This supernova, which at its peak was more than 200 times brighter than any other star in the

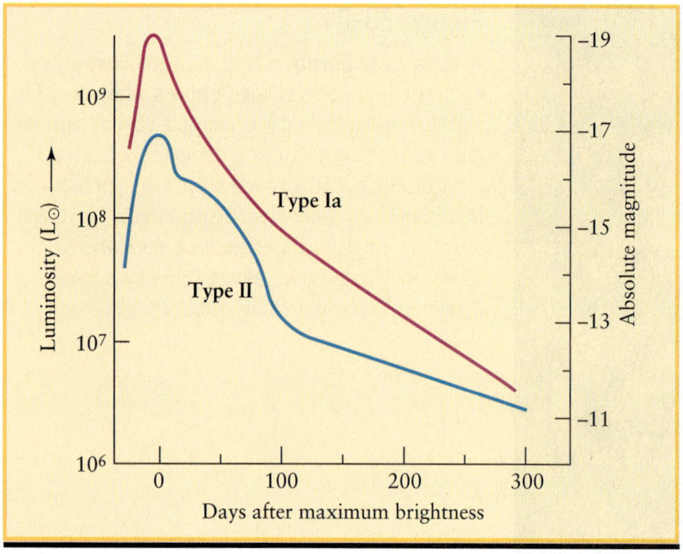

Figure 20-22

Supernova Light Curves A Type Ia supernova reaches maximum brightness in about a day, followed by a gradual decline in brightness. A Type II supernova reaches a maximum brightness only about one-fourth that of a Type Ia supernova and usually has alternating intervals of steep and gradual declines.

sky, took three years to fade into invisibility.) By contrast, the Type II light curve has a steplike appearance caused by alternating periods of steep and gradual declines in brightness.

For all supernova types, the energy source during the period of declining brightness is the decay of radioactive isotopes produced during the supernova explosion. Because a different set of thermonuclear reactions occurs for each type of supernova, each type produces a unique set of isotopes that decay at different rates. These decay rates help explain the distinctive light curves for different supernova types.

For the same reason, each type of supernova ejects a somewhat different mix of elements into the interstellar medium. As an example, Type Ia supernovae are primarily responsible for the elements near iron in the periodic table, because they generate these elements in more copious quantities than Type II supernovae.

A number of astronomers are now measuring the distances to remote galaxies by looking for Type Ia supernovae in those galaxies. This is possible because there is a simple relationship between the rate at which a Type Ia supernova fades away and its peak luminosity: The slower it fades, the greater its luminosity. Hence, by observing how rapidly a distant Type Ia supernova fades, astronomers can determine its peak luminosity. A measurement of the supernova's peak apparent brightness then tells us (through the inverse-square law) the distance to the supernova, and, therefore, the distance to the supernova's host galaxy. The tremendous luminosity of Type Ia supernovae allows this method to be used for galaxies more than 10^9 light-years distant. In Chapter 26 we will learn what such studies tell us about the size and evolution of the universe as a whole.

20-10 A supernova remnant can be detected at many wavelengths for centuries after the explosion

WEB LINK 20.12

Astronomers find the debris of supernova explosions, called **supernova remnants,** scattered across the sky. A beautiful example of a supernova remnant is the Veil Nebula, shown in Figure 20-23. The doomed star's outer layers were blasted into space with such violence that they are still traveling through the interstellar medium at supersonic speeds 15,000 years later. As this expanding shell of gas plows through space, it collides with atoms in the interstellar medium, exciting the gas and making it glow. We saw in Section 18-8 that the passage of a supernova remnant through the interstellar medium can trigger the formation of new stars, so the death of a single massive star (in a core-collapse supernova) or white dwarf (in a thermonuclear supernova) can cause a host of new stars to be born.

Many supernovae in our Galaxy are hidden from us by the obscuring interstellar medium

A few nearby supernova remnants cover sizable areas of the sky. The largest is the Gum Nebula, named after the astronomer Colin Gum, who first noticed its faint glowing wisps on photographs of the southern sky. Its astonishingly wide 40° angular diameter is centered on the constellation Vela (the Ship's Sail).

Figure 20-23 R I **V** U X G

The Veil Nebula—A Supernova Remnant This nebulosity is a portion of the Cygnus Loop, which is the roughly spherical remnant of a supernova that exploded about 15,000 years ago. The distance to the nebula is about 800 pc (2600 ly), and the overall diameter of the loop is about 35 pc (120 ly). (Palomar Observatory)

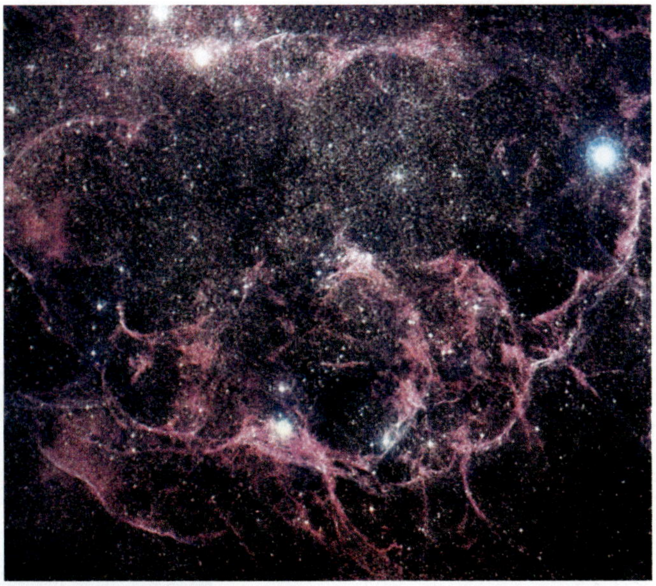

Figure 20-24 R I **V** U X G

The Vela Nebula—A Supernova Remnant This nebula has the largest angular size (40°) of any known supernova remnant. Only the central regions of the Gum Nebula are shown here. The supernova explosion occurred about 11,000 years ago, and the remnant now has a diameter of about 700 pc (2300 ly). (Royal Observatory, Edinburgh)

The Gum Nebula looks big because it's quite close to us and has had a long time to expand—its center is only about 1000 ly from Earth and it originated from a supernova explosion about a million years ago. Embedded in the Gum Nebula is another large supernova remnant called the Vela Nebula (Figure 20-24), which is about 16° wide. Studies of the nebula's expansion rate suggest that this supernova exploded around 9000 B.C. At maximum brilliance, the exploding star probably was as bright as the Moon at first quarter. Like the first quarter moon, it would have been visible in the daytime!

Many supernova remnants are nearly invisible at optical wavelengths. However, when the expanding gases collide with the interstellar medium, they radiate energy at a wide range of wavelengths, from X rays through radio waves. For example, Figure 20-25 shows a radio image of the supernova remnant Cassiopeia A. (Compare to Figure 18-24, which is a composite of observations of Cassiopeia A at X-ray, visible, and infrared wavelengths.) As a rule, radio searches for supernova remnants are more fruitful than optical searches. Only two dozen supernova remnants have been found in visible-light images, but more than 100 remnants have been discovered by radio astronomers.

From the expansion rate of Cassiopeia A, astronomers conclude that this supernova explosion occurred about 300 years ago. Although telescopes were in wide use by the late 1600s, no one saw the outburst (and no one today knows why). The last supernova seen in our Galaxy, which occurred in 1604, was observed by Johannes Kepler. In 1572, Tycho Brahe also recorded the sudden appearance of an exceptionally bright star in the sky. To find any other accounts of nearby bright supernovae, we must delve into astronomical records that are almost 1000 years old.

At first glance, this apparent lack of nearby supernovae may seem puzzling. From the frequency with which supernovae occur in distant galaxies, it's expected that we should have about two supernovae per century. Where have they been?

As we will learn when we study galaxies in Chapters 23 and 24, the plane of our Galaxy is where massive stars are born and supernovae explode. This disk-like region is so rich in interstellar dust, however, that we simply cannot see very far into space when looking in the plane of the disk, which is in the directions occupied by the Milky Way (see Section 20-2). In other words, supernovae probably do in fact erupt every few decades in remote parts of our Galaxy, but their detonations are hidden from our view by intervening interstellar matter.

Relics of the Fall: White Dwarfs, Neutron Stars, and Black Holes

A supernova remnant may be all that is left after some supernovae explode. But for core-collapse supernovae of Types II, Ib, and Ic, the core itself may also remain. If there is a relic of the core, it may be either a *neutron star* or a *black hole,* depending on the mass of the core and the conditions within it during the collapse. Neutron stars, as the name suggests, are made primarily of neutrons. Wholly unlike anything we have studied so far, these exotic objects are the subject of Chapter 21. We will study black holes, which are far stranger even than neutron stars, in Chapter 22.

CAUTION! Although neutron stars and black holes can be part of the debris from a supernova explosion, they are *not* called "supernova remnants." That term is applied exclusively to the gas and dust that spreads away from the site of the supernova explosion.

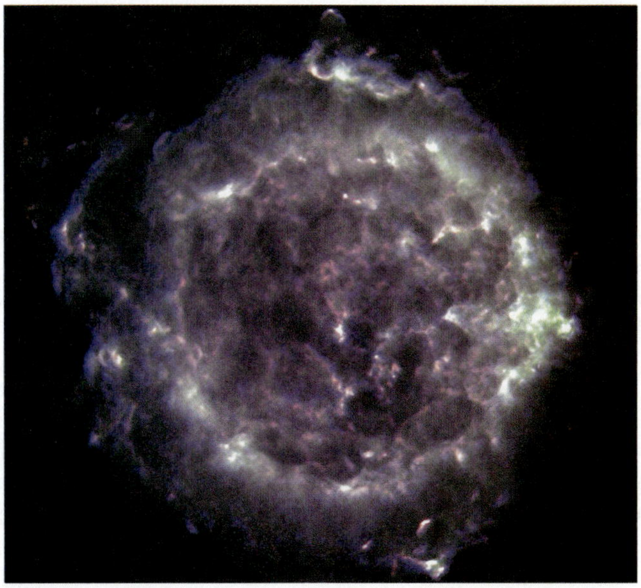

WEB LINK 20.13 ANIMATION 20.8

Figure 20-25 **R** I V U X G

Cassiopeia A—A Supernova Remnant This false-color radio image of Cassiopeia A was produced by the Very Large Array (see Figure 6-26). The impact of supernova material on the interstellar medium causes ionization, and the liberated electrons generate radio waves as they move. Cassiopeia A is roughly 3300 pc (11,000 ly) from Earth. (Image courtesy of NRAO/AUI)

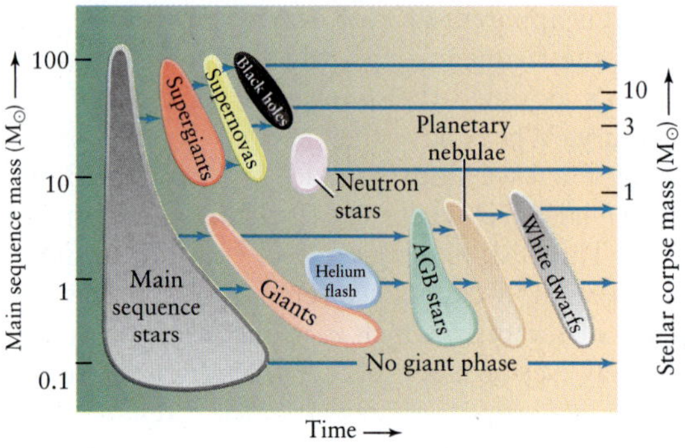

(a)

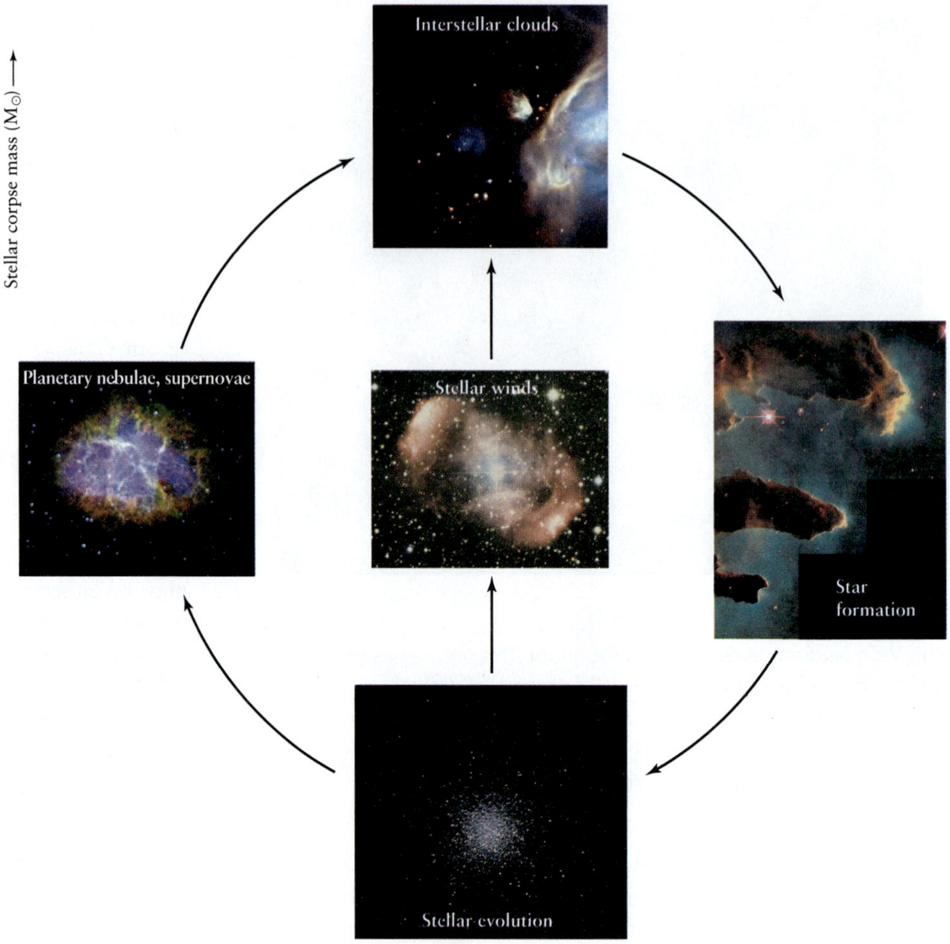

(b) R I V U X G

Figure 20-26

A Summary of Stellar Evolution **(a)** The evolution of an isolated star (one that is not part of a close multiple-star system) depends on the star's mass. The more massive the star, the more rapid its evolution. The scale on the left gives the mass of the star when it is on the main sequence, and the scale at the right shows the mass of the resulting stellar corpse. If the star's initial mass is less than about 0.4 $M_\odot$, it evolves slowly over the eons into an inert ball of helium. If the initial mass is in the range from about 0.4 $M_\odot$ to about 8 $M_\odot$, it ejects enough mass over its lifetime so that what remains is a white dwarf with a mass less than the Chandrasekhar limit of 1.4 $M_\odot$. If the star's initial mass is more than about 8 $M_\odot$ it ends as a core-collapse supernova, leaving behind a neutron star or black hole **(b)** This figure summarizes the key stages in the cycle of stellar evolution. (b: top inset, Infrared Space Observatory, NASA; right inset, Anglo-Australian Observatory/J. Hester and P. Scowen, Arizona State University/NASA; bottom inset, NASA; left inset, NASA; middle inset, Anglo-Australian Observatory)

We have seen in this chapter that only the most massive stars end their lives as core-collapse supernovae. We learned earlier that mass plays a central role in determining the speed with which a star forms and joins the main sequence (see Figure 18-10), the star's luminosity and surface temperature while on the main sequence (see Figure 17-21 and the *Cosmic Connections* figure on page 459 of Chapter 17), and how long a star can remain on the main sequence (see Table 19-1). Now we see that a star's initial mass also determines its eventual fate (Figure 20-26*a*).

Perhaps the most remarkable aspect of this rich and varied story of stellar evolution is that stars are wholly natural phenomena. Beginning with a handful of simple ingredients—hydrogen, helium, and perhaps a dash of heavier elements—stars evolve into immense, glowing orbs that shine because one chemical element naturally converts into another deep in the star's interior. As they evolve and eventually cease to shine, many stars naturally return much of their material to interstellar space, where over time new generations of stars emerge from this material (Figure 20-26*b*). To see an example of the magnificence and elegance of nature, one only needs to reflect on the processes of stellar evolution.

Key Words

asymptotic giant branch, p. 527
asymptotic giant branch star (AGB star), p. 527
carbon fusion, p. 533
carbon star, p. 528
Cerenkov radiation, p. 542
Chandrasekhar limit, p. 532
core-collapse supernova, p. 538
core helium fusion, p. 526
dredge-up, p. 528

helium shell flash, p. 529
horizontal branch, p. 526
mass-radius relation, p. 532
neon fusion, p. 535
neutron capture, p. 535
nuclear density, p. 537
oxygen fusion, p. 535
photodisintegration, p. 537
planetary nebula, p. 529
progenitor star, p. 539
red-giant branch, p. 526
shell helium fusion, p. 527
silicon fusion, p. 535
supergiant, p. 536
supernova (*plural* supernovae), p. 538
supernova remnant, p. 546
thermal pulse, p. 529
thermonuclear supernova, p. 545
Type I supernova, p. 543
Type Ia supernova, p. 543
Type Ib supernova, p. 543
Type Ic supernova, p. 543
Type II supernova, p. 543
white dwarf, p. 531

Key Ideas

Late Evolution of Low-Mass Stars: A star of moderately low mass (about 0.4 $M_\odot$ to about 4 $M_\odot$) becomes a red giant when shell hydrogen fusion begins, a horizontal-branch star when core helium fusion begins, and an asymptotic giant branch (AGB) star when the helium in the core is exhausted and shell helium fusion begins.

• As a moderately low-mass star ages, convection occurs over a larger portion of its volume. This takes heavy elements formed in the star's interior and distributes them throughout the star.

Planetary Nebulae and White Dwarfs: Helium shell flashes in an old, moderately low-mass star produce thermal pulses during which more than half the star's mass may be ejected into space. This exposes the hot carbon-oxygen core of the star.

• Ultraviolet radiation from the exposed core ionizes and excites the ejected gases, producing a planetary nebula.

• No further nuclear reactions take place within the exposed core. Instead, it becomes a degenerate, dense sphere about the size of Earth and is called a white dwarf. It glows from thermal radiation; as a white dwarf cools, it becomes dimmer.

Late Evolution of High-Mass Stars: Unlike a moderately low-mass star, a high-mass star (initial mass more than about 4 $M_\odot$) undergoes an extended sequence of thermonuclear reactions in its core and shells. These include carbon fusion, neon fusion, oxygen fusion, and silicon fusion.

• In the last stages of its life, a high-mass star has an iron-rich core surrounded by concentric shells hosting the various thermonuclear reactions. The sequence of thermonuclear reactions stops here, because the formation of elements heavier than iron requires an input of energy rather than causing energy to be released.

The Deaths of the Most Massive Stars: A star with an initial mass greater than 8 $M_\odot$ dies in a violent cataclysm in which its core collapses and most of its matter is ejected into space at high speeds. The luminosity of the star increases suddenly by a factor of around 10^8 during this explosion, producing a supernova.

• More than 99% of the energy from a core-collapse supernova is emitted in the form of neutrinos from the collapsing core.

• The matter ejected from the supernova, moving at supersonic speeds through interstellar gases and dust, glows as a nebula called a supernova remnant.

Other Types of Supernovae: An accreting white dwarf in a close binary system can also become a supernova when carbon fusion ignites explosively throughout such a degenerate star. This is called a thermonuclear supernova.

• A Type Ia supernova is produced by accreting white dwarfs in close binaries. A Type II supernova is the result of the collapse of the core of a massive star, as are supernovae of Type Ib and Type Ic; these latter types occur when the star has lost a substantial part of its outer layers before exploding.

• Most supernovae occurring in our Galaxy are hidden from our view by interstellar dust and gases.

Questions

Review Questions

1. What is the horizontal branch? Where is it located on an H-R diagram? How do stars on the horizontal branch differ from red giants or main-sequence stars?
2. Horizontal-branch stars are sometimes referred to as "helium main-sequence stars." In what sense is this true?
3. What is the asymptotic giant branch? Where is it located on an H-R diagram? How do asymptotic giant branch stars differ from red giants or main-sequence stars?
4. Is a carbon star a star that is made of carbon? Explain your answer.
5. What is the connection between dredge-ups in old stars and life on Earth?
6. What are thermal pulses in AGB stars? What causes them? What effect do they have on the luminosity of the star?
7. How is a planetary nebula formed?
8. How can an astronomer tell the difference between a planetary nebula and a planet?
9. What is the evidence that typical planetary nebulae are only a few thousand years old?
10. Why do we not observe planetary nebulae that are more than about 50,000 years old?
11. What is a white dwarf? Does it produce light in the same way as a star like the Sun?
12. How does the radius of a white dwarf depend on its mass? How is this different from other types of stars?
13. What is the significance of the Chandrasekhar limit?
14. On an H-R diagram, sketch the evolutionary track that the Sun will follow from when it leaves the main sequence to when it becomes a white dwarf. Approximately how much mass will the Sun have when it becomes a white dwarf? Where will the rest of the mass go?
15. What prevents thermonuclear reactions from occurring at the center of a white dwarf? If no thermonuclear reactions are occurring in its core, why doesn't the star collapse?
16. A white dwarf has a greater mass than either a red dwarf or a brown dwarf. Yet a white dwarf has a smaller radius than either a red dwarf or a brown dwarf. Explain why, in terms

of the types of pressure that keep the different kind of dwarfs from collapsing under their own gravity.

17. Why do you suppose that all the white dwarfs known to astronomers are relatively close to the Sun?
18. Why does the mass of a star play such an important role in determining the star's evolution?
19. Why is the temperature in a star's core so important in determining which nuclear reactions can occur there?
20. What is the difference between a red giant and a red supergiant?
21. Why does the evolutionary track of a high-mass star move from left to right and back again in the H-R diagram?
22. In what way does the structure of an aging supergiant resemble that of an onion?
23. What is nuclear density? Why is it significant when a star's core reaches this density?
24. Why is SN 1987A so interesting to astronomers? In what ways was it not a typical supernova?
25. Why are neutrinos emitted by core-collapse supernovae? How can these neutrinos be detected? How can they be distinguished from solar neutrinos?
26. What causes a thermonuclear supernova? How does a thermonuclear supernova compare with a core-collapse supernova?
27. What are the differences among Type Ia, Type Ib, Type Ic, and Type II supernovae? Which type is most unlike the other three, and why?
28. How can a supernova continue to shine for many years after it explodes?
29. How do supernova remnants produce radiation at nonvisible wavelengths?
30. There may have been recent supernovae in our Galaxy that have not been observable even though they are incredibly luminous. How is it this possible?
31. Is our own Sun likely to become a supernova? Why or why not?

Advanced Questions

Questions preceded by an asterisk (*) involve topics discussed in the Boxes in Chapter 1, Chapter 7, or Chapter 17.

Problem-solving tips and tools

The small-angle formula is given in Box 1-1. You may find it useful to review Box 17-4, which discusses stellar radii and their relationship to temperature and luminosity. Section 4-7 explains the formula for gravitational force, and Box 7-2 explains the concept of escape speed. Sections 5-2 and 5-4 describe some key properties of light, especially blackbody radiation. We discussed the relationship among luminosity, apparent brightness, and distance in Box 17-2. The relationship among absolute magnitude, apparent magnitude, and distance was the topic of Box 17-3. In our discussion of binary stars in Section 17-9 we saw how the masses of the stars, the orbital period, and the average distance between the two stars are all related.

32. Some blue main-sequence stars in our region of the Galaxy have the same luminosity and surface temperature as the horizontal-branch stars in the globular cluster M55 (see Figure 20-3). How do we know that the horizontal-branch stars in M55 are not main-sequence stars?
33. Stellar winds from an AGB star can cause it to lose mass at a rate of up to 10^{-4} $M_\odot$ per year. (**a**) Express this rate in metric tons per second. (One metric ton equals 1000 kilograms.) (**b**) At this rate, how long would it take an AGB star to eject an amount of mass equal to the mass of Earth? Express your answer in days.
34. The globular cluster M15 depicted in Figure 20-6*a* contains 30,000 old stars, but only one of these stars is presently in the planetary nebula stage of its evolution. Explain why planetary nebulae are not more prevalent in M15.
35. Explain why astronomers had to use infrared light to detect the faint wisps in planetary nebula NGC 7027 (Figure 20-6*c*).
36. The central star in a newly formed planetary nebula has a luminosity of 1000 $L_\odot$ and a surface temperature of 100,000 K. What is the star's radius? Give your answer as a multiple of the Sun's radius.
37. You want to determine the age of a planetary nebula. What observations should you make, and how would you use the resulting data?
38. Figure 20-6*b* shows the planetary nebula Abell 39. Use the information given in the figure caption to calculate the angular diameter of the nebula as seen from Earth.

*39. The Ring Nebula is a planetary nebula in the constellation Lyra. It has an angular size of 1.4 arcmin × 1.0 arcmin and is expanding at the rate of about 20 km/s. Approximately how long ago did the central star shed its outer layers? Assume that the nebula is 2,700 ly from Earth.

40. The accompanying image shows the planetary nebula IC 418 in the constellation Lepus (the Hare). (**a**) The image shows a small shell of glowing gas (shown in blue) within a larger glowing gas shell (shown in orange). Discuss how IC 418 could

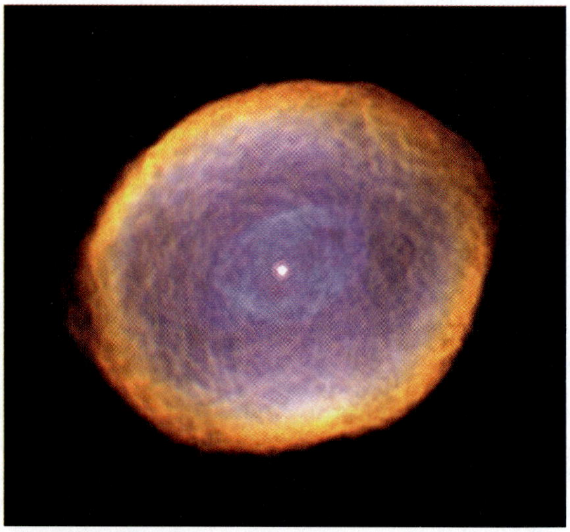

R I V U X G

(NASA and Hubble Heritage Team, STScI/AURA)

have acquired this pair of gas shells. (**b**) Explain why the outer shell looks thicker around the edges than near the middle.

41. (**a**) Calculate the wavelength of maximum emission of the white dwarf Sirius B. In what part of the electromagnetic spectrum does this wavelength lie? (**b**) In a visible-light photograph such as Figure 20-8, Sirius B appears much fainter than its primary star. But in an image made with an X-ray telescope, Sirius B is the brighter star. Explain the difference.

***42.** Sirius is 2.63 pc from Earth. By making measurements on Figure 20-8, calculate the distance between the centers of Sirius A and Sirius B at the time that this image was made in October 2003. Give your answer in astronomical units (AU). (Note that your result is the true distance only if Sirius A and Sirius B were exactly the same distance from Earth at the time the image was made. If one of the stars was closer to us than the other, the actual distance between them is greater than what you calculate.)

43. (**a**) Find the average density of a 1-$M_\odot$ white dwarf having the same diameter as Earth. (**b**) What speed is required to eject gas from the white dwarf's surface? (This is also the speed with which interstellar gas falling from a great distance would strike the star's surface.)

44. In the classic 1960s science-fiction comic book *The Atom,* a physicist discovers a basketball-sized meteorite (about 10 cm in radius) that is actually a fragment of a white dwarf star. With some difficulty, he manages to hand-carry the meteorite back to his laboratory. Estimate the mass of such a fragment, and discuss the plausibility of this scenario.

***45.** (**a**) Use the information in the caption to Figure 20-12 to calculate the diameter of the nebula SMC N76. Express your answer in parsecs. (**b**) How does your answer to part (a) compare to the diameters of the planetary nebulae depicted in Figure 20-6? Explain how this is consistent with the observation that gases ejected from a supergiant travel faster than gases ejected from an AGB star.

***46.** The supergiant star depicted in Figure 20-12 is actually one member of a binary star system. The masses of the two stars are 18 $M_\odot$ and 34 $M_\odot$, and the orbital period is 19.56 days. (**a**) What is the average separation between the two stars? Give your answer in AU. (**b**) Compare your answer in part (a) to the sizes of the orbits of Mercury, Venus, and Earth around the Sun.

47. (**a**) What kinds of stars would you monitor if you wished to observe a core-collapse supernova explosion from its very beginning? (**b**) Examine Appendices 4 and 5, which list the nearest and brightest stars, respectively. Which, if any, of these stars are possible supernova candidates? Explain your answer.

***48.** Consider a high-mass star just prior to a supernova explosion, with a core of diameter 20 km and density 4×10^{17} kg/m^3. (**a**) Calculate the mass of the core. Give your answer in kilograms and in solar masses. (**b**) Calculate the force of gravity on a 1-kg object at the surface of the core. How many times larger is this than the gravitational force on such an object at the surface of Earth, which is about 10 newtons? (**c**) Calculate the escape speed from the surface of the star's core. Give your answer in m/s and as a fraction of the speed of light. What does this tell you about how powerful a supernova explosion must be in order to blow material away from the star's core?

49. The shock wave that traveled through the progenitor star of SN 1987A took 3 hours to reach the star's surface. (**a**) Given the size of a blue supergiant star (see Section 20-7), estimate the speed with which the shock wave traveled through the star's outer layers. (The core of the progenitor star was very small, so you may consider the shock wave to have started at the very center of the star.) Give your answer in meters per second. (**b**) Compare your answer with the speed of sound waves in our atmosphere, about 340 m/s, and with the speed of light. (**c**) A shock wave traveling through a gas is a special case of a sound wave. In general, sound waves travel faster through denser, less easily compressed materials. Thus, sound travels faster through water (about 1500 m/s) than through our atmosphere, and faster still through steel (about 5900 m/s). Use this idea to compare the gases within the progenitor star of SN 1987A with the gases in our atmosphere in terms of their average density and how easily they are compressed.

50. The neutrinos from SN 1987A arrived 3 hours before the visible light. While they were en route to Earth, what was the distance between the neutrinos and the first photons from SN 1987A? Assume that neutrinos are massless and thus travel at the speed of light. Give your answers in kilometers and in AU.

***51.** Compared to SN 1987A (see Figure 20-17), the supernova SN 1993J (see Figure 20-16) had a maximum apparent brightness only 9.1×10^{-4} as great. Using the distances from Earth to each of these supernovae, determine the ratio of the maximum luminosity of SN 1993J to that of SN 1987A. Which of the two supernovae had the greater maximum luminosity?

***52.** Suppose that the brightness of a star becoming a supernova increases by 20 magnitudes. Show that this corresponds to an increase of 10^8 in luminosity.

***53.** Suppose that the red-supergiant star Betelgeuse, which lies some 425 light-years from Earth, becomes a Type II supernova. (**a**) At the height of the outburst, how bright would it appear in the sky? Give your answer as a fraction of the brightness of the Sun ($b_\odot$). (**b**) How would it compare with the brightness of Venus (about 10^{-9} $b_\odot$)?

***54.** In July 1997, a supernova named SN 1997cw exploded in the galaxy NGC 105 in the constellation Cetus (the Whale). It reached an apparent magnitude of +16.5 at maximum brilliance, and its spectrum showed an absorption line of ionized silicon. Use this information to find the distance to NGC 105. (*Hint:* Inspect the light curves in Figure 20-22 to find the *absolute* magnitudes of typical supernovae at peak brightness.)

55. Figure 20-23 shows a portion of the Veil Nebula in Cygnus. Use the information given in the caption to find the average speed at which material has been moving away from the site of the supernova explosion over the past 15,000 years. Express your answer in km/s and as a fraction of the speed of light.

56. The images that open this chapter show two kinds of glowing gas clouds: a planetary nebula and a supernova remnant. (**a**) Explain what makes the planetary nebula glow and what makes the supernova remnant glow. (*Hint:* The explanations are different for the two kinds of gas clouds.) (**b**) Which of these two kinds of gas clouds continues to glow for a longer time? Why?

57. The planetary nebula and supernova remnant shown in the images that open this chapter are both about the same age. Both objects consist of glowing gases that have expanded away from a central star. Based on these images, in which of these objects have the gases expanded more rapidly? Explain your reasoning.

Discussion Questions

58. Suppose that you discover a small, glowing disk of light while searching the sky with a telescope. How would you decide if this object is a planetary nebula? Could your object be something else? Explain.
59. Suppose the convective zone in AGB stars did *not* reach all the way down into their carbon-rich cores. How might this have affected the origin and evolution of life on Earth?
60. Imagine that our Sun was somehow replaced by a 1-$M_\odot$ white dwarf star, and that our Earth continued in an orbit of semimajor axis 1 AU around this star. Discuss what effects this would have on our planet. What would the white dwarf look like as seen from Earth? Could you look at it safely with the unaided eye? Would Earth's surface temperature remain the same as it is now?
61. The similar names *white dwarf*, *red dwarf*, and *brown dwarf* describe three very different kinds of objects. Suggest better names for these three kinds of objects, and describe how your names more accurately describe the objects' properties.
62. The major final product of silicon fusion is ^{56}Fe, an isotope of iron with 26 protons and 30 neutrons. This is also the most common isotope of iron found on Earth. Discuss what this tells you about the origin of the solar system.
63. SN 1987A did not agree with the theoretical picture outlined in Section 20-6. Does this mean that the theory was wrong? Discuss.

Web/eBook Questions

64. It has been claimed that the Dogon tribe in western Africa has known for thousands of years that Sirius is a binary star. Search the World Wide Web for information about these claims. What is the basis of these claims? Why are scientists skeptical, and how do they refute these claims?
65. Search the World Wide Web for recent information about SN 1993J. Has the shape of the supernova's light curve been adequately explained? Has the supernova produced any surprises?
66. Search the World Wide Web for information about SN 1994I, a supernova that occurred in the galaxy M51 (NGC 5194). Why was this supernova unusual? Was it bright enough to have been seen by amateur astronomers?
67. **Convection Inside a Giant Star.** Access and view the animation "Convection Inside a Giant Star" in Chapter 20 of the *Universe* Web site or eBook. Describe the motion of material in the interior of the star. In what ways is this motion similar to convection within the present-day Sun (see Section 16-2)? In what ways is it different? Is a dredge-up taking place in this animation? How can you tell?
68. **Types of Supernovae.** Access and view the animations "In the Heart of a Core-Collapse (Type II, Ib, or Ic) Supernova" and "A Thermonuclear (Type Ia) Supernova" in Chapter 20 of the *Universe* Web site or eBook. Describe how these two types of supernova are fundamentally different in their origin.

Activities

Observing Projects

Observing tips and tools

While planetary nebulae are rather bright objects, their brightness is spread over a relatively large angular size, which can make seeing them a challenge for the beginning observer. For example, the Helix Nebula (shown in the left-hand image on the page that opens this chapter) has the largest angular size of any planetary nebula but is also one of the most difficult to see. To improve your view, make your observations on a dark, moonless night from a location well shielded from city lights. Another useful trick, mentioned in Chapters 17 and 18, is to use "averted vision." Once you have the nebula centered in the telescope, you will get a brighter and clearer image if you look at the nebula out of the corner of your eye. The so-called Blinking Planetary in Cygnus affords an excellent demonstration of this effect; the nebula seems to disappear when you look straight at it, but it reappears as soon as you look toward the side of your field of view.

Another useful tip is to view the nebula through a green filter (a #58, or O III, filter available from telescope supply houses). Green light is emitted by excited, doubly ionized oxygen atoms, which are common in planetary nebulae but not in most other celestial objects. Using such a filter can make a planetary nebula stand out more distinctly against the sky. As a side benefit, it also helps to block out stray light from street lamps. The same tips also apply to observing supernova remnants.

69. Although they represent a fleeting stage at the end of a star's life, planetary nebulae are found all across the sky. Some of the brightest are listed in the accompanying table. Note that the distances to most of these nebulae are quite uncertain. Observe as many of these planetary nebulae as you can on a clear, moonless night using the largest telescope at your disposal. Note and compare the various shapes of the different nebulae. In how many cases can you see the central star? The central star in the Eskimo Nebula is supposed to be the "nose" of an Eskimo wearing a parka. Can you see this pattern?
70. Northern hemisphere observers with modest telescopes can see two supernova remnants, one in the winter sky and the other in the summer sky. Both are quite faint, however, so you should schedule your observations for a moonless night.

Planetary nebula	Distance (light-years)	Angular size	Constellation	Right ascension	Declination
Dumbbell (M27, NGC 6853)	490–3500	8.0 × 5.7	Vulpecula	19^h 59.6^m	+22° 43′
Ring (M57, NGC 6720)	1300–4100	1.4 × 1.0	Lyra	18 53.6	+33 02
Little Dumbbell (M76, NGC 650)	1700–15,000	2.7 × 1.8	Perseus	01 42.4	+51 34
Owl (M97, NGC 3587)	1300–12,000	3.4 × 3.3	Ursa Major	11 14.8	+55 01
Saturn (NGC 7009)	1600–3900	0.4 × 1.6	Aquarius	21 04.2	−11 22
Helix (NGC 7293)	450	41 × 41	Aquarius	22 29.6	−20 48
Eskimo (NGC 2392)	1400–10,000	0.5 × 0.5	Gemini	7 29.2	+20 55
Blinking Planetary (NGC 6826)	3300(?)	2.2 × 0.5	Cygnus	19 44.8	+50 31

The winter sky contains the Crab Nebula, which is discussed in detail in Chapter 23. The coordinates are R.A. = 5^h 34.5^m and Decl. = +22°00′, which places the object near the star marking the eastern horn of Taurus (the Bull). Whereas the entire Crab Nebula easily fits in the field of view of an eyepiece, the Veil or Cirrus Nebula in the summer sky is so vast that you can see only a small fraction of it at a time. The easiest way to find the Veil Nebula is to aim the telescope at the star 52 Cygni (R.A. = 20^h 45.7^m and Decl. = +30° 43′), which lies on one of the brightest portions of the nebula. If you then move the telescope slightly north or south until 52 Cygni is just out of the field of view, you should see faint wisps of glowing gas.

71. Use a telescope to observe the remarkable triple star 40 Eridani, whose coordinates are R.A. = 4^h 15.3^m and Decl. = −7° 39′. The primary, a 4.4-magnitude yellowish star like the Sun, has a 9.6-magnitude white dwarf companion, the most easily seen white dwarf in the sky. On a clear, dark night with a moderately large telescope, you should also see that the white dwarf has an 11th-magnitude companion, which completes this most interesting trio.

72. The red supergiant Betelgeuse in the constellation Orion will explode as a supernova at some time in the future. Use the *Starry Night Enthusiast*™ program to investigate how the supernova might appear if this explosion were to happen tonight. Click the **Home** button in the toolbar to show the sky as seen from your location at the present time. (If the program does not place you at your true location, use the **Viewing Location . . .** command in the **Options** menu.) Use the **Find** pane to locate Betelgeuse. If Betelgeuse is below the horizon, allow the program to reset the time to when it is visible. (**a**) At what time does Betelgeuse rise on today's date? At what time does it set? (**b**) If Betelgeuse became a supernova today, would it be visible in the daytime? How would it appear at night? Do you think it would cast shadows? (**c**) Are Betelgeuse and the Moon both in the night sky tonight? (Use the **Find** pane to locate the Moon.) If they are, and Betelgeuse were to become a supernova, what kinds of shadows might they both cast?

73. Use the ***Starry Night Enthusiast***™ program to show the location of Supernova 1987A. In the menu, select **Favorites > Deep Space > Local Universe** to display the Milky Way and other nearby galaxies, conveniently labeled, against the background of distant galaxies, from a distance of 0.282 Mly from the Sun. (If the Milky Way does not appear immediately, click once on either of the **Zoom** buttons.) Remove the image of the astronaut's feet by clicking on **View > Feet.** You can rotate the Milky Way Galaxy and its neighbor galaxies by holding down both the **Shift** button and the mouse button while moving the mouse. (On a two-button mouse, hold down the left mouse button.) (**a**) Use the **Find** pane to locate and center the Sun in the field of view. Describe the position of the Large Magellanic Cloud (LMC), within which SN 1987A lies, relative to the Milky Way Galaxy and to our solar system. (**b**) Use the **Find** pane to center on the LMC. You should be able to locate the Tarantula Nebula, shown in Figure 20-17. Is SN 1987A near to the center or the edge of the LMC? (Note that, although *Starry Night Enthusiast*™ depicts the LMC as being rather flat, it is thought to be an irregular blob of stars with some thickness.)

Collaborative Exercise

74. Imagine that a supernova originating from a close binary star system, both of whose stars have less than 4 solar masses, began (as seen from Earth) on the most recent birthday of the youngest person in your group. Using the light curves in Figure 20-22, what would its new luminosity be today and how bright would it appear in the sky (apparent magnitude) if it were located 10 parsecs (32.6 light-years) away? How would your answers change if you were to discover that the supernova actually originated from an isolated star with a mass 15 times greater than our Sun?

21 Neutron Stars

WEB LINK 21.1 RIVUXG

A composite image of the region around a neutron star within the Crab Nebula using X-ray (blue) and visible (red) wavelengths. (X-ray: NASA/CXC/ASU/J. Hester et al.; optical: NASA/CXC/ASU/J. Hester et al.)

Almost all the light we see from objects in the night sky is due to thermonuclear fusion reactions. Such reactions are what make stars shine, and in turn make any surrounding nebulae glow. Even the light we see from the Moon and planets—which is simply reflected sunlight—can be traced back to thermonuclear reactions in the Sun's core.

By contrast, the exotic object shown here—within the Crab Nebula in Taurus—is undergoing no thermonuclear fusion at all. Despite this, the object is so energetic that it emits copious amounts of X rays. This image shows tilted, glowing rings more than a light-year across, with oppositely directed jets flowing outward perpendicular to the plane of the rings. At the object's very center, too small to pick out in this image, is the source of all this dynamic activity: a neutron star, the leftover core of what was once a supergiant star.

Before 1967 neutron stars were considered pure speculation. But today we know of more than 1800 of these exotic stellar corpses, and hundreds of thousands more may be strewn across the Milky Way Galaxy. As we will discover in this chapter, neutron stars have powerful magnetic fields that are billions of times stronger than the Sun's field. As a neutron star rotates, its magnetic field sweeps beams of radiation across the sky. We detect these beams as pulsating radio signals.

Unlike normal stars, neutron stars actually have solid surfaces that can shift and fracture in a "starquake." When such a quake occurs on the surface of a neutron star with an unusually strong magnetic field, it releases a truly colossal burst of radiation that for a fraction of a second outshines an entire galaxy of stars.

We will find that neutron stars in close binary systems also display outrageous behavior, including bouts of explosive helium fusion that yield a vast outpouring of X rays. As we will see, a similar process involving a white dwarf in a close binary system produces a short-lived intense burst of visible light called a nova.

21-1 A neutron star is even more highly compressed than a white dwarf

On the morning of July 4, 1054, Yang Wei-T'e, the imperial astronomer to the Chinese court, made a startling discovery. Just a few minutes before sunrise, a new and dazzling object ascended above the eastern horizon. This "guest star," as Yang called it, was far brighter than Venus and more resplendent than any star he had ever seen.

Yang's records show that the "guest star" was so brilliant that it could easily be seen during broad daylight for the rest of July. Records from Constantinople (now Istanbul, Turkey) also

Learning Goals

By reading the sections of this chapter, you will learn

21-1 Why astronomers predicted the existence of neutron stars before they were discovered

21-2 What pulsars are and how they were discovered

21-3 The relationship between pulsars and neutron stars

21-4 Why pulsars slow down over time

21-5 What exotic states of matter are thought to exist inside neutron stars

21-6 How neutron stars with abnormally strong magnetic fields can produce amazingly intense bursts of radiation

21-7 How some pulsars are accelerated to truly dizzying rotation speeds

21-8 How pulsars in close binary systems can emit X-rays

21-9 How novae produced by white dwarfs are similar to X-ray bursters produced by neutron stars

21-10 What sets the upper limit on the mass of a neutron star

Figure 21-1 R I V U X G

A Supernova Pictograph? This drawing in an eleventh-century structure in New Mexico shows a ten-pointed star next to a crescent. It may depict the scene on the morning of July 5, 1054, when a "guest star" appeared next to the waning crescent moon. (Courtesy of National Parks Service)

describe this object, and works of art made by the Chacoans in the American Southwest suggest that they may have seen it as well (Figure 21-1). Over the next 21 months, however, the "guest star" faded to invisibility.

We now know that the "guest star" of 1054 was actually a remarkable stellar transformation: A massive star some 6500 light-years away perished in a supernova explosion, leaving behind both a supernova remnant (see Figure 1-8 and Section 20-10) and a bizarre object called a **neutron star**—an incredibly dense sphere composed primarily of neutrons. However, it took many centuries after Yang Wei-T'e's observations to understand what had happened; it was not until 1932 that the neutron itself was discovered by the English physicist James Chadwick. (We learned in Section 5-7 that the neutron has about the same mass as the positively charged proton, but has no electric charge.)

We saw in Section 20-6 that under very high pressure, a proton and an electron can combine to form a neutron (as well as a neutrino). Could such pressures be found within the core of a dying high-mass star? Within a year of Chadwick's discovery, astronomers Fritz Zwicky at the California Institute of Technology (Caltech) and his colleague Walter Baade at Mount Wilson Observatory made just this prediction—that a massive stellar core can transform into a sphere of neutrons. "With all reserve," Zwicky and Baade theorized, "we advance the view that supernovae represent the transition from ordinary stars into neutron stars, which in their final stages consist of extremely closely packed neutrons." In other words, there could be at least two types of stellar corpses—white dwarfs and neutron stars.

Zwicky and Baade realized that once a neutron star had formed in response to tremendous pressures, it would be able to resist any further compression. The neutrons themselves would provide a counterbalancing outward pressure. Motivated by the idea that white dwarfs are supported by degenerate electron pressure, Zwicky and Baade proposed that a highly compact ball of neutrons would similarly produce a **degenerate neutron pressure.** (Like electrons, neutrons obey the Pauli exclusion principle that we described in Section 19-3.) This powerful pressure could support a stellar corpse—perhaps one even more massive than a white dwarf, because degenerate neutron pressure can be much stronger than degenerate electron pressure. So, while a white dwarf collapses if its mass is above the Chandrasekhar limit (see Section 20-4), a neutron star might not.

Although neutron stars were first hypothesized in the 1930s, the idea seemed so strange that it was ignored for over thirty years

Although Zwicky and Baade's proposal would prove to be very close to the mark, most scientists politely ignored it for years. After all, a neutron star must be a rather weird object. If brought to Earth's surface, a single teaspoonful of neutron star matter would weigh 100 million tons. That weight is more than 10 million elephants!

A star compacted to such densities must be very small. A $1.4\text{-}M_\odot$ neutron star would have a diameter of only 20 km (12 mi), about the size of a moderate-sized city on Earth. Its surface gravity would be so strong that an object would have to travel at one-half the speed of light to escape into space. These conditions seemed outrageous until the late 1960s, when astronomers discovered pulsating radio sources.

21-2 The discovery of pulsars in the 1960s stimulated interest in neutron stars

As a young graduate student at Cambridge University, Jocelyn Bell spent many months helping construct an array of radio antennas covering $4\frac{1}{2}$ acres of the English countryside. The instrument was completed by the summer of 1967, and Bell and her colleagues in Anthony Hewish's research group began using it to scrutinize radio emissions from the sky. They were looking for radio sources that "twinkle" like stars; that is, they looked for random small fluctuations in brightness caused by the motion of gas between the source and the observer. What they discovered was something far more exotic.

The Discovery of Pulsars

While searching for random flickering, Bell noticed that the antennas had detected regular pulses of radio noise from one particular location in the sky. These radio pulses were arriving at regular intervals of 1.3373011 seconds—much more rapid than those of any other astronomical object known at that time. Indeed, they were so rapid and regular that the Cambridge team at first suspected that they might not be of natural origin. Instead, it was proposed that these pulses might be signals from an advanced alien civilization.

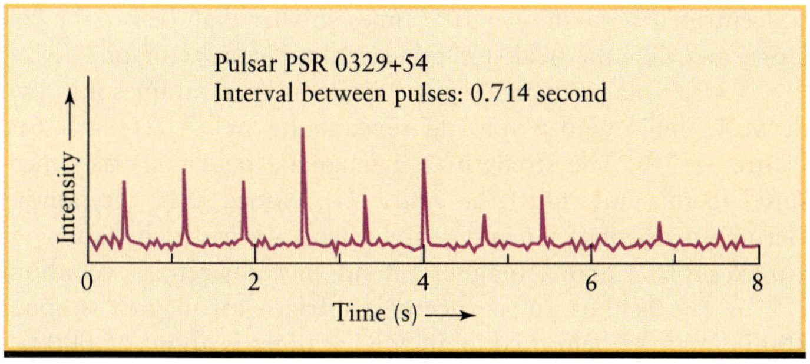

WEBLINK 21.4

Figure 21-2

A Recording of a Pulsar This chart recording shows the intensity of radio emission from one of the first pulsars to be discovered. (The designation means "pulsar at a right ascension of 03^h 29^m and a declination of $+54°$") Note that the interval between pulses is very regular, even though some pulses are weak and others are strong. (Adapted from R. N. Manchester and J. H. Taylor)

That possibility had to be discarded within a few months after several more of these pulsating radio sources, which came to be called **pulsars**, were discovered across the sky. In all cases, the periods were extremely regular, ranging from about 0.25 second for the fastest to about 1.5 seconds for the slowest (Figure 21-2).

Ruling Out Possibilities

What could these objects possibly be? As is often done in science, astronomers first decided what pulsars could *not* be. It was clear that pulsars were not ordinary stars or nebulae, because while these familiar objects emit some radio waves, their emissions do not pulsate (that is, their light output does not increase and decrease rhythmically). One type of star that *appears* to pulsate is an eclipsing binary star system, which appears dimmer when one star in the binary system passes in front of the other (see Section 17-11). However, in order to vary in brightness with a typical pulsar period of 1 second or less, an eclipsing binary would have to have an equally short orbital period. Newton's form of Kepler's third law (see Section 4-7) implies that the two stars would have to orbit only 1000 kilometers or so from each other. This distance is less than the diameter of any ordinary star, even a white dwarf star. Hence, the eclipsing binary hypothesis can be ruled out, because it leads to the highly implausible picture of two stars that overlap!

Astronomers also ruled out the idea that pulsars might be variable stars, which vary in luminosity as they increase and decrease in size (see Section 19-6). Pulsating variables have periods of days or weeks, while pulsars have periods of seconds or less. Just as the broad wings of an eagle could never flap as fast as a hummingbird's wings without breaking, an ordinary star or even a white dwarf is too large to pulsate in and out in less than a second.

Another possibility, however, was given serious consideration. Perhaps pulsars were rapidly rotating white dwarf stars with some sort of radio-emitting "hot spots" on their surfaces. But even this model had its limits, because a white dwarf that rotated on its axis in just 1 second would be on the verge of flying apart. What finally ruled out the spinning white dwarf model was the discovery of a pulsar in the middle of the Crab Nebula, the supernova remnant left behind by the supernova that Yang Wei-T'e saw in 1054 (see Figure 1-8). That discovery clarified the true nature of pulsars.

At the time of its discovery, the Crab pulsar was the fastest pulsar known to astronomers. Its period is 0.0333 second, which means that it flashes $1/0.0333 = 30$ times each second. It was immediately apparent to astrophysicists that white dwarfs are too big and bulky to generate 30 signals per second; calculations demonstrated that a white dwarf could not rotate that fast without tearing itself apart. Hence, the existence of the Crab pulsar means that the stellar corpse at the center of the Crab Nebula has to be much smaller and more compact than a white dwarf.

The rapid flashing of radiation from a pulsar gave evidence that white dwarfs are not the only endpoint of stellar evolution

This was a truly unsettling realization, because most astronomers in the mid-1960s thought that *all* stellar corpses are white dwarfs. The number of white dwarfs in the sky seemed to account for all the stars that must have died since our Galaxy was formed. It was generally assumed that all dying stars—even the most massive ones, which produce supernovae—somehow manage to eject enough matter so that their corpses do not exceed the Chandrasekhar limit. But the discovery of the Crab pulsar showed these conservative opinions to be in error.

It was now clear that pulsars had to be something far more exotic than rotating white dwarfs. As Thomas Gold of Cornell University emphasized, astronomers would now have to seriously consider the existence of neutron stars.

21-3 Pulsars are rapidly rotating neutron stars with intense magnetic fields

TUTORIAL 21.1

Science is, by its very nature, based on skepticism. If a scientist proposes a new and exotic explanation for an unexplained phenomenon, that explanation will be accepted only if it satisfies two demanding criteria. First, there must be excellent reasons why more conventional explanations cannot work. Second, the new explanation must help scientists understand a range of other phenomena.

A good example is the Copernican idea that Earth orbits the Sun, which gained general acceptance only after Galileo's observations (see Section 4-5). First, Galileo showed the older geocentric explanation of celestial motions to be untenable, and, second, it became clear that the Copernican model could also explain the motions of Jupiter's moons.

How the Neutron-Star Model Overcame Skepticism

We have seen how the discovery of the Crab pulsar ruled out the idea that pulsars were white dwarfs. But astronomers soon accepted the more exotic idea that pulsars were actually neutron stars. To understand why, we must assume the skeptical outlook of a scientist and ask some probing questions about this idea. Why should neutron stars emit radiation at all? In particular, why should they emit radio waves? Why should the emissions be pulsed? And how could the pulses occur as rapidly as 30 times per second, as in the Crab pulsar? The answers to all of these questions are intimately related to the small size of neutron stars.

LOOKING DEEPER 21.1

Because neutron stars are very small, they should also rotate rapidly. A typical star, such as our Sun, takes nearly a full month to rotate once about its axis. But just as a spinning ice skater speeds up when she pulls in her arms, a collapsing star also speeds up as its size shrinks. (We introduced this principle, called the *conservation of angular momentum,* in Section 8-4. See Figure 8–7.) If our Sun were compressed to the size of a neutron star, it would spin about 1000 times per second! Because neutron stars are so small and dense, they can spin this rapidly without flying apart.

The neutron-star model of pulsars had to pass several stringent tests to be accepted by astronomers

The small size of neutron stars also implies that they have intense magnetic fields. It seems safe to say that every star possesses some magnetic field, but typically the strength of this field is quite low. The magnetic field of a main-sequence star is spread out over billions of square kilometers of the star's surface. However, if such a star collapses down to a neutron star, its surface area (which is proportional to the square of its radius) shrinks by a factor of about 10^{10}. The magnetic field, which weaves through and is connected to the star's ionized gases (see Section 16-9), becomes concentrated onto an area 10^{10} times smaller than before the collapse, and thus the field strength increases by a factor of 10^{10}.

A star's magnetic field splits one of its spectral lines into two or more lines whose spacing reveals the field's strength (see Figure 16-20). The strength of a magnetic field is usually measured in the unit called the *gauss* (G), named after the famous German mathematician and astronomer Karl Friedrich Gauss. (As an example, Earth's magnetic field has a strength of about 0.5 G. The field at the surface of a refrigerator magnet is about 100 G, and the field used in an MRI scanner is about 15,000 G.) The magnetic splitting of lines in the spectra of main-sequence stars reveals field strengths in the range from about 1 G (the value for the Sun) to several thousand gauss. The splitting of lines in the spectra of certain white dwarfs corresponds to field strengths in excess of 10^6 G, comparable to the strongest fields ever produced in a laboratory on Earth. But the magnetic fields surrounding typical neutron stars are a million times stronger still, on the order of 1 trillion gauss (10^{12} G). As we will see in Section 21-6, some neutron stars may have fields as strong as 10^{15} G.

The magnetic field of a neutron star makes it possible for the star to radiate pulses of energy toward our telescopes. A number of different models have been proposed for how rotating neutron stars generate radiation; we'll discuss just one leading model. The idea behind this model is that the magnetic axis of a neutron star, the line connecting the north and south magnetic poles, is likely to be inclined at an angle to the rotation axis (Figure 21-3*a*). After

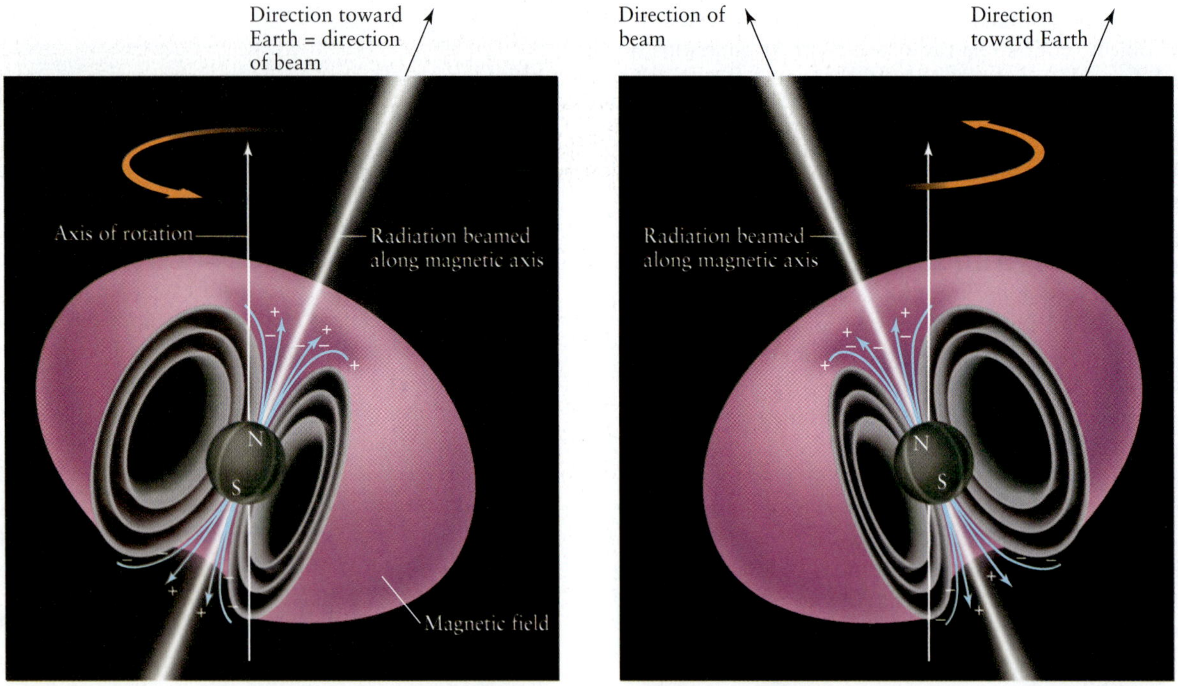

(a) One of the beams from the rotating neutron star is aimed toward Earth: we detect a pulse of radiation.

(b) Half a rotation later, neither beam is aimed toward Earth: we detect that the radiation is "off."

Figure 21-3

A Rotating, Magnetized Neutron Star Charged particles are accelerated near a magnetized neutron star's magnetic poles (labeled N and S), producing two oppositely directed beams of radiation. If the star's magnetic axis (a line that connects the north and south magnetic poles) is tilted at an angle from the axis of rotation, as shown here, the beams sweep around the sky as the star rotates. If Earth happens to lie in the path of one of the beams, we detect radiation that appears to pulse **(a)** on and **(b)** off.

all, there is no fundamental reason for these two axes to coincide. (Indeed, these two axes do not coincide for any of the planets of our solar system.)

In 1969, Peter Goldreich at Caltech pointed out that the combination of such a powerful magnetic field and rapid rotation would act like a giant electric generator, creating very strong electric fields near the neutron star's surface. Indeed, these fields are so intense that part of their energy is used to create electrons and positrons out of nothingness in a process called **pair production.** The powerful electric fields push these charged particles out from the neutron star's surface and into its curved magnetic field, as sketched in Figure 21-3*a*. As the particles spiral along the curved field, they are accelerated and emit energy in the form of electromagnetic radiation. (A radio transmission antenna works in a similar way. Electrons are accelerated back and forth along the length of the antenna, producing radio waves.) The result is that two narrow beams of radiation pour out of the neutron star's north and south magnetic polar regions.

ANALOGY A rotating, magnetized neutron star is somewhat like a lighthouse beacon. As the star rotates, the beams of radiation sweep around the sky. If at some point during the rotation one of those beams happens to point toward Earth, as shown in Figure 21-3*a*, we will detect a brief flash as the beam sweeps over us. At other points during the rotation the beam will be pointed away from Earth and the radiation from the neutron star will appear to have turned off (Figure 21-3*b*). Hence, a radio telescope will detect regular pulses of radiation, with one pulse being received for each rotation of the neutron star.

CAUTION! The name *pulsar* may lead you to think that the source of radio waves is actually pulsing. But in the model just described, this is not the case at all. Instead, beams of radiation are emitted continuously from the magnetic poles of the neutron star. The pulsing that astronomers detect here on Earth is simply a result of the rapid rotation of the neutron star, which brings one of the beams periodically into our line of sight, as Figure 21-3 shows. In this sense, the analogy between a pulsar and a lighthouse beacon is a very close one. If a pulsar's beams of radiation never point at Earth, we don't observe the pulses of radio emission directly. However, as we'll see with the Crab Nebula, there are cases where we can still see evidence of the pulsar.

The Crab Pulsar

WEB LINK 21.5

Soon after the rotating neutron star model of pulsars was proposed in the late 1960s, a team of astronomers at the University of Arizona began wondering if pulsars might also emit pulses at wavelengths other than the radio part of the spectrum. To investigate, they began a search for *visible-light*

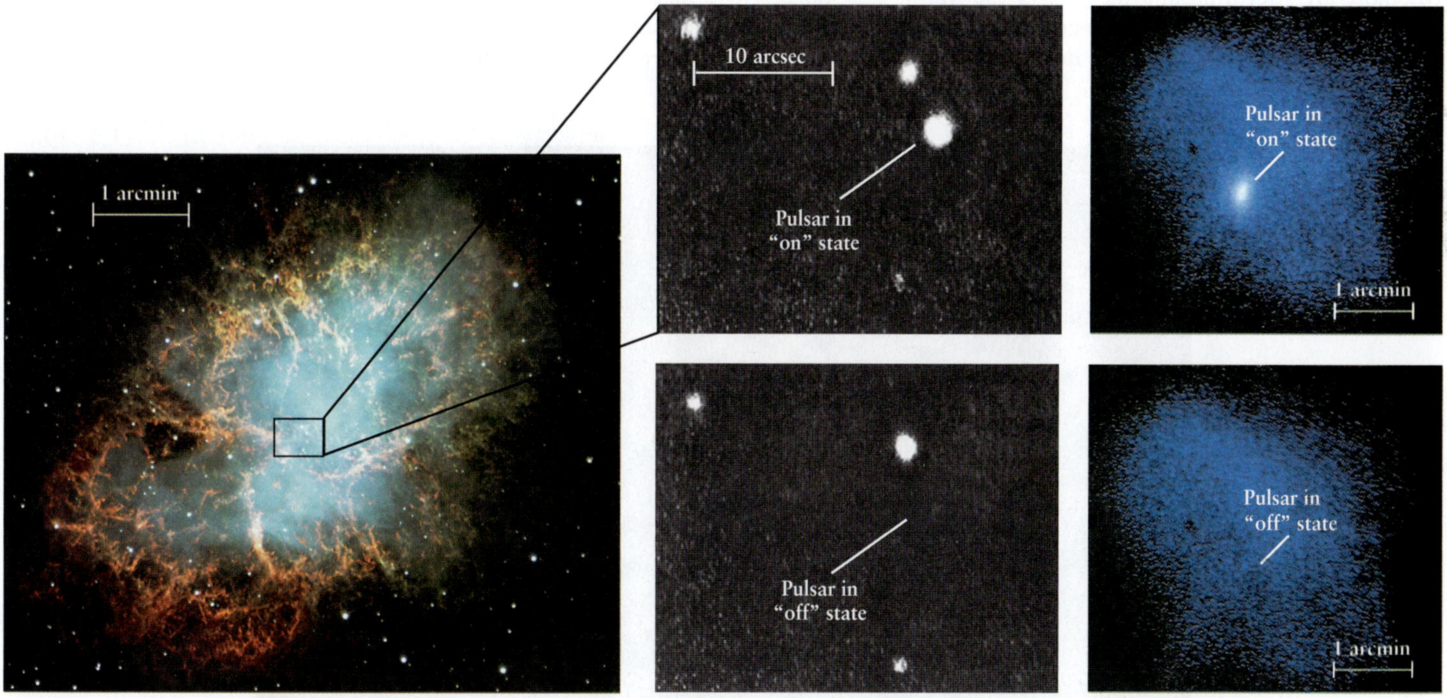

(a) The Crab Nebula R I **V** U X G

(b) The Crab pulsar in visible light R I **V** U X G

(c) The Crab pulsar in X rays R I V U **X** G

VIDEO 21.1

Figure 21-4

The Crab Pulsar (a) A pulsar is located at the center of the Crab Nebula, which is about 2000 pc (6500 ly) from Earth and about 3 pc (10 ly) across. The boxed area is shown in closeup in part (b). **(b)** We see a flash of visible light when the rotating pulsar's beam is directed toward us (the "on" state). The pulsar fades (the "off" state) when the beam is aimed elsewhere. **(c)** The Crab pulsar also pulses "on" and "off" at X-ray wavelengths. The radio pulses, visual flashes, and X-ray flashes all have a period of 0.033 s. (a: The FORS Team, VLT, European Southern Observatory; b: Lick Observatory; c: Einstein Observatory, Harvard-Smithsonian Center for Astrophysics) pulses from the Crab pulsar (Figure 21-4). They aimed a telescope at the center of the Crab Nebula (Figure 21-4*a*), then used a spinning disk with a slit in it to "chop," or interrupt, incoming light at rapid intervals. To their surprise and delight, they found that one of the stars at the center of the nebula appears to be flashing on and off 30 times each second (Figure 21-4*b*)—exactly the same pulse rate as observed with radio waves.

This observation was strong evidence in favor of pulsars being rotating, magnetized neutron stars, because charged particles accelerating in a magnetic field can radiate strongly over a very wide range of wavelengths. By contrast, the blackbody radiation from an ordinary star, which has a surface temperature in thousands of degrees, is very weak at radio wavelengths. (The Sun's weak radio emissions come not from its glowing surface but rather from charged particles spiraling in the solar magnetic field.)

Since the pioneering days of the 1960s, periodic flashes of radiation have been detected from the Crab pulsar at still other wavelengths, including X rays (Figure 21-4*c*). The pulsar period is the same at all wavelengths, just as we would expect if the emissions are coming from a portion of the neutron star, which comes into view periodically as the star rotates.

WEB LINK 21.6 The many successes of the neutron-star model of pulsars led to the rapid acceptance of the model. Since 1968 radio astronomers have discovered more than 1800 pulsars scattered across the sky, and it is estimated that perhaps 10^5 more are strewn around the disk of the Milky Way Galaxy. Each one is presumed to be the neutron star corpse of an extinct, massive star. Radio telescopes have detected pulsars with a wide variety of pulse periods, from a relatively sluggish 8.51 s to a blindingly fast 0.001396 s. In each case the pulse period is thought to be the same as the rotation period of the neutron star.

The Pulsar-Supernova Connection

The primary sources of pulsars are thought to be massive stars that become core-collapse supernovae (as described in Section 20-6). The supernova seen in 1054 that left behind the Crab Nebula and the Crab pulsar is thought to have been a Type II core-collapse supernova. However, not all supernovae leave pulsars amid their debris. For example, the supernovae observed by Tycho Brahe in 1572 and by Johannes Kepler in 1604 were probably thermonuclear supernovae in which a white dwarf in a binary system underwent a catastrophic explosion (see Section 20-9). It is difficult to imagine how the complete disruption of a white dwarf could produce a neutron star. Indeed, no pulsar has been found at the locations of either of those supernovae.

In many cases pulsars are found within the remnant of the supernova that spawned the pulsar. One example is the Crab pulsar, which is not far from the center of the Crab Nebula. This location is what we might expect: When the progenitor star of the supernova of 1054 exploded, its core remained at the center of the explosion and became a rotating neutron star, while the star's outer layers were blasted into space to form the nebula. However, Figure 21-5 shows a pulsar that lies *outside* its supernova remnant and is moving away from the remnant at hundreds of kilometers per second. It appears that when the progenitor star collapsed and exploded into a supernova, the explosion was much more powerful on one side of the core than the other. As a result,

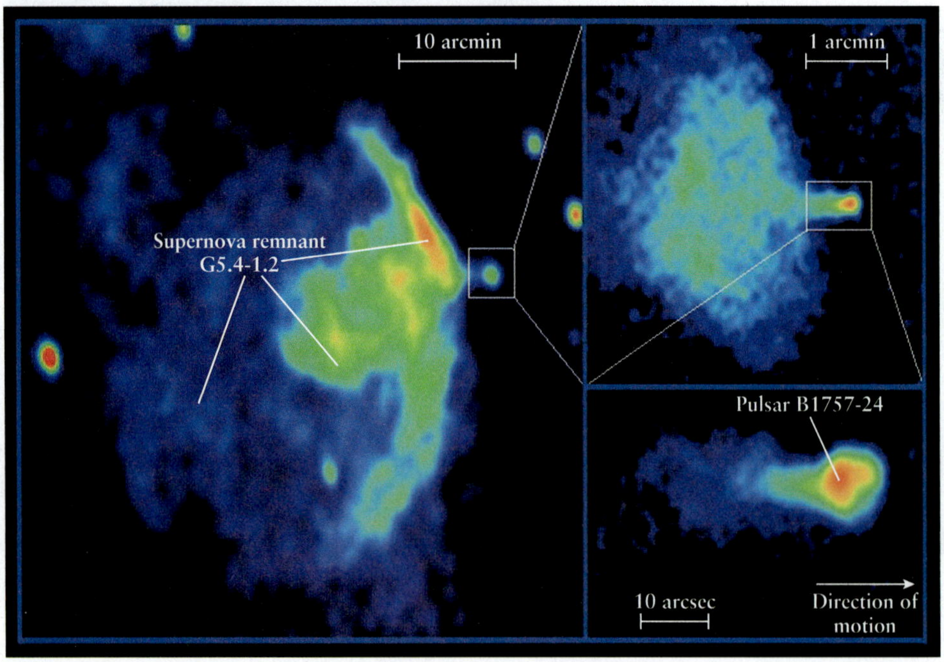

WEB LINK 21.7

Figure 21-5 R I V U X G

A Fast-Moving Pulsar When the progenitor star of this supernova remnant exploded, it ejected the star's core at such a high speed that the core overtook the blast wave and moved beyond the boundary of the supernova remnant. The ejected core became a pulsar, denoted B1757-24. Tens of thousands of years after the explosion, the pulsar is still moving at about 600 km/s (350 mi/s). In these false-color radio images, red denotes areas of strongest radio emission. (National Radio Astronomy Observatory) the core was sent flying at immense speed. Many such fast-moving pulsars have been observed, which suggests that supernova explosions are often asymmetrical. In this way studies of pulsars have given us new insights into the nature of supernovae.

The most intensely studied of all recent core-collapse supernovae is the great supernova of 1987 (SN 1987A, described in Section 20-7). Astronomers have been carefully observing its remains for signs of a pulsar. To date, however, no confirmed observations of pulses have been made, and the search for a pulsar in SN 1987A goes on.

21-4 Pulsars gradually slow down as they radiate energy into space

WEB LINK 21.8

As remarkable as the Crab pulsar is, the Crab Nebula that surrounds it is even more so. Although the nebula is 2000 pc (6500 ly) away, it can be spotted in dark skies with even a small telescope. It shines with a luminosity 75,000 times that of the Sun (and some 10^8 times the luminosity of the pulses from the Crab pulsar).

But why is the Crab Nebula so luminous? Certain other nebulae—the H II regions described in Section 18-2—can also be seen at great distances. But H II regions contain young stars that shine because of rapid thermonuclear reactions in their cores. These reactions heat the star's surfaces to such high temperatures that they give off prodigious amounts of ultraviolet radiation. This radiation ionizes the surrounding gas and provides the energy that the H II region emits into space. The Crab Nebula, however, is different from H II regions: It contains *no* young stars, and there are *no* thermonuclear reactions taking place within the neutron star that lies at the nebula's heart. What, then, is the source of this nebula's prodigious energy output? As we will see, the Crab pulsar provides the energy not only for radio pulses but for the entire Crab Nebula.

Rotational Energy as a Power Source

In 1966, two years before the discovery of pulsars, John A. Wheeler at Princeton University and Franco Pacini in Italy speculated that the ultimate source of the Crab Nebula's immense energy output might be the *spin* of a neutron star inside the nebula. In a kitchen blender, the energy of a rapidly spinning blade is used to chop food or mix beverages. Wheeler and Pacini proposed that the even greater energy of a rapidly spinning neutron star—which can rotate as rapidly as the blade in a blender, but has far more energy thanks to its much larger mass and radius—can provide the energy to light up the Crab Nebula.

The energy in a neutron star's rotation provides the power for its intense pulsar radiation

Wheeler and Pacini's speculation was quickly accepted as the correct explanation of the Crab Nebula's luminosity. What confirmed their idea were two discoveries: first, that there is a pulsar within the Crab Nebula, and second, that the Crab pulsar is slowing down.

Although pulsars were first noted for their very regular pulses, careful measurements by radio telescopes soon revealed that the rotation period of a typical pulsar—that is, the time for it to spin once on its axis—increases by a few billionths of a second each day. The Crab pulsar, which is one of the most rapidly rotating pulsars, is also slowing more quickly than most—its period increases by 3.8×10^{-8} second each day. Thirty billionths of a second may sound trivial, but there is so much energy in the rotation of a rapidly spinning neutron star that even the slightest slowdown corresponds to a tremendous loss of energy. In fact, the observed rate of slowing for the Crab pulsar corresponds to a rate of energy loss equal to the entire luminosity of the Crab Nebula.

Synchrotron Radiation

The interpretation is that the energy lost by the Crab pulsar as it slows down is transferred to the surrounding nebula. But *how* does this transfer of energy take place, and how is the energy radiated into space? The answer lies in the diffuse part of the Crab Nebula, which shines with an eerie bluish light (see Figure 21-4*a*). This type of light was first produced on Earth in 1947 in a particle accelerator in Schenectady, New York. This machine, called a *synchrotron,* was built by the General Electric Company to accelerate electrons to nearly the speed of light for experiments in nuclear physics. (The electrons are said to be *relativistic,* because their velocities are near the speed of light, and Einstein's theory of relativity—which we discuss in Chapter 22—must be applied to understand their motions.)

WEB LINK 21.9

As a beam of electrons whirled along a circular path within the cyclotron, held in orbit by powerful magnets, scientists noticed that it emitted a strange light. It is now known that this light, called **synchrotron radiation,** is emitted whenever high-speed electrons move along curved paths through a magnetic field (Figure 21-6*a*). Synchrotron radiation has a continuous spectrum, but one that is distinctively different from the continuous spectrum emitted by a heated blackbody (Figure 21-6*b*). This difference helped astronomers confirm that the Crab Nebula emits primarily synchrotron radiation, not blackbody radiation.

The total energy output of the Crab Nebula in synchrotron radiation is 3×10^{31} watts, compared to the relatively paltry 4×10^{26} watts emitted by the Sun. Thus, the Crab Nebula must contain quite a large number of relativistic electrons spiraling in an extensive magnetic field. Figure 21-7 shows where these electrons come from: They are ejected from the neutron star's vicinity, forming immense jets that emanate from the star's poles and rings that spread outward from its equator. (Figure 18-16 shows a related process that produces jets from young stars.) As the Crab pulsar slows, the energy of its rotation is transferred to these electrons through the magnetic field and is then emitted by them in the form of synchrotron radiation. So much energy is transferred in this way that the Crab Nebula shines very brightly.

This model for the luminosity of the Crab Nebula also provides a check on our ideas about the magnetic fields of neutron stars (Section 21-3). The stronger the magnetic field of a neutron star, the greater the rate at which it emits energy in the form of radiation and hence the more rapidly it will slow down. From the slowdown rate of the Crab pulsar, astronomers calculate that its magnetic field must be about 5×10^{12} G. This is very close to the estimate of 10^{12} G that we obtained in Section 21-3.

Aging Pulsars

Because a spinning neutron star slows down as it radiates away its rotational energy, it follows that an old neutron star should be

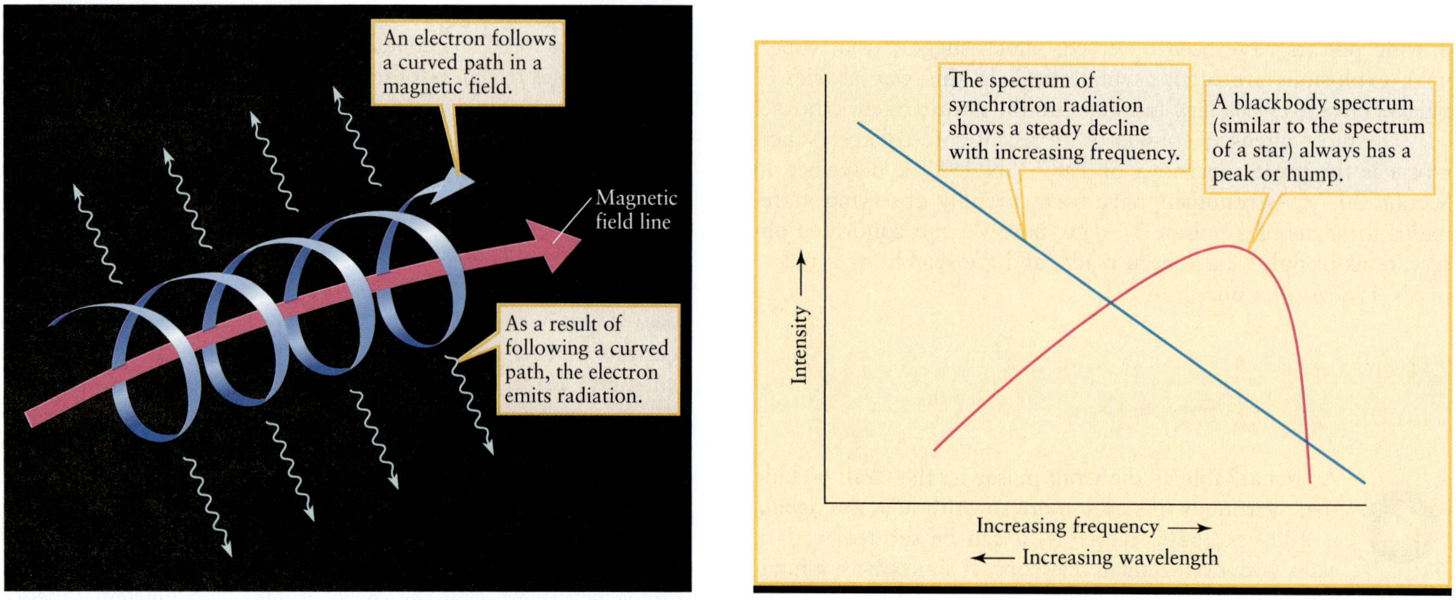

Figure 21-6

Synchrotron Radiation **(a)** Synchrotron radiation is emitted by electrons that move at high speed in a magnetic field. **(b)** Astronomers can distinguish synchrotron radiation from blackbody radiation because the two kinds of emission have very different spectra.

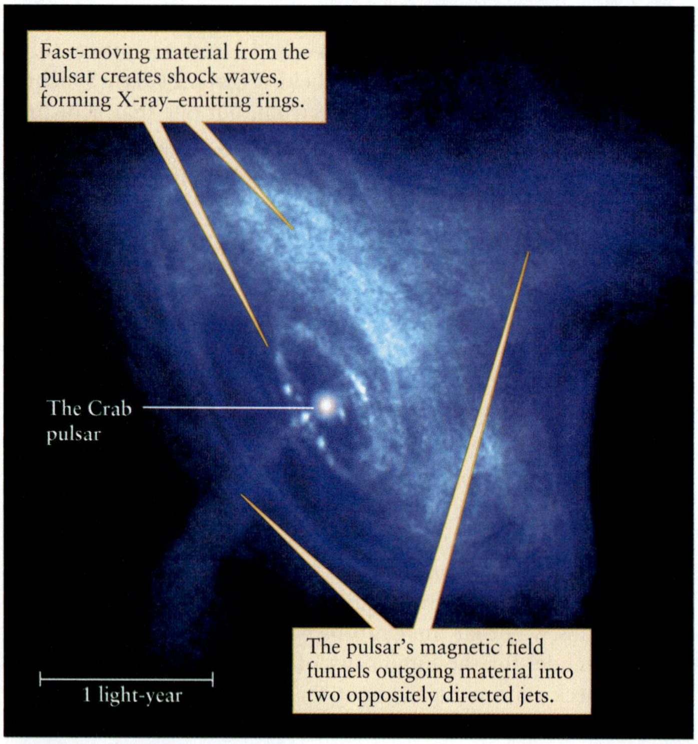

VIDEO 21.2 ANIMATION 21.1

Figure 21-7 RIVUXG

The Dynamic Heart of the Crab Nebula This image from the Chandra X-ray Observatory shows material being expelled from the Crab pulsar into the surrounding nebula. (The image that opens this chapter combines this X-ray view with a visible-light view.) Some of the ejected material travels at more than half the speed of light. (NASA/CXC/ASU/J. Hester et al.)

spinning more slowly than a young one. It is thought that slower pulsars were formed in supernova explosions that occurred hundreds of thousands of years ago. Over the ages, the supernova remnants have dispersed into the interstellar medium, and the pulsars left behind have slowed to periods of a second or more. (More than 400 such old pulsars are known.) Only relatively young, rapidly rotating pulsars like the Crab pulsar have enough energy to emit flashes of visible light along with radio pulses (see Figure 21-4*b*). We can summarize these ideas as a general rule:

An isolated pulsar slows down as it ages, so its pulse period increases.

From a pulsar's period (which increases as it ages) and the rate at which its period is increasing (which decreases as it ages), astronomers can estimate how old a pulsar is.

Like all general rules, the one relating a pulsar's age and its period has exceptions. In Section 21-7 we will see how some old pulsars that are *not* isolated—that is, are members of binary star systems—can be reaccelerated to truly dizzying rotation speeds.

21-5 Superfluidity and superconductivity are among the strange properties of neutron stars

The radiation from pulsars gives us information about the surroundings of neutron stars, as well as how they rotate. To deduce what conditions are like *inside* a neutron star, however, astrophysicists must construct theoretical models. We saw examples of such models for the Sun in Section 16-2 and for post–main-sequence stars in Sections 20-1 and 20-5.

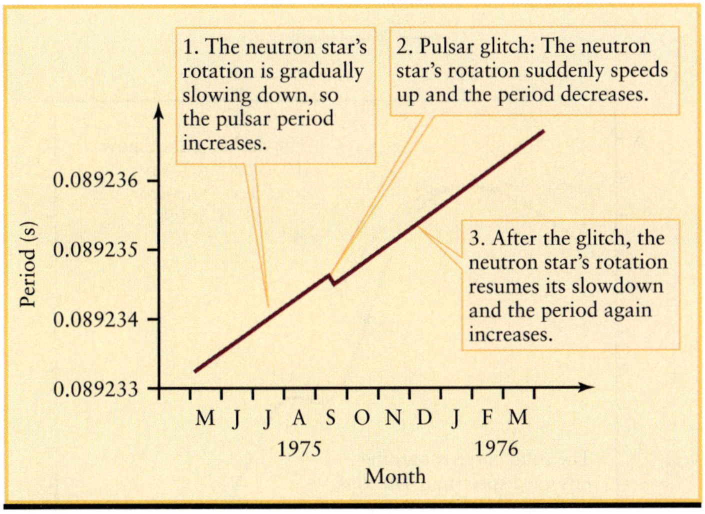

Figure 21-8

A Pulsar Glitch This graph shows how the period of the Vela pulsar varied during 1975 and 1976. Most of the time the neutron star's rotation slowed as its rotational energy was converted into radiation. In September 1975, however, a sudden speedup (glitch) occurred. A number of pulsars have been observed to undergo multiple glitches at intervals of a few years.

Superfluidity and Pulsar Glitches

The models for neutron stars are very different. For one thing, neutron stars are thought to have a solid crust on their surface. Furthermore, the interior of a neutron star is a sea of densely packed, degenerate neutrons, with properties quite unlike those of ordinary gases or even degenerate electrons. Detailed calculations strongly suggest that degenerate neutron matter can flow without any friction whatsoever, a phenomenon called **superfluidity.**

Material deep inside a neutron star may flow completely without friction

Superfluidity can be observed in the laboratory by cooling liquid helium to temperatures near absolute zero. Because it is frictionless, superfluid liquid helium exhibits strange properties such as being able to creep up the walls of a container in apparent defiance of gravity. Within a neutron star, elongated, friction-free whirlpools of superfluid neutrons may form. Their interaction with the star's crust may be the cause of sudden changes in a pulsar's rotation.

In addition to the general slowing down of pulsars, astronomers have found that they sometimes exhibit a sudden speedup, sometimes called a **glitch.** For example, Figure 21-8 shows measurements of the period of the Vela pulsar (named for the constellation in which it lies) during 1975 and 1976. On this graph, the pulsar's gradual slowdown is shown as a steady increase in its period. In September 1975, however, there was an abrupt speedup, after which the pulsar continued to slow down at its usual rate. Similar glitches are seen for other pulsars.

Current opinion among pulsar theorists is that glitches are caused by the superfluid neutrons within the neutron star. As a rotating neutron star radiates energy into space, the rotation of its crust slows down, but the neutron whirlpools in the star's interior continue to rotate with the same speed. Some of these whirlpools cling to the crust, as though they were bungee cords with one end attached to the crust and the other to the star's interior. As the crust slows down relative to the interior, these superfluid "bungee cords" stretch out. When the tension in them gets too great, they deliver a sharp jolt that makes the crust speed up suddenly. (An older model, in which pulsar glitches were caused by a "starquake" in the neutron star's settling crust, does not appear to agree with the accumulated observational data on glitches.)

The Ultimate Electric Conductor

Superfluidity is not the only exotic property of neutron star interiors. Models of the internal structure of a neutron star strongly suggest that the protons in the core can move around without experiencing any electrical resistance whatsoever. This phenomenon, called **superconductivity,** also occurs on Earth with certain substances at low temperatures. In a normal conductor of electricity such as the copper wires in a flashlight, an electric current only flows if there is a battery in the flashlight circuit to overcome the resistance of the wires. Once the battery goes "dead," the current stops and the flashlight turns off. Once you start an electric current moving in a superconducting material, however, it keeps moving forever.

CAUTION! You may have been surprised to read in the preceding paragraph about *protons* in a *neutron* star. While a neutron star is made up predominantly of neutrons, some protons and electrons must be scattered throughout the star's interior. Indeed, a pulsar's magnetic field must be anchored to the neutron star by charged particles. Neutrons are electrically neutral, so without the protons and electrons in its interior a neutron star would rapidly lose its magnetic field.

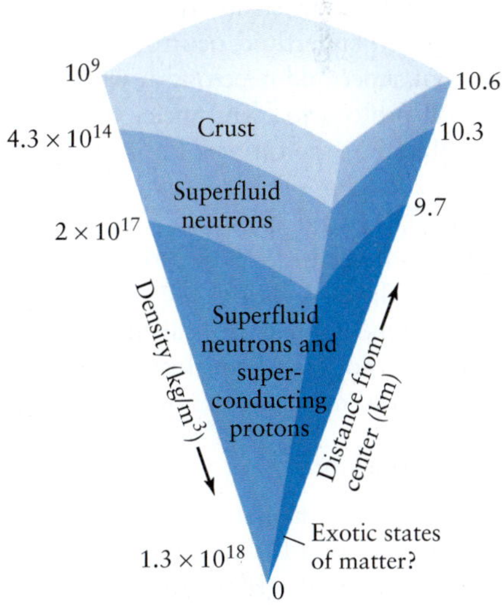

Figure 21-9

A Model of a Neutron Star This theoretical model for a 1.4-$M_\odot$ neutron star has a superconducting, superfluid core 9.7 km in radius. The core is surrounded by a 0.6-km thick mantle containing nuclei, electrons, and superfluid neutrons. The star's crust is only 0.3 km thick (the length of three football fields) and is composed of heavy nuclei (such as iron) and free electrons. (The thicknesses of the layers are not shown to scale.)

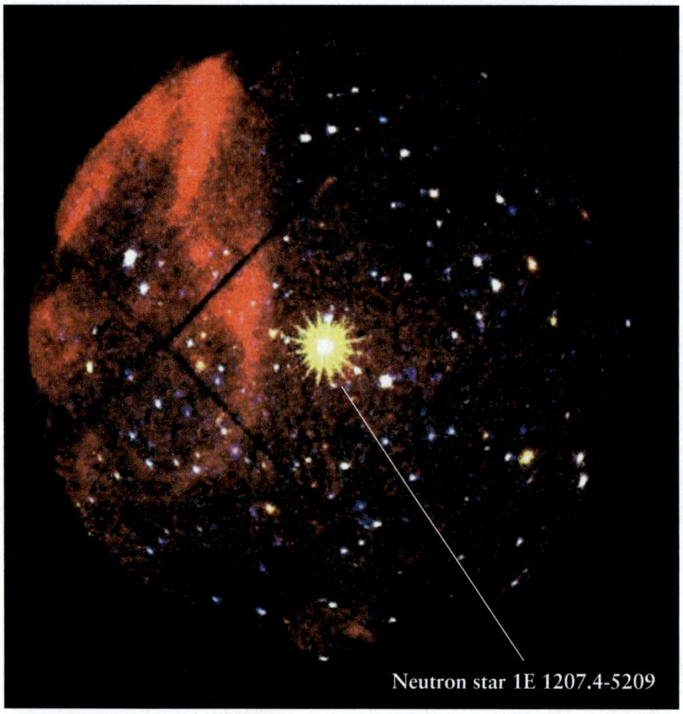
Neutron star 1E 1207.4-5209

(a) X-ray image R I V U X G

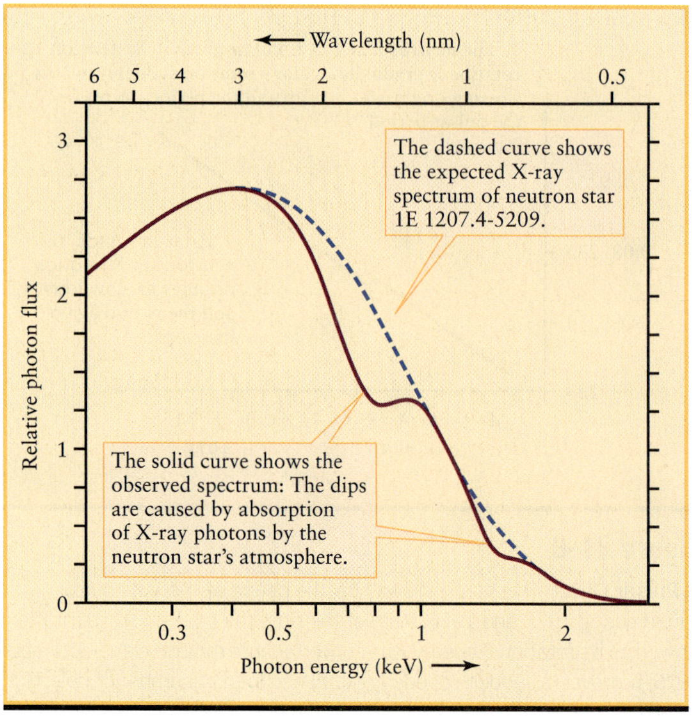
Wavelength (nm)
6 5 4 3 2 1 0.5
Relative photon flux
3
2
1
0
The dashed curve shows the expected X-ray spectrum of neutron star 1E 1207.4-5209.
The solid curve shows the observed spectrum: The dips are caused by absorption of X-ray photons by the neutron star's atmosphere.
0.3 0.5 1 2
Photon energy (keV)

(b) X-ray spectrum

WEB LINK 21.10

Figure 21-10

Evidence for a Neutron-Star Atmosphere **(a)** This image from XMM-Newton (see Figure 6-30*b*) shows a neutron star roughly 2000 pc (7000 ly) from Earth in the constellation Centaurus. Like the Crab pulsar (Figure 21-4*a*), this neutron star is an X-ray-emitting pulsar. **(b)** The dashed curve shows the X-ray spectrum that would be expected if the neutron star had no atmosphere. The observed dips are evidence that an atmosphere is present. (a: ESA/CESR/G. Bignami et al.)

As Figure 21-9 shows, the structure of a neutron star probably consists of a core with superfluid neutrons and superconducting protons, a mantle of superfluid neutrons in which some of the neutrons combine with protons to form nuclei, and a brittle crust less than a kilometer thick. At the center of a neutron star, the density is on the order of 10^{18} kg/m^3. This density is some 10 trillion (10^{13}) times greater than the central density of our present-day Sun and approximately twice the density of an atomic nucleus. It has been speculated that under these conditions the neutrons and protons dissolve into more fundamental particles called *quarks*. To date, however, astronomers have not found conclusive observational evidence of such extreme states of matter within neutron stars.

Neutron Star Atmospheres

While the deep interiors of neutron stars remain mysterious, astronomers have recently found evidence that the crusts of neutron stars are surrounded by a highly unusual atmosphere (Figure 21-10). The evidence comes from the neutron star 1E 1207.4-5209, which like the Crab pulsar emits X rays and lies at the center of a supernova remnant (Figure 21-10*a*). The dashed curve in Figure 21-10*b* shows the expected spectrum of the emitted X rays. The actual spectrum, shown by the solid curve in Figure 21-10*b*, has two pronounced dips. The interpretation of these dips is that they are caused by absorption of certain X-ray wavelengths by an atmosphere around the neutron star. (Compare Figure 21-10*b* with Figure 7-3.) The atoms that make up such an atmosphere must be dramatically different from those around you: In the intense magnetic field of a young neutron star, ordinarily spherical atoms are deformed into elongated cigar shapes and have radically altered spectral lines.

The bizarre properties of neutron star interiors and atmospheres are of great interest to astronomers and physicists precisely because they are so bizarre. Just as physicians learn about health by studying disease, scientists gain insight into the fundamental nature of matter by studying its behavior under extreme conditions.

21-6 The most highly magnetized pulsars can emit colossal bursts of radiation

WEB LINK 21.11

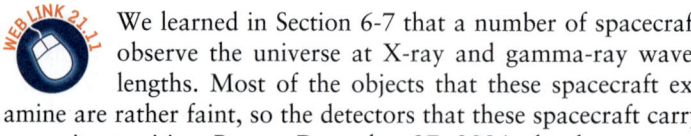

We learned in Section 6-7 that a number of spacecraft observe the universe at X-ray and gamma-ray wavelengths. Most of the objects that these spacecraft examine are rather faint, so the detectors that these spacecraft carry are quite sensitive. But on December 27, 2004, the detectors on seven such spacecraft were overloaded by an intense burst of radiation that lasted a mere 0.2 second.

By comparing the times at which spacecraft at different locations in space detected the burst, astronomers were able to triangulate the point in space from which the radiation emanated. The source proved to be a pulsar designated SGR 1806-20 that lies about 15,000 parsecs (50,000 light-years) from Earth in the constellation Sagittarius. For the spacecraft to detect such intense radiation from a source so far away, the pulsar's luminosity during the time it was emitting the burst must have been a staggering 10^{14} times greater than the luminosity of the Sun. It would take the Sun 300,000 years to radiate as much energy as SGR 1806-20 emitted in just 0.2 second!

How could a pulsar release such a stupendous amount of energy in such a brief amount of time? If the energy source were the rotation of a neutron star, as it is for the Crab Nebula, the neutron star's spin would have slowed appreciably after emitting the radiation burst. But observations of the pulsar made before and after the burst showed the *same* pulse period of 7.47 seconds. Since the pulse period and rotation rate are the same (see Figure 21-3), this means that the rotation did not slow at all. What is more, the total rotational energy of a neutron star spinning once every 7.47 seconds is only about 1% of the energy that was released as radiation in the 0.2-second burst. Hence, the neutron star's rotation could not have supplied the immense release of energy seen on December 27, 2004.

Short bursts of radiation that were nearly as intense as that seen from SGR 1806-20 have been detected from two other pulsars, one (SGR 0526-66) in 1979 and one (SGR 1900+14) in 1998. Like SGR 1806-20, neither of these pulsars showed any appreciable slowing after emitting their bursts, so their rotational energy could not have been the source of the bursts. To explain all three of these exotic events, astronomers have turned to another distinguishing characteristic of a pulsar: the neutron star's intense magnetic field. As we will see, the three pulsars in question appear to have the strongest magnetic fields of any object in the universe, far stronger than those of "ordinary" pulsars.

Magnetars and Starquakes

The magnetic field of an ordinary star such as the Sun is produced by the motion of its electrically conducting gas, or plasma (see Section 16-9). The same is true for a supergiant star. We saw in Section 21-3 that when thermonuclear reactions cease in a supergiant star and it begins to collapse, the star's magnetic field moves with the infalling material and becomes greatly amplified. This amplified magnetic field remains with the resulting neutron star, which typically ends up with a field strength of about 10^{12} G.

However, a neutron star can generate an additional magnetic field as it forms. For about the first 10 seconds after a neutron star forms from the core of a collapsing supergiant, it has an extremely high internal temperature—about 10^{11} K, as compared to a relatively frigid 1.55×10^7 K in the present-day Sun's core—and there is furious convection of the neutron star material. This material includes electrically charged protons and electrons, and the convective motions produce a magnetic field. If the newly formed neutron star spins relatively slowly, so that the time it takes a convection cell to rise and fall (about 0.005 s) is less than the time required for the neutron star to make a complete rotation, the magnetic fields caused by different convection cells remain uncorrelated and do not contribute much to the neutron star's magnetism. This seems to be what happens for the vast majority of neutron stars. But if a brand-new neutron star has a rotation period of less than 0.005 s, its rapid rotation causes the convective motions in different parts of the neutron star to organize together. When the neutron star has completely formed, it is left with a magnetic field that can be as strong as 10^{15} G, or about 1000 times stronger than in an ordinary neutron star. The magnetism of such a star could pull the keys out of your pocket and demagnetize your credit cards at a distance of 200,000 km, or half the distance from Earth to the Moon.

Astronomers Robert Duncan and Christopher Thompson developed the theory of highly magnetized neutron stars, or **magnetars**, in 1992. In their theory the magnetic field of a magnetar exerts tremendous stresses (or forces) on the electrically conducting material at the neutron star's surface (see Figure 21-9). From time to time these stresses can cause a localized fracture of the surface called a **starquake**. The magnetic field in the vicinity of the starquake responds by rearranging itself in a process that resembles magnetic reconnection in the Sun's magnetic field (see Figure 16-25) but is much more energetic. The energy released by such a localized starquake produces a moderate burst of gamma rays. But on rare occasions the crust can fracture simultaneously over much of the magnetar's surface. In such a catastrophic event the star's magnetic field undergoes a dramatic rearrangement that releases an immense amount of magnetic energy. The result is a short-lived but titanic blast of gamma rays and X rays.

A starquake on the surface of a magnetar releases magnetic energy in the form of an intense blast of X rays and gamma rays

Observing Magnetars

The theoretical model of a magnetar turns out to be an excellent description of the three pulsars that produced the energetic blasts observed in 1979, 1998, and 2004. Each of these pulsars emits moderate gamma-ray bursts at seemingly random intervals, which is why they are also called *soft gamma repeaters*. This name is the origin of the acronym "SGR" in the pulsars' designations. (The adjective "soft" means that the gamma rays emitted are of relatively low energy.) According to Duncan and Thompson's model, each such burst is caused by a localized starquake on a magnetar; the truly stupendous bursts are the result of rare quakes that span much of the magnetar's surface.

How can we test these radical ideas? As we saw in Section 21-4, the rate at which a pulsar slows down is a measure of its magnetic field. In 1998 the Greek astronomer Chryssa Kouveliotou determined that over the preceding 5 years the pulse period of SGR 1806-20 had increased by 0.008 seconds. (This change is small, but is about 40,000 times greater than the increase in the Crab pulsar's period over the same time interval.) Calculations show that the neutron star magnetic field needed in order to produce this rate of slowdown is 8×10^{14} G, in close agreement with the magnetar model.

Although magnetars begin their existence with dizzyingly fast rotation, their rapid slowdown soon makes them among the

COSMIC CONNECTIONS

Magnetars and Ordinary Pulsars

A magnetar is a neutron star with an extraordinarily strong magnetic field of 10^{15} gauss, a thousand times greater than the 10^{12}-gauss field of an ordinary neutron star. The magnetic energy stored in this intense field can be released in the form of powerful bursts of X rays and gamma rays. (After C. Kouveliotou, R.C. Duncan, and C. Thompson, "Magnetars", *Scientific American*, February 2003)

How magnetars form

1. Most neutron stars are thought to begin as massive but otherwise ordinary stars, between 8 and 20 times as massive as the Sun.

2. Massive stars die in a core-collapse supernova explosion, as the stellar core implodes into a ball of subatomic particles.

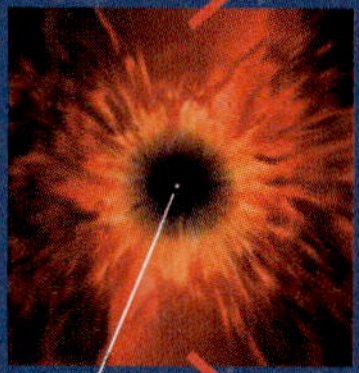

Newborn neutron star

Magnetar

Ordinary pulsar

3a. If the newborn neutron star spins fast enough, it generates an intense magnetic field. Field lines inside the star get twisted.

4a. The magnetar settles into neat layers, with twisted field lines inside and smooth lines outside. It might emit a narrow radio beam.

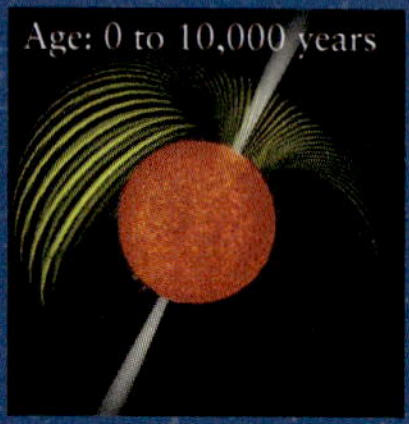

5a. The old magnetar has cooled off, and much of its magnetism has decayed away. It emits very little energy.

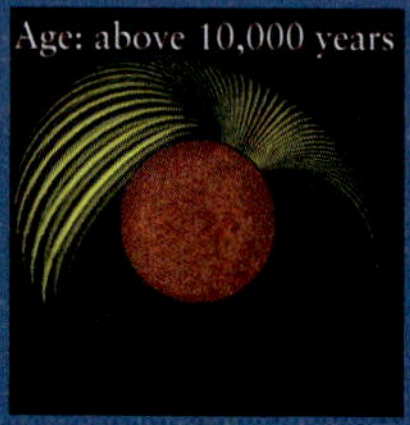

3b. If the newborn neutron star spins slowly, its magnetic field, though strong by everyday standards, does not reach magnetar levels.

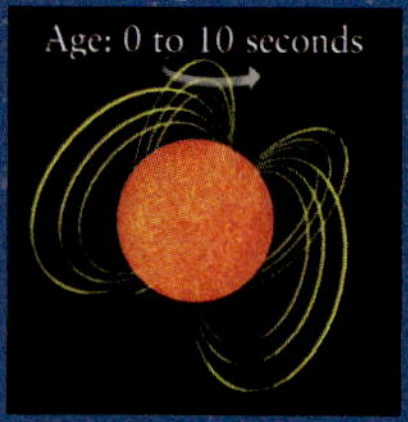

4b. The mature pulsar is cooler than a magnetar of equal age. It emits a broad radio beam, which radio telescopes can readily detect.

5b. The old pulsar has cooled off and no longer emits a radio beam.

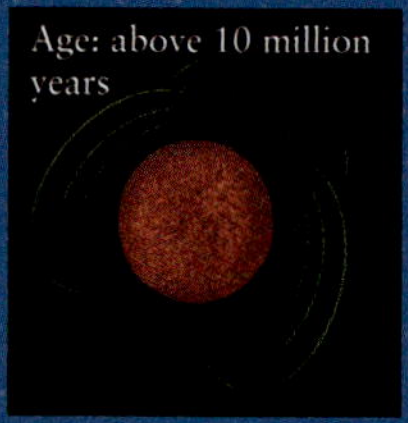

How magnetar bursts happen

The magentic field of a magnetar is so strong that the rigid crust sometimes breaks and crumbles, releasing a huge surge of energy.

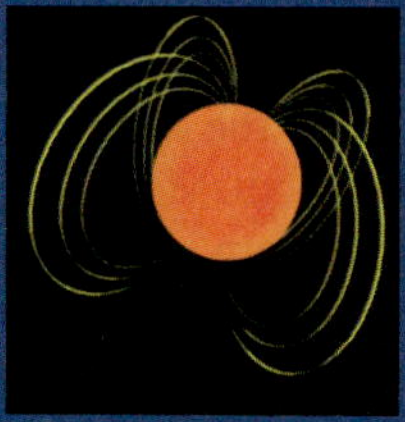

1. Most of the time the magnetar is quiet. But magnetic stresses are slowly building up.

2. At some point the solid crust is stressed beyond its limit. It fractures, probably into many small pieces.

3. This "starquake" creates a surging electric current, which decays and leaves behind a hot fireball.

4. The fireball cools by releasing X rays from its surface. It evaporates in minutes or less.

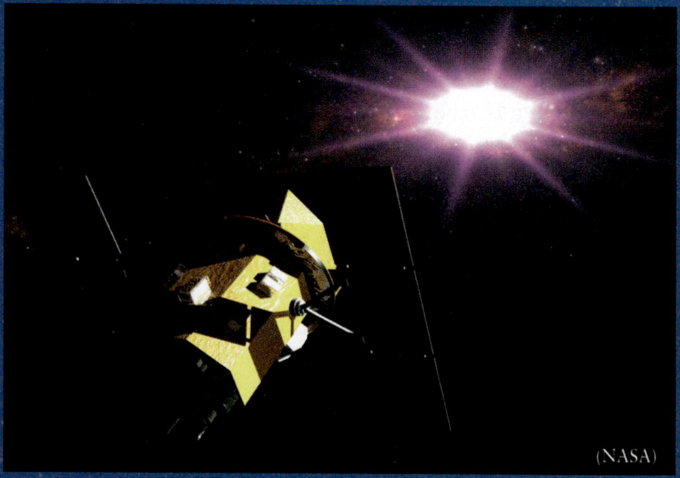

(NASA)

- A burst from a magnetar 15,000 parsecs away was observed on December 27, 2004.
- This burst was so intense that it caused a major disturbance of Earth's upper atmosphere.
- During the 0.2-second duration of the burst, the magnetar's luminosity was more than 100 times the total luminosity of the Milky Way Galaxy.
- It is estimated that this blast ripped away a layer of the magnetar's crust about 50 meters thick.

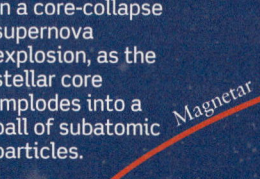

slowest of all pulsars. For example, in the few thousand years since SGR 1806-20 first formed from the core collapse of a supergiant star, its rotation period has increased from less than 0.005 s to 7.47 s. With the passage of time, a magnetar's rotation will continue to slow down and starquakes will eventually relieve the crust of all of its stresses. After perhaps 10,000 years, a magnetar will cease its activity altogether. It is estimated that there may be millions of these old, inert magnetars in our Galaxy alone.

As yet, only a handful of active magnetars are known. But because magnetars represent such an extreme state of matter, with magnetic effects stronger than those found anywhere else in the universe, astronomers are studying them with great interest. The *Cosmic Connections* figure depicts the key differences between magnetars and ordinary neutron stars.

Even neutron stars that lack the intense magnetic field of a magnetar can display exotic behavior. As we will see in the remainder of this chapter, this happens when a neutron star is one member of a binary star system in which the two stars orbit close to each other.

21-7 The fastest pulsars were probably created by mass transfer in close binary systems

In 1982, radio astronomers discovered a pulsar whose period is only 1.558 ms. (One millisecond, or 1 ms, equals a thousandth of a second). This remarkable pulsar, called PSR 1937+21 after its coordinates in the sky, is a neutron star spinning 642 times per second—about 3 times faster than the blades of a kitchen blender! Such incredibly rapid rotation is a clue to this pulsar's unusual history.

The fastest pulsars spin nearly a thousand times per second, more than twice as fast as a kitchen blender

As we have seen, pulsars should slow down as they age. The rapid spin rate of PSR 1937+21 suggests that it has hardly aged at all and so must be very young. A young pulsar should also be slowing down very rapidly. The faster it rotates, the more rapidly it can transfer rotational energy to its surroundings and the faster the slowdown. But PSR 1937+21 is actually slowing down at a very *gradual* rate, millions of times more gradually than the Crab pulsar. Such a gradual slowdown is characteristic of a pulsar hundreds of millions of years old. But if PSR 1937+21 is so old, how can it be spinning so rapidly?

PSR 1937+21 is not the only pulsar with unusually rapid rotation. Since 1982, astronomers have discovered more than 180 very fast pulsars, which are now called **millisecond pulsars.** All have periods between 1 and 10 ms, which means that these neutron stars are spinning at rates of 100 to 1000 rotations per second. (As of this writing the fastest known pulsar is PSR J1748-2446ad, which rotates 716 times a second.)

The Origin of Millisecond Pulsars

An important clue to understanding millisecond pulsars is that the majority are in close binary systems, with only a small separation between the spinning neutron star and its companion. (We know the separation must be small because the orbital periods of these binary systems are short, between 10 and 100 days. Kepler's third law for binary star systems, which we described in Section 17-9, tells us that the shorter the period, the smaller the distance between the two stars.) This fact suggests a scenario for how millisecond pulsars acquired their very rapid rotation.

Imagine a binary system consisting of a high-mass star and a low-mass star. The high-mass star evolves more rapidly than the low-mass star, and within a few million years becomes a Type II supernova that creates a neutron star. Like most newborn neutron stars, it spins several times per second, and, thus, initially we see a pulsar rather like the Crab or Vela pulsar.

Over the next few billion years, the pulsar slows down as it radiates energy into space. Meanwhile, the slowly evolving low-mass star begins to expand as it evolves away from the main sequence to become a red giant. When the red giant gets big enough to fill its Roche lobe, it starts to spill gas over the inner Lagrangian point onto the neutron star. (See Section 19-7 for a review of close binary systems.) The infalling gas strikes the neutron star's surface at high speed and at an angle that causes the star to spin faster. In this way, a slow, aging pulsar is "spun up" by mass transfer from its bloated companion.

What about the few millisecond pulsars, like PSR 1937+21, that are *not* members of close binaries? It may be that solitary millisecond pulsars were once part of close binary systems, but the companion stars have been eroded away by the high-energy particles emitted by the pulsar after it was spun up. The Black Widow pulsar—named for the species of spider whose females kill and eat their mates—may be caught in the act of destroying its companion in just such a process (Figure 21-11).

The Black Widow pulsar (a neutron star spinning 622 times per second) and its companion orbit each other with a period of 9.16 hours. The orbital plane of the system is nearly edge-on to our line of sight, so this system is an eclipsing binary (described in Section 19-11): The companion blocks out the pulsar's radio signals for 45 minutes during each eclipse. Just before and after an eclipse, however, the pulsar's signals are delayed substantially, as if the radio waves were being slowed as they pass through a cloud of ionized gas surrounding the companion star. This circumstellar material is probably the star's outer layers, dislodged by radiation and by fast-moving particles emitted from the pulsar. In a few hundred million years or so, the companion of the pulsar will have completely disintegrated, leaving behind only a spun-up, solitary millisecond pulsar.

It is also possible that both stars in a close binary system might evolve into neutron stars, forming a pair of pulsars. Several such *binary pulsars* are known. Because the two pulsars are so massive and move around each other in tight orbits, they lose energy through a process called *gravitational radiation.* (We will discuss this process, which is a prediction of Einstein's general theory of relativity, in more detail in Chapter 22.) As a result, the two neutron stars spiral toward each other and eventually collide. Computer simulations of such a collision show that some of the neutrons are ejected into space, where many of them decay into protons and combine to form various heavy elements. In addition to the heavy elements produced directly by supernova collisions,

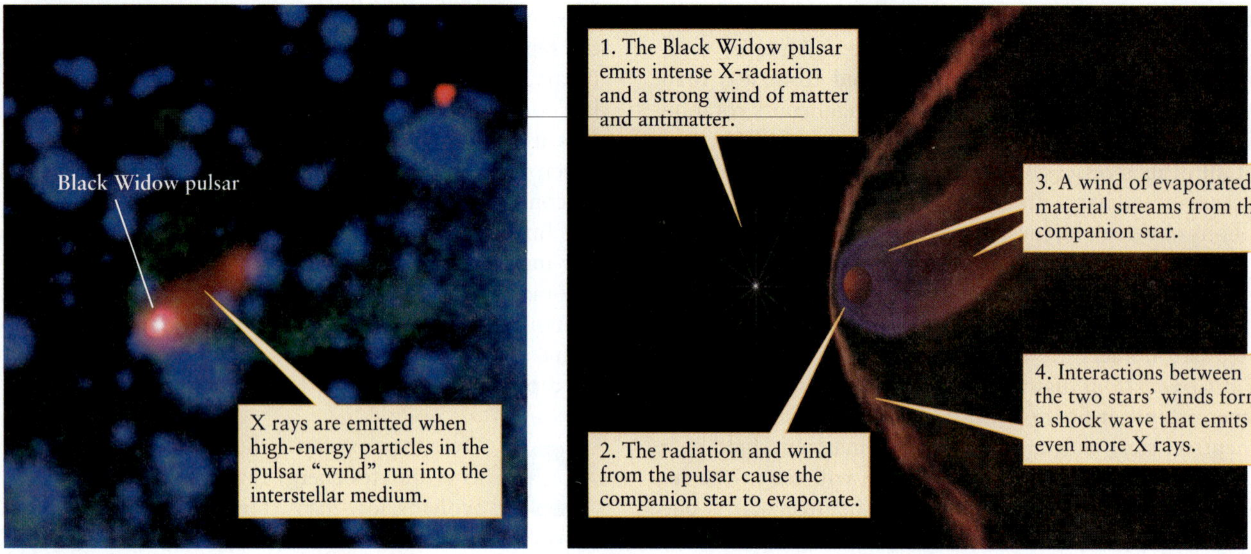

Black Widow pulsar
X rays are emitted when high-energy particles in the pulsar "wind" run into the interstellar medium.
1. The Black Widow pulsar emits intense X-radiation and a strong wind of matter and antimatter.
2. The radiation and wind from the pulsar cause the companion star to evaporate.
3. A wind of evaporated material streams from the companion star.
4. Interactions between the two stars' winds form a shock wave that emits even more X rays.

(a) The Black Widow pulsar R I V U X G

(b) An illustration of the pulsar and its companion

Figure 21-11

The Black Widow Pulsar (a) This false-color composite of optical and X-ray images shows pulsar B1957+20, called the Black Widow. As the pulsar moves through space at over 200 km/s, it emits a wind that streams behind it. **(b)** This artist's conception of the pulsar and its doomed companion shows an area far smaller than the white dot in (a) labeled "Black Widow pulsar." (a: NASA/CXC/ASTRON/B. Stappers et al. (X-ray), AAO/J. Bland-Hawthorn & H. Jones (visible); b: CXC/M. Weiss)

the elements produced by such neutron star collisions may play an important role in seeding the interstellar medium with the building blocks of planets (see Section 20-6).

21-8 Neutron stars in close binary systems can also be pulsating X-ray sources

Millisecond pulsars emitting radio waves are not the only result of having a neutron star in a close binary system with an ordinary star. Even more exotic are **pulsating X-ray sources,** in which material from the companion star is drawn onto the magnetic poles of the neutron star, producing intense hot spots that emit breathtaking amounts of X rays. As the neutron star spins and the X-ray beams are directed alternately toward us and away from us, we see the X rays flash on and off. Such pulsating X-ray sources, while similar in many ways to ordinary pulsars, are far more luminous.

Pulsating X-ray Sources and Binary Systems

In an X-ray binary pulsar, gas from a companion star is funneled to the magnetic poles of the neutron star, forming hot spots that emit X rays rather than radio waves

Pulsating X-ray sources were discovered in 1971 by the *Uhuru* spacecraft, the first to give astronomers a comprehensive look at the X-ray sky. (As we saw in Section 6-7, Earth's atmosphere is opaque to X rays.) The first X-ray source found to emit pulses was Centaurus X-3, the third X-ray source found in the southern constellation Centaurus (Figure 21-12). The X-ray pulses have a regular period of 4.84 s. A few months later, similar pulses were discovered coming from a source designated Hercules X-1, which had a period of 1.24 s. Because the periods of these two X-ray sources are so short, astronomers began to suspect that they were actually observing rapidly rotating neutron stars.

It soon became clear that Centaurus X-3 and Hercules X-1 are members of binary star systems. One piece of evidence is that every 2.087 days, Centaurus X-3 appears to turn off for almost 12 hours. Apparently, Centaurus X-3 is a neutron star in an eclipsing binary system, and it takes nearly 12 hours for the X-ray source to pass behind its companion star. Hercules X-1, too, appears to turn off periodically, corresponding to a 6-hour eclipse every 1.7 days by its visible companion, the star HZ Herculis. Moreover, careful timing of its X-ray pulses shows a periodic Doppler shifting every 1.7 days, which is direct evidence of orbital motion about a companion star. When the X-ray source is approaching us, the pulses from Hercules X-1 are separated by slightly less than 1.24 seconds. When the source is receding from us, slightly more than 1.24 seconds elapse between the pulses.

Putting all the pieces together, astronomers now realize that pulsating X-ray sources like Centaurus X-3 and Hercules X-1 are binary systems in which one of the stars is a neutron star. All these binary systems have very short orbital periods, which means that the distance between the two stars is quite small. The neutron star can therefore capture gases escaping from its ordinary companion (Figure 21-13). More than 20 such exotic binary systems, also called *X-ray binary pulsars,* are known. A typical rate

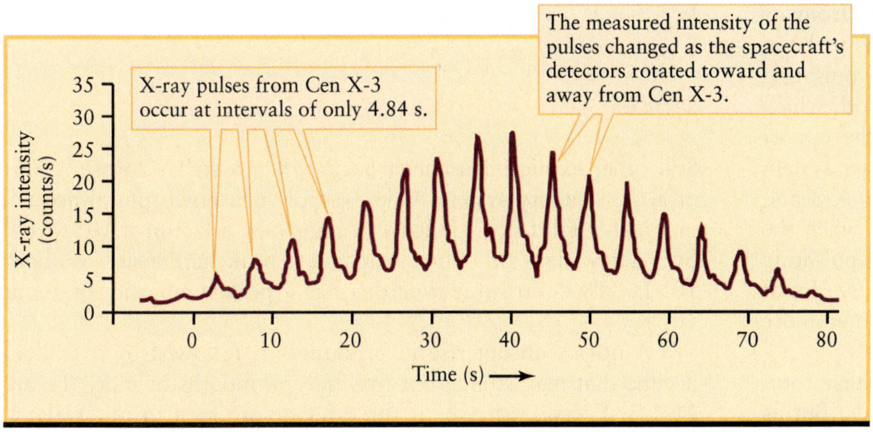

Figure 21-12

X-Ray Pulses from Centaurus X-3 This graph shows the intensity of X rays detected by the *Uhuru* spacecraft as Centaurus X-3 (often abbreviated as Cen X-3) moved across the satellite's field of view. Because the pulse period is so short, Centaurus X-3 is in all probability a rotating neutron star. (Adapted from R. Giacconi and colleagues)

of mass transfer from the ordinary star to the neutron star is roughly 10^{-9} solar masses per year.

Because of its strong gravity, a neutron star in a pulsating X-ray source easily captures much of the gas escaping from its companion. But like an ordinary pulsar, the neutron star is rotating rapidly and has a powerful magnetic field inclined to the axis of rotation (recall Figure 21-3). As the gas falls toward the neutron star, its magnetic field funnels the incoming matter down onto the star's north and south magnetic polar regions. The neutron star's gravity is so strong that the gas is traveling at nearly half the speed of light by the time it crashes onto the star's surface. This violent impact creates spots at both poles with searing temperatures of about 10^8 K. Material raised to these temperatures emits abundant X rays, and because so much material is dumped onto these hot spots the X-ray luminosity reaches roughly 10^{31} watts. This luminosity is much greater than the X-ray luminosity of isolated pulsars like the Crab (see Figure 21-4*b*) or 1E 1207.4-5209 (see Figure 21-10*a*), whose surface temperatures are a relatively chilly 10^6 K or so. It is also nearly 10^5 times greater than the combined luminosity of the Sun at all wavelengths.

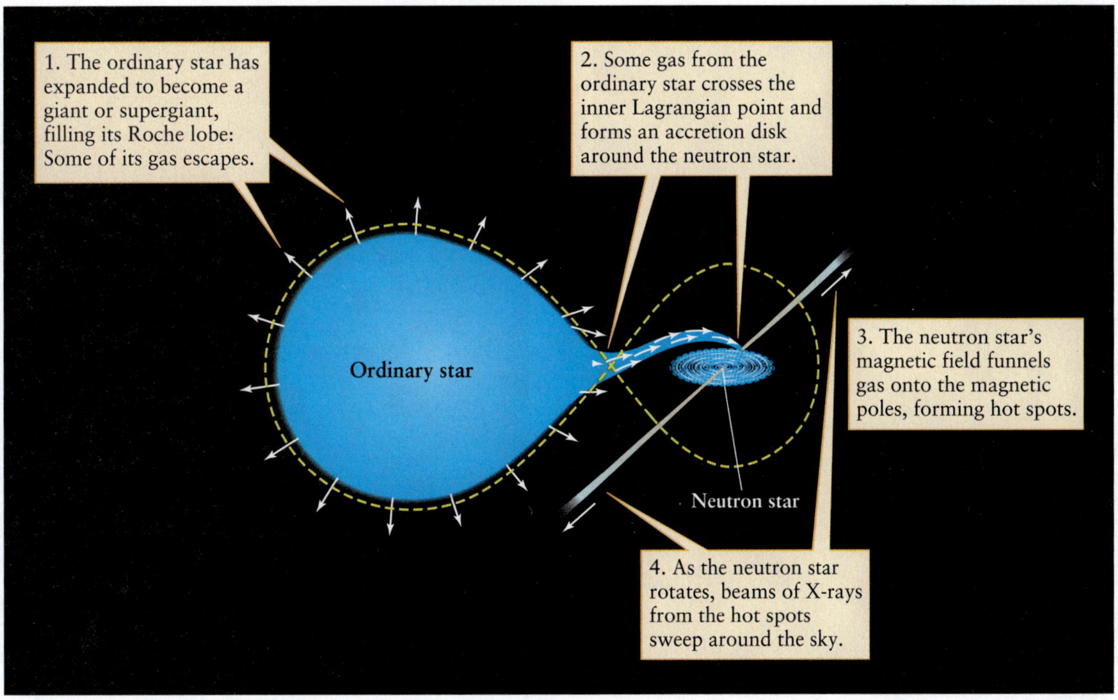

Figure 21-13

A Model of a Pulsating X-Ray Source Pulsating X-ray sources are close, semidetached binary systems (compare with Figure 19-20*b* and Figure 19-21*b*) in which one member is a rotating neutron star. Gas from the ordinary star flows onto the neutron star and creates hot spots that emit beams of X rays. If one of the beams is oriented so that it sweeps over Earth as the neutron star rotates, we see a pulse of X rays on each sweep.

As the neutron star rotates, the beams of X rays from its magnetic poles sweep around the sky. If Earth happens to be in the path of one of the two beams, we can observe a pulsating X-ray source. We detect a pulse once per rotation period, which means that the neutron star in Hercules X-1 spins at the rate of once every 1.24 seconds. If the orbital plane of the binary system is nearly edge-on to our line of sight, as for Centaurus X-3 and Hercules X-1, the pulsating X rays appear to turn off when the neutron star is eclipsed by the companion star. Other pulsating X-ray sources, such as Scorpius X-1, do not undergo this sort of turning off. The orbital planes of these systems are oriented more nearly face-on to our line of sight, so no eclipses occur.

The neutron stars in most pulsating X-ray sources have rotation periods (and hence pulse periods) of a few seconds. But as time passes, the neutron star accretes more and more mass from its companion. In the process, it will spin up and may eventually become a millisecond pulsar (Section 21-7). In 1998 the *Rossi X-ray Timing Explorer* spacecraft first observed an evolutionary "missing link" between pulsating X-ray sources and millisecond pulsars. This object, designated SAX J1808.4-3658, is an X-ray–emitting millisecond pulsar in a close binary system with a pulse period of just 2.5 ms. Apparently the pulsar has accreted a substantial amount of mass from its companion star. Like the Black Widow pulsar (see Figure 21-11), it has also blown much of the companion's mass into space. Within a billion years, the companion star may disappear altogether.

21-9 Explosive thermonuclear processes on white dwarfs and neutron stars produce novae and bursters

Still other exotic phenomena occur when a stellar corpse is part of a close binary system. One example is a **nova** (plural **novae**), in which a faint star suddenly brightens by a factor of 10^4 to 10^8 over a few days or hours, reaching a peak luminosity of about 10^5 $L_\odot$. By contrast, a *superova* has a peak luminosity of about 10^9 $L_\odot$.

A nova's abrupt rise in brightness is followed by a gradual decline that may stretch out over several months or more (Figure 21-14). Every year two or three novae are seen in our Galaxy, and several dozen more are thought to take place in remote regions of the Galaxy that are obscured from our view by interstellar dust.

Novae and White Dwarfs

In the 1950s, painstaking observations of numerous novae by Robert Kraft, Merle Walker, and their colleagues at the University of California's Lick Observatory led to the conclusion that all novae are members of close binary systems containing a white dwarf. Gradual mass transfer from the ordinary companion star, which presumably fills its Roche lobe, deposits fresh hydrogen

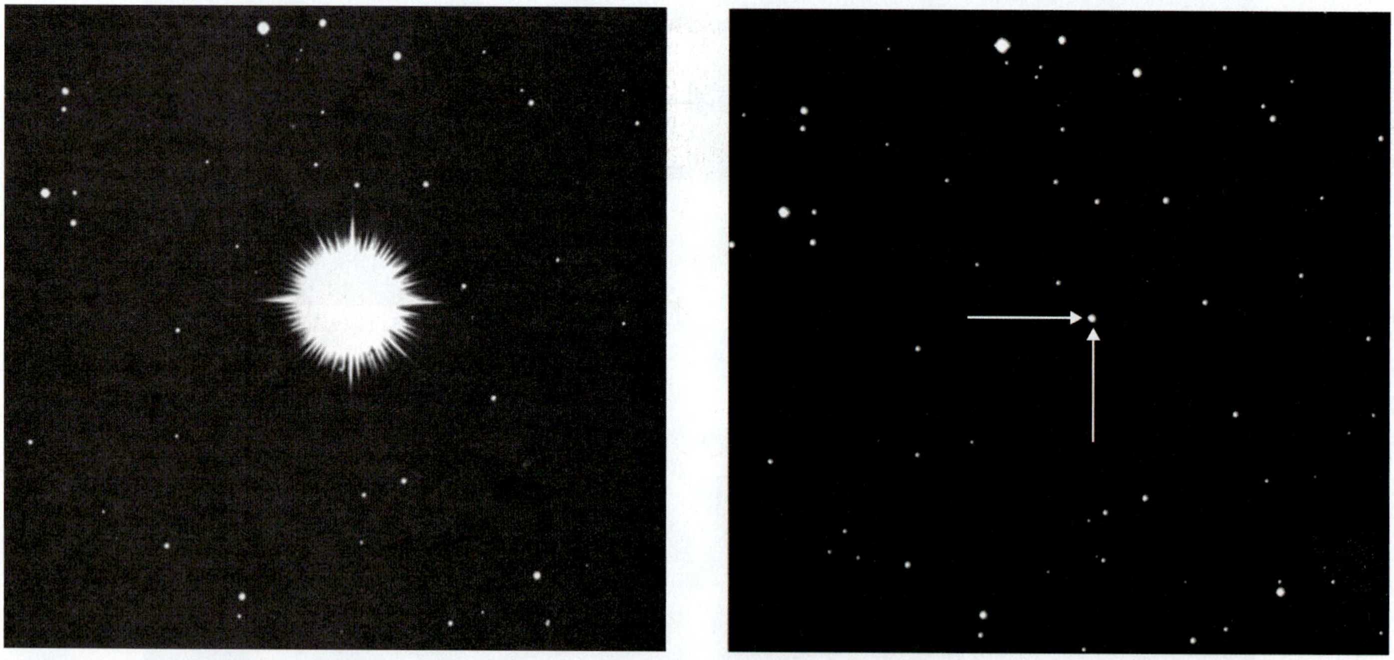

(a) Nova Herculis 1934 shortly after peak brightness

(b) Two months later

Figure 21-14 R I V U X G

Nova Herculis 1934 These two pictures show a nova **(a)** shortly after peak brightness as a star of apparent magnitude +3, bright enough to be seen easily with the naked eye, and **(b)** two months later, when it had faded by a factor of 4000 in brightness to apparent magnitude +12. (See Figure 17-6 for a description of the apparent magnitude scale.) Novae are named after the constellation and year in which they appear. (Lick Observatory)

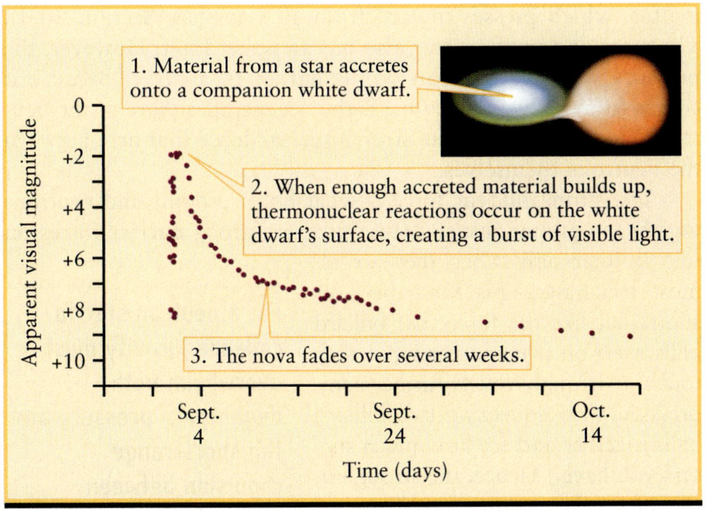

Figure 21-15

The Light Curve of a Nova This illustration and graph show the history of Nova Cygni 1975, a typical nova. Its rapid rise and gradual decline in apparent brightness are characteristic of all novae. This nova, also designated V1500 Cyg, was easily visible to the naked eye (that is, was brighter than an apparent magnitude of +6) for nearly a week. (Illustration courtesy CXC/M. Weiss)

onto the white dwarf (see Figure 19-21*b* for a schematic diagram of this sort of mass transfer).

Because of the white dwarf's strong gravity, this hydrogen is compressed into a dense layer covering the hot surface of the white dwarf. As more gas is deposited and compressed, the temperature in the hydrogen layer increases. When the temperature reaches about 10^7 K, hydrogen fusion ignites throughout the gas layer, embroiling the white dwarf's surface in a thermonuclear holocaust that we see as a nova (Figure 21-15).

CAUTION! It is important to understand the similarities and differences between novae and the thermonuclear (Type Ia) supernovae that we described in Section 20-9. Both kinds of celestial explosions are thought to occur in close binary systems where one of the stars is a white dwarf. But, as befits their name, supernovae are much more energetic. A Type Ia supernova explosion radiates 10^{44} joules of energy into space, while the corresponding figure for a typical nova is 10^{37} joules. (To be fair to novae, this relatively paltry figure is as much energy as our Sun emits in 1000 years.) The difference is thought to be that in a Type Ia supernova, the white dwarf accretes much more mass from its companion. This added mass causes so much compression that nuclear reactions can take place *inside* the white dwarf. Eventually, these reactions blow the white dwarf completely apart. In a nova, by contrast, nuclear reactions occur only within the accreted material. The reaction is more sedate because it takes place only on the white dwarf's surface (perhaps because the accretion rate is less or because the white dwarf had less mass in the first place).

Novae and thermonuclear supernovae both occur in close binary systems with a white dwarf, but a nova can recur while a supernova is a one-shot event

Because the white dwarf itself survives a nova explosion, it is possible for the same star to undergo more than one nova. As an example, the star RS Ophiuchi erupted as a nova in 1898, then put in repeat performances in 1933, 1958, 1967, 1985, and 2006. By contrast, a given star can only be a supernova once.

X-ray Bursters and Neutron Stars

A surface explosion similar to a nova also occurs with neutron stars. In 1975 it was discovered that some objects in the sky emit sudden, powerful bursts of X rays. Figure 21-16 shows the record of a typical burst. The source emits X rays at a constant low level until suddenly, without warning, there is an abrupt increase in X rays, followed by a gradual decline. An entire burst typically lasts for only 20 s. Unlike pulsating X-ray sources, there is a fairly long interval of hours or days between bursts, and the bursts don't repeat at regular intervals. Sources that behave in this fashion are known as **X-ray bursters.** Several dozen X-ray bursters have been discovered in our Galaxy.

X-ray bursters, like novae, are thought to involve close binaries whose stars are engaged in mass transfer. With a burster, however, the stellar corpse is a neutron star rather than a white dwarf. Gases escaping from the ordinary companion star fall onto the neutron star. The X-ray burster's magnetic field is probably not strong enough to funnel the falling material toward the magnetic poles, so the gases are distributed more evenly over the surface of the neutron star. The energy released as these gases crash down onto the neutron star's surface produces the low-level X rays that are continuously emitted by the burster.

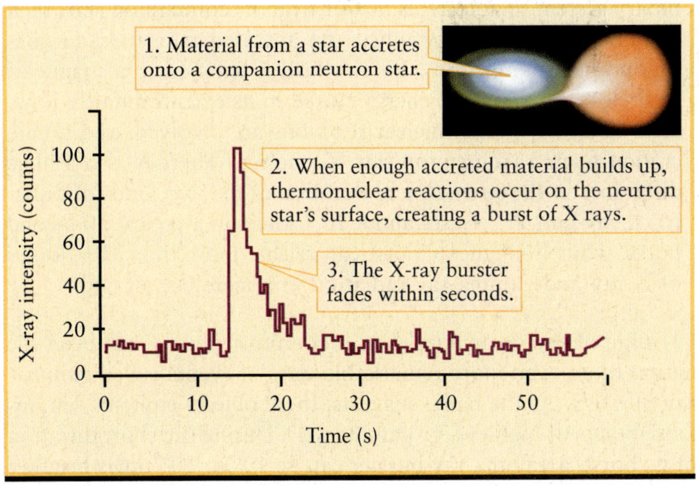

Figure 21-16

The Light Curve of an X-Ray Burster This illustration and graph show the history of a typical X-ray burster. A burster emits a constant low intensity of X rays interspersed with occasional powerful X-ray bursts. This burst was recorded on September 28, 1975, by an Earth-orbiting X-ray telescope. Contrast this figure with Figure 21-15, which shows a typical nova. (Data adapted from W. H. G. Lewin; illustration courtesy CXC/M. Weiss)

Most of the gas falling onto the neutron star is hydrogen, which the star's powerful gravity compresses against its hot surface. In fact, temperatures and pressures in this accreting layer become so high that the arriving hydrogen is converted into helium by hydrogen fusion. As a result, the accreted gases develop a layered structure that covers the entire neutron star, with a few tens of centimeters of hydrogen lying atop a similar thickness of helium. The structure is reminiscent of the layers within an evolved giant star (see Figure 20-2), although the layers atop a neutron star are much more compressed, thanks to the star's tremendous surface gravity.

When the helium layer is about 1 m thick, helium fusion ignites explosively and heats the neutron star's surface to about 3×10^7 K. At this temperature the surface predominantly emits X rays, but the emission ceases within a few seconds as the surface cools. Hence, we observe a sudden burst of X rays only a few seconds in duration. New hydrogen then flows onto the neutron star, and the whole process starts over. Indeed, X-ray bursters typically emit a burst every few hours or days.

Whereas explosive *hydrogen* fusion on a white dwarf produces a nova, explosive *helium* fusion on a neutron star produces an X-ray burster. In both cases, the process is explosive, because the fuel is compressed so tightly against the star's surface that it becomes degenerate, like the star itself. As with the helium flash inside red giants (described in Section 19-3), the ignition of a degenerate thermonuclear fuel involves a sudden thermal runaway. This is because an increase in temperature does not produce a corresponding increase in pressure that would otherwise relieve compression of the gases and slow the nuclear reactions.

CAUTION! Be careful not to confuse X-ray bursters with magnetars (Section 21-6), which are also neutron stars that emit powerful bursts of X rays. An X-ray burster is a member of a binary system that accretes matter from its companion star, then releases *nuclear* energy when the accreted matter undergoes thermonuclear reactions. By contrast, a burst from a magnetar is a release of *magnetic* energy stored in its extraordinarily powerful field; no thermonuclear reactions are involved, and no accretion from a companion star is required. There is also a huge difference in the strength and duration of the two kinds of burst. An X-ray burster releases about 10^{32} joules in a typical 20-second burst, while a magnetar burst can release more than 10^{39} joules of X-ray and gamma-ray radiation in a mere 0.2 second.

One of the great puzzles in modern astronomy has been the nature of an even more remarkable class of events called *gamma-ray bursters*. As the name suggests, these objects emit sudden, intense bursts of high-energy gamma rays. During the short duration of its burst, a gamma-ray burster can be 10^6 to 10^9 times brighter than a supernova at the same distance! We will discuss these exotic objects further in Chapter 22.

21-10 Like a white dwarf, a neutron star has an upper limit on its mass

A white dwarf will collapse if its mass is greater than the Chandrasekhar limit of 1.4 $M_\odot$. At that point, degenerate electron pressure cannot support the overpowering weight of the star's matter, which presses inward from all sides (see Section 20-4). The mass of a neutron star also has an upper limit. However, the pressure within a neutron star is harder to analyze, because it comes from *two* sources. One is the degenerate nature of the neutrons, and the other is the strong nuclear force that acts between the neutrons themselves.

The strong nuclear force is what holds protons and neutrons together in atomic nuclei. Neutrons exert strong nuclear forces on one another only when they are almost touching. This force behaves somewhat like the force that billiard balls exert on one another when they touch: It strongly resists further compression. (Try squeezing two billiard balls together and see how much success you have.) Hence, the strong nuclear force is a major contributor to the star's internal pressure. Unfortunately, there is a good deal of uncertainty about the details of this force. This uncertainty translates into uncertainties about how much weight the neutron star's internal pressure can support—that is, the neutron star's maximum mass. Theoretical estimates of this maximum mass range from 2 to 3 $M_\odot$.

For a neutron star to collapse, gravity must overwhelm both degeneracy pressure and the short-range repulsion between neutrons

Before pulsars were discovered, most astronomers believed all dead stars to be white dwarfs. Dying stars were thought to somehow eject enough material so that their corpses could be below the Chandrasekhar limit. The discovery of neutron stars proved this idea incorrect. Inspired by this lesson, astronomers soon began wondering what might happen if a dying massive star failed to eject enough matter to get below the upper limit for a neutron star. For example, what might a 5-$M_\odot$ stellar corpse be like?

The gravity associated with a neutron star is so strong that the escape speed from it is roughly one-half the speed of light. But if a stellar corpse has a mass greater than 3 $M_\odot$, so much matter is crushed into such a small volume that the escape speed actually *exceeds* the speed of light. Because nothing can travel faster than light, nothing—not even light—can leave this dead star. Its gravity is so powerful that it leaves a hole in the fabric of space and time. Thus, the discovery of neutron stars inspired astrophysicists to examine seriously one of the most bizarre and fantastic objects ever predicted by modern science, the black hole. We take up its story in the next chapter.

Key Words

Key Ideas

Neutron Stars: A neutron star is a dense stellar corpse consisting primarily of closely packed degenerate neutrons.

• A neutron star typically has a diameter of about 20 km, a mass less than 3 $M_\odot$, a magnetic field 10^{12} times stronger than that of the Sun, and a rotation period of roughly 1 second.

• A neutron star consists of a superfluid, superconducting core surrounded by a superfluid mantle and a thin, brittle crust.

• Intense beams of radiation emanate from regions near the north and south magnetic poles of a neutron star. These beams are produced by streams of charged particles moving in the star's intense magnetic field.

Pulsars: A pulsar is a source of periodic pulses of radio radiation. These pulses are produced as beams of radio waves from a neutron star's magnetic poles sweep past Earth.

• The pulse rate of many pulsars is slowing steadily. This reflects the gradual slowing of the neutron star's rotation as it radiates energy into space. Sudden speedups of the pulse rate, called glitches, may be caused by interactions between the neutron star's crust and its superfluid interior.

Magnetars: A magnetar is a pulsar with an extraordinarily strong magnetic field. This field is produced by convection inside the pulsar when it first forms.

• The solid crust of a magnetar is under tremendous magnetic stress. When the surface rearranges in a starquake, the released magnetic energy produces a powerful burst of X-rays and gamma rays.

Neutron Stars in Close Binary Systems: If a neutron star is in a close binary system with an ordinary star, tidal forces will draw gas from the ordinary star onto the neutron star.

• The transfer of material onto the neutron star can make it rotate extremely rapidly, giving rise to a millisecond pulsar.

• Magnetic forces can funnel the gas onto the neutron star's magnetic poles, producing hot spots. These hot spots then radiate intense beams of X rays. As the neutron star rotates, the X-ray beams appear to flash on and off. Such a system is called a pulsating X-ray variable.

Novae and Bursters: Material from an ordinary star in a close binary can fall onto the surface of the companion white dwarf or neutron star to produce a surface layer in which thermonuclear reactions can explosively ignite.

• Explosive hydrogen fusion may occur in the surface layer of a companion white dwarf, producing the sudden increase in luminosity that we call a nova. The peak luminosity of a nova is only 10^{-4} of that observed in a supernova.

• Explosive helium fusion may occur in the surface layer of a companion neutron star. This produces a sudden increase in X-ray radiation, which we call a burster.

Questions

Review Questions

1. What are neutron stars? What led scientists to propose their existence?
2. What is degenerate neutron pressure? How does it make it possible for a neutron star to be more massive than the Chandrasekhar limit for white dwarfs?
3. How were pulsars discovered? How do they differ from variable stars?
4. Why did astronomers rule out the idea that pulsars are rapidly rotating white dwarfs?
5. Why do astronomers think that pulsars are rapidly rotating neutron stars?
6. During the weeks immediately following the discovery of the first pulsar, one suggested explanation was that the pulses might be signals from an extraterrestrial civilization. Why did astronomers soon discard this idea?
7. Why do neutron stars rotate so much more rapidly than ordinary stars? Why do they have such strong magnetic fields?
8. How are rotating neutron stars able to produce pulses of radiation as seen by an observer on Earth?
9. Do all supernova remnants contain pulsars? Are all pulsars found within supernova remnants? For each question, explain why or why not.
10. Is our Sun likely to end up as a neutron star? Why or why not?
11. Why are some neutron stars seen moving through space at hundreds of kilometers per second?
12. What is synchrotron radiation? How is it involved in making the Crab Nebula glow?
13. Why does an isolated pulsar rotate more slowly as time goes by?
14. Astronomers have deduced that the Vela pulsar is about 11,000 years old. How do you suppose they did this?
15. What is the difference between superconductivity and superfluidity?
16. Does a neutron star contain only neutrons? If not, what else does it contain?
17. What is a pulsar glitch? How does a glitch affect a pulsar's period? What is thought to be the cause of glitches?
18. Compare the internal structure of a white dwarf to that of a neutron star. What are the similarities? What are the differences?
19. What is the evidence that neutron stars have atmospheres?
20. How does a magnetar differ from an ordinary pulsar? What determines whether a pulsar becomes a magnetar?
21. How are magnetars able to emit bursts of X-rays and gamma rays?
22. Why do astronomers think that millisecond pulsars are very old?
23. If millisecond pulsars are formed in close binary systems, why are some found without companion stars?
24. Describe a pulsating X-ray source like Hercules X-1 or Centaurus X-3. What produces the pulsation?
25. What is the connection between pulsating X-ray sources and millisecond pulsars?
26. What are the similarities between a nova and a Type Ia supernova? What are the differences?
27. What are the similarities between novae and X-ray bursters? What are the differences?
28. What are the similarities between pulsating X-ray sources and X-ray bursters? What are the differences?
29. (**a**) Rank the following explosive phenomena in order of the amount of energy released, from smallest to largest: (i) a nova; (ii) a Type Ia supernova; (iii) an X-ray burster; (iv) a major burst from a magnetar. (**b**) For each of the phenomena

listed in part (a), explain what the source of the released energy is.

30. Why is the maximum mass of neutron stars not known as accurately as the Chandrasekhar limit for white dwarfs?

Advanced Questions

Problem-solving tips and tools

The volume of a sphere of radius r is $4\pi r^3/3$. The small-angle formula is given in Box 1-1. Section 5-4 describes how to relate the temperature of a blackbody to its wavelength of maximum emission. Section 17-6 gives the formula relating a star's luminosity, surface temperature, and radius (see Box 17-4 for worked examples). Appendix 6 gives the conversion between seconds and years as well as the radius of the Sun.

31. Using a diagram like Figure 21-3, explain why the number of pulsars that we observe in nearby space is probably quite a bit less than the number of rotating, magnetized neutron stars in nearby space.

32. There are many more main-sequence stars of low mass (less than 8 $M_\odot$) than of high mass (8 $M_\odot$ or more). Use this fact to explain why white dwarf stars are far more common than neutron stars.

33. The distance to the Crab Nebula is about 2000 parsecs. In what year did the star actually explode? Explain your answer.

34. How do we know that the Crab pulsar is really embedded in the Crab Nebula and not simply located at a different distance along the same line of sight?

35. The Crab Nebula has an apparent size of about 5 arcmin, and this size is increasing at a rate of 0.23 arcsec per year. (**a**) Assume that the expansion rate has been constant over the entire history of the Crab Nebula. Based on this assumption, in what year would Earth observers have seen the supernova explosion that formed the nebula? (**b**) Does your answer to part (a) agree with the known year of the supernova, 1054 A.D.? If not, can you point to assumptions you made in your computations that led to the discrepancies? Or do you think your calculations suggest additional physical effects are at work in the Crab Nebula, over and above a constant rate of expansion?

36. Emission lines in the spectrum of the Crab Nebula exhibit a Doppler shift, which indicates that gas in the part of the nebula closest to us is moving toward us at 1450 km/s. (**a**) Assume that the expanding gas has been moving at the same speed since the original supernova explosion, observed in 1054 A.D., and calculate what radius and what diameter (in light-years) we should observe the nebula to have today. (**b**) Compare your result in part (a) to the actual size of the nebula, given in the caption to Figure 21-4.

37. The supernova remnant G5.4-1.2 shown in Figure 21-5 lies about 5000 pc (16,000 ly) from Earth in the constellation Sagittarius. (**a**) The green arc in the large left-hand image in Figure 21-5 represents part of the outer edge of this spherical supernova remnant. Estimate the diameter of this remnant in parsecs. (*Hint:* To calculate this, you will need to make measurements on the image.) (**b**) How far (in parsecs) did the neutron star travel from where it was formed (at the position of the supernova's progenitor star, presumably at the center of the present-day remnant) to the position shown in Figure 21-5? Explain your answer.

38. To determine accurately the period of a pulsar, astronomers must take into account Earth's orbital motion about the Sun. (**a**) Explain why. (**b**) Knowing that Earth's orbital velocity is 30 km/s, calculate the maximum correction to a pulsar's period because of Earth's motion. Explain why the size of the correction is greatest for pulsars located near the ecliptic.

39. If a pulsar has period P (in seconds) and its period is increasing at a rate R (in seconds per second), an approximate formula for the age T of the pulsar (in seconds) is $T = P/2R$. For the Crab pulsar, $P = 0.0333$ s and $R = 4.21 \times 10^{-13}$ s/s. (**a**) Calculate the approximate age of the Crab pulsar in years. (**b**) Based on the information given in Section 21-1, is your result in (a) an underestimate or an overestimate? Explain your answer.

40. (See Advanced Question 39.) (**a**) Magnetar SGR 1806-20 was the source of the intense burst observed on December 27, 2004. Based on the information given in Section 21-6, calculate the rate R (in seconds per second) at which the period of SGR 1806-20 is increasing. (**b**) The ordinary pulses from SGR 1806-20 have a period of 7.47 s. Calculate the approximate age of SGR 1806-20 in years. (**c**) Theory predicts that a magnetar becomes inactive after about 10,000 years. Is your result in part (b) consistent with this prediction? Explain why or why not.

41. A neutron has a mass of about 1.7×10^{-27} kg and a radius of about 10^{-15} m. (**a**) Compare the density of matter in a neutron with the average density of a neutron star. (**b**) If the neutron star's density is more than that of a neutron, the neutrons within the star are overlapping; if it is less, the neutrons are not overlapping. Which of these seems to be the case for average neutrons within the star? Which do you think is the case at the center of the neutron star, where densities are higher than average?

42. The total luminosity at all wavelengths of the magnetar burst observed on December 27, 2006, was approximately 10^{14} $L_\odot$. At what distance from the magnetar would the brightness of the burst have been equal to the brightness of the Sun as seen on Earth? Give your answer in AU and in parsecs.

43. X-ray pulsars are speeding up but ordinary (radio) pulsars are slowing down. Propose an explanation for this difference.

44. If the model for Hercules X-1 discussed in the text is correct, at what orientation of the binary system do we see its maximum optical brightness? Explain your answer.

45. Explain why heavy elements that are produced by neutron star collisions can still be regarded as having been processed through a supernova.

46. In an X-ray burster, the surface of a neutron star 10 km in radius is heated to a temperature of 3×10^7 K. (**a**) Determine the wavelength of maximum emission of the heated surface (which you may treat as a blackbody). In what part of the electromagnetic spectrum does this lie? (See Figure 5-7.) (**b**) Find the luminosity of the heated neutron star. Give your answer in watts and in terms of the luminosity of the Sun, given in Table 16-1. How does this compare with the peak luminosity of a nova? Of a Type Ia supernova?

47. The nearest neutron star, called RX J185635-3754, is just 60 pc (200 ly) from Earth. It is thought to be the relic of a star that underwent a supernova explosion about 1 million years ago. The explosion ejected the neutron star at high speed, so it is now moving through nearly empty space. (**a**) RX J185635-3754 is *not* a pulsar, that is, it does not emit pulses of radiation. Suggest why this might be so. (**b**) The neutron star has a surface temperature of 600,000 K. Find the wavelength at which it emits most strongly, and explain why the neutron star appears as a steady, nonpulsating object in an X-ray telescope. (**c**) RX J185635-3754 has a total luminosity at all wavelengths of about $0.046L_{\odot}$. Calculate its radius, and explain why astronomers conclude that it is a neutron star.

Discussion Questions

48. Imagine that we're somehow able to stand (and survive) at one of the magnetic poles of the Crab pulsar. Describe what you would see. How would the stars appear to move in the sky? What would you see if you looked straight up? What factors make this location a very unhealthy place to visit?
49. Accretion disks in close binary systems are too small to be seen directly with even the highest-resolution telescopes. How, then, can astronomers detect the presence of such accretion disks?
50. When neutrons are very close to one another, they repel one another through the strong nuclear force. If this repulsion were made even stronger, what effect might this have on the maximum mass of a neutron star? Explain your answer.

Web/eBook Questions

51. Search the World Wide Web for information about the latest observations of the stellar remnant at the center of SN 1987A. Has a pulsar been detected? If so, how fast is it spinning? Has the supernova's debris thinned out enough to give a clear view of the neutron star?
52. Search the World Wide Web for the latest information about magnetars and soft gamma repeaters (also known as soft gamma-ray repeaters). How many soft gamma repeaters are known? Have any of these produced an intense burst since December 27, 2004?
53. **Monitoring the Crab Pulsar.** Access the video "The Crab Pulsar" in Chapter 21 of the *Universe* Web site or eBook. View the video and use it to answer the following questions. For each part, explain how you determined your answer. (**a**) How many rotations does the neutron star complete during the duration of the video? (**b**) Is the neutron star visible at the beginning of the video? If not, explain why not. (**c**) How does the peak brightness of the Crab pulsar compare to the steady brightness of the nearby star? (**d**) What total amount of time is depicted in the video?

Activities

Observing Projects

54. If you did not take the opportunity to observe the Crab Nebula as part of the exercises in Chapter 20, do so now. The Crab Nebula is visible in the night sky from October through March. Its epoch 2000 coordinates are R.A. = 5^h 34.5^m and Decl. = 22° 00′, which is near the star marking the eastern horn of Taurus (the Bull). Be sure to schedule your observations for a moonless night. The larger the telescope you use, the better, because the Crab Nebula is quite dim.
55. Consult the World Wide Web to see if any novae have been sighted recently. If by good fortune one has been sighted, what is its apparent magnitude? Is it within reach of a telescope at your disposal? If so, arrange to observe it. Draw what you see through the eyepiece, noting the object's brightness in comparison with other stars in the field of view. If possible, observe the same object a few weeks or months later to see how its brightness has changed.
56. Use the *Starry Night Enthusiast™* program to observe the sky in July 1054, when the supernova that spawned the Crab Nebula was visible from the American Southwest. Select **Viewing Location . . .** in the **Options** menu, click on the **Latitude/Longitude** tab, enter 36° N for latitude and 109° W for longitude, and click on the **Set Location** button. You are now near the location of the pictograph shown in Figure 21-1. In the toolbar, set the **Date** to **July 5, 1054 A.D.** and the **Time** to **5:00 A.M.** Remove the artificial satellites from the view by clicking on **View > Solar System > Satellites,** since these would not have been in the sky at this time in history. Use the **Find** pane to find and center on the **Crab Nebula.** Zoom in or out until you can see both the position of the nebula and the Moon. You may find it helpful to turn daylight on or off (Select **Show Daylight** or **Hide Daylight** in the **View** menu). (**a**) What is the phase of the Moon? (**b**) Investigate how the relative positions of the Moon and the Crab Nebula change when you set the date to July 4, 1054, or July 6, 1054. On which date do the relative positions of the Moon and the Crab Nebula give the best match to the pictograph in Figure 21-1? (**c**) Zoom in on the Crab Nebula to see this supernova remnant. *Starry Night Enthusiast™* superimposes an X-ray image of this active region from the Chandra space telescope on to the visible light image. To compare the images from these two different wavelengths, you can remove the X-ray image by opening the **Options** pane, expanding the **Deep Space** layer and clicking in the box next to **Chandra.**
57. Use the *Starry Night Enthusiast™* program to locate the Small Magellanic Cloud, the site of a large concentration of X-ray pulsars. Open the **Favorites** menu and click on **Deep Space > Local Universe** to display the Milky Way and its nearby galaxies, suitably labeled, as seen from 0.282 Mly away. If the Milky Way does not appear immediately, click once on one of the Zoom buttons. Remove the image of the astronaut's feet by clicking on **View > Feet.** You can zoom in or out using the buttons at the upper right of the toolbar. You can rotate the view by holding down the **Shift** key while holding down the mouse button and moving the mouse. (On a two-button mouse, hold down the left mouse button.) (**a**) Open the **Find** pane and use the menu button to the left of the label for the **Sun** to **Centre** the field of view on the Sun. Describe the position of the Small Magellanic Cloud relative to the Milky Way Galaxy and to our solar system. (**b**) Use the **Find** pane to center on the Small Magellanic Cloud. Rotate this irregular galaxy to view it face-on and zoom in to see its component stars. Pulsars are produced by supernovae, and only certain types of stars be-

come supernovae. What evidence do you see that these types of stars are present in the Small Magellanic Cloud? (Note that *Starry Night Enthusiast*™ depicts the Small Magellanic Cloud as being rather flat but this galaxy is thought to be an irregular blob of stars with some thickness.)

Collaborative Exercises

58. Consider the graph showing a recording of a pulsar in Figure 21-2. Sketch and label similar graphs that your group estimates for: (1) a rapidly spinning, professional ice skater holding a flashlight; (2) an emergency signal on an ambulance; and (3) a rotating beacon at an airport.

59. As stars go, pulsars are tiny, only about 20 km across. Name three specific things or places that have a size or a separation of about 20 km.

60. If the Crab Nebula is slowing such that its period is increasing at a rate of 3×10^{-8} seconds per day, how much slower is it going today than on the day the youngest member of your group was born?

22 Black Holes

An artist's impression of a close binary star system that includes a black hole. The accretion disk around the black hole is made of material drawn from the blue companion star. (NASA/CXC/M. Weiss)

WEB LINK 22.1

Imagine a swirling disk of gas and dust, orbiting around an object that has more mass than the Sun but the object is so dark that it cannot be seen. Imagine the material in this disk being compressed and heated as it spirals into the unseen object, reaching temperatures so high that the material emits X rays. And imagine that the unseen object has such powerful gravity that any material that falls into it simply disappears, never to be seen again.

Such hellish maelstroms, like the one shown in the illustration, really do exist. The disks are called *accretion disks,* and the unseen objects with immensely strong gravity are called *black holes.*

The matter that makes up a black hole has been so greatly compressed that it violently warps space and time. If you get too close to a black hole, the speed you would need to escape exceeds the speed of light. Because nothing can travel faster than light, nothing—not even light—can escape from a black hole.

Black holes, whose existence was predicted by Einstein's general theory of relativity (our best description of what gravity is and how it behaves), are both strange and simple. The structure of a black hole is completely specified by only three quantities—its mass, electric charge, and angular momentum. Perhaps strangest of all, there is compelling evidence that black holes really exist. In recent years, astronomers have found that certain binary star systems contain black holes. They have also found evidence that extremely massive stars form black holes at their centers, producing immense bursts of gamma rays in the process. Even more remarkable is the discovery that black holes of more than a million solar masses lie at the centers of many galaxies, including our own Milky Way.

22-1 The special theory of relativity changes our conceptions of space and time

The idea of a black hole is a truly strange one. To appreciate it, we must discard some "commonsense" notions about the nature of space and time.

The Special Theory of Relativity

To understand black holes, we must first grasp the nature of space and time as described by Einstein's special theory of relativity

According to the classical physics of Newton, space is perfectly uniform and fills the universe like a rigid framework. Similarly, time passes at a monotonous, unchanging rate. It is always possible to know exactly how fast you are moving through this rigid fabric of space and time.

WEB LINK 22.2

Those ideas were upset forever in 1905 when Albert Einstein proposed his **special theory of relativity.** This theory describes how motion affects our measurements

Learning Goals

By reading the sections of this chapter, you will learn

22-1 The main ideas of Einstein's special theory of relativity

22-2 How Einstein's general theory of relativity describes the nature of gravitation

22-3 The evidence for black holes in binary star systems

22-4 How the sudden formation of black holes can explain the mysterious gamma-ray bursters

22-5 How astronomers have detected supermassive black holes at the centers of galaxies

22-6 The simple structure of a nonrotating black hole

22-7 How just three numbers completely describe the properties of a black hole

22-8 What it might be like to approach a black hole

22-9 How black holes evaporate over time

of distance and time. It is "special" in the sense of being specialized. In particular, it does not include the effects of gravity. The word "relativity" is used because one of the key ideas of the theory is that all measurements are made relative to an observer. Contrary to Newton, there is *no* absolute framework of space and time. In particular, the distance between two points is not an absolute, nor is the time interval between two events. Instead, the values that you measure for these qualities depend on how you are moving, and these values are relative to you. Someone moving in a different way would measure different values for these quantities.

Remarkably, Einstein's theory is based on just two basic principles. The first is quite simple:

Your description of physical reality is the same regardless of the constant velocity at which you move.

In other words, if you are moving in a straight line at a constant speed, you experience the same laws of physics as you would if you were moving at any other constant speed and in any other direction. As an example, suppose you were inside a closed railroad car moving due north in a straight line at 100 km/h. Any measurements you make inside the car—for example, how long your thumb is or how much time elapses between ticks of your watch—will give exactly the same results as if the car were moving in any other direction or at any other speed, or were not moving at all.

Einstein's second principle is much more bizarre:

Regardless of your speed or direction of motion, you always measure the speed of light to be the same.

To see what this implies, imagine that you are in a spaceship moving toward a flashlight. Even if you are moving at 99% of the speed of light, you will measure the photons from the flashlight to be moving at the same speed ($c = 3 \times 10^8$ m/s $= 3 \times 10^5$ km/s) as if your spaceship were motionless. This statement is in direct conflict with the Newtonian view that a stationary person and a moving person should measure different speeds (Figure 22-1).

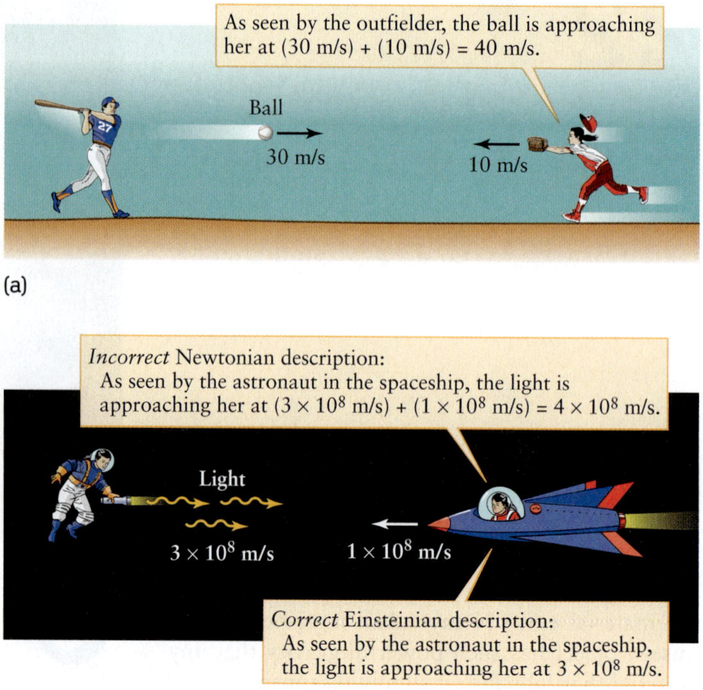

Figure 22-1

The Speed of Light Is the Same to All Observers **(a)** The speed you measure for ordinary objects depends on how you are moving. Thus, the batter sees the ball moving at 30 m/s, but the outfielder sees it moving at 40 m/s relative to her. **(b)** Einstein showed that this commonsense principle does not apply to light. No matter how fast or in what direction the astronaut in the spaceship is moving, she and the astronaut holding the flashlight will always measure light to have the same speed.

Speed involves both distance and time. Since speed has a very different behavior in the special theory of relativity than in Newton's physics, it follows that both space and time behave differently as well. Indeed, in relativity time proves to be so intimately intertwined with the three dimensions of space that we regard them as a single *four*-dimensional entity called **spacetime.**

Length Contraction

Einstein expressed his ideas about spacetime in a mathematical form and used this description to make a number of predictions about nature. All these predictions have been verified in innumerable experiments. One prediction is that the length you measure an object to have depends on how that object is moving; the faster it moves, the shorter its length along its direction of motion (Figure 22-2). This is called **length contraction.** In other words, if a railroad car moves past you at high speed, from your perspective on the ground you will actually measure it to be shorter than if it were at rest (Figure 22-2*a*). However, if you are on board the railroad car and moving with it, you will measure its length to be the same as measured on the ground when it was at rest. The word "relativity" emphasizes the importance of the relative speed between the observer and the object being measured.

If the idea of length contraction seems outrageous, it is because this effect is noticeable only at very high speeds, near the speed of light. But even the fastest spacecraft ever built by humans travels at a mere 1/25,000 of the speed of light. At this speed, a spacecraft 10 m long would be contracted in length by only 8 nm—a distance equal to the width of a single protein molecule! For moving cars, trains, and airplanes, length contraction is far too small to measure. As Box 22-1 describes, however, we can easily see the effects of length contraction for subatomic particles that do indeed travel at nearly the speed of light.

Time Dilation

A second result of relativity is no less strange: A clock moving past you runs more slowly than a clock that is at rest. Like length contraction, this **time dilation** is a very small effect unless the clock is moving at extremely high speeds (Figure 22-2*b*). Nonetheless,

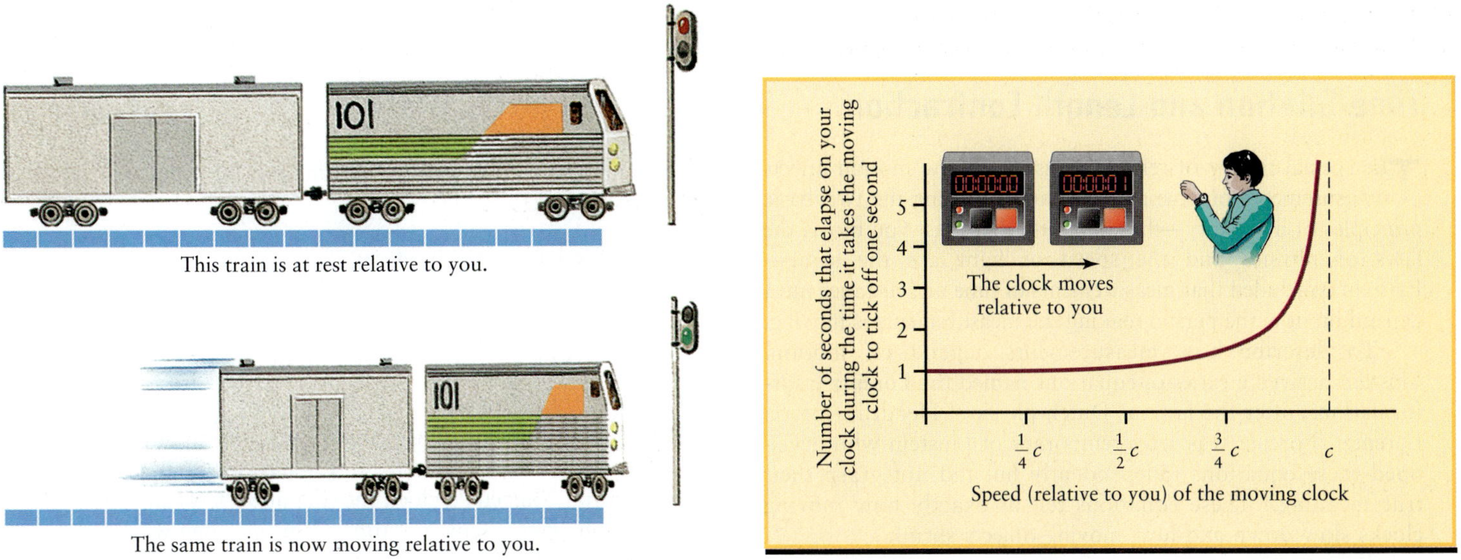

(a) Length contraction

(b) Time dilation

Figure 22-2

Length Contraction and Time Dilation **(a)** The faster an object moves, the shorter it becomes along its direction of motion. Speed has no effect on the object's dimensions perpendicular to the direction of motion. **(b)** The faster a clock moves, the slower it runs. This graph shows how many seconds (as measured on your clock) it takes a clock that moves relative to you to tick off one second. The effect is pronounced only for speeds near the speed of light *c*.

physicists have confirmed the existence of time dilation by using an extremely accurate atomic clock carried on a jet airliner. When the airliner landed, they found that the on-board clock had actually ticked off slightly less time than an identical, stationary clock on the ground. For the passengers on board the airliner during this experiment, however, time flowed at a normal rate: The on-board clock ticked off the seconds as usual, their hearts beat at a normal rate, and so on. You only measure a clock (or a beating heart) to be running slow if it is *moving* relative to you. Box 22-1 discusses time dilation in more detail.

The special theory of relativity predicts that the astronaut with the flashlight in Figure 22-1*b* sees the astronaut flying in the spaceship as shortened along the direction of motion and as having slowly ticking clocks. Remarkably, it also says that the *flying* astronaut sees the astronaut with the flashlight (who is moving relative to her) as being shortened and as having slowly ticking clocks! These observations may seem contradictory, but they are not: Each spaceship is moving relative to the other, and in the special theory of relativity only relative motion matters. Each astronaut's measurements are correct relative to his or her frame of reference.

CAUTION! You may have the idea that the special theory of relativity implies that there are *no* absolutes in nature and that everything is relative. But, in fact, the special theory is based on the principles that the laws of physics and the speed of light in a vacuum *are* absolutes. Only certain quantities, such as distance and time, depend on your state of motion. You may also have the idea that length contraction and time dilation are just optical illusions caused by high speeds. But these effects are *real.* A moving spaceship doesn't just look shorter or seem to be shorter—it really *is* shorter. Likewise, a moving clock doesn't merely look like it's ticking slowly, or seem to be ticking slowly—it really *does* tick more slowly. Einstein's theory is supported by every experiment designed to test it. Relativity is very strange, but it is also very real!

Mass and Energy

Einstein's special theory of relativity predicts another famous relationship: Any object with mass (m) has energy (E) embodied in its mass. This idea is expressed by the well-known formula $E = mc^2$, where c is the speed of light. In thermonuclear reactions, part of the energy embodied in the mass of atomic nuclei is converted into other forms of energy. This released energy is what makes the Sun and the stars shine (see Section 18-1). The sunlight we see by day and the starlight that graces our nights are a resounding verification of the special theory of relativity.

The special theory of relativity also explains why it is impossible for a spaceship to move at the speed of light. If it could, then a light beam traveling in the same direction as the spaceship would appear to the ship's crew to be stationary. But this would contradict the second of Einstein's principles, which says that all observers, including those on a spaceship, must see light traveling at the speed of light. Therefore, it cannot be possible for a spaceship, or indeed any material object, to travel at the speed of light. This conclusion too has been verified by experiments. Subatomic particles can be made to travel at 99.99999999% of the speed of light c, but they can never make it all the way to c.

BOX 22-1 Tools of the Astronomer's Trade

Time Dilation and Length Contraction

The special theory of relativity describes how motion affects measurements of time and distance. By using the two basic principles of his theory—that no matter how fast you move, the laws of physics and the speed of light are the same—Einstein concluded that measurements of time and distance must depend on how the person making the measurements is moving.

To describe how measurements depend on motion, Einstein derived a series of equations named the **Lorentz transformations,** after the famous Dutch physicist Hendrik Antoon Lorentz. (Lorentz was a contemporary of Einstein who developed these equations independently but did not grasp their true meaning.) These equations tell us exactly how moving clocks slow down and how moving objects shrink.

To appreciate the Lorentz transformations, imagine two observers named Sergio and Majeeda. Sergio is at rest on Earth, while Majeeda is flying past in her spaceship at a speed v. Sergio and Majeeda both observe the same phenomenon on Earth—say, the beating of Sergio's heart or the ticking of Sergio's watch—which appears to occur over an interval of time. According to Sergio's clock (which is not moving relative to the phenomenon), the phenomenon lasts for T_0 seconds. This time period is called the **proper time** of the phenomenon. But according to Majeeda's clock (which *is* moving relative to the phenomenon), the same phenomenon lasts for a different length of time, T seconds. These two time intervals are related as follows:

Lorentz transformation for time

$$T = \frac{T_0}{\sqrt{1 - (v/c)^2}}$$

T = time interval measured by an observer moving relative to the phenomenon

T_0 = time interval measured by a observer *not* moving relative to the phenomenon (proper time)

v = speed of the moving observer relative to the phenomenon

c = speed of light

EXAMPLE: Sergio heats a cup of water in a microwave oven for 1 minute. According to Majeeda, who is flying past Sergio at 98% of the speed of light, how long does it take to heat the water?

Situation: The phenomenon in question is heating the water in the microwave oven. Sergio is not moving relative to this phenomenon (so he measures the *proper* time interval T_0). Majeeda is moving at speed $v = 0.98c$ relative to this phenomenon (so she measures a different time interval T). Our goal is to determine the value of T.

Tools: We use the Lorentz transformation for time to calculate T.

Answer: We have $v/c = 0.98$, so

$$T = \frac{T_0}{\sqrt{1 - (v/c)^2}} = \frac{1 \text{ minute}}{\sqrt{1 - (0.98)^2}} = 5 \text{ minutes}$$

Review: A phenomenon that lasts for $T_0 = 1$ minute on Sergio's clock is stretched out to $T = 5T_0 = 5$ minutes as measured on Majeeda's clock moving at 98% of the speed of light. Other phenomena are affected in the same way: As measured by *Majeeda,* it takes 5 seconds for Sergio's wristwatch to tick off one second, and the minute hand on Sergio's wristwatch takes 5 hours to make a complete sweep. The converse is also true. As measured by *Sergio,* the minute hand on Majeeda's wristwatch will take 5 hours to make a complete sweep. This phenomenon, in which events moving relative to an observer happen at a slower pace, is called *time dilation.*

The Lorentz transformation for time is plotted in Figure 22-2*b*, which shows how 1 second measured on a stationary clock (say, Sergio's) is stretched out when measured using a clock carried by a moving observer (such as Majeeda). For speeds less than about half the speed of light, the mathematical factor $\sqrt{1 - (v/c)^2}$ is not too different from 1, and there is little difference between the recordings of the stationary and moving clocks. At the fastest speeds that humans have ever traveled (en route from Earth to the Moon), the difference between 1 and the factor $\sqrt{1 - (v/c)^2}$ is less than 10^{-9}. So, for any speed associated with human activities, stationary and slowly moving clocks tick at essentially the same rate. As the next example shows, however, time dilation is important for subatomic particles that travel at speeds comparable to c.

EXAMPLE: When unstable particles called *muons* (pronounced "mew-ons") are produced in experiments on Earth, they decay into other particles in an average time of 2.2×10^{-6} s. Muons are also produced by fast-moving protons from interstellar space when they collide with atoms in Earth's upper atmosphere. These muons typically move at 99.9% of the speed of light and are formed at an altitude of 10 km. How long do such muons last before they decay?

Situation: The phenomenon in question is the life of a muon, which lasts a time $T_0 = 2.2 \times 10^{-6}$ s as measured by an observer not moving with respect to the muon. Our goal is to find the time interval T measured by an observer on Earth, which is moving at $v = 0.999c$ relative to the muon (the same speed at which the muon is moving relative to Earth).

Tools: As in the previous example, we use the Lorentz transformation for time.

Answer: Using $v/c = 0.999$,

$$T = \frac{2.2 \times 10^{-6}\ \text{s}}{\sqrt{1 - (0.999)^2}} = 4.9 \times 10^{-5}\ \text{s}$$

Review: At this speed, the muon's lifetime is slowed down by time dilation by a factor of more than 22. Note that as measured by an observer on Earth, the time that it takes a muon produced at an altitude of 10 km = 10,000 m to reach Earth's surface is

$$\frac{\text{distance}}{\text{speed}} = \frac{10{,}000\ \text{m}}{0.999 \times 3.00 \times 10^8\ \text{m/s}} = 3.3 \times 10^{-5}\ \text{s}$$

Were there no time dilation, such a muon would decay in just 2.2×10^{-6} s and would never reach Earth's surface. But in fact, these muons *are* detected by experiments on Earth's surface, because a muon moving at $0.999c$ lasts more than 3.3×10^{-5} s. The detection at Earth's surface of muons from the upper atmosphere is compelling evidence for the reality of time dilation.

In the same terminology as "proper time," we say that a ruler at rest measures **proper length** or **proper distance** (L_0). According to the Lorentz transformations, distances perpendicular to the direction of motion are unaffected. However, a ruler of proper length L_0 held parallel to the direction of motion shrinks to a length L, given by

Lorentz transformation for length

$$L = L_0\sqrt{1 - (v/c)^2}$$

L = length of a moving object along the direction of motion

L_0 = length of the same object at rest (proper length)

v = speed of the moving object

c = speed of light

EXAMPLE: Again, imagine that Majeeda is traveling at 98% of the speed of light relative to Sergio. If Majeeda holds a 1-meter ruler parallel to the direction of motion, how long is this ruler as measured by Sergio?

Situation: The ruler is at rest relative to Majeeda, so she measures its *proper* length $L_0 = 1$ m. Our goal is to determine its length L as measured by Sergio, who is moving at $v = 0.98c$ relative to Majeeda and her ruler.

Tools: We use the Lorentz transformation for length.

Answer: With $v/c = 0.98$, we find

$$L = L_0\sqrt{1 - (v/c)^2} = (1\ \text{m})\sqrt{1 - (0.98)^2}$$
$$= (1\ \text{m}) \times (0.20) = 0.20\ \text{m} = 20\ \text{cm}$$

Review: We saw in the first example that according to *Sergio,* Majeeda's clocks are ticking only one-fifth as fast as his. This example shows that he also measures Majeeda's 1-meter ruler to be only one-fifth as long (20 cm) when held parallel to the direction of motion. Note that the converse is also true: If Sergio holds a 1-m ruler parallel to the direction of relative motion, *Majeeda* measures it to be only 20 cm long. This shrinkage of length is called *length contraction.*

EXAMPLE: Consider again the above example about muons created 10 km above Earth's surface. If a muon is traveling straight down, what is the distance to the surface as measured by an observer riding along with the muon?

Situation: Imagine a ruler that extends straight up from Earth's surface to where the muon is formed. This ruler is at rest relative to Earth, so its length of 10 km is the proper length L_0. Our goal is to find the length L of this ruler as measured by the observer riding with the muon.

Tools: As in the previous example, we use the Lorentz transformation for length.

Answer: With $v/c = 0.999$, we calculate

$$L = (1\ \text{km})\sqrt{1 - (0.999)^2} = 0.45\ \text{km} = 450\ \text{m}$$

Review: The distance is contracted tremendously as measured by an observer riding with the muon. This result gives us another way to explain how such muons are able to reach Earth's surface. As measured by the muon, the time required to travel the contracted distance is

$$\frac{\text{distance}}{\text{speed}} = \frac{450\ \text{m}}{0.999 \times 3.00 \times 10^8\ \text{m/s}} = 1.5 \times 10^{-6}\ \text{s}$$

This time is less than the 2.2×10^{-6} s that an average muon takes to decay. Hence, muons can successfully reach Earth's surface.

22-2 The general theory of relativity predicts black holes

Einstein's special theory of relativity is a comprehensive description of the behavior of light and, by extension, of electricity and magnetism. (Recall from Section 5-2 that light is both electric and magnetic in nature.) Einstein's next goal was to develop an even more comprehensive theory that also explained gravity. This was the **general theory of relativity,** which he published in 1915.

The Equivalence Principle

According to Newton's theory of gravity, an apple falls to the floor because the force of gravity pulls the apple down. But Einstein pointed out that the apple would appear to behave in exactly the same way in space, far from the gravitational influence of any planet or star, if the floor were to accelerate upward (in other words, if the floor came up to meet the apple).

In Figure 22-3, two famous gentlemen are watching an apple fall toward the floor of their closed compartments. They have no way of telling who is at rest on Earth and who is in the hypothetical elevator moving upward through empty space at a constantly increasing speed. This is an example of Einstein's **equivalence principle,** which states that in a small volume of space, the downward pull of gravity can be accurately and completely duplicated by an upward acceleration of the observer.

The equivalence principle is the key to the general theory of relativity. It allowed Einstein to focus entirely on motion, rather than force, in discussing gravity. A hallmark of gravity is that it causes the same acceleration no matter the mass of the object. For example, a baseball and a cannon ball have very different masses, but if you drop them side by side in a vacuum, they accelerate downward at exactly the same rate. To explain this observation, Einstein envisioned gravity as being caused by a *curvature of space.* In fact, his general theory of relativity describes gravity entirely in terms of the geometry of both space *and* time, that is, of spacetime. Far from a source of gravity, like a planet or a star, spacetime is "flat" and clocks tick at their normal rate. Closer to a source of gravity, however, space is curved and clocks slow down.

ANALOGY A useful analogy is to picture the spacetime near a massive object such as the Sun as being curved like the surface in Figure 22-4. Imagine a ball rolling along this surface. Far from the "well" that represents the Sun, the surface is fairly flat and the ball moves in a straight line. If the ball passes near the well, however, it curves in toward it. If the ball is moving at an appropriate speed, it might move in an orbit around the sides of the well. The curvature of the well has the same effect on a ball of any size, which explains why gravity produces the same acceleration on objects of different mass.

Testing the General Theory

Einstein's general theory of relativity and its picture of curved spacetime have been tested in a variety of different ways. In what follows we discuss some key experimental tests of the theory.

Careful experiments have verified the key ideas of Einstein's theory of gravity

Experimental Test 1: The gravitational bending of light. Light rays naturally travel in straight lines. But if the space through which the rays travel is curved, as happens when light passes near the surface of a massive object like the Sun, the paths of the rays will likewise be curved (Figure 22-5). In other words, gravity should bend light rays, an effect not predicted by Newtonian mechanics because light has no mass. This prediction was first tested in 1919 during

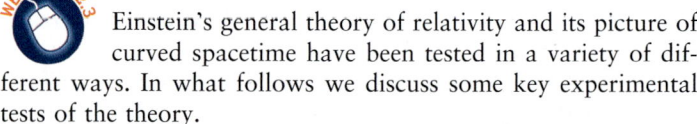

This compartment is at rest in Earth's gravitational field.
This compartment is moving in a gravity-free environment.

(a) The apple hits the floor of the compartment because Earth's gravity accelerates the apple downward.

(b) The apple hits the floor of the compartment because the compartment accelerates upward.

Figure 22-3

The Equivalence Principle The equivalence principle asserts that you cannot distinguish between **(a)** being at rest in a gravitational field and **(b)** being accelerated upward in a gravity-free environment. This idea was an important step in Einstein's quest to develop the general theory of relativity.

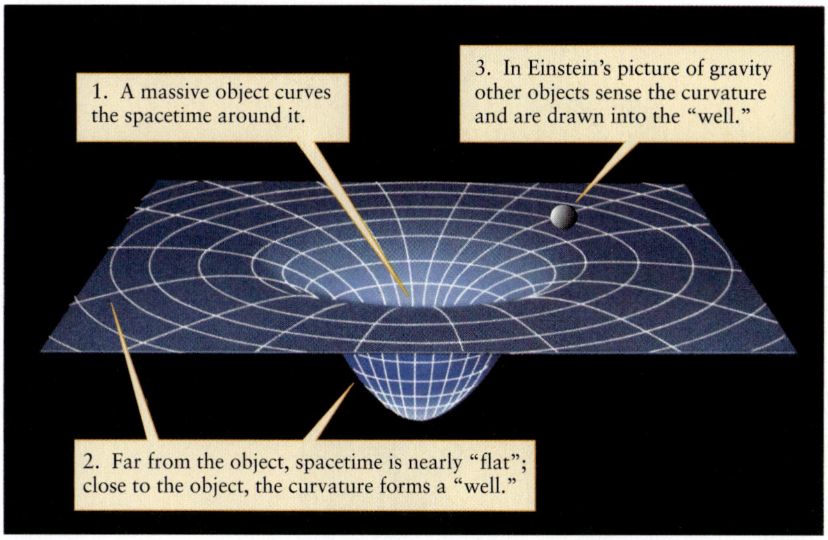

Figure 22-4

The Gravitational Curvature of Spacetime According to Einstein's general theory of relativity, spacetime becomes curved near a massive object. To help you visualize the curvature of four-dimensional spacetime, this figure shows the curvature of a *two*-dimensional space around a massive object.

a total solar eclipse. During totality, when the Moon blocked out the Sun's disk, astronomers photographed the stars around the Sun. Careful measurements afterward revealed that the stars around the Sun were shifted from their usual positions by an amount consistent with Einstein's theory.

Experimental Test 2: The precession of Mercury's orbit. As the planet Mercury moves along its elliptical orbit, the orbit itself slowly changes orientation or *precesses* (Figure 22-6). Most of Mercury's precession is caused by the gravitational pull of the other planets, as explained by Newtonian mechanics. But once the effects of all the other planets had been accounted for, there remained an unexplained excess rotation of Mercury's major axis of 43 arcsec per century. Although this discrepancy may seem very small, it frustrated astronomers for half a century. Some astronomers searched for a missing planet even closer to the Sun that might be tugging on Mercury; none has ever been found. Einstein showed that at Mercury's position close to the Sun, the general theory of relativity predicts a small correction to Newton's description of gravity. This correction is just enough to account for the excess precession.

Experimental Test 3: The gravitational slowing of time and the gravitational red shift (Figure 22-7). In the general theory of relativity, a massive object such as Earth warps time as well as space. Einstein predicted that clocks on the ground floor of a building should tick slightly more slowly than clocks on the top floor, which are farther from Earth (Figure 22-7*a*).

A light wave can be thought of as a clock; just as a clock makes a steady number of ticks per minute, an observer sees a steady number of complete cycles of a light wave passing by each

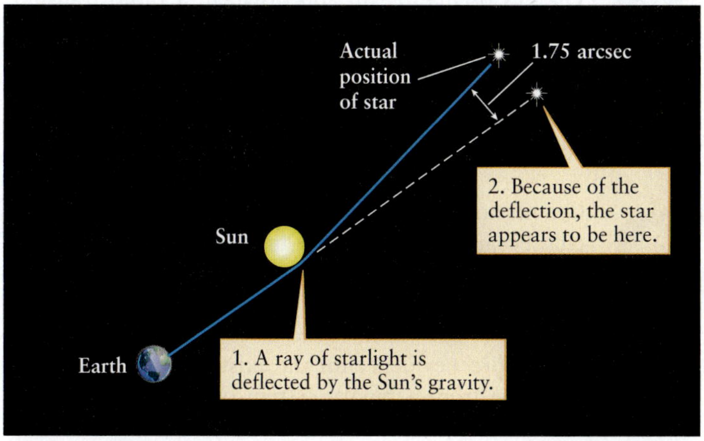

Figure 22-5

The Gravitational Deflection of Light Light rays are deflected by the curved spacetime around a massive object like the Sun. The maximum deflection is very small, only 1.75 arcsec for a light ray grazing the Sun's surface. By contrast, Newton's theory of gravity predicts *no* deflection at all. The deflection of starlight by the Sun was confirmed during a solar eclipse in 1919.

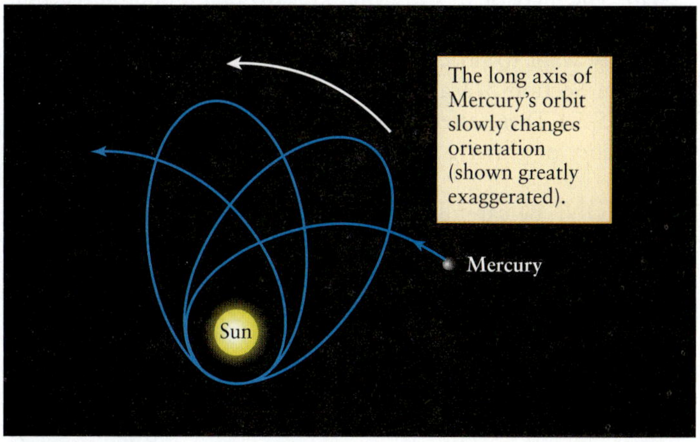

Figure 22-6

The Precession of Mercury's Orbit Mercury's orbit changes orientation at a rate of 574 arcsec (about one-sixth of a degree) per century. Newtonian mechanics predicts that the gravitational influences of the other planets should make the orientation change by only 531 arcsec per century. Einstein was able to explain the excess 43 arcsec per century using his general theory of relativity.

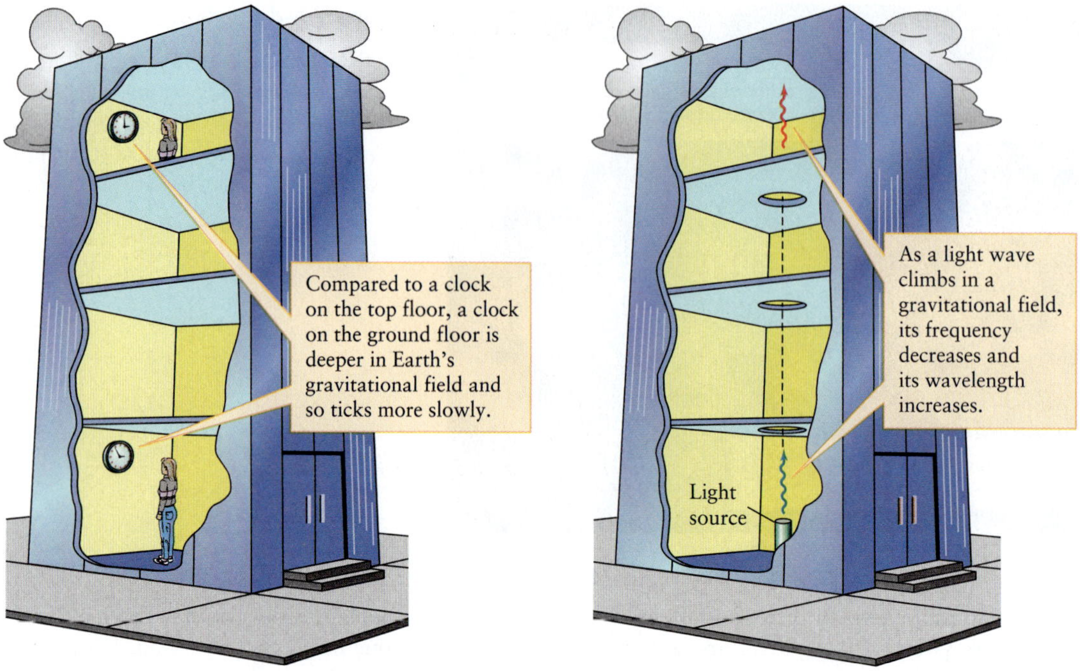

(a) The gravitational slowing of time

(b) The gravitational redshift

Figure 22-7

The Gravitational Slowing of Time and the Gravitational Redshift **(a)** Clocks at different heights in a gravitational field tick at different rates. **(b)** The oscillations of a light wave constitute a type of clock. As a light wave climbs from the ground floor toward the top floor, its oscillation frequency becomes lower and its wavelength becomes longer.

second. If a light beam is aimed straight up from the ground floor of a building, an observer at the top floor will measure a "slow-ticking" light wave with a lower frequency, and thus a longer wavelength, than will an observer on the ground floor (Figure 22-7*b*). The increase in wavelength means that a photon reaching the top floor has less energy than when it left the ground floor. (You may want to review the discussion of frequency and wavelength in Section 5-2 as well as the description of photons in Section 5-5.) These effects, which have no counterpart in Newton's theory of gravity, are called the **gravitational redshift.**

CAUTION! Be careful not to confuse the gravitational redshift with a Doppler shift. In the Doppler effect, redshifts are caused by a light source moving away from an observer. Gravitational redshifts, by contrast, are caused by time flowing at different rates at different locations. No motion is involved.

The American physicists Robert Pound and Glen Rebka first measured the gravitational redshift in 1960 using gamma rays fired between the top and bottom of a shaft 20 meters tall. Because Earth's gravity is relatively weak, the redshift that they measured was very small ($\Delta\lambda/\lambda = 2.5 \times 10^{-15}$) but was in complete agreement with Einstein's prediction.

Much larger shifts are seen in the spectra of white dwarfs, whose spectral lines are redshifted as light climbs out of the white dwarf's intense surface gravity. As an example, the gravitational redshift of the spectral lines of the white dwarf Sirius B (see Figure 20-8) is $\Delta\lambda/\lambda = 3.0 \times 10^{-4}$, which also agrees with the general theory of relativity.

LOOKING DEEPER 22.1

Experimental Test 4: Gravitational waves. Electric charges oscillating up and down in a radio transmitter's antenna produce electromagnetic radiation. In a similar way, the general theory of relativity predicts that oscillating massive objects should produce **gravitational radiation,** or **gravitational waves.** (Newton's theory of gravity makes no such prediction.)

WEB LINK 22.4

Gravitational radiation is exceedingly difficult to detect, because it is by nature much weaker than electromagnetic radiation. Although physicists have built a number of sensitive "antennas" for gravitational radiation, no confirmed detections have been made as of this writing (2009). But compelling *indirect* evidence for the existence of gravitational radiation has come from a binary system of two neutron stars. Russell Hulse and Joseph Taylor at the University of Massachusetts discovered that these stars are slowly spiraling toward each other and losing energy in the process (see Section 21-7). The rate at which they lose energy is just what would be expected if the two neutron stars are emitting gravitational radiation as predicted by Einstein. Hulse and Taylor shared the 1993 Nobel Prize in physics for their discovery.

The general theory of relativity has never made an incorrect prediction. It now stands as our most accurate and complete

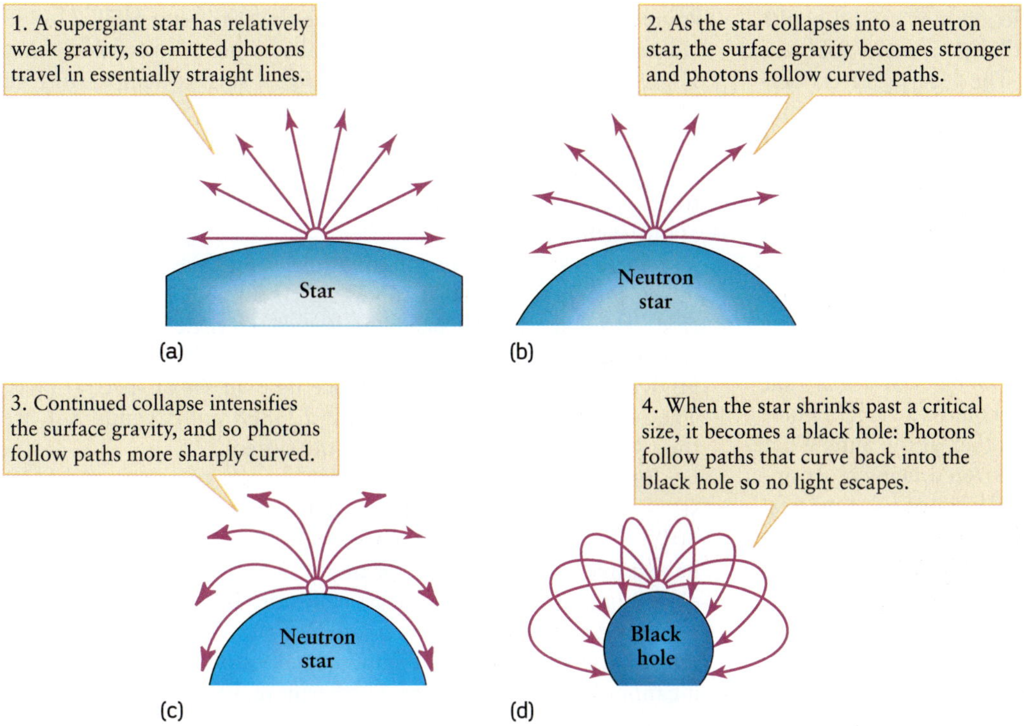

2. As the star collapses into a neutron star, the surface gravity becomes stronger and photons follow curved paths.

4. When the star shrinks past a critical size, it becomes a black hole: Photons follow paths that curve back into the black hole so no light escapes.

Figure 22-8

The Formation of a Black Hole
(a)–(c) These illustrations show four steps leading up to the formation of a black hole from a dying star. **(d)** When the star becomes a black hole, not even photons emitted directly upward from the surface can escape; they undergo an infinite gravitational redshift and disappear.

description of gravity. Einstein demonstrated that Newtonian mechanics is accurate only when applied to low speeds and weak gravity. If extremely high speeds or powerful gravity are involved, only a calculation using relativity will give correct answers.

Relativity and Black Holes

Perhaps the most dramatic prediction of the general theory of relativity concerns what happens when a large amount of matter is concentrated in a small volume. We have seen that if a dying star is not too massive, it ends up as a white dwarf star. If the dying star is more massive than the Chandrasekhar limit of about 1.4 $M_\odot$, it cannot exist as a stable white dwarf star and, instead, shrinks down to form a neutron star. But if the dying star has more mass than the maximum permissible for a neutron star, about 2 to 3 $M_\odot$, not even the internal pressure of neutrons can hold the star up against its own gravity, and the star contracts rapidly.

As the star's matter becomes compressed to enormous densities, the strength of gravity at the surface of this rapidly shrinking sphere also increases dramatically. According to the general theory of relativity, the space immediately surrounding the star becomes so highly curved that it closes on itself (Figure 22-8). Photons flying outward at an angle from the star's surface arc back inward, while photons that fly straight outward undergo such a strong gravitational redshift that they lose all their energy and cease to exist.

Ordinary matter can never travel as fast as light. Hence, if light cannot escape from the collapsing star, neither can anything else. An object from which neither matter nor electromagnetic radiation can escape is called a **black hole.** In a sense, a hole is punched in the fabric of the universe, and the dying star disappears into this cavity (Figure 22-9).

None of the star's mass is lost when it collapses to form a black hole, however. This mass gives the spacetime around the black hole its strong curvature. Thanks to this curvature, the black hole's gravitational influence can still be felt by other objects.

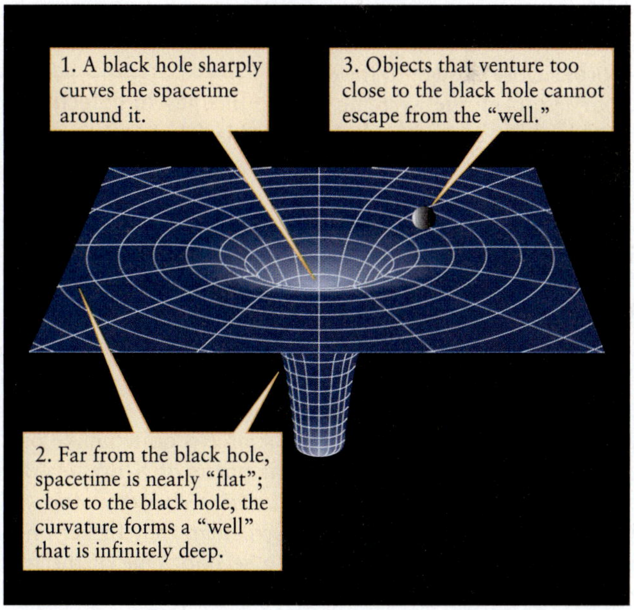

1. A black hole sharply curves the spacetime around it.
3. Objects that venture too close to the black hole cannot escape from the "well."
2. Far from the black hole, spacetime is nearly "flat"; close to the black hole, the curvature forms a "well" that is infinitely deep.

Figure 22-9

Curved Spacetime around a Black Hole This diagram suggests how spacetime is distorted by a black hole's mass. As in Figure 22-4, spacetime is represented as a two-dimensional surface. Unlike the situation in Figure 22-4, a black hole's gravitational "well" is infinitely deep.

CAUTION! Some low-quality science-fiction movies and books suggest that black holes are evil things that go around gobbling up everything in the universe. Not so! The bizarre effects created by highly warped spacetime are limited to a region quite near the hole. For example, the effects of the general theory of relativity predominate only within 1000 km of a 10-$M_\odot$ black hole. Beyond 1000 km, gravity is weak enough that Newtonian physics can adequately describe everything. If our own Sun somehow turned into a black hole (an event that, happily, seems to be quite impossible), the orbits of the planets would hardly be affected at all.

Perhaps the most remarkable aspect of black holes is that they really exist! As we will see, astronomers have located a number of black holes with masses a few times that of the Sun. What is truly amazing is that they have also discovered many truly immense black holes containing millions or billions of solar masses. These discoveries are a resounding confirmation of the ideas of the general theory of relativity. In the next three sections we will see how astronomers hunt down black holes in space.

22-3 Certain binary star systems probably contain black holes

TUTORIAL 22.2

Finding black holes is a difficult business. Because light cannot escape from inside the black hole, you cannot observe one directly in the same way that you can observe a star or a planet. The best you can hope for is to detect the effects of a black hole's powerful gravity.

Close binary star systems offer one way to find **stellar-mass black holes** (that is, ones with masses comparable to those of ordinary stars). Imagine that one of the stars in a binary system evolves into a black hole. If the black hole orbits close enough to the other, ordinary star in the system, tidal forces can draw matter from the ordinary star onto the black hole. If we can detect radiation coming from this "stolen" matter, we can infer the presence of the black hole.

One key to detecting black holes is to search for the radiation emitted by material as it falls into a hole

The Strange Case of Cygnus X-1

The first sign of such emissions from a binary system with a black hole came shortly after the launch of the *Uhuru* X-ray–detecting satellite in 1971. Astronomers became intrigued with an X-ray source designated Cygnus X-1. This source is quite unlike pulsating X-ray sources, which emit regular bursts of X rays every few seconds (see Section 21-8). Instead, the X-ray emissions from Cygnus X-1 are highly variable and irregular; they flicker on time scales as short as one-hundredth of a second. One of the fundamental concepts in physics is that nothing can travel faster than the speed of light (recall Section 22-1). Because of this limitation, an object cannot flicker in unison faster than the time required for light to travel across the object. Because light travels 3000 km in a hundredth of a second, Cygnus X-1 can be no more than 3000 km across, or about a quarter the size of Earth.

Cygnus X-1 occasionally radiates bursts of radio emission, and in 1971 radio astronomers used these bursts to show that the source was at the same location in the sky as the star HDE 226868 (Figure 22-10). Spectroscopic observations revealed that HDE 226868 is a B0 supergiant with a surface temperature of about 31,000 K. Such stars do not emit significant amounts of X rays, so HDE 226868 alone cannot be the X-ray source Cygnus X-1. Because binary stars are very common, astronomers began to suspect that the visible star and the X-ray source are in orbit about each other.

Further spectroscopic observations soon showed that the lines in the spectrum of HDE 226868 shift back and forth with a period of 5.6 days. This behavior is characteristic of a single-line spectroscopic binary (see Section 17-10); the companion of HDE 226868 is too dim to produce its own set of spectral lines. The clear implication is that HDE 226868 and Cygnus X-1 are the two components of a binary star system.

From what we know about the masses of other supergiant stars, HDE 226868 is estimated to have a mass of roughly 30 $M_\odot$. As a result, the unseen member of the binary system must have a mass of about 7 $M_\odot$ or more. Otherwise, it would not exert enough gravitational pull to make the B0 star wobble by the amount deduced from the periodic Doppler shifting of its spectral lines. Because the unseen companion does not emit visible light, it cannot be an ordinary star. Furthermore, because 7 solar masses is too large for either a white dwarf or a neutron star, Cygnus X-1 is likely to be a black hole.

The case for Cygnus X-1 being a black hole is not airtight. HDE 226868 might have a low mass for its spectral type, which would imply a somewhat lower mass for Cygnus X-1. In addition, uncertainties in the distance to the binary system could

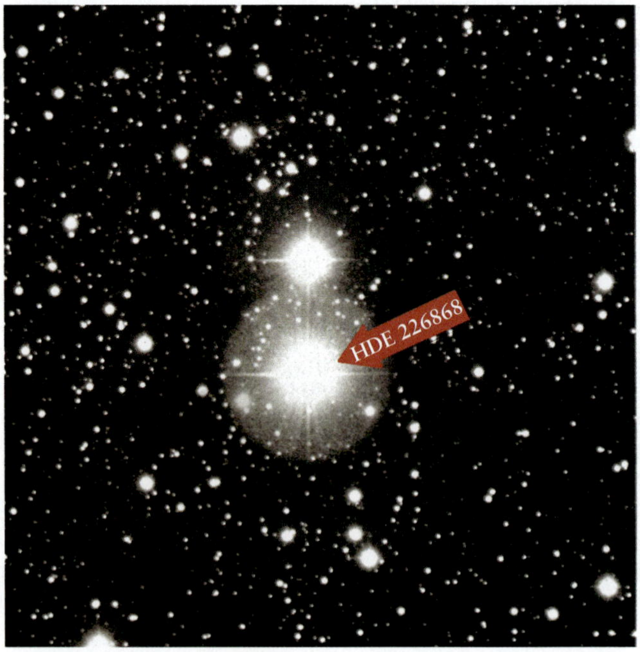

Figure 22-10 R I V U X G

HDE 226868 This photograph from the 5-m telescope on Palomar Mountain shows HDE 226868, the B0 blue supergiant star at the location of the X-ray source Cygnus X-1. This star is located about 2500 pc (87,000 ly) from Earth. The bright star directly above HDE 226868 is not part of the binary system. (Courtesy of J. Kristian, Carnegie Observatories)

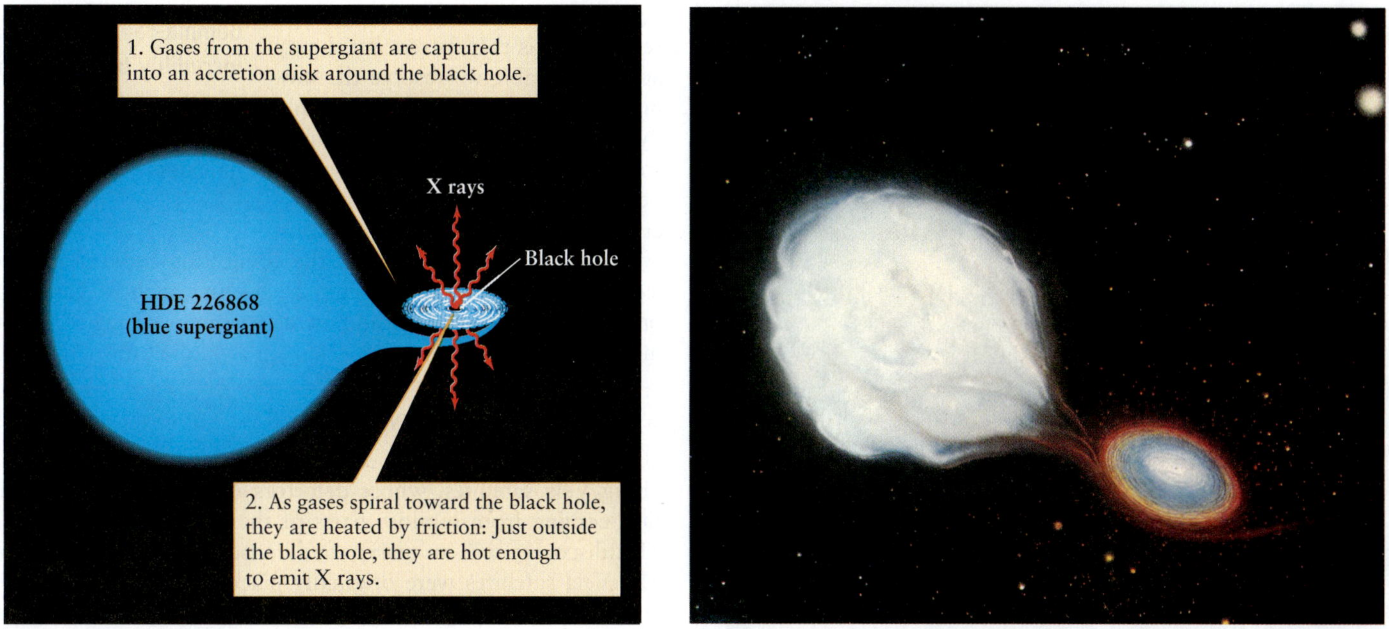

1. Gases from the supergiant are captured into an accretion disk around the black hole.
X rays
Black hole
HDE 226868 (blue supergiant)
2. As gases spiral toward the black hole, they are heated by friction: Just outside the black hole, they are hot enough to emit X rays.

(a) A schematic diagram of Cygnus X-1

(b) An artist's impression of Cygnus X-1

WEB LINK 22.5

Figure 22-11

The Cygnus X-1 System (a) The larger member of the Cygnus X-1 system is a B0 supergiant of about 30 $M_\odot$. The other, unseen member of the system has a mass of at least 7 $M_\odot$ and is probably a black hole. **(b)** This illustration shows how the Cygnus X-1 system might look at close range. (The illustration that opens this chapter depicts a similar system.) At even closer range, the black hole and its immediate surroundings might appear as shown in Figure 22-12. (Courtesy of D. Norton, Science Graphics)

further reduce estimates of the mass of Cygnus X-1. If all these uncertainties combined in just the right way, the estimated mass of Cygnus X-1 could be pushed down to about 3 $M_\odot$. Thus, there is a slim chance that Cygnus X-1 might contain the most massive possible neutron star rather than a black hole.

If Cygnus X-1 does contain a black hole, the X rays do not come from the black hole itself. Gas captured from HDE 226868 goes into orbit about the hole, forming an accretion disk about 4 million kilometers in diameter (Figure 22-11). As material in the disk gradually spirals in toward the hole, friction heats the gas to temperatures approaching 2×10^6 K. In the final 200 kilometers above the hole, these extremely hot gases emit the X rays that our satellites detect. Presumably the X-ray flickering is caused by small hot spots on the rapidly rotating inner edge of the accretion disk. In this way, the black hole's existence is announced by doomed gases just before they plunge to oblivion.

Other Black Hole Candidates

Astronomers have found more than 20 other black hole candidates like Cygnus X-1. All are compact X-ray sources orbiting ordinary stars in a spectroscopic binary system. One of the best candidates is V404 Cygni in the constellation Cygnus. Doppler shift measurements reveal that as the visible star moves around its 6.47-day orbit, its radial velocity (see Section 17-10) varies by more than 400 km/s. These data give a firm lower limit of 6.26 $M_\odot$ for the mass of the unseen companion. It is probably impossible for a neutron star to be more massive than 3 $M_\odot$, so V404 Cygni must almost certainly be a black hole.

Another particularly convincing case is the flickering X-ray source A0620-00 in the constellation Monoceros. (The A refers to the British satellite *Ariel 5*, which discovered the source; the numbers denote its position in the sky.) The visible companion to A0620-00 is an orange K5 main-sequence star called V616 Monocerotis, which orbits the X-ray source every 7.75 hours. Because this star is relatively faint, it is possible to observe the shifting spectral lines from *both* the visible companion and the X-ray source as they orbit each other. With this more complete information about the orbits, astronomers have determined the mass of A0620-00 to be more than 3.2 $M_\odot$, and more probably about 9 $M_\odot$.

WEB LINK 22.6

Astronomers have seen jets of hot, glowing material extending several light-years from some black hole candidates. The ejected material emerges from the vicinity of the black hole at speeds approaching the speed of light. It is thought that these jets are formed by strong electric and magnetic fields in the material around a *rotating* black hole, much as occurs for rotating protostars (see Figure 18-16) and for rotating neutron stars (see Figure 21-7). Figure 22-12 is an artist's impression of the immediate vicinity of such a rotating black hole. (We will discuss the curious features of rotating black holes in Section 22-7.)

If there really are black holes in close binary systems, how did they get there? One possibility is that one member of the binary explodes as a Type II supernova, leaving a burned-out core whose mass exceeds 3 $M_\odot$. This core undergoes gravitational collapse to form a black hole.

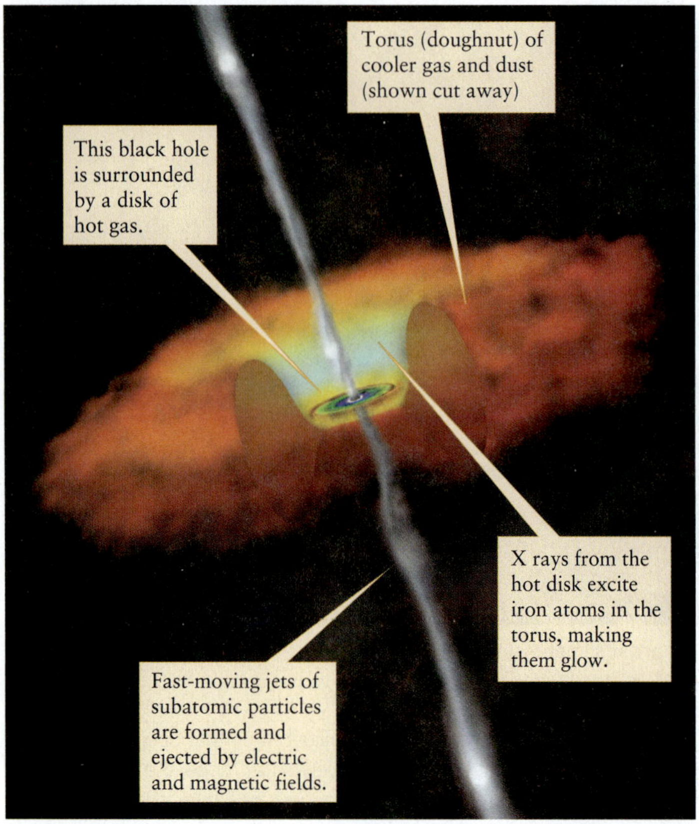

Figure 22-12

The Environment of an Accreting Black Hole If a black hole is rotating, it can generate strong electric and magnetic fields in its immediate vicinity. These fields draw material from the accretion disk around the black hole and accelerate it into oppositely directed jets along the black hole's rotation axis. This illustration also shows other features of the material surrounding such a black hole. (CXC/M. Weiss)

A white dwarf or a neutron star in a binary system could also become a black hole if it accretes enough matter from its companion star. This transformation can occur when the companion star becomes a red giant and dumps a significant part of its mass over its Roche lobe (see Figure 19-21*b*).

Finally, another possibility is two dead stars coalescing to form a black hole. For example, imagine two neutron stars orbiting each other, like the binary systems discussed in Section 21-7 and Section 22-2. Because the two stars emit gravitational radiation, they gradually spiral in toward each other and eventually merge. If their total mass exceeds 2 to 3 $M_\odot$, the entire system may become a black hole.

22-4 The most intense radiation bursts in the universe may be caused by the formation of black holes

WEB LINK 22.7

As we have seen, bursts of X rays from binary systems such as Cygnus X-1 lead us to the remarkable conclusion that black holes are real, and that certain stars can evolve into black holes. Even more remarkable is that astronomers may actually be able to observe the moment that a black hole forms from a dying star. The key to this comes from **gamma-ray bursters,** mysterious objects that emit the most powerful bursts of high-energy radiation ever measured. To understand the connection between these objects and black holes, we begin by looking at how gamma-ray bursters were discovered and how astronomers struggled to determine their true nature.

Gamma-ray bursters are incredibly bright flashes of radiation from remarkably distant galaxies

A Gamma-Ray Mystery

Gamma-ray bursters were discovered in the late 1960s by the orbiting Vela satellites, whose detectors noticed flashes of gamma rays coming from random parts of the sky at random intervals. This discovery was an unexpected consequence of the Cold War. The Vela satellites were originally placed in orbit by the United States to look for high-energy photons coming from above ground tests of nuclear weapons by the Soviet Union, tests that had been banned by treaty since 1963. (No such tests were ever detected.) More than 3000 of these cosmic flashes have been detected since 1967. With a new generation of gamma-ray telescopes currently in orbit (see Section 6-7), new gamma-ray bursters are being found at a rate of about one per day.

Gamma-ray bursters fall into two types. *Long-duration* gamma-ray bursters, which are more common, last from about 2 to about 1000 seconds before fading to invisibility. The less common *short-duration* gamma-ray bursters last from a few hundredths of a second to about 2 seconds, and tend to emit photons of shorter wavelength and hence higher energy. Unlike X-ray bursters (see Section 21-9), gamma-ray bursters of both types appear to emit only one burst in their entire history.

What are gamma-ray bursters, and how far away are they? These questions plagued astronomers for almost 30 years. One important clue is that gamma-ray bursters are seen with roughly equal probability in all parts of the sky, as Figure 22-13*a* shows. This suggests that they are not in the disk of our Galaxy, because then most gamma-ray bursters would be found in the plane of the Milky Way (Figure 22-13*b*). One idea was that gamma-ray bursters are relatively close to us and lie in a spherical halo surrounding the Milky Way. Alternatively, gamma-ray bursters could be strewn across space like galaxies, with some of them billions of light-years away.

Long-Duration Gamma-Ray Bursters and Supernovae

For many years there was no convincing way to decide between these competing models. This state of affairs changed dramatically in 1997, thanks to the Italian-Dutch BeppoSAX satellite. Unlike previous gamma-ray satellites, which could determine the position of a burst only to within a few degrees, BeppoSAX could pin down its position to within an arcminute. This made it far easier for astronomers using optical telescopes to search for a counterpart to the burster. Just 21 hours after BeppoSAX observed a long-duration gamma-ray burster on February 28, 1997, astronomers led by Jan van Paradijs of the University of Amsterdam located a visible-light afterglow of the burst. Since then,

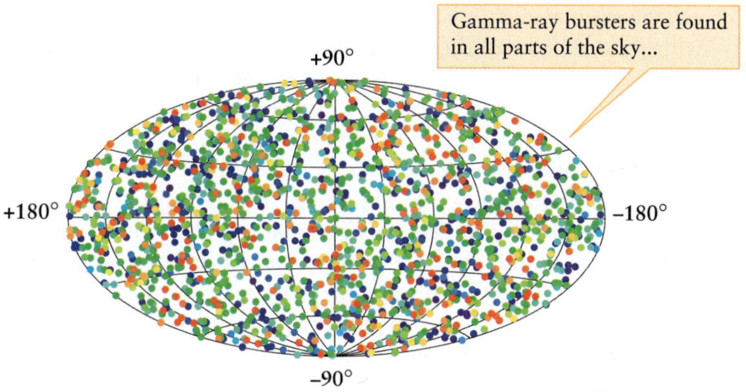
Gamma-ray bursters are found in all parts of the sky...
+90°
+180°
−180°
−90°

(a) Map of the entire sky showing the positions of gamma-ray bursters

R I V U X G

... and are *not* concentrated in the plane of the Milky Way Galaxy.
+90°
+180°
−180°
−90°

(b) Map of the entire sky at visible wavelengths

R I V U X G

Figure 22-13

Gamma-Ray Bursters **(a)** This map shows the locations of 1776 gamma-ray bursters detected by the Compton Gamma Ray Observatory (see Section 6-7). The entire celestial sphere is mapped onto an oval. The colors in the order of the rainbow indicate the total amount of energy detected from each burst; bright bursts appear in red and weak bursts appear in violet. **(b)** This map shows the same sky as in part (a) but at visible wavelengths. Comparing part (b) with part (a) shows that unlike X-ray bursters, which originate in the disk of the Milky Way Galaxy, gamma-ray bursters are seen in all parts of the sky.

astronomers using large visible-light and infrared telescopes have detected the afterglow of many gamma-ray bursters, and have found that they occur within galaxies (Figure 22-14). As of this writing (2009), the most distant gamma-ray burster yet seen lies within a galaxy some 13 billion (1.3×10^{10}) light-years away. (In Chapter 24 we will see how astronomers determine the distances to remote galaxies.) To be seen at such immense distances a gamma-ray burster must release a truly stupendous amount of energy in the form of radiation.

WEB LINK 22.8

The relationship between gamma-ray bursters and galaxies strongly suggests that a burster is a cataclysmic event involving one of a galaxy's stars. Observations of a relatively bright (and hence relatively close) long-duration gamma-ray burster have confirmed this idea. On March 29, 2003, the orbiting gamma-ray telescope HETE-2 (short for High Energy Transient Explorer) detected this burster—denoted GRB 030329 for the date it was observed—in the constellation Leo. (GRB 030329 was so intense that its high-energy photons temporarily ionized part of Earth's atmosphere—the only event of its kind ever caused by an object beyond our own Galaxy.) Within 90 minutes a visible-light afterglow of GRB 030329 was found by astronomers in Australia and Japan. The afterglow was bright enough that astronomers in the United States and Chile were able to measure its detailed spectrum, which turned out to be that of a Type Ic supernova at a distance of some 2.65 billion (2.65×10^9) light-years. We saw in Section 20-9 that a Type Ic supernova results when a massive star undergoes core collapse after having first shed the hydrogen and helium from its outer layers (see Figure 20-20*c*). Subsequent observations with the Hubble Space Telescope revealed an expanding shell of gas at the position of the gamma-ray burster. This shell is just what we would expect from a supernova explosion. Another long-duration gamma-ray burster with an associated Type Ic supernova was seen in 2006.

Beamed Radiation

A long-duration gamma-ray burster cannot simply be an ordinary Type Ic core-collapse supernova, however. Such supernovae release about 10^{46} joules of energy, of which only about 0.03% is

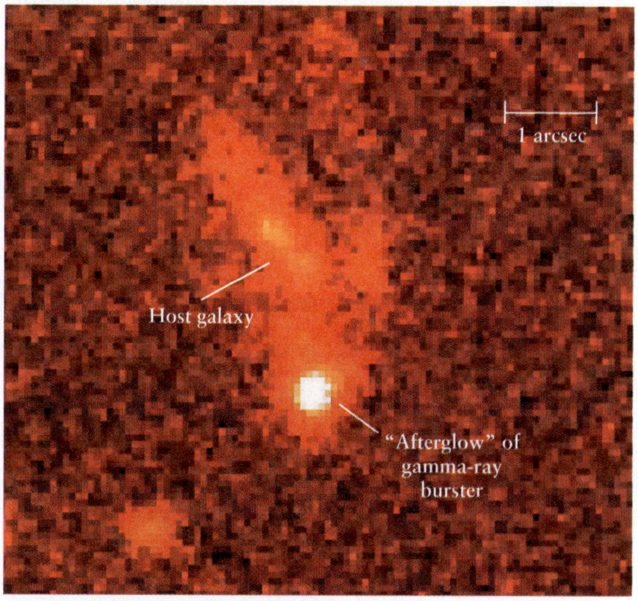
1 arcsec
Host galaxy
"Afterglow" of gamma-ray burster

Figure 22-14 R I V U X G

The Host Galaxy of a Gamma-Ray Burster This false-color image was made on February 8, 1999, 16 days after a gamma-ray burster was observed at this location. The host galaxy of the gamma-ray burster has a very blue color (not shown in this image), indicating the presence of many recently formed stars. The gamma-ray burst may have been produced when one of the most massive of these stars became a supernova. (Andrew Fruchter, STScI; and NASA) released as electromagnetic radiation. (Most of the remaining energy goes into neutrinos, while some goes into accelerating the debris that expands away from the supernova.) This radiation streams outward from the supernova in all directions equally, like light from the Sun. If we assume that a gamma-ray burster also emits light equally in all directions, the inverse-square law that relates brightness, luminosity, and distance (see Section 17-2) tells us that the most energetic bursters would have to emit about 3×10^{47} joules of radiation in less than a minute. It would take 100,000 ordinary supernovae going off simultaneously to release this much radiation!

This dilemma can be resolved if we imagine a type of supernova that emits most of its radiation in narrow beams. If such a beam happened to be aimed toward Earth, we would detect a far more intense burst of radiation than we would from an ordinary supernova.

ANALOGY An ordinary flashlight is an example of beamed radiation. The lightbulb in a flashlight is very small and produces only a weak light if you remove it from the flashlight housing. But the flashlight's curved mirror channels the bulb's light into a narrow beam. Thus, a flashlight produces an intense beam of light with only a small input of energy from the batteries.

Collapsars and the Birth of a Black Hole

One theoretical model that would produce beamed radiation invokes a special type of supernova called a **collapsar** (also called a *hypernova*). These objects are thought to result from progenitor stars that are very massive (more than about 30 $M_{\odot}$), have lost their outer layers of hydrogen and helium, and are rotating rapidly. The core of such a star is too massive to be a white dwarf or a neutron star, so when thermonuclear reactions cease the core will become a black hole. Figure 22-15 depicts what happens when thermonuclear reactions cease in the core of such a star. The black hole forms before the material outside the core has a chance to contract very much (Figure 22-15*a*). As a result, the black hole quickly forms an accretion disk from the surrounding stellar material as shown in Figure 22-15*b*. This process is aided because the progenitor star was rotating rapidly, which makes it easier for the infalling material to form a rotating disk around the black hole. Some of the infalling material does not fall into the black hole, but is ejected in powerful, back-to-back jets along the rotation axis of the accretion disk. (Much the same mechanism acts in close binary star systems in which one member is a black hole, as we described in Section 22-3.)

The jets are so energetic that they reach and break through the surface of the star within 5 to 10 seconds after being formed, as Figure 22-15*c* shows. (It helps that the star has already lost much of its outer layers, which makes it easier for the jets to break through what remains of the star.) As they travel outward, the jets produce powerful shock waves that blow the star to pieces.

If one of the jets from a collapsar happens to be aimed toward Earth, we see an intense burst of gamma rays. These are produced as the relativistic particles in the jet slow down and convert their kinetic energy (energy of motion) into radiation. The burst is short-lived because the jets have only a brief existence: It takes the black hole only about 20 seconds to accrete the entire inner core of the star, after which there is no longer an accretion disk to produce the jets. We do not see a burster if the jets are aimed away from Earth. Hence, for each gamma-ray burster that we observe, there may be hundreds or thousands of collapsars that go undetected.

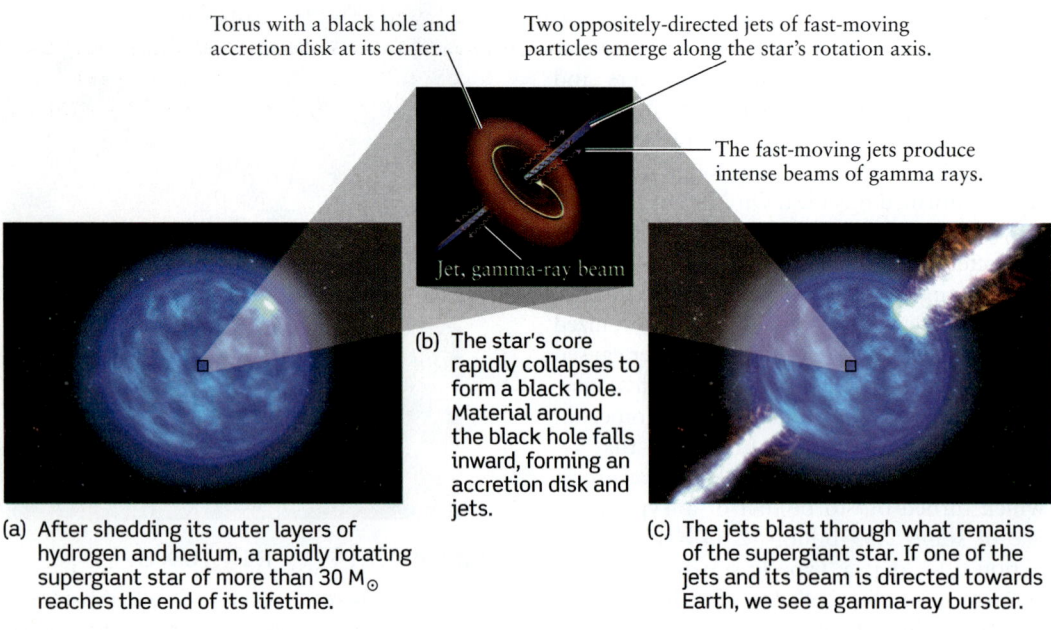

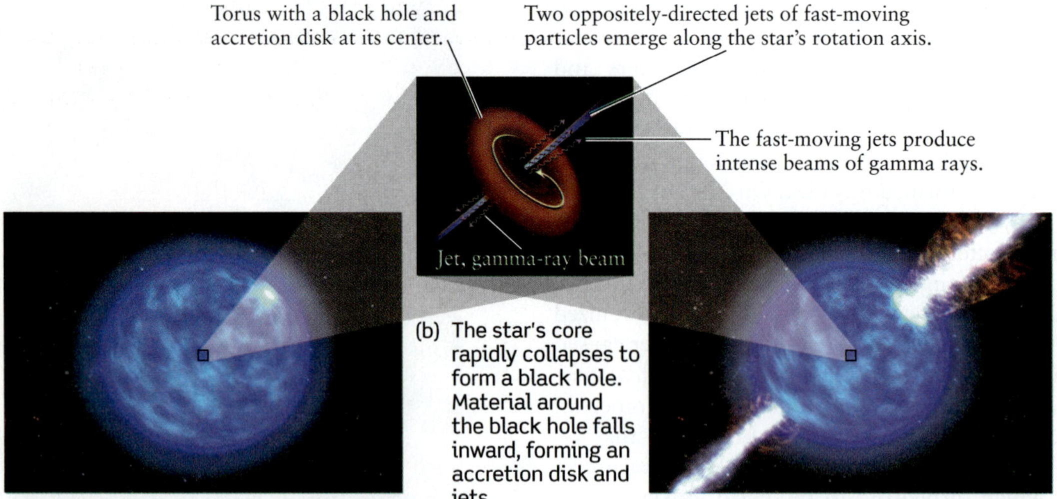

Torus with a black hole and accretion disk at its center.
Two oppositely-directed jets of fast-moving particles emerge along the star's rotation axis.
The fast-moving jets produce intense beams of gamma rays.
Jet, gamma-ray beam

(a) After shedding its outer layers of hydrogen and helium, a rapidly rotating supergiant star of more than 30 $M_{\odot}$ reaches the end of its lifetime.

(b) The star's core rapidly collapses to form a black hole. Material around the black hole falls inward, forming an accretion disk and jets.

(c) The jets blast through what remains of the supergiant star. If one of the jets and its beam is directed towards Earth, we see a gamma-ray burster.

ANIMATION 22.2

Figure 22-15

The Collapsar Model of a Long-Duration Gamma-Ray Burster These illustrations show the final few seconds in the life of a massive, rapidly rotating supergiant star that has lost its outer layers of hydrogen and helium. After the jets have subsided, what remains is a Type Ic supernova with a black hole at its center. (a, c: NASA; b: NASA and A. Field, STScI)

The collapsar model cannot explain short-duration gamma-ray bursters, whose lifetimes of less than 2 seconds are far shorter than those of a collapsar's jets. A number of models for short-duration bursters have been proposed, such as the energy released by the merger of two neutron stars or a particularly intense magnetar burst (see Section 21-6).

The newest tool in the search for answers about gamma-ray bursters is a NASA satellite called Swift, which was launched in 2004. In addition to sensitive gamma-ray detectors to identify bursters, Swift carries telescopes that are sensitive to X rays, ultraviolet light, and visible light. Thus, instead of having to wait for ground-based telescopes to make follow-up observations of gamma-ray bursters, Swift can make these observations itself as swiftly as possible (hence the spacecraft's name). Observations from Swift may help us better understand the nature of long-duration bursters, and provide important clues about the still enigmatic short-duration bursters.

22-5 Supermassive black holes exist at the centers of most galaxies

Stellar evolution makes it possible to form black holes with masses several times that of the Sun. But calculations suggest other ways to form black holes that are either extremely large or extraordinarily small.

Supermassive Black Holes

As we will discuss in Chapter 26, galaxies formed in the early universe by the coalescence of gas clouds. During this formation process, some of a galaxy's gas could have plunged straight toward the center of the galaxy, where it collected and compressed together under its own gravity. If this central clump became sufficiently dense, it could have formed a black hole. Typical galaxies have masses of 10^{11} $M_{\odot}$, so even a tiny percentage of the galaxy's gas collected at its center would give rise to a **supermassive black hole** with a truly stupendous mass.

The largest black holes have billions of times the mass of the Sun

Since the 1970s, astronomers have found evidence suggesting the existence of supermassive black holes in other galaxies. With the advent of a new generation of optical telescopes in the 1990s, it became possible to see the environments of these suspected black holes with unprecedented detail (Figure 22-16). Figure 22-16*a* shows radio emissions from two jets that extend some 15,000 pc (50,000 ly) from the center of the galaxy NGC 4261. These resemble an immensely magnified version of the jets seen emerging from some stellar-mass black holes (see Section 22-3). The highly magnified Hubble Space Telescope image in Figure 22-16*b* shows a disk some 250 pc (800 ly) in diameter at the very center of this galaxy. The jets are perpendicular to the plane of the disk, just as in Figure 22-12.

By measuring the Doppler shifts of light coming from the two sides of the disk, astronomers found that the disk material was orbiting around the bright object at the center of the disk at speeds of hundreds of kilometers per second. (By comparison, Earth orbits the Sun at 30 km/s.) Given the size of the material's orbit, and using Newton's form of Kepler's third law (see Section 4-7 and Box 4-4), they were able to calculate the mass of the central bright object. The answer is an amazing 1.2 *billion* (1.2×10^9) solar masses! What is more, the observations show that this object can be no larger than our solar system. The only possible explanation is that the object at the center of NGC 4261 is a black hole.

Dozens of other black holes in the centers of galaxies have been identified by their gravitational effect on surrounding gas and dust. Surveys of galaxies have shown that supermassive black holes are not at all unusual; most large galaxies appear to have them at their centers. As we will see in the next chapter, a black hole with several million solar masses lies at the center of our own Milky Way Galaxy, some 8000 pc (26,000 ly) from Earth.

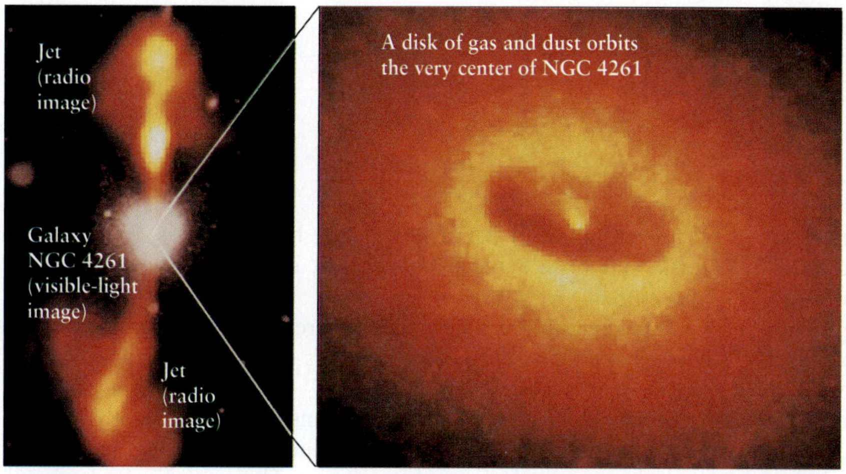

(a) Galaxy NGC 4261
R I V U X G

(b) Evidence for a supermassive black hole in NGC 4261
R I V U X G

Figure 22-16
A Supermassive Black Hole The galaxy NGC 4261 lies some 30 million pc (100 million ly) from Earth in the constellation Virgo. **(a)** This composite view superimposes a visible-light image of the galaxy (white) with a radio image of the galaxy's immense jets (orange). **(b)** This Hubble Space Telescope image shows a disk of gas and dust about 250 pc (800 ly) across at the center of NGC 4261. Observations indicate that a supermassive black hole is at the center of the disk. (NASA, ESA)

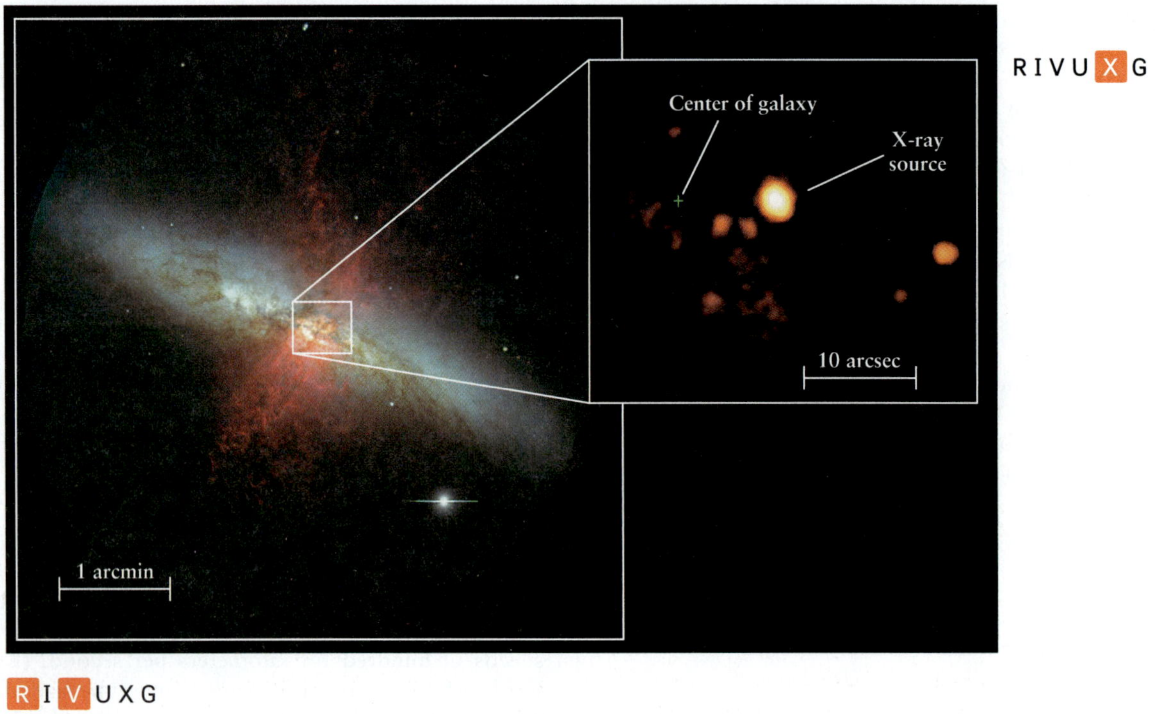

Figure 22-17

An Intermediate-mass Black Hole? M82 is an unusual galaxy in the constellation Ursa Major. (The cover of this book shows another image of M82.) The inset is an image of the central region of this galaxy from the Chandra X-ray Observatory. The bright, compact X-ray source varies in its light output over a period of months. The properties of this source suggest that it may be a black hole of roughly 500 solar masses or more. (Subaru Telescope, National Astronomical Observatory of Japan; inset: NASA/SAO/CXC)

Other Black Holes

Black holes seem to fall into two very different groups depending on their size. Supermassive black holes have masses from 10^6 to 10^9 $M_\odot$, while stellar-mass black holes like those discussed in Section 22-3 have masses around 10 $M_\odot$. In 2000 a team of astronomers from Japan, the United States, and Great Britain reported finding a "missing link" between these two groups.

The new discovery was a fluctuating X-ray source located some 200 pc (600 ly) from the center of the galaxy M82 (Figure 22-17). Images do not reveal material orbiting around this source, so we cannot determine the source's mass using Newton's form of Kepler's third law. But by relating the source's X-ray luminosity to the rate at which matter would have to fall onto the source to produce that luminosity, it can be shown that the source must have a mass of at least 500 $M_\odot$ or so. The rapid fluctuations of the source show that it has a very small diameter, which suggests that it is a black hole (see Section 22-3). Hence, the source in M82 may be an **intermediate-mass black hole** (also called a **mid-mass black hole**) Additional intermediate-mass black hole candidates have been found in other galaxies. Such black holes might have formed by the coalescence of many normal stars or by the direct merger of stellar-mass black holes.

During the early 1970s, Stephen Hawking of Cambridge University proposed the existence of yet another type of black hole. He suggested that during the Big Bang some 13.7 billion years ago, local pockets of the universe could have been dense enough and under sufficient pressure to form small black holes. According to his theory, these **primordial black holes** could have had masses as large as Earth or as small as 5×10^{-8} kg, about 1/40 the mass of a single raindrop. To date, however, no evidence has been found for the existence of primordial black holes in nature.

22-6 A nonrotating black hole has only a "center" and a "surface"

Understanding in detail how black holes form is a challenging task. But once a black hole forms, it has a remarkably simple structure.

Surrounding a black hole, where the escape speed from the hole just equals the speed of light, is the **event horizon.** You can think of this sphere as the "surface" of the black hole, although the black hole's mass all lies well inside this "surface." Once a massive dying star collapses to within its event horizon, it disappears permanently from the universe. The term "event horizon" is quite appropriate, because this surface is like a horizon beyond which we cannot see any events.

BOX 22-2 Tools of the Astronomer's Trade

The Schwarzschild Radius

The Schwarzschild radius (R_{Sch}) is the distance from the center of a nonrotating black hole to its event horizon. You can think of it as the "size" of the black hole. The Schwarzschild radius is related to the mass M of the black hole by

Schwarzschild radius

$$R_{Sch} = \frac{2GM}{c^2}$$

R_{Sch} = Schwarzschild radius of a black hole

G = universal constant of gravitation

M = mass of black hole

c = speed of light

When using this formula, be careful to express the mass in kilograms, not solar masses, because of the units in which G and c are commonly expressed ($G = 6.67 \times 10^{-11}$ N · m²/kg² and $c = 3.00 \times 10^8$ m/s). If you use M in kilograms, the answer that you get for R_{Sch} will be in meters.

EXAMPLE: Find the Schwarzschild radius (in kilometers) of a black hole with 10 times the mass of the Sun.

Situation: We are given the mass of the black hole (10 $M_\odot$) and wish to determine its Schwarzschild radius R_{Sch}.

Tools: We use the above formula for R_{Sch}, being careful to first convert the mass from solar masses to kilograms.

Answer: One solar mass is 1.99×10^{30} kg, so in this case $M = 10 \times 1.99 \times 10^{30}$ kg $= 1.99 \times 10^{31}$ kg. The Schwarzschild radius of this black hole is then

$$R_{Sch} = \frac{2GM}{c^2} = \frac{2(6.67 \times 10^{-11})(1.99 \times 10^{31})}{(3.00 \times 10^8)^2}$$

$$= 3.0 \times 10^4 \text{ m}$$

One kilometer is 10^3 meters, so the Schwarzschild radius of this 10-$M_\odot$ black hole is

$$R_{Sch} = (3.0 \times 10^4 \text{ m}) \times \frac{1 \text{ km}}{10^3 \text{ m}} = 30 \text{ km, or } 19 \text{ mi}$$

Review: The Schwarzschild radius is directly proportional to the mass of the black hole. Thus, a black hole with 10 times the mass (100 $M_\odot$) would have a Schwarzschild radius 10 times larger ($R_{Sch} = 10 \times 30$ km $= 300$ km); a black hole with half the mass (5 $M_\odot$) would have half as large a Schwarzschild radius ($R_{Sch} = 1/2 \times 30$ km $= 15$ km), and so on.

If the dying star is not rotating before it collapses, the black hole will likewise not be rotating. The distance from the center of a nonrotating black hole to its event horizon is called the **Schwarzschild radius** (denoted R_{Sch}), after the German physicist Karl Schwarzschild who first determined its properties. Box 22-2 describes how to calculate the Schwarzschild radius, which depends only on the black hole's mass. The more massive the black hole, the larger its event horizon.

Inside a Black Hole

Once a nonrotating star has contracted inside its event horizon, nothing can prevent its complete collapse. Equations describing the collapse indicate that the star's entire mass is crushed to zero volume—and hence infinite density—at a single point, known as the **singularity**, at the center of the black hole. We now can see that the structure of a nonrotating black hole is quite simple. As drawn in Figure 22-18, it has only two parts: a singularity at the center surrounded by a spherical event horizon.

All of the mass of a nonrotating black hole is concentrated in a point of zero size and infinitely high density

To understand why the complete collapse of such a doomed star is inevitable, think about your own life on Earth, far from any black holes. You have the freedom to move as you wish through the three dimensions of space: up and down, left and right, or forward and back. But you do not have the freedom to move at will through the dimension of time. Whether we like it or not, we are all carried inexorably from the cradle to the grave.

Inside a black hole, gravity distorts the structure of spacetime so severely that the directions of space and time become interchanged. In a sense, inside a black hole you acquire a limited ability to affect the passage of time. This seeming gain does you no good, however, because you lose a corresponding amount of freedom to move through space. Whether you like it or not, you will be dragged inexorably from the event horizon toward the singularity. Just as no force in the universe can prevent the forward march of time from past to future outside a black hole, no force in the universe can prevent the inward march of space from event horizon to singularity inside a black hole. Once an object dropped into a black hole crosses the event horizon, it is gone forever, like an object dropped into a bottomless pit.

The same is true for a light beam aimed at the black hole. Because all this light will be absorbed and none reflected back, a

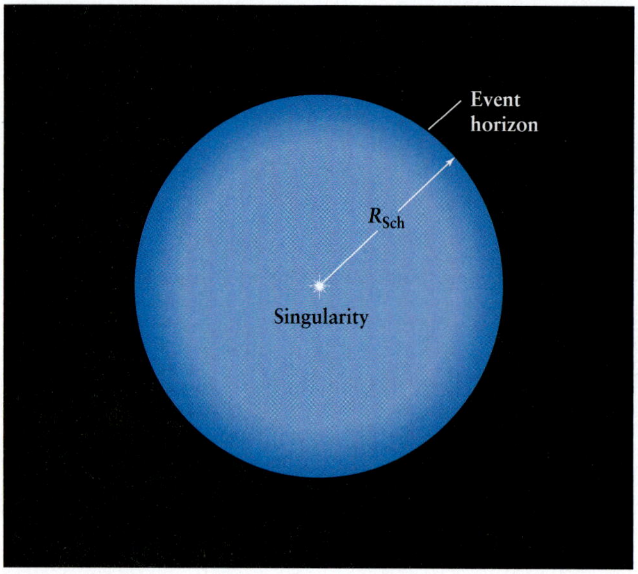

Figure 22-18

The Structure of a Nonrotating (Schwarzschild) Black Hole
A nonrotating black hole has only two parts: a singularity, where all of the mass is located, and a surrounding event horizon. The distance from the singularity to the event horizon is called the Schwarzschild radius (R_{Sch}). The event horizon is a one-way surface: Things can fall in, but nothing can get out.

black hole is totally and utterly black. (By contrast, even the blackest object on Earth reflects *some* light when you shine a flashlight on it. A black hole would reflect none.)

At a black hole's singularity, the strength of gravity is infinite, so the curvature of spacetime there is infinite. Space and time at the singularity are thus all jumbled up. They do not exist as separate, distinctive entities.

This confusion of space and time has profound implications for what goes on inside a black hole. All the laws of physics require a clear, distinct background of space and time. Without this identifiable background, we cannot clearly describe the arrangement of objects in space or the ordering of events in time. A description of the most central region of a black hole will require the development of additional laws of physics—this area of active research is called quantum gravity.

However, we are shielded from the inside of the black hole by the event horizon. In other words, although we don't know what happens at the singularity, none of the effects manage to escape beyond the event horizon. Consequently, the outside universe remains understandable.

Although black holes are really very simple objects, there are some common misconceptions about their nature. The *Cosmic Connections*: Black Hole "Urban Legends" exposes two of these misconceptions.

22-7 Just three numbers completely describe the structure of a black hole

Because light and matter can't escape, the event horizon prevents us from ever knowing much about anything that falls into a black hole, so a black hole would appear to be an "information sink." Many properties of matter's falling into a black hole, such as the matters chemical composition, texture, color, shape, and size, would appear to vanish as soon as it crosses the event horizon. The information might actually be preserved in microscopic quantum effects, but for all practical purposes, the information is unobservable through any existing techniques.

When an object falls into a black hole, all information about that object disappears except its mass, electric charge, and angular momentum

Because this information has seemingly vanished, it cannot affect the basic structure or properties of the hole. For example, consider two hypothetical black holes, one made from the gravitational collapse of 10 $M_\odot$ of iron and the other made from the gravitational collapse of 10 $M_\odot$ of peanut butter. Obviously, quite different substances went into the creation of the two holes. Once the event horizons of these two black holes have formed, however, both the iron and the peanut butter will have permanently disappeared from the universe. As seen from the outside, the two holes are identical, making it impossible for us to tell which ate the peanut butter and which ate the iron. A black hole is thus unaffected by the type of matter it consumes.

The Three Properties of a Black Hole

Because a black hole appears to be an information sink, it is reasonable to wonder whether we can determine anything at all about a black hole. What properties *do* characterize a black hole?

First, we can measure the *mass* of a black hole. One way to do this would be by placing a satellite into orbit around the hole. After measuring the size and period of the satellite's orbit, we could use Newton's form of Kepler's third law (recall Section 4-7 and Box 4-4) to determine the mass of the black hole. This mass is equal to the total mass of all the material that has gone into the hole.

Second, we can also measure the total *electric charge* possessed by a black hole. Like gravity, the electric force acts over long distances—it is a long-range interaction that is felt in the space around the hole. Appropriate equipment on a space probe passing near the hole could measure the strength of the electric force, and the electric charge could thus be determined.

In actuality, we would not expect a black hole to possess much, if any, electric charge. For example, if a hole did happen to start off with a sizable positive charge, it would vigorously attract vast numbers of negatively charged electrons from the interstellar medium, which would soon neutralize the hole's charge. For this reason, astronomers usually neglect electric charge when discussing real black holes.

Although a black hole might theoretically have a tiny electric charge, it can have no magnetic field of its own whatsoever. (We discussed in Section 22-3 how magnetic fields are involved in producing jets from black holes. However, these fields are associated with the accretion disk around the black hole, not the black hole itself.)

When a black hole is created, however, the collapsing star from which it forms may possess an appreciable magnetic field. The star must therefore radiate this magnetic field away before it

COSMIC CONNECTIONS

Black Hole "Urban Legends"

Black holes are such intriguing objects that a folklore has developed about them. As often is the case with folklore, myth and reality can become confused. These illustrations depict two misconceptions or "urban legends" about black holes.

Urban Legend #1:
When a star becomes a black hole, its gravitational pull becomes stronger.

Reality: At great distances from a black hole, the gravitational force that it exerts is exactly the *same* as that exerted by an ordinary star of the same mass. The truly stupendous gravitational effects of a black hole appear only if you venture close to the black hole's event horizon.

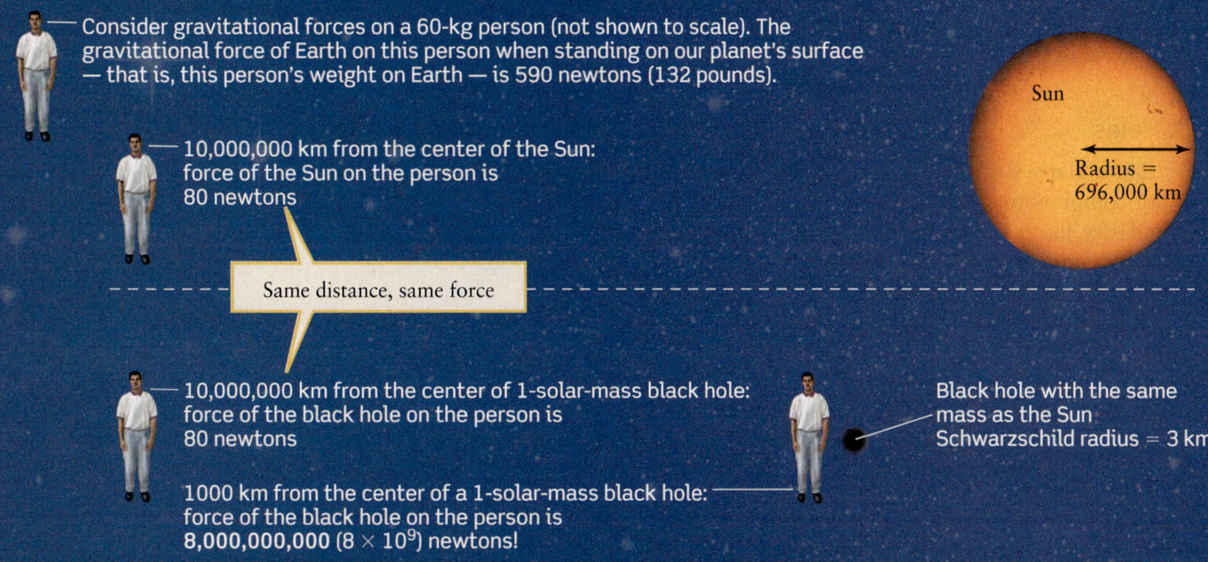

Consider gravitational forces on a 60-kg person (not shown to scale). The gravitational force of Earth on this person when standing on our planet's surface — that is, this person's weight on Earth — is 590 newtons (132 pounds).
10,000,000 km from the center of the Sun: force of the Sun on the person is 80 newtons
Sun
Radius = 696,000 km
Same distance, same force
10,000,000 km from the center of 1-solar-mass black hole: force of the black hole on the person is 80 newtons
Black hole with the same mass as the Sun Schwarzschild radius = 3 km
1000 km from the center of a 1-solar-mass black hole: force of the black hole on the person is **8,000,000,000** (8×10^9) newtons!

Urban Legend #2:
The larger a black hole, the more powerful the gravity near its Schwarzschild radius.

Reality: If you increase the mass of a black hole, its Schwarzschild radius increases by the same factor. Newton's law of universal gravitation, $F = Gm_1m_2/r2$, tells us that the force that an object of mass m_1 exerts on a second object of mass m_2 is directly proportional to mass m_1 but *inversely* proportional to the *square* of the distance r between the two objects. Hence the force that a black hole exerts on an object a given distance outside its Schwarzschild actually *decreases* as the black hole's mass and Schwarzschild radius increase.

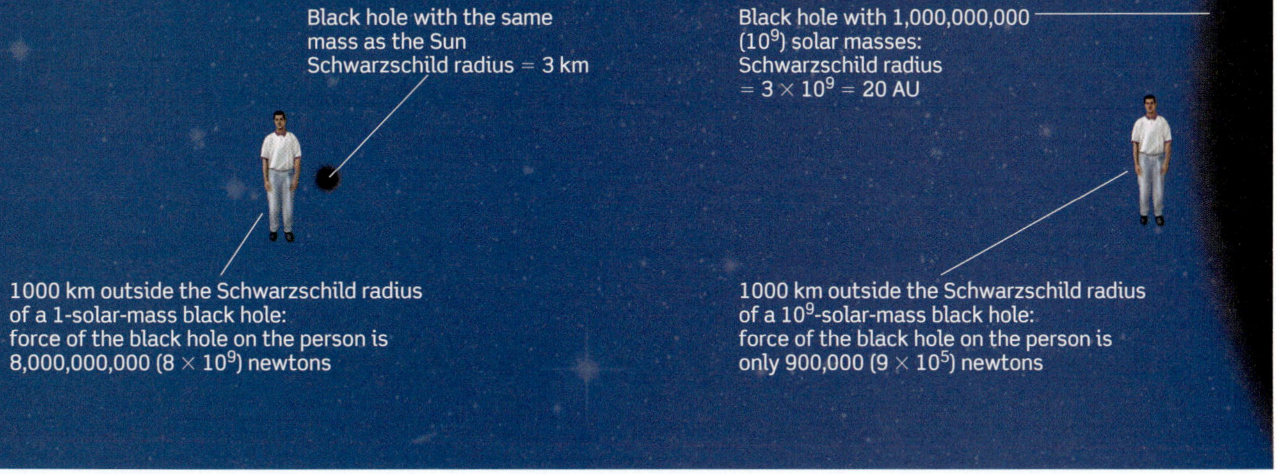

Black hole with the same mass as the Sun Schwarzschild radius = 3 km
Black hole with 1,000,000,000 (10^9) solar masses: Schwarzschild radius = 3×10^9 = 20 AU
1000 km outside the Schwarzschild radius of a 1-solar-mass black hole: force of the black hole on the person is 8,000,000,000 (8×10^9) newtons
1000 km outside the Schwarzschild radius of a 10^9-solar-mass black hole: force of the black hole on the person is only 900,000 (9×10^5) newtons

can settle down inside its event horizon. Theory predicts that the star does this by emitting electromagnetic and gravitational waves. As we saw in Section 22-2, gravitational waves are ripples in the overall geometry of space. Various experiments soon to be in operation may detect bursts of gravitational radiation emitted by massive stars as they collapse to form black holes.

Third, we can detect the effects of a black hole's rotation, that is, measure its *angular momentum.* An object's angular momentum is related to how fast it rotates and how the object's mass is distributed over its volume. As a dead star collapses into a black hole, its rotation naturally speeds up as its mass moves toward the center, just as a figure skater rotates faster when she pulls her arms and legs in. This same effect explains the rotation of the solar nebula (see Section 8-4), as well as why neutron stars spin so fast (see Section 21-3). We expect a black hole that forms from a rotating star to be spinning very rapidly.

Rotating Black Holes

When the matter that collapses to form a black hole is rotating, that matter does not compress to a point. Instead, it collapses into a ring-shaped singularity located between the center of the hole and the event horizon (Figure 22-19). The structure of such rotating black holes was first worked out in 1963 by the New Zealand mathematician Roy Kerr. Most rotating black holes should be spinning thousands of times per second, even faster than the most rapid pulsars (see Section 21-7).

WEB LINK 22.12

If a rotating black hole is surrounded by an accretion disk with a magnetic field, it may be possible for the magnetic field to steal some of the rotational energy and angular momentum from the black hole and transfer it to the disk. The magnetic field acts as a "brake" that retards the black hole's rotation and makes the disk's rotation speed up. (On Earth, magnetic braking is used to slow locomotives, amusement park rides, and hybrid automobiles to a smooth stop without generating excess heat.) Figure 22-20 shows an arching magnetic field that connects an accretion disk with a rotating black hole in just this way. This process may be taking place with the supermassive black hole at the center of the galaxy MCG–6-15-30. Observations with the XMM-Newton telescope (see Section 6-7) indicate that unusually intense X-rays are coming from the accretion disk around this black hole. The suspicion is that the energy to power this radiation may be extracted from the black hole's rotation. (This transfer of energy and angular momentum can also go the other way. The image that opens this chapter depicts a system in which magnetic fields are thought to transfer angular momentum *to* a black hole *from* its accretion disk. Robbing the accretion disk of its rotation makes it easier for its material to fall into the black hole.)

WEB LINK 22.13

Even a rotating black hole without an accretion disk can transfer energy to other objects. This energy transfer is possible according to Einstein's general theory of relativity, which makes the startling prediction that a rotating body drags spacetime around it. (Very precise spacecraft measurements indicate that the rotating Earth drags spacetime in just this way.) Surrounding the event horizon of every rotating black hole is a region where this dragging of space and time is so severe that it is impossible to stay in the same place. No matter what you do, you get pulled around the hole, along with the rotating geometry of space and time. This region, where it is impossible to be at rest, is called the **ergoregion** (see Figure 22-19).

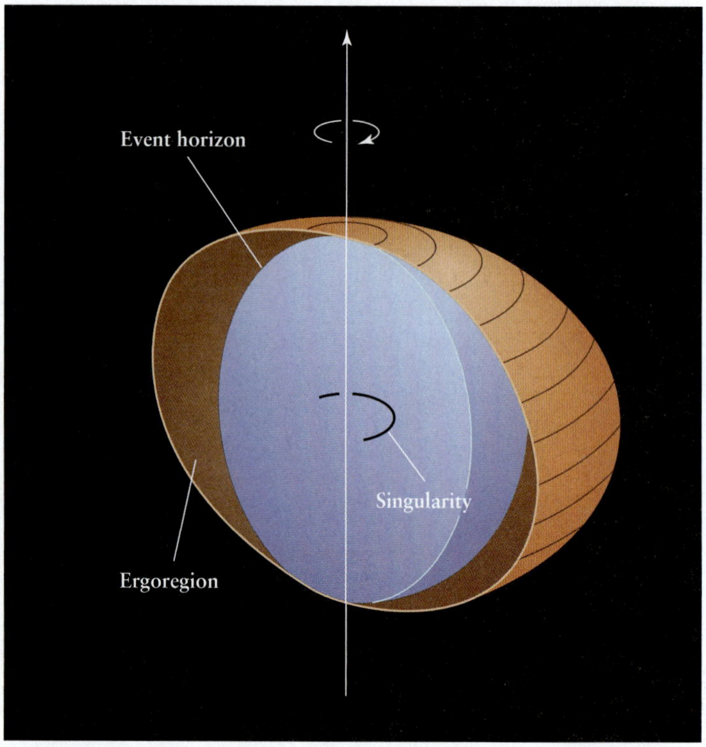

Figure 22-19

The Structure of a Rotating (Kerr) Black Hole The singularity of a rotating black hole is an infinitely thin ring centered on the geometrical center of the hole. (It appears as an arc in this cutaway diagram.) Outside the spherical event horizon is the doughnut-shaped ergoregion, where the dragging of spacetime around the hole is so severe that nothing can remain at rest. The tan-colored surface marks the outer boundary of the ergoregion.

To measure a black hole's angular momentum, we could hypothetically place two satellites in orbit about the hole. Suppose that one satellite circles the hole in the same direction the hole rotates and the other in the opposite direction. One satellite is thus carried along by the geometry of space and time, but the other is constantly fighting its way "upstream." The two satellites will thus have different orbital periods. By comparing these two periods, astronomers can deduce the total angular momentum of the hole.

Because the ergoregion is outside the event horizon, this bizarre region is accessible to us, and spacecraft could travel through it without disappearing into the black hole. According to detailed calculations, objects grazing the ergoregion could be catapulted back out into space at tremendous speeds. In other words, the ejected object could leave the ergoregion with more energy than it had initially, having extracted added energy from the hole's rotation. This process is called the *Penrose process,* after Roger Penrose, the British mathematician who proposed it.

Figure 22-20

A Rotating Supermassive Black Hole This artist's impression shows the accretion disk around the supermassive black hole at the heart of the galaxy MCG-6-15-30. The arching magnetic field allows the accretion disk to extract energy and angular momentum from the black hole. (XMM-Newton/ESA/NASA)

Mass, charge, and angular momentum are the only three properties that a black hole possesses. This simplicity is the essence of the **no-hair theorem**, first formulated in the early 1970s: "Black holes have no hair." In other words, "hair" would be some property of a black hole other than mass, charge, or angular momentum, yet another property is not possible.

22-8 Falling into a black hole is an infinite voyage

TUTORIAL 22.3 WEB LINK 22.14

Imagine that you are on board a spacecraft at a safe distance from a $5\text{-}M_\odot$ black hole. A distance of 1000 Schwarzschild radii, or 15,000 km, would suffice. You now release a space probe and let it fall into the black hole. To make the probe easier to track, it is coated with a phosphorescent paint that emits a blue glow. You also equip the probe with a video camera that transmits images of the view of the approaching black hole. What will you see as the probe falls?

As Figure 22-21*a* shows, at 1000 Schwarzschild radii from the black hole the video camera would send back a rather normal view of space. But as the probe approaches the black hole, the bending of light by the black hole (see Section 22-2) becomes more pronounced. Light rays passing close to the back hole are deflected so severely that at close range, the camera will actually see multiple images of the same stars (Figure 22-21*b*).

The gravitational field of a black hole distorts the image of other stars

How will the probe itself look from your vantage point at a safe distance? You might expect that as the probe falls, its speed should continue to increase. This expectation is true up to a point. But as the probe approaches the event horizon, where the black hole's gravity is extremely strong, the gravitational slowing of time becomes so pronounced that the probe will appear to slow down! From your point of view, the spacecraft takes an infinite amount of time to reach the event horizon, where it will appear to remain suspended for all eternity.

To watch these effects, however, you will need special equipment. The reason is the gravitational redshift: As the probe falls, the light reaching you from the probe is shifted to longer and longer wavelengths. The probe's blue glow will turn green, then yellow, then red, and eventually fade into infrared wavelengths that your unaided eye cannot detect.

If you could view the falling probe with infrared goggles, you would see it come to an unpleasant end (Figure 22-22). Near the event horizon, the strength of the black hole's gravity increases dramatically as the probe moves just a short distance closer to the hole. In fact, the side of the probe nearest the black hole feels a much stronger gravitational pull than the side opposite the hole. These *tidal forces* are like those that the Moon exerts on Earth (see Section 4-8), but tremendously stronger. Furthermore, the sides of the probe are pulled together, since the hole's gravity makes them fall in straight lines aimed at the center of the hole. The net effect is that the probe will be stretched out along the line pointing toward the hole, and squeezed together along the perpendicular directions. The stretching is so great that it will rip the probe into atoms, and even rip the atoms apart.

(a) Looking directly toward the black hole from a distance of 1000 Schwarzschild radii: Note positions of stars 1, 2, and 3.

(b) Looking directly toward the black hole from a distance of 10 Schwarzschild radii: Light bending causes multiple images.

Figure 22-21

The View Approaching a Black Hole These images are taken from a computer simulation of the appearance of the sky behind a black hole. **(a)** The bending of light by the black hole is negligible at large distances, but **(b)** becomes evident as you approach the black hole. (Courtesy Robert J. Nemiroff, Michigan Technological University)

Someone foolhardy enough to ride along with the probe would have a very different view of the fall than yours. From this astronaut's point of view, there is no slowing of time, and the probe continues its fall through the event horizon into the singularity. The astronaut is unlikely to enjoy the ride, however, since he will be disintegrated by tidal forces as he falls inward toward the singularity.

Could a black hole somehow be connected to another part of spacetime, or even some other universe? General relativity predicts that such connections, called **wormholes,** can exist for rotating black holes. Box 22-3 discusses the possibility that such wormholes really exist.

22-9 Black holes evaporate

With all the mass of a black hole hidden behind its event horizon and collapsed into a singularity, it may seem that there is no way of getting mass from the black hole back out into the universe. But in fact there is, as Stephen Hawking figured out in the 1970s. To understand how this is possible, we must look at the quantum mechanical behavior of matter at the submicroscopic scale of nuclei and electrons.

The **Heisenberg uncertainty principle** is a basic tenet of submicroscopic physics. This principle states that you cannot determine

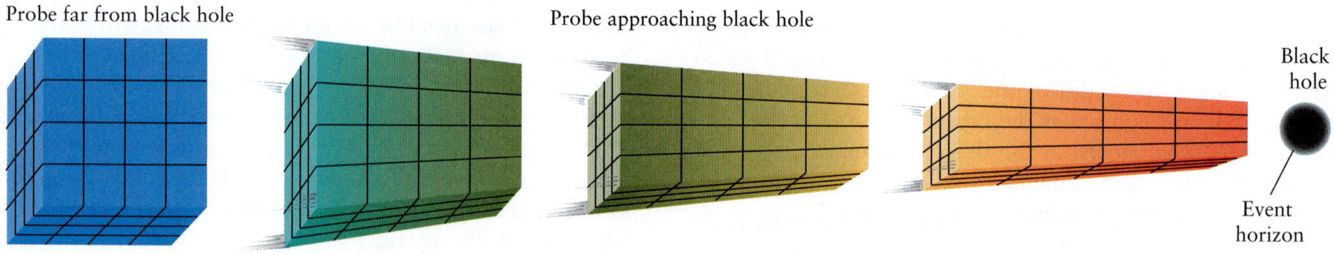

Figure 22-22

Falling into a Black Hole **(a)** A cube-shaped probe is dropped from a distance of 1000 Schwarzschild radii from a 5-$M_\odot$ black hole. **(b), (c), (d)** As the probe approaches the event horizon, it is distorted into a long, thin shape by the black hole's extreme gravity. A distant observer sees the probe change color as photons from the probe undergo a strong gravitational redshift.

BOX 22-3 Astronomy Down to Earth

Wormholes and Time Machines

Wormholes have become a staple of science fiction. They are certainly a convenient plot device: The gallant crew flies their spaceship into one end of a wormhole, and a short time later they emerge many light-years away. Time machines, too, have been part of science fiction ever since H. G. Wells published his classic novel *The Time Machine* at the end of the nineteenth century. What would it be like to go back in time and watch famous historical events—or even change history?

Wormholes and time machines challenge our normal ideas about space and time. So, too, does the general theory of relativity. Could it be that this theory really makes it possible to travel through wormholes and to travel in time?

In the 1930s, Einstein and his colleague Nathan Rosen discovered that a black hole could possibly connect our universe with a second domain of space and time that is separate from ours. The first diagram shows this connection, called an *Einstein-Rosen bridge.* You could think of the upper surface as our universe and the lower surface as a "parallel universe." Alternatively, the upper and lower surfaces could be different regions of our own universe. In that case, an Einstein-Rosen bridge would connect our universe with itself, forming a wormhole, shown in the second diagram.

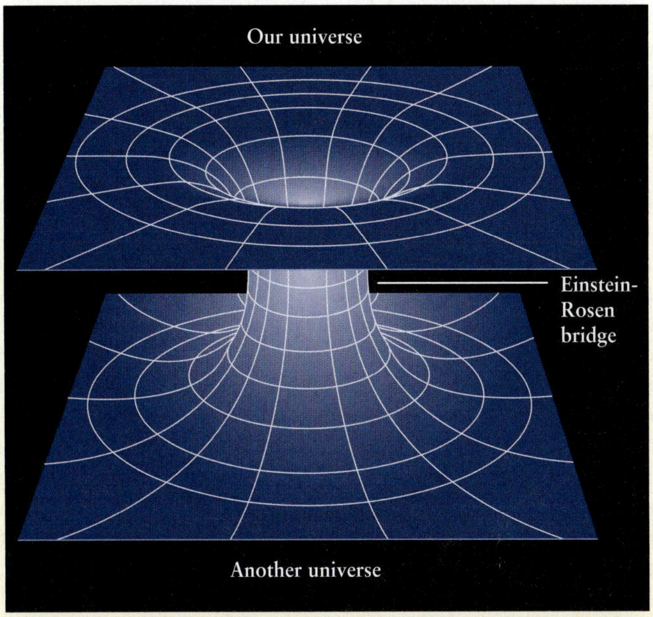

A wormhole may seem like a shortcut to distant places in our universe, but detailed calculations reveal a major obstacle. The powerful gravity of a black hole causes the wormhole to collapse almost as soon as it forms. As a result, to get from one side of a wormhole to the other, you would have to travel faster than the speed of light, which is not possible.

Caltech physicists Kip Thorne, Michael Morris, and Ulvi Yurtsever have proposed a scheme that might get around the difficulty of a collapsing wormhole. According to general relativity, they point out, pressure as well as mass can be a source of gravity. Normally we do not see the gravitational effects of pressure because they are so small. Thorne and his colleagues speculate that a technologically advanced civilization might someday be able to use pressure to produce *antigravity* strong enough to keep the wormhole open.

If a wormhole could be held open, it could also be a time machine. To see how, imagine you take one end of a wormhole and move it around for a while at speeds very near the speed of light. As we saw in Box 22-1, such motion causes clocks to slow down. Thus, when you stop moving that end of the wormhole, you find that it has not aged as much as the stationary end. In other words, one side of the wormhole has a different time from the other. As a result, you could go into one end of the wormhole at a late time and come out at an early time. For example, you might go in at 10 A.M. and come out at 9 A.M.

Time machines challenge ordinary logic. If you could get back from a trip an hour before you left, you could meet yourself and tell yourself what a nice journey you had. Then both of you could take the trip. If you and your twin return just before you both left, there would be four of you. And all four of you could take the trip again. And then all eight of you. Then all sixteen of you . . .

Making copies of yourself is an example of how time machines violate *causality,* the notion that effects must follow their causes. To date, we have never seen a violation of causality. Many scientists would therefore like to show that time machines cannot exist. The British astrophysicist Stephen Hawking points to one strong bit of observational evidence against time machines: We are not being visited by hordes of tourists from the future. If we could discover why nature seems to preclude time machines, we would have a much deeper understanding of the nature of space and time.

precisely both the position and the speed of a subatomic particle. Over extremely short distances or times, a certain amount of "fuzziness" is built into the nature of reality.

The Heisenberg uncertainty principle leads to the concept of **virtual pairs:** At every point in space, pairs of particles and antiparticles are constantly being created and destroyed. An antiparticle is quite like an ordinary particle except that it has an opposite electric charge and can annihilate an ordinary particle so that both disappear, usually leaving a pair of photons in their place. In the case of virtual pairs, the process of creation and annihilation occurs over such incredibly brief time intervals that these virtual particles and antiparticles cannot be observed directly.

Think about a tiny black hole. Furthermore, think about the momentary creation of a virtual pair of an electron and a positron (the antiparticle of an electron) just outside the hole's event horizon. It may happen that one of the virtual particles falls into the black hole. Its partner is then deprived of a counterpart with which to annihilate and must therefore become a real particle. To accomplish this conversion, some of the energy of the black hole's gravity must be converted into matter, according to $E = mc^2$. This decreases the mass of the black hole by a corresponding amount, and the real particle is free to escape from the hole. In this way, real particles can quantum-mechanically "leak" out from a microscopic region near the event horizon of a black hole, carrying some of the hole's mass with them (Figure 22-23).

The less mass a black hole has, the more easily particles can leak out through its event horizon. Jacob Bekenstein and Stephen Hawking proved mathematically that you can speak of the *temperature* of a black hole as a way of describing the amounts of energy carried away by particles leaking out of it. For example, a 1-trillion-ton (10^{15} kg) black hole emits particles and energy as if it were a blackbody with a temperature of nearly 10^9 K.

A fundamental uncertainty in physical knowledge makes it possible for black holes to emit particles and radiation into space

For stellar-mass black holes, such as Cygnus X-1, this effect is negligible over time spans of billions of years. The reason is that the temperature of a black hole is inversely proportional to the black hole's mass. Compared to a trillion-ton black hole, a 5-$M_\odot$ (10^{31} kg) hole has a mass that is 10^{16} times greater and a temperature T that is only 10^{-16} as much:

$$T = 10^{-16} \times 10^9 \text{ K} = 10^{-7} \text{ K}$$

This temperature is barely above absolute zero. In other words, ordinary black holes have such low temperatures that particles hardly ever manage to escape from them.

The story is different for low-mass black holes, such as the primordial black holes we discussed in Section 22-5. As particles escape from a small black hole, the mass of the black hole decreases, making its temperature go up. As its temperature rises, still more particles escape, further decreasing the hole's mass and forcing the temperature still higher. This runaway process of **black hole evaporation** finally causes the hole to vanish altogether. During its final seconds of evaporation, the hole gives up the last of its mass in a violent burst of energy equal to the detonation of a billion megaton hydrogen bombs.

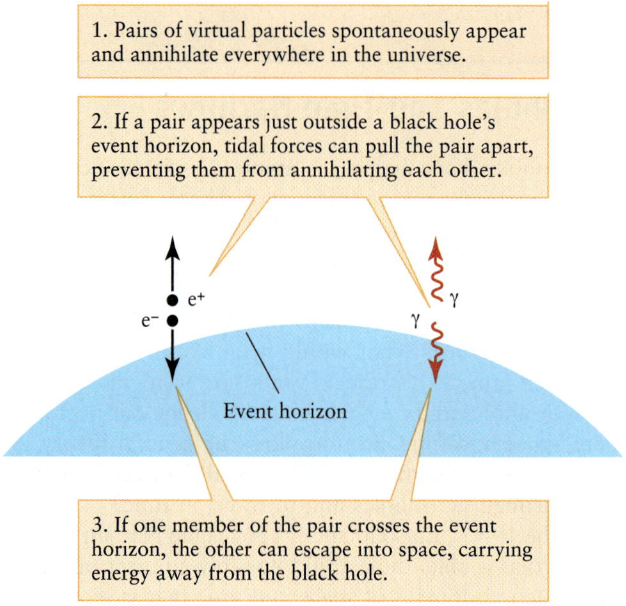

1. Pairs of virtual particles spontaneously appear and annihilate everywhere in the universe.
2. If a pair appears just outside a black hole's event horizon, tidal forces can pull the pair apart, preventing them from annihilating each other.
e+
e−
γ
γ
Event horizon
3. If one member of the pair crosses the event horizon, the other can escape into space, carrying energy away from the black hole.

Figure 22-23

Evaporation of a Black Hole This illustration shows two pairs of virtual particles—an electron (e^-) and a positron (e^+), and a pair of virtual photons (γ)—appearing just outside the event horizon of a black hole. If one member of the pair escapes and carries energy away from the black hole, the black hole decreases in mass and the event horizon shrinks.

A 10^{10}-kg primordial black hole (comparable to the mass of Mount Everest) would take about 15 billion years to evaporate. This period is close to the present age of the universe. If primordial black holes were formed in the Big Bang, we would expect to see some of them going through the explosive final stages of evaporation. To date, however, there is no compelling evidence that we have seen such evaporations taking place, and many astronomers suspect that there are no primordial black holes.

By contrast, a 5-$M_\odot$ black hole would take more than 10^{62} years to evaporate, and a supermassive black hole of 5 million solar masses would take more than 10^{80} years. These time spans are far longer than the age of the universe. We can safely predict that the black holes we have observed to date will remain as black holes for the foreseeable future.

Key Words

Terms preceded by an asterisk (*) are discussed in the Boxes.

black hole, p. 585
black hole evaporation, p. 600
collapsar, p. 590
equivalence principle, p. 582
ergoregion, p. 596
event horizon, p. 592
gamma-ray burster, p. 588
general theory of relativity, p. 582
gravitational radiation (gravitational waves), p. 584
gravitational redshift, p. 584
Heisenberg uncertainty principle, p. 598
intermediate-mass (mid-mass) black hole, p. 592
law of cosmic censorship, p. 594
length contraction, p. 578

*Lorentz transformations, p. 580
mid-mass (intermediate-mass) black hole, p. 592
no-hair theorem, p. 597
primordial black hole, p. 592
*proper length (proper distance), p. 581
*proper time, p. 580
Schwarzschild radius (R_{Sch}), p. 593
singularity, p. 593
spacetime, p. 578
special theory of relativity, p. 577
stellar-mass black hole, p. 586
supermassive black hole, p. 591
time dilation, p. 578
virtual pairs, p. 600
wormhole, p. 598

Key Ideas

The Special Theory of Relativity: This theory, published by Einstein in 1905, is based on the notion that there is no such thing as absolute space or time.

• The speed of light is the same to all observers, no matter how fast they are moving.

• An observer will note a slowing of clocks and a shortening of rulers that are moving with respect to the observer. This effect becomes significant only if the clock or ruler is moving at a substantial fraction of the speed of light.

• Space and time are not wholly independent of each other, but are aspects of a single entity called spacetime.

The General Theory of Relativity: Published by Einstein in 1915, this is a theory of gravity. Any massive object causes space to curve and time to slow down, and these effects manifest themselves as a gravitational force. These distortions of space and time are most noticeable in the vicinity of large masses or compact objects.

• The general theory of relativity is our most accurate description of gravitation. It predicts a number of phenomena, including the bending of light by gravity and the gravitational redshift, whose existence has been confirmed by observation and experiment.

• The general theory of relativity also predicts the existence of gravitational waves, which are ripples in the overall geometry of space and time produced by moving masses. Gravitational waves have been detected indirectly, and specialized antennas are under construction to make direct measurement of the gravitational waves from cosmic cataclysms.

Black Holes: If a stellar corpse has a mass greater than about 2 to 3 $M_{\odot}$, gravitational compression will overwhelm any and all forms of internal pressure. The stellar corpse will collapse to such a high density that its escape speed exceeds the speed of light.

Observing Black Holes: Black holes have been detected using indirect methods.

• Some binary star systems contain a black hole. In such a system, gases captured from the companion star by the black hole emit detectable X rays.

• Many galaxies have supermassive black holes at their centers. These are detected by observing the motions of material around the black hole.

Gamma-Ray Bursters: Short, intense bursts of gamma rays are observed at random times coming from random parts of the sky.

• By observing the afterglow of long-duration gamma-ray bursters, astronomers find that these objects have very large redshifts and appear to be located within distant galaxies. The bursts are correlated with supernovae, and may be due to an exotic type of supernova called a collapsar.

• The origin of short-duration gamma-ray bursters is unknown.

Properties of Black Holes: The entire mass of a black hole is concentrated in an infinitely dense singularity.

• The singularity is surrounded by a surface called the event horizon, where the escape speed equals the speed of light. Nothing—not even light—can escape from inside the event horizon.

• A black hole has only three physical properties: mass, electric charge, and angular momentum.

• A rotating black hole (one with angular momentum) has an ergoregion around the outside of the event horizon. In the ergoregion, space and time themselves are dragged along with the rotation of the black hole.

• Black holes can evaporate, but in most cases at an extremely slow rate.

Questions

Review Questions

1. You drop a ball inside a car traveling at a steady 50 km/h in a straight line on a smooth road. Does it fall in the same way as it does inside a stationary car? How does this question relate to Einstein's special theory of relativity?
2. In Einstein's special theory of relativity, two different observers moving at different speeds will measure the same value of the speed of light. Will these same observers measure the same value of, say, the speed of an airplane? Explain your answer.
3. A friend summarizes the special theory of relativity by saying "Everything is relative." Explain why this statement is inaccurate.
4. Serena flies past Michael in her spaceship at nearly the speed of light. According to Michael, Serena's clock runs slow. According to Serena, does Michael's clock run slow, fast, or at the normal rate? Explain your answer.
5. Ole flies past Lena in a spherical spaceship at nearly the speed of light. According to Lena, how does the distance from the front to the back of Ole's spaceship (that is, measured along the direction of motion) compare to the distance from the top to the bottom (that is, measured perpendicular to the direction of motion)? Explain your answer.
6. Why does the speed of light represent an ultimate speed limit?
7. Why is Einstein's general theory of relativity a better description of gravity than Newton's universal law of gravitation? Under what circumstances is Newton's description of gravity adequate?
8. Describe two different predictions of the general theory of relativity and how these predictions were tested experimentally. Do the results of the experiments agree with the theory?
9. How does a gravitational redshift differ from a Doppler shift?

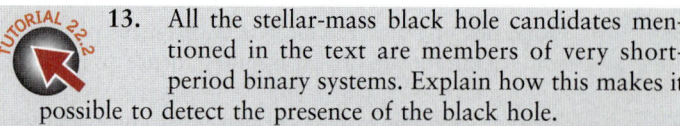

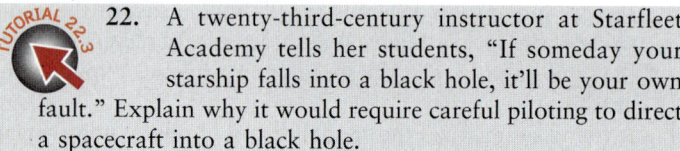

10. In what circumstances are degenerate electron pressure and degenerate neutron pressure incapable of preventing the complete gravitational collapse of a dead star?
11. Should we worry about Earth being pulled into a black hole? Why or why not?
12. How does rapid flickering of an X-ray source provide evidence that the source is small?
13. All the stellar-mass black hole candidates mentioned in the text are members of very short-period binary systems. Explain how this makes it possible to detect the presence of the black hole.
14. Astronomers cannot actually see the black hole candidates in close binary systems. How, then, do they know that these candidates are not white dwarfs or neutron stars?
15. Describe two ways in which a member of a binary star system could become a black hole.
16. What is a gamma-ray burster? What is the evidence that gamma-ray bursters are not located in the disk of our Galaxy or in a halo surrounding our Galaxy?
17. Summarize the evidence that gamma-ray bursters result from a process involving a star in a distant galaxy.
18. What is a collapsar? How does the collapsar model account for the existence of long-duration gamma-ray bursters?
19. How do astronomers locate supermassive black holes in galaxies?
20. What is an intermediate-mass black hole? How are such objects thought to form?
21. When we say that the Moon has a radius of 1738 km, we mean that this is the smallest radius that encloses all of the Moon's material. In this sense, is it correct to think of the Schwarzschild radius as the radius of a black hole? Why or why not?
22. A twenty-third-century instructor at Starfleet Academy tells her students, "If someday your starship falls into a black hole, it'll be your own fault." Explain why it would require careful piloting to direct a spacecraft into a black hole.
23. In what way is a black hole blacker than black ink or a black piece of paper?
24. If the Sun suddenly became a black hole, how would Earth's orbit be affected? Explain your answer.
25. According to the general theory of relativity, why can't some sort of yet-undiscovered degenerate pressure prevent the matter inside a black hole from collapsing all the way down to a singularity?
26. What is the law of cosmic censorship?
27. Is it possible to tell which chemical elements went into a black hole? Why or why not?
28. Why is it unlikely that a black hole has an electric charge?
29. What kind of black hole is surrounded by an ergoregion? What happens inside the ergoregion?
30. What is the no-hair theorem?
31. As seen by a distant observer, how long does it take an object dropped from a great distance to fall through the event horizon of a black hole? Explain your answer.
32. If even light cannot escape from a black hole, how is it possible for black holes to evaporate?

Advanced Questions

Questions preceded by an asterisk (*) involve topics discussed in the Boxes.

Problem-solving tips and tools

Remember that the time to travel a certain distance is equal to the distance divided by the speed, and that the density of an object is its mass divided by its volume. The volume of a sphere of radius r is $4\pi r^3/3$. Section 4-7 describes Newton's law of universal gravitation. Box 4-4 shows how to use Newton's formulation of Kepler's third law, which explicitly includes masses; when using this formula, note that the period P must be expressed in seconds, the semimajor axis a in meters, and the masses in kilograms. For another version of this formula, in which period is in years, semimajor axis in AU, and masses in solar masses, see Section 17-9.

*33. A spaceship flies from Earth to a distant star at a constant speed. Upon arrival, a clock on board the spaceship shows a total elapsed time of 8 years for the trip. An identical clock on Earth shows that the total elapsed time for the trip was 10 years. What was the speed of the spaceship relative to Earth?

*34. An unstable particle called a positive pion (pronounced "pie-on") decays in an average time of 2.6×10^{-8} s. On average, how long will a positive pion last if it is traveling at 95% of the speed of light?

*35. How fast should a meter stick be moving in order to appear to be only 60 cm long?

*36. An astronaut flies from Earth to a distant star at 80% of the speed of light. As measured by the astronaut, the one-way trip takes 15 years. (**a**) How long does the trip take as measured by an observer on Earth? (**b**) What is the distance from Earth to the star (in light-years) as measured by an Earth observer? As measured by the astronaut?

37. In the binary system of two neutron stars discovered by Hulse and Taylor (Section 22-2), one of the neutron stars is a pulsar. The distance between the two stars varies between 1.1 and 4.8 times the radius of the Sun. The time interval between pulses from the pulsar is not constant: It is greatest when the two stars are closest to each other and least when the two stars are farthest apart. Explain why this is consistent with the gravitational slowing of time (Figure 22-7*a*).

38. Find the total mass of the neutron star binary system discovered by Hulse and Taylor (Section 22-2), for which the orbital period is 7.75 hours and the average distance between the neutron stars is 2.8 solar radii. Is your result reasonable for a pair of neutron stars? Explain your answer.

39. Estimate how long it will be until the two neutron stars that make up the binary system discovered by Hulse and Taylor collide with each other. Assume that the distance between the two stars will continue to decrease at its present rate of 3 mm every 7.75 hours, and use the data given in Question 38. (You can assume that the two stars are very small, so they will collide when the distance between them is equal to zero.)

40. The orbital period of the binary system containing A0620-00 is 0.32 day, and Doppler shift measurements reveal that the radial velocity of the X-ray source peaks at 457 km/s (about 1 million miles per hour). (**a**) Assuming that the orbit of the X-ray source is a circle, find the radius of its orbit in kilometers. (This is actually an estimate of the semimajor axis of the orbit.) (**b**) By using Newton's form of Kepler's third law, prove that the mass of the X-ray source must be at least 3.1 times the mass of the Sun. (*Hint:* Assume that the mass of the K5V visible star—about 0.5 $M_\odot$ from the mass-luminosity relationship—is negligible compared to that of the invisible companion.)
41. Contrast gamma-ray bursters with X-ray bursters (discussed in Section 21-9). From our models of what causes these energetic phenomena, explain why X-ray bursters emit repeated pulses but gamma-ray bursters apparently emit just once.
42. Long-duration gamma-ray bursters are only observed in galaxies where there is ongoing star formation. Explain how this is consistent with the collapsar model of how these bursters occur.
43. The spectrum of a Type Ic supernova lacks absorption lines of hydrogen and helium. This means that when a black hole formed at the center of the progenitor star, the resulting jets were more easily able to escape into space. Explain why this is so.
44. Find the orbital period of a star moving in a circular orbit of radius 500 AU around the supermassive black hole in the galaxy NGC 4261 (Section 22-5).
*45. Find the Schwarzschild radius for an object having a mass equal to that of the planet Saturn.
*46. What is the Schwarzschild radius of a black hole whose mass is that of (**a**) Earth, (**b**) the Sun, (**c**) the supermassive black hole in NGC 4261 (Section 22-5)? In each case, also calculate what the density would be if the matter were spread uniformly throughout the volume of the event horizon.
*47. What is the mass in kilograms of a black hole whose Schwarzschild radius is 11 km?
*48. To what density must the matter of a dead 8-$M_\odot$ star be compressed in order for the star to disappear inside its event horizon? How does this compare with the density at the center of a neutron star, about 3×10^{18} kg/m^3?
*49. Prove that the density of matter needed to produce a black hole is inversely proportional to the square of the mass of the hole. If you wanted to make a black hole from matter compressed to the density of water (1000 kg/m^3), how much mass would you need?

Discussion Questions

50. The speed of light is the same for all observers, regardless of their motion. Discuss why this requires us to abandon the Newtonian view of space and time.
51. Describe the kinds of observations you might make in order to locate and identify black holes.
52. Speculate on the effects you might encounter on a trip to the center of a black hole (assuming that you could survive the journey).

Web/eBook Questions

53. Search the World Wide Web for information about a stellar-mass black hole candidate named V4641 Sgr. In what ways does it resemble other black hole candidates such as Cygnus X-1 and V404 Cygni? In what ways is it different and more dramatic? How do astronomers explain why V4641 Sgr is different?
54. Search the World Wide Web for information about supernova SN 2006aj, which was associated with gamma-ray burster GRB 060218. In what ways were this supernova and gamma-ray burster unusual? Are the observations of these objects consistent with the collapsar model?
55. Search the World Wide Web for information about the intermediate-mass black hole candidate in M82. Is this still thought to be an intermediate-mass black hole? What new evidence has been used to either support or oppose the idea that this object is an intermediate-mass black hole?

ANIMATION 22.1

56. **The Equivalence Principle** Access the animation "The Equivalence Principle" in Chapter 22 of the *Universe* Web site or eBook. View the animation and answer the following questions. (**a**) Describe what is happening as viewed from the frame of reference of the elevator. What causes the apple to fall to the floor of each elevator? (**b**) Describe what is happening as viewed from the frame of reference of the stars. What causes the apple to fall to the floor of each elevator? (**c**) Think of another experiment you could perform with the apple. Describe what would happen during this experiment as seen by Newton (in the left-hand box) and by Einstein (in the right-hand box).

Activities

Observing Projects

57. You cannot see a black hole with a telescope. Nevertheless, you might want to observe the visible companion of Cygnus X-1. The epoch 2000 coordinates of this ninth-magnitude star are R.A. = 19^h 58.4^m and Decl. = +35° 12′, which is quite near the bright star η (eta) Cygni. Compare what you see with the photograph in Figure 22-10.
58. Use the *Starry Night Enthusiast*™ program to plan observations of Cygnus X-1. Use the **Find** pane to **Centre** the view on Cygnus X-1. Select **View > Show Daylight** from the menu. (**a**) Using the time controls in the toolbar, determine when Cygnus X-1 rises and sets on today's date from your home location. (**b**) At approximately what time on today's date is Cygnus X-1 highest in the sky? Is tonight a good night for observing this star with a visible-light telescope? Would it be better placed in the sky for observation six months from now? Explain how you determined this.
59. Use the *Starry Night Enthusiast*™ program to examine X-ray images of galaxies with supermassive black holes at their centers. Open the **Options** pane and expand the **Deep Space** layer. Select **Chandra Images** and deselect all of the other op-

tions in this layer. Use the **Find** pane and Zoom controls to examine each of the following galaxies: (i) NGC 4261; (ii) Virgo A (M87); (iii) M31. Open the Options pane again and select Messier Objects and deselect Chandra Images and compare the visual images of Virgo A (M87) and M31. Suggest why supermassive black holes were discovered in these galaxies only after relatively recent advances were made in telescope and detector technology.

Collaborative Exercise

60. Using Einstein's theory of relativity, estimate (1) the length of your pencil or pen at constant speed at the speeds of a bicycle rider, a car traveling on the highway, and a commercial jet liner at cruising altitude; and (2) the speed of a light beam emitted by a spaceshift traveling at 200,000 kilometers per second toward another spaceship traveling at the same speed.

23 Our Galaxy

WEB LINK 23.1 Two views of the Milky Way: a wide-angle visible-light mosaic (upper) and a close-up infrared image from the Spitzer Space Telescope (lower). (upper: Dirk Hoppe; lower: NASA/JPL-Caltech/S. Stolovy, Spitzer Science Center/Caltech)

On a clear, moonless night, away from the glare of city lights, you can often see a hazy, luminous band stretching across the sky. This band, called the Milky Way, extends all the way around the celestial sphere. The upper of the two accompanying photographs is centered on the brightest part of the Milky Way, in the constellation Sagittarius.

Galileo, the first person to view the Milky Way with a telescope, discovered that it is composed of countless dim stars. Today, we realize that the Milky Way is actually a disk tens of thousands of parsecs across containing hundreds of billions of stars—one of which is our own Sun—as well as vast quantities of gas and dust. This vast assemblage of matter—our home—is collectively called the Milky Way Galaxy.

Just as Galileo's telescope revealed aspects of the Milky Way that the naked eye could not, modern astronomers use telescopes at nonvisible wavelengths to peer through our Galaxy's obscuring dust and observe what visible-light telescopes never could. For example, the lower of the two accompanying photographs is an infrared image that shows hundreds of thousands of stars near the center of the Galaxy. As we will see, radio, infrared, and X-ray observations reveal that at the very center of the Galaxy lies a black hole with a mass of 4.3 million Suns.

Modern astronomers have also discovered that most of the Milky Way's mass is not in its stars, gas, or dust, but in a halo of *dark matter* that emits no measurable radiation. What the character of this dark matter could be remains one of the greatest open questions in astronomy and physics.

The Milky Way is just one of myriad *galaxies,* or systems of stars and interstellar matter, that are spread across the observable universe. By studying our home galaxy, the Milky Way Galaxy, we begin to explore the universe on a grand scale. Instead of focusing on individual stars, we look at the overall arrangement and history of a huge stellar community of which the Sun is a member. In this way, we gain insights into galaxies in general and prepare ourselves to ask fundamental questions about the cosmos.

Learning Goals

By reading the sections of this chapter, you will learn

23-1 How astronomers discovered the solar system's location within the Milky Way Galaxy

23-2 The shape and size of our Galaxy

23-3 How the Milky Way's spiral structure was discovered

23-4 The evidence for the existence of dark matter in our Galaxy

23-5 What causes the Milky Way's spiral arms to form and persist

23-6 How astronomers discovered a supermassive black hole at the galactic center

23-1 The Sun is located in the disk of our Galaxy, about 8000 parsecs from the galactic center

TUTORIAL 23.1 Eighteenth-century astronomers were the first to suspect that because the Milky Way completely encircles us, all the stars in the sky are part of an enormous disk of stars—the **Milky Way Galaxy.** As we learned in Section 1-4, a **galaxy** is an immense collection of stars and interstellar matter, far larger than a star cluster. We have an edge-on view from inside the pancake-like disk of our own Milky Way Galaxy, which is why the Milky Way appears as a band around the sky (Figure 23-1).

Locating the Sun Within the Galaxy: Early Attempts

But where within this disk is our own Sun? Until the twentieth century, the prevailing opinion was that the Sun and planets lie at the Galaxy's center. One of the first to come to this conclusion was the eighteenth-century English astronomer William Herschel, who discovered the planet Uranus and was a pioneering cataloger of binary star systems (see Section 17-9). Herschel's approach to determining the Sun's position within the Galaxy was to count the number of stars in each of 683 regions of the sky. He reasoned that he should see the greatest number of stars toward the Galaxy's center and a lesser number toward the Galaxy's edge.

Herschel found approximately the same density of stars all along the Milky Way. Therefore, he concluded that we are at the center of our Galaxy (Figure 23-2). In the early 1900s, the Dutch astronomer Jacobus Kapteyn came to essentially the same conclusion by analyzing the brightness and proper motions of a large number of stars. According to Kapteyn, the Milky Way is about 17 kpc (17 kiloparsecs = 17,000 parsecs, or 55,000 light-years) in diameter, with the Sun near its center.

The Problem: Interstellar Extinction

Both Herschel and Kapteyn were wrong about the Sun being at the center of our Galaxy. The reason for their mistake was discerned in 1930 by Robert J. Trumpler of Lick Observatory. While studying star clusters, Trumpler discovered that the more remote clusters appear unusually dim—more so than would be expected from their distances alone. As a result, Trumpler concluded that interstellar space must not be a perfect vacuum: It must contain dust that absorbs or scatters light from distant stars.

Like the stars themselves, interstellar dust is concentrated in the plane of the Galaxy (see Section 18-2). As a result, it obscures our view within the plane and makes distant objects appear dim, an effect called **interstellar extinction.** Great patches of interstellar dust are clearly visible in wide-angle photographs such as the one that opens this chapter. Thanks to interstellar extinction, Herschel and Kapteyn were actually seeing only the nearest stars in the Galaxy. Hence, they had no idea of either the enormous size of the Galaxy or of the vast number of stars concentrated around the galactic center.

ANALOGY Herschel and Kapteyn faced much the same dilemma as a lost motorist on a foggy night. Unable to see more than a city block in any direction, he would have a hard time deciding what part of town he was in. If the fog layer were relatively shallow, however, our motorist would be able to see the lights from tall buildings that extend above the fog, and in that way he could determine his location (Figure 23-3*a*).

The same principle applies to our Galaxy. While interstellar dust in the plane of our Galaxy hides the sky covered by the Milky Way, we have an almost unobscured view out of the plane (that is, to either side of the Milky Way). To find our location in

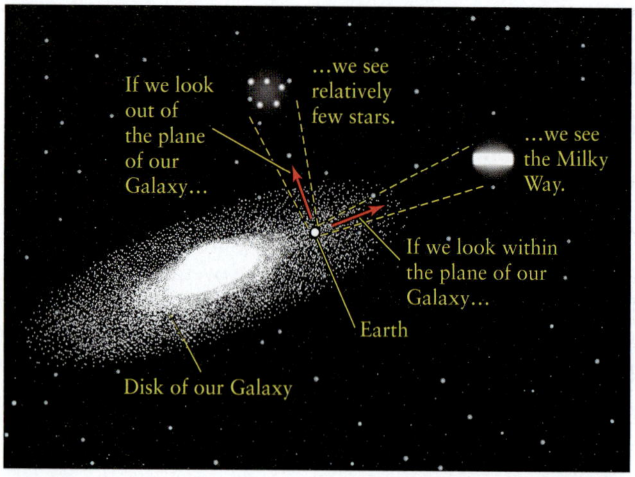

(a)

(b) RIVUXG

Figure 23-1

Our View of the Milky Way (a) The Milky Way Galaxy is a disk-shaped collection of stars. When we look out at the night sky in the plane of the disk, the stars appear as a band of light that stretches all the way around the sky. When we look perpendicular to the plane of the Galaxy, we see only those relatively few stars that lie between us and the "top" or "bottom" of the disk. **(b)** This wide-angle photograph shows a 180° view of the Milky Way centered on the constellation Sagittarius (compare with the photograph that opens this chapter). The dark streaks across the Milky Way are due to interstellar dust in the plane of our Galaxy. (Courtesy of Dennis di Cicco)

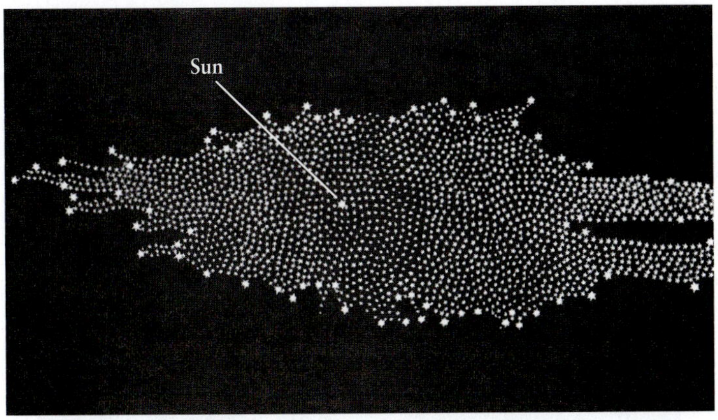

Figure 23-2

Herschel's Map of Our Galaxy In a paper published in 1785, the English astronomer William Herschel presented this map of the Milky Way Galaxy. He determined the Galaxy's shape by counting the numbers of stars in various parts of the sky. Herschel's conclusions were flawed because interstellar dust blocked his view of distant stars, leading him to the erroneous idea that the Sun is at the center of the Galaxy. (Yerkes Observatory)

the Galaxy, we need to locate bright objects that are part of the Galaxy but lie outside its plane in unobscured regions of the sky.

The Breakthrough: Globular Clusters and Variable Stars

WEB LINK 23.2

Fortunately, bright objects of the sort we need do in fact exist. They are the **globular clusters,** a class of star clusters associated with the Galaxy but which lie outside its plane (Figure 23-3*b*). As we saw in Section 19-4, a typical globular cluster is a spherical distribution of roughly 10^6 stars packed in a volume only a few hundred light-years across (see Figure 19-12).

However, to use globular clusters to determine our location in the Galaxy, we must first determine the distances from Earth to these clusters. (Think again of our lost motorist—glimpsing the lights of a skyscraper through the fog may be useful to the motorist, but only if he can tell how far away the skyscraper is.) Pulsating variable stars in globular clusters provide the distances, giving astronomers the key to the dimensions of our Galaxy.

Observations of pulsating variable stars revealed the immense size of the Milky Way

In 1912, the American astronomer Henrietta Leavitt reported her important discovery of the period-luminosity relation for Cepheid variables. As we saw in Section 19-6, Cepheid variables are pulsating stars that vary periodically in brightness (see Figure 19-18). Leavitt studied numerous Cepheids in the Small Magellanic Cloud (a small galaxy near the Milky Way) and found their periods to be directly related to their average luminosities. Figure 23-4 shows that the longer a Cepheid's period, the greater its average luminosity.

The period-luminosity law is an important tool in astronomy because it can be used to determine distances. For example, suppose you find a Cepheid variable in the sky. By measuring its period and using a graph like Figure 23-4, you can determine the star's average luminosity. Knowing the star's average luminosity, you can find out how far away the star must be in order to give the observed brightness. (Box 17-2 explains how this is done.)

Shortly after Leavitt's discovery of the period-luminosity law, Harlow Shapley, a young astronomer at the Mount Wilson Observatory in California, began studying a family of pulsating stars closely related to Cepheid variables called **RR Lyrae variables.** The light curve of an RR Lyrae variable is similar to that of a

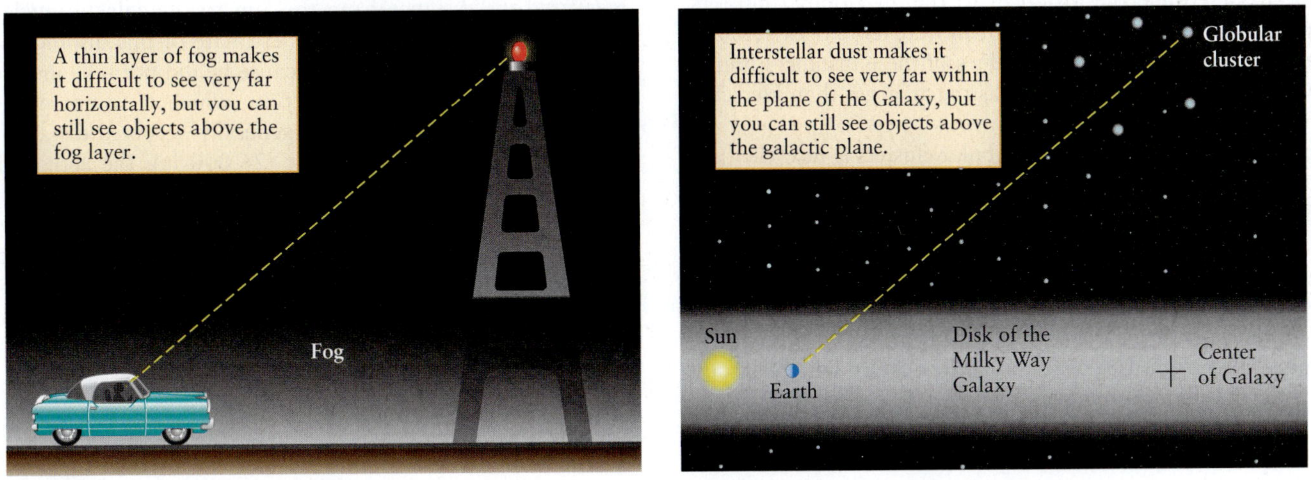

(a) Determining your position in the fog

(b) Determining your position in the Galaxy

Figure 23-3

Finding the Center of the Galaxy (a) A motorist lost on a foggy night can determine his location by looking for tall buildings that extend above the fog. **(b)** In the same way, astronomers determine our location in the Galaxy by observing globular clusters that are part of the Galaxy but lie outside the obscuring material in the galactic disk. The globular clusters form a spherical halo centered on the center of the Galaxy.

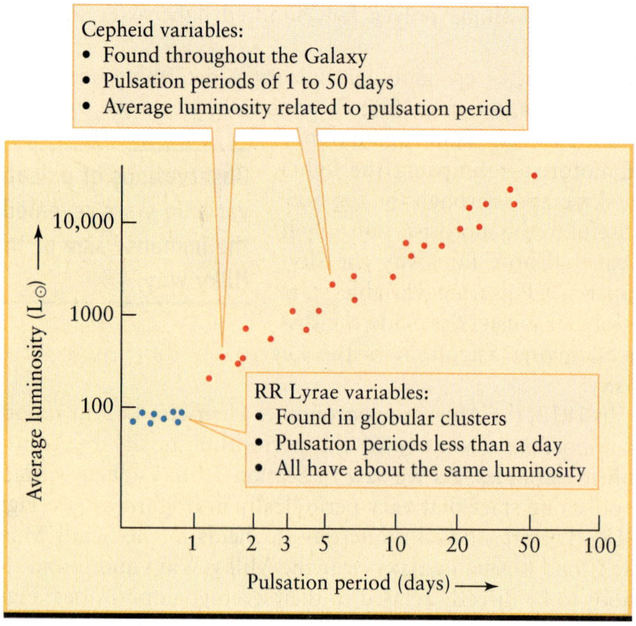

Figure 23-4

Period and Luminosity for Cepheid and RR Lyrae Variables This graph shows the relationship between period and average luminosity for Cepheid variables and RR Lyrae variables. Cepheids come in a broad range of luminosities: The more luminous the Cepheid, the longer its pulsation period. By contrast, RR Lyrae variables are horizontal branch stars that all have roughly the same average luminosity of about 100 $L_\odot$.

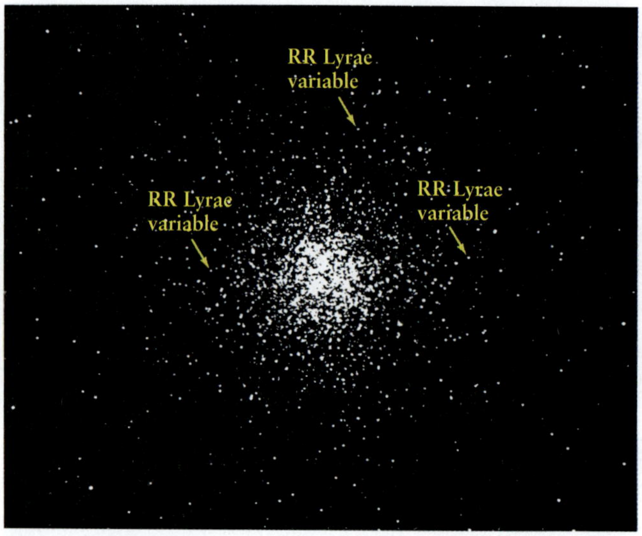

WEB LINK 23.3 VIDEO 23.1

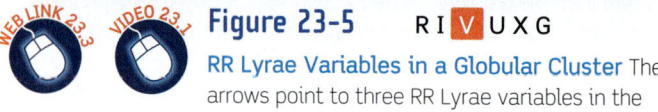

Figure 23-5 RIVUXG

RR Lyrae Variables in a Globular Cluster The arrows point to three RR Lyrae variables in the globular cluster M55, located in the constellation Sagittarius. From the average apparent brightness (as seen in this photograph) and average luminosity (known to be roughly 100 $L_\odot$) of these variable stars, astronomers have deduced that the distance to M55 is 6500 pc (20,000 ly). (Harvard Observatory)

Cepheid, but RR Lyrae variables have shorter pulsation periods and lower peak luminosities (see Figure 23-4).

The tremendous importance of RR Lyrae variables is that they are commonly found in globular clusters (Figure 23-5). By using the period-luminosity relationship for these stars, Shapley was able to determine the distances to the 93 globular clusters then known. He found that some of them were more than 100,000 light-years from Earth. The large values of these distances immediately suggested that the Galaxy was much larger than Herschel or Kapteyn had thought.

Another striking property of globular clusters is how they are distributed across the sky. Ordinary stars and open clusters of stars (see Section 18-6, especially Figures 18-18 and 18-19) are rather uniformly spread along the Milky Way. However, the majority of the 93 globular clusters that Shapley studied are located in one-half of the sky, widely scattered around the portion of the Milky Way that is in the constellation Sagittarius.

From the directions to the globular clusters and their distances from us, Shapley mapped out the three-dimensional distribution of these clusters in space. In 1920 he concluded that the globular clusters form a huge spherical distribution centered not on Earth but rather about a point in the Milky Way several kiloparsecs away in the direction of Sagittarius (see Figure 23-3*b*). This point, reasoned Shapley, must coincide with the center of our Galaxy, because of gravitational forces between the disk of the Galaxy and the "halo" of globular clusters. Therefore, by locating the center of the distribution of globular clusters, Shapley was in effect measuring the location of the galactic center.

Modern-Day Measurements

Since Shapley's pioneering observations, many astronomers have measured the distance from the Sun to the **galactic nucleus,** the center of our Galaxy. Shapley's estimate of this distance was too large by about a factor of 2, because he did not take into account the effects of interstellar extinction (which were not well understood at the time). Today, the generally accepted distance to the galactic nucleus is about 8 kpc (26,000 ly); the actual distance could be greater or less than that value by about 1 kpc (3300 ly).

Just as Copernicus and Galileo showed that Earth was not at the center of the solar system, Shapley and his successors showed that the solar system lies nowhere near the center of the Galaxy. We see that Earth indeed occupies no special position in the universe.

23-2 Observations at nonvisible wavelengths reveal the shape of the Galaxy

At visible wavelengths, light suffers so much interstellar extinction that the galactic nucleus is totally obscured from view. But the amount of interstellar extinction is roughly inversely proportional to wavelength. In other words, the longer the wavelength, the farther radiation can travel through interstellar dust without being scattered or absorbed. As a result, we can see farther into the plane of the Milky Way at infrared wavelengths than at visible wavelengths, and radio waves can travel all the way through the Galaxy. For this reason, telescopes sensitive to these nonvisible wavelengths are important tools for studying the structure of our Galaxy.

WEB LINK 23.4

Exploring the Milky Way in the Infrared

Infrared light is particularly useful for tracing the location of interstellar dust in the Galaxy. Starlight warms the dust grains to temperatures in the range of 10 to 90 K; thus, in accordance with Wien's law (see Section 5-4), the dust emits radiation predominately at wavelengths from about 30 to 300 μm. These are called **far-infrared** wavelengths, because they lie in the part of the infrared spectrum most different in wavelength from visible light (see Figure 5-7). At these wavelengths, interstellar dust radiates more strongly than stars, so a far-infrared view of the sky is principally a view of where the dust is. In 1983 the Infrared Astronomical Satellite (IRAS) scanned the sky with a 60-cm reflecting telescope at far-infrared wavelengths, giving the panoramic view of the Milky Way's dust shown in Figure 23-6*a*.

Our Galaxy's dust and stars—including the Sun—lie mostly in a relatively thin disk

In 1990 an instrument on the Cosmic Background Explorer (COBE) satellite scanned the sky at **near-infrared** wavelengths, that is, relatively short wavelengths closer to the visible spectrum. Figure 23-6*b* shows the resulting near-infrared view of the plane of the Milky Way. At near-infrared wavelengths, interstellar dust does not emit very much light. Hence, the light sources in Figure 23-6*b* are stars, which do emit strongly in the near-infrared (especially the cool stars, such as red giants and supergiants). Because interstellar dust causes little interstellar extinction in the near-infrared, many of the stars whose light is recorded in Figure 23-6*b* lie deep within the Milky Way.

Observations such as those shown in Figure 23-6, along with the known distance to the center of the Galaxy, have helped astronomers to establish the dimensions of the Galaxy. The **disk** of our Galaxy is about 50 kpc (160,000 ly) in diameter and about 0.6 kpc (2000 ly) thick, as shown in Figure 23-7. The center of the Galaxy is surrounded by a distribution of stars, called the **central bulge,** which is about 2 kpc (6500 ly) in diameter. This central bulge is clearly visible in Figure 23-6*b*. The spherical distribution of globular clusters traces the **halo** of the Galaxy.

This structure is not unique to our Milky Way Galaxy. Figure 23-8 shows another galaxy whose dust and stars lie in a disk and that has a central bulge of stars, just like the Milky Way.

Dust lies mostly in the plane of the Galaxy (seen edge-on)

(a) Infrared emission from dust at wavelengths of 25, 60, and 100 μm

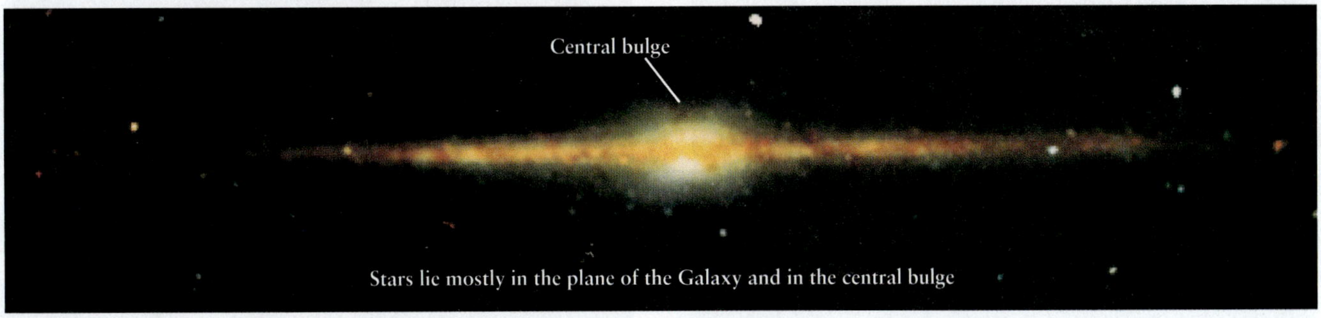
Central bulge
Stars lie mostly in the plane of the Galaxy and in the central bulge

(b) Infrared emission from dust at wavelengths of 1.2, 2.2, and 3.4 μm

WEB LINK 23.5

Figure 23-6 R I V U X G

The Infrared Milky Way **(a)** This view was constructed from observations made at far-infrared wavelengths by the IRAS spacecraft. Interstellar dust, which is mostly confined to the plane of the Galaxy, is the principal source of radiation in this wavelength range. **(b)** Observing at near-infrared wavelengths, as in this composite of COBE data, allows us to see much farther through interstellar dust than we can at visible wavelengths. Light in this wavelength range comes mostly from stars in the plane of the Galaxy and in the bulge at the Galaxy's center. (NASA)

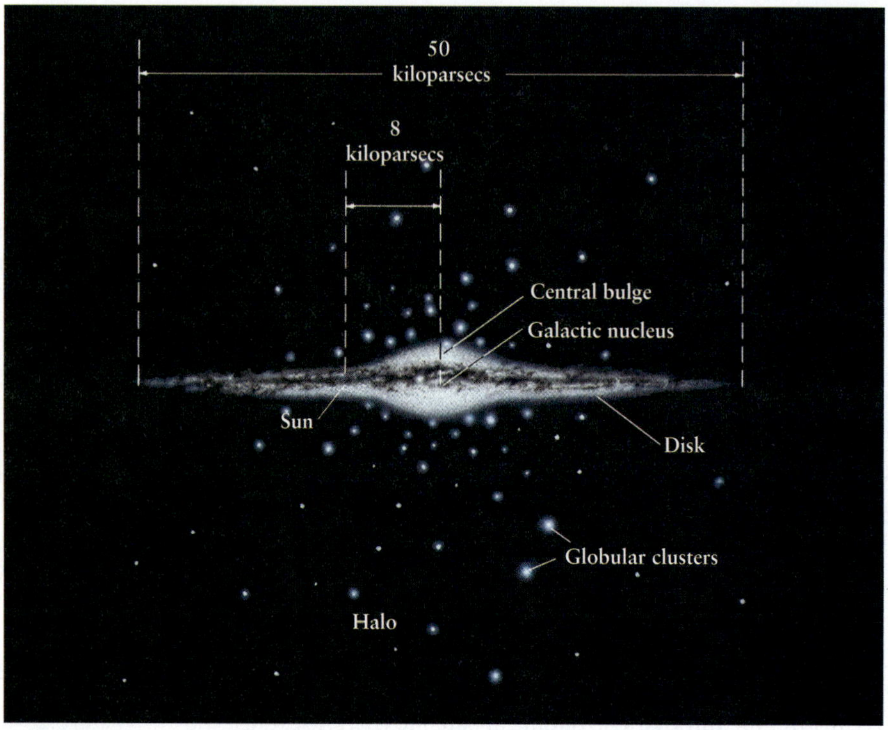

Figure 23-7

Our Galaxy (Schematic Edge-on View) There are three major components of our Galaxy: a disk, a central bulge, and a halo. The disk contains gas and dust along with metal-rich (Population I) stars. The halo is composed almost exclusively of old, metal-poor (Population II) stars. The central bulge is a mixture of Population I and Population II stars.

In the same way that our Sun is a rather ordinary member of the stellar community that makes up the Milky Way, the Milky Way turns out to be a rather common variety of galaxy.

The Milky Way's Distinct Stellar Populations

It is estimated that our Galaxy contains about 200 billion (2×10^{11}) stars. Remarkably, different kinds of stars are found in the various components of the Galaxy. The globular clusters in the halo are composed of old, metal-poor, Population II stars (see Section 19-5). Although these clusters are conspicuous, they contain only about 1% of the total number of stars in the halo. Most halo stars are single Population II stars in isolation, called **high-velocity stars** because of their high speed relative to the Sun. These ancient stars orbit the Galaxy along paths tilted at random

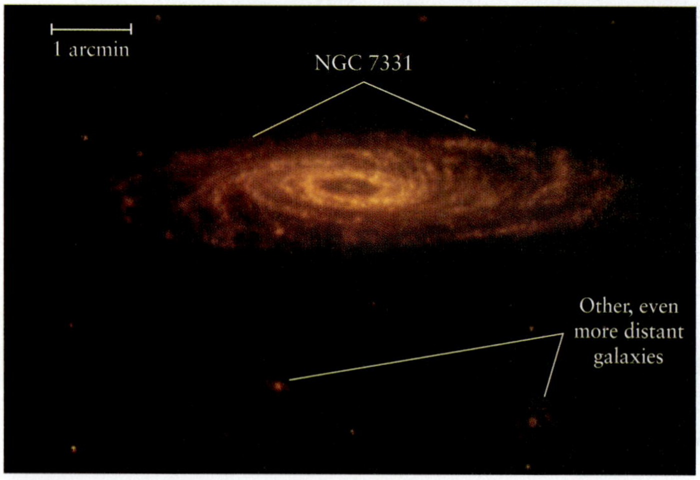

(a) Infrared emission from dust in NGC 7331 at 5.8 and 8.0 μm

(b) Infrared emission from stars in NGC 7331 at 3.6 and 4.5 μm

WEB LINK 23.6

Figure 23-8 R I V U X G

NGC 7331: A Near-Twin of the Milky Way If we could view our Galaxy from a great distance, it would probably look like this galaxy in the constellation Pegasus. As in Figure 23-6, the far-infrared image **(a)** reveals the presence of dust in the galaxy's plane, while the near-infrared image **(b)** shows the distribution of stars. These images of NGC 7331, which is about 15 million pc (50 million ly) from Earth, were made with the Spitzer Space Telescope (from the Spitzer Space Telescope (see Section 6-7, especially Figure 6-26). (NASA; JPL-Caltech; M. Regan (STScI); and the SINGS Team)

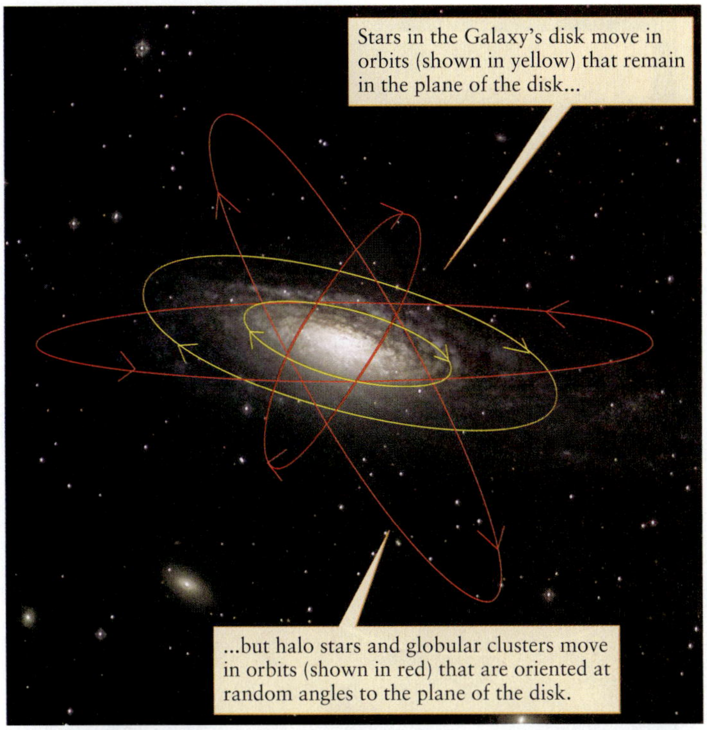

Figure 23-9 R I V U X G

Star Orbits in the Milky Way The different populations of stars in our Galaxy travel along different sorts of orbits. The galaxy in this visible-light image is the Milky Way's near-twin NGC 7331, the same galaxy shown at infrared wavelengths in Figure 23-8. (Daniel Bramich (ING) and Nik Szymanek)

angles to the disk of the Milky Way, as do the globular clusters. By contrast, stars in the disk travel along orbits that remain in the disk (Figure 23-9).

Unlike the halo, the stars in the disk are mostly young, metal-rich, Population I stars like the Sun. The disk of a galaxy like the Milky Way appears bluish because its light is dominated by radiation from hot O and B main-sequence stars. Such stars have very short main-sequence lifetimes (see Section 19-1, especially Table 19-1), so they must be quite young by astronomical standards. Hence, their presence shows that there must be active star formation in the galactic disk. By contrast, no O or B stars are present in the halo, which implies that star formation ceased there long ago.

The central bulge contains both Population I stars and metal-poor Population II stars. Since Population II stars are thought to have formed early in the history of the universe, this suggests that some of the stars in the bulge are quite ancient while others were created more recently. The central bulge looks yellowish or reddish because it contains many red giants and red supergiants (see Figure 1-7), but does *not* contain luminous, short-lived, blue O or B stars. Hence, there cannot be ongoing star formation in the central bulge. The same is true for other galaxies whose structure is similar to that of the Milky Way (Figure 23-10).

Why are there such different populations of stars in the halo, disk, and central bulge? Why has star formation stopped in some regions of the Galaxy but continues in other regions? The answers to these questions are related to the way that stars, as well as the gas and dust from which stars form, move within the Galaxy.

Only Population I (metal-rich) stars are found in the disk. The presence of hot, blue, young O and B stars indicates active star formation.
Both Population I (metal-rich) and Population II (metal-poor) stars are found in the central bulge — but there are *no* young blue stars. Hence star formation is not active there.

Figure 23-10 R I V U X G

Stellar Populations: Disk Versus Central Bulge The disk and central bulge of the Milky Way contain rather different populations of stars. The same is true for the galaxy NGC 1309, which has a similar structure to the Milky Way Galaxy and happens to be oriented face-on to us. NGC 1309 lies about 30 million pc (100 million ly) from us in the constellation Eridanus. (NASA; The Hubble Heritage Team; and A. Riess, STScI)

23-3 Observations of cold hydrogen clouds and star-forming regions reveal that our Galaxy has spiral arms

The galaxies shown in Figure 23-8 and Figure 23-10 both have **spiral arms,** spiral-shaped concentrations of gas and dust that extend outward from the center in a shape reminiscent of a pinwheel. Other galaxies' spiral arms would lead us to suspect that our own Milky Way Galaxy also has spiral arms. However, because interstellar dust obscures our visible-light view in the pancake-like plane of our Galaxy, a detailed understanding of the structure of our galactic disk had to wait until the development of radio astronomy. Thanks to their long wavelengths, radio waves can penetrate the interstellar medium even more easily than infrared light and can travel without being scattered or absorbed. As we shall see in this section, both radio and optical observations reveal that our Galaxy does indeed have spiral arms.

Mapping Hydrogen in the Milky Way

Hydrogen is by far the most abundant element in the universe (see Figure 8-4 in Section 8-2). Hence, by looking for concentrations of hydrogen gas, we should be able to detect important clues about the distribution of matter in our Galaxy. Unfortunately, ordinary visible-light telescopes are of little use in this quest, because hydrogen atoms can only emit visible light if they are first excited to high energy levels (see Section 5-8, and especially Figure 5-24). This excitation is quite unlikely to occur in the cold depths of interstellar space. Furthermore, even if there are some hydrogen atoms that glow strongly at visible wavelengths, interstellar extinction due to dust (see Section 23-1) would make it impossible to see this glow from distant parts of the Galaxy.

The magnetic nature of subatomic particles makes it possible to detect cold clouds of gas strewn throughout the galactic disk

What makes it possible to map out the distribution of hydrogen in our Galaxy is that even cold hydrogen clouds emit *radio* waves. As we saw in Section 23-2, radio waves can easily penetrate the interstellar medium, so we can detect the radio emission from such cold clouds no matter where in the Galaxy they lie. The hydrogen in these clouds is neutral—that is, not ionized—and is called **H I.** (This distinguishes it from ionized hydrogen, which is designated H II.) To understand how H I clouds can emit radio waves, we must probe a bit more deeply into the structure of protons and electrons, the particles of which hydrogen atoms are made.

In addition to having mass and charge, particles such as protons and electrons possess a tiny amount of angular momentum (that is, rotational motion) commonly called **spin.** Very roughly, you can visualize a proton or electron as a tiny, electrically charged sphere that spins on its axis. Because electric charges in motion generate magnetic fields, a proton or electron behaves like a tiny magnet with a north pole and a south pole (Figure 23-11).

If you have ever played with magnets, you know that two magnets attract when the north pole of one magnet is next to the south pole of the other and repel when two like poles (both north or both south) are next to each other (Figure 23-11*a*). In other words, the energy of the two magnets is least when opposite poles are together and highest when like poles are together. Hence, as shown in Figure 23-11*b,* the energy of a hydrogen atom is slightly different depending on whether the spins of the proton and electron are in the same direction or opposite directions. (According to the laws of quantum mechanics, these are the only two possibilities; the spins cannot be at random angles.)

If the spin of the electron changes its orientation from the higher-energy configuration to the lower-energy one—called a

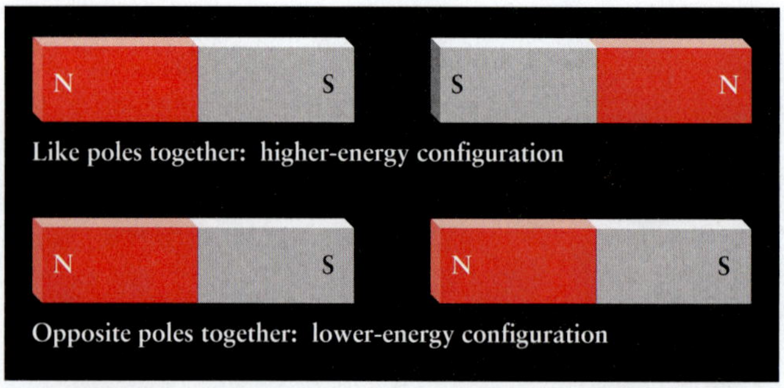

(a) The magnetic energy of two bar magnets depends on their relative orientation

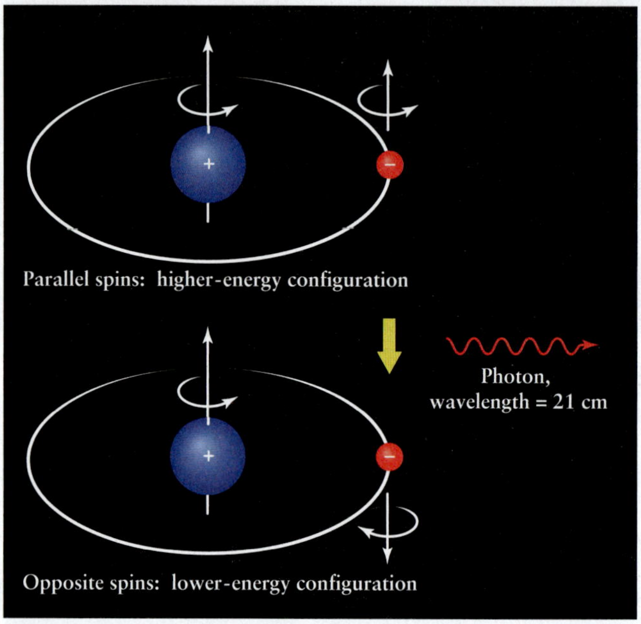

(b) The magnetic energy of a proton and electron depends on their relative spin orientation

Figure 23-11

Magnetic Interactions in the Hydrogen Atom **(a)** The energy of a pair of magnets is high when their north poles or their south poles are near each other, and low when they have opposite poles near each other. **(b)** Thanks to their spin, electrons and protons are both tiny magnets. When the electron flips from the higher-energy configuration (with its spin in the same direction as the proton's spin) to the lower-energy configuration (with its spin opposite to the proton's spin), the atom loses a tiny amount of energy and emits a radio photon with a wavelength of 21 cm.

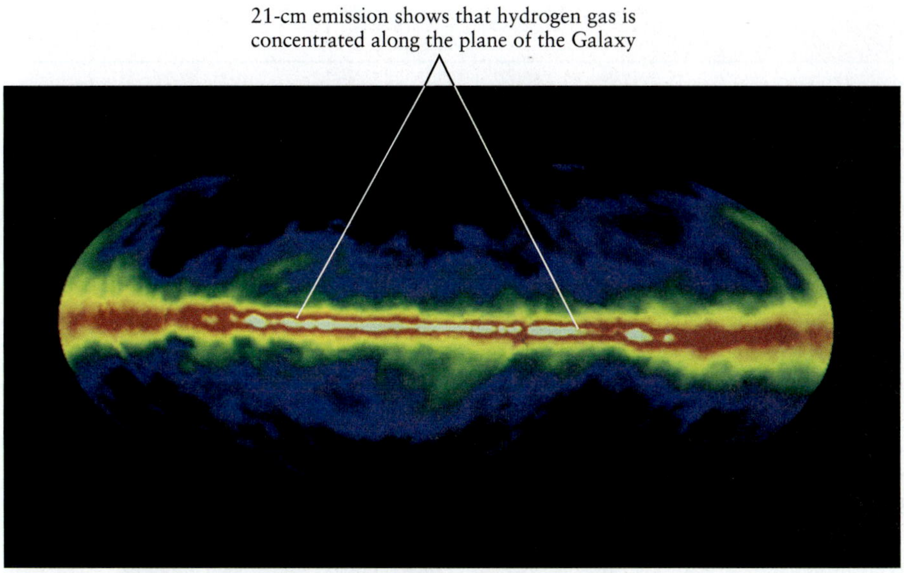
21-cm emission shows that hydrogen gas is concentrated along the plane of the Galaxy

Figure 23-12 R I V U X G

The Sky at 21 Centimeters This image was made by mapping the sky with radio telescopes tuned to the 21-cm wavelength emitted by neutral interstellar hydrogen (H I). The entire sky has been mapped onto an oval, and the plane of the Galaxy extends horizontally across the image as in Figure 23-6. Black and blue represent the weakest emission, and red and white the strongest. (Courtesy of C. Jones and W. Forman, Harvard-Smithsonian Center for Astrophysics)

spin-flip transition—a photon is emitted. The energy difference between the two spin configurations is very small, only about 10^{-6} as great as those between different electron orbits (see Figure 5-24). Therefore, the photon emitted in a spin-flip transition between these configurations has only a small energy, and thus its wavelength is a relatively long 21 cm—a radio wavelength.

The spin-flip transition in neutral hydrogen was first predicted in 1944 by the Dutch astronomer Hendrik van de Hulst. His calculations suggested that it should be possible to detect the **21-cm radio emission** from interstellar hydrogen, although a very sensitive radio telescope would be required. In 1951, Harold Ewen and Edward Purcell at Harvard University first succeeded in detecting this faint emission from hydrogen between the stars.

Figure 23-12 shows the results of a more recent 21-cm survey of the entire sky. Neutral hydrogen gas (H I) in the plane of the Milky Way stands out prominently as a bright band across the middle of this image.

LOOKING DEEPER 23.2 The distribution of gas in the Milky Way is not uniform but is actually quite frothy. In fact, our Sun lies near the edge of an irregularly shaped region within which the interstellar medium is very thin but at very high temperatures (about 10^6 K). This region, called the **Local Bubble,** is several hundred parsecs across. The Local Bubble may have been carved out by a supernova that exploded nearby some 300,000 years ago.

Remarkably, spin-flip transitions are used not only to map our Galaxy but also to map the internal structure of the human body. Box 23-1 on page 614 discusses this application, called magnetic resonance imaging.

Detecting Our Galaxy's Spiral Arms

The detection of 21-cm radio emission was a major breakthrough that permitted astronomers to reveal the presence of spiral arms in the galactic disk. Figure 23-13 shows how spiral arms were detected. Suppose that you aim a radio telescope along a particular line of sight across the Galaxy. Your radio receiver, located at *S* (the position of the solar system), picks up 21-cm emission from H I clouds at points 1, 2, 3, and 4. However, the radio waves from these various clouds are Doppler shifted by slightly different amounts, because the clouds are moving at different speeds as they travel with the rotating Galaxy.

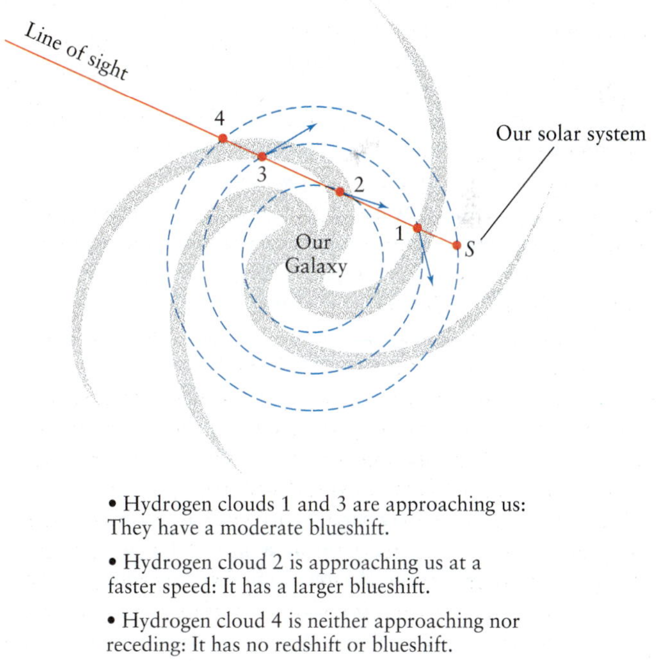
Line of sight
4
3
2
1
S
Our solar system
Our Galaxy
• Hydrogen clouds 1 and 3 are approaching us: They have a moderate blueshift.
• Hydrogen cloud 2 is approaching us at a faster speed: It has a larger blueshift.
• Hydrogen cloud 4 is neither approaching nor receding: It has no redshift or blueshift.

Figure 23-13

A Technique for Mapping Our Galaxy If we look within the plane of our Galaxy from our position at *S*, hydrogen clouds at different locations (shown as 1, 2, 3, and 4) along our line of sight are moving at slightly different speeds relative to us. As a result, radio waves from these various gas clouds are subjected to slightly different Doppler shifts. This permits radio astronomers to sort out the gas clouds and thus map the Galaxy.

BOX 23-1 Astronomy Down to Earth

Spin-Flip Transitions in Medicine

WEB LINK 23.7

Thanks to their spin, protons and electrons act like microscopic bar magnets. In a hydrogen atom, the interaction between the magnetism of the electron and that of the proton gives rise to the 21-cm radio emission. But these particles can also interact with outside magnetic fields, such as that produced by a large electromagnet. This physical principle is behind **magnetic resonance imaging (MRI)**, an important diagnostic tool of modern medicine.

Much of the human body is made of water. Every water molecule has two hydrogen atoms, each of which has a nucleus made of a single proton. If a person's body is placed in a strong magnetic field, the spins of the protons in the hydrogen atoms of their body can either be in the same direction as the field ("aligned") or in the direction opposite to the field ("opposed"). The aligned orientation has lower energy, and therefore the majority of protons end up with their spins in this orientation. But if a radio wave of just the right wavelength is now sent through the person's body, an aligned proton can absorb a radio photon and flip its spin into the higher-energy, opposed orientation. How much of the radio wave is absorbed depends on the number of protons in the body, which in turns depends on how much water (and, thus, how much water-containing tissue) is in the body.

In magnetic resonance imaging, a magnetic field is used whose strength varies from place to place. The difference in energy between the opposed and aligned orientations of a proton depends on the strength of the magnetic field, so radio waves will only be absorbed at places where this energy difference is equal to the energy of a radio photon. (This equality is called *resonance,* which is how magnetic resonance imaging gets its name.) By varying the magnetic field strength over the body and the wavelength of the radio waves, and by measuring how much of the radio wave is absorbed by different parts of the body, it is possible to map out the body's tissues. The accompanying false-color image shows such a map of a patient's head.

Unlike X-ray images, which show only the densest parts of the body, such as bones and teeth, magnetic resonance imaging can be used to view less dense (but water-containing) soft tissue. Just as the 21-cm radio emission has given astronomers a clear view of what were hidden regions of our Galaxy, magnetic resonance imaging allows modern medicine to see otherwise invisible parts of the human body.

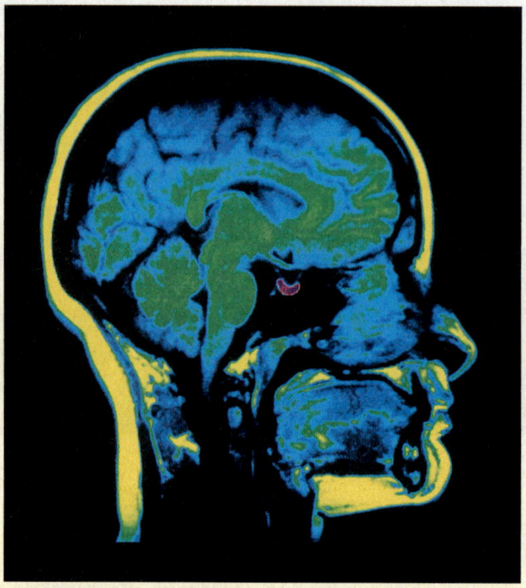

R I V U X G

(Scott Camazine, Photo Researchers)

It is important to remember that the Doppler shift reveals only motion along the line of sight (review Figure 5-26). In Figure 23-13, cloud 2 has the highest speed along our line of sight, because it is moving directly toward us. Consequently, the radio waves from cloud 2 exhibit a larger Doppler shift than those from the other three clouds along our line of sight. Because clouds 1 and 3 are at the same distance from the galactic center, they have the same orbital speed. The fraction of their velocity parallel to our line of sight is also the same, so their radio waves exhibit the same Doppler shift, which is less than the Doppler shift of cloud 2. Finally, cloud 4 is the same distance from the galactic center as the Sun. This cloud is thus orbiting the Galaxy at the same speed as the Sun, resulting in no net motion along the line of sight. Radio waves from cloud 4, as well as from hydrogen gas near the Sun, are not Doppler shifted at all.

These various Doppler shifts cause radio waves from gases in different parts of the Galaxy to arrive at our radio telescopes with wavelengths slightly different from 21 cm. It is therefore possible to sort out the various gas clouds and thus produce a map of the Galaxy like that shown in **Figure 23-14**.

Figure 23-14 shows that neutral hydrogen gas is not spread uniformly around the disk of the Galaxy but is concentrated into numerous arched lanes. Similar features are seen in other galaxies beyond the Milky Way. As an example, the galaxy in **Figure 23-15*a*** has prominent spiral arms outlined by hot, luminous, blue main-sequence stars and the red emission nebulae (H II regions) found near many such stars. Stars of this sort are very short-lived, so these features indicate that spiral arms are sites of active, ongoing star formation. The 21-cm radio image of this same galaxy, shown in Figure 23-15*b*, shows that spiral arms are also regions where neutral hydrogen gas is concentrated, similar to the structures in our own Galaxy visible in Figure 23-14. This similarity is a strong indication that our Galaxy also has spiral arms.

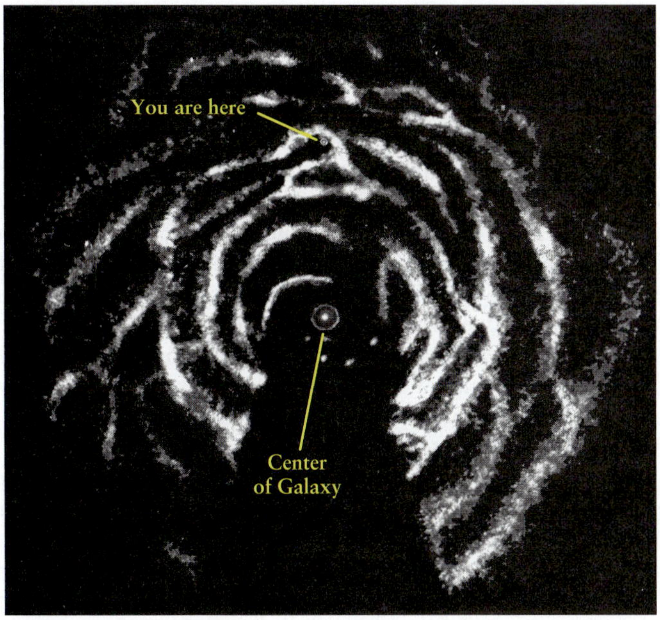

Figure 23-14 R I V U X G

A Map of Neutral Hydrogen in Our Galaxy This map, constructed from radio-telescope surveys of 21-cm radiation, shows the distribution of hydrogen gas in a face-on view of our Galaxy. The map suggests a spiral structure. Details in the blank, wedge-shaped region at the bottom of the map are unknown. Gas in this part of the Galaxy is moving perpendicular to our line of sight and thus does not exhibit a detectable Doppler shift. (Courtesy of G. Westerhout)

CAUTION Photographs such as Figure 23-15*a* can lead to the impression that there are very few stars between the spiral arms of a galaxy. Nothing could be further from the truth! In fact, stars are distributed rather uniformly throughout the disk of a galaxy like the one in Figure 23-15*a;* the density of stars in the spiral arms is only about 5% higher than in the rest of the disk. The spiral arms stand out nonetheless because they are where hot, blue O and B stars are found. One such star is about 10^4 times more luminous than an average star in the disk, so the light from O and B stars completely dominates the visible appearance of a spiral galaxy. An infrared image such as Figure 23-15*c* gives a better impression of how stars of all kinds are distributed through a spiral galaxy's disk.

Mapping the Spiral Arms and the Central Bulge

Figure 23-15*a* suggests that we can confirm the presence of spiral structure in our own Galaxy by mapping the locations of star-forming regions. Such regions are marked by OB associations, H II regions, and molecular clouds (see Section 18-7). Unfortunately, the first two of these are best observed using visible light, and interstellar extinction limits the range of visual observations in the plane of the Galaxy to less than 3 kpc (10,000 ly) from Earth. But there are enough OB associations and H II regions within this range to plot the spiral arms in the vicinity of the Sun.

Molecular clouds are easier to observe at great distances, because molecules of carbon monoxide (CO) in these clouds emit radio waves that are relatively unaffected by interstellar extinction. Hence, the positions of molecular clouds have been plotted even in remote regions of the Galaxy, as Figure 18-21 shows. (We saw in Section 18-7 that CO molecules in molecular clouds emit more strongly than the hydrogen atoms do, even though hydrogen is the principal constituent of these clouds.)

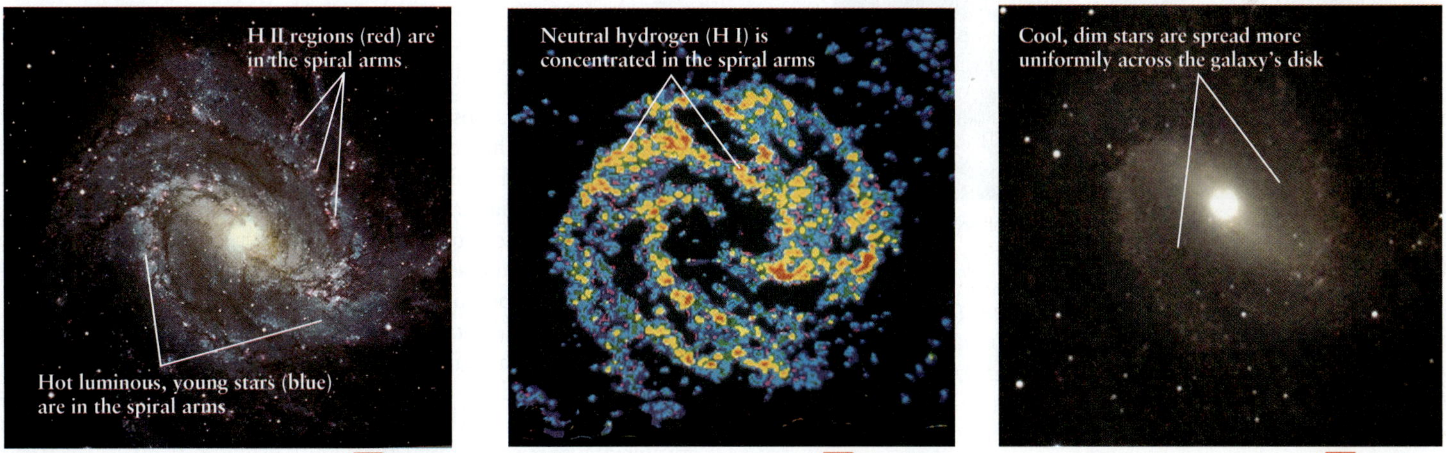

(a) Visible-light view of M83 R I V U X G

(b) 21-cm radio view of M83 R I V U X G

(c) Near-infrared view of M83 R I V U X G

WEB LINK 23.8

Figure 23-15

A Spiral Galaxy The galaxy M83 lies in the southern constellation Hydra about 5 million pc (15 million ly) from Earth. **(a)** This visible-light image clearly shows the spiral arms. The presence of young stars and H II regions indicates that star formation takes place in spiral arms. **(b)** This radio view at a wavelength of 21 cm shows the emission from neutral interstellar hydrogen gas (H I). Note that essentially the same pattern of spiral arms is traced out in this image as in the visible-light photograph. **(c)** M83 has a much smoother appearance in this near-infrared view. This shows that cooler stars, which emit strongly in the infrared, are spread more uniformly across the galaxy's disk. Note the elongated bar shape of the central bulge. (a: Anglo-Australian Observatory; b: VLA, NRAO; c: S. Van Dyk/IPAC)

(a) The structure of the Milky Way's disk

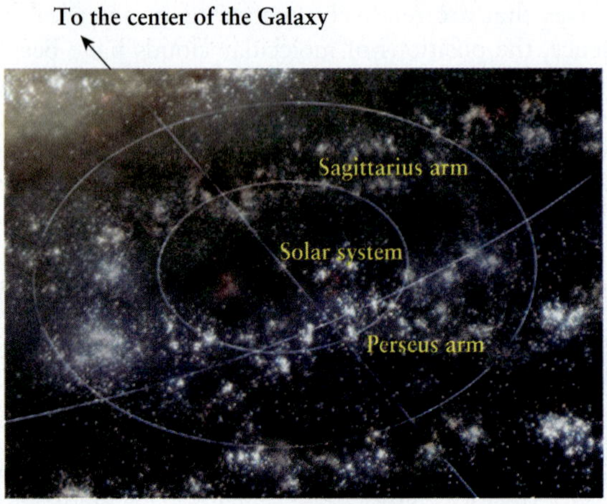

(b) Closeup of the Sun's galactic neighborhood

Figure 23-16

Our Galaxy Seen Face-on: Artist's Impressions (a) The Galaxy's diameter is about 50,000 pc (160,000 ly), and our solar system is about 8000 pc (26,000 ly) from the galactic center. The elongated central bulge is about 8300 pc (27,000 ly) long and is oriented at approximately 45° to a line running from the solar system to the galactic center. **(b)** Our solar system is located between the Sagittarius and Perseus arms, two of the major spiral arms in the Milky Way. (a: NASA/JPL-Caltech/R. Hurt, SSC; b: National Geographic)

Taken together, all these observations demonstrate that our Galaxy has at least four major spiral arms as well as several short arm segments (Figure 23-16). The Sun is located just outside a relatively short arm segment called the Orion arm, which includes the Orion Nebula and neighboring sites of vigorous star formation in that constellation.

Two major spiral arms border either side of the Sun's position. The Sagittarius arm is on the side toward the galactic center. You see this arm on June and July nights when you look at the portion of the Milky Way stretching across Scorpius and Sagittarius, near the center of the upper photograph on the first page of this chapter. In December and January, when our nighttime view is directed away from the galactic center, we see the Perseus arm. The other major spiral arms cannot be seen at visible wavelengths due to the obscuring effects of dust.

Figure 23-16*a* also shows that the central bulge of the Milky Way is not spherical, but is elongated like a bar. The Milky Way's bulge is unlike the bulge of the galaxy NGC 7331 shown in Figure 23-8, but similar to the bulge of the galaxy M83 shown in Figure 23-15. The elongated shape of the central bulge had been suspected since the 1980s; this shape was confirmed in 2005 using the Spitzer Space Telescope, which was used to survey the infrared emissions from some 3 million stars in the central bulge. Thus the artist's impression shown in Figure 23-16*a* is based on observations using both radio wavelengths (for the spiral arms) and infrared wavelengths (for the central bulge). We will see in Section 23-5 that this elongated shape may play a crucial role in sustaining the Galaxy's spiral structure.

Why are the young stars, star-forming regions, and clouds of neutral hydrogen in our Galaxy all found predominantly in the spiral arms? To answer this question, we must understand why spiral arms exist at all. Spiral arms are essentially cosmic "traffic jams," places where matter piles up as it orbits around the center of the Galaxy. This orbital motion, which is essential to grasping the significance of spiral arms, is our next topic as we continue our exploration of the Galaxy.

23-4 The rotation of our Galaxy reveals the presence of dark matter

The Doppler shifts observed in the disk of our Galaxy tell us that the disk rotates. This rotation shows that the stars, gas, and dust in our Galaxy are all orbiting the galactic center. Indeed, if this were not the case, mutual gravitational attraction would cause the entire Galaxy to collapse into the galactic center. In the same way, the Moon is kept from crashing into Earth and the planets from crashing into the Sun, because of their orbital motion (see Section 4-7).

Measuring the rotation of our Galaxy accurately is a difficult business. But such challenging measurements have been made, as we shall see, and the results lead to a remarkable conclusion: Most of the mass of the Galaxy is in the form of *dark matter,* a mysterious sort of material that emits no light at all.

Measuring How the Milky Way Rotates

Radio observations of 21-cm radiation from hydrogen gas provide important clues about our Galaxy's rotation. Doppler shift measurements of this radiation indicate that stars and gas all orbit in the same direction around the galactic center, just as the planets all orbit in the same direction around the Sun. Measurements also show that the orbital speed of stars and gas about the

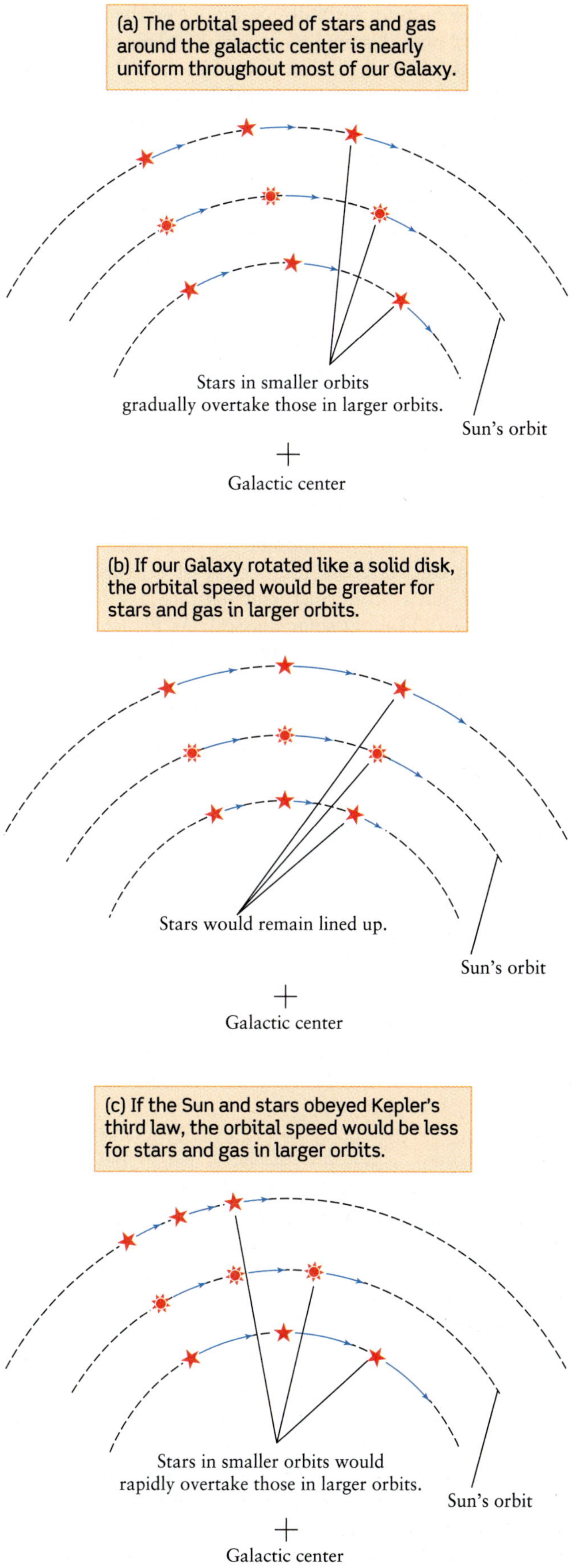
(a) The orbital speed of stars and gas around the galactic center is nearly uniform throughout most of our Galaxy.
Stars in smaller orbits gradually overtake those in larger orbits.
Sun's orbit
Galactic center
(b) If our Galaxy rotated like a solid disk, the orbital speed would be greater for stars and gas in larger orbits.
Stars would remain lined up.
Sun's orbit
Galactic center
(c) If the Sun and stars obeyed Kepler's third law, the orbital speed would be less for stars and gas in larger orbits.
Stars in smaller orbits would rapidly overtake those in larger orbits.
Sun's orbit
Galactic center

INTERACTIVE EXERCISE 23.1

Figure 23-17

The Rotation of Our Galaxy **(a)** This schematic diagram shows three stars (the Sun and two others) orbiting the center of the Galaxy. Although they start off lined up, the stars become increasingly separated as they move along their orbits. Stars inside the Sun's orbit overtake and move ahead of the Sun, while stars far from the galactic center lag behind the Sun. **(b)** The stars would remain lined up if the Galaxy rotated like a solid disk. This orientation is not what is observed. **(c)** If stars orbited the galactic center in the same way that planets orbit the Sun, stars inside the Sun's orbit would overtake us faster than they are observed to do.

galactic center is fairly uniform throughout much of the Galaxy's disk (Figure 23-17). As a result, stars inside the Sun's orbit complete a trip around the galactic center more quickly than the Sun, because the stars have a shorter distance to travel. Conversely, stars outside the Sun's orbit take longer to go once around the galactic center because they have farther to travel. As seen by Earth-based astronomers moving along with the Sun, stars inside the Sun's orbit overtake and pass us, while we overtake and pass stars outside the Sun's orbit (Figure 23-17*a*).

CAUTION Note that when we say that objects in different parts of the Galaxy orbit at the same speed, we do *not* mean that the Galaxy rotates like a solid disk. All parts of a rotating solid disk—a CD or DVD, for example—take the same time to complete one rotation. Because the outer part of the disk has to travel around a larger circle than the inner part, the speed (distance per time) is greater in the outer part (Figure 23-17*b*). By contrast, the orbital speed of material in our Galaxy is roughly the *same* at all distances from the galactic center.

The most familiar examples of orbital motion are the motions of the planets around the Sun. As we saw in Section 4-7, the farther a planet is from the Sun, the less gravitational force it experiences and the slower the speed it needs to have to remain in orbit. The same would be true for the orbits of stars and gas in the Galaxy *if* they were held in orbit by a single, massive object at the galactic center (Figure 23-17*c*). Hence, the 21-cm observations of our Galaxy, which show that the speed does *not* decrease with increasing distance from the galactic center, demonstrate that there is no such single, massive object holding objects in their galactic orbits.

Instead, what keeps a star in its orbit around the center of the Galaxy is the combined gravitational force exerted on it by *all* of the mass (including stars, gas, and dust) that lies within the star's orbit. (It turns out that the gravitational force from matter *outside* a star's orbit has little or no net effect on the star's motion around the galactic center.) This gives us a tool for determining the Galaxy's mass and how that mass is distributed.

The Sun's Orbital Motion and the Mass of the Galaxy

An important example is the orbital motion of the Sun (and the solar system) around the center of the Galaxy. If we know the semimajor axis and period of the Sun's orbit, we can use Newton's form of Kepler's third law (described in Section 4-7) to determine the mass of that portion of the Galaxy that lies within the orbit. We saw in Section 23-2 that the Sun is about 8000 pc (26,000 ly) from the galactic center. The orbit is in fact nearly circular, so we can regard 8000 pc as the radius r of the orbit. (This radius is also the semimajor axis of the Sun's orbit.) In one complete trip around the Galaxy, the Sun travels a distance equal to the circumference of its orbit, which is $2\pi r$. The time required for one orbit, or orbital period P, is equal to the distance traveled divided by the Sun's orbital speed v:

Period of the Sun's orbit around the galactic center

$$P = \frac{2\pi r}{v}$$

P = orbital period of the Sun

r = distance from the Sun to the galactic center

v = orbital speed of the Sun

Unfortunately, we cannot tell the Sun's orbital speed from 21-cm observations, since these reveal only how fast things are moving relative to the Sun. Instead, we need to measure how the Sun is moving relative to a background that is not rotating along with the rest of the Galaxy. Such a background is provided by distant galaxies beyond the Milky Way and by the globular clusters. (Since globular clusters lie outside the plane of the Galaxy, they do not take part in the rotation of the disk.) By measuring the Doppler shifts of these objects and averaging their velocities, astronomers deduce that the Sun is moving along its orbit around the galactic center at about 220 km/s—about 790,000 kilometers per hour or 490,000 miles per hour!

Using this information, we find that the Sun's orbital period is

$$P = \frac{2\pi \times 8000 \text{ pc}}{220 \text{ km/s}} \times \frac{3.09 \times 10^{13} \text{ km}}{1 \text{ pc}} = 7.1 \times 10^{15} \text{ s}$$
$$= 2.2 \times 10^{8} \text{ years}$$

Traveling at 790,000 kilometers per hour, it takes the Sun about 220 million years to complete one trip around the Galaxy. (In the 65 million years since the demise of the dinosaurs, our solar system has traveled less than a third of the way around its orbit.) The Galaxy is a very large place!

Box 23-2 shows how to combine the radius and period of the Sun's orbit to calculate the total mass of all the matter that lies inside the Sun's orbit. Such calculations give an answer of 9.0×10^{10} $M_\odot$ (90 billion solar masses). As Figure 23-7 shows, the Galaxy extends well beyond the Sun's orbit, so the mass of the entire Galaxy must be larger than this mass.

BOX 23-2 **Tools of the Astronomer's Trade**

Estimating the Mass Inside the Sun's Orbit

The force that keeps the Sun in orbit around the center of the Galaxy is the gravitational pull of all the matter *interior* to the Sun's orbit. We can estimate the total mass of all of this matter using Newton's form of Kepler's third law (see Section 4-7 and Box 4-4):

$$P^2 = \frac{4\pi^2 a^3}{G(M + M_\odot)}$$

In this equation P is the orbital period of the Sun, a is the semimajor axis of the Sun's orbit around the galactic center, G is the universal constant of gravitation, M is the amount of mass inside the Sun's orbit, and $M_\odot$ is the mass of the Sun.

Because the Sun is only one of more than 10^{11} stars in the Galaxy, the Sun's mass is minuscule compared to M. Hence, we can safely replace the sum $M + M_\odot$ in the above equation by simply M. If we now assume that the Sun's orbit is a circle, the semimajor axis a of the orbit is just the radius of this circle, which we call r. As we saw in Section 25-4, the period P of the orbit is equal to $2\pi r/v$, where v is the Sun's orbital speed. You can then show that

$$M = \frac{rv^2}{G}$$

(We leave the derivation of this equation as an exercise at the end of this chapter.)

Now we can insert known values to obtain the mass inside the Sun's orbit. Being careful to express distance in meters and speed in meters per second, we have v = 220 km/s = 2.2×10^5 m/s, $G = 6.67 \times 10^{-11}$ newton · m^2/kg^2, and

$$r = 26{,}000 \text{ light-years} \times \frac{9.46 \times 10^{12} \text{ km}}{1 \text{ light-year}} \times \frac{10^3 \text{ m}}{1 \text{ km}}$$
$$= 2.5 \times 10^{20} \text{ m}$$

Hence, we find that

$$M = \frac{2.5 \times 10^{20} \times (2.2 \times 10^5)^2}{6.67 \times 10^{-11}} = 1.8 \times 10^{41} \text{ kg}$$

or, in terms of the mass of the Sun,

$$M = 1.8 \times 10^{41} \text{ kg} \times \frac{1 \text{ M}_\odot}{1.99 \times 10^{30} \text{ kg}}$$
$$= 9.0 \times 10^{10} \text{ M}_\odot$$

This estimate involves only mass that is interior to the Sun's orbit. Matter outside the Sun's orbit has no net gravitational effect on the Sun's motion and thus does not enter into Kepler's third law. (This is strictly true only if the matter outside our orbit is distributed over a sphere rather than a disk. In fact, the dark matter that dominates our Galaxy seems to have a spherical distribution.)

Rotation Curves and the Mystery of Dark Matter

In recent years, astronomers have been astonished to discover how much matter may lie outside the Sun's orbit. The clues come from 21-cm radiation emitted by hydrogen in spiral arms that extend to the outer reaches of the Galaxy. Because we know the true speed of the Sun, we can convert the Doppler shifts of this radiation into actual speeds for the spiral arms. This calculation gives us a **rotation curve,** a graph of the speed of galactic rotation measured outward from the galactic center (Figure 23-18). We would expect that for gas clouds beyond the confines of most of the Galaxy's mass, the orbital speed should decrease with increasing distance from the Galaxy's center, just as the orbital speeds of the planets decrease with increasing distance from the Sun (see Figure 23-17*c*). But as Figure 23-18 shows, the Galaxy's rotation curve is quite flat, indicating roughly uniform orbital speeds well beyond the visible edge of the galactic disk.

To explain these nearly uniform orbital speeds in the outer parts of the Galaxy, astronomers conclude that a large amount of matter must lie outside the Sun's orbit. When this matter is included, the total mass of our Galaxy could exceed 10^{12} $M_\odot$ or more, of which about 10% is in the form of stars. This percentage implies that our Galaxy contains roughly 200 billion stars.

Unlike the Galaxy's stars and dust, its dark matter forms a roughly spherical halo

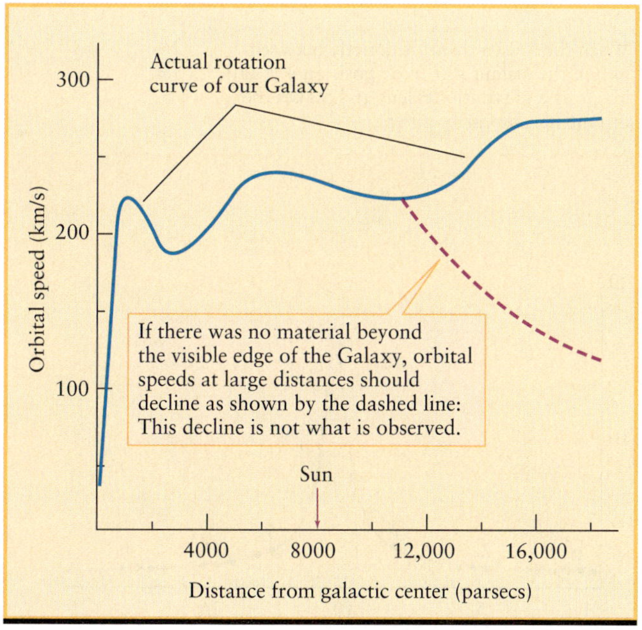

Figure 23-18

The Galaxy's Rotation Curve The blue curve shows the orbital speeds of stars and gas in the disk of the Galaxy out to a distance of 18,000 parsecs from the galactic center. (Very few stars are found beyond this distance.) The dashed red curve indicates how this orbital speed should decline beyond the confines of most of the Galaxy's visible mass. Because there is no such decline, there must be an abundance of invisible dark matter that extends to great distances from the galactic center.

WEB LINK 23.9 These observations lead to a profound mystery. Stars, gas, and dust account for only about 10% of the Galaxy's total mass. What, then, makes up the remaining 90% of the matter in our Galaxy? Whatever it's made of, it's dark. It does not show up on photographs, nor indeed in images made in any part of the electromagnetic spectrum. This unseen material, which is by far the predominant constituent of our Galaxy, is called **dark matter.** We sense its presence only through its gravitational influence on the orbits of stars and gas clouds.

CAUTION Be careful not to confuse dark *matter* with dark *nebulae*. A dark nebula like the one in Figure 18-4 emits no visible light, but does radiate at longer wavelengths. By contrast, no electromagnetic radiation of any kind has yet been discovered coming from dark matter.

Observations of star groupings outside the Milky Way suggest that our Galaxy's dark matter forms a spherical halo centered on the galactic nucleus, like the halo stars and globular clusters shown in Figure 23-7. However, the dark matter halo is much larger; it may extend to a distance of 100–200 kpc from the center of our Galaxy, some 2 to 4 times the extent of the visible halo (Figure 23-19). Analysis of the rotation curve in Figure 23-18 shows that the density of the dark matter halo decreases with increasing distance from the center of the Galaxy.

Dark Matter Speculations

WEB LINK 23.10 What is the nature of this mysterious dark matter? One proposal is that the dark matter halo is composed, at least in part, of dim objects with masses less than 1 $M_\odot$. These objects, which could include brown dwarfs, white dwarfs, or black holes, are called **massive compact halo objects,** or **MACHOs.** Astronomers have searched for MACHOs by monitoring the light from distant stars. If a MACHO passes between us and the star, its gravity will bend the light coming from the star. (In Section 24-2 we described how gravity can bend starlight.) As Figure 23-20 shows, the MACHO's gravity acts like a lens that focuses the light from the star. This effect, called **microlensing,** makes the star appear to brighten substantially for a few days.

Astronomers have indeed detected MACHOs in this way, but not nearly enough to solve the dark matter mystery. MACHOs with very low mass (10^{-6} to 0.1 $M_\odot$ each) do not appear to be a significant part of the dark matter halo. MACHOs of roughly 0.5 $M_\odot$ are more prevalent, but account for only about 10% to 20% of the dark matter halo.

The remainder of the dark matter is thought to be more exotic than the regular atoms (consisting of electrons, protons, and neutrons) which make known astrophysical objects. One idea proposed for dark matter is the neutrino. As we saw in Section 16-4, one type of neutrino can transform into another, and these transformations can only take place if neutrinos have mass. While we don't yet know the masses neutrinos have, we have come to understand that they are unlikely to account for halo dark matter. As we'll see in Chapter 27 (Section 27-6), fast-moving neutrinos can't "clump" enough to form a halo around our galaxy. The remainder of the dark matter is thought to be much more exotic.

Another possibility that has been proposed is a new class of subatomic particle called **weakly interacting massive particles,** or

Figure 23-19

The Galaxy and Its Dark Matter Halo The dark matter in our Galaxy forms a spherical halo whose center is at the center of the visible Galaxy. The extent of the dark matter halo is unknown, but its diameter is at least 100 kiloparsecs. The total mass of the dark matter halo is at least 10 times the combined mass of all of the stars, dust, gas, and planets in the Milky Way.

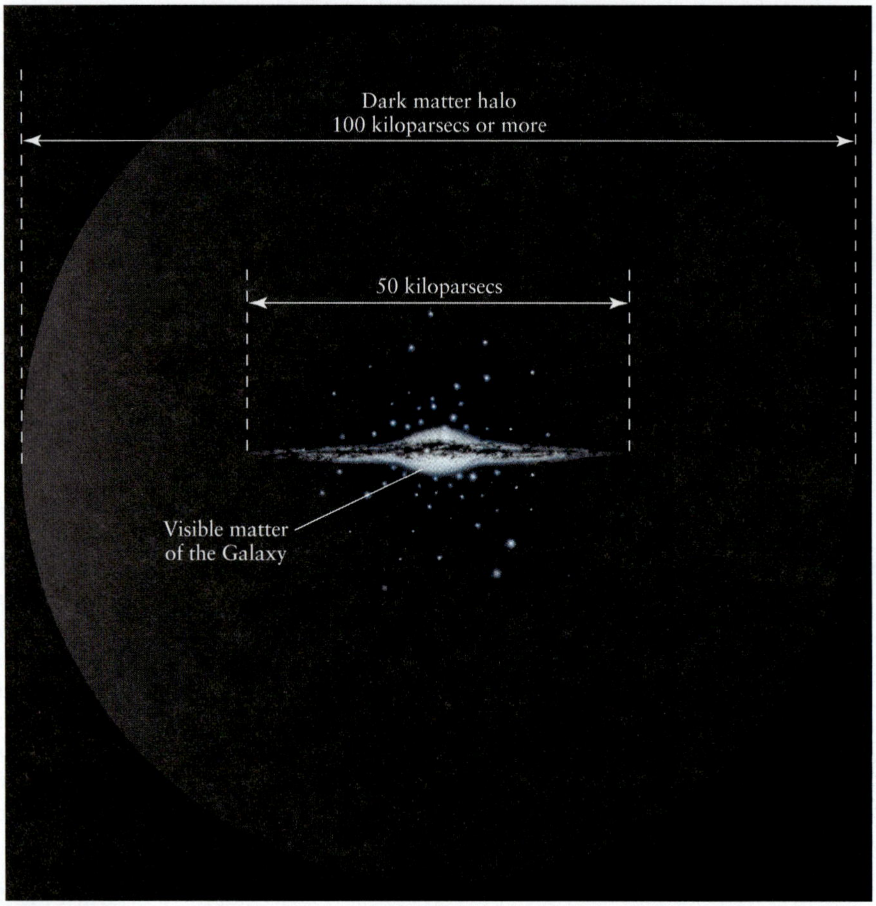

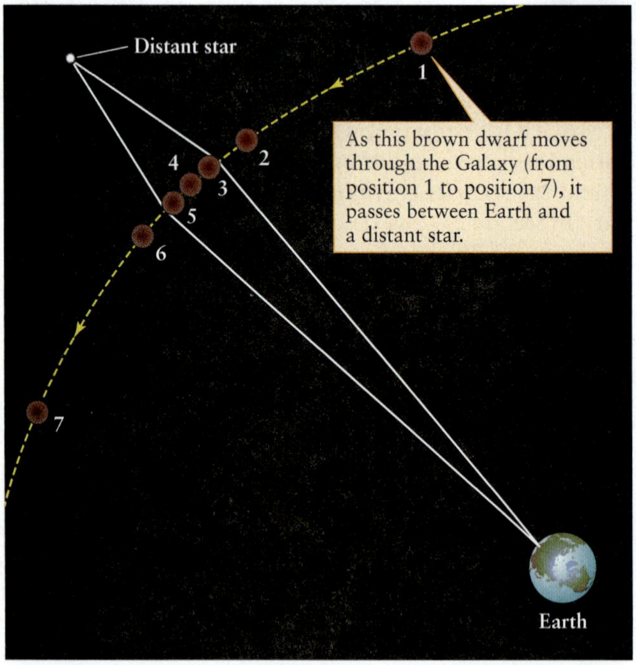

(a)

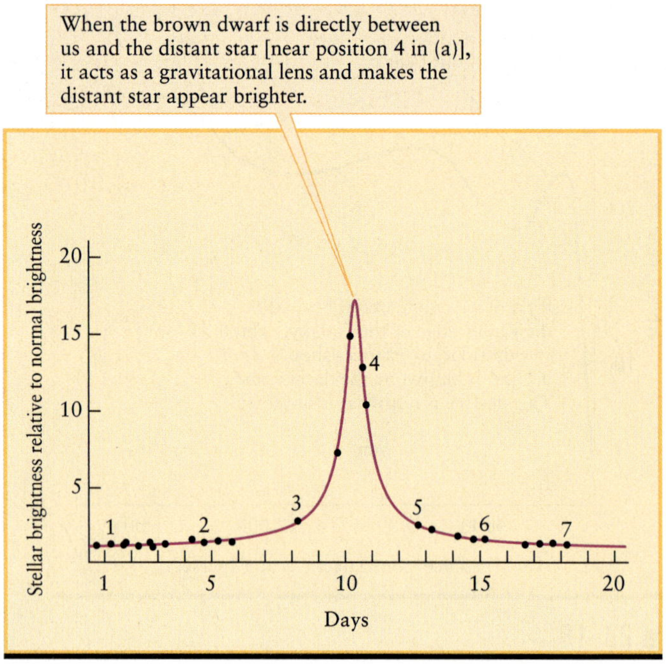

(b)

Figure 23-20

Microlensing by Dark Matter in the Galactic Halo **(a)** If a dense object such as a brown dwarf or black hole passes between Earth and a distant star, the gravitational curvature of space around the dense object deflects the starlight and focuses it in our direction. This effect is called microlensing. **(b)** This light curve shows the gravitational microlensing of light from a star in the Galaxy's central bulge. Astronomers do not know the nature of the object that passed between Earth and this star to cause the microlensing. (Courtesy of the MACHO and GMAN Collaborations)

WIMPs. These particles are suggested by certain theories which haven't yet been confirmed experimentally. In these theories, WIMPS don't emit or absorb electromagnetic radiation. Physicists are attempting to detect these curious particles, which would have masses 10 to 10,000 times greater than a proton or neutron, by using a large crystal cooled almost to absolute zero. If a WIMP should enter this crystal and collide with one of its atoms, the collision will deposit a tiny but measurable amount of heat in the crystal.

As yet, the true nature of dark matter remains a mystery. Furthermore, this mystery is not confined to our own Galaxy. In Chapter 24 we will find that other galaxies have the same sort of rotation curve as in Figure 23-18, indicating that they also contain vast amounts of dark matter. Indeed, dark matter appears to make up most of the mass in our universe. Hence, dark matter is one of the most important unsolved problems in physics and astronomy.

23-5 Spiral arms are caused by density waves that sweep around the Galaxy

The disk shape of our Galaxy is not difficult to understand. In Section 8-4 we described what happens when a large number of objects are put into orbit around a common center: Over time the objects tend naturally to orbit in the same plane. This is what happened when our solar system formed from the solar nebula. There a giant cloud of material eventually organized itself into planets, all of which orbit in nearly the same plane. In like fashion, the disk of our Galaxy, which is also made up of a large number of individual objects orbiting a common center, is very flat (see Figure 23-7). However, understanding why our Galaxy has spiral arms presents more of a challenge.

Our Sun and its solar system were spawned when a cloud of gas and dust passed through a spiral arm

The Winding Dilemma

One early explanation for the Galaxy's spiral structure was that the material in the Galaxy somehow condensed into a spiral pattern from the very start. In this view, once stars, gas, and dust had become concentrated within the spiral arms, the pattern would remain fixed. This fixed pattern would be possible only if the Galaxy rotated like a solid disk (see Figure 23-17*b*); the fixed pattern would be like the spokes on a rotating bicycle wheel. But the reality is that the Galaxy is not a solid disk. As we have seen, stars, gas, and dust all orbit the galactic center with approximately the same speed, as shown in Figure 23-17*a*. Let us see why this makes it impossible for a rigid spiral pattern to persist.

Imagine four stars, *A*, *B*, *C*, and *D*, that originally lie on a line extending outward from the galactic center (Figure 23-21*a*). In a given amount of time, each of the stars travels the same distance around its orbit. But because the innermost star has a smaller orbit than the others, it takes less time to complete one orbit. As a result, a line connecting the four stars is soon bent into a spiral (Figure 23-21*b*). Moreover, the spiral becomes tighter and tighter with the passage of time (Figures 23-21*c* and 23-21*d*). This "winding up" of the spiral arms causes the spiral structure to disappear completely after a few hundred million years—a very brief time compared to the age of our Galaxy, thought to be about 13.5 billion (1.35×10^{10}) years.

Figure 23-21 suggests that the Milky Way's spiral arms ought to have disappeared by now. The fact that they have not is called the **winding dilemma.** It shows that the spiral arms cannot simply be assemblages of stars and interstellar matter that travel around the Galaxy together, like a troop of soldiers marching in formation around a flagpole. What, then, can the spiral arms be?

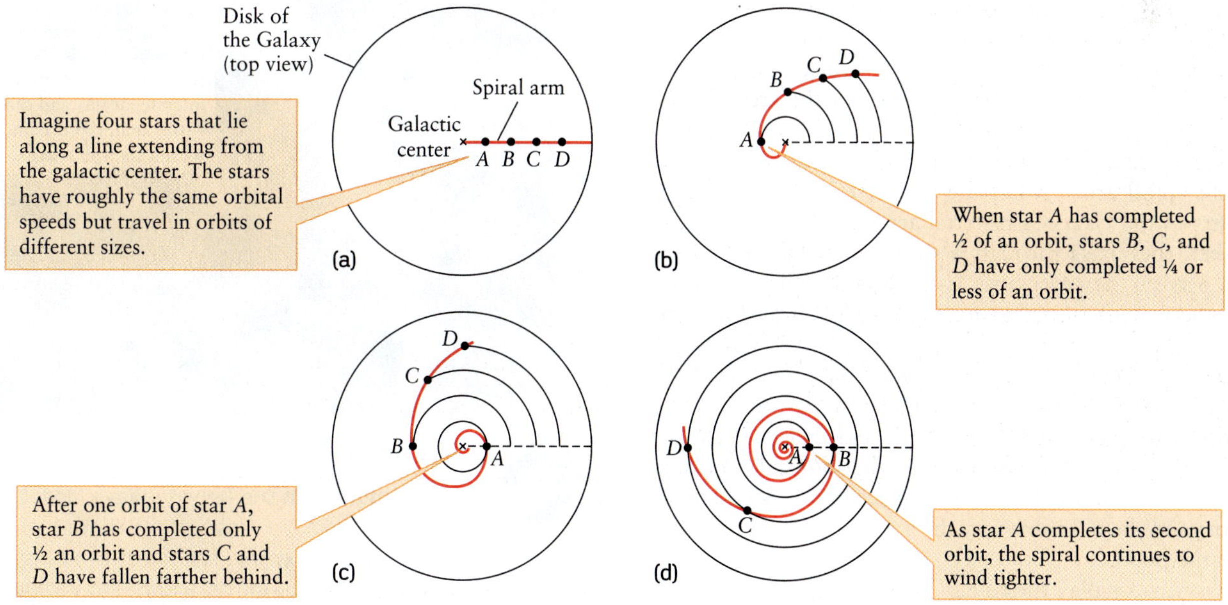

Figure 23-21

The Winding Dilemma This series of drawings shows that spiral arms in galaxies like the Milky Way cannot simply be assemblages of stars. If they were, the spiral arms would "wind up" and disappear in just a few hundred million years.

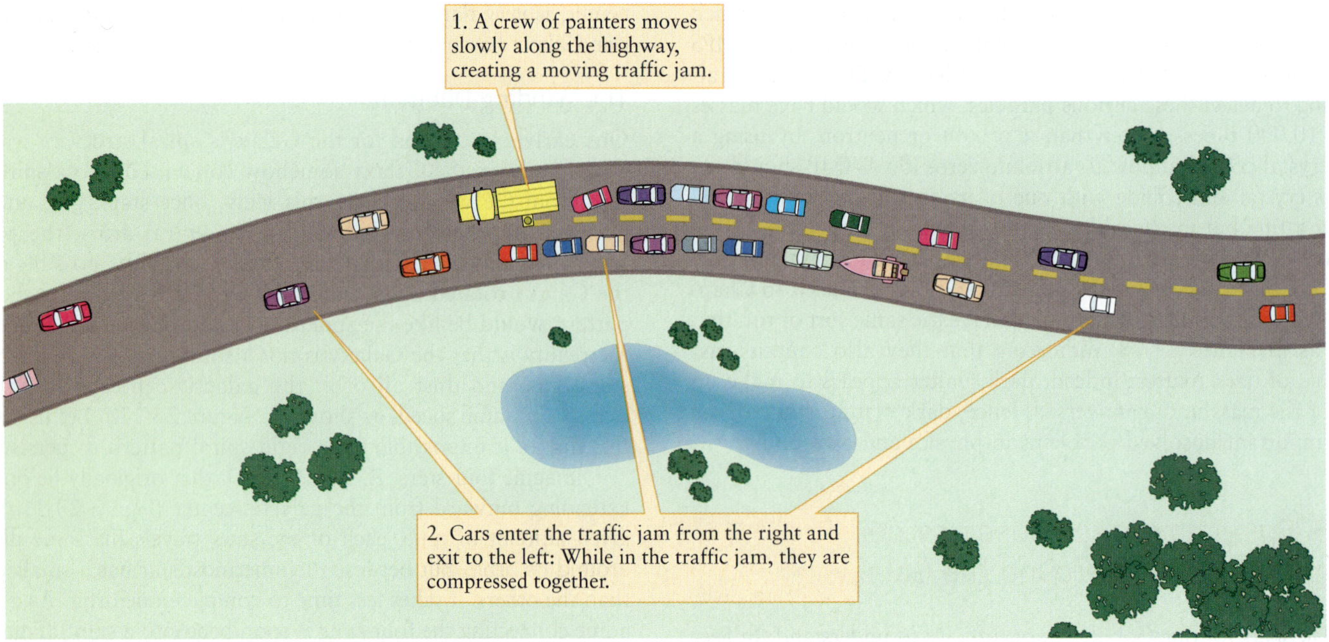
1. A crew of painters moves slowly along the highway, creating a moving traffic jam.
2. Cars enter the traffic jam from the right and exit to the left: While in the traffic jam, they are compressed together.

Figure 23-22

A Density Wave on the Highway A density wave in a spiral galaxy is analogous to a crew of painters moving slowly along the highway, creating a moving traffic jam. Like such a traffic jam, a density wave in a spiral galaxy is a slow-moving region where stars, gas, and dust are more densely packed than in the rest of the galaxy. As the material of the galaxy passes through the density wave, it is compressed. This triggers star formation, as Figure 23-23 shows.

The Density-Wave Model

In the 1940s, the Swedish astronomer Bertil Lindblad proposed that the spiral arms of a galaxy are actually a pattern that moves through the Galaxy like ripples on water. This idea was elaborated on greatly in the 1960s by the American astronomers Chia Chiao Lin and Frank Shu. In this picture, spiral arms are a kind of wave, like the waves that move across the surface of a pond when you toss a stone into the water. Water molecules pile up at a crest of the wave but spread out again when the crest passes. By analogy, Lindblad, Lin, and Shu pictured a pattern of **density waves** sweeping around the Galaxy. These waves make matter pile up in the spiral arms, which are the crests of the waves. Individual parts of the Galaxy's material are compressed only temporarily when they pass through a spiral arm. The pattern of spiral arms persists, however, just as the waves made by a stone dropped in the water can persist for quite awhile after the stone has sunk.

To understand better how a density wave operates in a galaxy, think again about a water wave in a pond. If one part of the pond is disturbed by dropping a stone into it, the molecules in that part will be displaced a bit. They will nudge the molecules next to them, causing those molecules to be displaced and to nudge the molecules beyond them. In this way the wave disturbance spreads throughout the pond.

In a galaxy, stars play the role of water molecules. Although stars and interstellar clouds of gas and dust are separated by vast distances, they can nonetheless exert forces on each other because they are affected by each other's gravity. If a region of above-average density should form, its gravitational attraction will draw

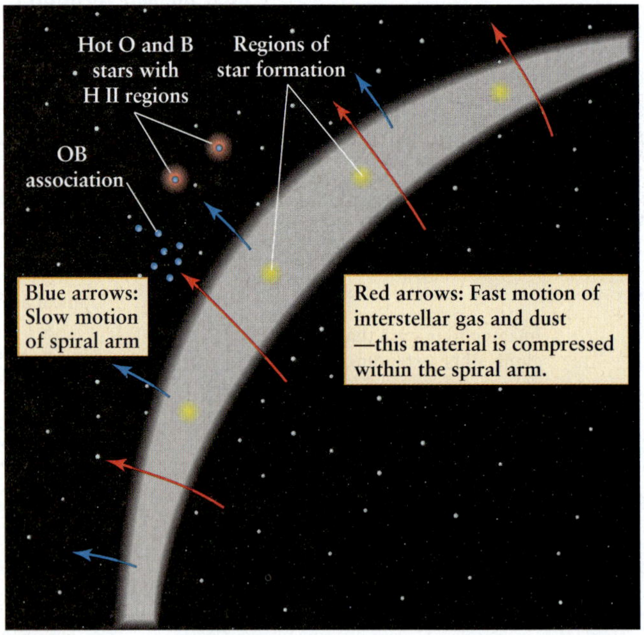
Hot O and B stars with H II regions
Regions of star formation
OB association
Blue arrows: Slow motion of spiral arm
Red arrows: Fast motion of interstellar gas and dust —this material is compressed within the spiral arm.

INTERACTIVE EXERCISE 23.2

Figure 23-23

Star Formation in the Density-Wave Model A spiral arm is a region where the density of material is higher than in the surrounding parts of a galaxy. Interstellar matter moves around the galactic center rapidly (shown by the red arrows) and is compressed as it passes through the slow-moving spiral arms (whose motion is shown by the blue arrows). This compression triggers star formation in the interstellar matter, so that new stars appear on the "downstream" side of the densest part of the spiral arms.

nearby material into it. The displacement of this material will change the gravitational force that it exerts on other parts of the galaxy, causing additional displacements. In this way a spiral-shaped density wave can travel around the disk of a galaxy.

ANALOGY A key feature of density waves is that they move more slowly around a galaxy than do stars or interstellar matter. To visualize this movement, imagine workers painting a line down a busy freeway (Figure 23-22). The cars normally cruise along the freeway at high speed, but the crew of painters is moving much more slowly. When the cars come up on the painters, they must slow down temporarily to avoid hitting anyone. As seen from the air, cars are jammed together around the painters. An individual car spends only a few moments in the traffic jam before resuming its usual speed, but the traffic jam itself lasts all day, inching its way along the road as the painters advance.

A similar crowding takes place when the matter between the stars, called the interstellar medium, enters a spiral arm. This crowding plays a key role in the formation of stars and the recycling of the interstellar medium. As interstellar gas and dust moves through a spiral arm, it is compressed into new nebulae (Figure 23-23). This compression begins the process by which new stars form, which we described in Section 18-3.

These freshly formed stars continue to orbit around the center of their galaxy, just like the matter from which they formed. The most luminous among these are the hot, massive, blue O and B stars, which may have emission nebulae (H II regions) associ-

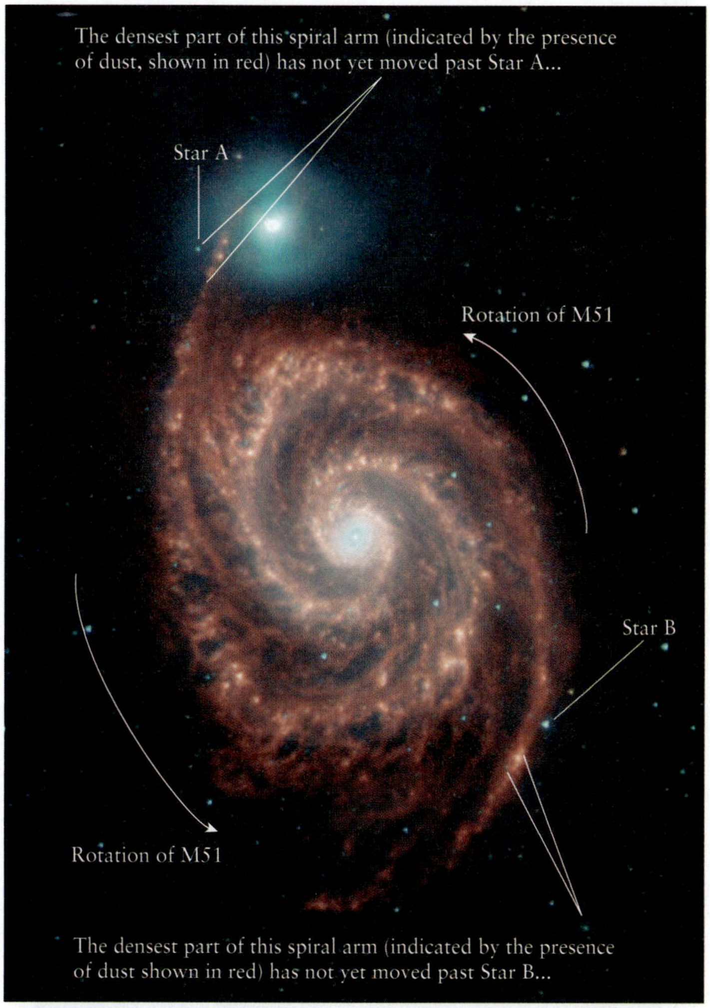

(a) An infrared view of M51 shows the locations of dust R I V U X G

(b) A visible-light view of M51 shows the locations of young stars R I V U X G

WEB LINK 23.1

Figure 23-24

Star Formation in the Whirlpool Galaxy The spiral galaxy M51 (called the Whirlpool) is a real-life example of the density-wave model illustrated in Figure 23-23. **(a)** This infrared image shows where dust has piled up as the material within M51 passes through its spiral arms. Radio images of M51 show that hydrogen gas also piles up in the same locations, thus beginning the formation of new stars. **(b)** By the time stars complete their formation process, their motion around the galaxy has swept them "downstream" of the positions of greatest dust density, just as depicted in Figure 23-23. (a: NASA, JPL-Caltech, and R. Kennicutt (Univ. of Arizona); b: DSS) ated with them. These stars have main-sequence lifetimes of only 3 to 15 million years (see Table 19-1), which is very short compared to the 220 million years required for the Sun to make a complete orbit around the Galaxy.

As a result, these luminous O and B stars can travel only a relatively short distance before dying off. Therefore, these stars, and their associated H II regions, are only seen in or slightly "downstream" of the spiral arm in which they formed. Figure 23-24 illustrates this for the spiral galaxy M51. Less massive stars have much longer main-sequence lifetimes, and thus their orbits are able to carry them all around the galactic disk. These less-luminous stars are found throughout the disk, including between the spiral arms (see Figure 23-15*c*).

The density-wave model of spiral arms explains why the disk of our Galaxy is dominated by metal-rich Population I stars. Because the material left over from the death of ancient stars is enriched in heavy elements, new generations of stars formed in spiral arms are likely to be more metal-rich than their ancestors. The *Cosmic Connections: Stars in the Milky Way* figure illustrates this cycle of star birth and death in the disk of our Galaxy.

The density-wave model is still under development. One problem is finding a driving mechanism that keeps density waves going in spiral galaxies. After all, density waves expend an enormous amount of energy to compress the interstellar gas and dust. Hence, we would expect that density waves should eventually die away, just as do ripples on a pond. The American astronomers Debra and Bruce Elmegreen have suggested that gravity can supply that needed energy. As mentioned in Section 23-3, the central bulge of our Galaxy is elongated into a bar shape, much like the central bulge of the galaxy M83 (see Figure 23-15*a* and Figure 23-16*a*). The asymmetric gravitational field of such a bar pulls on the stars and interstellar matter of a galaxy to generate density waves. Another factor that may help to generate and sustain spiral arms is the gravitational interactions *between* galaxies. We will discuss this in Chapter 24.

The Self-Propagating Star-Formation Model

Spiral density waves may not be the whole story behind spiral arms in our Galaxy and other galaxies. The reason is that spiral density waves should produce very well defined spiral arms. We do indeed see many so-called **grand-design spiral galaxies** (Figure 23-25*a*), with thin, graceful, and well-defined spiral arms. But in some galaxies, called **flocculent spiral galaxies** (Figure 23-25*b*), the spiral arms are broad, fuzzy, chaotic, and poorly defined. ("Flocculent" means "resembling wool.")

To explain such flocculent spirals, M. W. Mueller and W. David Arnett in 1976 proposed a theory of **self-propagating star formation.** Imagine that star formation begins in a dense interstellar cloud within the disk of a galaxy that does not yet have spiral arms. As soon as hot, massive stars form, their radiation and stellar winds compress nearby matter, triggering the formation of additional stars in that gas. When massive stars become supernovae, they produce shock waves that further compress the surrounding interstellar medium, thus encouraging still more star formation.

Although all parts of this broad, star-forming region have approximately the same orbital speed about the galaxy's center, the inner regions have a shorter distance to travel to complete one orbit than the outer regions. As a result, the inner edges of the star-forming region move ahead of the outer edges as the Galaxy rotates. The bright O and B stars and their nearby glowing nebulae soon become stretched out in the form of a spiral arm. These spiral arms come and go essentially at random across a galaxy. Bits and pieces of spiral arms appear where star formation has recently begun but fade and disappear at other locations where all the massive stars have

(a) Grand-design spiral galaxy

(b) Flocculent spiral galaxy

Figure 23-25 R I **V** U X G

Variety in Spiral Arms The differences from one spiral galaxy to another suggest that more than one process can create spiral arms. **(a)** NGC 628 is a grand-design spiral galaxy with thin, well-defined spiral arms. **(b)** NGC 4414 is a flocculent spiral galaxy with fuzzy, broken, and poorly defined spiral arms. (Courtesy of Gemini Observatory; Olivier Vallejo)

COSMIC CONNECTIONS

Stars in the Milky Way

Different populations of stars are found in different neighborhoods of our home galaxy. (The galaxy shown here is another spiral galaxy similar to our own.) The variations from one galactic region to another are due to the presence or absence of ongoing star formation.

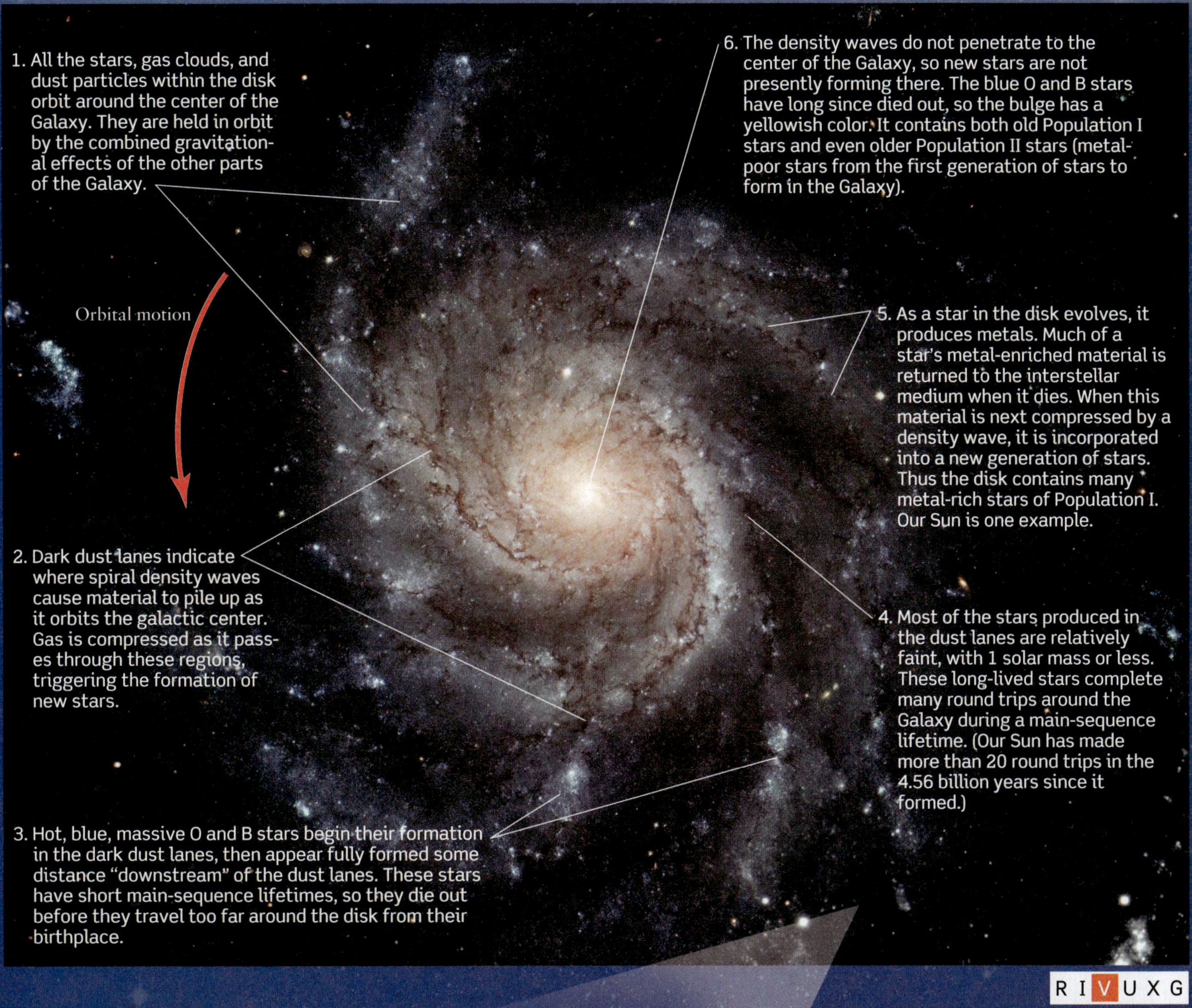

1. All the stars, gas clouds, and dust particles within the disk orbit around the center of the Galaxy. They are held in orbit by the combined gravitational effects of the other parts of the Galaxy.
Orbital motion
2. Dark dust lanes indicate where spiral density waves cause material to pile up as it orbits the galactic center. Gas is compressed as it passes through these regions, triggering the formation of new stars.
3. Hot, blue, massive O and B stars begin their formation in the dark dust lanes, then appear fully formed some distance "downstream" of the dust lanes. These stars have short main-sequence lifetimes, so they die out before they travel too far around the disk from their birthplace.
4. Most of the stars produced in the dust lanes are relatively faint, with 1 solar mass or less. These long-lived stars complete many round trips around the Galaxy during a main-sequence lifetime. (Our Sun has made more than 20 round trips in the 4.56 billion years since it formed.)
5. As a star in the disk evolves, it produces metals. Much of a star's metal-enriched material is returned to the interstellar medium when it dies. When this material is next compressed by a density wave, it is incorporated into a new generation of stars. Thus the disk contains many metal-rich stars of Population I. Our Sun is one example.
6. The density waves do not penetrate to the center of the Galaxy, so new stars are not presently forming there. The blue O and B stars have long since died out, so the bulge has a yellowish color. It contains both old Population I stars and even older Population II stars (metal-poor stars from the first generation of stars to form in the Galaxy).
R I V U X G

7. Also orbiting the Galaxy are the globular clusters. These star clusters have no mechanism to trigger star formation, so they contain only very old, metal-poor Population II stars. Studies of these clusters suggest that the first stars formed in them about 13.6 billion years ago — a mere 100 million years after the Big Bang.

8. Not visible to any telescope is the Galaxy's dark matter, which emits no electromagnetic radiation of any kind. The spherical halo of dark matter that envelops our Galaxy has at least 10 times the combined mass of all of the Galaxy's stars, planets, gas, and dust. Its fundamental nature remains a mystery.

(Image of spiral galaxy M101: NASA and ESA; image of globular cluster M3: S. Kafka and K. Honeycutt, Indiana University/WIYN/NOAO/NSF) died off. Self-propagating star formation therefore tends to produce flocculent spiral galaxies that have a chaotic appearance with poorly defined spiral arms, like the galaxy in Figure 23-25*b*.

The two theories presented here are very different in character. In the density-wave model, star formation is caused by the spiral arms; in the self-propagating star formation model, by contrast, the spiral arms are caused by star formation. A complete and correct description of spiral arms in our Galaxy remains a topic of active research.

23-6 Infrared, radio, and X-ray observations are used to probe the galactic center

TUTORIAL 23.2

The innermost region of our Galaxy is an active, crowded place. If you lived on a planet near the galactic center, you could see a million stars as bright as Sirius, the brightest single star in our own night sky. The total intensity of starlight from all those nearby stars would be equivalent to 200 of our full moons. In effect, night would never really fall on a planet near the center of the Milky Way. At the center of this empire of light, however, lies the darkest of all objects in the universe—a black hole millions of times more massive than the Sun.

At the very center of the Milky Way Galaxy is a maelstrom of activity centered on a supermassive black hole

Sagittarius A*: Heart of Darkness

Because of the severe interstellar absorption at visual wavelengths, most of our information about the galactic center comes from infrared and radio observations. Figure 23-26 shows three infrared views of the center of our Galaxy. Figure 23-26*a* is a wide-angle view covering a 50° segment of the Milky Way from Sagittarius through Scorpius. (The photograph that opens this chapter shows this same region at visible wavelengths, viewed at a different angle.) The prominent reddish band through the center of this false-color infrared image is a layer of dust in the plane of the Galaxy. Figure 23-26*b* is an IRAS view of the galactic center. It is surrounded by numerous streamers of dust (shown in blue). The strongest infrared emission (shown in white) comes from a grouping of several powerful sources of radio waves. One of these sources, **Sagittarius A*** (say "A star"), lies at the very center of the Galaxy. (Its position, pinpointed with simultaneous observations by radio telescopes scattered around the world, seems to be very near the gravitational center of the Galaxy.) The high-resolution infrared view in Figure 23-26*c*, made using adaptive optics (see Section 6-3), shows hundreds of stars crowded within 1 ly (0.3 pc) of Sagittarius A*. Compare this region to our region of the Galaxy, where the average distance between stars is more than a light-year.

Sagittarius A* itself does not appear in infrared images. Nonetheless, astronomers have used infrared observations to make truly startling discoveries about this object. Since the 1990s, two research groups—one headed by Reinhard Genzel of the Max

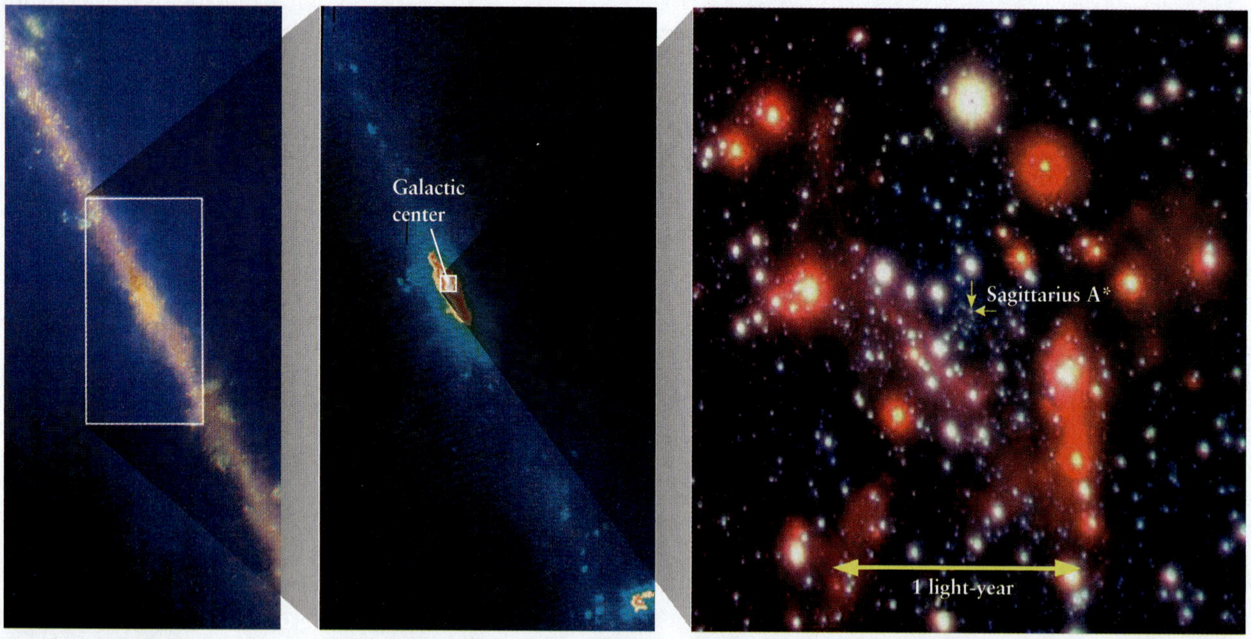

(a) A wide-angle (50°) infrared view

(b) A close-up view shows a more luminous region at the galactic center

(c) An extreme close-up view centered on Sagittarius A*, a radio source at the very center of the Milky Way Galaxy, shows hundreds of stars within 1 ly (0.3 pc)

Figure 23-26 R I V U X G

The Galactic Center (a) In this false-color infrared image, the reddish band is dust in the plane of the Galaxy and the fainter bluish blobs are interstellar clouds heated by young O and B stars. **(b)** This close-up infrared view covers the area outlined by the white rectangle in (a). **(c)** Adaptive optics reveals stars densely packed around the galactic center. (a, b: NASA; c: R. Schödel et al., MPE/ESO)

Planck Institute for Extraterrestrial Physics in Garching, Germany, and another led by Andrea Ghez at the University of California, Los Angeles—have been using infrared detectors to monitor the motions of stars in the immediate vicinity of Sagittarius A*. They have found a number of stars orbiting around Sagittarius A* at speeds in excess of 1500 km/s (Figure 23-27). (By comparison, Earth orbits the Sun at a lackadaisical 30 km/s.) In 2000 the UCLA group observed one such star, called SO-16, as its elliptical orbit brought it within a mere 45 AU from Sagittarius A* ($1\frac{1}{2}$ times the distance from the Sun to Neptune). At its closest approach, SO-16 was traveling at a breathtaking speed of 12,000 km/s, or 4% of the speed of light!

In order to keep stars like SO-16 in such small, rapid orbits, Sagittarius A* must exert a powerful gravitational force and hence must be very massive. By applying Newton's form of Kepler's third law to the motions of these stars around Sagittarius A*, the UCLA group calculates the mass of Sagittarius A* to be a remarkable 3.7 *million* solar masses ($3.7 \times 10^6\ M_\odot$). Furthermore, the small separation between SO-16 and Sagittarius A* at closest approach shows that Sagittarius A* can be no more than 45 AU in radius. An object this massive and this compact can only be one thing: a supermassive black hole (see Section 22-4).

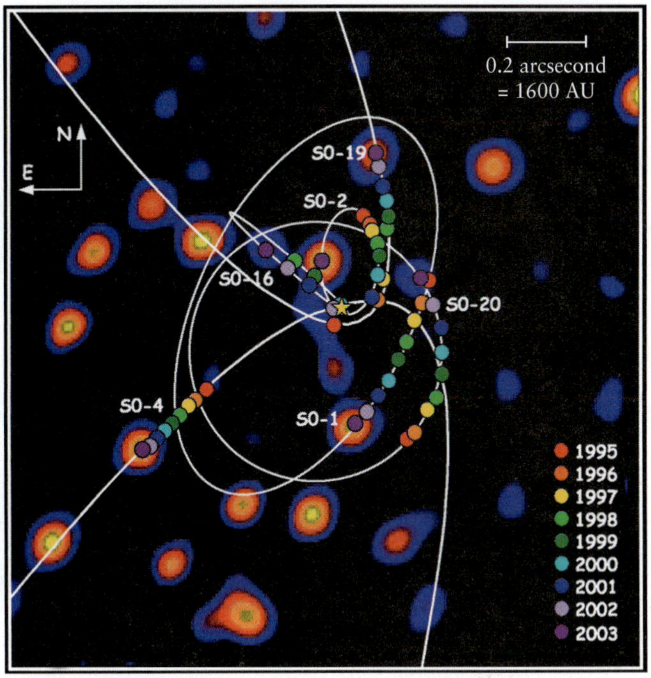

Figure 23-27 R I V U X G

Stars Orbiting Sagittarius A* The colored dots superimposed on this infrared image show the motion of six stars in the vicinity of the unseen massive object (denoted by the yellow five-pointed star) at the position of the radio source Sagittarius A*. The orbits were measured over an 8-year period. This plot indicates that the stars are held in orbit by a black hole of 3.7×10^6 solar masses. (UCLA Galactic Center Group)

X-rays from Around a Supermassive Black Hole

Evidence in favor of this picture comes from the Chandra X-ray Observatory, which has observed X-ray flares coming from Sagittarius A*. The flares brighten dramatically over the space of just 10 minutes, which shows that the size of the flare's source can be no larger than the distance that light travels in 10 minutes. (We used a similar argument in Section 22-3 to show that the flickering X-ray source Cygnus X-1 must be very small.) In 10 minutes light travels a distance of 1.8×10^8 km or 1.2 AU, and only a black hole could pack a mass of $3.7 \times 10^6\ M_\odot$ into a volume that size or smaller. The X-ray flares were presumably emitted by blobs of material that were compressed and heated as they fell into the black hole (see Section 22-3).

The X-ray flares from Sagittarius A* are relatively feeble, which suggests that the supermassive black hole is swallowing only relatively small amounts of material. But the region around Sagittarius A* is nonetheless an active and dynamic place. Figure 23-28*a* is a wide-angle radio image of the galactic center covering an area more than 60 pc (200 ly) across. Huge filaments of gas stretch for 20 pc (65 ly) northward of the galactic center (to the right and upward in Figure 23-28*a*), then abruptly arch southward (down and to the left in the figure). The orderly arrangement of these filaments is reminiscent of prominences on the Sun (see Section 18-10, especially Figure 18-28). This suggests that, as on the Sun, there is ionized gas at the galactic center that is being controlled by a powerful magnetic field. Indeed, much of the radio emission from the galactic center is synchrotron radiation: As we saw in Section 23-4, such radiation is produced by high-energy electrons spiraling in a magnetic field.

The false-color X-ray image in Figure 23-28*b* shows the immediate vicinity of Sagittarius A*. The black hole is flanked by lobes of hot, ionized, X-ray–emitting gas that extend for dozens of light-years. These lobes are thought to be the relics of immense explosions that may have taken place over the past several thousand years. Perhaps these past explosions cleared away much of the material around Sagittarius A*, leaving only small amounts to fall into the black hole. This could explain why the X-ray flares from Sagittarius A* are so weak.

There is nonetheless evidence that Sagittarius A* has been more active in the recent past. Between 2002 and 2005, Michael Muno and his colleagues at Caltech observed that certain nebulae within about 50 ly of Sagittarius A* suddenly became very bright at X-ray wavelengths. The character of the X-ray light was such that it could not have originated from the nebulae themselves. Muno and colleagues concluded that X-rays emitted from Sagittarius A* about 50 years earlier had struck and excited the nebulae, causing the nebulae to glow intensely at X-ray wavelengths. To produce such an intense X-ray glow, an object the size of the planet Mercury must have fallen into the supermassive black hole.

The supermassive black hole at the center of our Galaxy is not unique. Observations show that such titanic black holes are a feature of most large galaxies. In Chapter 25 we will see how black holes of this kind power *quasars*, the most luminous sustained light sources in the cosmos.

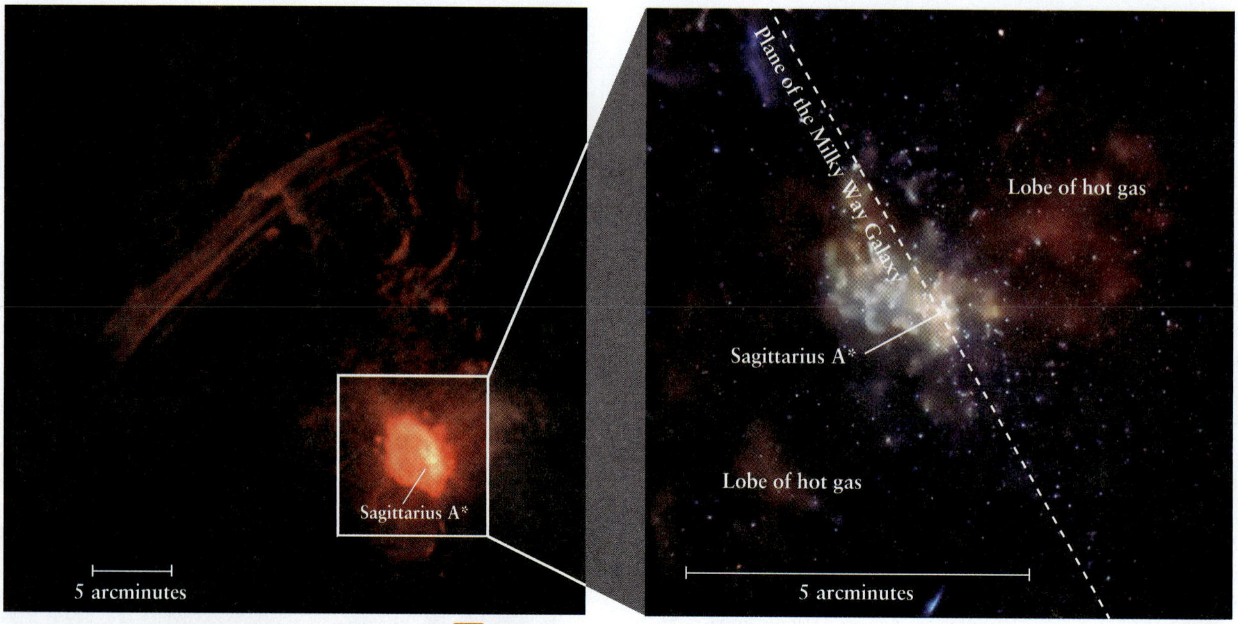

(a) A radio view of the galactic center R I V U X G

(b) An X-ray view of the galactic center R I V U X G

Figure 23-28

The Energetic Center of the Galaxy **(a)** The area shown in this radio image has the same angular size as the full moon. Sagittarius A*, at the very center of the Galaxy, is one of the brightest radio sources in the sky. Magnetic fields shape nearby interstellar gas into immense, graceful arches. **(b)** This composite of images at X-ray wavelengths from 0.16 to 0.62 nm shows lobes of gas on either side of Sagittarius A*. The character of the X-ray emission shows that the gas temperature is as high as 2×10^7 K. (a: NRAO/VLA/F. Zadeh et al.; b: NASA/CXC/MIT/F. K. Baganoff et al.)

WEB LINK 23.14

Astronomers are still groping for a better understanding of the galactic center. With future developments in very-long-baseline interferometry (described in Section 6-6), it may be possible to actually obtain a picture of the supermassive black hole lurking there. During the coming years, observations from Earth-orbiting satellites as well as from radio and infrared telescopes on the ground will certainly add to our knowledge of the core of the Milky Way.

Key Words

Word preceded by an asterisk (*) is discussed in Box 23-1.

central bulge (of a galaxy), p. 609
dark matter, p. 619
density wave, p. 622
disk (of a galaxy), p. 609
far-infrared, p. 609
flocculent spiral galaxy, p. 624
galactic nucleus, p. 608
galaxy, p. 606
globular cluster, p. 607
grand-design spiral galaxy, p. 624
H I, p. 612
halo (of a galaxy), p. 609
high-velocity star, p. 610
interstellar extinction, p. 606
Local Bubble, p. 613
*magnetic resonance imaging (MRI), p. 614
massive compact halo object (MACHO), p. 619
microlensing, p. 619
Milky Way Galaxy, p. 606
near-infrared, p. 609
rotation curve, p. 619
RR Lyrae variable, p. 607
Sagittarius A*, p. 626
self-propagating star formation, p. 624
spin (of a particle), p. 612
spin-flip transition, p. 613
spiral arm, p. 612
21-cm radio emission, p. 613
weakly interacting massive particle (WIMP), p. 619
winding dilemma, p. 621

Key Ideas

The Shape and Size of the Galaxy: Our Galaxy has a disk about 50 kpc (160,000 ly) in diameter and about 600 pc (2000 ly) thick, with a high concentration of interstellar dust and gas in the disk.

• The galactic center is surrounded by a large distribution of stars called the central bulge. This bulge is not perfectly symmetrical, but may have a bar or peanut shape.

• The disk of the Galaxy is surrounded by a spherical distribution of globular clusters and old stars, called the galactic halo.

• There are about 200 billion (2×10^{11}) stars in the Galaxy's disk, central bulge, and halo.

The Sun's Location in the Galaxy: Our Sun lies within the galactic disk, some 8000 pc (26,000 ly) from the center of the Galaxy.

• Interstellar dust obscures our view at visible wavelengths along lines of sight that lie in the plane of the galactic disk. As a result, the Sun's location in the Galaxy was unknown for many years. This dilemma was resolved by observing parts of the Galaxy outside the disk.

• The Sun orbits around the center of the Galaxy at a speed of about 790,000 km/h. It takes about 220 million years to complete one orbit.

The Rotation of the Galaxy and Dark Matter: From studies of the rotation of the Galaxy, astronomers estimate that the total mass of the Galaxy is about 10^{12} $M_\odot$. Only about 10% of this mass is in the form of visible stars, gas, and dust. The remaining 90% is in some nonvisible form, called dark matter, that extends beyond the edge of the luminous material in the Galaxy.

• Our Galaxy's dark matter may be a combination of MACHOs (dim, star-sized objects), massive neutrinos, and WIMPs (relatively massive subatomic particles).

The Galaxy's Spiral Structure: OB associations, H II regions, and molecular clouds in the galactic disk outline huge spiral arms.

• Spiral arms can be traced from the positions of clouds of atomic hydrogen. These can be detected throughout the galactic disk by the 21-cm radio waves emitted by the spin-flip transition in hydrogen. These emissions easily penetrate the intervening interstellar dust.

Theories of Spiral Structure: There are two leading theories of spiral structure in galaxies.

• According to the density-wave theory, spiral arms are created by density waves that sweep around the Galaxy. The gravitational field of this spiral pattern compresses the interstellar clouds through which it passes, thereby triggering the formation of the OB associations and H II regions that illuminate the spiral arms.

• According to the theory of self-propagating star formation, spiral arms are caused by the birth of stars over an extended region in a galaxy. Differential rotation of the galaxy stretches the star-forming region into an elongated arch of stars and nebulae.

The Galactic Nucleus: The innermost part of the Galaxy, or galactic nucleus, has been studied through its radio, infrared, and X-ray emissions (which are able to pass through interstellar dust).

• A strong radio source called Sagittarius A* is located at the galactic center. This marks the position of a supermassive black hole with a mass of about 3.7×10^6 $M_\odot$.

Questions

Review Questions

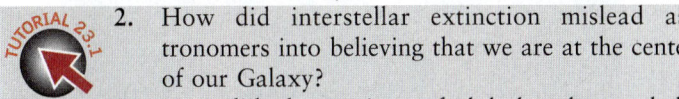

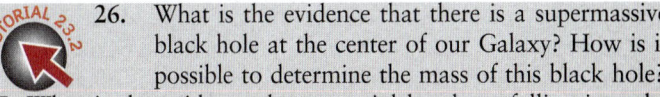

1. Why do the stars of the Galaxy appear to form a bright band that extends around the sky?
2. How did interstellar extinction mislead astronomers into believing that we are at the center of our Galaxy?
3. How did observations of globular clusters help astronomers determine our location in the Galaxy?
4. What are RR Lyrae stars? Why are they useful for determining the distance from our solar system to the center of the Galaxy?
5. Why are infrared telescopes useful for exploring the structure of the Galaxy? Why is it important to make observations at both near-infrared and far-infrared wavelengths?
6. The galactic halo is dominated by Population II stars, whereas the galactic disk contains predominantly Population I stars. In which of these parts of the Galaxy has star formation taken place recently? Explain your answer.
7. O or B main-sequence stars are found in the galactic disk but not in globular clusters. Why is this so?
8. What must happen within a hydrogen atom for it to emit a photon of wavelength 21 cm?
9. Most interstellar hydrogen atoms emit only radio waves at a wavelength of 21 cm, but some hydrogen clouds emit profuse amounts of visible light (see, for example, Figure 18-1*b* and Figure 18-2). What causes this difference?
10. How do astronomers determine the distances to H I (neutral hydrogen) clouds?
11. The radio map in Figure 23-14 has a large gap on the side of the Galaxy opposite to ours. Why is this?
12. In a spiral galaxy, are stars in general concentrated in the spiral arms? Why are spiral arms so prominent in visible-light images of spiral galaxies?
13. Many classic black-and-white photographs of spiral galaxies were taken using film that was most sensitive to blue light. Explain why the spiral arms were particularly prominent in such photographs.
14. What kinds of objects (other than H I clouds) do astronomers observe to map out the Galaxy's spiral structure? What is special about these objects? Which of these can be observed at great distances?
15. Why don't astronomers detect 21-cm radiation from the hydrogen in giant molecular clouds?
16. In what way are the orbits of stars in the galactic disk different from the orbits of planets in our solar system? What does this difference imply about the way that matter is distributed in the Galaxy?
17. How do astronomers determine how fast the Sun moves in its orbit around the Galaxy? How does this speed tell us about the amount of mass inside the Sun's orbit? Does this speed tell us about the amount of mass outside the Sun's orbit?
18. How do astronomers conclude that vast quantities of dark matter surround our Galaxy? How is this dark matter distributed in space?
19. Another student tells you that the Milky Way Galaxy is made up "mostly of stars." Is this statement accurate? Why or why not?
20. What is the difference between dark matter and dark nebulae?
21. What proposals have been made to explain the nature of dark matter? What experiments or observations have been made to investigate these proposals? What are the results of this research?
22. What is the winding dilemma? What does it tell us about the nature of spiral arms?
23. Do density waves form a stationary pattern in a galaxy? If not, do they move more rapidly, less rapidly, or at the same speed as stars in the disk?
24. In our Galaxy, why are stars of spectral classes O and B only found in or near the spiral arms? Is the same true for stars of other spectral classes? Explain why or why not.
25. Compare the kinds of spiral arms produced by density waves with those produced by self-propagating star formation. By examining Figure 23-16, cite evidence that both processes may occur in our Galaxy.
26. What is the evidence that there is a supermassive black hole at the center of our Galaxy? How is it possible to determine the mass of this black hole?
27. What is the evidence that material has been falling into the supermassive black hole at the galactic center?

Advanced Questions

Questions preceded by an asterisk (*) involve topics discussed in the Boxes.

Problem-solving tips and tools

Several of the following questions make extensive use of Newton's form of Kepler's third law, and you might find it helpful to review Box 4-4. Another useful version of Kepler's third law is given in Section 17-9. Box 19-2 discusses the relationship between luminosity, apparent brightness, and distance. We discussed the relationship between the energy and wavelength of a photon in Section 5-5. According to the Pythagorean theorem, an isosceles right triangle has a hypotenuse that is longer than its sides by a factor of $\sqrt{2} = 1.414$. The formula for the Schwarzschild radius of a black hole is given in Box 22-2. You will find the small-angle formula in Box 1-1 useful. It is also helpful to remember that an object 1 AU across viewed at a distance of 1 parsec has an angular size of 1 arcsecond. Remember, too, that the volume of a cylinder is equal to its height multiplied by the area of its base, the area of a circle of radius r is πr^2, and the volume of a sphere of radius r is $4\pi r^3/3$. You can find other geometrical formulae in Appendix 8.

28. Discuss how the Milky Way would appear to us if the Sun were relocated to (**a**) the edge of the Galaxy; (**b**) the galactic halo; (**c**) the galactic bulge.
29. Explain why globular clusters spend most of their time in the galactic halo, even though their eccentric orbits take them close to the galactic center.
30. The disk of the Galaxy is about 50 kpc in diameter and 600 pc thick. (**a**) Find the volume of the disk in cubic parsecs. (**b**) Find the volume (in cubic parsecs) of a sphere 300 pc in radius centered on the Sun. (**c**) If supernovae occur randomly throughout the volume of the Galaxy, what is the probability that a given supernova will occur within 300 pc of the Sun? If there are about three supernovae each century in our Galaxy, how often, on average, should we expect to see one within 300 pc of the Sun?

*31. An RR Lyrae star whose peak luminosity is 100 $L_\odot$ is in a globular cluster. At its peak luminosity, this star appears from Earth to be only 1.47×10^{-18} as bright as the Sun. Determine the distance to this globular cluster (**a**) in AU and (**b**) in parsecs.

32. A typical hydrogen atom in interstellar space undergoes a spin-flip transition only once every 10^7 years. How, then, is it at all possible to detect the 21-cm radio emission from interstellar hydrogen?
33. Calculate the energy of the photon emitted when a hydrogen atom undergoes a spin-flip transition. How many such photons would it take to equal the energy of a single H_α photon of wavelength 656.3 nm?
34. Suppose you were to use a radio telescope to measure the Doppler shift of 21-cm radiation in the plane of the Galaxy. (**a**) If you observe 21-cm radiation from clouds of atomic hydrogen at an angle of 45° from the galactic center, you will see the highest Doppler shift from a cloud that is as far from the galactic center as it is from the Sun. Explain this statement using a diagram. (**b**) Find the distance from the Sun to the particular cloud mentioned in (a).
35. Calculate approximately how many times our solar system has orbited the center of our Galaxy since the Sun and planets were formed 4.56×10^9 years ago.
36. Sketch the rotation curve you would obtain if the Galaxy were rotating like a rigid body.
37. The mass of our Galaxy interior to the Sun's orbit is calculated from the radius of the Sun's orbit and its orbital speed. By how much would this estimate be in error if the calculated distance to the galactic center were off by 10%? By how much would this estimate be in error if the calculated orbital velocity were off by 10%? Explain your reasoning.
38. A gas cloud located in the spiral arm of a distant galaxy is observed to have an orbital velocity of 400 km/s. If the cloud is 20,000 pc from the center of the galaxy and is moving in a circular orbit, find (**a**) the orbital period of the cloud and (**b**) the mass of the galaxy contained within the cloud's orbit.
39. According to the Galaxy's rotation curve in Figure 23-18, a star 16 kpc from the galactic center has an orbital speed of about 270 km/s. Calculate the mass within that star's orbit.
40. Speculate on the reasons for the rapid rise in the Galaxy's rotation curve (see Figure 23-18) at distances close to the galactic center.

*41. Show that the form of Kepler's third law stated in Box 23-2, $P^2 = 4\pi^2 a^3/G(M + M_\odot)$, is equivalent to $M = rv^2/G$, provided the orbit is a circle. (*Hint:* The mass of the Sun ($M_\odot$) is much less than the mass of the Galaxy inside the Sun's orbit (M).)

42. The image below shows the spiral galaxy M74, located about 55 million light-years from Earth in the constellation Pisces

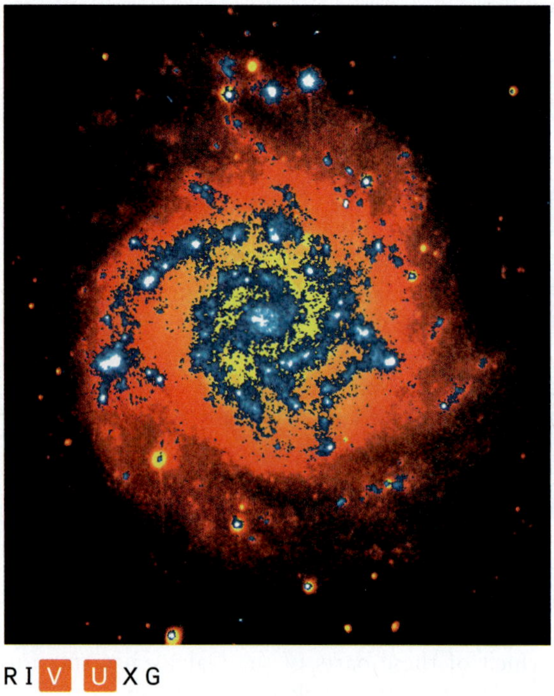

R I V U X G

(Astronomical Society of the Pacific)

(the Fish). It is actually a superposition of two false-color images: The red portion is an optical image taken at visible wavelengths, while the blue portion is an ultraviolet image made by NASA's Ultraviolet Imaging Telescope, which was carried into orbit by the space shuttle *Columbia* during the *Astro-1* mission in 1990. Compare the visible and ultraviolet images and, from what you know about stellar evolution and spiral structure, explain the differences you see.

43. The figure at the bottom of this page shows infrared images of two spiral galaxies. Explain which of these is a grand-design spiral galaxy and which is a flocculent spiral galaxy. Explain your reasoning.

44. (a) Calculate the Schwarzschild radius of a supermassive black hole of mass 3.7×10^6 $M_\odot$, the estimated mass of the black hole at the galactic center. Give your answer in both kilometers and astronomical units. (b) What is the angular diameter of such a black hole as seen at a distance of 8 kpc, the distance from Earth to the galactic center? Give your answer in arcseconds. Observing an object with such a small angular size will be a challenge indeed! (c) What is the angular diameter of such a black hole as seen from a distance of 45 AU, the closest that the star SO-16 comes to Sagittarius A? Again, give your answer in arcseconds. Would it be discernible to the naked eye at that distance? (A normal human eye can see details as small as about 60 arcseconds.)

45. (a) The scale bar in Figure 23-27 shows that at the distance of Sagittarius A, a length of 1600 AU has an angular size of 0.2 arcsecond. Use this information to calculate the distance to Sagittarius A*. (b) The star SO-16 moves around Sagittarius A* in an elliptical orbit with semimajor axis 1680 AU. Use this and the information given in the caption to Figure 23-27 to find the orbital period of SO-16. (c) Given the period and the semimajor axis of the star's orbit, is it possible to calculate the mass of SO-16 itself? If it is, explain how this could be done; if not, explain why not.

46. The stars S0-2 and S0-19 orbit Sagittarius A with orbital periods of 14.5 and 37.3 years, respectively. (a) Assuming that the supermassive black hole in Sagittarius A* has a mass of 3.7×10^6 $M_\odot$, determine the semimajor axes of the orbits of these two stars. Give your answers in AU. (b) Calculate the angular size of each orbit's semimajor axis as seen from Earth. (See Section 23-1 for the distance from Earth to the center of the Galaxy.) Explain why extremely high-resolution infrared images are required to observe the motions of these stars.

47. Consider a star that orbits around Sagittarius A* in a circular orbit of radius 530 AU. (a) If the star's orbital speed is 2500 km/s, what is its orbital period? Give your answer in years. (b) Determine the sum of the masses of Sagittarius A* and the star. Give your answer in solar masses. (Your answer is an estimate of the mass of Sagittarius A*, because the mass of a single star is negligibly small by comparison.)

Discussion Questions

48. From what you know about stellar evolution, the interstellar medium, and the density-wave theory, explain the

(a) (b)

R I V U X G

(NASA; JPL-Caltech; R. Kennicutt (University of Arizona); and the SINGS Team) appearance and structure of the spiral arms of grand-design spiral galaxies.

49. What observations would you make to determine the nature of the dark matter in our Galaxy's halo?

50. Describe how the appearance of the night sky might change if dark matter were visible to our eyes.

51. Discuss how a supermassive black hole could have formed at the center of our Galaxy.

Web/eBook Questions

52. Some scientists have suggested that the rotation curve of the Galaxy can be explained by modifying the laws of physics rather than by positing the existence of dark matter. Search the World Wide Web for information about these proposed modifications, called MOND (for *MO*dified *N*ewtonian *D*ynamics). Under what circumstances do MOND and conventional physics predict different behavior? What evidence is there that MOND might be correct?

VIDEO 23.2

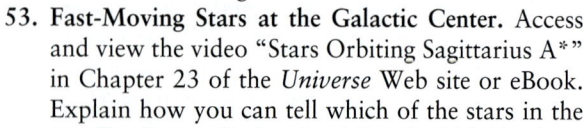

53. **Fast-Moving Stars at the Galactic Center.** Access and view the video "Stars Orbiting Sagittarius A*" in Chapter 23 of the *Universe* Web site or eBook. Explain how you can tell which of the stars in the video are actually close to Sagittarius A* and which just happen to lie along our line of sight to the center of the Galaxy.

Activities

Observing Projects

54. Use the *Starry Night Enthusiast*™ program to observe the Milky Way. (**a**) Display the entire celestial sphere by selecting **Favorites > Guides > Atlas.** Select **View > Stars > Milky Way** to display this galaxy. Select **Options > Stars > Milky Way,** move the **Brightness** slide-bar to the far right to brighten the Milky Way and click **OK.** In the **View** menu, ensure that the **Scrollbars** are activated and use them to look at different parts of the Milky Way. Can you identify the direction toward the galactic nucleus? In this direction the Milky Way appears broadest. Open the **Find** pane, enter **Sagittarius** in the Query box and press **Enter** to center on this constellation to check your identification. (**b**) Use this full-sky view to determine the orientation of the plane of the Galaxy with respect to the celestial sphere. Move the vertical scrollbar to its central position to display the Celestial Equator as a horizontal line across the lower part of the view. Move the horizontal scrollbar until the Milky Way is centered upon the view. Estimate the angle between the Milky Way and the celestial equator on the screen. How well aligned is the plane of the Milky Way with the plane of Earth's equator? (**c**) A third plane of interest is that of the ecliptic, which is shown as a green line. Use the scrollbars to adjust the view so that the ecliptic appears as a straight line rather than as a curve, thereby ensuring that you are viewing in a direction that lies in the ecliptic plane. Use the horizontal scrollbar to move the view to where you can see where the ecliptic crosses the Milky Way. Estimate the angle between the Milky Way and the ecliptic on the screen. How well aligned is the plane of the Milky Way to the ecliptic, the plane of Earth's orbit around the Sun? (**d**) Click on **Home** in the toolbar to return to your home view, stop **Time Flow** and set the local time to midnight (12:00:00 A.M.). Select **Options > Stars > Milky Way,** move the **Brightness** slide-bar to the far right to brighten the Milky Way and click **OK.** Adjust the date to January 1, then February 1, and so on. In which month is the galactic nucleus highest in the sky at midnight, so that it is most easily seen from your location?

55. Use the *Starry Night Enthusiast*™ program to measure the dimensions of the Milky Way Galaxy. Select **Favorites > Stars > Sun in Milky Way** to move to a position 0.15 Mly above the galactic plane, directly above the position of the Sun. (If the Milky Way does not appear immediately, click once on the + Zoom button.) Remove the image of the astronaut's feet by clicking on **View > Feet.** You can rotate the Milky Way Galaxy by putting the mouse cursor over the image, holding down the **Shift** key and the mouse button, and moving the mouse. (**a**) Rotate the Galaxy so that you are seeing it face-on. Use the Hand Tool to measure the distance on the screen from the position of the Sun to the center of the Galaxy, noting this value in both linear distance in light-years and in the subtended angle when viewed from the observer's position. This measurement will calibrate angular measurements in terms of linear distance when seen from this viewpoint. (**b**) Use the Hand Tool to measure several distances from galactic center to the fringes of the galaxy. (Note: These distances will often be displayed in both angular and linear dimensions.) Estimate the diameter of the galaxy in light-years. (**c**) Rotate the Galaxy so that you are seeing it edge-on. Use the Hand Tool to measure the angle subtended from the center of the galaxy to the upper edge of the central bulge of the Galaxy. Use the above measurements of the Sun-Galactic Center in both angle and linear dimension and simple proportions to calculate the half-width of the galactic bulge and multiply this number in light years by 2 to determine the bulge diameter. You can use the relationship between parsecs and light-years (1 pc = 3.26 light-years) to calculate this diameter in parsecs. (**d**) Calculate the thickness-to-diameter ratio of the Milky Way Galaxy.

Collaborative Exercises

56. Student book bags often contain a wide collection of odd-shaped objects. Each person in your group should rummage through their own book bags and find one object that is most similar to the Milky Way Galaxy in shape. List the items from each group member's belongings and describe what about the items is similar to the shape of our Galaxy and what about the items is not similar, then indicate which of the items is the closest match.

57. One strategy for identifying a central location is called *triangulation.* In triangulation, a central position can be pinpointed by knowing the distance from each of three different places. First, on a piece of paper, create a rough map showing where each person in your group lives. Second, create a circle around each person's home that has a radius equal to the distance that

each home is from your classroom. Label the place where the circles intersect as your classroom. Why can you not identify the position of the classroom with only two people's circles?

58. Figure 23-13 shows how emission spectra from hydrogen clouds would be shifted due to their motion around the Galaxy. Create a similar sketch showing an oval automobile racetrack with four cars moving on the track and a stationary observer outside the track at one end. Position and label the four moving cars all sounding their horns: (1) one that would have its horn sound shifted to longer wavelengths; (2) one that would have its horn sound shifted to shorter wavelengths; and (3) two cars moving in opposite directions so that their horn sounds would have no Doppler shift at all.

24 Galaxies

A century ago, most astronomers thought that the entire universe was only a few thousand light-years across and that nothing lay beyond our Milky Way Galaxy. One of the most important discoveries of the twentieth century was that this picture was utterly wrong. We now understand that the Milky Way is just one of billions of galaxies strewn across billions of light-years. The accompanying image shows two of them, denoted by rather mundane catalog numbers (NGC 1531 and NGC 1532) that give no hint to these galaxies' magnificence.

Some galaxies are spirals like NGC 1532 or the Milky Way, with arching spiral arms that are active sites of star formation. (The bright pink bands in NGC 1532 are H II regions, clouds of excited hydrogen that are set aglow by ultraviolet radiation from freshly formed massive stars.) Others, like NGC 1531, are featureless, ellipse-shaped agglomerations of stars, virtually devoid of interstellar gas and dust. Some galaxies are only one-hundredth the size and one ten-thousandth the mass of the Milky Way. Others are giants, with 5 times the size and 50 times the mass of the Milky Way. Only about 10% of a typical galaxy's mass emits radiation of any kind; the remainder is made up of the mysterious dark matter.

Just as most stars are found within galaxies, most galaxies are located in groups and clusters. These clusters of galaxies stretch across the universe, forming huge, lacy patterns. Remarkably, remote clusters of galaxies are receding from us; the greater their distance, the more rapidly they are moving away. This relationship between distance and recessional velocity, called the Hubble law, reveals that our immense universe is expanding. In Chapters 26 and 27 we will learn what this implies about the distant past and remote future of the universe.

WEB LINK 24.1

R I **V** U X G

The two galaxies NGC 1531 and NGC 1532 are so close together that they exert strong gravitational forces on each other. Both galaxies are about 17 million pc (55 million ly) from us in the constellation Eridanus. (Gemini Observatory/Travis Rector, University of Alaska, Anchorage)

Learning Goals

By reading the sections of this chapter, you will learn

24-1 How astronomers first observed other galaxies

24-2 How astronomers determined the distances to other galaxies

24-3 The basic types of galaxies

24-4 What techniques astronomers use to determine distances to remote galaxies

24-5 How the spectra of remote galaxies tell us that the universe is expanding

24-6 How galaxies are grouped into clusters and larger structures

24-7 What happens when galaxies collide

24-8 What observations indicate the presence of dark matter in other galaxies

24-9 How galaxies formed and evolved

24-1 When galaxies were first discovered, it was not clear that they lie far beyond the Milky Way

As early as 1755, the German philosopher Immanuel Kant suggested that vast collections of stars lie outside the confines of the Milky Way. Less than a century later, an Irish astronomer observed the structure of some of the "island universes" that Kant proposed.

William Parsons, the third Earl of Rosse, was a wealthy amateur astronomer who used his fortune to build immense telescopes. His largest telescope, completed in February 1845, had an objective mirror 1.8 meters (6 feet) in diameter (Figure 24-1). The mirror was mounted at one end of a 60-foot tube controlled by cables, straps, pulleys, and cranes (Figure 24-1*a*). For many years,

(a) Rosse's "Leviathan of Parsonstown" R I V U X G

(b) M51 as viewed through the "Leviathan"

Figure 24-1

A Pioneering View of Another Galaxy (a) Built in 1845, this structure housed the largest telescope of its day. The telescope itself (the black cylinder pointing at a 45° angle above the horizontal) was restored to its original state during 1996–1998. **(b)** Using his telescope, Lord Rosse made this sketch of spiral structure in M51. This galaxy, whose angular size is 8 × 11 arcminutes (about a third the angular size of the full moon), is today called the Whirlpool Galaxy because of its distinctive appearance. (a: Birr Castle Demesne; b: Courtesy of Lund Humphries)

this triumph of nineteenth-century engineering was the largest telescope in the world.

With this new telescope, Rosse examined many of the nebulae that had been discovered and cataloged by William Herschel. He observed that some of these nebulae have a distinct spiral structure. One of the best examples is M51, also called NGC 5194. (The "M" designations of galaxies and nebulae come from a catalog compiled by the French astronomer Charles Messier between 1758 and 1782. The "NGC" designations come from the much more extensive "New General Catalogue" of galaxies and nebulae published in 1888 by J. L. E. Dreyer, a Danish astronomer who lived and worked in Ireland.)

Lacking photographic equipment, Lord Rosse had to make drawings of what he saw. Figure 24-1*b* shows a drawing he made of M51, which compares favorably with modern photographs as shown in Figure 24-2. Views such as this inspired Lord Rosse to echo Kant's proposal that such nebulae are actually "island universes."

Many astronomers of the nineteenth century disagreed with this notion of island universes. A considerable number of nebulae are in fact scattered throughout the Milky Way. (Figures 18-2 and 18-4 show some examples.) The safest assumption was that "spiral nebulae," even though they are very different in shape from other sorts of nebulae, could also be components of our Galaxy.

As late as the 1920s it was unclear whether spiral nebulae were very remote "island universes" or simply nearby parts of our Galaxy

WEB LINK 24.3

The astronomical community became increasingly divided over the nature of the spiral nebulae. In April 1920, two opposing ideas were presented before the National Academy of Sciences in Washington, D.C. On one side was Harlow Shapley from the Mount Wilson Observatory, renowned for his recent determination of the size of the Milky Way Galaxy (see Section 23-1). Shapley thought the spiral nebulae were relatively small, nearby objects scattered around our Galaxy like the globular clusters he had studied. Opposing Shapley was Heber D. Curtis of the University of California's Lick Observatory. Curtis championed the island universe theory, arguing that each of these spiral nebulae is a rotating system of stars much like our own Galaxy.

The Shapley-Curtis "debate" generated much heat but little light. Nothing was decided, because no one could present conclusive evidence to demonstrate exactly how far away the spiral nebulae are. Astronomy desperately needed a definitive determination of the distance to a spiral nebula. Such a measurement became the first great achievement of a young man who studied astronomy at the Yerkes Observatory, near Chicago. His name was Edwin Hubble.

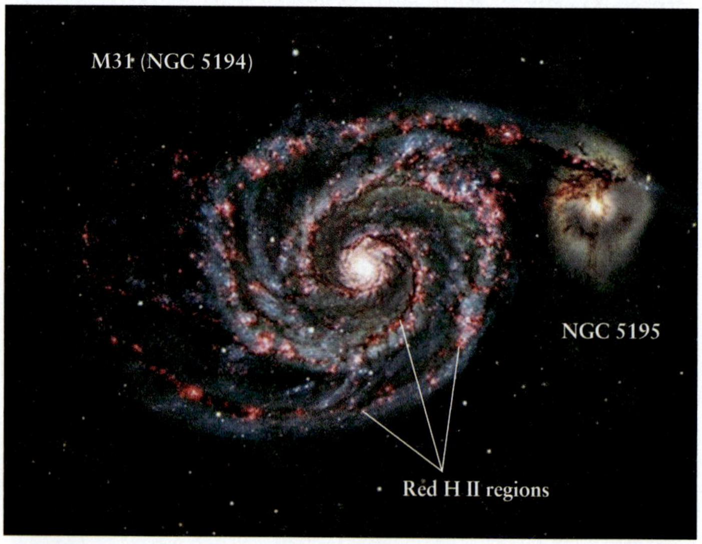

WEB LINK 24.2

Figure 24-2 R I V U X G

A Modern View of the Spiral Galaxy M51 This galaxy, also called NGC 5194, has spiral arms that are outlined by glowing H II regions. These regions reveal the sites of star formation (see Section 18-2). One spiral arm extends toward the companion galaxy NGC 5195. (CFHT)

Figure 24-3 R I **V** U X G
The Andromeda "Nebula" The Great Nebula in Andromeda, also known as M31, can be seen with even a small telescope. Edwin Hubble was the first to demonstrate that M31 is actually a galaxy that lies far beyond the Milky Way. M32 and M110 are two small satellite galaxies that orbit M31. (Bill & Sally Fletcher/Tom Stack & Associates)

24-2 Hubble proved that the spiral nebulae are far beyond the Milky Way

After completing his studies, Edwin Hubble joined the staff of the Mount Wilson Observatory in Pasadena, California. On October 6, 1923, he took a historic photograph of the Andromeda "Nebula," one of the spiral nebulae around which controversy raged. Figure 24-3 is a modern photograph of this object.)

Hubble carefully examined his photographic plate and discovered a *new* bright spot—what he at first thought to be a nova. Referring to previous plates of that region, he soon realized that the object was actually a Cepheid variable star. Further scrutiny of additional plates over the next several months revealed several more Cepheids. Figure 24-4 shows modern observations of a Cepheid in another "spiral nebula."

Just as RR Lyrae variable stars demonstrated the extent of the Milky Way, Cepheid variables revealed the immense distances to other galaxies

As we saw in Section 19-6, Cepheid variables help astronomers determine distances. An astronomer begins by carefully measuring the variations in apparent brightness of a Cepheid variable, then recording the results in the form of a plot of brightness versus time, or light curve (see Figure 19-18*a*). This graph gives the variable star's period and average brightness. Given the star's period, the astronomer then uses the period-luminosity relation shown in Figure 19-19 to find the Cepheid's average luminosity. (The astronomer must also examine the spectrum of the star to determine whether it is a metal-rich Type I Cepheid or a metal-poor Type II Cepheid. As Figure 19-19 shows, Type I and Type II Cepheids have somewhat different period-luminosity relations.) Knowing both the apparent brightness and luminosity of the Cepheid, the astronomer can then use the inverse-square law to calculate the distance to the star (see Box 17-2). Box 24-1 presents an example of this calculation. This procedure is very similar to that used by Harlow Shapley to measure the distances to the Milky Way's globular clusters using RR Lyrae variable stars (see Section 23-1).

Cepheid variables are intrinsically quite luminous, with average luminosities that can exceed 10^4 $L_{\odot}$. Hubble realized that for these luminous stars to appear as dim as they were on his

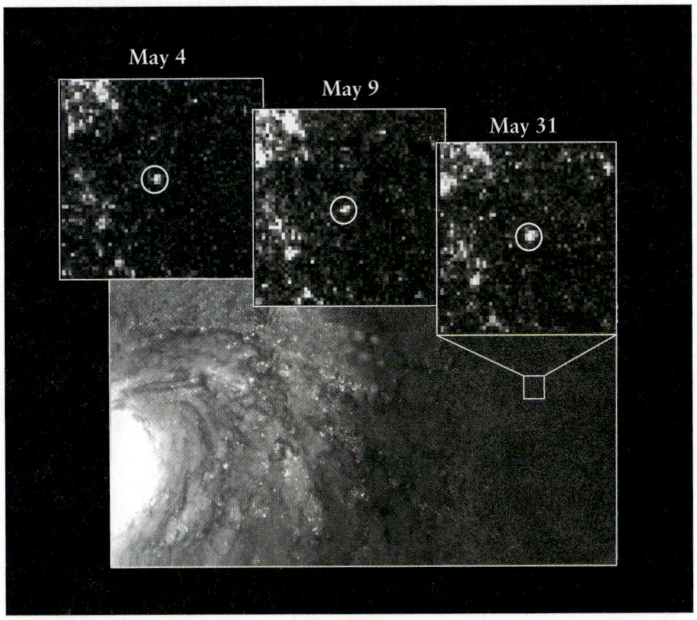

VIDEO 24.1

Figure 24-4 R I **V** U X G
Measuring Galaxy Distances with Cepheid Variables By observing Cepheid variable stars in M100, the galaxy shown here, astronomers have found that it is about 17 Mpc (56 million ly) from Earth. The insets show one of the Cepheids in M100 at different stages in its brightness cycle, which lasts several weeks. (Wendy L. Freedman, Carnegie Institution of Washington, and NASA)

BOX 24-1 Tools of the Astronomer's Trade

Cepheids and Supernovae as Indicators of Distance

Because their periods are directly linked to their luminosities, Cepheid variables are one of the most reliable tools astronomers have for determining the distances to galaxies. To this day, astronomers use this link—much as Hubble did back in the 1920s—to measure intergalactic distances. More recently, they have begun to use Type Ia supernovae, which are far more luminous and thus can be seen much farther away, to determine the distances to very remote galaxies.

EXAMPLE: In 1992 a team of astronomers observed Cepheid variables in a galaxy called IC 4182 to deduce this galaxy's distance from Earth. One such Cepheid has a period of 42.0 days and an average apparent magnitude (m) of +22.0. (See Box 19-3 for an explanation of the apparent magnitude scale.) By comparison, the dimmest star you can see with the naked eye has $m = +6$; this Cepheid in IC 4182 appears less than one one-millionth as bright. The star's spectrum shows that it is a metal-rich Type I Cepheid variable.

According to the period-luminosity relation shown in Figure 21-17, such a Type I Cepheid with a period of 42.0 days has an average luminosity of 33,000 $L_\odot$. An equivalent statement is that this Cepheid has an average absolute magnitude (M) of −6.5. (This compares to $M = +4.8$ for the Sun.) Use this information to determine the distance to IC 4182.

Situation: We are given the apparent magnitude $m = +22.0$ and the absolute magnitude $M = -6.5$ of the Cepheid variable star in IC 4182. Our goal is to calculate the distance to this star, and hence the distance to the galaxy of which it is part.

Tools: We use the relationship between apparent magnitude, absolute magnitude, and distance given in Box 19-3.

Answer: In Box 19-3, we saw that the apparent magnitude of a star is related to its absolute magnitude and distance in parsecs (d) by

$$m - M = 5 \log d - 5$$

This equation can be rewritten as

$$d = 10^{(m-M+5)/5} \text{ parsecs}$$

We have $m - M = (+22.0) - (-6.5) = 22.0 + 6.5 = 28.5$. (Recall from Box 19-3 that $m - M$ is called the *distance modulus.*) Hence, our equation becomes

$$d = 10^{(28.5+5)/5} \text{ parsecs} = 10^{6.7} \text{ parsecs} = 5 \times 10^6 \text{ parsecs}$$

(A calculator is needed to calculate the quantity $10^{6.7}$.)

Review: Our result tells us that the galaxy is 5 million parcsecs, or 5 Mpc, from Earth (1 Mpc = 10^6 pc). This distance can also be expressed as 16 million light-years.

EXAMPLE: Astronomers are interested in IC 4182 because a Type Ia supernova was observed there in 1937. All Type Ia supernovae are exploding white dwarfs that reach nearly the same maximum brightness at the peak of their outburst (see Section 22-9). Once astronomers know the peak absolute magnitude of Type Ia supernovae, they can use these supernovae as distance indicators. Because the distance to IC 4182 is known from its Cepheids, the 1937 observations of the supernova in that galaxy help us calibrate Type Ia supernovae as distance indicators.

At maximum brightness, the 1937 supernova reached an apparent magnitude of $m = +8.6$. What was its absolute magnitude at maximum brightness?

Situation: We are given the supernova's apparent magnitude m, and we know its distance from the previous example. Our goal is to calculate its absolute magnitude M.

Tools: We again use the relationship $m - M = 5 \log d - 5$.

Answer: We could plug in the value of d found in the previous example. But it is simpler to note that the distance modulus $m - M$ has the same value no matter whether it refers to a Cepheid, a supernova, or any other object, just so it is at the same distance d. From the Cepheid example we have $m - M = 28.5$ for IC 4182, so

$$M = m - (m - M) = 8.6 - (28.5) = -19.9$$

This absolute magnitude corresponds to a remarkable peak luminosity of 10^{10} $L_\odot$.

Review: Whenever astronomers find a Type Ia supernova in a remote galaxy, they can combine this absolute magnitude with the observed maximum apparent magnitude to get the galaxy's distance modulus, from which the galaxy's distance can be easily calculated (just as we did above for the Cepheids in IC 4182). This technique has been used to determine the distances to galaxies hundreds of millions of parsecs away (see Section 26-4).

photographs of the Andromeda "Nebula," they must be extremely far away. Straightforward calculations using modern data reveal that M31 is some 750 kiloparsecs (2.5 million light-years) from Earth. Based on its angular size, M31 has a diameter of 70 kiloparsecs—larger than the diameter of our own Milky Way Galaxy!

These results prove that the Andromeda "Nebula" is actually an enormous stellar system, far beyond the confines of the Milky Way. Today, this system is properly called the Andromeda Galaxy. (Under good observing conditions, you can actually see this galaxy's central bulge with the naked eye. If you could see the entire

Andromeda Galaxy, it would cover an area of the sky roughly 5 times as large as the full moon.) Galaxies are so far away that their distances from us are usually given in millions of parsecs, or *megaparsecs* (Mpc): 1 Mpc = 10^6 pc. For example, the distance to the galaxies in the image that opens this chapter is 17 Mpc.

Hubble's results, which were presented at a meeting of the American Astronomical Society on December 30, 1924, settled the Shapley-Curtis "debate" once and for all. The universe was recognized to be far larger and populated with far bigger objects than anyone had seriously imagined. Hubble had discovered the realm of the galaxies.

CAUTION! In everyday language, many people use the words "galaxy" and "universe" interchangeably. It is true that before Hubble's discoveries our Milky Way Galaxy was thought to constitute essentially the entire universe. But we now know that the universe contains literally billions of galaxies. A single galaxy, vast though it may be, is just a tiny part of the entire observable universe.

24-3 Galaxies are classified according to their appearance

Millions of galaxies are visible across every unobscured part of the sky. Although all galaxies are made up of large numbers of stars, they come in a variety of shapes and sizes.

Hubble classified galaxies into four broad categories based on their appearance. These categories form the basis for the **Hubble classification,** a scheme that is still used today. The four classes of galaxies are the spirals, classified S; barred spirals, or SB; ellipticals, E; and irregulars, Irr. Table 24-1 summarizes some key properties of each class. These various types of galaxies differ not only in their shapes but also in the kinds of processes taking place within them.

Spiral Galaxies: Stellar Birthplaces

M51 (Figure 24-2), M31 (Figure 24-3), and M100 (Figure 24-4) are examples of **spiral galaxies.** Figure 24-5 shows that spiral galaxies are characterized by arched lanes of stars, just as is our own Milky Way Galaxy (see Section 23-3). The spiral arms contain young, hot, blue stars and their associated H II regions, indicating ongoing star formation.

Thermonuclear reactions within stars create *metals,* that is, elements heavier than hydrogen or helium (see Section 17-5). These metals are dispersed into space as the stars evolve and die. So, if new stars are being formed from the interstellar matter in spiral galaxies, they will incorporate these metals and be Population I stars (see Section 19-5 and Section 23-2). Indeed, the visible-light spectrum of the disk of a spiral galaxy has strong metal absorption lines. Such a spectrum is a composite of the spectra of many stars and shows that the stars in the disk are principally of Population I. By contrast, there is relatively little star formation in the central bulges of spiral galaxies, and these regions are dominated by old Population II stars that have a low metal content. The lack of star formation also explains why the central bulges of spiral galaxies have a yellowish or reddish color; as a population of stars ages, the massive, luminous blue stars die off first, leaving only the longer-lived, low-mass red stars.

Hubble further classified spiral galaxies according to the texture of the spiral arms and the relative size of the central bulge. Spirals with smooth, broad spiral arms and a fat central bulge are called Sa galaxies, for spiral type *a* (Figure 24-5*a*); those galaxies with moderately well-defined spiral arms and a moderate-sized central bulge, like M31 and M51, are Sb galaxies (Figure 24-5*b*); and galaxies with narrow, well-defined spiral arms and a tiny central bulge are Sc galaxies (Figure 24-5*c*).

The differences between Sa, Sb, and Sc galaxies may be related to the relative amounts of gas and dust they contain. Observations with infrared telescopes (which detect the emission from interstellar dust) and radio telescopes (which detect radiation from interstellar gases such as hydrogen and carbon monoxide) show that about 4% of the mass of a Sa galaxy is in the form of gas and dust. This percentage is 8% for Sb galaxies and 25% for Sc galaxies.

Interstellar gas and dust is the material from which new stars are formed, so an Sc galaxy has a greater proportion of its mass involved in star formation than an Sb or Sa galaxy. Hence, a Sc galaxy has a large disk (where star formation occurs) and a small central bulge (where there is little or no star formation). By comparison, a Sa galaxy, which has relatively little gas and dust, and

Table 24-1 Some Properties of Galaxies

	Spiral (S) and barred spiral (SB) galaxies	Elliptical galaxies (E)	Irregular galaxies (Irr)
Mass ($M_\odot$)	10^9 to 4×10^{11}	10^5 to 10^{13}	10^8 to 3×10^{10}
Luminosity ($L_\odot$)	10^8 to 2×10^{10}	3×10^5 to 10^{11}	10^7 to 10^9
Diameter (kpc)	5 to 250	1 to 200	1 to 10
Stellar populations	Spiral arms: young Population I Nucleus and throughout disk: Population II and old Population I	Population II and old Population I	mostly Population I
Percentage of observed galaxies	77%	20%*	3%

**This percentage does not include dwarf elliptical galaxies that are as yet too dim and distant to detect. Hence, the actual percentage of galaxies that are ellipticals may be higher than shown here.*

(a) Sa (NGC 1357) (b) Sb (M81) (c) Sc (NGC 4321)

Figure 24-5 R I V U X G

Spiral Galaxies Edwin Hubble classified spiral galaxies according to the texture of their spiral arms and the relative size of their central bulges. Sa galaxies have smooth, broad spiral arms and the largest central bulges, while Sc galaxies have narrow, well-defined arms and the smallest central bulges. (a: Adam Block/Steve Mandel/Jim Rada and Students/NOAO/AURA/NSF; b: Giovanni Benintende; c: FORS Team, 8.2-meter VLT, ESO)

thus less material from which to form stars, has a large central bulge and only a small star-forming disk.

Barred Spiral Galaxies: Spirals with an Extra Twist

In **barred spiral galaxies,** such as those shown in Figure 24-6, the spiral arms originate at the ends of a bar-shaped region running through the galaxy's nucleus rather than from the nucleus itself. As with ordinary spirals, Hubble subdivided barred spirals according to the relative size of their central bulge and the character of their spiral arms. A SBa galaxy has a large central bulge and thin, tightly wound spiral arms (Figure 24-6*a*). Likewise, a SBb galaxy is a barred spiral with a moderate central bulge and moderately wound spiral arms (Figure 24-6*b*), while a SBc galaxy has lumpy, loosely wound spiral arms and a tiny central bulge (Figure 24-6*c*). As for ordinary spiral galaxies, the difference between SBa, SBb, and SBc galaxies may be related to the amount of gas and dust in the galaxy.

Bars appear to form naturally in many spiral galaxies. This conclusion comes from computer simulations of galaxies, which set hundreds of thousands of simulated "stars" into orbit around a common center. As the "stars" orbit and exert gravitational forces on one another, a bar structure forms in most cases. Indeed, barred spiral galaxies outnumber ordinary spirals by about two to one. (As we saw in Section 23-2, there is evidence that the Milky Way Galaxy is a barred spiral.)

Why don't all spiral galaxies have bars? According to calculations by Jeremiah Ostriker and P. J. E. Peebles of Princeton Uni-

(a) SBa (NGC 1291) (b) SBb (M83) (c) SBc (NGC 1365)

Figure 24-6 R I V U X G

Barred Spiral Galaxies As with spiral galaxies, Hubble classified barred spirals according to the texture of their spiral arms (which correlates to the sizes of their central bulges). SBa galaxies have the smoothest spiral arms and the largest central bulges, while SBc galaxies have narrow, well-defined arms and the smallest central bulges. (a: NASA/JPL-Caltech/CTIO; b: FORS Team, 8.2-meter VLT, ESO; c: FORS Team, 8.2-meter VLT Antu, ESO)

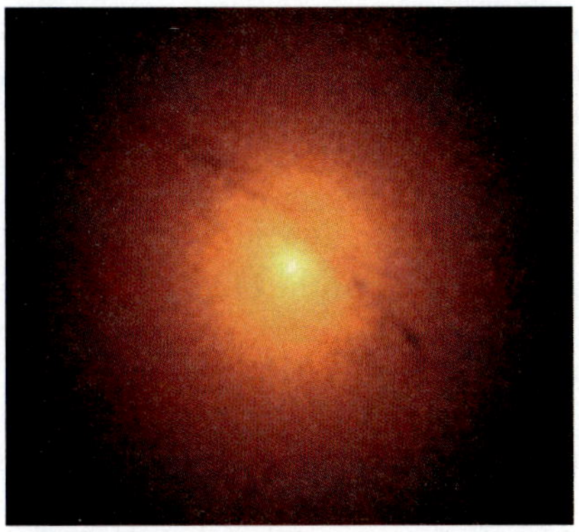

(a) E0 (M105)

(b) E3 (NGC 4406)

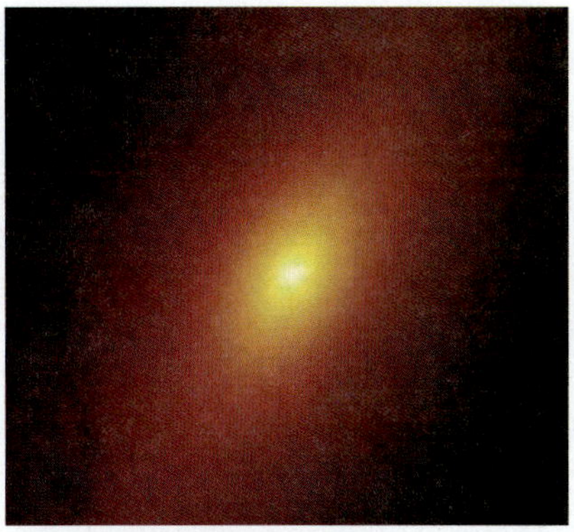

(c) E6 (NGC 3377)

Figure 24-7 R I V U X G

Elliptical Galaxies Hubble classified elliptical galaxies according to how round or flattened they look. A galaxy that appears round is labeled E0, and the flattest-appearing elliptical galaxies are designated E7. (a: Karl Gebhardt (University of Michigan), Tod Lauer (NOAO), and NASA; b: Jean-Charles Cuillandre, Hawaiian Starlight, CFHT; c: Karl Gebhardt (University of Michigan), Tod Lauer (NOAO), and NASA)

versity, a bar will not develop if a galaxy is surrounded by a sufficiently massive halo of nonluminous *dark matter*. (In Section 23-4 we saw evidence that our Milky Way Galaxy is surrounded by such a dark matter halo.) The difference between barred spirals and ordinary spirals may thus lie in the amount of dark matter the galaxy possesses. In Section 24-8 we will see compelling evidence for the existence of dark matter in spiral galaxies.

Elliptical Galaxies: From Giants to Dwarfs

Elliptical galaxies, so named because of their distinctly elliptical shapes, have no spiral arms. Hubble subdivided these galaxies according to how round or flattened they look. The roundest elliptical galaxies are called E0 galaxies and the flattest, E7 galaxies. Elliptical galaxies with intermediate amounts of flattening are given designations between these extremes (Figure 24-7).

CAUTION! Unlike the designations for spirals and barred spirals, the classifications E0 through E7 may not reflect the true shape of elliptical galaxies. An E1 or E2 galaxy might actually be a very flattened disk of stars that we just happen to view face-on, and a cigar-shaped E7 galaxy might look spherical if seen end-on. The Hubble scheme classifies galaxies entirely by how they *appear* to us on Earth.

Elliptical galaxies look far less dramatic than their spiral and barred spiral cousins. The reason, as shown by radio and infrared observations, is that ellipticals are virtually devoid of interstellar gas and dust. Consequently, there is little material from which stars could have recently formed, and indeed there is no evidence of young stars in most elliptical galaxies. For the most part, star formation in elliptical galaxies ended long ago. Hence, these galaxies are composed of old, red, Population II stars with only small amounts of metals.

Elliptical galaxies come in a wide range of sizes and masses. Both the largest and the smallest galaxies in the known universe are elliptical. Figure 24-8 shows two **giant elliptical galaxies** that are about 20 times larger than an average galaxy. These giant ellipticals are located near the middle of a large cluster of galaxies in the constellation Virgo. (We will discuss this and other clusters of galaxies in Section 24-6.)

Giant ellipticals are rather rare, but **dwarf elliptical galaxies** are quite common. Dwarf ellipticals are only a fraction the size of their normal counterparts and contain so few stars—only a few million, compared to more than 100 billion (10^{11}) stars in our Milky Way Galaxy—that these galaxies are completely transparent. You can actually see straight through the center of a dwarf galaxy and out the other side, as Figure 24-9 shows.

WEB LINK 24.5

Figure 24-8 R I V U X G

Giant Elliptical Galaxies The Virgo cluster is a rich, sprawling collection of more than 2000 galaxies about 17 Mpc (56 million ly) from Earth. Only the center of this huge cluster appears in this photograph. The two largest members of this cluster are the giant elliptical galaxies M84 and M86. These galaxies have angular sizes of 5 to 7 arcmin. (Jean-Charles Cuillandre, Hawaiian Starlight, CFHT)

Figure 24-9 R I V U X G

A Dwarf Elliptical Galaxy This diffuse cloud of stars is a nearby E4 dwarf elliptical called Leo I. It actually orbits the Milky Way at a distance of about 180 kpc (600,000 ly). Leo I is about 1 kpc (3000 ly) in diameter but contains so few stars that you can see through the galaxy's center. (Anglo-Australian Observatory)

The visible light from a galaxy is emitted by its stars, so the visible-light spectrum of a galaxy has absorption lines. But because a galaxy's stars are in motion, with some approaching us and others moving away, the Doppler effect smears out and broadens the absorption lines. The average motions of stars in a galaxy can be deduced from the details of this broadening.

For elliptical galaxies, studies of this kind show that star motions are quite random. In a very round (E0) elliptical galaxy, this randomness is **isotropic,** meaning "equal in all directions." Because the stars are whizzing around equally in all directions, the galaxy is genuinely spherical. In a flattened (E7) elliptical galaxy, the randomness of the stellar motions is **anisotropic,** which means that the range of star speeds is different in different directions.

Hubble also identified galaxies that are midway in appearance between ellipticals and the two kinds of spirals. These are denoted as S0 and SB0 galaxies, also called **lenticular galaxies.** Although they look somewhat elliptical, lenticular ("lens-shaped") galaxies have both a central bulge and a disk like spiral galaxies, but no discernible spiral arms. They are therefore sometimes referred to as "armless spirals." Figure 24-10 shows an example of an SB0 lenticular galaxy.

Edwin Hubble summarized his classification scheme for spiral, barred spiral, and elliptical galaxies in a diagram, now called the **tuning fork diagram** for its shape (Figure 24-11).

CAUTION! When Hubble first drew his tuning fork diagram, he had the idea that it represented an evolutionary sequence. He thought that galaxies evolved over time from the left to the right of the diagram, beginning as ellipticals and eventually becoming either spiral or barred spiral galaxies. We now understand that this evolution of galaxies is not the case at all! For one thing, elliptical galaxies have little or no overall rotation, while spiral and barred spiral galaxies have a substantial amount of overall rotation. There is no way that an elliptical galaxy could suddenly start rotating, which means that it could not evolve into a spiral galaxy.

A more modern interpretation of the Hubble tuning fork diagram is that it is an arrangement of galaxies according to their overall rotation. A rapidly rotating collection of matter in space tends to form a disk, while a slowly rotating collection does not. Thus, the elliptical galaxies at the far left of the tuning fork diagram have little internal rotation and hence no disk. Sa and SBa galaxies have enough overall rotation to form a disk, though their central bulges are still dominant. The galaxies with the greatest amount of overall rotation are Sc and SBc galaxies, in which the central bulges are small and most of the gas, dust, and stars are in the disk.

Irregular Galaxies: Deformed and Dynamic

Galaxies that do not fit into the scheme of spirals, barred spirals, and ellipticals are usually referred to as **irregular galaxies.** They are generally rich in interstellar gas and dust, and have both young and old stars. For lack of any better scheme, the irregular galaxies are sometimes placed between the ends of the tines of the Hubble tuning fork diagram, as in Figure 24-11.

Figure 24-10 R I V U X G

A Lenticular Galaxy NGC 2787 is classified as a lenticular galaxy because it has a disk but no discernible spiral arms. Its nucleus displays a faint bar (not apparent in this image), so NGC 2787 is denoted as an SB0 galaxy. It lies about 7.4 Mpc (24 million ly) from Earth in the constellation Ursa Major. (NASA and The Hubble Heritage Team, STScI/AURA)

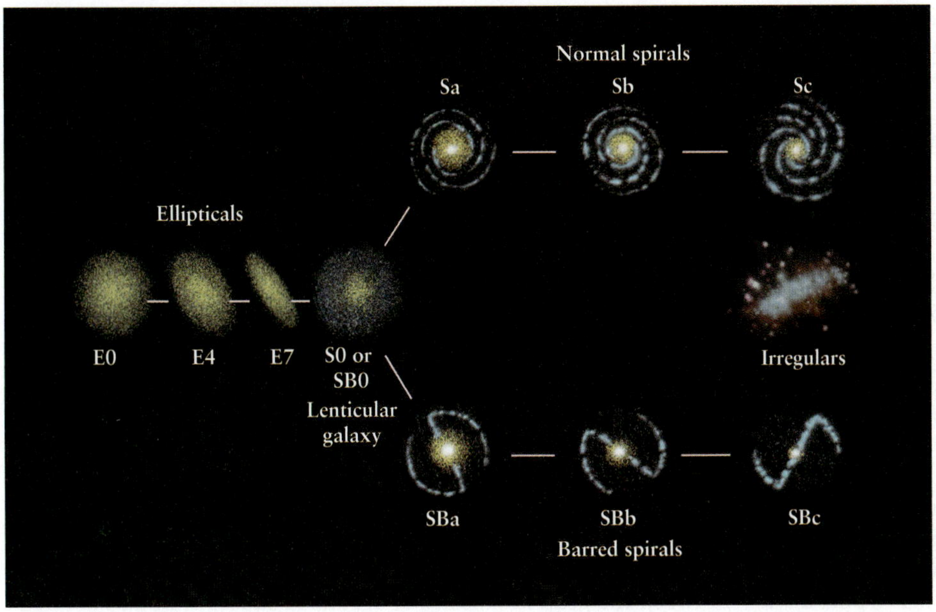

Figure 24-11

Hubble's Tuning Fork Diagram Edwin Hubble's classification of regular galaxies is shown in his tuning fork diagram. An elliptical galaxy is classified by how flattened it appears. A spiral or barred spiral galaxy is classified by the texture of its spiral arms and the size of its central bulge. A lenticular galaxy, is intermediate between ellipticals and spirals. Irregular galaxies do not fit into this simple classification scheme.

Hubble defined two types of irregulars. Irr I galaxies have only hints of organized structure, and have many OB associations and H II regions. The best-known examples of Irr I galaxies are the Large Magellanic Cloud (Figure 24-12) and the Small Magellanic Cloud. Both are nearby companions of our Milky Way and can be seen with the naked eye from southern latitudes. Both these galaxies contain substantial amounts of interstellar gas. Tidal forces exerted on these irregular galaxies by the Milky Way help to compress the gas, which is why both of the Magellanic Clouds are sites of active star formation.

WEB LINK 24.6

Figure 24-12 R I **V** U X G

The Large Magellanic Cloud (LMC) At a distance of only 55 kpc (179,000 ly), this Irr I galaxy is the third closest known companion of our Milky Way Galaxy. About 19 kpc (62,000 ly) across, the LMC spans 22° across the sky, or about 50 times the size of the full moon. One sign that star formation is ongoing in the LMC is the Tarantula Nebula, whose diameter of 250 pc (800 ly) and mass of $5 \times 10^6 M_\odot$ make it the largest known H II region (NOAO/AURA/NSF).

The other type of irregulars, called Irr II galaxies, have asymmetrical, distorted shapes that seem to have been caused by collisions with other galaxies or by violent activity in their nuclei. M82, shown in Figure 22-17 and in the image that opens Chapter 6, is an example of an Irr II galaxy.

24-4 Astronomers use various techniques to determine the distances to remote galaxies

WEB LINK 24.7

A key question that astronomers ask about galaxies is "How far away are they?" Unfortunately, many of the techniques that are used to measure distances within our Milky Way Galaxy cannot be used for the far greater distances to other galaxies. The extremely accurate parallax method that we described in Section 17-1 can be used only for stars within about 500 pc. Beyond that distance, parallax angles become too small to measure. Spectroscopic parallax, in which the distance to a star or star cluster is found with the help of the H-R diagram (see Section 17-8), is accurate only out to roughly 10 kpc from Earth; more distant stars or clusters are too dim to give reliable results.

The various methods of distance determination are interrelated because one is used to calibrate another

Standard Candles: Variable Stars and Type Ia Supernovae

To determine the distance to a remote galaxy, astronomers look instead for a **standard candle**—an object, such as a star, that lies within that galaxy and for which we know the luminosity (or, equivalently, the absolute magnitude, described in Section 17-3). By measuring how bright the standard candle appears, astronomers can calculate its distance—and hence the distance to the galaxy of which it is part—using the inverse-square law.

The challenge is to find standard candles that are luminous enough to be seen across the tremendous distances to galaxies. To be useful, standard candles should have four properties:

1. They should be luminous, so we can see them out to great distances.

2. We should be fairly certain about their luminosities, so we can be equally certain of any distance calculated from a standard candle's apparent brightness and luminosity.

3. They should be easily identifiable—for example, by the shape of the light curve of a variable star.

4. They should be relatively common, so that astronomers can use them to determine the distances to many different galaxies.

For nearby galaxies, Cepheid variable stars make reliable standard candles. These variables can be seen out to about 30 Mpc (100 million ly) using the Hubble Space Telescope, and their luminosity can be determined from their period through the period-luminosity relation depicted in Figure 19-19. Box 24-1 gives an example of using Cepheid variables to determine distances. RR Lyrae stars, which are Population II variable stars often found in globular clusters, can be used as standard candles in a similar way. (We saw in Section 25-1 how RR Lyrae variables helped determine the size of our Galaxy.) Because they are less luminous than Cepheids, RR Lyrae variables can be seen only out to 100 kpc (300,000 ly).

Beyond about 30 Mpc even the brightest Cepheid variables, which have luminosities of about $2 \times 10^4\ L_\odot$, fade from view. Astronomers have tried to use even more luminous stars such as blue supergiants to serve as standard candles. However, this idea is based on the assumption that there is a fixed upper limit on the luminosities of stars, which may not be the case. Hence, these standard candles are not very "standard," and distances measured in this way are somewhat uncertain.

WEB LINK 24.8

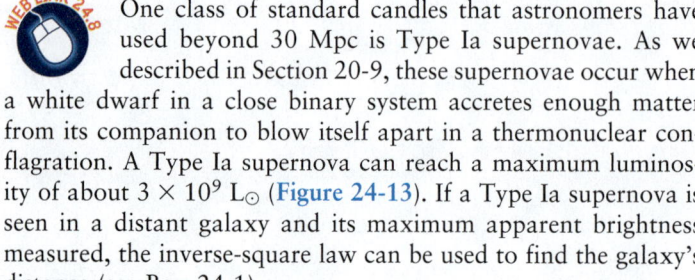

One class of standard candles that astronomers have used beyond 30 Mpc is Type Ia supernovae. As we described in Section 20-9, these supernovae occur when a white dwarf in a close binary system accretes enough matter from its companion to blow itself apart in a thermonuclear conflagration. A Type Ia supernova can reach a maximum luminosity of about $3 \times 10^9\ L_\odot$ (**Figure 24-13**). If a Type Ia supernova is seen in a distant galaxy and its maximum apparent brightness measured, the inverse-square law can be used to find the galaxy's distance (see Box 24-1).

One complication is that not all Type Ia supernovae are equally luminous. Fortunately, there is a simple relationship between the peak luminosity of a Type Ia supernova and the rate at which the luminosity decreases after the peak: The more slowly the brightness decreases, the more luminous the supernova. Using this relationship, astronomers have measured distances to supernovae more than 1000 Mpc (3 billion ly) from Earth.

Unfortunately, this technique can be used only for galaxies in which we happen to observe a Type Ia supernova. But telescopic surveys now identify many dozens of these supernovae every year, so the number of galaxies whose distances can be measured in this way is continually increasing.

(a) M100 in March 2002 R I V U X G

(b) M100 in February 2006, showing Supernova 2006X R I V U X G

Figure 24-13

A Supernova in a Spiral Galaxy These images from the Very Large Telescope show the spiral galaxy M100 **(a)** before and **(b)** after a Type Ia supernova exploded within the galaxy in 2006. (The two images were made with different color filters, which gives them different appearances.) Such luminous supernovae, which can be seen at extreme distances, are important standard candles used to determine the distances to faraway galaxies. The distance to M100 is also known from observations of Cepheid variables (see Figure 24-4), so this particular supernova can help calibrate Type Ia supernovae as distance indicators. (European Southern Observatory)

Distance Determination without Standard Candles

Other methods for determining the distances to galaxies do not make use of standard candles. One was discovered in the 1970s by the astronomers Brent Tully and Richard Fisher. They found that the width of the hydrogen 21-cm emission line of a spiral galaxy (see Section 23-3) is related to the galaxy's luminosity. This correlation is the **Tully-Fisher relation—**the broader the line, the more luminous the galaxy.

Such a relationship exists because radiation from the approaching side of a rotating galaxy is blueshifted while that from the galaxy's receding side is redshifted. Thus, the 21-cm line is Doppler broadened by an amount directly related to how fast a galaxy is rotating. Rotation speed is related to the galaxy's mass by Newton's form of Kepler's third law. The more massive a galaxy, the more stars it contains and thus the more luminous it is. Consequently, the width of a galaxy's 21-cm line is directly related to its luminosity.

Because line widths can be measured quite accurately, astronomers can use the Tully-Fisher relation to determine the luminosity of a distant spiral galaxy. By combining this information with measurements of apparent brightness, they can calculate the distance to the galaxy. This technique can be used to measure distances of 100 Mpc or more.

Elliptical galaxies do not rotate, so the Tully-Fisher relation cannot be used to determine their distances. But in 1987 the American astronomers Marc Davis and George Djorgovski pointed out a correlation between the size of an elliptical galaxy, the average motions of its stars, and how the galaxy's brightness appears distributed over its surface.

In geometry, three points define a plane, so the relationship among size, motion, and brightness is called the **fundamental plane.** By measuring the last two quantities, an astronomer can use the fundamental plane relationship to determine a galaxy's actual size. And by comparing this to the galaxy's apparent size, the astronomer can calculate the distance to the galaxy using the small-angle formula (Box 1-1). Ellipticals can be substantially larger and more luminous than spirals, so the fundamental plane can be used at somewhat greater distances than the Tully-Fisher relation.

The Distance Ladder

Figure 24-14 shows the ranges of applicability of several important means of determining astronomical distances. Because these ranges overlap, one technique can be used to calibrate another. As an example, astronomers have studied Cepheids in nearby galaxies that have also been host to Type Ia supernovae. The Cepheids provide the distances to these nearby galaxies, making it possible to determine the peak luminosity of each supernova using its maximum apparent brightness and the inverse-square law. Once the peak luminosity is known, it can be used to determine the distance to Type Ia supernovae in more distant galaxies. Because one measuring technique leads us to the next one like rungs on a ladder, the techniques shown in Figure 24-14 (along with others) are referred to collectively as the **distance ladder.**

ANALOGY If you give a slight shake to the bottom of a tall ladder, the top can wobble back and forth alarmingly. A change in distance-measuring techniques used for nearby objects can also have substantial effects on the distances to remote galaxies. For instance, if astronomers discovered that the distances to nearby Cepheids were in error, distance measurements using any technique that is calibrated by Cepheids would be affected as well. (As an example, the distance to the galaxy M100 shown in Figure 24-4 is determined using Cepheids. A Type Ia supernova

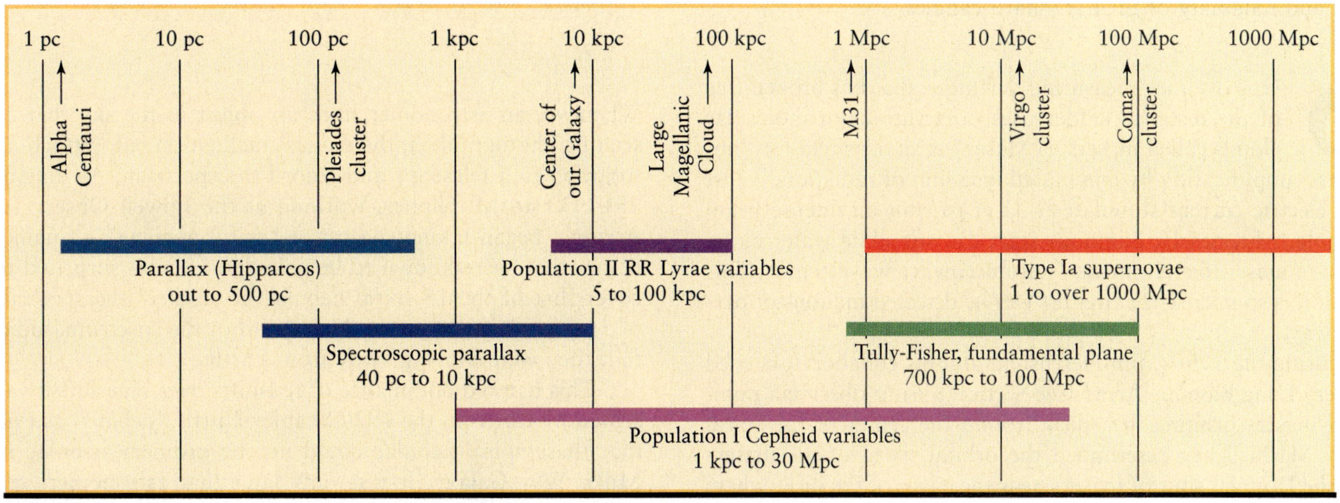

Figure 24-14

The Distance Ladder Astronomers employ a variety of techniques for determining the distances to objects beyond the solar system. Because their ranges of applicability overlap, one technique can be used to calibrate another. The arrows indicate distances to several important objects. Note that each division on the scale indicates a tenfold increase in distance, such as from 1 to 10 Mpc.

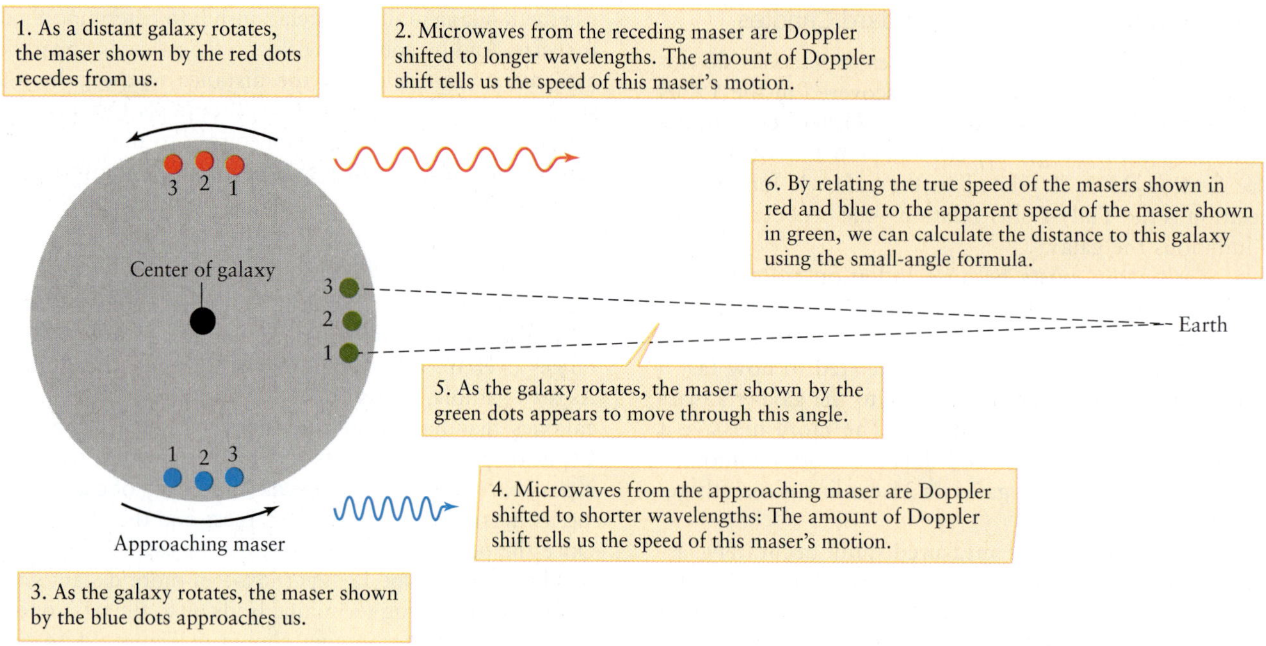

1. As a distant galaxy rotates, the maser shown by the red dots recedes from us.
2. Microwaves from the receding maser are Doppler shifted to longer wavelengths. The amount of Doppler shift tells us the speed of this maser's motion.
3 2 1
6. By relating the true speed of the masers shown in red and blue to the apparent speed of the maser shown in green, we can calculate the distance to this galaxy using the small-angle formula.
Center of galaxy
3
2
1
Earth
5. As the galaxy rotates, the maser shown by the green dots appears to move through this angle.
1 2 3
4. Microwaves from the approaching maser are Doppler shifted to shorter wavelengths: The amount of Doppler shift tells us the speed of this maser's motion.
Approaching maser
3. As the galaxy rotates, the maser shown by the blue dots approaches us.

Figure 24-15

Measuring the Distance to a Galaxy Using Masers This drawing shows interstellar clouds called masers (the colored dots) moving from position 1 to 2 to 3 as they orbit the center of a galaxy. The redshift and blueshift of microwaves from the masers shown in red and blue tell us their orbital speed. By relating this to the angle through which the masers shown in green appear to move in a certain amount of time, we can calculate the distance to the galaxy.

has been seen in M100, as Figure 24-13 shows, and its luminosity is determined using the Cepheid-derived distance to M100. Any change in the calculated distance to M100 would change the calculated luminosity of the Type Ia supernova, and so would have an effect on all distances derived from observations of how bright these supernovae appear in other galaxies.) For this reason, astronomers go to great lengths to check the accuracy and reliability of their standard candles.

WEB LINK 24.9

One distance-measuring technique that has broken free of the distance ladder uses observations of molecular clouds called **masers.** ("Maser" is an acronym for "microwave amplification by stimulated emission of radiation.") Just as an electric current stimulates a laser to emit an intense beam of visible light, nearby luminous stars can stimulate water molecules in a maser to emit intensely at microwave wavelengths. This radiation is so intense that masers can be detected millions of parsecs away.

During the 1990s, Jim Herrnstein and his collaborators used the Very Long Baseline Array (see Section 6-6) to observe a number of masers orbiting in a disk around the center of the spiral galaxy M106. They determined the orbital speed of the masers from the Doppler shift of masers near the edges of the disk, where they are moving most directly toward or away from Earth. They also measured the apparent change in position of masers moving across the face of M106. By relating this apparent speed to the true speed determined from the Doppler shift, they calculated that the masers and the galaxy of which they are part are 7.2 Mpc (23 million ly) from Earth (Figure 24-15).

The maser technique is still in its infancy. But because this technique is independent of all other distance measuring methods, it is likely to play an important role in calibrating the rungs of the distance ladder.

24-5 The Hubble law relates the redshifts of remote galaxies to their distances from Earth

Whenever an astronomer finds an object in the sky that can be seen or photographed, the natural inclination is to attach a spectrograph to a telescope and record the spectrum. As long ago as 1914, Vesto M. Slipher, working at the Lowell Observatory in Arizona, began taking spectra of "spiral nebulae"—a name used before they were known to be galaxies. He was surprised to discover that of the 15 spiral nebulae he studied, the spectral lines of 11 were shifted toward the red end of the spectrum, indicating that they were moving away from Earth.

This marked dominance of redshifts over blueshifts, was presented by Curtis in the 1920 Shapley-Curtis "debate" as evidence that these spiral nebulae could not be ordinary nebulae in our Milky Way Galaxy. It was only later that astronomers realized that the redshifts of spiral nebulae—that is, galaxies—reveal a basic law of our expanding universe.

WEB LINK 24.10

Redshift, Distance, and the Hubble Law

During the 1920s, Edwin Hubble and Milton Humason photographed the spectra of many galaxies with the

100-inch (2.5-meter) telescope on Mount Wilson in California. By observing the apparent brightnesses and pulsation periods of Cepheid variables in these galaxies, they were also able to measure the distance to each galaxy (see Section 24-2). Hubble and Humason found that most galaxies show a redshift in their spectrum. They also found a direct correlation between the distance to a galaxy and its redshift:

The greater the redshift of a distant galaxy, the farther away it is

The more distant a galaxy, the greater its redshift and the more rapidly it is receding from us.

In other words, nearby galaxies are moving away from us slowly, and more distant galaxies are rushing away from us much more rapidly. Figure 24-16 shows this relationship for five representative elliptical galaxies. This universal recessional movement is referred to as the **Hubble flow.**

Hubble estimated the distances to a number of galaxies and the redshifts of those galaxies. The **redshift,** denoted by the symbol z, is found by taking the wavelength (λ) observed for a given spectral line, subtracting from it the ordinary, unshifted wavelength of that line (λ_0) to get the wavelength difference ($\Delta\lambda$), and then dividing that difference by λ_0:

Redshift of a receding object

$$z = \frac{\lambda - \lambda_0}{\lambda_0} = \frac{\Delta\lambda}{\lambda_0}$$

z = redshift of an object

λ_0 = ordinary, unshifted wavelength of a spectral line

λ = wavelength of that spectral line that is actually observed from the object

From the redshifts, Hubble used the Doppler formula to calculate the speed at which these galaxies are receding from us. Box 24-2 describes this calculation. Plotting the data on a graph of distance versus speed, Hubble found that the points lie near a straight line. Figure 24-17 is a modern version of Hubble's graph based on recent data.

This relationship between the distances to galaxies and their redshifts was one of the most important astronomical discoveries of the twentieth century. As we will see in Chapter 28, this relationship tells us that we are living in an expanding universe. In 1929, Hubble published this discovery, which is now known as the **Hubble law.** The Hubble law is most easily stated as a formula:

The Hubble law

$$v = H_0 d$$

v = recessional velocity of a galaxy

H_0 = Hubble constant

d = distance to the galaxy

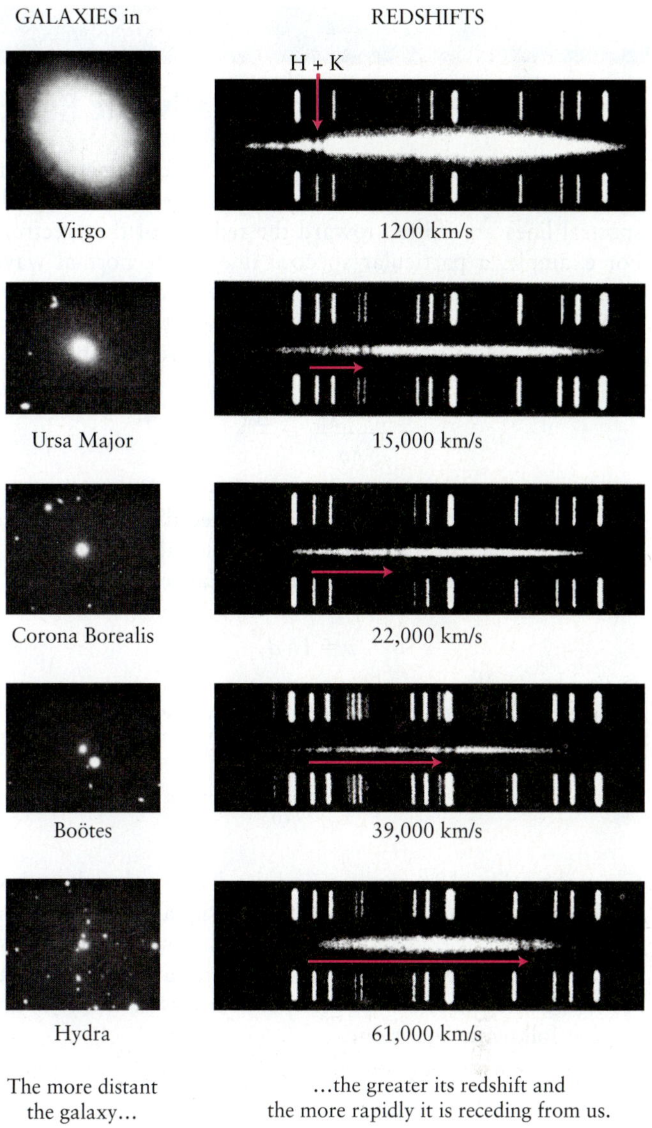

Figure 24-16 R I V U X G

Relating the Distances and Redshifts of Galaxies These five galaxies are arranged, from top to bottom, in order of increasing distance from us. All are shown at the same magnification. Each galaxy's spectrum is a bright band with dark absorption lines; the bright lines above and below it are a comparison spectrum of a light source at the observatory on Earth. The horizontal red arrows show how much the H and K lines of singly ionized calcium are redshifted in each galaxy's spectrum. Below each spectrum is the recessional velocity calculated from the redshift. The more distant a galaxy is, the greater its redshift. (Carnegie Observatories)

This formula is the equation for the straight line displayed in Figure 24-17, and the **Hubble constant** H_0 is the slope of this straight line. From the data plotted on this graph we find that H_0 = 73 km/s/Mpc (say "73 kilometers per second per megaparsec"). In other words, for each million parsecs to a galaxy, the galaxy's speed away from us increases by 73 km/s. For example, a galaxy located 100 million parsecs from Earth should be rushing away from us

BOX 24-2 Tools of the Astronomer's Trade

The Hubble Law and the Relativistic Redshift

Suppose that you aim a telescope at an extremely distant galaxy. You take a spectrum of the galaxy and find that the spectral lines are shifted toward the red end of the spectrum. For example, a particular spectral line whose normal wavelength is λ_0 appears in the galaxy's spectrum at a longer wavelength λ. The spectral line has thus been shifted by an amount $\Delta\lambda = \lambda - \lambda_0$. The redshift of the galaxy, z, is given by

$$z = \frac{\lambda - \lambda_0}{\lambda_0} = \frac{\Delta\lambda}{\lambda_0}$$

The redshift means that the galaxy is receding from us. According to the Hubble law, the recessional velocity v of a galaxy is related to its distance d from Earth by

$$v = H_0 d$$

where H_0 is the Hubble constant. We can rewrite this equation

$$d = \frac{v}{H_0}$$

Given the value of H_0, we can find the distance d to the galaxy if we know how to determine the recessional velocity v from the redshift z.

If the redshift and recessional velocity are not too great, we can ignore the effects of the special theory of relativity and use the following equation:

$$z = \frac{v}{c} \text{ (valid for low speeds only)}$$

where c is the speed of light. For example, a 5% shift in wavelength ($z = 0.05$) corresponds to a recessional velocity of 5% of the speed of light ($v = 0.05c$). For redshifts greater than about 0.1, however, we must use the relativistic equation for the Doppler shift:

$$z = \sqrt{\frac{c + v}{c - v}} - 1$$

A useful way to rewrite this relationship is as follows:

$$\frac{v}{c} = \frac{(z+1)^2 - 1}{(z+1)^2 + 1} \text{ (valid for all speeds)}$$

The accompanying graph displays this relationship between redshift z and speed v. Note that z approaches infinity as v approaches the speed of light.

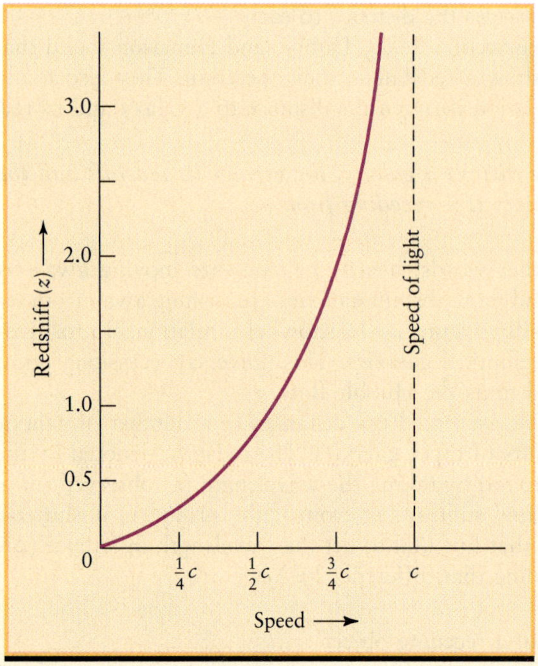

EXAMPLE: When measured in a laboratory on Earth, the so-called K line of singly ionized calcium has a wavelength $\lambda_0 = 393.3$ nm. But when you observe the spectrum of the giant elliptical galaxy NGC 4889, you find this spectral line at $\lambda = 401.8$ nm. Using $H_0 = 73$ km/s/Mpc, find the distance to this galaxy.

Situation: We are given the values of λ and λ_0 for a line in this galaxy's spectrum. Our goal is to determine the galaxy's distance d.

Tools: We use the relationship $z = (\lambda - \lambda_0)/\lambda_0$ to determine the redshift. We then use the appropriate formula to determine the galaxy's recessional velocity v, and finally use the Hubble law to determine the distance to the galaxy.

Answer: The redshift of this galaxy is

$$z = \frac{401.8 \text{ nm} - 393.3 \text{ nm}}{393.3 \text{ nm}} = 0.0216$$

This value is substantially less than 0.1, so we can safely use the low-speed relationship between recessional speed and redshift: $z = v/c$. So NGC 4889 is moving away from us with a speed

$$v = zc = (0.0216)(3 \times 10^5 \text{ km/s}) = 6480 \text{ km/s}$$

Using H_0 = 73 km/s/Mpc in the Hubble law, the distance to this galaxy is

$$d = \frac{v}{H_0} = \frac{6480 \text{ km/s}}{73 \text{ km/s/Mpc}} = 89 \text{ Mpc}$$

Review: This galaxy is receding from us at 0.0216 (2.16%) of the speed of light, and is 89 megaparsecs (290 million light-years) away. Thus the light we see from NGC 4889 today left the galaxy 290 million years ago, even before the first dinosaurs appeared on Earth.

EXAMPLE: In late 1997 astronomers observed a Type Ia supernova (called SN 1997ff) with a redshift $z = 1.7$. Use the Hubble law to find the distance to this supernova.

Situation: We are to use the redshift z to determine the distance d.

Tools: Since the redshift is large, we use the relativistic formula relating recessional velocity v and redshift z to calculate the value of v. We then use the Hubble law to determine the galaxy's distance d.

Answer: From the above relativistic formula,

$$\frac{v}{c} = \frac{(1.7+1)^2 - 1}{(1.7+1)^2 + 1} = \frac{(2.7)^2 - 1}{(2.7)^2 + 1} = \frac{6.29}{8.29} = 0.76$$

The supernova's recessional velocity is 76% of the speed of light, or $v = 0.76c = (0.76)(3 \times 10^5 \text{ km/s}) = 2.3 \times 10^5$ km/s. From the Hubble law, the distance to the supernova is

$$d = \frac{v}{H_0} = \frac{2.3 \times 10^5 \text{ km/s}}{73 \text{ km/s/Mpc}} = 3200 \text{ Mpc} = 10^{10} \text{ ly}$$

This supernova is a remarkable 10 *billion* light-years away.

Review: For this supernova the value of z is greater than 1. Had we used the low-speed relationship $z = v/c$, we would have come to the erroneous conclusion that $v = 1.7c$—that is, that the supernova is receding from us faster than the speed of light. Using the correct relativistic formula tells us that the recessional velocity is only 76% of the speed of light.

In Chapter 26 we will see how observations of very distant supernovae like SN 1997ff reveal important deviations from the Hubble law. In particular, these observations show that the expansion of the universe is not proceeding at a constant rate, but is actually accelerating.

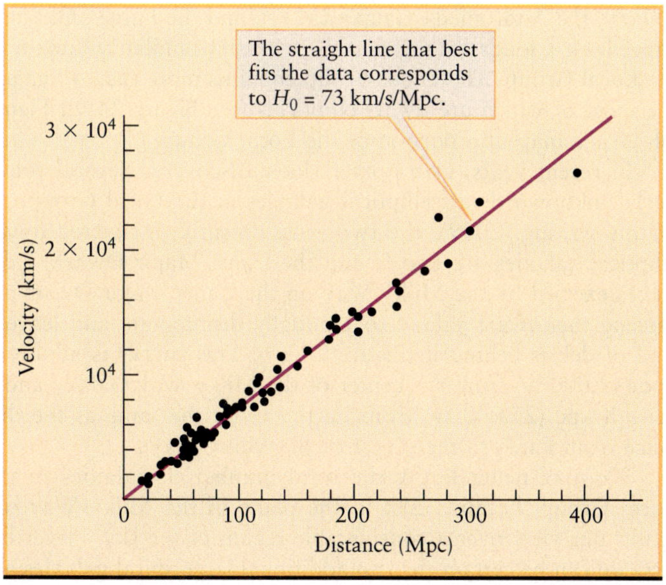

Figure 24-17

The Hubble Law This graph plots the distances and recessional velocities of a sample of galaxies. The straight line is the best fit to the data. This linear relationship between distance and recessional velocity is called the Hubble law.

with a speed of 7300 km/s. (In other books you may see the units of the Hubble constant written with exponents: 73 km s^{-1} Mpc^{-1}.)

CAUTION! A common misconception about the Hubble law is that *all* galaxies are moving away from the Milky Way. The reality is that galaxies have their own neighbor-induced motions relative to one another, due to their mutual gravitational attraction of neighboring galaxies. Astronomers call this neighbor-induced motion *intrinsic velocity*. For nearby galaxies, the speed of the Hubble flow is small compared to these intrinsic velocities. Hence, some of the nearest galaxies, including M31 (shown in Figure 24-3), are actually approaching us and have blueshifts rather than redshifts. But for distant galaxies, the Hubble speed $v = H_0 d$ is much greater than any intrinsic motion that the galaxies might have. Even if the intrinsic velocity of such a distant galaxy is toward the Milky Way, the fast-moving Hubble flow sweeps that galaxy away from us.

Pinning Down the Hubble Constant

A precise value of the Hubble constant has been a topic of heated debate among astronomers for several decades. The problem is that while redshifts are relatively easy to measure in a reliable way, distances to galaxies (especially remote galaxies) are not, as we saw in Section 24-4. Hence, astronomers who use different methods of determining galactic distances have obtained different values of H_0. To see why different values are measured, it is helpful to rewrite the Hubble law as

$$H_0 = \frac{v}{d}$$

This equation shows that if galaxies of a given recessional velocity (v) are far away (so d is large), the Hubble constant H_0 must be relatively small. But if these galaxies are relatively close (so d is small), then H_0 must have a larger value.

In the past, astronomers who used Type Ia supernovae for determining galactic distances found galaxies to be farther away than their colleagues who employed the Tully-Fisher relation. Therefore, the supernova adherents found values of H_0 in the range from about 40 to 65 km/s/Mpc, while the values from the Tully-Fisher relation ranged from about 80 to 100 km/s/Mpc.

In the past few years, the Hubble Space Telescope has been used to observe Cepheid variables with unprecedented precision and in galaxies as far away as 30 Mpc (100 million ly). These observations and others suggest a value of the Hubble constant of about 73 km/s/Mpc, with an uncertainty of no more than 10%. At the same time, reanalysis of the supernova and Tully-Fisher results have brought the values of H_0 from these techniques closer to the Hubble Space Telescope Cepheid value. We will adopt the value $H_0 = 73$ km/s/Mpc in this book.

Determining the value of H_0 has been an important task of astronomers for a very simple reason: The Hubble constant is one of the most important numbers in all astronomy. It expresses the rate at which the universe is expanding and, as we will see in Chapter 26, even helps give the age of the universe. Furthermore, the Hubble law can be used to determine the distances to extremely remote galaxies. If the redshift of a galaxy is known, the Hubble law can be used to determine its distance from Earth, as Box 24-2 shows. Thus, the value of the Hubble constant helps determine the distances of the most remote objects in the universe that astronomers can observe.

Because the value of H_0 remains somewhat uncertain, astronomers often express the distance to a remote galaxy simply in terms of its redshift z (which can be measured very accurately). Given the redshift, the distance to this galaxy can be calculated from the Hubble law, but the distance obtained in this way will depend on the particular value of H_0 adopted. Rather than going through these calculations, an astronomer might simply say that a certain galaxy is "at $z = 0.128$." From the Hubble law relating redshift and distance, this redshift makes it clear that the galaxy in question is more distant than one at $z = 0.120$ but not as distant as one at $z = 0.130$. When astronomers use redshift to describe distance, they are making use of the following general rule:

The greater the redshift of a distant galaxy, the greater its distance.

24-6 Galaxies are grouped into clusters and superclusters

WEB LINK 24.11

Galaxies are not scattered randomly throughout the universe but are found in **clusters.** Figure 24-18 shows one such cluster. Like stars within a star cluster, the members of a cluster of galaxies are in continual motion around one another. They appear to be at rest only because they are so distant from us.

WEB LINK 24.12

Figure 24-18 R I V U X G

The Hercules Cluster This irregular cluster of galaxies is about 200 Mpc (650 million ly) from Earth. The Hercules cluster contains many spiral galaxies, often associated in pairs and small groups. (NOAO)

Clusters of Galaxies: Rich and Poor, Regular and Irregular

A cluster is said to be either **poor** or **rich,** depending on how many galaxies it contains. Poor clusters, which far outnumber rich ones, are often called **groups.** For example, the Milky Way Galaxy, the Andromeda Galaxy (M31), and the Large and Small Magellanic Clouds belong to a poor cluster familiarly known as the **Local Group.** The Local Group contains more than 40 galaxies, most of which are dwarf ellipticals (see Figure 24-9). Figure 24-19 is a map of a portion of the Local Group.

In recent years, astronomers have discovered several previously unknown dwarf elliptical galaxies in the Local Group. As of this writing (2009), the two most recently discovered dwarf elliptical galaxies are Leo T and the Canis Major Dwarf. Tidal forces exerted by the Milky Way on the Canis Major Dwarf are causing this dwarf galaxy to gradually disintegrate and leave a trail of debris behind it (Figure 24-20). This galaxy is about 13 kpc (42,000 ly) from the center of the Milky Way Galaxy and a mere 8 kpc (25,000 ly) from Earth (about the same as the distance from Earth to the center of the Milky Way).

We may never know the total number of galaxies in the Local Group, because dust in the plane of the Milky Way obscures our view over a considerable region of the sky. Nevertheless, we can be certain that no additional large spiral galaxies are hidden by the Milky Way. Radio astronomers would have detected 21-cm radiation from them, even though their visible light is completely absorbed by interstellar dust.

The nearest fairly rich cluster is the Virgo cluster, a collection of more than 2000 galaxies covering a 10° × 12° area of the sky. Figure 24-8 shows a portion of this cluster. One member of this cluster not shown in Figure 24-8 is the spiral galaxy M100; mea-

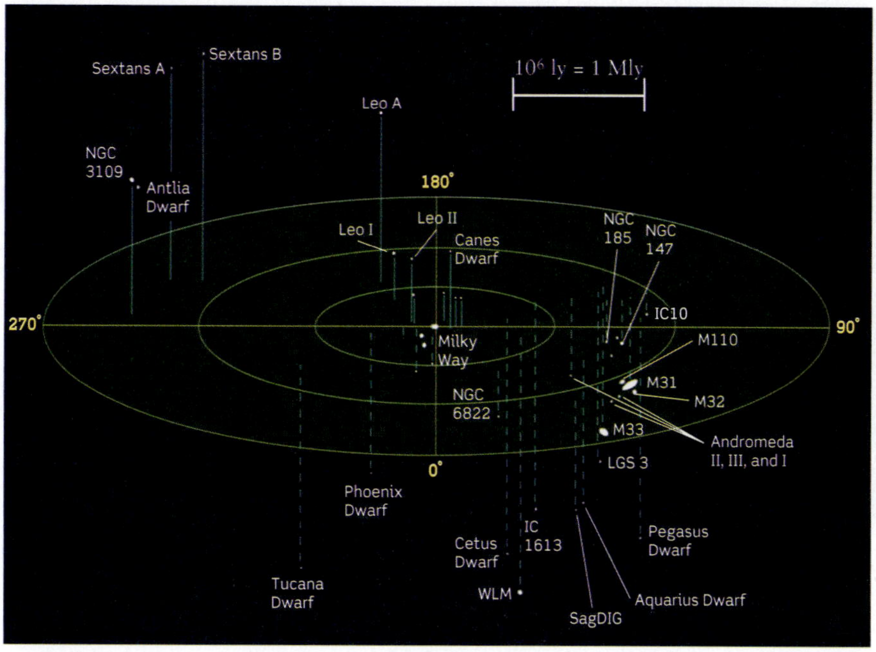

Figure 24-19

The Local Group This illustration shows the relative positions of the galaxies that comprise the Local Group, a poor, irregular cluster of which our Galaxy is part. (The blue rings represent the plane of the Milky Way's disk; 0° is the direction from Earth toward the Milky Way's center. Solid and dashed lines point to galaxies above and below the plane, respectively.) The largest and most massive galaxy in the Local Group is M31, the Andromeda Galaxy; in second place is the Milky Way, followed by the spiral galaxy M33. Both the Milky Way and M31 are surrounded by a number of small satellite galaxies. (© Richard Powell, www.atlasoftheuniverse.com)

surements of Cepheid variables in M100 (see Figure 24-4) give a distance of about 17 Mpc (56 million ly). The Tully-Fisher relation and observations of Type Ia supernovae (see Figure 24-13) give similar distances to this cluster. The overall diameter of the cluster is about 3 Mpc (9 million ly).

The center of the Virgo cluster is dominated by three giant elliptical galaxies. You can see two of these, M84 and M86, in Figure 24-8. The diameter of each of these enormous galaxies is comparable to the 750-kpc distance between the Milky Way and M31. In other words, one giant elliptical is approximately the same size as the entire Local Group!

Astronomers also categorize clusters of galaxies as regular or irregular, depending on the overall shape of the cluster. The Virgo cluster, for example, is called an **irregular cluster,** because its galaxies are scattered throughout a sprawling region of the sky. Our own Local Group is also an irregular cluster. In contrast, a **regular cluster** has a distinctly spherical appearance, with a marked concentration of galaxies at its center.

The nearest example of a rich, regular cluster is the Coma cluster, located about 90 Mpc (300 million light-years) from us toward the constellation Coma Berenices (Berenice's Hair) (Figure 24-21). Despite its great distance, telescopic images of this cluster show more than 1000 galaxies. The Coma cluster almost certainly contains many thousands of dwarf ellipticals, so the total membership of the cluster may be as many as 10,000 galaxies. The core of the Coma cluster is dominated by two giant ellipticals surrounded by many normal-sized galaxies.

The overall shape of a cluster is related to the dominant types of galaxies it contains. Rich, regular clusters contain mostly elliptical and lenticular galaxies. For example, about 80% of the brightest galaxies in the Coma cluster (see Figure 24-21) are ellipticals; only a few spiral galaxies are scattered around the cluster's outer regions. Irregular clusters, such as the Virgo cluster and the Hercules cluster shown in Figure 24-18, have a more even mixture of galaxy types.

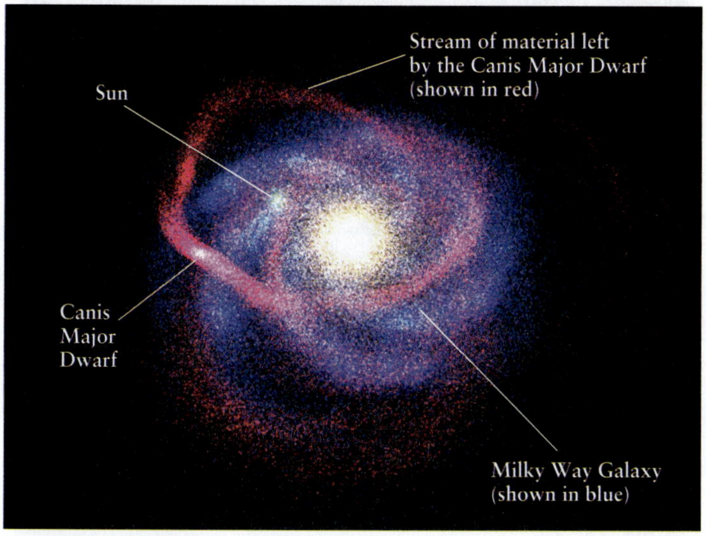

ANIMATION 24.1 WEB LINK 24.13

Figure 24-20

The Canis Major Dwarf Discovered in 2003, this dwarf elliptical galaxy is actually slightly closer to Earth than is the center of the Milky Way Galaxy. This illustration shows the stream of material left behind by the Canis Major Dwarf as it orbits the Milky Way. This material is torn away by the Milky Way's tidal forces (see Section 4-8). (R. Ibata et al., Observatoire de Strasbourg/Université Louis Pasteur; 2MASS; and NASA)

Superclusters: Clusters of Clusters of Galaxies

Clusters of galaxies are themselves grouped together in huge associations called **superclusters.** A typical supercluster contains dozens of individual clusters spread over a region of space up to 45 Mpc (150 million ly) across. Figure 24-22 shows the distribution of clusters in our part of the universe. The nearer ones out to the Virgo cluster, including our own Local Group, are

Figure 24-21 R I V U X G

The Coma Cluster This rich, regular cluster is about 90 Mpc (300 million light-years) from Earth. Almost all of the spots of light in this image are individual galaxies of the cluster. Two giant elliptical galaxies, NGC 4889 and NGC 4874, dominate the center of the cluster. The bright star at the upper right is within our own Milky Way Galaxy, a million times closer than any of the galaxies shown here. (O. Lopez-Cruz and I. K. Shelton, University of Toronto; KPNO)

members of the *Local Supercluster.* The other clusters shown in Figure 24-22 belong to other superclusters. The most massive cluster in the local universe is called the *Great Attractor;* its gravity is so great that the Milky Way as well as the rest of the Local Supercluster is moving toward it at speeds of several hundred kilometers per second.

Observations indicate that unlike clusters, superclusters are not bound together by gravity. That is, most clusters in each supercluster are drifting away from most of the other clusters in that same supercluster. Furthermore, the superclusters are all moving away from each other due to the Hubble flow.

Cosmic Voids and Sheets: The Distribution of Superclusters

Since the 1980s, astronomers have been working to understand how superclusters are distributed in space. Some of this structure is revealed by maps such as that shown in Figure 24-23, which displays the positions on the sky of 1.6 million galaxies. Such maps reveal that superclusters are not randomly distributed, but seem to lie along filaments. But to comprehend more fully the distribution of superclusters, it is necessary to map their positions in three dimensions. Their positions are mapped by measuring both the position of a galaxy on the sky as well as the galaxy's redshift. Using the Hubble law (see Section 24-5), astronomers can use each galaxy's redshift to estimate its distance from Earth and thus its position in three-dimensional space.

Superclusters of galaxies are not spread uniformly across the universe, but are found in vast sheets separated by truly immense voids

The first three-dimensional maps of this kind were made in the 1970s and included measurements of a few hundred galaxies. Collecting the data for such maps required many months or years of telescopic observations. Technology for astronomy has advanced tremendously since then, and it is now possible to measure the redshifts of 400 galaxies in a single hour!

WEB LINK 24.14 The most extensive galaxy maps available at this writing (early 2007) are those from the Sloan Digital Sky Survey, a joint project of astronomers from the United States, Japan, and Germany, and from the Two Degree Field Galactic Redshift Survey (2dFGRS), a collaboration between Australian, British, and U.S. astronomers. (The name refers to the 2° field of view of the telescope used for the observations, which is unusually wide for a research telescope.) Figure 24-24*a* shows a map made from 2dFGRS measurements of more than 60,000 galaxies. This particular map encompasses two wedge-shaped slices of the universe, one on either side of the plane of the Milky Way (Figure 24-24*b*). Earth (in the Milky Way) is at the apex of the wedge-shaped map; each dot represents a galaxy. The measurements used to create this map included galaxies with redshifts as large as $z = 0.25$, corresponding to a recessional velocity of 66,000 km/s. Using a Hubble constant of 73 km/s/Mpc, you can determine that the map in Figure 24-24*a* extends out to a distance of nearly 1000 Mpc, or 3 *billion* light-years from Earth.

Maps such as that shown in Figure 24-24*a* reveal enormous **voids** where exceptionally few galaxies are found. (These voids were first discovered in 1978 in a pioneering study by Stephen Gregory

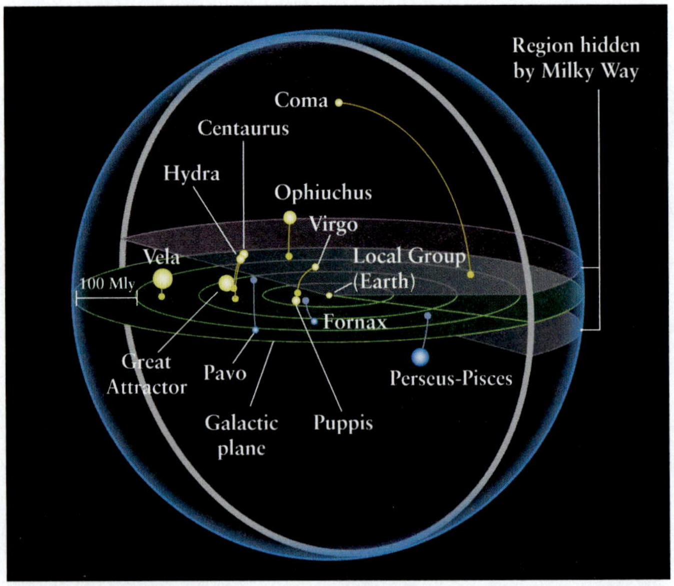

Figure 24-22

Nearby Clusters of Galaxies This illustration shows a sphere of space 800 million ly (250 Mpc) in diameter centered on Earth in the Local Group. Each spherical dot represents a cluster of galaxies. To better see the three-dimensionality of this figure, colored arcs are drawn from each cluster to the green plane, which is an extension of the plane of the Milky Way outward into the universe. Note that clusters of galaxies are unevenly distributed here, as they are elsewhere in the universe.

Figure 24-23 R I V U X G

Structure in the Nearby Universe This composite infrared image from the 2MASS (Two-Micron All-Sky Survey) project shows the light from 1.6 million galaxies. In this image, the entire sky is projected onto an oval; the blue band running vertically across the center of the image is light from the plane of the Milky Way. Note that galaxies form a lacy, filamentary structure. Note also the large, dark voids that contain few galaxies. (2MASS; IPAC/Caltech; and University of Massachusetts)

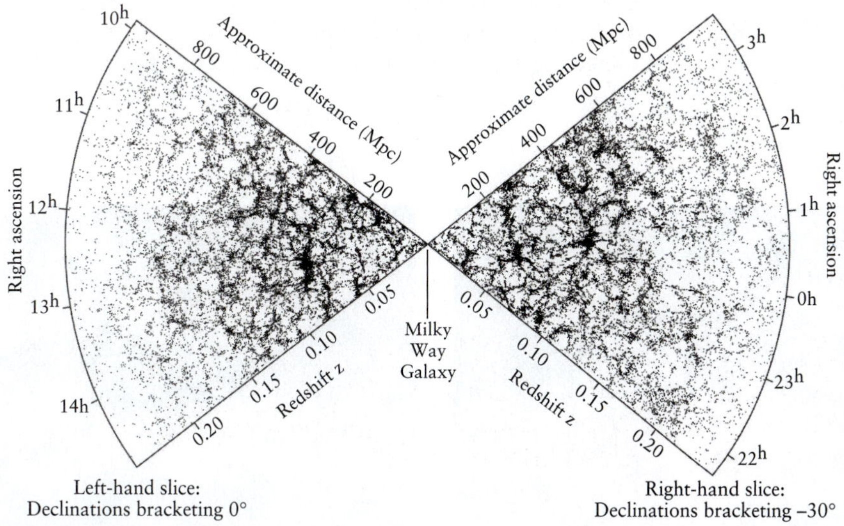

(a) The 2dF galaxy survey

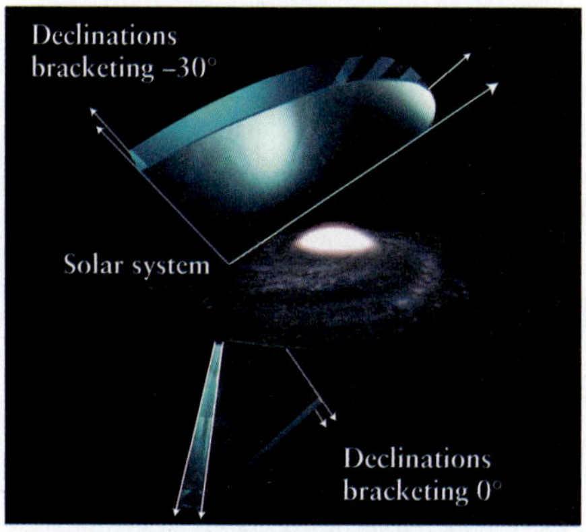

(b) Fields of view in the 2dF survey

ANIMATION 24.2

Figure 24-24

The Large-Scale Distribution of Galaxies **(a)** This map shows the distribution of 62,559 galaxies in two wedges extending out to redshift $z = 0.25$. (For an explanation of right ascension and declination, see Box 2-1.) Note the prominent voids surrounded by thin regions full of galaxies. **(b)** The two wedges shown in (a) lie roughly perpendicular to the plane of the Milky Way. These were chosen to avoid the obscuring dust that lies in our Galaxy's plane. (Courtesy the 2dF Galaxy Redshift Survey Team/Anglo-Australian Observatory) and Laird Thompson at the Kitt Peak National Observatory.) These voids are roughly spherical and measure 30 to 120 Mpc (100 million to 400 million ly) in diameter. They are not entirely empty, however. There is evidence for hydrogen clouds in some voids, while others may be subdivided by strings of dim galaxies.

Figure 24-24*a* shows that most galaxies are concentrated in sheets on the surfaces between voids. These sheets can be more than 100 Mpc long and several megaparsecs thick. This pattern is similar to that of soapsuds in a kitchen sink, with sheets of soap film (analogous to galaxies) surrounding air bubbles (analogous to voids). These titanic sheets of galaxies are the largest structures known in the universe: On scales much larger than 100 Mpc, the distribution of galaxies in the universe appears to be roughly uniform. As we will see in Chapter 27, this pattern of sheets and voids contains important clues about how clusters of galaxies formed in the early universe.

24-7 Colliding galaxies produce starbursts, spiral arms, and other spectacular phenomena

Occasionally, two galaxies within a cluster or from adjacent clusters can collide with each other. Past collisions have hurled vast numbers of stars into intergalactic space. In some cases, we can even observe a collision in progress, a cosmic catastrophe that gives birth to new stars. And astronomers can predict collisions that will not take place for billions of years, such as the collision that is fated to occur between the galaxy M31 and our own Milky Way Galaxy.

High-Speed Galaxy Collisions: Shredding Gas and Dust

When two galaxies collide at high speed, the huge clouds of interstellar gas and dust in the galaxies slam into each other—like two cars locking bumpers in a collision. In this way, two colliding galaxies can leave behind their interstellar gas and dust as the stars in each galaxy pass through the impact site.

The best evidence that such collisions take place is that many rich clusters of galaxies are strong sources of X rays (Figure 24-25). This emission reveals the presence of substantial amounts of hot **intracluster gas** (that is, gas within the cluster) at temperatures between 10^7 and 10^8 K. The only way that such large amounts of gas could be heated to such extremely high temperatures is in violent collisions between galaxies.

CAUTION! Although galaxies can and do collide, it is highly unlikely that the *stars* from two colliding galaxies actually run into each other. The reason is that the stars within a galaxy are very widely separated from one another, with a tremendous amount of space between them.

Gentle Galactic Collisions and Starbursts

In a less violent collision or a near-miss between two galaxies, the compressed interstellar gas may have more time to cool, allowing many protostars to form. Such collisions may account for **starburst galaxies** such as M82 (Figure 24-26), which blaze with the light of numerous newborn stars. These galaxies have bright centers surrounded by clouds of warm interstellar dust, indicating recent, vigorous star birth. Their warm dust is so abundant that starburst galaxies are among the most luminous objects in the universe at infrared wavelengths. (The right-hand image on the first page of Chapter 6 shows the infrared emission from M82's warm dust.)

The starburst galaxy M82 shown in Figure 24-26 also shows the effects of strong winds from young, luminous stars. It also contains a number of luminous globular clusters. Unlike the globular clusters in our Galaxy, whose stars are about 12.5 billion years old, those in M82 are no more than 600 million years old. These young globular clusters are another sign of recent star formation.

M82 is one member of a nearby cluster of galaxies that includes the beautiful spiral galaxy M81 and a fainter elliptical companion called NGC 3077 (Figure 24-27*a*). Radio surveys of that region of the sky reveal enormous streams of hydrogen gas connecting the three galaxies (Figure 24-27*b*). The loops and

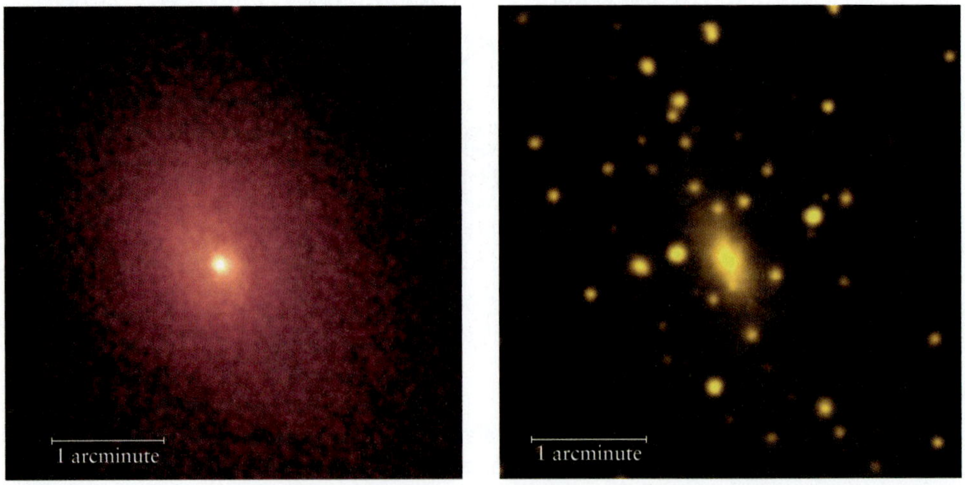

(a) An X-ray image of Abell 2029 shows emission from hot gas.
R I V U X G

(b) A visible-light image of Abell 2029 shows the cluster's galaxies.
R I V U X G

WEB LINK 24.15

Figure 24-25

X-ray Emission from a Cluster of Galaxies (a) An X-ray image of this cluster of galaxies shows emission from hot gas between the galaxies. The gas was heated by collisions between galaxies within the cluster. (b) The galaxies themselves are too dim at X-ray wavelengths to be seen in (a), but are apparent at visible wavelengths. This cluster, one of many cataloged by the UCLA astronomer George O. Abell, is about 300 Mpc (1 billion ly) from Earth in the constellation Serpens. (a: NASA, CXC, and University of California, Irvine/A. Lewis et al. b: Palomar Observatory DSS)

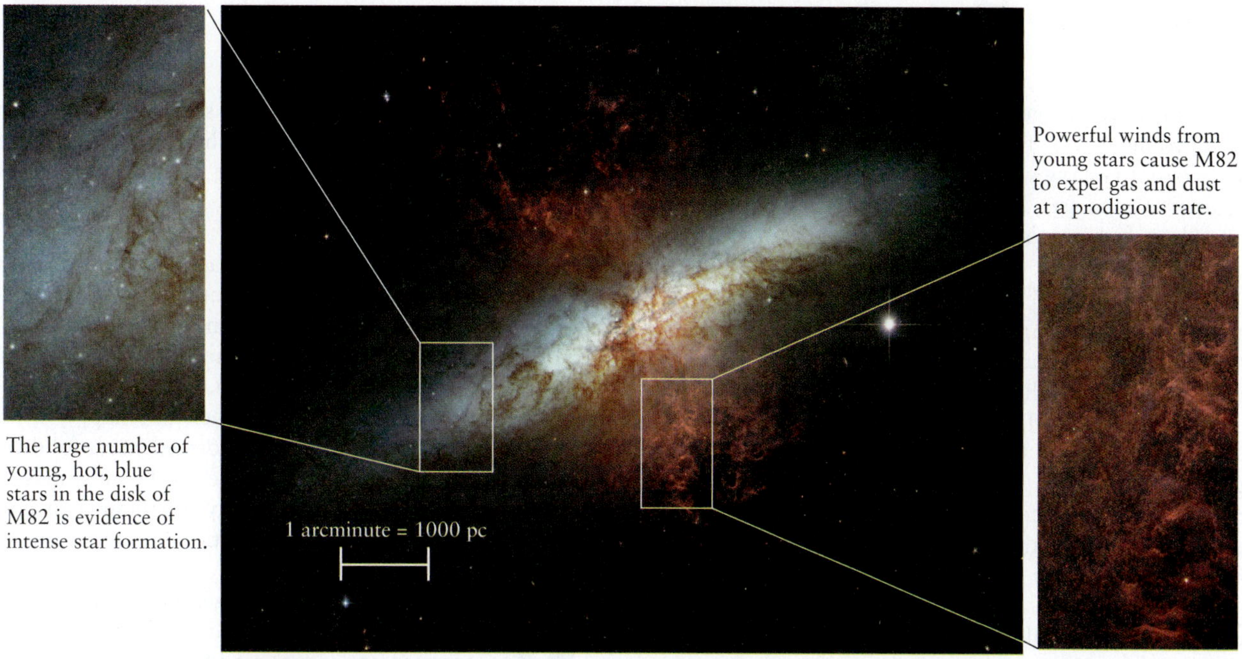

Powerful winds from young stars cause M82 to expel gas and dust at a prodigious rate.
The large number of young, hot, blue stars in the disk of M82 is evidence of intense star formation.
1 arcminute = 1000 pc

VIDEO 24.3

Figure 24-26 R I V U X G

A Starburst Galaxy Prolific star formation is occurring at the center of the irregular galaxy M82, which lies about 3.6 Mpc (12 million ly) from Earth in the constellation Ursa Major. M82 also contains an unusual X-ray source, shown in Figure 22-17. (NASA; ESA; and the Hubble Heritage Team, STScI/AURA)

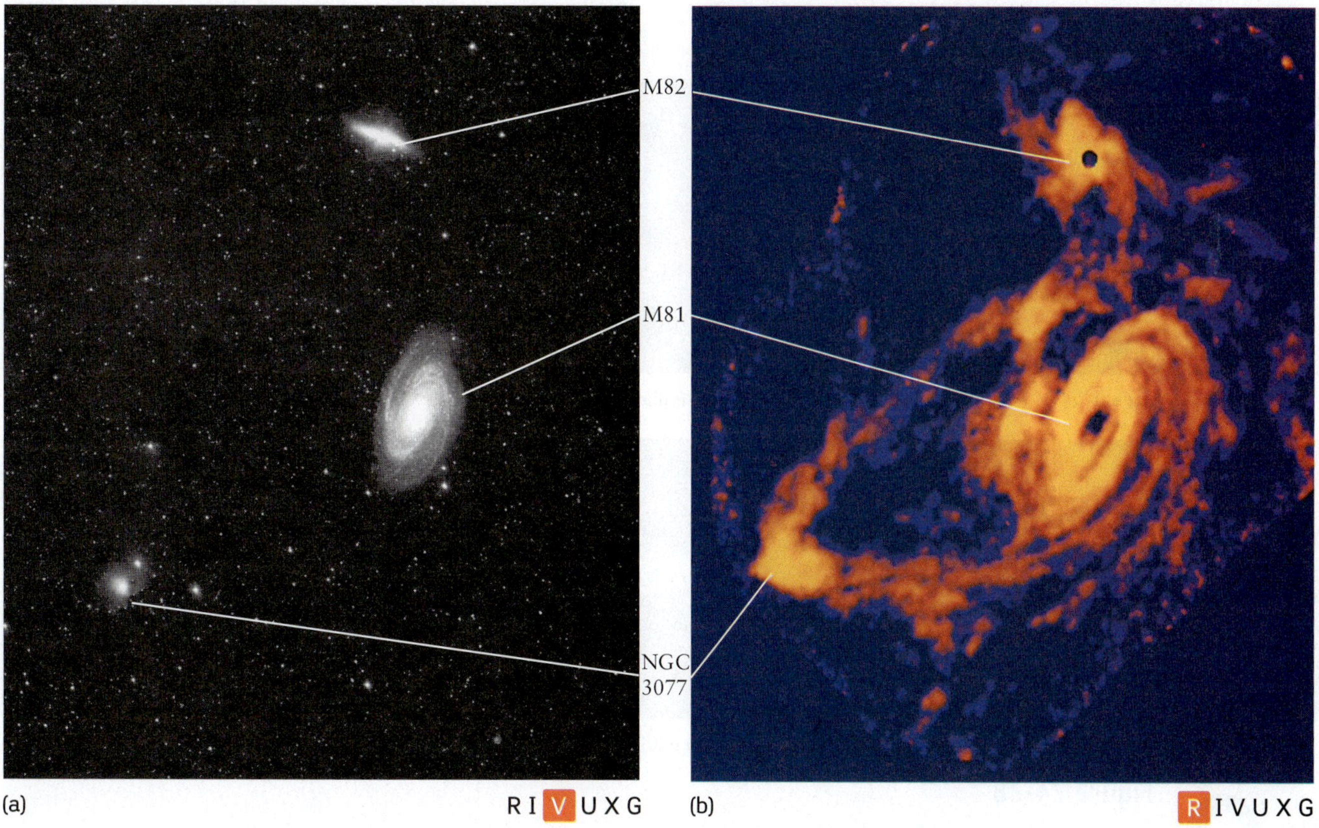

Figure 24-27

The M81 Group (a) The starburst galaxy M82 (see Figure 24-26) is part of a cluster of about a dozen galaxies. This wide-angle visible-light photograph shows the three brightest galaxies of the cluster. The area shown is about 1° across. **(b)** This false-color radio image of the same region, created from data taken by the Very Large Array, shows streamers of hydrogen gas that connect the three bright galaxies as well as several dim ones. (a: Palomar Sky Survey; b: M. S. Yun, VLA and Harvard) twists in these streamers suggest that the three galaxies have had several close encounters over the ages. A similar stream of hydrogen gas connects our Galaxy with its second nearest neighbor, the Large Magellanic Cloud (see Figure 24-12), suggesting a history of close encounters between our Galaxy and the LMC.

Tidal Forces and Galaxy Mergers

Tidal forces between colliding galaxies can deform the galaxies from their original shapes, just as the tidal forces of the Moon on Earth deform the oceans and help give rise to the tides (see Section 4-8, especially Figure 4-27). The galactic deformation is so great that thousands of stars can be hurled into intergalactic space along huge, arching streams. (This same effect has stripped material away from the Canis Major Dwarf galaxy as it orbits the Milky Way, as shown in Figure 24-20.) Supercomputer simulations of such collisions show that while some of the stars are flung far and wide, other stars slow down and the galaxies may merge.

Galaxies need not actually collide to exert strong forces on each other

Figure 24-28 shows one such simulation of stars (but no gas or dust). As the two galaxies pass through each other, they are severely distorted by gravitational interactions and throw out a pair of extended tails. The interaction also prevents the galaxies in the simulation from continuing on their original paths. Instead, they fall back together for a second encounter (at 625 million years). The simulated galaxies merge soon thereafter, leaving a single object. The *Cosmic Connections: When Galaxies Collide* explores a real-life example of two galaxies that are colliding in just this way.

Our own Milky Way Galaxy is expected to undergo a galactic collision like that shown in Figure 24-28. The Milky Way and the Andromeda Galaxy, shown in Figure 24-3, are actually approaching each other and should collide in another 6 billion years or so. (Recall that our solar system is only 4.56 billion years old.) When this happens, the sky will light up with a plethora of newly formed stars, followed in rapid succession by a string of supernovae, as the most massive of these stars complete their life spans. Any inhabitants of either galaxy will see a night sky far more dramatic and tempestuous than ours.

When two galaxies merge, the result is a bigger galaxy. If this new galaxy is located in a rich cluster, it may capture and devour additional galaxies, growing to enormous dimensions by **galactic cannibalism.** Cannibalism differs from mergers in that the galaxy that does the devouring is bigger than its "meal," whereas merging galaxies are about the same size.

Many astronomers suspect that galactic cannibalism is the reason that giant ellipticals are so huge. As we have seen, giant galaxies typically occupy the centers of rich clusters. In many cases, smaller galaxies are located around these giants (see Figure 24-8

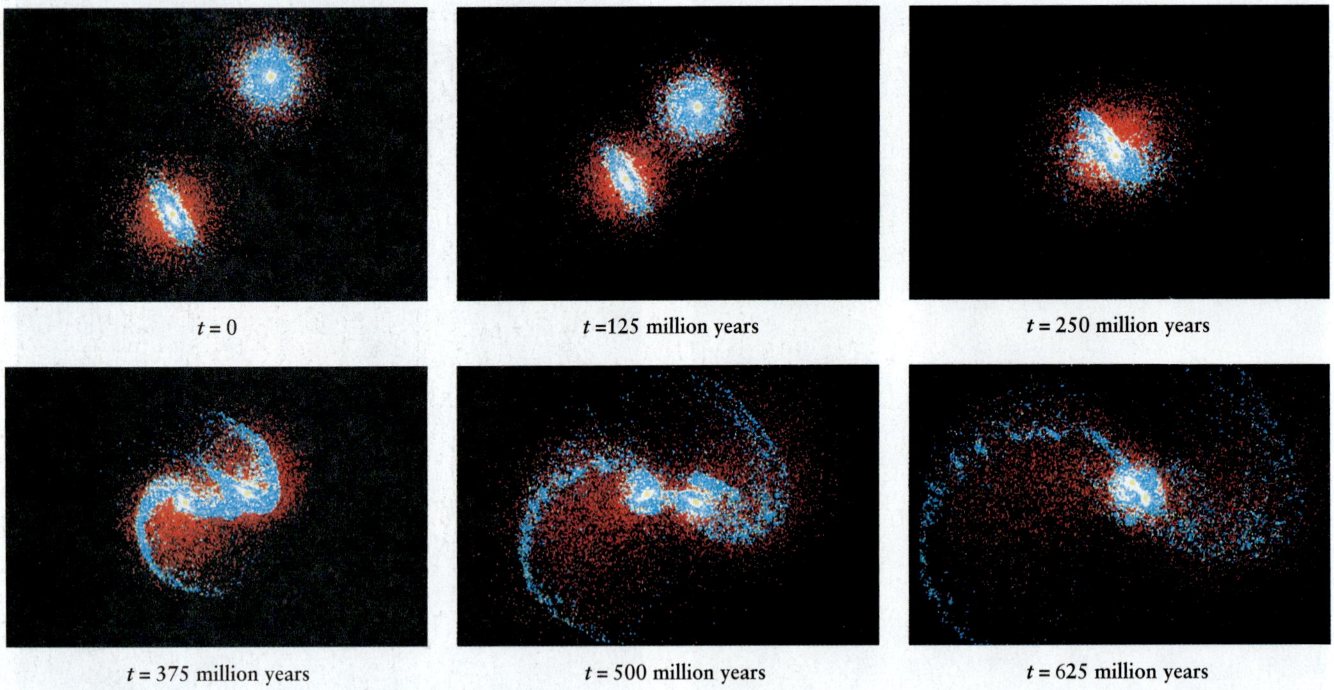

Figure 24-28

A Simulated Collision Between Two Galaxies These frames from a supercomputer simulation show the collision and merger of two galaxies accompanied by an ejection of stars into intergalactic space. Stars in the disk of each galaxy are colored blue, while stars in their central bulges are yellow-white. Red indicates dark matter that surrounds each galaxy. The frames progress at 125-million-year intervals. Compare the bottom frames with the image of the Antennae galaxies in the *Cosmic Connections: When Galaxies Collide.* (Courtesy of J. Barnes, University of Hawaii)

COSMIC CONNECTIONS

When Galaxies Collide

Although galaxies can collide at very high speeds by Earth standards, they are so vast that a collision can last hundreds of millions of years. Understanding what happens during a galactic collision requires ideas about tidal forces (Chapter 4), star formation (Chapter 18), and stellar evolution (Chapter 19).

1. One example of a galactic collision is the pair of galaxies called the Antennae, which lie 19 Mpc (16 million ly) from Earth in the constellation Corvus (the Crow). They probably began to interact several hundred million years ago.

Tidal forces between the galaxies pulled out these long "tidal tails" 200 to 300 million years ago.

R I V U X G

(Bob and Bill Twardy/Adam Block/NOAO/AURA/NSF)

2. As the gas and dust clouds of the two galaxies collide with each other, they are greatly compressed. This compression causes stars to form in tremendous numbers.

R I V U X G

(NASA; ESA; and the Hubble Heritage Team, STScI/AURA-ESA/ Hubble Collaboration)

R I V U X G

(NASA/JPL-Caltech/Z. Wang, Harvard-Smithsonian CfA)

3. This composite infrared and visible-light image of the Antennae allows us to see inside the two galaxies and reveals clouds of dust warmed by the light of hot young stars.

4. The globular clusters that orbit our Milky Way Galaxy contain only old stars; all of the short-lived blue stars have long since died. But some of the globular clusters that orbit the Antennae galaxies *do* have hot blue stars. Hence these clusters must be young. These, too, are a result of the compression of gas and dust that takes place in a collision between galaxies.

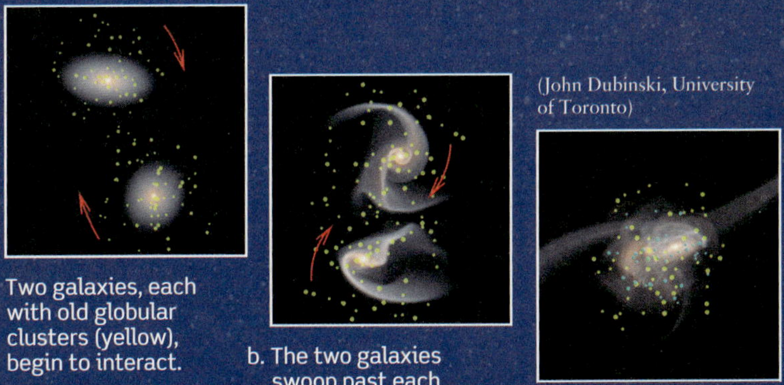

(John Dubinski, University of Toronto)

a. Two galaxies, each with old globular clusters (yellow), begin to interact.

b. The two galaxies swoop past each other before finally settling down as a single merged galaxy. Gas and dust is compressed in both galaxies in the process, creating new star clusters.

c. The combined galaxy has the original globular clusters (shown in yellow) as well as new ones (shown in blue).

VIDEO 24.4 WEB LINK 24.18

and Figure 24-21). As they pass through the extended halo of a giant elliptical, these smaller galaxies slow down and are eventually devoured by the larger galaxy.

Galaxy Interactions and Spiral Arms

ANIMATION 24.6

Close encounters between galaxies provide a third way of forming spiral arms (in addition to density waves and self-propagating star formation, discussed in Section 23-5). Supercomputer simulations clearly demonstrate that spiral arms can be created during a collision, either by drawing out long streamers of stars or by compressing clouds of interstellar gas. For example, the spiral arms of M51 (examine Figure 24-2) may have been produced by a close encounter with a second galaxy. The disruptive galaxy, NGC 5195, is now located at the end of one of the spiral arms created by the collision. The two galaxies shown in the image that opens this chapter are thought to be interacting in the same way.

The very fact of our existence may be intimately related to interactions between galaxies. Some astronomers argue that the spiral arms of our Milky Way Galaxy were produced by a close encounter with the Large Magellanic Cloud. As we saw in Section 23-5, spiral arms compress the interstellar medium in the Milky Way's disk to form Population I stars like our own Sun, which have enough heavy elements to produce Earthlike planets. Thus, the chain of events that led to the formation of our Sun, our solar system, and life on our planet may have been initiated by a long-ago interaction between two galaxies.

24-8 Most of the matter in the universe is mysterious dark matter

Galaxies in a cluster typically move at large speeds and must be held in the cluster by gravity. In other words, there must be enough matter in the cluster to prevent the galaxies from drifting away. Nevertheless, careful examination of a rich cluster, like the Coma cluster shown in Figure 24-21, reveals that the mass of the visually luminous matter (principally the stars in the galaxies) is not at all sufficient to bind the cluster gravitationally. The observed line-of-sight speeds of the galaxies, measured by Doppler shifts, are so large that the cluster should have broken apart long ago. Considerably more mass than is visible is needed to keep the galaxies bound in orbit about the center of the cluster.

We encountered a similar situation in studying our own Milky Way Galaxy in Section 25-4: The total mass of our Galaxy is more than the amount of visible mass. As for our Galaxy, we conclude that clusters of galaxies must contain significant amounts of nonluminous *dark matter*. If this dark matter were not there, the galaxies would have long ago dispersed in random directions and the cluster would no longer exist today. Analyses demonstrate that the total mass needed to bind a typical rich cluster is about 10 times greater than the mass of material that shows up on visible-light images.

The Dark-Matter Problem and Rotation Curves

As for our Galaxy, the problem is to determine what form the invisible mass takes. A partial solution to this **dark-matter problem,** which dates from the 1930s, was provided by the discovery in the late 1970s of hot, X-ray–emitting gas within clusters of galaxies (see Figure 24-25*a*). By measuring the amount of X-ray emission, astronomers find that the total mass of intracluster gas (mostly ionized hydrogen) in a typical rich cluster can be greater than the combined mass of all the stars in all the cluster's galaxies. The mass of intracluster gas is sufficient to account for only about 10% of the invisible mass, however. The remainder is dark matter of unknown composition.

Vast assemblages of dark matter reveal their presence by bending passing rays of light

Although we do not know what dark matter is made of, it is possible to investigate how dark matter is distributed in galaxies and clusters of galaxies. It appears that dark matter lies within and immediately surrounding galaxies, not in the vast spaces between galaxies. The evidence for this dark matter distribution comes principally from observations of the rotation curves of galaxies and of the gravitational bending of light by clusters of galaxies.

As we saw in Section 23-4, a rotation curve is a graph that shows how fast stars in a galaxy are moving at different distances from that galaxy's center. For example, Figure 23-18 is the rotation curve for our Galaxy. As Figure 24-29 illustrates, many other spiral galaxies have similar rotation curves that remain remarkably flat out to surprisingly great distances from each galaxy's center. In other words, the orbital speed of the stars remains roughly constant out to the visible edges of these galaxies.

This observation tells us that we still have not detected the *true* edges of these galaxies (and many similar ones). Near the true edge of a galaxy we should see a decline in orbital speed, in accordance with Kepler's third law (see Figure 23-18). Because this decline has not been observed, astronomers conclude that there must be a considerable amount of dark material that extends well beyond the visible portion of the disk.

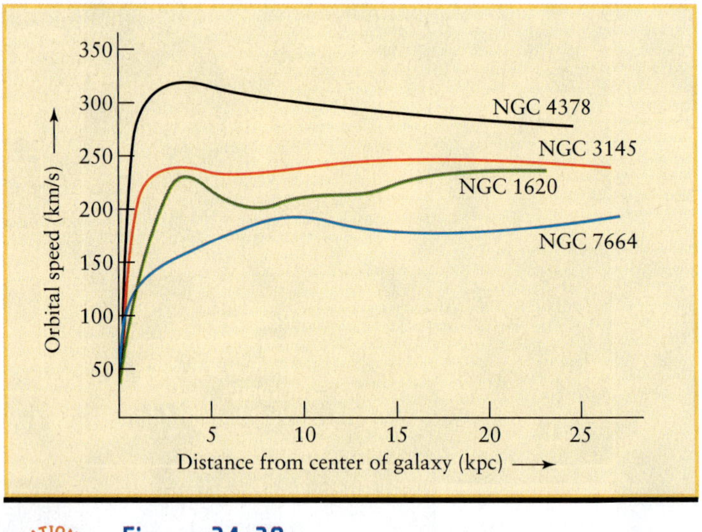

ANIMATION 24.7

Figure 24-29

The Rotation Curves of Four Spiral Galaxies This graph shows how the orbital speed of material in the disks of four spiral galaxies varies with the distance from the center of each galaxy. If most of each galaxy's mass were concentrated near its center, these curves would fall off at large distances. But these and many other galaxies have flat rotation curves that do not fall off. This indicates the presence of extended halos of dark matter. (Adapted from V. Rubin and K. Ford)

"Seeing" Dark Matter with Gravitational Lensing

Further evidence about how dark matter is distributed comes from the gravitational bending of light rays, which we described in Section 22-2. As Figure 22-5 shows, the gravity of a single star like the Sun can deflect light by only a few arcseconds. But a more massive object such as a galaxy can produce much greater deflections, and the amount of this deflection can be used to determine the galaxy's mass. For example, suppose that Earth, a massive galaxy, and a background light source (such as a more distant galaxy) are in nearly perfect alignment, as sketched in Figure 24-30*a*. Because of the warped space around the massive galaxy, light from the background source curves around the galaxy as it heads toward us. As a result, light rays can travel along two paths from the background source to us here on Earth. Thus, we should see two images of the background source.

A powerful source of gravity that distorts background images is called a **gravitational lens.** For gravitational lensing to work, the alignment between Earth, the massive galaxy, and a remote background light source must be almost perfect. Without nearly perfect alignment, the second image of the background star is too faint to be noticeable.

Beginning in 1979, astronomers have discovered a great number of examples of gravitational lensing. The example shown in Figure 24-30*b* is almost exactly like the ideal situation depicted in Figure 24-30*a*. If the alignment is very slightly off, the image of the distant galaxy is distorted into an arc as shown in Figure 24-30*c*. Figure 24-30*d* shows a more complicated example of lensing that results when the gravitational lens is not one but several massive galaxies.

Figure 24-31 shows a situation in which an entire cluster of galaxies acts as a gravitational lens. The image shows an ordinary-looking rich cluster of yellowish elliptical and spiral galaxies, but with a number of curious blue arcs. Reconstruction of the light paths through the cluster shows that all these blue arcs are actually distorted images of a single galaxy that lies billions of light-years beyond the cluster.

By measuring the distortion of the images of such background galaxies, J. Anthony Tyson of Bell Laboratories and his colleagues have determined that dark matter, which constitutes about 90% of the cluster's mass, is distributed much like the visible matter in the cluster. In other words, the overall arrangement of visible galaxies seems to trace the location of dark matter.

The Nature of Dark Matter

Many proposals have been made to explain the nature of dark matter. One reasonable suggestion was that clusters might contain

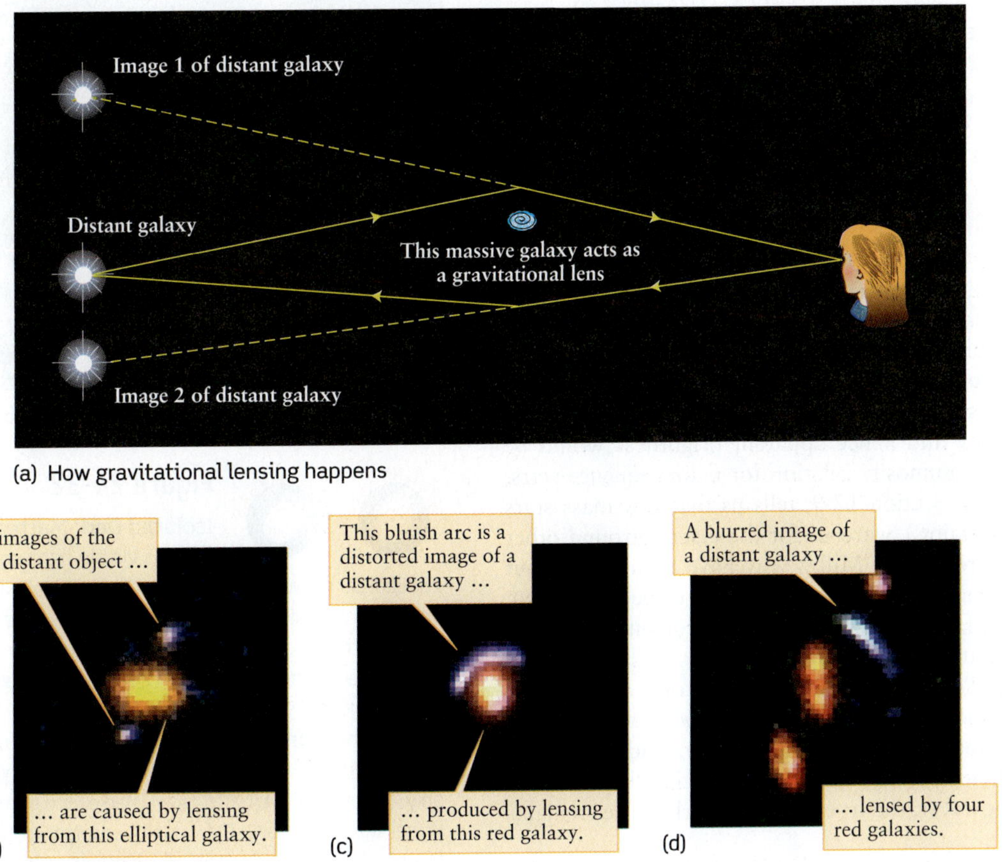
Image 1 of distant galaxy
Distant galaxy
This massive galaxy acts as a gravitational lens
Image 2 of distant galaxy
(a) How gravitational lensing happens
Two images of the same distant object ...
... are caused by lensing from this elliptical galaxy.
(b)
This bluish arc is a distorted image of a distant galaxy ...
... produced by lensing from this red galaxy.
(c)
A blurred image of a distant galaxy ...
... lensed by four red galaxies.
(d)

VIDEO 24.5

Figure 24-30 R I V U X G

Gravitational Lensing **(a)** A massive object such as a galaxy can deflect light rays like a lens so that an observer sees more than one image of a more distant galaxy. **(b), (c), (d)** Three examples of gravitational lensing. In each case a single, distant blue galaxy is "lensed" by a closer red galaxy or galaxies. (b, c, d: Kavan Ratnatunga, Carnegie Mellon University; ESA; and NASA)

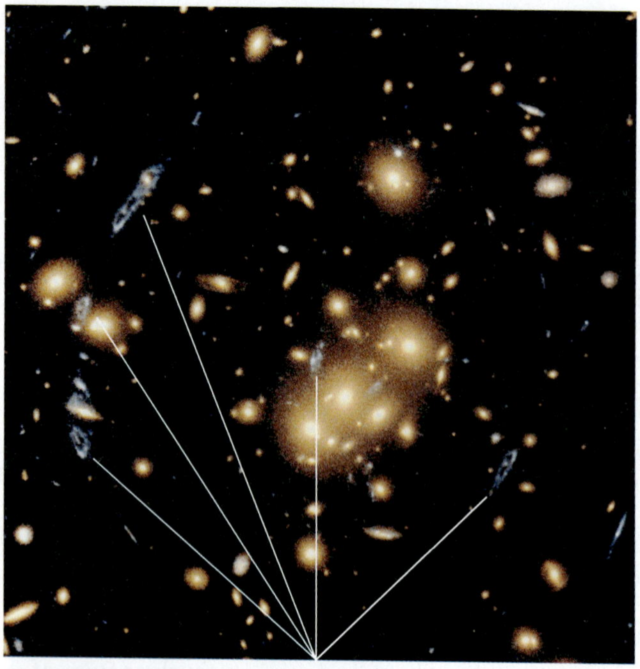

All of these blue arcs are images of the same distant galaxy.

Figure 24-31 R I V U X G

Gravitational Lensing by a Cluster of Galaxies The blue arcs in this image of the rich cluster CL0024+1654 are distorted multiple images of a single more distant galaxy. These images are the result of gravitational lensing by the matter in CL0024+1654. The cluster is about 1600 Mpc (5 billion ly) from Earth; the blue galaxy is about twice as distant. The blue color of the remote galaxy suggests that it is very young and is actively forming stars. (W. N. Colley and E. Turner, Princeton University; J. A. Tyson, Bell Labs, Lucent Technologies; NASA)

a large number of faint, red, low-mass (0.2 $M_\odot$ or less) stars. These faint stars could be located in extended halos surrounding individual galaxies or scattered throughout the spaces between the galaxies of a cluster. They would have escaped detection because their luminosity and hence apparent brightness would be very low. (The mass-luminosity relation for main-sequence stars, which we discussed in Section 17-9, tells us that low-mass stars are intrinsically very faint.) Searches for these stars around other galaxies as well as around the Milky Way have been carried out using the Hubble Space Telescope. None has yet been detected, so it is thought that faint stars are unlikely to constitute the majority of the dark matter in the universe.

As we described in Section 23-4, more exotic dark matter candidates include massive neutrinos, subatomic particles called WIMPs (weakly interacting massive particles), and MACHOs (massive compact halo objects, such as small black holes or brown dwarfs). To date, however, the true nature of dark matter remains unknown.

An important clue about the nature of dark matter was discovered in 2006 by examining a rich cluster of galaxies called 1E0657-56. Remarkably, the visible matter and dark matter in this cluster do *not* have the same distribution (Figure 24-32*a*).

(a) Composite image of galaxy cluster 1E0657-56 showing visible galaxies, X-ray-emitting gas (red) and dark matter (blue) R I V U X G

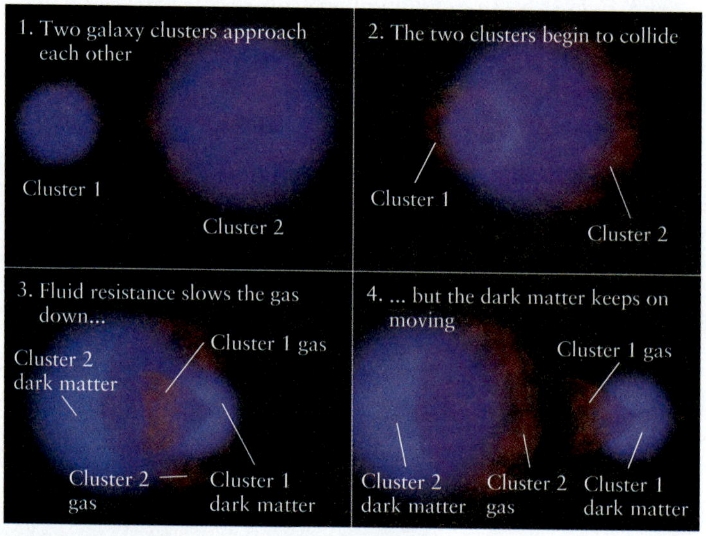

(b) A model of how the gas and dark matter in 1E0657-56 could have become separated

Figure 24-32

Isolated Dark Matter in a Cluster of Galaxies **(a)** This visible-light image of the galaxy cluster 1E0657-56 shows more than a thousand galaxies. The superimposed image in red shows the distribution of the cluster's hot, X-ray emitting gas, and the blue image shows the distribution of dark matter as determined by gravitational lensing (see Figure 24-30). **(b)** We can understand the separation of dark matter and gas in this cluster if we assume that dark matter does not feel any force of fluid resistance. This is what we would expect if dark matter responds to gravitational forces only. (a: X-ray: NASA/CXC/CfA/M. Markevitch et al.; Optical: NASA/STScI; Magellan/U. Arizona/D. Clew et al.; Lensing Map: NASA/STScI; SO FW; Magellan/U. Arizona/D. Clowe et al. b: NASA/CXC/M. Weiss)

The best explanation for how this could have come about is that 1E0657-56 is the result of a collision between two galaxy clusters, one larger than the other (Figure 24-32*b*). (In Section 24-7 we asked you to visualize collisions between entire galaxies; now you must imagine a vastly more immense collision between entire clusters of galaxies!) During the collision, the gas from one cluster slams into the gas from the other cluster and slows down due to *fluid resistance*. (You feel the force of fluid resistance pushing against you whenever you try to move through a liquid or gas—for example, when you swim in a lake or put your hand outside the window of a fast-moving car.) Fluid resistance is a consequence of the electric forces between adjacent atoms and molecules in a fluid. But if dark matter is made up of some curious material that responds only to *gravitational* forces, it is unaffected by fluid resistance. As a result, during the collision sketched in Figure 24-32*b* the gas is slowed by fluid resistance but the dark matter is not. The agreement of the simulation in Figure 24-32*b* with the observations in Figure 24-32*a* strongly reinforces the idea that dark matter, though mysterious, is quite real.

The study of dark matter forces us to reconsider our impression of the universe based on luminous matter alone. As we will see in Chapter 26, we now know that there is about 5 times as much dark matter in the observable universe as there is visible matter. Because dark matter is so dominant, "ordinary" visible matter—including this book, the air that you breathe, all the planets and stars, and your own body—is in fact relatively rare. Thus, the vast majority of mass in the universe is of completely unknown composition.

24-9 Galaxies formed from the merger of smaller objects

Over the eons, collisions and mergers have dramatically altered the population of galaxies

How do galaxies form and how do they evolve? Astronomers can gain important clues about galactic evolution simply by looking deep into space. The more distant a galaxy is, the longer its light takes to reach us. As we examine galaxies that are at increasing distances from Earth, we are actually looking further and further back in time. By looking into the past, we can see galaxies in their earliest stages.

Building Galaxies from the "Bottom Up"

The Hubble Space Telescope images in Figure 24-33 provide a glimpse of galaxy formation in the early universe. Figure 24-33*a* shows a number of galaxylike objects some 11 billion light-years (3400 Mpc) away and are thus seen as they were 11 billion years ago. These objects are smaller than even the smallest galaxies we see in the present-day universe and have unusual, irregular shapes (Figure 24-33*b*). Furthermore, these objects are scattered over an area only 600 kpc (2 million light-years) across—less than the distance between the Milky Way Galaxy and M31—making it quite probable that they would collide and merge with each other. These collisions would be aided by the dark matter associated with each subgalactic object, which increases the object's mass

(a) A portion of the constellation Hercules

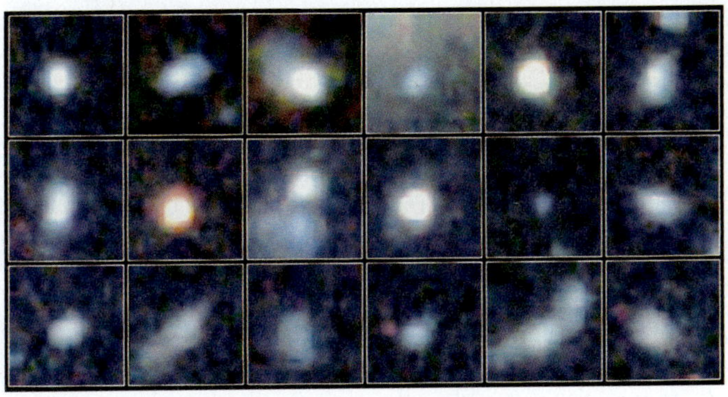

(b) Closeup images of the numbered objects in (a)

Figure 24-33 R I V U X G

The Building Blocks of Galaxies (a) In this Hubble Space Telescope image, the objects outlined by boxes are about 3400 Mpc (11 billion ly) from Earth and are only 600 to 900 pc (2000 to 3000 ly) across—larger than a star cluster but smaller than even dwarf elliptical galaxies like that shown in Figure 24-9. (b) If these objects were to coalesce, the result would be a full-sized galaxy such as we see in the nearby universe today. (Rogier Windhorst and Sam Pascarelle, Arizona State University; NASA) and, hence, the gravitational forces pulling the objects together. Such mergers would eventually give rise to a normal-sized galaxy.

Images of the young universe such as those in Figure 24-33 lead astronomers to conclude that galaxies formed "from the bottom up"—that is, by the merger of smaller objects like those in Figure 24-33*b* to form full-size galaxies. (These same images provide evidence against an older idea that galaxies formed "from the top down"—that is, fragmenting or breaking apart from immense, cluster-sized clouds of material.) The blue color of the objects in Figure 24-33*b* indicates the presence of young stars. Observations indicate that the very first stars formed about 13.5 billion years ago, when the universe was only about 200 million years old. We will see evidence in Chapter 26 that the matter in the universe formed "clumps" even earlier than this. These clumps evolved into objects like those shown in Figure 24-33*b*, which in turn merged to form the population of galaxies that we see today.

Forming Spirals, Lenticulars, and Ellipticals

Once a number of subgalactic units combine, they make an object called a *protogalaxy*. The rate at which stars form within a protogalaxy may determine whether this protogalaxy becomes a spiral or an elliptical. If stars form relatively slowly, the gas surrounding them has enough time to settle by collisions to form a flattened disk, much as happened on a much smaller scale in the solar nebula (see Section 8-4). Star formation continues because the disk contains an ample amount of hydrogen from which to make new stars. The result is a spiral or lenticular galaxy (Figure 24-34*a*). But if stars initially form in the protogalaxy at a rapid rate, virtually all of the available gas is used up to make stars before a disk can form. In this case what results is an elliptical galaxy (Figure 24-34*b*).

Figure 24-34*c* compares the stellar birthrate in the two types of galaxies. This graph helps us understand some of the differences between spiral and elliptical galaxies that we described in Section 24-3. Protogalaxies are thought to have been composed almost exclusively of hydrogen and helium gas, so the first stars were Population II stars with hardly any metals (that is, heavy elements). As stars die and form planetary nebulae or supernovae, they eject gases enriched in metals into the interstellar medium. In a spiral galaxy there is ongoing star formation in the disk, so these metals are incorporated into new generations of stars, making relatively metal-rich Population I stars like the Sun. By contrast, an elliptical galaxy has a single flurry of star formation when it is young, after which star formation ceases. Elliptical galaxies therefore contain only metal-poor Population II stars.

Figure 24-34*c* shows that both elliptical and spiral galaxies form stars most rapidly when they are young. This idea is borne out by the observation that very distant galaxies tend to be blue, which means that galaxies were bluer in the distant past than they are today. (Note the very blue colors of the distant, gravitationally lensed galaxies shown in Figure 24-30 and Figure 24-31, as well as those of the subgalactic objects shown in Figure 24-33*b*.) Spectroscopic studies of such galaxies in the 1980s by James Gunn and Alan Dressler demonstrated that most owe their blue color to vigorous star formation, often occurring in intense, episodic bursts. The hot, luminous, and short-lived O and B stars produced in these bursts of star formation give blue galaxies their characteristic color.

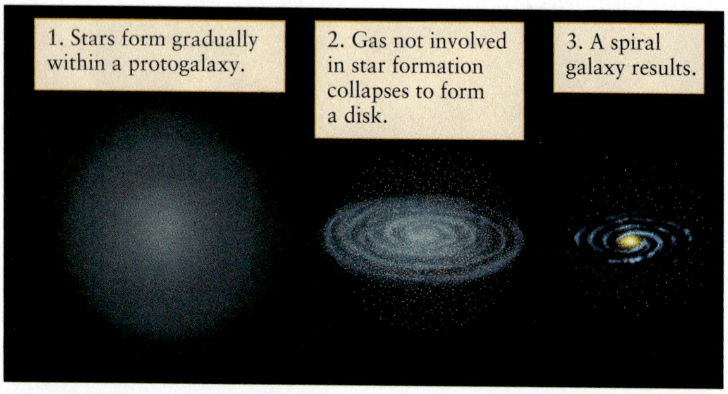

1. Stars form gradually within a protogalaxy.
2. Gas not involved in star formation collapses to form a disk.
3. A spiral galaxy results.

(a) Formation of a spiral galaxy

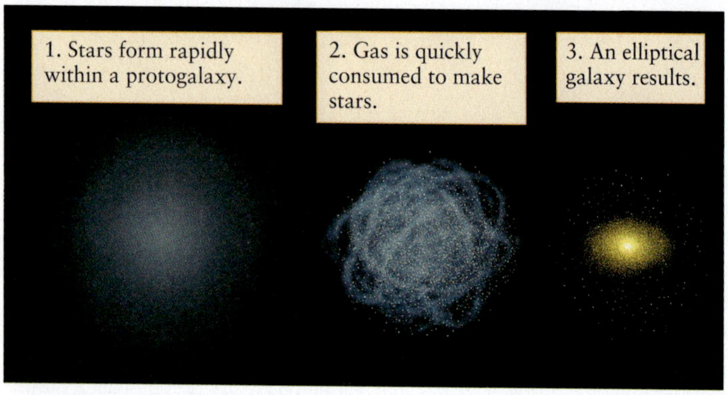

1. Stars form rapidly within a protogalaxy.
2. Gas is quickly consumed to make stars.
3. An elliptical galaxy results.

(b) Formation of an elliptical galaxy

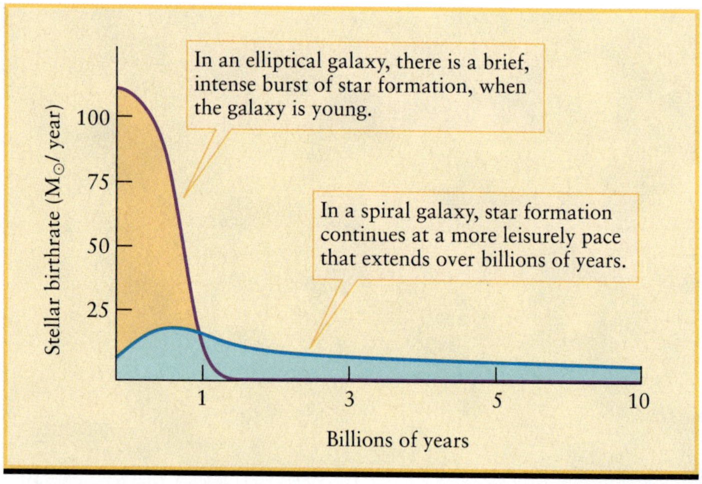

(c) The stellar birthrate in galaxies

Figure 24-34

The Formation of Spiral and Elliptical Galaxies **(a)** If the initial star formation rate in a protogalaxy is low, it can evolve into a spiral galaxy with a disk. **(b)** If the initial star formation rate is rapid, no gas is left to form a disk. The result is an elliptical galaxy. **(c)** This graph shows how the rate of star birth (in solar masses per year) varies with age in spiral and elliptical galaxies.

An Evolving Universe of Galaxies

In addition to changes in galaxy colors, the character of the galactic population has also changed over the past several billion years. In nearby rich clusters, only about 5% of the galaxies are spirals. But observations of rich clusters at a redshift of $z = 0.4$—which corresponds to looking about 4 billion years into the past—show that about 30% of their galaxies were spirals.

Why were spiral galaxies more common in rich clusters in the distant past? Galactic collisions and mergers are probably responsible. During a collision, interstellar gas in the colliding galaxies is vigorously compressed, triggering a burst of star formation (see the *Cosmic Connections: When Galaxies Collide* in Section 24-7). A succession of collisions produces a series of star-forming episodes that create numerous bright, hot O and B stars that become dispersed along arching spiral arms by the galaxy's rotation. Eventually, however, the gas is used up; star formation then ceases and the spiral arms become less visible. Furthermore, tidal forces tend to disrupt colliding galaxies, strewing their stars across intergalactic space until the galaxies are completely disrupted (see Figure 24-28).

A full description of galaxy formation and evolution must include the effects of dark matter. As we have seen, only about 10% of the mass of a galaxy—its stars, gas, and dust—emits electromagnetic radiation of any kind. As yet we have no idea what the remaining 90% looks like or what it is made of. The dilemma of dark matter is one of the most challenging problems facing astronomers today.

Key Words

anisotropic, p. 642
barred spiral galaxy, p. 640
clusters (of galaxies), p. 650
dark-matter problem, p. 658
distance ladder, p. 645
dwarf elliptical galaxy, p. 641
elliptical galaxy, p. 641
fundamental plane, p. 645
galactic cannibalism, p. 656
giant elliptical galaxy, p. 641
gravitational lens, p. 659
groups (of galaxies), p. 650
Hubble classification, p. 639
Hubble constant, p. 647
Hubble flow, p. 647
Hubble law, p. 647
intracluster gas, p. 654
irregular cluster, p. 651
irregular galaxy, p. 642
isotropic, p. 642
lenticular galaxy, p. 642
Local Group, p. 650
maser, p. 646
poor cluster, p. 650
redshift, p. 647
regular cluster, p. 651
rich cluster, p. 650
spiral galaxy, p. 639
standard candle, p. 643
starburst galaxy, p. 654
supercluster, p. 651
Tully-Fisher relation, p. 645
tuning fork diagram, p. 642
void, p. 652

Key Ideas

INTERACTIVE EXERCISE 24.2

The Hubble Classification: Galaxies can be grouped into four major categories: spirals, barred spirals, ellipticals, and irregulars.

• The disks of spiral and barred spiral galaxies are sites of active star formation.

• Elliptical galaxies are nearly devoid of interstellar gas and dust, and so star formation is severely inhibited.

• Lenticular galaxies are intermediate between spiral and elliptical galaxies.

• Irregular galaxies have ill-defined, asymmetrical shapes. They are often found associated with other galaxies.

Distance to Galaxies: Standard candles, such as Cepheid variables and the most luminous supergiants, globular clusters, H II regions, and supernovae in a galaxy, are used in estimating intergalactic distances.

• The Tully-Fisher relation, which correlates the width of the 21-cm line of hydrogen in a spiral galaxy with its luminosity, can also be used for determining distance. A method that can be used for elliptical galaxies is the fundamental plane, which relates the galaxy's size to its surface brightness distribution and to the motions of its stars.

The Hubble Law: There is a simple linear relationship between the distance from Earth to a remote galaxy and the redshift of that galaxy (which is a measure of the speed with which it is receding from us). This relationship is the Hubble law, $v = H_0 d$.

• The value of the Hubble constant, H_0, is not known with certainty but is close to 73 km/s/Mpc.

Clusters and Superclusters: Galaxies are grouped into clusters rather than being scattered randomly throughout the universe.

• A rich cluster contains hundreds or even thousands of galaxies; a poor cluster, often called a group, may contain only a few dozen.

• A regular cluster has a nearly spherical shape with a central concentration of galaxies; in an irregular cluster, galaxies are distributed asymmetrically.

• Our Galaxy is a member of a poor, irregular cluster called the Local Group.

• Rich, regular clusters contain mostly elliptical and lenticular galaxies; irregular clusters contain spiral, barred spiral, and irregular galaxies along with ellipticals.

• Giant elliptical galaxies are often found near the centers of rich clusters.

Galactic Collisions and Mergers: When two galaxies collide, their stars pass each other, but their interstellar media collide violently, either stripping the gas and dust from the galaxies or triggering prolific star formation.

• The gravitational effects during a galactic collision can throw stars out of their galaxies into intergalactic space.

• Galactic mergers may occur; a large galaxy in a rich cluster may tend to grow steadily through galactic cannibalism, perhaps producing in the process a giant elliptical galaxy.

The Dark-Matter Problem: The luminous mass of a cluster of galaxies is not large enough to account for the observed motions of the galaxies; a large amount of unobserved mass must also be present. This situation is called the dark-matter problem.

• Hot intergalactic gases in rich clusters account for a small part of the unobserved mass. These gases are detected by their X-ray emission. The remaining unobserved mass is probably in the form of dark-matter halos that surround the galaxies in these clusters.

• Gravitational lensing of remote galaxies by a foreground cluster enables astronomers to glean information about the distribution of dark matter in the foreground cluster.

Formation and Evolution of Galaxies: Observations indicate that galaxies arose from mergers of several smaller gas clouds.

• Whether a protogalaxy evolves into a spiral galaxy or an elliptical galaxy depends on its initial rate of star formation.

Questions

Review Questions

1. Why did many nineteenth-century astronomers think that the "spiral nebulae" are part of the Milky Way?
2. What was the Shapley-Curtis "debate" all about? Was a winner declared at the end of the "debate"? Whose ideas turned out to be correct?
3. How did Edwin Hubble prove that the Andromeda "Nebula" is not a nebula within our Milky Way Galaxy?
4. Are any galaxies besides our own visible with the naked eye from Earth? If so, which one(s)?
5. An educational publication for children included the following statement: "The Sun is in fact the only star in our galaxy. All of the other stars in the sky are located in other galaxies." How would you correct this statement?
6. What is the Hubble classification scheme? Which category includes the largest galaxies? Which includes the smallest? Which category of galaxy is the most common?
7. Which is more likely to have a blue color, a spiral galaxy or an elliptical galaxy? Explain why.
8. Which types of galaxies are most likely to have new stars forming? Describe the observational evidence that supports your answer.
9. Explain why the apparent shape of an elliptical galaxy may be quite different from its real shape.
10. Why do astronomers suspect that the Hubble tuning fork diagram does not depict the evolutionary sequence of galaxies?
11. Why are Cepheid variable stars useful for finding the distances to galaxies? Are there any limitations on their use for this purpose?
12. Why are Type Ia supernovae useful for finding the distances to very remote galaxies? Can they be used to find the distance to any galaxy you might choose? Explain your answers.
13. What is the Tully-Fisher relation? How is it used for measuring distances? Can it be used for galaxies of all kinds? Why or why not?
14. What are masers? How can they be used to measure the distance to a galaxy?
15. What is the Hubble law? How can it be used to determine distances?
16. How did the discovery of the Hubble Law reinforce the idea that the spiral "nebulae" could not be part of the Milky Way?
17. Why do you suppose it has been so difficult to determine the value of H_0?
18. Some galaxies in the Local Group exhibit blueshifted spectral lines. Why aren't these blueshifts violations of the Hubble law?
19. What are the differences between regular and irregular clusters?
20. What is the difference between a cluster and a supercluster? Are both clusters and superclusters held together by their gravity?
21. What measurements do astronomers make to construct three-dimensional maps of the positions of galaxies in space?
22. Describe what voids are and what they tell us about the large-scale structure of the universe.
23. Why is the intracluster gas in galaxy clusters at such high temperatures?
24. What are starburst galaxies? How can they be produced by collisions between galaxies?
25. Why do giant elliptical galaxies dominate rich clusters but not poor clusters?
26. What evidence is there for the existence of dark matter in clusters of galaxies?
27. What is gravitational lensing? Why don't we notice the gravitational lensing of light by ordinary objects on Earth?
28. How do observations of galaxy cluster 1E0657-56 help constrain the nature of dark matter?
29. What observations suggest that present-day galaxies formed from smaller assemblages of matter?
30. On what grounds do astronomers think that in the past, spiral galaxies were more numerous in rich clusters than they are today? What could account for this excess of spiral galaxies in the past?

Advanced Questions

Questions preceded by an asterisk (*) involve topics discussed in the Boxes.

Problem-solving tips and tools

Box 1-1 explains the small-angle formula, and Box 17-3 discusses the relationship among apparent magnitude, absolute magnitude, and distance. As Box 23-2 explains, a useful form of Kepler's third law is $M = rv^2/G$, where M is the mass within an orbit of radius r, v is the orbital speed, and G is the gravitational constant. Another form of Kepler's third law, particularly useful for two stars or two galaxies orbiting each other, is given in Section 17-9. The volume of a sphere of radius r is $4\pi r^3/3$. The mass of a hydrogen atom (^1H) is given in Appendix 7.

31. Hubble made his observations of Cepheids in M31 using the 100-inch (2.5-meter) telescope on Mount Wilson. Completed in 1917, this was the largest telescope in the world when Hubble carried out his observations in 1923. Why was it helpful to use such a large telescope?
32. The image on page 665 shows the Small Magellanic Cloud (SMC), an irregular galaxy that orbits the Milky Way. The SMC is 63 kpc (200,000 ly) from Earth and 8 kpc (26,000 ly) across, and can be seen with the naked eye from southern latitudes. What features of this image indicate that there has been recent star formation in the SMC? Explain your answer.

RIVUXG

(Anglo-Australian Observatory)

33. When the results from the Hipparcos mission were released, with new and improved measurements of the parallaxes of *nearby* stars within 500 pc, astronomers had to revise the distances to many *remote* galaxies millions of parsecs away. Explain why.

34. As Figure 19-19 shows, there are two types of Cepheid variables. Type I Cepheids are metal-rich stars of Population I, while Type II Cepheids are metal-poor stars of Population II. (**a**) Which type of Cepheid variables would you expect to be found in globular clusters? Which type would you expect to be found in the disk of a spiral galaxy? Explain your reasoning. (**b**) When Hubble discovered Cepheid variables in M31, the distinction between Type I and Type II Cepheids was not yet known. Hence, Hubble thought that the Cepheids seen in the disk of M31 were identical to those seen in globular clusters in our own Galaxy. As a result, his calculations of the distance to M31 were in error. Using Figure 19-19, explain whether Hubble's calculated distance was too small or too large.

*35. Astronomers often state the distance to a remote galaxy in terms of its distance modulus, which is the difference between the apparent magnitude m and the absolute magnitude M (see Box 17-3). (**a**) By measuring the brightness of supernova 1994I in the galaxy M51 (see Figure 24-2), the distance modulus for this galaxy was determined to be $m - M = 29.2$. Find the distance to M51 in megaparsecs (Mpc). (**b**) A separate distance determination, which involved measuring the brightnesses of planetary nebulae in M51, found $m - M = 29.6$. What is the distance to M51 that you calculate from this information? (**c**) What is the difference between your answers to parts (a) and (b)? Compare this difference with the 750-kpc distance from Earth to M31, the Andromeda Galaxy. The difference between your answers illustrates the uncertainties involved in determining the distances to galaxies!

*36. Suppose you discover a Type Ia supernova in a distant galaxy. At maximum brilliance, the supernova reaches an apparent magnitude of +10. How far away is the galaxy? (*Hint:* See Box 24-1.)

37. The masers that orbit the center of the spiral galaxy M106 travel at an orbital speed of about 1000 km/s. Astronomers observed these masers at intervals of 4 months. (**a**) What distance does a single maser move during a 4-month period? Give your answer in kilometers and in AU. (**b**) During this period, a maser moving across the line of sight (like the maser shown in green in Figure 24-15) appeared to move through an angle of only 10^{-5} arcsec. Calculate the distance to the galaxy.

38. The average radial velocity of galaxies in the Hercules cluster pictured in Figure 24-18 is 10,800 km/s. (**a**) Using $H_0 = 73$ km/s/Mpc, find the distance to this cluster. Give your answer in megaparsecs and in light-years. (**b**) How would your answer to (a) differ if the Hubble constant had a smaller value? A larger value? Explain your answers.

39. A certain galaxy is observed to be receding from the Sun at a rate of 7500 km/s. The distance to this galaxy is measured independently and found to be 1.4×10^8 pc. From these data, what is the value of the Hubble constant?

*40. In the spectrum of the galaxy NGC 4839, the K line of singly ionized calcium has a wavelength 403.2 nm. (**a**) What is the redshift of this galaxy? (*Hint:* See Box 24-2.) (**b**) Determine the distance to this galaxy using the Hubble law with $H_0 = 73$ km/s/Mpc.

*41. The galaxy RD1 has a redshift of $z = 5.34$. (**a**) Determine its recessional velocity v in km/s and as a fraction of the speed of light. (**b**) What recessional velocity would you have calculated if you had erroneously used the low-speed formula relating z and v? Would using this formula have been a small or large error? (**c**) According to the Hubble law, what is the distance from Earth to RD1? Use $H_0 = 73$ km/s/Mpc for the Hubble constant, and give your answer in both megaparsecs and light-years.

42. It is estimated that the Coma cluster (see Figure 24-21) contains about 10^{13} $M_\odot$ of intracluster gas. (**a**) Assuming that this gas is made of hydrogen atoms, calculate the total number of intracluster gas atoms in the Coma cluster. (**b**) The Coma cluster is roughly spherical in shape, with a radius of about 3 Mpc. Calculate the number of intracluster gas atoms per cubic centimeter in the Coma cluster. Assume that the gas fills the cluster uniformly. (**c**) Compare the intracluster gas in the Coma cluster with the gas in our atmosphere (3×10^{19} molecules per cubic centimeter, temperature 300 K); a typical gas cloud within our own Galaxy (a few hundred molecules per cubic centimeter, temperature 50 K or less); and the corona of the Sun (10^5 atoms per cubic centimeter, temperature 10^6 K).

*43. Two galaxies separated by 600 kpc are orbiting each other with a period of 40 billion years. What is the total mass of the two galaxies?

*44. Figure 24-29 shows the rotation curve of the Sa galaxy NGC 4378. Using data from that graph, calculate the orbital period of stars 20 kpc from the galaxy's center. How much mass lies within 20 kpc from the center of NGC 4378?

45. How might you determine what part of a galaxy's redshift is caused by the galaxy's orbital motion about the center of mass of its cluster?

46. The accompanying image shows the unusual elliptical galaxy NGC 5128. Explain how the properties of this galaxy seen in the infrared image can be explained if NGC 5128 is the result of a merger of an elliptical galaxy and a spiral galaxy.

R I V U X G

R I V U X G

(visible: Eric Peng, Herzberg Institute of Astrophysics and NOAO/AURA/NSF; infrared: Jocelyn Keene, NASA/JPL and Caltech)

47. Explain why the dark matter in galaxy clusters could not be neutral hydrogen.

48. According to Figure 24-34*c*, elliptical galaxies continue to form stars for about a billion years after they form. Give an argument why we might expect to find some Population I stars in an elliptical galaxy. (*Hint:* Table 19-1 gives the main-sequence lifetimes for stars of different masses.)

Discussion Questions

49. Earth is composed principally of heavy elements, such as silicon, nickel, and iron. Would you be likely to find such planets orbiting stars in the disk of a spiral galaxy? In the nucleus of a spiral galaxy? In an elliptical galaxy? In an irregular galaxy? Explain your answers.

50. Discuss what observations you might make to determine whether or not the various Hubble types of galaxies represent some sort of evolutionary sequence.

51. Discuss the advantages and disadvantages of using the various standard candle distance indicators to obtain extragalactic distances.

52. How would you distinguish star images from unresolved images of remote galaxies on a CCD?

53. Describe what sorts of observations you might make to search for as-yet-undiscovered galaxies in our Local Group. How is it possible that such galaxies might still remain to be discovered? In what part of the sky would these galaxies be located? What sorts of observations might reveal these galaxies?

Web/eBook Questions

54. When galaxies pass close to one another, as should happen frequently in a rich cluster, tidal forces between the galaxies can strip away their outlying stars. The result should be a loosely dispersed sea of "intergalactic stars" populating the space between galaxies in a cluster. Search the World Wide Web for information about intergalactic stars. Have they been observed? If so, where are they found? What would our nighttime sky look like if our Sun were an intergalactic star?

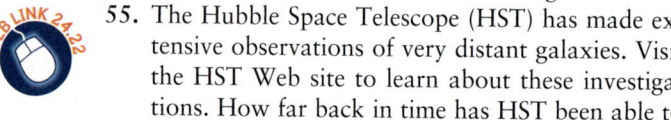

55. The Hubble Space Telescope (HST) has made extensive observations of very distant galaxies. Visit the HST Web site to learn about these investigations. How far back in time has HST been able to look? What sorts of early galaxies are observed? What is the current thinking about how galaxies formed and evolved?

56. **The Formation of "The Mice"** Access and view the animation "The Formation of 'The Mice'" in Chapter 24 of the *Universe* Web site or eBook. Based on this computer simulation, what will be the final fate of the two galaxies that make up the Antennae? (See the *Cosmic Connections: When Galaxies Collide* in Section 24-7.) Speculate on what might have happened if the two galaxies had collided at a much greater speed.

57. **Radiation from a Rotating Galaxy.** Access and view the animation "Radiation from a Rotating Galaxy" in Chapter 24 of the *Universe* Web site or eBook. Describe how the animation would have to be changed for a spiral galaxy of the same size but (**a**) greater mass and (**b**) smaller mass.

Activities

Observing Projects

58. Using a telescope with an aperture of at least 30 cm (12 in), observe as many of the spiral galaxies listed in the following table as you can. If your copy of the book comes with the *Starry Night Enthusiast*™ program, use it to help determine when these galaxies can best be viewed. Many of these galaxies are members of the Virgo cluster, which can best be seen from March through June. Because all galaxies are quite faint, be sure to schedule your observations for a moonless night. The best

view is obtained when a galaxy is near the meridian. While at the eyepiece, make a sketch of what you see. Can you distinguish any spiral structure? After completing your observations, compare your sketches with photographs found in an online catalog of Messier objects (the "M" in the galaxy designations stands for Messier).

Spiral galaxy	Right ascension	Declination	Hubble type
M31 (NGC 224)	0^h 42.7^m	+41° 16′	Sb
M58 (NGC 4579)	12 37.7	+11 49	Sb
M61 (NGC 4303)	12 21.9	+4 28	Sc
M63 (NGC 5055)	13 15.8	+42 02	Sb
M64 (NGC 4826)	12 56.7	+21 41	Sb
M74 (NGC 628)	1 36.7	+15 47	Sc
M83 (NGC 5236)	13 37.0	−29 52	Sc
M88 (NGC 4501)	12 32.0	+14 25	Sb
M90 (NGC 4569)	12 36.8	+13 10	Sb
M91 (NGC 4548)	12 35.4	+14 30	SBb
M94 (NGC 4736)	12 50.9	+41 07	Sb
M98 (NGC 4192)	12 13.8	+14 54	Sb
M99 (NGC 4254)	12 18.8	+14 25	Sc
M100 (NGC 4321)	12 22.9	+15 49	Sc
M101 (NGC 5457)	14 03.2	+54 21	Sc
M104 (NGC 4594)	12 40.0	−11 37	Sa
M108 (NGC 3556)	11 11.5	+55 40	Sc

Note: The right ascensions and declinations are given for epoch 2000.

59. Using a telescope with an aperture of at least 30 cm (12 in), observe as many of the following elliptical galaxies as you can. Six of these galaxies are in the Virgo cluster, which is conveniently located in the evening sky from March through June. If your copy of the book comes with the *Starry Night Enthusiast*™ program, use it to help determine when these galaxies can best be viewed. As in the previous exercise, be sure to schedule your observations for a moonless night, when the galaxies you wish to observe will be near the meridian. Do these elliptical galaxies differ in appearance from spiral galaxies?

Elliptical galaxy	Right ascension	Declination	Hubble type
M49 (NGC 4472)	12^h 29.8^m	+8° 00′	E4
M59 (NGC 4621)	12 42.0	+11 39	E3
M60 (NGC 4649)	12 43.7	+11 33	E1
M84 (NGC 4374)	12 25.1	+12 53	E1
M86 (NGC 4406)	12 26.2	+12 57	E3
M89 (NGC 4552)	12 35.7	+12 33	E0
M110 (NGC 205)	00 40.4	+41 41	E6

Note: The right ascensions and declinations are given for epoch 2000.

60. Using a telescope with an aperture of at least 30 cm (12 in), observe as many of the following interacting galaxies as you can. If your copy of the book comes with the *Starry Night Enthusiast*™ program, use it to help determine when these galaxies can best be viewed. As in the previous exercises, be sure to schedule your observations for a moonless night, when the galaxies you wish to observe will be near the meridian. While at the eyepiece, make a sketch of each galaxy. Can you distinguish hints of interplay among the galaxies? After completing your observations, compare your sketches with photographs found in an online catalog of Messier objects (the "M" in the galaxy designations stands for Messier).

Interacting galaxies	Right ascension	Declination
M51 (NGC 5194)	13^h 29.9^m	+47° 12′
NGC 5195	13 30.0	+47 16
M65 (NGC 3623)	11 18.9	+13 05
M66 (NGC 3627)	11 20.2	+12 59
NGC 3628	11 20.3	+13 36
M81 (NGC 3031)	9 55.6	+69 04
M82 (NGC 3034)	9 55.8	+69 41
M95 (NGC 3351)	10 44.0	+11 42
M96 (NGC 3368)	10 46.8	+11 49
M105 (NGC 3379)	10 47.8	+12 35

Note: The right ascensions and declinations are given for epoch 2000.

61. Use the *Starry Night Enthusiast*™ program to observe other galaxies. Select **Favorites > Guides > Atlas** to display the entire celestial sphere. Open the **Options** pane, expand the **Deep Space** list, and click the box for **Messier Objects** to display these objects. Open the **Find** pane and select each of the following galaxies in turn. Zoom in as necessary to see the shape of the galaxy to determine its Hubble classification as well as you can, and explain your reasoning: (i) Virgo A (M87); (ii) M105; (iii) M102; (iv) M104; (v) M109.

62. Use the *Starry Night Enthusiast*™ program to examine clusters of galaxies. Select **Favorites > Deep Space > Virgo Cluster** to center this collection of galaxies in the view, as seen from a distance of about 53 Mly from the Sun. You are looking at a three-dimensional view of the Tully Database. Open the **Find** pane and locate **Virgo A,** one of the galaxies examined in the previous question, which is close to the center of this cluster of galaxies. Right-click on this galaxy to open the contextual menu (Macintosh users Ctrl-click on this galaxy) and click on **Highlight "GA Virgo Cluster" Filament** to highlight this cluster in yellow. Click on the **"up"** arrow in the Viewing Location to move to about 30 Mly from the Sun. Hold down the Shift key while holding down the mouse button and move the mouse to use the location scroller to rotate this rich group of galaxies. (**a**) Describe the general shape of the Virgo cluster. (**b**) As you rotate the Virgo cluster, you should notice other groupings of galaxies. Stop this rotation at some position and make a sketch of the screen, circling what you believe are other groups on your sketch. Right-click (Macintosh users Ctrl-click) on one of the other clusters (and clouds and extensions) near to the Virgo Cluster to open the contextual menu and use the **Highlight** option to see how astronomers have grouped these other galaxies. Repeat this process until you have identified all of the clusters around Virgo. Outline and label these clusters on your drawing. (c) Choose three of these clusters, center on each in turn and right-click (Ctrl-click on a Macintosh) to open the contextual menu and use the **Centre** command. Use the **Zoom** facility and **location scroller** to move around these collections of galaxies, and describe their distribution compared to the Virgo cluster. For example, what are their shapes and relative sizes compared to Virgo and to each other? Are they rich spherical concentrations or walls of galaxies?

Collaborative Exercises

63. In the early twentieth century, there was considerable debate about the nature of spiral nebulae and their distance from us, but the debate was resolved by improvements in technology. As a group, list three issues that we, as a culture, did not understand in the past but understand today, and explain why we now have that understanding.

64. Even though there are billions of galaxies, there are not billions of different kinds. In fact, galaxies are classified according to their appearance. As a group, dig into your book bags and put all of the writing implements (pens, pencils, highlighters, and so on) you have in a central pile. Remember which ones are yours! Determine a classification scheme that sorts the writing implements into at least three to six piles. Write down the scheme and the number of items in each pile. Ask the group next to you to use your scheme and sort your materials. Correct any ambiguities before submitting your classification scheme.

65. Imagine your company, Astronomical Artistry, has been contracted by the local marching band to create a football halftime show about spiral galaxies. How exactly would you design the positions of the band members on the field to represent the different spiral galaxies of classes Sa, Sb, and Sc? Create two columns on your paper by drawing a line from top to bottom, drawing sketches in the left-hand column and writing a description of each sketch in the right-hand column. Also include what the band's opening formation and final formation should be.

25 Quasars and Active Galaxies

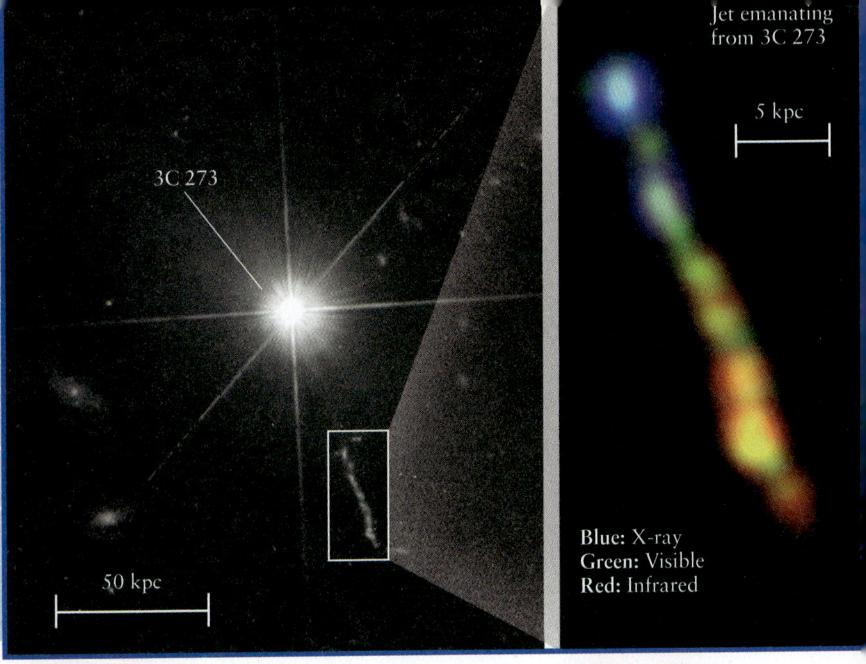

R I V U X G　　R I V U X G

WEB LINK 25.1 A powerful jet tens of thousands of parsecs in length blasts outward from quasar 3C 273. (Left: NASA and J. Bahcall, Institute for Advanced Study; right: NASA/JPL-Caltech/Yale University)

An ordinary star emits radiation primarily at ultraviolet, visible, and infrared wavelengths, in proportions that reflect the star's surface temperature. Ordinary galaxies, too, emit most strongly in these wavelength regions. But the object shown here, called 3C 273, is outrageously different: It emits strongly over an immense range of wavelengths from radio to X-ray. 3C 273 is also intensely luminous, with thousands of times the radiation output of an ordinary galaxy. And, as this X-ray image shows, some source of energy within 3C 273 causes it to eject a glowing, high-speed jet that extends outward for hundreds of thousands of light-years.

3C 273 is a *quasar*—one of many thousands of distant objects whose luminosity is far too great to come from starlight alone. Quasars are intimately related to *active galaxies,* whose tremendous luminosity can fluctuate substantially over a period of months, weeks, or even days.

What makes quasars and active galaxies so tremendously energetic? The quest to answer this question was one of the greatest scientific detective stories of the twentieth century. Like a detective story, in this chapter we will begin with the often startling evidence about quasars and active galaxies that confronted astronomers. The plot will thicken as we reveal the prime suspects—supermassive black holes that are thought to lie at the center of every quasar and active galaxy. As matter accretes onto such a black hole, it releases fantastic amounts of energy and can produce powerful jets. We will even catch the culprit in the act, with an image that shows a quasar about to devour material to feed its central black hole.

25-1 Quasars look like stars but have huge redshifts

Many of the most revolutionary discoveries in astronomy have been the result of advances in technology. So it was in 1610 when Galileo used a new device called the telescope to scan the sky, and in the process found evidence that shattered the ancient geocentric model of the universe (see Section 4-3). And so it was during the middle years of the twentieth century, when the new technique of radio astronomy revealed the existence of remote, dazzlingly luminous objects called *quasi-stellar radio sources* or *quasars*.

The Discovery of Quasars

WEB LINK 25.2 Grote Reber, a radio engineer and ham radio enthusiast, built the first true radio telescope in 1936 in his backyard in Illinois (see Section 6-6). By 1944 he had detected strong radio emissions from sources in the constellations Cassiopeia, Sagittarius, and Cygnus.

Two of these sources, named Cassiopeia A and Sagittarius A, happen to lie in our own Galaxy; they are a supernova remnant (see Figure 20-25) and the center of the Galaxy (see Section 23-6). The nature of the

Examining the spectra of quasars revealed that they are immensely distant

Learning Goals

By reading the sections of this chapter, you will learn

25-1 The distinctive features of quasars

25-2 The connection between quasars and distant galaxies

25-3 What Seyfert galaxies and radio galaxies are and how they compare to quasars

25-4 The properties of active galactic nuclei

25-5 How supermassive black holes can power active galactic nuclei

25-6 Why many active galaxies emit ultrafast jets of material

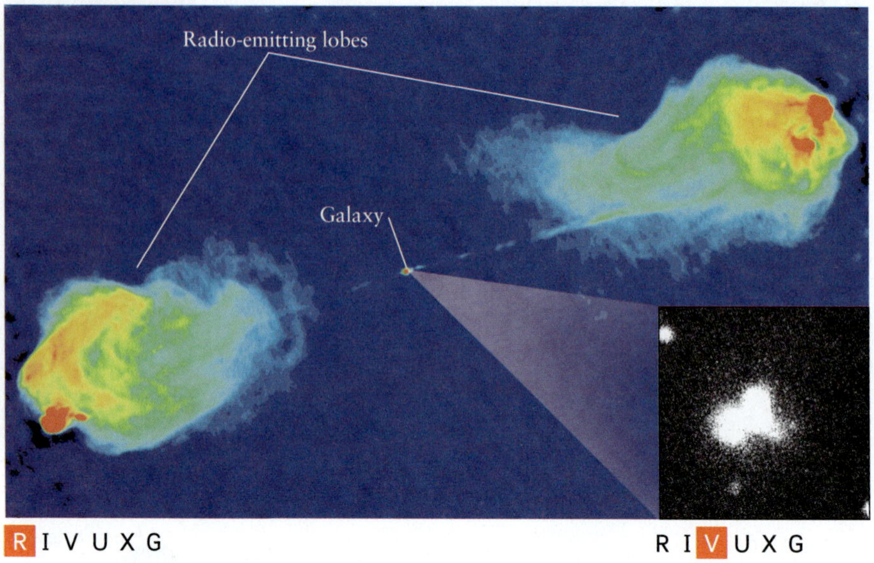

(a) Radio image of Cygnus A

(b) Visible-light close-up of the central galaxy

Figure 25-1

Cygnus A (3C 405) **(a)** This false-color radio image from the Very Large Array shows that most of the emission from Cygnus A comes from luminous radio lobes located on either side of a peculiar galaxy. (Red indicates the strongest radio emission, while blue indicates the faintest). Each lobe extends about 70 kpc (230,000 ly) from the galaxy. **(b)** The galaxy at the heart of Cygnus A has a substantial redshift, so it must be extremely far from Earth (about 230 Mpc, or 740 million ly). To be so distant and yet be one of the brightest radio sources in the sky, Cygnus A must have an enormous energy output. (a: R. A. Perley, J. W. Dreher, and J. J. Cowan, NRAO/AUI; b: Palomar Observatory)

third source, called Cygnus A (Figure 25-1*a*), proved more elusive. The mystery only deepened in 1951, when Walter Baade and Rudolph Minkowski used the 200-inch (5-meter) optical telescope on Palomar Mountain to discover a dim, strange-looking galaxy at the position of Cygnus A (see Figure 25-1*b*).

When Baade and Minkowski photographed the spectrum of Cygnus A, they were surprised to find a number of bright *emission* lines. By contrast, a normal galaxy has *absorption* lines in its spectrum (see Section 24-3 and Figure 24-16). This absorption takes place in the atmospheres of the stars that make up the galaxy, as we described in Section 17-5. In order for Cygnus A to have emission lines, something must be exciting and ionizing its atoms. Furthermore, the wavelengths of Cygnus A's emission lines are all shifted by 5.6% toward the red end of the spectrum. Astronomers use the letter z to denote redshift (see Section 24-5), and thus this object has $z = 0.056$, corresponding to a recessional velocity of 16,000 km/s.

If Cygnus A participates in the same Hubble flow as clusters of galaxies (described in Section 24-5), then this recessional velocity corresponds to a tremendous distance from Earth—about 230 Mpc (740 million ly) if the Hubble constant equals 73 km/s/Mpc. Yet despite its tremendous distance from Earth, radio waves from Cygnus A can be picked up by amateur astronomers with backyard equipment. The ease in detecting these radio waves means that Cygnus A must be one of the most luminous radio sources in the sky. In fact, its radio luminosity is 10^7 times as great as that of an ordinary galaxy like the Milky Way. The object that creates the Cygnus A radio emission must be something quite extraordinary.

Cygnus A was not the only curious radio source to draw the attention of astronomers. In 1960, Allan Sandage used the 200-inch telescope to discover a "star" at the location of a radio source designated 3C 48 (Figure 25-2). (The "3C" refers to the *Third Cambridge Catalogue*, a compendium of radio sources.) Ordinary stars are not strong sources of radio emission, so 3C 48 must be something unusual. Like Cygnus A, its spectrum showed a series of emission lines, but astronomers were unable to identify the chemical elements that produced these lines.

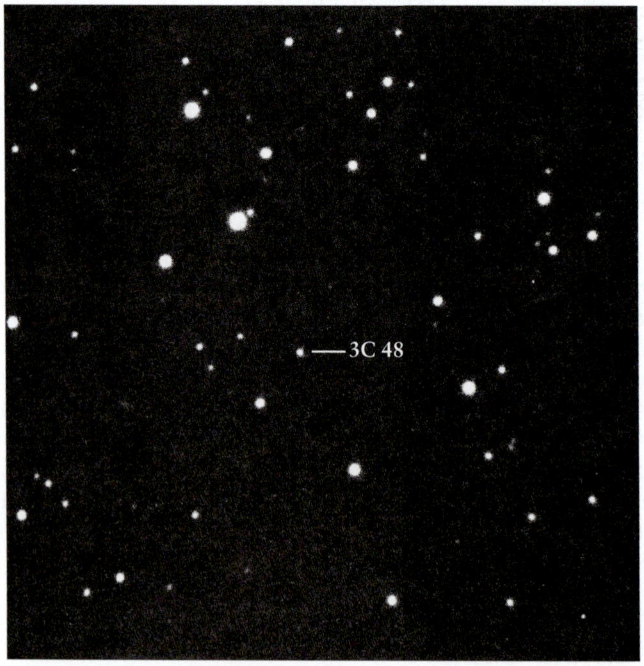

Figure 25-2 R I V U X G

The Quasar 3C 48 This object in the constellation Triangulum was thought at first to be a nearby star that happened to emit radio waves. In fact, the redshift of 3C 48 is so great ($z = 0.367$) that, according to the Hubble law, it must presently be approximately 1400 Mpc (4.6 billion ly) away. (Alex G. Smith, Rosemary Hill Observatory, University of Florida)

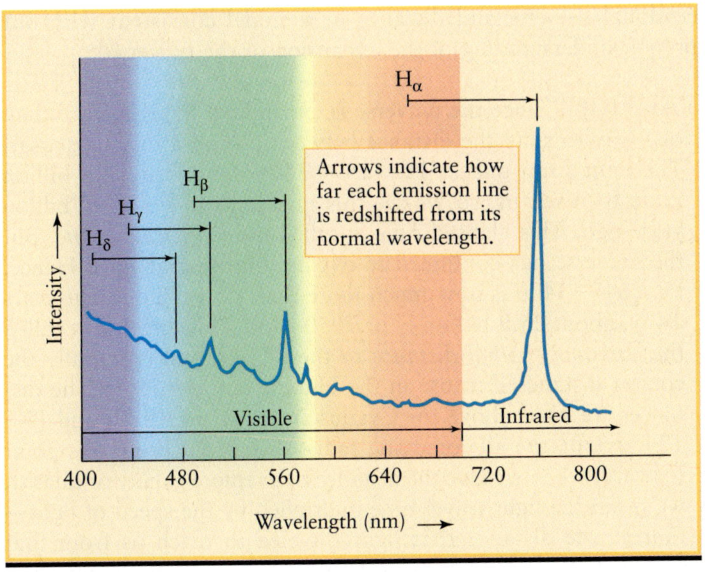

Figure 25-3

The Spectrum of 3C 273 The visible-light and infrared spectrum of this quasar is dominated by four bright emission lines of hydrogen (see Section 5-8). The redshift is $z = 0.158$, so the wavelength of each line is 15.8% greater than for a sample of hydrogen on Earth. For example, the wavelength of H_β is shifted from 486 nm (a blue-green wavelength) to $1.158 \times (486 \text{ nm}) = 563$ nm (a yellow wavelength).

Two years later, astronomers discovered a similar starlike optical counterpart to the radio source 3C 273. This "star" is even more unusual, with a luminous jet protruding from one side. (The image that opens this chapter shows 3C 273 and its accompanying jet.) Like 3C 48, its visible spectrum contains a series of emission lines that no one could explain.

Clues from Spectra

Although 3C 48 and 3C 273 were clearly oddballs, many astronomers thought they were just strange stars in our own Galaxy. A breakthrough occurred in 1963, when Maarten Schmidt at Caltech took another look at the spectrum of 3C 273. He realized that four of its brightest emission lines are positioned relative to one another in precisely the same way as four of the Balmer lines of hydrogen. However, these emission lines from 3C 273 were all shifted to much longer wavelengths than the usual wavelengths of the Balmer lines. Schmidt determined that 3C 273 has a redshift of $z = 0.158$, corresponding to a recessional velocity of 44,000 km/s (15% of the speed of light). No star could be moving this fast and remain within our Galaxy. Hence, Schmidt concluded that 3C 273 could not be a nearby star, but must lie outside the Milky Way.

According to the Hubble law, the recessional velocity of 3C 273 implies that its present distance from us is 630 Mpc (2 billion ly). To be detected at such distances, 3C 273 must be an extraordinarily powerful source of both visible light and radio radiation.

Figure 25-3 shows the visible and near-infrared spectrum of 3C 273. As we saw in Section 6-5 (review Figure 6-21), such a spectrum is a graph of intensity versus wavelength on which emission lines appear as peaks and absorption lines appear as valleys. Emission lines are caused by excited atoms, which emit radiation at specific wavelengths.

Upon learning how Schmidt deciphered the spectrum of 3C 273, two other Caltech astronomers, Jesse Greenstein and Thomas Matthews, found they could identify the spectral lines of 3C 48 as having suffered an even larger redshift of $z = 0.367$. That shift corresponds to a speed three-tenths that of light, which means that 3C 48 is presently twice as far away as 3C 273, or approximately 1400 Mpc (4.6 billion ly) from Earth.

Quasars: High Redshifts, Extreme Distances

WEB LINK 25.3

Because of their strong radio emission and starlike appearance, 3C 48 and 3C 273 were dubbed *quasi-stellar radio sources,* a term soon shortened to **quasars.** After the first quasars were discovered by their radio emission, many similar, high-redshift, starlike objects were found that emit little or no radio radiation. These "radio-quiet" quasars were originally called *quasi-stellar objects,* or *QSOs,* to distinguish them from radio emitters. Today, however, the term "quasar" is often used to include both types. Only about 10% of quasars are "radio-loud" like 3C 48 and 3C 273.

Thanks to sky surveys like those we discussed in Section 24-6, more than 100,000 quasars are now known. In images, they all look rather like stars (see Figure 1-10 for a side-by-side comparison), and all have large redshifts, ranging from 0.06 to at least 5.8 (Figure 25-4). Most quasars have redshifts of 0.3 or more, which

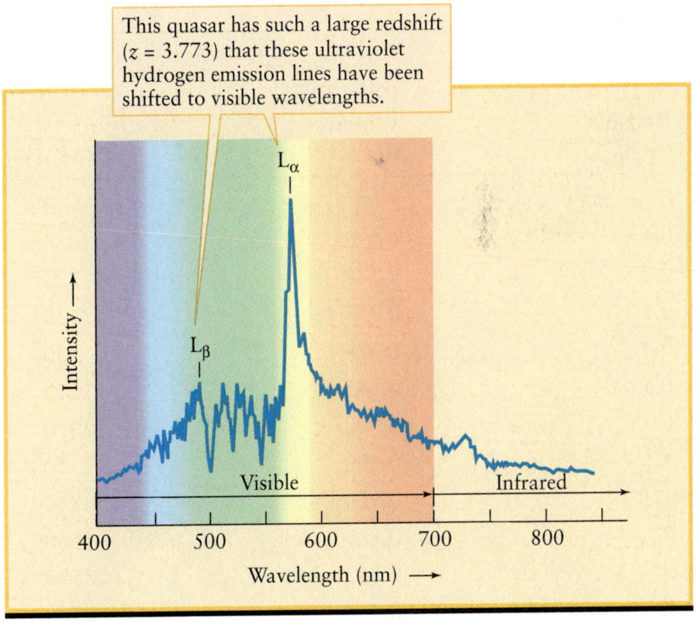

Figure 25-4

The Spectrum of a High-Redshift Quasar This quasar, known as PKS 2000-030, has a redshift $z = 3.773$. This redshift is so large that the Lyman spectral lines L_α and L_β of hydrogen, which are normally at ultraviolet wavelengths, have been shifted into the visible part of the spectrum. The designation of the quasar refers to its position on the celestial sphere and to the Parkes Observatory in Australia, where it was first discovered (see Figure 6-22).

implies that they are more than 1000 Mpc (3 billion light-years) from Earth. For example, the quasar PC 1247+3406 has a redshift of $z = 4.897$, which corresponds to a recessional velocity of more than 94% of the speed of light and (from the Hubble law) a distance of roughly 7950 Mpc (25.9 billion ly) from Earth.

CAUTION! A value of z greater than 1 does *not* mean that a quasar is receding from us faster than the speed of light. At high speeds, the relationship between redshift and recessional velocity must be modified by the special theory of relativity, as explained in Box 24-2. As the speed of a receding source approaches the speed of light, its redshift can become very much greater than 1. An infinite redshift ($z = \infty$) corresponds to a recessional velocity equal to the speed of light.

Light takes time to travel across space, so when we observe a very remote object, we are seeing it in the remote past. This means that for very remote objects, the relationship between redshift and distance from Earth depends on how the universe has evolved over time. As we will see in Chapter 26, the Hubble law reveals that the universe is expanding. In other words, if you could watch the motions of widely separated clusters of galaxies over millions of years, you would see them gradually moving away from one another. Furthermore, in Chapter 26 we will see evidence that the universe has expanded at different rates at different times in the past. Table 25-1 relates the redshift to the recessional velocity and distance in a model consistent with our present understanding of the expansion of the universe.

CAUTION! Since the universe is expanding, there is more than one way to state the distance between us and a distant quasar. The light from quasar PC 1247+3406 has taken 12.5 billion years to reach us, so we see this quasar as it was 12.5 billion years ago. This elapsed time is called the *light travel time*. But the universe has continued to expand during that time. Hence, PC 1247+3406 is now much more than 12.5 billion light-years away, about 25.9 billion ly or 7950 Mpc. This distance is called the *comoving radial distance* to the quasar, and is actually the correct distance d to use in the Hubble law $v = H_0 d$. (The distances we cited above for Cygnus A, 3C 273, 3C 48, and PKS 1247+3406 are all comoving radial distances.) From this point forward, when we give the distance to a remote galaxy or quasar, we mean the light travel time multiplied by the speed of light—that is, the distance that light traveled to reach us from that galaxy or quasar. Then the distance to a remote object expressed in light-years tells you how many years into the past you are looking when you view that object. (Table 25-1 lists both the distance calculated in this way and the comoving radial distance.) To avoid these issues about how best to define distances, many astronomers prefer simply to state the redshift z of a distant object; they then use the simple rule that the greater the redshift, the more distant the object (see Section 24-5).

WEB LINK 25.4

Table 25-1 Redshift and Distance

Redshift z	Recessional velocity v/c	Distance at which we see the object		Present distance to the object (comoving radial distance)	
		(Mpc)	(10^9 ly)	(Mpc)	(10^9 ly)
0	0	0	0	0	0
0.1	0.095	384	1.25	403	1.32
0.2	0.180	721	2.35	790	2.58
0.3	0.257	1020	3.32	1160	3.79
0.4	0.324	1280	4.17	1510	4.94
0.5	0.385	1510	4.93	1850	6.04
0.75	0.508	2070	6.48	2620	8.54
1	0.600	2350	7.65	3290	10.7
1.5	0.724	2840	9.26	4390	14.3
2	0.800	3160	10.3	5250	17.1
3	0.882	3520	11.5	6500	21.2
4	0.923	3710	12.1	7370	24.0
5	0.946	3830	12.5	8010	26.1
10	0.984	4060	13.2	9790	31.9
Infinite	1	4210	13.7	14500	47.4

This table assumes a Hubble constant $H_0 = 73$ km/s/Mpc, a matter density parameter $\Omega_m = 0.24$, and a dark energy density parameter $\Omega_\Lambda = 0.76$ (see Chapter 26). The distance at which we see the object (in light-years) is equal to the light travel time in years. The present distance to the object is the distance d to be used in the Hubble law, $v = H_0 d$.

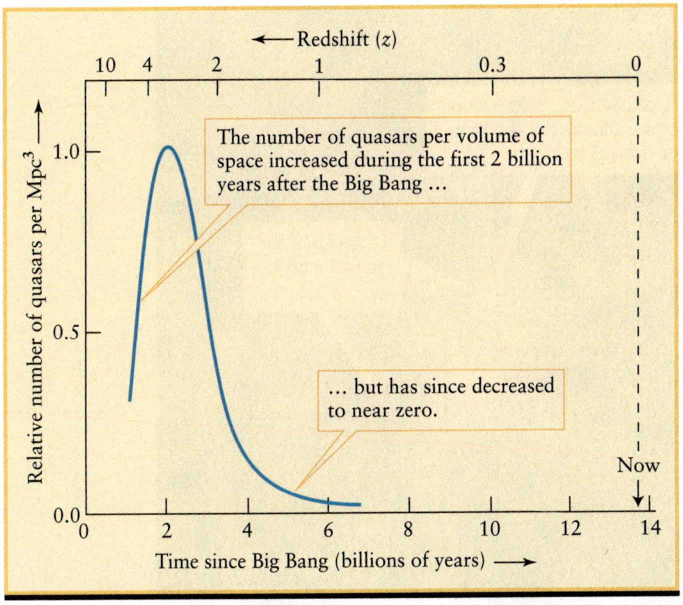

Figure 25-5

Quasars Are Extinct The greater the redshift of a quasar, the farther it is from Earth and the farther back in time we are seeing it. By observing the number of quasars found at different redshifts, astronomers can calculate how the density of quasars in the universe has changed over the history of the universe. The history of quasars is reminiscent of the history of the dinosaurs, which once populated the entire Earth but today are extinct. (Peter Shaver, European Southern Observatory)

Because there are no quasars with small redshifts, it follows that *there are no nearby quasars*. The nearest one is some 250 Mpc (800 million ly) from Earth. Hence, the absence of nearby quasars means that there have been no quasars for 800 million years. Indeed, the number of quasars began to decline precipitously roughly 10 billion years ago (Figure 25-5). Quasars were a common feature of the universe in the distant past, but there are none in the present-day universe.

25-2 Quasars are the ultraluminous centers of distant galaxies

If quasars are as distant as their redshifts show them to be, they must be extraordinarily luminous to be visible from Earth. What, then, are these strange objects? Quasars cannot simply be very large and luminous galaxies, because their spectra are totally different: The visible-light and infrared spectrum of a galaxy is dominated by absorption lines from the galaxy's stars, while the corresponding spectrum of a quasar is dominated by emission lines. Quasars also emit much more of their light at ultraviolet wavelengths than do stars or galaxies. Hence, whatever the origin of a quasar's radiation, it is not simply starlight. While quasars are not galaxies, we will see compelling evidence that the two are intimately related: Quasars turn out to be ultraluminous objects located at the centers of remote galaxies.

Quasars and Their Luminosities

A quasar's luminosity can be calculated from its apparent brightness and distance using the inverse-square law (see Section 17-2). For example, 3C 273 has a luminosity of about 10^{40} watts, which is equivalent to 2.5×10^{13} (25 trillion) Suns. (The Sun's luminosity is $L_{\odot} = 3.90 \times 10^{26}$ watts.) Generally, quasar luminosities range from about 10^{38} watts up to nearly 10^{42} watts. For comparison, a typical large galaxy, like our own Milky Way, shines with a luminosity of 10^{37} watts, which equals 2.5×10^{10} (25 billion) Suns. Thus, a bright quasar can be many thousands of times more luminous than the entire Milky Way Galaxy.

The most luminous quasars emit 100,000 times more radiation than the entire Milky Way Galaxy

WEB LINK 25.5

When quasars were first discovered, their energy output seemed so absurdly huge that a few astronomers began to question long-held ideas, such as the Hubble law. Traditionally, astronomers had used the Hubble law to calculate the distance to a galaxy from its redshift: the higher the redshift, the greater the distance. In the 1960s the American astronomer Halton C. Arp suggested that the Hubble law might not apply to quasars. If part of a quasar's redshift was caused by some yet-undiscovered phenomenon, then the quasar could be much closer to Earth than the Hubble law would have us believe. If so, a quasar's luminosity would not be so incredibly large.

In support of this theory, Arp and his collaborators drew attention to certain high-redshift quasars that seem to be located in or associated with low-redshift galaxies. These examples of "discordant redshifts" fueled a heated debate during the 1970s that was reminiscent of the Shapley-Curtis "debate" 50 years earlier (see Section 24-1). By the 1980s, however, the preponderance of evidence clearly favored the standard interpretation of redshifts according to the Hubble law. Most astronomers today regard Arp's discordant redshifts as simply a *projection effect,* whereby a distant quasar just happens to be in the same part of the sky as a nearby galaxy.

ANALOGY An analogy to the projection effect sometimes happens in photography on Earth. If a photographer poses a person with a telephone pole in the background, the photo can give the misleading impression that the unlucky person has a telephone pole growing out of his head.

What largely ended the redshift debate of the 1970s were observations showing that quasars are associated with remote galaxies. Long-exposure images of quasars showed that they are often found in groups or clusters of galaxies, and that the redshift of each of these quasars is essentially the same as the redshifts of the galaxies that surround it. Thus, the quasar's redshift indicates its distance from Earth, just as do the redshifts of galaxies.

Quasar "Fuzz" and Host Galaxies

A second important link between quasars and galaxies was established in the 1980s. In that decade astronomers first had the

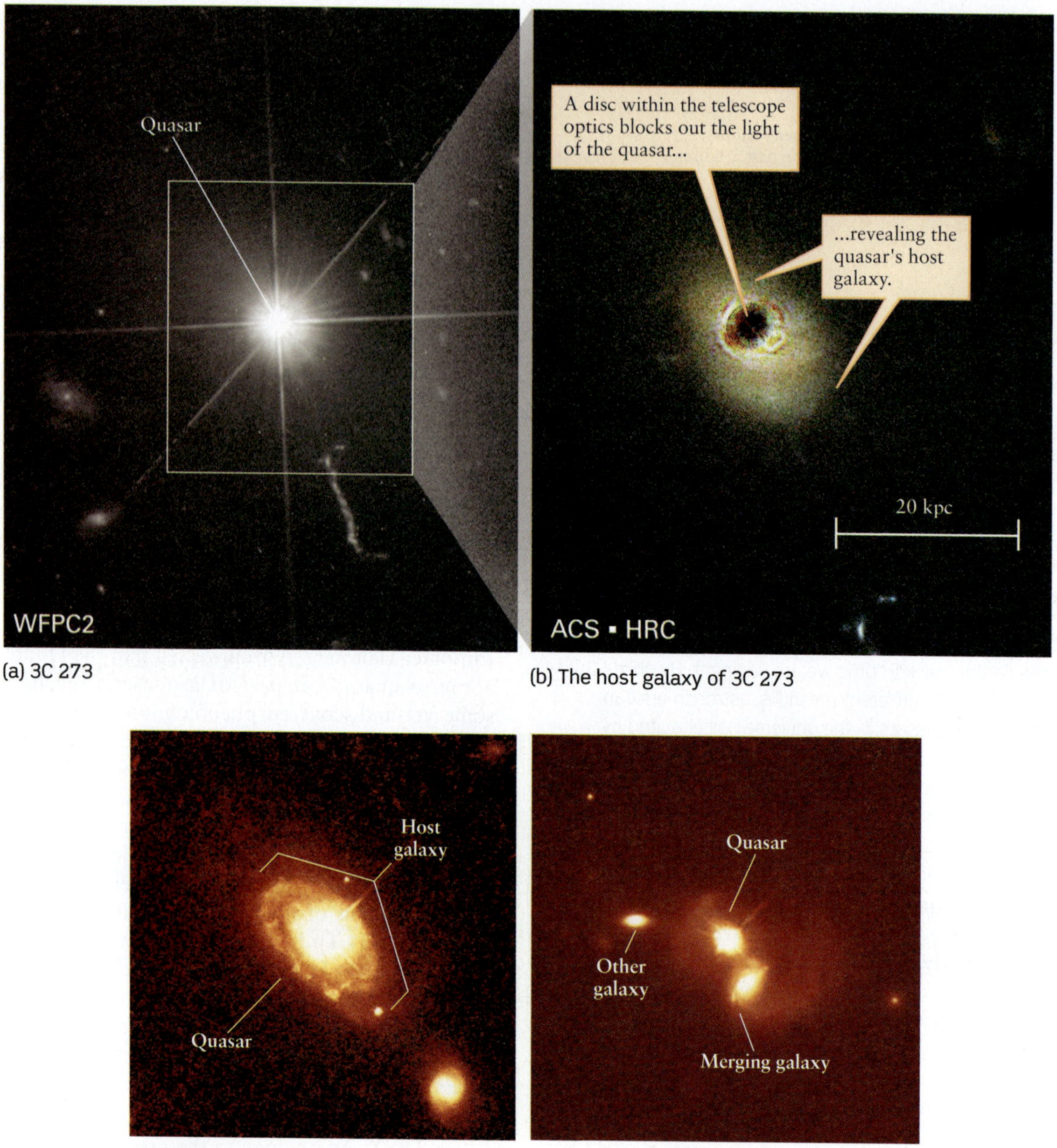
Quasar
WFPC2
A disc within the telescope optics blocks out the light of the quasar...
...revealing the quasar's host galaxy.
20 kpc
ACS ▪ HRC
Host galaxy
Quasar
Quasar
Other galaxy
Merging galaxy

(a) 3C 273

(b) The host galaxy of 3C 273

(c) PG 0052+251 and its host galaxy

(d) PG 1012+008 and its host galaxy

WEB LINK 25.6

Figure 25-6 R I V U X G

Quasars and Their Host Galaxies **(a)** In this 1994 image from the Hubble Space Telescope (HST), the glare of quasar 3C 273 hides its host galaxy. **(b)** This 2002 image that reveals the host galaxy was made using an upgraded camera aboard HST. **(c)** Quasar PG 0052+251 is located at the center of an apparently normal spiral galaxy at redshift $z = 0.155$. Other quasars are found at the centers of ordinary-looking elliptical galaxies. **(d)** The galaxy that hosts quasar PG 1012+008 (redshift $z = 0.185$) is in the process of merging with a second luminous galaxy. The wispy material surrounding the quasar may have been pulled out of the galaxies by tidal forces (see Figure 24-28). The two merging galaxies are just 9500 pc (31,000 ly) apart. Another small galaxy to the left of the quasar may also be merging with the others. (a: NASA and J. Bahcall (Institute for Advanced Study); b: NASA, A. Martel (JHU), H. Ford (JHU), M. Clampin (STScI), G. Hartig (STScI), G. Illingworth (UCO/Lick Observatory), the ACS Science Team and ESA; c, d: J. Bahcall (Institute for Advanced Study), M. Disney (University of Wales), and NASA)

technology to examine the faint "fuzz" seen around the images of some quasars. The spectrum of this fuzz shows stellar absorption lines, indicating that each of these quasars is embedded in a galaxy whose light is far fainter than that of the quasar itself. In each case, the absorption lines of the stars have the same redshift as the quasar's emission lines, further supporting the idea that quasars are at distances indicated by their redshifts and the Hubble law.

It is very difficult to observe the "host galaxy" in which a quasar is located because the quasar's light overwhelms light from the galaxy's stars (Figure 25-6*a*). Nevertheless, improved technology and painstaking observations have revealed some basic properties of these host galaxies (Figure 25-6*b*). Relatively nearby radio-quiet quasars (with redshifts of less than 0.2, corresponding to distances of less than about 750 Mpc) tend to be located in spiral galaxies, whereas radio-loud quasars as well as more distant radio-quiet quasars tend to be located in ellipticals (Figure 25-6*c*). However, a large percentage of these host galaxies have distorted shapes or are otherwise peculiar. Many have nearby companion galaxies, suggesting a link between collisions or mergers and the quasar itself (Figure 25-6*d*). We will see in Section 25-6 how the supermassive black hole model explains this link, and also helps explain why quasars are found only at great distances from Earth—and hence only in the distant past of the universe.

While observations like those in Figure 25-6 show that quasars lie at the centers of galaxies, they do not really explain what quasars *are*. As we will see, important clues come from a class of galaxies that are intermediate in luminosity between quasars and normal galaxies.

25-3 Seyferts and radio galaxies bridge the gap between normal galaxies and quasars

We have seen that some scientists in the 1970s preferred to challenge the Hubble law rather than accept the existence of highly luminous objects. One reason for their skepticism was the huge gap in energy output between normal galaxies and quasars. The gap was bridged, however, when astronomers realized that the centers of certain galaxies look like low-luminosity quasars.

Seyfert Galaxies and Radio Galaxies

The first "missing links" between quasars and ordinary galaxies were actually discovered before quasars themselves. In 1943, Carl Seyfert at the Mount Wilson Observatory made a systematic study of spiral galaxies with bright, compact nuclei that seem to show signs of intense and violent activity (Figure 25-7). Like quasars, the nuclei of these galaxies have strong emission lines in their spectra. These galaxies are now referred to as **Seyfert galaxies.**

Fast-moving jets emanate from quasars, radio galaxies, and Seyfert galaxies

A few percent of the most luminous spiral galaxies are Seyfert galaxies. These galaxies range in luminosity from about 10^{36} to 10^{38} watts, which makes the brightest Seyferts as luminous as faint quasars. Indeed, there is no sharp dividing line between the properties of Seyferts and those of quasars. Like radio-quiet quasars, Seyferts tend to have only weak radio emissions. And like quasars, some Seyferts are members of interacting pairs or exhibit the vestiges of mergers and collisions.

While Seyfert galaxies resemble dim, radio-quiet quasars, certain elliptical galaxies, called **radio galaxies** because of their strong radio emission, are like dim, radio-loud quasars. The energy output of radio galaxies covers approximately the same range as that of Seyferts. The first of these peculiar galaxies was discovered in 1918 by Heber D. Curtis. His short-exposure photograph of the giant elliptical galaxy M87 revealed a bright, starlike nucleus with a protruding jet like that of quasar 3C 273 (see the image that opens this chapter). Figure 25-8*a* shows an overall view of the entire galaxy. The Hubble Space Telescope image in Figure 25-8*b* shows the jet extending outward from the nucleus of M87.

Most of the light from the central regions of M87 is **thermal radiation,** with a spectrum like that of a blackbody. This radiation is caused by the random thermal motion of the atoms and molecules that make up the emitting object. The spectrum of thermal radiation depends only on the object's temperature. For a given temperature, an object emits the maximum amount of thermal radiation at one particular frequency and emits less at frequencies above or below that value. There are also absorption lines in the spectrum of thermal radiation from M87's nucleus, which indicates that this radiation is due to a profusion of stars crowded around the galaxy's center.

WEB LINK 25.7

Figure 25-7 R I V U X G

A Seyfert Galaxy This barred spiral galaxy, called NGC 1097, is a Seyfert galaxy that lies some 14 Mpc (45 million ly) from Earth in the southern constellation Fornax (the Furnace). Its nucleus glows much more brightly than that of a normal barred spiral galaxy. The small elliptical galaxy at upper left is a satellite of NGC 1097. (European Southern Observatory)

(a) The giant elliptical galaxy M87

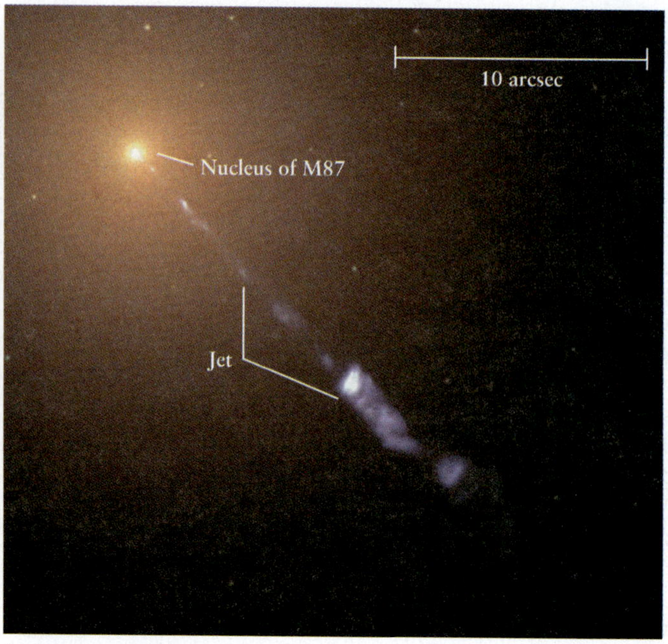

(b) A close-up of M87's jet

WEB LINK 25.8

Figure 25-8 R I V U X G

The Radio Galaxy M87 **(a)** M87 measures about 100 kpc (300,000 ly) across and lies in the Virgo cluster some 15 Mpc (50 million ly) from Earth. The numerous fuzzy dots that surround the galaxy are globular clusters. **(b)** From the galaxy's tiny, bright nucleus—less than 2 pc (6.5 ly) in diameter—a jet extends outward some 1500 pc (5000 ly). (a: NASA, ESA, Hubble Heritage Team (STScI/AURA); b: NASA and the Hubble Heritage Team, STScI/AURA)

By contrast, the light from M87's jet is **nonthermal radiation.** This radiation is *not* due to random thermal motion, and has a very different spectrum than does thermal radiation. In this case of M87's jet, the spectrum extends from radio to X-ray frequencies, which is a much broader range than thermal radiation.

The particular type of nonthermal radiation emitted from the jet is *synchrotron radiation.* As we saw in the discussion of the Crab Nebula in Section 21-4, synchrotron radiation is produced by relativistic electrons traveling in a strong magnetic field. (Recall that "relativistic" means "traveling near the speed of light.") As the electrons spiral around the magnetic field, they emit electromagnetic radiation with a distinctive spectrum very different from that of blackbody radiation (see Figure 21-6). The presence of synchrotron radiation coming from M87's jet indicates that relativistic particles are being ejected from the galaxy's nucleus and encountering a magnetic field.

One piece of evidence that the radiation from M87's jet is nonthermal is that some of it is **polarized radiation,** which means that the electric fields of the light waves are oriented in a specific direction. By contrast, thermal radiation—like light from the Sun, stars, or lightbulbs—is *unpolarized* radiation. Being unpolarized, thermal radiation consists of waves whose electric fields are oriented at random angles. (Most radio stations transmit signals that are polarized in a vertical plane. In order to detect these signals, the radio antennas on cars are oriented vertically rather than horizontally.)

Jets like those found with quasars and radio galaxies are also observed with Seyfert galaxies. One example is NGC 1097, which has jets that extend away from the galaxy's nucleus for more than 50 pc. However, these are too faint to appear in Figure 25-7.

Jets and Radio Lobes

During the 1960s and 1970s, astronomers found that many radio galaxies have a set of two **radio lobes,** one on each side of a parent galaxy. The radio lobes generally span a distance that is 5 to 10 times the size of the parent galaxy, although smaller and much larger examples are known. The parent galaxy—almost always a giant elliptical—sits midway between the radio lobes. Because of their overall shape, radio galaxies (and many similar quasars) are sometimes called **double radio sources.** Cygnus A, shown in Figure 25-1, is a fine example.

The spectrum and polarization of the radio-frequency emission from the lobes of a double radio source bear all the characteristics of synchrotron radiation. This suggests that radio galaxies should have jets of relativistic particles leading from the galaxy out to the radio lobes. A spectacular example of these jets is Centaurus A, one of the brightest radio sources in the southern sky (Figure 25-9). The peculiar parent galaxy of Centaurus A, shown in Figure 25-9*a*, has a broad dust lane studded with young, hot, massive stars. This galaxy, called NGC 5128, is thought to be an elliptical galaxy that has collided with a spiral galaxy. Radio waves pour from two lobes on opposite sides of the galaxy (Figure 25-9*b*). A second set of radio lobes farther from the galaxy (not shown in Figure 25-9) spans a volume 2 million light-years across. Figure 25-9*c* shows an X-ray–emitting jet emanating from

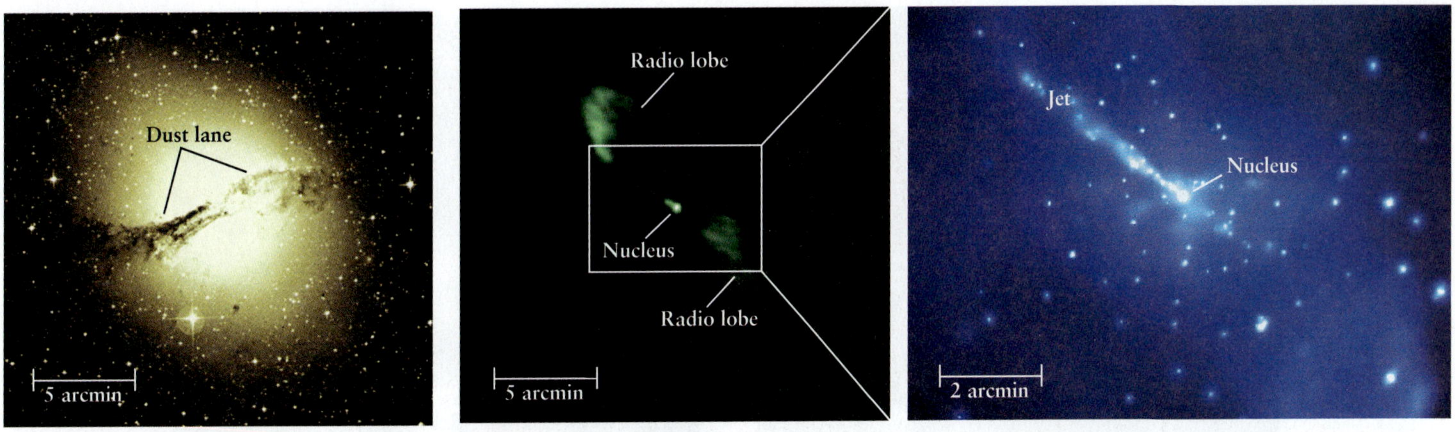

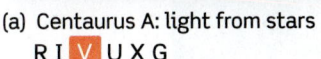

(a) Centaurus A: light from stars
R I **V** U X G

(b) Centaurus A: radio lobes
R I V U X G

(c) An X-ray-emitting jet emanates from the nucleus
R I V U **X** G

WEB LINK 25.9

Figure 25-9

The Radio Galaxy Centaurus A **(a)** This elliptical galaxy, called NGC 5128, lies about 4 Mpc (13 million ly) from Earth at the location of the radio source Centaurus A. The dust lane is evidence of a collision with another galaxy. **(b)** At radio wavelengths we see synchrotron radiation from two radio lobes centered on the galaxy's nucleus. **(c)** A luminous jet extends from the nucleus directly toward one of the lobes. (a: Digital Sky Survey/U. K. Schmidt Image/STScI; b: NRAO/VLA/J. Condon et al.; c: NASA/SAO/R. Kraft et al.)

the galaxy's nucleus. This jet, which is perpendicular to the galaxy's dust lane, is aimed toward one of the radio lobes.

How does a jet give rise to a radio lobe? To see the answer, note that Centaurus A is a member of a cluster of galaxies. We saw in Section 24-7 that the space between galaxies in such a cluster is filled with a substantial amount of hot intracluster gas whose presence can be detected in X-ray images (see Figure 24-25). As the particles in a jet plow through this gas, they are subjected to fluid resistance that slows them down. The particles lose energy in the process, and this energy is converted to electromagnetic radiation at radio wavelengths.

The idea that jets of particles are found in double radio sources is reinforced by the existence of **head-tail sources.** These sources are radio galaxies with a concentrated source of radio emission (the "head") to which are attached two long radio-emitting streams (the "tails"). Figure 25-10 shows one such head-tail source. The "head" coincides with the position of the elliptical galaxy NGC 1265, which is a member of the Perseus cluster of galaxies. NGC 1265 is known to be moving through the cluster at 2500 km/s. Figure 25-10 shows that the two "tails" are swept backward, like smoke emerging from a fast-moving steam locomotive. This orientation is just what we would expect if the "tails" are caused by jets emerging from the galaxy; as NGC 1265 moves through the intergalactic gas of the cluster, the jets feel a 2500-km/s wind blowing them back.

Like NGC 1265, many radio galaxies are found near the centers of rich clusters of galaxies. Hence, they are probably subjected to collisions and mergers like the one that NGC 5128 has experienced (see Figure 25-9*a*).

Seyfert galaxies and radio galaxies share many properties with the more remote, more luminous quasars. A key difference is that

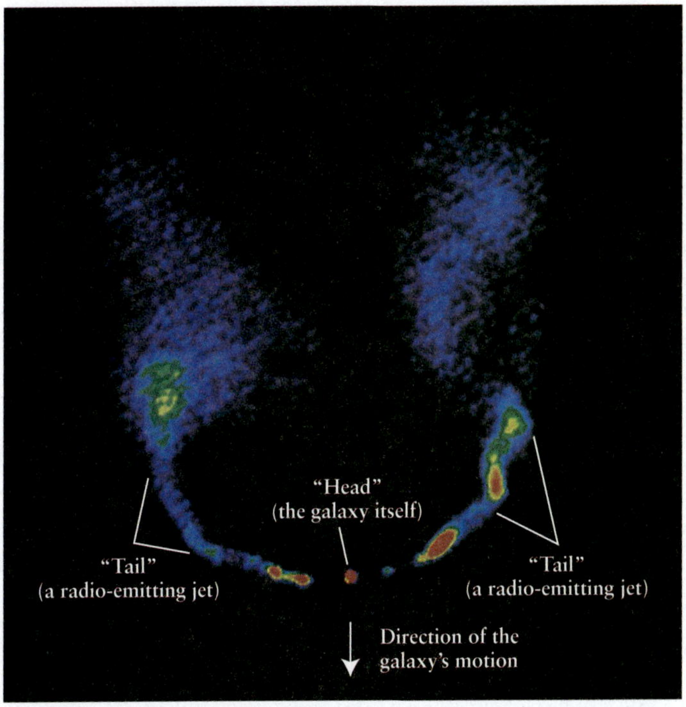

Figure 25-10 **R** I V U X G

The Head-Tail Source NGC 1265 The elliptical galaxy NGC 1265 would probably be an ordinary double radio source, except that it is moving at a high speed through the Perseus cluster of galaxies. Because of this motion, its two jets trail behind the galaxy, giving this radio source a distinctly windswept appearance. (NRAO)

(a) A quasar jet at radio wavelengths... R I V U X G

(b) ...and at X-ray wavelengths R I V U X G

WEB LINK 25.10

Figure 25-11

A Jet Emanating from a High-Redshift Quasar
Designated GB1508+5714, this quasar has a redshift $z = 4.3$, corresponding to a distance of 3800 Mpc (12.2 billion ly). We see this quasar as it was when the universe was only 10% of its present age. Quasars like this may have evolved into radio galaxies like those shown in Figures 25-8, 25-9, and 25-10. (a: NRAO/VLA/Brandeis/T. Cheung; b: NASA/CXC/SAO/A. Siemiginowska et al.)

we see many Seyferts and radio galaxies that are relatively close to us and thus in the recent past, whereas we only see quasars at great distances and hence in the remote past (Figure 25-11). Thus, quasar-like objects have not completely disappeared from the universe, but those that remain in the present-day universe—Seyferts and radio galaxies—are only a pale shadow of the intensely luminous quasars that populated the heavens when the universe was young.

Whatever lies at the centers of quasars, Seyferts, and radio galaxies, it must deliver energy at a prodigious rate to power not only these objects' tremendous luminosities but also their high-speed jets. As we will see in the next section, these energy sources are not only powerful; they are also highly variable.

25-4 Quasars, blazars, Seyferts, and radio galaxies are active galaxies

TUTORIAL 25.1

Inspired by the discovery of quasars, Seyferts, and luminous radio galaxies, astronomers during the 1960s and 1970s searched for clues to these powerful energy sources. One important clue turned out to be a new class of unusual objects called *blazars*. Like quasars, blazars are extraordinarily luminous objects that look like stars but prove to be the nuclei of distant galaxies. The difference is that unlike a quasar, the spectrum of a blazar is almost featureless, with hardly any absorption or emission lines at all!

Blazars and the Strange Case of Superluminal Motion

The first of this new class of objects to be discovered was BL Lacertae, or BL Lac for short, in the constellation Lacerta (the Lizard). When discovered in 1929, it was mistaken for a variable star, largely because its brightness varies by a factor of 15 within only a few months. Careful examination, however, revealed some fuzz (visible in Figure 25-12) around its bright, starlike core. In the early 1970s, Joseph Miller at the University of California's Lick Observatory blocked out the light from the bright center of BL Lac and managed to obtain a spectrum of the fuzz. This spectrum, which contains stellar absorption lines, strongly resembles the spectrum of an elliptical galaxy. In other words, BL Lac is an elliptical galaxy with an intensely luminous center.

Objects such as BL Lac are called **blazars.** They emit polarized light with a featureless, nonthermal spectrum typical of synchrotron radiation, as shown in Figure 21-6*b*. (Careful observations have revealed emission and absorption lines in blazar spectra, but these are swamped by the intense synchrotron radiation.) Detailed radio observations show that blazars are probably double radio sources—but oriented so that we see their jets pointing toward us. For example, high-resolution studies of blazars with the Very Large Array have revealed diffuse radio emission around a bright core in almost all cases. A faint radio halo around a bright radio core is exactly what you would expect if you were looking straight down the jet from a radio galaxy. As seen from this angle, the intense synchrotron radiation from the galaxy's bright nucleus is surrounded by weaker emission from its radio lobes.

WEB LINK 25.11

The idea that blazars are radio galaxies with their jets aimed toward Earth was supported by the surprising discovery of material moving away from the nucleus that *appeared* to be traveling faster than the speed of light. Such **superluminal motion** is also observed in some quasars, where very-long-baseline interferometry (described in Section 6-6) reveals a lumpy structure that changes with time. For example,

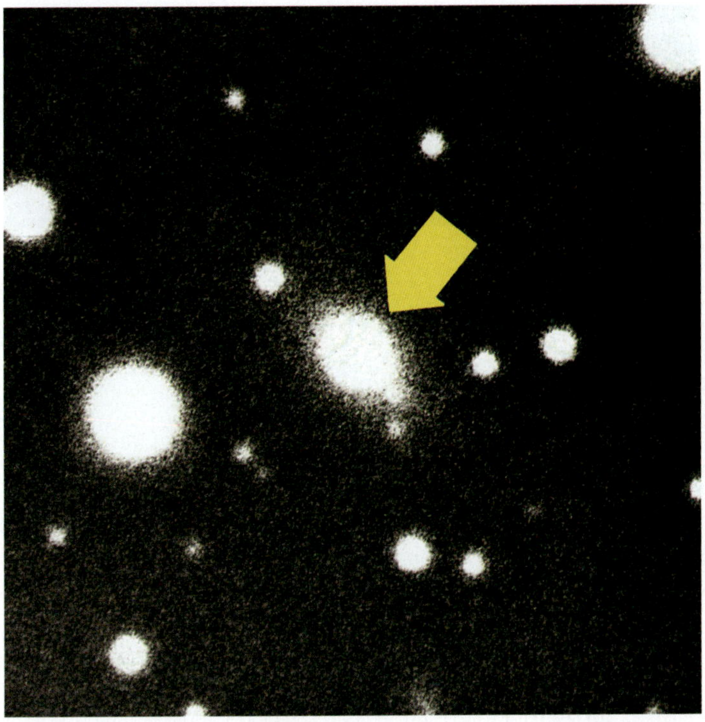

Figure 25-12 R I **V** U X G

BL Lac This photograph shows fuzz around the blazar BL Lac. The redshift of this fuzz indicates that BL Lac is about 280 Mpc (900 million ly) from Earth. Blazars are giant elliptical galaxies with bright, starlike nuclei that have many quasarlike properties. (Courtesy of T. D. Kinman; Kitt Peak National Observatory)

Figure 25-13 shows four high-resolution images of the quasar 3C 273 spanning three years. During this interval, a "blob" moved away from the quasar at a rate of almost 0.001 arcsec per year. Taking into account the distance to the source, this rate of angular separation corresponds to a speed 10 times that of light!

Superluminal motion was puzzling when it was first discovered, because it seemed to violate one of the basic tenets of the special theory of relativity: Nothing can move faster than light. Astronomers soon realized, however, that superluminal motion can be explained as movement slower than light—once we take into account the angle at which we view the radio source. If a relativistic beam of material is aimed close to your line of sight, it can appear to be moving faster than the speed of light. Figure 25-14 shows an example. Quasars generally exhibit lower superluminal speeds (from c to $5c$) than blazars (from $5c$ to $10c$), probably because the relativistic jets from quasars are not aimed as close to our line of sight as are blazar jets.

Active Galaxies and AGNs

Because of the many properties they share, quasars, blazars, Seyfert galaxies, and radio galaxies are now collectively called **active galaxies.** The activity of such a galaxy comes from an energy source at its center. Hence, astronomers say that these galaxies possess **active galactic nuclei,** or **AGNs.**

The rapid variability of active galactic nuclei tells us that they must be very small

Table 25-2 summarizes the properties of different kinds of active galactic nuclei. As this table shows, one of the features that distinguishes different types of AGNs from each other is the width of the emission lines in their spectra. Figures 25-3 and 25-4 show the spectra of typical quasars in the visible and infrared parts of the spectrum. The "spikes" in these spectra are emission lines from hydrogen gas as well as other elements. The widths of these emission lines indicate that individual light-emitting gas clouds are moving within the quasar at very high speeds (10,000 km/s or so). Some of the clouds are moving toward us, causing the emitted light to have a shorter wavelength and higher frequency, while other clouds are moving away from us and emit light with longer wavelength and lower frequency. This explains why the emission lines of quasars are quite broad. By contrast, the emission lines of

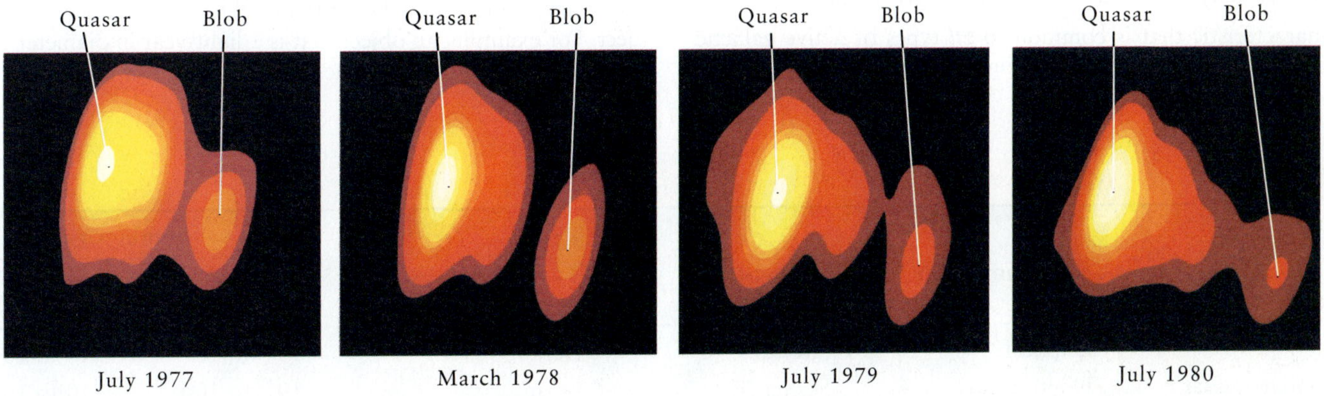

VIDEO 25.1

Figure 25-13 **R** I V U X G

Superluminal Motion in 3C 273 These four images are false-color, high-resolution radio maps of the quasar 3C 273 (shown in the image that opens this chapter). They show a blob that seems to move away from the quasar at 10 times the speed of light. In fact, a beam of relativistic particles from 3C 273 is aimed almost directly at Earth, giving the illusion of faster-than-light motion. Each image is about 7 milliarcseconds (0.007 arcsec) across. (NRAO)

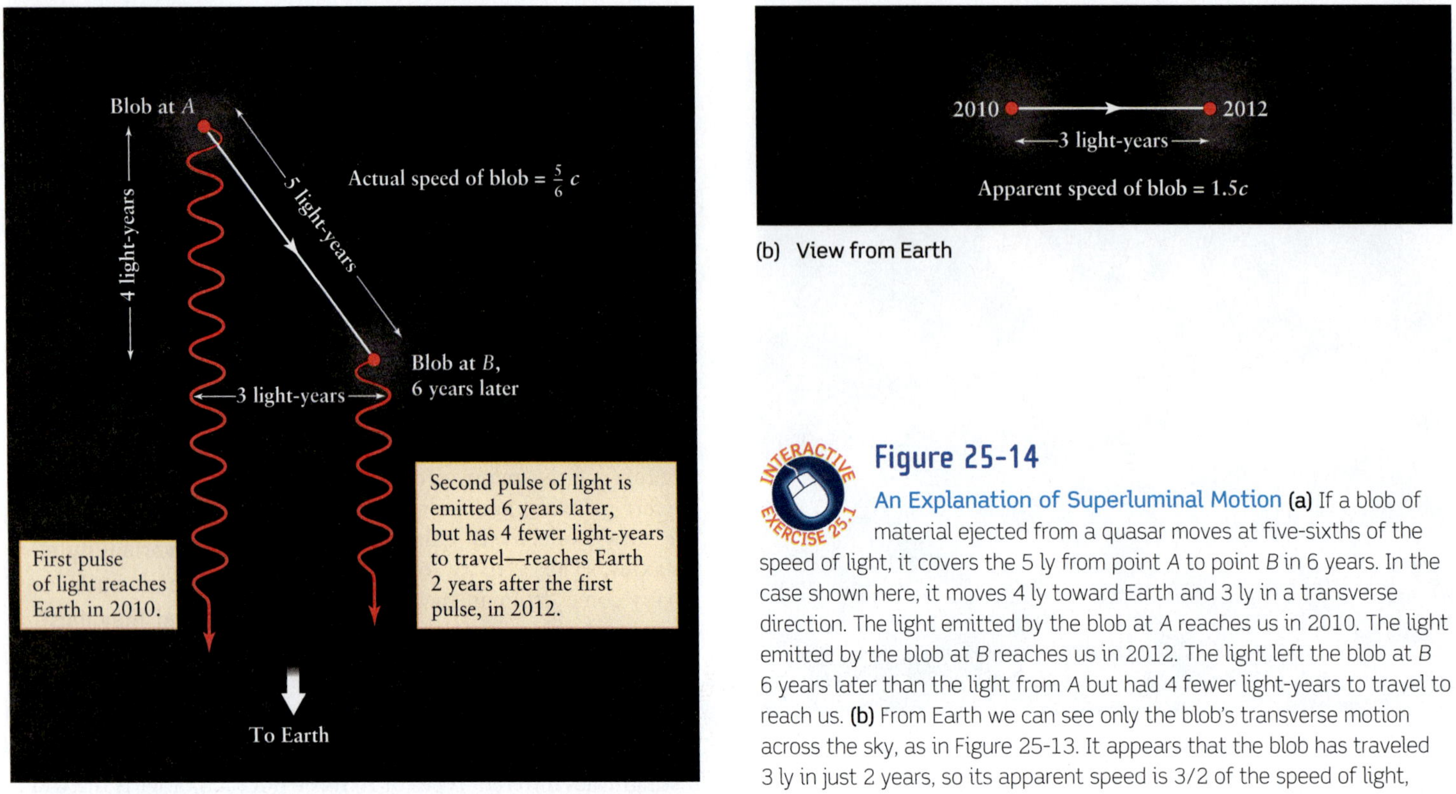

Blob at A
4 light-years
5 light-years
Actual speed of blob = $\frac{5}{6}c$
Blob at B, 6 years later
3 light-years
First pulse of light reaches Earth in 2010.
Second pulse of light is emitted 6 years later, but has 4 fewer light-years to travel—reaches Earth 2 years after the first pulse, in 2012.
To Earth
2010
2012
3 light-years
Apparent speed of blob = 1.5c

(a) View from above

(b) View from Earth

Figure 25-14

INTERACTIVE EXERCISE 25.1

An Explanation of Superluminal Motion (a) If a blob of material ejected from a quasar moves at five-sixths of the speed of light, it covers the 5 ly from point *A* to point *B* in 6 years. In the case shown here, it moves 4 ly toward Earth and 3 ly in a transverse direction. The light emitted by the blob at *A* reaches us in 2010. The light emitted by the blob at *B* reaches us in 2012. The light left the blob at *B* 6 years later than the light from *A* but had 4 fewer light-years to travel to reach us. **(b)** From Earth we can see only the blob's transverse motion across the sky, as in Figure 25-13. It appears that the blob has traveled 3 ly in just 2 years, so its apparent speed is 3/2 of the speed of light, or 1.5*c*.

radio galaxies do not show the same kind of broadening. If there are fast-moving gas clouds in these galaxies, they are hidden from our sight. (In Section 25-6 we will see why they are hidden from us.) In a similar way, astronomers distinguish between different subtypes of Seyfert galaxies (called *Seyfert 1* and *Seyfert 2*), depending on whether their dominant emission lines are broad or narrow.

AGN Variability and the Size of the Light Source

One characteristic that is common to *all* types of active galactic nuclei is variability. For example, Figure 25-15 shows brightness fluctuations of the quasar 3C 273 as determined from 29 years of observations. The brightness of 3C 273 increased by 60% from the beginning to the end of 1982, then declined to the starting value in just five months. Other AGNs undergo greater fluctuations in brightness (by a factor of 25 or more) that occur even more rapidly (X-ray observations reveal that some blazars vary in brightness over time intervals as short as 3 hours).

These fluctuations in brightness allow astronomers to place strict limits on the maximum size of a light source. An object cannot vary in brightness faster than light can travel across that object. For example, an object that is 1 light-year in diameter cannot vary significantly in brightness over a period of less than 1 year.

Table 25-2 Properties of Active Galactic Nuclei (AGNs)

Object	Found in which type of galaxy	Strength of radio emission	Type of emission lines in spectrum	Luminosity (watts)	Luminosity (Milky Way Galaxy = 1)
Blazar	Elliptical	Strong	None	10^{38} to 10^{42}	10 to 10^5
Radio-loud quasar	Elliptical	Strong	Broad	10^{38} to 10^{42}	10 to 10^5
Radio galaxy	Elliptical	Strong	Narrow	10^{36} to 10^{38}	0.1 to 10
Radio-quiet quasar	Spiral or elliptical	Weak	Broad	10^{38} to 10^{42}	10 to 10^5
Seyfert 1	Spiral	Weak	Broad	10^{36} to 10^{38}	0.1 to 10
Seyfert 2	Spiral	Weak	Narrow	10^{36} to 10^{38}	0.1 to 10

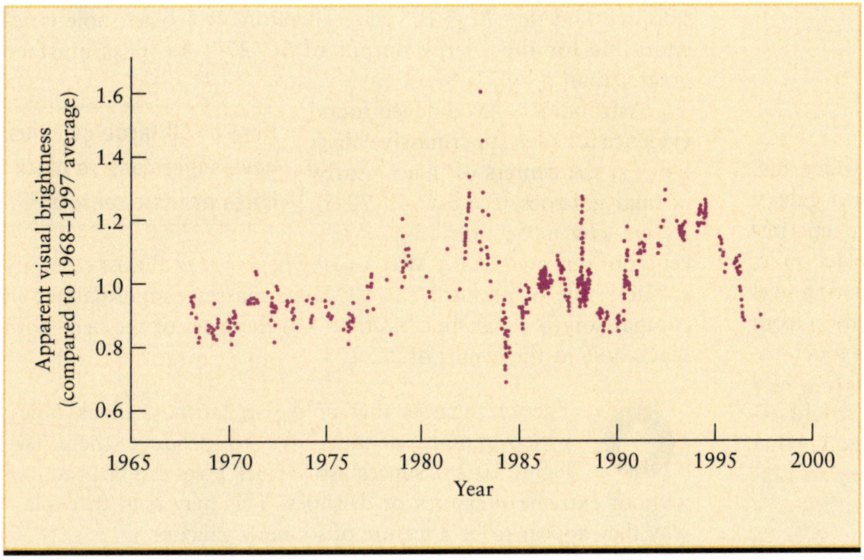

Figure 25-15

Brightness Variations of an AGN This graph shows variations over a 29-year period in the apparent brightness of the quasar 3C 273 (see Figure 25-6*a* and Figure 25-6*b*). Note the large outburst in 1982–1983 and the somewhat smaller ones in 1988 and 1992. (Adapted from M. Türler, S. Paltani, and T. J.-L. Courvoisier)

To understand this limitation, imagine an object that measures 1 light-year across, as in Figure 25-16. Suppose the entire object emits a brief flash of light. Photons from that part of the object nearest Earth arrive at our telescopes first. Photons from the middle of the object arrive at Earth 6 months later. Finally, light from the far side of the object arrives a year after the first photons. Although the object emitted a sudden flash of light, we observe a gradual increase in brightness that lasts a full year. In other words, the flash is stretched out over an interval equal to the difference in the light travel time between the nearest and farthest observable regions of the object.

The rapid flickering exhibited by active galactic nuclei means that they emit their energy from a small volume, possibly less than 1 light-day across. In other words, a region no larger than our solar system can emit more energy per second than a thousand galaxies! Astrophysicists therefore face the challenge of explaining how so much energy can be produced in such a very small volume.

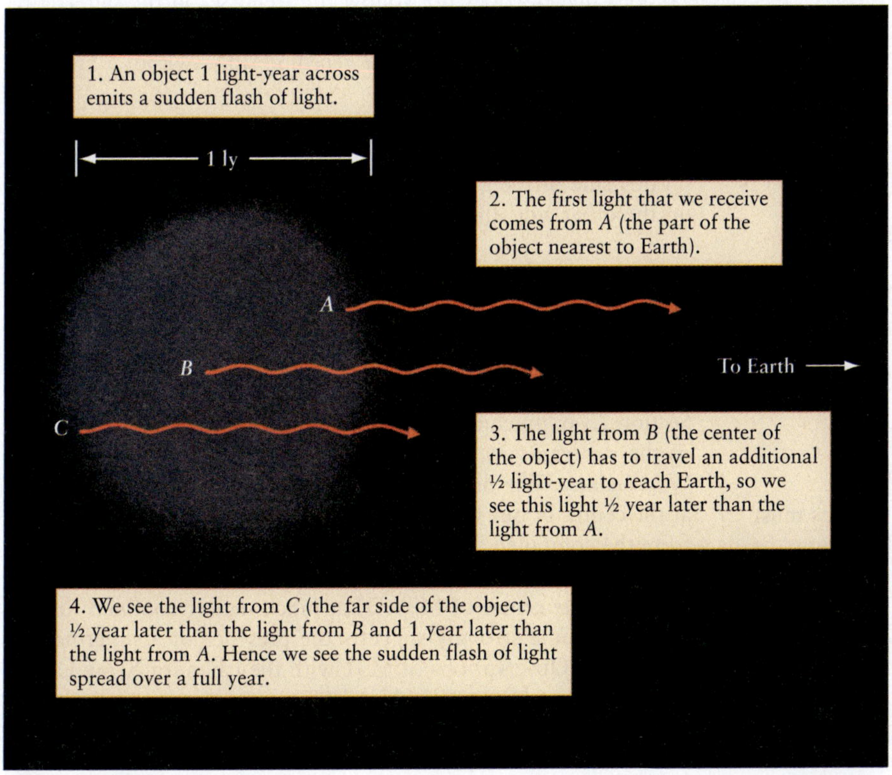

1. An object 1 light-year across emits a sudden flash of light.
1 ly
2. The first light that we receive comes from *A* (the part of the object nearest to Earth).
A
B
C
To Earth
3. The light from *B* (the center of the object) has to travel an additional ½ light-year to reach Earth, so we see this light ½ year later than the light from *A*.
4. We see the light from *C* (the far side of the object) ½ year later than the light from *B* and 1 year later than the light from *A*. Hence we see the sudden flash of light spread over a full year.

Figure 25-16

A Limit on the Speed of Variations in Brightness The rapidity with which the brightness of an object can vary significantly is limited by the time it takes light to travel across the object. If the object 1 light-year in size emits a sudden flash of light, the flash will be observed from Earth to last a full year. If the object is 2 light-years in size, brightness variations will last at least 2 years as seen from Earth, and so on.

25-5 Supermassive black holes are the "central engines" that power active galactic nuclei

As long ago as 1968, the British astronomer Donald Lynden-Bell pointed out that a black hole lurking at the center of a galaxy could be the "central engine" that powers an active galactic nucleus. Lynden-Bell theorized that as gases fall onto a black hole, their gravitational energy would be converted into radiation. (As we saw in Section 22-3, a similar process produces radiation from black holes in close binary star systems.) To produce as much radiation as is seen from active galactic nuclei, the black hole would have to be very massive. But even a gigantic black hole would occupy a volume much smaller than our solar system—exactly what is needed to explain how active galactic nuclei can vary so rapidly in brightness.

The Eddington Limit and Black Hole Sizes

How large a black hole would be needed to power an active galactic nucleus? You might think that what really matters is not the size of the black hole, but rather the amount of gas that falls onto it and releases energy. However, there is a natural limit to the luminosity that can be radiated by accretion onto a compact object like a black hole. This limit is called the **Eddington limit,** after the British astrophysicist Sir Arthur Eddington.

If the luminosity exceeds the Eddington limit, there is so much *radiation pressure*—the pressure produced by photons streaming outward from the infalling material—that the surrounding gas is pushed outward rather than falling inward onto the black hole. Without a source of gas to provide energy, the luminosity naturally decreases to below the Eddington limit, at which point gas can again fall inward. This limit allows us to calculate the minimum mass of an active galactic nucleus.

Numerically, the Eddington limit is

The Eddington limit

$$L_{Edd} = 30{,}000\left(\frac{M}{M_\odot}\right)L_\odot$$

L_{Edd} = maximum luminosity that can be radiated by accretion onto a compact object

M = mass of the compact object

$M_\odot$ = mass of the Sun

$L_\odot$ = luminosity of the Sun

The tremendous luminosity of an active galactic nucleus must be less than or equal to its Eddington limit, so this limit must be very high indeed. Hence, the mass of the black hole must also be quite large. For example, consider the quasar 3C 273 (shown in the image that opens this chapter and in Figure 25-6*a*), which has a luminosity of about 3×10^{13} $L_\odot$. To calculate the minimum mass of a black hole that could continue to attract gas to power the quasar, assume that the quasar's luminosity equals the Eddington limit. Inserting $L_{Edd} = 3 \times 10^{13}$ $L_\odot$ into the above equation, we find that $M = 10^9$ $M_\odot$. Therefore, if a black hole is responsible for the energy output of 3C 273, its mass must be greater than a billion Suns.

Astronomers have indeed found evidence for such **supermassive black holes** at the centers of many nearby normal galaxies (see Section 22-5). As we saw in Section 23-6, at the center of our own Milky Way Galaxy lies what is almost certainly a black hole of about 3.7×10^6 solar masses—supermassive in comparison to a star, but less than 1% the mass of the behemoth black hole at the center of 3C 273.

Most or all large galaxies have supermassive black holes at their centers

Theory suggests that unlike stellar-mass black holes, which require a supernova to produce them (see Figure 20-26), supermassive black holes can be produced without extreme pressures or densities. This may help to explain why they appear to be a feature of so many galaxies.

Measuring Black Hole Masses in Galaxies

One example of a galaxy that probably has a black hole at its center is the Andromeda Galaxy (M31), shown in Figure 24-3. M31 is only 750 kpc (2.5 million ly) from Earth, close enough that details in its core as small as 1 parsec across can be resolved under the best seeing conditions.

In the mid-1980s, astronomers made high-resolution spectroscopic observations of M31's core. By measuring the Doppler shifts of spectral lines at various locations in the core, we can determine the orbital speeds of the stars about the galaxy's nucleus. Figure 25-17 plots the results for the innermost 80 arcseconds of M31. (At M31's distance, this corresponds to a linear distance of 290 pc or 950 ly.) Note that the rotation curve in the galaxy's nucleus does not follow the trend set in the outer core. Rather, there are sharp peaks—one on the approaching side of the galaxy and the other on the receding side—within 5 arcsec of the galaxy's center.

The most straightforward interpretation is that the peaks are caused by the orbital motions of stars around M31's center. Stars on one side of the galaxy's center are approaching us while stars on the other side are receding from us.

The high speeds of stars orbiting close to M31's center indicate the presence of a massive central object. We can use Newton's form of Kepler's third law (described in Section 4-7 and Box 4-4) and our knowledge of these stars' orbital speeds to calculate the mass of this object. (As described in Section 23-6, much the same method is used to calculate the mass of the supermassive black hole at the center of our own Milky Way Galaxy. The difference is that we can track individual stars at the center of our Galaxy, while the data in Figure 25-17 comes from the combined light of many stars in M31.) Such calculations show that there must be about 3×10^7 solar masses within 5 pc (16 ly) of the center of M31. That much matter confined to such a small volume strongly suggests the presence of a supermassive black hole. Observations of M31 with the Chandra X-ray Observatory are consistent with this picture.

By applying high-resolution spectroscopy to the cores of other nearby galaxies, astronomers have discovered a number of

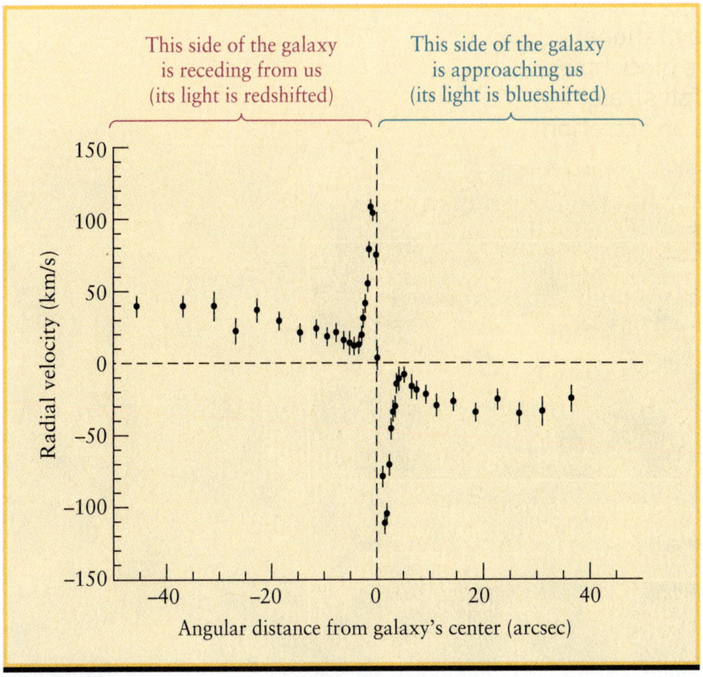

This side of the galaxy is receding from us (its light is redshifted)
This side of the galaxy is approaching us (its light is blueshifted)
Radial velocity (km/s)
150
100
50
0
-50
-100
-150
-40
-20
0
20
40
Angular distance from galaxy's center (arcsec)

Figure 25-17

The Rotation Curve of the Core of M31 This graph plots radial velocity of matter in the core of M31 versus the angular distance from the galaxy's center. Note the sharp peaks, one blueshifted and one redshifted, within 5 arcsec of the galaxy's center. This indicates the presence of a compact, very massive object at the center of the galaxy. At the distance of M31, 1 arcsec corresponds to 3.6 pc (12 ly). (Adapted from J. Kormendy)

supermassive black holes like the one in M31. Unfortunately, this technique for identifying black holes is difficult to apply to quasars, which are very distant and have small angular sizes. The evidence for supermassive black holes in quasars is therefore circumstantial, yet compelling: No other known energy source could provide enough power to sustain a quasar's intense light output.

If most galaxies have supermassive black holes at their centers, and supermassive black holes are the "central engines" of active galactic nuclei, why aren't all galaxies active? Why are there no nearby quasars? Why do radio galaxies and some quasars have jets? As we will see in the next section, the answers to all these questions may be provided by a model of what happens in an active galactic nucleus.

25-6 Quasars, blazars, and radio galaxies are probably the same kind of object seen from different angles

Accretion—the capture and consumption of material—onto a supermassive black hole is the most likely explanation of the immense energy output of active galactic nuclei. The challenge to astrophysicists is to understand how that accretion takes place. A successful model of the accretion process must also explain other properties of active galactic nuclei, including their unusual spectra, variable light output, and energetic jets.

Accretion Disks around Supermassive Black Holes

In the leading model of this process, at the heart of an active galaxy is a supermassive black hole surrounded by an **accretion disk,** an immense disk of matter captured by the hole's gravity and spiraling into it. We saw in Section 18-5 that accretion disks are found around stars in the process of formation. The accretion disks that astrophysicists envision around supermassive black holes are similar but far larger and far more dynamic. The *Cosmic Connections: Accretion Disks* depicts the physics of such accretion disks.

The power source for active galaxies is gravitational energy released by material falling toward a central black hole

Imagine a billion-solar-mass black hole sitting at the center of a galaxy, surrounded by a rotating accretion disk. According to Kepler's third law, the inner regions of this accretion disk would orbit the hole more rapidly than would the outer parts. Thus, the rapidly spinning inner regions would constantly rub against the slower moving gases in the outer regions. This friction, aided by magnetic forces within the disk, would cause the gases to lose energy and spiral inward toward the black hole.

As the gases move inward within the accretion disk, they are compressed and heated to very high temperatures. This causes the accretion disk to glow, thus producing the brilliant luminosity of an active galactic nucleus. The temperature of material in the accretion disk reaches 100,000 K or more, which helps explain why many active galactic nuclei emit far more X-ray and gamma-ray radiation than do ordinary galaxies.

In this model, the fundamental source of the energy output by an active galactic nucleus is *gravitational* energy released as thermal radiation by infalling material in the accretion disk. We saw in Section 20-6 that gravitational energy is also what powers the immense light output of a core-collapse supernova. The difference between such a supernova and an active galactic nucleus is that the infall of a supernova's outer layers is a one-time event, while gas in an AGN's accretion disk falls inward continuously. Any variations in the density of this infalling gas will cause the disk's luminosity to fluctuate, giving rise to the brightness variations observed by astronomers (see Figure 25-15).

Supercomputer simulations can help us visualize the flow of matter in such an accretion disk. These simulations combine general relativity with equations describing gas flow. At first, matter accelerates to supersonic speeds as it spirals inward toward a black hole. But because the matter is rotating around the hole as well as moving toward it, this inward rush stops abruptly near the hole. The reason is described by the law of conservation of angular momentum (described in Section 8-4 and Section 21-3). According to this law, rotating objects of all kinds tend to expand (like a spinning lump of dough that expands to make a pizza) rather than contract. The inward motion of gases stops where this tendency to expand outward balances the pull of the black hole's gravity. As a result, not all of the infalling gas reaches the black hole. Instead, part of the gas can become concentrated in high-speed orbits quite close to the hole. The sudden halt in the supersonic inflow creates a shock wave, which defines the inner edge of the accretion disk (see the *Cosmic Connections: Accretion*

COSMIC CONNECTIONS

Accretion Disks

Quasars and active galaxies are thought to be powered by gravitational energy released when gas falls toward a central supermassive black hole. Due to conservation of angular momentum, the gas does not fall straight into the black hole. Instead, it must negotiate its way through an accretion disk.

How an accretion disk forms around a black hole

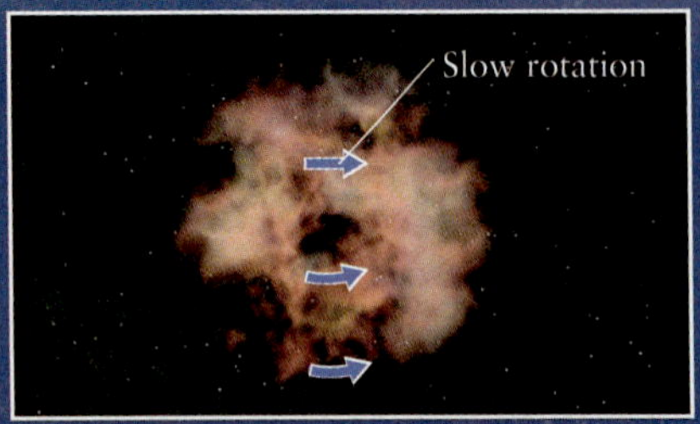

1. A slowly rotating cloud of gas coalesces around the black hole.

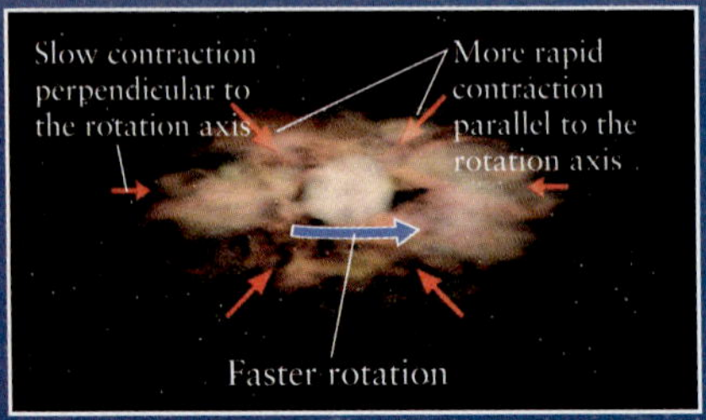

2. Due to conservation of angular momentum, the gas moves inward at different rates in different directions.

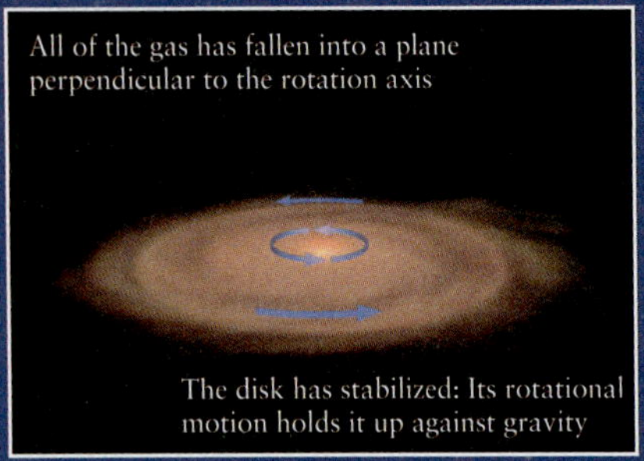

All of the gas has fallen into a plane perpendicular to the rotation axis
The disk has stabilized: Its rotational motion holds it up against gravity

3. The final configuration of the gas is a flattened disk.

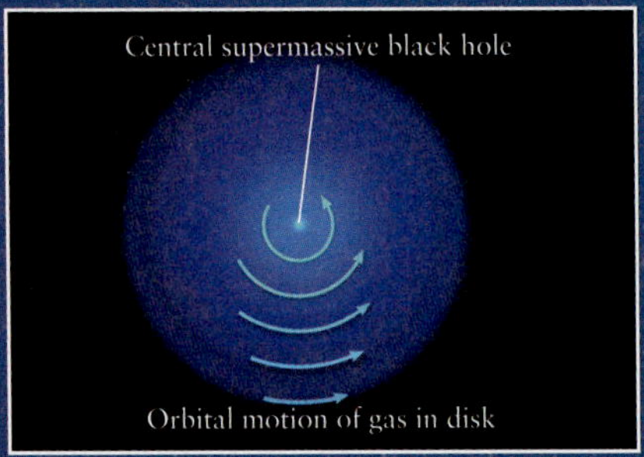

4. Like planets orbiting around the Sun, the closer a blob of gas is to the center of the disk, the faster its orbital speed.

Why gas in an accretion disk spirals in and radiates

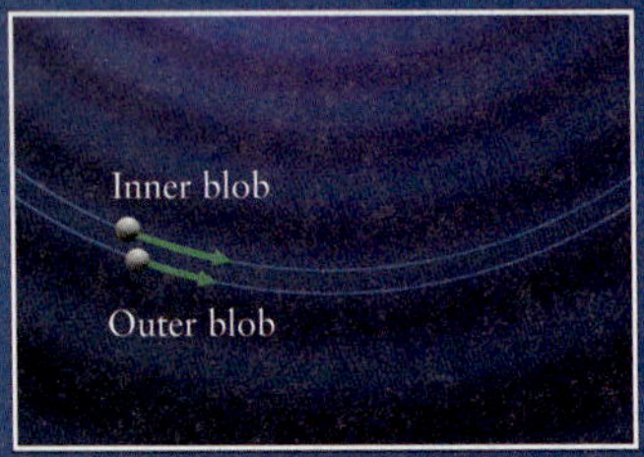

5. Consider two adjacent blobs of gas in the accretion disk. The inner blob moves a bit faster than the outer blob (see 4).

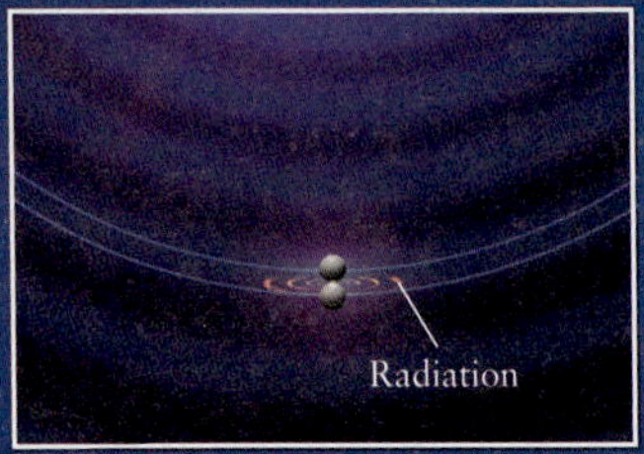

6. When the two blobs rub past each other, energy is transferred from the inner blob to the outer blob. The heated gas emits radiation.

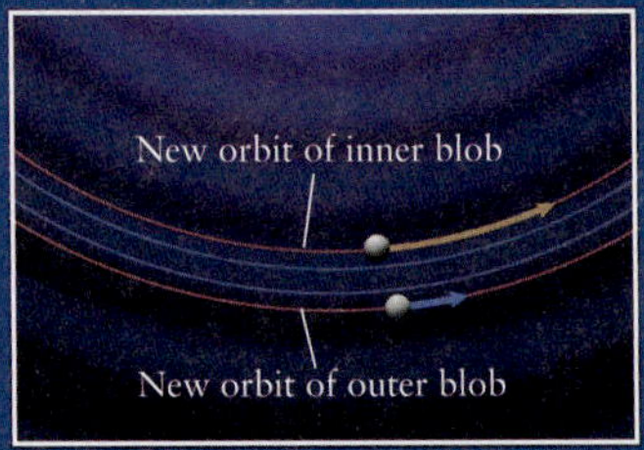

7. Having lost energy, the inner blob moves inward toward the center of the disk and speeds up. The outer blob moves outward and slows down.

8. Conservation of angular momentum slows the inward motion of the blobs of gas.

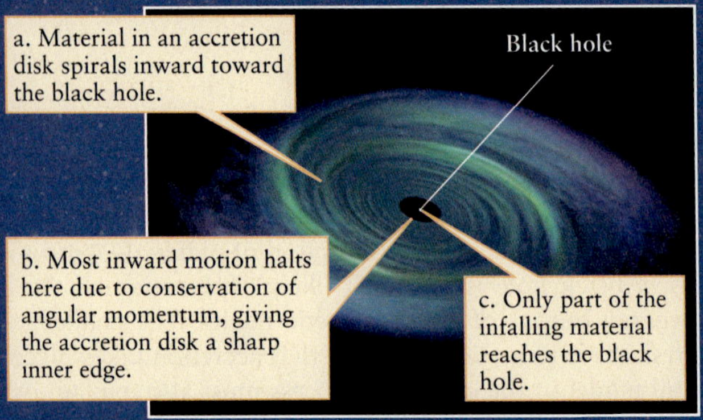

Upper images: Alfred T. Kamajian, from Omer Blaes, "A Universe of Disks," *Scientific American*, October 2004. Lower right image: Courtesy of Michael Owen and John Blondin, North Carolina State University.

Disks). Only a fraction of the material in the accretion disk can cross this inner edge and fall into the black hole.

Because of the constant inward crowding of hot gases, pressures climb rapidly in the inner accretion disk. These pressures relieve the congestion by expelling matter at extremely high speeds. This ejected material escapes moving at right angles to the plane of the accretion disk.

ANIMATION 25.1

Magnetic forces play a crucial role in steering these fast-moving particles. These forces arise because the hot gases in the accretion disk are ionized, forming a plasma, and the motions of this plasma generate a magnetic field (see Section 16-9). As the plasma in the disk rotates around the black hole, it pulls the magnetic field along with it. But because the disk rotates faster in its inner regions than at its outer rim, the magnetic field becomes severely twisted. This twisted field forms two helix shapes, one on either side of the plane of the disk. Relativistic particles flowing outward from the accretion disk tend to follow these magnetic field lines. The result is that the outflowing beams of particles are focused into two jets oriented perpendicular to the plane of the accretion disk, as Figure 25-18 shows. Figure 25-19 is an artist's conception of the accretion disk and jets as they might appear from a nearby planet.

Observations of active galaxies provide evidence in support of this model. One example is the radio galaxy NGC 4261, shown in Figure 22-16, whose radio lobes appear to emanate from a disk around a supermassive black hole. While the accretion disk itself is too small to see, we can observe a dusty ring, or torus, about 250 pc (800 ly) in diameter orbiting the central black hole (Figure 25-20). The motions of this torus reveal the mass of the central black hole to be 1.2×10^9 $M_\odot$.

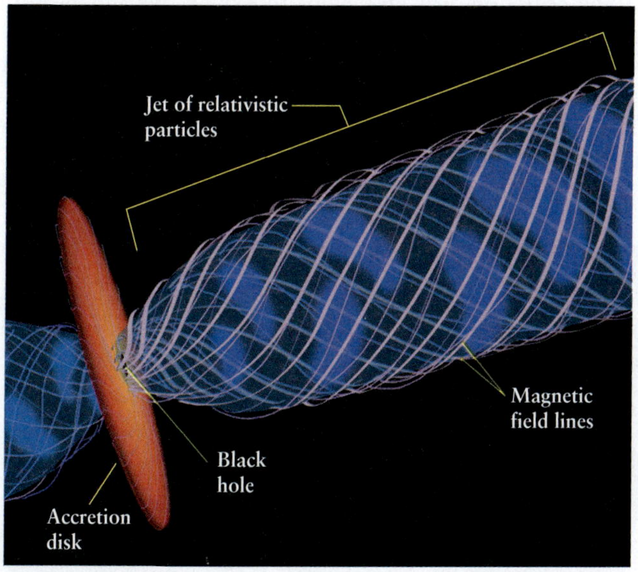

Figure 25-18

Jets from a Supermassive Black Hole The rotation of the accretion disk surrounding a supermassive black hole twists the disk's magnetic field lines into a helix. The field then channels the flow of subatomic particles pouring out of the disk. Over a distance less than a light-year, this channeling changes a broad flow into a pair of tightly focused jets, one on each side of the disk. Figure 18-16 shows a similar process that takes place on a much smaller scale in protostars. (NASA and Ann Feild, Space Telescope Science Institute)

ANIMATION 25.2

Figure 25-19

At the Core of an Active Galaxy This artist's rendering shows a supermassive black hole surrounded by an accretion disk. In the inner regions of the accretion disk, matter crowding toward the hole is diverted outward along two oppositely directed jets. As the relativistic particles move in spiral paths within the jets, they emit synchrotron radiation. (*Astronomy*)

The Unified Model of AGNs

If there were no dusty torus around an accreting supermassive black hole, an observer could view such a black hole from any angle and see the intense radiation from the accretion disk. But the presence of such a torus seems to be a natural result of the accretion process. As a result, from certain angles the torus blocks the view of the innermost part of the active galactic nucleus. This idea offers a single explanation for several types of active galaxies. Many astronomers suspect that the main difference between blazars, quasars, and radio galaxies is only the angle at which the black hole "central engine" is viewed. This idea is called the *unified model* of active galactic nuclei.

Figure 25-21 illustrates the unified model for a luminous active galactic nucleus with jets. If an observer looks straight down the axis of the jet, the observed radiation is dominated by synchrotron radiation from the jet. This radiation has a continuous spectrum with no emission or absorption lines (see Figure 21-6*b*). Hence, the observer sees a blazar, with a nearly featureless spectrum.

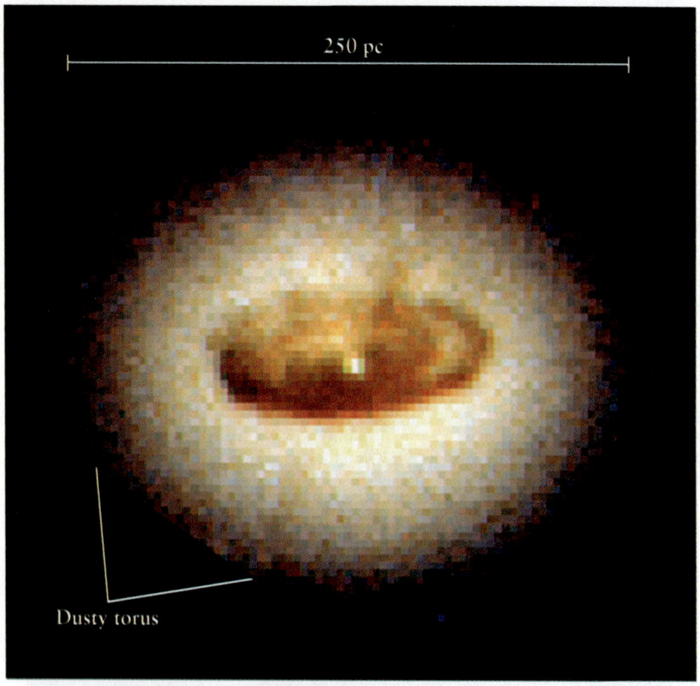

Figure 25-20 R I V U X G

A Dusty Torus Around a Supermassive Black Hole This immense doughnut of dust and gas orbits the black hole at the center of the radio galaxy NGC 4261. The radio lobes of this galaxy appear to be the endpoints of jets emerging perpendicular to the plane of this torus (see Figure 22-16). (L. Ferrarese/Johns Hopkins University, NASA)

At a more oblique angle, the observer gets a clear view of the luminous accretion disk and the turbulent region around the black hole. Because gases move at many different velocities in this region, the observer sees spectral lines that have been broadened by the Doppler effect. The observer also sees intense thermal radiation from the accretion disk and synchrotron radiation from the jets. From this angle, what the observer sees is a radio-loud quasar.

If the observer looks nearly edge-on at the torus, the accretion disk will be completely hidden. Some of the light reaching the observer comes from hot gas flowing out of the accretion disk, and this light has an emission-line spectrum. But this gas is not moving rapidly either toward or away from the observer, so there is little Doppler shift and the emission lines are narrow. The synchrotron emission from the jets is still visible, and so our observer reports seeing a radio galaxy.

It may happen that no jets are present, which means that the active galactic nucleus will lack the powerful radio synchrotron radiation that particles in a jet produce. In this case, an observer viewing the "central engine" either face-on or at an oblique angle will see a radio-quiet quasar.

Unlike our imaginary observer, we cannot move the vast distances through space that would be needed to view a given active galaxy from different angles. Instead, we may see a given active galaxy as either a blazar, a quasar, or a radio galaxy, depending on how the accretion disk and torus are oriented to our line of sight.

CAUTION! Note that we are really making use of *two* different but complementary models here. The unified model says that different types of active galactic nuclei are really different views of the same type of "central engine," while the accretion-disk model explains how the "central engine" works.

The accretion-disk idea helps to explain why there are no nearby quasars (see Figure 25-5). Over time, most of the available gas and dust surrounding a quasar's "central engine" is accreted onto the black hole. The "central engine" has less and less infalling matter to act as "fuel," and the quasar becomes much less active. The result is a relatively less luminous radio galaxy or

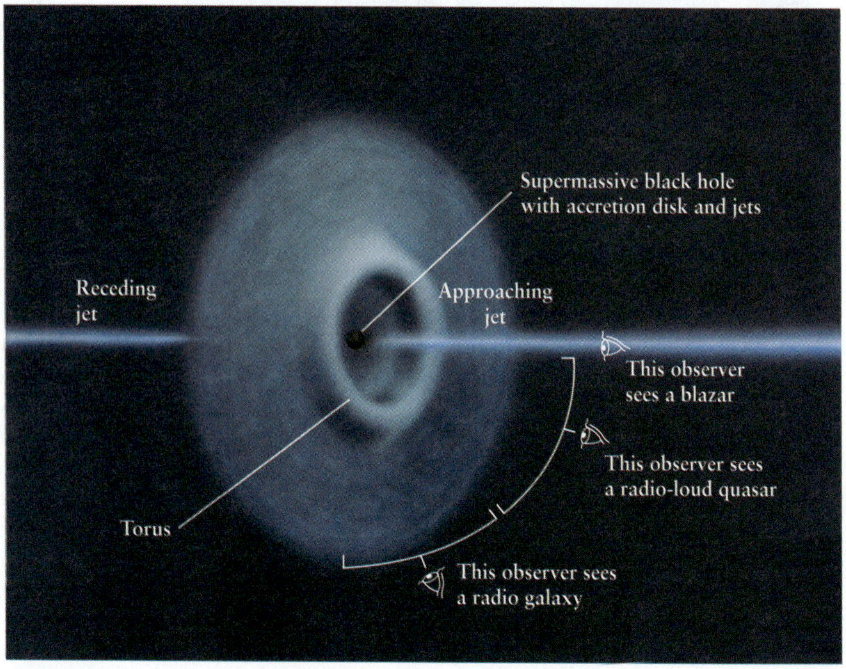

Figure 25-21

A Unified Model of Active Galaxies Blazars, radio-loud quasars, and radio galaxies may be the same type of object—a supermassive black hole, its accretion disk, and its relativistic jets—viewed at different angles. (Compare Figure 22-12, which shows the similar environment of a stellar-mass black hole.)

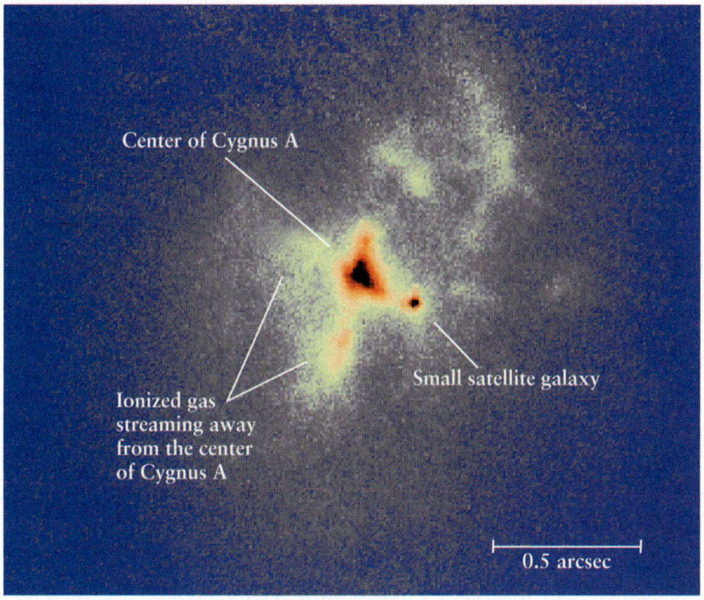

Figure 25-22 R I V U X G

Providing Fresh "Fuel" for a Supermassive Black Hole A small satellite galaxy has fallen into the central regions of the radio galaxy Cygnus A (see Figure 25-1). Its material may eventually accrete onto the supermassive black hole at Cygnus A's very center. This false-color extreme close-up was made using adaptive optics, described in Section 6-3. (Courtesy G. Canalizo, C. Max, D. Whysong, R. Antonucci, and S. E. Dahm)

Seyfert galaxy. (A Seyfert galaxy is essentially a low-luminosity, radio-quiet quasar. Whether it is a Seyfert 1 with predominantly broad emission lines or a Seyfert 2 with predominantly narrow lines may depend on whether we view its "central engine" more nearly face-on or edge-on.)

Finally, the accretion-disk idea can explain what happens when active galaxies collide or merge with other galaxies. Such collisions and mergers can transfer gas and dust from one galaxy to another, providing more "fuel" for a supermassive black hole. For example, the image in Figure 25-22 shows a close-up view of a supermassive black hole about to take in more "fuel" from a merger. Thus, it is not surprising that many of the most luminous active galaxies are also those that have recently undergone collisions, such as the radio galaxy Centaurus A shown in Figure 25-9.

Eventually an active galaxy will run out of "fuel" altogether, and its central supermassive black hole will become inactive. Most galaxies, including our own Milky Way, have supermassive black holes at their centers that are not presently active. Perhaps many of these galaxies once led much more dramatic lives as quasars or radio galaxies.

Key Words

accretion disk, p. 683
active galactic nucleus (AGN), p. 679
active galaxy, p. 679
blazar, p. 678
double radio source, p. 676
Eddington limit, p. 682
head-tail source, p. 677
nonthermal radiation, p. 676
polarized radiation, p. 676
quasar, p. 671
radio galaxy, p. 675
radio lobes, p. 676
Seyfert galaxy, p. 675
superluminal motion, p. 678
supermassive black hole, p. 682
thermal radiation, p. 675

Key Ideas

Quasars: A quasar looks like a star but has a huge redshift. These redshifts show that quasars are several hundred megaparsecs or more from Earth, according to the Hubble law.

- To be seen at such large distances, quasars must be very luminous, typically about 1000 times brighter than an ordinary galaxy.
- About 10% of all quasars are strong sources of radio emission and are therefore called "radio-loud"; the remaining 90% are "radio-quiet."
- Some of the energy emitted by quasars is synchrotron radiation produced by high-speed particles traveling in a strong magnetic field.

Seyfert Galaxies: Seyfert galaxies are spiral galaxies with bright nuclei that are strong sources of radiation. Seyfert galaxies seem to be nearby, low-luminosity, radio-quiet quasars.

Radio Galaxies: Radio galaxies are elliptical galaxies located midway between the lobes of a double radio source.

- Relativistic particles are ejected from the nucleus of a radio galaxy along two oppositely directed beams.

Blazars: Blazars are bright, starlike objects that can vary rapidly in their luminosity. They are probably radio galaxies or quasars seen end-on, with a jet of relativistic particles aimed toward Earth.

Active Galaxies: Quasars, blazars, and Seyfert and radio galaxies are examples of active galaxies. The energy source at the center of an active galaxy is called an active galactic nucleus.

- Rapid fluctuations in the brightness of active galaxies indicate that the region that emits radiation is quite small.

Black Holes and Active Galactic Nuclei: The preponderance of evidence suggests that an active galactic nucleus consists of a supermassive black hole onto which matter accretes.

- As gases spiral in toward the supermassive black hole, some of the gas may be redirected to become two jets of high-speed particles that are aligned perpendicularly to the accretion disk.
- An observer sees a radio galaxy when the accretion disk is viewed nearly edge-on, so that its light is blocked by a surrounding torus. At a steeper angle, the observer sees a quasar. If one of the jets is aimed almost directly at Earth, a blazar is observed.

Questions

Review Questions

1. When quasi-stellar radio sources were first discovered and named, why were they called "quasi-stellar"?
2. How were quasars first discovered? How was it discovered that they are very distant objects?
3. Explain why astronomers cannot use any of the standard candles described in Section 24-4 to determine the distances to quasars.

4. Suppose you saw an object in the sky that you suspected might be a quasar. What sort of observations might you perform to test your suspicion?
5. Quasar PC 1247+3406 is presently about 25.9 billion light-years from Earth. Explain how it is possible for astronomers to see this quasar, even though light travels at a speed of 1 light-year per year.
6. How does the spectrum of a quasar differ from that of an ordinary galaxy? How do spectral lines help astronomers determine the distances to quasars?
7. If quasars lie at the centers of galaxies, why don't we see strong absorption lines from the galaxy's stars when we look at the spectrum of a quasar (like those shown in Figures 25-3 and 25-4)?
8. It was suggested in the 1960s that quasars might be compact objects ejected at high speeds from the centers of nearby ordinary galaxies. Explain why the absence of blueshifted quasars disproves this hypothesis.
9. Why were some astronomers skeptical that the redshifts of quasars gave a true indication of their distance?
10. How do astronomers know that quasars are located in galaxies? In what sorts of galaxies are they found?
11. What is a Seyfert galaxy? Why do astronomers think that Seyfert galaxies may be related to radio-quiet quasars?
12. What is a radio galaxy? What is a double radio source? Why do astronomers think these objects may be related to radio-loud quasars?
13. How would you distinguish between thermal and nonthermal radiation?
14. What is the difference between polarized and unpolarized radiation? How does the polarization of radiation from M87's jet show that the radiation is nonthermal radiation?
15. What are head-tail sources? How do they provide evidence that double radio sources include jets of fast-moving particles?

TUTORIAL 25.1

16. What is a blazar? What is unique about its spectrum? How is it related to other active galaxies?
17. Some blazars or quasars appear to be ejecting material at speeds faster than light. Is the material really moving that fast? If so, how is this possible? If not, why does the material appear to be traveling so fast?
18. What do astronomers learn from the widths of the spectral lines of quasars?
19. What do the brightness fluctuations of a particular active galaxy tell us about the size of the energy-emitting region within that galaxy?
20. How could a supermassive black hole, from which nothing—not even light—can escape, be responsible for the extraordinary luminosity of a quasar?
21. What is the Eddington limit? Explain how it can be used to set a limit on the mass of a supermassive black hole, and explain why this limit represents a minimum mass for the black hole.
22. Explain how the rotation curve of a galactic nucleus can help determine whether a supermassive black hole is present.
23. How does matter falling inward toward a central black hole find itself being ejected outward in a high-speed jet?
24. Explain how the unified model of active galaxies suggests that quasars, blazars, and radio galaxies are the same kind of object viewed from different angles.
25. Why do you suppose there are no quasars relatively near our Galaxy?

Advanced Questions

Questions preceded by an asterisk (*) involve topics discussed in the Boxes in Chapters 4, 22, 24.

> **Problem-solving tips and tools**
>
> Table 25-1 gives the connection between redshift and distance, Box 24-2 discusses relativistic redshift, and Box 22-1 describes time dilation in the special theory of relativity. You may find it useful to recall from Section 1-7 that 1 light-year = 63,240 AU. A speed of 1 km/s is the same as 0.211 AU/yr. You can find Newton's form of Kepler's third law in Box 4-4 and the formula for the Schwarzschild radius in Box 22-2.

26. When we observe a quasar with redshift $z = 0.75$, how far into its past are we looking? If we could see that quasar as it really is right now (that is, if the light from the quasar could somehow reach us instantaneously), would it still look like a quasar? Explain why or why not.

*27. The quasar SDSS 1044–0125 has a redshift $z = 5.80$. At what speed does this quasar seem to be receding from us? Give your answer in km/s and as a fraction of the speed of light c.

*28. Suppose that an astronomer discovers a quasar with a redshift of 8.0. With what speed would this quasar seem to be receding from us? Give your answer in km/s and as a fraction of the speed of light.

29. In the quasar spectrum shown in Figure 25-4, there are many deep absorption lines to the left of the L_α emission line (that is, at shorter wavelengths). These lines, collectively called the *Lyman-alpha forest,* are due to remote gas clouds along our line of sight to the quasar. Hydrogen atoms in these clouds absorb L_α photons from the quasar. Explain why the L_α absorption lines due to these clouds are at shorter wavelengths than the L_α emission line from the quasar itself.
30. Explain how the existence of gravitational lenses involving quasars (review Section 24-8) constitutes evidence that quasars are located at the great distances inferred from the Hubble law.
31. The Seyfert galaxy NGC 1275 is actually two galaxies that are colliding. Images of NGC 1275 show a number of globular clusters with a distinctive blue color. Explain how this color shows that these clusters formed relatively recently, perhaps as a result of the collision.
32. In the image that opens this chapter, the close-up view of the Cygnus A jet shows that different wavelengths are preferentially emitted at different locations along the jet's length. Explain why, using the following principle: As an individual particle moves in a magnetic field, the greater its speed, the shorter the wavelength of the synchrotron radiation that it emits.
33. Suppose the distance from point *A* to point *B* in Figure 25-14*a* is 26 light-years and the blob moves at 13/15 of the speed of light. As the blob moves from *A* to *B,* it moves

24 light-years toward Earth and 10 light-years in a transverse direction. (**a**) How long does it take the blob to travel from *A* to *B*? (**b**) If the light from the blob at *A* reaches Earth in 2020, in what year does the light from *B* reach Earth? (**c**) As seen from Earth, at what speed does the blob appear to move across the sky?

*34. Suppose a blazar at $z = 1.00$ goes through a fluctuation in brightness that lasts one week (168 hours) as seen from Earth. (**a**) At what speed does the blazar seem to be moving away from us? (**b**) Using the idea of time dilation, determine how long this fluctuation lasted as measured by an astronomer within the blazar's host galaxy. (**c**) What is the maximum size (in AU) of the region from which this blazar emits energy?

35. (**a**) Calculate the maximum luminosity that could be generated by accretion onto a black hole of 3.7×10^6 solar masses. (This is the size of the black hole found at the center of the Milky Way, as described in Section 25-6.) Compare this to the total luminosity of the Milky Way, about 2.5×10^{10} $L_{\odot}$. (**b**) Speculate on what we might see if the center of our Galaxy became an active galactic nucleus with the luminosity you calculated in (a).

*36. Observations of a certain galaxy show that stars at a distance of 16 pc from the center of the galaxy orbit the center at a speed of 200 km/s. Use Newton's form of Kepler's third law to determine the mass of the central black hole.

*37. Calculate the Schwarzschild radius of a 10^9-solar-mass black hole. How does your answer compare with the size of our solar system (given by the diameter of Pluto's orbit)?

38. Figure 25-9 shows the double radio source Centaurus A. Is it possible that somewhere in the universe there is an alien astronomer who observes this same object as a blazar? Explain your answer with a drawing showing the relative positions of Earth, the alien astronomer, and Centaurus A.

Discussion Questions

39. The accompanying image from the Very Large Array (VLA) shows the radio galaxy 3C 75 in the constellation Cetus. This galaxy has several radio-emitting jets. High-resolution optical photographs reveal that the galaxy has two nuclei, which are the two red spots near the center of the VLA image. Propose a scenario that might explain the appearance of 3C 75.

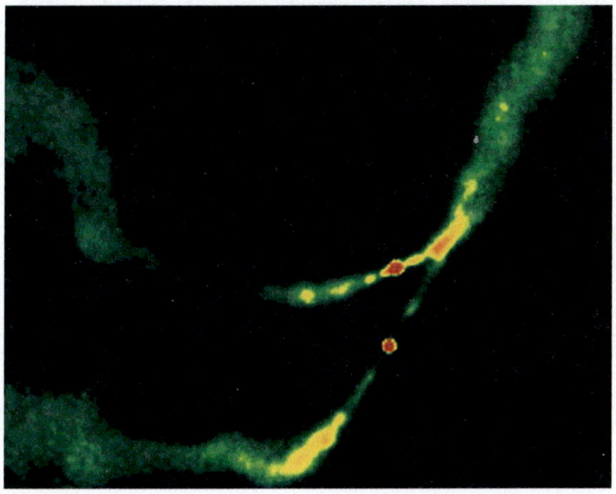

R I V U X G

(NRAO)

40. Some quasars show several sets of absorption lines whose redshifts are less than the redshifts of their emission lines. For example, the quasar PKS 0237-23 has five sets of absorption lines with redshifts in the range from 1.364 to 2.202, whereas the quasar's emission lines have a redshift of 2.223. Propose an explanation for these sets of absorption lines.

41. The Milky Way Galaxy is in the process of absorbing the satellite galaxy called the Canis Major Dwarf (see Section 24-6). Discuss whether this process could cause the Milky Way to someday become an active galaxy.

42. Figure 25-22 shows ionized gas streaming away from the "central engine" of the radio galaxy Cygnus A. Instead of spreading outward equally in all directions, the gas appears to be funneled into two oppositely directed cones. Discuss how this could be caused by a dusty torus surrounding a supermassive black hole at the center of Cygnus A.

Web/eBook Questions

43. Search the World Wide Web for information about "microquasars." These are objects that are found *within* the Milky Way Galaxy. How are they detected? What are the similarities and differences between these objects and true quasars? Are they long-lasting or short-lived?

44. The *Lockman hole* is a region in the constellation Ursa Minor where the Milky Way's interstellar hydrogen is the thinnest. By observing in this part of the sky, astronomers get the clearest possible view of distant galaxies and quasars. Search the World Wide Web for information about observations of the Lockman hole. What has been learned through X-ray observations made by spacecraft such as ROSAT and XMM-Newton?

Activities

Observing Projects

45. Use a telescope with an aperture of at least 20 cm (8 in) to observe the Seyfert Galaxy NGC 1068 (also known as M77). Located in the constellation Cetus (the Whale), this galaxy is most easily seen from September through January. The epoch 2000 coordinates are R.A. = $2^h\ 2.7^m$ and Decl. = $-0°\ 01'$. Sketch what you see. Is the galaxy's nucleus diffuse or starlike? How does this compare with other galaxies you have observed?

46. Use a telescope with an aperture of at least 20 cm (8 in) to observe the two companions of the Andromeda Galaxy, M32 and M110. Both are small elliptical galaxies, but only M32 is suspected of harboring a supermassive black hole. They are located on opposite sides of the Andromeda Galaxy and are most easily seen from September through January. The epoch 2000 coordinates are

Galaxy	Right ascension	Declination
M32 (NGC 221)	$0^h\ 42.7^m$	+40° 52′
M110 (NGC 205)	0 40.4	+41 41

Make a sketch of each galaxy. Can you see any obvious difference in the appearance of these galaxies? How does this difference correlate with what you know about these galaxies?

47. If you have access to a telescope with an aperture of at least 30 cm (12 in), you should definitely observe the Sombrero Galaxy, M104, which is suspected to have a supermassive black hole at its center. It is located in Virgo and can most easily be seen from March through July. The epoch 2000 coordinates are R.A. = $12^h\ 40.0^m$ and Decl. = $-11°\ 37'$. Make a sketch of the galaxy. Can you see the dust lane? How does the nucleus of M104 compare with the centers of other galaxies you have observed?

48. If you have access to a telescope with an aperture of at least 30 cm (12 in), observe M87 and compare it with the other two giant elliptical galaxies, M84 and M86, that dominate the central regions of the Virgo cluster. If your copy of the book comes with the *Starry Night Enthusiast*™ program, use it to help you plan your observations. The epoch 2000 coordinates are:

Galaxy	Right ascension	Declination
M84 (NGC 4374)	$12^h\ 25.1^m$	$+12°\ 53'$
M86 (NGC 4406)	12 26.2	+12 57
M87 (NGC 4486)	12 30.8	+12 24

49. If you have access to a telescope with an aperture of at least 40 cm (16 in), you might try to observe the brightest-appearing quasar, 3C 273, which has an apparent magnitude of nearly +13. It is located in Virgo at coordinates R.A. = $12^h\ 29^m\ 07^s$ and Decl. = $+2°\ 03'\ 07''$.

50. If the Milky Way had an active galactic nucleus, with an accretion disk around its central black hole, there might be a pair of relativistic jets emanating from its center. Use the *Starry Night Enthusiast*™ program to investigate how these jets might appear from Earth. On the toolbar, set the date and time to June 15 of this year at 12:00:00 A.M. (midnight), when the center of the Milky Way is prominent in the sky. Open the **Find** pane and center the field of view on the star HIP86919. The position of this star on the celestial sphere is less than 1° from the black hole at the center of the Milky Way. Select **View > Stars** and ensure that **Milky Way** is being displayed. Select **Options > Stars > Milky Way** to open the **Milky Way Options** dialog window, move the slidebar to the right to brighten the galaxy, and click the **OK** button. Close any open panes to ensure that that the entire window is again devoted to a view of the sky. Make a sketch of the Milky Way Galaxy and attempt to show how the night sky might appear on June 15 at 12:00:00 A.M. if our Galaxy had an active galactic nucleus. Label the Milky Way, the jets, the central black hole, and the accretion disk. Assume that the plane of the accretion disk is aligned with the plane of the Milky Way. Zoom in to a field of view of about 6°. An X-ray image of the Milky Way center is superimposed upon the galaxy. (If not, select **View > Deep Space** and click on **Chandra Images** to display this image.) Open the object contextual menu over this image and click on **Magnify** to enlarge this image to see the high temperature features of this violently active region of our galaxy.

51. Use the *Starry Night Enthusiast*™ program to examine the vicinity of the galaxy M87, shown in Figure 25-8. Select **Favorites > Deep Space > Virgo Cluster** to display this large cluster of galaxies. (**a**) You can use the upward and downward pointing triangles in the **Viewing Location** panel of the toolbar to move toward or away from the cluster. You can also rotate the Virgo Cluster by putting the mouse cursor over the image and, while holding down both the Shift key and the mouse button, move the mouse. (On a two-button mouse, hold down the left mouse button.) Use these controls to get a sense of the extent of the Virgo Cluster. Use the **Viewing Location** controls to move to a distance of about 30 Mly from the Sun. Open the **Find** pane and enter Virgo A. Click the menu button associated with Virgo A in the Find pane and select **Highlight "GA Virgo Cluster" Filament** to highlight the members of this cluster in yellow to see the extent of this huge grouping of galaxies. Describe where this active galaxy, also known as M87, is located in the cluster. (**b**) Discuss how the position of M87 in the Virgo Cluster might relate to its being an active galaxy.

Collaborative Exercises

52. Make a labeled sketch clearly showing how the spectrum of 3C 273, shown in Figure 25-3, would be different if the object was moving toward us at the same velocity. Compare your sketch to that of another group and resolve any inconsistencies.

53. Consider the relationship between redshift and distance as shown in Table 25-1. If an object has a recessional velocity (v/c) of 0.902, how many light-years away was the object when it emitted the light that we see today? How many light-years away is it now? Are these closer or farther than an object with a z of 3?

54. Two dramatic images of the radio galaxy M87 are shown in Figure 25-8. Invent an imaginary scenario that would be analogous to a local dance club where two pictures would show different aspects of the event and write a description of each. Be sure to include a description of how this is analogous to the images in Figure 25-8.

26 Cosmology: The Origin and Evolution of the Universe

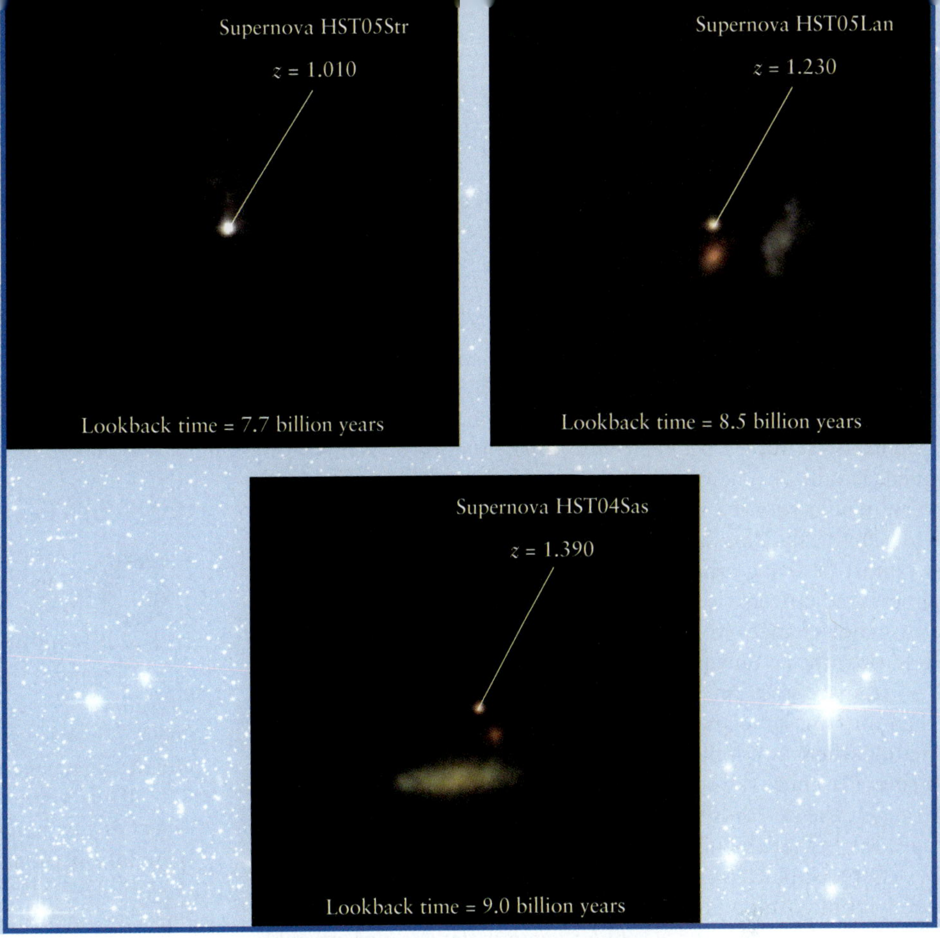

WEB LINK 26.1

R I V U X G

Very distant supernovae—which we see as they were billions of years ago—help us understand the evolution of the universe. (NASA; ESA; and A. Riess, STScI)

So far in this book we have cataloged the contents of the universe. Our scope has ranged from subatomic objects, such as atomic nuclei, to superclusters of galaxies hundreds of millions of light-years across. In between, we have studied planets, moons, and stars.

But now we turn our focus beyond the objects we find in the universe to the nature of the universe itself—the subject of the science called *cosmology*. How large is the universe? What is its structure? How long has it existed, and how has it changed over time?

In this chapter we will see that the universe is expanding. This expansion began with an event at the beginning of time called the *Big Bang*. We will see direct evidence of the Big Bang in the form of microwave radiation from space. This radiation is the faint afterglow of a primordial fireball that filled all space shortly after the beginning of the universe.

Will the universe continue to expand forever, or will it eventually collapse back on itself? We will find that to predict the future of the universe, we must first understand what happened in the remote past. To this end, astronomers study luminous supernovae like the example shown in the accompanying image. These can be seen across billions of light-years, and so can tell us about conditions in the universe billions of years ago. We will see how recent results from such supernovae, as well as from studies of the Big Bang's afterglow, have revolutionized our understanding of cosmology and given us new insights into our place in the cosmos.

26-1 The darkness of the night sky tells us about the nature of the universe

WEB LINK 26.2

Cosmology is the science concerned with the structure and evolution of the universe as a whole. One of the most profound and basic questions in cosmology may at first seem foolish: Why is the sky dark at night? This question,

Learning Goals

By reading the sections of this chapter, you will learn

26-1 What the darkness of the night sky tells us about the nature of the universe

26-2 What it means to say that the universe is expanding

26-3 How to estimate the age of the universe from its expansion rate

26-4 How astronomers detect the afterglow of the Big Bang

26-5 What the universe was like during its first 380,000 years

26-6 How the curvature of the universe reveals its matter and energy content

26-7 What distant supernovae tell us about the expansion history of the universe

26-8 How we are unveiling details of our universe

which haunted Johannes Kepler as long ago as 1610, was brought to public attention in the early 1800s by the German amateur astronomer Heinrich Olbers.

Olbers' Paradox and Newton's Static Universe

Olbers and his contemporaries pictured a universe of stars scattered more or less randomly throughout infinite space. Isaac Newton himself thought that no other model made sense. The gravitational forces between any *finite* number of stars, he argued, would in time cause them all to fall together, and the universe would soon be a compact blob.

Obviously, this has not happened. Newton concluded that we must be living amid a static, infinite expanse of stars. In this model, the universe is infinitely old, and it will exist forever without major changes in its structure. Olbers noticed, however, that a static, infinite universe presents a major puzzle.

If space goes on forever, with stars scattered throughout it, then any line of sight must eventually hit a star. In this case, no matter where you look in the night sky, you should ultimately see a star. The entire sky should be as bright as an average star, so, even at night, the sky should blaze like the surface of the Sun. **Olbers's paradox** is that the night sky is actually *dark* (Figure 26-1).

Olbers's paradox suggests that something is wrong with Newton's infinite, static universe. According to the classical, Newtonian picture of reality, space is like a gigantic flat sheet of inflexible, rectangular graph paper. (Space is actually three-dimensional, but it is easier to visualize just two of its three dimensions. In a similar way, an ordinary map represents the three-dimensional surface of Earth, with its hills and valleys, as a flat, two-dimensional surface.)

Figure 26-1 R I V U X G

The Dark Night Sky If the universe were infinitely old and filled uniformly with stars that were fixed in place, the night sky would be ablaze with light. In fact the night sky is dark, punctuated only by the light from isolated stars and galaxies, Hence, this simple picture of an infinite, static universe cannot be correct. (NASA; ESA; and the Hubble Heritage Team, STScI/AURA)

This rigid, flat, Newtonian space extends unchanged, on and on, totally independent of stars or galaxies or anything else. The same is true of time in Newton's view of the universe; a Newtonian clock ticks steadily and monotonously forever, never slowing down or speeding up. Furthermore, Newtonian space and time are unrelated, in that a clock runs at the same rate no matter where in the universe it is located.

Einstein's Revolution and His "Greatest Blunder"

WEB LINK 26.3

Albert Einstein overturned this view of space and time. His special theory of relativity (recall Section 22-1 and Box 22-1) shows that measurements with clocks and rulers depend on the motion of the observer. What is more, Einstein's general theory of relativity (Section 22-2) tells us that gravity curves the fabric of space. As a result, just as one object can bend the space around it, the matter that occupies the universe influences the overall shape of space throughout the universe.

If we represent the universe as a sheet of graph paper, the sheet is not perfectly flat but has a dip wherever there is a concentration of mass, such as a person, a planet, or a star (see Figure 22-4). Because of gravitational effects, clocks run at different rates depending on whether they are close to or far from a massive object, as Figure 22-7*a* shows.

Einstein narrowly missed predicting that our universe is not static

What does the general theory of relativity, with its many differences from the Newtonian picture, have to say about the structure of the universe as a whole? Einstein attacked this problem shortly after formulating his general theory in 1915. At that time, the prevailing view was that the universe was static, just as Newton had thought.

Einstein was therefore dismayed to find that his calculations could not produce a truly static universe. According to general relativity, the universe must be either expanding or contracting. In a desperate move to force his theory to predict a static universe, he added to the equations of general relativity a term called the **cosmological constant** (denoted by Λ, the capital Greek letter lambda). The cosmological constant was intended to represent a pressure that tends to make the universe expand as a whole. Einstein's idea was that this pressure would just exactly balance gravitational attraction, so that the universe would be static and not collapse.

ANALOGY Einstein's cosmological constant is analogous to the pressure of gas inside a balloon once the balloon has already been inflated. This pressure exactly balances the inward force exerted by the stretched rubber of the balloon itself, so the balloon maintains the same size.

Unlike other aspects of Einstein's theories, the cosmological constant did not have a firm basis in physics. He just added it to make the general theory of relativity agree with the prejudice that the universe is static.

Because Einstein doubted his original equations, he missed an incredible opportunity: He could have postulated that we live in an expanding universe. Einstein has been quoted as saying in his

later years that the cosmological constant was "the greatest blunder of my life." (In fact, the cosmological constant plays an important role in modern cosmology, although a very different one from what Einstein proposed. We will explore this in Section 26-7.)

Instead, the first hint that we live in an expanding universe came more than a decade later from the observations of Edwin Hubble. As we will see in Section 26-3, Hubble's discovery provides the resolution of Olbers's paradox.

26-2 The universe is expanding

TUTORIAL 26.1 WEB LINK 26.4

Hubble is usually credited with discovering that our universe is expanding. He found a simple linear relationship between the distances to remote galaxies and the redshifts of the spectral lines of those galaxies (review Section 24-5, especially Figures 24-16 and 24-17). This relationship, now called the *Hubble law,* states that the greater the distance to a galaxy, the greater is the galaxy's redshift. Thus, remote galaxies are moving away from us with speeds proportional to their distances. Now let us see how Hubble's law reveals an expanding universe.

The Hubble Law and the Expanding Universe

The Hubble law can be stated as a very simple equation that relates the recessional velocity v of a galaxy to its distance d from Earth:

$$v = H_0 d$$

where H_0 is the Hubble constant. Because remote galaxies are getting farther and farther apart as time goes on, astronomers say that the universe is expanding.

ANALOGY What does it actually mean to say that the universe is expanding? According to general relativity, space itself is not rigid. The amount of space between widely separated locations changes over time. A good analogy is that of baking a chocolate chip cake, as in Figure 26-2. As the cake expands during baking, the amount of space between the chocolate chips gets larger and larger. In the same way, as the universe expands, the amount of space between widely separated galaxies increases. The expansion of the universe *is* the expansion of space.

CAUTION! It is important to realize that the expansion of the universe occurs primarily in the vast spaces that separate clusters of galaxies. Just as the chocolate chips in Figure 26-2 do not expand as the cake expands during baking, galaxies themselves do not expand. Einstein and others have established that an object that is held together by its own gravity, such as a galaxy or a cluster of galaxies, is always contained within a patch of nonexpanding space. A galaxy's gravitational field produces this nonexpanding region, which is indistinguishable from the rigid space described by Newton. Thus, Earth and your body, for example, are not getting any bigger. Only the distance between widely separated galaxies increases with time. The *Cosmic Connections: Urband Legends about the Expanding Universe* has more to say about several misconceptions concerning the expanding universe.

The Hubble law is a direct proportionality—that is, a galaxy twice as far away is receding from us twice as fast. This property is just what we would expect in an expanding universe. To see why this property results from uniform expansion, imagine a grid of parallel lines (as on a piece of graph paper) crisscrossing the universe. Figure 26-3*a* shows a series of such gridlines 100 Mpc apart, along with five galaxies labeled A, B, C, D, and E that happen to lie where gridlines cross. As the universe expands in all directions, the gridlines and the attached galaxies spread apart. (This is just what would happen if the universe were a two-dimensional rubber sheet that was being pulled equally on all sides. Alternatively, you can imagine that Figure 26-3*a* depicts a very small portion of the chocolate chip cake in Figure 26-2, with galaxies taking the place of chocolate chips.)

Figure 26-3*b* shows the universe at a later time, when the gridlines are 50% farther apart (150 Mpc) and all the distances between galaxies are 50% greater than in Figure 26-3*a*. Imagine that A represents our galaxy, the Milky Way. The table accompanying Figure 26-3 shows how far each of the other galaxies have moved away from us during the expansion: Galaxies A and B and galaxies A and E were originally 100 Mpc apart, and have moved away from each other by an additional 50 Mpc; A and C, which were originally 200 Mpc apart, have increased their separation by an additional 100 Mpc; and the distance between A and D, originally 300 Mpc, has increased by an additional 150 Mpc. In other words, the increase in distance between any pair of galax-

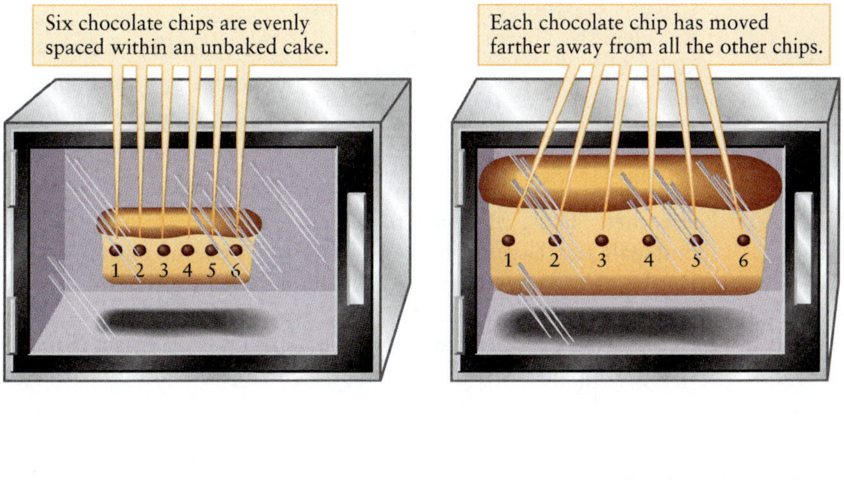

Six chocolate chips are evenly spaced within an unbaked cake.
1 2 3 4 5 6
Each chocolate chip has moved farther away from all the other chips.
1 2 3 4 5 6

Figure 26-2

The Expanding Chocolate Chip Cake Analogy The expanding universe can be compared to what happens inside a chocolate chip cake as the cake expands during baking. All of the chocolate chips in the cake recede from one another as the cake expands, just as all the galaxies recede from one another as the universe expands.

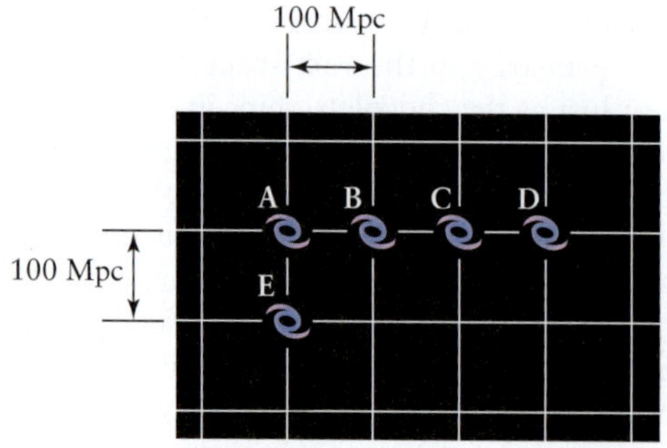

(a) Five galaxies spaced 100 Mpc apart

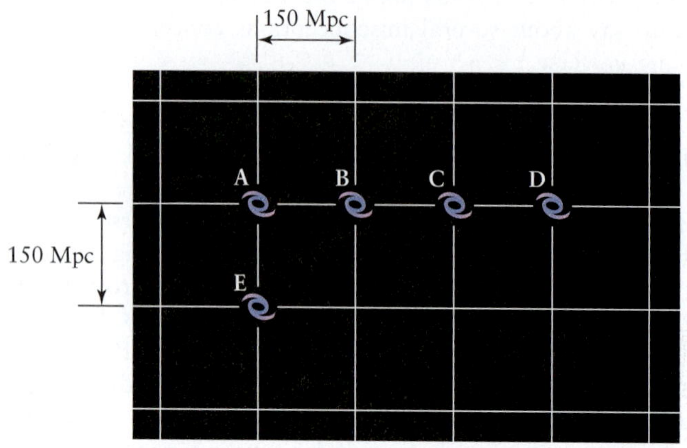

(b) The expansion of the universe spreads the galaxies apart

	Original distance (Mpc)	Later distance (Mpc)	Change in distance (Mpc)
A–B	100	150	50
A–C	200	300	100
A–D	300	450	150
A–E	100	150	50

ANIMATION 26.1

Figure 26-3

The Expanding Universe and the Hubble Law **(a)** Imagine five galaxies labeled A, B, C, D, and E. At the time shown here, adjacent galaxies are 100 Mpc apart. **(b)** As the universe expands, by some later time the spacing between adjacent galaxies has increased to 150 Mpc. The table shows that the greater the original distance between galaxies, the greater the amount that distance has increased. This agrees with the Hubble law.

ies is in direct proportion to the original distance; if the original distance is twice as great, the increase in distance is also twice as great.

To see what these distances tell us about the recessional velocities of galaxies, remember that velocity is equal to the distance moved divided by the elapsed time. (For example, if you traveled in a straight line for 360 kilometers in 4 hours, your velocity was (360 km)/(4 h) = 90 kilometers per hour.) Because the distance that each galaxy moves away from A during the expansion is directly proportional to its original distance from A, it follows that the velocity v at which each galaxy moves away from A is also directly proportional to the original distance d. This result is just the Hubble law, $v = H_0 d$.

CAUTION! It may seem that if the universe is expanding, and if we see all the distant galaxies rushing away from us, then we must be in a special position at the very center of the universe. In fact, the expansion of the universe looks the same from the vantage point of *any* galaxy. For example, as seen from galaxy D in Figure 26-3, the initial distances to galaxies A, B, and C are 300 Mpc, 200 Mpc, and 100 Mpc, respectively. Between parts *a* and *b* of the figure, these distances increase by 150 Mpc, 100 Mpc, and 50 Mpc, respectively. So, as seen from D as well, the recessional velocity increases in direct proportion with the distance, and in the same proportion as seen from A. In other words, no matter which galaxy you call home, you will see all the other galaxies receding from you in accordance with the same Hubble law (and the same Hubble constant) that we observe from Earth.

WEB LINK 26.5

Figure 26-2 also shows that the expansion of the universe looks the same from one galaxy as from any other. An insect sitting on any one of the chocolate chips would see all the other chips moving away. If the cake were infinitely long, it would not actually have a center; as seen from any chocolate chip within such a cake, the cake would extend off to infinity to the left and to the right, and the expansion of the cake would appear to be centered on that chip. Likewise, because every point in the universe appears to be at the center of the expansion, it follows that our universe has no center at all. (Later in this chapter we will see evidence that the universe, like our imaginary cake, is indeed infinite.)

CAUTION! "If the universe is expanding, what is it expanding into?" This commonly asked question arises only if we take our chocolate chip cake analogy too literally. In Figure 26-2, the cake (representing the universe) expands in three-dimensional space into the surrounding air. But the actual universe includes *all* space; there is nothing "beyond" it, because there is no "beyond." Asking "What lies beyond the universe?" is as meaningless as asking "Where on Earth is north of the North Pole?"

The ongoing expansion of space explains why the light from remote galaxies is redshifted. Imagine a photon coming toward us from a distant galaxy. As the photon travels through space, the space is expanding, so the photon's wavelength becomes stretched. When the photon reaches our eyes, we see an increased wavelength: The photon has been redshifted. The longer the photon's journey, the more its wavelength will have been stretched. Thus, photons from distant galaxies have larger redshifts than those of photons from nearby galaxies, as expressed by the Hubble law.

The redshifts of distant galaxies are not Doppler shifts; they are caused by the expansion of space itself

A redshift caused by the expansion of the universe is properly called a **cosmological redshift**. It is *not* the same as a Doppler shift. Doppler shifts are caused by an object's *motion through*

COSMIC CONNECTIONS

"Urban Legends" about the Expanding Universe

There are a number of common misconceptions or "urban legends" about what happens as the universe expands. The illustrations below depict the myth and the reality for three of these "urban legends."

Urban Legend #1:
The expansion of the universe means that as time goes by, galaxies move away from each other through empty space. In this picture, space is simply a background upon which the galaxies act out their parts .

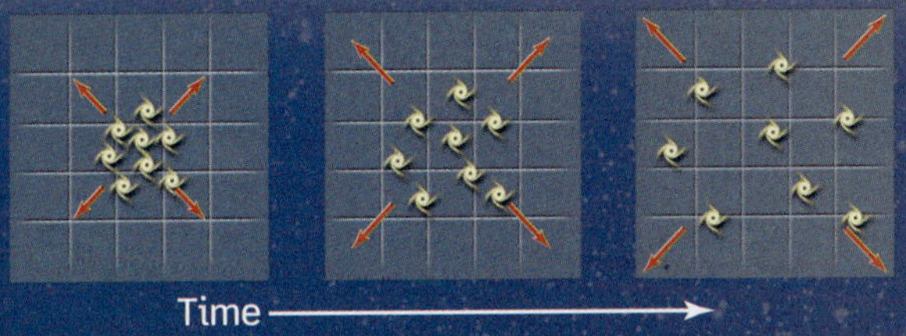

Reality:
The expansion of the universe means that as time goes by, *space itself* expands (shown here by the expanding grid). As it expands, it carries the galaxies along with it.

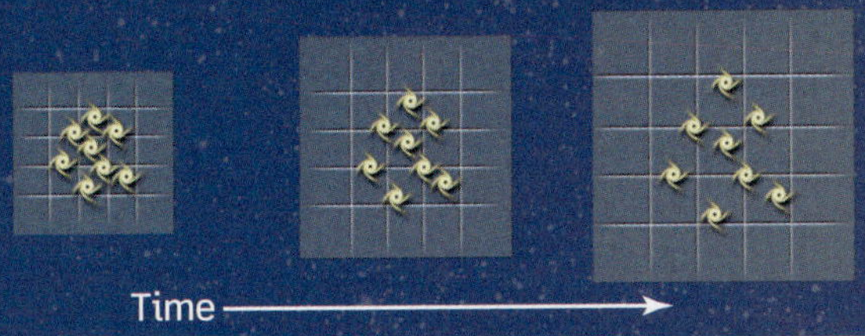

Urban Legend #2:
The redshift of light from distant galaxies is a Doppler shift. It occurs because these galaxies are moving away from us rapidly.

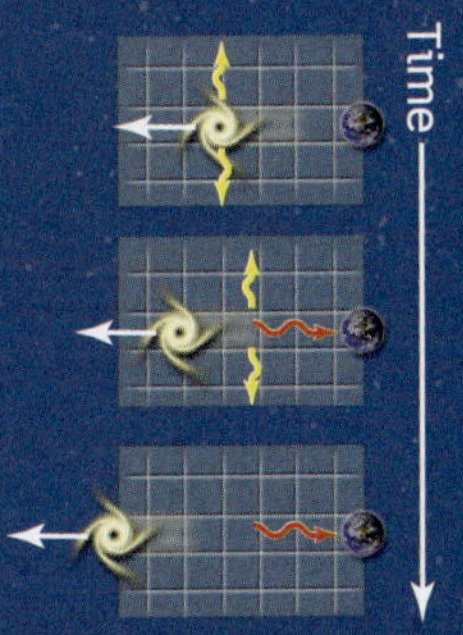

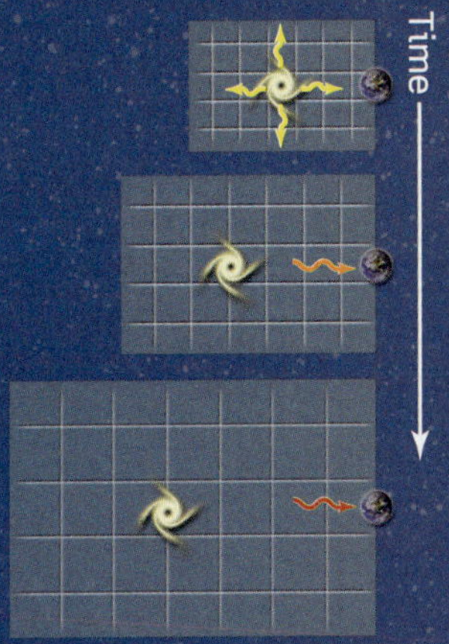

Reality:
As a photon travels through intergalactic space, its wavelength expands as the space through which it is traveling expands. This is called a *cosmological* redshift.

Urban Legend #3:
As the universe expands, so do objects within the universe. Hence galaxies within a cluster are now more spread out than they were billions of years ago.

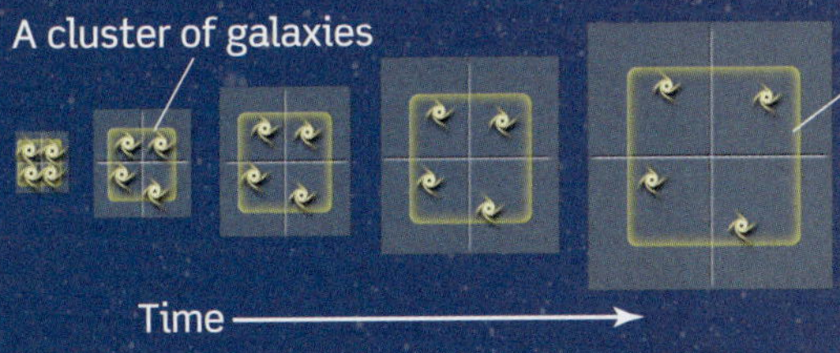

In this picture, the cluster expands as the universe expands.

Reality:
At first the expansion of the universe tends to pull the galaxies of a cluster away from each other. But the force of gravitational attraction binds the members of the clusters together, so the cluster stabilizes at a certain size.

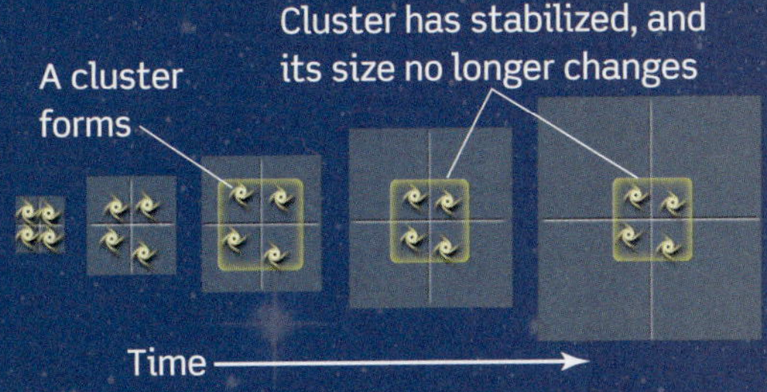

(Illustrations by Alfred T. Kamajian, from C. H. Lineweaver and T. M. Davis, "Misconceptions about the Big Bang," *Scientific American*, March 2005)

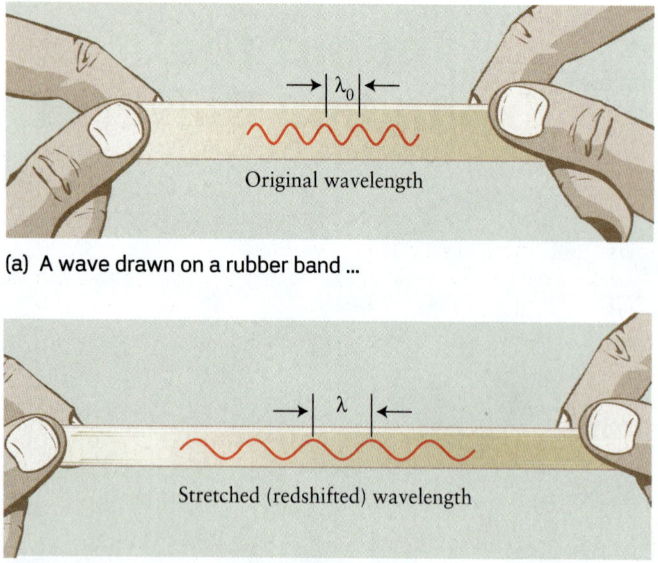

(a) A wave drawn on a rubber band ...

(b) ... increases in wavelength as the rubber band is stretched.

Figure 26-4

Cosmological Redshift A wave drawn on a rubber band stretches along with the rubber band. In an analogous way, a light wave traveling through an expanding universe "stretches," that is, its wavelength increases.

space, whereas a cosmological redshift is caused by the *expansion of space* (Figure 26-4).

Cosmological Redshift and Lookback Time

We can calculate the factor by which the universe has expanded since some ancient time from the redshift of light emitted by objects at that time. As we saw in Section 24-5, redshift (z) is defined as

$$z = \frac{\lambda - \lambda_0}{\lambda_0}$$

In this equation λ_0 is the unshifted wavelength of a photon and λ is the wavelength we observe. For example, λ_0 could be the wavelength of a particular emission line in the spectrum of light leaving a remote galaxy. As the galaxy's light travels through space, its wavelength is stretched by the expansion of the universe. Thus, at our telescopes we observe the spectral line to have a wavelength λ. The ratio λ/λ_0 is a measure of the amount of stretching. By rearranging terms in the preceding equation, we can solve for this ratio to obtain

$$\frac{\lambda}{\lambda_0} = 1 + z$$

For example, consider a galaxy with a redshift $z = 3$. Since the time that light left that galaxy, the universe has expanded by a factor of $1 + z = 1 + 3 = 4$. In other words, when the light left that galaxy, representative distances between widely separated galaxies were only one-quarter as large as they are today. A representative volume of space, which is proportional to the cube of its dimensions, was only $(1/4)^3 = 1/64$ as large as it is today. Thus, the density of matter in such a volume was 64 times greater than it is today.

If you know the redshift z of a distant object such as a remote galaxy, you can calculate that object's recessional velocity v (see Box 24-2 for this calculation). Then, using the Hubble law, you can determine the distance d to that object if you know the value of the Hubble constant H_0. This distance also tells you the **lookback time** of that object, that is, how far into the past you are looking when you see that object. For example, if the lookback time for a distant galaxy is a billion years, that means the light from the galaxy took a billion years to reach us and so we are seeing it as it was a billion years ago. The images that open this chapter show supernovae in distant galaxies with lookback times from 7.7 to 9.0 billion years. (Table 25-1 lists the distance at which we see objects with different redshifts. This distance measured in light-years is equal to the lookback time in years.)

Distances and lookback times determined in this way are somewhat uncertain, because there is some uncertainty in the value of the Hubble constant. Furthermore, as we will see in Section 26-8, the universe has not always expanded at the same rate. This variable rate of expansion means that the value of the Hubble constant H_0 was different in the distant past. (In other words, the Hubble "constant" is not actually constant in time.) Furthermore, the correct distance d to use in the Hubble law is not the distance at which we *see* the object, but rather the distance between us and the object *now*. The latter distance is larger because during the time that it takes to reach us from a distant object, that object has moved farther away due to the expansion of the universe.

To avoid dealing with these uncertainties and complications, astronomers commonly refer to times in the distant past in terms of redshift rather than years. For example, instead of asking, "How common were quasars 5 billion years ago?," an astronomer might ask, "How common were quasars at $z = 1.0$?" In this question, "at $z = 1.0$" is a shorthand way of saying "at the lookback time that corresponds to objects at $z = 1.0$." We will use this terminology later in this chapter. Remember that the greater the redshift, the greater the lookback time and hence the further back into time we are peering.

The Cosmological Principle

In cosmology, there is only one universe that we can observe. Unlike other sciences, we cannot carry out controlled experiments or even make comparisons between two similar systems, when the system is the universe itself. But then a question arises: Are we observing the universe from a special location, resulting in a unique appearance? Or, would observers anywhere see a universe with the same properties on large scales? The answer to this question determines how we interpret Hubble's law, because receding galaxies could be interpreted to mean that we are at the center of the universe. We reject this interpretation, however, because it violates a cosmological extension of Copernicus's argument that we do not occupy a special location in space. An expanding universe, on the other hand, has no center; at all locations in space, observers would see galaxies recede according to Hubble's law.

When Einstein began applying his general theory of relativity to cosmology, he made a daring assumption: Over very large distances the universe is **homogeneous,** meaning that every region is the same as every other region, and **isotropic,** meaning that the universe looks the same in every direction. In other words, if you could stand back and look at a very large region of space, any one part of the universe would look basically the same as any other part, with the same kinds of galaxies distributed through space in the same way. The assumption that the universe is homogeneous and isotropic constitutes the **cosmological principle.** It gives precise meaning to the idea that we do not occupy a special location in space.

Models of the universe based on the cosmological principle have proven remarkably successful in describing the structure and evolution of the universe and in interpreting observational data. More recently, direct observations reveal a homogeneous and isotropic universe on the largest scales—a validation of the cosmological principle.

26-3 The expanding universe emerged from a cataclysmic event called the Big Bang

The Hubble flow shows that the universe has been expanding for billions of years. Therefore, we can conclude that in the past the matter in the universe must have been closer together and therefore denser than it is today. Indeed, as we look farther into the past, we clearly see the density of the universe increasing. There is even strong evidence of increasing density back to the first moments of time. Therefore, some sort of tremendous event caused high-density matter to begin the expansion that continues to the present day. This event, called the **Big Bang,** marks the creation of the universe.

CAUTION! It is not correct to think of the Big Bang as an explosion. When a bomb explodes, pieces of debris fly off *into space* from a central location. If you could trace all the pieces back to their origin, you could find out exactly where the bomb had been. This process is not possible with the universe, however, because the universe itself always has and always will consist of all space. As we have seen, the universe logically cannot have an edge (see the discussion of the expanding chocolate chip cake in Section 26-2).

Estimating the Age of the Universe

How long ago did the Big Bang take place? To estimate an answer, imagine two galaxies that today are separated by a distance d and receding from each other with a velocity v. A movie of these galaxies would show them flying apart. If you run the movie backward, you would see the two galaxies approaching each other as time runs in reverse. We can calculate the time T_0 it will take for the reverse-run galaxies to collide by using the equation

$$T_0 = \frac{d}{v}$$

This expression says that the time to travel a distance d at velocity v is equal to the ratio d/v. (As an example, to travel a distance of 360 km at a velocity of 90 km/h takes (360 km)/(90 km/h) = 4 hours.) If we use the Hubble law, $v = H_0 d$, to replace the velocity v in this equation, we get

$$T_0 = \frac{d}{H_0 d} = \frac{1}{H_0}$$

Note that the distance of separation, d, has canceled out and does not appear in the final expression. Distance not appearing in the expression means that T_0 is the same for *all* galaxies. This period is the time in the past when all galaxies were crushed together, the time back to the Big Bang. In other words, the reciprocal (or inverse) of the Hubble constant H_0 gives us an estimate of the age of the universe, which is one reason why H_0 is such an important quantity in cosmology.

Observations suggest that $H_0 = 73$ km/s/Mpc to within a few percent, and this is the value we choose as our standard (see Section 24-5). Using this value, our estimate for the age of the universe is

$$T_0 = \frac{1}{73 \text{ km/s/Mpc}}$$

To convert this into units of time, we simply need to remember that 1 Mpc equals 3.09×10^{19} km and 1 year equals 3.156×10^7 s. Using the technique we discussed in Box 1-3 for converting units, we get

$$T_0 = \frac{1}{73}\ \frac{\text{Mpc s}}{\text{km}} \times \frac{3.09 \times 10^{19} \text{ km}}{1 \text{ Mpc}} \times \frac{1 \text{ year}}{3.156 \times 10^7 \text{s}}$$
$$= 1.34 \times 10^{10} \text{ years} = 13.4 \text{ billion years}$$

By comparison, the age of our solar system is only 4.56 billion years, or about one-third the age of the universe. Thus, the formation of our home planet is a relatively recent event in the history of the cosmos.

The value of H_0 has an uncertainty of about 5%, so our simple estimate of the age of the universe is likewise uncertain by at least 5%. Furthermore, the formula $T_0 = 1/H_0$ is at best an approximation, because in deriving it we assumed that the universe expands at a constant rate. In Section 26-8 we will discuss how the expansion rate of the universe has changed over its history. When these factors are taken into consideration, we find that the age of the universe is 13.7 billion years, with an uncertainty of about 0.2 billion years. This time period is remarkably close to our simple estimate.

Whatever the true age of the universe, it must be at least as old as the oldest stars. The oldest stars that we can observe readily lie in the Milky Way's globular clusters (see Section 19-4 and Section 23-1). The most recent observations, combined with calculations based on the theory of stellar evolution, indicate that these stars are about 13.4 billion years old (time period with an uncertainty of about 6%). Encouragingly enough, this time period is less than the calculated age of the universe: The oldest stars in our universe are younger than the universe itself!

Our Observable Universe and the Dark Night Sky

The Big Bang helps resolve Olbers's paradox, which we discussed in Section 26-1. We know that the universe had a definite beginning, and thus its age is finite (as opposed to infinite). If the universe is 13.7 billion years old, then the most distant objects that we can see are those whose light has traveled 13.7 billion years to reach us. (Due to the expansion of the universe, these objects are now more than 13.7 billion light-years away.) As a result, we can only see objects that lie within an immense sphere centered

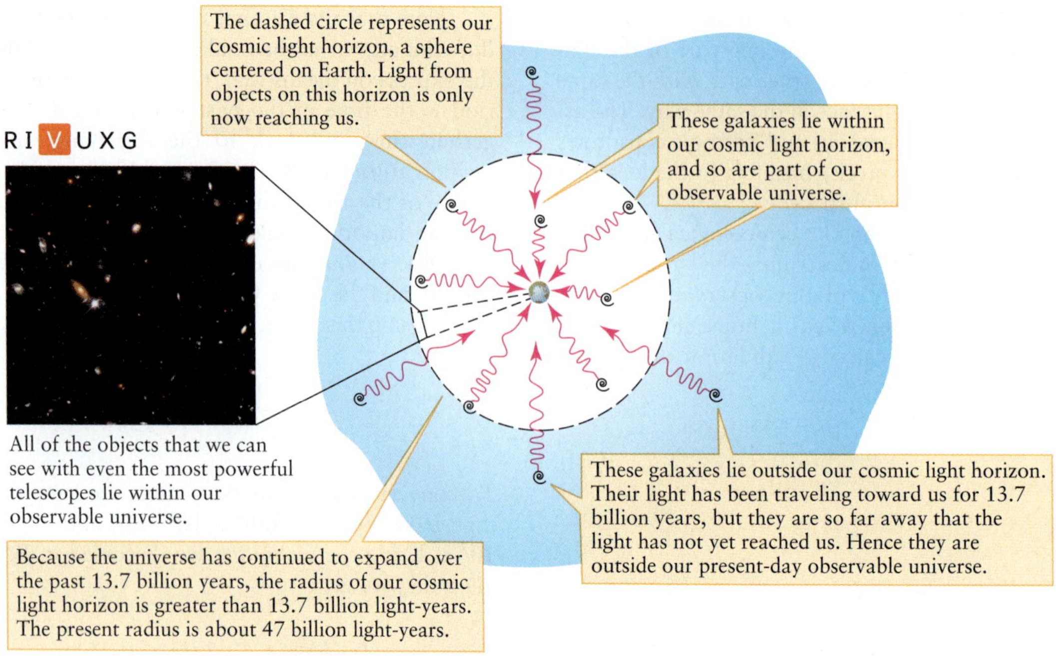

The dashed circle represents our cosmic light horizon, a sphere centered on Earth. Light from objects on this horizon is only now reaching us.
R I V U X G
These galaxies lie within our cosmic light horizon, and so are part of our observable universe.
All of the objects that we can see with even the most powerful telescopes lie within our observable universe.
These galaxies lie outside our cosmic light horizon. Their light has been traveling toward us for 13.7 billion years, but they are so far away that the light has not yet reached us. Hence they are outside our present-day observable universe.
Because the universe has continued to expand over the past 13.7 billion years, the radius of our cosmic light horizon is greater than 13.7 billion light-years. The present radius is about 47 billion light-years.

Figure 26-5

Our Observable Universe The part of the universe that we can observe lies within our cosmic light horizon. The galaxies that we can just barely make out with our most powerful telescopes lie inside our cosmic light horizon; we see them as they were less than a billion years after the Big Bang. We cannot see objects beyond our cosmic light horizon, because in the 13.7 billion years since the Big Bang their light has not had enough time to reach us. (Inset: Robert Williams and the Hubble Deep Field Team, STScI; NASA)

on Earth (Figure 26-5). This is true even if the universe is infinite, with galaxies scattered throughout its limitless expanse.

The surface of the sphere depicted in Figure 26-5 is called our **cosmic light horizon.** Our entire **observable universe** is located inside this sphere. We cannot see anything beyond our cosmic light horizon, because the time required for light to reach us from these incredibly remote distances is greater than the present age of the universe. As time goes by, light from more distant parts of the universe reaches us for the first time, and the size of the cosmic light horizon—the size of our observable universe—increases. The finite size of the observable universe, with a finite number of stars and galaxies, also helps to resolve Olber's paradox: Galaxies are distributed sparsely enough in our observable universe that there are no stars along most of our lines of sight. This sparse distribution of visible stars is one reason why the night sky is dark.

Measuring the recessional velocities of galaxies allows us to estimate the age of the universe

Besides the finite age of the universe, a second effect also contributes significantly to the darkness of the night sky—the redshift. According to the Hubble law, the greater the distance to a galaxy, the greater the redshift. When a photon is redshifted, its wavelength becomes longer, and its energy—which is inversely proportional to its wavelength (see Section 5-5)—decreases. Consequently, even though there are many galaxies far from Earth, they have large redshifts and their light does not carry much energy. A galaxy nearly at the cosmic light horizon has a nearly infinite redshift, meaning that the light we receive from that galaxy carries practically no energy at all. This decrease in photon energy because of the expansion of the universe decreases the brilliance of remote galaxies, helping to make the night sky dark.

The concept of a Big Bang origin for the universe is a straightforward, logical consequence of an expanding universe. In Section 26-4, we'll see direct evidence of the primordial fireball associated with the Big Bang and other confirmations of the Big Bang back to the first few moments. But how far back in time can our laws describe the universe? If you can just imagine far enough back into the past, you can arrive at a time 13.7 billion years ago, when the density throughout the universe was nearly infinite. As a result, throughout the universe space and time were jumbled up with nearly infinite curvature. A full description of this earliest instant requires (as do aspects of black holes), a valid mathematical theory of quantum gravity, which is a work in progress. But, when in the past did the known laws of physics begin to apply?

A very short time after the Big Bang, space and time began to behave in the way we think of them today. This short time interval, called the **Planck time** (t_P), is given by the following expression:

The Planck time

$$t_P = \sqrt{\frac{Gh}{c^5}} = 1.35 \times 10^{-43} \text{ s}$$

t_P = Planck time

G = universal constant of gravitation

h = Planck's constant

c = speed of light

LOOKING DEEPER 26.1

We do not yet understand how space, time, and matter behaved in that brief but important interval from the beginning of the Big Bang to the Planck time, about 10^{-43} seconds later. (Indeed, the laws of physics suggest that it might be impossible ever to know what happened during this extremely short time interval.) Hence, the Planck time represents a limit to our knowledge of conditions at the very beginning of the universe.

26-4 The microwave radiation that fills all space is evidence of a hot Big Bang

TUTORIAL 26.2

One of the major advances in twentieth-century astronomy was the discovery of the origin of the heavy elements. We know today that essentially all the heavy elements are created by thermonuclear reactions at the centers of stars and in supernovae (see Chapter 20). The starting point of all these reactions is the fusion of hydrogen into helium, which we described in Section 16-1. But as astronomers began to understand the details of thermonuclear synthesis in the 1960s, they were faced with a dilemma: There is far more helium in the universe than could have been created by hydrogen fusion in stars.

For example, the Sun consists of about 74% hydrogen and 25% helium by mass, and the universe as a whole is also about 25% helium. Some helium was produced by the thermonuclear fusion of hydrogen within stars, but calculations show that the amount of helium produced in this way is not nearly enough to account for the large fraction of helium observed in the universe. Because it was thought that the universe originally contained only hydrogen—the simplest of all the chemical elements—the presence of so much helium posed a major dilemma.

A Hot Big Bang and the Cosmic Microwave Background

Shortly after World War II, Ralph Alpher and Robert Hermann proposed that the universe immediately following the Big Bang must have been so incredibly hot that thermonuclear reactions occurred everywhere throughout space. Calculations for the amount of helium (and some other elements) produced by thermonuclear reactions in the first few minutes after the Big Bang are in extraordinary agreement with observations. This astounding prediction of the Big Bang was confirmed by detailed observations. But predictions coming from the Big Bang theory didn't stop there. Princeton University physicists Robert Dicke and P. J. E. Peebles calculated that in order for the entire universe to undergo the required thermonuclear reactions, the early universe must have been very hot and filled with many high-energy, short-wavelength photons. The properties of this radiation field depended on its temperature, as described by Planck's blackbody law (review Figure 5-11).

The universe has expanded so much since those ancient times that all those short-wavelength photons have had their wavelengths stretched by a tremendous factor. As a result, they have become low-energy, long-wavelength photons. The temperature of this cosmic radiation field is now only a few degrees above absolute zero. By Wien's law, radiation at such a low temperature should have its peak intensity at microwave wavelengths of approximately 1 millimeter. Hence, this radiation field, which fills all of space, is called the **cosmic microwave background** or **cosmic background radiation.** In the early 1960s, Dicke and his colleagues began designing an antenna to detect this microwave radiation.

Meanwhile, just a few miles from Princeton University, Arno Penzias and Robert Wilson of Bell Telephone Laboratories in New Jersey were working on a new microwave horn antenna designed to relay telephone calls to Earth-orbiting communications satellites (Figure 26-6). Penzias and Wilson were deeply puzzled when, no matter where in the sky they pointed their antenna, they detected faint background noise. Thanks to a colleague, they happened to learn about the work of Dicke and Peebles and came to realize that they had discovered the cooled-down cosmic background radiation left over from the hot Big Bang. Penzias and Wilson shared the 1978 Nobel Prize in Physics for their discovery.

The afterglow of the Big Bang was first discovered by a happy coincidence—and can be detected with an ordinary television

A TV using an antenna for signal reception (as opposed to cable or satellite dish) can actually detect cosmic background radiation. This radiation is responsible for about 1% of the random noise or "hash" that appears on the screen when you tune a television to a station that is off the air. Using far more sophisticated detectors than TV sets, scientists have made many measurements of the intensity of the background radiation at a variety of wavelengths. Unfortunately, Earth's atmosphere is almost totally opaque to wavelengths between about 10 μm and 1 cm (see Figure 6-25), which is just the wavelength range in which the back-

WEB LINK 26.7

Figure 26-6 R I V U X G

The Bell Labs Horn Antenna Using this microwave horn antenna, originally built for communications purposes, Arno Penzias and Robert Wilson detected a signal that seemed to come from all parts of the sky. After carefully removing all potential sources of electronic "noise" (including bird droppings inside the antenna) that could create a false signal, Penzias and Wilson realized that they were actually detecting radiation from space. This radiation is the afterglow of the Big Bang. (Bell Labs)

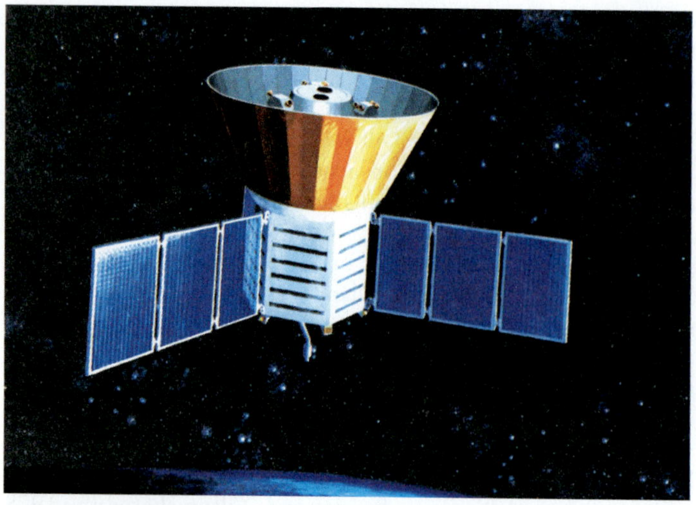

(a) The COBE spacecraft

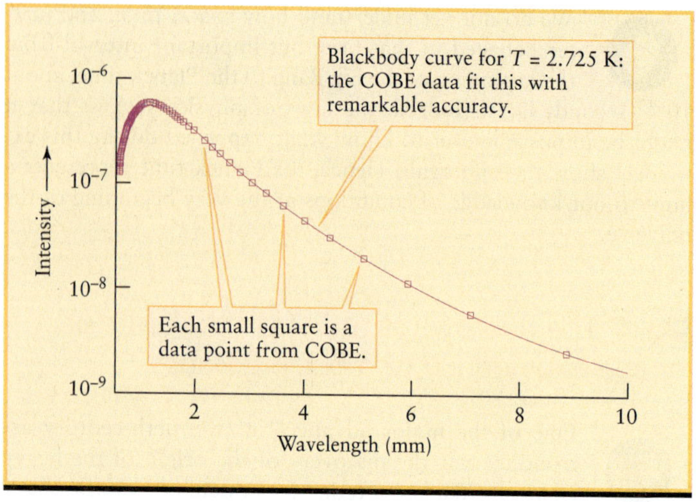

(b) The spectrum of the cosmic microwave background

WEB LINK 26.8

Figure 26-7

COBE and the Spectrum of the Cosmic Microwave Background **(a)** The Cosmic Background Explorer satellite (COBE), launched in 1989, measured the spectrum and angular distribution of the cosmic microwave background over a wavelength range from 1 μm to 1 cm. **(b)** A blackbody curve gives an excellent match to the COBE data. (a: Courtesy of J. Mather/NASA; b: Courtesy of E. Cheng/NASA COBE Science Team)

ground radiation is most intense. As a result, scientists have had to place detectors either on high-altitude balloons (which can fly above the majority of the obscuring atmosphere) or, even better, on board orbiting spacecraft.

A Detailed Look at the Cosmic Microwave Background

The first high-precision measurements of the cosmic microwave background came from the Cosmic Background Explorer (COBE, pronounced "coe-bee") satellite, which was placed in Earth orbit in 1989 (Figure 26-7*a*). Data from COBE's spectrometer, shown in Figure 26-7*b*, demonstrate that this ancient radiation has the spectrum of a blackbody with a temperature of 2.725 K. In recognition of this discovery, as well as others that we will discuss in Section 26-5, COBE team leaders John Mather of NASA and George Smoot of the University of California, Berkeley, were awarded the 2006 Nobel Prize in Physics.

An important feature of the microwave background is that its intensity is almost perfectly isotropic, that is, the same in all directions. In other words, we detect nearly the same background intensity from all parts of the sky. This is a striking confirmation of Einstein's assumption that the universe is isotropic (see Section 26-2). However, extremely accurate measurements first made from high-flying airplanes, and later from high-altitude balloons and from COBE, reveal a very slight variation in temperature across the sky. The microwave background appears slightly warmer than average toward the constellation of Leo and slightly cooler than average in the opposite direction toward Aquarius. Between the warm spot in Leo and the cool spot in Aquarius, the background temperature declines smoothly across the sky. Figure 26-8 is a map of the microwave sky showing this variation.

This apparent variation in temperature is caused by Earth's motion through the cosmos. If we were at rest with respect to the microwave background, the radiation would be even more nearly isotropic. Because we are moving through this radiation field, however, we see a Doppler shift. Specifically, we see shorter-than-average wavelengths in the direction toward which we are moving, as drawn in Figure 26-9. A decrease in wavelength corresponds to an increase in photon energy and thus an increase

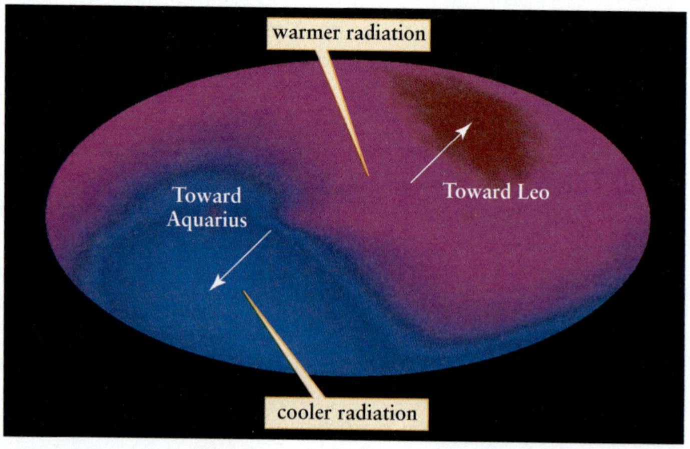

Figure 26-8 R I V U X G

The Microwave Sky In this map of the entire sky made from COBE data, the plane of the Milky Way runs horizontally across the map, with the galactic center in the middle. Color indicates temperature—red is warm and blue is cool. The small temperature variation across the sky—only 0.0033 K above or below the average radiation temperature of 2.725 K—is caused by Earth's motion through the microwave background. (NASA)

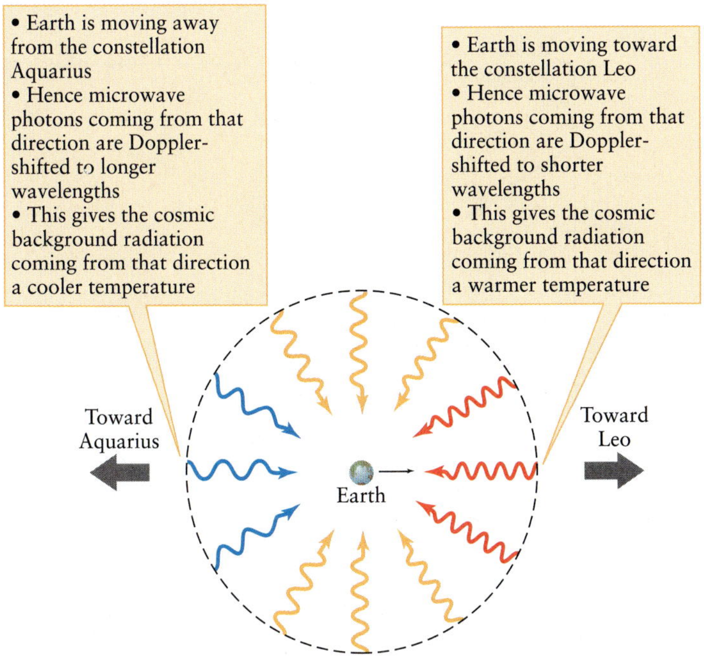
• Earth is moving away from the constellation Aquarius
• Hence microwave photons coming from that direction are Doppler-shifted to longer wavelengths
• This gives the cosmic background radiation coming from that direction a cooler temperature
• Earth is moving toward the constellation Leo
• Hence microwave photons coming from that direction are Doppler-shifted to shorter wavelengths
• This gives the cosmic background radiation coming from that direction a warmer temperature
Toward Aquarius
Toward Leo
Earth

Figure 26-9

Our Motion Through the Microwave Background Because of the Doppler effect, we detect shorter wavelengths in the microwave background and a higher temperature of radiation in that part of the sky toward which we are moving. This part of the sky is the area shown in red in Figure 26-8. In the opposite part of the sky, shown in blue in Figure 26-8, the microwave radiation has longer wavelengths and a cooler temperature.

in the temperature of the radiation. The slight temperature excess observed, about 0.00337 K, corresponds to a speed of 371 km/s. Conversely, we see longer-than-average wavelengths in that part of the sky from which we are receding. An increase in wavelength corresponds to a decline in photon energy and, hence, a decline in radiation temperature.

WEB LINK 26.9 Our solar system is thus traveling away from Aquarius and toward Leo at a speed of 371 km/s. Taking into account the known velocity of the Sun around the center of our Galaxy, we find that the entire Local Group of galaxies, including our Milky Way Galaxy, is moving at about 620 km/s toward the Hydra-Centaurus supercluster. Observations show that thousands of other galaxies are being carried in this direction, as is the Hydra-Centaurus supercluster itself. This tremendous flow of matter is thought to be due to the gravitational pull of an enormous collection of visible galaxies and dark matter lying in that direction. This immense object, dubbed the *Great Attractor,* lies about 50 Mpc (150 million ly) from Earth (see Figure 24-22).

The existence of such concentrations of mass, as well as the existence of superclusters of galaxies (see Section 22-6), shows that the universe is rather "lumpy" on scales of 100 Mpc or smaller. It is only on larger scales that the universe is homogeneous and isotropic.

26-5 The universe was a hot, opaque plasma during its first 380,000 years

Energy in the universe falls into one of two categories—radiation or matter. (We will encounter another type of energy in Section 26-7.) Photons are massless particles of light and are a form of radiation. There are many photons of starlight traveling across space, but the vast majority of photons in the universe belong to the cosmic microwave background. The matter in the universe is contained in such luminous objects as stars, planets, and galaxies, as well as in nonluminous dark matter. A natural question to ask is which plays a more important role in the universe, radiation or matter? As we will see, the answer to this question is different for the early universe from the answer for our universe today.

Radiation and Matter in the Universe

To make a comparison between radiation and matter, recall Einstein's famous equation $E = mc^2$ (see Section 16-1). We can think of the photon energy in the universe (E) as being equal to a quantity of mass (m) multiplied by the square of the speed of light (c). The amount of this equivalent mass in a volume V, divided by that volume, is the **mass density of radiation** (ρ_{rad}; say "rho sub rad").

We can combine $E = mc^2$ with the Stefan-Boltzmann law (see Section 5-4) to give the following formula:

Mass density of radiation

$$\rho_{rad} = \frac{4\sigma T^4}{c^3}$$

ρ_{rad} = mass density of radiation

T = temperature of radiation

σ = Stefan-Boltzmann constant $= 5.67 \times 10^{-8}\ \text{W m}^{-2}\ \text{K}^{-4}$

c = speed of light $= 3.00 \times 10^8$ m/s

For the present-day temperature of the cosmic background radiation, $T = 2.725$ K, this equation yields

$$\rho_{rad} = 4.6 \times 10^{-31}\ \text{kg/m}^3$$

The **average density of matter** (ρ_m; say "rho sub em") in the universe is harder to determine. To find this density, we look at a large volume (V) of space, determine the total mass (M) of all the stars, galaxies, and dark matter in that volume, and divide the volume into the mass: $\rho_m = M/V$. (We emphasize that this quantity is the *average* density of matter. It would be the actual density if all the matter in the universe were spread out uniformly rather than being clumped into galaxies and clusters of galaxies.) Determining how much mass is in a large volume of space is a challenging task. A major part of the challenge involves dark matter, which emits no electromagnetic radiation and can be detected only by its gravitational influence (see Section 23-4 and Section 24-8).

One method that appears to deal successfully with this challenge is to observe clusters of galaxies, within which most of the luminous mass in the universe is concentrated and the contribution from dark matter can also be estimated. Rich clusters are surrounded by halos of hot, X-ray–emitting gas, typically at temperatures of 10^7 to 10^8 K (see Figure 24-25). Such a halo should be in hydrostatic equilibrium, so that it neither expands nor contracts but remains the same size (see Section 16-2, especially Figure 16-2). The outward gas pressure associated with the halo's high temperature would then be balanced by the inward gravitational pull due to the total mass of the cluster. Thus, by measuring the temperature of the halo—which can be determined from the properties of the halo's X-ray emission—astronomers can infer the cluster's mass, including the contribution from dark matter!

From galaxy clusters and other measurements, the present-day average density of matter in the universe is estimated to be

$$\rho_m = 2.4 \times 10^{-27} \text{ kg/m}^3$$

with an uncertainty of about 15%. The mass of a single hydrogen atom is 1.67×10^{-27} kg. Hence, if the mass of the universe were spread uniformly over space, there would be the equivalent of $1\frac{1}{2}$ hydrogen atoms per cubic meter of space. By contrast, there are 5×10^{25} atoms in a cubic meter of the air you breathe! The very small value of ρ_m shows that our universe has very little matter in it.

Furthermore, by counting galaxies and other measurements, astronomers determine that the average density of *luminous* matter (that is, the stars and gas within clusters of galaxies) is about 4.2×10^{-28} kg/m^3. This density is only about 17% of the average density of matter of all forms. Thus, nonluminous dark matter is actually the predominant form of matter in our universe. The "ordinary" matter (protons and neutrons) of which the stars, the planets, and ourselves are made is only about one-fifth of the total matter!

When Radiation Held Sway Over Matter

Although the average density of matter in the universe is tiny by Earth standards, it is thousands of times larger than ρ_{rad}, the mass density of radiation. However, this ratio was not always the case. Matter prevails over radiation today only because the energy now carried by microwave photons is so small. Nevertheless, the number of photons in the microwave background is astounding. From the physics of blackbody radiation it can be calculated that there are today 410 million (4.1×10^8) photons in every cubic meter of space. In other words, the photons in space outnumber atoms by roughly a billion (10^9) to one. In terms of total number of particles, the universe thus consists almost entirely of microwave photons. This radiation field no longer has much effect on the universe however, because its photons have been redshifted to long wavelengths and low energies after 13.7 billion years of being stretched by the expansion of the universe.

In contrast, think back toward the Big Bang. The universe becomes increasingly compressed, and thus the density of matter increases as we go back in time. The photons in the background radiation also become more crowded together as we go back in time. But, in addition, the photons become less redshifted and thus have shorter wavelengths and higher energy than they do today. Because of this added energy, the mass density of radiation

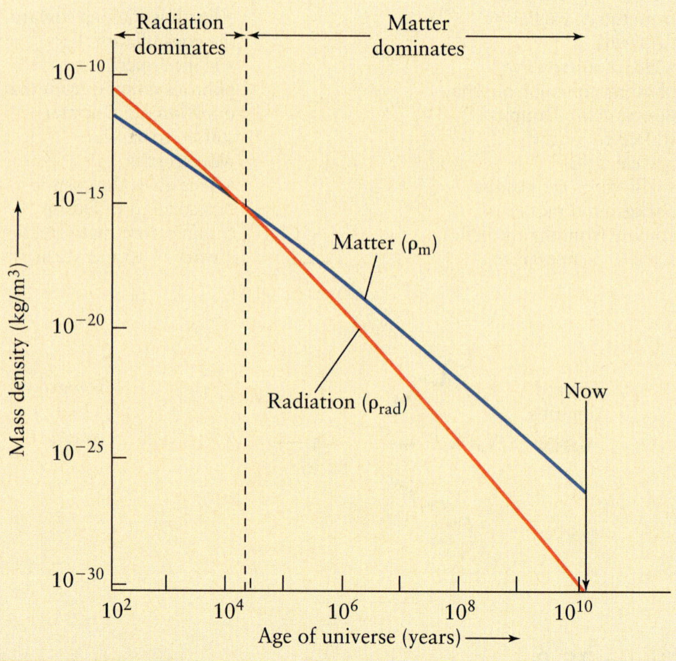

Figure 26-10

The Evolution of Density For approximately 24,000 years after the Big Bang, the mass density of radiation (ρ_{rad}, shown in red) exceeded the matter density (ρ_m, shown in blue), and the universe was radiation-dominated. Later, however, continued expansion of the universe caused ρ_{rad} to become less than ρ_m, at which point the universe became matter-dominated. (Graph courtesy of Clem Pryke, University of Chicago)

(ρ_{rad}) increases more quickly as we go back in time than does the average density of matter (ρ_m). In fact, as Figure 26-10 shows, there was a time in the ancient past when ρ_{rad} equaled ρ_m. Before this time, ρ_{rad} was greater than ρ_m, and radiation thus held sway over matter. Astronomers call this state a **radiation-dominated universe.** After ρ_m became greater than ρ_{rad}, so that matter prevailed over radiation, our universe became a **matter-dominated universe.**

This transition from a radiation-dominated universe to a matter-dominated universe occurred about 24,000 years after the Big Bang, at a time that corresponds to a redshift of about $z = 5200$. Since that time the wavelengths of photons have been stretched by a factor of $1 + z$, or about 5200. Today these microwave photons typically have wavelengths of about 1 mm. But when the universe was about 24,000 years old, they had wavelengths of about 190 nm, in the ultraviolet part of the spectrum.

The temperature of the background radiation has declined over the eons thanks to the expansion of the universe

To calculate the temperature of the cosmic background radiation at the time of this transition from a radiation-dominated universe to a matter-dominated one, we use Wien's law (see Section 5-4). This law says that the wavelength of maximum emission (λ_{max}) of a blackbody is inversely proportional to its

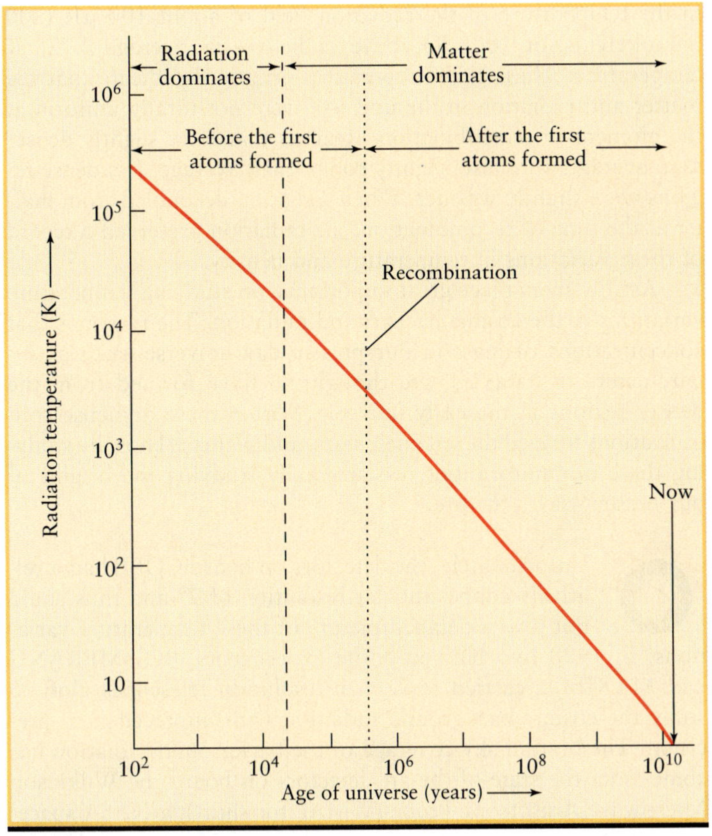

WEB LINK 26.10

Figure 26-11

The Evolution of Radiation Temperature As the universe expanded, the photons in the radiation background became increasingly redshifted and the temperature of the radiation fell. Approximately 380,000 years after the Big Bang, when the temperature fell below 3000 K, hydrogen atoms formed and the radiation field "decoupled" from the matter in the universe. After that point, the temperature of matter in the universe was not the same as the temperature of radiation. The time when the first atoms formed is called the era of recombination (see Figure 26-12). (Graph courtesy of Clem Pryke, University of Chicago)

temperature (T): A *decrease* of λ_{max} by a factor of 2 corresponds to an *increase* of T by a factor of 2.

The present-day peak wavelength of the cosmic background radiation corresponds to a blackbody temperature of 2.725 K. Hence, a peak wavelength 5200 times smaller corresponds to a temperature 5200 times greater: $T = 5200 \times 2.725 \text{ K} = 14{,}000 \text{ K}$. In other words, the radiation temperature at redshift z was greater than the present-day radiation temperature by a factor of $1 + z$. Therefore, the temperature of the radiation background was once much greater and has been declining over the ages, as Figure 26-11 shows.

When the First Atoms Formed

The nature of the universe changed again in a fundamental way about 380,000 years after the Big Bang, when z was roughly 1100 and the temperature of the radiation background was about $1100 \times 2.725 \text{ K} = 3000 \text{ K}$. To see the significance of this moment in cosmic history, recall that hydrogen is by far the most abundant element in the universe—hydrogen atoms outnumber helium atoms by about 12 to 1. A hydrogen atom consists of a single proton orbited by a single electron, and it takes relatively little energy to knock the proton and electron apart. In fact, ultraviolet radiation warmer than about 3000 K easily ionizes hydrogen. Thus, neutral hydrogen atoms could not survive in the universe that existed before $z = 1100$. That is, in the first 380,000 years after the Big Bang, the background photons had energies great enough to prevent electrons and protons from binding to form hydrogen atoms (Figure 26-12*a*). Only since $z = 1100$ (that is, since $t = 380{,}000$ years) have the energies of

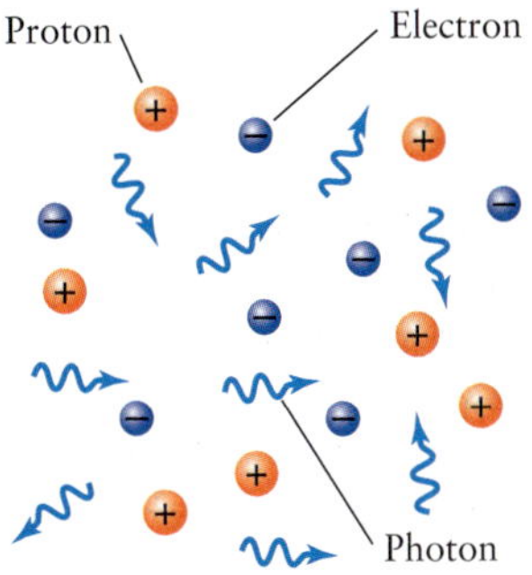

(a) Before recombination:
- Temperatures were so high that electrons and protons could not combine to form hydrogen atoms.
- The universe was opaque: Photons underwent frequent collisions with electrons.
- Matter and radiation were at the same temperature.

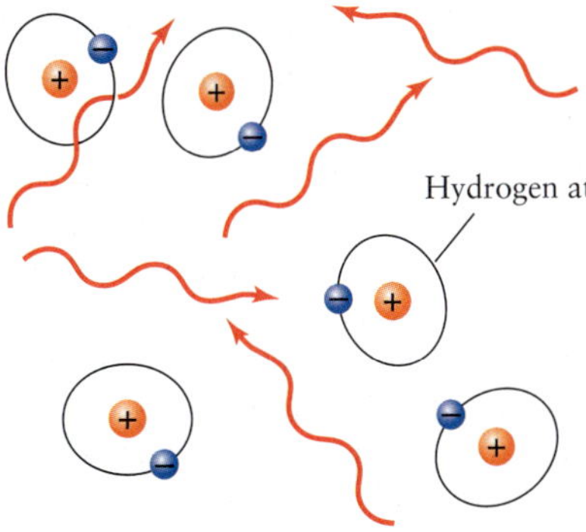

(b) After recombination:
- Temperatures became low enough for hydrogen atoms to form.
- The universe became transparent: Collisions between photons and atoms became infrequent.
- Matter and radiation were no longer at the same temperature.

Figure 26-12

The Era of Recombination **(a)** Before recombination, the energy of photons in the cosmic background was high enough to prevent protons and electrons from forming hydrogen atoms. **(b)** Some 380,000 years after the Big Bang, the energy of the cosmic background radiation became low enough that hydrogen atoms could survive.

these photons been low enough to permit hydrogen atoms to exist (Figure 26-12*b*).

The epoch when atoms first formed at $t = 380{,}000$ years is called the **era of recombination.** This name refers to electrons "recombining" to form atoms. (The name is a bit misleading, because the electrons and protons had never before combined into atoms.)

Prior to $t = 380{,}000$ years, the universe was completely filled with a shimmering expanse of high-energy photons colliding vigorously with protons and electrons. This state of matter, called a **plasma,** is opaque, just like the glowing gases inside a discharge tube (like a neon advertising sign). The surface and interior of the Sun are also a hot, glowing, opaque plasma (see Section 16-9). P. J. E. Peebles coined the term **primordial fireball** to describe the universe during its first 380,000 years of existence.

After $t = 380{,}000$ years, the photons no longer had enough energy to keep the protons and electrons apart. As soon as the temperature of the radiation field fell below about 3000 K, protons and electrons began combining to form hydrogen atoms. These atoms do not absorb low-energy photons, and so space became transparent! All the photons that heretofore had been vigorously colliding with charged particles could now stream unimpeded across space. Today, these same photons constitute the microwave background.

Before recombination, matter and the radiation field had the same temperature, because photons, electrons, and protons were all in continuous interaction with one another. After recombination, photons and atoms hardly interacted at all, and thus the temperature of matter in the universe was no longer the same as the temperature of the background radiation. Thus, $T = 2.725$ K is the temperature of the present-day background radiation field, *not* the temperature of the matter in the universe. Note that while the temperature of the background radiation is very uniform, the temperature of matter in the universe is anything but: It ranges from hundreds of millions of kelvins in the interiors of giant stars to a few tens of kelvins in the interstellar medium.

ANALOGY A good analogy is the behavior of a glass of cold water. If you hold the glass in your hand, the water will get warmer and your hand will get colder until both the water and your hand are at the same temperature. But if you set the glass down and do not touch it, so that the glass and your hand do not interact, their temperatures are decoupled: The water will stay cold and your hand will stay warm for much longer.

Because the universe was opaque prior to $t = 380{,}000$ years, we cannot see any further into the past than the era of recombination. In particular, we cannot see back to the era when the universe was radiation-dominated. The microwave background, whose photons have suffered a redshift of $z = 1100$, contains the most ancient photons we will ever be able to observe.

Nonuniformities in the Early Universe and the Origin of Galaxies

Careful analysis of COBE data showed that the cosmic background radiation is not completely isotropic. Even when the effects of Earth's motion are accounted for, there remain variations in the temperature of the radiation field of about 100 μK (100 microkelvins, or 10^{-4} K) above or below the average 2.725 K temperature. These tiny temperature variations indicate that the matter and radiation in the universe were not totally uniform at the moment of recombination. Regions that were slightly denser than average were also slightly cooler than average; less dense regions were slightly warmer. When radiation decoupled from matter at the time of recombination, the radiation preserved a record of these variations in temperature and density.

Astronomers place great importance on studying temperature variations in the cosmic background radiation. The reason is that concentrations of mass in our present-day universe, such as superclusters of galaxies, are thought to have formed from the denser regions in the early universe. Within these immense concentrations formed the galaxies, stars, and planets. Thus, by studying these nonuniformities, we are really studying the origins of our present-day structure.

WEB LINK 26.11

Unfortunately, the detectors on board COBE had a relatively coarse angular resolution of 7° and thus could not give a detailed picture of these temperature variations. In 1998 two balloon-borne experiments, BOOMERANG and MAXIMA, carried new, high-resolution telescopes aloft to study the cosmic background radiation with unprecedented precision. The best all-sky coverage of the background radiation has come from the state-of-the-art detectors on board the Wilkinson Microwave Anisotropy Probe (WMAP for short), a NASA spacecraft that was launched in 2001. Shown in Figure 26-13, the spacecraft is named for the late David Wilkinson of Princeton University, who was a pioneer in studies of the cosmic background

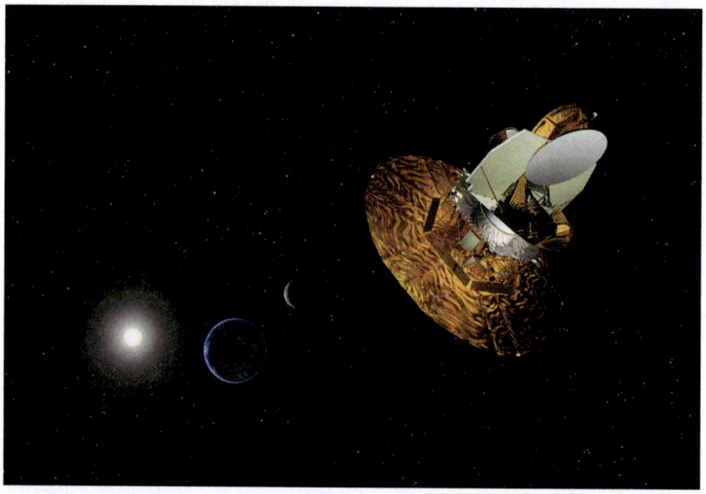

Figure 26-13

In Search of Ancient Photons This artist's impression shows the Wilkinson Microwave Anisotropy Probe (WMAP) en route to a location in space called L2, which lies about 1.5 million kilometers from Earth on the side opposite the Sun. At this position WMAP takes one year to orbit the Sun. The solar panels continually shade WMAP's detectors from sunlight, keeping them cold so that they can accurately measure the low-temperature photons ($T = 2.725$ K) of the cosmic background radiation. (NASA/WMAP Science Team)

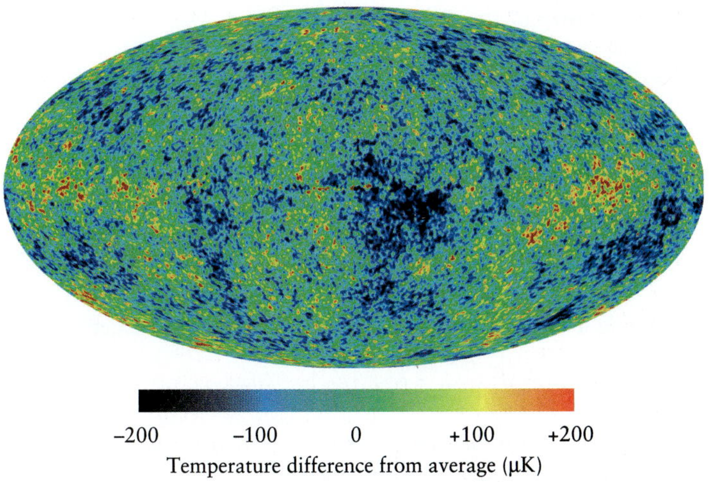

Figure 26-14 R I V U X G

Temperature Variations in the Cosmic Microwave Background This map from WMAP data shows small variations in the temperature of the cosmic background radiation across the entire sky. (The variations due to Earth's motions through space, shown in Figure 26-8, have been factored out.) Lower-temperature regions (shown in blue) show where the early universe was slightly denser than average. Note that the variations in temperature are no more than 200 μK, or 2×10^{-4} K. (NASA/WMAP Science Team)

radiation. Figure 26-14 shows a map of the sky based on data taken by the WMAP detectors. This map shows us the state of the universe when it was less than 0.003% of its present age.

Temperature variations in the cosmic background radiation do more than show us the origins of large-scale structure in the universe. As we will see in the next two sections, they actually reveal the shape of the universe as a whole.

26-6 The shape of the universe indicates its matter and energy content

We have seen that by following the mass densities of radiation (ρ_{rad}) and of matter (ρ_m), we can learn about the evolution of the universe. But it is equally important to know the combined mass density of *all* forms of matter and energy. (In an analogous way, an accountant needs to know the overall financial status of a company, not just individual profits or losses.) Remarkably, we can do this by investigating the overall shape of the universe.

The Curvature of the Universe

Einstein's general theory of relativity explains that gravity curves the fabric of space. Furthermore, the equivalence between matter and energy, expressed by Einstein's equation $E = mc^2$, tells that either matter or energy produces gravity. Thus, the matter and energy scattered across space should give the universe an overall curvature. The degree of curvature depends on the **combined average mass density** of all forms of matter and energy. This quantity, which we call ρ_0 (say "rho sub zero"), is the sum of the average mass densities of matter, radiation, and any other form of energy. Thus, by measuring the curvature of space, we should be able to determine the value of ρ_0 and, hence, learn about the content of the universe as a whole.

To see what astronomers mean by the curvature of the universe, imagine shining two powerful laser beams out into space so that they are perfectly parallel as they leave Earth. Furthermore, suppose that nothing gets in the way of these two beams, so that we can follow them for billions of light-years as they travel across the universe and across the space whose curvature we wish to detect. Figure 26-15 illustrates the only three possibilities:

1. We might find that our two beams of light remain perfectly parallel, even after traversing billions of light-years. In this case, space would not be curved: The universe would have **zero curvature,** and space would be **flat.**

2. Alternatively, we might find that our two beams of light gradually converge. In such a case, space would not be flat. Recall that lines of longitude on Earth's surface are parallel at the equator but intersect at the poles. Thus, in this case the three-dimensional geometry of the universe would be analogous to the two-dimensional geometry of a spherical surface. We would then say that space is **spherical** and that the universe has **positive curvature.** Such a universe is also called **closed,** because if you travel in a straight line in any direction in such a universe, you will eventually return to your starting point.

3. Finally, we might find that the two initially parallel beams of light would gradually diverge, becoming farther and farther apart as they moved across the universe. In this case, the universe would still have to be curved, but in the opposite sense from the spherical model. We would then say that the universe has **negative curvature.** In the same way that a sphere is a positively curved two-dimensional surface, a saddle is a good example of a negatively curved two-dimensional surface. Parallel lines drawn on a sphere always converge, but parallel lines drawn on a saddle always diverge. Mathematicians say that saddle-shaped surfaces are hyperbolic. Thus, in a negatively curved universe, we would describe space as **hyperbolic.** Such a universe is also called **open** because if you were to travel in a straight line in any direction, you would never return to your starting point.

Figure 26-15 summarizes the three cases of positive curvature, zero curvature, and negative curvature. Real space is three-dimensional, but we have drawn the three cases as analogous, more easily visualized two-dimensional surfaces. Therefore, as you examine the drawings in Figure 26-15, remember that the real universe has one more dimension. For example, if the universe is in fact hyperbolic, then the geometry of space must be the (difficult to visualize) three-dimensional analog of the two-dimensional surface of a saddle.

Note that in accordance with the cosmological principle, none of these models of the universe has an "edge" or a "center." This is clearly the case for both the flat and hyperbolic universes, because they are infinite and extend forever in all directions. A spherical universe is finite, but it also lacks a center and an edge.

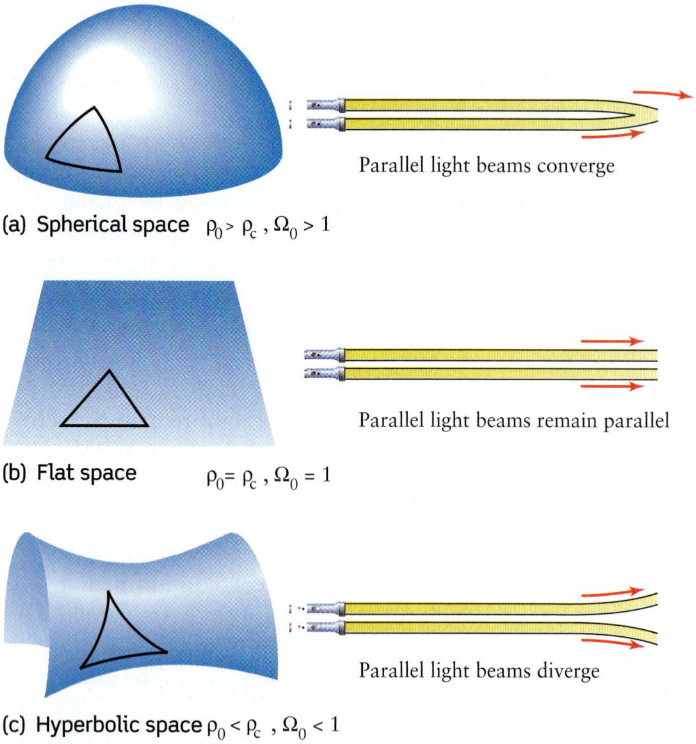

Figure 26-15

The Geometry of the Universe The curvature of the universe is either **(a)** positive, **(b)** zero, or **(c)** negative. The curvature depends on whether the combined mass density is greater than, equal to, or less than the critical density, or, equivalently, on whether the density parameter Ω_0 is greater than, equal to, or less than 1. In theory, the curvature could be determined by seeing whether two laser beams initially parallel to each other would converge, remain parallel, or spread apart.

You could walk forever around the surface of a sphere (like the surface of Earth) without finding a center or an edge.

The curvature of the universe is determined by the value of the combined mass density ρ_0. If ρ_0 is greater than a certain value ρ_c (say "rho sub cee"), the universe has positive curvature and is closed. If ρ_0 is less than ρ_c, the universe has negative curvature and is open. In the special case that ρ_0 is exactly equal to ρ_c, the universe is flat. Because of its crucial role in determining the geometry of the universe, ρ_c is called the **critical density.** It is given by the expression

Critical density of the universe

$$\rho_c = \frac{3H_0^2}{8\pi G}$$

ρ_c = critical density of the universe

H_0 = Hubble constant

G = universal constant of gravitation

Using a Hubble constant $H_0 = 73$ km/s/Mpc, we get

$$\rho_c = 1.0 \times 10^{-26}\ \text{kg/m}^3$$

A sample of hydrogen gas with this density would contain just 6 hydrogen atoms per cubic meter.

Many astronomers prefer to characterize the combined average mass density of the universe in terms of the **density parameter** Ω_0 (say "omega sub zero"). This parameter is just the ratio of the combined average mass density to the critical density:

Density parameter

$$\Omega_0 = \frac{\rho_0}{\rho_c}$$

Ω_0 = density parameter

ρ_0 = combined average mass density

ρ_c = critical density

An open universe (negative curvature) has a density parameter Ω_0 between 0 and 1, and a closed universe (positive curvature) has Ω_0 greater than 1. In a flat universe, Ω_0 is equal to 1. Thus, we can use the value of Ω_0 as a measure of the curvature of the universe (Table 26-1).

Measuring the Cosmic Curvature

How can we determine the curvature of space across the universe? In theory, if you drew an enormous triangle whose sides were each a billion light-years long (see Figure 26-15), you could determine the curvature of space by measuring the three angles of the triangle. If their sum equaled 180°, space would be flat. If the sum was greater than 180°, space would be spherical. And if the sum of the three angles was less than 180°, space would be hyperbolic. Unfortunately, this direct method for measuring the curvature of space is not practical.

Table 26-1 The Geometry and Average Density of the Universe

Geometry of space	Curvature of space	Type of universe	Combined average mass density (ρ_0)	Density parameter (Ω_0)
Spherical	positive	closed	$\rho_0 > \rho_c$	$\Omega_0 > 1$
Flat	zero	flat	$\rho_0 = \rho_c$	$\Omega_0 = 1$
Hyperbolic	negative	open	$\rho_0 < \rho_c$	$\Omega_0 < 1$

A way to determine the curvature of the universe that is both practical and precise is to see if light rays bend toward or away from each other, as shown in Figure 26-15. The greater the distance a pair of light rays has traveled, and hence, the longer the time the light has been in flight, the more pronounced any such bending should be. Therefore, astronomers test for the presence of such bending by examining the oldest radiation in the universe—the cosmic microwave background.

If the cosmic microwave background were truly isotropic, so that equal amounts of radiation reached us from all directions in the sky, it would be impossible to tell whether individual light rays have been bent. However, as we saw in Section 26-5, there are localized "hot spots" in the cosmic microwave background due to density variations in the early universe. The apparent size of these hot spots depends on the curvature of the universe (Figure 26-16). If the universe is closed, the bending of light rays from a hot spot will make the spot appear larger (Figure 26-16*a*); if the universe is open, the light rays will bend the other way and the hot spots will appear smaller (Figure 26-16*c*). Only in a flat universe will the light rays travel along straight lines, so that the hot spots appear with their true size (Figure 26-16*b*).

By calculating what conditions were like in the primordial fireball, astrophysicists find that in a flat universe, the dominant "hot spots" in the cosmic background radiation should have an angular size of about 1°. (In Section 26-9 we will learn how this is deduced.) This is just what the BOOMERANG and MAXIMA experiments observed, and what the WMAP observations have confirmed (see Figure 26-14). Hence, the curvature of the universe must be very close to zero, and the universe must either be flat or very nearly so.

As Table 26-1 shows, once we know the curvature of the universe, we can determine the density parameter Ω_0 and hence the combined average mass density ρ_0. By analyzing the data shown in Figure 26-14, astrophysicists find that $\Omega_0 = 1.0$ with an uncertainty of about 2%. In other words, ρ_0 is within 2% of the critical density ρ_c.

The flatness of the universe poses a major dilemma: Even if you include dark matter, there isn't enough matter to make the universe flat. We saw in Section 26-5 that the average mass density of matter in the universe, ρ_m, is 2.4×10^{-27} kg/m^3. This density is only 0.24 of the critical density ρ_c. We can express this ratio in terms of the **matter density parameter** Ω_m (say "omega sub em"), equal to the ratio of ρ_m to the critical density:

$$\Omega_m = \frac{\rho_m}{\rho_c} = 0.24$$

If matter and radiation were all there is in the universe, the combined average mass density ρ_0 would be equal to ρ_m (plus a tiny contribution from radiation, which we can neglect because the average mass density of radiation is only about 0.02% that of matter). Then the density parameter Ω_0 would be equal to Ω_m—that is, equal to 0.24—and the universe would be open. But the temperature variations in the cosmic microwave background clearly show that the universe is either flat or very nearly so. These variations also show that the density parameter Ω_0, which includes the effects of *all* kinds of matter and energy, is equal to 1.0. In other words, radiation and matter, including dark matter, together account for only 24% of the total density of the universe! The dilemma is to account for the rest of the density.

Dark Energy

The source of the missing density has evaded detection of the gravitational effects in the way astronomers use to detect dark matter. The missing energy density also doesn't appear to emit detectable radiation of any kind. With these properties in mind, we refer to this mysterious energy as **dark energy**.

The geometry of space reveals that the universe is filled with dark energy

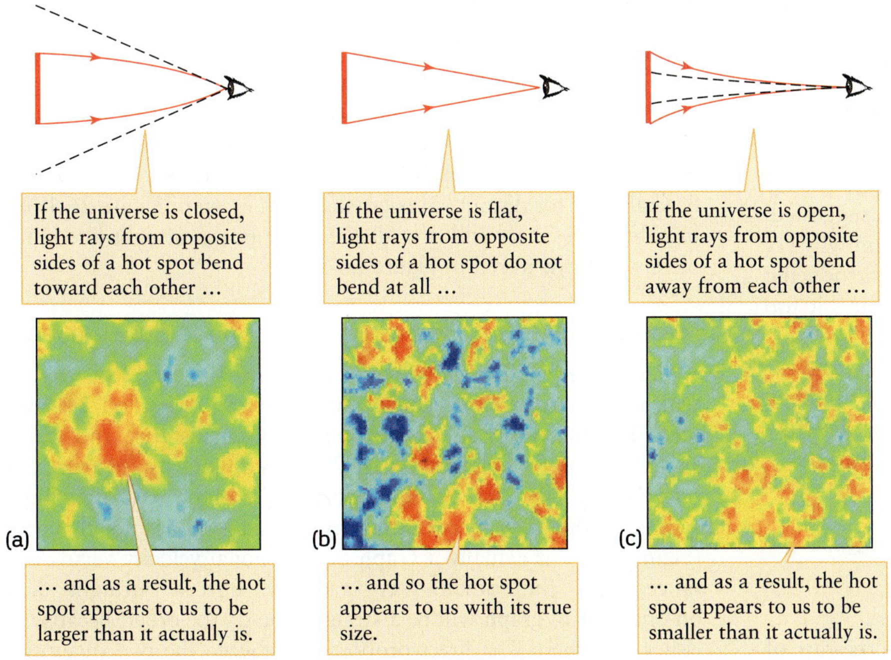
If the universe is closed, light rays from opposite sides of a hot spot bend toward each other ...
If the universe is flat, light rays from opposite sides of a hot spot do not bend at all ...
If the universe is open, light rays from opposite sides of a hot spot bend away from each other ...
(a)
(b)
(c)
... and as a result, the hot spot appears to us to be larger than it actually is.
... and so the hot spot appears to us with its true size.
... and as a result, the hot spot appears to us to be smaller than it actually is.

Figure 26-16 R I V U X G **The Cosmic Microwave Background and the Curvature of Space** Temperature variations in the early universe appear as "hot spots" in the cosmic microwave background. The apparent size of these spots depends on the curvature of space. (The BOOMERANG Group, University of California, Santa Barbara)

Just as we express the average density of matter and radiation by the matter density parameter Ω_m, we can express the average density of dark energy in terms of the **dark energy density parameter** Ω_Λ (say "omega sub lambda"). This parameter is equal to the average mass density of dark energy, ρ_Λ, divided by the critical density ρ_c:

$$\Omega_\Lambda = \frac{\rho_\Lambda}{\rho_c}$$

We can determine the value of Ω_Λ by noting that the combined average mass density ρ_0 must be the sum of the average mass densities of matter, radiation, and dark energy. As we have seen, the contribution of radiation is so small that we can ignore it, so we have

$$\rho_0 = \rho_m + \rho_\Lambda$$

If we divide this through by the critical density ρ_c, we obtain

$$\Omega_0 = \Omega_m + \Omega_\Lambda$$

That is, the density parameter Ω_0 is the sum of the matter density parameter Ω_m and the dark energy density parameter Ω_Λ. Solving for Ω_Λ, we find

$$\Omega_\Lambda = \Omega_0 - \Omega_m$$

Since Ω_0 is close to 1.0 and Ω_m is 0.24, we conclude that Ω_Λ must be $1.0 - 0.24 = 0.76$. Thus, whatever dark energy is, it accounts for 76% (about three-quarters) of the contents of the universe!

The concept of dark energy is actually due to Einstein. When he proposed the existence of a cosmological constant, he was suggesting that the universe is filled with a form of energy that by itself tends to make the universe expand (see Section 26-1). Unlike gravity, which tends to make objects attract, the energy associated with a cosmological constant would provide a form of "antigravity." Hence, it would not be detected in the same way as matter. (The subscript Λ in the symbol for the dark energy density parameter pays homage to the symbol that Einstein chose for the cosmological constant.)

If dark energy is in fact due to a cosmological constant, the value of this constant must be far larger than Einstein suggested. This change in the constant is needed if we are to explain why Ω_Λ has a large value of 0.76. If Einstein felt he erred by introducing the idea of a cosmological constant, his error was giving it too small a value!

These ideas concerning dark energy are extraordinary, and extraordinary claims require extraordinary evidence to confirm them. As we will see in the next section, a crucial test is to examine how the rate of expansion of the universe has evolved over the eons.

26-7 Observations of distant supernovae reveal that we live in an accelerating universe

We have seen that the universe is expanding. But does the rate of expansion stay the same? Because there is matter in the universe, and because gravity tends to pull the bits of matter in the universe toward one another, we would expect that the expansion should slow down with time. (In the same way, a cannonball shot upward from the surface of Earth will slow down as it ascends because of Earth's gravitational pull.) If there is a cosmological constant, however, its associated dark energy will exert an outward pressure that tends to accelerate the expansion. Which of these effects is more important?

Modeling the Expansion History of the Universe

To determine if the expansion of the universe is slowing down or speeding up, astronomers study the relationship between redshift and distance for extremely remote galaxies. We see these galaxies as they were billions of years ago. If the rate of expansion was the same in the distant past as it is now, the same Hubble law should apply to distant galaxies as to nearby ones. But if the rate of expansion has either increased or decreased, we will find important deviations from the Hubble law.

To see how astronomers approach this problem, first imagine two different parallel universes. Both Universe #1 and Universe #2 are expanding at constant rates, so for both universes there is a direct proportion between recessional velocity v and distance d as expressed by the Hubble law $v = H_0 d$. Hence, a graph of distance versus recessional velocity for either universe is a straight line, as Figure 26-17*a* shows. The only difference is that Universe #1 is expanding at a slower rate than Universe #2. Hence, a galaxy at a certain distance from Earth in Universe #1 will have a slower recessional velocity than a galaxy at the same distance from Earth in Universe #2. As a result, the graph of distance versus recessional velocity for slowly expanding Universe #1 (shown in blue) has a steeper slope than the graph for rapidly expanding Universe #2 (shown in green). Keep this observation in mind: A slower expansion means a steeper slope on a graph of distance versus recessional velocity.

Now consider *our* universe and allow for the possibility that the expansion rate may change over time. If we observe very remote galaxies, we are seeing them as they were in the remote past. If the expansion of the universe in the remote past was slower or faster than it is now, the slope of the graph of distance versus recessional velocity will be different for those remote galaxies. If the expansion was slower, then the slope will be steeper for distant galaxies (shown in blue in Figure 26-17*b*); if the expansion was faster, the slope will be shallower for distant galaxies (shown in green in Figure 26-17*b*). In either case, there will be a deviation from the straight-line Hubble law (shown in red in Figure 26-17*b*).

Measuring Ancient Expansion with Type Ia Supernovae

Which of the possibilities shown in Figure 26-17*b* represents the actual history of the expansion of our universe? In Section 24-5 we looked at the observed relationship between distance and recessional velocity for galaxies. Figure 24-17 is a plot of some representative data. The data points appear to lie along a straight line, suggesting that the rate of cosmological expansion has not changed. (Figure 24-17 is actually a graph of recessional velocity versus distance, not the other way around. But a straight line on one kind of graph will be a straight line on the other, because in either case there is a direct proportion between the two quantities being graphed.) However, the graph in Figure 24-17 was based on

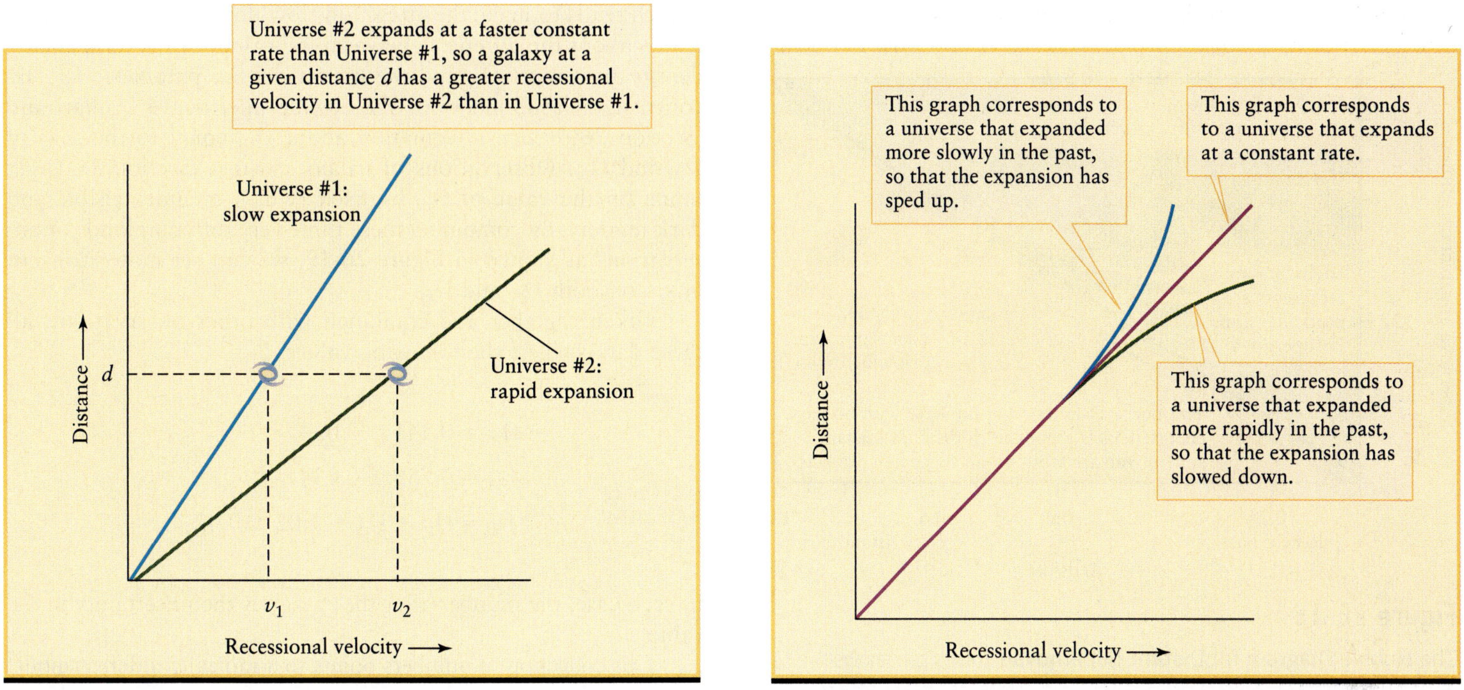
Universe #2 expands at a faster constant rate than Universe #1, so a galaxy at a given distance d has a greater recessional velocity in Universe #2 than in Universe #1.
Universe #1: slow expansion
Universe #2: rapid expansion
Distance
d
v_1
v_2
Recessional velocity
This graph corresponds to a universe that expanded more slowly in the past, so that the expansion has sped up.
This graph corresponds to a universe that expands at a constant rate.
This graph corresponds to a universe that expanded more rapidly in the past, so that the expansion has slowed down.
Distance
Recessional velocity

(a) Two universes with different expansion rates

(b) Possible expansion histories of the universe

Figure 26-17

Varying Rates of Cosmic Expansion **(a)** Imagine two universes, #1 and #2. Each expands at its own constant rate. For a galaxy at a given distance, the recessional velocity will be greater in the more rapidly expanding universe. Hence, the graph of distance *d* versus recessional velocity *v* will have a shallower slope for the rapidly expanding universe and a steeper slope for the slowly expanding one. **(b)** If the rate of expansion of our universe was more rapid in the distant past, corresponding to remote distances, the graph of *d* versus *v* will have a shallower slope for large distances (green curve). If the expansion rate was slower in the distant past, the graph will have a steeper slope for large distances (blue curve).

measurements of galaxies no farther than 400 Mpc (1.3 billion ly) from Earth, which means we are looking only 1.3 billion years into the past. The straightness of the line in Figure 24-17 means only that the expansion of the universe has been relatively constant over the past 1.3 billion years—only 10% of the age of the universe, and a relatively brief interval on the cosmic scale.

Now suppose that you were to measure the redshifts and distances of galaxies *several* billion light-years from Earth. The light from these galaxies has taken billions of years to arrive at your telescope, so your measurements will reveal how fast the universe was expanding billions of years ago. To determine the expansion, we need a technique that will allow us to find the distances to these very remote galaxies. We saw in Section 24-4 that one way to determine distances is to identify Type Ia supernovae in such galaxies. These supernovae are among the most luminous objects in the universe, and hence can be detected even at extremely large distances (see Figure 24-14). The maximum brightness of a supernova tells astronomers its distance through the inverse-square law for light, and the redshift of the supernova's spectrum tells them its recessional velocity. As an example, the image that opens this chapter shows Type Ia supernovae with redshifts $z = 1.010$, 1.230, and 1.390, corresponding to recessional velocities of 60%, 67%, and 70% of the speed of light. We see these supernovae as they were 7.7 to 9.0 billion years ago, when the universe was less than half of its present age.

WEB LINK 26.13 In 1998, two large research groups—the Supernova Cosmology Project, led by Saul Perlmutter of Lawrence Berkeley National Laboratory, and the High-Z Supernova Search Team, led by Brian Schmidt of the Mount Stromlo and Siding Springs Observatories in Australia—reported their results from a survey of Type Ia supernovae in galaxies at redshifts of 0.2 or greater, corresponding to distances beyond 750 Mpc (2.4 billion ly). Figure 26-18 shows some of their data, along with more recent observations, on a graph of apparent magnitude versus redshift. Recall that a greater apparent magnitude corresponds to a dimmer supernova (see Section 17-3), which means that the supernova is more distant. A greater redshift implies a greater recessional velocity. Hence, this graph is basically the same as those in Figure 26-17 (distance versus recessional velocity).

To interpret these results, we need guidance from **relativistic cosmology.** This field provides is a theoretical description of the universe and its expansion, based on Einstein's general theory of relativity and was developed in the 1920s by Alexander Friedmann in Russia, Georges Lemaître in France, and Willem de Sitter in the Netherlands. Given values of the mass density parameter Ω_m and the dark energy density parameter Ω_Λ, cosmologists can use the equations of relativistic cosmology to predict how the expansion rate of the universe should change over time. Such predictions can be expressed as curves on a graph of distance versus redshift such as Figure 26-18.

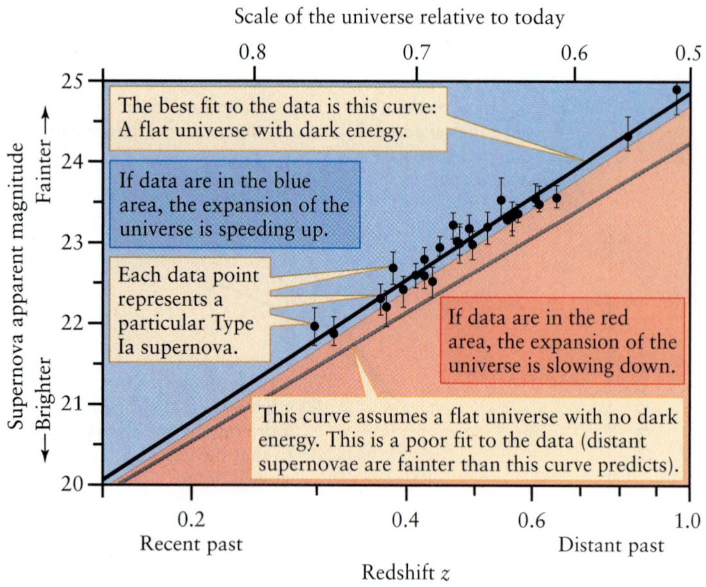

Figure 26-18

The Hubble Diagram for Distant Supernovae This graph shows apparent magnitude versus redshift for supernovae in distant galaxies. The greater the apparent magnitude, the dimmer the supernova and the greater the distance to it and its host galaxy. If the expansion of the universe is speeding up, the data will lie in the blue area; if it is slowing down, the data will lie in the red area. The data show that the expansion is in fact speeding up. (The Supernova Cosmology Project/R. A. Knop et al.)

The lower, gray curve in Figure 26-18 shows what would be expected in a flat universe with $\Omega_0 = 1.00$ but with no dark energy, so that $\Omega_\Lambda = 0$ and $\Omega_m = \Omega_0$ (that is, the density parameter is due to matter and radiation alone). In this model, and in fact in any model whose curve lies in the red area in Figure 26-18, the absence of dark energy means that gravitational attraction between galaxies would cause the expansion of the universe to slow down with time. Hence, the expansion rate would have been greater in the past (compare with the green curve in Figure 26-17*b*).

In fact, the data points in Figure 26-18 are almost all in the *blue* region of the graph, and agree very well with the curve shown in black. This curve also assumes a flat universe, but with an amount of dark energy consistent with the results from the cosmic microwave background ($\Omega_m = 0.24$, $\Omega_\Lambda = 0.76$, $\Omega_0 = \Omega_m + \Omega_\Lambda = 1.00$). In this model, and indeed in any model whose curve lies in the blue region of Figure 26-18, dark energy has made the expansion of the universe speed up over time. Hence, the expansion of the universe was slower in the distant past, which means that we live in an *accelerating* universe.

Just like the blue curve in Figure 26-17*b*, the data in Figure 26-18 show that supernovae of a certain brightness (and hence a given distance) have smaller redshifts (and hence smaller recessional velocities) than would be the case if the expansion rate had always been the same. These data provide compelling evidence of the existence of dark energy.

Roughly speaking, the data in Figure 26-18 indicate the relative importance of dark energy (which tends to make the expansion speed up) and gravitational attraction between galaxies (which tends to make the attraction slow down). Thus, these data tell us about the *difference* between the values of the dark energy density parameter Ω_Λ and the matter density parameter Ω_m. By contrast, measurements of the cosmic microwave background (Section 26-6) give information about Ω_0, equal to the *sum* of Ω_Λ and Ω_m. Observations of galaxy clusters (Section 26-5) set limits on the value of Ω_m by itself (which includes visible and dark matter). By combining these three very different kinds of observations as shown in Figure 26-19, we can set more stringent limits on both Ω_Λ and Ω_m.

Taken together and combined with other observations, all these data suggest the following values.

$$\Omega_m = 0.241 \pm 0.034$$

$$\Omega_\Lambda = 0.759 \pm 0.034$$

$$\Omega_0 = \Omega_m + \Omega_\Lambda = 1.02 \pm 0.02$$

In each case, the number after the ± sign is the uncertainty in the value.

This collection of numbers points to a radically different model of the universe from what was suspected just a few years ago. In the 1980s there was no compelling evidence for an accelerating expansion of the universe, so it was widely assumed that $\Omega_\Lambda = 0$. Evidence from distant galaxies suggested a flat universe, so it was presumed that $\Omega_m = 1$. Figure 26-19 shows that modern data rule out this model.

Dark energy became the dominant form of energy in the universe about the same time that our solar system formed

The model we are left with is one in which the universe is suffused with a curious dark energy due to a cosmological constant. Unlike matter or radiation, whose average densities decrease as the universe expands and thins out, the average density of this dark energy remains constant throughout the history of the universe (Figure 26-20). The dark energy was relatively unimportant over most of the early history of the universe. Today, however, the density of dark energy is greater than that of matter (Ω_Λ is greater than Ω_m). In other words, we live in a **dark-energy-dominated universe.**

Cosmic Expansion: From Slowing Down to Speeding Up

ANIMATION 26.4 ANIMATION 26.5

As Figure 26-20 shows, the dominance of dark energy is a relatively recent development in the history of the universe. Prior to about 5 billion years ago, the density of matter should have been greater than that of dark energy. Hence, we would expect that up until about 5 billion years ago, the expansion of the universe should have been slowing down rather than speeding up. Recently, astronomers have found evidence of this picture by using the Hubble Space Telescope to observe extremely distant Type Ia supernovae with redshifts z greater than 1. (The image that opens this chapter shows three of these supernovae.) When astronomers compare the distance to these supernovae (determined from their brightness) to their redshifts, they find that the redshifts are *greater* than would be the case if the expansion of the universe had always been at the same rate or had always been speeding up

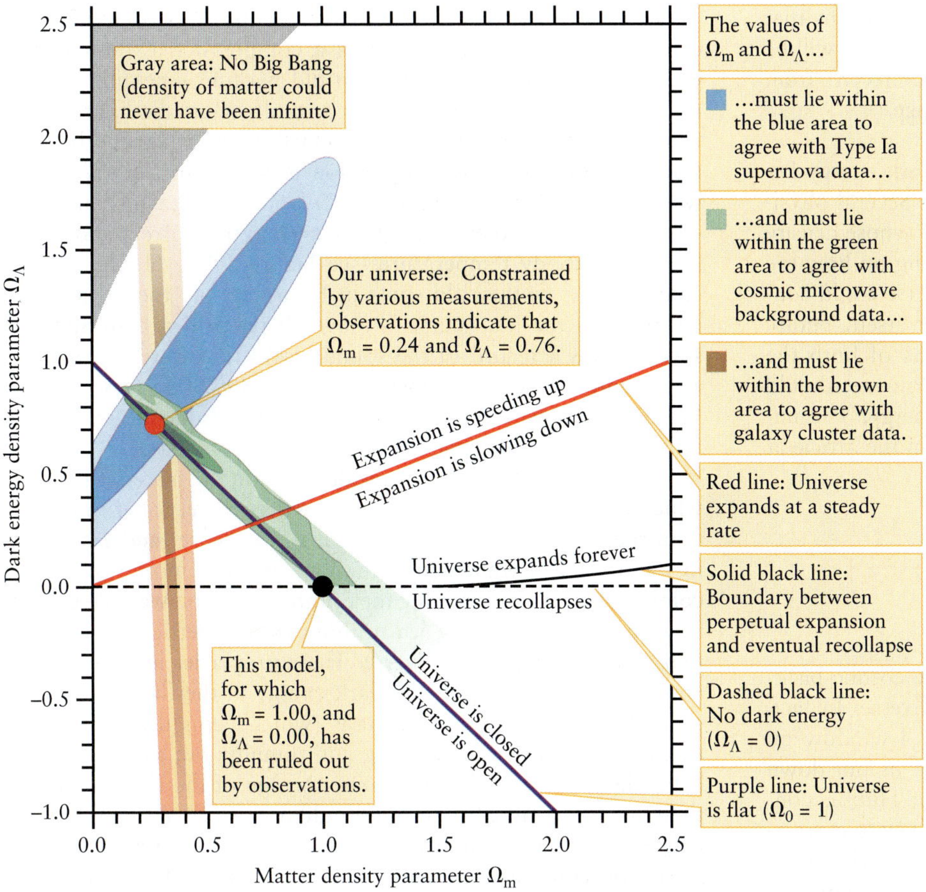

INTERACTIVE EXERCISE 26.2

Figure 26-19

Limits on the Nature of the Universe The three regions on this graph show values of the mass density parameter Ω_m and the dark energy density parameter Ω_Λ that are consistent with various types of observations. Galaxy cluster measurements (in brown) set limits on Ω_m. Observations of the cosmic microwave background (in green) set limits on the sum of Ω_m and Ω_Λ: A larger value of Ω_m (to the right in the graph) implies a smaller value of Ω_Λ (downward in the graph) to keep the sum the same, which is why this band slopes downward. Observations of Type Ia supernovae (in blue) set limits on the difference between Ω_m and Ω_Λ; this band slopes upward since a larger value of Ω_m implies a larger value of Ω_Λ to keep the difference the same. The best agreement to all these observations is where all three regions overlap (the red dot). (The Supernova Cosmology Project/R. A. Knop et al.)

(see Figure 26-17). This is just what would be expected if the expansion was slowing down in the very early universe. After about 5 billion years ago, the effects of dark energy became dominant and the expansion began to speed up.

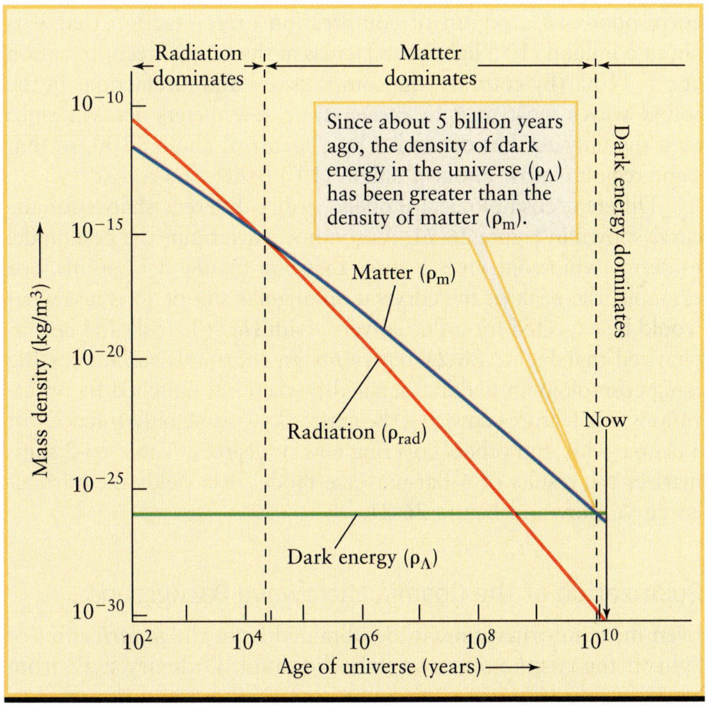

ANALOGY If you see a red light up ahead while driving, you would probably apply the brakes to make the car slow down. But if the light then turns green before your car comes to a stop, and the road ahead is clear, you would step on the gas to make the car speed up again. The expansion of the universe has had a similar history. The mutual gravitational attraction of all the matter in the universe means that "the brakes were on" for about the first 9 billion years after the Big Bang, so that the expansion slowed down. But for about the past 5 billion years, dark energy has "had its foot on the gas," and the expansion has been speeding up.

If dark energy truly is a cosmological constant, the density of dark energy will continue to remain constant, as shown in Figure 26-20. Due to the effects of this dark energy, the universe will keep on expanding forever, and the rate of expansion will continue to accelerate. Eventually, some 30 billion years from now, the universe will have expanded so much that only a thousand or so of the nearest galaxies will still be visible. The billions of other galaxies that

Figure 26-20

The Evolution of Density, Revisited The average mass density of matter, ρ_m, and the average mass density of radiation, ρ_{rad}, both decrease as the universe expands and becomes more tenuous. But if the dark energy is due to a cosmological constant, its average mass density ρ_Λ remains constant. In this model, our universe became dominated by dark energy about 5 billion years ago.

we can observe today will have moved so far away from us that their light will have faded to invisibility. Furthermore, they will be moving away from us so rapidly that what light we do receive from them will have been redshifted out of the visible range.

WEB LINK 26.14 There may be other explanations for dark energy besides a cosmological constant, however. Several physicists have proposed a type of dark energy whose density decreases slowly as the universe expands. Depending on how the density of dark energy evolves over time, the universe could continue to expand or could eventually recollapse on itself. Future observations, including space-based measurements of both the cosmic background radiation and of Type Ia supernovae, should help resolve the nature of the mysterious dark energy. The unknown nature of dark matter is discussed further in the article at the end of this chapter.

26-8 Primordial sound waves help reveal the character of the universe

We have seen how studying the "hot spots" in the cosmic background radiation reveals that we live in a flat universe. In fact, temperature variations reveal more: They give us a window on conditions in the early universe, and actually help us pin down the values of other important quantities such as the Hubble constant and the density of matter in the universe. The key to extracting this additional information from the cosmic background radiation is recognizing that the hot and cold spots in a map such as Figure 26-14 actually result from sound waves.

Sound Waves in the Early Universe

Sound waves can travel in fluids of all kinds. Sound waves in air are used in human speech and hearing (air is considered a fluid), while whales communicate using high-frequency underwater clicks and whistles. If you could take a snapshot of a sound wave, you would see that at any moment there are some regions, called **compressions,** where the fluid is squeezed together, and other regions, called **rarefactions,** where the fluid is thinned out or rarefied. (When a sound wave enters a human ear, the air next to the eardrum is alternately compressed and rarefied, which makes the eardrum flex back and forth. The ear detects this flexing and translates it into an electrical signal that is sent to the brain.)

Immense sound waves in the early universe left their imprint as variations in the cosmic microwave background

WEB LINK 26.15 There would also have been sound waves in the early universe before recombination. During the first 380,000 years after the Big Bang, the universe was filled with a fluid composed primarily of photons, electrons, and protons, with a density more than 10^9 times greater than that of our present-day universe. Just as water molecules in a glass of water collide with each other, photons and particles collided frequently with each other in this primordial fluid, triggering random sound waves with compressions and rarefactions.

Because there was more mass in a compression than in a rarefaction, photons emerging from a compression experienced a greater gravitational redshift than did photons emerging from a rarefaction (see Section 22-2). As a result, while the light we see from either a compression or a rarefaction has a blackbody spectrum, the light from a compression is shifted to slightly longer wavelengths. We saw in Sections 5-3 and 5-4 that a blackbody spectrum dominated by longer wavelengths corresponds to a lower blackbody temperature (see Figure 5-11). Hence, we see compressions as the cold spots in Figure 26-14, and we see rarefactions as the hot spots. The overall pattern of cold and hot spots is thus a record of the sound waves that were present just as the universe became transparent, some 380,000 years after the Big Bang. As we discussed in Section 26-5, from the cold compressions arose our present-day population of galaxies.

The nature of a sound wave depends on the material through which it passes. For example, sound waves travel faster in helium than they do in air (because helium is less dense) and faster still in water (which, while denser than air, is much more resistant to compression). So by studying the primordial sound waves recorded in Figure 26-14, we can learn about the properties of the fluid that made up the early universe. These properties include the average densities of matter and dark energy in the fluid, as well as the value of the Hubble constant (which helps determine how rapidly the fluid was expanding and thinning out as the universe expanded). We can also determine the age of the universe at the time that the cosmic background radiation was emitted, since this determines the maximum size to which a hot spot (rarefaction) or cold spot (compression) could have grown since the Big Bang.

Figure 26-21 shows an important way in which astronomers systematize their data about hot and cold spots in the cosmic background radiation. This graph shows the number of observed hot or cold spots of different angular sizes, with larger spots on the left and smaller spots on the right. The presence of peaks in the graph shows that spots of certain sizes are more common than others. The largest peak tells us that the predominant angular size is about 1°, which corresponds to a region of compression or rarefaction that was about a million (10^6) light-years across at the time of recombination at $z = 1100$. (By contrast, the compressions and rarefactions in the sound waves most used in speech are a few meters across.) Since then the universe has expanded by a factor of about 1100, so that same region is now about a billion (10^9) light-years across.

Different cosmological models predict different shapes for the curve shown in Figure 26-21. Astronomers determine the best model by seeing which one gives a curve that best fits the data points. For example, the peak of the curve at an angular size of 1° is just what would be expected for a flat universe with $\Omega_0 = 1$. It should be emphasized that Figure 26-21 represents an enormous success for Big Bang cosmology in making detailed predictions, matched by observations, for features covering the entire sky—some only fractions of a degree wide and others covering tens of degrees. Table 26-2 summarizes the results of a flat-universe model that yields the particular curve shown in Figure 26-21.

Polarization of the Cosmic Microwave Background

Even more information can be obtained from the *polarization* of light in the cosmic microwave background. Ordinary light from

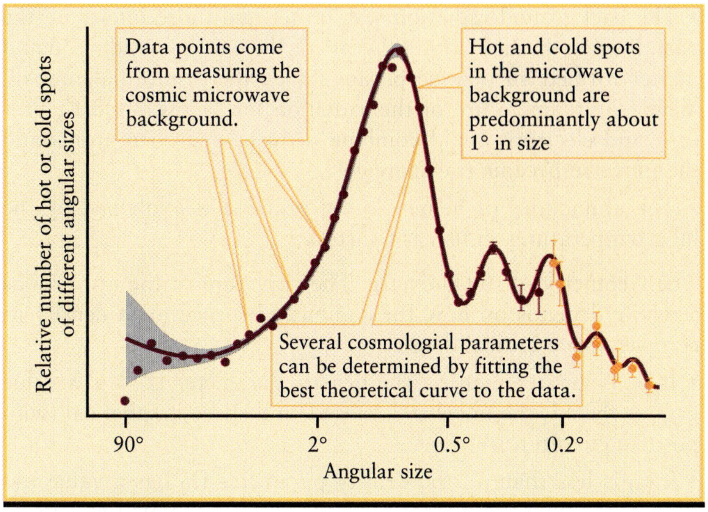

Figure 26-21

Sound Waves in the Early Universe Observations of the cosmic background radiation show that hot and cold spots of certain angular sizes are more common than others. A model that describes these observations helps to constrain the values of important cosmological parameters. Most of the data shown here is from the Wilkinson Microwave Anisotropy Probe; the data for the smallest angles (at the right of the graph) come from the CBI detector in the Chilean Andes and the ACBAR detector at the South Pole. (NASA/WMAP Science Team)

a lightbulb or from the Sun is *unpolarized,* which means that the electric fields of the light waves are oriented in random directions. But when light collides with and bounces off an object, it tends to become *polarized,* with its electric fields oriented in a specific direction. (As an example, the electric fields in sunlight reflected from the ground are mostly oriented horizontally. Polarizing sunglasses work by screening out electric fields of this orientation only, which helps eliminate distracting reflections.) In a similar way, the cosmic background radiation acquires a polarization when it scatters from material in a large hot spot. The amount of polarization turns out to be a very sensitive probe of conditions in the early universe, and, hence, of the nature of the universe itself.

In 2002 astronomers using the DASI microwave telescope at the South Pole reported the first detection of polarized light in the cosmic background radiation. Measurements made with the Wilkinson Microwave Anisotropy Probe have since provided additional information about polarization. In 2009, the Planck satellite began collecting data which will be ten times as sensitive as the Wilkinson Microwave Anisotropy Probe. The Planck satellite will also provide the first detailed polarization studies of the cosmic background radiation. Such measurements are very challenging to do. But the rewards for cosmology are very great, which is why researchers are devoting great effort to further studies of the polarization of the background radiation.

Our understanding of the universe as a whole has increased tremendously over the past several years. We have found compelling evidence that dark energy exists, and that it is the dominant form of energy in the universe. Studies of supernovae, galaxy clusters, and the cosmic background radiation have provided us with so much high-quality data that we can now express the key parameters of the universe (Table 26-2) with very high accuracy. When we look back to the situation in the 1980s, when the value of the Hubble constant was uncertain by at least 50%, it is no exaggeration to say that we have entered an age of *precision* cosmology.

Yet many questions remain unanswered. What is the nature of dark matter? What actually is dark energy? Can these mysterious entities be detected and studied in the laboratory? These and other questions will continue to occupy cosmologists for many years to come.

WEB LINK 26.16

Table 26-2 Some Key Properties of the Universe

Quantity	Significance	Value*
Hubble constant, H_0	Present-day expansion rate of the universe	$73.2^{+3.1}_{-3.2}$ km/s/Mpc
Density parameter, Ω_0	Combined mass density of all forms of matter *and* energy in the universe, divided by the critical density	1.02 ± 0.02
Matter density parameter, Ω_m	Combined mass density of all forms of matter in the universe, divided by the critical density	0.241 ± 0.034
Density parameter for ordinary matter, Ω_b	Mass density of ordinary atomic matter in the universe, divided by the critical density	0.0416 ± 0.001
Dark energy density parameter, Ω_Λ	Mass density of dark energy in the universe, divided by the critical density	0.759 ± 0.034
Age of the universe, T_0	Elapsed time from the Big Bang to the present day	$(1.373^{+0.016}_{-0.015}) \times 10^{10}$ years
Age of the universe at the time of recombination	Elapsed time from the Big Bang to when the universe became transparent, releasing the cosmic background radiation	$(3.79^{+0.08}_{-0.07}) \times 10^5$ years
Redshift z at the time of recombination	Since the cosmic background radiation was released, the universe has expanded by a factor $1 + z$	1089 ± 1

**Values for H_0, Ω_m, Ω_b, Ω_Λ, and T_0 are based on the three-year WMAP data with the assumption of a flat universe. Values for the time and redshift of recombination and for Ω_0 are from the first-year WMAP data. (NASA/WMAP Science Team)*

Key Words

average density of matter, p. 701
Big Bang, p. 697
closed universe, p. 705
combined average mass density, p. 705
compression, p. 712
cosmic background radiation, p. 699
cosmic microwave background, p. 699
cosmic light horizon, p. 698
cosmological constant, p. 692
cosmological principle, p. 697
cosmological redshift, p. 694
cosmology, p. 691
critical density, p. 706
dark energy, p. 707
dark energy density parameter, p. 708
dark-energy-dominated universe, p. 710
density parameter, p. 706
era of recombination, p. 704
flat space, p. 705
homogeneous, p. 696
hyperbolic space, p. 705
isotropic, p. 696
lookback time, p. 696
mass density of radiation, p. 701
matter density parameter, p. 707
matter-dominated universe, p. 702
negative curvature, p. 705
observable universe, p. 698
Olbers's paradox, p. 692
open universe, p. 705
Planck time, p. 698
plasma, p. 704
positive curvature, p. 705
primordial fireball, p. 704
radiation-dominated universe, p. 702
rarefaction, p. 712
relativistic cosmology, p. 709
spherical space, p. 705
zero curvature, p. 705

Key Ideas

The Expansion of the Universe: The Hubble law describes the continuing expansion of space. The redshifts that we see from distant galaxies are caused by this expansion, not by the motions of galaxies through space.

• The redshift of a distant galaxy is a measure of the scale of the universe at the time the galaxy emitted its light.

• It is meaningless to speak of an edge or center to the universe or of what lies beyond the universe.

The Cosmological Principle: Cosmological theories are based on the idea that on large scales, the universe looks the same at all locations and in every direction.

The Big Bang: The universe began as an infinitely dense cosmic singularity that began its expansion in the event called the Big Bang, which can be described as the beginning of time.

• The observable universe extends about 14 billion light-years in every direction from Earth. We cannot see objects beyond this distance because light from these objects has not had enough time to reach us.

• During the first 10^{-43} second after the Big Bang, the universe was too dense to be described by the known laws of physics.

Cosmic Background Radiation and the Evolution of the Universe: The cosmic microwave background radiation, corresponding to radiation from a blackbody at a temperature of nearly 3 K, is the greatly redshifted remnant of the hot universe as it existed about 380,000 years after the Big Bang.

• The background radiation was hotter and more intense in the past. During the first 380,000 years of the universe, radiation and matter formed an opaque plasma called the primordial fireball. When the temperature of the radiation fell below 3000 K, protons and electrons could combine to form hydrogen atoms and the universe became transparent.

• The abundance of helium in the universe is explained by the high temperatures in its early history.

The Geometry of the Universe: The curvature of the universe as a whole depends on how the combined average mass density ρ_0 compares to a critical density ρ_c.

• If ρ_0 is greater than ρ_c, the density parameter Ω_0 has a value greater than 1, the universe is closed, and space is spherical (with positive curvature).

• If ρ_0 is less than ρ_c, the density parameter Ω_0 has a value less than 1, the universe is open, and space is hyperbolic (with negative curvature).

• If ρ_0 is equal to ρ_c, the density parameter Ω_0 is equal to 1 and space is flat (with zero curvature).

Cosmological Parameters and Dark Energy: Observations of temperature variations in the cosmic microwave background indicate that the universe is flat or nearly so, with a combined average mass density equal to the critical density. Observations of galaxy clusters suggest that the average density of matter in the universe is about 0.24 of the critical density. The remaining contribution to the average density is called dark energy.

• Measurements of Type Ia supernovae in distant galaxies show that the expansion of the universe is speeding up. This may be due to the presence of dark energy in the form of a cosmological constant, which provides a pressure that pushes the universe outward.

Cosmological Parameters and Primordial Sound Waves: Temperature variations in the cosmic background radiation are a record of sound waves in the early universe. Studying the character of these sound waves, and the polarization of the background radiation that they produce, helps constrain models of the universe.

Questions

Review Questions

1. Why did Isaac Newton conclude that the universe was static? Was he correct?
2. What is Olbers's paradox? How can it be resolved?
3. What is a cosmological constant? Why did Einstein introduce it into cosmology?
4. TUTORIAL 26.1 What does it mean when astronomers say that we live in an expanding universe? What is actually expanding?
5. Describe how the expansion of the universe explains Hubble's law.
6. Would it be correct to say that due to the expansion of the universe, Earth is larger today than it was 4.56 billion years ago? Why or why not?
7. Using a diagram, explain why the expansion of the universe as seen from a distant galaxy would look the same as seen from our Galaxy.

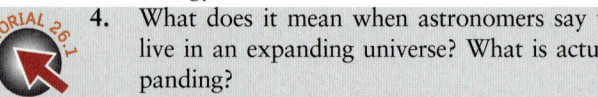

8. How does modern cosmology preclude the possibility of either a center or an edge to the universe?
9. Explain the difference between a Doppler shift and a cosmological redshift.
10. Explain how redshift can be used as a measure of lookback time. In what ways is it superior to time measured in years?
11. By what factor has the universe expanded since $z = 1$? Explain your reasoning.
12. What does it mean to say that the universe is homogeneous? That it is isotropic?
13. What is the cosmological principle? How is it justified?
14. How was the Big Bang different from an ordinary explosion? Where in the universe did it occur?
15. Some people refer to the Hubble constant as "the Hubble variable." In what sense is this justified?
16. What is meant by "the observable universe"?
17. (a) Explain why the radius of the observable universe is continually increasing. (b) Although the universe is 13.7 billion years old, the observable universe includes objects that are more than 13.7 billion light-years away from Earth. Explain why.
18. Imagine an astronomer living in a galaxy a billion light-years away. Is the observable universe for that astronomer the same as for an astronomer on Earth? Why or why not?

TUTORIAL 26.2

19. How did the abundance of helium in the universe suggest the existence of the cosmic background radiation?
20. Can you see the cosmic background radiation with the naked eye? With a visible-light telescope? Explain why or why not.
21. If the universe continues to expand forever, what will eventually become of the cosmic background radiation?
22. How can astronomers measure the average mass density of the universe?
23. What does it mean to say that the universe was once radiation-dominated? What happened when the universe changed from being radiation-dominated to being matter-dominated? When did this happen?
24. What was the era of recombination? What significant events occurred in the universe during this era? Was the universe matter-dominated or radiation-dominated during this era?
25. (a) Was there ever an era when the universe was radiation-dominated and matter and radiation were at the same temperature? If so, approximately when was this, and were there atoms during that era? If not, explain why not. (b) Was there ever an era when the universe was radiation-dominated and matter and radiation were *not* at the same temperature? If so, approximately when was this, and were there atoms during that era? If not, explain why not.
26. Describe two different ways in which the cosmic microwave background is not isotropic.
27. What is meant by the critical density of the universe? Why is this quantity important to cosmologists?
28. Describe how astronomers use the cosmic background radiation to determine the geometry of the universe.
29. Explain why it is important to measure how the expansion rate of the universe has changed over time. How is this rate measured?
30. What is dark energy? Describe two ways that we can infer its presence.
31. What does it mean to say that the universe is dark-energy–dominated? What happened when the universe changed from being matter-dominated to being dark-energy–dominated?
32. How can we detect the presence of sound waves in the early universe? What do these sound waves tell us?

Advanced Questions

> **Problem-solving tips and tools**
>
> We discussed Wien's law in Section 5-4. You may find it useful to recall that 1 parsec equals 3.26 light-years, that 1 Mpc equals 3.09×10^{19} km, and that a year contains 3.16×10^{7} seconds.

33. (a) For what value of the redshift z were representative distances between galaxies only 20% as large as they are now? (b) Compared to representative distances between galaxies in the present-day universe, how large were such distances at $z = 8$? Compared to the density of matter in the present-day universe, what was the density of matter at $z = 8$? (c) If dark energy is in the form of a cosmological constant, how does its present-day density compare to the density of dark energy at $z = 2$? At $z = 5$? Explain your answers.
34. The host galaxy of the supernova HST04Sas (see the image that opens this chapter) has a redshift $z = 1.390$. The light from this galaxy includes the Lyman-alpha (L_α) spectral line of hydrogen, with an unshifted wavelength of 121.6 nm. Calculate the wavelength at which we detect the Lyman-alpha photons from this galaxy. In what part of the electromagnetic spectrum does this wavelength lie?
35. Estimate the age of the universe for a Hubble constant of (a) 50 km/s/Mpc, (b) 75 km/s/Mpc, and (c) 100 km/s/Mpc. On the basis of your answers, explain how the ages of globular clusters could be used to place a limit on the maximum value of the Hubble constant.
36. Some so-called "creation scientists" claim that the universe came into being about 6000 years ago. Find the value of the Hubble constant for such a universe. Is this a reasonable value for H_0? Explain your answer.
37. The quasar HS 1946+7658 has a redshift $z = 3.02$. At the time when the light we see from HS 1946+7658 left the quasar, how many times more dense was the matter in the universe than it is today?
38. Use Wien's law (Section 5-4) to calculate the wavelength at which the cosmic microwave background ($T = 2.725$ K) is most intense.
39. If the mass density of radiation in the universe were 625 times larger than it is now, what would the background temperature be?
40. Suppose that the present-day temperature of the cosmic background radiation were somehow increased by a factor of 100, from 2.725 K to 272.5 K. (a) Calculate ρ_{rad} in this situation. (b) If the average density of matter (ρ_m) remained unchanged,

would it be more accurate to describe our universe as matter-dominated or radiation-dominated? Explain your answer.

41. Calculate the mass density of radiation (ρ_{rad}) in each of the following situations, and explain whether each situation is matter-dominated or radiation-dominated: (a) the photosphere of the Sun ($T = 5800$ K, $\rho_m = 3 \times 10^{-4}$ kg/m^3); (b) the center of the Sun ($T = 1.55 \times 10^7$ K, $\rho_m = 1.6 \times 10^5$ kg/m^3); (c) the solar corona ($T = 2 \times 10^6$ K, $\rho_m = 5 \times 10^{-13}$ kg/m^3).

42. If a photon from the cosmic microwave background had wavelength λ_0 when it was emitted at redshift z, its wavelength today is $\lambda = \lambda_0/(1 + z)$. (a) Let T be the symbol for the temperature of the cosmic microwave background today. Explain why the radiation temperature was $T_0 = T(1 + z)$ at redshift z. (b) What was the radiation temperature at $z = 1$? (c) At what redshift was the radiation temperature equal to 293 K (a typical room temperature)?

43. What would be the critical density of matter in the universe (ρ_c) if the value of the Hubble constant were (a) 50 km/s/Mpc? (b) 100 km/s/Mpc?

44. Consider the quasar HS 1946+7658 (see Advanced Question 37), which has $z = 3.02$. (a) Suppose that in the present-day universe, two clusters of galaxies are 500 Mpc apart. At the time that the light was emitted from HS 1946+7658 to produce an image on Earth tonight, how far apart were those two clusters? (b) What was the average density of matter (ρ_m) at that time? Assume that in today's universe, $\rho_m = 2.4 \times 10^{-27}$ kg/m^3. (c) What were the temperatures of the cosmic background radiation and the mass density of radiation (ρ_{rad}) at that time? (d) At this time in the remote past, was the universe matter-dominated, radiation-dominated, or dark-energy-dominated? Explain your answer.

45. Whether the expansion of the universe is speeding up or slowing down can be expressed in term of a quantity called the *deceleration parameter,* denoted by q_0. The expansion is slowing down if q_0 is positive and speeding up if q_0 is negative; if $q_0 = 0$, the expansion proceeds at a constant rate. If we assume that the dark energy is due to a cosmological constant, the deceleration parameter can be calculated using the formula

$$q_0 = \frac{1}{2}\Omega_0 - \frac{3}{2}\Omega_\Lambda$$

(Recall that the density parameter Ω_0 is equal to $\Omega_m + \Omega_\Lambda$.) (a) Show that if there is no cosmological constant, the expansion of the universe must slow down. (b) Using the values of Ω_m and Ω_Λ given in Table 26-2, find the value of the deceleration parameter for our present-day universe. Based on this, is the expansion of the universe speeding up or slowing down? (c) Imagine a universe that has the same value of Ω_m as our universe but in which the expansion of the universe is neither speeding up nor slowing down. What would be the value of Ω_Λ in such a universe? Which would be dominant in such a universe, matter or dark energy? Explain your answer.

46. In general, the deceleration parameter (see Advanced Question 45) is not constant but varies with time. For a flat universe, the deceleration parameter at a redshift z is given by the formula

$$q_z = \frac{1}{2} - \frac{3}{2}\left[\frac{\Omega_\Lambda}{\Omega_\Lambda + (1 - \Omega_\Lambda)(1 + z)^3}\right]$$

where Ω_Λ is the dark energy density parameter. Using the value of Ω_Λ given in Table 26-2, find the value of q_z for (a) $z = 0.5$ and (b) $z = 1.0$. (c) Explain how your results show that the expansion of the universe was actually decelerating at $z = 1.0$, but changed from deceleration to acceleration between $z = 1.0$ and $z = 0.5$.

47. The dark energy density parameter Ω_Λ is related to the value of the cosmological constant Λ by the formula

$$\Omega_\Lambda = \frac{\Lambda c^2}{3H_0^2}$$

where $c = 3.00 \times 10^8$ m/s is the speed of light. Determine the value of Λ if Ω_Λ and H_0 have the values given in Table 26-2. (*Hint:* You will need to convert units to eliminate kilometers and megaparsecs.)

Discussion Questions

48. Suppose we were living in a radiation-dominated universe. Discuss how such a universe would be different from what we now observe.

49. How can astronomers be certain that the cosmic microwave background fills the entire cosmos, not just the vicinity of Earth?

50. Do you think there can be "other universes," regions of space and time that are not connected to our universe? Should astronomers be concerned with such possibilities? Why or why not?

Web/eBook Questions

51. Before the discovery of the cosmic microwave background, it seemed possible that we might be living in a "steady-state universe" with overall properties that do not change with time. The steady-state model, like the Big Bang model, assumes an expanding universe, but does not assume a "creation event." Instead, matter is assumed to be created continuously everywhere in space to ensure that the average density of the universe remains constant. Search the World Wide Web for information about the steady-state theory. Explain why the existence of the cosmic microwave background was a fatal blow to the steady-state theory.

52. Search the World Wide Web for information on a European Space Agency mission called Planck. In what ways is Planck an improvement over the WMAP mission? Has it been launched? If yes, what have scientists learned from Planck? If no, what do they hope plan to learn?

Activities

Observing Projects

53. Use the *Starry Night Enthusiast*™ program to determine how the solar system moves through the cosmic microwave background. This motion appears to be taking us towards

the constellation Leo. First, select **Favorites > Guides > Atlas** to display the entire celestial sphere from the center of a transparent Earth. Open the **Find** pane and click on the magnifying glass symbol to display the Find categories and click on **Constellation.** Double-click on Leo to center on this constellation and click again on the **Find** pane tab to close this pane and display the full screen. Select **View > Constellations > Astronomical** and **View > Constellations > Labels** to display and label the constellations. (**a**) Draw a sketch showing the Sun, the plane in which Earth orbits the Sun, and the direction in which the solar system moves through the cosmic microwave background. (**b**) Use the date controls in the toolbar to step through the months of the year. In which month is the Sun placed most nearly in front of Earth as the solar system travels through the cosmic background radiation?

54. Use *Starry Night Enthusiast*™ to compare the distances of objects in the **Tully Database** with the radius of the Cosmic Light Horizon, the limit of our observable universe. As you will find, the most distant galaxies in this database are a long way away from Earth and yet these distances are only a small fraction of the distances from which we can see light in our universe. Select **Favorites > Deep Space > Tully Database** to display this collection of galaxies in their correct 3-dimensional positions in space around our position. Stop **Time** and click on **View > Feet** to remove the image of the astronaut's suit from the view. Select **Preferences** from the **File** menu (Windows) or the **Starry Night Enthusiast** menu (Macintosh). In the Preferences dialog, select **Cursor Tracking (HUD)** in the drop-down box and ensure that **Distance from observer, Name** and **Object type** are selected. The view shows the boundaries of the Tully database as a cube. Use the **location scroller** (hold down the **Shift** key and mouse button while moving the mouse) to rotate the cube to allow you to choose galaxies on the outer fringes of this space. Use the **Hand Tool** to examine a selection of the furthest objects from Earth, which is centered in the view, and write a list of 10–20 objects, noting the **Object type** and **Distance from observer.** (**a**) In your sample, is there a predominance of any one kind of galaxy? If so, what type of galaxy appears to be most common at these distances? (**b**) Select the furthest of these galaxies and compare their distances with the radius of the cosmic light horizon. What fraction of the radius of the observable universe is covered by the Tully database?

Collaborative Exercises

55. As a group, create a four- to six-panel cartoon strip showing a discussion between two individuals describing why the sky is dark at night.
56. Imagine your firm, Creative Cosmologists Coalition, has been hired to create a three-panel, folded brochure describing the principal observations that astronomers use to infer the existence of a Big Bang. Create this brochure on an $8\frac{1}{2} \times 11$ piece of paper. Be sure each member of your group supervises the development of a different portion of the brochure and that the small print acknowledges who in your group was primarily responsible for which portion.
57. The three potential geometries of the universe are shown in Figure 26-15. To demonstrate this, ask one member of your group to hold a piece of paper in one of the positions while another member draws two parallel lines that never change in one geometry, eventually cross in another geometry, and eventually diverge in another.

COSMOLOGY

Dark Forces at Work

Ten years ago two teams discovered that the universe will expand forever at an ever faster rate, thanks to an unseen energy. The leader of one of the groups, Saul Perlmutter, expects that new observations will soon illuminate the universe's dark side

BY DAVID APPELL

(from David Appell, "Dark Forces at Work," *Scientific American*, May 2008, 100, 102)

One of the chief astrophysicists behind the discovery of the acceleration of the expansion of the universe, among the most startling revelations in the history of cosmology, delights in the confusion about the observation. In fact, he wonders if the acceleration will end up being the most important feature in the ultimate explanation. "It might be something unexpected that looks like acceleration," says Saul Perlmutter, leader of the Supernova Cosmology Project (SCP), which first announced the astonishing fact in 1998. Ever the experimentalist, the 48-year-old Perlmutter is waiting, and planning, for more observations: "Until we go for a long run of more data, this just isn't a mature field."

Perlmutter philosophizes about the strangeness of the cosmos from his office at Lawrence Berkeley National Laboratory, high in the western hills of the San Francisco Bay Area. The room is the scientist's amalgam of too many computer screens, too many piles of papers and an equation-filled whiteboard that would have done Einstein proud. The spectacular view of the Golden Gate Bridge in the distance cannot help but promote lofty thinking.

It has been a decade since the science community learned of the shocking discovery made by Perlmutter's group and, independently, by the High-Z Supernova Search Team led by Brian Schmidt of the Australian National University (with analyses pioneered by Adam Riess of the Space Telescope Science Institute). The cosmos, the researchers found, is not just expanding; for unknown reasons, it is speeding up in its expansion.

The discovery took years of innovation and problem solving. The key was supernovae—specifically, those called type Ia. Such events are surprisingly invariable—the explosions have an intrinsic brightness that predictably fades over time, enabling astronomers to use them as "standard candles" and thus determine their distances from Earth. Perlmutter worked with Carl Pennypacker of the University of California, Berkeley, in the 1980s to robotically search for supernovae at relatively nearby distances. The field was then so young that their main competition came from Robert Evans, an amateur astronomer in Australia who identified supernovae with a backyard telescope.

In the beginning, the difficulty for Perlmutter's group lay in obtaining telescope time, always precious in the astronomical community. How would the researchers convince allocators to give them the chance to look for something—a supernova explosion—that had not yet taken place? So they worked out methods to predict and automatically detect supernovae in a given patch of the sky. But their goal of determining the universe's dynamics—then thought to be a decelerating expansion dominated by matter—still required additional observation to plot supernovae's brightness peaks and declines, which take place over a few weeks. Perlmutter twisted arms and begged colleagues for an hour or two on short notice, calling frantically around the world at all times of the day. Everyone knew him, he says, as half-annoying. "I was always worried about something that had to happen in the next 24 hours or sometimes the next two hours. It was a terrible way to lead an ordinary life," he recalls.

But persistence paid off. Observations of distant type Ia supernovae found them to be dimmer than expected. After eliminating the possibility of intergalactic dust and after years of painstaking data gathering and analysis at telescopes around the world (and in orbit), Perlmutter's team came to the conclusion that, incredibly, the universe is not only expanding, as Edwin Hubble discovered in 1929, but that its expansion rate is increasing. Some unknown force with negative pressure seems to be pushing the universe apart.

Subsequent balloon-borne observations of the cosmic microwave background made two years later showed that the universe is spatially flat—it was stretched out by an exponential expansion, called inflation, right after the big bang. The equations behind these experiments complemented those of the supernova teams taken a few years earlier, and together the results enabled scientists to calculate separately the density of dark energy in the universe and the density of matter.

But on the other hand, the discovery opened a mystery the size of, well, the universe. The simplest explanation is that dark energy is Einstein's famed "cosmological constant," an energy that permeates space but does not interact with any type of matter. Today astronomers have homed in on the details of this scenario; if true, then the universe consists of 72 percent antigravity dark energy, 23 percent dark matter (unseen and uncharacterized, but susceptible to gravity), and 5 percent normal matter (protons, neutrons, electrons). We would be just a small part of totality, surrounded by perplexity.

"It could well be that there's some big piece of reality that we don't fully understand," says astronomer Christopher Stubbs of Harvard University, who in a paper likened the new universe to "living in a bad episode of Star Trek." Physicist Steven Weinberg of the University of Texas at Austin calls it simply "a bone in the throat of theoretical physics."

Magic has not yet been proposed to explain the accelerating universe, but almost everything else has. In the past few years, physicists have widened their search beyond vacuum energy to include possible modifications to general relativity, spinless energy fields that vary with time and space, massive gravitons, brane worlds and extra dimensions. "All of them are so exciting, and any is going to rewrite the textbooks," says Eric Linder, a cosmologist at Lawrence Berkeley and U.C. Berkeley. The hypothetical repulsive dark energy field may well not survive in the final explanation.

"It's true the theorists right now are stuck," Perlmutter says. "But from an experimentalist's point of view, this is great: we have a mystery, and we have ways to get at it"—namely, in the form of new telescopes and satellites to look even farther across the universe (and, hence, farther back in time).

Ground-based projects are already gathering more data, looking for hundreds of type Ia events (instead of Perlmutter's and Schmidt's five dozen) to determine the relation between the pressure and density of the universe, akin to the ideal gas law. A galaxy like our Milky Way exhibits about one type Ia supernova every few hundred years, and its brightness fades in weeks, making the search for them quite a challenge. By observing the cosmic background radiation, the soon-to-launch Planck satellite will contribute more details about the universe's expansion.

Dark energy aficionados look especially to the Joint Dark Energy Mission, now in the planning stages in the U.S. for a possible launch in 2014. The probe will host a device that could find thousands of supernovae a year and provide far smaller error bars than anything done so far. One candidate is the SuperNova Acceleration Probe (SNAP), for which Perlmutter is the lead scientist and Linder the head theorist. It would host a telescope about two meters wide and have a gigapixel camera.

The discovery of cosmic acceleration will assuredly win a Nobel Prize, and over the years there has been some dispute over which team deserves priority. Perlmutter's SCP team announced the discovery first, but Schmidt's High-Z team beat the SCP group in publishing the finding. Both Perlmutter and Schmidt shared one fourth of the 2007 Gruber Cosmology Prize, with the remaining fraction going to their two teams collectively.

Gregarious and talkative, Perlmutter attributes his success to being able to convey his excitement and convince other researchers to join his team. An amateur violinist who also teaches an undergraduate physics and music course, he draws an orchestral analogy. "As a violinist, I always love the moments when a group of people are creatively tuned in together."

David Appell is based in Portland, Ore. A Q& A version of his interview is at **www.SciAm.com/sciammag.**

27 Exploring the Early Universe

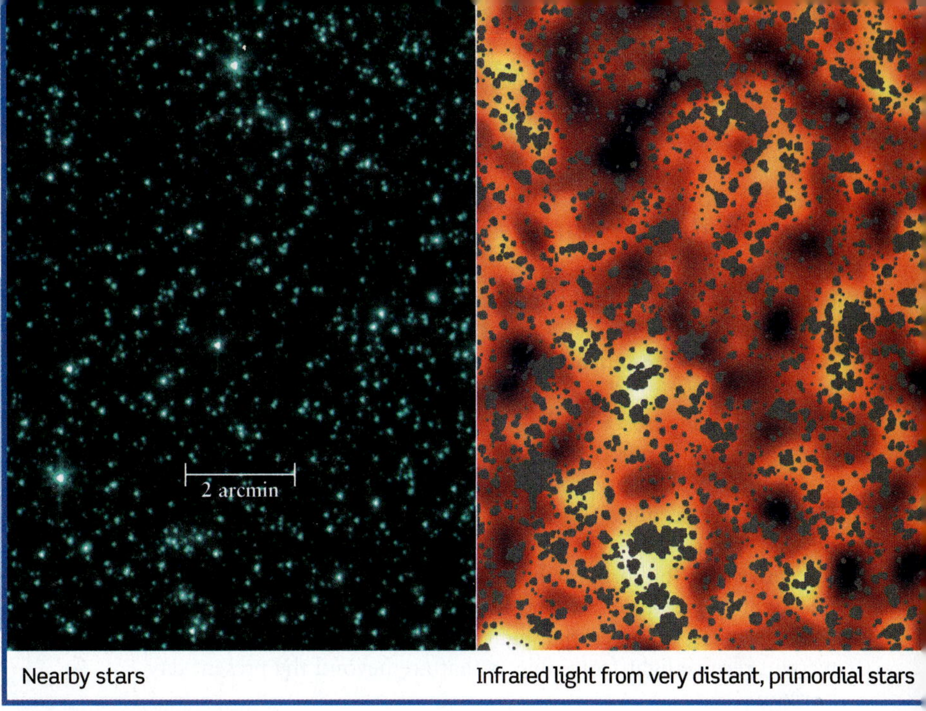

R I V U X G

These Spitzer Space Telescope images show (left) light from nearby stars and (right) light from a remote population of ancient stars in the same patch of sky with the nearby stars in our Galaxy removed. (NASA; JPL-Caltech; and A. Kashlinsky, Goddard Space Flight Center)

The two images shown here are of a patch of sky near the Big Dipper in the constellation Ursa Major. The left-hand image is dominated by relatively nearby stars in our Galaxy. But when these stars are removed digitally, what remains in the right-hand, false-color image is an intriguing pattern of highly redshifted light from objects much farther away. This radiation is thought to be some of the oldest starlight in the universe: It was emitted by members of the very first generation of stars, born when our universe was less than a billion years old.

Only recently have new telescopes begun to reveal the story of the first stars and first galaxies. But no telescope can ever hope to directly observe events during the first 380,000 years after the Big Bang, when the universe was so opaque that it blocked the free passage of light. We can nonetheless reconstruct some of the events of that hidden epoch, because many aspects of today's universe are relics of the earliest events in the cosmos.

In this chapter we will see evidence that during the first miniscule fraction of a second after the Big Bang, the universe inflated in size by a stupendously large factor of about 10^{50}. During the next 15 minutes after inflation came to an end, the universe was so dense and hot that particles were constantly colliding at high speeds. As we will see, the events of those 15 violent minutes set the stage for all that came afterward—from the formation of atoms 380,000 years later, to the appearance of the first stars and galaxies some 400 million years after the Big Bang, down to the diverse present-day universe of which we are part.

27-1 The newborn universe underwent a brief period of vigorous expansion

With the discovery of the cosmic microwave background, astronomers had direct evidence that the universe began with a hot Big Bang (see Section 26-4). Remarkably, the microwave background is incredibly uniform, or *isotropic*, across the sky. If we subtract the effects of our own motion through the microwave background (see Figure 26-8), we find that the temperature of the microwave background is the same in all parts of the sky to an accuracy of 1 part in 10,000. The small nonuniformities in the microwave background are also remarkable, in part because their angular sizes help indicate that our universe has a flat geometry (see Section 26-6, especially Figure 26-16). As striking as these observations are, they pose substantial challenges to the standard theory of the expansion of the universe.

Learning Goals

By reading the sections of this chapter, you will learn

27-1 How the very young universe expanded enormously in a brief instant of time

27-2 How the fundamental forces of nature and the properties of empty space changed during the first second after the Big Bang

27-3 How the physics of subatomic particles affected the evolution of the early universe

27-4 Why antimatter was once common in the universe, but is very rare today

27-5 Which chemical elements in today's universe are remnants of the primordial fireball

27-6 How the first stars and galaxies formed in the early universe

27-7 What steps scientists are taking in the quest toward an all-encompassing "theory of everything"

The Isotropy Problem: Why Is the Microwave Background So Uniform?

To appreciate why the uniformity of the microwave background poses a problem, think about two opposite parts of the sky, labeled *A* and *B* in Figure 27-1. Both of these points lie on a sphere centered on Earth; our **observable universe** lies within the sphere, and the surface is called our **cosmic light horizon** (see Section 26-3, especially Figure 26-5). We can see light coming from any object on or inside our cosmic light horizon, but we cannot yet see objects beyond this horizon. Even traveling at the speed of light—the ultimate speed limit—no information of any kind from objects beyond our cosmic light horizon has had time to reach us over the entire 13.7-billion-year history of the universe. Thus, the cosmic light horizon defines the limits of our observable universe. (As time passes, our cosmic light horizon expands, so eventually we will receive light from objects that are beyond the present-day horizon.)

Points *A* and *B* in Figure 27-1 lie near the distant edge of the observable universe, just inside our cosmic light horizon, so when we look at these points we are seeing about as far back into the past as possible—that is, the light we see from these points is the cosmic background radiation. In order for the radiation that reaches us from *A* and *B* to be nearly the same, the material of the early universe at *A* must have had the same temperature as at *B*. But for two objects to be at the same temperature, they should have been in contact and able to exchange heat with each other. (A hot cup of coffee and a cold spoonful of cream reach the same temperature only after you pour the cream into the coffee.)

The problem is that the widely separated regions at *A* and *B* have absolutely no connection with each other. As Figure 27-1 shows, point *A* lies outside the cosmic light horizon for point *B*, and point *B* lies outside the cosmic light horizon for point *A*. (Put another way, point *A* lies outside the observable universe of point *B* and vice versa.) Hence, no information has had time to travel between points *A* and *B* over the entire history of the universe. How is it possible, then, that these unrelated parts of the universe have almost exactly the same temperature? This dilemma is called the **isotropy problem** or the **horizon problem.**

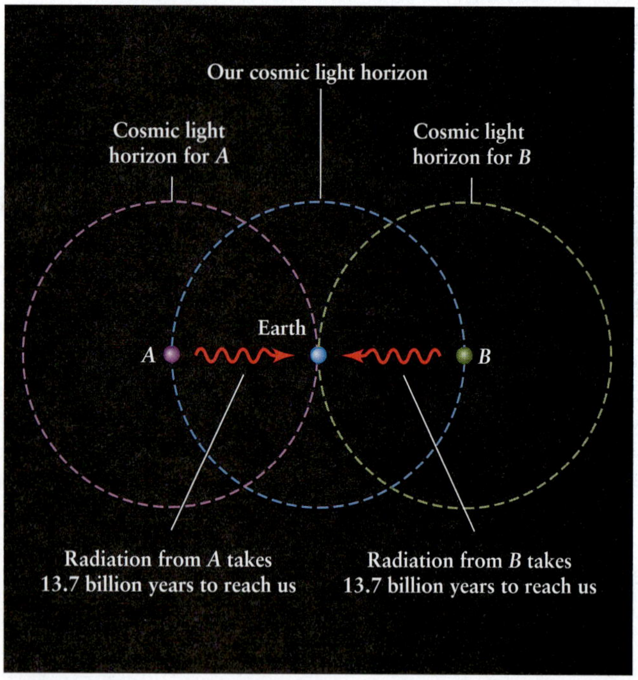

Figure 27-1

The Isotropy Problem Regions *A* and *B*, both of which lie on our cosmic light horizon, are so far apart that they seem never to have been in communication over the lifetime of the universe. Yet the cosmic background radiation from *A* and *B*, and from all other parts of the sky, shows that these disconnected regions have almost exactly the same temperature. The dilemma of why this should be is called the isotropy problem.

The Flatness Problem: Why Is $\Omega_0 = 1$?

The flatness of our universe presents us with a second enigma. Recall that the geometry of our universe depends on the density parameter Ω_0, which is the ratio of the combined average mass density in the universe (ρ_0) to the critical density (ρ_c) (see Section 26-6). Observations of temperature variations in the cosmic microwave background indicate that Ω_0 is very close to 1, which corresponds to a flat universe.

For the density parameter Ω_0 to be close to 1 today, it must have been *extremely* close to 1 during the Big Bang. In other words, the density of the early universe was almost exactly equal to the critical density. (The density was much higher than it is today, but the value of the critical density was also much higher. In a flat universe, the average mass density and the critical density decrease together as the universe expands, so that they remain equal at all times.)

The equations for an expanding universe show that any deviation from exact equality would have mushroomed within a fraction of a second. Had the average mass density been slightly less than ρ_c, the universe would have expanded so rapidly that matter could never have clumped together to form galaxies. Without galaxies, there would be no stars or planets, and humans would never have evolved. If, on the other hand, the density had been slightly greater than ρ_c, the universe would soon have become tightly packed with matter. Had this been the case, the gravitational attraction of this matter would long ago have collapsed the entire cosmos in a reversed Big Bang or "Big Crunch," and again humans would never have evolved.

In other words, immediately after the Big Bang the fate of the universe hung in the balance. The tiniest deviation from the precise equality $\rho_0 = \rho_c$ would have rapidly propelled the universe away from the special case of $\Omega_0 = 1$. Had there been such a deviation, we would not be here to contemplate the nature of the universe.

Happily, our universe is one in which galaxies, stars, planets, and humans do exist, a Big Crunch has not taken place, and Ω_0 is very close to 1. These conditions in the universe today tell us that the density of the universe immediately after the Big Bang must have been equal to the critical density to an incredibly high order of precision. In order for space to be as flat as it is today, the value of ρ_0 right after the Big Bang must have been equal to ρ_c to more than 50 decimal places!

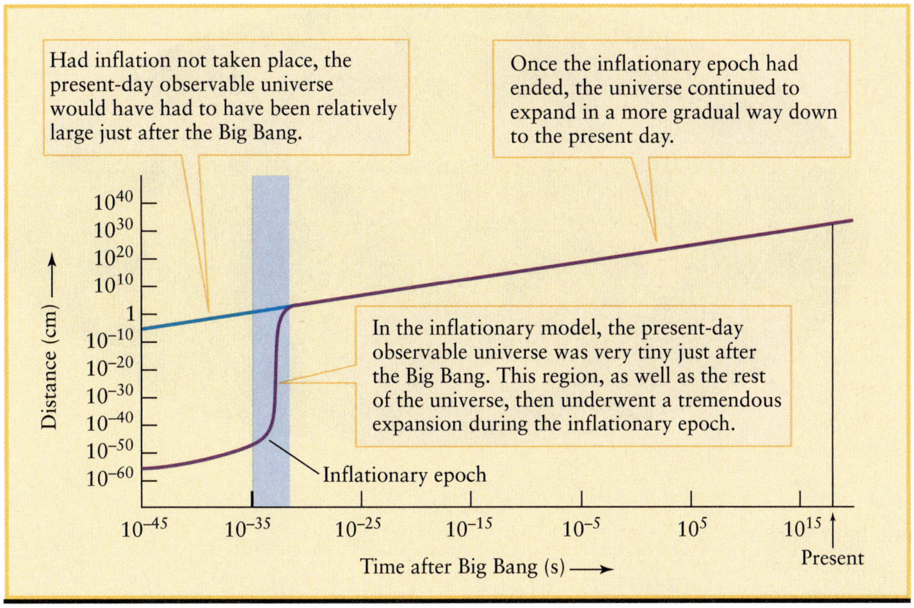

Had inflation not taken place, the present-day observable universe would have had to have been relatively large just after the Big Bang.
Once the inflationary epoch had ended, the universe continued to expand in a more gradual way down to the present day.
In the inflationary model, the present-day observable universe was very tiny just after the Big Bang. This region, as well as the rest of the universe, then underwent a tremendous expansion during the inflationary epoch.
Inflationary epoch
Distance (cm)
Time after Big Bang (s)
Present

Figure 27-2

The Observable Universe With and Without Inflation According to the inflationary model (shown in red), the universe expanded by a factor of about 10^{50} shortly after the Big Bang. This growth in the size of the present-day observable universe—that portion of the universe that lies within our present cosmic light horizon—occurred during a very brief interval, as indicated by the vertical shaded area on the graph. The blue line shows the projected size of the present-day observable universe soon after the Big Bang if inflation had not taken place. (Adapted from A. Guth)

What could have happened during the first few moments of the universe to ensure that $\rho_0 = \rho_c$ to such an astounding degree of accuracy? Because $\rho_0 = \rho_c$ means that space is flat, this enigma is called the **flatness problem.**

Solving the Problems: The Inflationary Model

WEB LINK 27.2

In the early 1980s, a remarkable solution was proposed to both the isotropy problem and the flatness problem. Working independently, Alexei Starobinsky at the L. D. Landau Institute of Theoretical Physics in Moscow and Alan Guth at Stanford University suggested that the universe might have experienced a brief period of **inflation,** or extremely rapid expansion, shortly after the Planck time. (As we saw in Section 26-3, before the Planck time the normal laws of physics do not properly describe the behavior of space, time, and matter.) During this **inflationary epoch,** the universe expanded outward in all directions by a factor of about 10^{50}. This epoch of dramatic expansion may have lasted only about 10^{-32} s (Figure 27-2).

The inflationary model explains not only why the cosmic microwave background is so uniform, but also how stars, galaxies, and humans can exist

Inflation accounts for the isotropy of the microwave background. During the inflationary epoch, much of the material that was originally near our location was moved out to tremendous

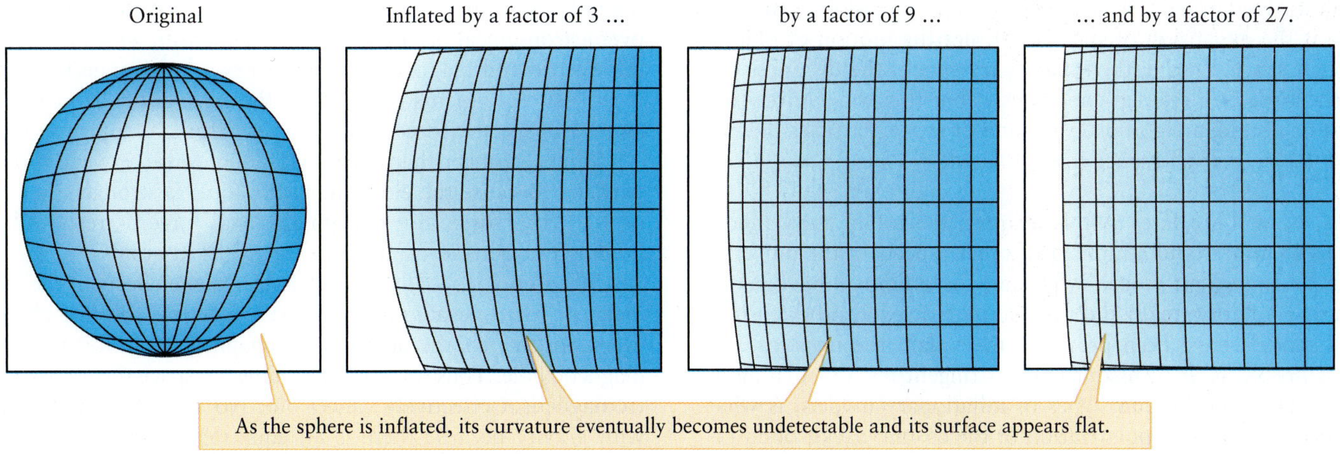

Original
Inflated by a factor of 3 ...
by a factor of 9 ...
... and by a factor of 27.
As the sphere is inflated, its curvature eventually becomes undetectable and its surface appears flat.

Figure 27-3

Inflation Solves the Flatness Problem This sequence of drawings shows how inflation can produce a locally flat geometry. In each successive frame, the sphere is inflated by a factor of 3 (and the number of grid lines on the sphere is increased by the same factor). Note how the curvature of the surface quickly becomes undetectable on the scale of the illustration. (Adapted from A. Guth and P. Steinhardt)

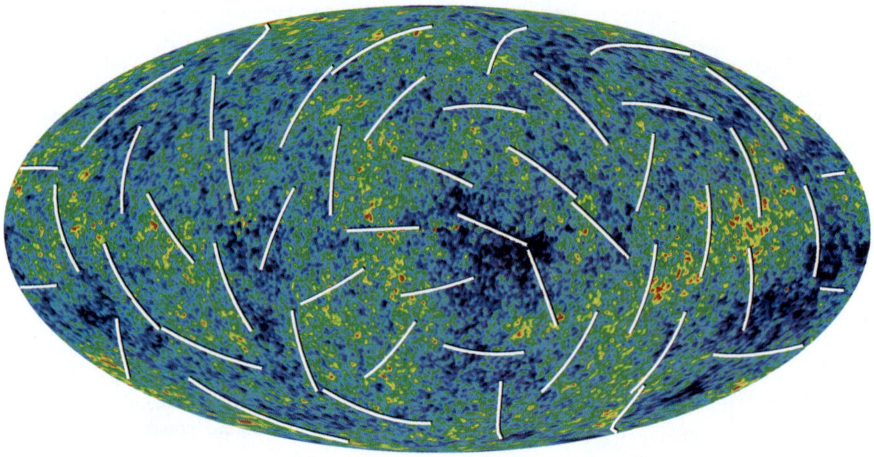

Figure 27-4 R I V U X G

Polarization of the Cosmic Microwave Background
The white lines on this map of the cosmic background radiation (compare Figure 26-14) indicate the directions in which radiation from different parts of the sky is polarized. This polarization was caused when photons in the background radiation scattered from electrons in the early universe. The distribution and density of these electrons depends on the earlier expansion history of the universe. Hence, polarization measurements can reveal whether inflation took place as well as the nature of the inflation. (NASA/WMAP Science Team)

distances. Over the past 13.7 billion years, our cosmic light horizon has expanded so that we can see radiation from these distant regions. Hence, when we examine microwaves from opposite parts of the sky, we are seeing radiation from parts of the universe that were originally in intimate contact with one another. This contact allowed for the exchange of heat and is why all parts of the sky have almost exactly the same temperature.

ANALOGY An inflationary epoch can also account for the flatness of the universe. To see why, think about a small portion of Earth's surface, such as a small lake. For all practical purposes, it is impossible to detect Earth's curvature over such a small area, and a small lake looks flat. Similarly, the observable universe is such a tiny fraction of the entire inflated universe that any overall curvature in it is virtually undetectable (Figure 27-3). Like a small lake, the segment of space we can observe looks flat.

CAUTION! It is important to note that the concept of inflation does not violate Einstein's dictum that nothing can travel faster than the speed of light. Remember that the expansion of the universe is the expansion of space itself, not the motion of objects through space. During the inflationary epoch, the distances between particles increased enormously, but this was entirely the result of a sudden vigorous *expansion of space.* Particles did not move through space; space itself inflated.

As we will see later in this chapter, the inflationary model helps us to understand a number of other aspects of the universe. This is an important reason why many astronomers have confidence that inflation really did take place. However, other cosmological models have been proposed to explain the isotropy and flatness problems. It turns out that a stringent test on such models, as well as on different types of inflationary models, is what they predict about the polarization of the cosmic background radiation. (We discussed this polarization in Section 26-8.) Figure 27-4 shows measurements of the background radiation's polarization from the Wilkinson Microwave Anisotropy Probe. While such measurements are very difficult and still in their infancy, the data obtained to date are entirely consistent with the inflationary model.

27-2 Inflation was one of several profound changes that occurred in the very early universe

If the universe went through an episode of extreme inflation, what could have triggered it? Our understanding is that inflation was one of a sequence of remarkable events during the first 10^{-12} seconds after the Big Bang. In each of these events there was a fundamental transformation of the basic physical properties of the universe. To understand what happened during that brief moment of time, when the universe was a hot, dense sea of fast-moving particles and energetic photons colliding with each other, we must first understand how particles interact at very high energies.

The Fundamental Forces of Nature

Just *four* fundamental forces—gravitation, electromagnetism, and the strong and weak forces—explain the interactions of everything in the universe. Of these forces, gravitation is the most familiar (Figure 27-5*a*). It is a long-range force that dominates the universe over astronomical distances. Electromagnetism, which accounts for the electric and magnetic forces, creates a long-range force, but it is intrinsically much stronger than the gravitational force. For example, the electric force between an electron and a proton is about 10^{39} times stronger than the gravitational force between those two particles. Because the electromagnetic force is stronger by a factor of 10^{39}, this force, not the gravitational force, is responsible for holding electrons in orbit about the nuclei in atoms.

We do not generally observe longer-range effects of the electromagnetic force, because there is usually a negative electric charge for every positive charge and a south magnetic pole for every north magnetic pole. Thus, over great volumes of space the effects of electromagnetism effectively cancel out. No similar canceling occurs with gravity because there is no equivalent "negative mass." Because the effects of gravity don't cancel out, the force that holds Earth in orbit around the Sun is gravitational, not electromagnetic.

The **strong force** holds protons and neutrons together to form the nuclei of atoms (Figure 27-5*b*). It is said to be a *short-range force:* Its influence extends only over distances less than about 10^{-15} m, about the diameter of a proton. Without the strong

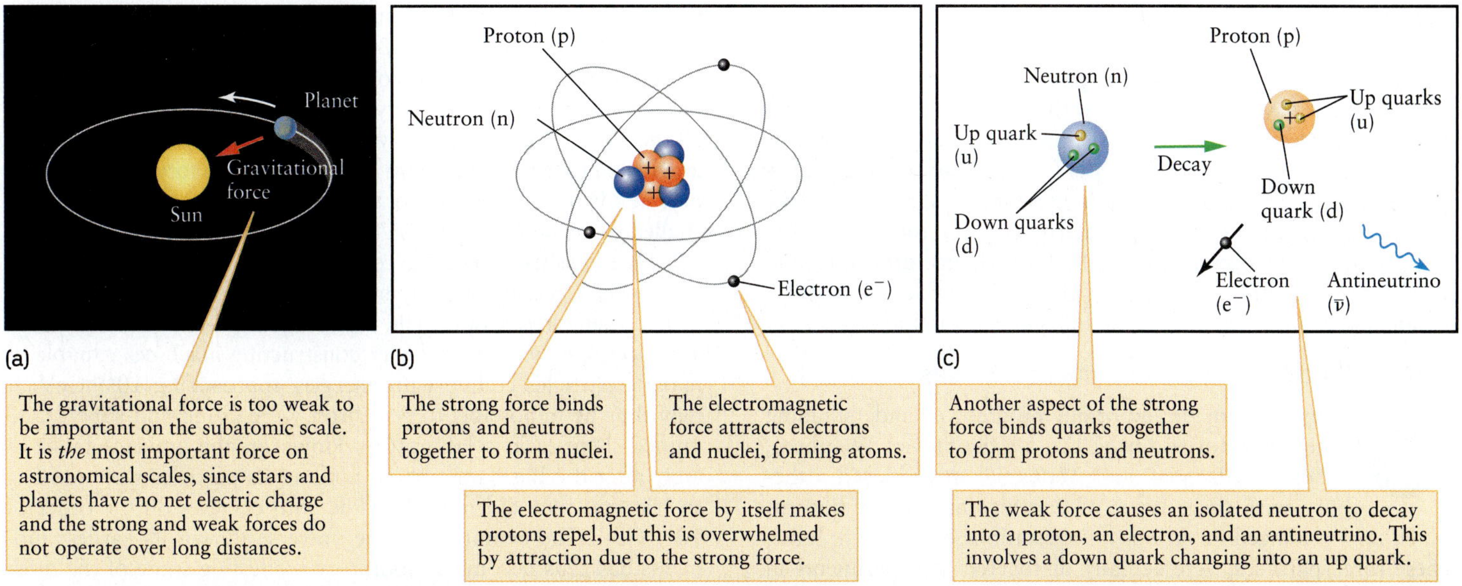
Planet
Gravitational force
Sun
(a)
The gravitational force is too weak to be important on the subatomic scale. It is *the* most important force on astronomical scales, since stars and planets have no net electric charge and the strong and weak forces do not operate over long distances.
Proton (p)
Neutron (n)
Electron (e^-)
(b)
The strong force binds protons and neutrons together to form nuclei.
The electromagnetic force attracts electrons and nuclei, forming atoms.
The electromagnetic force by itself makes protons repel, but this is overwhelmed by attraction due to the strong force.
Neutron (n)
Up quark (u)
Down quarks (d)
Decay
Proton (p)
Up quarks (u)
Down quark (d)
Electron (e^-)
Antineutrino ($\bar{\nu}$)
(c)
Another aspect of the strong force binds quarks together to form protons and neutrons.
The weak force causes an isolated neutron to decay into a proton, an electron, and an antineutrino. This involves a down quark changing into an up quark.

Force	Relative strength	Particles exchanged	Particles on which the force can act	Range	Example
Strong	1	gluons	quarks	10^{-15} m	holding protons, neutrons, and nuclei together
Electromagnetic	1/137	photons	charged particles	infinite	holding atoms together
Weak	10^{-4}	intermediate vector bosons	quarks, electrons, neutrinos	10^{-16} m	radioactive decay
Gravitational	6×10^{-39}	gravitons	everything	infinite	holding the solar system together

INTERACTIVE EXERCISE 27.1

Figure 27-5

The Four Forces **(a)** Gravitation is dominant on the scales of planets, star systems, and galaxies, while **(b), (c)** the strong, electromagnetic, and weak forces hold sway on the scale of atoms and nuclei.

force, nuclei would disintegrate because of the electric repulsion of the positively charged protons. In fact, the strong force overpowers the electric forces inside nuclei.

The **weak force,** which also is a short-range force, is at work in certain kinds of radioactive decay (Figure 27-5*c*). An example is the transformation of a neutron (n) into a proton (p), in which an electron (e^-) is released along with a nearly massless particle called an antineutrino ($\bar{\nu}$):

$$n \rightarrow p + e^- + \bar{\nu}$$

Numerous experiments in nuclear physics demonstrate that protons and neutrons are composed of more basic particles called **quarks,** the most common varieties being "up" (u) quarks and "down" (d) quarks. A proton is composed of two up quarks and one down quark, a neutron of two down quarks and one up quark.

In the 1970s the concept of quarks led to a breakthrough in our understanding of the strong and weak forces. The strong force holds quarks together, while the weak force is at work whenever a quark changes from one variety to another. For example, when a neutron decays into a proton, one of the neutron's down quarks changes into an up quark as shown in Figure 27-5*c*. Thus, the weak force is responsible for transformations such as

$$d \rightarrow u + e^- + \bar{\nu}$$

WEB LINK 27.3

Experimental and theoretical research into the basic forces of nature helps us understand the evolution of the cosmos

In the 1940s, physicists Richard P. Feynman and Julian S. Schwinger (working independently in the United States) and Sin-Itiro Tomonaga (in Japan) succeeded in developing a basic description of what we mean by force. Focusing their attention on the electromagnetic force, they tried to describe exactly what happens when two charged particles interact. According to their theory, now called **quantum electrodynamics,** charged particles interact by exchanging photons that cannot be directly observed. (We will explore such *virtual* photons in Section 27-3.)

Quantum electrodynamics has proved the most successful theory in modern physics. It describes with remarkable accuracy many details of the electromagnetic interaction between charged particles. Inspired by these successes, physicists have tried to develop similar theories for the other three forces. In these theories, the weak force occurs when particles exchange **intermediate vector bosons,** the gravitational force occurs when particles exchange **gravitons,** and quarks stick together by exchanging **gluons.** The table in Figure 27-5 summarizes these features of the four fundamental forces.

Unified Theories of the Fundamental Forces

Physicists made important progress in understanding the weak force during the 1970s. Steven Weinberg, Sheldon Glashow, and Abdus Salam proposed a theory with three types of intermediate vector bosons, which are exchanged in various manifestations of the weak force. These three particles were actually discovered in experiments in the 1980s, providing strong support for the theory.

A startling prediction of the Weinberg-Glashow-Salam theory is that the weak force and the electromagnetic force should be identical to each other for particles with energies greater than 100 GeV. (One GeV equals 10^9, or 1 billion, electron volts; see Section 5-5.) In other words, if particles are slammed together with a total energy greater than 100 GeV, then electromagnetic interactions become indistinguishable from weak interactions. We say that above 100 GeV the electromagnetic force and the weak force are "unified" into a single **electroweak force.**

This unification occurs because the three types of intermediate vector bosons behave just like photons above 100 GeV. At such high energies, these three particles actually lose their mass, and the weak force becomes a long-range force with the same intrinsic strength as electromagnetism. Physicists describe this similarity by saying that "symmetry is restored" above 100 GeV.

In the world around us, however, the typical energies with which particles interact are very much lower, on the order of 1 eV or less. Below 100 GeV, intermediate vector bosons behave like massive particles, but photons are always massless. Because intermediate vector bosons and photons are not similar at low energies, we say that "symmetry is broken" below 100 GeV, which is why the electromagnetic and the weak forces behave so differently in the world around us. In the language of physics, a **spontaneous symmetry breaking** occurs.

In the 1970s, Sheldon Glashow, Howard Georgi, Jogesh Pati, and Abdus Salam proposed **grand unified theories** (or **GUTs**), which predict that the strong force becomes unified with the weak and electromagnetic forces (but not gravity) at energies above 10^{14} GeV. In other words, if particles were to collide at energies greater than 10^{14} GeV, the strong, weak, and electromagnetic interactions would all be long-range forces and would be indistinguishable from each other.

Many physicists suspect that all four forces, including gravity, may be unified at energies greater than 10^{19} GeV (Figure 27-6*a*). That is, if particles were to collide at these colossal energies, there would be no difference between the gravitational, electromagnetic, and nuclear forces. However, no one has yet succeeded in working out the details of such a **supergrand unified theory,** which is sometimes called a **theory of everything** (or **TOE**). We've already seen with black holes and the Big Bang, that developing the answers to some questions requires a theory of quantum gravity (understanding the effects of strong gravity on microscopic scales). Combining all the forces (a theory of everything) would also describe quantum gravity, and the favored theory combining all the forces is **string theory.** While string theory looks very promising, many theoretical challenges remain, and it hasn't been confirmed experimentally.

Physicists use particle accelerators to examine the unification of the weak and electromagnetic forces by slamming particles together at energies around 100 GeV (Figure 27-6*b*). There is probably no hope, however, of ever constructing machines capable of making particles collide with energies approaching 10^{19} GeV. It may thus be impossible to test directly the grand and supergrand unified theories in a laboratory. However, the universe immediately after the Big Bang was so hot and its particles were moving so fast that they did indeed collide with energies on the order of 10^{19} GeV. The evolution of the universe during its earliest moments has thus become a laboratory for testing some of the most elegant, sophisticated theories in physics.

The Fundamental Forces in the Early Universe

Figure 27-6*a* shows how the various forces changed during the first fraction of a second after the Big Bang. Before the Planck time (from $t = 0$ s to $t = 10^{-43}$ s), particles collided with energies greater than 10^{19} GeV, and all four forces were unified. Because we do not yet have a TOE that properly describes the behavior of gravity, we remain ignorant of what was going on during the first 10^{-43} second of the universe's existence. We know, however, that by the end of the Planck time, the expansion and cooling of the universe had caused the energy of particles to fall to 10^{19} GeV. At energies below this, gravity is thought not to be unified with the other three forces.

In the language of physics, at $t = 10^{-43}$ s there was a spontaneous symmetry breaking in which gravity was "frozen out" of the otherwise unified hot soup that filled all space. In such a "soup," the typical energy of a particle (E) is related to temperature (T) by $E = kT$, where k is the Boltzmann constant (roughly 10^{-4} eV/K, or 10^{-13} GeV/K). Thus, the temperature of the universe was 10^{32} K when gravity emerged as a separate force.

As the universe expanded, its temperature decreased and the energy of particles decreased as well. (We discussed this property of gases in Box 19-1.) At $t = 10^{-35}$ s, the energy of particles in the universe had fallen to 10^{14} GeV, equivalent to a temperature of 10^{27} K. Below these energies and temperatures, the strong force is no longer unified with the electromagnetic and weak forces. Thus, at $t = 10^{-35}$ s, there was a second spontaneous symmetry breaking, at which time the strong force "froze out."

The inflationary epoch is thought to have begun at this point. Physicists hypothesize that before the strong force decoupled from the electroweak force, the universe was in an unstable state called a **false vacuum.** In this state, physicists hypothesize that the energy associated with a quantity called the *inflaton field* had a nonzero value (Figure 27-7*a*). (Just as the space around a magnet is permeated by a magnetic field, like that shown in Figure 7-13*a*, the entire universe is thought to be permeated by the inflaton field.) This state was unstable in the same sense as a ball perched atop a cone with a pointed top: The ball will stay there if left undisturbed, but will roll downhill if even slightly disturbed. In

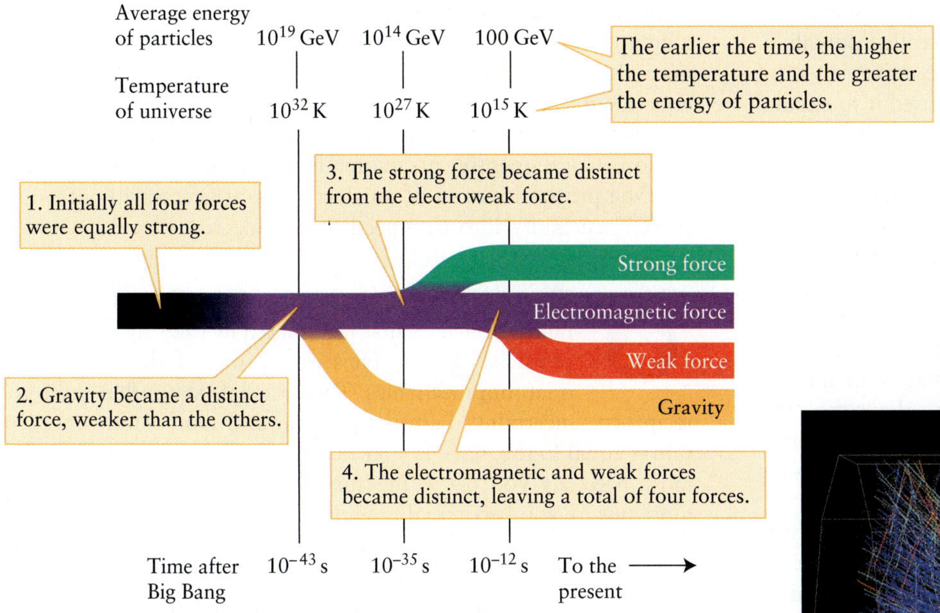

(a) How the four forces behave at different energies and temperatures

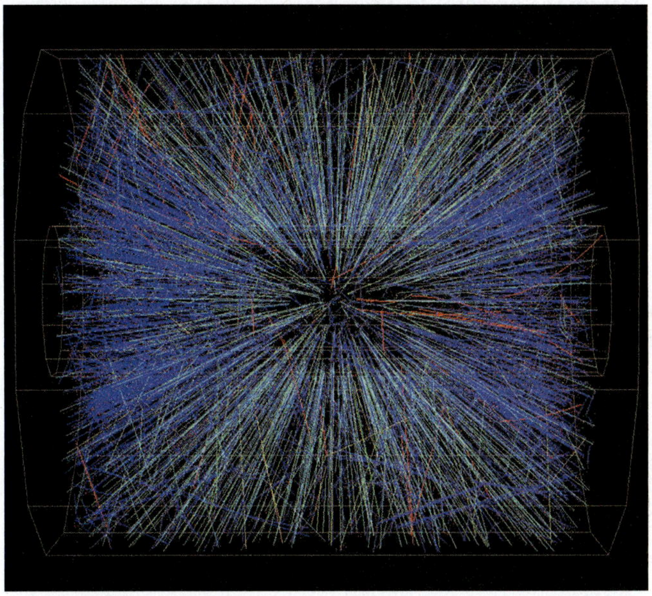

(b) The result of a high-energy collision between gold nuclei

WEBLINK 27.5

Figure 27-6

Unification of the Four Forces (a) The strength of the four fundamental forces depends on the energy of the particles that interact. As shown in this schematic diagram, the higher the energy, the more the forces resemble each other. Also included here are the temperature of the universe and the time after the Big Bang when the strengths of the forces are thought to have been equal. **(b)** This shower of subatomic particles was produced by a collision between two gold nuclei, each of which was traveling at 99.995% the speed of light. The energy per proton or neutron was 100 GeV, near the energy at which the electromagnetic and weak forces become unified. (Brookhaven National Laboratory)

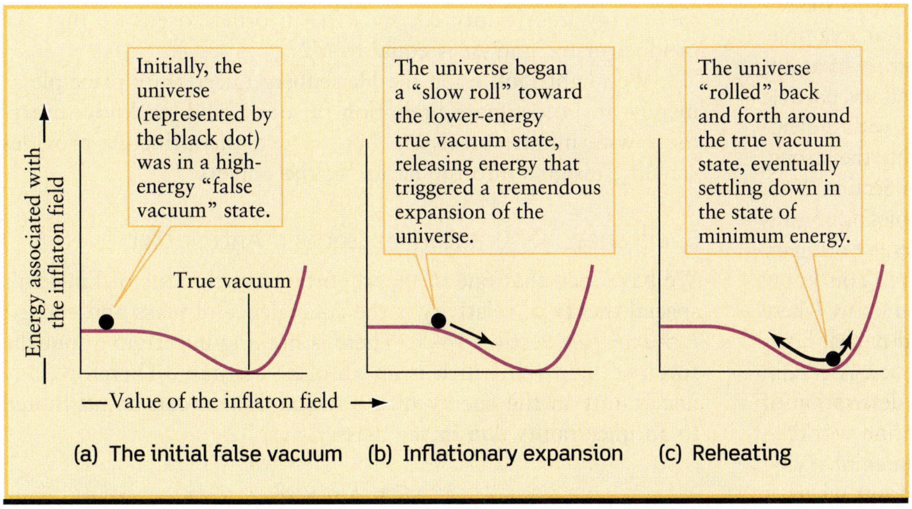

Figure 27-7

Inflation: Transitioning from the False Vacuum to the True Vacuum (a) The energy of the vacuum is thought to be determined by the value of a quantity called the inflaton field. As shown by the red curve in this diagram, this energy is at a minimum for a certain value of this field. The state of lowest energy is called the true vacuum. **(b)** Inflation is associated with the universe making a transition from its initial false-vacuum state to a true-vacuum state. This expansion took place over a period of about 10^{-32} s, during which the universe cooled to a temperature of about 3 K. **(c)** As the universe "settled in" to the true vacuum state, energy was released that reheated the universe to a temperature of 10^{27} K.

an analogous way, it is thought that at the time that the strong and electroweak forces decoupled, the universe "rolled downhill" to the *true vacuum,* a state of lower energy. The universe's transition to the true vacuum released energy that caused it to expand tremendously in a brief interval of time (Figure 27-6*b*). By the time the inflationary epoch had ended, about 10^{-32} s after the Big Bang, the universe had increased in scale by a factor of roughly 10^{50}.

The rapid expansion of the universe also gave rise to a rapid cooling. At the end of the inflationary epoch, the temperature of the universe may have dropped to about 3 K, about the same as the temperature that the cosmic background radiation has today. But as the universe finally settled into the vacuum-energy state, an additional amount of energy was released that went into *reheating* the universe to a temperature of 10^{27} K—about the same as it had before inflation began (Figure 27-7*c*). Thus, inflation caused the universe to expand tremendously while having no net effect on its temperature.

After the end of the inflationary epoch at $t = 10^{-32}$ s, the universe continued to expand and cool at a more sedate rate. At $t = 10^{-12}$ s, the temperature of the universe had dropped to 10^{15} K, the energy of the particles had fallen to 100 GeV, and there was a final spontaneous symmetry breaking and "freeze-out" that separated the electromagnetic force from the weak force. From that moment on, all four forces have interacted with particles essentially as they do today.

27-3 During inflation, all the mass and energy in the universe burst forth from the vacuum of space

WEB LINK 27.6

Inflation was a brief but stupendous expansion of the universe soon after the beginning of time. Physicists now realize that inflation helps explain where all the matter and radiation in the universe came from. To see how a violent expansion of space could create particles, we must first understand what quantum mechanics tells us about space.

Quantum Mechanics and the Heisenberg Uncertainty Principle

Quantum mechanics is the branch of physics that explains the behavior of nature on the atomic scale and smaller. For example, quantum mechanics tells us how to calculate the structure of atoms and the interactions between atomic nuclei. **Elementary particle physics** is the branch of quantum mechanics that deals with individual subatomic particles and their interactions, including the strong, weak, and electromagnetic forces that we discussed in Section 27-2.

The submicroscopic world of quantum mechanics is significantly different from the ordinary world around us. In the ordinary world we have no trouble knowing where things are. You know where your house is; you know where your car is; you know where this book is. In the subatomic world of electrons and nuclei, however, you can no longer speak with this same confidence. A certain amount of fuzziness, or uncertainty, enters into the description of reality at the incredibly small dimensions of the quantum world.

To appreciate the reasons for this uncertainty, imagine trying to measure the position of a single electron. To find out where it is located, you must observe it. And to observe it, you must shine a light on it. However, the electron is so tiny and has such a small mass that the photons in your beam of light possess enough energy to give the electron a mighty kick. As soon as a photon strikes the electron, the electron recoils in some unpredictable direction. Consequently, no matter how carefully you try to measure the precise location of an electron, you necessarily introduce some uncertainty into the speed of that electron.

WEB LINK 27.7

These ideas are at the heart of the **Heisenberg uncertainty principle,** first formulated in 1927 by the German physicist Werner Heisenberg, one of the founders of quantum mechanics. This principle states that there is a reciprocal uncertainty between position and momentum (momentum is equal to the mass of a particle multiplied by its velocity). The more precisely you try to measure the position of a particle, the more unsure you become of how the particle is moving. Conversely, the more accurately you determine the momentum of a particle, the less sure you are of its location. These restrictions are not a result of errors in making measurements; they are fundamental limitations inherent to all natural phenomena.

There is an analogous uncertainty involving energy and time. You cannot know the energy of a system with infinite precision at every moment in time. Over short time intervals, there can be great uncertainty about the amounts of energy in a subatomic system. Specifically, let ΔE be the smallest possible uncertainty in energy measured over a short interval of time Δt. (Astronomers and physicists often use the capital Greek letter delta, Δ, as a prefix to denote a small quantity or a small change in a quantity.)

Heisenberg uncertainty principle for energy and time

$$\Delta E \times \Delta t = \frac{h}{2\pi}$$

ΔE = uncertainty in energy

Δt = time interval over which energy is measured

h = Planck's constant = 6.625×10^{-34} J s

This equation says that the shorter the time interval Δt, the greater the energy uncertainty ΔE must be in order to ensure that the product of ΔE and Δt is equal to $h/2\pi$.

We might look upon the Heisenberg uncertainty principle as merely an unfortunate limitation on our ability to know everything with infinite precision. But, in fact, this principle provides startling insights into the nature of the universe.

Spontaneously Created Matter and Antimatter

We have seen that one of the important conclusions of Einstein's special theory of relativity is the equivalence of mass and energy: $E = mc^2$ (see Section 16-1). There is nothing uncertain about the speed of light (c), which is an absolute constant. Therefore, any uncertainty in the energy of a physical system can be attributed to an uncertainty Δm in the mass. Thus,

$$\Delta E = \Delta m \times c^2$$

Combining this expression with the previous equation, we obtain

Heisenberg uncertainty principle for mass and time

$$\Delta m \times \Delta t = \frac{h}{2\pi c^2}$$

Δm = uncertainty in mass

Δt = time interval over which mass is measured

h = Planck's constant = 6.625×10^{-34} J s

c = speed of light = 3.00×10^8 m/s

Quantum mechanics reveals that "empty" space is not empty, but is seething with particles and antiparticles that appear and then annihilate

This result is astonishing. It means that over a very brief interval Δt of time, we cannot be sure how much matter there is in a particular location, even in "empty space." During this brief moment, matter can spontaneously appear and then disappear. The greater the amount of matter (Δm) that appears spontaneously, the shorter the time interval (Δt) it can exist before disappearing into nothingness. This bizarre state of affairs is a natural consequence of quantum mechanics.

No particle can appear spontaneously by itself, however. For each particle created, so is a second, almost identical **antiparticle**; particles are made of matter, and antiparticles are made of antimatter. In other words, equal amounts of matter and **antimatter** come into existence and then disappear.

Despite its exotic name, there is actually nothing terribly mysterious about antimatter. A particle and an antiparticle are identical in almost every respect; their main distinction is that they carry opposite electric charges. For example, an ordinary electron (e^-) carries a negative charge; the corresponding antiparticle has the same mass but a positive charge, which is why it is called a **positron** (e^+). Because particles and antiparticles come and go in pairs, the total electric charge in the universe remains constant. Particles that have no electric charge can also have corresponding antiparticles. An example is the neutrino (ν); we met its antiparticle, the antineutrino ($\bar{\nu}$), in Section 27-2. The antineutrino is also electrically neutral, but differs from the neutrino in having opposite values of other, more subtle physical properties.

A spontaneously created particle-antiparticle pair lasts for only an incredibly brief time. For example, consider an electron and a positron, each with a mass of 9.11×10^{-31} kg. If we rewrite the Heisenberg uncertainty principle for mass and time to solve for Δt and then substitute the combined mass of $2 \times 9.11 \times 10^{-31}$ kg, we find that a spontaneously created electron-positron pair can last for a time

$$\Delta t = \frac{1}{\Delta m}\frac{h}{2\pi c^2} = \frac{1}{2 \times 9.11 \times 10^{-31}} \times \frac{6.625 \times 10^{-34}}{2\pi(3.00 \times 10^8)^2}$$
$$= 6.43 \times 10^{-22}\ \text{s}$$

In other words, an electron and a positron can spontaneously appear and then disappear without violating any laws of physics—but they can remain in existence for no longer than 6.43×10^{-22} s. The more massive the particle, the shorter the time interval it can exist. For example, the proton has about 2000 times more mass than the electron. Pairs of protons and their antiparticles, called **antiprotons,** can appear and disappear spontaneously, but they exist for only 1/2000 as long as pairs of electrons and positrons do.

Virtual Pairs

Spontaneous creation can and does happen absolutely anywhere and at any time, not just under the unusual conditions of the early universe. (It is happening right now in the space between this book and your eyes.) Quantum mechanics tells us that if a process is not strictly forbidden, then it must occur. Pairs of every conceivable particle and antiparticle are constantly being created and destroyed at every location across the universe. However, we have no way of observing these pairs directly without violating the uncertainty principle. For this reason, they are called **virtual pairs.** They do not "really" exist in the same sense as ordinary particles; they "virtually" exist. The particles that are exchanged in the four fundamental forces (see Section 27-2) are also virtual particles.

Although virtual pairs of particles and antiparticles cannot be observed directly, their effects have nonetheless been detected. Imagine, for example, an electron in orbit about the nucleus of an atom, such as a hydrogen atom. Ideally, the electron should follow its orbit in a smooth, unhampered fashion. However, because of the constant brief appearance and disappearance of pairs of particles and antiparticles, minuscule electric fields exist for extremely short intervals of time. These tiny, fleeting electric fields cause the electron to jiggle slightly in its orbit. This jiggling produces slight changes in the energies of different electron orbits in the hydrogen atom, which manifest themselves as a minuscule shift in the wavelengths of the hydrogen spectral lines. (We discussed the connection between the energies of electron orbits in hydrogen and the hydrogen spectrum in Section 5-8.)

This shift was first detected in 1947 by the American physicists Willis Lamb and R. C. Retherford and today is known as the **Lamb shift.** The Lamb shift provides powerful support for the idea that every point in space, all across the universe, is seething with virtual pairs of particles and antiparticles. In this sense, "empty space" is actually not empty at all. Figure 27-8 sketches the constant appearance and disappearance of virtual particles and antiparticles.

In some circumstances, virtual pairs can become *real* pairs of particles and antiparticles, a phenomenon called **pair production** (Figure 27-9*a*). In pair production, pairs of highly energetic gamma rays (photons) can convert their energy into pairs of particles and antiparticles. Quite simply, the gamma ray photons disappear upon colliding into each other, and a particle and an antiparticle appear in their place. These particles and antiparticles come from nature's ample supply of virtual pairs. The gamma rays provide a virtual pair with so much energy that the virtual particles can appear as real particles in the real world.

Pair production is routinely observed in high-energy nuclear experiments (see Figure 27-6*b*). Indeed, it is one of the ways in which physicists manufacture exotic species of particles and antiparticles. The only requirement is that nature's balance sheet be

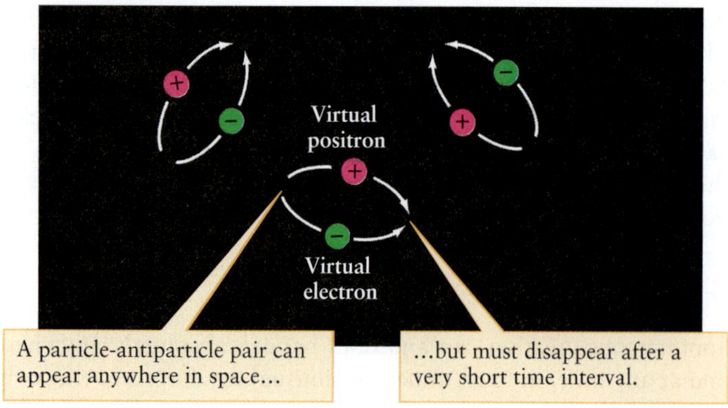

Figure 27-8

Virtual Pairs Pairs of particles and antiparticles can appear and then disappear anywhere in space provided that each pair exists only for a very short time interval, as dictated by the uncertainty principle. In this sketch, electrons are shown in green and positrons are shown in red.

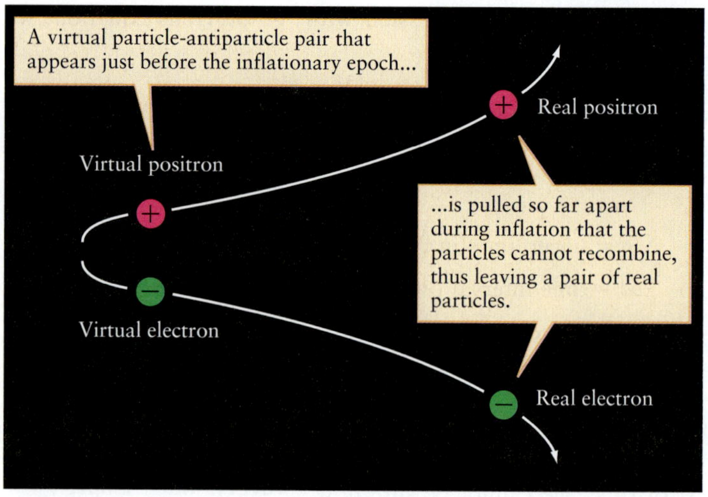

Figure 27-10

Inflation: From Virtual to Real Particles The universe expanded by such a tremendous factor during the inflationary epoch that the members of virtual particle-antiparticle pairs could no longer find each other. As a result, these virtual particles and antiparticles became real particles and antiparticles.

satisfied. To create a particle and an antiparticle having a total mass M, the incoming gamma-ray photons must possess an amount of energy E that is greater than or equal to Mc^2. If the photons carry too little energy (less than Mc^2), pair production will not occur. Likewise, the more energetic the photons, the more massive the particles and antiparticles that can be manufactured. While Figure 27-9*a* shows this process of pair production, Figure 27-9*b* shows the inverse process of **annihilation,** in which a particle and antiparticle collide with each other and are converted into high-energy gamma rays.

Around 1980 physicists began applying these ideas to their thinking about the creation of the universe. During the inflationary epoch, space was expanding with explosive vigor. As we have seen, however, all space is seething with virtual pairs of particles and antiparticles. Normally, a particle and an antiparticle have no trouble getting back together in a time interval (Δt) short enough to be in compliance with the uncertainty principle. During inflation, however, the universe expanded so fast that particles were rapidly separated from their corresponding antiparticles. Deprived of the opportunity to recombine and annihilate, these virtual particles became *real* particles in the real world. In this way, the universe was flooded with particles and antiparticles created by the violent expansion of space (Figure 27-10). For a "report card" on cosmology theories, including inflation, see the article at the end of this chapter.

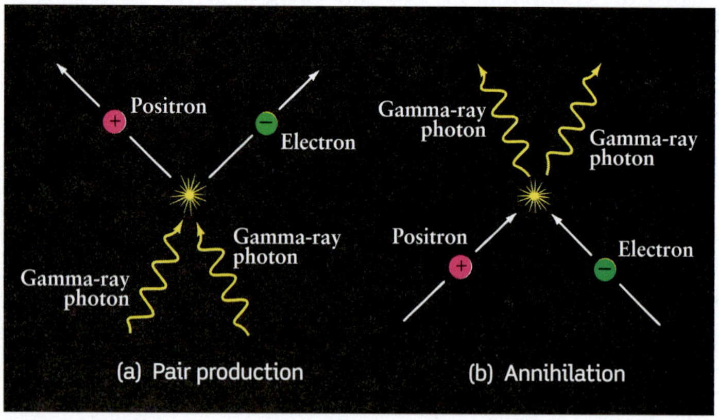

Figure 27-9

Pair Production and Annihilation **(a)** Pairs of virtual particles can be converted into real particles by high-energy gamma-ray photons. In this illustration, an electron (shown in green) and a positron (in red) are produced. This process can take place only if the combined energy of the two photons is no less than Mc^2, where M is the total mass of the electron and positron. **(b)** Conversely, a particle and an antiparticle can annihilate each other and be transformed into energy in the form of gamma rays.

27-4 As the early universe expanded and cooled, most of the matter and antimatter annihilated each other

As soon as the flood of matter and antimatter appeared in the universe, collisions between particles and antiparticles began to produce numerous high-energy gamma rays. As these gamma rays collided, they promptly turned back into the particles and antiparticles from which they came. As a result, the rate of pair production soon equaled the rate of annihilation. For example, for every electron and positron that annihilated each other to create gamma rays (Figure 27-9*b*), two gamma rays collided elsewhere to produce an electron and a positron (Figure 27-9*a*). In other words, annihilation and pair production reactions proceeded with equal vigor, and as many particles and antiparticles were being created as were being destroyed.

As the universe continued to expand, all the gamma-ray photons became increasingly redshifted. As a result, the temperature of the radiation field fell. Due to their frequent interaction, radiation and particles of all kinds were in **thermal equilibrium:** All particle species, including photons, were at the same temperature. Hence, as the radiation temperature decreased, the temperature of particles of different types decreased as well.

From Quark Confinement to Particle-Antiparticle Annihilation

The first change in the population of particles and antiparticles occurred at $t = 10^{-6}$ s, when the temperature was 10^{13} K and particles were colliding with energies of roughly 1 GeV. Prior to this moment, particles collided so violently that individual protons and neutrons could not exist, being constantly fragmented into quarks. After this time, appropriately called the period of **quark confinement,** quarks were finally able to stick together and became confined within individual protons and neutrons.

> **The cosmic background radiation we see today was spawned from a vast sea of particles and antiparticles in the early universe**

LOOKING DEEPER 27.1

As the universe continued to expand, temperatures eventually became so low that the gamma rays no longer had enough energy to create particular kinds of particles and antiparticles. We say that the temperature dropped below the particular particle's **threshold temperature.** Collisions between these types of particles and antiparticles continued to add photons to the cosmic-radiation background, but collisions between photons could no longer replenish the supply of particles and antiparticles.

At the same time that quark confinement became possible so that protons and neutrons appeared, the universe also became cooler than the 10^{13}-K threshold temperatures of both protons and neutrons. No new protons or neutrons were formed by pair production, but the annihilation of protons by antiprotons and of neutrons by antineutrons continued vigorously everywhere throughout space. This wholesale annihilation dramatically lowered the matter content (particles and antiparticles) of the universe, while simultaneously increasing the radiation (photon) content.

A little later, when the universe was about 1 second old, its temperature fell below 6×10^9 K, the threshold temperature for electrons and positrons. A similar annihilation of pairs of electrons and positrons further decreased the matter content of the universe while raising its radiation content. This radiation field, which fills all space, is the *primordial fireball* discussed in Section 26-5. This fireball, which dominated the universe for the next 380,000 years, therefore derived much of its energy from the annihilation of particles and antiparticles during the first second after the Big Bang.

Now we have a dilemma. If there had been perfect symmetry between particles and antiparticles, then for every proton there should have been an antiproton. For every electron, there should likewise have been a positron. Consequently, by the time the universe was 1 second old, every particle would have been annihilated by an antiparticle, leaving no matter at all in the universe.

Obviously, a total annihilation of all matter and antimatter never happened. The planets, stars, and galaxies we see in the sky are made of matter, not antimatter. If there were still substantial amounts of antimatter in the universe, it would eventually collide with ordinary matter. We would then see copious amounts of gamma rays being emitted from the entire sky. While we do observe gamma-ray photons from various locations in the universe, they are neither numerous enough nor of the right energy to indicate the presence of much antimatter. Thus, there must have been an excess of matter over antimatter immediately after the Big Bang, so that the particles outnumbered the antiparticles.

We can estimate the extent of this asymmetry between matter and antimatter. As noted in Section 26-5, there are roughly 10^9 photons today in the microwave background for each proton and neutron in the universe. Thus, for every 10^9 antiprotons, there must have been 10^9 plus one ordinary protons, leaving one surviving proton after annihilation. Similarly, for every 10^9 positrons, there must have been 10^9 plus one ordinary electrons. Theories of elementary particles and their interactions do indeed predict a slight preference for matter over antimatter of just this sort.

27-5 A background of neutrinos and most of the helium in the universe are relics of the primordial fireball

The early universe must have been populated with vast numbers of neutrinos (ν) and their antiparticles, the antineutrinos ($\bar{\nu}$). These particles have a very small mass, so their threshold temperature is quite low. These particles take part in the nuclear reactions that transform neutrons into protons and vice versa. For example, a neutron can decay into a proton by emitting an electron and an antineutrino:

$$n \rightarrow p + e^- + \bar{\nu}$$

This radioactive decay happens quickly (its half-life is about 10.5 minutes), which is why we do not find free neutrons floating around in the universe today. In the first 2 seconds after the Big Bang, however, neutrons were also created by collisions between protons and electrons:

$$p + e^- \rightarrow n + \nu$$

This reaction kept the number of neutrons approximately equal to the number of protons. This balance was maintained only as long as collisions between protons and electrons were frequent. But the number of electrons decreased precipitously as the temperature of the universe fell below 6×10^9 K and electrons and positrons annihilated each other without being replenished (Section 27-4). By the time the universe was about 2 seconds old, no new neutrons were being formed, the natural tendency for neutrons to decay into protons took over, and the number of neutrons began to decline.

Spawning the First Nuclei

WEB LINK 27.8

Before many of the neutrons could decay into protons, they began to combine with protons to form nuclei. Nuclei of helium, the first element more massive than hydrogen, consist of either two protons and two neutrons (^{4}He) or two protons and a single neutron (^{3}He). It is exceedingly improbable that two protons and one or two neutrons should all simultaneously collide with one another to form a helium nucleus. Instead, helium nuclei are built in a series of steps. The first step is to have a single proton and a single neutron combine to form deuterium (^{2}H), sometimes called "heavy hydrogen." A photon (γ) is emitted in this process, so we write this reaction as

The first atomic nuclei formed within a quarter-hour after the Big Bang

$$p + n \rightarrow {}^2H + \gamma$$

Forming deuterium does not immediately lead to the formation of helium, however. The problem is that deuterium nuclei are easily destroyed, because a proton and a neutron do not stick together very well. Indeed, in the early universe, high-energy gamma rays easily broke deuterium nuclei back down into independent protons and neutrons. As a result, the synthesis of helium could not get beyond the first step. This block to the creation of helium is called the **deuterium bottleneck.**

When the universe was about 3 minutes old, the background radiation had cooled enough that its photons no longer had enough energy to break up the deuterium. By this time, most of the neutrons had decayed into protons, and protons outnumbered neutrons by about 6 to 1. Because deuterium nuclei could now survive, the remaining neutrons combined with protons and rapidly produced helium. (The *Cosmic Connections: The Proton-Proton Chain* figure for Chapter 16 depicts a similar sequence of reactions that take place in the core of the present-day Sun.)

The result was what we find in the universe today—about one helium atom for every ten hydrogen atoms. In addition to helium, nuclei of lithium (Li, which has three protons) and beryllium (Be, which has four) were also produced in small numbers. The process of building up nuclei such as deuterium and helium from protons and neutrons is called **nucleosynthesis** (Figure 27-11).

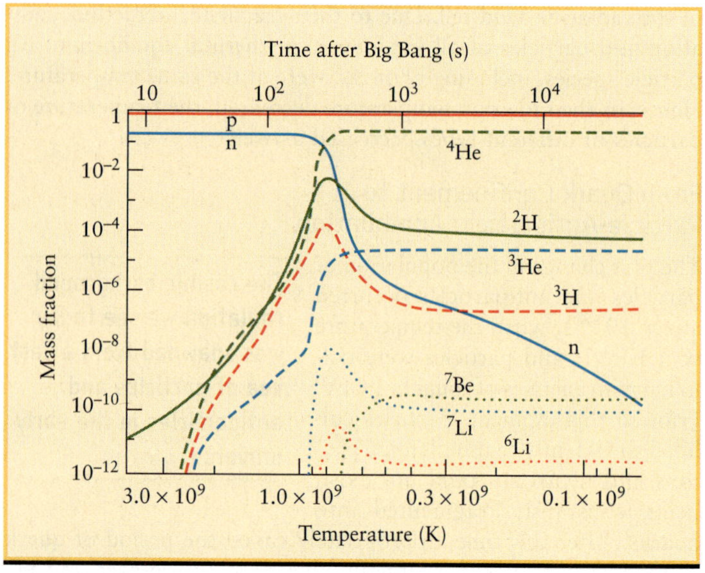

Figure 27-11

Nucleosynthesis in the Early Universe This graph shows how nuclei were produced between 10 seconds and 10 hours after the Big Bang. The vertical axis shows the fraction of the total mass that was in each type of particle or nucleus (p = proton, n = neutron, ^{2}H = deuterium, ^{3}He and ^{4}He = helium, ^{6}Li and ^{7}Li = lithium, ^{7}Be = beryllium). Very few nuclei were formed before the universe was 10 seconds old, due to the phenomenon of the deuterium bottleneck, which occurred at times earlier than those shown here. By about 10^3 seconds (roughly 15 minutes) after the Big Bang, the temperature had dropped below 4×10^8 K, no further nucleosynthesis was possible, and the relative amounts of different nuclei stabilized. The number of free neutrons declined rapidly as these particles decayed into protons, electrons, and antineutrinos. (Adapted from R. V. Wagoner)

Because nuclei have positive electric charges, bringing them together to form more massive nuclei requires that they overcome their mutual electric repulsion. They are unable to do so if they are moving too slowly, which will be the case if the temperature is too low. As a result, by about 15 minutes after the Big Bang the universe was no longer hot enough for nucleosynthesis to take place. Only the four lightest elements (hydrogen, helium, lithium, and beryllium) were present in appreciable numbers. The heavier elements would be formed only much later, once stars had formed and nuclear reactions within those stars could manufacture carbon, nitrogen, oxygen, and all the other elements.

CAUTION! Keep in mind that only *nuclei* formed in the first 15 minutes of the history of the universe. It would be another 380,000 years before temperatures became low enough for these nuclei to combine with electrons to form neutral atoms.

The Neutrino-Antineutrino Background

While nuclei were being formed in the early universe, what happened to all those primordial neutrinos and antineutrinos that had interacted so vigorously with the protons and neutrons before the universe was 2 seconds old? The answer is that by $t = 2$ s, matter was sufficiently spread out so that the universe became transparent to neutrinos and antineutrinos. From that time on, neutrinos and antineutrinos could travel across the universe unimpeded. (Recall from Section 16-4 that Earth itself is virtually transparent to neutrinos from the Sun.)

The neutrinos and antineutrinos that were liberated at $t = 2$ s should now fill the universe much as the cosmic microwave background does. Indeed, these ancient neutrinos and antineutrinos may be about as populous today as the photons in the microwave background (of which there are 4.1×10^8 per cubic meter). The neutrino-antineutrino background should be slightly cooler than the photon background, which received extra energy from electron-positron annihilations. Physicists estimate that the current temperature of the neutrino-antineutrino background is about 2 K, as opposed to 2.725 K for the microwave background. Unfortunately, because neutrinos and antineutrinos are so difficult to detect, we do not yet have direct evidence of the neutrino-antineutrino background.

27-6 Galaxies and the first stars formed from density fluctuations in the early universe

The distribution of matter in the universe today is quite lumpy. Stars are grouped together in galaxies, galaxies into clusters, and clusters into superclusters that stretch across 50 Mpc (150 million ly) or more (see Section 24-6). Furthermore, galaxies seem to be concentrated along enormous sheets, which in turn surround voids measuring 30 to 120 Mpc (100 million to 400 million ly) across. Figures 24-23 and 24-24 show these features, which characterize the large-scale structure of the universe. How did this large-scale structure arise from the chaos of the primordial fireball? When did stars first appear in the universe? And when and how did galaxies first form?

Density Fluctuations and the Jeans Length

At first glance the origin of large-scale structure seems puzzling, because the early universe must have been exceedingly smooth. To see why, think back to the era of recombination that occurred 380,000 years after the Big Bang (see Section 26-5). Before recombination, high-energy photons were constantly and vigorously colliding with charged particles throughout all space. After recombination, the universe became transparent, and these photons stopped interacting with the matter in the universe. Astronomers say that matter "decoupled" from radiation during the era of recombination. Because the cosmic microwave background is extremely isotropic, we can conclude that the matter with which these photons once collided so frequently must also have been spread smoothly across space.

The distribution of matter during the early universe could not have been *perfectly* uniform, however. If it had been, it would still have to be absolutely uniform today; there would now be only a few atoms per cubic meter of space, with no stars and no galaxies. Consequently, there must have been slight lumpiness, or **density fluctuations,** in the distribution of matter in the early universe. These fluctuations are thought to have originated in the very early universe, even before the inflationary epoch. Infinitesimally small quantum fluctuations in density, which are required by the Heisenberg uncertainty principle (see Section 27-3), were stretched during inflation to appreciable size. Through the action of gravity, these fluctuations initiated the clumping that eventually grew to become the galaxies and clusters of galaxies that we see today throughout the universe. As we saw in Section 26-5 and Section 26-8, the pattern of density fluctuations became imprinted on the cosmic background radiation during the era of recombination. Figure 26-14 shows a map of these fluctuations obtained from the WMAP microwave background spacecraft.

Our understanding of how gravity can amplify density fluctuations dates back to 1902, when the British physicist James Jeans solved a problem first proposed by Isaac Newton. Suppose that you have a gas with only very tiny fluctuations in density, as shown in Figure 27-12*a*. These regions of higher density will then gravitationally attract nearby material and thus gain mass. As the regions become more massive, however, the pressure of the gas inside these regions will also increase, which can make these regions expand and disperse. Jeans analyzed these opposing effects and asked the question: Under what conditions does gravity overwhelm gas pressure so that a permanent object can form?

Jeans proved that an object will grow from a density fluctuation provided that the fluctuation extends over a distance that exceeds the so-called **Jeans length** (L_J):

Jeans length for density fluctuations

$$L_J = \sqrt{\frac{\pi k T}{m G \rho_m}}$$

L_J = Jeans length

k = Boltzmann constant = 1.38×10^{-23} J/K

T = temperature of the gas (in kelvins)

m = mass of a single particle in the gas (in kilograms)

G = universal constant of gravitation = 6.67×10^{-11} N · m²/kg²

ρ_m = average density of matter in the gas

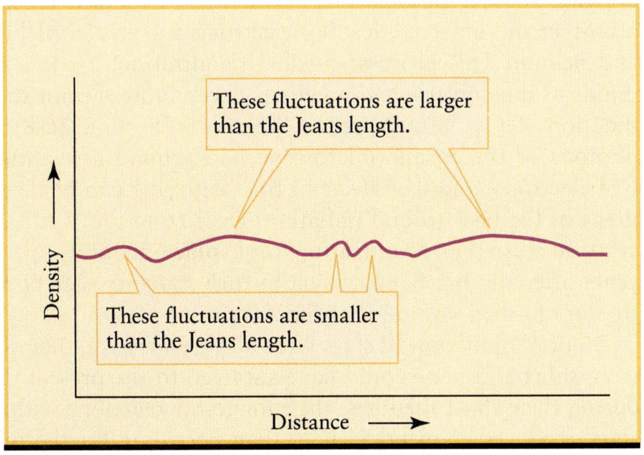

(a) At an early time

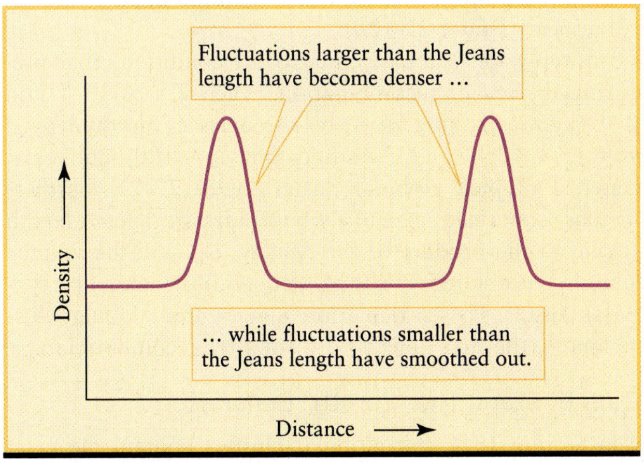

(b) At a later time

Figure 27-12

The Growth of Density Fluctuations (a) This conceptual illustration shows small density fluctuations in the distribution of matter shortly after the era of recombination. **(b)** If the size of a fluctuation is greater than the Jeans length (L_J), it becomes gravitationally unstable and can grow in amplitude.

Figure 27-13 RIVUXG

A Globular Cluster A typical globular cluster contains 10^5 to 10^6 stars, each with an average mass of about 1 $M_\odot$, so the total mass of a typical cluster is 10^5–10^6 $M_\odot$. Cluster diameters range from about 6 to 120 pc (20 to 400 ly). Because these masses and diameters are comparable to the Jeans length (L_J) during the era of recombination, astronomers suspect that globular clusters were among the first objects to form in the universe. (Hubble Heritage Team/AURA/STScI/NASA)

Density fluctuations that extend across a distance larger than the Jeans length tend to grow, while fluctuations smaller than L_J tend to disappear (Figure 27-12*b*).

We can apply the Jeans formula to the conditions that prevailed during the era of recombination, when $T = 3000$ K and $\rho_m = 10^{-18}$ kg/m^3. Taking m to be the mass of the hydrogen atom ($m = 1.67 \times 10^{-27}$ kg), we find that $L_J = 100$ light-years, the diameter of a typical globular cluster (Figure 27-13). Furthermore, the mass contained in a cube whose sides are 1 Jeans length in size (equal to the product of the density, ρ_m, and the volume of the cube, $L_J{}^3$) is about 5×10^5 $M_\odot$, equal to the mass of a typical globular cluster. These calculations suggest that globular clusters were among the first objects to form after recombination.

Population III Stars: The "Zeroth" Generation

We saw in Section 19-4 that globular clusters contain the most ancient stars we can find in the present-day universe. These are Population II stars with a low percentage of metals (elements heaver than hydrogen and helium), and are of an earlier stellar generation than metal-rich Population I stars like the Sun (see Section 19-5). However, the Population II stars in globular clusters *cannot* be the very first stars to have formed after the Big Bang. Those first stars could have contained only hydrogen, helium, and tiny amounts of lithium and beryllium; as Figure 27-11 shows, these were the only elements whose nuclei formed in the early universe. Hence, these original stars would have contained an even smaller percentage of metals than the Population II stars found in globular clusters. Such "zeroth-generation" stars are called **Population III stars.**

The "zeroth generation" of stars were much more massive and luminous than stars today

ANIMATION 27.1

Like stars in the present-day universe, Population III stars would have formed from clouds of gas. These ancient gas clouds were composed almost exclusively of hydrogen and helium atoms, and such clouds have higher internal pressures than do metal-rich clouds of the same temperature. A star can only form when the mutual gravitation of the various parts of a cloud (which tends to make the cloud collapse) overcomes the internal pressure of the cloud (which tends to prevent collapse). Hence, Population III stars could only form if their mass (and hence their mutual gravitation) was rather large. Calculations suggest that these stars had masses from 30 to 1000 $M_\odot$, compared to the range of 0.4 to roughly 120 $M_\odot$ for modern stars. Even the smallest Population III star would rank among the largest stars observed today.

Although no Population III stars have yet been observed directly, we have at least indirect evidence that they existed. Infrared images from the Spitzer Space Telescope like the one that opens this chapter reveal a distant infrared background of starlight that is what we would expect to see from these zeroth-generation stars.

Another bit of evidence follows from the tremendous energy output of these stars: A 1000-$M_\odot$ Population III star would have been millions of times more luminous than the Sun and have a surface temperature in excess of 10^5 K, causing it to emit a flood of short-wavelength, high-energy photons. The photons from even a small number of such stars would have ionized most of the atoms in the universe, leaving electrons and nuclei of hydrogen and helium. This process is called **reionization,** a name that reminds us the universe had previously been ionized prior to recombination at t = 380,000 years. We saw in Section 26-8 that the photons of the cosmic microwave background are scattered by free electrons, and the effects of this scattering can be detected in maps of the background radiation. Data from the WMAP spacecraft suggest that reionization took place around 400 million years after the Big Bang, which in turn suggests that Population III stars formed around that time.

Since Population III stars were all very massive, their lifetimes were short and none could have survived to the present day. But during their short lifetimes, thermonuclear reactions within these stars produced elements heavier than beryllium for the first time in the history of the universe. What is more, calculations suggest that when these stars exploded—and due to their great mass, all of them did—they did not leave a white dwarf, neutron star, or black hole behind. Instead, all of their mass was ejected into space to be incorporated into the next generation of stars. The presence

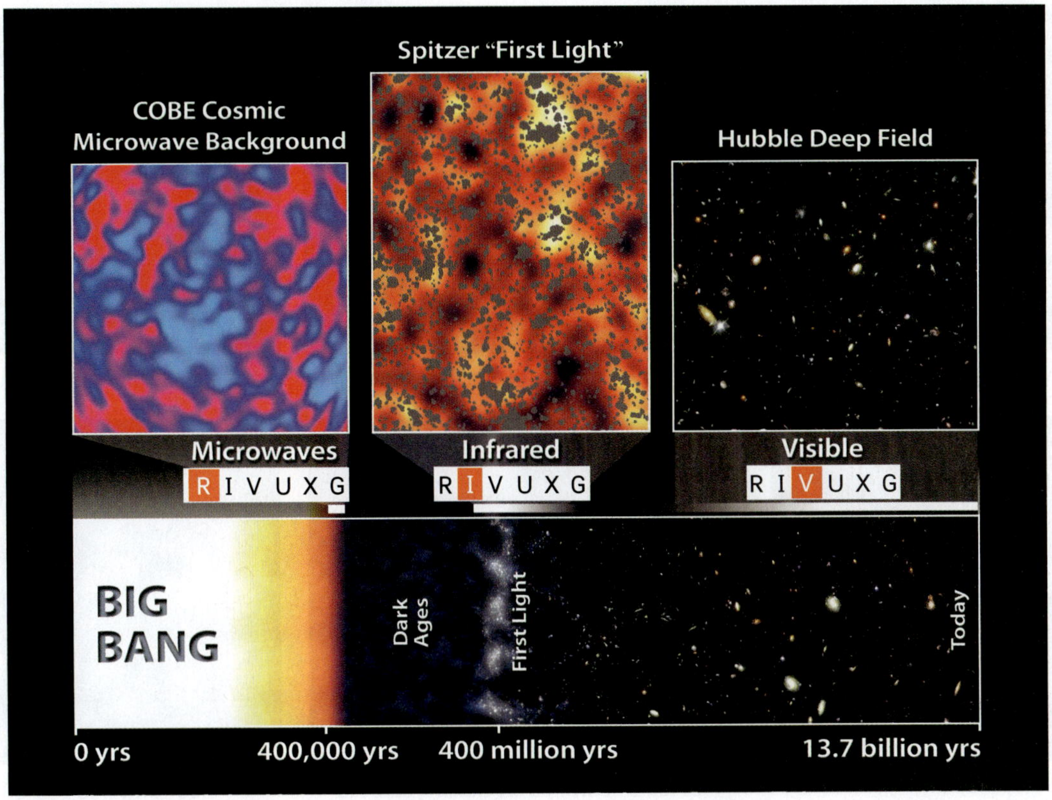

Figure 27-14

A Timeline of Light in the Universe The oldest light that we can see today is the cosmic background radiation, which comes from a time 380,000 years after the Big Bang when the universe first became transparent. This light has a redshift of about $z = 1100$ and appears in the microwave spectrum. Some 400 million years later at a redshift of about $z = 11$ the first stars appeared; their light is now redshifted to infrared wavelengths. Galaxies formed more recently and can be seen at visible wavelengths. (NASA; JPL-Caltech; and A. Kashlinsky, Goddard Space Flight Center)

of heavy elements in the ejected material meant that when this material subsequently formed into clouds, the internal pressure of these clouds was lowered substantially and it became possible for low-mass stars to form. This laid the foundation for today's universe, in which dim, low-mass stars are common and luminous, massive stars are the exception (see Figure 17-5).

The era from recombination at $t = 380{,}000$ years to the first stars at $t = 400$ million years is called the **dark ages.** The only photons present at that time were those that make up today's cosmic background radiation. The dark ages ended when the universe was filled for the first time with the light from stars (Figure 27-14).

Forming Large-Scale Structure

Once clumps the size of globular clusters had formed in the universe, how did they form into galaxies, clusters of galaxies, and larger structures? One issue that complicates this matter is the presence of dark energy, which acts to accelerate the expansion of the universe (see Section 26-6). This accelerated expansion pulls clumps of material away from each other and makes it more difficult for them to coalesce into larger structures. Another complication is that about 85% of the mass in the universe is in the form of dark matter, whose nature is not known (see Section 24-8). Researchers have hypothesized different types of dark matter in the hope of explaining the large-scale structure that we see. Neutrinos are an example of **hot dark matter,** so named because it consists of lightweight particles traveling at high speeds. **Cold dark matter,** on the other hand, consists of massive particles traveling at slow speeds. Examples include WIMPs (which we discussed in Section 23-4) as well as other exotic, speculative particles.

WEB LINK 27.9 Scientists use supercomputer simulations to see how different types of dark matter would influence the development of large-scale structure. Figure 27-15 shows the results of such a simulation for a flat universe with dark energy and *cold* dark matter. The simulation follows the motions of 2 million particles of cold dark matter in a box that expands as the universe expands. The box at the lower right of the figure, representing the present time (redshift $z = 0$) is 43 Mpc (160 million ly) a side. At earlier times, the box represents a volume whose side is smaller by a factor $1/(1 + z)$. For example, each side of the box for $z = 0.99$ is actually 1/1.99 as long as the box for $z = 0$.

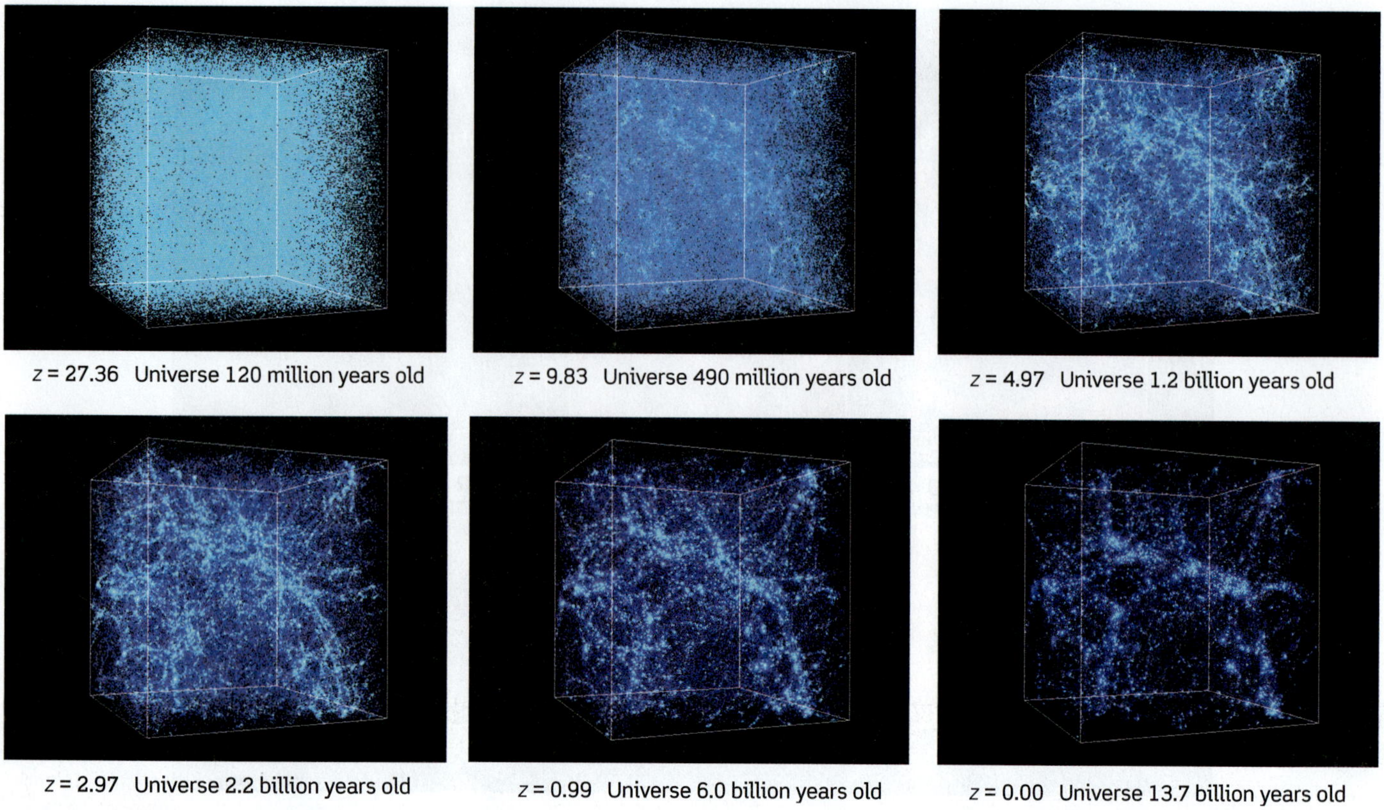

Figure 27-15

A Cold Dark Matter Simulation with Dark Energy These six views show the evolution of dark matter particles in a large, box-shaped volume of space. The box actually expands with time to follow the expansion of the universe and holds the same number of particles. In this figure, the boxes have been rescaled so they all appear at the same size. Small fluctuations in density are put into the simulation at the beginning (at upper left); these evolve over time to form structures that resemble those actually observed in our present-day universe ($z = 0.00$, shown at the lower right). (Simulations performed at the National Center for Supercomputer Applications by Andrey Kravtsov/University of Chicago and Anatoly Klypin/NMSU)

The simulation begins 120 million years after the Big Bang with an almost perfectly uniform distribution of particles, mimicking the tiny density fluctuations that must have been present just after inflation. A supercomputer then calculates how these particles move, based on Newton's laws in an expanding universe. As time goes on, the fluctuations grow into small, bright clumps whose sizes and masses are similar to those of galaxies. A large filament also forms, spanning the entire box from left to right. The simulation shows that no additional structures formed after the universe was about 6 billion years old, corresponding to redshift $z = 1$. The explanation is that after this time, the accelerating expansion of the universe becomes more important than gravitational attraction. The final frame of the simulation strongly resembles actual maps of galaxies in our present-day universe (see Figure 24-24).

Simulations similar to those in Figure 27-15 have also been carried out using *hot* dark matter in the form of neutrinos. A massless neutrino would always travel at the speed of light, just as a photon does. However, experiments show that neutrinos do have a small mass. (This nonzero mass is what allows one type of neutrino to transform into another. We saw in Section 16-4 that such transformations provided the explanation to the long-standing solar neutrino problem.) Hence, neutrinos travel slower than light, and slow down as the universe expands and cools. Slow-moving neutrinos would accumulate over time within density fluctuations, and the gravitational pull of these neutrinos on surrounding matter could eventually lead to the formation of clusters of galaxies.

A primary difference between simulations based on cold and hot dark matter is the way in which galaxies form. In calculations based on cold dark matter, the formation of galaxies takes place from the "bottom up." In these simulations, the densest gas undergoes collapse early in the history of the universe and stars begin to form. The regions of star formation stream along the filaments (Figure 27-16). When they meet at the intersections between filaments, they merge and group together into galaxies, then clusters of galaxies, then superclusters. But in calculations based on hot dark matter, galaxies form from the "top down."

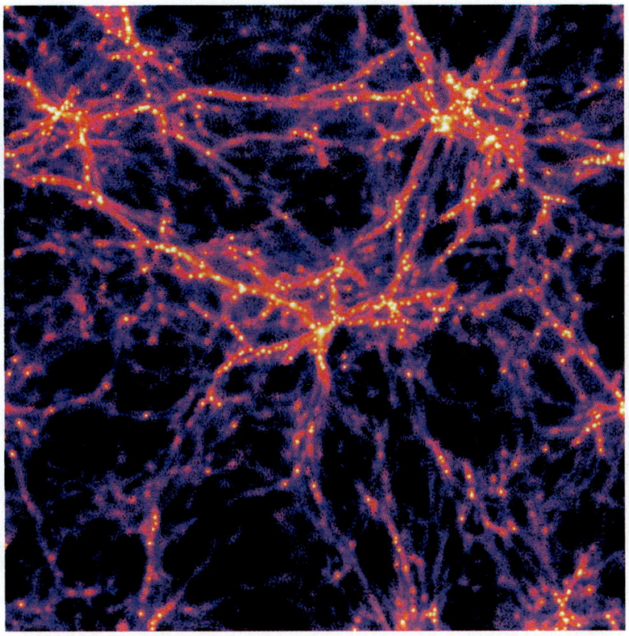

Figure 27-16

"Bottom-Up" Galaxy Formation: Simulation This image is taken from a cold dark matter simulation like that shown in Figure 27-15. A portion of the universe is shown at a time 2.2 billion years after the Big Bang, corresponding to redshift $z = 3.04$. The colors indicate the density of gas: Yellow is highest, red is medium, and blue is the lowest density. Over time, the gas tends to pile up at points where filaments intersect, forming galaxies and clusters of galaxies. (T. Theuns, MPA Garching/ESO)

Huge supercluster-sized sheets of matter form first and then fragment into galaxies. Observations of remote galaxies show that galaxies actually formed from the "bottom up" scenario. One piece of evidence for this is the image in Figure 27-17*a*, which shows a handful of "galaxy building blocks" at $z = 3.04$ (when the universe was 2.2 billion years old). These "building blocks," which have not yet coalesced into galaxies, lie within a long filament (see Figure 27-17*b*) that resembles those shown in the simulation of Figure 27-16. Figure 27-18 shows a collection of "building blocks" at a later stage in the process of merging into a galaxy. These observations strongly suggest that the dominant form of dark matter is cold, not hot.

What Large-Scale Structure Reveals

How might the universe have evolved if it had contained different amounts of cold dark matter and dark energy? Figure 27-19 shows some simulations designed to explore these possibilities. If the density of matter in the universe is kept constant for various simulations, the simulations predict approximately the same structure for different values of the dark energy density parameter Ω_Λ defined in Section 26-6 (see Figure 27-19*a* and Figure 27-19*b*). But if too large a matter density is used in the simulation, the voids between galaxies are smaller than what we actually observe in our universe (Figure 27-19*c*). Hence, observations of galaxy clustering coupled with supercomputer simulations of galaxy formation help determine the matter density of our universe. (We made use of this idea in Section 26-7. The brown band in Figure 26-19 shows the constraints on cosmological parameters from these observations and simulations of galaxies.)

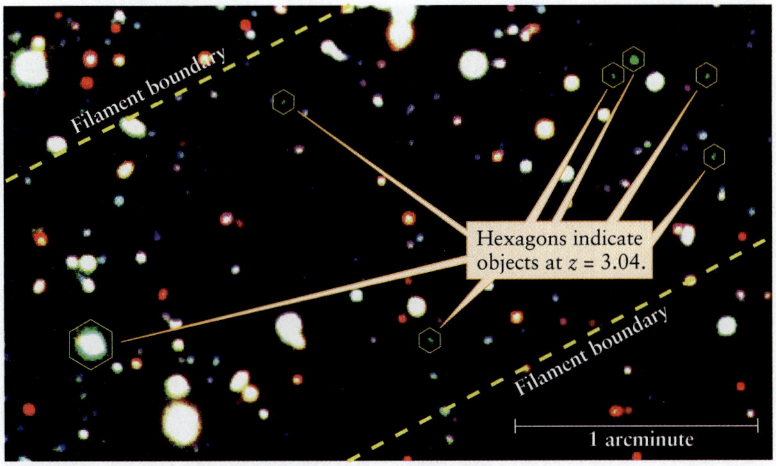

(a) High-redshift objects that lie within a filament

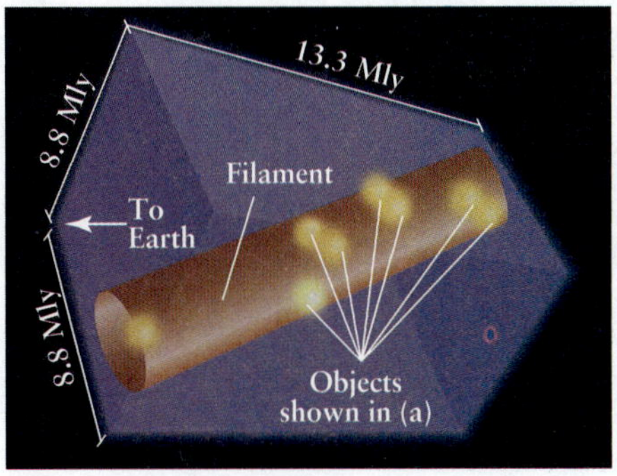

(b) Illustration of the filament

Figure 27-17 R I V U X G

"Bottom-Up" Galaxy Formation: Observation **(a)** The hexagons in this image from the Very Large Telescope show the positions of a number of sub-galaxy-sized objects at a redshift $z = 3.04$, the same as in the simulation shown in Figure 27-16. Excited hydrogen atoms in these objects emit ultraviolet photons, which are redshifted to visible wavelengths. This redshifted emission gives these objects a characteristic green color. (The object at lower left actually lies in front of a much brighter quasar.) **(b)** The objects in (a) all lie within an immense filament. The purple box shows the volume of space studied in this observing program. The dimensions are given in millions of light-years (Mly). (European Southern Observatory)

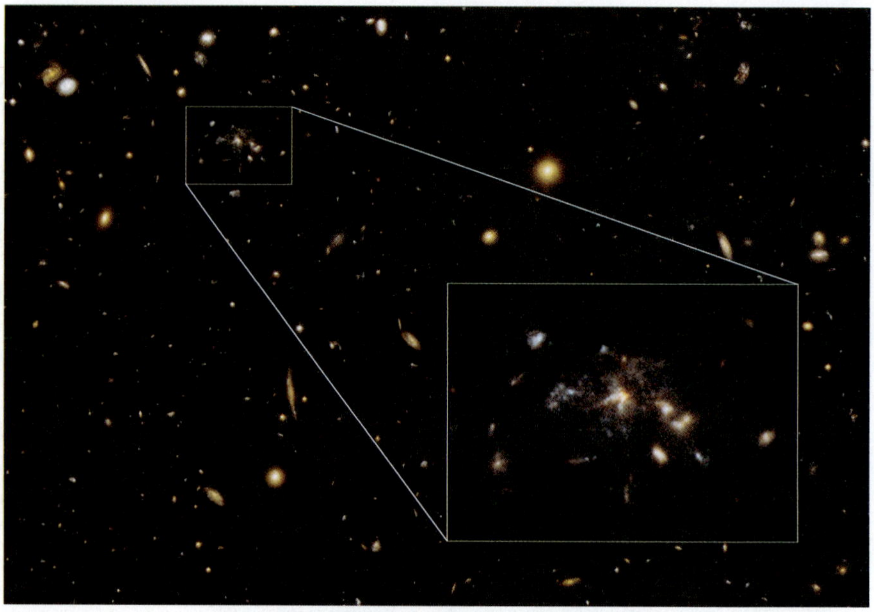

Figure 27-18 R I V U X G

A Galaxy Under Construction This Hubble Space Telescope image shows dozens of small galaxies in the process of merging into a single large galaxy. We see this galaxy at a redshift $z = 2.2$, corresponding to a time 3.1 billion years after the Big Bang. (NASA; ESA; G. Miley and R. Overzier, Leiden Observatory; and the ACS Science Team)

The best match to the observed distribution of galaxy clusters and to the cosmic background radiation data (see Figure 26-21) is a model like that shown in Figure 27-15, with dark energy and cold dark matter in the proportions listed in Table 26-2.

The *Cosmic Connections: The History of the Universe* summarizes the past history of our universe down to the present day. As we discussed in Section 26-7, the *future* of our universe is less certain, and depends on the detailed character of dark energy. More detailed data about galaxy clusters and the cosmic background radiation will be needed to pin down the future evolution of our universe.

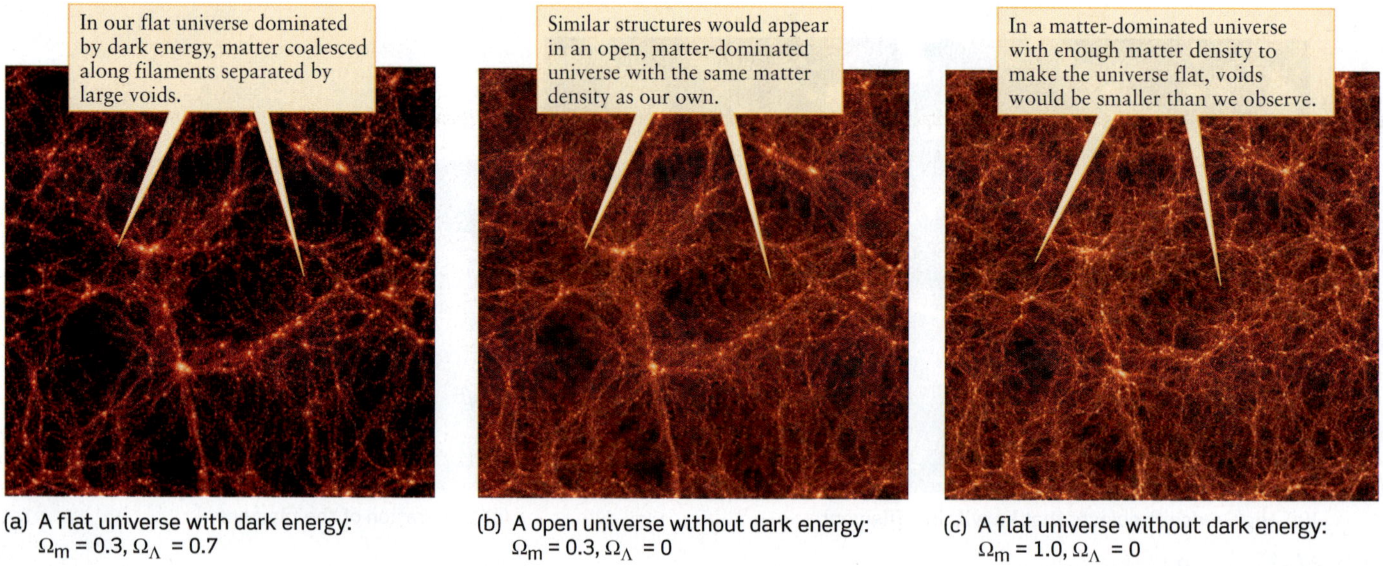
In our flat universe dominated by dark energy, matter coalesced along filaments separated by large voids.
Similar structures would appear in an open, matter-dominated universe with the same matter density as our own.
In a matter-dominated universe with enough matter density to make the universe flat, voids would be smaller than we observe.
(a) A flat universe with dark energy: $\Omega_m = 0.3, \Omega_\Lambda = 0.7$
(b) A open universe without dark energy: $\Omega_m = 0.3, \Omega_\Lambda = 0$
(c) A flat universe without dark energy: $\Omega_m = 1.0, \Omega_\Lambda = 0$

Figure 27-19

Using Simulations to Constrain the Matter Density of the Universe Cold dark matter simulations like those in Figures 27-15 and 27-16 help astronomers determine the value of the matter density parameter Ω_m. These three simulations show a portion of the universe at $z = 0$. **(a)** A simulation with $\Omega_m = 0.3$ and $\Omega_\Lambda = 0.7$, close to the values for our universe, gives a good match to the observed distribution of filaments and voids. **(b)** Nearly as good a match is obtained if we keep $\Omega_m = 0.3$, but eliminate dark energy so that $\Omega_\Lambda = 0$. **(c)** If we use a larger value of Ω_m, the distribution of matter in the simulation is a poor match to our universe. (Simulation by the Virgo Supercomputing Center using computers based at the Computer Center of the Max-Planck-Institute in Garching and at the Edinburgh Parallel Computer Centre)

COSMIC CONNECTIONS

The History of the Universe

Research in astronomy, elementary particle physics, and nuclear physics has allowed scientists to piece together the grand sweep of events over the first billion years of cosmic history. The greatest drama occurred in the first few minutes, and these graphs have been drawn to emphasize those earliest moments.

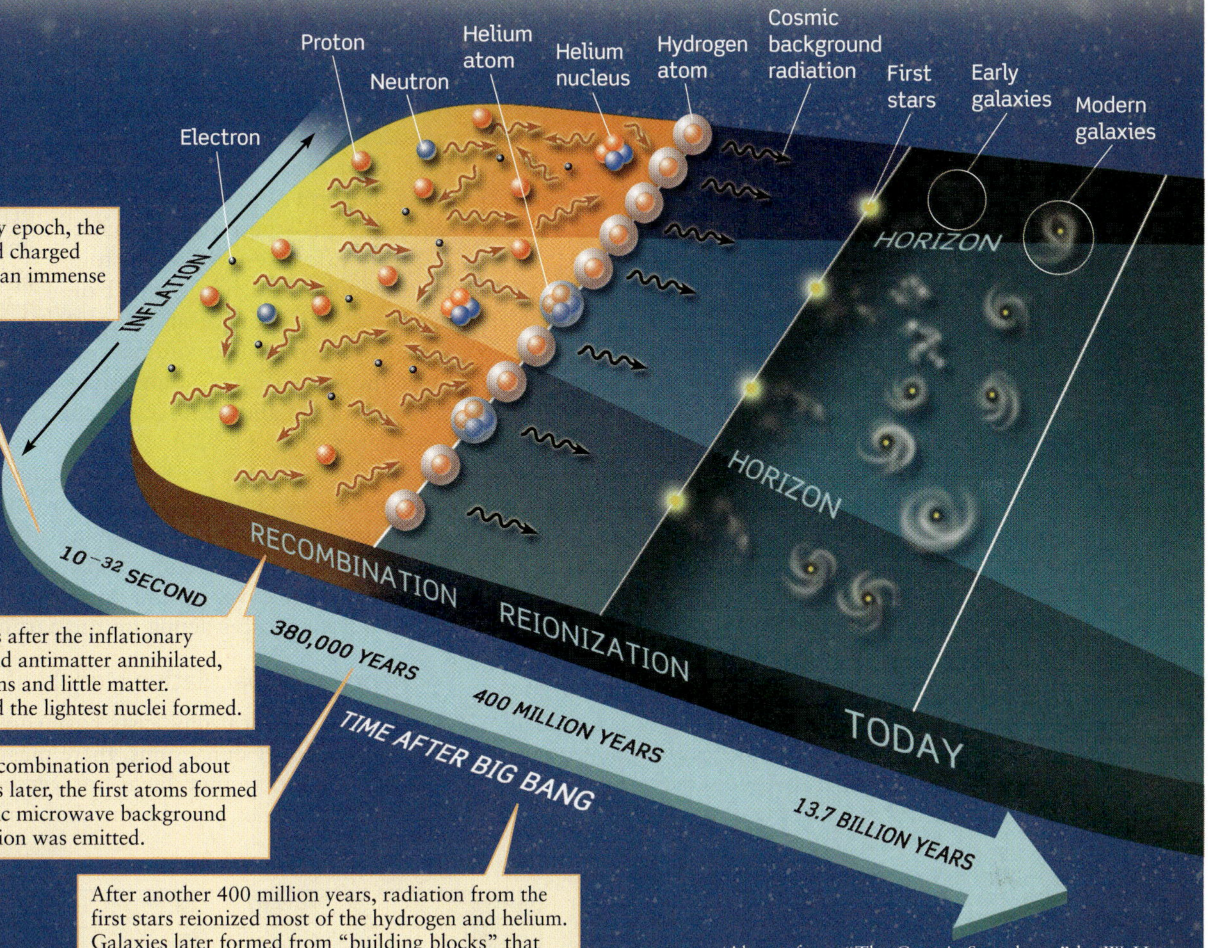

During the inflationary epoch, the plasma of photons and charged particles expanded by an immense factor.

During the 15 minutes after the inflationary epoch, most matter and antimatter annihilated, leaving a sea of photons and little matter. Protons, neutrons, and the lightest nuclei formed.

During the recombination period about 380,000 years later, the first atoms formed and the cosmic microwave background (CMB) radiation was emitted.

After another 400 million years, radiation from the first stars reionized most of the hydrogen and helium. Galaxies later formed from "building blocks" that were initially about the size of a globular cluster.

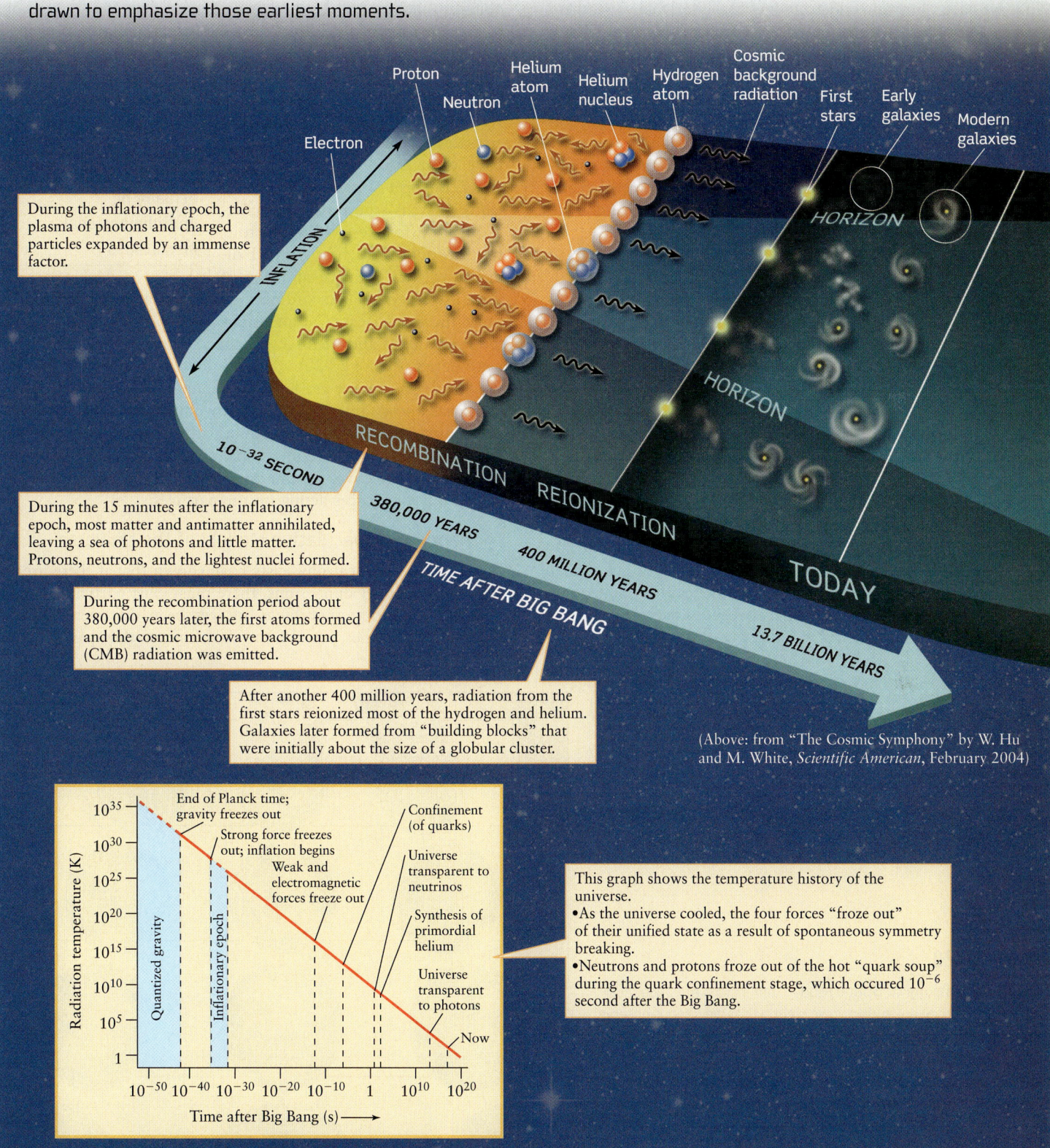

(Above: from "The Cosmic Symphony" by W. Hu and M. White, *Scientific American*, February 2004)

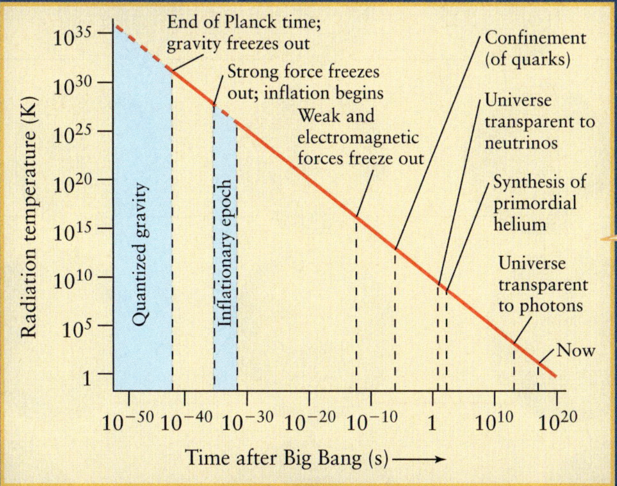

This graph shows the temperature history of the universe.

- As the universe cooled, the four forces "froze out" of their unified state as a result of spontaneous symmetry breaking.
- Neutrons and protons froze out of the hot "quark soup" during the quark confinement stage, which occured 10^{-6} second after the Big Bang.

27-7 Theories that attempt to unify the fundamental forces predict that the universe may have 11 dimensions

While we have a growing understanding of the early universe, there remains a veil obscuring the first 10^{-43} s after the Big Bang. These very *first* moments in the history of the universe, whose duration was the Planck time, set the stage for what would come after. To understand this brief interval, we need to construct a quantum-mechanical theory that unifies gravity with the other fundamental forces of nature and that reconciles quantum mechanics with gravity—a theory of quantum gravity. While unification remains an unfinished task, remarkable progress has been made in recent years. One major breakthrough is that physicists have had to abandon the idea that there are only three dimensions of space.

Beyond Four Dimensions: Kaluza-Klein Theories

In his special and general theories of relativity, Einstein combined time with the three dimensions of ordinary space, resulting in a four-dimensional combination called *spacetime* (see Section 22-1). In 1919, the Polish physicist Theodor Kaluza proposed the existence of a *fifth* dimension. Kaluza hoped to describe both gravity and electromagnetism in terms of the curvature of five-dimensional spacetime, just as Einstein had explained gravity by itself in terms of the curvature of four-dimensional spacetime (see Section 22-2).

A particle always follows the straightest possible path in the four space dimensions of Kaluza's theory. But in the three dimensions of ordinary space, the path appears curved. Hence, it appears to us that the particle has been deflected by gravitational and electromagnetic forces. Kaluza's hypothetical fifth dimension exists at every point in ordinary space but is curled up so tightly, like a very tiny loop, that it is not directly observable.

In 1926, the Swedish physicist Oskar Klein attempted to make Kaluza's five-dimensional theory compatible with quantum mechanics. While he was not successful, Klein discovered that particles of different masses could be identified with different vibrations of the tiny loop of Kaluza's fifth dimension. Today, any quantum-mechanical theory that uses more than four dimensions to provide a unified description of the forces of nature is called a **Kaluza-Klein theory.**

When Kaluza and Klein developed their theories, gravity and electromagnetism were the only known forces of nature. Today we know of four fundamental forces, which suggests that modern Kaluza-Klein theories should have even more than five dimensions. Edward Witten at Princeton University has argued that a geometric theory for describing all four forces would work best with 11 dimensions, ten of space and one of time. At every point in ordinary space and at every moment of time, the seven "extra" dimensions must be rolled up, like the loops of Kaluza and Klein, into compact structures far too tiny for us to detect (Figure 27-20).

Subatomic particles may actually be multidimensional membranes

While we do not yet have a definite 11-dimensional theory, physicists know something of how it would work. Particles would travel along the straightest possible paths in spacetime, but their paths in ordinary three-dimensional space would appear curved. From our perspective, we interpret these curved paths as the result of the four forces acting on the particles.

M-Theory and Speculative Models of the Universe

WEB LINK 27.10

Theoretical physicists have shown that if there are indeed 11 dimensions, there must exist particles so massive that they have not yet been discovered. These speculative particles may be the dark matter that pervades the

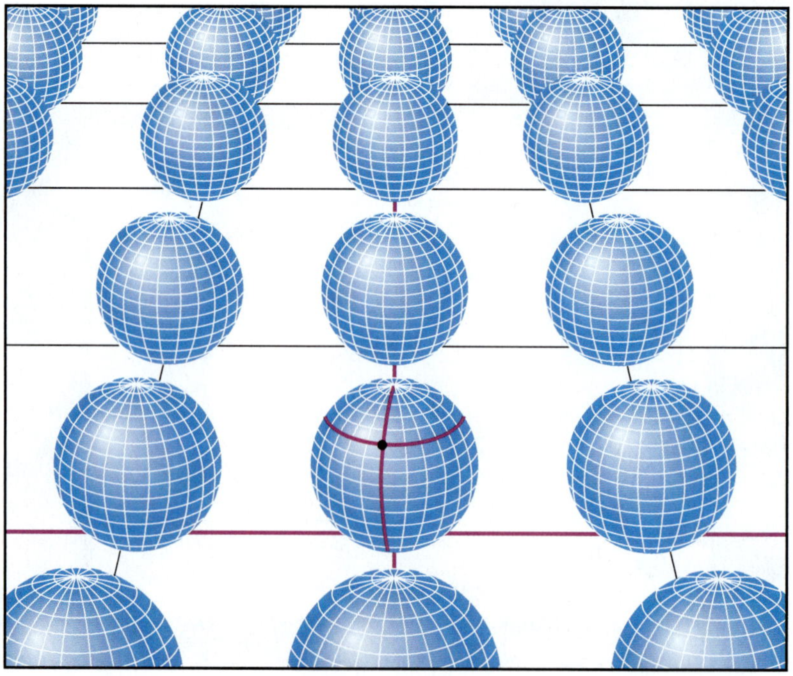

Figure 27-20

Hidden Dimensions of Space Hidden dimensions of space might exist provided they are curled up so tightly that we cannot observe them. This drawing shows how an ordinary two-dimensional plane might contain two additional dimensions. At every point on the plane, there is a very tiny sphere so small that it cannot be seen. To pinpoint a particular location, you need to give not only a position on the plane but also a position on the sphere that is tangent to the plane at that point, as indicated by the red lines. (Adapted from D. Freedman and P. Van Nieuwenhuizen) universe and holds together clusters of galaxies. Even more bizarre, the new theories no longer regard fundamental particles, such as electrons and quarks, as tiny points of mass. Instead, these particles may actually be multidimensional membranes, wrapped so tightly around the extra dimensions of space that they appear to us as points.

Andrew Strominger at Harvard University has shown that some of these membranes may fold on themselves in such a way that not even light can escape from them. In other words, these membranes may be black holes! This theory of membranes in 11 dimensions, or **M-theory**, may well explain the most exotic aspects of quantum mechanics, cosmology, and gravity. M-theory encompasses a number of related theories known as *superstring theories*, or referred to collectively as string theory.

One of the many speculative ideas inspired by M-theory is the *cyclic model* (known in an earlier version as the *ekpyrotic model*, from the Greek word for "conflagration"). In this model, our universe is one of two four-dimensional membranes (three space dimensions and one time dimension) that move with respect to each other along a fifth, "hidden" dimension. When the two membranes collide, a Big Bang results. The physics of the collision is such that each membrane automatically has the critical density, so that our universe is automatically flat. This provides a solution to the flatness problem that does not require inflation, so there is no inflationary epoch in the cyclic model. The cyclic model offers a natural explanation for dark matter: It is matter in the other membrane, the particles of which we cannot see but can detect through their gravitation. Dark energy in the cyclic model is a manifestation of the field that controls the interaction between the two membranes. The model is called *cyclic* because the membranes move apart and then back together in a rhythmic way. We would see a universe that expands for a time after the Big Bang, then eventually collapses back on itself in a Big Crunch. This is followed by another Big Bang, and the cycle repeats.

The cyclic model and the inflationary model are both *scientific* theories, because they both make predictions about things that can be observed. In particular, the two models make different predictions about the redshifts of very distant galaxies and about the polarization of the cosmic background radiation. Improved measurements will help us understand which of these models—each exotic in its own right—is a better description of our universe.

At present, M-theory is only the outline of a "theory of everything." We do not yet know how to describe a quark or an electron in terms of the higher dimensions of spacetime, nor can we yet explain the interactions among such particles. A full description of the first 10^{-43} s after the Big Bang is still far off. However, theoretical physicists around the world are devoting their efforts to these challenging problems. In league with astronomers and experimental physicists, they continue the age-old quest to understand the fundamental character of the universe.

We shall not cease from exploration
And the end of all our exploring
Will be to arrive where we started
And know the place for the first time.
T. S. Eliot, *Four Quartets*

Key Words

annihilation, p. 730
antimatter, p. 729
antiparticle, p. 729
antiproton, p. 729
cold dark matter, p. 735
cosmic light horizon, p. 722
dark ages, p. 735
density fluctuation, p. 733
deuterium bottleneck, p. 732
electroweak force, p. 726
elementary particle physics, p. 728
false vacuum, p. 726
flatness problem, p. 723
gluon, p. 726
grand unified theory (GUT), p. 726
graviton, p. 726
Heisenberg uncertainty principle, p. 728
hot dark matter, p. 735
inflation, p. 723
inflationary epoch, p. 723
intermediate vector boson, p. 726
isotropy problem (horizon problem), p. 722
Jeans length, p. 733
Kaluza-Klein theory, p. 740
Lamb shift, p. 729
M-theory, p. 741
nucleosynthesis, p. 732
observable universe, p. 722
pair production, p. 729
Population III star, p. 734
positron, p. 729
quantum electrodynamics, p. 725
quantum mechanics, p. 728
quark, p. 725
quark confinement, p. 731
reionization, p. 734
spontaneous symmetry breaking, p. 726
string theory, p. 726
strong force, p. 724
supergrand unified theory, p. 726
theory of everything (TOE), p. 726
thermal equilibrium, p. 731
threshold temperature, p. 731
virtual pairs, p. 729
weak force, p. 725

Key Ideas

Cosmic Inflation: A brief period of rapid expansion, called inflation, is thought to have occurred immediately after the Big Bang. During a tiny fraction of a second, the universe expanded to a size many times larger than it would have reached through its normal expansion rate.

• Inflation explains why the universe is nearly flat and the 2.725-K microwave background is almost perfectly isotropic.

The Four Forces and Their Unification: Four basic forces—gravity, electromagnetism, the strong force, and the weak force—explain all the interactions observed in the universe.

• Grand unified theories (GUTs) are attempts to explain three of the forces in terms of a single consistent set of physical laws. A supergrand unified theory would explain all four forces.

• GUTs suggest that all four fundamental forces were equivalent just after the Big Bang. However, because we have no satisfactory supergrand unified theory, we can as yet say nothing about the nature of the universe during this period before the Planck time ($t = 10^{-43}$ s after the Big Bang).

• At the Planck time, gravity froze out to become a distinctive force in a spontaneous symmetry breaking. During a second spontaneous symmetry breaking, the strong nuclear force became a distinct force; this transition triggered the rapid inflation of the universe. A final spontaneous symmetry breaking separated the electromagnetic force from the weak nuclear force; from that moment on, the universe behaved as it does today.

Particles and Antiparticles: Heisenberg's uncertainty principle states that the amount of uncertainty in the mass of a subatomic particle increases as it is observed for shorter and shorter time periods.

• Because of the uncertainty principle, particle-antiparticle pairs can spontaneously form and disappear within a fraction of a second. These pairs, whose presence can be detected only indirectly, are called virtual pairs.

• A virtual pair can become a real particle-antiparticle pair when high-energy photons collide. In this process, called pair production, the photons disappear, and their energy is replaced by the mass of the particle-antiparticle pair. In the process of annihilation, a colliding particle-antiparticle pair disappears and high-energy photons appear.

The Origin of Matter: Just after the inflationary epoch, the universe was filled with particles and antiparticles formed by pair production and with numerous high-energy photons formed by annihilation. A state of thermal equilibrium existed in this hot plasma.

• As the universe expanded, its temperature decreased. When the temperature fell below the threshold temperature required to produce each kind of particle, annihilation of that kind of particle began to dominate over production.

• Matter is much more prevalent than antimatter in the present-day universe, because particles and antiparticles were not created in exactly equal numbers just after the Planck time.

Nucleosynthesis: Helium could not have been produced until the cosmological redshift eliminated most of the high-energy photons. These photons created a deuterium bottleneck by breaking down deuterons before they could combine to form helium.

Density Fluctuations and the Origin of Stars and Galaxies: The large-scale structure of the universe arose from primordial density fluctuations.

• The first stars were much more massive and luminous than stars in the present-day universe. The material that they ejected into space seeded the cosmos for all later generations of stars.

• Galaxies are generally located on the surfaces of roughly spherical voids. Models based on dark energy and cold dark matter give good agreement with details of this large-scale structure.

The Frontier of Knowledge: The search for a theory that unifies gravity with the other fundamental forces suggests that the universe actually has 11 dimensions (ten of space and one of time), seven of which are folded on themselves so that we cannot see them. The idea of higher dimensions has motivated alternative cosmological models.

Questions

Review Questions

TUTORIAL 27.1

1. What is the horizon problem? What is the flatness problem? How can these problems be resolved by the idea of inflation?
2. The inflationary epoch lasted a mere 10^{-32} second. Why, then, is it worthy of so much attention by scientists?
3. In what ways is inflation similar to the present-day expansion of the universe? In what ways is it different?
4. Explain why the inflationary model does not violate the principle that the speed of light in a vacuum represents the ultimate speed limit.
5. Describe an example of each of the four basic types of interactions in the physical universe. Do you think it possible that a fifth force might be discovered someday? Explain your answer.
6. If gravity is intrinsically so weak compared to the strong force, why do we say that gravity rather than the strong force keeps the planets in orbit around the Sun?
7. Explain how changes in the energy of the vacuum can account for the rapid expansion during the inflationary epoch.
8. What is the Heisenberg uncertainty principle? How does it lead to the idea that all space is filled with virtual particle-antiparticle pairs?
9. What is the difference between an electron and a positron?
10. Is it possible for a single hydrogen atom, with a positively charged proton and a negatively charged electron, to be created as a virtual pair? Why or why not?
11. Which can exist for a longer time, a virtual electron-positron pair or a virtual proton-antiproton pair? Explain your reasoning.
12. Explain why antimatter was present in copious amounts in the early universe but is very rare today.
13. What is meant by the threshold temperature of a particle?
14. Explain the connection between the fact that humans exist and the imbalance between matter and antimatter in the early universe.
15. Explain the connection between particles and antiparticles in the early universe and the cosmic microwave background that we observe today.
16. What is the deuterium bottleneck? Why was it important during the formation of nuclei in the early universe?
17. Why were only the four lightest chemical elements produced in the early universe?
18. The first stars in the universe are thought to have appeared some 400 million (4×10^8) years after the Big Bang. Once these stars formed, thermonuclear fusion reactions began in their interiors. Explain why these were the first fusion reactions to occur since the universe was 15 minutes old.
19. Why is it reasonable to suppose that all space is filled with a neutrino background analogous to the cosmic microwave background?
20. What is the Jeans length? Why is it significant for the formation of structure in the universe?
21. Why do astronomers suspect that globular clusters were among the first objects to form in the history of the universe? Why not something larger and more massive?
22. What are Population III stars? How do they differ from stars found in the present-day universe? Why are they so difficult to detect directly?
23. Describe the large-scale structure of the universe as revealed by the distribution of clusters and superclusters of galaxies.
24. What is the difference between hot and cold dark matter? How do astronomers decide which was more important in the formation of large-scale structures such as clusters of galaxies?
25. How did the presence of dark energy help to "turn off" the process of structure formation in the universe?

26. Describe the observational evidence for (**a**) the Big Bang, (**b**) the inflationary epoch, (**c**) the confinement of quarks, and (**d**) the era of recombination.

Advanced Questions

Problem-solving tips and tools

We described Wien's law for blackbody radiation in Section 5-4. If light with wavelength λ_0 is emitted by an object at redshift z, the wavelength that we measure is $\lambda = \lambda_0(1 + z)$. As explained in Section 26-6, the critical density ρ_c is equal to $3H_0^2/8\pi G$; that is, ρ_c is proportional to the square of the Hubble constant H_0. If $H_0 = 73$ km/s/Mpc, the critical density is equal to about 1.0×10^{-26} kg/m^3. It is also useful to know that the mass of a proton is 1.67×10^{-27} kg, the mass of an electron is 9.11×10^{-31} kg, the mass of the Sun is 1.99×10^{30} kg, that 1 m^3 = 10^6 cm^3, that 1 light-year = 9.46×10^{15} m, and that 1 GeV = 10^3 MeV = 10^9 eV.

27. An electron has a lifetime of 1.0×10^{-8} s in a given energy state before it makes a transition to a lower state. What is the uncertainty in the energy of the photon emitted in this process?
28. How many times stronger than the weak force is the electromagnetic force? How many times stronger than the electromagnetic force is the strong force? Use this information to suggest one reason why the electromagnetic and weak forces can become unified at a lower energy than do the electroweak and strong forces.
29. How long can a proton-antiproton pair exist without violating the principle of the conservation of mass?
30. The mass of the intermediate vector boson W^+ (and of its antiparticle, the W^-) is 85.6 times the mass of the proton. The weak nuclear force involves the exchange of the W^+ and the W^-. (**a**) Find the rest energy of the W^+. Give your answer in GeV. (**b**) Find the threshold temperature for the W^+ and W^-. (**c**) From Figure 27-6, how long after the Big Bang did W^+ and W^- particles begin to disappear from the universe? Explain your answer.
31. Using the physical conditions present in the universe during the era of recombination ($T = 3000$ K and $\rho_m = 10^{-18}$ kg/m^3), show by calculation that the Jeans length for the universe at that time was about 100 ly and that the total mass contained in a sphere with this diameter was about 4×10^5 $M_\odot$.
32. (**a**) If a Population III star had a surface temperature of 10^5 K, what was its wavelength of maximum emission? In what part of the electromagnetic spectrum does this wavelength lie? (**b**) To ionize a hydrogen atom requires a photon of wavelength 91.2 nm or shorter. Explain how Population III stars caused reionization. (**c**) If reionization occurred at $z = 11$, what do we measure the wavelength of maximum emission of a Population III star to be? In what part of the electromagnetic spectrum does this wavelength lie? (**d**) The image that opens this chapter was made using infrared wavelengths. Suggest why these wavelengths were chosen.
33. (**a**) If the Hubble constant is 73 km/s/Mpc, the critical density ρ_c is 1.0×10^{-26} kg/m^3. The average density of dark matter is known to be about 0.20 times the critical density. Suppose that massive neutrinos constitute this dark matter, and the average density of neutrinos throughout space is 100 neutrinos per cubic centimeter. (In fact, the density of neutrinos is far less than this.) Under these assumptions, what must be the mass of the neutrino? Give your answers in kilograms and as a fraction of the mass of the electron. (**b**) Why do astronomers think that massive neutrinos are *not* the dominant type of dark matter in the universe?
34. A typical dark nebula (see Figure 18-4) has a temperature of 30 K and a density of about 10^{-12} kg/m^3. (**a**) Calculate the Jeans length for such a dark nebula, assuming that the nebula is mostly composed of hydrogen. Express your answer in meters and in light-years. (**b**) A typical dark nebula is several light-years across. Is it likely that density fluctuations within such nebulae will grow with time? (**c**) Explain how your answer to (b) relates to the idea that protostars form within dark nebulae (see Section 18-3).
35. At the time labeled $z = 4.97$ in Figure 27-15, how large was the length of each side of the box used in the simulation compared to its size in the present day ($z = 0$)? How much greater was the density at $z = 4.97$ than the present-day density?

Discussion Questions

36. If you hold an iron rod next to a strong magnet, the rod will become magnetized; one end will be a north pole and the other a south pole. But if you heat the iron rod to 1043 K (770°C = 1418°F) or higher, it will lose its magnetization and there will be no preferred magnetic direction in the rod. This demagnetization is an example of *restoring* a spontaneously broken symmetry. Explain why.
37. Some GUTs predict that the proton is unstable, although with a half-life far longer than the present age of the universe. What would it be like to live at a time when protons were decaying in large numbers?

Web/eBook Questions

38. Search the World Wide Web for information about the top quark. What kind of particle is it? How does it compare with the up and down quarks found in protons and neutrons? Why did physicists work so hard to try to find it?
39. Search the World Wide Web for information about primordial deuterium (that is, deuterium that was formed in the very early universe). Why are astronomers interested in knowing how abundant primordial deuterium is in the universe? What techniques do they use to detect it?
40. Search the World Wide Web for information about the South Pole Telescope. What is the purpose of this telescope? Why is it to be sited at the South Pole? How will it help us understand the early universe?

Activities

Observing Projects

41. Use the *Starry Night Enthusiast*™ program to observe globular clusters. First display the entire celestial sphere by selecting **Favorites > Guides > Atlas.** Select **View > Deep Space > Messier Objects** to display this set of diffuse objects in the sky. (**a**) Open the **Find** pane and locate and examine the following globular clusters. In each case, find the approximate

angular diameter of the cluster: (i) M3; (ii) M12; (iii) M13. (**b**) Speculate on how these clusters would appear if you could see them at the same distance at the time of recombination, before the first stars formed.

42. Use the *Starry Night Enthusiast*™ program to examine the distribution of galaxies in our local universe. Select **Favorites > Deep Space > Tully Database** to display the 3-dimensional distribution of the 28,000 galaxies nearest to the Milky Way. Stop **Time** and remove the image of the astronaut's feet by clicking on **View > Feet.** The Milky Way is at the center of the box. You can rotate the box by putting the mouse cursor over the image, holding down the mouse button and Shift key, and moving the mouse. (On a two-button mouse, hold down the left mouse button.) As you rotate this cube of galaxies, you will note the apparent lack of galaxies in one plane, the so-called Zone of Avoidance, caused by the obscuration of the light from distant galaxies in these directions by our own Milky Way Galaxy. You can zoom in or out using the buttons at the upper right of the toolbar. Note particularly the appearance of walls of galaxies, which surround voids in which few galaxies are found, and the clustering of galaxies at the interstices of these walls. Compare the box to the simulated present-day universe shown at the lower right of Figure 27-19. What are the similarities? What are the differences?

Collaborative Exercises

43. The four fundamental forces of nature are the strong force, the weak force, the gravitational force, and the electromagnetic force. List four things at your school that rely on one of these fundamental forces, and explain how each thing is dependent on one of the fundamental forces.

44. Consider the following hypothetical scenario adapted from a daytime, cable television talk show. Chris states that Pat borrowed Chris's telescope without permission. Tyler purchased balloons and a new telescope eyepiece without telling Chris. Sean borrowed star maps from the library, with the library's permission, but without telling Pat. Eventually, when the four met on Sunday evening, Chris was crying and speechless. Can you create a "grand unified theory" that explains this entire situation?

45. The *Cosmic Connection: The History of the Universe* shows the history of the universe in the form of a graph of the temperature versus the time after the Big Bang. Create a similar history of your class, starting with estimated outside temperature on the vertical axis and number of days since the beginning of the academic term on the horizontal axis. Include dates for major exams and assignments up through today. In different color ink, show your predictions for temperatures, days, and events from today until the end of the course.

Confused by all those theories? Good

Making Sense of Modern Cosmology

BY P. JAMES E. PEEBLES

(from P. James E. Peebles, "Making Sense of Modern Cosmology," *Scientific American*, January 2001, 54–55)

This is an exciting time for cosmologists: findings are pouring in, ideas are bubbling up, and research to test those ideas is simmering away. But it is also a confusing time. All the ideas under discussion cannot possibly be right; they are not even consistent with one another. How is one to judge the progress? Here is how I go about it.

For all the talk of overturned theories, cosmologists have firmly established the foundations of our field. Over the past 70 years we have gathered abundant evidence that our universe is expanding and cooling. First, the light from distant galaxies is shifted toward the red, as it should be if space is expanding and galaxies are pulled away from one another. Second, a sea of thermal radiation fills space, as it should if space used to be denser and hotter. Third, the universe contains large amounts of deuterium and helium, as it should if temperatures were once much higher. Fourth, galaxies billions of years ago look distinctly younger, as they should if they are closer to the time when no galaxies existed. Finally, the curvature of spacetime seems to be related to the material content of the universe, as it should be if the universe is expanding according to the predictions of Einstein's gravity theory, the general theory of relativity.

That the universe is expanding and cooling is the essence of the big bang theory. You will notice I have said nothing about an "explosion"—the big bang theory describes how our universe is evolving, not how it began.

I compare the process of establishing such compelling results, in cosmology or any other science, to the assembly of a framework. We seek to reinforce each piece of evidence by adding cross bracing from diverse measurements. Our framework for the expansion of the universe is braced tightly enough to be solid. The big bang theory is no longer seriously questioned; it fits together too well. Even the most radical alternative—the latest incarnation of the steady state theory—does not dispute that the universe is expanding and cooling. You still hear differences of opinion in cosmology, to be sure, but they concern additions to the solid part.

For example, we do not know what the universe was doing before it was expanding. A leading theory, inflation, is an attractive addition to the framework, but it lacks cross bracing. That is precisely what cosmologists are now seeking [see "Echoes from the Big Bang," on page 38]. If measurements in progress agree with the unique signatures of inflation, then we will count them as a persuasive argument for this theory. But until that time, I would not settle any bets on whether inflation really happened. I am not criticizing the theory; I simply mean that this is brave, pioneering work still to be tested.

More solid is the evidence that most of the mass of the universe consists of dark matter clumped around the outer parts of galaxies. We also have a reasonable case for Einstein's infamous cosmological constant or something similar; it would be the agent of the acceleration that the universe now seems to be undergoing. A decade ago cosmologists generally welcomed dark matter as an elegant way to account for the motions of stars and gas within galaxies. Most researchers, however, had a real distaste for the cosmological constant. Now the majority accept it, or its allied concept, quintessence [see "The Quintessential Universe," on page 46]. Particle physicists have come to welcome the challenge that the cosmological constant poses for quantum theory. This shift in opinion is not a reflection of some inherent weakness; rather it shows the subject in a healthy state of chaos around a slowly growing fixed framework. We are students of nature, and we adjust our concepts as the lessons continue.

The lessons, in this case, include the signs that cosmic expansion is accelerating: the brightness of supernovae near and far; the ages of the oldest stars; the bending of light around distant masses; and the fluctuations of the temperature of the thermal radiation across the sky [see "Special Report: Revolution in Cosmology," *Scientific American,* January 1999]. The evidence is impressive, but I am still uneasy about details of the case for the cosmological constant, including possible contradictions with the evolution of galaxies and their spatial distribution. The theory of the accelerating universe is a work in progress. I admire the architecture, but I would not want to move in just yet.

How might one judge reports in the media on the progress of cosmology? I feel uneasy about articles based on an inter-

continued

Report Card for Major Theories

Concept	Grade	Comments
The universe evolved from a hotter, denser state	A+	Compelling evidence drawn from many corners of astronomy and physics
The universe expands as the general theory of relativity predicts	A-	Passes the tests so far, but few of the tests have been tight
Darkmatter made of exotic particles dominates galaxies	B+	Many lines of indirect evidence, but the particles have yet to be found and alternative theories have yet to be ruled out
Most of the mass of the universe is smoothly distributed; it acts like Einstein's cosmological constant, causing the expansion to accelerate	B-	Encouraging fit from recent measurements, but more must be done to improve the evidence and resolve the theoretical conundrums
The universe grew out of inflation	Inc	Elegant, but lacks direct evidence and requires huge extrapolation of the laws of physics

view with just one person. Research is a complex and messy business. Even the most experienced scientist finds it hard to keep everything in perspective. How do I know that this individual has managed it well? An entire community of scientists can head off in the wrong direction, too, but it happens less often. That is why I feel better when I can see that the journalist has consulted a cross section of the community and has found agreement that a certain result is worth considering. The result becomes more interesting when others reproduce it. It starts to become convincing when independent lines of evidence point to the same conclusion. To my mind, the best media reports on science describe not only the latest discoveries and ideas but also the essential, if sometimes tedious, process of testing and installing the cross bracing.

Over time, inflation, quintessence and other concepts now under debate either will be solidly integrated into the central framework or will be abandoned and replaced by something better. In a sense, we are working ourselves out of a job. But the universe is a complicated place, to put it mildly, and it is silly to think we will run out of productive lines of research anytime soon. Confusion is a sign that we are doing something right: it is the fertile commotion of a construction site.

P. James E. Peebles is one of the world's most distinguished cosmologists, a key player in the early analysis of the cosmic microwave background radiation and the bulk composition of the universe. He has received some of the highest awards in astronomy, including the 1982 Heineman Prize, the 1993 Henry Norris Russell Lectureship of the American Astronomical Society and the 1995 Bruce Medal of the Astronomical Society of the Pacific. Peebles is currently an emeritus professor at Princeton University.

Further Information

The Evolution of the Universe. P. James E. Peebles, David N. Schramm, Edwin L. Turner and Richard G. Kron in *Scientific American,* Vol. 271, No. 4, pages 52–57; October 1994.

The Inflationary Universe: The Quest for a New Theory of Cosmic Origins. Alan H. Guth. Perseus Press, 1997.

Before the Beginning: Our Universe and Others. Martin Rees. Perseus Press, 1998.

The Accelerating Universe: Infinite Expansion, the Cosmological Constant, and the Beauty of the Cosmos. Mario Livio and Allan Sandage. John Wiley & Sons, 2000.

Concluding Remarks on New Cosmological Data and the Values of the Fundamental Parameters. P. James E. Peebles in *IAU Symposium 201: New Cosmological Data and the Values of the Fundamental Parameters,* edited by A. N. Lasenby, A. W. Jones and A. Wilkinson; August 2000. Preprint available at xxx.lanl.gov/abs/astro-ph/0011252 on the World Wide Web.

28 The Search for Extraterrestrial Life

WEB LINK 28.1 R I V U X G

An "alien" life-form on Earth: pink, eyeless worms in an underwater mound of yellow solid methane. (Dr. Charles Fisher, Eberly College of Science, Pennsylvania State University)

One of the most compelling questions in science is also one of the simplest: Are we alone? That is, does life exist beyond Earth? As yet, we have no definitive answer to this question. None of our spacecraft has found life elsewhere in the solar system, and radio telescopes have yet to detect signals of intelligent origin coming from space. Claims of aliens visiting our planet and abducting humans make compelling science fiction, but none of these stories has ever been verified.

Yet there are reasons to suspect that life might indeed exist beyond Earth. One is that biologists find living organisms in some of the most "unearthly" environments on our planet. An example (shown at the top of this page) is at the bottom of the Gulf of Mexico, where the crushing pressure and low temperature cause methane—normally a gas—to form solid, yellowish mounds. Amazingly, these mounds teem with colonies of pink, eyeless, alien-looking worms the size of your thumb. If life can flourish here, might it not also flourish in the seemingly hostile conditions found on other worlds?

In this chapter we will look for places in our solar system where life may once have originated, and where it may exist today. We will see how scientists estimate the chances of finding life beyond our solar system, and how they search for signals from other intelligent species. And we will learn how a new generation of telescopes may make it possible to detect the presence of even single-celled organisms on worlds many light-years away.

28-1 The chemical building blocks of life are found in space

WEB LINK 28.2

Suppose you were the first visitor to a new and alien planet. How would you recognize which of the strange objects around you were living, and which were inanimate? Questions such as these are central to **astrobiology,** the study of life in the universe. Most astrobiologists suspect that if we find living organisms on other worlds, they will be "life as we know it"—that is, their biochemistry will be based on the unique properties of the carbon atom, as is the case for all terrestrial life.

Organic Molecules in the Universe

Why carbon? The reason is that carbon has the most versatile chemistry of any element. Carbon atoms can form chemical bonds to create especially long and complex molecules (Figure 28-1). These carbon-based compounds, called **organic molecules,** include all the molecules of which living organisms are made. (Silicon has some chemical similarities to carbon, and it can also

Learning Goals

By reading the sections of this chapter, you will learn

28-1 How comets and meteorites could have aided the origin of life on Earth

28-2 On which other worlds of our solar system life might have evolved

28-3 About the controversy over proposed fossil life from Mars

28-4 How astronomers estimate the number of other civilizations in the Galaxy

28-5 Why radio telescopes may be a useful tool for contacting alien civilizations

28-6 How astronomers hope to detect Earthlike planets orbiting other stars

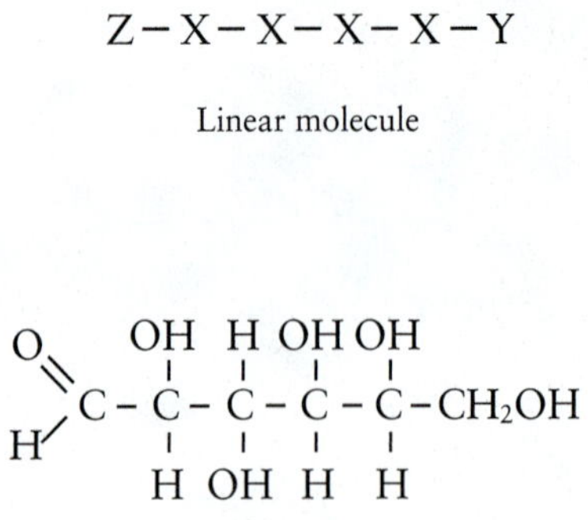

(a) Linear molecule

(b) Glucose

Figure 28-1

Complex Molecules and Carbon **(a)** Atoms that can bond to only two other atoms, like the atoms denoted X shown here, can form a chain of atoms called a linear molecule. The chain stops where we introduce an atom, such as those labeled Y and Z, that can bond to only one other atom. **(b)** A carbon atom (denoted C) can bond with up to *four* other atoms. Hence, carbon atoms can form more complex, nonlinear molecules like glucose. All organic molecules that are found in living organisms have backbones of carbon atoms.

form complex molecules. But as Figure 28-2 shows, complex silicon molecules do not have the right properties to make up complex systems such as living organisms.)

Organic molecules can be linked together to form elaborate structures, such as chains, lattices, and fibers. Some of these structures are capable of complex, self-regulating chemical reactions. Furthermore, the primary constituents of organic molecules—carbon, hydrogen, nitrogen, oxygen, sulfur, and phosphorus—are among the most abundant elements in the universe. The versatility and abundance of carbon suggest that extraterrestrial life is also likely to be based on organic chemistry.

If life is based on organic molecules, then these molecules must initially be present on a planet in order for life to arise from nonliving matter. We now understand that many carbon-based molecules originate from nonbiological processes in interstellar space. One such molecule is carbon monoxide (CO), which is made when a carbon atom and an oxygen atom collide and bond together. Carbon monoxide is found in abundance within giant interstellar clouds that lie along the spiral arms of our Milky Way Galaxy (see Figure 1-7) as well as in other galaxies (see Figure 1-9). Carbon atoms have also combined with other elements to produce an impressive array of interstellar organic molecules, including ethyl alcohol (CH_3CH_2OH), formaldehyde (H_2CO), methylcyanoacetylene (CH_3C_3N), and acetaldehyde (CH_3CHO). Radio astronomers have detected these molecules by looking for the telltale microwave emission lines of carbon-based chemicals in interstellar clouds.

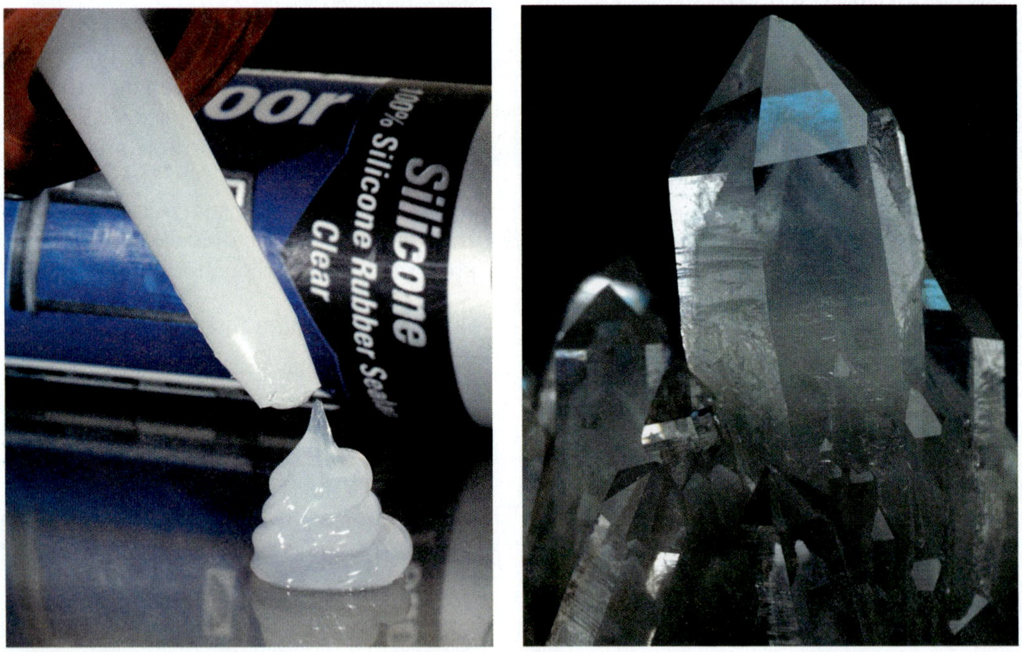

(a) (b)

Figure 28-2 RIVUXG

Why Silicon is Unsuitable for Making Living Organisms Like carbon, silicon atoms can bond with up to four other atoms. However, the resulting compounds are either too soft or too hard, or react too much or too little, to be suitable for use in living organisms. **(a)** Silicone has a backbone of silicon and oxygen atoms. The molecules form a gel or liquid rather than a solid, and react too slowly to undergo the rapid chemical changes required of molecules in organisms. **(b)** Molecules can also be made with a silicon-carbon-oxygen backbone, but the results (like this quartz crystal) are too rigid for use in organisms. (a: Richard Megna/Fundamental Photographs, New York; b: © Mark A. Schneider/Visuals Unlimited)

(a)

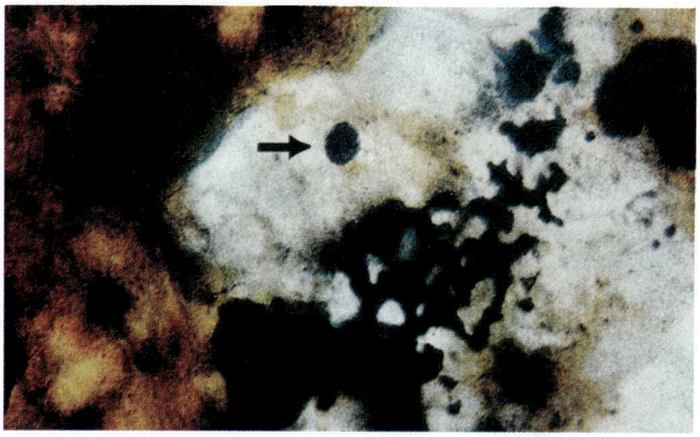

(b)

Figure 28-3 R I V U X G

A Carbonaceous Chondrite **(a)** Carbonaceous chondrites are primitive meteorites that date back to the very beginning of the solar system. This sample is a piece of the Allende meteorite, a large carbonaceous chondrite that fell in Mexico in 1969. **(b)** Chemical analyses of newly fallen specimens disclose that they are rich in organic molecules (arrow), many of which are the chemical building blocks of life. (a: From the collection of Ronald A. Oriti; b: Harvard-Smithsonian Center for Astrophysics)

The planets of our solar system formed out of interstellar material (see Section 8-5), and some of the organic molecules in that material must have ended up on the planets' surfaces. Evidence for this comes from meteorites called **carbonaceous chondrites,** like the one shown in Figure 28-3*a*. These meteorites are ancient, date from the formation of the solar system, and are often found to contain a variety of carbon-based molecules (Figure 28-3*b*). The spectra of comets (see Section 7-5)—which are also among the oldest objects in the solar system—show that they, too, contain an assortment of organic compounds.

Comets and meteoroids were much more numerous in the early solar system than they are today, and they were correspondingly more likely to collide with a planet. These collisions would have seeded the planets with organic compounds from the very beginning of our solar system's history. Organic compounds are also found in interplanetary dust particles (see Section 15-5), which continually rain down on the planets. Once meteorites, comets, and interplanetary dust particles bring simple organic chemicals to a planet's surface, additional chemical reactions can produce an even wider range of the complex organic compounds needed for life. Similar processes are thought to take place in other planetary systems, which are thought to form in basically the same way as did our own (see Figure 8-13 and the image that opens Chapter 8).

The Miller-Urey Experiment

WEB LINK 28.3

Comets and meteorites would not have been the only sources of organic material on the young planets of our solar system. In 1952, the American chemists Stanley Miller and Harold Urey demonstrated that under conditions that are thought to have prevailed on the primitive Earth, simple chemicals can combine to form the chemical building blocks of life. In a closed container, they prepared a sample of "atmosphere": a mixture of hydrogen (H_2), ammonia (NH_3), methane (CH_4), and water vapor (H_2O), the most common molecules in the solar system. Miller and Urey then exposed this mixture of gases to an electric arc (to simulate atmospheric lightning) for a week. At the end of this period, the inside of the container had become coated with a reddish-brown substance rich in amino acids and other compounds essential to life.

Since Miller and Urey's original experiment, most scientists have come to the conclusion that Earth's primordial atmosphere was composed of carbon dioxide (CO_2), nitrogen (N_2), and water vapor outgassed from volcanoes, along with some hydrogen. Modern versions of the Miller-Urey experiment (Figure 28-4) using these common gases have also succeeded in synthesizing a wide variety of organic compounds. The combination of comets and meteorites falling from space and chemical synthesis in the atmosphere could have made the chemical building blocks of life available in substantial quantities on the young Earth.

Organic molecules do not have to come from living organisms—they can also be synthesized in nature from simple chemicals

CAUTION! It is important to emphasize that scientists have *not* created life in a test tube. While organic molecules may have been available on the ancient Earth, biologists have yet to figure out how these molecules gathered themselves into cells and developed systems for self-replication. Nevertheless, because so many chemical components of life are so easily synthesized under conditions that simulate the primordial Earth, it seems reasonable to suppose that life could have originated as the result of chemical processes. Furthermore, because the molecules that combine to form these compounds are rather common, it seems equally reasonable that life could have originated in the same way on other planets.

Organic building blocks are commonplace throughout the universe, but their abundance does not guarantee that life is equally commonplace. If a planet's environment is hostile, life

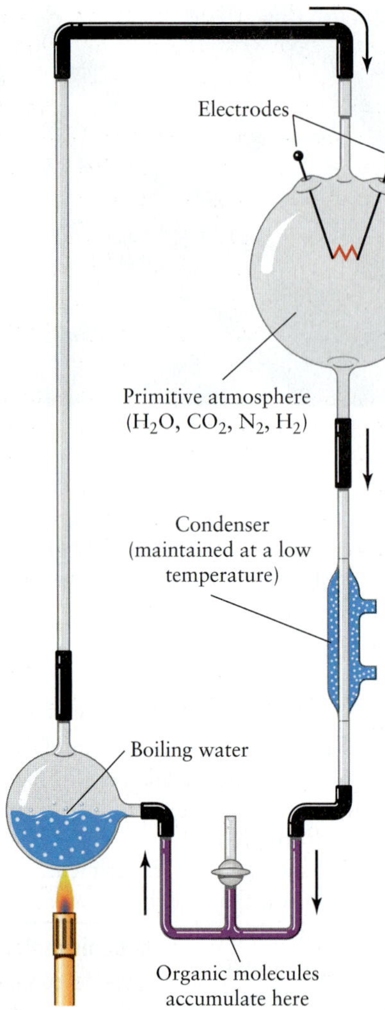

INTERACTIVE EXERCISE 28.1

Figure 28-4

An Updated Miller-Urey Experiment Modern versions of this classic experiment prove that numerous organic compounds important to life can be synthesized from gases that were present in Earth's primordial atmosphere. This experiment supports the hypothesis that life on Earth arose as a result of ordinary chemical reactions.

may never get started or may quickly be extinguished. But we now have evidence that Jupiter-sized planets and nearly Earth-sized planets orbit other stars (see Section 8-7), and that additional planetary systems are forming around young stars (see Section 8-4, especially Figure 8-8). It seems increasingly likely that Earthlike planets will be found orbiting other stars, and that conditions on some of these worlds may be suitable for life as we know it.

28-2 Europa and Mars have the potential for life to have evolved

If life evolved on Earth from nonliving organic molecules, might the same process have taken place elsewhere in our solar system? Scientists have carefully scrutinized the planets and satellites in an attempt to answer this question, and most of the answers have been disappointing.

The Importance of Liquid Water

One major problem is that liquid water is essential for the survival of life as we know it. The water need not be pleasant by human standards—terrestrial organisms have been found in water that is boiling hot, fiercely acidic, or ice cold (see the image that opens this chapter)—but it must be liquid. In order for water on a planet's surface to remain liquid, the temperature cannot be too hot or too cold. Furthermore, there must be a relatively thick atmosphere to provide enough pressure to keep liquid water from evaporating. Of all the worlds in our present-day solar system, only Earth has the right conditions for water to remain liquid on its surface.

WEB LINK 28.4

However, there is now compelling evidence that Europa, one of the large satellites of Jupiter (see Table 7-2 and Figure 7-4), has an ocean of water *beneath* its icy surface. As it orbits Jupiter, Europa is caught in a tug-of-war between gravitational forces from Jupiter and the other large satellites. These forces flex the interior of Europa, and this flexing generates enough heat to keep subsurface water from freezing. Chunks of ice on the surface can float around on this underground ocean, rearranging themselves into a pattern that reveals the liquid water beneath.

No one knows whether life exists in Europa's ocean. But interest in this exotic little world is great, and scientists have proposed several missions to explore Europa in more detail.

Searching for Life on Mars

The next best possibility for the existence of life is Mars. The present-day Martian atmosphere is so thin that water can exist only as ice or as a vapor. However, images made from Martian orbit show dried-up streambeds, flash flood channels, and sediment deposits. These features are evidence that the Martian atmosphere was once thicker and that water once coursed over the planet's surface. Could life have evolved on Mars during its "wet" period? If so, could life—even in the form of microorganisms—have survived as the Martian atmosphere thinned and the surface water either froze or evaporated?

Some areas on Mars were covered with liquid water for extended periods

In 1976, two spacecraft landed on different parts of Mars in search of answers to these questions. *Viking Lander 1* and *Viking Lander 2* each carried a scoop at the end of a mechanical arm to retrieve surface samples (Figure 28-5). These samples were deposited into a compact on-board biological laboratory that carried out three different tests for Martian microorganisms.

1. The *gas-exchange experiment* was designed to detect gases released due to chemical processes involving nutrients. For example, to get energy from food, we exhale carbon dioxide gas in our breath. The experiment involved a surface sample that was placed in a sealed container along with a controlled

WEB LINK 28.5

Figure 28-5 R I V U X G

Digging in the Martian Surface This view from the *Viking Lander 1* spacecraft shows the mechanical arm with its small scoop against the backdrop of the Martian terrain. The scoop was able to dig about 30 cm (12 in) beneath the surface. (NASA)

amount of gas and nutrients. The gases in the container were then monitored to see if their chemical composition changed.

2. The *labeled-release experiment* was designed to detect metabolic processes. A sample was moistened with nutrients containing radioactive carbon atoms. If any organisms in the sample consumed the nutrients, their waste products should include gases containing the telltale radioactive carbon.

3. The *pyrolytic-release experiment* was designed to detect photosynthesis, the biological process by which terrestrial plants use solar energy to help synthesize organic compounds from carbon dioxide. In the *Viking* experiments, a surface sample was placed in a container along with radioactive carbon dioxide and exposed to artificial sunlight. If plantlike photosynthesis occurred, microorganisms in the sample would take in some of the radioactive carbon from the gas.

The first data returned from these experiments caused great excitement, for in almost every case, rapid and extensive changes were detected inside the sealed containers. Further analysis of the data, however, led to the conclusion that these changes were due solely to nonbiological chemical processes. It appears that the Martian surface is rich in unstable chemicals that react with water to release oxygen gas. Because the present-day surface of Mars is bone-dry, these chemicals had nothing to react with until they were placed inside the moist interior of the *Viking Lander* laboratory.

At best, the results from the *Viking Lander* biological experiments were inconclusive. Perhaps life never existed on Mars at all. Or perhaps it did originate there, but failed to survive the thinning of the Martian atmosphere, the unstable chemistry of the planet's surface, and exposure to ultraviolet radiation from the Sun. (Unlike Earth, Mars has no ozone layer to block ultraviolet rays.) Another possibility is that Martian microorganisms have survived only in certain locations that the *Viking Lander*s did not sample, such as isolated spots on the surface or deep beneath the ground. And yet another option is that there is life on Mars, but the experimental apparatus on board the *Viking Lander* spacecraft was not sophisticated enough to detect it.

WEB LINK 28.6

An entirely different set of biological experiments were designed for the British spacecraft *Beagle 2*, which landed on Mars in December 2003. To search for organic building blocks of life, the spacecraft was to bore into the interiors of rocks to gather pristine, undisturbed samples, then heat these samples in the presence of oxygen gas. All carbon compounds decompose and form carbon dioxide (CO_2) when treated in this way, but biologically important molecules signal their presence by decomposing at a lower temperature. To test for past or present life on Mars, *Beagle 2* was to see how many of the CO_2 molecules contain the isotope ^{12}C (which appears preferentially in biological molecules) and how many contain ^{13}C (which does not). (See Box 5-5 for a description of isotopes.) Detecting an enhanced amount of ^{12}C would indicate that Martian rocks contain microorganisms that either survive to the present day or that died out at some point in the past.

Unfortunately, scientists on Earth were unable to establish contact with *Beagle 2* after its landing. However, similar experiments are planned for NASA's Mars Science Laboratory, to be launch in 2011. This robotic rover mission is designed to assess the past and present suitability of Mars for microbial life. The Mars Science Laboratory will also examine methane discovered on Mars. Microorganisms on Earth can gain energy by converting carbon dioxide to methane, and Martian microorganisms might do the same. Once it appears, methane rapidly decomposes in the Martian atmosphere, so we know that Mars presently has a source of methane. What we don't know is if this source of methane is geochemical or biological, and the Mars Science Laboratory has instruments that can potentially answer this question.

New Evidence for Martian Water

While results from the Mars Science Laboratory are years away, scientists searching for Martian life have been encouraged by other spacecraft that have provided new evidence of water on Mars. The European Space Agency's *Mars Express* spacecraft, which went into orbit around Mars in December 2003, used its infrared cameras to examine the ice cap at the Martian south pole (Figure 28-6). These cameras allowed scientists to see through the ice cap's surface layer of frozen carbon dioxide and reveal an underlying layer with the characteristic spectrum of water ice. (Figure 7-4 shows a similar spectrum obtained from Europa.) In January 2004, NASA successfully landed two robotic rovers named *Spirit* and *Opportunity* at two very different sites on opposite sides of Mars. While the terrain around the *Spirit* landing site appears to have been dry for billions of years, *Opportunity* landed in an area that appears to have been under water for extended periods (Figure 28-7*a*). Measurements made

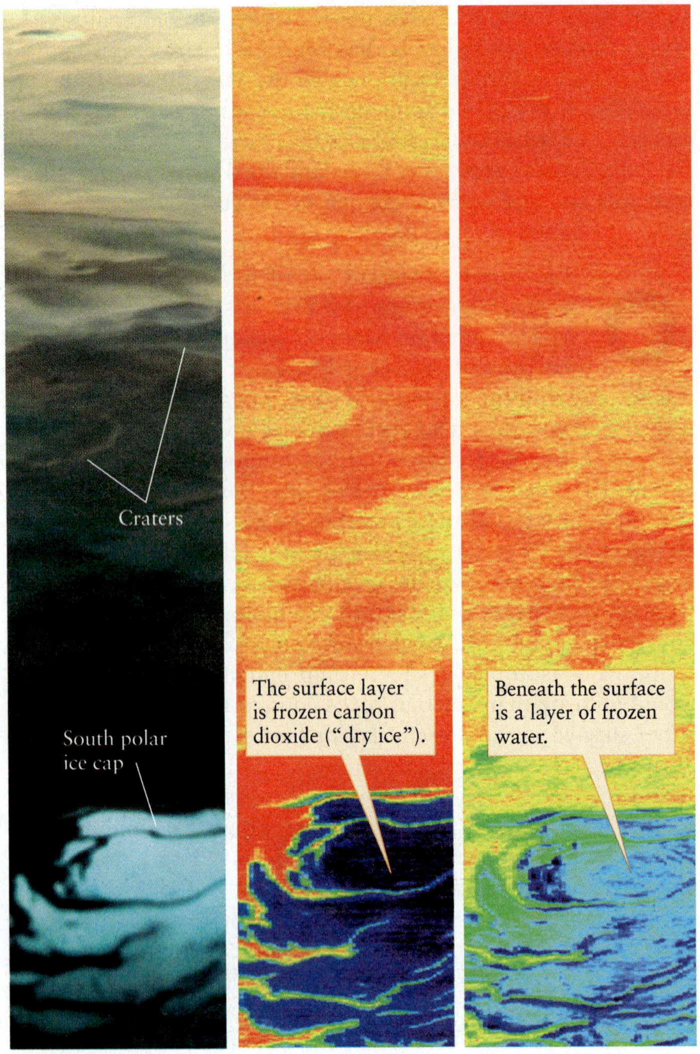

Craters
South polar ice cap
The surface layer is frozen carbon dioxide ("dry ice").
Beneath the surface is a layer of frozen water.

(a) Visible-light image (b) Infrared image of carbon dioxide ice (c) Infrared image of water ice

R I **V** U X G R **I** V U X G R **I** V U X G

Figure 28-6

Water at the Martian South Pole (a) This visible-light image shows the south polar cap of Mars, but does not indicate its chemical composition. But by using a camera tuned to different wavelengths of infrared light, the *Mars Express* spacecraft was able to identify the distinctive reflections of **(b)** an upper layer of carbon dioxide ice and **(c)** a deeper layer of water ice. Other observations have shown that there is also water ice at the Martian north pole. (ESA-OMEGA)

by *Opportunity* confirm that some of the very dark surface material at its landing site contains an iron-rich mineral called gray hematite (Figure 28-7*b*). On Earth, deposits of gray hematite are commonly found at the bottoms of lakes or mineral hot springs. The presence of gray hematite at the *Opportunity* site reinforces the argument that Mars once had liquid water on its surface, and helps hold open the possibility that living organisms could have evolved on Mars.

The Martian Civilization That Never Was

In 1976, while the *Viking Landers* were carrying out their biological experiments on the Martian surface, the companion *Viking Orbiter* spacecraft photographed some surface features that at first glance seemed to have been crafted by *intelligent* life on Mars. The *Viking Orbiter 1* image in Figure 28-8*a* shows what appears to be a humanlike face, perhaps the product of an advanced and artistic civilization. However, when the more

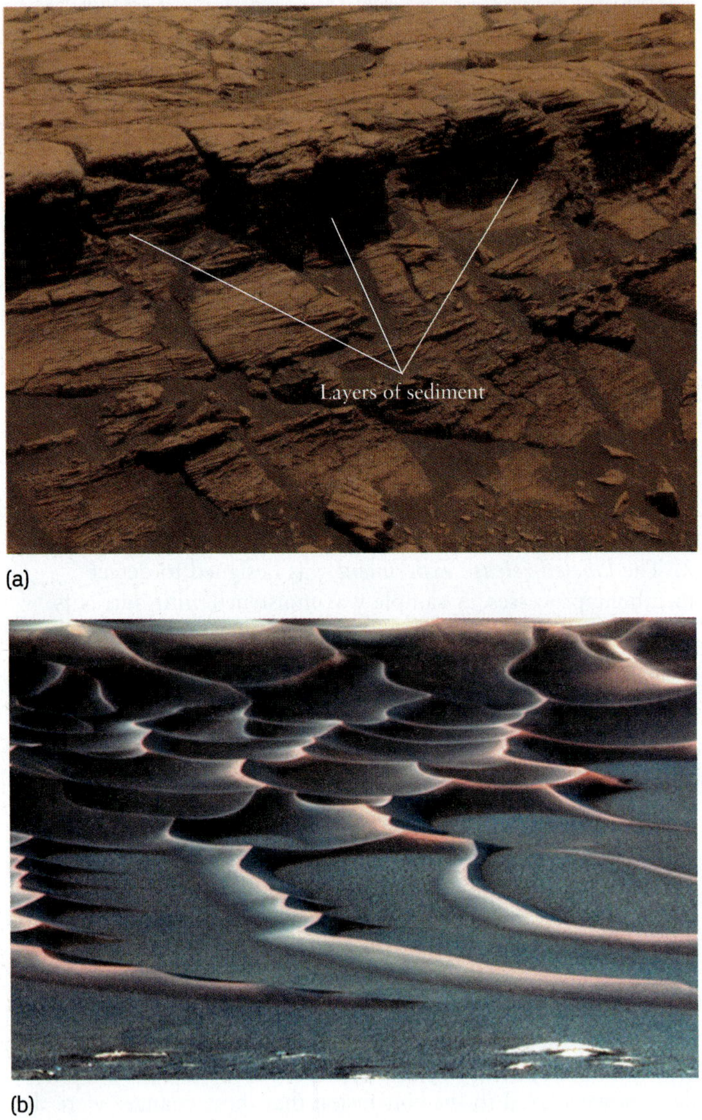

Layers of sediment
(a)
(b)

Figure 28-7 R I **V** U X G

Evidence of Ancient Martian Water **(a)** The Mars rover *Opportunity* photographed these sedimentary layers in a region called Meridiani Planum. Some of the layers are made of dust deposited by the Martian winds, but others were laid down by minerals that precipitated out of standing water. **(b)** In this false-color image from *Opportunity* of Martian sand dunes, the bluish color indicates the presence of millimeter-sized spheres of gray hematite. Such spheres naturally form in water-soaked deposits. (NASA/JPL/Cornell)

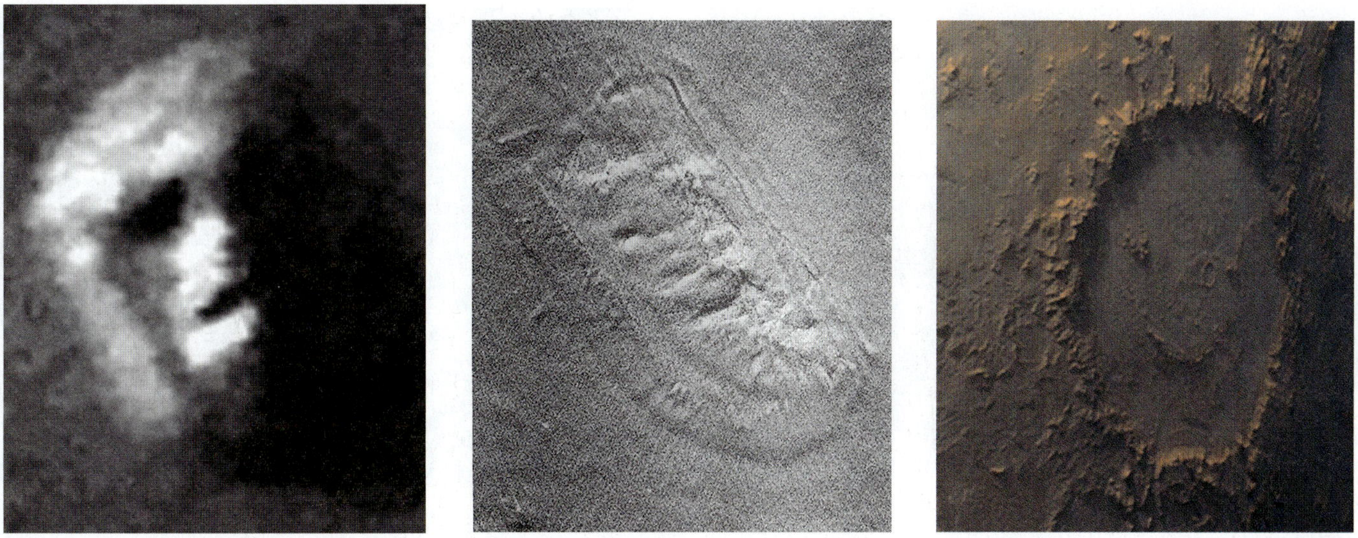

WEB LINK 28.9

Figure 28-8 RIVUXG

A "Face" on Mars? **(a)** This 1976 image from *Viking Orbiter 1* shows a Martian surface feature that resembles a human face. Some suggested that this feature might have been made by intelligent beings. **(b)** This 1998 *Mars Global Surveyor* (MGS) image, made under different lighting conditions with a far superior camera, reveals the "face" to be just an eroded hill. **(c)** This MGS image shows features of natural origin within a 215-km (134-mi) wide crater on Mars. Can you see this "face"? (a: NSSDC/NASA and Dr. Michael H. Carr; b, c: Malin Space Science Systems/NASA)

advanced *Mars Global Surveyor* spacecraft viewed the surface in 1998 using a superior camera (Figure 28-8*b*), it found no evidence for facial features.

Scientists are universally convinced that the "face" and other apparent patterns in the *Viking Orbiter* images were created by shadows on wind-blown hills. Microscopic life may once have existed on Mars, and may yet exist today, but there is no evidence that the red planet has ever been the home of intelligent beings.

28-3 Meteorites from Mars have been scrutinized for life-forms

While spacecraft can carry biological experiments to other worlds such as Mars, many astrobiologists look forward to the day when a spacecraft will return Martian samples to laboratories on Earth. Until that day arrives, we have the next best thing: Over thirty meteorites that appear to have formed on Mars have been found at a variety of locations on Earth.

What identifies meteorites as having come from Mars is the chemical composition of trace amounts of gas trapped within them. This composition is very different from that of Earth's atmosphere, but is a nearly perfect match to the composition of the Martian atmosphere found by the *Viking Landers*.

How could a rock have traveled from Mars to Earth? When a large piece of space debris collides with a planet's surface and forms an impact crater, most of the material thrown upward by the impact falls back onto the planet's surface. But some extraordinarily powerful impacts produce large craters—on Mars, roughly 100 km in diameter or larger. These tremendous impacts eject some rocks with such speed that they escape the planet's gravitational attraction and fly off into space.

There are numerous large craters on Mars, so a good number of Martian rocks have probably been blasted into space over the planet's history. These ejected rocks then go into elliptical orbits around the Sun. A few such rocks will have orbits that put them on a collision course with Earth, and these are the ones that scientists find as meteorites from Mars.

Using the radioactive age-dating technique (see Section 8-3), scientists find that most meteorites from Mars are between 200 million and 1.3 billion years old, much younger than the 4.56-billion-year age of the solar system. But one meteorite from Mars, denoted by the serial number ALH 84001 and found in Antarctica in 1984, was discovered in 1993 to be 4.5 billion years old (Figure 28-9*a*). Thus, ALH 84001 is a truly ancient piece of Mars. Analysis of ALH 84001 suggests that it was fractured by an impact between 3.8 and 4.0 billion years ago, was ejected from Mars by another impact 16 million years ago, and landed in Antarctica a mere 13,000 years ago.

ALH 84001 is the only known specimen of a rock that was on Mars during the era when liquid water existed on the planet's surface. Scientists have therefore investigated its chemical composition carefully, in the hope that this rock may contain clues to the amount of water that once flowed on the Martian surface. One such clue is the presence of rounded grains of minerals called *carbonates*, which can form only in the presence of water.

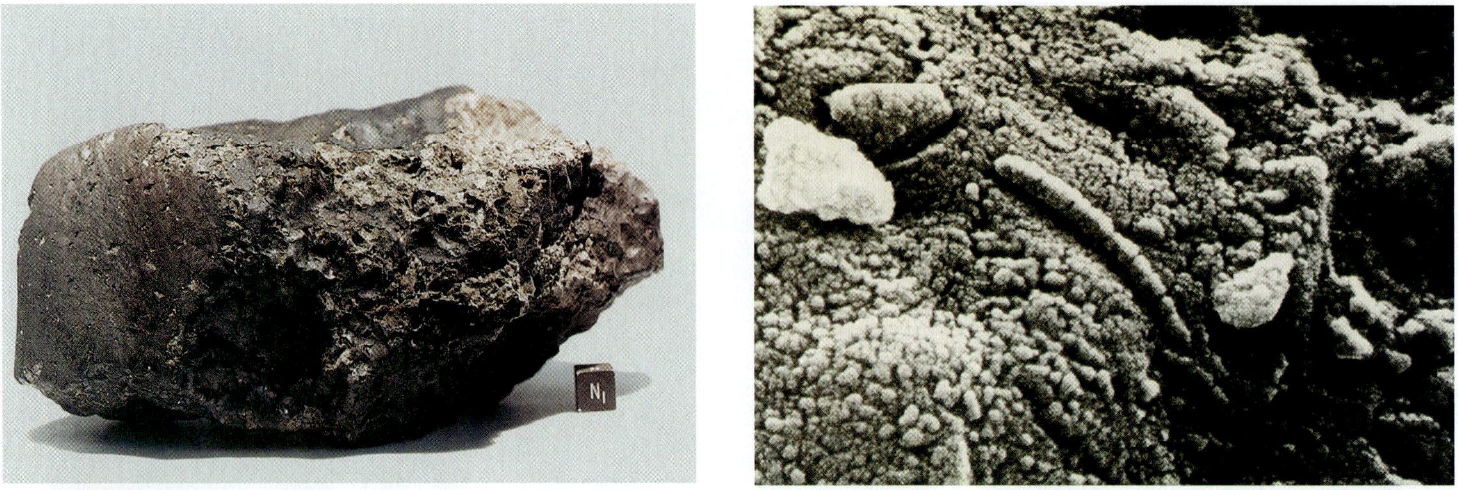

(a) R I V U X G (b)

WEB LINK 28.10

Figure 28-9

A Meteorite from Mars **(a)** This 1.9-kg meteorite, known as ALH 84001, formed on Mars some 4.5 billion years ago. About 16 million years ago a massive impact blasted it into space, where it drifted in orbit around the Sun until landing in Antarctica 13,000 years ago. The small cube at lower right is 1 cm (0.4 in) across. **(b)** This electron microscope image, magnified some 100,000 times, shows tubular structures about 100 nanometers (10^{-7} m) in length found within the Martian meteorite ALH 84001. One controversial interpretation is that these are the fossils of microorganisms that lived on Mars billions of years ago. (a: NASA Johnson Space Center; b: *Science,* NASA)

In 1996, David McKay and Everett Gibson of the NASA Johnson Space Center, along with several collaborators, reported the results of a two-year study of the carbonate grains in ALH 84001. They made three remarkable findings. First, in and around the carbonate grains were large numbers of elongated, tubelike structures resembling fossilized microorganisms (Figure 28-9*b*). Second, the carbonate grains contain very pure crystals of iron sulfide and magnetite. These two compounds are rarely found together (especially in the presence of carbonates) but can be produced by certain types of bacteria. Indeed, it appears that about one-fourth of the magnetite crystals found in ALH 84001 are of a type that on Earth are formed by bacteria. Third, the carbonates contain organic molecules—just the sort, in fact, that result from the decay of microorganisms.

McKay and Gibson proposed that the structures seen in Figure 28-9*b* are fossilized remains of microorganisms. If so, these organisms lived and died on Mars billions of years ago, during the era when liquid water was abundant.

Are McKay and Gibson's conclusions correct? Their claims of ancient life on Mars are extraordinary, and they require extraordinary proof. With only one rock like ALH 84001 known to science, however, such proof is hard to come by, and many scientists are skeptical. Skeptics point out that the tubelike structures and magnetite found in ALH 84001 could have been formed through geological processes on Mars. The question of a geological versus biological origin for the magnetite fuels an ongoing debate that will require more data to help settle the issue. Future spacecraft may help resolve the controversy by further examining rocks on the Martian surface. For now, the existence of microscopic life on Mars in the distant past remains an open question.

Claims that scientists have found Martian microorganisms are intriguing but very controversial

28-4 The Drake equation helps scientists estimate how many civilizations may inhabit our galaxy

TUTORIAL 28.1

We have seen that only a few locations in our solar system may have been suitable for the origin of life. But what about other planetary systems? The development of life on Earth seems to suggest that extraterrestrial life, including intelligent species, might evolve on terrestrial planets around other stars, given sufficient time and hospitable conditions. How can we learn whether such worlds exist, given the tremendous distances that separate us from them? This is the great challenge facing the **search for extraterrestrial intelligence,** or **SETI.**

Are we alone, or does the Galaxy teem with intelligent life? Or is the truth somewhere in between?

Close Encounters Versus Remote Communication

WEB LINK 28.11

A tenet of modern folklore is the belief that alien civilizations do exist, and that their spacecraft have visited Earth. Indeed, surveys show that between one-third and one-half of all Americans believe in unidentified flying objects (UFOs). A somewhat smaller percentage believes that aliens have landed on Earth. But, in fact, there is *no* scientifically verifiable evidence of alien visitations. As an example, many UFO proponents believe that the U.S. government is hiding evidence of an alien spacecraft that crashed near Roswell, New Mexico, in 1947. However, the bits of "spacecraft wreckage" found near Roswell turn out to be nothing more than remnants of an unmanned research balloon. To find real evidence of the presence or absence of intelligent civilizations on worlds orbiting other stars, we must look elsewhere.

With our present technology, sending even a small unmanned spacecraft to another star requires a flight time of tens of thousands of years. Speculative design studies have been made for unmanned probes that could reach other stars within a century or less, but these probes are prohibitively expensive. Instead, many astronomers hope to discover extraterrestrial civilizations by detecting radio transmissions from them. Radio waves are a logical choice for interstellar communication because they can travel immense distances without being significantly degraded by the interstellar medium, the thin gas and dust found between the stars (see Section 8-1).

Over the past several decades, astronomers have proposed various ways to search for alien radio transmissions, and several searches have been undertaken. In 1960, Frank Drake first used a radio telescope at the National Radio Astronomy Observatory in West Virginia to listen to two Sunlike stars, Tau Ceti and Epsilon Eridani, without success. More than 60 more extensive SETI searches have taken place since then, using radio telescopes around the world. Occasionally, a search has detected an unusual or powerful signal. But none has ever repeated, as a signal of intelligent origin might be expected to do. To date, we have no confirmed evidence of radio transmissions from another world.

Are They Out There?

Interactive Exercise 28.2

Should we be discouraged by this failure to make contact? What are the chances that a radio astronomer might someday detect radio signals from an extraterrestrial civilization? The first person to tackle this issue was Frank Drake, who proposed that the number of technologically advanced civilizations in the Galaxy could be estimated by a simple equation. This is now called the **Drake equation:**

Drake equation

$$N = R_* \, f_p \, n_e \, f_l \, f_i \, f_c \, L$$

N = number of technologically advanced civilizations in the Galaxy whose messages we might be able to detect

R_* = the rate at which solar-type stars form in the Galaxy

f_p = the fraction of stars that have planets

n_e = the number of planets per solar system that are Earthlike (that is, suitable for life)

f_l = the fraction of those Earthlike planets on which life actually arises

f_i = the fraction of those life-forms that evolve into intelligent species

f_c = the fraction of those species that develop adequate technology and then choose to send messages out into space

L = the lifetime of a technologically advanced civilization

The Drake equation is enlightening because it expresses the number of extraterrestrial civilizations in a simple series of terms. We can estimate some of these terms from what we know about stars, stellar evolution, and planetary orbits. The *Cosmic Connections* figure depicts some of these considerations.

For example, the first two factors, R_* and f_p, can be determined by observation. In estimating R_*, we should probably exclude stars with masses greater than about 1.5 times that of the Sun. These more massive stars use up the hydrogen in their cores in 3 billion (3×10^9) years or less. On Earth, by contrast, human intelligence developed only within the last million years or so, some 4.56 billion years after the formation of the solar system. If that is typical of the time needed to evolve higher life-forms, then a star of 1.5 solar masses or more probably fades away or explodes into a supernova before creatures as intelligent as we can evolve on any of that star's planets.

Although stars less massive than the Sun have much longer lifetimes they too seem unsuited for life because they are so dim. Only planets very near a low-mass star would be sufficiently warm for life as we know it, and a planet that close is subject to strong tidal forces from its star. We saw in Section 4-8 how Earth's tidal forces keep the Moon locked in synchronous rotation, with one face continually facing Earth. In the same way, a planet that orbits too close to its star would have one hemisphere that always faced the star, while the other hemisphere would be in perpetual, frigid darkness.

These considerations leave us with stars not too different from the Sun. (Like Goldilocks sampling the three bears' porridge, we must have a star that is not too hot and not too cold, but just right.) Based on statistical studies of star formation in the Milky Way, some astronomers estimate that roughly one of these Sunlike stars forms in the Galaxy each year in the galactic **habitable zone** (see the *Cosmic Connections* figure). This result sets R_* at 1 per year.

As we saw in Sections 8-4 and 8-5, the planets in our solar system formed as a natural consequence of the birth of the Sun. We have also seen evidence suggesting that planetary formation may be commonplace around single stars (see Figure 8-8). Many astronomers suspect that most Sunlike stars probably have planets, and so they give f_p a value of 1.

Unfortunately, the rest of the terms in the Drake equation are very uncertain. Let's play with some hypothetical values. The chances that a planetary system has an Earthlike world suitable for life are not known. Were we to consider our own solar system as representative, we could put n_e at 1. Let's be more conservative, however, and suppose that 1 in 10 solar-type stars is orbited by a habitable planet, making $n_e = 0.1$. Given the existence of life on Earth, we might assume that, given appropriate conditions, the development of life is a certainty, which would make $f_l = 1$. This is a topic of intense interest to astrobiologists.

For the sake of argument, we might also assume that evolution might naturally lead to the development of intelligence (a conjecture that is hotly debated) and also make $f_i = 1$. It's anyone's guess as to whether these intelligent extraterrestrial beings would attempt communication with other civilizations in the Galaxy, but were we to assume that they would, f_c would be put at 1 also.

The last variable, L, involving the longevity of a civilization, might be the most uncertain of all. Looking at our own example,

COSMIC CONNECTIONS

Habitable Zones for Life

Intelligent civilizations in our Milky Way Galaxy can evolve only in a certain region called the galactic habitable zone. In that zone, a suitable planet must lie within the planetary habitable zone of its parent star. (After C. H. Lineweaver, Y. Fenner, and B. K. Gibson)

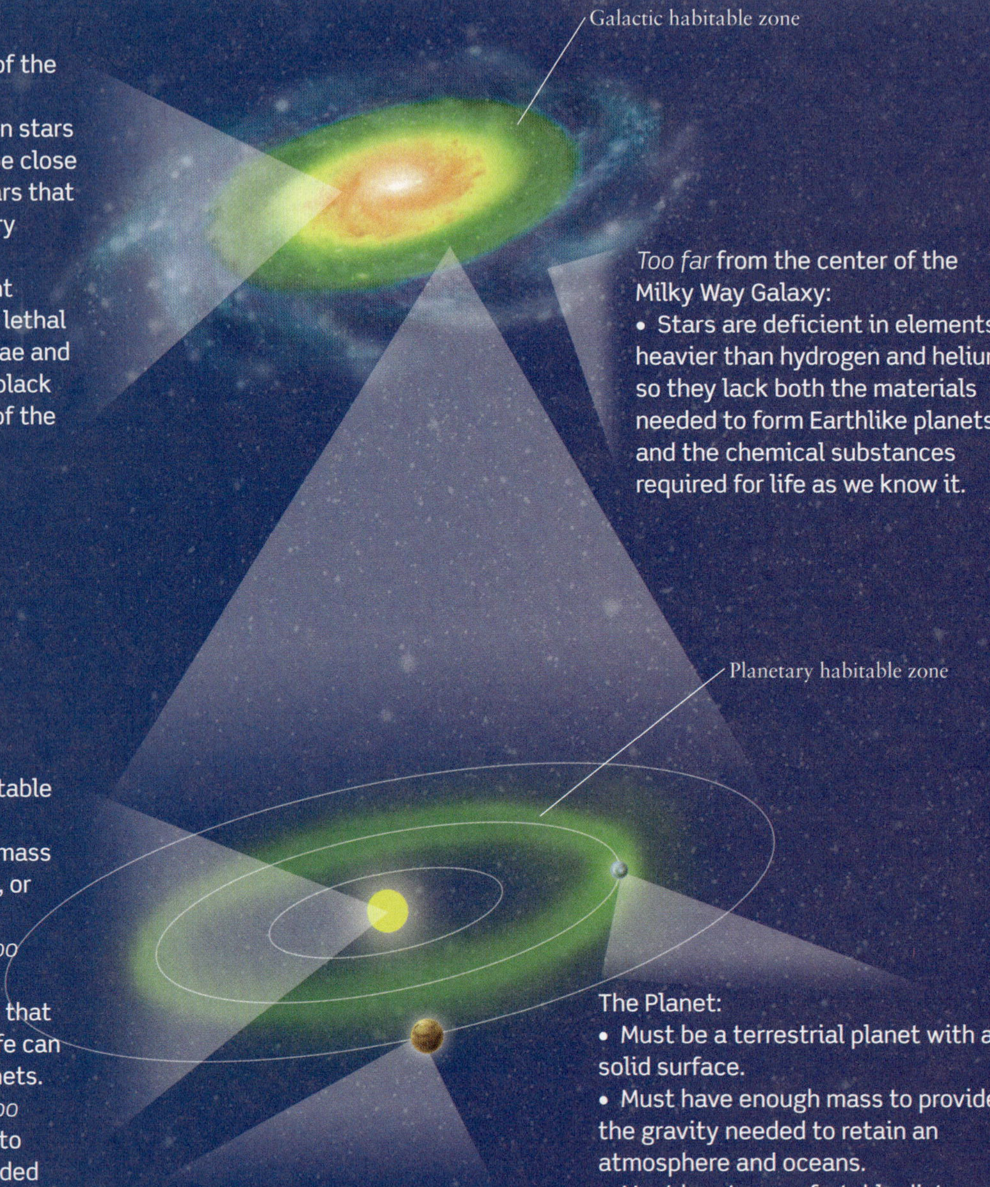

Too close to the center of the Milky Way Galaxy:
- The distances between stars are small, so there can be close encounters between stars that would disrupt a planetary system.
- There are also frequent outbursts of potentially lethal radiation from supernovae and from the supermassive black hole at the very center of the Galaxy.

Too far from the center of the Milky Way Galaxy:
- Stars are deficient in elements heavier than hydrogen and helium, so they lack both the materials needed to form Earthlike planets and the chemical substances required for life as we know it.

Inside the galactic habitable zone:
- The star must have a mass that is neither too large, or too small.
- If the star's mass is *too large*, it will use up its hydrogen fuel so rapidly that it will burn out before life can evolve on any of its planets.
- If the star's mass is *too small*, it will be too dim to provide the warmth needed for life.

The Neighborhood:
- There needs to be one or more large Jovian planets whose gravitational forces will clear away comets and meteors.

The Planet:
- Must be a terrestrial planet with a solid surface.
- Must have enough mass to provide the gravity needed to retain an atmosphere and oceans.
- Must be at a comfortable distance from the star so that the temperatures are neither too high nor too low.
- Must be in a stable, nearly circular orbit. (A highly elliptical orbit would cause excessively large temperature swings as the planet moved toward and away from the star.)

we see a planet whose atmosphere and oceans are increasingly polluted by creatures that possess nuclear weapons. If we are typical, perhaps L is as short as 100 years. Putting all these numbers together, we arrive at

$$N = 1/\text{year} \times 1 \times 0.1 \times 1 \times 1 \times 1 \times 100 \text{ years} = 10$$

In other words, out of the hundreds of billions of stars in the Galaxy, we would estimate that there are only 10 technologically advanced civilizations from which we might receive communications.

A wide range of values has been proposed for the terms in the Drake equation, and these various guesses produce vastly different estimates of N. Some scientists argue that there is exactly one advanced civilization in the Galaxy and that we are it. Others speculate that there may be hundreds or thousands of planets inhabited by intelligent creatures. If we wish to know whether our Galaxy is devoid of other intelligence, teeming with civilizations, or something in between, we must keep searching the skies.

28-5 Radio searches for alien civilizations are under way

Even if only a few alien civilizations are scattered across the Galaxy, we have the technology to detect radio transmissions from them. But if other civilizations are trying to communicate with us using radio waves, what frequency are they using? This is an important question, because if we fail to tune our radio telescopes to the right frequency, we might never know whether the aliens are out there.

A reasonable choice would be a frequency that is fairly free of interference from extraneous sources. SETI pioneer Bernard Oliver was the first to draw attention to a range of relatively noise-free frequencies in the neighborhood of the microwave emission lines of hydrogen (H) and hydroxide (OH) (Figure 28-10). This region of the spectrum can be called microwave or radio emission and is called the **water hole,** because H and OH together make H_2O, or water.

In 1989, NASA began work on the High Resolution Microwave Survey (HRMS), an ambitious project to scan the entire sky at frequencies spanning the water hole from 10^3 to 10^4 MHz. HRMS would have observed more than 800 nearby solar-type stars over a narrower frequency range in the hope of detecting signals that were either pulsed (like Morse code) or continuous (like the carrier wave for a TV or radio broadcast). The sophisticated signal-processing technology of HRMS would have been able to sift through tens of millions of individual frequency channels simultaneously. It would even have been able to detect the minute Doppler shifts in a signal coming from an alien planet as that planet spun on its axis and moved around its star.

WEB LINK 28.12

Sadly, just one year after HRMS began operation in 1992, the U.S. Congress imposed a mandate requiring that NASA no longer support HRMS or any other radio searches for extraterrestrial intelligence. This decision, which was made on budgetary grounds, saved a few million dollars—an entirely negligible amount compared to the total NASA budget. Ironically, the senator who spearheaded this move was from the state of Nevada, where tax dollars have been spent to signpost a remote desert road as "The Extraterrestrial Highway."

WEB LINK 28.13

Millions of personal computers have helped scan for alien radio transmissions

Even though NASA funding is no longer available, several teams of scientists remain actively involved in SETI programs. Funding for these projects has come from nongovernmental organizations such as the Planetary Society and from private individuals. Since 1995 the SETI Institute in California has been carrying out

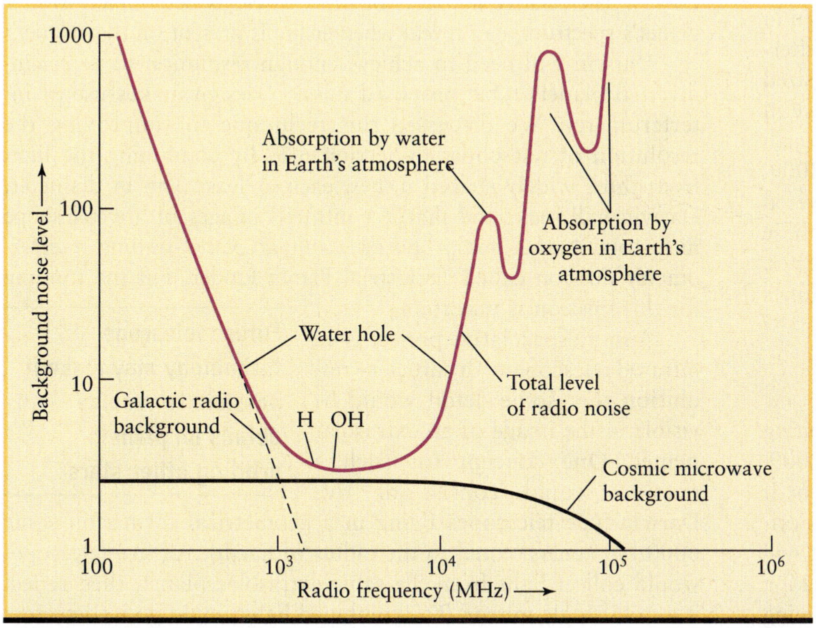

Figure 28-10

The Water Hole This graph shows the background noise level from the sky at various radio and microwave frequencies. The so-called water hole is a range of radio frequencies from about 10^3 to 10^4 megahertz (MHz) in which there is little noise and little absorption by Earth's atmosphere. Some scientists suggest that this noise-free region would be well suited for interstellar communication. Within the water hole itself, the principal source of noise is the afterglow of the Big Bang, called the cosmic microwave background. To put this graph in perspective, a frequency of 100 MHz corresponds to "100" on a FM radio, and 10^3 MHz is a frequency used for various types of radar. (Adapted from C. Sagan and F. Drake)

Project Phoenix, the direct successor to HRMS. When complete, this project will have surveyed a thousand Sunlike stars within 200 light-years at millions of radio frequencies. At Harvard University, BETA (the *B*illion-channel *E*xtra*T*errestrial *A*ssay) is scanning the sky at even more individual frequencies within the water hole. Other multifrequency searches are being carried out under the auspices of the University of Western Sydney in Australia and the University of California.

A major challenge facing SETI is the tremendous amount of computer time needed to analyze the mountains of data returned by radio searches. To this end, scientists at the University of California, Berkeley, have recruited nearly 5 million personal computer users to participate in a project called SETI@home. Each user receives actual data from a detector called SERENDIP IV (*S*earch for *E*xtraterrestrial *R*adio *E*missions from *N*earby, *D*eveloped, *I*ntelligent *P*opulations) and a data analysis program that also acts as a screensaver. When the computer's screensaver is on, the program runs, the data are analyzed, and the results are reported via the Internet to the researchers at Berkeley. The program then downloads new data to be analyzed. Since 1999, SETI@home users have provided the equivalent of two million years of computing time!

In 2007, the SETI Institute began operation of a large radio telescope dedicated to the search for intelligent signals. This telescope, called the Allen Telescope Array, is actually hundreds of relatively small and inexpensive radio dishes working together. Perhaps this new array will be the first to detect a signal from a distant civilization.

28-6 Telescopes in space will soon begin searching for Earthlike planets

Although no longer involved in SETI, NASA is planning a major effort to search for Earthlike planets suitable for the evolution of an advanced civilization. Such a search poses a major challenge. One problem is that Earth-sized planets are too dim to be seen in visible light against the glare of their parent star. Using a different approach, astronomers have discovered many Jupiter-sized planets by detecting the "wobble" that these planets produce in their parent star (see Section 8-6). But a planet the size of Earth exerts only a weak gravitational force on its parent star. In 2009, the first Earth-sized planet (at 1.9 Earth masses) was discovered through a tiny wobble. However, this planet, named Gliese 581 e, orbits too close to its parent star to allow for liquid water. Furthermore, the wobble is so tiny that few Earth-sized planets are expected to be found using this technique.

Searching for Planetary Transits

An alternative technique will be used by an orbiting telescope called Kepler, which was launched in 2009, and might discover hundreds of Earth-sized planets. If a star is orbited by a planet whose orbital plane is oriented edge-on to our line of sight, once per orbit the planet will pass in front of the star in an event called a *transit*. By blocking some of the star's light, this transit causes a temporary dimming of the light we see from that star. As we discussed in Section 8-6, astronomers have used Earth-based telescopes to observe dimming of this kind from Jupiter-sized planets transiting their parent stars. To detect the much slighter dimming caused by the transit of a small, Earth-sized planet, Kepler will use specialized detectors. Furthermore, by observing from an orbit around Earth and using a telescope with a wide field of view, Kepler will be able to continuously monitor thousands of stars at once.

If Kepler detects stars that dim slightly as expected, astronomers will have to make sure that the dimming really is due to a transiting planet and not some other cause. (One key test will be whether the dimming repeats with a definite period, as would be expected for a planet in a periodic orbit.) Once a transiting planet is confirmed, astronomers will be able to determine the transiting planet's size (inferred from how much dimming takes place), as well as the size of its orbit. Recall from Kepler's third law that the size of the orbit can be calculated using the orbital period of the planet, which is the same as the time interval between successive transits. Given the distance from the planet to its parent star and the star's luminosity, astronomers will even be able to estimate the planet's average temperature.

Searching Using Images and Spectra

The results from Kepler may help astronomers select stars to study in more detail using Darwin, a more advanced orbiting telescope under development by the European Space Agency. Planned for a 2018 launch, Darwin will search for Earthlike planets by detecting their infrared radiation. The rationale is that stars like the Sun emit much less infrared radiation than visible light, while planets are relatively strong emitters of infrared. Hence, observing in the infrared makes it less difficult (although still technically challenging) to detect planets orbiting a star.

Darwin will also analyze the infrared spectra of any planets that it finds, in the hope of seeing the characteristic absorption of atmospheric gases such as ozone, carbon dioxide, and water vapor (Figure 28-11). The relative amounts of these gases, as determined from a planet's spectrum, can reveal whether life is present on that planet.

Darwin will need to achieve enough resolution to detect individual planets. One proposed mission design makes use of interferometry. We discussed this technique for improving the resolution of telescopes in Section 6-6. By combining the light from three widely spaced dishes, each at least 3 m in diameter, Darwin will make the sharpest infrared images of any telescope in history. NASA has proposed a similar planet-finding interferometry mission called Terrestrial Planet Finder, but the funding for this mission is uncertain.

Future telescope technology may make it possible to resolve details on planets orbiting other stars

A more speculative project is an infrared telescope with sufficient resolution that some detail would be visible in the image of an extrasolar planet. One concept for such a mission would consist of five Darwin-type telescopes flying in a geometrical formation some 6000 km across (equal to the radius of Earth). All five telescopes would collect light from the same extrasolar planet, then reflect it onto a single mirror. The combined light would go to detectors

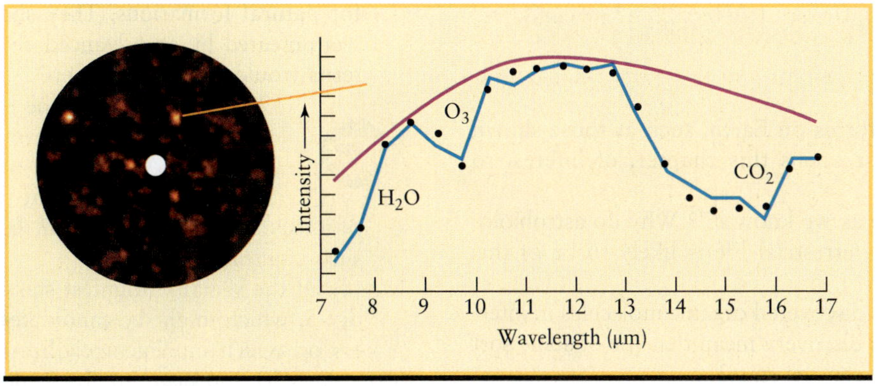

Figure 28-11

The Spectrum of a Simulated Planet The image on the left is a simulation of what the Terrestrial Planet Finder infrared telescope might see once it is launched. The white dot at the center is a nearby Sunlike star, and the smaller dots around it are planets orbiting the star. On the right is the simulated infrared spectrum of one of the planets, showing broad absorption lines of water vapor (H_2O), ozone (O_3), and carbon dioxide (CO_2). While all these molecules can be created by nonbiological processes, the presence of life will change the relative amounts of each molecule in the planet's atmosphere. Thus, the infrared spectrum of such planets will make it possible to identify worlds on which life may have evolved. (Jet Propulsion Laboratory)

on board a sixth spacecraft. The technology needed for such an ambitious mission does not yet exist, but may become available within a few decades.

Sometime during the twenty-first century, missions such as Darwin may answer the question "Are there worlds like Earth orbiting other stars?" If the answer is yes, radio searches for intelligent signals will gain even more impetus.

The potential rewards from such searches are great. Detecting a message from an alien civilization could dramatically change the course of our own civilization, through the sharing of scientific information with another species or an awakening of social or humanistic enlightenment. In only a few years our technology, industry, and social structure might advance the equivalent of centuries into the future. Such changes would touch every person on Earth. Mindful of these profound implications, scientists push ahead with the search for extraterrestrial intelligence.

Key Words

astrobiology, p. 747
carbonaceous chondrite, p. 749
Drake equation, p. 755
habitable zone, p. 755
organic molecules, p. 747
search for extraterrestrial intelligence (SETI), p. 754
water hole, p. 757

Key Ideas

Organic Molecules in the Universe: All life on Earth, and presumably on other worlds, depends on organic (carbon-based) molecules. These molecules occur naturally throughout interstellar space.

• The organic molecules needed for life to originate were probably brought to the young Earth by comets, meteorites, and interplanetary dust particles. Another likely source for organic molecules is chemical reactions in Earth's primitive atmosphere. Similar processes may occur on other worlds.

Life in the Solar System: Besides Earth, only two worlds in our solar system—the planet Mars and Jupiter's satellite Europa—may have had the right conditions for the origin of life.

• Mars once had liquid water on its surface, though it has none today. Life may have originated on Mars during the liquid water era.

• The *Viking Lander* spacecraft searched for microorganisms on the Martian surface, but found no conclusive sign of their presence. The Mars Science Laboratory is designed to assess the past and present suitability of Mars for microbial life.

• An ancient Martian rock that came to Earth as a meteorite shows circumstantial evidence that microorganisms once existed on Mars. Additional rock samples are needed to provide corroboration.

• Europa appears to have extensive liquid water beneath its icy surface. Future missions may search for the presence of life there.

Radio Searches for Extraterrestrial Intelligence: Astronomers have carried out a number of searches for radio signals from other stars. No signs of intelligent life have yet been detected, but searches are continuing and using increasingly sophisticated techniques.

• The Drake equation is a tool for estimating the number of intelligent, communicative civilizations in our Galaxy.

Telescope Searches for Earthlike Planets: A new generation of orbiting telescopes may be able to detect numerous terrestrial planets around nearby stars. If such planets are found, their infrared spectra may reveal the presence or absence of life.

Questions

Review Questions

1. Why are extreme life-forms on Earth, such as those shown in the photograph that opens this chapter, of interest to astrobiologists?
2. What is meant by "life as we know it"? Why do astrobiologists suspect that extraterrestrial life is likely to be of this form?
3. How have astronomers discovered organic molecules in interstellar space? Does this discovery mean that life of some sort exists in the space between the stars?
4. Mercury, Venus, and the Moon are all considered unlikely places to find life. Suggest why this should be.
5. Many science-fiction stories and movies—including *The War of the Worlds, Invaders from Mars, Mars Attacks!*, and *Martians, Go Home*—involve invasions of Earth by intelligent beings from Mars. Why Mars rather than any of the other planets?
6. Summarize the three tests performed by the *Viking Landers* to search for Martian microorganisms. Is the Mars Science Laboratory designed to repeat these test?
7. What arguments can you give against the idea that the "face" on Mars (Figure 28-8) is of intelligent origin? What arguments can you give in favor of this idea?
8. Suppose someone brought you a rock that he claimed was a Martian meteorite. What scientific tests would you recommend be done to test this claim?
9. Why are most searches for extraterrestrial intelligence made using radio telescopes? Why are most of these carried out at frequencies between 10^3 MHz and 10^4 MHz?
10. Explain why infrared telescopes like those proposed for Darwin and Terrestrial Planet Finder need to be placed in space.

Advanced Questions

Problem-solving tips and tools

The small-angle formula, discussed in Box 1-1, will be useful. Section 5-2 gives the relationship between wavelength and frequency, while Section 5-9 and Box 5-6 discuss the Doppler effect. Section 6-3 gives the relationship between the angular resolution of a telescope, the telescope diameter, and the wavelength used. You will find useful data about the planets in Appendix 1.

11. In 1802, when it seemed likely to many scholars that there was life on Mars, the German mathematician Karl Friedrich Gauss proposed that we signal the Martian inhabitants by drawing huge geometric patterns in the snows of Siberia. His plan was never carried out. (**a**) Suppose patterns had been drawn that were 1000 km across. What minimum diameter would the objective of a Martian telescope need to have to be able to resolve these patterns? Assume that the observations are made at a wavelength of 550 nm, and assume that Earth and Mars are at their minimum separation. (**b**) Ideally, the patterns used would be ones that could not be mistaken for natural formations. They should also indicate that they were created by an advanced civilization. What sort of patterns would you have chosen?
12. Assume that all the terms in the Drake equation have the values given in the text, except for N and L. (**a**) If there are 1000 civilizations in the Galaxy today, what must be the average lifetime of a technological civilization? (**b**) What if there are a million such civilizations?
13. (**a**) Of the visually brightest stars in the sky listed in Appendix 5, which might be candidates for having Earthlike planets on which intelligent civilizations have evolved? Explain your selection criteria. (**b**) Repeat part (a) for the nearest stars, listed in Appendix 4.
14. It has been suggested that extraterrestrial civilizations would choose to communicate at a wavelength of 21 cm. Hydrogen atoms in interstellar space naturally emit at this wavelength, so astronomers studying the distribution of hydrogen around the Galaxy would already have their radio telescopes tuned to receive extraterrestrial signals. (**a**) Calculate the frequency of this radiation in megahertz. Is this inside or outside the water hole? (**b**) Discuss the merits of this suggestion.
15. Imagine that a civilization in another planetary system is sending a radio signal toward Earth. As our planet moves in its orbit around the Sun, the wavelength of the signal we receive will change due to the Doppler effect. This gives SETI scientists a way to distinguish stray signals of terrestrial origin (which will not show this kind of wavelength change) from interstellar signals. (**a**) Use the data in Appendix 1 to calculate the speed of Earth in its orbit. For simplicity, assume the orbit is circular. (**b**) If the alien civilization is transmitting at a frequency of 3000 MHz, what wavelength (in meters) would we receive if Earth were moving neither toward nor away from their planet? (**c**) The maximum Doppler shift occurs if Earth's orbital motion takes it directly toward or directly away from the alien planet. How large is that maximum wavelength shift? Express your answer both in meters and as a percentage of the unshifted wavelength you found in (b). (**d**) Discuss why it is important that SETI radio receivers be able to measure frequency and wavelength to very high precision.
16. Astronomers have proposed using interferometry to make an extremely high-resolution telescope. This proposal involves placing a number of infrared telescopes in space, separating them by thousands of kilometers, and combining the light from the individual telescopes. One design of this kind has an effective diameter of 6000 km and uses infrared radiation with a wavelength of 10 mm. If it is used to observe an Earthlike planet orbiting the star Epsilon Eridani, 3.22 parsecs (10.5 light-years) from Earth, what is the size of the smallest detail that this system will be able to resolve on the face of that planet? Give your answer in kilometers.

Discussion Questions

17. Suppose someone told you that the *Viking Landers* failed to detect life on Mars simply because the tests were designed to detect terrestrial life-forms, not Martian life-forms. How would you respond?

18. Science-fiction television shows and movies often depict aliens as looking very much like humans. Discuss the likelihood that intelligent creatures from another world would have (**a**) a biochemistry similar to our own, (**b**) two legs and two arms, and (**c**) about the same dimensions as a human.

19. The late, great science-fiction editor John W. Campbell exhorted his authors to write stories about organisms that think as well as humans, but not *like* humans. Discuss the possibility that an intelligent being from another world might be so alien in its thought processes that we could not communicate with it.

20. If a planet always kept the same face toward its star, just as the Moon always keeps the same face toward Earth, most of the planet's surface would be uninhabitable. Discuss why.

21. How do you think our society would respond to the discovery of intelligent messages coming from a civilization on a planet orbiting another star? Explain your reasoning.

22. What do you think will set the limit on the lifetime of our technological civilization? Explain your reasoning.

23. The first of all Earth spacecraft to venture into interstellar space were *Pioneer 10* and *Pioneer 11,* which were launched in 1972 and 1973, respectively. Their missions took them past Jupiter and Saturn and eventually beyond the solar system. Both spacecraft carry a metal plaque with artwork (reproduced below) that shows where the spacecraft is from and what sort of creatures designed it. If an alien civilization were someday to find one of these spacecraft, which of the features on the plaque do you think would be easily understandable to them? Explain your reasoning.

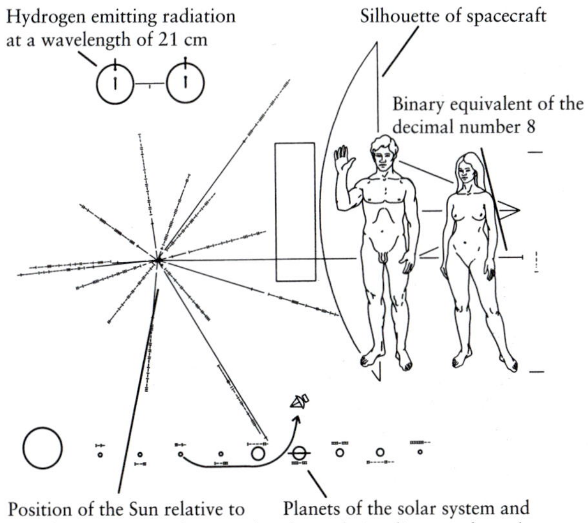

Web/eBook Questions

24. Any living creatures in the subsurface ocean of Europa would have to survive without sunlight. Instead, they might obtain energy from Europa's inner heat. Search the World Wide Web for information about "black smokers," which are associated with high-temperature vents at the bottom of Earth's oceans. What kind of life is found around black smokers? How do these life-forms differ from the more familiar organisms found in the upper levels of the ocean?

25. Search the World Wide Web for information about the *Mars Express* orbiter and the *Spirit* and *Opportunity* rovers. What discoveries have these missions made about water on Mars? Have they found any evidence that liquid water has existed on Mars in the recent past? Describe the evidence, if any.

26. Like other popular media, the World Wide Web is full of claims of the existence of "extraterrestrial intelligence"—namely, UFO sightings and alien abductions. (**a**) Choose a Web site of this kind and analyze its content using the idea of *Occam's razor,* the principle that if there is more than one viable explanation for a phenomenon, one should choose the simplest explanation that fits all the observed facts. (**b**) Read what a skeptical Web site has to say about UFO sightings. A good example is the Web site of the Committee for Skeptical Inquiry, or CSI. After considering what you have read on both sides of the UFO debate, discuss your opinions about whether intelligent aliens really have landed on Earth.

Activities

Observing Projects

27. Use the *Starry Night Enthusiast*™ program to view Earth as it might be seen by a visiting spacecraft. First, select **Viewing Location . . .** in the **Options** menu and set the viewing location to your city or town from the list of cities provided or click on the **Map** tab in the **Viewing Location** pane and use the mouse to click on your approximate position on the world map. Then click the **Set Location** button. Set the local time to 12:00:00 P.M. (noon). To see Earth from space, use the up and down elevation buttons on the toolbar to raise yourself above the surface until you can see the entire Earth. You can use the scrollbars (select **View > Show Scrollbars**) on the right side and bottom of the window to center Earth in your view. Earth can be rotated to allow you to see different locations by clicking and moving the mouse while its icon, a four-way arrow, is over Earth's image. (**a**) Describe any features you see that suggest life could exist on Earth. Explain your reasoning. (**b**) Using the controls at the right-hand end of the toolbar, zoom in to show more detail around your city or town. The amount of detail is comparable to the view from a spacecraft a few million kilometers away. Can you see any evidence that life does exist on Earth? (**c**) From a distance of a few million kilometers, are there any measurements that a spacecraft could carry out to prove that life exists on Earth? Explain your reasoning.

28. Use the *Starry Night Enthusiast*™ program to examine the planet Mars. Open the **Favorites** pane and double-click on **Solar System > Mars** to view this planet from about 6800 km above its surface. (Click on **View > Feet** to remove the astronaut's spacesuit from the view.) You can zoom in or out on Mars using the buttons in the Zoom section at the right of the toolbar. You can rotate Mars by placing the mouse cursor over the image and moving the mouse while holding down the mouse button. (On a two-button mouse, hold down the left mouse button.) Rotate Mars and zoom in and out to familiarize yourself with the different surface features. Based on what you observe, where on the Martian surface would you choose to land a spacecraft to search for the presence of life? Explain how you made your choice.

Collaborative Exercise

29. Imagine that astronomers have discovered intelligent life in a nearby star system. Your group is submitting a proposal for who on Earth should speak for the planet and what 50-word message should be conveyed. Prepare a maximum one-page proposal that states (**a**) who should speak for Earth and why; (**b**) what this person should say in 50 words; and (**c**) why this message is the most important compared to other things that could be said. Only serious responses receive full credit.

A Biologist's View of Astrobiology

by Kevin W. Plaxco

In the early 1960s, flush with the excitement of Sputnik and the first manned space missions, the Nobel Prize-winning biologist Joshua Lederberg coined the word "exobiology" to describe the scientific study of extraterrestrial life. But after a brief flurry of popularity in the 1960s and 1970s (due partly to the pioneering research and public outreach work of Carl Sagan), exobiology fell out of fashion among scientists; it was, after all, the only field of scientific study without any actual subject material to research.

In the last decade, however, the field of astrobiology has filled the void left by exobiology's demise by being simultaneously more encompassing and more practical. Astrobiology removes exobiology's distinction between life on Earth and life "out there" and focuses on broader, but more tractable, questions regarding the relationship between life (any life, anywhere) and the universe. In a nutshell, astrobiology uses our significant (if incomplete) knowledge of life on Earth to address three broad questions about life in the universe:

- Which of the physical attributes of our universe were necessary in order for it to support the origins and further evolution of life?
- How did life arise and evolve on Earth, and how might it have arisen and evolved elsewhere?
- Where else might life have arisen in our universe, and how do we best search for it?

Much of the worth and appeal of astrobiology lie in the fact that these questions, which are among the most fundamental posed by science, address the most profound issues regarding our origins and our nature.

Perhaps not surprisingly, some aspects of astrobiology are far better understood than others. We understand, for example, the broad details of how the almost metal-free hydrogen and helium left over after the Big Bang was enriched in the metallic elements by fusion reactions occurring in the more massive stars and how these elements—absolutely critical for biology because only they support the complex chemistry required for life—were then returned to the cosmos in supernova explosions. Similarly, we understand in detail how rocky planets—which we see as potentially habitable—form out of the dust and gas of prestellar nebulae. In contrast, however, much of the *biology* in astrobiology remains speculative at best. There is, for example, nothing even remotely resembling consensus regarding how life first arose on Earth much less how—or if—it might have arisen elsewhere in the cosmos.

Perhaps equally unsurprisingly, some branches of the astrobiology community are more "optimistic" than others. Astronomers generally view astrobiology in terms of the processes by which potentially habitable planets form. As these processes are reasonably well understood, astronomers tend to be rather upbeat regarding the events surrounding life's origins. The radio astronomer Frank Drake, who pioneered the search for extraterrestrial intelligence (SETI) in the 1960s, is a relatively extreme example of this hopefulness: When trying to estimate the number of intelligent, "communicating" civilizations in the Galaxy, he assumed that rocky, water-soaked planets almost invariably give rise to life. Biochemists, in contrast, tend to view the key issues in astrobiology as relating to the first steps in the origins of life—that is, the detailed processes by which inanimate chemicals first combine to form a living, evolving, self-replicating system. To date, about a half dozen theories have been put forth to explain how this might have occurred on Earth, but all are fraught with enormous scientific difficulties. Specifically, every single theory of the origins of life postulated to date either utterly lacks experimental evidence to back it up or requires events that are so improbable they make even the "astronomical" number of planets in the universe look small by comparison. When faced with what is, in effect, a complete unknown, biochemists tend to be much more cautious about estimating the probability of life arising and the chances of our being alone in the cosmos. To quote Francis Crick, co-discoverer of the structure of DNA, "We cannot decide whether the origin of life on Earth was an extremely unlikely event or almost a certainty, or any possibility in between these two extremes."

Irrespective of our degree of personal optimism or pessimism, though, no one would be more thrilled than us biologists by the detection of life that had arisen independently of life on Earth. Such a find would revolutionize our field by providing the first, tangible evidence of whether there are viable routes to the evolution of life other than the DNA/proteins/cells path that it took on Earth. But even barring such a momentous discovery as the detection of extraterrestrial life, astrobiology still provides a significant service to biology: By placing life into a broader context, astrobiology provides the perspective that we need in order to truly understand who we are and where we come from.

Prior to becoming a professor of biochemistry at the University of California, Santa Barbara, Kevin Plaxco earned his BS in chemistry from the University of California at Riverside and his Ph.D. in molecular biology from the California Institute of Technology. While his research focuses predominantly on the study of proteins, the "macromolecular machines" that perform the lion's share of the action in our cells, he has also performed NASA-funded research into methods for the detection of extraterrestrial life. In addition, he teaches a course in astrobiology and has published a popular science book on the subject.

Appendices

Appendix 1 The Planets: Orbital Data

Planet	Semimajor axis (10^6 km)	Semimajor axis (AU)	Sidereal period (years)	Sidereal period (days)	Synodic period (days)	Average orbital speed (km/s)	Orbital eccentricity	Inclination of orbit to ecliptic (°)
Mercury	57.9	0.387	0.241	87.969	115.88	47.9	0.206	7.00
Venus	108.2	0.723	0.615	224.70	583.92	35.0	0.007	3.39
Earth	149.6	1.000	1.000	365.256	—	29.79	0.017	0.00
Mars	227.9	1.524	1.88	686.98	779.94	24.1	0.093	1.85
Jupiter	778.3	5.203	11.86		398.9	13.1	0.048	1.30
Saturn	1429	9.554	29.46		378.1	9.64	0.053	2.48
Uranus	2871	19.194	84.10		369.7	6.83	0.043	0.77
Neptune	4498	30.066	164.86		367.5	5.5	0.010	1.77

Appendix 2 The Planets: Physical Data

Planet	Equatorial diameter (km)	Equatorial diameter (Earth = 1)	Mass (kg)	Mass (Earth = 1)	Average density (kg/m^3)	Rotation period* (solar days)	Inclination of equator to orbit (°)	Surface gravity (Earth = 1)	Albedo	Escape speed (km/s)
Mercury	4880	0.383	3.302×10^{23}	0.0553	5430	58.646	0.5	0.38	0.12	4.3
Venus	12,104	0.949	4.868×10^{24}	0.8149	5243	243.01^R	177.4	0.91	0.59	10.4
Earth	12,756	1.000	5.974×10^{24}	1.000	5515	0.997	23.45	1.000	0.39	11.2
Mars	6794	0.533	6.418×10^{23}	0.107	3934	1.026	25.19	0.38	0.15	5.0
Jupiter	142,984	11.209	1.899×10^{27}	317.8	1326	0.414	3.12	2.36	0.44	59.5
Saturn	120,536	9.449	5.685×10^{26}	95.16	687	0.444	26.73	1.1	0.47	35.5
Uranus	51,118	4.007	8.682×10^{25}	14.53	1318	0.718^R	97.86	0.92	0.56	21.3
Neptune	49,528	3.883	1.024×10^{26}	17.15	1638	0.671	29.56	1.1	0.51	23.5

**For Jupiter, Saturn, Uranus, and Neptune, the internal rotation period is given. A superscript* R *means that the rotation is retrograde (opposite the planet's orbital motion).*

Appendix 3 Satellites of the Planets

Planet	Satellite	Date of discovery	Average distance from center of planet (km)	Orbital (sidereal) period (days)*	Orbital eccentricity	Size of satellite (km)**	Mass (kg)
EARTH	Moon	—	384,400	27.322	0.0549	3476	7.349×10^{22}
MARS	Phobos	1877	9378	0.319	0.01	28 × 23 × 20	1.1×10^{16}
	Deimos	1877	23,460	1.263	0.00	16 × 12 × 10	1.8×10^{15}
JUPITER	Metis	1979	128,000	0.295	0.0012	44	1.2×10^{17}
	Adrastea	1979	129,000	0.298	0.0018	24 × 20 × 16	7.5×10^{15}
	Amalthea	1892	181,400	0.498	0.0031	270 × 200 × 155	2.1×10^{18}
	Thebe	1979	221,900	0.675	0.0177	98	1.5×10^{18}
	Io	1610	421,600	1.769	0.0041	3642	8.932×10^{22}
	Europa	1610	670,900	3.551	0.0094	3120	4.791×10^{22}
	Ganymede	1610	1,070,000	7.155	0.0011	5268	1.482×10^{23}
	Callisto	1610	1,883,000	16.689	0.0074	4800	1.077×10^{23}
	Themisto	2000	7,284,000	130.02	0.2426	9	6.9×10^{14}
	Leda	1974	11,165,000	240.92	0.1636	18	1.1×10^{16}
	Himalia	1904	11,461,000	250.56	0.1623	184	6.7×10^{18}
	Lysithea	1938	11,717,000	259.20	0.1124	38	6.3×10^{16}
	Elara	1905	11,741,000	259.64	0.2174	78	8.7×10^{17}
	Kallichore	2000	12,555,000	456.10	0.2480	4	1.5×10^{13}
	S/2003 J12	2003	15,912,000	489.52^{R}	0.6056	1	1.5×10^{12}
	Carpo	2003	16,989,000	456.10	0.4297	3	4.5×10^{13}
	Euporie	2001	19,304,000	550.74^{R}	0.1432	2	1.5×10^{13}
	S/2003 J3	2003	20,221,000	583.88^{R}	0.1970	2	1.5×10^{13}
	S/2003 J18	2003	20,514,000	596.59^{R}	0.0148	2	1.5×10^{13}
	Orthosie	2001	20,720,000	622.56^{R}	0.2808	2	1.5×10^{13}
	Euanthe	2001	20,797,000	620.49^{R}	0.2321	3	4.5×10^{13}
	Harpalyke	2000	20,858,000	623.31^{R}	0.2268	4	1.2×10^{14}
	Praxidike	2000	20,907,000	625.38^{R}	0.2308	7	4.3×10^{14}
	Thyone	2001	20,939,000	627.21^{R}	0.2286	4	9.0×10^{13}
	S/2003 J16	2003	20,963,000	616.36^{R}	0.2245	2	1.5×10^{13}
	Iocaste	2000	21,061,000	631.60^{R}	0.2160	5	1.9×10^{14}
	Mneme	2003	21,069,000	620.04^{R}	0.2273	2	1.5×10^{13}
	Hermippe	2001	21,131,000	633.90^{R}	0.2096	4	9.0×10^{13}
	Thelxinoe	2003	21,162,000	628.09^{R}	0.2206	2	1.5×10^{13}
	Helike	2003	21,263,000	634.77^{R}	0.1558	4	9.0×10^{13}
	Ananke	1951	21,276,000	629.77^{R}	0.2435	28	3.0×10^{16}
	S/2003 J15	2003	22,627,000	689.77^{R}	0.1916	2	1.5×10^{13}
	Eurydome	2001	22,865,000	717.33^{R}	0.2759	3	4.5×10^{13}
	Arche	2002	22,931,000	723.90^{R}	0.2588	3	4.5×10^{13}
	S/2003 J17	2003	23,001,000	714.47^{R}	0.2379	2	1.5×10^{13}
	Pasithee	2001	23,004,000	719.44^{R}	0.2675	2	1.5×10^{13}
	S/2003 J10	2003	23,042,000	716.25^{R}	0.4295	2	1.5×10^{13}
	Chaldene	2000	23,100,000	723.70^{R}	0.2519	4	7.5×10^{13}
	Isonoe	2000	23,155,000	726.25^{R}	0.2471	4	7.5×10^{13}

Appendix 3 Satellites of the Planets

Planet	Satellite	Date of discovery	Average distance from center of planet (km)	Orbital (sidereal) period (days)*	Orbital eccentricity	Size of satellite (km)**	Mass (kg)
JUPITER (continued)	Erinome	2000	23,196,000	728.51^R	0.2665	3	4.5×10^{13}
	Kale	2001	23,217,000	729.47^R	0.2599	2	1.5×10^{13}
	Aitne	2001	23,229,000	730.18^R	0.2643	3	4.5×10^{13}
	Taygete	2000	23,280,000	732.41^R	0.2525	5	1.6×10^{14}
	S/2003 J9	2003	23,384,000	733.29^R	0.2632	1	1.5×10^{12}
	Carme	1938	23,404,000	734.17^R	0.2533	46	1.3×10^{17}
	Sponde	2001	23,487,000	748.34^R	0.3121	2	1.5×10^{13}
	Megaclite	2000	23,493,000	752.88^R	0.4197	6	2.1×10^{14}
	S/2003 J5	2003	23,495,000	738.73^R	0.2478	4	9.0×10^{13}
	S/2003 J19	2003	23,533,000	740.42^R	0.2557	2	1.5×10^{13}
	S/2003 J23	2003	23,563,000	732.44^R	0.2714	2	1.5×10^{13}
	Kalyke	2000	23,566,000	742.03^R	0.2465	5	1.9×10^{14}
	S/2003 J14	2003	23,614,000	779.23^R	0.3439	2	1.5×10^{13}
	Pasiphaë	1908	23,624,000	743.63^R	0.4090	58	3.0×10^{17}
	Eukelade	2003	23,661,000	746.39^R	0.2721	4	9.0×10^{13}
	S/2003 J4	2003	23,930,000	755.24^R	0.3618	2	1.5×10^{13}
	Sinope	1914	23,939,000	758.90^R	0.2495	38	7.5×10^{16}
	Hegemone	2003	23,947,000	739.60^R	0.3276	3	4.5×10^{13}
	Aoede	2003	23,981,000	761.50^R	0.4322	4	9.0×10^{13}
	Kallichore	2003	24,043,000	764.74^R	0.2640	2	1.5×10^{13}
	Autonoe	2000	24,046,000	760.95^R	0.3168	4	9.0×10^{13}
	Callirrhoe	1999	24,103,000	758.77^R	0.2828	7	8.7×10^{14}
	Cyllene	2003	24,349,000	751.91^R	0.3189	2	1.5×10^{13}
	S/2003 J2	2003	29,541,000	979.99^R	0.2255	2	1.5×10^{13}
SATURN	Pan	1981	133,580	0.575	0	$35 \times 35 \times 23$	4.9×10^{15}
	Daphnis	2005	136,500	0.594	0	7	3×10^{14}
	Atlas	1980	137,670	0.602	0.0012	$46 \times 38 \times 19$	6.6×10^{15}
	Prometheus	1980	139,380	0.613	0.0022	$119 \times 87 \times 61$	1.6×10^{17}
	Pandora	1980	141,720	0.629	0.0042	$103 \times 80 \times 64$	1.4×10^{17}
	Epimetheus	1980	151,410	0.694	0.0098	$135 \times 108 \times 105$	5.3×10^{17}
	Janus	1980	151,460	0.695	0.0068	$193 \times 173 \times 137$	1.9×10^{18}
	Mimas	1789	185,540	0.942	0.0196	397	3.8×10^{19}
	Methone	2004	194,440	1.01	0.0001	3	2×10^{13}
	Pallene	2004	212,280	1.154	0.004	4	4×10^{13}
	Enceladus	1789	238,040	1.37	0.0047	504	1.1×10^{20}
	Tethys	1684	294,670	1.888	0.0001	1066	6.2×10^{20}
	Telesto	1980	294,710	1.888	0.0002	$29 \times 22 \times 20$	8×10^{15}
	Calypso	1980	294,710	1.888	0.0005	$30 \times 23 \times 14$	5×10^{15}
	Polydeuces	2004	377,200	2.737	0.0192	3.5	3×10^{13}
	Dione	1684	377,420	2.737	0.0022	1123	1.1×10^{21}
	Helene	1980	377,420	2.737	0.0071	$36 \times 32 \times 30$	2×10^{16}

Appendix 3 Satellites of the Planets

Planet	Satellite	Date of discovery	Average distance from center of planet (km)	Orbital (sidereal) period (days)*	Orbital eccentricity	Size of satellite (km)**	Mass (kg)
SATURN (continued)	Rhea	1672	527,070	4.518	0.001	1528	2.3×10^{21}
	Titan	1655	1,221,870	15.95	0.0288	5150	1.34×10^{23}
	Hyperion	1848	1,500,880	21.28	0.0274	360 × 280 × 225	5.7×10^{18}
	Iapetus	1671	3,560,840	79.33	0.0283	1472	2.0×10^{21}
	Kiviuq	2000	11,111,000	449.22	0.3288	16	3×10^{15}
	Ijiraq	2000	11,124,000	451.43	0.3163	12	1×10^{15}
	Phoebe	1898	12,947,780	550.31^{R}	0.1635	230 × 220 × 210	8.3×10^{18}
	Paaliaq	2000	15,200,000	686.93	0.3631	19	4×10^{15}
	Skathi	2000	15,541,000	728.21^{R}	0.2701	6	2×10^{14}
	Albiorix	2000	16,182,000	783.46	0.477	32	3×10^{16}
	S/2007 S2	2007	16,560,000	792.96^{R}	0.2418	6	1.5×10^{14}
	Bebhionn	2004	17,153,520	838.77	0.4691	6	2×10^{14}
	Erriapo	2000	17,343,000	871.18	0.4724	10	9×10^{14}
	Siarnaq	2000	17,531,000	895.55	0.2961	40	4×10^{16}
	Skoll	2006	17,473,800	862.37	0.418	6	2×10^{14}
	Tarqeq	2007	17,910,600	894.86	0.1081	7	2.3×10^{14}
	Tarvos	2000	17,983,000	926.23	0.5305	15	3×10^{15}
	Greip	2006	18,065,700	906.56	0.374	6	2×10^{14}
	S/2004 S19	2004	18,217,125	912^{R}	0.36	8	8×10^{14}
	S/2004 S13	2004	18,403,000	933.45^{R}	0.2586	6	2×10^{14}
	Jarnsaxa	2006	18,556,900	943.78	0.192	6	2×10^{14}
	Mundilfari	2000	18,685,000	952.67^{R}	0.21	7	3×10^{14}
	S/2006 S1	2006	18,930,200	972.41	0.13	6	2×10^{14}
	Narvi	2003	19,007,000	1003.93^{R}	0.4309	7	3×10^{14}
	S/2004 S15	2004	19,338,000	1005.93^{R}	0.1428	6	2×10^{14}
	S/2004 S17	2004	19,447,000	1014.7^{R}	0.1793	4	4×10^{13}
	Suttungr	2000	19,459,000	1016.67^{R}	0.114	7	3×10^{14}
	S/2004 S14	2004	19,856,000	1038.67^{R}	0.3715	6	2×10^{14}
	S/2004 S12	2004	19,878,000	1046.16^{R}	0.3261	5	2×10^{14}
	S/2004 S18	2004	20,129,000	1083.57^{R}	0.5214	7	3×10^{14}
	S/2004 S9	2004	20,390,000	1086.1^{R}	0.2397	5	2×10^{14}
	Thrymr	2000	20,474,000	1094.23^{R}	0.4652	7	3×10^{14}
	S/2007 S3	2007	20,518,500	1100^{R}	0.130	5	9×10^{14}
	S/2004 S10	2004	20,735,000	1116.47^{R}	0.252	6	2×10^{14}
	S/2004 S7	2004	20,999,000	1140.28^{R}	0.5299	6	2×10^{14}
	S/2006 S3	2006	21,076,300	1142.37	0.471	6	2×10^{14}
	Surtur	2006	22,288,916	1242.36	0.368	6	2×10^{14}
	Kari	2006	22,321,200	1245.06	0.341	7	3×10^{14}
	S/2004 S16	2004	22,453,000	1260.28^{R}	0.1364	4	4×10^{13}
	Loge	2006	22,984,322	1300.95^{R}	0.1390	6	1.5×10^{14}
	Ymir	2000	23,040,000	1315.21^{R}	0.335	18	3×10^{15}
	S/2006 S5	2006	23,190,000	1314^{R}	0.139	6	2×10^{14}
	S/2004 S8	2004	25,108,000	1490.87^{R}	0.2064	6	2×10^{14}

Appendix 3 Satellites of the Planets

Planet	Satellite	Date of discovery	Average distance from center of planet (km)	Orbital (sidereal) period (days)*	Orbital eccentricity	Size of satellite (km)**	Mass (kg)
Uranus	Cordelia	1986	49,800	0.335	0.0003	50 × 36	4.4×10^{16}
	Ophelia	1986	53,800	0.376	0.0099	54 × 38	5.3×10^{16}
	Bianca	1986	59,200	0.435	0.0009	64 × 46	9.2×10^{16}
	Cressida	1986	61,800	0.464	0.0004	92 × 74	3.4×10^{17}
	Desdemona	1986	62,700	0.474	0.0001	90 × 54	1.8×10^{17}
	Juliet	1986	64,400	0.493	0.0007	150 × 74	5.6×10^{17}
	Portia	1986	66,100	0.513	0.0001	156 × 126	1.7×10^{18}
	Rosalind	1986	69,900	0.558	0.0001	72	2.5×10^{17}
	Cupid	2003	74,800	0.618	0.0013	18	3.8×10^{15}
	Belinda	1986	75,300	0.624	0.0001	128 × 64	3.6×10^{17}
	Perdita	1986	76,420	0.638	0.003	26	1.3×10^{16}
	Puck	1985	86,000	0.762	0.0001	162	2.9×10^{18}
	Mab	2003	97,734	0.923	0.0025	24	1.0×10^{16}
	Miranda	1948	129,900	1.413	0.0013	471	6.59×10^{19}
	Ariel	1851	190,900	2.52	0.0012	1158	1.35×10^{21}
	Umbriel	1851	266,000	4.144	0.0039	1169	1.2×10^{21}
	Titania	1787	436,300	8.706	0.0011	1578	3.53×10^{21}
	Oberon	1787	583,500	13.46	0.0014	1522	3.01×10^{21}
	Francisco	2001	4,276,000	266.56	0.1459	22	1.4×10^{15}
	Caliban	1997	7,231,000	579.73^{R}	0.1587	72	7.4×10^{17}
	Stephano	1999	8,004,000	677.36^{R}	0.2292	32	6.0×10^{15}
	Trinculo	2001	8,504,000	749.24^{R}	0.22	18	7.5×10^{14}
	Sycorax	1997	12,179,000	1288.3^{R}	0.5224	150	5.4×10^{18}
	Margaret	2003	14,345,000	1687.01	0.6608	20	1.0×10^{15}
	Prospero	1999	16,256,000	1978.29^{R}	0.4448	25	2.1×10^{16}
	Setebos	1999	17,418,000	2225.21^{R}	0.5914	24	2.1×10^{16}
	Ferdinand	2003	20,901,000	2887.21^{R}	0.3682	21	4.4×10^{16}
Neptune	Naiad	1989	48,227	0.294	0.0004	96 × 60 × 52	1.9×10^{17}
	Thalassa	1989	50,075	0.311	0.0002	108 × 100 × 52	3.5×10^{17}
	Despina	1989	52,526	0.335	0.0002	180 × 150 × 130	2.1×10^{18}
	Galatea	1989	61,953	0.429	0	204 × 184 × 144	2.1×10^{18}
	Larissa	1989	73,548	0.555	0.0014	216 × 204 × 164	4.2×10^{18}
	Proteus	1989	117,647	1.122	0.0005	440 × 416 × 404	4.4×10^{19}
	Triton	1846	354,800	5.877^{R}	0	2706	2.15×10^{22}
	Nereid	1949	5,513,400	360.14	0.7512	340	3.1×10^{19}
	S/2002 N1	2002	15,728,000	1879.71^{R}	0.5711	62	1.8×10^{17}
	S/2002 N2	2002	22,422,000	2914.07	0.2931	44	6.3×10^{16}
	S/2002 N3	2002	23,571,000	3167.85	0.4237	42	5.5×10^{16}
	Psamathe	2003	46,695,000	9115.91^{R}	0.4499	24	1.0×10^{16}
	S/2002 N4	2002	48,387,000	9373.99^{R}	0.4945	60	1.6×10^{17}

This table was compiled from data provided by the Jet Propulsion Laboratory.

**A superscript* R *means that the satellite orbits in a retrograde direction (opposite to the planet's rotation).*

***The size of a spherical satellite is equal to its diameter.*

Appendix 4 The Nearest Stars

Name	Parallax (arcsec)	Distance (parsecs)	Distance (light-years)	Spectral Type	Proper motion (arcsec/yr)	Apparent visual magnitude	Absolute visual magnitude	Mass (Sun = 1)
Proxima Centauri	0.772	1.30	4.22	M5.5 V	3.853	+11.09	+15.53	0.107
Alpha Centauri A	0.747	1.34	4.36	G2 V	3.710	−0.01	+4.36	1.144
Alpha Centauri B	0.747	1.34	4.36	K0 V	3.724	+1.34	+5.71	0.916
Barnard's Star	0.547	1.83	5.96	M4.0 V	10.358	+9.53	+13.22	0.166
Wolf 359	0.419	2.39	7.78	M6.0 V	4.696	+13.44	+16.55	0.092
Lalande 21185	0.393	2.54	8.29	M2.0 V	4.802	+7.47	+10.44	0.464
Sirius A	0.380	2.63	8.58	A1 V	1.339	−1.43	+1.47	1.991
Sirius B	0.380	2.63	8.58	white dwarf	1.339	+8.44	+11.34	0.500
UV Ceti	0.374	2.68	8.73	M5.5 V	3.368	+12.54	+15.40	0.109
BL Ceti	0.374	2.68	8.73	M6.0 V	3.368	+12.99	+15.85	0.102
Ross 154	0.337	2.97	9.68	M3.5 V	0.666	+10.43	+13.07	0.171
Ross 248	0.316	3.16	10.32	M5.5 V	1.617	+12.29	+14.79	0.121
Epsilon Eridani	0.310	3.23	10.52	K2 V	0.977	+3.73	+6.19	0.850
Lacailee 9352	0.304	3.29	10.74	M1.5 V	6.896	+7.34	+9.75	0.529
Ross 128	0.299	3.35	10.92	M4.0 V	1.361	+11.13	+13.51	0.156
EZ Aquarii A	0.290	3.45	11.27	M5.0 V	3.254	+13.33	+15.64	0.105
EZ Aquarii B	0.290	3.45	11.27	—	3.254	+13.27	+15.58	0.106
EZ Aquarii C	0.290	3.45	11.27	—	3.254	+14.03	+16.34	0.095
Procyon A	0.286	3.50	11.40	F5 IV-V	1.259	+0.38	+2.66	1.569
Procyon B	0.286	3.50	11.40	white dwarf	1.259	+10.70	+12.98	0.500
61 Cygni A	0.286	3.50	11.40	K5.0 V	5.281	+5.21	+7.49	0.703
61 Cygni B	0.286	3.50	11.40	K7.0 V	5.172	+6.03	+8.31	0.630
GJ725 A	0.283	3.53	11.53	M3.0 V	2.238	+8.90	+11.16	0.351
GJ725 B	0.283	3.53	11.53	M3.5 V	2.313	+9.69	+11.95	0.259
GX Andromedae	0.281	3.56	11.62	M1.5 V	2.918	+8.08	+10.32	0.486
GQ Andromedae	0.281	3.56	11.62	M3.5 V	2.918	+11.06	+13.30	0.163
Epsilon Indi A	0.276	3.63	11.82	K5 V	4.704	+4.69	+6.89	0.766
Epsilon Indi B	0.276	3.63	11.82	T1.0	4.823			0.044
Epsilon Indi C	0.276	3.63	11.82	T6.0	4.823			0.028
DX Cancri	0.276	3.63	11.83	M6.5 V	1.290	+14.78	+16.98	0.087
Tau Ceti	0.274	3.64	11.89	G8 V	1.922	+3.49	+5.68	0.921
RECONS 1	0.272	3.68	11.99	M5.5 V	0.814	+13.03	+15.21	0.113
YZ Ceti	0.269	3.72	12.13	M4.5 V	1.372	+12.02	+14.17	0.136
Luyten's Star	0.264	3.79	12.37	M3.5 V	3.738	+9.86	+11.97	0.257
Kapteyn's Star	0.255	3.92	12.78	M1.5 V	8.670	+8.84	+10.87	0.393
AX Microscopium	0.253	3.95	12.87	M0.0 V	3.455	+6.67	+8.69	0.600

This table, compiled from data reported by the Research Consortium on Nearby Stars, lists all known stars within 4.00 parsecs (13.05 light-years).

**Stars that are components of multiple star systems are labeled A, B, and C.*

Appendix 5 The Visually Brightest Stars

Name	Distance: Designation	Distance (parsecs)	Distance (light-years)	Spectral type	Radial velocity (km/s)*	Proper motion (arcsec/year)	Apparent visual magnitude	Apparent visual brightness (Sirius = 1)**	Absolute visual magnitude
Sirius A	α CMa A	2.63	8.58	A1 V	−7.6	1.34	−1.43	1.000	+1.46
Canopus	α Car	95.9	313	F0 II	+20.5	0.03	−0.72	0.520	−5.63
Arcturus	α Boo	11.3	36.7	K1.5 III	−5.2	2.28	−0.04	0.278	−0.30
Alpha Centauri A	α Cen A	1.34	4.36	G2 V	−25	3.71	−0.01	0.270	+4.36
Vega	α Lyr	7.76	25.3	A0 V	−13.9	0.35	+0.03	0.261	+0.58
Capella	α Aur	12.9	42.2	G5 III	+30.2	0.43	+0.08	0.249	−0.48
Rigel	α Ori A	237	773	B8 Ia	+20.7	0.002	+0.12	0.240	−6.75
Procyon	α CMi A	3.50	11.4	F5 IV-V	−3.2	1.26	+0.34	0.196	+2.62
Achernar	α Eri	44.1	144	B3 V	+16	0.10	+0.50	0.169	−2.72
Betelgeuse	α Ori	131	427	M1 Iab	+21	0.03	+0.58	0.157	−5.01
Hadar	β Cen	161	525	B1 III	+5.9	0.04	+0.60	0.154	−5.43
Altair	α Aql	51.4	168	A7 V	−26.1	0.66	+0.77	0.132	−2.79
Aldebaran	α Tau A	20.0	65.1	K5 III	+54.3	0.20	+0.85	0.122	−0.65
Spica	α Vir	80.4	262	B1 III-IV	+1	0.05	+1.04	0.103	−3.49
Antares	α Sco A	185	604	M1.5 Iab	−3.4	0.03	+1.09	0.098	−5.25
Pollux	β Gem	10.3	33.7	K0 IIIb	+3.3	0.63	+1.15	0.093	+1.08
Fomalhaut	α PsA	7.69	25.1	A3 V	+6.5	0.37	+1.16	0.092	+1.73
Deneb	α Cyg	990	3230	A2 Ia	−4.5	0.002	+1.25	0.085	−8.73
Mimosa	β Cru	108	353	B0.5 IV	+15.6	0.05	+1.297	0.081	−3.87
Regulus	α Leo A	23.8	77.5	B7 V	+5.9	0.25	+1.35	0.077	−0.53

Data in this table were compiled from SIMBAD database operated at the Centre de Données Astronomiques de Strasbourg, France.

**A positive radial velocity means that the star is receding; a negative radial velocity means that the star is approaching.*

***This is a ratio of the star's apparent brightness to that of Sirius, the brightest star in the night sky.*

Note: Acrux, or α Cru (the brightest star in Crux, the Southern Cross) appears to the naked eye as a star of apparent magnitude +0.87, the same as Aldebaran. However, it does not appear in this table because Acrux is actually a binary star system. The blue-white component stars of this binary system have apparent magnitudes of +1.4 and +1.9, and so they are dimmer than any of the stars listed here.

Appendix 6 Some Important Astronomical Quantities

Astronomical Unit	$1\ \text{AU} = 1.4960 \times 10^{11}$ m
	$= 1.4960 \times 10^{8}$ km
Light-year	$1\ \text{ly} = 9.4605 \times 10^{15}$ m
	= 63,240 AU
Parsec	1 pc = 3.2616 ly
	$= 3.0857 \times 10^{16}$ m
	= 206,265 AU
Year	1 y = 365.2564 days
	$= 3.156 \times 10^{7}$ s
Solar mass	$1\ M_\odot = 1.989 \times 10^{30}$ kg
Solar Radius	$1\ R_\odot = 6.9599 \times 10^{8}$ m
Solar Luminosity	$1\ L_\odot = 3.90 \times 10^{26}$ W

Appendix 7 Some Important Physical Constants

Speed of light	$c = 2.9979 \times 10^{8}$ m/s
Gravitational constant	$G = 6.673 \times 10^{-11}$ N m^2/kg^2
Planck's constant	$h = 6.6261 \times 10^{-34}$ J s
	$= 4.1357 \times 10^{-15}$ eV s
Boltzmann constant	$k = 1.3807 \times 10^{-23}$ J/K
	$= 8.6173 \times 10^{-5}$ eV/K
Stefan-Boltzmann constant	$\sigma = 5.6704 \times 10^{-8}$ W m^{-2} K^{-4}
Mass of electron	$m_e = 9.1094 \times 10^{-31}$ kg
Mass of proton	$m_p = 1.6726 \times 10^{-27}$ kg
Mass of neutron	$m_n = 1.6749 \times 10^{-27}$ kg
Mass of hydrogen atom	$m_H = 1.6735 \times 10^{-27}$ kg
Rydberg constant	$R = 1.0973 \times 10^{7}$ m^{-1}
Electron volt	$1\ \text{eV} = 1.6022 \times 10^{-19}$ J

Appendix 8 Some Useful Mathematics

Area of a rectangle of sides a and b	$A = ab$
Volume of a rectangular solid of sides a, b, and c	$V = abc$
Hypotenuse of a right triangle whose other sides are a and b	$c = \sqrt{a^2 + b^2}$
Circumference of a circle of radius r	$C = 2\pi r$
Area of a circle of radius r	$A = \pi r^2$
Surface ares of a sphere of radius r	$A = 4\pi r^2$
Volume of a sphere of radius r	$V = 4\pi r^3/3$
Value of π	$\pi = 3.1415926536$

Glossary

You can find more information about each term in the indicated chapter or chapters. See the Key Words section in each chapter for the specific page on which the meaning of a given term is described.

A ring One of three prominent rings encircling Saturn. (Chapter 12)

absolute magnitude The apparent magnitude that a star would have if it were at a distance of 10 parsecs. (Chapter 17)

absolute zero A temperature of –273°C (or 0 K), at which all molecular motion stops; the lowest possible temperature. (Chapter 5)

absorption line spectrum Dark lines superimposed on a continuous spectrum. (Chapter 5)

acceleration The rate at which an object's velocity changes due to a change in speed, a change in direction, or both. (Chapter 4)

accretion The gradual accumulation of matter in one location, typically due to the action of gravity. (Chapter 8, Chapter 18)

accretion disk A disk of gas orbiting a star or black hole. (Chapter 25)

active galactic nucleus (AGN) The center of an active galaxy. (Chapter 25)

active galaxy A galaxy that is emitting exceptionally large amounts of energy; a Seyfert galaxy, radio galaxy, or quasar. (Chapter 25)

active optics A technique for improving a telescopic image by altering the telescope's optics to compensate for variations in air temperature or flexing of the telescope mount. (Chapter 6)

adaptive optics A technique for improving a telescopic image by altering the telescope's optics in a way that compensates for distortion caused by Earth's atmosphere. (Chapter 6)

aerosol Tiny droplets of liquid dispersed in a gas. (Chapter 13)

AGB star See *asymptotic giant branch star.*

AGN See *active galactic nucleus.*

albedo The fraction of sunlight that a planet, asteroid, or satellite reflects. (Chapter 9)

alpha particle The nucleus of a helium atom, consisting of two protons and two neutrons. (Chapter 19)

amino acids The chemical building blocks of proteins. (Chapter 15)

angle The opening between two lines that meet at a point. (Chapter 1)

angular diameter The angle subtended by the diameter of an object. (Chapter 1)

angular distance The angle between two points in the sky. (Chapter 1)

angular measure The size of an angle, usually expressed in degrees, arcminutes, and arcseconds. (Chapter 1)

angular momentum See *conservation of angular momentum.*

angular resolution The angular size of the smallest feature that can be distinguished with a telescope. (Chapter 6)

angular size See *angular diameter.*

anisotropic Possessing unequal properties in different directions; not isotropic. (Chapter 24)

annihilation The process by which the masses of a particle and antiparticle are converted into energy. (Chapter 27)

annular eclipse An eclipse of the Sun in which the Moon is too distant to cover the Sun completely, so that a ring of sunlight is seen around the Moon at mid-eclipse. (Chapter 3)

anorthosite Rock commonly found in ancient, cratered highlands on the Moon. (Chapter 10)

Antarctic Circle A circle of latitude $23\frac{1}{2}$° north of Earth's south pole. (Chapter 2)

antimatter Matter consisting of antiparticles such as antiprotons, antineutrons, and positrons. (Chapter 27)

antiparticle A particle with the same mass as an ordinary particle but with other properties, such as electric charge, reversed. (Chapter 27)

antiproton A particle with the same mass as a proton but with a negative electric charge. (Chapter 27)

aphelion The point in its orbit where a planet is farthest from the Sun. (Chapter 4)

apogee The point in its orbit where a satellite or the Moon is farthest from Earth. (Chapter 3)

apparent brightness The flux of a star's light arriving at Earth. (Chapter 17)

apparent magnitude A measure of the brightness of light from a star or other object as measured from Earth. (Chapter 17)

apparent solar day The interval between two successive transits of the Sun's center across the local meridian. (Chapter 2)

apparent solar time Time reckoned by the position of the Sun in the sky. (Chapter 2)

arcminute One-sixtieth (1/60) of a degree, designated by the symbol ′. (Chapter 1)

arcsecond One-sixtieth (1/60) of an arcminute or 1/3600 of a degree, designated by the symbol ″. (Chapter 1)

Arctic Circle A circle of latitude $23\frac{1}{2}$° south of Earth's north pole. (Chapter 2)

asteroid One of tens of thousands of small, rocky, planetlike objects in orbit about the Sun. Also called minor planets. (Chapter 7, Chapter 15)

asteroid belt A region between the orbits of Mars and Jupiter that encompasses the orbits of many asteroids. (Chapter 7, Chapter 15)

asthenosphere A warm, plastic layer of the mantle beneath the lithosphere of Earth. (Chapter 9)

astrobiology The study of life in the universe. (Chapter 28)

astrometric method A technique for detecting extrasolar planets by looking for stars that "wobble" periodically. (Chapter 8)

astronomical unit (AU) The semimajor axis of Earth's orbit; the average distance between Earth and the Sun. (Chapter 1)

asymptotic giant branch A region on the Hertzsprung-Russell diagram occupied by stars in their second and final red-giant phase. (Chapter 20)

asymptotic giant branch star (AGB star) A red giant star that has ascended the giant branch for the second and final time. (Chapter 20)

atmosphere (atm) A unit of atmospheric pressure. (Chapter 9)

atmospheric pressure The force per unit area exerted by a planet's atmosphere. (Chapter 9)

atom The smallest particle of an element that has the properties characterizing that element. (Chapter 5)

atomic number The number of protons that the nucleus of an atom of a particular element has. (Chapter 5, Chapter 8)

AU See *astronomical unit.*

aurora (plural **aurorae**) Light radiated by atoms and ions in Earth's upper atmosphere, mostly in the polar regions. (Chapter 9)

aurora australis Aurorae seen from southern latitudes; the southern lights. (Chapter 9)

aurora borealis Aurorae seen from northern latitudes; the northern lights. (Chapter 9)

autumnal equinox The intersection of the ecliptic and the celestial equator where the Sun crosses the equator from north to south. Also used to refer to the date on which the Sun passes through this intersection. (Chapter 2)

average density The mass of an object divided by its volume. (Chapter 7)

average density of matter In cosmology, the average density of all the material substance in the universe. (Chapter 26)

B ring One of three prominent rings encircling Saturn. (Chapter 12)

Balmer line An emission or absorption line in the spectrum of hydrogen caused by an electron transition between the second and higher energy levels. (Chapter 5)

Balmer series The entire pattern of Balmer lines. (Chapter 5)

Barnard object One of a class of dark nebulae discovered by E. E. Barnard. (Chapter 18)

barred spiral galaxy A spiral galaxy in which the spiral arms begin from the ends of a "bar" running through the nucleus rather than from the nucleus itself. (Chapter 24)

baseline In interferometry, the distance between two telescopes whose signals are combined to give a higher-resolution image. (Chapter 6)

belt A dark band in Jupiter's atmosphere. (Chapter 12)

Big Bang An explosion of all space that took place roughly 13.7 billion years ago and that marks the beginning of the universe. (Chapter 1, Chapter 26)

binary star (also **binary star system** or **binary**) Two stars orbiting about each other. (Chapter 17)

biosphere The layer of soil, water, and air surrounding Earth in which living organisms thrive. (Chapter 9)

bipolar outflow Oppositely directed jets of gas expelled from a young star. (Chapter 18)

black hole An object whose gravity is so strong that the escape speed exceeds the speed of light. (Chapter 1, Chapter 22)

black hole evaporation The process by which black holes emit particles. (Chapter 22)

blackbody A hypothetical perfect radiator that absorbs and re-emits all radiation falling upon it. (Chapter 5)

blackbody curve The intensity of radiation emitted by a blackbody plotted as a function of wavelength or frequency. (Chapter 5)

blackbody radiation The radiation emitted by a perfect blackbody. (Chapter 5)

blazar A type of active galaxy whose nucleus has a featureless spectrum without prominent spectral lines. (Chapter 25)

blueshift A decrease in the wavelength of photons emitted by an approaching source of light. (Chapter 5)

Bohr orbits In the model of the atom described by Niels Bohr, the only orbits in which electrons are allowed to move about the nucleus. (Chapter 5)

Bok globule A small, roundish, dark nebula. (Chapter 18)

bright terrain (Ganymede) Young, reflective, relatively crater-free terrain on the surface of Ganymede. (Chapter 13)

brightness See *apparent brightness.*

brown dwarf A starlike object that is not massive enough to sustain hydrogen fusion in its core. (Chapter 17)

brown ovals Elongated, brownish features usually seen in Jupiter's northern hemisphere. (Chapter 12)

C ring One of three prominent rings encircling Saturn. (Chapter 12)

capture theory The hypothesis that the Moon was gravitationally captured by Earth. (Chapter 10)

carbon fusion The thermonuclear fusion of carbon to produce heavier nuclei. (Chapter 20)

carbon star A peculiar red giant star whose spectrum shows strong absorption lines of carbon and carbon compounds. (Chapter 20)

carbonaceous chondrite A type of meteorite that has a high abundance of carbon and volatile compounds. (Chapter 15, Chapter 28)

Cassegrain focus An optical arrangement in a reflecting telescope in which light rays are reflected by a secondary mirror to a focus behind the primary mirror. (Chapter 6)

Cassini division An apparent gap between Saturn's A and B rings. (Chapter 12)

CCD See *charge-coupled device.*

celestial equator A great circle on the celestial sphere 90° from the celestial poles. (Chapter 2)

celestial poles The points about which the celestial sphere appears to rotate. See also *north celestial pole* and *south celestial pole.* (Chapter 2)

celestial sphere An imaginary sphere of very large radius centered on an observer; the apparent sphere of the sky. (Chapter 2)

center of mass The point between a star and a planet, or between two stars, around which both objects orbit. (Chapter 8, Chapter 10, Chapter 17)

central bulge A spherical distribution of stars around the nucleus of a spiral galaxy. (Chapter 23)

Cepheid variable A type of yellow, supergiant, pulsating star. (Chapter 19)

Cerenkov radiation The radiation emitted by a particle traveling through a substance at a speed greater than the speed of light in that substance. (Chapter 20)

Chandrasekhar limit The maximum mass of a white dwarf. (Chapter 20)

charge-coupled device (CCD) A type of solid-state device designed to detect photons. (Chapter 6)

chemical composition A description of which chemical substances make up a given object. (Chapter 7)

chemical differentiation The process by which the heavier elements in a planet sink toward its center while lighter elements rise toward its surface. (Chapter 8)

chemical element See *element.*

chondrule A glassy, roughly spherical blob found within meteorites. (Chapter 8)

chromatic aberration An optical defect whereby different colors of light passing through a lens are focused at different locations. (Chapter 6)

chromosphere A layer in the atmosphere of the Sun between the photosphere and the corona. (Chapter 16)

circumpolar A term describing a star that neither rises nor sets but appears to rotate around one of the celestial poles. (Chapter 2)

circumstellar accretion disk An accretion disk that surrounds a protostar. (Chapter 18)

close binary A binary star system in which the stars are separated by a distance roughly comparable to their diameters. (Chapter 19)

closed universe A universe with positive curvature, so that its geometry is analogous to that of the surface of a sphere. (Chapter 26)

cluster of galaxies A collection of galaxies containing a few to several thousand member galaxies. (Chapter 24)

cluster of stars A group of stars that formed together and that have remained together due to their mutual gravitational attraction. (Chapter 18)

CNO cycle A series of nuclear reactions in which carbon is used as a catalyst to transform hydrogen into helium. (Chapter 16)

cocoon nebula The nebulosity surrounding a protostar. (Chapter 18)

co-creation theory The hypothesis that Earth and the Moon formed at the same time from the same material. (Chapter 10)

cold dark matter Slowly moving, weakly interacting particles presumed to constitute the bulk of matter in the universe. (Chapter 27)

collapsar A proposed type of supernova in which a black hole forms at the center of a dying star before the outer layers of the star have time to collapse. (Chapter 22)

collisional ejection theory The hypothesis that the Moon formed from material ejected from Earth by the impact of a large asteroid. (Chapter 10)

color-magnitude diagram A plot of the apparent magnitudes (that is, apparent brightnesses) of stars in a cluster versus their color indices (a measure of their surface temperatures). (Chapter 19)

color ratio The ratio of the apparent brightness of a star measured in one spectral region to its brightness measured in a different region. (Chapter 17)

coma (of a comet) The diffuse gaseous component of the head of a comet. (Chapter 15)

coma (optical) The distortion of off-axis images formed by a parabolic mirror. (Chapter 6)

combined average mass density The average density in the observable universe of all forms of matter and energy, measured in units of mass per volume. (Chapter 26)

comet A small body of ice and dust in orbit about the Sun. While passing near the Sun, a comet's vaporized ices give rise to a coma and tail. (Chapter 7, Chapter 15)

compound A substance consisting of two or more chemical elements in a definite proportion. (Chapter 5)

condensation temperature The temperature at which a particular substance in a low-pressure gas condenses into a solid. (Chapter 8)

conduction The transfer of heat by directly passing energy from atom to atom. (Chapter 16)

compression A region in a sound wave where the medium carrying the wave is compressed. (Chapter 26)

conic section The curve of intersection between a circular cone and a plane; this curve can be a circle, ellipse, parabola, or hyperbola. (Chapter 4)

conjunction The geometric arrangement of a planet in the same part of the sky as the Sun, so that the planet is at an elongation of 0°. (Chapter 4)

conservation of angular momentum A law of physics stating that in an isolated system, the total amount of angular momentum—a measure of the amount of rotation—remains constant. (Chapter 8)

constellation A configuration of stars in the same region of the sky. (Chapter 2)

contact binary A binary star system in which both members fill their Roche lobes. (Chapter 19)

continuous spectrum A spectrum of light over a range of wavelengths without any spectral lines. (Chapter 5)

convection The transfer of energy by moving currents of fluid or gas containing that energy. (Chapter 9, Chapter 16)

convection cell A circulating loop of gas or liquid that transports heat from a warm region to a cool region. (Chapter 9)

convection current The pattern of motion in a gas or liquid in which convection is taking place. (Chapter 9)

convective zone The region in a star where convection is the dominant means of energy transport. (Chapter 16)

core (of Earth) The iron-rich inner region of Earth's interior. (Chapter 9)

core accretion model The hypothesis that each of the Jovian planets formed by accretion of gas onto a rocky core. (Chapter 8)

core-collapse supernova A supernova that occurs at the end of a massive star's lifetime when the star's core collapses to high density and the star's outer layers are expelled into space. (See Type Ib supernova, Type Ic supernova, and Type II supernova) (Chapter 20)

core helium fusion The thermonuclear fusion of helium at the center of a star. (Chapter 19, Chapter 20)

core hydrogen fusion The thermonuclear fusion of hydrogen at the center of a star. (Chapter 19)

corona (of the Sun) The Sun's outer atmosphere, which has a high temperature and a low density. (Chapter 16)

coronal hole A region in the Sun's corona that is deficient in hot gases. (Chapter 16)

coronal mass ejection An event in which billions of tons of gas from the Sun's corona is suddenly blasted into space at high speed. (Chapter 16)

cosmic background radiation See *cosmic microwave background.*

cosmic censorship, law of See *law of cosmic censorship.*

cosmic light horizon An imaginary sphere, centered on Earth, whose radius equals the distance light has traveled since the Big Bang. (Chapter 26, Chapter 27)

cosmic microwave background An isotropic radiation field with a blackbody temperature of about 2.725 K that permeates the entire universe. (Chapter 26)

cosmic rays Fast-moving subatomic particles that enter our solar system from interstellar space. (Chapter 11)

cosmological constant (Λ) In the equations of general relativity, a quantity signifying a pressure throughout all space that helps the universe expand; a type of dark energy. (Chapter 26)

cosmological principle The assumption that the universe is homogeneous and isotropic on the largest scale. (Chapter 26)

cosmological redshift A redshift that is caused by the expansion of the universe. (Chapter 26)

cosmology The study of the structure and evolution of the universe. (Chapter 26)

coudé focus An optical arrangement with a reflecting telescope. A series of mirrors is used to direct light to a remote focus away from the moving parts of the telescope. (Chapter 6)

crater See *impact crater.*

critical density The value of the combined average mass density throughout the universe for which space would be flat. (Chapter 26)

crust (of a planet) The surface layer of a terrestrial planet. (Chapter 9)

crustal dichotomy (Mars) The contrast between the young northern lowlands and older southern highlands on Mars. (Chapter 11)

crystal A material in which atoms are arranged in orderly rows. (Chapter 9)

current sheet A broad, flat region in Jupiter's magnetosphere that contains an abundance of charged particles. (Chapter 12)

D ring One of several faint rings encircling Saturn. (Chapter 12)

dark ages The era between the formation of the first atoms (when the universe was about 380,000 years old) and the formation of the first stars (when the universe was about 400 million years old). (Chapter 27)

dark energy A form of energy that appears to pervade the universe and causes the expansion of the universe to accelerate, but has no discernible gravitational effect. (Chapter 26)

dark energy density parameter (Ω_Λ) The ratio of the average mass density of dark energy in the universe to the critical density. (Chapter 26)

dark-energy-dominated universe A universe in which the mass density of dark energy exceeds both the average density of matter and the mass density of radiation. (Chapter 26)

dark matter Nonluminous matter that is the dominant form of matter in galaxies and throughout the universe. (Chapter 23)

dark-matter problem The enigma that most of the matter in the universe seems not to emit radiation of any kind and is detectable by its gravity alone. (Chapter 24)

dark nebula A cloud of interstellar gas and dust that obscures the light of more distant stars. (Chapter 18)

dark terrain (Ganymede) Older, heavily cratered, dark-colored terrain on the surface of Ganymede. (Chapter 13)

decametric radiation Radiation from Jupiter whose wavelength is about 10 meters. (Chapter 12)

decimetric radiation Radiation from Jupiter whose wavelength is about a tenth of a meter. (Chapter 12)

declination Angular distance of a celestial object north or south of the celestial equator. (Chapter 2)

deferent A stationary circle in the Ptolemaic system along which another circle (an epicycle) moves, carrying a planet, the Sun, or the Moon. (Chapter 4)

degeneracy The phenomenon, due to quantum mechanical effects, whereby the pressure exerted by a gas does not depend on its temperature. (Chapter 19)

degenerate electron (neutron) pressure The pressure exerted by degenerate electrons (neutrons). (Chapter 19, Chapter 21)

degree A basic unit of angular measure, designated by the symbol °. (Chapter 1)

degree Celsius A basic unit of temperature, designated by the symbol °C and used on a scale where water freezes at 0° and boils at 100°. (Chapter 5)

degree Fahrenheit A basic unit of temperature, designated by the symbol °F and used on a scale where water freezes at 32° and boils at 212°. (Chapter 5)

density See *average density.*

density fluctuation A variation of density from one place to another. (Chapter 27)

density parameter (Ω_0) The ratio of the average density of matter and energy in the universe to the critical density. (Chapter 26)

density wave In a spiral galaxy, a localized region in which matter piles up as it orbits the center of the galaxy; this is a proposed explanation of spiral structure. (Chapter 23)

detached binary A binary star system in which neither star fills its Roche lobe. (Chapter 19)

deuterium bottleneck The situation during the first 3 minutes after the Big Bang when deuterium (an isotope of hydrogen whose nucleus contains one proton and one neutron) inhibited the formation of heavier elements. (Chapter 27)

differential rotation The rotation of a nonrigid object in which parts adjacent to each other at a given time do not always stay close together. (Chapter 12, Chapter 16)

differentiated asteroid An asteroid in which chemical differentiation has taken place, so that denser material is toward the asteroid's center. (Chapter 15)

differentiation See *chemical differentiation.*

diffraction The spreading out of light passing through an aperture or opening in an opaque object. (Chapter 6)

diffraction grating An optical device, consisting of thousands of closely spaced lines etched in glass or metal, that disperses light into a spectrum. (Chapter 6)

direct motion The apparent eastward movement of a planet seen against the background stars. (Chapter 4)

disk (of a galaxy) The disk-shaped distribution of Population I stars that dominates the appearance of a spiral galaxy. (Chapter 23)

disk instability model The hypothesis that gases in the solar nebula coalesced rapidly to form the Jovian planets. (Chapter 8)

distance ladder The sequence of techniques used to determine the distances to very remote galaxies. (Chapter 24)

distance modulus The difference between the apparent and absolute magnitudes of an object. (Chapter 17)

diurnal motion Any apparent motion in the sky that repeats on a daily basis, such as the rising and setting of stars. (Chapter 2)

Doppler effect The apparent change in wavelength of radiation due to relative motion between the source and the observer along the line of sight. (Chapter 5)

double radio source An extragalactic radio source characterized by two large regions of radio emission, typically located on either side of an active galaxy. (Chapter 25)

double star A pair of stars located at nearly the same position in the night sky. Some, but not all, double stars are binary stars. (Chapter 17)

Drake equation An equation used to estimate the number of intelligent civilizations in the Galaxy with whom we might communicate. (Chapter 28)

dredge-up The transporting of matter from deep within a star to its surface by convection. (Chapter 20)

dust devil Whirlwinds found in dry or desert areas on both Earth and Mars. (Chapter 11)

dust grain A microscopic bit of solid matter found in interplanetary or interstellar space. (Chapter 18)

dust tail The tail of a comet that is composed primarily of dust grains. (Chapter 15)

dwarf elliptical galaxy A low-mass elliptical galaxy that contains only a few million stars. (Chapter 24)

dwarf planet A solar system body that is large enough to be spherical in shape and have a circular orbit around the Sun, but not large enough to clear its own path of other bodies. The term is used for Ceres, Pluto, Eris, Haumea, and Makemake. (Chapter 14)

dynamo The mechanism whereby electric currents within an astronomical body generate a magnetic field. (Chapter 7)

E ring A very broad, faint ring encircling Saturn. (Chapter 12)

earthquake A sudden vibratory motion of Earth's surface. (Chapter 9)

eccentricity A number between 0 and 1 that describes the shape of an ellipse. (Chapter 4)

eclipse The cutting off of part or all the light from one celestial object by another. (Chapter 3)

eclipse path The track of the tip of the Moon's shadow along Earth's surface during a total or annular solar eclipse. (Chapter 3)

eclipse year The interval between successive passages of the Sun through the same node of the Moon's orbit. (Chapter 3)

eclipsing binary A binary star system in which, as seen from Earth, stars periodically pass in front of each other. (Chapter 17)

ecliptic The apparent annual path of the Sun on the celestial sphere. (Chapter 2)

ecliptic plane The plane of Earth's orbit around the Sun. (Chapter 2)

Eddington limit A constraint on the rate at which a black hole can accrete matter. (Chapter 25)

electromagnetic radiation Radiation consisting of oscillating electric and magnetic fields. Examples include gamma rays, X rays, visible light, ultraviolet and infrared radiation, radio waves, and microwaves. (Chapter 5)

electromagnetic spectrum The entire array of electromagnetic radiation. (Chapter 5)

electromagnetism Electric and magnetic phenomena, including electromagnetic radiation. (Chapter 5)

electron A subatomic particle with a negative charge and a small mass, usually found in orbits about the nuclei of atoms. (Chapter 5)

electron volt (eV) The energy acquired by an electron accelerated through an electric potential of one volt. (Chapter 5)

electroweak force A unification of the electromagnetic and weak forces that occurs at high energies. (Chapter 27)

element A chemical that cannot be broken down into more basic chemicals. (Chapter 5)

elementary particle physics The branch of quantum mechanics that deals with individual subatomic particles and their interactions. (Chapter 27)

ellipse A conic section obtained by cutting completely through a circular cone with a plane. (Chapter 4)

elliptical galaxy A galaxy with an elliptical shape and no conspicuous interstellar material. (Chapter 24)

elongation The angular distance between a planet and the Sun as viewed from Earth. (Chapter 4)

emission line spectrum A spectrum that contains bright emission lines. (Chapter 5)

emission nebula A glowing gaseous nebula whose spectrum has bright emission lines. (Chapter 18)

Encke gap A narrow gap in Saturn's A ring. (Chapter 12)

energy flux The rate of energy flow, usually measured in joules per square meter per second. (Chapter 5)

energy level In an atom, a particular amount of energy possessed by an atom above the atom's least energetic state. (Chapter 5)

energy-level diagram A diagram showing the arrangement of an atom's energy levels. (Chapter 5)

epicenter The location on Earth's surface directly over the focus of an earthquake. (Chapter 9)

epicycle A moving circle in the Ptolemaic system about which a planet revolves. (Chapter 4)

epoch The date used to define the coordinate system for objects on the sky. (Chapter 2)

equal areas, law of See *Kepler's second law.*

equinox One of the intersections of the ecliptic and the celestial equator. Also used to refer to the date on which the Sun passes through such an intersection. (Chapter 2) See also *autumnal equinox* and *vernal equinox*.

equivalence principle In the general theory of relativity, the principle that in a small volume of space, the downward pull of gravity can be accurately and completely duplicated by an upward acceleration of the observer. (Chapter 22)

era of recombination The moment, approximately 380,000 years after the Big Bang, when the universe had cooled sufficiently to permit the formation of hydrogen atoms. (Chapter 26)

ergoregion The region of space immediately outside the event horizon of a rotating black hole where it is impossible to remain at rest. (Chapter 22)

escape speed The speed needed by an object (such as a spaceship) to leave a second object (such as a planet or star) permanently and to escape into interplanetary space. (Chapter 7)

eV See *electron volt.*

event horizon The location around a black hole where the escape speed equals the speed of light; the surface of a black hole. (Chapter 22)

evolutionary track The path on an H-R diagram followed by a star as it evolves. (Chapter 18)

excited state A state of an atom, ion, or molecule with a higher energy than the ground state. (Chapter 5)

exponent A number placed above and after another number to denote the power to which the latter is to be raised, as n in 10^n. (Chapter 1)

extinction (interstellar) See *interstellar extinction.*

extrasolar planet A planet orbiting a star other than the Sun. (Chapter 8)

eyepiece lens A magnifying lens used to view the image produced at the focus of a telescope. (Chapter 6)

F ring A thin, faint ring encircling Saturn just beyond the A ring. (Chapter 12)

false color In astronomical images, color used to denote different values of intensity, temperature, or other quantities. (Chapter 6)

false vacuum Empty space with an abnormally high energy density that is thought to have filled the universe immediately after the Big Bang. (Chapter 27)

far-infrared The part of the infrared spectrum most different in wavelength from visible light. (Chapter 23)

far side (of the Moon) The side of the Moon that faces perpetually away from Earth. (Chapter 10)

favorable opposition An opposition of Mars that affords good Earth-based views of the planet. (Chapter 11)

filament A portion of the Sun's chromosphere that arches to high altitudes. (Chapter 16)

first quarter moon The phase of the Moon that occurs when the Moon is 90° east of the Sun. (Chapter 3)

fission theory The hypothesis that the Moon was pulled out of a rapidly rotating proto-Earth. (Chapter 10)

flake tectonics A model of a planetary interior, particularly Venus, in which a thin crust remains stationary but wrinkles and flakes in response to interior convection currents. (Chapter 11)

flare See *solar flare.*

flat space Space that is not curved; space with zero curvature. (Chapter 26)

flatness problem The dilemma posed by the fact that the combined average mass density of the universe is very nearly equal to the critical density. (Chapter 27)

flocculent spiral galaxy A spiral galaxy with fuzzy, poorly defined spiral arms. (Chapter 23)

fluorescence A process in which high-energy ultraviolet photons are absorbed and the absorbed energy is radiated as lower-energy photons of visible light. (Chapter 18)

focal length The distance from a lens or mirror to the point where converging light rays meet. (Chapter 6)

focal plane The plane in which a lens or mirror forms an image of a distant object. (Chapter 6)

focal point The point at which a lens or mirror forms an image of a distant point of light. (Chapter 6)

focus (of an ellipse) (plural **foci**) One of two points inside an ellipse such that the combined distance from the two foci to any point on the ellipse is a constant. (Chapter 4)

focus (of a lens or mirror) The point to which light rays converge after passing through a lens or being reflected from a mirror. (Chapter 6)

force A push or pull that acts on an object. (Chapter 4)

frequency The number of crests or troughs of a wave that cross a given point per unit time. Also, the number of vibrations per unit time. (Chapter 5)

full moon A phase of the Moon during which its full daylight hemisphere can be seen from Earth. (Chapter 3)

fundamental plane A relationship among the size of an elliptical galaxy, the average motions of its stars, and how the galaxy's brightness appears distributed over its surface. (Chapter 24)

fusion crust The coating on a stony meteorite caused by the heating of the meteorite as it descended through Earth's atmosphere. (Chapter 15)

G ring A thin, faint ring encircling Saturn. (Chapter 12)

galactic cannibalism A collision between two galaxies of unequal mass and size in which the smaller galaxy seems to be absorbed by the larger galaxy. (Chapter 24)

galactic cluster See *open cluster.*

galactic nucleus The center of a galaxy. (Chapter 23)

galaxy A large assemblage of stars, nebulae, and interstellar gas and dust. (Chapter 1, Chapter 23)

Galilean satellites The four large moons of Jupiter. (Chapter 13)

gamma-ray bursters Objects found in all parts of the sky that emit a one-time intense burst of high-energy radiation. (Chapter 22)

gamma rays The most energetic form of electromagnetic radiation. (Chapter 5)

general theory of relativity A description of gravity formulated by Albert Einstein. It explains that gravity affects the geometry of space and the flow of time. (Chapter 22)

geocentric model An Earth-centered theory of the universe. (Chapter 4)

giant A star whose diameter is typically 10 to 100 times that of the Sun and whose luminosity is roughly that of 100 Suns. (Chapter 17)

giant elliptical galaxy A large, massive, elliptical galaxy containing many billions of stars. (Chapter 24)

giant molecular cloud A large cloud of interstellar gas and dust in which temperatures are low enough and densities high enough for atoms to form into molecules. (Chapter 18)

glitch A sudden speedup in the period of a pulsar. (Chapter 21)

global warming The upward trend of Earth's average temperature caused by increased amounts of greenhouse gases in the atmosphere. (Chapter 9)

globular cluster A large spherical cluster of stars, typically found in the outlying regions of a galaxy. (Chapter 19, Chapter 23)

gluon A particle that is exchanged between quarks. (Chapter 27)

grand-design spiral galaxy A galaxy with well-defined spiral arms. (Chapter 23)

grand unified theory (GUT) A theory that describes and explains the four physical forces. (Chapter 27)

granulation The rice grain–like structure found in the solar photosphere. (Chapter 16)

granule A convective cell in the solar photosphere. (Chapter 16)

grating See *diffraction grating.*

gravitational force See *gravity.*

gravitational lens A massive object that deflects light rays from a remote source, forming an image much as an ordinary lens does. (Chapter 24)

gravitational radiation (gravitational waves) Oscillations of space produced by changes in the distribution of matter. (Chapter 22)

gravitational redshift The increase in the wavelength of a photon as it climbs upward in a gravitational field. (Chapter 22)

graviton The particle that is responsible for the gravitational force. (Chapter 27)

gravity The force with which all matter attracts all other matter. (Chapter 4)

Great Red Spot A prominent high-pressure system in Jupiter's southern hemisphere. (Chapter 12)

greatest eastern elongation The configuration of an inferior planet at its greatest angular distance east of the Sun. (Chapter 4, Chapter 11)

greatest western elongation The configuration of an inferior planet at its greatest angular distance west of the Sun. (Chapter 4, Chapter 11)

greenhouse effect The trapping of infrared radiation near a planet's surface by the planet's atmosphere. (Chapter 9)

greenhouse gas A substance whose presence in a planet's atmosphere enhances the greenhouse effect. (Chapter 9)

ground state The state of an atom, ion, or molecule with the least possible energy. (Chapter 5)

group (of galaxies) A poor cluster of galaxies. (Chapter 24)

GUT See *grand unified theory.*

H I Neutral, unionized hydrogen. (Chapter 23)

H II region A region of ionized hydrogen in interstellar space. (Chapter 18)

habitable zone Regions of a galaxy or around a star in which conditions may be suitable for life to have developed. (Chapter 28)

half-life The time required for one-half of a quantity of a radioactive substance to decay. (Chapter 8)

halo (of a galaxy) A spherical distribution of globular clusters and Population II stars that surround a spiral galaxy. (Chapter 23)

head-tail source A radio source consisting of a bright "head" and a long, dimmer "tail." (Chapter 25)

Heisenberg uncertainty principle A principle of quantum mechanics that places limits on the precision of simultaneous measurements. (Chapter 22, Chapter 27)

heliocentric model A Sun-centered theory of the universe. (Chapter 4)

helioseismology The study of the vibrations of the Sun as a whole. (Chapter 16)

helium fusion The thermonuclear fusion of helium to form carbon and oxygen. (Chapter 19)

helium flash The nearly explosive beginning of helium fusion in the dense core of a red giant star. (Chapter 19)

helium shell flash A brief thermal runaway that occurs in the helium-fusing shell of a red supergiant. (Chapter 20)

Herbig-Haro object A small, luminous nebula associated with the end point of a jet emanating from a young star. (Chapter 18)

Hertzsprung-Russell (H-R) diagram A plot of the luminosity (or absolute magnitude) of stars against their surface temperature (or spectral type). (Chapter 17)

high-velocity star A star traveling at an exceptionally high speed relative to the Sun. (Chapter 23)

highlands (on Mars) See *southern highlands.*

highlands (on the Moon) See *lunar highlands.*

Hirayama family A group of asteroids that have nearly identical orbits about the Sun. (Chapter 15)

homogeneous Having the same property in one region as in every other region. (Chapter 26)

horizon problem See *isotropy problem.*

horizontal branch A group of stars on the color-magnitude diagram of a typical globular cluster that lie near the main sequence and having roughly constant luminosity. (Chapter 20)

horizontal-branch star A low-mass, post–helium-flash star on the horizontal branch. (Chapter 19)

hot dark matter Dark matter consisting of low-mass particles moving at high speeds. (Chapter 27)

hot spot (on Jupiter) An unusually warm and cloud-free part of Jupiter's atmosphere. (Chapter 12)

hot-spot volcanism Volcanic activity that occurs over a hot region buried deep within a planet. (Chapter 11)

H-R diagram See *Hertzsprung-Russell diagram.*

Hubble classification A method of classifying galaxies as spirals, barred spirals, ellipticals, or irregulars according to their appearance. (Chapter 24)

Hubble constant (H_0) In the Hubble law, the constant of proportionality between the recessional velocities of remote galaxies and their distances. (Chapter 24)

Hubble flow The recessional motions of remote galaxies caused by the expansion of the universe. (Chapter 24)

Hubble law The empirical relationship stating that the redshifts of remote galaxies are directly proportional to their distances from Earth. (Chapter 24)

hydrocarbon Any one of a variety of chemical compounds composed of hydrogen and carbon. (Chapter 13)

hydrogen envelope A huge, tenuous sphere of gas surrounding the head of a comet. (Chapter 15)

hydrogen fusion The thermonuclear conversion of hydrogen into helium. (Chapter 16)

hydrostatic equilibrium A balance between the weight of a layer in a star and the pressure that supports it. (Chapter 16)

hyperbola A conic section formed by cutting a circular cone with a plane at an angle steeper than the sides of the cone. (Chapter 4)

hyperbolic space Space with negative curvature. (Chapter 26)

hypothesis An idea or collection of ideas that seems to explain a specified phenomenon; a conjecture. (Chapter 1)

ice rafts (Europa) Segments of Europa's icy crust that have been moved by tectonic disturbances. (Chapter 13)

ices Solid materials with low condensation temperatures, including ices of water, methane, and ammonia. (Chapter 7)

ideal gas A gas in which the pressure is directly proportional to both the density and the temperature of the gas; an idealization of a real gas. (Chapter 19)

igneous rock A rock that formed from the solidification of molten lava or magma. (Chapter 9)

imaging The process of recording the image made by a telescope of a distant object. (Chapter 6)

impact breccia A type of rock formed from other rocks that were broken apart, mixed, and fused together by a series of meteoritic impacts. (Chapter 10)

impact crater A circular depression on a planet or satellite caused by the impact of a meteoroid. (Chapter 7, Chapter 10)

inertia, law of See *Newton's first law.*

inferior conjunction The configuration when an inferior planet is between the Sun and Earth. (Chapter 4)

inferior planet A planet that is closer to the Sun than Earth is. (Chapter 4)

inflation A sudden expansion of space. (Chapter 27)

inflationary epoch A brief period shortly after the Big Bang during which the scale of the universe increased very rapidly. (Chapter 27)

infrared radiation Electromagnetic radiation of wavelength longer than visible light but shorter than radio waves. (Chapter 5)

inner core (of Earth) The solid innermost portion of Earth's iron-rich core. (Chapter 9)

inner Lagrangian point The point between the two stars comprising a binary system where their Roche lobes touch; the point across which mass transfer can occur. (Chapter 19)

instability strip A region of the H-R diagram occupied by pulsating stars. (Chapter 19)

interferometry A technique of combining the observations of two or more telescopes to produce images better than one telescope alone could make. (Chapter 6)

intermediate-mass black hole See *mid-mass black hole*

intermediate-period comet A comet with an orbital period between 20 and 200 years. (Chapter 15)

intermediate vector boson The particle that is responsible for the weak force. (Chapter 27)

internal rotation period The period with which the core of a Jovian planet rotates. (Chapter 12)

interstellar extinction The dimming of starlight as it passes through the interstellar medium. (Chapter 18, Chapter 23)

interstellar medium Gas and dust in interstellar space. (Chapter 8, Chapter 18)

interstellar reddening The reddening of starlight passing through the interstellar medium as a result of blue light being scattered more than red. (Chapter 18)

intracluster gas Gas found between the galaxies that make up a cluster of galaxies. (Chapter 24)

inverse-square law The statement that the apparent brightness of a light source varies inversely with the square of the distance from the source. (Chapter 17)

Io torus A doughnut-shaped ring of gas circling Jupiter at the distance of Io's orbit. (Chapter 13)

ion tail The relatively straight tail of a comet produced by the solar wind acting on ions. (Chapter 15)

ionization The process by which a neutral atom becomes an electrically charged ion through the loss or gain of electrons. (Chapter 5)

iron meteorite A meteorite composed primarily of iron. (Chapter 15)

irregular cluster (of galaxies) A sprawling collection of galaxies whose overall distribution in space does not exhibit any noticeable spherical symmetry. (Chapter 24)

irregular galaxy An asymmetrical galaxy having neither spiral arms nor an elliptical shape. (Chapter 24)

isotope Any of several forms for the same chemical element whose nuclei all have the same number of protons but different numbers of neutrons. (Chapter 5)

isotropic Having the same property in all directions. (Chapter 24, Chapter 26)

isotropy problem The dilemma posed by the fact that the cosmic microwave background is isotropic; also called the horizon problem. (Chapter 27)

Jeans length The smallest scale over which a density fluctuation in a medium will contract to form a gravitationally bound object. (Chapter 27)

jet An extended line of fast-moving gas ejected from the vicinity of a star or a black hole. (Chapter 8)

joule (J) A unit of energy. (Chapter 5)

Jovian planet Low-density planets composed primarily of hydrogen and helium, including Jupiter, Saturn, Uranus, and Neptune. (Chapter 7)

Jupiter-family comet A comet with an orbital period of less than 20 years. (Chapter 15)

Kaluza-Klein theory A theory that attempts to describe the four physical forces in terms of the geometry of space-time with more than four dimensions. (Chapter 27)

kelvin (K) A unit of temperature on the Kelvin temperature scale, equivalent to a degree Celsius. (Chapter 5)

Kelvin-Helmholtz contraction The contraction of a gaseous body, such as a star or nebula, during which gravitational energy is transformed into thermal energy. (Chapter 8)

Kepler's first law The statement that each planet moves around the Sun in an elliptical orbit with the Sun at one focus of the ellipse. (Chapter 4)

Kepler's second law The statement that a planet sweeps out equal areas in equal times as it orbits the Sun; also called the law of equal areas. (Chapter 4)

Kepler's third law A relationship between the period of an orbiting object and the semimajor axis of its elliptical orbit. (Chapter 4)

kiloparsec (kpc) One thousand parsecs; about 3260 light-years. (Chapter 1)

kinetic energy The energy possessed by an object because of its motion. (Chapter 7)

Kirchhoff's laws Three statements about circumstances that produce absorption lines, emission lines, and continuous spectra. (Chapter 5)

Kirkwood gaps Gaps in the spacing of asteroid orbits, discovered by Daniel Kirkwood. (Chapter 15)

Kuiper belt A region that extends from around the orbit of Pluto to about 500 AU from the Sun where many icy objects orbit the Sun. (Chapter 7, Chapter 14, Chapter 15)

Lamb shift A tiny shift in the spectral lines of hydrogen caused by the presence of virtual particles. (Chapter 27)

lava Molten rock flowing on the surface of a planet. (Chapter 9)

law of cosmic censorship The hypothesis that all singularities must be surrounded by an event horizon. (Chapter 22)

law of equal areas See *Kepler's second law.*

law of inertia See *Newton's first law.*

law of universal gravitation A formula deduced by Isaac Newton that expresses the strength of the force of gravity that two masses exert on each other. (Chapter 4)

laws of physics A set of physical principles with which we can understand natural phenomena and the nature of the universe. (Chapter 1)

length contraction In the special theory of relativity, the shrinking of an object's length along its direction of motion. (Chapter 22)

lens A piece of transparent material (usually glass) that can bend light and bring it to a focus. (Chapter 6)

lenticular galaxy A galaxy with a central bulge and a disk but no spiral structure; an S0 galaxy. (Chapter 24)

libration An apparent rocking of the Moon whereby an Earth-based observer can, over time, see slightly more than one-half the Moon's surface. (Chapter 10)

light curve A graph that displays how the brightness of a star or other astronomical object varies over time. (Chapter 17)

light-gathering power A measure of the amount of radiation brought to a focus by a telescope. (Chapter 6)

light pollution Light from cities and towns that degrades telescope images. (Chapter 6)

light scattering The process by which light bounces off particles in its path. (Chapter 5, Chapter 12)

light-year (ly) The distance light travels in a vacuum in one year. (Chapter 1)

limb darkening The phenomenon whereby the Sun looks darker near its apparent edge, or limb, than near the center of its disk. (Chapter 16)

line of nodes The line where the plane of Earth's orbit intersects the plane of the Moon's orbit. (Chapter 3)

liquid metallic hydrogen Hydrogen compressed to such a density that it behaves like a liquid metal. (Chapter 7, Chapter 12)

lithosphere The solid, upper layer of Earth; essentially Earth's crust. (Chapter 9)

Local Bubble A large cavity in the interstellar medium within which the Sun and nearby stars are located. (Chapter 23)

Local Group The cluster of galaxies of which our Galaxy is a member. (Chapter 24)

local meridian See *meridian.*

long-period comet A comet that takes hundreds of thousands of years or more to complete one orbit of the Sun. (Chapter 15)

long-period variable A variable star with a period longer than about 100 days. (Chapter 19)

lookback time How far into the past we are looking when we see a particular object. (Chapter 26)

Lorentz transformations Equations that relate the measurements of different observers who are moving relative to each other at high speeds. (Chapter 22)

lower meridian The half of the meridian that lies below the horizon. (Chapter 2)

lowlands (on Mars) See *northern lowlands.*

luminosity The rate at which electromagnetic radiation is emitted from a star or other object. (Chapter 5, Chapter 16, Chapter 17)

luminosity class A classification of a star of a given spectral type according to its luminosity. (Chapter 17)

luminosity function The numbers of stars of differing brightness per cubic parsec. (Chapter 17)

lunar eclipse An eclipse of the Moon by Earth; a passage of the Moon through Earth's shadow. (Chapter 3)

lunar highlands Ancient, high-elevation, heavily cratered terrain on the Moon. (Chapter 10)

lunar month See *synodic month.*

lunar phase The appearance of the illuminated area of the Moon as seen from Earth. (Chapter 3)

Lyman series A series of spectral lines of hydrogen produced by electron transitions to and from the lowest energy state of the hydrogen atom. (Chapter 5)

M-theory A theory that attempts to describe fundamental particles as 11-dimensional membranes. (Chapter 27)

MACHO See *massive compact halo object.*

magma Molten rock beneath a planet's surface. (Chapter 9)

magnetar An unusual type of neutron star with a magnetic field about 1000 times stronger than that of an ordinary neutron star. (Chapter 21)

magnetic axis A line connecting the north and south magnetic poles of a planet or star possessing a magnetic field. (Chapter 14)

magnetic-dynamo model A theory that explains the solar cycle as a result of the Sun's differential rotation acting on the Sun's magnetic field. (Chapter 16)

magnetic reconnection An event where two oppositely directed magnetic fields approach and cancel, thus releasing energy. (Chapter 16)

magnetic resonance imaging (MRI) A technique for viewing the interior of the human body that makes use of how protons respond to magnetic fields. (Chapter 23)

magnetogram An image of the Sun that shows regions of different magnetic polarity. (Chapter 16)

magnetometer A device for measuring magnetic fields. (Chapter 7)

magnetopause That region of a planet's magnetosphere where the magnetic field counterbalances the pressure from the solar wind. (Chapter 9)

magnetosphere The region around a planet occupied by its magnetic field. (Chapter 9)

magnification The factor by which the apparent angular size of an object is increased when viewed through a telescope. (Chapter 6)

magnifying power See *magnification.*

magnitude scale A system for denoting the brightnesses of astronomical objects. (Chapter 17)

main sequence A grouping of stars on the Hertzsprung-Russell diagram extending diagonally across the graph from hot, luminous stars to cool, dim stars. (Chapter 17)

main-sequence lifetime The total time that a star spends fusing hydrogen in its core, and hence the total time that it will spend as a main-sequence star. (Chapter 19)

main-sequence star A star whose luminosity and surface temperature place it on the main sequence on an H-R diagram; a star that derives its energy from core hydrogen fusion. (Chapter 17)

major axis (of an ellipse) The longest diameter of an ellipse. (Chapter 4)

mantle (of a planet) That portion of a terrestrial planet located between its crust and core. (Chapter 9)

mare (plural **maria**) Latin for "sea"; a large, relatively crater-free plain on the Moon. (Chapter 10)

mare basalt A type of lunar rock commonly found in the mare basins. (Chapter 10)

maser An interstellar cloud in which water molecules emit intensely at microwave wavelengths. (Chapter 24)

mass A measure of the total amount of material in an object. (Chapter 4)

mass density of radiation The energy possessed by a radiation field per unit volume divided by the square of the speed of light. (Chapter 26)

mass loss A process by which a star gently loses matter. (Chapter 19)

mass-luminosity relation A relationship between the masses and luminosities of main-sequence stars. (Chapter 17)

mass-radius relation A relationship between the masses and radii of white dwarf stars. (Chapter 20)

mass transfer The flow of gases from one star in a binary system to the other. (Chapter 19)

massive compact halo object (MACHO) A dim star or low-mass black hole that may comprise part of the unseen dark matter. (Chapter 23)

matter density parameter (Ω_m) The ratio of the average density of matter in the universe to the critical density. (Chapter 26)

matter-dominated universe A universe in which the average density of matter exceeds both the mass density of radiation and the mass density of dark energy. (Chapter 26)

mean solar day The interval between successive meridian passages of the mean Sun; the average length of a solar day. (Chapter 2)

mean sun A fictitious object that moves eastward at a constant speed along the celestial equator, completing one circuit of the sky with respect to the vernal equinox in one tropical year. (Chapter 2)

medium (plural **media**) A material through which light travels. (Chapter 6)

medium, interstellar: See *interstellar medium.*

megaparsec (Mpc) One million parsecs. (Chapter 1)

melting point The temperature at which a substance changes from solid to liquid. (Chapter 9)

meridian (or **local meridian**) The great circle on the celestial sphere that passes through an observer's zenith and the north and south celestial poles. (Chapter 2)

meridian transit The crossing of the meridian by any astronomical object. (Chapter 2)

mesosphere A layer in Earth's atmosphere above the stratosphere. (Chapter 9)

metal In reference to stars and galaxies, any elements other than hydrogen and helium; in reference to planets, a material such as iron, silver, and aluminum that is a good conductor of electricity and of heat. (Chapter 17)

metal-poor star A star that, compared to the Sun, is deficient in elements heavier than helium; also called a Population II star. (Chapter 19)

metal-rich star A star whose abundance of heavy elements is roughly comparable to that of the Sun; also called a Population I star. (Chapter 19)

metamorphic rock A rock whose properties and appearance have been transformed by the action of pressure and heat beneath Earth's surface. (Chapter 9)

meteor The luminous phenomenon seen when a meteoroid enters Earth's atmosphere; a "shooting star." (Chapter 15)

meteor shower Many meteors that seem to radiate from a common point in the sky. (Chapter 15)

meteorite A fragment of a meteoroid that has survived passage through Earth's atmosphere. (Chapter 1, Chapter 8, Chapter 15)

meteoritic swarm A collection of meteoroids moving together along an orbit about the Sun. (Chapter 15)

meteoroid A small rock in interplanetary space. (Chapter 7, Chapter 15)

microlensing A phenomenon in which a compact object such as a MACHO acts as a gravitational lens, focusing the light from a distant star. (Chapter 23)

microwaves Short-wavelength radio waves. (Chapter 5)

mid-mass black hole A black hole with a mass of hundreds of Suns. (Chapter 22)

Milky Way Galaxy Our Galaxy; the band of faint stars seen from Earth in the plane of our Galaxy's disk. (Chapter 23)

millisecond pulsar A pulsar with a period of roughly 1 to 10 milliseconds. (Chapter 21)

mineral A naturally occurring solid composed of a single element or chemical combination of elements, often in the form of crystals. (Chapter 9)

minor planet See *asteroid.*

minute of arc See *arcminute.*

model A hypothesis that has withstood experimental or observational tests; or, the results of a theoretical calculation that gives the values of temperature, pressure, density, and so forth throughout the interior of an object such as a planet or star. (Chapter 1)

molecule A combination of two or more atoms. (Chapter 5)

moonquake Sudden, vibratory motion of the Moon's surface. (Chapter 10)

MRI See *magnetic resonance imaging.*

nanometer (nm) One billionth of a meter: 1 nm = 10^{-9} meter = 10^{-6} millimeter = 10^{-3} µm. (Chapter 5)

neap tide An ocean tide that occurs when the Moon is near first-quarter or third-quarter phase. (Chapter 4)

near-Earth object (NEO) An asteroid whose orbit lies wholly or partly within the orbit of Mars. (Chapter 15)

near-infrared The part of the infrared spectrum closest in wavelength to visible light. (Chapter 23)

nebula A cloud of interstellar gas and dust. (Chapter 1, Chapter 18)

nebular hypothesis The idea that the Sun and the rest of the solar system formed from a cloud of interstellar material. (Chapter 8)

nebulosity See *nebula.*

negative curvature The curvature of a surface or space in which parallel lines diverge and the sum of the angles of a triangle is less than 180°. (Chapter 26)

negative hydrogen ion A hydrogen atom that has acquired a second electron. (Chapter 16)

NEO See *near-Earth object.*

neon fusion The thermonuclear fusion of neon to produce heavier nuclei. (Chapter 20)

neutrino A subatomic particle with no electric charge and very little mass, yet one that is important in many nuclear reactions. (Chapter 16)

neutron A subatomic particle with no electric charge and with a mass nearly equal to that of the proton. (Chapter 5)

neutron capture The buildup of neutrons inside nuclei. (Chapter 20)

neutron star A very compact, dense star composed almost entirely of neutrons. (Chapter 21)

new moon The phase of the Moon when the dark hemisphere of the Moon faces Earth. (Chapter 3)

Newtonian mechanics The branch of physics based on Newton's laws of motion. (Chapter 1, Chapter 4)

Newtonian reflector A reflecting telescope that uses a small mirror to deflect the image to one side of the telescope tube. (Chapter 6)

Newton's first law of motion The statement that a body remains at rest, or moves in a straight line at a constant speed, unless acted upon by a net outside force; the law of inertia. (Chapter 4)

Newton's form of Kepler's third law A relationship between the period of two objects orbiting each other, the semimajor axis of their orbit, and the masses of the objects. (Chapter 4)

Newton's second law of motion A relationship between the acceleration of an object, the object's mass, and the net outside force acting on the mass. (Chapter 4)

Newton's third law of motion The statement that whenever one body exerts a force on a second body, the second body exerts an equal and opposite force on the first body. (Chapter 4)

no-hair theorem A statement of the simplicity of black holes. (Chapter 22)

noble gas An element whose atoms do not combine into molecules. (Chapter 12)

node See *line of nodes.*

nonthermal radiation Radiation other than that emitted by a heated body. (Chapter 12, Chapter 25)

north celestial pole The point directly above Earth's north pole where Earth's axis of rotation, if extended, would intersect the celestial sphere. (Chapter 2)

northern lights See *aurora borealis.*

northern lowlands (on Mars) Relatively young and crater-free terrain in the Martian northern hemisphere. (Chapter 11)

nova (plural **novae**) A star that experiences a sudden outburst of radiant energy, temporarily increasing its luminosity roughly a thousandfold. (Chapter 21)

nuclear density The density of matter in an atomic nucleus; about 4×10^{17} kg/m^3. (Chapter 20)

nucleosynthesis The process of building up nuclei such as deuterium and helium from protons and neutrons. (Chapter 27)

nucleus (of an atom) The massive part of an atom, composed of protons and neutrons, about which electrons revolve. (Chapter 5)

nucleus (of a comet) A collection of ices and dust that constitute the solid part of a comet. (Chapter 15)

nucleus (of a galaxy) See *galactic nucleus.*

OB association A grouping of hot, young, massive stars, predominantly of spectral types O and B. (Chapter 18)

OBAFGKM The temperature sequence (from hot to cold) of spectral classes. (Chapter 17)

objective lens The principal lens of a refracting telescope. (Chapter 6)

objective mirror The principal mirror of a reflecting telescope. (Chapter 6)

oblate Flattened at the poles. (Chapter 12)

oblateness A measure of how much a flattened sphere (or spheroid) differs from a perfect sphere. (Chapter 12)

observable universe That portion of the universe inside our cosmic light horizon. (Chapter 26, Chapter 27)

Occam's razor The notion that a straightforward explanation of a phenomenon is more likely to be correct than a convoluted one. (Chapter 4)

occultation The eclipsing of an astronomical object by the Moon or a planet. (Chapter 13, Chapter 14)

oceanic rift A crack in the ocean floor that exudes lava. (Chapter 9)

Olbers's paradox The dilemma associated with the fact that the night sky is dark. (Chapter 26)

1-to-1 spin-orbit coupling See *synchronous rotation.*

Oort comet cloud A presumed accumulation of comets and cometary material surrounding the Sun at distances of roughly 50,000 AU. (Chapter 7, Chapter 15)

open cluster A loose association of young stars in the disk of our Galaxy; a galactic cluster. (Chapter 18, Chapter 19)

open universe A universe with negative curvature, so that its geometry is analogous to a saddle-shaped hyperbolic surface. (Chapter 26)

opposition The configuration of a planet when it is at an elongation of 180° and thus appears opposite the Sun in the sky. (Chapter 4)

optical double star Two stars that lie along nearly the same line of sight but are actually at very different distances from us. (Chapter 17)

optical telescope A telescope designed to detect visible light. (Chapter 6)

optical window The range of visible wavelengths to which Earth's atmosphere is transparent. (Chapter 6)

organic molecules Molecules containing carbon, some of which are the molecules of which living organisms are made. (Chapter 28)

outer core (of Earth) The outer, molten portion of Earth's iron-rich core. (Chapter 9)

outgassing The release of gases into a planet's atmosphere by volcanic activity. (Chapter 9)

overcontact binary A close binary system in which the two stars share a common atmosphere. (Chapter 19)

oxygen fusion The thermonuclear fusion of oxygen to produce heavier nuclei. (Chapter 20)

ozone A type of oxygen whose molecules contain three oxygen atoms. (Chapter 9)

ozone hole A region of Earth's atmosphere over Antarctica where the concentration of ozone is abnormally low. (Chapter 9)

ozone layer A layer in Earth's upper atmosphere where the concentration of ozone is high enough to prevent much ultraviolet light from reaching the surface. (Chapter 9)

P wave One of three kinds of seismic waves produced by an earthquake; a primary wave. (Chapter 9)

pair production The creation of a particle and its antiparticle from energy. (Chapter 21, Chapter 27)

parabola A conic section formed by cutting a circular cone at an angle parallel to one of the sides of the cone. (Chapter 4)

parallax The apparent displacement of an object due to the motion of the observer. (Chapter 4, Chapter 17)

parsec (pc) A unit of distance; 3.26 light-years. (Chapter 1, Chapter 17)

partial lunar eclipse A lunar eclipse in which the Moon does not appear completely covered. (Chapter 3)

partial solar eclipse A solar eclipse in which the Sun does not appear completely covered. (Chapter 3)

Paschen series A series of spectral lines of hydrogen produced by electron transitions between the third and higher energy levels. (Chapter 5)

Pauli exclusion principle A principle of quantum mechanics stating that no two electrons can have the same position and momentum. (Chapter 19)

penumbra (of a shadow) (plural **penumbrae**) The portion of a shadow in which only part of the light source is covered by an opaque body. (Chapter 3)

penumbral eclipse A lunar eclipse in which the Moon passes only through Earth's penumbra. (Chapter 3)

perigee The point in its orbit where a satellite or the Moon is nearest Earth. (Chapter 3)

perihelion The point in its orbit where a planet or comet is nearest the Sun. (Chapter 4)

period (of a planet) The interval of time between successive geometric arrangements of a planet and an astronomical object, such as the Sun. (Chapter 4)

period-luminosity relation A relationship between the period and average density of a pulsating star. (Chapter 19)

periodic table A listing of the chemical elements according to their properties, invented by Dmitri Mendeleev. (Chapter 5)

photodisintegration The breakup of nuclei by high-energy gamma rays. (Chapter 20)

photoelectric effect The phenomenon whereby certain metals emit electrons when exposed to short-wavelength light. (Chapter 5)

photometry The measurement of light intensities. (Chapter 6, Chapter 17)

photon A discrete unit of electromagnetic energy. (Chapter 5)

photosphere The region in the solar atmosphere from which most of the visible light escapes into space. (Chapter 16)

photosynthesis A biochemical process in which solar energy is converted into chemical energy, carbon dioxide and water are absorbed, and oxygen is released. (Chapter 9)

physics, laws of See *laws of physics.*

pixel A picture element. (Chapter 6)

plage A bright region in the solar atmosphere as observed in the monochromatic light of a spectral line. (Chapter 16)

Planck time A fundamental interval of time defined by basic physical constants. (Chapter 26)

Planck's law The relationship between the energy of a photon and its wavelength or frequency; $E = hc/\lambda = h\nu$. (Chapter 5)

plane of the ecliptic The plane in which Earth orbits the Sun. (Chapter 3)

planetary nebula A luminous shell of gas ejected from an old, low-mass star. (Chapter 20)

planetesimal One of many small bodies of primordial dust and ice that combined to form the planets. (Chapter 8)

plasma A hot ionized gas. (Chapter 12, Chapter 16, Chapter 26)

plastic The attribute of being nearly solid yet able to flow. (Chapter 9)

plate A large section of Earth's lithosphere that moves as a single unit. (Chapter 9)

plate tectonics The motions of large segments (plates) of Earth's surface over the underlying mantle. (Chapter 9)

plutino One of about 100 objects in the Kuiper belt that orbit the Sun with nearly the same semimajor axis as Pluto. (Chapter 14)

polarized radiation Electromagnetic waves whose electric fields are aligned in a particular direction. (Chapter 25)

polymer A long molecule consisting of many smaller molecules joined together. (Chapter 13)

poor cluster (of galaxies) A cluster of galaxies with very few members; a group of galaxies. (Chapter 24)

Population I star A star whose spectrum exhibits spectral lines of many elements heavier than helium; a metal-rich star. (Chapter 19)

Population II star A star whose spectrum exhibits comparatively few spectral lines of elements heavier than helium; a metal-poor star. (Chapter 19)

Population III star One of the first stars to form after the Big Bang, composed of only hydrogen, helium, and tiny amounts of lithium and beryllium; a "zeroth-generation" star. (Chapter 27)

positional astronomy The study of the apparent positions of the planets and stars and how those positions change. (Chapter 2)

positive curvature The curvature of a surface or space in which parallel lines converge and the sum of the angles of a triangle is greater than 180°. (Chapter 26)

positron An electron with a positive rather than negative electric charge; the antiparticle of the electron. (Chapter 16, Chapter 27)

power of ten The exponent n in 10^n. (Chapter 1)

powers-of-ten notation A shorthand method of writing numbers, involving 10 followed by an exponent. (Chapter 1)

precession (of Earth) A slow, conical motion of Earth's axis of rotation caused by the gravitational pull of the Moon and Sun on Earth's equatorial bulge. (Chapter 2)

precession of the equinoxes The slow westward motion of the equinoxes along the ecliptic due to precession of Earth. (Chapter 2)

primary mirror See *objective mirror.*

prime focus The point in a telescope where the objective focuses light. (Chapter 6)

primitive asteroid See *undifferentiated asteroid.*

primordial black hole A type of black hole that may have formed in the very early universe. (Chapter 22)

primordial fireball The extremely hot gas that filled the universe immediately following the Big Bang. (Chapter 26)

principle of equivalence See *equivalence principle.*

progenitor star A star that later explodes into a supernova. (Chapter 20)

prograde orbit An orbit of a satellite around a planet that is in the same direction as the rotation of the planet. (Chapter 13)

prograde rotation A situation in which an object (such as a planet) rotates in the same direction that it orbits around another object (such as the Sun). (Chapter 11)

prominence Flamelike protrusions seen near the limb of the Sun and extending into the solar corona. (Chapter 16)

proper distance See *proper length.*

proper length A length measured by a ruler at rest with respect to an observer. (Chapter 22)

proper motion The angular rate of change in the location of a star on the celestial sphere, usually expressed in arcseconds per year. (Chapter 17)

proper time A time interval measured with a clock at rest with respect to an observer. (Chapter 22)

proplyd See *protoplanetary disk.*

proton A heavy, positively charged subatomic particle that is one of two principal constituents of atomic nuclei. (Chapter 5)

proton-proton chain A sequence of thermonuclear reactions by which hydrogen nuclei are built up into helium nuclei. (Chapter 16)

protoplanet A Moon-sized object formed by the coalescence of planetesimals. (Chapter 8)

protoplanetary disk (proplyd) A disk of material encircling a protostar or a newborn star. (Chapter 8, Chapter 18)

protostar A star in its earliest stages of formation. (Chapter 18)

protosun The part of the solar nebula that eventually developed into the Sun. (Chapter 8)

Ptolemaic system The definitive version of the geocentric cosmogony of ancient Greece. (Chapter 4)

pulsar A pulsating radio source thought to be associated with a rapidly rotating neutron star. (Chapter 1, Chapter 21)

pulsating variable star A star that pulsates in size and luminosity. (Chapter 19)

pulsating X-ray source An object that emits pulses of X rays at regular intervals of a few seconds. (Chapter 21)

quantum electrodynamics A theory that describes details of how charged particles interact by exchanging photons. (Chapter 27)

quantum mechanics The branch of physics dealing with the structure and behavior of atoms and their constituents as well as their interaction with light. (Chapter 5, Chapter 27)

quark One of several particles thought to be the internal constituents of certain heavy subatomic particles such as protons and neutrons. (Chapter 27)

quark confinement The permanent bonding of quarks to form particles like protons and neutrons. (Chapter 27)

quasar A very luminous object with a very large redshift and a starlike appearance. (Chapter 1, Chapter 25)

radial velocity That portion of an object's velocity parallel to the line of sight. (Chapter 5, Chapter 17)

radial velocity curve A plot showing the variation of radial velocity with time for a binary star or variable star. (Chapter 17)

radial velocity method A technique used to detect extrasolar planets by observing Doppler shifts in the spectrum of the planet's star. (Chapter 8)

radiant (of a meteor shower) The point in the sky from which meteors of a particular shower seem to originate. (Chapter 15)

radiation darkening The darkening of methane ice by electron impacts. (Chapter 14)

radiation-dominated universe A universe in which the mass density of radiation exceeds both the average density of matter and the mass density of dark energy. (Chapter 26)

radiation pressure Pressure exerted on an object by radiation falling on the object. (Chapter 15)

radiative diffusion The random migration of photons from a star's center toward its surface. (Chapter 16)

radiative zone A region within a star where radiative diffusion is the dominant mode of energy transport. (Chapter 16)

radio galaxy A galaxy that emits an unusually large amount of radio waves. (Chapter 25)

radio lobe A region near an active galaxy from which significant radio radiation emanates. (Chapter 25)

radio telescope A telescope designed to detect radio waves. (Chapter 6)

radio waves The longest-wavelength electromagnetic radiation. (Chapter 5)

radio window The range of radio wavelengths to which Earth's atmosphere is transparent. (Chapter 6)

radioactive dating A technique for determining the age of a rock sample by measuring the radioactive elements and their decay products in the sample. (Chapter 8)

radioactive decay The process whereby certain atomic nuclei spontaneously transform into other nuclei. (Chapter 8)

rarefaction A region in a sound wave where the medium carrying the wave is spread out or rarefied. (Chapter 26)

recombination The process in which an electron combines with a positively charged ion. (Chapter 18)

red dwarf A main-sequence star with a mass between about 0.08 $M_\odot$ and 0.4 $M_\odot$ which has a fully convective interior and which never goes through a red giant stage. (Chapter 19)

red giant A large, cool star of high luminosity. (Chapter 17, Chapter 19)

red-giant branch The region of an H-R diagram occupied by stars in their first red-giant phase. (Chapter 20)

reddening (interstellar) See *interstellar reddening.*

redshift The shifting to longer wavelengths of the light from remote galaxies and quasars; the Doppler shift of light from a receding source. (Chapter 5, Chapter 24)

reflecting telescope A telescope in which the principal optical component is a concave mirror. (Chapter 6)

reflection The return of light rays by a surface. (Chapter 6)

reflection nebula A comparatively dense cloud of dust in interstellar space that is illuminated by a star. (Chapter 18)

reflector A reflecting telescope. (Chapter 6)

refracting telescope A telescope in which the principal optical component is a lens. (Chapter 6)

refraction The bending of light rays when they pass from one transparent medium to another. (Chapter 6)

refractor A refracting telescope. (Chapter 6)

refractory element An element with high melting and boiling points. (Chapter 10)

regolith The layer of rock fragments covering the surface of the Moon. (Chapter 10)

regular cluster (of galaxies) A spherical cluster of galaxies. (Chapter 24)

reionization The process whereby high-energy photons emitted by the most massive of the first stars (Population III stars) ionized most of the atoms in the universe. (Chapter 27)

relativistic cosmology A cosmology based on the general theory of relativity. (Chapter 26)

residual polar cap An ice-covered polar region on Mars that does not completely evaporate during the Martian summer. (Chapter 11)

respiration A biological process that produces energy by consuming oxygen and releasing carbon dioxide. (Chapter 9)

retrograde motion The apparent westward motion of a planet with respect to background stars. (Chapter 4)

retrograde orbit An orbit of a satellite around a planet that is in the direction opposite to which the planet rotates. (Chapter 13)

retrograde rotation A situation in which an object (such as a planet) rotates in the direction opposite to which it orbits around another object (such as the Sun). (Chapter 11)

rich cluster (of galaxies) A cluster of galaxies containing many members. (Chapter 24)

rift valley A feature created when a planet's crust breaks apart along a line. (Chapter 11)

right ascension A coordinate for measuring the east-west positions of objects on the celestial sphere. (Chapter 2)

ring particles Small particles that constitute a planetary ring. (Chapter 12)

ringlet One of many narrow bands of particles of which Saturn's ring system is composed. (Chapter 12)

Roche limit The smallest distance from a planet or other object at which a second object can be held together by purely gravitational forces. (Chapter 12)

Roche lobe A teardrop-shaped volume surrounding a star in a binary inside which gases are gravitationally bound to that star. (Chapter 19)

rock A mineral or combination of minerals. (Chapter 9)

rotation curve A plot of the orbital speeds of stars and nebulae in a galaxy versus distance from the center of the galaxy. (Chapter 23)

RR Lyrae variable A type of pulsating star with a period of less than one day. (Chapter 19, Chapter 23)

runaway greenhouse effect A greenhouse effect in which the temperature continues to increase. (Chapter 11)

runaway icehouse effect A situation in which a decrease in atmospheric temperature causes a further decrease in temperature. (Chapter 11)

S wave One of three kinds of seismic waves produced by an earthquake; a secondary wave. (Chapter 9)

Sagittarius A* A powerful radio source at the center of our Galaxy. (Chapter 23)

saros A particular cycle of similar eclipses that recur about every 18 years. (Chapter 3)

scarp A line of cliffs formed by the faulting or fracturing of a planet's surface. (Chapter 11)

scattering of light See *light scattering.*

Schwarzschild radius (R_{Sch}) The distance from the singularity to the event horizon in a nonrotating black hole. (Chapter 22)

scientific method The basic procedure used by scientists to investigate phenomena. (Chapter 1)

seafloor spreading The separation of plates under the ocean due to lava emerging in an oceanic rift. (Chapter 9)

search for extraterrestrial intelligence (SETI) The scientific search for evidence of intelligent life on other planets. (Chapter 28)

second of arc See *arcsecond.*

sedimentary rock A rock that is formed from material deposited on land by rain or winds, or on the ocean floor. (Chapter 9)

seeing disk The angular diameter of a star's image. (Chapter 6)

seismic wave A vibration traveling through a terrestrial planet, usually associated with earthquake-like phenomena. (Chapter 9)

seismograph A device used to record and measure seismic waves, such as those produced by earthquakes. (Chapter 9)

self-propagating star formation The process by which the formation of stars in one location in a galaxy stimulates the formation of stars in a neighboring location. (Chapter 23)

semidetached binary A binary star system in which one star fills its Roche lobe. (Chapter 19)

semimajor axis One-half of the major axis of an ellipse. (Chapter 4)

SETI See *search for extraterrestrial intelligence.*

Seyfert galaxy A spiral galaxy with a bright nucleus whose spectrum exhibits emission lines. (Chapter 25)

shell helium fusion The thermonuclear fusion of helium in a shell surrounding a star's core. (Chapter 20)

shell hydrogen fusion The thermonuclear fusion of hydrogen in a shell surrounding a star's core. (Chapter 19)

shepherd satellite A satellite whose gravity restricts the motions of particles in a planetary ring, preventing them from dispersing. (Chapter 12)

shield volcano A volcano with long, gently sloping sides. (Chapter 11)

shock wave An abrupt, localized region of compressed gas caused by an object traveling through the gas at a speed greater than the speed of sound. (Chapter 9)

SI units The International System of Units, based on the meter (m), the second (s), and the kilogram (kg). (Chapter 1)

sidereal clock A clock that measures sidereal time. (Chapter 2)

sidereal day The interval between successive meridian passages of the vernal equinox. (Chapter 2)

sidereal month The period of the Moon's revolution about Earth with respect to the stars. (Chapter 3)

sidereal period The orbital period of one object about another as measured with respect to the stars. (Chapter 4)

sidereal time Time reckoned by the location of the vernal equinox. (Chapter 2)

sidereal year The orbital period of Earth about the Sun with respect to the stars. (Chapter 2)

silicon fusion The thermonuclear fusion of silicon to produce heavier elements, especially iron. (Chapter 20)

singularity A place of infinite space-time curvature; the center of a black hole. (Chapter 22)

small-angle formula A relationship between the angular and linear sizes of a distant object. (Chapter 1)

solar constant The average amount of energy received from the Sun per square meter per second, measured just above Earth's atmosphere. (Chapter 5)

solar corona Hot, faintly glowing gases seen around the Sun during a total solar eclipse; the uppermost regions of the solar atmosphere. (Chapter 3)

solar cycle See *22-year solar cycle.*

solar eclipse An eclipse of the Sun by the Moon; a passage of Earth through the Moon's shadow. (Chapter 3)

solar flare A sudden, temporary outburst of light from an extended region of the solar surface. (Chapter 16)

solar nebula The cloud of gas and dust from which the Sun and solar system formed. (Chapter 8)

solar neutrino A neutrino emitted from the core of the Sun. (Chapter 16)

solar neutrino problem The discrepancy between the predicted and observed numbers of solar neutrinos. (Chapter 16)

solar system The Sun, planets and their satellites, asteroids, comets, and related objects that orbit the Sun. (Chapter 1)

solar wind An outward flow of particles (mostly electrons and protons) from the Sun. (Chapter 8, Chapter 16)

south celestial pole The point directly above Earth's south pole where Earth's axis of rotation, if extended, would intersect the celestial sphere. (Chapter 2)

southern highlands (on Mars) Older, cratered terrain in the Martian southern hemisphere. (Chapter 11)

southern lights See *aurora australis.*

space velocity How fast and in what direction a star moves through space. (Chapter 17)

spacetime A four-dimensional combination of time and the three dimensions of space. (Chapter 22)

special theory of relativity A description of mechanics and electromagnetic theory formulated by Albert Einstein, which explains that measurements of distance, time, and mass are affected by the observer's motion. (Chapter 22)

spectral analysis The identification of chemical substances from the patterns of lines in their spectra. (Chapter 5)

spectral class A classification of stars according to the appearance of their spectra. (Chapter 17)

spectral line In a spectrum, an absorption or emission feature that is at a particular wavelength. (Chapter 5)

spectral type A subdivision of a spectral class. (Chapter 17)

spectrograph An instrument for photographing a spectrum. (Chapter 6)

spectroscopic binary A binary star system whose binary nature is deduced from the periodic Doppler shifting of lines in its spectrum. (Chapter 17)

spectroscopic parallax The distance to a star derived by comparing its apparent brightness to a luminosity inferred from the star's spectrum. (Chapter 17)

spectroscopy The study of spectra and spectral lines. (Chapter 5, Chapter 6, Chapter 7)

spectrum (plural **spectra**) The result of dispersing a beam of electromagnetic radiation so that components with different wavelengths are separated in space. (Chapter 5)

spectrum binary A binary star whose binary nature is deduced from the presence of two sets of incongruous spectral lines. (Chapter 17)

speed Distance traveled divided by the time elapsed to cover that distance. (Chapter 4)

spherical aberration The distortion of an image formed by a telescope due to differing focal lengths of the optical system. (Chapter 6)

spherical space Space with positive curvature. (Chapter 26)

spicule A narrow jet of rising gas in the solar chromosphere. (Chapter 16)

spin The intrinsic angular momentum possessed by certain particles. (Chapter 23)

spin-flip transition A transition in the ground state of the hydrogen atom, which occurs when the orientation of the electron's spin changes. (Chapter 23)

spin-orbit coupling See *1-to-1 spin-orbit coupling* and *3-to-2 spin-orbit coupling.*

spiral arms Lanes of interstellar gas, dust, and young stars that wind outward in a plane from the central regions of a galaxy. (Chapter 23)

spiral galaxy A flattened, rotating galaxy with pinwheel-like spiral arms winding outward from the galaxy's nucleus. (Chapter 24)

spontaneous symmetry breaking A process by which certain symmetries in the mathematics of particle physics are suddenly altered to produce new particles and forces. (Chapter 27)

spring tide An ocean tide that occurs at new moon and full moon phases. (Chapter 4)

stable Lagrange points Locations along Jupiter's orbit where the combined gravitational effects of the Sun and Jupiter cause asteroids to collect. (Chapter 15)

standard candle An astronomical object of known intrinsic brightness that can be used to determine the distances to other galaxies. (Chapter 24)

starburst galaxy A galaxy that is experiencing an exceptionally high rate of star formation. (Chapter 24)

star cluster See *cluster of stars.*

starquake A localized fracture of the surface of a highly magnetized neutron star, or magnetar. (Chapter 21)

stationary absorption line An absorption line in the spectrum of a binary star that does not show the same Doppler shift as other lines, indicating that it originates in the interstellar medium. (Chapter 18)

Stefan-Boltzmann law A relationship between the temperature of a blackbody and the rate at which it radiates energy. (Chapter 5)

stellar association A loose grouping of young stars. (Chapter 18)

stellar evolution The changes in size, luminosity, temperature, and so forth that occur as a star ages. (Chapter 18)

stellar-mass black hole A black hole with a mass comparable to that of a star. (Chapter 22)

stellar parallax The apparent displacement of a star due to Earth's motion around the Sun. (Chapter 17)

stony iron meteorite A meteorite composed of both stone and iron. (Chapter 15)

stony meteorite A meteorite composed of stone. (Chapter 15)

stratosphere A layer in Earth's atmosphere directly above the troposphere. (Chapter 9)

string theory The currently favored theory for combining all the known forces in nature. (Chapter 27)

strong force The force that binds protons and neutrons together in nuclei. (Chapter 27)

subduction zone A location where colliding tectonic plates cause Earth's crust to be pulled down into the mantle. (Chapter 9)

subtend To extend over an angle. (Chapter 1)

summer solstice The point on the ecliptic where the Sun is farthest north of the celestial equator. Also used to refer to the date on which the Sun passes through this point. (Chapter 2)

sunspot A temporary cool region in the solar photosphere. (Chapter 16)

sunspot cycle The semiregular 11-year period with which the number of sunspots fluctuates. (Chapter 16)

sunspot maximum/minimum That time during the sunspot cycle when the number of sunspots is highest/lowest. (Chapter 16)

supercluster A collection of clusters of galaxies. (Chapter 24)

superconductivity The phenomenon whereby a flowing electric current does not experience any electrical resistance. (Chapter 21)

superfluidity The phenomenon whereby a fluid flows without experiencing any viscosity. (Chapter 21)

supergiant A very large, extremely luminous star of luminosity class I. (Chapter 17, Chapter 20)

supergrand unified theory A complete description of all forces and particles, as well as the structure of space and time; a "theory of everything" (TOE). (Chapter 27)

supergranule A large convective feature in the solar atmosphere, usually outlined by spicules. (Chapter 16)

superior conjunction The configuration of a planet being behind the Sun as viewed from Earth. (Chapter 4)

superior planet A planet that is more distant from the Sun than Earth is. (Chapter 4)

superluminal motion Motion that appears to involve speeds greater than the speed of light. (Chapter 25)

supermassive black hole A black hole with a mass of a million or more Suns. (Chapter 22, Chapter 25)

supernova (plural **supernovae**) A stellar outburst during which a star suddenly increases its brightness roughly a millionfold. (Chapter 1, Chapter 15, Chapter 20)

supernova remnant The gases elected by a supernova. (Chapter 18, Chapter 20)

supersonic Faster than the speed of sound. (Chapter 18)

surface wave A type of seismic wave that travels only over Earth's surface. (Chapter 9)

synchronous rotation The rotation of a body with a period equal to its orbital period; also called 1-to-1 spin-orbit coupling. (Chapter 3, Chapter 10)

synchrotron radiation A type of nonthermal radiation emitted by charged particles moving through a magnetic field. (Chapter 12, Chapter 21)

synodic month The period of revolution of the Moon with respect to the Sun; the length of one cycle of lunar phases. Also called the lunar month. (Chapter 3)

synodic period The interval between successive occurrences of the same configuration of a planet. (Chapter 4)

T Tauri stars Young variable stars associated with interstellar matter that show erratic changes in luminosity. (Chapter 18)

T Tauri wind A flow of particles away from a T Tauri star. (Chapter 8)

tail (of a comet) Gas and dust particles from a comet's nucleus that have been swept away from the comet's head by the radiation pressure of sunlight and the solar wind. (Chapter 15)

tangential velocity That portion of an object's velocity perpendicular to the line of sight. (Chapter 17)

temperature See *degree Celsius, degree Fahrenheit,* and *kelvin.*

terminator The line dividing day and night on the surface of the Moon or a planet; the line of sunset or sunrise. (Chapter 10)

terrae Cratered lunar highlands. (Chapter 10)

terrestrial planet High-density worlds with solid surfaces, including Mercury, Venus, Earth, and Mars. (Chapter 7)

theory A hypothesis that has withstood experimental or observational tests. (Chapter 1)

theory of everything See *supergrand unified theory.*

thermal equilibrium A balance between the input and outflow of heat in a system. (Chapter 16, Chapter 27)

thermal pulse A brief burst in energy output from the helium-fusing shell of an aging low-mass star. (Chapter 20)

thermal radiation The radiation naturally emitted by any object that is not at absolute zero. Blackbody radiation is an idealized case of thermal radiation. (Chapter 12, Chapter 25)

thermonuclear fusion The combining of nuclei under conditions of high temperature in a process that releases substantial energy. (Chapter 16)

thermonuclear supernova A supernova that occurs when a white dwarf in a close binary system accretes so much matter from its companion star that the white dwarf explodes in a blast of thermonuclear fusion; a Type Ia supernova. (Chapter 20)

thermosphere A region in Earth's atmosphere between the mesosphere and the exosphere. (Chapter 9)

third quarter moon The phase of the Moon that occurs when the Moon is 90° west of the Sun. (Chapter 3)

3-to-2 spin-orbit coupling The rotation of Mercury, which makes three complete rotations on its axis for every two complete orbits around the Sun. (Chapter 11)

threshold temperature The temperature above which photons spontaneously produce particles and antiparticles of a particular type. (Chapter 27)

tidal force A gravitational force whose strength and/or direction varies over a body and thus tends to deform the body. (Chapter 4, Chapter 12)

tidal heating The heating of the interior of a satellite by continually varying tidal stresses. (Chapter 13)

time dilation The slowing of time due to relativistic motion. (Chapter 22)

time zone A region on Earth where, by agreement, all clocks have the same time. (Chapter 2)

TOE See *supergrand unified theory.*

total lunar eclipse A lunar eclipse during which the Moon is completely immersed in Earth's umbra. (Chapter 3)

total solar eclipse A solar eclipse during which the Sun is completely hidden by the Moon. (Chapter 3)

totality (lunar eclipse) The period during a total lunar eclipse when the Moon is entirely within Earth's umbra. (Chapter 3)

totality (solar eclipse) The period during a total solar eclipse when the disk of the Sun is completely hidden. (Chapter 3)

transit An event in which an astronomical body moves in front of another. See also *meridian transit* and *solar transit.* (Chapter 8)

transit method A method for detecting extrasolar planets that come between us and their parent star, dimming the star's light. (Chapter 8)

trans-Neptunian object Any small body of rock and ice that orbits the Sun within the solar system, but beyond the orbit of Neptune. (Chapter 7, Chapter 14)

triple alpha process A sequence of two thermonuclear reactions in which three helium nuclei combine to form one carbon nucleus. (Chapter 19)

Trojan asteroid One of several asteroids that share Jupiter's orbit about the Sun. (Chapter 15)

Tropic of Cancer A circle of latitude $23\frac{1}{2}°$ north of Earth's equator. (Chapter 2)

Tropic of Capricorn A circle of latitude $23\frac{1}{2}°$ south of Earth's equator. (Chapter 2)

tropical year The period of revolution of Earth about the Sun with respect to the vernal equinox. (Chapter 2)

troposphere The lowest level in Earth's atmosphere. (Chapter 9)

Tully-Fisher relation A correlation between the width of the 21-cm line of a spiral galaxy and the total luminosity of that galaxy. (Chapter 24)

tuning fork diagram A diagram that summarizes Edwin Hubble's classification scheme for spiral, barred spiral, and elliptical galaxies. (Chapter 24)

turnoff point The point on an H-R diagram where the stars in a cluster are leaving the main sequence. (Chapter 19)

21-cm radio emission Radio radiation emitted by neutral hydrogen atoms in interstellar space. (Chapter 23)

22-year solar cycle The semiregular 22-year interval between successive appearances of sunspots at the same latitude and with the same magnetic polarity. (Chapter 16)

Type I Cepheid A metal-rich Cepheid variable star. (Chapter 19)

Type I supernova A supernova whose spectrum lacks hydrogen lines. Type I supernovae are further classified as Type Ia, Ib, or Ic. (Chapter 20)

Type Ia supernova A supernova whose spectrum lacks hydrogen lines but has a strong absorption line of ionized silicon. (Chapter 20)

Type Ib supernova A supernova whose spectrum lacks hydrogen and silicon lines but has a strong helium absorption line. (Chapter 20)

Type Ic supernova A supernova whose spectrum is almost devoid of emission or absorption lines. (Chapter 20)

Type II Cepheid A metal-poor Cepheid variable star. (Chapter 19)

Type II supernova A supernova with hydrogen emission lines in its spectrum, caused by the explosion of a massive star. (Chapter 20)

UBV photometry A system for determining the surface temperature of a star by measuring the star's brightness in the ultraviolet (U), blue (B), and visible (V) spectral regions. (Chapter 17)

ultramafic lava A type of lava enriched in magnesium and iron. These give the lava a higher melting temperature. (Chapter 13)

ultraviolet radiation Electromagnetic radiation of wavelengths shorter than those of visible light but longer than those of X rays. (Chapter 5)

umbra (of a shadow) (plural **umbrae**) The central, completely dark portion of a shadow. (Chapter 3)

undifferentiated asteroid An asteroid within which chemical differentiation did not occur. (Chapter 15)

universal constant of gravitation (*G*) The constant of proportionality in Newton's law of gravitation. (Chapter 4)

upper meridian The half of the meridian that lies above the horizon. (Chapter 2)

Van Allen belts Two doughnut-shaped regions around Earth where many charged particles (protons and electrons) are trapped by Earth's magnetic field. (Chapter 9)

velocity The speed and direction of an object's motion. (Chapter 4)

vernal equinox The point on the ecliptic where the Sun crosses the celestial equator from south to north. Also used to refer to the date on which the Sun passes through this intersection. (Chapter 2)

very-long-baseline interferometry (VLBI) A method of connecting widely separated radio telescopes to make very high-resolution observations. (Chapter 6)

virtual pair A particle and antiparticle that exist for such a brief interval that they cannot be observed. (Chapter 22, Chapter 27)

visible light Electromagnetic radiation detectable by the human eye. (Chapter 5)

visual binary A binary star in which the two components can be resolved through a telescope. (Chapter 17)

VLBI See *very-long-baseline interferometry.*

void A large volume of space, typically 30 to 120 Mpc (100 to 400 million light-years) in diameter, that contains very few galaxies. (Chapter 24)

volatile element An element with low melting and boiling points. (Chapter 10)

waning crescent moon The phase of the Moon that occurs between third quarter and new moon. (Chapter 3)

waning gibbous moon The phase of the Moon that occurs between full moon and third quarter. (Chapter 3)

water hole A range of frequencies in the microwave spectrum suitable for interstellar radio communication. (Chapter 28)

watt A unit of power, equal to one joule of energy per second. (Chapter 5)

wavelength The distance between two successive wave crests. (Chapter 5)

wavelength of maximum emission The wavelength at which a heated object emits the greatest intensity of radiation. (Chapter 5)

waxing crescent moon The phase of the Moon that occurs between new moon and first quarter. (Chapter 3)

waxing gibbous moon The phase of the Moon that occurs between first quarter and full moon. (Chapter 3)

weak force The short-range force that is responsible for transforming certain particles into other particles, such as the decay of a neutron into a proton. (Chapter 27)

weakly interacting massive particle (WIMP) A hypothetical massive particle that may comprise part of the unseen dark matter. (Chapter 23)

weight The force with which gravity acts on a body. (Chapter 4)

white dwarf A low-mass star that has exhausted all its thermonuclear fuel and contracted to a size roughly equal to the size of Earth. (Chapter 17, Chapter 20)

white ovals Round, whitish feature usually seen in Jupiter's southern hemisphere. (Chapter 12)

Widmanstätten patterns Crystalline structure seen in certain types of meteorites. (Chapter 15)

Wien's law A relationship between the temperature of a blackbody and the wavelength at which it emits the greatest intensity of radiation. (Chapter 5)

WIMP See *weakly interacting massive particle.*

winding dilemma The problem of why the spiral arms of a galaxy like the Milky Way do not disappear. (Chapter 23)

winter solstice The point on the ecliptic where the Sun reaches its greatest distance south of the celestial equator. Also used to refer to the date on which the Sun passes through this point. (Chapter 2)

wormhole A conjectured connection between different regions of spacetime made possible by the gravitational effects of a black hole. (Chapter 22)

X-ray burster A nonperiodic X-ray source that emits powerful bursts of X rays. (Chapter 21)

X rays Electromagnetic radiation whose wavelength is between that of ultraviolet light and gamma rays. (Chapter 5)

ZAMS See *zero-age main sequence.*

Zeeman effect A splitting or broadening of spectral lines due to a magnetic field. (Chapter 16)

zenith The point on the celestial sphere directly overhead an observer. (Chapter 2)

zero-age main sequence The main sequence of young stars that have just begun to burn hydrogen at their cores. (Chapter 19)

zero-age main-sequence star A newly formed star that has just arrived on the main sequence. (Chapter 19)

zero curvature The curvature of a surface or space in which parallel lines remain parallel and the sum of the angles of a triangle is exactly 180°. (Chapter 26)

zodiac A band of 12 constellations around the sky centered on the ecliptic. (Chapter 2)

zonal winds The pattern of alternating eastward and westward winds found in the atmospheres of Jupiter and Saturn. (Chapter 12)

zone A light-colored band in Jupiter's atmosphere. (Chapter 12)

Answers to Selected Questions

CAUTION! Only mathematical answers are given in the following, not answers that require interpretation or discussion. Your instructor will expect you to show the steps required to reach each mathematical answer.

Chapter 1

24. 8.5×10^3 km **25.** 2.8×10^7 Suns **26.** About 3×10^{36} times larger **27.** 8.94×10^{56} hydrogen atoms **28.** **(a)** 1.581×10^{-5} ly **(b)** 4.848×10^{-6} pc **29.** 4.99×10^2 s **30.** 4.3×10^9 km **31.** **(a)** 1.59×10^{14} km **(b)** 16.8 years **32.** 4.32×10^{17} s **34.** **(a)** 1.5 m **(b)** 89 m **(c)** 5.4×10^3 m **35.** 6.9 m **36.** 3.4×10^3 km **37.** 3.7 km **38.** 0.320 arcmin

Chapter 2

28. Around 8:02 p.m. **33.** **(a)** About 9 hours **46.** October 25, 1917 **49.** 50° **50.** 3:50 a.m. local time **52.** **(a)** 6:00 p.m. **(b)** September 21

Chapter 3

31. **(a)** 0.91 hour **(b)** 11° **32.** 49 arcsec

Chapter 4

18. Semimajor axis = 0.25 AU, period = 0.125 year **19.** Average = 25 AU, farthest = 50 AU **23.** 6 newtons, 4 m/s² **26.** 1/9 as strong **33.** 87.97 days **38.** **(a)** 9 AU **(b)** 27 years **39.** **(a)** 16.0 AU **(b)** 0.5 AU **40.** **(a)** 1.26 AU **(b)** 258 days **41.** Earth exerts a force of 1.98×10^{20} newtons on the Moon **42.** Forces are approximately the same, Earth's acceleration is 100 times greater **43.** $\frac{1}{4}$ as much as on Earth **44.** 62 newtons, 0.13 **45.** 0.5 year **46.** **(a)** 24 hours **(b)** 43,200 km **47.** 119 minutes **49.** **(a)** 5 years **(b)** 2.92 AU **50.** **(a)** 3.43×10^{-5} newton **(b)** 3.21×10^{-5} newton **(c)** 2.2×10^{-6} newton

Chapter 5

2. 7.5 times **3.** 500 s **7.** 0.340 m **8.** 2.61×10^{14} Hz **14.** 135 nm **15.** 3400 K **24.** 9400 nm **25.** 9980°F **26.** About 10 μm **27.** 2.9 nm **28.** 3.9×10^{26} W **29.** 190 times more **30.** **(a)** 4890 nm = 4.89 μm **(b)** 540 times more **31.** **(a)** 5.75×10^8 W/m **(b)** 10,000 K **33.** 2.43×10^{-3} nm **34.** **(a)** 1005 nm **38.** Possible transitions: energy = 1 eV, λ = 1240 nm; energy = 2 eV, λ = 620 nm; energy = 3 eV, λ = 414 nm **39.** 656.6 nm **40.** Coming toward us at 13.0 km/s **41.** 8.6×10^4 km/s

Chapter 6

31. $\frac{1}{25}$ = 0.04 **33.** **(a)** 222× **(b)** 100× **(c)** 36× **(d)** 0.75 arcsec **36.** 300 km (Hubble Space Telescope, Jupiter's moons); 110 km (human eye, our Moon) **37.** **(a)** 34 ly **(b)** 37 km **40.** **(a)** 5.39×10^{-4} m **(c)** 0.56 m

Chapter 7

22. Mass = 6.4×10^{23} kg, average density = 3900 kg/m³ **24.** **(a)** 1.0×10^{13} kg **(b)** 1.2 m/s **25.** **(a)** 3.3×10^{21} J **(b)** Equivalent to 3.9×10^7 Hiroshima-type weapons **26.** **(a)** 2.02 km/s **(b)** 19.4 km/s **27.** 1.2 km/s **28.** **(a)** 618 km/s **30.** 1.76 years **32.** **(a)** 1000 years **(b)** 17 days **34.** In 100 years, probability is 8.3×10^{-8} (one chance in 12 million); in 10^6 years, probability is 8.3×10^{-4} (one chance in 1200)

Chapter 8

30. 0.40 kg after 1.3 billion years; 0.20 kg after 2.6 billion years; 0.10 kg after 3.9 billion years **31.** 2.6 billion years **33.** **(a)** About 180 AU **34.** 2.9×10^7 AU = 140 parsecs = 460 light-years **35.** **(a)** About 600 AU **(b)** About 5×10^{40} cubic meters **(c)** About 10^{55} atoms **(d)** About 3×10^{14} atoms per cubic meter **36.** 860 years **37.** 2.2×10^{30} kg = 1.1 times the mass of the Sun **38.** **(a)** 12 m/s **(b)** 1.3×10^{-3} arcsec **(c)** 9.0×10^{-5} arcsec

Chapter 9

29. **(a)** 6.8×10^{16} W **(b)** 1.07×10^{17} W **(c)** 209 watts per square meter **(d)** 246 K = −27°C **30.** 4 km **31.** Core: 17%; mantle: 82%; crust: 1% **32.** 0.020 (2% of the total mass) **33.** **(b)** About 15,000 kg/m³

Chapter 10

26. **(a)** 4671 km **(b)** 1707 km below the surface **(c)** 449 km **30.** 130 newtons on the Moon, 780 newtons on Earth **32.** **(a)** 5.78×10^{-5} newton **(b)** 4.15×10^{-5} newton **(c)** 1.39 **36.** 2.56 seconds **38.** **(b)** 10^6 times greater

Chapter 11

46. 13.0 arcsec **48.** **(a)** 0.16 AU **51.** **(a)** 3.03 m/s **(b)** 1.26 nm **52.** 0.615 AU **55.** For T = 460°C, λ_{max} = 4.0 μm **63.** **(a)** 3770 km **(b)** 370 km **64.** Difference between the round-trip times is 2.0×10^{-4} s **66.** 920 m **73.** Phobos: about 16 arcminutes; Deimos: about 2.7 arcminutes **74.** **(a)** Radius = 20,400 km, altitude = 17,000 km

Chapter 12

37. 12.7 km/s **40.** 8.5×10^{53} hydrogen atoms, 7.1×10^{52} helium atoms **41.** Roughly 600 km/h **43.** 127 K **45.** 59.5 km/s **46.** 8300 newtons **47.** 7.1×10^4 kg/m³ **51.** **(a)** 14.4 hours for inner edge of A ring, 7.9 hours for inner edge of B ring

Chapter 13

40. 760 km **44.** 2.8×10^{11} (280 billion) years **45.** About 4×10^{-11} (one part in 3×10^{10}) **46.** About 8.1 minutes **49.** Escape speed = 2.6 km/s; mass of molecule = 2.0×10^{-26} kg, corresponding to a molecular weight of 12 **52.** 0.09 arcsec **53.** **(a)** 65.7 hours **(b)** 27.6 arcsec

Chapter 14

29. Sun-Uranus force = 1.39×10^{21} newtons, Neptune-Uranus force = 2.24×10^{17} newtons; Neptune reduces the sunward gravitational pull on Uranus by 1.61×10^{-4}, or 0.0161% **31.** **(a)** 2900 kg/m³ **32.** **(a)** 8.4 hours **35.** **(a)** 1.13×10^{-3} as bright as on Earth **(b)** 4.10×10^{-4} as bright as on Earth **(c)** 2.77 times as bright **36.** About 7200 km **38.** 0.95 arcsec **43.** Without the boost, a one-way trip would take 30.5 years

Chapter 15

33. 2.82 AU **35.** 30 km **37.** **(a)** 3.6×10^{12} kg **(b)** 0.83 m/s **39.** About 4% **40.** 3.6×10^7 km = 0.24 AU **41.** **(a)** Period = 350 years, lifetime = 3.5×10^4 years **(b)** Period = 1.1×10^4 years, lifetime = 1.1×10^6 years **(c)** Period = 3.5×10^5 years, lifetime = 3.5×10^7 years **(d)** Period = 1.1×10^7 years, lifetime = 1.1×10^9 years **44.** **(a)** 10^{15} kg **(b)** 10^{-16} kg/m³

Chapter 16

29. **(a)** 1.8×10^{-9} J **(b)** 9.0×10^{16} J **(c)** 5.4×10^{41} J **30.** **(a)** 4.6×10^{-36} s **(b)** 2.3×10^{10} s **(c)** 1.4×10^{15} s = 4.4×10^7 years **31.** 0.048 (4.8%) of the Sun's mass will be converted from hydrogen to helium; chemical composition of the Sun (by mass) will be 69% hydrogen, 30% helium **32.** **(a)** 8.8×10^{29} kg of hydrogen consumed, 6.2×10^{26} kg lost **34.** **(a)** 1.64×10^{-13} J **(b)** 2.43×10^{-3} nm **35.** 1.4×10^{13} kg/s

39. 98,600 kg/m^3 **40.** 1.9 × 10^{-7} nm **42.** (a) 1700 nm **44.** For the photosphere, 500 nm; for the chromosphere, 58 nm; for the corona, 1.9 nm **47.** For the umbra, 670 nm; for the penumbra, 580 nm **48.** (a) (Flux from patch of penumbra)/(flux from patch of photosphere) = 0.55 (b) (Flux from patch of penumbra)/(flux from patch of umbra) = 1.8

Chapter 17

34. (a) 9.7 pc (b) 0.10 arcsec **35.** 6.54 pc **36.** (a) 161 km/s (b) 294 km/s **37.** Distance = 105 pc; tangential velocity would have to be about 5200 km/s **38.** 110 km/s **39.** (a) +59.4 km/s (c) 486.23 nm **45.** 37.0 AU **46.** 6.1 $L_\odot$ **47.** 0.38 pc = 7.9 × 10^4 AU **48.** (a) +13.8 (b) (Luminosity of HIP 72509)/(luminosity of Sun) = 2.4 × 10^{-4} **49.** +17 **50.** 6300 pc **51.** (a) Brightest star has $M = -2.37$; dimmest star has $M = +4.32$ (b) $M = +0.79$ **54.** (b) $m_B - m_V = -0.23$ (Bellatrix), +0.68 (Sun), +1.86 (Betelgeuse) **55.** 99 $R_\odot$ **59.** The radius of star X is 17 times larger than the radius of star Y **60.** Radius increases by a factor of 2, luminosity increases by a factor of 64 **62.** T = 10,000 K, $R = 3.8\ R_\odot$ **63.** $L = 35\ L_\odot$, so distance = 14 pc **64.** (b) 0.26 $R_\odot$ = 1.8 × 10^5 km **65.** 9700 years **66.** (a) 5.0 pc (b) 22.5 AU (c) 1.5 $M_\odot$ **67.** (a) 40 $M_\odot$ **68.** About 2500 pc, about 125 times greater volume

Chapter 18

30. 0.34% **32.** 3.4 × 10^4 atoms per cm^3 **35.** 100 $R_\odot$ = 7.4 × 10^8 km = 4.9 AU **37.** (a) 250 AU = 3.7 × 10^{10} km **39.** 1.3 × 10^{31} m^3 **41.** 2.2 × 10^3 km/s, or 7.5 × 10^{-3} (0.75%) of the speed of light

Chapter 19

29. 657,000 km **30.** (a) 618 km/s (b) 61.8 km/s **31.** (a) 12.0 km/s (b) 9.3 km/s 32. 2.3 × 10^{29} kg, or 0.15 (15%) of the original mass of hydrogen **36.** (a) 4.9 × 10^7 years (b) 3.8 × 10^{11} years **38.** About 1900 K (1600°C, or 2900°F) **39.** 3.4 × 10^7 years **44.** About 650 pc **45.** About 370 pc **49.** (a) 1.65 × 10^6 km

Chapter 20

36. 0.11 $R_\odot$ **39.** We see the nebula about 5900 to 8200 years after the central star shed its outer layers **41.** (a) 97 nm **43.** (a) 1.8 × 10^9 kg/m^3 (b) 6.5 × 10^3 km/s **44.** About 4 × 10^6 kg **46.** 0.53 AU **48.** (a) 1.7 × 10^{30} kg = 0.84 $M_\odot$ (b) 1.1 × 10^{12} newtons (c) 1.5 × 10^8 m/s, or 0.5 (50%) of the speed of light **49.** (a) 6.4 × 10^5 m/s **50.** 3.2 × 10^9 km = 22 AU **51.** (Maximum luminosity of SN 1993J)/(maximum luminosity of SN 1987A) = 4.5 **53.** (a) 7 × 10^{-7} $b_\odot$ (b) It would be about 700 times brighter than Venus **54.** 1.3 × 10^8 pc = 130 Mpc

Chapter 21

33. About 5500 B.C. **35.** (a) About 700 A.D. **36.** (a) Radius = 4.6 ly, diameter = 9.2 ly **37.** (b) about 40 pc **38.** (b) Maxiumum correction = 10^{-4} of the pulsar period **39.** (a) 1250 years **40.** (a) 5.07 × 10^{-11} seconds per second (b) 2330 years **41.** (a) Density of matter in a neutron = 4 × 10^{17} kg/m^3 **46.** (a) 0.097 nm (b) 5.8 × 10^{31} W = 1.5 × 10^5 $L_\odot$ **47.** (b) 48 nm (c) 14 km

Chapter 22

33. 0.6 of the speed of light **34.** 8.3 × 10^{-8} s **35.** 0.8 of the speed of light **36.** (a) 25 years (b) 20 light-years as measured by an Earth observer, 12 light-years as measured by the astronaut **38.** 2.8 $M_\odot$ **39.** 5.7 × 10^8 years **40.** (a) 2.01 × 10^6 km **44.** 0.32 year **45.** 0.84 m **46.** (a) R_{Sch} = 8.9 mm, density = 2.0 × 10^{30} kg/m^3 (b) R_{Sch} = 3.0 km, density = 1.8 × 10^{19} kg/m^3 (c) R_{Sch} = 3.5 × 10^9 km = 24 AU, density = 13 kg/m^3 **47.** 7.4 × 10^{30} kg **48.** 2.9 × 10^{17} kg/m^3 **49.** 2.7 × 10^{38} kg = 1.4 × 10^8 $M_\odot$

Chapter 23

30. (a) 1.2 × 10^{12} cubic parsecs (b) 1.1 × 10^8 cubic parsecs (c) Probability = 9.6 × 10^{-5}; we can expect to see a supernova with 300 pc once every 350,000 years **31.** (a) 8.3 × 10^9 AU (b) 4.0 × 10^4 pc **33.** 9.5 × 10^{-25} J; it takes 3.2 × 10^5 such photons to equal the energy of one H_α photon **34.** (b) 5700 pc **35.** 21 times **37.** A 10% error in radius results in a 10% error in mass; a 10% error in velocity results in a 20% error in mass **38.** (a) 3.1 × 10^8 years (b) 7.4 × 10^{11} $M_\odot$ **39.** 2.7 × 10^{11} $M_\odot$ **44.** (a) 1.1 × 10^7 km = 0.073 AU (b) 9.1 × 10^{-6} arcsec (c) 251 arcsec **47.** (a) 6.3 years (b) 3.7 × 10^6 $M_\odot$

Chapter 24

35. (a) 6.9 Mpc (b) 8.3 Mpc (c) 1.4 Mpc **36.** 9.5 Mpc **37.** (a) 1.1 × 10^{10} km = 70 AU (b) 7.0 Mpc **38.** (a) 152 Mpc = 4.96 × 10^8 ly **39.** 54 km/s/Mpc **40.** (a) z = 0.0252 (b) 106 Mpc **41.** (a) 2.85 × 10^5 km/s, or 0.951 (95.1%) of the speed of light (b) 1.60 × 10^6 km/s (c) 4020 Mpc = 1.31 × 10^{10} ly **42.** (a) 1.2 × 10^{70} atoms (b) 3.6 × 10^{-6} atom per cm^3 **43.** 1.2 × 10^{12} $M_\odot$ **44.** Period = 4.4 × 10^8 years; mass = 3.6 × 10^{11} $M_\odot$

Chapter 25

27. 2.87 × 10^5 km/s, or 0.958 (95.8%) of the speed of light **28.** 2.93 × 10^5 km/s, or 0.976 (97.6%) of the speed of light **33.** (a) 30 years (b) 2014 (c) 5/3 of the speed of light **34.** (a) 1.80 × 10^5 km/s, or 0.600 (60.0%) of the speed of light (b) 134 hours (c) 970 AU **35.** (a) 1.2 × 10^{11} $L_\odot$ **36.** 1.5 × 10^8 $M_\odot$ 37. 2.9 × 10^9 km = 20 AU

Chapter 26

33. $z = 4$ (b) Distances were 1/9 as great and the density of matter was 729 times greater **35.** (a) 20 billion years (b) 13 billion years (c) 9.7 billion years **36.** 1.6 × 10^8 km/s/Mpc **37.** 65.0 times denser **38.** 1.06 × 10^{-3} m = 1.06 mm **39.** 13.6 K **40.** (a) 4.6 × 10^{-23} kg/m^3 **41.** (a) 9.5 × 10^{-18} kg/m^3 (b) 4.8 × 10^{-18} kg/m^3 (c) 1.3 × 10^{-7} kg/m^3 **43.** (a) 4.7 × 10^{-27} kg/m^3 (b) 1.9 × 10^{-26} kg/m^3 **45.** (b) $q_0 = -0.595$ (c) $\Omega_\Lambda = 0.135$ **46.** (a) −0.17 (b) +0.12 **47.** 1.3 × 10^{-53} m^{-2}

Chapter 27

27. 1.1 × 10^{-26} J **29.** 3.5 × 10^{-25} s **30.** (a) 80.5 GeV (b) 9.34 × 10^{14} K **32.** (a) 29 nm (c) 348 nm **33.** (a) 2.0 × 10^{-35} kg = 2.2 × 10^{-5} electron mass **34.** (a) 1.1 × 10^{14} m = 0.011 ly **35.** Length was 0.168 of its present-day value; density was 213 times the present-day value

Chapter 28

11. (a) 3.7 cm **12.** (a) 10,000 years (b) 10 million years **14.** (a) 1430 MHz **15.** (a) 29.8 km/s (b) 0.10 m (c) 9.9 × 10^{-6} m, or 9.9 × 10^{-3}% of the unshifted wavelength **16.** 200 km

Index

Page numbers in *italics* indicate illustrations.

Star Charts

The following set of star charts, one for each month of the year, are from *Griffith Observer* magazine. They are useful in the northern hemisphere only. For a set of star charts suitable for use in the southern hemisphere, see the *Universe* Web site (**www.whfreeman.com/universe**).

To use these charts, first select the chart that best corresponds to the date and time of your observations. Hold the chart vertically as shown in the above illustration and turn it so that the direction you are facing is shown at the bottom.

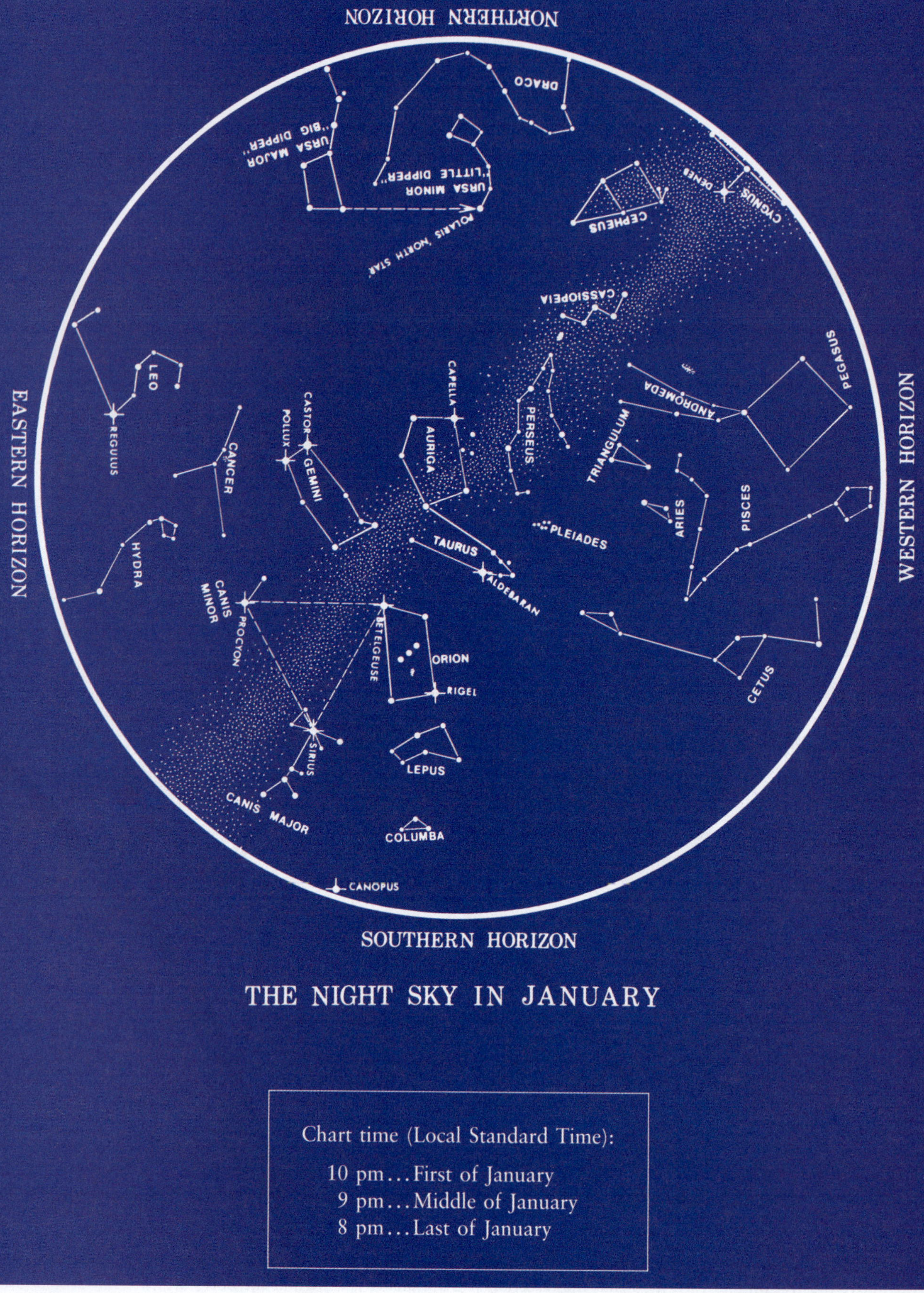

THE NIGHT SKY IN JANUARY

Chart time (Local Standard Time):

10 pm...First of January
9 pm...Middle of January
8 pm...Last of January

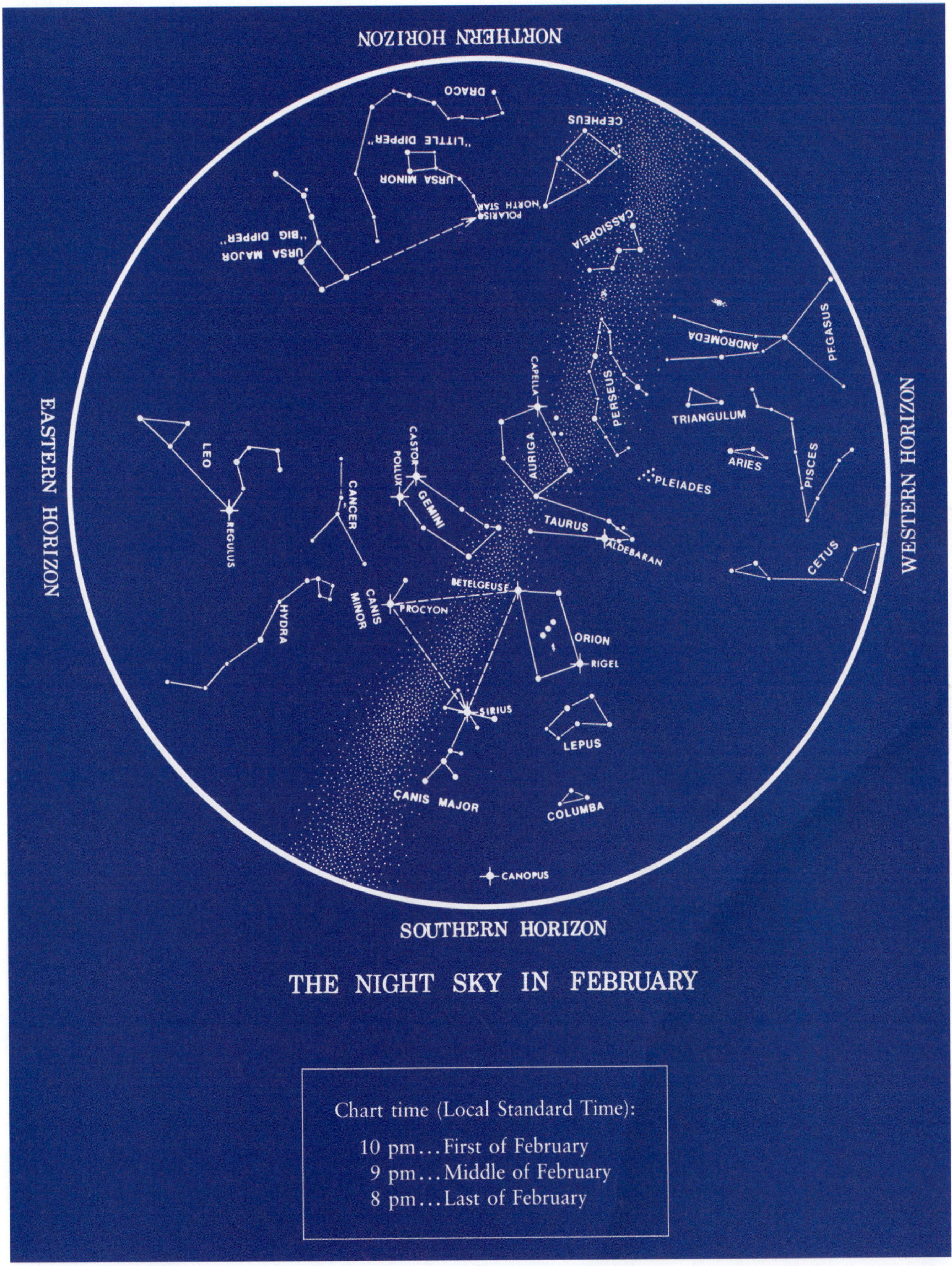

THE NIGHT SKY IN FEBRUARY

Chart time (Local Standard Time):

10 pm . . . First of February
9 pm . . . Middle of February
8 pm . . . Last of February

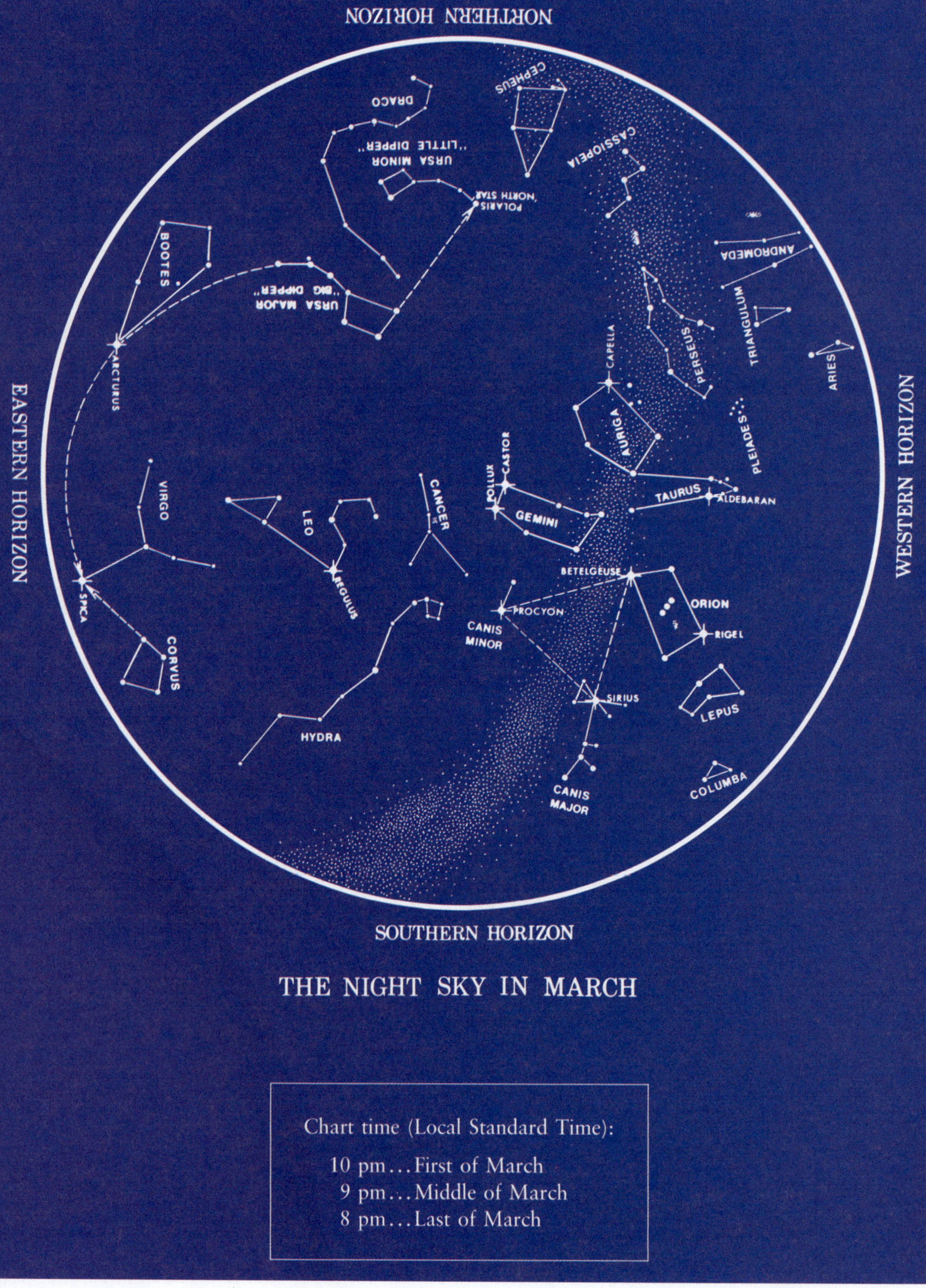

THE NIGHT SKY IN MARCH

Chart time (Local Standard Time):

10 pm... First of March
9 pm... Middle of March
8 pm... Last of March

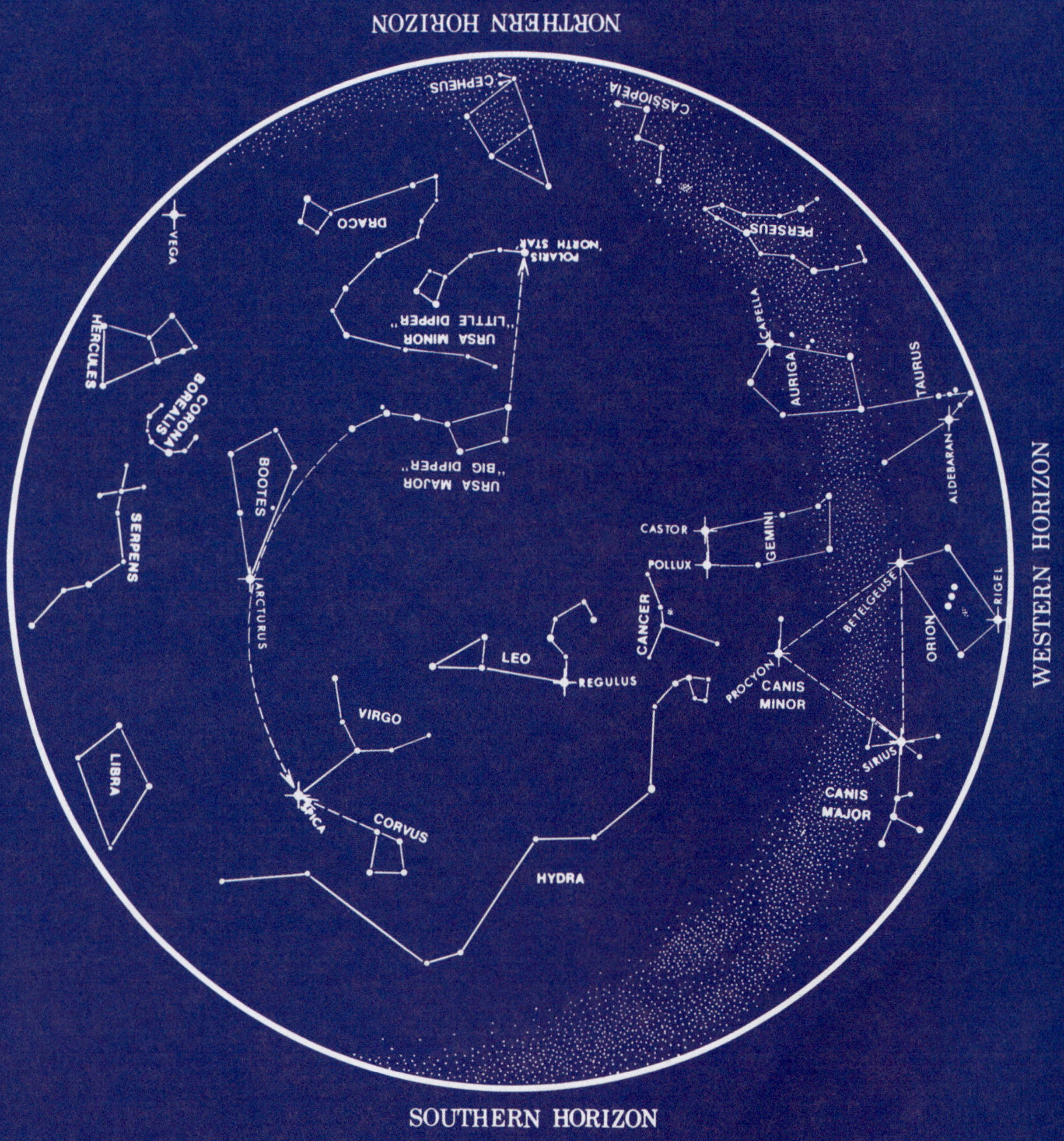

THE NIGHT SKY IN APRIL

Chart time (Daylight Saving Time):

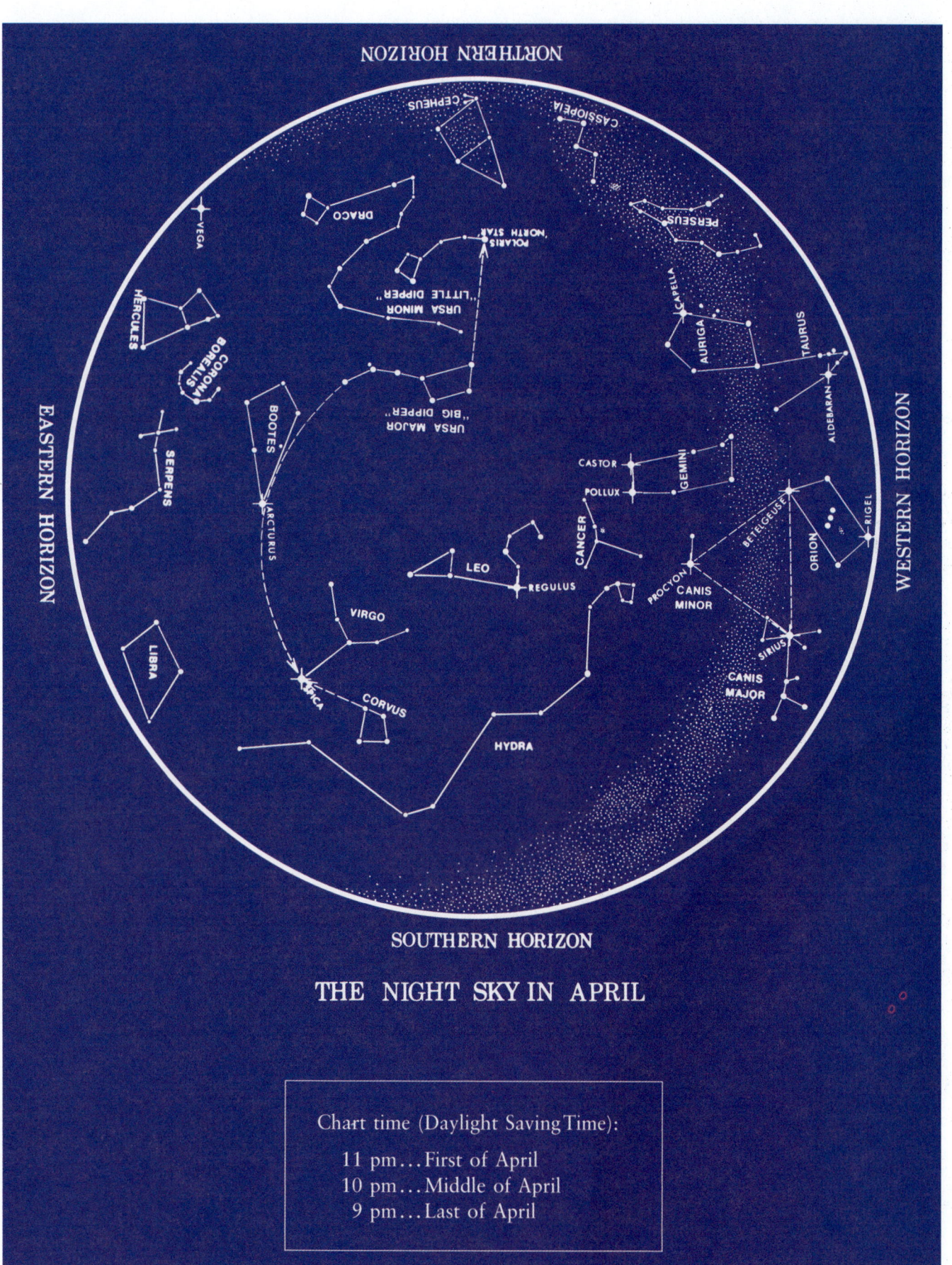

11 pm . . . First of April
10 pm . . . Middle of April
9 pm . . . Last of April

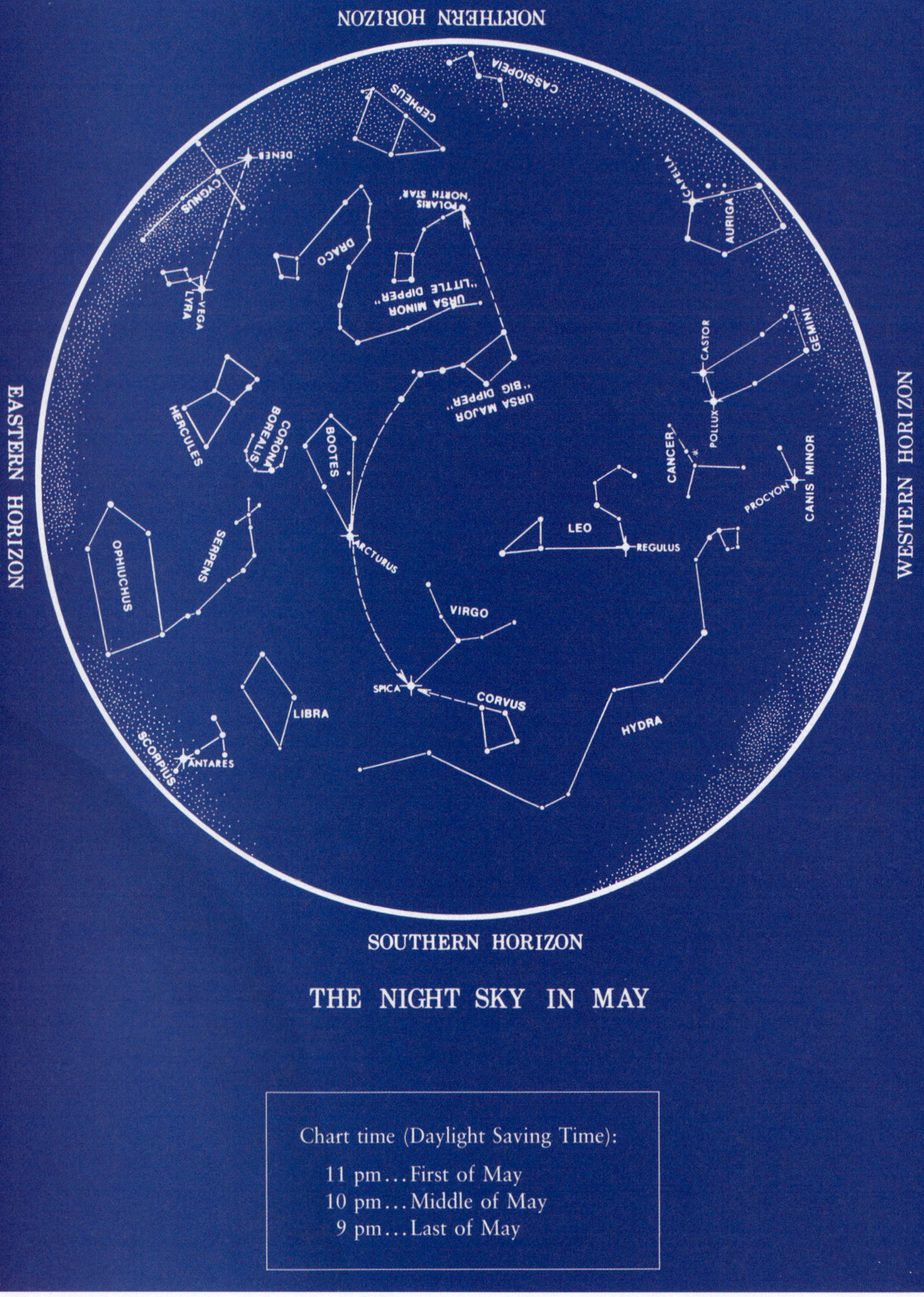

THE NIGHT SKY IN MAY

Chart time (Daylight Saving Time):

11 pm . . . First of May
10 pm . . . Middle of May
9 pm . . . Last of May

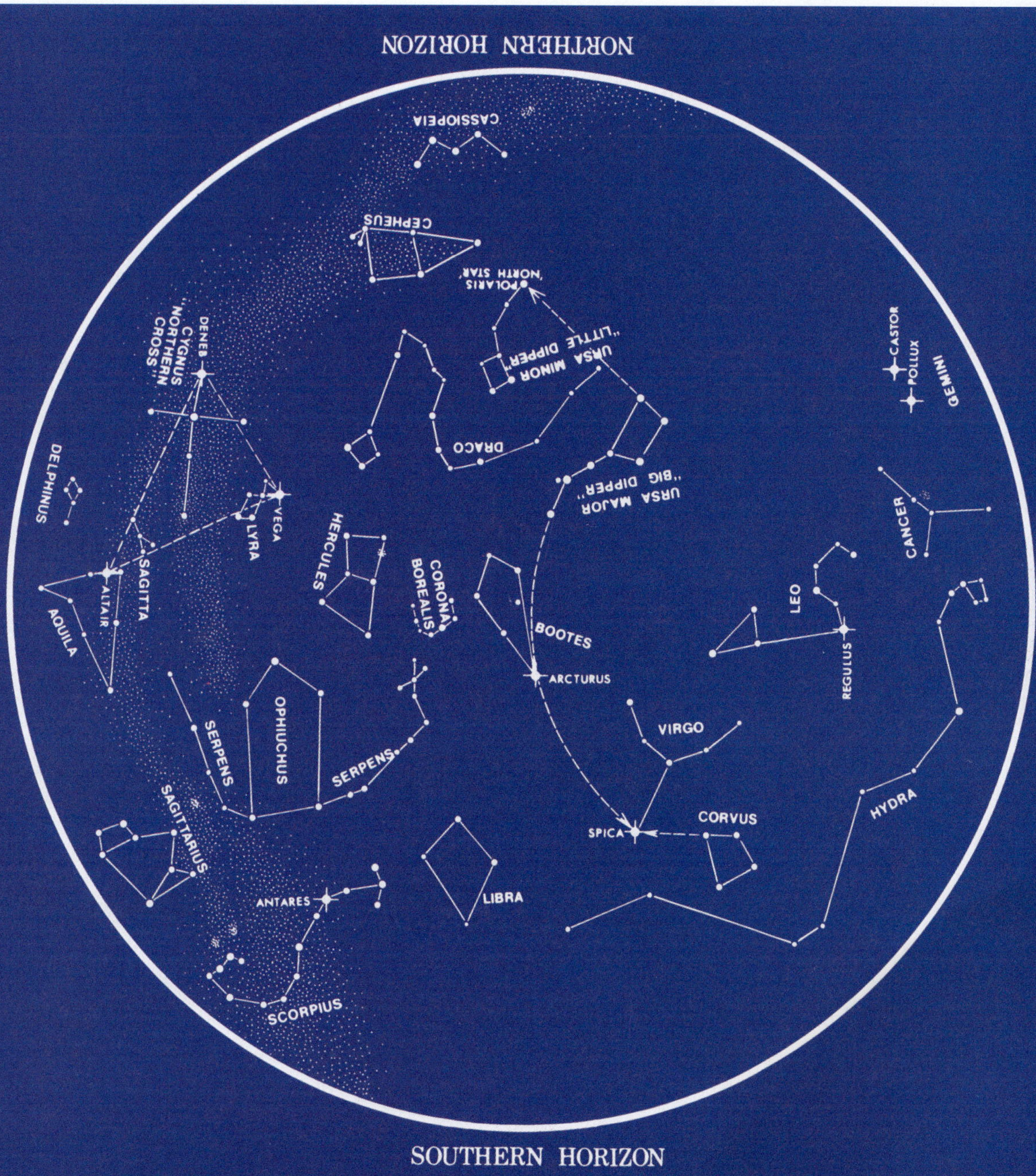

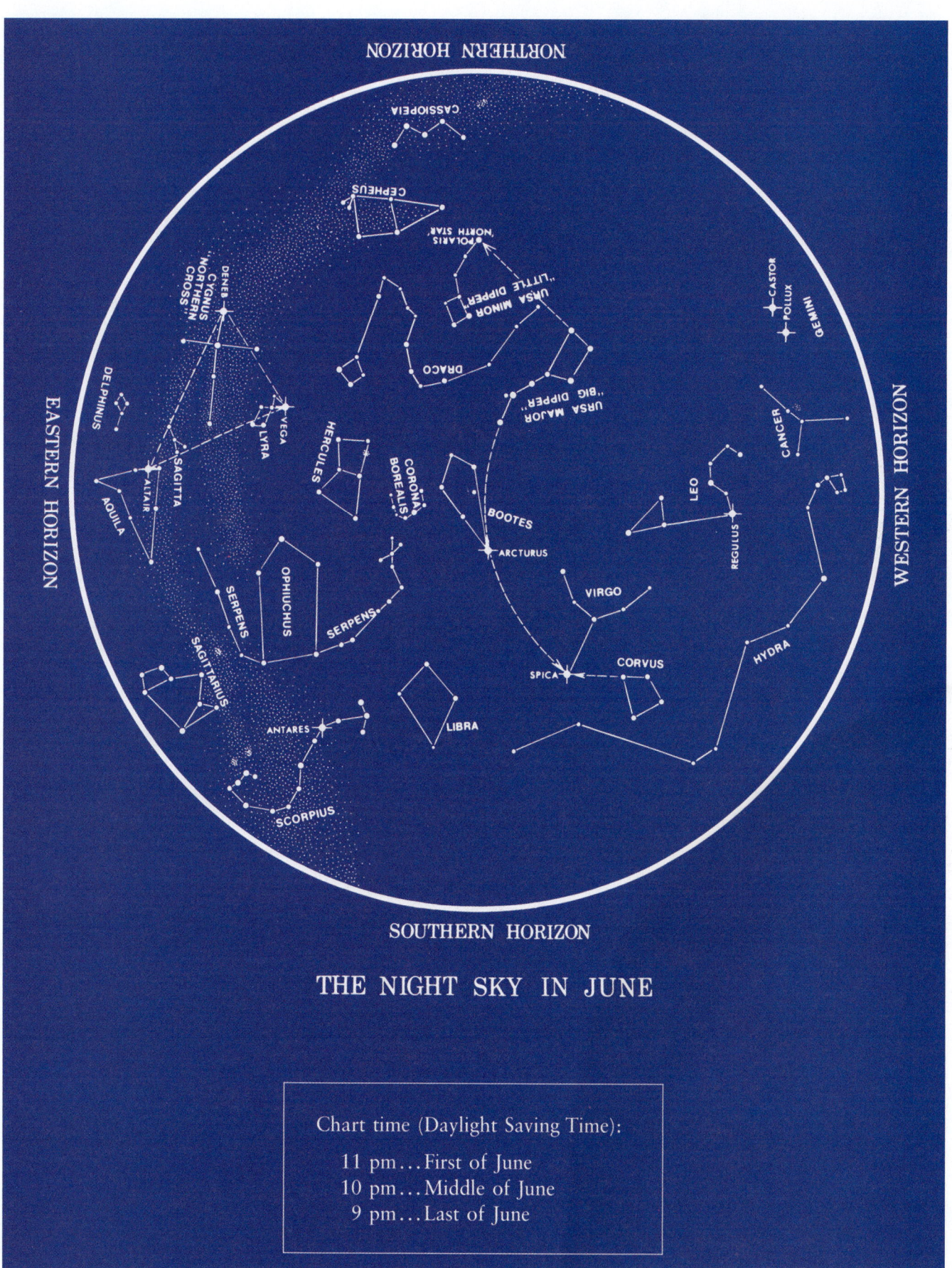

THE NIGHT SKY IN JUNE

Chart time (Daylight Saving Time):

11 pm... First of June
10 pm... Middle of June
9 pm... Last of June

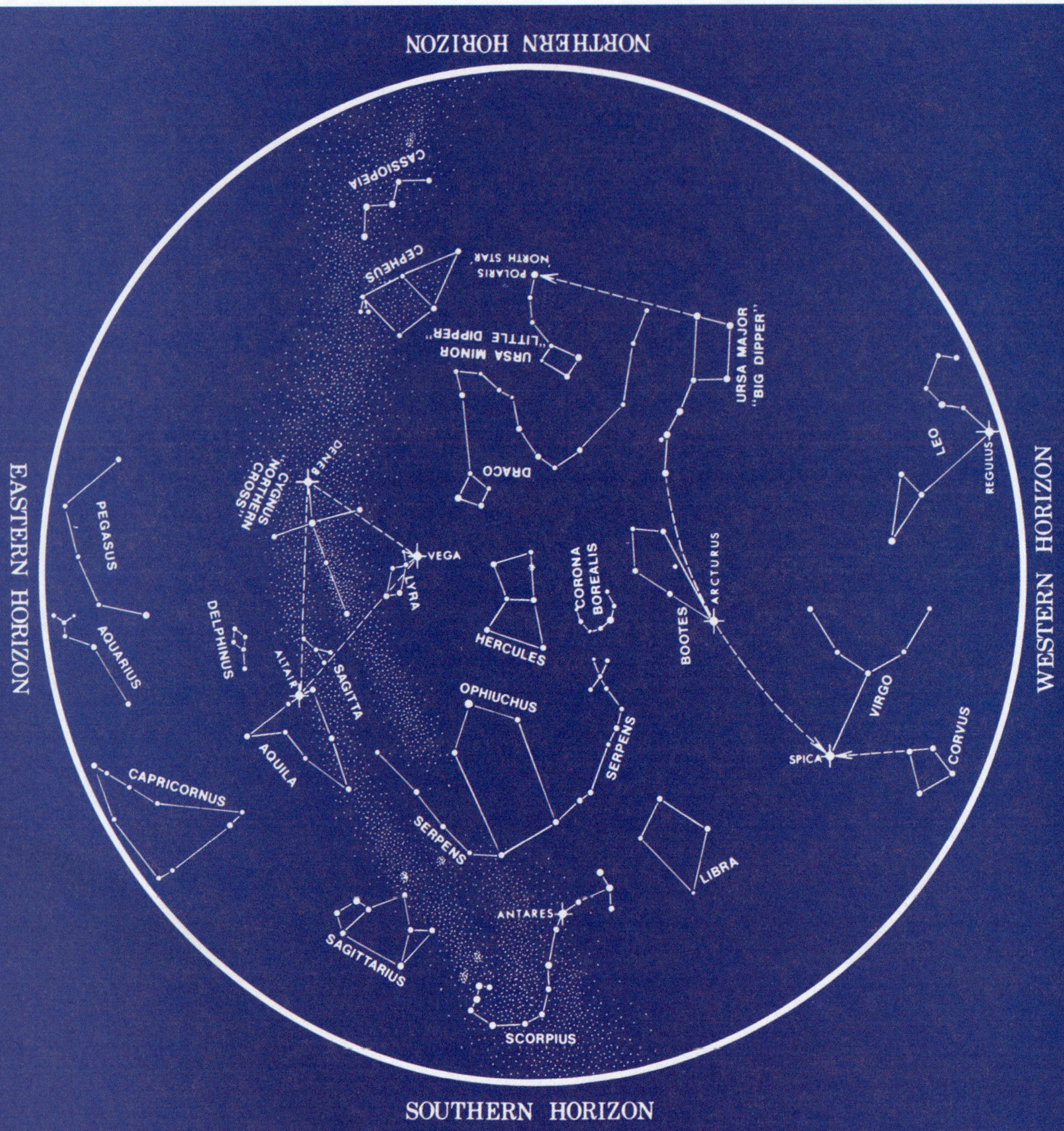

THE NIGHT SKY IN JULY

Chart time (Daylight Saving Time):

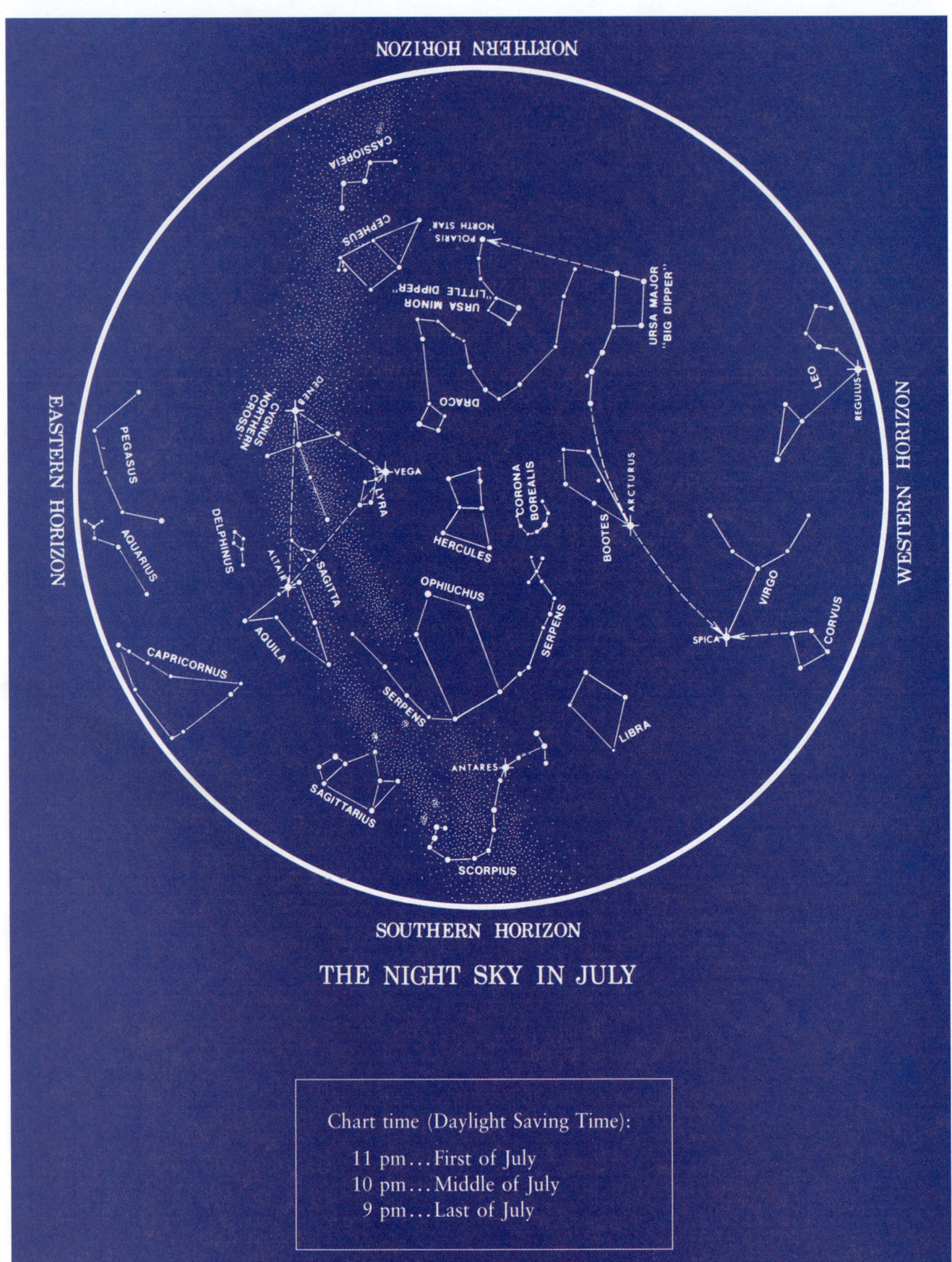

11 pm . . . First of July
10 pm . . . Middle of July
9 pm . . . Last of July

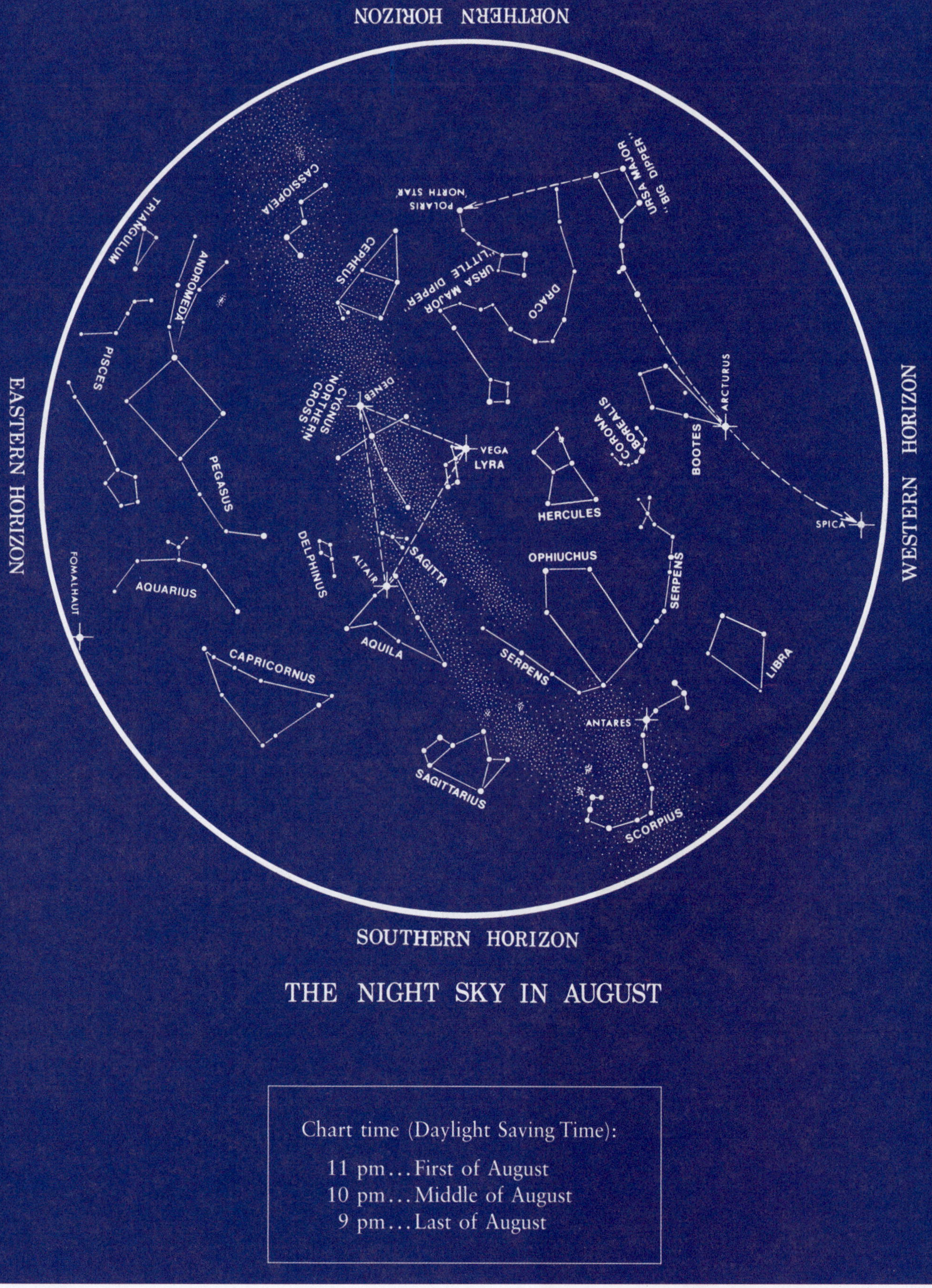

THE NIGHT SKY IN AUGUST

Chart time (Daylight Saving Time):

11 pm . . . First of August
10 pm . . . Middle of August
9 pm . . . Last of August

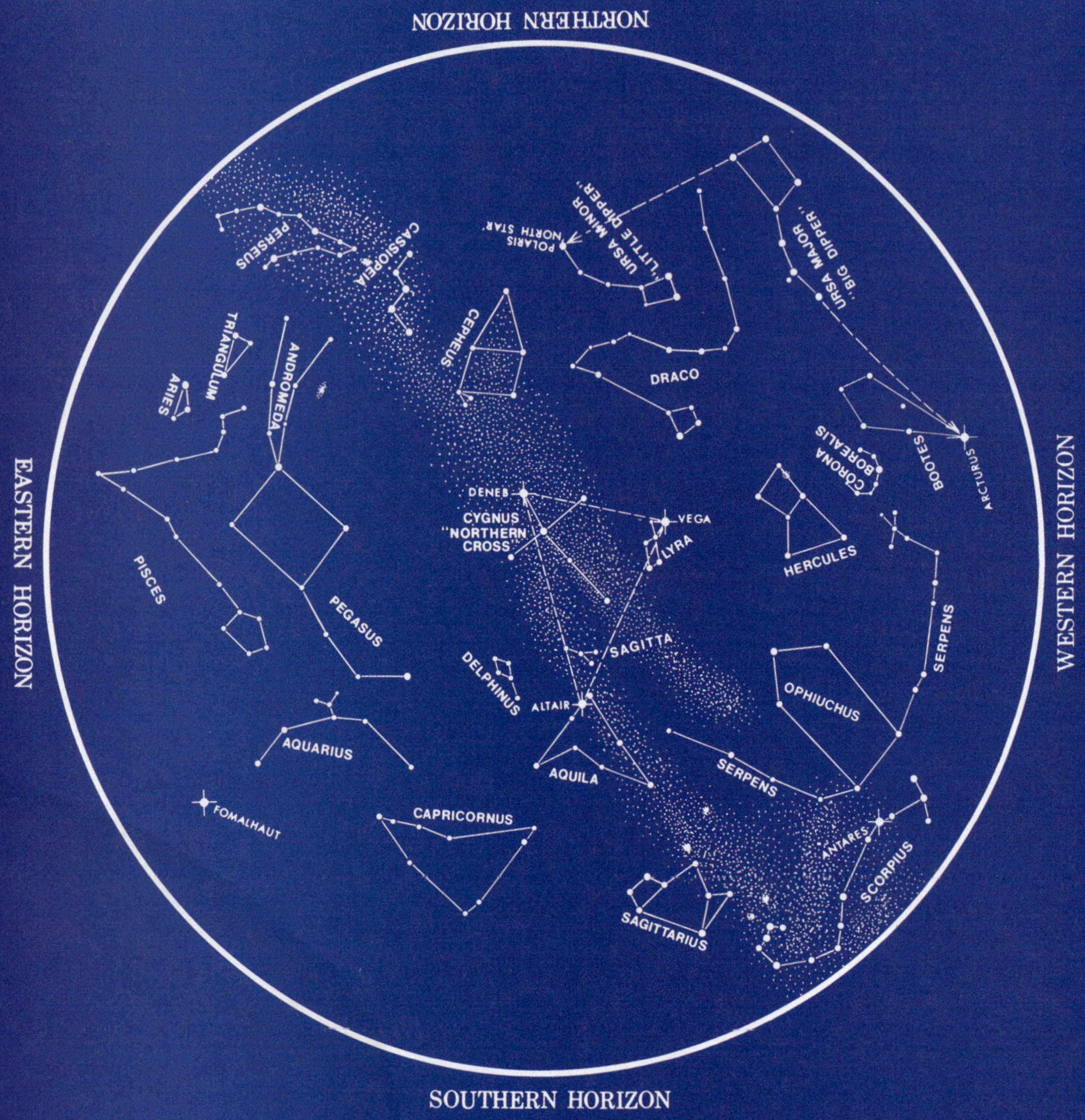

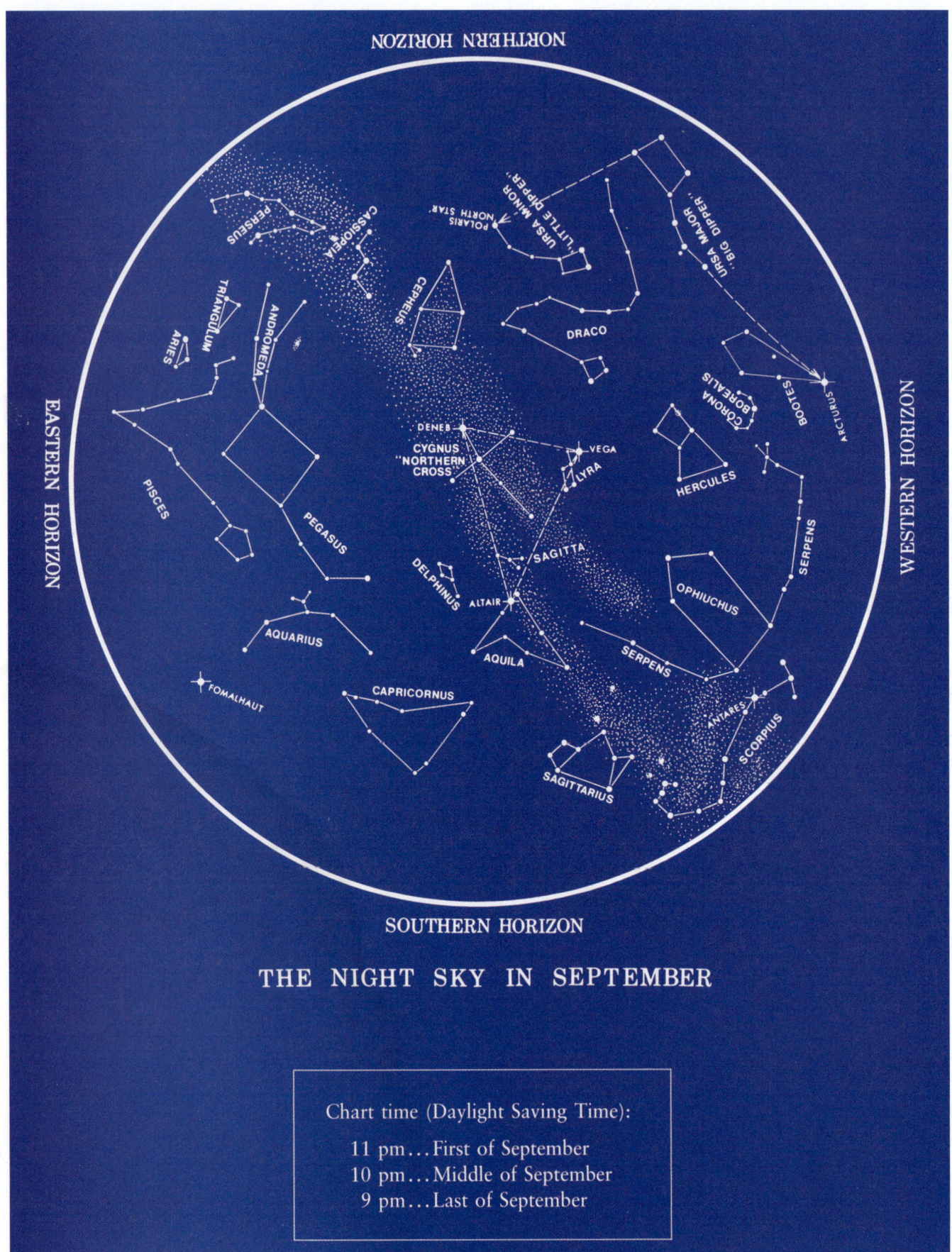

THE NIGHT SKY IN SEPTEMBER

Chart time (Daylight Saving Time):

11 pm . . . First of September
10 pm . . . Middle of September
9 pm . . . Last of September

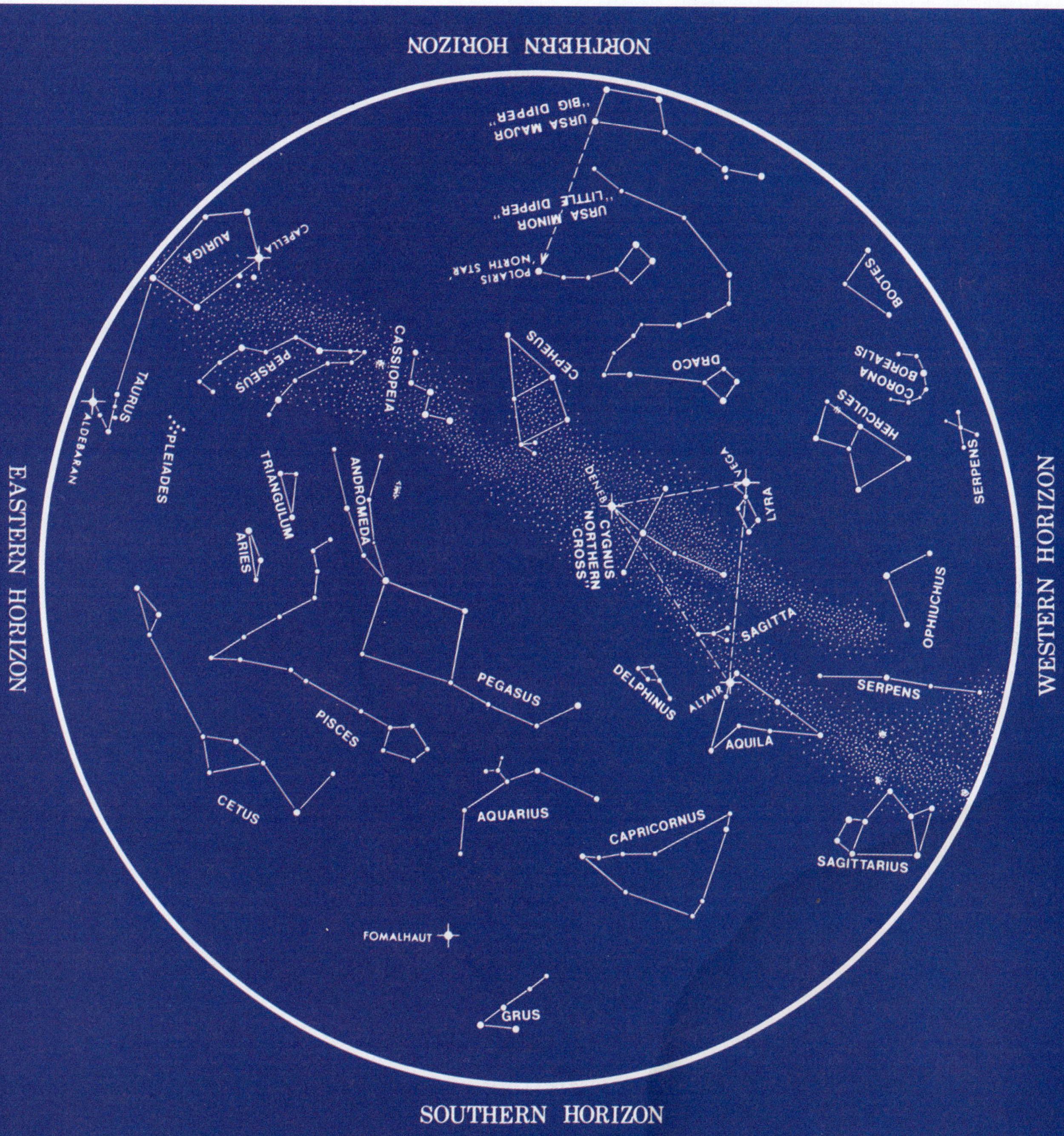

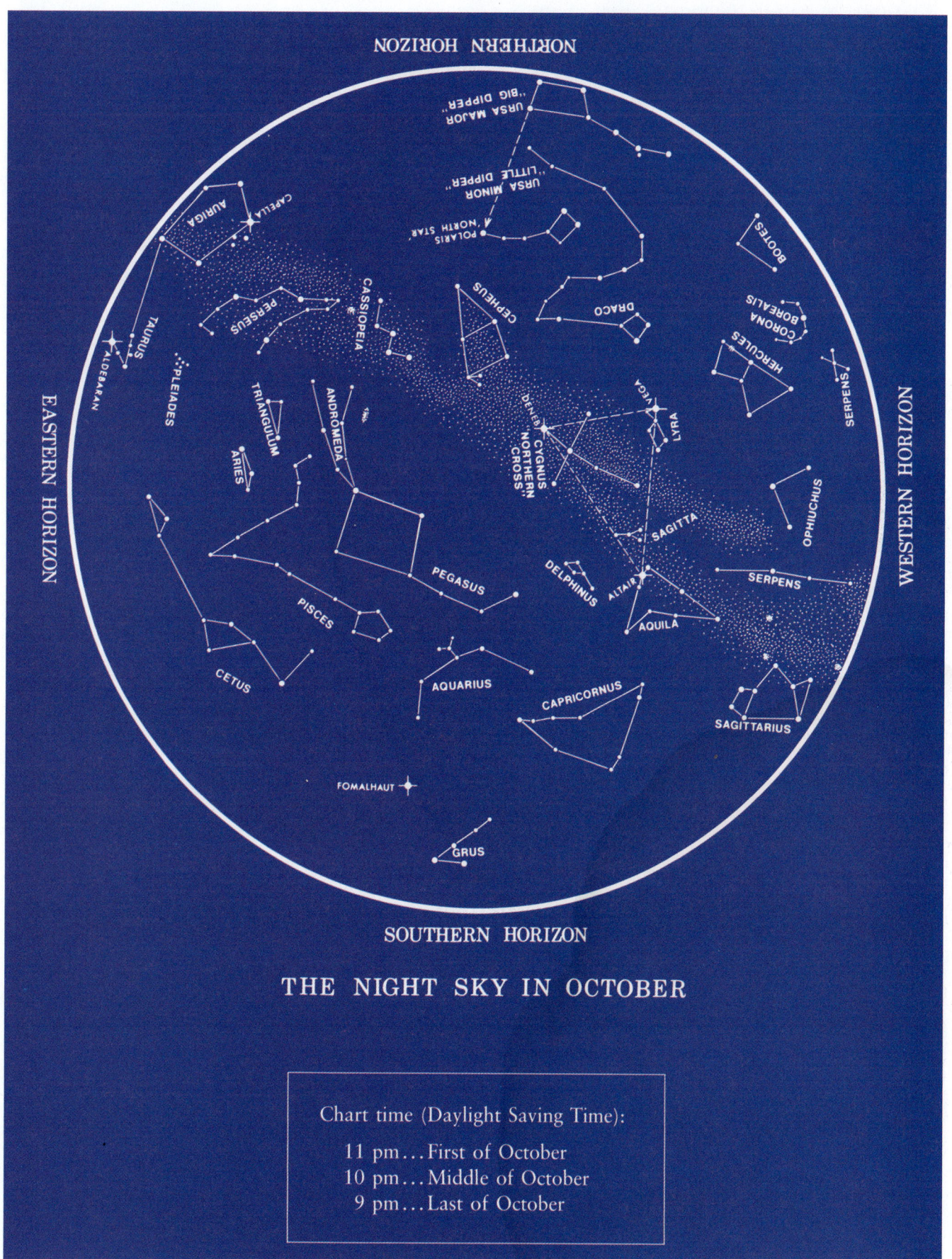

THE NIGHT SKY IN OCTOBER

Chart time (Daylight Saving Time):

11 pm . . . First of October
10 pm . . . Middle of October
9 pm . . . Last of October

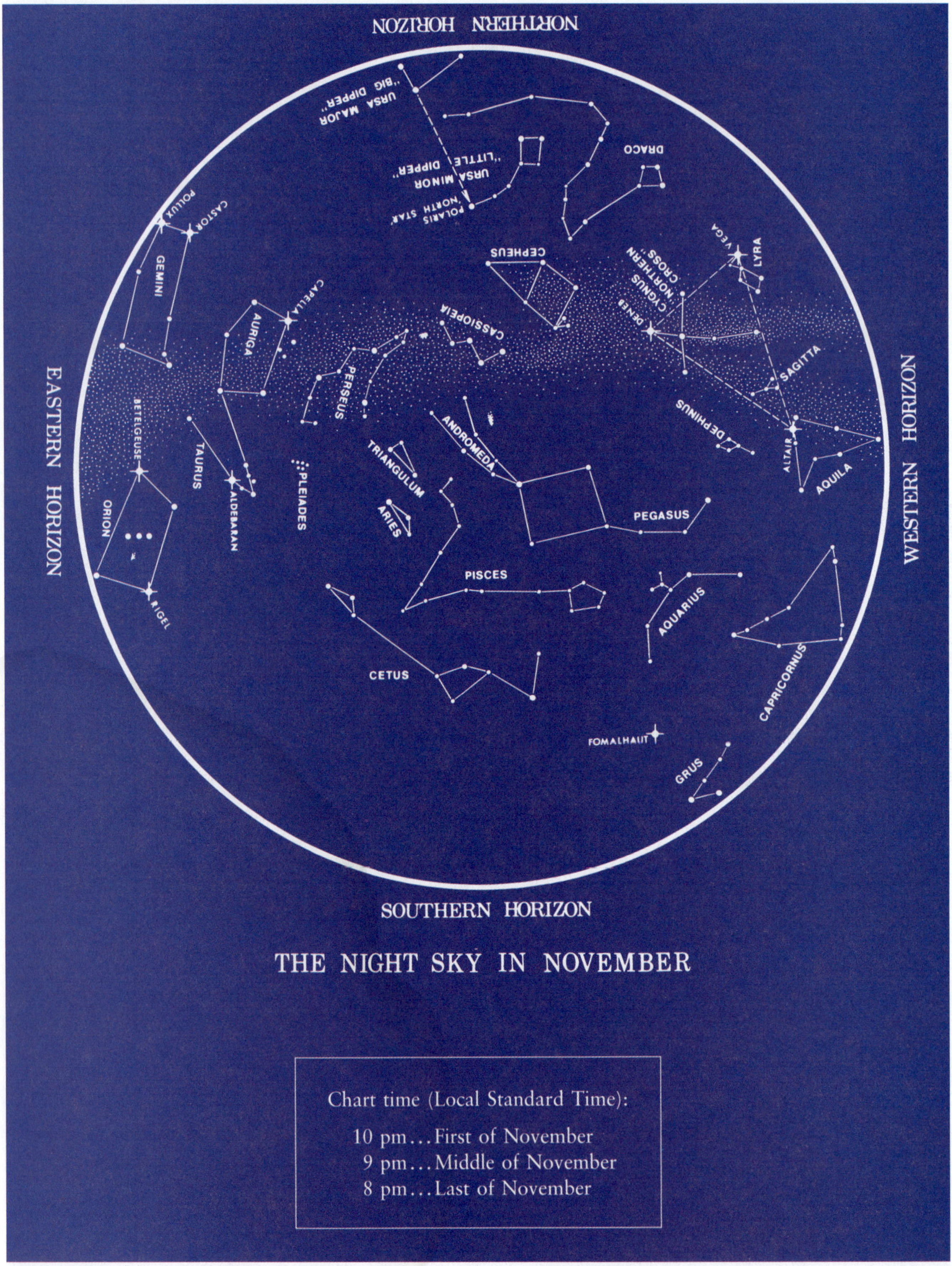

THE NIGHT SKY IN NOVEMBER

Chart time (Local Standard Time):

10 pm . . . First of November
9 pm . . . Middle of November
8 pm . . . Last of November

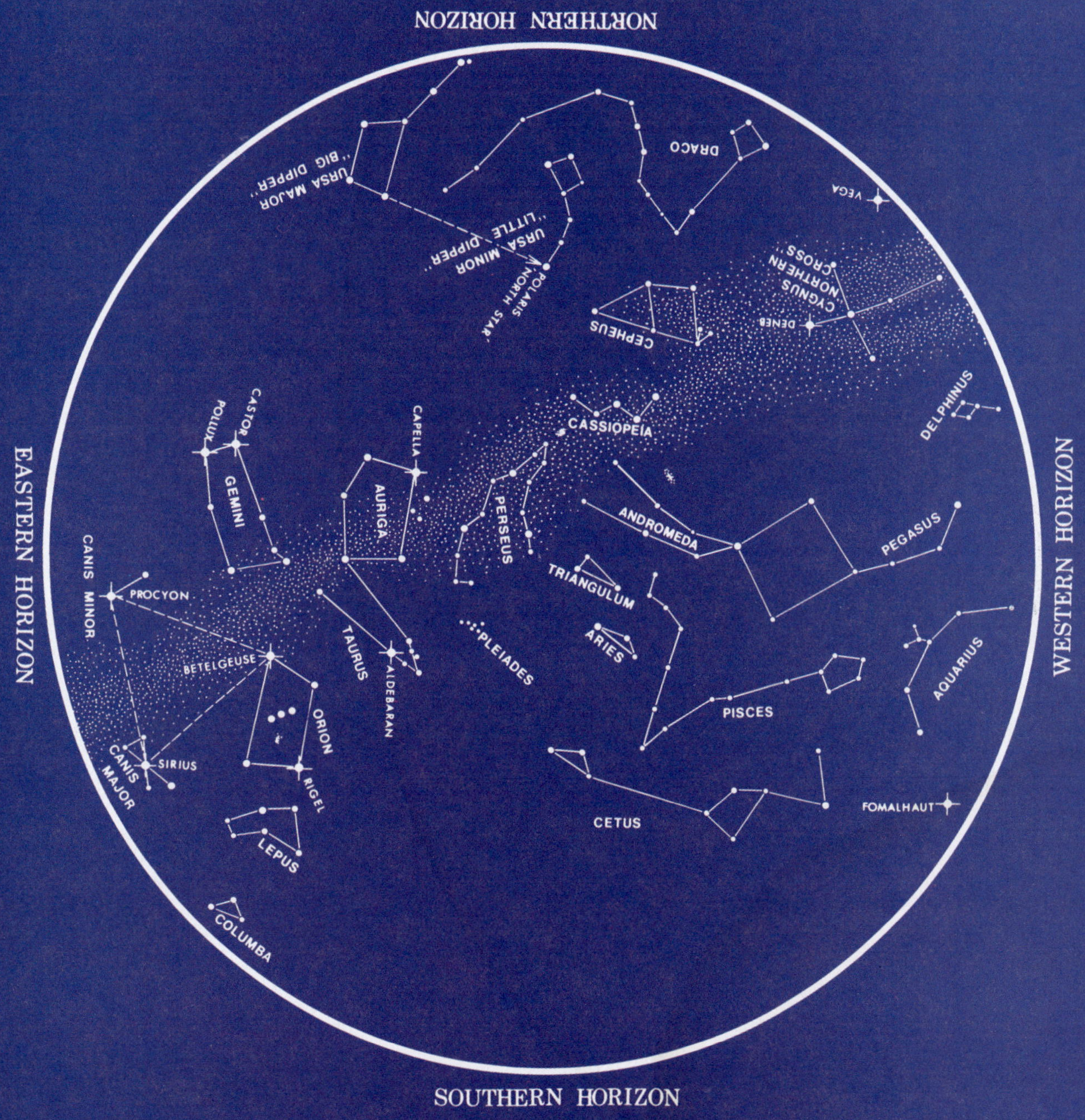

THE NIGHT SKY IN DECEMBER

Chart time (Local Standard Time):

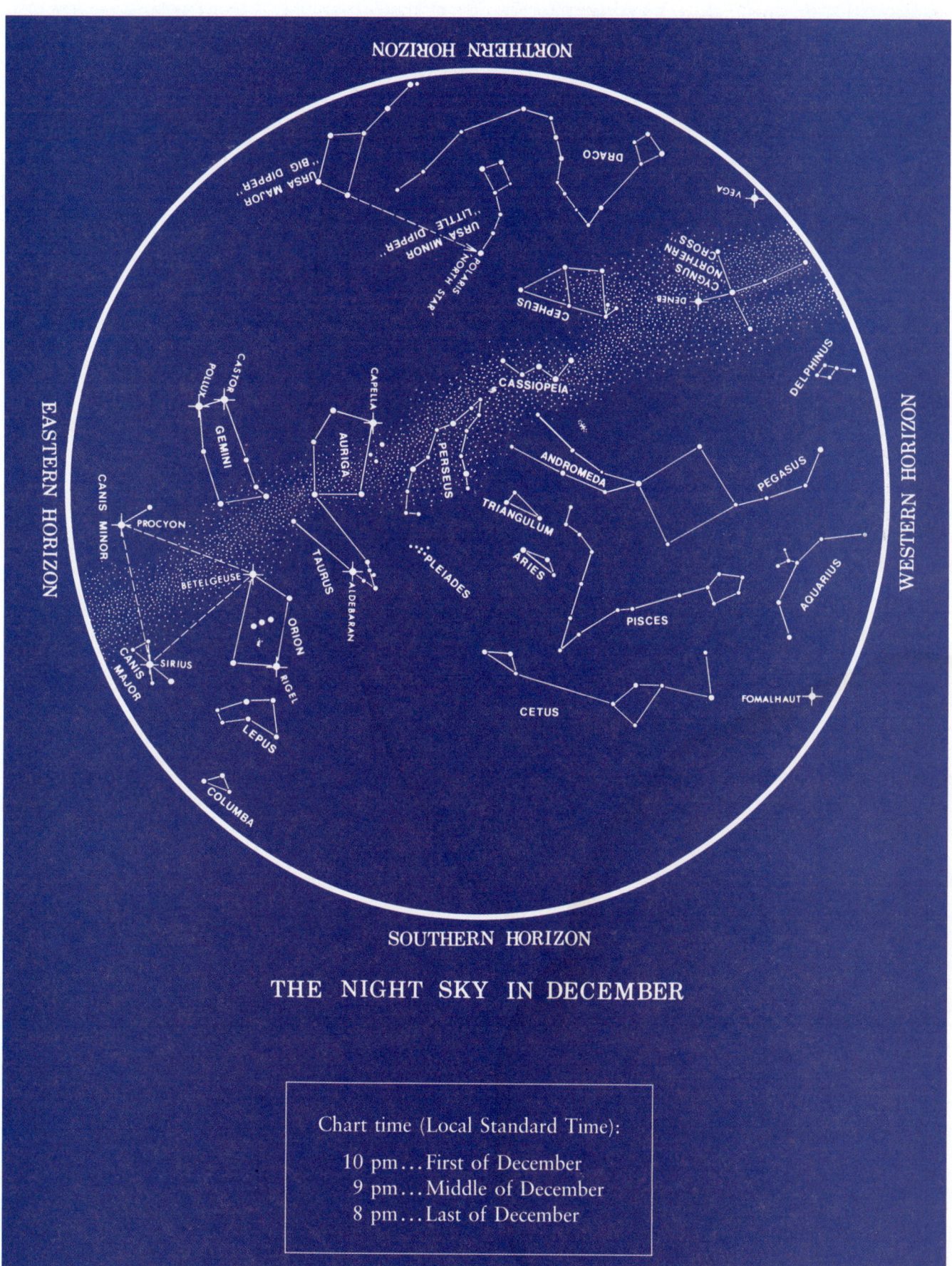

10 pm... First of December
9 pm... Middle of December
8 pm... Last of December